D-Amino Acids: A New Frontier in Amino Acid and Protein Research – Practical Methods and Protocols

D-AMINO ACIDS: A NEW FRONTIER IN AMINO ACID AND PROTEIN RESEARCH – PRACTICAL METHODS AND PROTOCOLS

RYUICHI KONNO
HANS BRÜCKNER
ANTIMO D'ANIELLO
GEORGE FISHER
NORIKO FUJII
HIROSHI HOMMA
EDITORS

Nova Biomedical Books
New York

For permission to use material from this book please contact us:
Telephone 631-231-7269; Fax 631-231-8175
Web Site: http://www.novapublishers.com

Library of Congress Cataloging-in-Publication Data
Available upon request.

ISBN 1-60021-075-9

Published by Nova Science Publishers, Inc. ✣ New York

To the memory of Drs. L. Pasteur, H. A. Krebs, A. Meister, J. J. Corrigan, E. H. Man, and the pioneers who devoted themselves to the research on D-amino acids.

CONTENTS

Preface

D-Amino acids are stereoisomers or optical isomers of naturally occurring L-amino acids. D Amino acids and L-amino acids have almost the same physical and chemical properties. D-Amino acids have been considered to be unnatural substances that are produced only when we synthesize amino acids in laboratories or factories. Indeed, equal amounts of D-amino acids and L-amino acids (racemic mixture) are produced by chemical synthesis.

Until recently, living organisms were thought to contain exclusively L-amino acids. Before the emergence of life, D- and L-amino acids should have existed on the primitive earth, because amino acids are considered to be produced under the conditions of high temperature, high pressure and lightning. It is an interesting and important question why D-amino acids were eliminated and organisms have evolved using only L-amino acids. This mystery has not been solved.

D-Amino acids have been considered exceptional amino acids that exist only in prokaryotes and a part of lower organisms. However, the development and improvement of analytical instruments and methods have revealed that D-amino acids are present in organisms far more in the amount and extent than previously considered. It has become clear that D-amino acids are present in invertebrates and vertebrates including humans and that some of D-amino acids have important physiological functions. Biological systems for synthesis and degradation of D-amino acids have also been found.

In the past few decades new evidence for D-amino acids has been uncovered and a vast new knowledge has been accumulated in the field of D-amino acid research. New experimental results have been published in a wide range of journals and have been reported in several conferences and symposia. This scattered information makes it difficult for the newcomers to the field to obtain a general oversight of current activity and new advances. Therefore, we launch out into a new publication of this book that includes comprehensive D-amino acid research. We have asked researchers worldwide who are actively studying D-amino acids to send us their contributions regarding their original methods and findings. Those contributions are compiled in this book. The readers will find what kind of research is conducted in the field of D-amino acids and what problems there are. This is a basic methodology book. We think this book is useful especially for young researchers and students to start their own researches in the field of D-amino acids.

Finally, we would like to thank all of the contributors who have agreed to participate in this project and are willing to spare their time to write the chapters and share their expertise.

Ryuichi Konno, Ph. D.
Department of Microbiology
Dokkyo Medical University

Hans Brückner, Ph. D.
Department of Food Sciences
University of Giessen

Antimo D'Aniello, Ph. D.
Laboratory of Neurobiology
Stazione Zoologica "A. Dohrn"

George Fisher, Ph. D.
Department of Physical Sciences
Barry University

Noriko Fujii, Ph. D.
Research Reactor Institute
Kyoto University

Hiroshi Homma, Ph. D.
School of Pharmaceutical Sciences
Kitasato University

SECTION 1. INTRODUCTION

In: D-Amino Acids: ISBN 1-60021-075-9
A New Frontier in Amino Acid and Protein Research, pp. 3-14

Chapter 1

AN ESSAY ON D-AMINO ACIDS: RETROSPECTION AND PERSPECTIVE

Kenji Soda*

Department of Biotechnology, Kansai University, Japan.

Keywords: amino acid racemase, D-amino acid aminotransferase, D-amino acid oxidase, D-amino acid metabolism, poly-γ-glutamate

1. INTRODUCTION - NATURAL OCCURRENCE OF D-AMINO ACIDS

D-Amino acids were first prepared by making use of the fungal asymmetric decomposition of the racemates in 1886 [1], and once regarded as unnatural compounds, because they were not naturally found in both free and bound forms in those days. However, that is not the case dually. First of all, amino acids quite likely occurred in the racemic form on the primitive Earth. Amino acids are, in a sense, stable mainly owing to the high dissociation energy of the α-carbon-hydrogen and α-carbon-carbon bonds. For example, it has been estimated that the racemization half-lives of usual amino acids, except for tryptophan and some other amino acids with unstable side chains, are 10^5 to 10^6 years at ambient temperatures [2]. However, amino acids are not so stable in another sense: it depends on the conditions and a timescale. Racemization rates are accelerated by metal ions, alkaline conditions, ionizing radiation [3], derivatization, high temperatures and some other conditions. Various compounds including amino acids reacted with one another on a timescale of ten million years and more. This time is more than enough for amino acids to be racemized under the condition of the primeval Earth [4]. Accordingly, it is presumed that amino acids occurred in a racemic form in the primordial soup even if L-amino acids were enantioselectively synthesized either terrestrially or extraterrestrially through any of various mechanisms proposed [5,6].

Only the L-enantiomers of racemic amino acids were specifically incorporated into proteins leading to the production of enzymes and eventually the birth of the first organisms. Accordingly, the first organisms were supposed to be certainly emerged in the presence of high concentrations of racemic and D-amino acids. The occurrence of free and peptidyl D-amino acids and D-amino acid dehydrogenase in hyperthermophilic archaea probably reflects the above situation [7].

* Correspondence concerning this article should be addressed to Kenji Soda, E-mail: soda@okurayama.mbox.media.kyoto-u.ac.jp, FAX:+81-774-32-7049.

Secondly, various free and peptidyl D-amino acids are now found not only in microorganisms, but in plants, insects, mammals, marine invertebrates such as sea shells, snails, and some others [8-10]. However, the physiological functions of D-amino acids and D-amino acid-containing peptides have hardly been elucidated. Fujii *et al.* [10] first demonstrated the occurrence of D-aspartyl and D-β-aspartyl residues in α-A- and α-B-crystallins of human lens, and elucidated the mechanism of post-translational stereoinversion of L-aspartyl residue in the proteins. They found the occurrence of D-β-aspartyl-containing protein in the elastic fibers of skin from elderly donors, and also showed that the formation of D-β-aspartyl residue is accelerated by exposure to sunlight and ultraviolet rays, and aging. Aspartyl residues are not uniformly converted into the D-counterparts, but inverted only in the specific structures found in α-crystallins and β-amyloid protein [11]. The amount of D-amino acid residues in particular proteins such as α-crystallins may serve as a measure of aging. The mechanism of post-translational inversion of the specific L-amino acid residues in particular proteins appears to be essentially analogous to that of the post-translational modification of the specific amino acid residue of the precursor enzyme to generate a catalytic center (a built-in coenzyme) in the quinoproteins [12,13].

Only recently, it was shown that D-histidine and D-lysine react with respective L-aminoacyl-tRNA synthetases of *Escherichia coli*: the stereospecificities of the enzymes are not as high as we have ever thought [14]. This suggests that D-amino acids may be very slowly incorporated into usual proteins to produce a slight amount of unusual proteins, though their functions are unknown. The research regarding the occurrence, the function and the incorporation mechanism of D-amino acids in proteins will be advanced with the development of more accurate analytical technique in enantioselective and site-specific determination of proteinaceous D-amino acids.

2. METABOLISM OF D-AMINO ACIDS

D-Amino acids are metabolized either directly or after their conversion into their L-enantiomers. The typical metabolic pathways are described here.

D-Amino acids are converted to the L-counterparts by a coupled reaction system of D-amino acid oxidase and an L-amino acid dehydrogenase such as alanine dehydrogenase and leucine dehydrogenase as follows.

$$\text{D-Alanine} + O_2 + H_2O \xrightarrow{\text{(D-Amino acid oxidase)}} \leftrightarrow \text{Pyruvate} + NH_3 + H_2O_2 \quad (1)$$

$$\text{Pyruvate} + NH_3 + NADH + H^+ \leftrightarrow \text{L-Alanine} + NAD^+ + H_2O \quad \text{(Alanine dehydrogenase)} \quad (2)$$

$$\text{D-Alanine} + O_2 + NADH + H^+ \leftrightarrow \text{L-Alanine} + NAD^+ + H_2O_2 \quad (3)$$

Alternatively, D-amino acids are interconverted to the L-counterpart by a coupling of D-amino acid aminotransferase (D-alanine aminotransferase, EC 2.6.1.21) and L-amino acid aminotransferase reactions. However, these reactions have not been shown *in vivo*.

$$\text{D-Alanine} + \text{2-Oxoglutarate} \leftrightarrow \text{Pyruvate} + \text{D-Glutamate} \quad \text{(D-Amino acid aminotransferase)} \quad (4)$$

(L-Amino acid aminotransferae)
L-Glutamate + Pyruvate ↔ 2-Oxoglutarate + L-Alanine (5)

D-Alanine + L-Glutamate ↔ L-Alanine + D-Glutamate (6)

The general metabolic pathways of D-amino acids are shown in Figure 1. There are four main metabolic pathways, which are characteristic of organisms.

1) Oxidative Deamination Pathway

Most of D-amino acids are first metabolized to the corresponding oxo analogues mainly by catalysis of D-amino acid oxidase in mammals, insects, other animals, and microorganisms as shown in the equation 1. The oxo acids formed are transaminated with L-amino acids to produce the L-amino acid counterparts by L-amino acid aminotransferases in mammals and microorganisms (equation 5). Alternatively, L-amino acids are reductively formed from the corresponding oxo acids by NAD(P)-dependent L-amino acid dehydrogenases such as glutamate dehydrogenase and alanine dehydrogenase in bacteria (equation 2). Glutamate dehydrogenase occurs in almost all organisms except for most *Bacillus* bacteria. D-Amino acids are indirectly converted to the corresponding L-enantiomers through these coupled enzyme reaction systems (equation 3).

Thus, several D-amino acids such as D-methionine are nutritionally utilized in many mammals. In fungi and mice, D-glutamate and D-aspartate, which are not the substrates of D-amino acid oxidase, are oxidatively deaminated to 2-oxoglutarate and oxaloacetate by FAD-dependent D-glutamate oxidase (EC 1.4.3.7) and D-aspartate oxidase (EC 1.4.3.1), respectively. Both oxo acids are metabolized directly, or after their enzymatic amination to the corresponding L-amino acids.

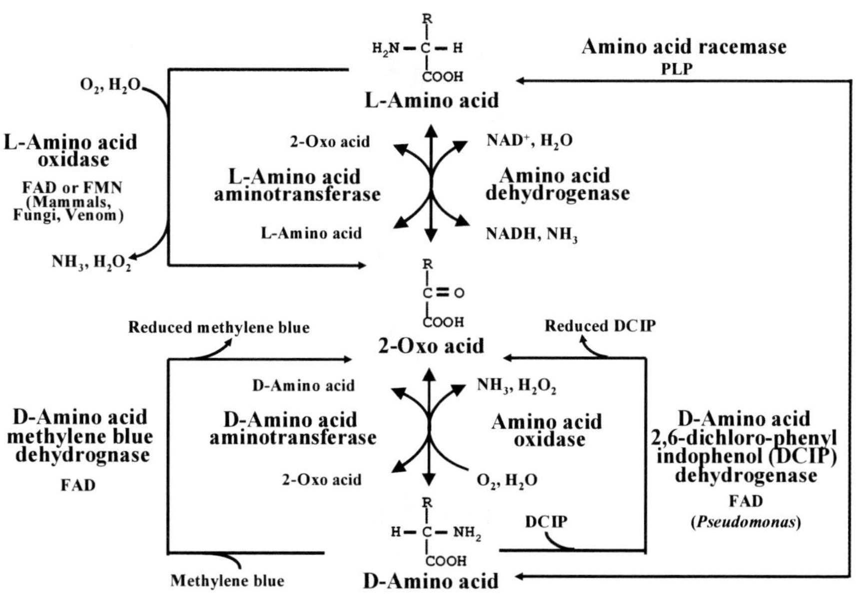

Figure 1. General metabolic pathways and enzymes participating in D-amino acid metabolism. Methalene blue and DCIP: artificial cofactors.

2) Dehydrogenation Pathway

D-Amino acid dehydrogenase (EC 1.4.00.1) irreversibly catalyzes the conversion of various D-amino acids to the corresponding oxo acids in the presence of methylene blue, 2,4-dichlorophenol indophenol and some other artificial electron acceptors. The oxo acids are aminated to the corresponding L-amino acids as described above (equations 2 and 5). This metabolic pathway is found mainly in bacteria.

3) Racemization Pathway

In bacteria, various D-amino acids are directly racemized or epimerized by amino acid racemases or amino acid epimerases, respectively, and the L-amino acids or *meso*-amino acids formed are metabolized by a variety of enzymes. Amino acid racemases are now found in fungi, plants and mammals, besides bacteria. Bacteria produce alanine racemase to supply D-alanine for the synthesis of peptidoglycan of cell walls. The activity is generally very low, but enough for the peptidoglycan biosynthesis on cell division. Most of amino acid racemases and amino acid epimerases show substantially absolute substrate specificity, but a few species of *Pseudomonas* bacteria produce racemases showing very low substrate specificities.

4) Transamination Pathway

D-Amino acid aminotransferase (D-alanine aminotransferase, EC 2.6.1.21), which is found in *Bacillus* species and some other bacteria, reversibly catalyzes transamination of various D-amino acids with oxo acids (equation 4). D-Amino acids formed by enzymatic racemization are transaminated with amino acceptor oxo acids to produce the corresponding oxo acids and D-amino acids.

Various D-amino acids are metabolized through their characteristic metabolic pathways. The metabolic pathways of a few D-amino acids are illustrated here. As shown in Figure 2, D-glutamate and D-glutamine are metabolized by various enzymes mainly in microorganisms. The incorporation of D-glutamate into the bacterial cell wall peptidoglycan through several enzyme reactions is found in every bacterium and is very important from the physiological viewpoint. D-Glutamate is metabolized in mammals in a confined way: it is intramolecularly dehydrated and cyclized to D-pyrolidone 5-carboxylate (D-oxoproline) by catalysis of D-glutamate cyclase.

D-Arginine is also metabolized chiefly in microorganisms (Figure 3), but hydrolyzed to D-ornithine and urea by mammalian D-arginase. Although the mammalian enzyme has not been purified, the enzyme was found also in bacteria, and characterized [15]. L-Arginase is the final enzyme of the urea cycle in ureotelic animals, and also plays an important role in regulation of arginine catabolism, affecting the metabolism of creatine, polyamines, proline and nitric oxide in various organisms. The bacterial D-arginase acts on not only D-arginine, but also 4-guanidinobutyrate, 3-guanidinopropionate and even L-arginine. Accordingly, the enzyme was regarded as a kind of guaninidobutyrase. D- and L-arginines are good substrates of D-amino acid oxidase and D-amino acid aminotransferase. 2-Oxoguanidinovalerate formed from D-arginine is transformed by Δ^1-pyrroline-2-carboxylase, and also by amino acid racemase of *Pseudomonas* bacteria to form the L-enantiomer.

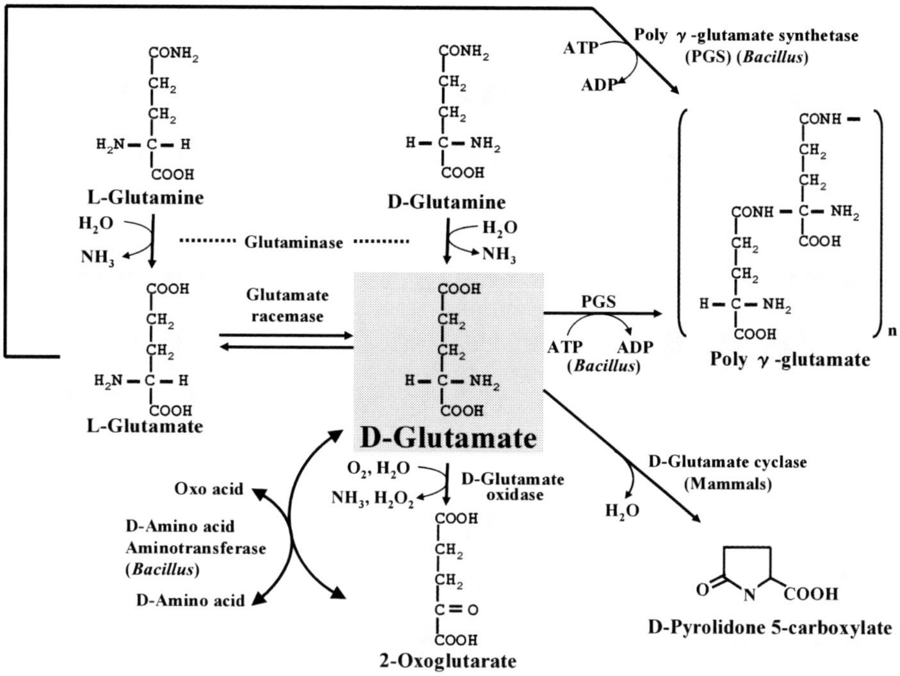

Figure 2. Metabolic pathways of D-glutamate.

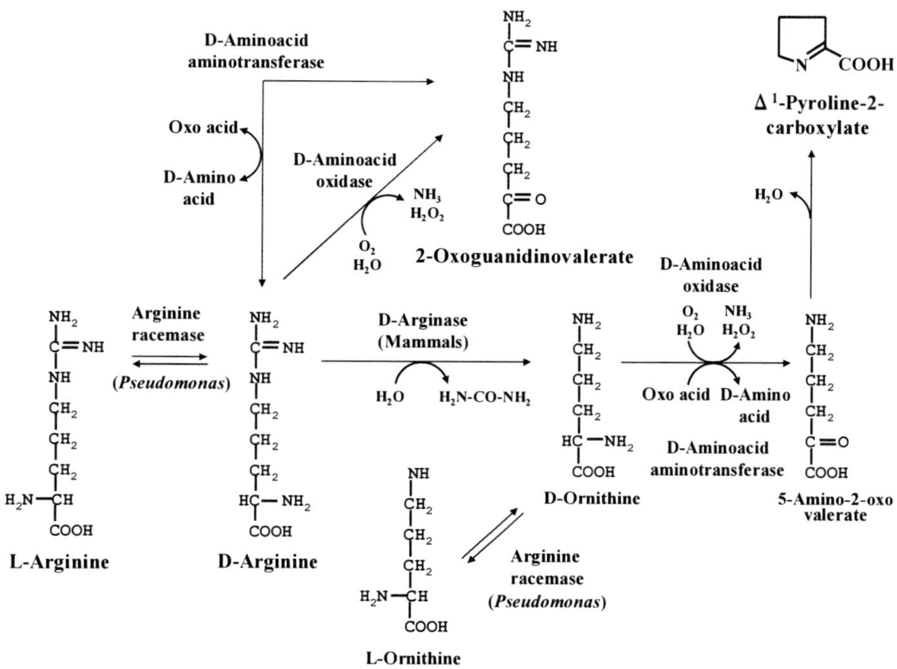

Figure 3. Metabolic pathways of D-arginine.

meso-2,6-Diaminopimelate is an essential constituent of cell wall peptidoglycan of some bacteria, and bacterially metabolized through lysine, L,L-2,6-diaminopimelate and L-2-amino-6-oxopimelate, which is spontaneously dehydrated to 2,3,4,5-tetrahydrodipicolate. However, the further metabolisms of L,L-2,6-diaminopimelate and 2,3,4,5-tetrahydrodipicolate are substantially unknown.

Figure 4 shows the metabolic pathways of D-lysine, which is converted to L-lysine, 6-amino-2-oxohexanate and D-2,5-diamainohexanoate by amino acid racemeases with low substrate specificity (or arginine racemase), D-amino acid aminotransferase and D-lysine 5,6-aminomutase, respectively. The information of further metabolisms of Δ^1-piperideine-2-carboxylate and D-Δ^1-pyrroline-2-methyl-carboxylate is not available. L-Lysine is decarboxylated to cadaverine by lysine decarboxylase (EC 4.1.1.18), transaminated with 2-oxoglutarate forming L-Δ^1-piperideine-6-caboxylate by lysine-6-aminotransferase (EC 2.6.1.36), which is oxidized to 5-aminovarelate by lysine 2-monooxygenase (1.13.12.2), or converted to β-lysine by lysine 2,3-aminomutase (EC 5.4.3.2) in bacteria.

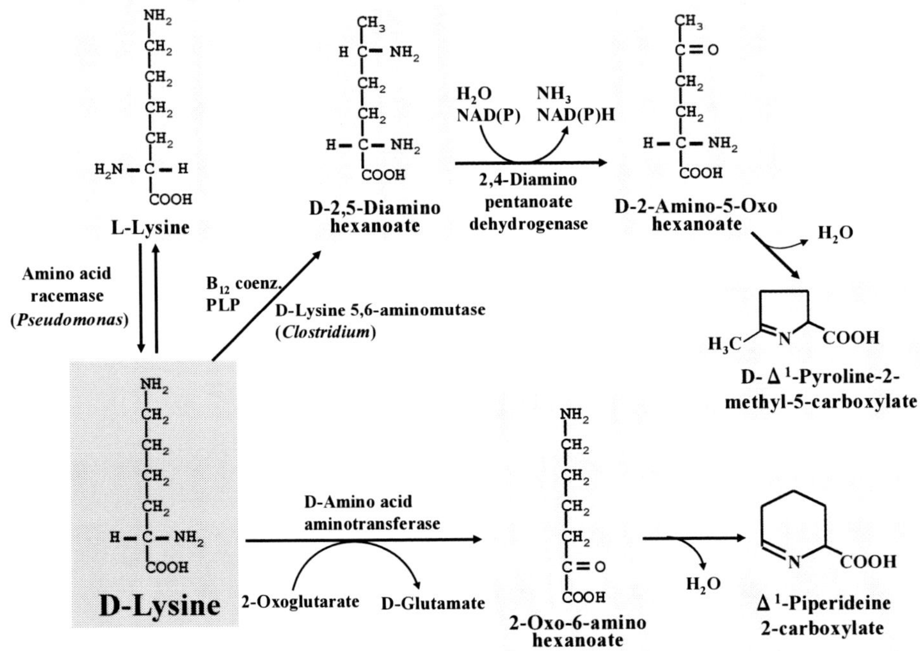

Figure 4. Metabolic pathways of D-lysine.

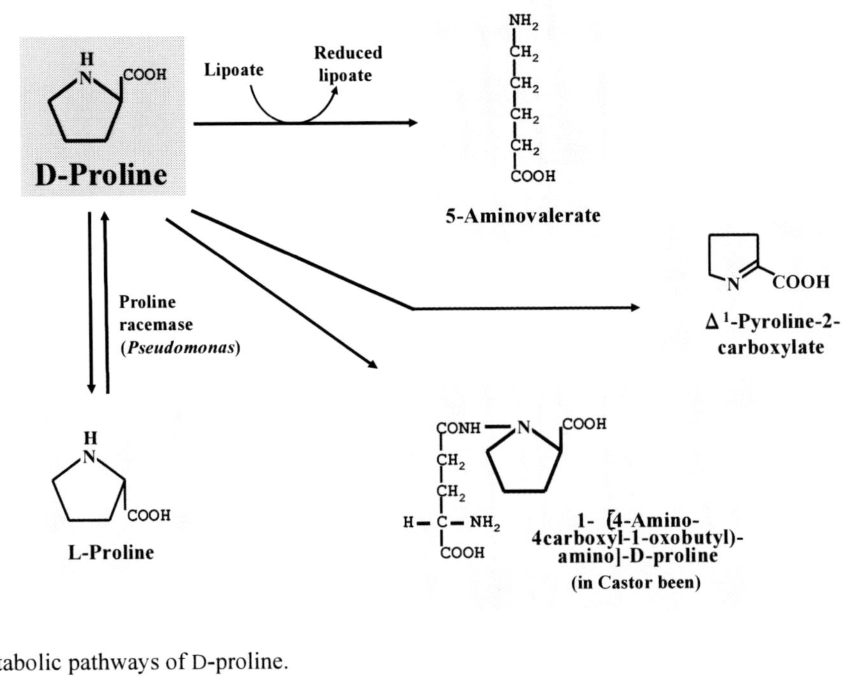

Figure 5. Metabolic pathways of D-proline.

Proline is not an amino acid, but a substituted amino acid from the chemical viewpoint. Accordingly, proline racemase is naturally independent of pyridoxal phosphate. D-Proline is metabolized by a specific enzyme, D-proline reductase to an achiral compound, 5-aminovalerate leading to glutarate (Figure 5).

D-Alanine plays a central role in the D-amino acid metabolism (Figure 6): it is a good substrate of two major enzymes participating in D-amino acid metabolism, alanine racemase and D-amino acid aminotransferase. Alanine racemase universally occurs in bacteria to supply D-alanine for the synthesis of peptidoglycan of bacterial cell walls. D-Amino acid aminotransferase is a sole enzyme to produce various D-amino acids from the corresponding oxo acids in the presence of an amino donor, and acts on D-alanine, D-glutamate or some other D-amino acids formed by amino acid racemases. In addition, D-alanine is an essential constituent of peptidoglycan of bacterial cell walls as described above.

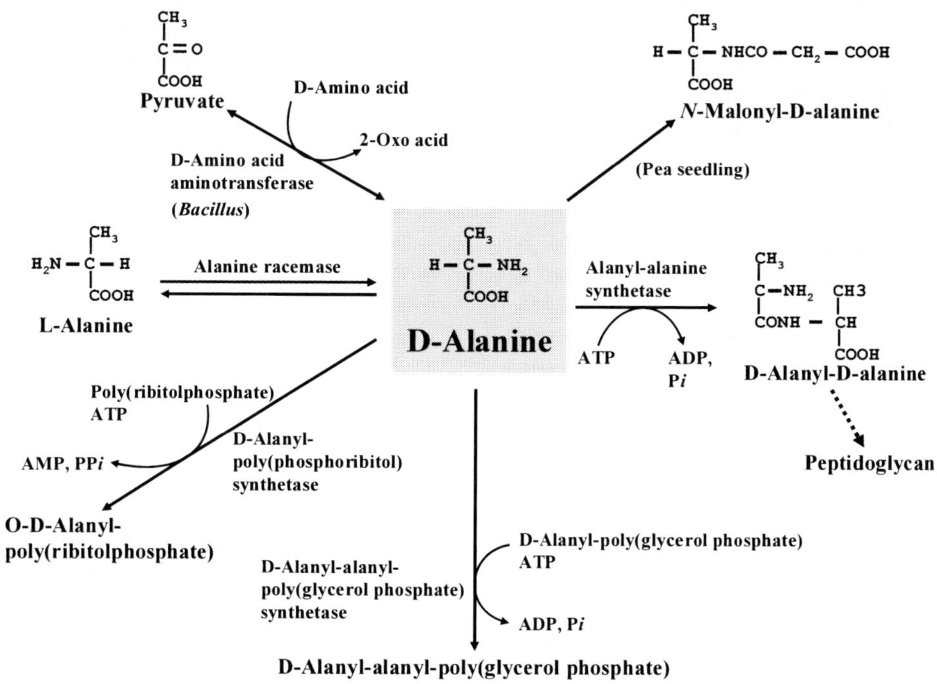

Figure 6. Metabolic pathways of D-alanine.

3. Enzymological Aspects of D-Amino Acid Metabolism

The typical enzymes participating in the D-amino acid metabolism are listed in Table 1. Only several metabolically important enzymes are described here, because it is beyond the scope of this article to review them in detail.

1) D-Amino acid oxidase requires FAD as a coenzyme, and catalyzes the following reaction. It has been studied in great detail. [16,17].

$$R\text{-}CH(NH_2)\text{-}COOH + O_2 + H_2O \rightarrow R\text{-}CO\text{-}COOH + NH_3 + H_2O_2 \quad (7)$$

However, hydrogen peroxide spontaneously reacts with oxo acids.

Table 1. D-Amino Acid Metabolizing Enzymes.

(1)	Amino Acid Racemases (Epimerases)
(2)	D-Amino Acid Aminotransferase
(3)	D-Serine Dehydratase
(4)	meso- α , ε -Diaminopimelate (DAP) Decarboxylase
(5)	meso- α , ε -DAP Epimerase
(6)	meso- α , ε -DAP D-Dehydrogenase
(7)	D-Amino Acid Oxidase
(8)	D-Glutamate Oxidase
(9)	D-Aspartate Oxidase
(10)	D -Glutamate Cyclase
(11)	D-Glutamyl Transpeptidase
(12)	D-Amino Acid Dehydrogenase
(13)	D-Arginase
(14)	D-Ala- D-Ala Synthetase
(15)	UDP-*N*-Acetylmuramoyl-L-Ala-D-Glu-L-Lys-D-Ala-D-Ala Synthetase
(16)	UDP-*N*^Acetylmuramoyl-L-Ala-D-Glu-*meso*-2,6-DAP Synthetase
(17)	UDP-*N*-Acetylmuramoyl-L-Ala-D-Glu-*meso*-2,6-DAP-D-Ala-D-Ala Synthetase
(18)	D-Ala-Ala-Poly(glycerophosphate) Synthetase
(19)	D-Ala-Poly(phosphoribitol)Synthetase
(20)	D-Amino Acid (D-Trp)Acetyltransferase
(21)	D-Lysine 5,6-Aminomutase
(22)	D-Ornithine 4,5-Aminomutase
(23)	D-Glutaminase
(24)	D-Proline Reductase (dithiol)
(25)	D-Proline Dehydrogenase (Dye-liked)

$$R\text{-}CO\text{-}COOH + H_2O_2 \rightarrow R\text{-}COOH + CO_2 + H_2O \quad (8)$$

(7) + (8)

$$R\text{-}CH(NH_2)\text{-}COOH + O_2 \rightarrow R\text{-}COOH + NH_3 + CO_2 \quad (9)$$

In the presence of catalase, which widely occurs in living systems, hydrogen peroxide is destroyed, and consequently the above oxidative deamination (equation 9) takes place *in vivo*. D-Amino acid oxidase is found in various eukaryotic organisms including mammals, insects, yeasts and fungi [18] and acts on various D-amino acids except acidic amino acids, D-lysine and a few others. The three-dimensional structure and reaction mechanism were intensively studied as described by Miura *et al.* [19]. The enzyme occurs in a variety of mammalian tissues, but the physiological substrate of the enzyme was unknown for a long time. Recently, it was shown that D-serine is probably the *in vivo* substrate of the enzyme, and is an endogenous ligand for the glycine site of the *N*-methyl-D-aspartate receptor to modulate neurotransmission [20]. Urai *et al.* [21] have shown the regional and cell type specific expression of D-amino acid oxidase gene in cultured rat astroocytes, and they give a detailed account of their study in this book.

D-Amino acid oxidase is produced also in eukaryotic microorganisms, in particular yeasts. Although it is supposed that the enzyme functions to catabolize D-amino acids in yeasts, the physiological role has not been clearly shown. *Candida lipolytica* and *C. tropoicalis* produce D-amino acid oxidase abundantly when they are cultivated in the presence of a large quantity of alkanes or alkanic acids, such as propionate, to form a peroxisome (microbody) showing a high activity of catalase [22]. However, the metabolic role of enzyme and the occurrence of D-amino acids in a peroxisome have not yet been studied.

2) D-Amino acid aminotransferase requires pyridoxal phosphate and pyridoxamine phosphate as a coenzyme, and occurs mainly in *Bacillus* bacteria. As stated above, its stereospecificity is absolute: D-amino acids are the exclusive amino donors and the products from amino acceptors. In contrast, the structural specificity is very low. D-Amino acid aminotransferase of *Bacillus* sp. YM-1 and branched-chain L-amino acid aminotransferase of *Escherichia coli* show a significant sequence homology, though are different from all other aminotransferases [22]. Both aminotransferases fall into the same subgroup of aminotransferases on the basis of the amino acid sequence [23]. The circular dichroism spectra of various aminotransferases show positive bands at the wavelengths where their internal Schiff bases absorb, but those of D-amino acid aminotransferases show negative ones.

We revealed that D-amino acid aminotransferase and branched-chain L-amino acid aminotransferase are unique in their stereospecific catalysis of *pro-R* hydrogen transfer between C-4' of the coenzyme moiety of ketimin Schiff base intermediate and C-1 of the amino acid moiety of aldimine Schiff base intermediate through the carbanionic intermediates (*pro-R* specific) in contrast to other various aminotransferases catalyzing the *pro-S* hydrogen transfer (Figure 7) [24]. Furthermore, the three dimensional structure of D-amino acid aminotransferase is unique: the amino acid residues including the catalytic active center base lysine 145 that are catalytically important are located on the *re*-face side of the plane of pyridoxal phosphate in contrast to aspartate aminotransferase whose catalytically important residues are found on the *si*-face side [25]. Accordingly, D-amino acid aminotransferae is peculiar in the substrate specificity, the circular dichroism, the amino acid sequence, the stereospecificity for the hydrogen transfer, and the stereostructure of the active center.

3) Amino acid racemases are unique in the substrate stereospecificity and mode of the reaction. They apparently catalyze a simple reaction, namely the removal of α-hydrogen from either enantiomer of a substrate amino acid, and subsequently a non-stereospecific reprotonation occurs spontaneously to attain equilibrium between both enantiomers. They are unique in lack of stereospecificity, because almost all the enzymes catalyze reactions with only one of an enantiomeric and diastereomeric pair of chiral compounds. Most require pyridoxal phosphate as a coenzyme, but some such as glutamate racemase and proline racemase require no coenzymes and no metals: they are diverse in the coenzyme requirement. Most of amino acid racemases show practically absolute substrate structural specificity. Alanine racemase acts on only alanine, and both enantiomers show substantially the same reactivity. In contrast, arginine racemase and amino acid racemase with low substrate specificity of *Pseudomonas* bacteria act on various amino acids except acidic and aromatic amino acids [26]. We revealed that alanine racemase shows the non-stereospecific hydrogen transfer: hydrogen is transferred on either *re*-face or *si*-face.

Glutamate racemase is a typical coenzyme-independent amino acid racemase. We characterized the enzyme from *Pediococcus pentosaceus*. The enzyme is composed of a single subunit with molecular weight of about 29,000, and acts on its γ-sulfinate analogue, homocysteine sulfinate as well, although very slowly. It contains an essential cysteine residue, which probably serves as a catalytic base to withdraw α-hydrogen from the substrate [27]. The enzyme shows no similarity to all other enzymes in the primary structure, but the structure of the active center is closely similar to that of the heme binding site of hemoglobin and myoglobin. Hemin stoichiometrically binds to the active center of the enzyme to inhibit the activity [28]. This finding is interesting from the viewpoint of molecular evolution of both proteins.

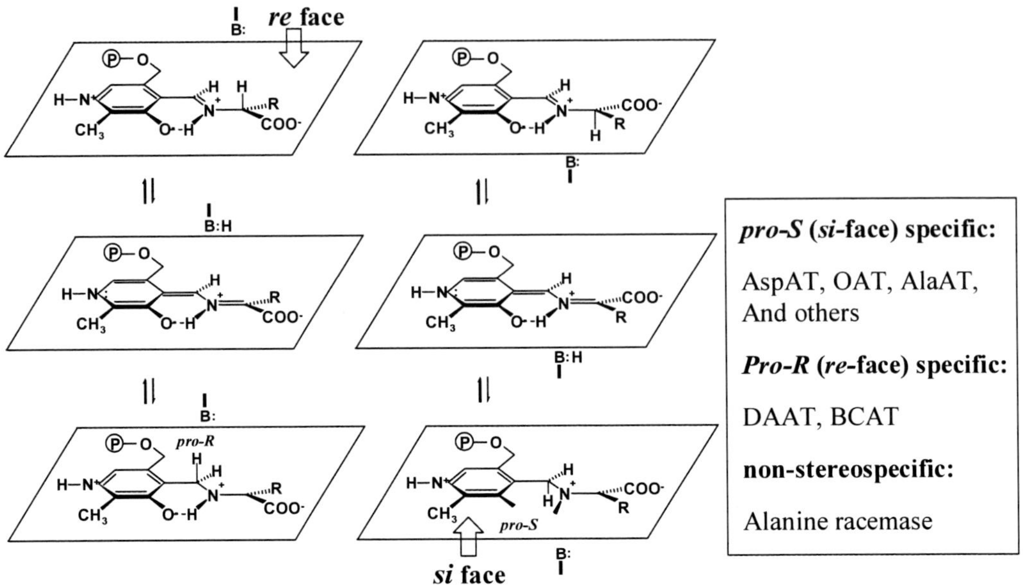

Figure 7. Stereochemistry of hydrogen transfer between the α-carbon of amino acid moiety of substrate-pyridoxal phosphate aldimine Schiff base and the carbon-4' of coenzyme moiety of substrate-pyridoxamine phosphate ketimine Schiff base. AspAT: aspartate aminotransferase; OAT: ornithine δ-aminotransferase; AlaAT: alanine aminotransferase; DAAT: D-amino acid aminotransferase; BCAT: branched-chain amino acid aminotransferase.

Poly-γ-glutamate is a unique γ-peptide composed of D-glutamate and L-glutamate and produced by several strains of *Bacillus subtilis,* and some others. The ratio of an amount of the D-enantiomer to that of the L-enantiomer ranges from 80 to 50% in the polymer produced by *B. subtilis* (natto), though it depends on the species and the growth conditions. We found the occurrence of two isozymes of glutamate racemase, Glr, which is involved in the poly-γ-glutamate synthesis, and YrpC participating in the synthesis of peptidoglycan of cell wall. Poly-γ-glutamate synthetase catalyzes synthesis of the polymer in the presence of ATP [29]. The γ-carboxyl group of a C-terminal glutamate residue of the growing poly-γ-glutamate is first activated to form a phosphate ester with hydrolysis of ATP. An amino group of a substrate glutamate attacks the activated carboxyl group as a nucleophile to elongate the polymer: this is an ATP and glutamate-dependent γ-ligase reaction. We showed that poly-γ-glutamate synthetase consists of three components. Component PgsB, which is located inside the membrane, catalyzes the ATP and glutamate-dependent γ-ligase reaction probably in association with PgsC. This occurs in the hydrophobic region of cell membrane. PgsA is not involved in the poly-γ-glutamate elongation directly, but serves as a kind of transporter and protects poly-γ-glutamate formed from the enzymatic and non-enzymatic depolymerization. The synthetase is peculiar in catalysis. The D,L-ratio of the product, poly-γ-glutamate is not regulated by the enzyme, but depends on the D,L-ratio of substrate glutamate controlled by glutamate racemase.

4. Perspective

As described here, D-amino acids have been extensively studied and significant information was obtained. However, the information was still fragmentary, and the biochemistry of D-amino acids has not been systematized. For example, even the occurrence of D-amino acids has not been fully examined in various organisms except the limited numbers of organisms, and the metabolism of D-amino acids has not been intensively studied in comparison with those of L-amino acids. Although several enzymes participating in D-amino acid metabolism have been studied in detail as stated above, other important enzymes such as mammalian D-glutamate cyclase and bacterial proline racemase have been little examined. There are many problems to be clarified in order to elucidate the biochemical and molecular biological aspects of D-amino acids. It is important to carry out the multidisciplinary study of D-amino acids. Advances in methodology will certainly give a clue to development of new aspects of D-amino acid biochemistry. More accurate and effective analytical procedures and instrumentation for specific determination of D-amino acids are needed to investigate D-amino acid incorporation into proteins more finely as reported by Masaki et al [14]. Commercially available D-amino acid preparations sometimes contain a trace amount of the L-enantiomers to prevent us from obtaining accurate results in the experiments with D-amino acids. Accordingly, cheaper supplies of the pure preparations of D-amino acids are needed.

The chemical, biochemical and astrobiochemical studies of amino acids and proteins certainly will stimulate the progress of D-amino acid biochemistry. The progress in D-amino acid research also will make important contributions to medicine, pharmaceutical science, molecular evolution and other related sciences.

REFERENCES

[1] Schulze, E., Bosshard, E. (1886) Untersuchungen Uber die Amidosauren, welche bei der Zersetzung der Eiweissstoffe durch Salzsaure und durch Barytwasser entstehen. *Z Physiol Chem, 10*, 134-145.

[2] Bada, J.L., Miller, S.L. (1987) Racemization and the origin of optically active organic compounds in living organisms. *Bio System, 20,* 21-26.

[3] Bonner, W. A., Blair, N.E., Lemmon, R. M. (1980) The radioracemization of amino acids by ionizing radiation: geochemical and cosmochemical implications. In: P.E. Hare, T.C., Hoering, K., King, Jr., *Biogeochemistry of amino acids.* Wiley, New York, 357-374.

[4] Bonner, W. A. (1998) Homochirality and Life: In: Jolles, P.(ed) *D-Amino Acids in Sequences of Secreted Peptides of Multicelullar Organisms.* Birkhauser Verlag, 159-188.

[5] Inoue, Y., Sugahara, N., Wada, T. (2001) Vital role of entropy in photochirogenesis. *Pure Appl Chem, 73,* 475-480.

[6] Kojo, S., Uchino, H., Yoshimura, M., Tanaka, K. (2004) Racemic D,L-asparagine causes enantiomeric excess of other coexisting racemic D,L-amino acids during recrystallization: a hypothesis accounting for the origin of L-amino acids in the biosphere. *Chem Commun, 19*, 2146-2147.

[7] Nagata, Y., Tanaka, K., Iida, T., Kera, Y., Yamada, R., Nakajima, Y., Fujiwara, T., Fukumori, Y., Yamanaka, T., Koga, Y., Tsuji, S., Kawaguchi-Nagata, K. (1999) Occurrence of D-amino acids in a few archaea and dehydrogenase activities in hyperthermophile *Pyrobaculum islandicum. Biochim biophys acta-protein struct molec enzymol,, 1435*, 160-166.

[8] Meister, A. (1965) *Biochemistry of the Amino Acids* (2nd edition) 113-118, New York: Academic Press.

[9] Jolles, P. (ed) (1998) *D-Amino Acids in Sequences of Secreted Peptides of Multicellular Organisms*, Basel, Birkhauser Verlag.

[10] Fujii N., Saito, T., (2004) Homochirality and life. *Chem Rec, 4,* 267-278.

[11] Roher, A. E., Lowenson, J. D., Clarke, S., Wolkow, C., Wang, R., Cotter, R. J., Reardon, I. M., Zurcherneely H. A., Heinrikson, R. L., Ball, M. J., Greenberg, B. D., (1993) Structural alterations in the peptide backbone of beta-amyloid core protein may account for its deposition and stability in Alzheimers-disease. *J Biol Chem, 268*, 3072-3083.

[12] Kim, M., Okajima, T., Kishishita, S., Yoshimura, M., Kawamori, A., Tanizawa, K., Yamaguchi, H. (2002) X-ray snapshots of quinone cofactor biogenesis in bacterial copper amine oxidase. *Nature Struct Biol, 9*, 591-596.

[13] Matsunami, H., Okajima, T., Hirota, S., Yamaguchi, H., Hori, H., Kuroda, S., Tanizawa, K. (2004) Chemical rescue of a site-specific mutant of bacterial copper amine oxidase for generation of the TOPA quinone cofactor. *Biochemistry, 43*, 2178-2187.

[14] Takayama, T., Ogawa, T., Hidaka, M., Shimizu, Y., Ueda, T., Masaki, H. (2005) Esterification of *Escherichia coli* tRNAs with D-histidine and D-lysine by aminoacyl-tRNA synthetases. *Biosci Biotechnol Biochem, 69*, 1040-1041.

[15] Arakawa, N., Igarashi, M., Kazuoka, T., Oikawa, T., Soda, K. (2003) D-Arginase of *Arthrobacter* sp KUJ 8602: characterization and its identity with Zn^{2+}-guanidinobutyrase. *J Biochem, 133*, 33-42.

[16] Yagi, K. (1971) D-Amino acid oxidase and its complexes *Methods Enzymol, 17B*, 608-622.

[17] Pilone, M.S. (2000) D-Amino acid oxidase new findings *CMLS, Cel. Life Sci, 57*, 1732-1747.

[18] Friedman, M. (1999) Chemistry, nutrition and microbiology of D-amino acids. *J Agric Food Chem, 47*, 3457-3479.

[19] Miura, R., Setoyama, C., Nishina, Y., Shiga, K., Mizutani, H., Miyahara, I, et al (1997) Structural and mechanistic studies on D-amino acid oxidase-substrate complex: implications of the crystal structure of enzyme-substrate analog complex. *J Biochem, 122*, 825-833.

[20] Mothet, J.-P., Parent, A. T., Wolosker, H., Brady, R. O., Linden, D. J., Ferris, C. D., Rogawski, M. A., Snyder, S. H. (2000) D-Serine is an endogenous ligand for the glycine site of the *N*-methyl-D-aspartate receptor *Proc Natl Acad Sci USA, 97*, 4926-4931.

[21] Urai, Y., Jinnouchi, O., Kwak, K.T., Suzue, A., Nagahiro, S., Fukui, K. (2002) Gene expression of D-amino acid oxidase in cultured rat astrocytes: regional and cell type specific expression. *Neurosci Lett, 324*, 101-104.

[22] Kawamoto, S., Kobayashi, M., Tanaka, A., Fukui, S. (1977) Studies on utilization of hydrocarbons by microorganisms. Production of D-amino-acid oxidase by Candida tropicalis. *J Ferment Technol, 55*, 13-18.

[23] Mehta, P.K., Hale, T.I., Christen, P. (1989) Evolutionary relationships among aminotransferases - tyrosine aminotransferase, histidinol-phosphate aminotransferase, and aspartate-aminotransferase are homologous proteins. *Eur J Biochem, 186*, 249-253.

[24] Yoshimura T., Nishimura, K., Ito, J., Esaki, N., Kagamiyama, H., Manning, J. M., Soda, K. (1993) Unique stereospecificity of D-amino acid aminotransferase and branched-chain L-amino acid aminotransferase for C-4'hydrogen transfer of the coenzyme. *J Amer Chem Soc, 115*, 3897-3900.

[25] Sugio, S., Petsko, G.A. Manning, J.M., Soda, K., Ringe, D. (1995) Crystal-structure of a D-amino-acid aminotransferase-how the protein controls stereoselectivity. *Biochemistry, 34*, 9661-9669.

[26] Esaki, N., Soda, K. (1995) Isomerizations in *"Enzymatic Catalysis in Organic Synthesis*" (Ed. Drauz, K., Waldaman, H.) Vol.II, 808-849.

[27] Choi, S. -Y, Esaki, N., Yoshimura, T., Soda, K. (1992) Reaction mechanism of glutamate racemase, a pyridoxal phosphate-independent amino acid racemase. *J Biochem, 112*, 139-142.

[28] Choi, S.-Y, Esaki, N., Ashiuchi, M., Yoshimura, T., Soda, K. (1994) Bacterial glutamate racemase has high sequence similarity with myoglobin and forms an equimolar inactive complex with hemin. *Proc Natl Acad Sci USA, 91*, 10144-10147.

[29] Ashiuchi, M., Soda, K, Misono, M. (1999) A poly-γ-glutamate synthetic system of *Bacillus subtilis* IFO 3336: gene cloning and biochemical analysis of poly-γ-glutamate produced by *Escherichia coli* clone cells. *Biochem Biophys Res Commun, 236*, 6-12.

SECTION 2. ANALYTICAL METHODS OF THE DETECTION OF D-AMINO ACIDS

In: D-Amino Acids:
A New Frontier in Amino Acid and Protein Research, pp.17-23
ISBN 1-60021-075-9

Chapter 2.1

CHROMATOGRAPHIC DETERMINATION OF FREE D- AND L-AMINO ACIDS IN MARINE ALGAE

Eizo Nagahisa* and Takehiko Yokoyama
School of Fisheries Sciences, Kitasato University, Japan.

Keywords: D-aspartate, D-alanine, marine algae, phytoplankton, diatom

1. INTRODUCTION

It has been generally accepted that D-amino acids are uncommon in higher organisms and that L-amino acids are of primary importance biologically. However, in recent years, evidence that D-amino acids also play an important role in higher organisms has increased. Furthermore, as a result of bacterial metabolism and chemically induced racemization, free D-amino acids are common constituents of microbially fermented foods, beverages, and various other processed foods. Microbially and chemically induced racemization of amino acids in food proteins and the nutritional utilization and safety of D-amino acids have been the focus of increasing attention.

The discovery of naturally occurring D-amino acids in organisms, in both free and combined forms, and the elucidation of their physiological functions are dependent on the availability of reliable, sensitive and rapid analytical methods. Numerous HPLC methods, more than can be cited here, have been used for the detection of D-amino acids in organisms and foods. However, it is generally believed that the main factor in the appearance of D-amino acids in higher organisms is racemization of L-amino acids by endogenous racemase enzymes. In fact, alanine racemase has been found in several higher organisms, including plants (Gamburg and Rekoslavskaya,1992) and marine invertebrates (Matsushima *et al.*,1982, Fujita *et al.*,1997, Shibata *et al.*,2000, Yoshikawa *et al.*,2002), in addition to microorganisms. Recently, we found an alanine racemase in the marine diatom *Thalassiosira* sp. (Yokoyama *et al.*,2005).

In this report, we introduce a new method for the detection of D-amino acids and present the results of this analysis of D-amino acids in marine macro- and micro-algae.

* Correspondence concerning this article should be addressed to Eizo Nagahisa, E-mail: nagahisa@kitasato-u.ac.jp, FAX: +81-0192-44-3930.

2. Materials and Methods

Materials

Instruments

High Performance Liquid Chromatograph (HPLC): with a fluorescence detector and an integrator (Jasco LC-800 series, Japan)

Column: Reversed phase column (TSK-ODS 80T, 4.6x250 mm, Tosoh, Japan)

Homogenizer: Polytron homogenizer (PT10135, Switzerland) and Ultrasonic homogenizer (Tomy UD200, Japan)

Centrifuge: High speed cooling centrifuge (KUBOTA 7930, Japan) and Table top centrifuge (Tomy LC-120, Japan)

Thermo-incubator: with illumination lamp (EYELA MTI-201B, Japan)

Reagents

The following reagent grade chemicals were purchased from Wako Pure Chem. Ind. (Japan): DL-amino acids, *o*-phthalaldehyde (OPA), *N*-acetyl-L-cysteine (NAC) and sodium borate. All other chemicals were analytical grade reagents.

Methods

Sources of Algae

Most of the macro-algae were collected at Okirai Bay (Iwate Prefecture, Japan). The freshwater algae, *Botrydiopsis alpina*, *Chlorella pyrenoidosa*, *Chlorella vulgaris* and *Scenedesmus obliquus*, were obtained from the Chlorella Kogyo Co. (Tokyo, Japan). Marine diatoms were isolated by the capillary method, and inoculated singly into each medium (Kotaki *et al.*,2000). Axenic strains were isolated by the method of Douglas and Bates (1992) and were cultured in germ-free conditions. *Asterionella* sp. and *Thalassiosira* sp. were cultured in Tl/seawater medium (Ogata *et al.* 1987) at 15°C, and *Nitzschia navis-varingica* and *Pseudonitzschia pungens* were cultured in an f/2/seawater medium (Guillard, 1983) at 20°C under an illumination intensity of 60-70 μE, with a 16 h light:8 h dark cycle, using cool-white fluorescent lamps. All species of microalgae were harvested after about two weeks in culture, when the population had reached the stationary phase.

Sample Preparation

The macro-alga samples (whole tissue wet weight 20 g) were rinsed well with sterilized artificial seawater (Aquamarine, Yasu Chem. Co., Japan). Then the macro-alga was finely sliced and homogenized with 3 volumes (w/v) of 99% ethanol using the Polytron homogenizer. The homogenate was centrifuged at 10,000×g at 4°C for 20 min. The precipitate was again extracted with 3 volumes of 80% ethanol and this process was repeated 3 times. The supernatants were combined and evaporated to dryness under reduced pressure at 40-45°C. The residue was dissolved in a small amount of distilled water and rinsed in diethyl ether to remove pigments and fatty materials. This aqueous extract was evaporated to dryness and dissolved in a small amount of deionized water.

Micro-algal cells were collected from the culture medium (500 mL) by centrifugation (400×g, 5 min), and the resulting cell pellet (about 10^9 cells) was rinsed three times with 5 volumes (w/v) of fresh medium. The pellet was then homogenized with 3 volumes of 99% ethanol using an ultrasonic homogenizer, 3 times for 2 min at 4°C. The homogenate was centrifuged at 10,000×g at 4°C for 20 min. The precipitate was then extracted using the same procedures described above for the macro-alga. Both extracts from macro- and micro-alga were stored at –30°C until the HPLC analysis was conducted.

Derivatization of Amino Acids

Each sample extract or standard DL-amino acid solution (12.5 μmol/L in 0.1 N HCl) was passed through a Millipore filter (Millex-LG; 0.20 μm pore size). Twenty μL of an OPA/NAC solution (8 mg of OPA and 10 mg of NAC in 1mL of methanol) and then 70 μL of 0.08 M sodium borate was mixed with 10 μL of the filtrate in a 1 mL vial for HPLC.

HPLC Analysis

After the derivatization reaction was incubated for 2 min, 10 μL of derivative solution was injected onto the column and analyzed by HPLC.

Chromatography was carried out automatically by the method of Nimura *et al.* (1986), with the alterations described here. Samples were eluted with a mixture of solvent A (50 mM sodium acetate buffer at pH 5.35) and solvent B (80% methanol/solvent A) at 40°C; a flow rate of 1.0 mL/min. The excitation and emission wavelengths for fluorescence detection were 348 nm and 450 nm, respectively. The elution gradient was set as follows: 0-50 min, 0-40% solvent B in solvent A; 50-60 min, 40-60% solvent B in solvent A; and 60-85 min, 60-100% solvent B in solvent A.

Calculations

$$C_{HW} = \frac{\text{peak area of standard amino acid (AA)}}{\text{content of the standard AA (pmol) injected onto column}}$$

$$\text{AA content } (\mu\,\text{mol/g}) = \frac{\text{peak area} \times \text{total derivative vol.} (100\,\mu\,\text{L}) \times \text{total extract vol.} (\text{mL}) \times \text{dilution}}{C_{HW} \times \text{injection vol.} \times \text{extract vol.} (10\,\mu\,\text{L}) \times \text{weight of sample alga (g)} \times 10^{6}}$$

Identification of D-Aspartate and D-Alanine Peaks

The peak fraction corresponding to D-aspartate isolated from the algal extracts by HPLC was collected and concentrated with an evaporator, and then analyzed again by HPLC. The peak corresponding to D-alanine was identified by treatment with D-amino acid oxidase. The reaction mixture consisted of 0.1 mL of algal extract, 100 μL of FAD, and 0.2 U of D-amino acid oxidase (from porcine kidney, Biozyme Lab., Blaenavon, UK) in 50 mM Tris-HCl buffer, pH 8.0, in a final volume of 1 mL. The reaction was started by adding the enzyme solution to the algal extract at 40°C. After incubating for 60 min, the reaction was stopped by heating for 3 min at 100°C, and the reaction mixture was then passed through a Millipore filter. The chromatographic conditions were the same as those described for Figure 1, with the exception that the elution gradient was: 0-10 min, 0-15% solvent B; 10-30 min, 15-25% solvent B; and 30-50 min, 25-30% solvent B in solvent A.

3. Results and Discussion

Chromatograms of Standard L- and D-Amino Acids and Sample Extract

Chromatograms of standard L- and D-amino acids are shown in Figure 1. The enantiomeric separation of some amino acids derivatives, such as cysteine, histidine, lysine, methionine, phenylalanine, proline and tryptophan was not possible by this analytic method.

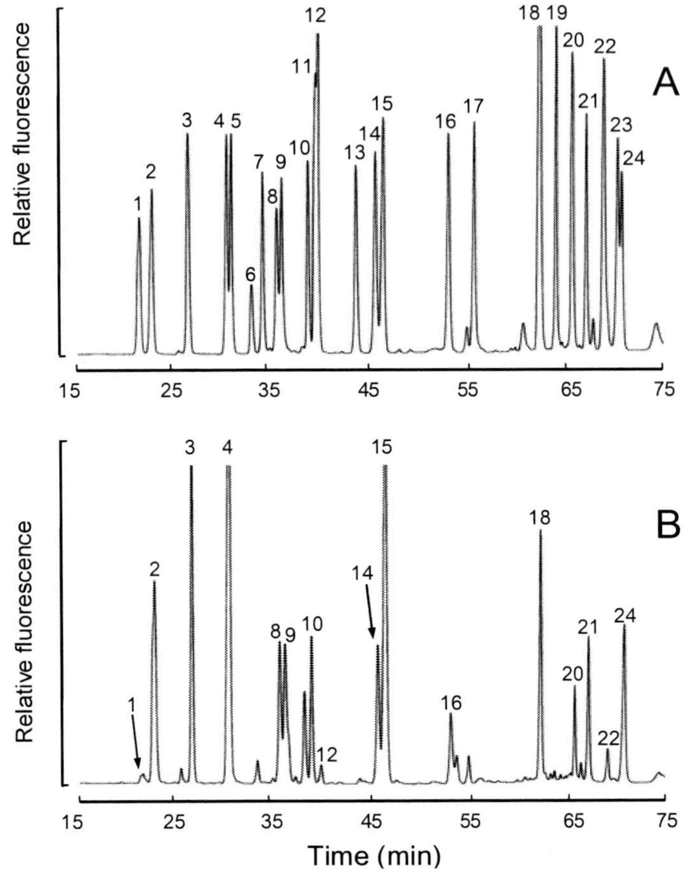

Figure 1. Typical HPLC Chromatograms of (A) the Converted D- and L-Enantiomers of Standard Amino Acids, and (B) an Extract of the Marine Diatom *Asterionella* sp. The number peaks correspond to: *1,* D-Asp; *2,* L-Asp; *3,* Ser; *4,* L-Glu; *5,* D-Glu; *6,* His; *7,* D-Thr; *8,* Gly; *9,* L-Thr; *10,* L-Arg; *11,* D-Arg; *12,* Tau; *13,* β-Ala; *14,* D-Ala; *15,* L-Ala; *16,* L-Tyr; *17,* D-Tyr; *18,* Met + L-Val; *19,* D-Val; *20,* Phe; *21,* L-Ile; *22,* D-Ile + Lys; *23,* D-Leu; *24,* L-Leu

The chromatographic pattern of the isolated, refractionated D-aspartate samples showed a single peak with the same retention time as the standard D-aspartate (Figure 2B). The chromatographic pattern of a sample extract treated with D-amino acid oxidase lacked the D-alanine peak, as shown in Figure 2D.

These results show that the peak 1 and peak 14 in Figure 1B is D-aspartate and D-alanine, respectively.

Applications

Results of the analysis of marine macro- and micro-algal extracts are shown in Table 1 and Table 2, respectively. Free D-aspartate and D-alanine were detected in both micro- and macro-algae, but other D-amino acids were not detected.

In macro-algae, D-aspartate was detected in small amounts, or not at all, in *Chlorophyta,* while the amino acid was strongly detected in most species of *Phaeophyta* and *Rhodophyta* (6 and 8 species, respectively). The content of D-aspartate was higher in *Phaeophyta* than in *Rhodophyta.* In contrast, the D-alanine content in *Chlorophyta* and *Rhodophyta* was generally higher than that of D-aspartate. The D-alanine content in most species belonging to the genus *Phaeophyta* was approximately the same level as D-aspartate.

Table 1. Distribution of D and L-Amino Acids in Marine Macroalgae*.

Species	Asp (nmol/g wet wt.)		(%)	Ala (nmol/g wet wt.)		(%)
	D	L	D/(D+L)	D	L	D/(D+L)
Chlorophyta						
Monostroma nitidum				62	550	10
Ulva pertusa	Tr	120	—	38	590	6.1
Enteromorpha compressa	2.2	12	15	40	581	6.4
Enteromorpha linza	2.5	12	17			
Enteromorpha intestinalis	ND	7.6	—	38	1085	3.4
Chaetomorpha moniligera	ND	98.6	—	15	39	28
Codiu fragile	ND	43.2	—			
Phaeophyta						
Heterochordaria abietina				83	733	10
Scytosiphon lomentarius				40	216	16
Lamiraria japonica	5.9	37	14	ND	460	—
Costaria costata	66	37	64	13	83	14
Undaria pinnatifida	12	30	29	ND	11500	—
Hizikia fusiformis	98	89	52	96	588	14
Sargassum fulvellum	1.2	6.6	15	65	183	26
Sargassum yezoensis	49	57	46			
Sargassum nigrifolium				78	255	23
Coccophora langsdorfii				68	2300	2.9
Rhodophyta						
Grateloupia filicina	3.4	38	8.2			
Gloiopeltis furcata	5.5	75	6.8	44	625	6.6
Chondrus ocellatu	ND	2.6	—	ND	731	—
Caroopeltis affinis	3	36	7.7			
Chondrus yendoi	9.0	38	19			
Rhodymenia palmata	Tr	106	—			
Lomentaria catenata	5.5	33	14			
Nemalion vermiculare				ND	203	—
Neodilsea yendoana				ND	652	—
Ahnfeltia paradoxu				39	80	33
Rhodomela larix	66	201	25			

Tr: Below 1.0 nmol/g wet wt.; ND: not detect; Blank column: Not determine.*: These data were quoted from our previous papers (Nagahisa *et al.*, 1992, 1995) with some alteration.

In micro-algae, D-aspartate was detected in all species of freshwater algae and marine diatoms examined, while D-alanine was detected only in the diatoms (Table 2). L-Isomers were detected in all specimens examined.

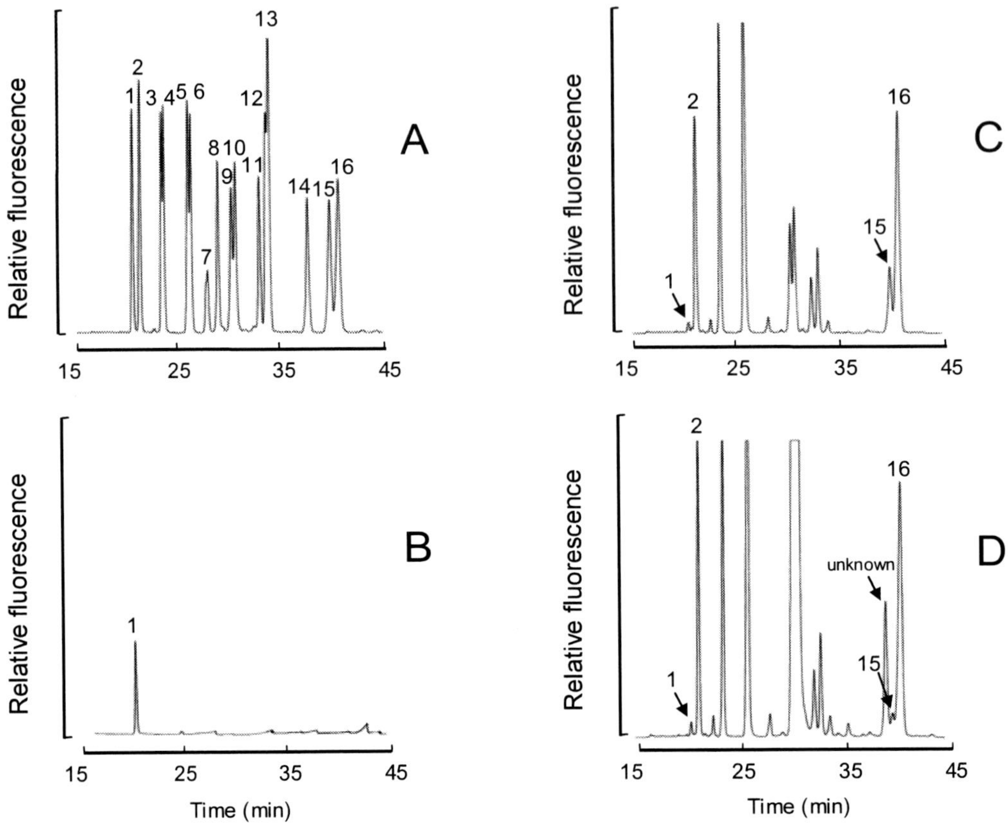

Figure 2. HPLC Chromatograms of (A) Standard DL-Amino Acids, (B) the Peak Corresponding to D-Aspartate Isolated from the Extract of *Asterionella* sp., and the Extract of *Asterionella* sp. before (C) and after (D) treatment with D-Amino Acid Oxidase. The number peaks correspond to *1*, D-Asp; *2*, L-Asp; *3*, D-Ser; *4*, L-Ser; *5*, L-Glu; *6*, D-Glu; *7*, His; *8*, D-Thr; *9*, Gly; *10*, L-Thr; *11*, L-Arg; *12*, D-Arg; *13*, Tau; *14*, β-Ala; *15*, D-Ala; *16*, L-Ala

Table 2. Distribution of D- and L-Amino Acids in the Microalgae*.

Species	Asp (nmol/10^6 cells)		(%)	Ala (nmol/10^6 cells)		(%)
	D	L	D/(D+L)	D	L	D/(D+L)
Freshwater alga						
*Botrydiopsis alpina**	Tr	0.08	—	ND	0.19	—
*Chlorella pyrenoidosa**	0.01	0.12	7.69	ND	0.77	—
*Chlorella vulgaris**	0.03	0.16	15.79	ND	2.11	—
*Scenedesmus obliquus**	0.01	0.10	9.09	ND	0.74	—
Marine diatom						
Asterionella sp.	0.09	4.77	1.85	0.40	5.19	7.16
Nitzschia navis-varingica	Tr	4.87	—	0.27	8.45	3.10
Pseudo-nitzschia pungens	0.10	2.05	4.65	0.24	3.61	6.23
Thalassiosira sp.	0.23	3.80	5.71	0.31	12.3	2.46

Tr, trace < 0.01; ND, not detected. ; These data were quoted from our previous paper (Yokoyama *et al.*, 2003)
*Obtained from Chlorella Kogyo Co. Ltd. (Tokyo Japan).
Each value for a marine diatom is the mean obtained from three or four samples taken from the respective cultures, and each value for a freshwater alga was obtained from one sample per culture. Each sample was collected after the alga had been cultured for 2 weeks (stationary phase).

REFERENCES

Douglas D.J, Bates S.S. (1992) Production of domoic acid, a neurotoxic amino acid, by an axenic culture of the marine diatom *Nitzschia pungens f. ultiseries* Hasle. *Can. J. Fish. Aquat. Sci., 49*:85-90

Fujita E., Okuma E, Abe H. (1997) Occurrence of alanine racemase in crustaceans and the changes of the properties during seawater acclimation of crayfish. *Comp. Biochem. Physiol.116A*:83-87

Gamburg K.Z.and Rekoslavskaya N.I.(1992) Formation and functions of D-amino acids in plants. *Siberian Institute of Plant Physiology and Biochemistry, 38(6)*:1236-1246

Guillard, R.R.L.(1983) Culture of phytoplankton for feeding marine invertebrates. *Culture of Marine Invertebrates*, 108-132

Kotaki Y., Koike K., Yoshida M., Chu VT, Nguyen TMH, Nguyen CH, Fukuyo Y., Kodama M. (2000) Domoic acid production in *Nitzschia* sp.(Bacillariophyceae) isolated from a shrimp-culture pond in Do Son, Vientnam. *J.Phycol.36*:1057-1060

Matsushima O., Katayama H., Yamada K., Kado Y.(1984) Occurrence of free D-alanine and alanine racemase activity in bivalve molluscs with special reference to intracellular osmoreguration. *Mar. Biol. Lett, 5*:217-225

Nagahisa E., Kan-no N., Sato M., Sato Y. (1992) Occurrence of free D-aspartic acid in marine macroalgae. *Biochem. Int., 28*:11-19

Nagahisa E., Kan-no N., Sato M., Sato Y. (1995) Occurrence of free D-alanine in marine macroalgae. *Biosci. Biotechnol. Biochem., 59*:2176-2177

Nimura N., Kinoshita T. (1986) *o*-Phthalaldehyde-*N*-acetyl-L-cysteine as a chiral derivatization reagent for liquid chromatographic optical resolution of amino acid enantiomers and its application to conventional amino acid analysis. *J. Chromatogr., 352*:169-177

Ogata T., Ishimaru T., Kodama M. (1987) Effect of water temperature and light intensity on growth rate and toxicity change in *Protogonyaulax tamarensis*. *Mar. Biol. 95*:217-220

Shibata K., Shirasuna K., Motegi K., Kera Y., Abe H., Yamada R. (2000) Purification and properties of alanine racemase from crayfish *Procambarus clarkii*. *Comp. Biochem. Physiol.126B*:599-608

Yokoyama T., Kan-no N., Ogata T., Kotaki Y., Sato M., Nagahisa E. (2003) Presence of free D-amino acids in microalgae. *Biosci. Biotechnol. Biochem*. 67(2):388-392

Yokoyama T., Tanaka Y., Sato M., Kan-no N., Nakano T., Yamaguchi T., Nagahisa E. (2005) Alanine racemase activity in the microalga *Thalassiosira* sp. *Fisheries Sci. 71*:924-930

Yoshikawa N., Dhome N., Takio K., Abe H. (2002) Purification, properties, and partial amino acid sequences of alanine racemase from the muscle of the blacktiger prawn *Penaeus monodon*. *Comp. Biochem. Physiol.*, 133B:445-453

In: D-Amino Acids:
A New Frontier in Amino Acid and Protein Research, pp. 25-32 ISBN 1-60021-075-9

Chapter 2.2

DETERMINATION OF ACIDIC D-AMINO ACIDS AND THEIR *N*-METHYL DERIVATIVES IN BIVALVES AND OTHER MOLLUSKS

Ryo-hei Yamada[1,*†], Yoshio Kera[1,‡], Shouji Takahashi[1], Takashi Watanabe[2] and Kimihiko Shibata[3]

[1]Department of Environmental Systems Engineering, Nagaoka University of Technology, Japan;
[2]Department of Chemical Engineering, Ichinoseki National College of Technology, Japan;
[3]Department of Chemistry and Biochemistry, Fukushima National College of Technology, Japan.

Keywords: (D and D)-aspartate, *N*-methyl-(D and L)-aspartate, *N*-methyl-(D and L)-glutamate, mollusks *Scapharca broughtonii.*

1. INTRODUCTION

Various bivalves as well as other invertebrates contain high levels of D-amino acids [1,2], of which D-aspartate is one of the most abundant. Besides, *N*-methyl-D-aspartate (NMDA), a well known neuroexitatory comopound that was regarded as an artificial compound in the past has been first isolated from the bivalve *Scapharca broughtonii* [3], demonstrating its natural occurrence. We have examined bivalves and other invertebrates for the presence of acidic D-amino acids including D-aspartate and their *N*-methyl derivatives, and also their L-enatiomers as well. Typical examples are shown here.

* Present address: 39-703 Kowata-Okurayama, Uji, Kyoto 611-0002, Japan. E-mail: ich19480@nifty.com; FAX: +81-774-32-4241.

† Correspondence concerning this article should be addressed to Ryo-hei Yamada at the present address.

‡ Reprint requests should be addressed to Yoshio Kera, Department of Environmental Systems Engineering, Nagaoka University of Technology, Japan. E-mail: yoshkera@vos.nagaokaut.ac.jp; FAX: +81-258-47-9634.

2. MATERIALS

2.1. Animals

Live specimens of the following animals were purchased at local fish markets in Niigata Prefecture, Japan: (a) mollusks Bivalvia: ark shell (or blood shell) *S. broughtonii*, half-crenate ark shell *Scapharca subcrenata*, asian hard clam *Meretrix lusoria*, blue mussel *Mytilus edulis*, sakhalin surf clam *Pseudocardium sachalinensis*, ezo giant scallop *Patinopecten yessoensis*, oyster *Crassostrea nippona* and tellins *Peronidia venulosa*, (b) mollusks Gastropoda: horned turban shell *Batillus cornutus*, whelk *Buccinum striatissimum*, ezo whelk *Buccinum middendorffi* and ezo neptune whelk *Neptunea polycostata*, (c) mollusks Cephalopoda: octopus *Paroctopus conispadiceus* and pacific flying squid *Todarodes pacificus*, (d) prochordates Urochordata: sea-squirt *Halocynthia roretzi*. Live specimens of echinoderms Echinoidea: sea urchin *Anthocidaris crassispina* were collected from a beach in Niigata Prefecture, Japan. The tissues of the animals were removed and washed with ice-cold saline, and were stored at -30°C until use. The foot and mantles of *S. broughtonii* were further divided: the foot was cut into two parts, the more intensely colored outside part (about 43% of the total mass) and the less intensely colored inside part [4]. The mantles were also divided into two parts, the circular fringe part and the other inner part.

2.2. Chemicals and Others

(+)-1-(9-Fluorenyl)ethyl chloroformate [(+)-FLEC] and (-)-1-(9-fluorenyl)ethyl chloroformate [(-)-FLEC] were purchased from Aldrich (Milwaukee, WI, USA). Solvents for high-performance liquid chromatography (HPLC) , such as acetonitrile, tetrahydrofuran and methanol of HPLC grade were obtained from Nacalai Tesque (Kyoto, Japan). NMDA, *N*-methyl-L-aspartate (NMLA), *N*-methyl-DL-glutamate and *N*-methyl-L-glutamate (NMLG) were purchased from Sigma (St. Louis, MO, USA). The anion-exchange resin, Dowex 1×2 (100-200 mesh, chloride form) was obtained from Muromachi Chemicals (Tokyo, Japan) and converted to acetate form just before use. A reverse-phase HPLC column (250×4.6 mm I.D.), packed with 4 μm-diameter J' sphere ODS-M80 was from YMC (Kyoto, Japan). Other chemicals were of analytical purity.

3. METHODS

3.1. Determination of Acidic D-Amino Acids Together with Neutral and Basic D-Amino Acids [4]

3.1.1. Preparation of Tissue Extracts

The frozen tissues were homogenized with a Potter-Elvehjem homogenizer first in 5 volumes of 5 mM potassium phosphate buffer (pH 7.4) containing 25 mM KCl, then further homogenized after addition of 5 volumes of 16% perchloric acid. The homogenate was centrifuged for 20 min at $12{,}000 \times g$ and the supernatant was neutralized with KOH. After centrifugation to remove the precipitated $KClO_4$, the supernatant was concentrated to dryness under reduced pressure at 40°C, and the residue was subjected to derivatization for HPLC analysis.

3.1.2. Derivatization

The residue, resulting from the procedure described above, was dissolved in 2-5 ml of 100 μM L-cysteic acid solution (adjusted to about pH 7.0 with KOH), which was employed as the internal standard. The derivatization of amino acids with *o*-phthaldialdehyde (OPA) and *N*-acetyl-L-cysteine (NAC) was performed by the method of Aswad [5] as follows. The OPA-NAC reagent was prepared by dissolving 4.0 mg OPA in 300 μl of methanol, adding the following in the order indicated; 250 μl of 0.4 M Na borate (pH 9.4), 390 μl of water, and 60 μl of 1.0 M NAC and finally adjusting to pH 5.0-6.0 with NaOH. Five μl of the reagent, which was stored at 4°C, was mixed with 20 μl of amino acid solution, and, after 2-3min, 475 μl of 50 mM Na acetate (pH 5.2) was added. The mixture was centrifuged at 10,000×g for 30 sec, and 250 μl of the upper portion was taken for direct injection into the HPLC system.

3.1.3. HPLC

The HPLC analysis was carried out with a Shimadzu (Kyoto, Japan) HPLC system consisting of two LC-6A pumps, an RF-10A fluorescence detector, a Chromatopac C-R4A data processor, a DGU-3A degasser and a J' sphere-ODS-M80 column (250×4.6 mm I.D.) (YMC, Kyoto, Japan). Excitation and emission wavelengths were 320 and 440 nm, respectively. Gradient elution was performed with 3.5-5.5% (0-45 min) methanol in 50 mM sodium acetate buffer (pH 5.53) at a flow rate of 1.0 ml / min. Peaks were identified by comparison with the behaviors of authentic compounds.

3.2. Determination of Acidic *N*-Methyl D-Amino Acids [6,7,8]

3.2.1. Preparation of Tissue Extracts

The frozen tissues (usually about 0.3 g) were thawed and homogenized with a high-speed microhomogenizer (Physcotron, Niti-on, Funabashi, Japan) or a Potter-Elvehjem homogenizer equipped with a Teflon pestle in 10 volumes of 8% perchloric acid under cooling in ice water. The homogenate was centrifuged for 20 min at 12,000×g, at 4°C, and the supernatant was neutralized with KOH and the precipitated $KClO_4$ was removed by centrifugation.

3.2.2. Pretreatment and Derivatization

The supernatant (2-5 ml) obtained was loaded on a column (2 ml volume) of Dowex 1×2 (100-200 mesh, acetate form), and the column was washed with 20 ml of water to remove neutral and basic substances. Acidic *N*-methyl amino acids and amino acids including aspartate and glutamate enantiomers were eluted with 20 ml of 1 M acetic acid. The eluate was concentrated to dryness under reduced pressure with a centrifugal evaporator (VC-960, Taitec, Saitama, Japan) at 40°C. The derivatization of acidic *N*-methyl amino acids were performed by the reaction with (+)- and (-)-FLEC shown in Fig. 1 after removal of primary amino acids with OPA as follows: The residue, resulting from the procedure described above, was dissolved in 2-3 ml of 0.1 M sodium borate buffer (pH 9.0). A 20 μl volume of this sample solution was mixed, in a 1.5 ml microcentrifuge tube, with 10 μl of OPA solution (5 mg/ml) in acetonitrile and was kept at 50°C for 15 min to remove primary amino acids. Then, 5 μl of 18 mM (+)- or (-)-FLEC in acetone and 5 μl of acetonitrile were added and, after reaction for 15 min at 50°C, a 10 μl volume of 100 mM aqueous L-cysteic acid in 0.1 M sodium borate buffer (pH 9.0) was added to remove the remaining FLEC. The mixture was allowed to react for another 7 min at the same temperature. After addition of 150 μl of 0.1 M sodium acetate buffer (pH 4.0), the mixture was filtered through a 0.45-μm filter (Nacalai Tesque).

3.2.3. HPLC Analysis

A portion (20-100 μl) of the filtrate obtained above was injected into a Shimadzu (Kyoto, Japan) HPLC system consisting of two LC-10AD pumps, a SIL-10A auto injector equipped with a sample cooler

S (3°C), a CTO-10A column oven, a SCL-10A system controller, a DUG-3A degasser, an RF-10A fluorescence detector and a Chromatopak C-R5A data processor. The analytical column was a reversed-phase J'sphere ODS-M80 (250×4.6 mm I.D.) (YMC, Kyoto, Japan). Unless otherwise stated, the mobile phase consisted of 86% (v/v) 0.1 M sodium acetate buffer (pH 5.59), 7% (v/v) acetonitrile and 7% (v/v) tetrahydrofuran, and isocratic elution was carried out at a flow-rate of 1.2 ml/min and at a column temperature of 40°C. For fluorometric detection of eluted FLEC derivatives the excitation and emission wavelengths were set at 260 and 315 nm, respectively.

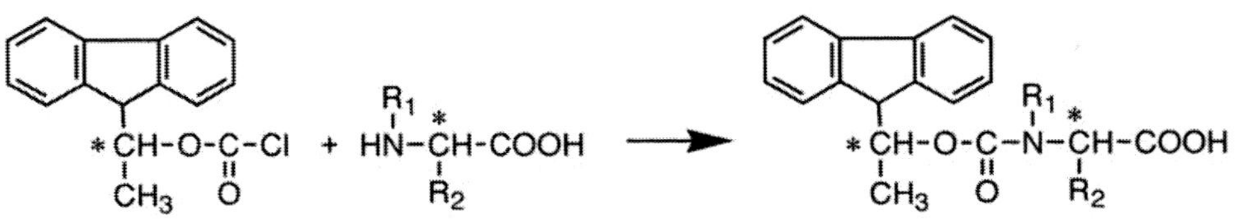

R_1=CH_3, and R_2=CH_2COOH for *N*-methyl-D,L-aspartic acid

R_1=H , and R_2=CH_2SO_3H for D,L-cysteic acid

Figure 1. Structure of (+)- and (-)-1-(9-fluorenyl)ethyl chloroformate (FLEC) and their reactions with enantiomers of *N*-methylaspartate and cysteic acid. Asterisks denote chiral carbon atoms. (From Todoroki et al. (1999), J. Chromatogr. 728B, 41-47, with permission)

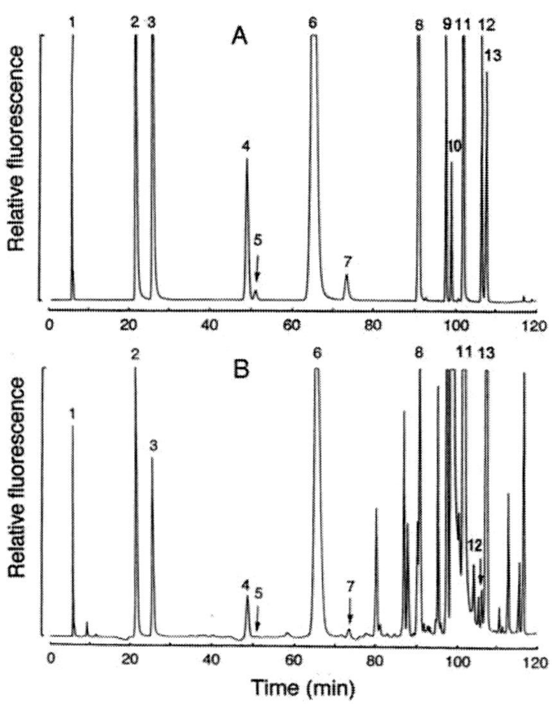

Figure 2. A typical HPLC resolution of the derivatized D- and L-enantiomers of some amino acids. HPLC chromatograms of (A) authentic amino acids, and (B) an extract from the outside part of foot of *S. broughtonii.* Peaks: 1= L-cysteic acid; 2=D-aspartate; 3=L-aspartate; 4=L-serine; 5=D-serine; 6=L-glutamate; 7=D-glutamate; 8=glycine; 9=L-arginine; 10=D-arginine; 11=β-alanine; 12=D-alanine; 13=L-alanine. (From Watanabe et al. (1998), Amino Acids *14*, 353-360, with permission)

While authentic NMDA, NMLA and NMLG were commercially available and served as HPLC standards, *N*-methyl-D-glutamate (NMDG) was not. We, therefore, used the NMDG in the commercially available racemic *N*-methyl-DL-glutamate as an HPLC standard, on assumption that the racemate is an exactly equimolar mixture of enantiomers. Comparison of the authentic NMLG and the racemate in the HPLC behavior enabled us to identifiy the peak of NMDG.

4. RESULTS

4.1. Determination of D-Aspartate and D-Glutamate together with Neutral and Basic D-Amino Acids [4]

The enantiomers of aspartate, glutamate, alanine and serine were all separated under the conditions employed for HPLC analysis as shown in Fig. 2, and were identified by comparison with the behaviors of authentic amino acids. Analysis thus conducted of the tissue extracts of *S. broughtonii* shows that most of the tissues examined contained D-aspartate at concentrations approximately equal to those of L-aspartate. On the other hand, D/(D + L) of glutamate, alanine and serine were generally at around 1% and at most a few percent, indicating that the high ratio of D-enantiomer was unique to aspartate in this bivalve. The highest concentration of D-aspartate was found in the intensely colored outside part of the foot and was about 4.8 μmol/g wet weght comparable to 4.4 μmol/g wet weght of L-aspartate. The D-aspartate concentration is lower than, but may be regarded as comparable to, the reported concentrations of the amino acid at around 10 μmol/g wet weight [1,2] in the foot or *Solemya reidi* and is at about the same level as those found in the foot and gill of *Mytilus edulis* and *Modiolus capax* [2]. The concentrations of D-glutamate, D-alanine and D-serine were about 0.11, 0.15 and 0.017 μmol/g wet weight, respectively.

4.2. Determination of *N*-Methyl-D-Aspartate and *N*-Methyl-D-Glutamate [6,7,8]

Fig. 3A shows that NMDA and NMLA derivatized with FLEC were completely resolved under the HPLC conditions employed. Similar resolution was obtained with the solvent system consisting of 84-90% (v/v) 0.1 M sodium acetate buffer (pH 4.4-5.8), 5-8% (v/v) acetonitrile and the same volume of tetrahydrofuran as acetonitrile. Fig. 3C shows that NMDA was detected under the same conditions for Fig. 3A even in the presence of D- and L-enantiomers of aspartate and glutamate, although NMLA was not separated from one of these acidic primary amino acids. By raising the pH of the solvent buffer from 5.6 to 5.8, NMLA became detectable, while NMDA, in turn, became unseparable from the acidic amino acids.

We tested various ways to remove primary amino acids from samples before derivatization with FLEC, and the pretreatment with OPA gave satisfactory results: When the same mixture of authentic *N*-methylaspartates and acidic amino acids as in Fig. 3C was treated with OPA before FLEC treatment as described in the 3.1.2 section, the chromatogram showed only the presence of *N*-methylaspartates with all the acidic amino acids disappeared (data not shown). When the extracts from tissues of *S. broughtonii* (3.1.1) underwent the FLEC treatment and HPLC under the same conditions as for Fig. 3A without the OPA pretreatment, the peak of NMDA could not be detected because of the interference by overwhelmingly large peaks of the acidic amino acids as shown in Fig. 4A. By removing primary amino acids with OPA, the peak of NMDA was clearly observable on the chromatogram as shown in Fig. 4B. The chromatogram also showed a peak of unknown substance that behaved as derivatized NMLA. However, derivatization with (-)-FLEC in place of (+)-FLEC employing otherwise the same procedure gave a chromatogram confirming the identity of NMDA and not of NMLA (data not shown).

To further confirm the identity of the peak of NMDA that became thus detectable, we treated the bivalve tissue extracts with pig kidney D-aspartate oxidase, which were reported to oxidize NMDA as one of the specific substrates [9]. The peak of NMDA was abolished by the enzyme treatment, confirming the identification, whereas the peak of unknown substance remained intact. The detection limit for NMDA was below 1 pmol, and the peak areas of their FLEC derivatives were linear up to 150 pmol with $r^2 \geq 0.995$. It should be added that the OPA treatment did not significantly affect the detection sensitivity for the secondary amino acids, as indicated by the experiments with authentic NMDA and NMLA.

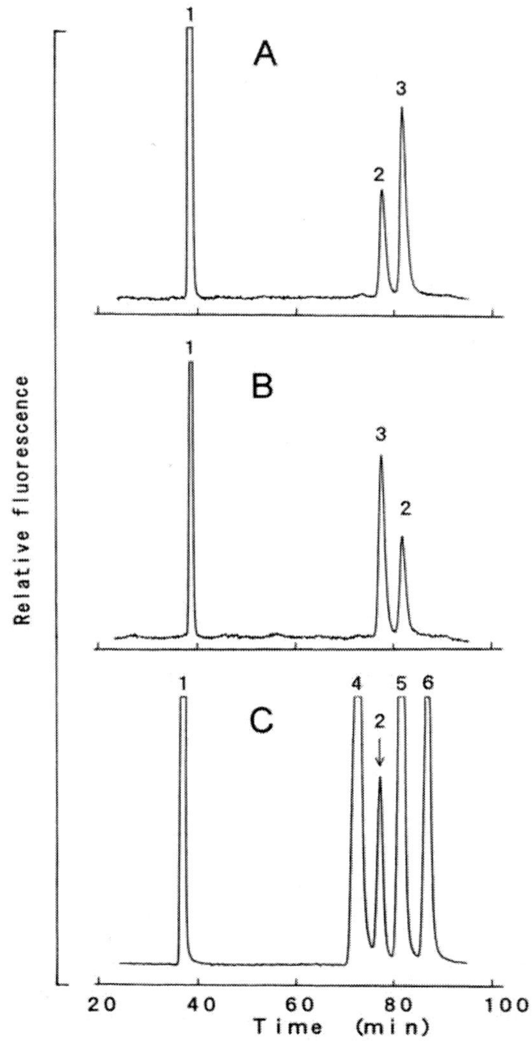

Figure 3. HPLC chromatogram of (A) (+)-FLEC derivatives of authentic *N*-methylaspartate enantiomers, (B) (-)-FLEC derivatives of authentic *N*-methylaspartate enantiomers, and (C) (+)-FLEC derivatives of authentic *N*-methylaspartate, aspartate and glutamate enantiomers. A mixture of 5 pmol of NMDA and 7.5 pmol of NMLA was treated with (+)- and (-)-FLEC for (A) and (B), respectively, and a mixture of 10 pmol each of L-enantiomers and 5 pmol each of D-enantiomers was treated with (+) FLEC for (C). HPLC was performed with isocratic elution with 86% 0.1 M sodium acetate buffer (pH 5.6): 7% acetonitrile: 7% tetrahydrofuran at a flow-rate of 1.2 ml/min and at 40°C. Peaks: 1= L-cysteic acid; 2= NMDA; 3=NMLA; 4=D- and L-aspartate; 5=D-glutamate and NMLA; 6= L-glutamate. (From Todoroki et al. (1999), J. Chromatogr. *728B*, 41-47, with permission)

The method thus established was applied to the determination of these compounds in the tissues of the blood shell, *S. broughtonii* and its closely related species, *Scapharca subcrenata*. In *S. broughtonii,* the outside part of mantle and the outside part of the foot contained about 110 nmol/g wet weight and 25 nmol/g wet weight of NMDA, respectively. The NMDA concentration in the foot and mantle of *S. subcrenata* was about 200 and 110 nmol/g wet weight, respectively. Comparable concentrations of NMDA were found also in the tissues of other bivalves: mantle of *Crassostrea nippona* and foot of *Meretrix lusoria* [7]. By the same method, *N*-methyl-D-glutamate and also *N*-methyl-L-glutamate were found in *S. broughtonii*, *S. subcrenata* and several other invertebrates [8], whereas *N*-methyl-L-aspartate was found only in two species [7].

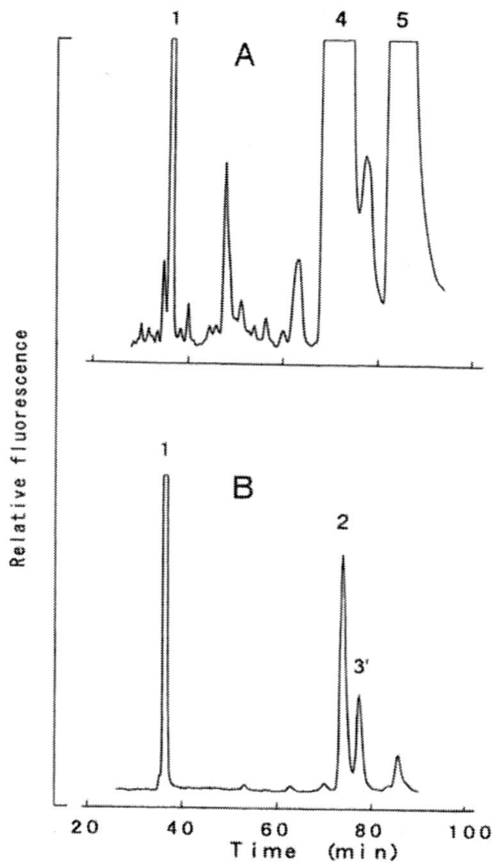

Figure 4. Effect of OPA treatment before (+)-FLEC derivatization of a tissue extract of *S. broughtonii* on HPLC chromatogram. The extract was prepared from the outside part of mantle of *S. broughtonii* and underwent the (+)-FLEC derivatization (A) without and (B) with the prior treatment with OPA, before injection into the HPLC. Peaks: 1=L-cysteic acid; 2=probably NMDA; 3'=unkown; 4=mainly D- and L-aspartate; 5=mainly D- and L-glutamate. (From Todoroki et al. (1999), J. Chromatogr. *728B*, 41-47, with permission)

5. Note

The buffer used for homogenization of tissues described in the section 3.1.1 contained KCl, while that described in the section 3.2.1 did not. KCl serves to prevent polymerization of myosin, but its addition is not always required.

References

[1] Felbeck, H (1985) Occurrence and metabolism of D-aspartate in the gutless bivalve *Solemya reidi*. *J. Exp. Zool. 234*, 145-149.

[2] Felbeck, H., and Wiley, S (1987) Free D-amino acids in the tissues of marine bivalves. *Biol. Bull. 173*, 252-259.

[3] Sato, M., Inoue, F., Kanno, N. and Sato, Y. (1987) The occurrence of *N*-methyL-D-aspartic acid in muscle extracts of the blood shell, *Scapharca broughtonii. Biochem. J. 241*, 309-311.

[4] Watanabe, T., Shibata, K., Kera, Y. and Yamada, R. (1998) Occurrence of free D-aspartate and aspartate racemase in the blood shell *Scapharca broughtonii. Amino Acids 14*, 353-360.

[5] Aswad, D.W. (1984) Determination of D- and L-aspartate in amino acid mixtures by high-performance liquid chromatography after derivatization with a chiral adduct of *o*-phthaldialdehyde. *Anal. Biochem. 137*, 405-409.

[6] Todoroki, N., Shibata, K., Yamada, T., Kera, Y. and Yamada, R. (1999) Determination of *N*-methyl-D-aspartate in tissues of bivalves by high-performance liquid chromatography. *J. Chromatogr. 728B*, 41-47.

[7] Shibata, K., Tarui, A., Todoroki, N., Kawamoto, S., Takahashi, S., Kera, Y., Yamada, R. (2001) Occurrence of *N*-methyl-L-aspartate in bivalves and its distribution compared with that of *N*-methyl-D-aspartate and D,L-aspartate. *Comp. Biochem. Physiol. 130B*, 493-500.

[8] Tarui, A., Shibata, K., Takahashi, S., Kera, Y., Munegumi, T., Yamada, R (2003) *N*-Methyl-D-glutamate and *N*-methyl-L-glutamate in *Scapharca broughtonii* (Mollusca) and other invertebrates. *Comp. Biochem. Physiol. 134B*, 79-87.

[9] Yamada, T., Hasegawa, A., Matsumura, H., Uchiyama, T., Kera, Y. and Yamada, R. (1996) Purification and properties of pig kidney D-aspartate oxidase and its utility in analysis of acidic D-amino acid. *Bull. Nagaoka Univ. Technol. 18*, 19-26.

In: D-Amino Acids:
A New Frontier in Amino Acid and Protein Research, pp.33-40
ISBN 1-60021-075-9

Chapter 2.3

PURIFICATION AND DETERMINATION OF *N*-METHYL-D-ASPARTIC ACID (NMDA) IN BIOLOGICAL TISSUES

Antimo D'Aniello*

Laboratory of Neurobiology, Stazione Zoologica, "A. Dohrn", Italy.

Keywords: *N*-methyl-D-aspartic acid (NMDA), D-aspartic acid (D-Asp), amino acids, HPLC.

1. INTRODUCTION

N-Methyl-D-aspartic acid (NMDA) is a molecule widely known for its property as an agonist of the L-glutamate receptor of NMDA-type and also for its activity in inducing hormone release in the hypothalamus and hypophysis. NMDA may be considered as a molecule derived from D-aspartic acid (D-Asp) in which a hydrogen atom of the amino group (NH_2) of D-Asp is substituted by a methyl group (CH_3) by a methyltransferase (NMDA synthetase):

D-Asp		NMDA
COOH		COOH
|		|
CH_2	NMDA synthetase	CH_2
|	——————————?	|
CH-NH_2		CH-$NHCH_3$
|		|
COOH		COOH

D–Asp **NMDA**

In fact, among of the three families of glutamate receptor proteins identified, one family of these receptors is activated by NMDA, and these receptors: NR1, NR2A, NR2B, NR2C and NR2D, are collectively referred to as glutamate receptors of the NMDA-type. Another family of receptors is activated by α-amino-3-hydroxyl-5-hydroxyl-4-isoxazole propionic acid (AMPA) and by kainate, and the third

* Correspondence concerning this article should be addressed to Antimo D'Aniello E-mail: daniello@szn.it. FAX: +39 081 7641 355.

family of glutamate receptors consist of a G-protein coupled receptors, the so-called matabotropic receptors (mGluR1-8) [for a review, see Refs 1 and 2]. Concerning the activity of NMDA in inducing hormones release, in 1978 Price et al. [4] and Ondo et al. [4] found that a very low amount of NMDA injected into the rat (1-10 nmol/g body weight of the animal) increased LH levels. Later, Gay et al. [5] and Downic et al. [6] also observed that a minute quantity of *N*-methyl- DL-aspartic acid (NMDLA) injected into Rhesus monkey (*Macaca mulatta*), induced the release of gonadotropin releasing hormone (GnRH) from the hypothalamus and growth hormone (GH) in ewes. More recently, we have found that NMDA occurs in the neuroendocrine tissue as an endogenous compound in vertebrates [7,8] and invertebrates [9], and that in both phyla of animals this amino acid is involved in neuroendocrine activity, particularly in the control of the hormone release of the hypothalamus-pituitary-gonad axis [7-9].

Because of the great interest in the determination of NMDA in biological tissues, here is described two specific methods for the determination of NMDA in animal tissue: an HPLC-enzymatic method and a TLC-enzymatic method. Both methods consist of a description of our previously described procedure [7,10], but opportunely improved in specificity and reliability.

2. Chemicals and Reagents

1. TCA (trichloroacetic acid).
2. Standard of amino acids and NMDA (Sigma Chemical Corp., St. Louis, MO, USA).
3. OPA (*o*-phthaldialdehyde) and 2-mercaptoethanol (Sigma Chemical Corp., St. Louis, MO, USA).
4. OPA-reagent for purifying NMDA: Prepared by mixing 1.34 g of OPA in 10 ml methanol (final concentration 1 M).
5. Sep-Pak® Cartridges, ODS-C_{18} (octadecylsilyl-C_{18}) (Waters Corporation, Milford, MA, USA).
6. Cation-exchange resin: AG 50W-X8, 200-400 mesh (Bio-Rad Laboratories, Hercules, CA 94547 U.S.A). Before use the resin is regenerated by treatment with 5 M NaOH, followed by distilled water until neutrality, then treated with 6 M HCl, followed by equilibration in 0.01 M HCl.
7. HPLC Gold system (Beckman, Milan, Italy).
8. HPLC column: ODS-C_{18} Supelcosil™ LC-18 column, 0.45 x 25 cm, 5 μm (Supelco, Inc., Belafonte, PA, U.S.A or Sigma-Aldrich, Supelco-Milan, Italy).
9. Buffers: Citrate-phosphate buffer (Mcllvain buffer), 1.0 M, pH 5.3, obtained by mixing a given volume of 1.0 M potassium phosphate dibasic (K_2HPO_4) with a given volume of 1.0 M citric acid ($C_6H_8O_6$) until pH 5.3. Sodium pyrophosphate buffer 0.5 M, pH 8.2, and pH 10. Tris-HCl buffer, 1 M pH 8.2, obtained by mixing a given volume of 1 M Tris [Tris(hydroxymethyl) aminomethane] with a given volume of 1 M HCl until pH 8.2 (Carlo Erba, Milan, Italy or Sigma Chemical Corp., St. Louis, MO, U.S.A).
10. Acetonitrile for HPLC (Carlo Erba, Milan, Italy)
11. TLC cellulose plate 20x20 cm, 0.5 mm thickness, PSC-Fertigplatten (art. 15275, Merck, Frankfurter, Germany)
12. Peroxidase (POD) 2000 U/ml (Sigma Chemical Corp., St. Louis, MO, USA)
13. Tyramine hydrochloride (Sigma Chemical Corp., St. Louis, MO, USA)
14. D-aspartate oxidase (D-AspO; EC 1.4.3.1), purified at 1-2 mg/ml, obtained from beef kidney by overexpression in *Escherichia coli* and then purified as described [11]. This enzyme reacts only with D-Asp and NMDA and at very low rate toward D-Glu. Any other D- or L-amino acid does not at all react with this enzyme. The enzyme catalyzes the oxidation of these three amino acids according the following reaction:

$$\text{D-Asp or NMDA or D-Glu} + O_2 + H_2O \xrightarrow{\text{D-AspO}} \alpha\text{-ketoacid} + H_2O_2 + NH_3 \text{ or } CH_3NH_2$$

(NH_3 is formed when D-Asp or D-Glu is used as substrate; CH_3NH_2 is formed when NMDA is used as substrate).

3. Methods for the Purification and Determination of NMDA

3.1. Purification

Since NMDA occurs at very low concentration in biological tissues (in order of 10^{-9}-10^{-6} M), in order to obtain reliable results, as first step, it is necessary to purify NMDA from other amino acids and amino compounds and also to concentrate it. The principle of purification is based on the fact that when free amino acids and amine compounds react with OPA (*o*-phthaldialdehyde), a strong complex "amino acids-OPA" occurs. If this complex is passed through a Sep-Pak of ODS-C_{18} resin (octadecylsilyl-C_{18} resin), it strongly binds to this resin, whereas NMDA passes through. In fact, since NMDA does not possess a primary amino group, it does not react with OPA-2-mercaptoethanol to form a complex, and consequently it does not bind to the ODS-C_{18} resin [10].

In brief the procedure id the follow:

Step 1. Homogenization: 200 mg of tissue (here is described a procedure starting from 200 mg of tissue, however, all the reagents will be used proportionally if other amounts of tissues will be used) is homogenized with 1.8 ml of 0.1 M TCA and centrifuged at 10-15,000 rpm.

Step 2. OPA treatment: The supernatant is brought to pH 9.5-10.0 with 1 M NaOH, then added 3 ml of distilled water, 0.5 ml of 0.5 M sodium pyrophosphate buffer, pH 8.5 (to fix the pH to about 8.5), and 200 µl of 1 M OPA reagent for purifying NMDA (reagent n. 4) and incubated at 37°C for 60 min. After incubation, the sample is brought to pH 2.5-3.0 with 1 M HCl and passed on a Sep-Pak® cartridges, ODS-C_{18} containing 800 mg resin (previously regenerated in 10 ml methanol and then washed in distilled water) follow by 2 ml 0.01 M HCl.

Note: If the sample becomes opalescent after HCl treatment it needs to be centrifuged before cation exchange resin purification.

Step 3. Purification on cation exchange resin: The sample as above obtained is passed on a column of cation exchange resin AG 50-WX-8 (200-400 mesh) containing 2 ml of settled resin following by 15-20 ml of 0.01 M HCl using 5 ml at a time. During this procedure the excess of OPA reagent will be completely eluted from the column, whereas NMDA remains bound to the resin). After washing with 0.01 M HCl, the column is eluted with 6 ml of 4 M NH_4OH, by using 3 ml at a time, and all the eluent (which contains NMDA), is collected and dried in a Petri dish placed on a plate under a hood, at 40-60 °C. The residue is dissolved in 400 µl of distilled water and used for the determination of NMDA by HPLC method.

Note: 1) It is important to verify that after wash of the cation exchange resin with 0.01 M HCl the OPA in excess is liberated from the resin. In order to test this, mix some drops of the eluate of the column with 200 µl of Nessler reagent (C. Erba, Italy or Merk, Germany), and if a yellow colour appears, it means that OPA is still present in the resin. In this case, continue to wash the column with 0.01 M HCl until all the OPA reagent is eliminated. 2) Before HPLC analysis, it is opportune to test if the sample contains ammonia (that came from ammonium hydroxide elution) because a high amount of this reagent can interferes with the determination of CH_3NH_2. For this purpose 20 µl of the sample are mixed with 200 µl of Nessler reagent and if a dense yellow colour appears, then ammonia is present. If ammonia is present in the sample, then add to the sample 10-20 μl of 1 M NaOH and let the ammonia evaporate (2-4 min in Petri dishes at 40-50°C). After that the pH of the sample is adjusted to about 8.2 with 5-10 μl of 1 M HCl.

3.2. Determination of NMDA by HPLC

This method is based on the measurement by HPLC of CH_3NH_2 (methylamine) which is generated by the oxidation of NMDA with D-aspartate oxidase (D-AspO) according to the following reaction:

COOH
|
CH_2
| + O_2 + H_2O —D-AspO—?
CH-$NHCH_3$
|
COOH
(NMDA)

COOH
|
CH_2
| + H_2O_2 + CH_3NH_2
C=O (methylamine)
|
COOH
(-oxaloacetate)

The methylamine formed is then derivatized with OPA-2-mercaptoethanol and determined by HPLC. This method is specific for NMDA since the CH_3NH_2 that is measured is generated only by NMDA treated with D-AspO (other methyl D- or L-amino acids are not oxidized by D-AspO [10].

Briefly, procedure is as follows:

In an Epperdorf tube is mixed 100 μl of purified sample with 50 μl of 0.5 M Na-phosphate buffer, pH 8.2, and 2 μl of purified D-AspO (1-2 mg/ml) and incubate at 37°C for 30 min. After that, mix 5 μl of OPA-2-mercaptoethanol reagent (prepared by mixing 20 mg OPA in 1 ml of methanol plus 40 μl of 2-mercaptoethanol). Mix for 2-5 sec and add 150 μl of 0.1 M pyrophosphate buffer, pH 10.0 (the final pH must be between 9.0-10) and mix. After 2 min, 100 μl of this mixture is injected into an ODS-C_{18} Supelcosil column. The column is connected to a Beckman-Gold HPLC system and a fluorescence detector (excitation wavelength 330 nm and emission wavelength 450 nm). The column is eluted at 1.2 ml/min with a gradient consisting of solution A (50 ml acetonitrile + 920 ml of H_2O + 30 ml of 1.0 M citrate-phosphate buffer, pH 5.3) and solution B (90% acetonitrile in distilled water). The gradient program is: 0-30% B over 3 min; 30-60% B over 11 min; 60-100 % B over 4 min and remaining at 100% for 3 min. The CH_3NH_2 is eluted, producing a sharp peak, at the retention time of 13.8-14.0 min (Fig. 2. Right panel). In the same way is carried out a blank sample, done as the sample, but without D-AspO and a standard consisting of 100 μl of NMDA at a concentration of 1.0 nmol/ml (dissolved in 0.2 M phosphate buffer, pH 7.8) instead of sample.

To calculate the quantity of NMDA in the 100 μl of the sample injected, the following formula is used: O.D. of the peak of CH_3NH_2 of the sample minus O.D. of the peak of CH_3NH_2 of the blank sample divided by the O.D. of the peak of CH_3NH_2 of the standard times 0.33 = nmol of NMDA in 100 μl of the purified sample (or in 8.33 mg of original tissue).

3.3.Thin-Layer Chromatography and Enzymatic Method for the Determination of NMDA

This method is based on the purification of NMDA by thin-layer chromatography (TLC) and its determination by using an enzymatic fluorimetric method. In brief, the method is the follows:

50 μl of the purified sample as obtained above (see: 3.1. Purification of NMDA) is applied to a TLC cellulose plate (20x20 cm, 0.5 mm thickness) at 2 cm from the origin of the plate. On the margin of the

plate is applied 50 μl of standard NMDA at a concentration of 1 μmol/ml (to identify the migration position of NMDA). The plate is developed in a solvent consisting of phenol-H_2O (100 g phenol:40 ml H_2O) using an amount of solvent in the tank to reach 1 cm high and left to run for about 10 hours. The plate is dried with a hair dryer and after that, in order know the position to which the NMDA sample has migrated, the TLC plate is covered with a paper sheet except for the channel where the NMDA standard was applied, and sprayed with ninhydrin (1 g in 100 ml methanol). The plate is dried at 50-60 °C to permit the visualization of the color of NMDA. In these conditions, the Rf for NMDA is about 0.6, that of Glu is 0.42 and Asp is 0.2 (Fig. 2, Ref. 7). The cellulose in which the sample has migrated, corresponding to the area in which the NMDA has migrated, is scraped off and mixed vigorously with 5 ml of 0.01 M HCl, centrifuged at 30,000 g for 20 min, and the supernatant is dried in a Petri dish. The residue is dissolved in 200 μl of 0.1 M Tris-HCl, pH 8.2, and used for the determination of NMDA by using an fluorescent enzymatic method as follows:

This method, is based on the fact that the H_2O_2 developed from the reaction between NMDA and D-AspO reacts with tyramine generating a highly fluorescent biphenyl derivative in the presence of POD (peroxidase) [11-12].

$$H_2O_2 + \text{tyramine} \xrightarrow{\text{POD}} \text{fluorescent biphenyl derivative}$$

Briefly, the procedure is the following:

a. 50 μl of sample as purified above (which corresponds to an amount of tissue of 4.16 mg of original tissue), are mixed with 100 μl of an assay mixture consisting of 20 mg tyramine hydrochloride in 5 ml of 0.1 M Tris-HCl, pH 8.2, plus 20 μl of peroxidase (POD, 2,000 U/ml) and 2 μl of D-AspO (1-2 mg/ml).
b. In another tube is carried out a blank sample consisting of sample, but without the D-AspO.
c. In a third tube is carried out a standard consisting of 50 μl of NMDA, 1 nmol/ml dissolved in 0.2 M Tris-HCl, pH 8.2, 100 μl of the assay mixture and 2 μl of D-AspO.
d. All tubes are incubated at 37°C for 30 min. Then 500 μl of 0.2 M Tris-HCl, pH 8.2, are added.
e. The fluorescence is read against distilled water at 330 nm excitation and 440 emission.
   To calculate the amount of NMDA in the sample using the following formulae: O.D. sample - O.D. of the blank sample divided by O.D. standard, times 0.05 = nmol in the 50 μl of sample. Multiply this value by 240 to obtain the nmoles of NMDA in 1g of original tissue (50 ul of sample used for the assay contains the NMDA from 4.16 mg of original tissue). The value 0.05 are the nmol of NMDA standard contained in 50 μl of standard.

4. RESULTS

In this study we have developed a system that permits us to eliminate all free amino acids and amino compounds from a biological sample (almost 90-95%), but not NMDA. Consequently NMDA can be detected (through the determination of CH_3NH_2, a molecule derived from the oxidation of NMDA by D-AspO) without interference by other amino acids. The CH_3NH_2 reacts quantitatively with OPA-2-mercaptoethanol giving a sharp peak that elutes by HPLC far from other amino acids if still present after purification. In fact, as is shown in Fig. 1 (left panel) if the sample is run by HPLC without purification, then the free amino acids cover completely the small peak of CH_3NH_2 that came from the oxidation of NMDA with D-AspO. Instead, if the sample is pretreated with OPA-2-mercaptoethanol, then almost all the amino acids have disappeared (Fig. 1, middle panel). However, when the purified and concentrated sample is treated with D-AspO, then the NMDA present in the sample will be oxidized into CH_3NH_2 giving a big peak at elution time 13.8-14.0 min.

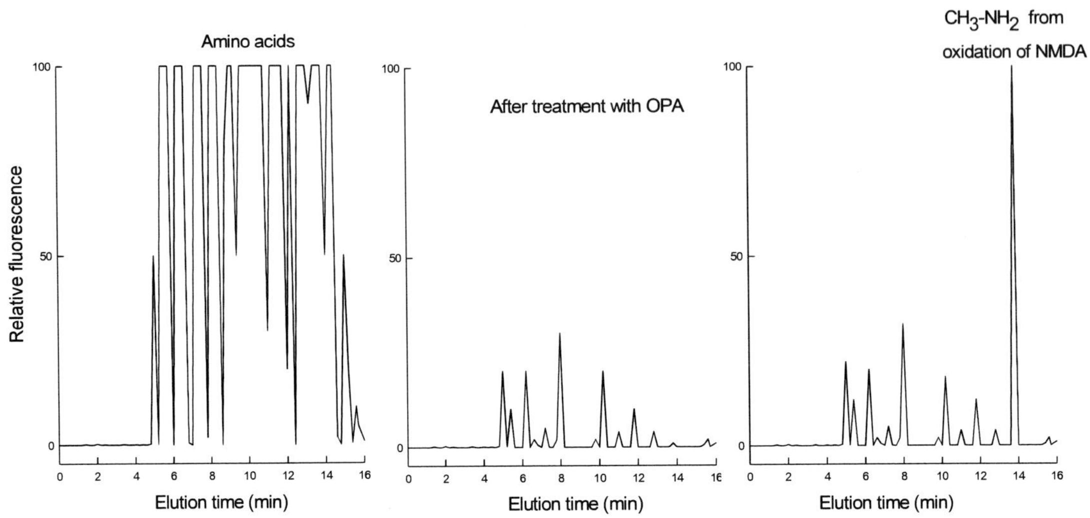

Figure 1: Example of purification and determination of NMDA in a biological sample (nervous ganglia of *Ciona intestinalis*). Left panel: a run of a sample consisting of a supernatant of a homogenate 1:10 in 0.2 M TCA of ganglia (10 μl of supernatant derivatized with 10 μl of OPA-2-merchaptoethanol, 100 μl of 0.5 M pyrophosphate buffer, pH 9.5, and water for a final volume of 1 ml). Middle panel: The same sample is purified with OPA-2-merchaptoethanol and cation exchange resin and 100 μl are derivatized as above and injected on HPLC. Right panel is the same sample as the middle panel, but after treatment with D-AspO.

The alternative method for the determination of NMDA in place of the HPLC method consists in the direct determination of the NMDA purified by thin-layer chromatography. In this case we measured the H_2O_2 produced from the reaction between NMDA and D-AspO. Also this method is sensitive (almost 10-20 pmol of NMDA can be determined) and is highly specific because NMDA is obtained completely separated from D-Asp and D-Glu (Fig 2), the only amino acids that can react with D-AspO.

5. Notes

1. The two methods described here for the determination of NMDA are specific, sensitive and precise and permit to detect a minimal amount of NMDA corresponding to 5-10 pmol for analysis. In order to obtain reliable results it is necessary to eliminate almost 95% of total amino acids and amine compounds. Since NMDA is present at very low concentration, other amino acids can interfere by overlapping the peak of methylamine (the target molecule that came from the oxidation of NMDA). The use of the D-AspO to oxidize NMDA giving CH_3NH_2 that is determined by HPLC constitutes a key procedure that ensures the reliability of the result.
2. The second method to determine NMDA, based on the determination of oxidized tyramine, also constitutes a precise method because we can to separate the NMDA from D-Asp and D-Glu if still present in the purified sample.

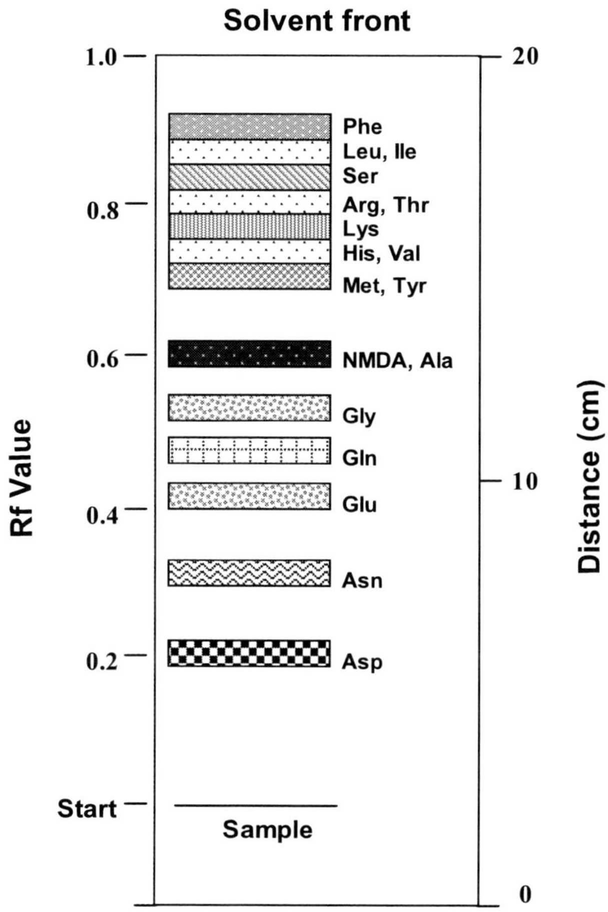

Figure 2. Separation of NMDA from D-Asp, D-Glu and other amino acids on TLC cellulose. The figure represents a separation of a mixture of standard amino acids (20 µl of the standard amino acids mixture in which each amino acid was at a concentration of 5 µmol/ml). The amino acids were visualized by sprayed with ninhydrin (1 g in 100 ml methanol).

REFERENCES

[1] Holman, M. & Heinemann, S. (1994). Cloned glutamate receptors. *Annu Rev Neurosci, 17,* 31-108.

[2] Danbolt Niels C. (2001). Glutamate uptake. *Progress in Neurobiology, 65,* 1-105.

[3] Price, M. T., Olney, J. W. & Cicero, T.J. (1978). Acute elevations of serum luteinizing hormone induced by kainic acid, *N*-methyl aspartic acid, or homocysteic acid. *Neuroendocrinology, 26*, 352-358

[4] Ondo, J., Wheeler, D. D., & Dom, R.M. (1988). Hypothalamic site of action for *N*-methyl-D-aspartate (NMDA) on LH secretion. *Life Sci, 43*, 2283-2286.

[5] Gay, V. L. & Plant, T. M. (1987). *N*-Methyl-D,L-aspartate elicits hypothalamic gonadotropin-releasing hormone release in prepubertal male rhesus monkeys (*Macaca mulatta*). *Endocrinology,120*, 2289-2296.

[6] Downing, J. A., Joss, J. & Scaramuzzi, R. J. (1996). The effects of *N*-methyl-D,L-aspartic acid and aspartic acid on the plasma concentration of gonadotropin, GH and prolactin in the ewe. *J Endocrinol,149*, 63-72.

[7] D'Aniello, A., Di Fiore, M. M., Fisher, G. H., Milone, A., Seleni, A., D'Aniello, S., Perna, A., & Ingrosso, D. (2000). Occurrence of D-Aspartic acid and *N*-methyl-D-aspartic acid in rat neuroendocrine tissues and their role in the modulation of luteinizing hormone and growth hormone release, *FASEB J, 14*, 699–714.

[8] D'Aniello, G., Tolino, A., D'Aniello, A., Fisher, G. H., & Di Fiore, M. M. (2000). The role of the aspartic acid and *N*-methyl-D-aspartic acid in the regulation of prolactin release. *Endocrinology, 141*, 3862–3870.

[9] D'Aniello, A., Spinelli, P., De Simone, A., D'Aniello, S., Branno, M., Aniello, F., Fisher, G., Di Fiore, M. M., & Rastogi, R. K. (2003). Occurrence and neuroendocrine role of D-Asp and NMDA in *Ciona intestinalis*. *FEBS Letters, 552*, 193-198.

[10] D'Aniello, A., De Simone, A., Spinelli, P., D'Aniello, S., Branno, M., Aniello, F., Rios, J., Tsesarkaja, M., & Fisher, G. (2002). A specific enzymatic high-performance liquid chromatography method to determine *N*-methyl-D-aspartic acid in biological tissues. *Anal Biochem, 308,* 42-51.

[11] D'Aniello, A., Spinelli, P., Ferrandino, P. G., Peterson, K., Tsesarskia, M., Fisher, G. H. & D'Aniello, A. (2005). Cephalopod vision involves dicarboxylic amino acids: D-aspartate, L-aspartate and L-glutamate. *Biochemical J, 386,* 331-340.

[12] Gross, A.J. & Sizer, I.W. (1959). The oxidation of tyramine, tyrosine, and related compounds by peroxidase. *J Biol Chem, 234,* 1611-1614.

In: D-Amino Acids:
A New Frontier in Amino Acid and Protein Research, pp. 41-47
ISBN 1-60021-075-9

Chapter 2.4

A SENSITIVE DIRECT HPLC METHOD FOR SIMULTANEOUS DETERMINATION OF *N*-METHYL-(D AND L)-ASPARTATE, *N*-METHYL-(D AND L)-GLUTAMATE AND (D AND L)-ASPARTATE IN BIOLOGICAL TISSUES

Mara Tsesarskaia[1], Erika Galindo[1], Gyula Szókán[2] and George Fisher[1,*]
[1]Department of Chemistry, Barry University, Miami Shores, FL, USA
[2]Institute of Organic Chemistry, Eötvös University, Budapest, Hungary

Keywords: *N*-methyl-(D and L)-aspartate, *N*-methyl-(D and L)-glutamate, (D and L)-aspartate, HPLC, NMDA, NMDG, *Scapharca broughtonii.*

1. INTRODUCTION

N-methyl-D-aspartate (NMDA) is a potent neuroexcitatory amino acid [1,2] that has specific action as an agonist for one of the glutamate sub-type NMDA receptors in the central nervous system of invertebrates and vertebrates [3-5]. NMDA receptors play an important role in learning and memory [6-7], and NMDA itself has been shown to improve social recognition potency in rats when administered systemically to adult rats [8].

Endogenous NMDA was first found (1987) in the muscle of the mollusk *Scapharca broughtonii* [9] and recently (2000) NMDA was found in rat nervous tissue and endocrine glands [10,11]. *N*-Methyl-L-aspartate (NMLA) was also shown to occur naturally in certain marine algae [12,13]. Subsequently, NMDA and NMLA as well as *N*-methyl-(D and L)-glutamate (NMDG and NMLG) were found in some mollusk bivalves [14,15].

Several methods have been developed for resolution of D- and L- enantiomers of primary amino acids by high performance liquid chromatography (HPLC) after preparation of diastereomeric derivatives of the

* Correspondence concerning this article should be addressed to George Fisher, e-mail: gfisher@mail.barry.edu; FAX: +1 305 899-3479.

amino acids with chiral reagents. However, very few chiral reagents are known to react with secondary amino acids such as *N*-methyl-(D and L)-aspartate and *N*-methyl-(D and L)-glutamate. In 2002 D'Aniello et al. [16] developed an indirect method for determination of NMDA based on the HPLC detection of the methylamine (CH_3NH_2) produced by the oxidation of NMDA in the presence of D-aspartate oxidase (D-AspO; EC 1.4.3.1), an enzyme that specifically catalyzes the oxidation of the acidic D-amino acids: D-Asp, D-Glu, NMDA, and NMDG. This method, however, would not distinguish between NMDA and NMDG since both of these amino acids would be oxidized by D-AspO to produce methylamine.

In 1999, Todoroki et al. [17] developed a direct HPLC method for determination of NMDA in biological tissues using precolumn derivatization with the chiral reagent (+)-1-(9-fluorenyl)ethyl chloroformate (FLEC) followed by isocratic resolution of the two diastereomers formed. This method, however, required the removal of neutral and basic substances by anion-exchange chromatography and removal of primary amino acids by treatment with *o*-phthaldialdehyde (OPA) before derivatization of the NMDA. Subsequently, Shibata et al. [14] and Tarui et al. [15] used (+)- and (–)-FLEC chiral derivatizing reagents to determine NMDA, NMLA, NMDG, and NMLG in bivalves. Again, these methods required prior removal of neutral and basic substances and primary amino acids, and the HPLC runs were extremely long (80 mins).

In 2002 the Homma group in Japan [18] developed an automated two step column-switching HPLC system for resolution of the D- and L-*N*-methyl aspartates. In the first step, the sample of amino acids is reacted with an achiral fluorescent reagent 4-fluoro-7-nitro-2,1,3-benzoxadiazole (NBD-F) and the NMDA and NMLA derivatives formed are separated from those of primary amino acids by HPLC. In the second step (column switching), the D- and L-*N*-methyl aspartates are resolved on a chiral column. The disadvantages of this method are the high cost and short life-time of the chiral column and long procedure times for the two step column-switching HPLC system.

In this chapter, we present a sensitive, one-step derivatization and HPLC method for simultaneous determination and resolution of NMDA, NMLA, NMDG, NMLG as well as D- and L-Asp. This method is based on a derivatization procedure with the chiral reagent *N*-α-(2,4-dinitro-5-fluorophenyl)-(D or L)-valine amide (FDNP-Val-NH_2), a close analog of Marfey's reagent 1-fluoro-2,4-dinitrophenyl-5-L-alanine amide [19]. The diastereomers formed are separated on an inexpensive reversed phase ODS-Hypersil column with elution by a 0.11% TFA/water – 0.11% TFA/MeCN gradient. UV absorption at 340 nm permits detection levels in the range of 5-10 picomoles.

2. Materials

1. Chemicals: Trifluoroacetic acid (TFA), acetonitrile (MeCN), NMDA, NMLA, NMDG, NMLG, (D and L)-Asp and other amino acids were purchased from Sigma-Aldrich (St. Louis, MO, USA). FDNP-(D or L)-Val-NH_2 was purchased from NovaBiochem (La Jolla, CA, USA). All solvents were HPLC grade. Water should be doubly distilled or deionized through a water purification system such as Millipore® or NANOpure®.
2. HPLC: The HPLC system should consist of at least a binary gradient pump, an in-line solvent degasser, a 10 μL injector loop, an ODS-C_{18} pre-column, connected to a 25 x 0.46 cm, Hypersil ODS-C_{18} (5μ) column (Column Engineering, Ontario, CA, USA), a UV-Vis detector, and a computing integrator data handling system.

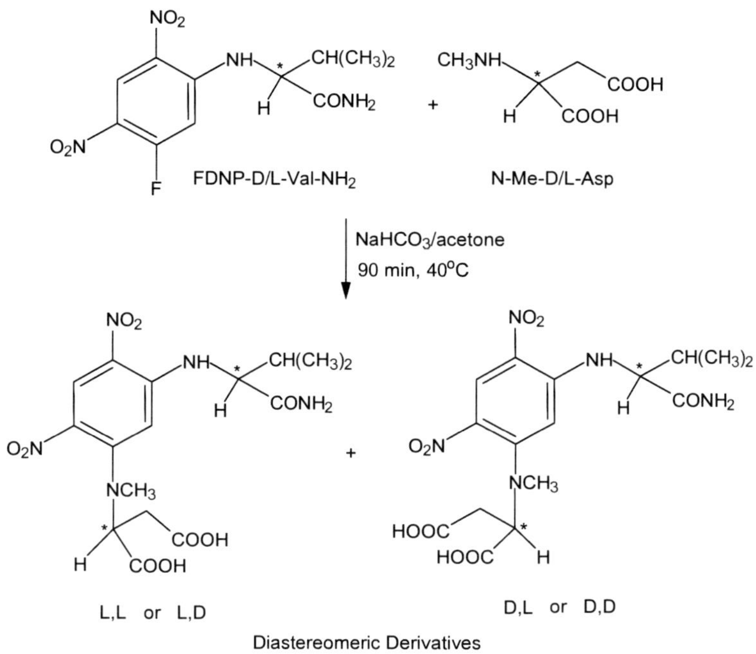

Figure 1. Derivatization scheme with the chiral reagent *N*-α-(2,4-dinitro-5-fluorophenyl)-(D or L)-valine amide (FDNP-Val-NH_2) to form a pair of diastereomers.

3. Methods

1. Derivatization: D,L-amino acids or tissue samples are derivatized with D- or L-FDNP-Val-NH_2, according to the procedure of Szókán et al. [19] (Fig. 1). 2.5 μmol of amino acid or a tissue homogenate from a biological sample are dissolved in 100 μL of 0.5 M $NaHCO_3$ to which 400 μL of a 1% solution of FDNP-Val-NH_2 in acetone are added. The mixture is incubated for 90 min at 40°C in a Thermomixer, and after cooling acidified with 2 M HCl to pH ~4. After 100-160 fold dilution with MeOH, 10 μL are injected into the HPLC.
2. HPLC analyses: The derivatized diastereomers formed are separated by HPLC on a 5μ ODS-Hypersil reversed phase column (25.0 x 0.46 cm) using a gradient pump, with detection at 340 nm, connected to a computing integrator. Two solvent mixtures are used: A) 0.11% TFA in water (v/v); B) 0.11% TFA in MeCN (v/v). The following HPLC program is used: isocratic for 2 min with 80% A:20% B, then a linear gradient to 62% A:38% B over the next 46 minutes, followed by a 4 min wash with 5% A:95% B, then back to 80% A:20% B. All flows rates are 1.1 mL/min.

4. Results

Figure 2 shows HPLC chromatograms for standards of (A) NMLA and NMDA, (B) D-Asp and L-Asp, and (C) NMDG and NMLG, derivatized with FDNP-D-Val-NH_2. The diastereomers of each pair are well resolved. The order of elution of the enantiomers is reversed when derivatized with FDNP-L-Val-NH_2. A large peak at 37.4 min is unreacted FDNP-Val-NH_2. Figure 3 shows an HPLC chromatogram of a standard mixture of all the amino acids that elute in the region from about 15 to 45 min. These other amino acids eluting in this region do not interfere with NMDA, NM(D/L)G or D/L-Asp. However, D-Thr coelutes with NMLA. This should not be a problem since most biological samples probably do not contain D-Thr. Also, when the sample is derivatized with FDNP-L-Val-NH_2 then L-Thr and NMDA coelute, and NMLA is resolved. Thus, NMLA can be identified.

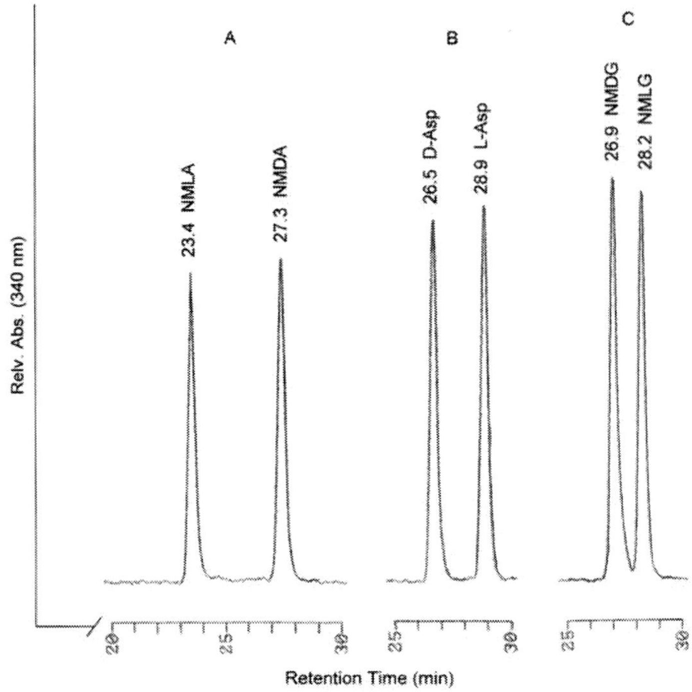

Figure 2. HPLC chromatograms for standards of (A) NMLA and NMDA, (B) D-Asp and L-Asp, and (C) NMDG and NMLG, each derivatized with FDNP-D-Val-NH_2.

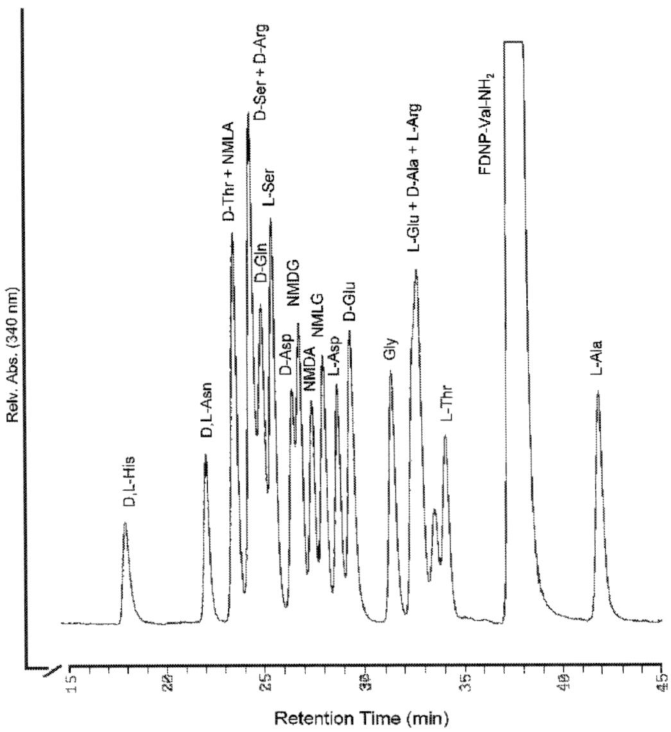

Figure 3. HPLC chromatogram of a standard mixture of amino acids derivatized with FDNP-D-Val-NH_2.

Peak areas as a function of concentration are linear in the range down to 5-10 pmoles, thus allowing calculation of the concentrations of these amino acids in biological tissue samples.

Figure 4 shows an HPLC chromatogram of a tissue extract of the mollusk bivalve *Scapharca broughtonii*, clearly indicating the presence of D-Asp, NMLA, NMDA, and NMDG, thus confirming results of previous researchers [13-15,17].

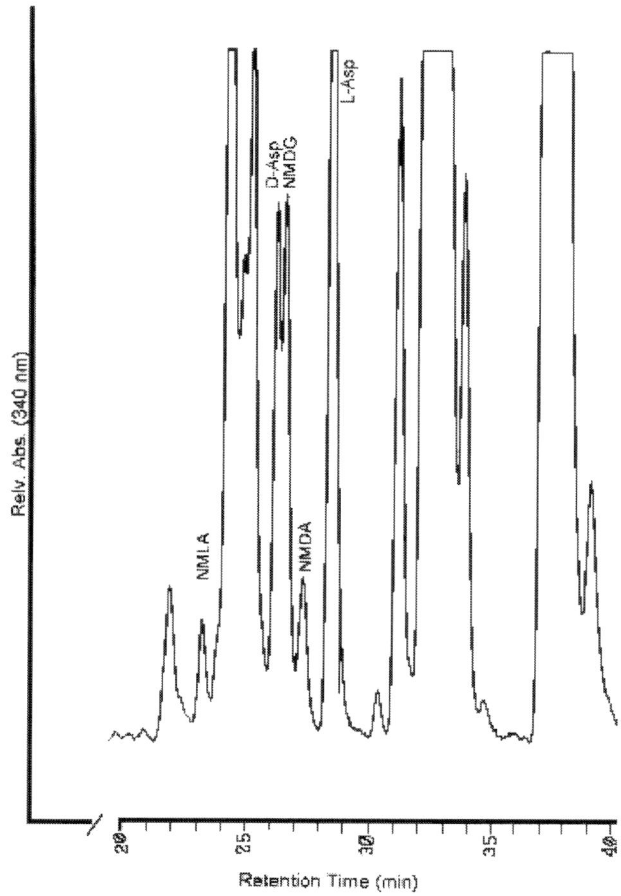

Figure 4. HPLC chromatogram of a tissue extract from the bivalve mollusk *Scapharca broughtonii* showing definite presence of D-Asp, *N*-methyl-(D & L)-Asp, and *N*-methyl-D-Glu.

5. Notes

1. This method is highly reliable and relatively fast (approximately 50 min HPLC run time). With this method the D- and L- enantiomers of *N*-methyl aspartic acid, *N*-methyl glutamic acid, and aspartic acid are well resolved. After extensive studies with all 20 natural amino acids plus GABA and taurine, only D-Thr partially overlaps with *N*-Me-L-Asp while *N*-Me-D-Asp stays well separated from any other amino acids. However, if one is interested in detecting specifically *N*-Me-L-Asp, this problem can be easily overcome by using the other enantiomer of the derivatizing agent (FDNP-L-Valine-NH_2). Then the order of elution will be reversed, and now *N*-Me-L-Asp will be resolved and *N*-Me-D-Asp will be obscured by L-Thr.
2. A definitive advantage to this method is that pretreatment of tissue samples with OPA to remove primary amino acids is not necessary since none of the diastereomers of these other amino acids overlap with D-Asp, *N*-Me-D-Asp or *N*-Me-D-Glu.

3. The order of elution of the D- and L- enantiomers of the derivatized amino acid can be proven by any of the following methods: a) derivatize and run each pure enantiomer and note its retention time; b) spike a racemic mixture of a D,L-amino acid with one of its pure enantiomers and note which HPLC peak increases; c) treat the amino acid with D-aspartate oxidase (D-AspO, EC 1.4.3.1) which specifically catalyzes oxidization of the D-amino acids NMDA, NMDG, and D-Asp such that the peak corresponding to the D-enantiomer is reduced or disappears.

ACKNOWLEDGEMENTS

This research was funded by an NIH grant MBRS SCORE: SO6 GM-45455. We wish to express sincerely appreciate to Dr. Hiroshi Homma and Dr. Ryo-Hei Yamada for generously providing samples of the mollusk bivalve *Scapharca broughtonii.*

REFERENCES

[1] Curtis, D. R. & Watkins, J. C. (1963). Acidic amino acids with strong excitatory actions on mammalian neurons. *J Physiol, 166,* 1-14.

[2] Watkins, J. C. & Evans, R. H. (1981). Excitatory amino acid transmitters. *Ann Rev Pharmacol Toxicol, 21*, 165-204.

[3] Monaghan, D. T. & Cotman, C. W. (1986). Identification and properties of *N*-methyl-D-aspartate receptors in rat brain synaptic plasma membranes. *Proc Natl Acad Sci USA, 83,* 176-179.

[4] Monaghan, D. T., Bridges, R. J., & Cotman, C. W. (1989). The excitatory amino acid receptors: their classes, pharmacology, and distinct properties in the function of the central nervous system. *Ann Rev Pharmacol Toxicol, 29*, 365-402.

[5] Hashimoto, A. & Oka, T. (1997). Free D-aspartate and D-serine in the mammalian brain and periphery. *Prog Neurobiol, 52,* 325-353.

[6] Mondadori, C., Weiskrantz, L., Buerki, H., Petschke, F. & Fagg, G. E. (1989). NMDA receptor antagonists can enhance or impair learning performance in animals. *Exp Brain Res, 75,* 449-456.

[7] Rison, R. A. & Stanton, P. K. (1995). Long-term potentiation and *N*-methyl-D-aspartate receptors: foundations of memory and neurologic disease? *Neurosci Biobehav Rev, 19*, 533-552.

[8] Hlinak, Z. & Krejci, I. (2002). *N*-Methyl-D-aspartate improved social recognition potency in rats. *Neurosci Letts, 330,* 227-230.

[9] Sato, M., Inoue, F., Kanno, N. & Sato, Y. (1987). The occurrence of *N*-methyl-D-aspartic acid in muscle extracts of the blood shell, *Scapharca broughtonii. Biochem J, 241*, 309-311.

[10] D'Aniello, A., Di Fiore, M. M., Fisher, G. H., Milone, A., Seleni, A., D'Aniello, S., Perna, A. & Ingresso, D. (2000). Occurrence of D-aspartic acid and *N*-methyl-D-aspartic acid in rat neuroendocrine tissues and their role in the modulation of luteinizing hormone and growth hormone release. *FASEB J, 14,* 699-714.

[11] Aniello, G. D., Tolino, A., D'Aniello, A., Fisher, G. H. & Di Fiore, M. M. (2000). The role of D-aspartic acid and *N*-methyl-D-aspartic acid in the regulation of prolactin release. *Endocrinol, 141,* 3862-3870.

[12] Sciuto, S., Piattelli, M. & Chillemi, R. (1979). *N*-methyl-L-aspartic acid from red alga *Halopytis incurvus*. *Phytochem, 18,* 1058.

[13] Sato, M., Nakano, T., Takeuchi, M., Kanno, N., Nagahisa, E. & Sato, Y. (1987). Distribution of neuroexcitatory amino acids in marine algae. *Phytochem, 42*, 1595-1597.

[14] Shibata, K., Tarui, A., Todoroki, N., Kawamoto, S., Takahashi, S., Kera, Y. & Yamada, R.-H. (2001). Occurrence of *N*-methyl-L-aspartic acid in bivalves and its distribution compared with that of *N*-methyl-D-aspartic acid and D,L-Asp. *Comp Biochem Physiol B, 130,* 493-500.

[15] Tarui, A., Shibata, K., Takahashi, S., Kera, Y., Munegumi, T. & Yamata, R.-H. (2003). *N*-methyl-D-glutamic acid and *N*-methyl-L-glutamic acid in *S. broughtonii* and other invertebrates. *Comp Biochem Physiol B, 134,* 79-87.

[16] D'Aniello, A., De Simone, A., Spinelli, P., D'Aniello, S., Branno, M., Aniello, F., Rios, J., Tsesarskaja, M. & Fisher, G. (2002). A specific high-performance liquid chromatography method to determine *N*-methyl-D-aspartic acid in biological tissues. *Anal Biochem, 308,* 42-51.

[17] Todoroki, N., Shibata, K., Yamada, T., Kera, Y. & Yamada, R.-H. (1999). Determination of *N*-methyl-D-aspartic acid in tissues of bivalves by HPLC. *J Chromatog B, 728,* 41-47.

[18] Skine, M., Fukuda, H., Nimura, N., Furuchi, T. & Homma, H. (2002). Automated column-switching high-performance liquid chromatography system for quantifying *N*-methyl-D- and -L-aspartate. *Anal Biochem, 310,* 114-121.

[19] Szókán, G., Mezö, G. & Hudecz, F. (1988). Application of Marfey's reagent in racemization studies of amino acids and peptides. *J Chromatog, 444,* 115-122.

In: D-Amino Acids:
A New Frontier in Amino Acid and Protein Research, pp.49-57
ISBN 1-60021-075-9

Chapter 2.5

Methods for Total Determination of D-Aspartic Acid, D-Glutamic Acid and *N*-Methyl-D-Aspartic Acid and for the Specific Determination of D-Aspartic Acid

Patrizia Spinelli[1], Enrico D'Aniello[1], Gabriele Farrandino[1], Enza Topo[1], Paola Pulzella[1], Tiziana Pollio, Angela Di Maio[1], George Fisher[2] and Antimo D'Aniello[1,*]

[1]Stazione Zoologica "A. Dohrn", Italy;
[2]Department of Chemistry, Barry University, USA.

Keywords: D-aspartic acid, D-glutamic acid, *N*-methyl-D-aspartic acid, methods

1. Introduction

Among various free D-amino acids present in animal tissue, D -alanine (D-Ala), D-serine (D-Ser) and D-aspartic acid (D-Asp) are those amino acids mostly diffuse in animals [1-3]. D-Ala is present in insects [1], crustaceans [4], echinoderms [5], molluscs [6] and mammals [7]. The biological role of D-Ala in mammals is not known. However, in marine animals, it has been hypothesized that this amino acid plays a role in the regulation of the animal's osmolarity with the external environment [6]. In fact, D-Ala is significantly higher concentrated in mutant mice lacking D-amino acid oxidase (D-AAO; EC 1.4.3.3; the enzyme that metabolises D-amino acids) than in normal mice, implying a possible role of D-AAO in metabolising D-amino acids [8]. Furthermore, if rats ingest large amounts of D-Ala (over the possibility to be destroyed by the D-AAO), a physiological damage occurs in the animal [9].

D-Ser has been found in various regions of rat brain and periphery nervous tissue [10-12] and in human brain [13-15]. D-Ser has been localized in astrocytes of grey-matter, hippocampus, amygdala and olfactory nucleus, has been defined as an endogenous synaptic modulator [11-12], and it has been hypothesized that D-Ser is an endogenous ligand for the glycine site of the NMDA-type glutamate

* Correspondence concerning this article should be addressed to Antimo D'Aniello, E-mail: daniello@szn.it, FAX: +39 081 7641 355.

receptors [15]. D-Ser also has been found in nervous tissues of mice along with D-proline (D-Pro) and D-Ala [16].

D-Asp was found for the first time in 1977 in the brain of *Octopus vulgaris* [17]. Since then, this amino acid has been widely studied. It is present in central and peripheral nervous tissues and in endocrine glands of many invertebrates and vertebrates [2-3], as well as in exocrine glands [18]. In the nervous system of chicken and mammals, during the embryonic stage and in the early postnatal life, the concentration of D-Asp increases very significantly in the brain and in the retina, implying a role of this amino acid in the development of the nervous system [19-20]. D-Asp also plays a role in neurotransmission acting as a neurotransmitter/neuromodulator in *Aplysia limacina* [3]. In this animal D-Asp has been detected in synaptosomes and in synaptic vesicles, and from these cell bodies it is released following chemical or electric stimulation [3]. A protein that transports D-Asp from synaptic clefts to presynaptic nerves has been reported [21]. In the endocrine system, D-Asp is involved in the regulation of luteinizing hormone and testosterone [22], and spermatogenesis [23]. In addition D-Asp also is capable to induce the release of gonadotropin releasing hormone (GnRH) from the hypothalamus [24] and the release of luteinizing hormone (LH), growth hormone (GH) [24] and prolactin (PRL) from the pituitary from pituitary [25]. The molecular mechanism by which D-Asp stimulates testosterone synthesis is regulated by a mRNA of a StAR gene [26]. Recently it has been demonstrated that D-Asp also is able to enhance oxytocin and vasopressin as well as capable to enhance the release of GABA and dopamine [27].

Thus, because D-Asp is implicated in multiple important functions, it consequently became very important to determine its presence in biological fluids/tissues using appropriate colorimetric or chromatographic methods. In this chapter thus we described some of these methods that have been used with success in our previously studies [2-3,18,23-25]. Specifically, we describe two colorimetric and one fluorimetric method for the total determination of D-Asp, D-Glu and NMDA and a HPLC method for the specific determination of D-Asp which are applicable in any biological tissue.

2. Materials

1. TCA (trichloroacetic acid), NaOH, H_2O_2, acetonitrile for HPLC (Carlo Erba, Milan, Italy)
2. Standards of D-Asp, D-Glu, *N*-methyl-D-aspartic acid (NMDA) and L-amino acids (Sigma Chemical Corp., St. Louis, MO, USA)
3. Sep-Pak® Cartridges, ODS-C_{18} (octadecylsilyl-C_{18}) (Waters Corporation, Milford, MA, USA).
4. Cation-exchange resin: AG 50W-X8, 200-400 mesh (Bio-Rad Laboratories, Hercules, CA 94547 USA). Before use the resin is regenerated by treatment with 5 M NaOH, followed by distilled water until neutrality, and then treated with 6 M HCl, followed by equilibration in 0.01 M HCl.
5. HPLC Gold system (Beckman, Milan, Italy).
6. HPLC column: ODS-C_{18}, Supelcosil™ LC-18 column, 0.45 x 25 cm, 5 μm (Supelco, Inc., Belafonte, PA, USA or Sigma-Aldrich, Supelco, Milan, Italy).
7. Buffers: Citrate-phosphate buffer for HPLC 1.0 M, pH 5.3, obtained by mixing a given volume of 1.0 M potassium phosphate dibasic (K_2HPO_4) with a given volume of 0.5 M citric acid ($C_6H_8O_6$) until pH 5.3. Sodium pyrophosphate buffer 1.0 M, pH 8.2 and pH 10, obtained by mixing a given volume of 0.5 M of sodium pyrophosphate ($Na_4P_2O_7$) with a given volume of 1.0 M of sodium phosphate monobasis (NaH_2PO_4) until pH 8.2 and 10, respectively. Tris-HCl buffer 1 M, pH 8.2, obtained by mixing a given volume of 1 M Tris [Tris(hydroxymethyl) aminomethane], with a given volume of 1 M HCl until pH 8.2 (Carlo Erba, Milan, Italy or Sigma Chemical Corp., St. Louis, MO, USA).
8. Acetonitrile for HPLC (Carlo Erba, Milan, Italy)

9. OPA-NAC (*o*-phthaldialdehyde-*N*-acetyl-L-cysteine) (Sigma Chemical Corp., St. Louis, MO, USA), prepared by dissolving 20 mg of OPA and 10 mg of NAC in 2 mL of methanol.
10. 2,4-dinitrophenylhydrazine (Sigma Chemical Corp. St. Louis, MO, USA), 5 mM solution in 5 M HCl.
11. POD (horseradish peroxidase), 2,500 U/mL (Sigma Chemical Corp., St. Louis, MO, USA).
12. 4-Aminoantipyrine and 2,4-dichlorophenol (Sigma Chemical Corp., St. Louis, MO, USA).
13. Tyramine hydrochloride (Sigma Chemical Corp., St. Louis, MO, USA)
14. D-Aspartate oxidase (D-AspO; EC 1.4.3.1), purified at 1-2 mg/mL, obtained from beef kidney by overexpression in *Escherichia coli* and then purified [1]. This enzyme reacts only with D-Asp, NMDA and at very low rate toward D-Glu. Any other D- or L-amino acid does not at all react with this enzyme. The enzyme catalyses the oxidation of these three amino acids according the following reaction [2]:

$$\text{D-Asp or NMDA or D-Glu} + O_2 + H_2O \xrightarrow{\text{D-AspO}} \alpha\text{-ketoacid} + H_2O_2 + NH_3 \text{ or } CH_3NH_2$$

(NH_3 is formed when D-Asp or D-Glu is used as substrate; CH_3NH_2 is formed when NMDA is used as substrate).

3. Methods

3.1. Extraction and Purification of free Amino Acids from Biological Tissues

Described here is a procedure starting with a given amount of tissue, however, the same can be accommodated by proportionally varying the volume of reagents. 200 mg tissue are homogenized with 1.8 mL of 0.1M TCA and centrifuged at 10-15,000 rpm. The supernatant is purified on a column of cation exchange resin AG 50W-X8 (200-400 mesh) containing 1 mL of settled resin (previously regenerated with 6 M HCl and equilibrated in 0.01 M HCl). After the sample is absorbed on the resin, the column is washed with 10 mL of 0.01 M HCl, by using 5 mL at a time. During this phase pigments, ketoacids, and all other organic compounds not containing an amino acid group are discarded. After washing the column is eluted with 4 mL of 4 M NH_4OH, using 2 mL at a time, and the eluent is collected and dried in a Petri dish placed on a plate under a hood, at 40-60 °C. The residue is dissolved in 1 mL of distilled water and then passed through a Sep-Pak ODS-C_{18} resin cartridge containing 400 mg resin, followed by 1 mL of distilled water and used for the determination of free D-amino acids.

The above procedure permits to one to obtain amino acids purified from proteins, peptides, organic pigments and other organic compounds and from salts in the original sample, including samples rich in pigments and many type of organic compounds such as liver or renal homogenates and urine and faeces of animals.

3.2. Colorimetric Method for total Determination of D-Asp, D-Glu and NMDA Based on the Measurement of the α-Ketoacids

This method is based on the determination of the α-ketoacids produced by a reaction between the above amino acids and D-aspartate oxidase (D-AspO). The α-ketoacids then react with 2,4-dinitrophenylhydrazine to yield a yellow hydrazone complex, which, in turn, reacts with NaOH to form an intense purple colour, measurable at 445 nm [2-3,17,22-25]. Since the D-AspO acts only on D-Asp, D-Glu and

NMDA [2], using this method we can determine only the total sum of these three amino acids. The reactions are the following:

1. D-Asp or D-Glu or NMDA + O_2 + H_2O $\xrightarrow{\text{D-AspO}}$ α-ketoacids + H_2O_2 + NH_3 (or CH_3-NH_2)
2. α-ketoacids + 2,4-dinitrophenylhydrazine → α-ketoacids-dinitrophenylhydrazone complex
3. α-ketoacids-dinitrophenylhydrazone complex + NaOH→dinitrophenylhydrazone (purple colour).

Briefly, the procedure is the following: In two Epperdorf tubes (sample and blank sample) are added 500 μL of purified sample and 50 μL of 1 M Tris-HCl, pH 8.2. Then, to the first tube (sample) is added 5 μL of purified D-AspO (1-2 mg/mL) obtained from beef kidney by overexpression [2] and mixed. The two tubes are incubated at 37°C for 30 min. After incubation, 50 μL of 5 mM 2,4-dinitrophenylhydrazine (5 mM dissolved in 5 M HCl) is added to the assay mixture, mixed and left at room temperature for 10 min. Then, 400 μL of 2 M NaOH is added and mixed. After 5 min, the absorbance of the sample is read at 445 nm against the blank sample.

In the same way, a standard and a general blanks are carried out. The standard is composed of 500 μL of D-Asp, 0.1 μmol/mL, 50 μL of 1 M Tris-HC, pH 8.2, and 5 μL of D-AspO; the general blank is the same as the standard, but without D-AspO. The standard and blank are incubated at 37°C and continued as sample and then the absorbance of the standard is read against the general blank at 445 nm.

The total quantity of D-Asp+D-Glu+NMDA in the sample is expressed as a D-Asp equivalent and is calculated as follows: O.D. sample divided by O.D. standard, times 0.1 = μmol in 500 μL of purified sample.

3.3. Colorimetric Method for total Determination of D-Asp, D-Glu and NMDA based on the H_2O_2 Measurement

This method is based on the determination of H_2O_2 produced by the reaction between the above amino acids and D-AspO. The H_2O_2 reacts with 4-aminoantipyrine and 2,4-dichlorophenol in the presence of POD (horseradish peroxidase) and turns into a red colour, measurable at 510 nm [2]. Because of the specificity of D-AspO [2], only D-Asp, D-Glu and NMDA can be determined. The reactions are the following:

D-Asp or D-Glu or NMDA + O_2 + H_2O $\xrightarrow{\text{D-AspO}}$ α-ketoacids + H_2O_2 + NH_3 (or CH_3NH_2)

H_2O_2 + 4-aminoantipyrine + 2,4-dichlorophenol $\xrightarrow{\text{POD}}$ red colour

Briefly, the procedure is the following: In two Epperdorf tubes (sample and blank sample) are added 500 μL of purified sample and 100 μL of an assay mixture prepared by mixing 1.0 mL of 1 M Tris-HCl buffer, pH 8.2, + 0.5 mL of 10 mM of 4-aminoantipyrine, + 0.5 mL of 8 mM 2,4-dichlorophenol, + 20 μL of POD, 2,500 U/mL. Then, to the first tube (sample) is added 5 μL of purified D-AspO (1-2 mg/mL). To the second tube blank is added 5 μL distilled water. Both tubes are mixed and incubated at 37°C for 30 min. Then, the absorbance of the sample is read at 510 nm against the blank sample. In the same way, is carried out a standard and a general blank. The standard consists of 500 μL of D-Asp, 0.1 μmol/mL, instead of sample, 50 μL of the assay mixture and 5 μL of D-AspO. The general blank consists of 500 μL of distilled water, 50 μL of the assay mixture and 5 μL of D-AspO. Both tubes are incubated as above and the absorbance of the standard is read at 510 nm against the general blank sample.

The total amount of D-Asp+D-Glu+NMDA in the sample is expressed as D-Asp equivalent and is calculated as follows: O.D. sample divided O.D. standard times 0.1 = μmol in 500 μL of purified sample.

3.4. Fluorimetric Method for Determination of D-Asp, D-Glu and NMDA Based on the Oxidized Tyramine Measurement

This method, is based on the fact that H_2O_2 develops from the reaction between one of above D-amino acids and D-AspO, and reacts with tyramine giving a highly fluorescent biphenyl derivative in the presence of POD (peroxidase) [28]. Because of the specificity of D-AspO only D-Asp, D-Glu and NMDA can be determined [2]. The reactions are the following:

$$\text{D-Asp or D-Glu or NMDA} + O_2 + H_2O \xrightarrow{\text{D-AspO}} \alpha\text{-ketoacids} + H_2O_2 + NH_3 \text{ (or } CH_3\text{-}NH_2\text{)}$$

$$H_2O_2 + \text{tyramine} \xrightarrow{\text{POD}} \text{fluorescent biphenyl derivative}$$

Briefly, the procedure is the following: In two Epperdorf tubes (sample and blank sample) are added 100 μL of purified sample and 100 μL of an assay mixture prepared by mixing 20 mg of tyramine hydrochloride in 10 mL of 0.2 M Tris-HCl, pH 8.2, plus 20 μL of POD (peroxidase, 2500 U/mL). To the first tube (sample) is added 2 μL of D-AspO (1-2 mg/mL). Both tubes are mixed and incubated at 37°C for 30 min. After that, 1.0 mL of 0.2 M Tris-HCl, pH 8.2, is added and mixed. The fluorescence is read at 325 nm excitation and 415 nm emission against the blank sample.

In the same way, is prepared a standard and a general blank. The standard consists of 100 μL of D-Asp at the concentration 0.1 μmol/mL, 100 μL of the assay mixture and 2 μL of D-AspO. The general blank consists of 100 μL of distilled H_2O, 100 μL of assay mixture and 2 μL of D-AspO. Both tubes are incubate at 37°C and continued as sample. The fluorescence of the standard is read against the general blank.

The total quantity of D-Asp+D-Glu+NMDA in the sample is expressed as D-Asp equivalent and is calculated as follows: O.D. sample divided by O.D. standard times 0.1 = μmol in 100 μL of purified sample.

3.5. Enzymatic-HPLC Method for the Specific Determination of D-Asp

The method is based on the chromatographic separation and quantification of D-Asp from L-Asp and other amino acids by HPLC. This method was pioneered by Aswad in 1984 [29] and has been used for a long time. Here we describe a modified procedure consisting of the introduction in the procedure the use of the D-AspO, the enzyme that oxidizes D-Asp, in order to measure specifically only D-Asp without any interference due to other compounds (if present in the sample) that can be eluted by HPLC at the same retention time as D-Asp. In this case, in fact, when the sample is analyzed by HPLC before and after treatment with D-AspO, it is possible to know specifically the peak area of D-Asp obtained by subtraction of the area of the sample after treatment with D-AspO from that of obtained before treatment with D-AspO.

The procedure is the following: In two Epperdorf tubes are mixed 20 μL of purified sample with 100 μL of 0.1 M sodium pyrophosphate, pH 8.2. Then, to the first tube is added 2 μL of distilled water and to the second tube is added 2 μL of purified D-AspO (1-2 mg/mL). Both tubes are incubated at 37°C for 30 min. After incubation to each tube is added 100 μL of 0.2 M sodium pyrophosphate, pH 10, and 20 μL of OPA-NAC reagent. After 2 min, distilled water is added to obtain a final volume of 1.0 mL. 100 μL of this mixture is injected into an ODS-C_{18} Supelcosil column (0.45 x 25 cm) connected to a Beckman-Gold HPLC system. The column is eluted at 1.2 mL/min, with a gradient consisting of solution A (50 mL acetonitrile + 920 mL of H_2O + 30 mL of 1.0 M citrate/phosphate buffer (McIlvaine buffer, pH 5.6) and solution B (90% acetonitrile in distilled water). The program gradients are the following: 0-10% solution B over 7 min; 7-100% solution B over next 2 min, 100% solution B for 6 min and returning to 100% A in 1 min. The fluorescence is read at an excitation wavelength of 330 nm and at an emission wavelength of 450 nm. D-Asp elutes with a peak at approximately 5.4 min, followed by L-Asp 0.7-0.8 min later and well separated from other amino acids. After the HPLC analysis of the sample contained in the first tube, the same HPLC analysis is carried out on the sample contained in the second tubes. In the same way is carried out a standard of D-Asp consisting of 10 µL of a mixture consisting of 17 L-amino acid standard, each at the concentration of 0.1 µmol/mL/amino acid, plus D-Asp at a concentration of 0.02 µmol/mL.

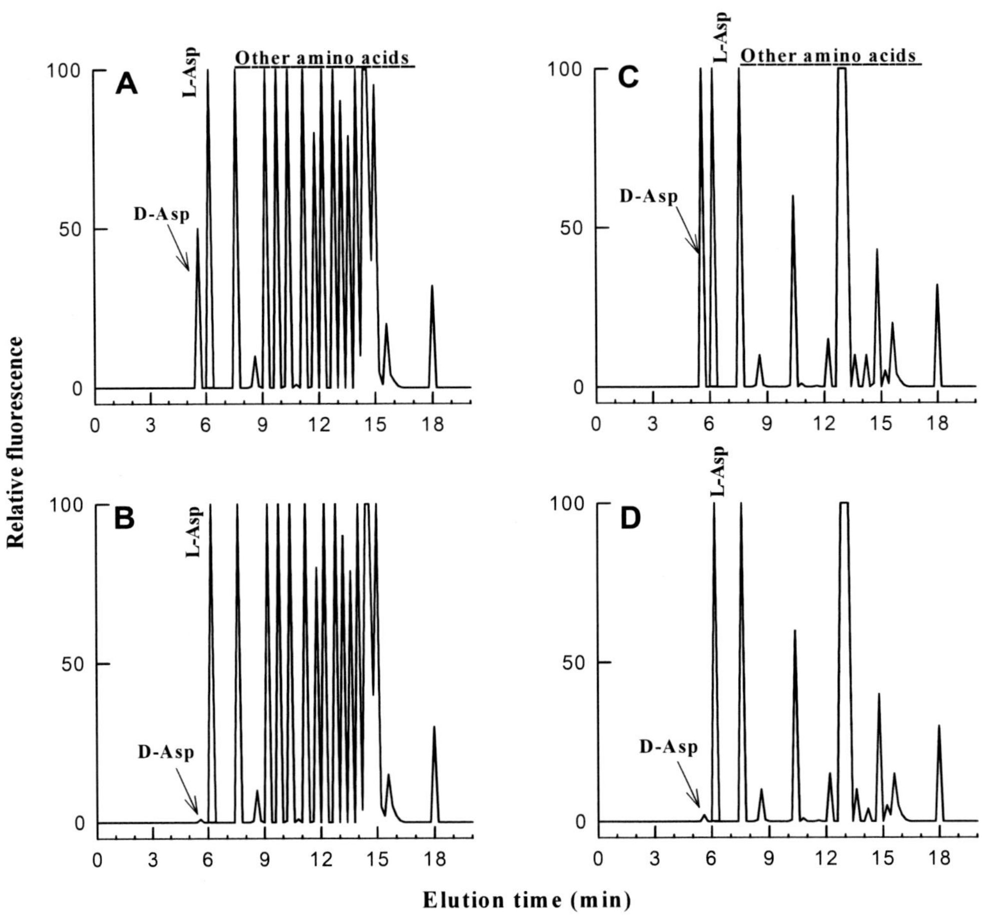

Figure 1. Typical example of determination of D-Asp in a standard mixture of amino acids and in a sample of *Octopus vulgaris* brain by HPLC. *Panel A*: HPLC profile of a standard mixture of amino acids (Sigma) consisting of 10 pmoles of D-Asp and 20 pmoles of each of other 17 standard amino acids. *Panel B*: HPLC analysis of the same sample as in the panel A, but after that the sample has been treated with D AspO. The peak corresponding to D-Asp is completely abolished by the action of D-AspO. *Panel C*: HPLC analysis of a sample consisting of a supernatant of a TCA homogenate of *Octopus vulgaris* brain. The peaks express amino acid compositions corresponding to 0.1 mg tissue. *Panel D*: HPLC analysis of the same sample as in the panel C, but after its incubation with D-AspO. The peak corresponding to D-Asp is completely abolished indicating the presence of D-Asp in this tissue.

In order to calculate the quantity of D-Asp in the sample, the following formula is used: Area of the peak of D-Asp of the sample (first tube) minus the area of peak corresponding to D-Asp of the blank sample (second tube) divided by the area of the peak of D-Asp of standard times 0.2 = nmol of D-Asp in 10 µL of sample.

4. RESULTS AND NOTES

In this chapter we described three methods for the determination of D-Asp plus NMDA plus D-Glu and a method for the determination of only D-Asp in any biological tissue. Among the methods for the total determination of D-Asp, D-Glu and NMDA two of them are colorimetric methods and measure the ketoacids or the H_2O_2 developed by one of the above D-amino acids and D-aspartate oxidase. These two methods permit us to determine an amount of D-Asp consistently of 1 nmol/assay. The third method is a fluorimetric assay which is highly sensitive and also based on the determination of H_2O_2 obtained by the enzymatic reaction between the D-amino acids and D-aspartato oxidase. The sensitivity of the method is such to determine consistently 0.1 nmol/assay. Lastly, we also described a sensitive and specific HPLC method for the direct determination of D-Asp which is based on the separation and quantification of D-Asp from all other amino acids. In addition, this last method is really specific for the determination of D-Asp since the sample is analysed before and after treatment with D-AspO, and then all other compounds that could elute at the same elution time as D-Asp will be considered (see Fig 1). This last method permits us to detect and amount of D-Asp of 10 pmoles/assay.

REFERENCES

[1] Maister A. (1965). Biochemistry of the Amino Acids (1965). In: A. Maister (2nd Ed), Vol. 1, Academic Press, New York, London, pp 113-139.

[2] D'Aniello, A., Spinelli, P., Ferrandino, G., Peterson, K., Tsesarskia, M., Fisher, G.H. & D'Aniello, A. (2005). Cephalopod vision involves dicarboxylic amino acids: D-aspartate, L-aspartate and L-glutamate. *Biochemical J, 386,* 331-340.

[3] Spinelli, P., Brown, E., Ferrandino,G., Branno, M., Montarolo, P.G. D'Aniello, E., Rastogi, R., D'Aniello B., Chieffi, G., Fisher, G. & D'Aniello A. (2006). D-Aspartic acid in the nervous system of *Aplysia limacina*: possible role in neurotransmission. *J. Cell Physiol, 206*, 672-681 (published on-line Oct. 2005).

[4] Okuma, E. & Abe, H. (1994). Simultaneous determination of D- and L-amino acids in the nervous tissues of crustaceans using precolumn derivatization with (+)-1-(9-fluorenyl)ethyl cloroformate and reversed-phase ion-pair high–performance liquid chromatography. *J Chromatog B, 660*, 243-250.

[5] D'Aniello, A., Nardi, G., Cipollaro, M., Piscetola, M., & Padula, L. (1990). Occurrence of D-alanine in the eggs and the developing embryo of the sea urchin *Paracentrotus lividus. Comp Biochem Physiol, 97B,* 291-290.

[6] Okuma, E., Watanabe, K., & Abe, H. (1998). Distribution of free D-amino acids in bivalve molluscs and the effects of physiological conditions on the levels of D- and L-alanine in the tissues of the hard clam *Meretrix lusoria. Fish Sci, 64*, 606-611.

[7] Nagata, Y., Yamamoto, K., Shimojo, T., Konno, R., Yasumura, Y., & Akino, T., (1992). The presence of D-alanine, D-proline and D-serine in mice. *Biochim Biophys Acta, 1115*, 208-211.

[8] Morikawa, A., Hamase, H., Inoue, T., Konno, R. Niwa, A. & Zaitsu, K. (2001). Determination of free D-aspartic acid, D-serine and D-alanine in the brain of mutant mice lacking D-amino acid oxidase activity. *J Chromatog B, 757*, 119-125.

[9] D'Aniello, A., D'Onofrio, G,. Pishetola, M, D'Aniello, G., Vetere, A., Petrucelli, L, & Fisher, G. (1993). Biological role of D-amino acid oxidase and D-aspartate oxidase. Effects of D-amino acids. *J Biol Chem, 268,* 26941-26949.

[10] Hashimoto, A., Oka, T., & Nishikawa, T. (1995). Anatomical distribution and postnatal changes in endogenous free D-aspartate and D-serine in rat brain and periphery. *Europ. J Neurosci, 7,* 1657-1663.

[11] Schell, M. J. Molliver, M. E., & Snyder, H. S. (1995). D-Serine, an endogenous synaptic modulator: Localization to astrocytes and glutamate-stimulated release. *Proc Natl Acad Sci USA, 92,* 3948-3952.

[12] Schell, M. J., Brady Jr, R. O., Molliver, M. E., & Solomon, H. S. (1997). D-Serine as a neuromodulator: Regional and developmental localizations in rat brain glia resemble NMDA receptors. *J Neurosci, 17,* 1604-1615.

[13] Hashimoto, A., Kumashiro, S., Nishikawa, T., Oka, T,. Takahashi, K,. Mito, T., Takashima, S., Doi, N., Mizutani, Y., Kaneco, T., & Ootomo, E. (1993). Embryonic development and postnatal changes in free D-aspartate and D-serine in the human prefrontal cortex. *J Neurochem, 61,* 348–351.

[14] Nagata. Y., Borghi, M., Fisher, G. H., & D'Aniello, A. (1995). Free D-Serine concentration in normal and Alzheimer human brain. *Brain Res Bu .l, 38,* 181-183.

[15] Mothet, J. P., Parent, A. T., Woloscker, H., Brady Jr, R. O., Linder, D. J., Ferris, C. D., Rogawski, M.A., & Snyder, S.H. (2000). D-Serine is an endogenous ligand for the glycine site of the N-methyl-D-aspartate receptor. *Proc Natl. Acad. Sci. USA, 97,* 4927-4931.

[16] Nagata, Y., Yamamoto, K., Shimojo, T., Konno, R., Yasumura, Y., & Akino, T. (1992). The presence of D-alanine, D-proline and D-serine in mice. *Biochim. Biophys. Acta., 1115,* 208-211.

[17] D'Aniello, A. & Giuditta, A. (1978). Presence of D-aspartate in squid axoplasm and in other regions of the cephalopod nervous system. *J Neurochem, 31,* 1107-1108.

[18] Raucci, F., Santillo, F., D'Aniello, A., Chieffi, P. & Chieffi Baccari, G. (2005). D-aspartate modulates trascriptional activity in harderian gland of frog, *Rana esculenta*: Morphological and Molecular Evidence. *J Cell Physiol 204,* 445-454.

[19] Neidle, A. & Dunlop, D. S. (1990). Developmental changes of free D-aspartic acid in the chicken embryo and in the neonatal rat. *Life Sci, 46,* 1517–1522.

[20] Hashimoto, A, Kumashiro, S, Nishikawa, T, Oka, T, Takahashi, K, Mito, T, Takashima, S, Doi N., Mizutani, Y., Kaneo, T, & Ootomo, E. (1993). Embryonic development and postnatal changes in free D-aspartate and D-serine in the human prefrontal cortex. *J Neurochem, 61,* 348–351.

[21] Kanai, Y. & Hediger, M. A. (1992). Primary structure and functional characterization of a high-affinity glutamate transporter. *Nature, 360,* 467-471.

[22] D'Aniello, A., Di Cosmo, A., Di Cristo, C., Annunziato, L., Petrucelli, L., & Fisher, G. H. (1996). Involvement of D-aspartic acid in the synthesis of testosterone in rat testes. *Life Sci, 59,* 97-104.

[23] D'Aniello, A., Di Fiore, M.M., D'Aniello, G., Colin, F.E., Lewis, G., & Setchell, B.P. (1998). Secretion of D-aspartic acid by the rat testis and its role in endocrinology of the testis and spermatogenesis. *FEBS Letters, 436,* 23-27.

[24] D'Aniello A., Di Fiore M.M., Fisher G.H., Milone A., Seleni A., D'Aniello S., Perna A. & Ingrosso D. (2000a). Occurrence of D-aspartic acid and *N*-methyl-D-aspartic acid in rat neuroendocrine tissues and their role in the modulation of luteinizing hormone and growth hormone release. *FASEB J, 14,* 699-714.

[25] D'Aniello G., Tolino A., D'Aniello A., Errico, F., Fisher G.H. & Di Fiore M.M. (2000b). The role of D-aspartic acid and *N*-Methyl-D-aspartic acid in the regulation of prolactin release. *Endocrinology, 141,* 3862-3870.

[26] Nagata, Y., Homma, H., Matsumoto, M. & Imai, K. (1999). Stimulation of steroidogenic acute regulatory protein (stAR) gene expression by D-aspartate in rat Leydig cells. *FEBS Letters, 454,* 317-320.

[27] Pampillo, M., Scimonelli, T., Bottin, M. C., Duvilanski, B. H., Rettori, V., Seilicovich, A., & Lasaga, M. (2002). The effect of D-aspartate on luteinizing hormone-releasing hormone, α-melanocyte-stimulating hormone, GABA and dopamine release. *Neuroreport, 13,* 2341–2344.

[28] Gross, A.J. & Sizer, I. W. (1959). The oxidation of tyramine, tyrosine, and related compounds by peroxidase. *J Bio Chem 234,* 1611-1614.

[29] Aswad, D.W. (1984). Determination of D- and L-aspartate in amino acid mixtures by high performance liquid chromatography after derivatization with a chiral adduct of *o*-phthaldialdehyde. *Anal Biochem, 118,* 405–407.

In: D-Amino Acids:
A New Frontier in Amino Acid and Protein Research, pp.59-64
ISBN 1-60021-075-9

Chapter 2.6

AN IMPROVED HPLC METHOD FOR DETERMINATION AND QUANTIFICATION OF D- AND L-ASPARTIC ACID

Erika Galindo[1], Mara Tsesarskaia[1], Susana Lopez[1], Collins Boston[1], Antimo D'Aniello[2], and George Fisher[1*]

[1]Department of Chemistry, Barry University, Miami Shores, FL, USA
[2]Department of Neurobiology, Stazione Zoologica 'A Dohrn', Naples, Italy

Keywords: D-aspartic acid (D-Asp), HPLC, *N*-methyl-D-Asp (NMDA), *o*-phthaldialdehyde (OPA), *N*-acetyl-L-cysteine (NAC)

1. INTRODUCTION

D-Aspartic acid (D-Asp) (Fig. 1) is an endogenous amino acid that occurs in the central nervous system and endocrine glands of vertebrates and invertebrates where it is involved in hormone release and synthesis [1,2]. D-Asp has also been shown to be involved in neurological disorders such as Alzheimer's disease, and debilitating diseases such as osteoarthritis [3-9]. The concentration of free D-Asp has been found to be lower in Alzheimer brains than in normal brains [3,4]. D-Asp is also the precursor molecule for the biosynthesis of *N*-methyl-D-aspartate (NMDA) [2], which is involved in the learning and memory processes. Thus, the decreased levels of D-Asp found in Alzheimer brains could contribute to the loss of memory associated with Alzheimer's disease. Other studies have shown that the concentration of protein-bound D-Asp is increased in phosphorylated tau protein of Alzheimer neurofibrillary tangles [5,6] and in β-amyloid protein of Alzheimer neuritic plaques [7]. D-Asp has also been found in cartilage of normal and osteoarthritic human knee joints [8,9].

With growing interest in D-Asp in biological/physiological tissues, an accurate and quick method for determination and quantification of D- and L-Asp has become of great importance. Since 1984 there have been several methods reported [10-14] for determination of D- and L-Asp by high performance liquid chromatography (HPLC) after derivatization with chiral fluorogens; e.g., D- and L-Asp react with *o*-phthaldialdehyde (OPA) and *N*-acetyl-L-cysteine (NAC) to form diastereomers (Fig. 2), which are

* Correspondence concerning this article should be addressed to George Fisher, E-mail: gfisher@mail.barry.edu; FAX: +1 305 899-3479.

separated on a reversed phase C_{18} column using isocratic or gradient elution with sodium acetate-methanol (NaOAc-MeOH), and fluorescence detection [9,10].

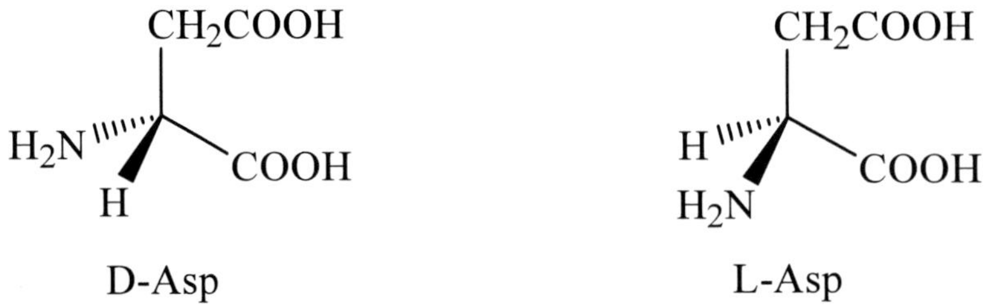

Figure 1. Stereochemical representations of D- and L-Asp.

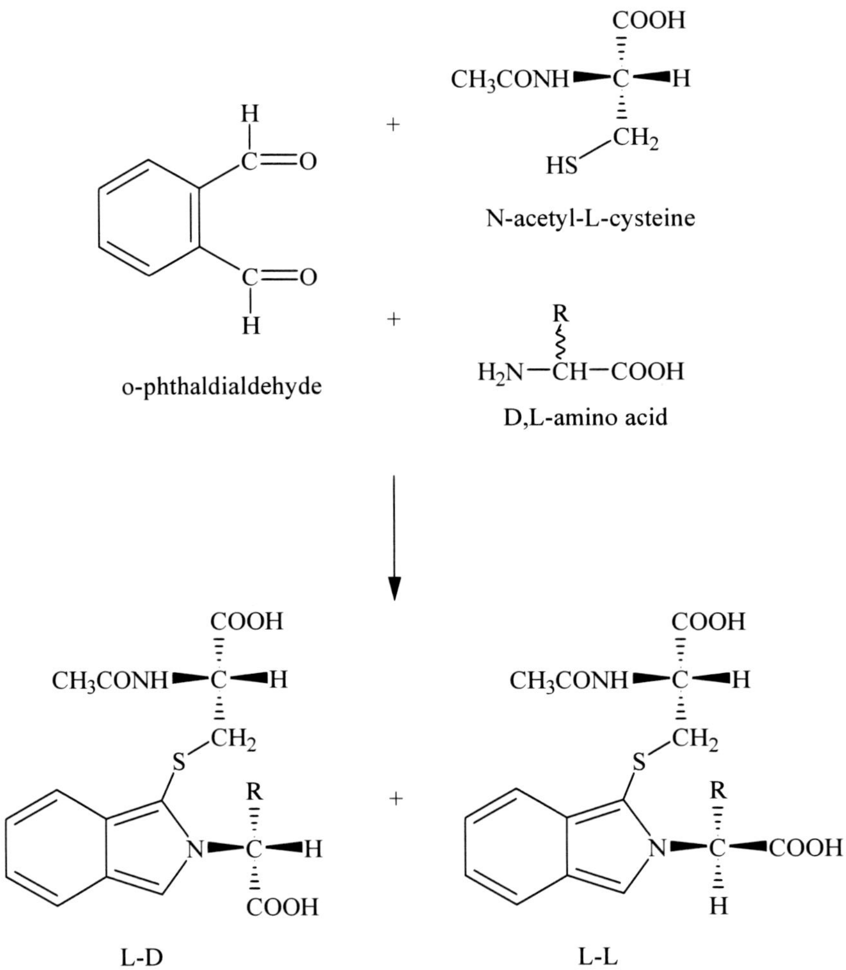

Figure 2. Derivatization of D/L-Asp with OPA-NAC to form a pair of fluorescent diastereomeric *N*-alkyl-2-thioalkyl isoindole derivatives.

We report here a method that gives better resolution of the D- and L- enantiomers of Asp in shorter time than those previous methods. We have developed an improved HPLC method to give maximum resolution and peak response within shorter retention times.

2. Materials

1. Chemicals: D,L-Asp, *o*-phthaldialdehyde (OPA, minimum 99% pure) and *N*-acetyl-L-cysteine (NAC) are from Sigma Chemicals. All solvents should be HPLC grade. Water should be doubly distilled or deionized through a water purification system such as Millipore® or NANOpure®.
2. Standards: A standard of 150 pmoles/μL of racemic D,L-Asp is prepared by dissolving 2.0 mg of D,L-Asp in water in a 100 mL volumetric flask.
3. Reagents: the OPA-NAC reagent is prepared by dissolving 4 mg OPA and 5 mg NAC in 0.5 mL of methanol.
4. Buffers: 0.1 M sodium borate buffer, pH 9.3-9.4, is prepared in water.
5. HPLC: The HPLC system should consist of at least a ternary or quaternary gradient pump, an in-line solvent degasser, a 10 μL injector loop, an ODS-C_{18} pre-column, connected to a 25 x 0.46 cm, Hypersil ODS-C_{18} (5μ) column, a fluorescence detector (325 nm excitation, 415 nm emission), and a computing integrator data handling system.

3. Methods

1. Derivatization: The D- and L- enantiomers of Asp are reacted with OPA-NAC reagent in sodium borate buffer to form a pair of fluorescent diastereomeric *N*-alkyl-2-thioalkyl isoindole derivatives (Fig. 2). Borate buffer is added to the appropriate amount of Asp standard, followed by the appropriate amount of OPA-NAC (see Table 1), mixed for 1 min, and then 10 μL are injected into the HPLC. A similar derivatization procedure can be done with a biological tissue sample extract instead of D,L-Asp standards.
2. The HPLC is calibrated with the racemic D,L-Asp standards, using the ratios shown in Table 1 of standard, borate buffer, and OPA-NAC to inject precise amounts of D- and L-Asp onto the column (using a 10 μL injector loop).

Table 1 – D,L-Asp Calibration.

D,L-Asp std pmoles/μL	μL Std	μL OPA-NAC	μL Borate	μL Total	μL injected	Total pmol Asp	pmoles D-Asp	pmoles L-Asp
150	5	5	365	375	10	20	10	10
150	10	5	285	300	10	50	25	25
150	15	5	205	225	10	100	50	50
150	30	10	185	225	10	200	100	100
150	50	10	90	150	10	500	250	250
150	100	20	30	150	10	1000	500	500

3. Mobile phases and HPLC gradient: The following mobile phases are used to elute the D- and L-Asp diastereomers from the column:

   A: 30 mM sodium citrate/10% MeOH (dissolve 8.823 g/L of $Na_3C_6H_5O_7 \cdot 2H_2O$ in water in a 1 L volumetric flask, adjust to pH 5.5-6.0 with solid citric acid, and add 111 mL of methanol to give a final concentration of 10% methanol; see Note 1).

   B: 100% methanol (MeOH).

   C. 100% acetonitrile (AcCN).

   The following gradient is used: 0-6 min, 100%A; 6-10 min linear gradient to 25%A, 40%B, and 35%C (to wash the column); returning to 100%A and equilibrating at 100%A over the next 5-6 min. All flow rates are 1.0 mL/min.

4. Quantification: A plot of peak areas vs. nmoles of Asp is made (Fig. 4). The areas under the peaks are used to quantify the picomoles of D- and L- Asp of the standards, which can then be used to quantify the picomoles of D- and L-Asp in biological samples.

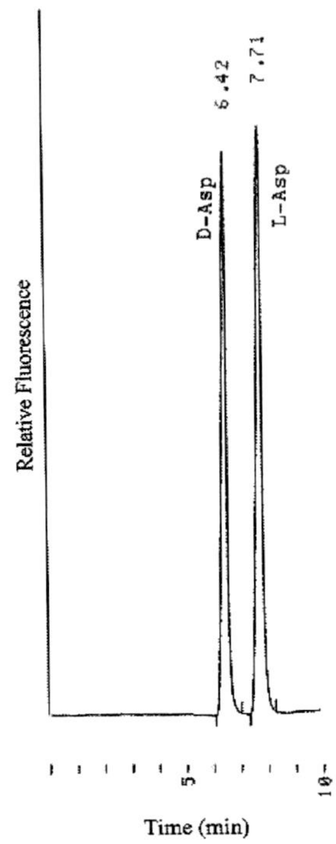

Figure 3. HPLC chromatogram of D- and L-Asp diastereomers.

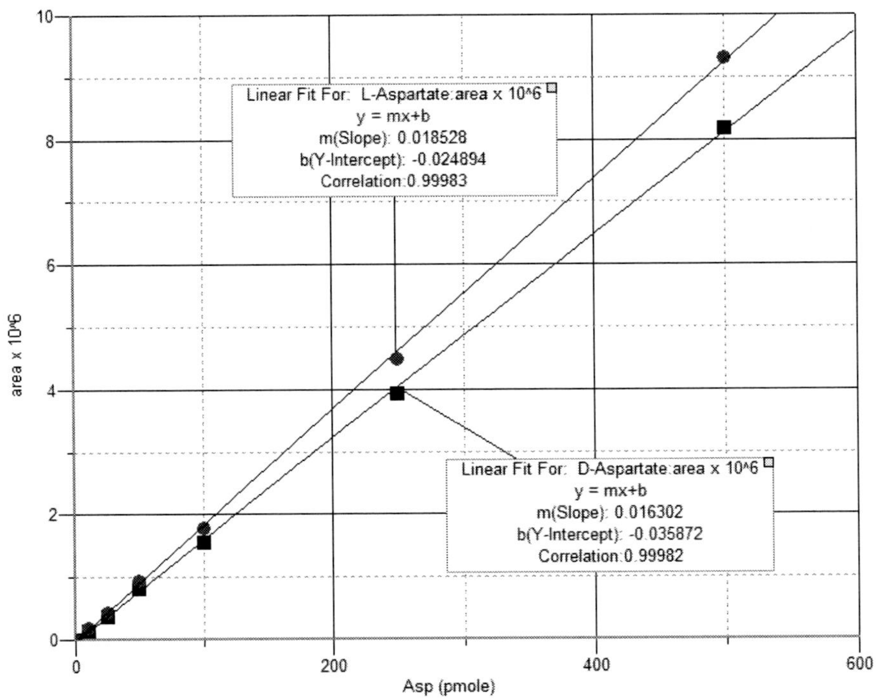

Figure 4. Fluorescence responses of D- and L-Asp. Linear correlation calibration of peak areas with picomoles of L-Asp (●) and D-Asp (■).

4. RESULTS

Under these conditions, D-Asp is well separated by 1.2 to 1.3 minutes from L-Asp (D-Asp elutes at ~6.3-6.5 min and L-Asp at ~7.6-7.8 min) (Fig. 3). Fluorescence responses (peak areas) of both the D- and L-Asp derivatives are linear over the range of 10 to 500 picomoles (correlation $r \geq 0.999$) (Fig. 4). The specific fluorescence of the D-Asp derivative is always about 12 ± 1% lower than that of L-Asp (Fig. 4). The D-Asp peak disappeares when incubated with D-aspartate oxidase (D-AspO; EC 1.4.3.1), which specifically catalyzes oxidation of D-aspartate [15], giving further proof that the first peak eluting at ~6.4 min is actually D-Asp.

Sodium citrate-methanol (NaCit-MeOH) gives better resolution in a shorter time than previously reported HPLC methods using sodium acetate-methanol gradients. In general, citrate tends to give better separation (resolution) than acetate. Since the molecular weight of citrate is much greater than that of acetate, an equimolar solution of citrate would tend to have a greater effect in its interaction with the OPA-NAC amino acid derivatives than does acetate. The organics (methanol and acetonitrile) tend to act as additional organic modifiers and decrease retention times as their concentration increases. Thus, this improved method is suitable for routine determination of small concentrations of D-Asp in multiple samples of biological tissues.

5. NOTES

1. The percent methanol in the citrate buffer may need to be adjusted slightly more or less in order to maximize the resolution and shorten the retention times. Instead of adding methanol directly to the citrate buffer, the HPLC pump may be programmed to add the methanol.
2. The OPA-NAC reagent is light sensitive and should be stored in a dark (light resistant) vial and prepared freshly at least every other day.
3. The HPLC should be calibrated with the D- and L-Asp standards each time a new batch of OPA-NAC reagent is prepared.
4. The picomoles of D- and L-Asp in 10 μL of a biological sample injected into the HPLC can be calculated by dividing the areas of D- and L-Asp sample peaks by the slope of the respective D- and L-Asp calibrations (e.g., as determined by calibration standards, Fig. 4).
5. If the D- and L-Asp are obtained from hydrolysis of a biological sample with 6 M HCl for 6 hr at 100°C, then the picomoles obtained from the HPLC should be multiplied by 0.995 to account for 5% hydrolysis-induced racemization.

ACKNOWLEDGEMENTS

This project was funded by grants from the National Institutes of Heath (NIH): MBRS SCORE S06 GM45455 and MARC T34 GM08021.

REFERENCES

[1] D'Aniello, A., DiFiore, M. M., and Fisher, G. (1998). Occurrence of D-aspartic acid in animal tissues and its role in the nervous and endocrine systems. *Trends in Comp Biochem Physiol, 4*, 1-24.

[2] D'Aniello, A., DiFiore, M. M., D'Aniello, G., Fisher, G. H., Milone, A., Seleni, A., D'Aniello, S., Perna, A.F., and Ingresso, D. (2000). Occurrence of D-aspartic acid and *N*-methyl-D-aspartic acid in rat neuroendocrine tissues and their role in the modulation of luteinizing hormone and growth hormone release. *FASEB J, 14*:699-714.

[3] Fisher, G. H., D'Aniello, A., Vetere, A. V., Padula, L., Cusano, G. P., and Man E. H. (1991). Free D-aspartate and D-alanine in normal and Alzheimer brain. *Brain Res Bull, 26*, 983-985.

[4] D'Aniello, A., Lee, J. M., Petrucelli, L., and Di Fiore, M. M. (1998). Regional decreases of free D-aspartate levels in Alzheimers's disease. *Neursci Letts, 250*, 131-134.

[5] Fisher, G. H., Payan, I. L., Chou, S.-J., Man, E. H., Cerwinski, S., Martin, T., Emory, C., and Frey, W. H. (1992). Racemized D-aspartate in Alzheimer neurofibrillary tangles. *Brain Res Bull, 28*, 127-131.

[6] Payan, I. L., Chou, S.-J., Fisher, G. H., Emory, C., and Frey, W. H. (1992). Altered aspartate in Alzheimer neurofibrillary tangles. *Neurochem Res, 17,* 187-191.

[7] Shapira R., Austin, G. E., and Mirra, S. S. (1988). Neuritic plaque amyloid in Alzheimer's disease is highly racemized. *J Neurochem, 50*, 69-74.

[8] Maroudas, A., Palla, G., and Gilav, E. (1992). Racemization of aspartic acid in human articular cartilage. *Connective Tissue Res, 28*, 161-70.

[9] Maroudas, A., Bayliss, M. T., and UchitelKaushansky, N. (1998). Aggregan turnover in human articular cartilage: Use of aspartic acid racemization as a marker of molecular age. *Archv Biochem Biophys, 350*, 61-71.

[10] Aswad, D.W. (1984). Determination of D- and L-aspartate in amino acid mixtures by high performance liquid chromatography after derivatization with a chiral adduct of *o*-phthaldialdehyde. *Anal Biochem, 137*, 405-407.

[11] Nimura, N. and Kinoshita, T. (1986). *o*-Phthaldialdehyde-*N*-acetyl-L-cysteine as a chiral derivatization reagent for liquid chromatographic optical resolution of amino acid enantiomers and its application to conventional amino acid analysis. *J Chromatog, 352*, 169-177.

[12] Nishikawa, T., Oka, T., Takahashi, K., and Hayashi, T. (1992). Determination of free amino acid enantiomers in rat brain and serum by HPLC after derivatization with *N-tert.*-butoxycarbonyl-L-cysteine and *o*-phthaldialdehyde. *J Chromatog, 582,* 41-48.

[13] Brüchner, H., Haasmann, S., Langer M., Westhauser, T., and Wittner, R. (1994). Liquid chromatographic determination of D- and L-amino acids by derivatization with *o*-phthaldialdehyde and chiral thiols. *J Chromatog, 666,* 259-273.

[14] Brüchner, H., Westhauser, T., and Godel, H. (1995). Liquid chromatographic determination of D- and L-amino acids by derivatization with *o*-phthaldialdehyde and *N*-isobutyl-L-cysteine. *J Chromatog, 711,* 201-215.

[15] D'Aniello, A., D'Onofrio, G., Pischetola, M., D'Aniello, G., Vetere, A., Petrucelli, L. and Fisher, G.H. (1993). Biological Role of D-Amino Acid Oxidase and D-Aspartate Oxidase: Effects of D-Amino Acids. *J Biol Chem, 268*, 26941-26949.

In: D-Amino Acids: ISBN 1-60021-075-9
A New Frontier in Amino Acid and Protein Research, pp.65-73

Chapter 2.7

Determination of Free D-Aspartate and D-Serine in Rat Brain and Periphery by Gas Chromatography and High-Performance Liquid Chromatography

Atsushi Hashimoto and Masanobu Yoshikawa*
Department of Clinical Pharmacology, Tokai University School of Medicine, Isehara, Japan.

Keywords: D-serine, D-aspartate, NMDA receptor, gas chromatography, high-performance liquid chromatography.

1. Introduction

A large body of evidence has been provided demonstrating that high levels of free D-aspartate and D-serine occur in the mammalian brain and periphery [1-8], although D-amino acids have long been assumed to be unnatural in mammals [7,9]. The transient emergence of a high concentration of D-aspartate is present in the brain and retina of chickens, rats and humans during early development [1]. D-Aspartate is widely observed in the adult rat periphery, such as the adrenals, testes, spleen and pituitary [6]. Because high concentrations of D-aspartate occur in the hormone-secreting organs such as the adrenals, testes, diencephalon, and pituitary, the D-amino acid could be involved in the regulation of proliferation and differentiation of the endocrine cells in mammals [6,7]. In contrast, D-serine is predominantly confined to the forebrain, where the N-methyl-D-aspartate (NMDA) type excitatory amino acid receptors exist [2,4,7,8]. In vivo microdialysis studies have also shown that the extracellular level of D-serine is similar to or is higher than that of glycine in the prefrontal cortex and striatum, respectively. Because D-serine potentiates the NMDA receptor-mediated transmission by selective stimulation of the glycine site of the NMDA receptor [11], D-serine has been proposed as an endogenous coagonist for an NMDA receptor-related glycine site in the mammalian brain [4,7].

Although enzymatic, gas chromatographic (GC) and high-performance liquid chromatographic (HPLC) methods have been successfully utilized to quantify D-amino acids, these techniques have some

* Correspondence concerning this article should be addressed to Atsushi Hashimoto, E-mail: hashimot@is.icc.u-tokai.ac.jp, FAX: +81-463-93-2896.

disadvantages, which include low specificity (e.g., enzymatic assay), time-consuming derivatization (e.g., GC analysis) or complicated extraction procedures. Further, they do not allow the simultaneous detection and determination of many amino acid enantiomers and some non-chiral amino acids such as glycine, β-alanine, O-phosphoethanolamine (O-PE), taurine and GABA. We show methods for the concurrent measurement of not only chiral but also non-chiral amino acids in the brain using GC and HPLC methods [2,3]. Especially, the HPLC method is rapid for sample preparation, shows high resolution and sensitivity and has revealed that considerable levels of free D-serine and D-aspartate are present in the brain and periphery [3,6].

2. MATERIALS

2.1. Materials for GC

1. Hewlett Packard 5890 equipped with nitrogen-phosphorus detector (Hewlett Packard, Japan).
2. SIC chromatocorder 11 integrator (System Instruments Co., Japan)
3. Chirasil-L-Val capillary column (25 m x 0.25 mm, film thickness 0.12 μm, Gasukuro Kogyo, Japan).
4. Pentafluoropropionic anhydride (PFPA, G.L.C. Science, Japan).
5. Trichloroacetic acid (TCA, Nacalai Tesque, Japan).
6. Isopropanol (Nacalai Tesque, Japan).
7. Methylene chloride (Nacalai Tesque, Japan).
8. Ethyl acetate (Nacalai Tesque, Japan).
9. Dry sodium sulfate (Nacalai Tesque, Japan).
10. All amino acids (Nacalai Tesque, Japan).
11. Nitrogen gas

2.2. Materials for HPLC

1. Waters 600E HPLC system (Waters, Japan).
2. 821-FPS spectrofluorometer (Jasco International, Japan)
3. Sample injector Model 231 (GILSON, Japan).
4. 4-μm Nova-Pak C18 column (300 mm x 3.9 mm I.D., Waters, Japan).
5. Precolumn (20 mm x 4.6 mm I.D., Eicom, Japan).
6. Waters 741 data module (Waters, Japan).
7. Acetate buffer: 0.1 M glacial acetic acid adjusted to pH 6.0 with sodium hydroxide.
8. Borate buffer: 0.4 M boric acid adjusted to pH 9.0 with sodium hydroxide.
9. N-tert-butyloxycarbonyl-L-cysteine (Boc-L-Cys, Novabiochem, Switzerland).
10. o-Phthaldialdehyde (OPA, Nacalai Tesque, Japan)
11. L-Homocysteic acid (L-HCA, Sigma, USA)
12. All solvents of HPLC grade (Kanto Chemicals, Japan)

3. METHODS

3.1. Animals and Tissue Preparation for GC and HPLC Analysis

1. Male Wistar rats (Clea Japan Inc., Japan) ranging in weight from 220-230 g at the time of the experiments.
2. The perfusion of rat brain by 0.9 % NaCl solution was carried out under pentobarbital anesthesia (40 mg/kg, intraperitoneally) to eliminate contamination by amino acids in the blood.
3. The perfused brain tissue was dissected out on ice and stored at -80 °C.
4. After the addition of L-norleucine for GC analysis or L-homocysteic acid for HPLC analysis as an internal standard, the brain tissue was homogenized in 10 volumes of 5 % TCA.
5. The homogenate was centrifuged at 16,000 X g for 30 min at 4 °C.
6. To remove TCA, the supernatant was washed three times with water-saturated diethyl ether.
7. The aqueous layer was then passed through a Millipore filter, HV (0.45 μm) and stored at -80 °C until derivatization.

3.2. Derivatization Procedure for GC

1. The 5 % TCA extract was evaporated under a gentle stream of nitrogen gas at room temperature. The last trace of waster was removed azeotropically using methylene chloride.
2. The residue was dissolved in 2 ml of isopropanol-2 N HCl.
3. After the mixture was heated at 100 °C for 1 h using a sea sand bath, the excess reagent and solvent were removed under a gentle stream of nitrogen gas at room temperature.
4. Four hundreds μl of methylene chloride and 100 μl of PFPA were added to the residue, which was heated at 100 °C for 20 min.
5. The excess reagent and solvent were then evaporated under a gentle stream of nitrogen gas at room temperature.
6. After addition of 2 ml distilled water, the derivatized sample was extracted three-times with ethyl acetate.
7. To remove any trace amount of water in the combined ethyl acetate extract, an appropriate amount of dry sodium sulfate was added to the combined extract and allowed to stand for 20 min.
8. After filtrating, the extract was carefully evaporated in the same way.
9. The derivatized sample was dissolved in an appropriate volume of ethyl acetate and stored at 4 °C until analysis.
10. 1-2 μl of the solution was injected into the gas chromatograph.

3.3. GC Conditions

1. Injection temperature: 250 °C
2. Detector temperature: 300 °C
3. Temperature program:
   80 °C, isothermal for 5 min
   80-140 °C at 2 °C/min
4. Split ratio: 1:10
5. Column head pressure: 110 kPa
6. Helium flow-rate on the column: 0.4 ml/min

7. Helium flow-rate (plus auxiliary gas): 30 ml/min
8. Hydrogen flow-rate at the detector: 3.5 ml/min
9. Air flow-rate at the detector: 110 ml/min

3.4. Derivatization Procedure for HPLC

1. The derivatization reagent was prepared daily by dissolving 20 mg of OPA and 20 mg of Boc-L-Cys in 2 ml of methanol.
2. A 20-µl volume of Boc-L-Cys-OPA reagent and 70 µl of the borate buffer were added to a vial containing 10 µl of the amino acid standard or sample.
3. After derivatization for 2 min at ambient temperature, an aliquot (5 µl) of the reaction mixture was introduced into the HPLC system.

3.5. HPLC Conditions

1. Excitation and emission wavelengths of the 821-FPS spectrofluorometer were 344 and 443 nm, respectively.
2. The Nova-Pak C18 column was operated at a constant flow-rate of 0.8 ml/min at 30 °C.
3. Data processing was performed using a Waters 741 data module.
4. Mobile phase (acetonitrile: acetate buffer: THF): A (7 %: 90 %: 3 %), B (22 %: 75 %: 3 %), C (50 %: 47 %: 3 %).
5. The amino acid separation was accomplished with a linear gradient from mobile phase A to B in 50 min. After that, the column was washed with mobile phase C for 15 min.

4. Calculations

The mean value of the ratios between the peak heights of the amino acid (AA) and the internal standard (I.S.) for the standard samples was used to calculate the AA concentration in the unknown samples according to the following equation:

AA (µmol/g wet weight) = (sample AA/I.S.) x (standard I.S./AA) x (µmol I.S. added/g wet weight)

5. Results

Figure 1a shows a representative gas chromatogram of N,O-pentafluoropropionyl (PFP)-isopropyl standard amino acid derivatives. The N,O-PFP-isopropyl amino acid derivatives provided good resolution for most of the amino acid enantiomers and non-chiral amino acids such as glycine, β-alanine, and GABA. Figure 1b indicates a typical gas chromatogram of N,O-PFP-isopropyl amino acid derivatives in the rat brain. A peak X at 11.5 min, which exhibited the same retention time as the N,O-PFP-isopropyl derivatives of authentic D-serine, was detected in the brain. Electron impact and positive chemical ionization mass spectra of the peak X of the brain were identical to those of authentic D-serine [2]. The level of free D-serine in the rat brain was calculated to be 0.27 µmol/g of tissue, which was 23 % of the total (D+L) serine level. A trace amount of D-alanine was also seen in the rat brain (<10 nmol/g of tissue).

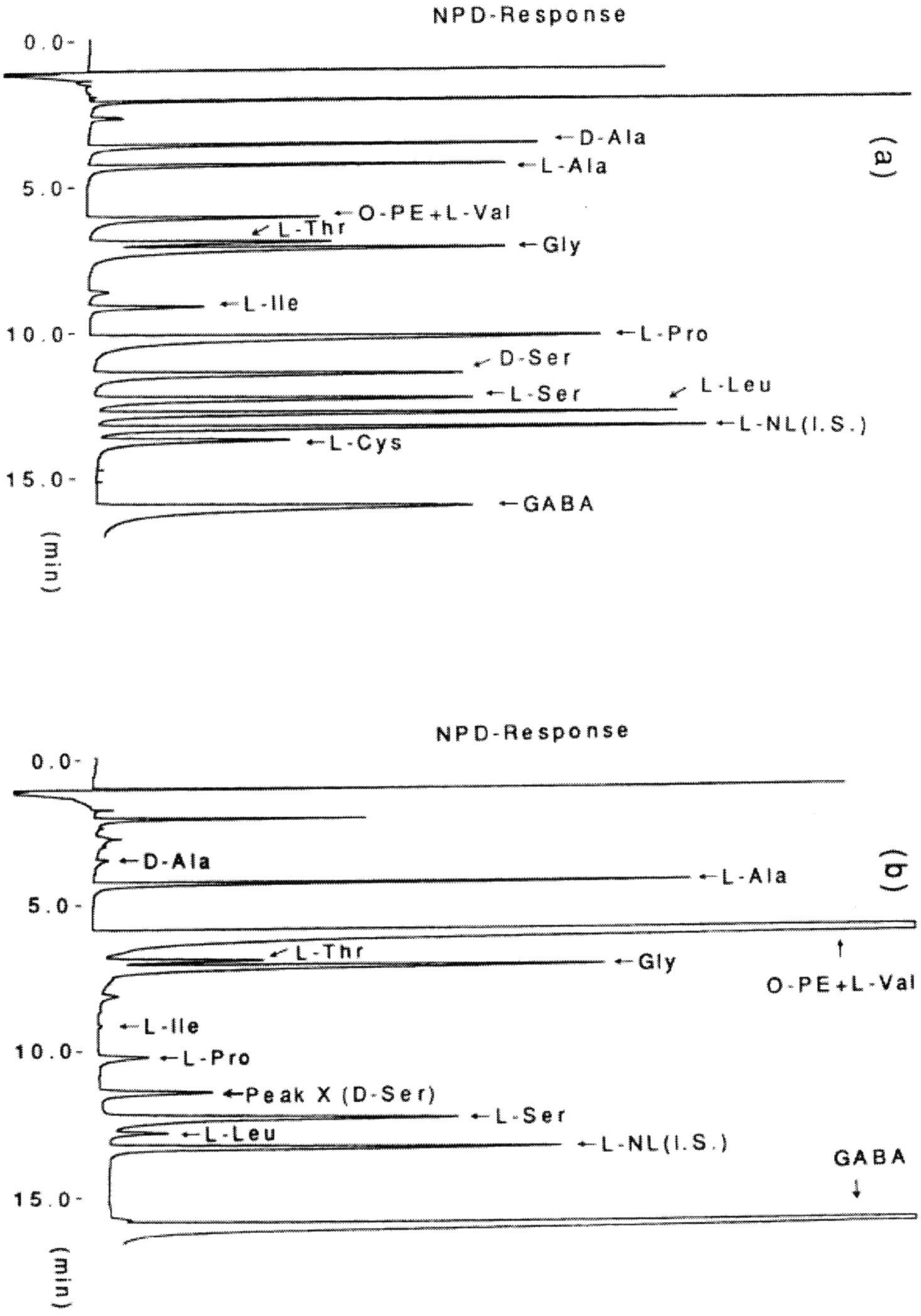

Figure 1. Gas chromatograms of N,O-PFP-isopropyl derivatives of standard amino acids (a) and free amino acids in the brain extracts (b) (From Hashimoto et al. (1992), FEBS Lett 296, 33-36, with permission).

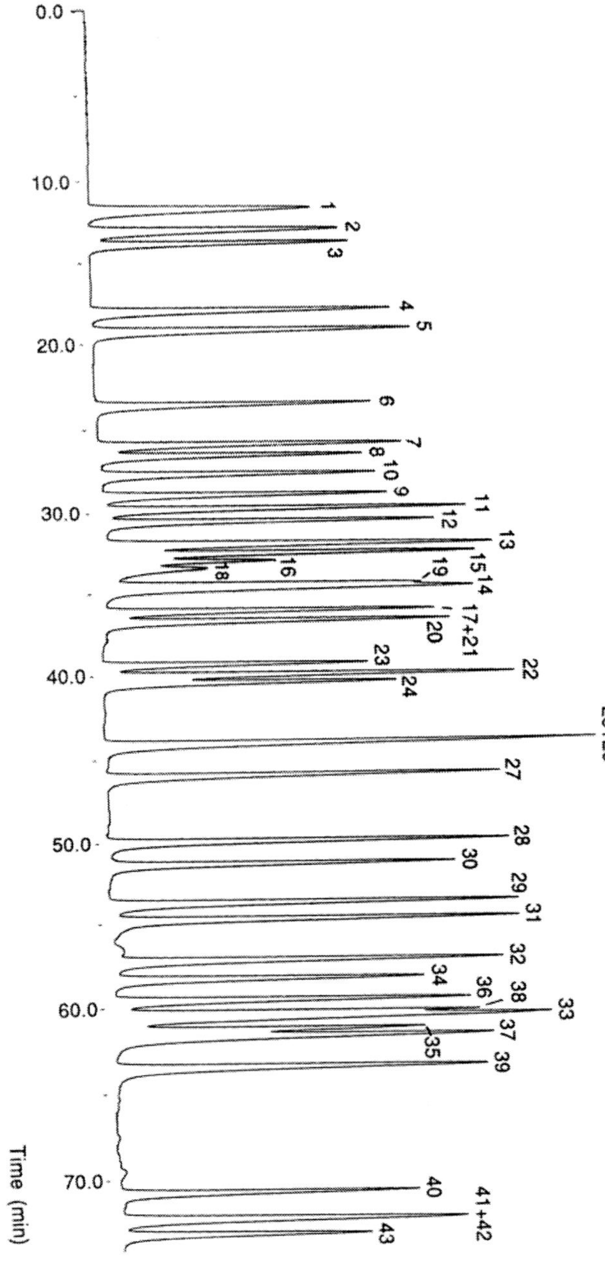

Figure 2. Chromatogram of Boc-L-Cys-OPA derivatives of standard amino acids. Each peak corresponds to 25 pmol of free amino acid. For chromatographic conditions, see Methods (From Hashimoto et al. (1992), J Chromatogr 582, 41-48, with permission). The peak numbers of amino acids are: 1; L-aspartate, 2; D-aspartate, 3; L-HCA, 4; L-glutamate, 5; D-glutamate, 6; L-asparagine, 7; D-asparagine, 8; L-serine, 9; D-serine, 10; L-glutamine, 11; D-glutamine, 12; citrulline, 13; L-threonine, 14; D-threonine, 15; glycine, 16; L-histidine, 17; D-histidine, 18; O-PE, 19; L-arginine, 20; D-arginine, 21; L-alanine, 22; D-alanine, 23; β-alanine, 24; taurine, 25; GABA, 26; L-tyrosine, 27; D-tyrosine, 28; L-valine, 29; D-valine, 30; L-methionine, 31; D-methionine, 32; L-isoleucine, 33; D-isoleucine, 34; L-tryptophan, 35; D-tryptophan, 36; L-phenylalanine, 37; D-phenylalanine, 38; L-leucine, 39; D-leucine, 40; L-ornithine, 41; D-ornithine, 42; L-lysine, 43; D-lysine.

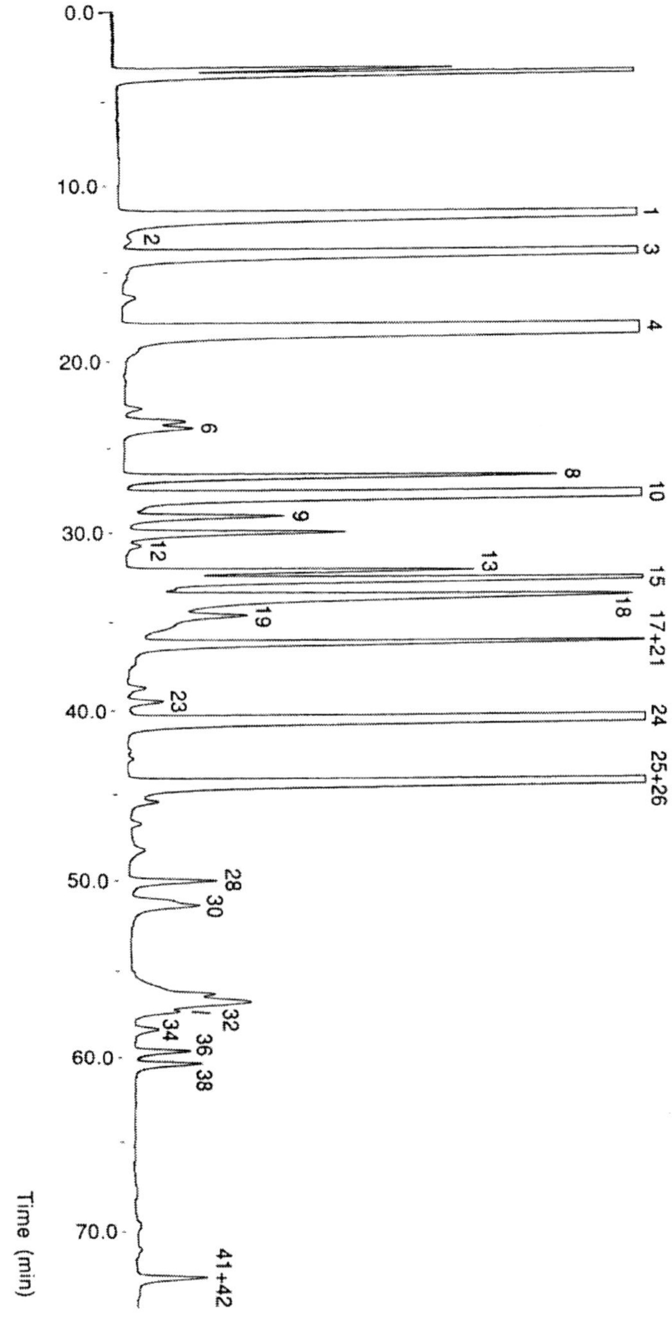

Figure 3. Chromatogram of Boc-L-Cys-OPA derivatives of standard amino acids. For chromatographic conditions and peak numbers of amino acids, see Methods and Figure 2, respectively (From Hashimoto et al. (1992), J Chromatogr 582, 41-48, with permission).

Figure 2 designates a representative HPLC chromatogram of Boc-L-Cys-OPA derivatives of authentic amino acids. The Boc-L-Cys-OPA derivative gave good resolutions for most of the amino acid enantiomers and non-chiral amino acids. The derivatives of D,L-aspartate, D,L-glutamate, D,L-asparagine, D,L-serine, D,L-glutamine, L-citrulline, L-threonine, glycine, L-histidine, O-PE, D-arginine, D-alanine, β-alanine, taurine, D,L-valine, D,L-methionine, L-isoleucine, D,L-tryptophan, D,L-phenylalanine, D-leucine, L-ornithine, and D-lysine were separated within 75 min. However, our chromatographic system failed to achieve good separations of the peaks corresponding to GABA, L-arginine, L-alanine, L-tyrosine, L-leucine, L-lysine, D-threonine, D-histidine, and D-ornithine. The HPLC method described here was applied to the determination of free amino acids in rat brain. As shown in Figure 3, a peak with a considerable height co-eluted with the Boc-L-Cys-OPA derivatives of authentic D-serine in the brain. The level of the D-amino acid was estimated to be 0.22 μmol/g of tissue, which was 24.6 % of the total (D+L) serine level. Moreover, a trace amount of D-aspartate was present in the brain (<10 nmol/g of tissue), whereas no other D-amino acid was found in the present HPLC analysis. The observation that the D-serine level determined by the present HPLC method is comparable to that given by GC with a completely different derivatization and detection process appears to support the accuracy of quantification by the HPLC-fluorescence detection technique. The present simple and reliable HPLC method for the detection of D-aspartate and D-serine will make a meaningful contribution to the investigation of the physiological and pathophysiological roles of D-serine in mammalian brain.

6. Notes

6.1. Notes for GC

1. Isopropanol-2 N HCl was prepared by bubbling anhydrous hydrogen chloride through a trap with concentrated sulfuric acid and then into isopropanol at 0 °C until the solution was 2 N in HCl as determined by weighting.
2. Injection of 25-500 pmols of each amino acid, dissolved in 0.5-1.0 μl of ethyl acetate, gave excellent resolution.
3. The NPD-response of the amino acid derivatives was linear over the range 25-1000 pmol per injection.
4. Amino acid stock solutions were prepared by dissolving amino acids in 0.01 M HCl to a concentration of 10 μmol/ml each. The stock solutions were used for calibration after appropriate dilution with 0.01 M HCl.

6.2. Notes for HPLC

1. A small amount of THF (3 %) in the mobile phase improved the resolutions of L-serine-D-asparagine and taurine-D-alanine without altering those of the other amino acids.
2. A higher column temperature was detrimental to the resolutions of amino acid enantiomers.
3. The fluorescence responses of the amino acid derivatives were linear over the range 2.5-500 pmol per injection.
4. Amino acid stock solutions were prepared by dissolving amino acids in 0.01 M HCl to a concentration of 0.1 μmol/ml with the exceptions of histidine, asparagine, glutamine, and tryptophan, which were dissolved in 20 % aqueous methanol. The stock solutions were used for calibration after appropriate dilution with 0.01 M HCl.

References

[1] Dunlop, D.S., Neidle, A., MaHale, D., Dunlop, D.M. & Lajtha, A. (1986) The presence of free D-aspartic acid in rodents and man. *Biochem. Biophys. Res. Commun. 141*, 27-32.

[2] Hashimoto, A., Nishikawa, T., Hayashi, T., Fujii, N., Harada, K., Oka, T. & Takahashi, K. (1992) The presence of free D-serine in rat brain. *FEBS Lett, 296*, 33-36.

[3] Hashimoto, A., Nishikawa, T., Oka, T., Takahashi, K. & Hayashi, T. (1992) Determination of free amino acid enantiomers in rat brain and serum by high-performance liquid chromatography after derivatization with N-tert.-butyloxycarbonyl-L-cysteine and o-phthaldialdehyde. *J Chromatogr, 582*, 41-48.

[4] Hashimoto, A., Nishikawa, T., Oka, T. & Takahashi, K. (1993) Endogenous D-serine in rat brain: N-methyl-D-aspartate receptor-related distribution and aging. *J Neurochem, 60*, 783-786.

[5] Hashimoto, A., Oka, T. & Nishikawa, T. (1995) Extracellular concentration of endogenous free D-serine in the rat brain as revealed by in vivo microdialysis. *Neuroscience, 66*, 635-643.

[6] Hashimoto, A., Oka, T. & Nishikawa, T. (1995) Anatomical distribution and postnatal changes in endogenous free D-aspartate and D-serine in rat brain and periphery. *Eur J Neurosci, 7*, 1657-1663.

[7] Hashimoto, A. & Oka, T. (1997) Free D-aspartate and D-serine in the mammalian brain and periphery. *Prog Neurobiol, 52*, 325-353.

[8] Schell, M.J., Brady, R.O., Molliver, M.E. & Snyder, S.H. (1997) D-Serine as a neuromodulator: regional and developmental localizations in rat brain glia resemble NMDA receptors. *J Neurosci, 17*, 1604-1615.

[9] Corrigan, J.J. (1969) D-Amino acids in animals. *Science, 164*, 142-149.

[10] Matsui, T., Sekiguchi, M., Hashimoto, A., Tomita, U., Nishikawa, T. & Wada, K. (1995) Functional comparison of D-serine and glycine in rodents: the effect on cloned NMDA receptors and the extracellular concentration. *J Neurochem, 65*, 454-458.

In: D-Amino Acids:
A New Frontier in Amino Acid and Protein Research, pp.75-83
ISBN 1-60021-075-9

Chapter 2.8

An LC Method for the Analysis of D-Amino Acids

Marianne Stenberg*
Department of Applied Nutrition and Food Chemistry, University of Lund, Sweden

Keywords: liquid chromatography, two-dimensional column liquid chromatography, chiral crown ether, chiral separation, racemization index

1. Introduction

Liquid chromatography is one of the most common methods of amino acid analysis described in the literature. It involves several separation mechanisms, and ion-exchange chromatography is the classical method, invented by Moore and Stein in the early 1950s [1]. The ionizable groups of the amino acids, which attract oppositely charged compounds form the underlying basis for this type of chromatography (Fig. 1). Cation-exchange separation is the preferred method. The resin used is saturated with counter-ions which, when the sample is injected, will be replaced by the amino acids. As each amino acid has a specific structure, it also has a specific dissociation constant, and by increasing the pH of the solutions, step by step, from acidic to basic, each amino acid reaches a specific pH at which the zwitterion predominates. At this so-called isoelectric point the affinity of the sample ion is weakened and it can be displaced by the eluent ion.

Analysis methods routinely used do not separate the D- and L-enantiomers of amino acids. Therefore, separation systems for D- and L-amino acids must include chiral selection of some kind. There are different ways of achieving this kind of chromatographic separation: gas chromatography with a chiral stationary phase, liquid chromatography with either chiral mobile phase additives or derivatization with chiral reagents, and capillary electrophoresis with chiral crown ether derivatives as the chiral selector [2-9]. Crown ethers are also an attractive alternative in liquid chromatography [10] where a complex is formed between an ammonium ion and the crown ether introduced by chemical reactions, e.g. naphthyl groups or carboxyl groups (Fig. 2).

* Correspondence concerning this article should be addressed to Marianne Stenberg, Department of Applied Nutrition and Food Chemistry, University of Lund, Sweden. E-mail: marianne-stenberg@spray.se; FAX: +46 46-222-4532.

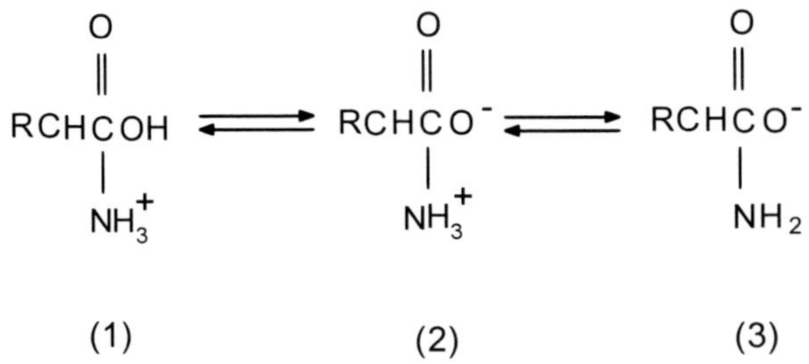

Figure 1. (1) Predominant form of the amino acid at pH 1; (2) predominant form at pH 7; (3) predominant form at pH 11.

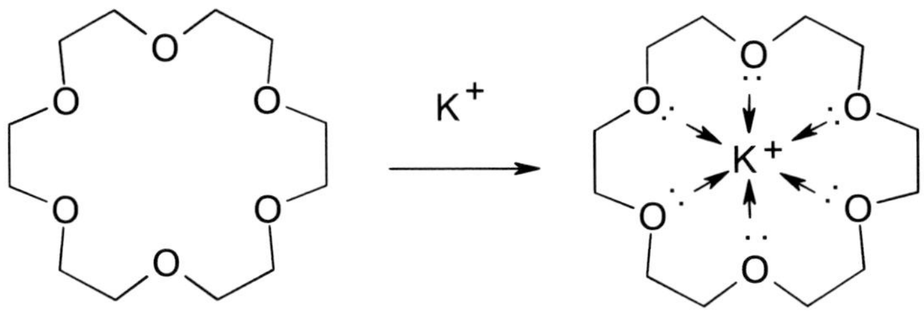

Figure 2. Crown ether; 18-crown-6.

Single amino acids and some combinations of amino acids can be separated on the chiral column alone, but since the chiral column has a limited capacity, regarding the amount of crown ether used for coating the C18 phase, it cannot separate all the enantiomeric amino acids simultaneously [11,12]. For the separation of D- and L-amino acids as constituents of a mixture, or a protein, a coupled column system with a conventional ion-exchange column connected to a chiral separation column is an alternative. This method is applicable in both HPLC systems and amino acid analysers.

Since the separation of the enantiomers is based on complex formation between the crown ether and the protonated amino group, a net positive charge on the amino acid is required for significant retention. It is thus important that the pH of not only the elution solution, but also the sample (= the eluate from the ion-exchange column), is sufficiently low to keep the amino acid positively charged. The practical consequences of the demand for low sample pH concerning the coupled system are evident. The pH of the eluate from ion-exchange columns varies with different systems and programmes, but by adding perchloric acid of a suitable pH to the eluate of the ion-exchange column prior to injection into the chiral column, the pH of the eluate can be adjusted to around 1 for most systems.

The analysis of amino acids in proteins requires hydrolysis, and this is often carried out with 6 M HCl at 110°C for 24 h [13]. Towards the end of the 1980s a new hydrolysis technique employing microwaves was introduced and evaluated [14]. Results have been reported from both commercially available and homemade equipment, and there are great variations in parameters such as the amount of sample, acid volume, temperature and hydrolysis time, but in summary the hydrolysis time can be reduced to minutes at temperatures around 150°C or higher [14,12,15,16]. Independent of the method, this will lead to partial racemization, most pronounced for the acidic amino acids, and it is obviously impossible to distinguish between the naturally formed D-amino acids and those arising from racemization during hydrolysis. Since free amino acids and peptide-bound ones become racemized at different rates and the degree of racemization depends not only on the amino acids involved, but also on their neighbours in the peptide, it is not possible to calculate the inversion rate of free amino acids for use as a correction factor [17]. An alternative to acid hydrolysis is enzymatic hydrolysis, in which amino acids sensitive to acid hydrolysis will be preserved, but this procedure is often complicated, lengthy and laborious and probably not an alternative for routine use [18,19]. Furthermore, enzymes are stereoselective by nature and may therefore not liberate D-amino acids from the peptide bonds [20,21]. However, by modifying normal acid hydrolysis and microwave hydrolysis regarding acid concentration and hydrolysis time, it is possible to reduce the hydrolysis-induced racemization to less than 1% [11].

2. Materials

1. HPLC system working in the mode for amino acid analysis, or an amino acid analyzer
2. Chiral column (Crownpac CR (+) 150x4 mm
3. Three-way selection valve
4. HPLC pump
5. Mixing block
6. Conventional hydrolysis oven or microwave oven
7. D- and L-amino acids to be used as standards
8. Perchloric acid
9. Methanol
10. Hydrochloric acid

3. Methods

For the quantitative separation of amino acids the routinely used method of LC is recommended. The following descriptions are intended for use in enantiomeric analysis and can be used in an LC gradient system, as well as in an amino acid analyser, and should be regarded as a chiral complement to an existing LC separation system. For this reason, details concerning primary separation and detection have been excluded. Neither are the practical aspects of conventional hydrolysis and sample preparation discussed, since there is no reason to change current working routines. However, the effect of various conditions of hydrolysis on racemization will be discussed.

In the methods described below it is assumed that an amino acid analyser is used, utilizing a programme for physiological samples.

I. Separation

One-Column System — Chiral Column for Separation of Single D- and L-Amino Acids

1. Replace the normal separation column with a chiral column, and install it outside any existing heating units in the equipment to make greater temperature variations possible.
2. Replace the routinely used separation buffers by perchloric acid (pH 1.5) and perchloric acid pH 1.5 containing 10% MeOH, and use a flow rate of 0.25 ml/min.
3. Dilute 100 μl of a single amino acid standard solution (500 μg/ml) in perchloric acid (pH 1.0) about 5 times.
4. Inject 50 μl of the standard solution into the column and obtain calibration and identification chromatograms.
5. Elute aspartic acid, threonine, serine, alanine, valine, histidine, arginine and lysine with perchloric acid at 0°C (ice-water), glutamic acid with perchloric acid at room temperature (23°C) and methionine, isoleucine, leucine, tyrosine and phenylalanine with the perchloric acid / MeOH mixture at room temperature (Table 1). Regarding the retention times in Table 1, it is clear that it is also possible to analyse combinations of certain amino acids (Fig. 3)

6. Inject 50 μl of the sample.
7. Calculate the amount of each D-amino acid as a percent of the total amount of amino acid: D-amino acid * 100 / (D-amino acid + L-amino acid)

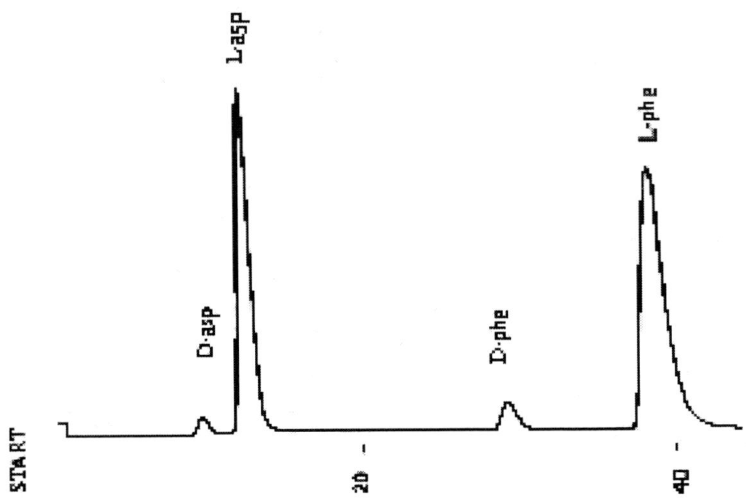

Figure 3. Enantiomeric separation of aspartic acid and phenylalanine. Separation was performed on a single-column system.

Table 1. Values employed in the separation of enantiomeric forms of single amino acids on a chiral column connected to an amino acid analyser.

Amino acid	$HClO_4$ pH	MeOH %	Temp °C	Retention time min D-AA	Retention time min L-AA	L-D min	Resolution
Aspartic acid	1,5	-	0	8.8	12,2	3.4	1.8
Threonine	1.5	-	0	10.0	13.9	3.9	2.1
Serine	1.5	-	0	9.8	11.2	1.4	0.7
Alanine	1.5	-	0	12.2	21.7	9.5	3.5
Valine	1.5	-	0	13.4	16.6	3.2	1.2
Histidine	1.5	-	0	11.7	14.8	3.1	1.3
Arginine	1.5	-	0	14.9	18.7	3.8	1.8
Lysin	1.5	-	0	10.9	14.7	3.8	1.2
Glutamic acid	1.5	-	23	9.5	16.5	7.0	3.5
Methionine	1.5	10	23	9.3	12.5	3.2	1.3
Isoleucine	1.5	10	23	12.4	14.3	1.9	0.5
Leucine	1.5	10	23	14.6	23.3	8.7	2.9
Tyrosine	1.5	10	23	15.8	21.7	5.9	1.8
Phenylalanine	1.5	10	23	22.6	31.2	8.6	2.0

Coupled-column — System for Separation of D- and L- Amino Acids as Constituents of a Mixture or a Protein

Figure 4 shows a schematic illustration of the coupled system employing an ion-exchange column and a chiral column. In this system, the chiral column is connected to the ion-exchange column with a three-way selection valve. Before entering the chiral column the eluate from the ion-exchange column has to be mixed with perchloric acid, delivered by an HPLC pump, to obtain the correct pH.

If all amino acids are continuously transported from the ion-exchange column to the chiral column, there will be high pressure over the column, creating considerable demands on the tightenings of the connections. Also, exposure of the chiral column to the analytical effluent buffers stresses the column and gradually results in loss of performance. Therefore, this procedure is not recommended [11].

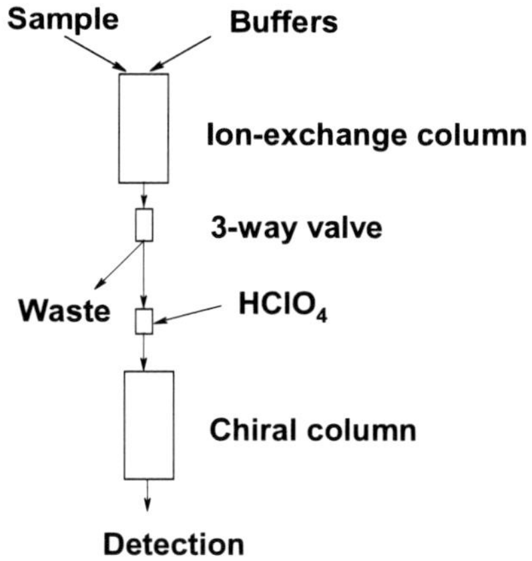

Figure 4. Schematic illustration of the coupled system employing an ion-exchange column and a chiral column.

An alternative method is to allow only selected amino acids to pass from the ion-exchange column into the chiral column and to lead the eluate not intended for analysis to the waste outlet. It is also possible to obtain good separation on the chiral column feeding groups of amino acids. This method may seem time consuming, but since aspartic acid is the acid most likely to racemize, it can be used as an index of racemization [22]. Stenberg et al. [11] proposed glutamic acid, which is also prone to racemization, as an additional racemization index. Glutamic acid often occurs in abundant quantities, and small amounts of the racemic forms will thus be easier to detect. Also, as will became clear in the following description, the elution parameters for D- and L-glutamic acid are favourable, sparing the column.

1. To determine the time at which the eluate from the ion-exchange column should be switched to the chiral column, disconnect the ion-exchange column and collect fractions of the eluate in glass test-tubes for one minute each. Analyse the content of each test-tube using the same technique as in the analysis procedure. If ninhydrin is the actual reagent, add half the volume of ninhydrin and heat the tubes about 5 minutes in boiling water. The eluate with amino acids turns purple.
2. Separate the sample on the analytical column using the normal routine. Effluent not intended for enantiomeric analysis can be led to the waste outlet and the eluate of interest directed to the chiral column. As soon as the amino acids of interest have passed into the chiral column, turn the three-way valve to the waste outlet.
3. Elute (from the chiral column) aspartic acid, threonine, serine, alanine, valine, histidine, arginine and lysine with perchloric acid, at pH 1 and a flow rate 0.25 ml/min at 0°C, glutamic acid with the same perchloric acid but at room temperature, and methionine, isoleucine, leucine, tyrosine and phenylalanine with perchloric acid, pH 2, mixed with 10% MeOH, at the same flow rate at room temperature (Table 2).

II. Hydrolysis

It is not possible to obtain both complete hydrolysis and minimal racemization of peptide-bound amino acids, so a compromise must be accepted. Therefore, analyse the sample quantitatively, following normal routines and use a modified method for enantiomeric analysis.

Exclude the internal standard to obtain a less complicated chromatogram.

Conventional hydrolysis for enantiomeric analysis: decrease hydrolysis-induced racemization by using 3 M HCl instead of 6 M HCl and shortening the time to 3 h –7 h.

Microwave hydrolysis for enantiomeric analysis: hydrolyse in 6 M HCl and avoid temperatures above 150°C.

Table 2. Values employed in the separation of enantiomeric forms of amino acids as constituents of a D L mixture on a chiral column connected to the ion-exchange column of an amino acid analyser.

Amino acid	$HClO_4$ pH	MeOH %	Temp °C	Retention time min		L-D min	Resolution
				D-AA	L-AA		
Aspartic acid	1.0	-	0	12.2	18.7	6.5	2.4
Threonine	1.0	-	0	12.4	17.3	4.9	2.4
Serine	1.0	-	0	8.3	10.6	2.3	1.4
Alanine	1.0	-	0	15.1	27.0	11.9	6.0
Valine	1.0	-	0	13.1	17.2	4.1	2.1
Histidine	1.0	-	0	11.0	16.3	6.3	1.6
Arginine	1.0	-	0	16.3	19.9	3.6	0.9
Lysine	1.0	-	0	11.0	15.0	4.0	0.8
Glutamic acid	1.0	-	23	9.7	19.2	9.5	3.8
Methionine	2.0	10	23	11.0	17.5	6.5	1.9
Isoleucine	2.0	10	23	11.5	12.0	0.5	0
Leucine	2.0	10	23	25.0	36.0	11.0	2.3
Tyrosine	2.0	10	23	15.0	24.0	8.5	3.2
Phenylalanine	2.0	10	23	28.0	37.5	9.5	4.2

4. Results

Linear response (r>0.996) was observed in the concentration range 0.005 to 3.8 μg / ml for all the amino acids studied [11]. The D-enantiomer always elutes before the L-enantiomer and the time lag between D- and L-elution varied for single amino acids between 1.4 min. (serine) and 9.5 min. (alanine) and for the amino acids isolated from a mixture, between 0.5 min. (isoleucine) and 37.5 min. (phenylalanine).

When analysing single amino acids, only serine and isoleucine had a resolution < 1, and the best resolution (3.5) was obtained with alanine and glutamic acid (Table 1). Among the amino acids separated by ion-exchange chromatography first, D- and L- isoleucine could not be separated on the chiral column, arginine and lysine had resolution factors just below 1 and the others >1 (Table 2).

It was also possible to obtain good separation on the chiral column when analysing groups of amino aicds. For example, aspartic acid, threonine, serine and glutamic acid cuold be injected into the chiral column as one fraction and still show the same resolution for D- and L-aspartic acid and D- and L-glutamic acid as for single fractions, but the two forms of threonine and serine was eluted as one peak. If aspartic acid and/or glutamic acid was used as a racemization index this would provide a fast method of determining racemization.

The method was developed using aspartame, a dipeptide consisting of aspartic acid and phenylalanine. As aspartame is an easily hydrolysed peptide, there was a need for an objective estimate of the efficiency of the different hydrolysis methods on protein. Therefore human serum albumin (HAS) was hydrolysed under the same conditions as aspartame.

With the conventional hydrolysis technique, 3 M and 6 M HCl had the same low racemization effect, 0.3%–0.5%, in the 3 h treatment of both D-aspartic acid and phenylalanine, but increasing the treatment time enhanced the effect more with 6 M HCl than with 3 M HCl. Hydrolysis with 3 M HCl for 24 h, gave 2.8% D-aspartic acid and 0.7% D-phenylalanine, compared with 4.7% D-aspartic acid and 1.6% D-phenylalanine with 6 M HCl (Table 3).

Table 3. Hydrolysis-induced racemization in aspartame using conventional hydrolysis for different periods of time. The general efficiency of hydrolysis was estimated by the degree of HSA hydrolysis. performed in the same way.

HCl	Duration h	Temp °C	n	Hydr. of Aspartame %	D-asp %	D-phe %	Hydr. of HSA % (n=2)
3 M	3	110	2	98	0.3	0.3	65
6 M	3	110	2	95	0.5	0.4	87
3 M	7	110	2	98	1.3	0.4	82
6 M	7	110	2	98	1.8	0.7	91
3 M	24	110	2	13	2.8	0.7	95
6 M	24	110	7	100	4.7 ± 0.6	1.6 ± 0.2	100

Table 4. Hydrolysis-nduced racemization in aspartame using microwave hydrolysis. The general efficiency of hydrolysis was estimated by the degree of HSA hydrolysis. performed in the same way.

HCl	Duration s	Final temp °C	n	Hydr.of aspartame %	D-asp %	D-phe %	Hydr.of HSA % (n=2)
3 M	60	185-190	2	100	~ 3	~ 3	
6 M	60	185-190	2	100	~ 3	~ 3	
3 M	120	150	5	93 ± 9	0.55 ± 0.11	0.12 ± 0.10	50
6 M	120	150	5	88 ± 6	0.29 ± 0.12	0.16 ± 0.06	52
3 M	180	160	3	100 ±1	4.83 ± 1.10	1.98 ± 0.50	70
6 M	180	160	3	93 ± 8	2.14 ± 0.58	1.43 ± 0.02	76
3 M	360	115	3	82 ± 4	0.19 ± 0.05	0.15 ± 0.03	47
6 M	600	115	2	77	0.64	0.36	60

The microwave hydrolysis was not performed in commercial vessels, so the principle, not the details should be applied. A temperature of 150°C seemed to be critical for racemization. Below this temperature racemization was between 0.12 and 0.64% for both the constituent amino acids, and already at 160°C it has increased to >1.4% (Table 4). The chiral analyses also show that 3 M HCl had a greater racemization effect on the substrate than 6 M HCl, especially in the case of aspartic acid.

5. Notes

1. Note that cautions must be exercised when handling perchloric acid.
2. The pH of the percloric acid, the flow rate and the temperatures employed here were found to be suitable for the amino acid analyser system we used. The number of combinations of parameters is

vast, and other values may be more suitable for other equipment. However, complete stereoselectivity will not be achieved if the pH of the sample is higher than 1–2.

3. Other combinations of parameter values could also be investigated in the hydrolysis process, however, the trend towards lower racemization with shorter hydrolysis time and modified HCl concentration is clear.

REFERENCES

[1] Moore, S. & Stein, W.H. (1951) Chromatography of amino acids on sulfonated polystyrene resins. *Journal of Biological Chemistry, 192*, 663

[2] Brückner, H. & Hausch, M. (1993). Gas chromatographic characterization of free D-amino acids in the blood serum of patients with renal disorders and of healthy volunteers. *Journal of Chromatography, 614*, 7-17

[3] Armani, E., Dossena, A. Marchelli, R. & Virgilli, R. (1988). Copper(II) complexes of diamino-diamido-type ligands as chiral eluents in the enantiomeric separation of D,L-dansylamino acids by reversed-phase high-performance liquid chromatography. *Journal of Chromatography, 441*, 275-286.

[4] Merino, I.M., Gonza´les, E.B. & Sanz-Mendel, A. (1990). Liquid chromatographic enantiomeric resolution of amino acids with β-cyclodextrin bonded phases and derivatization with o-phthaldehyde. *Analytica Chimica Acta, 234*, 127-131.

[5] Palla, G., Marchelli, R., Dossena, A., & Casnati, G. (1989). Occurrence of D-amino acids in food. Detection by capillary gas chromatography and by reversed-phase high-performance liquid chromatography with L-phenylalanineamides as chiral selectors. *Journal of Chromatography, 475*, 45-53.

[6] Tanaka Y., Otsuka, K. & Terabe, S. (2000). Separation of enantiomers by capillary electrophoresis – mass spectrometry employing a partial filling technique with a chiral crown ether. *Journal of Chromatography A, 875 1,2*, 323 - 330

[7] Pe'ter, A., Török, R. and Armstrong D. (2004). Direct high-performance liquid chromatographic separation of unusual secondary amino acids and a comparison of the performances of Chirobiotic T and TAG columns. *Journal of Chromatography A, 1057, 1-2*, 229-235.

[8] Piccinini, A-M., Schmid, M., Pajpanova, T. Pancheva, S., Grueva, E. and Gűbitz, G. (2004). Chiral separation of natural and unnatural amino acid derivatives by micro-HPLC on Ristocetin A stationary phase. *Journal of Biochemical and Biophysical Methods, 61, 1-2*, 11-21.

[9] Ding, G-S., Liu, Y. Cong R-Z. and Wang J-D. (2004). Chiral separation of enantiomers of amino acid derivatives by high-performance liquid chromatography on a norvancomycin-bonded chiral stationary phase. *Talanta, 62, 5*, 997-1003.

[10] Fanali, S. (1997). Controlling enantioselectivity in chiral capillary electrophoresis with inclusion – complexation *Journal of Chromatography A, 792, 1-2*, 227-267.

[11] Stenberg, M., Marko-Varga, G., & Öste, R. (2001). Racemization of amino acids during classical and microwave oven hydrolysis – application to aspartame and a Maillard reaction system. *Food Chemistry, 74*, 217-224.

[12] van de Merbel, N., Stenberg, M., Öste, R. Marko-Varga, G., Gorton, L., Lingeman, H. & Brinkman, U. (1995). Determination of D- and L-amino acids in biological samples by two-dimensional Column Liquid Chromatography. *Chromatographia, Vol 41, No ½, July* 6-14

[13] Hirs, C.H., Stein, W.H. & Moore, S. (1954). The amino acid composition of ribonuclease. *Journal of Biological Chemistry, 211*, 941-950.

[14] Chen, S.-T., Chiou, S.-H-., Chu, Y.-H. & Wang, K.-T. (1987). Rapid hydrolysis of proteins and peptides by means of microwave technology and its application to amino acid analysis. *International Journal of Peptide and Protein Research, 30*: 572-576.

[15] Acquistucci, R., Panfili, G., & Marconi, E. (1996). Application of the microwave hydrolysis to furosine determination in cereal and dairy foods. *Journal of Agricultural and Food Chemistry,* 44 (12) 3855-3857.

[16] Pe'ter, A., Laus, G., Tourwe', D., Gerlo, E. & van Binst, G. (1993). An evaluating of microwave heating for the rapid hydrolysis of peptide samples for chiral amino acid analysis. *Peptide Research,* 6:1, 48-52.

[17] Liardon, R. & Jost, R. (1981). Racemization of free and protein-bound amino acids in strong mineral acid. *International Journal of Peptide and Protein Research,* 18, 500-505.

[18] Liaset, B., Liel, E. & Espe M. (2000). Enzymatic hydrolysis of by-products from the fish-filleting industry; chemical characterisation and nutritional evaluation. *Journal of the Science of Food and Agriculture,* 80, 5, 581-589.

[19] Tsao & Otter, D.E. (1999). Quantification of glutamine in proteins and peptides using enzymatic hydrolysis and reverse-phase high-performance liquid chromatography. *Analytical Biochemistry,* 26, 1, 143-148.

[20] Hayashi, R.& Kameda, J. (1980). Decreased proteolysis of alkali-treated protein: Consequences of racemization in food processing. *Journal of Food Sciences,* 45, 5, 1430-1431.

[21] Masters, P.M. & Friedman, M (1979). Racemization of amino acids in alkali-treated food proteins. *Journal of Agricultural and Food Chemistry,* 27, (3) 507-511.

[22] Friedman, M., Zahnley, J. C. & Masters P. (1981). Relationship between *in vitro* digestibility of casein and its content of lysinoalanine and D-amino acids. *Journal of Food Science*, *46*, 127-131

In: D-Amino Acids:
A New Frontier in Amino Acid and Protein Research, pp.85-91
ISBN 1-60021-075-9

Chapter 2.9

SENSITIVE DETERMINATION OF D-AMINO ACIDS IN RAT BRAIN BY HPLC WITH A PIRKLE-TYPE CHIRAL COLUMN, FOLLOWING PRECOLUMN FLUORESCENCE DERIVATIZATION WITH NBD-F

Takeshi Fukushima[1,*], Junko Kawai[1], Kazuhiro Imai[2], and Toshimasa Toyo'oka[1]

[1]Division of Bio-Analytical Chemistry, School of Pharmaceutical Sciences, University of Shizuoka, Japan;
[2]Research Institute of Pharmaceutical Sciences, Musashino University, Japan.

Keywords: NBD-F, sumichiral OA-2500 (*S*), D-aspartate, pineal gland, microdialysis.

1. INTRODUCTION

The presence of D-amino acids has been confirmed in several mammalian tissues, such as the brain, testis, adrenal gland, etc., in the last decade [1-3], and research on their physiological significance has progressed. In the mammalian brain tissues, it is considered that D-aspartate could suppress the release of melatonin [4], while D-serine acts as a coagonist at the glycine site for *N*-methyl-D-aspartate (NMDA) receptor [5, 6]. Recent studies, in particular, have suggested that D-serine in the brain has a relationship with schizophrenia, a severe neuropsychiatric disorder, because the hypofunction of the NMDA receptor has been implicated in the pathophysiology of schizophrenia [7, 8].

With regard to the concentration, D-amino acids are found in traces in biological specimens in comparison with L-amino acids. In order to investigate the physiological functions of D-amino acids in mammals, a useful determination method with high sensitivity is required.

For the determination of D-amino acids in biological specimens, we herein introduce a sensitive high-performance liquid chromatography (HPLC) method, which includes precolumn derivatization with a fluorescence reagent, 4-fluoro-7-nitro-2,1,3-benzoxadiazole (NBD-F) (Fig. 1(a)), followed by enantiomeric

* Correspondence concerning this article should be addressed to Takeshi Fukushima, E-mail: fukushima@u-shizuoka-ken.ac.jp, Fax: +81-54-264-5655.

separation on a Pirkle-type chiral column, Sumichiral OA series (Fig. 1(b)) [9]. By utilizing this HPLC method, the occurrence of D-aspartate was first demonstrated in the rat pineal gland [10]. We report the determination of D-aspartate in the rat pineal gland [10] and of D-serine in the microdialysate from the prefrontal cortex [11] by the proposed HPLC method.

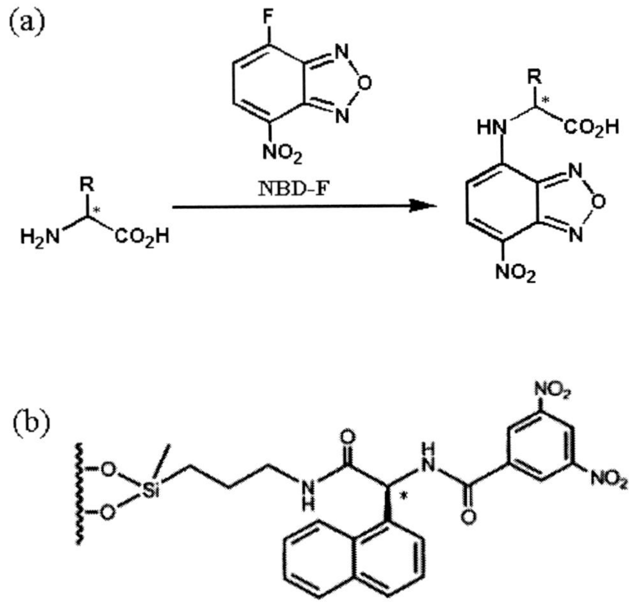

Figure 1. (a) Fluorescence derivatization scheme of amino acids with 4-fluoro-7-nitro-2,1,3- benzoxadiazole (NBD-F), (b) Chemical structure of the chiral moiety in a Pirkle-type chiral column, Sumichiral OA-2500 (*S*). *: asymmetric carbon.

2. Materials

2-1. D-Aspartate in the Pineal Gland

1. Standards of racemic aspartic acid and optically pure D- and L-aspartic acid
2. NBD-F (Tokyo Kasei Co., Ltd., Tokyo, Japan)
3. HPLC system with fluorescence detector
4. Chiral column: Sumichiral OA-2500 (*S*) (Sumika Chemical Analysis Service, Ltd., Osaka, Japan)
5. Mobile phase: 5 mM citric acid in methanol
6. Sprague-Dawley rat, and regular tools and instruments for dissection

2-2. D-Serine in the Microdialysate from Rat Brain

1. Standards of racemic serine and optically pure D- and L-serine
2. NBD-F (Tokyo Kasei Co., Ltd., Tokyo, Japan)
3. Column-switching HPLC system with fluorescence detector (Fig.2)
4. Octadecylsilica (ODS) column: TSKgel ODS-80TsQA (250 × 4.6 mm i.d., Tosoh Corporation, Tokyo, Japan)
5. Mobile phase 1: H_2O/CH_3CN (90/10) containing 0.1% trifluoroacetic acid, and H_2O/CH_3CN (10/90) containing 0.1% trifluoroacetic acid

6. Chiral column: Sumichiral OA-2500 (*S*) (Sumika Chemical Analysis Service, Ltd., Osaka, Japan) × 2
7. Mobile phase 2: 20 mM citric acid in methanol
8. Sprague-Dawley rat, and regular tools and instruments for dissection
9. Microdialysis system

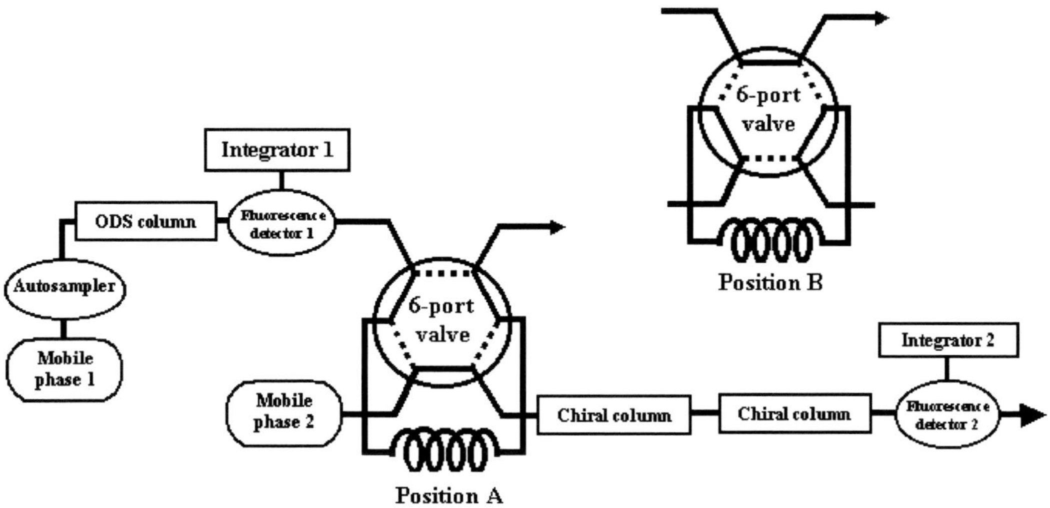

Figure 2. A block diagram of column-switching HPLC system. Time schedule of the 6-port valve position change is as follows: After the NBD-serine peak emerged in the chromatogram recorded by integrator 1, the 6-port valve was manually switched from the dotted line to the solid line pathway (Position B to A). Then, as soon as the peak top of the NBD-serine appeared in the chromatogram recorded by integrator 1, the 6-port valve was manually switched from the solid line to the dotted line pathway (Position A to B); thus, a portion of the peak fraction of NBD-serine was introduced into the chiral column through 100 μL of sample loop.

3. METHODS

3-1. D-Aspartate in the Pineal Gland

1. A Sprague-Dawley rat was sacrificed by exsanguination from the abdominal aorta under diethyl ether anesthesia. Then, the pineal gland was dissected out from the brain tissues and homogenized in 500 μL of methanol.
2. After centrifugation of the homogenate at 1,000 *g* for 5 min, an aliquot of the supernatant (10 μL) was transferred to another tube and evaporated to dryness in vacuo.
3. Next, 30 μL of 50 mM NBD-F, a fluorescent reagent, dissolved in CH_3CN, was added to the residue dissolved in 20 μL of borate buffer (pH 8.0). This mixture was heated at 60°C for 5 min, and then, added with 50 μL of 1% of acetic acid in methanol.
4. A standard of racemic or optically pure D- and L-aspartic acid was treated in a similar manner as described in step 3.
5. The resultant solution was injected onto the HPLC system equipped with a chiral column, namely, Sumichiral OA-2500 (*S*). The fluorescence peaks were monitored at 530 nm with 470 nm of excitation wavelength.

3-2. D-Serine in the Microdialysate from Rat Brain

1. A brain microdialysis surgery was carried out under anesthesia with pentobarbital.
2. After 2 days, the rat with the guide cannula recovered fully. The brain microdialysis probe was inserted, and sterilized sodium phosphate buffer in Ringer's solution (pH 7.0) was pumped at a flow rate of 1.0 μL/min. After stabilization, the sampling of microdialysate from the brain of the freely moving rat was initiated.
3. Twenty μL of 0.1 M borate buffer (pH 8.0), 20 μL of 10 μM internal standard (I.S., cysteinic acid) in H_2O, and 60 μL of 50 mM NBD-F in CH_3CN were serially added to the microdialysate (2.5-20 μL). The mixture was heated at 60°C for 2 min, and subsequently, 400 μL of H_2O/CH_3CN (90/10) containing 0.1% trifluoroacetic acid was added.
4. A standard of racemic or optically pure D- and L-serine was treated in a similar manner as described in step 3.
5. The final solution was injected onto a column-switching HPLC system equipped with an ODS column and a tandem of 2 chiral columns, namely, Sumichiral OA-2500 (*S*). The fluorescence peaks were monitored at 530 nm with 470 nm of excitation wavelength.

4. Results

4-1. D-Aspartate in the Pineal Gland

A standard of D, L-aspartic acid derivatized with NBD-F (NBD-aspartate) separated well on Sumichiral OA-2500 (*S*), enantiomerically (Fig. 3(a)). As shown in Fig. 3(b), the elution of NBD-D, L-aspartate can be reversed on Sumichiral OA-2500 (*R*). The Pirkle-type chiral column has a merit that the elution order of enantiomer can reverse easily, and this characteristic is useful for the confirmation of trace amounts of D-amino acid through the ratio of obtained peaks. Figs. 3(c) and (d) show the chromatograms of extracts from the rat pineal gland on Sumichiral OA-2500 (*S*) and (*R*), respectively. Based on the same ratio of the NBD-aspartate enantiomer, both chromatograms revealed that D-aspartate was present in the rat pineal gland.

The levels of D- and L-aspartate in the rat pineal gland were determined from a calibration curve wherein the obtained peak area was plotted against the concentration of standard samples of D- and L-aspartic acid. The detection limit of the proposed HPLC method was approx. 30 fmol on the column (a signal-to-noise ratio of 3). Since ion-exchange mode separation is performed between the carboxyl groups of NBD-aspartate and residual amino groups on the surface of Sumichiral OA-2500 (*S*) or (*R*), the retention times of acidic amino acids, such as aspartic acid, were exceptionally longer than those of other neutral NBD-amino acids. Therefore, the separation of NBD-aspartate from other neutral amino acids was possible when the prepared sample, after derivatization with NBD-F, was directly injected onto Sumichiral OA-2500 (*S*) or (*R*). Thus, the determination of D-aspartate in biological specimens, such as tissue homogenate, was easily performed by using the proposed HPLC method.

4-2. D-Serine in the Microdialysate from Rat Brain

In the case of D-serine, NBD-D-serine was coeluted with other neutral NBD-amino acids on Sumichiral OA-2500 (*S*). Therefore, a column-switching HPLC system consisting of an ODS column and a tandem series of two Sumichiral OA-2500 (*S*) (Fig. 2) needs to be employed. Briefly, NBD-D, L-serine was first separated from other neutral NBD-amino acids on the ODS column, and then, enantiomerically separated on two Sumichiral OA-2500 (*S*) columns by changing the 6-port valve position. The obtained chromatograms are shown in Figs. 4(a)-(d). Figures 4(a) and (c) were obtained by using the ODS column, while Figs. 4(b) and (d) were obtained by using the chiral columns. The peaks of NBD-D- and -L-serine were clearly found in the microdialysate from the rat prefrontal cortex. These NBD-D- and -L-serine peaks in the microdialysate were also identified using a tandem series of two Sumichiral OA-2500 (*R*).

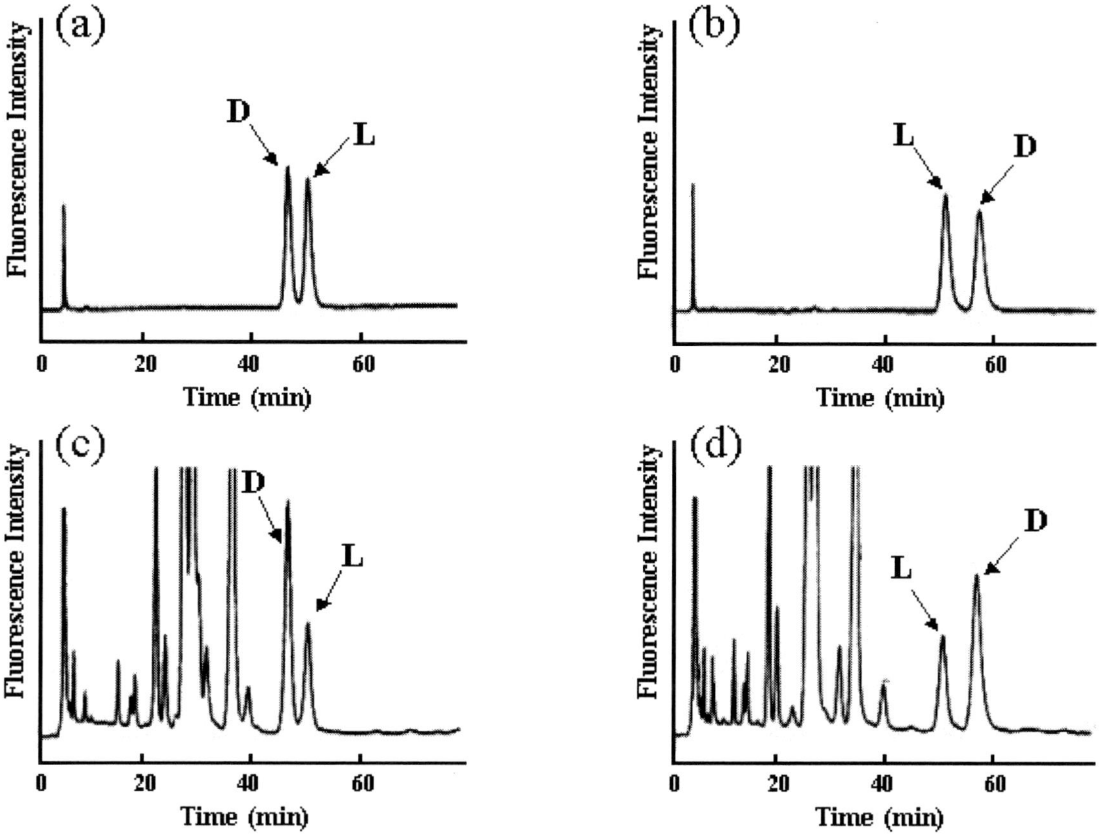

Figure 3. Chromatograms of a standard of D, L-aspartic acid derivatized with NBD-F (a and b) and of the extracts from the rat pineal gland on Sumichiral OA-2500 (*S*) and (*R*) (c and d, respectively). The chromatographic conditions are similar to those described in the paper by Imai *et al.* (10). (From Imai *et al.* (1996), *Biomed Chromatogr 9*, 106-109, 1996. Copyright John Wiley & Sons Limited. Reproduced from permission.).

The individual concentrations of D- and L-serine in the microdialysate were calculated as follows: the enantiomeric ratio of NBD-D- and -L-serine recorded by integrator 2 was multiplied with the total (D + L)-serine concentration, which was determined by the peak area ratio of serine derivative to I.S. derivative on the chromatogram recorded by integrator 1. Total (D + L)-serine concentration was determined by a calibration curve constructed from the standard samples of D- and L-serine.

A combination of the proposed HPLC method with *in vivo* microdialysis sampling is a beneficial technique for the research on the brain D-serine *in vivo* because it can provide data regarding any change in the brain D-serine levels in the awake, freely moving experimental animals. The time-course profile of D-serine concentration following intraperitoneal D-serine administration (50 mg/kg) is shown as an example in Fig. 5. Considering the rapid appearance of D-serine in the extracellular fluid, D-serine uptake through blood-brain barrier should occur readily. The D-serine concentration in the extracellular fluid decreases after 60 min. This might be due to its incorporation into the neuronal or glial cells, followed by metabolism by D-amino acid oxidase.

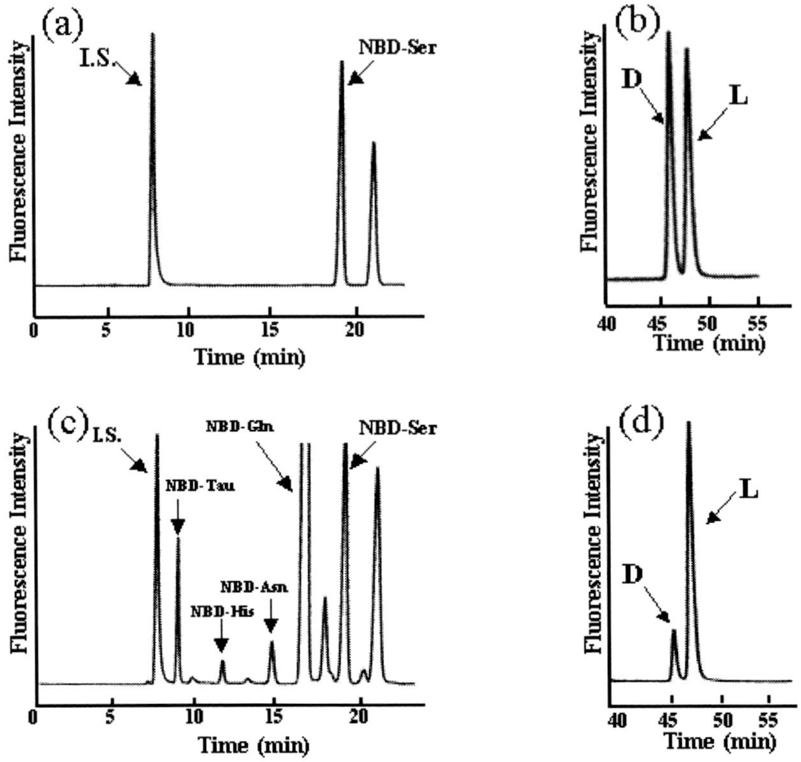

Figure 4. Chromatograms of a standard of racemic D, L-serine (a, b) and of a microdialysate from the rat prefrontal cortex (c, d), derivatized with NBD-F. (a) and (c) were obtained on the ODS column, while (b) and (d) were obtained on the chiral columns. The chromatographic conditions are similar to those described in the paper by Fukushima *et al.* (11). (From Fukushima *et al.* (2004), *Biomed Chromatogr 18*, 813-819, 2004. Copyright John Wiley & Sons Limited. Reproduced from permission.).

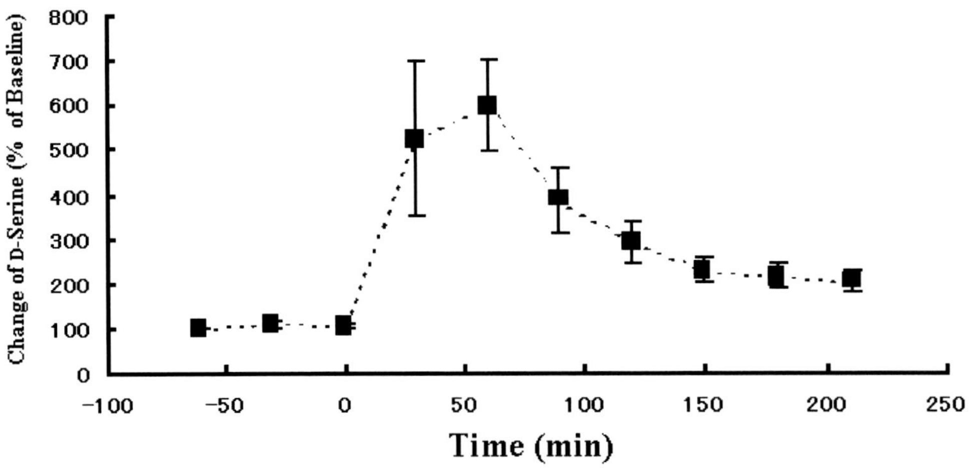

Figure 5. Time-course profile of D-serine concentration in the extracellular fluid of the rat prefrontal cortex, following intraperitoneal D-serine administration (50 mg/kg). D-Serine concentration recorded at –90- –60 min in each rat was considered to be 100%. The data are expressed as mean ±S.E., (n = 4). (From Fukushima *et al.* (2004), *Biomed Chromatogr 18*, 813-819, 2004. Copyright John Wiley & Sons Limited. Reproduced from permission.).

5. NOTES

The proposed HPLC method is applicable to other mammalian tissues as well as to tissue culture cells. The determination of D-amino acids in fermented foods and drinks is also possible by the proposed HPLC method [12].

REFERENCES

[1] Imai, K., Fukushima, T., Santa, T., Homma, H., Hamase, K., Sakai, K. & Kato, M. (1996). Analytical chemistry and biochemistry of D-amino acids. *Biomed Chromatogr 10*, 303-312.

[2] Imai, K., Fukushima, T., Santa, T., Homma, H., Huang, Y., Sakai, K. & Kato, M. (1997). Distribution of free D-amino acids in tissues and body fluids of vertebrates. *Enantiomer 2*, 143-145.

[3] Hashimoto, A. & Oka, T. (1997). Free D-aspartate and D-serine in the mammalian brain and periphery. *Prog Neurobiol 52*, 325-353.

[4] Takigawa, Y., Homma, H., Lee, J.A., Fukushima, T., Santa, T., Iwatsubo, T. & Imai, K. (1998). D-Aspartate uptake into cultured rat pinealocytes and the concomitant effect on L-aspartate levels and melatonin secretion. *Biochem Biophys Res Commun 248*, 641-647.

[5] Mothet, J.P., Parent, A.T., Wolosker, H., Brady, R.O., Linden, D.J., Ferris, C.D., Rogawski, M.A. & Snyder, S.H. (2000). D-Serine is an endogenous ligand for the glycine site of the *N*-methyl-D-aspartate receptor. *Proc Natl Acad Sci USA 97*, 4926-4931.

[6] Ribeiro, C.S., Reis, M., Panizzutti, R., Miranda, J.D. & Wolosker, H. (2002). Glial transport of the neuromodulator D-serine. *Brain Res 929*, 202-209.

[7] Tsai, G., Yang, P., Chung, L.C., Lange, N. & Coyle, J.T. (1998). D-Serine added to antipsychotics for the treatment of schizophrenia. *Biol Psychiatry 44*, 1081-1089.

[8] Hashimoto, K., Fukushima, T., Shimizu, E., Komatsu, N., Watanabe, H., Shinoda, N., Nakazato, M., Kumakiri, C., Okada, S., Hasegawa, H., Imai, K. & Iyo, M. (2003). Decreased serum levels of D-serine in patients with schizophrenia. Evidence in support of the *N*-methyl-D-aspartate receptor hypofunction hypothesis of schizophrenia *Arch Gen Psychiatry 60*, 572-576.

[9] Fukushima, T., Kato, M., Santa, T. & Imai, K. (1995). Enantiomeric separation and sensitive determination of D, L-amino acids derivatized with fluorogenic benzofurazan reagents on Pirkle type stationary phases. *Biomed Chromatogr 9*, 10-17.

[10] Imai, K., Fukushima, T., Hagiwara, K. & Santa, T. (1995). Occurrence of D-aspartic acid in rat brain pineal gland. *Biomed Chromatogr 9*, 106-109.

[11] Fukushima, T., Kawai, J., Imai, K. & Toyo'oka, T. (2004). Simultaneous determination of D- and L-serine in rat brain microdialysis sample using a column-switching HPLC with fluorimetric detection. *Biomed Chromatogr 18*, 813-819.

[12] Kato, M., Fukushima, T., Santa, T., Homma, H. & Imai, K. (1995). Determination of D-amino acids, derivatized with 4-fluoro-7-nitro-2,1,3-benzoxadiazole (NBD-F), in wine samples by high-performance liquid chromatography. *Biomed Chromatogr 9*, 193-194.

In: D-Amino Acids:
A New Frontier in Amino Acid and Protein Research, pp.93-102
ISBN 1-60021-075-9

Chapter 2.10

DETERMINATION OF SMALL AMOUNTS OF D-AMINO ACIDS IN MAMMALS USING COLUMN-SWITCHING HPLC

Kenji Hamase and Kiyoshi Zaitsu*
Graduate School of Pharmaceutical Sciences, Kyushu University, Japan.

Keywords: column-switching HPLC, fluorescence derivatization, D-alanine, D-leucine, D-proline.

1. INTRODUCTION

Various D-amino acids in the free form have been found in mammals, and they are now increasingly recognized as candidates of novel physiologically active substances or the marker molecules of physical conditions [1-3]. Especially, relatively large amounts of D-Ser and D-Asp were found in the brain and endocrine tissues, respectively, and there has been a significant number of reports concerning their functions and origins [4-6]. However, in most cases, the amounts of the free D-amino acids in mammalian tissues are extremely small, and their determination is troublesome because of the interference by large amounts of L-amino acids and by a large number of peptides. Therefore, a highly sensitive and selective method is needed for the determination of small amounts of D-amino acids in the tissue samples and physiological fluids. For this purpose, multi-dimensional chromatographic methods are useful, and various methods have been reported [7-10]. In this article, a powerful column-switching HPLC method [11-13] using a micro-ODS column and a chiral column is described. In this method, the amino acids are derivatized with a sensitive fluorescence derivatization reagent, 4-fluoro-7-nitro-2,1,3-benzoxadiazole [NBD-F, 14,15], prior to the HPLC determination. The resultant NBD-amino acids are separated as their D+L mixtures by the micro-ODS column, and the fraction of the target amino acid is collected on-line in a loop unit, then subsequently transferred into the chiral column in which the enantiomers are separated. This method enables the highly sensitive (detection limit of the NBD-amino acid is about 5 fmol per injection to the HPLC) and highly selective (no interfering substances are observed even when tissue samples are analyzed) determination of small amounts of D-amino acids in mammalian tissues and physiological fluids. The results obtained for D-Ala [11], D-Leu [12,13] and D-Pro [12] are described below.

* Correspondence concerning this article should be addressed to Kiyoshi Zaitsu, E-mail: zaitsu@phar.kyushu-u.ac.jp, FAX: +81-92-642-6596.

2. Materials

2.1. Reagents and Equipment for Tissue Treatment

1. diethyl ether
2. methanol (MeOH)
3. a tissue homogenizer
4. a centrifuge

2.2. Reagents and Equipment for NBD Derivatization of Amino Acids

1. 200 mM borate-NaOH buffer (pH 8.0)
2. 20 mM NBD-F in acetonitrile (MeCN)
3. 5% TFA in H_2O
4. a water bath

2.3. HPLC System

The column-switching HPLC system consists of the following units, and these units were connected as shown in Figure 1.

1. two HPLC pumps (P)
2. a sample injector (I)
3. a degasser (DG)
4. a column oven (CO)
5. two fluorescence detectors (D)
6. two integrators (IT)
7. a column selection unit with a loop of 400 μL (CS)
8. a guard column (GC)
9. a micro reversed-phase column (C1)
10. a chiral column (C2)

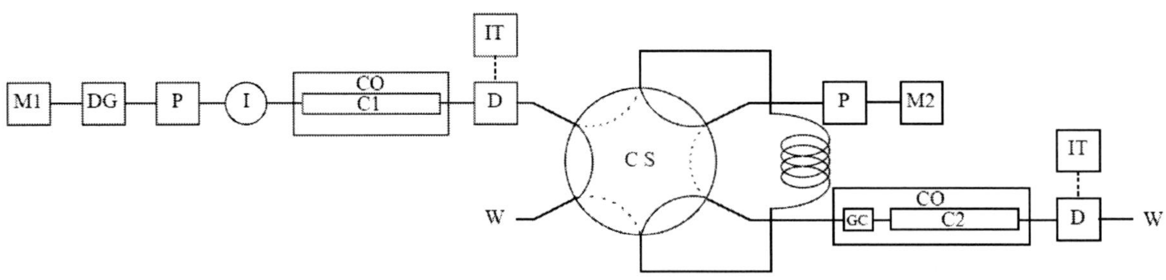

Figure 1. Column-switching HPLC system for the determination of D-amino acids. The analytical conditions are described in section 3.3. The abbreviations used are defined in the manuscript.

3. METHODS

3.1. Preparation of Tissue Samples

Animals were anesthetized with diethyl ether, and sacrificed by exsanguination from the abdominal aorta. The tissues (such as cerebrum, liver and kidney) were then immediately excised and stored at -80°C until analysis. The blood was centrifuged at 4500g for 15 min at 4°C, and the collected serum was also stored at -80°C. These tissues were homogenized on ice with 20X vol of MeOH (the serum was simply mixed with 20X vol of MeOH), then centrifuged at 4500g for 5 min, and the obtained supernatants were used for the derivatization of the amino acids.

3.2. Fluorescence Derivatization of Amino Acids

The tissue samples (supernatants of MeOH homogenates, 10 μL) were evaporated to dryness under reduced pressure at 40°C, and thoroughly dissolved again in 20 μL of 200 mM borate-NaOH buffer (pH 8.0). To the solution, 10 μL of 20 mM NBD-F in MeCN was added and heated at 60°C for 2 min. To this reaction mixture, 20 μL of 5% TFA in H_2O was added, and a portion of this reaction mixture (1 or 5 μL) was injected into the HPLC system. The reaction and the sample preparation procedure are shown in Figures 2 and 3.

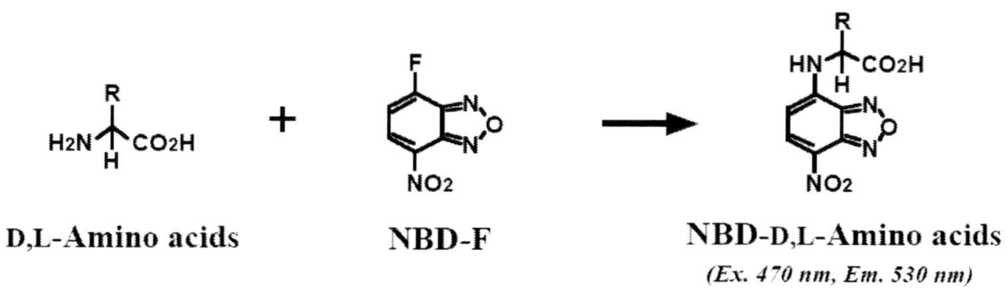

Figure 2. Fluorescence derivatization of amino acids with NBD-F.

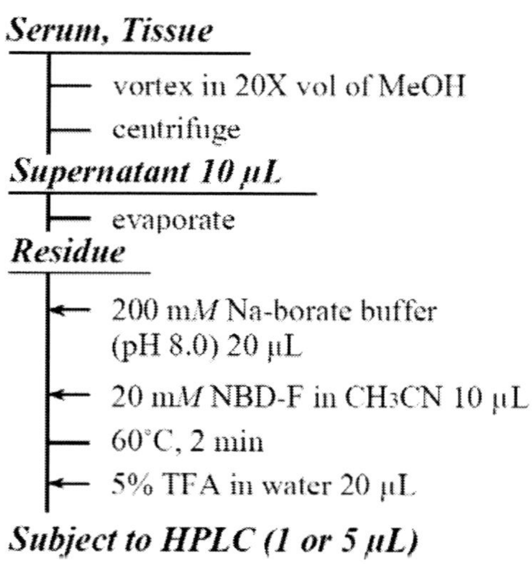

Figure 3. Sample treatment and derivatization procedure for the determination of D-Ala in the tissues or physiological fluids of mammals.

3.3. HPLC Determination of Amino Acids

For the determination of small amounts of D-amino acids, a column-switching HPLC system shown in Figure 1 was used. After the NBD derivatization, the tissue samples were injected to the HPLC system and separated by a micro-ODS column (first-step separation). The peaks of the NBD-amino acids were monitored by a fluorescence detector, and the fraction of the target amino acid was transferred into the loop *via* a column-selection unit. The enantiomers of the target amino acid were then separated by a chiral column (second-step separation), and both enantiomers were determined by a fluorescence detector (excitation wavelength is 470 nm, and the emission detection wavelength is 530 nm). We have already reported the separation conditions for the determination of D-Ala, D-Leu and D-Pro; the recommended conditions are summarized in Table 1. Other D-amino acids could also be determined by slightly changing the HPLC conditions.

Table 1. Separation conditions for the determination of NBD-amino acids using column-switching HPLC.

Amino acids	Micro-ODS column Mightysil RP-18 GP (1.0 mm i.d. x 100 mm, Kanto Chemical, Japan)		Column switching period (sec)	Chiral Column Sumichiral OA-2500S or 2500R (4.6 mm i.d. x 250 mm, Sumika Analytical Center, Japan)	
	Mobile phase MeCN / THF / TFA / H2O	Flow Rate (μl / min)		Mobile phase (citric acid in MeOH)	Flow Rate (ml / min)
D-Ala	10 / 1 / 0.02 / 89	50	210	5.0 mM	0.8
D-Leu	32.5 / 1 / 0.02 / 66.5	50	150	2.0 mM	0.8
D-Pro	15 / 1 / 0.02 / 84	50	150	5.0 mM	1.2

4. RESULTS

4.1. Determination of D-Ala in the Rat Tissues and Physiological Fluids [11]

D-Ala is the D-amino acid well known to be present in microorganisms. In mammals, the presence of a small amount of D-Ala in the blood and urine have been reported [1,2], however, the tissue distribution or the quantitative variation under various physiological conditions have hardly been clarified. In our recent report, we use the column-switching procedure for the determination of D-Ala, and small but significant amounts of D-Ala were found in the various tissues of mammals. Figure 4 shows the chromatograms obtained using a rat cerebrum. As shown in Figure 4A, NBD-Ala is eluted at 21 min as the D+L mixture on a micro-ODS column; the fraction is collected on-line to a loop (the flow diagram is shown in Figure 1) by changing the column-selection valve for 210 s (the HPLC conditions are shown in Table 1). This fraction is subsequently transferred to a chiral column by changing the column-selection valve again, and NBD-D-Ala and NBD-L-Ala were separated within 45 min (Figure 4B). The presence of D-Ala in the tissue samples is easily confirmed by changing the chiral column to the one having an opposite chiral center. Figure 5 shows the chromatograms obtained using the rat anterior pituitary gland. Replacement of OA-2500S (Figure 5A) by OA-2500R (Figure 5B) leads to a reversal in the elution order of NBD-D-Ala and NBD-L-Ala, and the presence of D-Ala is strongly supported when almost the same amount of D-Ala is determined by both chiral columns. The lower limit of determination of D-Ala by the present method is 5 fmol (per injection to the HPLC), which is about 1 / 100 that of the widely used OPA methods.

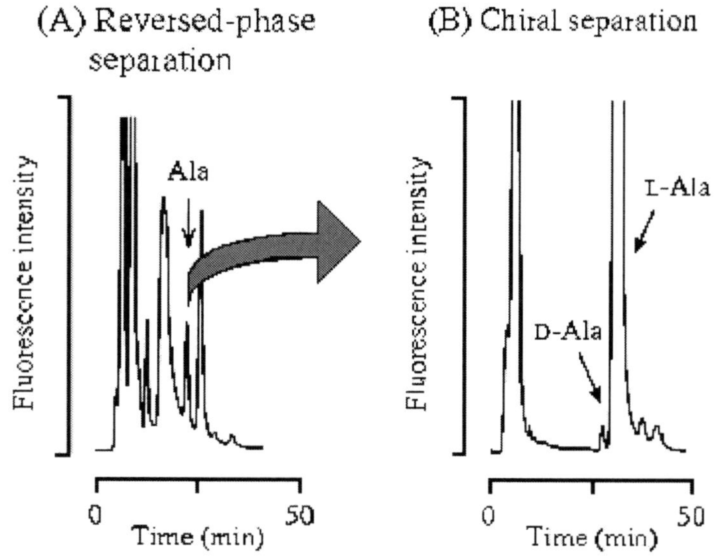

Figure 4. Separation of NBD-D- and L-Ala in the rat cerebrum on a micro-ODS column (A) and on a Sumichiral OA-2500S column using a column-switching system (B). The arrows show the peaks of NBD-Ala, NBD-D-Ala and NBD-L-Ala.

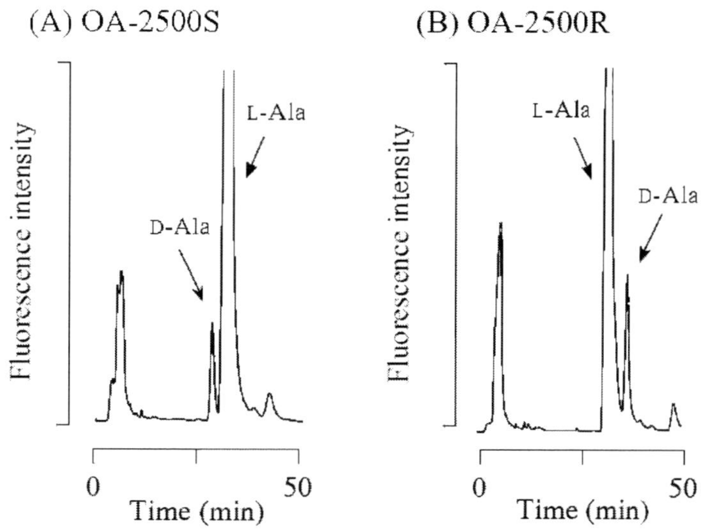

Figure 5. Separation of NBD-D- and L-Ala in the rat anterior pituitary gland using the column-switching HPLC system. A Sumichiral OA-2500S (A) and OA-2500R (B) were used as the chiral columns. Peaks: D-Ala, NBD-D-Ala; L-Ala, NBD-L-Ala. The HPLC conditions are described in Table1.

Using this method, the amounts of D-Ala were determined in 22 tissues including the central nervous system and periphery. In all the tissues tested, small but obvious peaks of D-Ala were found; which was the first time clearly showing the anatomical distribution of D-Ala in mammals. Figure 6 summarizes these results. Among the 22 tissues tested, the highest amount of D-Ala was found in the anterior pituitary gland (86.4 ± 9.9 nmol / g wet tissue), followed by the pancreas (29.2 ± 5.0 nmol / g wet tissue). All of the values described in Figure 6 are confirmed by changing the chiral column (for example, the value for anterior pituitary gland obtained using the opposite chiral column is 87.5 ± 9.4 nmol / g wet tissue). The postnatal changes of the D-Ala amount in several tissues (cerebrum, cerebellum, anterior pituitary, posterior pituitary, pineal gland, pancreas, adrenal gland and plasma) have also been investigated. As shown in

Figure 7, high amounts of D-Ala were observed in young-adult rats (6 weeks of age) in all tissues tested which gradually decreased thereafter. At any age, the amount of D-Ala was the highest in the anterior pituitary gland, and second highest in the pancreas. These results strongly suggest that this D-amino acid has some important roles in these endocrine glands, i.e., the anterior pituitary gland and pancreas. Surprisingly, the amount of D-Ala has a clear circadian rhythm in the anterior pituitary gland. The results obtained using the rats of 6 weeks of age (closed circles) and those of 9 weeks of age (open circles) are shown in Figure 8. In the rats of both ages, the amounts of D-Ala in the daytime are higher than those in the nighttime. Because the biosynthesis and secretion of various hormones have clear circadian rhythms, the present results also support the fact that D-Ala has some close relationships to pituitary hormones.

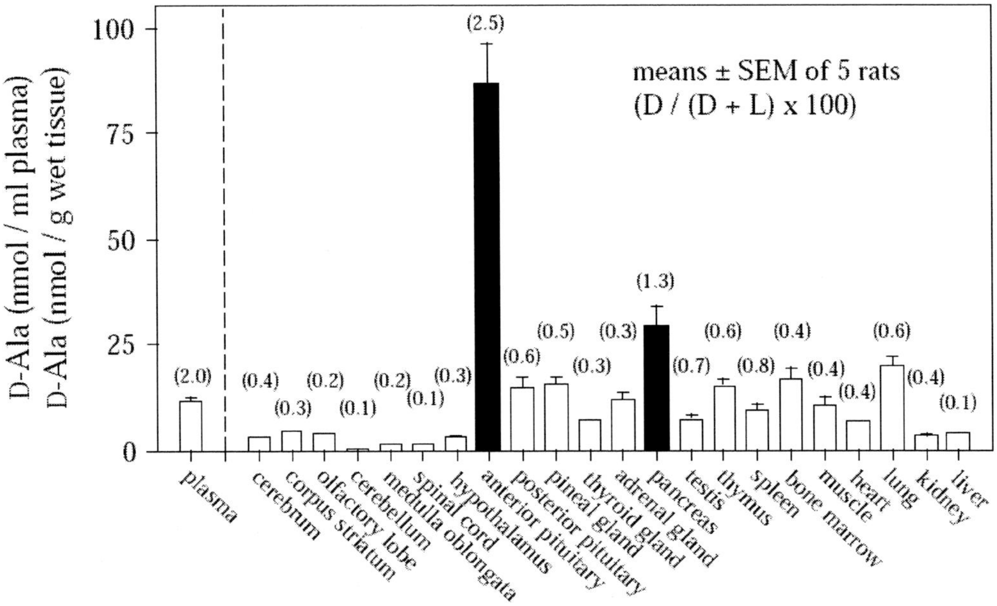

Figure 6. Anatomical distribution of D-Ala in the rat brain and periphery. The values are obtained using 5 animals (male Wistar rats, 9 weeks of age). Values in parentheses are the proportions of the D form (D/(L+D) x 100).

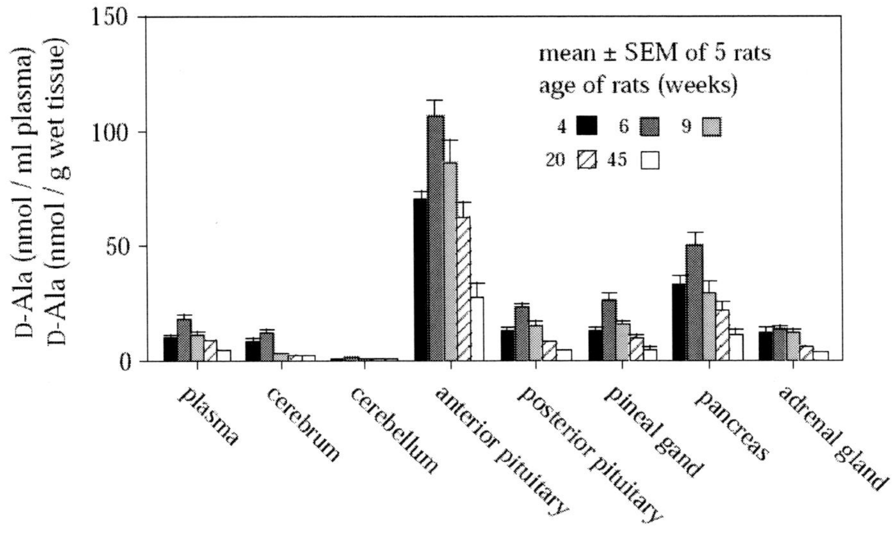

Figure 7. Age dependent changes in D-Ala amount in the rat brain and periphery. The values are obtained using 5 animals (male Wistar rats).

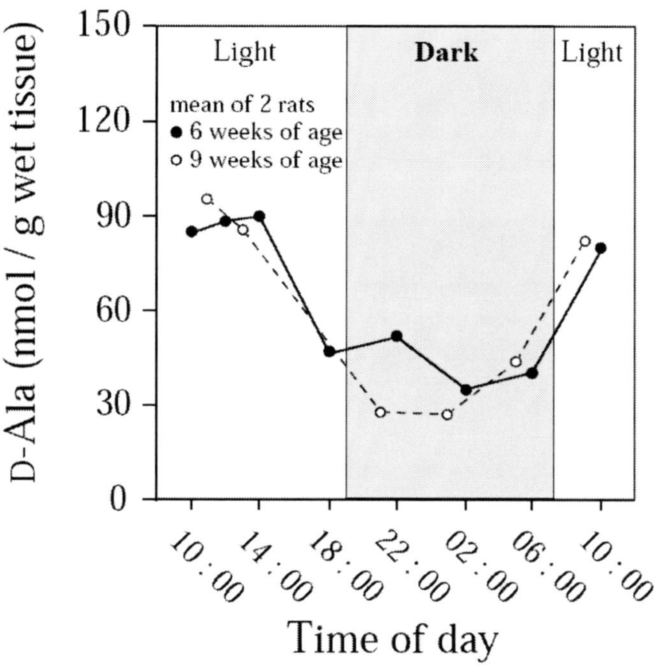

Figure 8. Circadian changes in D-Ala amount in the rat anterior pituitary gland. Open circles represent the values obtained using rats of 9 weeks of age, and closed circles are those for rats of 6 weeks of age. Values represent the average of 2 animals (male Wister rats).

4.2. Determination of D-Leu and D-Pro in the Brains of ddY/DAO$^+$ Mice and ddY/DAO$^-$ Mice [12,13]

D-Leu and D-Pro are also D-amino acids found in small amounts in mammals, and the regulation mechanisms of their tissue distribution and their quantity are of interest. We have determined these D-amino acids using the column-switching procedure, and the distributions of both D-Leu and D-Pro in the mouse brain have been demonstrated. Concerning the quantitative regulation mechanisms of these D-amino acids, the effect of D-amino acid oxidase [DAO, the enzyme which catalyzes the oxidative deamination of neutral D-amino acids to form the corresponding α-keto acids, 16] has been investigated using a mutant mouse strain, ddY/DAO$^-$ mouse [17,18]. These mice completely lack the DAO activity, and in these mice, relatively large amounts of some D-amino acids were observed. Figure 9 shows the chromatograms of NBD-D-Leu obtained using the mouse cerebrum. As shown in Figures 9A and 9B, the amount of D-Leu in the cerebrum of the ddY/DAO$^-$ mouse is about 10 times higher than that observed in the cerebrum of the ddY/DAO$^+$ mouse. These peaks of NBD-D-Leu are confirmed by the enzymatic degradation of D-Leu by DAO (commercially available DAO from hog kidney, purchased from Boehringar Mannheim Yamanouchi (Tokyo, Japan) was used) in addition to changing the chiral column. After DAO treatment, the peak of NBD-D-Leu almost completely disappeared (Figure 9C), which strongly supports the fact that the peaks are those for the genuine D-Leu. The distribution of D-Leu in the brains of ddY/DAO$^+$ and ddY/DAO$^-$ mice are summarized in Table 2. In the brains of ddY/DAO$^+$ mice, D-Leu was observed in all the brain areas tested, being slightly localized in the pineal gland. While, in the brains of ddY/DAO$^-$ mice, the amounts of D-Leu are significantly higher ($P<0.01$) than those in the ddY/DAO$^+$ mice in all brain areas tested except for the pineal gland. These results suggest that a large amount of D-Leu is originally present in the mouse tissues, which are metabolized by DAO in the ddY/DAO$^+$ mice (the normal mice having a strong DAO activity). The distribution of D-Pro in the brains of the ddY/DAO$^+$ and ddY/DAO$^-$ mice are summarized in Table 3. Different from the results for D-Leu, the amounts of D-Pro in the brains of the ddY/DAO$^-$ mice are almost the same as those in the brains of the ddY/DAO$^+$ mice; suggesting that D-Leu and D-Pro are regulated in the different manners in the mammalian brain.

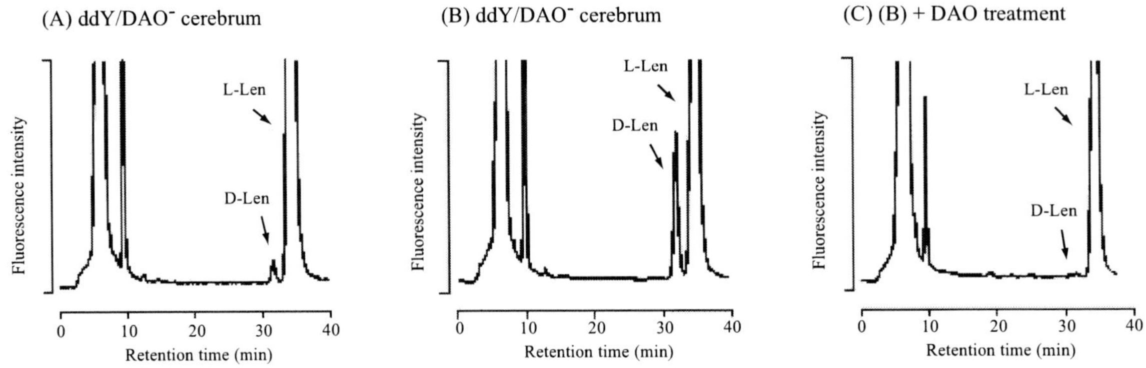

Figure 9. Chromatograms of NBD-D- and L-Leu in the cerebrum of a ddY/DAO$^+$ mouse (A) and a ddY/DAO$^-$ mouse (B) obtained by the column-switching HPLC system. The enantiomers of Leu in the cerebrum of ddY/DAO$^-$ mouse were determined after the treatment of DAO (C). Peaks: D-Leu, NBD-D-Leu; L-Leu, NBD-L-Leu. The HPLC conditions are described in Table 1.

5. Conclusions

The column-switching HPLC system described in the present article enables the sensitive and selective determination of D-amino acids in the complicated tissue samples or physiological fluids. This method is applicable to the determination of various D-amino acids by slightly changing the analytical conditions, which should be a powerful tool for the investigation of small amounts of D-amino acids in mammals.

Table 2. The amounts of D-Leu and L-Leu in the brains of ddY/DAO$^+$ and ddY/DAO$^-$ mice.

Brain regions	Mouse strain			
	ddY/DAO$^+$		ddY/DAO$^-$	
	D-Leu	L- Leu	D- Leu	L- Leu
Cerebrum	0.41±0.11	67.3±12.7	4.15±0.30**	62.8±14.3
Hippocampus	0.39±0.09	65.5±10.6	3.52±0.30**	57.9±11.9
Hypothalamus	0.55±0.09	82.1±11.9	4.75±0.25**	69.1±12.8
Pituitary gland	0.76±0.16	306±54.3	5.97±0.34**	267±19.8
Pineal gland	1.94±0.46	186±38.7	4.91±1.10	76.2±25.9
Cerebellum	0.24±0.04	55.5±10.9	5.02±0.23**	59.7±6.30
Medulla oblongata	0.27±0.05	58.2±13.7	4.05±0.53**	59.1±11.7

Values represent means ± SEM (nmol / g wet tissue) of 4 animals. The values for the pineal glands are expressed as pmol / whole gland. **$P < 0.01$, significant difference from the values of DAO$^+$ mice.

Table 3. The amounts of D-Pro and L-Pro in the brains of ddY/DAO$^+$ and ddY/DAO$^-$ mice.

Brain regions	Mouse strain			
	ddY/DAO$^-$		ddY/DAO$^-$	
	D-Pro	L-Pro	D-Pro	L-Pro
Cerebrum	0.31±0.15	123±30.6	0.12±0.04	72.3±12.1
Hippocampus	0.11±0.06	81.0±5.07	0.09±0.05	58.4±12.4
Hypothalamus	0.25±0.05	123±7.34	0.26±0.03	82.3±3.80
Pituitary gland	0.59±0.12	542±33.9	1.33±0.24	404±68.1
Pineal gland	1.11±0.26	194±42.5	1.52±0.22	72.6±20.6
Cerebellum	0.29±0.12	91.8±16.6	0.71±0.05	56.0±4.30
Medulla oblongata	0.16±0.10	67.3±6.90	0.38±0.01	51.7±10.4

Values represent means ± SEM (nmol / g wet tissue) of 4 animals. The values for the pineal glands are expressed as pmol / whole gland.

6. Notes

1. In case of small tissues (such as pineal gland and pituitary gland), the whole tissues should be homogenized in 500 μL of MeOH. Aliquots of the supernatant (100 μL) were then dried and used for the derivatization procedure.
2. The NBD-amino acids should be completely shielded from light. The use of the black tubes is strongly recommended for storing the derivatization reaction mixtures of the amino acids with NBD-F.
3. When the enantiomers of the NBD-amino acids are not sufficiently separated by the chiral column (Sumichiral OA-2500S or 2500R), the use of a lower concentration of citric acid in MeOH as the mobile phase would improve the separation.
4. The chiral column (Sumichiral OA-2500S and 2500R) must be washed with 50 mM ammonium acetate in MeOH for at least 20 min when a stable baseline is not obtained.

References

[1] Hamase, K., Morikawa, A. & Zaitsu, K. (2002). D-Amino acids in mammals and their diagnostic value. *J Chromatogr B*, *781*, 73-91.

[2] Fujii, N. (2002). D-Amino acids in living higher organisms. *Orig Life Evol Biosph*, *32*, 103-127.

[3] Imai, K., Fukushima, T., Santa, T., Homma, H., Hamase, K., Sakai, K. & Kato, M. (1996). Analytical chemistry and biochemistry of D-amino acids. *Biomed Chromatogr*, *10*, 303-312.

[4] Hashimoto, A. & Oka, T. (1997). Free D-aspartate and D-serine in the mammalian brain and periphery. *Prog Neurobiol*, *52*, 325-353.

[5] Wolosker, H., D'Aniello, A. & Snyder, S.H. (2000). D-Aspartate disposition in neuronal and endocrine tissues: Ontogeny, biosynthesis and release. *Neuroscience*, *100*, 183-189.

[6] Hashimoto, A., Oka, T. & Nishikawa, T. (1995). Anatomical distribution and postnatal changes in endogenous free D-aspartate and D-serine in rat brain and periphery. *Eur J Neurosci*, *7*, 1657-1663.

[7] Armstrong, D.W., Gasper, M.P., Lee, S.H., Ercal, N. & Zukowski, J. (1993). Factors controlling the level and determination of D-amino acids in the urine and plasma of laboratory rodents. *Amino Acids*, *5*, 299-315.

[8] Nagata, Y., Yamamoto, K. & Shimojo, T. (1992). Determination of D- and L-amino acids in mouse kidney by high-performance liquid chromatography. *J Chromatogr*, *575*, 147-152.

[9] Hamase, K., Homma, H., Takigawa, Y., Fukushima, T., Santa, T. & Imai, K. (1997). Regional distribution and postnatal changes of D-amino acids in rat brain. *Biochim Biophys Acta*, *1334*, 214-222.

[10] Zhao, H., Hamase, K., Morikawa, A., Qiu, Z. & Zaitsu, K. (2004). Determination of D- and L-enantiomers of threonine and allo-threonine in mammals using two-step high-performance liquid chromatography. *J Chromatogr B*, *810*, 245-250.

[11] Morikawa, A., Hamase, K. & Zaitsu, K. (2003). Determination of D-alanine in the rat central nervous system and periphery using column-switching high-performance liquid chromatography. *Anal Biochem*, *312*, 66-72.

[12] Hamase, K., Inoue, T., Morikawa, A., Konno R. & Zaitsu, K. (2001). Determination of free D-proline and D-leucine in the brains of mutant mice lacking D-amino acid oxidase activity. *Anal Biochem*, *298*, 253-258.

[13] Inoue, T., Hamase, K., Morikawa, A. & Zaitsu, K. (2000). Determination of minute amounts of D-leucine in various brain regions of rat and mouse using column-switching high-performance liquid chromatography. *J Chromatogr B*, *744*, 213-219.

[14] Imai, K. & Watanabe, Y. (1981). Fluorimetric determination of secondary amino acids by 7-fluoro-4-nitrobenzo-2-oxa-1,3-diazole. *Anal Chim Acta*, *130*, 377-383.

[15] Fukushima, T., Kato, M., Santa, T. & Imai, K., (1995). Enantiomeric separation and sensitive determination of D, L-amino acids derivatized with fluorogenic benzofurazan reagents on Pirkle type stationary phases. *Biomed Chromatogr*, *9*, 10-17.

[16] Pilone, M.S. (2000). D-Amino acid oxidase: new findings. *Cell Mol Life Sci*, *57*, 1732-1747.

[17] Konno, R. & Yasumura, Y. (1983). Mouse mutant deficient in D-amino acid oxidase activity. *Genetics*, *103*, 277-285.

[18] Konno, R. & Yasumura, Y. (1988). A simple and rapid method to screen for mutant mice lacking D-amino-acid oxidase activity. *Lab Anim Sci*, *38*, 292-295.

In: D-Amino Acids: ISBN 1-60021-075-9
A New Frontier in Amino Acid and Protein Research, pp.103-106

Chapter 2.11

SENSITIVE DETERMINATION OF AMINO ACID ENANTIOMERS BY CE/LIF

***Yi-Ming Liu*[1,*] *and Shulin Zhao*[2]**
[1]Department of Chemistry, Jackson State University, U.S.A;
[2]College of Chemistry and Chemical Engineering, Guangxi Normal University, China.

Keywords: amino acid enantiomers; quantification; capillary electrophoresis; laser induced fluorescence detection; biological samples.

1. INTRODUCTION

Unexpectedly high levels of both free and peptide-bound D-amino acids have been recently found in a variety of living systems [1,2]. In search for a better understanding of the biological significance of D-amino acids, enantiomeric determination of amino acids in biological samples has become highly desirable. Analytical methods based on high performance liquid chromatography (HPLC), capillary electrophoresis (CE), enzymatic and immunochemical biosensors, etc. were reported. A comprehensive review on analysis of amino acid enantiomers has been given [3]. A CE based method offers high separation efficiency, short run time, instrumentation simplicity, and compatibility with small volume/mass samples. In order to achieve the sensitivity needed for quantifying trace levels of amino acid enantiomers in biological samples, a highly sensitive detection must be used following CE separations. Laser induced fluorescence detection (LIF) is the most sensitive detection scheme available so far for CE. Prior to CE/LIF analysis, amino acids must be fluorescently tagged through precolumn derivatization. Among various tagging reagents, naphthalene-2,3-dicarboxaldehyde (NDA) has the advantage of reacting with amino acids in aqueous solutions producing stable and highly fluorescent derivatives. In addition, NDA reagent blank exhibits a very low fluorescence background. NDA-tagged amino acid enantiomers can be easily separated and sensitively detected by CE/LIF with β-cyclodextrin as chiral selector. Therefore, CE/LIF of NDA-tagged amino acid enantiomers is a method well suited for quantifying low levels of D-amino acids in biological samples. Determination of D/L-aspartic acid (Asp) enantiomers in rat brain tissue by using this method is described here as an example.

* Correspondence concerning this article should be addressed to: Yi-Ming Liu, E-mail: yiming.liu@jsums.edu., Shulin Zhao, E-mail: zhaoshulin001@163.com.

2. MATERIALS

1. The CE/LIF system consists of a high voltage supply (0-30 kV, Glassman High Voltage, Whitehouse Station, NJ, USA) and a lab-built LIF detector. Fused-silica capillaries with an internal diameter of 75 μm, an external diameter of 360 μm, and an effective length of 50 cm (Polymicro Technologies, Phoenix, AZ, USA) is used for CE separation. The buffer reservoir at the high-voltage end is enclosed in a plexiglass box. The 457.9 -nm line from an argon ion laser (INNOVA 90C FreD, Coherent, Santa Clara, CA, USA) is focused with a 25-mm focal length fused-silica lens onto the detection window of the separation capillary. The detection window is made by removing a 5-mm section of polyimide coating on the capillary. Fluorescence emission is collected from the detection window at an angle of 90° relative to the laser beam via a 40X microscope objective (Melles Griot, Irvine, CA, USA). The image of the collected fluorescence is focused on a spatial filter and passes through a GG495 cut-off filter (Melles Griot) before reaching a photomultiplier tube (R374 equipped with a C1556-50 DA-type socket assembly, Hamamatsu, Shizuoka, Japan). The photomultiplier tube is operated at –950V provided by a Hamamatsu C1309 regulated high voltage power supply. Output signals are recorded and processed with an IBM compatible computer using in-house written software.
2. Background electrolyte solution: 100 mM SDS, 20 mM β-cyclodextrin (β-CD) and 15% (v/v) methanol in 50 mM borate buffer (pH 9.0).
3. NDA solution: 2 mM in methanol.
4. KCN solution: 20 mM in water.
5. Borate buffer: 100 mM at pH 9.5.

3. METHODS

1. *Preparation of rat brain tissue samples.* Rat is anaesthetized with CO_2 and the brain is dissected out and ground with 0.1 M HCl solution (1 + 5) on ice using a tissue grinder. A portion of the homogenate is transferred into a 0.5-mL centrifuge tube and centrifuged (2,000 *g* for 10 min). The supernatant (100 μL) is transferred into another 0.5-ml centrifuge tube and 30% (w/v) trichloroacetic acid solution (30 μL) is added. The solution is vortexed and centrifuged (2,000 *g* for 10 min). The pH value of the supernatant is brought to about 9 with 1 M NaOH solution.
2. *Precolumn derivatization with NDA.* The sample solution prepared above (10 μL) is transferred into a 0.5-ml centrifuge tube, then 100 mM borate buffer solution (pH 9.5, 150 μL), 2 mM NDA solution (50μL), and 20 mM KCN solution (50 μL) are added in sequence. The solution is vortexed and kept at room temperature for 20 min before injection. The derivative solution is stable for at least 6 hours when kept in dark at 5 °C.
3. *Separation and quantification of D/L-Asp.* Separation is performed at 15 kV. All samples are injected into the capillary by hydrodynamic flow at a height differential of 20 cm for 20 S. D/L-Asp quantification is achieved by measuring the CE peak heights against a calibration curve prepared with D- and L-Asp enantiomeric solutions at concentrations ranging from 0.1 to 1.5 μM.

4. RESULTS

The method is very effective for resolving NDA-tagged amino acid enantiomers. Figure 1A shows the electropherogram from separating a mixture of four pairs of amino acid enantiomers. All of them are baseline separated. This CE/LIF method is well suited for the determination of D- and L-Asp in rat brain tissue samples. No interference from the endogenous compounds in this sample matrix is observed. The D- and L-Asp peaks are well identified in the electropherogram as shown in Figure 1B. After analyzing the rat brain samples obtained from rats at different growing stages, a relationship has been revealed between the contents of D-/L-Asp and the age [4]. The content of free D-Asp and the ratio of D-Asp to total Asp are found to be highest at 1 day before birth. D-Asp content and the D/L-Asp ratio diminish rapidly within one week after birth, and then remain very low during other postnatal stages. Analyses of tissue samples taken from different parts of rat brain show that the highest concentration of free D-Asp (81 nmol/g) is detected in cerebrum and the highest ratio of D-Asp to total Asp (19.5%) occurs in cerebellum.

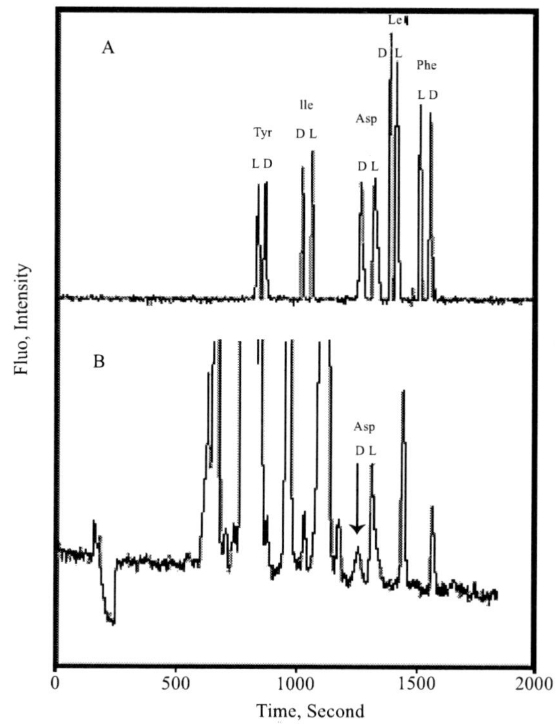

Figure 1. Electropherograms obtained from: (A) separating NDA-tagged amino acid enantiomers, and (B) separating a brain tissue sample dissected from a rat 1 day before birth. CE running buffer: 20 mM β-CD, 100 mM SDS, and 15% (v/v) methanol in 50 mM borate buffer (pH 9.0). Capillary: 50 cm (effective length) x 75 μm i.d. Applied voltage: 15 kV. LIF detection, $\lambda_{ex} = 457$ nm, λ_{em} =>495nm.

5. Notes

1. Capillary must be rinsed sequentially with 0.1 M NaOH, water, and the CE running buffer for approximately 3 min each between successive electrophoretic runs.
2. NDA solution is prepared weekly in methanol and kept in dark at 5 °C.
3. All electrolyte solutions must be filtered through a 0.2 μm syringe filter.
4. To achieve better resolutions of certain amino acid enantiomers, modified cyclodextrins can be used in place of β-CD. For example, D/L-Trp and D/L-Ser are better resolved with 15 mM hydroxypropyl-γ-cyclodextrin (HP-γ-CD) as chiral selector instead of 20 mM β-CD [5].
5. Better separation efficiencies can be obtained by using capillaries with smaller internal diameters such as 50 or 25 μm. However, these capillaries are more easily blocked (even partially) during the analysis of tissue samples.

6. This method is applicable to other biological samples including human urine and single neurons dissected from *Aplysia californica*, a widely used neuronal model.

References

[1] Fisher, G.H. (1998). Appearance of D-amino acids during aging: D-amino acids in tumor proteins. *EXS*, 85, 109 - 118.

[2] Imai, K., Fukushima, T., Santa, T., Homma, H., Huang, Y., Sakai, K., Kato, M. (1997). Distribution of free D-amino acids in tissues and body fluids of vertebrates. *Enantiomer*, 2, 143 - 145.

[3] Hamase, K., Morikawa, A., Zaitsu, K. (2002). D-Amino acids in mammals and their diagnostic value. *J. Chromatogr. B Analyt. Technol. Life Sci. 781*, 73-91.

[4] Zhao, S., Feng, Y., LeBlanc, M. H., Liu, Y.M. (2001). Determination of free aspartic acid enantiomers in rat brain by capillary electrophoresis with laser-induced fluorescence detection. *J. Chromatogr. B, 762*, 97–101.

[5] Zhao, S., Liu, Y.M. (2001). Electrophoretic separation of tryptophan enantiomers in biological samples. *Electrophoresis*, *22*, 2769-2774.

In: D-Amino Acids: ISBN 1-60021-075-9
A New Frontier in Amino Acid and Protein Research, pp.107-110

Chapter 2.12

METHODS OF DETECTION FOR AMINO ACID ENANTIOMERS IN THE VERTEBRATE RETINA

Robert F. Miller[1], Michael T. Bowser[2] and Kylie B. O'Brien[2,*]
[1]Department of Neuroscience,
[2]Department of Chemistry, University of Minnesota, Medical School Minneapolis, USA

Keywords: capillary electrophoresis, CE, D-serine, retina, enantiomer separation, NMDA receptors.

1. INTRODUCTION

There is now compelling evidence to support the idea that D-serine is an endogenous, functional amino acid in the vertebrate central nervous system [1] and that it plays a role as a coagonist for NMDA receptor function [2-4]. Although it was once thought that D-amino acids were present in trace amounts, accumulated from extraneous sources, studies have now demonstrated significant levels of D-serine in the brain [5]. Furthermore there is convincing evidence that D-serine is synthesized in the nervous system through serine racemase, a pyridoxyl-dependent enzyme that synthesizes D-serine from L-serine [6]. Both D-serine and serine racemase appear to be found primarily in astrocytes. Since D-serine can serve as a coagonist for NMDA receptors at the 'glycine binding site' an intriguing hypothesis of D-serine function has been proposed in which astrocytes, through synthesis and release of D-serine, provide the physiological basis for one form of glial-neuronal control in which a fast-acting glutamate receptor may have its efficacy controlled or partially controlled by glial cells [4,7].

While the localization of D-serine has been revealed in the brain and spinal cord, until recently, the vertebrate retina, another major division of the central nervous system, had not been explored for the presence of D-serine or its function. However, recent studies have demonstrated immunostaining for D-serine and serine racemase in the retinas of salamanders, mice and rats [8]. In addition, physiological studies have suggested that D-serine may serve as an endogenous coagonist for NMDA receptors of the inner retina.

The detection of amino acid racemates in the nervous system has been approached with HPLC methods [9]. However this separation technique typically involves fairly long separation times, usually

* Correspondence concerning this article should be addressed to: Robert F. Miller Email: rfm@umn.edu, FAX: +1 612 626-2136; Michael T. Bowser Email: bowser@chem.umn.edu; Kylie B. O'Brien Email: kylie.obrien@ucd.ie.

more than an hour. More recently, separation methods using capillary electrophoresis (CE) combined with laser-induced fluorescence (LIF) detection have been described which shortens the separation time considerably and one method, an on-line technique [10] resulted in racemate amino acid separation times of about 22 s or less, permitting continuous monitoring of amino acids released under various physiological states. More recently the on-line system was further refined and coupled to a brain microdialysis perfusion system with amino acid separations available every 12.5 s [11]. Unfortunately, technical problems with the on-line CE-LIF system using a perfused retina preparation did not provide reliable D-serine detection because of the low levels of D-serine released. For this reason, we developed a CE-LIF system to evaluate amino acid racemates from homogenized retinal tissue [12] using a commercial CE system and report on that method in this chapter.

2. Materials

1. Beckman Coulter P/ACE MDQ Capillary Electrophoresis unit with 488 nm laser module using Karat 32 software.
2. 4-fluoro-7-nitrobenz-2-oxa-1,3-diazole (NBD-F) obtained from Molecular Probes (Eugene, Oregon, USA)
3. Ultrasonic tissue homogenizer (XL2000, Misonix, Farmingdale, NY, USA)
4. BCA protein assay kit (Pierce Biotechnology, Rockford, IL, USA)
5. Hydroxypropyl-β-yclodextrin (HP-β-CD; Sigma Chemical, St. Louis, MO, USA)
6. Derivitization buffer consisting of 115 μL of perchloric acid (0.6 M) and 5 μL of D-glutamate (internal standard) neutralized with 37 μL of KOH (2M) (all reagents obtained from Sigma Chemical, St. Louis, MO, USA)
7. Silica capillary 50 μm inner diameter (Polymicro Technologies, Phoenix, AZ, USA)
8. Buffers:
   0.6 M perchloric acid (PCA)
   2 M KOH
   2 M NaOH
   1 mM o-phospho- L-threonine (in water)
   200 mM borate, pH 10.5
   60, 12, 6, 1.2, 0.6, 0.12, 0.06 mM D-serine in 200 mM borate, pH 10.5 (becomes 10, 5, 1, 0.5, 0.1, 0.05, 0.025 μM standards for constructing calibration curves).
   9.0 mg/mL NBD-F in acetonitrile
   34 mM HP-β-CD in 165 mM borate, pH 10.2
   Amphibian Ringer was made with the following composition (in mM): 111 NaCl; 2.5 KCl; 1.8 $CaCl_2$; 0.5 $MgCl_2$; 10 HEPES and 5 glucose; pH adjusted to 7.8 with NaOH

3. Methods

1. Salamander retinas were removed from the eye following decapitation and double pithing. Retinas were placed on filter paper (5 μ pore size, Millipore, Billerica, MA, USA). We used six retinas for each determination.
2. In a fume hood retinas were transferred from the filter paper to vials containing 115 μL PCA and 5 μL o-phospho- L-threonine on ice, followed by tissue sonication to fully homogenize the retinas using the XL 2000.
3. Prepare a blank sample without retinas for control

4. On ice, add 35 μL KOH to retinal samples and blank in 5 μL increments
5. Spin down at 14,000 rpm for five minutes
6. Remove supernatant
7. Re-suspend protein pellet in 120 μL 2M NaOH for protein determination. Store at –80°C
8. Add 120 μL borate buffer to supernatant from retinas (SR) and blank (SB)
9. Prepare the following samples for CE analysis:
   A. 50 μL (SR) + 5 mL 60 μM D-serine (in borate)
   B. 50 μL (SR) + 5 μL 12 μM D-serine (in borate)
   C. 50 μL (SR) + 5 μL 6 μM D-serine (in borate)
   D. 50 μL (SR) + 5 μL 1.2 μM D-serine (in borate)
   E. 50 μL (SR) + 5 μL 0 μM D-serine (in borate)
   F. 50 μL (SB) + 5 μL 0 μM D-serine (in borate)
10. Add 5 μL of NBD-F to each sample (A-F), 40 μM α-amino adipate (internal standard) and react 15 minutes at 60^0C
11. CE separation buffer
12. CE separation used a 40.2 cm, 50 μm inner diameter capillary, rinsed for one minute with 0.1 M NaOH and two minutes with the separation buffer at 20 psi; sample was loaded into the capillary by pressure injection (0.5 psi for 5 s) and a separation voltage of 15 kV was applied across the capillary for 20 minutes.

4. Results

Figure 1 illustrates an electropherogram of 25 NBD-F derivatized amino acid standards. Numbered peaks correlate with the indicated amino acids. D-Serine and glycine were omitted from the A sample, while GABA was eliminated in the B sample. The variations in peak area between runs arise from the differences in the injection between runs. When analyte peak areas were normalized for variability in the internal standard, the relative standard deviation between runs was about 3%.

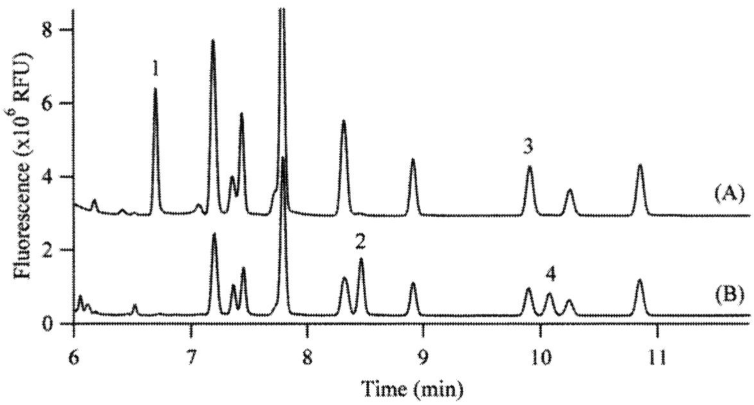

Figure 1. Electropherogram of 25 NBD-F labeled amino acids standards. Standard mixtures of L-asparagine, L-arginine, serotonin, dopamine, L-norepinephrine, GABA (1), L-β-ABA, D-β-ABA, β-alanine, taurine, L-citrulline, L-histidine, α-ABA, glycine (2), L-glutamine, L-serine (3), D-serine (4), L-alanine, L-threonine, L-tyrosine, phosphoethanolamine, glutathione, L-glutamate, D-glutamate, D-aspartate, and L-aspartate were prepared. D-serine and glycine were omitted from the reaction in (A), while GABA was omitted from the reaction in (B). Separation conditions were optimized to resolve GABA (1), glycine (2) and D-serine (4). D-and L-glutamate (not shown) are also resolved under these separation conditions (From *Monitoring D-serine Dynamics in the Central Nervous System by Capillary Electrophoresis*, Ph.D. thesis of K.B. O'Brien, University of Minnesota, 2005, p 92, with permission).

When bathing solution of a single retina is examined with the CE method outlined in this study, the D-serine level was too low to be reliably detected [12]. For that reason, we examined the D-serine levels in retina homogenates and found detectible levels when at least four retinas were included for each determination.

REFERENCES

[1] Hashimoto, A., Nishikawa, T., Hayashi, T., Fujii, N., Harada, K., Oka, T., and Takahashi, K. (1992). The presence of free D-serine in rat brain. *FEBS Lett, 296*, 33-36.

[2] Kumashiro, S., Hashimoto, A., and Nishikawa, T. (1995). Free D-serine in post-mortem brains and spinal cords of individuals with and without neuropsychiatric diseases. *Brain Res, 681*, 117-125.

[3] Snyder, S. H. and Kim, P. M. (2000). D-amino acids as putative neurotransmitters: focus on D-serine. *Neurochem Res, 25*, 553-560.

[4] Mothet, J. P., Parent, A. T., Wolosker, H., Brady, R. O., Jr., Linden, D. J., Ferris, C. D., Rogawski, M. A., and Snyder, S. H. (2000). D-serine is an endogenous ligand for the glycine site of the N-methyl-D-aspartate receptor. *Proc Natl Acad Sci USA, 97*, 4926-4931.

[5] Hashimoto, A. and Oka, T. (1997). Free D-aspartate and D-serine in the mammalian brain and periphery. *Prog Neurobiol, 52*, 325-353.

[6] Wolosker, H., Blackshaw, S., and Snyder, S. H. (1999). Serine racemase: a glial enzyme synthesizing D-serine to regulate glutamate-N-methyl- D-aspartate neurotransmission. *Proc Natl Acad Sci USA, 96*, 13409-13414.

[7] Schell, M. J., Molliver, M. E., and Snyder, S. H. (1995). D-serine, an endogenous synaptic modulator: localization to astrocytes and glutamate-stimulated release. *Proc Natl Acad Sci USA, 92*, 3948-3952.

[8] Stevens, E. R., Esguerra, M., Kim, P. M., Newman, E. A., Snyder, S. H., Zahs, K. R., and Miller, R. F. (2003). D-serine and serine racemase are present in the vertebrate retina and contribute to the physiological activation of NMDA receptors. *Proc Natl Acad Sci USA, 100*, 6789-6794.

[9] Hashimoto, A., Nishikawa, T., Oka, T., Takahashi, K., and Hayashi, T. (1992). Determination of free amino acid enantiomers in rat brain and serum by high-performance liquid chromatography after derivatization with N-tert.-butyloxycarbonyl- L-cysteine and *o*-phthaldialdehyde. *J Chromatogr, 582*, 41-48.

[10] O'Brien, K. B., Esguerra, M., Klug, C. T., Miller, R. F., and Bowser, M. T. (2003). A high-throughput on-line microdialysis-capillary assay for D-serine. *Electrophoresis, 24*, 1227-1235.

[11] Ciriacks, C. M. and Bowser, M. T. (2004). Monitoring D-serine dynamics in the rat brain using online microdialysis-capillary electrophoresis. *Anal Chem, 76*, 6582-6587.

[12] O'Brien, K. B., Miller, R. F., and Bowser, M. T. (2005). D-Serine uptake by isolated retinas is consistent with ASCT-mediated transport. *Neurosci Lett, 385*, 58-63.

In: D-Amino Acids: ISBN 1-60021-075-9
A New Frontier in Amino Acid and Protein Research, pp.111-113

Chapter 2.13

DETERMINATION OF N^G-NITRO-ARGININE ENANTIOMERS IN RAT PLASMA BY CAPILLARY ELECTROCHROMATOGRAPHY WITH A CHIRAL MOBILE PHASE

Yan-fei Xin, Xiang-jun Zhou and Yong-xiang Wang*
Laboratory of Pharmacology, School of Pharmacy, Shanghai Jiao Tong University, China.

Keywords: nitro-arginine, chiral separation, capillary electrochromatography.

1. INTRODUCTION

The conversion of some D-amino acids such as D-leucine and N^G-nitro-D-arginine (D-NNA) into L-enantiomers *in vivo* was verified in a number of studies [1,2]. The enzymatic process and organs responsible for the conversion remain unclear. The chiral separation of amino acid mixtures in plasma was the basis to study the mechanism of chiral inversions. The chiral mobile phase was a powerful and widely applicable analytical tool for this purpose [3]. Determination of N^G-nitro-arginine enantiomers in rat plasma by the chiral ligand exchange method via the capillary electrochromatography is shown below.

2. MATERIALS

1. Capillary electrochromatography system (Unimicro Technologies Inc., Pleasanton, CA, USA), equipped with PU 2080 plus transfusion pump, 10 nl sampling loop, UV detector, and TriSep-2000GV high voltage instrument.
2. Reverse phase C18 column (75 μm × 20 cm; Unimicro Technologies Co., Ltd., Shanghai, China).
3. N^G-nitro-D-arginine (D-NNA) and N^G-nitro-L-arginine (L-NNA)
4. Aspartame, cupric acetate, methanol, acetone, 50 mM acetate buffer solution (pH 5.0)
5. Rat plasma

* Correspondence concerning this article should be addressed to Yong-xiang Wang, Laboratory of Pharmacology, School of Pharmacy, Shanghai Jiao Tong University, China. E-mail: yxwang@sjtu.edu.cn; FAX: +86-21-3420-4772.

3. Methods

1. Standard rat plasma solutions of D-NNA and L-NNA were prepared by spiking a serial dilution of standard D-NNA and L-NNA to the rat blank plasma.
2. 100 μl plasma of D-NNA or L-NNA was deproteinated via mixing with methanol/acetone (1000 μl; v/v = 1:1) for 20 min.
3. After centrifugation at 10,000 rpm for 20 min, the organic layer was volatilized at -20°C and the residue was dissolved in 100 μl acetate buffer. After mixing for 1 min, an aliquot (10 nl) of the sample was injected into the capillary eletrochromatography system.
4. Cupric acetate (1 mM) and aspartame (2 mM) were dissolved in the methanol/acetate buffer (pH 5.0, v/v =1:20) to allow the formation of diastereomeric pairs with D-NNA or L-NNA.
5. The diastereomeric pairs were separated on a reverse phase C18 column at detection wavelength of 280 nm.
6. The linear relationship between the peak-area and the plasma concentration of D-NNA or L-NNA was demonstrated.
7. Five sets of plasma standards were prepared and analyzed for the within-day precision on the same day and the between-day precision over 5 days.

4. Results

As shown in Fig 1, D-NNA and L-NNA were separated on the baseline under the optimum chromatography conditions. The retention time was 16 and 19 min for L-NNA and D-NNA, respectively. The samples were quantified against standard curves that ranged from 0.025 to 0.75 mM ($r^2 > 0.98$). The between-day (n = 5) precision values measured by the time and area under the curve were 4.3 and 4.8% for L-NNA and L-NNA, respectively. The within-day precision of the quality control samples was 2% for both L-NNA and D-NNA [4].

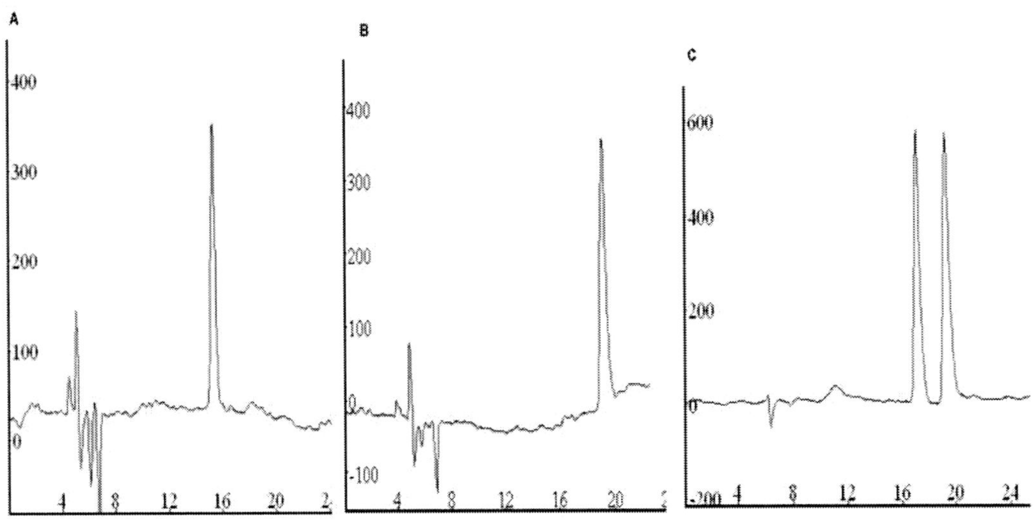

Figure 1. Typical chromatograms of 0.1 mmol/L *L*-NNA (A), 0.1 mmol/L *D*-NNA (B) and 0.2 mmol/L *DL*-NNA mixture (C) in rat plasma. Column: reverse phase C_{18} column (75 μm I.D. × 25 cm); operation pressure: 1000 p.s.i.; detection wavelength: 280 nm; mobile phase: acetate buffer (pH 5.0), 2 mM aspartame, 1 mM Cu (II), 5% (v/v) methanol.

5. Notes

1. This method is applicable for the separation of D-NNA and L-NNA in serum, plasma, and other biological samples.
2. Some other amino acids such as phenylalanine and proline may be used as chiral selectors for the separation of D-NNA and L-NNA.

References

[1] Hasegawa H, Matsukawa T, Shinohara Y, and Hashimoto T. (2000) Assessment of the metabolic chiral inversion of D-leucine in rat by gas chromatography-mass spectrometry combined with a stable isotope dilution analysis. *Drug Metab Dispos*, *28*, 920-924.

[2] Wang Q, Cwik M, Wright CJ, Cunningham F, and Pelligrino DA. (1999) The in vivo unidirectional conversion of nitro-D-arginine to nitro-L-arginine. *J Pharmacol Exp Ther, 288*, 270-273.

[3] Davankov VA. (2003) Enantioselective ligand exchange in modern separation techniques. *J Chromatogr A, 1000*, 891-915.

[4] Xin YF, Zhou XJ, Cheng X, Wang YX. (2005) Renal D-amino acid oxidase mediates chiral inversion of N^G-nitro-D-arginine. *J Pharmacol Exp Ther*, *312*, 1090-1096.

In: D-Amino Acids: ISBN 1-60021-075-9
A New Frontier in Amino Acid and Protein Research, pp.115-122

Chapter 2.14

ANALYSIS OF D-AMINO ACIDS IN PROTEIN HYDROLYSATES AND PEPTIDES SUBJECTED TO EDMAN DEGRADATION

*Anna Maria Salzano and Andrea Scaloni**
Proteomics and Mass Spectrometry Laboratory, ISPAAM, National Research Council, Italy.

Keywords: D-amino acids, sequence, Edman degradation, Marfey's reagent, chirality.

1. INTRODUCTION

Several procedures for the routine analysis of amino acid enantiomeric mixtures have been developed as result of need revealing and quantifying D-amino acids in natural extracts as well as checking the stereochemical purity of synthetic products [1]. Most methods available for characterization of amino acid racemates are based on general application of chiral stationary phases [2,3], chiral mobile phases [4,5] or precolumn derivatization with chiral reagents [6,7]; sometimes they have been used in combination. Precolumn derivatization with chiral reagents introduces an additional chiral center onto molecular structure of racemic analytes, thus determining formation of diasteromers with distinct physico-chemical properties, efficiently separable on conventional reversed-phase supports. In general, the derivatizing reagent is also used to introduce a fluorescent or UV absorptive mojety in enantiomers structure, allowing a sensitive detection of the resulting products. The amino group is the usual site for derivatization because modification of the carboxyl group could increase racemization risk. Substrate/reagent racemization should not take place during the derivatization; thus, mild reaction conditions are usually preferred (in water solution media). Moreover, precolumn derivatization has been routinely automated by using programmable autosamplers, now commercially available. Several pre-column chiral derivatizing reagents allowing an automated quantitative determination of D- and L-amino acids in protein hydrolysates are available on the market. Among them, molecules that found a wider application are *o*-phthaldialdehyde/N-isobutyryl-L-cysteine mixture [8], (+)-1-(9-fluorenyl)ethyl-chloroformate [9] and 1-fluoro-2,4-dinitrophenyl-5-L-alanine amide (L-FDAA), also known as Marfey's reagent [10]. Significant differences between these chemicals

* Correspondence concerning this article should be addressed to Andrea Scaloni, E-Mail: A.Scaloni@iabbam.na.cnr.it, FAX: 39-081-5965291.

were not observed in time requested for analysis. However, the first two reagents yield fluorescent derivatives easily detectable even in the low picomolar range.

When the same amino acid occurs repeatedly in peptide sequence, it is desiderable determining the chirality of each residue as a function of its position. Accordingly, several chiral reagents having an isothiocyanate group to be used in Edman degradation protocols have been proposed. Among them, two fluorescent chemicals proofed to be most useful, namely R(-)-4-(3-isothiocyanatopyrrolidin-1-yl)-7-(N,N-dimethylaminosulfonyl)-2,1,3-benzoxadiazole [11,12] and 7-methylthio-5-(2,1,3-benzoxadiazolyl)isothiocyanate [13]. However, these isothiocyanates showed that cleavage of thiocarbamoyl amino acids and their cyclization to thiazolinones did not occur without racemization (almost 20-30%). Accordingly, new sequencing procedures not affected by racemization have to be developed. In this manuscript, we describe the use of L-FDAA for detection of D-amino acids in protein hydrolysates as well as in the course of subtractive Edman degradation of peptides [14,15].

2. Materials

1. Nitrogen gas supply.
2. D- and L-amino acid standards (Sigma).
3. 1-fluoro-2,4-dinitrophenyl-5-L-alanine amide (Novabiochem).
4. 6 M HCl (Pierce).
5. Sequencing-grade phenylisothiocyanate (PITC), pyridine, trifluoroacetic acid, triethylamine, 1,5-difluoro-2,4-dinitrobenzene (Fluka).
6. Acetonitrile, acetone, 2-propanol are HPLC-grade solvents.
7. Pyrex vessel for vapour-phase acid hydrolysis with a vacuum/purge manifold and a Teflon screw cap, e.g., the Waters reaction vial and Mininert valve present in the Pico-Tag workstation (Waters).
8. Polypropylene microtubes (e.g., Gilson P/N 450831) or Pyrolyzed Pyrex glass (40 x 6 mm i.d.). Stoppered glass tubes for manual Edman degradation (3 ml).
9. Rotary evaporator, e.g., "Speed Vac" (Savant, Fullerton CA), equipped with a rotary vacuum pump and a cold trap kept at a temperature of -60 °C.
10. Vacuum desiccator containing phosphorus pentoxide.
11. Buffer for chiral derivatization: triethylamine/acetonitrile (2:3, v/v). Store at 4 °C. Use within 4 weeks.
12. Reagent for chiral derivatization: L-FDAA solution (1 mg/ml) in acetonitrile: acetone (1:2, v/v). Store at -20 °C (the reagent solution is stable for weeks).
13. Buffer for Edman degradation: pyridine/water (1:1, v/v). Pyridine should be freshly distilled. Use buffer within 24 h.
14. Reagent solution for Edman degradation: PITC/pyridine (1:9, v/v). Use it within 24 h.
15. Extraction solution after Edman degradation: ethylacetate/water (1:1, v/v). Mix under vigorous stirring and then let separate the two phases. Store the bottle closed at room temperature and use within a week.
16. An automated HPLC system for derivatization, separation and quantification of amino acid pairs consisting of two pumps, a column heater, an automated sample manipulator and injector with a rack thermostated at 40 °C (for the samples) and at 4 °C (for the reagents), a UV-detector set at 340 nm, a reversed-phase column and a database system for gradient control, quantitation of the chromatographic peaks and data analysis.
17. Column: Waters Spherisorb S5 ODS2 (2.0 x 250 mm, 5 μm) or Beckman Ultrasphere Octyl column (2.0 x 250 mm, 5 μm). Column flow: 0.2mL/min.

18. Chromatography injection solvent: water/ acetonitrile/ethylacetate (91:8:1, v/v/v).
19. Elution solvents: A: 0.2% trifluoroacetic acid in MilliQ water (filter through a 0.22 μm filter (Millipore); B. 0.1% trifluoroacetic acid in acetonitrile/2-propanol (4:1 v/v).

3. METHODS

3.1. Quantitative Determination of Amino acid Chirality in Protein Hydrolysates

1. Peptide or protein samples (0.5-1 nmol) are dissolved in a volatile solvent, placed into a Pyrex or polypropylene tube and dried in the rotary evaporator.
2. Hydrolyse samples by vapour-phase hydrolysis with 6 M HCl, under vacuum, placing the reaction vial at 110 °C for 24 h.
3. Cool down the vial, open it carefully and clean outside of the tubes from HCl excess.
4. Prepare tubes containing calibration standard mixture, with 1 nmol each of D- or L-amino acids, and blank samples containing water. Dry all samples in the rotary evaporator.
5. Add to each tube 10 μl of the chiral derivatization buffer and carefully vortex it. Add to each sample 10 μl of the chiral derivatization reagent solution and vortex again. Incubate at 40 °C for 60 min and then dry it.
6. Dissolve each sample in 25 μl of the injection solvent and analyse it by HPLC using the gradient shown in Table 1.
7. After the run, qualitative and quantitative analyses of unknown samples is given by the integrator, on the basis of the comparison with amino acid standard peaks areas. Blank samples are used to identify reagent-related peaks.

3.2. Determination of Amino acid Chirality During Subtractive Edman Degradation of Peptides

1. Place each peptide sample (5-25 nmol) to be sequenced into different glass tubes for manual Edman degradation and lyophilize (being the procedure time-consuming, the simultaneous processing of different samples is recommended).
2. Dissolve the samples in 100 μl of the Edman degradation buffer.
3. After addition of 50 μl of Edman degradation reagent to each tube, mix gently, flush with nitrogen and incubate for 30 min at 50 °C.
4. Dry the samples in the vacuum desiccator.
5. Add to each tube 100 μl of anhydrous trifluoroacetic acid, flush with nitrogen and incubate for 10 min at 50 °C, and then dry again.
6. Dissolve each sample in 250 μl of the aqueous phase from the Edman degradation extraction solution. Then, add 1 ml of the organic phase and vortex it. Let separate the two phases in each tube.
7. Remove carefully the organic phase from each sample and take a small aliquot of the aqueous phase for N-terminal analysis. Dry the remaining material and use it for the next cycle of Edman degradation, starting from step 2. Repeat for the suitable number of cycles.
8. For N-terminal analysis, place each sample into a Pyrex or polypropylene tube, carefully label it and dry in the rotary evaporator, as previously described. Similarly, dry in a separate tube 0.5-2 nmoles of the intact peptide.

9. Prepare other tubes containing each 1 nmol of D- and L-amino acid (calibration standards) or water (blank). Dry all samples in the rotary evaporator.
10. Add to each tube 10 μl of the chiral derivatization buffer and carefully vortex. Add to each sample 10 μl of the chiral derivatization reagent solution and vortex again. Incubate at 40 °C for 60 min, and then dry again.
11. Place all samples in the reaction vial for vapour-phase hydrolysis under vacuum and hydrolyse the samples at 110 °C for 5 h, as described above.
12. Cool down the vials, open carefully, clean the outside of the tubes from HCl excess and dry in the rotary evaporator.
13. Dissolve each sample in 25 μl of the injection solvent and analyse it as described above.
14. After the run, peaks are integrated and the identity of the N-terminal amino acid of the unknown is determined by comparison with peaks given by calibration standards. Blank samples are used to identify reagent-related peaks.

Table 1. Gradient parameters for the separation of amino acid derivatives obtained following derivatization with 1-fluoro-2,4-dinitrophenyl-5-L alanine amide. Gradient curves are linear.

Time (min)	% B (v/v)
0	15
3	15
110	50
111	85
117	85
118	15
125	end

4. Results

We set and optimized a fully automated system for the analysis of amino acid enantiomeric mixtures based on the complete automation of both derivatization and chromatography, using L-FDAA [14]. The reaction of L-FDAA with both primary and secondary amino acids is rapid and quantitative (Fig. 1). In our method, during the chromatographic analysis of a sample the next one is derivatized. Reagent and sample compartments were thermostated at different temperatures, allowing reagents storage without degradation. Elution profile of 15 pairs of enantiomeric amino acids plus Gly, after precolumn derivatization with L-FDAA, is shown in Fig. 2. Each D-amino acid derivative was well separated from the corresponding L-stereoisomer, except for the coelution of L-Leu and D-Val. Thus, each amino acid derivative can be detected and quantitated also in the presence of a large excess of the corresponding diastereomer. In our case each D-diastereoisomer eluted from the column after the corresponding L-diasteroisomer, except for His and Arg derivatives, probably because of the formation of an intramolecular hydrogen bonding [16]. When derivatization reaction was performed with 1-fluoro-2,4-dinitrophenyl-5-D-alanine amide, the elution order of diastereomers was reversed. Thus, the parallel use of L-FDAA and D-FDAA allowed an unequivocal determination, also in the case of coelution previously mentioned. The linearity of the response of all amino acids analyzed was verified in the range 50-1000 pmol. The use of L-norVal as internal standard, eluting 0.8 min later than L-Val, allowed calculating recovery for each sample.

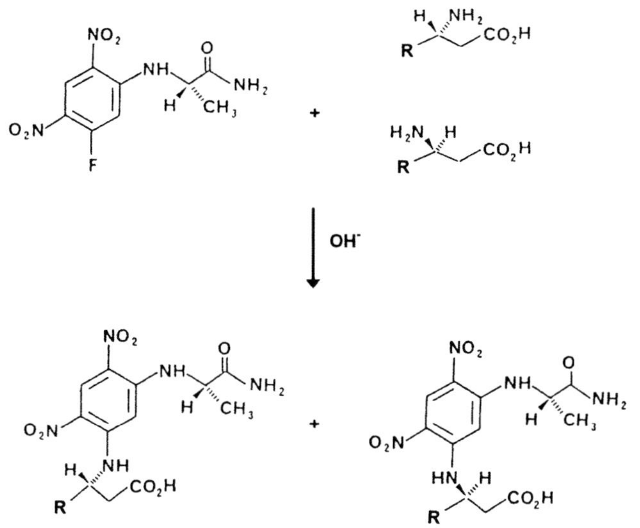

Figure 1. Reaction of 1-fluoro-2,4-dinitrophenyl-5-L-alanine amide with D- and L-amino acids.

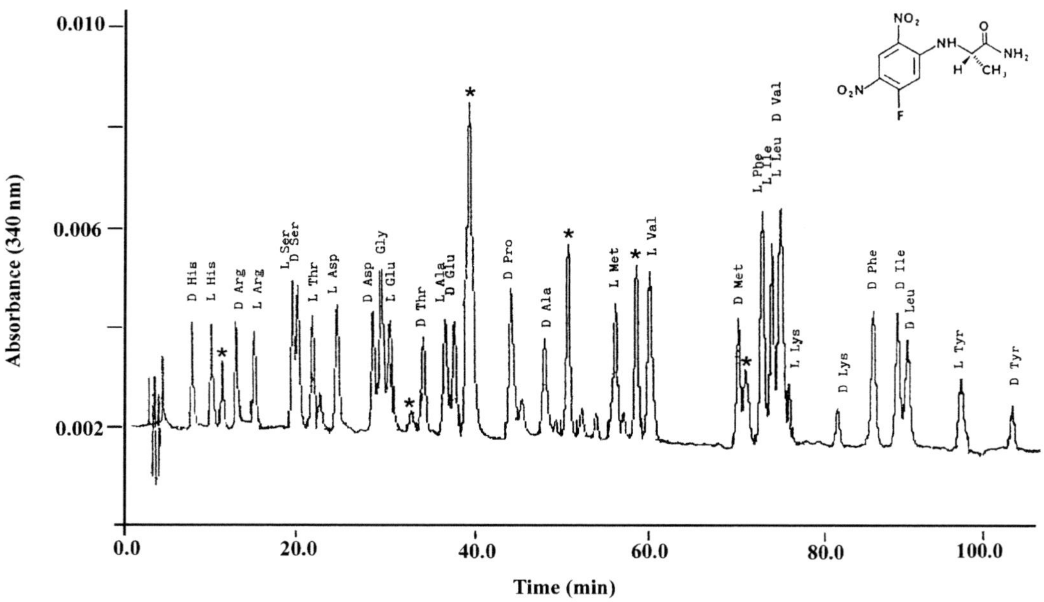

Figure 2. Elution profile of a standard mixture of D- and L-amino acids plus glycine (150 pmol each) derivatized with 1-fluoro-2,4-dinitrophenyl-5-L-alanine amide. Peaks denoted with asterisk are reagent related. Amino acids were derivatized and injected as described in Section 3.1. Injections were made using a Gilson autosampling injector consisting of a model 231 sample injector equipped with a code 9 modified rack, and a model 401 diluter, both controlled by a sample controller keypad. The HPLC system consisted of two LabFlow 4000 pumps (LabService Analytica, Italy), a model 430 UV-visible detector (Kontron AG, Switzerland) equipped with a 3 μL flow cell, a Gilson model 811C dynamic mixer (65 μL mixing chamber), and a Rheodyne injection valve (model 7125) equipped with a 20 μL sample loop. Gradient formation, quantitation of the peaks and data analysis was performed by a Kontron model 450 MT data system.

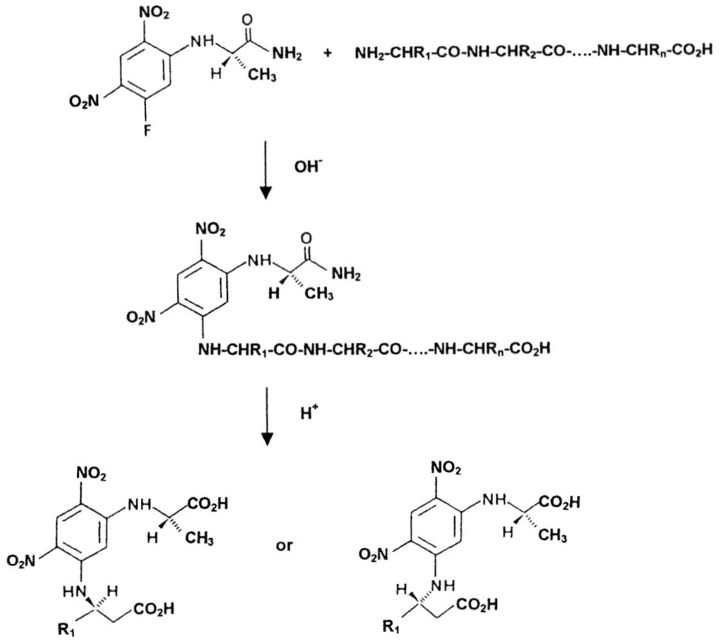

Figure 3. Labelling the N-terminal residue of a peptide with 1-fluoro-2,4-dinitrophenyl-5-L-alanine amide and release of its amino acid derivative following acid hydrolysis.

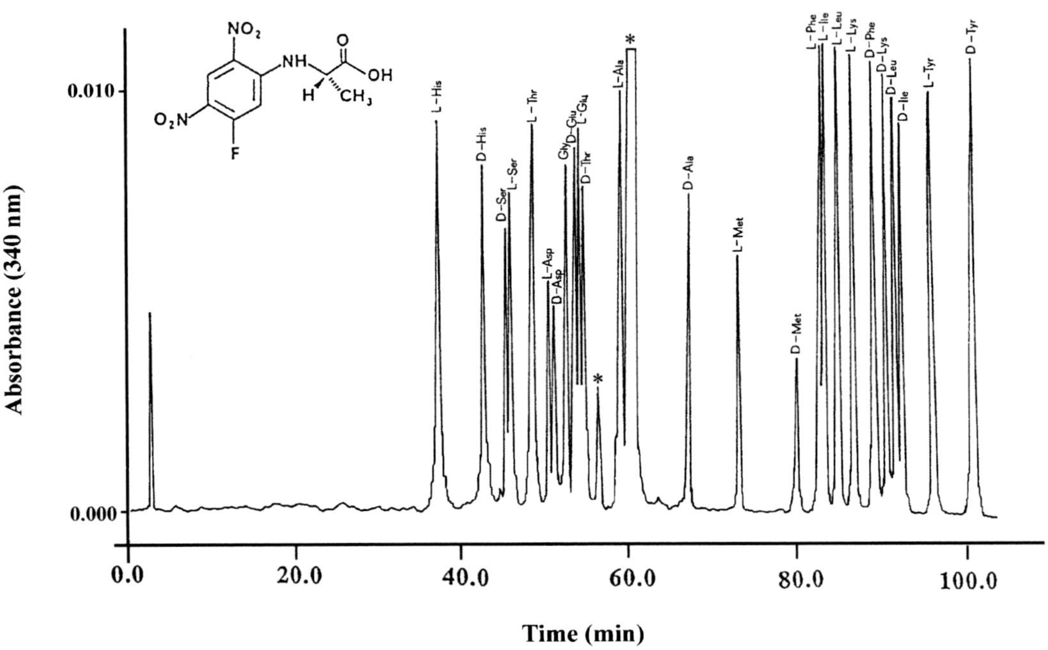

Figure 4. Elution profile of a mixture of D- and L-amino acids after derivatization with 1-fluoro-2,4-dinitrophenyl-5-L-alanine (L-FDA). Peaks denoted with asterisk are reagent related.

We also used L-FDAA reagent for the determination of the chirality of the amino acid residues in the course of sequence analysis of peptides. For this purpose, the Edman chemistry merely has been used as a subtractive procedure [15]: an aliquot of the shortened peptide was derivatized with the chiral reagent after each cycle of Edman degradation (Fig. 3). Under conditions needed to release the resulting derivatized N-terminal amino acid, the 2,4-dinitrophenyl-5-L-alanine amide group was hydrolysed to 2,4-dinitrophenyl-5-

L-alanine. An high recovery of each 2,4-dinitrophenyl-5-L-alanine amino acid derivative (L-DA-amino acid) was obtained after hydrolysis in vapour phase for 5 h, without racemization. Thus, analysis of resulting L-DA-amino acids allowed simultaneous identification and chirality determination of each N-terminal amino acid released following each cycle of Edman degradation. Resolution of diasteroisomer pairs of L-DA-amino acids plus Gly was obtained by reversed-phase HPLC analysis [15], as reported in Fig. 4. Separation of diasteroisomer L-DA-amino acids was optimized following derivatization of amino acid enantiomeric mixtures with L-FDA, an *ad hoc* synthesized reagent. The latter chemical quantitatively reacted with amino group of amino acids and peptides, similarly to L-FDAA, giving stable yellow derivatives, which absorb intensively at 340 nm ($\varepsilon = 4 \times 10^4\ M^{-1}cm^{-1}$).

5. Notes

1. The use of volatile eluants instead of the previously reported 40 mM triethylammine-phosphate buffer [14,15] will prevent damages to the HPLC system related to salt's corrosion and precipitation. Under these experimental conditions a column will allow at least a thousand of analysis without any loss of chromatographic resolution.
2. Samples should be dissolved in a solvent not containing salts or detergents.
3. Avoid performing acid hydrolysis of samples by using liquid-phase or microwave irradiation procedures. In the first case, background contamination is a frequent problem. In the second case, racemization of each amino acid during peptide/protein hydrolysis has been reported.
4. Reagents should be of the highest purity available. During Edman degradation, the use of freshly-open sequencing grade reagents is strongly recommended.
5. The volume removed at each Edman degradation cycle ranges from 1:10 to 1:1 of the total aqueous phase volume, depending on the number of cycles and the extent of degradation.
6. Highest recoveries of L-DA amino acids are obtained by adding to each acid hydrolysis vial a tube containing 1,5-difluoro-2,4-dinitrobenzene wetted with acetone. This compound has the best scavenger properties in preventing diastereoisomers destruction [15].
7. Alternatively to the methodology determining the chirality of the N-terminal amino acid, the different peptides present in the aqueous phase at each step of Edman degradation can be analysed in parallel for their chiral amino acid composition, as described in Section 3.1. The comparison of the amino acid content following each cycle will give indication on the occurrence of D-amino acids within the sequence [17].

References

[1] Hamase, K., Morikawa, A. & Zaitsu, K. (2002). D-amino acids in mammals and their diagnostic value. *J Chromatog B, 781,* 73-91.

[2] Davankov, V.A, Rogozhin, S.V., Semechkin, A.Y. & Sachkova, T.P. (1973). Ligand-exchange chromatography of racemates. Influence of the degree of saturation of the asymmetric resin by metal ions on ligand exchange. *J Chromatogr*, *82*, 359-365.

[3] Gubitz, G., Jellenz, W. & Santi W. (1981). Separation of the optical isomers of amino acids by ligand-exchange chromatography using chemically bonded chiral phases. *J Chromatogr, 203*, 377-384.

[4] Hare, P.E., Gil-Av, E. (1979). Separation of D and L amino acids by liquid chromatography: use of chiral eluants. *Science, 204*, 1226-1228.

[5] Weinstein, S., Engel, M.H. & Hare, P.E. (1982). The enantiomeric analysis of a mixture of all common protein amino acids by HPLC using a new chiral mobile phase. *Anal Biochem, 121*, 370-377.

[6] Manning, J.M. & Moore, S. (1968). Determination of D- and L-amino acids by ion exchange chromatography as L-D and L-L dipeptides. *J Biol Chem, 243*, 5591-5597.

[7] Mitchell, A.R., Kent, S.B.H., Chu, I.C. & Merrifield, R.B. (1978). Quantitative determination of D- and L-amino acids by reaction with tert-butyloxycarbonyl-L-leucine N-hydroxysuccinimide ester and chromatographic separation as L,D and L,L dipeptides. *Anal Chem, 50*, 637-640.

[8] Bruckner, H., Haasmann, S., Langer, M., Westhauser, T., Wittner, R. & Godel, H. (1994) Liquid chromatographic determination of D and L-amino acids by derivatization with o-phtaldialdehyde and chiral thiols. *J. Chromatogr, 666*, 259-273.

[9] Einarrson, S., Josefsson, B., Moller, P. & Sanchez, D. (1983). Separation of amino acid enantiomers and chiral amines using precolumn derivatization with 1-(9-fluorenyl)ethyl-chloroformate and reversed phase liquid chromatography. *Anal Chem, 59*, 1191-1195.

[10] Marfey, P. (1984). Determination of D-amino acids. II. Use of a bifunctional reagent, 1,5-difluoro-2,4-dinitrobenzene. *Carlsberg Res. Commun, 49,* 591-596.

[11] Imai, K., Matsunaga, H., Fukushima, T., Santa, T., Homma, H., Nakashima, K. & Akiyama, S. (1995). A novel method for the amino acid sequence/configuration determination of peptides containing D/L-amino acids utilizing a flurogenic Edman reagent, 7-N,N-dimethylaminosulphonyl-4-(2,1,3-benzoxadiazolyl)isothiocyanate. *Biomed Chromatogr, 9*, 152-154.

[12] Okiyama, N., Santa, T., Toriba, A., Nakagomi, K., Imai, K., Hiranuma, H. & Tanaka, H. (2002). An automated fluorescence protein sequencer using 7-methylthio-4-(2,1,3-benzoxadiazolyl) isothiocyanate (MTBD-NCS) as an Edman reagent. *Biomed Chromatogr, 16,* 183-186.

[13] Toyo'oka, T., Jin, D., Tomoi, N., Oe, T. & Hiranuma, H. (2001). R(-)-4-(3-Isothiocyanatopyrrolidin-1-yl)-7-(N,N-dimethylaminosulfonyl)-2,1,3-benzoxadiazole, a fluorescent chiral tagging reagent: sensitive resolution of chiral amines and amino acids by reversed-phase liquid chromatography. *Biomed Chromatogr, 15*, 56-67.

[14] Scaloni, A., Simmaco, M. & Bossa, F. (1995). D-L amino acid analysis using automated precolumn derivatization with 1-fluoro-2,4-dinitrophenyl-5-L-alanine amide. *Amino Acids, 8*, 305-313.

[15] Scaloni, A., Simmaco, M. & Bossa, F. (1991). Determination of the chirality of amino acid residues in the course of subtractive Edman degradation of peptides. *Anal Biochem, 197*, 305-310.

[16] Bruckner, H. & Keller-Hoehl, C. (1990). HPLC separation of DL-amino acids derivatized with N^2-(5-fluoro-2,4-dinitrophenyl)-L-amino acid amides. *Chromatography, 30*, 621-629.

[17] Mignogna, G., Simmaco, M., Kreil, G. & Barra, D. (1993). Antibacterial and haemolytic peptides containing D-alloisoleucine from the skin of *Bombina variegata. EMBO J, 12,* 4829-4832.

In: D-Amino Acids:
A New Frontier in Amino Acid and Protein Research, pp.123-133
ISBN 1-60021-075-9

Chapter 2.15

Quantitative Determination of the Protein of Bacterial Origin Based on D-Amino Acid Contents

János Csapó[1,2,*], Csilla Albert[2], Katalin Lóki[1], Zsuzsanna Csapó-Kiss[1] and Éva Varga-Visi[1]
[1]University of Kaposvár, Faculty of Animal Science, Hungary;
[2]Sapientia – Hungarian University of Transylvania, Csíkszereda Campus, Romania.

Keywords: D-aspartic acid, D-glutamic acid, diaminopimelic acid, bacterial protein synthesis, protein of bacterial origin.

1. Introduction

A common characteristic of new protein evaluation systems introduced in cattle nutrition in the past decade is that protein content of diets is evaluated on the basis of quantity of amino acids absorbed in the small intestine. In addition to endogenous amino acids, which are only a small proportion, there are two substantial sources of absorbed amino acids. These are microbial protein synthesised in the rumen and dietary protein not broken down in the rumen (by-pass protein). Approximately 70% of the protein content of the diet is broken down into amino acids in the rumen, which may be utilised for synthesis of amino acids or microbial proteins, or broken down still further to provide ammonia for the construction of microbial body proteins. Ammonia not utilised for synthesis of bacterial protein is absorbed from the rumen and converted into urea via the ornithine cycle in the liver. The quantity of microbial protein produced in the rumen must be established if we are to determine the proportion of by-pass protein in the diet.

In order to define the quantity of microbial protein synthesised in the rumen, it is essential that microbial protein in duodenal chyme be separable from the dietary by-pass protein and protein of endogenous origin. This is possible only if components characteristic solely of microbial protein can be identified in the total protein.

* Correspondence concerning this article should be addressed to János Csapó, Email: csapo@mail.atk.u-kaposvar.hu, FAX: +36 82 321 749.

Recently several methods have been developed for determination of the proportion which is of microbial origin of nitrogen-containing substances passed from the rumen into the abomasum. Attempts to estimate the proportion of microbial origin nitrogen-containing substances have been based on nucleic acid, vitamin B_{12} and sulphur 35 isotope. A critical evaluation and summary of these methods was given by Stern and Hoover [1].

Czerkawski [2] was successful in estimating protozoan nitrogen by means of the measurement of 2-amino-ethylphosphonic acid (AEP) and estimating nitrogen content of bacterial origin by measuring 2,6-diaminopimelic acid (DAPA). The AEP was found to occur almost exclusively in protozoa, while DAPA occurs exclusively in peptidoglycans in the bacterial cell wall. Despite the fact that the quantity of DAPA in the cell wall is strongly dependent on the species of bacterium, the ratio of DAPA to total bacterial protein does not vary under constant dietary conditions. Therefore, in comparative experiments, DAPA can be used effectively to estimate the proportion of protein of bacterial origin found in the contents of the intestine.

Schleifer and Kandler [3] discovered that, like DAPA, D-alanine (D-Ala) also occurs only in peptidoglycans in the bacterial cell wall. This compound was also suitable for use as a marker of proteins of bacterial origin and their quantitative determination. Garrett et al. [4] succeeded in determining nitrogen of bacterial origin by measuring the quantity of D-Ala in ruminal fluid. Garrett et al. [5] performed comparative experiments using DAPA and D-Ala to investigate the precision of estimating nitrogen of bacterial origin. They established that D-Ala was a better indicator of nitrogen of bacterial origin, since the coefficient of variation resulting from the data based on D-Ala was substantially lower than that for data based on DAPA. In addition, greater accuracy was achieved in determinations based on D-Ala than in those based on DAPA. In a series of experiments the authors [6] also established that both DAPA and D-Ala are suitable for use in the estimation of the quantity of protein of bacterial origin. No difference was observed between the two substances with respect to error either in analytical method (ion exchange column chromatography for DAPA and high performance liquid chromatography for D-Ala) or in determination of protein of bacterial origin.

Edols [7] determined DAPA in a hydrolysate from the ruminal fluid by the application of a two-column method using an automated amino acid analyser. By the optimization of the composition of buffers DAPA appeared between methionine and isoleucine in the chromatogram, well separated from these two amino acids, in the form of a sharp, easily quantifiable peak. Csapó et al. [8] oxidized samples with performic acid prior to protein hydrolysis, as a consequence of which – the disturbance effects of neighbouring amino acids having been eliminated – DAPA, even if present only in trace quantities, could be determined accurately. Subsequently to this, exploiting the similar representation on chromatograms of methionine and DAPA, the authors developed a fast method for the determination of DAPA alone by means of ion exchange column chromatography [9].

Analysis of data in the literature suggests that solutions have been found with respect to analytical methods for DAPA, allowing even trace quantities of the compound to be determined. The investigation of D-amino acid content of foodstuffs, particularly milk and milk products [10,11], revealed that, in addition to D-Ala, D-glutamic acid (D-Glu) and D-aspartic acid (D-Asp) could be detected in similar quantities. This was primarily true for products which were produced through bacterial activity. These findings suggested that D-Asp and D-Glu content of ruminal bacteria and duodenal chyme could provide additional markers for estimating the percentage of protein derived from bacteria.

2. MATERIAL AND METHOD

2.1. Methodology for the Animal Experiment

Chyme and ruminal bacteria samples were taken from two experiments using five growing bulls (crossbred Hungarian Simmental x Holstein Friesian) weighing 480–500 kg. Each was fitted with ruminal and duodenal fistulae. The main objectives of these experiments were to determine ruminal protein degradability of various diets, and to establish the effect of various types of feed additives on ruminal degradability of proteins. After a ten-day adaptation period duodenal chyme samples were taken through the duodenal fistula and ruminal fluid was taken through the ruminal fistula on several occasions during the same experimental period; the average chyme and ruminal fluid samples were analyzed for DAPA, D-Asp and D-Glu.

2.2. Preparation of Samples for Chemical Analysis

Ruminal fluid was centrifuged at 3000 g to separate the feed particles from infusorians. The bacterial mass was then separated by centrifugation of the fluid phase at 10 000 g. The bacterial mass was dried by lyophilization. Aliquot parts of the chyme samples taken from the duodenum were also lyophilised.

2.3. Chemical Analysis of Samples

DAPA content was determined, by means of the procedure developed by Csapó et al. [12], using Aminochrom-II (Labor MIM, Hungary) type amino acid analyzers. Protein was oxidized with performic acid followed by 24-hour hydrolysis with 6M hydrochloric acid containing 0.1% phenol.

Prior to determination of D-Asp and D-Glu, the protein was hydrolysed with 6M hydrochloric acid for 30 minutes at 170 °C in order to minimize the degree of racemization [13]. Separation and determination of the enantiomers was performed by means of high performance liquid chromatography in accordance with the method described by Einarsson et al. [14]. Derivatization and analysis were carried out with a MERCK-Hitachi (Darmstadt, Germany) HPLC containing the following modules: L-7250 programmable autosampler, L-7100 pump, L-7350 column thermostat, L-7480 fluorescence detector, AIA data conversion utility for D-7000 HPLC system manager.

For derivative formation, o-phthaldialdehyde (OPA) and 2,3,4,6-tetra-O-acetyl-1-thio-β-glucopyranoside (TATG) were purchased from Sigma (St. Louis, Mo). Enantiomer separation was performed on a reversed phase (250 x 4.6 mm internal diameter (i.d.), 5 μm particle size, Kromasil octyl (C_8)) analytical column. In order to extend the life of the column a guard column (RP_8, Newguard, 25 x 3.2 mm i.d., 7 μm particle size, Brownlee) was connected between the injector and the analytical column and a cleaning column (C_{18}, 36 x 4.5 mm i.d., 20 μm particle size, Rsil) was fitted between the pump and the injector. A gradient system consisting of two components was used for enantiomer separation, the composition of this system being the following: A = in 40 % methanol phosphate buffer (9.5 mM, pH = 7.05); B = acetonitrile. The rate of flow used was 1 ml per minute. The derivatives were detected with a fluorescence detector (ex.: 325 nm, em.: 420 nm).

Since, in this series of experiments, only the quantities of D-Asp and D-Glu were of interest to the authors, determination was restricted to the enantiomers of these two amino acids. By this method D-Asp and D-Glu present in very small quantities can be detected and determined alongside L-amino acid present in large quantities.

3. Results and Discussion

Table 1 presents the associations between DAPA and D-Asp, DAPA and D-Glu and D-Asp and D-Glu in both chyme and ruminal bacteria samples. Analysis of the data established that there were very high correlations between DAPA and D-Asp and between DAPA and D-Glu content. In both bacterial protein and chyme the r value observed between DAPA and D-Glu content of the chyme was the lowest, at 0.703, while the same correlation in ruminal bacteria was the highest, at 0.808. The highest correlation was that between D-Asp and D-Glu for chyme (r = 0.949) and for ruminal bacteria (r = 0.843). Figure 1 illustrates the relationships of D-Glu and D-Asp to DAPA content of ruminal bacteria, and Figure 2 shows the same relationships for chyme. Figure 3 demonstrates regression of D-Glu on D-Asp content of the chyme.

Table 1. Linear regression parameters and correlations among DAPA, D-Asp, and D-Glu contents of chyme and ruminal bacteria[a]

Parameter, statistical characteristic	Chyme			Ruminal bacteria		
	DAPA–D-Asp mg/kg	DAPA–D-Glu mg/kg	D-Asp–D-Glu mg/kg	DAPA–D-Asp mg/kg	DAPA–D-Glu mg/kg	D-Asp–D-Glu mg/kg
A	0.313	0.4398	-0.168	1.121	2.686	2.346
	0.089	0.142	0.074	0.560	0.436	0.437
B	0.655	0.830	1.331	0.840	0.771	0.729
	0.094	0.148	0.078	0.186	0.145	0.120
S.D.	0.118	0.188	0.083	0.238	0.185	0.169
n	34	34	34	17	17	17
r_{xy}	**0.78**	**0.70**	**0.95**	**0.76**	**0.81**	**0.84**
P	< 0.001	< 0.001	< 0.001	< 0.001	< 0.001	< 0.001

[a]Derived from regression equation Y=A+B*X, where Y is dependent and X is independent variable. In each column, first variable listed is Y and second is X. A = The intercept of the value of Y when X = 0. B = The slope of the line of the units change in Y per one unit change in X. sd = Standard deviation of A and B. S.D. = Standard deviation. n = number of observations. r_{xy} = correlation coefficient. P = Probability

Table 2 shows relationships between crude protein content of the ruminal bacteria and chyme and D-Asp, DAPA and D-Glu for the same samples. Figure 4 displays the relationships of DAPA, D-Asp and D-Glu with crude protein content of the ruminal bacteria. It was established that, for ruminal bacteria, the closest correlation was between DAPA and crude protein content (r = 0.737); the r value between D-Asp and crude protein content was only very slightly lower (r = 0.725), while the relationship between D-Glu and crude protein content was somewhat lower (r = 0.614). However, the same analysis of data on chyme samples did not produce a close correlation between the markers and crude protein content. In all three examinations, linear regression analysis demonstrated a very weak negative correlation (r values varying between -0.038 and -0.155). This lack of correlation could be explained by the fact that only part of the protein present in the chyme is derived from bacteria, the rest being comprised of dietary proteins which do not undergo bacterial degradation in the rumen, plus small amounts of endogenous protein. Ruminal degradability of dietary proteins is on average 70%. However, there is a very wide range of reported values. Results from experiments performed under *in vivo* and *in vitro* conditions have demonstrated very great differences among individual proteins with respect to ruminal degradability [15]. There are feedstuffs whose protein content undergoes almost total degradation in the rumen, while the ruminal degradability of other dietary proteins is only 15–20 %. The great deviations are attributed to protein structure, primarily

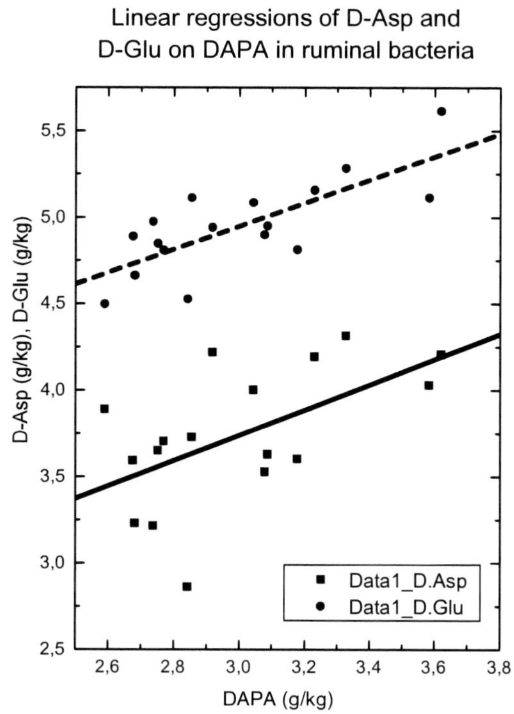

Figure 1. Linear regressions of D-Asp and D-Glu on DAPA in ruminal bacteria

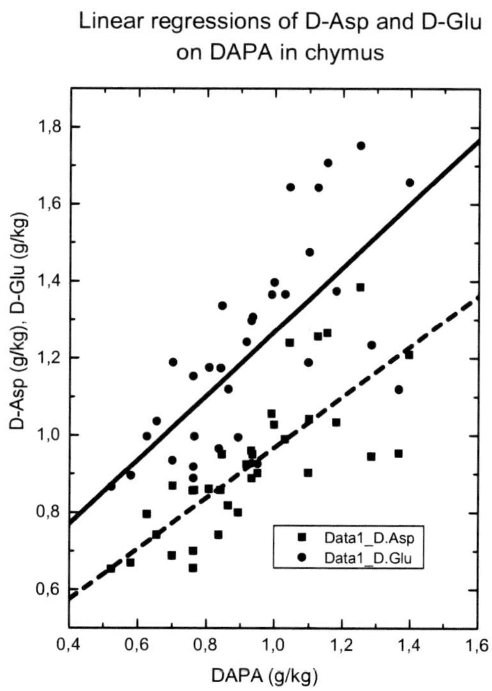

Figure 2. Linear regressions of D-Asp and D-Glu on DAPA in chymus.

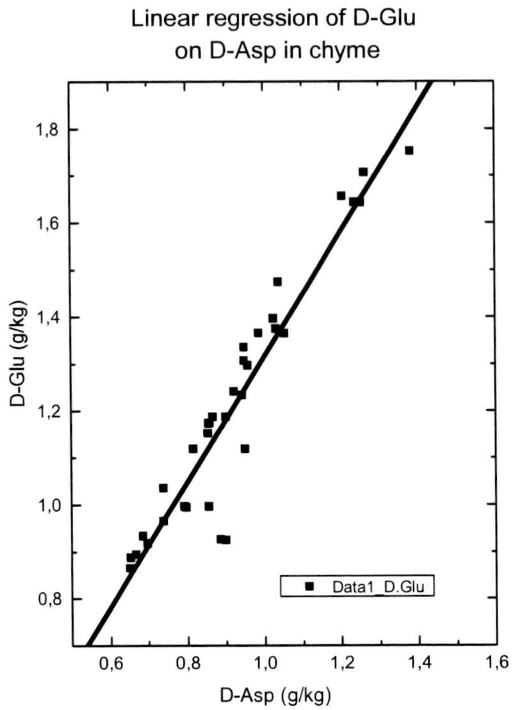

Figure 3. Linear regression of D-Glu on D-Asp in chime.

with the number of disulphide bonds present [16]. The quantity and ratio of protein fractions influence ruminal degradability [17] along with protein amino acid composition [18] and the chemical and heat treatment of protein [19, 20]. This implies that variations in the proportions of microbial protein in the chyme and non-degradable dietary protein in the rumen will be dependent on the dietary proteins. In the experiment from which the chyme and ruminal bacteria samples were taken the animals were fed diets of different ruminal degradability, which resulted in different ratios of microbial protein to by-pass dietary protein in the chyme for individual experimental periods. In the opinion of the authors, this is the fundamental reason for low associations of protein in the chyme with levels of DAPA and the two D-amino acids.

In the next step the average crude protein content and the average DAPA, D-Asp and D-Glu content were calculated from the analysis data for 17 ruminal bacteria produced by the method described in the section Material and method. The composition of the lyophilized ruminal bacteria (100% dry matter) was: 49.50% crude protein, 0.300% DAPA, 0.366% D-Asp and 0.494% D-Glu. By means of these results and with the use of crude protein content of the lyophilised bacteria, the DAPA, D-Asp and D-Glu content of the bacterial protein was calculated. The protein of the ruminal bacteria contained on average 0.606% DAPA, 0.739% D-Asp and 0.998% D-Glu. No possibility exists to compare values for D-Asp and D-Glu with literature values because, as far as the authors are aware, no other researchers have investigated these components of bacterial protein. The value of 0.606% obtained for DAPA is lower than the value of 1.0 ± 0.25% reported for bacterial protein by Orskov [21]. Estimates in the literature vary from 0.6 to 1.4%, and may be influenced by differences in the quality of diets fed to experimental livestock. Since the objective of these investigations was to evaluate new markers for bacteria, it was necessary to establish that the deviation of DAPA content from data in the literature did not influence the results obtained for D-Asp and D-Glu in this study.

Table 2. Linear regression parameters for crude protein regressed on DAPA, D-Asp and D-Glu for ruminal bacteria and chyme[a].

Parameter, statistical characteristic	Crude protein, %					
	Ruminal bacteria			Chyme		
	D-Asp mg/kg	DAPA mg/kg	D-Glu mg/kg	D-Asp mg/kg	DAPA mg/kg	D-Glu mg/kg
A	-0.698	-1.746	1.485	1.073	1.267	1.309
	1.704	1.125	1.152	0.322	0.379	0.453
B	0.088	0.096	0.070	-0.006	-0.014	-0.004
	0.035	0.023	0.023	0.013	0.015	0.018
S.D.	0.329	0.217	0.222	0.187	0.221	0.263
n	17	17	17	34	34	34
r	**0.73**	**0.74**	**0.61**	**-0.08**	**-0.16**	**-0.04**
P	< 0.01	< 0.001	< 0.01	0.644	0.381	0.832

[a]Derived from regression equation Y=A+B*X, where Y is dependent and X is independent variable. In each column, first variable listed is Y and second is X.

Abbreviations: Identical as at Table 1.

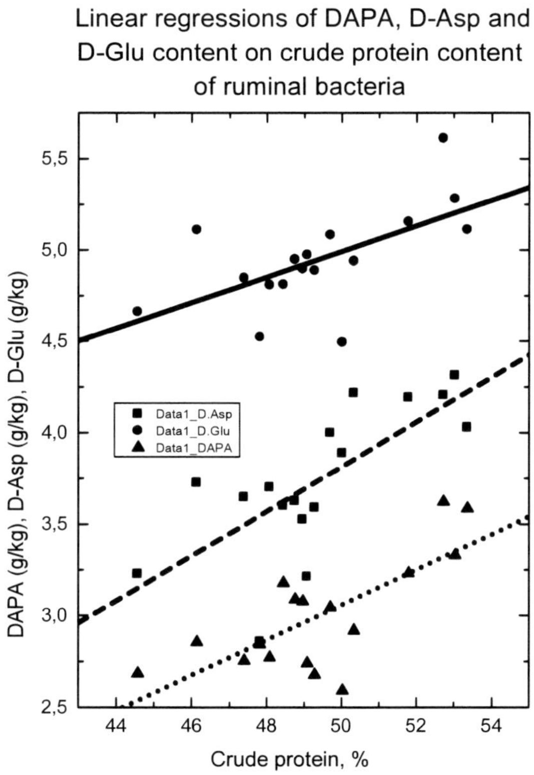

Figure 4. Linear regressions of DAPA, D-Asp and D-Glu content on crude protein content of ruminal bacteria.

After analysis of ruminal bacteria for DAPA, D-Asp, D-Glu and crude protein content, multiplying factors were calculated to enable the proportion of protein of bacterial origin in an unknown sample to be estimated on the basis of DAPA, D-Asp and D-Glu content. The multiplying factor used for DAPA was 100/0.606 = 165; for D-Asp, 100/0.739 = 135; and for D-Glu, 100/0.998 = 100. By the application of the multiplying factors the quantity of the protein of bacterial origin can be calculated very easily. In order to establish the applicability in practice of the multiplying factors calculated in this study, two experiments were performed. In the first, these multiplying factors were applied to various chyme samples. The results obtained are presented in Table 3. It can be seen that values estimated on the basis of DAPA content (the average of the 10 estimates being 12.774%) were on average 8% higher than the microbial protein quantities based on D-Glu (11.705%) or D-Asp (11.800%) content. The explanation for this may be that the DAPA content of ruminal bacteria was measured at a slightly lower value when the multiplying factors were calculated. When protein content determined on the basis of D-Glu was compared with that based on D-Asp, the level of concordance was obvious, and in most cases, the values obtained concurred almost exactly.

Table 3. Examples of the application of multiplying factors in the determination of protein of bacterial origin content in chyme samples.

Chyme sample	Analysis results*			Protein of bacterial origin, % calculated on the basis of		
	D-Asp %	DAPA %	D-Glu %	D-Asp	DAPA	D-Glu
1.	0.08546	0.07630	0.11513	11.54	12.59	11.67
2.	0.06681	0.05814	0.08935	9.03	9.59	9.06
3.	0.12546	0.11276	0.16415	16.96	18.61	16.64
4.	0.07402	0.06560	0.10342	10.01	10.82	10.49
5.	0.06519	0.05249	0.08644	8.81	8.66	8.77
6.	0.10546	0.09933	0.13637	14.26	16.38	13.83
7.	0.08564	0.07666	0.09952	11.58	12.64	10.09
8.	0.08671	0.07041	0.11865	11.72	11.61	12.03
9.	0.08591	0.08090	0.11730	11.61	13.34	11.89
10.	0.09230	0.08189	0.12402	12.48	13.50	12.58
			Mean	11.800	12.774	11.705

*Different chyme samples were analysed for D-Asp, DAPA and D-Glu, and the protein of bacterial origin was calculated by applying multiplying factors.

After the multiplying factors were tested for chyme samples, a mean sample was produced from 17 lyophilized ruminal bacteria samples, and the multiplying factors were also tested for ruminal bacteria. The crude protein content of this mean sample was 49.5%, its DAPA content was 0.325%, its D-Asp content was 0.364% and its D-Glu content was 0.492%. By application of the multiplying factors crude protein content values of 53.62% (0.325%x165), 49.14% (0.364%x135) and 49.20% (0.492x100) respectively were estimated.

Subsequently, the authors lyophilized beef to produce a meat meal of zero DAPA content. The D-Asp and D-Glu content was below 0.01% after hydrolysis of its protein. Racemization testing was performed during protein hydrolysis. A mixture of 1 g meat meal and 1 g bacterial sample was analyzed for DAPA, D-Asp and D-Glu content using 5 parallel analyses. This was then repeated using a mixture of 9 g meat meal and 1 g bacterial sample. The results are presented in Table 4. From the data given in Table 4 it can be established that the S.D.% for the first sample, containing a higher quantity of bacterial protein, are in each case below 5; thus, the standard deviation of the results meets the requirements for a reliable analytical method. The second sample contained only 20% of the quantity of bacterial protein contained by the first sample. S.D.% for D-Asp and D-Glu was below 5, but that calculated for DAPA slightly exceeded 5. When compared with the values calculated for bacterial origin protein content (24.76% and 4.95%), the only values to deviate significantly ($P < .05$) were three individual values and the two mean values based on DAPA content (Table 4.). DAPA estimates (26.58% and 5.54% protein of bacterial origin) averaged 9.5 and 13.5% respectively, higher than those for D-Asp (24.63 and 4.89%) and D-Glu (24.82 and 4.87%).

DAPA estimates averaged 7.3 and 11.9%, higher than the percentages calculated (24.76 and 4.95%) for samples one and two.

Table 4. Model experiment to investigate the accuracy of determination of protein of bacterial origin.

1:1 mixture of meat meal and bacterium with 24.76% calculated protein of bacterial origin						
Parallel analyses	Analysis results			Protein of bacterial origin, % calculated on the basis of		
	D-Asp %	DAPA %	D-Glu %	D-Asp	DAPA	D-Glu
1.	0.182	0.162	0.247	24.60	26.71	25.05
2.	0.181	0.159	0.239	24.47	26.22	24.24
3.	0.179	0.155	0.251	24.20	25.56	25.45
4.	0.184	0.164	0.241	24.87	27.04*	24.44
5.	0.185	0.166	0.246	25.01	27.37*	24.94
Mean	0.1822	0.1612	0.2448	24.63	26.58*	24.82
SE $\overline{X}$				0.144	0.318	0.217
9:1 mixture of meat meal and bacterium with 4.952% calculated protein of bacterial origin						
Parallel analyses	Analysis results			Protein of bacterial origin, % calculated on the basis of		
	D-Asp %	DAPA %	D-Glu %	D-Asp	DAPA	D-Glu
1.	0.037	0.037	0.047	5.001	6.101*	4.766
2.	0.035	0.035	0.049	4.731	5.772	4.969
3.	0.038	0.032	0.051	5.136	5.277	5.171
4.	0.034	0.031	0.046	4.596	5.112	4.664
5.	0.037	0.033	0.047	5.001	5.442	4.766
Mean	0.362	0.0336	0.0480	4.893	5.541*	4.867
SE $\overline{X}$				0.099	0.118	0.091

SE $\overline{X}$ = Standard error of mean.
*Differs from calculated protein of bacterial origin ($P < 0.05$).

4. Conclusions

These investigations have provided evidence that both D-Asp and D-Glu may be used to estimate percentage of protein of bacterial origin. The results obtained using these two markers proved to be approximately 8% lower than those obtained using DAPA. This deviation was not attributable to error associated with the new markers, but rather to unreliability of determination using DAPA. The analyses performed on samples of known bacterial protein content indicate that D-Asp and D-Glu produced practically identical values for bacterial protein content, which were very close to the theoretical (calculated) values.

The preparation process for DAPA analysis is lengthy due to the treatment with performic acid required. However, without performic acid treatment, the determination of DAPA present in small quantities is unreliable, due to interference of other amino acids in concentrations which may, in some cases, be higher by several orders of magnitude. In addition to the large amount of work required for the process, determination using DAPA also requires more time and more chemical agents, and is therefore considerably more expensive. An amino acid analyser operating on the principle of ion exchange column chromatography is suitable for analysis with DAPA, but separation of the D-amino acids cannot be performed with an amino acid analyser operating on the traditional principle.

Determination using D-Asp and D-Glu can be carried out with precision by means of high performance liquid chromatography. By the application of the fast procedure developed by the authors, determination of the quantity of D-Asp and D-Glu can be performed in 25% the amount of time required for ion exchange column chromatography analysis used for DAPA.

When determining D-amino acids, racemization may occur during protein hydrolysis, which can falsify the results obtained. Due to racemization, quantities of D-Asp and D-Glu greater than those actually present may be measured and lead to overestimation of the amount of protein of bacterial origin present. Two methods are recommended for the elimination of this source of error. One is the application of a hydrolysis procedure which allows only a very low degree of racemization to take place (for example, the protein hydrolysis method developed by the authors, performed at high temperature and for a short period, i.e. 160–170 °C for 30–45 minutes; Csapó et al., 1997b.). The other method is to determine, by means of the method to be applied, the D-Asp and D-Glu content of bacteria obtained from the rumen, and developing multiplying factors for use in estimating the quantity of protein of bacterial origin (this being our new method). In the latter case, racemization occurring during protein hydrolysis is regarded as a constant error, in both calculation of multiplying factors and analysis of real samples. As a result, it exerts no substantial influence on the accuracy of the determination procedure. Thus, the use of D-Asp or D-Glu is associated with minimum error if a protein hydrolysis procedure involving a low degree of racemization is used, and if multiplying factors are determined and applied.

REFERENCES

[1] Stern, M.D. & Hoover, W.H. (1979). Methods for determining and factors affecting rumen microbial synthesis: A review. *J Anim Sci, 49,* 1590-1603.

[2] Czerkawski, W.J. (1974). Methods for determining 2-6-diaminopimelic acid and 2-amino-ethylphosphonic acid in gut contents. *J Sci Food Agric, 25,* 45-55.

[3] Schleifer, K.H. & Kandler, O. (1972). Peptidoglycan type of bacterial cell walls and their taxonomic implications. *Bacteriol Rev, 36,* 407-477.

[4] Garrett, J.E., Goodrich, R.D. & Meiske, J.C. (1982). Measurement of bacterial nitrogen using D-alanine. Protein requirements for cattle. Symposium. Oklahoma State University, MP-109.

[5] Garrett, J.E., Goodrich, R.D. & Meiske, J.C. (1987). Measurement and use of D-alanine as a bacterial marker. *Can J Anim Sci, 67,* 735-743.

[6] Csapó, J. & Henics, Z. (1991). Quantitative determination of bacterial protein from the diaminopimelic acid and D-alanine content of rumen liquor and intestines. *Acta Agr Hung, 20,* 159-173.

[7] Edols, R.W. (1985). Simple method for the determination of diaminopimelic acid in rumen liquor hydrolysates. *J Chrom, 329,* 199-201.

[8] Csapó, J., Gombos, S., Henics, Z., & Tóth, L. (1986). Modified method of diaminopimelic acid determination in samples of biological origin. *Acta Aliment, 15,* 159-167.

[9] Csapó, J., Csapó-Kiss, Zs., Csordás, E., Martin, T.G., Folestad, S., Tivesten, A. & Némethy, S. (1995). Rapid method for the determination of diaminopimelic acid using ion exchange column chromatography. *Anal Let, 28,* 2049-2061.

[10] Csapó, J., Martin, T.G., Csapó-Kiss, Zs., Stefler, J. & Némethy, S. (1995). Influence of udder inflammation on the D-amino acid content of milk. *J Dairy Sci, 78,* 2375-2381.

[11] Csapó, J., Csapó-Kiss, Zs., Csordás, E., Fox, P.F., Wágner, L. & Tálos, T. (1997). Total free and free amino acid content of cheeses produced using various technological methods. *Hungarian Dairy J, 57,* 25-30.

[12] Csapó, J., Tóth-Pósfai. I. & Csapó-Kiss, Zs. (1986). Optimisation of hydrolysis at determination of amino acid content in food and feed products. *Acta Aliment, 15,* 3-21.

[13] Csapó, J., Csapó-Kiss, Zs., Folestad, S., Tivesten, A., Némethy, S., Wágner, L., Tálos, T. & Martin, T.G. (1997). Hydrolysis of proteins performed at high temperatures and for short times with reduced racemisation, in order to determine the enantiomers of D- and L-amino acids. *Anal Chim Acta, 339,* 99-107.

[14] Einarsson, S., Folestad, S. & Josefsson, B. (1987). Separation of amino acid enantiomers using precolumn derivatisation with o-phthalaldehyde and 2,3,4,6,-tetra-O-acetyl-1-thio-β-D-glucopyranoside. *J Liquid Chrom, 10,* 1589-1598.

[15] Chalupa, W. (1976). Degradation of amino acids by the mixed rumen microbial population. *J Anim Sci, 43,* 828-834.

[16] Mangan, J.L. (1972). Quantitative studies on nitrogen metabolism in the bovine rumen. 1. The rate of proteolysis of casein and ovalbumine and the release and metabolism of free amino acids. *Brit J Nutr, 27,* 261-283.

[17] Wohlt, J.E., Sniffen, C.J., Hoover, W.H., Johnson, L.L. & Walker, C.K. (1976). Nitrogen metabolism in weeters as affected by dietary protein solubility and amino acid profile. *J Anim Sci, 42,* 1280-1289.

[18] Tamminga, S. (1979). Relatio between different carbohydrates and microbial synthesis of protein. Kiel Group Seminar. Uppsala. Report No. 130.

[19] Kaufmann, W. & Lüpping, W. (1979). Zum Einfluss von Protected Protein und HMM-Ca auf die leistung von Milchkühen. *Z Tierphisiol Tierernähr Futtelmittelk, 41,* 202-217.

[20] Schmidt, J., Cenkvári, É., Sipőcz, J. & Gergácz, Z. (1993). Wirkung der Extrudierung auf den Abbau des Eiweisses im Pansen. *Acta Agronom Óvariensis, 35,* 147-154.

[21] Orskov, O.R. (1982). Protein Nutrition in Ruminants. Academic Press, London.

In: D-Amino Acids: ISBN 1-60021-075-9
A New Frontier in Amino Acid and Protein Research, pp.135-148

Chapter 2.16

DETECTION OF D-AMINO ACIDS BY D-AMINO ACID OXIDASE

Laura Caldinelli[1], Laura Motteran[1], Silvia Sacchi[1], Luciano Piubelli[1], Angelo Boselli[1], Jean-Pierre Mothet[2], Loredano Pollegioni[1,*] and Mirella S. Pilone[1]

[1] Department of Biotechnology and Molecular Sciences, University of Insubria, Italy;
[2] Laboratoire de Neurobiologie Cellulaire et Moleculaire, Centre National de la Recherche Scientifique Unite Propre de Recherche 9040,
Institut Federatif de Neurobiologie Alfred Fessard, France.

Keywords: analytical detection, biosensor, enzymatic assay, *in vivo* D-serine detection, D-amino acid oxidase.

1. INTRODUCTION

D-Isomers of amino acids are common constituents of bacterial cell walls; their presence has been reported also in marine invertebrates, even in appreciable concentrations, while only minute quantities of these compounds can be found in animal tissues [1].

The nutritional utilization of D-amino acids (D-aa) varies widely in animals and humans, and D-aa may be beneficial or deleterious. Thus, although D-phenylalanine in an all-amino-acid diet is utilized as a nutritional source of L-phenylalanine, high concentrations of D-tyrosine in such a diets inhibit the growth of mice [1].

Even though the toxicity of D-aa is still debated, the occurrence of D-aa in foods is normally associated with a decrease in proteins digestibility, thus affecting the bio-availability of essential amino acids, and ultimately impairing the nutritional quality of food.

Thus, both the D-aa content and the D/(D+L) ratio have been proposed as reliable molecular markers of ripening and as indexes for the assessment of food quality [2]. The presence of D-aa has been detected in various foods, mainly dairy products, through chromatographic methods for the enantiomeric separation of amino acids, *e.g.* gas chromatography, HPLC and thin-layer chromatography [3,4]. In particular, D-alanine content can be considered as an indicator of the presence of contaminants in milk [2], the relative amount

* Correspondence concerning this article should be addressed to Mirella S. Pilone, E-mail: mirella.pilone@uninsubria.it, FAX: +39 0332 421500.

of D-alanine, D-aspartate and D-glutamate allowing to discriminate between Parmigiano Reggiano and Grana Padano cheese and to establish the age of the cheese itself, and D-aa have been detected and quantified in reconstituted infant formula and in wine [1].

Finally, significant amounts of free D-serine and D-aspartic acid were found to occur in brain and peripheral tissues of mammals [5,6]. D-Aspartic acid has been found essentially in neurons of the neuroendocrine system and in neuroendocrine glands (pineal, adrenals...) and can be found in proteins where its amount increase with aging [5]. D-Serine has been recently identified as an astrocytic-derived messenger present in the forebrain where it serves as ligand for the *N*-methyl-D-aspartate subtype of glutamate receptors, thus modulating the glutamatergic neurotransmission [7]. In humans, its role has been related to the susceptibility to schizophrenia. In order to unravel the functional consequences of D-serine-mediated astrocyte-to-neuron signaling, it is essential to clarify the mechanism controlling the storage and the release pathway. Traditional analytical methods (GC, CE and HPLC) are not suitable for simple and accurate determination of D-aa. For this reasons, we have recently devised a D-amino acid oxidase-based bioassay for monitoring continuously the release of D-serine from cultured cells [8].

D-Amino acid oxidase (EC 1.4.3.3, DAAO) is a flavoenzyme that catalyzes the oxidative deamination of D-isomer of amino acids to the corresponding α-keto acid and ammonia. During this step, the FAD is reduced and the catalytic cycle is completed through its reoxidation by O_2 to yield hydrogen peroxide (Scheme 1).

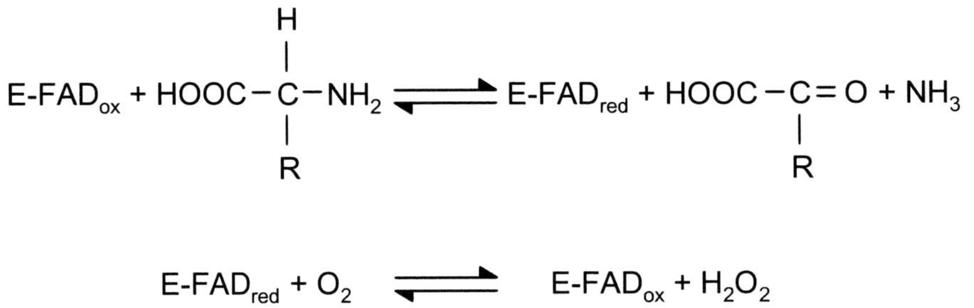

(1a)

(1b)

Scheme 1.

The reaction of DAAO with D-aa reported in Scheme 1 can be followed using different assays, namely D-aa or oxygen consumption, hydrogen peroxide, α-keto acid or ammonia production, or using artificial dyes as final indicator of the flavin redox reaction. Moreover, in recent years many research groups have proposed enzyme sensors for the specific detection of D-aa based on the detection of hydrogen peroxide or ammonia produced by the reaction catalysed by DAAO from pig kidney. However their application was limited by the low activity of the mammalian enzyme and the inactivity on acidic D-aa. The first problem was overcome by using DAAO from the yeast *Rhodotorula gracilis* (RgDAAO) [9], which exhibits a very high turnover number, and is mainly active on neutral and polar D-aa [for a review see 10]. The second one was solved by using the M213R RgDAAO mutant active on acidic substrates [11].

2. Materials

Recombinant wild-type RgDAAO was expressed and purified from *E. coli* cells as described in [9]; the enzyme concentration was determined using an extinction coefficient at 455 nm of 12.6 mM^{-1} cm^{-1}. The purified enzyme has a specific activity on D-alanine of 110 U/mg protein and a $K_{m,app}$ of 0.83 mM at pH 8.5 and 25 °C [9]. Recombinant M213R mutant RgDAAO was purified from *E. coli* cells as reported in [11]; the enzyme concentration was determined using an extinction coefficient at 455 nm of 13.5 mM^{-1} cm^{-1}. The purified M213R RgDAAO has a specific activity of 15.7 and 5.9 U/mg protein and a $K_{m,app}$ of 18 and 2 mM for D-alanine and D-aspartate, respectively, at pH 8.5 and 25 °C [11]. DAAO activity was assayed with an oxygen electrode in 75 mM disodium pyrophosphate, pH 8.5, at 25 °C and at air saturation ($[O_2]$ = 0.253 mM), using 28 mM D-alanine as substrate. One DAAO unit corresponds to the amount of enzyme that converts 1 μmol of D-aa per minute at 25 °C [10].

2.1. Materials for the Oxygen Consumption Assay

1. Hansatech oxygen electrode (Hansatech Instruments Ltd, Pentney, UK).
2. Disodium pyrophosphate, 100 mM, pH 8.5.
3. D-Alanine, solution in disodium pyrophosphate, 100 mM, pH 8.5.

2.2. Materials for the Spectrophotometric Assays

1. UV/Vis spectrophotometer (Jasco Europe, Cremella, Italy).
2. 2,4-Dinitrophenylhydrazine, 1 mM in 1 M HCl.
3. HCl.
4. NaOH.
5. D-Phenylglycine, 10 mM in disodium pyrophosphate, 75 mM, pH 8.5.
6. *o*-Dianisidine dihydrochloride, 10 mM in H_2O.
7. Peroxidase from horse radish (HRP, grade I, Roche Diagnostics, Basel Switzerland), 100 U/ml in disodium pyrophosphate, 100 mM, pH 8.5.
8. 4-Aminoantipyrine,15 mM in H_2O.
9. Phenol.
10. Glutamate dehydrogenase (GldH, Roche Diagnostics), 1500 U/ml in disodium pyrophosphate, 75 mM, pH 8.5.
11. NADH,15 mM in disodium pyrophosphate, 75 mM, pH 8.5.
12. α-Ketoglutaric acid, 100 mM in H_2O.

2.3. Materials for the Biosensor

1. Midas Pro amperometric biosensor (Sartorious A.G., Göttingen, Germany), equipped with a flow electrochemical cell containing a working electrode (graphite disk, 20 mm diameter, 2 mm thick, 100 μm microporosity) and a reference electrode (Ag/AgCl) (see Fig.1).
2. Calibration solution: 0.2 mg/ml hydroquinone.
3. Working buffer solution: disodium pyrophosphate, 100 mM, pH 8.5.
4. Substrate standard solutions: D-alanine 0.15-3 mM concentration range in working buffer and D-aspartate 0.5-10 mM concentration range in working buffer.
5. Enzymes: 1.5 units of wild-type RgDAAO (~ 14 μg protein); 0.17 units of M213R RgDAAO (~ 50 μg protein).
6. Washing solution: 1% sodium hypochlorite.

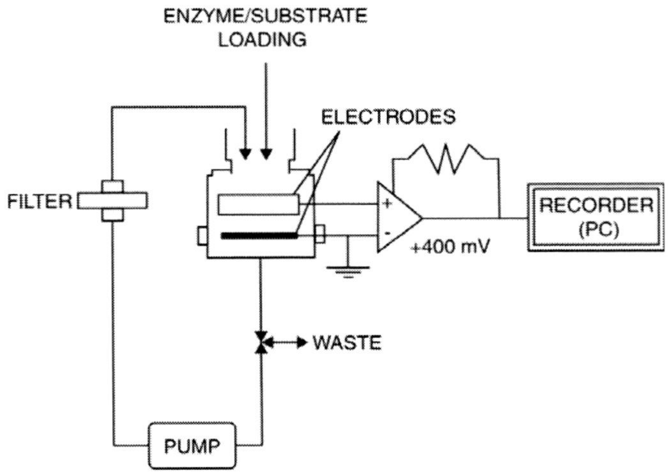

Figure 1. Schematic representation of the electrochemical flow-through cell for the amperometric detection of D-aa.

2.4. Materials for *in vivo* Determination of D-Serine

1. Rat glioma C6 cells were purchased from American Tissue Culture Collection (ATCC, Rockville, MD, USA).
2. Cell cultures were maintained in Dulbecco's modified Eagle's medium (DMEM) / nut mix F-12 (Invitrogen, Grand Island, NY, USA) supplemented with 2 mM glutamine and 10% fetal calf serum. Cells were grown in 75 cm^2 flasks at 37 °C in a humidified atmosphere of 5% CO_2 and 95% air. On day 2 or 4 in culture, C6 cells were proceeded for experiments. PBS was purchased from Invitrogen.
3. Experiments were performed using a laboratory made photon counter. In this device, the photons emitted during the bioassay are detected through a liquid-core optical fiber (Lumatec, Germany) by an ultra-low noise photomultiplier tube (PMT, R464SS, Hamamatsu, Japan; dark counts: 0.5 cps) running at 800 V. The anode signal is amplified and discriminated into brief logic pulses. Pulses are fed to an I/O port of a PC computer and counted over time by a laboratory made software.
4. The luminol-derived chemiluminescence (LDCL) signal was offline analyzed using Origin 7.0 software (Microcal Software Inc., Northampton, MA, USA); the Microcal LabTalk programming language was used to customize the analysis.

3. Methods

All assays were carried out at 25 °C, using a water circulating bath to keep the temperature at the correct value.

All the spectrophotometric assays are based on Lambert-Beer's law: the concentration was determined using the following equation and the known extinction coefficient of the compound of interest:

$$\mu mol/ml = \Delta Abs / (l \times \varepsilon) \quad \text{(eq. 1)}$$

where: $\Delta Abs = (Abs_{final} - Abs_{initial})$; l = cuvette path (cm); ε = extinction coefficient, expressed as $mM^{-1}\ cm^{-1}$.

3.1. Determination of Oxygen Consumption

This is a direct method employing a Hansatech Clark-type oxygen electrode. During DAAO reaction, the pO_2 decreases in the oxymeter chamber and its value is recorded by the CB1D electrode control box connected to a chart recorder. In order to achieve a fast equilibrium after the addition of DAAO, the assay solution in the oxymeter chamber is mixed by a magnetic stirrer. The assay is performed as follows:

1. Prepare the reaction mixture directly in the oxymeter chamber, mixing the solution containing the D-aa (or a standard solution of D-alanine), disodium pyrophosphate buffer, pH 8.5 and water to a final volume of 1-3 ml. The final concentration of disodium pyrophosphate is 75 mM, and the concentration range for the D-alanine is 0.1 - 4 mM.
2. Start the chart recorder, and wait until the signal is stable.
3. Add 0.1 μg of wild-type DAAO (~ 0.01 units in 5 – 10 μl) and follows the time course of the reaction for 3 – 5 minutes.
4. Measure the initial oxygen consumption (ΔpO_2/min), and calculate the enzymatic activity using the following equation:

$$\text{enzymatic activity} = \text{U/mg protein} = \frac{\Delta pO_2/\text{min x oxygen solubility (0.253 mM)}}{\text{mg of added DAAO}} \text{ x assay volume (1 or 3 ml)} \quad \text{(eq. 2)}$$

5. To determine the concentration of D-alanine, use the Michaelis – Menten equation and the known apparent kinetic parameters (see above).

3.2. Methods for the Spectrophotometric Assays

3.2.1. Determination of α-Keto Acid Production: Direct Method

The concentration of the α-keto acid produced by the DAAO activity on D-phenylglycine is evaluated following its characteristic absorption at 252 nm (see Scheme 2).

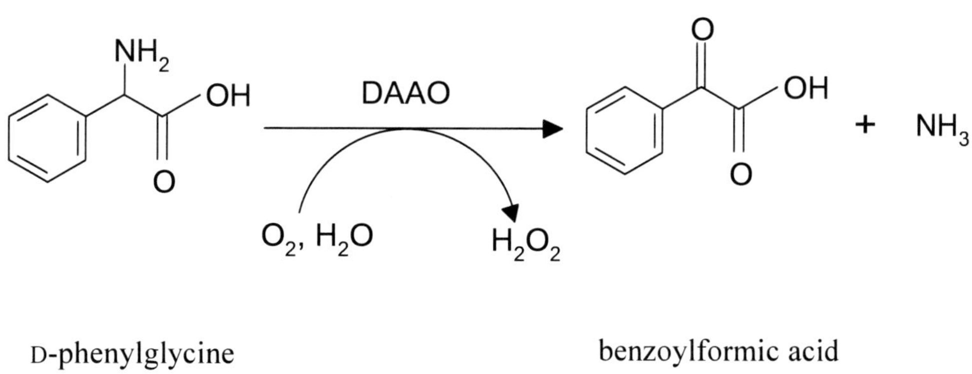

Scheme 2.

The concentration of the α-keto acid produced (corresponding to the amount of converted D-aa) is determined by a calibration curve set up using different known amounts of benzoylformic acid in the 1.5 – 80 μM concentration range. The assay is set up as follows:

1. Prepare the reaction mixture directly in a quartz microcuvette, mixing 10 – 80 μM D-phenylglycine, disodium pyrophosphate, pH 8.5 (final concentration 75 mM) and water.
2. Record the initial absorbance (Abs_i) at 252 nm of the mixture.
3. Add 10 units of wild-type DAAO to a final volume of 1 ml.
4. Record the absorbance after the addition of DAAO until its value is stable (Abs_f) (~ 20 min).
5. Determine the benzoyl formic acid produced using eq. 1 and the following extinction coefficients: $\varepsilon_{252\ nm} = 14.3\ mM^{-1}\ cm^{-1}$.

3.2.2. Determination of α-Keto Acid Production: Indirect Method with DNP

In this assay, the α-keto acid produced by the DAAO activity was evaluated as 2,4-dinitrophenylhydrazone derivative [12], obtained by the reaction with 2,4-dinitrophenylhydrazine (DNP) (Scheme 3).

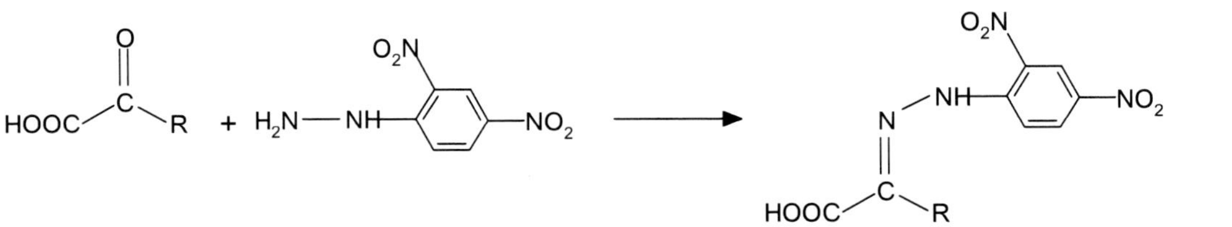

Scheme 3.

The 2,4-dinitrophenylhydrazone derivatives have a characteristic absorbance spectrum in the visible region, with an intensity maximum at 445 nm. The assay is set up as follows:

1. The reaction mixture is prepared in a microcentrifuge tube mixing 0.03 – 0.2 mM D-alanine, 75 mM disodium pyrophosphate, pH 8.5, and 1 unit of wild-type DAAO (all final concentrations) in a final volume of 300 μl.
2. Add 150 μl of 1 mM DNP, solubilized in 1 M HCl.
3. Incubate at 37 °C for at least 10 min.
4. Stop the reaction with 1050 μl of 0.6 M NaOH and incubate at room temperature for 5 min.
5. Read the absorbance value at 445 nm.
6. A blank solution is prepared using the same protocol, but replacing D-alanine with H_2O (Abs_{blank}).
7. Determine the concentration of the α-keto acid derivative produced using the eq. 1 and the extinction coefficient $\varepsilon_{445nm} = 3.5\ mM^{-1}\ cm^{-1}$, determined by a calibration curve prepared using pyruvate in the 0.01 – 0.2 mM concentration range.

3.2.3. Determination of Hydrogen Peroxide Production: Indirect Method with o-Dianisidine

The H_2O_2 produced by DAAO activity is decomposed by peroxidase (HRP, EC 1.11.1.7) whilst a colourless hydrogen donor (e.g., *o*-dianisidine) is oxidized to coloured compound (Scheme 4).

The compound *o*-dianisidine used in this method produces a red-brown dye with a broad absorption maximum centered at around 440 nm; the reaction is complete in three minutes and the colour is stable for several hours. The assay is set up as follows:

1. The assay mixture is prepared directly in a disposable plastic microcuvette mixing 0.02 – 0.1 mM D-alanine, 1 mM *o*-dianisidine, and 1 unit of HRP, in 75 mM disodium pyrophosphate, pH 8.5 (all final concentrations).

2. Record the initial absorbance (Abs_i) of the mixture at 440 nm.
3. Add 1 unit of DAAO to a final assay volume of 1 ml, and monitor the absorbance at 440 nm, until its value is stable (Abs_f) (~ 10 min).
4. The concentration of the coloured product (corresponding to that of D-alanine oxidized by DAAO) is determined using eq. 1 and an extinction coefficient of 13 mM^{-1} cm^{-1} [13].

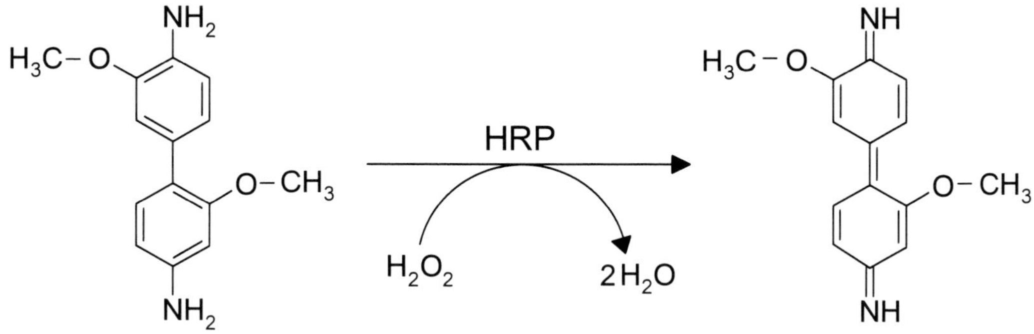

o-DNS colourless coloured product

Scheme 4.

3.2.4. Determination of Hydrogen Peroxide Production: Indirect Method with 4-Aminoantipyrine

This assay is based on the determination of H_2O_2 produced by the reaction catalyzed by DAAO coupled to the reduction of 4-aminoantipyrine (4-AAP) catalyzed by peroxidase in presence of phenol (see Scheme 5).

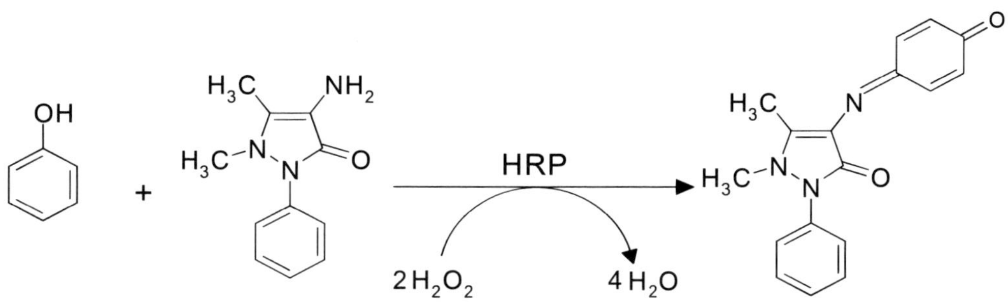

phenol 4-aminoantipyrine quinoeimine

Scheme 5.

The assay is set up as follows:

1. The reaction mixture is prepared directly in a disposable plastic microcuvette mixing 0.02 – 0.2 mM D-alanine, 2 mM phenol, 1.5 mM 4-AAP and 2.5 units of HRP, in 75 mM disodium pyrophosphate, pH 8.5 (all final concentrations).
2. Record the initial absorbance (Abs_i) at 505 nm.
3. Add 1 unit of DAAO to a final assay volume of 1 ml and monitor the absorbance at 505 nm until it is stable (Abs_f) (~ 10 min).
4. The concentration of produced quinoneimine is determined using eq. 1 and an extinction coefficient 6.58 $mM^{-1}cm^{-1}$ [14].

5. Two molecules of hydrogen peroxide are necessary to produce one molecule of quinoneimine (see Scheme 5); for this reason the concentration of D-alanine in the mixture is twice that of quinoneimine produced.

3.2.5. Determination of Ammonia Production

The enzyme glutamate dehydrogenase (GldH) in presence of NH_3 produced by DAAO catalyzes the reductive amination of α-ketoglutaric acid to L-glutamic acid. During this reaction, the cofactor NADH is oxidized to NAD^+ (see Scheme 6).

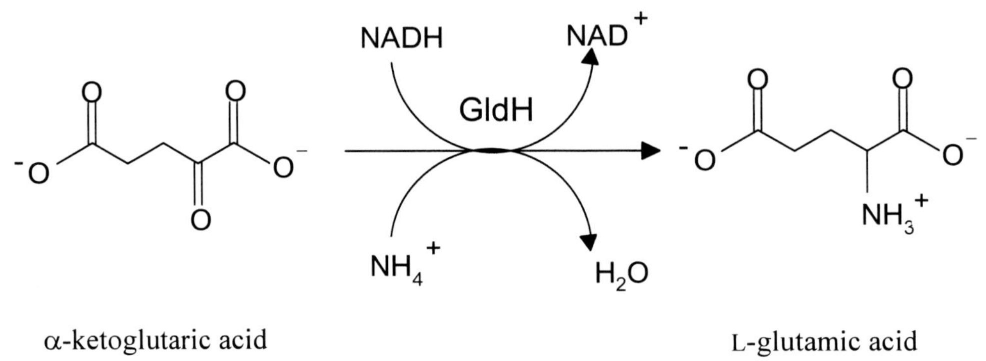

Scheme 6.

The assay is set up as follows:

1. The reaction mixture is prepared in a quartz micro-cuvette mixing 0.02 - 0.2 mM D-alanine, 5 mM α-ketoglutaric acid, 0.25 mM NADH, and 20 units of GldH, in 75 mM disodium pyrophosphate, pH 8.5 (all final concentrations).
2. Record the initial absorbance (Abs_i) of the mixture at 340 nm.
3. Add 1 unit of DAAO to a final volume of 1 ml and monitor the absorbance at 340 nm until its value is stable (Abs_f) (~ 20 min).
4. The concentration of NAD^+ enzymatically produced, corresponding to the converted D-aa, is determined measuring the decrease of absorbance at 340 nm using eq. 1 and an extinction coefficient of 6.3 mM^{-1} cm^{-1}.

3.3. Methods for the Biosensor

1. The electrochemical cell is filled with 5 ml of the working buffer solution and the anodic signal is monitored. When the recorded current is stable, 20 μl of the calibration solution were injected, and the corresponding calibration peak is measured.
2. The electrode is washed with 5 ml of washing solution and 20 ml H_2O.
3. A calibration curve is obtained using standard solutions of substrates; because of the different substrate specificity of the two RgDAAO variants employed in the assay, D-alanine and D-aspartate are used as standard substrates for wild-type and M213R mutant RgDAAO, respectively. The electrochemical cell is filled up with 5 ml of D-alanine (0.15 - 3 mM final concentration), or D-aspartate (0.5 - 10 mM final concentration) standard solutions. When the recorded current is stable, 1.5 units of wild-type or 0.17 units of M213R RgDAAO are injected.

4. The current values detected after 5 minutes with the standard solutions at different D-aa concentration (plateau values) are plotted as a function of substrate concentration and fitted by means of a classic Michaelis-Menten equation (see Fig. 2).
5. After the measurement, the cell is rinsed with 5 ml of washing solution.
6. For the analytical measurement, the electrochemical cell is filled with 4 ml of working buffer solution and 1 ml of a biological sample containing an unknown concentration of D-aa. When the recorded current is stable, 1.5 units of wild-type or 0.17 units of M213R mutant RgDAAO are injected.
7. The amount of neutral-basic and/or neutral-acidic D-aa is measured from the current values detected using the wild-type and M213R mutant RgDAAO, respectively, and the kinetic equations determined using the calibration curves (see above).

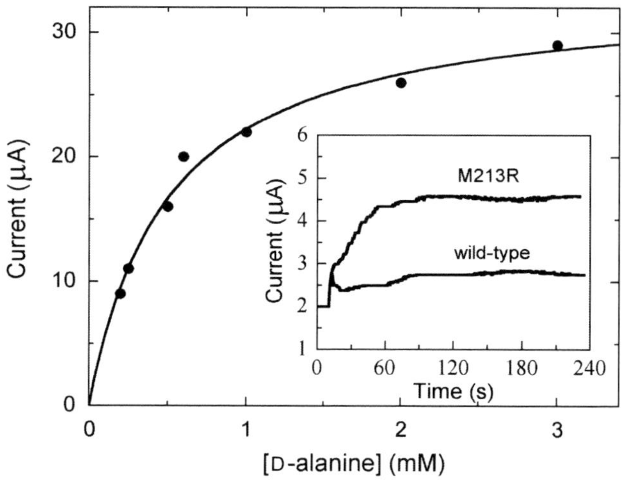

Figure 2. Calibration curve derived from the amperometric traces obtained for wild-type RgDAAO at different substrate concentrations. Inset: comparison of amperometric traces obtained for wild-type and M213R RgDAAO using 20 mM D-aspartate.

3.4. Determination of D-Serine from C6 Glioma Cells

D-Serine levels were detected using an enzymatic assay adapted from Wolosker et al. [15], based on the production of hydrogen peroxide by the reaction of RgDAAO; the produced H_2O_2 in turn is used by horseradish peroxidase (HRP) to oxidize luminol, a chemiluminescent probe. The latter compound is cleaved by HRP to produce one molecule of 3-aminophthalate ion which is in an electronically excited state and that, upon relaxation to the ground state provides light emission (at 425 nm, see Scheme 7).

Following the addition of RgDAAO and HRP/luminol to the saline solution in which astrocytes were bathed, any detectable levels of D-serine released in the medium can be visualized as an increase in emission light.

1. Cultured cells were first washed 2 times with PBS to remove any trace of medium and detached from the flasks. The cell suspension was then transferred to a Falcon tube and pelleted at 1000 x *g* for 10 min.
2. The supernatant was discarded and the cells were resuspended to 10^5 cells/ml in oxygenated Krebs-Henseleit (KH) buffer and maintained at 37 °C.

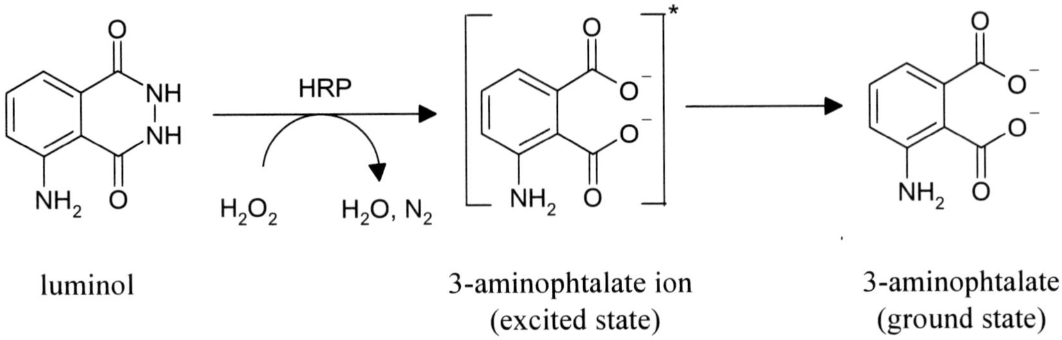

Scheme 7.

3. A 2-fold premix assay solution (2x pRM) was prepared in KH by addition of luminol, HRP and RgDAAO (200 μM, 3 units/100 μl and 0.6 units/100 μl final concentration, respectively). Continuous magnetic stirring was used to ensure a rapid mixing of the solution components in the dark.
4. 100 μl of the enzyme substrate solution (2x pRM) was mixed with 100 μl (i.e. 10,000 cells) of cells suspension in KH buffer to a final volume of 1.0 ml. Addition of the cells to the pRM produced an increase of light emission corresponding to unspecific signal; the latter decays over time and a stable background light emission (~ 1 cps) was observed after 5-10 min incubation. Then, various compounds can be added to the quartz cuvette using an Hamilton syringe, in the dark, and through a septum.
5. Recordings were made at room temperature (22-26 °C) by counting emitted photons on a 32 ms interval time period over a period of 30-60 s.
6. Baseline in luminol-derived chemiluminescence measurement was determined from the number of integrated photons during the period preceeding peaks and substracted to amplitude-time plots. LDCL spectra result from the average of *n* curves ($n \geq 4$).
7. To estimate the levels of D-serine released by cells, a simple calibration procedure in which fixed amounts of D-serine was added at the end of each test was used. The release was then normalized for protein content of each sample.
8. For measurement of D-serine content in cells, 2 or 4 days old C6 cells, were pelleted at low speed (1000 x *g*) for 10 min and surface-washed twice with ice-cold PBS. Free amino acids were extracted by addition of ice-cooled 5% trichloroacetic acid (TCA); TCA was removed with diethyl-ether, and the aqueous layer was stored at –80 °C until determination. D-Serine content was analyzed by the chemiluminescent assay by injection of 10 μl samples to the reaction chamber containing 1 ml pRM.

4. Results

4.1. Oxygen Consumption Assay

The first assay used to detect the D-aa content is based on the polarographic determination of the oxygen consumption during the reaction catalyzed by DAAO. The initial rate of the reaction, expressed as ΔpO_2/min, depends on the D-alanine concentration and the known kinetic parameters of wild-type RgDAAO [9]. From these values, the concentration of D-alanine can be determined.

4.2. Spectrophotometric Assays

1. In order to determine the D-aa content in a biological sample five different assays based on DAAO activity can be used. These assays are based on both direct or indirect methods and on the simple Lambert-Beer's law.
2. A direct spectrophotometric method is the one based on the extinction coefficient of the α-keto acid produced by the reaction. The concentration of the α-keto acid produced corresponds to the initial amount of the corresponding D-aa and this value is determined from the absorbance change and by means of appropriate calibration curve set up using different concentrations of the same α-keto acid.
3. Another method that allows to estimate the α-keto acid production is based on its reaction with 2,4-dinitrophenylhydrazine (Scheme 3). Also in this case, the determination of the concentration of the 2,4-dinitrophenylhydrazone derived from the reaction is based on the set up of a calibration curve obtained using different amounts of D-alanine.
4. Two indirect methods are instead based on the hydrogen peroxide produced by the reaction catalyzed by DAAO (Scheme 4 and 5). In both cases, H_2O_2 is used to oxidize an hydrogen donor, with formation of a coloured product, with a known extinction coefficient. In both cases, the coupling reaction is catalyzed by peroxidase.
5. The last method used is instead based on the enzymatic determination of the ammonia produced by DAAO reaction, that was estimated on the basis of a coupled reaction with the enzyme glutamate dehydrogenase.

4.3. RgDAAO Biosensor

1. The scheme of the amperometric biosensor is depicted in Fig. 1: the peristaltic pump allows an efficient mixing of the substrate/sample solutions. The signal is recorded at a fixed potential of +400 mV. This system shows an excellent sensitivity (up to 0.2 mM D-alanine concentration) and reproducibility (the average response for 20 repeated analysis of 2 mM D-alanine with the wild-type enzyme is 24.4 ± 2.8 μA) [16]. The calibration curve with D-alanine as substrate is shown in Fig. 2; in the same concentration range, no current is measured using L-alanine as substrate.
2. The system was used for the estimation of the overall content in neutral-basic D-aa in milk samples (1-30 days at 4 °C). The recovery of 1 mM D-alanine added to the crude milk samples is almost complete and no interference at an applied potential of +400 mV is detectable [16]. D-aa concentration is not detectable in fresh milk, but increases during samples aging, reaching a total concentration of 49 μM after 1 month at 4 °C, in good agreement with the values reported in literature using chromatographic methods (~45 μM) [2].
3. The M213R RgDAAO variant can be used for the detection of neutral-acidic D-aa in food samples, because the amperometric response of the mutant enzyme on D-aspartate is 100-fold higher with respect to value measured on D-alanine and no activity on basic D-aa is observed [11].

4.4. Enzyme-linked Assay to Monitor D-Serine Release from Cultured Cells

The study of the mechanisms subserving D-serine release from glial cells requires the sensitive and accurate direct measurement of D-serine on a high-resolution real-time scale. Such a goal is not accomplished by existing offline methods such as HPLC. To get new insights into the molecular mechanisms of D-serine release from astrocytes, we have adapted a previously offline enzyme-linked assay for D-serine [15] based on the DAAO and HRP/luminol systems to a new "on line" biological assay with pH, osmotic and ionic conditions compatible with cell viability.

1. Figure 3 shows the dose-response curves obtained with standard amounts of D-serine; the assay shows a limit of sensitivity of 2 pM and a dynamic response over 5 order magnitude.
2. The [D-serine] dependence of the observed chemiluminescence values follows a classic Michaelis-Menten behaviour (see Fig. 3).
3. Control experiments show that no light emission was detected in the absence of any components of the pRM (data not shown) or when the wild-type RgDAAO was replaced by a catalytic inactive mutant RgDAAO form (the Arg285Ala mutant).
4. Using our bioassay in offline mode, we estimated the C6 cells content in D-serine to be 196 ± 18 amoles/cell (n = 8 independent experiments) [8], a value in agreement with published data [15]. Time-course analysis during 24 hours of conditioned medium from different subconfluent C6 glioma cultures show large accumulation of D-serine in the extracellular medium from undetectable levels to 2.01 ± 0.63 nmol/mg of C6 cells protein which presupposes the release of D-serine from C6 cells [8].
5. We concluded that the new bioassay we designed is compatible with cell physiology and viability as no cell death was observed when C6 glioma cells were placed for 1 hour in the pRM [8].

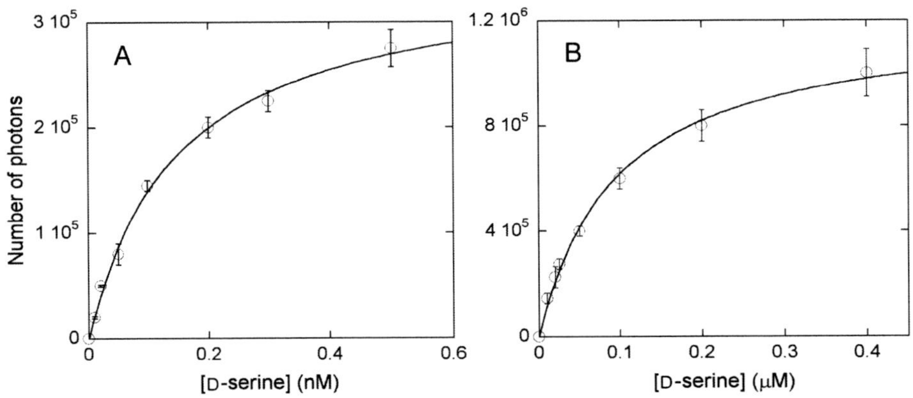

Figure 3. Standard curves of D-serine-induced LDCL. *A*) Dose-response curve for low D-serine quantities. *B*) Dose-response curve for high D-serine quantities. Each point is the mean (n ≥ 8 independent experiments) of light emission above background light production.

5. Notes

5.1. Oxygen Consumption Assay

Before the use, a calibration of the oxymeter is obtained considering as 100 % of oxygen content the amperometric value measured filling the oxymeter chamber with pure water (= 0.253 mM O_2 at 25 °C), and as 0 % the amperometric value measured following the addition of few crystals of dithionite.

5.2. Spectrophotometric Assays

1. The direct spectrophotometric method allows to detect a maximum D-aa concentration equal to the concentration of O_2 (*e.g.*, 0.253 mM at 25 °C). To determine higher concentrations of D-aa is necessary to insufflate O_2 in the reaction mixture. The calibration curve for pyruvate (the α-keto acid produced by D-alanine) gives a $\varepsilon_{340\ nm} = 0.017\ mM^{-1}\ cm^{-1}$.
2. The detection of D-aa content using the DNP method can be performed on different D-aa following the set up of a specific regression line (ε_{445nm} = 2.3 $mM^{-1}cm^{-1}$ for the 2,4-dinitrophenylhydrazone derivate of benzoylformic acid and of α-ketoglutaric acid).

5.3. RgDAAO Biosensor

1. The calibration procedure allows to compare the amperometric signal from different electrochemical cells or from different sessions of measurements.
2. The main limit of this system is represented by the operational stability. In order to have a good reproducibility in response, the electrode is washed with sodium hypochlorite after each measurement; for this reason an aliquot of the enzyme must be added for each measurement. Anyway, a very limited amount of RgDAAO is required for each single determination.
3. An intense anodic signal is detected after the addition of the milk samples in the electrochemical cell; however, the amperometric trace returns at the initial value after ~15 min.

5.4. Enzyme-linked Assay to Monitor D-Serine Release from Cultured Cells

1. The luminol/HRP coupling reaction does not significantly modify the kinetic properties of RgDAAO ($V_{max,app}$ = 70.5 U/mg protein and $K_{m,app}$ = 10 mM).
2. The selectivity and the sensitivity of the bioassay was not affected by others glial-derived neuroactive substances (*e.g.*, L-serine, L-glutamate, ATP, taurine, GABA and glycine) even at doses 500-fold times over [D-serine].

ACKNOWLEDGEMENTS

These works were supported by grants from FAR to L. Pollegioni, L. Piubelli and M. S. Pilone and from Fondazione Cariplo to L. Pollegioni.

REFERENCES

[1] Friedman, M. (1999). Chemistry, nutrition, and microbiology of D-amino acids. *J Agric Food Chem, 47,* 3457-3479.

[2] Gandolfi, I., Palla, G., Delprato, L., De Nisco, F., Marchelli, R. & Salvadori, C. (1992). D-Amino acids in milk as related to heat treatments and bacterial activity. *J Food Sci, 57*, 377–379.

[3] Palla, G., Dossena, A. & Marchelli, R. (1985). Oligo-oxaalkanoyl tetraamides derived from L phenylalanine as stationary phases in capillary gas chromatographic resolution of D,L-amino acids. *J Chromatogr, 349*, 9–15.

[4] Armani, E., Dossena, A., Marchelli, R. & Virgili, R. (1988). Copper(II) complexes of diamino-diamido-type ligands as chiral eluents in the enantiomeric separation of D,L-dansylamino acids by reversed-phase high-performance liquid chromatography. *J Chromatogr, 441,* 275–286.

[5] Fujii, N. (2002). D-Amino acids in living higher organisms. *Orig Life Evol Biosph, 32,* 103-127.

[6] Hashimoto, A. & Oka, T. (1997). Free D-aspartate and D-serine in the mammalian brain and periphery. *Prog Neurobiol, 52*, 325-353.

[7] Snyder, S.H. & Kim, P.M. (2000). D-Amino acids as putative neurotransmitters: focus on D-serine. *Neurochem Res, 25,* 553-560.

[8] Mothet, JP., Pollegioni, L., Ouanounou, G., Martineau, M., Fossier, P. & Baux, G. (2005). Glutamate receptor activation triggers a calcium-dependent and SNARE protein-dependent release of the gliotransmitter D-serine. *Proc Natl Acad Sci U S A, 102*, 5606-5611.

[9] Molla, G., Vegezzi, C., Pilone, M.S. & Pollegioni, L. (1998). Overexpression in *Escherichia coli* of a recombinant chimeric *Rhodotorula gracilis* D-amino acid oxidase. *Prot Expr Purif, 14*, 289–294.

[10] Pilone, M.S. (2000). D-Amino acid oxidase: new findings. *Cell Mol Life Sci, 57,* 1732–1747.

[11] Sacchi, S., Lorenzi, S., Molla, G., Pilone, M.S., Rossetti, C. & Pollegioni, L. (2002). Engineering the substrate specificity of D-amino-acid oxidase. *J Biol Chem, 277*, 27510-27516.

[12] Nagata, Y., Shimojo, T. & Akino, T. (1988). Two spectrophotometric assays for D-amino acid oxidase: for the study of distribution patterns. *Int J Biochem, 20*, 1235-1238.

[13] Chlumsky, L.J., Zhang, L., Rhamsey, A.J. & Schuman Jorns, M. (1993). Preparation and properties of recombinant corynebacterial sarcosine oxidase: evidence for posttranslational modification during turnover with sarcosine. *Biochemistry, 32*, 11132–11142.

[14] Mori, N., Sano, M., Tani, Y. & Yamada, H. (1980). Purification and properties of sarcosine oxidase from *Cylindrocarpon didymum* M-1. *Agric Biol Chem, 44*, 1391-1397.

[15] Wolosker, H., Blackshaw, S. & Snyder, S.H. (1999). Serine racemase: a glial enzyme synthesizing D-serine to regulate glutamate-*N*-methyl-D-aspartate neurotransmission. *Proc Natl Acad Sci U S A, 96*, 13409-13414.

[16] Sacchi, S., Pollegioni, L., Pilone, M.S. & Rossetti, C. (1998). Determination of D-amino acids using a D-amino acid oxidase biosensor with spectrophotometric and potentiometric detection. *Biotechnol Tech, 12,* 149–153.

Section 3. Free D-Amino Acids in Animal Tissues and their Physiological Role

In: D-Amino Acids: ISBN 1-60021-075-9
A New Frontier in Amino Acid and Protein Research, pp. 151-167

Chapter 3.1

A SYSTEMATIC APPROACH TO THE BRAIN D-SERINE SYSTEM

Toru Nishikawa*
Section of Psychiatry and Behavioral Sciences,
Tokyo Medical and Dental University Graduate School, Japan.

Keywords: brain, gene expression, metabolism and function, molecular mechanisms, D-serine.

1. INTRODUCTION

In mammals, the D-form of serine was initially described as a selective agonist for the strychnine-insensitive glycine binding site of the N-methyl-D-aspartate (NMDA) type glutamate receptor [1] (Fig. 1.) and has been used as an excellent tool for the functional study of the NMDA receptor in the central nervous system [2]. Moreover, the stimulation of the glycine modulatory site has been proven to be required for the NMDA receptor-mediated neurotransmission through experiments indicating that glutamate or NMDA failed to elicit an excitatory postsynaptic potential (EPSP) in the absence of the agonists for the glycine site including D-serine, glycine and D-alanine that by themselves lack the ability to produce EPSP at the glutamate receptor [2]. Based on these characteristic phenomena, D-serine and other glycine agonists are called co-agonists for the NMDA receptor [2]. These effects of D-serine and D-alanine were believed to be "non-physiological", because the protein and free amino acid pools of mammals had been presumed to be entirely composed of L-amino acids [3,4].

However, the present author and associates demonstrated by using gas chromatographic-mass spectrometric (GCMS) [5,6] and high performance liquid chromatographic (HPLC) methods [7,8,9,10] that a substantial amount of free D-serine is present in mammalian brains. The discovery of endogenous D-serine in this author's laboratory at the National Institute of Neuroscience, NCNP, in Tokyo was a consequence of this author's research project on the pathophysiology and novel pharmacotherapy of schizophrenia using D-serine and D-alanine, which facilitate the NMDA receptor function via its glycine site (Fig.1.), by noting the induction of schizophrenia-like psychosis by the glutamate receptor antagonists such as phencyclidine (PCP) [3,6,11]. According to this author's idea to overcome the low ability of these

* Correspondence concerning this article should be addressed to Toru Nishikawa, E-mail: tnis.psyc@tmd.ac.jp, FAX: +81-3-5803-0135.

polar D-amino acids to cross through the blood brain barrier (BBB) that their apolar compounds could easily permeate the BBB and improve the schizophrenic symptoms by their systemic administration, Dr. Hibino at Nippon Oil and Fats, Co., Ltd., designed and synthesized for our experiments, N-myristoyl-D-serine and N-myristoyl-D-alanine, which were shown to ameliorate an animal model of schizophrenia, PCP-induced abnormal behavior, following their intraperitoneal injection [3]. The present author further decided to try to prove the expected presence of free D-serine or D-alanine in the brain of animals treated with their fatty acid compounds by collaboration with the late Dr Hayashi at the National Institute of Neuroscience and Dr. Fujii at Tsukuba University, and provided the first evidence that free D-serine is constantly maintained at a high concentration in the mammalian brains [3, 5-8].

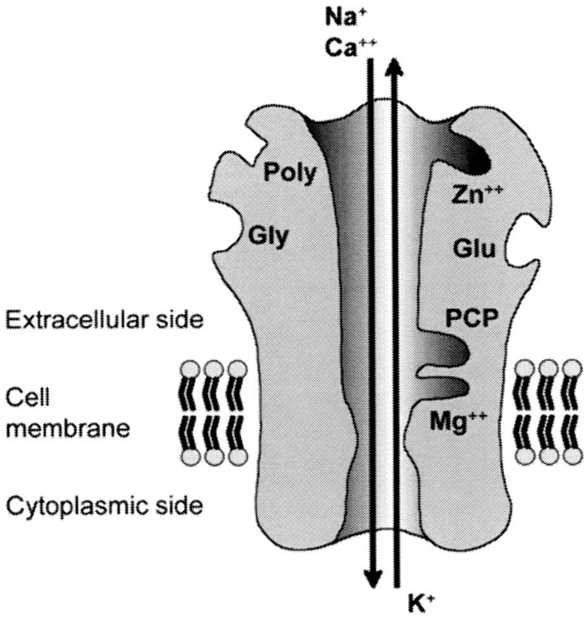

Figure 1. Schematic representation of the NMDA type glutamate receptor consisting of NR1 and NR2 subunits. D-Serine acts as a selective co-agonist of the NMDA receptor ion channel that has the multiple regulatory binding sites for glutamate (Glu), glycine and D-serine (Gly), phencyclidine (PCP), polyamine (Poly), magnesium ions (Mg^{++}), and magnesium ions (Zn^{++}).

Interestingly, endogenous D-serine was found to be exclusively concentrated in the brain with an NMDA receptor-related distribution during any postnatal stage of the rodents [8,12]. In terms of the neuroanatomical and functional correlations of D-serine, this author proposed that brain D-serine may, at least, act as an intrinsic ligand at the NMDA glycine site (Fig. 1.) [3,6,8,13]. According to this hypothesis, we have explored and revealed the processes of biosynthesis [14], release [13], binding to a specific receptor site [15,16], uptake [17,18], and degradation [10] for endogenous D-serine in the brain. To clarify the exact molecular cascades underlying these processes, we have been exploring by molecular cloning techniques D-serine-responsive transcripts [19,20] and molecules that modify D-serine accumulation in the cells [21].

Furthermore, the close relationships between D-serine and the NMDA receptor [8,13] and the anti-psychotomimetic [6,11,22,23,24,25], anti-schizophrenia [26,27], and anti-cerebellar ataxia [28,29] properties of D-serine suggest that endogenous D-serine could be involved in the higher brain functions and the pathophysiology of neuropsychiatric disorders. From this point of view, we have also measured the tissue contents of D-serine in the postmortem brains from patients with and without metabolic, neurodegenerative and mental disorders [30,31].

Based upon the above mentioned studies of D-serine in rats, mice and humans, we have applied behavioral, biochemical, molecular biological and histochemical methods. In this chapter, the methods required for our systematic approach to understand the metabolism and functions of brain D-serine are summarized (Table 1).

2. Quantitative Assay of D-Serine in the Central Nervous System

We accomplished the quantitative gas chromatographic (GC) analysis [5,6], and the GC-mass spectrometric (MS) analysis [5] to investigate the presence of D-serine and other D-amino acid in the brain and peripheral tissues of the rat, mouse and human. We have routinely performed the simultaneous quantitative measurement of D-serine and other chiral and non-chiral amino acids in mammalian tissues using high performance liquid chromatography (HPLC) with fluorometric detection [7,8].

2.1. Dissection of the Brain Areas

For the quantitative analysis of D-serine and other amino acids in the discrete brain areas of rats and mice, the cerebral neocortex (the dorsal part of the cerebral cortex divided along the rhinal fissure), pyriform cortex (the ventral part of the cerebral cortex divided along the rhinal fissure), striatum, hippocampus, limbic forebrain area (olfactory tubercle + nucleus accumbens + septum), thalamus, hypothalamus, midbrain, pons-medulla, cerebellum and spinal cord (cervical, thoracic and lumbo-sacral portions) were dissected out in the cold, frozen immediately, and stored at -80°C until biochemical assay [8,32,33].

2.2. GC and GC-MS Analysis

2.2.1. Reagents [5,6]

All amino acids and trichloroacetic acid (TCA) were obtained from Nakarai Tesque (Japan). All solvents were HPLC grade and purchased from Kanto Chemicals (Japan). A pentafluoropropionic anhydride (PFPA) was purchased from GL. C. Science (Japan). Amino acid stock solutions were prepared by dissolving amino acids in 0.01 M HCl to a concentration of 10 μmol/ml each. The standard amino acid stock solutions after appropriate dilution with 0.01 M HCI were used for calibration. L-Norleucine (L-NL) was used as an internal standard.

D-Serine may be present in the neurons and glial cells. D-Serine has been considered to be synthesized from L-serine or other amino acids and to be released to the extracellular fluid in a different way from classical neurotransmitters that depolarization of neurons failed to increase the extracellular D-serine contents. D-Serine acts as a selective co-agonist of the NMDA receptor ion channel (NR1/NR2 type) and is taken up into the neocortical P2 fraction. D-Amino acid oxidase can catalyze the breakdown of D-serine and mice lacking the activity of this enzyme exhibit increased D-serine concentrations in the brain areas.

Table 1. List of our research methods for the putative processes of D-serine metabolism and function in the mammalian brains.

Metabolic and functional procedures	Available methods	References
• Distribution	HPLC, GC, GC-MS	5-8, 12,13
• Synthesis	Differential cloning, DNA array, HPLC	14, 19, 20
• Storage	Differential cloning, DNA array, HPLC	19, 20
• Release	In vivo dialysis, Assay of release from the brain homogenates or Xenopus oocyte	13, 21
• Receptor or binding site	Electrophysiology, Binding assay	15, 16
• Uptake (Transmembrane transport)	Assay of uptake into the brain homogenates, Functional cloning	17, 18, 21
• Degradation	Differential cloning, DNA array, HPLC	10
• Behavior	Locomotor activity, Stereotypy, Ataxia	11, 22-24, 28
• Interaction with neurotransmitter systems	Measures of dopamine metabolism	25

2.2.2. GC Analysis [5,6]

The gas chromatographic analysis was carried out using a Hewlett Packard 5890 (Hewlett Packard.Japan) equipped with nitrogenphosphorus detector. Samples were analyzed on a Chirasil-L-Val capillary column (25 m x 0.25 mm. film thickness 0.12μm, Gasukuro Kogyo, Japan). The chromatographic conditions were: injection temperature, 250°C; detector temperature. 300°C; temperature program, 80°C, isothermal for 5 min, then 80-140°C at 2°C /min and split ratio, 1:10. The column head pressure was 110 kPa, and the helium flow-rate on the column was 0.4 ml/min; the flow-rate (plus auxiliary gas) was 30 ml/rnin. The hydrogen flow-rate and the air flow-rate at the detector were 3.5 and 1 10 ml/min, respectively. Peak area was determined using a SIC chromatocorder 11 integrator (System Instruments Co.).

After the addition of L-NL as an internal standard, the central nervous system or peripheral tissue was homogenized at 4°C in 5 vols of 5% TCA and the homogenate was centrifuged at 18000xg for 30 min at 4°C. To remove TCA, the supernatant was washed three-times with water-saturated diethylether. The aqueous layer was then passed through a Millipore filter, HV (0.45 mm) and stored at -80°C until derivatization.

The 5% TCA extract was evaporated under a gentle stream of nitrogen gas at room temperature. The last trace of water was removed azeotropically using methylene chloridep. The residue was dissolved in 2 ml of isopropanol-2N HCl. After the mixture was heated at l00°C for 1 h, the excess reagent and solvent were removed under a gentle stream or nitrogen gas at room temperature. Four hundred μl of methylene chloride and 100 μl of PFPA were added to the residue, which was heated at 100°C for 20 min. The excess reagent and solvent were then evaporated under a gentle stream of nitrogen gas at room temperature. After addition or 2 ml distilled water, the derivatized sample was extracted three-times with ethyl acetate. To remove any trace amount of water in the combined ethyl acetate extract, an appropriate amount of dry sodium sulfate was added to the combined extract and allowed to stand for 20 min. After filtrating the extract was carefully evaporated in the same way. Finally, the derivatized sample was dissolved in an appropriate volume or ethyl acetate and stored at 4°C until analysis. One or two μl of the solution was injected into the gas chromatograph.

2.2.3. MS Analysis [5]

The electron-impact (El)- and positive chemicalionization (PCI)-mass spectroscopic (MS) analysis in combination with the GC method (GC-MS analysis) was performed on a V.G. Masslab Trio-1 mass spectrometer (Jasco, International Co., Japan) equipped with a Hewlett Packard 5890A gas chromatograph. The GC conditions were the same as for the GC analysis with the exception of a splitless injection. The mass spectrometer conditions were: electron accelerating voltage, 70 eV; interface temperature, 170°C; photomultiplier voltage, 400 V; ion source temperature, 200°C (El) and 190°C (PCI); trap current, 150 μA (El); emission current. 150 μA (PCT). PCI-MS was performed using isobutane as a reactant gas.

2.3. HPLC Analysis

2.3.1. Reagents [7,8]

All amino acids except D-homocysteic acid (D-HCA), TCA, glacial acetic acid, and *o*-phthaldialdehyde were obtained from Nakarai Tesque (Japan). D-HCA, *N–tert*-butyloxycarbonyl-L-cysteine and all solvents of HPLC grade were purchased from Sigma (St. Louis, MO, USA), Novabiochem (Laufelfingen, Switzerland) and Kanto Chemicals (Japan), respectively. Amino acid stock solutions were prepared as described in the above 2.2.1 section.

2.3.2. HPLC Analysis [7,8]

The brain samples were homogenized in 2 ~ 10 volumes of 4% TCA after the addition of D-HCA as the internal standard, and the homogenate was centrifuged at 18,000 g for 30 min at 4°C. To remove the TCA, the supernatant was extracted three times with water-saturated diethyl ether. The aqueous layer was then passed through a Millipore filter (HV; pore size, 0.45 μm) and stored at -80°C until derivatization. The resultant sample was derivatized with N-*tert*-butyloxycarbonyl-L-cysteine and *o*-phthaldialdehyde for 2 min at room temperature. The derivatized sample was immediately applied to the HPLC system and analyzed using a 4-μm (particle size) Nova-Pak C18 column (300 X 3.9 mm) (Waters, Japan). The column was operated at a constant flow-rate of 1.0 ml/min at 35 °C. Mobile phase A was 0.1 M acetate buffer (pH 6.0) containing 10% acetonitrile and mobile phase B was the acetate buffer containing 20% acetonitrile. The separation of the amino acid derivatives was performed using a linear gradient from mobile phase A to B in 53 min. The fluorescent amino acid derivatives were detected using a Waters 2475 Multi λ fluorescence detector spectrofluorometer (Waters Co., Ltd., Japan). The excitation and emission wavelengths were 344 and 443 nm, respectively.

The present HPLC method was also applied for the measurements of the amounts of various amino acids in the dialysates obtained from the brain extracellular fluid (3.1.), the cultured cells and their medium, the Xenopus oocytes and their medium, and peripheral tissues (3.2.).

3. Release Mechanism

As predicted by the NMDA receptor co-agonist properties and NMDA receptor-like tissue distribution of D-serine, D-serine has been demonstrated to be present in the extracellular fluid of the brain regions [3,13]. For the studies of the control mechanisms and their temporal profiles of the extracellular D-serine dynamics, an in vivo microdialysis appears to be one of the most powerful techniques to estimate the extracellular D-serine contents in freely moving animals. We also observed the efflux of D-serine from the Xenopus oocytes preloaded with radiolabeled or non-radio-labeled D-serine.

3.1. In Vivo Microdialysis

For in vivo microdialysis studies [13,34,35], rats were anesthetized with pentobarbital (40 mg/kg, intraperitoneally) and mounted on a stereotaxic frame. A straight-shaped cellulose dialysis tube (3.0 mm in length, 0.16 mm internal diameter, molecular weight cutoff 50,000, EICOM Co. Ltd., Japan) was then implanted into the medial prefrontal cortex (mPFC: A +3.2 mm, V +5.2 mm, L -0.6 mm), striatum (A +0.5 mm, V +7.0 mm, L -3.0 mm) or cerebellum (A -11.3 mm, V +3.8 mm, L -0.6 mm) according to the atlas of Paxinos and Watson [33]. These dialysis probes and implantation procedures can also be applied to the mice [28] by adjusting the coordinates of the probe positions using the atlas of Paxinos and Franklin [36].

Two days after surgery, the dialysis probe was perfused with Ringer solution (NaCl, 147 mM; KCl, 4 mM; $CaCl_2$, 1.3 mM; pH 7.3) at the flow rate of 2 μl/min in a freely moving rat. After stabilizing for at least 80 min, the dialysate samples were collected every 20 min. The first three samples were used to determine the basal release of each amino acid. The addition of various agents to the perfusates allows us to analyze the mechanisms underlying the extracellular release of D-serine. For instance, veratrine (200 μg/ml) was included in the perfusion solution during the next 40-min perfusion period either in the presence or absence of TTX (2 μM) to investigate the influences of the nerve impulse flow on the extracellular release of D-serine [13]. The hypothetical requirements of Ca^{2+} for the possible neurotransmitter-like vesicular release of D-serine was tested by Ca^{2+}-free Ringer solution replacing $CaCl_2$ by NaCl with EGTA (2 mM) [13].

The collected samples are stored at -80°C until derivatization following the addition of D-HCA as an internal standard. The derivatized amino acids are quantified by HPLC with fluorometric detection (2.2.). After termination of the experiments, the location of the dialysis probe was macroscopically verified in each case on 150-μm-thick serial coronal slices.

3.2. Efflux of D-Serine Preloaded in the Xenopus Oocyte

In the experiments to test the possibility that the gene products expressed in the Xenopus oocytes might liberate or pump out the intraoocyte D-serine, the changes in the extra- and intra-oocyte concentrations of $[^3H]$D-serine (see the section 8) at various time after loading the radiolabeled D-serine [21]. Oocytes were preloaded with 1 mM of $[^3H]$ D-serine for 60 min at 26°C. Following removal of the nonspecific binding of the radiolabeled compound on the cell surface by a brief wash with FR solution at room temperature, 10 groups of 6 oocytes were transferred to the D-serine-free FR solution and incubated for 30-120 min at 26°C. The radioactivity in the medium and that remaining in the oocytes were then measured. The rate of efflux was calculated as the percentage of radioactivity in the medium to the total preloaded radioactivity [radioactivity in medium / (radioactivity in medium + radioactivity remaining in oocytes) X 100 %] [21].

4. Binding Site

Endogenous D-serine seems to act, at least, as a co-agonist for the NMDA receptor by binding to its glycine site with a high affinity around 2 or 3 X 10^{-7} M [2]. In in vitro experiments, D-serine has been shown to enhance the NMDA receptor-mediated depolarization, inward currents, Ca^{2+} influx, cyclic GMP production, and increase in radioligand binding to the phencyclidine (PCP) site within the NMDA ion channel in a stereoselective and NMDA antagonist-reversible manner [2]. The co-agonist potency of D-serine appears to be similar for any combination of NR2A (ε1 in mouse), NR2B(ε2), NR2C (ε3), or NR2D (ε4) and NR1 (ξ1) [15], although the distribution pattern of intrinsic D-serine is closely correlated with only that of the NR2B subunit mRNA [3,6,8]. Our further investigation [16] suggests that D-serine may gain access to a recognition site different from the glycine modulatory domain of the NMDA receptor in the presence of the selective glycine antagonist, 5,7-dichlorokynurenate (DCK) [2].

4.1. DCK-Insensitive D-Serine Biding Site

4.1.1. Membrane Preparation [16]

After the rats were killed by cervical dislocation, the cerebral cortex, cerebellum, liver, and kidney were dissected out in the cold, frozen immediately in liquid nitrogen, and stored at -80°C until preparation of the P2 fractions. These frozen tissues were used within 3 months. The synaptosomal P2 fractions were prepared from these brain or peripheral tissues on ice. Each tissue was homogenized in 20 volumes of ice-cold 0.32 *M* sucrose buffer containing 10 mM Tris-acetate (pH 7.4), 1 mM EDTA, 0.1 mM phenylmethylsulfonyl fluoride, and 5 μg/ml aprotinin using a glass-Teflon homogenizer. Following centrifugation at 1,000 *g* for 11 mm at 4°C, the supernatant was centrifuged at 14,000 *g* for 11 mm at 4°C.The resultant pellet (synaptosomal P2 fraction) was resuspended in a 50 mM Tris-acetate buffer (pH 7.4) containing 1 mM EDTA, 0.1 mM phenylmethylsulfonyl fluoride, and 5 μg/ml aprotinin using a Brinkmann Polytron homogenizer. The homogenate was centrifuged three times at 40,000 *g* for 20 mm at 4°Cand finally resuspended in the same buffer at a concentration of ~0.2 mg of the original wet weight/mi. The suspension was frozen at -80°C before the binding assay.

On the day of the binding assay, the stored homogenate was thawed and centrifuged at 40,000 *g* for 20 mm at 4°C.The final pellet was resuspended in 50 mM Tris-acetate buffer (pH 7.4) while the pH-related changes in the [^{3}H] D-serine binding were determined at pH 7.4 of the 50 mM Tris-acetale buffer. A part of the P2 fraction was withdrawn for the determination of the protein concentration.

4.1.2. [^{3}H]D-Serine Binding Experiments [16]

For the binding experiments, we used 50 mM Trisacetate buffer (pH 7.4) for the assay medium because preliminary studies indicated that specific [^{3}H]D-serine (see the section 8) binding to the cerebellar P2 preparation was greater in a 50 mM Tris-acetate buffer than in an equimolar Tris-HCI buffer and Tris-citrate buffer. The [^{3}H]D-serine binding assays were performed at 4°C with an aliquot of the washed P2 suspension (0.2 ~ 0.4 mg of protein) in a final volume of 0.5 ml of 50 mM Tris-acetate buffer (pH 7.4) containing appropriate concentrations of [3H]D-serine. To eliminate binding of the radioligand to the glycine modulatory site of the NMDA receptor (Fig. 1.), all binding reactions were carried out in the presence of DCK, a selective antagonist for the glycine site, at 100 μM [2]. The incubation was started by the addition of the P2 preparation, and the mixture was routinely kept at 4°C because the amount of the DCK-insensitive binding at 25°C was found to be lower than that at 4°C during the incubation time tested for 120 mm. Nonspecific binding was always defined in the presence of 1 mM unlabeled D-serine.

The binding reaction was terminated by centrifugation at 13,000 *g* for 10 min at 4°C. The supernatant was removed by aspiration. The resulting pellets were rapidly and superficially rinsed twice with 1 ml of ice-cold 50 mM Tris-acetate buffer and then solubilized in 100 p1 of 0.2 *M* NaOH. After neutralization of the solution by adding 80 μ1 of 0.2 M HCI, the bound radioactivity was quantified by liquid scintillation spectrometry using I ml of Ultima Gold XR (Packard, U.S.A.) at a counting efficacy of 35 ~ 38%. Each assay was performed in triplicate or quadruplicate. For the protein content measurement, the withdrawn P2 fractions were dissolved in 0.1 M NaOH. The protein content was assayed using the Bradford Reagent Kit (Bio-Rad, Richmond, CA, U.S.A.). The Scatchard plot was linear, indicating that [^{3}H]D-serine was most likely binding to a single population of recognition sites with a Kd of 614 ± 81 nM and a Bmax of 2.07 ± 0.39 pmol/mg protein in the presence of DCK.

5. UPTAKE ACTIVITY

The liberated D-serine into the extracellular fluid has been shown to be taken up into the brain P2 preparations [3,18] and C6 glioma cells [17] while neither the specific transporter nor the exact localization of the uptake activity are still unclear. The uptake activity of D-serine can be determined by the following methods [17,18] using the purified [^{3}H]D-serine (see section 8).

5.1. Uptake by Synaptosomal P2 Fraction Isolated from Fresh Rat Brain

5.1.1. Tissue Preparation [18]

After the animal was sacrificed by cervical dislocation, the cerebral cortex, cerebellum, liver and kidney were dissected out, and the P2 fractions were immediately prepared on ice from these tissues as described below. Each tissue was homogenized in 20 vol. of ice cold 0.32 M sucrose buffer containing 20 mM Tris-acetate (pH 7.5), 1 mM EDTA and 0.1 mM phenyl methyl sulfonyl fluoride (PMSF) using a glass-Teflon homogenizer. After centrifugation of the homogenate at 1,000 x g for 10 min, the supernatant was further centrifuged at 14,000 x g for 10 min. The resultant pellet (P2 fraction) was resuspended in ice-cold Tris-acetate buffer (20 mM Tris-acetate, pH 7.5, 120 mM NaCl, 5mM KCl, 1mM $MgSO_4$, 1mM NaH_2PO_4, 1 mM glucose and 0.75 mM $CaCl_2$), and used for the following uptake study.

5.1.2. Uptake Experiments [18]

An aliquot of the P2 suspension (1-1.3 mg protein) was preincubated in Tris-acetate buffer (1.8 ml in total) for 10 min at 32°C. The uptake reaction was then started by the addition of 0.2 ml of [^{3}H]D-serine at various concentrations in the Tris-acetate buffer to each assay medium and continued for 0.25 ~ 7 min at the same temperature. Blanks were treated in the same way but incubated at 0°C. Under this condition, the strychnine-insensitive glycine binding site of the NMDA receptor appears to be undetectable because there was no [^{3}H]D-serine binding at both 0°C and 32°C to this site as defined by the difference between the absence and presence of 100 μM of DCK, which is a selective antagonist for the glycine site [2]. For complete binding elimination of these radioligands to the glycine binding site, all uptake reactions were carried out in the presence of 100 μM of DCK. The uptake incubation was terminated by filtration on a glass fiber filter (Whatman GF/B) under vacuum and rinsing three times with 3 ml of Tris-acetate buffer. Each filter was then extracted in 1 ml of Ultima Gold XR scintillation cocktail (Packard Instrument, Meriden, CO, U.S.A.), and the accumulated radioactivity was measured by a liquid scintillation counter (LSC 6500, Beckman, Palo Alto, CA, U.S.A.) with a counting efficacy of 45 ~ 55%. Each assay was performed in triplicate or quadruplicate.

A non-linear least squares model fitting (GraphPad Prism 3, GraphPad Software, Inc., San Diego, CA) of the data from the saturation experiments ([^{3}H]D-serine: 10 to 1,100 μM) indicated that the cortical [^{3}H]D-serine uptake occurs by an apparent single-site transport system having a Km value of 283 ± 71 μM and a Vmax value of 207 ± 21 pmol/mg protein/min (n=3). The Na^+- and Cl^--dependent [^{3}H]D-serine accumulations in the P2 fraction of the cortex were markedly attenuated by glycine, the D- and L-forms of serine, but not by even high concentrations (1 mM) of β-alanine, γ-aminobutyric acid (GABA), and L- and D-glutamic acid. Both the cerebral cortex and cerebellum exhibited a high D-serine uptake (22.8 ± 1.8 and 13.6 ± 1.6 pmol/mg protein/3 min, respectively), whereas the liver and kidney accumulated only a small amount of [^{3}H]D-serine (1.8 ± 2.0 and 1.7 ± 1.3, respectively).

5.2. Uptake by C6 Glioma Cells

5.2.1. Culture of Rat Glioma C6BU-1 Cells [17]

Rat glioma C6BU-1 cells were cultured in Dulbecco's modified Eagle's medium containing 10% (v/v) fetal calf serum (Sanko Jyunyaku Co., Ltd., Japan) supplemented with penicillin (50 IU/ml) and streptomycin (50 μg/ml) at 37 °C in a humidified atmosphere of 90% air and 10% CO2. After a 2-day culturing, the cells were subcultured in multiwell culture dishes having a 22 mm diameter (Falcon, U.S.A.). The uptake study of the C6BU-1 cells was performed at their confluency.

5.2.2. Uptake experiments [17]

In the uptake study of [3H]D-serine, the cultured glioma cells attached to the dish (0.1 - 0.2 mg protein) are preincubated for 10 min in 1 ml of uptake buffer (pH 7.4) consisting of 5 mM HEPES, 5 mM KCl, 1 mM KH_2PO_4, 140 mM NaCl, 1.8 mM $CaCl_2$, 0.4 mM $MgSO_4$, 5 mM D-glucose and 2.5 mM NaOH. The uptake reaction was started by the addition of 0.4 ml of the uptake buffer containing 0.4 μCi [^{3}H]D-serine at various concentrations. After a 10-min incubation at 37 or 0 °C, the reaction was stopped by aspirating the buffer and washing the cells three times with 1 ml of ice-cold uptake buffer. Thereafter, 0.2 ml of 0.1% SDS was added to each dish and left for 1 h. The contents of the dish were then transferred into 1 ml of Ultimagold XR (Packard, USA) scintillation cocktail, and the accumulated radioactivity was measured by liquid scintillation spectroscopy (6500 Liquid Scintillation Counter: Beckman, USA). The specific uptake activity was estimated as the difference between the amino acid accumulation at 37 °C and that at 0 °C.

An Eadie-Scatchard plot of the data from the saturation experiments for the Michaelis-Menten equation indicates that the specific uptake of [^{3}H]D-serine into the glioma cells occurs by a single-site transport system with a Km of 2480.0±80.0 μM and a Vmax=6.7±0.1 nmol/min/mg protein (n=3) The sodium dependency and substrate specificity of the D-serine uptake into the C6 cells were found to resemble those of ASCT2 [37].

6. DIFFERENTIAL CLONING

Although the putative serine racemase synthesizing D-serine from the L-enantiomer has been reported in the rodent and human tissues [38] and a sodium-independent neutral amino acid transporter encoded by asc-1 displays a high affinity for D-serine as well as L-serine [39], little is known about the exact molecules involved in the metabolic or functional processes of the brain endogenous D-serine. One of the rationale approaches to clarify these molecules appears to identify transcripts that stereoselectively respond to D-serine because many other neuroactive substances influence the expression of the molecules specifically associated with their metabolism and function. To this end, we have successfully applied a differential cloning technique, an RNA arbitrarily primed PCR (RAP-PCR) [40], which can detect unknown transcripts, and recently cloned a novel D-serine-, but not L-serine-inducible transcript, *dsr-1* (D-serine-responsive transcript-1) [19], from the rat neocortex. Therefore, we have extended the exploration strategy to other candidates for the key genes for the brain D-serine system, which leads to the isolation of another D-serine regulated gene, *dsr-2* [20]. Now, we also use a DNA microarray technique to detect the expressed sequence tags which could be the parts of the gene transcripts encoding the as-yet-unrecognized D-serine-related proteins.

6.1. RNA Finger Printing by RAP-PCR

Rats were killed by cervical dislocation 3 h or 15 h after the i. p. administration of the D-, L-serine solution (9 mmol/kg body weight), or saline (vehicle) [19]. The neocortex (the dorsal part of the cerebral cortex divided along the rhinal fissure) was rapidly dissected out in the cold, frozen in liquid nitrogen, and stored at -80°C until use.

The total RNA was prepared from the neocortex using an RNeasy Midi Kit (Qiagen, GmbH, Hilden, Germany) and converted to single-stranded cDNA by reverse transcriptase (Superscript Preamplification system, GIBCO BRL, Rockville, MD) with random hexamer primers. In every experimental group, equal amounts of the resulting cDNAs of the individual rats were pooled. The pooled samples were serially diluted in TE buffer (pH 8.0) and used as a template for PCR using a rhodamine-labelled 12-mer arbitrary primer set. The PCR parameters were 94°C for 2 min, 40°C for 5 min, 72°C for 5 min for the first cycle, 94°C for 30 sec, 40°C for 2 min, 72°C for 1 min with 34 cycles, and then 72°C for 5 min for the extension. The PCR products were separated in a 5% denaturing polyacrylamide gel including 7 M urea and 1x TBE (90 mM Tris-borate and 2 mM EDTA). The gel display pattern of the cDNA bands was visualized and analyzed by an FMBIO II Multi-View fluorescence image analyzer (Hitachi, Tokyo, Japan). The cDNA bands that showed a higher intensity in the D-serine-treated group than in the saline-injected controls were cut from the gel. The extracted DNA fragments were re-amplified by a second PCR using the same primer set. The amplified DNA fragments were cloned using a TA cloning vector (pGEM-T Easy vector system, Promega, Madison, WI), and sequenced (ABI Prism 377, PE Biosystems, Foster City, CA).

To determine the full length structure from the isolated DNA fragments, we performed 5' and 3' rapid amplifications of the cDNA ends (RACE)-PCR with an aliquot of the oligo-dT selected RNA (1 μg) prepared from the neocortex using a SMART RACE cDNA amplification kit (Clontech, La Jolla, CA) according to the manufacturer's protocol. The expression and the copy number of the novel transcript were examined by Northern and Southern blot analyses, respectively.

6.2. Semiquantitave Co-Amplification RT-PCR [19,20]

In the experiments with the D- and L-serine injections, quantitative determinations of the target RNA levels in the neocortical samples from the individual animals were accomplished by co-amplification RT-PCR with an endogenous template as the internal standard to further control the variations in the sampling and processing between the samples [41,42]. As an internal control, 28S ribosomal RNA (rRNA) was employed because neither the D- nor L-serine affects the expression levels of 28S rRNA [41]. During the sample preparation step, a reverse transcription was performed using 0.2 μg of neocortical total RNA from an individual animal, and the resulting cDNA was suspended in 10 volumes of TE buffer (pH8.0) for the PCR template. For the assay for *dsr-2*, an appropriate portion (90.5% for the co-amplification with *dsr-2*) of the added primers specific for the 28S rRNA sequence is phosphorylated at their 3' ends so that the target and the control exponential phases will overlap. The following primers were used: for *dsr-2*: the 5'-primer, 5'-TGA GCC AGG AAT TTA GGA AGG TT-3' and the 3'-primer, 5'-AGC AAA TCT GGC CAA GTC TAA TG-3', size of PCR product = 530 bp; for r28S: the 5'-primer, 5'-CTC GCT GGC CCT TGA AAA TCC-3' and the 3'-primer, 5'-CCC AGC CCT TAG AGC CAA TCC TTA-3', the size of PCR = 173 bp. The PCR parameters used included 94°C for 2 min, 61°C for 3 min, 72°C for 5 min for the first cycle, 94°C for 45 sec, 61°C for 2 min, 72°C for 3 min with 2 cycles, 94°C for 30 sec, 61°C for 45 sec, 72°C for 1 min with 32 cycles and then 72°C for 5 min for extension. The PCR products were separated by electrophoresis of 3% NuSieve agarose 3:1 (FMC Bioproducts, Rockland, ME) gel in 1xTAE. The gel was stained with 0.5 μg/ml of ethidium bromide for 30min. The resultant cDNA bands were visualized by UV irraditation and quantitatively analyzed by an image analyzer (Lumi-Imager, Roche Diagnostics, Basel, Switzerland). The relative amount of the *dsr-2* transcript was evaluated as the ratio of the optical density of the *dsr-2* band to that of the 28S rRNA band.

7. Functional Cloning using Xenopus Oocytes

Because of the involvement of the NMDA receptor in divergent higher brain functions and the anti-schizophrenic and anti-ataxic properties of D-serine [see 2, 3, and 26 for reviews], the extracellular D-serine levels or translocation of the D-amino acid across the plasma and/or vesicular membrane should undergo a fine control by a specific molecular system. However, the exact molecules comprising the uptake and release machinery for the regulation of the extracellular D-serine and their relation to the D-serine metabolism in the intracellular structures are largely unknown. Therefore, to obtain insight into the D-serine carrier proteins or their regulators, we have explored molecules that modify the tritiated or non-radiolabeled D-serine accumulation into the *Xenopus* oocyte [21]. The measurements of the D-serine efflux from the oocytes and the quantitative HPLC analysis of the amino acids used in this cloning experiment are described in sections 3.2. and 2.2., respectively.

7.1. Construction and Screening of Rat Cerebral Neocortex cDNA Library [21]

From the male Wistar rat (PD 56) cerebral neocortex, poly(A)$^+$ RNA was purified and size-fractionated by 5-20% sucrose density gradient centrifuging using the SW41Ti rotor (Beckman, Palo Alto, CA) at 25,000 rpm for 12 h. Directional cDNA libraries were constructed from these fractions using the pSPORT1 SuperScript Plasmid System (Invitrogen, Carlsbad, CA) according to the manufacturer's instructions. Each library was subdivided into 50 pools, each containing 10,000 clones. The respective cDNA pool was transcribed *in vitro* by T7 RNA polymerase (mMESSAGE mMACHINE, Ambion, Austin, TX), and used for microinjection into the *Xenopus* oocytes.

7.2. Expression Screening in *Xenopus* Oocytes [21]

The expression screening assay was performed using collagenase-treated and manually defolliculated oocytes. The cells were microinjected with 50 nl (1 ng/ul dissolved in RNase-free H_2O) of poly(A)$^+$ RNA, or in vitro transcribed cRNA as mentioned above. The same volume of distilled H_2O was used for the microinjection as the control. The oocytes were incubated at 20°C in modified Barth's medium (MBM: 80 mM NaCl, 1 mM KCl, 0.33 mM $Ca(NO_3)_2$, 0.41 mM $CaCl_2$, 0.82 mM $MgSO_4$, 2.4 mM $NaHCO_3$, and 15 mM HEPES, pH 7.4) for 2 days before the assay.

7.3. Uptake Assay of D-Serine [21]

The uptake of D-serine was traced with [^{3}H]D-serine (see the section 8). Groups of 12-20 oocytes were incubated at 26°C for 60 min in 1.5 ml of Frog Ringer (FR: 120 mM NaCl, 2 mM KCl, 1.8 mM $CaCl_2$, and 5 mM HEPES, pH 7.4) containing 10 μM of [^{3}H]D-serine [39] based on the preliminary data that the uptake of [^{3}H]D-serine was found to be linear for at least 4 h and then approaches a plateau 8 h after the start of the reaction. At the end of the incubation, nonspecific binding of the radiolabeled compound on the cell surface was removed by six brief washings with 10 ml each of ice-cold FR solution. The individual oocytes were transferred to a microtube and completely lysed in 100 μl of 0.2% SDS in FR by intensive vortex mixing. One ml of scintillation fluid (Packard Ultima Gold XR, Perkin-Elmer, Boston, MA) was added to each lysate solution, and the radioactivity was determined by liquid-scintillation counting (LS6500, Beckman). Some fractions induced a 1.5-2 fold increase, while some other fractions had a 20-40% decrease in the uptake activity of D-serine into the oocytes. The positive fractions of the cDNA library were further sequentially subdivided into the next 50 pools, each of 500-1000 clones. After repeating the same step, one pool containing 100 clones exhibited a reproducible decreased D-serine uptake. Colonies from the positive pools were then individually seeded into the well of the 96-well plate to produce a grid system. From the combined testing of the rows and columns for the D-serine uptake activity, a single positive colony was isolated.

7.4. Semi-Quantitative RT-PCR (see also 6.3.)

The total RNA was extracted from tissues using an RNA extraction kit (RNeasy Midi Kit, Qiagen, Inc., Valencia, CA) and the complementary DNA (cDNA) synthesis was carried out using a SuperScript First Strand Synthesis System (Invitrogen) according to the manufacture's protocol. The following PCR parameters used were: *dsm-1*, 94°C for 30 sec, 68°C for 1 min, 72°C for 1 min with 5 cycles, and 94°C for 30 sec, 65°C for 45 sec, 72°C for 1 min with 31 cycles; GFAP, 94°C for 2 min, 60°C for 1 min, 72°C for 1 min for the first cycle, and 94°C for 30 sec, 60°C for 45 sec, 72°C for 1 min with 26 cycles, and; GAPDH, 94°C for 30 sec, 55°C for 45 sec, 72°C for 2 min with 25 cycles. The exact sequences of the specific primers of these experiments were described elsewhere [21].

7.5. Expression and Functional Assay of *dsm-1* in *Xenopus* Oocyte

The open reading frame (ORF) of *dsm-1* cDNA was subcloned into the pBScMxt vector, which contains 5' UTR and 3' UTR of the *Xenopus* β-globin gene for stable expression of cRNA in the oocytes [43]. The cRNA was synthesized by *in vitro* transcription using T3 RNA polymerase (mMESSAGE mMACHINE, Ambion) after the linearization of the plasmid by *Sal*I digestion. To examine the ion dependency for the uptake of D-serine, the oocytes were incubated in standard FR uptake solution, Na^+-free FR solution (sodium was displaced by choline), and Cl^--free FR solution (chloride was displaced by gluconate).

8. Behavioral Studies

The effects of the D-serine and related agents acting at the NMDA glycine site on the behavior of the rat and mouse under different conditions have been studied to obtain clues for the physiological and pathophysiological roles of the endogenous D-serine. The intraventricular infusion of D-serine or D-alanine has been shown to stereospecifically antagonize the ability of schizophrenomimetics, phencyclidine [3,6,11,22,23] and methamphetamine [24], to produce abnormal behavior including hyperlocomotion and/or stereotypy in a glycine site antabonist reversible manner. Moreover, the systemic administration of glycine site agonists, D-serine ethylester and D-cycloserine, caused a stereoselective and glycine site antagonist-sensitive inhibition of the cerebellar ataxia induced by neonatally injected cytosine arabinoside [28]. As sugested by these animal experiments, the agonists of the NMDA glycine site (Fig. 1.) including, the D-cycloserine, D-serine, glycine, D-alanine and glycine transporter inhibitor have been found to reduce the scores of antipsychotic-resistant symptoms of schizophrenia [26,27]. Also, D-cycloserine attenuates ataxia in the patients with cerebellospibnal degeneration [29].

8.1. Intraventricular Injection

Male Wistar rats (ST strain from Shizuoka Laboratory Animals, Shizuoka, Japan) weighing 200 to 250 g were used. The bilateral infusion of drugs into the lateral ventricle of conscious rats was performed via an indwelling cannula [11,22,23,32]. Stainless-steel guide cannulae (outer diameter, 0.55 mm) were stereotaxically implanted under pentobarbital (40 mg/kg i.p.) anesthesia 5 to 7 days before the manipulations. The lower end of the cannula was aimed at the cerebral cortex, 2.0 mm above the injection site. On the day of the experiment, an injection cannula (outer diameter, 0.33 mm) was inserted into the lateral ventricle through the guide cannula at the following stereotaxic coordinates [33]: AP, -0.8 mm; V,

+2.0 mm; and L, ± 1.5 mm. The injection cannula was connected to a motor driven 10 μl Hamilton syringe by polyethylene tubing. Drugs were infused in a volume of 5 μl for 4 min and the infusion cannulae were left in place for 5 min before removal. PCP (10 mg/Kg i.p.) was administered 10 min after the beginning of the local injection. At the end of the experimental session, the locations of the sites of the injection cannulae were macroscopically verified for every case on the 150-μm thick serial slices.

8.2. Behavioral Ratings

On each day of the experiment, rats were individually placed in the plastic observation cages (internal dimensions: 37 cm x 22 cm x 14 cm) and allowed for at least 3 hr to acclimate to the new environment, and then monitored for locomotor activity, stereotypy and ataxia. Behavioral ratings were made every 20-min from 10 to 90 min (or 150 min) after the i.p. injection of PCP. The cumulative behavioral rating for each animal was determined as the summation of each 20 min score for 90 min.

The intensity of these PCP-induced abnormal behaviors was evaluated according to the scoring system of Sturgeon *et al.* [44] with minor modifications as follows [11,22,23]: *locomotor activity rating scale*, 0) stationary, with little or no movement; 1) activity within a circumscribed area with the hindlimbs remaining stationary; 2) intermittent locomotor activity in less than one-half of the cage at a low-moderate rate; 3) continuous locomotor activity in less than one-half of the cage at a moderate-rapid rate; 4) intermittent locomotor activity in greater than one-half of the cage at a low-noderate rate; and 5) continuous locomotor activity in greater than one-half of the cage at a moderate-rapid rate; *stereotyped behavioral rating scale*, 0) no stereotyped behavior; 1) increased exploratory activity with occasional sniffing, grooming or rearing; 2) episodic nondirectional movement, weaving, reciprocal forepaw treading, higher frequency of sniffing, grooming or rearing greater than (1); 3) intermittent turning, backpeddling and weaving with or without bursts of frequent sniffing, grooming or rearing; 4) rapid and continuous turning, backpeddling, weaving, sniffing and gagging; and 5) dyskinetic extension and flexion of limbs, head and neck; *ataxia rating scale*, 0) inactive or coordinated movement; 1) awkward or jerky movenents or loss of balance while rearing; 2) moderate rate of falling while rearing or changing direction 3) frequent falling while rearing or changing direction, or partial impairment of antigravity reflexes; 4) inability to move beyond a small area or to support body weight on haunches or abdomen; and 5) inability to move except for twitching movement. In some experiments, the spontaneous locomotor activity was also quantified by automatically counting the number of interruptions of a magnetic field with a MK-ANIMEX AUTO instrument (Muromachi-kikai Co., Ltd., Tokyo, Japan).

In the drug interaction experiments, the rats were treated with the stereoisomers of amino acids acting on the strychnine-insensitive glycine receptor 10 min before i.p. injection of 10 mg/kg of PCP. D-Alanine was injected at doses ranging from 2 to 200 μg/rat, because higher doses of the amino acid by themslves caused ataxic movements. The maximal limit of the injectable amount of D-serine was 2000 μg/rat due to its solubitity in 10 μl of PBS. It was impossible to apply L-alanine and L-serine at doses greater than 1000 μg/rat due to their ability to induce apparent convulsions. Some animals received the i.c.v. injection of 7-chlorokynurenic acid or DCK (5,7-dichlorokynurenic acid), both of which are selective competitive antagonists of the glycine site associated with the NMDA receptor (Fig. 1.), 10 min before the application of D-alanine and D-serine.

8.3. Evaluation of Anti-Ataxic Effect by Falling Index

In the ataxia experiments, male mice of the ICR strain at 6–7 weeks of age were used. Ataxia was chemically induced in some animals by repeated subcutaneous injections of cytosine arabinoside (Ara-C; 50 mg/kg) on the first, second and third postnatal days [28].

To evaluate the severity of the ataxia, mice were tested using an open-field method and the falling index was calculated. The open-field apparatus is a 70-cm square field surrounded by a 15-cm high wall. The floor of the open field was marked out into 49 equal grids (each grid: 10 cm X 10 cm). The mice were tested individually. Thirty minutes after i.p. injection, each mouse was allowed to move for 30 min within the open field under observation. The movement score was assessed by counting the number of times the mouse crossed a gridline. During the same period, the number of falls, i.e., the falling score was determined. Falls were defined by rollover of the hind limbs. The ratio of the falling score to the movement score was calculated and designated as the falling index. Pre-drug baseline measurements of the falling index of the mutant mice were then determined.

9. Chemicals

PCP [1-(1-phenylcyclohexyl)piperidine] hydrochloride was kindly synthesized and donated by Sumitomo Pharmaceuticals, Co., Ltd. (Osakn, Japan) and by Yamanouchi Pharmaceutical, Co., Ltd. (Osakn, Japan). The other chemicals were obtained from commercially available sources.

[^{3}H] D-Serine (10—20 Ci/mmol) was purchased from Moravek Biochemicals (U.S.A.). The purchased o- [^{3}HJ serine should always be further purified using columns packed with Dowex 1-X 8 (100-200 mesh) resin as previously described [16-18,21]. In brief, the resin (1 ml) is first activated by 1 M HCl (4 ml) and subsequently washed with distilled water until its pH reached around 6. The purchased [^{3}H] D-serine solution (1 ml) is applied to the activated column, then eluted with distilled water. The eluted radioactivity was used for the various experiments.

10. Ethics

Our studies were performed in strict accordance with the guidance of the Ethical Committee (the Rinri Iinkai) of the National Institute of Neuroscience, National Center of Neurology and Psychiatry and with that of the Tokyo Medical and Dental University, and approved by their committees for animal experiment ethics.

Acknowledgements

The author thanks all my past and current collaborators concerned with the studies of D-serine described in this article, who are Ki. Takahashi, M. Takashima, A. Umino, H. Hibino, N. Hata, Y. Tanii, A. Hashimoto, A. Kashiwa, K. Nishijima, T. Hayashi, N. Fujii, K. Harada, T. Oka, R. Konno, S. Kumashiro, Ka. Takahashi, F. Hayashi, T. Matsui, M. Sekiguchi, U. Tomita, K. Wada, M. Matoba, N. Yamamoto, K. Kaneda, N. Kawaguchi, S. Sakurai, D. Shimazu, G. Taniguchi, H. Takebayashi, K. Saigoh, K. Matsui, H. Iwama, S. Kure, K. Narisawa, K. Tada, M. Mizoguchi, M. Ogawa, H. Shigeto, T. Yamamoto, Y. Oya, and M. Kawai, and listed as the co-authors of our papers in the reference section, Nos. 5 ~ 26, 29 ~ 33, 35 and 36.

REFERENCES

[1] Johnson, J.W., Ascher, P. (1987) Glycine potentiates the NMDA response in cultured mouse brain neurons. *Nature*, *325*, 529-531.

[2] Danysz, W,, Parsons, A.C. (1998) Glycine and N-methyl-D-aspartate receptors: physiological significance and possible therapeutic applications. *Pharmacol Rev. 50*, 597-664.

[3] Nishikawa, T. (2005) Metabolism and functional roles of endogenous D-serine in mammalian brains. *Biol Pharm Bull*, *28*, 1561-1565.

[4] Fujii, N. (2002) D-amino acids in living higher organisms. *Orig Life Evol Biosph. 32*:103-127.

[5] Hashimoto, A., Nishikawa, T., Hayashi, T., Fujii, N., Harada, K., Oka, T. & Takahashi, K. (1992). The presence of free D-serine in rat brain. *FEBS Lett*, *296*,33-36, 1992.

[6] Nishikawa, T., Hashimoto, A., Tanii, Y., Umino, A., Kashiwa, A., Kumashiro, S., Nishijima, K., Oka, T., Shirayama, Y. and Takahashi, K.: Disturbed neurotransmission via the N-methyl-D-aspartate receptor and schizophrenia: In *The Biology of Schizophrenia* (eds. T. Moroji and K. Yamamoto) Development of Psychiatry Series, Elsevier, Amsterdam, pp. 197-207, 1994.

[7] Hashimoto, A., Nishikawa, T., Oka, T., Takahashi, K. & Hayashi, T. (1992). Determination of free amino acid enantiomers in rat brain and serum by high performance liquid chromatography after derivatization with N-tert.-butyloxycarbonyl-L-cysteine and o-phthaldialdehyde. *J Chromatogr*, *582*, 41-48.

[8] Hashimoto, A., Nishikawa, T., Oka, T. & Takahashi, K. (1993). Endogenous D-serine in rat brain: N-Methyl-D-aspartate receptor-related distribution and aging. *J. Neurochem*, *60*, 783-786.

[9] Hashimoto, A., Kumashiro, S., Nishikawa, T., Oka, T., Takahashi, K., Mito, T., Takashima, S., Doi, N., Mizutani, Y., Yamazaki, T., Kaneko, T. & Ootomo, E. (1993). Embryonic development and postnatal changes in free D-aspartate and D-serine in the human prefrontal cortex. *J Neurochem*, *61*, 348-351.

[10] Hashimoto, A., Nishikawa, T., Konno R, Niwa A, Yasumura Y, Oka, T. & Takahashi, K. (1993). Free D-serine, D-aspartate and D-alanine in central nervous system and serum in mutant mice lacking D-amino acid oxidase. *Neurosci Lett*, *152*, 33-36.

[11] Nishikawa, T., Umino, A., Tanii, Y., Hashimoto, A., Hata, N., Takashima, M., Takahashi, K. and Toru, M. (1991) Dysfunction of excitatory amino acidergic systems and schizophrenic disorders. In: *Biological Basis of Schizophrenic Disorders* (ed Nakazawa T), Japan Scientific Societies Press (Tokyo) and Karger (Basel), pp65-76.

[12] Hashimoto, A., Oka, T. & Nishikawa, T. (1995). Anatomical distribution and postnatal changes in endogenous free D-aspartate and D-serine in rat brain and periphery. *Eur J Neurosci*, *7*, 1657-1663.

[13] Hashimoto, A., Oka, T. & Nishikawa, T. (1995). Extracellular concentration of endogenous free D-serine in the rat brain as revealed by in vivo dialysis. *Neuroscience*, *66*, 635-643, 1995.

[14] Takahashi, K., Hayashi, F. & Nishikawa, T. (1997). In vivo evidence for the link between L- and D-serine metabolism in rat cerebral cortex. *J Neurochem*, *69*, 1286-1290.

[15] Matsui, T., Sekiguchi, M., Hashimoto, A., Tomita, U., Nishikawa, T. & Wada, K. (1995). Functional comparison of D-serine and glycine in rodents: The effects on cloned NMDA receptors and the extracellular concentration. *J Neurochem*, *65*, 454-458.

[16] Matoba, M., Tomita, U. & Nishikawa, T. (1997). Characterization of 5,7-dichlorokynurenate-insensitive [^{3}H]D-serine binding to synaptosomal fraction isolated from rat brain tissues. *J Neurochem*, *69*, 399-405.

[17] Hayashi, F., Takahashi, K. and Nishikawa, T. (1997) Uptake of L- and D-serine in C6 glioma cells. *Neurosci Lett*, *239*, 85-88.

[18] Yamamoto, N., Tomita, U., Umino, A. & Nishikawa, T. (2001). Uptake of D-serine by synaptosomal P2 fraction isolated from rat brain. *Synapse*, *42*, 84-86.

[19] Tsuchida, H., Yamamoto, N., Kajii, Y., Umino, A. & Nishikawa, T. (2001). Cloning of a D-serine-regulated transcript dsr-1 from the rat cerebral cortex. *Biochem Biophys Res Commun, 280,* 1189-1196.

[20] Taniguchi, G., Yamamoto, N., Tsuchida, H., Umino, A., Shimazu, D., Sakurai, S., Takebayashi, H. & Nishikawa, T. (2005). Cloning of a D-serine-regulated transcript *dsr-2* from the rat cerebral neocortex. *J Neurochem, 95,* 1541-1549.

[21] Shimazu, D., Yamamoto, N., Umino, A., Ishii, S., Sakurai, S. & Nishikawa, T. (2006) Inhibition of D-serine accumulation to the Xenopus oocyte by expression of the rat ortholog of human 3'-phosphoadenosine 5'-phosphosulfate transporter gene isolated from the neocortex as D-serine modulator-1. *J Neurochem, 96*, 30-42

[22] Tanii, Y., Nishikawa, T., Hashimoto, A. & Takahashi, K. (1991). Stereoselective inhibition by D- and L-alanine of phencyclidineinduced locomotor stimulation in the rat. *Brain Res, 563*, 281-284.

[23] Tanii, Y., Nishikawa, T., Hashimoto, A. & Takahashi, K. (1994). Stereoselective antagonism by enantiomers of alanine and serine of phencyclidine-induced hyperactiviy, stereotypy and ataxia in the rat. *J Pharmacol Exp Ther*, *269*, 1040-1048.

[24] Hashimoto, A., Nishikawa, T., Oka, T. & Takahashi, K. (1991). D-Alanine inhibits methamphetamine-induced hyperactivity in rats. *Eur J Pharmacol*, *202*, 105-107.

[25] Umino, A., Takahashi, K. & Nishikawa, T. (1998) Characterization of the phencyclidine-induced increase in prefrontal cortical dopamine metabolism in the rat. *Br J Pharmacol*, *124*, 377-385.

[26] Javitt, D.C. (2004) Glutamate as a therapeutic target in psychiatric disorders. *Mol Psychiatry*, *9*, 984-997.

[27] Tsai, G.E., Yang, P., Chang, Y.C. & Chong, M.Y. D-Alanine added to antipsychotics for the treatment of schizophrenia. *Biol Psychiatry, 59*, 230-234.

[28] Saigoh, K., Matsui, K., Takahashi, K., Nishikawa, T. & Wada, K. (1998). The stereospecific effect of D-serine ethylester and D-cycloserine in ataxic mutant mice. *Brain Res*, *808*, 42-47.

[29] Ogawa, M., Shigeto, H., Yamamoto, T., Oya, Y., Wada, K., Nishikawa, T. & Kawai, M. (2003). D-Cycloserine for the treatment of ataxia in spinocerebellar degeneration. *J Neurol Sci*, *210*, 53-56.

[30] Kumashiro, S., Hashimoto, A. & Nishikawa, T. (1995). Free D-serine in post-mortem brains and spinal cords of individuals with and without neuropsychiatric diseases. *Brain Res*, *681*, 117-125.

[31] Iwama, H., Takahashi, K., Kure, S., Hayashi, F., Narisawa, K., Tada, K., Mizoguchi, M., Takashima, S., Tomita, U. & Nishikawa, T. (1997). Depletion of cerebral D-serine in non-ketotic hyperglycinemia: possible involvement of glycine cleavage system in control of endogenous D-serine. *Biochem Biophys Res Commun*, *231*, 793-796.

[32] Nishikawa, T. & Scatton, B. (1985): Inhibitory influence of GABA on central serotonergic transmission. Raphe nucleias the neuroanatomical site of the GABAergic inhibition of cerebral serotonergic neurons. *Brain Res, 331*, 91-103.

[33] Paxinos, G. and Watson, C. (2005). The rat brain in stereotaxic coordinates, the fifth edition, Elsevier Academic Press, Amsterdam, pp. 210.

[34] Tanii, Y., Nishikawa, T., Umino, A., Takahashi, K. (1990). Phencyclidine indreases extracellular dopamine metabolites in rat medial frontal cortex as measured by in vivo dialysis. *Neurosci Lett*, *112*, 318-323.

[35] Nishijima, K., Kashiwa, A. & Nishikawa, T. (1994) Preferential stimulation of extracellular release of dopamine in rat frontal cortex to striatum following competitive inhibition of the N-methyl-D-aspartate receptor. *J Neurochem*, *63*, 375-378.

[36] Paxinos, G. and Franklin, K.B.J. (2004). *The mouse brain in stereotaxic coordinates*, the compact second edition, Elsevier Academic Press, Amsterdam, pp. 142.

[37] Utsunomiya-Tate, N., Endou, H. & Kanai, Y. (1996) Cloning and functional characterization of a system ASC-like Na+-dependent neutral amino acid transporter. *J Biol Chem,* 271, 14883-14890.

[38] Wolosker, H., Blackshaw, S. & Snyder, S.H. (1999) Serine racemase: a glial enzyme synthesizing D-serine to regulate glutamate-N-methyl-D-aspartate neurotransmission. *Proc Natl Acad Sci USA, 96*, 13409-13414.

[39] Fukasawa, Y., Segawa, H., Kim, J.Y., Chairoungdua, A., Kim, D.K., Matsuo, H., Cha, S.H., Endou, H. & Kanai, Y. (2000) Identification and characterization of a Na(+)-independent neutral amino acid transporter that associates with the 4F2 heavy chain and exhibits substrate selectivity for small neutral D- and L-amino acids. *J Biol Chem*, *275*, 9690-9698.

[40] Welsh, J., Chada, K., Dalal, S.S., Cheng, R., Ralph, D. & McClelland, M. Arbitrarily primed PCR fingerprinting of RNA. (1992) *Nucleic Acids Res*, *20,* 4965-4970.

[41] Foley, K.P., Leonard, M.W. & Engel, J.D. (1993) Quantitation of RNA using the polymerase chain reaction. *Trends Genet, 9*, 380–385.

[42] Lombardo J. and Brown G. B. (1996) A quantitative and specific method for measuring transcript levels of highly homologous genes. *Nucleic Acids Res, 24*, 4812–4816.

[43] Ugawa, S., Sunouchi, Y., Ueda, T., Takahashi, E., Saishin, Y. & Shimada, S. (2001) Characterization of a mouse colonic system B(0+) amino acid transporter related to amino acid absorption in colon. *Am J Physiol, 281,* G365-370.

[44] Sturgeon, R.D., Fessler, R.G. & Meltzer, H.Y. (1979) Behavioral rating scales for assessing phencyclidine-induced locomotor activity, stereotyped behavior and ataxia in rats. *Eur J Pharmacol*, 59, 169-179

In: D-Amino Acids: ISBN 1-60021-075-9
A New Frontier in Amino Acid and Protein Research, pp.169-172

Chapter 3.2

IMMUNOHISTOCHEMICAL STUDY OF D-SERINE

Reiji Semba*

Institute for Developmental Research, Aichi Human Service Center, Japan.

Keywords: D-serine, antibody, affinity chromatography, immunohistochemistry, purification.

1. INTRODUCTION

After it was shown that glutamate and GABA could be demonstrated immunocytochemically [6], many studies have revealed immunohistochemical localization of various amino acids in various tissues. In case of D-serine, only a few laboratories could report the immunohistochemical localization of this amino acid. We got antibodies against D-serine in 1995 [10] and, through studies using the antibody, experienced that some special difficulties were present in immunohistochemical studies of this amino acid [7,8,9]. For example, the antibody was not easily recovered from D-serine affinity column by an acidic solution, the antibody recovered from the column was often still cross reactive with L-serine, and the D-serine antibody thus purified was not stable for days at 4 centigrade Celsius. Methods to overcome these difficulties are described here. For immunohistochemical studies of D-serine, preparation of tissue sections is also important [7].

2. MATERIALS

1. D-serine and other amino acids for tests
2. rabbit serum albumin (Calzyme laboratories, California)
3. glutaraldehyde, paraformaldehyde, ammonium sulfate, sodium borohydride
4. dialysis tubing, nitrocellulose membrane
5. Freund's complete adjuvant, Freund's incomplete adjuvant (Difco)
6. epoxyactivated Sepharose gel, prepacked gelfitration column
7. ActiSep Elution Medium (Sterogene Bioseparations Inc. California)
8. centrifuge

* Correspondence concerning this article should be addressed to Reiji Semba, E-mail: semba@inst-hsc.jp, FAX: +81-568-88-0829.

9. spectrophotometer

3. Methods

3.1. Preparation of Antigen (D-serine-glutaraldehyde-albumin conjugate) for Immunization

D-serine was conjugated to rabbit serum albumin via glutaraldehyde, which has two aldehyde groups in a molecule. In case of D- or L-serine, the chemically labile bond (Schiff base) between serine and glutaraldehyde was stabilized by reducing the bond by sodium borohydride, because we experienced amount of the antibody available from a rabbit was very small without the reduction.

3.1.1: Dissolve 10mg of D-serine in 1mL of 100mM sodium phosphate buffer pH 7.4 in a test tube.
3.1.2: Add 10mg of rabbit serum albumin to the D-serine solution and mix.
3.1.3: Add 25 microlitter of 25% glutaraldehyde solution to the tube, mix and leave for 2 hours at room temperature.
3.1.4: Measure 5mg of sodium borohydride in another test tube and dissolve it in 0.5mL of distilled water.
3.1.5: Add the fresh sodium borohydride solution into the tube with D-serine, glutaraldehyde, and rabbit serum albumin and leave for 1 hour.
3.1.6: Dialyze extensively against 10mM PBS (pH 7.4) and remove small molecules such as D-serine-glutaraldehyde and D-serine-glutaraldehyde- D-serine conjugates.
3.1.7: Recover the D-serine-glutaraldehyde-albumin conjugate and dilute it to make up 10 ml of PBS solution.
3.1.8: Divide into 9 sample tubes and store them frozen until use.
3.1.9: For the first time immunization, mix about 1.1 mL of the conjugate with an equal volume of Freund's complete adjuvant (Difco) and inject into multiple sites of the back of two rabbits.
3.1.10: For the second time and later immunizations, mix the conjugate with an equal volume of Freund's incomplete adjuvant and inject into the back of the rabbits. Immunizations were repeated monthly and the blood was sampled from the rabbits at 10-14 days after each booster injection.

3.2. Purification of D-Serine Antibody

Immunoglobulin fraction prepared from antisera is a mixture of various antibodies. Some of the antibodies, for example antibodies against glutaraldehyde, disturb the following immunohistochemical studies. So, we usually purify antibodies specific to D-serine by affinity chromatographies.

3.2.1: Making Affinity Columns

For immobilization of D-serine, we usually use epoxy gels (epoxyactivated Sepharose 6B or gels prepared by our hand after the method of Matsumoto [2]) following the instructioins of the manufacturer. The gels have spacer arms and do not adsorb antibodies against glutaraldehyde while they adsorb antibodies against amino acid-glutaraldehyde conjugate. When crossreactivities of the antibody was found in a dot immunobinding assay [1], we make affinity columns of the crossreactive amino acids for absorption of the crossreactive antibodies.

3.2.2: Immunoaffinity Chromatography

Crude immunoglobulin fraction is prepared from antiserum by 40% saturated ammonium sulfate and centrifugation. The IgG was dissolved in the original volume of physiological saline and applied onto the affinity column. After washing, antibody firmly bound to the D-serine column is recovered. D-serine antibody in the first bleeding can be recovered from the affinity column by an acidic elution buffer composed of glycine-HCl buffer pH 2.5 with 1M NaCl. However, from later bleedings, only a small amount of antibodies can be recovered by the acidic elution buffer. In those cases, ActiSep Elution Medium (Sterogene Bioseparations Inc, California) is useful, though separation of antibody from ActiSep by a gel filtration may be necessary.

3.2.3: Checking the Crossreactivities

Specificity of the eluted antibody can be cheked by a dot immunobinding assay [1]. Briefly, a small amount of various amino acid-albumin conjugates (20 microgram albumin/mL) are spotted on a nitrocellulose sheat, and stained by immunohistochemical method using antibodies to be tested.

3.2.4: Further Purification if Necessary

Antibody recovered from the D-serine column was often still reactive to L-serine or other amino acids in dot immunobinding assay in our experience. In those cases, the IgG fraction was repeatedly passed through the corresponding affinity columns to eliminate the crossreactive antibodies [7] and then applied on the D-serine affinity column. The method of making the affinity columns is the same as described in 3.2.1. Immuno-absorptions with overdose of L-serine-albumin or other amino acid-albumin conjugates are also used to eliminate the crossreactive antibodies [4]. For this purpose, amino acid- albumin conjugates made in steps 3.1.1 through 3.1.8 can be used.

3.2.5: Storage of Antibody

Antibodies are generally stable for years at 4 centigrade Celsius. However, in case of D-serine antibody, it loses 50% of its activity in a week at the temperature. For long term storage of the antibody, we are experimentally adding 15mg/mL of bovine serum albumin as a stabilizer and keeping them frozen at -80 centigrade Celsius.

3.3. Tissue preparation

For immunohistochemistry of amino acids, fixatives containing glutaraldehyde is essential [4]. The tissues should be fixed by perfusion fixation, or metabolism of amino acids may occur very quickly [3]. In addition, to stabilize the binding between D-serine and tissue proteins via glutaraldehyde, reduction of the fixed tissues with 0.5% of sodium borohydride or dimethylamine borane solution is advisable [7,8,9]. We experienced staining intensity of free floating sections without the reduction treatment decreased within a few weeks, while the intensity was not different in sections treated with the reducing agents [7].

4. Results

In the brainstem, strong D-serine immunoreactivity was found in a group of axons suggesting presence of D-serinergic neurons [7,9]. In the cerebral cortex, D-serine immunoreactivity was reported in apical dendrites and somata of many neurons [7] and in astrocytes [4,7].

5. Discussion

Localization of D-serine, which may control the function of NMDA receptors, is controversial. We found D-serine immunoreactivity mainly in neurons [7,9], while Dr. Snyder and his group claim that D-serine is localized only in astrocytes. This discrepancy may be partly caused by the difference of animals used for the examinations, because we used adults rats while Dr. Snyder used mainly young rats.

6. Notes

For examinations of the specificity of the D-serine antibody (step 3.2.3) by dot immunobinding assays [1], various amino acids conjugated to rabbit serum albumin were also prepared in similar manner (steps 3.1.1 through 3.1.7) keeping the molar ratio between amino acid and glutaraldehyde at around 3:2.

References

[1] Hawkes R, Niday E & Gordon J. (1982) A dot-immunobinding assay for monoclonal and other antibodies. *Anal Biochem 119*, 142-147.

[2] Matsumoto I, Mizuno Y & Seno N. (1979) Activation of Sepharose with epichlorohydrin and subsequent immobilization of ligand for affinity adsorbent. *J Biochem, 85*, 1091-1098.

[3] Saso M, Ma N, Yoshida S, Semba R & Uji Y. Immunocytochemical localization of glutamate in normal and detached cat retina. *Invest Opthal Visual Sci, 39*, 786-792, 1998.

[4] Schell MJ, Molliver ME & Snyder SH. (1995) D-serine, an endogenous synaptic modulator: Localization to astrocytes and glutamate-stimulated release. *Proc Natl Acad Sci, 92*, 3948-3952.

[5] Snyder SH & Kim PM. (2000) D-amino acids as putative neurotransmitters: Focus on D-serine, *Neurochem Res, 25*, 553-560.

[6] Storm-Mathisen J, Leknes AK, Bore A, Vaaland JL, Edminson P, Haug FMS & Ottersen OP. (1983) First visualization of glutamate and GABA in neurons by immunocytochemistry. *Nature, 301*, 517-520.

[7] Yasuda E, Ma N & Semba R. (2001) Immunohistochemical evidences for localization and production of D-serine in some neurons in the rat brain. *Neurosci Lett, 299*, 162-164.

[8] Yasuda E, Ma N & Semba R. (2001) Immunohistochemical demonstration of L-serine distribution in the rat brain. *Neuroreport, 12*, 1027-1030.

[9] Yasuda E, Ma N & Semba R. (2001) Comparative study of L-serine and D-serine immunoreactivities in the rat brain. *Mie Med J, 51*, 51-54.

[10] Wako K, Ma N, Shiroyama T & Semba R. (1995) Glial uptake of intracerebroventricularly injected D-serine in the rat brain: an immunocytochemical study. *Neurosci Lett, 185*, 171-174.

In: D-Amino Acids: ISBN 1-60021-075-9
A New Frontier in Amino Acid and Protein Research, pp.173-179

Chapter 3.3

IMMUNOHISTOCHEMISTRY FOR D-SERINE IN BRAIN

Michael J. Schell***

Department of Pharmacology, Uniformed Services University of the Health Sciences, USA.

Keywords: antibodies, astrocytes, fixation, glutaraldehyde, immunohistochemistry.

1. INTRODUCTION

Small molecule neurotransmitters are not immunogenic. To make antibodies against amino acids, the amino acids must first be reacted with other molecules and the product of the reaction is then injected into animals for antibody production. Although in principle a wide range of derivatizing reagents can be used, for immunohistochemical applications the most useful derivatives are made with commonly used fixatives, such as formaldehyde, glutaraldehyde or carbodiimide. Bifunctional molecules are ideal, since in one fixation step the molecules of interest are derivatized and cross-linked to the surrounding molecules in the tissue, thereby immobilizing them. Such fixatives have the further advantage of being compatible with electron microscopy, allowing localization at the ultrastructural level.

The key principle in small molecule immunohistochemistry is to recreate in the tissue an epitope resembling what was originally injected into the host animal. The successful production of antibodies against fixed biogenic amines was first reported in the 1970s [1], and the technique was perfected for amino acids during the 1980s [2]. An excellent methods paper describes the principles and techniques used to obtain specificity [3]. Today, good antibodies recognizing the fixed derivatives of most neurotransmitters are available commercially, and almost all of these show stereoselectivity.

To generate stereo-selective antibodies against D-serine [4], I coupled free D-serine to bovine serum albumin and then reduced the double bonds in the conjugate with sodium borohydride. This conjugate was used to immunize rabbits, according to standard methods by a commercial vendor. To further enhance the immunogenicity of the conjugate, I also coated this conjugate onto the surface of colloidal gold particles according to a published protocol [5], to create "artificial bacterial particles." These particles were injected intravenously into rabbits to further boost the immune response. The antiserum was collected after 5 months of immunization, and then purified by adsorbing away contaminating antibodies (negative purification). The production of the antibodies will not be described in greater detail here. The procedure

* Correspondence concerning this article should be addressed to Michael J. Schell, E-mail: mschell@usuhs.mil.

below assumes that the user has purchased a D-serine antibody from a commercial source (such as Abcam, Cambridge UK or Gemacbio, Cenon France) or obtained suitable antisera from a collaborator.

Parts 2.1 and 3.1 of the protocol describe how to negatively purify the antibody and then test it with a dot blot. Parts 2.2 and 3.2 describe how to use the antibody for immunohistochemistry in fixed brain slices.

2. MATERIALS

2.1. For Purifying the Antibody and Testing the Antibody by dot Blot

1. D-Serine and L-serine of the highest stereochemical purity obtainable
2. A collection of possible cross-reacting amino acids (such as glycine, L-alanine, and L-glutamate)
3. The carrier protein (usually bovine serum albumin (BSA) or keyhole limpet hemocyanin (KLH) that was used in the antibody generation step)
4. Glutaraldehyde (50% distilled solution, EM Sciences Hatfield PA)
5. Sodium borohydride ($NaBH_4$)
6. Dialysis tubing
7. CNBr-activated Sepharose beads (Pfizer-Pharmacia)
8. Alkaline phosphatase-coupled anti-rabbit (Jackson Immunoresearch, West Grove, PA)
9. Wash buffer (WB) 50 mM Tris, pH 7.4, 100 mM NaCl, 0.1% Tween-20, 0.05% NaN_3
10. Alkaline Phosphatase substrate kit (BCIP/NBT solution, Fluka)
11. Nitrocellulose

2.2. For Using the Purified D-Serine Antibody for Immunohistochemistry

1. Paraformaldehyde (prilled pellets, EM grade; EM Sciences)
2. Glutaraldehyde, distilled 50% solution, EM grade (EM Sciences, Hatfield, PA)
3. Krebs/Henzleit buffer: 118 mM NaCl, 4.7 mM KCl, 2 mM $CaCl_2$, 1.2 mM $MgSO4$, 1.2 mM KH_2PO_4, 25 mM $NaHCO_3$, and 11 mM glucose. Add the $CaCl_2$ last. Before use, warm to 37°C and bubble with 95% O_2/ 5% CO_2 for > 5 min.
4. Pentobarbitol or other suitable anesthetic
5. Fixative: 5% glutaraldehyde, 0.5% paraformaldehyde, 0.2% $Na_2S_2O_5$ (sodium metabisulfite) in 0.1 M NaPO4, pH 7.4. Notes: glutaraldehyde comes as a 50% stock from Electron Microscopy Sciences; paraformaldehyde is freshly depolymerized by heating and adding 1 drop of 10 M NaOH, and made as a 10% stock. To make 500 mL fixative, dilute glutaraldehyde and paraformaldehyde to 250 mL with 37°C water and add 250 mL of 37°C 0.2 M phosphate buffer containing freshly dissolved $Na_2S_2O_5$. Use within 30 min.
6. Cryoprotectant: 20% glycerol, 1% NaCl, 0.01% thymersol in 50 mM NaPO4, pH 7.4
7. Histochemistry TBS: 1.5% NaCl, 50 mM Tris-HCl, pH 7.4
8. Freezing stage sliding microtome (Leica)
9. Vectastain Elite ABC immunohistochemistry kit (universal species, Vector labs)
10. Diaminobenzidine (10 mg substrate tablets, Sigma)
11. 24-well disposable cell culture dishes
12. Rotating shaker
13. Fine, high-quality paintbrushes
14. Coated microscope slides (1% gelatin/ 0.07% Chrom alum)
15. Ethanol and clearing agent (Histoclear Xylene substitute)

16. DPX mounting media (Fluka)
17. Coverslips

3. METHODS

3.1. Purifying and Testing the Antibody

Most crude antisera do not have the necessary specificity to allow selective localization of endogenous fixed molecules of low abundance, so they must be purified first. However, many positive purification schemes (i.e. where the specific IgG molecules are adsorbed to the epitope, washed, and then eluted) lose or destroy the antibodies with the highest affinity. Therefore, an alternative technique is to "sponge away" the likely contaminating antibodies. This is done by coupling the glutaraldehyde-fixed carrier protein to a solid surface (Sepharose beads) without the amino acid. This protocol assumes that BSA was used, but it should be adjusted to reflect the carrier protein used to make the antibody being purified.

3.1.1. Making Conjugates and Beads for Purification and Specificity Testing

1. In a small glass vial containing a tiny stir bar, dissolve 15 mg BSA in 2.0 mL of 1.5 M Na acetate, pH 7.6.
2. Add 0.5 mL of 0.5 M glutaraldehyde, mix 20 min.
3. Add 0.5 mL of 0.1 M $NaBH_4$ (made fresh), mix 20 min.
4. Dialyze the solution overnight against 3 x 1 l of 0.5 M NaCl, 0.1 M $NaHCO_3$, pH 8.5.
5. The next day, wash about 5 mL (swollen, packed volume) of CNBr-activated Sepharose with dilute acid, according to the manufacturer's instructions, to remove protective groups. Then mix 3 mL of the dialyzed BSA with and equal packed volume of the Sepharose beads. Incubate on a slowly turning wheel overnight at 4°C.
6. The next day, wash the beads and chemically cap (quench) the excess reactive groups as described in the manufacturer's instructions.
7. Store the coupled, washed beads as a 1:1 slurry in TBS containing 0.05% NaN_3 at 4°C until use.
8. To negatively purify the antiserum, dilute 0.5 mL of serum 1:10 in TBS containing 0.05% NaN_3.
9. Mix the diluted antiserum with 2-3 mL of the packed BSA-Sepharose beads, and place on a slowly turning wheel or shaker for one hour at room temperature.
10. Centrifuge the beads, and collect the negatively purified antiserum in the supernatant (now diluted 1:10). Store at 4°C until testing or use.
11. To test the antiserum, make BSA conjugates of D-serine, L-serine, and possible contaminating amino acids, as described in the following steps:
12. In a glass vial containing a small stir bar, dissolve 5 mg of the amino acid in 1.0 mL of 1.5 M Na Acetate, pH 7.6.
13. Add 0.5 mL of 0.5 M glutaraldehyde, stir 10 min.
14. Add 1.0 mL of the 15 mg/mL BSA, stir 10 min.
15. Add 0.5 mL of 0.5% $NaBH_4$, stir 1 hr.
16. Add 150 μL of 1 M Tris-HCl, pH 7.4, stir 1 hr.
17. Add 200 μL 0.1 M triethanolamine pH 7.6.
18. Add NaN_3 to 0.5%.
19. Filter through a 0.45 syringe filter. Store at 4°C, stable for many months.
20. The screening and staining experiments will also require "liquid phase" conjugates of D-serine and L-serine. These are made by performing steps 12-19 above for the desired amino acid, but leaving

out the BSA in step 14 (i.e., just add 1 mL of acetate buffer instead). For serine liquid-phase conjugates, this procedure produces a final solution of 24.8 mM amino acid.

3.1.2. Testing specificity by Dot Blot

1. Using a soft pencil and ruler and wearing gloves, draw a 5 x 10 grid of 1 cm squares onto a few pieces of nitrocellulose.
2. In small tubes, pipet a dilution series for the various BSA-amino acid conjugates you wish to test. Start with a 1 mM solution of conjugate (i.e., 1:24 dilution of what you made using the protocol above), and then do serial 1:10, 1:100, and 1:1000 dilutions of that.
3. Spot 5 µL of each dilution onto a square on the paper and allow to air dry. At the end, you should have a rectangle where the X-axis consists of serial dilutions of each BSA-amino acid conjugates and the Y-axis is the collection of the various amino acid conjugates being tested (see Fig 1).
4. Block the nitrocellulose papers in: 3% BSA in WB >30 min at RT.
5. Incubate the nitrocellulose in primary antibody (diluted between 1:1000 and 1:10000 of the original antiserum in WB containing 0.3% BSA (remember that the purified antiserum has already been diluted 1:10). One might also include in this incubation liquid-phase conjugates of either L-serine or D-serine, at a concentration of 0.2-0.5 mM. These will improve antibody specificity (in the case of the L-serine conjugate) or block the reaction (in the case of the D-serine conjugate). Incubate 2 hrs at room temperature, or overnight in the cold.
6. Wash 4 x 5 min in WB
7. Incubate in secondary antibody (1:2000 dilution of alkaline phosphatase-coupled anti-rabbit) in WB/0.3% BSA. Incubate 1 hr room temperature.
8. Wash 5 x 5 min in WB
9. Develop dot blot in the BCIP/NBT substrate mixture, as described by the manufacturer. This takes between 2 and 30 min. The specificity of the antibody is demonstrated by a strong (high-affinity) reaction product across the D-serine dilution series, and a weak or absent signal against other amino acids. Also, the D-ser signal should be blocked if the primary incubation has been carried out in the D-serine conjugate but not the L-serine conjugate (see Fig. 1 for an example).

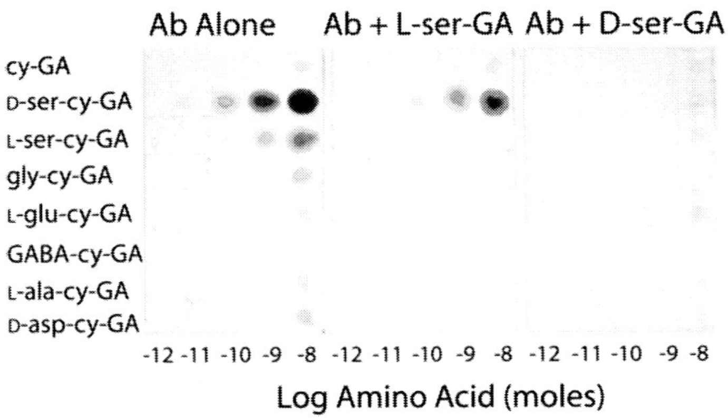

Fig. 1. Dot blot testing of the D-serine antibody. Various amino acid conjugates were serially diluted onto nitrocellulose and then probed with a 1:5000 dilution of purified antiserum. In the middle and right blots, the incubations were carried out in the presence of liquid-phase conjugates of either L-serine (*middle*) or D-serine (*right*). From reference [4]. Copyright 1995 Proceedings of the National Academy of Sciences, USA, used with permission.

3.2. Using the Purified Antibody for Immunohistochemistry

For this technique to work properly with the antibodies I made, the brain tissue must be well fixed with a high-glutaraldehyde-containing fixative, under reducing conditions. If other antibodies are used, the composition of the fixative should be adjusted to match the conditions originally used to generate the antigen. Perfusion through the aorta is a kind of simple surgery, and therefore it takes practice to perfect the technique. Also, maintaining a reducing condition until the blocking step (by including Na metabisulfite) in the buffers will improve selectivity. Because aldehyde fixatives are toxic, the perfusion procedure should be carried out in a fume hood and the researcher(s) performing the perfusion should wear a lab coat, protective gloves, a mask, and goggles.

1. Deeply anesthetize a rat with an I.P. injection of pentobarbitol or other anesthetic.
2. Open the chest cavity, cannulate the aorta, and perfuse at about 30-50 mL/ min with 37°C oxygen-saturated Krebs-Henseleit buffer. Perfuse until liver, eyes, and forepaws have cleared of blood (about 30 sec).
3. Rapidly switch to fixative, without producing air bubbles in the pump. The heart should continue to beat until the moment the bolus of fixative reaches the heart. Hold canula in place while rat is contracting. When animal has stiffened a bit, increase flow rate to about 80 mL/min. Perfuse 450 mL, making sure the cannula stays in aorta. Save the last 50 mL of fixative in a conical tube on ice for post-fixation.
4. After a >20 min. delay, remove brain and/or other organs. Cut tissues to increase penetration of fixative and then place in conical tubes for post-fixation for 2 hours at room temperature.
5. Cryoprotect for 2 days in the cryoprotection buffer at 4°C. Change buffer at least 3 times over the 2 days.
6. Cut sections frozen on a sliding microtome 20-40 μm. Collect in TBS containing 0.2% freshly dissolved sodium metabisulfite.
7. *NOTE: all incubation volumes are 2 mL and are performed at room temp on an orbital shaker set at high speed unless otherwise indicated.*
8. Reduce sections in 0.5% $NaBH_4$, 0.2% Na metabisulfite in TBS for 20 min.
9. Wash 4 x 10 min in TBS, 0.2% Na Metabisulfite.
10. Block / permeabilize 45 min. in TBS containing 4% normal goat serum (NGS), 0.2% TX-100, 0.02% NaN_3.
11. Primary antibody: dilute antibody (1:1000 to 1:5000) in TBS containing 2% NGS, 0.1% Tx-100, 0.02% NaN_3. Incubate 18-24 hours on orbital shaker at 4°C. In some experiments, liquid-phase glutaraldehyde conjugates of amino acids (200-500 μM amino acid) are preincubated with the diluted antibody overnight at 4°C or for 2 hr at room temperature before adding antibody to tissue--see protocol for making liquid-phase conjugates.
12. Day 2: Wash 3 x 10 min in TBS containing 1% NGS.
13. Secondary: Vector biotinylated anti-rabbit 1:200 in TBS 1.5% NGS, 45-60 min at room temperature starting with cold buffer.
14. Mix ABC reagent (Vectastain ABC *elite* kit). About 30 min after starting the secondary incubation, mix 1:50 of reagent A to TBS and vortex then add 1:50 of B and vortex. Make enough for 1.5 mL/ well. Set aside at room temperature.
15. Wash: 2 x 10 min in TBS, 1% NGS.
16. Wash 2 x 10 min in TBS.
17. Incubate in ABC reagent (1.5 mL/well) for 45 min.
18. Wash: 3 x 10 min in TBS.
19. Develop: 0.5 mg/mL DAB, 0.008% $NiCl_2$, 0.01% H_2O_2. Color develops in about 1-7 min.
20. Stop: TBS, 0.05% NaN_3.

21. Mount in TBS onto gelatin-coated slides; air-dry; wash under stream of distilled water and allow them to dry again. Dehydrate in graded alcohols, clear in xylenes, and coverslip in DPX plastic mounting media.

4. Results

Fig. 1 depicts and example dot blot using an antibody specific for D-serine. The incubation done in the presence of the L-serine liquid conjugate reduces the sensitivity of the antibody somewhat, but increases the specificity.

One of the best tests for whether the immunohistochemcal technique for D-serine is working with high sensitivity and specificity is to simply stain a saggital brain slice from adult rat [4,6]. Such an image is shown in Fig. 2. In adult rodents, levels of D-serine are high in the forebrain and very low in the hindbrain, and the staining pattern should reflect that. Another important control is to do the primary incubation in the presence of 0.5 mM of the D-serine liquid-phase conjugate, which should block the staining; an equivalent incubation in the presence of the L-serine liquid-phase conjugate should not block the staining.

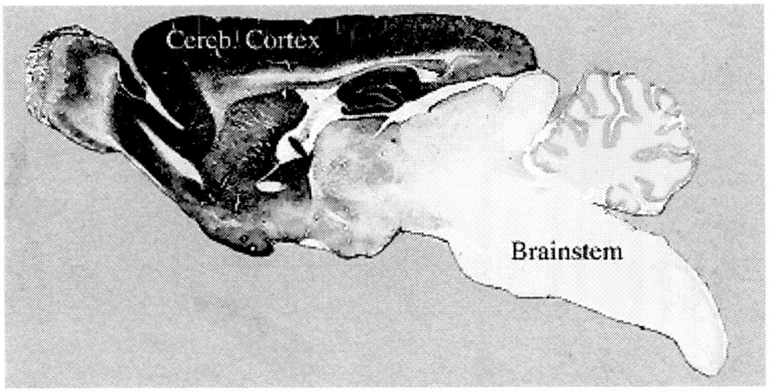

Fig 2. Staining pattern of D-serine in adult rat brain. The presence of strong label in the cerebral cortex but much lower label in the brainstem is an excellent test of antibody specificity. From reference [4]. Copyright 1995 Proceedings of the National Academy of Sciences, USA, used with permission.

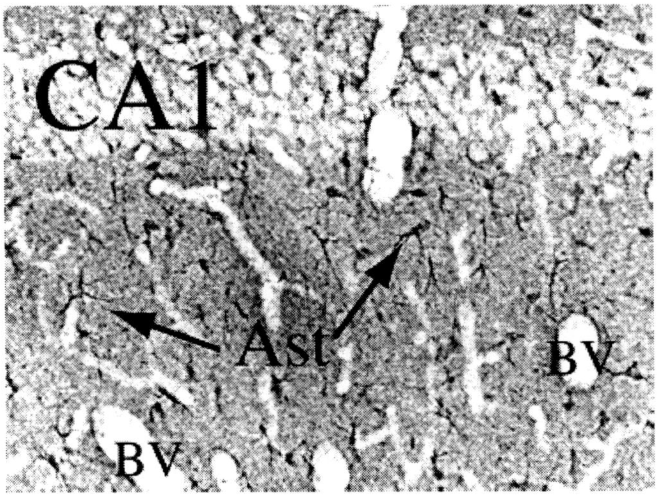

Fig 3. D-Serine at the cellular level. In the CA1 region of hippocampus, D-serine label is strong in gray matter astrocytes (Ast) with small, round cell bodies. Staining tends to be stronger in cells located near blood vessels (BV), and sometimes appears in puncta. D-Serine can also be observed in white matter glial cells (*not shown*). From reference [6]. Copyright 1997 Journal of Neuroscience, used with permission.

Under the microscope, the predominant cells staining for D-serine are astrocytes and oligodendrocytes, which have small cell bodies. An example of these cells in hippocampus is shown in Fig. 3. D-Serine glia have small, round cell bodies and tend to localize near blood vessels. Staining in neurons is much lower and may often be undetected. If one is staining cell cultures, another useful control is to add D-serine to the culture media and allow the cells to take up the amino acid prior to fixation; this should cause an increase in staining.

5. Notes

1. The high levels of glutaraldehyde required for creating the D-serine epitope preclude the use of most D-serine antibodies for immunofluorescence, because glutaraldeyde fixation produces a very high autofluorescence in the tissue. Also, immunofluorescence is a much less sensitive technique than the avidin-biotin peroxidase technique, decreasing the chance of detecting a specific signal. Recently, an antibody against paraformaldehyde-fixed D-serine has been reported [6]. This tool should facilitate double staining and allow for a more more detailed classification of D-serine cells. 2. To my knowledge, the immunohistochemical method for D-serine has only been applied successfully to mammalian brain, retina and kidney. However, because the D-serine antibodies recognize an amino acid conjugate, there is no reason why the technique could not also be applied to other species, such as flies or worms. It is not yet known if these species contain substantial levels of D-serine. 3. The high glutaraldehyde fixative allows for successful staining to be pursued to the ultrastructural level, especially if the immunoperoxidase technique with diaminobenzidine was used for the light level localizations. Our ultrastructural localization of D-serine suggested that D-ser was localized to the cytosol of glia [7]. A recent report suggests that, in some circumstances, D-serine release can occur via secretion from vesicles, which implies that there is a pool of stored D-ser inside some brain cells [8].

References

[1] Steinbusch, H. W., Verhofstad, A. A., & Joosten, H. W. (1978) *Neuroscience 3*, 811-819

[2] Storm-Mathisen, J., Leknes, A. K., Bore, A. T., Vaaland, J. L., Edminson, P., Haug, F. M., & Ottersen, O. P. (1983) *Nature 301*, 517-520

[3] Ottersen, O. P., Storm-Mathisen, J., Madsen, S., Skumlien, S., & Stromhaug, J. (1986) *Med Biol 64*, 147-158

[4] Schell, M. J., Molliver, M. E., & Snyder, S. H. (1995) *Proc Natl Acad Sci U S A 92*, 3948-3952

[5] Pow, D. V., & Crook, D. K. (1993) *J Neurosci Methods 48*, 51-63

[6] Williams, S. M., Diaz, C. M., Macnab, L. T., Sullivan, R. K., & Pow, D. V. (2006) Glia 53, 401-411

[7] Schell, M. J., Brady, R. O., Jr., Molliver, M. E., & Snyder, S. H. (1997) *J Neurosci 17*, 1604-1615

[8] Mothet, J. P., Pollegioni, L., Ouanounou, G., Martineau, M., Fossier, P., & Baux, G. (2005) *Proc Natl Acad Sci U S A 102*, 5606-5611

In: D-Amino Acids:
A New Frontier in Amino Acid and Protein Research, pp.181-186
ISBN 1-60021-075-9

Chapter 3.4

Brain Slice Preparation for Evaluation of the Pathophysiological Functions of D-Serine

Hiroshi Katsuki* and Akinori Akaike
Department of Pharmacology, Graduate School of Pharmaceutical Sciences, Kyoto University, Japan.

Keywords: cerebral cortex; Excitotoxicity; Ischemia; NMDA receptor.

1. Introduction

Ionotropic receptors for glutamate, the major excitatory neurotransmitter in the brain, include *N*-methyl-D-aspartate (NMDA) subtype and non-NMDA subtype. The unusual properties of NMDA receptor-associated channels such as voltage-dependency and high Ca^{2+} permeability enable this subtype to play an important role in neural activity-dependent processes. However, high levels of intracellular Ca^{2+} concentrations resulting from NMDA receptor over-activation are deleterious to neuronal cells.

A so-called 'glycine site' agonist in addition to glutamate is required for NMDA receptor-associated channels to be activated. D-Serine rather then glycine may be important as an endogenous 'glycine site' ligand in several regions of the central nervous system [1]. Here we describe experimental procedures using acute brain slice preparation, for examination of pathophysiological roles of endogenous D-serine in the cerebral cortex [2]. The advantage of acute brain slices is that they retain intact cellular architecture of brain tissues. Because D-serine is produced by and released from astrocytes and acts on NMDA receptors expressed on neurons, close apposition of astrocytes and neurons as observed in intact brain should be an important determinant of the functions of endogenous D-serine.

* Correspondence concerning this article should be addressed to Hiroshi Katsuki, E-mail: hkatsuki@pharm.kyoto-u.ac.jp, FAX: +81-75-753-4579.

2. Materials

1. Standard bathing solution: 124 mM NaCl, 3 mM KCl, 1.2 mM NaH_2PO_4, 2.4 mM $CaCl_2$, 1.3 mM $MgSO_4$, 26 mM $NaHCO_3$ and 10 mM glucose. Glucose was omitted from the solution in the case of ischemic treatment.
2. Tubes for supplies of 95% O_2 / 5% CO_2 and 95% N_2 / 5% CO_2 gasses.
3. Surgical scissors, forceps, scalpels and razor blades.
4. Beakers (50 mL and 300 mL), a petri dish and a pipette with a large tip diameter (~5-6 mm).
5. Vibrating blade microtome (VT1000S; Leica, Nussloch, Germany).
6. Cyanoacrylate glue and pieces of filter paper.
7. Nylon mesh beds [3]: We use culture plate inserts (Millicell-CM, 30 mm in diameter, Millipore, Bedford, MA, USA) in other experiments in our lab and routinely discard polystylene housing of the inserts. We utilized these housings by sticking nylon mesh to the bottom frame of the housings (see the photograph in Fig. 1).
8. Six-well culture plates (diameter of each well: 35 mm): Sterilization is not required. Small holes for insertion of gas tubes are made on the lid of the plate.
9. Water bath.
10. Phosphate-buffered saline (PBS).

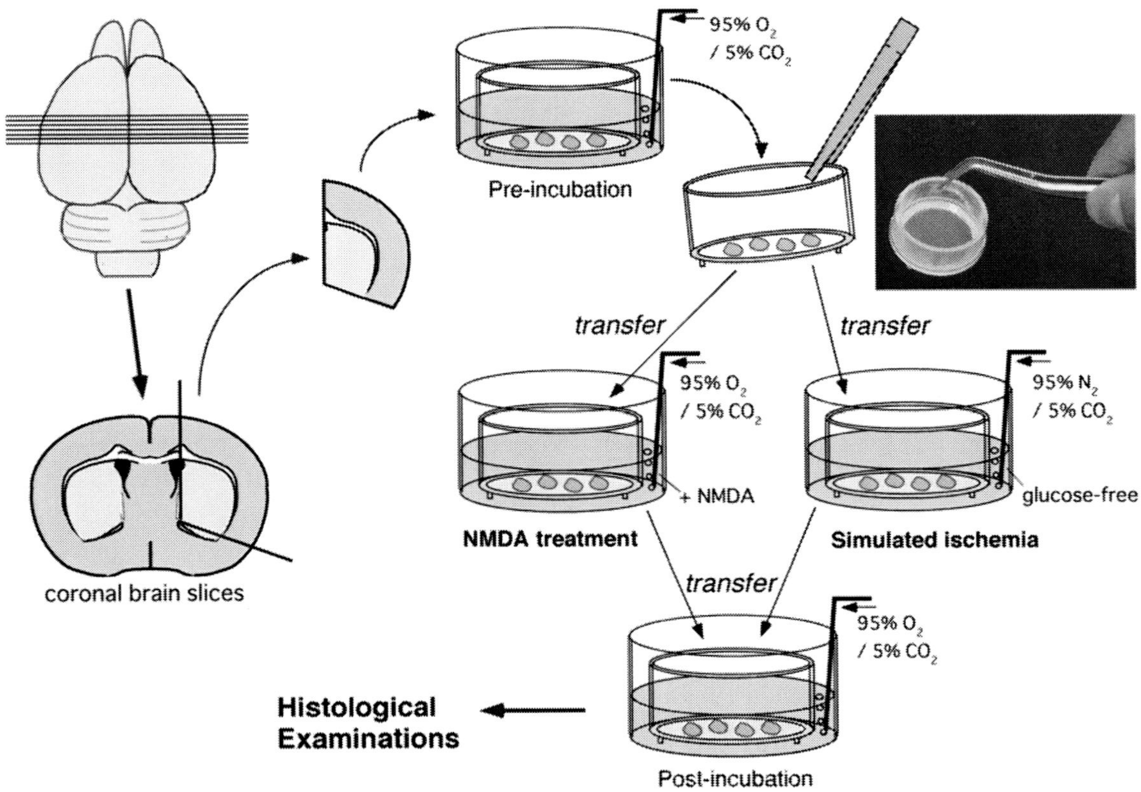

Figure 1. Schematic representation of experimental procedures. Coronal brain slices were prepared with a vibrating microtome. Cortico-striatal region of each slice was dissected out, and transferred into nylon mesh beds in a six-well plate containing the standard bathing solution saturated with 95% O_2 / 5% CO_2 at 34°C. NMDA or ischemic insult was applied to slices by transfer of nylon mesh beds to another six-well plate filled with NMDA-containing bathing solution or with glucose-free standard solution saturated with 95% N_2 / 5% CO_2, respectively. After treatment, slices were transferred into fresh standard bathing solution and incubated for 6 h, then processed for histological examinations. The inset photograph shows a nylon mesh bed (no slices are embedded) picked up by forceps.

11. Paraformaldehyde solution: 4% paraformaldehyde and 4% sucrose in 0.1 M phosphate buffer, pH 7.4.
12. Graded series of ethanol (70, 80, 90, 96 and 100%).
13. Xylene and paraffin.
14. Dry oven (60°C) and paraffin spreading apparatus.
15. Sliding microtome.
16. Silane-coated slide glasses, cover glasses, slide glass holders and staining jars.
17. Toluidine blue solution [4]: Dissolve 5.47 g Na_2HPO_4 and 7.06 g citric acid•H_2O in 500 mL H_2O, adjust pH to 4.0 with NaOH and add 0.5 g toluidine blue.

3. Methods

1. A Wistar rat (10-16 days old) under deep anesthesia was decapitated, the brain was rapidly removed from the skull and immersed into ice-cold standard bathing solution for 1-2 min in a 50 mL beaker. The solution was saturated with 95% O_2 / 5% CO_2.
2. The brain was moved onto a petri dish and cut into two pieces between the cerebral cortex and the cerebellum. After removal of excess moisture on the cut surface by a piece of filter paper, the cerebral part was fixed by cyanoacrylate glue onto a platform of a vibrating microtome, with the cut surface attached to the platform. Then, coronal brain slices of 400 μm thickness were prepared in the ice-cold standard bathing solution. Slices containing the parietal cortex and the striatum were collected, after other brain structures were removed with a scalpel (Fig. 1). Slices were picked up by a pipette, transferred to a 300 mL beaker filled with the standard bathing solution saturated with 95% O_2 / 5% CO_2 and were maintained at 4°C for 1-2 h.
3. Slices were transferred into nylon mesh beds (four slices per each bed) that were placed in a six-well plate, each well containing 6 mL of the standard bathing solution. Bathing solution was continuously bubbled with 95% O_2 / 5% CO_2 and maintained at 34°C by a water bath. After pre-incubation for 1 h, exposure of slices to NMDA was achieved by transfer of nylon mesh beds to another six-well plate filled with 6 mL of the bathing solution containing NMDA. After 30 min of treatment, slices were transferred into yet another plate filled with the standard bathing solution and incubated for further 6 h (Fig. 1).

3'. In the case of simulated ischemia, slices preincubated for 1 h in the standard oxygenated bathing solution were transferred into glucose-free bathing solution saturated with 95% N_2 / 5% CO_2 at 34°C. After oxygen / glucose deprivation for 20 min, slices were transferred into fresh, glucose-containing standard bathing solution bubbled with 95% O_2 / 5% CO_2 and incubated for further 6 h.

4. Slices were fixed with paraformaldehyde solution overnight at 4°C. After wash with PBS, slices were dehydrated with graded series of ethanol (70, 80, 90 and 96% ethanol, each for 20 min, and 100% ethanol overnight) and treated with xylene (three times for 30 min). Then, slices were immersed in xylene-paraffin for 1-2 h at room temperature and in paraffin at 60°C (three times for 30 min), and embedded in paraffin that was gradually cooled to room temperature.
5. Paraffin-embedded slices were cut into sections of 5 μm thickness with a sliding microtome. Cut sections were stretched on the surface of water, placed on silane-coated slide glasses and dried at 42°C overnight with a spreading apparatus.
6. Sections were immersed in xylene (twice for 3-5 min), in 100% ethanol (twice for 1 min) and in 70% ethanol for 3 min at room temperature. Then they were washed with water two or three times, and immersed into toluidine blue solution for 17 min (Nissl staining).

7. Sections were washed thoroughly with water, dehydrated with graded series of ethanol (70, 96 and 100%, each for 2 min), coverslipped and examined by bright-field microscopy.

4. RESULTS

The majority of cells inside cerebrocortical slices maintained in an oxygenated bathing solution displayed large round or oval cell bodies, which were judged as viable neurons (Fig. 2A). NMDA (10-300 μM) and simulated ischemia (20-30 min) caused a marked decrease in the number of viable cortical neurons.

We found a significant neuroprotective effect of 5,7-dichlorokynurenic acid (DCKA), a competitive antagonist acting at the glycine site of NMDA receptor complex, against NMDA cytotoxicity. DCKA was also effective in protecting cortical neurons from ischemic injury. Addition of glycine (10-30 μM) or D-serine (100-300 μM) in the bathing solution did not provide significant influences on NMDA- or ischemia-induced neuronal death. On the other hand, when D-amino acid oxidase (1-30 U/L; ICN Biomedicals, Irvine, CA, USA) was applied 1 h before and during 30 min treatment of cortical slices with 30 μM NMDA, the enzyme showed a marked protective effect in a concentration-dependent manner (Fig. 2B, C). D-Amino acid oxidase also protected cortical neurons from damage induced by 20 min of simulated ischemia (Fig. 2D).

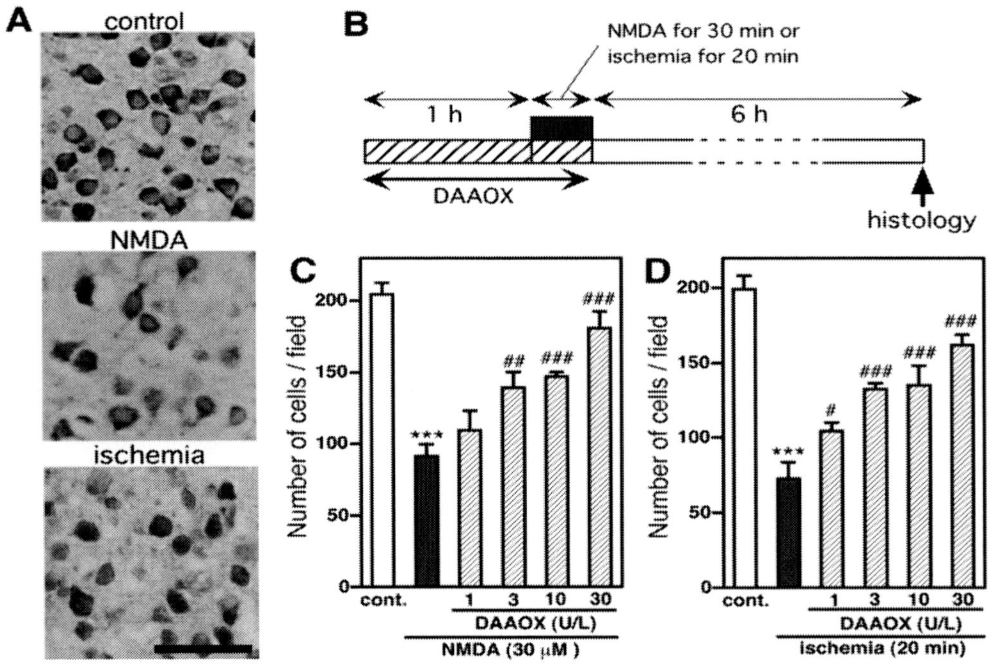

Figure 2. Neuronal damage induced by NMDA or simulated ischemia. A, Nissl-stained sections obtained from rat cerebrocortical slices. From top to bottom, a slice incubated under control conditions, a slice treated with 30 μM NMDA for 30 min, and a slice treated with oxygen/glucose-free solution for 20 min. Bar = 50 μm. B, The experimental schedule for evaluation of the effects of D-amino acid oxidase (DAAOX). Slices were treated with 30 μM NMDA for 30 min or with oxygen/glucose-free solution for 20 min, then incubated in oxygenated standard solution for further 6 h. DAAOX was applied 1 h before and during NMDA or ischemic treatment. C. Effect of DAAOX on NMDA-induced neuronal damage. *** $P < 0.001$ vs. control; ## $P < 0.01$, ### $P < 0.001$ vs. NMDA alone. D, Effect of DAAOX on ischemia-induced neuronal damage. *** $P < 0.01$ vs. control; # $P < 0.05$, ### $P < 0.001$ vs. ischemia alone. Data are reproduced from ref. 2, with permission from the American Society for Pharmacology and Experimental Therapeutics.

We provide two main conclusions. First, endogenous supply of glycine site agonist(s) is both necessary and sufficient for over-activation of NMDA receptors in the cerebral cortex. Second, D-serine plays an important role as an endogenous 'glycine site' agonist in determining excitotoxic consequences in the cerebral cortex.

5. Notes

1. Application of acute slice preparation for assessment of cell viability has been reported in slices obtained from various brain regions including hippocampus [5], midbrain substantia nigra [6] and spinal cord [7] as well as isolated retinal tissue [3]. Immunohistochemistry can be used for identification of specific phenotypes of neurons, such as midbrain dopaminergic neurons [6].
2. Data on cell number were obtained from the layer III of the forelimb / hindlimb area, the most dorsomedial parts of the parietal cortex [8], because this region contained a high and homogenous density of neurons.
3. Changes in morphology of cell nuclei can be examined by staining with fluorescent nuclear dyes [7]. Microtome sections of 5 μm thickness were treated with 2 μg/mL Hoechst 33342 in PBS at room temperature, washed three times in PBS, and viewed under fluorescence microscope with a DAPI filter set. Nuclear condensation and fragmentation were clearly observed in slices treated with 30 μM NMDA. Changes in nuclear morphology such as shrinkage and distorted appearance were evident in slices that received 20 min ischemia [2].
4. The activity of lactate dehydrogenase (LDH), which is released from damaged cells, is frequently used for determination of cell viability. However, we could not obtain reliable results by measuring LDH activity in the bathing solution. This may be attributable to the fact that LDH release is a slow event relative to morphological changes of cells. In this context, an improved method for LDH release assay on isolated tissue preparation has been reported [9].

References

[1] Hashimoto, A. & Oka, T. (1997). Free D-aspartate and D-serine in the mammalian brain and periphery. *Prog Neurobiol, 52*, 325-353.

[2] Katsuki, H., Nonaka, M., Shirakawa, H., Kume, T. & Akaike, A. (2004). Endogenous D-serine is involved in induction of neuronal death by *N*-methyl-D-aspartate and simulated ischemia in rat cerebrocortical slices. *J Pharmacol Exp Ther, 311*, 836-844.

[3] Izumi, Y., Benz, A.M., Kurby, C.O., Labruyere, J., Zorumski, C.F., Price, M.T. & Olney, J.W. (1995). An ex vivo rat retinal preparation for excitotoxicity studies. *J Neurosci Methods, 60*, 219-225.

[4] Noraberg, J., Kristensen, B.W. & Zimmer, J. (1999). Markers for neuronal degeneration in organotypic slice cultures. *Brain Res Protoc, 3*, 278-290.

[5] Izumi, Y., Ishii, K., Katsuki, H., Benz, A.M. & Zorumski, C.F. (1998). β-Hydroxybutyrate fuels synaptic function during development: histological and physiological evidence in rat hippocampal slices. *J Clin Invest, 101*, 1121-1132.

[6] Bywood, P.T. & Johnson, S.M. (2003) Mitochondrial complex inhibitors preferentially damage substantia nigra dopamine neurons in rat brain slices. *Exp Neurol, 179*, 47-59.

[7] Pizzi, M., Benarese, M., Boroni, F., Goffi, F., Valerio, A. & Spano, P.F. (2000). Neuroprotection by metabotropic glutamate receptor agonists on kainate-induced degeneration of motor neurons in spinal cord slices from adult rat. *Neuropharmacology, 39*, 903-910.

[8] Zilles, K. & Wree, A. (1995). Cortex: areal and laminar structure, in G. Paxinos (Ed.), The Rat Nervous System, Second Edition, (pp. 649-685). San Diego, CA: Academic Press.

[9] Izumi, Y., Izumi, M., Benz, A.M. & Zorumski, C.F. (2001). Lactate dehydrogenase release is facilitated by brief sonication of rat hippocampal slices and isolated retinas following acute neuronal damage. *J Neurosci Methods, 108*, 49-55.

In: D-Amino Acids: ISBN 1-60021-075-9
A New Frontier in Amino Acid and Protein Research, pp.187-190

Chapter 3.5

NEPHROTOXICITY OF D-SERINE IN RATS

Ryuichi Konno*
Department of Microbiology, Dokkyo Medical University, Japan.

Keywords: D-serine, nephrotoxicity, rat, D-amino-acid oxidase, glucosuria

1. INTRODUCTION

D-Serine is nephrotoxic in rats. Intraperitoneal injection of d-serine causes necrosis of the cells in the proximal tubules, resulting in glucosuria, aminoaciduria, proteinuria, and polyuria [1,2]. However, the precise mechanism by which D-serine induces nephrotoxicity is not well understood. D-Amino-acid oxidase (EC 1.4.3.3) is suspected to be involved in the D-serine-induced nephrotoxicity because (a) D-serine is reabsorbed in the straight part (pars recta) of the proximal tubules, where cell damage is observed [1-3], (b) D-amino-acid oxidase is localized in this region [4,5], (c) D-serine is a good substrate for this enzyme [6], and (d) one of the products of the D-amino-acid oxidase reaction, hydrogen peroxide, is a cytotoxin.

However, there are perplexing problems: (a) D-serine is not nephrotoxic in mice, guinea pigs, rabbits, dogs, hamsters, or gerbils that have D-amino-acid oxidase activity comparable to rats [7,8]; (b) D-alanine, which is also a good substrate for D-amino-acid oxidase, is not nephrotoxic in rats [8]. These facts cast doubt on the involvement of this enzyme in d-serine-induced nephrotoxicity.

Thus, D-serine-induced nephrotoxicity is a challenging problem. One example of the induction of the kidney damage by D-serine is described here.

2. MATERIALS

1. Male Fischer 344 rats (F344 rats, 6-10 weeks old, 100-250 g).
2. D-Serine (Sigma Chemical Co., St. Louis, MO, USA).
3. Metabolic cages (Model SN-781, Shinano Manufacturing Co., Ltd., Tokyo, Japan, or an equivalent).
4. Glucose assay kit (Wako Pure Chemicals, Osaka, Japan)

* Correspondence concerning this article should be addressed to Ryuichi Konno, E-mail: konno@dokkyomed.ac.jp, FAX: +81-282-86-5616.

3. Methods

1. Male F344 rats are housed in individual metabolic cages. They are allowed free access to the commercial diet and tap water.
2. Urine is collected for 24 h prior to D-serine administration.
3. D-Serine is dissolved in 0.85% NaCl at 80 mg/ml and is injected intraperitoneally into the rats at a dose of 0.8 g/kg.
4. Urine is collected daily at 24-h intervals for a week.
5. The volume of the urine is measured and the glucose content is measured.

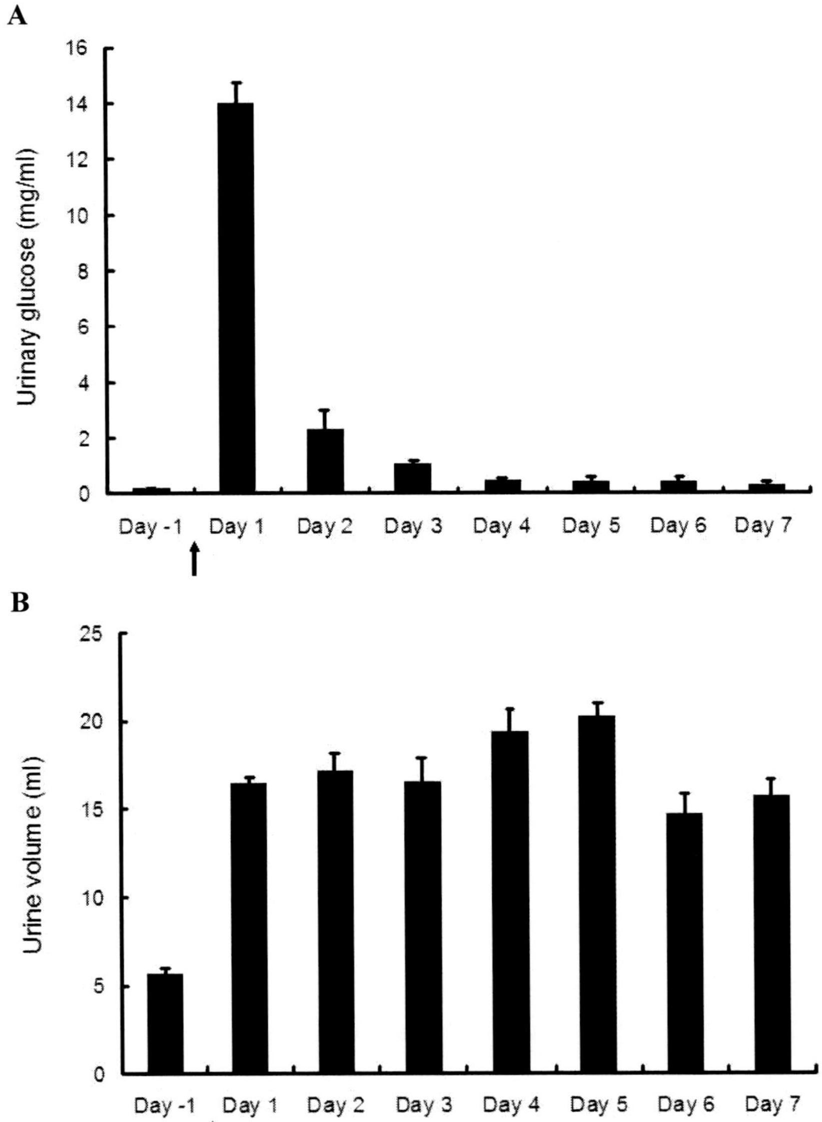

Figure 1. Glucosuria and polyuria induced by the administration of D-serine. D-Serine was injected intraperitoneally into adult male F344 rats at a dose of 0.8 g/kg (arrow). Their urine was collected daily at 24-h intervals for seven days. The glucose content in the urine (A) and the volume of the urine (B) were measured. The values represent the mean ± SEM in five rats. (Reprinted with permission from Maekawa *et al.*, *Chem Res Toxicol*, *18*, 1678-1682, Copyright (2005) American Chemical Society.)

4. RESULTS

Male adult F344 rats excreted approximately 6.0 ml of urine during 24 h (Day -1) and their urine contained approximately 0.2 mg of glucose/ml. Intraperitoneal injection of D-serine into the F344 rats caused glucosuria (Figure 1A). Urine collected for the first 24 h (Day 1) after the D-serine injection contained about seven times more glucose than that collected for the 24 h (Day -1) before the injection. However, the urinary glucose level dropped significantly on Day 2 and returned to the normal level on Day 4.

The high level of glucose in the urine was accompanied by an increase in the urinary volume (Figure 1B). The urinary volume after the D-serine injection was increased about three times on Day 1. In contrast to urinary glucose, the high level of the urine volume continued until Day 7.

5. NOTES

1. Inbred F344 rats are more sensitive to D-serine than Sprague-Dawley rats.
2. D-Serine doses of 0.5-1.0 g/kg are used to induce the obvious damage in the kidney.
3. Amino acid concentration in the urine is determined as follows [9]. Sulfosalicylic acid (20%) is added to the urine at a final concentration of 5% and vortexed. The mixture was centrifuged at 14,000 g for 10 min and the supernatant solution is filtered through a 0.22-μm membrane (Ultra-free C3GV, Nihon Millipore, Tokyo, Japan). The filtrate is applied to an automated amino acid analyzer (Model L-8500A, Hitachi, Tokyo, Japan). The amino acid concentration in the urine is usually expressed on the basis of the creatinine content in the urine. The creatinine content is determined using an assay kit (Wako Pure Chemicals).
4. The protein content in the urine is determined using a Non-Interfering Protein Assay kit (Geno Technology, St. Louis, Mo., USA) [9].
5. For histopathological examination, the kidney is fixed in 10% formalin and embedded in parfifin. Thin tissue sections are cut on a microtome and are stained with hematoxylin and eosin [9].
6. It has become clear that D-amino-acid oxidase is involved in D-serine-induced nephrotoxicity because D-serine injection did not cause glucosuria or polyuria in mutant rats lacking D-amino-acid oxidase [10]. The results that pretreatment of rats with sodium benzoate, a competitive inhibitor of D-amino-acid oxidase, resulted in a dose-dependent reduction of D-serine-induced nephrotoxicity [11] are consistent with this conclusion.

ACKNOWLEDGEMENTS

This work was supported in part by Research Grant from Asahi Glass Foundation.

REFERENCES

[1] Ganote, C. E., Peterson, D. R. and Carone, F. A. (1974). The nature of D-serine-induced nephrotoxicity. *Am J Pathol 77*, 269-282.

[2] Carone, F. A. and Ganote, C. E. (1975). D-Serine nephrotoxicity. *Arch Pathol 99*, 658-662.

[3] Silbernagl, S., Völker, K. and Dantzler, W. H. (1999). D-Serine is reabsorbed in rat renal pars recta. *Am J Physiol 276*, F857-F863.

[4] Usuda, N., Yokota, S., Hashimoto, T. et al. (1986). Immunocytochemical localization of D-amino acid oxidase in the central clear matrix of rat kidney peroxisomes. *J Histochem Cytochem 34*, 1709-1718.

[5] Perotti, M. E., Gavazzi, E., Trussardo, L. et al. (1987). Immunoelectron microscopic localization of D-amino acid oxidase in rat kidney and liver. *Histochem J 19*, 157-169.

[6] Dixon, M. and Kleppe, K. (1965). D-Amino acid oxidase. II. Specificity, competitive inhibition and reaction sequence. *Biochim Biophys Acta 96*, 368-382.

[7] Morehead, R. P., Poe, W. D., Williams, J. O. et al. (1946). The influence of age and species on the nephrotoxic action of DL-serine. *Am J Pathol 22*, 658.

[8] Kaltenbach, J. P., Ganote, C. E. and Carone, F. A. (1979). Renal tubular necrosis induced by compounds structurally related to D-serine. *Exp Mol Pathol 30*, 209-214.

[9] Konno, R., Ikeda, M., Yamaguchi, K. et al. (2000). Nephrotoxicity of D-propargylglycine in mice. *Arch Toxicol 74*, 473-479.

[10] Maekawa, M., Okamura, T., Kasai, N. et al. (2005). D-Amino-acid oxidase is involved in D-serine-induced nephrotoxicity. *Chem Res Toxicol, 18,* 1678-1682.

[11] Williams, R. E. and Lock, E. A. (2005). Sodium benzoate attenuates D-serine induced nephrotoxicity in the rat. *Toxicology 207*, 35-48.

In: D-Amino Acids: ISBN 1-60021-075-9
A New Frontier in Amino Acid and Protein Research, pp.191-201

Chapter 3.6

PREPARATION OF A POLYCLONAL ANTISERUM AGAINST D-ASPARTATE

Jen-Ai Lee[1,*] and Hiroshi Homma[2]

[1]Department of Pharmaceutical Analysis, School of Pharmacy, Taipei Medical University, Taiwan;
[2]Laboratory of Biomolecular Science, School of Pharmaceutical Sciences, Kitasato University, Tokyo 108-8641, Japan.

Keywords: Immunohistochemistry, localization, neuroendocrine, pineal gland, pituitary gland.

1. INTRODUCTION

It had long been thought that only L-forms of amino acids are present in mammalian tissues or body fluids; however, a significant quantity of the free form of D-Asp is present in various mammalian tissues [1-4]. In the rat pineal gland and testes, as much as 30% to 40% of the total Asp is in the D-form [5-8]. Recent investigations have revealed that this unique amino acid is involved in biological activities in these tissues [9-12]. In order to learn more about the physiological roles of D-Asp, we raised a polyclonal antibody against this amino acid. The raising of a high-titer anti– D-Asp antiserum is described below. The possibility of raising antibodies with high affinity for the target antigen is greatly reduced for haptens with a low molecular weight, such as D-Asp, and antibodies to such small-sized haptens can only be generated using the hapten conjugated to a carrier molecule [13]. Thus, a glutaraldehyde conjugate of D-Asp and BSA was prepared as the immunogen. The D-Asp/BSA conjugate (0.5 mg of protein) was emulsified with an equal volume of Freund's Complete Adjuvant for the first immunization and with Freund's Incomplete Adjuvant for the booster injections. The emulsion was injected subcutaneously into 2 sites on the back of a male New Zealand White rabbit. For achieving rapid and massive immune responses in the animals, another aliquot of the conjugate was mixed with muramyl dipeptide, adsorbed onto colloidal gold particles, and then administered simultaneously with each immunization to the rabbits via the marginal vein of the ear following the procedure of Pow and Crook [14]. Immunizations were continued every three weeks for

* Correspondence concerning this article should be addressed to:
Jen-Ai Lee, Department of Pharmaceutical Analysis, School of Pharmacy, Taipei Medical University, Taiwan. E-mail: jenai@tmu.edu.tw, Fax: +886-(2)-27361661 ext. 6125.
Hiroshi Homma, Laboratory of Biomolecular Science, School of Pharmaceutical Sciences, Kitasato University, 5-9-1 Shirokane, Minato-ku, Tokyo 108-8641, Japan. E-mail: hommah@pharm.kitasato-u.ac.jp Fax: +81-3-5791-6381.

24 weeks, and 10 days after the last injection the animal was bled and the antiserum was prepared. D-Asp antibodies were then purified using D-Asp-Sepharose affinity chromatography and a dot-blot assay [6,15]. The prepared antibodies were used to study the localization of D-Asp within various rat tissues, such as the pineal gland [15], adrenal gland [16,17], testis [6], embryonic brain [18], retina [19], pituitary gland [17,19], and cultured mammalian cells [20-22]. Here we describe the results for the pineal gland and pituitary gland.

2. Materials

2.1. Preparation of Immunogen:

2.1.1. Conjugation of D-Asp and BSA via Glutaraldehyde

1. D-Asp and other amino acids (Sigma Chemical Co., St. Louis, MO, USA).
2. Bovine serum albumin (BSA) as a carrier protein (Sigma Chemical Co., St. Louis, MO, USA).
3. 0.2 M sodium phosphate buffer, pH 7.4.
4. 50% glutaraldehyde solution (Sigma Chemical Co., St. Louis, MO, USA).
5. 4 mg/mL $NaBH_4$ solution (Kanto Kagaku, Tokyo, Japan).
6. Seamless cellulose tubing (Union Carbide Corp., Danbury, CT, USA) for dialysis.
7. Phosphate buffered saline (PBS): 8 g NaCl, 0.2 g KCl, 1.15 g Na_2HPO_4, and 0.2 g KH_2PO_4 dissolved in distilled water to 1000 mL.
8. Protein assay kit (Bio-Rad, Hercules, CA, USA).

2.1.2. Antigen and Adjuvant Peptide Co-Adsorbed on Gold Particles

1. Gold colloid: 20 nm, 0.01% as $HAuCl_4$ (Sigma Chemical Co., St. Louis, MO, USA).
2. 3 mM phosphate buffer, pH 5.3.
3. Muramyl dipeptide (Sigma Chemical Co., St. Louis, MO, USA).
4. 10% NaCl solution.
5. 0.9% NaCl solution.

2.2. Production of the Polyclonal Antiserum in Rabbit

2.2.1. Subcutaneous Immunization

1. Freund's Complete Adjuvant and Freund's Incomplete Adjuvant (Difco Laboratories, Detroit, MI, USA).
2. 0.5 mg antigen in PBS.
3. 70% ethanol.
4. 22-G needle.
5. 10-mL syringes.
6. Double-ended locking hub connector.

2.2.2. Bleeding from the Ear Artery

1. 18-G needle.
2. 50-mL plastic centrifuge tube.
3. 70% ethanol.

2.3. Purification of the D-Asp Antibody by Affinity Chromatography

2.3.1. Conjugation of D-Asp to CNBr-Activated Sepharose 4B

1. Cyanogen bromide (CNBr)-activated Sepharose 4B (Amersham Biosciences AB, Uppsala, Sweden).
2. 1 mM HCl solution.
3. Coupling buffer: 0.1 M $NaHCO_3$, 0.5 M NaCl solution, pH 8.3.
4. Acetate buffer: 2.45 g CH_3COONa, 4.75 mL CH_3COOH, and 29 g NaCl dissolved in distilled water to 1000 mL.
5. 25 mg/mL D-Asp aqueous solution.
6. Blocking buffer: 0.1 M Tris-HCl, pH 8.0
7. 0.02% NaN_3/PBS: 135 mM NaCl, 20 mM Na_2HPO_4, 1 mM KH_2PO_4, 2 mM KCl, 0.02% NaN_3, pH 7.4.
8. Column (~10 cm by 1 cm inner diameter).

2.3.2. Purification of D-Asp Antibody

1. TBS buffer: 0.15 M NaCl, 20 mM Tris-HCl, pH 7.5.
2. Washing buffer: 1% Tween 20 in TBS buffer.
3. 0.15 M NaCl solution.
4. Elution buffer: 0.1 M Gly-HCl, pH 2.5.
5. 1 M Tris buffer.

2.4. Dot-Blot Assay

1. Nitrocellulose membrane (Schleicher & Schuell BioScience).
2. Various amino acids conjugated to rabbit serum albumin (RSA) with glutaraldehyde.
3. Block Ace (Dainippon Pharmaceutical Co., Osaka, Japan).
4. 0.2% Tween 20 in PBS.
5. Diluted antibody.
6. Secondary antibody (HRP-conjugated goat anti-rabbit IgG) (DAKO, Denmark).
7. 3,3'-Diaminobenzidine tetrahydrochloride (Dojindo, Kumamoto, Japan).
8. H_2O_2 (Mitsubishi Gas Chemical Co., Tokyo, Japan).

2.5. Immunohistochemistry

1. Sodium pentobarbital (Abbott Laboratories, Chicago, IL, USA).
2. Ringer's solution: 4.55 g NaCl, 0.5 g KCl, 0.6 g HEPES, 0.6 g glucose, 4.0 mL of 0.5 M EGTA, 3.0 mL of 1 M $MgCl_2$, and 0.5 mL of 0.1 mM NaH_2PO_4, dissolved in distilled water to 500 mL, pH 7.4.
3. Heparin (1U/μL).
4. Glutaraldehyde and paraformaldehyde (EM Science, Pennsylvania, PA, USA) .
5. Sodium cacodylate (TAAB Laboratories, Reading, UK).
6. 50% Karnovsky's fixative: 2.5% glutaraldehyde, 2% paraformaldehyde, 0.1 M sodium cacodylate, pH 7.4.
7. 10%, 15%, and 20% sucrose in PBS.
8. OCT Compound (Miles Laboratories, Naperville, IL, USA).
9. Poly (L-lysine)-treated slides (Matsunami Glass Ind., Japan).

10. 0.5% $NaBH_4$ solution in PBS.
11. 10% calf serum, 0.1% NaN_3/PBS.
12. Anti– D-Asp antibody.
13. HRP-conjugated anti-rabbit IgG antibody (DAKO, Denmark).
14. Peroxidase-anti-peroxidase (PAP) complex (DAKO, Denmark).

3. Methods

3.1. Preparation of Immunogen:

3.1.1. Conjugation of D-Asp and BSA via Glutaraldehyde

1. Dissolve 100 μmol D-Asp and 12 mg BSA in 2 mL of 0.2 M sodium phosphate buffer, pH 7.4, in test tube and add 100 μmol glutaraldehyde solution (50%, 20 μL).
2. Incubate for 20 hrs at 20°C.
3. Add 0.1 mL of 4 mg/mL $NaBH_4$ solution. (Note 1)
4. Dialyze the conjugate for 48 hrs at 4°C against 2 liters of 0.1 M PBS, pH 7.4.
5. Determine the protein content with a Bio-Rad protein assay kit.
6. Divide the conjugate into aliquots and store at –20°C. Use 1.1 mg of total protein for each immunization.

3.1.2. Antigen and adjuvant peptide co-adsorbed on gold particles

1. Dialyze the antigen and 20 nm gold colloid separately overnight at 4°C against 2 L of 3 mM phosphate buffer, pH 5.3. (Note 2)
2. Add 20 μg muramyl dipeptide and 0.6 mg D-Asp-glutaraldehyde-BSA conjugate to 20 mL of gold colloid, then mix.
3. After 2 min, add 10% NaCl solution to a final concentration of 1%.
4. Centrifuge at 10,000 to 100,000 × *g* for 1.5 hrs at 4°C, then decant the supernatant.
5. Resuspend the residue in 1 mL of 0.9% NaCl solution.

3.2. Production of the Polyclonal Antiserum in Rabbit

3.2.1. Subcutaneous Immunization

1. Add an equal volume of Complete Freund's Adjuvant to 0.5 mg antigen. Emulsify the mixture with two 10-mL syringes that are connected with the double-ended locking hub connector. (Note 3).
2. Check the emulsion. (Note 4).
3. Attach a 22-G needle to one of the syringes and remove any air bubbles.
4. Place a rabbit into a basket and grasp the scruff of the neck with one hand and place the other hand over the rabbit's hindquarters.
5. Clean the areas to be injected with 70% ethanol, and inject 0.5 mL of the mixture into each site on the back of the rabbit.
6. Administer booster doses to the rabbit every three weeks with 0.5 mg antigen emulsified in an equal volume of Freund's Incomplete Adjuvant by following the above procedure.

3.2.2. Intravenous Immunization

1. Place the rabbit into a rabbit restrainer with its ears pulled out. Make sure that the rabbit is tightly restrained and cannot move freely.
2. Gently tap one of the ears with your hand and clean it with 70% ethanol.
3. Inject the prepared gold particle solution intravenously with 25G needle.

3.2.3. Bleeding from the Ear Artery

1. Place the rabbit into the rabbit restrainer with its ears pulled out. Make sure that the rabbit is tightly restrained and cannot move freely.
2. Gently tap one of the ears with your hand and clean it with 70% ethanol.
3. Insert an 18-G needle (without a syringe) into the arterial vessel, then push the needle 5 to 10 mm along inside the vessel.
4. Collect blood in a 50-mL plastic centrifuge tube.
5. Stop the blood flow. (Note 5)

3.2.4. Preparation of Antiserum from Whole Blood

1. Incubate the collected blood at 37°C for 40 min.
2. Scrape off the clot from the wall of centrifuge tube with a glass stirring rod.
3. Store the tube at 4°C overnight to shrink the clot.
4. Centrifuge at 3,000 rpm for 10 min at 4°C.
5. Store at –80°C before purification.

3.3. Purification of the D-Asp Antibody by Affinity Chromatography

3.3.1. Conjugation of D-Asp to CNBr-Activated Sepharose 4B

1. Gently add 3.5 g CNBr-activated Sepharose 4B to 30 mL of 1 mM HCl and stir slowly for 15 min.
2. Wash the Sepharose with 300 mL of 1 mM HCl on a glass filter.
3. Wash the Sepharose with coupling buffer.
4. Suspend the Sepharose in coupling buffer and add 2 mL of 25 mg/mL D-Asp solution.
5. Stir slowly for 1-2 hr at room temperature. (Note 6)
6. Wash with 5 volumes of coupling buffer.
7. Add 0.1 M Tris-HCl, pH 8.0, for blocking, and leave the suspension at room temperature for 2 hr.
8. Wash 3 times with 5 volumes each of acetate buffer and with coupling buffer.
9. Pack the Sepharose into a column (column size: approx. 10 cm x 1cm i.d.).
10. Wash the column with 10 times volume of 0.02% NaN_3/PBS.
11. Store at 4°C.

3.3.2. Purification of the D-Asp Antibody

1. Load 1 mL of the antiserum onto the affinity column (6-7 mL Sepharose).
2. Mix well.
3. Store at 4°C overnight.
4. Wash with 10 mL of TBS buffer.
5. Wash with 30 mL of washing buffer.
6. Wash with 30 mL of TBS buffer.
7. Wash with 10 mL of 0.15 M NaCl.
8. Add 2.5 mL of elution buffer.
9. Elute with elution buffer, monitoring the eluate at 254 nm with a UV monitor, and collecting fractions.

10. Neutralize the eluate with 1 M Tris buffer immediately.
11. Wash the column consecutively with elution buffer, distilled water, and 0.02% NaN_3/PBS buffer. Store the column at 4°C.
12. Before using the antiserum thus purified, complete the following procedure: Prepare liquid-phase conjugates of glutaraldehyde and L-Asp or D,L-Glu following the same procedure used to conjugate D-Asp to BSA. In order to minimize cross-reactivity against L-Asp and D,L-Glu, add the liquid-phase conjugates (0.5 mM each) to the antibody.

3.4. Dot-blot Assay

1. Conjugate various amino acids, for example D-serine (D-Ser), L-alanine (L-Ala), L-aspartate (L-Asp), L-glutamate (L-Glu), D-glutamate (D-Glu), γ-aminobutyrate (GABA), glycine (Gly), β-alanine (β-Ala) and taurine (Tau), to rabbit serum albumin (RSA) with glutaraldehyde as described above.
2. Spot serial dilutions of these conjugates onto a nitrocellulose membrane.
3. Wash the membrane twice with 0.2% Tween 20 in PBS for 5 min.
4. Block overnight with Block Ace (0.5 mL/cm^2 membrane) at 4°C.
5. Wash the membrane twice with 0.2% Tween 20 in PBS for 5 min.
6. Add D-Asp antibody that is appropriately diluted (0.1 mL/cm^2 membrane)
7. Incubate for 1 hr at 25°C.
8. Wash the membrane three times with 0.2% Tween 20 in PBS for 5 min.
9. Add secondary antibody (diluted 1:1000) (0.1 mL/cm^2 membrane).
10. Incubate for 1 hr at 25°C.
11. Wash the membrane four times with 0.2% Tween 20 in PBS for 5 min.
12. Visualize with 0.03% 3,3'-diaminobenzidine tetrahydrochloride and 0.015% H_2O_2.

3.5. Immunohistochemistry

1. Anesthetize rats by intraperitoneal injection of sodium pentobarbital (50 mg/kg).
2. Inject heparin solution (1 μL/g body weight) into the left cardiac ventricle.
3. Perfuse through the left ventricle of the heart for 1–2 min with Ringer's solution (pH 7.4) at 37°C to remove blood from the organs. (Note 7)
4. Perfuse immediately with 200 mL of 50% Karnovsky's fixative at a flow rate of 10 mL/min for 20 min at 37°C.
5. Isolate the tissues and postfix them overnight at 4°C in the same fixative.
6. Cryoprotect in successive solutions of 10%, 15%, and 20% sucrose in PBS for 4 hr each.
7. Freeze in embedding medium (OCT Compound) and cut at 10 μm on a cryostat.
8. Mount cryostat tissue sections on poly (L-lysine)-treated slides and air-dry for 30 min.
9. Wash the tissue sections with PBS three times for 5 min each.
10. Pretreat the sections for 20 min with 0.5% $NaBH_4$ in PBS to inactivate the glutaraldehyde.
11. Block with 10% calf serum, 0.1% NaN_3/PBS for 30 min.
12. Incubate overnight at 25°C with the anti–D-Asp antibody (at a dilution of 1:100 to 1:1000 in PBS containing 10% calf serum).
13. Wash the tissue sections three times with TS [50 mM Tris, 150 mM NaCl, pH 7.6] for 5 min.
14. Add diluted secondary antibody (1:50 in TS) for 2 hr.
15. Add the PAP complex. Here, the immunoreactivity was visualized by the peroxidase-anti-peroxidase method using the PAP complex (DAKO). To determine the immunohistochemical

specificity, a liquid-phase conjugate of glutaraldehyde and D-Asp (1 mM) was incubated overnight with diluted antibody (final dilution 1:15) before incubation with the brain sections.

4. Results

Figure 1 shows the specificity of the anti–D-Asp antiserum. Intense immunoreactivity was observed only with the D-Asp conjugate (Fig. 1a), and this immunoreactivity was abolished by preabsorption with a liquid-phase conjugate of glutaraldehyde and D-Asp (Fig. 1b) [6,15]. Immunohistochemical localization of D-Asp in a variety of rat tissues and cultured mammalian cells was carried out as previously described [3,4,6,15-22]. In Fig. 2, immunohistochemistry of rat pineal gland is depicted. The immunoreactivity was more evident in the distal region than in the proximal region of the gland (Fig. 2) [15]. Pinealocytes, the parenchymal cells of the gland in the distal region, are presumably involved in the synthesis and secretion of the pineal hormone melatonin. Therefore, the results of staining may indicate that D-Asp plays a role in the regulation of melatonin secretion. D-Asp is known to suppress norepinephrine-induced melatonin secretion in primary cultured rat pinealocytes [9].

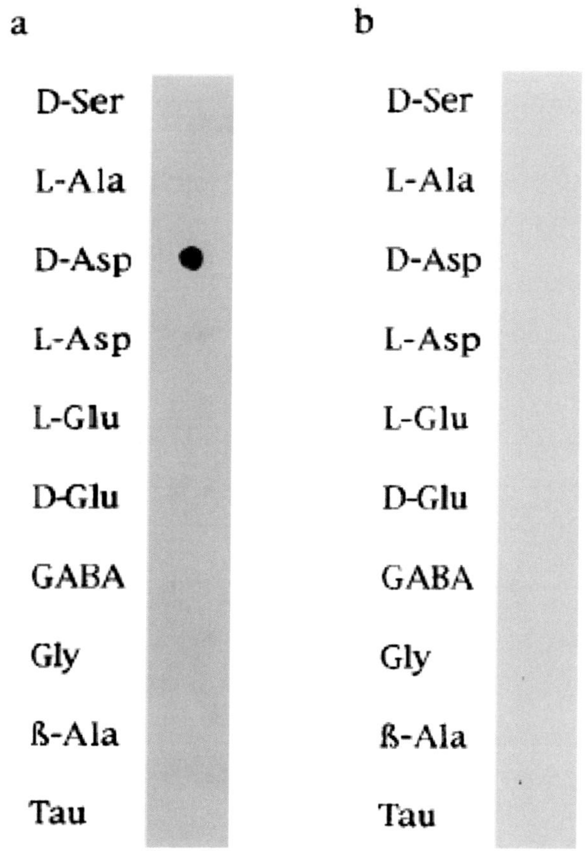

Figure 1. Results of dot-blot assays for determining the specificity of the antibody against D-Asp. Glutaraldehyde conjugates of rabbit serum albumin and various amino acids were spotted onto a nitrocellulose membrane, which was then incubated with the affinity-purified anti–D-Asp antibody at a final dilution of 1:500. The D-Asp antibody was preincubated with (a) liquid-phase conjugates of glutaraldehyde and L-Asp or D,L-Glu (0.5 mM each) or (b) with conjugates of the same amino acids plus D-Asp (1 mM). Reprinted from (6) with permission.

Figure 3 demonstrates the results in the pituitary gland of 6-week-old male rats. In this gland, D-Asp concentration is approximately 2%-4% of total Asp [7,8,23]. The gland consists of three regions: the anterior, intermediate, and posterior lobes, and D-Asp immunoreactivity was evident both in the anterior and posterior lobes but was not observed in the intermediate lobe (Fig. 3a). In the posterior lobe, immunoreactivity was spread homogeneously over an area presumably corresponding to the neuronal axons, while intense immunoreactivity was scattered throughout the anterior lobe (Fig. 3a). In the anterior lobe, at least six different types of cells are represented. Figures 3b-d are sections double stained with antibodies against D-Asp and prolactin, and the D-Asp–positive cells were found to be closely related to the prolactin-containing cells. In this case, we used a mouse monoclonal anti-rat prolactin antibody in addition to the rabbit polyclonal anti–D-Asp antibody. Glutaraldehyde in the fixative is necessary for immunostaining with the anti– D-Asp antibody, whereas it is deleterious for staining with the monoclonal anti-prolactin antibody. Therefore, the concentration of glutaraldehyde was reduced from 2.5% to 0.5% in the fixative for double staining with these antibodies, although this still resulted in weak immunostaining with the anti-prolactin antibody. Figures 3b-d demonstrate that D-Asp-positive cells (Fig. 3b, green in original figure in ref. 19) could be superimposed on prolactin-positive cells (Fig. 3c, red in original figure in ref. 19), as shown by the yellow color in Fig. 3d (yellow in original figure in ref.19). Cells can be seen that are D-Asp–positive and prolactin–negative in Fig. 3d (green in original figure in ref.19), presumably because staining with the monoclonal anti-prolactin antibody was weakened by the glutaraldehyde fixative. These results indicate that D-Asp–positive cells in the anterior lobe of the rat pituitary gland are prolactin-producing cells or closely related to these cells. Based on these results, we subsequently demonstrated that D-Asp increases thyrotropin releasing hormone (TRH)–induced prolactin secretion in GH3 cells, which are prolactin-secreting cultured cells [21].

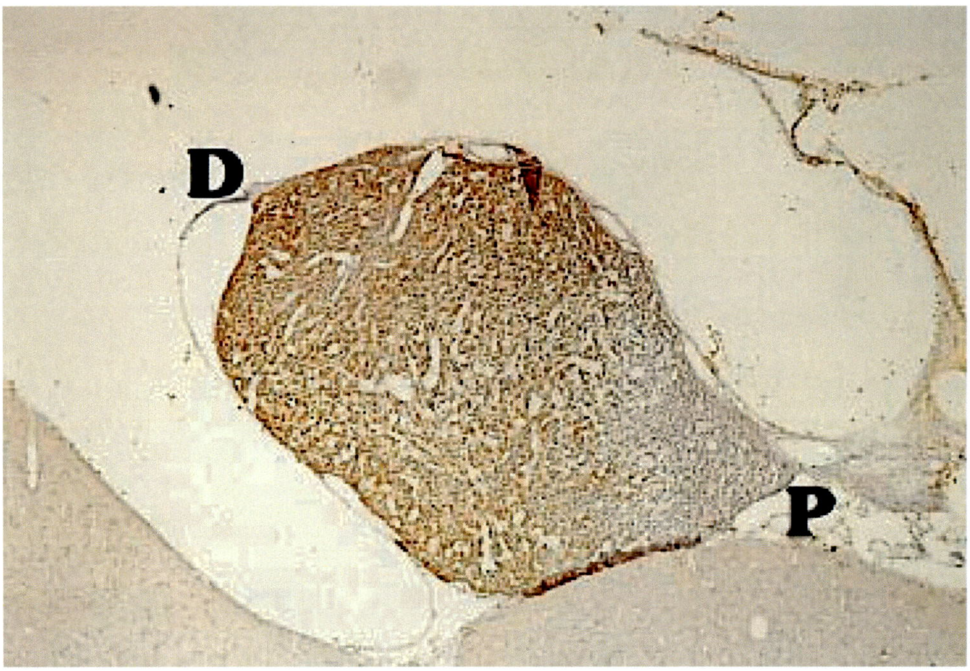

Figure 2. Immunohistochemical staining of the rat pineal gland. The pineal gland was prepared from an 8-week-old rat and incubated with the anti– D-Asp antibody. D and P represent the distal and proximal regions of the gland, respectively. Reprinted from (15) with permission.

Immunohistochemical studies with anti–D-Asp antiserum demonstrated that D-Asp was localized in specific types of cells in these endocrine tissues at distinct periods of development. Based on the results obtained, we have demonstrated D-Asp has biological activity in neuroendocrine and endocrine tissues such as the pineal gland, pituitary gland (see above), and testis [11,12].

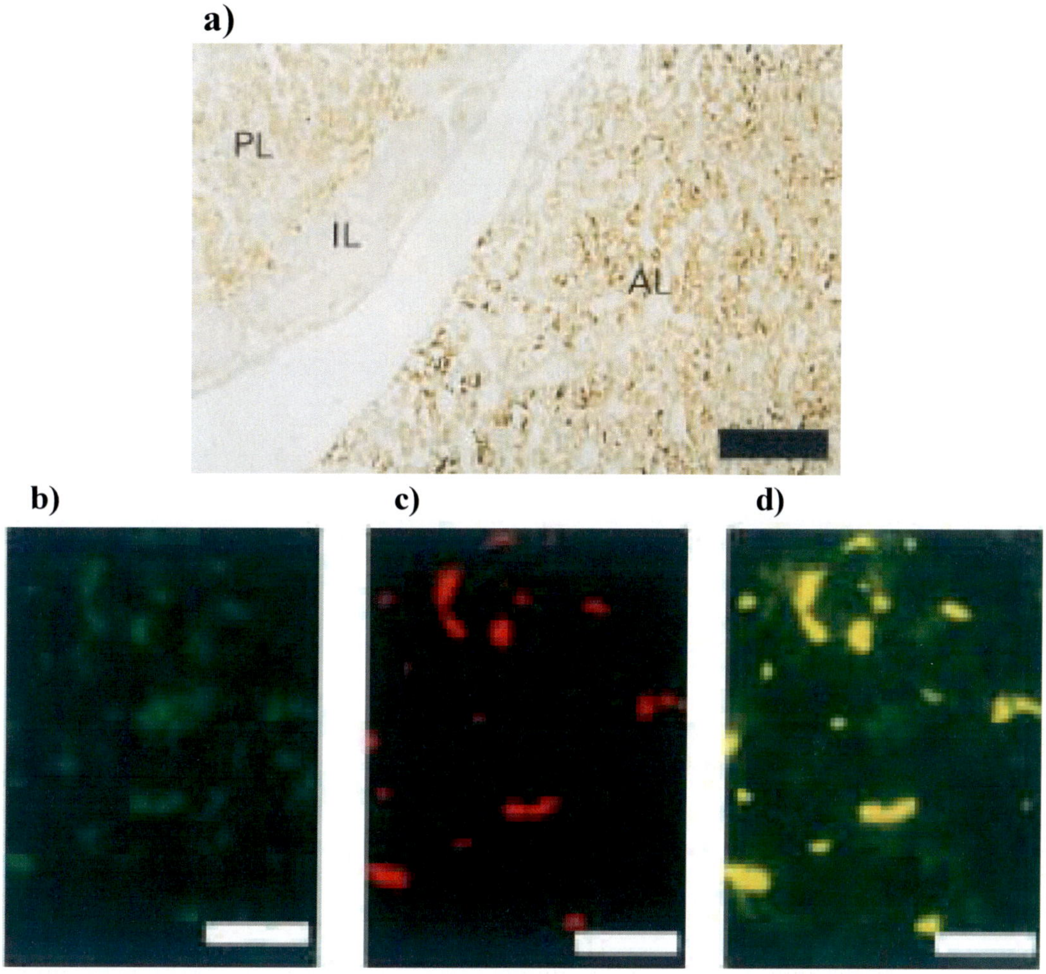

Figure 3. Immunohistochemical localization of D-Asp in the rat pituitary gland. (a) Pituitary gland of a 6-week-old rat. AL, anterior lobe; IL, intermediate lobe; PL, posterior lobe. The immunoreactivity was evident in the AL and PL, but not in the IL. Anti–D-Asp antiserum was diluted 1:300. (b–d) Anterior lobe of the pituitary gland (6-week-old rat) was probed with anti–D-Asp antiserum (b) and an anti-rat prolactin antibody (c). D-Asp-positive cells [(b), green in original figure in ref. 19] were superimposed on prolactin-positive cells [(c), red in original figure in ref. 19] giving a yellow color in (d) [yellow in original figure in ref.19]. D-Asp-positive cells appeared to be prolactin cells or some other closely related type of cell. Anti–D-Asp antiserum was diluted 1:100, monoclonal anti-prolactin antibody was diluted 1:1000. Reprinted from (19) with permission.

5. Notes

1. The conjugation of D-Asp to BSA via glutaraldehyde forms a Schiff base that lowers the specificity of the antiserum. Therefore, sodium borohydride is necessary to reduce the double bond.
2. The key factor in the adsorption of gold particles to protein is the pH value of the gold colloid solution. The pH value must be equal to or slightly higher than the pI value of the protein to make the protein electrically neutral. When this happens, the static force between the protein molecules and gold particles becomes low, with the surface tension of the protein molecule being high. This

makes the protein more easily adsorbed onto the gold particle's surface and stabilizes the colloid gold. If the gold colloid solution is acidic, the protein is positively charged, whereas the gold particles are negatively charged, and the protein and gold particles will form a complex with a markedly high molecular weight. If the pH value of the gold colloid is higher than the pI value of the protein, both the protein and the gold particles are negatively charged, and they will not form a complex.

3. Since a significant amount of the emulsion will be lost during emulsification, make 1 to 2 mL more emulsion than is required for immunization.
4. Drip a drop of the emulsion onto water. It should remain on the surface of the water.
5. Remove no more than 20 mL blood per kilogram per week.
6. Do not use a magnetic stirrer to mix this gel.
7. Stop the perfusion of Ringer's when the liver turns light in color.

ACKNOWLEDGEMENTS

The authors express their sincere appreciation to Profs. K. Imai and T. Iwatsubo and Drs. Z. Long and K. Sakai for invaluable discussions and significant experimental contributions.

REFERENCES

[1] Imai, K., Fukushima, T., Santa, T., Homma, H., Hamase, K., Sakai, K. & Kato, M. (1996) Analytical chemistry and biochemistry of D-amino acids. *Biomed. Chromatogr., 10*, 303-312.

[2] Hashimoto, A. & Oka, T. (1997) Free D-aspartate and D-serine in the mammalian brain and periphery. *Prog. Neurobiol., 52*, 325-353.

[3] Homma, H. (2002) D-Aspartate in the mammalian body. Viva Origino (freely available online at http://www.origin-life.gr.jp/3004/3004204/3004204.pdf), 30, 204-215.

[4] Furuchi, T. & Homma, H. (2005) Free D-aspartate in mammals. *Biol. Pharm. Bull., 28*, 1566-1570.

[5] Schell, M.J., Cooper, O.B. & Snyder, S.H. (1997) D-Aspartate localizations imply neuronal and neuroendocrine roles. *Proc. Natl. Acad. Sci. USA, 94*, 2013-2018.

[6] Sakai, K., Homma, H., Lee, J.-A., Fukushima, T., Santa, T., Tashiro, K., Iwatsubo, T. & Imai, K. (1998) Localization of D-aspartic acid in elongate spermatids in rat testis. *Arch. Biochem. Biophys., 351*, 96-105.

[7] Hamase, K., Homma, H., Takigawa, Y., Fukushima, T., Santa, T. & Imai, K. (1997) Regional distribution and postnatal changes of D-amino acids in rat brain. *Biochim. Biophys. Acta, 1334*, 214-222.

[8] Hashimoto, A., Oka, T. & Nishikawa, T. (1995) Anatomical distribution and postnatal changes in endogenous free D-aspartate and D-serine in rat brain and periphery. *Eur. J. Neurosci., 7*, 1657-1663.

[9] Takigawa, Y., Homma, H., Lee, J.-A., Fukushima, T., Santa, T., Iwatsubo, T. & Imai, K. (1998) D-Aspartate uptake into cultured rat pinealocytes and the concomitant effect on L-aspartate levels and melatonin secretion. *Biochem. Biophys. Res. Commun., 248*, 641-647.

[10] Ishio, S., Yamada, H., Hayashi, M., Yatsushiro, S., Noumi, T., Yamaguchi, A. & Moriyama, Y. (1998) D-Aspartate modulates melatonin synthesis in rat pinealocytes. *Neurosci. Lett., 249*, 143-146.

[11] Nagata, Y., Homma, H., Lee, J.-A. & Imai, K. (1999) D-Aspartate stimulation of testosterone synthesis in rat Leydig cells. *FEBS Lett., 444*, 160-164.

[12] Nagata, Y., Homma, H., Matsumoto, M. & Imai, K. (1999) Stimulation of steroidogenic acute regulatory protein (StAR) gene expression by D-aspartate in rat Leydig cells. *FEBS Lett., 454*, 317-320.

[13] Chappey, O., Debray, M., Niel, E. & Sherrmann, J.M. (1994) Association constants of monoclonal antibodies for hapten: heterogeneity of frequency distribution and possible relationship with hapten molecular weight. *J. Immunol., 172*, 219-225.

[14] Pow, D.V. & Crook, D.K. (1993) Extremely high titre polyclonal antisera against small neurotransmitter molecules: rapid production, characterisation and use in light- and electron-microscopic immunocytochemistry. *J. Neurosci. Methods, 48*, 51-63.

[15] Lee, J.-A., Homma, H., Sakai, K., Fukushima, T., Santa, T., Tashiro, K., Iwatsubo, T., Yoshikawa, M. & Imai, K. (1997) Immunohistochemical localization of D-aspartate in the rat pineal gland. *Biochem. Biophys. Res. Commun., 231*, 505-508.

[16] Sakai, K., Homma, H., Lee, J.-A., Fukushima, T., Santa, T., Tashiro, K., Iwatsubo, T. & Imai, K. (1997) D-Aspartic acid localization during postnatal development of rat adrenal gland. *Biochem. Biophys. Res. Commun., 235*, 433-436.

[17] Lee, J.-A., Long, Z., Nimura, N., Iwatsubo, T., Imai, K. & Homma, H. (2001) Localization, transport and uptake of D-aspartate in the rat adrenal and pituitary gland. *Arch. Biochem. Biophys., 385*, 242-249.

[18] Sakai, K., Homma, H., Lee, J.-A., Fukushima, T., Santa, T., Tashiro, K., Iwatsubo, T. & Imai, K. (1998) Emergence of D-aspartic acid in the differentiating neurons of the rat central nervous system. *Brain Res., 808*, 65-71.

[19] Lee, J.-A., Homma, H., Tashiro, K., Iwatsubo, T. & Imai, K. (1999) D-Aspartate localization in the rat pituitary gland and retina. *Brain Res., 838*, 193-199.

[20] Long, Z., Homma, H., Lee, J.-A., Fukushima, T., Santa, T., Iwatsubo, T., Yamada, R. & Imai, K. (1998) Biosynthesis of D-aspartate in mammalian cells. *FEBS Lett., 434*, 231-235.

[21] Long, Z., Lee, J.-A., Okamoto, T., Nimura, N., Imai, K. & Homma, H. (2000) D-Aspartate in a prolactin-secreting clonal strain of rat pituitary tumor cells (GH3). *Biochem. Biophys. Res. Commun., 276*, 1143-1147.

[22] Long, Z., Sekine, M., Nimura, N., Lee, J.-A., Imai, K., Iwatsubo, T. & Homma, H. (2001) Immunocytochemical study of D-aspartate in the 2068 rat pheochromocytoma cell line. *Bioimages, 9*, 61-67.

[23] Dunlop, D., S., Neidle, A., McHale, D., Dunlop, D., M. & Lajtha, A. (1986) The presence of free D-aspartic acid in rodents and man. *Biochem. Biophys. Res. Commun., 141*, 27-32.

In: D-Amino Acids:
A New Frontier in Amino Acid and Protein Research, pp.203-213
ISBN 1-60021-075-9

Chapter 3.7

VESICULAR STORAGE IN AND SECRETION OF D-ASPARTATE FROM NEUROENDOCRINE CELLS

Shunsuke Uehara[1,2]* and Yoshinori Moriyama[1]
[1]Laboratory of Membrane Biochemistry, Okayama University Graduate School of Medicine, Dentistry and Pharmaceutical Sciences, Okayama 700-8530, Japan.
[2]Laboratory of Biochemistry, Matsumoto Dental University, Shiojiri 399-0781, Japan.

Keywords: D-aspartate, neuroendocrine cell, PC12, exocytosis.

1. INTRODUCTION

In the central nervous system (CNS) and some neuroendocrine cells, L-glutamate acts as an excitatory amino acid, and exerts physiological and pharmacological effects [1]. To use L-glutamate as an intercellular transmitter, the CNS and peripheral tissues develop glutamatergic systems comprising input (receptors and subsequent intracellular signaling cascade), output (vesicular storage and exocytosis), and termination (sequestration of L-glutamate at the plasma membrane) L-glutamate signals (Fig. 1). The molecular components of the glutamatergic systems have been identified and characterized [1]. We have shown that the pineal gland, an endocrine organ for melatonin, possesses a prototype peripheral glutamatergic system [2]. Pinealocytes store L-glutamate in synaptic-like microvesicles, and secrete it through exocytosis. The released L-glutamate may bind to metabotropic glutamate receptor type 3 (mGluR3), and inhibit melatonin synthesis through a cAMP-mediated inhibitory cascade [3,4]. The glutamate signaling is terminated by a GLT-1-type glutamate transporter through sequestration of extracellular L-glutamate [5,6].

D-Aspartate is also known to be present in the CNS and peripheral neuroendocrine cells such as adrenal chromaffin cells and pituitary cells [7]. Localization of D-aspartate in the CNS and neuroendocrine cells suggests a role of D-aspartate as an intercellular chemical transmitter. Since pinealocytes contain an appreciable amount of D-aspartate [8], we hypothesized that D-aspartate acts as a transmitter in the pineal gland, and somehow affects melatonin synthesis. We have shown that D-aspartate seems to be de novo-synthesized and stored in the cytoplasm. The D-aspartate is gradually secreted from the cells via a Na^+-

* Correspondence concerning this article should be addressed to Shunsuke Uehara, Laboratory of Biochemistry, Matsumoto Dental University, Shiojiri 399-0781, Japan. E-mail: shuehara@po.mdu.ac.jp.

dependent glutamate transporter. The released D-aspartate in turn inhibits noradrenaline-stimulated melatonin synthesis in a dose-dependent and receptor-mediated manner [9]. Thus, D-aspartate should be regarded as an inhibitory modulator of melatonin synthesis [9]. Although our study revealed the role of D-aspartate as a modulator of endocrine function for the first time, the signaling cascade causing inhibition of melatonin synthesis dose not completely match that of classical neurotransmitters such as L-glutamate, as stated above.

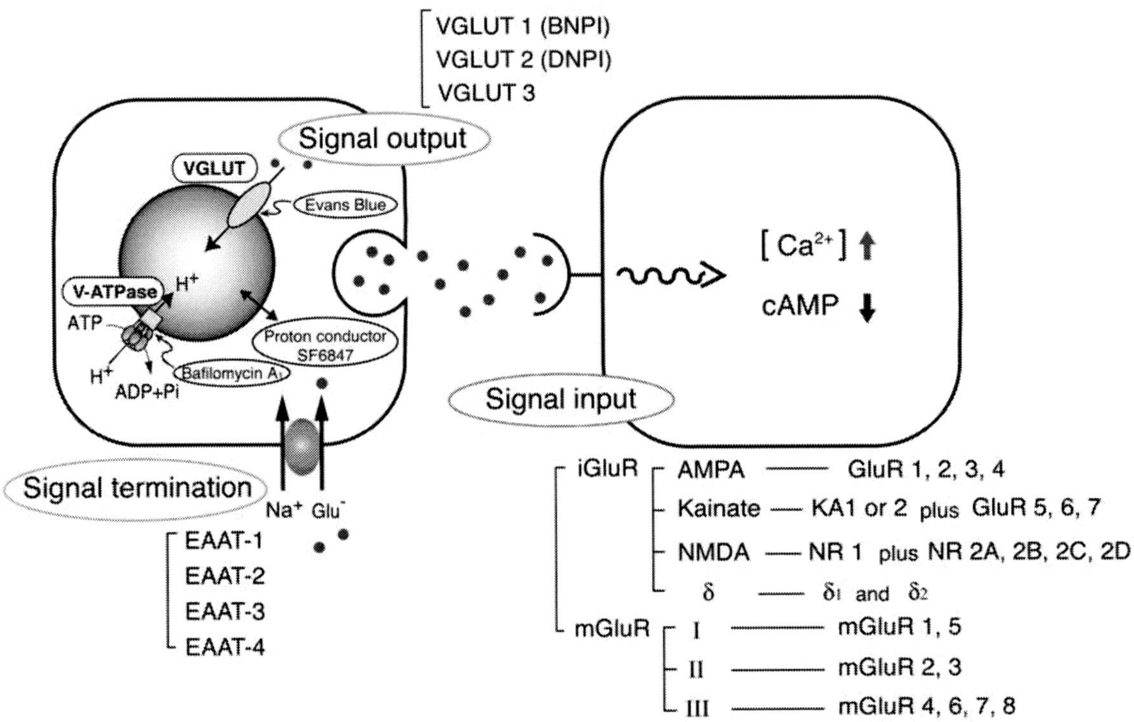

Fig. 1. Scheme of the glutamatergic system. Molecular components comprising the glutamatergic system are summarized.

Through extensive immunohistochemical studies in a search for D-aspartate containing cells involving specific polyclonal antibodies against D-aspartate, we found that various clonal neuroendocrine cells, such as human retinoblastoma Y79 cells [10] and rat pheochromocytoma PC12 cells [11], contain appreciable amounts of D-aspartate. To our surprise, D-aspartate is present punctuatedly in a subpopulation of PC12 cells (Fig. 2). More detailed studies involving immunohistochemical and biochemical techniques indicated that D-aspartate is confined to dopamine-containing secretory granules and not acetylcholine-containing synaptic-like microvesicles [12]. These results suggest that D-aspartate is accumulated in secretory granules through an active transport mechanism, and that the D-aspartate is co-secreted with dopamine through exocytosis from these cells. If this is the case, the PC12 cells may possess machinery for the output of D-aspartate, which allows vesicular storage and exocytosis of D-aspartate. In fact, we found that D-aspartate is accumulated in secretory granules in an ATP-dependent manner, and co-secreted with dopamine through regulated exocytosis. In this article, we describe in detail the procedures for the vesicular storage and exocytosis of D-aspartate.

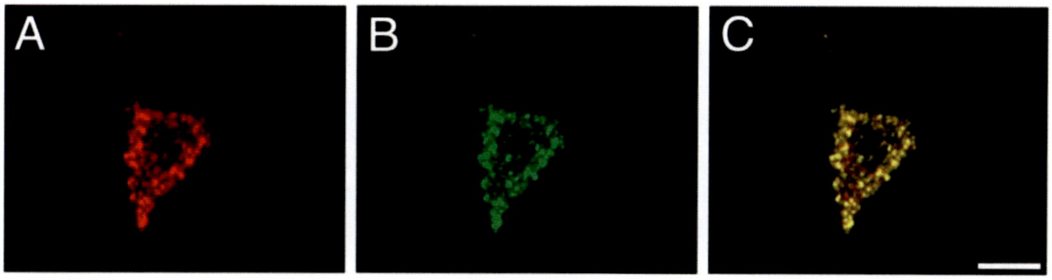

Fig. 2. Distribution of D-aspartate suggests its localization within secretory granules in PC12 cells. A subpopulation of PC12 cells contains the D-aspartate immunoreactivity, which is co-localized with dopamine, a marker of secretory granules. Yellow indicates co-localization. Bar, 10 μm.

2. MATERIALS

2.1. Reagents and Materials for Coating Coverslips with poly-L-lysine

1. The stock solution of poly-L-lysine (100 μg / ml), which was made with autoclaved ultra pure water, stored at -30°C and used within 6 months.
2. Poly-L-lysine hydrobromide (mol. wt. > 300,000) was purchased from Sigma (Saint Louis, Missouri). Coverslip was purchased from Matsunami (Osaka, Japan).

2.2. Reagents for Immunohistochemistry

1. Phosphate-buffered saline without Ca^{2+} and Mg^{2+} (PBS (-)) comprised 137 mM NaCl, 2.7 mM KCl, 8.1 mM Na_2HPO_4 and 1.5 mM KH_2PO_4 (pH 7.4). The PBS (-) was autoclaved, stored at 4 °C and used within 3 months.
2. Phosphate-buffered saline with Ca^{2+} and Mg^{2+} (PBS (+)) comprised 137 mM NaCl, 2.7 mM KCl, 8.1 mM Na_2HPO_4, 1.5 mM KH_2PO_4, 1 mM $CaCl_2$ and 0.5 mM $MgCl_2$ (pH 7.4). The stock solution, PBS (-), 100 mM $CaCl_2$ and 50 mM $MgCl_2$, were autoclaved, and added to PBS(-) solution to give 1 and 0.5 mM, respectively. The PBS (+) solution were stored at 4 °C and used within 3 months.
3. Bovine serum albumin (BSA), NaCl, KCl, Na_2HPO_4, KH_2PO_4 $CaCl_2$ and $MgCl_2$ were purchased from Sigma Aldrich Japan (SAJ) (Tokyo, Japan). Paraformaldehyde was purchased from Nacalai Tesque (Kyoto, Japan). Goat serum was purchased from Invitrogen (Carlsbad, California). Saponin and IMMUNOPEN were purchased from Wako (Osaka, Japan). Purified rabbit polyclonal antibodies against D-aspartate were made in our laboratory [see ref. 12]. Mouse monoclonal antibodies against dopamine were purchased from Biogenesis (Poole, UK). Alexa Flour 488 labeled anti-rabbit IgG and Alexa Flour 568 labeled anti-mouse IgG were purchased from Molecular Probes (Eugene, Oregon). Mounting Medium was purchased from Thermo Shandon (Pittsburgh, Pennsylvania).

2.3. Reagents for Cell Culture

1. The medium comprised Dulbecco's modified Eagle's medium (Invitrogen) supplemented with 5 % fetal bovine serum, 5 % horse serum, 55 mg/l sodium pyruvate, 4.5 g/l glucose, 100 units/ml

penicillin, 100 μg/ml streptomycin and 0.25 μg/ml amphotericin B. The medium was filtered, stored at 4 °C and used within 1 month.

2. Phosphate-buffered saline without Ca^{2+} and Mg^{2+} (PBS (-)). See section 2.2-1.
3. The cell culture dishes used were Biocoat poly-D-lysine cellware ones of 35 and 100 mm in diameter (Becton Dickinson Labware: Franklin Lakes, New Jersey). Filtration system: Stericup® vacuum-driven disposable filtration system (GP Express PLUS membrane; pore size, 0.22 μm) (Millipore; Molsheim, France).
4. Other reagents including sodium pyruvate, glucose, penicillin, streptomycin, amphotericin B, were purchased from SAJ.

2.4. Reagents for Measurement of Vesicular Storage of D-Aspartate

1. Buffer A comprised 20 mM MOPS-Tris (pH 7.0), 0.3 M sucrose, 2 mM Mg-acetate and 4 mM KCl. Buffer A was stored at 4 °C and used within 1 month.
2. Digitonin, 3,5-di-*tert*-butyl-4-hydroxybenzylidene-malononitrile (SF6847) and bafilomycin A1 were purchased from Wako.
3. D-[2,3-^{3}H]Aspartate (18 Ci/mmol) was from Perkin-Elmer Life Sciences (Wellesley, Massachusetts) and MP Biomedicals (Irvine, California).
4. MOPS (3-(N-Morpholino) propanesulfonic acid), tris (Tris[hydroxymethyl]aminomethane), sucrose, Mg-acetate and ATP-tris salt were purchased from SAJ.

2.5. Reagents for Measurement of Secretion of D-Aspartate

1. +Ca^{2+} Ringer's solution comprised 128 mM NaCl, 1.9 mM KCl, 1.2 mM KH_2PO_4, 2.4 mM $CaCl_2$, 1.3 mM $MgSO_4$, 26 mM $NaHCO_3$, 10 mM glucose, 10 mM HEPES/Tris (pH 7.4) and 0.2 % BSA.
2. 50 mM K^+ Ringer's solution comprised 80 mM NaCl, 50 mM KCl, 1.2 mM NaH_2PO_4, 2.4 mM $CaCl_2$, 1.3 mM $MgSO_4$, 26 mM $NaHCO_3$, 10 mM glucose, 10 mM HEPES/Tris (pH 7.4) and 0.2 % BSA.
3. –Ca^{2+} Ringer's solution comprised 128 mM NaCl, 1.9 mM KCl, 1.2 mM KH_2PO_4, 0.2 mM $CaCl_2$, 3.8 mM $MgSO_4$, 26 mM $NaHCO_3$, 10 mM glucose, 10 mM HEPES/Tris (pH 7.4), 1 mM EGTA/NaOH (pH 7.4) and 0.2 % BSA.
4. –Ca^{2+} 50 mM K^+ Ringer's solution comprised 80 mM NaCl, 50 mM KCl, 1.2 mM NaH_2PO_4, 0.2 mM $CaCl_2$, 3.8 mM $MgSO_4$, 26 mM $NaHCO_3$, 10 mM glucose, 10 mM HEPES/Tris (pH 7.4), 1 mM EGTA/NaOH (pH 7.4) and 0.2 % BSA. All Ringer's solutions were filtered, stored at 4 °C and used within 1 month.
5. For filtration, a Millex®-GV syringe-driven filter unit (PVDF membrane pore size; 0.22 μm; diameter, 33 mm) (Millipore) was used.
6. HEPES (2-[4-(hydroxyethyl)-1-piperazynyl]ethanesulfonic acid), EGTA (O,O'-bis(2-aminoethyl)ethyleneglycol-N,N,N',N'-tetraacetic acid) and EGTA/AM (acetoxymethylester) were purchased from Dojindo (Kumamoto, Japan), and $NaHCO_3$ was from Invitrogen. 5-(Methylamino)-2-[(2R,3R,6S,8S,9R,11R)-3,9,11-trimethyl-8-[(1S)-1-methyl-2-oxo-2-[(1H)-pyrrol-2-yl]ethyl]-1,7-[dioxaspiro[5,5]undec-2-yl]methyl]-4-benzoxazolecarboxylic acid (A23187) was purchased from Tocris (Avonmouth, UK). NaH_2PO_4, $MgSO_4$ and NaOH were purchased from SAJ

2.6. Reagents and Materials for Concentration of Acidic Amino Acids

1. Negative ion exchange resin (DOWEX 1X2; mesh size, 100-200; ionic form, Cl^-) (DOW Chemical Company: Midland, Michigan) in a column (1.22 cm in diameter) was used.

2.7. Reagents and Materials for Quantitative Analysis of D-Aspartate by HPLC [see Ref. 13]

1. The columns used were Cosmosil-packed-column [Column type: Waters] (Nacalai Tesque), 5C18AR-II (4.6 X 150 mm) and 5C18MS-II (4.6 X 150 mm). The two columns were connected by 10 cm of PEEK® tube.
2. The detector used was on FL detector L-7480 (Hitachi: Tokyo, Japan), the excitation and emission wavelength pair being 334 and 442 nm, respectively.
3. 50 mM Na-acetate buffer (pH 5.9) was used as mobile phase A. Mobile phase B comprised 200 ml of the same buffer made up with methanol to a final volume of 1 litter. Both buffers were filtered through cellulose acetate filters, stored at room temperature and used within 3 days. 0.4 M Sodium-Borate buffer (pH 9.4) (Na-B buffer) was also used, being stored at 4 °C and used within 6 months.
4. Reagents for detection. o-Phthalaldehyde / N-Acetyl-L-cysteine reagent (OPA-NAC): o-Phthalaldehyde (4 mg) was dissolved in 300 µl of methanol (solution X). N-Acetyl-L-cysteine (16 mg) was dissolved in 100 µl of Na-B buffer (solution Y). Then 300 µl of solution X and 60 µl of solution Y were mixed, and 640 µl of Na-B buffer was added. The mixture was stored at 4 °C and used within 3 days.
5. L-Homocysteic acid (HCA) was used as an internal standard.
6. The stock solutions of standards (D-aspartate and L-aspartate at 20mM) and internal standard (HCA at 20 mM) were stored at 4 °C and used within a year.
7. Cellulose acetate filters (pore size, 0.2 µm; diameter, 47 mm) were purchased from ADVANTEC (Tokyo, Japan). o-Phthalaldehyde, N-acetyl-L-cysteine and boric acid were from Wako. Na-acetate, acetic acid, methanol, L-homocysteic acid, D-aspartic acid and L-aspartic acid were purchased from SAJ.

3. METHODS

3.1. Coating Coverslips with Poly-L-lysine

1. Dilute poly-L-lysine stock solution to 20 µg / ml with autoclaved ultra pure water.
2. Put 10 coverslips (18 mm x 18 mm) on a dish (diameter, 100 mm).
3. Add 5 ml of poly-L-lysine solution.
4. Incubate for 2 hour at 37°C.
5. Wash coverslips three times with 5 ml of autoclaved ultra pure water.
6. Remove water from dish.
7. Put the dish in clean bench under UV lamp to dryness.

3.2. Immunohistochemistry

1. PC12 cells were cultured on poly-L-lysine coated coverslips.
2. Rinse the coverslips twice with 5 ml of PBS (-).
3. Cells were fixed with 15 ml of 4 % paraformaldehyde in PBS (+) for 30 min at room temperature.
4. Rinse the coverslips twice with 5 ml of PBS (-).
5. Cells were permeabelized and blocked with 10 ml of 0.4 % saponin, 1 % BSA and 2 % goat serum in PBS (-) for 15 min at room temperature in gently shaking.
6. Rinse the coverslips twice with 5 ml of PBS (-).
7. Write the edge of the coverslip with the IMMUNOPEN.
8. Place the coverslip on the dish (diameter, 35 mm).
9. React to the primary antibodies (100-diluted anti-D-aspartate antibodies and 500-diluted anti-dopamine antibodies) for 1 hour at room temperature.
10. Wash the coverslip four times with 3 ml of PBS (-).
11. React to the second antibodies (500-diluted Alexa Flour 488 labeled anti rabbit IgG and 1000-diluted Alexa Flour 568 labeled anti mouse IgG) for 1 hour at room temperature.
12. Wash the coverslip four times with 3 ml of PBS (-).
13. Drop Perma Flour Aquaous Mounting Medium on the slide glass.
14. Mount the coverslip on the mounting medium.

3.3. Cell Culture

1. PC12 cells were cultured on poly-D-lysine coated dishes (diameter, 100 mm) in 15 ml of medium, with incubation at 37 °C under 5 % CO_2.
2. When the cells became 80% confluent, passage treatment was performed. The cells were washed twice with 10 ml of PBS (-), and then dispersed by pipetting.
3. Then the cells were collected by centrifugation at 120 X g for 5 min at room temperature, suspended in the medium, and placed in poly-D-lysine coated dishes (diameter, 100 mm) to give 2 X 10^6 cells/dish. The medium was changed every 2 or 3 days.

3.4. Vesicular Storage of D-Aspartate

1. PC12 cells (1 X 10^6) on poly-D-lysine coated dishes (diameter, 35 mm) were washed with 500 µl of buffer A.
2. 500 µl of buffer A containing 10µM digitonin was added.
3. Incubation was performed for 10 min at 37 °C.
4. Then the buffer A was carefully removed by pipetting.
5. The cells were washed with 500 µl of buffer A.
6. 450 µl of buffer A, in the absence or presence of 2mM ATP, was added.
7. 50 µl of 1mM ^{3}[H] D-aspartate was added (final, 2.5 µCi)
8. Further incubation was performed for 10 min at 37 °C.
9. Then the supernatant was removed carefully, and the cells were washed twice with 1 ml of ice-cold buffer A.
10. 500 µl of 1% SDS solution was added to dissolve the cells.
11. Aliquots (200 µl) were taken to measure radioactivity with a liquid scintillation counter, and 50 µl ones were taken to assay protein.

3.5. Exocytosis of D-Aspartate

1. PC12 cells (1 X 10^7) on poly-D-lysine coated dishes (diameter, 100 mm) were washed three times with Ringer's solution or $-Ca^{2+}$ Ringer's solution.
2. Cells were incubated in Ringer's solution or $-Ca^{2+}$ Ringer's solution for 30 min at 37 °C under 5 % CO_2. In another experiment, the cells were incubated in Ringer's solution containing 50 μM EGTA/AM for 30 min at 37 °C under 5 % CO_2.
3. After preincubation, the cells were washed once with the same Ringer's solution.
4. Then the cells were stimulated with 15 ml of 50 mM K^+ Ringer's solution or $-Ca^{2+}$ 50 mM K^+ Ringer's solution. In another experiment, cells were stimulated with 15 ml of Ringer's solution containing 5 μM A23187 in the absence or presence of Ca^{2+}.
5. Immediately, half the supernatant was collected as a sample at 0 min.
6. Cells were incubated for the indicated times at 37 °C under 5 % CO_2.
7. Then the residual supernatant was collected carefully at the indicated time. All samples were stored at –30 °C.

3.6. Concentration of Acidic Amino Acids

1. 5 g resin (DOWEX) was swollen in 80 ml of ultra pure water for 30 min with occasional stirring.
2. Then a column was made with 3 ml swollen resin, and washed with 20 ml of 1 mol/l NaOH.
3. The column was further washed with ultra pure water until the pH of the eluate became neutral.
4. Then a sample (7.5 mL) was loaded.
5. The column was washed with ultra pure water until the pH of the eluate became neutral.
6. Then 2 mol/l acetic acid was loaded until the pH of the eluate became acidic (pH 6 confirmed with pH test paper), and the eluate (15 ml) was collected.
7. This solution was evaporated to give around 200 μl.
8. Then the sample was neutralized by adding 5 mol/l NaOH. The samples were stored at –30 °C until use.

3.7. Quantitative Analysis of D-Aspartate by HPLC

1. A sample (70 μl) was mixed with 20 μl of 0.4 μmol/l L-homocysteic acid.
2. 20 μl of OPA-NAC was added to start derivatization.
3. After 2 min incubation, 100 μl of the derivatization mixture was injected into a HPLC. (Note)

4. Results

Immunohistochemical evidence indicates that a subpopulation of PC12 cells (approximately 10 to 20 % cells) contains D-aspartate, which is associated with dopamine-containing secretory granules (Fig. 3). To elucidate the mechanism underlying vesicular storage of D-aspartate, we determined the ATP-dependent uptake of radiolabeled D-aspartate by digitonin-permeabilized PC12 cells. The treatment-exposed intracellular granules enabled us to detect the ATP-dependent uptake. Shown in Fig. 4, we found that such cells took up radiolabeled D-aspartate depending on ATP and Mg^{2+}. The ATP-dependent uptake was sensitive to bafilomycin A1, an inhibitor of vacuolar H^+-ATPase (V-ATPase) [14], and SF6847, a proton conductor that dissipates an electrochemical proton gradient. These results indicate that an electrochemical gradient of protons established by V-ATPase drove the uptake of D-aspartate in secretory granules.

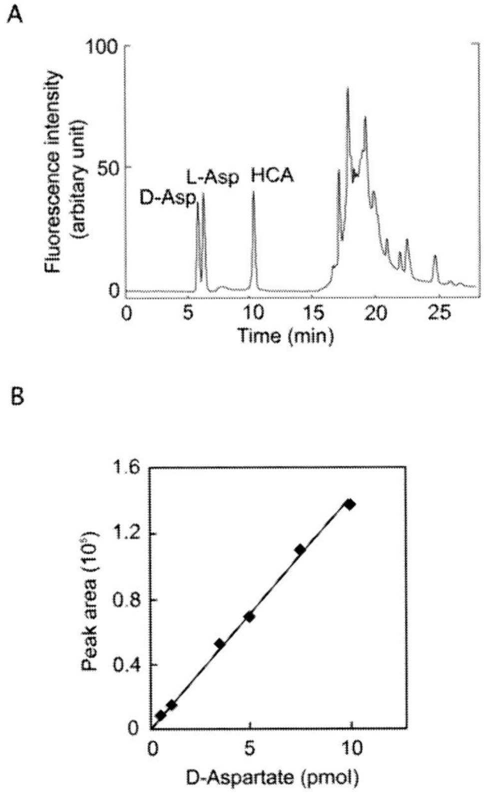

Fig. 3. Separation of D-aspartate by HPLC. A typical HPLC separation pattern is shown in (A). Linearity is shown in (B).

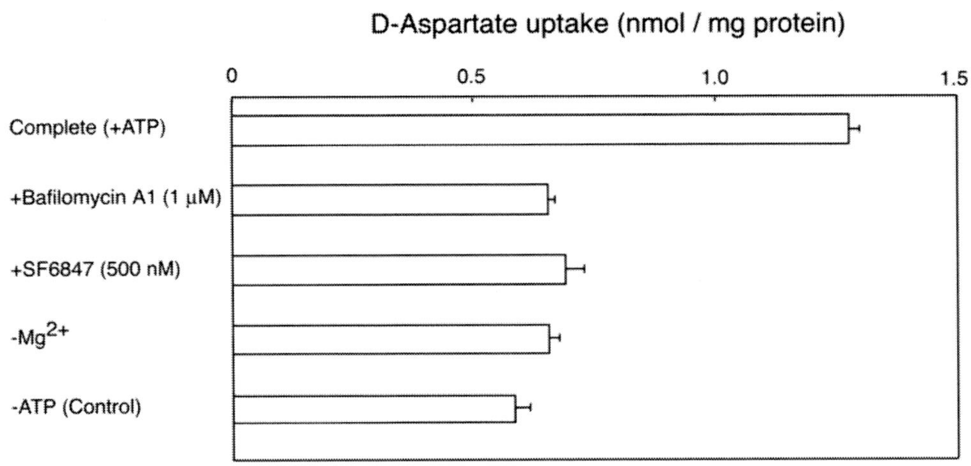

Fig. 4. ATP-dependent uptake of radiolabeled D-aspartate by digitonin-permeabilized cells. The uptake of radiolabeled D-aspartate by permeabilized cells was examined in the presence or absence of the listed compounds. In some experiments, magnesium acetate (-Mg^{2+}) or ATP (-ATP) was omitted. The results are the means ± S.E. for four independent experiments.

We further investigated whether or not the vesicular D-aspartate is released from the cells. It is well-established that dopamine is secreted through regulated exocytosis of secretory granules. The secretion was triggered upon depolarization. We depolarized PC12 cells by the addition of KCl. Fig. 5 indicates that KCl treatment stimulates secretion of D-aspartate from the cells. The amount of D-aspartate was 40.0 ± 7.1 pmol per 10^7 cells at 10 min, which corresponds to about 15.6 % of total D-aspartate. A Ca^{2+} ionophore, A23187, evoked D-aspartate release depended on the time and extracellular Ca^{2+}. Furthermore, EGTA/AM removed intracellular free Ca^{2+}, and inhibited KCl-evoked secretion of D-aspartate. Simultaneously, we detected the secretion of dopamine [see Ref. 15] from PC12 cells with similar kinetics and Ca^{2+} dependence to those in the case of D-aspartate. These results indicate that Ca^{2+} influx and elevation of the intracellular free Ca^{2+} concentration following depolarization are essential for D-aspartate secretion from PC12 cells, as in the case of dopamine.

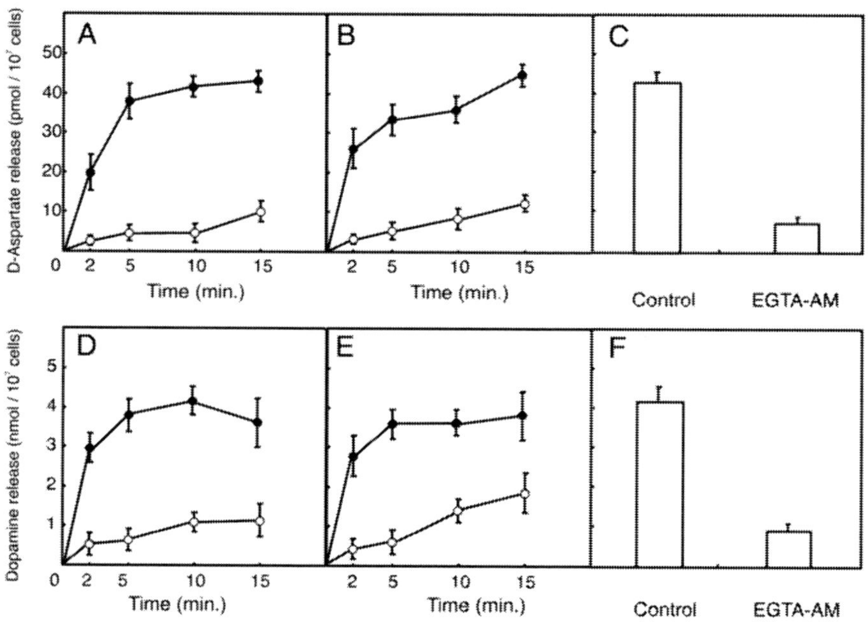

Fig. 5. Ca^{2+} dependent secretion of dopamine and D-aspartate from PC12 cells. A. Depolarization, and time-and Ca^{2+}-dependent D-aspartate release were examined as described under "Methods". The presence and absence of extracellular Ca^{2+} are indicated by closed and open circles, respectively. B. A23187, at 5 μM, was used instead of KCl, in the presence (closed circles) or absence (open circles) of extracellular Ca^{2+}. C. Cells were treated with 50 μM EGTA/AM for 30 min, and then stimulated with KCl. The dopamine contents in the same samples as above (A-C) were simultaneously measured and expressed, as shown in D-F, respectively (About method of dopamine quantitation, see ref. 15). All the results in the figure are the means ± S.E. for three independent experiments.

These results are consistent with idea that D-aspartate is stored in dopamine-containing secretory granules, and then secreted through exocytosis, as illustrated in Fig. 6. We concluded that a subpopulation of PC12 cells comprises D-aspartate-secreting cells. Our results suggest the presence of a vesicular transporter specific to D-aspartate. Furthermore, exocytosis of D-aspartate suggests the possibility that D-aspartate acts as an intercellular messenger in endocrine cells. To receive D-aspartate signaling, the presence of a D-aspartate receptor is expected. Identification of the membrane components responsible for D-aspartate signaling waits further study.

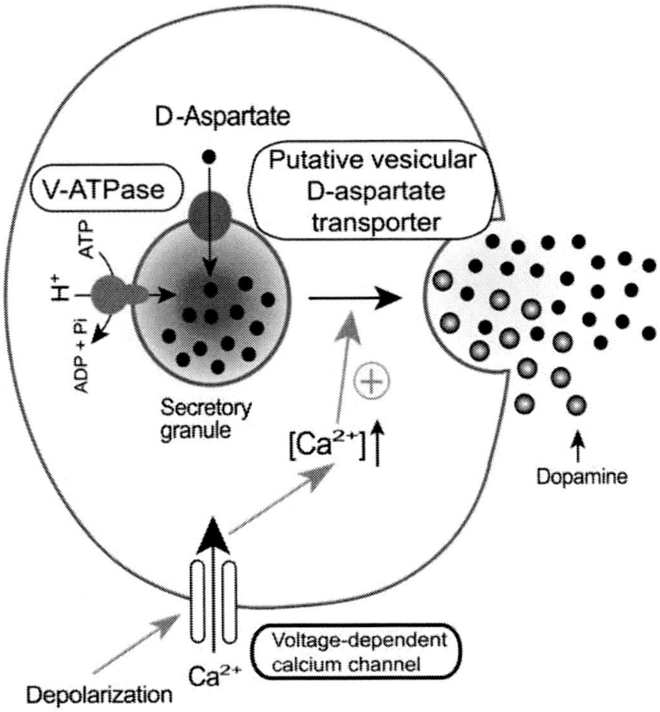

Fig. 6. Proposed mechanism for the storage and secretion of D-aspartate. Our results support the idea that D-aspartate is accumulated in secretory granules through a putative vesicular transporter for D-aspartate, and then co-secreted with dopamine through regulated exocytosis. Depolarization opens a voltage-gated calcium channel, and the resultant increase in intracellular Ca^{2+} triggers exocytosis.

5. Note

The linearity between the peak-area on the HPLC chromatogram and the amount of D-aspartate was always within the range of 0.1 pmol to 100 pmol (Fig. 3). The detection limit was below 0.01 pmol. A typical HPLC pattern is included in Fig. 3.

References

[1] Moriyama, Y. & Yamamoto, A. (2004). Glutamatergic chemical transmission: look! Here, there, and anywhere. *J Biochem*, *135*, 155-163.

[2] Moriyama, Y., Hayashi, M., Yamada, H., Yatsushiro, S., Ishio, S. & Yamamoto, A. (2000). Synaptic-like microvesicles, synaptic vesicle counterparts in endocrine cells, are involved in a novel regulatory mechanism for the synthesis and secretion of hormones. *J Exp Biol*, *203*, 117-125.

[3] Yamada, H., Yamamoto, A., Yodozawa, S., Kozaki, S., Takahashi, M., Morita, M., Michibata, H., Furuichi, T., Mikoshiba, K. & Moriyama, Y. (1996). Microvesicle-mediated exocytosis of glutamate is a novel paracrine-like chemical transduction mechanism and inhibits melatonin secretion in rat pinealocytes. *J Pineal Res*, *21*, 175-191.

[4] Yamada, H., Yatsushiro, S., Ishio, S., Hayashi, M., Nishi, T., Yamamoto, A., Futai, M., Yamaguchi, A. & Moriyama, Y. (1998). Metabotropic glutamate receptors negatively regulate melatonin synthesis in rat pinealocytes. *J Neurosci*, *18*, 2056-2062.

[5] Yamada, H., Yatsushiro, S., Yamamoto, A., Hayashi, M., Nishi, T., Futai, M., Yamaguchi, A. & Moriyama, Y. (1997). Functional expression of a GLT-1 type Na^+-dependent glutamate transporter in rat pinealocytes. *J Neurochem*, *69*, 1491-1498.

[6] Yatsushiro, S., Yamada, H., Kozaki, S., Kumon, H., Michibata, H., Yamamoto, A. & Moriyama, Y. (1997). L-Aspartate but not the D form is secreted through microvesicle-mediated exocytosis and is sequestered through Na^+-dependent transporter in rat pinealocytes. *J Neurochem*, *69*, 340-347.

[7] Schell, M. J., Cooper, O. B. & Snyder, S. H. (1997). D-Aspartate localizations imply neuronal and neuroendocrine roles. *Proc Natl Acad Sci*, *94*, 2013-2018.

[8] Imai, K., Fukushima, T., Hagiwara, K. & Santa, T. (1995). Occurrence of D-aspartic acid in rat brain pineal gland. *Biomed Chromatogr*, *9*, 106-109.

[9] Ishio, S., Yamada, H., Hayashi, M., Yatsushiro, S., Noumi, T., Yamaguchi, A. & Moriyama, Y. (1998). D-Aspartate modulates melatonin synthesis in rat pinealocytes. *Neurosci Lett*, *249*, 143-146.

[10] Hayashi, M., Itabashi, T. & Moriyama, Y. (1999). D-Aspartate is present in human retinoblastoma Y79 cells. *Neurosci Lett*, *267*, 37-40.

[11] Moriyama, Y., Yamada, H., Hayashi, M., Oda, T. & Yamaguchi, A. (1998). Identification of D-aspartate in rat pheochromocytoma PC12 cells. *Neurosci Lett*, *248*, 57-60.

[12] Nakatsuka, S., Hayashi, M., Muroyama, A., Otsuka, M., Kozaki, S., Yamada, H. & Moriyama, Y. (2001). D-Aspartate is stored in secretory granules and released through a Ca^{2+}–dependent pathway in a subset of rat pheochromocytoma PC12 cells. *J Biol Chem*, *276*, 26589-26596.

[13] Aswad, W. D. (1984). Determination of D- and L-aspartate in amino acid mixtures by high-performance liquid chromatography after derivatization with a chiral adduct of o- phthalaldehyde. *Anal Biochem*, *137*, 405-409.

[14] Bowman, E. J., Siebers, A. & Altendorf, K. (1988). Bafilomycins: a class of inhibitors of membrane ATPases from microorganisms, animal cells, and plant cells. *Proc Natl Acad Sci*, *85*, 7972-7976.

[15] Felice, L. J., Felice, J, D. & Kissinger, P. T. (1978). Determination of catecholamines in rat brain parts by reverse phase ion-pair liquid chromatography. *J Neurochem*, *31*, 1461-1465.

In: D-Amino Acids: ISBN 1-60021-075-9
A New Frontier in Amino Acid and Protein Research, pp.215-220

Chapter 3.8

D-ASPARTATE IN FROG HARDERIAN GLAND

Rossella Monteforte*[1]*, Alessandra Santillo*[1]*, Paolo Chieffi*[2] *and Gabriella Chieffi Baccari*[1]**

[1]Dipartimento di Scienze della Vita, Seconda Università di Napoli, Italy;
[2]Dipartimento di Medicina Sperimentale, Seconda Università di Napoli, Italy.

Keywords: Harderian gland, D-aspartate, Mallory stain, ERK 1, exocrine secretion

1. INTRODUCTION

In the green frog, *Rana esculenta*, a substantial amount of D-aspartate (D-Asp) is found endogenously within the Harderian gland (HG) following its synthesis from L-aspartate (L-Asp) by an aspartate racemase [1]. The frog HG is an orbital seromucoid gland that displays seasonal changes in secretory activity [2,3]. Injection of 2.0 μmol/g b.w. of D-Asp in frogs collected during the period of medium-low secretory activity revealed that HG can take up and accumulate D-Asp and that this amino acid rapidly induces activation of extracellular-signal regulated kinase (ERK1) and an increase in cells active in RNA synthesis [1]. Extracellular-signal regulated kinase belongs to the MAP kinases constituting a family of serine/threonine kinases that occupy a focal point in signal transduction, mainly by activating gene transcription via translocation into the nucleus [4,5]. The increase in transcriptional activity was followed by a significant increase in mucous secretion. By contrast, administration of exogenous D-Asp when HG was showing high activity rapidly induced inhibition of both ERK1 and transcriptional activity. Since D-Asp is known to be recognized by receptors for *N*-methyl- D-aspartic acid (NMDA), it is possible that in the HG, D-Asp mediated NMDA activation may enhance the kinase pathway. The above activation of opposing stimulatory and inhibitory processes could reflect different levels of NMDA-receptor activity, which could vary as a function of the level of gland activity. This study provides the first evidence of a role for this excitatory amino acid in exocrine secretion. The effects of D-Asp in HG appear to be specific since they were not seen in frogs treated with other D- or L-amino acids with known excitatory effects on neurosecretion.

---

* Correspondence concerning this article should be addressed to Gabriella Chieffi Baccari, E-mail: gabriella.chieffi@unina2.it, FAX: +39-0823 274571.

2. Materials

1. Small pieces of Harderian glands were fixed in Bouin's fluid, and paraffin sections (5 µm) were stained according to the Mallory's trichrome method (see Methods).
2. Nikon Eclipse E600 light microscope fitted with a JVC TK-C1381 photocamera connected to a Pentium III computer
3. Antibodies
   Polyclonal anti phospho-ERK1 (Thr202/Tyr204) antibody (#9101S, New England Biolab, MA, USA), raised in rabbit.
   Polyclonal anti ERK1 (#sc-94-G, Santa Cruz Biotechnology Inc., Santa Cruz, CA, USA), raised in rabbit.
   Horseradish-peroxidase-conjugated secondary antibody
4. Lysis buffer containing 50mM HEPES, 150 mM NaCl, 1 mM EDTA, 1 mM EGTA, 10% glycerol, 1% Triton-X-100 (1:2 weight/volume), 1 mM phenylmethylsulphonyl fluoride (PMSF) 1 µg aprotinin, 0.5 mM sodium orthovanadate and 20 mM sodium pyrophosphate, (Sigma Chemical Corporation, St. Louis, MO, USA).
5. Bradford assay (Bio-Rad, Melville, NY, USA).
6. SDS-PAGE (10% polyacrylamide)
7. Nitrocellulose membrane (Immobilon Millipore Corporation, Bedford, MA, USA)
8. Blocking solution (5% non-fat powdered milk in 25 mM Tris, pH 7.4; 200 mM NaCl; 0.5% Triton x-100, TBS/T)
9. Chemiluminescence (ECL) system (Amersham Life Science, UK).

3. Methods

Frogs of both sexes were divided into two groups: one group received i.p. 2.0 µmol/g b.w. of D-Asp; the control frogs were injected with solvent alone (distilled water). At various times after amino acid injection (from 1 to 72 h), frogs were anesthetized by immersion in a 1% solution of MS-222, then decapitated. HGs were dissected from the orbits and rapidly immersed in fixative for light microscopy. The contralateral glands were immediately stored at –20°C for biochemical and molecular determinations.

3.1. In this study cells active in RNA synthesis were detected using Mallory staining procedure [6]:

1. Dewax sections and bring to distilled water
2. Place slides in 1% Ruben S water solution for 1 min;
3. Differentiate the nuclei in fresh water;
4. Rinse in distilled water;
5. Place slides in 1% phosphomolibidic acid water solution. Leave 15 min at room temperature (RT);
6. Wash in distilled water;
7. Stain in Mallory mixture (0.5% aniline blue, 2% orange G, 2% oxalic acid, 100 mL distilled water). Leave for 20 min at RT;
8. Wash in distilled water;
9. Rapidly dehydrate, clear, mount.

This reaction is known to stain nuclei with orange G. Nuclei showing affinity for aniline blue, so-called "blue nuclei", have an increase in RNA indicative of an enhanced transcriptional activity [7,8]. To determine the number of cells active in RNA synthesis in D-Asp treated harderian glands, we calculated the percentage of blue nuclei at various times after the administration of the amino acid. The percentage of "blue nuclei" was quantified with the aid of an image analyser. Digitized transverse sections were viewed under a Nikon Eclipse E600 light microscope fitted with a JVC TK-C1381 photocamera connected to a Pentium III computer. Sections from each animal were analysed using Lucia 4.1 ScMeas on Mutech software. The "blue nuclei" index is expressed as the number of "blue nuclei" per total acinar cell-number count multiplied by 100. All data were compared using an analysis of variance followed either by Duncan's test for multi-group comparisons or by a Student's *t* test for between-group comparisons. All data are expressed as mean ± standard deviation. The levels of significance were taken as $P< 0.01$ and $P< 0.05$.

3.2. For protein extraction pool of frozen Harderian glands was.homogenised directly in lysis buffer (1:10) using a manual potter; then it was clarified by centrifugation at 14,000 x 10 min. Protein concentrations were estimated using a modified Bradford assay (Bio-Rad, Melville, NY).

3.3 Fifty micrograms of total protein extracts were boiled in Laemmli buffer for 5 min before electrophoresis. The samples were subjected to SDS-PAGE (10% polyacrylamide) under reducing conditions. After electrophoresis, proteins were transferred to a nitrocellulose membrane (Immobilon Millipore Corporation, Bedford, MA), complete transfer being assessed using prestained protein standards (Bio-Rad, Melville, NY). The membranes were treated for 2h with blocking solution (5 % non-fat powdered milk in 25 mM Tris, pH 7.4; 200 mM NaCl; 0.5% Triton x-100, TBS/T) and then the membranes were incubated for 1h at room temperature with the primary antibody, which was either: 1) against phospho-ERK1 (diluted 1: 2000), or 2) against ERK1 (diluted 1: 2000). After washing with TBS/T and TBS, membranes were incubated with the horseradish-peroxidase-conjugated secondary antibody (1:4000) for 45 min (at room temperature), and the reactions were detected using an enhanced chemiluminescence (ECL) system (Amersham Life Science, UK).

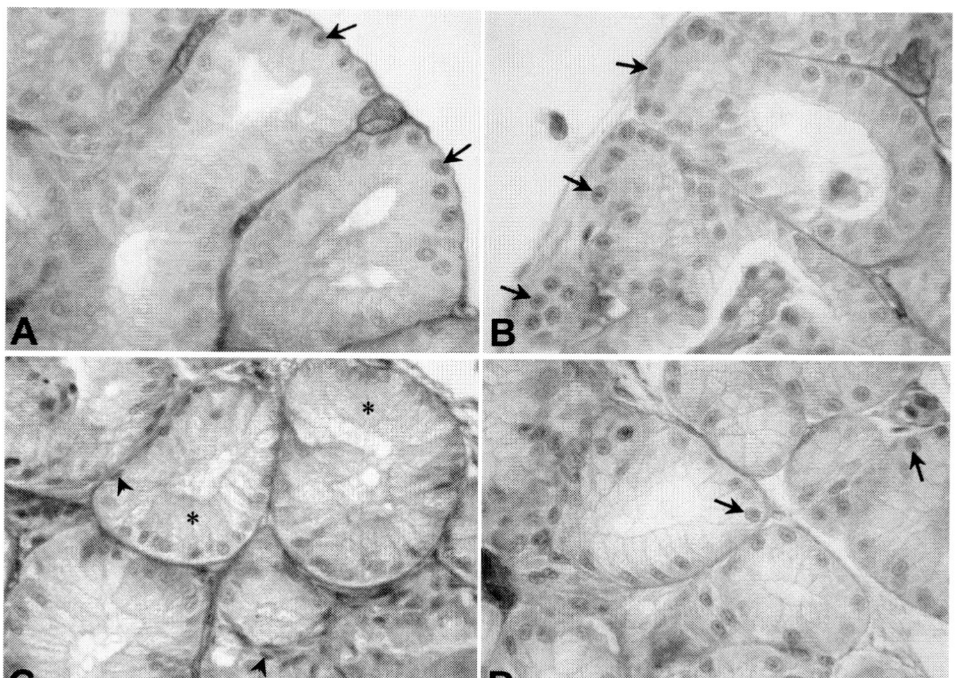

Figure 1. Paraffin sections of HG of *Rana esulenta*. A: The glandular cells of a March frog are columnar, with weakly basophilic cytoplasm. Note "blue nuclei" among orange-stained nuclei of glandular cells (arrows). B: Numerous blue nuclei (arrows) are present in the HG acini of a March frog at 2 h after D-Asp injection. C: Glandular cells of HG in a July frog are rich in secretory granules (asterisks). Nuclei, basally located, are apparently picnotic (arrowheads). D: At 6h after D-Asp treatment, the glandular cells of a July frog are weakly basophilic, and blue nuclei can be observed (arrows). Mallory stain; 500x.

4. Results

D-Asp is present at a high concentration (130 nmol/g tissue) in HG as an endogenous natural compound, and HG accumulates this amino acid following its injection In March frogs, the glandular cells of control HG were columnar in shape, with very few cytoplasmic secretory granules (Fig. 1A). When Mallory's trichrome stain was used, acinar nuclei displayed different affinities for the dye. Some orangiophilic nuclei (3%) showed an affinity for aniline blue (blue nuclei) (Fig. 1A). Such "blue nuclei" have an increase in nuclear RNA, indicative of an enhanced transcriptional activity [7,8]. At 2 and 6h after D-Asp injection in March frogs, a significantly (P<0.01) higher percentage of blue nuclei was present than at time 0 (Figs. 1B; 2). At 15h after D-Asp treatment, the percentage of blue nuclei was similar to that at time 0 (Fig. 2) and the glandular cells were filled with basophilic granules. At 24h after D-Asp injection, the cells appeared devoid of secretory granules and no blue nuclei were seen (Fig. 2).

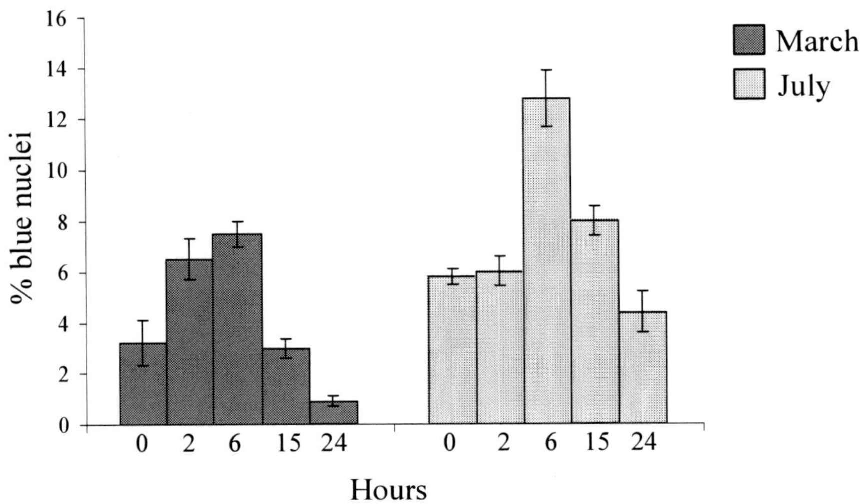

Figure 2. Percentage of "blue nuclei" in glandular cells at various times after D-Asp injection (2.0 μmol/g b.w.; n=5 animals for each data-point). *$P<0.01$ versus time 0.

In July frogs the control HG showed high secretory activity. Most acini were full of basophilic secretory granules and the acinar lumina contained secretory product (Fig. 1C). A higher percentage of blue nuclei was present than in March (Fig. 2). At 2h after D-Asp injection, the blue-nuclei index was unchanged but at 6h a significantly (P<0.01) higher percentage of blue nuclei was present than at time 0 (Figs. 1D, 2). At this time (6h), the glandular cells appeared devoid of secretory granules (Fig. 1D). At both 15 and 24h, the cells were filled with secretory granules and the percentage of blue nuclei had declined toward the basal level (Fig. 2).

Total proteins isolated from HG were analyzed by Western blot analysis. It is important to note that the basal level of ERK1 activity (phosphorylation of the Thr202/Tyr204 residues) in frog HG was higher in July than in March. The HG of March frogs injected with D-Asp showed an increase in ERK1 activity at both 2 and 6h (about 3- and 30-fold, respectively, versus control) (Fig. 3A, lanes 1,3,4 upper panel). At 15h, the activity was still elevated (Fig. 3A, lane 5 upper panel).

By contrast, the HG of July frogs injected with D-Asp showed a decrement in ERK1 activity after both 1 and 2h and then an increase at 6h, as compared with control (Fig. 3B, lanes 1,2,3,4 upper panel). By 15h, the activity had declined to the basal level (Fig. 3B, lane 5 upper panel).

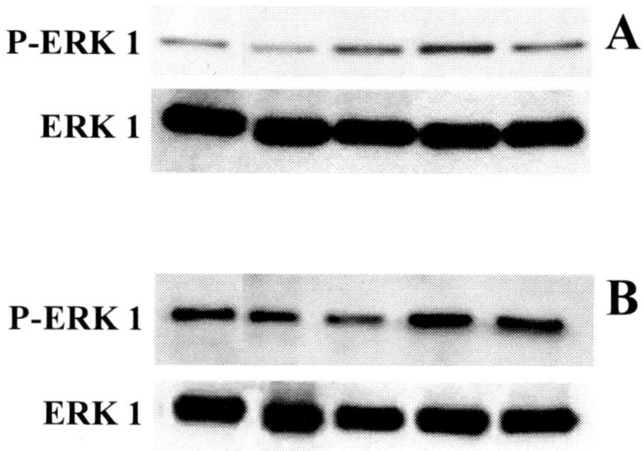

Figure 3. Western blot detection of ERK1 proteins in frog Harderian gland at various times after D-Asp injection (2.0 μmol/g body weight) in (A) March and (B) July. Total proteins (50 μg per sample) were resolved by SDS-PAGE, transferred to nitrocellulose membranes and then incubated with antibodies raised against phospho-ERK1 and ERK1 protein. A specific band was observed of size 44 kDa (by comparison with co-migrating size markers; Bio-Rad, Melville, NY). Values shown are each representative of three separate experiments.

D-Asp injections did not alter the expression of ERK1 itself, which appeared to be expressed to equal extents under all conditions and at all time-points after D-Asp injection (Fig. 3A,B lower panels).

5. Notes

Mallory trichrome procedure is applicable to paraffin sections. The dyes used, their color index number, suppliers and batch number are the follows:

Dyestuff	Supplier and batch number
Aniline blue (C.I.42755)	Merck (911976); Carlo Erba (235921)
Orange G (C.I.16230)	Carlo Erba (2785H100); Merck (4075775)
Oxalic acid	Carlo Erba (21734D5252)
Phosphomolybdic acid	Merck (936CC598732)
Ruben S (C.I.42685)	Merck (70110093)

References

[1] Raucci, F., Santillo, A., D'Aniello, A., Chieffi, P. & Chieffi Baccari, G. (2005). D-aspartate modulates transcriptional activity in Harderian gland of frog *Rana esculenta*: morphological and molecular evidence. *J. Cell Physiol., 204,* 445-454.

[2] Di Matteo, L., Minucci, S., Chieffi Baccari, G., Pellicciari, C., d'Istria & M., Chieffi, G. (1989). The harderian gland of the frog, *Rana esculenta*, during the annual cycle: Histology, histochemistry and ultrastructure. *Basic Appl Histochem, 33*, 93-112.

[3] Minucci, S., Chieffi Baccari, G., Di Matteo, L., Marmorino, C., d'Istria & M., Chieffi, G. (1990). Influence of light and temperature on the secretory activity of the Harderian gland of the green frog, *Rana esculenta*. *Comp Biochem Physiol A, 95(A),* 249-252.

[4] Cobb, MH. (1999). MAP kinase pathways. *Prog Biophys Mol Biol, 7,* 479-500.

[5] Friguls, B., Petegnief, V., Justicia, C., Pallas & M., Planas, AM. (2002). Activation of ERK and Akt signaling in focal cerebral ischemia: modulation by TGF-alpha and involvement of NMDA receptor. *Neurobiol Dis, 11*, 443-456.

[6] Mallory, F.B. (1900). A contribution to standing method. *J. Exp. Med, 5,* 15-20.

[7] Chieffi Baccari, G., Marmorino, C., Minucci, S., Di Matteo, L., Varriale, B., d'Istria, M. & Chieffi, G. (1992). Mallory stain may indicate differential rates of RNA synthesis: I. A seasonal cycle in the Harderian gland of the green frog (*Rana esculenta*). *Eur J Histochem, 36*, 81-90.

[8] Chieffi Baccari, G., Marmorino, C., Minucci, S., Di Matteo, L. & d'Istria, M. (1992). Mallory stain may indicate differential rates of RNA synthesis: II. Comparative observations in vertebrate nuclei. *Eur J Histochem, 36*, 187-196.

In: D-Amino Acids: ISBN 1-60021-075-9
A New Frontier in Amino Acid and Protein Research, pp.221-230

Chapter 3.9

The Role of D-Aspartic Acid in the Endocrine Regulation of the Testis: A Study in the Lizard *Podarcis S. Sicula*

Maria Maddalena Di Fiore* and Franca Raucci
Department of Life Sciences, Second University of Naples, Italy.

Keywords: D-Asp, testis, steroidogenesis, spermatogenesis, Reptiles.

1. Introduction

There is a growing body of evidence indicating free endogenous D-aspartic acid (D-Asp) in nervous and endocrine tissue in several animal phyla [1]. In the family of neuroexcitatory amino acids, D-Asp is now well known to play important neurotransmission and neurosecretion roles [2,3]. D-Asp is the precursor for the synthesis of N-methyl- D-aspartic acid (NMDA) and the enzyme catalyzing this reaction is the N-methyl transferase as reported in previous manuscripts carried out on rat [4,5]. D-Asp is localized in various neurons of rat brain, including the hippocampus, and in the hypothalamo-neurohypophyseal system suggesting its involvement in regulating the neurosecretory system [2,4,6,7]. In vertebrates, this amino acid occurs in the nervous tissues of chickens [8], rats [4,5,9-11] and humans [12,13]. An extremely high content of D-Asp was detected in human brain [12] as well as in the cerebrospinal fluid [13]. In addition, D-Asp may participate in brain development, differentiation and functioning [10,14-16] or may be a novel messenger in neuronal and neuroendocrine organs [2,3].

Most recent investigations have examined the endocrine system of mammals where D-Asp is well represented in the pineal gland [17-19], hypothalamus [7,13], pituitary [4,5], adrenals [20,21] and gonads [22-24]. Maximal content and transient emergence of D-Asp in these tissues correspond to their morphological and functional maturation [10,15]. Particularly, in the adenohypophysis, pineal gland and testis, D-Asp is involved in the hormone synthesis and release [4,5,7,9,10,12,14,19,24,25]. The relationship between D-Asp and endocrine activity has been well documented. Developmental changes in levels of D-Asp and testosterone in rat testis closely parallel each other: they increase to maximum levels at sexual maturity. As in testes, D-Asp also increases with age in the pituitary [23]. Intraperitoneal administration of D-Asp to adult male rats induces D-Asp accumulation in the pituitary gland and testis, followed by a

* Correspondence concerning this article should be addressed to MM Di Fiore, E-mail: MariaM.DiFiore@unina2.it.

significant increase in luteinizing hormone, testosterone, progesterone [4] and prolactin [5] in the blood. Moreover, D-Asp is contained in spermatids [22] and both Leydig [24,26] and Sertoli cells [24] and *in vitro* experiments, performed on isolated testis, demonstrated that this amino acid enhances the production of testosterone [23].

These data indicate that D-Asp may act as a novel putative regulator of hormonal synthesis. In support of findings in mammals, other studies on lower seasonal breeding vertebrates have shown the effects of D-Asp on gonads. In the amphibian *Rana esculenta,* D-Asp occurs in the ovary where it is involved in the control of testosterone release during the sexual cycle [27]. In the female lizard *Podarcis s. sicula,* D-Asp enhances follicular production of 17β-estradiol by up-regulating the local aromatase activity [28]. Raucci *et al.* [29] demonstrated that D-Asp was rapidly taken up by the testis of injected frogs and its rise was coupled with a significant increase of testosterone levels and a consequent increase of mitotic activity in the testis. On the other hand, D-aspartate may play a role in hormonal regulation, as it stimulates testosterone synthesis in the testis, increasing the mRNA level of a steroidogenic acute regulatory protein (StAR) [30]. These findings for D-Asp suggest a different target on sex steroid production machinery depending on the species and/or the sex, i.e., the amino acid operates differently, favouring androgen production in males and estrogen production in females.

Except for the data reported on male frogs, there are no studies yet available on D-Asp presence and effect on the testis of lower vertebrates. Seasonal breeders are good models to study the involvement of D-Asp on the testis because the effects of this molecule can be compared in response to the different phases of spermatogenesis. Therefore, to gather information on this aspect of D-Asp function, we investigated the occurrence of endogenous D-Asp in the testis of lizard, *Podarcis s. sicula.* To gain insight into the functional significance of D-Asp in this organ we studied the role of this amino acid in lizards collected during the main phases of their reproductive cycle. We studied the uptake of D-Asp in the testis and its putative role on the steroidogenesis (sex hormones in the testis and plasma evaluated during the cycle and on *in vivo* experiments). Finally, we attempted to determine whether the D-Asp present in the gonad could come from a local conversion of L-Asp by a specific racemase.

2. Materials

Animals

Adult male of Podarcis s. sicula were captured in the countryside (Caserta; Italy) during the pre-reproductive (November-February), reproductive (March-May) and post-reproductive (July) phases.

In Vivo Experimental Procedures on the Lizards:

1. D-aspartate (D-Asp), L-aspartate (L-Asp), D-glutamate (D-Glu), L-glutamate (L-Glu) (Sigma, Milan, Italy).

Sex Steroid Assay in Plasma and Testis

2. Enzyme immunoassay kits (EIA) (Adaltis Italia, spa, Bologna, Italy).
3. Ethyl ether.
4. Sodium phosphate buffer 0.05 M, pH 7.5, containing BSA at a concentration of 10 mg/mL.
5. UV spectrometer.

Specific Determination of D-Aspartic Acid

6. Perchloric acid (PCA) 0.5 M.
7. Column of cation exchange resin (AG 50 W-X8 resin, hydrogen ionic form, 200-400 mesh, Bio-Rad, Milan, Italy).
8. HCl 6.0 M and 0.001M.
9. NH_4OH 4M.
10. Sep-pak C-18 cartridge (300 mg; Waters, Milan, Italy).
11. Acetonitrile.
12. Savant centrifuge.
13. HPLC two-pump system and with a fluorimeter detecting excitation at 330 nm and emission at 450 nm. The column was a Supercosil C-18 (0.45 x 25 cm; 5 μm beads; Supelco, Bellefonte, PA, USA).
14. D-aspartate oxidase (D-AspO; EC 1.4.3.1). D-AspO enzyme was obtained by overexpression in *Escherichia coli* and was purified according to the procedure described previously [31], modified as follows: a single-stranded cDNA was synthesized from bovine kidney polyadenylated RNA, using reverse transcriptase and oligo (dT) as a primer cDNA, and was amplified by PCR using the single-stranded cDNA as a template and, as primers, BOVI (5′-ATGGATACAGTACGGATTGC-3′), complementary to the 5′ coding region, and BOVII (5′-TTGCTATCATTTCGGGTCAC-3′), complementary to the 3′ coding region of the cDNA of bovine kidney D-AspO. The resulting fragment was inserted into the expression vector pKK223-3, and the sequence was controlled by the deoxynucleotide termination procedure and inserted in *E. coli* JM105. Cells harbouring pKK223-3 D-AspO were grown in 1 litre of LB (Luria–Bertani) medium, containing 100 mg/mL ampicillin, at 37 °C. IPTG (isopropyl b-D-thiogalactoside) was added to a final concentration of 1 mM, and cells were grown for an additional 4–8 h. The bacterial suspension was centrifuged for 30 min at 3000 *g*, and the pellet was sonicated in 10 mL of 0.1 M citrate/phosphate buffer (McIlvaine buffer), pH 5.5, containing 30 mM sodium or potassium tartrate and 20 mM EDTA. The homogenate was heated at 55 °C for 5 min and centrifuged at 30000 *g* for 30 min. The supernatant was brought to 45% saturation with NH_4SO_4 and centrifuged as above. The precipitate was dissolved in 50 mL of 0.05 M Tris/HCl, pH 8.0, containing 10 mM sodium tartrate and 10 mM EDTA, and dialysed extensively against the same buffer. The dialysed sample was then added to a cation-exchange resin (DE-52 column, 2.5 cm×20 cm), equilibrated in 0.05 M Tris/HCl, pH 8.0, followed by 100 mL of the same buffer. The eluate was saturated to 45% with NH_4SO_4 and centrifuged as above. The pellet was dissolved in 5 mL of the same buffer and passed through a Sephacryl S-200 column (2.5 cm×80 cm) equilibrated in 0.1 M phosphate buffer, pH 7.0, containing 10 mM sodium or potassium tartrate. Fractions of 5 mL were collected, and those with the highest specific activity were pooled and concentrated to 2 mL by using an ultrafree filter membrane with cut-off of 30 kDa. The concentrated enzyme contained 5 mg/mL of protein and had an activity of 30 units/mg, where 1 unit is defined as the amount of the enzyme capable of oxidizing 1 mmol of D-Asp in 1 min at 37 °C.

Biosynthesis of D-Asp by a Racemase Activity

15. 0.05 M sodium phosphate buffer, pH 7.4, containing 10 mM of sodium EDTA.
16. Centrifuge (Beckman).
17. NaOH 1M.

3. Methods

Administration of D-Asp into Lizards

The D-Asp uptake by testis and the concomitant levels of testicular and plasma steroid hormones was studied. Lizards, belonging to main phases of their reproductive cycle, were distributed in two groups (25 animals each). Animals from the first group were injected intraperitoneally (i.p.) with 2.0 μmol of D-Asp/g body weight dissolved in 100 μL of reptilian saline. The lizards from second group received 100 μL saline solution and, therefore, were used as controls. Five injected lizards from each group were killed at set times within a period of 24 hrs (0, 3, 6, 15, 24 hrs after the injection, respectively).

The methods of capture and dissection and the captive rearing conditions were in accordance with Italian law (D. L.vo 116/92) and authorized by the appropriate Italian government administrative office (Servizio Veterinario della A.S.L. 44, Prot. Vet. 22/95).

Sex Steroid Assay in Plasma and Testis

1. Testis homogenized 1:5 (w/v) with distilled water.
2. Testis and plasma (100-200μL) samples vortexed with ethyl ether (1:10, v/v) for 5 min.
3. Centrifuged at 3,000 g for 10 min.
4. The extraction coming from two upper phase (ethyl ether) transferred to a glass tube.
5. Pooled ether phases left to evaporate on a hot plate at 40°-50°C under a hood.
6. The residue dissolved in a 0.5 mL sodium phosphate buffer and then utilized for the steroid assay performed by enzyme immunoassay kits purchased for laboratory research use. The assay is a colorimetric reaction which can be read at a wavelength between 405 and 420 nm. The following limits of detection were observed: for testosterone, sensitivity was 50 pg/mL (intra-assay variability 4.0 %, inter-assay variability 9.0 %) and for 17ß-estradiol, sensitivity was 6 pg/mL (intra-assay variability 6.0 %, inter-assay variability 7.5 %).

Specific Determination of D-Aspartic Acid

1. Testis and liver samples homogenized with 0.5 M PCA in a 1:10 ratio (w/v).
2. Centrifuged at 30,000 g for 20 min.
3. Supernatant was brought to pH 7.5-8.5 by the addition of 5 M KOH, cooled for 30 min at 0°C and the potassium perchlorate precipitate was removed by centrifugation.
4. The supernatant was adjusted to a pH of about 2.5 with 1 M HCl, and the amino acids were purified on a cation exchange column (AG 50W-X8 resin, hydrogen ionic form, 200-400 mesh) as described in step 5.
5. The sample was loaded on a column (1 x 3 cm) equilibrated with 0.01 M HCl, and, after a washing with 10 mL 0.01 M HCl, it was eluted with 8 mL of 4 M NH_4OH.
6. The eluates were dried by evaporation in small Petri dishes on a hot plate at 40-60 °C under a hood. The dry eluates were dissolved in 1 mL of 0.01 M HCl.
7. The samples prepared as in step 6 were further purified by slowly passing through a Sep-pak C-18 cartridge (300 mg) which had been previously activated with methanol (or acetonitrile) and washed with distilled water.
8. To recover the amino acids, the cartridge was eluted twice with 2 mL of 0.01 M HCl.

9. The resulting eluates were combined, and dried using a Savant centrifuge or left to evaporate in small Petri dishes at 40-50 °C under the hood.
10. The dry residues were then dissolved in 200 μL 0.01 M HCl.
11. The D-Asp was determined with a HPLC assay according to the method of OPA-NAC and using the D-aspartate oxidase (D-AspO) (EC 1.4.3.1) an oxidative enzyme that oxidizes D-Asp. This method has been widely described in a previous paper [27]. Briefly: 5–20 μL each sample were mixed with 20 μL 0.1 M NaOH (to obtain a final pH of about 9.5) and 10 μl *o*-phthaldialdehyde–*N*-acetyl-L-cysteine (OPA–NAC) reagent, prepared as follows: 4 mg OPA were dissolved in 300 μL methanol and then mixed with 500 μL 0·5 M borate buffer, pH 9.5, and 60 μL 1 M solution of NAC (previously brought to pH 5–6 with 4 M NaOH). After 2 min, 0.1 M sodium borate buffer, pH 9.5, was added to a final volume of 400 μL. Twenty microlitres of this mixture were injected onto a Supercosil C-18 HPLC column (0.45 x 25 cm; 5 μm beads; Supelco, Bellefonte, PA, USA) and the amino acids were eluted using a two-pump system. Buffer A consisted of 28 mL 1 M sodium acetate buffer, pH 5.5 (final 30 mM), 950 mL double-distilled water and 20 mL acetonitrile (2%). Buffer B consisted of 9.0 mL 1 M acetate buffer, pH 5.5 (final concentration 30 mM), 291 mL double-distilled water and 700 mL acetonitrile (70%). A linear gradient was developed from 0% B to 20% B over 20 min and then 100% B in 5 min. The flow rate was maintained at 1.2 mL/min throughout. The amino acid derivatives were detected fluorimetrically using excitation at 330 nm and emission at 450 nm. A standard curve was obtained using a mixture containing 17 different L-amino acids plus D-Asp, each at the concentration between 10-100 pmoles [4]. Recovery studies were performed by adding D-Asp a different concentrations to a sample homogenate and D-Asp was determined before and after addition of D-Asp. The percent (%) recovery obtained was between 94-108%. The inter-assay coefficient of variation (C.V.) was evaluated at several points along the standard curve in three different assays. The % C.V. was of 4.5-8.5% in the range between 10-100 pmol of D-Asp. The data were statistically analyzed using the STATVIEW version 4 program, ABASCUS Concepts.

Biosynthesis of D-Asp by a Racemase Activity

1. Testis and liver samples (five for each period) were homogenized (1:10 w/v) in 0.05 M sodium phosphate buffer, pH 7.4.
2. Centrifuged at 30,000 x *g* for 30 min.
3. 50 μl of the homogenate were mixed with 50 μL of 0.5 M L-Asp to the suspensions (or with other L-amino acids) in citrate buffer at different values of pH ranging from 4.0 to 8.0 and incubated at 37 °C for 120 min.
4. Control samples contained all components except L-amino acids. Incubations were stopped by immersion of the tubes in ice bath.
5. The amino acids present in the samples were extracted with 1.0 M perchloric acid (PCA).
6. Preformed D-Asp was determined using the HPLC as described in step 11.

Statistical Analysis

Data were compared by analysis of variance followed by Duncan's test for multi-group comparison and Student's *t* test for between-group comparison. All data were expressed as mean ± standard deviations. The level of significance was taken at $P < 0.01$ and $P < 0.05$. In addition, the correlation coefficient (r) between D-Asp content in the testis and both plasma and testicular concentrations of steroid hormones was calculated.

4. Results

Table 1 reports the profiles of endogenous levels of D-Asp and sex hormones (testosterone and 17ß-estradiol) in the testis of adult male lizards, *Podarcis s. sicula*, collected during the major phases of the sexual cycle. The plasma levels of sex hormones are also reported. Endogenous D-Asp in testis was significantly higher during the reproductive period than in pre- and post-reproductive. Likewise, testosterone concentration in the testis was higher in the reproductive phases and low in pre- and post-reproductive phases. In contrast, the 17ß-estradiol level was low in pre- reproductive and reproductive phases but significantly higher during the post-reproductive phase. The sex hormone levels in the plasma reflected those in the gonads, although in the plasma testosterone level was much higher in the reproductive phase. Comparison of free D-Asp content present in the testis with sexual steroid levels revealed a positive correlation between D-Asp and testosterone throughout the annual cycle ($r = 0.992$, $p<0.01$, in the testis and $r = 0.784$, $p<0.01$, in the plasma) and a negative correlation between D-Asp and 17ß-estradiol ($r = -0.932$, $p<0.01$, in the testis and $r = -0.999$, $p<0.01$, in the plasma).

In adult males *Podarcis s. sicula*, the i.p. injection of D-Asp (2.0 µmol/g body weight) was followed by its significant, although temporary, uptake by the testis in all sexual cycle stages (Fig. 1 A, B, C). In each phase of reproductive cycle D-Asp was rapidly taken up by the testis so that 3 hrs after injection its concentration in the tissue was greater than the value observed in animals injected with the saline alone. D-Asp levels were still high at 6 and 15 hrs after injection and then reached near baseline value within 24 hrs. The uptake of other D-/L- amino acids by the testis was evaluated, but was much lower than for D-Asp (data not shown). The D-Asp administration affected the levels of sex hormones (Fig. 1 Ai, Aii, Bi, Bii, Ci, Cii). A significant increase of testosterone was observed in testis at 3 hrs and in plasma at 6 hrs after D-Asp injection. The changes in 17β-estradiol after injection of D-Asp were opposed. Estradiol concentrations in testes decreased at 3 hrs and in the plasma at 6 hrs. However, this effect appeared to be reversible because plasma and testicular levels of 17β-estradiol returned to baseline values within 24 hrs of treatment.

Table 1. Endogenous D-Asp content in the testis and testosterone and 17ß-estradiol levels in the testis and plasma of the lizard *Podarcis s. sicula*, during the main phases of the reproductive cycle. Each value represents the mean ± SD of five determinations.

	Testis			Plasma	
Phases of the Reproductive cycle	*D-Asp (nmol/g)*	*Testosterone (ng/g)*	*17ß-estradiol (pg/g)*	*Testosterone (ng/ml)*	*17ß-estradiol (pg/ml)*
Pre-Reproductive	17.0 ± 1.2	35.0 ± 4.0	11.5 ± 0.9	8.0 ± 0.9	726 ± 212
Reproductive	30.0 ± 2.3	50.0 ± 4.9	7.5 ± 0.6	160.0 ± 9.7	217 ± 108
Post-Reproductive	3.5 ± 0.6	25.0 ± 1.4	33.5 ± 3.4	30.0 ± 3.1	1251 ± 324

Each value represents the mean ± SD of five determinations.

D-Asp: Pre-Reproductive versus Reproductive $p<0.01$; Reproductive versus Post-Reproductive $p<0.01$.

Testosterone in testis and plasma: Pre-Reproductive versus Reproductive $p<0.01$; Reproductive versus Post-Reproductive $p<0.01$.

17ß-estradiol in testis: Pre-Reproductive versus Post-Reproductive $p<0.05$; Reproductive versus Post-Reproductive $p<0.01$.

17ß-estradiol in plasma: Pre-Reproductive versus Reproductive $p<0.01$; Reproductive versus Post-Reproductive $p<0.01$

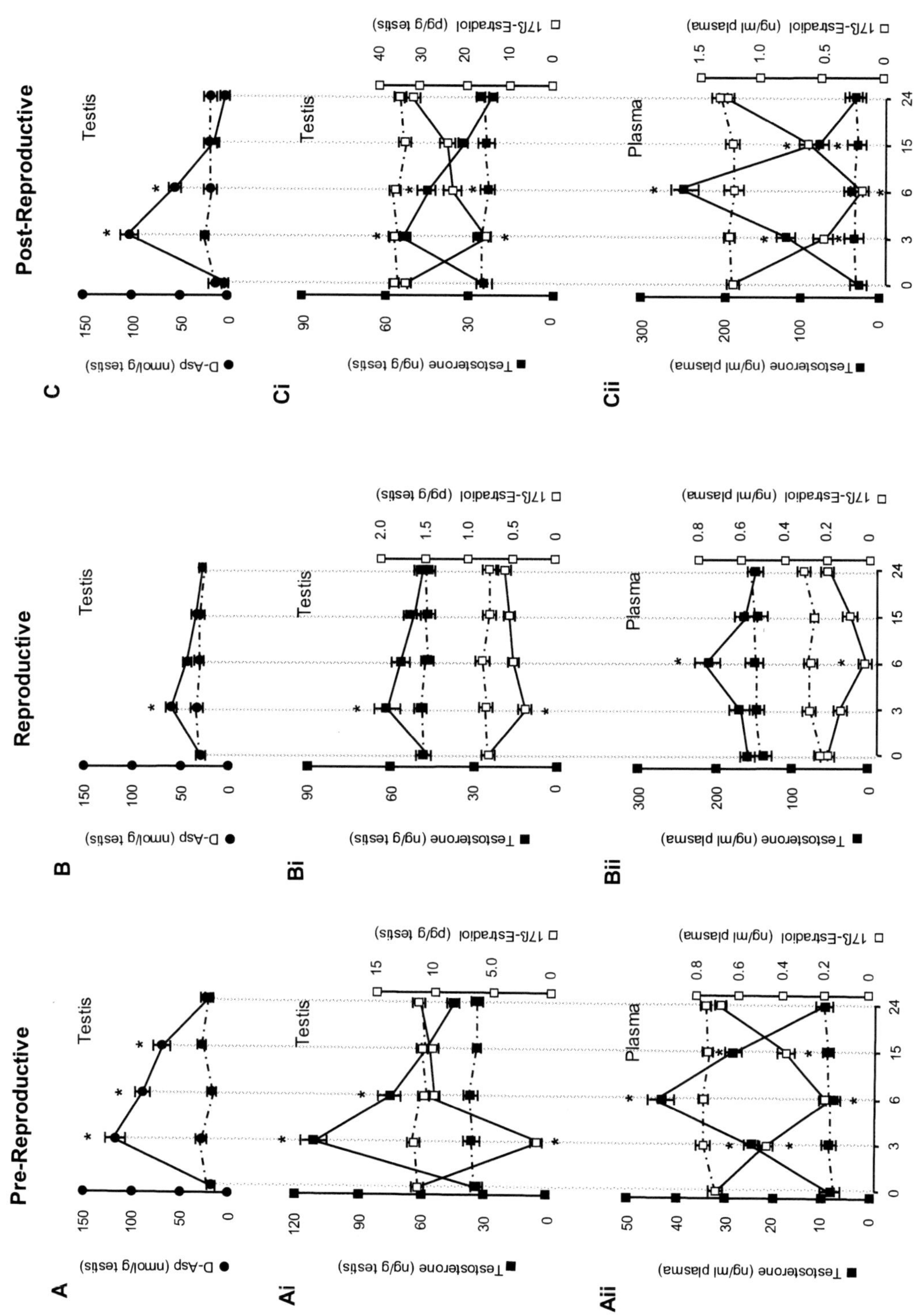

Figure 1. Uptake of D-Asp (A, B, C, —●—, - - - ● - - -) and concentrations of steroid hormones (testosterone = —■—, - - - ■ - - -; 17ß-estradiol = —□—, - - - □ - - -) in both testis (Ai, Bi, Ci) and plasma (Aii, Bii, Cii) during the reproductive cycle. Animals were injected at 0 time either with D-Asp or with saline solution. Solid lines represent D-Asp treated animals, broken lines represent control animals. Each value represents means ± SD of five determinations. *, P< 0.01 versus 0 time.

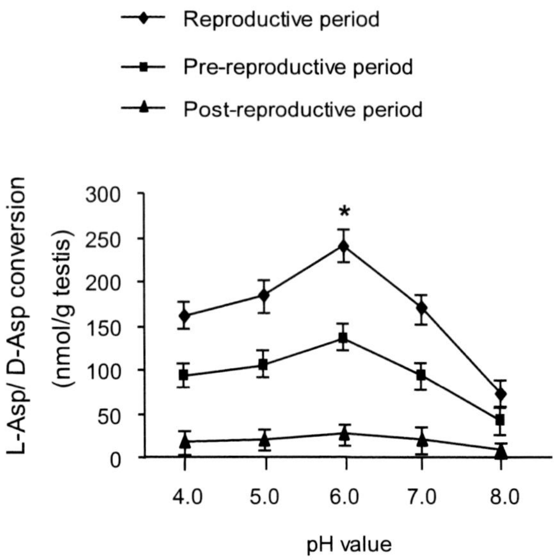

Figure 2. Aspartate racemase activity during reproductive cycle. Each value represents the mean ± SD of five determinations. *, P<0.01 versus pre- and post-reproductive periods.

To verify whether D-Asp is locally synthesized by L-Asp through an aspartate racemase we measured the racemase activity by evaluating the in vitro rate conversion of L-Asp into D-Asp during the reproductive cycle (Figure 2). In testis and liver tissue the conversion rate (L-Asp/D-Asp) was the highest when the in vitro incubation was carried out at pH 6.0. Testicular tissue converted L-Asp into D-Asp in all phases of the cycle and D-Asp biosynthesis significantly varied only during the reproductive period reaching its maximal level. In the liver (control tissue) racemase activity was observed but remained unchanged through the cycle (data not shown).

5. Notes

1. Sex steroid recovery was 85% from plasma and 80% from tissues. The addition of D-Asp to the standard curve did not modify the assay sensitivity.
2. HPLC assay method allowed us to determine a minimum amount of D-Asp corresponding to 10 pmol/assay mixture.

References

[1] D'Aniello, A., Di Fiore, M.M. & Fisher, G. (1998). Occurrence of D-aspartic acid in animal tissues and its role in the nervous and endocrine systems. *Trends in Comp Biochem and Physiol, 4*, 1-21.

[2] Schell, M.J., Cooper, O.B. & Snyder, S.H. (1997). D-Aspartate localizations imply neuronal and neuroendocrine roles. *PNAS, 94*, 2013-2018.

[3] Wolosker, H., D'Aniello, A. & Snyder, S.H. (2000). D-Aspartate disposition in neuronal and endocrine tissues: ontogeny, biosynthesis and release. *Neurosci, 100*, 183-189.

[4] D'Aniello, A., Di Fiore, M.M., Fisher, G.H., Milone, A., Seleni, A., D'Aniello, S., Perna, A.F. & Ingrosso, D. (2000a). Occurrence of D-aspartic acid and N-methyl- D-aspartic acid in rat

neuroendocrine tissues and their role in the modulation of Luteinizing hormone and Growth hormone release. *Faseb J, 14*, 699-714.

[5] D'Aniello, G., Tolino, A., D'Aniello, A., Errico, F., Fisher, G.H. & Di Fiore, M.M. (2000b). The role of D-aspartic acid and *N*-methyl-D-aspartic acid in the regulation of prolactin release. *Endocrinol, 141*, 3862-3870.

[6] Wang, H., Wolosker, H., Pevsner, J., Snyder, S.H. & Selkoe, D.J. (2000). Regulation of rat magnocellular neurosecretory system by D-aspartate: evidence for biological role(s) of a naturally occurring free D-amino acid in mammals. *J Endocrinol, 167*, 247-252.

[7] Wang, H., Wolosker, H., Morris, J.F., Pevsner, J., Snyder, S.H. & Selkoe, D.J. (2002). Naturally occurring free D-aspartate is a nuclear component of cells in the mammalian hypothalamo-neurohypophyseal-system. *Neurosci, 109*, 1-4.

[8] Neidle, A. & Dunlop, D.S. (1990). Developmental changes in free D-aspartic acid in the chicken embryo and in the neonatal rat. *Life Sci, 46,* 1517-1522.

[9] Dunlop, D.S., Neidle, A., McHale, D., Dunlop, D.M. & Lajtha, A. (1986). The presence of free D-aspartic acid in rodents and man. *Biochem Biophys Res Com, 141*, 27-32.

[10] Hashimoto, A., Nishikawa, T., Oka, T., Hayashi, T. & Takahashi, K. (1993). Widespread distribution of free D-aspartate in rat periphery. *FEBS Lett, 331*, 4-8.

[11] D'Aniello, A., Vetere, A. & Petrucelli, L. (1993). Further study on the specificity of D-amino acid oxidase and of D-aspartate oxidase and time course for complete oxidation of D-amino acids. *Comp Biochem Physiol, 105B,* 731-734.

[12] Fisher, G.H., D'Aniello, A., Vetere, A., Padula, L., Cusano, G.P. & Man, E.H. (1991). Free D-aspartate and D-alanine in normal and Alzheimer brain. *Brain Res Bull, 26,* 983-985.

[13] Fisher, G.H., Petrucelli, L., Gardner, C., Emory, C., Frey, W.H. 2nd, Amaducci, L., Sorbi, S., Sorrentino, G., Borghi, M., D'Aniello, A. (1994). Free D-amino acids in human cerebrospinal fluid of Alzheimer disease, multiple sclerosis, and healthy control subjects. *Mol Chem Neuropathol, 23,* 115-124.

[14] Hashimoto, A. & Oka, T. (1997). Free D-aspartate and D-serine in the mammalian brain and periphery. *Prog Neurobiol, 52,* 325-353.

[15] Imai, K., Fukushima, T., Santa, T., Homma, H., Sugihara, J., Kodama, H. & Yoshikawa, M. (1997). Accumulation of radioactivity in the brain and peripheral tissues including salivary gland after intravenous administration of C-14-aspartic acid. *Proc Japan Acad, 73 (B)*, 48-52.

[16] Sakai, K., Homma, H., Lee, J.A., Fukushima, T., Santa, T., Tashiro, K., Iwatsubo, T., Imai, K. (1998). Emergence of D-aspartic acid in the differentiating neurons of the rat central nervous system. *Brain Res,* 808, 65-71.

[17] Imai, K., Fukushima, T., Hagiwara, K. & Santa, T. (1995). Occurrence of D-aspartic acid in rat brain pineal gland. *Biomed Chromatog, 9,* 106-109.

[18] Lee, J.A., Homma, H., Sakai, K., Fukushima, T., Santa, T., Tashiro, K., Iwatsubo, T., Yoshikawa, M. & Imai, K. (1997). Immunohistochemical localization of D-aspartate in rat pineal gland. *Biochem Biophys Res Com, 231,* 505-508.

[19] Takigawa, Y., Homma, H., Lee, J.A., Fukushima, T., Santa, S., Iwatsubo, T. & Imai, K. (1998). D-aspartate uptake into cultured rat pinealocytes and the concomitant effect on L-aspartate levels and melatonin secretion. *Biochem Biophys Res Com, 248*, 641-647.

[20] Hamase, K., Homma, H., Takigawa, Y., Fukushima, T., Santa, T., Imai, K. (1997). Regional distribution and postnatal changes of D-amino acids in rat brain. *Biochim Biophys Acta, 1334,* 214-222.

[21] Lee, J.A., Long, Z., Nimura, N., Iwatsubo, T., Imai, K. & Homma, H. (2001). Localization, transport, and uptake of D-aspartate in the rat adrenal and pituitary glands. *Arch Biochem Biophys, 385,* 242-249.

[22] Sakai, K., Homma, H., Lee, J.A., Fukushima, T., Santa, T., Tashiro, K., Iwatsubo, T. & Imai, K. (1998). Localization of D-aspartic acid in elongate spermatids in rat testis. *Arch Biochem Biophys, 351,* 96-105.

[23] D'Aniello A, Di Cosmo A, Di Cristo C, Annunziato L, Petrucelli L, Fisher G (1996) Involvement of D-aspartic acid in the synthesis of testosterone in rat testes. *Life Sci, 59,* 97-104.

[24] D'Aniello, A., Di Fiore, M.M., D'Aniello, G., Colin, F.E., Lewis, G. & Setchell, B.P. (1998b). Secretion of D-aspartic acid by the rat testis and its role in endocrinology of the testis and spermatogenesis. F*EBS Lett, 436*, 23-27.

[25] Ishio, S., Yamada, H., Hayashi, M., Yatsushiro, S., Noumi, T., Yamaguchi, A. & Moriyama, Y. (1998). D-aspartate modulates melatonin synthesis in rat pinealocytes. *Neurosci Lett, 249,* 143-146

[26] Nagata, Y., Homma, H., Lee, J.A. & Imai, K. (1999a). D-aspartate stimulation of testosterone synthesis in rat Leydig cells. *FEBS Lett, 444,* 160-164.

[27] Di Fiore, M.M., Assisi, L., Botte, V. & D'Aniello, A. (1998). D-Aspartic acid is implicated in the control of testosterone production by the vertebrate gonad. Studies on the female green frog *Rana esculenta. J Endocrinol, 157,* 199-207.

[28] Assisi, L., Botte, V., D'Aniello, A. & Di Fiore. M.M. (2001). Enhancement of aromatase activity by D-aspartic acid in the ovary of the lizard *Podarcis s. sicula. Reprod, 121,* 803-808.

[29] Raucci F, Assisi L, D'Aniello S, Spinelli P, Botte V, Di Fiore MM (2004) Testicular endocrine activity is upregulated by D-aspartic acid in the green frog, *Rana esculenta. J Endocrinol, 182,* 365-376.

[30] Nagata, Y., Homma, H., Matsumoto, M. & Imai, K. (1999b). Stimulation of steroidogenic acute regulatory protein (StAR) gene expression by D-aspartate in rat Leydig cells. *FEBS Lett, 454,* 317-320.

[31] Negri, A., Tedeschi, G., Ceciliani, F. & Ronchi, S. (1999). Purification of beef kidney D-aspartate oxidase overexpressed in *Escherichia coli* and characterization of its redox potentials and oxidative activity towards agonists and antagonists of excitatory amino acid receptors. *Biochim Biophys Acta, 143,1* 212-222.

In: D-Amino Acids: ISBN 1-60021-075-9
A New Frontier in Amino Acid and Protein Research, pp.231-234

Chapter 3.10

D-Aspartic Acid in the Nervous System of a Mollusc: An Immunohistochemical Protocol

Biagio D'Aniello*[1]*, Luciano Luongo*[1]*, Antimo D'Aniello*[2]*,
***Claudia Pinelli*[3]*, and Rakesh K. Rastogi*[1,*]**
[1]Department of Structural and Functional Biology, University of Naples Federico II, Italy;
[2]Laboratory of Neurobiology, Stazione Zoologica "A. Dohrn", Italy;
[3]Department of Life Sciences, SUN, Italy.

Keywords: D-aspartic acid, immunohistochemistry, nervous system, mollusc, *Aplysia.*

1. Introduction

D-Aspartic acid (D-Asp) was first detected by biochemical methods in the nervous tissue of cephalopod mollusc [1,2]. Subsequently, this endogenous amino acid has been described in various tissues of the representatives of most vertebrate and invertebrate groups. The widespread distribution of this naturally occurring free amino acid is suggestive of its putative roles in neurotransmission and transcriptional activity [3-5]. Recently, D-Asp was described in the nervous system of *Aplysia limacina* (opisthobranch gasteropod mollusc) and its possible role in neurotransmission was suggested [6]. Here we describe an immunohistochemical protocol to demonstrate the intracellular distribution of D-Asp in the cerebral ganglia of *A. limacina.*

2. Materials

1. Fixative (2% glutaraldehyde plus 2% paraformaldehyde in 0.1M PBS).
2. Vacuum oven.
3. Rotary microtome.
4. Slide staining set with polypropylene dishes and racks plus Immuno slide staining trays.
5. Clearing agent, Bioclear (Bio Optica, Milan, Italy).

* Correspondence concerning this article should be addressed to Rakesh K. Rastogi, E-mail: rastogi@unina.it; FAX: +39-081-679210.

6. Anti-D-Asp antiserum (see below, Notes Section).
7. Biotinylated secondary antibody, goat anti-rabbit IgG (PIERCE, Rockford, IL, USA).
8. Avidin-biotin-horseradish peroxidase ABC Complex (Vectastain Kit, Vector Labs, Burlingame, CA, USA).
9. Chromogen, 3.3'-diaminobenzidine tetrahydrochloride (DAB; Sigma Chemical Co., St. Louis, MO, USA).
10. Mounting medium, Permount (Biomeda Corp., Foster City, CA, USA).
11. Professional research microscope.
12. Digital camera.

Adult *A. limacina* were collected in the gulf of Naples and acclimatized at Stazione Zoologica, in appropriate containers with circulating sea water at a temperature of 20±1°C. Laboratory grade standard reagents for histology and immunohistochemistry were used. Routine reagents were obtained from Carlo Erba Reagenti (Milan, Italy) and Merck (Darmstadt, Germany).

3. Methods

1. Animals were anaesthesized with one intraperitoneal injection of 0.034M $MgCl_2$ solution in PBS (ca. 5 mL/kg body weight).
2. Immediately after their surgical removal from the anaethesized animals, the cerebral ganglia were placed in the fixative for 24 h.
3. After dehydration, the ganglia were included in paraffin wax (56°C melting point), and 5-7 μm paraffin sections were placed on clean glass slides, air dried for 24 h and deparaffinized (two changes of 20 min each) in Bioclear.
4. Slides were then rehydrated through a graded series of ethanol (75%, 85%, 95%, absolute ethanol), with all steps repeated twice for 5 min each, rinsed in distilled water for 5 min and washed twice in 0.1 M PBS, pH 7.4, for a total of 20 min.
5. Non-specific binding sites were blocked by 20-min incubation in normal goat serum (Sigma), diluited 1:50 in PBS, at room temperature in the dark.
6. The sections were then rinsed in PBS and incubated with anti-D-Asp (diluted 1:10 after saturation with L-amino acid-glutaraldehyde-BSA) in a dark moist chamber overnight at 4°C.
7. Subsequently, the sections were rinsed in PBS for 20 min and incubated for 1 h with the secondary antibody (1:200), at room temperature in a dark moist chamber.
8. After a rinse in PBS, the sections were incubated for 1 h with ABC complex (1:200) at room temperature in a dark chamber.
9. Sections were then rinsed for 10 min in PBS followed by a rinse for 5 min in 0.1 M Tris-HCl, pH 7.4. The antigen-antibody complex was visualized by exposure at room temperature for 2 min to DAB (50 μg/mL in 0.05 M Tris buffer, pH 7.4) containing 0.0015% H_2O_2.
10. After a 10-min washing in distilled water, tissue sections were dehydrated using graded series of alcohols and Bioclear (two immersions of 10 min each) as clearing agent.
11. Slides were placed on a horizontal surface and small drops of Permount were applied to the tissue sections. A coverslip was carefully applied on the top of the Permount. Slides were placed horizontally for drying in an oven at 50°C for 45 min and allowed to cool down to room temperature for microscopic visualization.
12. Homologous and heterologous preabsorption controls and those omitting the primary antibody were carried out.

13. The specimens were examined in a Leica DMRB microscope and photographed using Canon Power Shot S50 digital camera. The digital images were adjusted for brightness and contrast using Adobe Photoshop 6, whereas panel montage and lettering were made using CorelDraw9 and Microsoft Office PowerPoint 2003.

4. Results

Using this technique, it has been possible to distinguish that D-Asp is localized in specific areas of the neuronal cell types within the *Aplysia* nervous system (Figure 1). Immunopositivity for D-Asp presented a substantial difference of distribution within the neuronal cell population. While the intensity of immunoreaction varies among neurons, there are immunonegative neurons as well. Frequently, either the cytoplasm has a more intense immunopositivity than the nucleus or *vice versa*. In some neurons, however, the immunopositivity is shown in the cytoplasm as well as in the nucleus. Tests for specificity of immunoreaction included omission of the primary antibody and such preparations did not show any immunopositivity in the neurons. Similarly, when the antibodies anti-D-Asp and L-amino acid-glutaraldehyde-BSA saturated antibody were preabsorbed either with D-Asp alone or with D-Asp-glutaraldehyde-BSA complex, no immunoreaction was visualized. Heterologous preabsorption of the primary antibody with immunogen complexes containing L-Asp, L-Glu or other amino acids including *N*-methyl- D-aspartic acid did not affect immunoreaction in the neurons. The homologous preabsorption test, however, showed that the immunopositivity was totally abolished.

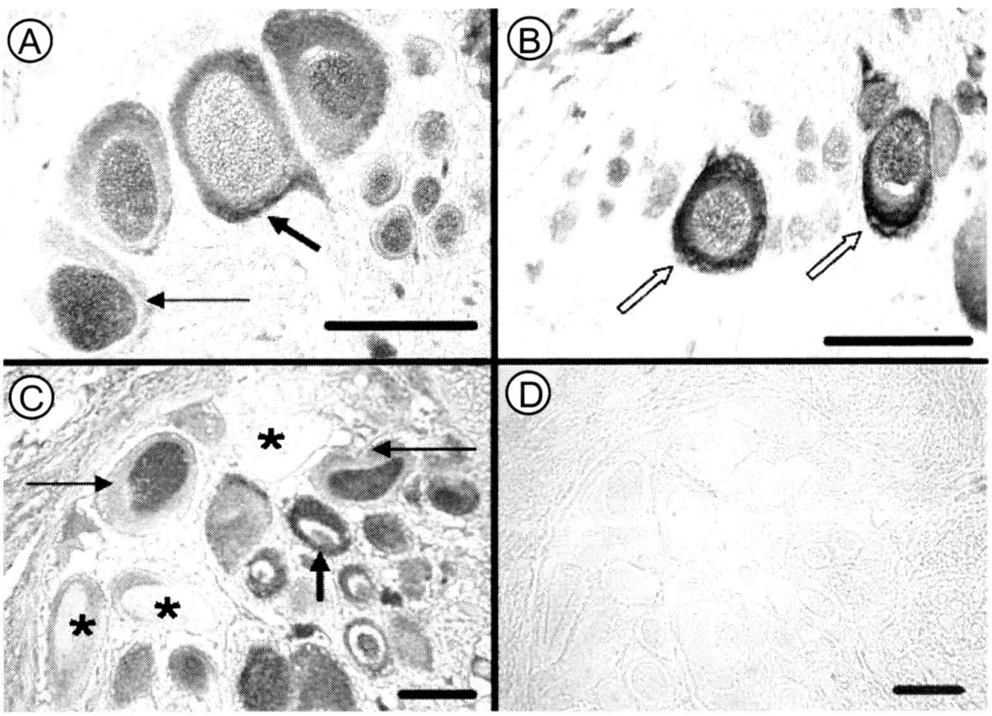

Figure 1. Microphotographs of cross sections of the cerebral ganglia of *Aplysia limacina*. Scale bars: A-D, 100 µm. Thin arrows indicate neurons with D-Asp immunoreactivity in the nucleus, thick arrows show similar immunoreactivity only in the cytoplasm whereas empty arrows point to neurons where such immunoreaction is present within the nucleus and the cytoplasm. Asterisks show immunonegative neurons. No immunoreactivity is observed in the control D.

5. Notes

Since D-Asp by itself is not immunogenic, the antibody against D-Asp was produced in rabbits in the laboratory of Neurobiology at the Stazione Zoologica by using D-Asp-glutaraldehyde-BSA complex as an immunogen. The methodology was substantially based upon that described earlier by Schell et al. [3]. Briefly, D-Asp-glutaraldehyde-BSA complex was dialyzed before being injected into 3-month old rabbit (six injections, fortnightly). ELISA was used to assess the immunogenicity of the antiserum. In order to avoid undesired and non-specific background staining as well cross reactivity with other immunogens, it is necessary to process the polyclonal antibody with affinity purification using nitrocellulose membranes (Hybond-C extra, nitrocellulose 0.45 μm; Cat. RPN203E Amersham Biosciences, Buckinghamshire, UK). Any cross reactivity of our antibody with BSA was removed by saturating the antibody with BSA prior to use. Use of L-amino acid-glutaraldehyde-BSA in specificity tests has been extremely important in verifying that what we see as immunostain was accounted by the presence of D-Asp. Despite the inherent limitations of immunohistochemistry, it is plausible that the immunohistochemical distribution profile of this amino acid may be important in future studies related to function.

Acknowledgements

Financial aid by the University of Naples Federico II is recognized.

References

[1] D'Aniello, A. & Giuditta, A. (1977). Identification of D-aspartic acid in the brain of *Octopus vulgaris*. *J Neurochem, 9*, 38-48.

[2] D'Aniello, A. & Giuditta, A. (1978). Presence of D-aspartate in squid axoplasm and in other regions of the cephalopod nervous system. *J Neurochem, 31*, 1107-1108.

[3] Schell, M.J., Cooper, O.B. & Snyder, S.H. (1997). D-Aspartate localizations imply neuronal and neuroendocrine roles. *Proc Natl Acad Sci USA, 94*, 2013-2018.

[4] D'Aniello, A., Spinelli, P., De Simone, A., D'Aniello, S., Branno, M., Aniello, F., Fisher, G.H., Di Fiore, M.M. & Rastogi, R.K. (2003). Occurrence and neuroendocrine role of D-Asp and NMDA in *Ciona intestinalis*. *FEBS Lett, 552*, 193-198.

[5] Raucci, F., Santillo, A., D'Aniello, A., Chieffi, P. & Chieffi Baccari, G. (2005). D-Aspartate modulates transcriptional activity in harderian gland of frog, *Rana esculenta*: morphological and molecular evidence. *J Cell Physiol, 204*, 445-454.

[6] Spinelli, P., Brown, E.R., Ferrandino, G., Branno, M., Montarolo, P.G., D'Aniello E., Rastogi, R.K., D'Aniello, B., Chieffi, G., Fisher, G. & D'Aniello, A. (2006). D-Aspartic acid in the nervous system of *Aplysia limacina*: possibile role in neurotransmission. *J Cell Physiol, 206*, 672-681.

In: D-Amino Acids: ISBN 1-60021-075-9
A New Frontier in Amino Acid and Protein Research, pp.235-247

Chapter 3.11

ANALYSIS, DISTRIBUTION, AND PHYSIOLOGICAL FUNCTIONS OF FREE D-ALANINE IN AQUATIC INVERTEBRATES

Hiroki Abe*
Department of Aquatic Bioscience, The University of Tokyo, Japan.

Keywords: alanine racemase, D-alanine, D-amino acid oxidase, invertebrate, physiological function.

1. INTRODUCTION

Research on free D-amino acids in invertebrates has rather a long history. In insect species, free D-alanine and D-serine have long been known to exist in milkweed bug and silkworm, respectively [1]. Serine racemase was recently purified partially from silkworm [2]. In marine invertebrates, D-amino acid research also dates back to 1970s. A large amount of free D-aspartate was firstly discovered in the brain of octopus *Octopus vulgaris* in 1977 [3] and thereafter free D-alanine was found in several crustacean muscles in 1980 [4]. In 1984, free D-alanine was also reported to exist in the mantle muscle of a small brackish water clam *Corbicula japonica*, foot muscles of short-necked clam *Tapes philippinarum* and hard clam *Meretrix lamarckii*, but not in mussel *Mytilus edulis* and oyster *Crassostrea gigas* [5]. Alanine racemase acitivity was also detected in the foot muscle of *C. japonica*. After that, D-amino acid research on bivalve mollusks preceded that on crustaceans. Free D-aspartate and aspartate racemase activity were detected in the foot muscle and gill of the bivalve *Solemya reidi* [6]. Although amino acid species were not specified, a survey of D-amino acid content in 43 species of 8 invertebrate phyla revealed that D-amino acids were found in 18 species in 6 phyla including bivalves, crustaceans, and annelids [7]. In these research works, D-amino acids were determined indirectly using D-amino acid and D-aspartate oxidases or other methods. On various species of bivalve mollusks, D-and L-enantiomers of alanine and aspartate were firstly determined directly by high-performance liquid chromatography (HPLC) in 1987. By using this method, Felbeck and Wiley [8] reported that the foot muscle of bivalves, *S. reidi*, *M. edulis*, and mantle muscle of *C. gigas* contained almost no D-alanine but the former two species contained a considerable amount of D-aspartate. Three lucinid clams, *Codakia* sp., represented the largest amount of D-alanine (22-187 μmol/g wet wt.) but no D-

* Correspondence concerning this article should be addressed to Hiroki Abe E-mail: aabe@mail.ecc.u-tokyo.ac.jp, FAX: +81-3-5841-8166.

aspartate. Thus, the distribution of these two D-amino acids is predicted to be exclusive to each other. One exception is a mytilid clam *Modiolus capax* which contained both D-amino acids in equal amounts.

From 1990s, several highly sensitive HPLC methods have been developed and made remarkable advances in D-amino acid research even in marine animals. In this chapter, I describe two HPLC methods for separating D-and L-amino acids, and the distribution and physiological functions of free D-alanine in aquatic invertebrates. I also describe alanine racemase and D-amino acid utilization in crustaceans and fish species, respectively.

2. Materials

1. Live invertebrate animals were obtained from local aquaculture farms or Tokyo Metropolitan Central Wholesale Fish Market.
2. High-performance liquid chromatograph (HPLC): A Jasco HPLC system (Tokyo, Japan) consisted of a PU-1580 pump, a DG-1580-53 line degasser, an LG-1580-02 ternary gradient unit, a CO-1565 column oven, an AS-2057 autosampler with cooling system, an FP-1520 fluorescence detector, and an 807-IT integrator, or a Shimadzu HPLC system (Kyoto, Japan) consisted of three LC-6A pumps, a CTO-6A column oven, an SCL-6B system controller, a GT-103 line degasser, a SIL-6B autosampler with cooling system, an RF-535 fluorescence detector, and a Chromatopack C-R7A data processor. A reversed-phase Shim-pack CLC-ODS (250 × 4.6 mm I.D.; Shimadzu) was employed with a guard column for both HPLC systems.
3. Amino acid analyzer: A Hitachi L-8500A amino acid autoanalyzer (Hitachi; Tokyo, Japan) for physiological amino acid was used for free amino acid determination.
4. Aquaria: For rearing invertebrate species, several 60 L volume glass aquaria with aerated circulating water were used with dechlorinated freshwater, filtrated natural seawater, or artificial seawater-salt mixture. Salinity was determined with a salinity refractometer S/Mill-E (Atago; Tokyo, Japan).
5. Homogenizer: A Polytron homogenizer with a PTA-10 blade (Kinematica; Littau-Lucerne, Switzerland) was used for tissue extraction.
6. Centrifuge: A high-speed centrifuge 50A-IV (Sakuma; Tokyo, Japan) with 1B rotor or a micro-centrifuge 1900 (Kubota; Tokyo, Japan) were employed.
7. Reagents: Chiral reagents, (+)-1-(9-fluorenyl)ethyl chloroformate (FLEC) and *N*-tert-butyloxycarbonyl-L-cysteine (Boc-L-Cys) were purchased from Eka Nobel (Surte, Sweden) and Novabiochem (Läufelfingen, Switzerland), respectively. Other reagents were of analytical grade and obtained from Sigma-Aldrich (St. Louise, MO, USA) or Wako Pure Chemical Industries (Osaka, Japan).

3. Methods

1. Acclimation of invertebrates to different salinity: Freshwater crayfish *Procambarus clarkii* were transferred from freshwater to 50% seawater and kept for 2 days. Subsequently, the animals were transferred to 66% seawater and then 75% after 2 days and kept for more 2 days. The other seawater species were transferred from 100% seawater to 125 or 75% seawater. After 2 days acclimation, the animals were further acclimated to 150 or 50% seawater, respectively, and kept for more 2 days. For preparing high salinity seawater over 100%, artificial seawater-salt mixture was added to the filtrated natural seawater. During acclimation periods, the animals were not fed.

2. Tissue extraction: Tissues were excised and quickly freeze-clamped using aluminum tong pre-cooled in liquid nitrogen or simply dipped into liquid nitrogen. One to two grams of tissues (minimum 0.5 g) were homogenized in 9 volumes of 8% perchloric acid. After centrifugation at 10,000 *g* for 10 min at 0°C, the supernatant was neutralized with solid potassium hydrogen carbonate and centrifuged again to eliminate the crystals of potassium perchlorate formed. The supernatant solution was used as tissue extract and served to free amino acid and D- and L-amino acid analyses.
3. Pre-column derivatization of amino acids: Pre-column derivatization of D- and L- amino acids with FLEC was performed essentially according to Einarsson et al. [9]. In a 2-mL Mixxor separation vial (Lidex; Technion, Israel), 0.1 mL of sample solution was mixed with 0.4 mL of 0.3 M borate-NaOH buffer, pH 11.0, and 0.5 mL of 18 mM FLEC in acetone. After reaction for 45 s at ambient temperature (Figure 1A), the reaction mixture was extracted twice with 2 mL of pentane to remove excess of reagent. The aqueous phase was filtered through a 0.45-μm filter (Toyo Roshi; Tokyo, Japan) and injected into the Shimadzu HPLC system. Another pre-column derivatization with Boc-L-Cys and *o*-phthaldialdehyde (OPA) was carried out according to Hashimoto et al. [10]. A 200-μL derivatization reagent (dissolving 50 mg OPA and 50 mg Boc-L-Cys in 5 mL of methanol) prepared daily and 700 μL of 0.1 M borate buffer, pH 9.2, were added to a vial containing 100 μL of amino acid standard or sample. After 2 min derivatization at ambient temperature, a 5 μL volume of the reaction mixture was injected into the Jasco HPLC system.
4. HPLC determination of D- and L-amino acids: The mobile phase for FLEC derivatized D- and L-amino acid separation consisted of 15 mM citric acid containing 10 mM tetramethylammonium chloride (TMA) as an ion-pair reagent, acetonitrile, and tetrahydrofuran (THF). After mixing these solvents, the apparent pH of the mobile phase was adjusted to a suitable value with 6 M NaOH or HCl (Figure 1B). Elution was carried out with a ternary gradient using high-pressure mixing with three pumps at a flow-rate of 0.7 mL/min and at a column temperature of 55°C. Separation was conducted with two separate runs differed in mobile phase composition and gradient profile (Figure 1B). FLEC-amino acids were monitored fluorimetrically with excitation and emission wavelengths of 260 and 310 nm, respectively. For the separation of Boc-L-Cys and OPA derivatized D-and L-amino acids, a mobile phase A consisted of 0.05 M monopotassium phosphate, pH 6.5, acetonitrile, and THF (92:5:3) and a mobile phase B consisted of the same reagents (45:50:5). A linear gradient elution was performed from 0%B to 49.8%B within 95 min at a flow rate of 0.7 mL/min at 32°C and then washed with 100%B for 12 min. The eluates were monitored with excitation and emission wavelengths of 344 and 443 nm, respectively.
5. Assay method for alanine racemase: Alanine racemase activity was assayed by determining D- and L-alanine with HPLC. Reaction mixture contained 0.1 M Tris-HCl, pH 8.5, 200 mM D- or L-alanine, and enzyme solution in a final volume of 1 mL. After 10-30 min incubation at 37°C, a 0.1-mL aliquot was taken and deproteinized with 0.1 mL of 0.6 M perchloric acid. After centrifugation at 10,000 *g* for 2 min, the supernatant was neutralized with 0.1 mL of 0.6 M potassium bicarbonate and centrifuged as above. The resulting supernatant was injected into HPLC for D- and L-alanine determination. Alanine enantiomers were separated by the above described Jasco HPLC system equipped with a chiral column, Sumichiral OA-5000 (150 × 4.6 mm I.D.; Sumika Chemical Analysis Center, Osaka, Japan). As the mobile phase, 1 mM copper sulfate was used at a flow rate of 1 mL/min. At an ambient temperature, D- and L-alanine eluted from the column were monitored at 254 nm as alanine-copper complex. Enzyme activity was calculated from the increase of D- or L-alanine.
6. Assay method for D-amino acid oxidase: D-Amino acid oxidase activity was determined essentially according to D'Aniello et al. [11]. In brief, the enzyme solution (200 μL) was incubated with 200 μL of 0.1 M D-alanine dissolved in 0.1 M Tris-HCl, pH 8.2, for 30 min at 37°C. After incubation, 200 μL of 25% trichloroacetic acid were added to terminate the reaction

and centrifuged. From the supernatant, 400 µL aliquots were mixed with 400 µL of 1 mM 2, 4-dinitrophenylhydrazine in 1 M HCl and incubated at 37°C for 10 min. Subsequently, 800 µL of 1.5 M NaOH was added. After further incubation for 10 min, the absorbance was read at 445 nm.

A

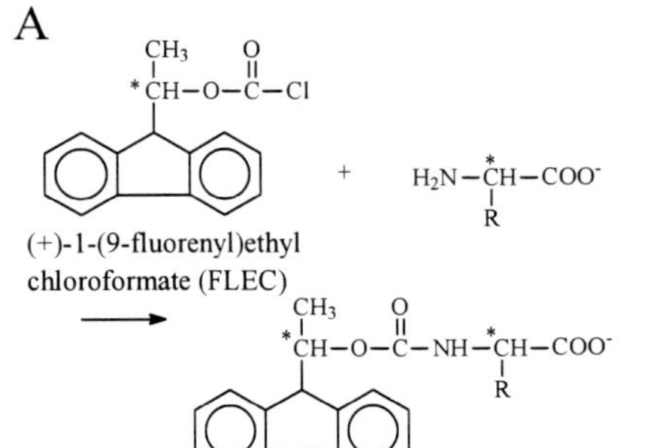

B

Mobile phase	Run 1 A	Run 1 B	Run 1 C%	Run 2 A	Run 2 B	Run 2 C%
Citrate-TMA	75	70	30	73	60	30
Acetonitrile	10	20	60	25	35	60
Tetrahydrofuran	15	10	10	2	5	10
pH	2.0	5.3	6.2	2.0	5.3	6.2
Gradient conditions						
0 min	100	0	0	100	0	0
3	20	80	0	60	40	0
10	20	80	0			
15				35	45	0
32				39	46	15
55				0	50	50
65				0	0	100
75				0	0	100
80	0	0	100			
90	0	0	100			

C D

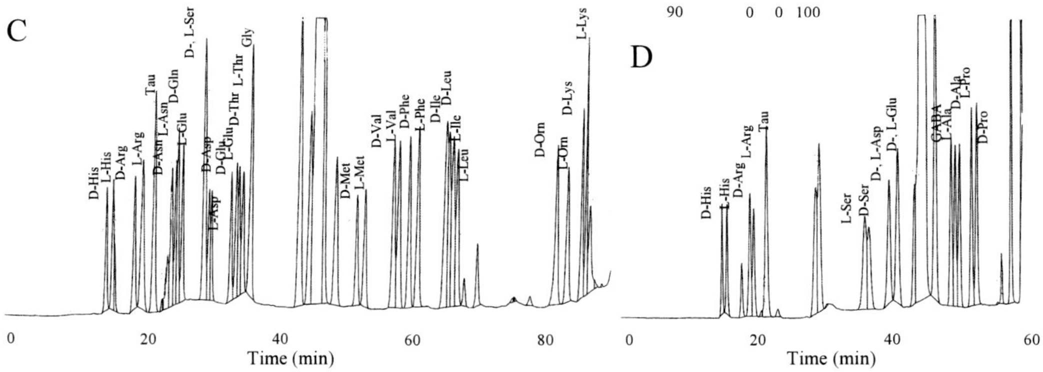

Figure 1. Determination of D- and L-enantiomers of amino acids pre-labeled with (+)-1-(9-fluorenyl) ethyl chloroformate (FLEC). A: reaction of amino acid with FLEC, B: mobile phase compositions and gradient conditions of two separate runs, C: a standard chromatogram of run 1, D: a chromatogram of run 2. Only major protein and non-protein amino acids are named in the chromatograms.

4. Results

1. Analyses of D- and L-Amino Acids in Invertebrate Tissues

An HPLC chromatogram of FLEC derivatives of D- and L-amino acids is shown in Figure 1C. In this run, however, D- and L-serine, -alanine, and -proline were overlapped with a large peak of hydrolyzed FLEC between 40 and 50 min. Thus, a second run (Figure 1D) is needed for the separation of these amino acids. With this FLEC method, 19 amino acids were separated into their D- and L-enantiomers and from other physiological amino compounds, such as taurine, γ-aminobutylic acid, ornithine, β-alanine, and so on [12]. Tyrosine, tryptophan, and cysteine did not react sufficiently with FLEC under the conditions described above. The detection limit for these D- and L-amino acids was below 1 pmol at a signal-to-noise ratio of 2.6:1.

FLEC is now available from Sigma-Aldrich but a very expensive reagent. Thus, we now use the Boc-L-Cys and OPA method (Figure 2). In this method, seven amino acids are separated into their D- and L-enantiomers and from glycine and taurine. The D-forms of these seven amino acids, aspartate, asparagine, glutamate, glutamine, serine, arginine, and alanine, have been found in animal tissues thus far examined and are practically sufficient for studying D-amino acids in animals. In most invertebrate tissues, especially in mollusks, however, there is a large amount of taurine and thus the separation of D-alanine from taurine is important for invertebrate research. For this reason, rather a long run time is needed in comparison with that of the original report [10]. The detection limit was also about 1 pmol for each D- and L-amino acid.

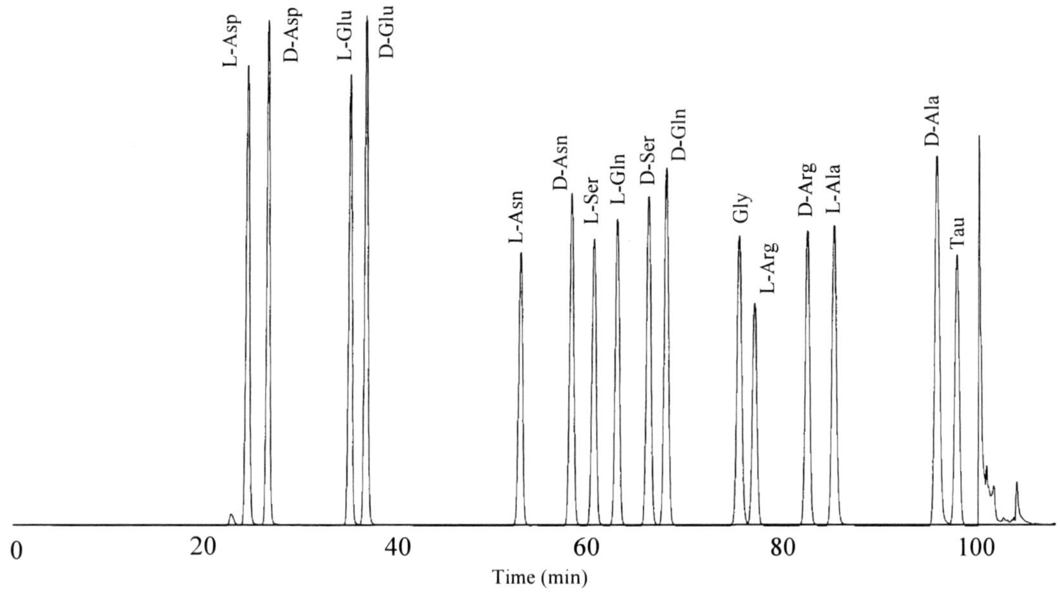

Figure 2. A chromatogram of major D- and L-amino acids pre-labeled with *N*-tert-butyloxycarbonyl-L-cysteine and *o*-phthaldialdehyde. Histidine and tyrosine appear within the analysis time but are separated from these amino acids.

2. Distribution of D-Alanine in Aquatic Invertebrate

A typical example for the distribution of free D-alanine in invertebrate tissues is shown in Figure 3. The data for most crustaceans and bivalves were obtained from the FLEC method. D-Alanine exists in all crustacean species and tissues, although it is restricted in amount in gill and hemolymph. The percentage of D-alanine to total alanine does not exceed 50% in crustaceans with some exceptions [13].

Bivalve mollusks belonging to subclass Heterodonta shown in Figure 3 contain an extremely large amount of D-alanine in all tissues. In particular, otter shell *Tresus keenae* contains over 50 μmol/g of D-alanine in siphon, of which content sometimes reaches 100 μmol/g and the percentage of D-alanine to total alanine exceeds 80% [14]. The other Heterodonta species also show higher percentage of D-alanine than crustacean species. In contrast, almost no D-alanine is found in bivalves belonging to subclass Pterimorphia, such as ark (blood) shell, oyster, and scallop (data not shown) [14]. In ark shell *Scapharca broughtonii*, however, a large amount of D-aspartate was reported to exist in the mantle and foot muscles [15] and aspartate racemase was isolated recently from foot muscle [16].

In these animals, there also exist some other D-amino acids such as D-arginine, D-aspartate, D-glutamate, D-proline in their tissues but their contents and percentages are low. When D-aspartate is detected in invertebrate tissues, however, the percentage is rather high, over 50% [13].

In invertebrates other than crustaceans and bivalves, sea urchin gonad has also been reported to contain D-alanine which decreased during development after fertilization and was suggested to involve in the embryonic development [17]. We also found over 30 μmol/g of D-alanine in the gonad of sea urchin *Strongylocentrotus nudus* (Figure 3). Several species of polychaete worms, phylum Annelida, contain a large amount of unspecified D-amino acids [7]. The distribution of D-alanine in polychaetes was highly species specific but we found 40 μmol/g of D-alanine in the polychaete *Perinereis aibuhitensis* (Figure 3, unpubl. data). D-Alanine was also detected in the sipunculid *Phascolosoma arcuatum* [18] and the percentage of D-alanine exceeded 50%. These data indicate that free D-alanine exists sporadically in various phyla of aquatic invertebrates.

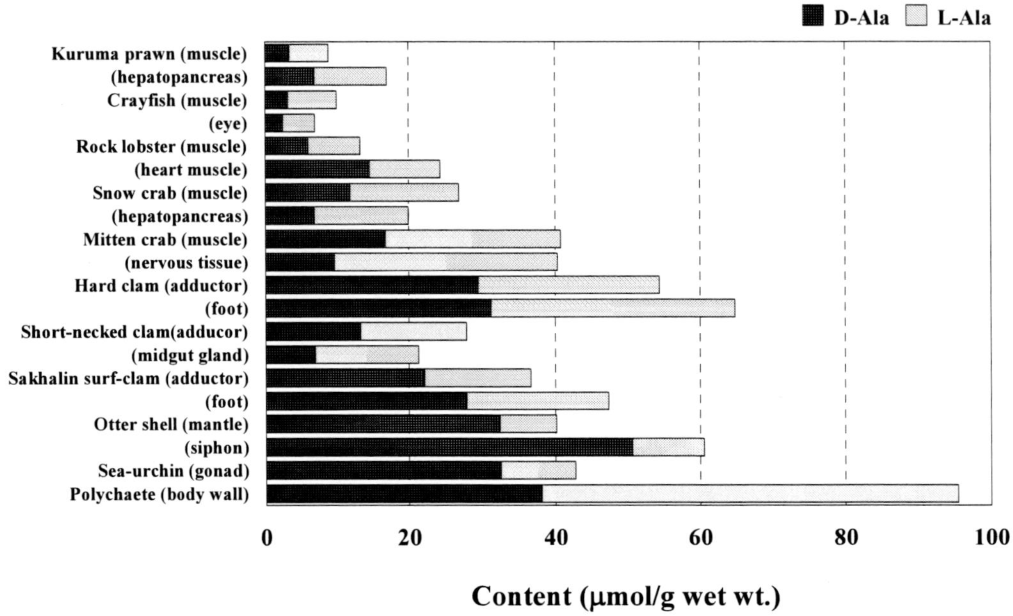

Figure 3. Distribution of free D- and L-alanine in the tissues of aquatic invertebrates. Only typical representatives are shown for crustaceans, bivalve mollusks (subclass Heterodonta), a sea-urchin, and a polychaete.

3. Physiological Function of D-Alanine as a Potent Osmolyte in Invertebrate Tissues

Special interest was focused on the physiological functions of free D-alanine in invertebrate tissues. In red-swamp crayfish *Procambarus clarkii* acclimated up to 75% seawater, total free amino acid increased over twice from freshwater to 75% seawater (Figure 4). The most responsible amino acids were glycine and D- and L-alanine, followed by L-glutamine and L-proline [19,20]. The increase of D- and L-alanine reached 6.7-fold and the percentage of D-alanine to total alanine also elevated from 38% in freshwater to 48% in 75% seawater. The increase of D- and L-alanine was also found in cardiac muscle, hepatopancreas, and nervous tissues except for gill and hemolymph. Thus, D-alanine, together with L-form, is concluded to be an effective osmolyte responsible for the intracellular isosmotic regulation or cell volume regulation.

Kuruma prawn *Penaeus* (*Marsupenaeus*) *japonicus* acclimated to 50 and 150% seawater also showed the twice increase of total free amino acid (Figure 4). In this species, glycine increase was the largest and responsible for half of total amino acid. Next to glycine, the increase of D- and L-alanine was large and the percentage of D-alanine kept a constant value of about 50%. Thus, glycine and D- and L-alanine are the major osmolytes in this species.

Japanese mitten crab *Eriocheir japonicus* is a euryhaline catadromous species and one of the most powerful osmoregulators in invertebrates. Post-larval juvenile crabs migrate upstream from brackish water area and grow up in freshwater river for several years. After maturation, adult crabs migrate downstream toward the sea for spawning. The crabs were captured in the river in the springtime (immature) and in the estuary and sea during spawning migration in October. Only D- and L-alanine levels significantly increased in estuary and the sea specimens and glycine increased only in the sea specimens [21,22]. Interestingly, during gonad maturation in the river, the crabs accumulate only D- and L-alanine and incorporate inorganic ions for preparing downstream migration toward the sea [21].

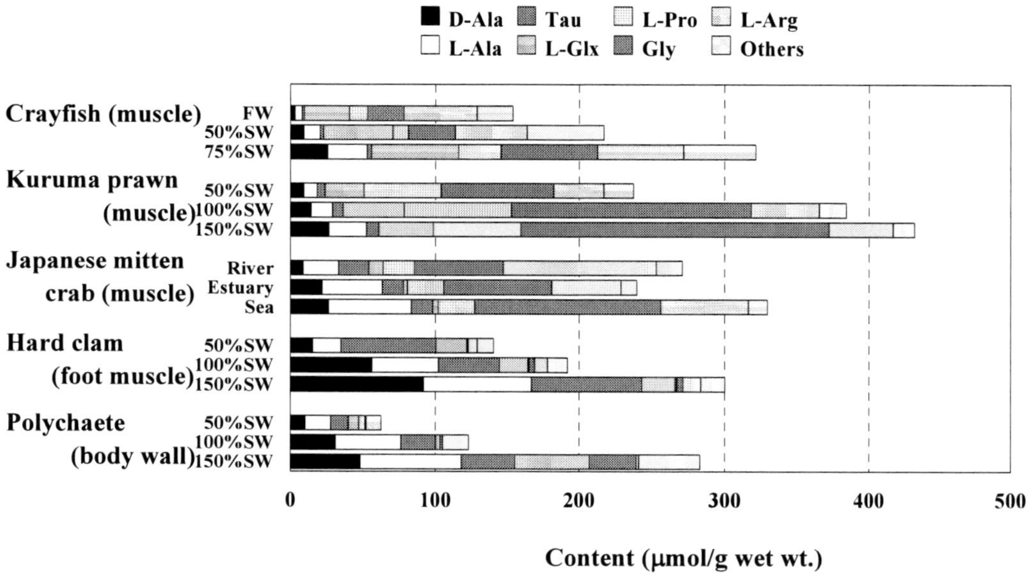

Figure 4. Changes of D- and L-alanine and other free amino acids in the muscle of several invertebrates under salinity stress. FW: freshwater, SW: seawater.

Hard clams *Meretrix lamarkii* and *M. lusoria* contain large amounts of D- and L-alanine which account for over half of the total free amino acid in 100% seawater. Under hyperosmotic stress in 150% seawater, only D- and L-alanine increased significantly. In contrast, they decreased largely in 50% seawater [14]. Thus, in this species, D- and L-alanine are the only responsible osmolytes for intracellular isosmotic regulation in their natural habitat.

Polychaete *Perinereis aibuhitensis* also contain D- and L-alanine which account for over half of the total free amino acid. In contrast to hard clam, however, taurine, L-glutamate, and glycine as well as D- and L-alanine also increased under hyperosmotic stress in 150% seawater. However, in their natural habitats in which salinity fluctuates from 50 to 100% seawater, only D- and L-alanine may work as the effective osmolytes for maintaining their body wall osmotic pressure.

4. Other Possible Physiological Functions of Free D-Alanine in Aquatic Invertebrates

Molecular oxygen is extremely restricted in aquatic habitats and fluctuates periodically and/or abruptly. Thus, aquatic animals well adapt to survive the severe environmental hypoxia or anoxia and harness the metabolic means to conquer the lack of oxygen, such as metabolic arrest and/or special anaerobic metabolism. During long lasted anaerobiosis, several bivalve mollusks have been well known to reduce metabolic rate and produce several anaerobic end products such as alanine, alanopine, succinate, and propionate [23]. Half of the alanine produced has been clarified to be D-form [14]. The crayfish acclimated to freshwater, 50, and 75% seawater as described above were exposed to anoxia ($PO_2<0.1$ mg/L) for 12h under constant bubbling of N_2 gas and recovered in normoxic water for 8h or 12h [20]. As is the case in mammals, crustacean species are known to produce only L-lactate as an anaerobic end product during anaerobiosis. Under anoxia, crayfish largely accumulated L-lactate in muscle along with the decrease of glycogen. The lactate content decreased to the initial value during recovery. Under anoxia, however, D- and L-alanine were also accumulated significantly in muscle and hepatopancreas of crayfish. The accumulation was largest in 75% seawater and lowest in freshwater. Thus, in crustaceans, D- and L-alanine as well as L-lactate are thought to be anaerobic end products under anoxia. The increase of D- and

L-alanine during anaerobiosis has also been reported in annelid *Arenicola marina* [24] and sipunculid *Sipunculus nudus* [25].

Just after the molting (ecdysis) of kuruma prawn, D- and L-alanine contents and alanine racemase activity were elevated significantly and declined thereafter [26]. Therefore, D- and L-alanine are possible to participate in the ecdysis of crustaceans. Judging from these fragmentary data, D- and L-alanine may be responsible to these and other important physiological events in invertebrates.

5. Biosynthesis of D-Alanine in Aquatic Invertebrates

The origin of free D-amino acids in bivalve mollusks had been controversial. Bivalves were postulated to incorporate exogenous D-amino acids from external media [7,27] or from symbiotic bacteria [6]. Alanine racemase [EC 5.1.1.1], however, has been found in bivalves [5,28] and crustaceans [29,30]. It is an enzyme catalyzing the specific interconversion between D- and L-alanine and well characterized in eubacteria. Eukaryotic alanine racemase has been purified to homogeneity from the fungus *Tolypocladium niveum* [31], crayfish muscle [32], mantle muscle of *C. japonica* [33], hepatopancreas [34] and muscle [35] of black tiger prawn *Penaeus monodon*. Alanine racemase from the muscle of *P. monodon* is a dimer with monomer molecular size of around 46 kDa which is almost similar to the other eukaryotic alanine racemases with some exceptions. The purified enzyme is highly specific to D- and L-alanine and do not catalyze the racemization of other amino acids.

All known alanine racemases require pyridoxal 5'-phosphate (PLP) as a cofactor. However, all the enzymes isolated from invertebrates showed activity even in the absence of PLP but are inhibited by several specific inhibitors to PLP-dependent enzymes, suggesting the tightly binding of PLP to the enzyme molecule in contrast to bacterial enzymes.

The most extraordinary feature of crustacean alanine racemase is an extremely low substrate affinity. Apparent *K*m values toward alanine for both L to D and D to L directions exceed 150 mM which is far beyond the concentrations of free D- and L-alanine in crustacean tissues (Figure 3). However, *V*max and *k*cat values of *P. monodon* enzyme are high and almost the same for both directions. The equilibrium constant (*K*eq) is 0.84 which is slightly lower than the theoretical one for racemization, i.e., 1.0. Judging from these data, crustacean alanine racemase is considered to regulate the subtle equilibrium between D- and L-alanine concentrations, in spite of the high *K*m values and low content in tissues.

By using primers designed from the partial amino acid sequences from the purified *P. monodon* alanine racemase, its cDNAs were cloned from the muscle and hepatopancreas of *P. japonicus* for the first time in animals [26]. The nucleotide sequence of 1,798 bp including polyadenylation signal and poly $(A)^+$ tail contained an open reading frame of 1,263 bp that encodes an identical protein of 421 amino acid residues both for muscle and hepatopancreas with a calculated molecular weight of 45,770. The amino acid sequence of *P. japonicus* enzyme shared about 25 and 27% identity with that of *Bacillus stearothermophilus* and *Schizosaccharomyces pombe* enzymes, respectively. In contrast to bacterial enzymes, *P. japonicus* enzyme contained a putative signal peptide at the N-terminus. However, a lysyl residue that binds PLP and a tyrosyl residue constituting the catalytic moiety, which are proposed in bacterial enzymes, were also conserved in *P. japonicus* enzyme. These data clearly indicate that alanine racemase gene is conserved from bacteria to crustaceans through a long evolutional time scale.

It is relevant to ask why several aquatic invertebrates maintain alanine racemase gene and utilize free D-alanine for some important physiological roles, especially for isosmotic regulation. As shown in Figure 4, several amino acids such as D- and L-alanine, glycine, L-glutamine, and L-proline serve as osmolytes in invertebrate tissues depending on the species and tissues. These amino acids are non-essential and easily biosynthesizable from glycogen via anaerobic glycolysis and tricarboxylic acid cycle. In particular, these amino acids are neutral ones which have no additional functional group interacting with proteins. For these reasons, these amino acids have a low chance to give harmful effects on protein functions or structures even if they are accumulated in considerable amounts in the cell. Such osmolytes have been referred to as 'compatible osmolyte' [36]. However, even compatible osmolyte will perturb the micro-environment of the cell if accumulated in large amounts. For instance, L-alanine, together with glycine, is the best compatible osmolyte but it is a substrate for some enzymes such as alanine aminotransferase and alanine dehydrogenase, and an inhibitor of pyruvate kinase, a regulatory enzyme in glycolysis. Therefore, when L-alanine is accumulated in large amounts in the cell, metabolic homeostasis will be collapsed. D-Alanine, however, does not inhibit pyruvate kinase and interact with other enzymes. For this reason, invertebrates may retain alanine racemase gene through a long evolutional period and transform L-alanine to D-form as an easily available and 'invisible (from enzymes)' [36] compatible osmolyte, although the cost is high.

6. Utilization of D-Amino Acids in Fish

As described up to this point, various invertebrate species contain a copious amount of free D-alanine in their tissues. Several fish species feed on these invertebrates (carnivorous and omnivorous fishes) or their larvae or juveniles (plankton feeders). However, fish contains only a trace amount of D-amino acids in tissues [37]. Thus, orally ingested D-alanine in these fish species is supposed to be dealt with some enzymes in fish tissues. D-Amino acid oxidase [EC 4.1.3.3] specific to neutral and basic D-amino acids and D-aspartate oxidase [EC 4.1.3.1] specific to acidic ones have been known to catabolize D-amino acids to corresponding 2-oxo acids, and distribute from bacteria to mammals. In aquatic animals, D-aspartate oxidase has been reported to exist in some fish liver [37] and several tissues of octopus [11,38,39] in which D-amino acid oxidase also exist [11]. Both enzyme activities are detected in intestine, liver (hepatopancreas), and kidney of various fishes [40] as well as in several organs of invertebrates [41]. The distribution of both enzyme activities, however, lacks any phylogenetic basis and is highly species specific. Careful inspection of the distribution pattern of the two oxidases revealed that both enzyme activities are higher in invertebrate-eaters than in fish-eaters. This tendency was much distinct in the distribution of D-amino acid oxidase activity [40].

To clarify the species specific distribution of both enzymes, common carp *Cyprinus carpio* which often feed on crustaceans and mollusks were reared with dry pellets containing D-alanine, D-aspartate, or D-glutamate for 4 weeks. The oral administration of D-alanine to carp increased the D-amino acid oxidase activity about 8-, 3-, and 1.5-fold in intestine, hepatopancreas, and kidney, respectively, but not in the brain [40]. In contrast, the ingestion of D-aspartate or D-glutamate gave no change of activities of both enzymes. These data suggest that D-amino acid oxidase is an inducible enzyme in carp with D-alanine ingestion. This may be a cause of species specific distribution of D-amino acid oxidase in fish.

To confirm the inducibility of D-amino acid oxidase in fish, a cDNA of the enzyme was cloned from hepatopancreas of carp fed D-alanine for the first time in animals other than mammals [42]. A 1,294 bp clone contained an open reading frame of 1,041 bp encoding 347 amino acids which was the same number as those of mammals except for rodents. This deduced amino acid sequence showed around 60% amino acid identity to mammalian enzymes and 21-29% to yeast and bacterial ones. The amino acid sequence contained an FAD binding consensus sequence, GXGXXG, which was GAGVIG commonly in the enzyme from animals including carp. The active site three residues, Tyr224, Tyr228, and Arg283 [43], were also conserved in carp enzyme. A conserved C-terminal sequence, Ser-(Lys/His/Arg)-Leu, is considered to be a peroxisomal signal [44] and also conserved in carp enzyme (Ser-Arg-Leu). The carp enzyme is confirmed to localize in peroxisomes [40].

D-Amino acid oxidase activity and its mRNA increased clearly in hepatopancreas of carp during oral administration of D-alanine (5 μmol/g body weight day) for 14 days. The enzyme activity increased rapidly to a plateau after 3 days and the level of mRNA also increased up to 7 days. As the increase of the activity and mRNA level did not coincide with each other, the enzyme induction may be regulated at a translation step or others as well as at a transcription step. After 14 days administration of D-alanine, a strong hybridization signal was detected in intestine followed by hepatopancreas and kidney but not in muscle. These data clearly indicate the inducible nature of carp D-amino acid oxidase with exogenous D-alanine. In mammals, D-amino acid oxidase is not an inducible enzyme [45], while D-aspartate oxidase is inducible in mouse liver after oral administration of D-aspartate [46]. This difference may reflect the fact that free D-alanine in carp feeds is wider in distribution and larger in amount than D-aspartate and D-glutamate. Therefore, D-amino acid oxidase in carp may be an important enzyme responsible for the effective utilization of the carbon skeletons of exogenous D-amino acids in intestine before absorption and after absorption in hepatopancreas and finally in kidney. It is clear from these data that D-alanine in crustaceans and bivalves is also utilized ingeniously by their predators.

For the first step to clarify the mechanisms of inducible nature of carp D-amino acid oxidase, it was overexpressed in *Escherichia coli* cells and purified homogeneity for the first time in animal tissues other than porcine kidney [47]. The purified preparation showed the highest activity toward D-alanine which is consistent with the inducible nature of the enzyme. In contrast, D-proline and D-methionine are the best substrate for mammalian enzymes. The apparent *K*m value of carp enzyme toward D-alanine (0.23 mM) is lower than that of mammalian and yeast counterparts. Carp D-amino acid oxidase has a k_{cat} value about 190 s^{-1} compared with 10 s^{-1} for porcine kidney enzyme and 300 s^{-1} for yeast one. The carp enzyme exhibited better thermal and pH stabilities than several yeast counterparts or the porcine kidney enzyme. Secondary structure topology consisted of 11 α-hilices and 17 β-strands that differed slightly from porcine kidney and *Rhodotorula gracilis* enzymes. A three-dimensional model of the carp enzyme constructed from a deduced amino acid sequence resembled that of porcine kidney enzyme but with a shorter active site loop and a longer C-terminal loop. Judging from these characteristics, carp D-amino acid oxidase is close to the porcine kidney enzyme structurally, but analogous to the *R. gracilis* enzyme enzymatically in turnover rate and pH and temperature stabilities. To unveil the induction mechanisms of carp D-amino acid oxidase, further research will be needed in future.

5. Conclusion

Aquatic habitat is very different from terrestrial one in the point of environmental factors such as salinity (0-35 ppt), oxygen concentration (below 1/30 in the air), and high pressure. Under these severe and daily-fluctuated environments, aquatic animals have adapted and radiated well during a long evolutional time scale. It became faintly visible now that these animals have developed to harness free D-amino acids as the effective means for their environmental adaptation. Thus, aquatic animals are dependent not only on L-amino acid biosystems but also on D-amino acid biosystems for maintaining life-support system. This may also be true for the other organisms on the earth. The research on D-amino acid biosystems has just started and many ingenious subsystems must be clarified in the future.

References

[1] Corrigan, J. J. (1969). D-Amino acid in animals. *Science*, *164*, 142-149.

[2] Uo, T., Yoshimura, T., Shimizu, S. & Esaki, N. (1998). Occurrence of pyridoxal 5'-phosphate-dependent serine racemase in silkworm, *Bombyx mori*. *Biochem Biophys Res Commun*, *246*, 31-34.

[3] D'Aniello, A. & Giuditta, A. (1977). Identification of D-aspartic acid in the brain of *Octopus vulgaris* LAM. *J Neurochem*, *29*, 1053-1057.

[4] D'Aniello, A. & Giuditta, A. (1980). Presence of D-alanine in crustacean muscle and hepatopancreas. *Comp Biochem Physiol*, *60B*, 319-322.

[5] Matsushima, O., Katayama, H., Yamada, K. & Kado, Y. (1984). Occurrence of free D-alanine and alanine racemase activity in bivalve mollusks with special reference to intracellular osmoregulation. *Mar Biol Lett*, *5*, 217-225.

[6] Felbeck, H. (1985). Occurrence and metabolism of D-aspartate in the gutless bivalve *Solemya reidi*. *J Exp Zool*, *234*, 145-149.

[7] Preston, R. L. (1987). Occurrence of D-amino acids in higher organisms: a survey of the distribution of D-amino acids in marine invertebrates. *Comp Biochem Physiol*, *87B*, 55-62.

[8] Felbeck, H. & Wiley, S. (1987). Free D-amino acids in the tissues of marine bivalves. *Biol Bull*, *173*, 252-259.

[9] Einarsson, S., Josefsson, B., Mōller, P. & Sanchez, D. (1987). Separation of amino acid enantiomers and chiral amines using precolumn derivatization with (+)-1-(9-fluorenyl)ethyl chloroformate and reversed-phase liquid chromatography. *Anal Chem*, *59*, 1191-1195.

[10] Hashimoto, A., Nishikawa, T., Oka, T., Takahashi, K. & Hayashi, T. (1992). Determination of free amino acid enantiomers in rat brain and serum by high-performance liquid chromatography after derivatization with *N-tert.*-butyloxycarbonyl-L-cysteine and *o*-phthaldialdehyde. *J Chromatogr*, *582*, 41-48.

[11] D'Aniello, A., D'Onofrio, G., Pischetola, M., D'Aniello, G., Vetere, A., Petrucelli, L. & Fisher, G. (1993). Biological role of D-amino acid oxidase and D-aspartate oxidase: effect of D-amino acids. *J Biol Chem*, *268*, 26941-26949.

[12] Okuma, E. & Abe, H. (1994). Simultaneous determination of D- and L-amino acids in the nervous tissues of crustaceans using precolumn derivatization with (+)-1-(9-fluorenyl)ethyl chloroformate and reversed-phase ion-pair high-performance liquid chromatography. *J Chromatogr*, *660*, 243-250.

[13] Okuma, E., Fujita, E., Amano, H., Noda, H. & Abe, H. (1995). Distribution of free D-amino acids in the tissues of crustaceans. *Fish Sci*, *61*,157-160.

[14] Okuma, E., Watanabe, K. & Abe, H. (1998). Distribution of free D-amino acids in bivalve mollusks and the effects of physiological conditions on the levels of D- and L-alanine in the tissues of the hard clam, *Meretrix lusoria*. *Fish Sci*, *64*, 606-611.

[15] Watanabe, T., Shibata, K., Kera, Y. & Yamada, R. (1998). Occurrence of free D-aspartate and aspartate racemase in the blood shell *Scapharca broughtonii*. *Amino Acids*, *14*, 353-360.

[16] Shibata, K., Watanabe, T., Yoshikawa, H., Abe, K., Takahashi, S., Kera, Y. & Yamada, R. (2003). Purification and characterization of aspartate racemase from the bivalve mollusk *Scapharca broughtonii*. *Comp Biochem Physiol*, *134B*, 307-314.

[17] D'Aniello, A., Nardi, G., Cipollaro, M., Pischetola, M. & Padula, L. (1990). Occurrence of D-alanine in the eggs and the developing embryo of the sea urchin *Paracentrotus lividus*. *Comp Biochem Physiol, 97B*, 291-294.

[18] Low, W. P., Ong, W. T. & Ip, Y. K. (1996). Different physiological functions of free D- and L-alanine in three body parts of the intertidal sipunculid *Phascolosoma arcuatum*. *J Comp Physiol*, *165B*, 558-564.

[19] Okuma, E. & Abe, H. (1994). Total D-amino and other free amino acids increase in the muscle of crayfish during seawater acclimation. *Comp Biochem Physiol*, *109A*, 191-197.

[20] Fujimori, T. & Abe, H. (2002). Physiological roles of free D- and L-alanine in the crayfish *Procambarus clarkii* with special reference to osmotic and anoxic stress responses. *Comp Biochem Physiol*, *131A*, 893-900.

[21] Abe, H., Okuma, E., Amano, H., Noda, H. & Watanabe, K. (1999). Role of free D- and L-alanine in the Japanese mitten crab *Eriocheir japonicus* to intracellular osmoregulation during downstream spawning migration. *Comp Biochem Physiol*, *123A*, 55-59.

[22] Abe, H., Okuma, E., Amano, H., Noda, H. & Watanabe, K. (1999). Effects of seawater acclimation on the levels of free D- and L-alanine and other osmolytes in the Japanese mitten crab *Eriocheir japonicus*. *Fish Sci*, *65*, 949-954.

[23] Hochachka, P. W. (1980). *Living without Oxygen: Closed and Open Systems in Hypoxia Tolerance*. Cambridge: Harvard University Press.

[24] Schöttler, U., Wienhause, G. & Westerman, J. (1984). Anaerobic metabolism in the lugworm *Arenicola marina* L.: the transition from aerobic to anaerobic metabolism. *Comp Biochem Physiol*, *79B*, 93-103.

[25] Pörtner, H. O., Vogeler, S. & Grieshaber, M. K. (1986). Recovery from anaerobiosis in the intertidal worm *Sipunculus nudus*. I. Restoration of aerobic steady-state energy metabolism. *J Exp Biol*, *122*, 37-50.

[26] Yoshikawa, N. (2003). Studies on alanine racemase in crustaceans. Ph D Thesis, The University of Tokyo, Tokyo.

[27] Bada, J. L. (1982). Racemization of amino acids in nature. *Interdisciplinary Sci Rev*, *7*, 30-46.

[28] Matsushima, O. & Hayashi, Y. S. (1992). Metabolism of D- and L-alanine and regulation of intracellular free amino acid levels during salinity stress in a brackish-water bivalve *Corbicula japonica*. *Comp Biochem Physiol*, *102A*, 465-471.

[29] Fujita, E., Okuma, E. & Abe, H. (1997). Occurrence of alanine racemase in crustaceans and the changes of the properties during seawater acclimation of crayfish. *Comp Biochem Physiol*, *116A*, 83-87.

[30] Fujita, E., Okuma, E. & Abe, H. (1997). Partial purification and properties of alanine racemase from the muscle of black tiger prawn *Penaeus monodon*. *Fish Sci*, *63*, 440-445.

[31] Hoffmann, K., Schneider-Scherzer, E., Kleinfauf, H. & Zocher, R. (1994). Purification and characterization of eukaryotic alanine racemase acting as key enzyme in cyclosporine biosynthesis. *J Biol Chem*, *269*, 12710-12714.

[32] Shibata, K., Shirasuna, K., Motegi, K., Kera, Y., Abe, H. & Yamada, R. (2000). Purification and properties of alanine racemase from crayfish *Procambarus clarkii*. *Comp Biochem Physiol*, *126B*, 599-608.

[33] Nomura, T., Yamamoto, I., Morishita, F., Furukawa, Y. & Matsushima, O. (2001). Purification and some properties of alanine racemase from a bivalve mollusc *Corbicula japonica*. *J Exp Zool*, *289*, 1-9.

[34] Uo, T., Ueda, M., Nishiyama, T., Yoshimura, T. & Esaki, N. (2001). Purification and characterization of alanine racemase from hepatopancreas of black-tiger prawn, *Penaeus monodon*. *J Mol Catal B*, *12*, 137-144.

[35] Yoshikawa, N., Dhomae, N., Takio, K. & Abe, H. (2002). Purification, properties, and partial amino acid sequences of alanine racemase from the muscle of the black tiger prawn *Penaeus monodon*. *Comp Biochem Physiol*, *133B*, 445-453.

[36] Hochachka, P. W. & Somero, G. N. (2002). *Biochemical Adaptation: Mechanism and Process in Physiological Evolution*. New York: Oxford University Press.

[37] Kera, Y., Hasegawa, S., Watanabe, T., Segawa, H. & Yamada, R. (1998). D-Aspartate oxidase and free acidic D-amino acids in fish tissues. *Comp Biochem Physiol*, *119B*, 95-100.

[38] Nagahisa, E., Kanno, N., Sato, M. & Sato, Y. (1993). Distribution of free D-aspartic acid and D-aspartate oxidase activity in tissues of *Octopus vulgaris*. *Nippon Suisan Gakkaishi*, *59*, 1897-1901.

[39] Tedeschi, G., Negri, A., Ceciliani, F., Ronchi, S., Vetere, A., D'Aniello, G. & D'Aniello, A. Properties of the flavoenzyme D-aspartate oxidase from *Octopus vulgaris*. *Biochim Biophys Acta*, *1207*, 217-222.

[40] Sarower, M. G., Matsui, T. & Abe, H. (2003). Distribution and characteristics of D-amino acid and D-aspartate oxidases in fish tissues. *J Exp Zool*, *295A,* 151-159.

[41] Sarower, M. G., Matsui, T. & Abe, H. (2004). Distribution and substrate specificity of D-amino acid and D-aspartate oxidases in marine invertebrates. *ScienceAsia*, *30*, 335-340.

[42] Sarower, M. G., Okada, S. & Abe, H. (2003). Molecular characterization of D-amino acid oxidase from common carp *Cyprinus carpio* and its induction with exogenous free D-alanine. *Arch Biochem Biophys*, *420*, 121-129.

[43] Todone, F., Vanoni, M. A., Mozzarelli, A., Bolognesi, M., Coda, A., Curti, B. & Mattevi, A. (1997). Active site plasticity in D-amino acid oxidase: a crystallographic analysis. *Biochemistry*, *36*, 5853-5860.

[44] Subramani, S. (1993). Protein import into peroxisomes and biogenesis of the organelle. *Ann Rev Cell Biol*, *9*, 445-478.

[45] Nagata, Y., Yamada, R., Nagasaki, H., Konno, R. & Yasumura, Y. (1991). Administration of D-alanine did not cause increase of D-amino acid oxidase activity in mice. *Experientia*, *47*, 835-838.

[46] Yamada, R., Nagasaki, H., Nagata, Y., Wakabayashi, Y. & Iwashima, A. (1989). Administration of D-aspartate increases D-aspartate oxidase activity in mouse liver. *Biochim Biophys Acta*, *990*, 325-328.

[47] Sarower, M. G., Okada, S. & Abe, H. (2005). Catalytic and structural characteristics of carp hepatopancreas D-amino acid oxidase expressed in *Escherichia coli*. *Comp Biochem Physiol*, *140B*, 417-425.

In: D-Amino Acids: ISBN 1-60021-075-9
A New Frontier in Amino Acid and Protein Research, pp.249-254

Chapter 3.12

D-GLUTAMIC ACID INDUCED MUSCLE CONTRACTION IN SILKWORM LARVA

***Hiroshi Hamamoto*[1], *Jorge Larranaga*[1], *Masae Sekine*[2], *Takemitsu Furuchi*[2], *Masumi Katane*[2], *Hiroshi Homma*[2], *Norio Matsuki*[1] *and Kazuhisa Sekimizu*[1,*]**
[1]Graduate School of Pharmaceutical Sciences, The University of Tokyo, Japan;
[2]School of Pharmaceutical Sciences, Kitasato University, Japan.

Keywords: silkworm, D-glutamic acid, muscle contraction, neurotransmitters.

1. INTRODUCTION

D- and L-amino acids have extremely similar chemical and physical properties, therefore it is not known why only L- amino acids were selected for polymerization and peptide formation at early evolutionary stages. While the biologic properties of L-amino acids have been extensively studied, D-amino acids have been systematically excluded from studies of living systems, until Corrigan reported the role of D-amino acids in the cell wall of microorganism [1], which opened a new field.

L-Glutamic acid functions as a neurotransmitter in the mammalian central nervous system [2-4], where it activates both ionotropic and G–protein–coupled metabotropic receptors, and has important roles in development [5,6], vision [7], olfactory memory [8,9], and spatial memory and locomotor control [10,11]. In invertebrates, L-glutamic acid also acts as a neurotransmitter in the central nervous system [12,13], where it has both excitatory and inhibitory actions [14-17]. L-Glutamic acid is also a neurotransmitter at the insect neuromuscular junction [18-20]. Transmission of excitatory stimuli in the brains of vertebrates and in the neuromuscular system of arthropods is mediated by the activation of postsynaptic ionotropic glutamate receptors, which are coupled with ion channels [21]. Ionotropic glutamate receptors are classified into three types based on their predominant sensitivity to one of three glutamate agonists: NMDA (N-methyl-D-aspartate), AMPA (α-amino-3-hydroxy-5-methyl-4-isoxazole propionate), or kainic acid [22].

Signal transmission at *Drosophila* neuromuscular junctions is primarily mediated by glutamatergic synapses localized on larval body wall muscles, which express at least two ionotropic glutamate receptor

* Correspondence concerning this article should be addressed to Sekimizu Kazuhisa, E-mail: sekimizu@mol.f.u-tokyo.ac.jp, FAX: +81-3-5684-2973.

subunits, DGluR-IIA [23] and DGluR-IIB [24]. These glutamate receptors have been cloned, and their ability to transfer information about glutamic acid binding and form ion channels has been demonstrated by electrophysiologic recordings of *Xenopus oocytes* expressing cloned receptors. Virtually nothing, however, is known about neuromuscular transmission in silkworm muscles.

The discovery of free D-alanine in the milkweed bug, *Oncopeltus fasciatus*, was the first report of free D-amino acids in animal tissues [25]. Later, D-serine was discovered in significant quantities in the hemolymph and tissues of *Bombyx mori* and two other silkworms [26], and other D-amino acids were also reported in other invertebrates [27], lower vertebrates [28], and rat tissues [29]. Free D-amino acids are related to developmental stages and aging, however, the physiologic role of D-glutamic acid remains unknown.

Therefore, the presence of free D-glutamic acid in *Bombyx mori* led us to establish a quantitative assay system to measure the contraction of silkworm larvae muscle, and to test the effect of L- and D- amino acids in muscle contraction.

2. Materials

1. 5th instar silkworm larvae (*Bombyx mori*, Hu·Yo x Tukuba·Ne)
2. 1-mL syringe attached to a 27-gauge needle
3. transducer to measure isotonic contraction with a load of 28 grams
4. analog recorder
5. nutrient saline (91 mM NaCl, 2.7 mM KCl, 1.8 mM $CaCl_2$, 1.05 mM $MgCl_2$, 11.9 mM $NaHCO_3$, and 5.6 mM glucose)
6. various L- or D- form amino acids and other neurotransmitters

3. Methods

1. Silkworm larvae were fed artificial food (Silkmate 2S, Nihon Nosan, Yokohama, Japan) at 27°C as described priviously [30].
2. Muscle specimens of silkworms (5th instar larval stage, 4-6 grams) were prepared as described below.
    i. The head of the silkworm was cut off with scissors.
    ii. Peritrophic membranes containing digesta and the silk glands were removed.
3. The specimen was tied with strings to a transducer.
4. The specimen was stabilized until autonomous vibration terminated.
5. Test samples were dissolved in nutrient saline.
6. The sample was injected into the body fluid of decapitated silkworms with a 1-mL syringe attached to a 27-gauge needle. Concentration values of injected materials were standardized according to the weight of the specimen.
7. The contraction process was very rapid; it reached maximum level within 1 s after injection and continued for more than 10 s thereafter.
8. The contraction ratio was calculated and expressed by measuring the maximum length of each individual specimen before (x cm) and after (y cm) injection using the formula (x-y)/x.

4. Results

After *in vivo* injection of L-glutamic acid into the hemolymph, the tension of the silkworm larva body decreased and locomotor activity was inhibited (similar to an anesthetized state) (Fig. 1). In larvae injected with D-glutamic acid, the muscles contracted and locomotor activity was inhibited (Fig.1). This lack of locomotor activity continued for 10 to 15 min and then the larvae recovered to the normal state, which indicates that D-glutamic acid is physiologically metabolized. Kainic acid produced effects similar to D-glutamic acid, although the larvae died after 10 to 15 min.

A transducer was used to quantitatively measure the effect of L- and D-forms of essential amino acids on muscle contraction in decapitated silkworm larvae. D-glutamic acid induced a strong contraction of silkworm larva specimens. D-aspartic acid also induced contractions, but had a weaker effect than D-glutamic acid (Fig. 2). Other amino acids, including L-glutamic acid, did not induce muscle contraction in this system. Muscle contraction of silkworm larva was also induced by kainic acid injection, but not by AMPA or NMDA; thus, D-glutamic acid might function via kainic-acid type glutamate receptors [31].

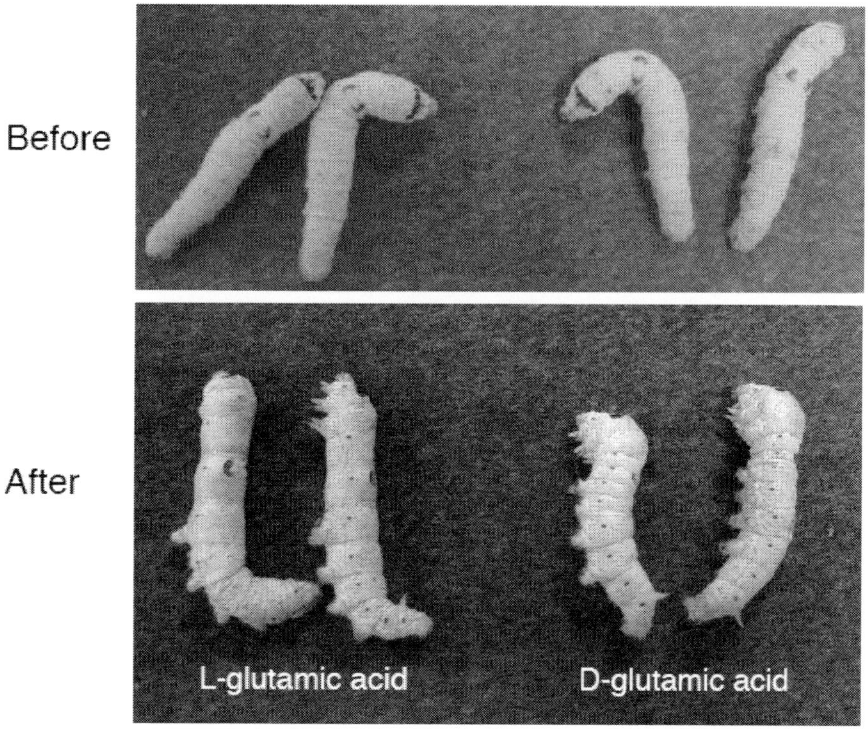

Figure 1. Effect of L- and D-glutamic acid on silkworm larva contraction. D-glutamic acid (1 μmol) (Right) or L-glutamic acid (1 μmol) (Left) was injected into silkworm hemolymph.

We previously reported that L-glutamic acid dose not have a contraction effect, but only pre-treatment of the muscle specimen with L-glutamic acid inhibited D-glutamic acid-induced contraction (Fig. 3). Neither NMDA nor AMPA had an inhibitory effect. Thus we conclude that the inhibitory effect was specific to L-glutamic acid.

These results strongly suggest that kainate-type glutamate receptors are responsible for the transfer of signals at the neuromuscular junctions of silkworm muscles, and that D-glutamic acid acts as a physiologic agonist to induce contraction. L-Glutamic acid is suggested to negatively regulate this process, although there are several hypotheses regarding the overall role of L-glutamic acid in the silkworm muscle contraction system [31].

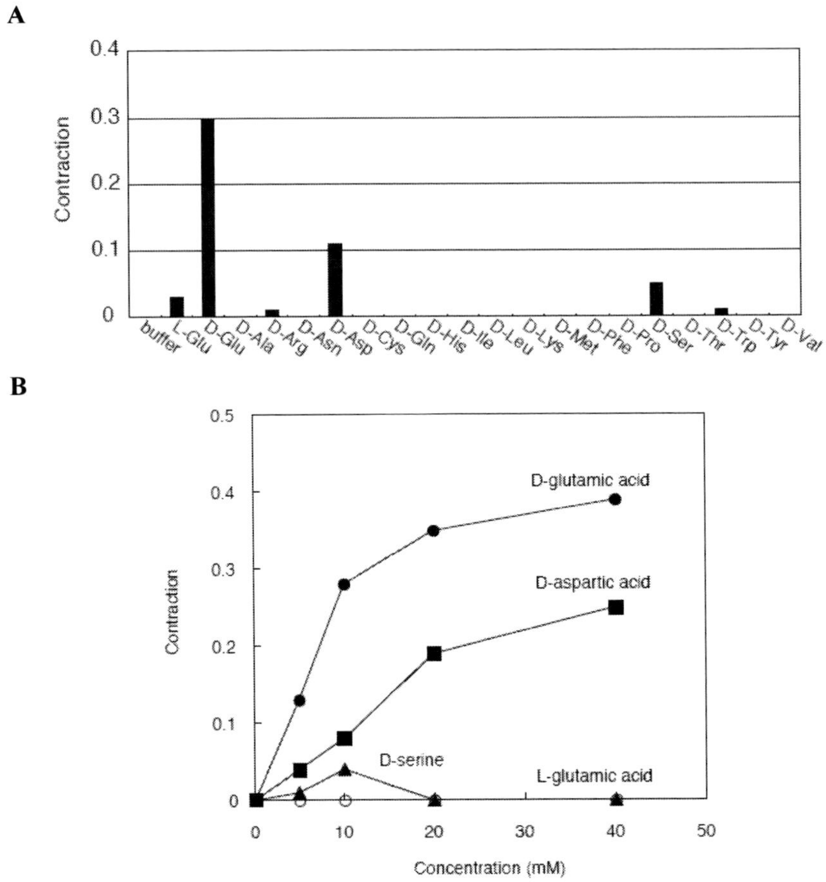

Figure 2. Muscle contraction induced by D-glutamic acid. A, Various D-amino acids (20 mM) dissolved in a nutrient buffer were injected (0.05 mL) into the silkworm specimen. B, Various concentrations of D-glutamic acid, D-aspartic acid, D-serine, and L-glutamic acid dissolved in nutrient buffer (0.05 mL) were injected into the silkworm specimens.

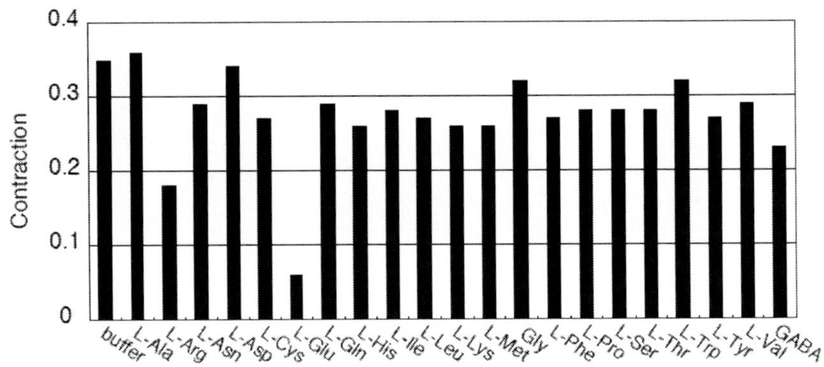

Figure 3. Inhibition of muscle contraction by L-glutamic acid. D-glutamic acid dissolved in nutrient buffer (0.05 mL) was injected into the silkworm specimen, followed by injection of various L-amino acids and GABA (20mM, 0.05mL).

5. Notes

1. The silkworm larva specimens were attached to the transducer within 1 min after decapitation in all cases.

2. After dissection, the specimen was already contracted in most cases, so weights were used to add tension and extend the specimen after attaching it to the transducer.
3. The optimum concentration of kainic acid induced a 40% contraction in this system.

REFERENCES

[1] Corrigan, J. J. (1969). D-amino acids in animals. *Science*, *164*, 142-149.

[2] Fonnum, F. (1984). Glutamate: a neurotransmitter in mammalian brain. *J Neurochem*, *42*, 1-11.

[3] Ottersen, O. P. & Storm-Mathisen, J. (1984) Neurons containing or accumulating transmitter amino acids. In classical transmitter and transmitter receptors in the CNS, Part II. In A. Bjorklund, T. Hokfelt, & M. J. Kuhar (Eds). *Handbook of chemical Neuroanatom*, (pp. 141–246). Elsevier, Amsterdam

[4] Collingridge, G. L. & Lester, R. A. (1989). Excitatory amino acid receptors in the vertebrate central nervous system. *Pharmacol Rev*, *41*, 143-210.

[5] Dudek, S. M. & Bear, M. F. (1989). A biochemical correlate of the critical period for synaptic modification in the visual cortex. *Science*, *246*, 673-675.

[6] Catania, M. V., Landwehrmeyer, G. B., Testa, C. M., Standaert, D. G., Penney, J. B., Jr. & Young, A. B. (1994). Metabotropic glutamate receptors are differentially regulated during development. *Neuroscience*, *61*, 481-495.

[7] Masu, M., Iwakabe, H., Tagawa, Y., Miyoshi, T., Yamashita, M., Fukuda, Y., Sasaki, H., Hiroi, K., Nakamura, Y., Shigemoto, R. & et al. (1995). Specific deficit of the ON response in visual transmission by targeted disruption of the mGluR6 gene. *Cell*, *80*, 757-765.

[8] Hayashi, Y., Momiyama, A., Takahashi, T., Ohishi, H., Ogawa-Meguro, R., Shigemoto, R., Mizuno, N. & Nakanishi, S. (1993). Role of a metabotropic glutamate receptor in synaptic modulation in the accessory olfactory bulb. *Nature*, *366*, 687-690.

[9] Kaba, H., Hayashi, Y., Higuchi, T. & Nakanishi, S. (1994). Induction of an olfactory memory by the activation of a metabotropic glutamate receptor. *Science*, *265*, 262-264.

[10] Aiba, A., Kano, M., Chen, C., Stanton, M. E., Fox, G. D., Herrup, K., Zwingman, T. A. & Tonegawa, S. (1994). Deficient cerebellar long-term depression and impaired motor learning in mGluR1 mutant mice. *Cell*, *79*, 377-388.

[11] Conquet, F., Bashir, Z. I., Davies, C. H., Daniel, H., Ferraguti, F., Bordi, F., Franz-Bacon, K., Reggiani, A., Matarese, V., Conde, F. & et al. (1994). Motor deficit and impairment of synaptic plasticity in mice lacking mGluR1. *Nature*, *372*, 237-243.

[12] Bicker, G., Schafer, S., Ottersen, O. P. & Storm-Mathisen, J. (1988). Glutamate-like immunoreactivity in identified neuronal populations of insect nervous systems. *J Neurosci*, *8*, 2108-2122.

[13] Horseman, B. G., Seymour, C., Bermudez, I. & Beadle, D. J. (1988). The effects of L-glutamate on cultured insect neurones. *Neurosci Lett*, *85*, 65-70.

[14] Kehoe, J. (1994). Glutamate activates a K+ conductance increase in Aplysia neurons that appears to be independent of G proteins. *Neuron*, *13*, 691-702.

[15] Bolshakov, V., Gapon, S. A. & Magazanik, L. G. (1991). Different types of glutamate receptors in isolated and identified neurones of the mollusc *Planorbarius corneus*. *J Physiol*, *439*, 15-35.

[16] Parker, D. (1994). Glutamatergic transmission between antagonistic motor neurones in the locust. *J. Comp. Physiol. [A]*, *175*, 737-748.

[17] Cleland, T. A. & Selverston, A. I. (1998). Inhibitory glutamate receptor channels in cultured lobster stomatogastric neurons. *J Neurophysiol*, *79*, 3189-3196.

[18] Jan, L. Y. & Jan, Y. N. (1976). L-Glutamate as an excitatory transmitter at the *Drosophila* larval neuromuscular junction. *J Physiol*, *262*, 215-236.

[19] Patlak, J. B., Gration, K. A. & Usherwood, P. N. (1979). Single glutamate-activated channels in locust muscle. *Nature*, *278*, 643-645.

[20] Delgado, R., Barla, R., Latorre, R. & Labarca, P. (1989). L-Glutamate activates excitatory and inhibitory channels in *Drosophila* larval muscle. *FEBS Lett*, *243*, 337-342.

[21] Magazanik, L. G., Bol'shakov, K. V., Buldakova, S. L., Gmiro, V. E., Dorofeeva, N. A., Lukomskaya, N. Y., Potap'eva, N. N., Samoilova, M. V., Tikhonov, D. B., Fedorova, I. M. & Frolova, E. V. (2002). Structural characteristics of ionotropic glutamate receptors as identified by channel blockade. *Neurosci Behav Physiol*, *32*, 173-182.

[22] Hollmann, M. & Heinemann, S. (1994). Cloned glutamate receptors. *Annu Rev Neurosci*, *17*, 31-108.

[23] Schuster, C. M., Ultsch, A., Schloss, P., Cox, J. A., Schmitt, B. & Betz, H. (1991). Molecular cloning of an invertebrate glutamate receptor subunit expressed in *Drosophila* muscle. *Science*, *254*, 112-114.

[24] Petersen, S. A., Fetter, R. D., Noordermeer, J. N., Goodman, C. S. & DiAntonio, A. (1997). Genetic analysis of glutamate receptors in *Drosophila* reveals a retrograde signal regulating presynaptic transmitter release. *Neuron*, *19*, 1237-1248.

[25] Auclair, J. L. & Patton, R. L. (1950). On the occurrence of D-alanine in the haemolymph of the milkweed bug, *oncopeltus fasciatus*. *Rev Can Biol*, *9*, 3-8.

[26] Srinivasan, N. G., Corrigan, J. J. & Meister, A. (1965). Biosynthesis of D-serine in the silkworm, *Bombyx mori*. *J Biol Chem*, *240*, 796-800.

[27] Tarui, A., Shibata, K., Takahashi, S., Kera, Y., Munegumi, T. & Yamada, R. H. (2003). *N*-Methyl-D-glutamate and *N*-Methyl-L-glutamate in *Scapharca broughtonii* (Mollusca) and other invertebrates. *Comp Biochem Physiol B Biochem Mol Biol*, *134*, 79-87.

[28] Kera, Y., Nagasaki, H., Iwashima, A. & Yamada, R. (1992). Presence of D-aspartate oxidase and free D-aspartate in amphibian (*Xenopus laevis*, *Cynops pyrrhogaster*) tissues. *Comp Biochem Physiol B*, *103*, 345-348.

[29] Kera, Y., Aoyama, H., Matsumura, H., Hasegawa, A., Nagasaki, H. & Yamada, R. (1995). Presence of free D-glutamate and D-aspartate in rat tissues. *Biochim Biophys Acta*, *1243*, 283-286.

[30] Hamamoto, H. & Sekimizu, K. (2005). Evaluation of the therapeutic effects of antibiotics using silkworm as an animal model. *Res. Adv. in Antimicrob. Agents Chemother.*, *5*, 1-23.

[31] Sekimizu, K., Larranaga, J., Hamamoto, H., Sekine, M., Furuchi, T., Katane, M., Homma, H. & Matsuki, N. (2005). D-Glutamic acid-induced muscle contraction in the silkworm, *Bombyx mori*. *J Biochem (Tokyo)*, *137*, 199-203.

In: D-Amino Acids: ISBN 1-60021-075-9
A New Frontier in Amino Acid and Protein Research, pp.255-258

Chapter 3.13

PRIMARY STUDY OF D-AMINO ACID ACCUMULATION SYSTEM

Wataru Masuda[*]
Department of Biosciences, Kyushu Dental College, Japan.

Keywords: salivary glands, D-aspartic acid, transport

1. INTRODUCTION

It has long been believed that only L-enantiomers of amino acids occur in nature, and D-amino acids (D-AAs) have been regarded as "unnatural isomers" or laboratory artifacts. However, due to progress in development of analytical methods, various D-AAs have been found in higher as well as lower animals, plants, and foods [reviewed in 1,2]. The amounts of some D-AAs indicate tissue-specific expression, development-related expression, and circadian rhythms of expression. These findings suggest that D-AAs play some important physiological roles and that their quantities are carefully regulated. Elucidation of the origins of D-AAs in each organ is thus a very important problem. Two origins are candidates: one is *de novo* synthesis by endogenous enzyme(s), while the other is transportation from the blood, in which D-AAs are dissolved, following absorption from the intestine.

In initiating study of the D-AA accumulation system, it must be determined whether D-AAs are transported into specific (or target) organs *in vivo*. Here, I present an easy non-RI method and the results for D-aspartic acid (D-Asp) administration [3]. In this method, D-Asp is compatible with other D-AAs. If it appeared that D-AAs were transported into specific organs or cells, this experimental system could be used not only to study the D-AAs accumulation/transport systems, but also for other experimental purposes, *e.g.* searches for the specific gene/protein expression associated with D-AAs.

[*] Correspondence concerning this article should be addressed to Wataru Masuda, E-mail: wmasuda@kyu-dent.ac.jp, FAX: +81-93-561-7086.

2. Materials

1. Experimental animals (Wistar male rats).
2. Disposable syringe equipped with 25 G needle.
3. Dissecting tools.
4. Physiological saline (0.9 % NaCl solution).
5. D- and L-aspartic acid (D- and L-Asp).
6. Pentobarbital.
7. Trichloroacetic acid (TCA).
8. L-homocysteic acid (L-HCA).
9. Refrigerated centrifuge.
10. Water-saturated diethyl ether.
11. Millipore filter LCR13-LH (0.5 μm).
12. Nova-Pak C18 (4 μm) column (Waters).
13. High-performance liquid chromatography (HPLC) system equipped with a fluorometric detector.
14. N-*tert.*-butyloxycarbonyl-L-cysteine (BOC-L-Cys).
15. *o*-phthalaldehyde (OPA).
16. Acetic acid.
17. Acetonitrile.
18. Tetrahydrofuran (THF).
19. Sodium hydroxide (NaOH).

3. Methods

1. Rats of 3 and 7 weeks of age were intraperitoneally administered either 1 mmol/kg D-Asp, 1 mmol/kg L-Asp, or physiological saline.
2. After 1 hour, animals were perfused with physiological saline under pentobarbital anesthesia (50 mg/kg, intraperitoneally) to eliminate contamination of amino acids in the blood.
3. Salivary glands (parotid gland (PG), submandibular gland (SMG), sublingual gland (SLG)) were removed, weighed, homogenized in 10 vols. of 5 % TCA after the addition of L-HCA, and centrifuged at 18,000 xg for 30 min at 4 °C. The supernatants were collected.
4. To remove TCA, the supernatants were extracted three times with water-saturated diethyl ether. The aqueous layers were then passed through a LCR13-LH Millipore filter (0.5 μm).
5. The TCA extracts were derivatized with Boc-L-Cys and OPA for 2 min at room temperature, and the derivatives were immediately applied to the HPLC system.
   Analytical conditions were:
   - Column: Nova-Pak C18 (4 μm) column, 3.9 x 300 mm (Waters).
   - Detection: Excitation and emission wavelengths of 344 and 443 nm, respectively.
   - Column temperature: 30 °C.
   - Flow rate: 0.8 ml/min.
   - Mobile phase A: Acetate buffer with 7 % acetonitrile and 3 % THF. An acetate buffer was prepared using 0.1 M acetic acid adjusted to pH 6.0 with NaOH.
   - Mobile phase B: Acetate buffer with 47 % acetonitrile and 3 % THF.
   - Separation conditions: Linear gradient from mobile phase A to B in 2 hours.
6. D- and L-Asp concentrations in each organ were calculated from internal standard L-HCA peak areas, amounts of injected samples, and tissue weights.

4. Results

At 3 weeks of age (Fig. 1, left side), when D-Asp was administered, D-Asp concentration was dramatically increased and L-Asp slightly decreased in all salivary glands compared with saline-treated rats. When L-Asp was administered, neither D- nor L-Asp concentration changed in any salivary gland. In contrast, at 7 weeks of age (Fig. 1, right side), D-Asp administration did not affect D-Asp level in any gland, while L-Asp concentration was slightly decreased in PG and SMG. When L-Asp was administrated, L-Asp level increased nearly twice in PG and SMG, while D-Asp level was changed in none of the glands. These results indicate that a D-Asp transport system from blood into cells exists in immature rat salivary glands. However, it appears that this system is no longer present in adults.

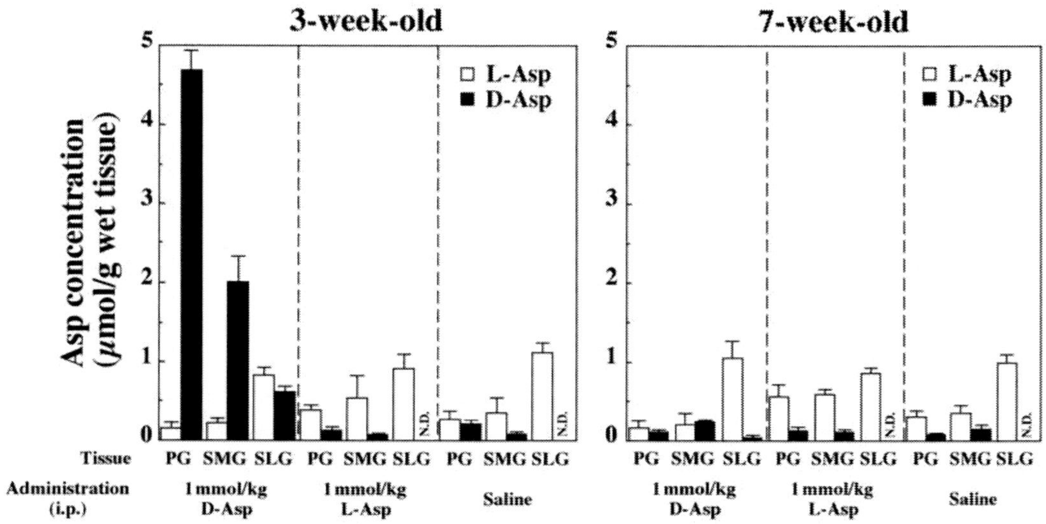

Figure 1. Changes in D- and L-Asp concentrations after intraperitoneal administration of D- or L-Asp. Rats of 3 and 7 weeks of age were intraperitoneally administered 1 mmol/kg D-Asp, 1 mmol/kg L-Asp, or physiological saline (0.9 % NaCl solution), and amounts of D- and L-Asp in salivary glands were determined after 1 hour as described in "Methods". Values are means with S.E. of four rats. (Reprinted from *Arch Biochem Biophys*, 420, Masuda, W., *et al.*, Free D-aspartic acid in rat salivary glands, 46-54, Copyright (2003), with permission from Elsevier)

5. Notes

1. This method is applicable to other experimental animals, tissues or culture cells as well. Experimental conditions (*e.g.* types of amino acids, concentrations of amino acids, duration after administration, *etc.*) are also changeable.
2. Tissue preparation (Step 1 to 4 in Methods) was carried out according to the method of Hashimoto *et al.* [4].
3. The TCA extracts could be stored at -70 °C until derivatization.
4. The determination of amino acid enantiomers and non-chiral amino acids by HPLC (Step 5 in Methods) was performed as previously described [5].
5. Since several methods and conditions for determination of amino acids have been reported [6,7,8], whichever method is preferred can be used.

REFERENCES

[1] Imai, K., Fukushima, T., Santa, T., Homma, H., Hamase, K., Sakai, K. & Kato, M. (1996). Analytical chemistry and biochemistry of D-amino acids. *Biomed Chromatogr, 10,* 303-312.

[2] Hashimoto, A. & Oka, T. (1997). Free D-aspartate and D-serine in the mammalian brain and periphery. *Prog Neurobiol, 52,* 325-353.

[3] Masuda, W., Nouso, C., Kitamura, C., Terashita, M. & Noguchi, T. (2003). Free D-aspartic acid in rat salivary glands. *Arch Biochem Biophys, 420,* 46-54.

[4] Hashimoto, A., Nishikawa, T., Oka, T., Hayashi, T. & Takahashi, K. (1993). Widespread distribution of free D-aspartate in rat periphery. *FEBS Lett, 331,* 4-8.

[5] Hashimoto, A., Nishikawa, T., Oka, T., Takahashi, K. & Hayashi, T. (1992). Determination of free amino acid enantiomers in rat brain and serum by high-performance liquid chromatography after derivatization with N-*tert.*-butyloxycarbonyl-L-cysteine and *o*-phthaldialdehyde. *J Chromatogr, 582.* 41-48.

[6] Hamase, K., Homma, H., Takigawa, Y., Fukushima, T., Santa, T. & Imai, K. (1997). Regional distribution and postnatal changes of D-amino acids in rat brain. *Biochim Biophys Acta, 1334,* 214-222.

[7] Nagata, Y., Yamamoto, K. & Shimojo, T. (1992). Determination of D- and L-amino acids in mouse kidney by high-performance liquid chromatography. *J Chromatogr, 575.* 147-152.

[8] D'Aniello, A., Di Fiore, M. M., D'Aniello, G., Colin, F. E., Lewis, G. & Setchell, B. P. (1998). Secretion of D-aspartic acid by the rat testis and its role in endocrinology of the testis and spermatogenesis. *FEBS Lett, 436,* 23-27.

In: D-Amino Acids: ISBN 1-60021-075-9
A New Frontier in Amino Acid and Protein Research, pp.259-269 2008 Nova Science Publishers, Inc.

Chapter 3.14

D-AMINO ACIDS AND IMMUNE RESPONSE

Einat Zisman[1] and Michael Sela[2,*]
[1]Yeda Research & Development Company at The Weizmann Institute of Science, Israel;
[2]Department of Immunology, The Weizmann Institute of Science, Israel.

Keywords: D-amino acids, immunogenicity, polypeptides, specificity

1. INTRODUCTION

Peptides and polypeptides either fully or partially built of D-amino acids interest researchers because of their pharmacokinetic advantages over all L peptides and polypeptides. The renaissance in research and applications of D-amino acids in biomedicine is based on their contribution to the extended biological half-life of the product [1], as was exemplified for several synthetic analogs of neuropeptides and peptide hormones, such as: desmopressin (vasopressin analog), goserelin and leuprolin [gonadorelin (GnRH) analogs], and octreotide (somatostatin analog). However, the dogma that D-peptides would probably possess little or no immunogenicity (see, for example, ref 2), since they could not be processed and presented to the major histocompatibility complex (MHC) class 2 molecules for recognition by helper T cells or cytotoxic lymphocytes is a misconception, which grew despite the early published studies of the immunogenicity and antigenic specificity of these isomeric peptide analogs, D-peptides and peptides containing a mixture of L- and D-amino acids [3-5].

Recently, Hudecz *et al.* showed that substitution of the N- and C-terminal flanking regions of MUC2 peptide by 2-3 D-amino acids not only increased dramatically its resistance against proteolytic degradation, but also retained antibody recognition [6]. Thus, in exploiting the use of D-amino acids, one should realize that the resulting molecules are nonetheless not inert, but rather may induce a unique immune response, which is not necessarily cross-reactive with the L-enantiomer.

* Correspondence concerning this article should be addressed to Michael Sela, Department of Immunology, The Weizmann Institute of Science, Israel. E- mail: Michael.sela@weizmann.ac.il; FAX: +972-8-946-9713.

2. IMMUNOGENICITY

D-Amino acids are present in prokaryotes and eukaryotes. For example, D-alanyl-D-alanine sequence is a component of the peptidoglycan layer of the cell wall of many bacterial species; D-amino acid peptides are found in amphibia and in the nervous system of invertebrates such as cephalopods (for review, see refs 7-8); and D-amino acids are present in many living tissues, including human teeth, eye lenses and brain, where they may have accumulated during aging (see ref 9-12).

The first report of the immunological properties of a macromolecule composed exclusively of D-amino acids concerns poly-γ-D-glutamic acid, a natural polymer isolated from the cell wall of *Bacillus anthracis* (anthrax;13). In this polymer the peptide bonds are between the α-amino and γ-carboxyl functions of glutamic acid. Whereas most of the antibodies produced against *anthrax* react with poly-γ-D-glutamic acid, it could not elicit antibody formation when administered into experimental animals. Thus, it is a cross-reacting antigen but is not immunogenic. Immunogenicity was defined as the capacity to provoke an immune response and is independent of the specificity of the antibodies, or T cells, formed. Recently, in view of the renewed interest in immunity against *anthrax*, Robbins et al. aimed at developing a safe and effective vaccine based on synthetic peptides [14]. The researchers elicited antibody response to poly-γ-D-glutamic acid by immunizing mice with conjugates of poly-γ-D-glutamic acid or its peptides to BSA or recombinant anthrax Protective Protein, respectively. The antibodies obtained by immunizing with these carriers induced opsonophagocytic killing of anthrax. Thus the carriers altered the immunogenicity of poly-γ-D-glutamic acid while preserving the specificity of the immune response to anthrax. In another example, attachment of tyrosine peptides, whether L- [15] or D- [16], to gelatin enhances its immunogenicity, but if only a small amount of tyrosine is attached it does not change gelatin specificity [17].

Early studies of polymers composed exclusively of D-amino acids led to the conclusion that they were not immunogenic in experimental animals such as rabbits and guinea pigs [5]. Only later was it realized that D-amino acid polymers are immunogenic [18-19]; but in the case of D-amino acid polymers, the dose administered is of crucial importance. The amount of antigen injected should be small; otherwise immunological paralysis occurs as measured by antibody formation [20], similar to bacterial polysaccharides. Also, a booster injection, which increases the antibody amount formed in the case of L-amino acid polymers, may cause paralysis for D-amino acid polymers [19]. Indeed, metabolic studies suggested that the apparent lack of immunogenicity of D-amino acid polymers was due to their prolonged retention in the organs and gradual release over a long period of time [18]. In addition, Janeway and Humphrey [21] found that if mice were left for a long time before receiving a booster injection, they displayed an anamnestic response.

Ultimately, studies in monkeys [5], mice [19], guinea pigs [22], and rats [23] showed that these species all responded to the D-amino acid polymers. At low doses, the D-enantiomer is as immunogenic as the L-enantiomer [19]. Moreover, immunization with D-polypeptides elicited antibodies of the IgM and IgG isotypes [24]; the predominance of the IgG subclass in the immune response implies the formation of memory B cells.

In accordance with these results, Benkirane et al. [25] described the generation of antibodies against an all-D-hexapeptide and against single D-amino acid modifications of sequence IRGERA corresponding to the carboxyl-terminal residues 130-135 of histone H3 after immunization of mice. The difference between the immune responses to the all-L- and the D-amino acid containing peptides was in the IgG subtypes: In the case of L-peptide, antibodies belonged to the IgG1, IgG2a, IgG2b, and IgG3 subclasses, whereas in the case of the D-peptides the IgG3 antibody response, which is common for thymus-independent antigens (e.g., polysaccharides, see reference 26) was predominant. Interestingly, the immunogenicity of hexapeptides containing single D-amino acid modifications was reduced markedly compared with their all-L- or all-D- counterparts [25].

In addition to the reports on the immunogenicity of D-peptides in other mammals, IgG and IgA antibodies with binding specificity for the carboxyl-terminal R-D-Ala-D-Ala sequence of the precursor peptide from bacterial cell wall peptidoglycan were detected in several human sera [27-28], especially in correlation with rheumatic diseases [29].

In view of the above accumulated data on the immunogenicity of D-enantiomers, the lack of immune response to an all-D-amino acid protein, rubredoxin (a 45 amino acid protein, ref. 30) might be due to the following: *1)* multiple administrations (injections) of the antigen, previously shown to induce tolerance [19], *2)* use of alum as adjuvant whereas, as stated by Dintzis et al. [30], the use of complete Freund's adjuvant induced "a small but significant IgG response," and *3)* the immune response (Ir) background of the particular strain used in that study.

3. Antigenic Specificity at the B Cell Level

Epitopes built of peptides of L-amino acid residues generally do not cross-react at the antibody level with their D-amino acid enantiomers, i.e., they are strictly stereospecific. The size of the antigenic determinants for both L-alanine peptides and D-alanine peptides is of three to four amino acid residues [31,32]. The region of the antigenic epitope furthest removed from the protein carrier, to which the alanine peptides were attached, is of paramount importance in determining the specificities of the antibodies formed [33]. In initial studies, D-peptides inhibited the homologous D-system (poly-D-alanyl protein reacting with anti-poly-D-alanine antibodies) much more efficiently than were the L-peptides in inhibiting the L-system. This was due to the presence of proteolytic enzymes in rabbit serum capable of splitting peptides of L-alanine, but not those of D-alanine [31]. A comparison of the size of combining sites of IgG and IgM antibodies with poly-D-alanyl specificity [34] showed that in both cases the efficiency of inhibition by D-alanine peptides increased up to that of tetrapeptide whereas the penta-D-alanine showed an extent of inhibition similar to the tetrapeptide. Thus, the antigen-combining sites of both IgG and IgM antibodies are similar in size and capable of accommodating a peptide composed of four alanine residues, which is not surprising since their variable regions are developed from the same germ line genes. A detailed analysis of the combining sites of anti-poly-D-alanine has been given by Schechter [35]. Using his mapping procedure based on the number and spatial orientation of the subsites involved in the binding site, quantitative data on the free energies of binding in the various subsites were obtained and enthalpies and entropies of binding were determined.

It has long been known that D-amino acids play a prominent role in bacterial physiology. Reports in animals have been sporadic, mostly in invertebrates, but in recent years have also included mammals. One such study [12] described the localization of endogenous D-serine in rat brain, making use of an antibody highly specific for D-serine conjugated to glutaraldehyde. Thus, thanks to the exquisite discrimination between the optical enantiomers, antibodies to D-amino acid residues may serve as useful tools in biological studies. Stereospecific importance of D-amino acid residues is exemplified by the exchange within a peptide epitope of an L-amino acid residue with a D-amino acid residue [36]. The exchange of an L-amino acid with its D-analog usually diminishes the cross-reaction, but different positions may affect it to various degrees [37].

Of special interest is the puzzling observation [25] that, whereas changes with D-amino acid residues within a peptide derived from the sequence of histone H3 led to losses of cross-reactions to different degrees as expected, when the antibodies were of the IgG3 isotype (whether induced against the D- or the L-peptides) they were able to bind both the homologous peptide and its mirror enantiomer. However, anti-L-peptide antibodies of the IgG1, IgG2a, and IgG2b isotypes did not cross-react with the D-peptide and vice versa [25,38]. This implied that the specificity of the IgG3 antibodies was against the peptide backbone.

4. Preferential Formation of Antibodies to D-Amino Acid Sequences

Polypeptidyl proteins are proteins to which peptide chains are attached. They have been used extensively in studies concerned with immunogenicity, antigenic specificity, and immunological tolerance [3,4]. An investigation of the specificity of the anti-polyalanyl antibodies formed upon immunization of rabbits and goats with poly-DL-alanyl protein showed that the specificity of the antibodies formed was directed predominantly to peptide sequences of D-alanine [39]. The sequence of D-alanine and L-alanine residues within the poly-DL-alanine chains is random, permitting sequences composed either mostly of D-alanine or mostly of L-alanine. The phenomenon observed seems to be a general one, since poly-DL-phenylalanyl and poly-DL-tyrosyl proteins elicited in rabbits anti-polypeptide antibodies with specificity directed mainly to D-amino acid sequences [39].

The above observation may serve as an example of the phenomenon of antigenic competition between sequences composed of L-, D-, and DL-amino acid residues, the D sequences being the most efficient. When rabbits were immunized with a mixture of poly-L-alanyl human serum albumin and poly-D-alanyl human serum albumin, similar amounts of antibodies were formed against both optical isomers, suggesting that in this case there was no antigenic competition [40]. Earlier studies of poly-DL-alanyl proteins had already implied the preferential response to D-alanine peptides [41,42]. In studies of catalytic antibodies, racemic transition-state analogs, i.e., mixtures of enantiomers that correspond to both the L- and D-amino acid substrates led to formation of monoclonal antibodies capable of catalyzing only the D-enantiomer and were specifically active toward the respective D-ester or amide [43,44].

5. Thymus Independence

Multichain polypeptides built of D-amino acids were originally defined as thymus-independent (Ti) on the basis of their ability to activate B lymphocytes and mount an in vivo immune response in thymectomized or nude mice, i.e., in the relative absence of T lymphocytes [45-46]. Other Ti antigens are polysaccharides (e.g., pneumococcal SIll, Ficol, and dextran), several proteins in their nondenatured form (e.g., flagellin and collagen), and synthetic polymers (e.g., polyvinylpyrrolidone). A common feature of Ti antigens is the presence of repeating identical antigenic determinants. However, this requirement is insufficient for thymus independence because multichain polypeptides built of L-amino acids, although different from the D-polypeptide by their chirality, are thymus-dependent. Another important feature of Ti antigens is a slow rate of metabolism, which results in persistence of detectable amounts of antigens in macrophages throughout the body even several weeks after immunization [5,30,45]. As mentioned earlier, the immune response to many Ti antigens was reported to be predominantly the IgG3 isotype of antibodies [26]; this is also true for the multichain-D-polypeptide (D-Phe,D-Glu)-D-Pro--D-Lys [47] and for short D-amino acid peptides [38].

It was shown that covalent attachment of a peptide from heat shock protein-60 (HSP-60) could shift the immune response of classical Ti antigens, such as polysaccharides and polymers composed of D-amino acids to become thymus-dependent [48]. The shift from Ti to thymus dependence was accompanied by a shift in the IgG isotype from predominantly IgG3 to IgG1 [47].

Despite the Ti characteristics of D-peptides and polypeptides, T cells specific to these antigens do exist in vivo. For example, a T cell line specific for the chemically well-defined peptidoglycan of bacterial cell wall, disaccharide tetrapeptide, was established from Lewis rats immunized with the antigen covalently linked to the autologous rat serum albumin. Katsuki et al. [49] examined antigen specificity with various analogs or derivatives of peptidoglycan. The cell line was reactive to analogs with the carboxyl-terminal D-amino acid, but least reactive to those with L-amino acid as carboxyl terminus, suggesting that this site is part of the T cell epitope. Also, Maillere et al. [50] could detect specific stimulation of T cells by L-peptides containing single D-amino acid substitutions after a single immunization of BALB/c mice with the analogs in CFA. Moreover, a T cell hybridoma specific to an all-L sequence could not respond to single D-amino acid substitutions unless the substitutions were distant to the MHC class II or T cell receptor contacts [50].

6. T-Cell Immunity

In agreement with these reports, we have shown that after immunization of mice with the Ti multichain polypeptide (D-Phe,D-Glu)-D-Pro--D-Lys [51], which is composed exclusively of D-amino acids, their lymph node cells could be stimulated to proliferate in vitro in response to the immunizing Ti antigen [51]. Moreover, we have described the establishment and characterization of T cell hybridomas specific to this Ti polypeptide and compared these hybridomas to T cell hybridomas specific to the polymer of the same composition and structure, but built exclusively of L-amino acids and thus thymus-dependent. Most of the D-polypeptide specific hybridomas that were established from splenocytes of BALB/c mice immunized with (D-Phe,D-Glu)-D-Pro--D-Lys could be stimulated by the original polypeptide, but also by its mirror image, the L-enantiomer. In contrast, less than half of the L-specific hybridomas tested could respond to the D-enantiomer [52]. The crossreactivity at the T cell level between the D-specific hybridomas and their corresponding L-enantiomers, shown by us [51], is conceptually similar to the observation by Guichard et al. [38] that, at the antibody level, IgG3 antibodies could recognize both the immunizing peptide and its mirror enantiomer. On the other hand, the less frequent ability of a D-peptide to stimulate an L-specific T cell line is supported by the report of Lamb et al. [53], who failed to either stimulate or tolerate an influenza haemagglutinin peptide specific line by its mirror image enantiomer.

Using a fluorescent, flow-cytometry based direct binding assay (described in refs 54-55), we could demonstrate that the Ti D-polymer is presented to T cells by MHC class II molecules on antigen-presenting cells [51]. In this assay, splenic antigen-presenting cells were incubated with biotinylated (D-Phe,D-Glu)-D-Pro--D-Lys for 0.5-18 h. Any appearance of the biotinylated antigen at the cell surface was detected using a fluorophore (e.g., phycoerythrin) conjugated to streptavidin. Moreover, by the ELISA technique, using avidin-coated plates, we have shown that the surface-bound biotinylated (D-Phe,D-Glu)-D-Pro--D-Lys was directly associated with MHC class II molecules [51]. The affinity of interaction of (D-Phe,D-Glu)-D-Pro--D-Lys with the MHC class II molecules is similar to that of the thymus-dependent L-polymer. Furthermore, the incubation with the D-polymer induces increased expression of membranal MHC class II molecules, similar to the reports on other Ti antigens, namely, trinitrophenylated *Brucella abortus* [56] and antibody-conjugated dextran [57]. The D-polymer has a faster kinetics of presentation than the L-polymer. This faster kinetics may explain the antigenic competition between stretches of D- and L- peptides within poly (DL-peptidyl) proteins described earlier in this review that elicited predominantly antibodies toward D- isomers after immunization with poly-DL-alanyl, poly-DL-tyrosyl, or poly-DL-phenylalanyl proteins [39]. The association of the D-polymer with MHC class II on murine splenic antigen-presenting cells is not altered by pretreatment of the splenocytes with chloroquine, whereas the association of the L-polymer with the MHC class II gene products is reduced significantly *($P<0.001$)*, reflecting the importance of processing for the L-polypeptide. Thus, Ti D-polymers can engage the T cell receptors and stimulate or tolerate T cells if they are presented to them as a complex with MHC class II molecules, in a manner similar to the tertiary complexes formed with "classical-conventional" L-peptides. The ability of MHC class II molecules to present in their groove not only "classical" L-peptides had already been speculated by Demotz et al. [58] and Rudensky et al. [59], who found the presence of small quantities of substance, that could not be digested by proteases in the elution profiles of material extracted from purified MHC class II molecules.

Among these substances, direct interactions with MHC molecules were revealed for the heavy metals gold and nickel [60-61], for carbohydrates [62], for lipids [63], and by us for DNA [64].

7. Retro-Inverso Peptides

Retro-inverso peptides are peptides made of reversed D-amino acids, so they are mirror images of a mirror image. The L to D shift results in inverted chirality. The reversed order of the amide bonds (i.e., -NH-CO- instead of -CO-NH-) regenerates the shape and space of the original L-peptide. Thus, the retroinverso peptides hold the biological activity of the original L-peptide, but with a longer half-life. These characteristics make them excellent candidates for pharmacological research. Since retro-inverso peptides have the chirality of D-amino acids, but the overall structure of the L-peptide, analysis was done concerning their immunogenicity and antigenic specificity [38,65-66]. For example, the retro-inverso analog of the IRGERA sequence from histone H3 elicited primarily cross-reactive IgG3 antibodies, while IgG1, IgG2a, and IgG2b antibodies that were obtained from immunization with the L- and the retro-inverso peptides recognized both the L- and the retro-inverso peptides, but not the D-peptides. It was shown that certain retro-inverso peptide analogs mimic successfully the antigenic and immunogenic properties of the parent peptides and elicit high levels of long-lasting neutralizing and protective antibodies and T-cell responses (e.g., IL-2 secretion) against a viral challenge [67-70]. More recently, Arnon et al. immunized mice with a retro-inverso peptide of the protective B-cell epitope hemagglutinin (HA) 91-108 from influenza HA [71]. Nasal immunization with the retro-inverso HA peptide induced a strong mucosal and systemic immune response which protected the mice against challenge with the virus [71], suggesting a possible novel peptide-based vaccine for influenza.

8. An L-Polymer, but not its D-Enantiomer, as a Drug against MS

Of special interest is the story of the biological differences between an L-amino acid copolymer, denoted Cop 1, an FDA approved drug for treatment of multiple sclerosis, and its D-amino acid analog. Cop 1 is a synthetic random polymer of L-alanine, L-glutamic acid, L-lysine, and L-tyrosine in a residue molar ratio of 6.0/1.9/4.7/1.0. The polymer is quite effective in preventing or healing experimental allergic encephalomyelitis (EAE) [72,73] as well as the exacerbating-remitting type of multiple sclerosis [74,75], most probably because of its immunological cross-reaction both at the antibody and the T cell levels, with myelin basic protein (MBP), the putative cause of the autoimmune phenomenon in both the above-mentioned diseases. The suppressive activity of Cop 1 is specific and is limited to EAE and multiple sclerosis, having no effect on other autoimmune diseases. In total contrast, the D-amino acid analog of Cop 1 has no capacity whatsoever to prevent or suppress EAE [76] probably because of the total lack of immunological cross-reaction with MBP. Apparently, the complex of class II MHC antigen with Cop 1 can compete for the MBP-specific T cell receptor with the complex of class II antigen with peptides derived from MBP. On the other hand, because of the lack of immunological similarity between D-Cop 1 and MBP, the complex of class II antigen with D-COP 1 will not be able to compete with class II MBP complex. As stated above, the first step is the formation of complex with class II MHC antigen. This is formed very efficiently by Cop 1 and MBP; actually, Cop 1 is able to displace MBP peptides from the groove of class II antigen. Moreover, Cop 1 and MBP do not require processing before binding to class II MHC molecules on living antigen-presenting cells [77]. This is the basis for our finding that Cop 1 efficiently prevents murine graft-vs.-host disease in vivo [78]. D-Cop 1 is as efficient as Cop 1 in reacting with the groove of class II antigen; indeed, we found that D-Cop 1 is even more efficient than Cop I in preventing the graft-vs.-host disease [79]. The preferential binding of D-Cop 1 over Cop 1 in binding to the MHC resembles our observation with the multichain polymers described earlier in this review [51]. The above findings stress the difference in the mechanisms of action of Cop 1 and its D-analog in the two diseases. Whereas the competition between MBP and either Cop 1 or D-Cop 1 for the groove of class II MHC antigen is what matters for the prevention of graft-vs.-host disease, in the case of EAE this is only the first, necessary but not sufficient, step. Subsequent specific steps that result from cross-reactivity with the autoantigen MBP are of crucial importance.

9. Concluding Remarks

This review summarizes our knowledge about the role of D-amino acids in immunological phenomena. Some of the results presented here might have been anticipated whereas others are unexpected. As D-amino acids begin to play an important role in the preparation of therapeutic agents, including for cancer therapy and vaccination [80-83], an awareness of their immunological potential definitely becomes relevant.

References

[1] Pappenheimer, J. R., Dahl, C. E., Karnovsky, M. L., and Maggio. J. E. (1994). Intestinal absorption and excretion of octapeptides composed of D-amino acids. *Proc Natl Acad Sci USA, 91*, 1942- 1945.

[2] Milton, R. C., Milton, S. C., and Kent, S. B. (1992). Total chemical synthesis of a D-enzyme: the enantiomers of HIV-1 protease show reciprocal chiral substrate specificity. *Science, 256*, 1445-1448.

[3] Sela, M. (1966). Immunological studies with synthetic polypeptides. *Adv Immunol, 5*, 29-129.

[4] Sela, M. (1969). Antigenicity: Some molecular aspects. *Science, 166*, 1365-1374.

[5] Gill, T. J., III (1972). The chemistry of antigens and its influence on immunogenicity. In F. Borek (Ed.), *Immunogenicity,* (pp. 1-14). North-Holland Publ. Co., New York.

[6] Tugyi, R., Uray, K., Ivan, D., Fellinger, E., Perkins, A., and Hudecz, F. (2005). Partial D-amino acid substitution: Improved enzymatic stability and preserved Ab recognition of a MUC2 epitope peptide. *Proc Natl Acad Sci USA, 102,* 413-418.

[7] Kreil, G. (1994). Peptides containing a D-amino acid from frogs and molluscs. *J Biol Chem, 269,* 10967-10970.

[8] Mor, A., Amiche, M., and Nicolas, P. (1992). Enter a new posttranslational modification: D-amino acids in gene-encoded peptides. *Trends Biochem Sci, 17,* 481-485.

[9] Fisher, G. H., Petrucelli, L., Gardner, C., Emory, C., Frey, W. H., Amaducci, L., Sorbi, S., Sorrentino, G., Borghi, M., and D'Aniello, A. (1994). Free D-amino acids in human cerebrospinal fluid of Alzheimer disease, multiple sclerosis, and healthy control subjects. *Mol Chem Neuropathol, 23,* 115-124.

[10] Fisher, G. H., Torres, D., Bruna, J., Cerwinski, S., Martin, T., Bergljung, C., Gruneiro, A., Chou, S. J., Man, E. H., and Pappatheodorou, S. (1995). Presence of D-aspartate and D-glutamate in tumor protein. *Cancer Biochem Biophys, 15,* 79-82.

[11] Nagata, Y., Borghi, M., Fisher, G. H. and D'Aniello, A. (1995). Free D-serine concentration in normal and Alzheimer human brain. *Brain Res Bull, 38,* 181-183.

[12] Schell, M. J., Molliver, M. E., and Snyder, S. H. (1995). D-Serine, an endogenous synaptic modulator: localization to astrocytes and glutamate-stimulated release. *Proc Natl Acad Sci USA, 92,* 3948-3952.

[13] Ivanovics, G., and Bruckner, V. (1937). Chemische und immunologische Studien uber den Mechanismus der Milzbrandinfektion und Immunitat. *Z Immunitatsforsch, 91,* 175-176.

[14] Schneerson, R., Kubler-Kielb, J., Liu, T. Y., Dai, Z. D., Leppla, S. H., Yergey, A., Backlund, P., Shiloach, J., Majadly, F., Robbins, J. B. (2003). Poly(γ-D-glutamic acid) protein conjugates induce IgG antibodies in mice to the capsule of Bacillus anthracis: a potential addition to the anthrax vaccine *Proc Natl Acad Sci USA, 100,* 8945-50.

[15] Sela, M., and Arnon, R. (1960). Studies on the chemical basis of the antigenicity of proteins. I. Antigenicity of polypeptidyl gelatins. *Biochem J, 75,* 91-102.

[16] Sela, M., and Fuchs, S. (1965). On the role of charge and optical configuration in antigenicity. In *Proceedings of a Symposium on Molecular Cellular Basis of Antibody Formation,* (pp. 43-56). Publishing House of the Czechoslovak Acad Sci, Prague

[17] Arnon, R., and Sela, M. (1960). Studies on the chemical basis of the antigenicity of proteins. II. Antigenic specificity of polytyrosyl gelatin. *Biochem J, 75,* 103-109.

[18] Gill, T. J., III, Kunz, H. W., Gould, H. J., and Doty, P. (1964). Studies on synthetic polypeptide antigens. XI. The antigenicity of optically isomeric synthetic polypeptides. *J Biol Chem, 239,* 1107-1113.

[19] Janeway, C. A., Jr., and Sela, M. (1967). Synthetic antigens composed exclusively of L- or D-amino acids. *Immunology, 13,* 29-38.

[20] Mozes, E., Sela, M., and Taussig, M. J. (1974). Tolerance to thymus- independent antigens: characteristics of induction of tolerance to thymus-independent synthetic polypeptides. *Immunology, 27,* 64 1-646.

[21] Janeway, C. A., Jr. and Humphrey, J. I-I. (1968). Synthetic antigens composed exclusively of L- or D-amino acids. II. Effect of optical configuration on the metabolism and fate in mice. *Immunology, 14,* 225-234.

[22] Stupp, Y. and Sela, M. (1967). Further studies on the immunogenicity of D-amino acid polymers. *Biochim Biophys Acta, 140,* 349-359.

[23] Simonian, S. J., Gill, T. J., III, and Gershoff, S. N. (1968). Studies on synthetic polypeptide antigens. *J Immunol, 101,* 730-742.

[24] Schmitt-Verhulst, A.-M., Mozes, E., and Sela, M. (1974). Control of the immune response to a thymus-independent synthetic polypeptide. *Immunogenetics, 1,* 357-369.

[25] Benkirane, N., Friede, M., Guichard, G., Bnand, J. P., Van Regenmortel, M. H., and Muller, S. (1993). Antigenicity and immunogenicity of modified synthetic peptides containing D-amino acid residues. Antibodies to a D-enantiomer do recognize the parent L-hexapeptide and reciprocally. *J Biol Chem, 268,* 26279-26285

[26] Mond, J. J., Lees, A., and Snapper, C. P. (1995). T cell-independent antigens type 2. *Annul. Rev. Immunol, 13,* 666-692.

[27] Franken, N., Seidl, P. H., Kuchenbauer, T., Kolb, H. J., Schleifer, K. H., Weiss, L., and Tympner, K. D. (1984). Specific immunoglobulin A antibodies to a peptide subunit sequence of bacterial cell wall peptidoglycan. *Infect Immunol, 44,* 182-187.

[28] Franken, N., Seidl, P. H., Zauner, E., Kolb, H. J., Schleifer, K. H., and Weiss, L. (1985). Quantitative determination of human IgG antibodies to the peptide subunit determinant of peptidoglycan by an enzyme-linked immunosorbent assay. *Mol Immunol, 22,* 573-579.

[29] Todome, Y., Ohkuni, H., Mizuse, M., Furuya, M., Fujikawa, S., Tanaka, S., Watanabe, N., Fujii, K., and Zabriskie, J. B. (1992). Detection of antibodies against streptococcal peptidoglycan and the peptide subunit (synthetic tetra-D-alanyl-bovine serum albumin complex) in rheumatic diseases. *Int Arch Allergy Immunol, 97,* 301-307.

[30] Dintzis, H. M., Symer, D. E., Dintzis, R. Z., Zawadzke, L. E., and Berg, J. M. (1993). A comparison of the immunogenicity of a pair of enantiomeric proteins. *Proteins, 16,* 306-308.

[31] Schechter, I., Schechter, B., and Sela, M. (1966). Combining sites of antibodies with L-alanine and D-alanine peptide specificity and the effect of serum proteolytic activity on their estimation. *Biochim Biaphys Acta, 127,* 438-456.

[32] Schechter, B., Schechter, I., and Sela, M. (1970). Specific fractionation of antibodies to peptide determinants. *Immunochemistry, 7,* 587-597.
[33] Schechter, B., Schechter, I., and Sela, M. (1970). Antibody combining sites to a series of peptide determinants of increasing size and defined structure. *J Biol Chem, 245,* 1438-1447.
[34] Haimovich, J., Schechter, I., and Sela, M. (1969). Combining sites of IgG and IgM antibodies of poly-D-alanyl specificity. *Eur J Biochem, 7,* 537-543
[35] Schechter, I. (1972). Mapping of the combining sites of antibodies specific to polypeptide chains. *Ann NY Acad Sci, 190,* 394- 419.
[36] Ananyeva, N. M., Tischenko, V. A., Bespalova, J. D., Pekelis, B. L., Yakubov, L. Z., Vinogradov, V. A., Rokhlin, O. V., and Titov, M. I. (1988). Monoclonal antibodies to dalargin-a synthetic analog of enkephalins. *Immunol Lett, 18,* 269-273.
[37] Geysen, H. M., Rodda, S. J., and Mason, T. J. (1981). A prior delineation of a peptide which mimics a discontinuous antigenic determinant. *Mol Immunol, 23,* 709-715.
[38] Guichard, G., Benkirane, N., Zeder-Lutz, G., van Regenmortel, M. H., Briand,J. P., and Muller, S. (1994). Antigenic mimicry of natural L-peptides with retro-inverso-peptidomimetics. *Proc Natl Acad Sci USA, 91,* 9765-9769.
[39] Schechter, I. and Sela, M. (1967). Preferential formation of antibodies specific toward D-amino acid residues upon immunization with poly-DL-peptidyl proteins. *Biochemistry, 6,* 897-905.
[40] Schechter, I. (1968). Antigenic competition between polypeptidyl determinants in normal and tolerant rabbits. *J Exp Med, 127,* 237-250
[41] Brown, R. K., Trzpis, M. A., Sela, M., and Anfinsen, C. B. (1963). Studies on the antigenic structure of ribonuclease. IV. Polyalanyl ribonuclease. *J Biol Chem, 238,* 3876-3883
[42] Sage, H. J., Deutsch, G. F., Fasman, G. D., and Levine. L. (1964). The serological specificity of the poly-alanine immune system. *Immunochemistry 1,* 133-144.
[43] Pollack, S. J., Hsuin, P., and Schultz, P. G. (1989). Stereospecific hydrolysis of alkyl esters by antibodies. *J Am Chem Soc, 111,* 5961-5962.
[44] Martin, M. T., Angeles, T. S., Sugasawala, R., Ainan, N. I., Napper, A. D., Darsley, M.., Sanchaz, R. I.. Booth, P., and Titmas. R. C. (1994). Antibody-catalyzed hydrolysis of an unsubstituted amide. *J Am Chem Soc. 116,* 6508-6512.
[45] Sela, M., Mozes, E., and Shearer, G. M. (1972). Thymus-independence of slowly metabolized immunogens. *Proc Natl Acad Sci USA, 69,* 2696-2700.
[46] Mosier, D. E., and Subbarao, B. (1982). Thymus independent antigens: complexity of B-lymphocyte activation revealed. *Immunol Today, 3,* 217-222.
[47] Konen-Waisman, S. (1996). HSP6O T cell epitopes convert T independent into T dependent antigens. Thesis dissertation, The Weizmann Institute of Science, Rehovot, Israel.
[48] Konen-Waisman, S., Fridkin, M., and Cohen, I. R. (1995). Self and foreign 60 kD heat-shock protein T-cell epitope peptides serve as immunogenic carrier for a T-cell independent sugar antigen. *J Immunol 154,* 5977-5985.
[49] Katsuki, M., Kakimoto, K., Kawata, S., Kotani, S., and Koga, T. (1987). Induction of delayed-type hypersensitivity by the T cell line specific to bacterial peptidoglycans. *J Immunol, 139,* 3570- 3572.
[50] Maillere, B., Mourier, G., Cotton, J., Herve, M., Leroy, S., and Menez, A. (1995). Probing immunogenicity of a T-cell epitope by L-alanine and D-amino acid scanning. *Mol Immunol, 32,* 1073-1080.
[51] Zisman, E., Dayan, M., Sela, M., and Mozes, E. (1993). Ia-antigen-T-cell interactions for a thymus-independent antigen composed of D-amino acids. *Proc. Natl Acad Sci USA, 90,* 994-998.
[52] Sela, M., and Zisman, E. (1997). Different roles of D-amino acids in immune phenomena. *FASEB J, 11,* 449-56.

[53] Lamb,J. R., Feldmann, M., Green, N., and Lerner, R. A. (1986). Influence of antigen structure on the activation and induction of unresponsiveness in cloned human T lymphocytes. *Immunology, 57,* 331-335.

[54] Mozes, E., Dayan, M., Zisman, E., Brocke, S., Licht, A., and Pecht, I. (1989). Direct binding of a myasthenia gravis related epitope to MHC class II molecules on living murine antigen-presenting cells. *EMBO J, 8,* 4049-4052.

[55] Zisman, E., Sela, M., and Mozes, E. (1991). Direct binding of a synthetic multichain polypeptide to class II major histocompatibility complex molecules on antigen-presenting cells and stimulation of a specific T-cell line require processing of the polypeptide. *Proc Nat. Acad Sci USA, 88,* 9738-9742

[56] Vitetta, E. S., Bossie, A., Fernandez-Botran, R., Myers, C. D., Oliver, K. G., Sander, V. M., and Stevens, T. L. (1987). Interaction and activation of antigen-specific T and B cells. *Immunol Rev, 99,* 193-239.

[57] Brunswick, M., June, C. H., Finkelman, F. D., Dintzis, H. M., Inman, J. K., and Mond, J.J. (1989). Surface immunoglobulin mediated B-cell activation in the absence of detectable elevations in intracellular ionized calcium: A model for T-cell-independent B-cell activation. *Proc Natl Acad Sci USA, 86,* 6724-6728.

[58] Demotz, S., Grey, H. M., Appella, E., and Sette, A. (1989). Characterization of a naturally processed MHC class II-restricted T-cell determinant of hen egg lysozyme. *Nature (London), 342,* 682-684

[59] Rudensky, A. Y., Preston Hurlburt, P., Hong, S. C., Barlow, A., and Janeway, C. A., Jr. (1991). Sequence analysis of peptides bound to MHC class II molecules. *Nature (London), 353,* 622-627

[60] Romagnoli, P., Labhardt, A. M., and Sinigaglia, F. (1991). Selective interaction of Ni with an MHC-bound peptide. *EMBO J, 10,* 1303-1306.

[61] Romagnoli, P., Spinas, G. A., and Sinigaglia, F. (1992). Gold-specific T cells in rheumatoid arthritis patients treated with gold. *J Clin Invest, 89,* 254-258.

[62] Jackson, S., Folks, T. M., Wetterskog, D. L., and Kindt, T. J. (1984). A rabbit helper T cell clone reactive against group specific streptococcal carbohydrate. *J Immunol, 133,* 1553- 1557.

[63] Beckman, E. M., Porcelli, S. A., Morita, C. T., Behar, S. M., Furlong, S. T., and Brenner, M. B. (1994). Recognition of a lipid antigen by CD1-restricted alpha beta+ T cells. *Nature (London), 372,* 691-694

[64] Waisman, A., Zisman, E., and Mozes, E. (1996). MHC class II mediated T cell response to DNA. *Cell Immunol, 173,* 7-14.

[65] Muller, S., Guichard, G., Benkirane, N., Brown, F., Van Regenmortel, M. H. V., and Briand, J.-P. (1995). Enhanced immunogenicity and cross-reactivity of retro-inverso peptidomimetics of the major antigenic site of foot-and-mouth disease virus. *Peptide Res. 8,* 138-144.

[66] Briand,J.-P., Guichard, G., Dumortier, H., and Muller, S. (1995). Retro-inverso peptidomimetics as new immunological probes. *J Biol Chem, 270,* 20686-20691.

[67] Briand, J. P., Benkirane, N., Guichard, G., Newman, J. F., Van Regenmortel, M. H., Brown, .F, Muller, S. (1997). A retro-inverso peptide corresponding to the GH loop of foot-anD-mouth disease virus elicits high levels of long-lasting protective neutralizing antibodies. *Proc Natl Acad Sci USA, 94,* 12545-12550.

[68] Van Regenmortel, M. H., Guichard, G., Benkirane, N., Briand, J. P., Muller, S. and Brown, F. (1998). The potential of retro-inverso peptides as synthetic vaccines. *Dev Biol Stand, 92,* 139-143.

[69] Nargi, F., Kramer, E., Mezencio, J., Zamparo, J., Whetstone, C., Van Regenmortel, M. H., Briand, J. P., Muller, S. and Brown, F. (1999). Protection of swine from foot-and-mouth disease with one dose of an all-D retro peptide. *Vaccine 17,* 2888-2893.

[70] Meziere, C., Viguier, M., Dumortier, H., Lo-Man, R., Leclerc, C., Guillet, J. G., Briand, J. P. and Muller, S. (1997). In vivo T helper cell response to retro-inverso peptidomimetics. *J Immunol, 159,* 3230-3237.

[71] Ben-Yedidia, T., Beignon, A.S., Partidos, C.D., Muller, S. and Arnon, R. (2002). A retro-inverso peptide analogue of influenza virus hemagglutinin B-cell epitope 91-108 induces a strong mucosal and systemic immune response and confers protection in mice after intranasal immunization. *Mol Immunol, 39,* 323-31.

[72] Sela, M., Arnon, B., and Teitelbaum, D. (1990). Suppressive activity of Cop 1 in EAE and its relevance to multiple sclerosis. *Bull Inst Pasteur, 88,* 303-314.

[73] Sela, M., and Arnon, R. (1992). Synthetic approaches to vaccines for infectious and autoimmune diseases. *Vaccine, 10,* 991-999.

[74] Bornstein, M. B., Miller, A., Slagle, S., Weitzman, M., Crystal, H., Drexler, E., Keilson, M., Merriam, A., Wassertheil-Smoller, S., Spada, V., Weiss, W., Arnon, R., Jacobsohn, I., Teitelbaum, D., and Sela, M. (1987). A pilot trial of Cop I in exacerbating-remitting multiple sclerosis. *N Engl J Med, 317,* 408-414.

[75] Johnson, K. P., Brooks, B. R., Cohen, J. A., Ford, C. C., Goldstein, J., Lisak, R. P., Myers, L. W., Panitch, H. S., Rose,J. W., Schiffer, R. B., Voilmer, T., Weiner, L. P., Wolinsky, J. S., and the Cop 1 MS Study Group. (1995). Copolymer 1 reduces relapse rate and improves disability in relapsing-remitting multiple sclerosis. *Neurology, 45,* 1268-1276.

[76] Webb, C., Teitelbaum, D., Hers, A., Arnon, R., and Sela, M. (1976). Molecular requirements involved in suppression of EAE by synthetic basic copolymers of amino acids. *Immunochemistry, 13,* 333-337

[77] Fridkis-Hareli, M., Teitelbaum, D., Arnon, R., and Sela, M. (1995). Synthetic copolymer 1 and myelin basic protein do not require processing prior to binding to class II region histocompatibility complex molecules on living antigen-presenting cells. *Cell. Immunol, 163,* 229-236.

[78] Schlegel, P. G., Aharoni, R., Chen, *Y.,* Chen, J., Teitelbaum, D., Arnon, R., Sela, M., and Chao, N. H. (1996). A synthetic random basic copolymer with promiscous binding to class II major histocompatibility complex molecules inhibits T-cell proliferative responses to major and minor histocompatibility antigens in vitro and confers the capacity to prevent murine graft-versus-host disease in vivo. *Proc Natl Acad Sci USA, 93,* 5061-5066.

[79] Aharoni, R., Scheiegel, P. G., Teitelbaum, D., RoikheL-Karper, O., Arnon, R., Sela, M., and Chao, N. J. (1997). Studies on the mechanism and specificity of the effect of the synthetic random copolymer GLAT on graft-versus-host disease. *Immunol Lett, 58,* 79-87.

[80] Papo, N., Braunstein, A., Eshhar, Z. and Shai, Y. (2004). Suppression of human prostate tumor growth in mice by a cytolytic D-, L-amino Acid Peptide: membrane lysis, increased necrosis, and inhibition of prostate-specific antigen secretion. *Cancer Res, 64,* 5779-5786.

[81] Papo, N. and Shai, Y. (2003). New lytic peptides based on the D,L-amphipathic helix motif preferentially kill tumor cells compared to normal cells. *Biochemistry, 42,* 9346-9354.

[82] Papo, N., Shahar, M., Eisenbach, L. and Shai, Y. (2003). A novel lytic peptide composed of DL-amino acids selectively kills cancer cells in culture and in mice. *J Biol Chem, 278,* 21018-21023.

[83] Nieddu, E., Melchiori, A., Pescarolo, M. P., Bagnasco, L., Biasotti, B., Licheri, B., Malacarne, D., Tortolina, L., Castagnino, N., Pasa, S., Cimoli, G., Avignolo, C., Ponassi, R., Balbi, C., Patrone, E., D'arrigo, C., Barboro, P., Vasile, F., Orecchia, P., Carnemolla, B., Damonte, G., Millo, E., Palomba, D., Fassina, G., Mazzei, M. and Parodi, S. (2005). Sequence specific peptidomimetic molecules inhibitors of a protein-protein interaction at the helix 1 level of c-Myc. *FASEB J, 19,* 632-634.

In: D-Amino Acids: ISBN 1-60021-075-9
A New Frontier in Amino Acid and Protein Research, pp.271-276

Chapter 3.15

CONSIDERATIONS FOR THE USE OF 'RAPID' CELL SUPERFUSION AND VOLTAGE CLAMP TO INVESTIGATE THE ROLE OF RARE AMINO ACIDS IN SYNAPTIC TRANSMISSION

Euan R Brown** and Stefania Piscopo
Neurobiology Laboratory, Stazione Zoologica "A.Dohrn", Naples, Italy

Keywords: whole-cell voltage clamp, D-aspartic acid, neurotransmission, neuromodulation, glutamate receptors.

1. INTRODUCTION AND THEORETICAL CONSIDERATIONS

D- and L-aspartic acids have been found to be present in the majority of nervous systems examined and the most significant levels have been reported in mollusc brains, particularly cephalopods [1]. As a result, it has been proposed that these D-amino acids could act as neurotransmitters or neuromodulators [2]. However, it has not proved possible so far to establish a precise role in synaptic transmission because D-amino acids are present in low concentration in these nervous systems (when compared with 'classical' transmitters). Furthermore, functional identification requires that a number of strict criteria are met, some of which are/will be difficult to achieve with certainty when considering such 'rare' substances (see Table 1). Here we highlight a protocol we have developed to investigate the possible role of these amino acids at the membrane level using a solenoid valve perfusion system and whole-cell voltage clamp. This protocol has established a possible role for and L- and D- aspartic acid in cloned L-glutamate ligand- gated ion channel modulation [3].

* Correspondence concerning this article should be addressed to Euan R Brown and Stefania Piscopo brown@szn.it, piscopo@szn.it; FAX: +39 081 7641355.

Table 1. Generally accepted 'classical' criteria for defining a substance as a neurotransmitter.

- The amino acid must be synthesized or stored in the presynaptic cell.
- The amino acid must be secreted by the presynaptic cell.
- The amino acid must have a postsynaptic effect that duplicates the effect of activating the synapse.
- The amount released must be adequate to account for the entire postsynaptic effect.
- An antagonist that blocks the action of the substance must also block the postsynaptic potential.

Amino acids that satisfy some but not all of the above criteria are often referred to as "transmitter candidates".

2. Materials

Standard external solution for patching whole cells: modified Tyrode solution containing in bidistilled water 5 mM KCl, 135 mM NaCl, 5 mM Hepes-Na, 2 mM $CaCl_2$, 1 mM $MgCl_2$, 10 mM glucose, and 25 mM sucrose (pH 7.4).

Standard internal solution for patching whole cells: Internal solution in bidistilled water 120 mM KCl, 20 mM NaCl, 10 mM Hepes-Na, 1 mM $MgCl_2$, and 0.5 mM EGTA (pH 7.2).

Perfusion solutions: 200 μM D-aspartate (Sigma-Aldrich, St. Louis, MO, U.S.A.) in Tyrode solution; 200 μM L-aspartate (Sigma-Aldrich) in Tyrode solution; 20 mM D-glutamate (Sigma-Aldrich) in Tyrode solution; and 20 mM L-glutamate (Sigma-Aldrich) in Tyrode solution.

Cells: HEK-293 human embryonic kidney cells were transfected with a subunit of a squid glutamate receptor (SqGluR) (NCBI accession number *AJ242600.1, http://www.ncbi.nlm.nih.gov/entrez/viewer.fcgi?db=protein&val=6688642)*. HEK-293 cells were cultured by standard methods. Lipofectamine 2000 reagent (Life Technologies, Rockville, MD, USA) was used to transfect cells at 60-95% confluence. The cells were re-plated on no. 1 glass coverslips (22 x 22 mm , Knittel glaser, Germany) and used within 72 h.

3. Methods

Whole-Cell Voltage Clamp

Cover slips wilth HEK cells expressing the squid glutamate receptor were removed from the culture dishes and placed directly into a chamber on an inverted microscope. The chamber was continuously perfused at 2 ml per minute with modified Tyrode solution. Cells were subjected to whole-cell voltage clamp. Briefly, 3-5 Mohm pipettes were brought into contact with the cell surface by micromanipulation, GΩ resistance cell-attached seals obtained by gentle suction and whole–cell access achieved by rupturing the patch with direct suction. Using an Axoclamp 2B amplifier and a 0.1 Ul headstage, capacity transients and feedback resistance were adjusted optimally and the cell clamped to –60 mV. Current and voltage outputs from the amplifier were digitized by a Digidata 1200 digital acquisition system and stored on an IBM PC. Amplifier, data acquisition system, and software were all from Axon instruments Inc. (Foster City, CA, USA, part of Molecular Devices Corporation, Union City, CA, USA).

Application of Amino Acids

Multiple solution applications are required to compare the effects of D- and L-amino acids along with classical ligands. Some results from a typical experiment are shown in Figure 1. The upper panel shows two applications of L-glutamate (20 mM) and in the second application 200 μM D-aspartic acid and 20 mM L-glutamate are added simultaneously. We originally used a laboratory- made two-channel pinch-valve system that gave a time resolution for solution changes of around 0.5 s. However more channels were required (at least 3 in Figure 1) and switching between channels caused a transient pressure pulse on valve closure that often disrupted seal integrity resulting in loss of whole-cell clamp. We then opted for a commercially available multi-valve system using TTL (Transistor-Transistor Logic) switchable solenoid valves. The 'Valvelink 8' valve system (Automate Scientific Inc, San Francisco, CA, USA), allows 8 solenoid valves to be remotely switched using TTL logic control pulses. Control pulses were generated using Clampex software (Axon Instruments Inc.) through the digital outputs of the Digidata Acquisition system, were amplified to 12 V and delivered to the Valvelink control panel. Control steps are shown in the lower panel of Figure 1 (0=off, 1=on). 10 PSI pressure was applied to the solution reservoirs. The multiple output tubes from the valves led to a 'perfusion pencil' (Automate Scientific Inc., San Francisco, CA, USA), which was an 8 way manifold with a dead- space of around 10 μl. Channel one of the Valvelink system perfused the target cells continuously with Tyrode solution (Figure 1). A suitable target cell was located under x 40 objective and 100 μM diameter perfusion pipette fitted to the perfusion manifold was micro- manipulated to within 300 μm of the cell under voltage clamp in such a way that the flow from the pipette was against the prevailing current generated by the bath perfusion. It was found necessary to supply a continuous flow of control Ringer over the cells from the perfusion pipette under resting conditions. This was because sudden changes in perfusion flow (i.e. Zero to 10 PSI) caused a pressure artefact on the surface of the cells similar to that experienced with the lab-built pinch valve system.

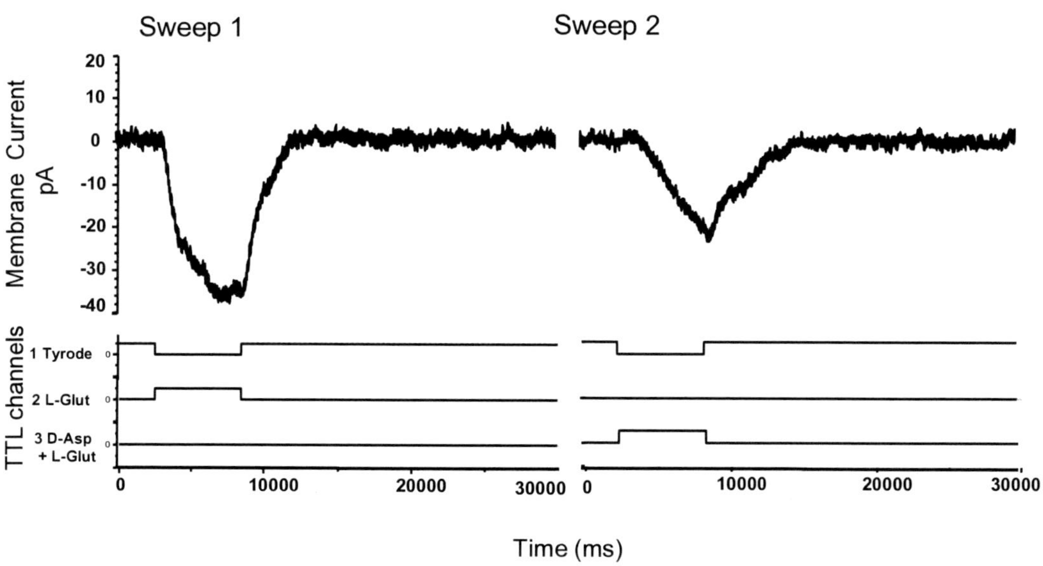

Figure 1. Two recordings from a single HEK cell transfected with SqGluR. The upper panel shows the currents evoked by the application of L-glutamate (20mM) at a holding potential under voltage clamp of –60 mV. Bottom panel shows the TTL control steps that open and close the solenoid valves. The content of the tubes supplying the valves are shown to the left with the valve number. In the second sweep, there is a significant slowing of the activation of the glutamate receptor when D-aspartic acid is applied together with L-glutamate.

4. Results

As shown in Figure 1, it was possible to make high resolution whole-cell voltage clamp recordings from single cells transfected with the squid glutamate receptor (SqGlurR) to investigate ligand-gated currents evoked by amino acids. The advantage of this system was that a single cell could be exposed to a series of amino acid additions in a rapid and reproducible way. For example, the trace in Figure 1 shows that application of 200 μM D-aspartic acid applied at the same time as 20 mM L-glutamate slows significantly the ligand-gating of the glutamate receptor: whereas application of D or L-aspartic acid was ineffective at gating a current (data not shown). The results suggest that this squid glutamate receptor may be modulated by D- and L-aspartic acid [3].

Table 2. Comparison of methods used to apply amino acids to cells with the time-course of postsynaptic activation of glutamate receptors during synaptic transmission. Other time-courses are activation time constants of post synaptic receptors when exposed to the various application techniques.

Method	Speed of 'application ' (fastest step)	'perfusion channels' (number)	Ease of use	Drawbacks
Synaptic transmission	Activation time constant of postsynaptic receptors around 3.5 ms (ref 4)	-	Easy (requires nerve stimulation and postsynaptic recording).	Interpretation difficult
Iontophoresis	50 -60 ms (ref 5)	Up to 6	Difficult	Only 1 or at maximum 2 application channels possible at any one time, concentration of applied substance difficult to determine.
Pressure ejection	1-2 seconds	Up to 16	Moderate	Needs the pulling of specialized multichannel pipettes. Difficult to equalize the pressure between channels. Tendency for the pressure pulse to cause artefacts.
Concentration jump	fast	1 channel per time	Easy	Impossible to determine the time course of application and the concentration of the drug difficult to reproduce because of large pressure artefacts.
Valve perfusion	100 ms	8 -16 channels depending on the system	Easy	Needs constant local perfusion so tends to destabilize whole cell seals.

5. Notes

Synaptic Complexity

One of the complicating factors in determining the effect of rare amino acids in the nervous system is that synapses are complex elements. They consist of not only pre and postsynaptic domains but also glial and spatial elements, while the location and number of membrane proteins (ligand-gated ion channels and transporters) that may respond to the amino acid of interest render a clear interpretation difficult. As a result we have concentrated on establishing the third point in Table 1 and to do this are using a synapse free system with defined receptors and whole-cell voltage clamp. Although this is a very "reductionist" step, it is essential one to accurately determine the role of these amino acids in the nervous system. For the above reasons, we used cultured cells that expressed cloned AMPA-like glutamate receptors from squid. We used this approach because the highest concentrations of D-amino acids have been reported in cephalopod brains [1], and have reasoned that if they are active on ligand-gated ion channels, the effects should be most obvious in channels from these animals. In the case of the squid AMPA-like receptor used in the present experiments, D and L- aspartic acid have significant modulatory effects on the ligand gated currents but are unable to gate the channel [3]. However it is worth pointing out that this technique could also be applied to enzymatically isolated neurons or neurons in primary culture where receptor expression is well defined.

Application of Amino Acids

The time course of application of the amino acids should approximate at least the real time-course of activation of native receptors during synaptic transmission, which may be considered to be around 3-4 msec (see Table 2). Thus, we have sought to use techniques that allow us to apply drugs rapidly that have minimal artefacts. Table 2 shows a comparison of the *pros* and *cons* of some of these systems compared to synaptic transmission. The most effective combination we have found is to use a solenoid valve switching system as this allows us to apply up to 8 different solutions with a time resolution of around 10 msec. However, the fastest application times are achievable usefully with iontophoresis. There is a danger that if a channel is able to inactivate on exposure to the ligand that the real time course of channel activation will be obscured by combination of inactivation, activation and deactivation. These artifacts are possible when using a valve operated perfusion system. In this case, a sensible approach is to use a multi-channel perfusion system to screen for effects, and then look in more detail at the kinetics using iontophoresis.

AKNOWLEDGEMENTS

SP was supported by a FIRB program grant (RBAU018RWY) to ERB.

REFERENCES

[1] D'Aniello, S., Spinelli, P., Ferrandino, G., Peterson, K., Tsesarskia, M., Fisher, G. and D'Aniello A. (2005). Cephalopod vision involves dicarboxylic amino acids: D-aspartate, L-aspartate and L-glutamate. *Biochem J 386*, 331–340.

[2] Spinelli, P., Brown, E. R., Ferrandino, G., Branno, M., Montarolo, P. G., D'Aniello, E., Rastogi, R. K., D'Aniello. B., Baccari, G. C., Fisher, G., and D'Aniello A. (2006). D-Aspartic acid in the nervous system of *Aplysia limacina*: possible role in neurotransmission. *J Cell Physiol 206,* (3):672-81.

[3] Brown ER, Piscopo S, Chun JT, Francone M, Mirabile I, D'Aniello A. (2006) Modulation of an AMPA-like glutamate receptor (SqGluR) gating by L- and D-aspartic acids. *Amino Acids.* Jun 1; [Epub ahead of print]

[4] Forsythe, I. D. and Westbrook, G. L. (1988) Slow excitatory postsynaptic currents mediated by *N*-methyl-D-aspartate receptors on cultured mouse central neurones. *J Physiol (Lond), 396*, 515–533.

[5] Lima, P. A, Nardi, G, and Brown, E. R. (2003) AMPA/kainate and NMDA-like glutamate receptors at the chromatophore neuromuscular junction of the squid: role in synaptic transmission and skin patterning. *Eur J Neurosci*, *17*, 507–516.

In: D-Amino Acids: ISBN 1-60021-075-9
A New Frontier in Amino Acid and Protein Research, pp.277-282

Chapter 3.16

ARE D-AMINO ACIDS PREVALENT AMONG EUKARYOTES?

Yoko Nagata*

Department of Materials and Applied Chemistry, College of Science and Technology, Nihon University, Kanda-Surugadai, Chiyoda-Ku, Tokyo, 101-8308 Japan.

Keywords: D-amino acid, D-amino acid oxidase, D-aspartate oxidase, peptide.

It seems that the ratio of occurrence of D- to L-amino acids was 1:1 when Earth was created since meteorites contain both of the amino acids at almost an equal ratio, and the amino acids synthesized *in vitro* in a laboratory are racemized. Today, the occurrence of D-amino acids is not generalized in higher organisms, whereas many more D-amino acids have been detected in primitive organisms such as bacteria (Table 1). The homochirality of amino acids in organisms, the so called L-amino-acid world, is likely to be established being accompanied by evolution of life on Earth. However, occurrences of D-amino acids in higher animals are increasingly reported these days, although the concentrations of D-amino acids are usually low, and the number of reported animals and tissues are few. Is it true that animals contain only small amounts of D-amino acids?

1. FREE AMINO ACIDS

De novo synthesis of D-amino acids was predicted as normal metabolism in mice by an experiment using SPF and germ-free mice [1], and serine racemase was purified from rat brain [2] as the first mammalian enzyme that catalyzes conversion of a free L-amino acid to the corresponding D-amino acid, and vice versa. Then, a serine racemase gene was cloned from mice [3] and human [4] brains. Expression of serine racemase mRNA was shown by Northern blot analysis not only in the human brain, but also expressed in heart, muscle, kidney and liver. The expression of the enzyme protein was confirmed by an immunohistochemical study [5]. Recently, mRNA for the enzyme was detected in human placenta, and D-serine transport to the fetus was reported [6]. In the report, D-serine concentrations in maternal and the fetal

* Correspondence concerning this article should be addressed to Yoko Nagata,
E-mail: nagata@chem.cst.nihon-u.ac.jp.

sera were high, *i.e.*, 5.8 μM and 14.6 μM, respectively, while in normal adult plasma, it is 0.42-1.76 μM [7]. These results may be explained well by the fact that D-amino acid oxidase activity in nephron of neonatal rats is 4.5-fold lower than that in adult rats [8].

Table 1. Free D-amino acid ratios (%D=D/D+L).

	Ser	Ala	Pro	Glu	Asp
Streptococcus mutans [a]	16.2	54.7	1.8	0.4	3.8
Paracoccus denitrificans [b]	1.6	32.1	42.5	11.7	5.0
Escherichia coli [b]	1.2	31.8	2.2	7.6	2.9
Pyrobaculum islandicum [c]	12.7	0.9	3.5	1.2	3.6
Physarum polycephalum [d]	3.1	3.4	0	4.4	0.7
Chlamydomonas reinhardtii [e]	0.15	0.15	0	0.12	0
Tokunagayusurika akamusi [d]	0.94	0.94	0	0.12	0.30
Radish (sprouts) [d]	0.18	0.18	0	0	0
Soy bean (sprouts) [d]	0	0	0	0	0

[a] Nagata, Y. *et al.* (2006) Life Sci., 78, 1677.
[b] Nagata, Y. *et al.*(2001) *J. Mol. Catal. B: Enzymatic, 12,* 109.
[c] Nagata, Y. *et al.*(1999) *Biochim. Biophys. Acta, 1435*, 160.
[d] Higashi, M.& Nagata, Y., unpublished.
[e] Ishii, Y. *et al.*, unpublished.

Considerably higher concentrations of D-amino acids were detected in the serum [9] of a mutant mouse [10] lacking D-amino acid oxidase, and plasma from patients with renal diseases such as glomerulonephritis [7]. This was more than those of normal mice and healthy individuals, respectively. D-Amino acid oxidase (EC 1.4.3.3) is an enzyme that catalyzes the oxidative deamination of free neutral D-amino acids [11]. Most of the enzyme is localized in renal proximal tubules, cerebellum and liver in mammals [12]. Hence, both of the above mutant mice sera, and the renal disease patients' plasma, had not received the catabolic function of D-amino acid oxidase. In the brain, D-serine was detected at a 6.4-fold higher concentration in the cerebrum than in the cerebellums of mice, and conversely, D-amino acid oxidase activity is 6.7-fold higher in the cerebellum than in the cerebrum in mice [13]. High levels of D-alanine and D-proline were observed in human saliva [39]and gastric juice [40] where the enzyme activity is negligible.

D-amino acid oxidase is an inducible enzyme. In carp, the enzyme is induced by feeding them a D-alanine supplemented diet [14]. In mammals, the enzyme is always expressed. It may mean that substrates of the enzyme, free neutral D-amino acids are continuously present in the mammal bodies. The physiological role of this enzyme has clearly been shown by oral administration of D-alanine to the mutant mice lacking D-amino acid oxidase [15]. That is, in the experiment, the serum D-alanine level increased 114-fold in the mutant mice contrary to the increase of 13-fold in the normal mice. The role has also been demonstrated in a plant, *Arabidopsis thaliana*, by introducing the yeast D-amino acid oxidase gene into the plant [16].

D-Aspartate oxidase (EC 1.4.3.1) catalyzes the oxidative deamination of free D-aspartate [17]. The relation between D-aspartate and D-aspartate oxidase is similar to that between D-amino acid and D-amino acid oxidase. *De novo* synthesis of D-aspartate has been strongly suggested by the occurrence and increase of the amino acid in a chicken eggs [18], and PC12 cultured cells [19]. Inverse brain localization of D-aspartate and D-aspartate oxidase is reported to be in the rat brain [20]. D-aspartate oxidase is also an inducible enzyme, and the enzyme activity was increased in mouse liver by administration of D-aspartate

[21]. A highly active D-aspartate oxidase was purified from a yeast strain, that was selected from soil cultures carried out in a medium containing D-aspartate as a nitrogen source [22].

All these results conclusively point to the fact that neutral and acidic D-amino acids are biosynthesized in higher animals, and not by habitant bacteria but solely by the animals. If it were not D-amino acid oxidase and D-aspartate oxidase in the animals, high levels of D-amino acids would be observed in the animals.

2. Peptide Amino Acids

D-Amino-acid containing neuropeptides and peptide toxins have been found in eukaryotes (Table 2), such as a natriuretic peptide from the venom of a platypus, a mammal, [23], and potent opioid heptapeptides, dermorphin [24], dermenkephalin [25] and deltorphin [26] in the skin of a South American frog. Bombinin H [27] is a bactericidal peptide secreted from an European frog. In molluscs, neuropeptides are found such as achatin I [28], and fulicin [29] in a giant land snail, and a decapeptide from the bivalve, *Mytilus* [30]. A D-tryptophan containing octapeptide, contryphan [31], and a large peptide, r11a [32], grouped to the I-superfamily conotoxin, are isolated from the venom of the fish-hunting marine snail, *Conus radiatus*. A cardioactive peptide [33], and a hyperglycemic hormone [34], were isolated from the octopus brain and crustacean sinus glands, respectively. Many of these peptides are small, and are composed of less than ten amino acids. In most cases, the D-configuration of the amino acid residue in the peptides (Table 2) is important for the activity.

Posttranslational formation of D-amino acid in these peptides, except the contryphan, has strongly been suggested based on the presence of mRNA encoding the precursor peptide of dermorphin [35]. ω-Agatoxin IV B [36] is a spider venom which is a calcium channel inhibitor that possesses D-serine at position 46 of the 48-amino acid peptide. The isomerase that convert the precursor L-Ser^{46} to D-Ser^{46} was purified from the spider [37]. Most of the peptides in Table 2 contain the D-enantiomers in position 2 or 3 from the N- terminus, and in position 3 from the C-terminus. There may be favored positions near the N or C terminus that are preferential sites for isomerization [33]. Recently, the isomerase that produces D-*allo*-isoleucine from the L-isoleucine residue in the bombinin H precursor was purified and the cDNA was cloned [38]. Of interest, the homologous DNA sequences to bombinin H were found in human, chicken and fugu fish [38].

Up to date, reports on the D-amino acid peptides are few. It may be ascribed to the fact that there is no simple and accurate method to distinguish a D-amino acid residue *in situ* in a peptide. Usually, a D-amino acid peptide is found by bioassays that detect differences in activity between the peptides containing a D-amino acid, and those composed solely of L-amino acids. Determination of the D-amino acid and the position in the peptide may need hydrolysis of the peptide followed by analysis using high performance liquid chromatography *et al.* Furthermore, the D-amino acid residue and the position in the peptide must be confirmed by comparison of the natural peptide with the synthesized peptide. If elaborative studies were to be carried out in many laboratories, 'it would not be surprising that many proteins/peptides containing D-amino acids will be discovered in other higher organisms in the future [23]'.

Table 2. D-Amino-acid containing peptides.

		Positions
1.Platypus venom	Leu-<u>Leu</u>-His-Asp-His-Phe· · ·Cys	2, 39
2.Dermorphin	Tyr-<u>Ala</u>-Phe-Gly-Tyr-Pro-Ser	2, 7
3.Dermenkephalin	Tyr-<u>Met</u>-Phe-His-Leu-Met-Asp	2, 7
4.Deltorphin I	Tyr-<u>Ala</u>-Phe-Glu-Val-Val-Gly	2, 7
5.Bombinin H	Ile-<u>*a*Ile</u>-Gly-Pro-Val-Leu-Gly	2, 7
6.Achatin I	Gly-<u>Phe</u>-Ala-Asp	2, 4
7.Fulicin	Phe-<u>Asn</u>-Glu-Phe-Val	2, 5
8.FMRMamide	Ala-<u>Leu</u>-Ala-Gly-Asp-His-Phe-Phe-Arg-Phe	2, 10
9.Contryphan	Gly-Cys-Hyp-<u>Trp</u>-Glu-Pro-Trp-Cys	4, 8
10.Conotoxin r11a	-Thr-Ser-Ser-Phe-<u>Phe</u>-Lys-Ile	44, 46
11.Cardioactive peptide	Gly-<u>Phe</u>-Gly-Asp	2, 4
12.Hyperglycemic peptide	Glu-Val-<u>Phe</u>-Asp-Glu-Ala-Lys-	3
13.ω -Agatoxin IVB	-Met-Glu-Gly-Leu-<u>Ser</u>-Phe-Ala	46, 48

—, D-enantiomer; Hyp, hydroxyproline

REFERENCES

[1] Nagata, Y. & Akino, T. (1990) D-Amino acids in mouse tissues are not of microbial origin. *Experientia, 46*, 466-468.

[2] Wolosker, H., Sheth, K.N., Takahashi, M., Mothet, J.-P., Brady Jr., R.O., Ferris, C.D. & Snyder, S.H. (1999) Purification of serine racemase: biosynthesis of the neuromodulator D-serine. *Proc. Natl. Acad. Sci. U.S.A. 96*, 721-725.

[3] Wolosker H., Blackshaw, S. & Snyder, S.H. (1999) Serine racemase: a glial enzyme synthesizing D-serine to regulate glutamate-*N*-methyl-D-aspartate neurotransmission. *Proc. Natl. Acad. Sci. U.S.A. 96*, 13409-13414.

[4] De Miranda, J., Santoro, A., Engelender, S. & Wolosker, H. (2000) Human serine racemase: molecular cloning, genomic organization and functional analysis. *Gene 256*, 183-188.

[5] Xia, M., Liu, Y., Figueroa, D.J., *et al.* (2004) Characterization and localization of a human serine racemase. *Mol. Brain Res. 125*, 96-104.

[6] Chen, Z., Huang, W., Srinivas, S.R., Jones, C.R., Ganapathy, V. & Prasad, P.D. (2004) Serine racemase and D-serine transport in human placenta and evidence for a transplacental gradient for D-serine in humans. J. Soc. *Gynecol. Investig. 11*, 294-303.

[7] Nagata, Y., Masui, R. & Akino, T. (1992) The presence of free D-serine, D-alanine and D-proline in human plasma. *Experientia 48*, 986-988.

[8] Chan, A.W.K., Perry, S.G., Burch, H.B., Fagioli, S., Alvey, T.R. & Lowry, O.H. (1979) Distribution of two aminotransferases and D-amino acid oxidase within the nephron of young and adult rats. *J. Histochem. Cytochem. 27*, 751-755.

[9] Nagata, Y., Yamamoto, K., Shimojo, T., Konno, R., Yasumura, Y. & Akino, T. (1992) The presence of free D-alanine, D-proline and D-serine in mice. *Biochim. Biophys. Acta 1115*, 208-211.

[10] Konno, R., Yasumura, Y. (1983) Mouse mutant deficient in D-amino acid oxidase activity. *Genetics 103*, 277-285.

[11] Bright, H.J. & Porter, J.J.T. (1975) Flavoproteinoxidases. In: Edtitor, P. D. Boyer, *The Enzyme*, 3rd Edition, Vol 12,. New York: Academic Press, 421-505.

[12] Dunn, J.T. & Perkoff, G.T. (1963) D-Amino acid oxidase activity in human tissues. *Biochim. Biophys. Acta 73*, 327-331.

[13] Nagata, Y. (1992) Involvement of D-amino acid oxidase in elimination of D-serine in mouse brain. *Experientia 48*, 753-755.

[14] Sarower, M.G., Okada, S. & Abe, H. (2003) Molecular characterization of D-amino acid oxidase from common carp *Cyprinus carpio* and its induction with exogenous free D-alanine. *Arch. Biochem. Biophys. 420*, 121-129.

[15] Nagata, Y., Konno, R. & Niwa, A. (1994) Amino acid levels in D-alanine-administered mutant mice lacking D-amino acid oxidase. *Metabolism 43*, 1153-1157.

[16] Erikson, O., Hertzberg, M. & Näsholm, T. (2004) A conditional marker gene allowing both positive and negative selection in plants. *Nature Biotec. 22*, 455-458.

[17] Hamilton, G.A. (1985) Peroxisomal oxidases and suggestions for the mechanism of action of insulin and other hormones. *Adv. Enzymol. Relat. Areas Mol. Biol. 57*, 85-178.

[18] Nidle, A. & Dunlop, D.S. (1990) Developmental changes in free D-aspartic acid in the chicken embryo and in the neonatal rat. *Life Sci. 46*, 1517-1522.

[19] Long, Z.Q., Homma, H., Lee, J.A., Fukushima, T., Santa, T., Iwatsubo, T., Yamada, R. & Imai, K. (1998) Biosynthesis of D-aspartate in mammalian cells. *FEBS Lett. 434*, 231-235.

[20] Schell, M.J., Cooper, D.B. & Snyder, S.H. (1997) D-aspartate localizations imply neuronal and neuroendocrine roles. *Proc. Natl. Acad. Sci. 94*, 2013-2018.

[21] Yamada, R., Nagasaki, H., Nagata, Y. & Wakabayashi, Y. (1989) Administration of D-aspartate increases D-aspartate oxidase activity in mouse liver. *Biochim. Biophys. Acta 990*, 325-328.

[22] Yamada, R., Ujiie, H., Kera, R., Nakase, T., Kitagawa, K., Imasaka, T., Arimoto, K., Takahasi, M. & Matsumura Y, (1996) Purification and properties of D-aspartate oxidase form *Cryptococcus humicolus* UJ1. *Biochim. Biophys. Acta 1294*, 153-158.

[23] Torres, A.M., Menz, I., Alewood, P.F., Bansal, P., Lahnstein, J., Gallagher, C.H. & Kuchel, P.W. (2002) D-Amino acid residue in the C-type natriuretic peptide from the venom of the mammal, *Ornithorhynchus anatinus*, the Australian platypus. *FEBS Lett. 524*, 172-176.

[24] Montecucchi, P.C., de Castiglione, R., Piani, S., Gossini, L. & Erspamer, V. (1981) Amino acid composition and sequence of dermorphin, a novel opiate-like peptide from the skin of *Phyllomedusa savagei. Int. J. Pept. Protein Res. 17*, 275-283.

[25] Mor, A., Delfour, A., Sagan, S., Amiche, M., Pradelles, P., Rossier, J. & Nicolas, P. (1989) Isolation of dermenkephalin from a amphibian skin, a high-affinity delta-selective opioid heptapeptide containing a D-amino acid residue. *FEBS Lett. 255*, 269-274.

[26] Kreil, G., Barra, D., Simmaco, M., Erspamer, V., Erspamer, G.F., Negri, L., Severini, C., Corsi, R. & Melchiorri, P. (1989) Deltorphin, a novel amphibian skin peptide with high selectivity and affinity for delta opioid receptors. *Eur. J. Pharm. 162*, 123-128.

[27] Mignogna, G., Simmaco, M., Kreil, G. & Barra, D. (1993) Antibacterial and haemolytic peptides containing D-*allo*isoleucine from the skin of *Bombina variegate. EMBO J. 12*, 4829-4832.

[28] Kamatani, Y., Minakata, H., Kenny, P.T.M. *et al.* (1989) Achatin-I, an endogenous neuroexcitatory tetrapeptide from *Achatina fulica* Férussac containing a D-amino acid residue. *Biochem. Biophys. Res. Commun. 160*, 1015-1020.

[29] Ohta, N., Kubota, I., Takao, T., Shimonishi, Y., Yasuda-Kamatani, Y., Minakata, H., Nomoto, K., Muneoka, Y. & Kobayashi, M. (1991) Fulicin, a novel neuropeptide containing a D-amino acid residue isolated from the ganglia of *Achatina fulica. Biochem. Biophys. Res. Commun. 178*, 486-493.

[30] Fujisawa, Y., Ikeda, T., Nomoto, K., Yasuda-Kamatani, Y., Minakata, H., Kenny, P.T.M., Kubota, I. & Muneoka, Y. (1992) The FMRFamide-related decapeptide of *Mytilus* contains a D-amino acid residue. *Comp. Biochem. Physiol. 102C*, 91-95.

[31] Jimenéz, E.C., Olivera, B.M., Gray, W.R. & Cruz, L.J. (1996) Contryphan is a D-tryptophan-containing *Conus* peptide. *J. Biol. Chem. 271*, 28002-28005.

[32] Buezek, O., Yoshikami, D., Bulaj, G., Jimenez, E.C. & Olivera, B.M. (2005) Post-translational amino acid isomerization. *J. Biol. Chem. 280*, 4247-4253.

[33] Iwakoshi, E., Hisada, M. & Minakata, H. (2000) Cardioactive peptides isolated from the brain of a Japanese octopus, *Octopus minor. Peptides 21*, 623-630.

[34] Soyez, D., van Herp, F., Rossier, J., Le Caer, J.P., Tensen, C.P. & Lafon, R. (1994) Evidence for a conformational polymorphism of invertebrate neurohormones. *J. Biol. Chem. 269*, 18295-18298

[35] Richter, K., Egger, R. & Kreil, G. (1987) D-Alanine in the frog skin peptide dermorphin is derived from L-alanine in the precursor. *Science 238*, 200-202.

[36] Heck, S. D., Kelbaugh, P.R., Kelly, M.E., Thadeio, P.F., Saccomano, N.A., Stroh, J.G. & Volkman, R.A. (1994) Disulfide bond assignment of ω-Agatoxins IVB and IVC: Discovery of a D-serine residue in ω-Agatoxins IVB. *J. Am. Chem. Soc. 116*, 10426-10436.

[37] Shikata, Y., Watanabe, T., Teramoto, T., Inoue, A., Kawakami, Y., Nishizawa, Y., Katayama, K. & Kuwada, M. (1995) Isolation and characterization of a peptide isomerase from funnel web spider venom. *J. Biol. Chem. 270*, 16719-16723.

[38] Jilek, A., Mollay, C., Tippelt, C. *et al.* (2005) Biosynthesis of a D-amino acid in peptide linkage by an enzyme from frog skin secretions. *Proc. Nat. Acad. Sci. U.S.A. 102*, 4235-4239.

[39] Nagata, Y., Higashi, M., Ishii, Y. et al. (2006) The presence of high concentrations of free D-amino acids in human saliva. Life Sci. 78, 1677-1681.

[40] Nagata, Y., Sato, T., Enomoto, N. et al., High concentration of D-amino acids in human gastric juice. Amino Acids, in press.

Section 4. Free d-Amino Acid in Food Stuffs and their Nutritional Aspects

In: D-Amino Acids: ISBN 1-60021-075-9
A New Frontier in Amino Acid and Protein Research, pp.285-298

Chapter 4.1

NUTRITIONAL EVALUATION OF D-AMINO ACIDS

Mendel Friedman*

Western Regional Research Center, Agricultural Research Service,
U. S. Department of Agriculture, U.S.A.

Keywords: D-amino acid nutrition; mouse bioassay; protein digestibility; antinutritional effects; toxic effects; food processing; food safety; racemization.

1. INTRODUCTION

Most amino acids of importance in nutrition exist as L-isomers. During food processing, the L-amino acids may be may be transformed to their mirror image configuration D-isomers. D-Amino acids can also be synthesized by microorganisms. As part of a program to evaluate the chemistry and the nutritional and toxicological potential of novel amino acids, including D-amino acids, formed during food processing, we compared the weight gain in mice fed free amino acid diets in which the test D-amino acid was substituted for the L-isomer. The results obtained reflect the ability of mice to utilize D-amino acids in the complete absence of the L-form. In the case of essential amino acids, the mice must meet the entire metabolic demand from the D-isomeric forms.

The nutritional effectiveness protein-bound essential D-amino acids depends on the amino acid composition, digestibility, and physiological utilization of released amino acids. Since an amino acid must be liberated by digestion before nutritional assimilation can occur, the decreased susceptibility to proteolytic enzymes of D-D, D-L, and L-D peptide bonds in D-amino acid-containing proteins to digestion by enzymes is a major factor adversely affecting the bioavailability of protein-bound D-amino acids; reviewed in [1-4].

Mice provide a good animal model to study the biological utilization and biological effects of D-amino acids. A major advantage of mouse bioassays is that they require about one-fifth of the test substance needed for rats and can be completed in 14 days [5-10].

* Correspondence concerning this article should be addressed to Mendel Friedman, E-mail: mfried@pw.usda.gov, FAX: ; +1 510 559-5963.

2. Methods

Growth Assay in Mice

Biological utilization by weanling male mice (Swiss Webster strain, Simonsen Laboratories, Gilroy, California) of D-amino acids was tested in a 14-day growth assay by using an all free amino acid diet in which each L-amino acid was replaced by the D-isomer (Table 1). Mice were housed singly or two per cage. The cages were polycarbonate with stainless steel wire tops and pine shavings for litter. Feed and water were provided *ad libitum*. The temperature of the animal room was 22 ± 1 °C. Humidity was maintained at 50 ± 10%. The diurnal light cycle was 0600-1800h light and 1800-0600 h dark, as regulated by an automatic timer. Animals were assigned so that all treatment groups had nearly the same initial mean body weight.

Table 1. Composition of amino acid diet used to determine nutritional values of D-amino acids in mice.

Ingredient	%
L-Ala	0.35
L-Arg • HCl	1.35
L-Asn	0.60
L-Asp	0.35
L-Glu	3.50
Gly	2.33
L-His • HC1	0.41
L-Ile	0.82
L-Leu	1.11
L-Lys • HC1	1.35
L-Met	1.17
L-Phe	1.51
L-Pro	0.35
L-Ser	0.35
L-Trp	0.174
L-Thr	0.82
L-Val	0.82
Alphacel[a]	3.00
Corn oil	8.00
Cornstarch	20.00
Dextrose	38.33
Salts USP XIV[b]	5.00
Sodium acetate	1.31
Water (added)	5.00
Complete vitamin mixture[c]	2.00
Total	100.00

[a]Nutritional Biochemicals, Cleveland, OH. [b]Salts USP XIV provided in g/kg diet: cupric sulfate, 3.88 mg; ferric ammonium citrate, 764 mg; manganese sulfate, 10.0 mg; ammonium alum, 4.62 mg; potassium iodide, 2.03 mg; sodium fluoride, 25.4 mg; calcium carbonate, 3.43; calcium citrate, 15.4; calcium biphosphate, 5.64; magnesium carbonate, 1.76; magnesium sulfate, 1.92; potassium chloride, 6.24; dibasic potassium phosphate, 10.9; and sodium chloride, 3.86. The diet was supplemented with zinc, 6.25 mg, and cobalt, 0.29 mg/kg diet. [c]Vitamin mixture provided in diet: per g diet: vitamin A acetate, 18 IU; cholecalciferol, 2.0 IU; DL-α-tocopherol acetate, 0.11 IU; per kg diet: choline chloride, 2.0 g; (in mg) menadione, 20; nicotinic acid, 90; riboflavin, 20; pyridoxine • HC1, 20; thiamin•HC1, 20; calcium pantothenate, 60; D-biotin, 0.40; folic acid, 4.0; inositol, 100; vitamin C, 900; vitamin B-12 crystalline, 27 μg/kg.

Table 2. Relative Growth Response to D-Amino Acids and Peptides in Mice Fed Amino-Acid Diets. Adapted from [5-10].

D-Amino acids and peptides	Relative nutritional value to L-form, %
D-Methionine (first experiment)	79.5
D-Phenylalanine	51.6
D-Tryptophan	24.7
D-Leucine	12.4
D-Histidine	8.5
D-Valine	5.1
D-Threonine	3.1
D-Isoleucine	1.2
D-Lysine	-8.2
L-L + D,L-Lysinoalanine	3.8
D-Methionine (second experiment)	87.5
N-Acetyl-L-methionine	89.6
N-Acetyl-D-methionine	22.9
L-Methionyl-L-methionine	99.2
L-Methionyl-D-methionine	102.9
D-Methionyl-L-methionine	80.4
D-Methionyl-D-methionine	41.0
L-Methionine sulfoxide	87.5
D-Methionine sulfoxide	26.8
N-Acetyl-L-methionine sulfoxide	58.7
N-Acetyl-D-methionine sulfoxide	1.7
L-Methionine hydroxy analogue	55.4
D-Methionine hydroxy analogue	85.7

The growth assay was standardized with six mice per group, with potency estimation being based of response of two to seven groups (growth data from 12 to 42 mice). Statistical evaluation of potencies of the D-amino acids was calculated as the slopes of growth curves (weight gain after 14 days per unit of dietary concentration) where linearity was approximated. Slope ratio analyses of growth data were performed as described by Finney with SAS using the general linear model (GLM). Confidence intervals of potencies relative to the corresponding L-amino acid were estimated using Fieller's theorem. Mean body weight gains were compared by Duncan's multiple range test using individual values.

3. RESULTS

Figures 1 and 2 illustrate weight gains in mice fed increasing dietary amounts of D-amino acids. Table 2 lists the nutritional values (relative potencies) of specific D-amino acids and D-peptides as sources of the corresponding L-isomers. To place our findings in proper perspective, we integrate the results of our studies and related studies by other investigators into the following brief summary of nutritional and health- related aspects of D-amino acids listed alphabetically.

D-Alanine

D-Alanine induced cytotoxic oxidative stress in brain tumor cells [11].

D-Arginine

Both L- and D-arginine protected against oxygen-radical-induced injury of rat heat tissue [12] and against endotoxin shock in rabbits [13]. D-arginine and other D-amino acids inhibited cell proliferation and tumor growth in rats [14].

D-Aspartic Acid

D-Aspartic acid aggravated the pyeolnephritis in rats induced by *Staphylococcus aureus* bacteria [15]. D-Aspartate prevented potassium and magnesium depletion in rats induced by cardiac drugs and diuretics [16].

D-Cysteine

Although L-CysSH had a sparing effect of L-Met consumed by mice, D-CysSH did not [8]. In fact, D-CysSH imposed a metabolic burden as indicated by depressed growth when fed to mice with less than optimal levels of D-Met. The 24% decrease in weight gain of the D-CysSH plus L-Met amino-acid diet compared to L-Me alone implies that D-CysSH is nutritionally antagonistic or toxic. D-Cysteine but not N-acetyl-D-cysteine lowered rat blood cyanide levels derived from acrylonitrile [17]. L-Cysteine-glutathione disulfide but not the D-cysteine analog protected mice against acetaminophen-induced liver damage [18].

D-Cystine

Although L-Cys is not an essential amino acid for rodents, less L-Met is needed for growth if the diet contains L-Cys [8]. Our results show that L-Cys is somewhat more efficient in sparing D-Met than in sparing L-Met in diets containing low levels (0.29%) of the two isomers. Supplementation of D-Met with an equal sulfur equivalent of L-Cys doubled growth. Thus, the overall response was equal to that produced by L-Met in the presence of L-Cys. In contrast, supplementation of suboptimal levels of L-Met with increasing concentrations of D-Cys reduced the growth rate of mice. This implies that excess D-Cys in the diet is toxic.

D-Histidine

D-His enhanced zinc accumulation, but reduced the fraction of zinc that was retained and absorbed by fish [19]. Both D- and L-His enhance the DNA degradation by hydrogen peroxide and ferric ions [20].

Lanthionine (LAN) Isomers

LAN isomers are formed during the biosynthesis of microbially-derived peptide antibiotics [21] and during exposure of proteins to alkali and heat. Reaction of the SH group of CysSH and the double bond of dehydroalanine gives rise to one pair of optically active D- and L-isomers and one diastereomeric (*meso)* form of LAN [22]. The mixture of DL + *meso*-LAN has a sparing effect on L-Met, as evidenced by a 27% greater weight gain when the two amino acids were fed together, than when fed suboptimal L-Met [8,10].

D-Lysine and Lysinoalanine (LAL) Isomers

In contrast to sulfur-containing amino acids, D-Lys is not utilized as a nutritional source of L-Lys by chicks, dogs, mice, rats, or humans, presumable because D-amino oxidase does not act on D-lysine. The amino acid LAL, formed concurrently with D-amino acids in alkali-treated proteins, has two asymmetric C-atoms, making possible four separate diastereoisomeric forms: DD, DL, LD, and LL. The nutritional value of the individual LAL isomers as a source of L-Lys is not known. Table 2 shows that a mixture of LL and LD isomers has a nutritional value for the mouse equivalent on a molar basis to 3.8% of L-Lys. For comparison, the table also lists the nutritional values of other D-amino acids we determined.

The four LAL isomers differ in their ability to chelate metal ions such as copper [23]. The transformation of even a small fraction of L-Lys to D-Lys and to LAL adversely affects the nutritional quality of cereal proteins to a greater extent than would be the case for legume (soy) and animal (casein) proteins, whose L-Lys content is much higher [1,24,25]. Because of its low toxicity, D-Lys may be a better candidate to reduce radioactivity uptake by the kidneys during therapy with radionuclides than is L-Lys [26].

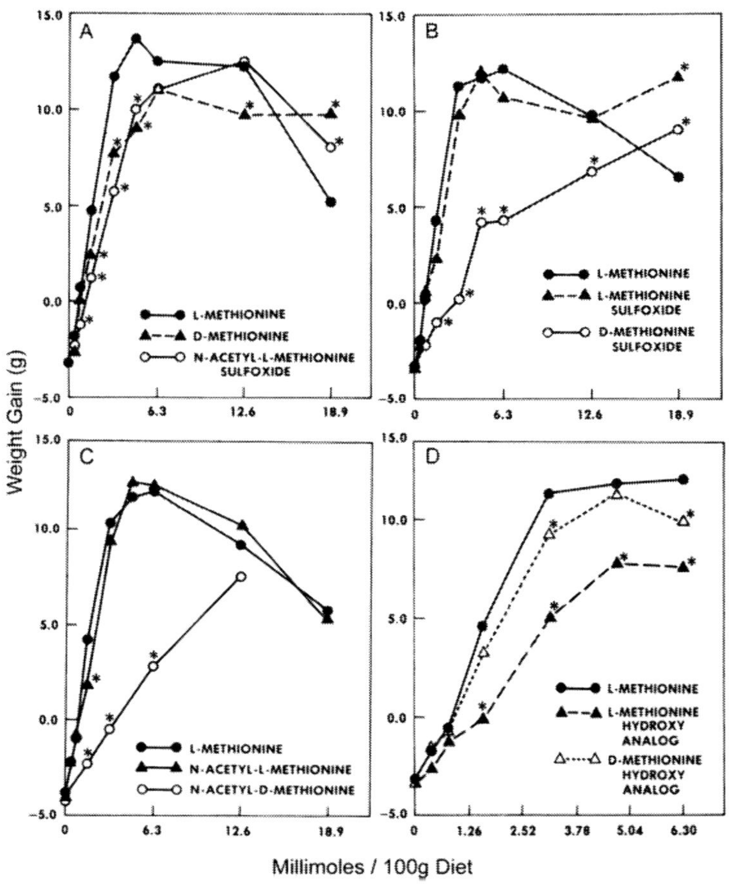

Figure 1. Weight gain in mice fed increasing dietary levels of L-Met, D-Met, and isomeric methionine derivatives and analogs for 14 days. Plotted are mean values; asterisk indicates significant difference from L-Met at the same dietary concentration [8,10].

D-Methionine

The nutritional value of D-Met in mice approaches that of the L-Met (Table 2). D-Met appears to be poorly utilized by humans when consumed either orally or during total parenteral nutrition. One factor giving rise to inconsistencies in the utilization of D-Met is the dose dependency of the apparent potency of D-Met relative to its L-isomer, *i.e.* the dietary level of the D-form for any given growth response relative to that of the L-form which would produce the same growth response. This dose-dependency is a result of the non-linear nature of the dose-response curves (Figure 1). This complicates attempts to compare results from mice with those of other animal species. The latter studies often report data based on a single substitution of the D- for the L-isomer. Figure 1 shows that high levels of L-Met (but not of D-Met) are toxic since they inhibited growth of mice. D-Met containing solutions inhibited tumor cell growth *in vitro* [27] and protected against cisplatin ototoxicity [28,29].

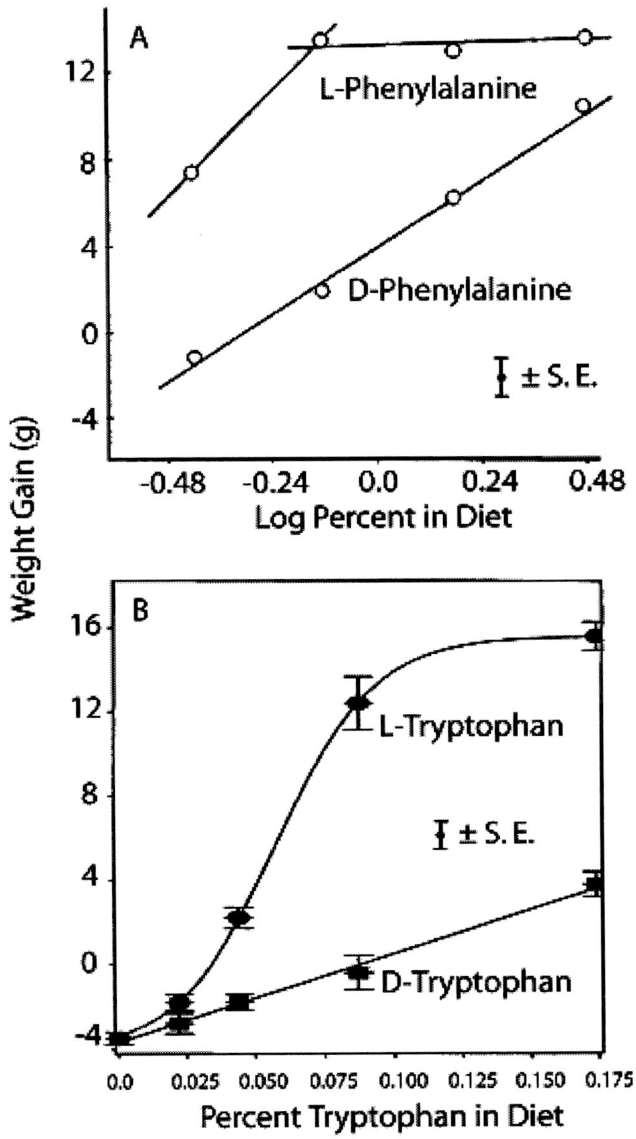

Figure 2. Relationships of weight gains to percent of L- and D-amino isomers in amino acid diets fed to mice: A. Phe isomers; B. Trp isomers [9,38,43].

D-Phenylalanine

The relative growth rate of mice fed D-Phe replacing the L-isomer in an free-amino-acid diet is concentration-dependent, ranging from 28.3 to 81.3% when compared to control diets containing the same amounts of L-Phe (Figure 2)[9]. The data suggest the absence of any antinutritional effects or toxicity from feeding either Phe isomer at twice the optimum dietary level.

D-Proline

L-Pro is a non-essential amino acid. Oral feeding of an aqueous solution of D-Pro (50 mg/kg body weight/day) for one month to rats induced fibrosis and necrosis of kidney liver cells and elevation of serum enzymes [30]. No evidence was found that orally fed D-Pro and D-Asp induced acute toxicity in rats [31].

Selenomethionine Isomers

Selenomethionine is a major source of selenium in the diets of both animals and humans. Comparison of the effects of oral consumption of seleno-L-methionine, seleno-DL-methionine, and selenized yeast on the reproduction of mallard ducklings revealed that although both seleno-methionine preparations were of similar toxicity, their potency was greater than that of selenium present in yeast [32]. Heinz et al. [32] found that the survival of day-old ducklings consuming L-selenomethionine after two weeks was significantly lower (36%) than that of ducklings consuming the DL-isomer (100%). D-Selenomethionine protected against adverse effects induced by space radiation [33].

D-Serine

Protein-bound L-Ser racemizes faster to the D-isomer than any of the other amino acids (Figure 4) [34]. D-Ser has been reported to enlarge rat kidney cells (cytomegaly) similar to that observed with LAL. D-Serine-induced tubular kidney necrosis in rats may be alleviated by a protein-deficient diet [35]. The amino acid may also be involved in neurotransmission in the mammalian brain [36]. Because brain uptake of D-Ser was significantly higher than of L-Ser, exogenous (dietary?) D-Ser may affect glutamatergic neurotransmission and associated diseases [37].

D-Threonine

L-Thr is second-limiting amino acid in maize (corn) proteins. The utilization of D-Thr by the chick, rat, mouse, or human as a nutritional source of the L-isomer is insignificant [3].

D-Tryptophan

L Trp contributes to protein synthesis and regulates numerous physiological mechanisms. These include serving as a precursor of the neurotransmitter serotonin and of the vitamin niacin. The relative nutritional potency of D-Trp compared to the L-Trp in mice is strongly dose-dependent, being related to the dietary concentration and ranging from 29 to 64% (Figure 2) [38]. The maximum growth obtainable for L-

Trp occurred at 0.174% in the diet. By increasing the dietary concentration of D-Trp up to 0.52%, growth passed through a maximum at 82% of that achieved with the L-isomer. This occurred at 0.44% of D-Trp.

Considerable species variation exists for the nutritive value of D-Trp [3]. In chick's diets, relative potency of the D- to the L-isomer has been reported to be 20%. D-Trp was well-utilized by growing pigs [39]. The value for humans is about 10%. Rats utilize D-Trp as efficiently as L-Trp. D-Trp can serve as a niacin precursor in rats to the same extent as does the L-form [40,41].

D-Tyrosine

Nutritionally, L-Tyr is classified as a semi-essential amino acid [42]. Combinations of L-Tyr and L-Phe are complementary in supporting growth of mice [9]. Thus, under conditions where L-Phe may be limiting, L-Tyr may supply half the requirement of L-Phe alone for chicks, mice, rats, and humans. Our feeding studies showed that with D-Tyr in an amino acid diet, growth inhibition was severe at a D-Tyr/L-Tyr ratio of 2:1, but was less so when the ratio was 1:1. Similar results were obtained with a casein diet supplemented with D-Tyr. The antimetabolic manifestation of D-Tyr may be ascribed to interference with the biosynthesis of vital neurotransmitters and proteins *in vivo* [9,43].

Our findings demonstrate that D-Tyr, unlike L-Tyr, has no sparing effect for L-Phe in mice. In fact growth inhibition may become evident when D-Tyr is present in the diet at level equal or greater than L-Phe. This effect remains to be defined toxicologically, and the potential for chronic toxicity following exposure to lower levels of D-Tyr remains unknown. Formation of D-tyrosyl-$tRNA^{Tyr}$ may be responsible for the toxicity of D-Tyr toward *Escherichia coli* [44].

D-Valine

Administration of D-valine-containing parenteral solutions to rats inhibited tumor growth without adversely affecting nutritional status [45].

4. Notes

A better understanding of the dietary significance of D-amino acids requires knowledge about the factors that induce racemization of L- to D-amino acids in food proteins during food processing. To facilitate such an understanding, Table 3 shows the D-amino acid content of eight alkali-treated proteins, Figure 3 offers derivations of the kinetic equations governing racemization, and Figure 4 illustrates racemization rates and related aspects of peptide-bound D-amino acids [37].

There is a need to standardize methods designed to ascertain the role of D-amino acids in nutrition. Because the organisms is forced to use the D-amino acid as the sole source of the L-form, the use of all-amino-acid diets in which the L-isomer is completely replaced with different levels of the corresponding D-amino acid may be preferable to supplementation of proteins with D-amino acids. D-Amino acids along a peptide chain may be less assimilable than the L-forms. The utilization of any D-amino acid may be affected by the presence of other D-amino acids in the diet. Largely unresolved are the following questions. (a) How do biological effects of D-amino acids vary, depending on whether they are consumed in the free state or as part of a food protein? (b) Can poorly digestible racemized food proteins serve as dietary fiber in the digestive tracts of humans? (c) Do D-amino acids and D-peptides alter the normal microflora of the intestine. (d) Do metabolic interactions, antagonisms, or synergisms among free and protein-bound D-amino acids occur *in vivo*? (e) Does racemization alter protein conformations and charge distributions (isoelectric points) resulting in beneficial protective effects against protein-induced allergy, celiac disease, and bacterial, plant, and venom toxic proteins? (f) Does food processing-induced formation of D and other modified amino acids [46-49] adversely or beneficially affect the safety of concurrently formed potentially toxic acrylamide [50,51]?

$$\text{L-amino acids} \underset{k'}{\overset{k}{\rightleftarrows}} \text{D-amino acids}$$

$$-dL/dt = kL - k'D$$

$$\text{if } D_{t=0} \ll L_{t=0}, \text{ then } L_{t=0} = L + D \text{ and } L_{t=0} - L = D$$

$$-dL/dt = kL - k'(L_{t=0} - L) = kL - k'L_{t=0} + k'L$$

$$= (k + k')L - k'L_{t=0}$$

$$dL/dt = -(k + k')L + k'L_{t=0}$$

$$dL/dt + (k + k')L = k'L_{t=0}$$

$$dL/dt\, e^{(k+k')t} + (k + k')L\, e^{(k+k')t} = k'L_{t=0}\, e^{(k+k')t}$$

$$d/dt\,(Le^{(k+k')t}) = k'L_{t=0}\, e^{(k+k')t}$$

$$\int d(Le)(Le^{(k+k')t}) = \int (k'L_{t=0}\, e^{(k+k')t}) \cdot dt$$

$$Le^{(k+k')t} = [k/(k+k')]L_{t=0}\, e^{(k+k')t} + \text{constant}$$

$$L = [k'/(k+k')]L_{t=0} + \text{constant} \cdot e^{-(k+k')t}$$

$$\text{At } t = 0,\ L = L_{t=0} \text{ and } e^{-(k+k')t} = 1$$

$$L_{t=0} = [k'/(k+k')]L_{t=0} + \text{constant}$$

$$L_{t=0} - [k'/(k+k')]L_{t=0} = \text{constant}$$

$$L_{t=0}[1 - k'/(k+k')] = \text{constant}$$

$$L = [k'/(k+k')]L_{t=0} + L_{t=0}\, k/(k+k')e^{-(k+k')t}$$

$$\text{if } L_{t=0} = L + D$$

$$L = [k'/(k+k')](L + D) + (L + D)\, k/(k+k')e^{-(k+k')t}$$

$$L/(L + D) = k'/(k+k') + [k/(k+k')]e^{-(k-k')t}$$

$$\frac{L(k+k') - k'(L+D)}{(L+D)(k+k')} = [k/(k+k')]e^{-(k+k')t}$$

$$= \frac{Lk - Dk'}{(L+D)(k+k')}$$

$$\left[\frac{L - D(k'/k)}{L + D}\right] = e^{-(k+k')t}$$

$$= \left[\frac{L[(1 - (D/L)(k'/k)]}{L[(1 + (D/L)]}\right]$$

$$= \left[\frac{1 - (D/L)(k'/k)}{1 + D/L}\right]$$

$$\ln\left[\frac{1 - (D/L)(k'/k)}{1 + (D/L)}\right] = -(k + k')t$$

$$\ln\left[\frac{1 + (D/L)}{1 - (D/L)(k'/k)}\right] = -(k + k')t$$

$$\text{if } k'/k = K' = 1/K_{eq};\ K_{eq} = D/L \text{ ratio at equilibrium} = 1$$

$$\ln\left[\frac{1 + D/L}{1 - K'(D/L)}\right] = (1 + K') \cdot k \cdot t \qquad \text{(eq. 1)}$$

An alternate mathematical treatment of the first-order kinetics of racemization gives equation 2. Equation 1 is operationally more efficient than equation 2 because we can measure D/L ratios with a 1-3% error compared to a 10-15% error when concentrations of D only are measured.

$$\ln(D_e/D_e - D_t) = (k + k')t = k_{(obs)}t \qquad \text{(eq. 2)}$$

where D_e = equilibrium value of D and D_t = D at time t

Figure 3. Derivations of first order kinetic equations for reversible amino acid racemization. Adapted from [55,56].

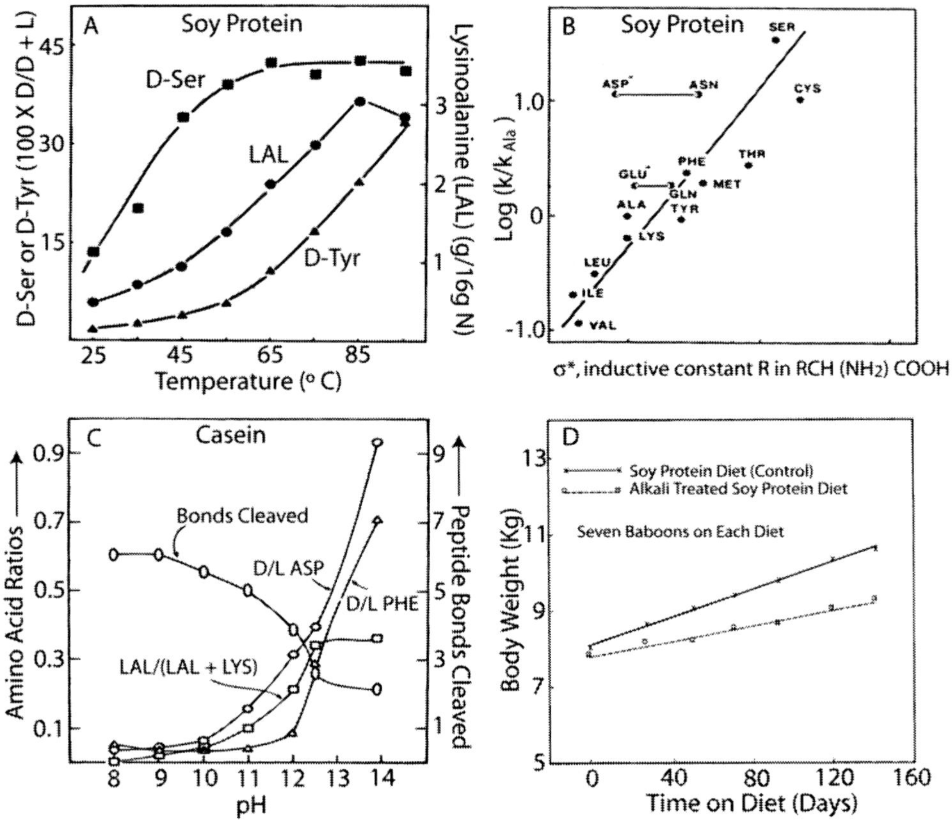

Figure 4. A. Effect of temperature on content D-Ser, D-Tyr, and LAL of soybean protein [34,53,57]; B. Relationship between racemization rates (k) relative to Ala and the inductive constant (σ*) of the amino acid side chain [34,52]. Analogous linear free-energy relationships probably operate for most racemized proteins. See also Table 3; C. Inverse relationship among D-amino acid and LAL content of casein and extent of digestibility of peptide bonds by trypsin [58]; D. Weight gain of baboons fed a 20% toasted and alkali-treated soy protein diets [59].

Table 3. D-Amino acid content [(D/D + L) x 100)] of eight alkali treated proteins[a].

Amino Acid	casein	lactalbumin	wheat gluten	zein protein	fish protein	soybean protein	bovine albumin	hemoglobin
Ala	15.2	14.4	18.6	22.2	19.3	15.8	22.1	17.1
Val	2.6	2.7	4.0	4.9	3.1	2.5	3.5	4.0
Leu	7.4	5.0	7.2	7.8	6.8	6.3	8.2	6.6
Ile	3.3	3.1	4.0	5.5	3.6	3.9	5.7	5.0
Cys	--	32.1	32.0	43.7	22.8	21.0	23.0	30.0
Met	24.7	32.3	33.1	29.8	29.2	24.3	30.0	26.2
Phe	24.4	24.3	24.4	32.4	28.0	25.5	28.1	30.0
Lys	8.1	7.2	9.4	8.0	11.5	11.3	13.3	9.9
Asp	29.2	22.6	25.6	41.6	25.0	30.8	27.0	18.9
Glu	19.7	19.5	32.3	35.0	18.9	21.1	18.4	19.8
Ser	41.0	47.1	42.2	44.0	42.1	44.2	43.0	44.5
Thr	29.3	29.1	30.0	36.3	32.8	27.8	28.3	31.2
Tyr	15.0	18.9	19.5	35.5	16.3	13.7	15.3	22.6
LAL[b]	4.4	5.4	0.9	0.3	2.8	3.2	8.5	4.4

[a] Adapted from (34, 52-54). [b] Mixture of (LD + LL) LAL (lysinoalanine) isomers in g/16g N.

Acknowledgments

I am most grateful to Patricia M. Masters (Scripps Institution of Oceanography, La Jolla, CA), Remy Liardon (Nestle Research Centre, Lausanne, Switzerland), Kevin Pearce (New Zealand Dairy Research Center), and Michael R. Gumbmann (this laboratory) for excellent scientific collaboration and to Carol E. Levin for assistance with the figures.

References

[1] Friedman, M. (1991). Formation, nutritional value, and safety of D-amino acids. in: Friedman, M. (Ed.) *Nutrtional and Toxicological Consequences of Food Processing* (New York: Plenum Press); *Adv Exp Med Biol, 289*, 447-81.

[2] Friedman, M. (1999). Chemistry, nutrition, and microbiology of D-amino acids. *J Agric Food Chem, 47*, 3457-3479.

[3] Borg, B. S. & Wahlstrom, R. C. (1989) Species and isomeric variation in the utilization of amino acids, in: Friedman, M. (Ed.) *Absorption and Utilization of Amino Acids*, pp. Volume II, pp. 155-171 (Boca Raton: CRC Press).

[4] Man, E. H. & Bada, J. L. (1987). Dietary D-amino acids. *Annu Rev Nutr, 7*, 209-25.

[5] Friedman, M. & Gumbmann, M. R. (1979). Biological availability of ε-N-methyl-L-lysine, 1-N-methyl-L-histidine, and 3-N-methyl-L-histidine in mice. *Nutrition Reports International, 19*, 437-443.

[6] Friedman, M. & Gumbmann, M. R. (1981). Bioavailability of some lysine derivatives in mice. *J Nutr, 111*, 1362-1369.

[7] Friedman, M. & Gumbmann, M. R. (1982). The nutritional value of lysinoalanine as a source of lysine for mice. *Nutrition Reports International, 26*, 937-943.

[8] Friedman, M. & Gumbmann, M. R. (1984). The utilization and safety of isomeric sulfur-containing amino acids in mice. *J Nutr, 114*, 2301-2310.

[9] Friedman, M. & Gumbmann, M. R. (1984). The nutritive value and safety of D-phenylalanine and D-tyrosine in mice. *J Nutr, 114*, 2089-96.

[10] Friedman, M. & Gumbmann, M. R. (1988). Nutritional value and safety of methionine derivatives, isomeric dipeptides and hydroxy analogs in mice. *J Nutr, 118*, 388-397.

[11] Stegman, L. D., Zheng, H., Neal, E. R., Ben-Yoseph, O., Pollegioni, L., Pilone, M. S. & Ross, B. D. (1998). Induction of cytotoxic oxidative stress by D-alanine in brain tumor cells expressing *Rhodotorula gracilis* D-amino acid oxidase: a cancer gene therapy strategy. *Hum Gene Ther, 9*, 185-93.

[12] Suessenbacher, A., Lass, A., Mayer, B. & Brunner, F. (2002). Antioxidative and myocardial protective effects of L-arginine in oxygen radical-induced injury of isolated perfused rat hearts. *Naunyn Schmiedebergs Arch Pharmakol, 365*, 269-76.

[13] Wiel, E., Pu, Q., Corseaux, D., Robin, E., Bordet, R., Lund, N., Jude, B. & Vallet, B. (2000). Effect of L-arginine on endothelial injury and hemostasis in rabbit endotoxin shock. *J Appl Physiol, 89*, 1811-8.

[14] Szende, B. (1993). The effect of amino acids and amino acid derivatives on cell proliferation. *Acta Biomed Ateneo Parmense, 64*, 139-45.

[15] Koyuncuoglu, H., Gungor, M., Ang, O., Inanc, D., Ang-Kucuker, M., Sagduyu, H. & Uysal, V. (1988). Aggrevation by morphine and D-aspartic acid of pyelonephritis induced by i.v. inoculation of *Staphylococcus aureus* in rats. *Infection, 16*, 42-5.

[16] Iezhitsa, I. N., Spasov, A. A., Zhuravleva, N. V., Sinolitskii, M. K. & Voronin, S. P. (2004). Comparative study of the efficacy of potassium magnesium L-, D- and DL-aspartate stereoisomers in

overcoming digoxin- and furosemide-induced potassium and magnesium depletions. *Magnes Res, 17*, 276-92.

[17] Benz, F. W., Nerland, D. E., Pierce, W. M. & Babiuk, C. (1990). Acute acrylonitrile toxicity: studies on the mechanism of the antidotal effect of D- and L-cysteine and their N-acetyl derivatives in the rat. *Toxicol Appl Pharmacol, 102*, 142-50.

[18] Berkeley, L. I., Cohen, J. F., Crankshaw, D. L., Shirota, F. N. & Nagasawa, H. T. (2003). Hepatoprotection by L-cysteine-glutathione mixed disulfide, a sulfhydryl-modified prodrug of glutathione. *J Biochem Mol Toxicol, 17*, 95-7.

[19] Glover, C. N. & Hogstrand, C. (2002). Amino acid modulation of in vivo intestinal zinc absorption in freshwater rainbow trout. *J Exp Biol, 205*, 151-8.

[20] Tachon, P. (1990). DNA single strand breakage by H_2O_2 and ferric or cupric ions: its modulation by histidine. *Free Radical Res Commun, 9*, 39-47.

[21] Sahl, H. G. & Bierbaum, G. (1998). Lantibiotics: biosynthesis and biological activities of uniquely modified peptides from gram-positive bacteria. *Annu Rev Microbiol, 52*, 41-79.

[22] Friedman, M., Noma, A. T. & Wagner, J. R. (1979). Ion-exchange chromatography of sulfur amino acids on a single-column amino acid analyzer. *Anal Biochem, 98*, 293-304.

[23] Friedman, M. & Pearce, K. N. (1989). Copper(II) and cobalt(II) affinities of LL- and LD-lysinoalanine diastereomers: implications for food safety and nutrition. *J Agric Food Chem, 37*, 123-127.

[24] Friedman, M. & Finot, P. A. (1990). Nutritional improvement of bread with lysine and γ-glutamyl-lysine. *J Agric Food Chem, 38*, 2011-2020.

[25] Smith, G. A. & Friedman, M. (1984). Effect of carbohydrates and heat on the amino acid composition and chemically available lysine content of casein. *J Food Sci, 49*, 817-820, 843.

[26] Bernard, B. F., Krenning, E. P., Breeman, W. A. et al. (1997). D-lysine reduction of indium-111 octreotide and yttrium-90 octreotide renal uptake. *J Nucl Med, 38*, 1929-33.

[27] Sasamura, T., Matsuda, A. & Kokuba, Y. (1999). Effects of D-methionine-containing solution on tumor cell growth in vitro. *Arzneimittelforschung, 49*, 541-3.

[28] Campbell, K. C., Meech, R. P., Rybak, L. P. & Hughes, L. F. (2003). The effect of D-methionine on cochlear oxidative state with and without cisplatin administration: mechanisms of otoprotection. *J Am Acad Audiol, 14*, 144-56.

[29] Wimmer, C., Mees, K., Stumpf, P. Welsch, U., Reichel, O. & Suckfull, M. (2004). Round window application of D-methionine, sodium thiosulfate, brain-derived neurotrophic factor, and fibroblast growth factor-2 in cisplatin-induced ototoxicity. *Otol Neurotol, 25*, 33-40.

[30] Kampel, D., Kupferschmidt, R. & Lubec, G. (1990) Toxicity of D-proline, in: Lubec, G. & Rosenthal, G. A. (Eds.) *Amino Acid: Chemistry, Biology and Medicine*, pp. 1164-1171 (Leiden, Netherlands: Escom Science Publishers).

[31] Schieber, A., Brückner, H., Rupp-Classen, M., Specht, W., Nowitzki-Grimm, S. & Classen, H. G. (1997). Evaluation of D-amino acid levels in rat by gas chromatography-selected ion monitoring mass spectrometry: no evidence for subacute toxicity of orally fed D-proline and D-aspartic acid. *J Chromatogr B Biomed Sci Appl, 691*, 1-12.

[32] Heinz, G. H., Hoffman, D. J. & Lecaptain, L. J. (1996). Toxicity of seleno-L-methionine, seleno-DL-methionine, high selenium wheat, and selenized yeast to mallard ducklings. *Arch Environ Contam Toxicol, 30*, 93-9.

[33] Kennedy, A. R., Ware, J. H., Guan, J., Donahue, J. J., Biaglow, J. E., Zhou, Z., Stewart, J., Vazquez, M. & Wan, X. S. (2004). Selenomethionine protects against adverse biological effects induced by space radiation. *Free Radic Biol Med, 36*, 259-66.

[34] Friedman, M. & Liardon, R. (1985). Racemization kinetics of amino acid residues in alkali-treated soybean proteins. *J Agric Food Chem, 33*, 666-672.

[35] Levine, S. & Saltzman, A. (2003). Acute uremia produced in rats by nephrotoxic chemicals is alleviated by protein deficient diet. *Renal Failure, 24*, 517-523.

[36] Kappor, R. & Kapoor, V. (1997). Distribution of D-amino acid oxidase (DAO) activity in the medulla and thoracic spinal cord of the rat: implications for a role for D-serine in autonomic function. *Brain Res, 771*, 351-5.
[37] Bauer, D., Hamacher, K., Bröer, S., Pauleit, D., Palm, C., Zilles, K., Coenen, H. H. & Langen, K. J. (2005). Preferred stereoselective brain uptake of D-serine - a modulator of glutamatergic neurotransmission. *Nucl Med Biol, 32*, 793-7.
[38] Friedman, M. & Cuq, J. L. (1988). Chemistry, analysis, nutritional value, and toxicology of tryptophan in food. A review. *J Agric Food Chem, 36*, 1079-1093.
[39] Arentson, B. E. & Zimmerman, D. R. (1985). Nutritive value of D-tryptophan for the growing pig. *J Anim Sci, 60*, 474-9.
[40] Shibata, K., Swabe, M., Fukuwatari, T. & Sugimoto, E. (2000). Efficiency of D-tryptophan as niacin in rats. *Biosci Biotechnol Biochem, 64*, 206-9.
[41] Carter, E. G. A., Carpenter, K. J. & Friedman, M. (1982). The nutritional value of some niacin analogs for rats. *Nutrition Reports International, 25*, 389-397.
[42] Mercer, L. P., Dodds, S. J. & Smith, D. L. (1989) Dispensable, indispensable, and conditionally indispensable amino acids ratios in the diet, in: Friedman, M. (Ed.) *Absorption and Utilization of Amino Acids*, Volume I, pp. 2-13 (Boca Raton, FL: CRC Press).
[43] Anonymous (1985). Dietary D-tyrosine as an antimetabolite in mice. *Nutrition Reviews, 43*, 156-158.
[44] Soutourina, O., Soutourina, J., Blanquet, S. & Plateau, P. (2004). Formation of D-tyrosyl-tRNATyr accounts for the toxicity of D-tyrosine toward *Escherichia coli*. *J Biol Chem, 279*, 42560-5.
[45] Sasamura, T., Matsuda, A. & Kokuba, Y. (1998). Nutritional effects of a D-methionine-containing solution on AH109A hepatoma-bearing rats. *Biosci Biotechnol Biochem, 62*, 2418-20.
[46] Finot, P. (2005). The absorption and metabolism of modified amino acids in processed foods. *J AOAC Int, 88*, 894-903.
[47] Gilani, G., Cockell, K. & Sepehr, E. (2005). Effects of antinutritional factors on protein digestibility and amino acid availability in foods. *J AOAC Int, 83*, 967-987.
[48] Friedman, M. (2003). Nutritional consequences of food processing. *Forum Nutr, 56*, 350-352.
[49] Friedman, M. (2004) Effects of food processing, in: Corke, H. & Walker, C. (Eds.) *Encyclopedia of Grain Science*, Volume 2, pp. 328-340 (Oxford, UK: Elsevier).
[50] Friedman, M. (2003). Chemistry, biochemistry, and safety of acrylamide. A review. *J Agric Food Chem, 51*, 4504-4526.
[51] Friedman, M. (2005) Biological effects of Maillard browning products that may affect acrylamide safety in food, in: Friedman, M. & Mottram, D. S. (Eds.) Chemistry and Safety of Acrylamide in Food (New York: Springer); *Adv. Exp. Med. Biol. 561*, 135-155.
[52] Friedman, M. & Masters, P. M. (1982). Kinetics of racemization of amino acid residues in casein. *J Food Sci, 47*, 760-764.
[53] Liardon, R. & Friedman, M. (1987). Effect of peptide bond cleavage on the racemization of amino acid residues in proteins. *J Agric Food Chem, 35*, 661-667.
[54] Liardon, R., Friedman, M. & Philippossian, G. (1991). Racemization kinetics of free and protein-bound lysinoalanine (LAL) in strong acid media. Isomeric composition of bound LAL in processed proteins. *J Agric Food Chem, 39*, 531-537.
[55] Masters, P. M. & Friedman, M. (1979). Racemization of amino acids in alkali-treated food proteins. *J Agric Food Chem, 27*, 507-511.
[56] Masters, P. M. & Friedman, M. (1980) Amino acid racemization in alkali-treated food proteins - chemistry, toxicology, and nutritional consequences, in: Whitaker, J. R. & Fujimaki, M. (Eds.) *Chemical Deterioration of Proteins*, pp. 165-194 (Washington, D.C.: American Chemical Society).
[57] Friedman, M., Levin, C. E. & Noma, A. T. (1984). Factors governing lysinoalanine formation in soy proteins. *J Food Sci, 49*, 1282-1288.

[58] Friedman, M., Zahnley, J. C. & Masters, P. M. (1981). Relationship between in vitro digestibility of casein and its content of lysinoalanine and D -amino acids. *J Food Sci, 46*, 127-131.

[59] Friedman, M. (1999). Chemistry, biochemistry, nutrition, and microbiology of lysinoalanine, lanthionine, and histidinoalanine in food and other proteins. *J Agric Food Chem, 47*, 1295-1319.

In: D-Amino Acids: ISBN 1-60021-075-9
A New Frontier in Amino Acid and Protein Research, pp.299-315

Chapter 4.2

D-AMINO ACIDS IN FOOD

Rosangela Marchelli*, Gianni Galaverna, Arnaldo Dossena, Gerardo Palla, Anna Bobbio, Salvatore Santaguida, Kalina Grozeva, Roberto Corradini, and Stefano Sforza
Department of Organic and Industrial Chemistry, University of Parma, Italy.

Dedicated to the memory of F. De Nisco

Keywords: D-amino acids, cheese, yoghurt, milk, chocolate

1. INTRODUCTION

Natural macromolecules are generally believed to be homochiral, that is molecules which contain stereogenic centers of the same configuration. Thus, proteins are composed of only L-amino acids and carbohydrates mainly of D-monosaccharides. The existence of the other enantiomer (the D-amino acid in the case study) was neglected or considered as the result of artificial transformations. The tremendous amount of research on chiral separations carried out since 1970 led to the development of many methods for enantiomeric separations based on different principles [1-3], which allowed the detection of the presence of D-amino acids in biological systems and in foods. Actually, D-amino acids are quite spread in marine invertebrates and molluscs, in virtually all species of bacteria, which use them to build up the peptidoglycan layer of their cell wall, and also in mammalians and humans as well [4,5]. D-Serine and D-Aspartate were found in plasma and human brain where they fulfill specific biological functions (neurotransmission, development and endocrine regulation) [6]. In addition to naturally present D-amino acids, our diet also contains those formed during food processing (high temperature, alkaline treatment) [7-9] or produced by microorganisms in fermented foods [10]. The metabolic fate of D-amino acids in humans is controversial [11,12]. On the other hand, altered levels of D-amino acids were shown to induce serious injury especially to kidneys [6,13]. The biological effects are different whether they are ingested as free amino acids or as proteins. Indeed, peptide bonds with D-L, L- D or D- D configurations are resistant to proteolytic enzymes, thus inducing a reduced digestibility and eventually the formation of oligopeptides

* Correspondence concerning this article should be addressed to Rosangela Marchelli, E-mail: rosangela.marchelli@unipr.it, Fax: +39-0521-905472.

with unknown biological activity [14,15]. No toxic effects were found in trials with parenteral nutrition of adults and children with free D, L-amino acids [16]. For D-amino acids two metabolic pathways are known: a) conversion to the L-enantiomer by epimerases or racemases; b) oxidative deamination to ketoacids, catalysed by D-amino acid oxidase [17], and subsequent transamination to L-amino acids. Both mechanisms were proved to occur in microorganisms, whereas only the latter was found to be operative in mammalians. Whether the human organism is able to fully utilize D-amino acids is still unclear, although D-amino acid oxidase was found in human liver and kidneys [18,19].

In a general scheme aiming at studying the mechanisms of chiral separation, we devised many years ago several methods for the enantiomeric separation of amino acids: TLC [20], HPLC [21-24], GC [25], CE [26], a chemical sensor [27] and a biosensor [28]. By using the original methods thus developed, we investigated the presence and the role of D-amino acids in several foodstuffs and their metabolic fate.

2. Materials

All reagents and solvents used were of analytical grade. Milk, yoghurt and fruit samples were purchased by or produced in collaboration with Parmalat S.p.A. (Collecchio, Parma, Italy). Cheese samples were obtained by Consorzio del Formaggio Parmigiano Reggiano (Reggio Emilia, Italy) or by Istituto Lattiero-Caseario di Lodi (Lodi, Milano, Italy) or purchased directly from the market. *In vivo* experiments were performed in collaboration with the Dipartimento di Medicina Sperimentale (University of Parma, Parma, Italy).

3. Methods

3.1. GC

The GC analyses were performed on a Dani 8500 instrument (DANI, Monza, Italy) equipped with a flame ionization detector (FID) and an open tubular (WCOT) fused silica column (L = 25 m, I.D.= 0.25 mm) statically wall-coated with a dichloromethane solution of the chiral tetraamide L-Phe-3-O-TA (0.15%) and OV-101, according to the method previously reported [25]. GC-MS analyses were performed on a Finnigan 1020 instrument at 70 eV, using the same chiral columns. Single ion monitoring (SIM) was performed at m/z 69 and 140 for Alanine, m/z 184 and 212 for Aspartic acid and m/z 198 and 226 for Glutamic acid.

3.2. HPLC

HPLC analyses were performed on a Waters chromatographic system (Waters, Milford, MA, USA) equipped with a fluorescence detector and post-column derivatization. The system was composed of two parts: an ion exchange apparatus for the separation of amino acids and a reversed-phase column for the chiral separation [23,24].

3.3. CE

A P/ACE 5000 apparatus (Beckman, Fullerton, CA, USA), equipped with an on-column UV detector set at 200 nm, was used for the electrophoretic experiments [26].

3.4. Sensor and Biosensor

The chemical sensor is based on the use of a modified β-cyclodextrin bearing a metal binding site and a dansyl fluorophore for the discrimination of enantiomers of amino acids, according to a ligand exchange mechanism and fluorescence "switch-on - switch-off". A fast protocol using fluorescence quenching by the copper(II)/amino acid complexes in a fluorescence microplate reader (FLUOstar, BMG LabTechnologies, Offenburg, Germany) was developed [27].

The biosensor is based on D-amino acid oxidase covalently bound to a preactivated Immunodyne membrane (PALL Filtration Corporation, New York, USA) and coupled to an Amperometric Biosensor Detector (Universal Sensors, Inc., Los Angeles, CA, USA) to detect the hydrogen peroxide developed in the catalytic conversion of D-amino acid to the corresponding ketoacid [28].

3.5. Yoghurt Production

The experiments were performed both on a laboratory scale and on a pilot scale. During the processes, samples of yoghurt were drawn, immediately refrigerated at –20°C and analyzed within the following 24 h. The shelf life studies were performed on the same samples maintained at controlled temperatures (4° or 25°C) until the deadline (30 days).

Laboratory scale fermentations were performed with 2 L of regular milk inoculated with 2% (w/w) of starter, homogenized and incubated at 42°C for 4h (final pH≈4.2).

Pilot scale fermentations were performed in a bioreactor similar to those normally used in industrial yoghurt production with 200 L of regular milk heated at 85°C for 1 min, cooled to 42°C, inoculated with 2% (w/w) of starter and incubated at 42°C for 4 h (final pH≈4.2). Different fermentation studies were performed: (a) milk fermented with *S. thermophilus*, (b) milk fermented with *L. bulgaricus*, (c) milk fermented with a commercial mixture of *S. thermophilus* and *L. bulgaricus* about 1:1, (d) milk fermented with a commercial mixture of *S. thermophilus* and *L. bulgaricus* about 1:10.

Eight portions of each mixture were prepared and analyzed at different times during the biotechnological process (from t=0 to t=240 min) and during the shelf life (until the deadline). Each sample was analyzed to determine:

a) *S. thermophilus* microbial counts in a M17 (OXOID) medium. Inoculated plates were incubated 48 h at 40°C: the results were recorded as colony forming units (CFU)/mL.
b) *L. bulgaricus* microbial count by using M.R.S. acidified medium (OXOID, pH=5.4). Inoculated plates were incubated in anaerobiosis for 48 h at 40°C and the results were recorded as colony forming units (CFU)/mL.
c) pH (Beckmann pH-meter with temperature compensation).
d) Total acidity, expressed as °Dornic/10g of yoghurt, measured by titration with N/9 NaOH (phenolphthalein as indicator).
e) Free amino acid concentration (by ion exchange HPLC).
f) D-amino acid concentration (by GC-MS and HPLC single amino acid analysis).

3.6. Yoghurt Sample Preparation for Amino Acid Analysis

To 5 g of yoghurt, diluted to 50 mL with double distilled water, 7 µmol of D,L-Nleu and 3 mL of trifluoroacetic acid were added. The mixture was stirred for 10 min and centrifuged at 5000 rpm for 10 min; the surnatant was extracted twice with 50 mL of ethyl ether to remove lipids and filtered through a 0.45 µm filter (Millipore). The aqueous phase, reduced to small volume and adjusted to pH≈5-6, was eluted through an Amberlite 200 column and washed with 10 mL of double distilled water. The cationic resin was subsequently washed with 60 mL of 2N HCl. The eluate was divided in two equal parts, one for chiral analyses in GC, the other for free amino acids and enantiomeric analyses in HPLC.

3.7. Cheese Sample Preparation for Amino Acid Analysis

To 1 g of finely grated cheese suspended in 50 mL of double distilled water 40 μmol of D,L-NLeu were added as internal standard. The resulting suspension, added of 5 mL of trifluoroacetic acid to induce the denaturation and the precipitation of proteins, was stirred for 10 min and centrifuged at 5000 rpm for 10 min; the surnatant was filtered on paper filter and extracted twice with 50 mL of ethyl ether to remove lipids. The aqueous phase, reduced to small volume and adjusted to pH≈5-6, was treated as above.

3.8. *In Vivo* Experiments

Twenty subjects were chosen among healthy volunteers as confirmed by urine and blood analyses to exclude hepatic and renal diseases, diabetes mellitus, dyslipoproteinemia and hemopoietic disorders. A total amount of 62 g of cheese was administered to every person. Blood samples (20 mL) were drawn at different time intervals: before consuming cheese (basal value – t0), and after 30 min (t30), 1 h (t1h), 2 h (t2h), and 3 h (t3h) they had consumed cheese. Serum samples, obtained by centrifugation at 1500 rpm for 20 min at 4°C, were stored at –20°C until the analyses were performed (within 3 days).

3.9. Serum Sample Preparation for Amino Acid Analysis

To 5 mL of serum sample 1.5 μmol of D,L-NLeu were added as internal standard. After dilution with a methanol:water mixture (9:1 v/v), to precipitate proteins, and centrifugation at 3000 rpm for 10 min, the surnatant was taken, reduced to 1 mL, added of 25 mL of double distilled water and extracted twice with 50 mL of ethyl ether to remove lipids. The aqueous phase was treated as for yoghurt and cheese samples.

4. Results and Discussion

4.1. Methods for Chiral Separation of D,L-Amino Acids

Several chromatographic and electrophoretic methods are currently available to perform the enantiomeric separation of amino acids [1-3]. Some original methods were devised by our group and applied to different foodstuffs [10,21-28].

Today, Chirasil-Val or cyclodextrin-based chiral stationary phases are commercially available and generally used for the enantiomeric separation of D,L-amino acids by GC [1].

More or less simultaneously to the development of Chirasil-Val, an original chiral chromatographic method was developed by our group, based on capillary columns coated with the chiral stationary phases constituted by oligo-oxaalkanoyl tetraamides which were obtained by linking two L-Phenylalanine via an oxaalkanoyl bridge: the selector was then mixed with Carbowax 20 M or a methyl silicone phase (OV-101) in a 1:1 ratio (w/w) and statically coated on the wall column [25]. The best selector turned out to be Phe-3-O-TA [25,29]. Amino acids were analysed after derivatization to *N*-trifluoroacetyl-amino acid isopropyl esters. Hydrogen bond formation was at the basis of the recognition mechanism. This phase allowed to resolve ca. 20 amino acids in one single run, with a very good limit of detection (0.1%) (Fig. 1).

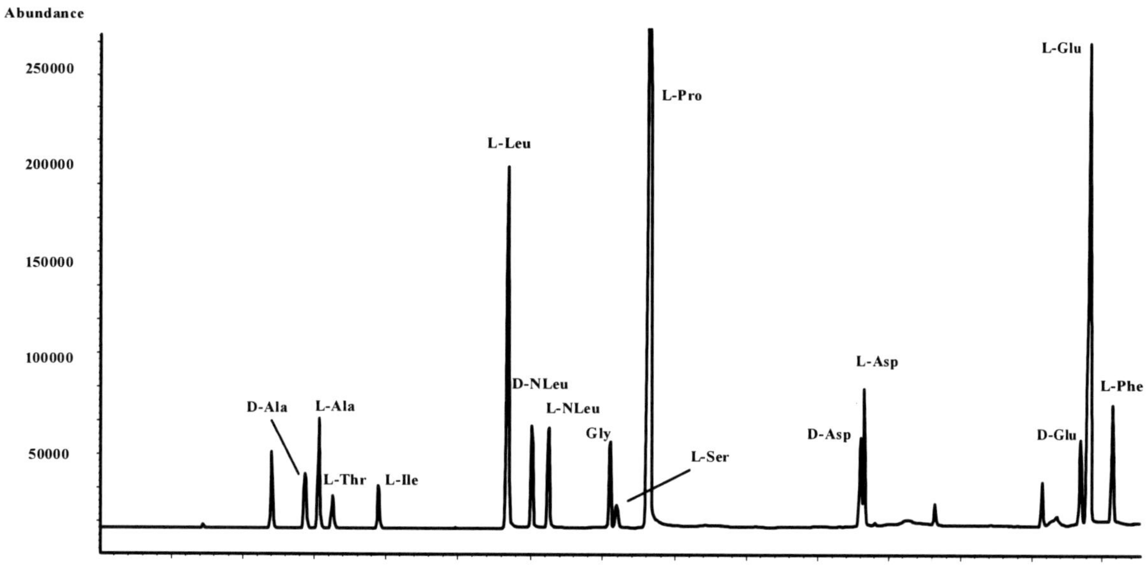

Fig. 1. GC enantiomeric separation of D,L-amino acids in Parmigiano-Reggiano cheese.

Chiral ligand exchange chromatography is one of the most powerful tool to perform the enantiomeric separation of bidentate analytes such as amino acids in HPLC [2]. We successfully used chiral copper(II) complexes of amino acid amides added to the eluent (chiral eluents) to perform the enantiomeric discrimination of D,L-Dns-amino acids [21,22] and unmodified D,L-amino acids [23,24]. These chiral complexes were particularly effective and allowed to obtain the enantiomeric separation with high enantioselectivity: the mechanism is consistent with a ligand exchange between the initial binary copper(II) complex and the amino acid enantiomers, leading to the formation of diastereomeric complexes of different stability and/or different affinity for the column stationary phase. The method is particularly simple as it involves the use of commercially available amino acid amides (L-Phe-A) and normal reversed phase C18 column, thus avoiding the use of expensive and dedicated chiral columns. On the other hand, the direct application to complex mixtures such as those present in natural systems, biological fluids, foods and beverages is difficult as both chemoselectivity and enantioselectivity are highly affected also by the presence of other components (amino alcohols, biogenic amines, dipeptides and oligopeptides). To overcome these problems, an improvement was set up via a two-dimensional HPLC system [24]: first, amino acid separation was achieved by means of an ion-exchange column by elution with a lithium chloride-lithium citrate buffer, then each peak corresponding to an individual amino acid was switched by an automatic valve to a reversed phase C18 column and eluted with an aqueous solution containing chiral copper(II) complexes which perform chiral discrimination via a ligand-exchange mechanism (Fig. 2). By using post-column derivatization with *o*-phthalaldehyde and fluorimetric detection it was possible to evaluate D-amino acids up to 0.1% D/(D+L) ratio in the nanomolar range.

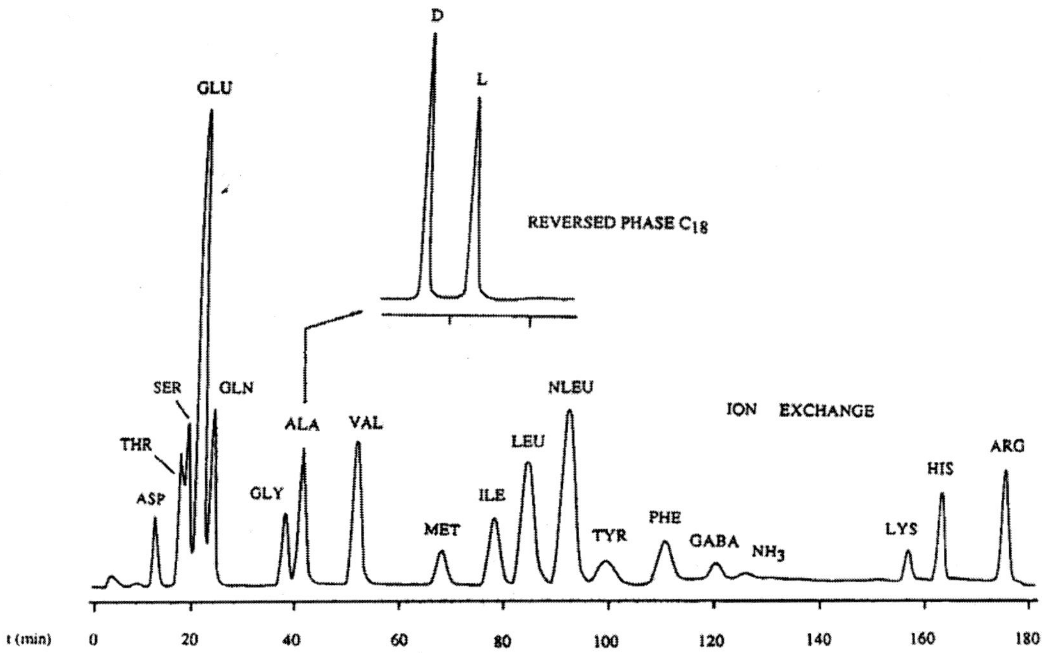

Fig. 2. HPLC single amino acid analysis in Parmigiano-Reggiano cheese. Lower trace: amino acid separation on an ion exchange column. Insert: enantiomeric separation of D,L-Ala switched to a reversed phase column with a chiral eluent [24].

Later, capillary electrophoresis became quite popular and showed high potential, especially because of its efficiency, speed of analysis, the small volumes and quantities of chiral selector and analytes needed, easiness and flexibility of the method. Chiral separations were performed using different chiral additives to the electrophoretic buffer in CE, e.g. natural and modified cyclodextrins, chiral surfactants, crown ethers, metal complexes and even antibiotics [30]. We successfully used positively charged β-CDs, bearing a Histamine moiety linked to the upper rim, either through the amino or the imidazole nitrogen (CD-hm and CD-mh), to perform the enantiomeric separation of dansyl-amino acids [26]. Ion pair interaction coupled to inclusion complexation is at the base of the discrimination mechanism. Inversion of the migration order was observed with the two CDs, depending on the relative position of the charged moieties on the upper rim, thus allowing for a reliable confirmation of the enantiomers.

Since a fast screening for the presence of D-amino acids was required from industry to check microbial contamination of fruit juices, we devised a biosensor based on D-amino acid oxidase (D-AAO) [28]. The biosensor is based on the catalytic activity of D-AAO (which catalyzes the reaction of oxidative deamination of the D-amino acid to the corresponding ketoacid with production of ammonia and hydrogen peroxide), coupled to an amperometric transduction of the hydrogen peroxide produced. This biosensor is able to detect the presence of D-Ala in a very short time (the whole operation required 3 min).

Later, a chemical sensor was devised by combining ligand exchange and inclusion mechanisms based on a modified β-cyclodextrin bearing a copper binding site and a dansyl fluorophore. It allowed the chiral discrimination of a single amino acid at a time (D,L-Pro or D,L-Val) simultaneously in 50 samples, by fluorescence "switch on" detection [27].

4.2. D-Amino Acids in Technological Processes (Heat Treatments)

Factors inducing amino acid racemization in proteins under food processing conditions have been extensively studied [7-9].

An investigation was performed on amino acids and proteins in milk to determine whether racemization was induced by commonly used industrial processes such as pasteurization and ultra-high temperature (UHT) processes [31]. It was demonstrated that heat treatments at a relatively high temperature but for a short duration (pasteurization, UHT) did not induce racemization in proteins and free amino acids.

Higher racemization degrees were measured for sterilized or powder milk, as a consequence of the technological process (Table 1).

Table 1. Variation of the %D/(D+L) of free amino acids in different milk samples.

Milk sample	Ala	Asp	Glu
Raw	3-4	2-3	2-3
Pasteurized	3-4	1-3	3-4
UHT (dir. treatm.)	4-6	2-3	3-4
UHT (ind. treatm.)	6-8	2-3	2-3
Sterilized	6-7	2-5	3-4
Powder	4-12	4-6	3-4
High digestibility	5-6	3-5	2-3
Infant formula	10-12	3-5	2-3
Human milk	< 0.2	< 0.2	< 0.2

However, a certain amount of D-Alanine (3-4 % as D/D+ L%), D-Glutamic acid (2-3 % as D/D+L%) and D-Aspartic acid (2-3 % as D/D+L%) was present in the free amino acid pool which could be considered physiological, as they were present in all cow milk samples examined, including raw milk samples immediately frozen after milking. They were found to derive from the lactic acid bacteria cell walls and from the activity of racemases and epimerases present in cow rumen microorganisms.

In fact, lactic acid bacteria contain D-Alanine (D-Ala), D-Glutamic acid (D-Glu) and D-Aspartic acid (D-Asp) in the peptidoglycan moiety of the cell wall, whereas other bacteria contain only D-Ala and D-Glu. Raw milk is usually contaminated with rumen microorganisms, mainly anaerobic bacteria of the genera *Bacteroides*, *Ruminococcus* and *Butyrivibrio*. Thus, the low, but significant, levels of D-amino acids found in milk are the results of the digestion and autolysis of rumen bacteria [31].

A dramatic international controversy arose as whether microwave heating of reconstituted milk for infants (infant formula) could induce racemization and, in particular, the formation of D-Proline and *cis*- D-Hydroxyproline, supposed to be toxic in chicken [32]. Indeed, the experimental conditions used in this work were quite extreme and far from the normal culinary practice: 10 min in a closed vessel, temperature unknown, high pressure. Thus, an extensive study was performed for comparing the racemization of proteins and free amino acids in samples of infant formula heated in a domestic micro-wave oven, a conventional oil bath and in a waveguide system at controlled temperatures and for specific periods of time [33]. In particular, the special wave-guide device allowed to carefully monitor the thermal kinetics of milk heating. No significant racemization was found to occur under the conditions studied. In particular, D-Proline and *cis*- D-Hydroxyproline were never found both in proteins and in the free amino acid pool. Also the toxicity tests were criticized by other authors [34].

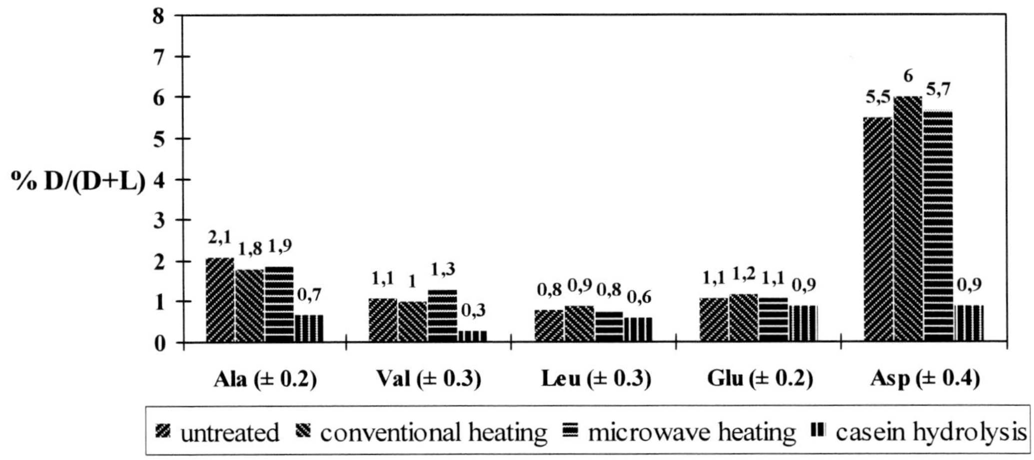

Fig. 3. % D/(D+L) ratio for protein amino acids in instant chocolate prepared with different heating devices (comparison with the racemization induced by protein hydrolysis previous to amino acid analysis).

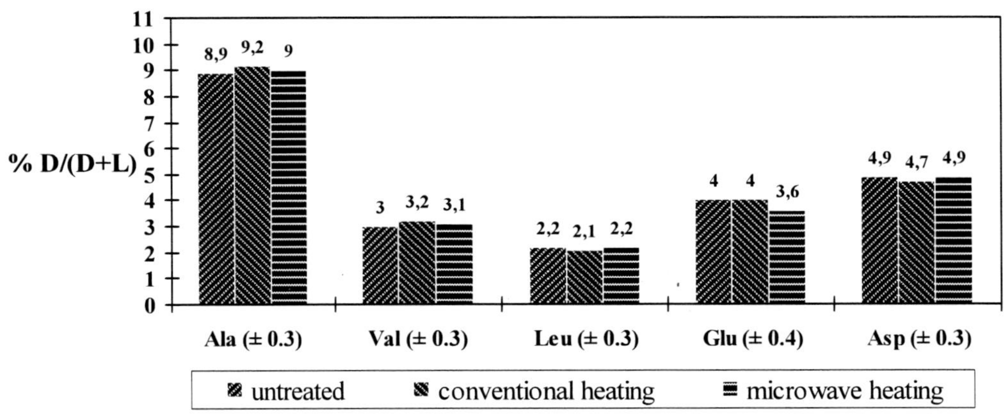

Fig. 4. % D/(D+L) ratio for free amino acids in instant chocolate prepared with different heating devices.

Instant chocolate was also prepared by heating with a microwave oven or oil bath: although the amounts of D-amino acids both in the free amino acid pool and in protein amino acids were higher than those measured in milk samples, no significant differences were observed between the untreated samples and the different heating systems. In particular, 6% of D-Asp was found in protein amino acids (Fig. 3) and 9% D-Ala among free amino acids (Fig. 4) in the cocoa powder. Actually, the partial racemization of amino acids with apolar side chains as Val and Leu can be justified by harsh chemical (basic pH) or high thermal treatments. Indeed, to obtain cocoa powder, cocoa beans are fermented for some days, toasted at 100°C-150°C up to 120 min and treated with an alkaline aqueous solutions.

4.3. D-Amino Acids as Markers of Contamination in Unfermented Foods

Fruit juices may contain D-amino acids on account of fortification or adulteration, whereas the presence of only D-Ala, D-Glu or D-Asp is indicative of microbial contamination.

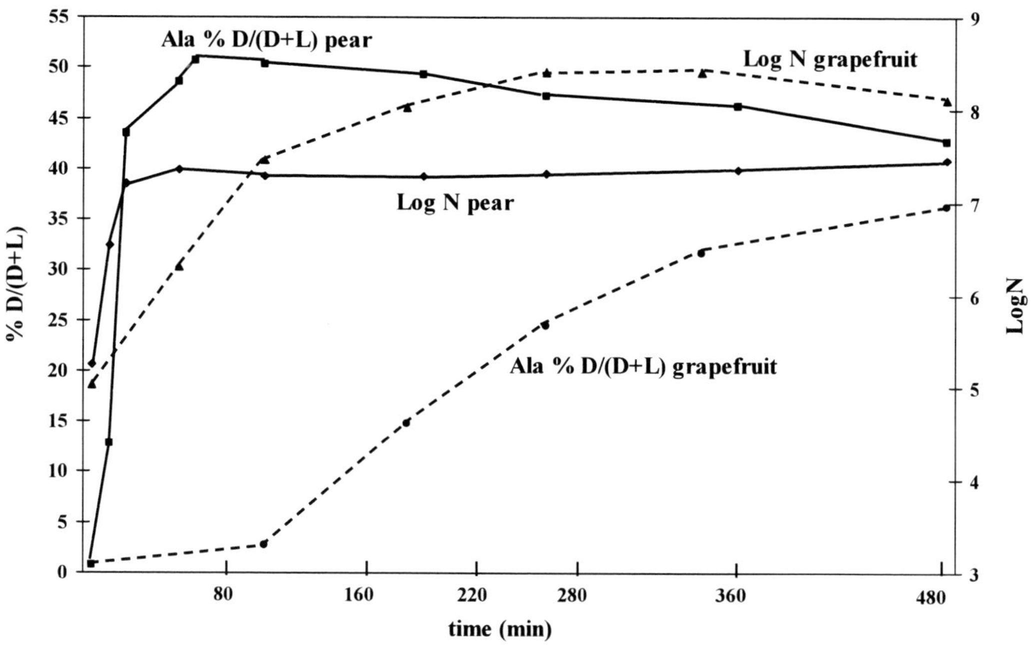

Fig. 5. Correlation between microbial growth and % D/(D+L) Ala in pear and grapefruit juice inoculated with *L. plantarum* [35].

In a screening investigation of commercially available fruit juices, several samples (in particular, grapefruit and pear juices) were found to contain significant amount of D-Ala (8-31 ppm) [35]. In these products the presence of D-Ala could not be related to adulteration, rather to the bacterial contamination and to the poor quality of the fruit used to make the juice. In spontaneously deteriorating grapefruits significant amounts of D-Ala (5-20 mg/L) were found after microbial contamination with *Lactobacillus plantarum*, one of the main microorganisms responsible for raw and finished juice spoilage, reached a concentration > 10^7 CFU/mL. Moreover, sterile juices inoculated with the same microorganism showed a high concentration of D-Ala (20 ppm ca. after 40 days, with a D/(D+L) ratio = 40%) with a good correlation between the content of D-Ala and the microbial growth (Fig. 5). In contrast, D-Ala was never found in sterile juices inoculated with yeasts (*Saccharomyces cerevisiae*) or moulds. As the content of D-Ala is not influenced by conventional thermal treatments (pasteurization, sterilization, UHT) and is not linked to other fruit characteristics (geographical origin, color, species), it can be considered a good marker of bacterial contamination occurring before or during juice processing, particularly when microbial counts is ineffective (sterilized juices).

4.4. D-Amino Acids in Fermented Foods

In the case of fermented foods the presence of D-amino acids is to be considered "natural". Indeed, D-Ala, D-Glu and D-Asp have been reported as widespread in cheese, yoghurt, wine, vinegar, sausages, etc. [10,36,37].

4.4.1. Cheese

Fairly large amounts of D-Ala, D-Glu and D-Asp (up to 5 g/Kg) were found in long aged cheeses such as Parmigiano-Reggiano and Grana Padano [38], which are manufactured from raw milk by addition of rennet and bacterial starter cultures: during ripening, the lysis of bacterial cell walls releases racemases and epimerases in the medium, which continue to be active as long as the water activity (a_w) allows it.

Analysis of 20 samples from 10 different cheese factories showed a significant presence of D-Ala, D-Glu and D-Asp in all samples (Fig. 1 and Fig. 2). In the oldest samples (18-24 months) the % content of D-Ala was around 31% (± 6), D-Asp 32% (± 7) and D-Glu 15% (± 6), with a total average amount of 0.7 g of D-Ala, 1.20 g of D-Asp and 2.3 g of D-Glu per kg of Parmigiano-Reggiano cheese.

During ageing the D/(D+L) ratios showed different trends: a gradual increase was observed for Glu (Fig. 6), a decrease for Ala and no significant variations for Asp. This behaviour was due to the different kinetics of formation for the L- and the D-enantiomers of the amino acids. Indeed, for all of them a substantial increase of the amount of the D-enantiomer was observed during ageing, but at a different rate.

Thus, for Ala the D-enantiomer actually increased during ageing, but less than the L-enantiomer so that the % D/(D+L) ratio decreased (Fig. 7). In the case of Glu, both the D- and L-enantiomers increased during ageing up to 14-15 months, after that the L-enantiomers reached a plateau, whereas the D-enantiomer still increased thus inducing a constant increase of the % D/(D+L) ratio. In the case of Asp the two enantiomers were produced at a comparable rate, thus the D/(D+L) ratio showed no significant variations.

For the Grana Padano cheese the trends of the D/(D+L) ratio was practically the same as observed in the case of Parmigiano Reggiano cheese, although the D/(D+L) values were lower than for Parmigiano Reggiano.

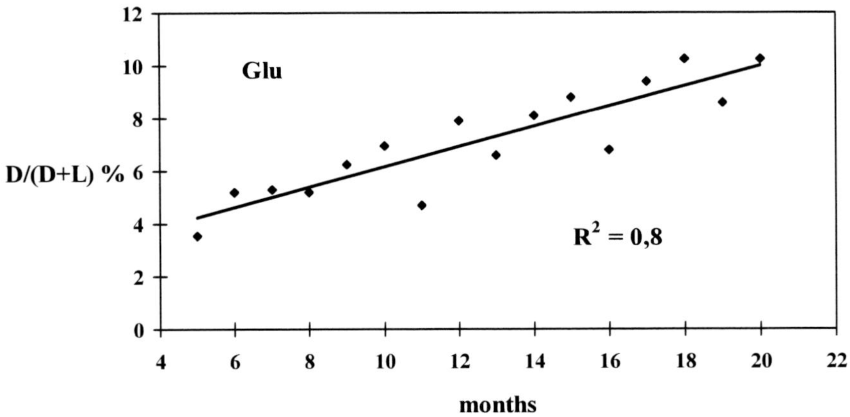

Fig. 6. Correlation between the % D/(D+L) Glu with ageing time in Parmigiano-Reggiano cheese [38].

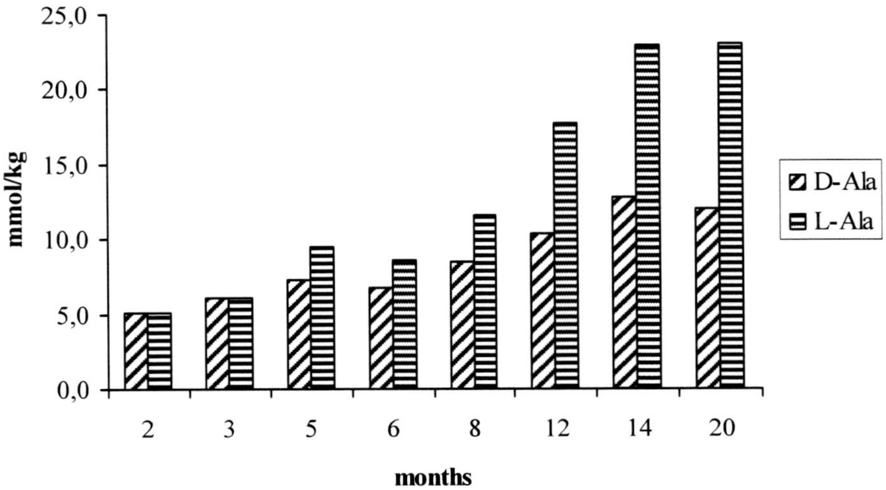

Fig. 7. Increase of D- and L-Ala during ripening in Parmigiano-Reggiano cheese [38].

The proteolysis process and the free amino acids content are well known markers of the ageing process in Parmigiano Reggiano cheese [39]: also D-amino acids can be considered as markers of ageing, since they are produced by the enzymatic activity linked to the ageing process. A least square analysis of the amount of these amino acids showed a good correlation with the age of the cheese. The 87% of variance was explained and the uncertainty of the recalculated value was around 1.3 months until 15-18 months of ageing. Moreover, a statistical PCA analysis on 30 samples of Parmigiano Reggiano and 32 samples of Grana Padano, based on the quantitative determination of D- and L-Ala, D- and L-Glu, D- and L-Asp, allowed to differentiate the two products to a good extent (markers of tipicity) [38].

During these studies, an interesting and unexpected phenomenon arose about D-Proline, which was found occasionally in some samples of hard cheeses also in very high amounts (up to 40% D/(D+ L) ratio), but with no correlation with the age or the cheese or the kind of starter used (Table 2). Extending the investigation to other hard-type cheeses, such as Bagoss and Val Trompia cheeses, two products of high alpine valleys in Lombardia (Italy) produced according to an artisan cheesemaking technology (without the use of starter cultures), D-Pro was found to be always present in huge amounts and it increased significantly with ageing.

Table 2. Maximum % D/(D+L) ratio and amount of D-Pro occasionally found in hard cheeses.

Sample	% D/(D+L)	mg/100g
Emmentaler	15	8
Limburger	26	1
Swiss Emmental	17	41
French Emmental	40	10
Bagoss	48	268
Pecorino	24	72
Grana Padano	22	340
Parmigiano Reggiano	37	158
Asiago	7	16
Fontina	7	11
Val Trompia	45	391

The origin of D-Pro in cheese is not known: it could be the product of enzymatic activity linked to the autochthonous microflora of milk or to bacterial contamination from the environment. According to the literature [40], D-Pro could be formed 1) from L-Pro via Proline racemase; 2) from L-Pro via formation of 1-Pirrolyn-2-carboxylate (P2C) by reductase and subsequent transformation via D-amino acid oxidase; 3) from L-Arg, via L-Orn, D-Orn, 5-Amino-2-oxovalerate and P2C; 4) from D-Pyroglutamic acid (5-Oxo-proline), which could be formed during ageing from D-Glutamic acid. Actually, these cheeses showed high amounts of Citrulline, γ-Aminobutyric acid, 5-Aminovaleric acid and Ornithine, all linked to the metabolic cycle of Arg and Pro, whereas Pyroglutamic acid was always present particularly in Parmigiano Reggiano and Grana Padano and increased with ageing up to 6-7 mg/g [41]. In the analysed samples, no correlations were observed beetwen the D-Pro content and 5-Aminovaleric acid. Moreover, only L-Pyroglutamic acid was found and not its D-enantiomer from which D-Proline could derive [41]. Thus, the origin of D-Pro in cheese to our knowledge is still mysterious.

4.4.2. Yoghurt

A good correlation was obtained between the amount of D-Ala, D-Glu and D-Asp in yoghurt and the microbial growth of *Streptococcus thermophilus* and *Lactobacillus delbrueckii subs. bulgaricus* [42]. The

D-amino acid content is strictly related to bacterial activity, in particular D-Asp is produced only by *Lactobacillus bulgaricus*.

Four different series of fermentation experiments were studied on a laboratory scale: one with only *Streptococcus thermophilus* as a starter (a), one with only *Lactobacillus bulgaricus* (b), and two with both microorganisms in a 1:1 or 1:10 ratio (c and d). The last two fermentation experiments were also reproduced on an industrial pilot plant.

During the fermentation experiments a general increase of the free amino acid concentration was observed: the amount of amino acids was moderate at the beginning of the fermentation, whereas it increased sharply at the end, in particular in the presence of both microorganisms. These findings were consistent with the synergic action of the two starters, *S. thermophilus* working in the early stages to promote the optimal acidity in the medium for the subsequent growth of *L. bulgaricus* which displays a larger proteolytic activity.

A general increase of both enantiomers was observed during the incubation period and the D/(D+L) ratio gradually increased for Ala, Asp and Glu during fermentation for yoghurt (Table 3).

Table 3. Variation of the amount of lactic bacteria and of the % D/(D+L) of free amino acids in yoghurt at different fermentation times.

Sample at:	Lactic flora (c.f.u./g)	% D-Ala D/(D+L)	% D-Asp D/(D+L)	% D-Glu D/(D+L)
start	3.1×10^{7}	21	7	4
60 min	5.5×10^{7}	49	24	9
120 min	4.7×10^{8}	47	25	15
180 min	8.5×10^{8}	55	27	14

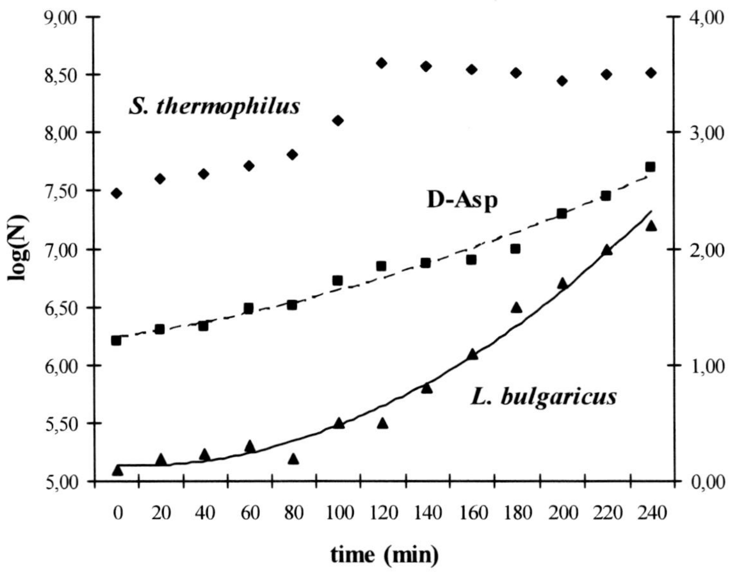

Fig. 8. Correlation between the total amount of D-Asp and *L. bulgaricus* growth during yoghurt fermentation [42].

In the fermentation experiment (a), where only *S. thermophilus* was present, an increase for D-Glu and D-Ala (D/(D+L)%) was observed, whereas in (b), where only *L. bulgaricus* was present, D-Asp was produced with a trend similar to the growth curve of the microorganism. Therefore, D-Asp can be regarded as a marker of the presence of *L. bulgaricus* and is strictly related to its amount.

In fact, as shown in the fermentation test with both microorganisms (c) and (d), the amount of D-Asp showed a remarkable increase during the exponential growth of *L. bulgaricus*, reaching values around 2 μmol/100g of yoghurt when the microorganism is present at least at $1 \cdot 10^7$ CFU/mL (Figure 8).

Although at the end of the fermentation the yoghurt was refrigerated at 4°C, it was possible to observe, during the first 24 h, a further increase of the % D/ D+L ratio for D-Ala, D-Asp and D-Glu. This is probably due to the residual enzymatic activity in the product.

We analysed also the fate of these enantiomers during the shelf-life of the product (0-40 days): at 4°C the total amount and the enantiomeric excesses of amino acids did not change significantly, whereas at 25°C both values increased significantly following a logarithmic curve, probably on account of accelerated proteolytic processes and cell lysis phenomena. In this case, the amount of D-Asp, D-Ala and D-Glu may be used as a marker of uncorrect preservation conditions.

4.5. Metabolic Fate of Dietary D-Amino Acids

The presence of high amounts of D-amino acids in the human diet aroused great interest about their metabolic fate which is linked to the activity of D-amino acid oxidases, present in human liver and kidneys. The localization of these enzymes implies that D-amino acids can enter the blood stream and be transported to these organs. *In vitro* experiments on human fibroblasts have shown that amino acid transport through cell membranes is regulated by specific transport systems [43]. It has been recently demonstrated that while the Glutamic acid transport is stereospecific (only the L-enantiomer is transported), the Aspartic acid transport is not [44]. The receptor could probably interact unselectively with either carboxylic group of Aspartic acid, whereas in the case of Glutamic acid, which has another $-CH_2-$ in the side chain, the interaction occurs only with the α carboxylic group, thus inducing stereoselectivity.

In order to study the metabolic fate of D-amino acids in the human diet, we designed different experiments based on the human consumption of the hard cheeses previously analysed as source of D-amino acids: 20 healthy volunteers were asked to eat 62 g of cheese (either Parmigiano Reggiano or Bagoss) and amino acid analysis was performed on blood samples taken at different time intervals (0, 30, 60, 120, 180 min after ingestion of the cheese). The presence of D- and L-Ala, D- and L-Glu, D- and L-Asp and D- and L-Pro was investigated in serum.

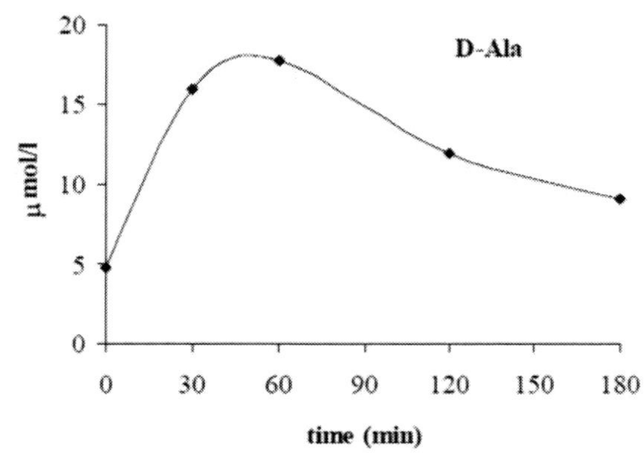

Fig. 9. Uptake curve of D-Ala in human serum after cheese consumption.

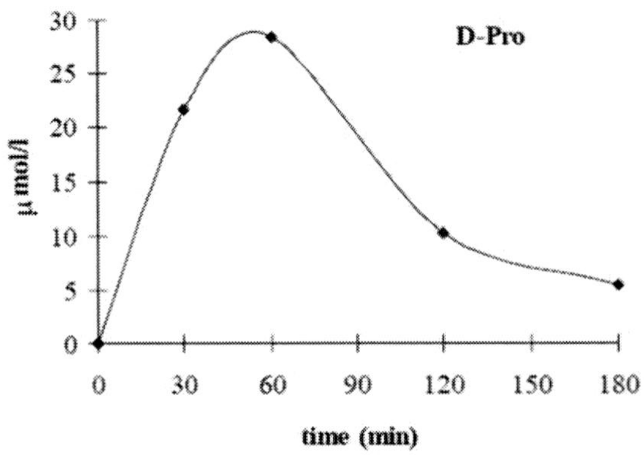

Fig. 10. Uptake curve of D-Pro in human serum after cheese consumption.

In particular for D- and L-Ala and D- and L-Pro a similar trend was observed for all subjects: both enantiomers were found in serum with the maximum absorption between 30 and 60 min after cheese consumption. After 3 hours a significant amount of the D-amino acid was still present. The data was best fitted by a cubic equation (correlation coefficient close to 1). An example for D-Ala and D-Pro is reported in Figures 9 and 10. Thus, both D-Ala and D-Pro were adsorbed in the blood stream and slowly metabolized. D-Asp and D-Glu were never found in human serum during these experiments. For D-Glu this could be due to the stereospecificity of the transport system, whereas for D-Asp a very fast metabolic pathway could be operative.

5. Conclusions

In conclusion, D-amino acids in foods turned out to be valuable markers of the process [45]. The formation of D-Amino acids in foods is in general promoted by harsh technological processes, as they are frequently produced by high temperature or extreme pH values. They can be also found as a consequence of adulteration, as in the case of hydrolyzed proteins added to foodstuffs to mask low nutrient content. On the contrary, D-Ala, D-Asp and D-Glu are natural components of bacterial cell walls: thus, they are valuable markers of fermentation and ageing in fermented foods (yoghurt, cheese, wine, vinegar, sausages, ecc.), but they can be also useful markers of undesired fermentations such as those occurring during microbial spoilage of fruit juices and purees. The occasionally occurrence of D-Pro in fermented products has probably a microbial/enzymatic origin and is probably due to the growth and development of an autochthonous microflora when starter cultures are not used. D-Ala and D-Pro are rapidly absorbed and metabolized by humans.

Acknowledgements

Many other people have participated to the research lines here described which are gratefully acknowledged: Prof. E. Neviani and Prof. G. Mucchetti (University of Parma), Dr. I. Gandolfi and Prof. C. Salvadori (Parmalat S.p.A.), together with many students who performed their "Laurea thesis" on this subject.

REFERENCES

[1] Schurig, V. (2002). Chiral separations using gas-chromatography. *Trends in Analytical Chemistry, 21*, 647-661.

[2] Davankov, V. A. (2003). Enantioselective ligand exchange in modern separation techniques. *J. Chromatogr. A, 1000*, 891-915.

[3] Wan, H., Blomberg, L. G. (2000). Chiral separation of amino acids and peptides by capillary electrophoresis. *J. Chromatogr. A, 875*, 43-88.

[4] Corrigan, J. J. (1969). D-Amino acids in animals. *Science, 164*, 142-149.

[5] Preston, R. L. (1987). Occurrence of D-amino acids in higher organisms: a survey of the distribution of D-amino acids in marine invertebrates. *Comp. Biochem. Physiol., 87B*, 55-62.

[6] Fuchs, S. A., Berger, R., Klomp, L. W. J., de Koning, T. J. (2005). D-Amino acids in the central nervous system in health and disease. *Molecular Genetics and Metabolism, 85*, 168-180.

[7] Masters, P. M., Friedman, M. (1979). Racemization of amino acids in alkali-treated food proteins. *J. Agric. Food Chem., 27*, 507-511.

[8] Hayase, F., Kato, H., Fujimaki, M. (1973). Racemization of amino acid residues in proteins during roasting. *Agric. Biol. Chem., 37*, 191-192.

[9] Liardon, R., Hurrell, R. F. (1983). Amino acid racemization in heated and alkali-treated proteins. *J. Agric. Food Chem., 31*, 432-437.

[10] Palla, G., Dossena, A., Marchelli, R., Casnati, G. (1989). Occurrence of D-amino acids in food. Detection by capillary gas chromatography and by reversed-phase high-performance liquid chromatography with L-Phenylalaninamides as chiral selectors. *J. Chromatogr., 475*, 45-53.

[11] Man, E. H., Bada, J. L. (1987). Dietary D-amino acids. *Annual Rev. Nutr., 7*, 209-225.

[12] Friedman, M. (1991). Formation, nutritional value, and safety of D-amino acids. *Adv. Exp. Med. Biol., 289*, 447-481.

[13] Kaltenbach, J. P., Ganote, C. E., Carone, F. (1979). Renal tubular necrosis induced by compounds structurally related to D-Serine. *Exp. Molec. Pathol., 17*, 209-214.

[14] Bunjapamai, S., Mahoney, R. R., Fagerson, I. S. (1982). Determination of D-amino acids in some processed foods and effect of racemization on *in vitro* digestibility of casein. *J. Food Sci., 47*, 1229-1234.

[15] de Vrese, M., Frik, R., Roos, N., Hagemeister, H. (2000). Protein-bound D-amino acids, and to a lesser extent Lysinoalanine, decrease true ileal protein digestibility in minipigs as determined by ^{15}N-labeling. *Nutrient Interactions and Toxicity*, 2026-2031.

[16] Bruckner, H., Jeak, P., Langer, M., Godel, H. (1992). Liquid chromatographic determination of D-amino acids in cheese and cow milk. Implication of starter cultures, amino acid racemases, and rumen microorganisms on formation, and nutritional consideration. *Amino acids, 2*, 271-284.

[17] Konno, R., Yasumura, Y. (1992). D-amino-acid oxidase and its physiological function. *Int. J. Biochem., 24*, 519-524.

[18] Bender, A. E., Krebs, H. A. (1950). The oxidation of various synthetic α-amino acids by mammalian D-amino acid oxidase, L-amino acid oxidase of cobra venom, and the L- and D-amino acid oxidases of Neurospora crassa. *Biochem. J., 46*, 210-219.

[19] Dixon, M., Kenworthy, P. (1967). D-Aspartate oxidase of kidney. *Biochem. Biophys. Acta, 146*, 54-76.

[20] Marchelli, R., Virgili, R., Armani, E., Dossena, A. (1986). Enantiomeric separation of D, L-Dns-amino acids by one- and two-dimensional thin-layer chromatography. *J. Chromatogr., 355*, 354-357.

[21] Armani, E., Dossena, A., Marchelli, R., Virgili, R. (1988). Copper(II) complexes of diamino-diamido-type ligands as chiral eluents in the enantiomeric separation of D,L-dansylamino acids by reversed-phase high-performance liquid chromatography. *J. Chromatogr., 441*, 275-286.

[22] Armani, E., Barazzoni, L., Dossena, A., Marchelli, R. (1988). Bis(L-amino acid amidato) copper(II) complexes as chiral eluents in the enantiomeric separation of D, L-dansylamino acids by reversed-phase high-performance liquid chromatography. *J. Chromatogr.*, *441*, 287-298.

[23] Galaverna, G., Corradini, R., de Munari, E., Dossena, A., Marchelli, R. (1993). Chiral separation of unmodified amino acids by ligand exchange high-performance liquid chromatography using copper(II) complexes of L-amino acid amides as additives to the eluent. *J. Chromatogr.*, *657*, 43-54.

[24] Dossena, A., Galaverna, G., Corradini, R., Marchelli, R. (1993). Two-dimensional high-performance liquid chromatographic system for the determination of enantiomeric excess in complex amino acid mixtures. Single amino acid analysis. *J. Chromatogr.*, *653*, 229-234.

[25] Palla, G., Dossena, A., Marchelli, R. (1985). Oligo-oxaalkanoyl tetraamides derived from L-Phenylalanine as stationary phase in capillary gas chromatographic resolution of D, L-amino acids. *J. Chromatogr.*, *349*, 9-15.

[26] Galaverna, G., Corradini, R., Dossena, A., Marchelli, R., Vecchio, G. (1997). Histamine-modified β-cyclodextrins for the enantiomeric separation of dansyl-amino acids in capillary electrophoresis. *Electrophoresis*, *18*, 905-911.

[27] Corradini, R., Paganuzzi, C., Marchelli, R., Pagliari, S., Sforza, S., Dossena, A., Galaverna, G., Duchateau, A. (2005). Fast parallel enantiomeric analysis of unmodified amino acids by sensing with fluorescent β-cyclodextrins. *J. Mater. Chem.*, *15*, 2741-2746.

[28] Venturi, E., Labadini, L., Marchelli, R., Dossena, A., Mori, G., Pilone, M. (1996). A biosensor for food quality control. In *Sensors and Microsystems*, Di Natale, C. and D'Amico, A. eds., World Scientific, Singapore, 187-191.

[29] Corradini, R., Marchelli, R., Palla, G. (1994). Properties of L-Phenylalanine tetraamides as chiral selectors for D,L-amino acids by capillary gas chromatography. *Chromatographia*, *38*, 173-176.

[30] Chankvetadze, B. (1997). *Capillary electrophoresis in chiral analysis.* John Wiley & Sons, eds., Chichester, UK.

[31] Gandolfi, I., Palla, G., Delprato, B., De Nisco, F., Marchelli, R., Salvadori, C. (1992). D-Amino acids in milk as related to heat treatments and bacterial activity. *J. Food Sci.*, *57*, 377-379.

[32] Lubec, G., Wolf, C., Bartosch, B. (1989). Amino acids isomerisation and microwave exposure. *Lancet*, *334*, 1392-1393.

[33] Marchelli, R., Dossena, A., Palla, G., Audhuy-Peaudecerf, M., Lefeuvre, S., Carnevali, P., Freddi, M. (1992). D-Amino acids in reconstituted infant formula: a comparison between conventional and microwave heating. *J. Sci. Food Agric.*, *59*, 217-226.

[34] Segal, W. (1990). D-Amino acids and microwaves. *Lancet*, *336*, 49.

[35] Gandolfi, I., Palla, G., Marchelli, R., Dossena, A., Puelli, S., Salvadori, C. (1994). D-Alanine in fruit juices: a molecular marker of bacterial activity, heat treatments and shelf-life. *J. Food Sci.*, *59*, 152-154.

[36] Chiavaro, E., Caligiani, A., Palla, G. (1998). Chiral indicators of ageing in balsamic vinegars of Modena. *Ital. J. Food Sci.*, *10*, 329-337.

[37] Innocente, N., Palla, G. (1999). Occurrence of D-amino acids in a typical semi-hard cheese. *J. Dairy Res.*, *66*, 633-637.

[38] Marchelli, R., Palla, G., Dossena, A., Galaverna, G., Corradini, R., Clementi, S. (1997). D-Amimo acids: molecular markers of ageing and authenticity for Parmigiano Reggiano and Grana Padano cheese. *Sc. Tecn. Latt.-Casearia*, *48*, 21-32.

[39] Resmini, P., Pellegrino, L., Pazzaglia, C., Hogenboom, J. A. (1985). Free amino acids for the quality control of Parmigiano-Reggiano cheese and particularly for the grated products. *Sci. Tecn. Latt.-Casearia*, *36*, 557-592.

[40] Aral, B., Kamoun, P. (1997). The Proline biosynthesis in living organisms. *Amino acids*, *13*, 189-217.

[41] Mucchetti, G., Locci, F., Gatti, M., Neviani, E., Addeo, F., Dossena, A., Marchelli, R. (1999). Pyroglutamic acid in cheese: presence, origin, and correlation with ripening time of Grana Padano cheese. *J. Dairy Sci.*, 83, 659-665.

[42] Pantò, F., De Nisco, F., Pinardi, M., Marchelli, R., Palla, G., Dossena, A., Mori, G. (1993). Molecular markers of yoghurt fermentation for quality control and sensory evaluation. In: *Progress in Food Fermentation*, Proceedings of FECS VII, IATA-CSIC, Valencia, *2*, 410-415.

[43] Bannai, S., Christensen, H. N., Vadgama, J. V., Ellory, J. C., Englesberg, E., Guidotti, G. G., Gazzola, G. C., Kilberg, M. S., Lajtha, A., Sacktor, B. (1984). Amino acid transport systems. *Nature*, *311*, 308.

[44] Gazzola, G. C., Dall'Asta, V., Bussolati, O., Makowska, M., Cristensen, H. N. (1981). A stereoselective anomaly in dicarboxylic amino acid transport. *J. Biol. Chem.*, *256*, 6054-6059.

[45] Marchelli, R., Dossena, A., Palla, G. (1996). The potential of enantioselective analysis as a quality control tool. *Trends Food Sci. Tech.*, *7*, 113-119.

In: D-Amino Acids: ISBN 1-60021-075-9
A New Frontier in Amino Acid and Protein Research, pp.317-336

Chapter 4.3

D-AMINO ACID DETERMINATION IN FOODS, BEVERAGES, AND BIOLOGICAL SAMPLES

Molly Warnke and Daniel W. Armstrong*
Department of Chemistry & Biochemistry
University of Texas at Arlington, U.S.A.

Keywords: enantioselective analysis, HPLC, GC, CE, column switching, macrocyclic glycopeptides, cyclodextrins

1. INTRODUCTION

The revolution in high efficiency enantiomeric separations and analysis has led to the possibility of determining small amounts of D-amino acids within complex sample matrices such as food, beverages, and physiological fluids and tissues. Advances in the development of chiral stationary phases (CSPs) have led to routine enantiomeric separations by high performance liquid chromatography (HPLC) and HPLC coupled with mass spectrometry detection (HPLC-MS) along with gas chromatography (GC) and capillary electrophoresis (CE).

Many components of foods and beverages are chiral. The predominate taste sensations in humans are sweet, sour, salty, bitter and umami, which is a brothy or "meaty" taste [1-5]. It is known that enantiomers can have completely different tastes. Many D-amino acids are sweet as opposed to the bitter taste of many L-amino acids [5,6]. The D-enantiomers of hydrophobic amino acids are especially sweet [6]. D-tryptophan, D-phenylalanine, D-histidine, D-tyrosine, and D-leucine are 35, 7, 7, 6, and 4 times sweeter than sucrose, respectively [4]. Certain amino acids are desired for their savory flavors. L-glutamic acid and L-aspartic acid are bitter in their undissociated states, but as part of a sodium salt, they have umami tastes [5]. Stereospecific flavor and fragrance components are important to researchers, as are stereochemical differences in the nutritional value of many components of foods and beverages.

In the past twenty years, stereochemical analysis of amino acid content in food has drawn more and more attention as methods to determine the D- and L-enantiomers in complex samples have improved. Research in this area through 1996 was reviewed in a book chapter by Ekborg-Ott and Armstrong [5]. The

* Correspondence concerning this article should be addressed to Daniel W. Armstrong, E-mail: sec4dwa@uta.edu; FAX 817-272-0619

ability to determine D-amino acids in foods and beverages can be important for the following reasons: i) identifying adulterated foods or beverages, ii) possibility to monitor fermentation more exactly, iii) evaluation and identification of age, iv) treatment and storage effects, v) more exact evaluation of some flavor components, and vi) fingerprinting complex mixtures [7].

The identification of adulterated food and beverage products is an important consequence of the ability to perform stereochemical analysis. Flavors can change at many steps in production: while stored as raw materials, during processing, or with storage of the finished product. Altered flavor components may lead to an improvement in flavor or a ruined product. Although flavor components can change unintentionally through time and/or processing, they can also be changed deliberately. Intentional changes to flavors can include syntheses of new flavor compounds, hiding flavors which are already present, changing one flavor compound into another, removing flavor compounds, and hindering the formation of a certain flavor compound [5,8]. An example is coffee. Spray- and freeze-drying can cause coffee to lose volatile flavor and aroma compounds. Replacement compounds are added as closely to packaging as possible because they are sensitive to oxygen and moisture. The age, storage, shipping, and handling of the coffee or other food and beverage products can sometimes be determined by the enantiomeric purity of certain chiral components [5].

Since food is an important route of exposure to D-amino acids, it is important to note that they are found more often in foods that i) have been highly processed by exposure to extreme heat or pH, etc. [5,9-13], ii) are exposed to microbial activity via fermentation, aging, etc. [5,9-12,14-17], and iii) naturally contain D-amino acids (certain seafoods) [18-22]. Foods that are commonly known to contain D-amino acids due to processing are beer, mead, wine, sake, vinegar, cocoa powder, coffee, dried tea leaves, fermented dairy products, lactic acid fermented vegetables, fermented sausage, and soy sauce, among others [5].

Natural proteins found in foods and beverages are often changed by manufacturing processes to affect taste, texture, and shelf life. L-amino acids may undergo acid- or base-catalyzed racemization in processing and preparation involving acid, alkali, or heat [5,13,23-31]. The nutritional value and safety of foods may change because of racemization since D-amino acids that are formed can be nutritional antagonists or act as antimetabolites [5,17,28,29,31,32].

Recent studies have shown that some D-amino acids may be synthesized in mammalian tissue [33]. Dunlop and Niedle used radiolabeled L-serine to study racemization within mouse brains [34]. Wolosker et al. discovered a mammalian [35,36] and human serine racemase in rat brains [37]. The racemization studies and serine racemases suggest that D-serine naturally occurs in the mammalian brains. Berg [38] and Neuberger [39] reviewed early studies on essential amino acid racemization and showed that D-amino acid isomers in mammals are poorly utilized, sometimes inhibit growth, and are mainly excreted in the urine [5]. Other studies confirmed Berg and Neuberger's conclusions [33,41-44]. The determination of D-amino acids found in mammals can be indicative of their origin and their function in the body. Some D-isomers have been found to play roles in the central nervous system (i.e., D-Ser) and in endocrine organs (i.e., D-Asp) [33].

2. Determination of Native Amino Acids

There are four main ways to separate native, underivatized amino acids into their enantiomeric pairs: 1) ligand-exchange methods [45-51], 2) chiral crown ether stationary phases for primary amino acids [52], 3) teicoplanin macrocyclic antibiotic CSP [53-62], and 4) an α-cyclodextrin bonded phase for aromatic amino acids [63]. The separation mechanism for method four will be discussed later with the discussion on derivatized amino acids.

Ligand-exchange chromatography, first introduced by Davankov and co-workers [64,65], involves attaching a chiral bidentate ligand to a support. Common ligands are proline and hydroxyproline. When copper (II) is added to the mobile phase, the copper coordinates with the chiral selector and the amino acid. This forms a ternary complex and if the free energies of the diastereomeric complexes of each enantiomer differ sufficiently, there will be an enantiomeric separation. The ligand-exchange method can be used in thin-layer chromatography [5,66].

The use of chiral crown ethers for enantiomeric separations provided an alternative to ligand-exchange methods. Cram and co-workers [67-69] first introduced the chiral crown ether stationary phases. Inclusion complexes between ions the size of potassium and ammonium can form with [18]-crown-6-polyethers. Primary amines, hence most amino acids, can be separated with this method. Mobile phases must be acidic to keep the primary amine group protonated in its ammonium form [5,66].

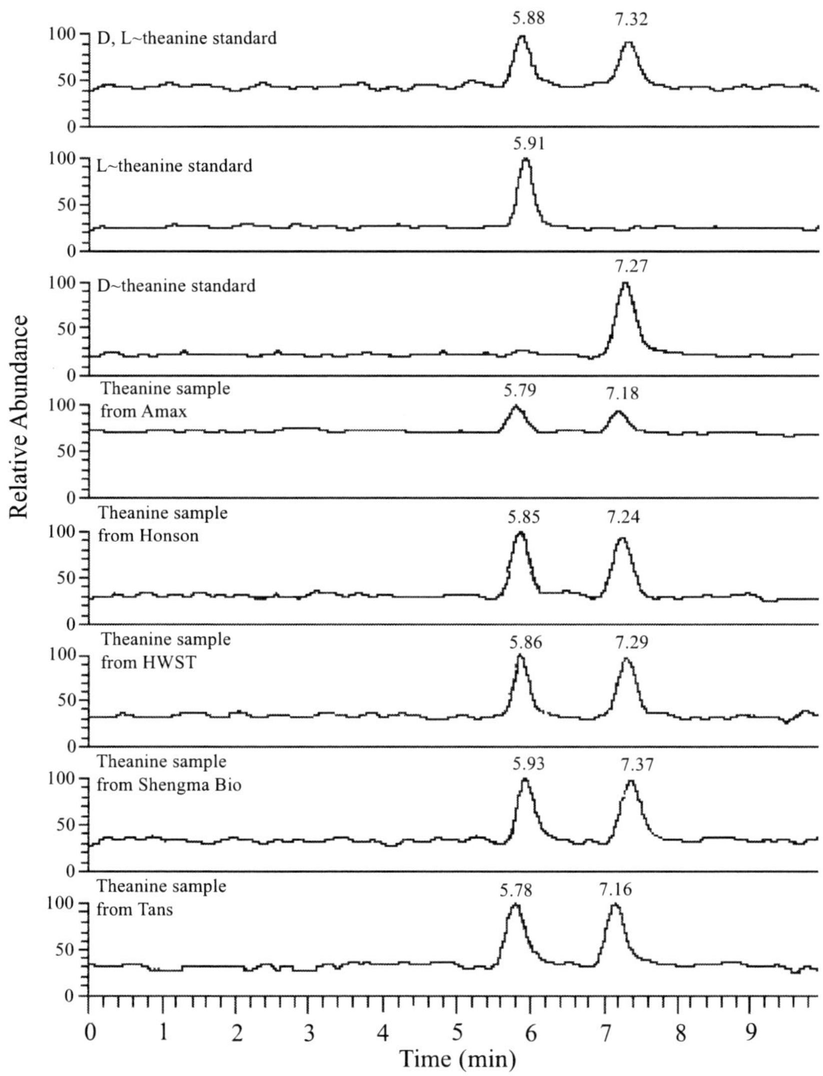

Figure 1. Chromatograms of five of six theanine samples containing both L- and D-theanine compared with theanine standards. The retention times are slightly shifted due to fluctuations in the ambient temperature from day-to-day. Separation conditions for all standards and samples: LC/APCI-MS, SIM: m/z 175, mobile phase conditions: 80:20 (1.0% ammonium trifluoroacetate in methanol/0.1% formic acid in H_2O) 0.8 mL/min. Reproduced from Ref. 74 with permission.

The macrocyclic glycopeptide (or antibiotic) teicoplanin CSP, introduced by Armstrong, has proven to be a very useful chiral selector [66,70-72]. Teicoplanin can separate all of the chiral proteinic amino acids (primary and secondary) in the reverse-phase mode [72]. The main interaction for enantioselectivity between amino acids and teicoplanin is that between the amine group on the aglycone basket of the antibiotic and the carboxylate moiety of the amino acid. When the carbohydrate moieties are removed from teicoplanin, the remaining molecule is referred to as the teicoplanin aglycone. This moiety has been used to make a related CSP that is even more selective for amino acid enantiomers [73].

While native amino acids can be determined without derivatization, there are still detection sensitivity issues with UV/visible detection because only the aromatic amino acids contain a chromophore. Mass spectrometric detection has become an alternative detection method for native amino acids [59,74-76].

Recently, Desai and Armstrong demonstrated that all 19 chiral amino acids could be separated on a teicoplanin or teicoplanin aglycone (teicoplanin without its carbohydrate group) CSP and detected with atmospheric pressure chemical ionization-MS (APCI-MS) without fluorescent derivatizing agents [76]. Using this method, D-theanine (N-ethyl glutamic acid) was detected in five of six products labeled as L-theanine as shown in Figure 1 [74]. Pharmacokinetic studies of D-theanine in rats were performed [75]. D-theanine was determined in rat urine and plasma with a teicoplanin CSP by APCI-MS.

3. Derivatization Methods for HPLC Analysis

Derivatization of amino acids often is done in order to enhance sensitivity with UV/visible, fluorescence, or electrochemical detection [77-83]. Derivatization also can enhance sample clean-up and pre-concentration as well as increase interactions with the CSP [5,84-86]. The sensitivity and selectivity for amino acids can be enhanced by the use of achiral derivatizing agents. Achiral derivatizing agents are often used in HPLC and CE, along with the use of a chiral selector, either as a part of a stationary phase (HPLC) or run buffer additive (CE). Though derivatization adds another step in the procedure, the improved sensitivity is important as are possible enhanced interactions between the tagged amino acid and the chiral selector [87]. Table 1 shows many of the achiral derivatizing agents commonly used.

Some derivatizing agents have disadvantages when used to increase the sensitivity of D-amino acid determination. For example, when using phenyl isothiocyanate (PITC), the excess reagent must be evaporated before HPLC separation. Dansyl chloride (DNS-Cl) derivatization is difficult with trace analyte levels in complex mixtures, such as food, beverage, or biological samples [100]. *o*-Phthalaldehyde (OPA) with appropriate chiral thiol agents is much faster and more sensitive than the DNS-Cl reagent, but only reacts with primary amines [101,102]. Both OPA and DNS-Cl have been used in labs for many years, however neither is stable in solution for an extended period of time [5].

FMOC (9-fluorenylmethyl chloroformate), FMOC-Gly-Cl (9-fluorenylmethoxy-carbonylglycine chloride), and AQC (6-aminoquinolyl-*N*-hydroxysuccinimidyl carbamate) seem to be better choices as derivatizing agents for the detection and resolution of amino acids in complex food, beverage, biological or other "real world" samples. These agents have been shown not to cause racemization if used properly and are compatible for use with aqueous samples. They also tend to stabilize the amino acid during sample preparation and analysis. Another advantage is that these reagents can derivatize both primary and secondary amino acids [85].

Detection interference by an excess of derivatizing reagent that may fluoresce or absorb at a similar wavelength to the analyte is another disadvantage to derivatization. AQC does not suffer from this problem because the byproducts from its derivatization reaction do not interfere chromatographically with the amino acids or other analytes of interest [5]. In the polar organic mode, AQC amino acids can be separated easily on native cyclodextrin CSPs [86]. The use of cyclodextrin CSPs and the polar organic mobile phase leads to better sensitivity, quicker equilibration, and a more stable baseline [5,86]. All FMOC protein amino acids are easily separated on a ristocetin-A (Chirobiotic R) LC stationary phase [54].

Table 1. Achiral derivatizing agents used in HPLC and CE.

Structure	Compound Name (Abbreviation)	Reference
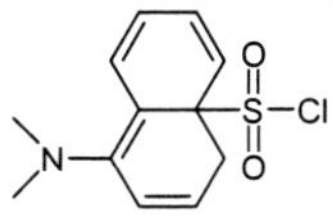	dansyl chloride (DNS-Cl)	5, 88
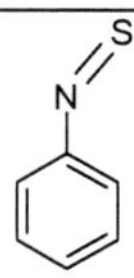	phenyl isothiocyanate (PITC)	5, 81
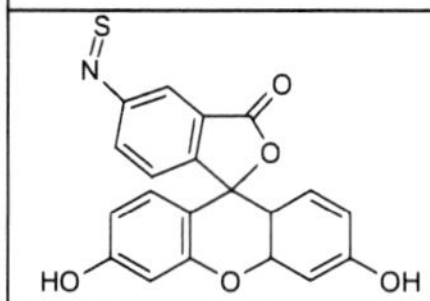	fluorescein isothiocyanate (FITC)	88
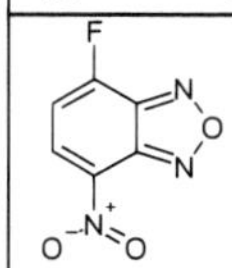	4-fluoro-7-nitrobenzofuran (NBD-F)	89-92
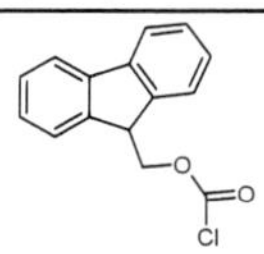	9-fluorenylmethyl chloroformate (FMOC)	73, 75, 93
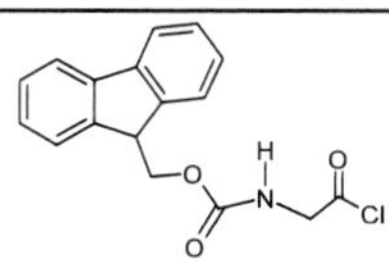	9-fluorenylmethoxy-carbonylglycine chloride (FMOC-Gly-Cl)	93, 94
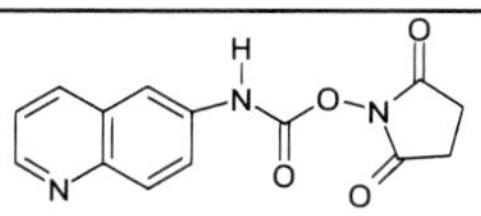	6-aminoquinolyl-*N*-hydroxysuccinimidyl carbamate (AQC)	95, 120
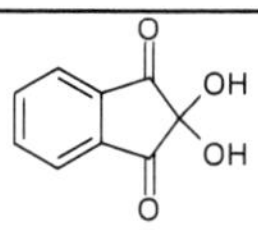	ninhydrin	96
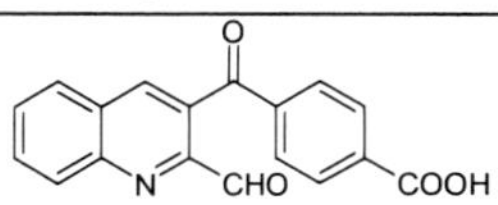	3(4-carboxybenzoyl)-2-quinoline-carboxaldehyde (CBQCA)	97
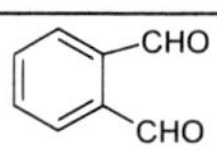	*o*-phthalaldehyde (OPA) *use with chiral thiol reagent	77, 98, 99

Chiral derivatizing agents are also used to enhance detection sensitivity. When derivatizing amino acids, diastereomeric pairs are formed and can be separated on achiral columns (e.g. C_{18}). It is important that the agent is optically pure and the reaction conditions will not lead to racemization. Some of the chiral reagents that have been used in this manner are shown in Table 2.

Table 2. Chiral derivatizing agents used in HPLC and CE.

Structure	Name (Abbreviation)	Reference
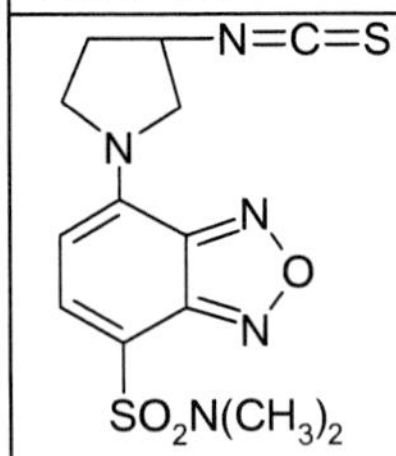	R(-)-4-(3-isothiocyanatopyrrolidin-1-yl)-7-(N,N-dimethylaminosulfonyl)-2,1,3-benzoxadiazole (DBD-PyNCS)	104, 105
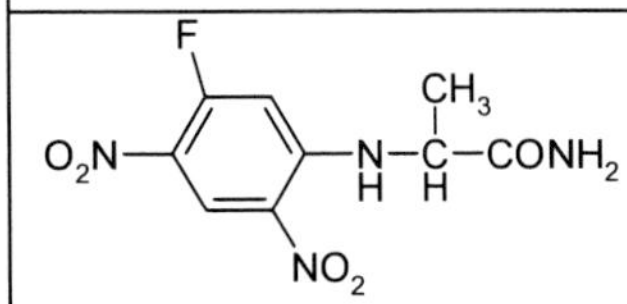	1-fluoro-2,4-dinitrophenyl-5-L-alanine amide (Marfey's reagent)	5, 103
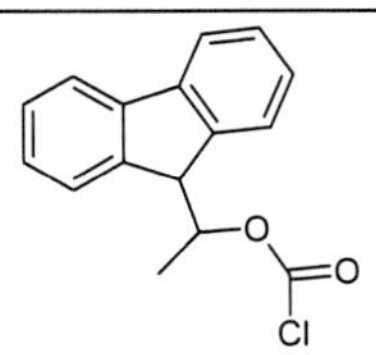	(+)-1-(9-Fluorenyl)ethyl chloroformate (FLEC)	5
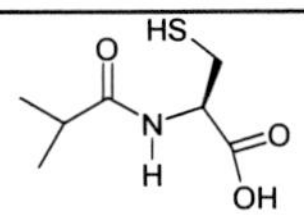	*Chiral thiol reagent used with OPA *N*-isobutyryl-L-cysteine	98
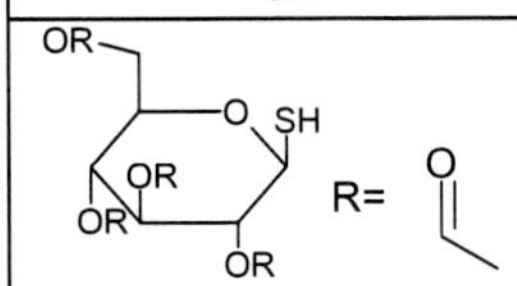	*Chiral thiol reagent used with OPA 2,3,4,6-tetra-*O*-acetyl-1-thio-β-D-glucopyranose (TATG)	99

4. HPLC Analysis of Derivatized Amino Acids

Amino acids derivatized with achiral agents must be separated on CSPs. The most common CSPs used for amino acid analysis of derivatized amino acids are cyclodextrins [93,94,106-108], macrocyclic glycopeptide [54,70,72], and π-complex type stationary phases [89-92,109-112].

Cyclodextrins are macrocyclic compounds formed by glucopyranose units. The native cyclodextrins, α, β, and γ, have 6, 7, and 8 units, respectively. In the reverse phase mode of operation, cyclodextrin CSPs separate enantiomers by forming inclusion complexes [106,113]. The hydrophobic part of the analyte molecule complexes with the hydrophobic interior cavity of the cyclodextrin. Additionally, interactions with the hydroxyl groups or derivative groups on the rim of the cyclodextrin are required to achieve enantioselective complexation [113]. Derivatized cyclodextrin CSPs, which provide additional sites for chiral recognition interactions, may also be used to separate amino acids [108,112,114]. Cyclodextrin CSPs can also be operated in polar organic mode where acetonitrile is generally the main component of the mobile phase [66].

Macrocyclic glycopeptide CSPs also can resolve amino acid enantiomers that have been derivatized with achiral agents. Teicoplanin was discussed in a previous section. Other macrocyclic glycopeptides that have been shown to separate amino acids are based on vancomycin and ristocetin-A in both reversed-phase mode and polar organic mode [54,70,115].

Pi-complex CSPs contain a pi-acidic group, pi-basic group, or both. In order for enantioselective resolution, the analyte must contain a pi-acid group when used with a pi-basic CSP or vice versa [66]. Most amino acids lack pi-acid or pi-base groups. In the case of pi-complex CSPs, derivatization not only improves detection sensitivity but also leads to increased interactions with the chiral selector. Although these CSPs were used in the early days of LC enantiomeric separations, they are not used as much today, especially for amino acid separations.

Column switching is a common technique used in the HPLC analysis of analytes in complex matrices (Fig. 2). In column switching, an achiral column is coupled with a chiral column. The amino acids can be separated from each other and other matrix interferences by the achiral column. The amino acid racemate is then introduced onto the column containing an appropriate CSP, where it is separated into its enantiomers. Generally the achiral system is a C_{18} stationary phase operated in reversed-phase mode, and the chiral system is a cyclodextrin or macrocyclic glycopeptide stationary phase operated in reversed-phase or polar organic mode [40,85,93,94,116-120].

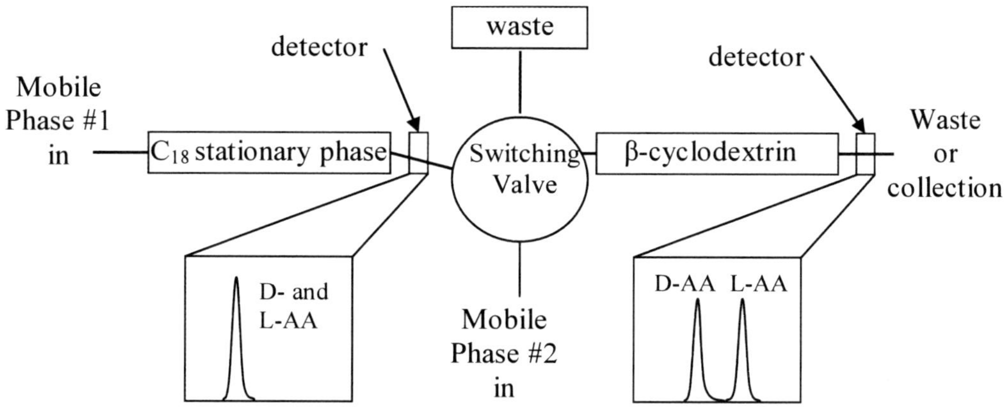

Figure 2. Box diagram of a column switching apparatus. The amino acid enantiomers co-elute on the C18 column. The eluent containing the amino acid can then be "switched" onto the chiral stationary phase (β-cyclodextrin in this example). The CSP resolves the amino acid into its enantiomeric pair.

In the last few years, many studies have reported the presence of D-amino acids in both food products and biological samples. FMOC-Cl was used as a fluorescent tagging agent for D-Pro, while FMOC-Gly-Cl provided better enantiomeric separations for primary amino acids and was used to tag D-Leu and D-Phe in a study of the free D-amino acid content in beer [93]. Column switching from a C_{18} stationary phase to RN-β-cyclodextrin (Cyclobond I-RN) column in polar organic mode was used to determine D-Pro. D-Leu and D-Phe were determined by column switching from a C_{18} to a β-cyclodextrin (Cyclobond I) stationary phase in polar organic mode. FMOC-Gly-Cl was also used to determine D-theanine in various types of tea. Column switching from a C_{18} column to a γ-cyclodextrin (Cyclobond II) column operated in polar organic mode was used [94].

D-Isomers of proline, alanine, asparagine, aspartic acid, valine, methionine, leucine, and phenylalanine were found in tobacco [120]. The amino acids were derivatized with AQC. Column switching from a C_{18} stationary phase to various native or derivatized cyclodextrin CSPs was used. Interestingly, alanine was shown to have racemized completely during processing in one of the smokeless tobacco samples tested.

The highly fluorescent achiral derivatizing agent NBD-F has also been used to determine D-amino acids in mouse and rat brain tissue [89-92]. Column switching was used to improve selectivity and sensitivity from the mammalian tissue samples. The free D-Leu and D-Pro content in different parts of a mouse brain were determined using a microbore C_{18} stationary phase followed by a 1-naphthylglycine pi-complex type chiral stationary phase [90]. D-Ala was determined in rat cerebellum by similar methods [91]. NBD-F was also used to derivative D-Thr and D-allo-Thr found in various rat tissues [89]. In this case, fractions containing the NBD-derivatized amino acids were collected after elution on the C_{18} stationary phase. The collected fractions were then injected onto the pi-complex type chiral stationary phase. This same method was used to determine 17 D-amino acids in rat brain tissues [92]. Also, OPA, along with *N*-isobutyryl-L-cysteine, was used by Brückner in HPLC experiments to determine D-Asp and D-Asn in beer samples [98].

Amino acids derivatized with chiral agents are predominately separated using C_{18} columns. The D-amino acid content in various food products and human urine was determined by Jin et al. using DBD-PyCNS(R(-)-4-(3-isothiocyanatopyrrolidin-1-yl)-7-(*N,N*-dimethylaminosulfonyl)-2,1,3-benzoxadiazole) as a chiral derivatizing agent [104]. The thiocarbamoyl derivatives were analyzed under reverse phase mode conditions on a C_{18} column with fluorescence detection. The dairy products tested contained D-Glu, D-Asp, and D-Ala, while tomato products, wine, and beer were found to contain just D-Glu and D-Ala. Human urine contained D-Ser and D-Ala. Electrospray ionization-MS along with UV detection was used to elucidate the structures of the D-amino acids in this study. Toyo'oka et al. also used DBD-PyCNS and similar methods to determine D-amino acids in human urine, tomato juice, wine, and beer [105]. Marfey's reagent (1-fluoro-2,4-dinitrophenyl-5-L-alanine amide) has also been used as a chiral derivatizing agent to determine D-amino acids in eubacteria, archaea and eukaryotes [103]. The samples were derivatized; fractions were collected from TLC plates, and then analyzed on a C_{18} stationary phase.

Ninhydrin was used to determine D-Glu and D-Asp in soy products by Stenberg and coworkers [96]. A two-column system was used. The amino acids were first separated at varying pHs with an ion-exchange column and then passed onto a crown ether CSP. The amino acids were derivatized with ninhydrin after elution through the crown ether CSP.

5. Derivatization Methods for GC Analysis

For GC analysis, amino acids must be derivatized. Native amino acids are not volatile and must be derivatized twice before separation in GC. First is the esterification of the carboxylic acid moiety, followed by reaction of the amine group with a desired agent (trifluoroacetic anhydride, for example) to form an amide [5].When using a hydrogen bonding GC chiral stationary phase such as Chirasil- L-Val, analytes need to be derivatized not only to increase volatility but to possibly add a hydrogen bonding group for increased interactions [121]. The most prevalent achiral derivatization scheme is perfluoroacylation using trifluoroacetyl derivatives [98,122-125] and pentafluoropropionyl derivatives [98,126-131]. The acylation can be performed by substitution of a trifluoroacetyl, pentafluoropropionyl, or heptafluorobutyryl group for the amino or hydroxyl moieties [5,122,130,131]. Representative GC derivatizing agents are shown in Table 3.

Table 3. Derivatizing agents for GC.

Structure	Name (Abbreviation)	Reference
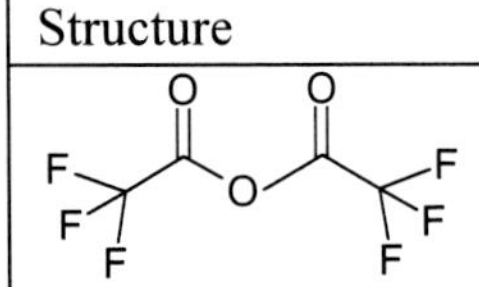	Trifluoroacetic anhydride (TFAA)	5, 98,125,129,130
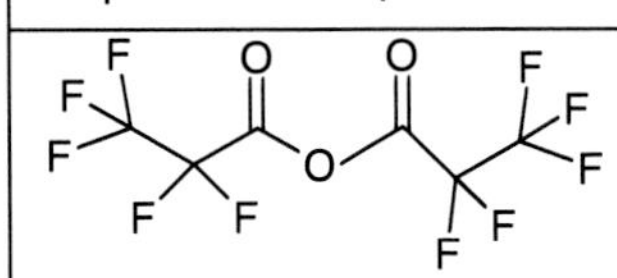	Pentafluoropropionic anhydride (PFPA)	5, 98,129,130
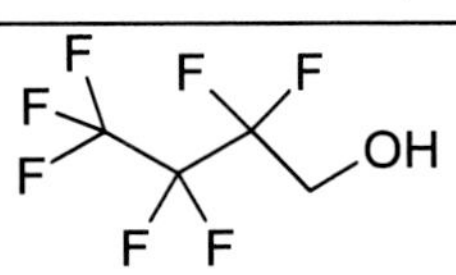	2,2,3,3,4,4,4-heptafluoro-1-butanol	5,132

6. GC Analysis of Derivatized Amino Acids

Chiral separations in gas chromatography rely on CSPs (for a review on GC CSPs, see ref. 133). The first GC stationary phase to resolve amino acid enantiomers was introduced in 1966 and uses amino acids and peptides as chiral selectors [134]. The CSP used most frequently in food and biological samples in the past few years is an *N-tert.*-butyl-L-valine-polysiloxane fused silica capillary column (Chirasil-L-Val) introduced by Frank, Nicholson, and Bayer [111,135,136]. The mechanism for chiral selectivity with the Chirasil-Val column is based on hydrogen bonding.

The other important CSP for GC analysis of amino acids is based on derivatized cyclodextrins [106,137-140]. A wide range of chiral compounds could be separated on the cyclodextrin GC stationary phases. Many compounds that could not be resolved by HPLC, such as nonaromatic and nonderivatizable compounds could easily be resolved by GC on the cyclodextrin based CSPs [5].

Sterochemical analysis by GC offers several advantages over HPLC including speed of analysis, low detection limits for enantiomeric impurities, universal detectors, easy resolution of most volatile compounds, durability, and high absolute sensitivity that is independent of functional groups [5,141].

GC has been used with various derivatizing agents, CSPs, and detection methods to determine D-amino acids in foods, beverages, and biological samples in the past few years. PFPA (pentafluoropropionic anhydride) along with 2-propanol and hydrochloric acid, converts amino acids to their *N(O)*-pentafluoropropionyl 2-propyl esters. Brückner et al. used this derivitization method along with the Chirasil- L-Val CSP to determine D-amino acid content in beer and its raw materials (Fig. 3) [98], fortified wines [114], and rat tissues [115]. Flame ionization detection (FID) as well as selected ion monitoring-MS (SIM-MS) detection was used to detect and quantify the D-amino acids present in the various samples. In SIM-MS, characteristic fragments are preselected for monitoring. SIM-MS is advantageous over FID because PFPA contains some trifluoroacetic acid, which forms an *N(O)*-trifluoroacetyl amino acid 1-propyl ester that can be an interference in analysis [98]. Similar methods were used by Brückner et al. to determine D-amino acid content in various physiological fluids and tissues in higher vertebrates [142].

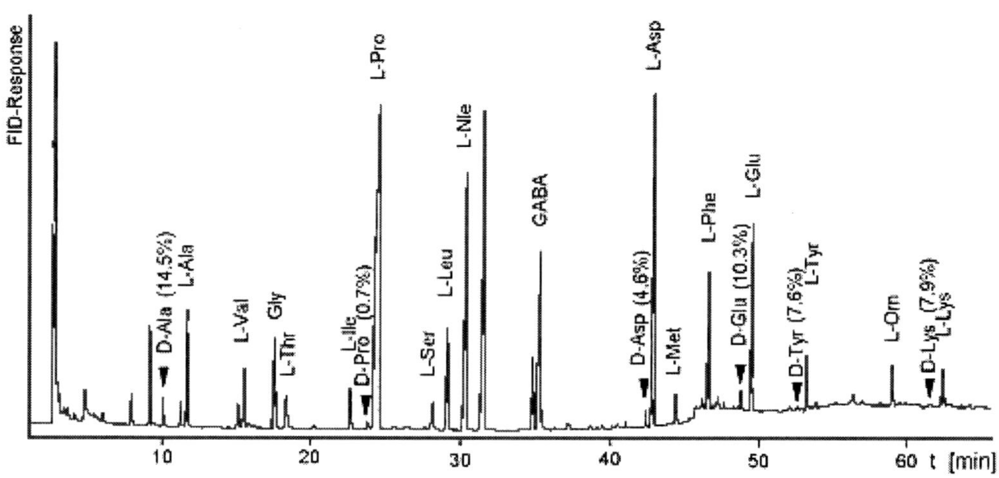

Figure 3. GC of a Kriek-lambic beer; amino acids were separated on Chirasil-L-Val as trifluoroacetyl amino acid 1-propyl esters. Reproduced from Ref. 98 with permission.

D-Amino acids in some dairy products, beer, and coffee were determined by derivatizing with ethylchloroformate and 2,2,3,3,4,4,4-heptafluoro-1-butanol to make *N(O,S)*-ethoxycarbonyl heptafluorobutyl esters [117]. The derivatives were analyzed on the Chirasil-L-Val chiral stationary phase with flame ionization detection (FID).

Nokihara and Gerhardt developed an automated GC chiral analysis system. They used hydrochloric acid and 1-propyl alcohol to esterify the amino acid, followed by acylation with TFAA in trifluoroacetylethylester for derivatization [110]. Enantiomers were separated with either the Chirasil- L-Val CSP or a 3-*O*-butanoyl-2,6-di-*O*-n-pentyl)-β-cyclodextrin CSP. D-Amino acids were determined in various milk samples as well as tissue proteins. The automated instrument hydrolyzed and analyzed the samples without purifying or separating before GC analysis with FID and MS detection.

7. CE Analysis of Derivatized Amino Acids

Trace amounts of amino acids can be determined by CE [143]. Many types of polar and nonpolar molecules can be separated by CE, making it an alternative to HPLC and GC. CE has the advantage of short separation times and the use of small amounts of solvent, but is unpractical for preparative-scale separations and has relatively poor detection sensitivity [5]. Some CE methods require derivatization, while others separate underivatized amino acids. Many of the same derivatizing agents that are used in HPLC are also used in CE analysis.

In order to achieve an enantioselective separation in CE, a chiral agent must be added to the run buffer. Common chiral additives for the enantioseparation of amino acids are cyclodextrins and their derivatives [88,97,144-154], chiral surfactants [99,152,155-160], ligand-exchange compounds (metal complexes) [158,161], proteins [162-165], crown ethers [166,167], oligosaccharides [168], and macrocyclic antibiotics [169-172].

A popular technique for resolving amino acid enantiomers in CE is micellar capillary electrophoresis (MCE) [97,99,153,173,174]. The addition of surfactants creates micelles which act as a pseudophase and can incorporate nonpolar molecules. This approach was "borrowed" from earlier LC methods [175,176]. Chiral surfactants can be used for enantiomeric separations. Additionally, chiral selectors such as cyclodextrins can be added to the run buffer [97,153,154]. The presence of surfactant helps to increase the solubility of some cyclodextrins [5].

D-Asp and D-Glu were determined in beer and yogurt after derivatization with OPA and TATG (2,3,4,6-tetra-*O*-acetyl-1-thio-β-D-glucopyranose) [99]. The tagged derivatives were separated by MCE with octylglucoside, a neutral surfactant. Sodium dodecyl sulfate (SDS) and β-cyclodextrins were used in the run buffer to separate amino acids found in rat vitreous perfusates [97]. The amino acids were derivatized with CBQCA (3-(4-carboxybenzoyl)-2-quinoline-carboxaldehyde). A MCE method with FMOC derivatized amino acids and β-cyclodextrins confirmed the absence of D-amino acids in a sport nutritional supplement containing branched amino acids [153].

D-Asp derivatized with NBD-F was separated by the addition of modified cyclodextrins to the run buffer without the addition of surfactant (Fig. 4) [154]. DNS-Cl and FITC (fluorescein isothiocyanate) were used as derivatizing agents to determine D-amino acid content in orange juice as a method to detect product adulteration. The separations were carried out with cyclodextrins as run buffer additives, but MS was used for detection [88].

8. Conclusion

The determination of D-amino acids in foods and beverages is important to detect adulteration, analyze flavor components, and investigate handling and storage processes. Diet exposes mammals to the D-amino acids present in fermented foods and products that are exposed to extremes in heat or pH. Further studies may indicate what type of role ingested D-amino acids play in physiological function. The determination of

D-amino acids in biological samples may also lead to more information on the function of naturally occurring enantiomers.

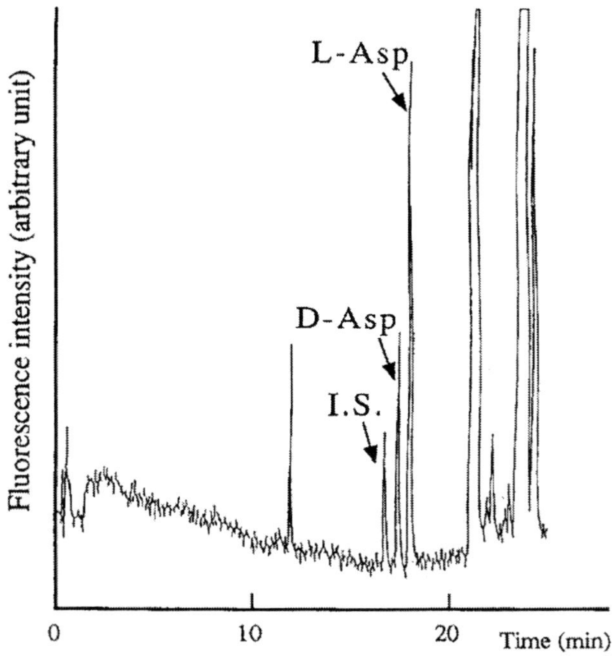

Figure 4. Electropherogram obtained from rat pineal gland (male Sprague-Dawley rat, 7 weeks old) after derivatization with NBD-F. Conditions: running buffer, 30mM heptakis(2,3,6-tri-*O*-methyl)-β-cyclodextrin in 50 mM phosphate buffer (pH 4.0). The internal standard (I.S.) is N-methyl- D-aspartate. Reproduced from Ref. 152 with permission.

The most widely used HPLC method for determining D-amino acids in complex samples is with derivatization by achiral agents and column-switching. The derivatization method offers the most sensitivity available with UV or fluorescence detection. Column-switching can enhance sensitivity as well since the samples are essentially "cleaned up" in the process. HPLC-MS with the use macrocyclic glycopeptide CSPs will become more widespread because the amino acids do not need to be derivatized.

GC methods for D-amino acid determination always involve a two-step derivatization to volatilize the amino acids. Chirasil-L-Val is the most widely used GC chiral stationary phase. The use of mass spectrometric detection may become more common because it can reduce interference depending on the derivatization scheme.

Capillary electrophoretic determinations of D-amino acids are dominated by micellar CE with the addition of cyclodextrins as chiral selectors. The amino acids are almost always derivatized to improve sensitivity. As methods to couple CE with MS improve, there is a chance that CE-MS may gain in popularity for D-amino acid analysis.

References

[1] Kawamura, Y. (1987) Y. Kawamura & M.R. Kare (Eds.),*Umami: A Basic Taste: Biochemistry, Nutrition, Food Science.* (p. v, vi) New York: Marcel Dekker.

[2] Teranishi, R. (1989) In R. Teranishi, R.G. Buttery & F. Schahidi (Eds.) *Flavor Chemistry—Trends and Developments,* (pp 1-6). ACS Symposium Series 388. Washington, DC: Amercian Chemical Society.

[3] Kirimura, J., Shimizu, A., Kimizuka, A., Ninomiya, T. & Katsuya, N. (1969) Contribution of peptides and amino acids to the taste of foods. *J Agric Food Chem*, 17, 689-695.

[4] Solms, J. (1969) Taste of amino acids and proteins. *J Agric Food Chem*, 17, 684-688.

[5] Ekborg-Ott, K.H. & Armstrong, D.W. (1997) Stereochemical Analyses of Food Components. In S. Ahuja (Ed.), *Chiral Separations: Applications and Technology*, (pp. 201-270). Washington, D.C.: American Chemical Society.

[6] Wieser, H., Jugel & H., Belitz, H.-D. (1977) *Z Lebensm Unters Forsch, 164,* 277-282.

[7] Armstrong, D.W., Chang, C.D. & Li, W.L. (1990) Relevance of enantiomeric separations in food and beverage analyses. *J Agric Food Chem, 38,* 1674-1677.

[8] Eriksson, C.E. (1978) Flavor modification. In G. Charalambous & G.E. Inglett (Eds.), *Flavor of Foods and Beverages,* (pp. 1-14). Orlando, FL: Academic.

[9] Liardon, R. & Hurrell, R.F. (1983) Amino acid racemization in heated and alkali-treated proteins. *J Agric Food Chem, 31,* 432-437.

[10] Bunjapamai, S., Mahoney, R.R. & Fagerson, I.S. (1982) Determination of D-amino acids in some processed foods and effect of racemization on in vitro digestibility of casein. *J Food Sci, 47,* 1229-1234.

[11] Friedman, M., Zahnley, J.C. & Masters, P.M. (1981) Relationship between in vitro digestibility of casein and its content of lysoalanine and D-amino acids. *J Food Sci, 46,* 127-131, 134.

[12] Hayashi, R. & Kameda, I. (1980) Racemization of amino acid residues during alkali-treatment of protein and its adverse effect of pepsin digestibility. *Agric Biol Chem, 44,* 891-895.

[13] Friedman, M. (1991) Formation, nutritional value, and safety of D-amino acids. In M. Friedman (Ed.) *Nutritional and Toxicological Consequences of Food Processing,* (pp 447-481). New York: Plenum.

[14] Bruckner, H. & Hausch, M. (1990) In G. Lubec & G.A. Rosenthal (Eds.) *Amino Acids: Chemistry, Biology and Medicine,* (pp. 1172-1182). Leiden, Netherlands: ESCOM Scientific.

[15] Palla, G., Marchelli, R., Dossena, A. & Casnati, G. (1989). Occurrence of D-amino acids in food: Detection by capillary gas chromatography and by reversed-phase high performance liquid chromatography with L-phenylalaninamides as chiral selectors. *J Chromatogr A, 475,* 45-53.

[16] Gandolfi, I., Palla, G., Delprato, L., De Nisco, F., Marchelli, R. & Salvadori, C. (1992) D-Amino acids in milk as related to heat treatments and bacterial activity. *J Food Sci, 57,* 377-379.

[17] Bruckner, H. & Hausch, M. (1989) Gas chromatographic detection of D-amino acids as common constituents of fermented foods. *Chromatographia, 28,* 487-492.

[18] Preston, R.L. (1987) Occurrence of D-amino acids in higher organisms: a survey of the distribution of D-amino acids in marine invertebrates. *Comp Biochem Physiol, 87B,* 55-62.

[19] Corrigan, J.J. (1969) D-Amino acids in animals. *Science, 164,* 142-149.

[20] D'Aniello, A. & Giuditta, A. (1978) Presence of D-aspartate in squid axplasm and in other regions of the cephalopod nervous system. *J Neurochem, 31,* 1107-1108.

[21] Felbeck, H. (1985) Occurrence and metabolism of D-aspartate in the gutless bivalve solemya reidi. *J Exp Zool, 234,* I145-149.

[22] Matsushima, O., Katayama, H., Yamada, K. & Kado, Y. (1984) Occurrence of free D-alanine and alanine racemase activity in bivalve mollusks with special reference to intracellular osmoregulation. *Mar Biol Lett, 5,* 217-225.

[23] Whitaker, J.R. & Feeney, R.E. (1983) Chemical and physical modification of proteins by the hydroxide ion. *CRC Crit Rev Food Sci Nutr, 19,* 173-212.

[24] Masters, P.M & Friedman, M. (1979) Racemization of amino acids in alkali-treated food proteins. *J Agric Food Chem, 27,* 507-511.

[25] Friedman, M. & Liardon, R. (1985) Racemization kinetics of amino acid residues in alkali-treated soybean protein. *J Agric Food Chem, 33,* 666-672.

[26] Blackburn, S. (1978) Sample preparation and hydrolytic methods [for amino acids]. In S. Blackburn (Ed.), *Amino Acid Determinations: Methods and Techniques* (2nd edition), (pp. 7-37). New York: Marcel Dekker.
[27] Liardon, R. & Ledermann, S. (1986) Racemization kinetics of free and protein-bound amino acids under moderate alkaline treatment. *J Agric Food Chem, 34,* 557-565.
[28] Friedman, M., Gumbmann, M.R. & Masters, P.M. (1984) Protein-alkali reactions: chemistry, toxicology, and nutritional consequences. In M. Friedman (Ed.) *Nutritional and Toxicological Aspects of Food Safety,* Vol. 177, Chapter 18, (pp. 367-412). New York: Plenum.
[29] Tovar, L.R. & Schwass, D.E. (1983) In *D-Amino Acids in Processed Proteins: Their Nutritional Consequences,* ACS Symposium Series 234, (pp. 169-185). Washington, DC: American Chemical Society.
[30] Man, E.H. & Bada, J.L. (1987) Dietary D-amino acids. *Ann Rev Nutr, 7,* 209-225.
[31] Calendar, R. & Berg, P. (1967) D-tyrosylRNA: Formation, hydrolysis and utilization for protein synthesis. *J Mol Biol, 26,* 39-54.
[32] Cavallini, D., DeMarco, D. & Mondovi, B. (1958) Experiments with D-cysteine in the rat. *J Biol Chem, 230,* 25-30.
[33] Hamase, K., Morikawa, A. & Zaitsu, K. (2002) D-Amino acids in animals and their diagnostic value. *J Chromatogr B, 781,* 73-91.
[34] Dunlop, D.S. & Niedle, A. (1997) The origin and turnover of D-serine in brain, *Biochem Biophys Res Commun, 246,* 26-30.
[35] Wolosker, H., Sheth, K.N., Takahashi, M., Mothet, J.P., Brady Jr., R.O., Ferris, C.D. & Snyder, S.H. (1999) Purification of serine racemase: biosynthesis of the neuromodulator D-serine. *Proc Natl Acad Sci USA, 96,* 721-
[36] Wolosker, H., Blackshaw, S. & Snyder, S.H. (1999) Serine racemase: a glial enzyme synthesizing D-serine to regulate glutamate-N-methyl- D-aspartate neurotransmission. *Proc Natl Acad Sci USA, 96,* 13409-
[37] Miranda, J.D., Santoro, A., Engelender, S. & Wolosker, H. (2000) Human serine racemase: molecular cloning, genomic organization and functional analysis. *Gene, 256,* 183-188.
[38] Berg, C.P. (1959) Utilization of D-amino acids. In A.A. Albanese (Ed.) *Protein and Amino Acid Nutrition,* (pp. 57-96). Orlando, FL: Academic.
[39] Neuberger, A. (1948) The metabolism of D-amino acids in mammals. *Biochem Soc Symp, 1,* 20-32.
[40] Armstrong, D.W., Gasper, M.P., Lee, S.H., Ercal, N., & Zukowski, J. (1993). Factors controlling the level and determination of D-amino acids in the urine and plasma of laboratory rodents. *Amino Acids, 5,* 299-315.
[41] Friedman, M. & Gumbmann, M.R. (1984) The nutritive value and safety of D-phenylalanine and D-tyrosine in mice. *J Nutr, 114,* 2089-2096.
[42] Friedman, M. & Gumbmann, M.R. (1984) The utilization an safety of isomeric sulfur-containing amino acids in mice. *J Nutr, 114,* 2301-2310.
[43] Kies, C., Fox, H. & Aprahamian, S. (1975) Comparative value of L-, DL -, and D-methionine supplementation of an oat-based diet for humans. *J Nutr, 105,* 809-814.
[44] Stegink, L.D., Bell, E.F., Filer Jr, L.J., Ziegler, E.E., Andersen, D.W. & Seligson, F.H. (1986) Effects of equimolar doses of L-methionine, D-methinonine and L-methionine-dl-sulfoxide on plasma and urinary amino acid levels in normal adult humans. *J Nutr, 116,* 1185-1192.
[45] Gubitz, G. & Jellenz, W. (1981) Separation of the optical isomers of amino acids by ligand exchange chromatography using chemically bonded chiral phases. *J Chromatogr, 203,* 377-384.
[46] Charmot, D., Audebert, R. & Quivoron, C. (1985) High efficiency ligand exchange chromatography of amino acid enantiomers. *J Liq Chromatogr, 8,* 1769-1781.

[47] Davankov, V.A., Bochkov, A.S. & Belov, Yu. P. (1981) Ligand exchange chromatography of racemates: XV. Resolution of alpha-amino acids on reversed-phase silica gels coated with n-decyl-L-histidine. *J Chromatogr, 218,* 547-557.

[48] Davankov, V.A. & Zolotarev, Yu. A. (1978) Ligand exchange chromatography of racemates: VII. Separation of optical isomers of amino acids on a polystyrene resin containing L-allo-hydroxyproline as the fixed ligand. *J Chromatogr, 155,* 303-310.

[49] Gubitz, G., Juffman, F. & Jellenz, W. (1982) Direct separation of amino enantiomers by high performance ligand exchange chromatography on chemically bonded chiral phases. *Chromatographia, 16,* 103-106.

[50] Kiniwa, H., Baba, Y., Ishida, T. & Katoh, H. (1989) General evaluation and application to trace analysis of a chiral column for ligand exchange chromatography. *J Chromatogr, 461, I397-405.*

[51] Galushko, S.V., Shishkina, I.P. & Soloshonok, V.A. (1992) High-performance ligand exchange chromatography of some amino acids containing two chiral centers. *J Chromatogr, 592,* 345-348.

[52] Hilton, M.L. & Armstrong, D.W. (1991) Evaluation of the enantiomeric separation of dipeptides using a chiral crown ether liquid chromatography column. *J Liq Chromatogr, 14,* 9-28.

[53] Berthod, A., Liu, Y., Bagwill, C. & Armstrong, D.W. (1996) Facile liquid chromatographic enantioresolution of native amino acids and peptides using a teicoplanin chiral stationary phase. *J Chromatogr A, 731,* 123-137.

[54] Ekborg-Ott, K., Liu, Y. & Armstrong, D.W. (1998) Highly enantioselective HPLC separations using the covalently bonded macrocyclic antibiotic, ristocetin A, chiral stationary phase. *Chirality, 10,* 434-483.

[55] Berthod, A., Yu, T., Kullman, J.P., et al (2000) Evaluation of the macrocyclic glycopeptide A-40,926 as a high-performance liquid chromatographic chiral selector and comparison with teicoplanin chiral stationary phase. *J Chromatogr A, 897,* 113-129.

[56] Berthod, A., Chen, X. & Kullman, J.P., et al. (2000) Role of the carbohydrate moieties in chiral recognition on teicoplanin-based LC stationary phases. *Anal Chem, 72,* 1767-1780.

[57] Tesarova, E., Bosakova, A. & Pacakova, V. (1999) Comparison of enantioselective separation of N-tert-butyloxycarbonyl amino acids and their non-blocked analogues on teicoplanin-based chiral stationary phase. *J Chromatogr A, 838,* 121-129.

[58] Risley, D.S. & Strege, M.A. (2000) Chiral separations of polar compounds by hydrophilic interaction chromatography with evaporative light scattering detection. *Anal Chem, 72,* 1736-1739.

[59] Petritis, K., Valleix, A., Elfakir, C. & Dreux, M. (2001) Simulataneous analysis of underivatized chiral amino acids by liquid chromatography-ionspray tandem mass spectrometry using a teicoplanin chiral stationary phase. *J Chromatogr A, 913,* 331-340.

[60] Jandera, P., Backovska, V. & Felinger, A. (2001) Analysis of the band profiles of the enantiomers of phenylglycine in liquid chromatography on bonded teicoplanin columns using the stochastic theory of chromatography. *J Chromatogr A, 919,* 66-77.

[61] Jandera, P., Skavrada, M., Klemmova, K., Backovska, V. & Guichon, G. (2001) Effect of the mobile phase on the retention behaviour of optical isomers of carboxylic acids and amino acids in liquid chromatography on bonded teicoplanin columns. *J Chromatogr A, 917,* 123-133.

[62] Berthod, A., Valleix, A., Tizon, V.L., Estelle, X.X., Caussignac, C. & Armstrong, D.W. (2001) Retention and selectivity of teicoplanin stationary phases after copper complexation and isotopic exchange. *Anal Chem, 73,* 5499-5508.

[63] Armstrong, D.W., Yang, X., Han, S.M.& Menges, R.A. (1987) Direct liquid chromatographic separation of racemates with an alpha-cyclodextrin bonded phase. *Anal Chem, 59,* 2594-2596.

[64] Davankov, V.A. & Rogozhin, S.V. (1971) The chromatography and electrophoresis of the perbromate ion. *J Chromatogr, 60,* 280-283.

[65] Davankov, V.A., Kurganov, A.A. & Bochkov, A.S. (1983) Resolution of racemates by high-performance liquid chromatography. In J.C. Giddings, E.Grushka, J. Cazes, & P.R. Brown (Eds.) *Advances in Chromatography,* vol. 22, (p. 71). New York: Marcel Dekker.
[66] Armstrong, D.W. (1997) The evolution of chiral stationary phases for liquid chromatography. *LC-GC, 59 (Supplemental Issue),* S20-S28.
[67] Helgeson, R., Timko, J., Moreau, P., Peacock, S., Mayer, J. & Cram, D.J. (1974) Models for chiral recognition in molecular complexation. *J Am Chem Soc, 96,* 6762-6763.
[68] Sogah, G.D.Y & Cram, D.J. (1976) Total chromatographic optical resolution of alpha-amino acids and ester salts through chiral recognition by a host covalently bound to polystyrene resin. *J Am Chem Soc, 98,* 3038-3041.
[69] Newcomb, M., Toner, J., Helgeson, R. & Cram, D.J. (1979) Host-guest complexation. 20. Chiral recognition in transport as a molecular basis for a catalytic resolving machine. *J Am Chem Soc, 101,* 4941-4947.
[70] Armstrong, D.W., Tang, Y., Chen, S., Zhou, Y., Bagwill, C. & Chen, J.-R. (1994) Macrocyclic antibiotics as a new class of chiral selectors for LC. *Anal Chem, 66,* 1473-1484.
[71] Chen, S., Liu, Y., Armstrong, D.W., Borrell, J.I., Matrinez-Teipel, B. & J.L. Matallana. (1995) Enantioresolution of substituted 2-methoxy-6-oxo-1,4,5,6-tetrahydropyridine-3-carbon on macrocyclic antibiotic and cyclodextrin stationary phases. *J Liq Chromatogr, 18,* 1495-1507.
[72] Armstrong, D.W., Liu, Y., & Ekborg-Ott, K.H. (1995) A covalently bonded teicoplanin chiral stationary phase for high performance liquid chromatography enantioseparations. *Chirality, 7,* 474-497.
[73] Berthod, A., Chen, X., Kullman, J. P., Armstrong, D. W., Gasparrini, F., D'Acquarica, I., Villani, C. & Carotti, A. (2000) Role of the Carbohydrate Moieties in Chiral Recognition on Teicoplanin-Based LC Stationary Phases. *Anal Chem, 72*, 1767-1780.
[74] Desai, M.J. & Armstrong, D.W. (2004) Analysis of derivatized and underivatized theanine enantiomers by high-peformance liquid chromatography/atmospheric pressure ionization-mass spectrometry. *Rapid Commun Mass Spectrom, 18,* 251-256.
[75] Desai, M.J., Gill, M.S., Hsu, W.H. & Armstrong, D.W. (2005) Pharmokinetics of theanine enantiomers in rats. *Chirality, 17,* 154-162.
[76] Desai, M.J. & Armstrong, D.W. (2004) Analysis of native amino acid and peptide enantiomers by high performance liquid chromatography/atmospheric pressure chemical ionization mass spectrometry. *J Mass Spectrom, 39,* 177-187.
[77] Jegerov, A., Triska, J., Trnka, T. & Cerny, M. (1988) Separation of alpha-amino acid enantiomers by reversed-phase high performance liquid chromatography after derivitization with o-phthaldialdehyde and a sodium salt of 1-thio-beta-D-glucose. *J Chromatogr, 434,* 417-
[78] Schuster, R. (1988) Determination of amino acids in biological, pharmaceutical, plant and food samples by automated precolumn derivatization and high-performance liquid chromatography. *J Chromatogr, 431,* 271-
[79] Stocchi, V., Piccoli, G., Magnani, M., Palma, F., Biagiarelli, B. & Cucchiarini, L. (1989) Reversed phase high-performance liquid chromatography separation of dimethylaminoazobenzen sulfonyl- and dimethylaminoazobenzene thiohydantoin- amino acid derivatives for amino acid analysis and microsequencing studies at the picomole level. *Anal Biochem, 178,* 107
[80] Praminik, B.C., Moomaw, C.R., Evans, C.T., Cohen, S.A. & Slaughter, C.A. (1989) Identification of phenylthiocarbamyl amino acids by high-performance liquid chromatography/mass spectrometry. *Anal Biochem, 176,* 269-
[81] Sherwood, R.A., Titherage, A.C. & Richards, D.A. (1990) Measurement of plasma and urine amino acids by high-performance liquid chromatography with electrochemical detection using phenylisothiocynate dervitization. *J Chromatogr, 528,* 293-

[82] Pronce, T. & Tilquin, B. (1996) Trace analysis in chiral separation of selected amino enantiomers. *J Pharm Biomed Anal, 14,* 1175.

[83] Tcherkas, Y.V. & Denisenko, A.D. (2001) Simultaneous determination of several amino acids, including homocysteine, cysteine and glutamic acid, in human plasma by isocratic reversed-phase high-performance liquid chromatography with fluorimetric detection.

[84] Zukowski, J., Pawlowska, M. & Armstrong, D.W. (1992) Efficient enantioselective separation and determination of trace impurities in secondary amino acids (i.e. imino acids). *J Chromatogr, 33,* 33-41.

[85] Zukowski, J., Pawlowska, M., Nazatkina, M. & Armstrong, D.W. (1993) High-performance liquid chromatography enantioseparation of glycyl di- and tri-peptides on native cyclodextrin bonded phases: mechanistic considerations. *J Chromatogr, 629,* 169-179.

[86] Pawlowska, M., Chen, S., & Armstrong, D.W. (1993) Enantiomeric separation of fluorescent, 6-aminoquinolyl-N-hydroxysuccinimidyl carbamate, tagged amino acids. *J Chromatogr, 641,* 257-265.

[87] Rizzi, A.M., Cladrowa-Runge, S., Jonsson, H. & Osla, S. (1995) Enantiomeric resolution of derivatized DL-amino acids by high-performance liquid chromatography using a β-cyclodextrin chiral stationary phase: A comparison between derivatization labels. *J Chromatogr A, 710,* 287-295.

[88] Simo, C., Rizzi, A., Barbas, C. & Cifuentes, A. (2005) Chiral capillary electrophoresis-mass spectrometry of amino acids in foods. *Electrophoresis, 26,* 1432-1441.

[89] Zhao, H., Hamase, K., Morikawa, A., Qiu, Z. & Zaitsu, K. (2004) Determination of D- and L-enantiomers of threonine and allo-threonine in mammals using two-step high-performance liquid chromatography. *J Chromatogr B, 810,* 245-250.

[90] Hamase, K., Inoue, T., Morikawa, A., Konno, R. & Zaitsu, K. (2001) Determination of free D-proline and D-leucine in the brains of mutant mice lacking D-amino acid oxidase activity. *Anal Biochem, 298,* 253-258.

[91] Morikawa, A., Hamase, K. & Zaitsu, K. (2003) Determination of D-alanine in the rat central nervous system and periphery using column-switching high-performance liquid chromatography. *Anal Biochem, 312,* 66-72.

[92] Hamase, K., Homma, H., Takigawa, Y., Fukushima, T., Santa, T. & Imai, K. (1997) Regional distribution and postnatal changes of D-amino acids in rat brain. *Biochim Biophys Acta, 1334,* 214-222.

[93] Ekborg-Ott, K.H. & Armstrong, D.W. (1996) Evaluation of the concentration and enantiomeric purity of selected free amino acids in fermented malt beverages(beers). *Chirality, 8,* 49-57.

[94] Ekborg-Ott, K.H., Taylor, A. & Armstrong, D.W. (1997) Varietal differences in the total and enantiomeric composition of theanine in tea. *J Agric Food Chem, 45,* 353-363.

[95] Armstrong, D.W., Kullman, J.P., Chen, X. & Rowe, M. (2001) Composition and chirality of amino acids in aerosol/dust from laboratory and residential enclosures. *Chirality, 13,* 153-158.

[96] Stenberg, M., Marko-Varga, G. & Öste, R. (2002) Enantioseparation of D- and L-amino acids by a coupled system consisting of an ion-exchange column and a chiral column and determination of D-aspartic acid and D-glutatmic acid in soy products. *Food Chem, 79,* 507-512.

[97] Thongkhao-On, K., Kottegoda, S., Pulido, J. & Shippy, S. (2004) Determination of amino acids in rat vitreous perfusates by capillary electrophoresis. *Electrophoresis, 25,* 2978-2984.

[98] Erbe, T. & Brückner, H. (2000) Chromatographic determination of amino acid in beers and raw materials used for their manufacture. *J Chromatogr A, 881,* 81-91.

[99] Tivetsen, A., Lundqvist, A. & Folestad, S. (1997) Selective chiral determination of aspartic and glutamic acid in biological samples by capillary electrophoresis. *Chromatographia, 44,* 623-633.

[100] Neadle, D.J. & Pollitt, R.J. (1965) Formation of 1-dimethylaminonaphthalene-5-sulfonamide during the preparation of 1-dimethylaminonaphthalene-5-sulfonylamino acids. *J Biol Chem, 192,* 607-608.

[101] Roth, M. (1971) Fluorescence reaction for amino acids. Anal Chem, 43, 880-882.

[102] Joseph, M.H. & Davies, P. (1983) Electrochemical analysis of o-phthalaldehyde-mercaptoethanol derivatives of amino acids. Application to high-performance liquid chromatographic determination of amino acids in plasma and other biological materials. *J Chromatogr, 277,* 125-136.

[103] Nagata, Y., Fujiwara, T., Kawaguchi-Nagata, K., Fukimori, Y. & Yamanaka, T. (1998) Occurrence of peptidyl D-amino acids in soluble fractions of several eubacteria, archaea, and eukaryotes. *Biochim Biophys Acta, 1379,* 76-82.

[104] Jin, D., Miyahara, T., Oe, T. & Toyo'oka, T. (1999) Determination of D-amino acids labeled with fluorescent chiral reagents, R(-) and S(+)-4-(3-isothiocyanatopyrrolidin-1-yl)-7-(N,N-dimethylaminosulfonyl)-2,1,3-benzoxadiazoles, in biological and food samples by liquid chromatography. *Anal Biochem, 269,* 124-132.

[105] Toyo'oka, T., Jin, D. Tomoi, N., Oe, T. & Hiranuma, H. (2001) R(-)-4-(3-isothiocyanatopyrrolidin-1-yl)-7-(N,N-dimethylaminosulfonyl)-2,1,3-benzoxadiazole, a fluorescent chiral tagging reagent: sensitive resolution of chiral amines and amino acids by reversed-phase liquid chromatography. *Biomed Chromatogr, 15,* 56-67.

[106] Armstrong, D.W. & DeMond, W. (1984) Cyclodextrin bonded phases for the liquid chromatographic separation of optical, geometrical, and structural isomers. *J Chromatogr Sci, 22,* 411-415.

[107] Hinze, W.L., Riehl, T.E., Armstrong, D.W., DeMond, W., Alak, A. & Ward, T.J. (1985) Liquid chromatographic separation of enantiomers using a chiral beta-cyclodextrin bonded stationary phase and conventional aqueous-organic mobile phases. *Anal Chem, 57,* 237-242.

[108] Pawlowska, M., Chen, S. & Armstrong, D.W. (1993) Enantiomeric separation of fluorescent, 6-aminoquinolyl-N-hydroxysuccinimidyl, tagged amino acids. *J Chromatogr, 641,* 257-265.

[109] Pirkle, W.H, Pochapsky, T.C., Mahler, G.C., Corey, D.E., Reno, D.S. & Alessi, D.M. (1986) Useful and easily prepared chiral stationary phases for the direct chromatographic separation of the enantiomers of a variety of derivatized amines, amino acids, alcohols, and related compounds. *J Org Chem, 51,* 4991-5000.

[110] Kato, M., Fukushima, T., Santa, T., Imai, K., Nakashima, K. & Nishioka, R. (1998) Preparation and evaluation of new Pirkle type chiral stationary phases with long alkyl chains for the separation of amino acid enantiomers derivatized with NBD-F. *Analyst, 123,* 2877-2882.

[111] Kato, M., Fukushima, T., Shimba, N., Shimada, I., Kawakami, Y. & Imai, K. (2001) A study of chiral recognition for NBD-derivatives on a Pirkle-type chiral stationary phase. *Biomed Chromatogr, 15,* 227-234.

[112] Armstrong, D.W., Gasper, M., Lee, S.H., Zukowski, J. & Ercal, N. (1993) D-Amino acid levels in human physiological fluids. *Chirality, 5,* 375-378.

[113] Armstrong, D.W., Ward, T.J., Armstrong, R.D. & Beesley, R.E. (1986) Separation of drug stereoisomers by the formation of β -cyclodextrin inclusion complexes. *Science, 232,* 1132-1135.

[114] Chang, S.C., Wang, L.R. & Armstrong, D.W. (1992) Facile resolution of N-tert-butoxycarbonyl amino acids: the importance of enantiomeric purity in peptide synthesis. *J Liq Chromatogr, 15,* 1411-1429.

[115] Peter, A., Vekes, E. & Armstrong, D.W. (1994) Effects of temperature on retention of chiral compound on a ristocetin A chiral stationary phase. *J Chromatogr A, 958,* 89-107.

[116] Pawloska, M. & Armstrong, D.W. (1994) Evaluation of enantiomeric purity of selected amino acids in honey. *Chirality, 6,* 270-276.

[117] Armstrong, D.W., Zukowski, J., Ercal, N. & Gasper, M. (1993) High-performance liquid chromatographic enantioseparation of glycyl di- and tripeptides on native cyclodextrin bonded phases. Mechanistic considerations. *J Pharm Biomed Anal, 11,* 881-886.

[118] Stalcup, A.M., Faulkner Jr., J.R., Tang, Y., Armstrong, D.W., Levy, L.W. & Regalado, E. (1991) Determination of the enantiomeric purity of scopolamine isolated from plant extract using achiral/chiral coupled column chromatography. *Biomed Chromatogr, 5,* 3-7.

[119] Armstrong, D.W., Duncan, J.D. & Lee, S.H. (1991) Evaluation of D-amino acid levels in human urine and in commercial L-amino acid samples. *Amino Acids, 1,* 97-106.

[120] Kullman, J.P., Chen, X. & Armstrong, D.W. (1999) Evaluation of the enantiomeric composition of amino acids in tobacco. *Chirality, 11,* 669-673.

[121] Schurig, V. (1994) Enantiomer separation by gas chromatography on chiral stationary phases. *J Chromatogr A, 666,* 111-129.

[122] König, W.A.(1987) In Bertsch, W., Jennings, W.G. & Kaiser, R.E. (Eds.) *The Practice of Enantiomeric Separation by Capillary Gas Chromatography.* Heidelberg, Germany: Hüthig.

[123] König, W.A. (1984) In Schreier, P. (Ed.), *Analysis of Volatiles—Methods and Applications,* (pp. 77-91). Berlin, Germany: de Gruyter.

[124] König, W.A., Sievers, S. & Benecke, I. (1981) Enantiomer analysis by complexation gas chromatography : Scope, merits and limitations. In Kaiser, R.E. (Ed.) *Proceedings of the 4th International Symposium on Capillary Chromatography,* (pp. 703-725). Bad Dürkheim, Germany: Institute of Chromatography.

[125] Nokihara, K. & Gerhardt, J. (2001) Development of an improved automated gas-chromatographic chiral analysis system: application to non-natural amino acids and natural protein hydrosylates. *Chirality, 13,* 431-434.

[126] Frank, H., Nicholoson, G.J. & Bayer, E. (1977) Rapid gas chromatographic separation of amino acid enantiomers with a novel chiral stationary phase. *J Chromatogr Sci, 15,* 174-176.

[127] Schurig, V. (1988) Enantiomer analysis by complexation gas chromatography: Scope, merits and limitations. *J Chromatogr, 441,* 135-153.

[128] Frank, H., Woiwode, W., Nicholson, G. & Bayer, E. (1981) Determination of the rate of acid-catalyzed racemization of protein amino acids. *Liebigs Ann Chem, 3,* 354-365.

[129] Pätzold, R., Nieto-Rodriguez, A. & Brückner, H. (2003) Chiral gas chromatographic analysis of amino acids in fortified wines. *Chromatographia Supplement, 57,* S207-S212.

[130] Schieber, A. Brückner, H., Rupp-Classen, Specht, W., Nowitzki-Grimm, S. & Classen, H.G. (1997) Evaluation of D-amino acid levels in rat by gas chromatography-selected ion monitoring mass spectrometry: no evidence of subacute toxicitiy of orally fed D-proline and D-aspartic acid. *J Chromatogr B, 691,* 1-12.

[131] Armstrong, D.W., Li, W. & Pitha, J. (1990) Reversing enantioselectivity in capillary gas chromatography with polar and nonpolar cyclodextrin derivative phases. Anal Chem, 62, 214-217.

[132] Casal, S. Oliveria, M. & Ferreira, M. (2000) Gas chromatographic quantification of amino acid enantiomers in food matrices by their *N(O,S)*-ethoxycarbonyl heptafluorobutyl ester derivatives. *J Chromatogr A, 886,* 221-230.

[133] Schurig, V. (2001) Separation of enantiomers by gas chromatography. *J Chromatogr A, 906,* 275-299.

[134] Gil-Av, E., Feibush, B. & Charles-Sigler, R. (1966) Separation of enantiomers by gas liquid chromatography with an optically active stationary phase. *Tetrahedron Lett, 10,* 1009-1015.

[135] Frank, H., Nicholoson, G.J. & Bayer, E. (1978) Gas chromatographic--mass spectrometric analysis of optically active metabolites and drugs on a novel chiral stationary phase. *J Chromatogr, 146,* 197-206.

[136] Frank, H., Nicholoson, G.J. & Bayer, E. (1978) Chiral polysiloxanes for resolution of optical antipodes. *Angew Chem Int Ed, 17,* 363-365.

[137] Schurig, V. & Nowotny, H.-P. (1990) Gas chromatographic separation of enantiomers on cyclodextrin derivatives. *Angew Chem Int Ed, 29,* 939-957.

[138] Keim, W., Köhnes, A. Meltzow, W. & Römer, H. (1991) Enantiomer separation by gas chromatography on cyclodextrin chiral stationary phases. *J High Resolut Chromatogr, 14,* 507-529.

[139] König, W.A. (1992) *Enantioselective Gas Chromatography with Modified Cyclodextrins.* Heidelberg, Germany: Hüthig.

[140] Snopek, J., Smolková-Keulemansová, E., Cserháti, Gahm, K. & Stalcup, A. (1996) Cyclodextrins in analytical separation methods. In J. Szejtli & T.Osa (Eds.), *Comprehensive Supramolecular Chemistry, Vol 3, Chapter 18,* (pp. 515-571).
[141] König, W. (1993) Enantioselective gas chromatography. *Trends Anal Chem, 12,* 130-137.
[142] Pätzold, R., Schieber, A. & Brückner, H. (2005) Gas chromatographic quantification of free D-amino acids in higher vertebrates. *Biomed Chromatogr, 19,* 466-473.
[143] Bergman, T. & Joernvll, H. (1992) Capillary electrophoresis for preparation of peptides and direct determination of amino acids. In R.H. Angeletti (Ed.), *Techniques in Protein Chemistry 3,* (pp. 129-134). Orlando, FL: Academic.
[144] Armstrong, D.W., Tang, Y., Ward, T. & Nichols, M. (1993) Derivatized cyclodextrins immobilized on fused-silica capillaries for enantiomeric separations via capillary electrophoresis, gas chromatography, or supercritical fluid chromatography. *Anal Chem, 65,* 1114-1117.
[145] Nowotny, H.-P., Schmalzing, D., Wistuba, D. & Schurig, V. (1989) Extending the scope of enantiomer separation on diluted methylated β -cyclodextrin derivatives by high-resolution gas chromatography. *J High Resolut Chromatogr, 12,* 383-393.
[146] Werner, A., Nassauer, T., Kiechle, P. & Erin, F. (1994) Chiral separation by capillary zone electrophoresis of an optically active drug and amino acids by host—guest complexation with cyclodextrins. *J Chromatogr A, 666,* 375-379.
[147] Fanali, S. (1991) Use of cyclodextrins in capillary zone electrophoresis: Resolution of terbutaline and propranolol enantiomers *J Chromatogr, 545,* 437-444.
[148] Shibukawa, A., Lloyd, D.K & Wainer, I.W. (1993) Simultaneous chiral separation of leucovorin and its major metabolite 5-methyltetrahydrofolate by capillary electrophoresis using cyclodextrins as chiral selectors: estimation of the formation constant and mobility of the solute-cyclodextrin complexes. *Chromatographia, 35,* 419-429.
[149] Snopek, J., Soini, H., Novotny, M., Smolková-Keulemansová, E. & Jelinek, I. (1991) Selected applications of cyclodextrin selectors in capillary electrophoresis. *J Chromatogr, 559,* 215-222.
[150] Guttman, A., Paulus, A. Cohen, A.S., Grinberg, N. & Karger, B.L. (1988) Use of complexing agents for selective separation in high-performance capillary electrophoresis : Chiral resolution via cyclodextrins incorporated within polyacrylamide gel columns. *J Chromatogr, 448,* 41-53.
[151] Ward, T.J., Nichols, M., Sturdivant, L. & King, C.C. (1995) Use of organic modifiers to enhance chiral selectivity in capillary electrophoresis. *Amino Acids, 8,* 337-344.
[152] Otsuka, K. & Terabe, S. (1990) Enantiomeric resolution by micellar electrokinetic chromatography with chiral surfactants. *J Chromatogr, 515,* 221-226.
[153] Boniglia, C., Carratu, B. & Sanzini, E. (2002) Enantiomer separation of D- L branched amino acids by capillary electrophoresis in sport nutritional supplements. *J Food Sci, 67,* 1352-1355.
[154] Tsunoda, M., Kato, M., Fukushima, T., Santa, T., Homma, H., Yanai, H., Soga, T. & Imai, K. (1999) Determination of aspartic acid enantiomers in bio-samples by capillary electrophoresis. *Biomed Chromatogr, 13,* 335-339.
[155] Fanali, S. (1989) Separation of optical isomers by capillary zone electrophoresis based on host-guest complexation with cyclodextrins. *J Chromatogr, 474,* 441-446.
[156] Bushey, M.M. & Jorgenson, J.W. (1989) Separation of dansylated methylamine and dansylated methyl-d3-amine by micellar electrokinetic capillary chromatography with methanol-modified mobile phase. *Anal Chem, 61,* 491-493.
[157] Cole, R.O., Sepaniak, M.J. & Hinze, W.L. (1990) Optimization of binaphthyl enantiomer separations by capillary zone electrophoresis using mobile phases containing bile salts and organic solvent. *J High Resolut Chromatogr, 13,* 579-582.
[158] Snopek, J., Jelinek, I. & Smolková-Keulemansová, E. (1988) Micellar, Inclusion and metal-complex enantioselective pseudophases in high-performance electromigration methods. *J Chromatogr, 452,* 571-590.

[159] Terabe, S., Shibata, M. & Miyashita, Y. (1989) Chiral separation by electronkinetic chromatography while bile salt micelles. *J Chromatogr, 480,* 403-411.

[160] Dobashi, A., Ono, T., Hara, S. & Yamaguchi, J. (1989) Optical resolution of enantiomers with chiral mixed micelles by electrokinetic chromatography. *Anal Chem, 61,* 1984-1986.

[161] Gassmann, E., Kuo, J.E. & Zare, R.N. (1985) Electrokinetic Separation of Chiral Compounds. *Science, 230,* 813-814.

[162] Barker, G.E., Russo, P. & Hartwick, R.A. (1992) Chiral separation of leucovorin with bovine serum albumin using affinity capillary electrophoresis. *Anal Chem, 64,* 3024-3028.

[163] Sun, P., Wu, N., Barker, G. & Hartwick, R.A. (1993) Chiral separations using dextran and bovine serum albumin as run buffer additives in affinity capillary electrophoresis. *J Chromatogr, 648,* 475-480.

[164] Sun, P., Barker, G.E., Hartwick, R.A., Grinberg, N. & Kaliszan, R. (1993) Chiral separations using an immobilized protein—dextran polymer network in affinity capillary electrophoresis. *J Chromatogr, 652,* 247-252.

[165] Voltcheva, L., Mohammad, J., Pettersson, G. & Hjerten, S. (1993) Chiral separation of β-blockers by high-performance capillary electrophoresis based on non-immobilized cellulose as enantioselective protein. *J Chromatogr, 638,* 263-267.

[166] Kuhn, R., Erni, F., Bereuter, T. & Hausler, J. (1992) Chiral recognition and enantiomeric resolution based on host-guest complexation with crown ethers in capillary zone electrophoresis. *Anal Chem, 638,* 263-267.

[167] Kuhn, R., Stoecklin, R. & Erni, F. (1992) Chiral separations by host-guest complexation with cyclodextrin and crown ether in capillary zone electrophoresis *Chromatographia, 33,* 32-36.

[168] D'Hulst, A. & Verbeke, N. (1992) Chiral separation by capillary electrophoresis with oligosaccharides. *J Chromatogr, 608,* 275-287.

[169] Armstrong, D.W., Rundlett, K.L, Chen, J.-R. (1994) Evaluation of the macrocyclic antibiotic vancomycin as a chiral selector for capillary electrophoresis. *Chirality, 6,* 496-509.

[170] Armstrong, D.W., Gasper, M.P. & Rundlett, K.L. (1994) Highly enantioselective capillary electrophoretic separations with dilute solutions of the macrocyclic antibiotic ristocetin A. *J Chromatogr A, 689,* 285-304.

[171] Rundlett, K.L., Gasper, M.P., Zhou, E. & Armstrong, D.W. (1996) Capillary electrophoretic enantiomeric separations using the glycopeptide antibiotic, teicoplanin. *Chirality, 8,* 88-107.

[172] Armstrong, D.W., Rundlett, K. & Reid III, G.L. (1994) Use of a macrocyclic antibiotic, Rifamycin B, and indirect detection for the resolution of racemic amino alcohols by CE. *Anal Chem, 66,* 1690-1695.

[173] Terabe, S. (2004) Micellar Electrokinetic Chromatography. *Anal Chem A-Pages, 76,* 240A-246A.

[174] Otsuka, K. & Terabe, S. (2004) Chiral micellar electrokinetic chromatography. In G. Gublitz & M.G. Schmid., *Methods in Molecular Biology, vol. 243,* (pp. 355-363). Totowa, NJ: Humana Press, Inc.

[175] Armstrong, D.W. & Nome, F. (1981) Partitioning behavior of solutes eluted with micellar mobile phases in liquid chromatography. *Anal Chem, 53,* 1662-1666.

[176] Armstrong, Daniel W. (1980) Pseudophase liquid chromatography: applications to TLC. *J Liq Chromatogr, 3,* 895-900.

In: D-Amino Acids: ISBN 1-60021-075-9
A New Frontier in Amino Acid and Protein Research, pp.337-352

Chapter 4.4

GAS CHROMATOGRAPHIC DETERMINATION OF D-AMINO ACIDS IN FOOD AND BEVERAGES

Ralf Pätzold and Hans Brückner*
Department of Food Sciences, Interdisciplinary Research Center, University of Giessen, Germany.

Keywords: derivatization protocols, food stuffs (plant juices, honey, cacao, cheese), authenticity and quality control, Chirasil-Val, racemization mechanism, Maillard reaction

1. INTRODUCTION

In the past formation and analysis of D-amino acids in proteins or foodstuffs treated with alkali or subjected to otherwise harsh conditions attracted priority attention [1-3]. In the last two decades it was fully realized, however, that bacterial fermented foods and beverages are major sources of D-amino acid uptake with the diet [4-10]. Free D-amino acids have also been detected as natural constituents of plants, algae, bivalves, crustaceae and wild or domesticated animals [11-16]. Consequently, food of plant or animal origin, are additional sources of dietary D-amino acid uptake. Furthermore, we have recently shown that D-amino acids are formed on heating saccharides with L-amino acids, a reaction referred to as the Maillard reaction. It was postulated that D-amino acids are formed on decay of relatively stable intermediates of this reaction named the Amadori compounds [17-19].

For the analysis of D-amino acids in foodstuffs preferably high performance liquid chromatographic (HPLC) and gas chromatographic (GC) methods are used. For HPLC analysis precolumn derivatization of amino acids with Marfey's reagent (2,4-dinitrofluorobenzene-L-alanine amide) [20] and UV detection of the diastereoisomers formed were used for the investigation of proline isomers in wine [21]. For analyzing of a broad range of foodstuffs (cheese, honey, yeast spread) and beverages (wine, beer, fruit juices) derivatization of amino acids with o-phthaldialdehyde together N-isobutyryl-L-cysteine and resolution of the resulting fluorescent diastereomeric isoindol derivatives on ODS-stationary phases were used [12,22,23]. Ligand exchange chromatography and use of L-phenylalaninamide together with copper (II) acetate as chiral additive were employed for the HPLC analysis of food D-amino acids [8]. Capillary electrophoresis was used for the determination of DL-Asp in white and red wine [24].

* Correspondence concerning this article should be addressed to Hans Brückner, E-mail: Hans.Brueckner@ernaehrung.uni-giessen.de, FAX:+49/641 99 39149.

The most frequently used chiral selector for gas chromatographic chiral amino acid analysis is Chirasil®-L-Val representing dimethylpolysiloxane functionalized with L-valine tert. butylamide. This is attributed to the thermal stability and commercial availability of this stationary phase [25-27]. The resulting fused silica capillary column was and still is intensively used for the investigation of fermented and nonfermented foods as well as alcoholic and nonalcoholic fermented beverages such as beer, wine, vinegars, fruit juices and plant products [5-7,11,28]. For GC analysis conversion of non-volatile amino acids into volatile derivatives is required. Trifluoroacyl or pentafluoropropionyl amino acid alkyl esters are usually employed. Recently the separation of *N(O,S)*-ethoxycarbonyl heptafluorobutyl esters of food DL-amino acids on Chirasil®-L-Val was reported [29].

Other commercially available chiral columns for GC such as functionalized γ-cyclodextrin (Lipodex® E) have also been used for food analysis, albeit much less frequently [7,10]. Finally it should be mentioned that laboratory designed trioxaundecanoyl tetramide (Phe-3-O-TA) was intensively used for the analysis of DL-amino acids in foods and drinks [8,30,31].

In the following we provide protocols and demonstrate the versatility of Chirasil®-L-Val for the detection of D-amino acids in food and beverages. Furthermore, we propose a mechanism for the generation of D-amino acids in the course of the Maillard reaction which is of major importance in the food industry.

2. Materials

2.1. Equipment

1. Shimadzu 14 A gas chromatograph with flame ionization detector (FID) or GC 17 A coupled to a Model QP5000 mass spectrometer (MS) (Shimadzu, Kyoto, Japan).
2. Fused silica capillary column Chirasil®-L-Val (*N*-propionyl-L-valine *tert.* butylamide polysiloxane) (25 m x 0.25 mm id; film thickness 0.12 µm) (Varian-Chrompack, Darmstadt, Germany)
3. Reactitherm Heating/Stirring Module with 'Reacti-Vap' Evaporator Model 18780 (Pierce, Rockford, IL, USA) or Heating Module (Model BT 2000, Kleinfeld Labortechnik, Gehrden, Germany) and nitrogen distributor device.
4. Ultra-Turrax® homogenizer (Model T25, IKA Labortechnik, Staufen, Germany).
5. Vials (1 or 5 mL) with teflon-lined screw caps (ReactiVials®, Wheaton, NJ, USA).
6. Rotatory evaporator with vacuum pump (Model Laborata 4000; Heidolph Instruments, Schwabach, Germany).
7. Quartz distill for generating doubly distilled water (Model Destamat Bi 18^E; Heraeus Instruments, Osterode, Germany).
8. Centrifuge (Model Labofuge 400, Heraeus Instruments).
9. Microsyringes (Mikroliter # 701; Hamilton, Reno, NV, USA).
10. Pasteur capillary pipettes (15 cm x 0.5 cm id), outlet closed with glass wool plug.
11. Nitrogen supply.

2.2. Chemicals and Reagents

1. Solvents: methanol (MeOH), 1-propanol (1-PrOH), 2-propanol (2-PrOH), dichloromethane (DCM), petroleum ether (b.p. 40–70°C).

2. Derivatizing reagents: trifluoroacetic acid anhydride (TFAA), pentafluoropropionic acid anhydride (PFPAA), acetyl chloride (AcCl), 10% aqueous trichloroacetic acid (TCA).
3. AcCl/1-PrOH: 1-PrOH (9 mL) and AcCl (1 mL) are mixed with stirring and chilling.
4. BHT (10%, *g/v*) in 1-PrOH: BHT (50 mg) dissolved in 1-PrOH (5 mL).
5. Internal standard 10 mM L-Nle: Nle (13.1 mg) dissolved in 0.1 M HCl (10 mL).
6. Amino acid standard solution (AA-S-18; from Sigma Chemicals, St. Louis, MO, USA) containing 2.5 M L-amino acids: Ala, Arg, Asp, Cys (1.25 M), Glu, Ile, Leu, Lys, Met, Phe, Pro, Ser, Thr, Tyr, Val and achiral Gly; GABA and L-Nle or special amino acids have to be added.
7. A qualitative standard composed of D- and L-amino acids in the ratio 1:2 is prepared by mixing appropriate quantities of enantiomers (or addition of L-amino acids to DL-amino acids).
8. Dowex 50W X8-400 cation exchanger (200 – 400 mesh from Supelco Park, Bellefonte, Pennsylvania, USA).
9. Doubly distilled water from a quartz distill was used exclusively.

2.3. Sources of Foodstuffs

Cocoa beans: 'Forastero', Ivory Coast, roasted at 150°C for 120 min, provided by a major chocolate producer, Munich, Germany.
Forest honey: 'Waldhonig Auslese', commercial product, Germany.
Maple syrup: 'Spring Tree', Grad A, from Quebec, Canada.
Grape juice concentrate: Turkish 'Pekmez', imported product.
Raisin: 'Sweet Valley' product of USA.
Sherry: Jerez Fino 'Tio Pepe', from Jerez, Spain.
Cheddar: 'Farmhouse' cheddar, matured for 12 months, Somerset, UK.

3. METHODS

Treatment and analyses are described exemplarily in the following. Procedures can be applied to solid, syrupy and liquid foodstuffs.

3.1. Cacao

Cacao beans (10 g) were grinded in an electric coffee mill. To aliquots (0.5 g) petroleum ether, b.p. 50-70°C (3 x 4 mL) was added, the mixture was stirred for about 30 min and centrifuged (1200 *g*, 15 min). The procedure was repeated twice. The organic solvents were discarded and the remaining residues extracted with 0.1 M HCl (3 x 3 mL). The combined extracts were evaporated to dryness using a rotatory evaporator at 10 mbar and a bath temperature of 40°C. To the remaining residue 0.1 M HCl (2 mL) was added. To aliquots (100 µl) bidistilled water (2 mL) and the internal standard L-Nle (10 mM in 0.01 M HCl, 20 µl) was added and the pH was adjusted to 2.3 by addition of 0.5 M HCl. The solution was passed through a Pasteur pipette filled with activated Dowex 50W X8 cation exchanger (bed volume 0.5 cm x 5 cm length). The resin was washed with water (about 10 mL) until neutral and amino acids were eluted with 4 M aqueous ammonia (3 mL). The effluent was evaporated to dryness *in vacuo* in a conical flask (20 mL), then 0.1 M HCl (1 mL) was added and the solution transferred into a ReactiVial®. The solvent was removed in a stream of nitrogen. To the remaining residue 2.5 M HCl in 1-PrOH (0.5 mL) and BHT in 1-PrOH (20 µl) were added and the mixture heated in the tightly closed vial at 100 °C for 1 hour. Then reagents were removed at ambient in a stream of nitrogen, DCM (300 μl) and PFPAA (50 μl) were added and the mixture heated for 20 min at 100 °C and excess of reagents removed in a stream of nitrogen. Then DCM (100 μl) was added, solvent removed in a stream of nitrogen, the procedure repeated once and finally the remaining residue was dissolved in DCM (200 μl) and 0.5 to 1.0-μl aliquots were injected onto the GC column.

3.2. Honey, Maple Syrup and Grape Juice Concentrate

To aliquots (1 g) of foodstuffs water (5 mL) was added and resulting solutions were adjusted to pH 2.3 by addition of 0.01 M HCl. Analytes were passed through cation exchanger columns (bed volume 0.5 x 3.5 cm) and amino acids eluted, derivatized and analyzed as described above. After washing with water (10 mL) the effluent was evaporated to dryness, derivatized as described above and analyzed by GC-SIM-MS.

3.3. Raisins

Amounts (1 g) of raisins were briefly washed with ethanol in order to clean the surface and then minced with a lancet. To aliquots (1 g) 10% aqueous TCA (4 mL) was added and the mixture was blended using of an Ultra-Turrax® homogenizer. Then the mixture was centrifuged, the precipitate was discarded and the supernatant transferred into a separatory funnel. The aqueous phase was extracted with diethyl ether (3 x 10 mL) in order to remove excess of TCA. Then the aqueous phase was adjusted to pH 2.3 by addition of 0.1 M HCl and passed through a Dowex 50W X8 cation exchanger. The effluent was evaporated to dryness and amino acids converted into derivatives and analyzed as described above for cacao.

3.4. Sherry Wine

An aliquots of the fortified wine (5 mL) was adjusted to pH 2.3 by addition of 0.01 M HCl. The resulting solution was passed without further treatment through the cation exchanger and amino acids were eluted, derivatized and analyzed as described above.

3.5. Cheese

The interior of Cheddar cheese (about 200 g) was finely grinded and to an aliquot (100 mg) water (5 mL) and 50 mM L-Nle in TFA (40 μl) were added with magnetic stirring. Additional TFA (1 mL) was added in order to precipitate proteins. The mixture was stirred for 15 min and then filtered through fluted paper. The filter was washed with water, the filtrate transferred to a separatory funnel and extracted with petroleum ether (2 x 20 mL). The organic phase containing lipids was discarded. Then the aqueous phase was evaporated to dryness, 5 mM HCl (2 mL) was added and the solution passed through the cation exchanger. The effluent was treated and analyzed as described above.

4. RESULTS

Chromatograms resulting from analyses of a standard and food samples are compiled in Figure 1. The GC-SIM-MS on Chirasil®-L-Val of derivatives of an amino acid standard composed of D- and L-amino acids (ratio about 1:2), including achiral Gly and GABA and the internal standard L-Nle, together with a peak resulting from the antioxidant BHT, is presented in Figure 1a. Under the conditions used D-Met and L-Asx (the latter actually representing L-Asp) are not resolved. Using SIM-MS this does not cause problems since co-eluting amino acid derivatives can be distinguished owing to their characteristic fragment ions.

Enantioselective GC-SIM-MS of roasted cacao beans is shown in Figure 1b. The high relative quantities of D-amino acids and their diversity are worth of noting and have been recognized previously [7,32]. On roasting of cocoa beans Amadori compounds (fructose amino acids) are formed and proceeding of the Maillard Reaction is indicated by colour change and formation of characteristic flavour compounds. Amadori compounds are precursors of D-amino acids (see Figure 2).

The chromatogram of a forest honey is shown in Figure 1c. Honey is produced from plant nectars by *Apis mellifera* L. Since honey is not subject of severe technological treatment the presence of D-amino acids is somewhat unexpected. However, honey might be considered to represent a very concentrated solution of glucose and fructose in water which contains also free amino acids besides other minor constituents. Consequently, the Maillard reaction proceeds to a certain extent and presence of D-amino acids in honey is attributed to racemization of L-amino acids in the course of this reaction.

This result is supported by analysis of a maple syrup (Fig. 1d), representing a plant sap from the maple tree *Acer saccharum* L. which is commonly concentrated by boiling. Relative high amounts of D-amino acids can be seen in the chromatogram. Maple sap contains saccharose and amino acids. On heating and proton catalysis saccharose is converted partly into glucose and fructose. The former yields fructose amino acids (Amadori compounds), the latter glucose amino acids (Heyns compounds) on heating with amino acids.

The chromatogram of concentrated grape juice (Turkish 'Pekmez'), representing a dark brown syrup, indicating intensive Maillard reaction, is shown in Figure 1e. Large quantities of D-amino acids can be seen.

The chromatogram of amino acid enantiomers detected by GC-FID of raisins (dried grape fruits) is presented in Figure 1f. Here again large quantities of D-amino acids can be seen. Grapes are rich in glucose, fructose and amino acids. On drying concentrations increase and proceeding of the Maillard reaction is indicated by the typical brown colour of raisins.

The chromatogram of a Sherry wine, representing an alcoholic fermented beverage, is presented in Figure 1g. Sherry, like fortified wines in general, contains very high amounts of D-amino acids which approach or exceed 50% of the corresponding L-enantiomer in many cases [33].

Finally, the GC-FID of free amino acids isolated from an English cheddar cheese is presented in Figure 1h. Here, the large quantities of D-Pro are conspicuous. Cheese is a typical bacterial fermented food produced with the aid of lactobacilli and lactic acid streptococci. This explains the abundance of D-AAs in cheddar or fermented sour milk products in general [5].

These representative analyses demonstrate that D-amino acids are quite common in food and beverages and their presence, structural variety and formation requires explanations.

Formation of D-amino acids in foodstuffs that had been the subject of severe technological treatments (alkali, heat, pressure etc.) is well known for many years and has been reviewed in detail [3]. It has also been fully realized in recent years that all kinds of bacterial fermented food, i.e. food that attributes its appearance, texture, flavor and nutritional acceptance to the action of bacteria, contain D-amino acids. Bacteria have an abundance of racemases and contain D-amino acids in their cytoplasm and cell walls (peptidoglycan) [9].

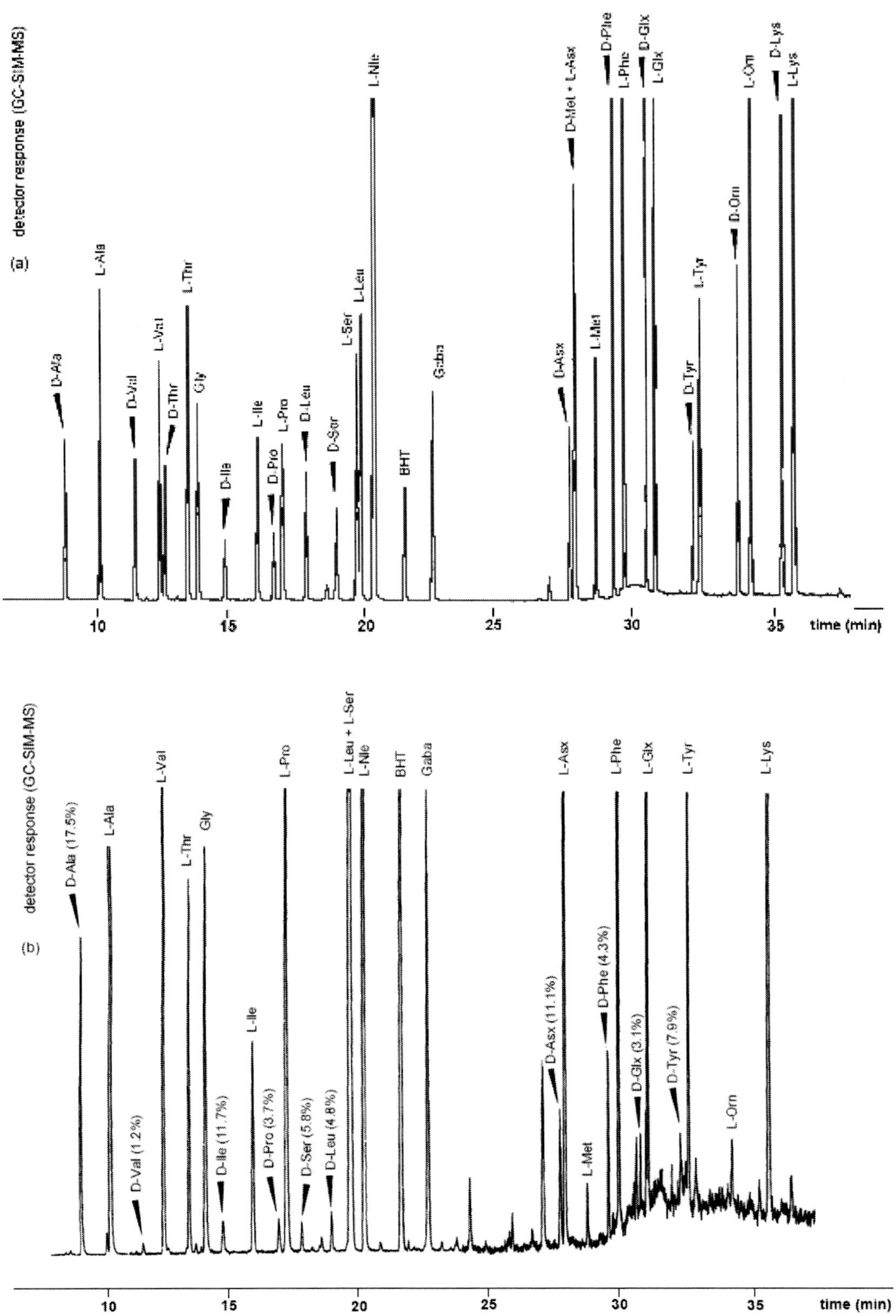

Figure 1. GC-SIM-MS of PFP-amino acid-(2)-Prp esters resolved on Chirasil®-L-Val of (a) standard of DL-amino acids (ratio D:L ca. 1:2), and amino acids extracted from (b) roasted cocoa from Ivory Coast.

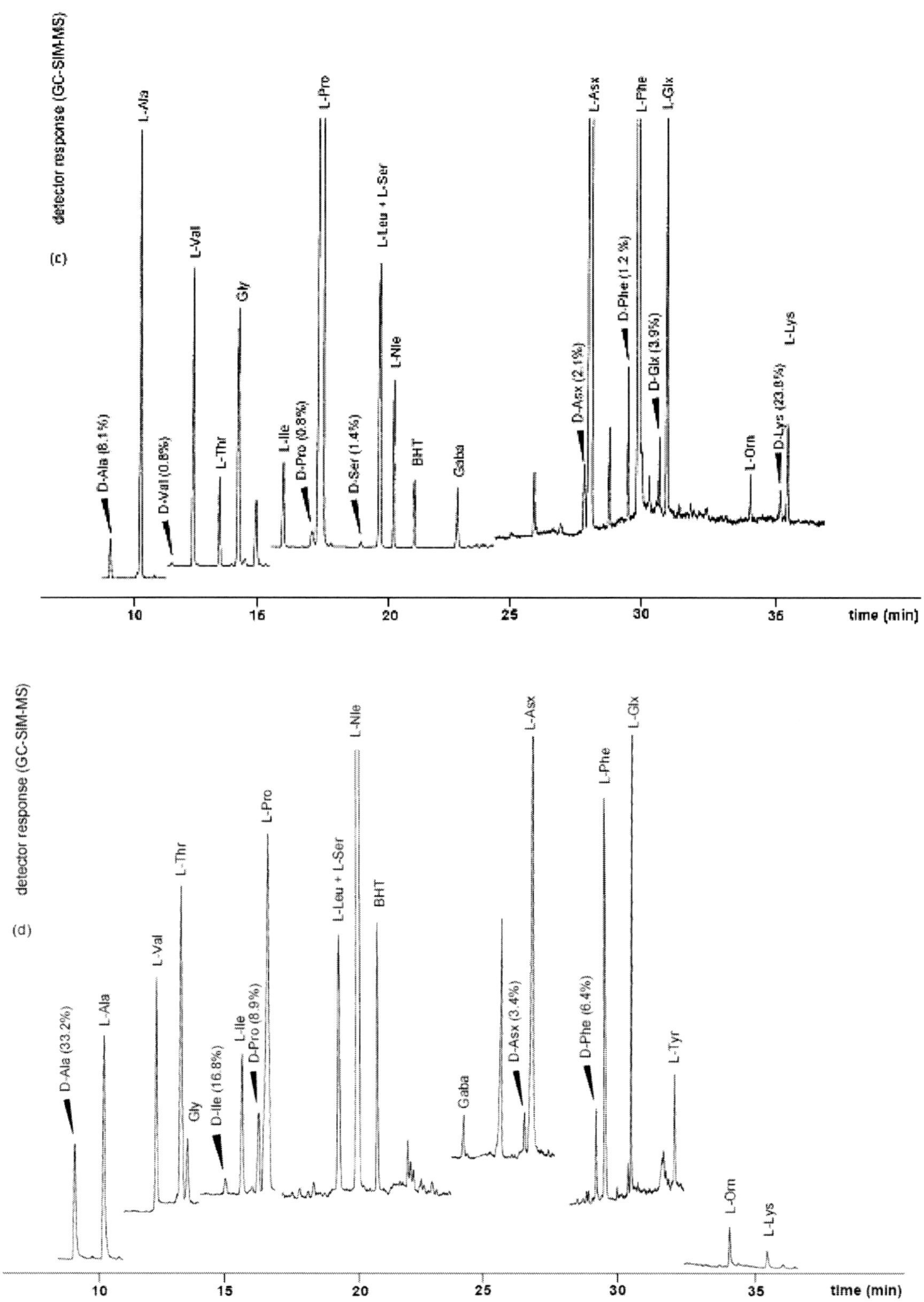

Figure 1. GC-SIM-MS of PFP-amino acid-(2)-Prp esters resolved on Chirasil®-L-Val of (c) forest honey, (d) maple syrup.

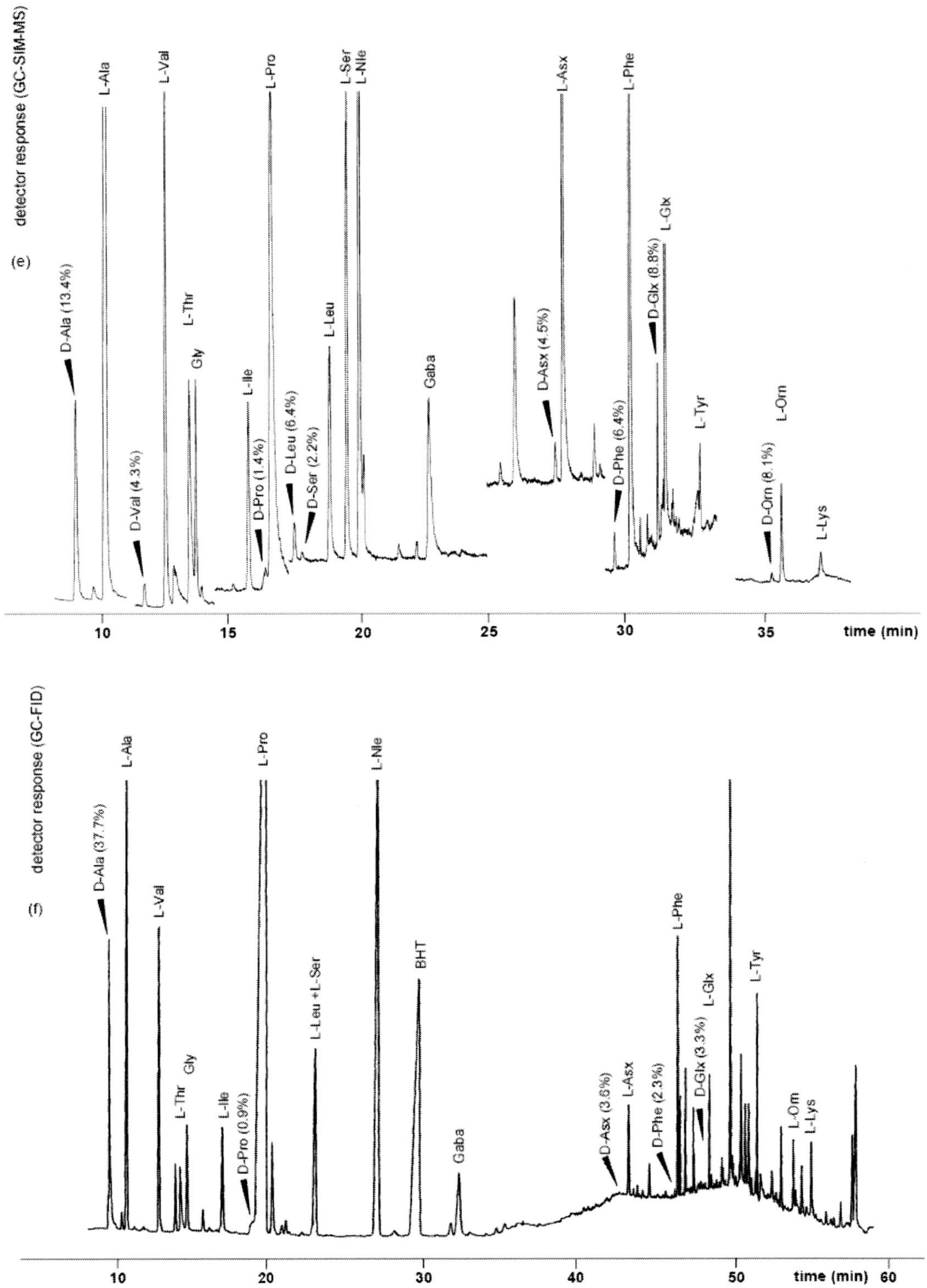

Figure 1. GC-SIM-MS of PFP-amino acid-(2)-Prp esters resolved on Chirasil®-L-Val of (e) Pekmez, and GC-FID of PFP-amino acid-(2)-Prp esters from (f) raisin 'Sweet Valley'.

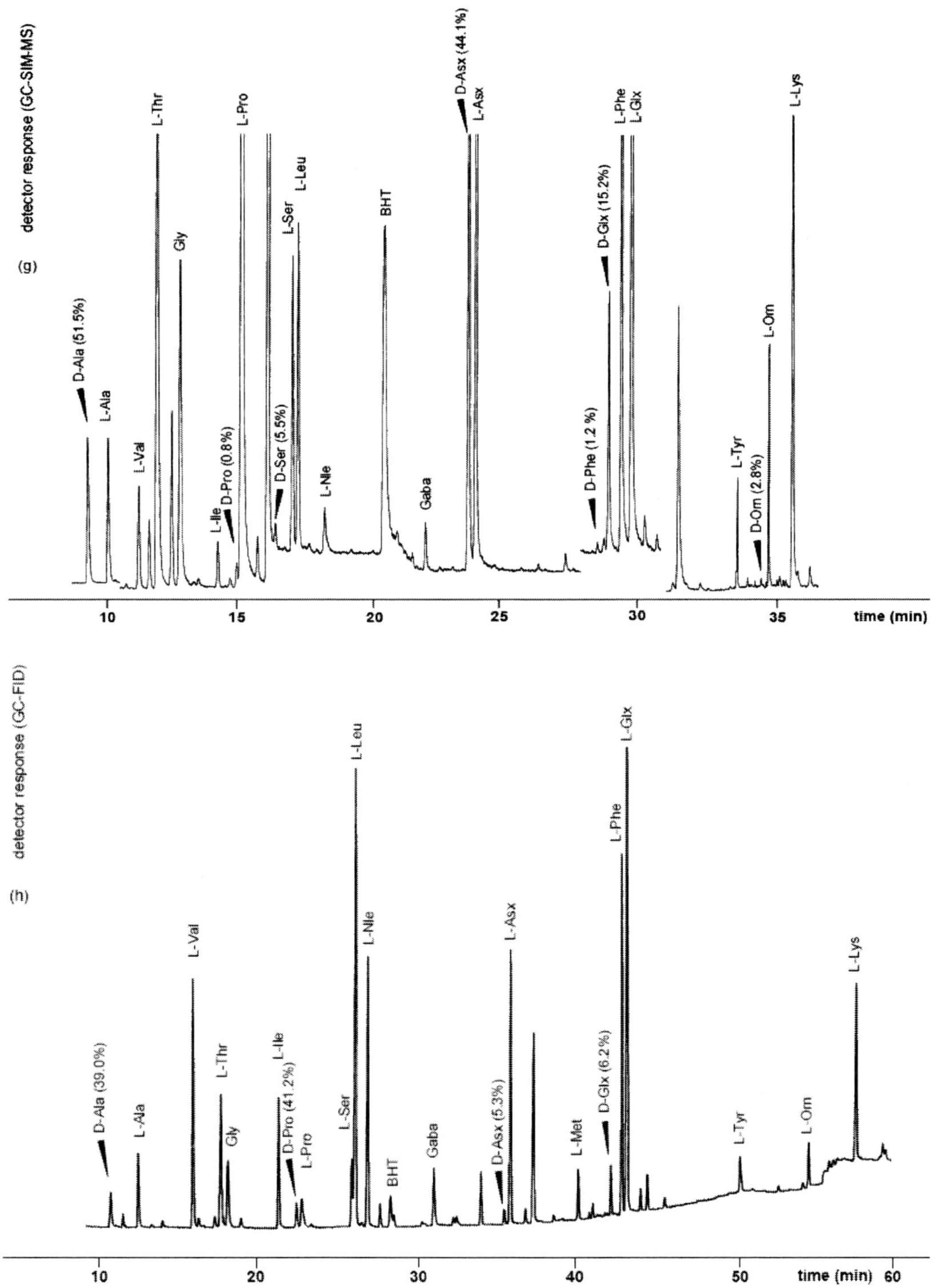

Figure 1. GC-FID of PFP-amino acid-(1)-Prp esters from (g) Sherry, and (h) cheddar cheese. For chromatographic conditions, see note 7 and 8.

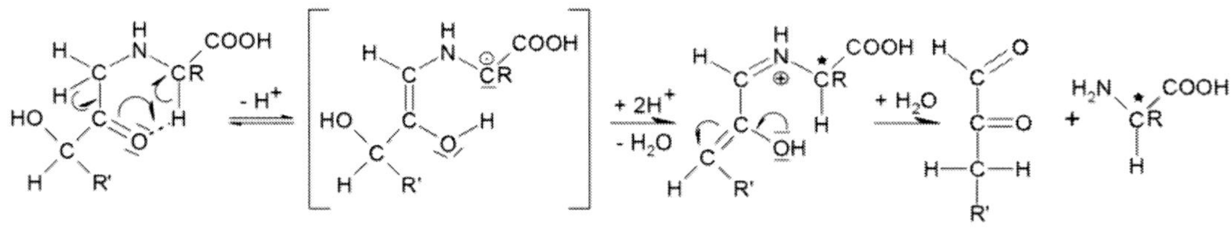

Figure 2. Proposed racemization mechanism of amino acids in Amadori compounds 1 formed in the course of the Maillard reaction. C^{α}–proton abstraction from bonded amino acids is favored by formation of an intramolecular hydrogen bridge as indicated in 1, followed by 1,2-enaminolisation and generation of an intermediate carbanion 2 that is capable of being racemized on reprotonation. Dehydratisation yields the imin cation 3 which decomposes on heating with formation of 3-deoxyosone 4 and release of (partially) racemized amino acid 5. R = amino acid side chain; R' = sugar residue; asterisk refers to racemized amino acid.

However, neither harsh technological treatment nor bacteria are satisfactory explanations for the occurrence of D-amino acids in honey and cocoa, or dried plant products such as raisins or fruit syrups.

We had recognized in previous studies that heating of saccharides together with L-amino acids leads to the formation of D-amino acids [17]. This interaction is well known as non-enzymic browning or the Maillard reaction. Consequently, it is reasonable to assume that D-amino acids are generated in the course of the Maillard reaction. All conditions favoring the Maillard reaction will lead to the formation of D-amino acids. Kinetics of formation are dependent on concentrations and structures of reactants, temperature, pH, water activity, and presence of catalysts. Notably, the Maillard reaction proceeds also at ambient temperature or in physiological fluids [34].

Relative stable intermediates of the Maillard reaction are the Amadori and Heyns compounds (fructose amino acids and glucose amino acids). Indeed, Amadori compounds (fructose amino acids) are formed and have been detected in many foodstuffs such as milk powder, roasted cacao beans, fruit and vegetable powders, dried apricots and raisins, fruit juice concentrates and tobacco [32,35-37]. Heating of synthetic Amadori compounds containing L-amino acids leads to the release of D-amino acids and *vice versa*. A tentative racemization mechanism via intermediate formation of a planar carbanion has been suggested [18,19] and is presented in Figure 2. Proceeding of the Maillard reaction in foodstuffs will be accompanied inevitably with the formation of D-amino acids [38,39].

Relevance of D-amino in foodstuffs and uptake therefrom is still a matter of debate and also discussed in detail in this volume by M. Friedman. Peptide bonded D-amino acids without doubt prevent or hamper digestibility of food proteins and thus reduce nutritional value. With regard to free D-amino acids corresponding to the protein L-amino acids and under conditions of realistic food intake we do not expect adverse health effects on normal human beings. This view is corroborated by data on clinical administration of large quantities of racemic amino acid mixtures to infants. Owing to the expenses of pure amino acid enantiomers such clinical administration was common at the beginning of parenteral nutrition [9,37,40]. Reports on nephrotoxic effects of D-Ser administered in high doses to rat should be discussed in the light of the natural occurrence of D-Ser in man and animal brain [16,41,42] and pharmaceutical use of DL-Ser as chelating agent for iron or use of the magnesium salt of DL-Asp as cardioprotective drug [43]. Further, D-amino acids are converted by mammalian D-amino acid oxidases to α-keto acids which are further metabolized. Notably, an abundance of D-amino acids are excreted with the urine. Consequently, research on physiological D-amino acids has changed to their endogenous biosynthesis and the specific roles which D-amino acids may play in living organisms [41,44,45].

5. Notes

Based on many years of experience concerning gas and liquid chromatographic determination of D-amino acids in biosamples we provide some notes related to analytical methodology.

1. Sample Treatment

- Samples should be adjusted to pH 2.3 before subjected to ion exchange.
- Loss of aspartic acid might occur at pH<2 or on loading large analytical volumes
- Triglycerides and lipids from analytes should be removed by extraction with diethyl ether *n*-hexane or ethyl acetate.
- In some cases it is of advantage to extract amino acids with 70% ethanol or methanol. In these solvents proteins and polysaccharides are not or sparingly soluble.
- Proteins can be precipitated and removed by centrifugation using trichloroacetic acid, picric acid or sulfosalicylic acid.

2. Ion Exchange

- Batches of the cation exchanger should be carefully washed successively with doubly distilled water, 2 M sodium hydroxide, water, 2 M HCl, 5 M aqueous ammonia and finally with water until neutral. The ion exchanger should be stored in 0.1 M HCl and washed with water until neutral prior to use. Ion exchangers should be stored protected from light. Used ion exchanger should be discarded and not regenerated.
- Water should be of highest purity in order to avoid contamination with amino acids. We prefer doubly distilled water from a quartz distill as we have observed that traces of D-amino acids may occur even in commercially available water 'for chromatography' produced by reversed osmosis. D-amino acids may result from bacterial biofilms.

3. Glasware and Derivatization Procedure

- Precaution: reagents and solvents should be checked regularly for contamination with D-amino acids in a blank run.
- All kinds of bacterial contamination (including dust adjacent to glassware) will lead to the presence D-amino acids.
- If trace amounts of amino acids have to be determined glassware should be wrapped in aluminum folia and heated in a muffle furnace at 500 °C.
- A small crystal of BHT (or 25 μl of a 0.5% solution in the alcohol used for esterification) should be added to the analyte prior to derivatization.
- It is recommended to prepare a qualitative standard for calibration by mixing appropriate amounts of solid D- and L-amino acids (or addition of L-amino acids to the corresponding DL-amino acid) and fine grinding of the resulting standard mixture in a mortar. Thus gram amounts of a solid standard of almost infinite lifetime can be produced and aliquots handed over to coworkers and students in order to avoid contamination. A skilled analyst will use about 2 mg aliquots for derivatization and calibration.
- To aliquots of this standard special amino acids can be added, if required (e.g. hydroxyproline stereoisomers, Orn etc.).

- For quantitative purposes commercially available standard mixtures of L-amino acids can be used and L-Nle added since GC response factors for derivatives of L- and D-enantiomers are equal (see Experimental).
- We have realized that change of the splitting ratio of the GC injector influences response factors Therefore, concentration or dilution of samples by otherwise unchanged analytical conditions are recommended.

4. Chirasil®-L-Val and Chirasil®-D-Val

- Manufactures columns are tested with TFA/OMe esters of few DL-amino acids using hydrogen as carrier gas and isothermal temperature. We use PFP/2-Prp esters in most cases owing to the best overall separation in particular with regard to DL-Pro and enantiomers of DL-Glu and DL-Phe.
- Various companies offer Chirasil®-L-Val columns but only Macherey-Nagel the Chirasil®-D-Val having D-Val as chiral selector in place of L-Val. This reverses the elution order of enantiomers [11,10,27]. If using an FID detector use of both columns makes results very reliable. If SIM is used, this is not necessary. Furthermore, overlapping peaks can be distinguished according to specific fragment ions.
- The critical amino acid that is worst separated on Chirasil®-L-Val is DL-Pro. We have observed that this might lead to problems using the Permabond® column (i.e. crosslinked) Chirasil®-L-Val column from Machery-Nagel. Notably, DL-Pro is best resolved on γ-cyclodextrins such as Lipodex® E [7,10].
- Stereoisomers of special collagen amino acids such as hydroxyproline or hydroxylysine, bacterial 2,4-diaminopimelic acid or lanthionine and enantiomers of dipeptides are also resolvable on Chirasil®-Val [46-48].
- It should be noted that many data published by the inventors of the Chirasil®-L-Val and Chirasil®-D-Val columns have been obtained from laboratory made and coated glass (not quartz) columns using hydrogen as carrier gas [25]. These columns were also fine-tuned and selected with regard to best enantioresolution. Commercially available columns show somewhat different elution behavior of enantiomers.

5. Problematic Amino Acids

- Problematic amino acids are Arg (not derivatized), His (imidazol group requires special derivatization using e.g. ethyloxycarbonylchloride); Asn, Gln (amides are hydrolyzed to acids), Ser/Thr (low stability of esters), Cys_2 (reduction to cystein required); Ile, Thr (*allo*–forms may lead to problems owing to assignment of stereoisomers); Trp (partly destroyed).

6. Relative and Absolute Quantification of Amino Acid Enantiomers

- Relative amounts of D-amino acids were calculated according to equation (1)

$$\%D = 100\, A_D / (A_D + A_L) \tag{1}$$

where %D is the relative amount of the D-amino acid to be determined, and A_D and A_L are the peak areas (ion abundances) of the D- or L-enantiomer, respectively, determined by GC-SIM-MS. For quantification response factors of amino acids in a standard mixture (for composition see above) were determined in relation to the internal standard (IS) L-Nle. Equimolar amounts of amino acids of the standard mixture, including L-Nle, were injected into the GC-MS system.

- Response factors were calculated according to equation (2)

$$f_R = A_{LAA}/A_{IS} \quad (2)$$

where f_R is the response factor of the amino acid to be determined, A_{LAA} the peak area of amino acid to be determined, and A_{IS} the peak area of the IS obtained from the standard amino acid mixture. Amino acids in samples were quantified according to equation (3)

$$c_{LAA} = (1/f_R \, x \, A_{LAA})/(A_{IS} \, x \, c_{IS}) \quad (3)$$

where c_{LAA} is the amount of L-amino acid, f_R the response factor, A_{LAA} the peak area of the L-amino acid in the sample, A_{IS} the peak area of the IS added to the sample and c_{IS} the amount of the IS added to the sample. If required, from the relative quantities of D-amino acids *(%D)* and the quantities of L-amino acids c_{LAA} presented in Tables 2-4, the absolute quantities of D-amino acid c_{DAA} can be calculated according to equation (4).

$$c_{DAA} = c_{LAA} \, (\%D) \, / \, (100 - \%D) \quad (4)$$

7. Conditions for GC-SIM-MS

- The conditions described here for GC-MS and GC-FID have been used for foodstuffs shown in Figure 1.
- Gas chromatographic conditions are dependent on the instruments and chiral columns used, together with the carrier gas and amino acid derivatives employed.
- -Carrier gas helium at an inlet pressure of 5 kPa, purge flow 3 mL /min and flow rate 0.5 mL /min. Injector and interface temperatures 250°C and split ratio 1:30. The temperature program 70 °C for 1 min, then 2.5 °C/min to 100 °C, 2 min isothermal, then 3.5°C /min to 135°C, then 5 °C/min to 150 °C, then 20°C /min to 190°C, and 8 min isothermal.
- Pressure of carrier gas 5 kPa for 1 min, then 0.2 kPa/min to 7.0 kPa, 2 min isobaric; then 0.3 kPa/min to 10.8 kPa, then 1.4 kPa/min to 13.0 kPa, then 2.4 kPa/min to 15.0 kPa, 8 min isobaric.
- For SIM appropriate ion sets of the PFP/2-Prp esters of the amino acids were selected and characteristic mass fragments (*m/z*) were used: Ala (190, 191, 235), Val (203, 218, 219), Thr (202, 203), Gly (176, 177), Pro (216), Leu (190, 232, 233), Ser (188, 189), Asx (189, 234, 235, 262), Met (203, 221, 263), Phe (91, 148, 190, 266), Glx (202, 230, 248, 276), Tyr (253, 266), Orn (216), Lys (176, 230), GABA, (176, 204, 232, 249).

8. Conditions for GC-FID

- Carrier gas helium at an inlet pressure of 46.8 kPa, constant flow 17.9 mL /min and split flow 15.0 mL /min; injector and detector temperatures 250 °C and split ratio 1:30. Hydrogen flow adjusted to 30.0 mL /min and synthetic air flow to 300.0 mL /min.
- Temperature program 2.5 °C/min from 85 °C to 100 °C, 2 min isothermal, then 0.6 °C/min to 115 °C, then 9.0 °C/min to 130 °C, then 6.0°C/min to 180°C, then 5.0°C/min to 190°C and 12.0 min isothermal.

References

[1] Liardon, R. & Hurrell, R. F. (1983). Amino acid racemization in heated and alkali-treated proteins. *J Agric Food Chem, 31*, 432-437.

[2] Zagon, J., Dehne, L. I. & Bögl, K. W. (1994). D-Amino acids in organisms and food. *Nutr Res, 14*: 445-463.

[3] Friedman, M. (1999). Chemistry, nutrition, and microbiology of D-amino acids. *J Agric Food Chem, 47*: 3457-3479.

[4] Abe, H., Park, J. N., Fukumoto, Y., Fujita, E., Tanaka, T., Washio, T., Otsuka, S., Shimizu, T. & Watanabe, K. (1999). Occurence of D-amino acids in fish sauces and other fermented fish products. *Fisheries Sci, 65*: 637-641.

[5] Brückner, H. & Hausch, M. (1989). Gas chromatographic detection of D-amino acids as common constituents of fermented foods. *Chromatographia, 28*: 487-492.

[6] Brückner, H. & Hausch, M. (1989). Detection of free D-amino acids in food by chiral phase capillary gas chromatography. *J High Res Chromatogr, 12*: 680-684.

[7] Brückner, H. & Hausch, M. D-Amino acids as ubiquitous constituents in fermented foods. (1990) In Lubec, G. (Ed.) *Amino Acids: Chemistry, Biology and Medicine,* Escom, Leiden, The Netherlands, pp. 1172-1177.

[8] Palla, G., Marchelli, R., Dossena, A. & Casnati, G. (1989). Occurrence of D-amino acids in food. *J Chromatogr, 475*, 45-53.

[9] Brückner, H., Jaek, P., Langer, M. & Godel, H. (1992). Liquid chromatographic determination of D-amino acids in cheese and cow milk. Implication of starter cultures, amino acid racemases, and rumen microorganisms on formation, and nutritional considerations. *Amino Acids, 2*, 271-284.

[10] Brückner, H., Becker, D. & Lübke, M. (1993). Chirality of amino acids of microorganism used in food biotechnology. *Chirality, 5*, 385-392.

[11] Brückner, H. & Lüpke, M. (1991). Determination of amino acids enantiomers in orange juices by chiral phase gas chromatography. *Chromatographia, 31*, 123-128.

[12] Brückner, H. & Westhauser, T. (1994). Chromatographic determination of D-amino acids as native constituents of vegetables and fruits. *Chromatographia, 39*, 419-426.

[13] Brückner, H. & Westhauser, T. (2003). Chromatographic determination of L- and D-amino acids in plants. *Amino Acids, 24*, 43–55.

[14] Okuma, E. & Abe, H. (1994). Simultaneous determination of D- and L-amino acids in the nervous tissues of crustaceans using precolumn derivatisation with (+)-1-(9-fluorenyl)ethyl chloroformate and reversed-phase ion-pair high-performance liquid chromatography. *J Chromatogr B, 660*, 243-250.

[15] Okuma, E., Watanabe, K. & Abe, H. (1998). Distribution of free D-amino acids in bivalve mollusks and the effect of physiological conditions on the level of D- and L-alanine in the tissue of the hard clam, *Metrix lusoria. Fisheries Sci, 64*, 606-611.

[16] Pätzold, R., Schieber, A. & Brückner, H. (2005). Gas chromatographic quantification of free D-amino acids in higher vertebrates. *Biomed Chromatogr, 19*, 466-473.

[17] Brückner, H., Justus, J. & Kirschbaum, J. (2001). Saccharide induced racemization of amino acids in the course of the Maillard reaction. *Amino Acids, 21*, 429-433.

[18] Brückner, H., Kirschbaum, J. & Pätzold, R. (2002) A new racemization mechanism for amino acids, in E. Benedetti, C. Pedone (Eds.), *Peptides 2002. Proceedings of the Twenty-Seventh European Peptide Symposium*, Sorrento, August 31- September 6. 2002, Edizione Ziino, Napoli, Italy, pp. 54-55.

[19] Pätzold, R. & Brückner, H. (2004) Studies on the racemization (epimerization) of amino acids and peptides in the course of the Maillard Reaction, in M. Flegel, M. Fridkin, C. Gilon, J. Slaninova

(Eds.), *Peptides 2004. Proceedings of the Third International and Twenty-Eighth European Peptide Symposium*, Prague, September 5-10, 2004, Kenes International, Geneva, Switzerland, pp 997-998.

[20] Bhushan, R. & Brückner, H. (2004). Marfey's reagent for chiral amino acid analysis: A review. *Amino Acids, 27*, 231-247.

[21] Calabrese, M. & Stancher, B. (1999). A study of the proline isomerisation in typical Italian wines. *J Sci Food Agric, 79*, 1357-1360.

[22] Brückner, H., Langer, M., Lüpke, M., Westhauser, T. & Godel, H. (1995). Liquid chromatographic determination of amino acid enantiomers by derivatisation with *o*-phthaldialdehyd and chiral thiols. Application with reference to food science. *J Chromatogr A, 697*, 229-245.

[23] Erbe, T. & Brückner, H. (2000). Chromatographic determination of amino acid enantiomers in beers and raw materials used for their manufacture. *J Chromatogr A, 881*, 81–91.

[24] Tsunoda, M., Kato, M., Fukushima, T., Santa, T., Homma, H., Yanai, H., Soga, T. & Imai, K. (1999). Determination of aspartic acid enantiomers in bio-samples by capillary electrophoresis. *Biomed Chromatogr 13*, 335-339.

[25] Frank, H., Nicholson, G. J. & Bayer, E. (1977). Rapid gas chromatographic separation of amino acid enantiomers with a novel chiral stationary phase. *J Chromatogr Sci, 15*, 174-176.

[26] Schurig, V. (2001). Separation of enantiomers by gas chromatography. *J Chromatogr A, 906*, 275-299.

[27] Pätzold, R. & Brückner, H. (2005) Chiral separation of amino acids by gas chromatography, in I. Molnar-Perl (Ed.), *Quantitation of Amino Acids and Amines by Chromatography*, Journal of Chromatography Library, vol. 70, Elsevier, Amsterdam, The Netherlands, pp. 98-118.

[28] Erbe, T. & Brückner, H. (2000). Studies on the optical isomerization of dietary amino acids in vinegar and aqueous acetic acid. *Eur Food Res Technol, 211*, 6-12.

[29] Casal, S., Oliveira, M. B. & Ferreira, M. A. (2000). Gas chromatographic quantification of amino acid enantiomers in food matrices by their *N(O,S)*-ethoxycarbonyl heptafluorobutyl ester derivatives. *J Chromatogr A, 866*, 221-230.

[30] Marchelli, R., Dossena, A. & Palla, G. (1996). The potential of enantioselective analysis as a quality control tool. *Trends Food Sci Tech, 7*, 113-119.

[31] Gandolfi, I., Palla, G., Marchelli, R., Dossena, A., Puelli, S. & Salvadori, C. (1994). D-alanine in fruit juices: a molecular marker of bacterial activity, heat treatments and shelf-life. *J Food Sci, 59*, 152-154.

[32] Pätzold, R. & Brückner, H. (2006). Gas chromatographic determination and mechanism of formation of D-amino acids occurring in fermented and roasted cocoa beans, cocoa powder, chocolate and cocoa shell. *Amino Acids,* in press, DOI10.1007/s00726-006-0330-1.

[33] Pätzold, R., Nieto-Rodriguez, A. & Brückner, H. (2003). Chiral gas chromatographic analysis of amino acids in fortified wines. *Chromatographia Suppl, 57*, S 207-S 211.

[34] Ledl, F. & Schleicher, E. (1990). New aspects of the Maillard reaction in Foods and in the human body. *Angew Chem Int Edit, 29*, 565-594.

[35] Cremer, D. R., Vollenbroeker, M. & Eichner, K. (2000). Investigations of the formation of Strecker aldehydes from the reaction of Amadori rearrangement products with α-amino acids in low moisture model systems. *Eur Food Res Technol, 211*, 400-403.

[36] Anet, E. F. L. J. & Reynolds, T. M. (1956). Reactions between amino acids, organic acids and sugars in freeze dried apricots. *Nature, 177*, 1082.

[37] Ali, H., Pätzold, R. & Brückner, H. (2005). Determination of L- and D-amino acids in smokeless tobacco products and tobacco. *Food Chem, 99*, 803-812.

[38] Pätzold, R. & Brückner, H. (2005) Mass spectrometric detection and formation of D-amino acids in processed plant saps, syrups, and fruit juice concentrates. *J Agric Food Chem 53*, 9722-9729.

[39] Pätzold, R., & Brückner H. (2006) Gas chromatographic detection of D-amino acids in natural and thermally treated bee honeys and studies on the mechanism of their formation as result of the Maillard reaction. *Eur Food Res Technol* 223, 347-354.

[40] Schieber, A., Brückner, H., Rupp-Classen, M., Specht, W., Nowitzki-Grimm, S. & Classen, H. G. (1997). Evaluation of D-amino acid levels in rat by gas chromatography-selected ion monitoring mass spectrometry: no evidence for subacute toxicity of orally fed D-proline and D-aspartic acid. *J Chromatogr B, 691*, 1-12.

[41] Nagata, Y., Horiike, K. & Maeda, T. (1994). Distribution of free D-serine in vertebrate brains. *Brain Res, 634*, 291-295.

[42] Wolosker, H., Blackshaw, S. & Snyder, S. H. (1999). Serine racemase: a glial enzyme synthesizing D-serine to regulate glutamate-*N*-methyl-D-aspartate neurotransmission. *Proc Natl Acad Sci USA, 96*, 13409-13414.

[43] Brückner, H., Westhauser, T. & Godel, H. (1995). Liquid chromatographic determination of D- and L-amino acids by derivatization with *o*-phthaldialdehyde and *N*-isobutyryl-L-cysteine. Applications with reference to the analysis of peptidic antibiotics, toxins, drugs and pharmaceutically used amino acids. *J of Chromatogr A, 711*, 201-215.

[44] Konno, R. & Yasumura, Y. (1992). D-Amino-acid oxidase and its physiological function. *Int J of Biochem, 24*, 519-524.

[45] Fujii, N. (2002). D-Amino Acids in Living Higher Organisms. *Origins of Life and Evolution of the Biosphere, 32*, 103-127.

[46] Erbe, T. & Brückner, H. (1999). Microwave treatment of dietary gelatin does not generate *cis*-4-hydroxy-L-proline, an inhibitor of collagen biosynthesis. *Z Lebensm Untersuch Forsch A, 208*, 424-428.

[47] Schieber, A., Brückner, H. & Ling, J. R. (1999). GC-MS analysis of diaminopimelic acid stereoisomers and amino acid enantiomers in rumen bacteria. *Biomed Chromatogr, 13*, 46-50.

[48] Pätzold, R., Theis, C., & Brückner, H. (2006). Gas chromatographic separation of stereoisomers of dipeptides, *Chirality 18*, 551-557.

Section 5. Protein-Bound D-Amino Acids

In: D-Amino Acids: ISBN 1-60021-075-9
A New Frontier in Amino Acid and Protein Research, pp.355-367

Chapter 5.1

DETECTION OF THE SPECIFIC D-ASPARTIC ACID RESIDUES IN PROTEIN

Noriko Fujii* and Takumi Takata
Research Reactor Institute, Kyoto University, Japan.

Keywords: D-aspartate, isomerization, racemization, lens, aging, skin.

1. INTRODUCTION

We have believed that proteins consist exclusively of L-amino acids, however, recently, D-aspartic acid (D-Asp) has been detected in various tissues such as tooth [1-4], bone [5-7], aorta [8], brain [9-11], erythrocyte [12], eye lens [13-16], skin [17,18], ligament [19], and lung [20] from elderly individuals. D-Serine (D-Ser) was found in beta-amyloid protein of Alzheimer's disease [11,21]. It has been explained that the presence of D-Asp is the result of racemization of Asp residues in the protein during the life span, inasmuch as the proteins in these tissues are metabolically inert. The accumulation of D-amino acid in protein will change the higher order structure of the protein and subsequently will decrease the function of the protein. D-Asp residues are observed in cataract lens, in the brain of Alzheimer disease, and in the aorta of arteriosclerosis from elderly donors. The earlier studies documented almost exclusively simply the presence of D-Asp in whole tissues. Recently, the specific sites were identified in alpha A-crystallin [15] and alpha B-crystallin [14] in lens, and a beta-amyloid protein in brain [11].

At first, we here describe the biochemical method of identification of the specific sites in alpha A- and alpha B-crystallin from lenses of elderly donors (section A).

Secondary, we present the comprehensive analysis to detect the D-Asp containing protein in any tissues (section B) or cells (section C) by immunohistochemical method or proteomics analysis using the anti-D-Asp containing peptide antibody.

---

* Correspondence concerning this article should be addressed to Noriko Fujii, E-Mail: nfujii@HL.rri.kyoto-u.ac.jp, FAX: +81-724-51-26.

A. Lens: To Search the Specific Asp Residues in Lens Protein by Biochemical Analysis

A-1. Materials

1. Human lens
2. DEAE-Toyopearl column TSK-650S
3. Sephacryl S300HR column
4. TSK gel-ODS-80 TM, 4.6 x 250 mm, Tosoh, Tokyo
5. Nova-Pak ODS, 3.9 x 300 mm, Waters Japan
6. HPLC (UV and fluorescence detection are necessary)
7. SDS-PAGE
8. Protein sequencer (Applied Biosystems 477A/120A, Foster City, CA, USA)
9. PicoTag Work Stations (Waters, Tokyo)
10. Matrix-assisted Laser Desorption/Ionization Time-of-flight Mass Spectrometry (MALDI-TOF-MS, Kompact MALDI IV, Shimadzu, Kyoto, Japan)

A-2. Methods

A-2-1. The Identification of D-Asp Containing Protein in Lens

The methods were summarized in Schemes 1 and 2.

1. Human lenses were homogenized in 50 mM Tris-HCl buffer (50mM NaCl, 1mM EDTA, pH 7.4).
2. Centrifuge the homogenate samples at 15000g for 20 min at 4°C.
3. Apply the supernatant (water-soluble fractions) to a DEAE-Toyopearl column equilibrated with 50 mM Tris-HCl buffer (50mM NaCl, 1mM EDTA, pH 7.4).
4. The elution was carried out with 0, 0.15, and 2M NaCl in the same buffer.
5. Each fraction was collected, dialyzed against water, and lyophilized.
6. After the hydrolysis of the sample, the D/L ratio of amino acid was measured by the method described in the section A-2-4.
7. As a result, high D/L ratios of Asp (D/L=0.1) were detected in the proteins eluted with 2M NaCl fraction in the DEAE chromatography.
8. The 2M NaCl fraction was applied on SDS-PAGE and every bands were eluted by electrical elution.
9. The No. 4 band in the Figure 1 had the highest D/L ratio of Asp [22].
10. **(Result)** This D-Asp-containing protein was identified to be alpha-crystallin [23]. Alpha-crystallin exsists as a large aggregate (600-1000 kDa) which is comprised of two types of subunits, alpha A (173 amino acids) and alpha B (175 amino acids) in human lenses.

A-2-2. Purification of Alpha A- and Alpha B-Crystallins from Lenses of Eldery Donors

Since the D-Asp containing protein was identified to be alpha-crystallin, we extracted large amounts of alpha-crystallin in human lens.

1. The homogenate of human (mean age: 80 years as aged sample, and 11 months as young sample) lenses were separated into water-soluble and water-insoluble fractions by centrifugation at 15000g for 20 min at 4°C.

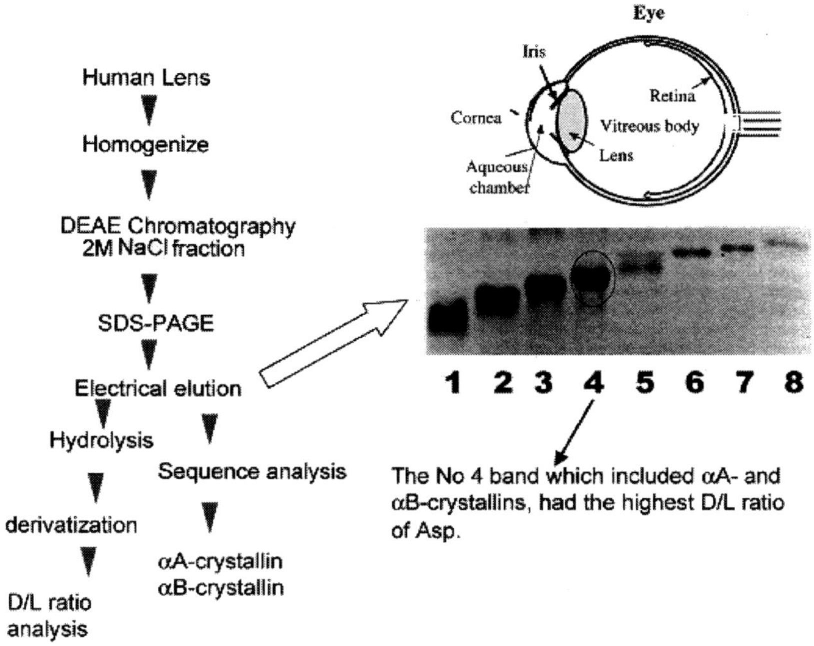

Scheme 1.

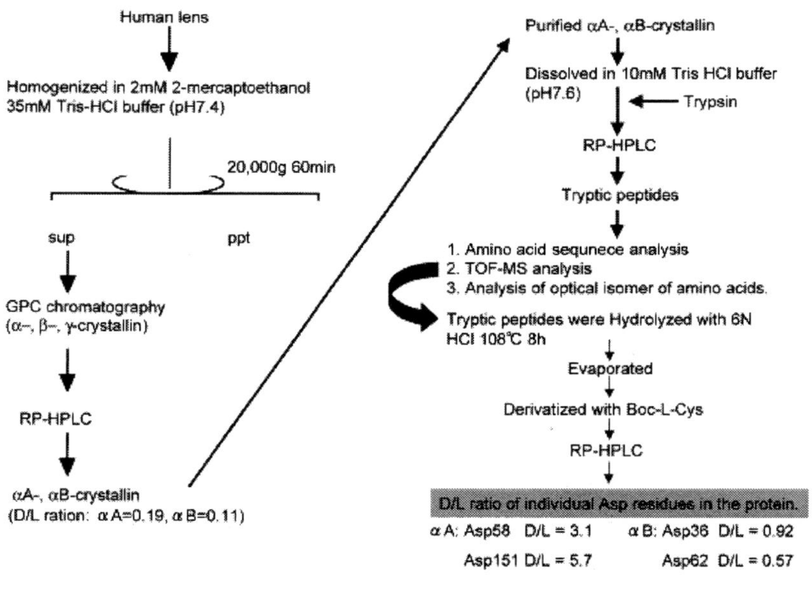

Scheme 2.

2. The water-soluble fractions were applied to a Sephacryl S300HR column which had been equilibrated with the same buffer and fractions containing alpha-crystallin were collected.
3. Alpha A- and B-crystallins were separated by reverse-phase (RP)-HPLC according to the procedure of Swamy and Abraham [24].
4. After the hydrolysis of the alpha A- and B-crystallins, the D/L ratios of amino acids were measured by the method described in the section A-2-4.
5. **(Result):**In aged human lenses, the D/L ratios of Asp residues in alpha A- and alpha B-crystallin were 0.19 and 0.11, respectively [25].

A-2-3. The Determination of the D-Asp Sites in Human Alpha A-or Alpha B-crystallin

Human alpha A-crystallin has 15 Asp and 2 Asn residues while alpha B- crystallin contains 11 Asp and 3 Asn residues (Fig 1a and 1b). As discussed in the previous section, the D/L ratios of total Asp in alpha A- and alpha B-crystallin of aged human lenses were estimated to be 0.19 and 0.11, respectively. Therefore, these values are considered to be the average value of 17 Asp (15 plus 2) residues of alpha A-crystallin and 14 Asp (11 plus 3) residues in alpha B-crystallin. The indivisual D/L ratios of Asp residues were determined as follows:

1. The alpha A- and alpha B-crystallin samples were digested with trypsin at an enzyme-to-substrate ratio of 1:50 (mol/mol) in 0.1 M Tris-HCl buffer (pH 7.6) for 20 h at 37°C.
2. The resulting peptides were separated by RP-HPLC (LC-10A, Shimadzu, Kyoto) using a C18 column (TSK gel-ODS-80 TM, 4.6 x 250 mm, Tosoh, Tokyo) with a linear gradient of 0-40% acetonitrile in the presence of 0.1% trifluoroacetic acid, at a flow rate of 0.8 ml/min, with monitoring at 215 nm (Fig. 2).
3. The fractions containing the desired peptides were collected into tubes using a fraction collector (FRC10 Shimadzu, Kyoto).
4. The tryptic peptides were identified by sequence analysis, and mass spectrometry.
5. The D/L ratios of the individual Asp and Asn residues in the peptides were measured by the method described in section A-2-4.

A-2-4. Method for the Determination of D/L Ratio of Amino Acids in Protein or Peptides

This method enables accurate analysis of amino acid enantimoers at the picomole level. We can determine the D/L ratios of the individual Asp residues in alpha A- and alpha B-crystallin from only one lens of donor.

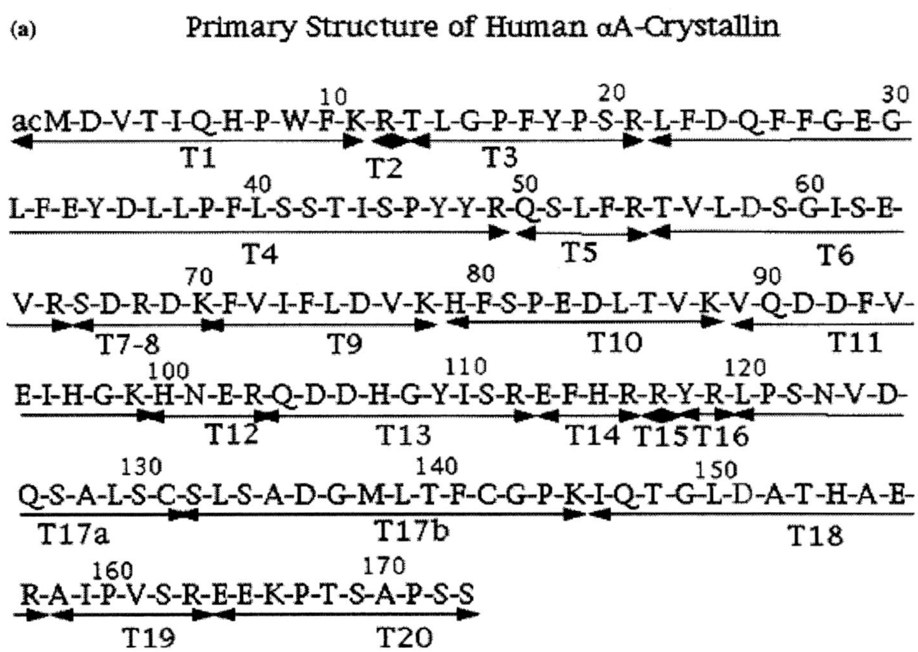

Figure 1a. Primary structure of human alpha A-crystallin.

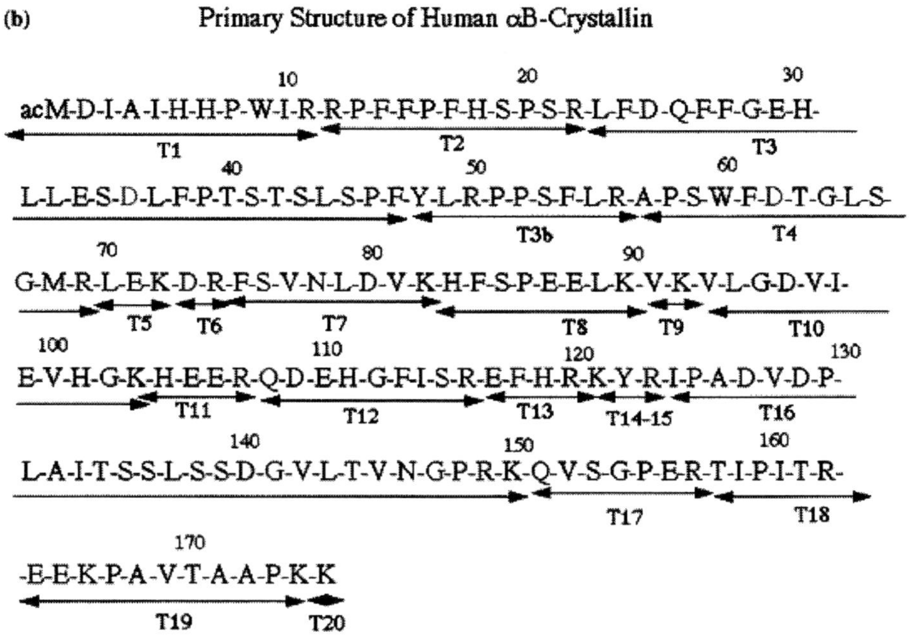

Figure 1b. Primary structure of human alpha B-crystallin.

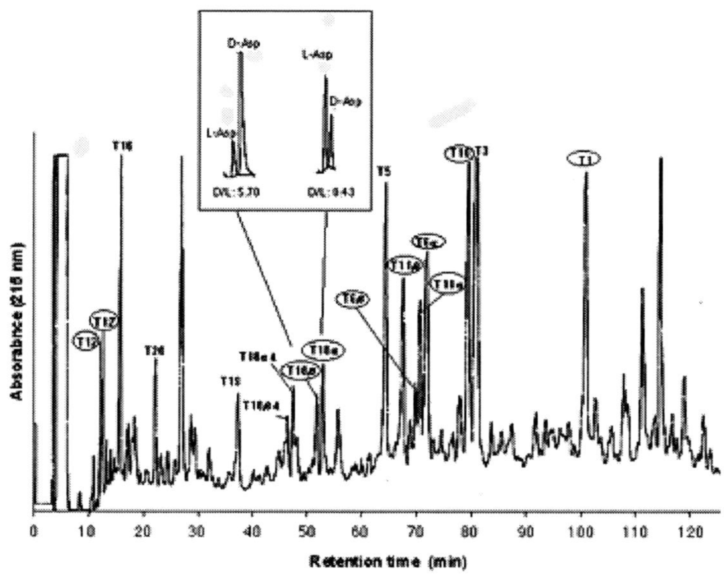

Figure 2. Elution profiles of tryptic (T) peptides of alpha A-crystallin from the lenses of elderly donors (80 years old). The peptides were separated by RP-HPLC using a C18 column (TSK gel-ODS-80 TM, 4.6 x 250 mm, Tosoh, Tokyo) with a linear gradient of 0-40% acetonitrile, 0.1% trifluoroacetic acid, at a flow rate of 1 ml/min over 120 min. Peptides were detected by measuring their absorbance at 215 nm. The assignment of each peak was based on amino acid composition, sequence analysis, and mass analysis. The tryptic peptides containing Asp/Asn residues are circled.

1. All glassware was baked at 500°C for 3 h.
2. The samples were hydrolyzed with gas-phase 6 N HCl in vacuo at 108°C for 24 h (PicoTag Work Stations, Waters, Tokyo).
3. The hydrolysates were evaporated under a reduced pressure environment.

4. The samples were derivatized with *o*-phthalaldehyde (OPA) and *N-tert*-butyloxycarbonyl-L-cysteine (Boc-L-Cys) to form diastereomers.
5. The determination of the D/L ratio of amino acids was performed by RP-HPLC with a C18 column (Nova-Pak ODS, 3.9 x 300 mm, Waters Japan) using fluorescence detection (344 nm as excitation and 433 nm as emission wavelength). The samples were eluted with linear gradient of 5-47 % acetonitrile plus 3% tetrahydrofuran in 0.1 M acetate buffer (pH 6.0) over 120 min at flow rate of 0.8 ml/min, at 30°C.

A-3. Results

A-3-1. The Specific Sites of D-Asp Residues in Alpha A- or Alpha B-Crystallins

As shown in Table 1, we found that Asp-58 and Asp151 residues in aged human alpha A-crystallin were highly inverted to D-isomers (D/L ratio of Asp-58, 3.1; Asp-151, 5.7) [14] and Asp-36 and Asp-62 were highly racemized (D/L ratio of Asp-36, 0.92; Asp-62, 0.57) [15]. Other Asp residues in the proteins were not racemized.

Table 1. The specific sites of D-beta-aspartic acid residues in alpha A-and alpahB-crystallins from lenses of 80-year-old donors

Crystallin	Site	D/L ratio of Asp	Linkage of Asp
alpha A	Asp-58	3.10	beta
alpha A	Asp-151	5.70	beta
alpha B	Asp-36	0.92	beta
alpha B	Asp-62	0.57	beta

We also clarify the mechanism why D-Asp and beta-Asp spontaneously form in protein.

A-3-2. Mechanism of D-Asp and Beta-Asp Formation in Protein

We also clarify the mechanism why D-Asp spontaneously forms in protein. As shown in Fig. 3, the simultaneous formation of beta- and D-Asp residues in the protein could be explained as follows: i) When the carbonyl group of the side chain of the L-alpha-aspartyl residue is attacked by the nitrogen of the amino acid residue following the Asp residue, L-succinimide is formed by intramolecular cyclization. ii) L-succinimide may be converted to D-succinimide through an intermediate [I] that has the prochiral alpha-carbon in the plane of the ring. iii) Protonation of the intermediate [I] would occur from the upper or lower side of the plane in an ordinary peptide or protein. In the case of the Asp 151 and 58 residues in alpha A-crystallin, since the area surrounding these Asp residues might form a chiral reaction field (Figure 3, shaded parts), D-succinimide more predominantly formed than L-succinimide [26]. iv) D- and L-succinimide are hydrolyzed at either side of their two carbonyl groups, yielding both beta- and alpha-Asp residues, respectively. The rate of succinimide formation is expected to depend on the neighboring residue of the Asp residue. When the neighboring amino acid of the Asp residue has a small side chain, such as glycine, alanine, or serine, the formation of succinimide occurs easily because there is no steric hindrance [27-29].

D-Asp formation was also accompanied by isomerization from the natural alpha-Asp to the biologically uncommon beta-Asp (isoaspartate). Therefore, four isomers, which are normal L-alpha-Asp, biologically uncommon L-beta-Asp, D-alpha-Asp, and D-beta-Asp are formed in alpha A-crystallins. The formation of these isomers at Asp-151 and Asp-58 of human alpha A-crystallin started shortly after birth, increased with age and the amount of D- beta-Asp was more than that of normal L-alpha-Asp in the alpha A-crystallin of the human lenses of 80 years range [30].

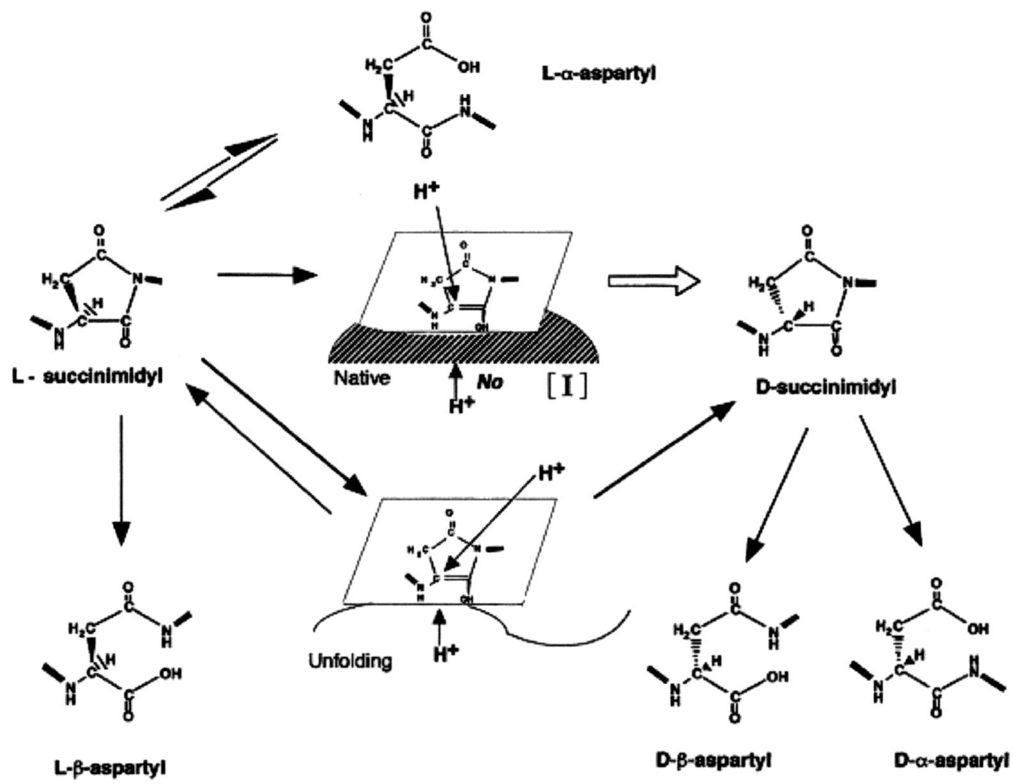

Figure 3. Reaction pathways for spontaneous inversion and isomerization of aspartyl residues in protein.

B. Skin: Detection of D-beta-Asp Containing Protein in Skin by Immunohistochemical Analysis Using Anti D-beta-Asp Containing Protein Antibody

As described in the previous section, D-beta-Asp forms much more easily in proteins than was thought. Therefore, we prepared a highly specific anti D-beta-Asp containing protein antibody in order to find D-beta-Asp containing protein in any tissues by immunohistochemical studies [31].

B-1. Materials

1. Paraffin-embedded skin specimens from elderly donors
2. Anti-D-beta-Asp containing protein antibody (antibody for peptide 3R) [31].
3. Anti-rabbit immunoglobulin antibody
4. Avidin-biotin complex system

B-2. Methods

1. Skin specimens were embedded in paraffin and cut into 5 μ thick sections.
2. The sections were deparaffinized, pretreated with 0.05% pronase (Dako, Glostrup, Denmark) in Tris-buffered saline (TBS) for 20 min at room temperature, and blocked with 10% goat serum.

3. The sections were incubated with the antibody for peptide 3R at 1: 100 dilution for 24 h at 4°C then incubated with biotin-conjugated anti-rabbit Ig antibody (Dako) at 1: 500 dilution for 1 h at room temperature.
4. The location of antigen-antibody was detected by avidin-biotin complex system (Dako). 3-Amino-9-ethyl carbasol (Dako) was used as a substrate.

B-3. Results

The antibody recognized integrated or disintegrated elastic fibers in the sun-exposed skin but not in the sun-protected skin of the elderly donors and in the face skin from young donor (Figure 4). These results suggest that UV irradiation and aging are closely related to the formation of D-beta-Asp in the elastic fibers of skin. The antibody could be a useful indicator for sun damage of the skin.

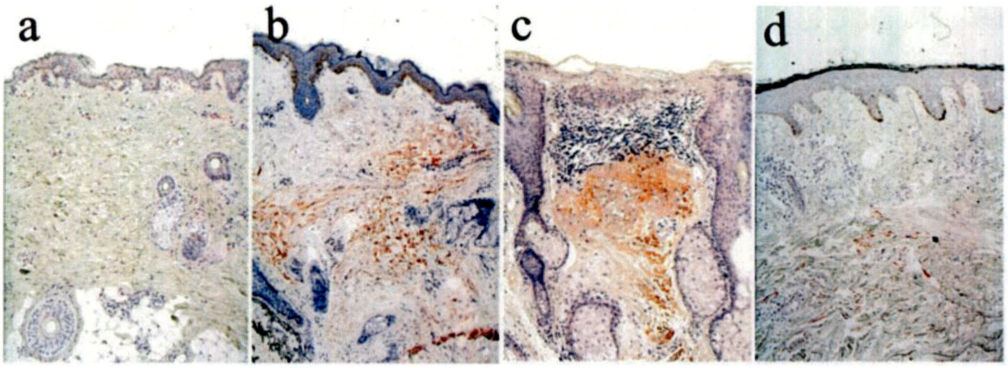

Figure 4. Immunohistochemistry of the antibody for D-beta-Asp containing peptide in the skin. Skin specimens obtained from the sun-exposed area (face) of 9-year-old (a), 59-year-old (b), and 86-year-old (c), or sun-protected area (buttock) of 85-year-old (d).

C. Detection of D-beta-Asp Containing Protein in Cell

We here present D-beta-Asp containing proteins in a cultured N/N1003A cell line derived from lens epithelial cell, which is commonly used in lens research. Recently, we detected the several different of D-beta-Asp containing proteins in the cell by 2D-PAGE and Western blot analysis using an anti-peptide 3R antibody. Subsequently, we identified these proteins by a combination of 2D-PAGE, Matrix-assisted Laser Desorption/Ionization Time-of-Flight Mass Spectrometry (MALDI-TOF-MS) analysis, computer searching algorithm and the MS-Fit online database searching algorithm [32] (scheme 3).

C-1. Materials

1. The N/N1003A rabbit cell line
2. Dulbecco's modified Eagle's medium (DMEM)
3. Two-dimensional gel electrophoresis
4. Antibodies: (a) Anti peptide 3R antibody, (b) A goat anti-rabbit IgG conjugated to horseradish peroxidase (HRP) (Biosource International, Camarillo, CA)
5. Trypsin (Promega, Madison, WI)
6. αcyano-4-hydroxycinnamic acid (WAKO)

7. Matrix-assisted Laser Desorption/Ionization Time-of-flight Mass Spectrometry (Voyager DE Pro (Applied Biosystems, Foster City, CA)

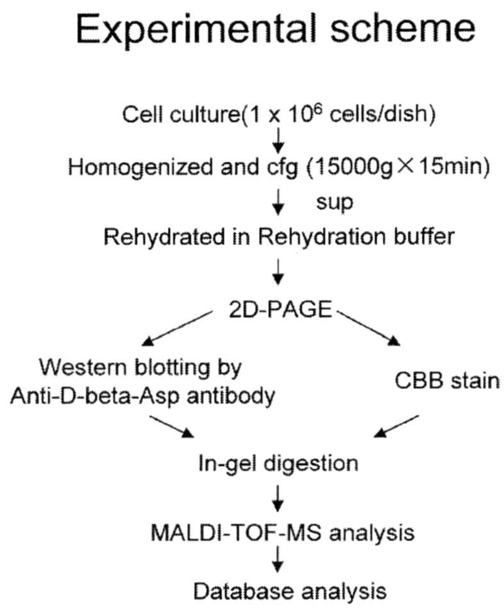

Scheme 3.

C-2. Methods

C-2-1 The Culture of the N/N1003A Cell Line

1. The N/N1003A cells were cultured in Dulbecco's modified Eagle's medium (DMEM) containing 1% antimycotic antibiotic and 10% fetal calf serum (FCS) that was heat-inactivated (56°C, 30 min) before use.
2. The N/N1003A cell line was seeded at approximately 1.0 x 10^6 cells in 100 mm plates in culture medium.
3. Cells were harvested at 80-90% confluency using a cell scraper and then washed with Dulbecco's phosphate buffered saline (DPBS).
4. Cells were re-suspended in extraction buffer (8 M urea, 2 M thiourea, 4% 3-[(3-cholamidopropyl) dimethylammonio]-1-propanesulfonate (CHAPS), 65 mM dithiothreitol (DTT)) and then disrupted by homogenization on ice.
5. The cell-free extract was clarified by centrifugation (15,000 g, 15 min) and the supernatant was collected.
6. A 2% immobilized pH gradient (IPG) buffer and a trace of bromophenol blue were added to the supernatant. Total protein concentration was determined using the Bradford assay [33].

C-2-2. Two-Dimensional Gel Electrophoresis

1. Each preparation was subjected to two-dimensional gel electrophoresis (isoelectric focusing (IEF) in the first dimension and SDS-PAGE in the second dimension).
2. The IEF separation was carried out using Immobiline Dry Strips (18 cm, pH 3-10) by following the manufacturer's suggested method (Amersham Biosciences, Piscataway, NJ). Prior to the first IEF analysis, the supernatant was incubated overnight with dry strips at 4°C. IEF analysis was performed using an EPS 3501XL power supply (Amersham Biosciences) for 18 hours at 15°C.

3. After IEF, the IPG strips were equilibrated for 20 min in equilibration buffer (50 mM Tris-HCl, pH 6.8, containing 6 M urea, 2% SDS, 30% glycerol, and 200 mM DTT). Strips were then re-equilibrated for 20 min in the same buffer containing 135 mM iodoacetamide instead of DTT.
4. In all cases, molecular weight separation was achieved using an ExcelGel SDS XL gradient 12-14 (Amersham Biosciences). The protein spots on the gel were visualized by staining with Coomassie blue.

C-2-3. Western Blot

1. Extracted proteins from the cells were separated by 2D-PAGE and then transferred to Immobilon™ transfer PVDF membrane (0.45 μm) (Millipore, Bedford, MA).
2. The membrane was blocked with 5% non-fat milk, 0.1% BSA in 0.1 % polyoxyethylenesorbitan monolaurate (Tween-20) containing Tris-buffered saline (TBST pH 7.4) overnight at 4°C.
3. The membrane was washed three times with TBST containing 0.1% Tween-20 and then incubated for 1 hour at room temperature with the first antibody (1:1000).
4. After three washes with TBST containing 0.1% Tween-20, the membrane was incubated (1 hour, room temperature) with horseradish peroxidase-conjugated secondary antibody (1:1000).
5. Following another three washes with TBST containing 0.1% Tween-20, labeled proteins were visualized using enhanced chemiluminescence (ECL Plus Western blotting detection kit, Amersham Biosciences) on BioMax MS film (Kodak, Rochester, NY).

C-2-4. Image Analysis

1. Expression levels were determined from the relative spot intensity of each protein versus other non-relative spots on a CBB stained gel and Western blotting membrane.
2. Differentially expressing proteins detected by Western blotting were selected for excision and MALDI-TOF-MS analysis.

C-2-5. In-Gel Digestion

The enzyme digestion of proteins in the CBB stained gel was performed as follows.

1. Protein spots excised from the CBB stained gel were de-stained with 50% MeOH/50 mM ammonium bicarbonate and cut into 1-2 mm slices using a spot cutter.
2. The gel pieces were then washed with 50% acetonitrile/50 mM ammonium bicarbonate and incubated with acetonitrile.
3. The gel pieces were dried in a SpeedVac vacuum concentrator and then re-hydrated using 25 μl of trypsin solution (Promega, Madison, WI). Digestion was continued at 37°C for 12 hours.
4. The tryptic peptides were first extracted using 0.1% TFA/50% acetonitrile and then re-extracted twice using the same solution after 15 min agitation.
5. The extracted samples were mixed in an Eppendorf tube and then dried in a vacuum concentrator prior to MALDI-TOF-MS analysis.

C-2-6. Identification of D-beta-Asp Containing Proteins by MALDI-TOF-MS

All MALDI-TOF-MS experiments were performed using a Voyager DE Pro (Applied Biosystems, Foster City, CA) instrument.

1. The matrix (α-cyano-4-hydroxycinamic acid) was made up to a saturated solution with 0.1% trifluoroacetic acid/70% acetonitrile.
2. The sample solution and the matrix solution was mixed and applied in a 100 well target using Zip-Tip c-μ18 (Millipore) and air-dried.
3. External calibration was achieved with a peptide mixture provided by ABI.

4. Protein identification was conducted using Ms-Fit software (available on the internet) against the Swiss-Plot Database.

C3 Results

One of D-beta-Asp containing proteins was an enolase showing homology with Tau-crystallin. Other D-beta-Asp containing proteins were lamin A/C, cytoplasmic NADP$^+$-dependent isocitrate dehydrogenase, fructose-bisphosphate aldolase A, aldose reductase, L-lactate dehydrogenase A or calponin H2, phosphoglycerate mutase 1, phosphatidylethanolamine-binding protein, alpha-B-crystallin and peptidyl-prolyl cis-trans isomerase A (PPlase) (Figure 5). The results indicate that the inversion and isomerization of an Asp-residue occurs during the posttranslational modification of specific proteins in actively proliferating cells.

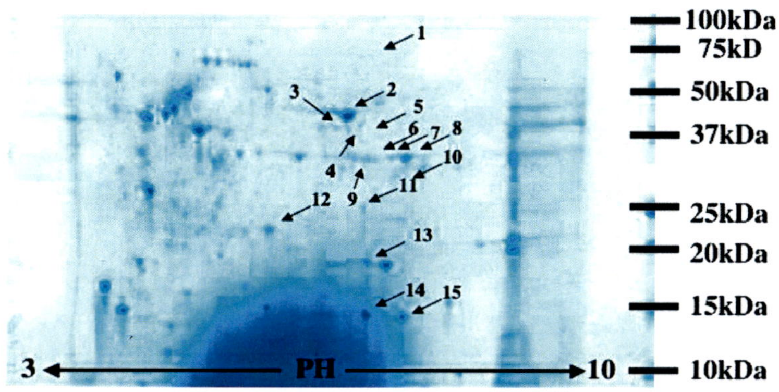

Figure 5. Two-dimensional gel electrophoretic separation of proteins derived from a rabbit lens epithelium cell line (N/N1003A) and identification of D-beta-Asp containing proteins in the pH 3-10 range. Spots that indicated by arrows, were immunoreactive proteins with anti D-beta-Asp containing protein antibody. They were excised and analyzed by in-gel trypsin digestion and MALDI-TOF-MS.

Prospects

Most researchers have held the idea that L-amino acids in proteins could never change to D-isomers under the physiological conditions in the living body. This is because proteins are not easily modifiable chemically as a result of the selection process during evolution which has worked to ensure very stable properties of such molecules. This general idea had no real scientific basis but became an accepted because D-amino acids could not in fact be found.However, recent improvements in analytical techniques now enable accurate analysis of amino acid enantiomers at the picomole level. Therefore, we were able to identify a very small quantity of D-aspartic acids at specific sites in lens proteins comprised almost entirely of L-amino acids, similar to looking for a needle in a haystack. Aspartic acid residues were not racemized uniformly but the racemization proceeded at specific sites of proteins. We were also able to propose a mechanism for D- and beta-aspartic acid formation in alpha-crystallin. The formation of D-aspartic acids in protein depends on primary structure and higher order structure surrounding Asp residues. We propose that a chiral reaction field exists in the native higher order structure of human alpha A-crystallin which induces the inversion of L-Asp to D-Asp residues at Asp-151.

D-Amino acid formation with age partially proceeds in proteins which contained only one handed structures comprised of L-amino acids in a process of evolution opposite to the evolution of life.

When we try to find D-Asp containing protein in the tissue or cell where we don't know the presence of the protein, we can analyze the immunohistochemistry or 2D-PAGE and Western blotting as described in section B and C, respectively.

If the D-beta-Asp containing proteins were detected, the spot was identified by the MALDI-TOF-MS analysis. Subsequently, the large amounts of the protein will be purified from the tissues or cells and the specific sites of D-beta-Asp residues in the protein can be determined by biochemical analysis as described in section A-2-3.

References

[1] Helfman, P.M.and Bada, J.L. (1975). Aspartic acid racemization in tooth enamel from living humans. *Proc Natl Acad Sci U S A, 72*, 2891-2894.

[2] Helfman, P.M.and Bada, J.L. (1976). Aspartic acid racemisation in dentine as a measure of ageing. *Nature, 262*, 279-281.

[3] Masters, P.M. (1983). Stereochemically altered noncollagenous protein from human dentin. *Calcif Tissue Int, 35*, 43-47.

[4] Masuda, W., Nouso, C., Kitamura, C., et al. (2002). D-Aspartic acid in bovine dentine non-collagenous phosphoprotein. *Arch Oral Biol, 47*, 757-762.

[5] Ritz, S., Turzynski, A., Schutz, H.W., et al. (1996). Identification of osteocalcin as a permanent aging constituent of the bone matrix: basis for an accurate age at death determination. *Forensic Sci Int, 77*, 13-26.

[6] Ohtani, S., Yamamoto, T., Matsushima, Y., et al. (1998). Changes in the amount of D-aspartic acid in the human femur with age. *Growth Dev Aging, 62*, 141-148.

[7] Cloos, P.A.and Fledelius, C. (2000). Collagen fragments in urine derived from bone resorption are highly racemized and isomerized: a biological clock of protein aging with clinical potential. *Biochem J, 345 Pt 3*, 473-480.

[8] Powell, J.T., Vine, N., Crossman, M. (1992). On the accumulation of D-aspartate in elastin and other proteins of the ageing aorta. *Atherosclerosis, 97*, 201-208.

[9] Fisher, G.H., Garcia, N.M., Payan, I.L., et al. (1986). D-aspartic acid in purified myelin and myelin basic protein. *Biochem Biophys Res Commun, 135*, 683-687.

[10] Shapira, R.and Chou, C.H. (1987). Differential racemization of aspartate and serine in human myelin basic protein. *Biochem Biophys Res Commun, 146*, 1342-1349.

[11] Roher, A.E., Lowenson, J.D., Clarke, S., et al. (1993). Structural alterations in the peptide backbone of beta-amyloid core protein may account for its deposition and stability in Alzheimer's disease. *J Biol Chem, 268*, 3072-3083.

[12] McFadden, P.N.and Clarke, S. (1982). Methylation at D-aspartyl residues in erythrocytes: possible step in the repair of aged membrane proteins. *Proc Natl Acad Sci U S A, 79*, 2460-2464.

[13] Masters, P.M., Bada, J.L., Zigler, J.S., Jr. (1977). Aspartic acid racemisation in the human lens during ageing and in cataract formation. *Nature, 268*, 71-73.

[14] Fujii, N., Satoh, K., Harada, K., et al. (1994). Simultaneous stereoinversion and isomerization at specific aspartic acid residues in alpha A-crystallin from aged human lens. *J Biochem, 116*, 663-669.

[15] Fujii, N., Ishibashi, Y., Satoh, K., et al. (1994). Simultaneous racemization and isomerization at specific aspartic acid residues in alpha B-crystallin from the aged human lens. *Biochim Biophys Acta, 1204*, 157-163.

[16] Groenen, P.J.T.A., van den IJssel, P.R.L.A., Voorter, C.E.M., et al. (1990). Site-specific racemization in aging alpha A-crystallin. *FEBS Lett, 269*, 109-112.

[17] Fujii, N., Tajima, S., Tanaka, N., et al. (2002). The presence of D-beta-aspartic acid-containing peptides in elastic fibers of sun-damaged skin: a potent marker for ultraviolet-induced skin aging. *Biochem Biophys Res Commun, 294*, 1047-1051.

[18] Ritz-Timme, S., Laumeier, I., Collins, M.J. (2003). Aspartic acid racemization: evidence for marked longevity of elastin in human skin. *Br J Dermatol, 149*, 951-959.

[19] Ritz-Timme, S., Laumeier, I., Collins, M. (2003). Age estimation based on aspartic acid racemization in elastin from the yellow ligaments. *Int J Legal Med, 117*, 96-101.

[20] Stephenson, R.C.and Clarke, S. (1989). Succinimide formation from aspartyl and asparaginyl peptides as a model for the spontaneous degradation of proteins. *J Biol Chem, 264,* 6164-6170.

[21] Kaneko, I., Yamada, N., Sakuraba, Y., et al. (1995). Suppression of mitochondrial succinate dehydrogenase, a primary target of beta-amyloid, and its derivative racemized at Ser residue. *J Neurochem, 65*, 2585-2593.

[22] Fujii, N., Muraoka, S., Harada, K. (1989). Purification and characterization of a protein containing D-aspartic acid in bovine lens. *Biochim Biophys Acta 999*, 239-242.

[23] Muraoka, S., Fujii, N., Ueda, Y., et al. (1991). Racemization of aspartic acid in alpha A-crystallin of bovine lenses. *Biomed Res, 12*, 61-64.

[24] Swamy, M.S.and Abraham, E.C. (1991). Reverse-phase HPLC analysis of human alpha crystallin. *Curr Eye Res, 10*, 213-220.

[25] Fujii, N., Muraoka, S., Satoh, K., et al. (1991). Racemization of aspartic acids at specific sites in alpha A-crystallin from aged human lens. *Biomed Res, 12*, 315-321.

[26] Fujii, N., Harada, K., Momose, Y., et al. (1999). D-Amino acid formation induced by a chiral field within a human lens protein during aging. *Biochem Biophys Res Commun, 263*, 322-326.

[27] Geiger, T.and Clarke, S. (1987). Deamidation, isomerization, and racemization at asparaginyl and aspartyl residues in peptides. Succinimide-linked reactions that contribute to protein degradation. *J Biol Chem, 262*, 785-794.

[28] Tyler-Cross, R.and Schirch, V. (1991). Effects of amino acid sequence, buffers, and ionic strength on the rate and mechanism of deamidation of asparagine residues in small peptides. *J Biol Chem, 266*, 22549-22556.

[29] Fujii, N., Momose, Y., Harada, K. (1996). Kinetic study of racemization of aspartyl residues in model peptides of alpha A-crystallin. *Int J Peptide Protein Res, 48*, 118-122.

[30] Fujii, N., Takemoto, L.J., Y., M., et al. (1999). Formation of four isomers at Asp-151 residue of aged human alpha A-crystallin by natural aging. *Biochem Biophys Res Comm , 265*, 746-751.

[31] Fujii, N., Shimo-Oka, T., Ogiso, M., et al. (2000). Localization of biologically uncommon D-beta-aspartate-containing alphaA-crystallin in human eye lens. *Mol Vis, 6*, 1-5.

[32] Takata, T., Shimo-Oka, T., Miki, K., et al. (2005). Characterization of new D-beta-aspartate-containing proteins in a lens-derived cell line. *Biochem Biophys Res Commun, 334*, 1022-1031.

[33] Bradford, M. (1976). A rapid and sensitive method for the quantitation of microgram quantities of protein utilizing the principle of protein-dye binding. *Anal. Biochem, 72*, 248-254

In: D-Amino Acids: ISBN 1-60021-075-9
A New Frontier in Amino Acid and Protein Research, pp.369-374

Chapter 5.2

DETECTION OF AMYLOID β PEPTIDES WITH L-ISOASPARTATE IN ALZHEIMER'S DISEASE

Takahiko Shimizu and Takuji Shirasawa*
Research Team for Molecular Biomarkers, Tokyo Metropolitan Institute of Gerontology, Japan.

Keywords: isoaspartate, amyloid β, Alzheimer's disease, antibody, immunohistochemistry

1. INTRODUCTION

Alzheimer's disease (AD) is one of the most common neurodegenerative disorders characterized by the extracellular deposition of amyloid β peptides (Aβ) in the brain parenchyma and cortical blood vessels [1]. In the amyloid deposits of senile plaques, there are some peptides with different C-termini among which Aβ40 and Aβ42 representing the major Aβ species [2]. Biochemical studies on Aβ isolated from AD brains revealed that they were posttranslationally modified by isomerization, racemization, pyroglutamylation, and truncation [3,4]. In general, the isomerization and racemization of aspartyl residues are formed by spontaneous nonenzymatic chemical reactions that give rise to L- and D-isoaspartate, and D-aspartate.

To investigate in details isomerization of Aβ in AD brains, we generated anti-isoAsp7 and anti-isoAsp23 antibodies. We confirmed that isomerized Aβ was dominantly deposited in senile plaques and amyloid-bearing vessels. Moreover, we showed that Aβ isomerized at either position 7 or 23, was differentially deposited on senile plaques and vascular amyloids of AD brains. We therefore postulate that spontaneous modifications of Aβ such as isomerization and racemization are potentially responsible for the development of sporadic AD [5].

In this text, we introduce the method of the peptide synthesis, antibody preparation and immunohistochemistry on AD brain sections.

* Correspondence concerning this article should be addressed to Takuji Shirasawa, E-mail: sirasawa@tmig.or.jp, FAX: +81-3-3579-4776.

2. Materials

1. Fmoc amino acids, Fmoc-L-Asp(OH)-OtBu for L-isoaspartic acid, Fmoc-D-Asp(OH)-OtBu for D-isoaspartic acid, Fmoc-D-Asp(OtBu)-OH, *N*-[(Dimethylamino)-1*H*-1,2,3-triazolo[4,5-*b*]pyridin-1-ylmethylene]-*N*-methylmethanaminium hexafluorophosphate *N*-oxide (HATU) and N,N-diisopropylethylamine (DIPEA) (Applied Biosystems, Foster City, USA)
2. Pioneer Peptide Synthesizer (Applied Biosystems, Foster City, USA)
3. High performance liquid chromatography (HPLC) or equivalent
4. Reverse-phased columns, ODS-HG-5 and ODS-UG-5 (Nomura Chemicals, Seto, Japan)
5. Maleimide-activated KLH (Pierce, Rockford, IL, USA).
6. FMP-Cellulofine (Seikagaku Kogyo, Tokyo, Japan).
7. Nitrocellulose membrane (BA83; Schleicher & Schuell, Dassel, Germany)
8. Block Ace (Dainihon Seiyaku, Tokyo, Japan).
9. Horseradish peroxidase-coupled goat anti-rabbit IgG antibody (Kirkegaard & Perry Laboratories, Gaithersburg, MD, USA)
10. Enhanced chemiluminescence (ECL) system (Amersham Pharmacia Biotech, Buckinghamshire, UK)
11. LAS-1000 Luminoimage analyzer (Fuji film, Tokyo Japan)
12. Hippocampal sections of AD brains (kindly provided by Dr. S. Murayama of Tokyo Metropolitan Institute of Gerontology)
13. Formic acid (specific grade)
14. Anti-Aβ11-28 monoclonal antibody (12B2, IBL, Takasaki, Japan)
15. Biotinylated secondary antibody and ABC reagent kit (Vector Laboratories, Burlingame, CA, USA)
16. 3,3'-Diaminobenzidine (Dojindo, Kumamoto, Japan).

3. Methods

3-1. Peptide Synthesis [6,7]

1. Peptides were synthesized using the Fmoc-solid phase strategy with a Pioneer Peptide Synthesizer.
2. Fmoc-L-Asp(OH)-OtBu and Fmoc-D-Asp(OH)-OtBu were used for peptides containing L-isoaspartic acid and D-isoaspartic acid, respectively. Fmoc-D-Asp(OtBu)-OH was used for the peptides containing D-aspartic acid.
3. The N-terminus Fmoc group was deblocked with 20% peperidine in dimethylformamide (DMF) for 5 min, and the coupling reaction was carried out using Fmoc amino acid (4 equv), HATU (4 equv), and DIPEA (8 equv) in DMF for 60 min.
4. After completion of the chain elongation, the peptide resin was treated with a cocktail containing tifluoroacetic acid (TFA), phenol, triisopropylsilane, and water (8.8:0.5:0.2:0.5) for final deprotection and cleavage from the resin.
5. After 2 hr of shaking at room temperature, the crude peptides were precipitated by diethyl ether and dried.
6. The crude peptides were dissolved in 0.1% TFA and purified by reversed-phase HPLC on Develosil ODS-HG-5 under acidic conditions (0.1% TFA-acetonitrile solution)

7. The crude Aβ1-42 and Aβ1-40 peptides were each dissolved in 0.1% NH_4OH and purified by reversed-phase HPLC on Develosil ODS-UG-5 under basic conditions (0.1% NH_4OH-acetonitrile solution).
8. Aβ peptides were dissolved in 0.02% NH_4OH at 250 μM and stored at –80°C until use.
9. The purity of these synthetic peptides was checked by reversed-phase HPLC, mass spectrometry, and amino acid composition analysis.

3-2. Preparation of Anti-isoAsp Aβ Antibodies [8,9]

1. The antibodies specific to Aβ(isoAsp7) or Aβ(isoAsp23) were produced using synthetic peptides, DAEFRH-(L-isoAsp)-SGYEVC [Aβ1-12(L-isoAsp7) + C] and acetyl-AE-(L-isoAsp)-VGSNKC [Aβ21-28(L-isoAsp23) + C], respectively.
2. Immunogens peptides were conjugated to maleimide-activated KLH according to the manufacturer's instructions.
3. The immunogens, 100-150 μg peptide/rabbit, together with complete or incomplete Freund's adjuvant, were subcutaneously injected into rabbits over every 3 weeks.
4. A week after the third booster, antisera were collected.
5. The specific antibodies were affinity-purified using the immunogen peptides immobilized on FMP Activated Cellulofine.
6. Finally, the anti-isoAsp antibodies were applied to Aβ1-12 or Aβ21-28 peptide- immobilized resin and flow-through fractions were collected.

3-3. Specificity of Anti-isoAsp Aβ Antibodies on Dot Blot Analysis [9]

1. Synthetic peptides (1 μg) Aβ1-42, Aβ1-42 with L-isoAsp7 [Aβ1-42(L-isoAsp7)], and Aβ1-42 with L-isoAsp23 [Aβ1-42(L-isoAsp23)] were applied to nitrocellulose membranes and dried in air.
2. Membranes were blocked for 1 hr in Block Ace.
3. Membranes were incubated for 1 hr with anti-isoAsp antibodies [0.4 μg/ml in 0.05% Tween 20, 150 mM NaCl, 50 mM Tris-HCl, pH 7.6 (TTBS), and 1% bovine serum albumin].
4. Filters were washed 3 times for 5 min in TTBS.
5. Horseradish peroxidase-coupled goat anti-rabbit IgG antibody was incubated in a 1:10,000 dilution in TTBS for 1 hr.
6. Filters were washed 3 times for 5 min in TTBS.
7. Immunoreactive proteins were visualized with an ECL system and Luminoimage analyzer LAS-1000.
8. To confirm the specificity of anti-isoAsp antibodies, antibodies were preincubated with or without the various isomerized or racemized Aβ peptides (200 molar excess) as indicated in Figure 1.

3-4. Immunohistochemistry [8-10]

1. We used the brains and sections with the informed consent obtained from the patient's family by paper, and the consent was approved by the ethical committee of Tokyo Metropolitan Institute of Gerontology and Tokyo Metropolitan Geriatric Hospital.
2. Hippocampal sections of AD patients were deparafinized and blocked with 20% goat serum for 1 hr.

3. Sections were treated with 99% formic acid for 10 sec and washed 3 times in TBS before immunostaining.
4. Sections were incubated with anti-isoAsp antibodies (1:200) or anti-Aβ11-28 monoclonal antibody (1:50) at 4°C overnight.
5. After being washed, sections were incubated with biotinylated secondary antibody for 90 min at room temperature.
6. Sections were then incubated with ABC reagent mix for 1 hr, and the immunoreactivity was visualized using 3,3'-diaminobenzidine.
7. After being washed in distilled water, sections were counterstained with hematoxylin.

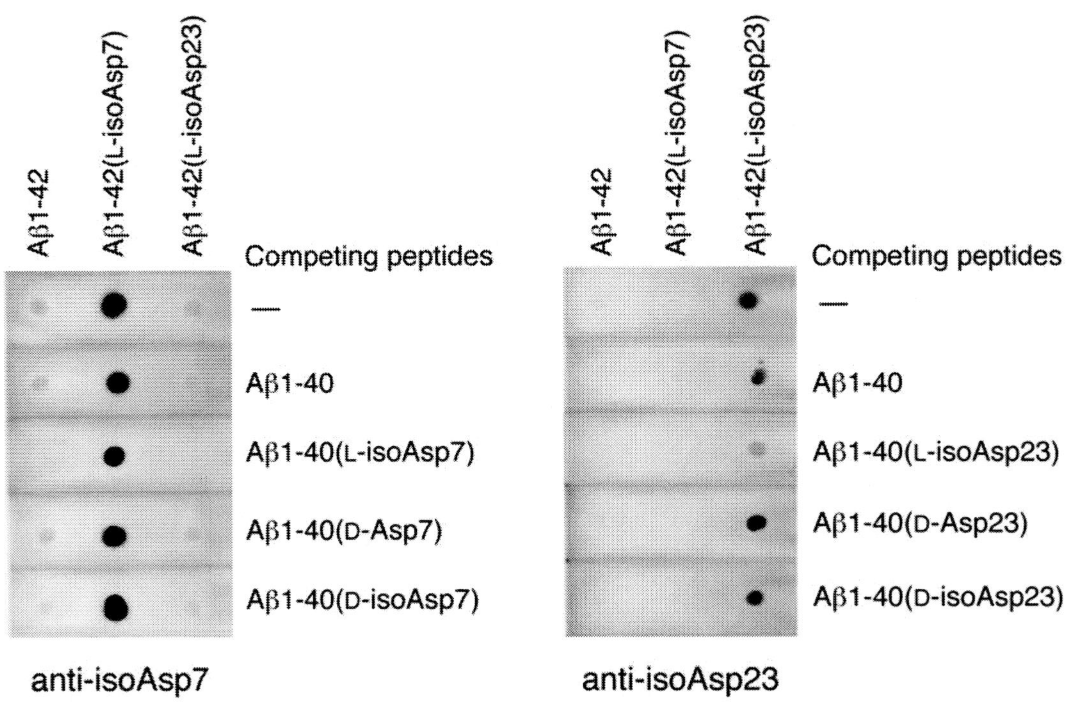

Figure 1. Characterization of anti-isoAsp antibodies. The peptide (1 μg) of Aβ1-42, Aβ1-42(L-isoAsp7), and Aβ1-42(L-isoAsp23) were subjected to dot blot analysis using anti-isomerized Aβ antibodies as indicated. Antibody binding was performed in the absence (-) or presence of the various isomerized or racemized Aβ1-40 peptides (200 molar excess) as indicated.

4. RESULTS

Characterization of Anti-isoaspartyl Aβ Antibodies

To investigate the specificity of anti-Aβ with L-isoasparte antibodies, we performed a dot blot analysis using the competitive Aβ peptides (Figure 1). Binding specificity was confirmed in an inhibition assay with either isomerized or racemized Aβ peptides. Anti-isoAsp7 and anti-isoAsp23 antibodies specifically recognized Aβ1-42(L-isoAsp7) and Aβ1-42(L-isoAsp23), respectively. Furthermore, we confirmed that anti-isoAsp7 and anti-isoAsp23 antibodies specifically reacted to Aβ1-40(L-isoAsp7) and Aβ1-40(L-isoAsp23) but not Aβ1-40, Aβ1-40(D-Asp7), or Aβ1-40(D-Asp23) and Aβ1-40(D-isoAsp7) and Aβ1-40(D-isoAsp23) in dot blot (data not shown). These results were confirmed by western blot analysis (data not shown). The recognition of anti-isoAsp23 antibody was specifically inhibited by Aβ1-40 with L-isoAsp23

[Aβ1-40(L-isoAsp23)] but not by other Aβ isomers. However, Aβ1-40 with L-isoAsp7 [Aβ1-40(L-isoAsp7)] completely failed to interfere with the binding of anti-isoAsp7 antibody to Aβ 1-42(L-isoAsp7). The result suggests that anti-isoAsp7 antibody recognizes Aβ(L-isoAsp7) fixed on the membrane, but fails to bind the soluble form of isomerized Aβ. In western blot analysis, we also confirmed the specificity of these antibodies [9]. Both antibodies specifically recognized each isomerized Aβ1-42. The results indicated that these anti-isoAsp antibodies recognized the specific structure of Aβ with site-specific isomerization at position 7 or 23.

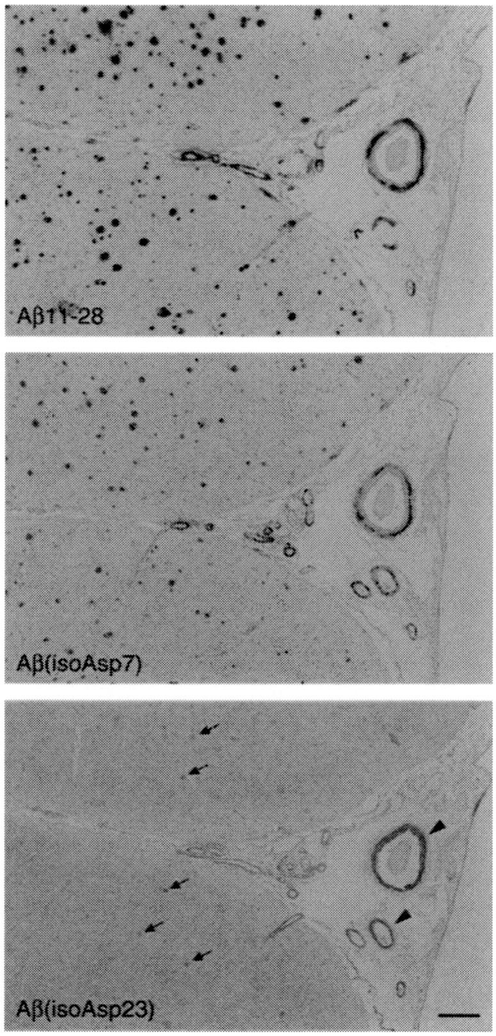

Figure 2. Immunohistochemical analysis of AD brains. Hippocampal sections were stained with anti- Aβ11-28 (upper panel), anti-isoAsp7 (middle panel), and anti-isoAsp23 (lower panel) antibodies. Conventional anti-Aβ and anti-isoAsp7 antibodies stained almost all plaques and amyloid-bearing vessels. Amyloid-bearing vessels (arrowheads) and senile plaques (arrows) were characteristically stained by anti-isoAsp23 antibody. The scale bar indicates 100 μm.

Isoaspartate Formation of Aβ in Senile Plaques and Vascular Amyloid

We performed an immunohistochemical analysis using anti-isoAsp antibodies on autopsied brains from AD patients. In the hippocampal sections in an AD case, conventional Aβ antibody that recognized both Aβ1-40 and Aβ1-42 stained almost all the plaques and amyloid-bearing vessels (Figure 2, upper panel). In adjacent sections, almost all the plaques and amyloid-bearing vessels were also positively stained with anti-isoAsp7 antibody (Figure 2, middle panel). In other adjacent sections, senile plaques and amyloid-bearing vessels showed, albeit less intensive, positive staining with anti-isoAsp23 antibody whereas the immunoreactivity was more obvious in vessel lesions (Figure 2, lower panel). Our findings show that isomerized Aβ, either at Asp7 or Asp23 was differentially deposited in plaques and amyloid-bearing vessels in AD. We therefore postulate that spontaneous modifications of Aβ such as isomerization and racemization are potentially responsible for the development of sporadic AD.

References

[1] Selkoe, D. J. (1999). Translating cell biology into therapeutic advances in Alzheimer's disease. *Nature, 399*, A23-31.

[2] Iwatsubo, T., Odaka, A., Suzuki, N., et al. (1994). Visualization of Abeta 42(43) and Abeta 40 in senile plaques with end-specific Abeta monoclonals: evidence that an initially deposited species is Abeta 42(43). *Neuron, 13*, 45-53.

[3] Saido, T. C., Iwatsubo, T., Mann, D. M., et al. (1995). Dominant and differential deposition of distinct beta-amyloid peptide species, Abeta N3(pE), in senile plaques. *Neuron, 14*, 457-466.

[4] Roher, A. E., Lowenson, J. D., Clarke, S., et al. (1993). Structural alterations in the peptide backbone of beta-amyloid core protein may account for its deposition and stability in Alzheimer's disease. *J Biol Chem, 268*, 3072-3083.

[5] Shimizu, T., Matsuoka, Y. & Shirasawa, T. (2005). Biological significance of isoaspartate and its repair system. *Biol Pharm Bull, 28*, 1590-1596.

[6] Fukuda, H., Shimizu, T., Nakajima, M., et al. (1999). Synthesis, aggregation, and neurotoxicity of the Alzheimer's Abeta1-42 amyloid peptide and its isoaspartyl isomers. *Bioorg Med Chem Lett, 9*, 953-956.

[7] Gonzalez, L. J., Shimizu, T., Satomi, Y., et al. (2000). Differentiating alpha- and beta-aspartic acids by electrospray ionization and low-energy tandem mass spectrometry. *Rapid Commun Mass Spectrom, 14*, 2092-2102.

[8] Shimizu, T., Watanabe, A., Ogawara, M., et al. (2000). Isoaspartate formation and neurodegeneration in Alzheimer's disease. *Arch Biochem Biophys, 381*, 225-234.

[9] Shimizu, T., Fukuda, H., Murayama, S., et al. (2002). Isoaspartate formation at position 23 of amyloid beta peptide enhanced fibril formation and deposited onto senile plaques and vascular amyloids in Alzheimer's disease. *J Neurosci Res, 70*, 451-461.

[10] Shin, Y., Cho, H. S., Fukumoto, H., et al. (2003). Abeta species, including IsoAsp23 Abeta, in Iowa-type familial cerebral amyloid angiopathy. *Acta Neuropathol (Berl), 105*, 252-258.

In: D-Amino Acids: ISBN 1-60021-075-9
A New Frontier in Amino Acid and Protein Research, pp.375-378

Chapter 5.3

DETECTION OF D-AMINO ACID IN PEPTIDES BY RP-HPLC AND MASS SPECTROMETRY

Céline Ollivaux* and Daniel Soyez
Biogenèse des Peptides Isomères - FRE CNRS 2852, Protéines: biochimie structurale et fonctionnelle, Université Paris 6, France.

Keywords: Crustacea, D-amino acid, mass spectrometry, neuropeptides, RP-HPLC.

1. INTRODUCTION

Post-translational epimerization of a L-amino acid residue to its D-counterpart in a peptide chain is a subtle modification, transparent to classical methods of molecular biology and analytical biochemistry. To date, this original phenomenon has been shown only in a few Vertebrates and in some Invertebrates, including crustaceans (review in [1]). In the American lobster, two neurohormones have been shown to be synthesized and secreted as stereoisomers: the CHH (Crustacean Hyperglycemic Hormone) with two forms differing only by the third residue, a phenylalanine, that is either in L- or D- configuration (CHH and D-Phe[3] CHH) and the VIH (Vitellogenesis Inhibiting Hormone) with a L- or a D- tryptophane in position 4 (VIH and D-Trp[4]VIH) [2,3]. Detection of D-amino acid containing peptides is complicated since both isomers present identical molecular mass, p*I* and primary sequence. In addition, the nature and the position of the D-residue as well as the nature of the neighbouring residues vary according to the peptides identified so far and therefore cannot be predicted, contrary to others post-translational modifications such as glycosylation or phosphorylation which occur at well defined sequence sites [4]. The only clues available to date are – i) that D-residue has been always found near peptide termini, most frequently at N-terminus (until aa^{4}) or, in a few instances, at C-terminus (aa^{n-2}), and – ii) that the D-amino acid containing peptide has generally been found to be more hydrophobic than its L-counterpart. Consequently, when both stereoisomers are present in a tissue preparation, they can be separated by RP-HPLC.

In this report, we describe a method with complementary techniques of analytical biochemistry and physicochemistry, which have been used to characterize putative stereoisomer peptides. The example of VIH is shown here.

* Correspondence concerning this article should be addressed to Céline Ollivaux, E-mail: celine.ollivaux@snv.jussieu.fr, FAX: +33-1-44-27-23-61.

2. Materials

1. Solution of acetic acid (10 %) at room temperature and heated at 80°C.
2. Glass-glass microgrinder to homogenize tissues.
3. Microcentrifuge (Eppendorf, Hamburg, Germany).
4. RP-HPLC columns: 1- an analytical 250 mm length x 4.6 mm internal diameter column, filled with Kromasil SI, 3.5 μm particle size (AIT Chromato, Le Mesnil-le-roi, France), 2- a microbore 250 mm length x 2.0 mm internal diameter column filled with Nucleosil C18, 5 μm particle size (Machery-Nagel EURL, Hoerdt, France) connected to a SpectraSYSTEM P4000 pumping gradient system (Thermo Separation Products, Fremont, CA, USA) with a TSP SpectroSeries UV100 spectrophotometer (set at 220 nm).
5. Acetonitrile plus 0.08% trifluoroacetic acid.
6. Water plus 0.1% trifluoroacetic acid.
7. Speed-Vac (Savant Instruments Inc. Holbrook, NY, USA).
8. Endoproteinase Asp-N sequencing grade (Roche Diagnostics, Mannheim, Germany).
9. Incubator at 37°C.
10. Standard peptides supplied by Neosystem (Strasbourg, France) or synthesized by the solid phase FastMoc chemistry on an Applied Biosystems 433A Automated Peptide Synthesizer (Applera, Courtaboeuf, France). Standard heptapeptides are synthesized with the sequence corresponding to the N-terminus of VIH (A-S-A-W-F-T-N) with all L-residues (standard Hep-L) or with a D-residue in position 2, 3, 4 or 5 (standards Hep-DS^2, Hep-DA^3, Hep-DW^4 and Hep-DF^5, respectively).
11. Mass spectrometer: reflectron matrix-assisted laser desorption/ionization time-of-flight (MALDI-TOF) mass spectrometer (Voyager DE RP, Perseptive Biosystems Inc., Framingham, MA).

3. Methods

1. Homogenization of tissues containing the peptides (30 sinus glands dissected from lobster eyestalks) in 80°C 10% acetic acid, centrifugation (10 000 g 5 min), then collection of the supernatant, and two complementary extractions of the pellet with 10% acetic acid at room temperature.
2. Fractionation of the pooled supernatants by RP-HPLC (column 1) to purify VIHs (VIH I and II), collection of the fractions containing the two hormones, and speed-vac drying of the fractions.
3. Cleavage of each VIH by endoproteinase Asp-N for 22 hours at 37°C.
4. Fractionation of digests by RP-HPLC (column 2)[1].
5. Screening of the digest fractions by MALDI-TOF MS.
6. MS sequencing of the peptide with the mass of the standard heptapeptides (795.9 Da).
7. Determination of the retention times of synthetic peptides by RP-HPLC using the same conditions as in 4 above.
8. Comparison of retention times of standard peptides and digests.

[1] One of the fragments corresponding to the primary sequence of the N-terminal heptapeptide of the VIH (A-S-A-W-F-T-N) should be contained in the digest mixture.

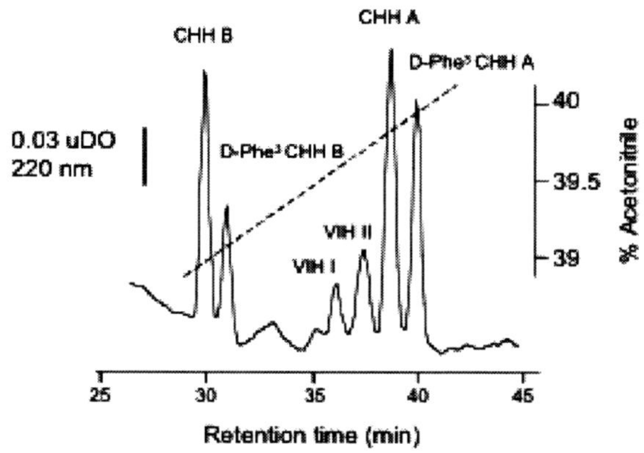

Figure 1: RP-HPLC analysis of an acetic acid extract of 30 lobster sinus glands. Only the part of the chromatogram where CHHs and VIHs neuropeptides are eluted is shown. The nature of the ultraviolet absorbance peaks was assessed by comparison with previous published similar analysis [5].

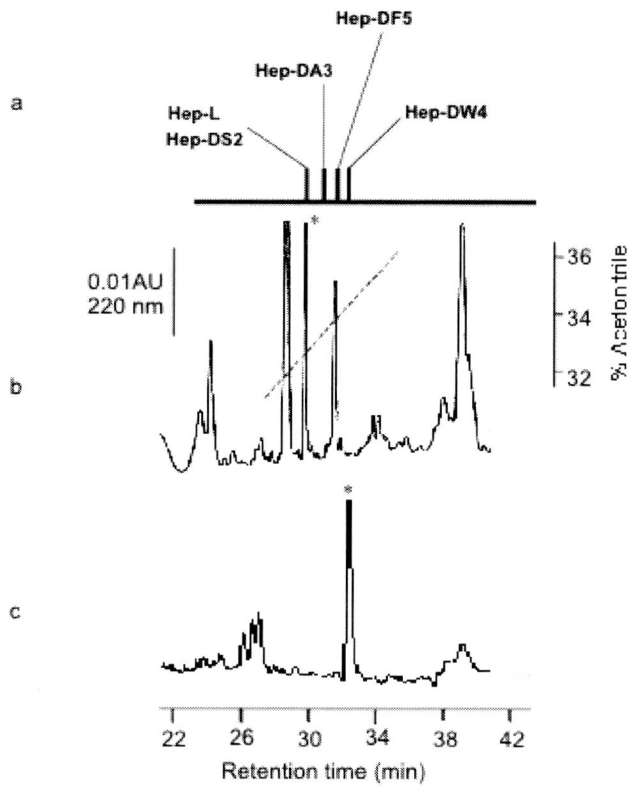

Figure 2: a: Retention times of standards Hep-L, Hep-DS2, Hep-DA3, Hep-DW4 and Hep-DF5. b: RP-HPLC of a VIH I digest (10 sinus gland equivalents). It is worth to note that the more hydrophobic peak contains uncleaved VIH. The nature of the ultraviolet absorbance peaks was assessed by comparison with HPLC profile of standard, i.e. heptapeptides Hep-L, Hep-DS2, Hep-DA3, Hep-DW4 and Hep-DF5, coupled to MALDI-TOF mass analysis. The asterisk indicates peptides with a mass of 795.9 Da c: RP-HPLC of a VIH II digest (10 sinus gland equivalents).

4. RESULTS

Two peptides were identified from sinus gland extracts as VIH I and II, respectively (Fig.1), on basis of their retention time in RP-HPLC (by comparison with previous similar analysis [5]). This assumption was further confirmed by MALDI-TOF mass spectrometry (VIH has a mass of 9135 Da). RP-HPLC of VIH I digest showed that one peptide was eluted with the same retention time as the standard peptides Hep-L and Hep-DS2, which comigrated in our HPLC system (Fig. 2.a and b). This fragment was found by MS to display identical mass (795.9 Da) and sequence (A-S-A-W-F-T-N) as VIH N-terminal heptapeptide[2].

In the same way, analysis of VIH II digest showed that a peptide with the correct mass and sequence was eluted with the same retention time as the standard Hep-DW4 (Fig. 2.c).

Therefore, VIH I corresponds to all L-residue neuropeptide whereas VIH II displays a D-tryptophane in position 4. Our results have demonstrated that the crustacean neuroendocrine system synthesizes a mixture of stereoisomers of two different neuropeptides (CHH and VIH), where the D-residues differ by their nature and their position (D-Phe^3 CHH and D-Trp^{4}VIH).

REFERENCES

[1] Kreil G. (1997) D-amino acids in animal peptides, *Annu. Rev. Biochem. 66*, 337-345.

[2] Soyez D., Van Herp F., Rossier J., *et al.* (1994) Evidence for a conformational polymorphism of invertebrate neurohormones. D-amino-acid residue in crustacean hyperglycemic hormone, *J. Biol. Chem. 269*, 18295-18298.

[3] Ollivaux C., Amiche M., Gallois D. *et al.* A novel D-amino acid containing peptide of the Crustacean Hyperglycemic Hormone family: biochemical and immunochemical characterization, *submitted.*

[4] Bennet H.P., Bradbury A.F., Huttner W.B. & Smyth D.G. (1993) Processing of propeptides: glycosylation, phosphorylation, sulfation, acetylation, and amidation. In *Mechanisms of intracellular trafficking and processing of proproteins* (Peng Loh, Y., ed) pp. 251-288, CRC, London, Tokyo.

[5] Soyez D., Lecaer J.P., Noel P.Y., *et al.* (1991) Primary Structure of Two Isoforms of the Vitellogenesis Inhibiting Hormone from the Lobster Homarus-Americanus, *Neuropeptides. 20*, 25-32.

---

[2] The presence of a D-Ser^2 residue in VIH I fragment was later invalidated by immunoassays using a specific antiserum against Hep-L.

In: D-Amino Acids: ISBN 1-60021-075-9
A New Frontier in Amino Acid and Protein Research, pp.379-387

Chapter 5.4

DETERMINATION OF D-AMINO-ACID RESIDUES IN PEPTIDES FROM ANIMAL VENOM BY NMR: APPLICATION TO PLATYPUS VENOM PEPTIDE OVCNPS

Allan M. Torres[1], Paramjit S. Bansal[2], Paul F. Alewood[2], Dominic P. Geraghty[3] and Philip W. Kuchel[1]*

[1]School of Molecular and Microbial Biosciences, University of Sydney, Australia;
[2]Institute for Molecular Bioscience, University of Queensland, Australia;
[3]School of Human Life Sciences, University of Tasmania, Australia.

Keywords: D-amino acid, HPLC, natriuretic peptides, NMR, platypus venom.

1. INTRODUCTION

The discoveries of D-amino-acid-containing biologically active peptides in various complex organisms have refuted the longstanding notion that such unusual molecules only exist in lower forms of organisms. To date, D-amino-acid-containing bioactive peptides have been found in frogs [1,2], snails [3,4], crustaceans [5], spiders [6,7], and even a mammal [8]. It is now accepted that such unusual biomolecules are produced post-translationally from standard peptides that contain all amino acids in the L-form through a modification process that is mediated by a specific enzyme referred to as a peptidylaminoacyl-L/-D-isomerase or simply L- to D-isomerase [9,10].

D-amino-acid-containing peptides or proteins are not easily detected in protein analysis because common procedures, such as Edman degradation and mass spectrometry, do not discriminate between the L- and D-forms of the amino acid. D-amino acid residues are usually discovered only after confusing experimental results are obtained. This can arise for any of the following reasons: (1) two forms of the peptide exist with identical primary structures are observed in chromatography; (2) the synthetic version of the peptide fails to elute with the native peptide in chromatography; and (3) the activity of the synthetic peptide is different from that of the native peptide.

* Correspondence concerning this article should be addressed to Philip W. Kuchel, Email: p.kuchel@mmb.usyd.edu.au, FAX: +61-2-9351-4726.

D-amino acids in peptides are commonly determined biochemically through the use of proteolytic enzymes that can discriminate between the L- and D-forms of the amino acid. For example, digestion of a peptide by amino peptidase or carboxy peptidase is inhibited by the presence of a D-amino acid in the sequence. D-amino oxidase, on the other hand, only acts on the D-form of free amino acids so that the disappearance of an amino acid "signal" in a protein digest is taken as a strong indication of its presence [1]. The use of proteolytic enzymes solely to determine D-amino acids in peptides is, however, a tedious process that can take several steps, especially if the suspected peptide has many residues.

In this Chapter, we present how we have used NMR spectroscopy as an alternative method to detect the presence of D-amino acid residues in peptides. In particular, we applied this technique to two forms of OvCNP from platypus venom [8,11]. The procedure described here may be readily applied to other sources of peptides and proteins whenever one or more D-amino acid residues are suspected. The method assumes that the primary structures of the two peptides are known and are proven to be identical, and that about 1 mg of each peptide is available for the NMR experiments. The NMR results may be used in conjunction with biochemical results to conclusively establish the presence of a D-amino acid residue. Alternatively, synthetic versions of a peptide can be made to conclusively confirm the hypothesis provided by NMR.

2. Materials

Venom Samples

Samples used in the experiments consisted of either crude venom or venom-gland extracts. Crude venom samples were obtained from trapped live adult male platypuses that were later released unharmed or from deceased animals that had been accidentally killed by motor vehicles. Extracts were obtained from excised venom glands from deceased platypuses. A 1% (w/v) trehalose solution was used to extract soluble components in the venom gland. Trehalose is routinely used in protein extraction to stabilize or protect components from denaturation, especially during freezing and drying.

Reverse-Phase HPLC

A GBC HPLC system (Gibco, Dandenong, Vic., Australia) with an LC 11250 pump and LC 1210 uv/vis dual wavelength detector was used in the chromatographic analyses. The C18 analytical column (250 mm × 4.60 mm) used was either VYDAC (Hesperia, CA, USA) 218TP54 or Phenomenex (Torrance, CA, USA) Synergi 4 μ Hydro-RP. The solvent system consisted of 0.1 % (v/v) trifluoroacetic acid in MilliQ water (Buffer A) and 0.1 % (v/v) trifluoroacetic acid in acetonitrile (HPLC grade, Buffer B). Acetonitrile must be handled with care as it is poisonous and flammable; proper ventilation is needed when working with it and skin contact must be avoided.

Synthetic OvCNPa and OvCNPb

Chemical syntheses of L-2Leu-OvCNP(OvCNPa) and D-2Leu-OvCNP(OvCNPb) were performed manually using 2-(1-*H*-Benzotriazol-1-yl-)1,1,3,3 tetramethyluronium hexafluorophosphate (HBTU) activation of Boc-amino acids with in-situ neutralization chemistry, as previously described [12]. Details of the procedure are beyond the scope of this Chapter and are described in Torres et al. [8].

NMR Spectroscopy

All NMR experiments were performed on a Bruker (Karlsruhe, Germany) AVANCE-600 DRX spectrometer using a 5-mm ^{1}H-inverse probe operating at a sample temperature of 25°C. The spectrometer was equipped with XWINNMR software (Bruker) for data acquisition and processing. A Silicon Graphics Workstation with XEASY software [13] was used in the spectral analysis.

3. Methods

Preparation of Crude Venom and Venom Gland Extracts

1. Crude venom was obtained from captured platypuses by directly holding the spurs onto a small collection container (Eppendorf centrifuge tube). Additional venom samples were also from the spurs by direct aspiration. Note that the amount of venom material (~50-100 μL) that was obtained from the spurs was relatively small compared to the amount of material obtained from venom glands.
2. Venom glands were processed by first dissecting-out the two glands from the deceased animals after tying the ducts with ligatures. The glands were finely sliced, the ligatures removed, and the tissue was then placed in a container with 50 mL of cold 1% (w/v) trehalose solution. The material was then homogenized with a Bamix cutter using short bursts (~5 s) to ensure that the temperature remained low and thus avoided denaturation.
3. The resulting mixture was centrifuged for 5-10 min at 5 °C; the liquid in the extract was then decanted and passed through a 0.2 μm Sartorius Minisart filter (Göttingen, Germany) to avoid contaminating the HPLC system during chromatography. The solid material that was separated by centrifugation was discarded.
4. The resulting filtrate in step (3) was dispensed into 1 mL Eppendorf tubes and then frozen for later use as starting material for the HPLC and NMR experiments described below.

Reverse-phase HPLC

1. The frozen samples were defrosted and then centrifuged at high speed for ~1 min to ensure that no solid material was introduced to the HPLC system. This was done because the venom samples still displayed, or formed, cloudiness or precipitates after some time even though they had already been filtered.
2. Depending on the amount of dissolved material in the venom sample, 10-500 μL of the liquid mixture was loaded onto the HPLC column. Components were eluted at a flow rate of 1 mL min^{-1} and were detected at either 215 nm or 280 nm. Chromatograms recorded at 215 nm usually had a higher signal-to-noise ratio than those obtained at 280 nm; however the baselines of the chromatograms were flatter at 280 nm.
3. Two linear gradient profiles with different durations were usually implemented.
   a. For a long acquisition run of ~60 min, the solvent gradient in % of B were 5-20 for 5 min, 20-45 for 40 min, 45-60 B for 5 min, 60 for 5 min and 60-5 for 5 min.
   b. For a short acquisition run of ~35 min, the solvent gradient in % of B were 5-25 for 5 min, 25-35 for 16 min, 35-60 for 2 min, 60 for 2 min, and 60-5 for 2 min.
4. OvCNP peptide samples were collected manually in 1 mL Eppendorf tubes, pooled in 15 mL centrifuge tubes, and then lyophilized.

NMR Spectroscopy

1. NMR samples were prepared by dissolving 1-2 mg of the lyophilized native or synthetic OvCNP peptides in 0.350 mL of 90% H_2O/10% D_2O in a 5-mm magnetic-susceptibility matched Shigemi (Allison Park, PA, USA) NMR tube. Regular NMR tubes can also be used but this special NMR tube is preferable if only limited small amounts are available.
2. One-dimensional NMR experiments were first performed for each peptide sample at ambient temperature. Significant differences between the chromatogram of each sample may be evident at this stage, especially in the chemical shift region between 6.5 to 10 ppm, which corresponds to the backbone amide resonances. A small difference in pH of the sample may also lead to small shifts (less than 0.05 ppm) in this region.
3. To pinpoint which part of a sequence contains the D-amino acid(s), the identity of the NMR signals must be established or assigned; and this requires performing at least two 2D NMR experiments, a TOCSY [14] and a NOESY [15] (see Notes). In the present study, a TOCSY experiment with a mixing time of 90 ms and a NOESY with a mixing time of 200 ms were performed at 25° C (ambient temperature).
4. The 2D data were processed in XWINNMR and then analyzed using XEASY. Knowing the amino-acid sequence, the cross-peaks in the TOCSY and NOESY spectra were assigned using standard NMR procedures [16] (see Notes). A table of ^{1}H chemical shift assignments for each form of the OvCNP was then obtained (see Table 1).
5. The NH and Hα chemical shift differences between the two forms of the peptide were plotted as a function of the residue number (see Figure 3). This plot provided information on the possible presence of a D-amino acid residue, and its location in the sequence.
6. 2D TOCSY maps of the two forms were plotted and carefully compared to confirm conclusively the location of the suspected region of altered structure. Once determined, three OvCNP synthetic peptides containing D-amino acids at positions 1, 2 and 3 were synthesized.
7. NMR (standard 1D and 2D TOCSY) and RP-HPLC experiments of the synthetic versions were performed using identical experimental conditions as employed earlier. The resulting data were then compared with the previous data. The 2D TOCSY spectrum provided the most conclusive result that showed that a D-amino acid at position 2 was present in OvCNPb (see Figure 4).

Table 1. Partial ^{1}H NMR chemical shifts of the first five amino-acid residues in OvCNPa and OvCNPb at pH 3 and 298 K.

Residue	OvCNPa			OvCNPb		
	NH	Hα	Hβ	NH	Hα	Hβ
Leu1	-	4.00	1.66, 1.61	-	4.00	1.71, 1.60
Leu2	8.60	4.32	1.58, 1.47	8.80	4.20	1.48, 1.34
His3	8.67	4.69	3.25, 3.16	8.95	4.84	3.11, 3.38
Asp4	8.36	4.60	2.73	8.27	4.57	2.71
His5	8.51	4.99	3.13, 3.25	8.59	4.99	3.13, 3.25

4. RESULTS

Typical RP-HPLC chromatograms of the crude venom and venom gland extract are given in Figure 1. The venom contains several proteins and peptides many of which are not found in any other source [17]. There are also variations in composition between samples depending on season of sample collection and age of the animal. Clearly, OvCNPa and OvCNPb in both chromatograms eluted at significantly different times although their amino-acid sequences are identical. DLP-2 and DLP-4 share the same characteristics as OvCNPs, so that one of them might also contain a D-amino-acid residue.

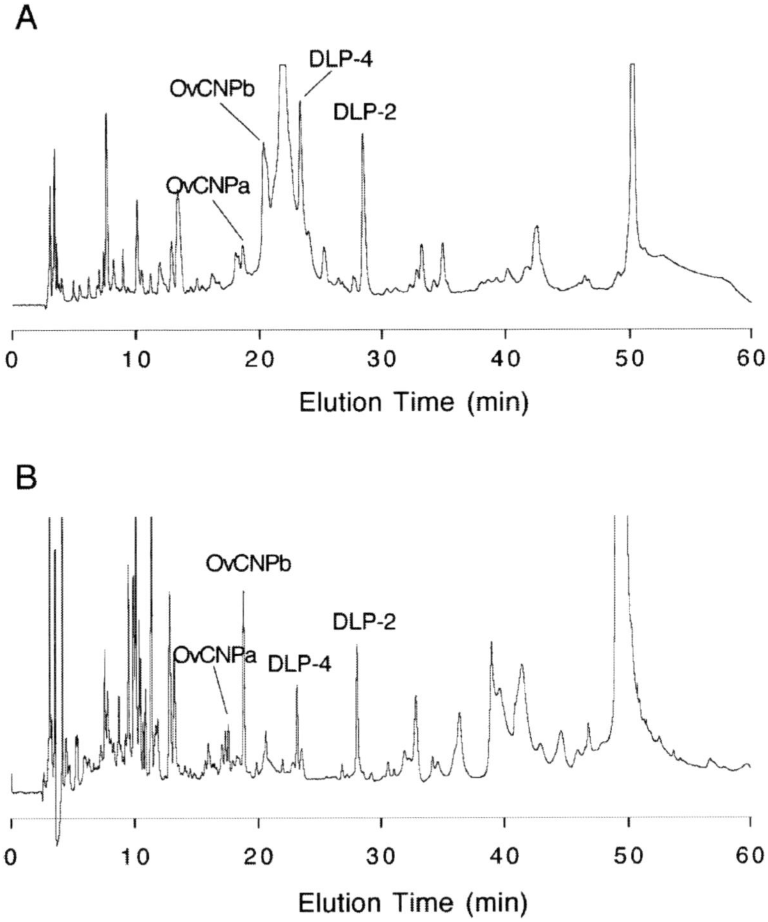

Figure 1. RP-HPLC chromatograms of (A) crude venom and (B) venom gland extracts from platypus. The components were eluted using the 60 min acquisition as described in Methods and were detected by using light absorption at a wavelength of 215 nm. Labeled peaks represent peptides that were suspected of containing D-amino acid residues.

It had been established previously that a synthetic sample of OvCNP which contained all L- amino acids co-eluted with OvCNPa [11]. This implied that OvCNPa contained all L-amino acids while OvCNPb was suspected to contain at least one D-amino acid residue. Sufficient amounts of native OvCNPb for NMR investigation were obtained by one-step RP-HPLC purification; however, the amount of native OvCNPa collected was insufficient. This was due to fact that the amounts of OvCNPa in the samples as shown in the RP-HPLC traces in Figure 1 were considerably less than those for OvCNPb. NMR experiments were therefore performed on the synthetic sample of OvCNPa.

Visual comparison of the 1D NMR spectra of OvCNPa and OvCNPb, as shown in Figure 2, already displayed significant differences between the two forms of OvCNP. The ^{1}H chemical shift assignments (see Table 1) and the appropriate section of the 2D TOCSY spectra of the two peptides clearly showed that the differences lay near the N-terminus whereby NH cross-peaks of residues 2-5 of the two forms were different. The plot of the NH and Hα NMR chemical shift differences of the two peptide forms, as shown in Figure 3, further confined the difference to residues 2 and 3, since the differences were greater for these residues than those for the rest.

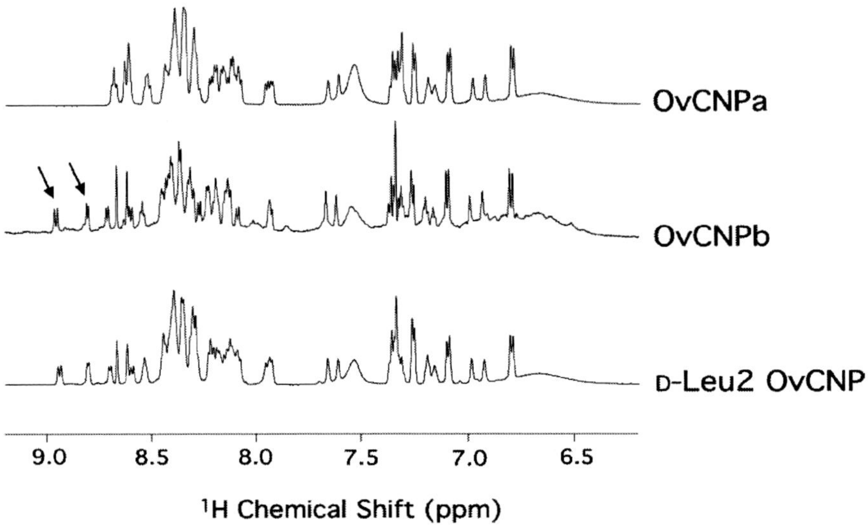

Figure 2. Parts of 1D ^{1}H NMR spectra of OvCNPs. The peaks indicated by arrows in the spectrum of OvCNPb represent backbone amide signals that were visually different in OvCNPa and OvCNPb. Those two peaks were clearly present in the synthetic peptide, D-Leu2 OvCNP. The spectral plots are patterned on similar plots in [8].

It would be easy to establish biochemically at this stage if a D-amino acid near the N-terminus of OvCNPb were present simply by subjecting both peptides to an amino peptidase. This has been done and has been reported elsewhere [8]. The experiment showed that the rate of digestion of OvCNPb was considerably less than that for OvCNPa; and this further supported the presence of a D-amino acid in OvCNPb. However, additional experiments would be required to definitively pin-point the location of the D-amino acid in the sequence. Alternatively, chemical synthesis of a D-amino-acid-containing analogue of OvCNP, followed by RP-HPLC and NMR spectroscopy can be performed without performing the peptidase digestion to determine conclusively if a D-amino acid is present. These two methods combined to establish the exact location of the D-amino-acid residue in the OvCNP sequence. Three forms of OvCNP were chemically synthesized with the D-amino acid introduced at either positions 1, 2 or 3. NMR spectral comparisons as shown in Figures 2 and 4 conclusively identified the synthetic peptide with D-Leu at position 2 (D-leu2 OvCNP) as having essentially identical 1D and 2D TOCSY spectra to OvCNPb. RP-HPLC experiments (results not shown) supported the conclusions, since D-leu2 OvCNP and OvCNPb eluted at the same time. These results showed beyond reasonable doubt that the difference between OvCNPa and OvCNPb was due to an unusual amino acid (D-Leu) at position 2.

The results of this study have important biological implications in that they showed for the first time that a biologically active D-amino-acid-containing peptide exists in a mammal. It is likely that other D-amino-acid-containing peptides and the requisite L- to D-isomerase are also present in the platypus venom. Thus, there is the possibility that other higher animals, including humans may be using D-amino-acid-containing proteins that are accordingly altered to have 'desirable' properties. D-amino acids have been discovered in the human lens but they are due to aging or disease and are apparently non-functional [18]. The search for biologically active peptides or proteins that contain D-amino acid residues is a difficult task. Perhaps, it would be prudent to concentrate the search for such residues in small proteins or peptides such as hormones, neuropeptides and anti-microbial peptides as has been suggested recently [10]. The NMR methods describe here would certainly be valuable in detecting and establishing their presence.

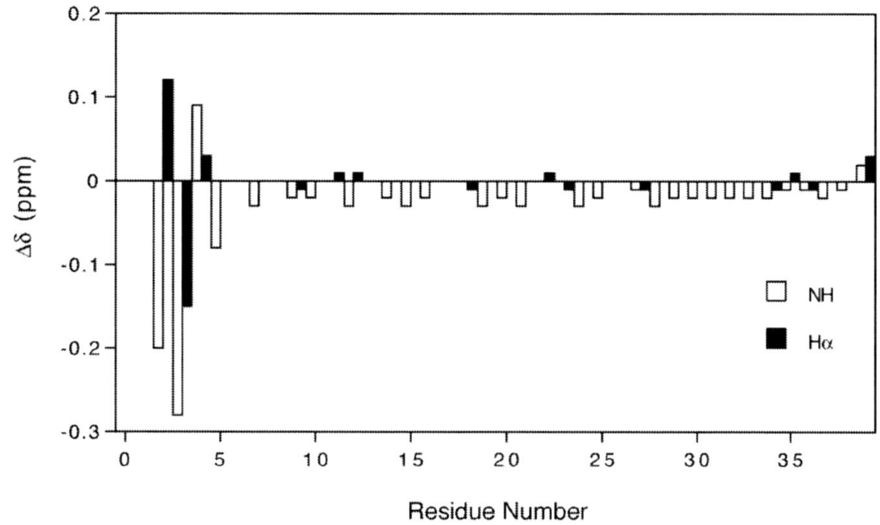

Figure 3. ¹H NMR chemical shift differences between OvCNPa and OvCNPb. (■) Hα and (□) NH ¹H NMR chemical shift differences. Values were obtained by subtracting OvCNPb chemical shifts from those of OvCNPa. The plot is patterned on a similar one in [8].

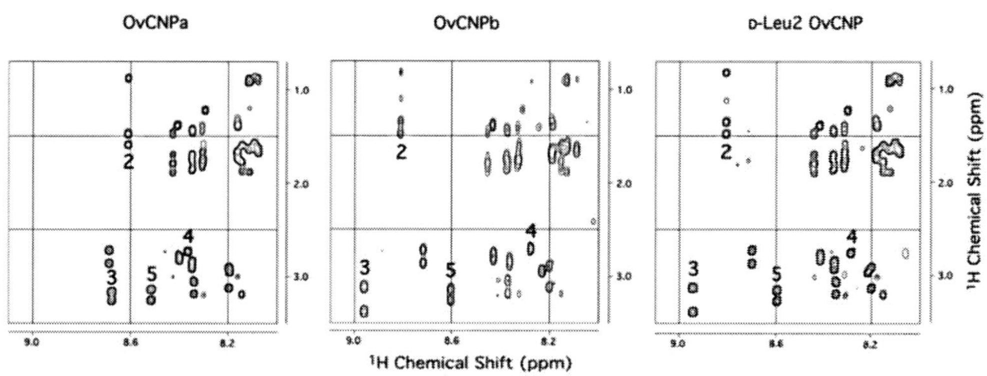

Figure 4. Comparsion of 2D TOCSY spectra of OvCNPs. Backbone amide (NH) cross-peaks that were different in OvCNPa and OvCNPb are indicated by sequence numbers placed either above (3, 4, 5) or below (2) the cross-peaks. The 2D maps are patterned on a figure in [8].

5. Notes

Determination of ¹H chemical shift assignments in proteins and peptides usually requires implementing several 2D NMR experiments; and the number of experiments needed depends on the solution behavior of the peptide and the length of the amino-acid sequence. Large polypeptides or proteins with molecular weights greater than 5 k may also require labelling with the magnetic isotopes ^{15}N and ^{13}C, hence the cost may be prohibitive.

It is assumed that the researcher is at least familiar with NMR methods of protein structure determination. Ideally, only two 2D experiments, a TOCSY, and a NOESY for larger peptides (> 2 kDa) or a ROESY for smaller peptides (< 2 kDa), are needed. These provide sufficient information for this type of study. A complete ^{1}H chemical shift assignment is not necessary but a sequential NMR assignment that only includes the backbone amide (NH) and Hα resonances is required. A TOCSY obtained with a 90 ms mixing time and a NOESY with a 200-300 ms mixing time are usually a prudent starting point with peptides of size greater than 2 kDa.

Other 2D experiments, such as a double-quantum filtered (DQF)-COSY [19] may be used to aid in the assignment of resonances to spin systems in amino-acid residues. Performing 2D experiments at different temperatures and mixing times is often useful for resolving overlapping peaks and in eliciting more peaks, especially those where there has been co-incidence with the water signal. The duration of a typical TOCSY experiment for OvCNP is ~12 h while a NOESY experiment typically takes ~24 h to achieve useful signal-to-noise in the spectra.

The spectral analysis may be straightforward or complicated depending on the molecular properties, as already described above. Analysis of short peptide chains (less than 10 amino acid residues) requires less time and may be completed within a day while longer peptides may require several days.

It is important to note that NMR spectra obtained at this stage provide information about the secondary and tertiary structures of the two peptide forms. Thus, it should be possible to calculate the three-dimensional fold of each peptide if the ^{1}H chemical shifts are completely determined. Structure calculations from NMR data are, however, tedious requiring more time and effort than what was needed in the present study of platypus venom peptides.

The key to the usefulness of NMR spectroscopy in detecting D-amino acids is through discrimination of NMR spectral properties of the two sorts of peptides. The application is similar to the use of RP-HPLC, however NMR provides considerably more structural information and is more structurally specific. This is because it is theoretically possible that two isoforms can have identical elution times in RP-HPLC, but it is very unlikely that their NMR spectra will be equivalent. However, NMR spectroscopy has two basic drawbacks: (1) substantial amounts of sample are needed (mgs); and (2) the experiments and spectral analysis require a significant amount of time to run, and expertise to analyse. But if the peptide were small, or if peptide fragments were available then this method is easy to perform. In addition, structural information on conformations of the peptides in solution is a welcome offshoot of this approach.

REFERENCES

[1] Montecucchi, P. C., de Castiglione, R., Piani, S., Gozzini, L. & Erspamer, V. (1981). Amino acid composition and sequence of dermorphin, a novel opiate-like peptide from the skin of *Phyllomedusa sauvagei*. *Int J Pept Prot Res, 17*, 275-283.

[2] Richter, K., Egger, R. & Kreil, G. (1987). D-Alanine in the frog-skin peptide dermorphin is derived from L-alanine in the precursor. *Science, 238*, 200-202.

[3] Kamatani, Y., Minakata, H., Kenny, P. T., Iwashita, T., Watanabe, K., Funase, K., Sun, X. P., Yongsiri, A., Kim, K. H. & Novales-Li, P. (1989). Achatin-I, an endogenous neuroexcitatory tetrapeptide from *Achatina fulica* Ferussac containing a D-amino acid residue. *Biochem Biophys Res Commun, 160*, 1015-1020.

[4] Jacobsen, R. B., Jimenez, E. C., de la Cruz, R. G. C., Gray, W. R., Cruz, L. J. & Olivera, B. M. (1999). A novel D-leucine-containing Conus peptide: diverse conformational dynamics in the contryphan family. *J Pept Res, 54*, 93-99.

[5] Soyez, D., Van Herp, F., Rossier, J., Le Caer, J. P., Tensen, C. P. & Lafont, R. (1994). Evidence for a conformational polymorphism of invertebrate neurohormones. D-amino acid residue in crustacean hyperglycemic peptides. *J Biol Chem, 269*, 18295-18298.

[6] Heck, S. D., Siok, C. J., Krapcho, K. J., Kelbaugh, P. R., Thadeio, P. F., Welch, M. J., Williams, R. D., Ganong, A. H., Kelly, M. E. & Lanzetti, A. J. (1994). Functional consequences of posttranslational isomerization of Ser46 in a calcium channel toxin. *Science, 266*, 1065-1068.

[7] Shikata, Y., Watanabe, T., Teramoto, T., Inoue, A., Kawakami, Y., Nishizawa, Y., Katayama, K. & Kuwada, M. (1995). Isolation and characterization of a peptide isomerase from funnel web spider venom. *J Biol Chem, 270*, 16719-16723.

[8] Torres, A. M., Menz, I., Alewood, P. F., Bansal, P., Lahnstein, J., Gallagher, C. H. & Kuchel, P. W. (2002). D-Amino acid residue in the C-type natriuretic peptide from the venom of the mammal, *Ornithorhynchus anatinus*, the Australian platypus. *FEBS Lett, 524*, 172-176.

[9] Kreil, G. (1997). D-amino acids in animal peptides. *Annu Rev Biochem, 66*, 337-345.

[10] Jilek, A., Mollay, C., Tippelt, C., Grassi, J., Mignogna, G., Mullegger, J., Sander, V., Fehrer, C., Barra, D. & Kreil, G. (2005). Biosynthesis of a D-amino acid in peptide linkage by an enzyme from frog skin secretions. *Proc Natl Acad Sci, USA 102*, 4235-4239.

[11] de Plater, G. M., Martin, R. L. & Milburn, P. J. (1998). A C-type natriuretic peptide from the venom of the platypus (*Ornithorhynchus anatinus*): structure and pharmacology. *Comp Biochem Physiol Part C Toxicol Pharmacol, 120*, 99-110.

[12] Schnolzer, M., Alewood, P., Jones, A., Alewood, D. & Kent, S. B. (1992). In situ neutralization in Boc-chemistry solid phase peptide synthesis. Rapid, high yield assembly of difficult sequences. *Int J Pept Prot Res, 40*, 180-193.

[13] Bartels, C., Xia, T. H., Billeter, M., Güntert, P. & Wüthrich, K. (1995). The program XEASY for computer-supported NMR spectral-analysis of biological macromolecules. *J Biomol NMR, 6*, 1-10.

[14] Bax, A. & Davis, D. G. (1985). MLEV-17-based two-dimensional homonuclear magnetization transfer spectroscopy. *J Magn Reson, 65*, 355-360.

[15] Kumar, A., Ernst, R. R. & Wüthrich, K. (1980). A two-dimensional nuclear Overhauser enhancement (2D nOe) experiment for the elucidation of complete proton-proton cross-relaxation networks in biological macromolecules. *Biochem Biophys Res Commun, 95*, 1-6.

[16] Wüthrich, K. (1986). *NMR of Proteins and Nucleic Acids*. New York: John Wiley and Sons.

[17] Torres, A. M., Wang, X., Fletcher, J. I., Alewood, D., Alewood, P. F., Smith, R., Simpson, R. J., Nicholson, G. M., Sutherland, S. K., Gallagher, C. H., King, G. F. & Kuchel, P. W. (1999). Solution structure of a defensin-like peptide from platypus venom. *Biochem J, 341*, 785-794.

[18] Fujii, N., Satoh, K., Harada, K. & Ishibashi, Y. (1994). Simultaneous stereoinversion and isomerization at specific aspartic acid residues in alpha A-crystallin from human lens. *J Biochem, 116*, 663-669.

[19] Rance, M., Sorensen, O. W., Bodenhausen, G., Wagner, G., Ernst, R. R. & Wüthrich, K. (1983). Improved spectral resolution in COSY ^{1}H NMR spectra of proteins via double quantum filtering. *Biochem Biophys Res Commun, 117*, 479-485.

In: D-Amino Acids: ISBN 1-60021-075-9
A New Frontier in Amino Acid and Protein Research, pp.389-397

Chapter 5.5

PROTEIN L-ISOASPARTYL-*O*-METHYLTRANSFERASE CATALYZES *IN SITU* FORMATION OF D-ASPARTATE AND D-ISOASPARTATE IN PROTEINS

Dana W. Aswad*
Department of Molecular Biology and Biochemistry, University of California, USA

Keywords: aspartate, isoaspartate, isomerization, racemization, methyltransferase.

1. INTRODUCTION

Although D-amino acids are rare in eukaryotic cellular proteins, significant levels of D-aspartate have been documented at specific sites in certain long-lived proteins such as the crystallins of the eye lens, elastins, and β-amyloid [1]. The unusual susceptibility of aspartate to racemization is attributable to the tendency of aspartyl and asparaginyl residues to undergo the transpeptidation reaction shown in Fig. 1. This non-enzymatic damage reaction involves the formation of a metastable succinimide (cyclic imide) intermediate that hydrolyzes to a mixture of mainly L-aspartyl (15-30%) and L-isoaspartyl (60-85%) forms. D-aspartyl and D-isoaspartyl forms (not shown in Fig. 1) are also generated as minor by-products because the succinimide, with a somewhat acidic α-proton, is prone to forming a non-chiral planar carbanion that can flip into the D- or L- configuration when the proton is recaptured [2]. In long-lived proteins, this mechanism can give rise to significant accumulation of D-aspartyl sites at specific succinimide-prone locations.

In recent years it has become apparent that the accumulation of D-aspartyl sites is not limited to long-lived proteins with low metabolic turnover. The protein repair enzyme, protein L-isoaspartyl-*O*-methyltransferase (PIMT) appears to make a major contribution to this more rapid appearance of D-aspartate in proteins via the pathway shown in Fig. 2. In its attempts to repair atypical L-isoaspartyl linkages, PIMT promotes the repetitive formation of the metastable cyclic imide allowing greatly increased opportunity for partial racemization to occur.

* Correspondence concerning this article should be addressed to Dana W. Aswad E-mail: dwaswad@uci.edu, FAX: +01-949-824-8551.

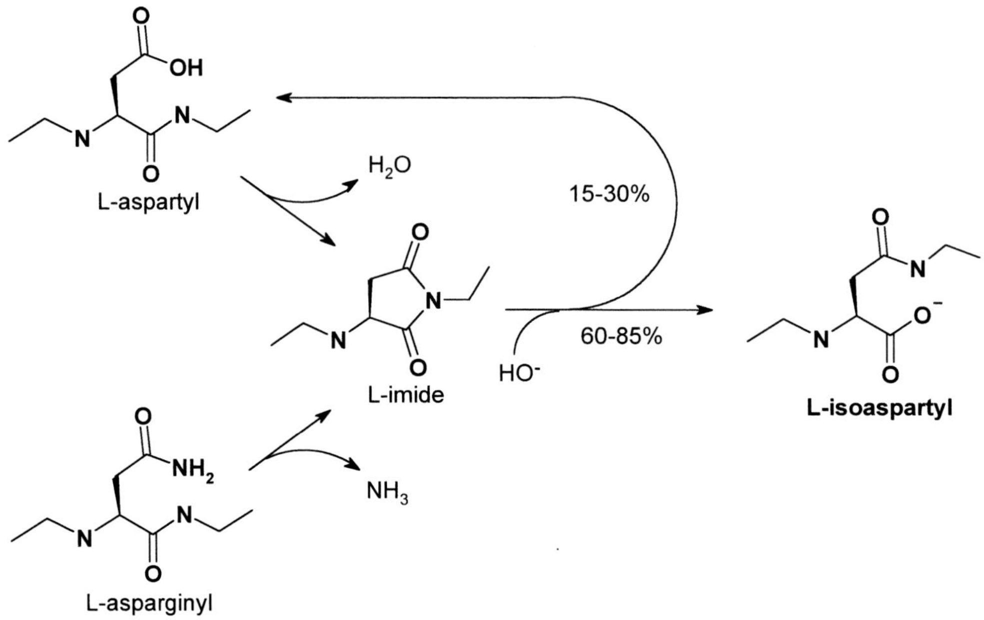

Figure 1. Mechanism by which L-isoaspartyl sites arise from L-aspartyl and L-asparaginyl sites in peptides and proteins. This spontaneous intramolecular rearrangement occurs most readily at Asn-Gly, Asn-Ser and Asp-Gly sequences in flexible regions of polypeptides. The L-isoaspartyl form typically accounts for 60-85% of the succinimide hydrolysis product, while 15-30% goes to a normal L-aspartyl form.

Several lines of experimental evidence provide compelling support for the idea that PIMT catalyzes the accumulation of D-aspartate in cellular proteins as a side-reaction to its main repair function. [1] D-Asp and D-isoAsp sites have been observed as the major by-products of PIMT-dependent *in vitro* repair of synthetic L-isoaspartyl peptides [3,4]. [2] Inhibition of PIMT activity in red blood cells, occurring as a consequence of kidney disease, results in decreased levels of D-aspartate in red cell proteins [5]. [3] The N-terminus of the catalytic subunit of protein kinase A contains a deamidation-prone Asn-Gly sequence. The *in vivo* deamidation products were found to be -L-Asp-Gly-, -D-Asp-Gly-, and -D-isoAsp-Gly-, with little or no -L-isoAsp-Gly [6]. This spectrum of products is exactly as predicted from the mechanism shown in Fig. 2. PIMT most likely prevented accumulation of -L-isoAsp-Gly- (the expected major product of deamidation) but promoted racemization via repetitive imide formation. [4] One known target of PIMT-dependent repair *in vivo* in mammals is aspartate-25 in histone H2B [7]. Approximately 11-14% of this site is present as the D-enantiomer (D-isoAsp and D-Asp) regardless of whether the H2B is isolated from 5 week-old mouse brain or 15 year old dog brain [8]. Other Asx residues in H2B are not prone to L-isoAsp formation *in vivo* and do not accumulate significant levels of D-aspartate. A catalytic role for PIMT in racemization at Asp-25 is evident from the fact that the amount of D-Asp/D-isoAsp found at those sites is greatly reduced in PIMT knockout (KO) mice.

This chapter reviews the key experimental data obtained by our lab that supports a role for PIMT in promoting the accumulation of D-Asp and D-isoAsp sites in cellular proteins.

2. Materials

1. OPA-NAC chiral fluorogen reagent. Dissolve 0.4 mg of *o*-phthalaldehyde (OPA) in 0.3 mL methanol and combine with 0.25 mL of 0.4 M Na-borate (pH 9.4), 0.39 mL water, and 0.06 mL of 1.0 M *N*-acetyl-L-cysteine (adjusted to pH 7 with NaOH). The final pH of the reagent should be 9.3-9.4.

2. Amino acid standards. Separate stock solutions of 1.0 mM L-aspartic acid and 1.0 mM D-aspartic acid are made up in 10 mM Na-phosphate (pH 7-7.2) and stored frozen when not in use.

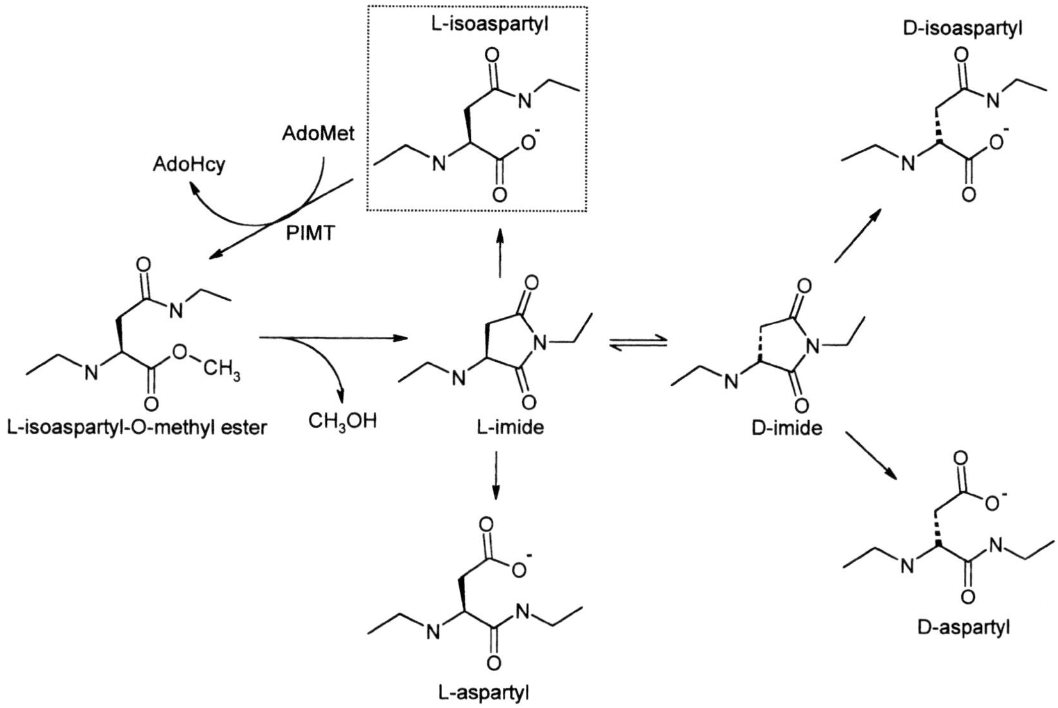

Figure 2. Mechanism for PIMT-dependent repair and racemization of L-isoaspartyl sites. PIMT catalyzes the methylation of L-isoaspartyl sites (dotted box) to form α-aspartyl *O*-methyl esters (far left). At physiological pH and temperature, the methyl esters spontaneously demethylate with a half-life of ca. 5-15 min to form the more stable L-succinimide (L-imide, center) which has a half-life of several hours. Hydrolysis of the L-succinimide generates a mixture of L-aspartyl and L-isoaspartyl peptides, the former representing completion of one repair cycle. Several additional cycles of methylation and demethylation convert nearly all of the original L-isoaspartyl sites to L-aspartyl sites. In the succinimide form, the acidity of the aspartate α-carbon markedly increases, thereby promoting production of D-imide (center right) via spontaneous racemization. Both the L-imide and the D-imide hydrolyze to form the corresponding isoaspartyl and aspartyl peptides. Although succinimide racemization is slow compared to its hydrolysis, its relative stability, combined with its constant replenishment during repeated repair cycles, provide ample opportunity for accumulation of D-isoaspartyl (upper right) and D-aspartyl sites (lower right), which are poor substrates for PIMT.

3. Methods

1. Acid Hydrolysis of Proteins and Peptides

The analyte (peptide or protein) should be HPLC pure and substantially free of salt and small molecules containing primary amines that can react with OPA. The analyte (2-20 µg) is subjected to acid hydrolysis in 0.2 mL of 6 N HCl for 18 h at 110° C in an evacuated side-arm 5 mL hydrolysis tube (Pierce #29560) using a Reacti-Therm Heating Module (Pierce #18790 or equivalent). After cooling, the hydrolysate is reduced to dryness in a vacuum centrifuge. The residue is taken up in 0.2 mL of triethylamine:ethanol:water (1:2:2), re-dried, and finally dissolved in 75 µL of water. The hydrolysate can be stored frozen until use.

2. Derivatization of Amino Acids with OPA-NAC Reagent

A 20.0 µL sample of acid hydrolysate is thoroughly mixed with 5.0 µL of OPA-NAC reagent in a polyethylene microfuge tube. After 2-3 min at room temperature the reaction is stopped by adding 255 µL of 0.1 M Na-acetate (pH 5.2). From this 285 µL stopped reaction solution, a sample of 250 µL is taken immediately for HPLC analysis as described below.

3. HPLC-Fluorescence Analysis of derivatized Amino Acids

250 µL of derivatized sample (described above) is injected into a 4.6 x 100-mm Hypersil 3 µm ODS column (Thermo Electron) fitted with a 10-mm guard cartridge. Isocratic elution is carried out at 1.0 mL/min in 50 mM Na-acetate (pH 5.9), containing 3.2% (v/v) methanol. Elution of derivatized D- and L-aspartate is monitored with an on-line fluorescence detector set for excitation at 340 nm and emission at 455 nm. Aspartate D/L ratios are calculated from peak areas after correction for the appropriate standards and blanks, and for acid hydrolysis-induced racemization, as described below. This protocol is an update to our 1984 paper which describes the first use of OPA in combination with a chiral thiol (*N*-acetyl-L-cysteine) to separate D- and L-amino acids [9]. A example of an original HPLC fluorescence trace from that earlier work is shown in Fig. 3. An amino acid standard curve indicates that this method is linear over a range of 0-100 pmol or more (Fig. 4). It is noteworthy that the specific fluorescence of L-aspartate is approximately 12% greater than that of D-aspartate.

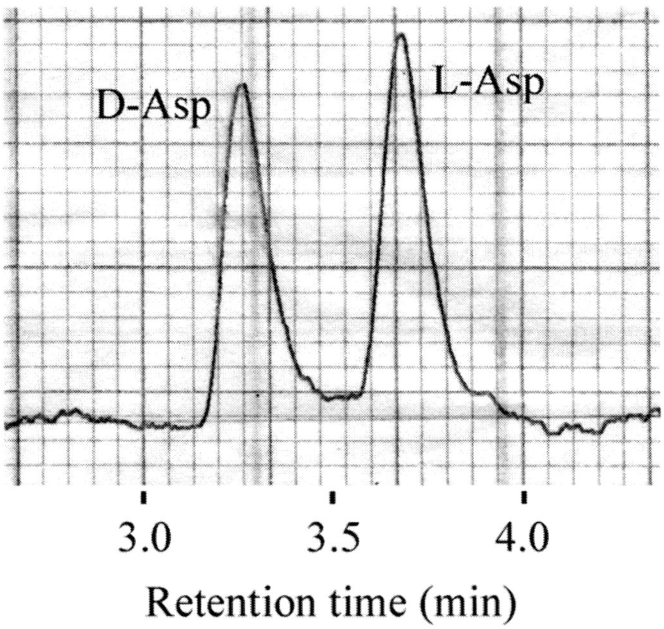

Figure 3. Separation of D- and L-aspartate by reversed-phase HPLC after derivatization with OPA-NAC reagent. A mixture of D- and L-aspartate standards, 10 pmol each, was injected onto a reversed-phase column as described in our 1984 publication [9]. Note that the D-aspartate peak is somewhat smaller than the L-aspartate peak due to the lower specific fluorescence of the former.

4. Recommendations Regarding Blanks

It is well established that amino acids are ubiquitous trace contaminants in laboratory solutions, containers, and even "purified water." We have found that acid hydrolysate "blanks" consisting of only 6 N HCl, can yield anywhere from 2-20 pmol of L-aspartate and lower, but detectable levels of D-aspartate. If one is analyzing the aspartate D/L ratio in a peptide purified by HPLC, a typical situation in our lab, we recommend the following. During the HPLC peptide purification run, collect two or three "blank" fractions in regions near the peptide peak of interest - but in regions that have no peptide as judged by the absence of UV absorbance at 214 nm. The volume of the blank fractions should be the same as the volume of the peptide fractions collected. These blanks should then be analyzed in parallel with the peptide samples, using all the same solutions, acid hydrolysis time, etc. The average aspartate D and L peak areas from the blanks are subtracted directly from the aspartate peak areas from the peptide samples prior to converting sample peak areas to pmol.

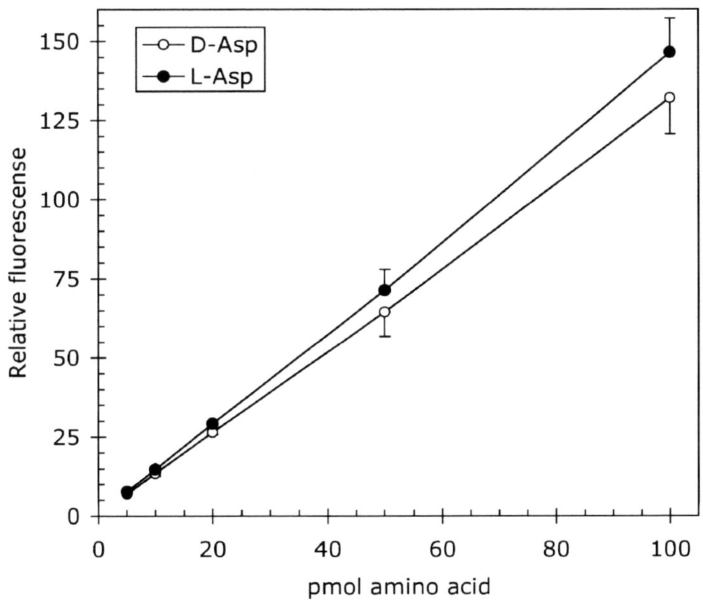

Figure 4. Standard curve relating HPLC fluorescence peak area to pmol of L-Asp and D-Asp standards over the range of 0-100 pmol. Linearity extends up to at least 250 pmol.

5. Correcting for Acid Hydrolysis-Induced Racemization

Using the protocols described above, we found that the synthetic peptide Ac-KKDGKKY-NH_2, consisting of only L-amino acids, yielded significant amounts of D-aspartate after acid hydrolysis. After acid hydrolysis times of 4, 12, 18 and 20 h, we found that 1.1%, 2.8%, 4.1% and 4.5% respectively of the total free aspartate was recovered in the D-configuration. To correct experimental data for this acid hydrolysis artifact, we use the following relationships:

$$D_o = D_i - D_i{*}R + L_i{*}R$$

$$L_o = L_i - L_i{*}R + D_i{*}R,$$

where

D_o = observed value of D-Asp/(D-Asp+L-Asp)
L_o = observed value of L-Asp/(D-Asp+L-Asp)
D_i = initial (true) value of D-Asp/(D-Asp+L-Asp)
L_i = initial (true) value of L-Asp/(D-Asp+L-Asp)
R = fraction of L-Asp converted to D-Asp (e.g., 0.041 at 18 h) during acid hydrolysis.

4. Results

In 1987 we reported the results of a study designed to test the idea that PIMT can repair atypical β-aspartyl peptide bonds (isoaspartyl sites) in proteins [3]. We incubated a synthetic L-isoaspartyl 12-mer peptide (KQVVD^SAYEVIK) with purified bovine brain PIMT and the methyl donor, *S*-adenosyl-L-methionine, for 0-24 h at 37 °C, pH 7.4. At various times during the incubation, samples were removed and subjected to HPLC to characterize the products formed and the kinetics of the reaction. The results of this experiment are shown in Fig. 5. After 24 h, the major product was a normal L-aspartyl peptide indicating that PIMT had indeed catalyzed substantial repair of the atypical β-aspartyl peptide bond. Interestingly, the major by-products of this repair reaction were the D-isoaspartyl and D-aspartyl versions of the original peptide. Partial racemization of this aspartate is consistent with the known susceptibility of cyclic imides to slow racemization (Fig. 2). The accumulation of D-isoaspartyl peptide over 24 h is significant because the repair reaction is inefficient; PIMT must drive the L-isoaspartyl peptide through the cyclic imide intermediate multiple times in order to achieve substantial overall repair. Each cycle through the cyclic imide provides an additional opportunity for this metastable intermediate to racemize.

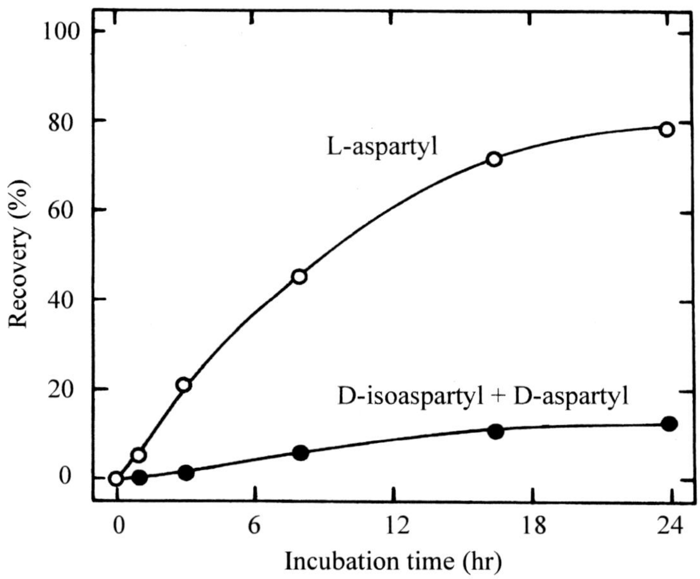

Figure 5. Kinetics of PIMT-dependent repair and racemization of a synthetic L-isoaspartyl peptide. Methylation reactions of varying duration were subjected to reversed-phase HPLC analysis with UV detection to separate and quantitate products and intermediates. The peak areas for the two major products (L-aspartyl and D-aspartyl/D-isoaspartyl peptides) is expressed as a percentage of the area of the starting L-isoaspartyl 12-mer peptide. Each point is the mean of duplicate determinations.

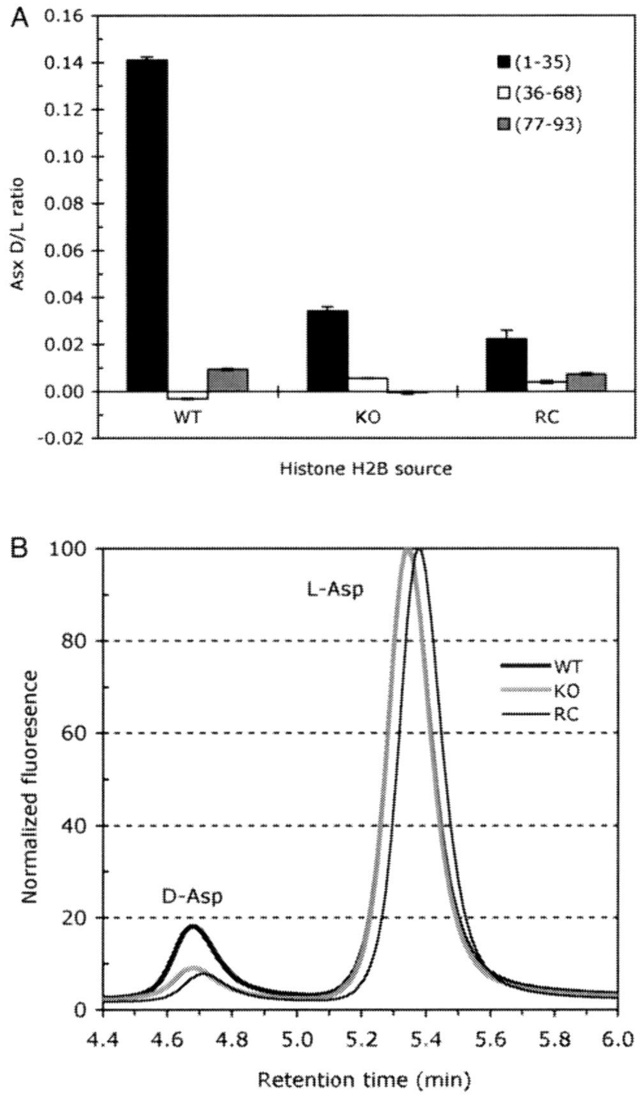

Figure 6. The D/L ratio of Asp-25 is dramatically reduced in mice deficient in PIMT. The bar graph shows a comparison of Asx D/L ratios in H2B peptides 1-35 (gray; containing Asp-25 as the only Asx residue), 36-68 (black; containing 4 Asx residues), and 77-93 (white; containing Asn-84 as the only Asx residue). Data are shown for histone H2B peptides isolated from wild type mice (WT), PIMT deficient mice (KO), and from recombinant chick H2B expressed in *E. coli* (RC). The recombinant chick H2B was included as an additional control to determine the Asx D/L ratio expected for the 1-35 peptide in the absence of *in vivo* racemization. Error bars represent standard deviation around the mean of triplicate D/L-Asp determinations from a given acid hydrolysate.

Mammalian histone H2B is an important substrate for PIMT *in vivo* [7]. The site of isoaspartate formation has recently been localized to Asp-25 which lies in the sequence -KKDGKK- in the flexible N-terminal domain of H2B (W.G. Carter, K.J. Reissner, M.J. Mamula, H.A Doyle, and D.W. Aswad; manuscript in preparation). We thought histone H2B would be an excellent test case to see if PIMT catalyzes significant D-Asp accumulation *in vivo*. To carry out this test, we isolated H2B from PIMT-KO mice and from litter mates that were wild type (WT) with regard to PIMT expression, digested the H2B with proteases, and then analyzed the aspartate D/L ratio in these peptides [8]. If PIMT catalyzes D-Asp formation *in vivo*, as it does *in vitro*, we would expect to find levels of D-Asp significantly above baseline in the PIMT wild-type mice but not in knockout mice. Moreover, the increased D-Asp should appear only at Asp-25, and not at other Asx sites, since only Asp-25 is the target of PIMT-dependent repair. An important caveat in this prediction is that D-Asp accumulation could be limited by turnover of the D-Asp-containing forms of H2B. Indeed, if the *in vivo* function of PIMT is to target age-damaged, isoaspartyl proteins for degradation, rather than repair, then we would not expect to see significant accumulation of D-Asp at any sites in H2B.

Results of this study are shown in Fig. 6. The H2B 1-35 peptide (containing Asp-25 as the only Asx residue) from wild-type mouse brain (WT) yielded an aspartate D/L ratio of 0.14. In contrast, the aspartate D/L ratio at the same site from the KO mice was only 0.035, a value similar to that found in a recombinant form of chicken H2B (RC) which had never in been in contact with PIMT. The 36-76 and 77-93 peptides (which contain 4 and 1 Asx residues, respectively) exhibited aspartate D/L ratios of 0.01 or less regardless of H2B source. These results are in complete agreement with the hypothesis that Asp-25 is a hot-spot for racemization and peptide-bond racemization, and that PIMT plays a major role in the accumulation of D-aspartate/D-isoaspartate at this location.

5. Notes

1. Preparation OPA-NAC reagent.
   When making the NAC stock solution, one typically uses solid N-acetyl-L-cysteine-HCl. Because of the HCl content, it is important to neutralize this solution (as indicated in the Methods section) before combining in with the OPA and borate buffer. If the NAC stock solution is not neutralized, it may acidify the reagent enough to render it ineffective. The final pH of OPA-NAC should be dictated by the borate buffer; i.e, it should be in the range of 9.3-9.4, a pH necessary for efficient reaction of primary amines with the reagent.
2. Water purity.
   We recommend using commercial bottled HPLC grade water for preparing all solutions involved in amino acid analysis. We have found that even a well-maintained laboratory water purification system (a "Milli-Q" cartridge system fed with reversed-osmosis supply water) was inferior to commercial HPLC grade water with respect to free amino acid content.
3. Separation of D- and L-aspartate by reversed-phase HPLC.
   The conditions for separation of D- and L-aspartate by HPLC are recommendations based on what has worked recently in the author's lab. The exact solvent conditions (% methanol) for column equilibration and analyte elution will need to be fine-tuned in each laboratory and may vary with brand of column. It is recommended that each HPLC run (or at least every third run) end with a column washing step, such as a steep linear gradient to 80% methanol, to remove amino acids that are more retentive than aspartate. This should be followed by an extensive period of column equilibration to the initial starting conditions. Ideally, blanks and standards should be run both before and after the experimental samples. Great care should also be taken to flush the injection port well between runs, especially after injecting an aspartate standard or any sample with high levels of aspartate.

References

[1] Fujii, N. (2002) D-amino acids in living higher organisms. *Orig Life Evol Biosph 32*, 103-27.

[2] Radkiewicz, J. L., Zipse, H., Clarke, S., and Houk, K. N. (1996) Accelerated racemization of aspartic acid and asparagine residues via succinimide intermediates: an ab initio theoretical exploration of mechanisms. *J Am Chem Soc 118*, 9148-9155.

[3] Johnson, B. A., Murray, E. D., Jr., Clarke, S., Glass, D. B., and Aswad, D. W. (1987) Protein carboxyl methyltransferase facilitates conversion of atypical L-isoaspartyl peptides to normal L-aspartyl peptides. *J Biol Chem 262*, 5622-5629.

[4] McFadden, P. N., and Clarke, S. (1987) Conversion of isoaspartyl peptides to normal peptides: Implications for the cellular repair of damaged proteins. *Proc Natl Acad Sci USA 84*, 2595-2599.

[5] Perna, A. F., D'Aniello, A., Lowenson, J. D., Clarke, S., De Santo, N. G., and Ingrosso, D. (1997) D-aspartate content of erythrocyte membrane proteins is decreased in uremia: implications for the repair of damaged proteins. *J Am Soc Nephrol 8*, 95-104.

[6] Kinzel, V., Konig, N., Pipkorn, R., Bossemeyer, D., and Lehmann, W. D. (2000) The amino terminus of PKA catalytic subunit--a site for introduction of posttranslational heterogeneities by deamidation: D-Asp2 and D-isoAsp2 containing isozymes. *Protein Sci 9*, 2269-77.

[7] Young, A. L., Carter, W. G., Doyle, H. A., Mamula, M. J., and Aswad, D. W. (2001) Structural integrity of histone H2B in vivo requires the activity of protein L-isoaspartyl O-methyltransferase, a putative repair enzyme. *J Biol Chem 276*, 37161-37165.

[8] Young, G. W., Hoofring, S. A., Mamula, M. J., Doyle, H. A., Bunick, G. J., Hu, Y., and Aswad, D. W. (2005) Protein L-isoaspartyl methyltransferase catalyzes in vivo racemization of aspartate-25 in mammalian histone H2B. *J Biol Chem 280*, 26094-8.

[9] Aswad, D. W. (1984) Determination of D- and L-aspartate in amino acid mixtures by high-performance liquid chromatography after derivatization with a chiral adduct of o-phthaldialdehyde. *Anal Biochem 137*, 405-409.

In: D-Amino Acids: ISBN 1-60021-075-9
A New Frontier in Amino Acid and Protein Research, pp.399-401

Chapter 5.6

Purification of a Novel Mammalian Proteinase for D-Aspartate-Containing Protein, D-Aspartyl Endopeptidase (DAEP)

Tadatoshi Kinouchi*

Department of Radiation Life Science and Radiation Medical Science, Research Reactor Institute, Kyoto University, Japan.

Keywords: D-aspartate, proteinase, peptidase, mitochondria.

1. Introduction

D-Aspartyl endopeptidase (DAEP) is a novel proteinase found in mammal, specifically degrading D-aspartate-containing protein [1]. DAEP is mainly localized in the inner mitochondrial membrane of liver, kidney, spleen, and also brain. The existence of DAEP in mitochondria would be explained by ROS (reactive oxygen species), because they damage and modify various molecules in mitochondria. In fact, D-Asp is detected in rat liver mitochondria [2]. Interestingly, DAEP activity is not detected in *E. coli*, *S. cerevisiae*, and *C. elegans*. Therefore DAEP may serve as a scavenger against accumulation of racemized proteins in aging in the long life span of mammals.

In order to analyze the physiological functions of DAEP, it is necessary to purify it from mammalian tissue. However, it is quite difficult to purify DAEP which keeps high specific activity. Because DAEP is a kind of the membrane-integrated protein and has a multi-complex structure (MW: 600 kDa), I suppose that DAEP would lose some components in purifying steps. Actually, the highly purified DAEP has a half life of 2-3 days at 4°C.

Here, let me introduce how to purify DAEP from mammalian liver. This is the most effective way to purify DAEP now.

* Correspondence concerning this article should be addressed to Tadatoshi Kinouchi, E-mail: kinouchi@HL.rri.kyoto-u.ac.jp, TEL/FAX: +81-724-51-2630.

2. MATERIALS

1. Centrifuge and ultracentrifuge
2. Spectrofluorometer (ex. Hitachi F-2000, Hitachi High-Technologies Co.) or fluorescence microplate reader (ex. Wallac ARVOsx multilabel counter, PerkinElmer Life & Analytical Sciences)
3. Spectrophotometer (ex. DU-650, Beckman Coulter, Inc.) or microplate reader (ex. Model 3550, Bio-Rad Laboratories, Inc.)
4. Column Chromatograph system (ex. ÄKTAFPLC™, GE Healthcare)
5. Inhibitor and substrate of DAEP (Peptide Institute, Inc.)
6. DC protein assay kit (Bio-Rad Laboratories, Inc.)
7. Other chemicals were purchased from Sigma-Aldrich Co., Nakarai Tesque, Inc., and Kanto Chemical Co. Inc.

3. METHODS

1. Proteolytic Assay for DAEP

The DAEP activities are determined using a synthesized substrates: Succinyl-D-Aspartic acid α-(4-methyl-coumaryl-7-amides) (Suc-D-Asp-MCA: newly released by Peptide Institute, Inc., http://www.peptide.co.jp). Suc-D-Asp-MCA is hydrolyzed and the fluorescence of liberated aminomethylcoumarin (AMC) can be measured at $\lambda ex = 380$ nm and $\lambda em = 460$ nm. This substrate (1 mM in DMSO) is added to the assay buffer (10 mM Tris-HCI, pH 8.5, 200 mM NaCl, 3 mM $MnCl_2$) with the enzyme at final concentration of 100 μM. Generally, 90 μl of the assay mixture is mixed to 10 μl of sample and incubated at 37°C. If you use to measure the fluorescence with a fluorescence microplate reader, you should measure in every two minutes for 15 min. If you use a spectrofluorometer, 100 μl of 10% SDS solution is immediately added to the reaction mixture after incubation for 15 min to stop the reaction, 1.3 ml of 0.5 M sodium acetate buffer (pH 4.5) is added following that, and then the fluorescence should be measured. AMC is used as a standard to measure the fluorescence (usually 10 nM-100 μM). The protein concentration of samples is determined with a DC protein assay kit using bovine serum albumin as a standard.

2. Purification of DAEP from Mouse Liver

All subsequent operations are carried out at 4°C and column works in the purification step are performed by a column chromatography system, ex. ÄKTAFPLC™. Mouse livers are homogenized in a Potter-Elvehjem homogenizer in homogenization buffer (20 mM Tris-HCl (pH 8.5), 250 mM sucrose, 0.5 mM EDTA, 10 volumes) and are centrifuged at 100 x g for 10 min to remove debris. The supernatant is centrifuged, first at 600 x g for 10 min (precipitate=nuclei), then at 9,000 x g for 10min (precipitate=mitochondria). The precipitate is resuspended in homogenization buffer and centrifuged again at 9,000 x g for 10min to remove contaminated lysosome. The mitochondorial fraction is suspended in osmotic shock buffer (20 mM Tris-HCI, pH 8.5, 1 mM EDTA, 1 mM DTT, 14 μM E-64, 21 μM leupeptin, 15 μM pepstatin A, 36 μM bestatin, 0.8 μM aprotinin and 1.4 mM AEBSF: these inhibitors are premixed and obtainable as "Protease Inhibitor Cocktail for use with mammalian cell and tissue extracts (P 8340)" from Sigma-Aldrich) for 60 min. After the osmotic shock, the fraction is mixed with equal volume of 1M LiCl solution, sonicated and centrifuged at 100,000 x g for 60 min. The precipitate (= mitochondrial membrane fraction) was suspended in the extracting buffer (20 mM Tris-HCI, pH 8.5, 1 mM EDTA, 1% CHAPS, 20% glycerol, 0.5 M $NaSO_4$,1 mM DTT, 14 μM E-64, 21 μM leupeptin, 15 μM pepstatin A, 36 μM bestatin, 0.8 μM aprotinin and 1.4 mM AEBSF) and stirred by a tube rotator for 12 hours. This is an extracting step of DAEP from the mitochondrial membrane, therefore, CHAPS is included in all buffers in the following purification steps. Next, the CHAPS extract after centrifugation at 100,000 x g for 60 min was loaded onto the centrifugal filter device: MacroSep 100 (molecular cut-off limit: 100 kDa) to concentrate the extract and to remove small molecules included in the extract. The concentrated extract is diluted at 50 times by buffer A (20 mM Tris-HCI, pH 8.5, 1 mM EDTA, 0.25 % CHAPS, 20% glycerol, 1 mM DTT, 14 μM E-64, 21 μM leupeptin, 15 μM pepstatin A, 36 μM bestatin, 0.8 μM aprotinin and 1.4 mM AEBSF) and is loaded onto DEAE Affi-Gel® blue column (Bio-Rad, Richmond, CA, USA) and finally eluted with a liner gradient of 0-0.6 M NaCl in buffer A. The active fractions as detected by the DAEP assay are collected and diluted at three times by buffer A, and then loaded onto RESOURCE™ Q column (6 ml). In this column chromatography, the fractions were eluted with a liner gradient of 0-0.6 M NaCl in buffer A. The active fractions was dialyzed to 10 mM potassium phosphate in buffer B (1 mM EDTA, 0.25 % CHAPS, 20% glycerol, 1 mM DTT, pH 8.0) and placed on a column of hydroxyapatite: Bio-Scale CHT2-I (Bio-Rad), and bounded protein is then eluted with a liner gradient of 10-600 mM potassium phosphate in buffer B. The active fractions are collected and concentrated by MacroSep 100, and then loaded onto a gel filtration column: Superose® 6 HR 10/30 (GE) and finally eluted with buffer C (20 mM Tris-HCI, pH 8.5, 1 mM EDTA, 150 mM NaCl, 0.25 % CHAPS, 20% glycerol,1 mM DTT). If you need to concentrate the active fractions from Superose® 6, those are loaded onto Mono-Q and eluted with a liner gradient of 0-0.6 M NaCl in buffer A.

4. Results

Using these purification steps, DAEP was finally purified a 271-fold compared to the original mitochondorial fraction, and the purified sample was used for followed experiment. Specific inhibitor is one of the necessaries to analyze the physiological functions of enzymes. However, DAEP is not sensitive to popular protease inhibitors as described in method. Therefore you should use i-DAEP (benzoyl-Arg-His-D-Asp-CH_2Cl), which covalently bind to the active site of DAEP [1]. Zn^{2+} (~3 mM) is also a strong inhibitor for DAEP.

5. References

[1] Kinouchi T, Ishiura S, Mabuchi Y, et al. (2004) Mammalian D-aspartyl endopeptidase: a scavenger for noxious racemized proteins in aging. *Biochem. Biophys. Res. Commun.*, *314*; 730-736.

[2] Nagata Y, Fukuda A, Sakai M, et al. (2001) D-Amino acid contents mitochondria and some purple bacteria. *J. Mol. Catal. B Enzymat.*, *12*; 109-113.

In: D-Amino Acids: ISBN 1-60021-075-9
A New Frontier in Amino Acid and Protein Research, pp.403-407

Chapter 5.7

MICRO-PURIFICATION AND STRUCTURAL ASSAY OF POLY-γ-GLUTAMATE, A D-AMINO ACID-CONTAINING BIOPOLYMER

Makoto Ashiuchi* and Haruo Misono
Department of Bioresources Science, Kochi University, Japan

Keywords: poly-γ-glutamate, purification, stereochemistry, molecular size, D-glutamate.

1. INTRODUCTION

Poly-γ-glutamate (PGA) is an anionic γ-isopolypeptide consisted only of glutamate [1]. There are stereochemically different three types of PGA: the homo-polymer of D-glutamate (D-PGA), the homo-polymer of L-glutamate (L-PGA), and the co-polymer in that D- and L-glutamyl residues are randomly lined (DL-PGA). *Bacillus anthracis* produces capsular D-PGA that functions in evading mammalian immune defense mechanisms [1,2]. Deep understanding of D-PGA synthesis, therefore, helps to develop pharmaceuticals against the pathogen. In contrast, according to the review of Shih *et al.* [3], D-PGA will serve as a delivery tool for drugs, *e.g.* chemotherapeutic agents against human malignancies. As the *natto* mucilage, containing D-glutamate-rich DL-PGA [1], increases Ca^{2+} solubility and intestinal Ca^{2+} absorption [4], the polymer may be useful for a therapeutic tool for osteoporosis. On the other hand, many literatures indicate that L-PGA functions as a multi-functional adaptation agent in various environments [1]. Such the findings provide insights into novel biochemical applications of PGA [1].

PGA now draws attention as a most promising material in industrial, pharmaceutical, and environmental uses. In view of the wide variety of uses of PGA, it is urgent to achieve the safe and effective synthesis of a set of PGAs bearing the molecular structures suitable to practical purposes, and thermoplastic functionality of biopolymers generally increases with an increase in the enantiomeric homogeneity and the degree of polymerization [1]. Gel permeation chromatography is often applied for PGA study [1]; however, further improvement about the purification techniques and structural assays of PGA is required in order to characterize the polymer molecule in more details. Here we present some new methods for such investigation, *i.e.* the anion-exchange chromatography and micro-purification of PGA,

* Correspondence concerning this article should be addressed to Makoto Ashiuchi, E-mail: ashiuchi@cc.kochi-u.ac.jp, FAX.: +81-88-864-5200.

the determination and stereochemical assay of PGA, and the molecular-size estimation of PGA with 1-fluoro-2,4-dinitrobenzene (FDNB).

2. Materials

1. Authentic PGAs with the average molecular masses of 50, 250, 500, 1,000, 2,000, 5,000, and 7,000 kDa (BioLeaders Co., Daejeon, Korea).
2. Sep-Pak Plus Waters Accell™ Plus QMA anion-exchange cartridge (bed volume, 2 mL; Millipore, Bedford, MA).
3. Vivapure IEX spin column Mini-M with a Q-type anion-exchange membrane (bed volume, 60 µL; Sartorius, Göttingen, Germany).
4. CHIRALPAK MA(+) column (4.6 by 50 mm; DAICEL, Tokyo, Japan).
5. Mini-ProteanII Ready Gel J (linear gradient from 5 to 15%; Bio-Rad, Richmond, CA).
6. HMW marker kit (Amersham Pharmacia Biotech, Little Chalfont, UK).

3. Methods

3.1. Anion-exchange Chromatography of PGA

1. Crude PGA (20 mg) is subjected to the QMA cartridge.
2. The PGA-charged cartridge is washed with 10 mL of distilled water.
3. The cartridge is stepwise developed with 5 mL of NaCl solutions (0 to 1.0 M).

3.2. Micro-purification of PGA

1. A small amount (in µg order) of PGA sample (total volume, ~100 µL) is subjected to the Q-type membrane of the spin column Mini-M.
2. By centrifugation at 400 *g* at 4°C, the PGA-charged membrane is 4-times washed with 0.2 mL of distilled water.
3. The cartridge is stepwise developed with 0.1 mL of NaCl solutions (0 to 1.0 M).

3.3. Determination and Stereochemical Assay of PGA

1. PGA is hydrolyzed with distilled HCl (6 M) at 105°C for 8 h *in vacua* with a Hydrolysis Station AHST-1 (Shimadzu, Kyoto, Japan).
2. The hydrolysate is lyophilized, dissolved in 0.2 mL of distilled water, and analyzed using HPLC with a CHIRALPAK MA(+) column as the chiral carrier.
3. The total amount and DL ratio of glutamate in the hydrolysate are determined from the area of its corresponding peaks [1,5].

3.4. Molecular-Size Estimation of PGA (FDNB method)

1. PGA is lyophilized and dissolved in 0.1 mL of 0.1 M borate buffer (pH 8.5).
2. The PGA sample is centrifuged at 12,000 *g* for 10 min at 25°C.
3. The supernatant is mixed with 10 µl of 10 mM FDNB solution (in acetone) [6].
4. The mixture is incubated at 65°C for 45 min in the dark in order to modify the amino group of PGA with FDNB.
5. Termination of the chemical modification and hydrolysis of FDNB-modified PGA are performed by the use of distilled HCl (3.3 to 4.0 M) at 105°C for 12 h.
6. The resultant is lyophilized and dissolved in 0.1 mL of distilled water.
7. Glutamate monomers and the dinitrophenyl glutamate thus formed are determined by the chiral HPLC described above and by monitoring an increase in the absorbance at 356 nm with an Ultrospec 2100 Pro spectrophotometer (Amersham Bioscience, Uppsala, Sweden) [7].

4. Results

PGA was easily purified and fractionated by anion-exchange chromatography. Figure 1 presents the result of the micro-purification of 1,000-kDa PGA. Besides, glutamate, 50-, 250-, 500-, 2,000-, 5,000-, and 7,000-kDa PGAs were eluted in the 0.1, 0.4, 0.5-0.6, 0.6-0.7, 0.7-0.8, 0.7-0.9, and 0.8-1.0 M NaCl fractions, respectively; this profile was almost identical to that with the QMA cartridge. PGA could be obtained in the µg order (or in pmol order for the 1,000-kDa polymer).

The FDNB method showed a high sensitivity; even when 1,000-kDa PGA was used, only a small amount of the sample, such as 0.15 mg, was needed for the average-molecular-mass estimation.

In usual, D-glutamate content of PGA is changeable, depending on the used strains [1]. Figure 2 showed the stereochemistry (DL-ratio) of PGA produced by *Bacillus subtilis* under the different culture conditions. The result indicates that ammonium sulfate may affect the structural feature of *B. subtilis* PGA.

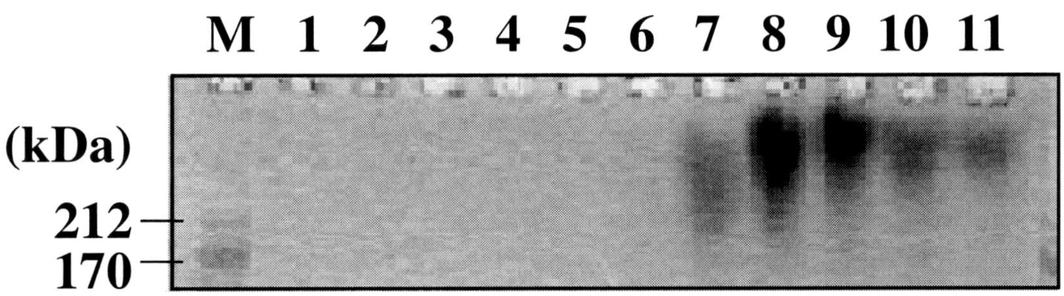

Figure 1. Micro-purification of PGA. Ten µg of 1,000-kDa PGA was purified as described in the Methods section. Ten-µl aliquots of the eluates (lane 1, 0 M NaCl fraction; 2, 0.1 M; 3, 0.2 M; 4, 0.3 M; 5, 0.4 M; 6, 0.5 M; 7, 0.6 M; 8, 0.7 M; 9; 0.8 M; 10, 0.9 M; and 11, 1.0 M) were subjected to SDS-PAGE using a Mini-ProteanII Ready Gel J with an HMW marker kit (lane M) and PGA was visualized by methylene-blue staining [5].

5. Notes

1. Highly purified PGA is needed for the precise structural analysis of the polymer.

2. Anion-exchange chromatography is a most useful technique for PGA purification, since the acidity of the polymer increases with an increase in molecular mass corresponding to the number of γ-glutamyl residues.
3. The micro-purification of PGA, in particular, serves to the detailed analysis of enzymatic PGA synthesis [5].
4. The FDNB method is established on the basis of the structural feature of PGA, namely that every molecule of PGA has one terminal free amino group irrespective of molecular size. Therefore, the ratio of numbers between the amino group and γ-glutamyl residues of the polymer is parallel to the molecular mass.
5. The FDNB method will be applicable for the molecular-size estimation of polyamino acids other than PGA, *e.g.* polyaspartate, as well.
6. The FDNB method is useful for assessment of the purity of PGA sample besides the molecular-size estimation of the polymer [7].
7. The presented methods will contribute to the assessment and improvement of PGA applicability as well.

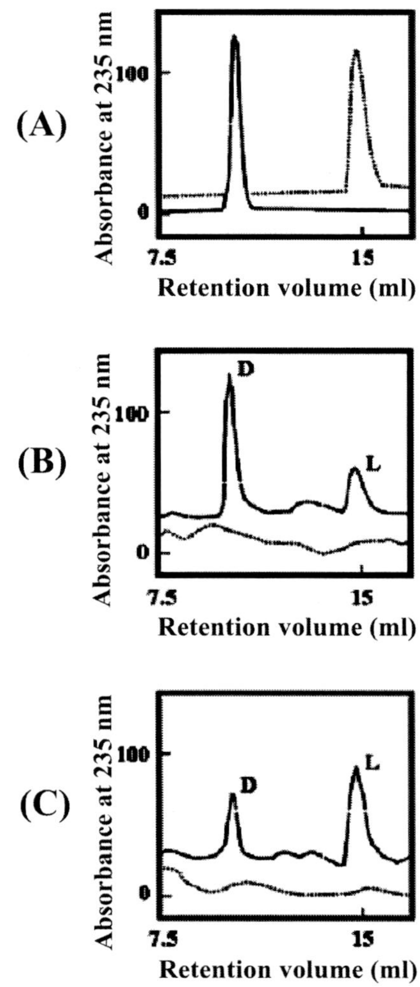

Figure 2. Stereochemical analysis of PGA using chiral HPLC. (A) Authentic D-glutamate (elution volume, 10.3 mL; *solid line*) and L-glutamate (14.9 mL; *dotted line*) (each 1 nmol). High-molecular-mass PGAs (average 1,000 kDa) produced by *B. subtilis* subsp. *chungkookjang* in a GS medium [5] (B) and in a GS medium supplemented with 1% $(NH_4)_2SO_4$ (C); with no treatment (*dotted lines*) and after hydrolysis (*solid lines*).

REFERENCES

[1] Ashiuchi, M. & Misono, H. (2002). Poly-γ-glutamic acid. *Biopolymers, vol.7* (S.R. Fahnestock, S.R. & Steinbüchel, A., Eds.), pp.123-174, Weinheim: Wiley-VCH.

[2] Drysdale, M., Heninger, S., Hutt, J. et al. (2005). Capsule synthesis by *Bacillus anthracis* is required for dissemination in murine inhalation anthrax. *EMBO J. 24*, 221-227.

[3] Shih, I.L. & Van, Y.T. (2001). The production of poly-(γ-glutamic acid) from microorganisms and its various applications. *Bioresour Technol 79*, 207-225.

[4] Tanimoto, H., Mori, M., Motoki, M. et al. (2001). Natto mucilage containing poly-γ-glutamic acid increases soluble calcium in the rat small intestine. *Biosci Biotechnol Biochem 65*, 516-521.

[5] Ashiuchi, M., Shimanouchi, K., Nakamura, H. et al. (2004). Enzymatic synthesis of high-molecular-mass poly-γ-glutamate and regulation of its stereochemistry. *Appl Environ Microbiol 70*, 4249-4255.

[6] Paraskevas, G., Atta-Politou, J. & Koupparis, M (2002). Spectrophotometric determination of lisinopril in tablets using 1-fluoro-2,4-dinitrobenzene reagent. *J. Pharm. Biomed. Anal. 29*, 865-872.

[7] Park, C., Choi, J.C., Choi, Y.H., et al. (2005). Synthesis of super-high-molecular-weight poly-γ-glutamate from *Bacillus subtilis* subsp. *chungkookjang*. *J. Mol. Catal. B: Enzym. 35*, 128-133.

In: D-Amino Acids:
A New Frontier in Amino Acid and Protein Research, pp.409-414
ISBN 1-60021-075-9

Chapter 5.8

Estimation of Chronological Age using the Aspartic Acid Racemization Method on Dentin Samples

Susumu Ohtani*
Institute for Frontier Oral Science and Department of Forensic Medicine,
Kanagawa Dental College, Japan.

Keywords: forensic medicine, age estimation, dentin, racemization, D-aspartic acid

1. Introduction

At present, one of the most accurate methods for the estimation of chronological age uses the racemization reaction of aspartic acid (the degree of racemization) in dentin. This was discovered in 1975, when Helfman and Bada [1], using aspartic acid, which has the highest racemization reaction rate, found a very high correlation (r=0.921) between its degree of racemization (D/L ratio) in enamel and chronological age in 19 cases. Subsequently, they reported that using dentin from the crown yielded better results (r=0.979) than enamel dentin in 20 cases [2]. Similarly, the relationship between the degree of racemization of teeth and age was also studied by Shimoyama and Harada [3] and Ogino et al. [4], and in agreement with Helfman & Bada, they also reported a high degree of correlation. Furthermore, these results have been double-checked and confirmed by Masters and many others [5-9]. Thus, the high reliability of the racemization method has been well established. In 1987 the current author and Yamamoto studied the relationship between age and degree of aspartic acid racemization in detail with respect to dentin from various parts of the tooth, and applied this racemization method to actual appraisal of age [10].

In this article, the author will describe the outline of his technique for analysis by the racemization method and the results obtained using this technique [11,12].

* Correspondence concerning this article should be addressed to Susumu Ohtani, E-mail: ohtanisu@kdcnet.ac.jp, FAX: +81-46-822-8801.

2. Materials

1. Teeth of known age extracted due to alveolar pyorrhea and other medical indications. Any caries did not extend into the dentin.
2. Isomet low speed saw (BUEHLER, Chicago, IL), diamond wafering blade, Series 15 LC diamond (102mm × 0.3mm).
3. Vibratory Sieve Shaker (FRITSCH, Idar-Oberstein, Germany).
4. Gas chromatograph (GC-17A, Shimadzu, Kyoto, Japan)
5. Capillary column (length: 25m, inner diameter: 0.3mm) comprising the fixing phase of the optical activity is lined with Chirasil-Val (GL Science, Tokyo, Japan), which is commercially available.

3. Methods [13]

1. Separate dentin from all other tooth constituents using the cutter (Fig. 1.).
2. Wash samples with 0.2 M HCl for 5 min.
3. Wash samples three times with distilled water for 5 min. each.
4. Wash samples with ethanol for 5 min.
5. Wash samples with ether for 5 min.
6. Pulverize samples with a pulverizer.
7. Weigh 10 mg from each pulverized sample.
8. Hydrolyze powdered samples with 6N HCl for 6 hrs at 100 °C.
9. Dry the powdered samples by evaporation.
10. Add 5 ml of distilled water.
11. Apply samples to ion-exchange resin (Dowex, 50W-X8, 50-100 mesh, Dow Chemical Company, MI) to collect amino acids.
12. Wash resin with 10 ml of distilled water.
13. Elute amino acids from resin with 10 ml of 2N NH_4OH.
14. Dry eluted fractions by evaporation.
15. Add 2 ml of a mixture of isopropyl alcohol and acetyl chloride (8:2 v/v).
16. Let samples stand for 30 min. at 100 °C.
17. Dry samples by nitrogen gas aeration.
18. Add 800 μl of dichloromethane.
19. Add 200 μl of Trifluoroacetic anhydride.
20. Let samples stand for 30 min. at room temperature.
21. Dry samples by nitrogen gas aeration.
22. Add 50 μl of ethyl acetate.
23. Apply samples to a gas chromatograph (to separate L- and D-aspartic acids) (Fig. 2).
24. Calculate D/L ratios and estimate age from a test sample by substitution to the regression line deduced from standard samples.

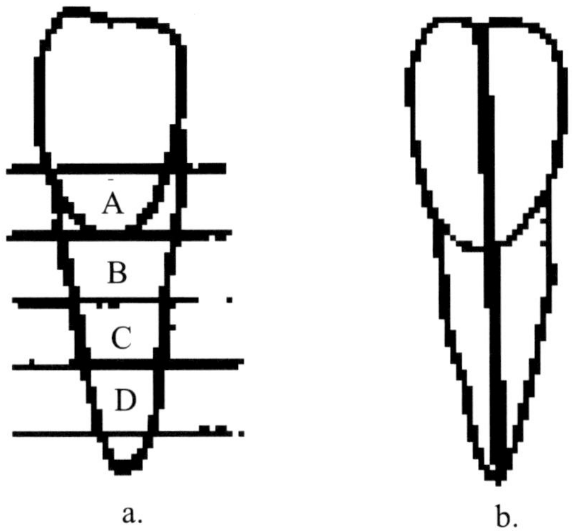

Figure 1. Sites used for determining the aspartic acid D/L ratio in dentin. a, positions of transverse sections cut from root apex to crown of central incisor; b, position of a longitudinal section.

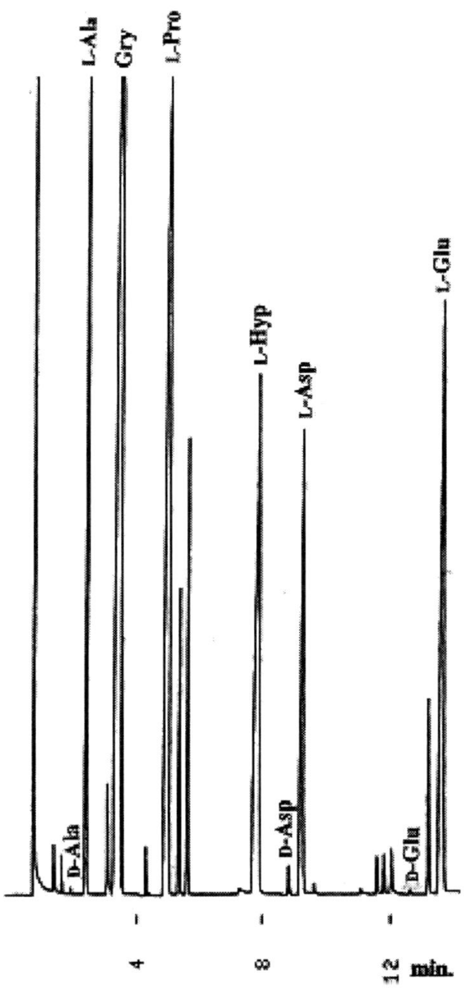

Figure 2. Gas chromatogram of N(0)-TFA isopropyl esters of amino acids in a single dentin specimen. Column temperature: 90°C, 4 min hold followed by an increase to 180°C at 4°C /min; Injection temperature: 280°C; Carrier gas: He; Split ratio: 40/1.

4. RESULTS

Differences in the Degree of Racemization in Dentin from Different Parts of the Tooth (Upper Central Incisor)

Formation of homonymous teeth of the same jaw (the left and right teeth) occurs in an individual at approximately the same time, resulting in a similar degree of racemization [14]. Hence, in homonymous teeth of the same jaw, whole dentin specimens were prepared from transverse sections were made in one tooth (the tooth was split into 4 blocks from A to D, Fig. 1a) and from longitudinal sections made in the other tooth (Fig. 1b). The degree of racemization was then determined, to compare the correlation with age. As shown in Fig. 3, the highest correlation (r=0.996), and the highest racemization rate were both found in the whole dentin samples [15].

Estimation of Age from Whole Dentin

The correlation coefficient between the degree of racemization of a sample of dentin and age (221 cases) was r=0.971-0.995 [10]. The correlation coefficient between the degree of racemization of a sample of whole dentin of the lower teeth and age in 166 cases (ranging from 13 to 82 years of age) was r=0.991-0.997. In the central incisor and first and second premolars it was possible to estimate age with an accuracy of ±3 years (Table 1) [10]. Thus, to accurately estimate age it is preferable to use whole dentin rather than only a portion of dentin of the test tooth.

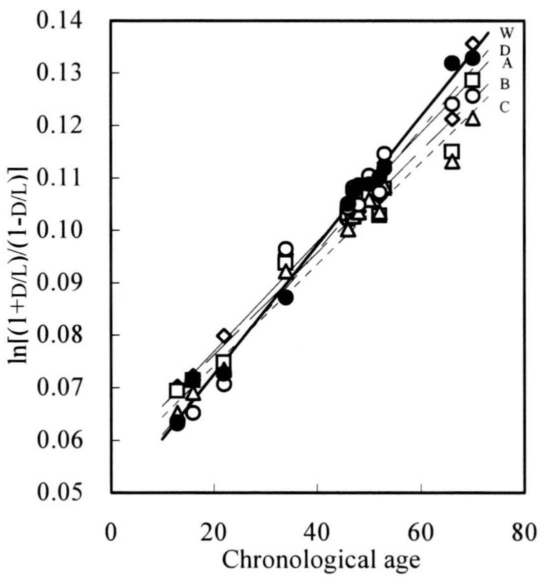

Figure 3. Comparison of the rates of racemization in the different blocks of dentin. ◇ (A), D/L=0.001043t+0.0560 r=0.986 k=5.2157×10^{-4}; □(B), D/L =0.000976t+0.0566 r=0.984 k=4.8830×10^{-4}; △(C), D/L =0.000970t+0.0547 r=0.987 k=4.8495×10^{-4}; ○(D), D/L =0.001163t+0.0494 r=0.984 k=5.8148×10^{-4}; ●(W, Whole dentin), D/L =0.001230t+0.0479, r=0.996, k=6.1480×10^{-4}; D/L, ln[(1+ D/L)/(1- D/L)]; t, age (years); k, racemization reaction rate constant (years).

Table 1. The correlation coefficient between the degree of racemization of whole dentin and chronological age.

Materials	Age calculation formula	Correlation coefficient	Error region samples						Total
			±0	±1	±2	±3	±4	±5	
Central incisor	Y=0.001078X+0.0393	r=0.996	8 (21)	19 (50)	8 (21)	3 (8)	0	0	38
Lateral incisor	Y=0.001130X+0.0327	r=0.994	13 (21)	36 (68)	8 (13)	5 (8)	0	0	62
Canine	Y=0.001242X+0.0261	r=0.991	4 (14)	12 (43)	5 (18)	4 (14)	3 (11)	0	28
First premolar	Y=0.001265X+0.0245	r=0.997	8 (57)	4 (29)	2 (14)	0	0	0	14
Second premolar	Y=0.001252X+0.0208	r=0.992	5 (21)	8 (33)	9 (38)	2 (8)	0	0	24
Total			38 (23)	79 (48)	32 (19)	14 (8)	3 (2)	0	166

Y, Age (years); X, ln[(1+D/L)/(1-D/L)]; ( ), %

5. Notes

1. In the appraisal of age from teeth using the racemization method, the degree of racemization of total amino acids should be used [13].
2. Longitudinal sections of all the teeth to be tested should be cut. All the pieces including dentin only, should be powdered and mixed, and 10 mg of the powder is used as the specimen (whole dentin) [10,15].
3. As the sample tooth for the estimation of age, the lower incisor or premolar are most suitable, if possible, as both are single-rooted, small in size, and whole dentin is easily obtainable [10].
4. The racemization reaction rarely proceeds after death, because body temperature quickly drops. Accordingly, even from the teeth of bodies kept in the air for 10-odd years it is generally possible to estimate the age at the time of death [4,16,17].
5. The age of burned bodies can be accurately estimated as long as the surface of the teeth is the same color as vital teeth [18].
6. In the appraisal of age of more than 100 cases conducted so far by this author, the error has not exceeded ±3 years in most of the cases [19].

References

[1] Helfman, PM. & Bada, JL. (1975). Aspartic acid racemisation in tooth enamel from living humans. *Proc Natl Acad Sci USA, 72*, 2891-2894.

[2] Helfman, PM. & Bada, JL. (1976). Aspartic acid racemisation in dentine as a measure of ageing. *Nature,* 262, 279-281.

[3] Shimoyama, A. & Harada, K. (1984). An age determination of an ancient burial mound man by apparent racemization reaction of aspartic acid in tooth dentin. *Chem Lett,* 1984, 1661-1664.

[4] Ogino, T., Ogino, H. & Nagy, B. (1985). Application of aspartic acid racemization to forensic odontology: post mortem designation of age at death. *Forensic Sci Int, 29*, 259-267.

[5] Masters, PM. (1986). Age at death determinations for autopsied remains based on aspartic acid racemization in tooth dentin: importance of postmortem conditions. *Forensic Sci Int, 32*, 179-184.

[6] Ritz, S., Schütz, HW. & Peper, C. (1993). Postmortem estimation of age at death based on aspartic acid racemization in dentin: its applicability for root dentin. *Int J Legal Med, 105*, 289-293.

[7] Saleh, N., Deutsch, D. & Gil-Av, E. (1993). Racemization of aspartic acid in the extracellular matrix proteins of primary and secondary dentin. *Calcif Tissue Int, 53*, 103-110.

[8] Fu, SJ., Fan, CC., Song, HW. & Wei, FQ. (1995). Age estimation using a modified HPLC determination of ratio of aspartic acid in dentin. *Forensic Sci Int, 73*, 35-40.

[9] Carolan, VA., Gardner, MLG., Lucy, D. & Pollard, AM. (1997). Some considerations regarding the use of amino acid racemization in human dentine as an indicator of age at death. *J Forensic Sci, 42*, 10-16.

[10] Ohtani, S. & Yamamoto, K. (1987). Age estimation using the racemization of amino acid in human dentin. *Nippon Hoigaku Zasshi, 41*,181-190 (in Japanese).

[11] Ohtani, S., Ito, R. & Yamamoto, T. (2003). Differences in the D/L aspartic acid ratios in dentin among different types of teeth from the same individual and estimated age. *Int J Legal Med, 117*, 149-152.

[12] Ohtani, S., Ito, R., Arany, S. & Yamamoto, T. (2005). Racemization in enamel among different types of teeth from the same individual. *Int J Legal Med, 119*, 66-69.

[13] Ohtani, S. & Yamamoto, K. (1991). Age estimation using the racemization of amino acid in human dentin. *J Forensic Sci, 36*, 792-800.

[14] Ohtani, S. (1997). Different racemization ratios in dentin from different locations within a tooth. *Growth Develp Aging, 61*, 93-99.

[15] Ohtani, S. (1995). Estimation of age from dentin by using the racemization reaction of aspartic acid. *Am J Forensic Med Pathol, 16*, 158-161.

[16] Ohtani, S. (1995). Estimation of age from dentin by utilizing the racemization of aspartic acid: influence of pH. *Forensic Sci Int, 75*, 181-187.

[17] Ohtani, S., Ohhira, H., Watanabe, A., Ogasawara A., Sugimoto, H. (1997). Estimation of age from teeth by amino acid racemization: influence of fixative. *J Forensic Sci, 42*, 137-139.

[18] Ohtani, S., Sugeno, H., Marumo, T. & Yamamoto, K. (1989). Two cases of age estimation from teeth of burned body using amino acid racemization. *Nippon Hoigaku Zasshi, 43*, 191-197 (in Japanese).

[19] Ohtani, S. (1995). Estimation of age from the teeth of unidentified corpses using the amino acid racemization method with reference to actual cases. *Am J Forensic Med Pathol, 16*, 238-242.

In: D-Amino Acids: ISBN 1-60021-075-9
A New Frontier in Amino Acid and Protein Research, pp.415-430

Chapter 5.9

THE USE OF D-AMINO ACIDS IN PEPTIDE DESIGN

Radhakrishnan Mahalakshmi and Padmanabhan Balaram*
Molecular Biophysics Unit, Indian Institute of Science, India.

Keywords: peptide hairpins, α-aminoisobutyric acid, Schellman motif, ambidextrous helices, LD peptides

1. INTRODUCTION

Proteins and most naturally occurring peptides are composed of amino acids of the L-configuration. D-amino acids are found as constituents of natural peptides produced primarily, by microorganisms, using a non-ribosomal mechanism of synthesis. Research in this field dates back to over 60 years ago when Lipmann *et al* noted the presence of D-amino acids in tyrocidines and gramicidins [1]. Post-translational epimerization is an infrequently used mechanism in higher organisms for the introduction of D-amino acids into polypeptide chains [2]. Some notable examples are the production of opioid peptides, namely dermorphins and deltorphins form the skin of various frog species belonging to the Phyllomedusinae sub-family [3] and the contryphans produced by cone snails via post-translational epimerization [4]. Recently, total chemical synthesis of proteins containing all-D residues has been carried out and the characteristics and enzymatic activity of such enantiomeric proteins has been analyzed [5-7]. The incorporation of D-amino acids into polypeptide chains imposes local conformational constraints [8]. The purpose of this review is to examine the use of D-amino acids in the design of peptides adopting well-defined folded conformations.

2. CONFORMATIONAL CHARACTERISTICS OF D-AMINO ACIDS

The effects of residue configuration on polypeptide chain conformation are readily understood by considering the Ramachandran maps for N-acetyl-L-Ala-NHMe and N-acetyl-D-Ala-NHMe shown in Fig. 1A & 1B [9-12]. The map for the D-residue is derived by inverting the classical L-Ala map through the origin. Most of the stereochemically allowed regions for the L-residues correspond to negative values of ϕ, while for D-residues, the allowed regions correspond to positive values of ϕ. The classical 3_{10} and α-helical

* Correspondence concerning this article should be addressed to Padmanabhan Balaram, E-mail: pb@mbu.iisc.ernet.in, Fax: +91-80-23600535.

structures formed by all-L polypeptides adopt a right-handed twist ($3_{10\text{-}R}/\alpha_R$), while the corresponding D-polypeptides form helices with a left-handed twist ($3_{10\text{-}L}/\alpha_L$). Notably, there are small sterically allowed regions, which correspond to positive ϕ values for L-residues and negative ϕ values for D-residues. L-Proline is the most constrained of the amino acids occurring in proteins. The restraints imposed by the formation of the pyrrolidine ring restrict the allowed range of ϕ values in L-Pro to -60 ± 20° (Fig. 1C) [11,12]. By extension, the allowed values for D-Pro are +60 ± 20° (Fig. 1D). D-Pro can be used to advantage in restricting the local conformation of peptide chains, facilitating the formation of prime β-turns. The only achiral amino acid occurring in proteins is glycine, a residue for which considerably larger regions of ϕ-ψ space are sterically allowed, as a consequence of the absence of a substituent at the C^α atom (Fig. 2A). The Gly residue can adopt conformations on either side of the ϕ-ψ map. The observed distribution of Gly residues in proteins (Fig. 2B) establishes an approximately symmetric distribution about the origin [13]. α-Aminoisobutyric acid (Aib) is an achiral residue, which contains a pair of methyl groups at the C^α carbon atom. This residue may be formally viewed as a combination of L-Ala and D-Ala residues. The stereochemically allowed regions of Ramachandran space for Aib are readily derived by considering only the regions of overlap between the L-Ala and D-Ala ϕ-ψ maps (Fig. 2C). Clearly, the Aib residue is largely limited to the right (α_R) and left (α_L) handed helical structures. The observed distribution of crystallographically obtained conformations for Aib residues in peptides provides overwhelming support for this expectation (Fig. 2D) [11,12]. Achiral amino acids can, therefore, also be viewed as surrogate D-amino acids in peptide design because they can adopt conformations on both sides of the ϕ-ψ map. While the achiral Gly and Aib residues can support conformations with both positive and negative values of ϕ, D-amino acids may be used in design when positive ϕ values are required to form a desired structure.

3. β-HAIRPIN DESIGN

β-Hairpins are widely present in proteins with antiparallel β-sheet formation being facilitated by a nucleating reverse turn. Two-residue β-turns belonging to the type II' and I' category are most often observed as nucleating features for hairpins in proteins [14]. In the prime turns, the residue at position i+1 necessarily adopts the conformation of $\phi \sim +60°$, while ψ values of ~ -120° and ~ +30° are observed for type II' and I' turns, respectively [15]. Fig.3A illustrates a view of an ideal hairpin formed by a centrally positioned nucleating type II' β-turn. When the stereochemistry of the β-turn is type II (ϕ_{i+1} = -60°, ψ_{i+1} = +120°), registry of the antiparallel strands is not readily achieved. Although chain reversal occurs, the antiparallel strands drift away from one another (Fig. 3B).

In the design of β-hairpins, ^{D}Pro-Xxx sequences can provide a site for nucleation of prime turns. Indeed, insertion of ^{D}Pro residues has proved to be a successful strategy for peptide hairpin design [8,11,12,16-21]. Fig. 4 illustrates the superposition of the crystal structures of six designed peptide hairpins containing a central ^{D}Pro-Xxx type II' β-turn. ^{D}Pro residues provide a strong conformational determinant, when inserted into L-amino acid sequences. In ongoing studies, centrally positioned ^{D}Pro-Gly sequences have been used to generate well-defined β-hairpin scaffolds which permit spectroscopic probing of cross-strand interactions between aromatic side chains (Fig. 5A & 5B) [22]. Multiple ^{D}Pro-Xxx segments may be used to nucleate multistranded antiparallel β-sheet structures as exemplified by the design of three, four and five stranded β-sheets (Fig. 5C & 5D) [23-26].

In principle, hairpin nucleation may be achieved by using sequences with a high propensity to form prime β-turns. Specifically, the Aib-^{D}Ala segment has been used to nucleate the formation of a β-hairpin in the octapeptide Boc-Leu-Phe-Val-Aib-^{D}Ala-Leu-Phe-Val-OMe (Fig. 6A) [27]. In this case, crystallographic studies establish that the Aib-^{D}Ala β-turn adopts a type I' conformation, undoubtedly as a consequence of the tendency of the Aib residue to form α_R or α_L conformations. Thus, D-amino acid containing dipeptide segments may be used to generate both type II' and I' structures. While both turn types can sustain β-hairpin formation, the orientations of the strands are significantly different, as illustrated in Fig. 6C & 6D.

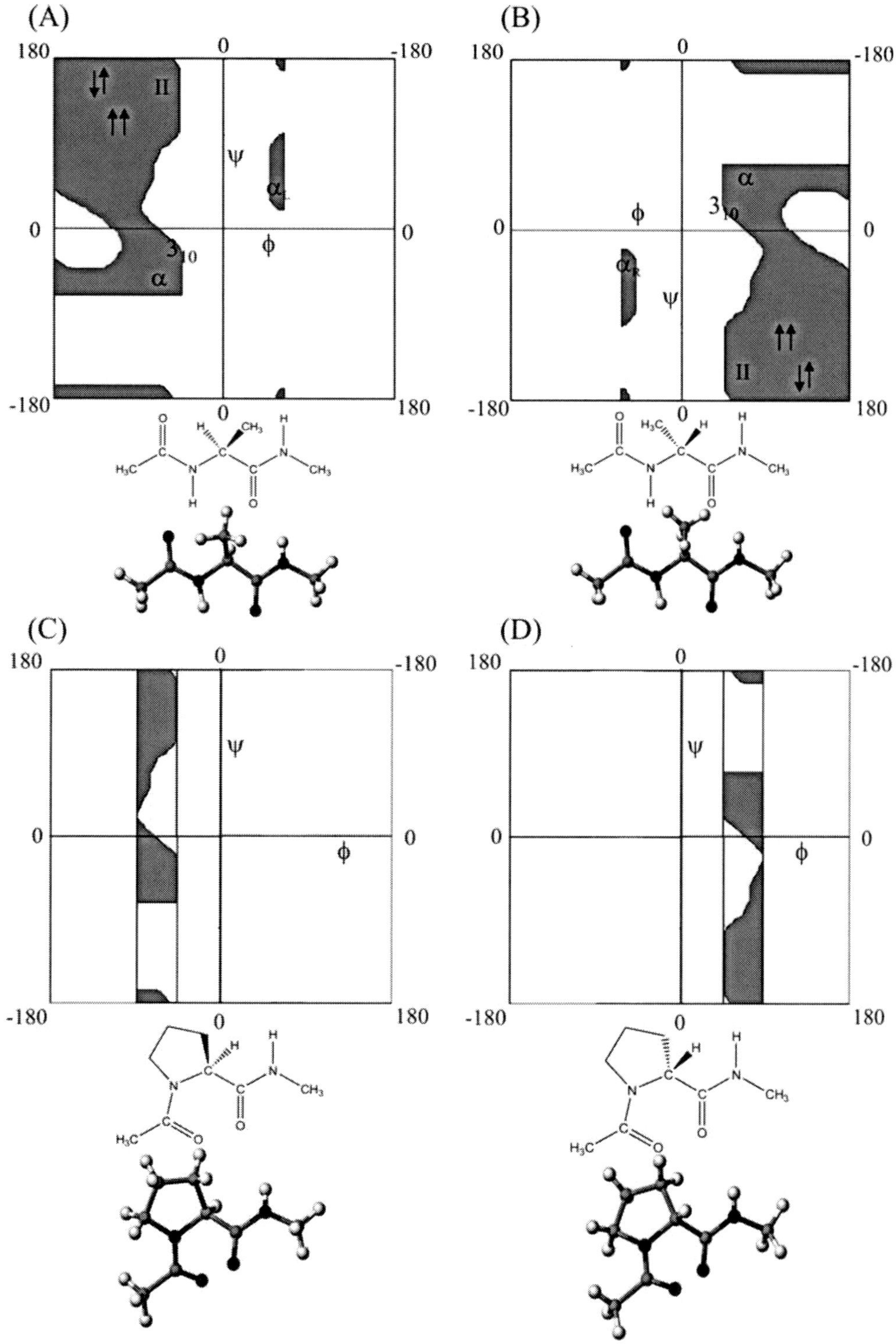

Figure 1. Ramachandran maps showing allowed regions for (A) L-Ala (B) D-Ala (C) L-Pro (D) D-Pro residues, respectively. The chemical formula and a ball-and-stick notation of N-acetyl-amino acyl-NHMe (fully extended conformation) for each amino acid are also shown below their respective maps. Figures have been adapted from [11]. Map in Figure (A) reprinted with permission from [11]; Copyright (2001) American Chemical Society.

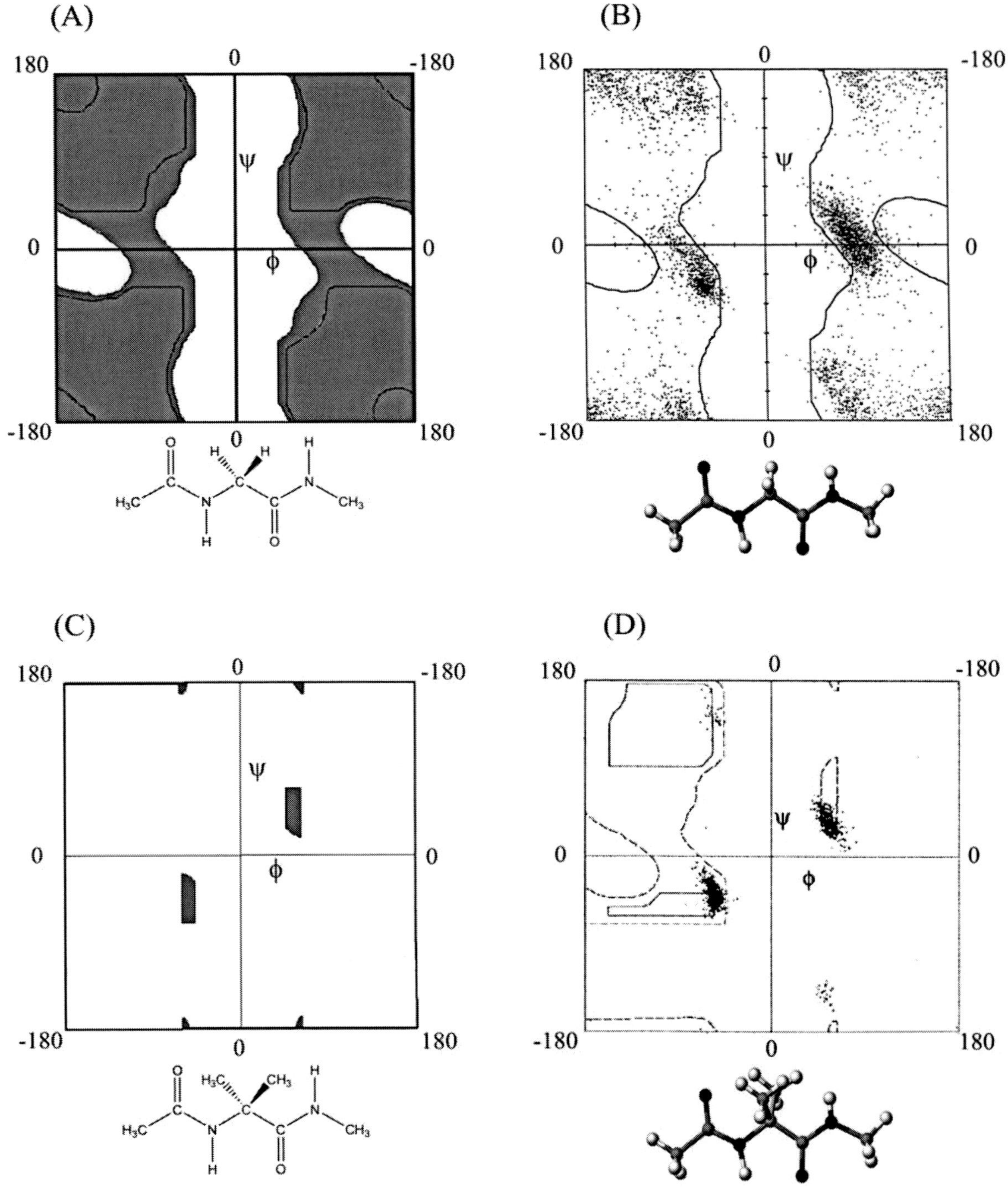

Figure 2. Allowed regions in the Ramachandran plot for Gly (A) and Aib (C) residues compared with ϕ-ψ values obtained from crystal structures of proteins (B) for Gly and peptides (D) for Aib. Also indicated are the chemical structures and ball-and-stick notations for these residues. Figures (A) and (C) reprinted with permission from [11]; Copyright (2001) American Chemical Society. Figures (B) and (D) reprinted with permission from [12]; Copyright (2003) Indian Academy of Sciences.

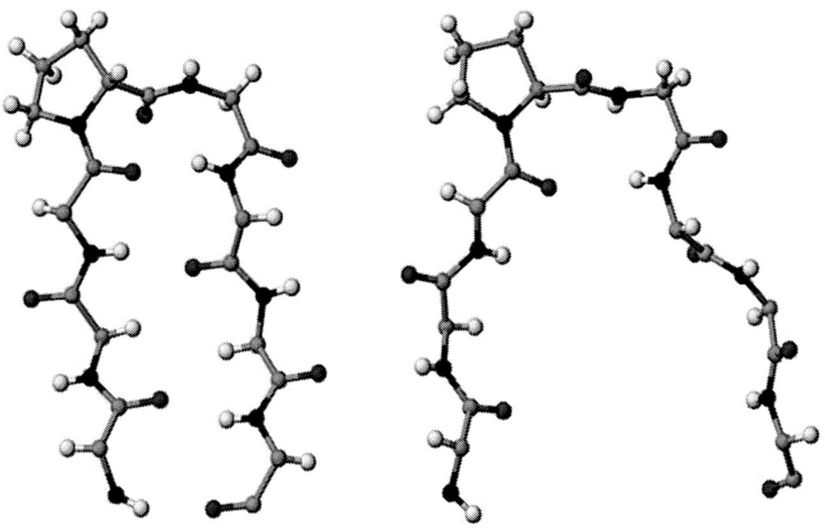

Figure 3. (A) Peptide hairpin with a type II' turn formed using ^{D}Pro-Gly as a turn nucleator. Shown is the crystal structure of the peptide Boc-Leu-Val-Val-^{D}Pro-Gly-Leu-Val-Val-OMe [17]. Only the peptide backbone and pyrrolidine ring of ^{D}Pro residue are shown. (B) ^{D}Pro in the previous peptide was replaced by ^{L}Pro and the ϕ-ψ values of the turn segment were altered to ideal type II values. Although turn nucleation does occur, it is evident that the two strand segments fray away from each other. This clearly establishes the superiority of ^{D}Pro to nucleate tight β-turns compared to its L-analog.

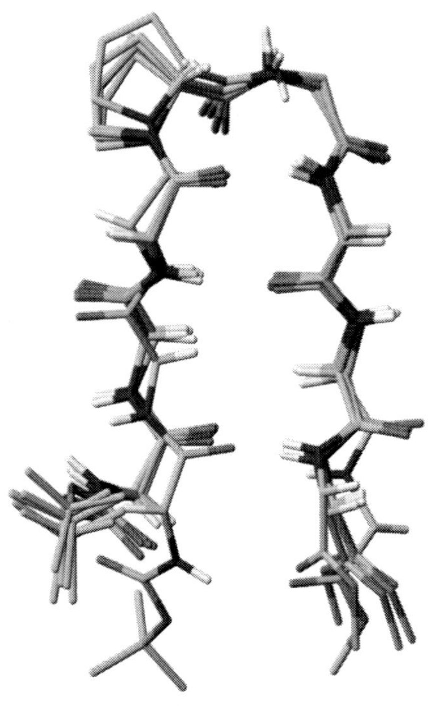

Figure 4. Superposition of crystal structures of six peptide hairpins obtained using a nucleating ^{D}Pro-Xxx segment. All the peptides shown, form type II' turns. Sequences: Boc-Leu-Val-Val-^{D}Pro-Gly-Leu-Val-Val-OMe [17], Boc-Leu-Val-Val-^{D}Pro-Ala-Leu-Val-Val-OMe (2 molecules) [20], Boc-Leu-Phe-Val-^{D}Pro-Ala-Leu-Phe-Val-OMe [21], Boc-Leu-Val-Val-^{D}Pro-Gly-Leu-Phe-Val-OMe [21], Boc-Leu-Val-Val-^{D}Pro-Aib-Leu-Val-Val-OMe [21]. These crystal structures provide experimental evidence for the turn-nucleating ability of ^{D}Pro-Xxx segments, which subsequently propagates to well-registered strand segments.

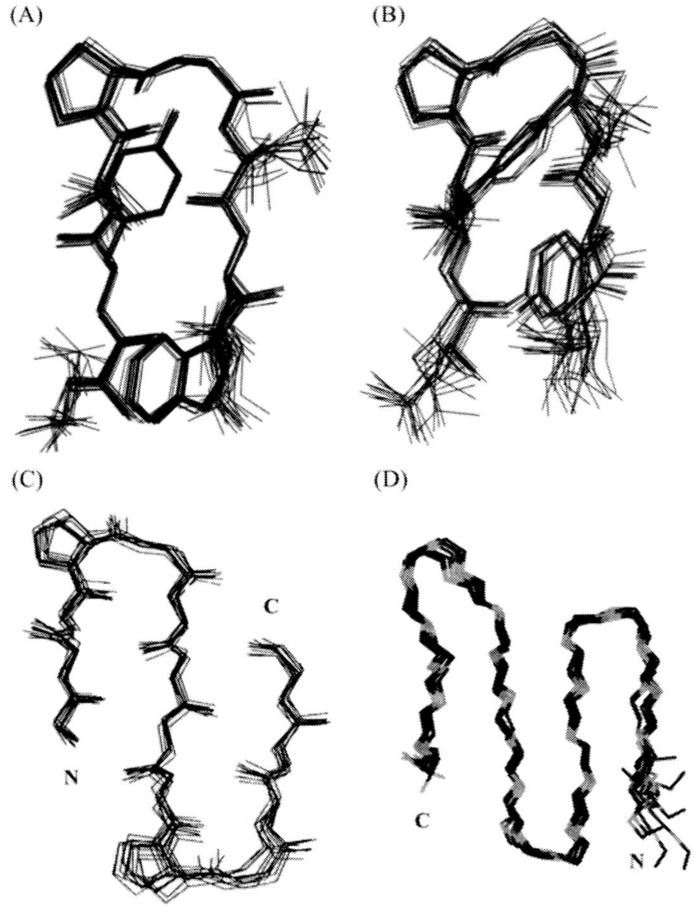

Figure 5. (A and B) Solution NMR structures of octapeptide hairpins formed using ^{D}Pro-Gly as a turn nucleator [22]. The tight turn forming ability of ^{D}Pro-Gly segments can be exploited for the construction of predefined hairpin scaffolds, which can then be made use of, for the study of tertiary interactions, such as aromatic interactions, in solution, as is demonstrated in these examples. (C and D) The ability of ^{D}Pro-Xxx segments to nucleate tight turns can be exploited for the construction of multistranded β-sheet structures. Shown are the solution NMR structures of a 14-residue three-stranded β-sheet [23] and a 26-residue four-stranded β-sheet [24]. Figures C and D reprinted with permission from [11]; Copyright (2001) American Chemical Society.

4. D-Amino Acid as Helix Termination Signals

Polypeptide helices in proteins are characterized by a succession of residues, which adopt ϕ-ψ values that lie in the right handed helical (α_R) region of the Ramachandran map. Helix termination at the N- and C-terminal ends occurs when residues drift away to other regions of ϕ-ψ space, most notably, the extended (β-sheet (β)/ polyproline (P_{II})) and left handed helical (α_L) regions (Fig. 1A). In proteins, a widely observed structural feature at the C-terminal end of helices is the Schellman motif, which is formed when the helix terminating residue adopts an α_L conformation [28,29]. In proteins structures this mode of termination predominantly involves the occurrence of Gly or Asn residues at the C-terminal position. This is a consequence of the fact that the achiral Gly residue can be readily accommodated in α_L conformations, while Asn has the greatest α_L propensity of the remaining chiral residues [30]. The sequence of residue conformations α_R α_R α_R α_R α_L leads to helix termination with the simultaneous formation of a pair of 4→1 and 6→1 hydrogen bonds (Fig. 7A).

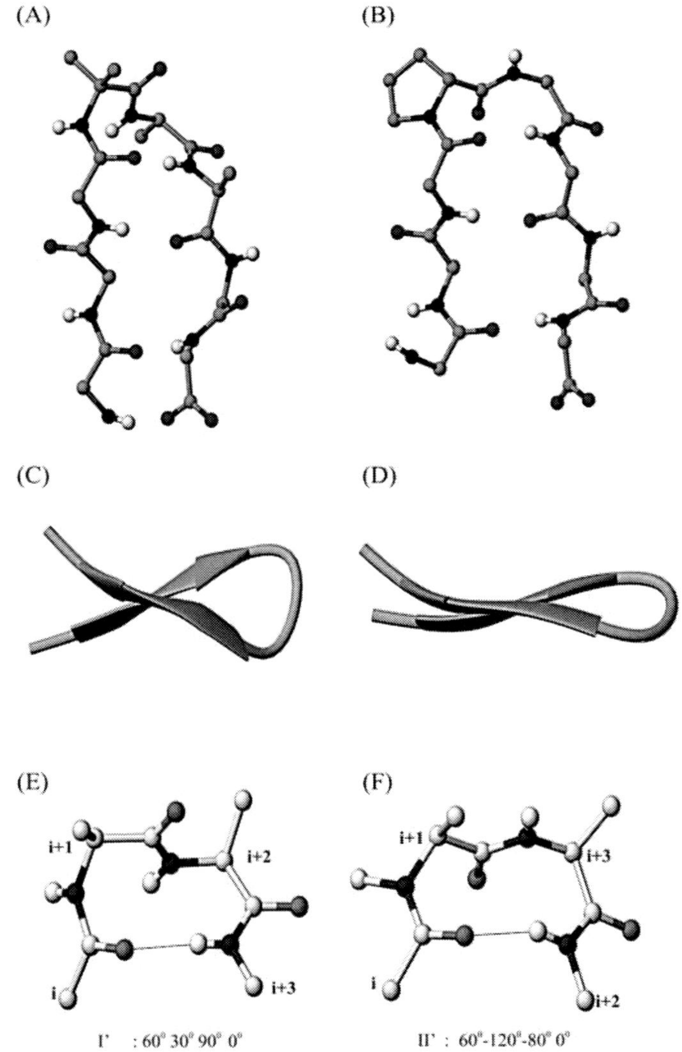

Figure 6. (A) Peptide hairpin nucleated by Aib-^{D}Ala segment (Boc-Leu-Phe-Val-Aib-^{D}Ala-Leu-Phe-Val-OMe) (type I' turn) [27] compared with a peptide hairpin formed using ^{D}Pro-Gly (Boc-Leu-Val-Val-^{D}Pro-Gly-Leu-Val-Val-OMe) (type II' turn) [17] (B). Side chains of only the turn residues have been included in the figure. (C and D) Strand orientations in both peptides are however different, as illustrated by a side view of the ribbon diagram of the two hairpins. Also shown are ball-and-stick notations of the two turn types: I' (E) and II (F).

In synthetic peptides, helix termination by a Schellman motif can be readily achieved by placement of an achiral residue or a D-residue towards the C-terminal of a potentially helical segment consisting of L-amino acids. Crystallographic characterization of Schellman motifs in peptides have been achieved in sequences containing the achiral Aib residue at the penultimate position from the C-terminus [31]. In the heptapeptide Boc-Leu-Aib-Val-Ala-Leu-Aib-Val-OMe, the molecule forms a right handed helix over the segment residues 1-5. Aib6 adopts an α_L conformation resulting in helix termination (Fig. 7B) [32]. Extension of this sequence to the level of a 10-residue peptide by addition of the tripeptide sequence ^{D}Ala-^{D}Leu-Aib-OMe results in an extension of the N-terminal helical segment with termination occurring by Schellman motif formation with ^{D}Ala8 adopting an α_L conformation (Fig. 7C) [33]. This example illustrates the energetic advantage of propagating the helix with concomitant formation of intramolecular hydrogen bonds and the propensity of the ^{D}Ala residue to adopt α_L conformation. Unexpectedly, the structure of this peptide also revealed an interesting chain reversal stabilized by the formation of a C-H...O hydrogen bond between the Ala4 $C^{\alpha}H$ and ^{D}Leu9 C=O groups. This structural feature is a consequence of the extended conformation at the ^{D}Leu9 residue with a positive value of ϕ and a negative value of ψ. The generality of this structural feature has been confirmed in the crystal structure of the decapeptide analog, where Ala4 is replaced by Gly. Most importantly, replacement of ^{L}Ala4 by ^{D}Ala4 or Aib4 results in a completely different peptide conformation, establishing the importance of the C-H...O interaction in stabilizing this conformational feature [34]. A similar structural motif has also been identified in proteins [35].

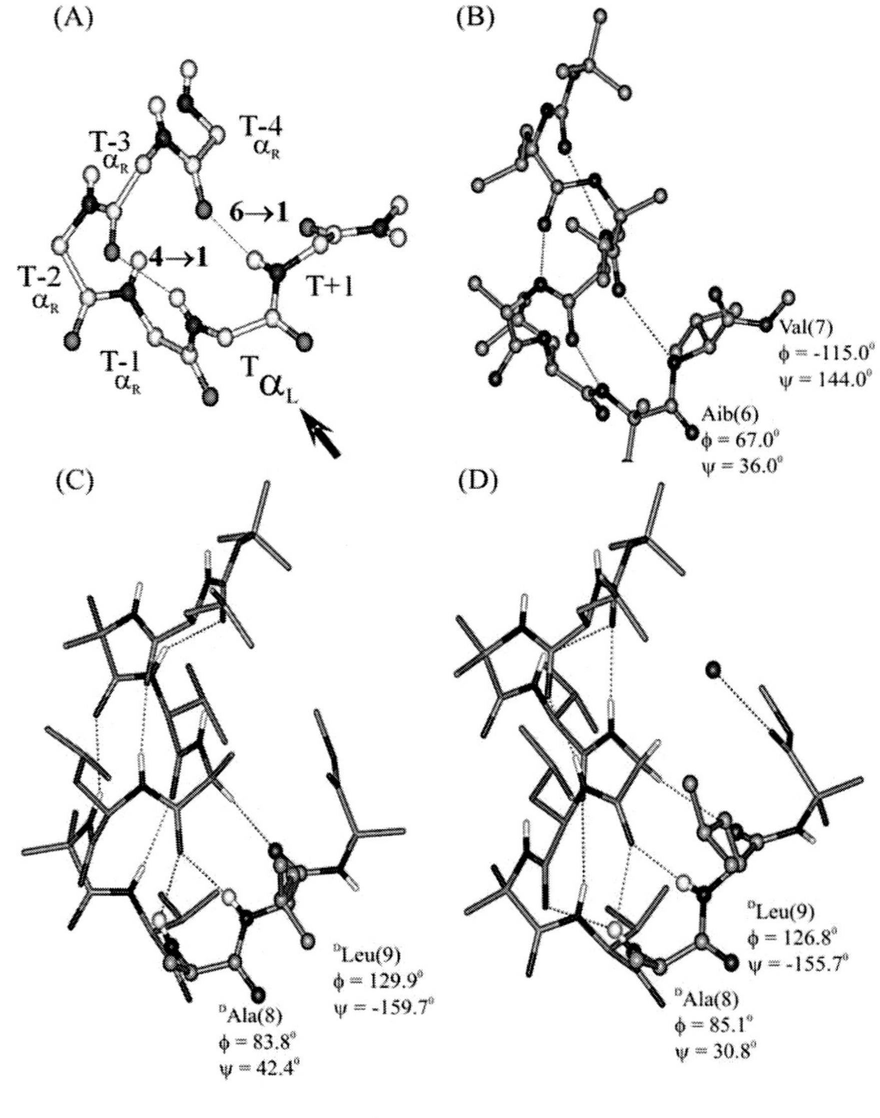

(B) Boc-Leu-Aib-Val-Ala-Leu-Aib-Val-OMe
(C) Boc-Leu-Aib-Val-Ala-Leu-Aib-Val-^{D}Ala-^{D}Leu-Aib-OMe
(D) Boc-Leu-Aib-Val-Gly-Leu-Aib-Val-^{D}Ala-^{D}Leu-Aib-OMe

Figure 7. (A) Helix-terminating Schellman motif obtained when the terminal residue (T) adopts an α_L conformation (indicated by an arrow). A 4→1 and 6→1 hydrogen bond formation stabilizes this motif [32]. (B) Crystal structure of a heptapeptide forming the Schellman motif with Aib6 adopting an α_L conformation [32]. (C and D) Decapeptides containing D-amino acid residues, which function to signal helix termination [33, 34]. Figures (A), (B), (D) reprinted with permission from [12]; Copyright Indian Academy of Sciences. Figure (C) reprinted with permission from [34]; Copyright (2003) American Chemical Society.

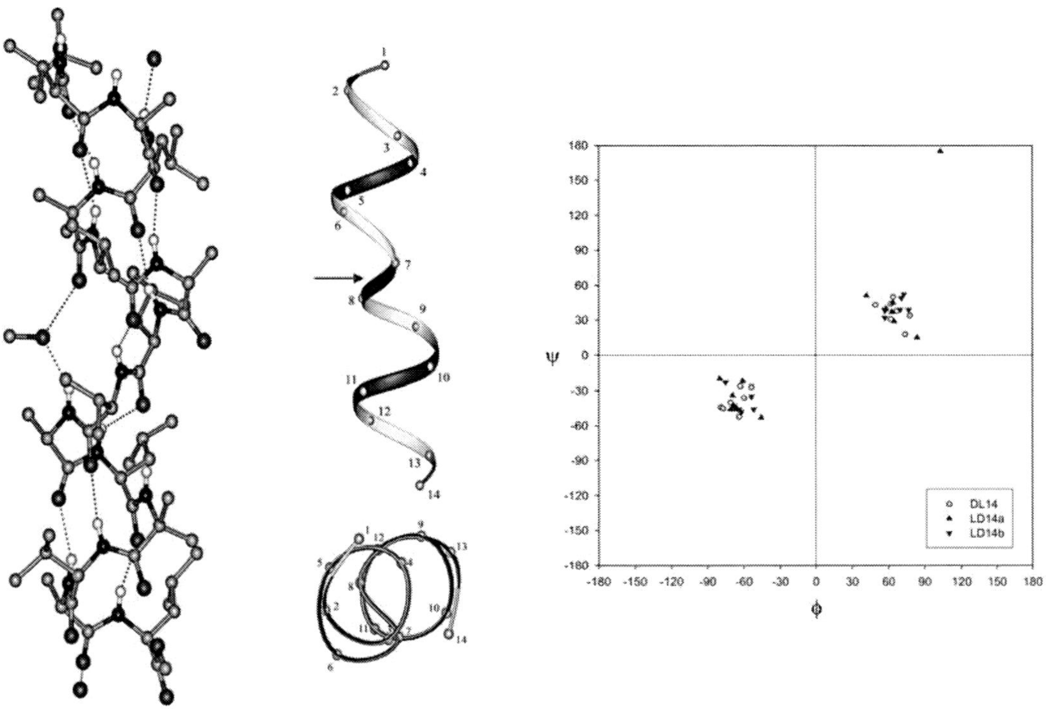

(A) (B) (C)

Figure 8. Ambidextrous helices formed by fusing left- and right-handed helical segments. (A) Ball-and-stick notation of a 14-residue peptide Boc-^{D}Val-^{D}Ala-^{D}Leu-Aib-^{D}Val-^{D}Ala-^{D}Leu-Val-Ala-Leu-Aib-Val-Ala-Leu-OMe (DL14) [36]. The DL junction, indicated by an arrow bears the Schellman motif. (B) Ribbon diagram of the peptide viewed along the helix axis and down the helix axis. (C) Plot of the ϕ-ψ values obtained from the crystal structures of two ambidextrous helices. One of the peptides (Boc-Val-Ala-Leu-Aib-Val-Ala-Leu-^{D}Val-^{D}Ala-^{D}Leu-Aib-^{D}Val-^{D}Ala-^{D}Leu-OMe, LD14) crystallized in two polymorphic forms [37]. Figures (A) and (B) reprinted from [12]; Copyright (2003) Indian Academy of Sciences.

The ability of D-amino acid residues to facilitate Schellman motif mediated helix termination may be used to advantage in creating ambidextrous structures, which consist of directly fused right-handed and left-handed helical segments. Fig. 8 illustrates the structure of a 14-residue peptide (Boc-^{D}Val-^{D}Ala-^{D}Leu-Aib-^{D}Val-^{D}Ala-^{D}Leu-Val-Ala-Leu-Aib-Val-Ala-Leu-OMe) (DL14), which consists of two distinct heptapeptide segments containing amino acid sequences of opposite chirality [36]. The achiral Aib residue at positions 4 and 11 in the sequence nucleates helix formation with the handedness of the helix being determined by the configuration of the flanking amino acid sequences. Careful inspection of the crystal structure in Fig. 8 reveals that the molecule is composed of two distinct helical segments; a left-handed helical segment for residues 1-7 and a right handed helical segment for residues 8-14. Residue 8, which adopts an α_R conformation, acts as the helix terminating residue with respect to the N-terminal *left handed* helical (α_L) segment. The Schellman motif is distorted by the insertion of a methanol molecule into the 6→1 hydrogen bond. The robustness of this fold is further confirmed by the crystal structure of the enantiomeric peptide (Boc-Val-Ala-Leu-Aib-Val-Ala-Leu-^{D}Val-^{D}Ala-^{D}Leu-Aib-^{D}Val-^{D}Ala-^{D}Leu-OMe) (LD14), which curiously crystallized in a polymorphic form containing two molecules in the crystallographic asymmetric unit [37]. Comparison of the ϕ-ψ values (Table AII of [37]) confirms the similarity of the folding pattern of these sequences containing separate blocks of L- and D-amino acids.

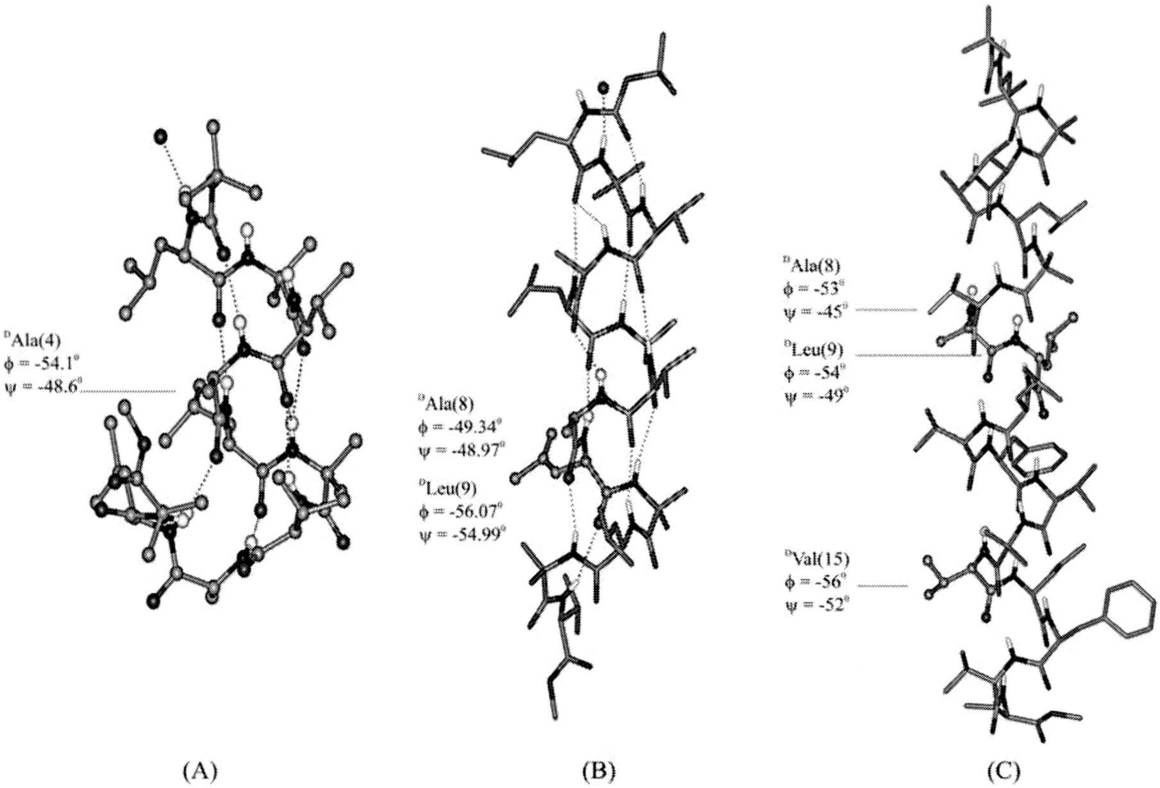

Figure 9. Examples of peptide helices containing D-amino acids incorporated in the right handed helical scaffold. (A) One D-amino acid in the right-handed helix (Boc-Leu-Aib-Val-^{D}Ala-Leu-Aib-Val-^{D}Ala-^{D}Leu-Aib-OMe) [34]. (B) Two D-amino acids in the right-handed helix (Boc-Leu-Aib-Val-Ala-Leu-Aib-Val-^{D}Ala-^{D}Leu-Aib-Leu-Aib-Val-OMe) [40]. (C) Three D-amino acids in the right-handed helix (Boc-Leu-Aib-Val-Ala-Leu-Aib-Val-^{D}Ala-^{D}Leu-Leu-Val-Phe-Val-Aib-^{D}Val-Leu-Phe-Val-Val-OMe) [41]. Figures (A) and (B) reprinted from [12]; Copyright (2003) Indian Academy of Sciences.

5. Guest D-Amino Acids in Right Handed Helical Segments

In the synthesis of analogs of biologically active peptides, it is often desirable to introduce D-amino acids into host sequences composed of L-amino acids. The site of D-amino acid insertion becomes resistant to proteolytic cleavage, thereby resulting in greater *in vivo* stability of such analogs [38]. In designing analogs, it is necessary to retain the conformational features, which are critical for biological activity. Thus, the site of D-amino acid replacement must be carefully chosen in order to minimize the possibility of major conformational changes. The helix is one of the two most widely distributed secondary structures in peptides and proteins. It is therefore instructive to examine the consequences of inserting D-amino acids into the center of potential helical sequences. Inspection of the conformational maps in Fig. 1B clearly establishes that there is a small 'allowed window' for D-amino acids in the α_R region of conformational space, suggesting that, in principle, D-amino acids can be accommodated into *right-handed* helical segments. The relatively small energy penalty for adopting positive ϕ values can be readily offset by compensating interactions. Fairman *et al* have estimated the helix-forming propensity of D-residues in right handed α-helices and find a penalty of 0.95kcal/mol for D-Ala when compared to the L-Ala standard [39]. Fig. 9 shows examples of synthetic peptide helices in which D-amino acids have been incorporated without any disruption of the right handed twist of the polypeptide chain. In the decapeptide (Boc-Leu-Aib-Val-^{D}Ala-Leu-Aib-Val-^{D}Ala-^{D}Leu-Aib-OMe), the ^{D}Ala residue at position 4 adopts an α_R conformation and is incorporated into the right-handed helix formed by the segments 1-7. In this case, the observed ϕ-ψ values for the ^{D}Ala residue in the two crystallographically independent molecules are -46.6, -52.9 and -54.1, -48.6 respectively [34]. The 13-residue peptide (Boc-Leu-Aib-Val-Ala-Leu-Aib-Val-^{D}Ala-^{D}Leu-Aib-Leu-Aib-Val-OMe) provides an interesting example of a double D-segment (^{D}Ala8, ^{D}Leu9) incorporated into a right-handed helix [40]. The crystal structure of the 19-residue peptide (Boc-Leu-Aib-Val-Ala-Leu-Aib-Val-^{D}Ala-^{D}Leu-Leu-Val-Phe-Val-Aib-^{D}Val-Leu-Phe-Val-Val-OMe) is a dramatic example of the accommodation of as many as three D-amino acids into a right-handed helical scaffold [41]. These examples clearly emphasize the fact that right-handed helices can readily accommodate D-residues. Selective D-amino acid substitution in strongly helical sequences may provide a convenient means of enhancing proteolytic stability, without significant disruption of overall secondary structure.

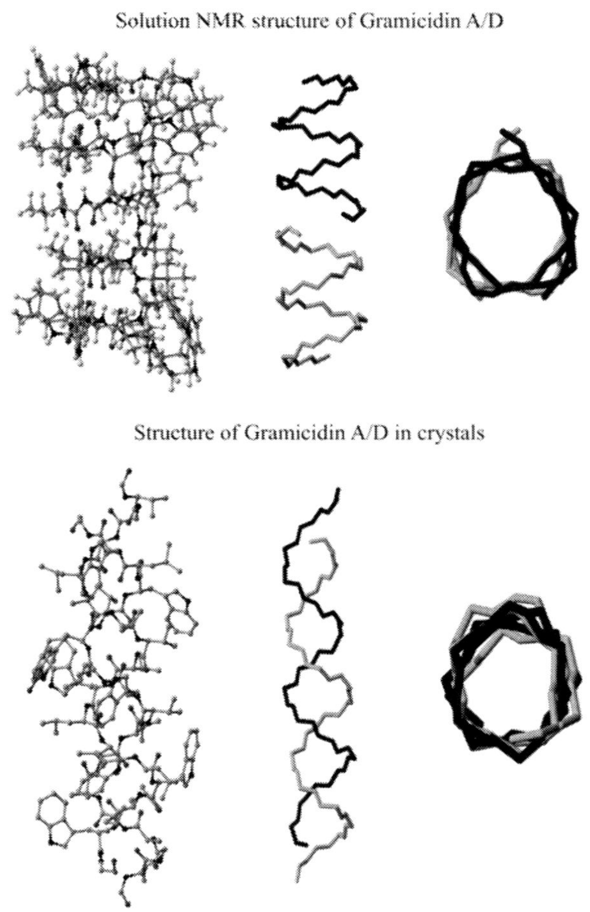

Figure 10. LD peptide helices of gramicidin. Sequence: For-^{L}Val-Gly-^{L}Ala-^{D}Leu-^{L}Ala-^{D}Val-^{L}Val-^{D}Val-^{L}Trp-^{D}Leu-^{L}Trp-^{D}Leu-^{L}Trp-^{D}Leu-^{L}Trp-Eta. Shown are the NMR- and X-ray derived structures of gramicidin A/D (PDB coordinates 1JNO [45] and 1ALZ [47], respectively). Note that although the gramicidin structures have been named 'A' and 'D', their sequences are essentially the same. Interestingly, the NMR structure reveals a π_{LD} helical structure for the peptide, whereas the crystal structure is a double helix.

(A)

H_2N-^{D}Mpg-^{D}Dpg-^{L}Val-^{D}Dpg-^{L}Mpg-^{D}Dpg-^{L}Mpg-^{D}Dpg-^{L}Val-^{D}Dpg-^{L}Mpg-^{L}Phe-^{L}Asp-COOH

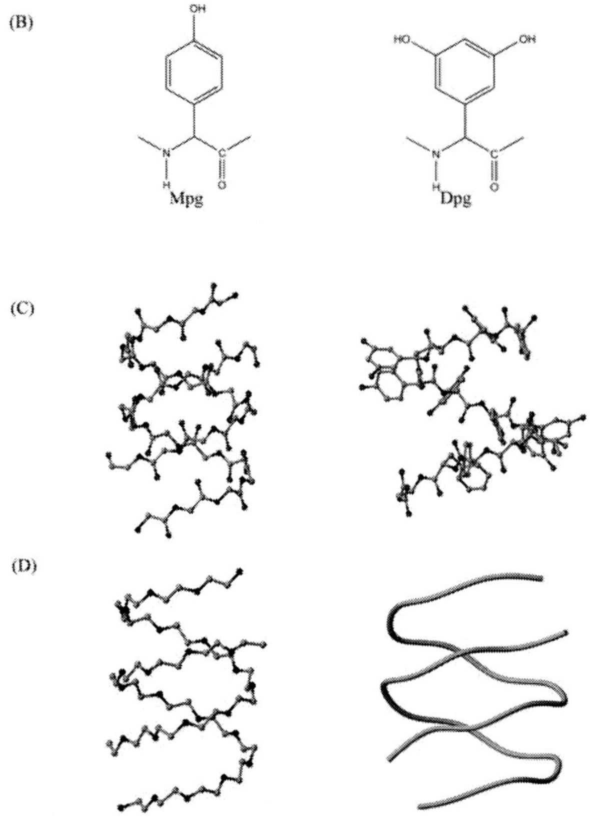

Figure 11. Feglymycin (A) sequence (B) chemical structure of the unnatural amino acids (C) different views of the double helix. Also shown is the ball-and-stick notation of the single molecule. PDB code: 1W7Q [48].

6. Alternating LD Sequences

Any discussion of the conformational effects of D-residues in peptide sequences will be incomplete without a consideration of LD polypeptides. Regular alternation of configuration is observed in the linear gramicidins, which are membrane channel forming polypeptides produced by *Bacillus brevis*. Early theoretical studies suggested that alternating LD sequences would adopt wide-bore helical structures in which an approximately extended conformation of the polypeptide was wound around a helical axis. The π_{LD} (LD_n helices, where n=3,4,5) helices so generated may be viewed as helices formed by repeating LD dipeptides, in which the two residues adopt β-sheet conformations that lie on opposite sides of the ϕ-ψ map. The helices so formed are stabilized by parallel β-sheet hydrogen bonds between segments of the polypeptide chain on adjacent helical turns [42]. The NMR study of Gramicidin A provides confirmatory evidence for the π_{LD} helix (Fig. 10A) [43-45]. In crystals, an interesting variant of the β-helical conformation is observed with two antiparallel chains forming a double helical structure (Fig. 10B) [46, 47]. The unique structures formed by alternating LD sequences are also highlighted by a recent structure determination of the 13-residue peptide Feglymycin [48]. In this case also, an antiparallel double helical structure is formed (Fig. 11). Fig. 12 summarizes the distribution of ϕ-ψ values observed for the alternating LD segments. The LD helices are characterized by the formation of a central pore, which can permit passage of cations, a property that is of functional significance in these natural peptides [49].

7. Conclusions

D-amino acids can be effectively used in peptide design since their conformational preferences can facilitate formation of specific folded structures. β-Turn formation is favored in dipeptide sequences of alternating configuration (LD peptides). ^{D}Pro, a constrained residue, can be used to force the formation of prime turns (type II'/I'), which inturn serves as nuclei for antiparallel β-hairpin formation in designed peptides. Multiple ^{D}Pro-Xxx segments may be used to promote the formation of multistranded β-sheets. D-amino acids can also be used to mimic helix termination signals like the Schellman motif. Peptides containing building blocks of L- and D- amino acids can be induced to form ambidextrous helices in which right- and left-handed segments are fused. Single D-amino acids and double D-segments may be incorporated into synthetic peptides forming right-handed helical structures. The energy penalties for insertion of a D-residue into a right-handed helix are small and readily compensated by minor structural alterations. Alternating LD sequences provide a special example of the formation of β-helices, which may be viewed as structures generated by winding an extended polypeptide chain around a cylinder. Defining the role of D-amino acids on peptide conformations is of importance in the rational design of analogs of biologically active peptides, which may possess a high degree of resistance to proteolysis.

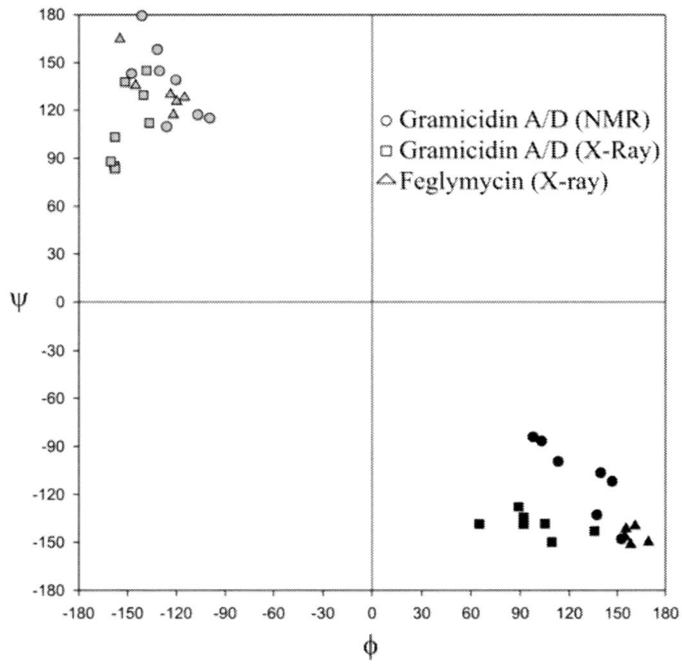

Figure 12. Plot of the observed ϕ-ψ values for the two structures of gramicidin A/D and the crystal structure of feglymycin. It is evident that the dihedral angles of the individual residues populate the extended region of the ϕ-ψ map, although the overall structure formed is helical.

8. Acknowledgments

RM acknowledges the Council of Scientific and Industrial Research (CSIR), India for the award of a Senior Research Fellowship (SRF). The structural analysis of peptides in crystals has been made possibly by long-standing collaboration with Dr. I. L. Karle, Naval Research Laboratory, USA and Prof. N. Shamala, Department of Physics, Indian Institute of Science, Bangalore, India. Work in this area at

Bangalore has been supported by grants from the Department of Science and Technology and the Department of Biotechnology.

REFERENCES

[1] Lipmann, F., Hotchkiss, R. D. & Dubos, J. (1941). The occurrence of D-amino acids in gramicidin and tyrocidine. *J Biol Chem, 141*, 163-169.

[2] Kleinkauf, H. & von Dohren, H. (1996). A nonribosomal system of peptide biosynthesis. *Eur J Biochem, 236*, 335-351.

[3] Kreil, G. (1997). D-amino acids in animal peptides. *Annu Rev Biochem, 66*, 337-345.

[4] Craig, G., Bandyopadhyay, P. & Olivera, B. M. (1999). Post-translationally modified neuropeptides from conus venoms. *Eur J Biochem, 264*, 271-275.

[5] Milton, R. C. L., Milton, S. C. F. & Kent, S. B. H. (1992). Total chemical synthesis of a D-enzyme: The enantiomers of HIV-1 protease show demonstration of reciprocal chiral substrate specificity. *Science, 256*, 1445-1448.

[6] Zawadzke, L. E. & Berg, J. M. (1992). A racemic protein. *J Am Chem Soc, 114*, 4002-4003.

[7] Fitzgerald, M. C., Chernuschevich, I., Standing, K. G., Kent, S. B. H. & Whitman, C. P. (1995). Total chemical synthesis and catalytic properties of the enzyme enantiomers L- and D-4-oxalocrotonate tautomerase. *J Am Chem Soc, 117*, 11075-11080.

[8] Balaram, P. (1999). De novo design: Backbone conformational constraints in nucleating helices and β-hairpins. *J Peptide Res, 54*, 195-199.

[9] Ramachandran, G. N. & Sasisekharan, V. (1968). Conformation of polypeptides and proteins. *Adv Protein Chem, 23*, 283-438.

[10] Ramachandran, G. N. & Chandrasekaran, R. *Progress in Peptide Research.* Lande, S., Ed.; Gordon & Breach: New York, 1972; Vol. II (Proceedings of the Second American Peptide Symposium, Cleveland, 1970), p 195.

[11] Venkatraman, J., Shankaramma, S. C. & Balaram, P. (2001). Design of folded peptides. *Chem Rev, 101*, 3131-3152.

[12] Aravinda, S., Shamala, N., Roy, R. S. & Balaram, P. (2003). Non-protein amino acids in peptide design. *Proc Indian Acad Sci (Chem Sci), 115*, 373-400.

[13] Ramakrishnan, C. & Srinivasan, N. (1990). Glycyl residues in proteins and peptides: An analysis. *Curr Sci, 59*, 851-861.

[14] Wilmot, C. M. & Thornton, J. M. (1988). Analysis and prediction of the different types of β-turn in proteins. *J Mol Biol, 203*, 221-232.

[15] Creighton, T. E. (1997) *Proteins: Structure and molecular properties* (2nd edition), New York, NY: Freeman Press.

[16] Awasthi, S. K., Raghothama, S. & Balaram, P. (1995). A designed β-hairpin peptide. *Biochem Biophys Res Commun, 216*, 375-381.

[17] Karle, I. L., Awasthi, S. K. & Balaram, P. (1996). A designed β-hairpin peptide in crystals. *Proc Natl Acad Sci USA, 93*, 8189-8193.

[18] Stanger, H. E. & Gellman, S. H. (1998). Rules for antiparallel β-sheet design: D-Pro-Gly is superior to L-Asn-Gly for β-hairpin nucleation. *J Am Chem Soc, 120*, 4236-4237.

[19] Karle, I. L., Das, C. & Balaram, P. (2000). De novo protein design: Crystallographic characterization of a synthetic peptide containing independent helical and hairpin domains. *Proc Natl Acad Sci USA, 97*, 3034-3037.

[20] Das, C., Naganagowda, G. A., Karle, I. L. & Balaram, P. (2001). Designed β-hairpin peptides with defined tight turn stereochemistry. *Biopolymers, 58*, 335-346.

[21] Aravinda, S., Harini, V. V., Shamala, N., Das, C. & Balaram, P. (2004). Structure and assembly of designed β-hairpin peptides in crystals as models for β-sheet aggregation. *Biochemistry, 43*, 1832-1846.

[22] Mahalakshmi, R., Raghothama, S. & Balaram, P. (2006). NMR analysis of aromatic interactions in designed peptide β-hairpins. *J Am Chem Soc, 128*, 1125-1138.

[23] Das, C., Raghothama, S. & Balaram, P. (1998). A designed three stranded β-sheet peptide. *J Am Chem Soc, 120*, 5812-5813.

[24] Das, C., Raghothama, S. & Balaram, P. (1999). A four stranded β-sheet structure in a designed synthetic polypeptide. *Chem Commun*, 967-968.

[25] Venkataraman, J., Naganagowda, G. A., Sudha, R. & Balaram, P. (2001). *De novo* design of a five-stranded β-sheet anchoring a metal-ion binding site. *Chem Commun*, 2660-2661.

[26] Venkatraman, J., Nagana Gowda, G. A. & Balaram, P. (2002). Design and construction of an open multistranded β-sheet polypeptide stabilized by a disulfide bridge. *J Am Chem Soc, 124*, 4987-4994.

[27] Aravinda, S., Shamala, N., Rajkishore, R., Gopi, H. N. & Balaram, P. (2002). A crystalline β-hairpin peptide nucleated by a type I' Aib-D-Ala β-turn: Evidence for cross-strand aromatic interactions. *Angew Chem Int Ed Engl, 41*, 3863-3865.

[28] Schellman, C. (1980) The α_L conformation at the ends of helices, in *Protein folding* (Jaenicke, R., ed.), pp. 53-61, Elsevier/North-Holland, New York.

[29] Nagarajaram, H. A., Sowdhamini, R., Ramakrishnan, C. & Balaram, P. (1993). Termination of right handed helices in proteins by residues in left-handed helical conformations. *FEBS Lett, 321*, 79-83.

[30] Srinivasan, N., Anuradha, V. S., Ramakrishnan, C., Sowdhamini, R. & Balaram, P. (1994). Conformational characteristics of asparginyl residues in proteins. *Int J Peptide Protein Res, 44*, 112-122.

[31] Karle, I. L., Flippen-Anderson, J. L., Uma, K. & Balaram, P. (1993). Peptide mimics for structural features in proteins. Crystal structures of three heptapeptide helices with a C-terminal 6→1 hydrogen bond. *Int J Pept Protein Res, 42*, 401-410.

[32] Banerjee, A., Datta, S., Pramanik, A., Shamala, N. & Balaram, P. (1996). Heterogeneity and stability of helical conformations in peptides: Crystallographic and NMR studies of a model heptapeptide. *J Am Chem Soc, 118*, 9477-9483.

[33] Aravinda, S., Shamala, N., Pramanik, A., Das, C. & Balaram, P. (2000). An unusual C-H...O hydrogen bond mediated reversal of polypeptide chain direction in a synthetic peptide helix. *Biochem Biophys Res Commun, 273*, 933-36.

[34] Aravinda, S., Shamala, N., Bandyopadhyay, A. & Balaram, P. (2003). Probing the role of the C-H...O hydrogen bond stabilized polypeptide chain reversal at the C-terminus of designed peptide helices. Structural characterization of three decapeptides. *J Am Chem Soc, 125*, 15065-15075.

[35] Babu, M. M., Singh, S. K. & Balaram, P. (2002). A C-H triple bond O hydrogen bond stabilized polypeptide chain reversal motif at the C terminus of helices in proteins. *J Mol Biol, 322*, 871-880.

[36] Banerjee, A., Raghothama, S. R., Karle, I. L. & Balaram, P. (1996). Ambidextrous molecules: Cylindrical peptide structures formed by fusing left- and right-handed helices. *Biopolymers, 39*, 279-285.

[37] Karle, I. L. (2001). Controls exerted by the Aib residue: Helix formation and helix reversal. *Biopolymers (Pept Sci), 60*, 351-365.

[38] van Regenmortel, M. & Muller, S. (1998). D-peptides as immunogens and diagnostic reagents. *Curr Opin Biotechnol, 9*, 377-382.

[39] Fairman, R., Anthony-Cahill, S. J. & DeGrado, W. F. (1992). The helix-forming propensity of D-alanine in a right handed α helix. *J Am Chem Soc, 114*, 5458-5459.

[40] Aravinda, S., Shamala, N., Desiraju, S. & Balaram, P. (2002). A right-handed peptide helix containing a central double D-amino acid segment. *Chem Commun*, 2454-2455.

[41] Karle, I. L., Gopi, H. N. & Balaram, P. (2003). Crystal structure of a hydrophobic 19-residue peptide helix containing three centrally located D amino acids. *Proc Natl Acad Sci USA*, *100*, 13946-13951.

[42] Ramachandran, G. N. & Chandrasekaran, R. (1972). Conformation of peptide chains containing both L- and D-residues: Part I-helical structures with alternating L- and D-residues with special reference to the LD-ribbon and the LD-helices. *Indian J Biochem Biophys*, *9*, 1-11.

[43] Arseniev, A. S., Barsukov, I. L., Lomize, A. L., Orekhov, V. Y. & Bystrov, V. F. (1992). Refinement of the spatial structure of the gramicidin A ion channel. *Bioorg Khim*, *18*, 182-200.

[44] Ketchem, R. R., Lee, K. C., Huo, S. & Cross, T. A. (1996). Macromolecular structural elucidation with solid-state NMR-derived orientational constraints. *J Biomol NMR*, *8*, 1-14.

[45] Townsley, L. E., Tucker, A., Sham, S. & Hinton, J. F. (2001). Structures of gramicidins A, B and C incorporated in sodium dodecyl sulfate micelles. *Biochemistry*, *40*, 11676-86.

[46] Langs, D. A. (1988). Three-dimensional structure at 0.86Å of the uncomplexed form of the transmembrane ion channel peptide gramicidin A. *Science*, *241*, 188-191.

[47] Burkhart, B. M., Gassman, R. M., Langs, D. A., Pangborn, W. A. & Duax, W. L. (1998). Heterodimer formation and crystal nucleation of gramicidin D. *Biophys J*, *75*, 2135-2146.

[48] Bunkoczi, G., Vertesy, L. & Sheldrick, G. M. (2005). The antiviral antibiotic feglymycin: First direct-methods solution of a 1000+ equal-atom structure. *Angew Chem Int Ed*, *44*, 1340-1342.

[49] Hladky, S. B. & Haydon, D. A. (1970). Discreteness of conductance change in bimolecular lipid membranes in the presence of certain antibiotics. *Nature*, *225*, 451-53.

In: D-Amino Acids: ISBN 1-60021-075-9
A New Frontier in Amino Acid and Protein Research, pp.431-440

Chapter 5.10

CELLULAR APPROACH OF THE BIOGENESIS OF D-AMINO-ACID CONTAINING PEPTIDES IN EUKARYOTES: THE CRUSTACEAN MODEL

Daniel Soyez [*]

Groupe Biogenèse des Peptides Isomères, FRE CNRS 2852: Protéines: Biochimie Structurale et Fonctionnelle, Université Pierre et Marie Curie, France.

ABSTRACT

The presence of a D-amino acid residue was observed several years ago in isoforms of a major Crustacean neuropeptide, the Crustacean Hyperglycemic Hormone (CHH) from lobster and crayfishes. To study the cellular distribution of the different stereoisomers, hapten-specific antisera have been raised against synthetic peptides with sequences corresponding to the N-terminal octapeptide of CHH isoforms. Then, crayfish and lobster neuroendocrine complexes were explored by immunohistochemistry (IHC) and immunocytochemistry (ICC) as well. Very recently, a CHH-related neurohormone, the Vitellogenesis-Inhibiting Hormone (VIH) was shown to be present in lobster neurohemal organs as two stereoisomers, like CHH. However, the N-ter amino-acid sequence of both neuropeptides is very different and the D-residue is a phenylalanine in position 3 in CHH but a tryptophan in position 4 in VIH. Specific conformational antibodies were developed, allowing localization of VIH stereoisomers by IHC and ICC, in conjunction with CHH detection.

These studies, which constitute first characterizations of D-amino acid containing peptide producing cells in Eukaryotes, raise intriguing questions regarding the cellular mechanism of posttranslational isomerization.

Keywords: D-amino-acid, neuropeptides, neuroendocrine cells, Crustacean hyperglycaemic hormone, Crustacea.

---

[*] Correspondence concerning this article should be addressed to Daniel Soyez, E-mail: daniel.soyez@snv.jussieu.fr.

INTRODUCTION

Since the early publication by Montecuchi et al. [1] describing the presence of a D-Ala residue in second position of the sequence of dermorphin, an octapeptide isolated from the skin of the frog *Phyllomedusa sauvagei*, a number of different opioid and antimicrobial peptides containing a D-amino acid (aa) residue have been characterized from frog skin serous glands [reviews in 2,3]. D-aa containing peptides have been evidenced in peptides from other animal groups as well, especially in toxins present in different animal venoms [reviews in 4,5]. D-aa containing peptides were also identified in endogenous neuropeptides such as in achatin, fulicin and FFRFamide from Molluscs [review in 6], and in Crustacean neurohormones, which are discussed in the present article. The different D-aa containing peptides identified to date in Eucaryotes are presented in Table 1.

Table 1: Peptides with a D-amino acid residue isolated from eukaryotic tissues. The list of opioid peptides from frog skin is not exhaustive. The D-residue is in bold italics.

Peptide	Sequence	Organism
Opïoid and antibacterial peptides from frog skin		
Dermorphin	Tyr-***Ala***-Phe-Gly-Tyr-Pro-SerNH_2	*Phyllomed*
Dermenkephalin	Tyr-***Met***-Phe-His-Leu-Met-AspNH_2	"
Deltorphins I/II	Tyr-***Ala***-Phe-Asp/Glu-Val-Val-Gly	
Leu-deltorphin	Tyr-***Leu***-Phe-Ala-Asp-Val-Ala-....	
Bombinin H	Ile-***Alle***-Gly-Pro-Val-Leu-Gly-	*Bombina*
Toxin from spider venom	46	
ω-Agatoxin	...-Gly-Leu-***Ser***-Phe-Ala	*Agelenopsis*
Toxins from Mollusc venom	46	
Conotoxine r11a:......	...-Ser-Phe-***Phe***-Lys-Ile	*Conus*
Contryphan	Gly-Cys-Hyp-***Trp***-Glu-Pro-Trp-CysNH2	*Conus*
Toxins from mammal venom		
OvCNP:	Leu-***Leu***-His-Asp-His-Pro-Asn-....	*Ornithorhyncus*
DLP-2	Ile-***Met***-Phe-Phe-Glu-Met-	"
Molluscan neuroexcitatory peptides		
Achatin	Gly-***Phe***-Ala-Asp	*Achatina*
Fulicin	Phe-***Asn***-Glu-Phe-ValNH2	"
Fulyal	Tyr-***Ala***-Glu-Phe-LeuNH2	"
FFRFamid	Ala-***Leu***-Ala-Gly-Asp-His-Phe-Phe-Arg-PheNH_2	*Mytilus*
NdWFamid	Asp-***Trp***-PheNH2	*Aplysia*
Crustacean neurohormones		
CHH	pGln-Val-***Phe***-Asp-Glu-Ala-Cys....	Astacidea
VIH	Ala-Ser-Ala-***Trp***-Phe-Thr-Asn...	*Homarus*

Considering that all cloned D-aa containing peptides do exhibit a precursor encoded by a mRNA with a classical codon, it is now generally recognized that biosynthesis of D-aa containing peptides is a posttranslational mechanism [7] catalyzed by a specialized enzyme, a peptide-isomerase [8].

As stressed above, most D-aa containing peptides have been isolated from acellular system such as exudates from frog skin glands or venom from Molluscs or Spiders and the investigations previously reported deal mainly with peptide characterization and mode of action. By contrast, D-aa containing neuropeptides are elaborated by neuroendocrine cells; though several of them such as achatin, fulicin, fulyal were isolated from molluscan nervous tissue, it is only in Crustaceans that cellular studies were developed, probably because D-aa containing neurohormones are elaborated in discrete, well-identified neuroendocrine cells located in an easily accessible organ. The crustacean model constitutes to date a unique and irreplaceable tool to study the biogenesis of these peculiar peptides at the cellular level.

The present mini-review aims at summarizing the major results obtained these last years in our laboratory on D-aa containing peptides biosynthesis in the crustacean neuroendocrine system.

THE CRUSTACEAN HYPERGLYCEMIC HORMONE FAMILY AND THE X-ORGAN-SINUS GLAND COMPLEX

The Crustacean Hyperglycemic Hormone (CHH) family constitutes an original neuropeptide group found primarily in Crustaceans, later in different Arthropods and even in the genome of a Nematode. This family comprises, to date, more than 50 members which are 72-78 amino acid peptides exhibiting common structural features such as a conserved position of 6 cysteine residues paired in three constant disulfide bridges (Table 2). In Crustaceans, these peptides were demonstrated to constitute key physiological regulators. They control energetic metabolism (CHH), but also molting and growth via the molt-inhibiting hormone (MIH) and reproduction via the Vitellogenesis-Inhibiting hormone (VIH, also called Gonad inhibiting Hormone, GIH) and the Mandibular-Organ Inhibiting Hormone (MOIH) [review in 9].

CHH-related peptides are mostly synthesized by a peculiar neuroendocrine structure located in the eyestalks of Decapods: the X-organ-sinus gland complex (XO-GS) (Fig 1). The X-organ, situated at the surface of a nervous ganglion on the optic nerve (*medulla terminalis*), is a cluster of a few hundreds of neurosecretory cells projecting their axons into the sinus gland, a storage and release site of the neuropeptides within the surrounding hemolymph. Among these cells, a number that is variable with the species and development stage, elaborate CHH-related peptides while others are devoted to synthesis and release of other neuropeptides such as chromatophorotropins involved in the control of pigment dispersion in the epidermal pigment cells and the retina [review in 10].

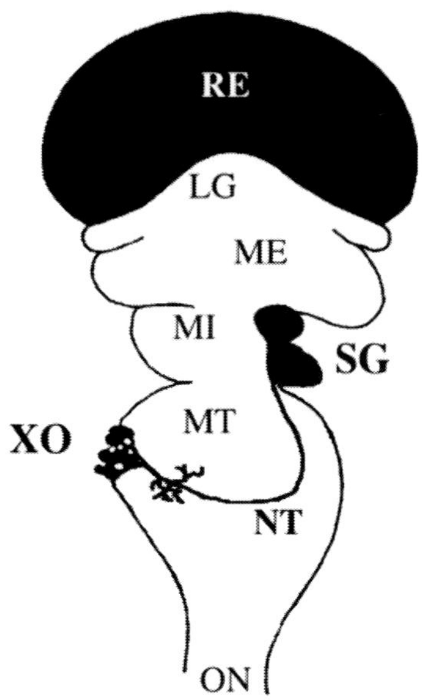

Figure 1. Schematic representation of the X-organ-sinus gland complex in a crustacean eyestalk. LG: *lamina ganglionaris*, ME: *medulla externa*, MI: *medulla interna*, MT: *medulla terminalis*, NT: nervous tract, ON: optic nerve, RE: retina, SG: sinus gland, XO: X-organ.

Characterization of a D-AA Residue in Crustacean Hyperglycemic Hormone Sequences

In the late 1980's several publications have pointed to the presence of several CHH peaks after fractionation of lobster sinus gland extract by RP-HPLC [11,12,13]. Cloning experiments have demonstrated that the CHH was present as two variants differing by 6 residues [14] (Table 2). In addition, each variant was consistently present as doublet peaks having identical amino acid composition, molecular mass and p*I* [12]. Using peptide mapping after limited proteolysis, mass spectrometry and chiral analysis, the presence of a D-Phe residue in the third position of the more hydrophobic isoforms of each doublet was demonstrated [15]. Using similar methodology, the presence of a D-Phe3 was found in CHH isoforms from the crayfish *Procambarus clarkii* [16]. For sake of clarity, CHH isoforms with only L residues will be called L-CHH while D-CHH will be used for D-Phe^{3}CHH in the subsequent text.

Table 2. Amino acid sequence of the Crustacean Hyperglycemic Hormones (CHHA and CHHB) and Vitellogenesis Inhibiting Hormone (VIH) from the lobster *Homarus americanus*. Cysteine residues are bolded.

CHHA: pQVFDQA**C**KGVYDRNLFKKLDRV**C**ED**C**YNLYRKPFVATT**C**REN**C**YSNWVFRQ**C**LDDLLLSDVIDEYVSNVQMVNH2
CHHB: pQVFDQA**C**KGVYDRNLFKKLNRV**C**ED**C**YNLYRKPFIVTT**C**REN**C**YSNRVFRQ**C**LDDLLMIDVIDEYVSNVQMVNH2
VIH: ASAWFTNDE**C**PGVMGNRDLYEKVAWV**C**ND**C**ANIFRNNDVGVM**C**KKD**C**FHTMDFLW**C**VYATERHGEIDQFRKWVSILRANH2

Antibodies were produced against synthetic octapeptides with the sequence of the N-terminal end of CHH (pGlu-Val-Phe-Asp-Gln-Ala-Cys-Lys) and D-Phe^{3}CHH (pGlu-Val-D-Phe-Asp-Gln-Ala-Cys-Lys) respectively, coupled to bovine thyroglobulin. The conjugates were injected in rabbits. When assayed by ELISA with CHHs separated from sinus glands, these antibodies (that will be called r-anti-L-CHH and r-anti-D-CHH respectively) were able to discriminate efficiently L- and D-CHH [17].

Consequently these antibodies were used to screen CHHs extracted from several species with identical N-terminal sequence and thus the presence of D-CHH was evidenced in the lobster *Homarus gammarus* [18] and in the following crayfish species: *Procambarus bouvieri* [19], *Orconectes limosus* [17], *Astacus leptodactylus* [20] and *Cherax quadriatus* [21]. In this last animal, the existence of a D-phenylalanyl residue was also checked by chiral amino acid analysis and results were in total agreement with immunological data.

To date, the question remains open whether or not D-aa residues are present in CHH with different sequences, in other species that Astacidea.

D-Phe isomerization is important for peptide function since it was established that the D-CHH displays a stronger and longer hyperglycemic activity than its L-conterpart. In addition D-CHH exclusively is involved in the control of molt and osmoregulation [review in 9].

Cellular Study of PHE3 Isomerization

As stated above, a major advantage of the crustacean model to study D-aa containing peptide biogenesis at cellular level is that the X-organ-sinus gland complex may be easily dissected from animal eyestalks and have been extensively studied over the years, for physiology and morphology as well [10].

In a first study on organ sections of *Orconectes limosus* and *Homarus americanus*, staining with r-anti-D-CHH was observed in certain cells only while every CHH cell was stained with r-anti-L-CHH antiserum [17]. This leads to the hypothesis that isomerization was occuring in specialized cells only (called D-cells in the subsequent text).

To develop this study with confocal methodology, double immunolabeling was necessary; therefore new antisera were prepared by injecting conjugate of bovine thyroglobulin with L-octapeptide to guinea pigs. The resulting antibodies will be called gp-anti L-CHH.

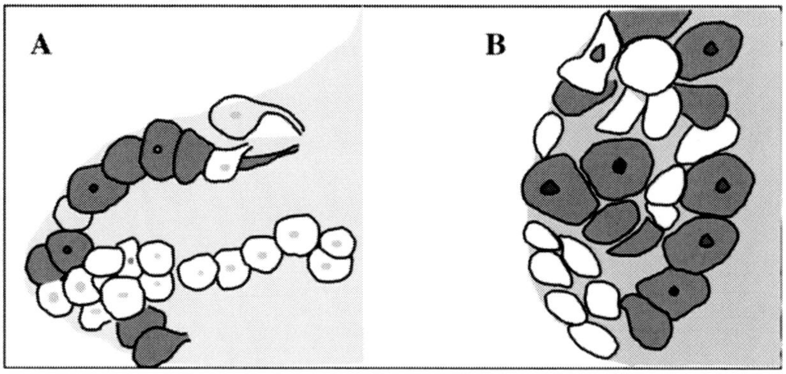

Figure 2. Schematic drawing of a confocal view of X-organ from the crayfish *Orconectes limosus* (A) and the lobster *Homarus gammarus* (B) after double immunolabeling with r-anti-D-CHH and gp-anti-L-CHH. Cells labeled with both antisera (D-cells) are in dark grey while cells labeled only with gp-anti-L-CHH are white. The grey area figures the limit of the X-organs.

Whole mounts of crayfish X-organ were incubated with gp-anti L-CHH and r-anti D-CHH and then with appropriate fluorescent secondary antisera. The resulting images confirmed the views obtained from organ sections. A constant number of 8 cells was found to contain a mixture of L- and D-CHH [22] (Fig 2A).

Interestingly, the images suggested that L- and D-CHH colocalization was occurring mostly in the perikarya, while axons and sinus gland terminals appeared to be labeled almost exclusively by the r-anti-D-CHH antiserum.

This confirms the previous observations that D-CHH synthesis was confined to specialized cells only but was not indicative of the mechanism involved in D-CHH synthesis. As a matter of fact, the question could be addressed whether L-CHH was an intermediate toward D-CHH synthesis in D-cells or if L- and D-CHH were coexpressed in D-cells, D-Phe epimerization occurring at the precursor level, as suggested for dermorphin [23].

This was further investigated by metabolic labeling experiments and peptide mapping [24,22]. It was demonstrated that no D-aa residue could be detected at the CHH precursor level, Phe3 isomerization being therefore likely to occur as a late maturation step, after propeptide cleavage.

In the lobsters *H. americanus* or *H. gammarus* [18], confocal analysis after labeling with the specific antisera revealed that the number of D-cells was greater than in crayfishes (Fig 2B). This was in total agreement with the chromatographic analysis that showed that the D-CHH was more abundant in the lobster SG extract than in crayfishes [17].

To deepen the immunohistochemical observations, it was decided to investigate the D-CHH producing system at the ultrastructural level. Therefore, double immunogold labeling followed by electron microscopy (EM) was performed on crayfish XO-GS ultrathin sections [25]. Four categories of secretory granules were observed consistently in CHH cells: granules not labeled, labeled by gp-anti-L-CHH only, labeled by r-anti-D-CHH only and labeled by both antisera (Fig 3). Two cell populations could be distinguished according to their granule content: L-cells with the two first categories exclusively and D-cells with the four granule types. Other granule type combinations were never observed within a cell.

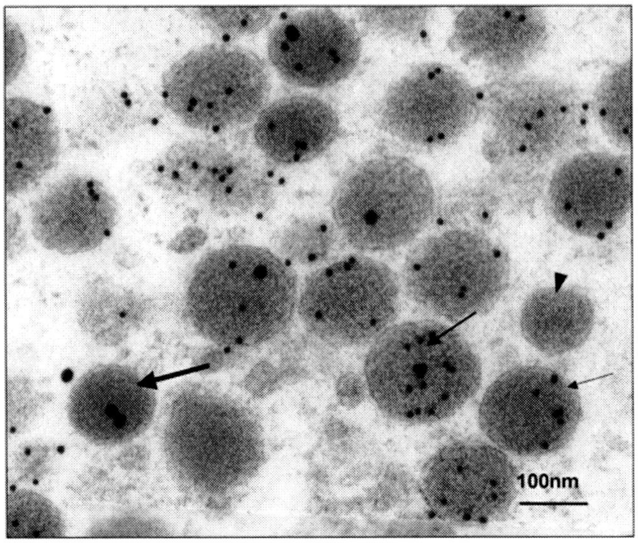

Figure 3. Electron micrography of a crayfish D-cell body after double immunogold labeling. Four types of granules are visible: non-labeled granules: arrowhead; granules with 10nm gold particles (r-anti-D-CHH): thin arrow; granules with both 10 nm and 20 nm (gp-anti-L-CHH) gold particles: medium arrow; granules with 20 nm gold particles only: thick arrow. From [25].

Confocal observation that D-cells axons and SG terminals contained only D-CHH was not confirmed since, by EM, the same granule population was observed in every part of D-cells. To examine the possibility that a cross-reaction between gp-anti-L-CHH and D-CHH could generate this image, a number of experiments were performed, including affinity purification of the specific IgGs and dilution experiments. Even when cross-reactivity (tested by ELISA) was totally abolished, a labeling of D-cells tracts and terminals with gp-anti-L-CHH was always observed, which indicated that L-CHH was always present in D-cells

Interestingly, quantification by image analysis after double immunogold labeling of D-cells revealed that D-labeling was significantly increased from the cell body to the axon terminal. This was interpreted as a progress in Phe3 isomerization process during granule migration within the neurosecretory cell. Accordingly, the proportion of granules containing only L-CHH was lower in terminals (5%) than in cell bodies (40%). This low proportion may explain the absence of detection of L-CHH in D-cells axons and terminals during confocal analysis. In addition, detection of L-CHH may have been reduced by the combination of fluorochromes selected in the confocal work: the fluoresceine isothiocyanate used for L-CHH exhibits a higher background and is less perceptible than the rhodamine used for D-CHH. Moreover, immunoglod detection at the ultrastructural level has a greater sensitivity than fluorescent labeling and is not so strictly dependant upon image analysis as confocal imaging.

More generally, our experience of these techniques clearly indicates that confocal images should be interpreted with great care.

Another interesting observation after immunogold labeling was the large heterogeneity in the granule population and the great variability in the proportion of L- and D-labeling in granules, in every cell region. This indicated that not all the CHH synthesized was isomerized at a given time and suggested a heterogeneous distribution of the enzyme involved in Phe3 isomerization within the secretory pathway. In addition, one may consider that isomerization occurs only after cleavage of the proCHH by a prohormone convertase [22,26]. Therefore, the proportion of L- and D-CHH in a given granule may reflect the dynamics of two successive enzymatic steps, which are unfortunately still poorly known in our model.

Characterization of a D-AA Residue in Vitellogenesis Inhibiting Hormone Sequence

Is has been shown since earlier studies [27] that after RP-HPLC fractionation of *H. americanus* SG extract, VIH is eluted, like CHHs, as two peaks with identical physicochemical parameters such as amino acid composition, mass, p*I* (Fig 2). Therefore, it was tempting to hypothesize that these two peaks may represent stereoisomers. This hypothesis was recently demonstrated [28] by using peptide mapping of both isoforms followed by mass spectrometry and comparison with synthetic peptides with a D-aa residue in different positions of the sequence. Thus, the presence of a D-tryptophan residue in pos. 4 of the sequence of the more hydrophobic VIH isoform was evidenced, while a L-tryptophan residue was present in the hydrophilic isoform.

It may be noted that D-VIH was always found more abundant than L-VIH in sinus gland extracts, though a great number of different preparations from different batches of animals has been processed over years. This is different from D-CHH that was always in smaller quantity than L-CHH.

Cellular Study of TRP4 Isomerization

In a next step, antibodies were prepared in rat and mice using decapeptides with the sequence Ala-Ser-Ala-Trp/D-Trp-Phe-Thr-Asn-Asp-Glu-Cys coupled to keyhole lumpet hemocyanin (KLH). The resulting antibodies were tested by ELISA and showed a relatively high level of cross-reactivity with the two VIH stereoisomers. After affinity purification, cross-reaction was drastically reduced. Unfortunately, despite their satisfactory behavior in ELISA, utilization of these antibodies for immunohisto/cytochemistry appeared very difficult, with a rather weak signal when performing either fluorescent labeling and confocal microscopy or immunogold labeling and EM. Nevertheless, as for CHH, the existence in the lobster X-organ of cells expressing only L-VIH and others expressing a mixture of L- and D-VIH was observed [28]. Among these cells, important variations of coloration were observed, indicating a variable proportion of L- and D-VIH in the different D-VIH containing cells.

A fascinating question is how the D-CHH and D-VIH producing sytem are intricated in the XO-GS. In a previous article [29], the colocalization of CHH and VIH (without any distinction of stereoisomers) was observed in several cell bodies of the lobster X-organ, by IHC. However, in the same article, both peptides were mostly found in different axonal endings, by EM after immunogold labeling on successive sections. In another study, colocalization of CHH and VIH in several XO perikarya was confirmed at the molecular level by *in situ* hybridization [30].

At the present time, the question of whether D-isomers of CHH and VIH are produced in the same cells or not is not definitely answered, due to the difficulty of using the available antibodies, however, recent results have significantly clarified this problem. When double immmunogold labeling with rabbit-anti-D-CHH and mice-anti-D-VIH was realized on lobster ultrathin sections, it was observed that every D-CHH terminal in the sinus gland contained also D-VIH. Interestingly, this double labeling was observed in the same secretory granules [28]. This clearly indicates that the D-isomers of CHH and VIH are elaborated in the same neurosecretory cells.

Concluding Remarks

The observation that D-stereoisomers of CHH and VIH are coexpressed in the same neurosecretory cells addresses urgently the question of the nature of the mechanisms involved in the biogenesis of these peptides. It is known from molecular studies that CHH and VIH precursors are very different: in the VIH precursor, the mature hormone segment directly flanks the signal peptide while the CHH preprohormone is composed of three regions: a signal peptide, a cryptic peptide named CPRP (CHH Precursor Related Peptide) and the hormone. After propeptide cleavage, further maturation steps are still necessary to get fully mature CHH such as C-terminal amidation and N-terminal cyclisation [31] while peptide signal excision generates VIH, which is probably C-ter amidated [32]. Consequently, posttranslational processing is very different for the VIH and CHH precursors. Biochemical studies combined with ICC as described above have established that the inversion from the L-Phe to the D-Phe in CHH biosynthesis was occurring as a late step in precursor processing, after propeptide cleavage and amidation but before or simultaneously with glutamine N-terminal cyclization [24,22]. The fact that isomerization may occur, in the same granule, on the CHH Phe3 and VIH Trp4 suggests that this mechanism is probably not linked to the preceding maturation steps.

Post-translational isomerization is likely to involve a specialized enzyme, a peptide-isomerase. There is only a few examples of such enzymes described in the literature: a peptide-isomerase which inverts the configuration of a D-serine residue located in pos. 46 of the 48-residue peptide ω-agatoxin, was isolated from the venom of the black widow spider *Agenelopsis aperta* [33]. More recently an enzyme was isolated and cloned, that may be responsible for the isomerization of the second residue (Ile) of the frog antibacterial peptide bombinin [34]. These two enzymes are totally unrelated to each other in terms of sequence, but both displayed a rather high substrate-specificity.

In this context, the Crustacean model again appears to be original since our observations address the intriguing question whether one single enzyme may isomerize Phe3 in CHH and Trp4 in VIH or if two specialized enzymes are present in the D-peptide producing cells of the X-organ. This question will remain unanswered until *in vitro* assays with purified enzyme preparations are available, which is not the case at the moment.

Over the last decade, studies of the Crustacean neuroendocrine system have resulted in a number of original data on the biogenesis of D-aa containing peptides in Eukaryotes. One question that may be addressed is whether the mechanisms unraveled here are applicable to other organisms or if the modalities of isomerization are different according to the species considered. Further studies will answer this question but undoubtedly, the story of the D-aa containing peptides is only at its infancy.

Aknowledgments

Many thanks to Prof. Ernest Chang, Drs Céline Ollivaux and Jean-Yves Toullec for their valuable criticisms and suggestions on this manuscript.

References

[1] Montecucchi, P. C., De Castiglione, R., Piani, S., Gozzini, L. & Erspamer, V. r. (1981) Amino acid composition and sequence of dermorphin, a novel opiate-like peptide from the skin of Phyllomedusa sauvagei. *Int J Peptide Protein Res, 17*, 275-283.

[2] Mignogna, G., Simmaco, M. & Barra, D. (1998) Occurence and function of D-amino acid containing peptides and proteins: antimicrobial peptides. In Jollès, P. (Ed.) *D-amino acids in sequence of secreted peptides of multicellular organisms*, (pp. 29-36), Basel, Boston, Berlin.Birkhäuser Verlag.

[3] Amiche, M., Delfour, A. & Nicolas, P. (1998) Opioid peptides from frog skin. In Jollès, P. (Ed.) *D-amino acids in sequences of secreted peptides of multicellular organisms*, (pp. 57-72), Basel, Boston, Berlin.Birkhäuser Verlag.

[4] Craig, A. G., Bandyopadhyay, P. & Olivera, B. M. (1999) Post-translationally modified neuropeptides from Conus venoms. *Eur J Biochem, 264*, 271-275.

[5] Torres, A. M., Menz, I., Alewood, P. F., Bansal, P., Lahnstein, J., Gallagher, C. H. & Kuchel, P. W. (2002) D-amino acid residue in the C-type natriuretic peptide from the venom of the mammal, Ornithorhynchus anatinus, the Australian platypus. *FEBS Let, 524*, 172-176.

[6] Yasuda-Kamatani, Y. (1998) Molluscan neuropeptides. In Jollès, P. (Ed.) *D-Amino Acids in Sequences of Secreted Peptides of Multicellular Organisms*, (pp. 37-56), Basel.Birkhäuser.

[7] Mor, A., Amiche, M. & Nicolas, P. (1992) Enter a new post-translational modification: D-amino-acids in gene encoded peptides. *TIBS, 17*, 481-485.

[8] Heck, S. D., Siok, C. J., Krapcho, K. J., Kelbaugh, P. R., Thadeio, P. F., Welsh, M. J., Williams, R. D., Ganong, A. H., Kelly, M. E., Lanzetti, A. J., Gray, W. R., Phillips, D., Parks, T. N., Jackson, H., Ahlijanian, M. K., Saccomano, N. A. & Volkman, R. A. (1994) Functional consequences of posttranslational isomerization of Ser46 in a calcium channel toxin. *Science, 266*, 1065-1068.

[9] Soyez, D. (2003) Recent data on the crustacean hyperglycemic hormone family. In M. Fingerman and R. Nagabhushanam (Eds) *Recent Advances in Marine Biotechnology*, Vol 10: Molecular genetics of marine organisms, (pp 280-301) Enfield (NH) USA, Plymouth, UK. Science Publishers

[10] Van Herp, F. & Kallen, J. (1990) Neuropeptides and neurotransmitters in the X-organ sinus gland complex, an important neuroendocrine integration center in the eyestalk of crustacea. In G.B. Stephano and E.. Florey (Eds) *Comparative Aspects of Neuropeptides*, (pp. 351-364.), Manchester. Manchester University Press.

[11] Chang, E. S., Bruce, M. J. & Newcomb, R. W. (1987) Purification and aminoacid composition of a peptide with molt-inhibiting activity from the lobster Homarus americanus. *Gen Comp Endocrinol, 65*, 56-64.

[12] Soyez, D., Noel, P. Y., Van Deijnen, J. E., Martin, M., Morel, A. & Payen, G. G. (1990) Neuropeptides from the sinus gland of the lobster Homarus americanus: characterization of hyperglycemic peptides. *Gen Comp Endocrinol, 76*, 261-274.

[13] Tensen, C. P., Janssen, K. P. C., Soyez, D. & Van Herp, F. (1991) Comparative characterization of hypergycemic neuropeptides from the lobster Homarus americanus. *Peptides, 12*, 241-249.

[14] Tensen, C. P., Dekleijn, D. P. V. & Vanherp, F. (1991) Cloning and Sequence Analysis of cDNA Encoding Two Crustacean Hyperglycemic Hormones from the Lobster Homarus-Americanus. *Eur J Biochem, 200*, 103-106.

[15] Soyez, D., Van Herp, F., Rossier, J., Le Caer, J. P., Tensen, C. P. & Lafont, R. (1994) Evidence for a conformational polymorphism of invertebrate neurohormones. D-amino-acid residue in crustacean hyperglycemic hormone. *J Biol Chem, 269*, 18295-18298.

[16] Yasuda, A., Yasuda, Y., Fujita, T. & Naya, Y. (1994) Characterization of crustacean hyperglycemic hormone from the crayfish (Procambarus clarkii): Multiplicity of molecular forms by stereoinversion and diverse functions. *Gen Comp Endocrinol, 95*, 387-398.

[17] Soyez, D., Laverdure, A. M., Kallen, J. & Van Herp, F. (1998) Demonstration of a cell-specific isomerization of invertebrate neuropeptides. *Neuroscience, 82*, 935-942.

[18] Ollivaux, C., Vinh, J., Soyez, D. & Toullec, J. Y. (2005) Crustacean Hyperglycemic Hormone family peptides in the lobster Homarus gammarus: occurrence of genetic variants and posttranslationally modified isoforms. *submitted,*

[19] Aguilar, M. B., Soyez, D., Falchetto, R., Arnott, D., Shabanowitz, J., Hunt, D. & Huberman, A. (1995) Aminoacid sequence of the minor isomorph of the crustacean hyperglycemic hormone (CHH-II) of the Mexican crayfish Procambarus bouvieri (Ortmann): presence of a D-amino acid. *Peptides, 16*, 1375 1383.

[20] Serrano, L., Charmantier, G., Soyez, D., Grousset, E. & Spanings-Pierrot, C. (2003) Putative involvement of crustacean hyperglycemic hormone isoforms in the neuroendocrine mediation of osmoregulation in the crayfish Astacus leptodactylus. *J Exp Biol, 206*, 979-988.

[21] Bulau, P., Meisen, I., Reichwein-Roderburg, B., Peter-Katalinic, J. & Keller, R. (2003) Two genetic variants of the crustacean hyperglycemic hormone (CHH) from the Australian crayfish, Cherax destructor: detection of chiral isoforms due to posttranslational modification. *Peptides, 24*, 1871-9.

[22] Soyez, D., Toullec, J. Y., Ollivaux, C. & Geraud, G. (2000) L- to D-amino-acid isomerization in a peptide hormone is a late post-translational event occurring in specialized neurosecretory cells. *J Biol Chem, 275,* 37870-37875.

[23] Mor, A., Delfour, A. & Nicolas, P. (1991) Identification of a D-alanine containing polypeptide precursor for the peptide opioid, Dermorphin. *J Biol Chem, 266*, 6264-6270.

[24] Ollivaux, C. & Soyez, D. (2000) Dynamics of biosynthesis and release of crustacean hyperglycemic hormone isoforms in the X-organ-sinus gland complex of the crayfish Orconectes limosus. *Eur J Biochem, 267*, 5106-5114.

[25] Gallois, D., Brisorgueil, M. J., Conrath, M., Mailly, P. & Soyez, D. (2003) Posttranslational isomerization of a neuropeptide in crustacean neurosecretory cells studied by ultrastructural immunocytochemistry. *Eur J Cell Biol, 82*, 431-40.

[26] Toullec, J. Y., Kamech, N., Gallois, D., Papon, V., Maïbèche, M., Boscameric, M. & Soyez, D. (2002) Molecular cloning and cellular expression of crustacean PC2-like prohormone convertase. *Biochim. Biophys. Acta, 1574*, 145-151.

[27] Soyez, D., Lecaer, J. P., Noel, P. Y. & Rossier, J. (1991) Primary Structure of Two Isoforms of the Vitellogenesis Inhibiting Hormone from the Lobster Homarus-Americanus. *Neuropeptides, 20*, 25-32.

[28] Ollivaux, C., Amiche, M., Gallois, D., Boscameric, M., Nicolas, P. & Soyez, D. (2006) A novel D-amino acid containing peptide of the Crustacean Hyperglycemic Hormone family: biochemical and immunochemical characterization. *submitted,*

[29] Kallen, J. & Meusy, J. J. (1989) Do the neurohormones VIH (vitellogenesis inhibiting hormone) and CHH (crustacean hyperglycemic hormone) of the crustacean have the same precursor? Immunolocalization of VIH and CHH in the X-organ sinus gland complex of the lobster Homarus americanus. *Invert Reprod Dev, 16*, 43-52.

[30] De Kleijn, D. P. V., Coenen, T., Laverdure, A. M., Tensen, C. P. & Vanherp, F. (1992) Localization of messenger RNAs encoding crustacean hyperglycemic hormone and gonad inhibiting hormone in the X-Organ sinus gland complex of the lobster Homarus americanus. *Neuroscience, 51*, 121-128.

[31] De Kleijn, D. & Van Herp, F. (1995) Molecular biology of neurohormone precursors in the eyestalk of crustacea. *Comp Biochem Physiol, 112B*, 573-579.

[32] De Kleijn, D. P. V., Sleutels, F. J. G. T., Martens, G. J. M. & Van Herp, F. (1994) Cloning and expression of mRNA encoding prepro-gonad-inhibiting hormone (GIH) in the lobster Homarus americanus. *FEBS Lett, 353*, 255-258.

[33] Heck, S. D., Faraci, W. S., Kelbauch, P. R., Saccamo, N. A., Thadeio, P. F. & Volkmann, R. A. (1996) Posttranslational amino acid epimerization: enzyme-catalyzed isomerization of amino acid residues in peptide chain. *Proc Natl Acad Sci, USA, 93*, 4036-4039.

[34] Jilek, A., Mollay, C., Tippelt, C., Grassi, J., Mignogna, G., Mullegger, J., Sander, V., Fehrer, C., Barra, D. & Kreil, G. (2005) Biosynthesis of a D-amino acid in peptide linkage by an enzyme from frog skin secretions. *Proc Natl Acad Sci, U S A, 102*, 4235-9.

Section 6. Enzymes Involved in Synthesis and Metabolism of D-Amino Acids

In: D-Amino Acids: ISBN 1-60021-075-9
A New Frontier in Amino Acid and Protein Research, pp.443-445

Chapter 6.1

PURIFICATION OF ALANINE RACEMASE FROM *HELICOBACTER PYLORI*

Katsushi Nishimura* and Yoko Nagata
Department of Materials and Applied Chemistry, College of Science and Technology, Nihon University, Japan.

Keywords: *Helicobacter pylori*, racemase, enzyme purification.

1. INTRODUCTION

Helicobacter pylori is a Gram-negative bacterium that colonizes the gastric mucosa of humans and is implicated in a wide range of gastroduodenal diseases [1-4]. *H. pylori* is a microaerophilic bacterium exhibiting a strict respiratory form of metabolism and oxidizing organic acids as an energy source. Many organisms utilize glucose as the energy source, but for *H. pylori* it is equivocal that glucose be used as the energy source. A high rate of utilization of L-serine and L-proline, followed by D-alanine and D-proline, as respiratory substrates has been observed in this organism [5]. Therefore, these amino acids are considered to be possible candidates for respiratory substrates and energy sources, and alanine racemase seems to be an important enzyme for *H. pylori.*

We describe herein a purification procedure for an alanine racemase that is responsible for D-alanine utilization in *H. pylori.*

2. MATERIALS

1. *H. pylori* NCTC 11637 cells (from Dr. K. Nagata, Department of Bacteriology, Hyogo College of Medicine, Nishinomiya, Hyogo 663-8501, Japan)
2. A French press (Ohtake Works, Tokyo, Japan)
3. A HPLC system (Shimadzu, Kyoto, Japan)

* Correspondence concerning this article should be addressed to Katsushi Nishimura, E-mail: nisimura@chem.cst.nihon-u.ac.jp, FAX: +81-3-3259-0432.

4. A HiTrap-DEAE-Sepharose column and a HiTrap-Phenyl-Sepharose column (Amersham Biosciences, Piscataway, NJ, USA)
5. A Ceramic Hydroxyapatite, CHT column (Bio-Rad, Hercules, CA, USA)
6. A KW-2003-gel-filtration column (Showa Denko, Tokyo, Japan)

3. Methods

1. Cultured cells of *H. pylori* were harvested by centrifugation at 8,000 *g* for 10 min. The cells (60 g) were suspended in 50 mM sodium phosphate buffer (pH 7.0) containing 1 mM EDTA, 0.1 mM phenylmethanesulfonyl fluoride, and 50 μM pyridoxal 5'-phosphate (PLP).
2. The cell suspension was passed twice through a French pressure cell at 135 megapascals.
3. The experimental procedures described hereafter were carried out at 4°C unless otherwise indicated. After removal of the unbroken cells by centrifugation at 8,000 *g* for 10 min, the supernatant solution was further centrifuged at 140,000 *g* for 1 h in order to eliminate the membrane fraction and was used as a cell-free extract.
4. The cell-free extract was brought to 45% saturation with ammonium sulfate, and the resultant precipitate was removed by centrifugation. Similarly, ammonium sulfate fractionation was carried out with 80% ammonium sulfate. Precipitate of the 45-80% fraction was dissolved in buffer A (20 mM Tris-HCl, 50 μM PLP, pH 8.5) and dialyzed against the same buffer.
5. The dialyzed solution was applied to 7 HiTrap-DEAE-Sepharose columns (5 ml) in series equilibrated with buffer A. A gradient elution was performed at 25°C with buffer B (20 mM Tris-HCl, 300 mM NaCl, 0.05 mM PLP, pH 8.5) by increasing the concentration linearly from 0 to 100%. The active fractions were pooled, concentrated, and dialyzed against buffer C (10 mM sodium phosphate, 50 μM PLP, pH 7.2).
6. The dialyzed solution was applied to a Ceramic Hydroxyapatite column (20 x 200 mm) equilibrated with buffer C. A gradient elution was performed at 25°C with buffer D (350 mM sodium phosphate, 50 μM PLP, pH 7.2) by increasing the concentration linearly from 0 to 100%. The active fractions were pooled and dialyzed against buffer E (50 mM Tris-HCl, 40%-saturated ammonium sulfate, 50 μM PLP, pH 7.5).
7. The dialyzed solution was applied to a HiTrap-Phenyl-Sepharose column (5 ml) equilibrated with buffer E. A gradient elution was performed at 25°C with buffer F (50 mM Tris-HCl, 50 μM PLP, pH 7.5) by increasing the concentration linearly from 0 to 100%. The active fractions were pooled and concentrated.
8. The concentrated solution was further purified by gel filtration with a KW-2003 column (20 x 300 mm) equilibrated with buffer G (50 mM sodium phosphate, 200 mM NaCl, 50 μM PLP, pH 7.2). An elution was performed at 25°C with buffer G at a flow rate of 0.8 ml/min. The active fractions were pooled and desalted by ultrafiltration.

4. Results

The purification magnification and recovery rate of the final active fraction were 327 and 5.35%, respectively, as shown in Table 1. The molecular weight of the enzyme was determined to be approximately 65,000 by gel filtration. This enzyme preparation can be used for characterization of enzymatic properties and the amino acid sequence analysis.

Table 1. Purification of alanine racemase.

	Total protein mg	Total activity µmol/min	Specific activity µmol/min/mg	Yield %
Cell-free extract	1640	310	0.189	100
Ammonium sulfate fractionation	1010	252	0.250	81.3
DEAE sepharose	109	120	1.10	38.7
Ceramic Hydroxyapatite	7.10	86.4	12.2	27.9
Phenyl sepharose	1.47	58.1	39.5	18.7
Gel filtration with KW-2003	0.269	16.6	61.8	5.35

5. Notes

1. Protein concentrations were determined by the method of Lowry et al. [6] using bovine serum albumin as the standard.
2. Enzyme activity was assayed as follows. The standard reaction mixture was composed of 50 µg of enzyme sample, 25 mM D- or L-amino acid, 10 µM PLP, and 50 mM Tris-HCl buffer (pH 8.3) in a total volume of 100 µL. The mixture was incubated at 37°C for 20 min, and the reaction was stopped by the addition of 5% (final) trichloroacetic acid. D- And L-amino acids were assayed by using 1-fluoro-2,4-dinitrophenyl-5-L-alanineamide (FDAA) and HPLC [7]. One enzyme unit was defined as the amount catalyzing the isomerization of 1 µmol of alanine for 1 min.

References

[1] Blaser, M. J. (1990). *Helicobacter pylori* and the pathogenesis of gastroduodenal inflammation. *J Infect Dis, 161*, 626–633.

[2] Parsonnet, J., Friedman, G. D., Vandersteen D. P., Chang, Y., Vogelman, J. H., Orentreich, N. & Sibley, R. K. (1991). *Helicobacter pylori* infection and the risk of gastric carcinoma. *N Engl J Med, 325*, 1127–1131.

[3] Parsonnet, J., Hansen, S., Rodriguez, L., Gelb, A. B., Warnke, R. K., Jellum, E., Orentrich, N., Vogelman, J. H. & Friedman, G. D. (1994). *Helicobacter pylori* infection and gastric lymphoma. *N Engl J Med, 330*, 1267–1271.

[4] Rabeneck, L. & Ransohoff, D. F. (1991). Is *Helicobacter pylori* a cause of duodenal ulcer? A methodologic critique of current evidence. *Am J Med, 91*, 566–572.

[5] Nagata, K., Nagata, Y., Sato, T., Fujino, M. A., Nakajima, K. & Tamura, T. (2003). L-Serine, D- and L-proline and alanine as respiratory substrates of *Helicobacter pylori*: correlation between *in vitro* and *in vivo* amino acid levels. *Microbiology, 149,* 2023-2030.

[6] Lowry, O. H., Rosebrough, N. J., Farr, A. L. & Randall, R. J. (1951). Protein measurement with the Folin phenol reagent. *J Biol Chem, 193,* 265-275.

[7] Nagata, Y., Yamamoto, K. & Shimojo, T. (1992). Determination of D- and L-amino acids in mouse kidney by high-performance liquid chromatography. *J Chromatogr 575*, 147-152.

In: D-Amino Acids: ISBN 1-60021-075-9
A New Frontier in Amino Acid and Protein Research, pp.447-451

Chapter 6.2

DISRUPTION OF ALANINE RACEMASE GENE OF *SCHIZOSACCHAROMYCES POMBE*

Takayuki Kazuoka, Takuma Uo and Nobuyoshi Esaki***
Institute for Chemical Research, Kyoto University, Japan.

Keywords: alanine racemase, *Schizosaccharomyces pombe*, gene disruption, D-amino acid metabolism

1. INTRODUCTION

Attention has been paid to clarification of the physiological role of free D-amino acids in higher organisms such as mammals [1]. However, except D-serine in mammalian brain [1,2,3], physiological functions and metabolic pathways of free D-amino acids in eukaryotes including yeasts remain unsettled. A budding yeast, *Saccharomyces cerevisiae* and a fission yeast, *Schizosaccharomyces pombe* are useful model eukaryotes for the studies of biological functions of D-amino acids. Toxic effects of free D-amino acids on *S. cerevisiae* cells have been studied in some detail [4]. For example, D-alanine inhibits the growth of *S. cerevisiae* cells [5]. In the whole genome sequence, *S. cerevisiae* does not have the gene homologous with alanine racemase, which is responsible for biosynthesis and catabolism of D-alanine in bacteria [6,7]. On other hands, the growth of *S. pombe* cells is not inhibited by D-alanine, and *S. pombe* has a gene encoding an alanine racemase (*alr1*$^+$ gene) [8]. Certain features of *S. pombe* such as cell cycle, chromosome structure and RNA splicing are similar to those of mammalian cells [9]. Therefore, the studies of *S. pombe* will provide us with information about the physiological roles and metabolism as well as toxicity of free D-amino acids in eukaryotes.

The construction of disruption mutant and investigation of its phenotypic alteration are useful to determine physiological roles of the genes participating in D-amino acid metabolism.

2. MATERIALS

1. *S. pombe* SP1 strain (*h*$^+$ *leu1-32 ura4-D18*).

* Correspondence concerning this article should be addressed to Nobuyoshi Esaki, E-mail: esaki@SCL.kyoto-u.ac.jp, FAX: +81-774-38-3248.

2. Pure gene, yeast and gram-positive bacteria DNA isolation kit (Funakoshi, Tokyo).
3. PCR primers.
   Primer pair A (5'-CAG TTT CAT TTC AAA ACG AAT GTT ATA GAA CAT C-3' (sense primer) and 5'-CGA ATT CCT GCA GCC CGG GGG ATC CAA TTG CTA AAA CAT GAC CGC TTG G-3' (antisense primer)).
   Primer pair B (5'-GCT TAT CGA TAC CGT CGA CCG GAG AGT TCC TTT ACA GTA CAC-3' (sense primer) and 5'-CAG ACT ACA TAG ATT TAC ACA TCC TTC TTT ACC-3' (antisense primer)).
4. Thermal cycler.
5. Agarose.
6. Tris/Acetate/EDTA (TAE) buffer: 40 mM Tris-Acetate, 1 mM EDTA.
7. Electrophoresis tank.
8. Ethidium bromide dye solution (10 mg/mL).
9. UV-Transilluminator.
10. EASYTRAP™ Ver. 2 Kit (TAKARA BIO, Shiga).
11. Transformation regents for lithium acetate method [10].
12. Minimal medium for *S. pombe*: 2% glucose, 0.05% KH_2PO_4, 0.036% CH_3COOK, 0.05% $MgSO_4 \cdot 7H_2O$, 0.01% NaCl, 0.01% $CaCl_2 \cdot 2H_2O$, 0.1% (vol/vol) trace element solution (0.05% H_3BO_4, 0.004% $CuSO_4 \cdot 5H_2O$, 0.01% KI, 0.02% $FeCl_3 \cdot 6H_2O$, 0.04% $MnSO_4 \cdot H_2O$, 0.02% $Na_2MoO_4 \cdot 2H_2O$, and 0.04% $ZnSO_4 \cdot 7H_2O$), 1% (vol/vol) biotin (10 mg/ml), 1% (vol/vol) vitamin solution (0.1% pantothenic acid calcium salt, 1% nicotinic acid, and 1% myoinositol), 10 mM ammonium sulfate.

3. Methods

1. Genomic DNA was extracted from *S. pombe* SP1 cells using the extraction kit.
2. A *LEU2* deletion cassette for *alr1*$^+$ was prepared by PCR.
   2-1. DNAs corresponding to the upstream region (0.6 kbp) and the downstream region (0.8 kbp) of *alr1*$^+$ were obtained by PCR with the genomic DNA of *S. pombe* as a template and primer sets A and B.
   2-2. DNA (2.2 kbp) containing the *LEU2* gene was prepared from pART1 plasmid by digestion with HindIII and SalI and purification from 1% agarose using EASYTRAP™ Ver. 2 Kit.
   2-3. PCR was carried out with the DNA of 2-2 as a template and those of 2-1 as primers.
3. The resulting PCR product (3.6 kbp) containing the *LEU2* gene flanked by the upstream and downstream regions of *alr1*$^+$ was used for the transformation of the haploid *S. pombe* cells by the lithium acetate method [10].
4. Positive clones were selected on the minimal medium not supplemented with L-leucine.
5. pART2 plasmid was prepared by the exchange of the *LEU2* marker in the pART1 plasmid with the *ura4*$^+$ marker derived from the pBSUra4 plasmid at the HindIII and SalI sites.
6. The SP1 strain was transformed with the plasmids pART1 and pART2 and then used for examination of nitrogen sources utilized.
7. The *alr::LEU2* strain was transformed with the plasmid pART2 and then used for examination of nitrogen sources utilized.
8. *S. pombe* cells were harvested at a logarithmic growth phase, washed three times with the minimal medium from which ammonium sulfate was omitted, and then examined whether it can utilize D- or L-alanine by incubation at 32°C with aeration in the minimal medium or the alanine-

supplemented medium, in which ammonium sulfate of the minimal medium was replaced by D- or L-alanine as a sole nitrogen source.

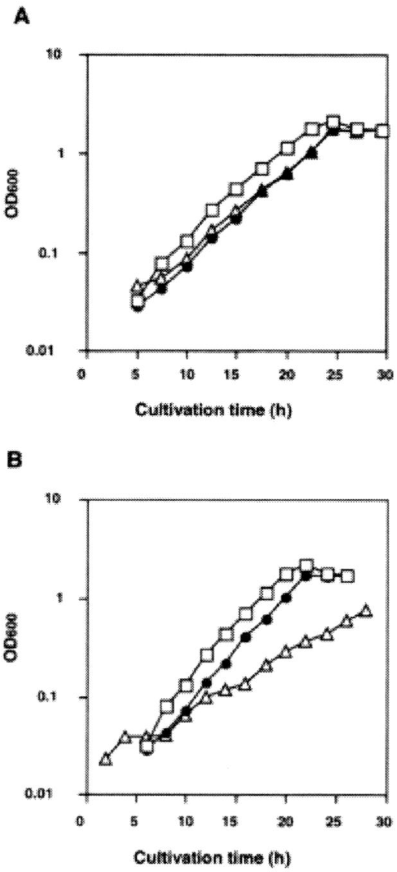

Figure 1. Growth of wild-type (A) and a deletion mutant (*alr1*Δ) (B) of *S. pombe* with D- or L-alanine as a sole nitrogen source. (A) *S. pombe* SP1 (Leu$^-$ Ura$^-$) was made Leu$^+$ Ura$^+$ by introduction of both pART1 and pART2 (see Methods). (B) *alr1::LEU2* strain (Alr1Δ Leu$^+$ Ura) derived from *S. pombe* SP1 was transformed with pART2 to be converted to Alr1Δ Leu$^+$ An Ura$^+$. Each recombinant was precultured in the minimal medium and transferred to a culture containing, as a sole nitrogen source, 5 mM ammonium sulfate (squares), 10 mM D-alanine (triangles), or L-alanine (circles). The culture was grown at 32°C with reciprocal shaking.

4. Results

S. pombe grows on either enantiomer of alanine as a sole nitrogen source (Figure 1A). Since *S. pombe* has Alr1p, which catalyzes the interconversion between the enantiomers, it is reasonable to assume that the yeast survives on either enantiomer through racemization with only a single catabolic pathway for L- or D-alanine. Alternatively, the yeast may have a separate pathway for each enantiomer. A deletion mutant of *S. pombe* lacking the *alr1*$^+$ gene (Figure 1B) was constructed, and the resulting *alr1*Δ strain grew as well as the original strain on a complete medium such as YPD, indicating that the *alr1*$^+$ gene is not essential for *S. pombe*. However, the deletion mutant grows about 50% less slowly than the original strain on a medium containing D alanine as a sole nitrogen source (Figure 1B). The response to the antipode was different: both strains grew equally well on L-alanine as a sole nitrogen source. Moreover, the growth rate on L-alanine was twice that on D-alanine. Therefore, *S. pombe* has catabolic pathways for each enantiomer of alanine, and the pathway for L-alanine plays a major role in the catabolism of D-alanine by coupling with racemization. L-Alanine is utilized in *Candida maltosa* by L-alanine aminotransferase, which is inducibly formed by L-alanine [11]. *S. pombe* also has two putative homologs of this enzyme (GenBank accession no. CAB46671.1 and CAA93786.1). D-Alanine is utilized by D-amino acid oxidase in *Rhodotorula gracilis* and *Candida boidinii* [12,13]. *S. pombe* grown on D-alanine as a sole nitrogen source also shows D-amino acid oxidase activity toward D-alanine as a substrate. Thus, alanine racemase participates in the catabolism of D-alanine by providing L-alanine aminotransferase with its substrate L-alanine and thereby supports the growth of the yeast, which would otherwise not be efficient with D-amino acid oxidase alone.

5. Notes

1. *S. pombe* was grown in YPD medium (1% yeast extract, 2% peptone, 2% glucose).
2. The 3' end of the upstream region and the 5' end of the downstream region are complementary to the 5' and 3' ends of the *LEU2* gene, respectively. KOD DNA polymerase (TOYOBO, Osaka) was used for PCR because the fidelity of the enzyme is high. In place of KOD DNA polymerase, Z-taq (Takara, Ohtsu), AmpliTaq Gold (PE Applied Biosystems, Foster City, CA), Expand High Fidelity (Roche, Mannheim), LA *Taq* (Takara) were also used satisfactorily.
3. Disruption of $alr1^{+}$ gene was verified by Southern blotting and PCR analysis.
4. The resulting *alr1*Δ strain (h^{+} *leu1-32 ura4-D18 alr1::LEU2*) and the SP1 strain are not suitable for examination of nitrogen sources; at low concentrations of uracil and leucine, which must be added to supplement the defects in mutated *ura4* and *leu1* genes, respectively, can support the growth of the yeast without addition of nitrogen sources. Therefore, nitrogen sources for the deletion mutant were examined by introduction of the $ura4^{+}$ gene on a plasmid, pART2. In the same manner as *alr1*Δ strain, the SP1 strain was introduced the $ura4^{+}$ and *LEU2* genes on plasmids pART1 and pART2.

References

[1] Atsushi, H. & Oka, T. (1997). Free D-aspartate and D-serine in the mammalian brain and periphery. *Prog Neurobiol, 52,* 323-353.

[2] Mothet, J. P., Parent, A. T., Wolosker, H., Brady R. O. Jr, Linden, D. J., Ferris, C. D., Rogawski, M. A. & Snyder, S. H. (2000). D-Serine is an endogenous ligand for the glycine site of the *N*-methyl-D-aspartate receptor. *Proc Natl Acad Sci U S A, 97,* 4926-4931.

[3] Wolosker, H., Blackshaw, S. & Snyder, S. H. (1999). Serine racemase: a glial enzyme synthesizing D-serine to regulate glutamate-*N*-methyl-D-aspartate neurotransmission. *Proc Natl Acad Sci U S A, 96,* 13409-13414.

[4] Soutourina, J., Blanquet, S. & Plateau, P. (2000). D-Tyrosyl-tRNA(Tyr) metabolism in *Saccharomyces cerevisiae. J Biol Chem, 275,* 11626-11630.

[5] Rytka, J. (1975). Positive selection of general amino acid permease mutants in *Saccharomyces cerevisiae. J Bacteriol, 121,* 562-570.

[6] Walsh, C. T. (1989). Enzymes in the D-alanine branch of bacterial cell wall peptidoglycan assembly. *J Biol Chem, 264,* 2393-2396.

[7] Soda, K. & Esaki, N. (1994). Pyridoxal enzymes acting on D-amino acids. *Pure Appl Chem, 66,* 709-714.

[8] Uo, T., Yoshimura, T., Tanaka, N., Takegawa, K. & Esaki, N. (2001). Functional characterization of alanine racemase from *Schizosaccharomyces pombe*: a eucaryotic counterpart to bacterial alanine racemase. *J Bacteriol, 183,* 2226-2233.

[9] Moreno, S., Klar, A. & Nurse, P. (1991). Molecular genetic analysis of fission yeast *Schizosaccharomyces pombe*. *Methods Enzymol, 194,* 795-823.

[10] Okazaki, K., Okazaki, N., Kume, K., Jinno, S., Tanaka, K. & Okayama, H. (1990). High-frequency transformation method and library transducing vectors for cloning mammalian cDNAs by trans-complementation of *Schizosaccharomyces pombe*. *Nucleic Acids Res, 18,* 6485-6489.

[11] Umemura, I., Yanagiya, K., Komatsubara, S., Sato, T. & Tosa, T. (1991). Characteristics of alanine aminotransferase from *Candida maltosa*, p. 229-231. *In* T. Fukui, H. Kagamiyama, K. Soda, and H. Wada (ed.), Enzymes dependent on pyridoxal phosphate and other carbonyl compounds as cofactors. Pergamon Press, Oxford, England.

[12] Goodman, J. M., Trapp, S. B., Hwang, H. & Veenhuis, M. (1990). Peroxisomes induced in *Candida boidinii* by methanol, oleic acid and D-alanine vary in metabolic function but share common integral membrane proteins. *J Cell Sci, 97,* 193-204.

[13] Perotti, M. E., Pollegioni, L. & Pilone, M. S. (1991). Expression of D-amino acid oxidase in *Rhodotorula gracilis* under induction conditions: a biochemical and cytochemical study. *Eur J Cell Biol, 55,* 104-113.

In: D-Amino Acids: ISBN 1-60021-075-9
A New Frontier in Amino Acid and Protein Research, pp.453-458

Chapter 6.3

ASSAY METHOD FOR ASPARTATE RACEMASE ACTIVITY BY HPLC

***Noriyuki Nimura*[1,*] *and Hiroshi Homma*[2]**
[1]Faculty of Pharmaceutical Sciences, Josai International University, Japan;
[2]School of Pharmaceutical Sciences, Kitasato University, Japan.

Keywords: chiral derivatization, enantio-separation, fluorometric detection, *o*-phthalaldehyde, reversed-phase high performance liquid chromatography.

1. INTRODUCTION

High performance liquid chromatographic (HPLC) chiral separation has been widely used to search for biological D-amino acids and related enzyme activities [1-3]. Chiral derivatization, a chromatographic enantio-separation technique, is useful for determining target compounds in complex biological samples. The chiral derivatization of enantiomeric amino acids, using *o*-phthalaldehyde (OPA) with an optically active thiol reagent, converts amino acids into pairs of fluorescent diastereomers that are readily separable on a reversed-phase column by HPLC [4-8]. The use of N-(*tert*-butylthiocarbamoyl)-L-cysteine ethyl ester (BTCC) as the optically active thiol reagent (Fig. 1) is particularly useful [9]. Derivatized D-aspartate (D-Asp) is eluted before the L-aspartate derivative (Fig. 2), and therefore can be quantified accurately and with high sensitivity, even in the presence of excess amounts of L-Asp. Therefore this system can be used for assaying aspartate racemase, which synthesizes a small amount of D-Asp in the presence of a large amount of the substrate L-Asp. Furthermore, since the chiral derivatization reaction proceeds rapidly at room temperature, a fully-automated system has been developed for derivatization and sample injection. This automated method is practical and has been applied successfully to an archaeal aspartate racemase assay.

* Correspondence concerning this article should be addressed to:
Noriyuki Nimura, E-mail: nimura@jiu.ac.jp, FAX: +81-475-53-4554.
Hiroshi Homma, E-mail: hommah@pharm.kitasato-u.ac.jp, FAX: +81-3-5791-6381.

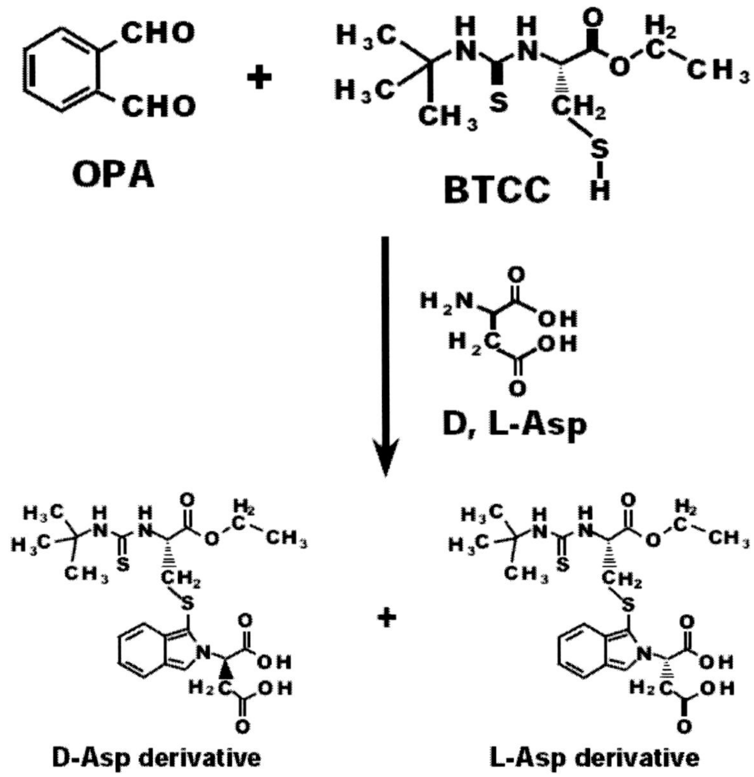

Figure 1. Derivatization of amino acids with BTCC and OPA. (From Nimura et al. (2003), *Anal. Biochem. 315*, 262-269, with permission)

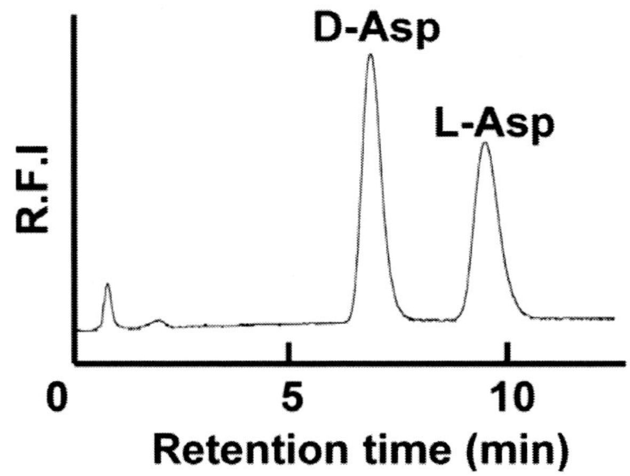

Figure 2. Chromatograms of the diastereomers formed from standard D,L-Asp and OPA/BTCC. Fluorescence intensities were detected at 445 nm with excitation at 335 nm. (From Nimura et al. (2003), *Anal. Biochem. 315*, 262-269, with permission)

2. Materials

1. High performance liquid chromatograph (HPLC) equipped with an octadecylsilyl (ODS) silica gel column and a fluorescence detector
2. Chiral derivatization reagent: N-(*tert*-butylthiocarbamoyl)-L-cysteine ethyl ester (BTCC) and *o*-phthalaldehyde (OPA)

BTCC was prepared from commercially available L-cysteine ethyl ester hydrochloride and *tert*-butyl isothiocyanate [9] as follows. L-Cysteine ethyl ester hydrochloride (13.3 g), *tert*-butyl isothiocyanate (7.0 g), and triethylamine (7.2 g) were mixed and kept overnight at room temperature with gentle stirring in 300 mL of dry acetonitrile. The acetonitrile was evaporated and the resulting residue was redissolved in ethyl acetate. The ethyl acetate solution was washed consecutively with 1 M hydrochloric acid, 5% sodium hydrogen carbonate, and distilled water. The organic layer was dried over anhydrous sodium sulfate. After evaporation of the ethyl acetate, a syrup of the chiral reagent BTCC (approximately 9.93 g, reaction yield 65%) was obtained.

A mixture of BTCC (158.4 mg) and commercially available *o*-phthalaldehyde (OPA) (80.5 mg) in 10 mL of acetonitrile was used as the chiral derivatization reagent. (The reagent solution was stable for at least one month in a refrigerator.)

3. Borate buffer (0.2 M, pH 10.2)
4. HPLC-grade methanol and acetonitrile
5. Phosphate buffer (50 mM, pH 6.8)
6. D- and L-aspartic acids and D,L(racemic)-aspartic acid
7. Pyridoxal 5'-phosphate (PLP)
8. Phospahte buffer (100 mM, pH 8.0)
9. Homogenized sample(s) containing amino acid racemase activity.

2. Methods

1. Chiral derivatization was carried out automatically with an autosample injector (GILSON model 231) as follows. A sample solution of the racemase assay mixture containing D,L-Asp (10 μL) was added to a sample tube placed on a rack of the sample injector. The tube was sealed and 30 μL of 0.2 M borate buffer and 20 μL of the derivatization reagent were added automatically by the autosample injector. The automatic diluter of the autosample injector mixed the solution by repeated suction and ejection. These processes required approximately 1.5 min, during which chiral derivatization was almost completed. An aliquot of the mixture (5 μL) was injected automatically into the chromatograph.
2. HPLC conditions:
   Column: Mightysil RP-18 GP (75 x 3.0 mm, i.d. 5 μm) maintained at 20°C
   Mobile phase: 50 mM phosphate buffer (pH 6.8):methanol:acetonitrile (70:10:20)
   Flow rate: 0.5 mL/min
   Fluorescence detector: Ex, 335 nm; Em, 420 nm
   Time program: Following injection of the sample, the mobile phase was delivered for 12 min for the analysis and the column was washed with 80% aqueous acetonitrile for 4 min. The column was equilibrated with the mobile phase for the next injection. These steps were carried out by the chromatograph system controller. The time program was repeated every 23 min.
3. The calibration curve of diastereomers formed from standard D- and L-Asp was prepared prior to analysis of the racemic samples. To prepare samples with different D-Asp/L-Asp ratios, various amounts of D-Asp were added to standard sample solutions containing given amounts of L-Asp. The correlation between the D-Asp peak area and D-Asp/L-Asp ratio was evaluated.
4. The reaction mixture (200 μL) for the aspartate racemase assay contained 100 mM sodium phosphate buffer (pH 8.0), 10 mM L-Asp as substrate, and 50 μM PLP as the coenzyme. The mixture was incubated at 70°C for 60 min. The reaction was stopped by chilling at 4°C and the addition of 10 μL of 100 % TCA. The precipitated proteins were removed by centrifugation at 15000 rpm (20,600 x g) for 10 min and the supernatant was subjected to automated HPLC analysis.

5. Frozen cells of *Thermoplasma acidophilum* HO-62 [10] were disrupted at 4°C by sonication in ice-cold 10 mM sodium phosphate buffer (pH 7.4) containing 0.9% NaCl. The homogenate was centrifuged at 3000 rpm (900 x g) for 15 min to remove large cellular debris and the supernatant was centrifuged again at 33000 rpm (100,000 x g) for 60 min. The resultant supernatant was assayed for aspartate racemase activity.
6. The protein concentration was determined by the Lowry method using BSA as the standard.

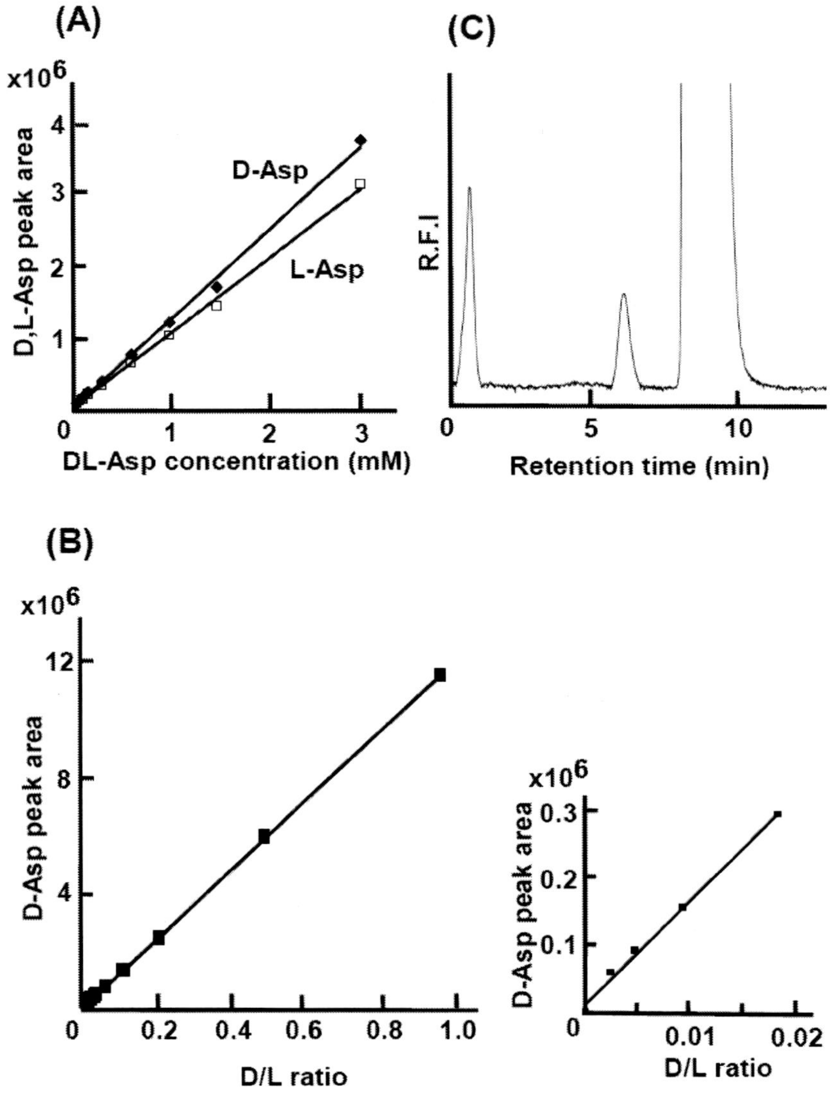

Figure 3. Calibration curve of the diastereomers formed from D,L-Asp and OPA/BTCC. A; Peak areas of the diastereomers formed from D-,L-Asp and OPA/BTCC were determined by HPLC. Correlations between the areas and the Asp concentrations were evaluated. The averaged ratio of the peak areas of D- and L-Asp was approximately 1.2. B; Various levels of D-Asp were added to sample solutions containing given levels of L-Asp, in order to prepare samples with various ratios of D-Asp/L-Asp. The correlation between the D-Asp peak area and the D-Asp/L-Asp ratio was evaluated. Right panel: the correlation is indicated for small D/L ratios. C; Chromatogram of the diastereomers formed from standard D, L-Asp at a ratio of 1/400. (From Nimura et al. (2003), *Anal. Biochem. 315*, 262-269, with permission)

3. RESULTS

Linearities for D- and L-Asp were evaluated by measuring the correlations between peak areas and concentrations ranging from 0.03 to 3.0 mM (Fig. 3A). These correlations were satisfactory, since the regression coefficients (r^2) were 0.9975 and 0.9977 for D- and L-Asp, respectively. The correlations of variance of the peak areas of 0.3 mM of D- and L-Asp were 3.3 % and 4.5 % (n=10), respectively, and those of the retention times for each isomer under continuous running were 1.5 % (D-Asp) and 1.1 % (L-Asp) (n=10). Fluorescence quantum yield of the diastereomer formed from D-Asp was approximately 1.2-fold higher than that from L-Asp [9]. Consistent with this result, the peak area of D-Asp was higher than that of L-Asp over a wide range of the concentrations (Fig. 3A) and the averaged ratio of the peak areas of D- and L-Asp was approximately 1.2. Figure 3B shows the correlation between the D-Asp peak area and the D-Asp/L-Asp ratios in the premixed samples. The correlation was linear over a wide range of ratios from 2.5 x 10^{-3} (1:400) (Fig 3C) to 1.0.

Figure 4 shows a chromatogram of the reaction mixture for aspartate racemase in *Thermoplasma acidophilum*, as described in a previous report [11]. The small amount of D-Asp produced was detected even in the presence of a large amount of L-Asp. This result confirmed that D-Asp could be quantified accurately with high sensitivity, even in the presence of excess amounts of L-Asp. The detection limit for D-Asp was approximately 1 pmol.

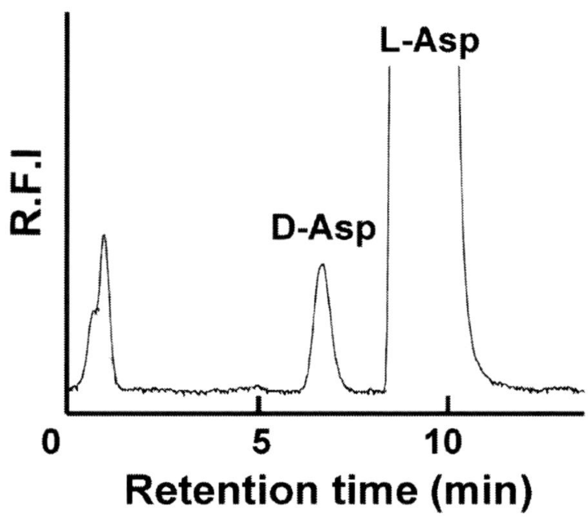

Figure 4. Chromatogram of the reaction mixture for the aspartate racemase assay in *Thermoplasma acidophilum*. (From Nimura et al. (2003), *Anal. Biochem. 315*, 262-269, with permission)

4. NOTES

1. BTCC is readily prepared from commercially available reagents and the product can be used directly for chiral derivatization without any special refining.
2. Since the reaction of the reagent with amino acids proceeds rapidly under mild conditions, the derivatization procedure can be automated by using a programmable robotic autosampling injector.
3. This method has enabled us to detect a small amount of D-Asp, even in the presence of an excess amount of L-Asp. The automated procedure allows a large number of samples to be assayed easily, rapidly, and accurately.

4. This method is of practical use, since only a change of the guard column after approximately 500 injections is required for maintaining efficiency.
5. The derivatization and injection procedures can be carried out manually when continuous sample processing is not required.
6. This HPLC method is applicable to the measurement of various amino acid racemase activities and suitable for the enantio-eparation of numerous amino acids, amino alcohols, and catecholamines.

REFERENCES

[1] Imai,K., Fukushima,T., Santa,T., Homma,H., Hamase,K., Sakai,K. & Kato,M. (1996). Analytical chemistry and biochemistry of D-amino acids. *Biomed. Chromatogr. 10*, 303-312.

[2] Bhushan, R, & Brückner, H. (2004) Marfey's reagent for chiral amino acid analysis: A review. *Amino Acids 27*, 231-247.

[3] Hamase, K., Morikawa, A. & Zaitsu, K. (2002) D-Amino acids in mammals and their diagnostic value. *J Chromatogr B Analyt Technol Biomed Life Sci. 781*, 73-91.

[4] Kinoshita, T. & Nimura, N. (1983). *Japanese Patent Application*, 58- 547147079.

[5] Nimura, N. & Kinoshita, T. (1986). *o*-Phthalaldehyde-N-acetyl-L-cysteine as a chiral derivatization reagent for liquid chromatographic optical resolution of amino acid enantiomers and its application to conventional amino acid analysis. *J. Chromatogr. 352,* 169-552.

[6] Aswad, D.W. (1984). Determination of D- and L-aspartate in amino acid mixtures by high-performance liquid chromatography after derivatization with a chiral adduct of *o*-phthalaldehyde. *Anal. Biochem. 137*, 405-409.

[7] Jegorov, A., Trika, J. & Tranka, T. (1994). 1-Thio-β-D-galactose as a chiral derivatization agent for the resolution of D,L-amino acid enantiomers. *J. Chromatogr. A, 67*, 3286-290.

[8] Buck, R. H. & Krummen, K. (1984). Resolution of amino acid enantiomers by high-performance liquid chromatography using automated pre-column derivatization with a chiral reagent. *J. Chromatogr. 315*, 279-285.

[9] Nimura, N., Fujiwara, T., Watanabe, A., Sekine, M., Furuchi, T., Yohda, M., Yamagishi, A., Oshima, T. & Homma, H. (2003). A novel chiral thiol reagent for automated precolumn derivatization and high-performance liquid chromatographic enantioseparation of amino acids and its application to the aspartate racemase assay. *Anal Biochem* 315, 262-269.

[10] Yasuda, M., Oyaizu, H., Yamagishi, A. & Oshima, T. (1995). Morphological variation of new *Thermoplasma acidophilum* isolates from Japanese hot springs. *Appl. Environ. Microbiol. 61*, 3482–3485.

[11] Long, Z., Lee, J.-A., Okamoto, T., Sekine, M., Nimura, N., Imai, K., Yohda, M., Maruyama, T., Sumi, M., Kamo, N., Yamagishi, A., Oshima, T. & Homma, H. (2001). Occurrence of D-amino acids and a pyridoxal 5'-phosphate-dependent aspartate racemase in the acidothermophilic archaeon, *Thermoplasma acidophilum. Biochem. Biophys. Res. Commun. 281*, 317–321.

In: D-Amino Acids: ISBN 1-60021-075-9
A New Frontier in Amino Acid and Protein Research, pp.459-462

Chapter 6.4

SERINE RACEMASE OF SILKWORM, *BOMBYX MORI*

Tohru Yoshimura*
Graduate School of Bioagricultural Sciences, Nagoya University, Japan.

Keywords: serine racemase, D-serine, pyridoxal 5'-phosphate, silkworm, *Bombyx mori.*

1. INTRODUCTION

D-Serine occurs in forbrain structures of mammals and acts as a co-agonist at the glycine site of *N*-acetyl D-aspartate (NMDA) receptor [1]. Wolosker et al. have demonstrated the serine racemase, which is probably involved in D-serine biosynthesis, in an extract of rat brain [2]. They also identified the gene encoding the mouse serine racemase, and cloned its c-DNA [3]. Mouse serine racemase is composed of 339 amino acids with a predicted molecular mass of 36.3 kDa, and contains pyridoxal 5'-phosphate (PLP) as a cofactor. The primary structure of the enzyme is different from that of bacterial serine racemase, but resembles that of bacterial L-threonine dehydratase. Genes homologous to that of mouse serine racemase are distributed in various eukaryotes such as *Schizosaccharomyces pombe*, *Caenohabiditis elegans*, and *Arabidopsis thaliana*. However, no homologs have been found in insects so far, though D-serine has been demonstrated at late larval and pupal stages in various lepidopteran species including silkworm [4]. We cannot deny the possibility that insects contain serine racemase that differs from the mammalian enzyme. I here describe the method for the partial purification of serine racemase from the pupae of silkworm, *Bombyx mori,* by ammonium sulfate fractionation, and Butyl-TOYOPEARL and DEAE-TOYOPEARL column chromatographies [5].

2. MATERIALS

1. Pupae of silkworm, *Bombyx mori* (type, Kinshu x Showa) reared on mulberry leaves
2. Phenylmethyl sulfonyl fluoride, *p*-toluene sulfonyl-L-phenylalanine chloromethyl ketone, phenylthiourea, leupeptin, pepstatin A, and chymostatin (Nacalai Tesque, Kyoto, Japan)

* Correspondence concerning this article should be addressed to Tohru Yoshimura, Graduate School of Bioagricultural Sciences, Nagoya University, Japan. E-mail: yosimura@agr.nagoya-u.ac.jp; Fax: +81-52-789-4132.

3. L-Serine freed from contaminated D-serine
4. *N*-tert-Butyloxycarbonyl L-cysteine (BOC-L-Cys) (Novabiochem, Switzerland)
5. Butyl-TOYOPEARL and DEAE-TOYPPEARL (Tosoh, Japan)
6. HPLC equipped with a fluorescence detector {two CCPD dual pumps, a CCP solvent programmer, a FS-8000 spectrofluorometer (Tosoh, Tokyo, Japan), an ERC-3310 degasser (Erma Optical Works, Tokyo, Japan), and a 4-mm Nova-Pack C18 column (3.9 x 300 mm) (Waters)}
7. Spectrophotometer
8. Standard buffer consisting of 10 mM potassium phosphate (pH 7.2), 50 mM KCl, 10 mM PLP, 1 mM dithiothreitol (DTT) and 1 mM ethylenediamine-tetraacetic acid (EDTA)
9. Potter-Elvenhjem glass equipped with a motor-driven Teflon pestle

3. METHODS

Purification of Serine Racemase from Silkworm Pupae

1. All procedures were carried out at 4°C. Three-days old pupae of silkworm, *Bombyx mori* (about 100 g), were homogenized with 5 volume of the standard buffer supplemented with 0.1 mM phenylmethyl sulfonyl fluoride, 0.1 mM *p*-toluene sulfonyl-L-phenylalanine chloromethyl ketone, 50 μg/ml phenylthiourea, 1 μg/ml leupeptin, 0.5 μg/ml pepstatin A, and 2 μg/ml chymostatin, in a Potter-Elvenhjem glass equipped with a motor-driven Teflon pestle (3,000 rpm).
2. The homogenates were centrifuged at 20,000 x g for 20 min. The supernatant solution was brought to 45% saturation with ammonium sulfate. The solution was stirred for 40 min, and centrifuged at 20,000 x g for 20 min. The resulting pellet was dissolved in 75 ml of the standard buffer, and dialyzed against the standard buffer overnight.
3. The supernatant solution obtained by centrifugation at 20,000 x g for 10 min was brought to 20% saturation with ammonium sulfate. After standing for 1 h, the solution was centrifuged to remove insoluble materials. The supernatant solution was loaded onto a Butyl-TOYOPEARL column (150 ml) equilibrated with the standard buffer supplemented with ammonium sulfate (20% saturation). A stepwise elution was performed with 300 ml of the buffer supplemented with 20, 10, 5.0, 2.5, 1.0, and 0% saturation of ammonium sulfate in this order.
4. Each fraction was brought to 45% saturation with ammonium sulfate. The pellet was dissolved in a small volume of the buffer and dialyzed overnight against the standard buffer. The active fraction, which was eluted with 10% saturation of ammonium sulfate, was loaded onto a DEAE-TOYOPEARL column (50 ml) equilibrated with the standard buffer. The column was washed with 100 ml of the standard buffer, and the enzyme was eluted with 100 ml of the standard buffer supplemented with 50 mM KCl. The enzyme was precipitated with 45% saturation of ammonium sulfate, and dialyzed against the standard buffer overnight.

Assay of Serine Racemase

1. Serine racemase was assayed by measuring the rate of conversion of L- or D-serine to the corresponding antipode. Enantioselective determination of D, L-serine was carried out according to the method of Hashimoto *et al* [6]. The reaction mixture (100 μl) contained 100 mM 1,3-bis[tris(hydroxymethyl)-methylamino]-propane (Bis-tris Propane) buffer (pH 8.5), 40 mM L- or D-serine, 7 μM PLP, 0.7 mM DTT, 0.7 mM EDTA, and the enzyme. After incubation at 37°C for 8 h, a 20 μl-aliquot was withdrawn from the reaction mixture and diluted 10-fold with 0.05 M

HCl. After centrifugation, 50 μl of the supernatant solution was used for the enantioselective determination of D, L-serine [6].

2. The supernatant solution was incubated for 3 min at room temperature with 350 μl of 0.4 M borate buffer (pH 9.0) and 100 μl of BOC-L-Cys-OPA reagent, which was prepared by dissolving 50 mg of OPA and 50 mg of Boc-L-Cys in 5 ml of methanol. A 10 μl-aliquot of the reaction mixture was subjected to HPLC, whose system comprised two CCPD dual pumps, a CCP solvent programmer, a FS-8000 spectrofluorometer, an ERC-3310 degasser, and a Nova-Pack C18 column. Other conditions were the same as described by Hashimoto *et al* [6]. One unit of the enzyme was defined as the amount of enzyme that catalyzes the formation of 1 nmol of the product enantiomer per minute. Protein was assayed according to the method of Bradford with bovine serum albumin as a standard [7].

4. Results

Silkworm serine racemase was partially purified by ammonium sulfate fractionation, and Butyl-TOYOPEARL and DEAE-TOYOPEARL column chromatographies. A summary of the purification is shown in Table 1. The partially purified enzyme catalyzed the production of D-serine from L-serine without production of other amino acids. Apparent *K*m and *V*max values were estimated: L-serine, 4.6 mM and 0.80 nmol/min/mg; D-serine, 16.5 mM and 2.1 nmol/min/mg.

Table 1. Partial purification of serine racemase from silkworm pupae.

Step	Total protein (mg)	Specific act. (units/mg)		Total act. (units)	
		L-Ser	D-Ser	L-Ser	D-Se
Pupae homogenate	4420			N.D.[a]	N.D.[a]
$(NH_4)_2SO_4$ Precipitation	1275	0.16	0.31	200	400
Butyl-TOYOPEARL	725	0.26	0.51	190	370
DEAE-TOYOPEARL	125	1.1	2.4	140	300

a. N.D., not detectable.

Serine racemase activity was inhibited strongly by sodium cyanoborohydride, sodium borohydride, phenylhydrazine and hydroxylamine. The enzyme was also inactivated by dialysis against the buffer containing hydroxylamine, and full activity was recovered by subsequent dialysis against the PLP-supplemented buffer. These results indicate that the enzyme depends on PLP.

5. Notes

1. Commercially available L-serine contains about 1% (w/w) D-serine. The contaminated D-serine could be removed with D-serine dehydratase, whose gene (*dsdA*) of *Escherichia coli* JM109 was cloned by polymerase chain reaction (PCR) according to the reported sequence of the gene (accession number of GenBank, JO 1603). The enzyme was purified from the recombinant cells

as reported previously [8]. A mixture (40 ml) containing 400 mM potassium phosphate buffer (pH 7.8) and 466 units of the enzyme was incubated at 37°C for 12 h, and the product was purified with Dowex 50 X 8 (H^+). The optical purity of the L-serine obtained was more than 99.7% enantiomeric excess (*ee*).

References

[1] Boehning, D. & Snyder, S.H. (2003). Novel neural modulators. *Annu Rev Neurosci*, *26*, 105-131.

[2] Wolosker, H., Sheth, K.N., Takahashi, M., Mothet J.P., Brady, R.O. Jr, Ferris, C.D., Snyder, S.H. (1999). Purification of serine racemase: biosynthesis of the neuromodulator D-serine. *Proc Natl Acad Sci, 96*, 721-725.

[3] Wolosker, H., Blackshaw, S., & Snyder S.H. (1993) Serine racemase: a glial enzyme synthesizing D-serine to regulate glutamate-*N*-methyl-D-aspartate neurotransmission. *Proc Natl Acad Sci, 96*, 13409-13414.

[4] Corrigan, J.J. & Srinivasan, N.G. (1966) The occurrence of certain D-amino acids in insects. *Biochemistry*, *5*, 1185-1190.

[5] Uo, T., Yoshimura, T., Shimizu, S. & Esaki, N. (1998) Occurrence of pyridoxal 5'-phosphate-dependent serine racemase in silkworm, *Bombyx mori*. *Biochem Biophys Res Commun, 246*, 31-34.

[6] Hashimoto, A., Nishikawa, T., Oka, T., Takahashi, K., & Hayashi, T. (1992) Determination of free amino acid enantiomers in rat brain and serum by high-performance liquid chromatography after derivatization with *N-tert*-butyloxycarbonyl-L-cysteine and *o*-phthaldialdehyde. *J Chromatogr*, *582,* 41-84.

[7] Bradford, M.M. (1976) A rapid and sensitive method for the quantitation of microgram quantities of protein utilizing the principle of protein-dye binding. *Anal Biochem, 72*, 248-252.

[8] Marceau M., McFall E., Lewis, S.D. & Shafer, J.A. (1988) D-Serine dehydratase from *Escherichia coli*. DNA sequence and identification of catalytically inactive glycine to aspartic acid variants. *J Biol Chem, 263,* 16926-16933.

In: D-Amino Acids: ISBN 1-60021-075-9
A New Frontier in Amino Acid and Protein Research, pp.463-466

Chapter 6.5

RECOMBINANT SERINE RACEMASE PREPARATION

Veronika N. Foltyn and Herman Wolosker*
Department of Biochemistry, B. Rappaport Faculty of Medicine, Technion-Israel Institute of Technology, Israel.

Keywords: serine racemase, recombinant, racemization, D-serine, pyridoxial 5-phosphate.

1. INTRODUCTION

D-Serine is an endogenous coagonist of NMDA receptors in the brain and it is synthesized from L-serine by serine racemase enzyme [1-4]. Understanding the regulation of D-serine synthesis will shed light on mechanisms regulating NMDA receptor neurotransmission. Serine racemase is a member of the fold type II of pyridoxal 5' phosphate (PLP) dependent family of enzymes and was the first cloned eukaryotic racemase. A major problem encountered in the study of serine racemase properties was to obtain large amounts of purified protein suitable for biochemical characterization, antibody production and crystallization studies. Early attempts to produce the enzyme in bacteria failed due to insolubility of the protein, which was packed in bacterial inclusion bodies. To overcome this caveat, we previously established a method to purify small quantities of GST fusion serine racemase from transfected HEK293 cells [5]. We now describe a short protocol on purification of recombinant histidine-tagged mouse serine racemase from *E. coli*. The method gives a high yield and careful induction of protein expression minimizes protein packing into inclusion bodies. By using this method, we recently described new properties of serine racemase, such as its ability to catalyze α,β elimination of water from L- and D-serine, generating pyruvate and ammonia [6,7].

2. MATERIALS

1. Incubator shaker
2. UV/Vis spectrophotometer
3. Preparative centrifuge (Sorvall RC5C)

* Correspondence concerning this article should be addressed to Herman Wolosker, E-mail:hwolosker@tx.technion.ac.il, FAX: +972-4-829-5384.

4. Probe sonicator
5. Table centrifuge
6. Luria-Bertani (LB) medium
7. Antibiotics kanamycin (50 mg/ml stock solution) and chloramphenicol (30 mg/ml stock solution)
8. Isopropyl β-D thiogalactopyranoside (IPTG) (Sigma)
9. Pyridoxal 5'-phosphate (PLP) (Sigma)
10. Imidazole (1 M solution) (Sigma)
11. Phosphate buffered saline (PBS) solution
12. Triton X-100 (Sigma)
13. Phenyl methyl sulfonyl fluoride (PMSF) (Sigma)
14. Talon Metal Affinity Resin (BD Bioscience)
15. Sodium chloride (Sigma)
16. Dialysis membrane (10,000 MW cutoff, Spectrapor)
17. Mouse serine racemase cDNA subcloned into pET 28c+ vector (Novagen)
18. BL21 codon plus bacteria (Novagen)

3. Methods

- BL21 codon plus bacteria are first transformed with mouse serine racemase plasmid construct encoding for N-terminal poly-histidine tag according to the manufacturer's instructions (Novagen).
- On the first day, inoculate 10 ml of LB medium containing 50 µg/ml kanamycin and 30 µg/ml chloramphenicol with an isolated colony of transformed BL21 cells and grow the inoculum overnight on a shaker set at 250 rpm and 37°C.
- Next morning dilute the overnight inoculum into 500 ml LB containing 50 µg/ml kanamycin and return to shaker.
- When the absorbance at 600 nm reaches about 0.6-0.7, decrease the shaker temperature to 30°C and add IPTG to a final concentration of 1 mM.
- Incubate under shaking for additional 16 hours.
- Pellet the bacteria by centrifugation at 4°C (9,000 x g for 10 minutes in a preparative centrifuge).
- Resuspend the pellet in 20 ml of freshly prepared lysis medium containing 15 µM PLP, 10 mM imidazole, 0.4 mM PMSF dissolved in PBS supplemented with 500 mM NaCl.
- Disrupt the cells by sonication on ice.
- Add triton X-100 to a final concentration of 1.2% and rotate the suspension for 10 minutes in a cold room.
- Centrifuge at 35,000 x g for 20 minutes.
- Transfer the supernatant to 50 ml tubes and add 0.5 ml of a 50% slurry Talon resin beads previously washed with PBS.
- Rotate in cold room for 2 hours.
- Collect the beads by brief centrifugation (1 min 4,000 x g on a table centrifuge).
- Wash the beads 5 times with 10 ml of buffer A (15 µM PLP, 20 mM imidazole, 1% triton X-100 dissolved in PBS supplemented with 400 mM NaCl).
- Wash the beads 3 times with 10 ml buffer B (15 µM PLP dissolved in PBS).
- Batch elute the protein by sequential additions of a 20 mM Tris-HCl (pH 8.5) solution containing 15 µM PLP and 200 mM imidazole.
- Dialyze against PBS containing 15 µM PLP with four buffer changes.
- If desired, purified protein can be concentrated with membrane concentrators and stored at -70°C.

4. Results

Gel electrophoretic analysis of the preparation reveals a major band corresponding to recombinant serine racemase that migrates at 37 kDa (Fig. 1). The yield of the preparation is about 700 μg purified protein.

5. Notes

The preparation can be scaled up and adapted according to individual needs. The yield of the elution step is variable since not all the protein can be removed from the beads. Occasionally a portion of the protein may precipitate during dialysis. We found that serine racemase is fully active when bound to the resin and enzyme activity can be monitored with beads, without the need to elute the protein. In this case, the protein can be stored as 50% bead slurry in PBS containing 15 μM PLP for a few days at 4°C.

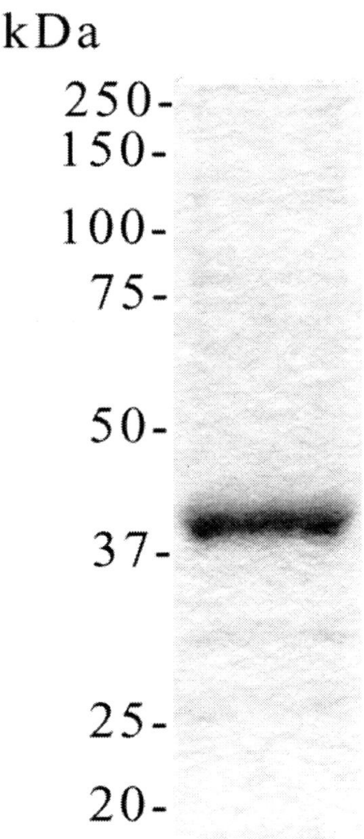

Figure 1. SDS-gel analysis of purified recombinant serine racemase. The lane contains 1 mg protein.

References

[1] De Miranda, J., Santoro, A., Engelender, S. & Wolosker, H. Human serine racemase: molecular cloning, genomic organization and functional analysis. *Gene 256*, 183-8. (2000).

[2] Wolosker, H. et al. Purification of serine racemase: biosynthesis of the neuromodulator D-serine. *Proc Natl Acad Sci U S A 96*, 721-5. (1999).

[3] Wolosker, H., Blackshaw, S. & Snyder, S. H. Serine racemase: a glial enzyme synthesizing D-serine to regulate glutamate-*N*-methyl-D-aspartate neurotransmission. *Proc Natl Acad Sci U S A 96*, 13409-14. (1999).

[4] Wolosker, H., Panizzutti, R. & De Miranda, J. Neurobiology through the looking-glass: D-serine as a new glial-derived transmitter. Neurochem Int 41, 327. (2002).

[5] Panizzutti, R., De Miranda, J., Ribeiro, C. S., Engelender, S. & Wolosker, H. A new strategy to decrease *N*-methyl-D-aspartate (NMDA) receptor coactivation: inhibition of D-serine synthesis by converting serine racemase into an eliminase. *Proc Natl Acad Sci U S A 98*, 5294-9. (2001).

[6] De Miranda, J., Panizzutti, R., Foltyn, V. N. & Wolosker, H. Cofactors of serine racemase that physiologically stimulate the synthesis of the *N*-methyl-D-aspartate (NMDA) receptor coagonist D-serine. *Proc Natl Acad Sci U S A 99*, 14542-7 (2002).

[7] Foltyn, V. N. et al. Serine racemase modulates intracellular D-serine levels through an alpha,beta-elimination activity. *J Biol Chem 280*, 1754-63 (2005).

In: D-Amino Acids:
A New Frontier in Amino Acid and Protein Research, pp.467-476
ISBN 1-60021-075-9

Chapter 6.6

D-Amino Acid Oxidase Activity Assays

Maria A. Vanoni* and Bruno Curti
Dipartimento di Scienze Biomolecolari e Biotecnologie, Universita' degli Studi di Milano, Italy.

Keywords: D-amino acid oxidase, activity assay, peroxidase, D-α-phenylglycine, isoelectrophocusing.

1. Introduction

D-Amino acid oxidase (E.C. 1.4.3.3, DAAO) catalyzes the oxidative deamination of a variety of D-α-amino acids according to Eqs. 1-4

$$\text{D-}\alpha\text{-amino acid} + O_2 \leftrightarrow \alpha\text{-keto acid} + H_2O_2 \quad (1)$$

$$\text{D-}\alpha\text{-amino acid} + \text{E-FAD}_{ox} \leftrightarrow \text{E-FAD}_{ox}\bullet\ \text{D-}\alpha\text{-amino acid} \leftrightarrow \text{E-FAD}_{hq}\bullet\alpha\text{-imino acid} \quad (2)$$

$$\text{E-FAD}_{hq}\bullet\alpha\text{-imino acid} + O_2 \leftrightarrow \text{E-FAD}_{ox}\bullet\alpha\text{-imino acid} + H_2O_2 \quad (3)$$

$$\alpha\text{-imino acid} + H_2O \leftrightarrow \alpha\text{-keto acid} + NH_4^+ \quad (4)$$

DAAO has been the object of a large number of structural, mechanistic and computational studies since its discovery in the 1930's (see [1-11] and references therein). Beside being the prototype of flavin-dependent oxidases, DAAO is of wide interest for its medical [12-15] and biotechnological (see [16] for a recent review) implications. In order to facilitate studies on this enzyme, we here report on the three main methods used to assay for DAAO activity exploiting: (i) oxygen consumption; (ii) the non-physiological substrate D-α-phenylglycine; (iii) hydrogen peroxide evolution. For sake of example, we show how these assays have been used to characterize two pig kidney DAAO isoforms, the effect of viscosity on the enzyme kinetics and on the stability of the holoenzyme. Finally, we show how the peroxidase-coupled assay has been used to carry out a preliminary characterization of the reaction of pig kidney DAAO reactivity with D-dihydroxyphenylalanine (DOPA), a compound related to the metabolism of neurotransmitters [17].

* Correspondence concerning this article should be addressed to Maria A. Vanoni, E-mail: maria.vanoni@unimi.it; bruno.curti@unimi.it; FAX: +39-0250314895.

2. MATERIALS

1. Chemicals

All chemicals were from Sigma-Aldrich Chemical Co. (www.sigmaaldrich.com; St. Louis, MO, USA) or Merck (www.merck.de; Darmstadt, Germany). All solutions were made up in ultrapure water prepared with a MilliQ apparatus (www.millipore.com; Millipore, Billerica, MA, USA).

2. Enzymes

The DAAO-benzoate (DAAO-Bz) complex was prepared as described in [18,19] and stored in 20 mM Hepes/KOH buffer, pH 8.3 containing 0.2 mM sodium benzoate at -80 °C [20]. The apoprotein form of DAAO (Apo-DAAO) was obtained by extensive dialysis against 20 mM Hepes/KOH, pH 8.3, 3 mM EDTA and 1 M KBr followed by dialysis against the same buffer lacking KBr [21,20]. The holoenzyme (DAAO) and the DAAO-Bz complex were obtained by incubating Apo-DAAO with 40-fold molar excess FAD in the absence or presence of 500–molar excess sodium benzoate for 30 min at 25°C [22]. DAAO isoforms were isolated by chromatofocusing by loading 0.1 – 10 mg DAAO-benzoate complex or Apo-DAAO on a MonoP column (AP Biotech[Note1]) connected to an FPLC apparatus. The column was equilibrated with 25 mM imidazole/acetate, pH 7.6 and eluted with solvent formed by 7.7% (by volume) Polybuffer 96 (AP Biotech [Note1]) brought to pH 6.0 with acetic acid. The flow rate was 1 mL/min and 0.5 mL fractions were collected and analyzed. Ampholines (AP Biotech [Note1]) were removed by repeated dilution/concentration in Centricon 30 (Amicon-Millipore, www.millipore.com; Billerica, MA, USA). Isoelectrofocusing (IEF) was carried out on 5% polyacrylamide gels using a flat-bed apparatus from Amersham[Note 1] according to [23] to analyse the fractions corresponding to the acidic and basic DAAO form[Note 2].

3. Solutions

1. Polarographic activity assay and spectrophotometric activity assay with D-α-phenylglycine.
   - Buffer A: 0.1 M sodium pyrophosphate buffer, pH 8.5 (stored at 0-5°C)
   - D-alanine stock: 280 mM D-α-alanine solution in water (stored frozen).
   - D-α-phenylglycine stock: 25 mM D-α-phenylglycine in Buffer A (stored for up to 1 week at 0-5°C)
   - FAD stock: 0.1 mM FAD in Buffer A (stored frozen).
2. Peroxidase-coupled assay
   - *Buffer B:* 100 mM Hepes/KOH, pH 8.0 (stored at 0-5°C);
   - *FAD stock:* 1 mM FAD in water (stored frozen);
   - *HRP solution*: 1 mg/mL horse radish peroxidase (freshly diluted in 50 mM Hepes/KOH buffer, pH 8.0 from the commercial stock provided as an ammonium sulphate suspension (Catalog no. P6140);
   - *o-dianisidine solution*: 8 mM o-dianisidine in 20% Triton X-100 (stored for up to 1 week at 0-5°C)
   - *D-amino acid solution:* D- or DL- amino acid solutions should be prepared as concentrated stock solutions [depending on the corresponding K_m value (Table 2) and solubility] in water and neutralized as needed.

4. Equipment

Oxygen consumption was measured using a Clark-type liquid-phase oxygen electrode from Hansatech Instruments Ltd. (www-hansatech-instruments.com, Norfolk, UK). Spectrophotometric assays were carried out with double beam spectrophotometers (Cary 219 or Cary 100, Varian, Palo Alto, CA, USA, www.varianinc.com). All instruments were connected to water baths.

3. Methods

1. Polarographic Assay with D-α-Amino Acids

Oxygen consumption can be measured in a Clark-type oxygen electrode in 2 mL reaction mixtures equilibrated at 25°C [Note 3]. For assay set up using D-alanine as the substrate[Note 4], 1 mL Buffer A, 0.8 mL D-alanine stock and 0.2 mL FAD stock are mixed. After equilibration at 25°C for 5 min, 1-10 μg DAAO (in 1-50 μL) are added to start the reaction.

In this assay activity is expressed as μmol O_2 consumed per minute per mL enzyme solution (or mg enzyme). The rate of oxygen consumption (in μmol/min) is calculated with the following formula:

$$(\Delta P_{O2}/\text{min}) * \alpha * 2 \text{ mL}$$

where: $(\Delta P_{O2}/\text{min})$ is the fractional change in partial oxygen pressure per minute during measurements of initial velocities of reactions and α is a conversion factor that takes into account oxygen solubility in water as a function of temperature (see Table 1 for the most commonly used values of α).

Table 1. Values of dissolved oxygen concentrations in water at 1 atm.

Temperature (°C)	α (mM)
20	0.276
25	0.253
30	0.230
37	0.214

2. Spectrophotometric Assay with D-α-Phenylglycine

A modification of the method of Fonda and Anderson [24] is widely employed for the measurement of DAAO activity and it exploits the increase of absorbance in the UV region that accompanies the oxidative deamination of D-α-phenyglycine to give the corresponding α-keto acid (benzoylformic acid) and ammonia[Note5, Note 6]. Assays are set up in 3 mL quartz cuvettes by mixing 1.2 mL Buffer A, 1.5 mL D-α-phenylglycine stock and 0.3 mL FAD stock.

After equilibration at 25°C for 5 min, 1-10 μg DAAO (in 1-50 μL) is added to start the reaction that is continuously monitored at 243 nm.

Activity is typically expressed as ΔA_{243}/min and referred to 1 mL or 1 mg of enzyme solution.

Table 2. Steady-state kinetic parameters of the reaction of pig kidney DAAO with a series of D-amino acids and the polarographic or peroxidase-coupled assays.

D-Amino acid	Assay, pH[a]	K_M [a]	K_M (pH 8.5)[b]	V_{max} [a] (U/mg)	V_{max} [a] (%)	V_{max}, pH 8.5 [b] (%)
D-Ala (1.4-28)	O_2, 8.0	2.6 ± 0.4	2.3	10.9 ±0.5	100	100
D-Ala (1.8 – 28)	H_2O_2, 8.0	2.5 ±0.4		10.5±0.5	100	
D-Phe (3.0 – 50.6)	H_2O_2, 8.0	2.1 ± 0.6	1.6	12.6 ± 0.9	119	90
D-Ser (3.2 – 9.1)	H_2O_2, 8.0	41.8 ± 2.7	4.6	13.6 ± 0.4	127	48
D-Thr (4.2 – 157)	H_2O_2, 8.0	9.0 ± 1.5	2.3	0.4 ± 0.02	4.0	27
D-Arg (5-42.4)	H_2O_2, 8.0	26.9 ± 4.8	NR	4.6 ± 0.4	6.2	NR
D-Lys (7.5 – 60)	H_2O_2, 8.0	6.7 ± 1.1	0	0.13 ± 0.01	1.2	0
D-Pro (0.9- 17)	O_2, 8.5	2.4 ± 0.3	2.3	32.0 ± 1.1	163	302
D-Val (1.3 –17.1)	O_2, 8.5	0.8 ± 0.2	1.2	11.1 ± 0.3	50	65
D-Leu (0.1 – 6.9)	O_2, 8.5	0.2 ± 0.02	0.2	4.9 ± 0.1	23.8	22
D-α-Phenylglycine, (2.5- 25)	H_2O_2, 8.0	10.8 ± 3.8		10.8 ± 2.1		
D-DOPA (0.87 – 4.34)	H_2O_2, 8.0	1.6 ± 0.3		0.9 ± 0.06		
DL-DOPA (1.6- 8.4)	H_2O_2, 8.0	4.2 ± 0.76		1.1 ± 0.1		

[a]The V_{max} and K_M values were determined with the oxygen electrode (O_2) or the peroxidase-coupled assay (H_2O_2) in 50 mM Hepes/KOH buffer, pH 8.0 or 0.1 M sodium pyrophosphate, pH 8.5 in the presence of the indicated concentration range of the given D-amino acid (in mM).

[b]Values derived from the histogram reported in [27] and obtained in 0.1 sodium pyrophosphate buffer, pH 8.5. NR, not reported.

3. Peroxidase-Coupled Assay

The method for assaying glycolate oxidase [25] or D-aspartate oxidase [26] has been adapted for measuring DAAO activity [22].

Assays are carried out in semi-micro glass or quartz cuvettes (optical path: 1 cm) in 1 mL reaction mixtures containing: 500 μl buffer B, 10 μl FAD stock solution and varying amounts of the amino acid solution. Water is added to reach the final volume of 1 mL. After equilibration at 25°C, the reaction is started by adding 50 μL *o*-dianisidine and 15 μL HRP solutions. The instrument is zeroed and 1-10 μL (0.5-5 μg) DAAO is added. The increase of absorbance is monitored continuously at 436 nm. For calculation of enzyme activity an extinction coefficient (ε) of 11.6 $mM^{-1}cm^{-1}$ is used for oxidation of *o*-dianisidine, which is stoichiometric with D-amino acid oxidation and H_2O_2 production.

4. Determination of Steady-State Kinetic Parameters of the Reactions

In all cases the steady-state kinetic parameters have been determined by varying the concentration of the D-amino amino acid (see Tables 2, 3) in the presence of constant concentrations of all other reagents, and by fitting the data to the Michaelis-Menten equation (eq. 5):

$$v = (V_{max}*S)/(K_M + S) \quad (5)$$

where v, is the initial reaction velocity; V_{max} is the apparent maximal velocity and K_M is the apparent K_M value for the varied substrate.

The apparent dissociation constant (K_d) of the DAAO-FAD complex was also determined kinetically using Apo-DAAO, constant concentration of D-phenylglycine or D-alanine and varying concentrations of

FAD (see Table 3), using eq. 5 and assuming that the dissociation constant for FAD is equal to the apparent K_M of the reaction.

Data analysis was carried out using the Grafit software package (Erithacus Software Ltd, Horley, UK, www.erithacus.com).

Table 3. Comparison of the kinetic properties of Apo-DAAO isoforms as determined with the D-α-phenylglycine spectrophotometric assay.

Enzyme	V_{max} (ΔA_{243}/(min*mg)	K_M (mM)	K_{FAD} (μM)
Apo-DAAO	267.4 ± 5.7	4.8 ± 0.2	0.7 ± 0.04
Apo-DAAO-A	264.5 ± 5.4	4.3 ± 0.2	0.8 ± 0.05
Apo-DAAO-B	259.1 ± 7.9	4.9 ± 0.3	0.7 ± 0.03

Apparent V_{max} and K_M values for D-α-phenylglycine were determined by varying the substrate concentration between 1.25 and 12.5 mM in the presence of 10 μM FAD in 0.1 M sodium pyrophosphate buffer, pH 8.5 at 25°C. 1 μg of protein was used to start the assays. Care was taken to calculate activity values during the linear portion of the trace to allow for reconstitution of the holoenzyme under the given condition. Velocities are expressed as ΔA243/(min*mg). The apparent dissociation constant of the enzyme-FAD complex (K_{FAD}) was determined with 12.5 mM D-α-phenylglycine and varying concentrations of FAD (0.5 – 10 μM).

4. Results

Activity of DAAO with D-Amino Acids as Measured with the Polarographic and Peroxidase-Coupled Assays

In order to substitute the traditional polarographic assay of DAAO with a continuous spectrophotometric method, we have shown that the peroxidase-coupled assay of DAAO provides an efficient and simple tool to study DAAO-catalysed D-amino acid oxidation [22]. During the set-up of the assay, the results obtained with the polarographic and peroxidase-coupled methods were compared with each other and with those derived from the early work of Dixon and Kleppe ([27], Table 2). These authors reported the steady-state kinetic parameters measured for the different amino acids in a histogram from which we calculated the corresponding values. In order to check if our calculations were correct, we determined V and K values with the polarographic method under the same conditions as those used in [27]. Values similar to the published ones were indeed found. With D-alanine, V and K values determined with the peroxidase-coupled assay were also indistinguishable from those determined with the polarographic method when the assays were carried out in the same 50 mM Hepes/KOH buffer, pH 8.0 and at the same temperature (25°C) demonstrating that the peroxidase-coupled assays is a valid method to assay for DAAO. The values are also similar to those reported in [27] at pH 8.5 in agreement with the small changes in the reaction kinetics at the two pH values as shown in [28]. The peroxidase-coupled assay (at pH 8.0) was also used to redetermine the V and K values associated with DAAO-catalysed oxidation of other D-amino acids. The differences between the values shown in Table 2, and those calculated from [26] are to be ascribed to the different pH value chosen for these assays as shown by the pH-dependence of the reaction of DAAO with various amino acids reported, for example, in the detailed kinetic analyses carried out in the Fitzpatrick laboratory [28]. Therefore, care should be taken to choose experimental conditions (buffer, pH, temperature) in studies aimed to establish DAAO activity levels in tissues and cell lines. Furthermore, modeling studies aimed to evaluate the *in vivo* metabolic role of DAAO should be accompanied by careful evaluation of the catalytic behaviour of the enzyme during reaction with a given substrate under conditions of pH, temperature and, most likely, oxygen concentration that actually mimic those of the peroxisomal compartment in which DAAO resides. For example, the K_m value for D-serine is about 4 mM in 0.1 M sodium pyrophosphate buffer, pH 8.5, but it was found to be as high as 40 mM when assays were run in sodium Hepes buffer, pH 8.0 in the coupled peroxidase assay.

The peroxidase-coupled assay has also been used to assay DAAO activity in the presence of D-α-phenylglycine and D- or DL-DOPA (Table 2). L-DOPA is the precursor of dopamine while D-DOPA is supposed to be biologically inactive. However, it has been reported [17] that administration of D-DOPA to rats induced the same circling behaviour as that obtained with L-DOPA, and that the effect of D-DOPA could be abolished by co-injection of benzoate. These results were taken as evidence that DAAO may participate in dopamine production by converting D-DOPA in the corresponding α-ketoacid. The product could be aminated to L-DOPA, with the latter compound being the actual dopamine precursor. As shown in Table 2, D-DOPA is a substrate of DAAO, while the presence of L-DOPA in the assays does not affect the V value, but only doubles the observed K value, as expected for the strict specificity of DAAO with respect to the D-isomer of the amino acid substrate. While the V value observed with D-DOPA is only 10 % that measured with D-alanine, the K value is of the same order of magnitude making V/K with D-DOPA similar to that measured, for example, with D-serine. These results may be of interest within the context of studies aimed to establish the role of DAAO in the metabolism of neurotransmitters.

Use of the Spectrophotometric Assay with D-α-Phenylglycine to Characterise the Isoforms found in DAAO Preparations.

The assay has been used in a variety of instances to monitor DAAO activity since its design by Fonda and Anderson [24] [Note 7]. To provide an example of its application, we will here show the results of the preliminary characterization of two isoforms of DAAO found in enzyme preparations from pig kidney[Note 8].

The D-α-phenylglycine spectrophotometric assay was used to compare the apo-DAAO isoforms with respect to their kinetic behavior with this substrate and their apparent affinity for the FAD cofactor (Table 3). No significant difference between the isoforms was found. While we do not know if the isoforms arise during purification or storage of DAAO (perhaps due to limited and catalytically irrelevant deamidation) or if they are present, initially, in the kidneys used as the protein source, their potential presence will have to be taken into account during the evaluation of DAAO protein levels in tissues and cells exploiting, for example, proteomics approaches.

Table 4. Effect of glycerol on V, $K_{D\text{-}\alpha\text{-phenylglycine}}$, and K_{FAD}.

Glycerol, %	V (ΔA_{243}/(min*mg)	$K_{D\text{-}\alpha\text{-phenylglycine}}$, mM	K_{FAD}, μM
0	188 ± 7.5	4.3 ± 0.5	
10	145 ± 5.8	3.0 ± 0.4	
0	127 ± 1.3		0.4 ± 0.03
1	132 ± 1.3		0.4 ± 0.03
5	128 ± 1.7		0.3 ± 0.03
10	119 ± 2.0		0.2 ± 0.03
20	109 ± 1.7		0.2 ± 0.04

Assays were carried out at 25 °C in 0.1 M sodium pyrophosphate. The K for D-α-phenylglycine was calculated by varying D-α-phenylglycine between 1.25 and 12.5 mM in the presence of 10 μM FAD. The K_d for the enzyme-FAD complex (K_{FAD}) was calculated at 12.5 mM D-α-phenylglycine and varying FAD between 1 and 10 μM. Glycerol was added from a 50 % solution.

IEF also showed a striking effect of glycerol in preventing FAD release from the DAAO-Bz complex: when 10 or 20% glycerol was included in the IEF gel, bands corresponding to Holo-DAAO isoforms appeared with pI of 5.8 (Holo-DAAO-B) and 5.4 (Holo-DAAO-A) with a minor component at pI 5.3, as judged by the fact that activity staining could be obtained in the absence of added FAD. At 20% glycerol only the holoDAAO species were found. Again, the spectrophotometric D-α-phenylglycine assay could be used to study the effect of glycerol on DAAO kinetics (Table 4). In agreement with the IEF results, glycerol was found to decrease the dissociation constant of the holoenzyme to yield apoDAAO and free FAD although only by a factor of 2 with a less pronounced effect on V_{max} and K_M for D-α-pheylglycine (Table 2). The latter result is in full agreement with the results of Fitzpatrick at al. [30]. While chromatofocusing may provide a novel method to prepare DAAO apoprotein, the stabilization afforded by glycerol on DAAO holoenzyme may be of practical relevance in studies aimed to isolate DAAO forms from tissues and to preserve the integrity of Holo-DAAO in samples.

In conclusion, we here summarized three different methods to assay DAAO activity, with the peroxidase-coupled assay being the most versatile and accessible one. Variations of this assay may be implemented in order to adapt it to high-throughput assays or to increase its sensitivity by using fluorescent substrates of the peroxidase. However, it should be mentioned here that HPLC assays have also been developed (e.g.: [31]) as well as several methods for the direct detection of DAAO activity in tissues (e.g.: [32]).

We also highlighted how studies of *in vivo* levels of the enzyme through activity measurements and modeling of metabolic fluxes require a careful evaluation of the actual steady-state kinetic parameters exhibited by the enzyme with a given substrate under well-defined conditions with pH being an important factor to be taken into account together with solvent viscosity. The latter strongly influences the stability of the holoenzyme, and to a lesser extent the steady-state kinetic parameters of DAAO catalyzed reaction. Finally, we demonstrated that the peroxidase-coupled assay may be used to study DAAO-catalyzed oxidation of D-DOPA, which has been shown to be a DAAO substrate as efficient as, for example, D-serine. This finding may provide an additional tool to study the *in vivo* role of DAAO, in particular its involvement in neurotransmission.

5. NOTES

1. Amersham and Pharmacia, also known as AP Biotech, are now part of GE Healthcare (www.GEHealthcare.com; Little Chalfont, UK).
2. After focusing, the gel was sliced in order to detect protein bands by Coomassie Blue staining and DAAO activity in the absence or presence of added FAD. The gel strip was incubated at 30-37°C in 0.05 M sodium pyrophosphate buffer, pH 8.5, containing 40 mM D-alanine, 0.1 mg/mL iodonitrotetrazolium and 50 μM FAD. FAD was omitted to assay for the presence of holo-DAAO forms. After color development the gels were fixed by immersion in 7% acetic acid.
3. An essential step in the polarographic assay of DAAO is the calibration of the oxygen electrode, which is accomplished according to the manufacturer's instructions by (i) recording the signal of an assay mixture equilibrated at the given temperature to obtain the signal corresponding to 100% dissolved oxygen (in air), and (ii) adding a few crystals of dithionite and recording the signal corresponding to 0% oxygen concentration. The observed change yields the signal amplitude for 100% change in oxygen concentration and will be used to calculate the fractional change in partial oxygen pressure (ΔP_{O2}) in a given time-interval during measurements of initial velocities of reactions.
4. D-Alanine can be substituted by other amino acids, which can also be prepared in water or diluted buffer.
5. The assay can be also carried out in semi-micro cuvettes in 1 mL reaction mixtures by reducing the volumes of each solution accordingly. 0.5- 1 μg of DAAO is typically used to start the reaction.

6. For measurement of steady-state kinetic parameters in 1 mL assays, if needed, a 0.2 M sodium pyrophosphate, pH 8.5, buffer stock is prepared and D-α-phenylglycine and FAD are made up in water. In this case the assay is set up using 500 μL buffer, and varying amounts of substrate, cofactors or other ligands and water to reach the final volume of 1 mL.
7. The limit of this assay is the high background absorbance due to D-α-phenylglycine and FAD, as well as that due to the enzyme-containing solution. Thus, it is typically used to characterize homogeneous preparations of the enzyme during, for example, chemical modification (e.g.: [20]) or limited proteolysis (e.g.: [29]) studies.
8. Isoelectrophocusing (IEF) experiments have been carried out on several DAAO preparations under native conditions. The experiments revealed that during IEF in the absence of glycerol, the DAAO-benzoate complex is converted into the apoprotein form and gives rise to two protein bands with pI of 6.5 (Apo-DAAO-B) and 6.25 (Apo-DAAO-A), respectively. A minor component with a pI value of 6.1 was also sometimes present. A similar result was obtained by chromatofocusing, which led to separation of the isoforms. The two isoforms were not in equilibrium with each other since re-chromatography (or IEF) of each protein species yielded a single protein peak (or protein band on IEF) and their structural differences must be small since no heterogeneity in the starting preparations was found by mass spectrometry-time of flight. The analysis yielded signals consistent with the presence of one protein species of 39312, which matches well the calculated mass of Apo-DAAO (39330).

ACKNOWLEDGMENTS

Experimental work has been carried out with financial support from Ministero dell'Istruzione e della Ricerca Scientifica, Rome, and Consiglio Nazionale delle Ricerche.

REFERENCES

[1] Bright, H.J., Porter, D.J.T. (1963) Flavoprotein oxidases. *Enzymes*, *12*, 421-505.

[2] Curti, B., Ronchi, S., Pilone, M.S. (1992) D- and L-amino acid oxidases. in "*Chemistry and biochemistry of flavoenzymes*" (Mueller, F., Ed.), Vol. III, pp. 69-94, CRC Press, Boca Raton.

[3] Mattevi, A., Vanoni, M.A., Curti, B. (1997) Structure of D-amino acid oxidase: new insights from an old enzyme. *Curr Op Str Biol, 7*, 804-810.

[4] Palfey, B.A., Massey, V. (1998) Flavin-dependent enzymes. in "*Comprehensive biological catalysis*", Ch. 29, pp. 82-154.

[5] Pilone, M.S. (2000) D-Amino acid oxidase: new findings. *Cell Mol Life Sci*, *57*, 1732-1747.

[6] Mattevi, A., Vanoni, M.A., Todone, F., Rizzi, M, Teplyakov, A., Coda, A., Bolognesi, M., Curti B. (1996) Crystal structure of D-amino acid oxidase: a case of mirror-image convergent evolution with flavocytochrome b2. *Proc Natl Acad Sci USA, 93*, 7496-7501.

[7] Mizutani, H., Miyahara, I., Hirotsu, K., Nishima, Y., Shiga, K., Setoyama, C., and Mikura, R. (1996) Three-dimensional structure of porcine kidney D-amino acid oxidase at 3.0 Å resolution. *J Biochem, 120*, 14-17.

[8] Umhau, S., Pollegioni, L., Molla, G., Diederichs, K., Welte, W., Pilone, M.S., Ghisa, S. (2000) The x-ray structure of D-amino acid oxidase at bery high resolution identifies the chemical mechanism of flavin-dependent substrate dehydrogenation. *Proc Natl Acad Sci USA, 97*, 12463-12468.

[9] Kawazoe, T., Iwana, A., Ono, K., Park, H. K., Yorita, K., Tomita, Y, Tsuge, H., Fukui, K. (2005) Purification and crystal structure of human D-amino acid oxidase. in "*Flavins and Flavoproteins 2006*" (Nishino, T., Miura, R.,, Tanokura, M., Fukui, K., Eds.) pp. 33-36, ARchiTect Inc., Tokyo.

[10] Fitzpatrick, P.F. (2004) Carbanion versus hydride transfer mechanisms in flavoprotein-catalysed dehydrogenations. *Bioorg Chem, 32*, 125-139.

[11] Tilocca, A., Gamba, A., Vanoni, M.A., Fois, E. (2002) First-principles molecular dynamics investigation of the D-amino acid oxidative half reaction catalyzed by the flavoenzyme D-amino acid oxidase. *Biochemistry, 41*, 14111-14121.

[12] Snyder, S.H., Kim, P.M. (2000) D-Amino acids as putative neurotransmitters: focus on D-serine. *Neurochem Res*, 25, 553-560.

[13] Chumakov, I., Blumenfeld, M., Guerassimenko, O., Cavarec, L., Palicio, M., et al. (2002) Genetic and physiological data implicating the new human gene G72 and the gene for D-amino acid oxidase in schizophrenia. *Proc Natl Acad Sci USA, 99*, 13675-13680.

[14] Konno, R., Ikeda, M., Yamaguchi, K., Ueda, Y., Niwa, A. (2000) Nephrotoxicity of D-propargylglycine in mice. *Arch Toxicol, 74*, 473-479.

[15] Williams, R.E., Lock, E.A. (2005) Sodium benzoate attenuates D-serine induced nephrotoxicity in the rat. *Toxicology, 207*, 35-48.

[16] Tishkov, V.I., Khoronenkova, S.V. (2005) D-Amino acid oxidae: structure, catalytic mechanism, and practical application. *Biochemistry (Moscow), 70*, 40-54.

[17] Moses, J., Siddiqui, A., Silverman, P.B. (1996) Sodium benzoate differentially blocks circling induced by D- and L-DOPA in the hemi-parksonian rat. *Neurosci Lett, 218*, 145-148.

[18] Brumby, P.E., Massey, V. (1968) D-Amino acid oxidase from pig kidney in "*Biochemical preparations*" (Lands, W.E.M., Ed.), Vol 12, pp. 31-41, Wiley, New York.

[19] Curti, B., Ronchi, S., Branzoli, U., Ferri, G., Williams, C.H. (1973) Improved purification, amino acid analysis and molecular weight of homogeneous D-amino acid oxidase from pig kidney. *Biochim Biophys Acta, 327*, 266-273.

[20] Vanoni, M.A., Pilone Simonetta, M., Curti, B. Negri, A. and Ronchi, S. (1987) Phenylglyoxal modification of arginines in mammalian D-amino acid oxidase. *Eur J Biochem, 167*, 261-267.

[21] Massey, V., Curti, B. (1966) A new method of preparation of D-amino acid oxidase apoprotein and a conformational change after its combination with flavin adenine dinucleotide. *J Biol Chem*, 241, 3417-3423.

[22] Vanoni, M.A., Cosma, A., Mazzeo, D., Mattevi, A., Todone, F., Curti B. (1997) Limited proteolysis and X-ray crystallography reveal the origin of substrate specificity and of the rate limiting product release during oxidation of D-amino acids catalysed by mammalian D-amino acid oxidase. *Biochemistry, 36*, 5624-5632.

[23] O' Farrell, P.H. (1975) High resolution two-dimensional electrophoresis of proteins. *J Biol Chem, 250*, 4007-4021

[24] Fonda, M.L., Anderson, B.M. (1969) D-Amino acid oxidase. I. Spectrophotometric studies. *J Biol Chem, 244*, 666-674

[25] Macheroux, P., Mulrooney, S.B., Williams, C.H., Massey, V. (1992) Direct expression of active spinach glycolate oxidase in *Escherichia coli*. *Biochim Biophys Acta, 1132*, 11-16.

[26] Negri, A., Massey, V., Williams, C.H. (1987) D-Aspartate oxidase from beef kidney. Purification and properties. *J Biol Chem, 262*, 10026-10034.

[27] Dixon, M., Kleppe, K. (1965) D-amino acid oxidase. II. Specificity, competitive inhibition and reaction sequence. *Biochim Biophys Acta, 96*, 368-382.

[28] Denu, J.M., Fitzpatrick, P.F. (1992) pH and kinetic isotope effects on the reductive half-reaction of D-amino acid oxidase. Biochemistry, 31, 8207-8215.

[29] Torri Tarelli, G., Vanoni, M. A., Negri, A., Curti B. (1990) Characterization of a fully active N-terminal 37-kDa polypeptide obtained by limited tryptic cleavage of pig kidney D-amino acid oxidase. *J Biol Chem, 265*, 21242 - 21246.

[30] Fitzpatrick, P.F., Denu, J.M. and Emanuele, J.J. (1996) Evidence for rate-limiting conformational changes during the D-amino acid oxidase reaction. In "*Flavins and Flavoproteins 1996*" (Stevenson, K., Williams, C.H. & Massey, V., Eds.), pp. 195-198, Calgary University Press.

[31] Biondi, P.A., Guidotti, L., Negri, A., Secchi, C. (1991) High-performance liquid chromatographic determination of D-amino acid oxidase activity. *J Chromatogr, 566*, 377-382.

[32] van den Munckhof, R.J. (1996) In situ heterogeneity of peroxisomal oxidase activities: an update. *Histochem J, 28*, 401-429.

In: D-Amino Acids:
A New Frontier in Amino Acid and Protein Research, pp.477-480
ISBN 1-60021-075-9

Chapter 6.7

HISTOCHEMICAL DETECTION OF D-AMINO-ACID OXIDASE ACTIVITY IN RAT BRAIN

Hiroyuki Tanaka**, *Atsushi Yamamoto, Tetsuo Ishida and Kihachiro Horiike
Division of Molecular Physiological Chemistry, Department of Biochemistry and Molecular Biology, Shiga University of Medical Science, Japan.

Keywords: D-amino-acid oxidase, D-serine, Bergmann glial cells, astrocyte, diaminobenzidine.

1. INTRODUCTION

D-Amino-acid oxidase (DAO, a flavoenzyme, EC1.4.3.3) catalyzes the oxidation of D-amino acids to 2-imino acids. In the case of noncyclic amino acids except cyclic amino acids such as D-proline, the 2-imino acid is non-enzymatically hydrolyzed to the corresponding 2-oxo acid and ammonia [1]:

$$R - CH(NH_3^+) - COO^- + O_2 + H_2O \rightarrow R - CO - COO^- + NH_4^+ + H_2O_2$$

DAO is expressed in various tissues including the kidney, liver and brain of a large number of animal species [2,3].

We are particularly interested in the physiological role of DAO in the central nervous system, because DAO may be involved in the pathogenesis of schizophrenia through the metabolism of D-serine, an endogenous agonist of the NMDA receptor [4]. To elucidate the role of DAO, using H_2O_2 produced from DAO reaction we developed a sensitive procedure coupled with peroxidase for histochemical detection of DAO activity [5]. This method is based on the intensification and color modification of diaminobenzidine-based oxidized products by nickel ion. Using this method, the histological localization of DAO and the identification of the type of cells containing DAO have been successfully carried out [2,5,6,7]. Here we show histochemical detection of DAO activity in rat brain [8].

* Correspondence concerning this article should be addressed to Hiroyuki Tanaka, Division of Molecular Physiological Chemistry, Department of Biochemistry and Molecular Biology, Shiga University of Medical Science, Japan. E-mail: tanakah@belle.shiga-med.ac.jp; FAX: +81-77-548-2157.

2. Materials

1. Adult male rats
2. Sodium pentobarbital
3. Peristaltic pump
4. Fixative
   2% paraformaldehyde plus 2% glutaraldehyde in 0.1 M sodium phosphate (pH 7.4)
5. Vibratome (DOSAKA EM, Kyoto, Japan)
6. Incubation medium
   0.1% horseradish peroxidase (Grade III, Toyobo, Osaka, Japan), 0.005% 3,3'-diaminobenzidine tetrahydrochloride (Grade II, Sigma, St. Louis), 0.065% sodium azide, 0.02% nickel ammonium sulfate and D-amino acid (20 mM D-proline, 20 mM D-alanine or 50 mM D-serine) as substrate in 0.05 M Tris-HCl, pH 7.6.
7. Gelatin-coated glass slides
8. Graded ethanol series (70%-100%) and xylene
9. Entellan (Merck, Dermstad, Germany)
10. Light microscope

3. Methods

1. Adult male rats, weighing 200-300 g, were anesthetized with an intraperitoneal injection (*ca.* 50 mg/kg) of sodium pentobarbital.
2. The whole bodies of rats were perfused through the left ventricle after incision of the right auricle, initially with 150 mL of 0.9% NaCl in 0.01 M sodium phosphate (pH 7.4) at a flow rate of 50 mL/min with a peristaltic pump to wash out the blood, and subsequently with 100 mL of the ice-cold fixative at a flow rate of 50 mL/min and then 400 mL of the same fixative at a rate of 30 mL/min.
3. After removal of the fixed tissues, the areas of interest of the removed tissues were dissected out into small blocks.
4. The tissue blocks of tissues were cut into 20-μm or 50-μm thin sections with a vibratome.
5. The tissue sections were floated in the incubation medium with agitation at 4°C or 20°C for 12-16 hr.
6. After incubation, the samples were rinsed with 0.1 M sodium phosphate buffer containing 0.9% NaCl, pH 7.4, to terminate the reaction.
7. The sections were mounted on gelatin-coated glass slides, dried with air, dehydrated through a graded ethanol series and finally cleared with xylene.
8. The sections were coverslipped with Entellan for observation under a light microscope.

4. Results

Figure 1a-c show the sagittal sections of the rat whole brain stained with D-serine, D-alanine and D-proline, respectively. The locations of the brain stained with the three substrates were identical, but the intensity in the staining decreased in the following order: D-proline > D-alanine > D-serine. D-Proline was the best substrate for the detection of DAO activity by the present method. This difference in staining intensity was consistent with the substrate specificity of DAO for these substrates. In the absence of D-amino acids (Fig. 1d) or peroxidase in the incubation medium, no staining was developed. In the presence of sodium benzoate (10 mM), a competitive inhibitor, in the incubation medium, histochemical staining was significantly suppressed (Fig. 1e).

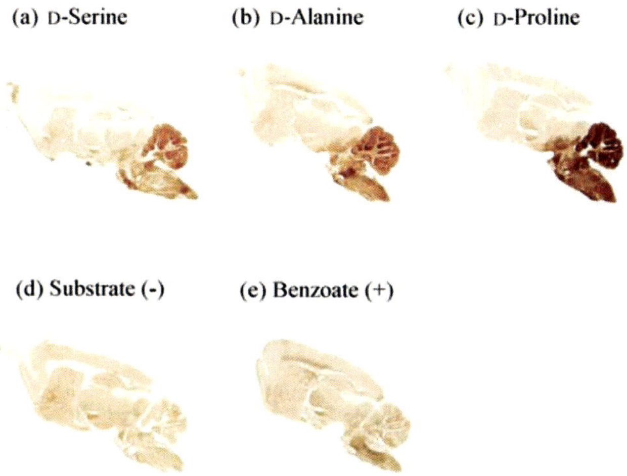

Figure 1. Histochemical staining of D-amino-acid oxidase activity in the sagittal sections (20 μm) of rat brain. The substrates used were (a) 50 mM D-serine, (b) 20 mM D-alanine and (c) 20 mM D-proline. Control experiments were carried out in the absence of D-proline (d) and in the presence of 10 mM sodium benzoate (e) in the reaction medium.

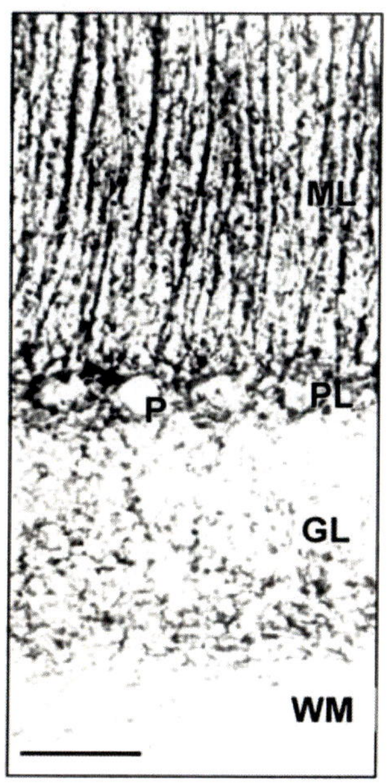

Figure 2. Histochemical staining of D-amino-acid oxidase activity in rat cerebellum (50-μm section). In this case, 20 mM D-proline and 0.6% nickel ammonium sulfate were used. In the molecular layer (ML), deposit-containing ascending processes of Bergmann glial cells cross the layer parallel to each other and perpendicularly to the pial surface. In the Perkinje cell layer (PL), deposit-containing processes of Bergmann glial cells envelop Purkinje cell bodies (P), but Purkinje cells are unstained. In the granular layer (GL) and the white matter (WM), astrocytes were stained. The scale bar denotes 50 μm.

The activity of DAO was completely absent in the telencephalon, scarce in the diencephalon, and was almost completely confined to the brain stem (the midbrain, pons and medulla oblongata), cerebellum and spinal cord. In the brain stem, the oxidase activity was mainly distributed in the tegmentum, characteristically in the reticular formation. In the cerebellum, the enzyme activity was observed in all gross regions (Fig. 2). Both grey and white matter showed DAO activity.

On light microscopic observation, the oxidase activity was present only in astrocytes including Bergmann glial cells (Fig. 2), and not in neuronal components, endothelial cells, oligodendrocytes or ependymal cells [2,5,6,7].

5. Notes

1. In the present method, the color-formation of diaminobenzidine was intensified by using nickel ions. The optimum concentration of nickel ion was determined to be 0.02%-0.6% by judging both the positive effect on the color-formation and the negative effect on the DAO activity. When D-serine was used as substrate, staining intensity with 0.6% nickel ammonium sulfate was weak.
2. This method is sensitive enough to allow histochemical identification of the type of DAO-positive cells in brain sections.
3. Other peroxisomal enzymes such as D-aspartate oxidase [9] and urate oxidase [5] can be histochemically detected by the present method.
4. The present method is applicable to electron microscopic investigation [7].

References

[1] Krebs, H. A. (1935). Metabolism of amino-acids. III. Deamination of amino-acid. *Biochem J, 29*, 1620-1644.

[2] Horiike, K., Ishida, T., Tanaka, H. & Arai, R. (2001). Distribution of D-amino-acid oxidase and D-serine in vertebrate brains. *J Mol Catal 12B*, 37-41.

[3] Nagata, Y., Horiike, K. & Maeda, T. (1994). Distribution of free D-serine in vertebrate brains. *Brain Res, 634*, 291-295.

[4] Chumakov, I., Blumenfeld, M., Guerassimenko, O. et al. (2002). Genetic and physiological data implicating the new human gene G72 and the gene for D-amino acid oxidase in schizophrenia. *Proc Natl Acad Sci USA, 99*, 13675-13680.

[5] Horiike, K., Arai, R., Tojo, H., Yamano, T., Nozaki, M. & Maeda, T. (1985). Histochemical staining of cells containing flavoenzyme D-amino acid oxidase based on its enzymatic activity: Application of a coupled peroxidation method. *Acta Histochem Cytochem, 18*, 539-550.

[6] Horiike, K., Tojo, H., Arai, R., Nozaki, M. & Maeda, T. (1994). D-Amino-acid oxidase is confined to the lower brain stem and cerebellum in rat brain: Regional differentiation of astrocytes. *Brain Res, 652*, 297-303.

[7] Horiike, K., Tojo, H., Arai, R., Yamano, T., Nozaki, M. & Maeda, T. (1987). Localization of D-amino acid oxidase in Bergmann glial cells and astrocytes of rat cerebellum. *Brain Res Bull, 19*, 587-596.

[8] Tanaka, H., Yamamoto, A., Ishida, T. & Horiike K. (2003) Distribution of D-amino-acid oxidase activity and D-serine metabolism in rat brain. *Seikagaku, 75*, 717-717.

[9] Schell, M. J., Cooper, O. B. & Snyder, S. H. (1997). D-Aspartate localizations imply neuronal and neuroendocrine roles. *Proc Natl Acad Sci USA, 94*, 2013-2018.

In: D-Amino Acids: ISBN 1-60021-075-9
A New Frontier in Amino Acid and Protein Research, pp.481-490

Chapter 6.8

EVALUATION OF CHIRAL INVERSION OF D-AMINO ACID BY STABLE ISOTOPE METHODOLOGY

Hiroshi Hasegawa**, *Yoshihiko Shinohara and Takao Hashimoto
Department of Pathophysiology, Tokyo University of Pharmacy and Life Science, Japan.

Keywords: chiral inversion, stable isotope, GC-MS

1. INTRODUCTION

Utilization of D-amino acids as nutrients was repeatedly confirmed in different species of animals, and it was found that several D-amino acids were utilized for growth [1]. The utilization of exogenous D-amino acid depends on whether it can be efficiently transformed into the L-enantiomer. However, little information is available on the extent to which D-amino acid is converted into the L-enantiomer because the L-amino acid formed is indistinguishable from endogenous L-amino acid.

The use of gas chromatography-mass spectrometry (GC-MS) and stable-isotope labeled compounds as tracers has enjoyed broad applications in metabolic studies. One of the major advantages of this technique is that an endogenous compound and its exogenously administered labeled analog are separately measurable by using GC-MS.

The objective of this chapter is to evaluate the kinetic behavior of amino acid enantiomers and the extent of conversion of D-amino acid into the corresponding α-keto acid or L-enantiomer in vivo by a stable isotope methodology. To achieve the objective, the following steps are required: (a) preparation of stable isotope labeled amino acid enantiomers and α-keto acid for tracers and analytical internal standards; (b) development of analytical method for chiral separation of amino acid enantiomers; (c) estimation of the extent of conversion from plasma concentration data after administration of tracers.

The present chapter describes a procedure for determining the extent that D-leucine is converted into the corresponding α-keto acid, α-ketoisocaproic acid (KIC), and L-leucine in rats.

* Correspondence concerning this article should be addressed to Hiroshi Hasegawa, E-mail: hasegawa@ps.toyaku.ac.jp, FAX: +81-42-676-5686.

2. MATERIALS

1. Amino acids and α-keto acid. D-Leucine (Peptide Institute, Osaka, Japan), L-leucine (Wako Pure Chemicals, Osaka, Japan), and sodium 2-oxo-4-methylpentanoate (KIC: Aldrich, St. Louis, MO, U.S.A.).
2. Commercially available stable-isotope labeled compounds. DL-[2,3,3-$^2H_3$]Leucine (DL-[$^2H_3$]leucine: Isotec, Miamisburg, OH, U.S.A.), DL-[4,5,5,5,6,6,6-$^2H_7$]leucine (DL-[$^2H_7$]leucine: Isotec), and sodium [5,5,5-$^2H_3$]2-oxo-4-methylpentanoate ([$^2H_3$]KIC: Isotec).
3. Derivatizing reagents. (*S*)-(+)-α-Methoxy-α-trifluoromethylphenylacetyl chloride (MTPA-Cl: Tokyo Kasei, Tokyo, Japan), *N*-phenyl-1,2-diaminobenzene (Aldrich), 10% HCl in methanol (Tokyo Kasei).
4. Enzymes. Acylase (from porcine kidney, Tokyo Kasei), L-amino acid oxidase (crude dried venom from Crotalus artox, Sigma, St. Lous, MO, U.S.A.), catalase (from bovine liver, Sigma).
5. Strong cation exchange solid-phase extraction column BondElut SCX (H^+ form, size 1 ml/100mg: Varian, Harbor City, CA, U.S.A.). Before use, the column was pre-washed and activated with methanol, a mixture of methanol and 0.1 M HCl (1 : 1, v/v) and 0.1 M HCl.
6. Solvents. Chloroform stabilized with amylene was used (Note 1). All other chemicals and solvents were of analytical reagent grade and used without further purification.
7. Animals. Male Sprague-Dawley rats (7 weeks old) were used. They were housed in stainless-steel cages in an air-conditioned room maintained at 23 ± 1 °C and 55 ± 5% humidity with a 12-h dark/light cycle. All rats were acclimated for at least 7 days, during which time they were allowed free access to water and food.

3. METHODS

3.1. Resolution of DL-[$^2H_7$]leucine

3.1.1. Preparation of N-acetyl-DL-[$^2H_7$]leucine

1. DL-[$^2H_7$]Leucine (300 mg) was dissolved in acetic acid (10 mL) and then the solution was refluxed for 5 min.
2. Acetic anhydride (1 mL) was added and the solution was allowed to reflux for 5 min.
3. The solution was evaporated under reduced pressure.
4. The residual syrup was treated with water followed by evaporation to yield *N*-acetyl-DL-[$^2H_7$]leucine (390 mg) as a colorless solid.

3.1.2. Hydrolysis of N-acetyl-DL-[$^2H_7$]leucine by acylase

1. *N*-Acetyl-DL-[$^2H_7$]leucine (390 mg) was suspended in water (20 mL).
2. To the solution was added 1M NH_3 until *N*-acetyl-DL-[$^2H_7$]leucine was dissolved and a pH of the solution reached 7.2-7.8.
3. Acylase was added in two portions (2 mg each) at 24 h intervals, being incubated at 37°C for 48 h.
4. The reaction was terminated by acidification with 1M HCl.
5. After the protein was filtered off with the aid of charcoal, the filtrate was concentrated under reduced pressure.
6. The concentrate was applied to cation-exchange Dowex 50W X8 (H^+ form) column.
7. Because *N*-acetyl-D-[$^2H_7$]leucine was not retained to the column, the *N*-acetyl derivative was eluted in the waste and water washing. After evaporation of the solvent under reduced pressure,

the residual syrup was diluted with acetone followed by evaporation to yield *N*-acetyl-D-[$^2H_7$]leucine (190 mg) as a colorless solid.

8. L-[$^2H_7$]Leucine was eluted with 1M NH_3. After evaporation of the solvent under reduced pressure, the residue was dissolved in a small portion of water and then the pH was adjusted to 5 with 1M acetic acid. The solvent was evaporated under reduced pressure and then the residue was recrystallized with water-ethanol to yield L-[$^2H_7$]leucine (50 mg) as a colorless solid.

3.1.3. Acid Hydrolysis of N-acetyl-D-[$^2H_7$]leucine

1. A solution of *N*-acetyl-D-[$^2H_7$]leucine (190 mg) in 2M HCl (10 mL) was refluxed for 3 h.
2. After evaporation of the solvent under reduced pressure, the residue was dissolved in a small portion of water and then the pH was adjusted to 5 with 1M LiOH.
3. The solvent was evaporated under reduced pressure and the residue was recrystallized with water-ethanol to yield D-[$^2H_7$]leucine (50 mg) as a colorless solid.

3.2. Preparation of Sodium [4,5,5,5,6,6,6-$^2H_7$]2-oxo-4-methylpentanoate ([$^2H_7$]KIC)

1. DL-[$^2H_7$]Leucine (500 mg) was dissolved in water (100 mL) and the pH was adjusted to 7.4-7.8 with 1M NH_3.
2. Catalase (5 mg) and L-amino acid oxidase (50 mg) were added and then the solution was incubated at 37°C under a gentle bubbling of oxygen for 6 h.
3. The reaction was terminated by acidification with 1M HCl and the protein was filtered off with the aid of charcoal.
4. After concentration of the filtrate under reduced pressure, the concentrate was applied to Dowex 50W X8 (H^+ form) column.
5. Because [$^2H_7$]KIC was not retained to the column, the α-keto acid was eluted in the waste and water washing. After evaporation of the solvent under reduced pressure, the residual syrup was diluted in a small portion of water and then the pH was adjusted to 4 with 1M NaOH. The solvent was evaporated under reduced pressure and the residue was recrystallized with water-ethanol to yield sodium [$^2H_7$]KIC (190 mg) as a colorless solid.
6. D-[$^2H_7$]Leucine was eluted with 1M NH_3. After evaporation of the effluent under reduced pressure, the residue was dissolved in a small portion of water and then the pH was adjusted to 5 with 1M acetic acid. The solvent was evaporated under reduced pressure and the residue was recrystallized with water-ethanol to yield D-[$^2H_7$]leucine (200 mg) as a colorless solid.

3.3. Dose Experiments

1. Rats were anesthetized with sodium pentobarbital (50 mg/kg body weight, i.p.) after an overnight fast.
2. D-[$^2H_7$]Leucine, [$^2H_7$]KIC or L-[$^2H_7$]leucine (35 μmol/kg body weight, each) dissolved in saline (0.5 mL of the dosing solution/kg body weight) was administered into the femoral vein.
3. Blood samples were collected from the jugular vein at several time periods after administration and the samples were centrifuged to separate plasma at 3000 rpm for 10 min.
4. The plasma samples were stored at -20°C until analysis.

3.4. Sample Preparation for GC-MS

1. To a 50 μL of plasma were added DL-[$^2H_3$]leucine (1 nmol) and [$^2H_3$]KIC (0.1 nmol) as analytical internal standards (Note 2).
2. The plasma sample was deproteinized and extracted with ethanol (0.5 mL x 2).
3. The supernatant was evaporated at 40°C under a stream of nitrogen.
4. The residue was dissolved in 0.5 mL of 40 mM HCl and then applied to a BondElut SCX cartridge.
5. KIC was eluted with water.
   5-1. To the combined waste and water effluent (KIC fraction) were added 1 mL of 6M HCl and 3 mg of *N*-phenyl-1,2-diaminobenzene.
   5-2. The solution was heated at 100°C for 1 h.
   5-3. After extraction with 1 mL of a mixture of diethyl ether and hexane (1:1, v/v), the solvent was evaporated to dryness at room temperature under a stream of nitrogen.
   5-4. The residue was dissolved in 20 μL of diethyl ether and then 1-2 μL of the solution was subject to GC-MS.
6. After washing the BondElut SCX cartridge with 1 mL of methanol, leucine enantiomers were eluted with 0.5 mL of 10% HCl in methanol (Note 3).
   6-1. The effluent (leucine fraction) was heated at 60°C for 1 h.
   6-2. After removal the solvent under a stream of nitrogen, the residue was reconstituted in 100 μL of 2% MTPA-Cl in chloroform and the solution was left at room temperature for 1 h.
   6-3. After washing the solution with water, the solvent was evaporated at room temperature under a stream of nitrogen.
   6-4. The residue was dissolved in 20 μL of ethyl acetate and then 1-2 μL of the solution was subject to GC-MS.

3.5. GC-MS-SIM

3.5.1. Apparatus

GC-MS with selected ion monitoring (SIM) were conducted on a Shimadzu (Kyoto, Japan) QP1000EX gas chromatograph-mass spectrometer equipped with a data processing system. A methylsilicone bonded-phase fused-silica capillary column SPB-1 (15 m x 0.25 mm I.D.) with a 0.25 μm thin film thickness (Supelco, Bellefonte, PA, U.S.A.) was connected directly into the ion source. Helium was used as the carrier gas. A split-splitless injection system Shimadzu SPL-G9 operating in the splitless mode was used. The purge activation time was 2 min after injection.

3.5.2. Determination of Labeled and Non-Labeled Leucine Enantiomers

The conditions of the GC-MS-SIM were as follows [2]. The initial column temperature was set at 120°C. After the sample injection, temperature was maintained for 2 min, increased at 15°C/min to 190°C and held at 190°C for 3 min. The mass spectrometer was operated in chemical ionization mode with isobutane as the reactant gas. The ionization voltage and ionization current were 200 eV and 150 μA, respectively. SIM was performed on the molecular-related ions of the MTPA methyl ester (-OMe) derivatives of leucine species ($^2H_0$, m/z 362; $^2H_3$, m/z 365; $^2H_7$, m/z 369) and the respective peak area values were determined. The peak area ratios (m/z 362/365 for leucine) and (m/z 369/365 for [$^2H_7$]leucine) were calculated. The plasma concentrations were calculated by comparing the peak area ratios obtained from the unknown samples with those obtained from the standard mixtures.

3.5.3. Determination of Labeled and Non-Labeled KIC

The conditions of the GC-MS-SIM were as follows [3]. The initial column temperature was set at 120°C. After the sample injection, temperature was maintained for 2 min, increased at 30°C/min to 250°C and held at 250°C for 2 min. The mass spectrometer was operated in electron ionization mode at energy of 70 eV (Note 4). SIM was performed on the molecular ions of the *N*-phenyl-quinoxalinone derivative of KIC species ($^2H_0$, m/z 278; $^2H_3$, m/z 281; $^2H_7$, m/z 285) and the respective peak area values were determined. The peak area ratios (m/z 278/281 for KIC) and (m/z 285/281 for [$^2H_7$]KIC) were calculated. The plasma concentrations were calculated by comparing the peak area ratios obtained from the unknown samples with those obtained from the standard mixtures.

3.6. Data Analysis

3.6.1. Pharmacokinetic Parameters

Pharmacokinetic parameters were calculated by model-independent analysis using a macro-program MOMENT(EXCEL) [4] running on Microsoft Excel. The elimination half-life ($t_{1/2}$) was obtained by least squares regression of the terminal log linear portion of the plasma concentration-time curve. The area under the plasma concentration-time curve (AUC) was calculated by the trapezoidal method, and was extrapolated to infinity using the last detectable plasma concentration and the terminal elimination rate constant. Total clearance (CL) was calculated by CL=Dose/AUC.

3.6.2. Extent of Conversion

The extent of conversion of compound A into metabolite B ($F_{A\to B}$) is calculated as:

$$F_{A\to B} = \frac{\text{amount of metabolite B formed}}{\text{amount of compound A administered}} = \frac{CL_{AB} \cdot AUC_A}{Dose_A} \tag{1}$$

where CL_{AB} is the metabolic clearance associated with conversion of compound A to metabolite B and AUC_A is the AUC of compound A. According to mass balance, the rate of change of metabolite B in the body is represented as follows:

$$\text{Rate of change of metabolite B} = CL_{AB} \cdot C_A - CL_B \cdot C_{B(\leftarrow A)} \tag{2}$$

where CL_B is the total clearance of metabolite B and C_A and $C_{B(\leftarrow A)}$ are the respective plasma concentrations of compound A and metabolite B. Because no metabolite B is present in the body at zero or at infinite time, integrating equation (2) between these time limits yields

$$CL_{AB} = CL_B \cdot AUC_{B(\leftarrow A)} / AUC_A \tag{3}$$

where $AUC_{B(\leftarrow A)}$ is the AUC of metabolite B after administration of compound A. When the compound B is administered, the total clearance CL_B is defined as the following equation:

$$CL_B = Dose_B / AUC_B \tag{4}$$

Substitution for CL_{AB}, according to equation (3) and (4), into equation (1) yields

$$F_{A \to B} = \frac{AUC_{B(\leftarrow A)} / Dose_A}{AUC_B / Dose_B} \tag{5}$$

This method does not depend on where metabolite B is formed and whether or not compound A is eliminated by other routes, although intravenous administrations of compound A and B are necessary.

4. Results

4.1. Synthesis of D-[$^2H_7$]leucine, L-[$^2H_7$]leucine and [$^2H_7$]KIC

D-[$^2H_7$]Leucine, L-[$^2H_7$]leucine and [$^2H_7$]KIC (Note 5) were prepared from commercially available DL-[$^2H_7$]leucine (Fig. 1) [5]. Stereoselective hydrolysis of the *N*-acetyl derivative of DL-[$^2H_7$]leucine with porcine kidney acylase gave L-[$^2H_7$]leucine. The remaining *N*-acetyl-D-[$^2H_7$]leucine was hydrolyzed with diluted HCl to give D-[$^2H_7$]leucine. On the other hand, stereoselective de-amination of DL-[$^2H_7$]leucine with venom L-amino acid oxidase gave [$^2H_7$]KIC and D-[$^2H_7$]leucine. These methods will be applicable to prepare other labeled amino acid enantiomers and α-keto acids.

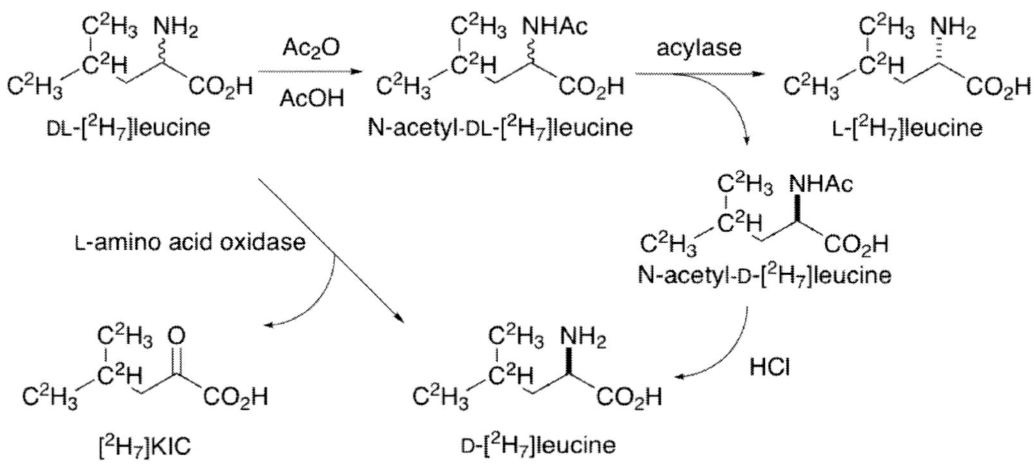

Figure 1. Synthetic scheme for D-[$^2H_7$]leucine, L-[$^2H_7$]leucine and [$^2H_7$]KIC.

4.2. GC-MS

4.2.1. Simultaneous Determination of the D- and L-enantiomers of [$^2H_7$]leucine and leucine

Leucine enantiomers are converted into methyl ester by HCl in methanol and subsequent *N*-acylation with MTPA-Cl to give the diastereomeric MTPA-OMe derivative with high yield (>80%) (Note 6). Because the respective MTPA-OMe derivatives of leucine, [$^2H_3$]leucine and [$^2H_7$]leucine produce strong molecular-related ions $[M+H]^+$ at m/z 362, 365 and 369 in chemical ionization mass spectrometry, the molecular-related ions have been used as the selected-ion monitoring ions.

Figure 2A shows the representative SIM profiles for MTPA-OMe derivatives of labeled and non-labeled leucine enantiomers in plasma. There was no interference from endogenous compounds in the vicinity of the peaks of analytes in the mass fragmentgrams. MTPA-OMe derivatives of D-leucine and L-leucine were observed at retention time 7.07 min and 7.20 min, respectively. In tracing at m/z 362, a peak for MTPA-OMe derivative of L-isoleucine was also observed at 7.31 min because of their identical mass numbers (Note 7). No detectable amount of endogenous D-leucine has been found in rat plasma.

4.2.2. Determination of [$^2H_7$]KIC and KIC

When KIC, [$^2H_3$]KIC and [$^2H_7$]KIC were converted into the *N*-phenyl-qunoxalinone derivatives by condensation with *N*-phenyl-1,2-diaminobenzene, mass spectra for the respective derivatives at electron ionization mode showed the molecular ions at m/z 278, 281 and 285 in relative high intensities. Figure 2B shows the representative SIM profiles of *N*-phenyl-quinoxalinone derivatives of labeled and non-labeled KIC in plasma. Although the *N*-phenyl-quinoxalinone derivative of α-keto-β-methylvaleric acid (KMV), the α-keto acid of isoleucine, has the identical mass number with that of KIC, their derivatives were well separated.

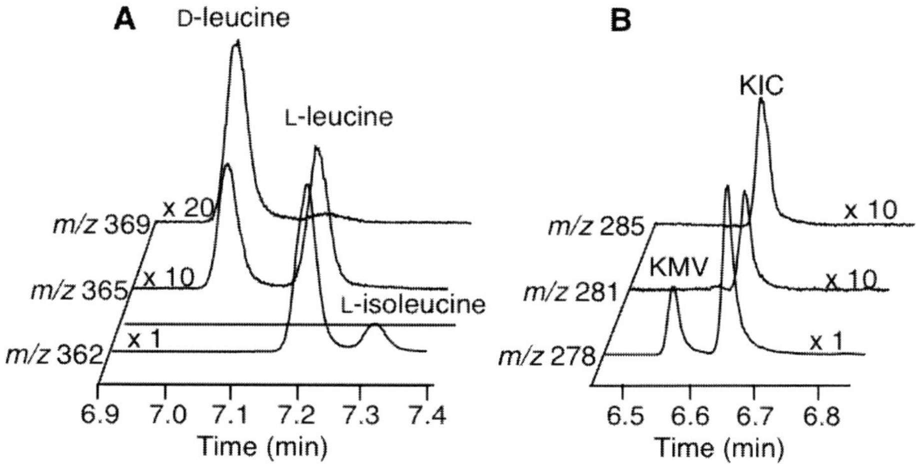

Figure 2. Representative SIM profiles for (A) MTPA-OMe derivatives of labeled and non-labeled leucine enantiomers in plasma and (B) *N*-phenyl quinoxalinone derivatives of labeled and non-labeled KIC in plasma. (A) Masses at *m/z* 362, 365 and 369 were monitored for MTPA-OMe derivatives of leucine, [$^2H_3$]leucine and [$^2H_7$]leucine, respectively. (B) Masses at *m/z* 278, 281 and 285 were monitored for *N*-phenyl-quinoxalinone derivatives of KIC, [$^2H_3$]KIC and [$^2H_7$]KIC, respectively.

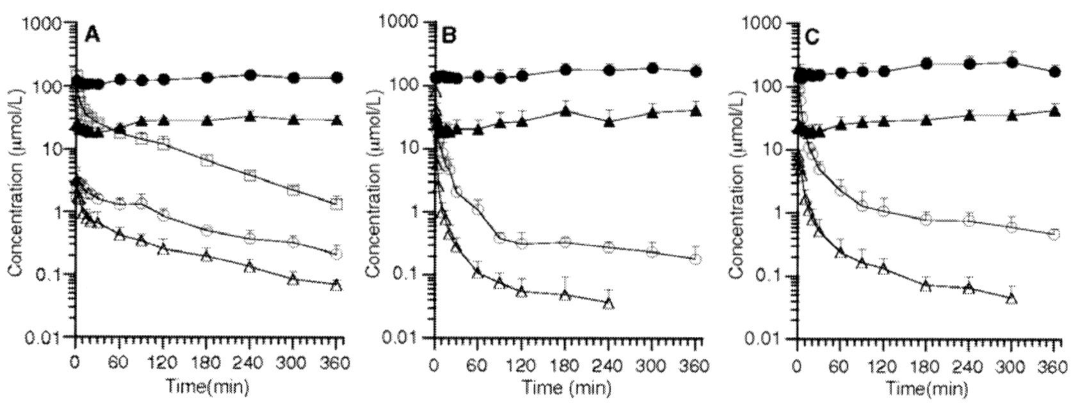

Figure 3. Plasma concentration-time profiles for D-[$^2H_7$]leucine (□), [$^2H_7$]KIC (△), L-[$^2H_7$]leucine (○), endogenous L-leucine (●) and endogenous KIC (▲) in rats after i.v. administration of (A) D-[$^2H_7$]leucine, (B) [$^2H_7$]KIC and (C) L-[$^2H_7$]leucine. Reproduced, with permission, from Hasegawa et al. [6].

4.3. Estimation of the Extent of Conversion

Figure 3 shows the plasma concentration-time profiles for D-[$^2H_7$]leucine, L-[$^2H_7$]leucine or [$^2H_7$]KIC together with endogenous L-leucine and KIC after intravenous administrations of D-[$^2H_7$]leucine, L-[$^2H_7$]leucine or [$^2H_7$]KIC to rats [6]. No detectable amount of endogenous D-leucine has been found. Following D-[$^2H_7$]leucine administration (Fig. 3A), the substrate disappeared biexponentially from the plasma. The appearances of [$^2H_7$]KIC and L-[$^2H_7$]leucine in the plasma were very rapid. After administration of [$^2H_7$]KIC (Fig. 3B), plasma concentration of [$^2H_7$]KIC was observed to decline in apparently triexponential manner. L-[$^2H_7$]Leucine was quickly appeared in the plasma whereas no detectable amounts of D-[$^2H_7$]leucine were produced. After administration of L-[$^2H_7$]leucine (Fig. 3C), plasma concentration of L-[$^2H_7$]leucine was observed to decline in apparently triexponential manner and [$^2H_7$]KIC was quickly appeared. However, administration of L-[$^2H_7$]leucine produced no detectable amounts of D-[$^2H_7$]leucine.

Table 1. AUC values for D-[$^2H_7$]leucine, [$^2H_7$]KIC and L-[$^2H_7$]leucine after i.v. administration of D-[$^2H_7$]leucine, [$^2H_7$]KIC or L-[$^2H_7$]leucine to rats.

Administered	Measured		
	D-[$^2H_7$]leucine	[$^2H_7$]KIC	L-[$^2H_7$]leucine
D-[$^2H_7$]leucine	3998 ± 406	111 ± 31	314 ± 40
[$^2H_7$]KIC		168 ± 45	488 ± 48
L-[$^2H_7$]leucine		105 ± 32	1190 ± 288

Each value (min · µmol/L) is expressed as mean ± SD (n=6).
Reproduced, with permission, from Hasegawa et al. (6).

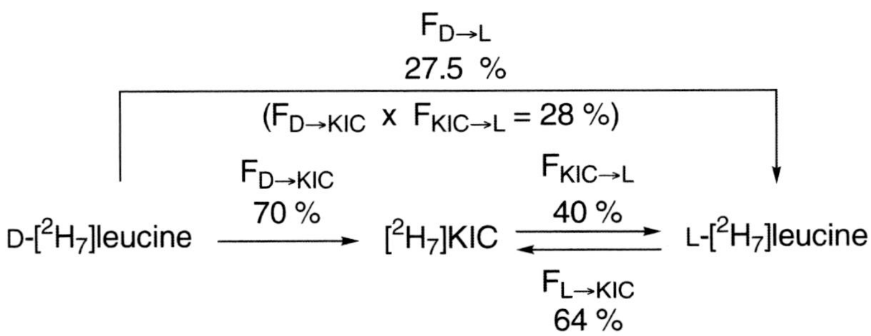

Figure 4. Fraction of conversion of D-[$^2H_7$]leucine into L-[$^2H_7$]leucine via [$^2H_7$]KIC. Reproduced, with permission, from Hasegawa et al. [6].

The AUC values for D-[$^2H_7$]leucine, [$^2H_7$]KIC and L-[$^2H_7$]leucine after administrations of D-[$^2H_7$]leucine, [$^2H_7$]KIC or L-[$^2H_7$]leucine were calculated. The results were shown in Table 1. The fraction of D-[$^2H_7$]leucine that converted to [$^2H_7$]KIC ($F_{D \to KIC}$) was estimated according to equation (5) using the mean AUC value of [$^2H_7$]KIC on the D-[$^2H_7$]leucine administration ($AUC_{KIC(\leftarrow D)}$) and that of [$^2H_7$]KIC on the [$^2H_7$]KIC administration (AUC_{KIC}) to yield 70 % (Fig. 4) (Note 8). Similarly, the fraction of conversion from [$^2H_7$]KIC to L-[$^2H_7$]leucine ($F_{KIC \to L}$) and that from L-[$^2H_7$]leucine to [$^2H_7$]KIC ($F_{L \to KIC}$) were estimated to be 40 % and 64 %, respectively. Since the fraction that D-[$^2H_7$]leucine is converted into L-[$^2H_7$]leucine ($F_{D \to L}$) was considered to the product of $F_{D \to KIC}$ and $F_{KIC \to L}$, the value was 28%. Regard of [$^2H_7$]KIC, on the other hand, $F_{D \to L}$ was also calculated directly using $AUC_{L(\leftarrow D)}$ and AUC_L to yield 27.5%.

5. NOTES

1. Analytical grade of chloroform contains methanol as a stabilizer. Since MTPA-Cl reacts with methanol, MTPA-Cl was diluted with chloroform containing amylene as a stabilizer.
2. Commercially available DL-[2,3,3-$^2H_3$]leucine and [5-$^2H_3$]KIC were used as analytical internal standards. The $^2H_3$-labeled internal standards mimic D-[$^2H_7$]leucine, L-[$^2H_7$]leucine, L-leucine, [$^2H_7$]KIC and KIC in plasma. Therefore, if the extraction and/or the derivatization are not 100% complete, the leucine and KIC species measurement will not be affected because the $^2H_3$-labeled internal standards will also undergo losses to the same extent as D-[$^2H_7$]leucine, L-[$^2H_7$]leucine, D-leucine, L-leucine, [$^2H_7$]KIC and KIC.
3. Elution of leucine species from BondElut SCX column was well carried out by 1M NH_3 and the recover yield was satisfactory. Replacement of water with methanol improved recovery of leucine (>90%) and the effluent was evaporated with ease. Moreover, replacement of NH_3 with HCl also gave satisfactory recovery yield (>90%). The advantage for using HCl in methanol was that methyl esterification was achieved by heating the effluent directly.
4. GC-MS analysis of *N*-phenyl-quinoxalinone derivative of KIC on CI mode with isobutane as a reactant gas also gave satisfactory results. In this case, the molecular-related ions at m/z 279, 282, 286 were used as SIM ions for *N*-phenyl-quinoxalinone derivative of KIC, [$^2H_3$]KIC and [$^2H_7$]KIC, respectively.
5. Successful application of stable isotope dilution methodology to the present procedure is dependent upon the availability of compounds labeled at predesigned positions that are chemically and biologically inert. We have chosen D-[$^2H_7$]leucine, L-[$^2H_7$]leucine and [$^2H_7$]KIC labeled at C-4, C-5 and C-6 positions for tracers because the deuterium labels were placed at sufficient enough distance from the amino group of leucine to avoid loss of the label under both de-amination and transamination process.
6. Because amino acids are not volatile and do not permit direct analysis on GC, they must be converted into suitable derivatives. Separation of DL-amino acids on GC is achieved by the two methods: (a) derivatization of achiral reagents followed by direct separation on a chiral column [7,8]; (b) derivatization of chiral reagents followed by diastereomeric separation on an achiral column [2,9]. The latter method requires that the chiral reagent has enough optical purity and no racemization of both amino acid and the chiral reagent occurs under derivatization. We have chosen MTPA-Cl [10] as the chiral reagent because the reagent is highly resistant to racemization due to the lack of α-hydrogen and is commercially available in its resolved form with high enantiomeric purity.
7. When (*R*)-(-)-MTPA-Cl was used as a chiral reagent, the elution order of the MTPA-OMe derivative of leucine enantiomers was reversed and the MTPA-OMe derivatives of D-leucine and L-isoleucine was not resolved on GC.
8. There are two approaches for determining the extent of conversion of compound A into metabolite B. Method (A) is based on estimation of the ratio between the rate constant for conversion of compound A into the metabolite B and the overall elimination rate constant of compound A. It requires a fitting of the observed set of plasma concentration data of compound A and the metabolite B to a kinetic compartment model with first-order rate constant. Method (B) is based on estimation of the ratio between AUC of metabolite B formed after administration of compound A and AUC of compound B after administration of compound B itself, as shown in equation (5). In this study, method (B) is used.

REFERENCES

[1] Berg, C.P. (1959). Utilization of D-amino acids. In Protein and Amino Acid Nutrition (Albanese, A.A. ed.), pp57-96. New York: Academic Press.

[2] Hasegawa, H., Matsukawa, T., Shinohara, Y. & Hashimoto, T. (1999). Simultaneous determination of the enantiomers of leucine and [$^2H_7$]leucine in plasma by capillary gas chromatography-mass spectrometry. *J Chromatogr B, 735*, 141-149.

[3] Matsukawa, T., Hasegawa, H., Shinohara, Y. & Hashimoto, T. (2001). Gas chromatographic-mass spectrometric determination of α-ketoisocaproic acid and [$^2H_7$]α-ketoisocaproic acid in plasma after derivatization with *N*-phenyl-1,2-phenylenediamine. *J Chromatogr B, 751*, 213-220.

[4] http://www.pharm.kyoto-u.ac.jp/byoyaku/Kinetics/download.html

[5] Hasegawa, H., Matsukawa, T., Shinohara, Y. & Hashimoto, T. (2000). Assessment of the metabolic chiral inversion of D-leucine in rat by gas chromatography-mass spectrometry combined with a stable isotope dilution analysis. *Drug Metab Dispos, 28,* 920-924.

[6] Hasegawa, H., Matsukawa, T., Shinohara, Y. & Hashimoto, T. (2002). Kinetics of sequential metabolism from D-leucine to L-leucine via α-ketoisocaproic acid in rat. *Drug Metab Dispos, 30,* 1436-1440.

[7] Liardon, R. Ledermann, S. & Ott, U. (1981). Determination of D-amino acids by deuterium labelling and selected ion monitoring. *J Chromatogr, 203*, 385-395.

[8] Brückner, H. & Schieber, A. (2001). Determination of amino acid enantiomers in human urine and blood serum by gas chromatography-mass spectrometry. *Biomed Chromatogr, 15*, 166-172.

[9] Bonner, W. A., Van Dort, M. A. & Flores, J. J. (1974). Quantitative gas chromatographic analysis of leucine enantiomers. A comparative study. *Anal Chem, 46*, 2104-2107.

[10] Dale, J. A. Dull, D. L. & Mosher, H. S. (1969). α-Methoxy-α-trifluoromethylphenylacetic acid, a versatile reagent for the determination of enantiomeric composition of alcohols and amines. *J Org Chem, 34*, 2543-2549.

In: D-Amino Acids: ISBN 1-60021-075-9
A New Frontier in Amino Acid and Protein Research, pp.491-501

Chapter 6.9

TECHNIQUES FOR THE ENGINEERING OF D-AMINO ACID OXIDASE SUBSTRATE SPECIFICITY

Gianluca Molla, Elena Rosini, Giorgia Letizia Marcone, Simona Lorenzi, Mirella S. Pilone and Loredano Pollegioni*
Department of Biotechnology and Molecular Sciences, University of Insubria, Italy.

Keywords: protein engineering, rational design, directed evolution, D-amino acid oxidase, substrate specificity

1. INTRODUCTION

D-Amino acid oxidase (DAAO, EC 1.4.3.3) is a strictly stereoselective flavoenzyme which catalyses the oxidative deamination of D-amino acids to produce the corresponding α-keto acids and ammonia.

Because of its peculiar characteristics, i.e. high turnover number and broad substrate specificity, the enzyme from the yeast *Rhodotorula gracilis* (RgDAAO) can be employed as biocatalyst in a number of biotechnological applications [1]. The two-step conversion of cephalosporin C to 7-aminocephalosporanic acid and the biosensor for detection of whole D-aa content in biological samples represent the most interesting applications. However, regarding the latter application, the very low activity of DAAO towards acidic D-aa represents the major obstacle.

In order to obtain DAAO mutants with a broader substrate specificity, two different and complementary approaches can be used: the "rational design", whose major tool is represented by the site-directed mutagenesis (SDM) technique, and the "irrational design (directed evolution)" in which mutants are generated "randomly" by site-saturation mutagenesis (SSM) or error-prone PCR (EP-PCR) techniques.

1.1. Rational Design

Since its first application more then 20 years ago [2], site-directed mutagenesis technique has been widely employied for the specific alteration of protein functional and structural properties. The engineering

* Correspondence concerning this article should be addressed to Loredano Pollegioni, E-mail: loredano.pollegioni@uninsubria.it.

of an enzyme to make it more suitable for biotechnological applications represents one of the most important applications of this technique. Regarding this topic, the choice of the target amino acid(s) that has to be changed (or deleted) in order to generate the mutant protein possessing the desired properties requires a careful *a priori* planning procedure (hence the definition "rational design"). This preliminary work consists in the analysis of the whole body of data regarding the functional and structural properties of the target protein and, in particular, the detailed analysis of its 3D structure (if available). When the 3D structure of the protein of interest has not yet been resolved, it is possible to build a theoretical model of the protein by homology modelling (if a suitable template is available) [3].

As stated above, we have tackled the alteration of the substrate specificity of yeast DAAO to render it more suitable for biotechnological applications (specifically to develop a biosensor for the determination of the D-aa concentration in food specimens) [4]. The determination of the structural determinants that modulate the substrate specificity of the enzyme by a detailed analysis of the 3D structure of the enzyme-substrate complex represents a fondamental starting point. Since the crystallographic structure of the "Michaelis complex" is seldom available, the interaction between the protein and the desired substrate-ligand is usually predicted by a computational approach. This can be achieved using the software package AutoDock that allows the prediction of ligand-protein interactions based on a Montecarlo simulated annealing algorithm [5]. The same kind of analysis can be performed using a model of the designed mutant(s) as the target (starting) protein in the docking simulation; this with the final aim to theoretically predict the effect of the designed amino acid substitution(s) on the binding properties of the enzyme.

The designed amino acid substitutions are then inserted by mutating the corresponding nucleotides in the cDNA coding RgDAAO. This latter part of the project is quite straightforward since several commercial site-directed mutagenesis kits are available. Precautions should be kept that additional (non desired) mutations could be inserted during the mutagenesis procedure: for this reason the whole gene is usually sequenced after the site-directed mutagenesis reaction.

1.2. Directed Evolution (Site-Saturation Mutagenesis and Error-Prone PCR)

The optimization of binding-site geometry for new or improved enzymatic properties can be performed by a combinatorial site-saturation mutagenesis approach on the key residues identified by docking analysis (see above). This method is used to replace a selected codon by an alternative set of codons that, upon translation, should yield a set of mutant proteins containing all 20 amino acids at this specific site [6]. These mutations can be introduced employing the same kits used for the site-directed mutagenesis reaction and using oligonucleotides degenerated in appropriate key positions. For each altered codon, ~ 200 clones are screened, a number that allows to have ~ 96 % of probability to introduce each single amino acid at the desired position.

On the other hand, in order to obtain amino acid changes along the entire protein sequence, the EP-PCR approach is used. This method allows the introduction of random mutations through misincorporation of nucleotides by DNA polymerase during the PCR amplification [7,8]. In general, DNA polymerases lacking proofreading activity are employed in EP-PCR experiments; the error rate may be further increased by mixing unequal concentrations of the four dNTPs, by manganese-induced misinsertion as well as reduced concentration of each dNTP, and/or by increasing the number of PCR cycles [9,10]. The EP-PCR approach shows a higher variability in the number and type of introduced mutations, and larger mutant libraries may be screened compared to the SSM approach.

The directed enzyme evolution approach requires a method for the rapid and sensitive selection of the property of interest. The development of an efficient method to screen large mutant libraries for the desired property is an important element in determining the success of the overall protein design strategy. If possible, the mutant libraries can be screened using an *in vivo* approach on Petri's plates (e.g., when the presence of an altered catalytic activity provides a growth advantage to the microorganism: genetic selection). When such a selection is not possible (in the majority of cases), the individual microorganism clones of the mutant library must be physically separated and individually assayed, *e.g.* using a 96-well microplate. Simple colorimetric assays are usually used when the evolved property generates a visible signal. As an example, here we developed a colorimetric assay based on the coupling of the reaction catalyzed by DAAO with the reaction catalyzed by horseradish peroxidase.

2. MATERIALS

2.1. Molecular Biology

1. pT7.7 expression vector carrying the cDNA coding for RgDAAO [11].
2. Agarose gel: 1% LE agarose in 1X TAE buffer (40 mM Tris-acetate, 1 mM EDTA) and 1.4 μg/ml ethidium bromide.
3. DNA size marker (New England Biolabs 1 kb, Beverly, MA, USA).
4. UV-Transilluminator equipped with a gel documentation system (GelDoc 2000, BioRad, Hercules, CA, USA).
5. Restriction enzymes (New England Biolabs; Promega Life Sciences, Madison, WI, USA).
6. Plasmid DNA purification kit (Nucleospin Plasmid, Macherey-Nagel GmbH, Düren, Germany).
7. Kit for DNA extraction from agarose gel (Nucleospin Extract, Macherey-Nagel).
8. JM109 *E. coli* competent cells (for cloning).
9. BL21(DE3)pLysS *E. coli* competent cells (for protein expression).
10. Ampicilline (100 μg/ml final concentration) and chloramphenicol (34 μg/ml final concentration).
11. LB (Luria Bertani) and LB-agar plates containing antibiotics.
12. Shaking, thermostatic incubator for cells growth.
13. Thermocycler (GeneAmp PCR System 2400, Perkin Elmer, Norwalk, CT, USA).

2.2. SDM (Rational Design)

1. Amino acid sequence (P80324) and 3D structure coordinates (PDB code: 1C0L) of D-amino acid oxidase from *R. gracilis*.
2. Personal computer (Pentium IV processor equivalent or higher, Unix (Linux) operative system).
3. Deep view - SwissPDBviewer (http://www.expasy.org/spdbv/). A software for 3D structure visualization, analysis and generation of models of mutant protein [12].
4. Structure model of the ligands of interest generated:
   - using a chemistry modeling software (i.e., Chem3D, CambrigdeSoft) or the on-line facility "Dundee PRODRG2 Server" (http://davapc1.bioch.dundee.ac.uk./programs/prodrg/prodrg.html)
   - searching a database of 3D structures of small molecules (e.g., the macromolecular structure database of EMBL-EBI) (http://www.ebi.ac.uk/msD-srv/chempdb/cgi-bin/cgi.pl).
5. AutoDock 3.0 [5]: a suite of automated docking tools designed to predict the binding mode of small molecules to a protein of known 3D structure.
6. Site-directed mutagenesis kit, such as "Quick-change site-directed mutagenesis kit" (Stratagene, La Jolla, CA, USA) or "Altered Sites® II *in vitro* Mutagenesis System" (Promega). These kits allow to introduce single/multiple point mutations or to introduce/delete short sequences in the cDNA encoding for the protein of interest.

7. Mutagenic primer(s) carrying the desired mutation(s). In general, a ~25 bases oligonucleotide with the mismatch located in the center is sufficient for single-base mutations. For mutations involving more than one mismatch, oligonucleotides longer than 25 bases are needed (at least 12-15 perfectly matched nucleotides on either sides of the mismatch are required). Larger deletions or multiple, non adjacent, mutations require an oligonucleotide having 20-30 matched bases on either side of the mismatched region. Furthermore, the primer should possess a ≥ 40% GC content and should terminate with one or more G/C bases. In order to quickly screen the clones carrying the mutant cDNA by restriction analysis, it is also useful to introduce (or remove), in addition to the desired mutation(s), a specific restriction site.

2.3. SSM (Irrational Design of Specific Positions)

1. A mixture of appropriate PCR primers differing for one random codon. These primers allow to combinatorially generate protein variants containing all 20 amino acids at the desired position, bearing all or a subset of the 64 codons in a specific position depending on the numbers and type of the desired amino acidic mutation. The primers should possess the characteristics described at the point 2.2.7.
2. Site-directed mutagenesis kit such as "Quick-change site-directed mutagenesis kit" (Stratagene) or "Altered Sites® II *in vitro* Mutagenesis System" (Promega).

2.4. EP-PCR (Irrational Design)

1. Two appropriate PCR amplification primers designed to amplify the cDNA region subjected to random mutagenesis. In the case of RgDAAO, the two primers allow the amplification of the entire RgDAAO cDNA (1.2 kb).
2. dNTPs mixtures (2.5 mM each): each dNTP is used at different final concentrations.
3. Taq DNA polymerase (Promega) and AmpliTaq DNA polymerase (Applied Biosystems, Foster City, CA, USA) with appropriate 10X buffer.
4. 25 mM $MgCl_2$, used at different final concentrations.
5. 25 mM $MnCl_2$, used at different final concentrations.
6. T4 DNA ligase (Roche Diagnostics, Basel, Switzerland).

2.5. Screening of Mutants Libraries

1. 96-Well microplates.
2. Plate reader (Biotrak II plate reader, Amersham Biosciences, Piscataway, NJ, USA).
3. Horseradish peroxidase (1 U/μl, Roche) dissolved in 100 mM phosphate buffer, pH 7.0.
4. *o*-Dianisidine (3.2 mg/ml, Sigma) dissolved in water.
5. Solutions of D-aa as substrates (Sigma, St. Louise, MO, USA).
6. Lysis buffer: 50 mM sodium pyrophosphate buffer pH 8.5, 1 mM EDTA, 100 mM NaCl, 40 μg/ml lysozyme and 1 μg/ml DNase I.

3. METHODS

3.1. SDM (Rational Design)

1. 3D coordinates of the ligands of interest were obtained from the macromolecular structure database of EMBL-EBI. D-Aspartate was modelled applying a mirror simmetry operator to L-aspartate coordinates.
2. The mode of binding of the ligands at the active site of the proteins (RgDAAO and D-aspartate oxidase, DASPO) was predicted by AutoDock software. The correct definition of the 3D space that will be explored by AutoDock during docking analysis is a critical point. For this reason, we decided to use a 3D grid encompassing the whole active site. In order to avoid interactions between the ligand and the protein surface that could lead to incorrect results, the grid may contain the minimal part of the protein surface. The complexes with the lowermost free energy were selected for subsequent analysis.
3. In order to identify the active site residues critical for the determination of the substrate specificity, the selected enzyme-ligand complexes were visualized using SwissPDBviewer software. As shown in Fig. 1A, the side chain of the substrate CF_3-D-alanine in RgDAAO points towards M213. On the other hand, in DASPO model at the same position R216 is instead present (Fig. 1B). The presence of a positively charged side chain in this protein position was thus proposed to facilitate the binding of acidic D-amino acids.
4. A theoretical model of the M213R RgDAAO mutant was generated using SwissPDBviewer. The model was used to carry out the same docking analysis performed with the wild-type RgDAAO in order to verify if the planned mutation increases the affinity of the mutant for the ligand(s) of interest.
5. The mutation was introduced in the cDNA coding for the RgDAAO using the above cited site-directed mutagenesis kit and the following oligonucleotide: 5'-GCAAGCGATGCACGCGTGAC TCGTCCGACCC-3'.
6. The PCR mixture was used to transform JM109 *E. coli* competent cells.
7. Preliminary screening of the mutant clones was made by restriction analysis with *Mlu*I and further confirmed by nucleotidic sequence of specific clones.
8. In order to express the mutant protein, the plasmid carrying the cDNA coding for the mutant M213R enzyme was transferred into BL21(DE3)pLysS *E. coli* cells. The main advantage of the "Quick-change site-directed mutagenesis kit" (Stratagene) is represented by the possibility to perform the mutagenesis reaction on the same plasmid used for protein expression (i.e., it does not require to subclone the mutant cDNA in a specific plasmid). On the other hand, if the "Altered Sites® II *in vitro* Mutagenesis System" (Promega) is used, the gene coding for the mutant enzyme must be subcloned into the pAlter-1 vector.
9. Mutant proteins are usually expressed in the BL21(DE3)pLysS *E. coli* cells transformed with the plasmid carrying the cDNA coding for the M213R mutant and purified following the protocol optimized for the wild-type RgDAAO [11].

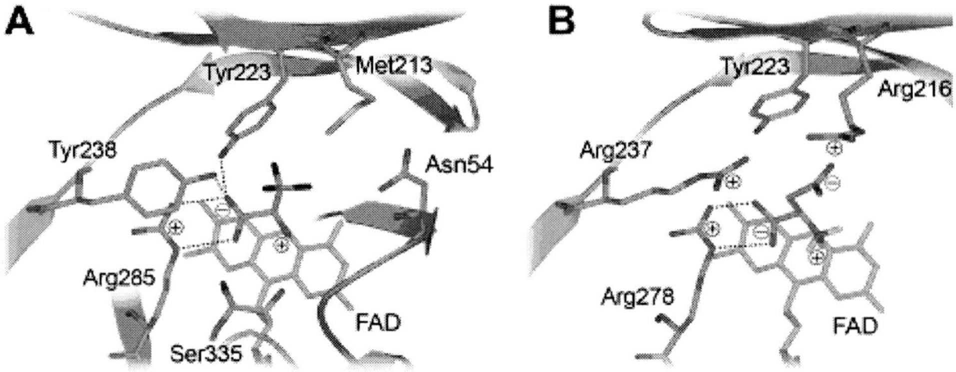

DAAO-CF_3-D-alanine complex DASPO-D-aspartate complex

Figure 1. A. Active site of wild-type RgDAAO in complex with trifluoro-D-alanine. The side chain of M213 is ~ 3.8 Å from the β-carbon of the bound substrate. B. Proposed model of the active site of wild-type DASPO in complex with D-aspartate. The model has been generated by the Swiss-Model server [3] using the protein sequence of bovine kidney DASPO (SwissProt accession number: P31228) and DAAO atomic coordinates (PDB code: 1C0L). R216 in DASPO seems to be located in a position resembling that of M213 in RgDAAO.

3.2. SSM (Irrational Design of Specific Positions)

1. Plasmid DNA containing the cDNA was purified by Nucleospin Plasmid kit (Macherey-Nagel) and its concentration was determined by electrophoresis on agarose gel.
2. 20 ng of DNA template was used for each PCR reaction.
3. The genetic variability was introduced using the "Quick-change site-directed mutagenesis" kit (Stratagene) protocol, appropriate primers (see point 2.3.1) and the following amplification program: 30 s at 95 °C; 14 cycles of 30 s denaturation at 95 °C, 1 min annealing at 57 °C and 14 min extension at 68 °C.
4. The PCR mixture was used to transform JM109 *E. coli* competent cells.
5. JM109 *E. coli* recombinant cells were collected for DNA extraction.
6. The plasmid DNA from JM109 *E. coli* cells was extracted using the Nucleospin Plasmid kit (Macherey-Nagel) and 20 ng of extracted DNA was used to transform BL21(DE3)pLysS *E. coli* cells (random mutants library).
7. The random mutants library was finally screened as described at point 3.4.

3.3. EP-PCR (Irrational Design)

1. Plasmid DNA containing the RgDAAO cDNA was purified by Nucleospin Plasmid kit (Macherey-Nagel) and its concentration was determined by electrophoresis on agarose gel.
2. Different conditions can be used to modulate the mutational frequency. In the case of RgDAAO the following conditions were used:
   - Condition 1: the reaction mixture (100 μl) contained 20 ng of template DNA, 0.1 μM each of the 5' and 3' primers, 20 μM each dNTPs, 2.5 U AmpliTaq, 1X buffer, 2 mM $MgCl_2$ and 0.25 mM $MnCl_2$.
   - Condition 2: the reaction mixture (100 μl) contained 20 ng of template DNA, 0.5 μM each of the 5' and 3' primers, 50 μM each dNTPs, 2.5 U Taq DNA polymerase, 1X buffer, 10 mM 2-mercaptoethanol, 10% DMSO and 1.5 mM $MgCl_2$.

- Condition 3: the reaction mixture (100 μl) contained 20 ng of template DNA, 0.1 μM each of the 5' and 3' primers, 200 μM each dNTPs, 2.5 U Taq DNA polymerase, 1X buffer and 7.5 mM $MgCl_2$.
- Condition 4: the reaction mixture (25 μl) contained 20 ng of template DNA, 0.4 μM each of the 5' and 3' primers, 0.2 mM dGTP and dATP, 1 mM dCTP and dTTP, 2.5 U AmpliTaq, 1X buffer, 2 mM $MgCl_2$ and 0.25 mM $MnCl_2$.

3. The PCR reaction was performed for 40 cycles (conditions 1-3) or 25 cycles (condition 4) and the following program: 40 s denaturation at 94 °C, 30 s annealing at 60 °C, and 90 s extension at 72 °C (a ten minutes extension time was used in the last cycle).
4. The PCR-generated DAAO fragments were purified and digested with *Not*I-*Xba*I restriction enzymes.
5. Digested fragments were separated by gel electrophoresis, purified and ligated to a pT7.7 vector linearized using the same restriction enzymes.
6. The ligation reaction mixture was used to transform JM109 *E. coli* competent cells.
7. The plasmid DNA from JM109 *E. coli* cells was extracted, and (20 ng) was used to transform BL21(DE3)pLysS *E. coli* cells (random mutants library).
8. The random mutants library was finally screened as described at point 3.4.
9. The plasmid carrying the cDNA coding for improved mutant(s) was extracted from *E. coli* cells and was used as a template for further rounds of mutagenesis or for large scale production of the "evolved" RgDAAO variants.

3.4. Screening of Mutants Libraries

1. The expression of the recombinant RgDAAO proteins was induced by addition of 1 mM IPTG to *E. coli* cultures grown at saturation and further incubation at 30 °C for 2 hours.
2. 50 μl of the induced cell broth was transferred to three wells of a 96-well microplate (one for each substrate tested).
3. Cells were lysed adding 150 μl of lysis buffer and incubating the plates at 37 °C for 30 min.
4. The oxidase activity was assayed on the crude extracts by adding 100 μl of a solution containing 90 mM substrate (D-alanine, D-aspartate or D-arginine), 0.3 mg/ml *o*-dianisidine and 1 U horseradish peroxidase in 100 mM sodium pyrophosphate buffer pH 8.5.
5. The time course of the absorbance change at 440 nm was followed at room temperature using a microtiter plate reader.
6. The clones showing an improved activity on the substrates tested were subsequently used for a larger-scale protein production (2-10 liters of fermentation broth) for their biochemical characterization.

4. Results

Engineering of RgDAAO by:

4.1. SDM (rational design)

Rational design has been used to broaden the substrate specificity of yeast DAAO in order to render it active also on acidic D-aa (D-aspartate and D-glutamate); these latters compounds are substrates of the flavoprotein DASPO [1]. The model of the DASPO active site was built by comparative modelling using the 3D structure of DAAOs from pig kidney and yeast as template (40% and 27% of sequence identity with DASPO, respectively). The mode of binding of the substrate D-aspartate at the active site of DAAO and DASPO was then predicted using AutoDock. Detailed analysis of the DASPO-D-aspartate complex reveals the presence of two additional arginines in its active site compared to DAAO, one of which can make a salt bridge interaction with the carboxylic group of the substrate side chain (Fig. 1B). We thus decided to place an arginine in the corresponding position of the RgDAAO active site (i.e., in the position occupied by methionine 213 in wild-type RgDAAO) [13] generating a mutant possessing an active site more similar to that of DASPO.

The designed mutant was produced by site-directed mutagenesis, expressed in *E. coli* and purified as reported in [14]. The kinetic characterization of the M213R RgDAAO mutant shows that it acquired the ability to efficiently oxidize negatively charged D-aa with a catalytic efficiency (k_{cat}/K_m ratio) even higher than DASPO (i.e., 118 $mM^{-1}min^{-1}$ vs. 104 $mM^{-1}min^{-1}$ on D-aspartate as substrate, respectively). One of the most interesting characteristics of the mutant RgDAAO is that it retains the ability to oxidize its "original" substrates, i.e. the neutral and polar D-aa (although with a lower efficiency compared to the wild-type RgDAAO) [14].

The newly acquired substrate specificity renders the M213R mutant RgDAAO suitable to develope a biosensor for the determination of the content of D-amino acids in food specimens. In fact, a current response was obtained for neutral and acidic D-aa using the M213R RgDAAO (Fig. 2), while no current was measured using the wild-type DAAO with D-aspartate as substrate [14].

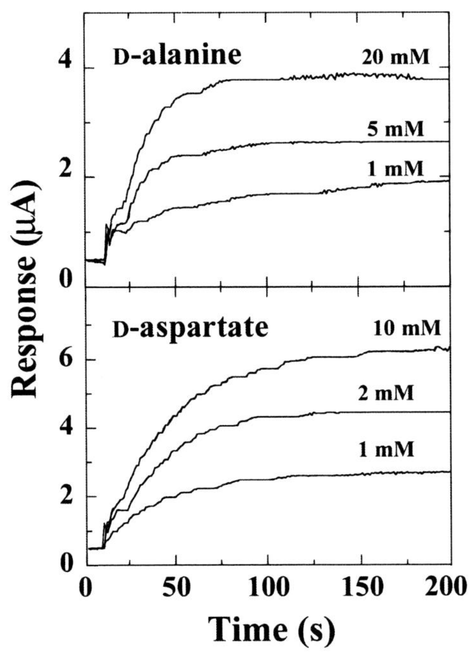

Figure 2. Amperometric traces of D-alanine (top) and D-aspartate (bottom) detection using the M213R RgDAAO biosensor [14].

4.2. SSM (Irrational Design of Specific Positions)

Site-saturation mutagenesis was performed focusing on the active site residues involved in substrate binding and product release using the "Quick-change site-directed mutagenesis" kit and starting from the wild-type RgDAAO cDNA. Preliminary results indicate that a large fraction of the libraries did not contain variants with evolved properties compared to the starting RgDAAO enzyme. Anyway, the few most active variants were purified and their kinetic parameters are under investigation to be compared to those of wild-type RgDAAO.

4.3. EP-PCR (Irrational Design)

In all cases, the amount of amplification product recovered by mutagenic PCR was lower than that obtained under optimal (non-mutagenic) conditions. In particular, under condition 1 a yield of ~ 25% in comparison with native conditions was obtained, whereas under conditions 2-4 a 50-80% amplification yield was obtained [15]. For each set of tested conditions, a library of ~ 10000 clones was generated and up to 1200 independent clones for each PCR condition were screened for DAAO activity on different substrates (D-alanine, D-aspartate and D-arginine). Frequencies of mutation of 0.15-0.77% were obtained, that gave rise to 91, 27, 17 and 63% of inactive mutants for conditions 1-4, respectively.

Two mutants possessing interesting features were identified: namely L118H and T60A/Q144R/K152E [15]. These mutants showed a higher activity on at least two out the three substrates tested compared with wild-type RgDAAO. In addition, they exhibited the lowest variability of response as function of D-aa composition (Fig. 3) and therefore are useful for the analytical determination of the total content of D-aa in biological samples [15].

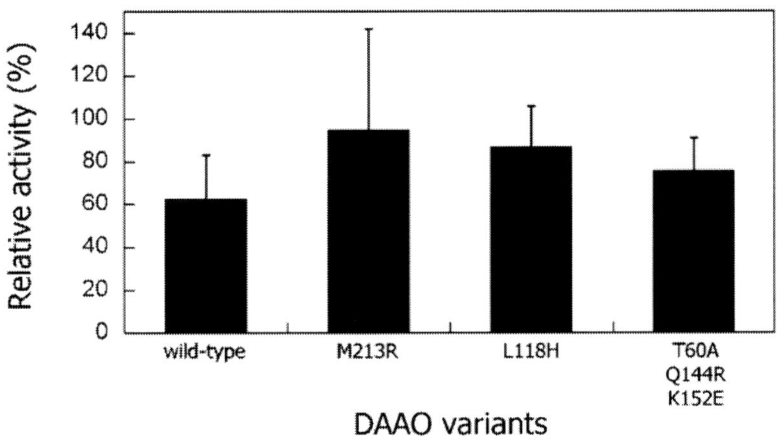

Figure 3. Comparison of the average of the relative activity of wild-type, M213R and random mutants of DAAO on nine D-aa mixtures (10 mM final concentration) containing different ratios of D-Ala, D-Glu, D-Lys, D-Gln and D-Met [15]. The values determined on 10 mM D-alanine with each enzyme were taken as 100 %. The error bars represent the variability of response of each enzyme on all mixtures.

5. Notes

1. The substrate specificity of DAAO from *Rhodotorula gracilis* has been modified either by "rational" and by "directed evolution" approach but, intriguingly, acting on different structural determinants: in the rational design approach (by site-directed mutagenesis and by site-saturation mutagenesis) the active site residues involved in substrate recognition were subjected to mutagenesis, while in the irrational design approach (error-prone PCR), residues far from the active site were mutated [14,15].
2. Concerning the effect of these substitutions, and in the absence of detailed structural data for the mutants, it is difficult to explain the mechanism(s) by which some of these amino acid mutations modulate the substrate specificity of evolved DAAOs. Substitutions on the protein surface (not far

from the active site entrance) could synergistically generate small structural changes that affect the dynamic and/or stability of the protein in a way that enhances substrate binding or catalytic turnover.

3. The combination of structural studies [13], rational design [14], and directed evolution experiments [15] allowed the modulation of substrate specificity in this flavoenzyme and the production of RgDAAO variants suitable for biotechnological applications.
4. The activity of each single clone was assayed in double on the three substrates tested.
5. The time course of the absorbance change at 440 nm during the screening procedure was followed up to 16 hours depending on the substrate used.
6. During the screening procedure, the plates are maintained at room temperature and under shaking.

Acknowledgements

This work was supported by grants from Italian MIUR *(Prot 2004058243)* to M.S. Pilone, from Fondo di Ateneo per la Ricerca to L. Pollegioni, G. Molla and M. S. Pilone, and from Fondazione Cariplo to L. Pollegioni.

References

[1] Pilone, M.S. & Pollegioni, L. (2002). D-Amino acid oxidase as an industrial biocatalyst. *Biocatal Biotransf, 20*, 145-159.

[2] Winters, G., Fersht, A.R., Wilkinson, A.J., Zoller, M. & Smith, M. (1982). Redesigning enzyme structure by site-directed mutagenesis: tyrosil tRNA synthetase and ATP binding. *Nature, 299*, 756-758.

[3] Schwede, T., Kopp, J., Guex, N. & Peitsch, M.C. (2003). SWISS-MODEL: an automated protein homology-modeling server. *Nucleic Acids Res, 31*, 3381-3385.

[4] Sacchi, S., Pollegioni, L., Pilone, M. S. & Rossetti, C. (1998). Determination of D-amino acids using a D-amino acid oxidase biosensor with spectrophotometric and potentiometric detection. *Biotechnol Tech, 12*, 149-153.

[5] Morris, G M., Goodsell, D.S., Halliday, R.S., Huey, R., Hart, W.E., Belew, R.K. & Olson, A.J. (1998). Automated docking using a Lamarckian genetic algorithm and empirical binding free energy function. *J Computl Chem, 19*, 1639-1662.

[6] Miyazaki, K. & Arnold, F.H. (1999). Exploring nonnatural evolutionary pathways by saturation mutagenesis: rapid improvement of protein function. *J Mol Evol, 49*, 716-720.

[7] Eckert, K.A & Kunkel, T.A. (1990). High fidelity DNA synthesis by the *Thermus aquaticus* DNA polymerase. *Nucleic Acids Res, 18*, 3739-3744.

[8] Lehtovaara, P.M., Koivula, A.K., Bamford, J. & Knowles, J.K.C. (1998). A new method for random mutagenesis of complete genes: enzymatic generation of mutant libraries *in vitro*. *Protein Eng, 2*, 63-68.

[9] Fromant, M., Blanquet, S. & Plateau, P. (1995). Direct random mutagenesis of gene-sized DNA fragments using polymerase chain reaction. *Anal Biochem, 224*, 347-353.

[10] Lin-Goerke, J.L., Robbins, D.J. & Burczak, J.D. (1997). PCR-based random mutagenesis using manganese and reduced dNTP concentration. *Biotechniques, 23*, 409-412.

[11] Molla, G., Vegezzi, C., Pilone M.S. & Pollegioni, L. (1998). Overexpression in *Escherichia coli* of a recombinant chimeric *Rhodotorula gracilis* D-amino acid oxidase. *Protein Expres Purif, 14*, 289-294.

[12] Guex, N. & Peitsch, M.C. (1997). SWISS-MODEL and the Swiss-PdbViewer: an environment for comparative protein modeling. *Electrophoresis*, *18*, 2714-2723.

[13] Umhau, S., Pollegioni, L., Molla, G., Diederichs, K., Welte, W., Pilone, M.S. & Ghisla, S. (2000). The X-ray structure of D-amino acid oxidase at very high resolution identifies the chemical mechanism of flavin-dependent substrate dehydrogenation. *Proc Natl Acad Sci USA*, *97*, 12463-12468.

[14] Sacchi, S., Lorenzi, S., Molla, G., Pilone, M.S., Rossetti, C. & Pollegioni, L. (2002). Engineering the substrate specificity of D-amino-acid oxidase. *J Biol Chem*, *277*, 27510-27516.

[15] Sacchi, S., Rosini, E., Molla, G., Pilone, M.S. & Pollegioni, L. (2004). Modulating D-amino acid oxidase substrate specificity: production of an enzyme for analytical determination of all D-amino acids by directed evolution. *Protein Eng Des Sel*, *17*, 517-525.

In: D-Amino Acids: ISBN 1-60021-075-9
A New Frontier in Amino Acid and Protein Research, pp.503-508

Chapter 6.10

TRANSFORMATION OF PLANTS WITH D-AMINO ACID RESISTANCE SELECTABLE MARKERS

Oskar Forsum and Torgny Näsholm*
Umeå Plant Science Centre, Department of Forest Genetics and Plant Physiology, Swedish University of Agricultural Sciences, Sweden.

Keywords: selectable markers, D-serine, D-alanine, D-serine dehydratase, D-amino acid oxidase.

1. INTRODUCTION

Production of transgenic plants usually includes a selection procedure due to the fact that the fraction of cells that are transformed is generally low. To enable successful selection, a marker gene giving a selectable trait is therefore usually incorporated along with the gene(s) for the desired trait(s). The marker genes used today are usually either conferring antibiotic or herbicide resistance, such as *nptII* (neomycine phosphotransferase II), *hpt* (hygromycin phosphotransferase) and *bar* (phosphinothricin acetyltransferase), which confer resistance to the antibiotics kanamycin, hygromycin and the herbicide L-phosphinothricin, respectively [1,2]. The use of antibiotic and herbicide resistance selectable markers has been an issue of concern and this has stimulated the development of new markers, and several less controversial alternatives, such as *manA* [phosphomannose isomerase; 3] mediating selection on mannose, *xylA* [xylose isomerase; 4], mediating selection on xylose, and galT [UDP-glucose:galactose-1-phosphate uridyltransferase; 5] mediating selection on galactose have been developed recently. We have recently shown that the *dao1* gene from the yeast *Rhodotorula gracilis*, encoding D-amino acid oxidase (DAAO), can be used as a conditional marker allowing selection either for or against transgenic plants, depending on the selective agent chosen [6]. Similarly, the *E. coli* gene *dsdA* encoding D-serine ammonia lyase may be used as a marker gene together with D-serine as the selection agent [7].

The potential for using D-amino acids and their metabolism in biotechnology is based on the variation in toxicity exerted by different D-amino acids on plants, and the possibility to modify this plant-specific toxicity by introducing genes encoding enzymes specific for the D-enatiomers. The ability to metabolise D-amino acids is widespread among both prokaryotes and eukaryotes, but the metabolism of D-amino acids in plants is poorly understood. Oxidative deamination by DAAO is one of the most studied pathways for D-

* Correspondence concerning this article should be addressed to Torgny Näsholm, Email: torgny.nasholm@genfys.slu.se, Fax: +46-90-7868165.

amino acid metabolism, and has been found in numerous eukaryotic organisms, including yeast, fungi, insects, amphibians, reptiles, birds and mammals [8], but there is no report of the enzyme in plants. Furthermore, there are several reports of D-amino acid toxicity towards plants [9-11]. Our studies have shown that D-alanine and D-serine are toxic to plants at relatively low concentrations, while the D-enantiomers of valine and isoleucine are non-toxic, even at high concentrations. Furthermore, transgenic plants expressing *dao1* are resistant to D-serine and D-alanine, and *dao1* may therefore be used to select for transgenic plants. In contrast to wild-type plants, *dao1*-transgenic plants are sensitive to exposure to D-valine and D-isoleucine, most likely because DAAO activity against these substances results in formation of the toxic 3-methyl-2-oxobutanoic acid and 3-methyl-2-oxopropanoic acid, respectively. Hence *dao1* may also function as a counter selectable marker enabling selection against transformed plants [6].

The dual function of *dao1*, enabling both negative selection and counter selection, makes this marker gene ideal for selection and later removal of the selectable marker. When the intention is to retain, rather than delete, the marker gene, a gene encoding an enzyme with high specificity for the selective agent may be preferred. One such candidate is D-serine ammonia lyase (DSD, EC 4.2.1.14, formerly D-serine dehydratase), encoded by the *dsdA* gene from *Escherichia coli.* DSD is a pyridoxal 5'-phosphate-requiring enzyme which catalyses the deamination of the D-enantiomers of serine, threonine and allothreonine to the corresponding keto acids, ammonium and water. The *DAD*-catalyzed decomposition of the enantiomers of serine and threonine indicates that the enzyme exhibits considerable kinetic specificity for the D-forms; the K_m for D-serine is 0.086, compared to 11.04 mM for L-serine [12]. Thus, DSD has a high specificity for a few D-amino acids, suggesting that the risk of interference with plant endogenous amino acid metabolism of the marker gene should be low.

The genes encoding D-amino acid oxidase and D-serine ammonia lyase are two examples of genes encoding D-amino acid metabolising enzymes that may be used as selectable markers in plant biotechnology. This chapter provides protocols for the use of *dao1* and *dsdA* as selectable markers in plant biotechnology. The protocols describe the use of the markers for studies on *Arabidopsis thaliana.* With suitable amendments of the protocol, the methods may be applied to other plant species as well.

2. MATERIALS AND METHODS

D-Amino Acid Oxidase

1. The yeast *Rhodotorula gracilis* (strain ATCC 26217, available from stock center DSMZ, http://www.dsmz.de) was grown in liquid culture containing 30 g/L malt extract, 3 g/L soya peptone and 30 mM D–alanine to induce the D–amino acid oxidase gene *dao1*.
2. Total RNA was isolated from the yeast using the hot acid phenol extraction [13] and used for cDNA synthesis.
3. The PCR primers (forward) 5′–TACTACTCGAAGGACGCCATG and (reverse) 5′–ACAGCCA CAATTCCCGCCCTA, and with appropriate cloning sites added suited for the vector, were used to amplify the *dao1* gene from the cDNA template by PCR. PCR was performed for 30 cycles with 20 sec denaturation at 98°C, 20 sec annealing at 60°C and 20 sec extension at 72°C.
4. The PCR fragments were sub–cloned into a sequencing vector and the vectors were subjected to restriction analysis and sequencing to check that they contained the correct constructs.
5. The fragment containing the *dao1* gene was sub-cloned into an expression cassette of a binary vector system.
6. The binary vector was electroporated into electrocompetent cells of *Agrobacterium tumefaciens.*
7. *Arabidopsis thaliana* plants were transformed by the floral dip method [14] with *A. tumefaciens* carrying the vector.

8. Transgenic T_1 plants were selected on half-strength MS medium [15] containing 3 mM D-alanine or 1 mM D-serine (Figure 1). Alternatively, selection may be performed by spraying young (soil-grown) seedlings with an aqueous solution of 50 mM D-serine or 50 mM D-alanine. The spraying procedure should be repeated at least three times to ensure no escapes.

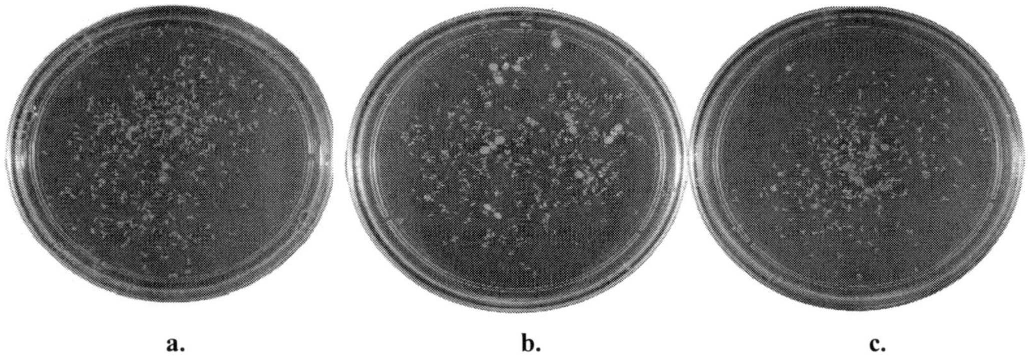

Figure 1. Selection of transformants using either dsdA (a) or dao1 (b, c) as selectable marker genes and D-serine (a, c) or D-alanine (b) as selection agent. D-amino acids at 3 mM concentration were included in 1/2 strength MS media. Plates are photographed 5 days after germination of seedlings. (Color picture is available on request.)

9. Lines containing a single T–DNA insertion locus were selected by statistical analysis of T–DNA segregation in the T_2 population that germinated on D-alanine or D-serine–containing medium. Plants with a single locus of inserted T–DNA were grown and self–fertilized. Homozygous T_3 seed stocks were then identified by analysing T–DNA segregation in T_3 progenies and confirmed to be expressing the introduced gene by northern blot analyses.
10. *dao1* can also be used as a negative marker when D–isoleucine or D–valine are used as selectable agents (Figure 2). Unambiguous negative selection was achieved when *dao1* transgenic seeds were sown on either D–isoleucine or D–valine at 10 mM and 15 mM respectively.

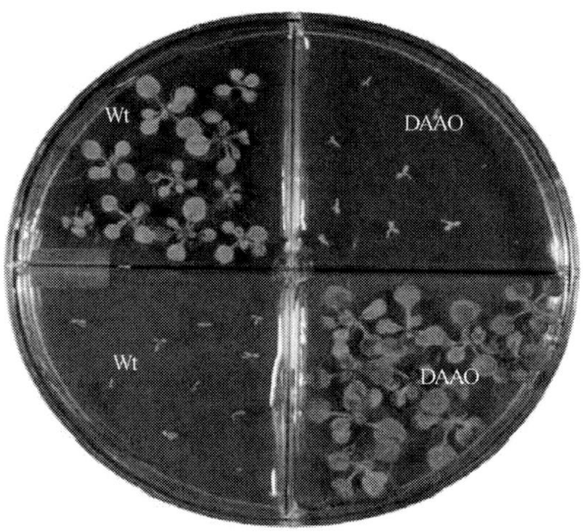

Figure 2. Growth of wild-type (Wt) and *dao1* expressing (DAAO) Arabidopsis thaliana on 1/2 strength MS media amended with either D-isoleucine (top left and right) or D-alanine (bottom left and right). D-isoleucine is indifferent to Wt plants but is toxic to *dao1* expressing plants while the reverse is true for D-alanine. (Color picture is available on request.)

D-Serine Dehydratase

1. *E. coli* (strain K12) was grown on LB plates.
2. Bacterial cells were picked from single colonies and used as template for the PCR amplification of the *dsdA* gene.
3. The PCR primers (forward) 5′-TCATCTAAGCGCAAAGAGACGTACTATGG and (reverse) 5′-ATGCTGCGTTGAAACGTTATTAACGG and with appropriate cloning sites added suited for the vector, were used to amplify the *dsdA* gene. PCR was performed for 30 cycles with 20 sec denaturation at 98°C, 20 sec annealing at 58°C and 20 sec extension at 72°C.
4. The PCR fragments were sub–cloned into a sequencing vector and the vectors were subjected to restriction analysis and sequencing to check that they contained the correct constructs.
5. The fragment containing the *dsdA* gene was subcloned into an expression cassette of a binary vector system.
6. The binary vector was electroporated into electrocompetent cells of *Agrobacterium tumefaciens.*
7. *Arabidopsis thaliana* plants were transformed by the floral dip method [14] with *A. tumefaciens* carrying the vector.
8. Transgenic T_1 plants were selected on half-strength MS medium [15] containing 3 mM D-serine (Figure 1). Alternatively, selection may be performed by spraying young (soil-grown) seedlings with an aqueous solution of 30 mM D-serine (Figure 3). The spraying procedure should thorough and the solution must cover all leaves. Spraying should be repeated at least three times to ensure no escapes. As seedlings develop and get older, sensitivity to D-serine spraying decreases significantly.
9. Lines containing a single T–DNA insertion locus were selected by statistical analysis of T–DNA segregation in the T_2 population that germinated on D-serine–containing medium. Plants with a single locus of inserted T–DNA were grown and self–fertilized. Homozygous T_3 seed stocks were then identified by analysing T–DNA segregation in T_3 progenies and confirmed to be expressing the introduced gene by northern blot analyses.

Figure 3. *dsdA* transgenic plants and wild type plants grown on soil photographed following selection by spraying with D-serine. Two rows of transgenic plants, with a row of wild type plants between them, were grown for seven days following germination before applying the selective treatment, which consisted of spraying with an aqueous 30 mM solution of D-serine with 0.05 % Tween 80 on three consecutive days. (Color picture is available on request.)

3. Results

Using the *dsdA* and *dao1* genes as markers in the floral dip protocol usually results in a transformation efficiency of 1-2% in *A. thaliana* which is similar to frequencies achieved with other selectable markers. Transformants can be identified at a very early stage, usually just after the cotyledons have emerged. Growth of transformants is not affected by the expression of *dsdA* or *dao1*. Using other selectable markers, a lag-phase in growth is often found after the selection step. This lag-phase does not occur when *dao1* or *dsdA* are used.

4. NOTES

Several D-amino acids may function as selectable agents for plants expressing *dao1*. Our experience is that D-alanine is the preferred agent because the concentration range within which transgenes and wild type plants can be separated is realtively wide. For D-serine, fine tuning of the concentration used during selection is more critical.

REFERENCES

[1] Kuiper, H. A, Kleter, G. A., Noteborn, H. P. J. M. and Kok, E. J. (2001) Assessment of the food safety issues related to genetically modified foods. *Plant J, 27,* 503-528.

[2] Miki, B. and McHugh, S. (2004). Selectable marker genes in transgenic plants: applications, alternatives and biosafety. *J Biotechnol, 107,* 193–232.

[3] Joersbo, M., Donaldson, I., Kreiberg, J., Guldager Petersen, S., Brunstedt, J. and Okkels, F. T. (1998) Analysis of mannose selection used for transformation of sugar beet. *Mol Breed, 4*, 111-117.

[4] Haldrup, A., Guldager Petersen, S. and Okkels, F. T. (1998) The xylose isomerase gene from *Thermoanaerobacterium thermosulfurogenes* allows effective selection of transgenic plant cells using D-xylose as the selection agent. *Plant Mol Biol, 37*, 287-296.

[5] Joersbo, M., Joregensen, K. and Brunstedt, J. (2003). A selection system for transgenic plants based on galactose as selective agent and a UDP-glucose:galactose-1-phosphate uridyltransferase gene as selective gene. *Mol Breed, 11*, 315-323.

[6] Erikson, O., Hertzberg, M. & Näsholm, T. (2004). A conditional marker gene allowing both positive and negative selection in plants. *Nat Biotech, 22*, 455-458.

[7] Erikson, O., Hertzberg, M. and Näsholm, T. (2005). The *dsdA* gene from *Escherichia coli* provides a novel selectable marker for plant transformation. *Plant Mol Biol, 57*, 425-433.

[8] Pilone, M.S. (2000). D–Amino acid oxidase: new findings. *Cell Mol Life Sci, 57*, 1732–1747.

[9] Aldag, R. W. and Young, J. L. (1970) D-Amino acids in soils. I. Uptake and metabolism by seedling maize and ryegrass. *Agron J, 62*, 184-189.

[10] Valdovinos, J. G. and Muir, R. M. (1965). Effects of D and L amino acids on foliar abscission. *Plant Physiol, 40*,335-339.

[11] Manabe, H. and Ohira, K. (1981). Effects of D- and L-alanine on the growth of suspension-cultured rice, soybean and tobacco cells. *Soil Sci Plant Nutr, 27*, 383-386.

[12] Federiuk, C. S., Bayer, R. and Shafer, J. A. (1983) Characterization of the catalytic pathway for D-serine dehydratase. *J Biol Chem, 258*, 5379-5385.

[13] Sambrook, J. and Russell, D. W. (2001). Molecular Cloning: A Laboratory Manual 3rd ed. Cold Spring Harbor Laboratory Press, New York.

[14] Clough, S. J. & Bent, A. F. (1998). Floral dip: a simplified method for *Agrobacterium*–mediated transformation of *Arabidopsis thaliana*. *Plant J, 16*, 735–743.

[15] Murashige, T. & Skoog, F. (1962). A revised medium for rapid growth and bioassay with tobacco tissue cultures. *Physiol Plant, 15*, 310–313.

In: D-Amino Acids: ISBN 1-60021-075-9
A New Frontier in Amino Acid and Protein Research, pp.509-512

Chapter 6.11

SCREENING FOR MUTANT MICE LACKING D-AMINO-ACID OXIDASE ACTIVITY

Ryuichi Konno*
Department of Microbiology, Dokkyo Medical University, Japan.

Keywords: D-amino-acid oxidase, mutant mouse, screening, urine, TLC

1. INTRODUCTION

D-Amino-acid oxidase (DAO, EC 1.4.3.3) catalyzes oxidative deamination of D-amino acids to the corresponding 2-oxo acids [1]. In the course of the reaction, hydrogen peroxide and ammonia are produced. This enzyme is present in the kidney, liver and brain in higher animals. Since D-amino acids have been considered very rare in higher animals, the physiological function of D-amino-acid oxidase has been enigmatic [2,3]. However, improvements and development of analytical methods have shown that D-amino acids are present in higher animals as well. Therefore, it has become clear that the metabolism of D-amino acids of intrinsic and extrinsic origin is one of the physiological roles of D-amino-acid oxidase. Recently Chumakov et al. [4] have shown that this enzyme is involved in the schizophrenia through the metabolism of D-serine in the brain.

Mutant animals having disorders in D-amino-acid oxidase would be useful for the study of schizophrenia and also for elucidation of yet discovered physiological functions of this enzyme. A simple and rapid method to screen for mutant mice lacking D-amino-acid oxidase activity is described here. This screening method is based on the fact that a small amount of ingested D-amino acid is completely metabolized by D-amino-acid oxidase in normal mice whereas it is not metabolized and eventually excreted into urine in mutant mice lacking D-amino-acid oxidase.

2. MATERIALS

1. Mice (non-inbred, adult, both male and female).

* Correspondence concerning this article should be addressed to Ryuichi Konno, E-mail: konno@dokkyomed.ac.jp, FAX: +81-282-86-5616.

2. D-Methionine (Sigma Chemical Co., St. Louis, USA).
3. Silica gel thin-layer plate (20 x 20 cm, TLC plates Silica gel 60, Merck, Darmstadt, Germany).
4. TLC-developing chamber.
5. Drying oven.
6. n-Butanol, acetic acid, ethanol.
7. Ninhydrin (Sigma).
8. An atomizer (a sprayer).

3. Methods

1. Mice are numbered and housed in cages.
2. They are given bottled distilled water containing 0.02 % (w/v) D-methionine. They are allowed free access to the water and commercial diet.
3. The following day, fresh urine is collected by letting the mice urinate spontaneously or by massaging them on the bladder.
4. Five μl of the urine is spotted on a TLC plate approximately 1.5 cm from the bottom of the plate.
5. The plate is placed in a TLC-developing chamber containing a layer of solvent (n-butanol/acetic acid/water, 3:1:1, w/w) approximately 0.5 cm deep.
6. The chamber is closed and ascending development is carried out for approximately five h.
7. The plate is dried at 100°C for 10 min in a drying oven.
8. A solution of 0.2% ninhydrin in ethanol is sprayed onto the plate.
9. The plate is heated at 100°C until the color develops (approximately 5 min).

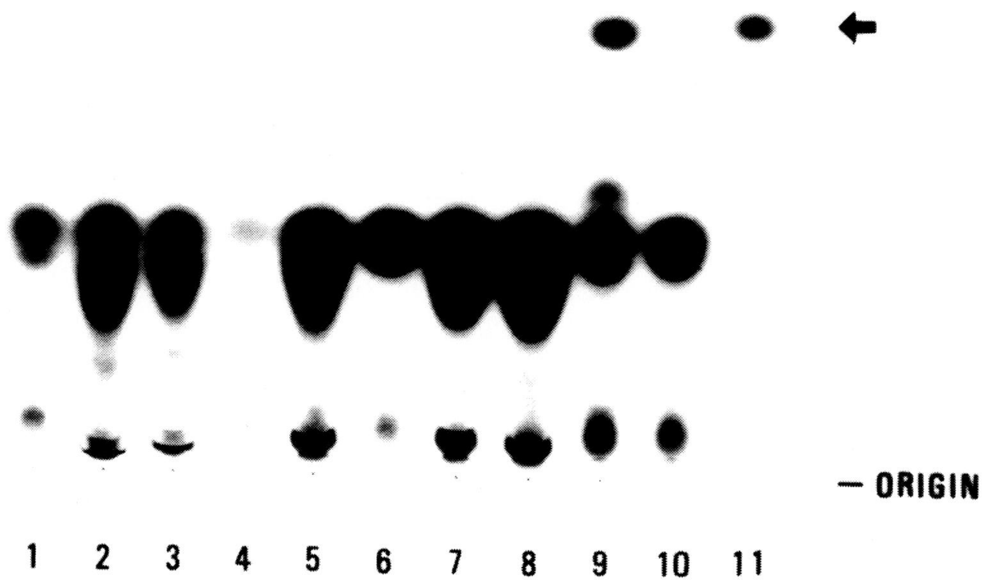

Figure 1. Thin-layer chromatogram of urine of mice given water containing 0.02% D-methionine. Urine was developed with n-butanol/acetic acid/water on a silica gel plate. Direction of the development was from the bottom to the top. Numbers 1 through 10 represent the urine samples of the mice numbered as such. Number 11 is authentic D-methionine, 1.5 μg. Arrow indicates the spot of methionine (From Konno and Yasumura. (1988), *Lab. Anim. Sci. 38*, 292-295, with permission).

4. RESULTS

Non-inbred ddY mice were given water containing 0.02% D-methionine overnight. Their fresh urine was correct the following day. Five ul of the urine was applied onto a silica gel thin-layer plate and developed with n-butanol/acetic acid/water. The plate was dried and sprayed with 0.2% ninhydrin.

A representative chromatogram is shown in Figure 1. One urine sample (No. 9) out of 10 had a spot that corresponds to authentic methionine. These results suggested that the No. 9 mouse was a mutant mouse that lacked D-amino-acid oxidase activity. Post-mortem assay for D-amino-acid oxidase activity in the kidney, indeed, indicated that the No. 9 mouse did not have the enzyme activity whereas the other 9 mice had a various level of activity (Table 1).

Table 1. D-Amino-acid oxidase activity in the kidney of mice used for screening.

Mouse No.*	Sex	Activity (nmol/min per mg protein)
1	Female	1.3
2	Female	1.9
3	Female	1.6
4	Female	2.5
5	Male	1.4
6	Male	2.6
7	Male	3.8
8	Male	3.4
9	Male	<0.1
10	Male	4.6

*Mouse number corresponds with the urine number shown in Figure 1.

5. NOTES

1. This screening method is simple and quick. Intact mutant mice are obtained in two days. It is also economical because special instruments or equipments are not required. It is possible to screen a large number of mice at a time.
2. This screening method is applicable to other animal species.
3. A cellulose thin-layer plate can be used in place of silica gel plate.
4. D-Phenylalanine can be used in place of D-methionine.
5. Phenol/water (3:1, w/w) can be used in place of n-butanol/acetic acid/water.
6. The solvent (n-butanol/acetic acid/water) smells badly. Therefore, the drying of the TLC plate and the color development should be carried out in a fume hood or in a well-ventilated room.
7. Using this screening method, mutant mice lacking D-amino-acid oxidase were isolated from the population of non-inbred ddY mice [5]. These mutant mice have been used for various studies, such as genetics [6], nutrition [7], molecular biology [8], physiology [9,10], toxicology [11], and neuroscience [12-14].

References

[1] Krebs, H. A. (1935). CXCVII. Metabolism of amino-acids. III. Deamination of amino-acids. *Biochem J, 29*, 1620-1644.

[2] Meister, A. (1965). *Biochemistry of the Amino Acids* (2nd edition) New York: Academic Press.

[3] Konno, R. and Yasumura, Y. (1992). D-Amino-acid oxidase and its physiological function. *Int J Biochem, 24*, 519-524.

[4] Chumakov, I., Blumenfeld, M., Guerassimenko, O. et al. (2002). Genetic and physiological data implicating the new human gene G72 and the gene for D-amino acid oxidase in schizophrenia. *Proc Natl Acad Sci USA, 99*, 13675-13680.

[5] Konno, R. and Yasumura, Y. (1988). A simple and rapid method to screen for mutant mice lacking D-amino-acid oxidase activity. *Lab Anim Sci, 38*, 292-295.

[6] Konno, R. and Yasumura, Y. (1983). Mouse mutant deficient in D-amino acid oxidase activity. *Genetics, 103,* 277-285.

[7] Konno, R. and Yasumura, Y. (1984). Involvement of D-amino-acid oxidase in D-amino acid utilization in the mouse. *J Nutr, 114,* 1617-1621.

[8] Sasaki, M., Konno, R., Nishio M. et al. (1992). A single-base-pair substitution abolishes D-amino-acid oxidase activity in the mouse. *Biochim Biophys Acta, 1139,* 315-318.

[9] Konno, R., Isobe, K., Niwa A. et al. (1988). Lack of D-amino-acid oxidase activity causes a specific renal aminoaciduria in the mouse. *Biochim Biophys Acta, 967,* 382-390.

[10] Hasegawa, H., Matsukawa, T., Shinohara, Y. et al. (2004). Role of renal D-amino-acid oxidase in pharmacokinetics of D-leucine. *Am J Physiol Endocrinol Metab, 287,* E160-E165.

[11] Konno, R., Ikeda, M., Yamaguchi, K. et al., (2000). Nephrotoxicity of D-propargylglycine in mice. *Arch Toxicol, 74,* 473-479.

[12] Wake, K., Yamazaki, H., Hanzawa, S. et al. (2001). Exaggerated responses to chronic nociceptive stimuli and enhancement of *N*-methyl-D-aspartate receptor-mediated synaptic transmission in mutant mice lacking D-amino-acid oxidase. *Neurosci Lett, 297,* 25-28.

[13] Haradahira, T., Okauchi, T., Maeda, J. et al. (2003). Effects of endogenous agonists, glycine and D-serine, on in vivo specific binding of [^{11}C]L-703,717, a PET radioligand for the glycine-binding site of NMDA receptors. *Synapse, 50,* 130-136.

[14] Hashimoto, A., Yoshikawa M., Niwa, A. et al. (2005). Mice lacking D-amino acid oxidase activity display marked attenuation of stereotypy and ataxia induce by MK-801. *Brain Res, 1033,* 210-215.

In: D-Amino Acids: ISBN 1-60021-075-9
A New Frontier in Amino Acid and Protein Research, pp.513-521

Chapter 6.12

ASTROGLIAL EXPRESSION OF D-AMINO ACID OXIDASE: REGIONAL AND CELL-TYPE SPECIFIC EXPRESSION

Hwan Ki Park, Yoshiteru Urai, Osamu Jinnouchi, Atsuhiko Suzue, Tokujiro Kanamori, Kyung Tak Kwak, Oded Ben-Yoseph, Shinji Nagahiro, and Kiyoshi Fukui*

Department of Gene Regulatorics, Institute for Enzyme Research, The University of Tokushima, Japan.

Keywords: D-amino acid oxidase, D-serine, schizophrenia, type-1 astrocytes, type-2 astrocytes.

1. INTRODUCTION

D-Amino acid oxidase (DAO) is a flavoenzyme that catalyzes the oxidative deamination of D-amino acids [1]. Histochemical studies revealed that DAO activity was located to some types of astrocytes [2]. It is notable that the novel human gene G72 has recently been implicated in schizophrenia and the gene product has been shown to bind with DAO and up-regulate its catalytic activity. DAO was itself associated with schizophrenia, and the combination of G72/DAO genotype had a synergistic effect on disease risk [3]. Although the existence of in vivo substrate of DAO was not known for many years, recently it had been demonstrated that substantial amounts of free D-serine are present in mammalian brain [4]. D-Serine was found to be localized to type-2 astrocytes in culture, and released by glutamate stimulation. Finally, D-serine was shown to be an endogenous agonist of the *N*-methyl-D-aspartate (NMDA) receptor in mammalian brain [5].

The major types of glial cells in the central nervous system are type-1 astrocytes, type-2 astrocytes, and oligodendrocytes [6]. They are important for maintaining the environment around neurons. Based on morphology and expression of certain antigen, astrocytes are divided into type-1 and type-2 astrocytes. Type-1 astrocytes were originally defined as polygonal, glial fibrillary acidic protein (GFAP) positive cells

* Correspondence concerning this article should be addressed to Kiyoshi Fukui, E-mail: kiyo@ier.tokushima-u.ac.jp, FAX: +81-88-633-7431.

that were A2B5 negative, while type-2 astrocytes were defined as process bearing, GFAP positive cells that were A2B5 positive [7]. In addition, Ran-2 is known as a specific marker for type-1 astrocytes [8]. Type-2 astrocytes develop in culture from a progenitor cell that was found to be the same progenitor cell as that for oligodendrocytes and was thus called the O-2A progenitor [9]. Commitment of O-2A progenitors to differentiate in culture to either oligodendrocytes or type-2 astrocytes depends on the culture condition. High serum in the media leads to type-2 astrocytes development while a serum-free media causes oligodendrocytes development to occur.

To investigate the physiological role of DAO in the brain, we established new method for purified cultures of type-1 and type-2 astrocytes from the any rat brain region and examined the DAO gene expression in cultured astrocytes.

2. Materials

2.1. Cell Cultures

1. Neonatal Sprague-Dawley rats
2. Clean bench, centrifuge, CO_2 incubator and incubator shaker (Innova 4000; New Brunswick Scientific Co., Edison, NJ, USA)
3. Sterile Pasteur pipets
4. Cell strainer (Falcon, no. 2340, Bedford, MA, USA)
5. Tissue culture plasticware or glassware including pipets and 75-cm^2 flasks
6. Fine surgical instrument
   Forceps, extra fine points (INOX, no.5, Natsume, Tokyo, Japan)
   Forceps, special fine points, curves (INOX, no. 7)
   Disposable Scalpel (Feather, no. 11, Osaka, Japan)
   Scissors
7. Basic working reagent
   DMEM (Sigma, no. D5796, St. Louis, MO, USA)
   Leibovitz's L-15 medium (L-15) (Gibco, no. 11415-064, Grand Island, NY, USA)
   Poly-D-lysine (PDL) (Sigma, no. P6407)
   Fetal bovine serum (Sigma, no. F2442)
   DNase 1 Type IV (Sigma, no. D5025)
   Bovine serum albumin (Sigma, no. A4161)
   Horse serum (Gibco, no. 16050-122)
   Cytosine arabinoside (Ara-C) (Sigma, no. C1768)
   0.25 % trypsin without Ca^{2+}, Mg^{2+} (Gibco, no. 15050-057)
   M-CSF (Sigma, no. M6518)
8. Reagents for panning dish
   100 x 15 mm petri dish (Falcon, no. 351029, not tissue culture grade)
   Tris-HCl buffer (50 mM), pH 9.5
   Goat IgG fraction to mouse IgG (ICN Pharmaceuticals, Inc., Aurora, OH, USA)
   Affinipure goat anti-mouse IgG (H+L) (Jackson Immuno Research Laboratories, West Grove, PA, USA)
   Ran-2 hybridoma supernatant (ATCC, Manassas, VA, USA)
9. Chemically defined medium (DMEM-BS) with growth factors
   DMEM (Gibco, no. 11995-065)
   100 μg/ml transferrin (Sigma, no. T1147)

100 μg/ml BSA (Sigma, no. A4161)
16 μg/ml putrescine dihydrochloride (Sigma, no. P5780)
60 ng/ml progesterone (Sigma, no. P8783)
40 ng/ml sodium selenite (Sigma, no. S5261)
630 ng/ml *N*-acetyl-L-cysteine (NAC) (Nacalai, Osaka, Japan)
Penicillin-Streptomycin (Gibco, no. 15140-122)
Growth factors
  PDGF (PeproTech, no. 100-13A, London, UK)
  NT-3 (PeproTech, no. 450-03)
  bFGF (PeproTech, no. 100-18B)
  Insulin (Sigma, no. I6634)

2.2. Immunocytochemistry

1. Slide flasks (Nalge Nunc International, no. 170920, Roskilde, Denmark)
2. Fluorescence microscope
3. Basic working reagents and antibodies
   PBS without Ca^{2+}, Mg^{2+}; PBS (-)
   Goat serum
   Acid-alcohol solution (5% glacial acetic acid plus 95 % ethanol)
   4', 6-diamidino-2-phenylindole (DAPI)
   Monoclonal A2B5 antibody (Ascites fluid)
   Monoclonal Ran-2 antibody (hybridoma supernatant)
   Fluorescein isothiocyanate (FITC)-conjugated goat anti-mouse IgG (Zymed Laboratories, Inc., San Francisco, CA, USA)
   Rabbit anti-GFAP serum (InRo Biomedtec, Ured, Sweden)
   Cy-5-conjugated goat anti-rabbit immunoglobulin (Jackson Immuno Research Laboratories, USA)

2.3. Reverse Transcriptase-PCR (RT-PCR)

1. RNA extraction reagent (Isogen, Nippon gene, Toyama, Japan)
2. Spectrophotometer
3. Electrophoresis tank
4. Thermocycler (PC-700; ASTEC, Fukuoka, Japan)
5. UV-transilluminator equipped with Charge-coupled device (CCD) camera (Printgraph AE-6910; ATTO, Tokyo, Japan)
6. SuperScript Preamplification System for cDNA synthesis (Invitrogen Corp., Carlsbad, CA, USA)
7. AmpliTaq Gold enzyme for PCR (Applied Biosystems, Foster City, CA, USA)
8. PCR primers
   rat DAO (F: 5'-GTGGCATCTTTCGAAGAGGT-3' and R: 5'-AGGCGTTCTCTTTCTA GCCG-3')
   rat β-actin (F: 5'-TTGTAACCAACTGGGACGATATGG-3' and R: 5'-GATCTTGATCT TCATGGTGCTAGG-3')
9. Basic working reagent
   Agarose
   Tris/borate/EDTA (TBE) buffer

DNA size marker
Ethidium bromide (EtBr)

3. Methods

3.1. Cell Cultures

3.1.1. Preparation of PDL-Coated Culture Flask

1. Add 8 ml of PDL solution (100 μg/ml) to coat 75-cm^2 flask and place it at clean bench for 8 h.
2. Rinse the culture flask 3 times with sterile deionized water.
3. Dry the culture flask on clean bench.

3.1.2. Mixed Glial Cells

1. Dissect brain from newborn or 1-day old Sprague-Dawley rats and place it into L-15 medium.
2. Divide cerebral cortex and cerebellum onto 90 mm glass plate containing L-15 medium.
3. Peel off the meningeal tissue.
4. Wash 3 times with L-15 medium.
5. Mince tissue into small pieces with a scalpel.
6. Transfer them to a 50-ml centrifuge tube.
7. Add 10 ml of 0.25% trypsin without Ca^{2+}, Mg^{2+} and incubate at 37°C for 10 min with 2 min-interval shaking.
8. Add 200 μl of DNase I type IV solution (10 mg/ml), and then add 5 ml of horse serum to 50-ml centrifuge tube.
9. Fill it up to 50 ml with PBS (-).
10. Centrifuge at 1,000 rpm for 5 min at 4 °C.
11. Remove supernatant and tap the tube well.
12. Add 15 ml of DMEM/FBS (10%) to 50-ml centrifuge tube.
13. Triturate into cell suspension by pipetting up and down, first in a 5-ml pipet and then in a Pasteur pipet.
14. Add 15 ml of DMEM/FBS (10%).
15. Continue pipetting until no lumps of tissue and add 20 ml of DMEM/FBS (10%).
16. Filter the cell suspensions through 40-μm cell strainer.
17. Rinse the filter with 10 ml of DMEM/FBS (10%).
18. Count cells and transfer cells suspension to PDL-coated 75-cm^2 flask with a density of 1×10^7 cells for cerebral cortex and 5×10^6 cells for cerebellum.
19. Incubate the cells in a humidified 37°C, 5 % CO_2 incubator.
20. Change the culture medium (DMEM/FBS 10%) two times a week.

3.1.3. Type-1 Astrocytes

Continue mixed glial cell culture (3.1.2) to step 20.

21. When the culture is confluent at 7 to 10 days, seal lid firmly, wrap with parafilm and transfer the flask to a shaking incubator.
22. Shake the flask for 10 min at 80 rpm to dislodge microglia from the surface of the culture. The shaker should be kept at 37°C.
23. Transfer culture medium to 100-mm culture dish (see chapter "microglia").
24. Add fresh DMEM/FBS (10%) to flask and place it at CO_2 incubator for 30 min to recovery the cells.

25. Transfer the flask to shaking incubator and shake the flask at 275 rpm for cultures derived form cortical cortex or at 240 rpm for those from cerebellum to dislodge the population of oligodendrocyte-type-2 astrocyte (O-2A) progenitors overnight.
26. The following day, recover the supernatant in 50-ml centrifuge tube (see chapter "O-2A progenitors").
27. For purification of type-1 astrocytes, add 10 ml of fresh DMEM/FBS (10%) medium containing 20 μM Ara-C to 75-cm^2 flask and place the cultures in CO_2 incubator for 48 h. This treatment will kill off dividing the cells (such as meningeal contaminants, O-2A progenitors and microglia).
28. Change the culture medium with fresh DMEM/FBS (10%) and place the cells in CO_2 incubator for 24 h to recover the cells.
29. Repeat steps (27 and 28) to remove further any remaining dividing cells.
30. You can get pure type-1 astrocyte cultures.

3.1.4. Microglia

Continue Type-1 astrocyte culture (3.1.3) to step 23.

24. Place the culture dish in CO_2 incubator for 30 min.
25. Discard supernatant and add DMEM/FBS (10%) containing M-CSF (10 ng/ml) to culture dish.
26. Incubate in a humidified 37°C, 5% CO_2 incubator and change the medium with twice a week.

3.1.5. Purification of O-2A Progenitor Cells by Panning

Preparation of panning dish: Incubate two petri dishes overnight at 4°C in 10 ml of 50 mM of Tris-HCl buffer, pH 9.5 containing goat IgG fraction to mouse IgG at 42 μg/ml and affinity purified anti-mouse IgG (H+L) at 10 μg/ml. Wash each dish three times with PBS. Add Ran-2 antibody (1.5 ml of the Ran-2 hybridoma supernatant plus 4.5 ml of 0.2% BSA/PBS) to affinity purified goat anti-mouse IgG (H+L)-coated dish, incubate for at least one hour, and then wash the plate three times with PBS. Add enough PBS to coat the dish and discard the PBS just before use.

Continue Type-1 astrocyte culture (3.1.3) to step 26.

27. Centrifuge at 1,000 rpm for 5 min, discard supernatant and suspend pellet in 10 ml of L-15 medium containing insulin (50 μg/ml) and BSA (0.02 %).
28. Place cell suspension on the goat IgG-coated dish for 30 min (agitate gently at 15 min to ensure access of all cells to the panning surface area) to eliminate microglia.
29. Transfer the non-adherent cells to the Ran-2 antibody coated dish for 30 min to trap meningeal and astrocytes contamination.
30. Transfer the non-adherent cells to 15-ml centrifuge tube.
31. Centrifuge at 1,000 rpm for 5 min at room temperature.
32. Discard supernatant and resuspend pellet in 15 ml of DMEM-BS containing 10 ng/ml PDGF, 10 ng/ml bFGF, 6.3 ng/ml NT-3 and 5 μg/ml insulin.
    O-2A progenitor cells must be grown in chemically defined medium (for example, DMEM-BS), because fetal serum will induce these cells to differentiate into type-2 astrocytes.
33. Transfer cell suspension to PDL-coated 75-cm^2 flask and incubate the cells.
34. You can get O-2A progenitors.

3.1.6. Type-2 Astrocytes

Continue O-2A progenitor cell culture (3.1.5) to step 34.

35. When the cultures become semi-confluent, change a half volume of culture medium with DMEM-BS/FBS (30%) to induce differentiation into type-2 astrocytes.
36. Incubate cells in CO_2 incubator.
37. Change culture medium with DMEM-BS/ FBS (15%) twice a week.

3.2. Immunocytochemistry

1. Plate the purified type-1 and type-2 astrocytes in slide flasks and incubate the cells in CO_2 incubator.
2. When the cultures show desirable cell density, discard the culture medium.
3. Block the cells with PBS (-)/BSA (3%) for 3 min.
4. Remove the blocking solution.
5. Incubate the cells with either monoclonal A2B5 antibody (for type-2 astrocytes) (1:100) or monoclonal Ran-2 antibody (for type-1 astrocytes) for 30 min at 4°C.
6. Wash 3 times with PBS (-).
7. Incubate cells with the FITC-conjugated goat anti-mouse antibody (1:100).
8. Wash the cells with PBS(-).
9. Fix the cells in acid-alcohol solution (5% glacial acetic acid plus 95% ethanol) for 10 min at -30°C.
10. Wash serially 3 times with PBS (-).
11. Add PBS/goat serum (10%) to the cells to block the non-specific binding sites.
12. Briefly, wash the cells with PBS (-).
13. Incubate the both cells with rabbit anti-GFAP serum (1:250) for 30 min at 4°C, and then with Cy-5-conjugated goat anti-rabbit antibodies (1:100) for 30 min at room temperature.
14. Wash serially 3 times with PBS (-).
15. Incubate the cells with 300 nM of DAPI for 5 min.
16. Obtain images using fluorescence microscope.

3.3. RT- PCR

1. Extract total RNA from each cultures using Isogen according to the supplier's instructions.
2. Measure the concentration of RNA with spectrophotometer.
3. Synthesize cDNA from 5 μg of total RNA primed with oligo (dT) primers using a SuperScript Preamplification System.
4. Prepare reaction mixture (50 μl) containing 2 μl of cDNA sample as template, 0.2 μM each of the forward and reverse primers, dNTP mix (0.2 mM each), 1 mM $MgCl_2$, 2.5 units of AmpliTaq Gold enzyme.
5. Program following PCR profile and run the thermocycler: DAO, denaturation for 60 s at 95°C, annealing for 120 s at 57°C and extension for 150 s at 72°C with 33 cycles; β-actin, denaturation for 60 s at 95°C, annealing for 120 s at 57°C and extension for 150 s at 72°C with 20 cycles.
6. Run ten μl of PCR products on a 2% agarose gel.
7. Obtain the images of EtBr-stained DNA band using UV-transilluminator equipped with CCD camera.

4. Results

1. We established new method for purified cultures of type-1 and type-2 astrocytes derived from cerebral cortex and cerebellum as shown in Figure 1 [10]. This method utilizes combination of cell type specific separation by standard shaking method [11] and subsequent purification by immunopanning (for type-2 astrocytes) [12] or Ara-C treatment (for type-1 astrocytes) [13].

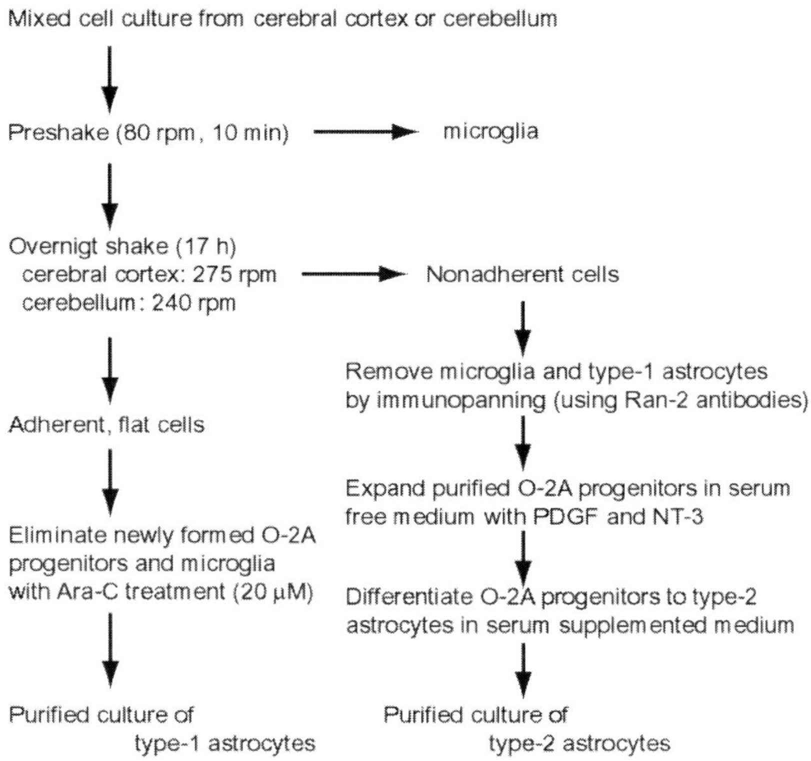

Figure 1. Schematic representation of the protocol used to prepare type-1 and type-2 astrocyte cultures.

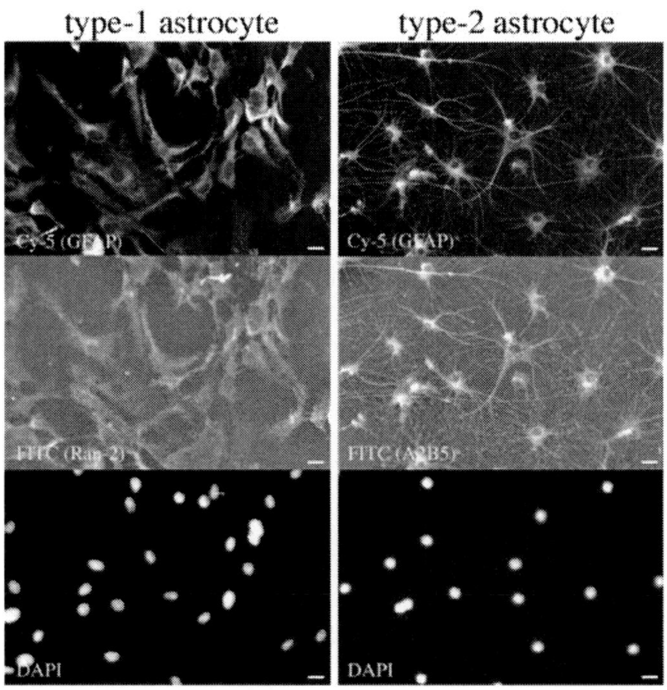

Figure 2. Immnocytochemical identification of purified type-1 (left column) and type-2 (right column) astrocytes from rat cerebellum. Cells cultured on slide flasks were stained with Ran-2 or A2B5 monoclonal antibody (FITC). After acid alcohol fixation, they were incubated with anti-GFAP serum (Cy-5), and then counterstained with DAPI. Scale bar, 10 μm.

2. We confirmed high purity of the cultures using immunocytochemical analysis as shown in Figure 2 [10]. Type-1 astrocytes are characterized by the expression of GFAP and Ran-2, whereas type-2 by GFAP and A2B5. Over 95% cells in our preparation of each type of astrocytes expressed the markers in simultaneously prepared cells in the flask.
3. We detected the RT-PCR products of predicted size (DAO: 398 bp, β-actin: 764 bp) as single ban and observed higher DAO expression in type-1 astrocytes cultures from cerebellum than that in type-2 astrocytes by RT-PCR as shown in Figure 3 [10]. In addition, the expression level was higher in cerebellum than th at in cerebral cortex.

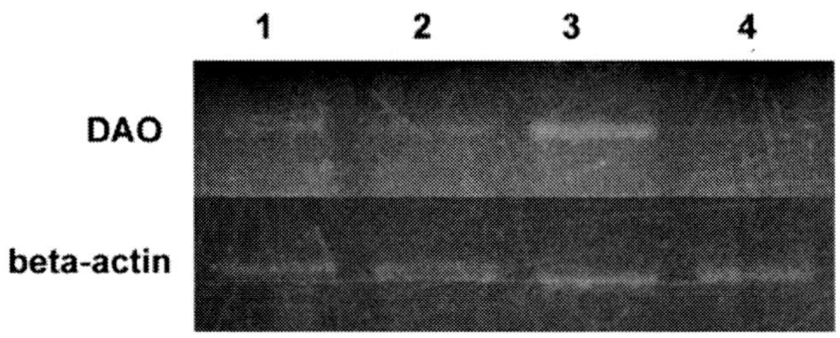

Figure 3. PCR amplication of DAO cDNA and β-actin. Total RNAs extracted from purified astrocyte cultures of cerebral cortex and cerebellum were reverse-transcribed to cDNA. Lane 1; type-1 astrocyte cultures from cerebral cortex, Lane 2; type-2 astrocyte cultures from cerebral cortex, Lane 3; type-1 astrocyte cultures from cerebellum, Lane 4; type-2 astrocyte cultures from cerebellum.

5. Notes

1. This protocol of primary cell cultures corresponds to the experiment with the use of about 12 neonatal rats.
2. When triturating, do not pass air bubbles through the medium because air bubble will kill many cells.
3. When you shake culture flask overnight (3.1.3 step 25), make sure that cells are not exposed to light.
4. For the amplification of rat DAO cDNA, a primer pair (F: 5'-GTGGCATCTTTCG AAGAGGT-3' and R: 5'-AGGCGTTCTCTTTCTAGCCG-3') produced a 398 bp fragment. For rat β-actin cDNA, a primer pair (F: 5'-TTGTAACCAAC TGGGACGATATGG-3' and R: 5'-GATCTTGAT CTTCATGGTGCTAGG-3') amplified a 764 bp fragment.
5. Figures are reproduced from Neuroscience Letters, Vol. 324, Y. Urai, O. Jinnouchi, T.K. Kwak, A. Suzue, S. Nagahiro, K. Fukui, Gene expression of D-amino acid oxidase in cultured rat astrocytes: regional and cell type specific expression, pp 101–104, Copyright 1994, with permission from Elsevier.

References

[1] Krebs, H. A. (1935). Metabolism of amino-acids: deamination of amino-acids. *Biochem J, 29*, 1620–1644.

[2] Horiike, K. Tojo, H., Arai, R., Nozaki, M. & Maeda, T. (1994). D-Amino acid oxidase is confined to the lower brain stem and cerebellum in rat brain: regional differentiation of astrocytes. *Brain Res, 652*, 297–303.

[3] Chumakov, I., Blumenfeld, M., Guerassimenko, O. et al. (2002). Genetic and physiological data implicating the new human gene G72 and the gene for D-amino acid oxidase in schizophrenia. *Proc Natl Acad Sci USA 99*, 13675–13680.

[4] Hashimoto, A., Nishikawa, T., Hayashi, T., Fujii, N., Harada, K., Oka, T. & Takahashi, K. (1992). The presence of free D-serine in rat brain. *FEBS Lett, 296*, 33–36.

[5] Matsui, T., Sekiguchi, M., Hashimoto, A., Tomita, U., Nishikawa, T. & Wada, K. (1995). Functional comparison of D-serine and glycine in rodents: the effect on cloned NMDA receptors and the extracellular concentration. *J Neurochem, 65*, 454–458.

[6] Fedoroff, S. (1985). Macroglial cell lineages. In G. M. Edelman, W. E. Gall, W. M. Cowan (Eds), *Molecular Bases of Neural Development*, (pp. 91-117). New York, NY: John Wiley.

[7] Raff, M. C., Miller, R. H. & Noble, M. (1983). A glial progenitor cell that develops *in vitro* into an astrocyte or an oligodendrocyte depending on culture medium. *Nature 303*, 390–396.

[8] Raff, M.C., Abney, E. R., Cohen, J., Lindsay, R. & Noble, M. (1983). Two types of astrocytes in cultures of developing rat white matter: differences in morphology, surface gangliosides, and growth characteristics. *J Neurosci, 2,* 1289–1300.

[9] Raff, M.C., Abney, E. R. & Miller, R. H. (1984). Two glial cell lineages diverge prenatally in rat optic nerve. *Develop Biol 106*, 53–60.

[10] Urai, Y., Jinnouchi, O., Kwak, K. T., Suzue, A., Nagahiro, S. & Fukui, K. (2002). Gene expression of D-amino acid oxidase in cultured rat astrocytes: regional and cell type specific expression. *Neurosci Lett* 324, 101–104.

[11] McCarthy, K.D. & de Vellis, J. (1980). Preparation of separate astroglial and oligodendroglial cell cultures from rat cerebral tissue. *J Cell Biol*, 85, 890–902.

[12] Barres, B.A., Hart, I.K., Coles, H.S.R., Burne, J.F., Voyvodic, J.T., Richardson, W.D. & Raff, M.C. (1992). Cell death and control of cell survival in the oligodendrocyte lineage. *Cell*, *70*, 31–46.

[13] Noble, M. & Murray, K. (1984). Purified astrocytes promote the in vitro division of a bipotential glial progenitor cell. *EMBO J*, 3, 2243–2247.

In: D-Amino Acids: ISBN 1-60021-075-9
A New Frontier in Amino Acid and Protein Research, pp.523-533

Chapter 6.13

D-Amino Acid Oxidase: A Versatile Tool to Study the Function of the Glial-Derived Neuromodulator D-Serine at Glutamatergic Synapses

Jean-Pierre Mothet*

Institut de Neurobiologie Alfred Fessard, Laboratoire de Neurobiologie Cellulaire et Moléculaire, CNRS UPR 9040, France.

Keywords: D-serine, NMDA receptors, D-amino acid oxidase, synapse

1. Introduction

It has long been thought that D-amino acids only occur in substantial quantity in bacteria or invertebrates [1]. Recently, techniques to distinguish isomers of amino acids in biological samples have led to the identification in some mammalian tissues of large amounts of at least two D-amino acids, D-serine and D-aspartate [2].

D-Serine and its biosynthetic enzyme serine racemase occur primarily in the brain, with highest concentrations in regions enriched in *N*-methyl-D-aspartate receptors (NMDAR). Interestingly, functional activity of NMDAR requires the binding of glutamate and the binding of an endogenous coagonist. Because exogenous glycine and inhibitors of glycine transporters increases NMDA synaptic responses [3,4], it has long been assumed that glycine is the endogenous coagonist. However, several lines of evidence accumulated over the last decade have led to the hypothesis that another amino acid, D-serine, is the endogenous ligand for the glycine site of NMDAR [5]. First, D-serine is localized to protoplasmic astrocytes, which ensheathe glutamatergic synapses bearing NMDAR. Secondly, stimulation of astrocytic AMPA/kainate and metabotropic glutamate receptors led to the release of D-serine [6,7]. Finally, treating brain slices or neuron-astrocyte cultures with purified D-amino acid oxidase (DAAO) reduces NMDAR-driven glutamatergic neurotransmission in hippocampus by reducing extracellular D-serine [8]. Different groups have recently used DAAO to investigate the function of D-serine in different brain areas. By

* Correspondence concerning this article should be addressed to Jean-Pierre Mothet, Email: mothet@nbcm.cnrs-gif.fr, Fax : +33 1 69 82 41 41.

applying DAAO in mixed neurons-astrocytes cultures or in brain slices, Yang and colleagues have clearly shown that D-serine is required for the induction of long term potentiation in hippocampus [9]. In addition, endogenous D-serine is a necessary component of full NMDA receptor activation in the retina. Indeed DAAO added to the bathing medium significantly decreased currents evoked by exogenously applied NMDA [10] and the NMDA receptor-mediated component of the proximal negative response recorded in amphibian eyecup preparation [10]. In the developing cerebellum, DAAO blocks the migration of granule cells, a phenomenon depending on NMDAR activity and controlled by Bergmann glia where D-serine is localized [11]. Pretreatment of the conditioned medium from amyloid β-peptide-activated primary microglia with DAAO also blocked the effects on neuronal $[Ca^{2+}]_i$ and neurotoxicity [12]. Finally, neuronal cell death induced by application of NMDA or by oxygen-glucose deprivation (simulated ischemia) was markedly suppressed by treatment of cerebrocortical slices with DAAO [13].

Thus, DAAO represents a versatile tool for studying the modulation of NMDARs by D-serine in many assays using brain slices or mixed cell cultures. The use of DAAO is a powerful approach to delineate the physiopathological function of D-serine in the brain by *in vitro* and *in vivo* assays. Here, I describe the use of DAAO in combination with electrophysiological recordings as originally designed [8] and employed in my laboratory.

2. MATERIALS

2.1. Animals

Rats were housed under a 12-h light/12-h dark cycle with food and water *ad libitum*. All animal use procedures are in accordance with the French Government *Guide for the Care and Use of Laboratory Animals* (law: 2001-464).

2.2. Cell Cultures

2.2.1. Media and Chemicals

-Bovine serum albumin

-Completed Eagle Modified medium for astrocytes cultures:

Fetal bovine serum (*Biowest, Distributor: AbCys, Paris, France*)		50 mL
Glucose 20%		15 mL
Penicillin-streptomycin 100 x (*Invitrogen, Cergy-Pontoise, France*)		5 mL
Sodium pyruvate 100 mM (*Invitrogen*)		5 mL
L-Glutamine 200 mM (*Invitrogen*)		5 mL
Fungizone 100x (250 µg/ml, *Invitrogen*)		5 mL
MEM 1x (*Invitrogen*)	qsp	500 mL

-Completed HBSS

HBSS 10x (*Invitrogen*)		2.5 mL
Hepes 300 mM		825 µL
Penicillin-streptomycin 100 x		250 µL
Sterile H_2O	qsp	25 mL

-Completed Neurobasal-B27 medium:

1 ml B27 50x (*Invitrogen*) stored at -20°C

125 µl Glutamine 400x (*Invitrogen*) aliquoted and stored at -20°C

500 µl Penicillin-streptomycine 100x (*Invitrogen*) aliquoted and stored at -20°C

500 µl Fungizone 100x (*Invitrogen*) aliquoted and stored at -20°C
50µl Glutamate (*Invitrogen*) dissolved in neurobasal at 25 mM, aliquoted and stored at -20°C
Qsp 50 ml with neurobasal (*Invitrogen*)

-Cytosine arabinoside (*Calbiochem, Distributor: Merck Biosciences Ltd., Nottingham, UK*)
Prepare a 100x stock solution (500µM) by resuspending 12.16 mg in 100 mL of neurobasal medium. The solution was stored at -20°C.

-1 mM L-Cysteine

-DMEM/Hepes solution
Dulbecco's modified Eagle medium (D-MEM) (1X)
with L-Glutamine, 4500 mg/L D-Glucose, 25 mM HEPES, without sodium pyruvate

-DNase I (specific activity, >2000 U/mg, *Roche Applied Science, Mannheim, Germany*)

-Earle's balanced salt solution

-0.5 mM EDTA

-Low molecular weight (53 kDa) poly-D-lysine (*Sigma-Aldrich, St quentin Fallavier, France*)

-Trypsine-DNaseI dissociation solution:

Trypsine 10x (*Invitrogen*)	750 µL
DNaseI	750 µL
Completed HBSSqsp	7.5 mL

2.2.2. Small Material and Equipment for Cell Cultures

70 *µ*m Falcon nylon cell strainer
75 cm² flasks
Six- well culture plates
Glass coverslips
Dissecting instruments
37°C water bath
Binocular loop with back illumination
Cell hemacytometer
Incubator, 5% CO_2, humidified
Dissecting tools (fine forceps, scissors)

3. METHODS

3.1. Preparation of Hippocampal Slices for Patch-Clamp Recordings

Brain slices are widely used for electrophysiological studies [14,15], and the methods used to prepare brain slices are widely documented. Different labs have their own individual variants; below is a brief description of the procedures that we use and which can be applied for a variety of brain areas.

(a) Postnatal day 20-25 male Sprague-Dawley rats were anesthetized lightly with halothane and decapitated. The brain was removed and placed in oxygenated (95% O_2/5% CO_2) ice-cold Ca^{2+}-free solution A (Table 1) within 60 seconds of decapitation. *NB: The solution should be prepared in advance and place for 2hrs in the freezer.*

(b) Pause for 3-5 minutes while the tissue cools down.

(c) Trim or block off the tissue using clean cuts with a sharp scalpel blade in preparation for gluing to the stage of the tissue slicer. Remove the hippocampi. *NB: Avoid squeezing or otherwise deforming the tissue at this stage.*

(d) Apply a thin layer of cyanoacrylate glue (Super-glue) to the stage of the tissue slicer and then gently place the tissue at the correct orientation. In the case of hippocampi, we use transverse slices (Figure 1). Immediately, add Ca^{2+} free ice-cold solution (A) over the tissue until it is submerged.

(e) Cut slices (400-500 μm thick) with vibrating slicer.

(f) Using a Pasteur pipette cut and fire-polished to an opening of 3-5 mm across, when produced, transfer each slice to the holding chamber which should be in a water bath at 32-35°C with a good steady flow of O_2/CO_2 bubbling through the solution (A) containing 2 mM $CaCl_2$.

(g) Incubate the slices at 32-35°C for at least 1 h before beginning recordings.

3.2. Neuron-Astrocyte Cultures

Co-cultures of neuron and astrocyte were obtained from rat hippocampi according to the procedures of Liu that we have adapted [16]. The overall scheme of the procedure consists in growing cells in separating cultures with optimized media for each cell type and then to plate the neurons onto confluent astrocytes monolayer. All steps were performed in sterile room.

Table 1. Composition of extracellular solutions.

	Hippocampal slices (A)	Neuron-astrocytes culture (B)
Compounds	Standard pH 7.4	Standard pH 7.4
NaCl	126	140
KCl	1.5	5
$CaCl_2$	2	3
Mg^{2+}	1.5	-
$MgSO_4$	1.5	-
KH_2PO_4	1.25	-
$NaHCO_3$	26	-
Glucose	10	-
Dextrose	-	10

The concentrations mentioned in mM take into account the ionic addition due to pH adjustment.

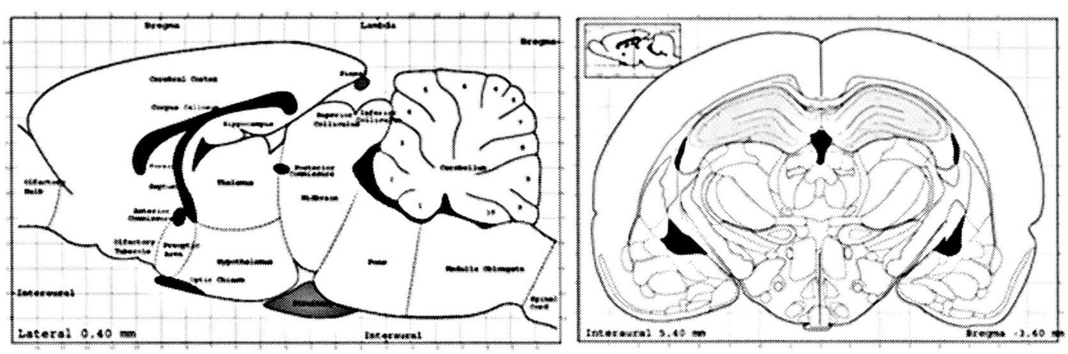

Figure 1. Schematic representation of longitudinal (left) and coronal section (right) of rat brain. The hippocampus is localized between the thalamus and the corpus callosum. Lateral 0.40 mm (left) and Bregma -3.60 mm (Paxinos and Watson, The rat brain, 1996).

3.2.1. Dissection of Hippocampi for Neurons Cultures

We are here describing the procedures used to dissect the hippocampi for astrocytes and neuron cultures. This method can be applied to E16-E19 rat embryos and to 1-5 postnatal rat pups.

1. Prepare the area for dissection by spraying with 70% ethanol and wiping dry.
2. Sterilize dissection instruments by immersion in 70% ethanol for 10 min. Air-dry dissection instruments by propping on Petri dish covers.
3. Place two 100-mm petri dishes on ice and add 15 ml of DMEM/Hepes to each.
4. Euthanize a pregnant rat with CO_2. When newborn rats are used, they are quickly decapitated with scissors.
5. Remove the embryos from the uterus and place in the petri dish containing DMEM/Hepes on ice. The head of newborn rats are also placed in ice-cold DMEM/Hepes.
6. Use fine scissors or forceps to remove bone (cranium) overlying the brain.
7. Transfer brains to the second 100-mm petri dish containing DMEM/Hepes.
8. Remove the pia, starting at the bottom of the brain.
9. Cut the brain along the midline. Pinch off thalamus at the ventral medial surface.
10. With the medial surface of the brain facing you, remove the hippocampus by pinching with a pair of fine forceps.
11. Transfer the hippocampus into a new petri dish containing fresh DMEM/Hepes and cut it with fine forceps into pieces.

3.2.2. Dissociation of Hippocampi for Astrocytes Cultures

Eight to ten hippocampi from 2- to 4-day-old rat pups cleaned of meninges, are transferred to a 50 ml Falcon tube with a cut and flamed Pasteur pipette or a P1000 micropipette, and rinsed twice with 5-10 ml completed Hanks' Balanced Salt Solution (HBSS). Maximum of HBSS is discarded and 5 ml Trypsine/DNaseI solution is added (*NB: minimizing the amount of HBSS prevents dilution of the enzyme solution*). The tissue is then incubated in enzyme solution at 37°C for 20 min in rocking water bath (200 rpm) and mechanically dissociated twice or three times. After decantation, the supernatant is transferred to a new 50 ml Falcon tube. Ice-cold plating medium (completed MEM with 10% FBS) is added (vol/vol) and the tube is kept on ice. Three ml Trypsine/DNaseI solution is added to the cell pellet and the dissociated tissue is incubated at 37°C for 20 min in rocking water bath (200 rpm). After further mechanical dissociation with a P1000 pipette, ice-cold plating medium is added (vol/vol). The contents of the two 50 ml Falcon tubes are pooled and passed through a 70 μm Falcon nylon cell strainer. The dissociated, single-cell type suspension is centrifuged at low speed (~80×*g*) for 15 min and resuspended in plating medium. Viable cells are counted using a hemacytometer. Cells are plated at 0.5-1$x10^6$ cells per flask in 75 cm^2 flasks. Medium is changed completely after 48 h and twice weekly thereafter. Once a confluent monolayer is obtained (after ~10-15 days) the flasks are vigorously shaken overnight on a rotary shaker at 180 rpm at 37°C followed by two changes of medium to remove microglia, O2A progenitor cells and any surviving neurons. For co-culture experiments of astrocytes and neurons, astrocytes are first trypsinized, then re-plated onto glass coverslips pre-coated with low molecular weight (53 kDa) poly-D-lysine. When the cells reached confluence again they are transiently exposed to 5 μM cytosine arabinoside for 2 days and then maintained for <7 days in completed neurobasal-B27 medium before being used for co-cultures. Cultures prepared in this way contain ≥ 95% type-1 astrocytes as determined by glial-fibrillary acid protein positive ($GFAP^+$) immunocytochemistry and morphological examination.

3.2.3. Preparation and Culture of Hippocampal Neurons

The hippocampi when removed, and cut into little pieces as described in section above are dissociated in 5 ml Earle's balanced salt solution (EBSS) containing 20 U/ml papain, 0.01% DNaseI, 0.5 mM EDTA and 1 mM L-cysteine and rocked in a 37°C incubator for 35–40 min. Single-cell type suspensions are obtained by triturating the tissue with a cut and flamed Pasteur pipette, then re-suspended in EBSS with 1 mg/mL trypsin inhibitor (TI) and 1 mg/mL bovine serum albumin (BSA). The suspension is layered over 5 ml of EBSS with 10 mg/mL TI and 10 mg/mL BSA in a 15-mL centrifuge tube. The gradient is spun at ~80×*g* for 5 min to remove dead cells and debris which remain in the supernatant, while live cells compose the pellet. The cell pellet is then re-suspended in neurobasal-B27 medium and plated at a density of 1.5–2×10^5 cells/dish on poly-D-lysine-coated glass coverslips. The cultures are kept at 37°C in a humidified atmosphere without any added CO_2. Most of the cells plated in this manner are immunostained with MAP2, a monoclonal antibody to neuron-specific microtubules.

3.2.4. Mixed Cultures

Mixed culture were prepared in the same manner as pure neuronal cultures except that neurons are plated at 10 000 cells/cm^2 on the monolayer of primary astrocytes (10–14 d) cultured on poly-D-lysine-coated glass coverslips. In mixed cultures, neurons were found to grow on top of a layer of astrocytes. Astrocytes are easily distinguished from neurons by their morphological characteristics under a phase-contrast microscope.

3.3. Electrophysiological Recordings

3.3.1. Recording Apparatus

Electrophysiological experiments are conducted under voltage-clamp conditions, using a computer-controlled MultiClamp 700A microelectrode patch-clamp amplifier (*Axon Instruments, Distributor: Dipsi, Chatillon, France*) in the whole-cell recording mode. The amplifier is operated using a control panel program (MultiClamp Commander host software, *Axon Instruments*) which runs on a Pentium IV Dell personal computer and communicates with the amplifier via a serial cable. Recording electrodes are pulled on a horizontal flaming/Brown micropipette puller (*Sutter Instrument* Model P-97, *Distributor: Dipsi, Chatillon, France*) using thin-walled filamented borosilicate glass (TF150, *Clark Biomedical*). The electrode when filled with the internal pipette solution (Table 2), has a 2-5 MΩ nominal resistance. The series resistance are compensated prior to the recordings. Whole-cell currents are recorded from individual pyramidal cells in 12- to 15-day old mixed cultures under voltage-clamp at room temperature. Whole-cell currents are filtered at 2 kHz and sampled at 5 kHz using a Digidata 1322A interface (AD/DA converter; *Axon Instruments*) driven by the pClamp 8 software package (*Axon Instruments*) installed on a pentium IV Dell personal computer. All recordings are carried out at room temperature (22–25°C) on the stage of a Zeiss inverted microscope (Figure 2). An ungrounded stimulus isolation unit (SIU) is used to supply voltage pulses to the stimulating electrode. SIUs are triggered by a pulse generator.

Table 2. Composition of the intracellular solution.

	Stock solution	Final concentration
HEPES	1000	40
$MgCl_2$	1000	5
Cesium-BAPTA	1000	10
Na_2-ATP	1000	5
Na_3-GTP	1000	0.5
Lidocaine N-ethyl bromide	1000	5

The pH and osmolarity of the solution are adjusted as mentioned in the text. The concentrations are mentioned in mM.

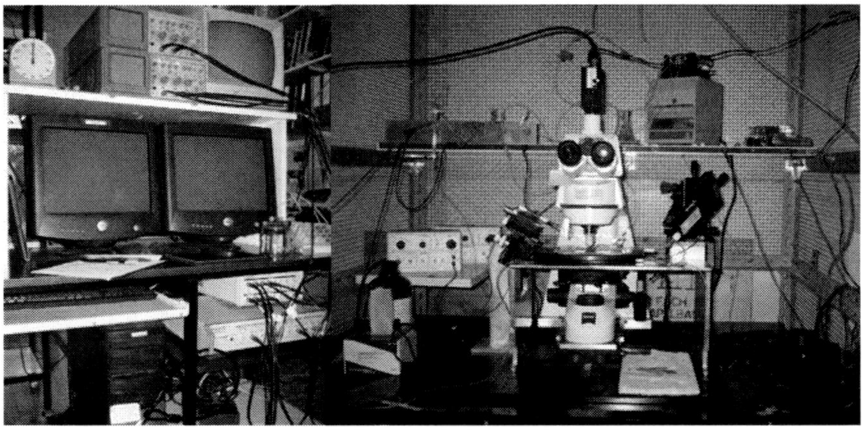

Figure 2. Example of an electrophysiological setup for brain slices/cultures recordings.

3.3.2. The Whole-Cell Configuration

After the formation of a 'giga-ohm seal' in the cell-attached mode, the whole-cell configuration is obtained by applying a short and strong suction to rupture the membrane patch. With this method, low resistance pathway is obtained (access resistance ≈ 5 MΩ), and therefore an improved voltage-clamp and a nearly complete dialysis of the intracellular compartment can be achieved. However, this means also that soluble endogenous components can be lost due to dialysis. In order to allow full equilibration of the intracellular content, experiments in the whole-cell mode are always started 10-15 min after rupture of the membrane patch.

3.4. Perfusion Media and Intracellular Solutions

3.4.1. Composition of the Extracellular Solutions

Experiments were performed at room temperature and cells are continuously superfused with the extracellular medium. The standard perfusion solution (A, see Table1) is a modified Krebs' solution, whose pH is buffered with Hepes and adjusted to 7.4 with sodium hydroxide (NaOH). The osmolarity of the extracellular solution is determined by the freezing point depression method using an automatic osmometer and set at 285 mosmol/L by adding mannitol when necessary (Table 1).

In whole-cell recordings experiments of NMDAR-driven synaptic currents from neurons in mixed hippocampal neuron-astrocyte cultures or from brain slices, the standard perfusion solutions are completed with NBQX (10 μM), picrotoxin (50 μM), strychnine (5 μM) and SCH50911 (10 μM) to block the AMPA/Kainate currents, the $GABA_A$ currents, the glycine inhibitory sensitive currents and the $GABA_B$ currents, respectively (extracellular solution C).

Tetrodotoxin (TTX, *Tocris*) is added at 1 μM (final concentration) to standard perfusion solutions to eliminate the sodium currents (action potentials) in experiments where synaptic responses are evoked by application of agonists.

3.4.2. Perfusion System and Application of Test Substances

The recording chamber is gravity fed at 0.5-1 mL/min with the extracellular solutions contained in 50 mL syringes. Test substances are dissolved in external solution to the desired final concentration and are applied by using the multibarrel gravity-fed perfusion system. Generally, changes of the extracellular medium and drug applications are realised by switching between syringes. Thus, because of the small volume of the recording chamber (≈ 500 μl), the bath solution can be completely exchanged within 15-20 sec. However, in some experiments, drugs are directly applied by air-pressure ejection (5-15 psi) onto recorded cells from a glass micropipette. The technique of Puffer accelerates drug delivery or medium changes and then limits the exposure to the recorded cell. The perfusion solutions are maintained at room temperature (22-25°C).

3.4.3. Composition of the Intracellular Solution

The intracellular solution is mainly caesium-based to block most of the potassium currents and therefore unmask Ca^{2+} currents and improve the space-clamp. The pH of the pipette solutions is adjusted to 7.3 with CsOH and the osmolarity set at ≈ 280 mosm/L with caesium gluconate (CsGl). Lidocaine N-ethyl bromide (QX 314) is used to block Na^{+} channels. The composition of the intracellular solution used is described in more detail in Table 2.

3.5. Treatment of Brain Slices or Cultures with D-Amino Acid Oxidase

D-Amino acid oxidase (DAAO) is a flavoprotein oxidase, flavin being the only prosthetic group. In the presence of molecular oxygen, DAAO oxidatively deaminates D-amino acids to corresponding α-keto acids.

In our initial work, we purchased DAAO from Boehringer Mannheim [8]. This enzyme is not anymore commercially available from this company. Different sources of pig kidney DAAO can be used as substitutes provided enough purity and high activity. We and colleagues have tested the effects of the enzyme from Worthington Biochemical Corporation (http://www.worthington-biochem.com) or from Sigma and found that these enzymes worked well. Other commercial DAAOs can be purchased from USB Corporation (http://www.usbweb.com, Ohio, USA) or ABCR (http://www.abcr.de, Karlsruhe, Germany) but we have no experience with these enzymes. All commercially DAAOs are partially purified enzyme provided has a lyophilized powder or as a suspension in 3.2M ammonium sulfate solution. Each batch of enzyme should be dialyzed (after re-suspension when starting with lyophilized powder) for 8-10 h at 4°C against 20 mM phosphate buffer (pH 7.4) containing flavin adenine dinucleotide (FAD) in order to remove contaminants and ammonium sulphate. The dialyzed enzyme is stored at -20°C until use. Another source for DAAO is provided by recombinant protein purified from microbial sources. We are now using in routine the DAAO purified from the yeast *Rhodotorula gracilis* (RgDAAO). RgDAAO displays peculiar features in comparison with mammalian enzyme, such as the tight binding of the coenzyme FAD (K_d=2.0 x 10^{-8}M), a strict stereospecificity for D-amino acids, and a significant higher catalytic efficiency for its substrates [17]. Furthermore, for control experiments we are using catalytic inactive mutant.

Active or inactive DAAO (mutant enzyme or apoenzyme) is diluted in extracellular solutions at 0.1-0.2U/mL. FAD is added at 100 μM when mammalian enzyme is used. *Note: The yellow coloration observed after addition of FAD to the extracellular solution is normal. In control experiments we have found no effect of FAD at 100 μM on electrophysiological recordings.*

3.6. Protocols of Electrophysiological Recordings and Brain/Cells Treatment with DAAO

3.6.1. Basic Considerations for Recording Synaptic Activity from Brain Slices or Cultures

1. Position the stimulus electrode on the tissue slice near, but not necessarily on, the region to be activated before the patch pipet forms a seal.

NB: After seal formation, more precise positioning will be made. With visual patch recording (as opposed to "blind patching"), one can, with practice, visually identify cells that will likely produce stable whole-cell recordings.

2. Establish whole-cell recording.

NB: At this stage, compensate series resistance and decide if it is low enough and stable enough for the recording.

3. Identify presynaptic fibers.

NB: Placement of the stimulus pipet on a presynaptic cell or axon can be a trial-and-error process. It is important to not physically dislodge the patch pipet during the process.

4. Evaluate stability response amplitude. Record with low-frequency stimulation (0.1 to 1 Hz) for several minutes to establish that the synaptic response is relatively stable.

NB: Amplitude fluctuation is normal, but systematic "run-down" (decline of response by >10 to 15% during the recording period) may mean the recording should be abandoned. If run-down is unavoidable, then record long enough to determine the rate of run-down before applying a drug.

5. Change from standard bath to recording solution.

NB: The standard extracellular solution A was used to establish the recording. Once the usable cell is identified, we switched the solution to one containing the pharmacological cocktail needed to block unwanted ion channels (Solution C).

6. Begin recording responses, monitoring the cell continually. Check series resistance and recompensate it every few minutes. Check for run-down. If testing the effects of a drug, make certain to demonstrate reversibility after washing out the drug (i.e., response returns to within 90% of the control value).

4. Results

4.1. Brain Slices Experiments

A single slice is transferred to the recording chamber and continuously perfused with normal extracellular solution completed with inhibitors of non-NMDA currents (extracellular solution C). Stimuli (50 μsec; 50-500 μA) are delivered through surface bipolar stainless steel electrodes in CA1 stratum radiatum to stimulate the Schaffer collateral fibers that contact the pyramidal neurons (Figure 3). Evoked postsynaptic currents (EPSCs) are recorded in whole-cell configuration every 15-20 sec interval.

After a baseline period 15-30 min, hippocampal slices are perfused with extracellular solution C containing DAAO for at least 30 min while recording the evoked EPSCs as before. This treatment should reduce the NMDAR-driven EPSCs. Figure 3B shows two representations of electrophysiological recordings obtained in control conditions and after treatment with DAAO when postsynaptic neuron is maintained at + 40mV. The NMDARs generate a relatively slow synaptic current. When the postsynaptic cell is maintained at low potentials (-60 to -70 mV), Mg^{2+} is removed from the extracellular solution and stimulation will induce an inward current.

NB: In the initial work we added catalase (1000U/ml) to extracellular DAAO containing medium to circumvent the possible effect of H_2O_2 produced during DAAO-driven degradation of D-serine. This could be easily omitted as we found no difference with or without catalase.

4.2. Electrophysiological Recordings Experiments in Cultures

Mixed astrocyte-neuron cultures are preferentially used in order to provide enhanced access to synaptic sites during DAAO perfusion. Electrophysiological recordings of synaptic events in mixed astrocyte-neuron cultures are performed by maintaining recorded neurons at -70 mV. For assaying synaptic connectivity, each neuron is stimulated at a low frequency (0.02 Hz) by 1-msec-step depolarization from -70 to + 30 mV in voltage-clamp mode. When spontaneous and evoked NMDAR-mediated inward synaptic events are recorded, the preparation is exposed to active DAAO for 5 min. To directly access the effect and the time course of DAAO on postsynaptic NMDARs, NMDAR currents are evoked by application of NMDARs' agonists (Figure 4).

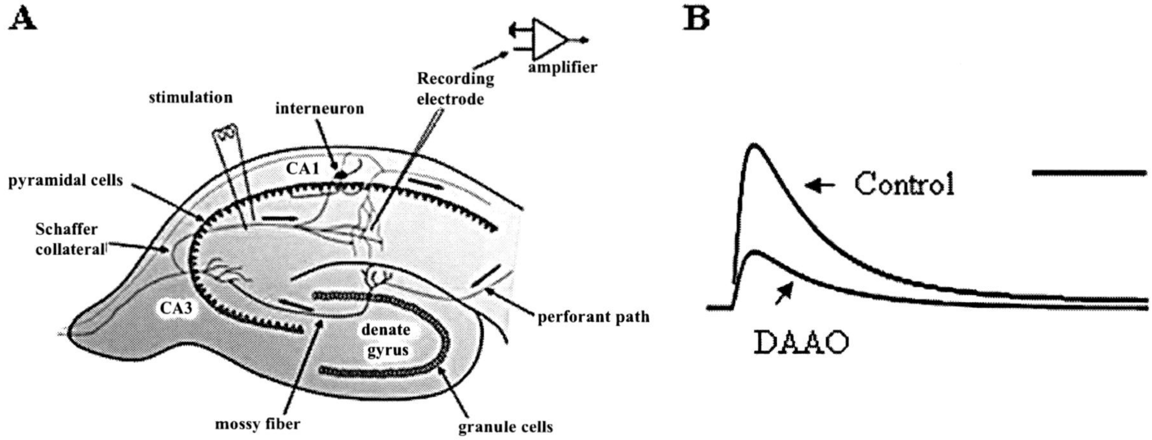

Figure 3. A) Schematic representation of a hippocampus slice with the emplacement of stimulation and recording electrodes. B) Two EPSCs are shown under control condition and after DAAO application in the bath medium. Bar: 200 msec.

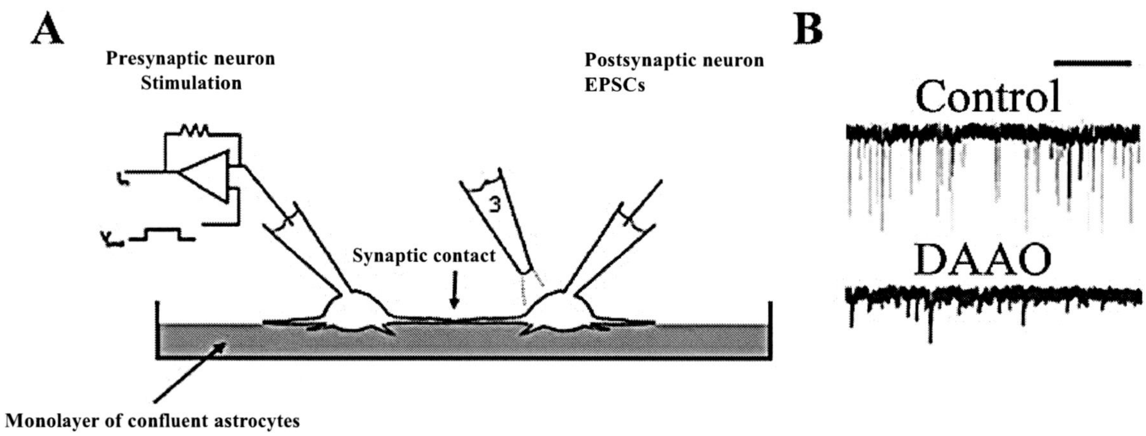

Figure 4. A) Schematic representations of recording configuration used in mixed cell cultures. The presynaptic neuron is depolarized while postsynaptic NMDAR currents are recorded in the postsynaptic neuron. In some experiments, NMDA was directly applied on the soma of the recorded cells (3). B) An example of spontaneous current is shown on right. After treatment with DAAO, the amplitude and the frequency of synaptic currents are reduced. Bar: 200 msec.

ACKNOWLEDGEMENTS

I am grateful to Magalie Martineau and Gilles Ouanounou for critical reading and their helpful suggestions during the redaction of the manuscript. I thank Dr. Gérard Baux for his continual support. I'd like to thank also Dr. Loredano Pollegioni for kindly providing us with RgDAAO.

REFERENCES

[1] Fujii, N. and Saito, T. (2004). Homochirality and life. *Chem Rec*, *4*, 267-278.

[2] Hashimoto, A. and Oka, T. (1997). Free D-aspartate and D-serine in the mammalian brain and periphery. *Prog Neurobiol*, *52*, 325-353.

[3] Berger, A. J., Dieudonne, S. and Ascher, P. (1998). Glycine uptake governs glycine site occupancy at NMDA receptors of excitatory synapses. *J Neurophysiol*, *80*, 3336-3340.

[4] Eulenburg, V., Armsen, W., Betz, H. et al. (2005). Glycine transporters: essential regulators of neurotransmission. *Trends Biochem Sci*, *30*, 325-333.

[5] Miller, R. F. (2004). D-Serine as a glial modulator of nerve cells. *Glia*, *47*, 275-283.

[6] Schell, M.J., Molliver, M. E. and Snyder S. H. (1995). D-Serine, an endogenous synaptic modulator: localization to astrocytes and glutamate-stimulated release. *Proc Natl Acad Sci USA*, *92*, 3948-3952.

[7] Mothet, J. P., Pollegioni, L., Ouanounou, G. et al. (2005). Glutamate receptor activation triggers a calcium-dependent and SNARE protein-dependent release of the gliotransmitter D-serine. *Proc Natl Acad Sci USA*, *102*, 5606-5611.

[8] Mothet, J. P., Parent, A .T., Wolosker, H. et al. (2000). D-Serine is an endogenous ligand for the glycine site of the *N*-methyl-D-aspartate receptor. *Proc Natl Acad Sci USA*, *97*, 4926-4931.

[9] Yang, Y., Ge, W., Chen, Y. et al. (2003). Contribution of astrocytes to hippocampal long-term potentiation through release of D-serine. *Proc Natl Acad Sci USA*, *100*, 15194-15199.

[10] Stevens, E. R., Esguerra, M., Kim, P. M. et al. (2003). D-Serine and serine racemase are present in the vertebrate retina and contribute to the physiological activation of NMDA receptors. *Proc Natl Acad Sci USA*, *100*, 6789-6794.

[11] Kim, P. M., Aizawa, H., Kim, P. S. et al. (2005). Serine racemase: activation by glutamate neurotransmission via glutamate receptor interacting protein and mediation of neuronal migration. *Proc Natl Acad Sci USA*, *102*, 2105-2110.

[12] Wu, S. Z., Bodles, A. M., Porter, M. M. et al. (2004). Induction of serine racemase expression and D-serine release from microglia by amyloid beta-peptide. *J Neuroinflammation*, *1*, 1-11.

[13] Katsuki, H., Nonaka, M., Shirakawa, H. et al. (2004). Endogenous D-serine is involved in induction of neuronal death by *N*-methyl-D-aspartate and simulated ischemia in rat cerebrocortical slices. *J Pharmacol Exp Ther*, *311*, 836-844.

[14] Piet, R., Vargova, L., Sykova, E. et al. (2004). Physiological contribution of the astrocytic environment of neurons to intersynaptic crosstalk. *Proc Natl Acad Sci USA*, *101*, 2151-2155.

[15] Bollmann, J.H. and Sakmann, B. (2005). Control of synaptic strength and timing by the release-site Ca^{2+} signal. *Nat Neurosci*, *8*, 426-434.

[16] Liu, Q. Y., Schaffner, A. E., Li, Y. X. et al. (1996). Upregulation of GABAA current by astrocytes in cultured embryonic rat hippocampal neurons. *J Neurosci*, *16*, 2912-2923.

[17] Pilone, M. S. (2000). D-Amino acid oxidase: new findings. *Cell Mol Life Sci*, *57*, 1732-1747.

In: D-Amino Acids: ISBN 1-60021-075-9
A New Frontier in Amino Acid and Protein Research, pp.535-543

Chapter 6.14

RESONANCE RAMAN STUDY ON CHARGE-TRANSFER COMPLEXES OF PORCINE D-AMINO ACID OXIDASE

***Yasuzo Nishina**,1, Kyosuke Sato*[2] *and Kiyoshi Shiga*[2]**
[1]Department of Physiology, School of Health Sciences and [2]Department of Molecular Physiology, Graduate School of Medical Sciences, Kumamoto University, Japan.

Keywords: D-amino acid oxidase, charge-transfer complex, flavoenzyme, Raman spectroscopy, resonance Raman spectroscopy.

1. INTRODUCTION

Porcine kidney D-amino acid oxidase [D-amino acid:O_2 oxidoreductase (deaminating), EC 1.4.3.3] (DAO), which has FAD as the prosthetic group, is one of the most extensively investigated flavoenzymes. DAO catalyzes the dehydrogenation of the substrate D-amino acid to the corresponding imino acid; and the imino acid is spontaneously hydrolyzed to 2-ketoacid and ammonia. In enzyme reactions, specific interactions between enzyme and substrate take place during the course of the reaction. The interactions are responsible for the substrate recognition/specificity and the remarkable acceleration of the reaction. Thus, information about the interactions, e.g., the origin and strength of the interacting force, is essential for elucidation of the molecular mechanism of the catalytic reaction. Since flavin can exist in three redox states (Fig. 1) [1], i.e., oxidized, semiquinoid (one-electron reduced), and reduced (two-electron reduced) states, information about the interaction of substrate/ligand with DAO in three redox states is valuable for investigation of the molecular mechanism. The charge-transfer (CT) complexes of DAO in the three redox states are known [2,3], and are attractive materials for elucidation of the molecular mechanism; the charge-transfer complexes are characterized by a broad long wavelength (500–700 nm) absorption band. Figure 2 shows the absorption spectra of typical CT complexes of DAO in the three redox states.

* Correspondence concerning this article should be addressed to Yasuzo Nishina, E-mail: nishina@hs.kumamoto-u.ac.jp, FAX: +81-96-373-5490.

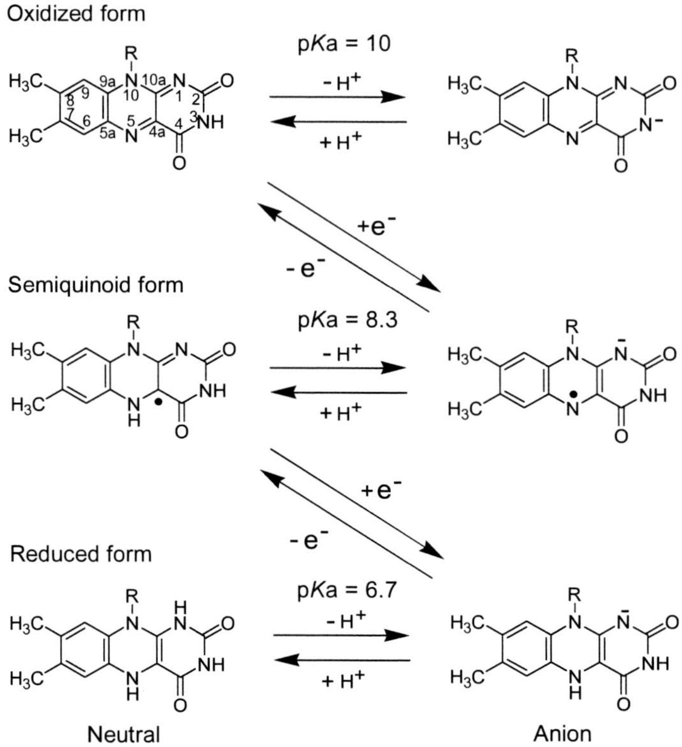

Figure 1. The structures of neutral and anionic flavin species in the three redox states. Modified from Ref. [1].

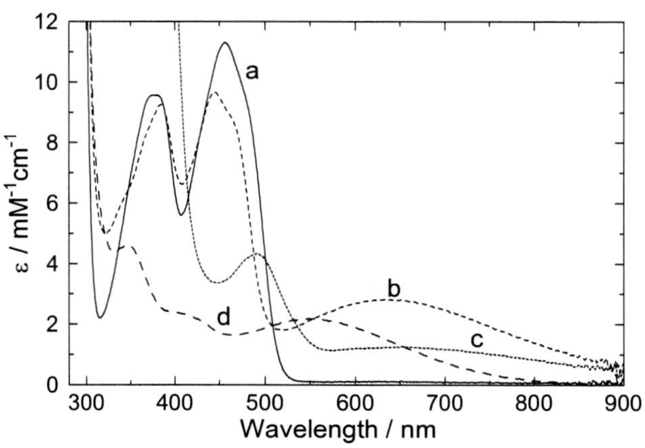

Figure 2. Absorption spectra of the CT complexes of DAO in three redox states. (a) oxidized DAO free in solution; (b) the complex of oxidized DAO with Δ^2-piperideine-2-carboxylate (D2PC); (c) the complex of semiquinoid DAO with Δ^1-piperideine-2-carboxylate (D1PC); (d) the complex of reduced DAO with D1PC. Modified from Ref. [21].

Vibrational (infrared and Raman) spectroscopies bring out the information about molecular vibration. Molecular vibration is sensitive for the chemical nature of a molecule, e.g., configuration of atoms and chemical bonds, thus the spectroscopies are useful for not only identification of molecule but also the elucidation of molecular structure. Furthermore, since molecular vibration is affected by the interaction between the molecule and the environment, the spectroscopies are also powerful for the investigation on the interaction. Among them, resonance Raman (RR) spectroscopy has provided a practical method for observing selectively the vibrational bands of chromophoric groups of biomolecules without the interference of Raman bands of the surrounding amino acid residues of apoprotein, because the bands of the chromophore are selectively enhanced with excitation within the absorption band. RR spectroscopy on flavoproteins has been shown to be a powerful tool for studying the structural as well as mechanistic details [4,5] and a practical method has been shown [6]. In particular, RR studies on CT complexes of flavoenzymes are useful for structural determination of the enzyme-bound ligand and for elucidation of the interaction between the flavin and the ligand, because Raman scattering excited at the CT absorption band selectively reveals the vibration of chemical groups involved in the complex formation [7–17].

In RR studies, correct assignments of the RR bands are very important for the above purpose. The isotope effects of the bands are extremely useful for band assignment. Furthermore, comparison of the vibrational spectra of a substrate/ligand bound in an active site with the spectra of the same species free in solution is also important to yield information not only on the structure but also on the electron distribution in substrate functional groups and on the molecular interaction between the substrate/ligand and the active site.

In this article, a way of preparing solution for RR studies about charge-transfer complexes of DAO, assignments of observed Raman bands, analysis of interactions among flavin, substrate/ligand, and apoDAO are described.

2. Materials

2.1. Enzymes

1. DAO purified in the form of its benzoate complex has been stored in the condition of ammonium sulfate precipitates at 5 °C. The enzyme was stable at least for several months.
2. DAO stock suspension is dissolved by a dialysis against 50 mM sodium pyrophoshate buffer pH 8.3. Benzoate is removed from the enzyme by passing the enzyme after reduction with D-alanine through a Sephadex G-50 column.
3. The apoenzyme of DAO is prepared by repeated ultrafiltration of DAO with a Centricon YM-30 (Millipore) at 5 °C. FAD was removed with 50 mM sodium pyrophosphate buffer, pH 8.3, containing 0.3 mM EDTA and 2 M KBr, and then KBr is removed with the same buffer, but containing no KBr.
4. The holoenzyme is reconstituted with isotope-labeled FAD by incubation of FAD samples in slight molar excess with the apoenzyme at room temperature for 30 min in 50 mM pyrophosphate buffer. Excess FAD is removed by gel filtration or ultrafiltration.
5. The concentrations of the holoenzyme and the apoenzyme are determined using molar absorption coefficients of 11,300 M^{-1} cm^{-1} at 455 nm and 72,600 M^{-1} cm^{-1} at 280 nm, respectively.

2.2. Chemicals

1. Synthesis of isotopically enriched FAD: ^{13}C- and ^{15}N-enriched riboflavin is essentially according to Yagi et al [18]. Isotope-enriched diethyl malonate, urea, and sodium nitrite are used as starting materials. [4-carbonyl-^{18}O]riboflavin is prepared by incubating riboflavin in 1 M $Na^{18}OH$ ($H_2{}^{18}O$) [19]. Each enriched riboflavin is converted to FAD with FAD synthetase from *Brevibacterium ammoniagenes* [20].
2. Various isotope-enriched amino acids are commercially available. To prepare isotope-labeled ligands, various commercial available compounds are used as starting materials, e.g., [carboxyl-^{13}C]*o*-aminobenzoate (OAB) can be prepared by using [α-^{13}C]toluene as starting material [21].

3. Method

3.1. Measurements of Raman and Resonance Raman Spectra

1. Raman spectra are measured in 50 mM sodium pyrophosphate or potassium phosphate buffer at room temperature (ca. 25 °C). Sample solution should be clarified by the membrane filter of 0.45 μm pore size before use.
2. Raman spectra shown here were obtained with a JASCO NR-1800 spectrophotometer (Japan Spectroscopic) with a He-Ne laser (NEC GLG 5900)(632.8 nm) or an Ar ion laser (INNOVA 70-2)(488.0 and 514.5 nm) as a light source. Kr ion (647.1 nm, 752.6 nm) and He/Cd (441.6 nm) lasers are also used in general.
3. The enzyme concentration of charge-transfer complexes for RR spectra are ca. 0.5 mM and the laser power (632.8 nm) at the sample point is ca. 30 mW.
4. The concentration for Raman spectra of ligands free in solution are ca. 300 mM and the laser powers at the sample point are ca. 30 and ca. 50 mW for lights of 632.8 nm and 488.0 nm (or 514.5 nm), respectively.
5. For the measurements of Raman spectra, ca. 30 μl of sample is put into a cell (quartz or pyrex, 2×2 mm internal width and 50 mm height) and excitation light is irradiated from the bottom of the cell: but the alignment of irradiation and detection is different in each apparatus.
6. The wavenumber axis of the Raman spectra is calibrated against known Raman peaks of standard material such as indene.

3.2. Sample Preparation

1. Complexes of oxidized DAO: The complex of DAO with a ligand such as OAB for measurement of RR spectra is prepared by concentration of DAO solution containing a ligand with a Centricon YM-30. The high concentration sample can be prepared, for the solubility of the DAO complex is higher than that of ligand-free DAO.
2. Complexes of reduced DAO: The purple complex of DAO with D-proline is prepared by mixing concentrated DAO (0.3–0.8 mM) with D-proline (ca. 50 mM) and the complex with D-alanine is obtained by addition of pyruvate (ca. 50 mM), ammonium sulfate (ca. 150 mM), and D-alanine (ca. 100 mM) to DAO. The complex of reduced DAO with a ligand such as picolinate is prepared by reduction with D-arginine (20–60 mM) of DAO solution containing a ligand.
3. Complexes of semiquinoid DAO: The complex of semiquinoid DAO with a ligand such as picolinate is prepared by the addition of dithionite (10–30 mM) to DAO solution containing ligands (100–300 mM, the concentration depends on the ligand), and kept in an N_2 atmosphere. The Raman spectra can also be obtained even under aerobic conditions for at least 2 h.

4. Results

Complexes of oxidized DAO——When OAB binds to DAO, a very broad absorption band appears in the long-wavelength region extending beyond 700 nm. In the RR spectra (Fig. 3) of the complex with excitation at 632.8 nm within the broad absorption band, Raman bands at 1,583 and 570 cm^{-1} are derived from oxidized flavin and OAB, respectively. These results verify that the broad absorption band has a CT nature of the interaction between oxidized flavin and OAB [21]. The 1,583-cm^{-1} band is known to involve the vibrational displacements of the N(5) and C(4a) atoms of isoalloxazine [22] and the 570-cm^{-1} band is derived from a ring deformation mode of OAB. This suggests that OAB lies along the N(5)=C(4a) bond and parallel to the flavin face [21]; the positioning was confirmed by X-ray crystallography [23]. The 1,366-cm^{-1} band shifts to 1,355 cm^{-1} with [carboxyl-^{13}C]OAB. The 1,383-cm^{-1} band of OAB free in solution shifted to 1,362 cm^{-1} upon [carboxyl-^{13}C]-labeling. The bands associated with C=O stretching of the COOH and COO$^-$ symmetric stretching of carboxylate appear around 1,700 cm^{-1} and in the 1,340–1,440 cm^{-1} region, respectively. The difference spectrum in the inset shows the 1,368-cm^{-1} band shifts to 1,355 cm^{-1} upon [carboxyl-^{13}C]-labeling. Thus the 1,368- and 1,383-cm^{-1} bands in the complexed state and in the free state, respectively, are clearly assigned to the COO$^-$ symmetric stretching, but not to COOH. This indicates that the carboxylate group of OAB remains unprotonated when it binds to DAO. Data on COO$^-$ stretching are summarized in Table 1. The band-shift from 1,383 to 1,368 cm^{-1} upon the complexation (15 cm^{-1} downward shift) shows that the bond strength of the carboxyl group in OAB is weakened by the complexation. This suggests the existence of the interaction of the carboxylate group in the active site; the interaction of the carboxylate with Arg283 and the hydroxy moiety of Tyr228 has been clearly shown by studies of crystal structure [23].

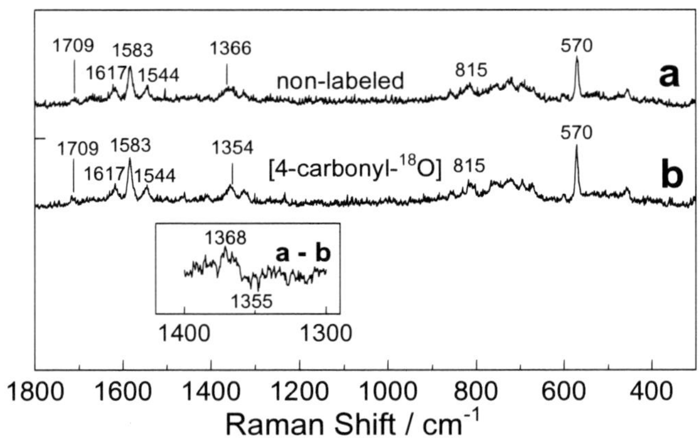

Figure 3. RR spectra excited at 632.8 nm of the complexes of oxidized DAO with nonlabeled and isotopically labeled OAB. The spectra were recorded in 50 mM sodium pyrophosphate buffer, pH 8.3. The inset shows the difference spectrum (a – b). Modified from Ref. [21].

Table 1. COO$^-$ Raman frequencies of ligands free in solution and bound to DAO (cm^{-1}).

	Free (m)	Complex (n)	m – n
Oxidized DAO-AECK	1,376	1,362	17
Oxidized DAO-OAB	1,383	1,368	15
Reduced DAO-picolinate	1,393	1,376	17
Reduced DAO-D1PC	1,387	1,369	18
	1,352	1,339	13
Semiquinoid DAO-D1PC	1,387	1,371	16
	1,352	ca. 1,339	ca. 13

Reproduced from Ref. 21. AECK, aminoethylcysteine-ketimine; D1PC, Δ^1-piperideine-2-carboxylate

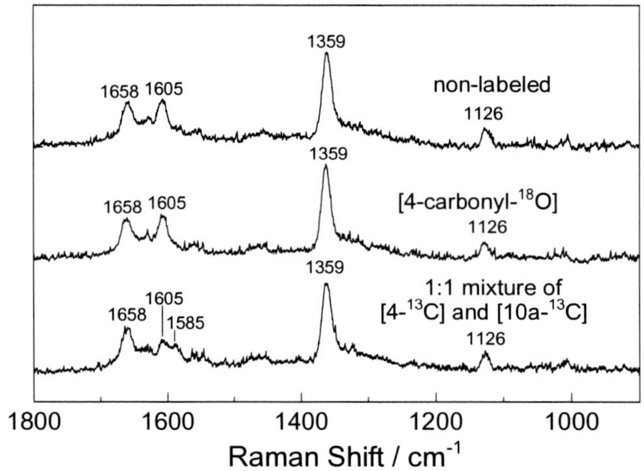

Figure 4. RR spectra of the purple intermediate with D-proline and native DAO or reconstituted with FADs isotopically enriched in the flavin moiety. Positions enriched in the flavin moiety are indicated in the figure. The spectra were recorded in 50 mM sodium pyrophosphate buffer, pH 8.3. Reproduced from Ref. [19].

Complexes of reduced DAO——RR studies for purple intermediates of DAO produced in an anaerobic reaction of DAO clarified that the purple intermediate has the charge-transfer nature between anionic reduced flavin and an imino acid in a zwitterionic form derived from D-amino acid as described below. Figure 4 shows the RR spectra of the purple intermediates prepared with D-proline as a substrate and the DAO reconstituted with FAD, where designated positions of flavin are enriched. Various isotope shifts are summarized in Table 2. The 1,658-cm^{-1} band with D-proline is assigned to the C=N stretching mode of an imino acid derived from D-proline, i.e., Δ^1-pyrrolidine-2-carboxylate, since the band shifts to 1,633 cm^{-1} upon [^{15}N]-substitution and its stretching frequency is generally found in this frequency region. Since the band shifts to low frequency in D_2O, the imino nitrogen of the imino acid is in a protonated cationic form. The 1,359-cm^{-1} band with D-proline shifts to 1,341 cm^{-1} upon [carboxyl-^{13}C]-labeling; thus the band involves a contribution of COO^- symmetric stretching vibration. The band around 1,605 cm^{-1} does not shift upon any isotope substitution of substrates, but downshifted by 15–20 cm^{-1} and 4 cm^{-1} upon [4,10a-$^{13}C_2$]- or [4a-^{13}C]-labeling, respectively. The 1,605-cm^{-1} band splits into two bands at 1,605 and 1,585 cm^{-1} in the 1:1 mixture of [4-^{13}C]- and [10a-^{13}C]-FAD. This indicates that only one of the labels affects the band frequency. The band is downshifted by 4 cm^{-1} upon [4a-^{13}C]-labeling, but is not shifted upon [4-carbonyl-^{18}O]-labeling or [1,3-$^{15}N_2$]-labeling. Furthermore, the band frequency is too high to be associated with a single bond. Therefore, the band is mainly due to the C(4a)=C(10a) stretching mode and the 1,585-cm^{-1} band observed in the mixture is from the enzyme reconstituted with [10a-^{13}C]-FAD. The bands derived from oxidized or semiquinoid flavin observed near 1,605 cm^{-1} have different isotope effects from the 1,605-cm^{-1} band. Therefore, it is concluded that the band around 1,605 cm^{-1} is derived from fully reduced flavin. Furthermore, the flavin is in the anionic reduced form as shown by NMR [24]. The spectra of the ligand free in solution are also important for comparison with the bound form. Piperideine-2-carboxylate (DPC, an oxidative product of D-pipecolate), a cyclic imino acid, is stable even in aqueous solution; it is the ligand in the purple complexes of DAO prepared by anaerobic mixing with D-lysine or D-pipecolate (Fig. 2). DPC in solution is mainly in the imino form (D1PC), not in the enamine form (D2PC). DPC binds to reduced and oxidized DAOs in the forms of zwitterionic D1PC and monoanionic D2PC, respectively. C=N and COO^- stretching frequencies of D1PC free in solution and bound to reduced native DAO or DAO reconstituted with 7,8-Cl_2-FAD are summarized in Tables 1 and 3.

Table 2. Isotopic frequency shifts of RR bands of DAO purple complexes with proline.
In parentheses are the values of isotopic shifts.

FAD							
Non-labeled	[4a-$^{13}C_2$]	[4,10a-$^{13}C_2$]	[2-^{13}C]	1:1 mixture of [4-^{13}C] and [10a-^{13}C]	[5-^{15}N]	[1,3-$^{15}N_2$]	[4-carbonyl-^{18}O]
1,605	1,601 (−4)	1,585 (−20)	1,606 (1)	1,605 (0) 1,585 (−20)	1,606 (1)	1,606 (1)	1,605 (0)
Non-labeled	carboxyl-^{13}C	^{15}N	Non-labeled, D_2O				
1,658	1,658 (0)	1,633 (−25)	1,635 (−23)				
1,359	1,341 (−18)	1,359 (0)	1,371 1,334				

Reproduced from Refs. 11 and 19.

Table 3. C=N Raman Frequencies of D1PC free in solution and bound to DAO (cm^{-1}).
In parentheses are differences from the frequency in free state.

Free	1,704 (0)
Reduced DAO	1,675 (29)
Reduced DAO reconstituted with 7,8-Cl_2-FAD	1,680 (24)
Semiquinoid DAO	1,681 (23)

Reproduced from Ref. 21.

Complexes of semiquinoid DAO——The C=N stretching and COO^- stretching bands of zwitterionic D1PC appear in RR spectra of the CT complex of semiquinoid DAO with D1PC with excitation at 632.8 nm (Tables 1 and 3).

The COO^- bands for the complexed state are observed around 1,370 cm^{-1} as summarized in Table 1, and the frequencies in the complex are lower than those of the ligands free in solution. The magnitude of the shift is almost the same, i.e., about 17 cm^{-1}, regardless of redox state of DAO. This suggests that the interaction modes in the three redox states are similar to each other.

The low shift of the C=N band in the complex is qualitatively explained by weakening of the C=N bond by electron donation from the HOMO of reduced flavin or the SOMO of semiquinoid flavin to the LUMO, which has a node in the C=N, of D1PC. The shift of C=N stretching of D1PC bound to reduced DAO is smaller in the case of 7,8-Cl_2-FAD-DAO than the case of native DAO; electronegative chlorine atom weakens electron donating power of reduced FAD. The smaller shift of C=N stretching of D1PC in the case of semiquinoid DAO than in the case of reduced DAO (Table 3) suggests that the electron-donating ability of semiquinoid flavin is weaker than that of reduced flavin.

CT complexes of many other enzymes besides DAO have been investigated by RR spectroscopy and the spectroscopy is useful for elucidation of the molecular mechanism of the catalytic reaction.

5. Notes

1. A photomultiplier was used to detect the Raman scattered light in the Raman spectra shown here. At present a charge-coupled device (CCD) camera in general is used as a detector for its high

sensitivity. Therefore we can obtain Raman spectra of sample solution at low concentration in a short time.

2. RR spectra are observed without interference by the fluorescence of flavin, for fluorescence does not occur from CT absorption band.
3. This method is applicable to CT complexes of various flavoenzymes.
4. The observed bands have been assigned by experimental and theoretical methods, e.g., on the basis of the band shifts with chemically and isotopically substituted flavins or ligands, and normal mode analysis [25,26].

REFERENCES

[1] Müller, F. (1991) Free flavins: Synthesis, chemical and physical properties. In *Chemistry and Biochemistry of Flavoenzymes Vol. I*, F. Müller (ed.), (pp. 1–71), Florida, CRC Press

[2] Massey, V. & Ghisla, S. (1974) Role of charge-transfer interactions in flavoprotein catalysis. *Ann. NY Acad. Sci. 227*, 446-465

[3] Nishina, Y., Tojo, H., & Shiga, K. (1988) Resonance Raman spectra of anionic semiquinoid form of a flavoenzyme, D-amino acid oxidase. *J. Biochem. 104*, 227–231

[4] Morris, M.D. & Bienstock, R.J. (1986) Resonance Raman spectroscopy of flavins and flavoproteins. In *Spectroscopy of Biological Systems.* R.J.H. Clark & R.E. Hester (Eds.), (pp. 395–442). New York, John Wiley & Sons.

[5] McFarland, J.T. (1987) Flavins. In *Biological Applications of Raman Spectroscopy Vol. II* T.G. Spiro (Ed.), (pp. 221–302). New York, John Wiley & Sons.

[6] Macdonald, I.D.G. (1999) Vibrational spectroscopy of flavoproteins. In *Method in Molecular Biology 13,* S.K. Chapman & G.A. Reid (Eds.), (pp. 125–138). Totowa, NJ: Humana.

[7] Kitagawa, T., Nishina, Y., Shiga, K., Watari, H., Matsumura, Y., & Yamano, T. (1979) Resonance Raman evidence for charge-transfer interactions of phenols with the flavin mononucleotide of old yellow enzyme. *J. Am. Chem. Soc. 101*, 3376–3378

[8] Schmidt, J., Reinsch, J.R., & McFarland, J. T. (1981) Mechanistic studies on fatty acyl-CoA dehydrogenase. *J. Biol. Chem. 256*, 11667–11670

[9] Nishina, Y., Shiga, K., Tojo, H., Miura, R., Watari, H., & Yamano, T. (1981) Resonance Raman study of D-amino acid oxidase-inhibitor complexes. *J. Biochem. 90*, 1515–1520

[10] Williamson, G., Engel, P.C., Nishina, Y., & Shiga, K. (1982) A resonance Raman study on the nature of charge-transfer interactions in butyryl CoA dehydrogenace. *FEBS Lett. 138*, 29–32

[11] Nishina, Y., Shiga,K., Miura, R., Tojo, H., Ohta, M., Miyake, Y., Yamano, T., & Watari, H. (1983) On the structures of flavoprotein D-amino acid oxidase purple intermediates: A resonance Raman study. *J. Biochem. 94*, 1979–1990

[12] Nishina, Y., Tojo, H., Miura, R., Miyake, Y., & Shiga, K. (1988) Complex formation between anionic semiquinoid form of a flavoenzyme D-amino acid oxidase and ligands: Stabilization mechanism of anionic semiquinoid flavoemzyme. *J. Biochem. 104*, 727–733

[13] Suzuki, H., Koyama, H., Nishina, Y., Sato, K., & Shiga, K. (1991) A resonance Raman study on a reaction intermediate of *Pseudonomas* L-phenylalanine oxidase (deaminating and decarboxylating). *J. Biochem. 110*, 169–172

[14] Nishina, Y., Sato, K., Hazekawa, I., & Shiga, K. (1995) Structural modulation of 2-enoyl-CoA bound to reduced acyl-CoA dehydrogenase: A resonance Raman study of a catalytic intermediate. *J. Biochem. 117*, 800–808

[15] Hazekawa, I., Nishina, Y., Sato, K., Shichiri, M., & Shiga, K. (1995) Substrate activating mechanism of short-chain acyl-CoA, medium-chain acyl-CoA, long-chain acyl-CoA, and isovaleryl-

CoA dehydrogenase from bovine liver: A resonance Raman study on the 3-ketoacyl-CoA complexes. *J. Biochem. 118*, 900–910

[16] Tamaoki, H., Setoyama, C., Miura, R., Hazekawa, I., Nishina, Y., & Shiga, K. (1997) Spectroscopic studies of rat liver acyl-CoA oxidase with reference to recognition and activation of substrate. *J. Biochem. 121*, 1139–1146

[17] Zheng, Y., Wagner, M.A., Jorns, M.S., & Carey, P.R. (2001) Selective enhancement of ligand and flavin Raman modes in charge-transfer complexes of sarcosine oxidase. *J. Raman Spectrosc. 32*, 79–92

[18] Yagi, K., Ohishi, N., Takai, A., Kawano, K., & Kyogoku, Y. (1976) ^{15}N Nuclear magnetic resonance of flavins. *Biochemistry 15*, 2877–2880

[19] Nishina, Y., Sato, K., Miura, R., Matsui, K., & Shiga, K. (1998) Resonance Raman study on reduced flavin in purple intermediate of flavoenzyme: Use of [4-cabonyl-^{18}O]-enriched flavin. *J. Biochem. 124*, 200–208

[20] Manstein, D.J. & Pai, E.F. (1986) Purification and characterization of FAD synthetase from *Brevibacterium ammoniagenes*. *J. Biol. Chem. 261*, 16169–16173

[21] Nishina, Y., Sato, K., Shi, R., Setoyama, C., Miura, R., & Shiga, K. (2001) On the ligands in charge-transfer complexes of porcine kidney flavoenzyme D-amino acid oxidase in three redox states: A resonance Raman study. *J. Biochem. 130*, 637–647

[22] Kitagawa, T., Nishina, Y., Kyogoku, Y., Yamano, T., Ohishi, N., Takai-Suzuki, A., & Yagi, K. (1979) Resonance Raman spectra of carbon-13- and nitrogen-15-labeled riboflavin bound to egg-white flavoprotein. *Biochemistry 18*, 1804–1808

[23] Miura, R., Setoyama, C., Nishina, Y., Shiga, K., Mizutani, H., Miyahara, I., & Hirotsu, K. (1997) Structural and mechanistic studies on D-amino acid oxidase-substrate complex: Implications of the crystal structure of enzyme-substrate analog complex. *J. Biochem. 122*, 825–833

[24] Miura, R. & Miyake, Y. (1987) ^{13}C-NMR studies on the reaction intermediates of porcine D-amino acid oxidase reconstituted with ^{13}C-enriched flavin adenine dinucleotide. *J. Biochem. 102*, 1345–1354

[25] Abe, M. & Kyogoku, Y. (1987) Vibrational analysis of flavin derivatives: Normal coordinate treatments of lumiflavin. *Spectrochim. Acta 43A*, 1027–1037

[26] Zheng, Y., Dong, J., Palfey, B.A., & Carey, P.R. (1999) Using Raman spectroscopy to monitor the solvent-exposed and "buried" forms of flavin in *p*-hydroxybenzoate hydroxylase. *Biochemistry 38*, 16727–16732

In: D-Amino Acids: ISBN 1-60021-075-9
A New Frontier in Amino Acid and Protein Research, pp.545-548

Chapter 6.15

IMMUNODETECTION OF D-ASPARTATE OXIDASE

Gabriella Tedeschi, Severino Ronchi and Armando Negri *
Dipartimento di Patologia Animale Igiene e Sanità Pubblica Veterinaria, Sezione di Biochimica, Università di Milano, Italy.

Keywords: D-aspartate oxidase; anti-D-aspartate oxidase; immunodetection; immunoprecipitation; D-aspartate.

1. INTRODUCTION

Increasing evidences suggest that D-amino acids are present in animals and humans in high concentrations and fulfil specific biological functions [1]. In particular, D-aspartate appears to play a role in development and differentiation of cells in tissues, such as CNS, the retina, the adrenals and the testes [2,3]. D-aspartate oxidase (DASPO) (EC 1.4.3.1), a flavoprotein of 40 kDa catalyzing the oxidative deamination of D-aspartate by O_2 to oxaloacetate, ammonia and hydrogen peroxide [4], is the only enzyme known to metabolize D-aspartate in animal tissues. Therefore, it is suggested that the amount of D-aspartate in tissues can be regulated by the levels of DASPO [5,6]. In order to investigate expression at the protein level, localization and distribution of the enzyme in different cell types it is necessary to complement researches based on the detection of the catalytic activity of the enzyme with more sensitive immunological methods. For such purpose, strictly specific affinity-purified antibodies can be obtained following the protocols reported in this communication.

2. MATERIALS

1. Purified beef kidney DASPO, either from natural source [4] or overexpressed in E. coli [7]
2. Serum rabbit antibodies against purified DASPO (PRIMM, Milano, Italy)
3. Anion exchange membrane Q-100 (Sartorius AG, Goettingen, Germany)
4. CNBr-activated Sepharose 4B (Amersham, Uppsala, Sweden)

* Correspondence concerning this article should be addressed to Armando Negri, Dipartimento di Patologia Animale Igiene e Sanità Pubblica Veterinaria, Sezione di Biochimica, Università di Milano, Milano, Italy. E.mail: armando.negri@unimi.it.

5. Buffers: (A) 0.16 M KPi, pH 6.5; (B) 0.01 M KPi, pH 7.2, 0.14 M NaCl; (C) 0.1 M KPi pH 8.5, 0.5 M NaCl, 2 μM FAD; (D) 0.1 M borate pH 8.5; (E) 3 M KSCN in 0.01 M KPi pH 6.8; (F) 60 mM Tris/HCl pH 7.4, 1.25 % Triton X-100, 190 mM NaCl, 1 mM EDTA, 0.2 % SDS; (G) 50 mM Tris/HCl pH 7.5, 0.1 % Triton X-100, 150 mM NaCl, 5 mM EDTA; (H) buffer G plus 0.02 % SDS; (I) 50 mM Hepes pH 8.0 plus protease inhibitor cocktail (Complete Mini, Roche, Milano, Italy)
6. Protein A-Sepharose CL 4 B packed beads (Amersham, Uppsala, Sweden)
7. Iodoacetamide
8. Apparatus and reagents for SDS-PAGE electrophoresis and electroblotting onto nitrocellulose membranes
9. Secondary antibody: horseradish peroxidase-linked calf anti-rabbit IgG (Sigma Aldrich, Milano, Italy)
10. Chemiluminescence detection kit and films (Amersham, Uppsala, Sweden)

3. Methods

3.1. Affinity Purification of Anti-DASPO

1. 1 Preparation of the affinity column: DASPO was immobilized on CNBr-activated Sepharose 4B by incubating 7 mg of purified enzyme with 10 mL gel in buffer C for 2 h at room temperature under gentle rotation. The following steps of the procedure followed the manufacturer's instructions.
2. Serum rabbit antibodies against DASPO were initially purified by ammonium sulfate fractionation (50%) at 4 °C.
3. After dialysis against buffer A, the material was loaded on an anion exchange membrane Q-100 equilibrated in buffer A and eluted in buffer B.
4. 20 mg of the partially purified antibodies were incubated with the gel-immobilized DASPO for 2 h under gentle rotation. The suspension was washed with buffer D followed by elution with buffer E

3.2. Immunoblotting Using Anti-DASPO

1. Sample preparation: eggs, embryos and ovaries from *Xenopus laevis* were suspended in 500 μL buffer I on ice. Mechanical lysis was done by a Potter homogenizer and the material was centrifuged at 4 °C for 15 min in a microcentrifuge. The supernatant was extracted by freon in order to remove yolk proteins. Following centrifugation, the upper phase was added to 4 volumes of cold acetone on ice and the precipitated proteins centrifuged and resuspended in the appropriate buffer.
2. Samples were subjected to SDS-PAGE or, if needed, two-dimensional electrophoresis, and electroblotted on nitrocellulose following classical procedures, essentially as described in [8,9].
3. Incubation with anti-DASPO (1:200) was performed overnight at 4 °C. The following steps, incubation with secondary antibodies and chemiluminescence detection, were performed following classical procedures (see Notes 1).

3.3. Immunoprecipitation of DASPO from Tissues

1. The acetone precipitated tissue homogenate was resuspended in 6 mL of buffer F and incubated with 200 μL of the primary affinity purified antiboby at 4 °C for 12 h.
2. The sample was spun for 5 min, the supernatant was added to 350 μL of a 1:1 suspension of protein A-Sepharose CL 4 B (0.1 g of packed beads) and incubated with end-over-end mixing at room temperature for 2 h.
3. The beads were pelleted by centrifugation and washed 4 times in buffer H at room temperature and once in buffer G.
4. 400 μL SDS-PAGE sample buffer containing 50 mM DTT were added to the beads and the sample was heated at 100 °C. Free SH-groups were blocked by adding 1 mM iodoacetamide and incubated for 45 min at 37 °C. After centrifugation the supernatant was fractionated by 12% SDS-PAGE for further analyses.

4. Results

The availability of suitable amounts of recombinant bovine DASPO in *E.* coli allowed obtaining a highly specific polyclonal antibody following affinity chromatography using recombinant DASPO immobilised on Sepharose. Lack of cross-reactivity between purified anti-DASPO and D-amino acid oxidase, which shows 41 % similarity with DASPO [10], was checked by immunodecoration following dot-blotting of the two proteins on nitrocellulose membranes.

The importance of using purified antibodies in order to detect the presence of DASPO in tissues is highlighted in Fig. 1, which shows the immunodecoration of protein extracts from *Xenopus* embryo, eggs and ovaries using either anti-serum (Fig. 1A) or affinity-purified anti-DASPO (Fig. 1B). Such absolute specificity can be exploited to detect DASPO in very complex mixtures separated by 2D-electrophoresis and to immunoprecipitate the protein from tissues homogenates [9].

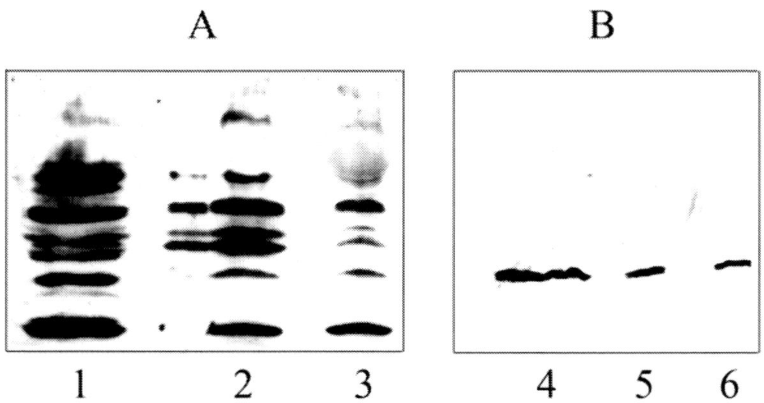

Figure 1. SDS-PAGE of homogenate from *Xenopus* ovaries (lanes 1 and 4), eggs (lanes 2 and 5) and embryos (lanes 3 and 6) immunodecorated with anti-DASPO following blotting onto nitrocellulose using rabbit anti-serum (Fig. 1A, lanes 1-3) or affinity-purified antibodies (Fig. 1B, lanes 4-6). Bands were visualised by chemiluminescence.

5. Notes

1. In the present case, the following detailed protocol was used following blotting onto nitrocellulose: the membrane was shaken in PBS-Tween-20 (0.2%) (PBS-T) containing 10 % powder milk for 1 h at room temperature; after washing with PBS-T with 5% powder milk, the membrane was incubated overnight at 4 °C in PBS-T containing the appropriate amount of primary antibody. Following the washing step (5 x 3 min with TBS-T and 3 x 3 min with PBS-T), the membrane was incubated for 1 h with the secondary antibody (horseradish peroxidase-linked calf anti-rabbit IgG, dilution 1 : 4000 in PBS-T with 5% powder milk). After the same washing step detailed above, the chemiluminescence detection was performed according to the manufacturer's instructions.

References

[1] Fuchs, S.A., Berger, R., Klomp, L.W.J. & de Koning, T.J. (2005). D-Amino acids in the central nervous system in health and disease. *Mol Genet Metab, 85,* 168-180.

[2] Wolosker, H., D'Aniello, A. & Snyder, S.H. (2000). D-Aspartate disposition in neuronal and endocrine tissues: ontogeny, biosynthesis and release. *Neuroscience, 100,* 183-189.

[3] Homma, H. (2002). D-Aspartate in the mammalian body. *Viva Origino, 30,* 204-215.

[4] Negri, A., Massey, V. & Williams, C.H., Jr. (1987). D-aspartate oxidase from beef kidney. *J Biol Chem, 262*, 10026- 10034.

[5] Zaar, K., Kost, H.-P., Schad, A., Volkl, A., Baumgart, E. & Fahimi, H.D. (2002). Cellular and subcellular distribution of D-aspartate oxidase in human and rat brain. *J Comp Neurol, 450,* 272-282.

[6] Shell, M.J., Cooper, O.B. & Snyder, S.H. (1997). D-aspartate localizations imply neuronal and neuroendocrine roles. *Proc Natl Acad* Sci, *94,* 2013-2018.

[7] Negri, A., Tedeschi, A., Ceciliani F. & Ronchi, S. (1999). Purification of beef kidney D-aspartate oxidase overexpressed in *E. coli* and characterization of its redox potentials and oxidative activity towards agonists of excitatory amino acid receptors. *Biochim Biophys Acta, 1431*, 212-222.

[8] Matsudaira, P. (1987) Sequence from picomole quantities of proteins electroblotted onto polyvinylidene difluoride membranes. *J Biol Chem, 262,* 10035-10038.

[9] Tedeschi, G., Bernardini, G., Mortarino, M., Negri, A., Ceciliani, F. & Ronchi, S. (1998) Two-dimensional protein maps of *Xenopus* eggs and embryos at different developmental stages. *Cell Biol Int, 22*, 517-525.

[10] Negri, A., Ceciliani, F., Tedeschi, G., Simonic T. & Ronchi, S. (1992) The primary structure of the flavoenzyme D-aspartate oxidase from beef kidney. *J Biol Chem, 267,* 11865-11871.

In: D-Amino Acids: ISBN 1-60021-075-9
A New Frontier in Amino Acid and Protein Research, pp.549-553

Chapter 6.16

HIGH-PERFORMANCE LIQUID CHROMATOGRAPHIC DETERMINATION OF D-ASPARTATE OXIDASE ACTIVITY IN RAT TISSUES

Atsushi Yamamoto**, *Hiroyuki Tanaka, Tetsuo Ishida and Kihachiro Horiike
Department of Biochemistry and Molecular Biology, Shiga University of Medical Science, Japan.

Keywords: D-aspartate oxidase, D-aspartate, flavoenzyme, oxalacetate, HPLC.

1. INTRODUCTION

D-Aspartate oxidase (DDO, EC.1.4.3.1) catalyzes the oxidative deamination of D-aspartate to oxalacetate [1]. Mammalian neuronal and endocrine tissues contain near-millimolar concentrations of D-aspartate [2]. In these tissues, DDO plays an important role in maintaining the physiological levels of D-aspartate [3].

To determine DDO activity in rat tissues, we developed a high-performance liquid chromatographic (HPLC) analysis of oxalacetate and other 2-oxo acids following derivatization with 3-methyl-2-benzothiazolinone hydrazone (MBTH) [4,5]. In aqueous solution, oxalacetate exists in the keto and enol forms, and is easily subjected to nonenzymatic decarboxylation [6]. To circumvent possible interference in the quantification of oxalacetate due to these properties, oxalacetate was completely decarboxylated to pyruvate during its incubation with MBTH, and quantified as the MBTH-azine of pyruvate.

2. MATERIALS

1. Adult male Sprague-Dawley rats (6 weeks old, weighting 200-250 g)
2. Sodium pentobarbital
3. Peristaltic pump
4. micro-glass homogenizer (100 μL)

* Correspondence concerning this article should be addressed to Atsushi Yamamoto, Department of Biochemistry and Molecular Biology, Shiga University of Medical Science, Japan. E-mail: atsu@belle.shiga-med.ac.jp. FAX: +81-77-548-2157.

5. Homogenization buffer
   50 mM sodium pyrophosphate (pH 8.3)
6. Assay mixture
   Air-saturated assay mixture (200 μL) consisting of 50 mM sodium pyrophosphate (pH 8.3), 50 μM FAD, 12.5 μg catalase, and 50 mM D-aspartate
7. Derivatization with MBTH
   0.1% 3-methyl-2-benzothiazolinone hydrazone (MBTH)
   12.5% trichloroacetic acid
   1M sodium acetate (pH 5.0)
   1 mM 2-oxoglutaric acid
8. High-performance liquid chromatography
   5C18 column
   HPLC pump
   Gradient controller
   UV–VIS detector
   Trifluoroacetic acid (HPLC grade)
   Acetonitrile (HPLC grade)

3. Methods

1. An adult rat was anesthetized by intraperitoneal injection of sodium pentobarbital (50 mg/kg body weight).
2. After incision of the right auricle, the whole body was perfused for 3 min through the left ventricle with 10 mM sodium phosphate (pH 7.4) containing 0.9% NaCl at a flow rate of 50 mL/min using a peristaltic pump.
3. The rat tissues were excised and homogenized with 9 volumes of 50 mM sodium pyrophosphate (pH 8.3) with a glass homogenizer at 4°C.
4. An aliquot of the homogenate (2–20 μL) was added to an assay mixture (200 μL). As a control, the same amount of the homogenate was incubated with the assay mixture without D-aspartate.
5. After a 0–60 min incubation at 37°C, the reaction was stopped by the addition of 40 μL of 12.5% trichloroacetic acid.
6. After the addition of 10 μL of 1 mM 2-oxoglutarate (10 nmol) as an internal standard, the reaction mixture was centrifuged at 5700 × *g* for 5 min.
7. An aliquot of the supernatant (100 μL) was transferred to an Eppendorf tube. To this supernatant, 200 μL of 1 M sodium acetate buffer (pH 5.0) and 80 μL of 0.1% MBTH (aqueous solution) was added in this order.
8. The mixture was incubated at 50°C for 30 min. After incubation, it was cooled to room temperature.
9. A 5-μL aliquot of the MBTH-treated sample was applied to a Cosmosil 5C18 column (4.6 × 50 mm, Nacalai Tesque). The azine-derivatives were eluted at a flow rate of 0.8 mL/min and 25°C with a linear gradient of acetonitrile from 0% to 80% in 8 min in the presence of 0.1% trifluoroacetic acid. Elution of the azine derivatives was monitored spectrophotometrically at 350 nm.

4. Results

The reaction scheme of the MBTH treatment is shown in Figure 1. When oxalacetate was treated with MBTH according to the present method, the two azine derivatives were detected (Fig. 2 a), and they were identical to those from pyruvate (Fig. 2 b). This result indicates that oxalacetate is completely decarboxylated to pyruvate during the derivatization [6]. The two peaks of MBTH-azine derivatives from oxalacetate (or pyruvate) correspond to *Z*- and *E*-form of the azine derivatives (Fig. 1) [5]. In contrast, those of 2-oxoglutarate were detected as a single peak with a slightly large half peak width in the present chromatographic conditions (Fig. 2 c).

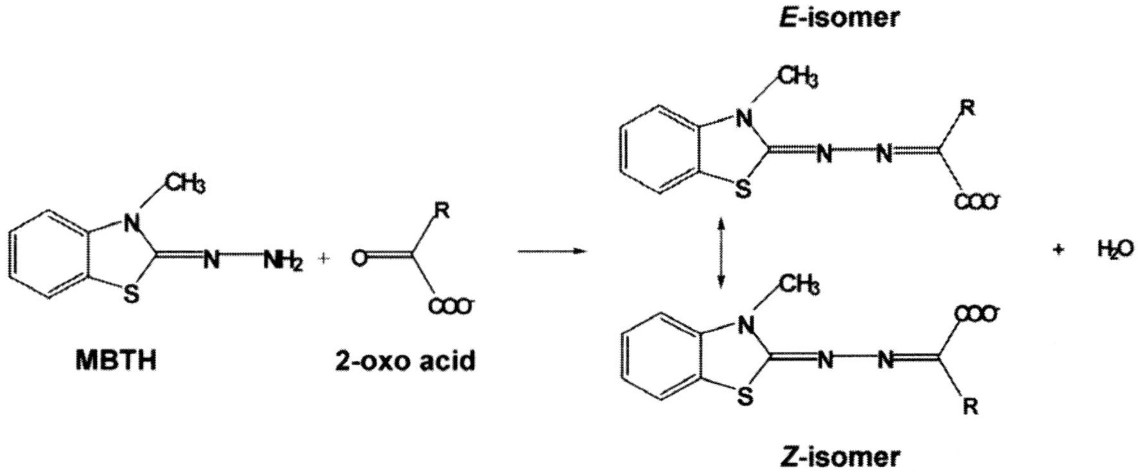

Fig. 1. A reaction scheme of 2-oxo acid with MBTH. Two isomers of azine-derivatives (*E*- and *Z*-) are produced for each 2-oxo acid. R: $-CH_3$ (pyruvate), $-CH_2COO^-$ (oxalacetate) and $-CH_2CH_2COO^-$ (2-oxoglutarate).

The sum of the peak areas of the two derivatives of oxalacetate was plotted against the amount of oxalacetate as the ratio to that of the internal standard (2-oxoglutarate) (Fig. 3). The result shows that the present method can quantify 0.1–50 nmol levels of oxalacetate present in 200-μL assay mixtures.

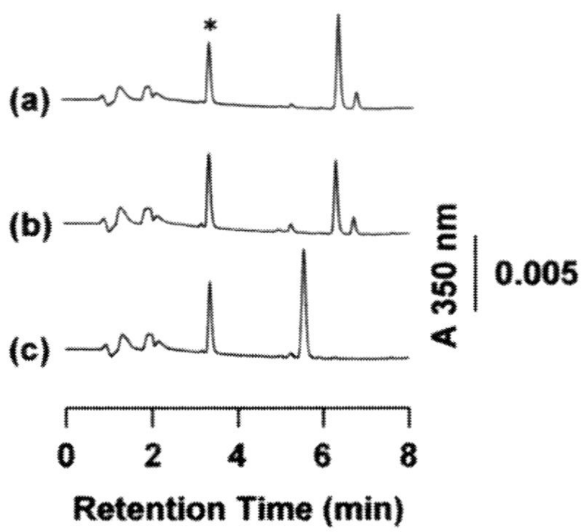

Fig. 2. Elution profiles of authentic azine-derivatives of 2-oxo acids. A reaction mixture (200 μL), containing 10 nmol of authentic 2-oxo acid (a; oxalacetate, b; pyruvate and c; 2-oxoglutarate) was analyzed according to the present method. The retention time of the two isomers of the MBTH-azine of pyruvate were 6.3 min and 6.7 min, respectively, and that of 2-oxoglutarate was 5.6 min. *: peak deriving from the MBTH reagent.

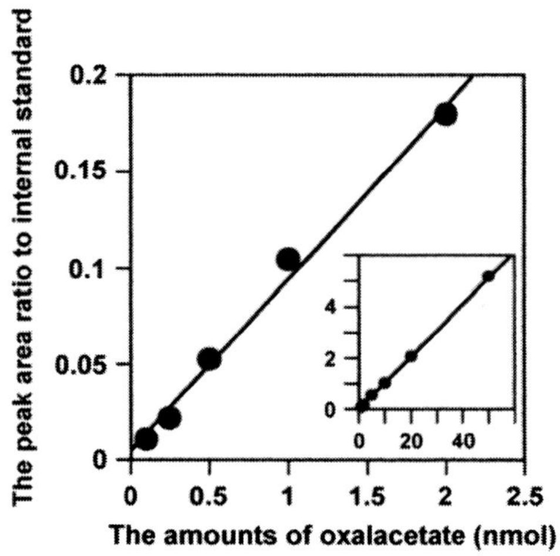

Fig. 3. Calibration curve for the quantification of oxalacetate. The ratio of the combined area of the two peaks from oxalacetate and the peak area of the 2-oxoglutarate derivative (relative peak area) was plotted against the amount of oxalacetate. Inset shows the calibration curve in higher concentration range.

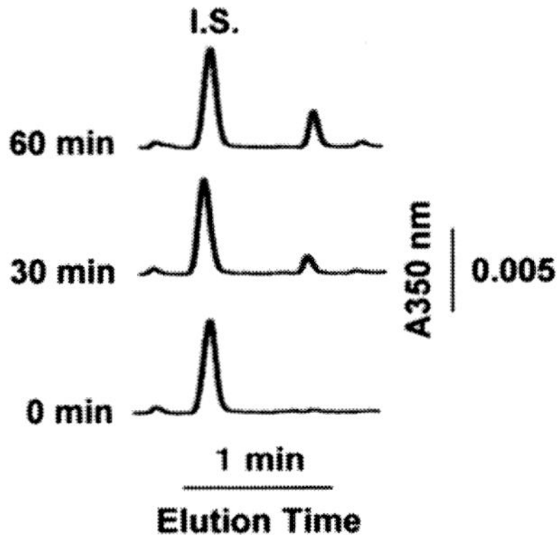

Fig. 4. Typical chromatographs obtained from the assay for DDO activity in rat cerebellum. A 20-μL aliquot of the homogenate was used for the assay. The reaction was stopped after 0-, 30- and 60-min incubation. I.S. is the MBTH azine from the internal standard (2-oxoglutarate, 10 nmol).

Table 1 summarizes the DDO activity in various rat tissues. The kidney and liver are rich sources of DDO, like D-amino-acid oxidase [7,8]. No activities of DDO were found in lung, heart and spleen.

Table 1. The D-Aspartate oxidase activity in rat tissues.

Tissues	Activity (nmol/min/mg protein)
Cerebrum	0.26 ± 0.08
Cerebellum	0.21 ± 0.05
Kidney	1.39 ± 0.13
Liver	0.93 ± 0.11
Lung	<0.01
Heart	<0.01
Spleen	<0.01

The values are means ± S.D. (n = 5)

Using this method, the DDO activity in rat cerebellum was measured (Fig. 4). To confirm that the oxalacetate detected was actually produced by the DDO reaction, the incubation was carried out using argon-saturated assay solutions (i.e., oxygen-free solution) or in the presence of D-tartrate, a competitive inhibitor. In these conditions, the formation of oxalacetate was almost completely inhibited (data not shown).

5. Notes

1. This method is applicable to determine the activity of other enzymes that catalyze the production (or consumption) of 2-oxo acid.
2. We used 2-oxoglutarate as internal standard based on its structural similarity to oxalacetate and its low abundance in rat tissues. Other 2-oxo acid should be used as an internal standard to quantify 2-oxo acids structurally different from oxalacetate.
3. The present assay method can be done using only 25-μL reaction mixtures to determine DDO activity when only a small volume of tissue sample is available.

References

[1] Still, J. L., Buell, M. V., Knox, W. E. & Green, D. E. (1949). Studies on the cyclophorase system VII. D-Aspartic oxidase, *J Biol Chem 179*, 831–837.

[2] Hashimoto, A. & Oka, T. (1997). Free D-aspartate and D-serine in the mammalian brain and periphery, *Prog Neurobiol 52*, 325–353.

[3] Schell, M. J., Cooper, O. B. & Snyder, S. H. (1997). D-Aspartate localizations imply neuronal and neuroendocrine roles, *Proc Natl Acad Sci USA 94*, 2013–2018.

[4] Soda, K. (1967). A spectrophotometric microdeterminetion of keto acids with 3-methyl-2 benzothiazolone hydrazone, *Agr Biol Chem 31*, 1054–1060.

[5] Chiavari, G. & Facchini, M. C. (1987). Behaviour of 3-methyl-2-benzothiazolone azines of carbonyl compounds in high-performance liquid chromatography, *J Chromatogr 387*, 459–466.

[6] Flint, D. H., Nudelman, A., Calabrese, J. C. & Gottlieb, H. E. (1992) Enol oxalacetic acid exists in the *Z* form in the crystalline state and in solution, *J Org Chem 57*, 7270–7274.

[7] Horiike, K., Ishida, T., Tanaka, H. & Arai, R. (2001) Distribution of D-amino-acid oxidase and D-serine in vertebrate brains, *J Mol Catal 12B*, 37–41.

[8] Horiike, K., Tojo, H., Arai, R., Nozaki, M. & Maeda, T. (1994) D-Amino-acid oxidase is confined to the lower brain stem and cerebellum in rat brain: regional differentiation of astrocytes, *Brain Res 652*, 297–303.

In: D-Amino Acids:
A New Frontier in Amino Acid and Protein Research, pp.555-560
ISBN 1-60021-075-9

Chapter 6.17

Cloning of the Human Brain D-Aspartate Oxidase cDNA and its Expression in *Escherichia coli*

Chiaki Setoyama* and Retsu Miura

Department of Molecular Enzymology, Graduate School of Medical Sciences, Kumamoto University, Japan.

Keywords: D-aspartate oxidase, flavoenzyme, RT-PCR, protein expression.

1. Introduction

D-Aspartate oxidase (DDO, EC 1.4.3.1) (the abbreviation is based on the one-letter code D for aspartate) is a flavoprotein that catalyzes the oxidative deamination of dicarboxylic D-amino acids to give the corresponding α-ketoacids and ammonia. The *in vivo* function of DDO was considered enigmatic because D-amino acids had not been found in higher animals, with the exception of some classes of bacteria and some invertebrates [1-3]. However, advances in detection methods for chiral amino acids have led to the discovery that free D-amino acids are present in significant quantities in mammals [4-6]. Although the biological role of DDO is still unclear, DDO is phylogenetically conserved in mammalian central nervous system and periphery [7,8]. These discoveries have raised the possibility that DDO is involved in the regulation of the central nervous system and/or the development of the brain in mammals.

To understand the catalytic mechanism and biological role of mammalian DDO, we isolated human brain DDO cDNA and developed an expression system for DDO in *E. coli*.

* Correspondence concerning this article should be addressed to Chiaki Setoyama, E-mail: setoyama@gpo.kumamoto-u.ac.jp, Fax: +81-96-373-5064.

2. Materials and Methods

Isolation of Human Brain D-Aspartate Oxidase cDNA

1. Materials

1) Human brain RNA (Invitrogen, Carlsbad, CA, USA)
2) High Fidelity RNA PCR kit (Takara Shuzo, Otsu, Japan)
3) PCR primers
   One corresponding to the sequence from 3 to 20 of DDO-1 cDNA (5'-pGACACAGCACGGATTGCA-3'), and the other corresponding to the anti-sense strand sequence from 1,014 to 1,033 (5'-pATGTCATCTACAGGTTTGAC-3') [9]
4) Thermal cycler
5) Low melting point gel agarose (Takara Shuzo)
6) Electrophoresis apparatus
7) Tris/Borate/EDTA (TBE) buffer

2. Methods

1) cDNA was synthesized from human brain RNA with oligo(dT) primer using High Fidelity RNA PCR kit. The reaction mixture contained 1 µg human brain RNA, 5mM $MgSO_4$, 0.5 mM dNTP mixture, 0.5 U RNase inhibitor, 2.1 U *Bca* PLUS RTase, 0.25 µM oligo(dT) primer and 1 x buffer. Reaction was performed for one cycle of: 60°C for 1 min, 30°C for 1 min, gradual rise to 65°C (15 min), 65°C for 15 min, 98°C for 5 min and 5°C for 5 min.
2) The DDO cDNA was amplified by PCR using the human brain cDNA as a template. The reaction mixture contained 20 µl of cDNA solution, 0.3 mM dNTP mixture, 1 µM each of the forward and reverse primers, *Pyrobest* DNA polymerase and 1 x buffer. After 30 cycles of denaturation (1 min, 94°C), annealing (1 min, 65°C) and elongation (30 s, 72°C), two kinds of PCR products of approximately 1 kb (DDO-1) and 850 bp (DDO-2), were obtained. These fragments were purified from the 1% agarose gel electrophoresis.

Construction of Expression Plasmid

1. Materials

1) pET-11d plasmid (Novagen, Madison, WI, USA)
2) Restriction enzyme *Nco* I (Takara Shuzo)
3) Klenow fragment (Takara Shuzo)
4) Calf intestinal alkaline phosphatase (CIP) (Takara Shuzo)
5) T4 Ligase (Takara Shuzo)

2. Methods

1) The expression plasmid pET-11d was digested at the unique *Nco* I site.
2) CIP was added and the mixture was incubated for 30 min at 37°C.
3) The linearized fragment was purified through a 0.8% low melting point agarose gel.
4) To convert a 5'-protruding end to a blunt end, the DNA fragment was treated with Klenow flagment.
5) The amplified D-aspartate oxidase cDNA was inserted into the pET-11d.
6) Finally, the expression plasmids pETDDO-1 and pETDDO-2 were obtained.

Expression and Purification of D-Aspartate Oxidase

1. Materials

1) Expression plasmids pETDDO-1and pETDDO-2
2) *E. coli* BL21(DE3) (Novagen)
3) LB medium (Wako Pure Chemical Industries, Osaka, Japan)
4) Carbenicillin (Nakarai Tesque, Kyoto, Japan)
5) Isopropyl-β-D-thiogalactopyranoside (IPTG) (Wako)
6) Autoclave
7) Incubator
8) Lysozyme (Wako)
9) Sodium potassium tartrate (Wako)
10) Potassium phosphate buffer, pH 8.0
11) Phenyl Sepharose CL-4B (Amersham Biosciences Corp., Piscataway, NJ, USA)
12) DEAE Sepharose CL-6B (Amersham)

2. Methods

1) *E. coli* BL21(DE3) cells harboring pETDDO-1 or pETDDO-2 were grown in LB medium containing 50 μg/ml carbenicillin at 37°C to an absorbance of 0.6 at 600 nm.
2) IPTG was added to a final concentration of 0.01 mM and cells were grown for additional 20 h.
3) The cells were harvested by centrifugation.
4) The cell pellets were suspended in 50 mM potassium phosphate, pH 8.0, containing 5 mM sodium potassium tartrate, 20 μM FAD and 0.3 mM EDTA and treated with 1 mg/ml lysozyme for 1 h in an ice bath.
5) The cells were then disrupted by sonication.
6) After centrifugation, the supernatant was brought to pH 5.2 with acetic acid and was heated at 55°C for 15 min, then cooled rapidly to 4°C in an ice bath, and the precipitate formed was removed by centrifugation.
7) The supernatant was brought to pH 6.7 by the addition of 1 M potassium phosphate, pH 8.0, and then applied to a Phenyl Sepharose CL-4B column (3.2 x 12.5 cm) equilibrated with 200 mM potassium phosphate, pH 8.0, containing 5 mM sodium potassium tartrate.
8) After washing of the column with 10 column volumes of the equilibration buffer, proteins were eluted with 5 mM potassium phosphate buffer, pH 8.0.
9) Yellow fractions were pooled, dialyzed against 10 mM potassium phosphate, pH 8.0, containing 5 mM sodium potassium tartrate and applied to a DEAE Sepharose CL-6B column (3.2 x 20 cm) equilibrated with the same buffer.
10) Elution of proteins was performed using the same buffer.
11) Fractions containing DDO and showing a single protein band on SDS-PAGE were pooled.

4. RESULTS

We cloned the human DDO cDNA by the RT-PCR method. Interestingly, two types of DDO cDNA (DDO-1 and DDO-2) were amplified. Sequence analysis revealed that DDO-1 consists of 341 amino acid residues. In several peroxisomal enzymes, the *C*-terminal sequence (SKL- or SHL-motif) is known to be a minimal peroxisome-targeting signal [10]. The SNL sequence of human DDO-1, in which the lysine (or histidine) residue is substituted by aspargine, is also functional peroxisome-targeting signal [11]. On the other hand, DDO-2 encoded a 282-amino acid peptide whose sequence was identical to that of DDO-1 except for the absence of 59 amino acids covering from residue 95 to residue 153 of DDO-1 (Fig. 1). The difference between the two types (DDO-1 and DDO-2) of DDO mRNA is limited to a small region. Comparison of the nucleotide sequences between the DDO cDNAs and the DDO gene revealed that DDO-1 and DDO-2 mRNAs are produced by alternative splicing from a single gene.

A. Amino acid sequence of human brain D-aspartate oxidase

MDTARIAVVGAGVVGLSTAVCISKLVPRCSVTIISDKFTPDTTSDVAAGMLIPHTYPDTPIHTQKQWFRE

TFNHLFAIANSADAGDAGVHLVSGLQIFQSTPTEEVPFWADVVLGFRKMTEAELKKFPQYVFGQAFTTLK

CECPAYLPWLEKRIKGSGGWTLTRRIEDLWELHPSFDIVVNCSGLGSRQLAGDSKIFPVRGQVLQVQAPW

VEHFIRDGSGLTYIYPGTSHVTLGGTRQKGDWNLSPDAENSREILSRCCALEPSLHGACNIREKVGLRPY

RPGVRLQTELLARDGQRLPVVHHYGHGSGGISVHWGTALEAARLVSECVHALRTPIPKSNL

B. Comparison of the structure between DDO-1 and DDO-2

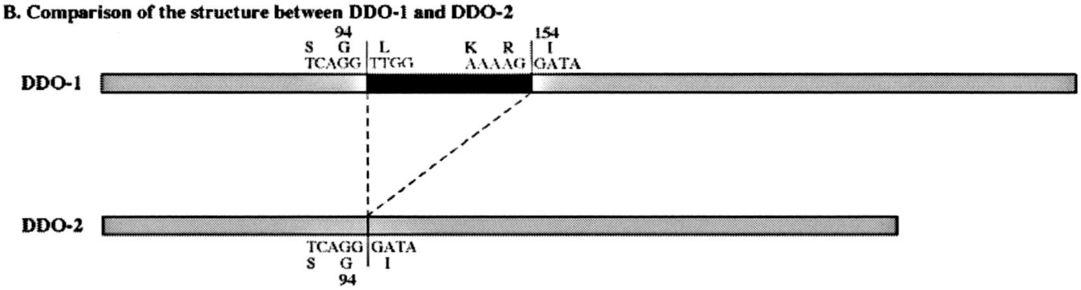

Figure 1. Construction of the expression vector of human D-aspartate oxidase.

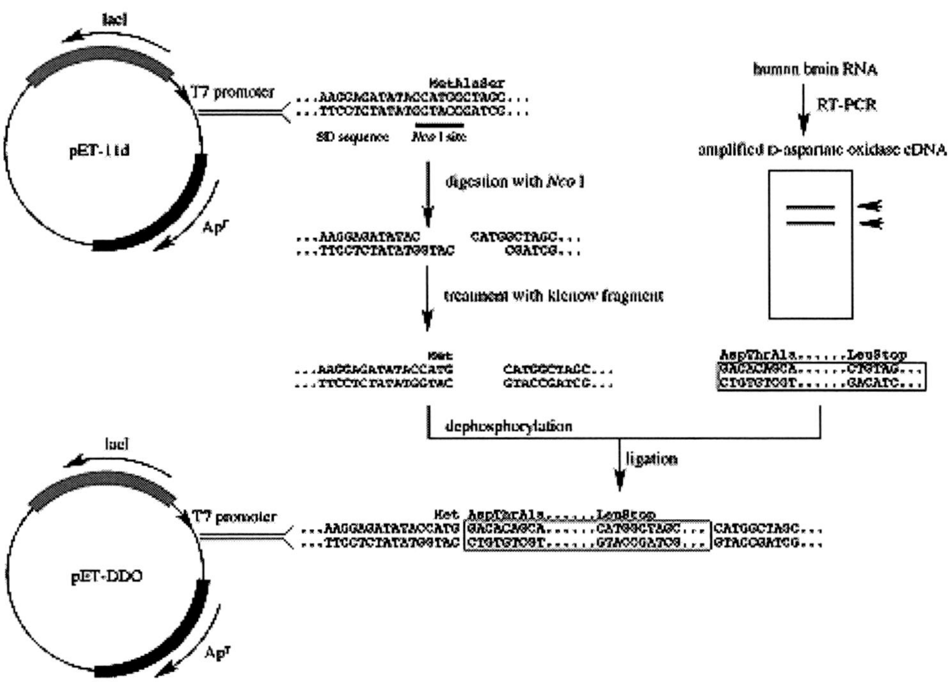

Figure 2. Two forms of the human brain D-aspartate oxidase. The underline in (A) and the filled region in (B) indicate the amino acid sequence (amino acid residues 95-153) specific for the DDO-1 cDNA, which is absent in the DDO-2 sequence.

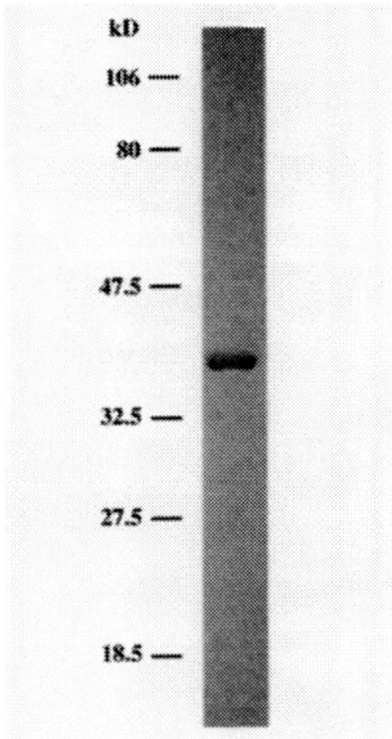

Figure 3. SDS-PAGE of recombinant human DDO-1. Approximately 1 μg of the purified enzyme was analyzed by SDS-PAGE with 12.5% gel. The values of molecular-mass standards in kD are shown to the left side.

To obtain homogeneous preparations of human DDO-1 and DDO-2, we developed an expression system of each type of DDO in *E. coli* cells as summarized in Fig. 2. Recombinant DDO-1 expressed in *E. coli* BL21(DE3) cells was purified by a three-step procedure as described under "Methods". The DDO-1 preparation obtained was homogeneous as judged by SDS-PAGE (Fig. 3), which showed a single protein band corresponding to the molecular mass of 37 kDa.

5. Notes

1. Using our expression system for DDO in *E. coli*, about 200 mg of DDO-1 was obtained from 72 liters of IPTG-induced culture. In the case of DDO-2, most of the expressed protein was accumulated in an insoluble form.
2. D-Aspartate and its *N*-methylated derivative, *N*-methyl-D-aspartate (NMDA), were good substrates for human DDO-1, but D-glutamate turned out to be a poor substrate.

References

[1] Corrigan, J.J. and Srinivasan N.G. (1996) The occurrence of certain D-amino acids in insects. *Biochemistry, 5*, 1185-1190.

[2] Corrigan J.J. (1969) D-Amino acids in animals. *Science, 164,* 142-149.

[3] Neims, A.H. and Hellerman L. (1970) Flavoenzyme catalysis. *Annu Rev Biochem, 39,* 867-888.

[4] Nagata, Y., Akino, T., and Ohno, K. (1989) The presence of free D-amino acids in mouse tissues. *Experimentia, 45*, 330-332.

[5] Fisher, G. H., D'Aniello, A., Vetere, A., Padula, L., Cusano, G. P., and Man, E. H. (1991) Free D-aspartate and D-alanine in normal and Alzheimer brain. *Brain Res Bul, 26*, 983-985.

[6] Nagata, Y., Horiike, K., and Maeda, T. (1994) Distribution of free D-serine in vertebrate brains. *Brain Res, 634*, 291-295.

[7] Kera Y., Nagasaki, H., Iwashima, A., and Yamada, R. (1993) Gender dependence of D-aspartate oxidase activity in rat tissues. *Comp Biochem Physiol, 104*, 739-742.

[8] Zaar, K., Köst, H.-P., Schad, A., Völkl, A., Baumgart, E. and Fahimi, H. D. (2002) Cellular and subcellular distribution of D-aspartate oxidase in human and rat brain. *J Comp Neurol, 450*, 272-282.

[9] Setoyama, C. and Miura, R. (1997) Structural and functional characterization of the human brain D-aspartate oxidase. *J Biochem*, *121*, 798-803.

[10] Gould, S.J., Keller, G.A., Hosken, N., Wilkinson, J. and Subramani, S. (1989) A conserved tripeptide sorts proteins to peroxisomes. *J Cell Biol, 108*, 1657-1664.

[11] Amery, L., Brees, C., Baes, M., Setoyama, C., Miura, R., Mannaerts, G.P. and Van Veldhoven, P. (1998) C-terminal tripeptide Ser-Asn-Leu (SNL) of human D-aspartate oxidase is a functional peroxisome-targeting signal. *Biochem J, 336*, 367- 371.

In: D-Amino Acids: ISBN 1-60021-075-9
A New Frontier in Amino Acid and Protein Research, pp.561-571

Chapter 6.18

METHODS FOR *N*-ACYL-D-AMINO ACID AMIDOHYDROLASE RESEARCH

***Mamoru Wakayama*[1,*], *Kazuaki Yoshimune*[2] *and Mitsuaki Moriguchi*[3]**
[1]Department of Bioscience and Biotechnology, Faculty of Science and Engineering, Ritsumeikan University, Japan;
[2]Research Institute of Genome-based Biofactory, National Institute of Advanced Industrial Science and Technology, Japan;
[3]Department of Food and Bioscience, Faculty of Food and Nutrition, Beppu University, Japan.

Keywords: *N*-acyl-D-amino acid amidohydrolases, screening, assay methods, gene cloning.

1. INTRODUCTION

Among amino acids, the L-form is by far the predominant enantiomer found in nature. Recently, however, the sensitivity and efficiency of the separate determination of D- and L-amino acids have improved dramatically due to the development of analytical procedures. As a result, an increasing number of studies on the physiological role of D-amino acids in plants, microbes, and mammals have been reported. Furthermore, D-amino acids have been found and used intentionally in physiologically active compounds such as antibiotics and therapeutic drugs. D-Amino acids are important chiral intermediates for pharmaceuticals, agrochemicals, and food additives. To produce various D-amino acids effectively, the application of *N*-acyl-D-amino acid amidohydrolase to D-amino acid production has recently received increased attention. *N*-Acyl-D-amino acid amidohydrolase (EC. 3.5.1) catalyzes the hydrolysis of *N*-acyl derivatives of various D-amino acids to D-amino acids and fatty acids [1,2]. As far as we know, studies on the microbial *N*-acyl-D-amino acid amidohydrolase are limited to some bacterial genera of *Streptomyces* [3,4], *Pseudomonas* [5-8], *Alcaligenes* [9-14], *Variovorax* [15], and *Defluvibacter* [16] and to one fungus genus of *Trichoderma* [17]. No *N*-acyl-D-amino acid amidohydrolase has been reported in mammals. In this report, we describe the methods for isolating a microorganism producing *N*-acyl-D-amino acid amidohydrolase and its enzyme assay. Methods of gene cloning are also briefly discussed.

* Correspondence concerning this article should be addressed to Mamoru Wakayama, E-mail: wakayama@se.ritsumei.ac.jp; Fax: +81-77-561-2659.

2. Materials

a. Screening and Assay

N-Acetyl-D-amino acid, D-amino acid, cetyltrimethylammonium bromide (CTAB), acetic anhydride, 2,4,6-trinitrobenzenesulfonic acid (TNBS), D-amino acid oxidase, catalase, peroxidase, chromophore (*o*-dianisidine, *o*-phenylene diamine, 4-aminoantipyrine-phenol, ninhydrin, *o*-phthaldialdehyde (OPA), *N*-ethyl-*N*-(2-hydroxy-3-sulfopropyl-)3-methylaniline), *N*-tert butyloxycarbonyl-L-cysteine (Boc), enantio-selective column (Sumichiral OA series, Sumika Chemical Analysis Service, Tokyo, Japan), spectrophotometer, incubator. All these chemicals were from Wako Pure Chemicals, Osaka, Japan.

b. Gene Cloning

Ampicillin, IPTG, and X-gal were purchased from Wako Pure Chemicals, Osaka, Japan. Toyo No.5 filter paper was obtained from Advantec, Tokyo, Japan. Cloning vector pUC18, *Escherichia coli* JM109, DNA-modifying enzymes such as restriction enzymes, DNA ligase, PCR in vitro Cloning Kit, and thermal cycler were purchased from Takatra shuzo, Kyoto, Japan.

3. Methods

3-1. *N*-Acetyl-D-amino Acid Synthesis

D-Leucine (9.93g) is dissolved in 100 ml distilled water. Acetic anhydride is added gradually to the D-leucine solution to a total volume of 20 ml, and the pH is maintained at 7-8 with NaOH solution. The solution is mixed well for 1 h and then is concentrated to a volume of about 10 ml by evaporation. The *N*-acetyl-D-leucine is crystallized by adjusting the pH to 2.6 ~ 2.7 with HCl. The solution is kept at 4 °C for 2 h and then is filtrated with a glass filter. The product is dried with P_2O_5 and stored at –20 °C.

3-2. Culture Condition

a. Screening for Microorganisms Producing N-Acyl-D-Amino Acid Amidohydrolase

To isolate microorganisms producing *N*-acyl-D-amino acid amidohydrolase from the natural environment by enrichment culture, a medium containing *N*-acyl-D-amino acid is used, with the target substrate as the sole carbon and nitrogen source (Table 1). A source of microorganisms, such as soil, is inoculated into the medium, and the culture is incubated with shaking at 30°C. The cultivation temperature is changed depending on the purpose in each case. The *N*-acyl-D-amino acid assimilating microorganisms in the mixed population are further enriched by two or three consecutive daily transfers in the same medium. In the case of the selective isolation of fungus, the addition of an antibiotic such as streptomycin to the medium is effective [17]. Bacteria are purely isolated on an agar plate containing *N*-acyl-D-amino acid as the sole carbon and nitrogen source. For the selection of microorganisms producing *N*-acyl-D-amino acid amidohydrolase among type cultures, stock cultures or type cultures preserved in the laboratory are cultivated in the best medium for the growth of each microorganism or in a medium containing *N*-acyl-D(L)-amino acid as an inducer in addition to carbon and nitrogen sources [4]. A cell-free extract, CTAB-treated cells, or toluene treated cells are prepared for the enzyme assay [18]. *N*-Acyl-D-amino acid amidohydrolase is measured according to the assay procedure described in Section 3-3.

Table 1. Medium for the isolation of a microorganism producing *N*-acyl-D-amino acid amidohydrolase.

Contents	Concentration (%)
N-Acetyl-D(L)-amino acid[1)]	0.1-0.5
Carbon source[2)]	0-0.2
Nitrogen source[3)]	0-0.2
KH_2PO_4	0.1
K_2HPO_4	0.1
$MgSO_4 \cdot 7H_2O$	0.01
Yeast extract	0.01

1) As an inducer, *N*-acetyl-D(L)-amino acid or D(L)-amino acid, described in Table 2, is useful.
2) Glucose, glycerol, maltose, fructose, or citric acid.
3) $(NH_4)_2SO_4$, peptone, casamino acids, or urea.

b. Enzyme Production

When the enzyme producer is cultivated in a medium containing *N*-acyl-D-amino acid or D-amino acid as the sole carbon and nitrogen source, the resulting enzyme is categorized as an inducible enzyme. To our knowledge, all bacterial *N*-acyl-D-amino acid amidohydrolases are inducible enzymes (Table 2). On the other hand, the enzyme from *Trichoderma harzianum* is not inductively produced by *N*-acyl-D-amino acid or D-amino acid but is produced regularly in the medium often used for microbial cultivation. The specific activities of fungus enzymes are lower than those of bacterial ones.

Table 2. Induction of bacterial *N*-acyl-D-amino acid amidohydrolase.

Enzyme	Substrate	Inducer	Strain	References
D-Aminoacylase	*N*-Ac-D-Met *N*-Ac-D-Leu *N*-Ac-D-Phe, etc.	*N*-Ac-γ-Methyl-D-Leu *N*-Ac-D-Leu, etc.	*Alcaligenes xylosoxydans* subsp. *xylosoxydans* A-6	14
	N-Ac-D-Met *N*-Ac-D-Leu *N*-Ac-D-Phe, etc.	γ-Methyl-D-Leu *N*-Ac-D-Leu, etc	*Alcaligenes denitrificans* subsp.*xylosoxydans* MI-4	26, 27
	N-Ac-D-Met *N*-Ac-D-Leu *N*-Ac-D-Phe, etc.	*N*-Ac-DL-Leu *N*-Ac-D-Val, etc	*Alcaligenes faecalis* DA1	11, 20, 28
	N-Ac-D-Met *N*-Ac-D-Leu *N*-Ac-D-Phe, etc.	*N*-Ac-DL-Leu *N*-Ac-D-Val, etc.	*Alcaligenes denitrificans* DA181	29
	N-Ac-D-Met *N*-Ac-D-Leu *N*-Ac-D-Phe, etc.	DL-Val *N*-Ac-DL-Val	*Streptomyces olivaceus* 62-3	3
	N-Ac-D-Met *N*-Ac-D-Leu *N*-Ac-D-Ala, etc.		*Variovorax paradoxus* Iso1	15
N-Acyl-D-Glutamate amidohydrolase	*N*-Ac-D-Glu (specific)	*N*-Ac-D-Glu	*Alcaligenes xylosoxydans* subsp. *xylosoxydans* A-6	12, 30
	N-Ac-D-Glu (specific)	*N*-Ac-D-Glu D-Glu, D-Asp, D-Asn	*Pseudomonas* 5f-1	8
N-Acyl-D-Aspartate amidohydrolase	*N*-Ac-D-Asp (specific)	*N*-Ac-D-Asp	*Alcaligenes xylosoxydans* subsp. *xylosoxydans* A-6	13, 30

3-3. Assay Methods for *N*-acyl-D-amino Acid Amidohydrolase

N-Acyl-D-amino acid amidohydrolase catalyzes the hydrolysis of *N*-acyl derivatives of D-amino acids to the corresponding D-amino acids and fatty acids. For an enzyme assay, in which *N*-acetyl-D-amino acid is usually used as a substrate, the amount of D-amino acid released by the reaction can be determined. Some methods often used for the determination of D-amino acid are described below. In general, one unit of enzyme activity is defined as the amount of enzyme which catalyzes the formation of 1 μmol of D-amino acid per min in the hydrolysis reaction.

a. TNBS Method

2,4,6-Trinitrobenzenesulfonic acid (TNBS), which specifically reacts with a free amino group of amino acids, gives trinitrophenyl (TNP) derivative its yellow color. This yellow product has a maximum absorbance at 340 ~ 350 nm. In the standard method, the detectable concentrations of D-amino acid range from 0.4 to 4 mM. Scheme 1 shows the procedure for the TNBS reaction. After the enzyme reaction, by adding TNBS into the reaction mixture and measuring the absorbance at 520 nm (wavelength providing maximum absorbance in the condition), the amount of D-amino acid produced in the reaction can easily be determined [19]. Since this method is convenient and highly sensitive, it is widely used for the assay of *N*-acyl-D-amino acid amidohydrolase.

Reaction mixture (RM)	
Buffer	100 mM
N-Acetyl-D-amino acid	20 mM
Enzyme	

Total 0.2 ml
- incubate at 30 °C for 10 min
- add 100 μl of 0.25 N NaOH to stop the reaction
- incubate at 30 °C for 10 min
- add 100 μl of 0.125 M NaB_4O_7 to RM
- add 40 μl of 0.2 M TNBS to RM
- incubate at 30 °C for 10 min
- add 800 μl of NaH_2PO_4•Na_2SO_3* solution to RM
- measure the absorbance at 520 nm

*freshly prepare by mixing 0.1 M NaH_2PO_4 with 0.1 M Na_2SO_3 in a 19.7 to 0.3 ratio.

Scheme 1. TNBS method.

b. D-Amino Acid Oxidase Method

D-Amino acid oxidase catalyzes the deamination of D-amino acid to α-keto acids, hydrogen peroxide, and ammonia. The enzyme does not oxidize acidic D-amino acids and shows little activity on L-proline. In this method, the amount of D-amino acid released by *N*-acyl-D-amino acid amidohydrolase can be determined indirectly by measuring α-keto acids, hydrogen peroxide, or ammonia. The concentration of ammonia can be measured by the Conway-Indophenol method, Nestler's reagent, and glutamate dehydrogenase. In general, the amounts of α-keto acid and hydrogen peroxide are measured. 2,4-Dinitrophenyl hydrazine (DNPH), which specifically reacts with keto acid, gives a hydrazone derivative having a maximum absorbance at 445 nm. The amount of α-keto acid corresponding to D-amino acid can easily be determined by adding DNPH to a D-amino acid reaction mixture and measuring the absorbance at 445 nm. On the other hand, in measuring hydrogen peroxide, the peroxidase-coupling method is convenient. In this method, peroxidase and a reduced-form chromophore such as 4-aminoantipyrine and phenol are mixed in the D-amino acid oxidase reaction solution. The D-amino acid oxidase-peroxidase coupling reaction is begun by adding the *N*-acyl- D-amino acid amidohydrolase reaction mixture. After 10 – 20 min, the absorbance is measured at 500 nm. The detectable range is from 0.02 to 0.1 mM in the standard method. Other chromophores often used are *o*-phenylenediamine, *o*-dianisidine, and *N*-ethyl-*N*-(2-hydroxy-3-sulfopropyl-)3-methylaniline [16]. Scheme 2 shows the procedures for the DNPH method and the peroxidase coupling method.

Reaction mixture (RM)

Buffer	100 mM
N-Acetyl-D-amino acid	20 mM
Enzyme	
	Total 0.2 ml

- incubate at 30 °C for 10 min
- add 100 μl of 0.25 N NaOH to stop the reaction
- incubate at 30 °C for 5 min
- add 100 μl of 0.25 N HCl to neutralize RM
- centrifuge at 12,000 rpm for 10 min
- transfer 100 μl of the supernatant to D-amino acid oxidase reaction mixture

Reaction mixture (peroxidase coupling method)

Buffer (HEPES, pH 8.0)	100 mM
D-Amino acid oxidase	0.5 U
Peroxidase	3.0 U
4-Aminoantipyrine/phenol	20 μg/0.4 μl
Supernatant	
	Total 1.0 ml

- incubate at 30 °C for 10 min
- add 100 μl of 6 N HCl to stop the reaction
- measure the absorbance at 500 nm

Reaction mixture (DNPH method)

Buffer (HEPES, pH 8.0)	100 mM
D-Amino acid oxidase	0.5 U
Catalase	2.0 U
Supernatant	100 μl
	Total 0.5 ml

- incubate at 30 °C for 10-60 min
- add 250 μl of 0.1 % DNPH* solution
- add 250 μl of water
- incubate at 30 °C for 3 min
- add 500 μl of 4 N NaOH
- measure the absorbance at 470 nm

*dissolve DNPH into 2 N HCl by heating.

Scheme 2. Methods using D-amino acid oxidase.

c. HPLC

There are two ways to determine the amount of D-amino acid released in the *N*-acyl-D-amino acid amidohydrolase reaction by HPLC. The D-amino acid released in the reaction can be separated by enantioselective HPLC using a Sumichiral OA series column. Its concentration is obtained by measuring the absorbance at a wavelength ranging from 200 to 260 nm, derived directly from the D-amino acid produced in the reaction. Another way is to measure the fluorescence of the D-amino acid-*o*-phthaldialdehyde (OPA) derivative by HPLC using reverse-phase column. This method is highly sensitive, and the resolution of DL-amino acid becomes possible by combining *N*-tert butyloxycarbonyl-L-cysteine (Boc) with OPA. Fluorescence is measured at an excitation wavelength of 344 nm to detect emissions at 433 nm. The D-amino acid detection process by HPLC is illustrated in Scheme 3.

d. Ninhydrin Method

Ninhydrin has been widely used for the detection of amino acid and peptide, but there are a few reports that the ninhydrin method has been used for the assay of *N*-acyl-D-amino acid amidohydrolase. An explanation of the ninhydrin method is not given here.

Reaction mixture
↓
Stop the reaction by boiling
↓
Supernatant
↓

Directly applied to the chiral column

Column: SUMICHIRAL OA-5000
Eluent: 2 mM CuSO4:isopropylalcohol=85:15
Reactor: UV254nm

Preparation of the derivative with OPA-Boc

OPA-Boc	10 mg/ml
Borate buffer (pH 9.0)	0.3 M
Supernatant	
	Total 0.5 ml

--- incubate at room temperature for 3 min
--- apply to column

Column: COSMOSIL 5C18-MS-II (4.6x250 mm)
Eluent: 0.2 M acetate buffer (pH 6.0):AcCN:THF
50 : 7 : 3 (A buffer)
50 : 47 : 3 (B buffer) arbitrary gradient
Detector Excitation: 344 nm
Emission: 433 nm

Scheme 3. HPLC method.

3-4. Gene Cloning of *N*-acyl-D-amino Acid Amidohydrolase

a. Shotgun Cloning by the D-Amino Acid Oxidase Method

To obtain the *N*-acyl-D-amino acid amidohydrolase gene, the D-amino acid oxidase-peroxidase coupling method is used, as described in "Assay method" [18]. This method is outlined in Scheme 4. A gene library is prepared by shotgun cloning from a microorganism with a cloning vector such as pUC18. *Escherichia coli* transformants are spread on an LB/ampicillin agar plate containing IPTG and X-gal. The resulting white colonies, indicating clones carrying a vector with a genomic insert, are transferred to an LB/ampicillin agar plate containing IPTG and 0.2% *N*-acetyl-D-amino acid. To detect the *N*-acyl-D-amino acid amidohydrolase activity of each transformant, the replica printing method is used. Colonies grown on an LB/ampicillin agar plate containing IPTG and 0.2% *N*-acetyl-D-amino acid are transferred to a replica paper. The paper is immersed in an *N*-acyl-D-amino acid amidohydrolase reaction mixture composed of 100 mM buffer (pH 7.0), 20 mM *N*-acyl-D-amino acid, and 0.1 mg/ml CTAB. After a 30-min reaction, the paper is dried and then immersed in a D-amino acid-peroxidase coupling reaction mixture composed of 100 mM buffer (pH 7.0), 0.2 units/ml D-amino acid oxidase (as the substrate is *N*-acetyl-D-aspartate or *N*-acetyl-D-glutamate, D-aspartate oxidase or D-glutamate oxidase is used, respectively.), 10 units/ml peroxidase, and 0.5 mg/ml *o*-dianisidine. As a consequence of these series of reactions, the surroundings of a recombinant *E. coli* cell in which the *N*-acyl-D-amino acid amidohydrolase gene is expressed will show a deep red color due to oxidized *o*-dianisidine. This detection method is simple and useful for the cloning of *N*-acyl-D-amino acid amidohydrolase.

b. Gene Amplification by PCR Using Mixed Primers Designed Based on the Conserved Amino Acid Sequence

An increasing amount of available sequence information for D-aminoacylase has facilitated the isolation of a fragment of a gene by PCR. The crystal structure of D-aminoacylase from *Alcaligenes faecalis* DA1 shows the conserved amino acids that locate in its active site [20]. The two peptide regions from Pro60 to Asp72 and from Gly365 to Phe383 locate near the active center of D-aminoacylase from *Alcaligenes faecalis* DA1. This conserved amino acid sequence is reverse translated, and a mixture of oligonucleotides representing all possible DNA sequences coding for that particular amino acid sequence can be synthesized. These two degenerated primers are employed in a PCR reaction. The amplified PCR products can be extended by cassette-ligation-mediated PCR [21] to determine the D-aminoacylase gene with the TaKaRa PCR in vitro Cloning Kit as recommended by the manufacturer. Alternatively, the PCR products can be used as a probe to screen a genomic library by colony hybridization [22].

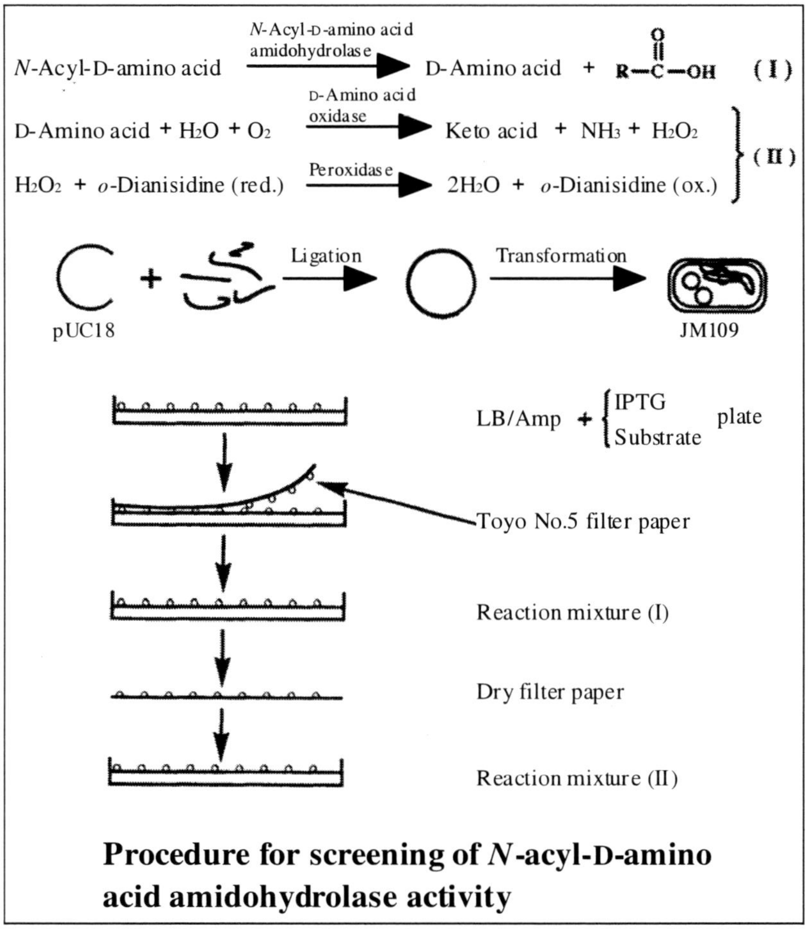

Scheme 4. Screening method for a transformant exhibiting *N*-acyl-D-amino acid amidohydrolase activity.

4. Results

The properties of microbial *N*-acyl-D-amino acid amidohydrolases reported to date are summarized in Table 3. Most of these are metallo-enzymes containing zinc ion(s) and are induced by substrates, substrate analogues, and D-amino acids, except some enzymes. All enzymes prefer *N*-acetyl or *N*-formyl derivatives of D-amino acids as substrates. These data are helpful in the preparation of the media for the isolation of microorganisms having *N*-acyl-D-amino acid amidohydrolase and the reaction mixture for the enzyme assay. Recently, we performed a screening of the microorganisms exhibiting *N*-acyl-D-amino acid amidohydrolase activity. (Refer to data in Table 2.) Among 260 isolates, one bacterium showing relatively high *N*-acyl-D-amino acid amidohydrolase activity for *N*-acetyl-D-leucine was found. The cell-free extract prepared from the isolate exhibited 0.67 U/mg for *N*-acetyl-D-leucine as a specific activity.

Table 3. Characteristics of microbial *N*-acyl-D-amino acid amidohydrolase.

Enzyme	Strain	Best substrate	Optimal pH	Optimal Temperature	Metal	Inducer (1 mM)	Inhibitor	pI
D-Aminoacylase	*Alcaligenes xylosoxydans* A-6	*N*-Ac-D-Leu	7	40° C	Zn^{2+}	*N*-Ac- -methyl-D-Leu	Cu^{2+}, Zn^{2+}, EDTA	5.2
	Alcaligenes denitrificans MI-4	*N*-Ac-D-Met	7. 4	50° C	-	-Methyl-D-Leu	Cu^{2+}, Hg^{2+}	-
	Alcaligenes faecalis DA1	*N*-Ac-D-Met	7.0-8.0	45° C	Zn^{2+}	*N*-Ac-DL-Leu	-	5.4
	Alcaligenes denitrificans DA181	*N*-Ac-D-Met	7. 5	45° C	Zn^{2+}	*N*-Ac-D-Met	Cu^{2+}, Zn^{2+}, EDTA	4.4
	Streptomyces olivaceus 62-3	*N*-Ac-D-Leu	7	50° C	-	D-Phenylglycine	Ag^{2+}, Hg^{2+}, EDTA	-
	Streptomyces tuirus	*N*-Ac-D-Leu	7.0-7.5	-	-	DL-Val	Ag^{2+}, Hg^{2+}(0.01 mM)	-
	Variovorax paradoxus	*N*-Ac-D-Met	7. 5	50° C	Co^{2+}, Mg^{2+}	-	Cu^{2+}, Fe^{2+}, Hg^{2+}, Zn^{2+}	-
	Defluvibacter sp. 131-3	*N*-Ac-D-Val	8	37° C	-	*N*-Ac-D-Val	Co^{2+}, Mn^{2+}, Ni^{2+}, Zn^{2+}	5.3
	Trichoderma harzianum	*N*-Ac-D-Phe	8	55° C	-	-	Co^{2+}, Fe^{2+}, Ni^{2+}, Zn^{2+}	4.5
N-Acyl-D-aspartate amidohydrolase	*Alcaligenes xylosoxydans* A-6	*N*-Ac-D-Asp	7.5-8.0	50° C	Zn^{2+}	*N*-Ac-DL-Asp	Co^{2+}, Fe^{2+}, Hg^{2+}, Mn^{2+}, Zn^{2+}	4.8
N-Acyl-D-glutamate amidohydrolase	*Alcaligenes xylosoxydans* A-6	*N*-Ac-D-Glu	7.5-8.0	55° C	Zn^{2+}	*N*-Ac-DL-Glu	Cu^{2+}, Co^{2+}	5.5
	Pseudomonas	*N*-Ac-D-Glu	6.5-7.5	45° C	Zn^{2+}	DL -Glu	Cu^{2+}, Hg^{2+}	8.8
sp	. 5f-1						Fe^{2+}, Zn^{2+}, EDTA	

5. Notes

1. The enrichment culture is considered to be effective for the production of *N*-acyl-D-amino acid amidohydrolase because its substrate, *N*-acyl-D-amino acid, or product, D-amino acid, could often be a strong inducer of the bacterial enzyme. For example, *N*-acyl-D-glutamate amidohydrolase and *N*-acyl-D-aspartate amidohydrolase from *Alcaligenes* A-6 were induced by *N*-acyl-D-glutamate and *N*-acyl-D-aspartate, respectively. D-Aminoacylase from *Alcaligenes* A-6 was strongly induced by the substrate analogue, *N*-acetyl -γ-methyl-D-leucine [12,13,14].

2. D-Amino acid oxidase from pig kidney is commercially available, but it has a substrate preference. Its activity on some D-amino acids, in particular, on D-aspartate and D-glutamate, is ineffective.
3. In this D-amino acid oxidase reaction, a catalase must be added to the reaction mixture because the α-keto acid produced in the reaction is readily degraded hydrogen peroxide.
4. Since D-amino acid oxidase acts specifically on D-amino acid, the enzyme can be assayed even if *N*-acyl-DL-amino acid is used as a substrate when *N*-acyl-L-amino acid amidohydrolase is included in the enzyme solution.
5. D-Glutamate oxidase and D-aspartate oxidase cannot be purchased commercially. D-Aspartate oxidases and D-glutamate oxidases from yeasts and fungi have been reported [23,24,25].
6. TNBS is relatively easy to oxidize at room temperature (turning from yellow to deep red), and it causes an increase in the background absorbance during the D-amino acid detection reaction. Before TNBS is used, the color of the solution should be checked. If it exhibits a deep red color, it must be freshly prepared.

REFERENCES

[1] Wakayama, M. & Moriguchi, M. (2001). Comparative biochemistry of bacterial *N*-acyl-D-amino acid amidohydrolase. *J Mol Catal B: Enzyme, 12,* 15-25.

[2] Wakayama, M., Yoshimune, K., Hirose, Y., & Moriguchi, M. (2003). Production of D-amino acids by *N*-acyl-D-amino acid amidohydrolase and its structure and function. *J Mol Catal: Enzyme, 23,* 71-85.

[3] Sugie, M. & Suzuki, H. (1978). Purification and properties of D-aminoacylase of *Streptomyces olivaceus*. *Agr Biol Chem, 42,* 107-113.

[4] Sugie, M. & Suzuki, H. (1980). Optical resolution of DL-amino acids with D-aminoacylase of *Streptomyces*. *Agric Biol Chem, 44,* 1089-1095.

[5] Kameda, Y., Toyoura, E., Yamazoe, H., Kimura, Y., & Yasuda, Y. (1952). Hydrolysis and metabolism by soil bacteria of benzoyl derivatives of D- and L-forms of some amino acids. *Nature, 170,* 888-889.

[6] Kameda, Y., Hase, H., Kanamoto, S., & Kita, Y. (1978). Studies on acylase activity and microorganisms. XXVI. Purification and properties of D-acylase (*N*-acyl-D-amino acid amidohydrolase) from AAA 6029 (*Pseudomonas* sp.). *Chem Pharm Bull (Tokyo), 26*, 2698-2704.

[7] Kubo, K., Ishikura, T., & Fukagawa, Y. (1980). Deacetylation of PS-5, a new β-lactam compound. II. Separation and purification of L- and D-amino acid acylases from *Pseudomonas* sp. 1158. *J Antibiot*, *43*, 550-555.

[8] Sakai, K., Oshima, K., & Moriguchi, M. (1991). Production and characterization of *N*-acyl-D-glutamate amidohydrolase from *Pseudomonas* sp. Strain 5f-1. *Appl Environ Microbiol*, *57*, 2540-2543.

[9] Moriguchi, M. & Ideta, K. (1988). Production of D-aminoacylase from *Alcaligenes denitrificans* subsp. *xylosoxydans* MI-4. *Appl Environ Microbiol*, *54*, 2767-2770.

[10] Tsai, Y. C., Tseng, C. P., Hsiao, K. M., & Chen, L. Y. (1988). Production and purification of D-aminoacylase from *Alcaligenes denitrificans* and taxonomic study of the strain. *Appl Environ Microbiol*, *54*, 984-989.

[11] Yang, Y. B. Y., Lin, C. S., Tseng, C. P., Wang, Y. J., & Tsai, Y. C. (1991). Purification and characterization of D-aminoacylase from *Alcaligenes faecalis* DA1. *Appl Environ Microbiol*, *51*, 1259-1260.

[12] Sakai, K., Imamura, K., Sonoda, Y., Kido, H., & Moriguchi, H. (1991). Purification and characterization of *N*-acyl-D-glutamate deacylase from *Alcaligenes xylosoxydans* subsp. *xylosoxydans* A-6. *FEBS Lett*, *289*, 44-46.

[13] Moriguchi, M., Sakai, K., Katsuno, Y., Maki, T., & Moriguchi, M. (1993). Purification and characterization of novel *N*-acyl-D-aspartate amidohydrolase from *Alcaligenes xylosoxydans* subsp. *xylosoxydans* A-6. *Biosci Biotech Biochem, 57*, 1145-1148.

[14] Moriguchi, M., Sakai, K., Miyamoto, Y., & Wakayama, M. (1993). Production, purification, and characterization of D-aminoacylase from *Alcaligenes xylosoxydans* subsp. *xylosoxydans* A-6. *Biosci Biotech Biochem*, *57*, 1149-1152.

[15] Lin, P. H., Su, S. C., Tsai, Y. C., & Lee, C. Y. (2002). Identification and characterization of a new gene from *Variovorax paradoxus* Isol encoding *N*-acyl-D-amino acid amidohydrolase responsible for D-amino acid production. *Eur J Biochem*, *269*, 4868-4878.

[16] Kumagai, S., Kobayashi, M., Yamaguchi, S., Kanaya, T., Motohashi, R., & Isobe, K. (2004). A new D-aminoacylase from *Defluvibacter* sp. A 131-3. *J Mol Catal B: Enzyme*, *30*, 159-165.

[17] Wakayama, M., Kitahata, S., Manoch, L., Tachiki, T., Yoshimune, K., & Moriguchi, M. (2004). Production, purification and properties of D-aminoacylase from a newly isolated *Trichoderma* sp. SKW-36. *Process Biochem*, *39*, 1119-1124.

[18] Wakayama, M., Katsuno, Y., Hayashi, S., Miyamoto, Y., Sakai, K., & Moriguchi, M. (1995). Cloning and sequencing of a gene encoding D-aminoacylase from *Alcaligenes xylosoxydans* subsp. *xylosoxydans* A-6 and expression of the gene in *Escherichia coli*. *Biosci Biotech Biochem*, *59*, 2115-2119.

[19] Fields, R. (1972). The rapid determination of amino groups with TNBS. *Methods Enzymol*, *25*, 464-468.

[20] Liaw, S. H., Chen, S. J., Ko, T. P., Hsu, C. S., Chen, C. J., Wang, A. H., & Tsai, Y. C. (2003). Crystal structure of D-aminoacylase from *Alcaligenes faecalis* DA1. A novel subset of amidohydrolases and insights into the enzyme mechanism. *J Biol Chem*, *278*, 4957-4962.

[21] Isegawa, Y., Sheng, J., Sokawa. Y., Yamanishi, K., Nakagomi, O., & Ueda, S. (1992). Selective amplification of cDNA sequence from total RNA by cassette-ligation mediated polymerase chain reaction (PCR): application to sequencing 6.5 kb genome segment of hantavirus strain B-1. *Mol Cell Probes*, *6*, 467-475.

[22] Sambrook, J. & Russell, D. W. Molecular cloning: A laboratory manual, 3rd ed. Cold Spring Harbor Laboratory, Cold Spring Harbor, N.Y., 2001.

[23] Yamada, R., Ujiie, H., Kera, Y., Nakase, T., Kitagawa, K., Imasaka, T., Arimoto, K., Takahashi, M., & Matsumura, Y. (1996). Purification and properties of D-aspartate oxidase from *Cryptococcus humicolus* UJ1. *Biochim Biophys Acta*, *1294*, 153-158.

[24] Wakayama, M., Nakashima, S., Sakai, K., & Moriguchi, M. (1994). Isolation, enzyme production and characterization of D-aspartate oxidase from *Fusarium sacchari* var. *elongatum* Y-105. *J Ferment Bioeng*, *78*, 377-379.

[25] Wakayama, M., Takashima, K., Tau, Y., Nakashima, S., Sakai, K., & Moriguchi, M. (1997). Spectrophotometric assay of D-aspartate and D-glutamate using D-aspartate oxidase with malate dehydrogenase and glutamate dehydrogenase. *Anal Biochem*, *250*, 252-253.

[26] Sakai, K., Obata, T., Takano, S., & Moriguchi, M. (1989). A novel inducer, γ-methyl-D-leucine, of D-aminoacylase from *Alcaligenes denitrificans* subsp. *xylosoxydans* MI-4. *Agric Biol Chem*, *53*, 2285-2286.

[27] Sakai, K., Obata, T., Ideta, K., & Moriguchi, M. (1991). Purification and properties of D-aminoacylase from *Alcaligenes denitrificans* subsp. *xylosoxydans* MI-4. *J Ferment Bioeng*, *71*, 79-82.

[28] Tsai, Y. C., Lin, C. S., Tseng, T. H., Lee, H., & Wang, Y. J. (1992). Production and immobilization of D-aminoacylase of *Alcaligenes faecalis* DA1 for optical resolution of *N*-acyl-DL-amino acids. *Enzyme Microbiol Technol*, *14*, 384-389.

[29] Yang, Y. B., Hsiao, K. M., Li, H., Yano, H., Tsugita A., & Tsai Y. C. (1992). Characterization of D-aminoacylase from *Alcaligenes denitrificans* DA181. *Biosci Biotech Biochem*, *56*, 1392-1395.

[30] Sakai, K., Imamura, K., Goto, M., Hirashiki, I., & Moriguchi, M. (1990). Occurrence of novel enzymes, *N*-acetyl-D-glutamate deacetylase and *N*-acetyl-D-aspartate deacetylase, in *Alcaligenes xylosoxydans* subsp. *xylosoxydans* A-6. *Agric Biol Chem*, *54*, 841-844.

In: D-Amino Acids: ISBN 1-60021-075-9
A New Frontier in Amino Acid and Protein Research, pp.573-577

Chapter 6.19

ACTIVITY MEASUREMENT OF HYDANTOIN RACEMASE ENZYME: A KEY FOR THE PRODUCTION OF OPTICALLY PURE D-AMINO ACIDS

Sergio Martínez-Rodríguez, Josefa María Clemente-Jiménez, Felipe Rodríguez-Vico and Francisco Javier Las Heras-Vázquez*
Departamento de Química-Física, Bioquímica y Química Inorgánica.
Edificio C.I.T.E. I. Universidad de Almería, Spain.

Keywords: hydantoin racemase, hydantoinase process, 5-monosubstituted hydantoins synthesis

1. INTRODUCTION

Optically pure D-amino acids can be enzymatically obtained from D,L-5-monosubstituted hydantoins, a process that is cheaper and less contaminating than chemoenzymatic production. In this enzymatic reaction, called "hydantoinase process" [1], firstly the chemically synthesised D,L-5-monosubstituted hydantoin ring is hydrolysed by a stereoselective hydantoinase enzyme. Further hydrolysis of the resulting enantiospecific *N*-carbamoyl-D-amino acid to the free D-amino acid is catalysed by highly enantiospecific *N*-carbamoyl-D-amino acid amidohydrolase (D-carbamoylase). At the same time as D-hydantoinase hydrolyses the enantiospecific D-5-monosubstituted hydantoin, the chemical and/or enzymatic racemization of L-5-monosubstituted hydantoin starts.

Hydantoin racemase plays an important role in the production of optically pure D-amino acids because it allows the racemization of the 5-monosubstituted hydantoins under physiological conditions where chemical racemization is not favoured [2]. The substrates of the hydantoin racemase enzyme are optically pure D- and L-5-monosubstituted hydantoins which can be synthesized from their corresponding D- and L-amino acids. This enzyme together with a stereoselective hydantoinase and a stereospecific D-carbamoylase guarantee the total conversion from D,L-5-monosubstituted hydantoins with a low velocity of racemization to optically pure D-amino acids [3]. Knowledge of the biochemical, physical and thermodynamic characteristics of the hydantoin racemase enzyme, after its purification, will allow the

* Correspondence concerning this article should be addressed to Francisco Javier Las Heras-Vázquez, E-mail: fjheras@ual.es, FAX: + 34 950 015008.

better design of biocatalysts, such as immobilized enzymes or whole cell biocatalysts. For this purpose we describe an appropriate substrate synthesis method together with its fast and sensitive chiral HPLC separation to characterise the enzyme.

2. MATERIALS

1. Substrate Synthesis

1.1. Optically pure D- and L-amino acids and potassium cyanate (Fluka, Madrid, Spain). Short neck round bottom flask (100 mL) (POBEL, Madrid, Spain).
1.2. Magnetic stirrer with heating (IKAMAG RCT basic, IKA, Staufen, Germany), electronic contact thermometer (ETS-D 4 fuzzy IKATRON, IKA, Staufen, Germany) and air condenser (POBEL, Madrid, Spain).
1.3. Hydrochloric acid (Fluka, Madrid, Spain).
1.4. Buchner funnel and water jet filter pump (Brand, Wertheim, Germany).
1.5. Ethanol (Fluka, Madrid, Spain).
1.6. Crystallizing dish (900 mL) (POBEL, Madrid, Spain).
1.7. Desiccator (POBEL, Madrid, Spain).
1.8. UV spectrophotometer (Spectronic Helios β, Thermo, Bonsai Technologies S.A., Madrid, Spain).

2. Activity Assay

2.1. Purified recombinant hydantoin racemase [4,5]. Optically pure 5-monosubstituted hydantoins synthesized from the previous section. Sodium dihydrogen phosphate anhydrous and disodium hydrogen phosphate anhydrous (Merck Farma y Química S.A., Barcelona, Spain).
2.2. Thermomixer compact (Eppendorf Iberica, Madrid, Spain) and ortho-phosphoric acid (H_3PO_4) (Merck Farma y Química S.A., Barcelona, Spain).
2.3. The HPLC system (Waters 1515 isocratic HPLC pump, Waters 2487 UV detector and Waters 717plus autosampler). Chiral Chirobiotic T column (4.6×250 mm, Astec Inc.). Methanol Lichrosolv (Merck Farma y Química S.A., Barcelona, Spain) and MiliQ water, both for liquid chromatography.

3. METHODS

1. Substrate Synthesis

1.1. Add 20 mmols of optically pure D- or L- amino acid and 25 mmols of potassium cyanate to a short neck round bottom flask with 15 mL of H_2O.
1.2. Heat reaction mixture to reflux and stir slightly at 60-80°C for 1 hour.
1.3. Add 5 mL of hydrochloric acid (37 %) to the stirred solution and continue heating for 15 min more.
1.4. Cool the reaction mixture on ice until the product is crystallized.
1.5. Vacuum filter the product in a Buchner funnel.
1.6. Add 20 mL of cold 20 % ethanol (v/v) solution to the Buchner funnel containing the product to wash it. Repeat this step twice.

1.7. Recrystallize the 5-monosubstituted hydantoin from hot 20 % ethanol (v/v).
1.8. Dry the powder in a filter paper bag, placed in a desiccator for one day.
1.9. Measure the maximum wavelength of each optically pure D- and L-5-monosubstituted hydantoin in a spectrophotometer.

2. Activity Assay

2.1. Mix purified hydantoin racemase (end concentration 1-5 μM) together with 5 mM of optically pure 5-monosubstituted hydantoin dissolved in 100 mM sodium phosphate buffer (pH 7.5) in 1000 μL of reaction volume in a 1.5 mL Eppendorf tube.
2.2. Incubate the reaction mixture at 40 °C in a thermomixer and take out aliquots (50 μL each) at different intervals (e.g. 0, 15, 30 min, etc.). Stop the reaction of each aliquot by adding it to 400 μL of 1% H_3PO_4 and centrifuge at 10,000 r.p.m. for 10 min in a microcentrifuge.
2.3. Transfer the supernatant to an HPLC autosampler vial. Inject 20 μL of sample into the HPLC with a chiral column and run it for 30 min. The mobile phase used in the analysis, under isocratic conditions, is 80 % water and 20 % methanol, with a flow of 0.5 mL min^{-1}.
2.4. Repeat steps 2.1., 2.2. and 2.3. at least three times.

4. RESULTS

Kinetic parameters of hydantoin racemases 1 and 2 from *Agrobacterium tumefaciens* C58 (AtHyuA1 and AtHyuA2, respectively) have been obtained for D- and L- isomers of 5-monosubstituted hydantoins (Table 1). Both enzymes have shown higher affinity for hydantoins with aliphatic substituents than those with aromatic ones at the 5-position. However, in a hypothetical application of both hydantoin racemase enzymes in a multienzymatic system for production of optically pure D-amino acids, AtHyuA1 [4] would be more viable for industrial application than AtHyuA2 [5], due to its higher substrate affinity and racemization velocity.

Table 1. Kinetic parameters of *A. tumefaciens* hydantoin racemase 1 and 2 (AtHyuA1 and AtHyuA2)[a].

Substrate[b]	AtHyuA1			AtHyuA2		
	K_m (mM)	k_{cat} (s^{-1})	k_{cat}/K_m (s^{-1} * mM^{-1})	K_m (mM)	k_{cat} (s-1)	k_{cat}/K_m (s^{-1} * mM^{-1})
L-EH	4.45 ± 0.60	13.64 ± 1.53	3.06 ± 0.63	19.42 ± 2.34	1.81 ± 0.01	0.09 ± 0.01
D-EH	4.26 ± 0.70	16.47 ± 1.47	3.87 ± 0.59	12.54 ± 1.81	1.80 ± 0.10	0.14 ± 0.03
L-BH	5.56 ± 1.45	1.07 ± 0.23	0.19 ± 0.04	18.42 ± 4.80	0.18 ± 0.02	0.01 ± 0.00
D-BH	4.71 ± 0.62	2.36 ± 0.43	0.50 ± 0.10	20.77 ± 5.47	0.46 ± 0.07	0.02 ± 0.00
L-MTEH	5.41 ±1.06	1.57 ± 0.37	0.35 ± 0.08	10.90 ±1.48	0.78 ± 0.07	0.07 ± 0.01
D-MTEH	6.31 ± 0.31	0.50 ± 0.01	0.08 ± 0.01	1.23 ± 0.20	2.05 ± 0.53	1.66 ± 0.13
L-IBH	1.23 ± 0.20	2.05 ± 0.53	1.66 ± 0.13	3.02 ± 0.78	0.48 ± 0.02	0.16 ± 0.05
D-IBH	4.58 ±0.51	4.16 ± 0.67	0.90 ± 0.08	6.79 ± 0.60	0.83 ± 0.03	0.12 ± 0.01

[a] The kinetic parameters of hydantoin racemases were determined at 40 °C for 15 min at pH 8.5. The k_{cat} was defined as the mmol of D- or L-5-monosubstituted hydantoin racemized per s and mmol of enzyme at 40 °C.
[b] D- and L-EH: D- and L-5-ethyl-hydantoin; D- and L-BH: D- and L-5-benzylhydantoin; D- and L-MTEH: D- and L-5-methylthioethylhydantoin; D- and L-IBH: D- and L-5-isobutylhydantoin.

5. NOTES

1. Substrate Synthesis

1.1. From the following amino acids, D- and L-2-aminobutyric acid (≥99 %), D- and L-methionine (≥99 %), D- and L-leucine (≥99 %) and D- and L-phenylalanine (≥99 %), the 5-monosubstituted hydantoins obtained are D- and L-5-ethylhydantoin (D- and L-EH), D- and L-5-methylthioethyl-hydantoin (D- and L-MTEH), D- and L-5-isobutylhydantoin (D- and L-IBH) and D- and L-5-benzylhydantoin (D- and L-BH), respectively. For less soluble amino acids (e.g. phenylalanine), it is preferable to pre-heat water before adding the potassium cyanate.

1.2. As only slight stirring is required, a little piece of porous plate may be used in the round bottom flask instead of a magnet.

1.3. The HCl solution must be added very slowly in order to avoid producing corrosive hydrochloric acid vapours. In this step the cyclization process of the carbamoyl-amino acid to 5-monosubstituted hydantoin takes place.

1.4. The duration of this step can vary from a few minutes to several days depending on the solubility of the 5-monosubstituted hydantoin synthesized.

1.5. Use the cold 20 % ethanol (v/v) solution to empty out and clean the round bottom flask. In this way we ensure that the entire product formed in the previous steps is filtered in the Buchner funnel.

1.6. The maximum absorbance of the hydantoin ring is normally between 195 and 215 nm. Indeed, the maximum wavelengths were 210 nm for D- and L-BH and D- and L-MTEH, 200 nm for D- and L-EH, and 205 nm for D- and L-IBH.

2. Activity Assay

2.1. Due to the different kinetics of the enzyme, the optimal enzyme concentration, pH and temperature reaction have to be determined previously. Pre-warm the substrate and the buffer in the tubes at reaction temperature (40 °C for these enzymes) for at least 10 min before adding the enzyme. In this way, we ensure that the reaction starts at the desired temperature. Likewise, in order to avoid variations in temperature, the stock concentration of the enzyme has to be high enough for the volume added to the reaction tube to be less than 5 % of the whole volume reaction. In order to check the spontaneous racemization of each optically pure 5-monosubstituted hydantoin, a control assay should be made without the enzyme.

2.2. The reaction time finishes when the optically pure 5-monosubstituted hydantoin (100 %) has been racemized to 50 % of each isomer (Figure 1A and B). The 1 % H_3PO_4 guarantees that the reaction has been stopped because the protein has been completely denaturalised. 1 % HCl could also be used to stop the reaction. If biochemical constants (K_m, k_{cat}) need to be determined, carry out the reaction with different substrate concentrations (from 0 to 100 mM, if the substrate is soluble enough).

2.3. The solvents have to be filtered daily through a 0.45-μm cellulose membrane filter before use, and degassed with helium both before and during use. The UV detector should be set to the maximum wavelength of each substrate (see point 1.9.).

2.4. A calibration line is carried out by injecting several known concentrations (from 0 to 1 mM) of both optically pure forms (D- and L-). This enables us to find the relationship between peak areas in the chromatogram and concentrations for each compound. Kinetic parameters of each substrate can be obtained from hyperbolic saturation curves by least-squares fit of the data to the Michaelis-Menten equation.

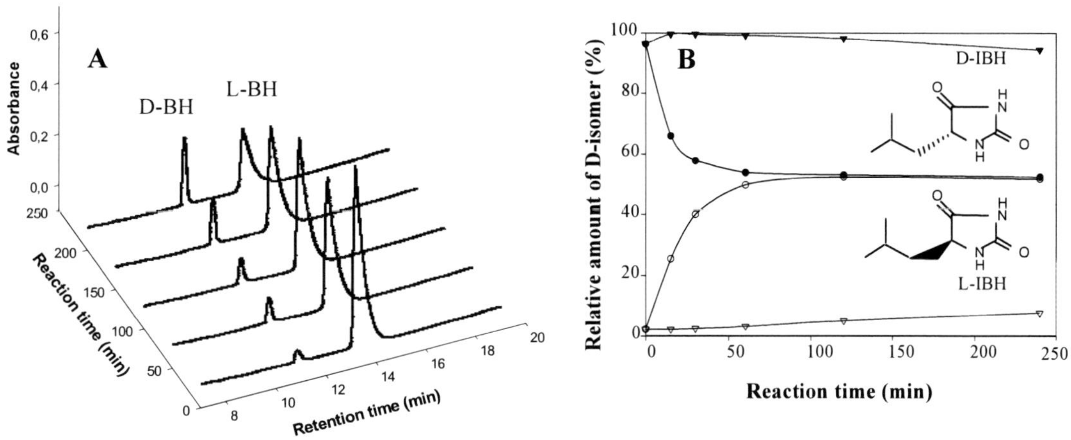

Figure 1. (A) HPLC separation of enantiomers D- and L-5-benzylhydantoin (D- and L-BH) produced during the enzymatic racemization of L-BH by hydantoin racemase 1 of *Agrobacterium tumefaciens* C58 (AtHyuA1). (B) Diagram of the evolutionary progression of the AtHyuA1 activity from the D-isobutylhydantoin (D-IBH) (•) and L-isobutylhydantoin (L-IBH) (○) enantiomers in comparison with the chemical racemization for the same substrates (D-IBH, ▼; L-IBH,▽).

REFERENCES

[1] Pietzsch, M. & Syldatk, C. (2002). Hydrolysis and formation of hydantoins. In K. Drauz, H. Waldmann (Eds.), *Enzyme Catalysis in Organic Synthesis*, (pp. 761-799), Weinheim, Wiley-VCH.

[2] Martinez-Rodriguez, S., Las Heras-Vazquez, F. J., Clemente-Jimenez, J. M., Mingorance-Cazorla, L. & Rodriguez-Vico, F. (2002). Complete conversion of D,L-5-monosubstituted hydantoins with a low velocity of chemical racemization into D-amino acids using whole cells of recombinant *Escherichia coli*. *Biotechnol Prog*, *18*, 1201-1206.

[3] Altenbuchner, J., Siemann-Herzberg, M. & Syldatk, C. (2001). Hydantoinases and related enzymes as biocatalysts for the synthesis of unnatural chiral amino acids. *Curr Opin Biotech*, *12*, 559-563.

[4] Las Heras-Vazquez, F. J., Martinez-Rodriguez, S., Mingorance-Cazorla, L., Clemente-Jimenez, J. M. & Rodriguez-Vico, F. (2003). Overexpression and characterization of hydantoin racemase from *Agrobacterium tumefaciens* C58. *Biochem Biophys Res Commun*, *303*, 541-547.

[5] Martinez-Rodriguez, S., Las Heras-Vazquez, F. J., Clemente-Jimenez, J. M. & Rodriguez-Vico, F. (2004). Biochemical characterization of a novel hydantoin racemase from *Agrobacterium tumefaciens* C58. *Biochimie*, *86*, 77-84.

In: D-Amino Acids: ISBN 1-60021-075-9
A New Frontier in Amino Acid and Protein Research, pp.579-589

Chapter 6.20

ENZYMES ACTING ON D-AMINO ACID AMIDES

Yasuhisa Asano*

Biotechnology Research Center, Toyama Prefectural University, Japan.

Keywords: D-amino acid amide, D-aminopeptidase, alkaline D-peptidase, *Ochrobactrum anthropi*, *Bacillus cereus*.

1. INTRODUCTION

D-Amino acids are important starting materials for various pharmaceuticals, herbicides and food additives: they act as intermediates for the preparation of β-lactam antibiotics such as semisynthetic cephalosporins and penicillins [1] or, like D-alanine, constitute an important part of a synthetic sweetener such as alitame [2].

We needed D-stereospecific amino acid amidases, speculating that such enzymes would be useful in the D-stereospecific amide synthesis and D-stereospecific kinetic resolution of amino acid amides. Knowing that few study has been done on this class of enzymes, we started screening for the enzymes, and discovered three new bacterial enzymes, D-aminopeptidase (EC 3.4.11.19), D-amino acid amidase, and alkaline D-peptidase. These enzymes were used for the chiral synthesis of amino acid derivatives. We developed a new method of D-stereoselective kinetic resolution of racemic amino acid amides and the synthesis of D-amino acid *N*-alkyl amides in organic solvents.

Some peptidases act on peptides containing D-amino acids. Soluble *Streptomyces* carboxypeptidase DD catalyzes not only the transpeptidation reaction on the peptide intermediate in peptidoglycan biosynthesis, but also the hydrolysis of *N,N*-diacetyl-L-lysyl-$(D\text{-}Ala)_2$ in water [3]. In *Enterococcus*, the *van*X gene product, $(D\text{-}Ala)_2$ hydrolase plays a role in vancomycin resistance [4] (Table 1). D-Amino acid containing peptides are found in the bacterial peptidoglycan, D-alanine containing dipeptides in rice plant, the peptide antibiotics gramicidin S, tyrosidine and bacitracin, and dermorphin in the frog [5].

In this report, the isolation procedures of the microorganisms producing D-stereospecific amide hydrolases, and the properties of D-aminopeptidase (DAP) produced by *Ochrobactrum anthropi* C1-38, and alkaline D-peptidase (ADP) from *Bacillus cereus* DF4-B, are described.

* Correspondence concerning this article should be addressed to Yasuhisa Asano, E-mail: asano@pu-toyama.ac.jp, FAX: +81-766-56-2498.

Table 1. D-Amino acid containing peptide hydrolases of microbial origin.

Enzyme	Origin	Active Center	Reference
Carboxypeptidase DD	microorganisms	Ser	[21]
D-Aminopeptidase (DAP)	*Ochrobactrum anthropi*	Ser	[12,15]
Alkaline D-peptidase (ADP)	*Bacillus cereus*	Ser	[16]
D-Amino acid amidase (DaaA)	*Ochrobactrum anthropi*	Ser	[24]
DD-Dipeptidase (VanX)	*Enterococcus*	Zn^{2+}	[4]
VanX homolog (DdpX)	*Escherichia coli*	Zn^{2+}	[25]
D-Aminopeptidase (DppA)	*Bacillus subtilis*	Zn^{2+}	[26]

2. Materials

1. DEAE-Toyopearl 650 M, Butyl-Toyopearl 650 M, HPLC columns G-3000 SW and DEAE-5PW were purchased from Tosoh Corp. (Tokyo, Japan), Sephacryl S-300 from Pharmacia (Sweden), marker proteins for molecular weight determination from Oriental Yeast (Tokyo, Japan).
2. D-Alanylglycine, D-alanyl-D-alanine, D-alanyl-D-alanyl-D-alanine, D-alanyl-L-alanyl-L-alanine, D-alanylglycylglycine, glycine amide HCl were purchased from Sigma (St. Louis, USA).
3. D-Alanyl-L-alanine, D-alanyl-D-alanyl-D-alanyl-D-alanine, D-alanine-*p*-nitroanilide HCl, L-alanyl-D-alanine, L-alanyl-L-alanine, and L-alanylglycine amide HCl were from Bachem (Switzerland).
4. D-Alanine methylester AHCl, D-alanine amide HCl, glycine methylester AHCl, $(Boc)_2O$ (di-*t*-butyl dicarbonate) [6], Boc-ON (2-(*t*-butoxycarbonyloxyimino)-2-phenylacetonitrile) [7], and 4 *N* hydrogen chloride in ethylacetate were purchased form Kokusan Chemicals (Tokyo, Japan). For microbial screening, a large amount of D-alanine methylester AHCl, and D-alanine amide HCl were synthesized in house as described below.
5. Thionylchloride, carbodiimide, isobutyl chloroformate, absolute methanol, and other chemicals were purchased from Wako Pure Chemicals Co. (Osaka, Japan).
6. Absolute methanol was saturated with dry ammonia gas by bubbling ammonia gas into absolute methanol cooled at about -10 °C by ice, methanol and NaCl.
7. *N*-Nitrosomethylurea was from ICN K & K Laboratories, Inc. (England).
8. Palladium-carbon (10%) was purchased from Nippon Engelhard Inc. (Tokyo, Japan).
9. All other chemicals were from commercial sources and used without further purification.
10. DNA manipulations were carried out under standard procedures [8].

3. Methods

1. Synthesis of D-Alanine Amide, $(D\text{-}Phe)_4$ and other Substrates for the Characterization of the Enzymes

The methylesters of amino acids were synthesized by slowly dropping 1.2 to 5.0 equivalents of thionylchloride into amino acids suspended in methanol at -20 °C [9]. Boc-Amino acids were synthesized from free amino acids by the use of $(Boc)_2O$ [6] or Boc-ON [7]. Methylesters of the Boc-D-amino acids were synthesized by methylation of the Boc-D-amino acids by ethereal diazomethane, which was evolved from *N*-nitrosomethylurea. Boc-D-Glutamic acid dimethylester, Boc-D-aspartic acid dimethylester, Boc-D-aspartic acid-β-benzylester-α-methylester, Boc-D-glutamic acid-γ-benzylester-α-methylester, Boc-D-serine methylester, Boc-D-threonine methylester, Boc-D-*allo*-threonine methylester, Boc-D-valine methylester, Boc-D-isoleucine methylester, Boc-DL-α-aminobutyric acid methylester, Boc-L-threonine methylester, and *N, N*-diBoc-D-lysine methylester were synthesized by this method. Deprotection of Boc and Z moieties were carried out in 4 *N* hydrogen chloride in ethylacetate, and by catalytic hydrogenation in the presence of 10 % Palladium-carbon, respectively.

Amino acid amides were synthesized by ammonialysis [9] of amino acid methylesters AHCl or Boc-amino acid methylesters, in absolute methanol saturated with dry ammonia gas at room temperature. The amino acid amides thus synthesized included Boc-D-alanine amide, D-alanine amide HCl, D-α-aminobutyric acid amide HCl, D-serine amide HCl, D-threonine amide HCl, D-*allo*-threonine amide HCl, D-methionine amide HCl, D-norvaline amide HCl, D-norleucine amide HCl, D-phenylglycine amide HCl, D-proline amide HCl, D-glutamine amide HCl, D-lysine amide HCl, D-asparagine amide HCl, D-aspartic acid amide HCl, D-glutamic acid amide, D-valine amide HCl, D-isoleucine amide HCl, D-leucine amide HCl, D-histidine amide HCl, D-phenylalanine amide HCl, D-tyrosine amide HCl, D-tryptophan amide HCl, D-phenylglycine amide HCl, L-α-aminobutyric acid amide HCl, L-threonine amide HCl, and L-methionine amide HCl. The syntheses of D-alanine benzylamide HCl, D-alanine-3-aminopentane amide HCl, D-alanine n-butylamide HCl, D-alanine n-laurylamide HCl, and D-alanine anilide HCl are described elsewhere [18]. Some of these amides were actually used for determining the substrate specificities of the enzymes, although some do not appear in this text.

Peptide substrates used to screen and test the substrate specificity were synthesized from D- and L-Phe. NH_2 and COOH termini were protected by Boc [6] and methyl groups, respectively. Isobutyl chloroformate [10] and carbodiimide [11] condensed the monomer to a dimer and the dimer to a tetramer of Phe, respectively. The following peptide derivatives were synthesized: Boc-D-Phe, D-Phe *tert*-butyl ester, $(D\text{-}Phe)_2$ HCl, Boc-$(D\text{-}Phe)_2$, $(D\text{-}Phe)_2$ methyl ester HCl, Boc-$(D\text{-}Phe)_2$ methyl ester, $(D\text{-}Phe)_3$ HCl, Boc-$(D\text{-}Phe)_3$, Boc-$(D\text{-}Phe)_3$ methyl ester, Boc-$(D\text{-}Phe)_3$ *tert*-butyl ester, $(D\text{-}Phe)_4$ HCl, Boc-$(D\text{-}Phe)_4$, Boc-$(D\text{-}Phe)_6$, Boc-$(D\text{-}Phe)_4$ methyl ester, L-Phe methyl ester HCl, $(L\text{-}Phe)_2$ methyl ester HCl, $(L\text{-}Phe)_3$ HCl, $(L\text{-}Phe)_4$ HCl, Boc-$(L\text{-}Phe)_4$, Boc-$(L\text{-}Phe)_4$ methyl ester, $(D\text{-}Phe)_2$-D-Tyr HCl, D-Tyr-$(D\text{-}Phe)_2$ HCl, D-Phe-$(L\text{-}Phe)_2$ HCl, L-Phe-$(D\text{-}Phe)_2$ HCl, $(D\text{-}Phe)_2$-L-Phe HCl, $(L\text{-}Phe)_2$-D-Phe HCl, D-Phe-L-Phe-D-Phe HCl, L-Phe-D-Phe-L-Phe HCl, D-Phe-L-Phe HCl, and L-Phe- D-Phe HCl.

2. Screening for D-Aminopeptidase (DAP) [12,13]

An enrichment culture technique [14] in a medium containing D-alanine amide as the sole source of nitrogen, was applied for the selection of a desirable microorganism. A number of microorganisms utilizing D-alanine amide have been isolated from soil samples using the screening medium composed of 5 g glycerol, 2 g D-alanine amide HCl, 2 g K_2HPO_4, 1 g NaCl, 0.2 g $MgSO_4$ $7H_2O$, 0.5 g yeast extract, and vitamin mixture in 1 liter of tap water, pH 7.0. The amidases hydrolyzing D-alanine amide were purified from each of the strains. Some of the cell-free extracts of the strains were active toward both of the enantiomers of alanine amide. Therefore, the enzymes responsible for the hydrolysis of D-alanine amide were purified to homogeneites from the likely source of the microorganisms, since some cell-free extracts contained an enzyme with low stereospecificity, some contained several kinds of enzymes acting on the enantiomers of D-amino acid amide. After careful examination of the stereospecificities of the enzymes, the one from *Ochrobactrum anthropi* C1-38 was chosen, because it showed strict stereoselectivity toward D-alanine amide, etc.

3. Activity Measurement of DAP

The enzyme activity was assayed by the hydrolysis of D-alanine amide or by the formation of *p*-nitroaniline from D-alanine-*p*-nitroanilide at 405 nm as follows. A reaction mixture (1.0 ml) containing 5 mM D-alanine-*p*-nitroanilide, 100 mM Tris-HCl, pH 8.0, and the enzyme was monitored by the change in absorbance at 405 nm. A linear change in absorbance for at least 1 min was employed in the kinetic study. Typically, the amount of the enzyme (as a homogeneous enzyme of the native structure) in the reaction mixture was 0.004 ~ 0.2 μg, causing a change of 0.01 to 0.5 absorbance units/min.

4. Preparation of DAP

A Gram-negative, motile, obligatory aerobic, non-fermenting rod, isolated from a soil sample was chosen as a likely source of the enzyme. The strain was identified as *Ochrobactrum anthropi*. The enzyme was formed intracellularly and constitutively formed. The enzyme was purified about 2,800-fold with a 17 % yield from the cell-free extracts with a procedure of ammonium sulfate fractionation, column chromatographies on DEAE-Toyopearl, Butyl-Toyopearl, Sephacryl S-300, and DEAE-5PW (HPLC). The gene for the enzyme was cloned and the amount of the enzyme in the cell-free extract of an *Escherichia coli* transformant with an expression plasmid was elevated about 3,600-fold over the wild strain [15].

5. Screening $(D\text{-Phe})_4$ Degrading Microorganisms for Alkaline D-Peptidase (ADP)

We screened the ability of microorganisms to hydrolyze $(D\text{-Phe})_4$ in LB medium in enriched cultures at 30 °C [16]. $(D\text{-Phe})_4$ was dissolved in DMSO (10 % w/v), and added to 2 ml of LB medium containing soil samples. The mixture was then aerobically shaken for 2 days. A loopful of the culture broth was transferred to the same medium and aerobically incubated under the same conditions. A small portion of the culture broth was streaked onto a plate of the same medium containing 1.5% agar and incubated at 30 °C overnight. Strains forming clear zones around the colonies were isolated and $(D\text{-Phe})_4$ degradation in the liquid culture was monitored by TLC (chloroform : methanol : acetic acid = 8 : 2 : 1 or 10 : 2 : 1) and visualized with ninhydrin. $(D\text{-Phe})_n$ (n = 1 - 4) were well separated on TLC with the solvent. A bacterial strain isolated from Kanagawa Prefecture, Japan, completely degraded the substrate by forming $(D\text{-Phe})_2$.

6. Activity Measurement of ADP

Alkaline D-peptidase (ADP) activity was routinely assayed at 30°C by measuring the production of $(D\text{-Phe})_2$ from $(D\text{-Phe})_4$ by HPLC. ADP activity was routinely assayed at 30 °C by measuring the production of $(D\text{-Phe})_2$ from $(D\text{-Phe})_4$ [15]. The reaction mixture was composed of 1 mmol of $(D\text{-Phe})_4$, 10 ml of DMSO, 50 mmol of Tris-HCl, pH 9.0, and 1 mmol of $MgSO_4$, and the assay was started by addition of the enzyme in a total volume of 500 ml. The reaction was terminated with 50 ml of 2 N $HClO_4$, and the amount of $(D\text{-Phe})_2$ formed was estimated with a Waters 600E HPLC apparatus equipped with a Cosmosil 5C18-MS reverse-phase column (4.6 x 150 mm) at a flow rate of 1.0 ml/min, using 35% methanol in 5 mM KH_2PO_4/ H_3PO_4 buffer, pH 2.9. Absorbance of the eluate was monitored at 254 nm. One unit of enzyme activity is defined as the amount of enzyme that catalyzes the formation of 2 mmol of $(D\text{-Phe})_2$ from 1 mmol of $(D\text{-Phe})_4$. The substrate specificity was examined qualitatively by thin-layer chromatography first and then was quantitatively assayed by the following methods. (*a*) The enzyme activity toward peptide substrates was measured as described above with 2.5 units of enzyme. The amounts of $(D\text{-Phe})_2$, D-Phe, L-Phe- D-Phe, Boc-D-Phe, and Boc-$(D\text{-Phe})_2$ were quantitatively assayed by HPLC on a Cosmosil 5C18-MS reverse-phase column (4.6 x 150 mm) at a flow rate of 1.0 ml/min, using the following solvent system of 5 mM KH_2PO_4/H_3PO_4 buffer, pH 2.9: methanol = 13 : 7 or 9 : 11 (v/v). To determine the kinetic constants of the enzyme for the peptide substrate, a reaction mixture containing 0.1–12.5 μmol of the substrate, 50 μmol of Tris-HCl, pH 9.0, 1 μmol of $MgSO_4$, 10 μl of Me_2SO, and 100 μl of the enzyme solution in a total volume of 500 μl was used. (*b*) The enzyme activity toward β-lactam compounds was determined by measuring the consumption of the substrate in a reaction mixture containing 100 μmol of potassium phosphate buffer, pH 7.0, 2.5 μmol of β-lactam compound, and 0.46 units of the enzyme in a total volume of 1 ml. After incubation at the same temperature, 100 ml of the reaction mixture was added to 900 ml of 100 mM KH_2PO_4/H_3PO_4 buffer, pH 2.9, to stop the reaction. After centrifugation (15,000 x *g*, 10 min), the amount of β-lactam compounds in the supernatant was determined by HPLC on an ODS-80Ts column (46 x 150 mm) with a solvent of 25% methanol in 5 mM KH_2PO_4/H_3PO_4 buffer, pH 2.9, and of 50 μM potassium phosphate buffer, pH 7.0, for ampicillin and penicillin G, respectively. The column was eluted by fast liquid protein chromatography (Pharmacia) at 1.0 ml/min, and the active fractions were collected and concentrated by ultrafiltration.

7. Cloning of the *adp* Gene

B. cereus genomic DNAs were partially digested with *Mbo*I and fractionated by sucrose density gradient ultracentrifugation (5–25%; 100,000 x *g*, 16 h). DNA fragments of 3–6 kb were purified and ligated into *Bam*HI-digested and dephosphorylated pUC118 by T4 ligase. Ampicillin-resistant transformants expressing ADP activity were directly selected by monitoring halo formation from $(D\text{-Phe})_4$ because the host *E. coli* JM109 cells did not show ADP activity. One of 17,000 transformants exhibited ADP activity, and it harbored a plasmid designated pBDP2 with a 5-kb DNA insert. For subcloning, pBDP2 was digested with *Eco*RI and *Sal*I, and the resulting 1.3-kb fragment was introduced into the *Eco*RI and *Sal*I sites of pUC118 to yield pBDP22 [17].

4. RESULTS

1. Substrate Specificity and Distinguishing Features of D-Aminopeptidase (DAP)

The enzyme acted on D-alanine amide (relative velocity: 100%, K_m value: 0.65 mM), glycine amide (44%, 22.3 mM), D-α-aminobutyric acid amide (30%, 18.3 mM), D-serine amide (29%, 27.0 mM), D-alanine 3-aminopentane amide (32%, 2.27 mM), D-alanine anilide (73%), D-alanine benzylamide (72%, 0.51 mM), D-alanine-*p*-nitroanilide (96%, 0.51 mM), D-alanine n-butylamide (66%, 0.73 mM), D-alanine methylester (75%), glycine methylester (229%), D-alanylglycine (95%, 0.98 mM), D-alanylglycylglycine (45%, 0.37 mM), D-alanyl-D-alanine (21%, 10.2 mM), D-alanyl-D-alanyl-D-alanine (92%, 0.57 mM), D-alanyl-D-alanyl-D-alanyl-D-alanine (89%, 0.32 mM), D-alanyl-L-alanine (46%, 1.03 mM), and D-alanyl-L-alanyl-L-alanine (100%, 0.65 mM), *etc.* These results show that the enzyme has higher affinity toward peptide substrates rather than amino acid amides. The enzyme showed neither endopeptidase, nor carboxypeptidase activity. When the release of amino acids from D-alanylglycylglycine was followed with time, the enzyme exhibited a mode of action typical of aminopeptidases (EC 3.4.11). Thus, the enzyme was named as "D-aminopeptidase" (DAP). There has been no report of an aminopeptidase catalyzing the stereospecific hydrolysis of D-amino acid containing peptides, although the peptides do exist in nature.

DAP, carboxypeptidase DD, and β-lactamases catalyze the hydrolytic cleavage of the amide bond between D-amino acids or the same configuration. The former two catalyze the transpeptidation reaction; carboxypeptidase DD catalyzes the cross-linkage of the peptidoglycan, and DAP catalyzes the transpeptidation and aminolysis of D-amino acid ester and amides, respectively, in water and organic solvents. The aminolysis reaction catalyzed by DAP in butylacetate is efficient (*kcat*: 7,700 /min) [18]. The transpeptidation reaction can also be catalyzed in water, although the product D-alanine-3-aminopentane amide was easily hydrolyzed in water. Inheritance of the carboxypeptidase-like tertiary structure in the DAP would have made the transpeptidation reaction in organic solvents possible. The enzyme was actually inhibited by β-lactam compounds, such as 6-aminopenicillamic acid, 7-aminocephalosporic acid, benzylpenicillin and ampicillin, although they are not the substrate for the enzyme [15]. The enzyme is a new member of the "penicillin-recognizing enzymes", because of: (i) similarities in the primary structure by gene sequencing, (ii) similarities in the reactions catalyzed in water and organic solvents, and (iii) the findings obtained by kinetic studies of the mutants generated by the site-directed mutagenesis and (iv) the inhibition by β-lactam compounds, although the enzyme does not appear to be a selective target of the β-lactam compounds. The enzyme is the first example of an aminopeptidase with Ser at the active site [19].

2. Structure of DAP

The *Mr* of the native enzyme was calculated to be approximately 122,000 by gel filtration on HPLC. The molecular weight of the subunits was calculated to be 59,000 by SDS-PAGE. The *dap* gene coding for its product DAP consisted of an open reading frame of 1,560 nucleotides which specifies a protein of *Mr* of 57,257. The enzyme had an isoelectric point of pH 4.2. The native enzyme has a dimeric structure consisting of two identical subunits.

The sequence of the enzyme showed similarities with the *Streptomyces* R61 extracellular carboxypeptidase DD (EC 3.4.16.-) and β-lactamases (EC 3.5.2.6), which are involved in the synthesis of bacterial peptidoglycan and the hydrolysis of its analog, β-lactam compounds, respectively. Seven mutants of the enzyme were generated by site-specific mutagenesis to explore the roles of the residues of interest, around the sequence Ser61-X_{aa}-X_{aa}-Lys64, where X_{aa} is any amino acid, since the identical sequences also appear in the penicillin-recognizing peptide hydrolases with Ser at the active sites. Replacements of the site at Ser61 and Lys64 yielded mutants showing significantly reduced V_{max} values, while most of the K_m values remained unchanged.

The crystal structure of DAP has been determined to 1.9 Å resolution. The enzyme folds into three domains, A, B and C. Domain A, which contains conserved catalytic residues, has the classical fold of serine β-lactamases, whereas domains B and C are both antiparallel eight-stranded β barrels [19].

3. Features of Alkaline D-Peptidase (ADP)

An enrichment culture in LB medium containing a synthetic substrate $(D\text{-Phe})_4$ led to an isolation of a bacterial strain *Bacillus cereus* DF4-B. The extracellular enzyme hydrolyzing $(D\text{-Phe})_4$ was purified and characterized [16].

4. Substrate Specificity of ADP

As summarized in Table 2, the enzyme was active toward $(D\text{-Phe})_4$ (relative activity: 100%, K_m value: 0.398 mM) and $(D\text{-Phe})_3$ (90%, 0.127 mM), forming $(D\text{-Phe})_2$ and D-Phe. The enzyme is also active toward tripeptides L-Phe-$(D\text{-Phe})_2$ (119%, 0.455 mM), L-Phe-D-Phe-L-Phe (28.1 %, 1.63 mM), D-Tyr-$(D\text{-Phe})_2$ (83.6 %), $(D\text{-Phe})_2$-D-Tyr (83.6 %), and on Boc-$(D\text{-Phe})_n$ (n = 2 - 4) forming Boc-D-Phe, $(D\text{-Phe})_2$ and D-Phe. The enzyme had esterase activity toward D-Phe methyl ester and $(D\text{-Phe})_2$ methyl ester. The products from Boc-(D-Phe)$_3$ *tert*-butyl ester were Boc-D-Phe, D-Phe and D-Phe *tert*-butyl ester. A dimer was formed when D-Phe methyl ester and D-Phe-NH_2 were the substrates. The enzyme also showed poor β-lactamase activity toward ampicillin (8.9 %, 73.1 mM) and penicillin G (9.7 %, 48.9 mM). On the other hand, carboxypeptidase DD [21] and DAP [12] activities were undetectable.

Table 2. Substrate Specificity of ADP [16].

Substrate	Relative Activity (%)	K_m (mM)	V_{max} (units/mg)	V_{max}/K_m (units/mg/mM)
(D-Phe)$_6$	1.8			
(D-Phe)$_4$	100.0	0.4	199.0	500.0
(D-Phe)$_3$	90.0	0.1	130.0	1020.0
(D-Phe)$_2$	0.2	50.1	13.7	0.3
D-Phe-L-Phe	< 0.1			
(D-Phe)$_2$-L-Phe	14.9	0.5	30.6	59.0
L-Phe-(D-Phe)$_2$	119.0	0.5	154.0	346.0
L-Phe-D-Phe-L-Phe	28.1	1.6	66.0	41.0
D-Tyr-(D-Phe)$_2$	83.6			
(D-Phe)$_2$-D-Tyr	83.6			
D-Phe-OMe	15.0	1.8		
D-Phe-NH_2	0.1	0.1		
D-Phe-*p*NA	4.2			
Boc-(D-Phe)$_4$	1.8	0.8		
Boc-(D-Phe)$_3$	3.2	1.1		
Boc-(D-Phe)$_2$	7.0			
Boc-(D-Phe)$_3$-O^{t}Bu	1.2	0.3		
Boc-(D-Phe)$_4$-OMe	0.5	0.2		
Boc-(D-Phe)$_3$-OMe	0.7	0.3		
Boc-(D-Phe)$_2$-OMe	1.4			
Ampicillin	8.9	73.1	262.0	3.6
Penicillin G	9.7	48.9	250.0	5.1

The following compounds were inert as a substrate: (L-Phe)$_4$, (L-Phe)$_3$, D-Phe-L-Phe-D-Phe, D-Phe-(L-Phe)$_2$, (L-Phe)$_2$-D-Phe, (L-Phe)$_2$, L-Phe-D-Phe, L-Phe methyl ester, L-Phe amide, Boc-(L-Phe)$_2$, Boc-(L-Phe)$_4$ methyl ester, D-Leu *p*-nitroanilide, D-Ala *p*-nitroanilide, D-Phenylglycine amide, (D-Ala)$_5$, (D-Ala)$_4$, (D-Ala)$_3$, (D-Ala)$_2$, (D-Val)$_3$, (D-Leu)$_2$, and DL-Ala-DL-Phe.

As shown in Fig. 1, time course of the (D-Phe)$_4$ degradation was measured. (D-Phe)$_4$ was hydrolyzed to (D-Phe)$_2$ and D-Phe. No (D-Phe)$_3$ was detected. These results coincide with the kinetic properties of the enzyme described above. The mode of action of the enzyme was examined with the synthetic substrates D-Tyr-(D-Phe)$_2$ and (D-Phe)$_2$-D-Tyr. When D-Tyr-(D-Phe)$_2$ was the substrate, D-Phe was released first, then D-Tyr was slowly formed. When (D-Phe)$_2$-D-Tyr was used as a substrate, D-Tyr was released first, then D-Phe was slowly formed. In both the reactions, the second peptide bond from the *N*-terminus of the substrate was hydrolyzed first (Fig. 2). These results showed that the enzyme acts as a D-stereospecific dipeptidylendopeptidase.

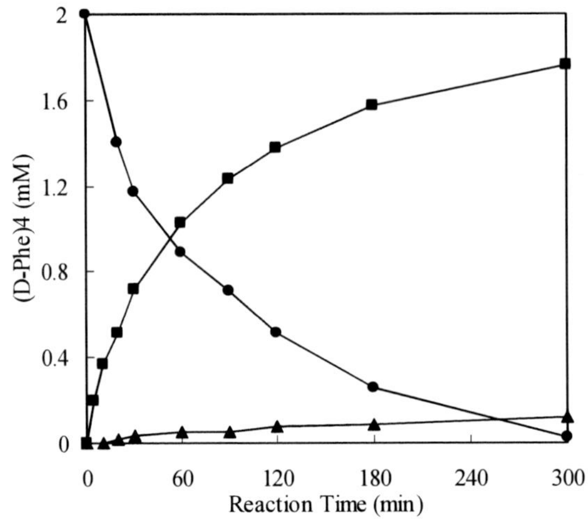

Figure 1. The Time Course of (D-Phe)$_4$ Hydrolysis by ADP. The reaction mixture was consisted of 2 mM of (D-Phe)$_4$, 100 mM of Tris-HCl, pH 9.0, 2 mM $MgSO_4$, 2% (v/v) of DMSO, and 0.038 units of the enzyme solution in a total volume of 500:l. The reaction was carried out at 30°C and terminated with $HClO_4$. The amount of (●) (D-Phe)$_4$, (■) (D-Phe)$_2$, and (▲) D-Phe in the supernatant was measured by a Waters HPLC equipped with a Cosmosil 5C18-MS at a flow rate of 1.0 ml/min with a mixture of methanol (55%) and 5 mmol KH_2PO_4 -H_3PO_4, pH 2.9 (45 %).

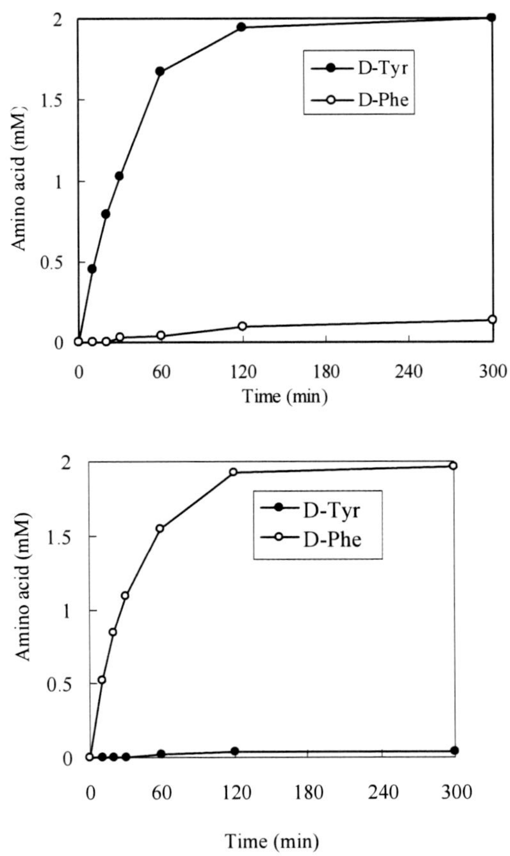

Figure 2. The Mode of Action of ADP toward (A) (D-Phe)$_2$-D-Tyr and (B) D-Tyr-(D-Phe)$_2$. The reaction mixture (500:l) containing 2 mM of the substrate (D-Phe)$_2$-D-Tyr or D-Tyr-(D-Phe)$_2$, 50 mM of Tris-HCl, pH 9.0, 2 mM of $MgSO_4$, 2% (v/v) of DMSO and 0.038 units of the enzyme was incubated at 30°C. After the reaction was terminated, the reaction mixture was analyzed with a Hitach L-8500 amino acid analyzer. Symbols: (●) D-Tyr, (○) D-Phe.

The enzyme is an endopeptidase that acts D-stereospecifically upon peptides composed of aromatic D-amino acids, recognizing the D-configuration of the amino acid whose caroboxy-terminal peptide bond is hydrolyzed. The enzyme had an optimum pH at around 10.3. Thus, the enzyme was named "Alkaline D-peptidase" (D-stereospecific peptide hydrolase (EC 3.4.11.-)). Although the enzyme showed poor β-lactamase activity toward ampicillin and penicillin G, carboxypeptidase DD [21] and DAP [12,15] activities were not detected. The enzyme can be categorized as a member of the "penicillin-recognizing enzymes" [20].

5. Structure of ADP

The gene coding for ADP (*adp*) was cloned into plasmid pUC118, and a 1,164 bp open reading frame consisting of 388 codons with an *Mr* of 42,033 was identified as the *adp* gene [15]. ADP would be synthesized with a signal peptide. The *Mr* of the subunit calculated was about 36,000 by SDS-PAGE, 37,000 by HPLC, and 37,952 by mass spectrometry. The absorption of the purified enzyme in 0.01 M potassium phosphate buffer, pH 7.0, was maximal at 281 nm.

The deduced primary structure of ADP is similar to carboxypeptidase DD from *Streptomyces* R61 (35.0 % identical over 346 a.a.) [21], class C β-lactamases of *Enterobacter cloacae* (25.1% identical over 191 amino acids) [22], DAP from *O. anthropi* (27.5% over 182 a.a.) [15], and esterase from *Pseudomonas* sp. (30.5% over 154 a.a.) [23], etc. The sequence Ser-Xaa-Xaa-Lys is perfectly conserved among the "penicillin-recognizing enzymes" [20], which include penicillin-binding proteins, β-lactamases and DAP.

The enzyme showed D-stereospecific dipeptidyl aminopeptidase and endopeptidase activities. The calculated V_{max}/K_m values against $(D\text{-Phe})_3$, and $(D\text{-Phe})_4$ were about 1000 times higher than that for $(D\text{-Phe})_2$. The enzyme was active toward $(D\text{-Phe})_n$, Boc-$(D\text{-Phe})_n$, $(D\text{-Phe})_n$ methyl ester, D-Phe-NH_2, Boc-$(D\text{-Phe})_n$ methyl ester, Boc-$(D\text{-Phe})_n$ *tert*-butyl ester, but not toward $(D\text{-Ala})_n$ (n = 2 -4), $(D\text{-Val})_3$, and $(D\text{-Leu})_2$. The enzyme digested $(D\text{-Phe})_2$-D-Tyr into $(D\text{-Phe})_2$ and D-Tyr, and D-Tyr-$(D\text{-Phe})_2$ into D-Tyr-D-Phe and D-Phe. The enzyme attacked $(D\text{-Phe})_4$, forming $(D\text{-Phe})_2$, which was then slowly hydrolyzed to D-Phe. The product $(D\text{-Phe})_3$ was not formed. The enzyme had an optimum pH at around 10.3, and it was stable in a relatively wide pH region. The enzyme was inhibited by PMSF, indicating that a serine residue is responsible for exerting the enzyme activity. Thus, the enzyme was named "Alkaline D-peptidase" (D-stereospecific peptide hydrolase (EC 3.4.11.-)). Although the enzyme showed β-lactamase activity toward ampicillin and penicillin G, carboxypeptidase DD [21] and DAP activities were not detected. The kinetic measurements have cleary shown that the enzyme prefers $(D\text{-Phe})_3$ and $(D\text{-Phe})_4$ to the penicillins.

REFERENCES

[1] Yamada, H. & Shimizu, S. (1988). Microbial and enzymatic processes for the production of biologically and chemically useful compounds. *Angew. Chem. Int. Ed. Engl. 27*, 622–642.

[2] Glowaky, R. C., Hendrick, M. E., Smiles, R. E., & Torres, A. (1991) Development and uses of alitame. A novel dipeptide amide sweetener; ACS Symp. Ser. *450*, Sweeteners, 57-67.

[3] Frère, J.-M., & Joris, B. (1985). Penicillin-sensitive enzymes in peptidoglycan biosynthesis; *Crit. Rev. Microbiol., 11*, 299-396.

[4] Reynolds, P. E., Depardieu, F., Dutka-Malen, S., Arthur, M., & Courvalin, P. (1994). Glycopeptide resistance mediated by enterococcal transposon Tn 1546 requires production of VanX for hydrolysis of D-alanyl-D-alanine, *Mol. Microbiol, 13*, 1065-1070.

[5] Krell, G. (1994) Conversion of L- to D-amino acids: posttranslational reaction, *Science, 266*, 996-997.

[6] Tarbell, D. S., Yamamoto, Y., & Pope, B. M. (1972). New method to prepare *N-t*-butoxycarbonyl derivatives and the corresponding sulfur analogs from di-*t*-butyl dicarbonate or di-*t*-butyl dithiol dicarbonates and amino acids, *Proc. Natl. Acad. Sci. USA, 69*, 730-732.

[7] Ito, M., Hagiwara, D., & Kamiya, T. (1977). Peptides. VI. Some oxime carbonates as novel *t*-butoxycarbonylating reagents, *Bull. Chem. Soc. Jpn., 50*, 718-721.

[8] Maniatis, T., Fritsch, E. F., & Sambrook, J. (1982). *Molecular Cloning: A Laboratory Manual*, Cold Spring Harbor Laboratory, Cold Spring Harbor, NY.

[9] Altman, J., Shoef, N., Wilchek, M., & Warshawsky, A. (1983). Bifunctional chelating agents. Part 1. 1-(*p*-Aminophenethyl)-ethylenediaminetetra-acetic acid, *J. Chem. Soc. Perkin Trans I,* 365-368.

[10] Vaughan, J. R., Jr., & Osato, R. L. (1952). The preparation of peptides using mixed carbonic-carboxylic acid anhydrides, *J. Am. Chem. Soc., 74,* 676–678.

[11] Sheehan, J. C., Preston, J., & Cruikshank, P. A. (1965). A rapid synthesis of oligopeptide derivatives without isolation of intermediates, *J. Am. Chem. Soc., 87,* 2492–2493.

[12] Asano, Y., Nakazawa, A., Kato, Y. & Kondo, K. (1989). Properties of a novel D-stereospecific aminopeptidase from *Ochrobactrum anthropi. J. Biol. Chem., 264*, 14233-14239.

[13] Asano, Y., Nakazawa, A., Kato, Y., & Kondo, K. (1989) Isolierung einer D-stereospezifischen Aminopeptidase und ihre Anwendung als Katalysator in der Organischen Synthese. *Angew. Chem. 101*, 511-512.

[14] Asano, Y. (2002). Overview of screening for new microbial catalysts and their uses in organic synthesis -Selection and optimization of biocatalysts-, *J. Biotechnol., 94*, 65-72.

[15] Asano, Y., Kato, Y., Nakazawa, A., & Kondo, K. (1992). Structural similarity of D-aminopeptidase to carboxypeptidase DD and β-lactamase, *Biochemistry, 31*, 2316-2328.

[16] Asano, Y., Ito, H., Dairi, T., & Kato, Y. (1996). An alkaline D-stereospecific endopeptidase with beta-lactamase activity from *Bacillus cereus, J Biol Chem., 271,* 30256-30262.

[17] Komeda, H., & Asano, Y. (2003). Genes for an alkaline D-stereospecific endopeptidase and its homologue are located in tandem on *Bacillus cereus* genome, *FEMS Microbiol. Lett., 228,* 1-9.

[18] Kato, Y., Asano, Y., Nakazawa, A., & Kondo, K. (1989). First synthesis of D-amino acid *N*-alkyl amide catalyzed by D-aminopeptidase, *Tetrahedron, 45*, 5743-5754.

[19] Bompard-Gilles, C., Remaut, H., Villeret, V., Prangé, T., Fanuel, L., Delmarcelle, M., Joris, B., Frère, J.-M., & Van Beeumen, J. (2000). Crystal structure of D-aminopeptidase from *Ochrobactrum anthropi*, a new member of the 'penicillin-recognizing enzyme' family, *Structure, 8*, 971-980.

[20] Frère, J.-M., Joris, B., Dideberg, O., Charlier, P. & Ghuysen, J.-M. (1988). Penicillin-recognizing enzymes., *Biochem. Soc. Trans., 16*, 934-938.

[21] Duez, C., Piron-Fraipont, C., Joris, B., Dusart, J., Urdea, M. S., Martial, J. A., Frère, J.-M. & Ghuysen, J.-M. (1987). Primary structure of the *Streptomyces* R61 extracellular DD-peptidase. *Eur. J. Biochem., 162*, 509-518.

[22] Galleni, M., Lindberg, F., Normark, S., Cole, S., Honore, N., Joris, B., & Frère, J.-M. (1988). Sequence and comparative analysis of three *Enterobacter cloacae* ampC beta-lactamase genes and their products, *Biochem J., 250*, 753-760.

[23] McKay, D. B., Jennings, M. P., Godfrey, E. A., MacRae, I. C., Rogers, P. J., & Beacham, I. R. (1992). Molecular analysis of an esterase-encoding gene from a lipolytic psychrotrophic pseudomonad, *J. Gen. Microbiol., 138*, 701-708.

[24] Komeda, H. & Asano, Y. (2000). Gene cloning, nucleotide sequencing, and purification and characterization of the D-stereospecific amino acid amidase from *Ochrobactrum anthropi* SV3., *Eur. J. Biochem., 267,* 2028-2036.

[25] Lessard, I. A., & Walsh, C. T. (1999). VanX, a bacterial D-alanyl-D-alanine dipeptidase: resistance, immunity, or survival function?, *Proc. Natl. Acad. Sci. USA., 96*,11028-11032.

[26] Remaut, H., Bompard-Gilles, C., Goffin, C., Frère, J.-M., & Van Beeumen, J. (2001). Structure of the *Bacillus subtilis* D-aminopeptidase DppA reveals a novel self-compartmentalizing protease. *Nat. Struct. Biol., 8*, 674-678.

In: D-Amino Acids: ISBN 1-60021-075-9
A New Frontier in Amino Acid and Protein Research, pp.591-607

Chapter 6.21

ROLE OF AMMONIUM PHOSPHATES IN TRYPTOPHANASE ACTIVITY TOWARD D-TRYPTOPHAN

Akihiko Shimada*
Life Sciences and Bioengineering, Graduate School of Life and Environmental Sciences, University of Tsukuba, Japan.

Keywords: tryptophanase, D-tryptophan, ammonium phosphate, enzyme stereoselectivity, origin of homochirality.

1. INTRODUCTION

The problem of chiral breaking in biological amino acids has intrigued scientists since Louis Pasteur first discovered the chirality of molecules over 150 years ago. While we do not know how the homochirality of biological molecules (the use in living organisms of L-amino acids and D-sugars) originated, we do know that enzymes strongly support the current homochiral world. Known enzymes thus far number 4,361, and it is likely that many more await discovery [1]. The homochirality of biological molecules cannot be imagined without these enzymes. The origins of chiral breaking have been actively discussed as abiotic chemical or physical processes [2,3], but a general consensus has yet to be reached, suggesting that not all origins of homochirality are involve in them [4].

Enzyme stereospecificity on L-amino acids is absolute and stable enough to maintain the contemporary homochiral biological environment. If this balance becomes unstable, the exclusive selection of L-amino acids would no longer be possible with the homochiral environment immediately disappearing and life on earth as we know it dying out. Its immutability is essential to life and its survival — without homochirality, there would be no life [5]. Perhaps life itself began simultaneously with birth of homochirality [6]. We know we're repeating ourselves, but the enzyme stereospecificity is intimately linked with chiral homogeneity. But for some reason, there hasn't been the kind of emphasis on this that we are putting on it, and I think we should discuss about the enzyme stereospecificity that is believed absolutely immutable.

* Correspondence concerning this article should be addressed to Akihiko Shimada, E-mail: ashimada@kankyo.envr.tsukuba.ac.jp.

Although it remains debatable how many amino acids are required to have stereospecificity in oligo- or polypeptide, we can imagine that early polypeptides might be enantiometrically-mosaic polypeptides with mixed L- and D-amino acids as building blocks [7]. Their specificity in addition to activity might thus be very low. They presumably continued to select only necessary amino acids repeatedly, abandoning the rest to obtain higher specificity and activity. By the time catalytically active early polypeptides, i.e., primitive enzymes, were present, the switch to the exclusive use of L-amino acids might already have been made. If early polypeptides were thus synthesized by spontaneous abiotic processes in a primitive racemic environment, the mechanism of homochirality may have already been incorporated into them, and their descendants may be traced to present enzymes. Current high stereoselectivity reflects the historical consequences of enzyme evolution, and should be involved in the stereospecificity of current enzymes. In this context, clues to elucidate the origin of homochirality should lie in the mechanism of stereospecificity in extant enzymes [8]. However, no global picture of this has been presented. Before attempting to solve this problem, we require a better understanding of enzyme stereospecificity (Note 1).

Recent progress in the analysis of D-amino acids has shown that D-amino acids are biologically more widely distributed than previously expected. D-alanine, D-aspartic acid and D-serine, for example, are present in eubacteria, archaea, bivalves, and crayfish as well as in the human eye and brain [9-16]. To our astonishment, D-alanine in a neuropeptide secreted by frog skin is shown to play an important role in expressing physiological activity [17]. Enzyme activity on biological amino acids is known to occur only in the L enantiomer and to be entirely inactive for the opposite enantiomer. The widespread distribution of D-amino acids, however, raises doubts regarding the absolute exclusive selectivity of enzyme stereospecificity, suggesting that enzyme stereospecificity may be more flexible than previously imagined.

Tryptophanase is an enzyme with extremely tight stereospecificity, cleaving L-tryptophan into indole, having no activity on D-tryptophan under ordinary conditions (Note 2). However, it becomes active toward D-tryptophan in highly concentrated ammonium phosphate solutions — quite different from what was expected. The only salts inducing the reaction are ammonium phosphates and ammonium sulfate, although different salts have been tested. Free tryptophan is more readily influenced by alkaline pH or strong ion strengths than other biological amino acids. If ammonium phosphates affect chemical racemization on D-tryptophan, the enzymological significance of this reaction vanishes, so it is important to confirm that D-tryptophan is not racemized by these salts. We used an HPLC column appropriate for tryptophan resolution to analyze free D-tryptophan, demonstrating that this reaction is enzymatic. Because the flexible stereospecificity of tryptophanase is probably due to small conformational changes [18-20], we applied circular dichroism and fluorescence spectrophotometric analyses to tryptophanase in an ammonium phosphate solution. D-tryptophan is inactive in the absence of ammonium phosphates, so it could be concluded that it does not interact with D-tryptophan, but kinetic studies show competitive interaction between them. A binding form of D-tryptophan at the active site of tryptophanse is assumed based on spectrophotometric analysis, offering insights into how tryptophanase excludes only the D form.

2. Materials

1. Salts, Reaction Mixture and Enzyme Assay

1. NH_4Cl, NH_4NO_3, $(NH_4)_2SO_4$, ammonium dihydrogenphosphate ($NH_4H_2PO_4$, MAP), diammoniumhydrogen phosphate ($(NH_4)_2HPO_4$, DAP), triammonium phosphate ($(NH_4)_3PO_4$, TAP), $(NH_4)_2CO_3$, CH_3COONH_4, $(COONH_4)_2$, ammonium citrate, KCl, K_2HPO_4, NaCl, Na_2HPO_4, Na_2SO_4, sodium citrate, $MgSO_4$, seawater
2. 100 mM of phosphate buffer with a pH of 8.3 (PB)
3. Pyridoxal 5′-phosphate (PLP)

4. Apotryptophanase (apoTPase) (Sigma-Aldrich Co., St. Louis, MO, USA)
5. D-tryptophan (D-Trp) (Peptide Institute Inc., Osaka, Japan)
6. L-tryptophan (L-Trp) (Peptide Institute Inc., Osaka, Japan)
7. Dry Thermo Unit (DTU-1B) (Taitec, Tokyo, Japan)
8. n-Butanol
9. *p*-DMAB (*p*-dimethylaminobenzaldehyde)
10. conc. HCl
11. Ethanol
12. Clean 99 CL (Clean Chemical Co. Ltd., Osaka, Japan)

2. Resolution of free D-Trp

1. Crownpack CR (+) Resolution HPLC column (Daicel Industry Ltd., Tokyo, Japan).
2. Ultrafree-MC 5000 NMWL Filter Unit (Millipore Co. Bedford, MA, USA)
3. Perchloric acid
4. UV monitor (1040A HPLC Detection System, Hewlett-Packard, CA, USA)
5. CD detector (CD-1595 detector, Jasco, Tokyo, Japan)
6. HPLC pump (635 Liquid Chromatogram, Hitachi, Ltd., Tokyo, Japan)
7. Degasser (DG-980-50, Jasco, Tokyo, Japan)

3. Fluorescence and Circular Dichroism Spectrophotometers

1. Bis-ANS (4,4'-dianilino-1,1'-binaphthalene-5,5'-disulfonic acid, $\varepsilon = 23\times10^3$ $cm^{-1}M^{-1}$ at 395 nm, a major fluorescence probe of hydrophobic protein surfaces)
2. Fluorescence spectrophotometer F4500 (Hitachi, Tokyo, Japan)
3. Circular dichroism spectrophotometer J805 (Jasco, Tokyo, Japan)

3. METHODS

1. Effect of Different Salts on D-tryptophan Activity

1. See Note 3 for reagents and glasswares.
2. Each salt was dissolved completely in 80-85 °C water and slowly cooled to room temperature. This concentrated salt solution was left at room temperature for about a week until crystals were deposited (Note 4). The saturated solution thus prepared was defined as a 100% saturation concentration.
3. Each salt solution was diluted to the required saturation concentration with PB before experiments, e. g., 50% saturation of TAP was prepared by adding 1 volume of PB to 1 volume of TAP saturated solution.
4. To determine the effect of different salts on TPase activity toward D-Trp, you must prepare reaction mixtures to include 1.1 mM of PLP, 245 μM of D-Trp and 0.2 μM of apoTPase in a 50% saturation concentration solution of each salt. The total volume of the reaction mixture was 2 mL per tube.
5. Reactions were conducted at 37 °C for 2 hr in Dry Thermo Unit. The reaction was stopped by adding 2 mL of n-butanol, vigorously mixed, then immediately centrifuged at about 1,000 G for

10 min. Centrifuging caused 3 layers to form, — a butanol layer (supernatant), a denatured enzyme layer (middle), and an aqueous layer (bottom).

6. Tryptophanase-catalyzed tryptophan was degraded into indole, which was equally distributed in the butanol and aqueous layers. An aliquot of 1.5 mL from the supernatant of a centrifuge tube was withdrawn for indole assay, vigorously mixed with 1.5 mL of Ehrlich reagent (Note 5), then stained red at 60 °C for 30 min in Dry Thermo Unit. TPase activity was spectrophotometrically determined by measurement at λ=570 nm.

2. Dependence on Reaction Temperature

1. Saturated solutions of MAP, DAP, and TAP were diluted to 50 % saturation with 100 mM of PB; the pH of MAP was 3.6, that of DAP 7.8, and that of TAP 9.2. After 1.1 mM of PLP and 0.2 μM of apoTPase were added, they were then warmed to the required temperature (30 - 70 °C).
2. The enzyme reaction was initiated by adding 245 μM of D-Trp. After 30 min, the reaction was stopped by adding n-butanol equal in volume to the reaction mixture. As in Method 1.6, the resulting indole was stained red with Ehrlich's reagent, and the absorbance was determined at 570 nm.
3. As a reference, D-Trp was reacted with TPase in 100 mM of potassium phosphate buffer with a pH of 3.6, 7.8, and 9.2 under the same conditions except for excluding ammonium phosphate salts. D-Trp activity was not detected in their absence.

3. Analysis of free D-Trp

1. Perchloric acid was added dropwise to a few liters of distilled water with vigorous stirring at room temperature until pH decreased to 2.0. This was used as the eluting buffer.
2. D-Trp was exposed to a $(NH_4)_2CO_3$, K_2HPO_4 or TAP solution of 50 % saturation concentration at 55 °C for 30 min, immediately filtered with Ultrafree-MC 5000, and resolved at a flow of 1.0 mL/min with the eluting buffer on a Crownpack CR(+) Resolution HPLC column (Note 6).
3. The eluant was monitored with an UV monitor for UV absorbance at a wavelength of λ=280 nm or a CD detector for ellipticity at a wavelength of λ=233 nm. UV absorbance of the eluting buffer was set to zero. The ellipticity of D-Trp was detected on the negative side and that of L-Trp on the positive side when that of the eluent was set to zero. The retention time of D-Trp was 24 min and that of L-Trp 32 min. Units of absorbance and ellipticity are arbitrary.

4. Measurement of Fluorescence

1. Bis-ANS is a hydrophobic fluorescent probe indicating the existence of solvent-exposed hydrophobic domains in a protein's structure. A stock solution of bis-ANS was prepared to a final concentration of 0.5 μM by dissolving 0.3 mg in 1 mL of PB before experiments (Note 7); 20 μL of bis-ANS stock was added to a 1 mL solution of PB or MAP, DAP and TAP solutions of 50 % saturation concentration including 0.5 μM of apoTPase and 1.1 mM of PLP. These aliquots were left to stand at room temperature for at least 15 min before measurement, then placed in a cuvette in a fluorescence spectrophotometer.
2. Emission from 420 to 650nm was measured at an excitation wavelength of 395 nm. The fluorescence of bis-ANS bound to TPase was calculated by subtracting a blank prepared identically but without TPase. Fluorescence intensity is measured in arbitrary units.

5. Measurement of Circular Dichroism

1. A 200 μL solution including 50 % saturation MAP, DAP, or TAP with 0.5 μM of apoTPase and 1.1 mM of PLP was placed in a 0.1 cm path length cell in a circular dichroism (CD) spectrophotometer.
2. Spectra were recorded at wavelengths from 200 to 350nm at room temperature (Note 8). Five scans were accumulated per spectrum, averaged, and expressed as molar ellipticity in degrees cm^2 $dmol^{-1}$.
3. The CD spectra were analyzed with a CDPro software package and calculated by subtracting a blank prepared identically but without TPase.

6. D-Trp as a Competitive Inhibitor

1. For kinetic analyses, reaction mixtures were prepared by combining required L-Trp (98-980 μM) in the absence of TAP with fixed apoTPase (0.08 units) and PLP concentrations (1.1 mM) in 10 mL of PB, adjusted to a pH of 8.3.
2. The reaction was conducted at 37 °C for 30 min in Dry Thermo Unit with 0, 9.8 or 19.6 mM of D-Trp, and stopped by adding n-butanol. The reaction mixture was vigorously mixed and immediately centrifuged at about 1,000 *g*. Assays of resulting indole were conducted the same as in Methods 1.6.
3. For kinetic analysis, initial velocities (μmoles/min/mg) represented indole that formed within 1 minute after 1 mg of TPase was added. TPase activity was assayed in triplicate at the same substrate concentration. Initial velocities were averaged and Lineweaver-Burk plots drawn based on the least squares method.
4. The reciprocal of initial velocity (ordinate) versus the reciprocal of substrate concentration (abscissa) was plotted for each concentration of D-Trp. Inhibition constant Ki was determined from the slopes of these double-reciprocal plots (Note 9).

4. RESULTS

Tryptophanse has absolute stereospecificity with L-tryptophan and is utterly inactive on D-tryptophan (Note 2). Most salts could not induce TPase activity on D-Trp (Table 1), but several salts made it positive, e.g., K_2HPO_4 showed positive activity, but the reaction was not enzymatic. Figure 1B shows an HPLC resolution chromatograph of D-Trp exposed to 50% saturation K_2HPO_4 solution at room temperature for 30 min, indicating a small peak at a retention time of L-Trp in addition to an unidentified by-product also seen in D-Trp exposed to 50% saturation $(NH_4)_2CO_3$ solution (Figure 1A). K_2HPO_4 thus racemized D-Trp into L-Trp through a chemical process. Positive activity with K_2HPO_4 meant that chemically racemized L-Trp reacted with TPase.

TPase was also converted to active form for D-Trp in a highly concentrated ammonium sulfate solution [21] and had higher activity in ammonium phosphates than in ammonium sulfate [19]. Ammonium phosphates have 3 species — ammonium dihydrogenphosphate (MAP), diammoniumhydrogen phosphate (DAP) and triammonium phosphate (TAP). DAP and TAP produced TPase activity on D-Trp, but MAP denatured to inactivate TPase. TPase was reversibly restored to its native state when DAP or TAP was removed from the reaction mixture. This reaction depended on the ammonium phosphate concentration, PLP, and the reaction temperature. Note that tryptophan is more labile in an alkaline, salty or high-temperature environment than other amino acids [22,23]. We carefully studied the possibility that D-Trp was racemized into L-Trp due to a chemical reaction with concentrated ammonium phosphates. D-Trp exposed to a 50% saturation concentration of TAP with PLP at 55 °C for 30 min was resolved through a Crownpack CR (+) HPLC column and monitored with a UV detector (Figure 1C) and a CD detector (Figure 1D). The same chromatograms were obtained from D-Trp exposed to DAP and ammonium sulfate solutions (not shown), demonstrating that neither L-Trp nor a by-product was produced. We thus eliminated the possibility that D-Trp was chemically racemized by concentrated ammonium phosphates. We studied the chemical decomposition of D-Trp into indole in another preliminary experiment, but found it was not possible (data not shown). We demonstrated that TPase activity on D-Trp was enzymatic. Racemization is still potentially possible by reprotonation of the quinonoid intermediate from the face opposite the PLP complex at the active site of TPase (Note 10). If the quinonoid intermediate is formed from D-Trp and it alternatively partitions D-Trp between elimination and racemization, tryptophanase-catalyzed racemization may be possible [24], so we analyzed D-Trp exposed to TPase and PLP at 55 °C for 30 min. The same chromatograms as Figure 1C and D were obtained (data not shown), indicating that this, too, was not possible. These results thus indicated that D-Trp reacted directly with TPase in the presence of ammonium phosphates.

Table 1. Effect of different salts on TPase activity toward D-Trp.

Salt species	Activity on D-Trp	By-product
NH_4Cl	–	–
NH_4NO_3	–	–
$(NH_4)_2SO_4$	+	–
$NH_4H_2PO_4$	–	–
$(NH_4)_2HPO_4$	++	–
$(NH_4)_3PO_4$	+++	–
$(NH_4)_2CO_3$	–	+
CH_3COONH_4	–	–
$(COONH_4)_2$	–	–
Ammonium citrate	–	–
KCl	–	–
K_2HPO_4	++	+
NaCl	–	–
Na_2HPO_4	–	–
Na_2SO_4	–	–
Sodium citrate	–	–
$MgSO_4$	–	–
Seawater	–	–

– no activity, + active, ++ more active, +++ most active.

Tryptophanase activity on D-Trp depended on the reaction temperature, increasing sharply with it (Figure 2), but decreasing quickly above 55°C. Higher activity was seen in TAP than in DAP at between 30 and 55°C. Activity at 55°C was about 3 to 4 times that at 37°C, a standard temperature in enzyme reactions. Optimal reaction conditions were a PLP concentration of 1.1 mM, a TAP solution with 50% saturation concentration, and a reaction temperature of 55°C. Environmental factors such as salt or temperature could have influenced the TPase stereostructure, in turn inducing to change in its stereospecificity.

Ammonium phosphates actually acted on TPase, inducing a conformational change in TPase and enabling it to react with D-Trp. Fluorescence and circular dichroism were examined to detect this. Fluorescence intensity increases when the fluorescent probe bis-ANS is bound to hydrophobic surfaces of TPase. The increased fluorescence of bis-ANS is a useful probe for exposing of hydrophobic sites and enables the protein's conformational states to be compared. The effect of increasing exposure of the hydrophobic region is related to conformational changes in proteins [25]. We measured the degree of exposure of hydrophobic sites as an indication of the conformational change of TPase among different ammonium phosphates. When the fluorescence intensity of bis-ANS-bound TPase was measured at room temperature in 50% saturation solutions of MAP, DAP, and TAP (Figure 3), it was highest in the MAP solution, showing that such conformational changes were too great to maintain TPase activity. The increasing exposure of hydrophobic amino acids, i.e., high fluorescence intensity, is connected with the unfolding or denaturing with which TPase is inactivated. Intensity was slightly higher in a DAP or TAP solution than in a PB solution, but spectra were almost the same. Compared to the MAP solution, their fluorescence intensity was very low, indicating that appropriate exposure to hydrophobic amino acids is very important in D-Trp activity. TAP was therefore ideal for providing TPase with the optimal exposure of hydrophobic amino acids.

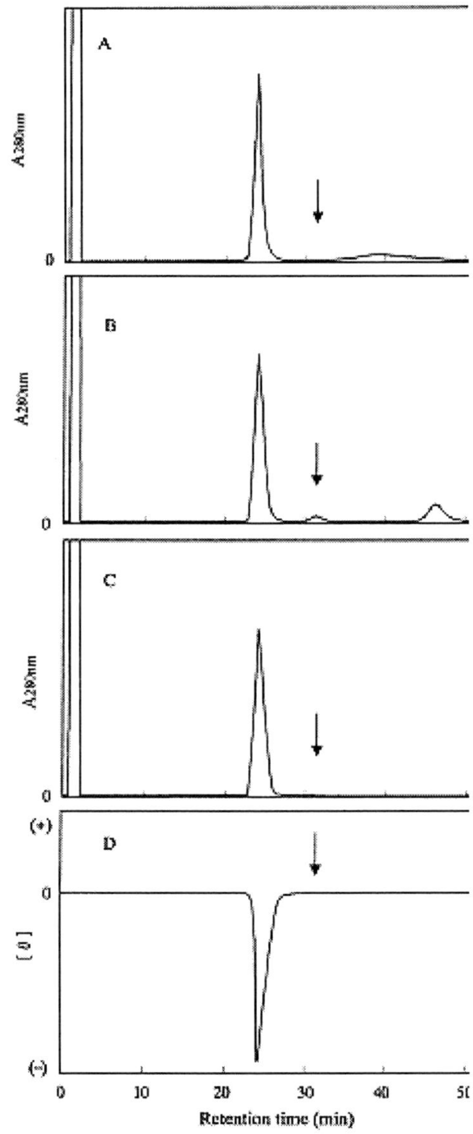

Figure 1. HPLC chromatographic analysis of D-Trp exposed to different salt solutions at 55 °C for 30 min. A: D-Trp in a 50% saturation solution of $(NH_4)_2CO_3$; B: D-Trp in a 50% saturation solution of K_2HPO_4; C: D-Trp in a 50% saturation solution of TAP; D: CD measurement of Fig. 1C. The arrow shows the retention time of L-Trp. A, B, and C show absorbance at λ=280 nm and D ellipticity at λ=233 nm.

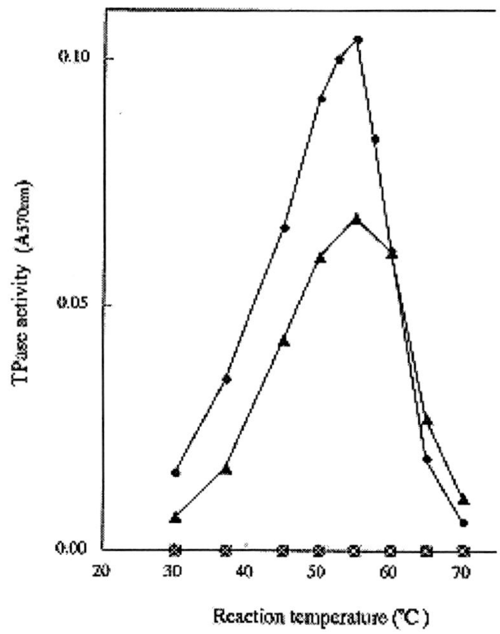

Figure 2. Reaction-temperature-dependent TPase activity on D-Trp in a 50% saturation concentration of MAP, DAP, and TAP solutions. TPase activity was determined from absorbance at λ=570nm. ■ MAP, ▲ DAP, ● TAP, × PB.

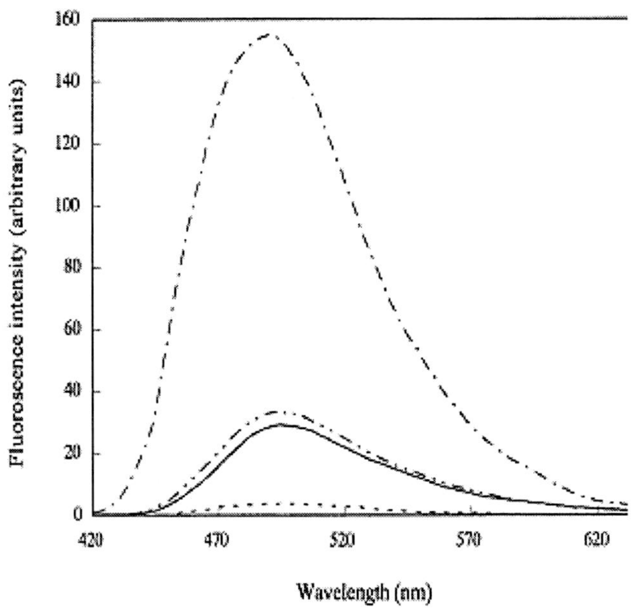

Figure 3. Fluorescence spectra of bis-ANS bound to TPase in MAP, DAP, TAP, and PB solutions. PB (– – –), MAP (– · –), DAP (– · · · –), TAP (——).

Figure 4 shows CD spectra of TPase for PB, MAP, DAP, and TAP solutions at wavelengths of 200-350 nm (Note 8). CD spectra in the near-uv region (250-350 nm) did not produce any difference among PB, TAP, DAP, and MAP, indicating that TPase in highly concentrated ammonium phosphate solutions had a tertiary structure similar to its natural structure. CD spectra in the far-uv region (200-250 nm) produced detectable differences, however, indicating that ammonium phosphates produced changes in the secondary structure of TPase. Theses are useful for assessing the degree to which ammonium phosphates change it. Choosing λ=220 nm as the single wavelength for monitoring specific features of the protein structure, we compared the signal at this wavelength among TAP, DAP, and MAP. When the degree of conformational change was defined as 100% unfolding in the MAP solution, it was 21% in DAP and 7 % in TAP. Measurement of the CD spectra showed that a limited secondary structural change was required for TPase activity to appear on D-Trp. Judging from fluorescence and CD measurements, the degree of conformational change is very small.

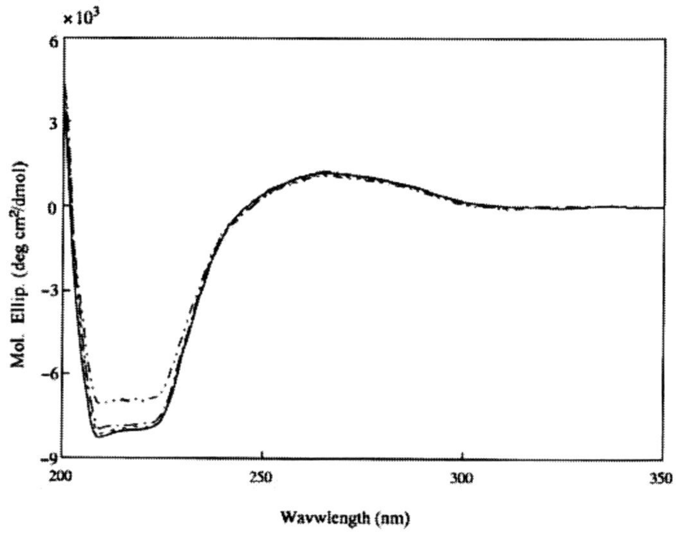

Figure 4. Circular dichroism spectra of TPase in PB, MAP, DAP, and TAP solutions. PB (——), MAP (– · · –), DAP (– · –), TAP (– – –).

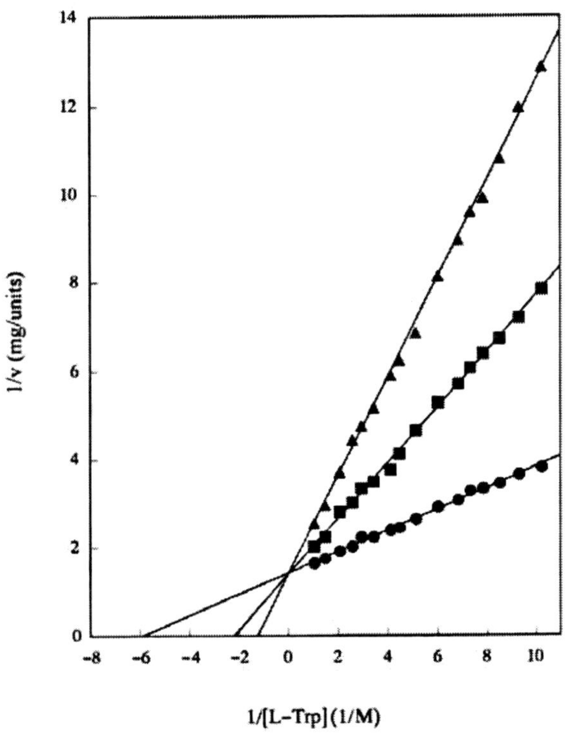

Figure 5. Competitive inhibition of L-Trp decomposition in the absence of ammonium phosphates. 0 mM of D-Trp (●), 9.8 mM of D-Trp (■), 19.6 mM of D-Trp (▲).

D-tryptophan is inactive under ordinary reaction conditions without ammonium phosphates, which would suggest that D-Trp does not interact with TPase. Figure 5, however, indicates the inhibitory effect of D-Trp on the reaction of tryptophanase with L-Trp in the absence of ammonium phosphates. D-Trp served as a competitive inhibitor with 5 mM of inhibition constant Ki (Note 9). The competitive inhibitor usually shares a highly structural resemblance with its substrate. The D-Trp stereostructure fits into the active site structure of TPase, so D-Trp can bind at the active site and L-Trp activity is inhibited. The relationship between structural features of L-amino acid inhibitors and the effective inhibition constant is reported elsewhere for tryptophanase from *E. coli*, showing a good linear correlation between –RTlnKi and the hydrophobicity of the side chain from which negative deviations are interpreted as being a consequence of steric interactions and the positive deviations as being evidence for favorable specific interactions [26]. We took an analogous approach to a competitive inhibitor D-Trp to analyze the interaction with TPase. The deviation from this linearity was about -3 kcal/mol. The negative value was lower than would be expected, accounting for its hydrophobicity. D-Trp belongs to a group of amino acids with a branched side chain as L-leucine or L-valine. Lower affinity to D-Trp than to L-Trp suggests there is a steric problem between D-Trp and the active site structure of TPase. D-Trp causes steric hindrance similar to L-leucine with a hydrophobic long chain, possibly attributable to the heterocyclic moiety of D-Trp. When D-Trp binds to the active site of TPase, it is invariably at a standstill at the beginning due to steric hindrance, which is why tryptophanase excludes D-Trp from the L-Trp metabolic pathway.

D-Trp serves as a competitive inhibitor in the absence of ammonium phosphates, indicating that D-Trp can bind at the active site of TPase. L-Trp is degraded into indole through gem-diamine, external aldimine, quinonoid intermediate, iminoquinonoid intermediate and α-aminoacrylate (Note 10). It remains open as to which intermediate is a binding type for D-Trp in this figure, but the possibility is greater that the binding type is gem-diamine or external aldimine because D-Trp may stop at the first step in the series of the reaction since D-Trp behaves similarly to L-leucine with a long branched side chain [27]. The presence of ammonium phosphates produces a trivial conformational change in TPase to fit D-Trp partly into the active site structure, helping D-Trp to shift to the next reaction intermediate. Subtle steric interactions, whose details are still being investigated, result in the advent of TPase activity toward D-Trp. The present reaction system is appropriate for investigating the flexible stereospecificity of TPase. This mechanism remains ambiguous, but homochiral origins in early metabolism will be elucidated once this becomes clear.

5. Notes

1. For a better understanding of this study, it is significant to describe the marked distinctive specificity of enzymes differing from chemical catalysts. Enzymes are catalysts and governed by the same rules that apply to all catalysts, including chemical catalysts, that speed reactions up. However, they have a number of unique properties not available to any great extent in chemical catalysts, i.e., reaction range, control of enzyme activity, catalytic power, and specificity. Specificity is especially interesting because it is very closely connected to chiral homogeneity.

   Chemical catalysts display only limited selectivity, whereas enzymes show 3 levels of specificity — for the susceptible bond involved in enzyme reaction, for the type of product produced, and for the reactant. A typical example of susceptible bond specificity is provided by proteases. Different proteases show specificity for peptide bonds located between particular amino acid residues [28-31]. Product specificity ensures that the final product is not contaminated with by-products [32-34]. The formation of by-products by side reactions is a troublesome problem associated with most of the less specific chemical catalysts. The degree of specificity for a substrate varies from fairly broad to absolute. Alkaline phosphatase, carboxylesterase, and carboxypeptidase, for example, are hydrolytic with wide substrate specificity [35-37], while hexokinase is less specific and shows group specificity for a small set of related sugar molecules [38]. In contrast, urease is absolutely specific for its substrate urea, and structural analogs are not hydrolyzed at all by the enzyme [39]. A useful advantage of enzyme specificity is that, in a number of cases, it extends to selective discrimination between stereoisomers of a substrate molecule, and is also closely related to homochirality. Stereospecificity is shown by the enzyme D-amino acid oxidase, which is specific for D-amino acids alone, and does not catalyze oxidation of L-amino acid stereoisomers [40]. Aspartase, which deaminates only an L-aspartate, also has very strict stereospecificity [41]. Generally speaking, enzyme stereospecificity to optically active substances is absolute except for racemases.

A limited number of amino acid residues of protein sequences participate in this enzyme specificity because an enzyme reaction takes in a particular region of the enzyme or the active site designed to accommodate specific participants involved in the reaction. This is normally a small pocket, cleft, or crevice on the surface of the enzyme. Upon substrate binding, an enzyme-substrate (ES) complex is formed, involving noncovalent interactions between the substrate and site amino acid residues. During any reaction, reactants briefly enter a state in which the susceptible substrate is not completely broken and new bonds in the product are not completely formed. This transient condition is called the transition state, and the substrate in the transition state the intermediate. It is energy dependent because it requires energy to break chemical bonds — about 350 kJ /mol for a carbon-carbon single bond [42]. This represents an energy barrier to successful reaction and is why the vast majority of reactions proceed extremely slowly in the absence of external aid. The high catalytic power of enzymes is due to precise molecular interactions occurring at the active site that lower the energy barrier and enable formation of the transition state. There are at least 4 types of interaction that accomplish this effect, and they may operate singly or in combination. First, the active site in many enzymes provides a nonpolar microenvironment [43], and the removal of a substrate molecule from an aqueous polar solution to a nonpolar phase may alter the conformation of the substrate toward the transition state. Second, the precise alignment of substrate molecules in the active site presents susceptible bonds at the correct angle so that a collision between reactants will result in the formation of a transition state. Third, the substrate molecule is normally held firmly in the active site by a number of noncovalent interactions and small movements in the conformation of the enzyme molecule can be transmitted to the active site, distorting the substrate structure [44,45], weakening the susceptible bond, and reducing the amount of energy required to form a transition state. Fourth, site amino acid residues contribute catalytic functional groups to participate directly in the reaction. Amino acid side chains such as hydroxyl, sulphydryl, carboxyl, and amino function as proton donors or acceptors, depending on the state of ionization of the side chain [46]. By adding and/or withdrawing electrons, protons, or other group, catalytic functional groups at the active site push substrate molecules toward the transition state, thereby, increasing a reaction. The substrate molecule must thus have the correct shape and/or functional groups to fit into the active site and participate in interactions. Enzymes with absolute stereospecificity exclusively demand precise shape/interaction requirements that are only found in a particular enantiomer. A rough outline of the mechanism of enzyme stereospecificity is thus given based on the interactions above.

2. TPase is a pyridoxal 5′-phosphate (PLP)-dependent enzyme that catalyzes the hydrolytic β-elimination of L-tryptophan to indole and ammonium pyruvate as shown below, consisting of 4 identical 52 kDa momomers. Each monomer contains 1 molecule of PLP, which forms an aldimine bond with a lysine residue. TPase is an enzyme with absolute stereospecificity for optically active stereoisomers [47].

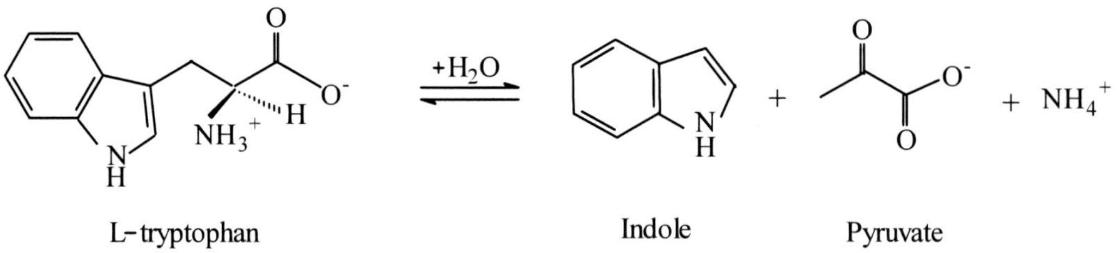

Tryptophanase activity on D-tryptophan was unexpectedly discovered when I purified to isolate tryptophanase from *E. coli* by use of adsorption chromatography. I thought in the beginning it was an erroneous result led by wrong chromatographic operation because enzyme stereospecificity was generally believed to remain unchanged forever. However, the activity was established because the same results were reproduced by repeated experiments.

3. Pure crystalline apotryptophanase purchased from Sigma-Aldrich Co. is easily soluble in an aqueous solution. Prior to experiments, SDS-PAGE shows a single band at MW 55,000, indicating that apotryptophanase is completely pure; 1 mg of it released 75 to 150 μg of indole from L-Trp in 10 min at a pH of 8.3 at 37°C in the presence of 0.04 mM of PLP. All chemicals are high reagent grade, purchased from Wako Pure Chemical Industries, Ltd., unless otherwise noted. All glassware is washed by soaking for more than 3 days in a detergent, Clean 99 CL, thoroughly rinsed, and then dried in an oven. All aqueous solutions are prepared from deionized and ceramics-distilled water.
4. The hot water that salts are dissolved in must not be boiling. When ammonium phosphates, for example, are dissolved while water is boiling, no tryptophanase activity on D-tryptophan appears. All processes should be conducted at temperatures lower than the boiling point of water. Great attention should also be paid to dissolve salts as slowly as possible and to cool their solutions at room temperature. It took about a week to complete crystallization. Use of a saturated solution should be avoided until crystallization finishes.
5. Erlich reagent is prepared as follows: 2 g of *p*-DMAB is completely dissolved in 16 mL of conc. HCl, then 200 mL of ethanol is added. Erlich reagent should be kept in a black or deep brown glass bottle with a cap. It can be used for a year if tightly capped. Old or deteriorated ethanol reacts naturally with Erlich reagent, so its use must be avoided.
6. The Crown pack CR column has 2 types of (+) and (–). The former first elutes the L form of tryptophan, and the latter D form. The appropriate column is chosen based on the application. These columns should be used at a pressure not exceeding 100 kg/cm^2 with a guard column because they are vulnerable to high pressure. Both a guard column, designed to trap chemical and physical debris that can cause problems at the head of the analytical column and a degasser should be used to stabilize the baseline.
7. Dissolving bis-ANS is time-consuming and should be done immediately before fluorescence is measured. It should be used only after confirming that it has completely dissolved in water. A bis-ANS stock solution should be made each time measurement is done.
8. Circular dichroism (CD) spectroscopy measures differences in the absorption of left-handed polarized light, versus right-handed polarized light, that arise due to structural asymmetry. The absence of a regular structure results in a CD intensity of zero, while an ordered structure results in a spectrum that can contain both positive and negative signals. The secondary structure, an α helix or β-sheet, is determined by CD spectroscopy in the 190-250 nm spectral region (far-uv). At these wavelengths, the chromophore is the peptide bond and the signal arises when it is located in a regular, folded environment. The CD spectrum of a protein in the 250-350 nm spectral region (near-uv) is sensitive to certain aspects of a tertiary structure. At these wavelengths, chromophores are aromatic amino acids (tryptophan, tyrosine, or phenylalanine) and disulfide bonds, and the CD signals they produce are sensitive to the overall tertiary structure of the protein. If a protein retains a secondary structure but no

defined 3-dimensional structure, signals in the near-uv region will be nearly zero. The presence of significant signals is a good indication that the protein is folded into a well-defined tertiary structure. The near-uv CD spectrum is sensitive to small changes in the tertiary structure due to conformational changes in solvent conditions.

9. Competitive inhibitors usually share a structural resemblance with a substrate. An inhibitor (I) competes with a substrate (S) at an active site on an enzyme (E) in competitive inhibition as shown below. The EI complex does not react to form E + P, and the enzyme is unable to bind both S and I simultaneously.

$$E + S \underset{k_{-1}}{\overset{k_{+1}}{\rightleftharpoons}} ES \underset{k_{-2}}{\overset{k_{+2}}{\rightleftharpoons}} E + P$$

$$E + I \overset{K_i}{\rightleftharpoons} EI$$

Reaction velocity v is $v = d[P]/dt = k_{+2}[ES] - k_{-2}[E][P]$. Initial velocity v is equal to $k_{+2}[ES]$, however, because [P] is zero at t = 0.

$v = k_{+2}[ES]$ ($[E]_o = [E] + [ES] + [EI]$)

$[E]_o$ is the total concentration of the enzyme. At equilibrium,

$K_s = [E][S]/[ES] = k_{-1}/k_{+1}$, $K_i = [E][I] / [EI]$

K_s is the substrate constant for the ES complex and K_i the dissociation constant (=inhibitor constant) for the EI complex. Solving this gives

$v = V_m/\{1+K_s(1 + [I]/K_i)/[S]\}$

Here, V_m ($= k_{+2}[E]_o$) is the maximum velocity. In a steady state ($k_{+2} \ll k_{-1}$), Michaelis constant $K_m = (k_{-1} + k_{+2})/k_{+1}$ is given, related to K_s.

$$K_m = (k_{-1} + k_{+2})/k_{+1} \approx k_{-1}/k_{+1} = K_s$$

Therefore,

$$1/v = K_m(1 + [I]/K_i)/V_m \cdot 1/[S] + 1/V_m$$

K_m (app) $= K_m(1 + [I]/K_i)$ and Vm is unchanged, i.e., the effect of a competitive inhibitor is to decrease K_m by $1 + [I]/K_i$. An inhibitor constant, K_i, which is quantitative measure of the inhibitory strength of reagent, can be estimated. K_i is determined by first conducting a standard K_m analysis.

10. L-Trp is degraded into indole, pyruvate and ammonia at the active site of TPase *via* several intermediates as shown below [48].

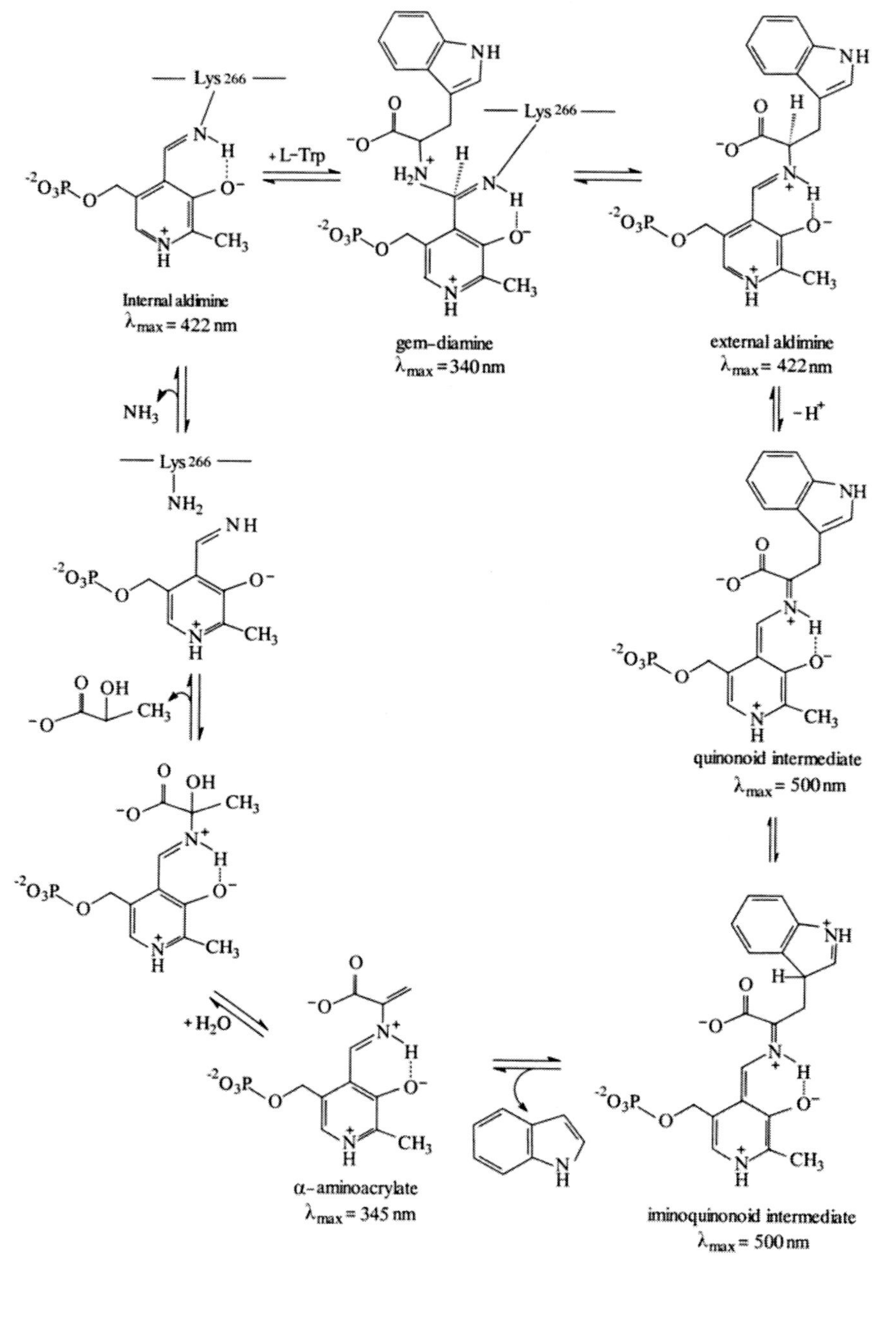

REFERENCES

[1] Enzyme Nomenclature Database, updates up to 31 May 2005.

[2] Miyamura, O. (1995). Aberian dominance of chiral-symmetry braking on SU(2) lattice. *Nucl Phys B, Suppl. 42*, 538-540.

[3] Laforgue, A. (1995). From the deadened oscillation to the spontaneous chiral breaking. *Theochem-J Mol Struct, 330*, 89-95.

[4] Shimada, A., Shishido H., Kogure, H., Nakamura, I., Fujii, N. & Akaboshi, M. (1998). Evolutionary significance of active site for D-tryptophan on tryptophanase. *Viva Origino*, *26*, 219-227.

[5] Yuasa, S. (1994). Molecular chirality and the origin of life. *J Biol Phys*, *20*, 229-233

[6] Bonner, W. A. (1995). Chirality and life. *Origins Life Evol B*, *25*, 175-190.

[7] Shimada, A., Shishido H., Kogure, H., Nakamura I., Fuji, N. & Akaboshi, M. (2000). Tryptophanase-catalyzed D-tryptophan degradation reaction and its significance for chiral homogeneity: The Role of Radiation in the Origin and Evolution of Life, Eds. Akaboshi, M., Fuji, N. & Gonzalez, R. N., Kyoto University Press, Kyoto, pp.243-257.

[8] Babincova, M., Babinec, P. (1994). Evolution of homochirality in a system of coupled chiral biomolecules. *Czech J Phys*, *44*, 165-167.

[9] Nagata, Y., Fujiwara, T., Kawaguchi-Nagata, K. et al. (1998). Occurrence of peptidyl D-amino acids in soluble fractions of several eubacteria, archaea and eukaryotes. *BBA-Gen Subjects*, *1379*, 76-82.

[10] Long, Z. Q., Lee, J. A., Okamoto, T. et al. (2001). Occurrence of D-amino acids and a pyridoxal 5 '-phosphate-dependent aspartate racemase in the acidothermophilic archaeon, *Thermoplasma acidophilum*. *Biochem Biophys Res Comm*, *281*, 317-321.

[11] Sarower, M. G., Matsui, T. & Abe, H. (2003). Distribution and characteristics of D-amino acid and D-aspartate oxidases in fish tissues. *J Exp Zoo*, *295A*, 151-159.

[12] Abe, H. (2002). Distribution, metabolism and physiological functions of free D-amino acids in aquatic invertebrates. *Nippon Suisan Gakk*, *68*, 516-525.

[13] Bulau, P., Meisen, I., Reichwein-Roderburg, B. et al. (2003). Two genetic variants of the crustacean hyperglycemic hormone (CHH) from the Australian crayfish, Cherax destructor: detection of chiral isoforms due to posttranslational modification. *Pepides*, *24*, 1871-1879.

[14] Yang, H. B., Yuan, J. G. & Peng, X. Z. (2003). Progress in the function of D-serine in mammalian central nervous system. *Prog Biochem Biophys*, *30*, 852-854.

[15] Fujii, N. (2002). D-amino acids in living higher organisms. *Origins Life Evol B*, *32*, 103-127.

[16] Lazarus, L. H. & Attila, M. (1993). The toad, ugly and venomous, wears yet a precious jewel in his skin. *Prog Neurobiol*, *41*, 473-507.

[17] Amiche, M., Delfour, A. & Nicolas, P. (1998). Opioid peptides from frog skin. *EXS*, *85*, 57-71.

[18] Shimada, A., Kogure, H., Shishido, H. et al. (1997). Reaction pathway of tryptophanase degrading D-tryptophan. *Amino Acids*, *12*, 379-383.

[19] Shimada, A., Shishido, H. & Nakamura, I. (1996). Tryptophanase-catalysed degradation of D-tryptophan in highly concentrated diammonium hydrogen phosphate solution. *Amino Acids*, *11*, 83-89.

[20] Shimada, A., Fujii, N. & Saito, T. (2004). Tryptophanase on D-tryptophan. In Paliyi, G., Zucchi, C., Caglioti, L. (Eds.), (pp. 321-327), Elsevier, (Oxford).

[21] Shimada, A. and Nakamura, I. (1992). Degradation of D-tryptophan by tryptophanase under high salt concentration. *Viva Origino*, *20*, 147-162.

[22] Zhao, S.L. & Liu, Y.M. (2001). Electrophoretic separation of tryptophan enantiomers in biological samples. *Electrophoresis*, *22*, 2769-2774.

[23] Yokoyama, Y., Hikawa, H. & Murakami, Y. (2001). Does water suppress the racemization and decomposition of amino acids? *J Chem Soc Perkin Transact-1*, *12*, 1431-1434.

[24] Phillips, R. S., Sundararaju B. & Faleev, N. G. (2000). Proton transfer and carbon-carbon bond cleavage in the elimination of indole catalyzed by *Escherichia coli* tryptophan indole-lyase. *J Am Chem Soc*, *122*, 1008 -1014.

[25] Sheluho, D. & Ackerman, S. H. (2001). An accessible hydrophobic surface is a key element of the molecular chaperone action of Atp11p. *J Biol Chem*, *276*, 39945-39949.

[26] Zakomirdina, L. N., Kulikova, V. V., Gogoleva, O. I., Dementieva, I. S., Faleev, N. G. & Demidkina, T. V. (2002) Tryptophan indole-lyase from *Proteus vulgaris*: kinetic and spectral properties. *Biochemistry (Moscow)*, *67*, 1189-1196.

[27] Gogoleva, O. I., Zakomirdina, L. N. & Demidkina, T. V. (2003) Tryptophanase in aqueous methanol: the solvent effects and a probable mechanism of the hydrophobic control of substrate specificity. *Enzyme Microb Tech*, 32, 843-850.

[28] Wu, T. P., Yee, V., Tulinsky, A., Chrusciel, R.A., Nakanishi, H., Shen, R., Priebe, C. & Kahn, M. (1993). The structure of a designed peptidomimetic inhibitor complex of alpha-thrombin. *Protein Eng*, *6*, 471-478.

[29] Oshikawa, K., Aoki, K., Yoshino, Y. & Terada, S. (2000). Purification and characterization of a basic amino acid-specific peptidase from seeds of jack bean (Canavalia ensiformis). *Biosci Biotech Bioch*, *64*, 2186-2192.

[30] Bhagya, S., Srinivasan K. S., & Prakash V. (1993). Effect of proteases on groundnut proeins (*Arachis hypogaea* L.) . *Nahrung*, *37*, 528-537.

[31] Csizmok, V., Bokor, N., Banki, P., et al. (2005). Primary contact sites in intrinsically unstructured proteins: The case of calpastatin and microtubule-associated protein 2. *Biochemistry*, *44*, 3955-3964.

[32] Lee P. C., Mijts, B. N., Petri R., et al. (2004). Alteration of product specificity of Aeropyrum pernix farnesylgeranyl diphosphate synthase (Fgs) by directed evolution. *Protein Eng Design Selection*, *17*, 771-777.

[33] Collins, R. E., Tachibana, M., Tamaru, H., et al. (2005). In vitro and in vivo analyses of a Phe/Tyr switch controlling product specificity of histone lysine methyltransferases. *J Biol Chem*, *280*, 5563-5570.

[34] Boucher, Y., Kamekura, M., & Doolittle, W. F. (2004). Origins and evolution of isoprenoid lipid biosynthesis in archaea. *Mol Microbiol*, *52*, 515-527.

[35] Bingham, E. W., Garver, K. & Powlen, D. (1992). Purification and proteins of alkaline-phosphatase in the lactating bovine mammary-gland. *J Daily Sci*, *75*, 3394-3401.

[36] Maxwell, D. M. & Brecht, K. M. (2001). Carboxylesterase: Specificity and spontaneous reactivation of an endogenous scavenger for organophosphorus compound. *J Appl Toxicol*, *21*, S103-S107 Suppl.

[37] Raksakulthai, R. & Haard, N. F. (2001). Purification and characterization of a carboxypeptidase from squid hepatopancreas (Illex illecebrosus). *J Agr Food Chem*, *49*, 5019-5030.

[38] Jang, J. C. & Sheen, J. (1994). Sugar sensing in higher plants. *Plant Cell*, *6*, 1665-1679.

[39] Greenwood, J. A., Mills, J., Tyler, P. D., et al. (1998). Physiological regulation, purification and properties of urease from Methylophilus methylotrophus. *FEMS Microbiol Lett*, *160*, 131-135.

[40] Gabler, M., Hensel, M. & Fischer, L. (2000). Detection and substrate selectivity of new microbial D-amino acid oxidases. *Enzyme Microb Tech*, *27*, 605-611.

[41] Kato, Y. & Asano, Y. (1995). Purification and properties of crystalline 3-methylaspartate from 2 facultative anaerobes, *Citrobacter* sp. strain YG-0504 and *Morganella Morganii* strain YG-0601. *Biosci Biotech Bioch*, *59*, 93-99.

[42] Smith, D. W. (2001). Carbon-carbon pi antibonding effects on the thermochemistry of alkanes, elucidated by angular overlap and MO calculations. *Phys Chem Chem Phys*, *3*, 3562-3568.

[43] Grossman, S. H. (1994). An equilibrium study of the dependence of secondary and tertiary structure of creatinine-kinase on subunit association. *BBA-Protein Struct M*, *1209*, 19-23.

[44] Andre, G. & Tran, V. (2004). Putative implication of alpha-amylase loop 7 in the mechanism of substrate binding and reaction products release. *Biopolymers*, *75*, 95-108.

[45] Synstad, B., Gaseidnes, S., van Aalten, D. M. F., et al. (2004). Mutational and computational analysis of the role of conserved residues in the active site of a family 18 chitinase. *Eur J Biochem*, *271*, 253-262.

[46] Gelpi, J. L., Jackson, R. M. & Holbrook, J. J. (1993). Electrostatic interaction energies in lactate dehydrogenase catalysis. *J Chem Soc Faraday T*, *89*, 2707-2716.

[47] Snell, E. E. (1975). Tryptophanase: structure, catalytic activities, and mechanism of action. *Adv Enxymol Relat Area Mol Biol*, *42*, 287-333.

[48] Kulikova, V. V., Zakomirdina, L. N., Bazhulina, N. P., Dementieva, I. S., Faleev, N. G., Gollnick, P. D. & Demidkina, T. V. (2003) Role of arginine 226 in the mechanism of tryptophan indole-lyase from *Proteus vulgaris*. *Biochemistry (Moscow)*, *68*, 1181-1188.

INDEX

A

B

C

D

E

F

G

H

N

O

P

Q

R

S

T

U

V

W

X

Y

Z

Entomology in Focus

Volume 2

Series editor
Fernando L. Cônsoli, Piracicaba, Brazil

Insects are the most common and widespread organisms on Earth, where they colonize the most diverse habitats and are part of our everyday life. Insects are important organisms in nature as they constitute an important source of nutrients in the diet of a number of invertebrates and vertebrates, having a direct impact on many food chains. Their contributions to natural environments are recognized as they are important in recycling nutrients and maintaining reproduction in several plants due to their role as pollinators. Insects are also recognized by their beneficial and prejudicial interactions with humans. The beneficial interactions involve the direct or indirect exploitation of insects as a source of food or secondary products, for example, while the prejudicial interactions involve the damage they cause to our cultivated plants and to the diseases they vector to humans, plants and animals.

The book series *Entomology in Focus* is intended to accelerate our understanding on the diversity of insects and on their mode of life and ecology in ways to provide a broad understanding on how these organisms can affect our lives and how we can diminish their noxious effects and exploit the benefits they provide to nature and humans. Therefore, books in this series are expected to provide a comprehensive synthesis of basic and/or applied topics in the field of Entomology.

More information about this series at http://www.springer.com/series/10465

Antônio R. Panizzi • Jocélia Grazia
Editors

True Bugs (Heteroptera) of the Neotropics

Springer

Sociedade Entomológica do Brasil · 1972 ·

Editors
Antônio R. Panizzi
Laboratório de Entomologia
Embrapa Trigo, Passo Fundo
RS, Brazil

Jocélia Grazia
Departamento de Zoologia
Instituto de Biociências
Universidade Federal do Rio Grande do Sul (UFRGS)
Porto Alegre, RS, Brazil

ISSN 2405-853X ISSN 2405-8548 (electronic)
Entomology in Focus
ISBN 978-94-017-9860-0 ISBN 978-94-017-9861-7 (eBook)
DOI 10.1007/978-94-017-9861-7

Library of Congress Control Number: 2015950651

Springer Dordrecht Heidelberg New York London

Printed on acid-free paper

Springer Science+Business Media B.V. Dordrecht is part of Springer Science+Business Media (www.springer.com)

C.W. Schaefer in Antônio's office –
photo by Patrick De Clercq

On April 29, 2015, we were shocked to learn the sad news that our friend and colleague Prof. Carl W. Schaefer had passed away. He will not see his nice chapter on Pyrrhocoridae/Largidae nor the one he co-authored on Alydidae published in this book. However, his legacy as a great heteropterist and as a great man will remain among us. We dedicate this book to Carl's memory for all his contribution to the knowledge we have on true bugs.

Foreword

In entomology, the appearance of group specialists and comprehensive treatments often occurs at very long intervals, and the Heteroptera are no exception. The reasons for this phenomenon are easy to understand: too many insects and too few specialists to study them. This volume represents a real contribution in filling that inevitable gap.

After receiving my PhD in 1971, I had the opportunity to teach in the Brazilian Federal University system for two academic years in the city of Belém, Pará, Brazil. Part of my responsibility was to work with a small group of students—*bolsistas*—on in-depth studies of insects. We often focused on Heteroptera, because that is the group that I knew best, and they served as excellent models for addressing broader questions in entomology. The search for suitable texts and teaching aids led me to two works in particular. On a general level it became clear that the single-volume, 1970 edition of the *Insects of Australia*, recently published by Melbourne University Press, was the reference of choice. It offered far and away the best family-level coverage of the Neotropical fauna because of the many shared faunal similarities of these two tremendously diverse Gondwana fragments. Nonetheless, at a more detailed level discussions of Australian species cited in the work were of little use in understanding the fauna of Brazil in general, or of Amazonia in particular.

The work of recourse for understanding the details of Brazilian Heteroptera was the 351-page 1940 volume on Hemiptera, part of the classical *Os Insetos do Brasil*, a set of 12 volumes authored by Angelo Moreira da Costa Lima. Despite its tremendous popularity among entomologists in Brazil, the volume on Hemiptera (Heteroptera) was of limited utility. This is understandable considering Costa Lima is not a heteropterist per se but rather had a background as an applied entomologist and as a medical sanitarist working at the Instituto Oswaldo Cruz, the center for the study of the triatomine vectors of Chagas disease. He concentrated on few family groups, particularly those of greatest economic and human-health importance, doubtless a sign of the types of entomological studies he conducted and also the main priority at that time. Although it was more comprehensive than the enduring *Hemiptera Argentina* of Carlos Berg from the late 1880s, both works understandably presented far from what we now know to be the actual diversity of the

Neotropical Heteroptera fauna. Also, at the time Costa Lima's work was published, family-level nomenclature in the Heteroptera was in a state of flux as was our understanding of relationships within the suborder; both of these issues inevitably led to the work becoming out of date.

Therefore, I take great pleasure in welcoming the publication of *True Bugs of the Neotropics*, organized and edited by Antônio R. Panizzi and Jocélia Grazia. This impressive volume benefits from events that have taken place since the time of Costa Lima and of which we can now enjoy the full advantage. Standing out prominently is the cadre of young specialists who have contributed to this volume, many from the Neotropical Region and many inspired and trained by coeditor Grazia. Several of them have made real contributions in their own right, and much of that knowledge is summarized as part of their authorship contributions to this volume. The current authors have also had the benefit of contributions made by authorities likely directly influenced by *Os Insetos do Brasil*. Three of those pioneering workers—heteropteran titans—who stand out in my mind are José C. M. Carvalho (Miridae), Herman Lent (Reduviidae, Triatominae), and Pedro Wygodzinsky (Dipsocoromorpha, Reduviidae).

Users of this volume will benefit from the uniformity of treatment across family groups, the digital revolution in photography and photocomposition, and the greatly improved knowledge of the Neotropical fauna that has resulted from extensive field work, local and international collection development, and extensive revisionary work in almost every family of Heteroptera over the last 75 years. The current volume provides the reader with up-to-date knowledge concerning the classification, nomenclature, and biology of Neotropical taxa and offers a solid foundation for specialists, mentors, and students alike in their studies of the true bugs.

My congratulations to the entire team whose knowledge and dedication have brought this volume to fruition.

New York, NY, USA Randall T. Schuh
January 2015

Preface

The idea of making a book on the Heteroptera of the neotropics matured with time. Along the years, the two editors got together in several entomological congresses in Brazil and elsewhere, and, every time they met, they discussed about the need to publish a book on such matter. The initial idea was to publish a text on the stink bugs (Pentatomidae) of Brazil. This is understandable considering the background of the editors, ARP as a biologist and JG as a taxonomist, both working on pentatomids.

Back in 2010, ARP was the president of the Entomological Society of Brazil (SEB), and the society started negotiations with Springer to publish its journal *Neotropical Entomology*. When the agreement was signed, another idea came out, now to start an *Entomology Series Books* with Springer, through the SEB, which would focus on any aspect related to entomology in the neotropics. With that on mind, we editors had a meeting with the senior publishing editor of Springer, Ing. Zuzana Bernhart. She welcomed our plan to publish a more comprehensive book to cover not only the stink bugs but all (most) families of Heteroptera, not only from Brazil but from the whole Neotropical Region.

Not many books have been published recently on true bugs (Heteroptera). Of general interest, two texts are worth of mention: *True Bugs of the World (Hemiptera: Heteroptera) Classification and Natural History* by R. T. Schuh and J. A. Slater, Cornell University Press, 1995, and *Heteroptera of Economic Importance* edited by C. W. Schaefer and A.R. Panizzi, CRC Press, 2000. Fifteen years from this later publication, we believe that it was time to produce a book on Heteroptera from the Neotropical Region, doubtless the most diverse and complex biogeographic zone of the world.

The book is divided in seven parts. Part I is devoted to general aspects, and includes three chapters. Chapter 1 is an introduction to the true bugs of the neotropics; Chap. 2 covers general aspects of the morphology, ontogeny, reproduction, and feeding of true bugs; and Chap. 3 focuses on the classification and biogeography of the Neotropical true bugs. Parts II–VII are dedicated to specific aspects. We decided to divide these parts considering each of the seven infraorders (except the first two that were merged in one part) as follows: Enicocephalomorpha (the unique-headed bugs, Chap. 4) and the Dipsocoromorpha (the minute litter bugs, Chap. 5),

Gerromorpha (the semiaquatic bugs, Chap. 6), Nepomorpha (the true aquatic bugs, Chap. 7), Leptopodomorpha (the shore bugs, Saldidae, Chap. 8), Cimicomorpha (minute pirate bugs, Anthocoridae and Lyctocoridae, Chap. 9; plant bugs, Miridae, Chap. 10; damsel bugs, Nabidae, Chap. 11; assassin bugs, Reduviidae excluding Triatominae, Chap. 12; hematophagous bugs, Reduviidae and Triatominae, Chap. 13; and lace bugs, Tingidae, Chap. 14), and Pentatomomorpha (flat bugs, Aradidae, Chap. 15; the big-eyed bugs, chinch bugs, and seed bugs, Lygaeoidea, Chap. 16; the cotton strainers, Pyrrhocoridae, and bordered plant bugs, Largidae, Chap. 17; the broad-headed bugs, Alydidae, Chap. 18; the leaf-footed bugs, Coreidae, Chap. 19; the scentless plant bugs, Rhopalidae, Chap. 20; the burrower bugs, Cydnidae, Chap. 21; the stink bugs, Pentatomidae, Chap. 22; the shield bugs, Scutelleridae, Chap. 23; the negro bugs, Thyreocoridae, Chap. 24, and the less diverse pentatomoid families, Acanthosomatidae, Canopidae, Dinidoridae, Megarididae, Phloeidae, and Tessaratomidae, Chap. 25).

The chapters are written mostly by experts on the different taxonomic groups of heteropterans of the neotropics, considering their knowledge of the literature. To elaborate the chapters, we passed to authors a general outline. We asked them to introduce the theme, to provide general characteristics and diagnosis of the representative of the family (group) covered, and to discuss the biology and ecology, eggs and oviposition, nymph feeding and development, adult feeding and reproduction, and classification and diversity. We also asked authors to list the main and secondary species of each family and to highlight their economic importance. Of course, the layout was a suggestion because we did not want to put authors in a "forced frame," and we allowed them some leeway. Most authors followed the original layout, and we believe that the final product met our expectations, and we hope those of the readers interested in the true bugs of the neotropics.

Passo Fundo, RS, Brazil — Antônio R. Panizzi
Porto Alegre, RS, Brazil — Jocélia Grazia

Acknowledgements

The editing of this book was only possible with the help of many persons. We want to express our gratitude to Zuzana Bernhart, senior publishing editor who coaxed us from the very beginning when the publication contract was signed with Springer to the end of the project. We are also grateful to the editorial assistant Mariska van der Stigchel who reminded us gently of deadlines and other details in the process of having all materials set in time to meet the chronology of the project. We want to thank the gentle manner with which we, editors and authors, were treated by Nagarajan Paramasivam and Rajeswari Sathiamoorthy of Springer, SPi Global, India, in handling proofs of cover, front material, chapters and index. Without their help we surely could not have avoided the many mistakes that usually occur in the production of such a voluminous book.

Special thanks go to the authors of the chapters. Of course, they are the reason by which we ended up with such a nice and complete volume on the true bugs of the neotropics. All of them were very kind and cooperative to meet the deadlines and to follow all the instructions we editors passed to them in the beginning of this endeavour.

Finally, we want to thank Fernando Cônsoli, editor of the Entomology in Focus series, for the invitation to contribute this volume.

Contents

Contributors

Kim R. Barão Programa de Pós-Graduação em Biologia Animal, Instituto de Biociências, Universidade Federal do Rio Grande do Sul (UFRGS), Porto Alegre, RS, Brazil

Julianna F. Barbosa Instituto de Biologia, Departamento de Zoologia, Laboratório de Entomologia, Universidade Federal do Rio de Janeiro, Rio de Janeiro, RJ, Brazil

Aline Barcellos Museu de Ciências Naturais, Fundação Zoobotânica do Rio Grande do Sul, Porto Alegre, RS, Brazil

Filipe M. Bianchi Programa de Pós-Graduação em Biologia Animal, Instituto de Biociências, Universidade Federal do Rio Grande do Sul (UFRGS), Porto Alegre, RS, Brazil

S. Kristine Braman Department of Entomology, UGA Griffin Campus, Griffin, GA, USA

Luiz A. Campos Departamento de Zoologia, Instituto de Biociências, Universidade Federal do Rio Grande do Sul (UFRGS), Porto Alegre, RS, Brazil

Diego L. Carpintero Museo Argentino de Ciencias Naturales Bernardino Rivadavia, Buenos Aires, Argentina

Scott P. Carroll Department of Entomology, University of California, Davis, CA, USA

Lívia A. Coelho Museu Regional de Entomologia, Universidade Federal de Viçosa (UFV), Viçosa, MG, Brazil

Eugenia F. Contreras Instituto de Biología de la Altura, Universidad Nacional de Jujuy, San Salvador de Jujuy, Argentina

Marcela Cornelis Facultad de Ciencias Exactas y Naturales, Universidad Nacional de La Pampa, Santa Rosa, La Pampa, Argentina

Maria del C. Coscarón División Entomología, Museo de La Plata, La Plata, Buenos Aires, Argentina

Alan L. de Melo Departamento de Parasitologia, Instituto de Ciências Biológicas, Universidade Federal de Minas Gerais, Belo Horizonte, MG, Brazil

Alexandre S. de Paula Departamento de Ciências Biológicas, Instituto de Ciências Exatas e Biológicas, Universidade Federal de Ouro Preto (UFOP), Ouro Preto, MG, Brazil

Pablo M. Dellapé División Entomología, Museo de La Plata, La Plata, Buenos Aires, Argentina

Joseph Eger Jr. Dow AgroSciences LLC, Tampa, FL, USA

José Antônio Marin Fernandes Instituto de Ciências Biológicas, Universidade Federal do Pará, Belém, PA, Brazil

Augusto Ferrari Instituto de Ciências Biológicas (ICB), Universidade Federal do Rio Grande - FURG, Rio Grande, RS, Brazil

Paulo S.F. Ferreira Departamento de Entomologia, Universidade Federal de Viçosa (UFV), Viçosa, MG, Brazil

Dimitri Forero Laboratorio de Entomología, Unidad de Ecología y Sistemática, Pontificia Universidad Javeriana, Bogotá, Colombia

Trevor M. Fowles Department of Entomology, University of California, Davis, CA, USA

Cleber Galvão Laboratório Nacional e Internacional de Referência em Taxonomia de Triatomíneos, Instituto Oswaldo Cruz, Fiocruz, Rio de Janeiro, RJ, Brazil

Thereza de A. Garbelotto Programa de Pós-Graduação em Biologia Animal, Instituto de Biociências, Universidade Federal do Rio Grande do Sul (UFRGS), Porto Alegre, RS, Brazil

Hélcio R. Gil-Santana Laboratório de Diptera, Fundação Instituto Oswaldo Cruz (Fiocruz), Rio de Janeiro, RJ, Brazil

Jocélia Grazia Departamento de Zoologia, Instituto de Biociências, Universidade Federal do Rio Grande do Sul (UFRGS), Porto Alegre, RS, Brazil

Caroline Greve Departamento de Zoologia, Instituto de Biociências, Universidade Federal do Rio Grande do Sul (UFRGS), Porto Alegre, RS, Brazil

Marcus Guidoti Programa de Pós-graduação em Biologia Animal, Instituto de Biociências, Universidade Federal do Rio Grande do Sul (UFRGS), Porto Alegre, RS, Brazil

Éric Guilbert UMR7205 MNHN/CNRS/EPHE/UPMC, Muséum National d'Histoire Naturelle, Paris, France

Thomas J. Henry Systematic Entomology Laboratory, Agricultural Research Service, United States Department of Agriculture, c/o Smithsonian Institution, National Museum of Natural History, MRC 168, Washington, DC, USA

José Jurberg Laboratório Nacional e Internacional de Referência em Taxonomia de Triatomíneos, Instituto Oswaldo Cruz, Fiocruz, Rio de Janeiro, RJ, Brazil

Laurence Livermore Biodiversity Informatics, Insects Division, The Natural History Museum, London, UK

Viviana C. Matesco Programa de Pós-Graduação em Biologia Animal, Instituto de Biociências, Universidade Federal do Rio Grande do Sul (UFRGS), Porto Alegre, RS, Brazil

Paula Levin Mitchell Department of Biology, Winthrop University, Rock Hill, SC, USA

Sara I. Montemayor División Entomología, Museo de La Plata, La Plata, Buenos Aires, Argentina

Felipe F.F. Moreira Laboratório Nacional e Internacional de Referência em Taxonomia de Triatomíneos, Instituto Oswaldo Cruz, Fiocruz, Rio de Janeiro, RJ, Brazil

Cristiane Nardi Departamento de Agronomia (DEAGRO), Universidade Estadual do Centro-Oeste (UNICENTRO), Guarapuava, PR, Brazil

Malin Nikunlassi Biodiversity Informatics, The Natural History Museum, London, UK

Antônio R. Panizzi Laboratório de Entomologia, Embrapa Trigo, Passo Fundo, RS, Brazil

Higor D.D. Rodrigues Museu de Zoologia da Universidade de São Paulo, São Paulo, SP, Brazil

Carl W. Schaefer Author deceased at the time of publication

Cristiano F. Schwertner Departamento de Ciências Biológicas, Instituto de Ciências Ambientais, Químicas e Farmacêuticas, Universidade Federal de São Paulo (UNIFESP), Diadema, SP, Brazil

Felipe L. Simões Departamento de Zoologia, Instituto de Biociências, Universidade Federal do Rio Grande do Sul (UFRGS), Porto Alegre, RS, Brazil

Luciana Weiler Programa de Pós-Graduação em Biologia Animal, Instituto de Biociências, Universidade Federal do Rio Grande do Sul (UFRGS), Porto Alegre, RS, Brazil

Christiane Weirauch Department of Entomology, University of California, Riverside, CA, USA

About the Editors

Antônio R. Panizzi is a senior research entomologist at the Wheat Research Center of Embrapa (Brazilian Enterprise of Agricultural Research), in Passo Fundo, Rio Grande do Sul, Brazil. He earned his BS in Agronomy in 1972 from the University of Passo Fundo; his MS in Entomology in 1975 from the Federal University of Paraná, both in Brazil; and his PhD in 1985 from the University of Florida. He is the recipient of the Alexandre Rodrigues Ferreira Prize given by the Brazilian Society of Zoology and editor in chief of the *Annals of the Entomological Society of Brazil* (1993–1998). He served as president of the Entomological Society of Brazil (2008–2012) and in 2014 became Honorary Member of the society. He has been an invited speaker at several congresses and symposia in different parts of the world and an invited scientist at the National Institute of Agro-Environmental Sciences, Tsukuba, Japan (1991). From 2002 to 2004 worked as a visiting scientist at the USDA laboratory at Peoria, IL, Dr. Panizzi coedited *Heteroptera of Economic Importance* (CRC Press, 2000) and *Insect Bioecology and Nutrition for Integrated Pest Management* (CRC Press, 2012) and served as guest coeditor of *True Bugs (Heteroptera) Chemical Ecology of Invasive and Emerging Pest Species* (special issue of Psyche, 2012). He has published extensively on Hemiptera (Heteroptera), over 150 peer-reviewed publications, including an *Annual Review of Entomology* article on wild hosts of Pentatomidae. He teaches a course on insect nutritional ecology at the Federal University of Paraná and is an invited professor at the Universidad de la Republica in Uruguay, where he serves as advisor for MSc and PhD students. His current research focuses on the interactions of heteropterans (mostly Pentatomidae) with their wild and cultivated host plants and the management of pest species on field crops.

Jocélia Grazia is a full professor of the Department of Zoology at the Federal University of Rio Grande do Sul and a researcher of National Research Council of Brazil for the last 50 years. She earned her PhD in Zoology in 1976 from the State University of Campinas (UNICAMP), São Paulo, Brazil, and between October 1994 and April 1995 obtained a postdoc fellowship to develop research at the Heteroptera Laboratory of the American Museum of Natural History, New York, under the supervision of RT Schuh. She is recipient of the Jesus Santiago Moure Prize (2010) and Alexandre Rodrigues Ferreira Prize (2013) given by the Brazilian Society of Zoology for the best paper published in taxonomy and the best book in zoology (senior author of the Hemiptera chapter), respectively. She served as editor in chief of the *Annals of the Entomological Society of Brazil* (1978–1984) and as president of the Entomological Society of Brazil (1984–1990), and in 2014 she received the Edilson Bassoli de Oliveira Prize given by the Brazilian Entomological Society in recognition to the members who have contributed to the development of entomology in Brazil. She served as president of the International Heteropterist's Society (2002–2006). She has published extensively on Hemiptera (Heteroptera), 180 peer-reviewed publications. She advised more than 70 students (undergraduate, MSc, and PhD) in four universities in Brazil (Federal University of Paraná, Federal University of Rio Grande do Sul, State University of Campinas, and Federal University of Alagoas). She was the director of the Natural Science Museum, Zoo-botanical Foundation, Porto Alegre, Brazil (1972–1975), and vice rector of graduate courses, Federal University of Rio Grande do Sul (2003–2004). Her current research focuses on the systematic and phylogeny of heteropterans (mostly Pentatomoidea).

Part I
General Aspects

Chapter 1
Introduction to True Bugs (Heteroptera) of the Neotropics

Antônio R. Panizzi and Jocélia Grazia

Abstract True bugs (Heteroptera) are a diverse and complex group of insects, particularly in the neotropics. The fauna of these bugs has been investigated through time, but our knowledge of the species living in the Neotropical Region is limited. In this introductory chapter, we give a general view on true bugs classification and biogeography, with concise comments on their general characteristics and bioecology of each major taxon that comprise each of the seven infraorders of Heteroptera.

1.1 Introduction

The true bugs (Heteroptera) constitute a very interesting widely distributed group of insects, which is greatly diversified in tropical zones. Considered the largest group of insects with incomplete metamorphosis, heteropterans have been studied on both basic and applied aspects worldwide.

Along the years, several books have been published on Heteroptera, the majority on specific aspects, such as certain groups (taxa) of particular areas, and others on more general comprehensive issues. Of more broad interest, two books about the latter were published relatively recently. The first was dedicated to the classification and natural history of true bugs in particular, with insights on the history of the study of Heteroptera, how to collect and preserve true bugs, historical biogeogra-

A.R. Panizzi (✉)
Laboratório de Entomologia, Embrapa Trigo, Caixa Postal 3081,
Passo Fundo, RS 99001-970, Brazil
e-mail: antonio.panizzi@embrapa.br

J. Grazia
Departamento de Zoologia, Instituto de Biociências, Universidade Federal
do Rio Grande do Sul (UFRGS), Av. Bento Gonçalves 9500, prédio 43435,
Bairro Agronomia, Porto Alegre, RS 91501-970, Brazil
e-mail: jocelia@ufrgs.br

A.R. Panizzi, J. Grazia (eds.), *True Bugs (Heteroptera) of the Neotropics*,
Entomology in Focus 2, DOI 10.1007/978-94-017-9861-7_1

phy, and comments of species of economic importance (Schuh and Slater 1995). The second book is a massive review of the world literature on species of economic importance, ranging from pest species of crops to species of medical importance (e.g., transmitters of Chagas disease) and to species that are of "nuisance" to humans (Schaefer and Panizzi 2000).

Considering the Neotropical Region, not a general book devoted to Heteroptera has been published in the last 70+ years. There are texts dealing with Hemiptera (Heteroptera) such as the one by Costa Lima (1940) on the true bugs of Brazil and the one by Berg (1879) devoted to true bugs of Argentina. Recently, Grazia and Fernandes (2012) made an overview of the true bugs of Brazil, updating what was published before on many scattered publications.

In this introductory chapter, we give a concise and fast overview of the true bugs of the neotropics. This includes a general view on their classification and biogeography, with concise comments on their general characteristics and bio-ecology of each major taxon that comprise each of the seven infraorders of Heteroptera.

1.2 Classification, Biogeography, General Characteristics, and Bioecology of Neotropical True Bugs

The order Hemiptera, suborder Heteroptera, is divided into seven infraorders: Enicocephalomorpha, Dipsocoromorpha, Gerromorpha, Nepomorpha, Leptopodomorpha, Cimicomorpha, and Pentatomomorpha. These seven infraorders were recognized by Schuh (1979) based on the information on the evolutionary tendencies of heteropterans mouthparts and feeding habits described by Cobben (1978). Recently, Weirauch and Schuh (2011) revised the actual knowledge on the systematics and evolution of the heteropterans.

1.2.1 Enicocephalomorpha

This infraorder, the unique-headed bugs, is divided in two families, Aenictopecheidae and Enicocephalidae. The first family is subdivided into four subfamilies with ten genera and 20 species. These bugs live on or underneath the soil debris or under tree barks showing omnivorous feeding habits (Grazia and Fernandes 2012). The Enicocephalidae has 33 genera and over 400 species divided into five subfamilies. Members of this family show swarming behavior with sexual dimorphism, females being much bigger than males. Wygodzinsky and Schmidt (1991) presented a monography of the fauna of the New World.

1.2.2 Dipsocoromorpha

This infraorder includes the smallest heteropterans with 47 genera and over 200 species, divided into five families, of which Ceratocombidae and Schizopteridae are better studied in the New World. The fauna in the neotropics is greatly diverse and poorly studied (Grazia and Fernandes 2012). Weirauch and Štys (2014) performed a phylogenetic analysis of this infraorder supporting its monophyly.

These small-sized heteropterans, very inconspicuous, live within fallen leaves, in natural small vegetation, or in tree canopy. They are hard to intercept, and many resemble small beetles.

1.2.3 Gerromorpha

Known as the semiaquatic bugs (water striders), the infraorder Gerromorpha comprises over 2,000 species in 11 families, with the majority of species in the neotropics belonging to the families Gerridae, Hebridae, Hydrometridae, Mesoveliidae, and Veliidae (Grazia and Fernandes 2012). The most comprehensive references for Gerromorpha, including morphology, phylogeny, biogeography, and keys for identification, were published by Andersen (1982). Great habitats dominated by freshwater, such as the Amazonia and the Pantanal in the neotropics of mostly Brazil, include a great number of species, many undescribed and waiting to be studied.

The Gerridae include bugs that live mostly on the water surface, with median and posterior legs elongated. Eight subfamilies are recognized (Andersen 1982), of which Trepobatinae contains a greater number of genera, at least in Brazil (Grazia and Fernandes 2012).

The Hebridae are small (1.3–3.7 mm) cryptic bugs living on herbaceous vegetation along waterways. Over 150 species are known and included in seven genera and two subfamilies (Hebrinae and Hyrcaninae) (Andersen 1981, 1982).

The Hydrometridae are long bugs (2.7–22 mm) found on vegetation or on rocks along the side of creeks and rivers. The over 100 species described are included in three subfamilies (Hydrometrinae, Heterocleptinae, and Limnobatodinae) (Andersen 1977).

The Mesoveliidae are small (1.2–4.2 mm), with variable wing development and shape. They live associated with floating vegetation or on open areas over the water. There are about 40 species included in two subfamilies (Mesoveliinae and Madeoveliinae). Very little is known about the Neotropical mesoveliids.

The Veliidae is a fairly large family (38 genera and about 600 species) (Andersen 1982). These bugs posses a truncated body, 1–10 mm long, well adapted to live on the water surface. There are six subfamilies (Ocelloveliinae, Rhagoveliinae, Perittopinae, Veliinae, Microveliinae, and Haloveliinae). They are easily found and captured and a well-studied group in many parts of the world, but not in the Neotropical Region.

1.2.4 *Nepomorpha*

The Nepomorpha, or true water bugs, comprises about 2,000 species placed on 11 families. Their size varies from 1 mm (Pleidae and Helotrephidae) to 130 mm (Belostomatidae). They are, in general, predators, with some Corixidae feeding on algae and decaying vegetal matter. In the neotropics, Pereira et al. (2007) presented keys for family and genera identification of species occurring in the Central Amazon in Brazil. Species of the following families are found in the neotropics: Belostomatidae, Corixidae, Gelastocoridae, Naucoridae, Nepidae, Notonectidae, Ochteridae, Pleidae, and Potamocoridae. The most speciose families are Naucoridae; Belostomatidae, which include the giant water bugs, *Lethocerus* spp.; Nepidae; and Notonectidae (Grazia and Fernandes 2012).

1.2.5 *Leptopodomorpha*

About 400 species distributed in four families comprise the infraorder Leptopodomorpha, or shore bugs. The major reference to the literature and classification of the group is by Schuh et al. (1987). In the neotropics only Saldidae is represented. Usually ovoids, these dark fast-moving bugs are found in nearby water in humid open areas. They are predators, even able to dig in order to capture their preys (Grazia and Fernandes 2012).

1.2.6 *Cimicomorpha*

This infraorder is the largest infraorder of Heteroptera including about 20,000 species in 16 families (Schuh and Slater 1995). The Cimicomorpha is a well-studied infraorder, especially because of its economic and health importance. Of the 16 families referred to by Schuh and Štys (1991), the most common and important ones will be covered in this book.

Minute Pirate Bugs (Anthocoridae)

These small to minute flattened, dark, oval or elongated insects are predaceous, but may feed on pollen and other parts of plants. Schuh and Slater (1995) presented an overview of the family worldwide, presenting a key for the seven tribes, and Carpintero (2002) wrote the first catalog devoted to Neotropical species. Lattin (2000) reviewed the species that are known to occur in managed ecosystems. Apparently, they are important as biological control of pests in agroecosystems. In Brazil there are about 100 species described (Grazia and Fernandes 2012).

Plant Bugs (Miridae)

This is, by far, the most diverse and speciose family within the Heteroptera, with about 1,300 genera and over 11,000 described species worldwide and over 500 genera and more than 3,000 species in the neotropics. In the neotropics, the majority of the species belong to the subfamilies Orthotylinae, Mirinae, and Deraeocorinae. The main taxonomic work for the Neotropical mirids was carried out by José C. M. Carvalho, who described about 400 genera and over 2,000 species and published the world catalog from 1957 to 1960 in a series of five publications (references in Chap. 10).

Mirids, known as plant bugs or capsids, are small- to medium-sized, elongated to ovoid bugs with triangular heads; they show variable coloration, being either phytophagous or carnivorous (predators).

The biology of mirids including habits and hosts has been published in several different publications and were synthesized by Wheeler (2000a, b, 2001). They are important pests in agricultural crops, and the predatory species show potential to be used in biological control programs. Among the most common species, we can cite those in the genera *Lygus*, *Monalonion*, *Pseudatomoscelis*, *Tenthecoris*, *Microtechnites*, and *Pycnoderes* as pests and *Phytocoris*, *Blepharidopterus*, *Campylomma*, *Campyloneura*, and *Engytatus* as potential predators.

Damsel Bugs (Nabidae)

The nabids, known as damsel bugs, are small- to medium-sized, elongated and robust bugs, with pale coloration. They are active predators, found in agricultural and urban areas. Schuh and Slater (1995) presented a key to the two subfamilies, Prostemmatinae and Nabinae. In the neotropics there are about 11 genera and 83 species described, with the genus *Arachnocoris*, *Praecarthasis*, *Hoplistoscelis*, *Nabis*, *Alloeorhynchus*, and *Pagasa* the most speciose (see Chap. 11). The majority of the literature refers to species of the Old World, and the New World fauna is treated by Harris (1930, 1931, 1939) and Kerzhner (1986).

Assassin Bugs (Reduviidae Excluding Triatominae)

The so-called assassin bugs, the reduviids, are voracious terrestrial predators with about 7,000 described species in over 900 genera and 25 subfamilies; it is considered one of the three most abundant in a number of species within the Heteroptera (Forero 2008; Henry 2009). Putchkov and Putchkov (1985) and Maldonado (1990) published catalogs of the species. For the Neotropical Region, over 200 genera and about 1,400 species included in 21 subfamilies are recorded (Forero 2004, and personal communication to JG).

Reduviids show a great morphological diversity with a range in size that varies from few millimeters to very large insects (Schuh and Slater 1995). Excluding the Triatominae that feed on vertebrate blood, all others live on preying.

The most common species belong to the following subfamilies: Chryxinae, Ectrichodiinae (the most abundant in the genus *Brontostoma* Kirkaldy), Elasmodeminae (exclusively from the neotropics, living under tree bark), Emesinae, Hammacerinae, Harpactorinae (with the most speciose genus *Apiomerus* Hahn and the common *Arilus* Hahn and *Zelus* (F.)), Peiratinae, Phymatinae, Reduviinae (which include the most studied genus, at least in Brazil, *Zelurus* Hahn), Saicinae, Sphaeridopinae, and Stenopodaine (Grazia and Fernandes 2012).

Hematophagous Bugs (Reduviidae, Triatominae)

The Triatominae are blood-sucking bugs known in Portuguese as "barbeiros." This subfamily includes about 140 species within 15 genera and five tribes. The actual taxonomic status of this subfamily that recognizes five tribes is based mainly on the revision by Lent and Wygodzinsky (1979), the most important systematic study of the group. This study has been updated by several other inclusions of new taxa (Carcavallo et al. 1998/1999; Schofield and Galvão 2009; see also Chap. 13).

These bugs are fairly big, in general from 20 to 28 mm length in body size, although some species may have about 5 mm or can reach over 40 mm; body color is mainly brownish black with spots in yellow, orange, or red (Jurberg et al. 2012). They can be found living on bird nests, underneath the bark of trees, and in crevices of various species of trees such as bromeliads, palms, and others and feeding on animals.

They are known worldwide as transmitters of the Chagas disease, caused by the flagellate *Trypanosoma cruzi* (Chagas), a major disease affecting human population in several regions of the world, and very important in the neotropics (see Chap. 13 for details). The most common species in the neotropics belong to the genera *Panstrongylus* Berg, *Triatoma* Laporte, and *Rhodnius* Stål, and they have been illustrated by Jurberg et al. (2004, 2014).

Lace Bugs (Tingidae)

The tingids, known as lace bugs, are delicate and small bugs that show outgrowth of the pronotum and forewings with lacy aspect. The family is divided into three subfamilies (Tinginae, Cantacaderinae, and Vianaidinae), all of them represented in the Neotropical Region (Schuh et al. 2006). There are around 300 genera and 2,500 species in this family, with 70 genera and 600 species in the neotropics.

Very few species have been studied in the neotropics with regard to their biology, such as *Corythaica cyathicollis* (Costa) (Kogan 1960) and *Leptopharsa heveae* Drake and Poor (Cividanes et al. 2004), both species of economic importance. The first species is pest of several solanaceous plants, and the second damages rubber plants. Various other genera (over 15) include species of economic importance (see details in Chap. 14). Neal and Schaefer (2000) reviewed the species of economic importance of the world.

1.2.7 Pentatomomorpha

A variable number of superfamilies have been recognized within the infraorder Pentatomomorpha, with five currently accepted: Aradoidea, Coreoidea, Lygaeoidea, Pentatomoidea, and Pyrrhocoroidea (see details in Chap. 3). In this book, we will discuss in details the families most important of this infraorder, as follows.

Flat Bugs (Aradidae)

These bugs, commonly known as "flat bugs" or "bark bugs," encompass a great number of species worldwide, with over 200 genera and about 1,800 species included in eight subfamilies (Schuh and Slater 1995). For the Neotropical Region, there are 80 genera and 509 species (Coscarón and Contreras 2012).

Aradids are small to medium in length. Many tropical species are wingless and the dorsal surface appears granular or rugose. They have the mouth stylets extremely elongated and coiled; they feed on juices of fungi (mycetophagous) and also can feed on phloem, cambium, and xylem of living trees. Some species might be associated with termites and occasionally can live on bird nests. Detailed information on their biology and ecology and a list of species from the Neotropical Region and their geographical distribution are presented in Chap. 15.

Lygaeoidea Families (Berytidae, Colobathristidae, Lygaeidae, and Piesmatidae)

The Lygaeoidea is the second largest superfamily in the infraorder Pentatomomorpha, with about 700 genera and more than 4,200 species in the world. In the neotropics, there are about 184 genera and 836 species (see Chap. 16). Sweet (2000) made a comprehensive world review of the Lygaeoidea, stating their distribution, life history, biology, and damage and control of the major and minor species of economic importance.

Berytidae

The so-called stilt bugs have a slender elongated body, with long antennae and legs, measuring 2.5–11 mm, dull yellowish or reddish brown; two subfamilies are recognized: Berytinae and Metacanthinae (Schuh and Slater 1995). Berytids are phytophagous, being polyphagous on several species of plants of different families; however, they do have strong predatory tendencies (Henry 2000); he reviewed the most important species of the world, considering their importance either as pests, predators, or pollinators. In Brazil there are over 20 species according to the literature (Grazia and Fernandes 2012).

Colobathristidae

These bugs, elongated with slender legs, 6–20 mm long, feed exclusively on grasses, are well represented in the Neotropical Region; it is divided in two subfamilies, Colobathristinae and Dayakiellinae (Schuh and Slater 1995). In Brazil, there are 13 species (Grazia and Fernandes 2012).

Lygaeidae

These bugs known as "seed bugs," 1.2–12 mm long, are variable in shape and color, brown or black, showing aposematic coloration (red or yellow). Baranowski and Slater (2005) provided the key for the 10 subfamilies, the most important being Lygaeinae, Blissinae, Geocorinae, and Rhyparochrominae. In Brazil, 120 species and 58 genera are registered (Grazia and Fernandes 2012).

Sweet (2000) in his world review of the major and minor species of economic importance pointed out the main pest species, which belong to the following genera: *Blissus*, *Cavalerius*, *Spilostethus*, *Nysius*, *Oxycarenus*, *Chauliops*, and *Elasmolomus*. Of these, species of *Nysius* and *Elasmolomus* (pest of peanuts probably introduced) are of economic concern in the neotropics (see Chap. 16).

Piesmatidae

Called the ash-gray leaf bugs, the piesmatids are small (<5 mm) bugs. Very little is known about these bugs in the neotropics, with about only five species recorded for Brazil (Grazia and Fernandes 2012). Narisu (2000) reviews the world species of economic importance, referring to species of *Piesma* as transmitters of virus to sugar beet.

Cotton Stainers (Pyrrhocoridae) and Bordered Plant Bugs (Largidae)

The cotton stainers (pyrrhocorids) are medium to large bugs, frequently red or yellow and black. About 30 genera and 300 species are known. They occur in all zoogeographic regions, mostly in tropical and subtropical areas (Schuh and Slater 1995). They feed on fruits and seeds, mostly on Malvales plants. Schaefer and Ahmad (2000) made a world review of the species of economic importance, particularly of those in the genus *Dysdercus*, which include at least 17 species, of which *Dysdercus peruvianus* Guérin-Méneville, *D. maurus* Distant, and *D. ruficollis* (L.) are major pests in the neotropics.

The bordered plant bugs (largids), commonly colored bugs, medium to big size, are divided into two subfamilies, Physopeltinae and Larginae. A key to the subfamilies and tribes of Largidae for New and Old World is presented in Chap. 17. They feed on seeds and on plant extracts. Several species mimic ants. About 60 species included in seven genera were described from South America (Grazia and Fernandes 2012).

Broad-Headed Bugs (Alydidae)

The broad-headed bugs (alydids) are small to medium, slender, with a triangular head. They are divided into two subfamilies, Alydinae and Micrelytrinae, each divided into two tribes, Daclerini and Alydini and Micrelytrini and Leptocorisini, respectively. A key for the two subfamilies, including the tribes for the last presented by Schaefer (2004), is included in Chap. 18, with slight modifications.

The family Alydidae has over 50 genera with about 250 species; in the neotropics there are over 20 genera. The most common species belong to the genera *Neomegalotomus* Schaffner and Schaefer, *Hyalymenus* Amyot & Serville, *Stenocoris* Burmeister, *Cydamus* Stål, and *Trachelium* Herrich-Schäffer (see Chap. 18). The most studied species in the neotropics is the alydine *Neomegalotomus parvus* (Westwood), usually associated with legumes, and may be a pest of soybean. The alydids of economic importance in the world were reviewed by Panizzi et al. (2000b).

Leaf-Footed Bugs (Coreidae)

The coreids, known as leaf-footed bugs or squash bugs, are, in general, medium to very large, strong robust bugs and may be strikingly colorful, showing expansion of femora, tibiae, humeral angles, or antennae. They are distributed worldwide, but are more abundant in the tropics.

The New World Coreidae catalog lists 158 genera with species from the neotropics (Packauskas 2010). The keys to the subfamilies of the Neotropical Region coreids are presented in Chap. 19, adapted from Packauskas (1994). These include the three subfamilies, Pseudophloeinae, Meropachyinae, and Coreinae.

Coreids are polyphagous plant feeders on gymnosperms and angiosperms, monocots and dicots. However, dicots are far more common and few coreid species are exclusively gymnosperms feeders (Schaefer and Mitchell 1983). Mitchell (2000) reviewed the coreids of economic importance of the world. In the neotropics, the most common species include *Crinocerus sanctus* (F.), *Anisoscelis foliaceus* (F.), *Holhymenia histrio* (F.), *Leptoglossus gonagra* (F.), *Leptoglossus zonatus* (Dallas), *Phthia picta* (Drury), *Anasa tristis* (De Geer), and *Spartocera dentiventris* Berg (see details in Chap. 19).

Scentless Plant Bugs (Rhopalidae)

Rhopalidae are plant feeders bugs, mostly associated with reproductive tissues and seeds. The rhopalids are divided into two subfamilies, Rhopalinae and Serinethinae, with about 200 species in over 20 genera (Henry 1988, 2009). Its taxonomy has been studied over time by several authors (e.g., Schaefer 1965; Chopra 1967) and a world catalog was published (Göllner-Scheiding 1983). Several specific catalogs were produced for particular areas of the neotropics (e.g., Froeschner 1981, 1985, 1999; Maes and Göllner-Scheiding 1993; Pall and Coscarón 2012).

Rhopalids are small to medium and have a narrow body that resembles coreids, with brown dull coloration. They are called "scentless" plant bugs, which is an inappropriate name, since they have scent glands that produce volatile compounds (Aldrich et al. 1990).

In the neotropics, the Serinethinae of the genus *Jadera* are the most common species, although little studied regarding their biology and ecology. For example, for Brazil, there are nine species belonging to this genus (Grazia and Fernandes 2012). The species *Jadera choprai* Göllner-Scheiding is an exception, and its biology and its association with the balloon vine, *Cardiospermum halicacabum* L. (Sapindaceae), a weed plant in soybean fields in southern Brazil, have been investigated in detail (Panizzi and Hirose 2002; Panizzi et al. 2002, 2005).

Rhopalids, despite their wide distribution and occurrence in natural vegetation and on cultivated plants, are in general overseen and their role in these habitats is underestimated.

Burrower Bugs (Cydnidae)

The cydnids, called burrower bugs, are small- to median-sized ovoid bugs, black or brown, with flattened head and legs with strong and long spines; tibia and tarsi are modified in species with fossorial habitats that feed on roots. Some cydnids live above ground on vegetation, feeding on falling seeds or plant tissues (Schuh and Slater 1995; Schaefer 2009).

Cydnids are worldwide distributed, being well represented in tropical and temperate regions (Lis 1999, 2002). It includes more than 750 species in 93 genera and is divided into six subfamilies (Grazia et al. 2008). The cydnids in the neotropics are poorly known; Froeschner (1981) presented a key to the subfamily Cydninae of South America, the biggest and more diverse with 90 genera and 300 species.

Lis et al. (2000) made a review of the species of economic importance of the world. In the neotropics, the most common species are *Scaptocoris carvalhoi* Becker, *Scaptocoris castanea* Perty, *Cyrtomenus bergi* Froeschner, and *Pangaeus bilineatus* (Say) (see Chap. 21).

Stink Bugs (Pentatomidae)

The pentatomids, called stink bugs, are in general broad and ovoid, 4–20 mm in length, the fourth largest family in Heteroptera, with about 800 genera and 4,700 species in the world; in the neotropics about 230 genera and 1,400 species are included in seven subfamilies of the world's nine subfamilies (Schuh and Slater 1995; see Chap. 22).

Species in the subfamily Asopinae are predacious and are important as biological control agents. The most common species in the neotropics include those in the genera *Alcaeorrhynchus* Bergroth, *Apateticus* Dallas, *Euthyrhynchus* Dallas, *Brontocoris* Thomas, *Perillus* Stål, *Podisus* Herrich-Schäffer, *Stiretrus* Laporte, *Supputius* Distant, and *Tylospilus* Stål. Two species, *Podisus maculiventris* (Say) and *Podisus nigrispinus* (Dallas), show potential in biological control programs (De Clercq 2000).

The exclusively Neotropical subfamily Cyrtocorinae is small with four genera and 11 species (Packauskas and Schaefer 1998); they are cryptically colored black to brown and are small to medium. Cyrtocorines are rare and of minor economic importance and little is known about their biology. Nymphs and adults of *Cyrtocoris egeris* Packauskas & Schaefer were observed feeding on immature soybean pods in Argentina (Schaefer et al. 2005).

The subfamily Discocephalinae includes two tribes, Discocephalini and Ochlerini, with more than 70 genera and over 300 species; almost all are medium sized and dark colored and restricted to the neotropics. The most known species of Discocephalini belong to the genus *Antiteuchus* Dallas, and in Ochlerini, the genera *Lincus* Stål and *Macropygium* Spinola include species of economic importance by vectoring *Phytomonas* spp. (Trypanosomatidae) in palms in South America (see details in Chap. 22).

The subfamily Edessinae is one of the largest with about 300 species (da Silva et al. 2013) distributed mostly in the Neotropical Region. It includes colored and large stink bugs. They feed on plants and may show economic importance, such as *Edessa meditabunda* (F.), particularly on soybean and on tobacco, and *Edessa rufomarginata* (De Geer) that feeds on Leguminosae and on Solanaceae causing damage to tomato and potato (Panizzi et al. 2000a).

The subfamily Pentatominae is the most diverse with an estimated number between 400 to 600 genera and 2,700 to 3,300 species; the exact number of groups (tribes, genera, and species) is uncertain because the definition of the subfamily is still under discussion (Rider 2013).

They are all phytophagous and many species are economically important as pests of cultivated crops. Main species include *Nezara viridula* (L.), *Piezodorus guildinii* (Westwood), and *Euschistus heros* (F.) and species of *Arvelius* Spinola, *Loxa* Amyot & Serville, *Agroecus* Dallas, *Dichelops* Spinola, and *Chinavia* Orian as important pests of soybean and other crops, and *Mormidea v-luteum* (Lichtenstein), *Oebalus poecilus* (Dallas), *Oebalus ypsilongriseus* (De Geer), and *Tibraca limbativentris* Stål may cause damage to rice and wheat (Panizzi et al. 2000a). Pentatominae are, in general, highly polyphagous, and their life history includes several host plant sequences (Panizzi 1997).

The subfamilies Podopinae and Stirotarsinae are represented by only one species each in the Neotropical Region.

The review of stink bugs of economic importance in the world was carried out by Panizzi et al. (2000a).

Shield Bugs (Scutelleridae)

The scutellerids are known as shield bugs or, sometimes, jewel bugs. These common names are based on their enlarged scutellum that covers the abdomen and their striking vivid color, with body length 5–20 mm long (Schuh and Slater 1995). These bugs are included in 80 genera and about 500 species worldwide (Tsai et al. 2011). In the neotropics, 25 genera and 112 species are registered, although these numbers are much probably underestimated (see Chap. 23).

Scutellerids are included in eight subfamilies, of which three occur in the neotropics, Pachycorinae with 23 genera and Scutellerinae and Elvisurinae with one genus each. The most common species in the neotropics are *Pachycoris torridus* (Scopoli), *Pachycoris klugii* Burmeister, *Augocoris illustris* (F.), *Augocoris gomesii* Burmeister, *Agonosoma flavolineata* (Laporte), *Agonosoma trivittata* (Panzer), *Symphylus deplanatus* (Herrich-Schaeffer), and *Tetyra pinguis* Germar (see details in Chap. 23).

Shield bugs are phytophagous feeding on an array of plants, but they might be necrophagous and be attracted to carrion. Javahery et al. (2000) reviewed the species of economic importance of the world.

Negro Bugs (Thyreocoridae)

The so-called negro bugs, Thyreocoridae Amyot & Serville (Corimelaenidae Uhler), include two subfamilies, Thyreocorinae and Corimelaeninae, this last represented in the Neotropical Region. They are relatively small and dark colored, with oval-elongated body and large scutellum (Grazia and Fernandes 2012). *Galgupha* Amyot & Serville is the most diverse Neotropical genus with 155 species in 15 subgenera; about 160 species are recorded in the neotropics, their biology poorly known (see Chap. 24).

Negro bugs are phytophagous feeding on flowers and on developing or ripe fruits. Despite their occurrence on many crops, they are regarded as minor pests on orchards and flowers and are not mentioned among the Heteroptera of economic importance (Schaefer and Panizzi 2000).

Pentatomoid Families (Acanthosomatidae, Canopidae, Dinidoridae, Megarididae, Phloeidae, and Tessaratomidae)

These are pentatomoid bugs that belong to small families, which include less than 70 species in less than 30 genera in the neotropics; species may show maternal care and have cryptic coloration; despite information on their taxonomic status, and on their biology and ecology, in general, they are poorly studied (see Chap. 25).

Acanthosomatidae

These are medium to large bugs, distributed mainly in temperate regions or at high altitudes in the subtropics. They are phytophagous, and some species show maternal care (Tallamy and Schaefer 1997). Few Neotropical species had their biology studied (Faúndez and Osorio 2010a, b). No acanthosomatids with economic importance are recorded from the neotropics (Schaefer et al. 2000).

Canopidae

Exclusively distributed in the neotropics, canopids are small- to medium-sized bugs and are shiny black. Due to their appearance, they are usually mistaken as coleopterans. Eight known species belong to the genus *Canopus* F. Their biology and ecology are less known. They feed on fungi (McHugh 1994).

Dinidoridae

Dinidorids are large, aposematic colored bugs, brown or black with red/yellow stripes, and included in two subfamilies Dinidorinae and Megymeninae, mostly distributed in the Afrotropical and Oriental regions. In the neotropics, only the genus *Dinidor* Latreille is recorded, with six species distributed in South America (see Chap. 25).

Species studied are phytophagous and polyphagous. Schaefer et al. (2000) reviewed the species of economic importance; none were referred to from the neotropics. Few data are available on their biology and ecology in the neotropics; *Dinidor mactabilis* (Dinidorinae) was recorded on *Smilax japecanga* Grisebach (Smilacaceae) in southern Brazil, which deposit large number of eggs in clusters on branches of the host plant (Grazia and Fernandes 2012).

Megarididae

These are small, dark, and shiny bugs and have ovoid and convex shape. They are distributed exclusively in the neotropics, in which 16 extant species are known all in the genus *Megaris* Stål. They are considered phytophagous and their biology and ecology are mostly unknown (see Chap. 25 for more details).

Phloeidae

The phloeids are large, flat depressed bugs with global expansions around the body which makes them look peculiar; they form colonies on the tree trunks where they live, being strongly mimetic (Grazia and Fernandes 2012). They are distributed mostly in the Neotropical Region, with four species in three genera: one species in the genus *Serbana* Distant, which occurs only in Borneo (Leston 1953), two species in *Phloea* Spinola, and one species in *Phloeophana* Leston, the latter two restricted to Brazil (Lent and Jurberg 1965).

Tessaratomidae

Tessaratomids are robust, ovoid-elongated, medium to large bugs; Neotropical species are in general dark green; they are mostly distributed in the Old World tropics with over 40 genera and over 200 species worldwide; the genus *Piezosternum* Amyot & Serville has worldwide distribution and includes three endemic species to the neotropics (see Chap. 25).

The biology and ecology of tessaratomids were studied in the Old World, and few data are available from the neotropics. They are polyphagous on plants; the Neotropical species *Piezosternum subulatum* (Thunberg) was recorded on some cultivated plants in Nicaragua (Maes 1994). However, there are no records on species causing economic damage in the neotropics; the bronze orange bug, *Musgraveia sulciventris* Stål, is referred to as a pest of citrus in the Australian region (Schaefer et al. 2000).

1.3 Concluding Remarks

Heteropterans are indeed a diverse and complex group of insects. They live on several different habitats (terrestrial, aquatic, intertidal and in association with other insects and spiders) as phytophagous (from seed, fruit, leaf, root to mycelia of fungi feeding), or as carnivorous (predators on other insects and arthropods and even feeding on blood of vertebrates); they have great ability to disperse and to adapt to different habitats through mimicry and protective coloration and shape (Schuh and Slater 1995). As a consequence of their success, the clash with humans' interests is inevitable, and many species of heteropterans became with time of economic and/or medical concern. Efforts are focused on research to better understand their biology and ecology in order to develop control strategies to manage pest species to crops or others of medical importance (Schaefer and Panizzi 2000).

Despite the many years of research, heteropterans in the Neotropical Region are still poorly known, as the readers of this book will rapidly verify. However, we hope that the information provided by each chapter will facilitate to learn the knowledge we actually have on the fauna of the heteropterans of the neotropics.

References

Aldrich JR, Carroll SP, Oliver JE, Lusby WR, Rudmann AA, Waters RM (1990) Exocrine secretions of scentless plant bugs; *Jadera, Boisea, Niesthrea* species (Hemiptera: Heteroptera: Rhopalidae). Biochem Syst Ecol 18:369–376

Andersen NM (1977) A new and primitive genus and species of Hydrometridae (Hemiptera, Gerromorpha) with a cladistic analysis of relationships within the family. Entomol Scand 8:301–316

Andersen NM (1981) Semiaquatic bugs: phylogeny and classification of the Hebridae (Heteroptera: Gerromorpha) with revisions of *Timasius, Neotimasius* and *Hyrcanus*. Syst Entomol 6:377–412

Andersen NM (1982) The semiaquatic bugs (Hemiptera, Gerromorpha). Phylogeny, adaptations, biogeography and classification. Entomonograph 3:1–455

Baranowski RM, Slater JA (2005) The Lygaeidae of the West Indies, Florida Agric Experiment Station. Bull 402. Agricultural Experiment Station, Gainesville, pp 1–266

Berg C (1879) Hemiptera Argentina enumeravit speciesque novas. Pauli E.Coni, Bonariae

Carcavallo RU, Galíndez-Girón I, Jurberg J, Lent H (eds) (1998/1999) Atlas of Chagas disease vectors in the Americas. Editora Fiocruz, Rio de Janeiro

Carpintero DL (2002) Catalogue of the Neotropical Anthocoridae (Heteroptera). Rev Soc Entomol Arg 61:25–44

Chopra NP (1967) The higher classification of the family Rhopalidae (Hemiptera). Trans R Entomol Soc Lond 119:363–399

Cividanes FJ, Fonseca FS, Santos TM (2004) Distribuição de *Leptopharsa heveae* em seringal no Estado de São Paulo. Pesq Agropec Brasil 39:1053–1056

Cobben RH (1978) Evolutionary trends in Heteroptera, Part II. Mouthpart-structures and feeding strategies. Centre for Agricultural Publishing and Documentation, Wageningen

Coscarón MC, Contreras EF (2012) Catalog of Aradidae (Hemiptera: Heteroptera) for the Neotropical Region. Zootaxa 3466:1–103

Costa Lima AM (1940) Insetos do Brasil. Hemipteros. Tomo II. Escola Nacional de Agronomia, Rio de Janeiro

da Silva VJ, Nunes DM, Fernandes JAM (2013) *Paraedessa*, a new genus of Edessinae (Hemiptera: Heteroptera: Pentatomidae. Zootaxa 3716:395–416

De Clercq P (2000) Predaceous stinkbugs (Pentatomidae: Asopinae). In: Schaefer CW, Panizzi AR (eds) Heteroptera of economic importance. CRC Press, Boca Raton, pp 737–789

Faúndez EI, Osorio GA (2010a) Contribution to the knowledge of *Cylindrocnema plana* Mayr, 1864 (Hemiptera: Heteroptera: Acanthosomatidae). An Ins Patagonia 38:159–161

Faúndez EI, Osorio GA (2010b) New data on the biology of *Sinopla perpunctatus* Signoret, 1864 (Hemiptera: Heteroptera: Acanthosomatidae). Bol Biodiversidad Chile 3:24–31

Forero D (2004) Diagnosis de los géneros neotropicales de la família Reduviidae (Hemiptera: Heteroptera), y su distribución en Colombia (excepto Harpactorinae). In: Fernández F, Andrade G, Amat G (eds) Insectos de Colombia, vol 3. Academia Colombiana de Ciencias Exactas, Físicas y Naturales, Bogotá, pp 128–275

Forero D (2008) The systematics of the Hemiptera. Rev Colomb Entomol 34:1–21

Froeschner RC (1981) Heteroptera or true bugs of Ecuador: a partial catalog. Smithsonian Contrib Zool 322:1–147

Froeschner RC (1985) Synopsis of the Heteroptera or true bugs of the Galápagos Islands. Smithson Contrib Zool 407:39–40

Froeschner RC (1999) True bugs (Heteroptera) of Panama: a synoptic catalog as a contribution to the study of Panamanian biodiversity. Am Entomol Inst 61:1–393

Göllner-Scheiding U (1983) General-kataloge der Familie Rhopalidae (Heteroptera). Mitt Zool Mus Berlin 59:37–189

Grazia J, Fernandes JAM (2012) Subordem Heteroptera Linnaeus, 1758. In: Rafael JA, Melo GAR, Carvalho CJKB, Casari AS, Constantino R (eds) Insetos do Brasil. Diversidade e Taxonomia. Ed. Holos, Ribeirão Preto, pp 369–405

Grazia J, Schuh RT, Wheeler WC (2008) Phylogenetic relationships of family groups in Pentatomoidea based on morphology and DNA sequences (Insecta: Heteroptera). Cladistics 24:932–976

Harris HM (1930) Notes on some South American Nabidae, with descriptions of new species (Hemiptera). Ann Carnegie Mus 19:241–248

Harris HM (1931) Nabidae from the state of Paraná. Ann Mus Zool Polonici 9:179–185

Harris HM (1939) Miscelánea sobre Nabidae sudamericanos (Hemiptera). Not Mus La Plata 4:367–377

Henry TJ (1988) Family Rhopalidae Amyot and Serville, 1843. In: Henry TJ, Froeschner RC (eds) Catalog of the Heteroptera, or true bugs, of Canada and the continental United States. Brill, Leiden, pp 652–664

Henry TJ (2000) Stilt bugs (Berytidae). In: Schaefer CW, Panizzi AR (eds) Heteroptera of economic importance. CRC Press, Boca Raton, pp 725–735

Henry TJ (2009) Biodiversity of Heteroptera. In: Foottit RG, Adler PH (eds) Insect biodiversity: science and society. Blackwell Publishing Ltd., Oxford, pp 223–263

Javahery M, Schaefer CW, Lattin JD (2000) Shield bugs (Scutelleridae). In: Schaefer CW, Panizzi AR (eds) Heteroptera of economic importance. CRC Press, Boca Raton, pp 475–503

Jurberg J, Galvão C, Noireau F, Carcavallo RU, Rocha DS, Lent H (2004) Uma iconografia dos triatomíneos. Entomol Vect 11:457–494

Jurberg J, Galvão C, Rocha DS, Dale C, Cunha V (2012) Vetores da doença de Chagas no Brasil. Instituto Oswaldo Cruz, Rio de Janeiro

Jurberg J, Rodrigues JMS, Moreira FFF, Dale C, Cordeiro IRS, Lamas VD Jr, Galvão C, Rocha DS (2014) Atlas iconográfico dos triatomíneos do Brasil (Vetores da doença de Chagas no Brasil). Instituto Oswaldo Cruz, Rio de Janeiro

Kerzhner IM (1986) Neotropical Nabidae (Heteroptera), 1: A new genus, some new species, and notes on synonymy. J NY Entomol Soc 94:180–193

Kogan M (1960) *Corythaica cyathicollis* (Costa, 1864), aspectos sistemáticos, biológicos e econômicos (Hemiptera, Tingidae). Mem Instituto Oswaldo Cruz 58:59–88

Lattin JD (2000) Minute pirate bugs (Anthocoridae). In: Schaefer CW, Panizzi AR (eds) Heteroptera of economic importance. CRC Press, Boca Raton, pp 607–637

Lent H, Jurberg J (1965) Contribuição ao conhecimento dos Phloeidae Dallas, 1851, com um estudo sôbre genitália (Hemiptera, Pentatomoidea). Rev Bras Biol 25:123–144

Lent H, Wygodzinsky P (1979) Revision of Triatominae (Hemiptera, Reduviidae) and their significance as vectors of Chagas disease. Bull Am Mus Nat Hist 163:123–529

Leston D (1953) "Phloeidae" Dallas: systematics and morphology, with remarks on the phylogeny of "Pentatomoidea" leach and upon the position of "*Serbana*" distant (Hemiptera). Rev Bras Biol 13:121–140

Lis JA (1999) Burrower bugs of the Old World – a catalogue (Hemiptera: Heteroptera: Cydnidae). Genus 10:165–249

Lis JA (2002) Burrower bugs described after the Old World catalogue of the family (Hemiptera: Heteroptera: Cydnidae). Pol Pis Entomol 71:7–17

Lis JA, Becker M, Schaefer CW (2000) Burrower bus (Cydnidae). In: Schaefer CW, Panizzi AR (eds) Heteroptera of economic importance. CRC Press, Boca Raton, pp 405–419

Maes J-M (1994) Catálogo de los Pentatomoidea (Heteroptera) de Nicaragua. Rev Nicar Entomol 28:1–29

Maes PJ, Göllner-Scheiding U (1993) Catálogo de los Coreoidea (Heteroptera). Rev Nica Entomol 25:1–19

Maldonado CJ (1990) Systematic catalogue of the Reduviidae of the world. Carib J Sci Spec Publ 1:1–694

McHugh JV (1994) On the natural history of Canopidae (Heteroptera: Pentatomoidae). J New York Entomol Soc 102:112–114

Mitchell PL (2000) Leaf-footed bugs. In: Schaefer CW, Panizzi AR (eds) Heteroptera of economic importance. CRC Press, Boca Raton, pp 337–403

Narisu (2000) Ash-gray leaf bugs (Piesmatidae). In: Schaefer CW, Panizzi AR (eds) Heteroptera of economic importance. CRC Press, Boca Raton, pp 265–270
Neal JW Jr, Schaefer CW (2000) Lace bugs (Tingidae). In: Schaefer CW, Panizzi AR (eds) Heteroptera of economic importance. CRC Press, Boca Raton, pp 85–137
Packauskas RJ (1994) Key to the subfamilies and tribes of the New World Coreidae (Hemiptera) with a checklist of published keys to genera and species. Proc Entomol Soc Wash 96:44–53
Packauskas R (2010) Catalog of the Coreidae, or leaf-footed bugs, of the new world. Fort Hays Stud 5:1–270
Packauskas RJ, Schaefer CW (1998) Revision of the Cyrtocoridae (Hemiptera: Pentatomoidea). Ann Entomol Soc Am 91:363–386
Pall JL, Coscarón MC (2012) The Rhopalidae (Hemiptera: Heteroptera) of Argentina. J Nat Hist 46:1441–1465
Panizzi AR (1997) Wild hosts of pentatomids: ecological significance and role in their pest status on crops. Annu Rev Entomol 42:99–122
Panizzi AR, Hirose E (2002) Seed-carrying and feeding behavior of *Jadera choprai* Göllner-Scheiding (Heteroptera: Rhopalidae). Neotrop Entomol 31:327–329
Panizzi AR, McPherson JE, James DG, Javahery M, McPherson RM (2000a) Stink bugs (Pentatomidae). In: Schaefer CW, Panizzi AR (eds) Heteroptera of economic importance. CRC Press, Boca Raton, pp 421–474
Panizzi AR, Schaefer CW, Natuhara Y (2000b) Broad-headed bugs (Alydidae). In: Schaefer CW, Panizzi AR (eds) Heteroptera of economic importance. CRC Press, Boca Raton, pp 321–336
Panizzi AR, Hirose E, Chocorosqui VR (2002) Unusual oviposition behavior by a seed feeding bug (Heteroptera: Rhopalidae). Neotrop Entomol 31:477–479
Panizzi AR, Schaefer CW, Hirose E (2005) Biology and descriptions of nymphal and adult *Jadera choprai* (Hemiptera: Rhopalidae). Ann Entomol Soc Am 98:515–526
Pereira DLV, de Melo AL, Hamada N (2007) Chaves para identificação para famílias e gêneros de Gerromorpha e nepomorpha (Insecta: Heteroptera) na Amazônia central. Neotrop Entomol 36:210–228
Putchkov VG, Putchkov PV (1985) A catalog of assassin-bugs genera of the world (Heteroptera: Reduviidae). Published by Authors, Kiev
Rider DA (2013) Pentatomoidea home page. North Dakota State University, Fargo
Schaefer CW (1965) The morphology and higher classification of the Coreoidea (Hemiptera-Heteroptera). Part III. The families Rhopalidae, Alydidae, and Coreidae. Misc Publ Entomol Soc Am 5:1–76
Schaefer CW (2004) Key to the genera of New World Alydidae (Hemiptera: Heteroptera). Proc Entomol Soc Wash 106:280–287
Schaefer CW (2009) *Prosorrhyncha* (Heteroptera and Coleorrhyncha). In: Resh VH, Cardé RT (eds) Encyclopedia of insects, 2nd edn. Academic, Amsterdan, pp 839–855
Schaefer CW, Mitchell PL (1983) Food plants of the Coreoidea (Hemiptera: Heteroptera). Ann Entomol Soc Am 76:591–615
Schaefer CW, Ahmad I (2000) Cotton strainers and their relatives (Pyrrhocoroidea: Pyrrhocoridae and Largidae). In: Schaefer CW, Panizzi AR (eds) Heteroptera of economic importance. CRC Press, Boca Raton, pp 271–308
Schaefer CW, Panizzi AR (2000) Heteroptera of economic importance. CRC Press, Boca Raton
Schaefer CW, Panizzi AR, James DG (2000) Several small pentatomoid families (Cyrtocoridae, Dinidoridae, Eurostylidae, Plataspidae, and Tessaratomidae). In: Schaefer CW, Panizzi AR (eds) Heteroptera of economic importance. CRC Press, Boca Raton, pp 505–512
Schaefer CW, Panizzi AR, Coscarón MC (2005) New records of plants fed upon by the uncommon heteropterans *Cyrtocoris egeris* Packauskas & Schaefer and *C. trigonus* (Germar) (Hemiptera: Cyrtocoridae) in South America. Neotrop Entomol 34:127–129
Schofield CJ, Galvão C (2009) Classification, evolution and species groups within the Triatominae. Acta Trop 110:88–100

Schuh RT (1979) Evolutionary trends in Heteroptera. Part II. Mouthpart-structures and feeding strategies. Syst Zool 28:653–656

Schuh RT, Slater JA (1995) True bugs of the world (Hemiptera: Heteroptera). Classification and natural history. Cornell University Press, Ithaca

Schuh RT, Štys P (1991) Phylogenetic analysis of Cimicomorphan family relationships (Heteroptera). J NY Entomol Soc 99:298–350

Schuh RT, Galil B, Polhemus JT (1987) Catalog and bibliography of Leptopodomorpha (Heteroptera). Bull Am Mus Nat Hist 185:243–406

Schuh RT, Cassis G, Guilbert É (2006) Description of the first recent macropterous species of Vianaidinae (Heteroptera: Tingidae) with comments on the phylogenetic relationships of the family within the Cimicomorpha. J NY Entomol Soc 114:38–53

Sweet MH II (2000) Seed and chinch bugs (Lygaeoidea). In: Schaefer CW, Panizzi AR (eds) Heteroptera of economic importance. CRC Press, Boca Raton, pp 143–264

Tallamy DW, Schaefer CW (1997) Maternal care in the Hemiptera: ancestry, alternatives, and current adaptive value. In: Choe JC, Crespi BJ (eds) The evolution of social behavior in insects and arachnids. Cambridge University Press, Cambridge/New York, pp 94–115

Tsai J-F, Rédei D, Yeh GF, Yang MM (2011) Jewel bugs of Taiwan (Heteroptera: Scutelleridae). National Chung Hsing University, Taichung

Weirauch C, Schuh RT (2011) Systematics and evolution of Heteroptera: 25 years of progress. Annu Rev Entomol 56:487–510

Weirauch C, Štys P (2014) Litter bugs exposed: phylogenetic relationships of Dipsocoromorpha (Hemiptera: Heteroptera) based on molecular data. Insect Syst Evol 45:1–20

Wheeler AG (2000a) Plant bugs (Miridae) as plant pests. In: Schaefer CW, Panizzi AR (eds) Heteroptera of economic importance. CRC Press, Boca Raton, pp 37–83

Wheeler AG (2000b) Predacious plant bugs (Miridae). In: Schaefer CW, Panizzi AR (eds) Heteroptera of economic importance. CRC Press, Boca Raton, pp 657–693

Wheeler AG (2001) Biology of the plant bugs (Hemiptera: Miridae). Pests, predators, opportunists. Cornell University Press, Ithaca

Wygodzinsky PW, Schmidt K (1991) Revision of the New World Enicocephalomorpha (Heteroptera). Bull Am Mus Nat Hist 200:1–265

Chapter 2
Morphology, Ontogeny, Reproduction, and Feeding of True Bugs

Jocélia Grazia, Felipe L. Simões, and Antônio R. Panizzi

Abstract In this chapter we present an overview of the morphology, ontogeny, reproduction, and feeding of the 28 higher-level categories (infraorder/superfamily/family/subfamily) of true bugs occurring in the neotropical region, which are treated in greater detail in Sect. 2.2.

2.1 Introduction

The Heteroptera can be diagnosed by the presence of mouthparts typically hemipteran with mandibular stylets concentric and surrounding maxillary stylets. The labium is inserted anteriorly on the head and a distinct gular area is always present, often closed behind to form a buccular bridge. Scent-gland structures are often paired, located and exiting ventrally in the metathoracic pleuron; nymphs with paired scent glands which are placed at the junction of one or more of the following abdominal terga: III/IV, IV/V, V/VI, and VI/VII. A detailed description of the adult morphology of heteropterous insects is found in Schuh and Slater (1995). The eggs may be elongate, cylindrical, curved, or barrel-shaped; they are inserted into a plant tissue, cemented to a substratum, or laid free. A distinct operculum may be present, the same applying to a true egg burster. The cephalic pole of the egg of most bugs bears one or two micropyles or aero-micropyles (apparently with gaseous exchange function). Aero-micropylar processes ornament the egg pole; in all bugs except Cimicoidea, the sperm enters the egg during fertilization through the micropyles. There are ordinarily five nymphal instars, which typically resemble adults and live in similar environments. Aside from the lack of ocelli, wings, and genitalia, nymphs are primarily distinguished by the presence of one

J. Grazia (✉) • F.L. Simões
Departamento de Zoologia, Instituto de Biociências,
Universidade Federal do Rio Grande do Sul (UFRGS),
Av. Bento Gonçalves 9500, prédio 43435, Bairro Agronomia,
Porto Alegre, RS 91501-970, Brazil
e-mail: jocelia@ufrgs.br; felipe.lorenz@ufrgs.br

A.R. Panizzi
Laboratório de Entomologia, Embrapa Trigo, Caixa Postal 3081,
Passo Fundo, RS 99001-970, Brazil
e-mail: antonio.panizzi@embrapa.br

A.R. Panizzi, J. Grazia (eds.), *True Bugs (Heteroptera) of the Neotropics*,
Entomology in Focus 2, DOI 10.1007/978-94-017-9861-7_2

less tarsal segment than those found in adults. Most heteropterous males apparently deposit sperm in the female in the shape of a spermatophore; the sperm is then stored by the female in the spermatheca. Fertilization and oviposition are simultaneous. Plant-feeding species make up the majority of the Heteroptera. The phytophagous habit has been acquired independently at least twice from predatory precursors over the long millennia in which heteropterans have evolved (Schuh and Slater 1995). Phytophagy is found mostly in the Pentatomomorpha and Cimicomorpha. Considering the heteropteran families, the majority are predaceous upon insects and other arthropods. Some heteropterans have also developed the ability to utilize the blood of vertebrates, a characteristic that has arisen at least three and possibly four times independently in the taxon's evolutionary history.

A brief description of the morphology (item 2), ontogeny and reproduction (item 3), and feeding (item 4) of true bugs is presented. Those issues are discussed in more detail in the chapters of Session II. The taxa are listed in alphabetical order, aiming to make more practical the finding of the information.

2.2 Morphology

2.2.1 Alydidae

The Alydidae are 8–20 mm long, usually elongate and slender with disproportionately large heads. The broad-headed bugs as they are known are very often myrmecomorphic, particularly as nymphs. Additional diagnostic characters include: bucculae very short; antennae dorsally inserted with segment 1 not constricted at base; ocelli not placed on elevations; corium elongated on costal margin; metathoracic scent-gland auricles well developed; and tibia nonsulcate (Schuh and Slater 1995).

2.2.2 Anthocoridae

The anthocorids vary from 1.15 to 5 mm long, with a porrect head and ocelli always present in macropterous forms. The labium is four segmented (even though it appears to be three segmented), and the labrum is wide and flap-like. The metathoracic scent gland is deeply bilobed behind with a median accessory gland present. The forewing has an elongate median corial groove, with a short ventral fold, which does not reach the cuneus along the costal margin. The hind wing can have or lack a hamus. Both macropterous and brachypterous forms occur. The tibiae have spongy fossae, either well developed or vestigial; foretibiae in males are, in most of genera, armed with a row of teeth or spines in the inner margin; and the tarsi have three segments. The male genitalia are always asymmetrical. The ovipositor is usually developed or sometimes reduced.

2.2.3 Aradidae

The Aradidae are 3–11-mm-long insects mainly diagnosed by the mandibular and maxillary stylets, which are extremely elongate and coiled within, and the absence of ocelli. Their head is of the prognathous type, highly specialized to accommodate the Mycetophagus sucking mouthparts, and the labium is usually short and stout with 4 distinct segments. The wings, when present, are usually dimorphic, well developed, and exceedingly diverse in development and venation. Metathoracic scent glands are well developed, and the canals and evaporating areas vary in shape with striking development. The legs in Aradidae are considerably variable and sometimes modified for stridulation; the trochanters are commonly and usually distinct from the femora; the front tibiae have an apical or periapical comb on the inner side; and the tarsi are bisegmented. In the abdomen stridulatory structures are present in the ventral surface; the connexivum, even lateroventrally, is morphologically a part of the tergum, since the spiracles are considered to mark the lower limit of the tergum. The connexiva of the first segment or the lateral parts of the first tergite are completely lost. The modification of abdominal segments into reproductive systems begins in the seventh segment for females and in the eighth for males.

2.2.4 Coreidae

Coreids are usually heavy-bodied insects, elongate or elliptical in shape, ranging from 6 to 40 mm. The head is smaller than the pronotum and the antennae have four segments. Their hind femora may be enlarged, with spines and tubercles, and frequently incrassate, while the hind tibiae are curved, with spines or leaflike expansions; apterous, brachypterous, and macropterous forms occur in the family. Abdominal spiracles ventral; three trichobothria on abdominal segments III–VI, two on segment VII; nymphal dorsal abdominal scent-gland openings between terga 4/5 and 5/6. The genital plates are flattened and the spermatheca has only a proximal pump flange. In several coreid tribes, a pronounced sexual dimorphism occurs in the size and armament of the hind femur (Miller and Emlen 2010).

2.2.5 Cydnidae

With ovoid and convex bodies, the cydnids range from 2 to 25 mm. Their head is either quadrate or semicircular, wide, and often explanate; the antennae have five segments. The scutellum never surpasses the middle of the abdomen. The development of wings is variable with the possibility of brachypterous and macropterous forms occurring in the same species; the last three nymphal instars have wing pads, which become more evident in the 5th. The distal margins of the coxae have a row of setae or bristles, known as coxal combs; the tibia has strong and long spines in all extension, and the anterior tibia is usually compressed; and the tarsi have three segments. Female genital plates: Laterotergites 8 fused; gonocoxites 8 large and broad; the spermatheca is relatively small usually with two flanges.

2.2.6 *Dipsocoromorpha*

Of small size (from 1 to 4 mm) and often strongly declivous head, the Dipsocoromorpha port antennae are with scape and a short pedicel, flagellomeres long, thin and typically with numerous stout setae; the proepisternum is often inflated. One of the most prominent characteristics of the group is wing morphology. While four wing types are found in the taxon (macropterous, submacropterous, brachypterous, and apterous, some of these occurring concomitantly in the same species), wing morphology is even more complex since the forewing is typically not clearly divided into corium and membrane, ranging from completely membranous to entirely elytrous; the hind wing is often seen with deeply separated lobes. The male genitalia are either symmetrical, with appendage-like laterotergites, or strongly asymmetrical.

2.2.7 *Enicocephalomorpha*

Enicocephalomorpha are small to midsize with body length typically ranging between 2 and 15 mm, are often dull colored, and generally resemble small Reduviidae. Wings are membranous and well developed, but micropterous, brachypterous, and wingless forms, especially in females, also occur (Wygodzinsky and Schmidt 1991). They are characterized by the elongate and porrect head that is subdivided into anterior and postocular lobes by a usually conspicuous postocular constriction (Štys 1995).

2.2.8 *Gerromorpha*

Semiaquatic bugs are rather diversified in general appearance and structurally very diverse (Andersen 1979), which makes morphologically defining the group somewhat difficult (Nieser and Melo 1997). The head of the gerromorphans is typically elongate, with prominent eyes; contrary to the Nepomorpha, the antennae of Gerromorpha are always long and exposed (Heckman 2011); and the rostrum is typically long and slender. The pronotum is a broad plate with a short anterior lobe, while the scutellum is sometimes reduced. The wings in Gerromorpha are very particular, since pterygopolymorphism is common in the infraorder, with some species showing continuous variation (Schuh and Slater 1995). The forewings are usually not differentiated into the anterior coriaceous and posterior membranous portions; on the other hand, the hind wings are membranous and shorter but much broader than the forewing. Brachypterous and apterous forms also occur, the latter being more common than macropterous. The scent apparatus of semiaquatic bugs belongs to the "type omphalion" of Carayon (1971), with a single median scent orifice on the metasternum. The legs of Gerromorpha generally have modifications that facilitate movement across the surface of the water, so as to stay over the tension layer, and which can even facilitate copulation. The pretarsi have dorsal and ventral arolia (a synapomorphy of the group); the adult tarsus has three segments, but these can be fused sometimes (Andersen 1982). The abdomen has eight dorsal nonreproductive segments in females and seven in males. Nymphs of semiaquatic bugs differ from adults in their usually softer integument; the metathoracic scent apparatus is not formed until the adult stage.

The male genitalia consist basically of a cylindrical abdominal segment eight, a boat-shaped segment nine (the pygophore), and a lid-shaped segment ten (the proctiger), which covers the posterior end of the pygophore. A relatively simple phallic organ is internally attached to the pygophore, and a pair of parameres arises laterally from it. As for the female genitalia, it is formed by the eighth and ninth abdominal segments. The eighth segment has a basal sclerite or first gonocoxa ventrolaterally on either side, and this sclerite carries ventrally and caudally the first gonapophyses. The second pair of gonocoxae is found on segment nine, to which are attached the second pair of gonapophyses. The male genital segments are often more conspicuous than those of females, which may be concealed.

2.2.9 Lygaeoidea

Being the second largest superfamily of Pentatomomorpha, the Lygaeoidea have a plethora of variable morphological characteristics which can hardly be summarized in such a short manner. Nevertheless, the synapomorphies of the taxon can be highlighted: reduced venation nearly always lacking closed cells on the hemelytral membrane and incrassate fore femora found in all basal taxa but lost in a number of distal groups. Also, the Lygaeoidea can be apterous, sub-brachypterous, brachypterous, and macropterous.

2.2.10 Miridae

Mirids can stretch from one to more than 15 mm in ovoid or elongate bodies, although these are variable in adults. The head is triangular and usually prognathous; the eyes are usually large; and the antennae and labium have four segments. Even though most mirids are macropterous, it is common to find submacroptery, brachyptery, microptery, and coleoptery. The hemelytra are distinctive, with the corium divided distally into a triangular cuneus and an apical membrane with one or two unequal cells divided by a short longitudinal vein. Trochanters of all legs are divided. The tarsi are most often three segmented. The metathoracic scent glands are paired and the external efferent or evaporative system is usually well developed. The male genitalia are asymmetrical, with a partially membranous endosoma, often inflatable, and confined by a sclerotized phallotheca. The female ovipositor is laciniate; the sperm-storage organ is a large, saclike structure.

2.2.11 Nabidae

The Nabidae range from 5 to 15 mm with elongate or stout bodies. The head is also more or less elongated, with a cylindrical posterior portion; the labium is flexible and usually curving, reaching onto the pro- or mesothorax, with four distinct segments; antennal prepedicellite is present, although variable in length. Being bell shaped, the thorax itself has no particular features; however, wing polymorphism is common in many species, particularly in those living at higher latitudes; depending on the type of prey, legs can either be simple or enlarged and armed with heavy spines, forming a formidable opposable grasping apparatus. Worth noticing on the abdomen is the presence of trichobothria in some taxa. The male genitalia are usually symmetrical, with eventual asymmetrical parameres or phallus; Ekblom's organ is also usually present. In females, a reduced laciniate ovipositor is found in some cases.

2.2.12 Nepomorpha

The bodies of the nepomorphans can be ovate and flattened dorsoventrally, globular and rounded, convex on the dorsal surface and straight ventrally, elongated, or elongate-ovoid. Besides the variation of forms and colors, aquatic heteropterans are widely diverse in total body length, with some species measuring a little more than 1 mm (Pleidae), whereas others are more than 100 mm long (Belostomatidae). Particularly, the Nepomorpha have reduced antennae (with one to four segments) and a generally small and robust rostrum (three or four segments); the pronotum is long, covering the meso- and metanotum, and the scutellum is always exposed. According to Schuh and Slater (1995), nepomorphans generally have the following six types of wings: micropterous, staphylinid, brachypterous, coleopteroid, submacropterous, and macropterous, although the former five can be grouped into the brachypterous group. The forelegs are generally modified and may be raptorial with the fore femora enlarged or scoop shaped with the fore tarsi unsegmented. The hind legs are generally fringed with golden setae and in some groups can be slightly flattened. The dorsal and ventral arolia are usually present in adults and nymphs

(Schuh and Slater 1995). Sexual dimorphism is observed in some families: the male genital segments vary from symmetrical to asymmetrical, including modifications, and the female ovipositor is generally not developed.

2.2.13 Pentatomidae

The bodies of most pentatomids are of moderate to large size, ranging in length from 4 to 20 mm, and generally ovoid or broadly elliptical. There are, however, some relatively elongate slender species, most of which are associated with grasses. The antennae are usually five segmented, although some species have four. The scutellum is large and usually triangular or subtriangular, in a shield-like form. The forewings are half coriaceous and half membranous, while the hind wings are completely membranous. The tarsi have three segments (two segmented in Cyrtocorinae). Segments VIII, IX, and X of the abdomen are modified into genital structures: whereas the males possess a pygophore with other internal structures like the phallus, ejaculatory reservoir, and spermatheca, females have external plates (laterotergites and gonocoxites) with a generally platelike ovipositor, never truly laciniate.

2.2.14 Pentatomoid Small Families

Acanthosomatidae

Species included in this family resemble median to large pentatomids and may be recognized by the shorter scutellum, two-segmented tarsi, male abdominal segment 8 large and exposed, and most of the females with Pendergrast's organ present.

Canopidae

Species included in this family are medium-sized (5–7 mm), almost rounded, dorsally convex and ventrally flat convex and ventrally flat, totally black, and shiny, with purple and green reflexes. The scutellum is enlarged, covering entirely the abdomen and most part of the hemelytra (Grazia et al. 2012).

Dinidoridae

Adults have large size (10–30 mm), an ovoid and robust body, and a dark brown or black general color, with red/yellow stripes and maculae usually present. The head is carinate laterally, the buccula short and elevated, and the rostrum reaching the meso- and metacoxa. The humeral angles never developed. The scutellum is triangular, almost as wide as long, never covering the corium, and the apex usually blunt. Hemelytral membrane venation is reticulate. *Dinidor* Latreille is the only genus that occurs in the neotropics; it has antennae with five segments and tarsi with three segments.

Megarididae

Megaridids are small (5 mm or less in length) and coleopteroid in shape. Strongly convex dorsally, the scutellum is enlarged and covering the abdomen and wings. They are dark in color and polished. The anterior margins of the head and pronotum are carinate and the bucculae undeveloped. The antennae are four segmented, with many setae as long as the diameter of segments in females is much longer than in males. The tibia is without spines, and the tarsi are two segmented.

Phloeidae

Their bodies are depressed; the external margins of juga and pronotum and the base of corium and abdomen broadly foliate. Their eyes are divided into dorsal and ventral portions. Its antenna is three segmented, only partially visible dorsally, and hidden below the expanded juga; segment 1 is very long and segment 3 curved. They have peritreme opening near the lateral margin of the pleuron. Its hemelytral membrane is reticulate. The tarsi are three segmented. The abdominal sterna 3–7 with trichobothria arranged longitudinally to the spiracular line. Spiracle 2 is present and partially exposed.

Tessaratomidae

Piezosternum Amyot and Serville, the only genus representative of the family in the neotropical region, has worldwide distribution and includes three endemic species to the neotropics. Similar to large pentatomids, tessarotomids can be distinguished from them by its very small head; usually four-segmented antenna; short rostrum; metasternum produced anteriorly, generally into a large bladelike projection reaching the anterior coxae; and pronotum extending over the base of the scutellum.

2.2.15 Pyrrhocoridae

As the only known neotropical genus of Pyrrhocoridae, *Dysdercus* serves as the reference in this section (some of the characteristics may be shared with other pyrrhocorids).

The body length of *Dysdercus* range from 8 to 18 mm, with an elongate to oval, somewhat flattened, body. The slender and elongated sucking mouthparts reach the middle of the thorax and occasionally the abdomen. The pronotum is clearly divided into an anterior third and a posterior two-thirds, these sometimes of different colors. As in other heteropterans, the hind wings are membranous, but the forewings are partly sclerotized and partly membranous, the latter porting two closed cells. The fore femora are, at most, slightly swollen (usually slender), and the metathoracic scent gland lacks an auricle.

Pyrrhocoridae can be distinguished from Largidae most readily by the presence in the females of the latter of a median split on the seventh sternum, indicative of an elongate or laciniate ovipositor; the ovipositor of Pyrrhocoridae is platelike, and the female's seventh sternum is entire.

2.2.16 Reduviidae

Most Reduviidae are medium to large bugs and often show elongate or ovoid body shapes. Overall, Reduviidae is one of the most morphologically diverse groups of Heteroptera, ranging from delicate and elongate to large and robust. Reduviidae can also mimic other groups of Heteroptera (Stride 1954), as well as other insects, such as Hymenoptera, with a number of species participating in Müllerian mimicry complexes (Haviland 1931; Hogue 1993). Some of the most distinctive characteristics of the assassin bugs are the necklike shape of the head behind the eyes and the labium, which is short, strongly curved, and inflexible (Schuh and Slater 1995). Other relevant body structures include the membrane of hemelytra, usually with two or three elongated cells; the presence of a fossula spongiosa at the apex of the fore and mid tibiae in many taxa; and the presence of Brindley's glands between the metathorax and the first abdominal segment. The female genitalia have a lateral spermathecae; males with the eighth abdominal segment telescoped largely into the seventh segment and usually with symmetrical genitalia (Schuh and Slater 1995; Weirauch 2008).

Triatominae

The subfamily is diagnosed, among other features, by the slender and straight labium of its members and their blood-feeding habits (Lent and Wygodzinsky 1979; Clayton 1990; Weirauch 2008). Their general appearance is similar to that of other reduviids. The length of most species varies from 20 to 28 mm, but *Dipetalogaster maxima* (Uhler) can reach 44 mm in length, and the smallest species is *Alberprosenia goyovargasi* (Martínez & Carcavallo), with 5 mm. The color pattern varies, with an overall black or piceous color and spotted patterns of yellow, brown, orange, or red (Jurberg et al. 2004, 2012, 2013). Opposite to the females, in which the genital plates are simple and extremely homogeneous, the male genital complex of the Triatominae is composed of approximately 15 features, which are highly variable and can be used for generic and specific differentiation (Singh-Pruthi 1925; Dupuis 1955, 1963; Lent and Jurberg 1969, 1975).

2.2.17 Rhopalidae

With a body length ranging from 4 to 15 mm, rhopalids vary greatly in both shape and color. The main characteristics of the head which can be highlighted are the four-segmented antennae, clypeus that surpasses the mandibular plates, and ocelli based on low tubercles. As for the thorax, it is notable that the metathoracic scent-gland openings are commonly, but not always, obsolete or obsolescent, hence the common name "scentless plant bugs"; the corium frequently has large hyaline areas, and the membrane of the forewings always has numerous slender parallel veins arising from a transverse basal vein. Females have inner laterotergites with a platelike ovipositor, whereas the pygophore in males has lateral, median, and paralateral lobes; internally, the spermatheca consists of a round bulb, small pump, and long, generally coiled duct.

2.2.18 Saldidae

Most shore bugs are typically small, ranging from 2 to 8 mm, with an ovate body covered by fine setae that repel water and bind layers of air to the body, which indicates that these insects are able to adapt to periods of submergence. Usually, the eyes of the Saldidae are very large, occupying the side of the head that is usually relatively short and broad with visible antennae. All three basic wing forms (apterous, brachypterous, and macropterous) have been recorded for Saldidae (Polhemus and Chapman 1979). Its legs are usually relatively short, slender, or longer; the hind coxae are large, transverse, and broad coxal cavity in the thoracic pleurae and adapted for hopping or jumping; the femora sometimes are armed with spines; and all tarsi are usually three segmented with apical claws. The male genitalia are symmetrical, and the parameres are hook shaped with a distinct processus. In females, the valvulae of the ovipositor are developed or reduced and platelike; also, a spermatheca is present.

2.2.19 Scutelleridae

Shield bugs represent a diverse pentatomoid family, which are mainly recognized by the beetle-like appearance, due to the great development of the mesoscutellum (scutellum), which covers almost the entirety of the abdomen. Scutellerids vary greatly in shape, color, and size, but they are usually medium-sized pentatomoids (10–20 mm). The main thoracic characteristics are the well-developed prosternal keel, weakly sclerotized hemelytra, ostiolar peritreme (which can be present or absent), and usually unarmed legs. The external genitalia are either visible or concealed by sternite VII; thus, most of the diagnostic genital characters (apart from the genital plates) are internal, such as in females with the spermathecal bulb with a simple wall and the genital capsule with a ringlike 8th segment. In males, the phallus often has two or three pairs of well-developed conjunctival projections, and the parameres are commonly uncinate (Schuh and Slater 1995; Tsai et al. 2011; Barcellos et al. 2014).

2.2.20 *Thyreocoridae*

Thyreocorids are 3–8-mm-long bugs with oval-elongate, dorsally convex, and ventrally flat bodies, which are characteristically dark colored (usually shiny). The head is declivent and subtriangular in dorsal view, and the antennae are five segmented. The main thoracic features include a strongly convex scutellum, which covers most of the hemelytra, a reduced corium, tibiae with setae, and spines in all their extension and three-segmented tarsi.

2.2.21 *Tingidae*

Tingids range from 2 to 8 mm and are poorly colored, generally varying in shades of brown. The head is armed with up to nine spines, which can be distributed in variable ways; the antennae have four segments. The pronotum is convex or flat; composed by a collar, a pronotal disk with or without carinae, and a pair of lateral expansions (paranota); and can have a posterior projection covering the scutellum. The collar can be developed, forming a hood, which can fully or partially cover the head. An important morphological characteristic of most tingids is the lacelike network of areolae on the wings, which is very often used in species identification. The male genitalia are symmetrical and the phallus has some sclerotized eversible structures (Lee 1969); the parameres are usually directed backward. The ovipositor is laciniate and the connection between the first valvula and first valvifer is lost, as in Miridae (Schuh and Slater 1995). The spermatheca is absent and the pseudospermatheca is present, but it seems to be unrelated with the sperm-storage function (Marchini et al. 2010).

2.3 Ontogeny and Reproduction

2.3.1 *Alydidae*

For the neotropical alydids, not much data is available on their biology, with the exception of a few species with economic interest. *Neomegalotomus parvus* (Westwood) eggs are laid singly or in groups; on soybean, eggs are laid on the upper third of the plant, preferably on the lower side of leaves, close to the midrib (Panizzi et al. 1996). On pigeon pea, *Cajanus cajan* (L.) eggs are laid preferably in-between seeds (crevices) of mature pods (Ventura and Panizzi 2000, 2003). Nymphs mimic ants and are darkish.

2.3.2 *Anthocoridae*

The duration of embryonic development after spawning depends greatly on the temperature, while the total duration of the hatching takes a few minutes. The nymphs do not differ greatly from the adult stage, and growth after hatching has no metamorphosis; it is only driven by shedding or ecdysis. The last one, called fledging, is accompanied by some structural completions, including the full development of the wings (for winged forms) and sexual organs. The normal number of nymphal stages is five, but it is possible that some species evolve in just four steps. The sexual organs are externally invisible except at the fifth instar, where the ovipositor of the female is apparent.

Insemination is done through "traumatic insemination," i.e., injection in the hemocoel or in secondarily developed organs and that is not associated with the genital vias, which are only subsequently reached. The way to oviposit varies according to the different development levels of the ovipositor of the females of the different taxa. Fertilization takes place in the vitellarium, where there are two to seven testes follicles.

2.3.3 *Aradidae*

Aradids generally have one or two generations per year with five nymphal stages: eggs are laid in May, and nymphs hatch in June (the time of incubation depends on temperature and species), reach the fourth instar during the first summer, become adults from July to August, and after which they hibernate. It is not until the following spring that the bugs mate and begin laying eggs.

After copulating (when the male lies beneath the female) in their natural habitats, in fungi, and under tree barks, the eggs are deposited in frass beneath the eggs, and are found partially embedded in softwood in the galleries of termites. A single female can lay 14–45 eggs.

2.3.4 *Coreidae*

The number of generations per year varies with latitude and as a function of crop cycles, host plant phenology, and host plant specificity. The eggs are cemented to the substrate by the female and laid either in long chains; singly; in small groups, adjacent rows, or large clusters; or even atop one another. Most eggs are laid on host plants, generally on the undersides of leaves, but also along a linear surface or on the backs of males. Five nymphal stadia occur between hatching and adult emergence. Adults live long, reaching more than 50 days (Mitchell 2006).

Courtship and mating involve antennation, touching with the legs, or other body contact prior to copulation (Eberhard 1998; Tatarnic and Spence 2013; Miller 2011), and males and females may mate multiple times for as long as 6–7 h (McLain et al. 1993; Tatarnic and Spence 2013). Mating often occurs on fruits, and feeding by one or both partners during copulation is common. The most frequent mating position is end to end, with the male facing the opposite direction from the female. Copulatory success appears to correlate with the size of the male (McLain et al. 1993; Tatarnic and Spence 2013).

2.3.5 Cydnidae

The females of Cydnidae can lay a single egg underground (García and Bellotti 1980; Riis et al. 2005), egg masses in shallow cracks on the soil surface (Sites and McPherson 1982), or inside the fruits of *Ficus* (Moraceae) (Mayorga and Cervantes 2001). After that, incubation time can vary from 1 to 4 weeks. Newly hatched nymphs measure from 1.0 to 2.5 mm in length (García and Bellotti 1980; Sites and McPherson 1982; Riis et al. 2005), and their development also occurs in five instars, and, even though the time spent in these stages is mostly unknown, studies found they can last from 54 to as much as 180 days.

Although little is known about the reproduction of neotropical cydnids, for species that live in the soil, mating occurs in this habitat. During copulation, males and females remain connected by the genitalia and in opposite directions (Willis and Roth 1962). Whereas Cydnidae present solitary behavior, mating occurs in the soil and finding sexual partners is difficult, since the environment where these bugs live restricts large-scale communication between individuals. Like other Heteroptera, it is likely that copulation in Cydnidae is mediated by chemicals and sound signals.

2.3.6 Dipsocoromorpha

Not much is known about the development and reproduction of neotropical Dipsocoromorpha, with sparse works mentioning, or dealing with, the five nymphal stages and morphology/coupling of the genitalia.

2.3.7 Enicocephalomorpha

Swarming is a biological phenomenon that is rather common in this group and has been documented for species in five genera and both families in the New World (Wygodzinsky and Schmidt 1991). Štys (1981) speculated that all species that are capable of flight might also swarm. Swarms are possibly formed for the purpose of mating and are composed of large numbers of males and much smaller numbers of females (Kritsky 1977; Štys 1981). Some enicocephalids have the unusual ability to autotomize, or shed, their wings. This behavior may be related to the cryptic way of life of most species. In females, it could be important when individuals retreat, after swarming and copulation, into the soil, where wings might be a hindrance, similar to wing shedding in termites and ants (Wygodzinsky and Schmidt 1991).

2.3.8 *Gerromorpha*

The eggs of gerromorphan bugs are laid in a wide variety of places, such as on moss and algae, in holes of stones, inserted in plant tissue, on floating objects and living or dead leaves, and many other possibilities. One to 500 eggs can be laid at once or through several days, with incubation periods ranging from 6 to 23 days. Only a few works describing the life cycle of nymphs of neotropical Gerromorpha exist, but it is known that they have four to five instars; the development time takes from 14 to 33 days, during which the two wing pairs start developing before the final molt, where the teneral development takes places, hardening and adding pigmentation to the cuticle, resorption of the now useless muscles, and, in macropterous adults, with the growth of the endoskeleton and flight muscles (Andersen 1982).

The reproductive behavior of semiaquatic bugs has been extensively studied, but the vast majority of the experiments are restricted to temperate species. Four basic mating systems have been described among gerromorphans (Spence and Andersen 1994), but it consists of the male jumping on the back of the female and then engaging in the copulation, usually by attaching its legs to the female's thorax subsequently transferring the sperm in a process that can last as quick as minutes or as long as weeks.

2.3.9 *Lygaeoidea*

As mentioned in the morphology section, lygeoids are a diverse, highly successful group of true bugs and there have been a significant number of important reviews of the habits, life history, and reproduction. For further details on one of the 12 neotropical families of Lygaeoidea, please refer to the respective chapter.

2.3.10 *Miridae*

The life histories of most Miridae, especially tropical ones, are poorly studied or are unknown (Schuh and Slater 1995; Wheeler 2001). Much of what we know about courtship and oviposition behavior, fecundity, longevity, and voltinism is the result of studies involving species of agricultural importance. However, it is known that they often have life cycles that last no more than 6 weeks. Firstly, the eggs are inserted, usually, into plant tissue. After eclosion, the transition from the first stage to the final, which can be from three to six (Schuh and Slater 1995; Wheeler 2001), takes 15–30 days.

Plant bugs are remarkable for their genitalic evolution (Kelton 1959; Cassis and Schuh 2012), including traumatic insemination and sexual antagonistic coevolution (Tatarnic et al. 2006; Tatarnic and Cassis 2010; Cassis and Schuh 2012). Parthenogenesis occurs in a few mirid species, such as the predatory *Campyloneura virgula* (Herrich-Schaeffer), where males are rare (Wheeler and Henry 1992; Henry 2012).

2.3.11 Nabidae

Most details provided in this section are related to *Nabis capsiformis* (Germar), which is representative of the family (Hormchan et al. 1976), even though some information on other species is also available (however, very little is known about the development of neotropical nabid species).

Generations can be as long lived as 59 days, depending on the conditions provided. After being laid in plant tissue, eggs (more than 110 can be laid) hatch into nymphs after 7–8 days of development. Then, they proceed through the five instars, a process which can last up to 18 days. Adults can live for 21.6 days under laboratory conditions.

Nabidae reproduction is done mainly through vaginal copulation, with fertilization taking place in the mesodermal oviducts near the base of the ovarioles or near the pedicels. The other possibility is through traumatic intravaginal insemination (Carayon 1977).

2.3.12 Nepomorpha

Nepomorphan eggs are deposited individually or grouped, submerged or slightly submerged above the water surface, or even distant from the water. They can be deposited over the substrate (plant and algae masses, rocks, logs, and the hemelytra of males) or inside it (usually plant tissues). After eclosion, the nymphs develop across four to five instars, which have a similar lifestyle of the adults. The development time and number of generations are not known for most neotropical species, but in the tropics and/or subtropics, the species or populations can be multivoltine, with three or more generations annually (Saulich and Musolin 2007).

Reproduction in nepomorphans generally takes place in water and includes various methods across the families of the infraorder. Stridulation is quite common in these bugs and plays an important role in mating. The males generally make use of different leg pairs to hold the female together for copulation; however, the methods are too variable to be explained in this section (please refer to the Nepomorpha chapter for further details).

2.3.13 Pentatomidae

The pentatomid barrel-shaped eggs are laid on the underside of leaves in clusters with tight rows; in early spring, overwintered adult females seek out suitable hosts and typically deposit their eggs on wild host plants. Often, these overwintering populations are found along field borders, particularly along tree lines near their overwintering sites. Later-developing cultivated plants become more attractive when these initial wild hosts dry down, and their proximity allows easy access for stinkbug colonization in crops; emerging nymphs are gregarious and remain on/near the egg mass and then begin to feed and disperse as they grow. Eggs with translucent chorion, through which it is possible to observe their contents, are more common in Pentatomidae than pigmented chorion; the spinose pattern of chorion sculpture seems to be the most frequent, although several studied species show reticulated chorion and a few others a granulated chorion. The number of aero-micropylar processes is rarely used for specific recognition, due to its overlapping in congeneric species; the number varies between 10 and 86. The number of eggs per mass is relatively constant within Pentatomidae species, related to the number of ovarioles in females and the oviposition strategy of each species. Small egg masses are laid, usually with 14 eggs (Matesco et al. 2007), the number of ovarioles more frequently found in Pentatomidae. Among pentatomids, several life history traits are strongly influenced by temperature, humidity, and food quality. The duration in days of the development from the egg to adulthood is about 48–57.

2.3.14 Pentatomoid Small Families

Acanthosomatidae

Faúndez (2007a) gives a brief description of the life cycle of *D. punctiventris*, providing details of the mating behavior (Faúndez 2007a; Faúndez et al. 2009). Females lay eggs in clutches, corresponding more or less with the number of ovarioles. Egg and nymphal development time, prereproductive females, and adult longevity have not been studied in any of the neotropical species. Faúndez and Osorio (2010) described maternal care behavior for the neotropical species *Sinopla perpunctatus* Signoret. The authors also reported coloration change in the female associated to the reproductive period and guard of eggs and nymphs.

Canopidae

The biology of the canopids is poorly known.

Dinidoridae

Eggs or nymphs of any species of the neotropical Dinidoridae have not been formally described up to date. Knowledge is restricted to species from other regions (Malipatil and Kumar 1975; Danielczok and Kocorek 2003). Eggs and nymphs of *D. mactabilis* were illustrated in Schwertner and Grazia (2014). The unique characteristics of the immature include the shape of the eggs and of the aero-micropylar processes, aspect of the chorion surface, and the development of some nymph structures. Eggs are cylindrical, deposited in rows, horizontally positioned to the substrate.

Megarididae

The general body of nymphs is similar to adults in shape and color; they are heavily sclerotized, with abdominal terga not distinguished; and there is no formal description of eggs.

Phloeidae

The life cycle of these bugs occurs entirely on the trunks of the host (Salomão et al. 2012). Females protect eggs and early nymphs; the latter attach themselves to the abdomen of the mother (Lent and Jurberg 1965; Guilbert 2003).

Tessaratomidae

Eggs are spherical, deposited perpendicular to the substrate. The chorion is smooth and thinner than in Dinidoridae. Aero-micropylar processes are reduced, knob-like, and circulating the middle region of the egg, in variable numbers (30–60). The operculum is not apparent, with the line of the operculum dorsolateral extending basally. Nymphs of the 1st, 2nd, and 3rd instars have oval, flattened body; late instars have elongated and more robust body. All species are univoltine and usually use more than one host plant throughout its life cycle (Schaefer et al. 2000; Dzerefos et al. 2009), with biological aspects very similar to Dinidoridae. Depending on the species, females can lay in masses of four rows (3-4-4-3 formula), with an average of 14–28 eggs per clutch. Fertility is considered low (Schaefer et al. 2000). Maternal care has been described for some tessaratomids of the subfamily Oncomerinae in the Australian region (Gogala et al. 1998; Monteith 2006).

2.3.15 *Pyrrhocoridae*

As mentioned above, *Dysdercus* serves as the model genus for this section. These bugs' biology and ecology remain poorly known.

2.3.16 *Reduviidae*

Through their mostly platelike ovipositors, reduviids usually glue their eggs to the plant, often in a group, and sometimes cover them a gelatinous material. However, few species are known to guard their eggs. Ground-dwelling species may insert their eggs into the soil, or the eggs may be laid loose. Some reduviines have more well-developed ovipositors, suitable for the insertion of eggs into cracks or crevices.

Triatominae

Triatominae species are found in almost any habitat offering a degree of climatic stability and access to a blood source. Most of them tolerate a range of air humidity between 30 and 80 %, and temperatures of 24–28 °C are satisfactory. Their development is usually slow at temperatures below 16 °C, whereas temperatures above 40 °C are lethal. During the rigorous winters of Patagonia, they undergo lethargy, and on strong summers they increase the number of blood meals in order to compensate the higher loss of water. In general, most triatomines are nocturnal, and during the day they remain in their resting places, although they may sometimes go out to suck blood during the day under adverse conditions. In general, the eggs of the majority of triatomine species are deposited free in the environment, although some species have an adhesive substance that makes the eggs stick to the substrate. The triatomines are parasites with slow development, whose life cycle from egg to adult includes five nymphal stages, ranging from 3 to 4 months in *Rhodnius prolixus* up to 2 years in *Panstrongylus megistus*. This life span can vary considerably according to the temperature to which the bugs are submitted. The first information on copulation in triatomines was published by Neiva (1914), who noted that *Panstrongylus megistus* females appeared to copulate only once, maintaining the eggs fertile throughout their life span. Courting is not complex in this group of insects, and copulation in some species has been observed in the laboratory by some authors. Prior to copulation, the male approaches the female, attempting to immobilize her with the three legs on one side of the body in a dorsolateral position (Abalos and Wygodzinsky 1951; Hack and Bar 1979; Lent and Wygodzinsky 1979; Lima et al. 1986; Rojas et al. 1990; Manrique and Lazzari 1994).

2.3.17 *Rhopalidae*

Since detailed studies of eggs and oviposition have not been made for most neotropical rhopalids, this section is based mainly in temperate and subtropical congeners and conspecifics. Rhopalids typically lay their eggs on or near host plants, more specifically on sepals, nearby stems, leaf midribs, floral pappi, the undersides of leaves, and flower pedicels, but they can also be buried in the soil or simply dropped to the ground; while as many as 1,000 eggs can be laid by a single female (Wheeler 1977), they require about 8 days to hatch over an average monthlong oviposition period (Yonke and Walker 1970a, b). As with the eggs, not much is known about neotropical rhopalid nymphs, but the relative genera from other regions have a mature period through the five instars that ranges from 22 to 40 days (Yonke and Walker 1970a, b; Paskewitz and McPherson 1983). Adult males have been reported to live for as long as the average life span of 86,6 days, while females lived for up to 47,7 days.

Mostly, what is known of the reproduction of rhopalids focuses on the consequences of low food availability to reproductive life span and diapause; however, this is not the main focus of this section. Apart from that, Carroll (1988, 1991) reported two different mating systems: promiscuous mating, when females are abundant, and a guarding system when males are much more abundant than females.

2.3.18 *Saldidae*

Even though the life cycles of Saldidae are not yet completely understood, some generalizations can be made. The eggs are laid in over the vegetation or other substrates. Their maturation requires approximately 1 week, and the hatched nymphs will develop across the five instars in around 2–3 weeks (Wiley 1922; Jordan and Wendt 1938). There are some cold-adapted species and others that can overwinter in the egg or adult stages (Wroblewski 1966).

Saldidae usually copulate side by side (Polhemus and Chapman 1979), and according to Cobben (1957) the female's forewing and male abdominal segments show some modifications to facilitate this copulatory position.

2.3.19 *Scutelleridae*

Even though the length of life cycles in Scutelleridae is quite variable, some model studies do exist, where total cycles lasted for as long as 60 days. Scutellerid eggs are laid in two or more rows, on several parts of a host plant. These egg masses vary in number, averaging from 14 to almost a hundred eggs (around 75–92); the incubation time is not very well known, but a 7-day period has been recorded for one species (Paleari 1992). After eclosion, the nymphs proceed through five instars until reaching the adult stage.

2.3.20 *Thyreocoridae*

The life cycle of the Thyreocoridae generally corresponds to that of the Pentatomidae (Grazia and Schwertner 2008). They can be uni-, bi-, or multivoltine: in low latitudes, a higher number of generations per year occur, whereas in higher latitudes only one generation per year is possible (McPherson 1972; Lung and Goeden 1982; Bundy and McPherson 1997, 2009). Eggs are laid singly and glued laterally to the substrate, often in the host plants' reproductive parts and less frequently in branches or leaves; the development varies between 8 and 11 days. After eclosion, the first instar nymphs are active, differing from the nymphs of other Pentatomoidea families, feeding and not showing gregarious behavior; the subsequent development through the five nymphal stages takes from 30 to 45 days.

Not much can be said about the reproduction of thyreocorids, but a pre-copulatory behavior has been described by Bundy and McPherson (1997), and it consists on the male standing behind the female and starting the contact through antennation. After the female's acceptance, the copulation takes place with the individuals positioned in opposite directions, just touching genitalic structures; this process can last for hours.

2.3.21 *Tingidae*

In general, tingid eggs have been relatively poorly studied; thus, not much is known about their egg development. However, some studies delved into the oviposition sites and strategies. Generally, there are one or two generations per year (univoltine or bivoltine), although multivoltine species are known (Neal and Douglas 1990). Oviposition sites are variable, as species drop their eggs on roots, stems, bud flowers, and/or leafs. The eggs can be inserted on the spongy mesophyll, partially inserted in the vegetal tissue (pseudo-endophytic oviposition), or placed at the surface of the vegetal tissue (exophytic oviposition). They are usually oviposited in the abaxial surface of the leaf and can be arranged in small or big groups, or they can be isolated. Although most species have five nymphal instars, four-instar cycles have already been reported (Johnson 1936). Apart from that, tingid nymphs have yet to be further explored, since only recently they started to be featured in the specialized literature (e.g., Guilbert and Montemayor 2010; Guidoti and Barcellos 2013; Montemayor 2009; Montemayor and Dellapé 2010; Montemayor et al. 2011), and their function and evolution were just studied in a few occasions (Guilbert 2004; Scholze 1992).

As with the life cycle, the reproductive behavior of the Tingidae has been seldom explored, with no courtship behavior reported.

2.4 Feeding

2.4.1 Alydidae

The alydids are phytophagous, feeding on noncultivated and cultivated legumes, solanaceas, and other plant families. Research work on their biology, particularly on their life history in nature, to reveal their association with host plants either cultivated or not is needed.

2.4.2 Anthocoridae

Anthocoridae are chiefly predaceous bugs that occur in a variety of habitats where they feed on different arthropods, including mites and insects. Other examples of food source not so often found in Anthocoridae are plants (forbs, shrubs, and trees), ants, and litter.

Nymphal food does not differ substantially from that of adults. The young Anthocoridae, after hatching, are voracious predators of small arthropods and they attack the same prey as that of imagos, preferring only individuals proportional to their size (eggs, young larvae). The number of prey consumed can reach or exceed several hundred mites or a hundred aphids during the nymphal development of a single individual. It seems at least one meal is needed between two molts, and the most hearty meal precedes the ecdysis. Growth is sometimes compatible with a not predatory feeding habitus (Péricart 1972).

2.4.3 Aradidae

Most aradid species are subcortical and mycetophagous, mainly feeding on the bark of dead trees and fungi juices. They are mostly found on branches that have died recently and are not totally dessicated. When the tree becomes too dry, they apparently abandon it for another branch. They have elongated maxillary and mandibular stylets that are coiled within the head; in the resting position, both stylets are coiled up to form a compact double spiral (Weber 1930), and when the bug is feeding, the length of the uncoiled stylets may reach as much as five to six times that of the insect. Nymphs are found with the adults and feed from the same sources.

2.4.4 Coreidae

Coreids are phytophagous, feeding on gymnosperms and angiosperms, through the phloem, buds, fruits, developing seeds, and mature seeds (Mitchell 2000); no coreids feed exclusively from xylem, but hydration from xylem occurs. Opportunistic feeding on fecal matter, carrion, and insect eggs has been reported (Adler and Wheeler 1984; Menezes-Netto et al. 2012) but is never the primary source of nutrients. Plant preference spans a wide spectrum from specialization on a single genus to extreme polyphagy. Nymphs of the first instar often remain in the vicinity of the egg mass and appear to need only a source of water for successful molt to the second stadium (Cook and Neal 1999). Probing into plant tissue may occur, but the tissue selected is always vegetative (tendrils, leaves, terminal buds) (Rodrigues et al. 2007; Wheeler and Miller 1990). The second through fifth instars require a nutrient source and feed similarly to the adults in most cases, although the shorter stylets may prevent access to deeper plant tissues.

2.4.5 Cydnidae

Cydnid nymphs are typically polyphagous and likely feed on plants closer to the site of hatching. Feeding of host plants through suction, for both nymphs and adults, includes roots, ground pods, aboveground structures, fallen seeds, and fruits. Plants of several families have been reported as hosts (Timonin 1958; Becker 1967; Riis et al. 2005; Mayorga and Cervantes 2001; Chapin et al. 2006), ranging from small-sized plants, such as *A. hypogea*, to larger-sized ones, such as *Eucalyptus* spp. Very often these insects are reported as agricultural pests and, depending on the population level, the symptoms of their occurrence can be checked on the shoots, whose leaves turn yellow and dry (Gallo et al. 2002; Riis et al. 2005; Oliveira and Malaguido 2004; Pereira et al. 2012).

2.4.6 Dipsocoromorpha

Given the small size and cryptic habits of most Dipsocoromorpha, direct observations of their biology are limited. However, it is known that they can feed on chironomid larvae (Esaki and Miyamoto 1959), Collembola, mites, and other Dipsocoromorpha.

2.4.7 Enicocephalomorpha

Enicocephalids are predaceous and appear to feed on soft-bodied insects and other arthropods, but few direct observations are available. In captivity, they were fed with larvae and freshly killed adult flies, symphylids (Symphyla), and campodeids (Diplura), as well as ant larvae and pupae.

2.4.8 *Gerromorpha*

In general, semiaquatic bugs are intermediate opportunistic stage predators (polyphagous carnivores and scavengers) in the food chains of their communities, normally attacking small invertebrates that fall on the water surface and become trapped or those who swim up to it (Menke 1979). Many species of Gerromorpha are also known to be cannibalistic, feeding upon weaker individuals of their own kind (Butler 1923; Sprague 1956; Andersen 1982). The main sources of food for nymphs and adults are dead or half-dead midges, mosquito larvae and pupae, ostracods, cladocerans, ephemeropterans, and collembolans, while the adults can also feed on pelagic cnidarians, sea anemones, planktonic crustaceans, and fish larvae.

As mentioned above, cannibalism is very common in the infraorder, to the point where some (if not most) experiments with live cultures can be completely lost due to this behavior.

2.4.9 *Lygaeoidea*

Most lygaeids feed on seeds or plant sap, whereas only a few are predatory. They have piercing-sucking mouthparts, which are made up of the mandibles and maxillae modified to form needlelike stylets lying within a grooved labium (Schuh and Slater 1995). Feeding methods can be divided into two "types": "stylet-sheath" feeders and "lacerate-flush" feeders (Schuh and Slater 1995; Sweet 2000). The majority of Lygaeoidea are lacerate-flush feeders, a method commonly used by species that feed on portions of the plant rich in nutrients, such as seeds (Schuh and Slater 1995), whereas the families Blissidae, Malcidae, and Colobathristidae are predominantly sap feeders. The lygaeids are often related as pests of agriculture crops (Sweet 2000).

Specific food sources and modified strategies for the neotropical families can be found in the Lygaeoidea chapter in this book.

2.4.10 *Miridae*

Most plant bugs are oligophagous, occurring on members of one plant family or on plants belonging to a few closely related families (Wheeler 2000), but their piercing-sucking mouthparts and the presence of powerful salivary secretions also allow access to different nutrient resources, such as dried carrion and feces (Wheeler 2001). Even though mirids are often not recognized as predators, specially due to the lack of specialized morphological characteristics for preying, plant bugs are frequently causing severe injuries in crops and have recently invaded several areas across the world due to changes in agricultural practices that lead to the increase of the host range of many plant bugs (Wheeler 2001).

2.4.11 Nabidae

The nabids are usually described as pests of cotton or other important agroecosystems. However, they can get their nutrients from other insects as well. Unfortunately, there's not much information on their feeding sources.

2.4.12 Nepomorpha

Both nymphs and adults are predaceous. They feed on crustaceans, dipteran larvae, and other insects, sometimes even on small vertebrates, such as fishes, tadpoles, salamander and freshwater snakes (Menke 1963), but usually on matter found in deposits on the bottom of the pool and upon the dead leaves that have lodged in its shallow waters, such as diatoms, desmids, *Oscillatoria*, and spores of various algae. Cannibalism is sometimes observed in the taxon, specially from starved individuals feeding on nymphs.

2.4.13 Pentatomidae

The majority of the Pentatomidae is herbivorous, but members of one subfamily (Asopinae) are predaceous on other insects. Both adults and nymphs of plant-feeding species may damage plants, mostly by piercing the plant tissues and thus opening a path for pathogens to enter the plant. Independent of the food source, most species are generalists.

2.4.14 Pentatomoid Small Families

Acanthosomatidae

All known acanthosomatids are phytophagous, usually recorded in more than one host, which includes trees and shrubs (Kumar 1974; Schaefer and Ahmad 1987; Faúndez 2007b, 2009). The feeding sites of acanthosomatids include the young tissues of the host and reproductive parts (Schaefer and Ahmad 1987; Faúndez 2007b, c). Casual records of feeding on decaying organic matter and predation including cannibalism (Miller 1971) are known, probably related to the shortage of suitable host plants (Schaefer and Ahmad 1987). Polyphagy seems to be the rule, but specializations can occur at different levels (Schaefer and Ahmad 1987; Faúndez 2007c, 2009). For instance, *Sinopla perpunctatus* is registered only in *Nothofagus* spp. in Chile and is considered a specialized species (Faúndez 2007c; Osorio 2009). Otherwise, *Acrophyma cumingii* (Westwood) and *Ditomotarsus punctiventris* Spinola are considered generalists (Faúndez 2007a, 2009).

Canopidae

Nymphs and adults of *Canopus* spp. were registered on fungi (McHugh 1994).

Dinidoridae

All species studied so far are exclusively phytophagous, feeding both in reproductive and vegetative parts of their hosts. The data indicate that polyphagy is widespread, but monophagy in some species is likely. Some species may exhibit gregarious behavior during a part of the year, being found in large quantities together on the host, i.e., *Dinidor mactabilis* which was recorded in *Smilax japecanga* Grisebach (Smilacaceae) in southern Brazil (Grazia et al. 2012).

Megarididae

The biology and ecology of the megaridids are completely ignored; it is accepted that all species are exclusively phytophagous.

Phloeidae

The unique cryptic morphology of the phloeids makes these bugs to be confused with the bark of trees where they live (Lent and Jurberg 1965; Grazia et al. 2012). Nymphs and adults feed on the vascular system of the hosts (Bernardes et al. 2005).

Tessaratomidae

All species studied so far are exclusively phytophagous, feeding on both reproductive and vegetative parts of their host. Polyphagy seems to be the rule, but as for Dinidoridae, some species may show host preference [e.g., *Musgraveia sulciventris* (Stål) feeds primarily on plants of the family Rutaceae and is considered pest in the Australian region] (Schaefer and Ahmad 1987; Schaefer et al. 2000).

2.4.15 Pyrrhocoridae

Pyrrhocoridae often feed on the ripening seeds, or secretions on the seeds, of common, fairly low plants (mostly members of the order Malvales). Since they basically feed on a restricted number of sources, some genera or species can be of economic importance, like *Dysdercus* that become serious pests of cotton. Other food sources include wheat, water and energy from the nectar of various flowers and from probing fruits (Myers 1927; Van Doesburg 1966), and even other insects (Schouteden

1912; Kamble 1971). Citrus seems to be such a favored plant (Moizant and Téran 1970, Riley & Howard in Myers 1927), but it may only reflect the likelihood that citrus groves often lay near cotton fields.

2.4.16 Reduviidae

As the common name implies, the assassin bugs are strictly predaceous. Although many Reduviidae appear to feed on a wide variety of arthropods, they sometimes show prey preferences or even prey specializations (see summary in Hwang and Weirauch 2012). Some of these preferences include millipedes (Cachan 1952; Forthman and Weirauch 2012), termite predators (McMahan 1983), ants (Jacobson 1911), spiders (Wignall and Taylor 2008; Jackson et al. 2010), and other reduviids (Bérenger and Pluot-Sigwalt 2009).

Triatominae

Triatomine bugs are obligatory hematophagous insects in all stages of their development and in both sexes, requiring numerous blood meals to complete their development. The amount of blood ingested varies according to the species as well as in relation to the life stage, and generally the 4th and 5th instar nymphs are the ones that feed the most. The vast majority of species feed on the blood of mammals or birds, but some can feed on reptile or amphibian blood (Carcavallo et al. 1998/1999). Coprophagy (ingestion of feces), kleptohematophagy (sucking blood already ingested by another triatomine), and hemolymphagy (sucking hemolymph from other arthropods) have also been reported by various authors (Lafont 1912; Brumpt 1914; Lent and Martins 1940; Wood 1941; Ryckman 1951; Sandoval et al. 2000). Resistance to long periods of fasting and the fact that many species are generalists favor their survival in nature.

2.4.17 Rhopalidae

The rhopalids are exclusively plant-feeding bugs, which specialize on particular host plant taxa as food resources, mainly herbs and woody plants. However, even specialists may take water or nutriment from other sources, including flowers, sweet fruits, and dead insects, which may sometimes lead to confusion about the definitive host on which they depend for growth and reproduction. Usually, Rhopalidae are found hosting on the seeds of plants of the family Sapindaceae. Often, groups of nymphs are found feeding on a single seed, forming a cluster around the seed; feeding on seeds of sapindaceous canopy and light gap lianas in aggregations may exceed a million individuals in number (Carroll and Loye 1987; Wolda and Tanaka 1987). Only some genera, like *Boisea*, have been reported as nuisance pests (e.g., Schowalter 1986).

2.4.18 Saldidae

Most members of Saldidae are predaceous, feeding on small insects (Brooks and Kelton 1967), mainly Diptera larvae, other saldids (Wroblewski 1966), and even their own eggs (Rimes 1951). Some species can adapt to scavenger or carnivore habits (Merritt et al. 2008). After catching the prey, the rostrum is inserted and proteases enzyme are injected to immobilize and liquefy the body contents before they can be sucked up as in several other predaceous Heteroptera (Picado 1937, 1939; Cheng 1967; Pereira and Melo 1998).

2.4.19 Scutelleridae

Scutellerids are generally considered to be exclusively phytophagous, although some authors consider these bugs generalists (Javahery et al. 2000), with extreme examples, such as when some were found feeding on a decomposing snake (Chérot et al. 1998) and on putrefied fish carrion baits (Eger, personal observation). Stems, leaves, and flowers may be used as food although fruit and seeds seem to be most commonly used. Despite their mostly polyphagous habit, several species have been reported to prefer plants from the family Euphorbiaceae. Other known host plant families include Malvaceae and Lamiaceae. The first instar nymphs do not feed, while the other instars feed on the very same plants where their eggs were deposited.

2.4.20 Thyreocoridae

Being exclusively herbivorous, thyreocorids are generally associated to bushes, feeding and living in the reproductive parts of the host plants. Even though most species are restricted to one or two plants (Biehler and McPherson 1982; Bundy and McPherson 1997), up to ten different plant families hosting a single species have been recorded.

2.4.21 Tingidae

These exclusively phytophagous bugs are usually found on the undersides of leaves (with some exceptions), where they feed on the sap of living plants by piercing the epidermis with their very slender stylets. These stylets are protrusile and retractile, and can easily penetrate the cellular tissue to extract the sap. Their feeding activities may cause great injury and plasmolysis of the foliage. Many cultivated and wild plants of prime importance in agriculture and horticulture are seriously affected by the feeding activities of these insects (Drake and Ruhoff 1965). Some non-neotropical genera have been reported to induce galls and feed on pollen and moss (Péricart 1983).

2.5 Concluding Remarks

Morphological studies on the neotropical Heteroptera are increasing in the last four decades specially for the groups with economic interest. Considering the neotropical fauna, the knowledge of the ontogeny, reproduction, and feeding is still incipient.

References

Abalos JW, Wygodzinsky P (1951) Las Triatominae Argentinas (Reduviidae, Hemiptera). Publ Inst Med Reg 601:1–179

Adler PH, Wheeler AG Jr (1984) Extra-phytophagous food source of Hemiptera-Heteroptera: bird droppings, dung, and carrion. J Kans Entomol Soc 57:21–27

Andersen NM (1979) Phylogenetic inference as applied to the study of evolutionary diversification of semiaquatic bugs (Hemiptera: Gerromorpha). Syst Zool 28:554–578

Andersen NM (1982) The semiaquatic bugs (Hemiptera, Gerromorpha). Phylogeny, adaptations, biogeography and classification. Entomonograph 3:1–455

Barcellos A, Eger J Jr, Grazia J (2014) Scutelleridae. In: Roig-Juñent S, Claps LE, Morrone JJ (eds) Biodiversidad de Artrópodos Argentinos, vol 3. Sociedad Entomologica Argentina, Mendoza (in press)

Becker M (1967) Estudos sobre a família Scaptocorinae na região neotropical (Hemiptera: Cydnidae). Arq Zool 15:291–325

Bérenger J-M, Pluot-Sigwalt D (2009) Notes sur *Micrauchenus lineola* (Fabricius 1787), espèce termitophile et termitophage (Heteroptera: Reduviidae: Harpactorinae, Apiomerini). Ann Soc Entomol Fr (n s) 45:129–133

Bernardes JL, Grazia J, Barcellos A, Salomão AT (2005) Descrição dos estágios imaturos e notas sobre a biologia de *Phloea subquadrata* (Heteroptera, Phloeidae). Iheringia Sér Zool 95:415–420

Biehler JA, McPherson JE (1982) Life history and laboratory rearing of *Galgupha ovalis* (Hemiptera: Corimelaenidae), with descriptions of immature stages. Ann Entomol Soc Am 75:465–470

Brooks AR, Kelton LA (1967) Aquatic and semiaquatic Heteroptera of Alberta, Saskatchewan and Manitoba (Hemiptera). Mem Ent Soc Can 51:3–92

Brumpt E (1914) Importance du cannibalisme et coprophagie chez les Réduvidés hématophàges (*Rhodnius* et *Triatoma*) pour la conservation des trypanosomas pathogenes en dehors de l'hote vertebré. Bull Soc Pathol Exot 7:702–705

Bundy CS, McPherson JE (1997) Life history and laboratory rearing of *Corimelaena obscura* (Heteroptera: Thyreocoridae) with descriptions of immature stages. Ann Entomol Soc Am 90:20–27

Bundy CS, McPherson JE (2009) Life history and laboratory rearing of *Corimelaena incognita* (Hemiptera: Heteroptera: Thyreocoridae), with descriptions of immature stages. Ann Entomol Soc Am 102:1068–1076

Butler EA (1923) A biology of British Hemiptera-Heteroptera. Witherby, London

Cachan P (1952) Etude de la prédation chez les Réduvides de la région Éthiopienne. 1. La prédation en groupe chez *Ectrichodia gigas* H-Sch. Phys Comp Oecol 2:378–385
Carayon J (1971) Notes et documents sur l'appareil odorant métathoracique des hémiptères. Ann Soc Entomol Fr (N S) 7:737–770
Carayon J (1977) Insémination extra génitale traumatique. In: Grassé PP (ed) Traité de Zoologie, Anatomie, Systématique, Biologie, T. VIII, fasc. V-A. Masson, pp 351–390
Carcavallo RU, Galíndez-Girón I, Jurberg J, Lent H (eds) (1998/1999) Atlas of Chagas disease vectors in the Americas. Editora Fiocruz, Rio de Janeiro
Carroll SP (1988) Contrasts in reproductive ecology between temperate and tropical populations of *Jadera haematoloma*, a mate-guarding hemipteran (Rhopalidae). Ann Entomol Soc Am 81:54–63
Carroll SP (1991) The adaptive significance of mate guarding in the soapberry bug, *Jadera haematoloma* (Hemiptera: Rhopalidae). J Insect Behav 4:509–530
Carroll SP, Loye JE (1987) Specialization of *Jadera* (Hemiptera: Rhopalidae) on the seeds of Sapindaceae (Sapindales), and coevolutionary responses of defense and attack. Ann Entomol Soc Am 80:373–378
Cassis G, Schuh RT (2012) Systematics, biodiversity, biogeography, and host associations of the Miridae (Insecta: Hemiptera: Heteroptera: Cimicomorpha). Annu Rev Entomol 57:377–404
Chapin JW, Sanders TH, Dean LO, Hendrix KW, Thomas JS (2006) Effect of feeding by a burrower bug, *Pangaeus bilineatus* (Say) (Heteroptera: Cydnidae), on peanut flavor and oil quality. J Entomol Sci 41:33–39
Cheng L (1967) Studies on the biology of Gerridae (Hem., Heteroptera). I: Observations on the feeding of *Limnogonus fossarum* (F.). Entomol Mon Mag 102:121–129
Chérot F, Pauwels OSG, Eger JE, Chimsunchart C (1998) Unusual feeding behavior in a scutellerid: a case of scavenging on snake. Nat Hist Bull Siam Soc 46:207–208
Clayton RA (1990) A phylogenetic analysis of the Reduviidae (Hemiptera: Heteroptera) with redescription of the subfamilies and tribes. Doctor thesis, The George Washington University, Washington, DC, USA
Cobben RH (1957) Beitrag zur Kenntnis der Uferwanzen (Hem. Het. Fam. Saldidae). Ent Berichten 17:245–257
Cook CA, Neal JJ (1999) Feeding behavior of larvae of *Anasa tristis* (Heteroptera: Coreidae) on pumpkin and cucumber. Environ Entomol 28:173–177
Danielczok T, Kocorek A (2003) External morphology of four African species of *Coridius* Ill. (Heteroptera: Pentatomoidea: Dinidoridae). Pol J Entomol 72:63–73
Drake CJ, Ruhoff FA (1965) Lacebugs of the world: a catalog (Hemiptera: Tingidae). US Nat Mus Bull 243:1–634
Dupuis C (1955) Les genitalia des Hemipteres. Mem Mus Nat Hist Nat Ser A Zool 6:183–278
Dupuis C (1963) Progrés récents de l'étude des genitalia des Hétéroptères (etude bibliographique critique). Doctor thesis, Museum National d'Histoire Naturelle
Dzerefos CM, Witkowski ETF, Toms R (2009) Life-history traits of the edible stinkbug, *Encosternum delegorguei* (Hem., Tessaratomidae), a traditional food in southern Africa. J Appl Entomol 133:749–759
Eberhard WG (1998) Sexual behavior of *Acanthocephala declivis guatemalana* (Hemiptera: Coreidae) and the allometric scaling of their modified hind legs. Ann Entomol Soc Am 91:863–871
Esaki T, Miyamoto S (1959) A new or little known *Hypselosoma* from Amami-Oshima and Japan, with the proposal of a new tribe for the genus (Hemiptera). Sieboldia 2:109–120
Faúndez EI (2007a) Lista de las especies de Acanthosomatidae (Insecta: Hemiptera) de la región de Magallanes (Chile). An Ins Patagonia 35:75–78
Faúndez EI (2007b) Notes on the biology of *Ditomotarsus punctiventris* Spinola, 1852 (Hemiptera: Acanthosomatidae) in the Magellan region, and comments about the crypsis in Acanthosomatidae. An Ins Patagonia 35:67–70

Faúndez EI (2007c) Asociación críptica entre *Sinopla perpunctatus* Signoret, 1863 (Acanthosomatidae: Hemiptera) y el Ñirre *Nothofagus antarctica* (G. Forster) Oersted (Fagaceae) en la Región de Magallanes (Chile). Bol Soc Entomol Aragonesa 40:563–564

Faúndez EI (2009) Contribution to the knowledge of the genus *Acrophyma* Bergroth, 1917 (Hemiptera: Heteroptera: Acanthosomatidae). Zootaxa 2137:57–65

Faúndez EI, Osorio GA (2010) Contribution to the knowledge of *Cylindrocnema plana* Mayr, 1864 (Hemiptera: Heteroptera: Acanthosomatidae). An Ins Patagonia 38:159–161

Faúndez EI, Osorio GA, Bahamondez CP, Monsalve EA (2009) Comportamiento gregario en los hábitos sexuales de *Ditomotarsus punctiventris* Spinola, 1852 (Hemiptera: Heteroptera: Acanthosomatidae). Bol Biodiversidad Chile 1:15–18

Forthman M, Weirauch C (2012) Toxic associations: a review of the predatory behaviors of millipede assassin bugs (Hemiptera: Reduviidae: Ectrichodiinae). Eur J Entomol 109:147–153

Gallo D, Nakano O, Silveira-Neto S, Carvalho RPL, Baptista GC, Berti Fo E, Parra JRP, Zucchi RA, Alves SB, Vendramim JD, Marchini LC, Lopes JRS, Omoto C (2002) Entomologia agrícola. Fealq, Piracicaba

García CA, Bellotti AC (1980) Estudio preliminar de la biología y morfología de *Cyrtomenus bergi* F. Nueva plaga de la yuca. Rev Colomb Entomol 6:55–61

Gogala M, Yong H-S, Brühl C (1998) Maternal care in *Pygoplatys* bugs (Heteroptera: Tessaratomidae). Eur J Entomol 95:311–315

Grazia J, Cavicchioli RR, Wolff VRS, Fernandes JAM, Takiya DM (2012) Hemiptera. In: Rafael JA, Melo GAR, Carvalho CJB, Casari S (eds) Os insetos do Brasil: Diversidade e taxonomia. Editora Holos, Ribeirão Preto, pp 348–405

Guidoti M, Barcellos A (2013) On the nymphs of lantana lace bug *Teleonemia scrupulosa* Stål (Hemiptera: Heteroptera: Tingidae: Tinginae): ontogenetic features of integumentary structures highlighted. Zootaxa 3613:289–296

Guilbert É (2003) Habitat use and maternal care of *Phloea subquadrata* (Hemiptera: Phloeidae) in the Brazilian Atlantic forest (Espirito Santo). Eur J Entomol 100:61–63

Guilbert É (2004) Do larvae evolve the same way as adults in Tingidae (Insecta: Heteroptera)? Cladistics 20:139–150

Guilbert É, Montemayor SI (2010) Tingidae (Insecta, Heteroptera) from the Argentinan Yungas: new records and descriptions of selected fifth instars. Zoosystema 32:549–565

Hack WH, Bar ME (1979) Comportamento sexual de *Triatoma infestans* en laboratório. Facena 3:105–117

Haviland MD (1931) The Reduviidae of Kartabo Bartica District, British Guiana. Zoologica 7:129–154

Heckman CW (2011) Encyclopedia of South American aquatic insects: Hemiptera – Heteroptera. Illustrated keys to known families, genera, and species in South America. Springer, New York

Henry TJ (2012) First eastern North American records of *Campyloneura virgula* (Hemiptera: Heteroptera: Miridae: Bryocorinae). Proc Entomol Soc Wash 114:159–163

Hogue CL (1993) Latin American insects and entomology. University of California Press, Berkeley

Hormchan P, Schuster MF, Hepner LW (1976) Biology of *Tropiconabis capsiformis*. Ann Ent Soc Am 69:1016–1018

Hwang WS, Weirauch C (2012) Evolutionary history of assassin bugs: insights from divergence dating and ancestral state reconstruction. PLoS ONE 7:e45523

Jackson RR, Salm K, Nelson XJ (2010) Specialized prey selection behavior of two East African assassin bugs, *Scipinnia repax* and *Nagusta* sp. that prey on social jumping spiders. J Insect Sci 10(82):1–19

Jacobson E (1911) Biological notes on the hemipteron *Ptilocerus ochraceus*. Tijdschr Ent 54:175–179

Javahery M, Schaefer CW, Lattin JD (2000) Shield bugs (Scutelleridae). In: Schaefer CW, Panizzi AR (eds) Heteroptera of economic importance. CRC Press, Boca Raton, pp 475–503

Johnson CG (1936) The biology of *Leptobyrsa rhododendri* Horvath (Hemiptera, Tingitidae), the Rhododendron lacebug. 1. Introduction, bionomics and life history. Ann Appl Biol 23:342–368

Jordan KHC, Wendt A (1938) Zur Biologie von *Salda litoralis* L. (Hem. Het.). Stettiner Ent Zeitng 99:273–292

Jurberg J, Galvão C, Noireau F, Carcavallo RU, Rocha DS, Lent H (2004) Uma Iconografia dos Triatomíneos. Entomol Vect 11:457–494

Jurberg J, Galvão C, Rocha DS, Dale C, Cunha V (2012) Vetores da doença de Chagas no Brasil. Instituto Oswaldo Cruz, Rio de Janeiro

Jurberg J, Cunha V, Cailleaux S, Raigorodschi R, Lima MS, Rocha DS, Moreira FFF (2013) *Triatoma pintodiasi* sp. nov. do subcomplexo *T. rubrovaria* (Hemiptera, Reduviidae, Triatominae). Rev Pan-Amaz Saúde 4:43–56

Kamble ST (1971) Bionomics of *Dysdercus koenigii* Fabr. (Hemiptera: Pyrrhocoridae). J NY Entomol Soc 79:154–157

Kelton LA (1959) Male genitalia as taxonomic characters in the Miridae (Hemiptera). Can Entomol Suppl 11:1–72

Kritsky G (1977) Observations on the morphology and behavior of the Enicocephalidae (Hemiptera). Entomol News 88:105–110

Kumar R (1974) A revision of world Acanthosomatidae (Heteroptera: Pentatomidae). Key to and description of subfamilies, tribes and genera, with designation of types. Aust J Zool Supl Ser 34:1–60

Lafont A (1912) Trypanosomide d'un reduvide (*Conorhenus rubrofasciatus*) inoculable au rat et à la sourris. Ann Inst Pasteur 26:893–922

Lee CE (1969) Morphological and phylogenic studies on the larvae and male genitalia of the East Asiatic Tingidae (Heteroptera). J Fac Agric 15:138–256

Lent H, Jurberg J (1965) Contribuição ao conhecimento dos Phloeidae Dallas, 1851, com um estudo sôbre genitália (Hemiptera, Pentatomoidea). Rev Bras Biol 25:123–144

Lent H, Jurberg J (1969) O gênero *Rhodnius* Stål, 1859 com um estudo sobre a genitália das espécies (Hemiptera, Reduviidae, Triat.). Rev Bras Biol 29:487–560

Lent H, Jurberg J (1975) O gênero *Panstrongylus* Berg, 1879 com um estudo sobre a genitália externa das espécies (Hemiptera, Reduviidae, Triatominae). Rev Bras Biol 35:379–438

Lent H, Martins AV (1940) Estudo sobre os Triatomideos do Estado de Minas Gerais com descrição de uma espécie nova. Rev Entomol 11:877–886

Lent H, Wygodzinsky P (1979) Revision of Triatominae (Hemiptera, Reduviidae) and their significance as vectors of Chagas disease. Bull Am Mus Nat Hist 163:123–529

Lima MM, Jurberg P, Almeida JR (1986) Behavior of triatomines (Hemiptera: Reduviidae) vectors of Chagas disease. I. Courtship and copulation of *Panstrongylus megistus* (Burm., 1835) in the laboratory. Mem Inst Oswaldo Cruz 81:1–5

Lung KYH, Goeden RD (1982) Biology of *Corimelaena extensa* on tree tobacco, *Nicotiana glauca*. Ann Entomol Soc Am 75:177–180

Malipatil MB, Kumar R (1975) Biology and immature stages of some Queensland Pentatomomorpha (Hemiptera: Heteroptera). J Aust Entomol Soc 14:113–128

Manrique G, Lazzari CR (1994) Sexual behaviour and stridulation during mating in *Triatoma infestans* (Hemiptera: Reduviidae). Mem Inst Oswaldo Cruz 89:629–633

Marchini D, Del-Bene G, Dallai R (2010) Functional morphology of the female reproductive apparatus of *Stephanitis pyrioides* (Heteroptera, Tingidae): a novel role for the pseudospermathecae. J Morphol 271:473–482

Matesco VC, Grazia J, Campos LA (2007) Description of new genus and species of Ochlerini from Central America (Hemiptera: Pentatomidae: Discocephalinae). Zootaxa 1565:63–68

Mayorga MC, Cervantes PL (2001) Life cycle and description of a new species of *Amnestus* Dallas (Hemiptera-Heteroptera: Cydnidae) associated with the fruit of several species of *Ficus* (Moraceae) in Mexico. J NY Entomol Soc 109:392–402

McHugh JV (1994) On the natural history of Canopidae (Heteroptera: Pentatomoidea). J NY Entomol Soc 102:112–114

McLain DK, Burnette LB, Deeds DA (1993) Within season variation in the intensity of sexual selection on body size in the bug *Margus obscurator* (Hemiptera Coreidae). Ethol Ecol Evol 5:75–86

McMahan EA (1983) Adaptations, feeding preferences, and biometrics of a termite-baiting assassin bug (Hemiptera: Reduviidae). Ann Entomol Soc Am 76:483–486

McPherson JE (1972) Life history of *Corimelaena lateralis lateralis* (Hemiptera: Thyreocoridae) with descriptions of immature stages and list of other species of Scutelleroidea found with it on wild carrot. Ann Entomol Soc Am 65:906–911

Menezes-Netto AC, Varella AC, Fernandes OA (2012) Maize-dwelling insects omnivory in *Spodoptera frugiperda* (J.E. Smith) (Lepidoptera: Noctuidae) egg masses. Braz Arch Biol Technol 55:97–100

Menke AS (1963) A review of the genus *Lethocerus* in North and Central America, including the West Indies (Hemiptera: Belostomatidae). Ann Entomol Soc Am 56:261–267

Menke AS (1979) The semiaquatic and aquatic Hemiptera of California (Heteroptera: Hemiptera). Bull Calif Insect Surv 21:1–166

Merritt RW, Cummins KW, Berg MB (2008) Aquatic insects of North America. Kendall/Hunt Publ. Co., Dubuque

Miller CW (2011) The heliconia bug, Leptoscelis tricolor. Featured creatures. http://entnemdept.ufl.edu/creatures/misc/bugs/heliconia_bug.htm

Miller CW, Emlen DJ (2010) Across and within population differences in the size and scaling relationship of a sexually-selected trait in *Leptoscelis tricolor* (Hemiptera: Coreidae). Ann Entomol Soc Am 103:209–215

Miller NCE (1971) The biology of the Heteroptera, 2nd edn. E.W. Classey Ltd., Hampton

Mitchell PL (2000) Leaf-footed bugs. In: Schaefer CW, Panizzi AR (eds) Heteroptera of economic importance. CRC Press, Boca Raton, pp 337–403

Mitchell PL (2006) Polyphagy in true bugs: a case study of *Leptoglossus phyllopus* (L.) (Hemiptera, Heteroptera, Coreidae). Denisia 19:1117–1134

Moizant RC, Téran JB (1970) *Dysdercus maurus* Distant (Hemiptera, Pyrrhocoridae) sobre *Citrus* spp. Agron Trop (Maracay) 20:267–269

Monteith GB (2006) Maternal care in Australian oncomerine shield bugs (Insecta, Heteroptera, Tessaratomidae). Denisia 19:1135–1152

Montemayor SI (2009) Description of a new *Corythucha* from Argentina (Hemiptera: Heteroptera: Tingidae), with description of its life cycle. Zootaxa 2170:61–68

Montemayor SI, Dellapé PM (2010) On the identity of *Gargaphia subpilosa* Berg, 1879, *G. bergi* Monte, 1940 and *G. penningtoni* Drake, 1928 (Insecta, Hemiptera, Heteroptera, Tingidae), with the description of immatures of *G. bergi*. Zoosystema 32:155–162

Montemayor SI, González-Herrera A, Villalobos K (2011) Description of a new *Pleseobyrsa* (Heteroptera: Tingidae) from Costa Rica. Rev Mex Biodiversidad 82:475–480

Myers JG (1927) Ethological observations on some Pyrrhocoridae of Cuba (Hemiptera-Heteroptera). Ann Entomol Soc Am 20:279–300

Neal JW Jr, Douglas LW (1990) Seasonal dynamics and the effect of temperature in *Corythucha cydoniae* (Heteroptera: Tingidae). Environ Entomol 19:1299–1304

Neiva A (1914) Revisão do gênero Triatoma Lap. Doctor thesis, Faculdade de Medicina, Rio de Janeiro, Brasil

Nieser N, Melo AL (1997) Os heterópteros aquáticos de Minas Gerais: guia introdutório com chave de identificação para as espécie de Nepomorpha e Gerromorpha. Editora UFMG, Belo Horizonte

Oliveira LJ, Malaguido AB (2004) Flutuação populacional dos percevejos castanhos da raiz, *Scaptocoris castanea* Perty (Hemiptera: Cydnidae), no perfil do solo em áreas produtoras de soja nas regiões centro-oeste e sudeste do Brasil. Neotrop Entomol 33:283–291

Osorio GA (2009) Primer registro para la región del Maule de *Sinopla perpunctatus* Signoret, 1864 (Hemiptera: Heteroptera: Acanthosomatidae). Bol Biodiversidad Chile 1:35–37

Paleari LM (1992) Biologia de *Agonosoma flavolineata* Laporte, 1832 (Hemiptera, Scutelleridae). Rev Bras Entomol 36:521–526

Panizzi AR, Hirose E, Oliveira EDM (1996) Egg allocation by *Megalotomus parvus* (Westwood) (Heteroptera: Alydidae) on soybean. An Soc Entomol Bras 25:537–543

Paskewitz SM, McPherson JE (1983) Life history and laboratory rearing of *Arhyssus lateralis* (Hemiptera: Rhopalidae) with descriptions of immature stages. Ann Entomol Soc Am 76:477–482

Pereira MH, Melo AL (1998) Influência do tipo de presa no desenvolvimento e na preferência alimentar de *Belostoma anurum* (Herrich-Schaffer, 1848) e *B. plebejum* (Stal, 1858) (Heteroptera, Belostomatidae). Oecol Bras 5:41–49

Pereira MF, Peres RM, Borges RS (2012) Population of *Scaptocoris castanea* Perty (Hemiptera: Cydnidae) in a crop-livestock integration system. Neotrop Entomol 41:409–413

Péricart J (1972) Anthocoridae, Cimicidae et Microphysidae de l'Ouest- Paléarctique. 7. Hemiptera. In: Masson (ed) Fauna de l'Europe et du Bassin Méditerranéen. Paris

Péricart J (1983) Hémiptères Tingidae Euro-Méditerranéens. Faune Fr 69:1–620

Picado TC (1937) Estudo experimental sobre o veneno de *Lethocerus delpontei* (De Carlo) (Hemiptera-Belostomidae). Mem Inst Butantan 10:305–310

Picado TC (1939) Étude experimentale du venin de *Lethocerus delpontei* (De Carlo). Trav Stn Zool Wimereux 13:533–562

Polhemus JT, Chapman HC (1979) Family Saldidae/shore bugs. In: Menke AS (ed) The semi-aquatic and aquatic Hemiptera of California (Heteroptera: Hemiptera), Bulletin of the California insect survey 21. University of California Press, Berkeley, pp 16–33

Riis L, Belotti AC, Arias B (2005) Bionomics and population growth statistics of *Cyrtomenus bergi* (Hemiptera: Cydnidae) on different host plants. Fla Entomol 88:1–10

Rimes GD (1951) Some new and little-known shore-bugs (Heteroptera-Saldidae) from the Australian region. Trans R Soc S Aust 74:135–145

Rodrigues D, Sampaio DS, Isaias RMDS, Moreira GRP (2007) Xylem and seed feeding by two passion vine leaffooted bugs, *Holymenia clavigera* and *Anisoscelis foliacea marginella* (Hemiptera: Coreidae: Anisoscelini), with notes on mouthpart morphology. Ann Entomol Soc Am 100:907–913

Rojas JC, Malo EA, Gutierrez-Martínez A (1990) Mating behavior of *Triatoma mazzottii* Usinger (Hemiptera: Reduviidae) under laboratory conditions. Ann Entomol Soc Am 83:598–602

Ryckman RE (1951) Recent observations of cannibalism in *Triatoma* (Hemiptera, Reduviidae). J Parasitol 37:43–437

Salomão AT, Postali TC, Vasconcellos-Neto J (2012) Bichos-cascas na Serra do Japi: história natural dos percevejos Phloeidae (Hemiptera). In: Vasconcellos-Neto J, Polli PR, Penteado-Dias AM (eds) Novos olhares, novos saberes sobre a Serra do Japi: ecos de sua biodiversidade. Editora CRV, Curitiba, pp 321–337

Sandoval CM, Joya M, Gutierrez R, Angulo VM (2000) Cleptohaemathophagia of the triatome bug *Belminus herreri*. Med Vet Entomol 14:100–101

Saulich AH, Musolin DL (2007) Seasonal development of aquatic and semiaquatic true bugs (Heteroptera). St. Petersburg University Press, St. Petersburg (in Russian, with extended 6p English Summary). ISBN 978-5-288-04332-1

Schaefer CW, Ahmad I (1987) The food plants of four pentatomoid families (Hemiptera: Acanthosomatidae, Tessaratomidae, Urostylidae, and Dinidoridae). Phytophaga 1:21–34

Schaefer CW, Panizzi AR, James DG (2000) Several small pentatomoid families (Cyrtocoridae, Dinidoridae, Eurostylidae, Plataspidae, and Tessaratomidae). In: Schaefer CW, Panizzi AR (eds) Heteroptera of economic importance. CRC Press, Boca Raton, pp 505–512

Scholze W (1992) Sekretschwitzen_ bei Netzwanzenlarven: Zur Morphologie, Chemie und biologischen Bedeutung von Integumentbildungen mit sekretorischer Funktion bei Netzwanzenlarven (Heteroptera, Tingidae). Inaugural-dissertation zur Erlangung des Doktorgrades der Fakultät Biologie, Chemie und Geowissenschaften. Universität Bayreuth, Bayreuth

Schouteden H (1912) Les hémiptères parasites des cotonniers en Afrique. Rev Zool Africaine 1:297–321

Schowalter TD (1986) Overwintering aggregation of *Boisea rubrolineatus* (Heteroptera: Rhopalidae) in western Oregon. Environ Entomol 15:1055–1056

Schuh RT, Slater JA (1995) True bugs of the world (Hemiptera: Heteroptera): classification and natural history. Cornell University Press, Ithaca

Schwertner CF, Grazia J (2014) Dinidoridae, Megarididae y Tessaratomidae. In: Roig-Juñent S, Claps LE, Morrone JJ (eds) Biodiversidad de artrópodos Argentinos, vol 3. Sociedad Entomologica Argentina, Mendoza (in press)

Singh-Pruthi H (1925) The morphology of the male genitalia in Rhynchota. Trans R Entomol Soc 1:127–267

Sites RW, McPherson JE (1982) Life history and laboratory rearing of *Sehirus cinctus cinctus* (Hemiptera: Cydnidae), with descriptions of immature stages. Ann Entomol Soc Am 75:210–215

Spence JR, Andersen NM (1994) Biology of water striders: interactions between systematics and ecology. Annu Rev Entomol 39:101–128

Sprague IB (1956) The biology and morphology of *Hydrometra martini* Kirkaldy. Univ Kans Sci Bull 38:579–693

Stride GO (1954) On the specific status of *Phonoctonus subimpictus* Stål. Rev Zool Bot Afr 50:13–16

Štys P (1981) Unusual sex ratios in swarming and light-attracted Enicocephalidae (Heteroptera). Acta Entomol Bohemoslov 78:430–432

Štys P (1995) Dipsocoromorpha. In: Schuh RT, Slater JA (eds) True bugs of the world (Hemiptera: Heteroptera): classification and natural history. Comstock Publishing Associates, Cornell University Press, Ithaca, pp 67–73

Sweet MH II (2000) Seed and chinch bugs (Lygaeoidea). In: Schaefer CW, Panizzi AR (eds) Heteroptera of economic importance. CRC Press, Boca Raton, pp 143–264

Tatarnic NJ, Cassis G (2010) Sexual coevolution in the traumatically inseminating plant bug genus *Coridromius*. J Evol Biol 23:1321–1326

Tatarnic NJ, Spence JR (2013) Courtship and mating in the crusader bug *Mictis profana* (Fabricius). Aust J Entomol 52:151–155

Tatarnic NJ, Cassis G, Hochuli DF (2006) Traumatic insemination in the plant bug genus *Coridromius* Signoret (Heteroptera: Miridae). Biol Lett 2:58–61

Timonin MI (1958) *Scaptocoris talpa* Champ. on the roots of banana and other plants. Plant Prot Bull FAO 6:1–2

Tsai J-F, Rédei D, Yeh GF, Yang MM (2011) Jewel bugs of Taiwan (Heteroptera: Scutelleridae). National Chung Hsing University, Taichung

Van Doesburg PH Jr (1966) Heteroptera of Suriname: I. Largidae and Pyrrhocoridae. Stud Fauna Suriname Guyanas 9:1–60

Ventura MU, Panizzi AR (2000) Oviposition behavior of *Neomegalotomus parvus* (West.) (Hemiptera: Alydidae): daily rhythm and site choice. An Soc Entomol Bras 29:391–400

Ventura MU, Panizzi AR (2003) Population dynamics, gregarious behavior and oviposition preference of *Neomegalotomus parvus* (Westwood) (Hemiptera: Heteroptera: Alydidae). Braz Arch Biol Technol 46:33–39

Weber H (1930) Biologie der Hemipteren. Biologische Studienbücher, XI. J. Springer, Berlin

Weirauch C (2008) Cladistic analysis of Reduviidae (Heteroptera: Cimicomorpha) based on morphological characters. Syst Ent 33:229–274

Wheeler AG (1977) Life history of *Niesthrea louisianica* (Hemiptera: Rhopalidae) on rose of Sharon in North Carolina. Ann Entomol Soc Am 70:631–634

Wheeler AG (2000) Plant bugs (Miridae) as plant pests. In: Schaefer CW, Panizzi AR (eds) Heteroptera of economic importance. CRC Press, Boca Raton, pp 37–83

Wheeler AG (2001) Biology of the plant bugs (Hemiptera: Miridae). Pests, predators, opportunists. Cornell University Press, Ithaca

Wheeler AG, Henry TJ (1992) A synthesis of the Holarctic Miridae (Heteroptera): distribution, biology, and origin, with emphasis on North America. Thomas Say Foundation, Entomological Society of America, Lanham

Wheeler AG Jr, Miller GL (1990) *Leptoglossus fulvicornis* (Heteroptera: Coreidae), a specialist on magnolia fruits: seasonal history, habits, and descriptions of immature stages. Ann Entomol Soc Am 83:753–765

Wignall AE, Taylor PW (2008) Biology and life history of the araneophagic assassin bug Stenolemus bituberus including a morphometric analysis of the instars (Heteroptera, Reduviidae). J Nat Hist 42:59–76

Wiley GO (1922) Life history notes on two species of Saldidae (Heteroptera) found in Kansas. Kans Univ Sci Bull 14:301–311

Willis ER, Roth LM (1962) Soil and moisture relations of *Scaptocoris divergens* Froeschner (Hemiptera: Cydnidae). Ann Entomol Soc Am 55:21–33

Wolda H, Tanaka S (1987) Dormancy and aggregation in a tropical insect *Jadera obscura* (Hemiptera: Rhopalidae). Proc Koninklijke Nederlandse Akademie van Wetenschappen Ser C Biol Med Sci 90:351–356

Wood SF (1941) Notes on the distribution and habits of reduviid vectors of Chagas'disease in the Southwestern United States. Pan-Pac Entomol 17:85–94

Wroblewski A (1966) Shorebugs (Heteroptera, Saldidae) of Poland. Polskie Pismo Ent 36:219–302

Wygodzinsky PW, Schmidt K (1991) Revision of the New World Enicocephalomorpha (Heteroptera). Bull Am Mus Nat Hist 200:1–265

Yonke TR, Walker DL (1970a) Field history, parasites, and biology of *Harmostes reflexulus* (Say) (Hemiptera: Rhopalidae). J Kans Entomol Soc 4:444–450

Yonke TR, Walker DL (1970b) Description of the egg and nymphs of *Harmostes reflexulus* (Hemiptera: Rhopalidae). Ann Entomol Soc Am 63:1749–1754

Chapter 3
Classification and Biogeography of Neotropical True Bugs

Augusto Ferrari, Kim R. Barão, Filipe M. Bianchi, Luiz A. Campos, and Jocélia Grazia

Abstract A review of Heteroptera classification, biogeography, and phylogeography is presented. The use of molecular data significantly expanded the knowledge of phylogenetic relationships among and within heteropteran infraorders. However, taxa historically less studied continue to receive little attention. Promising, new molecular approaches with increased genetic markers and broader taxon sampling, as well as new morphological approaches (e.g., microtomography), are the future for more stable classifications and a better comprehension of the heteropteran evolutionary history, but their application is still incipient. A non-exhaustive overview of studies about Neotropical heteropteran biogeography is made and discussed, including those about intercontinental connections and regional distribution patterns. The most comprehensible studies, and more promising area, seem to be those focused on distribution patterns, especially employing macroecological methods, and trying to elucidate what are the major factors responsible for the distribution of the group in the Neotropics. Finally, we present an overview of phylogeographic studies involving Neotropical Heteroptera. It is clear that the best biogeographic and phylogeographic studied groups are those with medical and economical importance (e.g., Reduviidae and Pentatomidae).

A. Ferrari (✉)
Instituto de Ciências Biológicas (ICB), Universidade Federal do Rio Grande - FURG, Campus Carreiros - Av. Itália, km 8, Prédio 6, Sala 22g, Rio Grande, RS 96203-900, Brazil
e-mail: ferrariaugusto@gmail.com

K.R. Barão • F.M. Bianchi
Programa de Pós-Graduação em Biologia Animal, Instituto de Biociências, Universidade Federal do Rio Grande do Sul (UFRGS), Av. Bento Gonçalves 9500, prédio 43435, Bairro Agronomia, Porto Alegre, RS 91501-970, Brazil
e-mail: kbarao@gmail.com; fmichels2@gmail.com

L.A. Campos • J. Grazia
Departamento de Zoologia, Instituto de Biociências, Universidade Federal do Rio Grande do Sul (UFRGS), Av. Bento Gonçalves 9500, prédio 43435, Bairro Agronomia, Porto Alegre, RS 91501-970, Brazil
e-mail: luiz.campos@ufrgs.br; jocelia@ufrgs.br

A.R. Panizzi, J. Grazia (eds.), *True Bugs (Heteroptera) of the Neotropics*, Entomology in Focus 2, DOI 10.1007/978-94-017-9861-7_3

3.1 Introduction

The knowledge of the Heteroptera goes back to the eighteenth century, during the 1750s to 1770s. This period was called "Classical" (Slater 1974) and began with C. Linnaeus, followed by J.C. Fabricius and others, which studied the Heteroptera in general and provided the first major higher classifications of the suborder. Carl Stål, from the mid-1850s to the end of the 1870s, represents the watershed of the revisional studies in many heteropteran families (Stål 1870–1876). The appearance of specialized works started in 1870, primarily in Europe, favored by the enrichment of the European museums by major collecting expeditions in all parts of the world. It was a period of dominance for taxonomy, not only with descriptive works but also with faunistic (e.g., Distant 1880–1893) and comprehensive catalogs (e.g., Lethierry and Severin 1893–1896, a general catalog, and Kirkaldy 1909, a Rhynchota catalog). From the seventeenth to the early twentieth centuries, not so many heteropterists were dedicated to describe new taxa in the New World, and mostly faunistic works were done (Mayr 1864, 1866a, b; Uhler 1869, 1894; Berg 1879, 1884 – Hemiptera Argentina; Bergroth 1891, 1893, 1894, 1905, 1908, 1914, 1918; Breddin 1903a, b, c, d, 1904a, b, c, d, 1907, 1908a, b, 1909, 1910, 1912a, b, c, 1914).

Reuter (1910) pioneered in evaluating the classification of Heteroptera, making explicit interpretations of the characters used by previous authors, presenting objective arguments for definition of groups (Schuh 1986), and producing a phylogenetic scheme for the Hemiptera. As in many other works, character polarity was the weakest aspect in Reuter's work (Schuh and Slater 1995). Although Reuter (1910) proposed subordinal names, most were never consistently adopted, but he already had a close concept of the current Gerromorpha, Nepomorpha, and Cimicomorpha. For a more complete view of Reuter's work, see Schuh and Slater (1995).

From 1900 to the early 1940s, several workers have devoted to the description of new genera and species. For their influence in the knowledge about classification of Neotropical Heteroptera, worth to mention is the world catalog of part of the families' currently classified in Pentatomoidea (Kirkaldy 1909) and the world revision of Thyreocoridae (McAtee and Malloch 1933). Other relevant taxonomists in this period were G. Breddin, E.E. Bergroth, R. Jeannel, H.G. Barber, G. Horvath, and E.P. Van Duzee.

From the early 1950s to the beginning of the 1990s, taxonomists focused mainly on the revision and description of taxa within families. Outstanding exceptions are the world catalogs for genera and species of Miridae (Carvalho 1952; Schuh 1995, 2002–2013), Tingidae (Drake and Ruhoff 1960, 1965; Froeschner 1996, 2001), Lygaeidae (Slater 1964; Slater and O'Donnell 1995), and the world classification reviews of Aradidae (Usinger and Matsuda 1959) and Cimicidae (Usinger 1966). Other relevant taxonomists in this period were T. Esaki and H.B. Hungerford for Nepomorpha and Gerromorpha, respectively; D. Leston for Cimicoidea; R. Cobben for Leptopodomorpha; R. Matsuda for Gerridae; P.W. Wygodzinsky for Enicocephalomorpha, Dipsocoromorpha, and Reduviidae; P.D. Ashlock for

Lygaeidae; and R. Ruckes, L.H. Rolston, M. Becker, and A.A. Pirán for Pentatomidae. For an extensive review of the world major taxonomists in Heteroptera, see Schuh and Slater (1995).

In a wider classification approach within the Heteroptera, the work of Leston et al. (1954) was a keystone to heteropteran modern classification. Leston et al. introduced the terms Cimicomorpha and Pentatomomorpha in the first attempt to recognize natural groups within the polyphyletic Geocorisae, based on accumulated evidence from comparative studies of internal anatomy and external morphology of the Heteroptera (Schuh and Slater 1995). The influence of Leston et al. work was widely felt and stimulated other authors into attempting to document the monophyly of higher groups within Heteroptera. As outlined by Štys and Kerzhner (1975), such attempts resulted in the recognition of seven infraorders, which at the time were not proposed based on cladistic analysis: Enicocephalomorpha Stichel; Dipsocoromorpha Miyamoto; Gerromorpha Popov; Nepomorpha Popov; Leptopodomorpha Popov; Cimicomorpha Leston, Pendergrast, and Southwood; and Pentatomomorpha Leston, Pendergrast, and Southwood.

3.2 Phylogenetic Systematics and Modern Higher Classifications of Heteroptera

Hennig's ideas on defining and recognizing monophyletic groups based on shared derived characters (i.e., synapomorphies) took some time to influence the classification of Heteroptera, even after the publication of its English edition (Hennig 1950, 1966). The first studies to apply the concepts proposed by Hennig were approximations of what is now traditionally called "cladistic analysis" – based on characteristics usually employed in taxonomy and comparative morphology, such works aimed to identify equivalence between phylogenetic hypotheses and taxonomic classifications (Schuh 1979, 1986). At higher taxonomic ranks, such hypotheses of relationships were presented as hierarchical branching diagrams, depicting the distributional congruence of attributes (characters) between taxa. At the time of Hennig's work, important aspects of phylogenetic analysis, such as broad sampling of outgroups and questions about rooting, character polarity, and character coding had not yet been developed. Nevertheless, Hennig's legacy laid the foundations to produce a general reference system for biology based exclusively on monophyletic groups and a way to summarize the knowledge contained in classifications and taxa diagnoses (Schuh 1986).

The 1970s and 1980s saw the rapid development of phylogenetic methods and an arising dispute between schools of thought (evolutionary systematics, numerical taxonomy or numerical phenetics, and systematic cladistics) for the establishment of the most suitable classification method (Hull 1988). It was not clear which were the most appropriate methods for assessing phylogenetic relationships neither if this would be a task that could be achieved in the near future.

On his seminal work on adult and immature heteropteran morphology, Cobben (1968, 1978) provided cues for the definition of characters within Heteroptera and enormously expanded the knowledge about features never studied before. Cobben's sampling of the Neotropical fauna was notable, by sampling species from the Antilles, Costa Rica, Brazil, and Chile in its first volume (Cobben 1968), then, on its second volume, by extending the Neotropical sampling to species occurring in Honduras, Paraguay, Peru, and Trinidad (Cobben 1978). Cobben's studies on heteropteran morphology would be later compiled in morphological matrices (Schuh 1979; Mahner 1993; Wheeler et al. 1993), but many characters proposed by Cobben are yet to be coded in a formal cladistic analysis (Weirauch and Schuh 2011).

Cobben (1978) also ventured into proposing phylogenetic relationships within Heteroptera, even though not using formal cladistic analysis. Cobben recognized Gerromorpha as paraphyletic and at the base of Heteroptera. Schuh (1979) reaccessed Cobben's work employing cladistic analyses, proposing the first phylogenetic analysis using parsimony criterion and outgroup comparison to polarize characters, in order to recognize higher-level groups in Heteroptera. Schuh (1979) demystifies wrong concepts implicit in Cobben's work, such as the impossibility to cladistically analyze higher taxa of Heteroptera because of the enormous amounts of parallelisms. Schuh (1979) found the Enicocephalomorpha to be the sister group to the remaining Heteroptera, Dipsocoromorpha as sister group to Neoheteroptera, and Gerromorpha as sister group to Nepomorpha + Leptopodomorpha and Cimicomorpha + Pentatomomorpha.

Since Schuh's reanalysis of Cobben's work, phylogenetic analyses based on morphological and molecular datasets have increasingly influenced the understanding of relationships within Heteroptera, resulting in explicit cladistic hypotheses of relationships among the seven infraorders (Weirauch and Schuh 2011). Wheeler et al. (1993) hypothesis treats the Enicocephalomorpha as the sister group to the remaining Heteroptera, Dipsocoromorpha as sister group to the Neoheteroptera, Gerromorpha as sister group to the Panheteroptera, and Nepomorpha as sister taxon to the Leptopodomorpha (Cimicomorpha + Pentatomomorpha).

Other competing hypotheses, based on morphological data of fossil and extant taxa, include Nepomorpha as the sister taxon to the remaining Heteroptera (Mahner 1993; Scherbakov and Popov 2002), without resolving the relationships among the other infraorders. Li et al. (2012), based on molecular data, agree with the Nepomorpha as sister taxon to the remaining Heteroptera, but find the Dipsocoromorpha + Gerromorpha as sister taxon to Enicocephalomorpha + Leptopodomorpha and Cimicomorpha + Pentatomomorpha. Yet, others agree with Wheeler et al. (1993) with the Enicocephalomorpha as the sister taxon to the remaining Heteroptera, but disagree on the relationships of other infraorders (Xie et al. 2008).

The only sister-group relationship that is not contradicted by any of these hypotheses is the Cimicomorpha + Pentatomomorpha. This clade was also supported in combined morphological and molecular analysis by Schuh et al. (2009) and in morphological analysis focusing on the rhabdom structure of the ommatidia (Fischer et al. 2000).

Spangenberg et al. (2013) provided new data on head morphology (including musculature, tentorium, cephalic nervous system, and alimentary tract) based on serial sectioning and computer-based 3D reconstruction for representatives of the potentially basal heteropteran lineages (Enicocephalomorpha, Dipsocoromorpha, and Gerromorpha). Spangenberg et al. cladistically analyzed 71 cephalic characters scored for 16 heteropteran terminals. Heteropteran monophyly was strongly supported, but some of the recovered relationships between infraorders are not consistent with previous studies, such as the paraphyly of Nepomorpha and Dipsocoromorpha – the authors thus suggest that cephalic features alone are apparently insufficient for a clarification of the relationships of the major lineages of Heteroptera. They also mapped cephalic character data on three alternative hypotheses, being Wheeler et al. (1993) the most parsimonious one, requiring 201 steps for their data. The topologies of Xie et al. (2008) and Mahner (1993) required 202 and 212 steps, respectively.

There has been a recent increase in studies employing DNA data, especially those with a more comprehensive taxon sampling. The availability of new sequencing techniques promises a revolution, by reducing costs and increasing the availability of molecular data. Phylogenomics and proteomics approaches are making possible a better understanding of the relationships within Insecta (e.g., http://www.1kite.org/index.html). At the same time, the availability of molecular data is increasing, as is the refinement of morphology-driven studies with the diffusion of new approaches. New and improved technology for image capture and treatment, such as confocal microscopes and 3D-reconstructions by scanning electron microscopy and X-ray microtomography, are making possible to revisit and investigate underexplored morphological characters (Friedrich et al. 2014). The new challenge will be to integrate new molecular and morphological datasets enabling its combined analysis, as has been recently done by some researchers (Friedrich et al. 2014).

3.2.1 Phylogenetic Relationships Within the Heteropteran Infraorders

Nothing is known about the phylogenetic relationships within Enicocephalomorpha. Therefore, phylogenetic relationships will be presented for the remaining six infraorders of Heteroptera.

Dipsocoromorpha

According to Weirauch and Štys (2014), the Dipsocoromorpha are the least documented heteropteran infraorder, comprising about 300 species in five families. The monophyly of the group has been considered to be controversial because its families show a great morphological variability (Weirauch and Schuh 2011). Weirauch and Štys (2014) were the first authors to comprehensively sample the Dipsocoromorpha and perform a phylogenetic analysis of this infraorder. Their results support a monophyletic Dipsocoromorpha, as well as the families included in the analysis (i.e., Ceratocombidae, Dipsocoridae, and Schizopteridae). However, the position of this infraorder in Heteroptera still needs to be settled, since the dataset alone (ribosomal data) is not sufficient to resolve deep relationships among the Heteroptera.

Gerromorpha

The infraorder Gerromorpha, commonly called the semiaquatic bugs, comprises about 2,100 species in eight families (Damgaard 2008). Although the monophyly of this taxon has been extensively corroborated (Andersen 1982; Damgaard 2008), the monophyly and relationship among its superfamilies have been recently disputed (Damgaard 2012). The pioneerism of Andersen's contributions (Andersen 1981, 1982) is still the most comprehensive comparative study of this infraorder (Damgaard 2012). Damgaard (2008) proposed a phylogenetic hypothesis for Gerromorpha using morphological and molecular data summarized from several studies (Muraji and Tachikawa 2000; Andersen and Weir 2004; Damgaard et al. 2005; Damgaard and Cognato 2003) addressing the systematics of major lineages within Gerromorpha. Damgaard (2008) corroborated the monophyly of the infraorder, as well as the relationship of the Mesoveliidae as the sister group to all other gerromorphan families, and the relationship between Gerridae and Veliidae. Gerroidea, Hydrometroidea, Vellinae, and Cylindrostethinae were recovered as paraphyletic.

Nepomorpha

The Nepomorpha, or true water bugs, comprises about 2,000 species placed on 11 families. The monophyly of Nepomorpha is widely accepted (Wheeler et al. 1993; Mahner 1993; Scherbakov and Popov 2002; Xie et al. 2008), but Hua et al. (2009) found contrasting results, suggesting Nepomorpha was paraphyletic and the Pleidae should be elevated to infraordinal level, i.e., Pleomorpha; given their small taxon sample and lack of a combined effort including morphology, their taxonomic decisions have not been accepted (Weirauch and Schuh 2011).

Within Nepomorpha, there is an agreement in placing the Nepidae + Belostomatidae as the sister group to the remaining Nepomorpha and in treating the Helotrephidae, Notonectidae, and Pleidae as a clade (Rieger 1976; Mahner 1993; Hebsgaard et al. 2004).

Leptopodomorpha

About 400 species distributed in four families comprise the Leptopodomorpha, or shore bugs. Family-level relationships within the infraorder have not changed since the publication of Schuh and Polhemus (1980) that, based on morphological data, proposed Saldidae + Aepophilidae to represent the sister group to Omaniidae + Leptopodidae. Recently, Zhang et al. (2014) found the same relationships among the extant leptopodomorphan families, placing the fossil Palaeoleptidae as sister group to Omaniidae + Leptopodidae, and the fossil Archegocimicidae more closely related to Aepophilidae. Other phylogenetic contributions to the Leptopodomorpha are the cladistic revisions of *Saldula* Van Duzee and of *Pseudosaldula* Cobben (Lindskog and Polhemus 1992; Schuh and Polhemus 2009).

Cimicomorpha

Cimicomorpha is the largest infraorder of Heteroptera, comprising about 20,000 species in 16 families (Schuh and Slater 1995). The Cimicomorpha is a well-studied infraorder, especially because of its economic and health importance; however, few studies were concerned with the phylogenetic relationships within the family level (Schuh et al. 2009). Ford (1979) and Kerzhner (1981) were the first to propose hypotheses of phylogenetic relationships within Cimicomorpha, but the phylogenetic relationships proposed by Schuh and Štys (1991) were the most accepted. Since this publication, knowledge about cimicomorphan morphology has increased for many families. Schuh and Slater (1995) made important new observations about the Cimicomorpha morphology; other important contributors to the morphology of the infraorder are Weirauch (2003, 2005, 2006), Schuh (2006), Popov (2008), and Cassis et al. (1999).

Tian et al. (2008) analyzed the phylogenetic relationships of 12 cimicomorphan families based on molecular dataset and different methods of phylogenetic reconstruction, finding small differences from Schuh and Štys (1991), but did not propose taxonomic changes. Analyzing molecular and morphological data, Schuh et al. (2009) found a paraphyletic Cimicomorpha, with the Thaumastocoridae as sister group to the Pentatomomorpha, but the authors avoided taking taxonomic decisions at the infraordinal level because of ambiguities between the datasets results. Within the Cimicomorpha, Schuh et al. (2009) expanded the Cimiciformes to include the Joppeicidae, Microphysidae, Velocipedidae, and Curaliidae; the monophyly of Miroidea, Cimicoidea, Reduviidae, Tingidae, and Miridae were corroborated. At family level or lower levels, families such as the Miridae (Weirauch and Schuh 2010), Reduviidae (Paula et al. 2005, 2007; Weirauch 2008; Weirauch and Munro 2009; Hwang and Weirauch 2012), and Tingidae (Montemayor and Costa 2009; Guilbert 2012a, b; Guilbert et al. 2014) were analyzed phylogenetically, and the Reduviidae has been the focus of studies on comparative morphological and behavioral evolution (Berniker and Weirauch 2012; Weirauch et al. 2011; Zhang and Weirauch 2013).

Pentatomomorpha

Influenced by the work of Tullgren (1918) and Singh-Pruthi (1925) and by their own observations, Leston et al. (1954) coined the Pentatomomorpha to include the Aradoidea and Trichophora. As conceived by Tullgren (1918), the Trichophora gathered the taxa presenting ventral abdominal trichobothria and a certain type of pulvilli (*sic* arolia). As conceived by Leston et al. (1954), the Pentatomomorpha could be recognized by wing venation, pretarsal structure, salivary glands, internal genitalia, and egg morphology; the family composition of Pentatomomorpha little changed after their work, with the exception of the Thaumastocoridae and Saldidae, and Leptopodidae being transferred to Cimicomorpha and Leptopodomorpha, respectively (Štys and Kerzner 1975).

Four to six superfamilies have been recognized within the Pentatomomorpha (Štys 1961; Schaefer 1993; Carver et al. 1991; Schuh 1986; Henry and Froeschner 1988), illustrating the uncertainty of relationships in the infraorder. Currently, five superfamilies are accepted: Aradoidea, Coreoidea, Lygaeoidea, Pentatomoidea, and Pyrrhocoroidea. So far, seven works have addressed the phylogeny of the Pentatomomorpha, all finding the infraorder to be monophyletic and only the Aradoidea and Pentatomoidea are consistently recognized as monophyletic (Henry 1997; Li et al. 2005, 2006; Xie et al. 2005; Grazia et al. 2008; Hua et al. 2008; Tian et al. 2011; Yao et al. 2012). The Aradoidea seems to be the sister group of Trichophora, and the relationships within Trichophora are far from being understood. Some phylogenetic hypotheses agree in Coreoidea and Lygaeoidea probably being polyphyletic (Li et al. 2005, 2006; Xie et al. 2005) and the Pyrrhocoroidea sometimes appearing as sister group to Coreoidea+Lygaeoidea (Henry 1997; Xie et al. 2005; Hua et al. 2009) or as sister group to the Pentatomoidea (Li et al. 2005, 2006) or to the Coreoidea (Tian et al. 2011).

3.3 Biogeography of Neotropical Heteroptera

Since the contributions of Augustin P. de Candolle and Friedrich W. H. A. von Humboldt in the early nineteenth century, the Neotropical region has been the focus of studies on how the geophysical processes influence the distribution of living organisms (Llorente et al. 2000). From the very beginning, naturalists studying the Neotropical biota were amazed by its richness and diversity and were puzzled on how to explain the biogeographical evolution of such a diverse ecosystem. Most of the questions made by the first naturalists remain unanswered: Are there distributional patterns? And how can they be explained? Are distributional patterns similar to all organisms or each taxon responds differently to it? How is the Neotropical biota related to the biota of other biogeographical areas?

Methodological approaches and the understanding of biogeographic patterns were developed concomitantly, resulting in a wide variety of explanations to the current distributional patterns of the Neotropical biota. During the 1850s and 1860s, Joseph Hooker hypothesized the existence of intercontinental land bridges (e.g., between South America and Africa) and posteriorly changed his opinion and accepted the idea of centers of diversity, agreeing with the dispersal views of C. Darwin and A.F. Wallace (Llorente et al. 2000; Morrone 2007). Matthew's (1915) work was the foundation for the, subsequently named, "New York School of Zoogeography" (Croizat 1958; Morrone 2009), a school of thought reconciling biogeography, natural selection, and dispersal theories – a paradigm to neo-Darwinists. In this context, dispersal explanations for distributional patterns state that ancient species live in remote areas, whereas more recently derived species are located in the centers of origin where they speciated. Even though dispersalism was not a unified research program, Simpson (1940), Darlington (1957), and Mayr (1942) followed W.D. Matthew's ideas (Morrone 2002). Simpson (1940) proposed concepts such as *filter bridges* and *sweepstake routes*, stating that any taxon can be originated in a center of origin and expands its distribution in every direction, until it reaches an impassable barrier. Thus, the biota would have random distributions, because the distribution of each taxon would be determined by its dispersal capabilities.

Panbiogeography was developed by L. Croizat to test dispersalist ideas. By comparing the distribution of plant and animal species, Croizat (1958, 1964) found a limited number of distributional patterns (Craw et al. 1999; Morrone 2007), somewhat contrary to the dispersalist assumption of random distribution patterns. Croizat's work had either a low impact or was negatively criticized at its time (Llorente et al. 2000).

The development of cladistics, providing rigorous means of recognizing groups and its relationships and the wide acceptance of plate tectonics, promoted a shift on biogeography, from dispersal, centers of origin, and identification of subregions to the establishment of areas of endemism and degrees of distributional concordance between different taxa and the areas taxa occupy (Schuh and Slater 1995). Biogeographic approaches became methodologically explicit only after Nelson and Platnick's (1981) formalization of the cladistic biogeography method. The method's assumption is that there is evidence of shared common distributional patterns if congruence between phylogenetic and biogeographic patterns of two or more taxa is found.

Biogeographic studies on Heteroptera face old and new impediments, such as limited knowledge on phylogenetic relationship for most groups, limited taxon sampling, lack of dating in phylogenetic hypotheses, and limited knowledge on the distribution of taxa. Most biogeographical discussions are drawn from comparisons between distributional patterns of studied taxa and other taxa, and geological information. Also, biogeographical discussions frequently are neither based on phylogenetic information, nor dating nodes, nor on assumption of ancestral areas from related fossils. Such approaches have serious methodological flaws (Heads 2005), because congruence or incongruence between non-dated biogeographical events may be derived from pseudo-congruence or pseudo-incongruence patterns (Page 1990; Hunn and Upchurch 2001; Donoghue and Moore 2003).

A non-exhaustive overview of studies discussing the biogeography of Heteroptera on the Neotropics, including those about intercontinental connections and local patterns, is presented below. This section is intended as a starting point for new biogeographic studies on Heteroptera occurring in the Neotropics, facilitating literature review, and is organized by infraorder. To avoid the flaws exposed above, we do not assume generalized patterns for the studies discussed.

3.3.1 Continental Biogeographical Connections

Gerromorpha

Andersen (1999) reviewed the phylogenetic, ecological, and geographical aspects of species diversity of the Gerromorpha, finding that about 10 % of the 152 gerromorphan genera are marine and confined to the subtropical and tropical regions of the world. According to Andersen, most marine water striders of the Neotropical region have rather restricted distributional areas, such as the *Rhagovelia* (*Trochopus*) (Carpenter) (Veliidae) and *Rheumatobates* Bergroth (Gerridae), occurring in the Caribbean and Pacific coasts of Central and South America, while other genera, such as *Husseyella* Herring (Veliidae) and *Telmatometroides* Polhemus (Gerridae), are confined to one of these two areas. The fauna of the Atlantic, Caribbean, and East Pacific regions is mainly composed of marine lineages that most likely evolved multiple times from limnic ancestors. Andersen (1999) hypothesized that *Darwinivelia* Andersen and Polhemus, *Rheumatobates*, and *R.* (*Trochopus*) evolved during the Late Pliocene, before the emergence of the Isthmus of Panama. During the Cenozoic, different shallow sea connections (e.g., Atlantic/Caribbean/East Pacific and Indo/West Pacific) were available as dispersal routes for marine water striders and other organisms occurring in the Pacific Ocean, the Tethys Sea, and Southern Africa.

Andersen (2000) described and discussed phylogenetic, paleontological, and biogeographic aspects of eight gerromorphan species from Eocene Baltic amber. Andersen hypothesized that these gerromorphan species seem to have their closest living relatives either in the Palearctic region or in the Neotropical region or belong to an ancestral lineage, which has left few living relict species in the Neotropical region and on isolated islands of the Central Pacific.

Based on two putatively sibling genera, the South American *Mesoveloidea* Hungerford and the West African *Mesovelia* Poisson, Andersen and Polhemus (2003) suggested mesoveliids have had an ancient, wide tropical distribution since the Cretaceous and could, therefore, have diversified in response to global tectonic events. According to Andersen and Polhemus (2003), the restricted distributions of many extant species may be an artifact of the lack of knowledge and sampling effect. Andersen supports his hypothesis on the Neotropical genera *Cryptovelia* Andersen and Polhemus, a monotypic genus known only to Brazil with an allied species found in Borneo, and the *Darwinivelia*, described to contain a species found on the Galapagos archipelago, with other species subsequently recorded from the Atlantic and Pacific coasts of South America.

Polhemus and Polhemus (2008) reviewed the global diversity and taxonomic richness of freshwater aquatic and semiaquatic Heteroptera (Gerromorpha, Nepomorpha, and Leptopodomorpha). Modern gerromorphan lineages can be traced back to the Mesozoic by fossil evidence and can be further supported from Gondwanan distributions, assuming vicariant explanations (South America/Africa, *Platyvelia* Polhemus and Polhemus/*Angilia* Stål (Veliidae); South America/Australasia, *Metrobates* Uhler/*Metrobatoides* Polhemus and Polhemus (Gerridae)).

Based on their own and other colleagues studies, Polhemus and Polhemus (2008) assumed that the main areas of endemicity for aquatic Heteroptera in the world are (1) Madagascar, (2) New Guinea, (3) Indochina, (4) the Malay Archipelago, (5) Australia, (6) tropical Central and Western Africa, (7) the Guiana Shield of northern South America, and (8) the Atlantic rainforests of eastern South America. The northern South American Guiana Shield area is species-rich for aquatic Heteroptera, according to Polhemus and Polhemus (2008) due to its ancient geological age, proximity to the Equator, and topographic complexity. The Guiana Shield has many endemic species, but still most of its genera occur in other areas of tropical South America.

Leptopodomorpha

Before the work of Grimaldi et al. (2013), the Leptosaldinae (Leptopodidae) were known from two Neotropical species, *Leptosalda chiapensis* Cobben, a fossil from Mexican Miocene amber, and the extant *Saldolepta kistnerorum* Schuh and Polhemus, from Ecuador and Colombia. Grimaldi et al. described two new fossil species of *Leptosalda* Cobben from Miocene amber of the Dominican Republic and a remarkable new fossil genus and species, *Archaesalepta schuhi* Grimaldi and Engel, from Early Eocene Cambay amber from western India. They believe *A. schuhi* is the sister taxon to the New World leptosaldines, suggesting leptosaldines were more widespread and that their present-day distribution is relict. Their conclusions should be taken cautiously, since the proposed relationships and biogeographic hypothesis are not based on a phylogenetic analysis.

Cimicomorpha

The understanding of Triatominae (Reduviidae) evolution still puzzles the knowledge about its ancestral areas and its geographical distribution patterns. Trying to answer some biogeographical questions, Schofield (1988, 2000) proposed a polyphyletic origin of triatomines based on biogeographic arguments, with Asian fauna being composed of at least two independent lineages (Hypsa et al. 2002). In a phylogenetic analysis of the New and Old World, including species of the Rhodniini, Linshcosteini, and Triatomini (Reduviidae: Triatominae), Hypsa et al. (2002) found conflicting results (dependent of rooting) for the origin of Triatominae, suggesting either an origin in the northern areas of South America, in Central America, or in southern North America. Hypsa et al. also found a clade of Asian species within the Triatominae.

Studying the Tingidae *sensu stricto* in a phylogenetic framework, Lis (1999) erected the Cantacaderini to family status and applied Bremer's (1992) ancestral areas to explore its biogeography. The ancestral area of Cantacaderidae (*sensu* Lis 1999) was smaller than the current distribution, included the Australian continent, and the taxon probably originated around 140 million years ago (mya). Along a dispersalist narrative, Lis (1999) argues that the sister taxa of the Neotropical genera *Stenocader* Hambleton and *Nectocader* Drake are from New Zealand and Australia, respectively, suggesting the Cantacaderidae colonization of the Neotropics via two dispersal events through a connection between Antarctica and South America, about 50 mya. However, according to Guilbert (2012a), the Vianaidinae (an exclusively South American taxon) is sister to all other Tingidae, including the Cantacaderinae (as a subfamily of Tingidae). This relationship hypothesis would imply that the common ancestor to Vianaidinae and Cantacaderinae could not be restricted to Australia, but should rather have a Gondwanan distribution (Guilbert 2012b). However, this author agrees with Lis (1999) and Wappler (2006) that the Cantacaderinae colonized the Neotropics by ancient vicariance followed by several recent dispersal events.

Pentatomomorpha

Based on evidence of Gondwanan origin to Aradoidea, Sweet (2006) proposed that the Aradidae is an old taxon and its worldwide distribution most likely results from vicariance. The Isoderminae (Aradidae) show an "Antarctic distribution," occurring in Chile, New Zealand, and Australia, while other subfamilies (Aradinae, Calisiinae, Mezirinae, Carventinae, and Aneurinae) have cosmopolitan distributions, more diverse in the tropics.

Mecidea Dallas comprises a group of stink bugs (Pentatomidae) occurring in subtropical and adjacent temperate areas, apparently coinciding with xerophytic or semi-xerophytic environments (Sailer 1952). To Sailer, *Mecidea*'s distribution is similar to countless taxa and is evidence of an ancient faunistic and floristic relationship between the semidesert and desert regions of the Mediterranean Basin, South Africa, southern South America, and southwestern North America. According to a dispersalist point of view, the Ethiopia, Eritrea, and Uganda highlands were suggested as the center of distribution of *Mecidea*, because five of the 14 species of *Mecidea* occur in or are adjacent to this area.

Through Bremer's (1992) ancestral area analysis, Lis (1999) suggests the Cephalocteinae (Cydnidae) originated on the Indian continental block (as part of the Gondwanaland) more than 125 mya. From the absence of paleontological data for the taxa, he assumed that only migratory events could explain distribution patterns and uses Gondwanan age for the group to substantiate the events of migration between land masses without resorting to long-dispersal events.

Scaptocoris Perty (Cydnidae: Scaptocorini) reached the Neotropics through colonization from Antarctica, between 125 and 110 mya (Lis 1999); after, it radiated to South America, giving rise to the endemic *Atarsocoris* Becker in Brazil and dispersing to Central America. The distribution of the single species of Scaptocorini known to Africa (Angola) was explained by dispersion of a "*Scaptocoris* ancestral species," from South America to West Africa, and evolved into a separate genus, which displayed the same evolutionary trends in some characters as those in the South American *Atarsocoris*.

3.3.2 Biogeography of the Neotropical Region

Enicocephalomorpha

Štys (2008) reviewed the geographic distribution of the unique-headed bugs' genera of Aenictopecheidae and Enicocephalidae, emphasizing distributional patterns based on zoogeographic regions. According to Štys, the proposition of historical biogeography hypotheses for the group would be premature, given the lack of knowledge for some faunas, particularly of the islands and archipelagos of the Atlantic and Indian Oceans, and the lack of phylogenies at the generic level.

Five enicocephalomorphan genera have been described for the Neotropical region, occurring from the Caribbean islands and southern Mexico to as south as the Tucuman province and Santa Catarina state, in Argentina and Brazil, respectively. Of these genera, four are assigned to the Enicocephalidae and one to the Aenictopecheidae. About 40 unique-headed bug species have been described to the Neotropical region, but it is estimated that at least 100 more have yet to be described (Štys 2008). Additionally, a monotypic genus, *Gamostolus* Bergroth (Aenictopecheidae), occurs on the South American Subantarctic region in southern Chile (Magallanes and Osorno provinces) and Argentina (Staten Island).

Gerromorpha and Nepomorpha

Based on track analysis of 60 species of Belostomatidae, Corixidae, Micronectidae, and Gerridae from Chaco province, Morrone et al. (2004) found five generalized tracks and three panbiogeographic nodes: Tracks include (1) Bolivia and northwestern and central Argentina (11 spp.); (2) southern Brazil, eastern Bolivia, Paraguay, and northeastern Argentina (four spp.); (3) southeastern Brazil and northeastern Argentina (eight spp.); (4) southeastern Brazil, Uruguay, and central western Argentina (five spp.); and (5) southern Argentina (three spp.). Nodes include (1) northeastern Argentina, (2) central Argentina, (3) and central south Argentina. The authors stated that the Chaco province appears to be a natural biogeographical area that shares taxa with the Amazonian, Parana, and Patagonian biogeographic provinces.

Leptopodomorpha

Trying to answer whether the Andean distribution of *Pseudosaldula* (Saldidae) is part of a broader Austral distributional pattern (including New Zealand, Australia, and part of Antarctica), Schuh and Polhemus (2009) discuss the genus' biogeography based on a morphological phylogenetic hypothesis. Using a qualitative approach to analyze the genus' distribution, they recognized five areas of endemism: Northern Andes, Northern Peru, Puna, Central Chile, and Subantarctic. Using two different cladograms obtained by different optimization criteria (equal weights and implied weighting), area cladograms were constructed, and they found either the Northern and Southern Andes being commingled or a great hierarchical structuring of the Andes, with southern areas basal on the cladogram. The following hypothesis for the biogeography of the group, is proposed (*sic* Schuh and Polhemus 2009): (1) Subantarctic is the basal area on the cladogram; (2) Central Chile is the next area on the cladogram, and either *P. penai* Schuh and Polhemus or *P. pilosa* Schuh and Polhemus are endemic to it; (3) Puna is the next area on the cladogram; and (4) Northern Peru and Northern Andes are sister areas.

Cimicomorpha

Neotropical species of *Rahasus* Amyot and Serville (Reduviidae: Peiratinae) and other three related genera (*Eidmania* Teuber, *Melanolestes* Stål, and *Thymbreus* Stål) were studied by Morrone and Coscarón (1998), using a cladistic biogeographic approach. General area cladograms were constructed using two methods (Component 2.0 by Page 1993 and paralogy-free subtree, Nelson and Ladiges 1996) and resulted in scenarios previously proposed by Morrone and Coscarón (1996): "open vegetated" provinces (Desierto, Caatinga, Cerrado, and Chacoan), found basally on the trees, and "forest" provinces (Caribbean, Amazonian, Paraná, and Atlantic). The general area cladogram was in agreement with the previous scenario proposed by Morrone and Coscarón (1996) and resulted in "open vegetated" provinces (Desierto, Caatinga, Cerrado, and Chacoan) at the base, whereas the "forest" provinces (Caribbean, Amazonian, Paraná, and Atlantic) formed a group. Morrone and Coscarón (1998) relate the formation of the South American arid diagonal with the uplift of the Andes and hypothesized it as a vicariant event that separated the forested areas into two portions (Caribbean + Amazonian and Paranaense + Atlantic) during the middle Miocene. This can be considered one of the most comprehensive cladistic biogeography analyses searching for general patterns on Neotropical Heteroptera.

Abad-Franch and Monteiro (2007) presented an approximation to the historical biogeography and evolution of the main triatomine lineages (*Panstrongylus* Berg and *Triatoma* Laporte) that occur in the greater Amazon. They found the distributional pattern of the Rhodniini in the Amazon forest to be associated with two biogeographically well-defined groups of species ("*pictipes* group" and "*robustus* group," with trans- and cis-Andean distribution, respectively), invoking a basal evolutionary split within the tribe in this area. The origin of Amazonian triatomine species was suggested to be associated with historical, ecological, and anthropogenic ecological disturbances.

Paula et al. (2007) proposed a biogeographic hypothesis to species of Rhodniini (Reduviidae: Reduviinae), based on the historical background of the Neotropical region proposed by Amorim (2009). The authors employed a reconciled tree method to deduce taxon-area associations, implemented in TreeMap (Charleston and Page 2001). Twelve optimal solutions were found to explain Rhodniini biogeographical scenarios, with six vicariance events, 20 duplications (sympatry), at least three dispersals, and one extinction event. However, Paula et al. (2007) decided not to use any of these reconstructions, instead opting to discuss specific scenarios for some lineages that are not necessarily present in all optimal solutions. The main events evoked to explain biogeographic patterns of the group are the uplift of the Central Andes and Andes breakup into three separate cordilleras, emergence of the Isthmus of Panama, and uplift of the Serra do Mar and Serra da Mantiqueira (Brazil). The lack of resolution in the biogeographical history of the group may be the result from poor taxon sampling and inappropriate areas included in the study.

The bee assassin *crassipes* and *pictipes* species groups of *Apiomerus* Hahn (Reduviidae: Harpactorinae) were studied by Berniker and Weirauch (2012). They used the phylogeny to explore biogeographic patterns of 12 species and determine the boundary between Nearctic and Neotropical areas of endemism. Their results support the limits between the Neotropical and Nearctic regions along the Isthmus of Tehuantepec (Mexico), congruent with previous biogeographical analyses for New World insects (e.g., Halffter 1987; Morrone and Marquez 2001). Under a Brooks's parsimony analysis (BPA), Berniker and Weirauch (2012) corroborate the relationship between Chiapan-Guatemalan Highlands, Talamancan Cordillera, and northwest South American areas and together are sister to the remainder of the Neotropics; this pattern was found by previous studies, such as Amorim and Pires (1996) and Amorim (2001). However, this assertion must be taken cautiously, since the authors did not use a BPA area classification with the intent of testing the assumptions of Amorim and Pires (1996) of an Amazonia composed of two distinct biotic components.

Pentatomomorpha

Brochymena Amyot and Serville and *Parabrochymena* Larivière are the only representatives of Halyini (Pentatomidae: Pentatominae) occurring in the New World: *Brochymena* all over North and Central America between 50°N and 15°N and *Parabrochymena* only in the eastern United States, southeastern Canada, and Central America. Generally, Larivière (1994) explains the distribution of both genera based on major geological events and ecological aspects of North America, particularly assumptions concerning climate change. Under a dispersalist framework, he asserts that areas of greatest diversity of *Brochymena* and *Parabrochymena* correspond well with areas occupied by hypothesized oldest lineages of the genera. He speculates there is a general trend of diversity reduction towards northern latitudes, perhaps due to Pleistocene glaciation, but also pinpoint there is an overall pattern observed in pentatomids to not extend far beyond warm temperate conditions.

Grazia (1997) explored the biogeographical relationships of the *Evoplitus* genus group (*Evoplitus* Amyot and Serville, *Pseudevoplitus* Ruckes, and *Adevoplitus* Grazia and Becker; Pentatomidae: Pentatominae). *Evoplitus* occurs on the Atlantic Forest, including Paraguay and northern Argentina, and its sister group, *Adevoplitus* + *Pseudevoplitus*, is distributed in Central America, northern Colombia, and eastern Venezuela and Amazonian Basin. The contact zone between the two clades is along the Amazonas, Madeira, and Mamoré rivers. Under the same reasoning employed by Grazia (1997), the biogeography of the Neotropical *Brachysthetus* Laporte was studied by Barcellos and Grazia (2003). In both works, the authors find distributional patterns congruent with the hypotheses of Amorim and Pires (1996) and Amorim (2001) of two distinct Amazonias, the formation of an epicontinental sea in the Maracaibo region, after the Late Cretaceous separated an area in Mesoamerica from the remaining Amazonian elements.

Fortes and Grazia (2005) and Simões et al. (2012) studied *Serdia* Stål (Pentatomidae: Pentatominae), a Neotropical genus distributed from Costa Rica to southern Brazil and northern Argentina, with highest diversity in southern and southeastern Brazil. Reanalyzing the morphological matrix of Fortes and Grazia (2005), Simões et al. (2012) proposed biogeographic hypotheses to *Serdia* based on vicariance analysis (Hovenkamp 1997, 2001), which consists of finding the disjunct (allopatric/vicariant) distributions in the nodes of the cladogram. They identified three nodes sharing disjunct distributions, which might be associated to vicariant events related to the development of the Chacoan subregion (Morrone 2006) that, during the Paleogene and Neogene (former Tertiary), divided the once continuous Amazonian-Parana Forest.

The ancestral area of *Schraderiellus* group (Pentatomidae: Discocephalinae) is the Amazonian subregion, from where the species successively dispersed (Campos and Grazia 2006). The southern distributional limits of the *Schraderiellus* group, the Amazonas, Madeira, and Mamoré river basins, can also be explained by the independent evolution of biota from the northwest and southeast of Amazonia proposed by Amorim (2001). The *Schaefferella* group (Pentatomidae: Discocephalinae), widely distributed in Central and South Americas, is believed to have originated in the Paraná subregion with further expansions to the north, towards southeastern Amazonian subregion, and to the south, towards the Chacoan subregion. The distributions of *Uvaldus* Rolston and *Clypona* Rolston (Pentatomidae: Discocephalinae), restricted to the *Araucaria* Forest and Chaco, respectively, could be the result of vicariant events separating the Parana and Chacoan subregions.

Procleticini (Pentatomidae: Pentatominae) is a small tribe (about 30 species in 11 genera) found only in the New World. According to Bernardes et al. (2009), the group *Lobepomis* Berg + *Neoderoploa* Pennington + *Procleticus* Berg and basal species of *Thoreyella* Spinola are restricted to open vegetation formations of Neotropical region (the Chacoan subregion), and the most apical *Thoreyella* species are more related to forested habitats in the Parana Forest subregion.

3.3.3 Distributional Patterns of Neotropical Heteroptera

Recently, studies of distribution patterns within Heteroptera have increased but are still largely biased towards better-known taxa. In most of those studies, authors point out to the lack of phylogenetic hypotheses and outdated phylogenies for the group and stress their studies are early approaches to the understanding of distributional patterns of the taxa. Macroecological and biogeographic aspects of Heteroptera are hampered from being further understood because of the lack of accessible online databases of biological collections and more research on phylogenetic relationships.

One of the best studied groups of Heteroptera is the Triatominae (Reduviidae). The group has been employed in macroecological studies in attempts to understand its richness and distribution patterns on the Neotropics (e.g., Rodriguero and Gorla 2004; Diniz-Filho et al. 2013; Fergnani et al. 2013).

Rodriguero and Gorla (2004) examined how New World Triatominae species richness responds to latitudinal gradients and explored the relevance of geographical area and available energy. Based on the distribution of 118 triatomine species, the authors found for the first time in obligatory hematophagous organisms an evident increase in the number of species of Triatominae from higher to lower latitudes. Such a pattern is not expected to insects (Rodriguero and Gorla 2004), but is found in some mammalian taxa. Habitable area effect on species richness is different in each hemisphere: on the Southern Hemisphere, an increase in species richness is correlated with an increase in habitable area from the poles towards the Equator, whereas on the Northern Hemisphere, there is an increase of species richness towards low latitudes; habitable area does not affect this relationship. There is a significant correlation of species richness with habitable area, mainly in the Southern Hemisphere and eastern longitudes.

Diniz-Filho et al. (2013) tested different environmental hypotheses to understand geographical patterns of richness and the distribution of 115 Neotropical triatomine species. Testing seven different hypotheses of climatic effects as drivers of species richness, using spatial eigenvector mapping and nonspatial ordinary least squares multiple regression models, they found water, energy, temperature, and temperature seasonality to be the environmental variables with relatively higher explanatory power of species richness and elevation variables having minor effects. The unique effects of these variables are quite difficult to disentangle because of the collinearity among variables and residual autocorrelation and may also be due to low taxonomic sampling.

Fergnani et al. (2013) conducted a complementary study to understand large-scale spatial patterns of morphological diversity and species richness of triatomine species in terms of environmental gradients. Studying 91 species and 12 morphological attributes of Neotropical triatomine, they found the latitudinal gradient of species richness to be in agreement with previous studies (Rodriguero and Gorla 2004; Diniz-Filho et al. 2013). The morphological diversity, by means of Gower index, also followed a latitudinal gradient, suggesting an overall spatial congruence between species richness and morphological diversity. However, the overall correlation pattern between species richness and morphological diversity is not homogeneous throughout taxon distribution (e.g., Brazil and Argentina have regions of high species diversity, but low morphological diversity). According to Fergnani et al. (2013), their's was the first study where this kind of relationship was demonstrated for insects on a continental scale, demonstrating that for the Triatominae, species richness and morphological diversity cannot be considered substitutes for one another.

One of the few studies that looked for distribution patterns exclusively on Neotropical Pentatomidae (Hemiptera) was done by Ferrari et al. (2010), analyzing the distribution of 222 species belonging to 14 genera using analysis of endemicity (NDM – Szumik et al. 2002; Szumik and Goloboff 2004) and identifying hierarchical endemic areas in the Atlantic Forest. The Amazonian region was identified as a single area on the consensus, its southeastern portion share elements with the Chacoan and Paraná subregions, while the Cerrado and the Caatinga were not identified as unique areas of endemism.

3.3.4 Phylogeography

Almost 30 years after Avise et al. (1987) coined the term phylogeography, two major papers reviewed the subject, including works dealing with the Neotropics. As emphasized by Beheregaray (2008), out of the 3049 phylogeographic papers published around the world until 2006, only 15 % explored Southern Hemisphere organisms; the numbers are even lower if we consider only published articles about terrestrial invertebrates on South America (8 % out of 313 papers). Focusing on South America, Turchetto-Zolet et al. (2013) indicate that, from 1987 to 2011, only 29 studies on invertebrates were published, most emphasizing in vectors of human diseases, such as Chagas disease (Hemiptera: Reduviidae) and malaria (Diptera: Culicidae). When compared to other biogeographical regions and continents, phylogeographical studies in South America are still incipient, the same happening in the rest of the Neotropical Region.

The first effort on phylogeography of Heteroptera was populational studies about species of economic or medical importance (e.g., Monteiro et al. 1999; Sosa-Gomez et al. 2004), triggering studies which focused on Triatominae vectors of *Trypanosoma cruzi* (Chagas), a parasitic euglenoid protozoan that causes Chagas disease. The Triatominae (Reduviidae), known as kissing bugs due their hematophagous feeding habit, are distributed from Argentina to the United States, between the latitudes 46°S and 46°N (García et al. 2013). Of the 15 known genera of Triatominae, 14 occur in the New World (Gourbière et al. 2012), and almost all of the 145 species recognized for Triatominae (Gonçalves et al. 2013) seem to be potential vectors of trypanosomatids. Three species are of greatest epidemiological relevance: *Triatoma infestans* Klug, *Triatoma brasiliensis* Neiva, and *Triatoma dimidiata* (Latreille).

Triatoma infestans is the most studied species because, alone, it is responsible for half of the Chagas disease cases in the Neotropics. The evolutionary lineage of *T. infestans* originated between 2.7 and 1.0 mya as a derivation from *T. platensis* Neiva (Bargues et al. 2006). Bargues et al. (2006) suggest *T. infestans* originated from Bolivian highlands, and the current lineages allows the identification of clear differences between Andean and non-Andean populations. Two initial dispersal routes of *T. infestans* were hypothesized, one through the Andean highlands of Bolivia and Peru and another through the lowlands of Argentina, Paraguay, Uruguay, Brazil, and Chile. Both these primary lineages are corroborated by independent nuclear, mitochondrial, and multilocus microsatellite studies (Monteiro et al. 1999; Piccinali et al. 2009; Pérez-de-Rosas et al. 2011). Torres-Pérez et al. (2010) suggested that *T. infestans* would have originated outside of Bolivia, but they also recognized the Andean and non-Andean lineages with estimated divergence time around 0.388–0.588 mya, agreeing globally with the estimate dates found by Bargues et al. (2006). On the other hand, the widespread distribution of *T. infestans* in South America is attributed to its domiciliation, a probable recent event following the New World colonization by humans around 12,000 years ago (González-José et al. 2003) and main range expansion in post-Colombian times, after the seventeenth century until today (Panzera et al. 2004).

It is believed that the contemporary population structure has been dramatically influenced by human-vector interactions, as indicated by the complex biogeographic pattern of *T. infestans*, at both local and regional scales (Torres-Pérez et al. 2010). The relationship between human dwellings and *T. infestans* has elicited studies aiming to understand: insecticide effectiveness through haplotype diversity/persistence, the fine-scale migration ability, offspring viability between populations/morphs, capability of house infestation, and reinfestation of sylvatic and peridomestic populations (e.g., Marcet et al. 2008; Pérez-de-Rosas et al. 2008, 2013; Piccinali et al. 2009, 2011), as resources for developing rational vector control strategies (Piccinali et al. 2009).

In Northeastern Brazil, *T. brasiliensis* is the most important Chagas disease vector. Monteiro et al. (2004) analyzed the populations of *T. brasiliensis* across its distribution and indicated that each chromatic form is genetically different from each other, suggesting the existence of a species complex: the forms *juazeiro*, *melanica*, and *brasiliensis* + *macromelasoma* should be treated as separate species. Nowadays, *T. juazeirensis* Costa & Felix and *T. melanica* Neiva & Lent are considered species of the *T. brasiliensis* complex. The population structure of *brasiliensis* does not show strong geographic segregation, suggesting recent range expansion. The divergence time between *brasiliensis* and *melanica* is estimated to have occurred around 5.2 mya, which would place their common ancestor in the early Pliocene. On the other hand, *brasiliensis* + *macromelasoma* population structure does not show strong geographic segregation, suggesting recent range expansion.

Triatoma dimidiata is the main vector of *T. cruzi* throughout Central America to Ecuador. The diversity of habitats explored by *T. dimidiata* involves domestic and peridomestic dwellings and non-domiciliated and sylvatic populations, act as sources of reinfestation. Along the distribution of *T. dimidiata*, some populations/subspecies are suggested by phenotypical, genotypical, and ethological traits. Bargues et al. (2008) analyzed its intraspecific variability, finding high haplotypic diversity supporting three groups: group 1, widespread (Colombia, Ecuador, Guatemala, Honduras, Mexico, Nicaragua, and Panama) and highly variable; group 2, geographically restricted to Guatemala and Mexico; and group 3, occurring in Guatemala, Honduras, and Mexico. The high haplotypic variability found in *T. dimidiata* seems to be remarkably outside the limits of the intraspecific variability range known for *Triatoma* species (Bargues et al. 2008; Monteiro et al. 2013).

Monteiro et al. (2013) refined the analyses of Bargues et al. (2008) by adding different molecular markers and finding four genetic groups for *T. dimidiata*. Both groups of authors found a complex of monophyletic cryptic species including the subspecies of *T. dimidiata* plus *T. hegneri* Mazzotti. The *Triatoma dimidiata* species complex is believed to have originated 5.9–10.5 mya. Hypotheses to each population/subspecies origin, expansion, and distribution are explored on the literature (Bargues et al. 2006; Monteiro et al. 2013; Gómez-Palacio and Triana 2014) and include distributional expansions by relatively recent human activity, mountain uplifts, and the connection of South and North Americas through the Isthmus of Panama during the Pliocene (3–5 mya).

Other Neotropical reduviid genera have been studied under phylogeographical approaches. On the Amazonian region, the phylogeography of *Rhodnius prolixus* Stål and *R. robustus* Larrousse was studied by Monteiro et al. (2003), sampling in seven Latin American countries. Five clades were found, being *R. prolixus* homogeneous. *Rhodnius robustus* is represented by four paraphyletic clades, one of them (*R. robustus* Clade I) more closely related to the *R. prolixus* clade. The *R. prolixus* and *R. robustus* Clade I is believed to have originated around 1.4 mya, when an ancestral stock was spread in different refugia at the Orinoco lowland forest (Venezuela). The other *R. robustus* clades are from the Amazon forest region. Extremely low nucleotide diversity of *R. prolixus* suggests a recent bottleneck and posterior dispersion facilitated by human activity. The clade *R. prolixus*+*R. robustus senso lato* suggests an origin in between 3.7 and 2.4 mya. Dating within the Pleistocene, the pattern of phylogenetic discontinuity with geographical distribution of haplotypes could involve long-term biogeographical barriers to gene flow (Avise et al. 1987) and could be explained by the refugium theory (Monteiro et al. 2003).

Maia-da-Silva et al. (2007) compared the phylogeny of *Rhodnius* with *Trypanosoma rangeli* Tejera, 1920 populations, showing a significant overlap in the distribution and demographic correspondence between the *Rhodnius* spp. and *T. rangeli* lineages. The pattern is consistent with a hypothetical long parasite-vector coexistence with and is supported by a high congruence between the phylogeographical analysis of both *T. rangeli* lineages and *Rhodnius* species.

Mepraia Mazza, Gajardo, and Jörg, comprised by three species, is an important endemic Chilean vector of *T. cruzi* in the sylvatic cycle. Campos et al. (2013) found three lineages of *Mepraia*, congruent with the current recognized species. All populations tested were highly structured, suggesting that they have not been affected by strong bottlenecks and/or experienced sudden demographic changes due to repeated climatic fluctuations. The origin of *Mepraia* is suggested to have occurred around 3.6 mya, during the pre-Pleistocene. *Mepraia spinolai* (Porter) is recognized as the oldest lineage, followed by *M. gajardoi* Frias, Henry, and Gonzales (originating 0.99 mya) and *M. parapatrica* Frías-Lasserre (originating 0.66 mya).

The southern green stink bug *Nezara viridula* (L.) (Pentatomidae) is a polymorphic and worldwide pentatomid pest, causing economic damage to many crops (Panizzi et al. 2000). To elucidate its origin and dispersion routes, Karvar et al. (2006) sampled specimens from four continents (Africa, America, Europe, and Asia), and Li et al. (2010) added populations from China and Iran, both agreeing in a basal phylogeographic position of *N. viridula* from Africa and suggesting a deep division between African and non-African populations. Using different molecular clocks, Karvar et al. (2006) suggested the division of African and non-African populations of *N. viridula* since the Pliocene, while Li et al. (2010) suggested the split during the Miocene. Both groups of authors concluded that the New World populations of *N. viridula* are more closely related to European populations. Karvar et al. (2006) hypothesize different routes for the American colonization: one from the eastern Mediterranean (Greece, Italy) to Central America and from there to eastern United States and the western coastal areas of South America and another originating in the western Mediterranean (Iberian Peninsula) and dispersing to the eastern coastal areas of South America. The relationships among Neotropical populations should be better investigated, as well as recent human activity-mediated dispersion.

As showed above, most phylogeographic studies are biased to species with medical relevance and economic impact, and the aim of these studies are often intertwined with applied interests. Nevertheless, as side results, speciation hypotheses, taxa coalescence, population demography, and geographical processes have been improved to those taxa on the Neotropical region. Furthermore, the integrative nature of phylogeography and increased access to molecular evidences are helping elucidate intraspecific relationships, influencing phylogenetic hypotheses and taxonomic decisions.

Phylogeographic studies on the Neotropics are incipient, especially to terrestrial invertebrates (Turchetto-Zolet et al. 2013), and most Neotropical biomes, such as the Amazonia, Cerrado, Atlantic Forest, Pampa, and Caribbean islands, are poorly known. The current systematic, ecological, and distributional knowledge about Neotropical heteropterans make them a useful source to phylogeography. The phylogeography could support the identification of refugia and contact zones, enlightening the biogeographical processes and improving the understanding of lineages history and processes of diversification.

3.4 Concluding Remarks

Much remains to be studied regarding the systematics and biogeography of Heteroptera, especially on the Neotropics. Despite the high diversity of true bugs, most families occurring in the Neotropical region are neglected. In fact, the best known taxa are those of economical or medical importance (e.g., Reduviidae, Pentatomidae), although Pentatomidae lacks an updated Neotropical catalog. In the last decades, though, the training of new systematists has improved the knowledge on some families, as will be presented in the following chapters. However, it is time to expand the scope of revisionary, systematic, and biogeographic studies.

There is a general and global lack of phylogenetic studies for Neotropical taxa, especially with those taxa of broad taxonomic coverage and consistent out-group sampling. The search for heteropteran biogeographic patterns is hampered by the lack of phylogenies and a clear use of biogeographic methodologies. Phylogeographic studies are incipient, focused on taxa of economic importance, even though the integrative nature of the subject can help elucidate the evolutionary history on the Neotropics.

Overall, future works should incorporate new techniques for data gathering and new methodologies for data analysis and become taxonomically broader and comprehensive. All of it would facilitate the understanding of the evolutionary history of the different groups.

Acknowledgments We thank M. Guidotti (UFRGS) for his help with the Cimicomorpha phylogenetic section, the financial support from the Conselho Nacional de Desenvolvimento Científico e Tecnológico (CNPq) (Ed. Universal 470796/2012-0), a postdoctoral fellowship from the Coordenação de Aperfeiçoamento de Pessoal de Nível Superior (CAPES) to A. Ferrari (PNPD 02637/09-0 and CAPES/FAPERGS 07/2012), and doctoral fellowships to K.R. Barão (CNPq 142447/2011-0, CAPES/PDSE BEX5641/13-6) and F.M. Bianchi (CAPES/PDSE), and research fellowships from CNPq to J. Grazia and L.A. Campos are also greatly acknowledged.

References

Abad-Franch F, Monteiro FA (2007) Biogeography and evolution of Amazonian triatomines (Heteroptera: Reduviidae): implications for Chagas disease surveillance in humid forest ecoregions. Mem Inst Oswaldo Cruz 102:57–70

Amorim DS (2001) Dos Amazonias. In: Llorente-Bousquets J, Morrone JJ (eds) Introducción a La biogeografia em Latinoamérica: teórias, conceptos, métodos y aplicaciones. Facultad de Ciencias, UNAM, México, pp 245–255

Amorim DS (2009) Neotropical Diptera biogeography. In: Pape R, Bickel D, Meier R (eds) Diptera diversity: status, challenges and tools. Brill, Boston

Amorim DS, Pires MRS (1996) Neotropical biogeography and a method for maximum biodiversity estimation. In: Bicudo CEM, Menezes NA (eds) Biodiversity in Brazil, a first approach. CNPq, São Paulo, pp 183–219

Andersen NM (1981) A fossil water measurer (Insecta, Hemiptera, Hydrometridae) from the Paleocene/ Eocene of Denmark and its phylogenetic relationships. Bull Geol Soc Den 30:91–96

Andersen NM (1982) The semiaquatic bugs (Hemiptera, Gerromorpha). Phylogeny, adaptations, biogeography, and classification. Entomonography 3:1–455

Andersen NM (1999) The evolution of marine insects: phylogenetic, ecological and geographical aspects of species diversity in marine water striders. Ecography 22:98–111

Andersen NM (2000) Fossil water striders in the Eocene Baltic amber (Hemiptera, Gerromorpha). Insect Syst Evol 31:257–284

Andersen N, Polhemus DA (2003) A new genus of terrestrial mesoveliidae from the Seychelles (Hemiptera: Gerromorpha). J N Y Entomol Soc 111(1):12–21

Andersen NM, Weir TA (2004) Mesoveliidae, Hebridae, and Hydrometridae of Australia (Hemiptera: Heteroptera: Gerromorpha), with a reanalysis of the phylogeny of semiaquatic bugs. Invertebr Syst 18:467–522

Avise JC, Arnold J, Ball RM, Bermingham E (1987) Intraspecific phylogeography: the mitochondrial DNA bridge between population genetics and systematic. Annu Rev Ecol Syst 18:489–522

Barcellos A, Grazia J (2003) Cladistic analysis and biogeography of Brachystethus Laporte (Heteroptera, Pentatomidae, Edessinae). Zootaxa 256:1–14

Bargues MD, Klisiowics DR, Panzera F, Noireau F, Marcilla A, Perez R, Rojas MG, O'Connor JE, Gonzales-Candelas F, Galvão C, Juberg J, Carcavallo RU, Dujardin JP, Mas-Coma S (2006) Origin and phylogeography of the Chagas disease main vector *Triatoma infestans* based on nuclear rDNA sequences and genome size. Infect Genet Evol 6:46–62

Bargues MD, Klisiowicz DR, Gonzalez-Candelas F, Ramsey JM, Monroy C, Ponce C, Salazar-Schettino PM, Panzera F, Abad-Franch F, Sousa OE, Schofield CJ, Dujardin JP, Guhl F, Mas-Coma S (2008) Phylogeography and genetic variations of *Triatoma dimidiata*, the main Chagas disease vector in Central America, and its position within the genus *Triatoma*. Plos Negl Trop D2:e233

Beheregaray L (2008) Twenty years of phylogeography: the state of the field and the challenges for the Southern Hemisphere. Mol Ecol 17:3754–3774

Berg C (1879) Hemiptera Argentina enumeravit speciesque novas descripsit. Pauli E. Coni, Bonariae

Berg C (1884) Addenda et emmendanda ad Hemiptera Argentina. Paul E. Coni, Bonariae

Bergroth E (1891) Contributions a l'etude des pentatomides. Rev Entomol 10:200–235

Bergroth E (1893) Descriptions of some Rhynchota of geographical interest. Entomol Mon Mag 29:61–63

Bergroth E (1894) Liste de quelques Hémiptres de la Plata. Rev Entomol 13:164–167

Bergroth E (1905) Rhynchota neotropica. Rev Entomol 24:104–112

Bergroth E (1908) Enumeratio Pentatomidarum post Catalogum bruxellensem descriptarum. Mem Soc Entomol Belg 15:131–200

Bergroth E (1914) Pentatomides nouveaux de la Guyane Française (Hémipt. Pentatomidae). Ann Soc Entomol Fr 83:423–441

Bergroth E (1918) Hendecas generum Hemipterorum novorum vel subnovorum. Ann Hist Nat Mus Nat Hung 16:298–314

Bernardes JL, Schwertner CF, Grazia J (2009) Cladistic analysis of *Thoreyella* and related genera (Hemiptera: Pentatomidae: Pentatominae: Procleticini). Zootaxa 2310:1–23

Berniker L, Weirauch C (2012) New World biogeography and the evolution of polychromatism: evidence from the bee assassin genus Apiomerus (Heteroptera: Reduviidae: Harpactorinae). Syst Entomol 37:32–54

Breddin G (1903a) Beiträge zur Hemipteren-fauna der Anden. Sitz Ges Freunde Berlin 1903:366–383

Breddin G (1903b) Neue Rhynchoten aus den Anden. Soc Entomol 18:122–125

Breddin G (1903c) Neue Arten der Gattung Edessa Fab. Soc Entomol 18:131–132

Breddin G (1903d) Neue Arten der Gattung Edessa Fab. (Schluss.). Soc Entomol 18:139–140

Breddin G (1904a) Neue Rhynchotenausbeute aus Süs-America. Soc Entomol 18:147–149

Breddin G (1904b) Neue Rhynchotenausbeute aus Süd-Amerika (Fortsetzung). Soc Entomol 18:153–154

Breddin G (1904c) Neue Rhynchotenausbeute aus Süd-Amerika (Fortsetzung). Soc Entomol 18:177–178

Breddin G (1904d) Neue Rhynchotenausbeute aus Süd-Amerika (Fortsetzung). Soc Entomol 19:49–50, 58, 147–149

Breddin G (1907) Beiträge zur Systematik der Pentatomiden Südamerikas. Erstes Stück. Sitz Ges Freunde Berl 1907:327–348

Breddin G (1908a) Zwei neue Nezara-Arten (Rhynchota Heteroptera). Soc Entomol 23:68–69

Breddin G (1908b) Beiträge zur Systematik der Pentatomiden Südamerikas. Zweites Stück. Sitz Ges Freunde Berl 1908:24–36

Breddin G (1909) Beiträge zur Systematik der Pentatomiden Südamerikas. Drittes Stück. Sitz Ges Freunde Berl 1909:154–161

Breddin G (1910) Beiträge zur Systematik der Pentatomiden Südamerikas. Viertes Stück. Sitz Ges Freunde Berl 1909:615–631

Breddin G (1912a) Beschreibung einer Art der Pentatomiden-Gattung Edessa Fabr. (Hem.). Arch Naturgesch 78:89–90

Breddin G (1912b) Zwei neue neotropische Pentatomiden-Gattungen. (Hem.). Arch Naturgesch 78:90–93

Breddin G (1912c) Hemiptera tria neotropica. Entomol Mitteilungen 1:351–354

Breddin G (1914) Neue odor wenig gekannte neotropische Hemiptera. Abh Senckenb Naturforsch Ges 36:53–59

Bremer K (1992) Ancestral areas: a cladistic reinterpretation of the center of origin concept. Syst Biol 41:436–445

Campos LA, Grazia J (2006) Cladistic analysis and biogeography of Ochlerini (Heteroptera, Pentatomidae, Discocephalinae). Iheringia Sér Zool 96:147–163

Campos R, Torrez-Pérez F, Botto-Mahan C, Coronado X, Solani A (2013) High phylogeographic structure in sylvatic vectors of Chagas disease of the genus *Mepraia* (Hemiptera: Reduviidae). Infect Genet Evol 19:280–286

Carvalho JCM (1952) On the major classification of the Miridae (Hemiptera). (With keys to subfamilies and tribes and a catalogue of the world genera.). An Acad Bras Cienc 24:31–110
Carver M, Gross GF, Woodward TE (1991) Hemiptera (bugs, leafhoppers, cicadas, aphids, scale insects, etc.). In: Naumann ID, Carne PB, Lawrence JF, Nielsen ES, Spradbery JP (eds) The insects of Australia. Melbourne University Press, Carlton, pp 429–509
Cassis G, Schuh RT, Brailovsky H (1999) A review of *Onymocoris* (Heteroptera: Thaumastocoris), with a new species, and notes on hosts and distributions of other thaumastocorid species. Acta Soc Zool Bohemoslov 63:19–36
Charleston MA, Page RDM (2001) TreeMap for Macintosh. Version 2.0. http://taxonomy.zoology.gla.ac.uk/rod/treemap.html. Accessed 30 Jul 2014
Cobben RH (1968) Evolutionary trends in Heteroptera. Part I. Eggs, architecture of the shell, gross embryology and eclosion. Centre for Agricultural Publishing and Documentation, Wageningen
Cobben RH (1978) Evolutionary trends in Heteroptera, Part II. Mouthpart-structures and feeding strategies. Centre for Agricultural Publishing and Documentation, Wageningen
Craw RC, Grehan JR, Heads MJ (1999) Panbiogeography: tracking the history of life. Oxford University Press, New York
Croizat L (1958) Panbiogeography. Published by the Author, Caracas
Croizat L (1964) Space, time, form: the biological synthesis. Published by the Author, Caracas
Damgaard J (2008) Phylogeny of the semiaquatic bugs (Hemiptera-Heteroptera, Gerromorpha). Insect Syst Evol 39:431–460
Damgaard J (2012) What do we know about the phylogeny of the semi-aquatic bugs (Hemiptera: Heteroptera: Gerromorpha)? Entomol Am 118:81–98
Damgaard J, Cognato AI (2003) Sources of character conflict in a clade of water striders (Heteroptera: Gerridae). Cladistics 19:512–526
Damgaard J, Andersen NM, Meier R (2005) Combining molecular and morphological analyses of water strider phylogeny (Hemiptera-Heteroptera, Gerromorpha): effects of alignment and taxon sampling. Syst Entomol 30:289–309
Darlington PJ Jr (1957) Zoogeography. The geographical distribution of animals. Wiley, New York
Diniz-Filho JAF, Ceccarelli S, Hasperué W, Rabinovick J (2013) Geographical patterns of Triatominae (Heteroptera: Reduviidae) richness and distribution in the Western Hemisphere. Insect Conserv Divers 6(6):704–714. doi:10.1111/icad.12025
Distant WL (1880) Insecta. Rhynchota. Hemiptera-Heteroptera. Biol Cent Am 1:1–88
Distant WL (1893) Insecta. Rhynchota. Hemiptera-Heteroptera. Biol Cent Am 1:369–462
Donoghue MJ, Moore BR (2003) Toward an integrative historical biogeography. Integr Comp Biol 43:261–270
Drake CJ, Ruhoff FA (1960) Lace-bug genera of the world. (Hemiptera: Tingidae). Proc US Natl Mus 112:1–105
Drake CJ, Ruhoff FA (1965) Lace-bugs of the world: a catalogue. (Hemiptera: Tingidae). Bull US Natl Mus 243:1–643
Fergnani PN, Ruggiero A, Ceccarelli S, Menu F, Rabinovich J (2013) Large-scale patterns in morphological diversity and species assemblages in Neotropical Triatominae (Heteroptera: Reduviidae). Mem Inst Oswaldo Cruz 108:997–1008
Ferrari A, Paladini A, Schwertner CF, Grazia J (2010) Endemism analysis of Neotropical Pentatomidae (Hemiptera, Heteroptera). Iheringia Sér Zool 100:449–462
Fischer C, Mahner M, Wachmann E (2000) The rhabdom structure in the ommatidia of the Heteroptera (Insecta), and its phylogenetic significance. Zoomorphology 120:1–13
Ford L (1979) The phylogeny and biogeography of the Cimicoidea (Insecta: Hemiptera). Masters thesis, University of Connecticut, Storrs, CT, USA
Fortes NDFD, Grazia J (2005) Review and cladistic analysis of *Serdia* Stål (Heteroptera, Pentatomidae, Pentatomini). Rev Bras Entomol 49:294–339
Friedrich F, Matsumura Y, Pohl H, Bai M, Hornschemeyer T, Beutel RG (2014) Insect morphology in the age of phylogenomics: innovative techniques and its future role in systematics. Entomol Sci 17:1–24

Froeschner RC (1996) Lace bug genera of the world, I: Introduction, subfamily Canthacaderinae (Heteroptera: Tingidae). Smithson Contrib Zool 574:1–43
Froeschner RC (2001) Lace bug genera of the world, II: subfamily Tinginae: tribes Litadeini and Ypsotingini (Heteroptera; Tingidae). Smithson Contrib Zool 611:1–28
García BA, Pérez de Rosas AR, Blariza MJ, Grosso CG, Fernandez CJ, Stroppa MM (2013) Molecular population genetics and evolution of the Chagas' disease vector *Triatoma infestan* (Hemiptera: Reduviidae). Curr Genomic 14:316–323
Gómez-Palacio A, Triana O (2014) Molecular evidence of demographic expansion of the chagas disease vector *Triatoma dimidiata* (Hemiptera, Reduviidae, Triatominae) in Colombia. Plos Negl Trop D8:e2734
Gonçalves TCM, Tevez-Neves SC, dos Santos-Mallet JR, Carbajal-de-la-fuente AL, Lopes CM (2013) *Triatoma jatai* sp. *nov*. in the state of Tocantins, Brazil (Hemiptera: Reduviidae: Triatominae). Mem Inst Oswaldo Cruz 108:429–437
González-José R, Gonzalez-Martin A, Hernandez M, Pucciarelli HM, Sardi M, Rosales A, Van-der-Molen S (2003) Craniometric evidence for Palaeoamerican survival in Baja California. Nature 425:62–65
Gourbière S, Dorn P, Tripet F, Dumonteil E (2012) Genetics and evolution of triatomines from phylogeny to vector control. Heredity 108:190–202
Grazia J (1997) Cladistic analysis of the Evoplitus genus group of Pentatomini (Heteroptera: Pentatomidae). J Comp Biol 2:43–48
Grimaldi DA, Engel MS, Singh H (2013) Bugs in the biogeography: Leptosaldinae (Heteroptera: Leptopodidae) in amber from the Miocene of Hispaniola and Eocene of India. J Kans Entomol Soc 86:226–243
Guilbert E (2012a) Phylogeny of Cantacaderinae (Heteroptera: Tingidae) revisited after the description of a new genus and new species from New Caledonia. Eur J Entomol 109:111–116
Guilbert E (2012b) Biogeography of the Cantacaderinae Stål (Insecta: Heteroptera: Tingidae) revisited. Invertebr Syst 6:316–322
Guilbert E, Damgaard J, D'haese CA (2014) Phylogeny of the lacebugs (Insecta: Heteroptera: Tingidae) using morphological and molecular data. Syst Entomol 39:431–441
Halffter G (1987) Biogeography of the montane entomofauna of Mexico and Central-America. Annu Rev Entomol 32:95–114
Heads M (2005) Dating nodes on molecular phylogenies: a critique of molecular biogeography. Cladistics 21:62–78
Hebsgaard MB, Andersen NM, Damgaard J (2004) Phylogeny of the true water bugs (Nepomorpha: Hemiptera-Heteroptera) based on 16S and 28S rDNA and morphology. Syst Entomol 29:488–508
Hennig W (1950) Grundzüge einer Theorie der phylogenetischen Systematik. Deutscher Zentralverlag, Berlin
Hennig W (1966) Phylogenetic systematic. University of Illinois Press, Urbana
Henry TJ (1997) Phylogenetic analysis of family groups within the infraorder Pentatomomorpha (Hemiptera: Heteroptera), with emphasis on the Lygaeoidea. Ann Entomol Soc Am 90:275–301
Henry TJ, Froeschner RC (1988) Catalog of the Heteroptera or true bugs of Canada and the Continental United States. Brill, New York
Hovenkamp P (1997) Vicariance events, not areas, should be used in biogeographical analysis. Cladistics 13:67–79
Hovenkamp P (2001) A direct method for the analysis of vicariance patterns. Cladistics 17:260–265
Hua J, Li M, Dong P, Cui Y, Xie Q, Wenjun B (2008) Comparative and phylogenomic studies on the mitochondrial genomes of Pentatomomorpha (Insecta: Hemiptera: Heteroptera). BMC Genomics 9:610
Hua J, Li M, Dong P, Cui Y, Xie Q, Wenjun B (2009) Phylogenetic analysis of the true water bugs (Insecta: Hemiptera: Heteroptera: Nepomorpha): evidence from mitochondrial genomes. BMC Evol Biol 9:134

Hull D (1988) Science as a process: an evolutionary account of the social and conceptual development of science. University of Chicago Press, Chicago

Hunn CA, Upchurch P (2001) The importance of time/space in diagnosing the causality of phylogenetic events: towards a "chronobiogeographical" paradigm? Syst Biol 50:1–17

Hwang WS, Weirauch C (2012) Evolutionary history of assassin bugs (Insecta: Hemiptera: Reduviidae): insights from divergence dating and ancestral state reconstruction. PLoS ONE 7:e45523

Hypša V, Tietz DF, Zrzavý J, Rego ROM, Galvão C, Jurberg J (2002) Phylogeny and biogeography of Triatominae (Hemiptera: Reduviidae): molecular evidence of a New World origin of the Asiatic clade. Mol Phylogenet Evol 23:447–457

Karvar T, Pavlovcic P, Susnik S, Meglic V, Virant-Doberlet M (2006) Genetic differentiation of geographically separated populations of the southern green stink bug *Nezara viridula* (Hemiptera: Pentatomidae). Bull Entomol Res 96:117–128

Kerzhner IM (1981) Bugs of the family Nabidae. In: Fauna of the USSR Rhynchota, vol 13(2). Nauka, Leningrad, Russia

Kirkaldy GW (1909) Catalogue of the Hemiptera (Heteroptera) with biological and anatomical references, lists of foodplants and parasites, etc, vol I. Prefaced by a discussion on Nomenclature, and an analytical table of families Cimicidae. Berlim, Germany

Larivière MC (1994) *Parabrochymena* Larivière (Hemiptera: Pentatomidae): systematics, natural history, chorological affinities, and evolutionary relationships, with a biogeographic analysis of *Parabrochymena* and *Brochymena* Amyot and Audinet-Serville. Can Entomol 126:1193–1250

Leston D, Pendergrast JG, Southwood TER (1954) Classification of the terrestrial Heteroptera (Geocorisae). Nature 174:91–92

Lethierry L, Severin G (1893) Catalogue Général des Hémiptères. Hétéroptères Pentatomidae, vol I. Hayez, Bruxelles

Lethierry L, Severin G (1894) Catalogue Général des Hémiptères. Hétéroptères Coreidae, Berytidae, Lygaeidae, Pyrrhocoridae, vol II. Hayez, Bruxelles

Lethierry L, Severin G (1896) Catalogue Général des Hémiptères. Hétéroptères Tingidae, Phymatidae, Aradidae, Hebridae, Hydrometridae, Henicocephalidae, Reduviidae, Ceratocombidae, Cimicidae, Anthocoridae, vol III. Berlin, p 275

Li HM, Deng RQ, Wang JW, Cheng ZY, Jia FL, Wang XZ (2005) A preliminary phylogeny of the Pentatomomorpha (Hemiptera: Heteroptera) based on nuclear 18S rDNA and mitochondrial DNA sequences. Mol Phylogenet Evol 37:313–326

Li HM, Deng RQ, Wang XZ (2006) Phylogenetic relationships of the Pentatomomorpha (Hemiptera: Heteroptera) inferred from nuclear 18S rDNA sequences. Zool Res 27:307–316

Li M, Xi L, Fan ZH, Hua JM, Niu CJ, Li CX, Bu WJ (2010) Phylogeographic relationships of the Green stink bug *Nezara viridula* (Hemiptera: Heteroptera). Insect Sci 17:448–458

Li M, Tian Y, Zhao Y, Bu W (2012) Higher level phylogeny and the first divergence time estimation of Heteroptera (Insecta: Hemiptera) based on multiple genes. PLoS ONE 7:e32152

Lindskog P, Polhemus JT (1992) Taxonomy of Saldula: revised genus and species group definitions, and a new species of the pallipes group from Tunisia (Heteroptera: Saldidae). Entomol Scand 23:63–88

Lis B (1999) Phylogeny and classification of Cantacaderini (=Cantacaderidae stat. nov.) (Hemiptera: Tingidae). Ann Zool 49:157–196

Llorente J, Morrone JJ, Bueno A, Pérez R, Viloria A, Espinosa-Organista D (2000) Historia Del desarrollo y La recepción de las ideas panbiogeográficas de Léon Croizat. Rev Acad Colomb Cienc 24:549–577

Mahner M (1993) Systema Cryptoceratorum Phylogeneticum (Insecta, Heteroptera). Zoologica 48:1–302

Maia-da-Silva F, Junqueira ACV, Campaner M, Rodrigues AC, Crisante G, Ramirez LE, Caballero ZCG, Monteiro FA, Coura JR, Anez N, Teixeira MMG (2007) Comparative phylogeography of *Trypanosoma rangeli* and *Rhodnius* (Hemiptera: Reduviidae) supports a long coexistence of parasite lineages and their sympatric vectors. Mol Ecol 16:3361–3373

Marcet PL, Mora MS, Cutrera AP, Jones L, Gurtler RE, Kitron U, Dotson EM (2008) Genetic structure of Triatoma infestans populations in rural communities of Santiago del Estero, northern Argentina. Infect Genet Evol 8:835–846
Matthew WD (1915) Climate and evolution. Ann NY Acad Sci 24:171–318
Mayr GL (1864) Diagnosen neuer Hemipteren. Verh Zool-Bot GesWien 14:903–914
Mayr GL (1866a) Diagnosen neuer Hemipteren III. Verh Zool-Bot GesWien 16:361–366
Mayr GL (1866b) Hemiptera. Reise der Österreichischen Freggate Novara um die Erde in den Jahren 1857, 1858, 1859 unter den Befehlen des Commodore B. von Wüllerstorf-Urbair, II. Zoologischer Theil, Wien, Austria
Mayr E (1942) Systematics and the origin of species, from the viewpoint of a zoologist. Harvard University Press, Cambridge
McAtee WL, Malloch JR (1933) Revision of the subfamily Thyreocoridae of the Pentatomidae (Hemiptera-Heteroptera). Ann Carnegie Mus 21:191–411
Monteiro FA, Pérez R, Panzera F, Dujardin JP, Galvão C, Rocha D, Noireau F, Schofield CJ, Beard CB (1999) Mitochondrial DNA variation of *Triatoma infestans* populations and its implication on the specific status of *T. melanosoma*. Mem I Oswaldo Cruz 94:229–238
Monteiro FA, Tabey V, Barrett TV, Fitzpatrick S, Cordon-Rosales C, Feliciangeli D, Beard CB (2003) Molecular phylogeography of the Amazonian Chagas disease vectors *Rhodnius prolixus* and *R. robustus*. Mol Ecol 12:997–1006
Monteiro FA, Donnelly MJ, Beard CB, Costa J (2004) Nested clade and phylogeographic analyses of the Chagas disease vector *Triatoma brasiliensis* in Northeast Brazil. Mol Phylogenet Evol 32:46–56
Monteiro FA, Peretolchina T, Lazoski C, Harris K, Dotson EM, Abad-Franch F, Tamayo E, Pennington PM, Monroy C, Cordon-Rosales C, Salazar-Schettino PM, Gomez-Palacio A, Grijalva MJ, Beard CB, Marcet PL (2013) Phylogeographic pattern and extensive mitochondrial DNA divergence disclose a species complex within the Chagas disease vector *Triatoma dimidiata*. PLoS ONE 8:e70974
Montemayor SI, Costa LAA (2009) Systematic revision of *Macrotingis* and phylogenetic analysis of the genera *Macrotingis* and *Ceratotingis* (Heteroptera: Tingidae). Eur J Entomol 106:631–642
Morrone JJ (2002) El espectro del dispersalismo: de los centros de origem de las áreas ancestrales. Rev Soc Entomol Argent 61:1–14
Morrone JJ (2006) Biogeographic areas and transition zones of Latin America and the Caribbean islands based on panbiogeography and cladistic analysis of the entomofauna. Annu Rev Entomol 51:467–494
Morrone JJ (2007) Hacia una biogeografia evolutiva. Rev Chil Hist Nat 80:509–520
Morrone JJ (2009) Evolutionary biogeography: an integrative approach with case studies. Columbia University Press, New York
Morrone JJ, Coscarón C (1996) Distributional patterns of the American Peiratinae (Heteroptera: Reduviidae). Zool Med Leiden 70:1–15
Morrone JJ, Coscarón MC (1998) Cladistics and biogeography of the assassin bug genus Rasahus Amyot & Serville (Heteroptera: Reduviidae: Peiratiinae). Zoologische Mededelingen Leiden 72:73–87
Morrone JJ, Marquez J (2001) Halffter's Mexican transition zone, beetle generalized tracks, and geographical homology. J Biogeogr 28:635–650
Morrone JJ, Mazzucconi SA, Bachmann AO (2004) Distributional patterns of chacoan water bugs (Heteroptera: Belostomatidae, Corixidae, Micronectidae and Gerridae). Hydrobiologia 523:159–173
Muraji M, Tachikawa S (2000) Phylogenetic analysis of water striders (Hemiptera: Gerroidea) based on partial sequences of mitochondrial and nuclear ribosomal RNA genes. Entomol Sci 3:615–626
Nelson G, Ladiges PY (1996) Paralogy in cladistic biogeography and analysis of paralogy-free subtrees. Am Mus Novit 3167:1–58

Nelson GJ, Platnick NI (1981) Systematics and biogeography: cladistics and vicariance. Columbia University Press, New York

Page RDM (1990) Temporal congruence and cladistic analysis of biogeography and cospeciation. Syst Zool 39:205–226

Page RDM (1993) Component Ver 2.0. http://taxonomy.zoology.gla.ac.uk/rod/cpw.html. Accessed 30 Jul 2014

Panizzi AR, McPherson JE, James MJ, Javahery M, McPherson RM (2000) Stink bugs (Pentatomidae). In: Schaefer C, Panizzi AR (eds) Heteroptera of economic importance. CRC Press, Boca Raton, pp 421–474

Panzera F, Dujardin JP, Nicolini P, Caraccio MN, Rose V, Tellez T, Bermudez H, Bargues MD, Mas-Coma S, O'Connor JE, Perez R (2004) Genomic changes of Chagas disease vector, South America. Emerg Infect Dis 10:438–446

Paula AS, Diotaiuti L, Schofield CJ (2005) Testing the sister-group relationship of the Rhodniini and Triatomini (Insects: Hemiptera: Reduviidae: Triatominae). Mol Phylogenet Evol 35:712–718

Paula AS, Diotaiuti L, Galvao C (2007) Systematics and biogeography of Rhodniini (Heteroptera: Reduviidae: Triatominae) based on 16S mitochondrial rDNA sequences. J Biogeogr 34:699–712

Pérez-de-Rosas AR, Segura EL, Fichera L, Garcia BA (2008) Macrogeographic and microgeographic genetic structure of the Chagas' disease vector *Triatoma infestans* (Hemiptera: Reduviidae) from Catamarca, Argentina. Genetica 133:247–260

Pérez-de-Rosas AR, Segura EL, García BA (2011) Molecular phylogeography of the Chagas' disease vector Triatoma infestans in Argentina. Heredity 107:71–79

Pérez-de-Rosas AR, Segura EL, Fusco O, Bareiro ALG, Garcia BA (2013) Fine-scale genetic structure in populations of the Chagas disease vector *Triatoma infestans* (Hemiptera, Reduviidae). Genetica 141:107–117

Piccinali RV, Marcet PL, Noireau F, Kitron U, Gürtler RE, Dotson EM (2009) Molecular population genetics and phylogeography of the Chagas disease vector *Triatoma infestans* in South America. J Med Entomol 46:796–809

Piccinali RV, Marcet PL, Ceballos LA, Gürtler RE, Dotson EM (2011) Genetic variability, phylogenetic relationships and gene flow in Triatoma infestans dark morphs from the Argentinean Chaco. Infect Genet Evol 11:895–903

Polhemus JT, Polhemus DA (2008) Global diversity of true bugs (Heteroptera; Insecta) in freshwater. In: Balian EV, Lévêque C, Segers H et al (eds) Freshwater animal diversity assessment. Springer, Dordrecht, pp 379–391

Popov YA (2008) Pavlostysia wunderlichi gen. nov. and sp. nov., the first fossil spider-web bug (Hemiptera: Heteroptera: Cimicomorpha: Plokiophilidae) from the Baltic Eocene amber. Acta Entomol Mus Nat Pragae 48:497–502

Reuter OM (1910) Neue beiträge zur phylogenie und systematik der Miriden nebst einleitenden Bemerkungen über die Phylogenie der Heteropteren-familien. Acta Soc Sci Fenn 37:1–171

Rieger C (1976) Skeleton and musculature of the head and prothorax of Ochterus marginatus contribution towards clarification of the phylogenetic relationships of the Ochteridae Insecta Heteroptera. Zoomorphologie 82:109–191

Rodriguero MS, Gorla DE (2004) Latitudinal gradient in species richness of the New World Triatominae (Reduviidae). Glob Ecol Biogeogr 13:75–84

Sailer RI (1952) A review of the stink bugs of the genus Mecidea. Proc US Natl Mus 102:471–505

Schaefer CW (1993) The Pentatomomorpha (Hemiptera: Heteroptera): an annotated outline of its systematic history. Eur J Entomol 90:105–122

Scherbakov DE, Popov YA (2002) Superorder Cimicidae Laicharting, 1781. Order Hemiptera Linné, 1758. The bugs, cicadas, plantlice, scale insects, etc. In: Rasnitsyn AP, Quicke DLJ (eds) History of insects. Kluwer, Dordrecht, pp 143–157

Schofield CJ (1988) Biosystematics of the Triatominae. In: Service MW (ed) Biosystematics of haematophagous insects. Clarendon, Oxford, pp 285–312

Schofield CJ (2000) Biosystematics and evolution of the Triatominae. Cad Saúde Públ 16:89–92

Schuh RT (1979) Evolutionary trends in Heteroptera. Part II. Mouthpart-structures and feeding strategies. Syst Zool 28:653–656

Schuh RT (1986) The influence of cladistics on heteropteran classification. Annu Rev Entomol 31:67–93

Schuh RT (1995) Plant bugs of the world (Insecta: Heteroptera: Miridae). Systematic catalog, distributions, host list, and bibliography. The New York Entomological Society, New York

Schuh RT (2002–2013) Online systematic catalog of plant bugs (Insecta: Heteroptera: Miridae). http://research.amnh.org/pbi/catalog/. Accessed 30 Jul 2014

Schuh RT (2006) *Heissophila macrotheleae*, a new genus and new species of Poliophilidae from Thailand (Hemiptera, Heteroptera), with comments on the family diagnosis. Denisia 19:637–645

Schuh RT, Polhemus JT (1980) Analysis of taxonomic congruence among morphological, ecological, and biogeographic data sets for the Leptopodomorpha (Hemiptera). Syst Zool 29:1–26

Schuh RT, Polhemus JT (2009) Revision and analysis of *Pseudosaldula* Cobben (Insecta: Hemiptera: Saldidae): a group with a classic Andean distribution. Bull Am Mus Nat Hist 323:1–102

Schuh RT, Slater JA (1995) True bugs of the world (Hemiptera: Heteroptera): classification and natural history. Cornell University Press, New York

Schuh RT, Stys P (1991) Phylogenetic analysis of cimicomorphan family relationships (Heteroptera). J NY Entomol Soc 99:298–350

Schuh RT, Weirauch C, Wheeler WC (2009) Phylogenetic relationships within the Cimicomorpha (Hemiptera: Heteroptera): a total-evidence analysis. Syst Entomol 34:15–48

Simões FL, Ferrari A, Grazia J (2012) Is *Elsiella* Froeschner, 1981 a valid genus? (Hemiptera: Heteroptera: Pentatomidae: Pentatominae). Zootaxa 3238:39–48

Simpson GG (1940) Mammals and land bridges. J Wash Acad Sci 30:137–163

Singh-Pruthi H (1925) The morphology of male genitalia in Rhynchota. Trans Entomol Soc Lond 1:127–267

Slater JA (1964) A catalogue of the Lygaeidae of the world. University of Connecticut, Storrs

Slater JA (1974) Class Insecta. Order Hemiptera. Suborder Heteroptera. In: Coaton WGH (ed) Status of the taxonomy of Hexapoda of Southern Africa. Entomol Mem 38:66–74

Slater JA, O'Donnell JE (1995) A catalogue of the Lygaeidae of the world (1960–1994). New York Entomological Society, New York

Sosa-Gomez DR, Delpin KE, Almeida AMR, Hirose E (2004) Genetic differentiation among Brazilian populations of *Euschistus heros* (Fabricius) (Heteroptera: Pentatomidae) based on RAPD analysis. Neotrop Entomol 33:179–187

Spangenberg R, Friedemann K, Weirauch C, Beutel RG (2013) The head morphology of the potentially basal heteropteran lineages Enicocephalomorpha and Dipsocoromorpha (Insecta: Hemiptera: Heteroptera). Arthropod Syst Phyl 71:103–136

Stål C (1870) Enumeratio Hemipterorum. Bidrag till en förtickning öfver aller hittills kända Hemiptera, jemte systematiska meddelanden, 1. Köngl Svenska Vet-Akad Handl 9:1–232

Stål C (1872) Enumeratio Hemipterorum. Bidrag till en förtickning öfver aller hittills kända Hemiptera, jemte systematiska meddelanden, 2. Köngl Svenska Vet-Akad Handl 10:1–159

Stål C (1873) Enumeratio Hemipterorum. Bidrag till en förtickning öfver aller hittills kända Hemiptera, jemte systematiska meddelanden, 3. Köngl Svenska Vet-Akad Handl 11:1–163

Stål C (1874) Enumeratio Hemipterorum. Bidrag till en förtickning öfver aller hittills kända Hemiptera, jemte systematiska meddelanden, 4. Köngl Svenska Vet-Akad Handl 12:1–186

Stål C (1876) Enumeratio Hemipterorum. Bidrag till en förtickning öfver aller hittills kända Hemiptera, jemte systematiska meddelanden, 5. Köngl Svenska Vet-Akad Handl 14:1–162

Stys P (1961) Morphology of the abdomen and female ectodermal genitalia of the trichophorous Heteroptera and bearing on their classification. In: Proceedings of the 11th international congress of entomology, Vienna, 17–25 August 1960

Stys P (2008) Zoogeography of Enicocephalomorpha (Heteroptera). Bull Insectol 61:137–138

Stys P, Kerzhner I (1975) The rank and nomenclature of higher taxa in recent Heteroptera. Acta Entomol Bohemosl 72:65–79

Sweet MH (2006) Justification for the Aradimorpha as an infraorder of the suborder Heteroptera (Hemiptera, Prosorrhyncha) with special reference to the pregenital abdominal structure. Denisia 19:225–248

Szumik C, Goloboff PA (2004) Areas of endemism an improved optimality criterion. Syst Biol 53:968–977

Szumik C, Cuezzo F, Goloboff PA, Chalup A (2002) An optimality criterion to determine areas of endemism. Syst Biol 51:806–816

Tian Y, Zhu W, Li M, Xie Q, Bu W (2008) Influence of data conflict and molecular phylogeny of major clades in Cimicomorphan true bugs (Insecta: Hemiptera: Heteroptera). Mol Phylogenet Evol 47:581–597

Tian X, Xie Q, Li M, Gao C, Cui Y, Xi L, Bu W (2011) Phylogeny of pentatomomorphan bugs (Hemiptera-Heteroptera: Pentatomomorpha) based on six Hox gene fragments. Zootaxa 2888:57–68

Torres-Pérez F, Acuna-Retamar M, Cook JA, Bacigalupo A, Garcia A, Cattan PE (2010) Statistical phylogeography of Chagas disease vector *Triatoma infestans*: testing biogeographic hypotheses of dispersal. Infect Genet Evol 11:167–174

Tullgren A (1918) Zur morphologie und systematik der Hemipteren I. Entom Tidskr Arg 38:115–133

Turchetto-Zolet AC, Pinheiro F, Salgueiro F, Palma-Silva C (2013) Phylogeographical patterns shed light on evolutionary process in South America. Mol Ecol 22:1193–1213

Uhler PR (1869) Notices of the Hemiptera obtained by the expedition of Prof. James Orton in Ecuador and Brazil. Proc Boston Soc Nat His 12:321–327

Uhler PR (1894) On the Hemiptera-Heteroptera of the Island of Grenada, West Indies. Proc Zool Soc Lond 1894:167–224

Usinger RL (1966) Monograph of Cimicidae (Hemiptera – Heteroptera). The Thomas Say Foundation, vol 7. Entomological Society of America, Lanham, p 585

Usinger RL, Matsuda R (1959) Classification of Aradidae (Hemiptera-Heteroptera). British Museum, London

Wappler T (2006) Lutetiacader, a puzzling new genus of cantacaderid lace bugs (Heteroptera: Tingidae) from the Middle Eocene Messel Maar, Germany. Palaeontology 49:435–444

Weirauch C (2003) Glandular areas associated with the male genitalia in *Triatoma rubrofasciata* (Triatominae, Reduviidae, Hemiptera) and other Reduviidae. Mem Inst Oswaldo Cruz 98:73–776

Weirauch C (2005) Pretarsal structures in Reduviidae (Heteroptera, Insecta). Acta Zool Stockh 86:91–110

Weirauch C (2006) Evaporator structures associated with the metathoracic glands in Reduviidae (Heteroptera: Cimicomorpha). Eur J Entomol 103:97–108

Weirauch C (2008) Cladistic analysis of Reduviidae (Heteroptera: Cimicomorpha) based on morphological characters. Syst Entomol 33:229–274

Weirauch C, Munro JB (2009) Molecular phylogeny of the assassin bugs (Hemiptera: Reduviidae), based on mitochondrial and nuclear ribosomal genes. Mol Phylogenet Evol 53:287–299

Weirauch C, Schuh RT (2010) Southern hemisphere distributional patterns in plant bugs (Hemiptera : Miridae : Phylinae): *Xiphoidellus*, gen. nov. from Australia and *Ampimpacoris*, gen. nov. from Argentina, show transantarctic relationships. Invertebr Syst 24:473–508

Weirauch C, Schuh RT (2011) Systematics and evolution of Heteroptera: 25 years of progress. Annu Rev Entomol 56:487–510

Weirauch C, Stys P (2014) Litter bugs exposed: phylogenetic relationships of Dipsocoromorpha (Hemiptera: Heteroptera) based on molecular data. Insect Syst Evol 45(4):351–370. doi:10.1163/1876312X-45012109

Weirauch C, Forero D, Jacobs DH (2011) On the evolution of raptorial legs – an insect example (Hemiptera: Reduviidae: Phymatinae). Cladistics 27:138–149

Wheeler WC, Schuh RT, Bang R (1993) Cladistic relationships among higher groups of Heteroptera: congruence between morphological and molecular data sets. Entomol Scand 24:121–137

Xie Q, Bu W, Zheng L (2005) The Bayesian phylogenetic analysis the 18S rRNA sequences from the main lineages of Trichoptera (Insecta: Heteroptera: Pentatomorpha). Mol Phylogenet Evol 34:448–451

Xie Q, Tian Y, Zheng L, Bu W (2008) 18S rRNA hyper-elongation and the phylogeny of Euhemiptera (Insecta: Hemiptera). Mol Phylogenet Evol 47:463–471

Yao Y, Ren D, Rider DA, Cai W (2012) Phylogeny of the infraorder Pentatomomorpha based on fossil and extant morphology, with description of a new fossil family from China. PLoS ONE 7:e37289

Zhang W, Weirauch W (2013) Molecular phylogeny of Harpactorini (Insecta: Reduviidae): correlation of novel predation strategy with accelerated evolution of predatory leg morphology. Cladistics 30:339–351

Zhang W, Engel MS, Yao Y, Ren D, Shih C (2014) The Mesozoic family Archegocimicidae and phylogeny of the infraorder Leptopodomorpha (Hemiptera). J Syst Palaeontol 12:93–111

Part II
Infraorders Enicocephalomorpha and Dipsocoromorpha

Chapter 4
The Unique-Headed Bugs (Enicocephalomorpha)

José Antônio Marin Fernandes and Christiane Weirauch

Abstract Enicocephalomorpha, the unique-headed bugs, is one of the more basal taxa within Heteroptera. The small size, unique morphology, and cryptic habits of most unique-headed bugs are the main reasons that most genera and species are poorly known from a biological and taxonomical point of view. Few experts have focused on this group in recent times and the boundaries of several taxa from species to family level are still unclear. In the Neotropical Region, Enicocephalidae and Aenictopecheidae are represented by ca. 70 described species in seven genera, but the true diversity is certainly much larger than these numbers would indicate.

4.1 Introduction

The infraorder Enicocephalomorpha, the unique-headed bugs, has long been considered to be the most basal clade of Heteroptera and the sister group of the Euheteroptera (Schuh 1979; Štys 1995). However, recent analyses have questioned this basal position (Li et al. 2012) and additional phylogenetic investigation is required to firmly establish the sister group of the Enicocephalomorpha. This taxon comprises only two small and still severely understudied families with little more than 320 described species (Štys 2008). The present classification of the taxon was proposed by Štys (1989), in which Aenictopecheidae were raised from the level of a subfamily to family status (see also Štys 1970; Wygodzinsky and Schmidt 1991). The suprageneric classification of both families is in flux. However, Štys (2008) indicated that a cladistic analysis is in preparation that together with the description of new genera will likely change the present suprageneric classification. Aenictopecheidae and Enicocephalidae occur worldwide, but most of the genera

J.A.M. Fernandes (✉)
Instituto de Ciências Biológicas, Universidade Federal do Pará,
Rua Augusto Corrêa n°1, Belém, PA 66075-110, Brazil
e-mail: joseamf@ufpa.br

C. Weirauch
Department of Entomology, University of California, Riverside, CA 92521, USA
e-mail: christiane.weirauch@ucr.edu

A.R. Panizzi, J. Grazia (eds.), *True Bugs (Heteroptera) of the Neotropics*,
Entomology in Focus 2, DOI 10.1007/978-94-017-9861-7_4

and described species are restricted to the Old World (Štys 2008). Identification of the described Neotropical genera is possible using the key to New World taxa by Wygodzinsky and Schmidt (1991) or the one to the world genera by Štys (2002b). All Enicocephalomorpha are small to midsized with body length typically ranging between 2 and 15 mm, often dull colored, and generally resemble small Reduviidae. Wings are membranous and well developed, but micropterous, brachypterous, and wingless forms, especially in females, also occur. Some species in both families swarm, a behavior that is rare among true bugs. Also unusual is the fact that swarms may contain more than one species (e.g., Wygodzinsky and Schmidt 1991).

The greatest species diversity is seen in humid tropical and subtropical forests, but species also occur in cool-temperate climates and arid zones where they are found in cryptic microhabitats with high humidity (Wygodzinsky and Schmidt 1991; Štys 1995).

4.2 General Characteristics and Diagnosis

According to the most recent and comprehensive diagnosis (Štys 1995), Enicocephalomorpha are characterized, among other features, by: the elongate and porrect head that is subdivided into anterior and postocular lobes by a usually conspicuous postocular constriction (Fig. 4.1); the ocelli, if present, on the posterior lobe and well removed from the eyes; the long gula; the four-segmented labium short and straight to arcuate, never exceeding the length of the head; the moderately long antennae flagelliform to cylindrical or slightly tapering; the forewings always completely membranous, i.e., without a distinct corium and wing only subdivided into a remigium and clavus; the ambient vein (the vein encircling the margin of the wing) in the remigium marginal or submarginal, if venation reduced, then the vein represented by a row of macrotrichia; the forewing sometimes reduced or absent, occasionally prone to autotomy; the medial fracture located anterior to R; basal crossvein connecting marginal veins (C, Sc, and/or R) with M + Cu; foreleg usually with enlarged femur and tibia, thus presumably raptorial; tibia usually with distoventral process carrying one or two clusters of spiniform setae; opposing ventral

Fig. 4.1 A species of Enicocephalidae of the genus *Enicocephalus* (Courtesy of F Carvalho-Filho)

surface of one- or two-segmented foretarsus typically also with spiniform setae; the symmetrical male genitalia either with typical heteropteran phallus (Aenictopecheidae) or with paired genital plates that can be fused into a racquet-shaped guide (Enicocephalidae); parameres small, either movable (Aenictopecheidae) or not (Enicocephalidae); ovipositor present (Aenictopecheidae) to strongly reduced or absent (Enicocephalidae); and subgenital plate of males formed by sternum 8 rather than sternum 7 as in other Heteroptera.

4.3 General Biology and Ecology

The small size and cryptic habits of Enicocephalomorpha make direct observations of their biology and ecology rare events. Biological information is therefore typically scattered in the taxonomic literature. Wygodzinsky and Schmidt (1991) summarized and cited most of the relevant literature. Most sources indicate that Enicocephalomorpha are found in leaf litter, under bark, in rotten logs, under stones at the edges of streams, in superficial but probably also deeper layers of soil, in moss and liverwort mats, suspended leaf litter and soil, and in nests of pack rats (e.g., Wygodzinsky and Schmidt 1991; Usinger 1939, 1945; Schuh 1970; Villiers 1961; Kritsky 1976, 1977). Villiers (1960) reported that species of *Systelloderes* Blanchard may also be collected in bee tunnels and associated with ants.

Enicocephalids are predaceous and appear to feed on soft-bodied insects and other arthropods, but few direct observations are available. In captivity, they were fed on larvae and freshly killed adult flies, symphylids (Symphyla) and campodeids (Diplura), as well as ant larvae and pupae.

Swarming is a biological phenomenon that is rather common in Enicocephalomorpha and has been documented for species in five genera and both families in the New World (Wygodzinsky and Schmidt 1991). Štys (1981) speculated that all species that are capable of flight might also swarm. Swarms are possibly formed for the purpose of mating and are composed of large numbers of males and much smaller numbers of females (Kritsky 1977; Štys 1981). As an example, Schuh (1970) counted 441 males and only seven females in a swarm of *Systelloderes* in Costa Rica. Interestingly, swarms that comprise more than one species of Enicocephalidae are fairly common (Wygodzinsky and Schmidt 1991). Swarming is a complex phenomenon and its behavioral and morphological aspects in Enicocephalomorpha are poorly documented, making this a wide open field for future research.

Some enicocephalids have the unusual ability to autotomize, or shed, their wings. This behavior may be related to the cryptic way of life of most species. In females, it could be important when individuals retreat, after swarming and copulation, into the soil, where wings might be a hindrance, similar to wing shedding in termites and ants (Wygodzinsky and Schmidt 1991).

4.4 Classification and Diversity

Both families Aenictopecheidae and Enicocephalidae are known to occur in the New World. Their taxonomic history and classification, with emphasis on the Neotropical fauna, is outlined below. Štys (1989) proposed the classification used today (Štys 1995). Wygodzinsky and Schmidt (1991) taxonomically revised the New World fauna of Enicocephalomorpha with exception of the speciose genus *Systelloderes* that in the New World alone contains more than 40 described and undescribed species, with many more that await discovery. This monograph also includes substantial morphological information including many scanning electron micrographs and line drawings as well as biological data and a phylogenetic scheme of relationships that could serve as a basis for future cladistic investigations into the group. Štys (2002b) presented a key to world genera and subgenera and a list of valid names and synonymies. Additional information on the two families is found in Štys (1981, 1989, 1995, 2002b, 2008).

4.4.1 Aenictopecheidae

Aenictopecheids are small (2.6–10 mm long) and rarely collected true bugs. Worldwide, they comprise four subfamilies, 11 genera, and about 20 extant species (Štys 1995, 2002b). The four subfamilies show fairly divergent morphologies and the family is mostly diagnosed by the absence of presumably derived characters that occur in the Enicocephalidae. In addition, some species have been reported to display characters commonly found in Auchenorrhyncha but not normally seen in other Heteroptera, a situation that further complicates homology assessment and systematic conclusions (Štys 1995).

According to Štys (1995), Aenictopecheidae are most easily distinguished from Enicocephalidae by the fairly typical heteropteran genitalia in both males and females compared to the strongly modified and reduced genitalia in Enicocephalidae. In the males, these include a typical heteropteran phallus, movable parameres, and a pygophore that is never subdivided into tergum, laterotergites, and sternum; females are characterized by a well-developed ovipositor. Other characters to differentiate this family from the Enicocephalidae include the poorly demarcated posterior lobe of the pronotum, a short costal fracture (absent in Enicocephalidae), and nymphs with wing pads that are not contiguous along the midline.

Among the four subfamilies, only Aenictopecheinae that are subdivided into two tribes are represented in the Neotropical Region. Aenictopecheini with the two monotypic genera *Aenictipechys* Breddin and *Lomagostus* Villiers are restricted to the Oriental Region and Madagascar. Gamostelini comprises four genera, only two of which occur in the Neotropical Region: *Gamostolus* Berg with a single described species, *G. subantarcticus* (Berg), was described from Southern Argentina and Chile. Wygodzinsky and Schmidt (1991) examined nymphs from Venezuela and

Colombia that they classified as *Gamostolus*, but Štys (2008) indicated that these nymphs likely belong to an undescribed genus. *Gamostolus* is diagnosed by the distinct arrangement of fore tibial spines, the foretarsus with four spines, as well as several genitalic features (see Wygodzinsky and Schmidt 1991 for details).

In addition, the Neotropical fauna comprises the genus *Tornocrusus* Kritsky with eight described, and several undescribed, species that are distributed in Central and South America. Among other New World Aenictopecheinae, *Tornocrusus* is most readily identified by the reduced proepimera and the apical mid- and hind tibial armature consisting of only one large spine and one bristle comb (illustrated in Wygodzinsky and Schmidt 1991). Comprehensive diagnoses, drawings, and a key the species can be found in Wygodzinsky and Schmidt (1991). Species of *Tornocrusus* are difficult to identify due to intraspecific variation and similarity among species. Štys (2008) reported the discovery of two new genera from the New World, but both remain undescribed.

4.4.2 Enicocephalidae

This family contains the great majority of described enicocephalomorphan genera and species and the bulk of specimens deposited in natural history collections. Species are small (2–15 mm long) and mostly dull colored, ranging from uniformly yellow or brown to blackish (Štys 1995). Worldwide, Enicocephalidae comprises about 300 described extant species that are classified into 47 genera and five subfamilies (Štys 1995, 2002b). Štys (2008) indicated that at least 30 genera and hundreds of new species have so far remained undescribed. A key to the subfamilies of Enicocephalidae was provided by Štys (1995).

According to Štys (1995), Enicocephalidae are distinguished from Aenictopecheidae by the pronotum that is usually subdivided into three distinct lobes (except in *Alienates* Barber), the common occurrence of wing polymorphism featuring macropterous males and brachypterous to apterous females, the absence of a costal fracture, the strongly modified male and female genitalia, and the large and contiguous wing pads of later nymphal instars large that are sometimes slightly overlapping along the midline. The most prominent feature of the male genitalia is a distally perforated "guide" of uncertain homology that is associated with the ventral margin of the posterior foramen of the pygophore. The parameres are always immobile, fused at the bases or reduced to flat sclerites. The external female genitalia are absent or retained as remnants in the Systelloderini and the female genital opening is covered by an extensive subgenital plate formed by the sternum 8.

In the Neotropical Region, Enicocephalidae are represented by two subfamilies, the Alienatinae and the Enicocephalinae. Alienatinae only contain the genus *Alienates* that can be recognized by very distinctive characters such as the small body size, the one-segmented tarsi of the adult, the males with the posterior pronotal lobe virtually absent, the strongly reduced forewing venation, and the females being largely desclerotized, apterous, and with strongly reduced eyes (Wygodzinsky and

Schmidt 1991; Štys 1995). Species in the genus occur in the Southwestern United States, Central America, and the Caribbean, with 12 out of 13 species being Neotropical (Wygodzinsky and Schmidt 1991). Štys (2008) indicated that about ten species await description. Wygodzinsky and Schmidt (1991) pointed out that morphological variation among examined specimens renders the formulation of species hypotheses difficult.

Enicocephalinae are represented in the Neotropical Region by 48 described species that are classified in six genera. These taxa are divided in two tribes: Systelloderini with the single genus *Systelloderes* and the Enicocephalini that contain the five genera *Chiricocoris* Wygodzinsky and Schmidt, *Enicocephalus* Westwood, *Hymenocoris* Uhler, *Neoncylocotis* Wygodzinsky and Schmidt, and *Xenicocephalus* Wygodzinsky and Schmidt. Three additional genera of Enicocephalini occur in the New World but are restricted to the Nearctic Region (*Brevidorus* Kristsky, *Lysenicocephalus* Wygodzinsky and Schmidt, and *Urnacephala* Wygodzinsky and Schmidt).

Two publications are fundamental for the identification of Neotropical Enicocephalidae: Wygodzinsky and Schmidt (1991) diagnosed and described or redescribed all genera except *Systelloderes*, and Štys (2002b) added diagnostic characters for several genera in his key. *Chiricocoris* is known only from a few female specimens collected in Panama, which represent the single known species *C. dybasi* Wygodzinsky and Schmidt. This genus is distinguished from other New World Enicocephalidae by the very robust overall aspect, the extremely thick forelegs, the great length of the pronotal midlobe and reduction of the hind lobe, the posteriorly closed anterior coxal cavities, and the presence of a large conical projection at the anteroventral angle of the foretarsus. The forewings lack basal and discal cells.

Enicocephalus is diagnosed by features that include the extremely small or obsolete outer claw of the foreleg, one of the foretarsal spines being half-moon shape, and the tibia with five apical spines, with the innermost spine being long, slender, and curved toward the remaining spines. Males are winged and females are either winged or capable of autotomizing their wings. This genus comprises 17 species, keyed by Wygodzinsky and Schmidt (1991), which range from Southern Mexico to Southern Brazil, but undescribed species were mentioned by Wygodzinsky and Schmidt (1991) and Štys (2008).

Males of *Hymenocoris* are delicate and have foretarsal claws that are nearly equal in length, simple setae on the dorsal surface of the abdomen, and apices of the mid- and hind tibiae with two spines, and they frequently display a red forewing stigma. Females are wingless and possess a membranous abdomen and closed foreacetabula, the spine VI of the foretibia is shaped like a molar, and their forefemur features a basal projection. Among the three known species, only *H. hintoni* occurs in the Neotropical Region, with the remaining two species being Nearctic. The delineation of species is considered to be problematic due to substantial intraspecific variation and overlapping species-diagnostic characters (Wygodzinsky and Schmidt 1991).

Neoncylocotis can be identified by the scale-like setae on the abdominal dorsum, the subequal foretarsal claws, the conical to peg-shaped foretarsal spines, the emarginated posterior margin of the pronotum, and the presence of at least a few macrotrichia within the forewings cells. Among New World genera, *Neoncylocotis* is most similar to *Enicocephalus* but can be distinguished by the above characters. All 17 species are Neotropical and range from Southern Mexico to Southern Brazil and Northern Argentina.

Xenicocephalus is recognized by the large body size of the adults (at least 8 mm) and the dense tuberculation on much of the cuticular surface. In addition, distinctive foreleg characters include a heavily sclerotized, ridgelike projection ventrally on the trochanter; the inner apical angle of the tibia shaped into a pointed, strongly sclerotized projection with 3–5 straight and slender, subapically inserted spines; and the foretarsus bearing only one stout claw. This genus was described based on one damaged female and a few nymphs; the male is unknown. The authors acknowledge that this is not an ideal situation but decided to describe this genus based on the unique characteristics, especially the size of the specimens. The single species, *X. giganteus* Wygodzinsky and Schmidt, is known from Colombia, and nymphs were collected in Costa Rica, Panama, and Guyana.

Systelloderes is diagnosed by the autapomorphic fusion of the ninth and tenth abdominal terga, which form the roof of the pygophore, and is most easily distinguished by the presence of a well-developed supradistal plate. Additional diagnostic features comprise: the moderately sized eyes; the distinctly trilobate pronotum; the posterior margin of the pronotal midlobe entire, without Y-shaped impressions and the lateral parts of the midlobe at most with a pair of pits; the apex of the foretibia with none of its spiniform setae spatulate; and the fore claws normally developed and subequal. This genus comprises 11 Neotropical described species ranging from Southern Mexico to Northern Argentina. *Systelloderes* spp. occur in moderate to high altitudes and have small endemic ranges and apparently low mobility. Consequently, dozens or hundreds of undescribed species await discovery and/or description (Wygodzinsky and Schmidt 1991; Štys 2008). Wygodzinsky and Schmidt (1991) suggested that *Systelloderes* is likely not monophyletic and that the Neotropical and Old World species are not congeneric. Štys (1970) divided the genus into three groups: the *moschatus* group from the Americas, the Ethiopian Region, Australia (except New Zealand), and the Asiatic Archipelago; the *maclachlani* group from New Zealand; and the *uvarovi* group from the Ethiopian Region and the Middle East. Štys (2002a) more recently reexamined the latter group and divided it into four new genera.

4.5 Concluding Remarks

Enicocephalomorpha, the unique-headed bugs, is a basal clade of Heteroptera comprising only two families and little more than 320 species. Species are small and usually dull colored resembling small reduviids. Biological and ecological data are scarce due to their cryptic habits, also responsible by the small number of specimens deposited in entomological collections. Better sampling technics, dedicated to soil and suspended soil, will improve considerably the number of species captured in the Neotropical Region. Descriptions of new species and genus will contribute to more robust suprageneric definitions, helping to better characterize families and subfamilies. This suprageneric taxonomical arrangement certainly will be influenced or changed by future cladistic analyses of the infraorder.

References

Kritsky G (1976) The Enicocephalidae (Hemiptera) in Illinois. Trans Ill State Acad Sci 69:192–193

Kritsky G (1977) Observations on the morphology and behavior of the Enicocephalidae (Hemiptera). Entomol News 88:105–110

Li M, Tian Y, Zhao Y, Bu W (2012) Higher level phylogeny and the first divergence time estimation of Heteroptera (Insecta: Hemiptera) based on multiple genes. PLoS ONE 7:e32152. doi:10.1371/journal.pone.0032152

Schuh RT (1970) Swarming in an undescribed Enicocephalid (Hemiptera) from Costa Rica. Entomol News 81:66–68

Schuh RT (1979) Review of evolutionary trends in Heteroptera. Part II. Mouthpart-structures and feeding strategies, by R. H. Cobben. Syst Zool 28:653–656

Štys P (1970) On the morphology and classification of the family Dipsocoridae s. lat., with particular reference to the genus *Hypsipteryx* Drake (Heteroptera). Acta Entomol Bohemoslov 67:21–46

Štys P (1981) Unusual sex ratios in swarming and light-attracted Enicocephalidae (Heteroptera). Acta Entomol Bohemoslov 78:430–432

Štys P (1989) Phylogenetic systematics of the most primitive true bugs (Heteroptera: Enicocephalomorpha, Dipsocoromorpha). Prace Slov Entomol Spol SAV 8:69–85

Štys P (1995) Dipsocoromorpha. In: Schuh RT, Slater JA (eds) True bugs of the world (Hemiptera: Heteroptera): classification and natural history. Comstock Publishing Associates, Cornell University Press, Ithaca, pp 67–73

Štys P (2002a) New enicocephaline genera similar to *Systelloderes* (Heteroptera: Enicocephalidae). Acta Univ Carolinae Biol 45:319–338

Štys P (2002b) Key to the genus-group taxa of the extant Enicocephalomorpha of the world, their list, and taxonomic changes (Heteroptera). Acta Univ Carolinae Biol 45:339–368

Štys P (2008) Zoogeography of Enicocephalomorpha (Heteroptera). Bull Insectol 61:137–138

Usinger RL (1939) A new genus of Pacific Island Enicocephalidae with new species from the Hawaiian and Philippine Islands (Hemiptera). Proc Hawaii Entomol Soc 10:267–270

Usinger RL (1945) Classification of the Enicocephalidae (Hemiptera, Reduvioidea). Ann Entomol Soc Am 38:321–342

Villiers A (1960) Revision des *Systelloderes* Africains. Rev Francaise Entomol 27:155–162

Villiers A (1961) Un Henicocephalid myrmecophile de l'Afrique austral: *Composederes stuckenbergi*, n. sp. (Hemiptera). Ann Natal Mus 15:69

Wygodzinsky PW, Schmidt K (1991) Revision of the New World Enicocephalomorpha (Heteroptera). Bull Am Mus Nat Hist 200:1–265

Chapter 5
The Minute Litter Bugs (Dipsocoromorpha)

Christiane Weirauch and José Antônio Marin Fernandes

Abstract Dipsocoromorpha, the minute litter bugs, are the putatively least known infraorder of true bugs. Despite their astounding morphology, the small size and cryptic habits of Dipsocoromorpha have resulted in limited curated museum collections. Few specialists have focused on this group. In the Neotropical and elsewhere region, Ceratocombidae, Dipsocoridae, and Schizopteridae are represented by ca. 120 described species in 22 genera. The true diversity in the Neotropics, but also the tropics of the Old World, is clearly much larger than this number.

5.1 Introduction

The infraorder Dipsocoromorpha, the minute litter bugs, comprise five small and poorly studied families that together amount to little more than 320 described species (Štys 1995; Henry 2009; Weirauch and Štys 2014). Ceratocombidae, Dipsocoridae, and Schizopteridae are distributed worldwide (Fig. 5.1), whereas Hypsipterygidae (Indo-Pacific, Afrotropics, Baltic amber) and Stemmocryptidae (Oriental) are restricted to the Old World. The greatest species diversity is seen in humid tropical forests, but species occur from temperate rain forests (Scudder 2010) and other cool temperate climates (Hill 1980) to the deserts of the Western United States (Uhler 1904). All Dipsocoromorpha are small (typically 1–2 mm, rarely up to 4 mm), many are dull colored or black, the antennal flagellomeres are typically strongly setose, and male genitalia range from symmetrical to strongly asymmetrical and include appendages not observed in other infraorders of true bugs. Far from being uniform despite their small size, Dipsocoromorpha show tremendous morphological variation in the male genitalia, which frequently expand to abdominal segments considered as "pregenital" in other Heteroptera (Fig. 5.1). The second

C. Weirauch (✉)
Department of Entomology, University of California, Riverside, CA 92521, USA
e-mail: christiane.weirauch@ucr.edu

J.A.M. Fernandes
Instituto de Ciências Biológicas, Universidade Federal do Pará, Rua Augusto Corrêa n°1, Belém, PA 66075-110, Brazil
e-mail: joseamf@ufpa.br

A.R. Panizzi, J. Grazia (eds.), *True Bugs (Heteroptera) of the Neotropics*,
Entomology in Focus 2, DOI 10.1007/978-94-017-9861-7_5

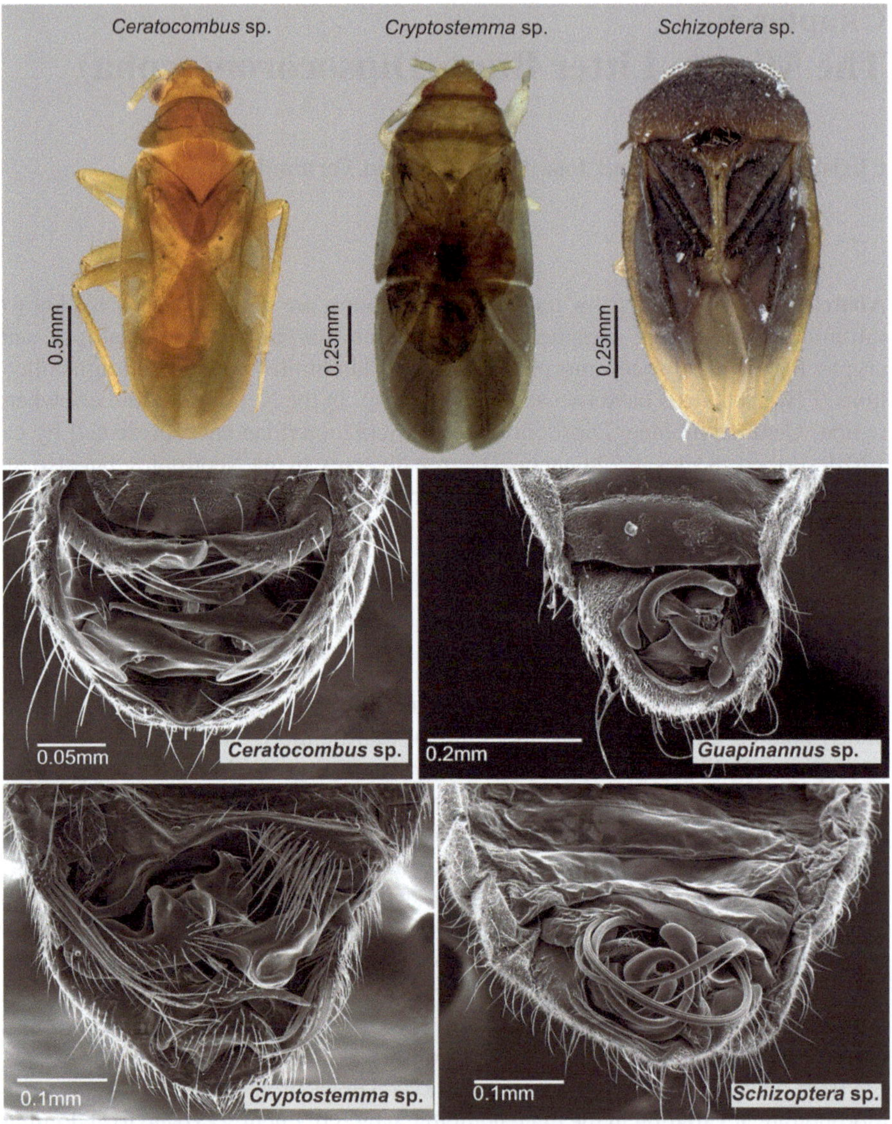

Fig. 5.1 Habitus of species of *Ceratocombus*, *Cryptostemma*, and *Schizoptera*. Scanning electron micrographs showing dorsal views of the distal part of abdomen and pygophore of species of *Ceratocombus*, *Guapinannus*, *Cryptostemma*, and *Schizoptera*

character complex with astounding variation among Dipsocoromorpha is the hemelytra, better referred to as forewings, since a clear division of the corium and membrane is frequently absent. Macropterous, submacropterous, brachypterous, and apterous wing types exist, and sometimes multiple types co-occur in the same species. The most intriguing wings are found among females (and less frequently males) of certain Ceratocombidae and many Schizopteridae: they are shortened, convex, and sclerotized and give these true bugs a beetle-like appearance. This wing type is therefore called "coleopteroid."

The small size of Dipsocoromorpha, together with their cryptic life style, poses several obstacles to taxonomic and systematic investigations. Dipsocoromorpha occur in leaf litter, among gravel along streams, in low herbaceous vegetation, and even in the tree canopy. However, few specialists are capable of recognizing them in the field, and given the oddly beetle-like appearance of many species, specimens may even be overlooked in samples from passive traps, such as Berlese funnels as well as flight intercept and pan traps. Most natural history collections therefore hold relatively few curated specimens of Dipsocoromorpha. In addition, their small size makes thorough examination for taxonomic and phylogenetic purposes difficult. Taxonomic work before the middle of the twentieth century was therefore often hampered by poor illustrations and the lack of detailed morphological descriptions. Our current understanding of the morphology and systematics of the group is mostly based on the excellent work conducted by P. Štys, P. Wygodzinsky, L. Hill, and M. Emsley and a few other authors who have contributed less comprehensively to the study of Dipsocoromorpha.

5.2 General Characteristics and Diagnosis

According to the most recent and comprehensive diagnosis, by Štys (1995), Dipsocoromorpha are characterized, among other features, by: the small size; the often strongly declivous head; the antenna with short scape and pedicel and long and thin flagellomeres and typically with numerous stout setae; the proepisternum often inflated; the forewing typically not clearly divided into the corium and membrane, ranging from completely membranous to entirely elytrous; the hind wing often with deeply separated lobes; and the male genitalia either symmetrical with appendage-like laterotergites (some Ceratocombidae) or strongly asymmetrical (some Ceratocombidae, Dipsocoridae, and Schizopteridae), with the asymmetry being distinctly different in the three groups.

5.3 General Biology and Ecology

Given the small size and cryptic habits of most Dipsocoromorpha, direct observations of their biology are limited. Esaki and Miyamoto (1959) kept specimens of the schizopterid *Hypselosoma hirashimai* Esaki & Miyamoto in the laboratory and fed them on chironomid larvae. Emsley's (1969) monograph on the fauna of Schizopteridae also includes some observations of specimens of *Hoplonannus* and *Schizoptera* that he kept in captivity. Lattin (2000) focused on *Ceratocombus vagans* McAtee & Malloch, and this life history study is the most comprehensive account published to date on the biology of any species of Dipsocoromorpha. Specimens were observed to feed on Collembola and mites but also to engage in cannibalism. The sparse knowledge on the ecology of Dipsocoromorpha, including habitats and microhabitats, is mostly derived from specimen records and few published observations. Literature data and some personal observations are summarized in the family accounts below.

5.4 Phylogeny and Evolution

Our current understanding of the phylogeny and evolution of Dipsocoromorpha is mostly based on the comparative morphological studies by Štys (1970, 1982, 1985), supplemented by publications by P. Wygodzinsky and L. Hill. However, Štys also pointed out that the monophyly of the group is far from being well established (Štys 2014; Weirauch and Schuh 2011). Molecular data were virtually unavailable until recently (Wheeler et al. 1993; Schuh et al. 2009) and are still scarce after the first published molecular study focusing on relationships among the three larger families of Dipsocoromorpha (Weirauch and Štys 2014). Weirauch and Štys (2014), based on molecular cladistic analyses of 87 taxa (35 of them Dipsocoromorpha) and two gene regions, found strong support for the monophyly of Dipsocoromorpha and the three included families. The three recognized subfamilies of Schizopteridae were included in the analyses: Hypselosomatinae were supported as the sister taxon to a clade comprising the Ogeriinae and Schizopterinae. More comprehensive molecular analyses clearly need to follow up on this first effort: only one genus was included for both Ceratocombidae and Dipsocoridae; Schizopteridae were represented by ca.15 genera. Combined morphological and molecular analyses will allow for including rarely collected taxa, such as the Hypsipterygidae and Stemmocryptidae, as well as fossil species, and at the same time test diagnostic characters for recognized families.

5.5 Classification and Diversity of Neotropical Dipsocoromorpha

Only the families Ceratocombidae, Dipsocoridae, and Schizopteridae are known to occur in the New World. The taxonomic history for each of the three families, with emphasis on the Neotropical fauna, is outlined below. A summary on the scarce natural history data is provided.

5.5.1 Ceratocombidae

Ceratocombidae contain two subfamilies, the Ceratocombinae with worldwide distribution and the Trichotonanninae that are restricted to the tropics of the Old World. The current classification is based on Štys (1982). Ceratocombinae are further subdivided into two tribes, the Ceratocombini and Issidomimini. Ceratocombini include three valid genera, *Ceratocombus* Signoret and *Leptonannus* Reuter, which are represented in the Neotropical region, and *Feshina* Štys that is only documented from Africa. *Leptonannus* is distinguished from *Ceratocombus*, among other characters, by the trilobed hind wing, although Wygodzinsky (1953) pointed out that the status of the two genera is in need of revision. Issidomimini comprise three valid genera that are restricted to the Old World and not discussed here.

Ceratocombus comprises 25 extant species worldwide, only seven of which were described from the Neotropical region (Reuter 1891; Poppius 1910; McAtee and Malloch 1925; Usinger 1946; Štys 1995). Species are small and typically dull colored (Fig. 5.1), although some species, especially in the Old World, can have striking, contrasting color patterns. Species are difficult to identify, mostly due to the fact that species descriptions by Poppius, Reuter, Usinger, and McAtee and Malloch were rather generic and omitted reference to genitalic features. Based on observations by Štys (1995), hundreds of undescribed species of *Ceratocombus* await description, including dozens of species from the Neotropics (C. Weirauch personal observation). *Ceratocombus cuneatus* McAtee & Malloch that was described from Southeastern Brazil has recently also been reported from Argentina (Dellapé et al. 2010).

The genus *Leptonannus*, first treated as a subgenus of *Ceratocombus* and later raised to genus rank (Reuter 1891; McAtee and Malloch 1925; Wygodzinsky 1953), contains three extant species known from Africa (Reuter 1891; Wygodzinsky 1953), North America (Uhler 1904), and the Caribbean (Uhler 1894) and one Chiapas amber fossil from Mexico (Azar and Nel 2010).

Species of *Ceratocombus* are most commonly associated with leaf litter and moss and therefore frequently collected using Berlese or Tullgren funnels (Štys 1959; Crowson 1972; Lattin 2000). Specimens of *Ceratocombus* can also be abundant in Malaise traps, as recently documented for an Asian species (Jung et al. 2011). Data derived from surveys in 12 countries in the Neotropical region indicate that species of *Ceratocombus* are frequently found in Malaise traps also in the New World, with 244 being the highest number of individuals recorded from a single trap in Amazonian lowland rainforest (C. Weirauch 2014). The Malaise trap data are an indication that, although ground dwelling, species of *Ceratocombus* are also relatively mobile and thus collected in flight intercept traps that rarely sample other Dipsocoromorpha such as Schizopteridae and Dipsocoridae.

5.5.2 Dipsocoridae

Dipsocoridae in the modern sense comprise only the genera *Alpagut* Kiyak (Palaearctic), *Cryptostemma* Herrich-Schäffer (worldwide), and *Pachycoleus* Fieber (Palaearctic) and about 30 described species. According to Štys (1995), *Pachycoleus* that has often been considered as a subgenus of *Cryptostemma* is represented in the Neotropical region by many undescribed species. Nine species of *Cryptostemma* were described from the Neotropical region (Fig. 5.1), five of which from the Caribbean including Trinidad (Uhler 1894; McAtee and Malloch 1925; China 1946; Nieser 1973). *Cryptostemma pratti* Unsiger is known from Puerto Rico (Usinger 1945), *C. haywardi* Wygodzinsky from Argentina (Wygodzinsky 1952), *C. pedunculatum* McAtee & Malloch from Panama (McAtee and Malloch 1925), and *C. uhleri* from Mexico (McAtee and Malloch 1925). Similar to the situation in Ceratocombidae, species are difficult to tell apart partly due to poorly illustrated original descriptions that lack detail on important genitalic features. Wygodzinsky's (1952) exceptionally well-illustrated description of *C. haywardi* is an exception.

Species of *Cryptostemma* are typically found on or underneath gravel and stones along streams (Leston 1954; Hill 1987; Heckmann and Rieger 2001) but have also been collected using suction (Halbert and Brambila 2002) and light traps (Péricart and Matocq 2003). *Pachycoleus* has been collected in moss (Champion 1908; Golub 2004; Whitehead 2005) but is also found in leaf litter extracts (Costas et al. 1993).

5.5.3 Schizopteridae

All three recognized subfamilies of Schizopteridae, the Hypselosomatinae, Ogeriinae, and Schizopterinae, occur in the Neotropical region, although Schizopterinae contribute the greatest proportion of the genus- and species-level diversity (Emsley 1969; Weirauch and Štys 2014). The family comprises about 48 genera and more than 200 described species (Weirauch and Štys 2014). Nineteen of these genera occur in and are restricted to the New World, and all except four are only documented from the Neotropical region. Hypselosomatinae are diagnosed, among other characters, by large eyes, a well-developed ovipositor (Esaki and Miyamoto 1959), and typically a labium with four segments; Ogeriinae and Schizopterinae share a reduced ovipositor and smaller eyes. According to Emsley (1969), Ogeriinae and Schizopterinae can be separated based on wing venation: in Schizopterinae, four free veins originate from the trapezoidal cell; Ogeriinae are characterized by five free veins emanating from the trapezoidal cell.

Genus-level identification of Schizopteridae mostly relies on original literature and reference collections, given that available keys are incomplete. McAtee and Malloch (1925) keyed the nine genera of Neotropical Schizopteridae described at the time, whereas Emsley (1969) provided a key to the four genera that according to his studies occurred in Trinidad, and Weirauch (2012) keyed the five genera that constitute the *Corixidea* genus group.

Hypselosomatinae, the "big-eyed bugs," comprise about 14 genera worldwide that are most diverse in the Old World (Hill 1984, 2013). They are represented in the Neotropical region by described and undescribed species in the genera *Ommatides* Uhler and *Williamsocoris* Carpintero & Dellapé and undescribed species of *Glyptocombus* Heidemann (Fig. 5.2; Hill 1980; C. Weirauch 2014). Uhler created the monotypic genus *Ommatides* to accommodate *O. insignis* Uhler from St. Vincent that based on Emsley's (1969) examination is closely related to species in the genus

Glyptocombus. The monotypic *Williamsocoris* was described from Formosa in Argentina and is distinguished from other Hypselosomatinae by unique paired processes on the third and fourth labial segments. Undescribed Hypselosomatinae are known from Colombia, Ecuador, Peru, and Trinidad (Hoey-Chamberlain and Weirauch 2014) indicating that the actual species-level diversity of this subfamily in the Neotropics is distinctly higher.

The subfamily Ogeriinae was created by Emsley (1969) to accommodate the New World genus *Chinannus* (Fig. 5.2) together with several Old World genera (*Ogeria* Distant, *Pachyplagia* Gross, *Luachimonannus* Wygodzinsky, and *Kokeshia* Miyamoto; *Kaimon* Hill was later included). Given the characters shared between *Chinannus* and the monotypic genus *Itagunannus* Wygodzinsky (outlined by Wygodzinsky 1948), the latter genus is here treated as part of the subfamily Ogeriinae, despite the fact that Emsley (1969) placed it as *incertae sedis*. The two described species of *Chinannus*, *C. trinitatis* (China) and *C. bierigi* Wygodzinsky, are known from Trinidad and Costa Rica, respectively, but undescribed species range as far south as Bolivia (Knyshov 2014). *Itagunannus itaguaiensis*

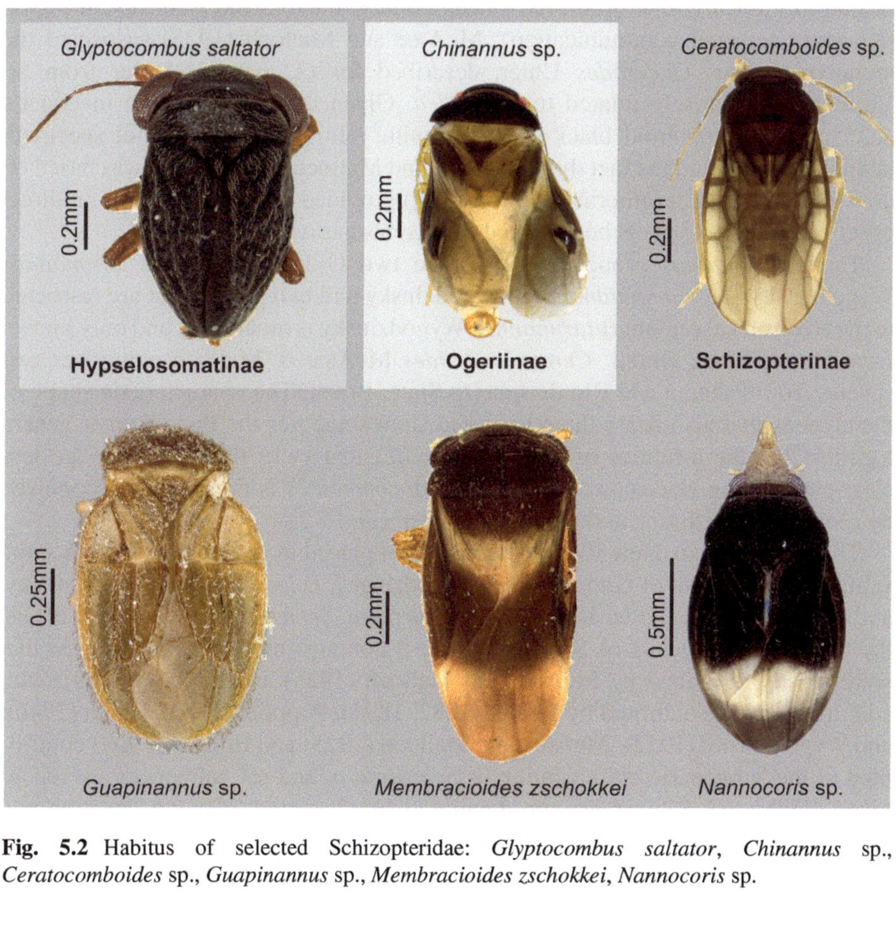

Fig. 5.2 Habitus of selected Schizopteridae: *Glyptocombus saltator*, *Chinannus* sp., *Ceratocomboides* sp., *Guapinannus* sp., *Membracioides zschokkei*, *Nannocoris* sp.

Wygodzinsky, described from the state of Rio de Janeiro in Brazil, is known only from the male holotype. Wygodzinsky (1948) speculated that this species could be sexually dimorphic similar to the situation in *Chinannus*, where males are macropterous, but females coleopteroid.

Emsley (1969) organized Schizopterinae in a number of "associations" of genera that he thought to be closely related. Two of these groups occur in the New World, the *Corixidea* group and the *Biturunannus* group. Emsley recognized the *Corixidea* group of genera based on the uniquely truncate apex of the labium that is shared by *Corixidea* Reuter, *Hoplonannus* McAtee & Malloch, *Membracioides* McAtee & Malloch (Fig. 5.2), *Voccoroda* Wygodzinsky, and the recently described *Voragocoris* Weirauch (Emsley 1969). *Corixidea* includes seven described species from Central America including the Caribbean; *Hoplonannus* three described species from Guatemala and Trinidad; and *Membracioides* two species described from Mexico and Panama. *Voccoroda* is monotypic and only known from Rio de Janeiro State, Brazil. *Voragocoris* now includes two species from Peru and Suriname (Weirauch 2012; Makhan 2013). A key to genera in this group was provided by Weirauch (2012). Numerous undescribed species, some likely representing new genera, that belong to the *Corixidea* group have been collected across the Neotropical region (Knyshov personal communication). McAtee and Malloch (1925) suspected the monotypic genus *Oncerodes* Uhler, described for *O. robustus* Uhler from St. Vincent, to be closely related to *Corixidea.* Given the lack of detail in Uhler's description (coleopteroid black specimen with yellowish legs, sex not specified, labium tapering) and the fact that McAtee's and Malloch's assessment was based on drawings (the head, prothorax, and forewing) provided by China and not on direct observation, *Oncerodes* is here treated as *incertae sedis*.

The *Biturunannus* group comprises the two Old World genera *Bironannus* Wygodzinsky and *Humpatanannus* Wygodzinsky and two genera that are restricted to the Neotropical region: *Biturunannus* Wygodzinsky is monotypic and only known from Parana State, Brazil. *Ceratocomboides* McAtee & Malloch comprises two species from Panama and Rio de Janeiro State, Brazil. The characteristic shape of the clypeus that also bears three long and stout setae ties the three former genera together, but the affinities of *Ceratocomboides* that lacks these features are less clear-cut. Among Neotropical genera, *Ceratocomboides* is most easily recognized by a small cell posterior to the trapezoidal cell (Fig. 5.2).

The remaining six New World genera are not placed in genus groups. In fact, the subfamily placement of *Guapinannus* Wygodzinsky, *Peloridinannus* Wygodzinsky, and *Tropistotrochus* Reuter is tentative at this point. *Schizoptera* Fieber (Fig. 5.1) is the by far largest genus of Schizopterinae. It contains more than 60 species that were mostly described by McAtee and Malloch (1925) and Emsley (1969), with additional species authored by Reuter (1882, 1891), Poppius (1910), China (1946), and Wygodzinsky (1952). McAtee and Malloch (1925) and Emsley (1969) contributed to the subgeneric classification of *Schizoptera*, and six subgenera are recognized. Species of *Schizoptera* are diagnosed by characters of wing venation and male genitalia, especially the uniquely modified and asymmetrical subgenital plate in the males. *Nannocoris* Reuter is likely the most easily recognized and charismatic among all Neotropical schizopterine genera: the majority of species are recognized by the elongate and pointed anteocular head region (Fig. 5.2). Twelve valid species range from Central America and the Caribbean to Argentina. As true for the most Neotropical schizopterine genera, a number of new species await description (Weirauch 2014). *Ptenidiophyes* Reuter is a rather enigmatic monotypic, female-based genus, described by Reuter (1891) for *P. mirabilis* Reuter from Santa Catarina State, Brazil. As speculated by Weirauch and Štys (2014), *P. mirabilis* may represent the female of a sexually dimorphic species. In the molecular analyses by Weirauch and Štys (2014), two specimens tentatively identified as *Ptenidiophyes* grouped with *Schizoptera*.

Some of the most stunning looking Schizopteridae in the New World are found in the genera *Guapinannus*, *Peloridinannus*, and *Tropistotrochus*. The holotype of the monotypic genus *Tropistotrochus*, *T. ampliatipennis* (Santa Catarina State, Brazil), is lost, but Reuter illustrated the wing venation that closely resembles the one seen in *Guapinannus*. *Guapinannus* (Fig. 5.2) was described by Wygodzinsky (1950) to accommodate *G. bierigi* from Costa Rica. The explanate forewings and in particular the very long fracture of the forewing that appears to divide the wing into two sections distinguish *Guapinannus* from other Neotropical Schizopterinae. *Peloridinannus* is also monotypic and was described by Wygodzinsky (1950) for *P. margaritatus* Wygodzinsky from Costa Rica. Its generic name emphasizes the fact that species in this genus resemble Coleorrhyncha in the family Peloridiidae in having flat and explanate wings with a complex pattern of wing venation. Undescribed species of *Peloridinannus* are known from Peru, Ecuador, and Panama, and species of *Guapinannus* also occur in Ecuador and Brazil.

The biology and ecology of Neotropical Schizopteridae are largely unknown, except the information provided by Emsley (1969) and notes on specimen labels recorded in Wygodzinsky's publications. Specimens are typically thought to be best collected using leaf litter extraction methods or light traps (Emsley 1969). Recent data (Weirauch Lab 2014) from sorting Schizopteridae from passive trapping samples show that out of 3,488 schizopterid specimens (35 countries; sampling all biogeographic regions), 822 specimens were collected using leaf extraction methods, only 17 in light traps, but 171 in pan traps and 1,396 specimens in flight intercept and Malaise traps (sampling method unknown for ca. 400; the remainder are other methods including hand collecting, pit fall traps, and canopy fogging). This indicates that many Schizopteridae may be more mobile than typically assumed.

5.6 Concluding Remarks

The minute litter bugs are clearly understudied in the Neotropical region, as they are elsewhere. The existing classification of subfamilies and genera will almost certainly undergo substantial changes as new genera are described and phylogenies produced that will put the current classification to the test. Similarly exciting as these changes above the genus level is the fact that a plethora of new species remain to be discovered and described from the Neotropical region.

Acknowledgments We thank Rochelle Hoey-Chamberlain, Aleksandr Knyshov, and Stephanie Leon for contributing some of the habitus images and acknowledge NSF DEB 1257702 (ARTS: Litter Bugs: revisionary and phylogenetic research on the least studied true bug infraorder (Insecta: Hemiptera: Dipsocoromorpha) for financial support.

References

Azar D, Nel A (2010) The earliest fossil schizopterid bug (Insecta: Heteroptera) in the Lower Cretaceous amber of Lebanon. Ann Soc Entomol Fr 46:193–197

Champion GC (1908) *Pachycoleus rufescens* Sahlb. Fam. Ceratocombidae in Devonshire. Entomol Mag Lond 44:8

China WE (1946) New Cryptostemmatidae Hemiptera from Trinidad, British West Indies. Proc R Entomol Soc Lond (B) 15:148–154

Costas M, Vázquez MA, López T (1993) *Pachycoleus pusillimus* (Sahlberg, 1870) (Heteroptera: Dipsocoridae), primera cita para la Península Ibérica. Bol Asoc Esp Entomol 17:77–81

Crowson RA (1972) A macropterous *Ceratocombus coleoptratus* Zett Hem, Dipsocoridae in Scotland. Entomol Mon Mag 108:247

Dellapé PM, Carpintero DL, Melo MC (2010) New records of *Dipsocoromorpha*, *Cimicomorpha* and *Pentatomomorpha* (Hemiptera: Heteroptera) from Argentina. Zootaxa 2436:57–64

Emsley MG (1969) The Schizopteridae (Hemiptera: Heteroptera) with the description of new species from Trinidad. Mem Am Entomol Soc 25:1–154

Esaki T, Miyamoto S (1959) A new or little known Hypselostoma from Amani-Oshima and Japan, with proposal of a new tribe for the genus (Hemiptera). Sieboldia. Acta Biolo 2:109–120

Golub VB (2004) Record of *Pachycoleus pusillimus* from Voronezh Province of Russia (Heteroptera: Dipsocoridae). Zoosyst Rossica 12:216

Halbert SE, Brambila J (2002) Dipsocoridae (Heteroptera) found for the first time in Florida. Insecta Mundi 16:24

Heckmann R, Rieger C (2001) Wanzen aus Baden-Württemberg – Ein Beitrag zur Faunistik und Ökologie der Wanzen in Baden-Württemberg (Insecta, Heteroptera). Carolinea 59:81–98

Henry TJ (2009) Biodiversity of the Heteroptera. In: Foottit RG, Adler PH (eds) Insect biodiversity: science and society. Wiley-Blackwell, Oxford, pp 223–263

Hill L (1980) Tasmanian Dipsocoroidea (Hemiptera: Heteroptera). J Aust Entomol Soc 19:107–127

Hill L (1984) New Genera of Hypselosomatinae (Heteroptera: Schizopteridae) from Australia. Aust J Zool Suppl Ser 103:1–55

Hill L (1987) Four new Australian species of *Hypselosoma* Reuter (Heteroptera: Schizopteridae). Aust J Entomol 26:265–278

Hill L (2013) A revision of *Hypselosoma* Reuter (Heteroptera: Schizopteridae) from New Caledonia. Mem Qld Mus Nat 56:407–455

Jung S, Kerzhner IM, Lee S (2011) A new record of the infraorder Dipsocoromorpha (Hemiptera: Heteroptera) from South Korea. J Asia-Pac Entomol 14:163–165

Lattin JD (2000) Life history of *Ceratocombus vagans* (Hemiptera: Heteroptera: Ceratocombidae), with notes on the immature stages. Great Lakes Entomol 33:131–139

Leston D (1954) Some aquatic and semi-aquatic Heteroptera (Hem.) from West Riding, Yorks. (V.C.64): with a note on the morphology of Cryptostemmatidae. Entomol Mon Mag 90:115–117

Makhan D (2013) *Soekhnandanius aschne* gen. et sp. nov., Schizopterinae (Hemiptera: Heteroptera: Schizopteridae) from Suriname. Calodema 291:1–3

McAtee WL, Malloch JR (1925) Revision of bugs of the family Cryptostemmatidae in the collection of the United States National Museum. Proc US Nat Mus 67:1–42

Nieser N (1973) Semiaquatic Heteroptera from Cuba. Res Exped Biospeleol Cubano-Roumaines Cuba 1:347–350

Péricart J, Matocq A (2003) Two species of Dipsocoridae new for Algeria, one new for science. Nouv Rev D'Entomol 20:255–257

Poppius B (1910) Neue Ceratocombiden. Ofversigt af Finska Vetenskapssocietetens Forhandligar Helsingfors 52A(1):1–14

Reuter OM (1882) Sur le genre *Schizoptera* Fieber. Rev d'Entomol 1:162–164

Reuter OM (1891) Monographia Ceratocombidarum orbis terrestres. Acta Soc Scient Fennicae 19:1–28

Scudder GGE (2010) The Schizopteridae (Hemiptera), a family new to Canada

Schuh RT, Weirauch C, Wheeler WC (2009) Cimicomorphan relationships (Insecta: Heteroptera): combining morphological and DNA sequence data (Insecta: Heteroptera). Syst Entomol 34:15–34

Štys P (1959) The 5th stage of *Ceratocombus (Ceratocombus) coleoptratus* (Zetterstedt, 1819) and notes on the morphology and systematics of Dipsocoridae (Heteroptera). Act Entomol Mus Natur Pragae 33:377–388

Štys P (1970) On the morphology and classification of the family Dipsocoridae s. lat., with particular reference to the genus *Hypsipteryx* Drake (Heteroptera). Acta Entomol Bohemoslov 67:21–46

Štys P (1982) A new genus and species of Schizopteridae with porrect head from Papua New Guinea Heteroptera. Acta Entomol Bohemoslov 796:450–456

Štys P (1985) Two new species of *Kokeshia* (Heteroptera, Schizopteridae) from Nepal and appraisal of alleged synapomorphies of Paraneoptera. Acta Entomol Bohemoslov 82:187–205

Štys P (1995) Dipsocoromorpha. In: Schuh RT, Slater JA (eds) True bugs of the world (Hemiptera: Heteroptera): classification and natural history. Comstock Publishing Associates, Cornell University Press, Ithaca, pp 74–83

Uhler PR (1894) On the Hemiptera-Heteroptera of the Island of Grenada, West Indies. Proc Zool Soc Lond 167–224

Uhler PR (1904) List of Hemiptera-Heteroptera of Las Vegas Hot Springs, New Mexico, collected by Messrs. E.A. Schwarz and Herbert S. Barber. Proc US Nat Mus 27:349–364

Usinger RL (1945) Notes on the genus *Cryptostemma* with a new record for Georgia and a new specie from Puerto Rico (Hemiptera, Cryptostemmatidae). Entomol News 56:238–241

Usinger RL (1946) Notes and descriptions of *Ceratocombus* (Hemiptera: Cryptostemmatidae). Proc Hawaiian Entomol Soc 12:633–636

Weirauch C (2012) *Voragocoris schuhi*, a new genus and species of Neotropical Schizopterinae (Hemiptera: Schizopteridae). Entomol Am 118:285–294

Weirauch C, Schuh RT (2011) Systematics and evolution of Heteroptera: 25 years of progress. Annu Rev Entomol 56:487–510

Weirauch C, Štys P (2014) Litter bugs exposed: phylogenetic relationships of Dipsocoromorpha (Hemiptera: Heteroptera) based on molecular data. Insect Syst Evol 45:1–20

Wheeler WC, Schuh RT, Bang R (1993) Cladistic relationships among higher groups of Heteroptera: congruence between morphological and molecular data sets. Entomol Scand 24:121–137

Whitehead P (2005) *Pachycoleus waltli* Fieber 1860 (Hem., Dipsocoridae) new to Worcestershire (VC37). Entomol Mon Mag 141:166

Wygodzinsky P (1948) On two new genera of "Schizopterinae" (Cryptostemmatidae) from the Neotropical region (Hemiptera). Rev Bras Biol 8:143–155

Wygodzinsky P (1950) Contribution to the knowledge of the family Cryptostemmatidae. Rev Bras Biol 10:377–392

Wygodzinsky P (1952) Sobre algunos Cryptostemmatidae, principalmente de la Argentina. Acta Zool Lilloana 10:51–74

Wygodzinsky P (1953) Cryptostemmatinae from Angola (Cryptostemmatidae, Hemiptera). Publ Cult Co Diam Angola 16:27–48

Part III
Infraorder Gerromorpha

Chapter 6
The Semiaquatic Gerromorphans

Felipe F.F. Moreira

Abstract The Gerromorpha comprise a group of mainly semiaquatic predatory bugs that play important roles in freshwater ecosystems. Despite its current low economic importance, there are evidences that certain groups may be utilized in the biological control of mosquitoes and agricultural pests or as potential indicators of biological quality in aquatic habitats. They are rather diversified in general appearance, structurally very diverse, and their two-dimensional habitat makes them ideal subjects for ecological and behavioral analysis. The Neotropical fauna of Gerromorpha is relatively well known, but more than a hundred undescribed species are known to exist. This fact, allied to the existence of large under-collected areas, makes taxonomic and faunistic studies concerning the semiaquatic bugs from tropical America urgent.

6.1 Introduction

Heteroptera constitutes a very interesting worldwide distributed group of insects, which is more diversified in tropical zones (Mazzucconi et al. 2009). Most of the species is terrestrial, but numerous others are aquatic or semiaquatic. Out of the seven infraorders of the suborder, the true aquatic bugs (most of which live in the water) constitute the Nepomorpha, and the mainly semiaquatic species constitute the Gerromorpha (water striders) and Leptopodomorpha (shore bugs) (Nieser and Melo 1997; Polhemus and Polhemus 2008).

Water bugs in general play an important role in freshwater ecosystems, and knowledge about them is essential for the study of water biology and the proper management of aquatic habitats. Furthermore, there is increasing evidence that

This chapter is dedicated to the memory of Dr. John T. Polhemus (1929–2013), one of the greatest researchers who worked with Gerromorpha and other aquatic Heteroptera.

F.F.F. Moreira (✉)
Laboratório Nacional e Internacional de Referência em Taxonomia de Triatomíneos,
Instituto Oswaldo Cruz, Fiocruz, Av. Brasil, 4365, Manguinhos,
Rio de Janeiro, RJ 21040-360, Brazil
e-mail: felipento@hotmail.com

A.R. Panizzi, J. Grazia (eds.), *True Bugs (Heteroptera) of the Neotropics*,
Entomology in Focus 2, DOI 10.1007/978-94-017-9861-7_6

certain groups may be utilized in the biological control of mosquitoes (Menke 1979; Armúa de Reyes et al. 2005; Rodríguez-Castro et al. 2006; Shaalan and Canyon 2009) and agricultural pests (Nakasuji and Dyck 1984; Ignacimuthu 2002) or as potential indicators of biological quality in aquatic habitats (Mazzucconi et al. 2009; Cabette et al. 2010).

The Gerromorpha, or semiaquatic bugs, have been recognized as a unique taxonomic entity since the time of Dufour (1833) and were treated for a long period as "Amphibicorisae" due to their ability to walk on the surface of the water, distinctly from the true aquatic bugs "Hydrocorisae" and the terrestrial bugs "Geocorisae" (Schuh and Slater 1995; Damgaard 2008b). The name Gerromorpha itself was proposed only by Popov (1971).

In his view, the Nepomorpha, Gerromorpha, and Leptopodomorpha arose in the Triassic, and the ancestral stocks of the modern superfamilies all had diverged by the end of the Jurassic. However, while Nepomorpha has a very extensive fossil record dating back into the Triassic, the Gerromorpha has by comparison left much fewer fossils (Grimaldi and Engel 2005). So far, only a few more than 30 fossil species belonging to six families have been described or recorded, spanning more than 120 million years of geological history. The evolutionary history of water striders probably began in the early Mesozoic, and the oldest members of the infraorder are Lower Cretaceous age, belonging to the families Mesoveliidae, Veliidae, and Hydrometridae (Andersen 1998; Damgaard 2008b).

As for the extant fauna, the Gerromorpha comprises more than 2,100 species distributed all over the World's continents, except for Antarctica (Andersen 1982; Polhemus and Polhemus 2008). Most of the known species occupy an adaptative zone defined by water surfaces (the pleustonic zone), and their habitats range from the few square centimeters surface area of water accumulated in tree holes to the millions of square kilometers of the world's great oceans (Andersen 1982, 1998). Some species have also become secondarily terrestrial, being found, for example, on forest litter or moss covering fallen trees (Andersen and Polhemus 1980; Malipatil and Monteith 1983).

Some gerromorphans, like most members of the family Gerridae and many of the Veliidae, spend nearly the entire active period of their lives on the water surface (Schuh and Slater 1995), and their two-dimensional habitat makes them ideal subjects for ecological and behavioral analysis, with many northern temperate species being extensively used as model organisms in such studies (Spence and Andersen 1994; Andersen 1998). Studies of the systematics, functional morphology, and ecological diversity of semiaquatic bugs (Andersen 1973, 1975, 1976, 1978) have also demonstrated a striking variation in habitat preference and ways of life among the Gerromorpha.

The Neotropical fauna of Gerromorpha is relatively well known (Polhemus and Polhemus 2007), especially because of the works of Champion (1898), and C. J. Drake, H. M. Harris, and H. B. Hungerford from the 1920s to the 1950s. Taxonomic studies tended to decline since then, with the exception of the studies of N. Nieser on the Caribbean, Suriname, and northern South America, J. T. Polhemus on Mexico, and the monograph on Neotropical *Rhagovelia* Mayr (1865) by D. A.

Polhemus (1997). During the last two decades, however, there has been increasing interest concerning Neotropical Gerromorpha and their taxonomy and diversity, especially by workgroups from Colombia, Brazil, and Argentina. However, detailed studies on ecology, behavior, genetics, and evolution are still very scarce when compared to those based on Palearctic and Nearctic species.

6.2 General Characteristics and Diagnosis

Semiaquatic bugs are rather diversified in general appearance and structurally very diverse (Fig. 6.1) (Andersen 1979), which makes morphologically defining the group somewhat difficult (Nieser and Melo 1997). Besides that, pterygopolymorphism is common in the infraorder, with some species showing continuous variation, whereas others are known only from the macropterous and/or micropterous or apterous morphs (Schuh and Slater 1995). Nevertheless, Andersen (1982) gives an excellent account of the structural features common to all or at least most of the families of Gerromorpha.

Out of the characters mentioned by him, the following, among others, are considered synapomorphies of the group: presence of three pairs (rarely four) of cephalic trichobothria inserted in deep cuticular pits in the adult; epipharynx with a long, narrow, external projection; mandibular levers quadrangular; maxillary levers absent; forewings usually not differentiated into anterior coriaceous and posterior membranous portions; pretarsus with dorsal and ventral arolium, the ventral one being sometimes modified; body covered by a 2-layered hair pile; and female gynatrial complex with long, tubular, entirely glandular spermatheca and secondary fecundation canal (Fig. 6.2) (Schuh and Slater 1995).

The cuticular surface of semiaquatic bugs is covered by dense hair layers, which makes these insects appear dull or velvety, with a silvery or golden sheen in certain angles of illumination (Andersen 1982). The greater part of the head and body of most species is covered by hair layers which have the same basic structure: a macrohair layer of long, tapering and pointed macrotrichia, flexibly inserted in sockets, and a micro-hair layer of much smaller, stiff, usually filiform cuticular outgrowths or macrotrichia (Fig. 6.3). The surface of the antennae, rostrum, and leg segments, however, has only a layer of macrotrichia interspersed with other setae. These elaborate surface structures comprise without a doubt a system designed to prevent wetting and also function secondarily as protection against UV radiation from the sun (Cheng et al. 1978).

Another peculiar characteristic of the body surface of gerromorphans is the presence of peg-plates or "sieve-pores" (Fig. 6.4) (Cobben 1978). Under the stereomicroscope, they appear as minute, shiny dots, and in the scanning electron microscope, they are seen as ovate or circular plates bordered by a rim surrounding minute subconical pegs. Peg-plates are distributed all over the surface of the head, thorax (except posterior lobe of pronotum), and abdomen, but they are usually more numerous on the pleural and ventral parts of the body. They are commonly absent

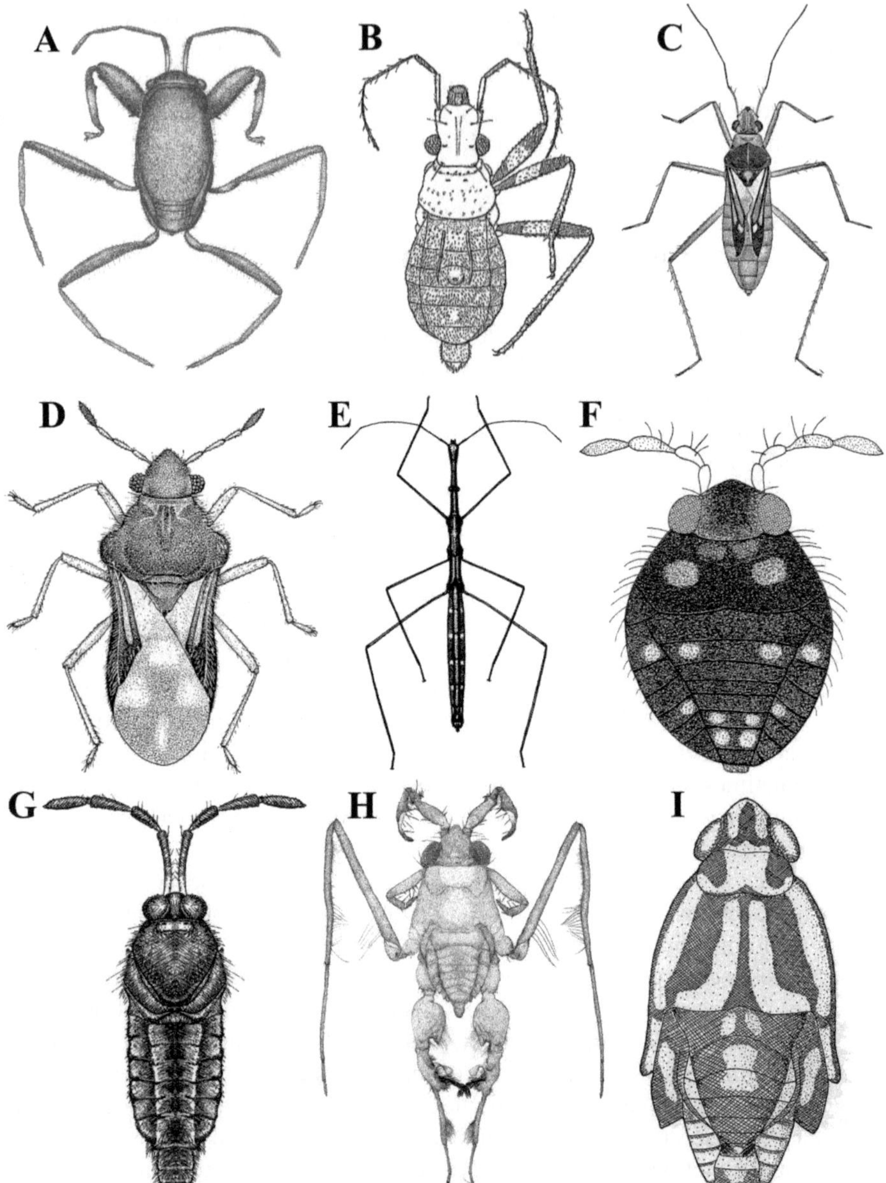

Fig. 6.1 Morphological diversity among the Gerromorpha. (**a**) *Hermatobates* (Hermatobatidae). (**b**) *Chepuvelia* (Macroveliidae). (**c**) *Mesovelia* (Mesoveliidae). (**d**) *Merragata* (Hebridae). (**e**) *Hydrometra* (Hydrometridae). (**f**) *Euvelia*, legs omitted (Veliidae). (**g**) *Rhagovelia*, legs omitted (Veliidae). (**h**) *Rheumatobates* (Gerridae). (**i**) *Halobatopsis*, antennae and legs omitted (Gerridae) (Modified from **a** – Andersen and Weir 2000; **b** – China 1963; **c** – Cobben 1960; **d** – Drake and Cobben 1960; **e** – Menke 1979; **f** – Polhemus and Polhemus 1984a; **g** – Moreira and Ribeiro 2009; **h** – Padilla-Gil and Pacheco-Chaves 2012; **i** – Nieser and Melo 1999)

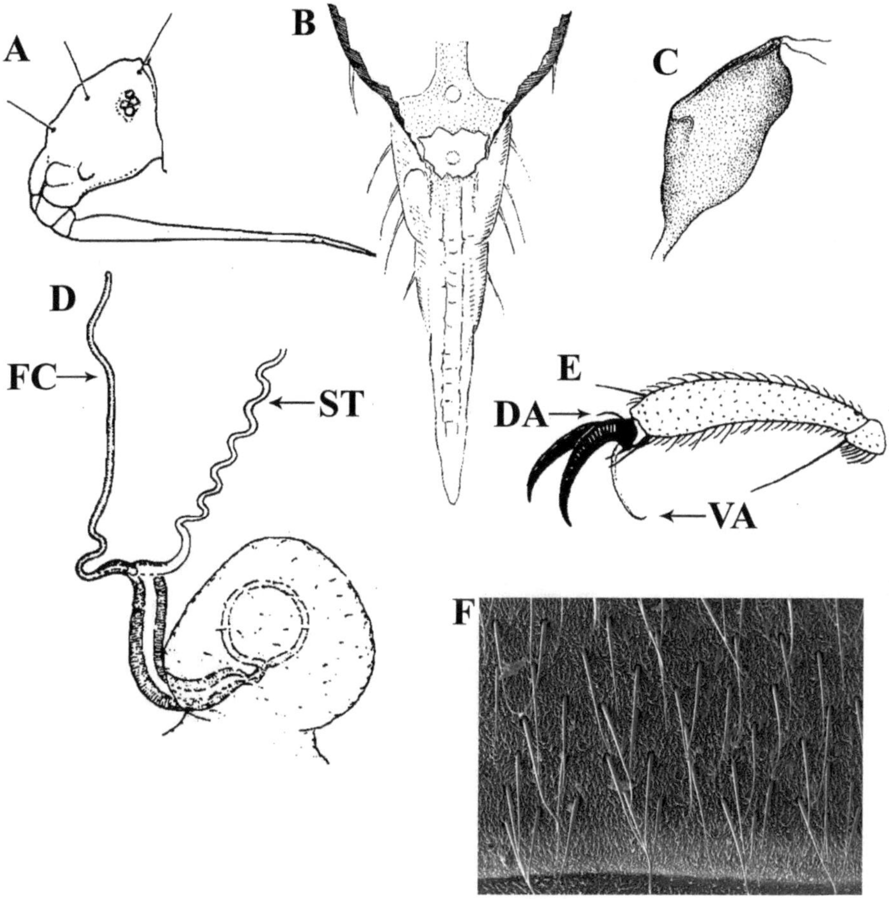

Fig. 6.2 Some of the synapomorphies of the Gerromorpha. (**a**) *Cryptovelia* (Mesoveliidae), head in lateral view showing three cephalic trichobothria. (**b**) *Hydrometra* (Hydrometridae), epipharynx in ventral view. (**c**) *Hydrometra* (Hydrometridae), quadrangular mandibular lever in plane view. (**d**) *Hyrcanus* (Hebridae), female gynatrial complex (*FC* fecundation canal, *ST* spermathecal tube). (**e**) *Hyrcanus* (Hebridae), fore tarsus (*DA* dorsal arolium, *VA* ventral arolium). (**f**) *Velia* (Veliidae), hind portion of fourth abdominal sternite showing layers of macro- and micro-hairs (Modified from **a**, **d**–**e** – Andersen 1982; **b, c** – Cobben 1978; **f** – Cianferoni and Santini 2013)

from leg segments (except coxae) and have been recorded in all families of semi-aquatic bugs except Hermatobatidae and Gerridae (Andersen 1982).

As mentioned above, water striders also typically bear three pairs of cephalic trichobothria, which were first noticed by China (1955) and that are very likely to be mechanoreceptors (Andersen 1976). The distribution of these long setae on the dorsal head surface is practically the same in all semiaquatic bugs, with the posterior pair inserted at the base of the head and usually quite close to the hind corners of the eyes and the two anterior pairs inserted on the frontal surface of the head. An additional pair of trichobothria is inserted laterally to the two anterior pairs in most

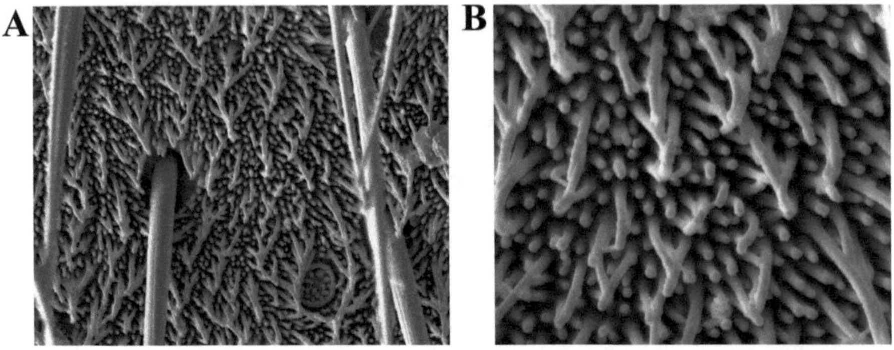

Fig. 6.3 Cuticular hair layers of the gerromorphans. (**a**) *Velia* (Veliidae), fourth abdominal sternite, detail of the socket of a macro-hair (micro-hairs also visible). (**b**) *Velia* (Veliidae), fourth abdominal sternite, double layer of micro-hairs (Modified from Cianferoni and Santini 2013)

Fig. 6.4 *Velia* (Veliidae), fourth abdominal sternite, peg-plate (Modified from Cianferoni and Santini 2013)

Gerridae. While cephalic trichobothria-like setae are recorded in several Heteropterans, the type of insertion found in semiaquatic bugs is unique. The long and slender seta arises from a deep pitlike depression in the cuticle, and this socket is located on a domelike elevation (Andersen 1982).

The head of the gerromorphans (Fig. 6.5) is typically elongate, with prominent eyes, usually with a pair of ocular setae. A pair of ocelli can be found near inner eye margins, but is reduced along with wing reduction in polymorphic species and totally lost in some Hydrometridae, most Veliidae, and all Gerridae. The surface of the head has very few sutures and is usually longitudinally impressed or has longitudinal rows of shiny punctures. A pair of dorsal indentations is also commonly present (Andersen 1982).

Contrary to the Nepomorpha, the antenna of Gerromorpha is always long and exposed (Heckman 2011). They are inserted medio-dorsally on the antennal tuber-

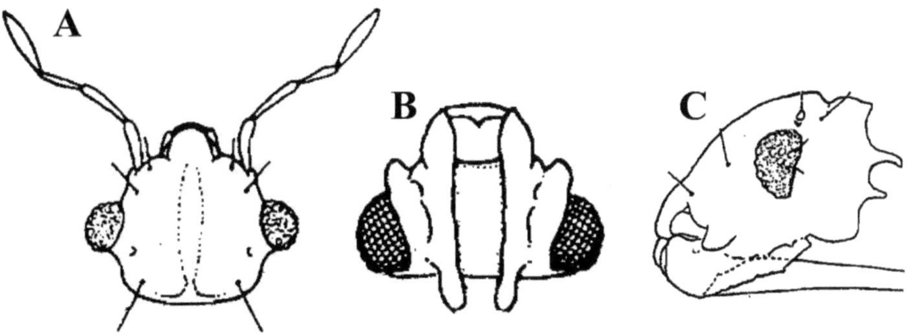

Fig. 6.5 General aspect of the head of the gerromorphans. (**a**) *Merragata* (Hebridae), dorsal view. (**b**) *Timasius* (Hebridae), ventral view. (**c**) *Hebrus* (Hebridae), lateral view (Modified from Andersen 1982)

cles, usually well in front of eyes, and always have four articles, besides short intersegmental pieces which are usually found between antennomeres II–III and III–IV. A superficial ring on antennomere IV of some genera of Hebridae gives the impression of a five-segmented antenna, which was misinterpreted in the past. The rostrum of semiaquatic bugs is typically long and slender, reaches far backward beneath the body when not in use, and is always four-segmented (Andersen 1982).

Most of the externally visible parts of the prothorax of gerromorphan bugs are composed of marginal evaginations or lobes. The pronotum is a broad plate with a short anterior lobe that surrounds the postocciput of the head, sometimes forming a narrow anterior collar. The lateral and, in particular, the posterior margin of the pronotum are expanded into a large posterior lobe, which in most semiaquatic bugs with winged adults extends far behind as the level of the wing bases, covering the whole mesonotum and usually also the median part of the metanotum (Andersen 1982).

The mesonotum is always the largest part of the pterothorax tergum in semiaquatic bugs. It consists of a broad anterior scutum and, in Mesoveliidae and Hebridae only, of a posterior scutellum. A large triangular scutellar lobe, typically found in other Heteroptera, is only present in *Mesoveloidea* Hungerford and *Madeovelia* Poisson of the family Mesoveliidae. In all other gerromorphan bugs, the scutellum is reduced (*Mesovelia* Mulsant & Rey), rudimentary (Hebridae), or absent (other families) (Andersen 1982).

The above notes on the thoracic skeleton apply only to the macropterous adult form (Fig. 6.6a). Most semiaquatic bugs commonly occur in a form which has more or less reduced wings (brachypterous or micropterous) or without recognizable wings (apterous). In these forms, the posterior lobe of the pronotum is usually reduced in length and width or completely absent, the mesonotum is thickened and sclerotized, and the pterothoracic endoskeleton is more or less reduced (Fig. 6.6b) (Andersen 1982).

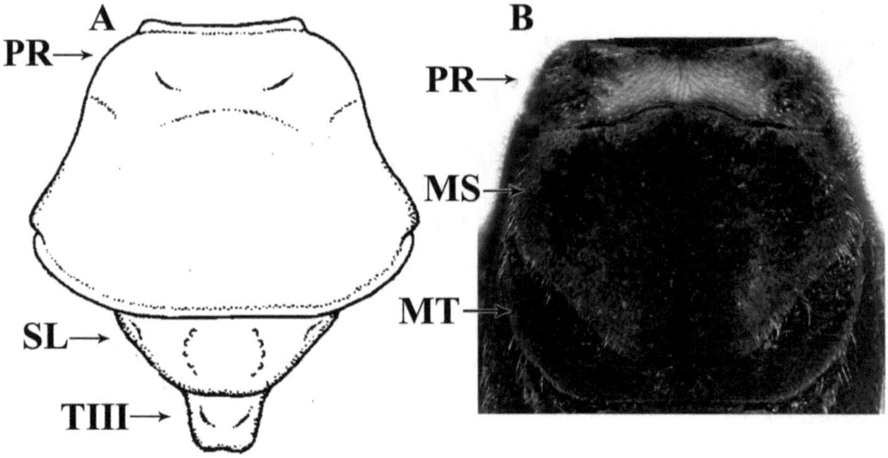

Fig. 6.6 Variation of the structure of the thorax in Gerromorpha. (**a**) *Mesovelia* (Mesoveliidae), macropterous thorax in dorsal view. (**b**) *Rhagovelia* (Veliidae), apterous thorax in dorsal view (*MS* mesonotum, *MT* metanotum, *PR* pronotum, *SL* scutellum, *TIII* metanotal lobe; modified from: **a** – Andersen 1982; **b** – Moreira et al. 2010)

There is a great diversity within the Gerromorpha concerning the functional morphology of the legs (Andersen 1976). In the Mesoveliidae, they are inserted close to the ventral midline of the body, and the long metacoxae almost touch each other, whereas in most other semiaquatic bugs, the coxae of each thoracic segment are widely separated (Andersen 1982). The plesiomorphic state of the inclination of the coxal axis in Gerromorpha is oblique in all legs, but it can range from a vertical inclination on fore and middle legs of *Hydrometra* Latreille to an almost horizontal inclination on middle and hind legs of Gerridae (Andersen 1976).

Members of some genera of the families Veliidae and Gerridae show very long legs with characteristic modifications that facilitate movement across the surface tension of the water. Others, like the Hydrometridae, have legs that do not appear markedly different from those of terrestrial bugs, but which nevertheless easily support the insect on the surface tension layer (Andersen 1976; Heckman 2011).

The trochanter of the Gerromorpha has proximal groups of cuticular sense organs, which are most likely proprioceptive and more or less uniform in constitution and number among most families, except for Gerridae and Veliidae, in which they are more numerous (Andersen 1982). Some or all of their tibiae have distal grooming structures composed by modified macrotrichia, usually arranged in combs, and used as cleaning devices to keep the hair layers of the legs and body tidily arranged and free of dust or water drops (Andersen 1976). In most families of Gerromorpha, the adult tarsus has three segments, whereas in Hebridae, some Veliidae, and all Gerridae, the number of tarsomeres is reduced due to the fusion of articles (Andersen 1982).

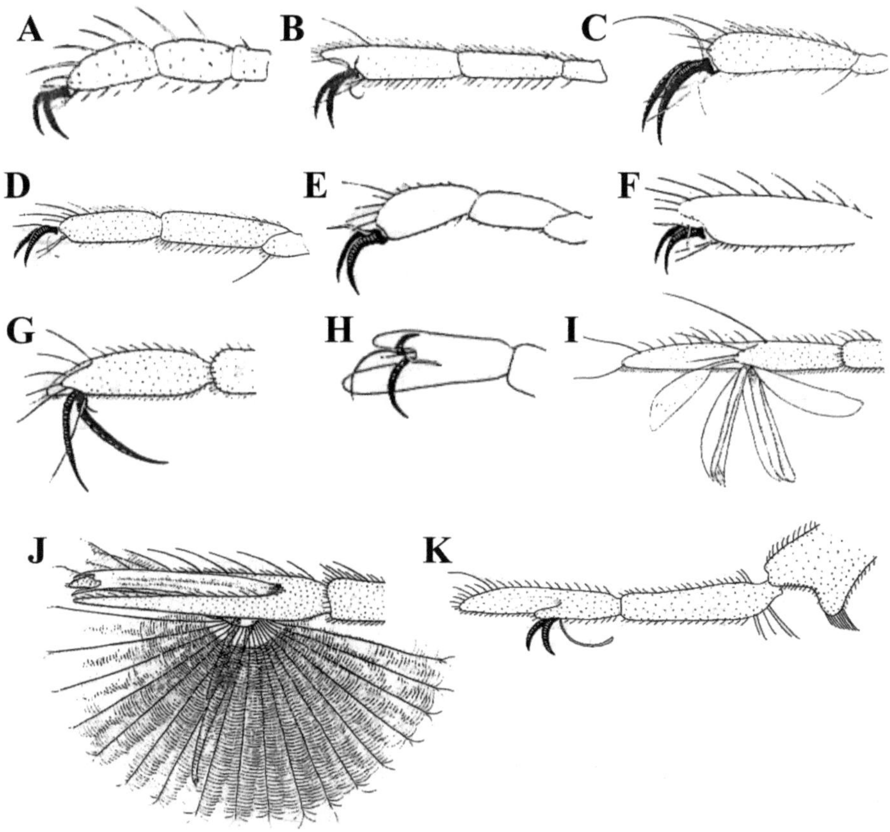

Fig. 6.7 Variation of the structure of the pretarsus in Gerromorpha. (**a**) *Cryptovelia* (Mesoveliidae). (**b**) *Mesoveloidea* (Mesoveliidae). (**c**) *Timasius* (Hebridae). (**d**) *Macrovelia* (Macroveliidae). (**e**) *Veliometra* (Hydrometridae). (**f**) *Limnobatodes* (Hydrometridae). (**g**) *Platyvelia* (Veliidae). (**h**) *Oiovelia* (Veliidae). (**i**) *Veloidea* (Veliidae). (**j**) *Rhagovelia* (Veliidae). (**k**) *Halobates* (Gerridae) (Modified from Andersen 1982)

The pretarsus structure in Gerromorpha is complex and highly variable (Fig. 6.7) (Cobben 1968, 1978). On the plesiomorphic condition, it consists of two long, thick, and hook-shaped claws articulating to the tarsus and connected to a small posterior plate, the unguiractor, which extends to the distal part of the last tarsal segment. From the distal end of the unguiractor arises a pair of setiform parempodia, and between the claws originate two vertically oriented arolia, one dorsal and one ventral (Andersen 1982). The dorsal arolium is typically setiform, but may be horizontally flattened like the ventral arolium, which is usually the longer (Andersen 1976).

The position of the point of articulation between the pretarsus and the tarsal segments varies within semiaquatic bugs, being apical or preapical. The apical type is undoubtedly the plesiomorphic condition among the Gerromorpha, being found in almost all species of the basal families. The preapical type is typical of the Veliidae and Gerridae, produced by a prolongation of the anterior part of the last tarsal segment, and with parempodia usually modified, asymmetrically developed or reduced (Andersen 1982).

The gerromorphan fore wing is not divided into a leathery basal part (corium and clavus) and membranous apical part (membrane) like the typical hemielytron of Heteroptera. Most of the wing is sclerotized but relatively thin, and only the proximal parts of the longitudinal veins are thickened. The hind wing, on its turn, is membranous and shorter but much broader than the fore wing. The two wings of each side of the thorax are coupled during flight, and the hind wing is moved passively. The coupling apparatus consists of a hook-shaped protuberance on the lower side of the fore wing and an upturned fold along the anterior margin of the hind wing. The number of muscles in the pterothorax of gerromorphan bugs is much reduced as compared with other Heteroptera, and in adults with reduced wings, all muscles involved in their operation are reduced or lost (Andersen 1982).

The scent apparatus of semiaquatic bugs belongs to the "type omphalien" of Carayon (1971), with a single median scent orifice on the metasternum, which is a feature shared with Dipsocoromorpha, Enicocephalomorpha, Leptopodomorpha, and Nepomorpha. The scent orifice is overgrown by integument, so that the secretion has to flow each side through two short channels, and a distinct evaporative apparatus as found in terrestrial bugs is not present. Some Gerridae and Veliidae, however, have a more complex apparatus than other families, with prolonged lateral channels that reach the pleural regions of the metathorax. An abdominal scent gland is found in nymphs and adults of all semiaquatic families except for some Hydrometridae, the Veliidae and the Gerridae, and is situated between tergum III and IV (Andersen 1982).

An abdomen with eight plate-shaped segments on the abdominal dorsum of the female and seven on the male is the plesiomorphic condition among the Gerromorpha (Fig. 6.8). The eighth segment of females is divided ventrally along the midline, and in males it is cylindrical and usually withdrawn into the pregenital abdomen. A typical gerromorphan abdominal segment is divided dorsally into mediotergites and a pair of laterotergites, with endotergites being present in some females. The laterotergites form the abdominal connexives, with usually six segments, and the pleural and sternal parts of the abdominal segments are fused to each other to form a sternopleuron. The first abdominal sternopleuron is fused with the metasternum, except in the subfamily Rhagadotarsinae of the Gerridae, where it is secondarily retained (Andersen 1982).

The male genitalia of Gerromorpha were considered the most plesiomorphic among the Heteroptera by Cobben (1968, 1978). It consists basically of a cylindrical abdominal segment eight, a boat-shaped segment nine (the pygophore), and a lid-shaped segment ten (the proctiger), which covers the posterior end of the pygophore (Fig. 6.9). A relatively simple phallic organ is internally attached to the pygophore, and a pair of parameres arises laterally from it. The phallus is typically composed of two parts: the sclerotized proximal phalloteca and a membranous dis-

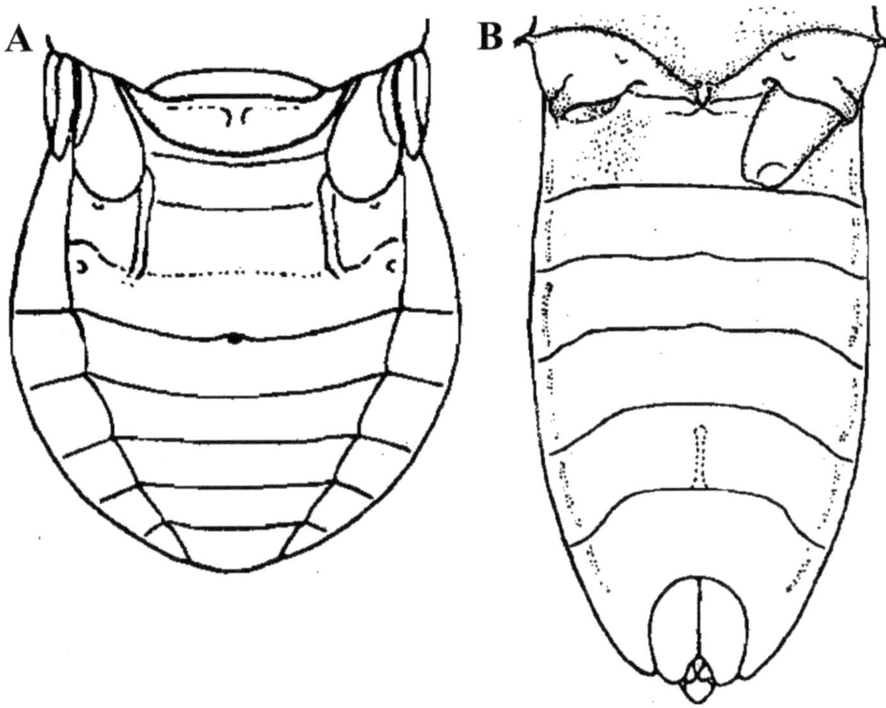

Fig. 6.8 General aspect of the abdomen of the gerromorphans. (**a**) *Merragata* (Hebridae), dorsal view of the micropterous form. (**b**) *Ocellovelia* (Veliidae), ventral view of the female (Modified from Andersen 1982)

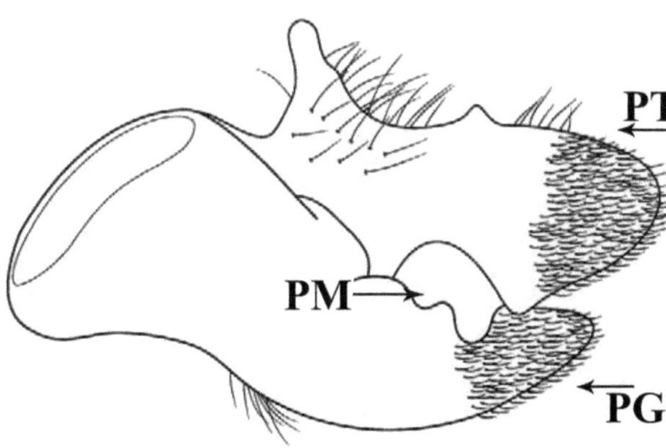

Fig. 6.9 General aspect of the external male genitalia of *Paravelia* (Veliidae) in lateral view, with parameres omitted (*PG* pygophore, *PM* place of insertion of the paramere, *PT* proctiger; modified from Moreira and Barbosa 2012)

tal endosoma, which is usually subdivided into a membranous conjunctivum and a vesica normally armed with sclerotized pieces (Andersen 1982).

As for the female genitalia (Fig. 6.10), it is formed by the eighth and ninth abdominal segments. The eighth segment has a basal sclerite or first gonocoxa

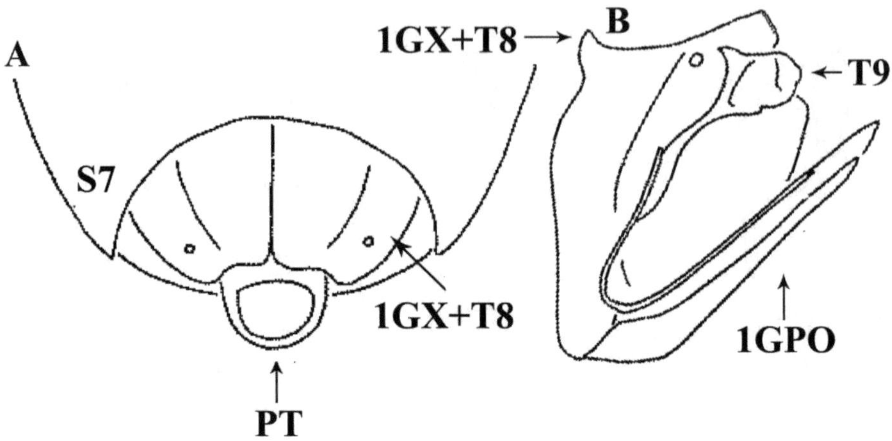

Fig. 6.10 General aspect of the external female genitalia in Gerromorpha. (**a**) *Rhagovelia* (Veliidae), apex of abdomen in ventral view. (**b**) *Hebrus* (Hebridae), dissected genitalia in lateral view (*1GPO* first gonapophysis, *1GX* first gonocoxa, *S7* seventh abdominal sternite, *PT* proctiger, *T8* paratergite 8, *T9* paratergite 9; modified from Scudder 1959)

ventrolaterally on either side, and this sclerite carries ventrally and caudally the first gonapophyses. The second pair of gonocoxae is found on segment nine, to which are attached the second pair of gonapophyses. Besides those, there is a cone-shaped structure named proctiger, which represents the tenth abdominal segment, and apically a plate that flanks the lower edge of the anal opening. The female ovipositor of most Gerromorpha is plate-shaped, with flap-like gonapophyses, adapted for depositing the egg on the surface of a substrate, embedded in a gelatinous mass. In Mesoveliidae, however, the ovipositor is laciniate and used to embed the eggs in the substrate (Andersen 1982).

6.3 General Biology and Ecology

The Gerromorpha occupy a wide variety of habitats from saltwater pools to mountain lakes, from hot springs to large rivers, and are the only group of insects found on open seas (Menke 1979; Cheng 1985). Andersen (1982) divided the kinds of habitats occupied by them in a few categories, as follows: (1) humid terrestrial habitats, not necessarily close to free water; (2) marginal aquatic habitats comprising wet soil, tufts of plants, or carpets of moss; (3) plant-covered water surface; (4) free surface of stagnant freshwater bodies; (5) free surface of flowing freshwater bodies; (6) surface of intertidal marine water; and (7) the surface of oceanic water. Representatives of most species of Gerromorpha will perish within a short time if kept dry (Andersen 1976), and although the vast majority of them is found above the water surface, the genus *Nieserius* Zettel from Asia includes subaquatic species of the family Hebridae (Zettel 1999).

In terms of habitat preferences, the better-known species are almost invariably European and North American, whereas tropical species are unfortunately very poorly studied. Among Neotropical Mesoveliidae, specimens of *Mesoveloidea williamsi* Hungerford can be found on wet moss close to streams or walking on top of fallen logs or boulders, similarly to *Mesovelia amoena* Uhler, which also dwells crevices at the edges of ponds and rivers. *Mesovelia mulsanti* White, by its turn, is a more active species commonly encountered running on floating leaves of aquatic plants on ponds (Andersen 1982; Damgaard et al. 2012). Other Neotropical mesoveliids inhabit environments very different from those commonly associated with the family (Damgaard et al. 2012). For example, *Speovelia mexicana* Polhemus was collected under rocks atop a reef-like coastal area, running about on the sea water trapped by the reef after the tide receded; species of *Darwinivelia* Andersen and Polhemus are associated with mangroves; and *Cryptovelia terrestris* Andersen and Polhemus is a semiterrestrial species, living on soil litter in primary or high secondary forests (Polhemus 1975; Andersen and Polhemus 1980; Andersen 1982).

As for the family Hebridae, the Neotropical species are very rarely collected or studied, and the habits of most of them are unknown. Representatives of the genus *Hebrus* Curtis commonly hide among the marginal vegetation alongside pools, lakes, and streams, occasionally being found on brackish water; *Merragata* White is typically associated with ponds completely covered by algal mats or heavy grown by plants with floating leaves; and specimens of *Lipogomphus* Berg have been collected on the water's edge of ponds, slow-moving parts of streams, and quiet secluded coves of lakes (Drake and Chapman 1958; Andersen 1982).

Among Neotropical Hydrometridae, the common and widespread genus *Hydrometra* is found on the reedy margins of rivers and streams where the current is subdued and at edges of lakes, ponds, and pools (Andersen 1982). However, species of the less diverse and rarer genera of the family are found in conditions very different from those. *Chepuvelia usingeri* China is endemic to forest leaf-litter in Chile; *Veliometra schuhi* Andersen was collected at marginal vegetation in the impounded area of a forest stream in the Amazon; and species of *Bacillometra* Esaki and *Bacillometroides* Polhemus and Polhemus are confined to swift rocky upland streams in the Guiana Plateau and the foothills of the Andes (China 1963; Andersen 1977; Polhemus and Polhemus 2010).

An even wider variety of habitats is found among the Veliidae and Gerridae. In the former family, most *Microvelia* Westwood are found near shore, on the plant-covered zone of stagnant waters. Some, however, can be found on open water surface of puddles and pools filled with rainwater or on the water pockets of terrestrial and arboreal bromeliads (an habitat shared only with *Paravelia* Breddin), inside crab holes or tree holes (Andersen 1982; Polhemus and Polhemus 1991). Species of *Husseyella* Herring are found on tidal streams of mangrove swamps; those of the poorly studied genus *Aegilipsovelia* Polhemus live hygropetrically on wet rock surfaces and in wet moss on rocks; *Platyvelia* Polhemus and Polhemus occupies ponds with overhanging vegetation and floating leaves or emergent plants along the margins; *Stridulivelia* Hungerford dwells among emergent vegetation or close to the shore under overhanging plants in both small and large rivers; and *Oiovelia* Drake and Maldonado-Capriles is common on foam masses in blackwater streams (Herring 1955; Wilson 1958; Drake and Menke 1962; Polhemus 1970; Andersen 1982; Spangler 1986). Finally, representatives of the highly diverse *Rhagovelia* are found almost exclusively on running freshwater, with the exception of those of the *salina* group *sensu* Polhemus (1997) from marine habitats, although they can be caught considerably distant from the sea (Drake and van Doesburg 1966; Andersen 1982).

Some genera of Neotropical Gerridae occupy a wide variety of habitats, like the *Rheumatobates* Bergroth, that can be caught in ponds, lakes, quiet pools of larger streams, artificial canals, estuaries, and mangrove swamps; *Eurygerris* Hungerford and Matsuda from margins of small streams, ponds of varied sizes, and lakes; and *Limnogonus* Stål and *Neogerris* Matsumura, which are a common sight in stagnant waters such as ponds, pools, and sometimes lakesides and backwaters of streams (Herring 1949; Polhemus 1969; Andersen 1982; Molano-Rendón et al. 2008). Others, however, are stricter in their environmental preferences. Species of *Metrobates* Uhler inhabit turbulent streams and rivers free of vegetation; specimens of *Brachymetra* Mayr and *Tachygerris* Drake are usually collected in running water, in or near the shaded areas of the habitat; those of *Cylindrostethus* Fieber can be found in streams, always avoiding both turbulent and still waters; *Potamobates* Champion and *Platygerris* White are exclusive of swift currents; and *Charmatometra* Kirkaldy is common in streams and rivers covered by dense vegetation (Hynes 1948; Andersen 1982; Molano-Rendón et al. 2008). The sea skaters of the genus *Halobates* Eschscholtz inhabit both coastal and oceanic marine waters, and *Hermatobates breddini* Herring, the only representative of the family Hermatobatidae from the Western Hemisphere, is associated with coral reefs and rubble (Andersen 1982; Schuh and Slater 1995).

In general, semiaquatic bugs are intermediate stage predators in the food chains of their communities, normally attacking small invertebrates that fall on the water surface and become trapped or those who swim up to it (Menke 1979). On the other hand, gerromorphans are prey to hunting spiders, damselflies, dragonflies, back-swimmers, beetles, fishes, frogs, lizards, turtles, crocodiles, birds, and bats (Warren 1915; Riley 1925; Anderson 1932; Williams 1944; Corbet 1959; Haverschmidt 1962; Ashmole and Ashmole 1967; Hoyt 1970; Harrington and Harrington 1972; Callahan 1974; Andersen 1982; Cooper 1984; Cheng 1985; Teran et al. 1995; Svensson et al. 2002; Frick et al. 2009). Many species of Gerromorpha are also known to be cannibalistic, feeding upon weaker individuals of their own kind (Butler 1923; Sprague 1956; Andersen 1982).

Antipredator adaptations in gerromorphan bugs comprise structural features (camouflage) as well as behavioral means. Cryptic coloration and countershading, obscuring the contours of the bugs, are common phenomena. The slender body shape with stilt legs possibly also affords some camouflage protection against predators (Andersen 1982). *Hydrometra* is often seen to rise and lower its body rhythmically, which tends to obscure the outline of the insect (Rensing 1962). Another strategy observed in *Hydrometra* is death feigning. When specimens are roughly handled, they become immobile, the antennae are stretched forward, and the legs pressed against the body (catalepsy). The insect stays in this condition for several minutes and may be able to get a predator to lose interest (Steiniger 1933).

Although the function of the scent glands of gerromorphan bugs is rather obscure, they produce substances with a distinctly unpleasant smell and may function as a chemical defense against predators. However, fast escape movements are probably the most effective antipredator behavior in semiaquatic bugs and can be associated with gregarious behavior in many veliids and gerrids, reducing predation by larger animals (Andersen 1982). In addition, at least some species of *Rhagovelia* and *Halobates* can dive and swim underwater if disturbed, with movements similar to those used when skating on the surface (Wharton 1981).

Besides being attacked by numerous predators, semiaquatic bugs are also target to different groups of parasites. Ectoparasitic fungi of the order Laboulbeniales occur on mesoveliids, hebrids, macroveliids, and veliids (Poisson 1957; Benjamin 1967). Immature stages of water mites belonging to several families can attach to the exoskeleton of the gerromorphan host, being sometimes very numerous and interfering with its development (Menke 1979; Lanciani 1985; Smith 1989). A variety of endoparasites and commensals is also known to be associated with semiaquatic bugs, including cnidosporids, nematodes, trypanosomatids, and flagellates (Poisson 1957; Arnqvist and Mäki 1990). Hymenopterous parasitoids of the families Scelionidae, Mymaridae, and Trichogrammatidae attack the eggs of semiaquatic Hemiptera (Poisson 1957; Henriquez and Spence 1993), and Cobben (1965) described a microorganism found in eggs and nymphs of *Mesovelia* that he believed to be a symbiont, a condition unknown in other Gerromorpha.

Wing polymorphism is common in the water strider families, and apterous forms are often much more common than winged individuals. The evolutionary significance of flightless forms is complex, but the essential consequence of brachyptery is that it insures continued usage of the immediate resource. Generally, flightless morphs are found in permanent habitats, where brachyptery insures that at least a part of the population will survive and also allows more energy to be allocated on the process of reproduction instead of having to share it with flight muscle and wing developmental processes. On the other hand, the ability of at least a part of the population to migrate by flight is favored in temporary habitats (Menke 1979). Seasonal changes in wing morph proportions within a population are usually well defined in temperate species, but in one of the few studies of polymorphism regarding Neotropical species, Nummelin (1997) could not detect this effect on a population of *Limnogonus franciscanus* (Stål) from Barbados.

An interesting behavior involving the wings occurs in some macropterous rhagadotarsine and trepobatine Gerridae and some Mesoveliidae which break off a portion of their wings (de la Torre-Bueno 1908; Jordan 1951). The break occurs along a transverse cleavage line in gerrids. Two theories were proposed by Torre-Bueno to explain this autotomy. One is that after bugs have migrated to a new habitat, they no longer need their wings. It is well documented that macropters in *Gerris* F. resorb their flight muscle after spring populations are established (Brinkhurst 1959; Andersen 1973), and this fact adds credence to his theory. Torre-Bueno's second hypothesis is that wings hinder copulation, and a mechanism has developed enabling them to be broken off.

As for the locomotion of semiaquatic bugs, Andersen (1976) established that movement on surface film of the water by means of ordinary limb action is achieved in three different ways: walking on alternating tripods in a way comparable to ordinary locomotion in terrestrial insects (*Mesovelia, Hebrus, Hydrometra, Microvelia*), rowing by simultaneous strokes of the middle legs while at least the hind legs slide on the water (*Rhagovelia*), and jumping, where the simultaneous power stroke of the middle legs causes the insect to take off the water surface and glide through the air (Gerridae).

Besides these basic modes, Brinkhurst (1960) described an intriguing phenomenon in *Velia* Latreille: after struggling to make headway against a strong current without progress, bugs would stop all movement, but instead of being swept downstream, they suddenly shot forward. Linsenmair and Jander (1963) found that in *Velia* the saliva discharged from the rostrum acts as a detergent, lowering the surface tension. The insect is thus carried forward (or sideways or in an arc) by the contracting surface film, with direction of movement apparently being controlled by the highly mobile rostrum (Andersen 1976). This mode of locomotion was called "expansion skating" and is also found in adults and nymphs of *Microvelia* and *Rhagovelia*.

Living on the water surface, gerromorphans developed means of using superficial waves for prey location and for communicating with each other. This phenomenon is documented in the Gerridae and Veliidae and might also play a role in the life of other semiaquatic bugs (Polhemus 1990; Schuh and Slater 1995). Southwood and Leston (1959), for example, suggested that hydrometrids locate their prey by sensing vibrations caused by the struggling animal trapped on the superficial tension of the water. Veliids, by their turn, apparently perceive prey by a combination of visual and mechanical means. Meyer (1971a, b) performed experiments on *Velia caprai* Tamanini with various kinds of visual stimuli and concluded that this species locates prey by means of a visual "central flickering detector" in combination with a water surface vibration detector.

Besides ripple communication, stridulation might also occur in gerromorphans, particularly in those of the genus *Stridulivelia*, which have a row of pegs on the paratergites and a file or roughened patch on the hind femur. A similar apparatus occurs in *Paravelia stenoptera* Polhemus and Polhemus, with an arcuate row of pegs on the abdominal venter and a rasplike area on the hind trochanter (Andersen 1982; Polhemus and Polhemus 1984b). Polhemus (1994) also recorded a possible stridulatory apparatus in *Metrobates* (family Gerridae), in which males and females of some species have fine ridges on mesoacetabulum and a sharp ridge on hind trochanter.

6.3.1 Eggs and Oviposition

The eggs of gerromorphan bugs are laid in a wide variety of places, each usually characteristic for a particular species or genus (Menke 1979). In the most plesiomorphic state, the eggs are deposited freely or partly hidden, attached lengthwise by a gelatinous substance. Deposition of the egg in an upright position by means of a pedicel (*Hydrometra*) or without one (*Hermatobates* Carpenter), or embedding the egg in plant tissue (*Mesovelia*) are inferred to be derived conditions (Cobben 1968; Andersen 1982).

A thin, entirely solid egg shell, marked with a hexagonal pattern from the follicular cells, is considered plesiomorphic among the Gerromorpha. Porous, aerostatic layers have probably developed several times independently in the Heteroptera as well as in semiaquatic bugs. The evolution of the micropylar system has most likely started with a single micropyle in the center of the anterior pole of the egg. This primitive condition is shared by all families of Gerromorpha except the Macroveliidae, which may have up to five or seven micropyles, and the Veliidae, which usually have two to six micropyles. Primitively, the micropyle forms a single perpendicular canal through the chorion. Usually this canal has an internal projection pointing toward the fore side of the egg. Those species having many, displaced micropyles usually have the inner projections of these oriented in a clockwise direction (Cobben 1968; Andersen 1982).

Hebrus species lay their eggs on moss, usually secreting them in leaf axils or between closely spaced leaves. They are elongate, oval, about twice as long as wide, with rounded ends, and pearly white when laid, but yellowish white as the embryo develops. Incubation takes from 6 to 12 days, being shorter at higher temperatures. In the laboratory, *Merragata* species lay their eggs on moss leaves or under algae, with incubation taking about the same time as in *Hebrus* (Cobben 1968; Menke 1979). In nature, Porter (1950) found their eggs under filaments of algae on stones and in surface holes of the stones.

As for *Mesovelia*, eggs are inserted in plant tissue at the waters' edge by means of the well-developed, sawlike ovipositor which pierces stems by means of rocking motion of the body. The egg is elongate, oval, and the micropylar end curves up to a flat operculum or lid which is exposed when the egg is deposited. When laid, the egg is white, but it becomes transparent as the embryo develops (Menke 1979). *Mesovelia mulsanti* lays about 100 eggs, which have an incubation period of 7–9 days, while *M. amoena* lays from 100 to 200, which incubate for 12 days (Hoffmann 1932). Unlike most Gerromorpha which have a sharp egg burster, the mesoveliid embryo develops a bladderlike outgrowth in the frontal region which pushes off the egg cap. The nymph then struggles free and swims to the surface, breaking through the surface film easily (Ekblom 1930).

The eggs of *Hydrometra* are spindle-shaped, superficially sculptured, stalked, and pearly white when deposited, but turning brown upon exposure to moist air or wet surfaces. They are laid singly on almost any object, usually just above the surface of the water, although Takahashi (1921) often found *Hydrometra* eggs under water in Japan. The female exudes a glistening drop of a gummy substance from the genital opening, which she then presses against the oviposition site where it hardens rapidly. The base of the egg stalk follows this fluid, and as the egg emerges, the female lifts her body, drawing the egg out perpendicular to the support (Menke 1979). A single female can produced up to 11 eggs per day, with a total of 173 over 37 days, and the incubation time of studied species varies from 7 to 23 days, depending on temperature, with the average being 11 to 13 days. Hydrometrid embryos split the shell with a sclerotized egg burster and then thrust the head through the slit (Hungerford 1920; Menke 1979).

Among the Veliidae, eggs can be laid on floating objects, moss, duckweed, stones, and living or dead leaves either just above or below the water surface. Eggs are laid singly or in clusters and are attached with gelatinous glue that covers much of the egg (Cobben 1968). They are white when laid, but turn amber as development progresses, and are often covered with sparse lines, cubical projections, or a honeycomb mesh structure. The eggs are oval and usually slightly flattened lengthwise on one side. de la Torre-Bueno (1917) found only two eggs at a time being deposited in *Microvelia pulchella* Westwood, but Miyamoto (1953) found that *M. diluta* Distant laid an average of 7.4 eggs/day, and Frick (1949) noted that *M. pulchella* laid up to 505 eggs.

Finally, gerrids are known to normally lay their eggs singly or in rows, just under the surface on floating sticks, vegetation, or other material. The eggs can be white when first laid and then turning amber, and the incubation time usually ranges from 8 to 12 days. Both Hungerford (1920) and Hoffmann (1924) studied the life cycle of *Trepobates pictus* (Herrich-Schäffer) and found that it lays three to ten eggs at a time in solid masses of gelatinous material on the underside of leaves and sticks in the water, with an incubation time of 5–12 days. As for members of the subfamily Rhagadotarsinae, they are unique among gerrids by the presence of an ovipositor used to insert the eggs in plant tissue (Silvey 1931). Finally, in the marine *Halobates*, eggs of coastal species are laid in clusters on substrates near shore, being olive green when laid, but turning bright orange in 2–3 days. In the open ocean species, eggs are laid on floating substrates, including, among others, feathers, small pieces of wood, mollusk shells, seeds, tar lumps, and insect carcasses (Cheng 1985).

6.3.2 Nymph Feeding and Development

Nymphs of semiaquatic bugs differ from adults in their usually softer integument, which is more or less distinctly divided into darker sclerites and lighter membranes. The ecdysial line is usually visible as a pale, longitudinal, median line through the thoracic nota and first abdominal tergum; on the head, the line branches in front of the eyes, thus assuming the shape of a Y. The number of antennal segments is the same as in adults, but there is only one segment in each tarsus. The metathoracic scent apparatus is not formed until the adult stage, but differentiation of the genitalia begins to become apparent in the fourth instar, and fifth instar nymphs are easily separated by sex (Andersen 1982).

The postembryonic stages of semiaquatic bugs have several features which may be used in comparisons between higher taxa, but which are less useful than adult characters at species level since they are often reduced or otherwise modified during the ontogenetic development. Such features include the structure of the eyes and the chaetotaxy of the head and body, especially in the earlier nymphal stages (Cobben 1968, 1978).

It was suggested by Cobben (1978) that the most plesiomorphic number of ommatidia in the eye of the first instar nymph of Heteroptera is probably five. The gerromorphan bugs show a progressive increase in ommatidial number: *Hebrus* (10), *Microvelia* (28), *Hermatobates* (30), *Macrovelia* Uhler (35), *Hydrometra* (38), *Mesovelia* (55), *Hebrovelia* Lundblad (about 70), and *Gerris* (about 130) (Cobben 1978; Andersen 1982). The low number of ommatidia in *Hebrus* and the small increase from the first instar nymph (10) to adult (35) may be representative of the plesiomorphic condition among semiaquatic bugs (Andersen 1982).

The eye of the early instar nymphs has a pair of ocular setae inserted between the ommatidia of the posterior surface of the eye, except in *Hydrometra*. These setae are usually preserved through the postembryonic development, but are however absent from adults of Macroveliidae, Hydrometridae (except *Veliometra* Andersen), and some Veliinae (Andersen 1982; Polhemus and Polhemus 1993).

As for the chaetotaxy, the first instar nymph of *Hebrus* and *Mesovelia* has relatively few macrotrichia on the dorsal surface of the head, thorax, and abdomen. This primary hair pattern can be traced in first instar nymphs of *Hydrometra* and *Hermatobates*, but is lost or obscured by a dense macro-hair layer in first instar nymphs of Veliidae and Gerridae. The number of these macro-hairs increases gradually through the nymphal instars of all semiaquatic bugs, including *Mesovelia* and *Hebrus* (Andersen 1982).

Some works describing the nymphs of Neotropical Gerromorpha exist (Estévez and Schnack 1980; Mazzucconi and Bachmann 1997; Konopko and Mazzucconi 2005, 2011; Padilla-Gil and Arcos 2011), but life histories studies are available only for species with wide geographical distributions (*Mesovelia mulsanti*, *M. amoena*, *Microvelia pulchella*, *M. hinei* Drake), and the experiments were almost always conducted in temperate environments (Hungerford 1917; de la Torre-Bueno 1917; Lanciani 1987; McPherson 1988; Taylor and McPherson 1999, 2000, 2003; McPherson and Taylor 2006). The few exceptions are the works of Frick (1949), which took place in Panama, and Iglesias and Crespo (2003, 2008) about the growth pattern in exclusively Neotropical species of the genera *Brachymetra* and *Rheumatobates*.

The usual number of nymphal instars in the heteropterous bugs, including the Gerromorpha, is five. However, species with less than five instars are found throughout the suborder (Southwood and Leston 1959). In Gerromorpha, Hoffmann (1932) found only four instars in *Mesovelia cryptophila* Hungerford, and Cheng and Fernando (1971) recorded the same number in *Rhagovelia obesa* Uhler. However, it is in the genus *Microvelia* that this phenomenon has been studied most carefully (Andersen 1982).

When rearing *Microvelia borealis* Torre-Bueno (= *M. pulchella*), de la Torre-Bueno (1917) found only four nymphal stages. Hungerford (1920) reared the same species and found more than four stages in some individuals; he suggested that variation might exist or that apterous forms have fewer stages than macropterous. Contrary to the latter theory, Frick (1949), in rearings of what was probably *M. pulchella* (his *M. capitata* Guérin-Méneville), obtained nine apterous adults developing through four instars. The majority of apterous and macropterous adults, however, developed through five nymphal stages, and it is so far unsettled which factors determine the variation in number of instars in *Microvelia* and other Gerromorpha (Andersen 1982).

The process of molting in Gerromorpha may take place both on a firm substrate and on the surface of the water. Sprague (1956) described this process with details in *Hydrometra*, as follows: in the hour before molting, the nymph imbibes water more frequently than before, as the gut is distended. During the later part of the period, the nymph becomes less active. The actual molting process takes place rather rapidly. The fore and hind legs are extended forward and backward, respectively, and arranged so that they are parallel with the body axis; the middle legs are planted firmly on the substrate at right angles with the body. When the nymph has placed its legs securely, it starts to suck air rather than water. The cuticle of the thorax then begins to cleave longitudinally, in the median ecdysial line. Small bubbles of air pass through the digestive tube and distend and lengthen the abdomen; in this way, alternate waves of contraction and dilation move up and down the body. The head, which is strongly bent ventrad, splits along its Y-shaped ecdysial line. As the body stretches upward, the proximal parts of the legs emerge from the exuvium; first the prothoracic and then the meso- and metathoracic legs are pulled away; with the legs touching the substrate, the antennae and finally the rostrum pull free.

The growth of the two wing pairs starts relatively early during the postembryonic development. In the second and third instar nymphs, the posterolateral angles of the meso- and metatergum are produced, indicating the formation of the wing pads. The wing pads of those third instar nymphs that will become macropterous are normally much more conspicuous than those of nymphs that will give rise to apterous adults. After the final molt, the adult individual passes through the so-called teneral development. The teneral stage is largely taken up with hardening and pigmentation of the cuticle, resorbation of muscles which only function during nymphal life, and, in macropterous adults, with growth of endoskeleton and flight muscles (Andersen 1982).

Although temperature clearly is important to the rate of nymphal development (Spence et al. 1980), it is certainly not the only factor. Photoperiod, food, and crowding are known to affect the development of many insects, including Heteroptera. In species that occur in the Neotropics, total time of development is known for *Mesovelia mulsanti* (20.5 days at 21–27 °C, 12 days at 28–33 °C), *Microvelia pulchella* (14.2–20.7 days), *M. hinei* (25.0 days), and *Platyvelia brachialis* (Stål) (27 days) (Frick 1949; Wilson 1958; Galbreath 1973; Taylor and McPherson 2003).

Feeding, in particular in nature, has only been observed and described in relatively few species of semiaquatic bugs. All of them, including the nymphs, are polyphagous carnivores or scavengers, feeding upon dead or half-dead arthropods, mostly other insects. However, the food preferences and behavior during prey location seem to vary considerably among the Gerromorpha (Andersen 1982). Hungerford (1920) studied the food preferences of *Mesovelia mulsanti* and suggested that microcrustaceans may form the staple food for this species, especially the small nymphs. He reared *Mesovelia* in aquarium without other food resources than an abundant population of microcrustaceans and the weaker individuals of their own species.

Hungerford (1920) and Sprague (1956) recorded *Hydrometra* feeding upon adult midges, mosquito larvae and pupae, and various microcrustaceans. The former author observed a third instar nymph catching an ostracod resting on the surface film. The nymph stalked slowly up to the prey, its body aquiver and weaving from side to side. The tips of its antennae were turned down to the surface, and the rostrum directed down and slightly forward. When close enough, the prey was caught and carried upon the tip of the rostrum with a sudden move. Cannibalistic behavior, where adult hydrometrids feed upon their own nymphs and the older nymphs upon the younger ones, has frequently been observed in laboratory (Andersen 1982).

Among the Veliidae, Hungerford (1920) observed first instar nymphs of what was probably *Microvelia pulchella* eat ostracods larger than themselves. First they felt around twisting the distal article of their rostrum in order to find a vulnerable spot to insert their stylets. After succeeding in doing so, the prey was lifted from the surface upon the upturned rostrum. *Microvelia* also often attack individuals of their own kind, but usually without success (Andersen 1982). Cheng and Fernando (1971) managed to rear *Rhagovelia obesa* using commercial fish food containing pulverized insects.

In Gerridae, smaller species and nymphs of the larger species may feed on various microcrustaceans (Hungerford 1920; Ekblom 1926; Silvey 1931; Liche 1936; Wesenberg-Lund 1943; Wilson 1958; Southwood and Leston 1959). They are known to have strong cannibalistic tendencies, which add to the difficulties in rearing these insects in the laboratory, especially because nymphs are particularly vulnerable when molting (Andersen 1982). In marine species, attempts to feed newly hatched nymphs of *Halobates* in the laboratory have not been successful, and in one rearing study, they all died within 5 days. The young nymphs occasionally preyed on one another, but they did not feed on any marine organism or on freshly killed *Drosophila* Fallén offered to them, although older nymphs (from third instar onward) and adults readily take such food organisms. How first instar nymphs survive in the open ocean is still a mystery. They may be able to utilize nutrients from the organic-rich surface film, but this has not been substantiated (Cheng 1985).

6.3.3 Adult Feeding and Reproduction

As explained for the nymphs, most semiaquatic bugs are opportunistic predators and scavengers, although small- or medium-sized arthropods with a soft integument are preferred as food items in nature. Two basic types of predator strategies can be recognized among them: the searching strategy and the waiting strategy, with some overlapping occurring between them, as well as the same species changing its strategies from time to time (Andersen 1982).

Species of *Mesovelia*, *Hebrus*, *Hydrometra*, and probably also *Microvelia* seem to be largely searching predators which move around examining crevices and cavities in the vegetation or other floating objects on the water surface looking for something edible. The prey is probably recognized by the tactile or olfactorial sensory apparatus rather than by vision (Andersen 1982). *Mesovelia* have been observed feeding upon dead or half-dead midges of Culicidae and Chironomidae, ostracods, and cladocerans (Hungerford 1920; Ekblom 1930).

Individuals of the genus *Hydrometra* have been observed feeding upon ephemeropterans, adult chironomids which were floating on the surface of the water, larvae and pupae of ceratopogonids, microcrustaceans, and smaller individuals of their own kind (Hungerford 1920; Ekblom 1926; Sprague 1956; Maier 1977). *Microvelia* can feed upon ostracods, collembolans, mosquito eggs, Anophelinae larvae, and also their own nymphs (Hungerford 1920; Jordan 1932; Frick 1949; Laird 1956; Miyamoto 1953). Some individuals can even attack adults of their own kind, but usually without success (Andersen 1982).

In *Rhagovelia* and almost certainly most gerrids, the waiting strategy is the prevailing one. They can detect and locate live prey trapped in the surface film of water solely by the surface vibrations generated by the struggling animal. Visual stimuli are probably significant for close-range location of prey, whether moving or not, when the surface bug patrols the water (Jamieson and Scudder 1979; Andersen 1982). Representatives of the genus *Rhagovelia* usually aggregate in places where the water current may concentrate floating objects including potential food items, such as microcrustaceans and small insects (Bacon 1956; Andersen 1982). A similar feeding behavior is seen in *Oiovelia*, where individuals would gather around a dying adult ephemeropteran that was trapped on foam retained on the edge of a stream (personal observation).

The prey taken by freshwater gerrids consist of various soft-bodied arthropods, including microcrustaceans, terrestrial insects that fall on the water or emerging insects with aquatic larvae (Andersen 1982). Among the oceanic species of *Halobates*, food items include pelagic cnidarians, sea anemones, planktonic crustaceans, and fish larvae (Andersen and Polhemus 1976). Besides that, it was shown by Riley (2006) that *Limnogonus luctuosus* (Montrouzier), a freshwater gerrid, preferred mobile prey, whereas *Halobates hawaiiensis* Usinger preferred immobile prey.

The reproductive behavior of semiaquatic bugs has been extensively studied, but as mentioned for other aspects of their biology, the vast majority of the experiments are sadly restricted to temperate species. For example, the possible role of the stridulatory devices found in *Stridulivelia* in acoustic communication and reproduction has not yet been investigated.

Many temperate species of Gerromorpha, on the other hand, are models for the study of mating systems, sexual selection, and evolution, especially those of the genera *Aquarius* Schellenberg, *Gerris*, and *Limnoporus* Stål (e.g., Vepsäläinen et al. 1985; Kaitala 1991; Vermette and Fairbairn 2002; Rowe and Arnqvist 2011). One of the few studies of such type based on a species from the Neotropics is that of Villagra et al. (2001), who analyzed sexual dimorphism and behavior in the water strider *Aquarius chilensis* (Berg).

According to Spence and Andersen (1994), four basic mating systems have been described among gerromorphans: (1) a kind of monogamy resulting from extended postcopulatory guarding or monopolization of the female's sperm-storage apparatus; (2) resource defense polygyny through male territoriality and defense of oviposition sites, which has evolved independently in several gerrid genera; (3) pure dominance or lek polygyny, charactering the mating system of *Halobates robustus* Barber; and (4) the scramble competition polygyny associated with a prolonged search for receptive females and variable postcopulatory guarding, seen in *Microvelia* spp. and many gerrids.

The location of a mate itself apparently presents no problem for gerromorphans, at least in gregarious species like most Gerridae and Veliidae. Sexually active males seem to be attracted by any moving object, and attempts to copulate with females of a different species or even with other males have been observed (Andersen 1982). However, it has been shown that males of *Gerris* can accurately detect the sex of other adults by the type of surface wave they generate on the water, even without visual cues (Wilcox 1979).

The mating in semiaquatic bugs is usually initiated by the male, which can use a variety of strategies to conquer the female. The most simple copulation strategy observed in Gerromorpha is that of the *Mesovelia*, where the male jumps on and engages the female very quickly and leaves her almost immediately after the actual sperm transfer (Andersen 1982). Individuals of *Mesovelia mulsanti* often mate when resting on the water surface, with the male clasping its fore legs around the mesothorax of the female, resting the middle legs on the water or over the female, and holding the hind legs in the air (Hungerford 1920; Hoffmann 1932).

An interesting fact among the Mesoveliidae is the apparently parthenogenetic condition seen in *Mesovelia amoena*. Males of this species have only been collected in North America, Mexico, and Hispaniola and only females from other parts of the Neotropical Region and from Hawaii, where the species was introduced (Menke 1979).

Different copulating positions are seen in veliids, where the male jumps on the female and grasps her thorax with a firm grip, assisted by the grasping comb present on the fore tibia. Unlikely the *Mesovelia*, however, the hind legs are also used to secure his position on the female's back (Andersen 1982). In *Microvelia* several matings can take place, each lasting a couple of minutes, and between which the male follows the female closely (Frick 1949).

However, it is not only during reproduction that males ride on top of females; some males of *Microvelia* and *Rhagovelia* "piggyback" on the females' dorsum whether copulating or not, and female's thorax can show modifications that facilitate this habit (Polhemus 1974; Moreira et al. 2010). In *Platyvelia brachialis*, on the other hand, the two sexes have reversed their roles so that the female mounts the male and stimulates him to copulate by rubbing his abdomen with her hind legs (Wilson 1958).

Reproduction in gerrids usually occurs with the male passively riding on the back of the female during copulation, securing his position by clasping her thorax with his fore legs just in front of the mesoacetabula. During this period, which can last from days to weeks, they copulate several times; the male feeds on female preys and submerges when she is laying eggs beneath the surface.

From these observations on the sexual behavior of semiaquatic bugs, it appears that there is a distinct trend toward a prolongation of the time the male spends with the same female, whether actual copulation takes place or not. This way, the male protects his own gametes from competition with those of other males and also protects the female from possible risks involved in new mating attempts. Females, in their turn, have a longer period of time to feed and lay eggs without being pursued by other sexually active males (Andersen 1982).

In the Gerromorpha, primary sexual dimorphism involves the reproductive organs and terminal segments of abdomen. Male genital segments are often more conspicuous than those of females, which may be concealed. The secondary sexual dimorphism, however, involves many different body parts, varying from group to group (Andersen 1982).

Several modifications on the legs of Veliidae may enhance male's ability to copulate, like the grasping comb present on for tibiae and the many spines seen on the enlarged hind legs of some species of *Rhagovelia*. More striking forms of secondary dimorphism occur, for example, in *Microvelia longipes* Uhler, in which only the males have hind legs much longer than their own body, and in many species of *Rheumatobates* that have males with extremely modified antennae and legs, which probably evolved as a response to sexual conflict (Westlake 1998; Rowe et al. 2006).

6.4 Classification and Diversity

The more than 2,100 known species of Gerromorpha are currently divided into eight families and approximately 160 genera. Most of the known diversity of the infraorder was described from the Oriental Region, but more than 500 species and 45 genera have been recorded from the Neotropical Region, representing roughly 24 and 29 % of the global specific and generic diversities (Polhemus and Polhemus 2008).

The current classification of the semiaquatic bugs follows almost entirely that summarized by Andersen (1982), with the exception of the progressive abandonment of the superfamilies Mesovelioidea (Mesoveliidae), Hebroidea (Hebridae), Hydrometroidea [Paraphrynoveliidae + (Macroveliidae + Hydrometridae)], and Gerroidea [Hermatobatidae + (Veliidae + Gerridae)]. This rejection is especially true after the proposition of the phylogeny of the Gerromorpha by Damgaard (2008a) (Fig. 6.11), which showed that both Hydrometroidea and Gerroidea were paraphyletic entities.

Out of the eight currently recognized families of Gerromorpha, only Paraphrynoveliidae is not found on the Neotropical Region. This monotypic family holds only two species from South Africa, which have rarely been collected (Schuh and Slater 1995; Polhemus and Polhemus 2008). The also monotypic marine family Hermatobatidae has a single Neotropical representative, *Hermatobates breddini* Herring, which occurs along the West Indian islands. The eight remaining species

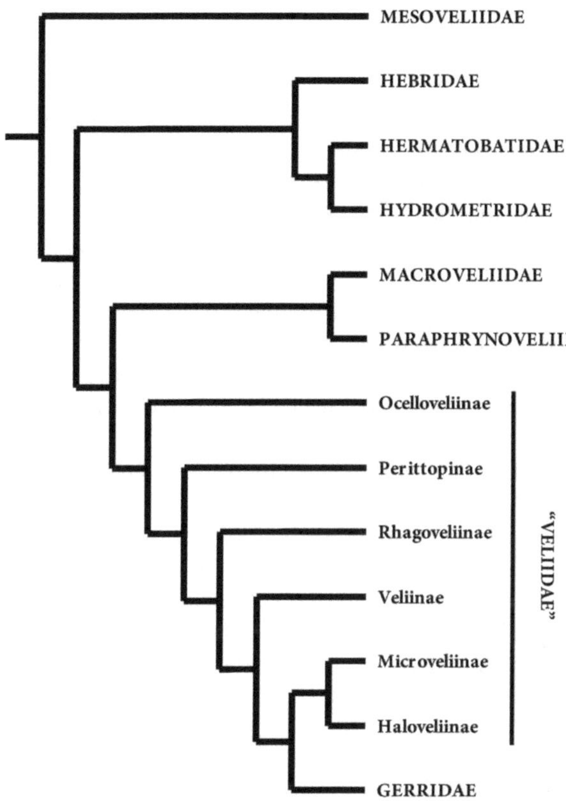

Fig. 6.11 Phylogeny of the Gerromorpha proposed by Damgaard (2008a)

of the family have been collected on the Indian and Pacific oceans (Polhemus and Polhemus 2008; Muñoz et al. 2010).

Macroveliidae is another family of semiaquatic bugs with very low diversity. Only three species, from three monotypic American genera, have been included in the family: *Macrovelia hornii* Uhler, *Oravelia pege* Drake and Chapman, and *Chepuvelia usingeri*. The last species was described from few specimens collected on southern Chile and is the only representative of the family to occur on the Neotropics.

A classification of the Mesoveliidae, which includes a few more than 45 extant species and subspecies, was presented by Andersen and Polhemus (1980) and later reanalyzed by Andersen (1999), with the acceptance of the subfamilies Madeoveliinae, including *Madeovelia* Poisson, and *Mesoveloidea*; and Mesoveliinae, with the remaining nine genera of the family. After that, the terrestrial genus *Seychellovelia* Andersen and Polhemus was described in the Mesoveliinae (Andersen and Polhemus 2003), and the status of the subfamilies and relationships among the genera were questioned by Damgaard et al. (2012). The phylogeny presented by the later authors showed both subfamilies of Mesoveliidae to be non-monophyletic, as well as some genera; however, no classificatory changes have been so far proposed.

On the Neotropical Region, Mesoveliidae is represented mainly by the common and widespread genus *Mesovelia*, with nine species ranging from Mexico to Argentina. Two of them, namely, *M. amoena* and *M. mulsanti*, also occur on the United States and Canada, besides being introduced in the Hawaiian Islands. The other genera of the family are less diverse and more rarely collected, like *Mesoveloidea* with just *M. williamsi* occurring from Mexico to southeastern Brazil and *M. peruviana* Drake, known only from the type series from Peru. *Darwinivelia* contains only three species from the coasts of Colombia, northern Brazil, and the Galápagos Islands; and both *Cryptovelia* and *Speovelia* have a single Neotropical representative each, respectively, from northern Brazil and Mexico (Damgaard et al. 2012).

According to the classification presented by Andersen (1981), Hebridae is subdivided into the Oriental Hyrcaninae and the cosmopolite Hebrinae. Probably because of the difficulties to collect and study the hebrids, they are very poorly represented in most collections and in taxonomic works, and no updated phylogenies or propositions of classification have been published after Andersen's study.

Out of the approximate 220 species described in the family (Polhemus and Polhemus 2008), 32 have been so far recorded from the Neotropical Region, included in three genera. *Lipogomphus* gathers four species, which are almost exclusively Neotropical except for *L. brevis* (Champion), which extends from southern United States to Panama (Polhemus and Polhemus 1988). The other species of the genus are known from Guatemala to Panama (*L. leucostictus* (Champion) and *L. acola* (Drake and Chapman)) and from central-southern South America (*L. lacuniferus* Berg) (Drake and Chapman 1958; Moreira et al. 2011b).

Merragata, on its turn, is represented on the tropical America by four of its five species, ranging from Mexico to Argentina. Out of the Neotropical species, *M. hebroides* White also occurs on Canada and United States and is the only species of the genus recorded from other zoogeographical regions, on the Hawaiian and Canary Islands, probably as a result of unintentional introduction. Finally *Hebrus*, which gathers the majority of the species of the family, is represented on the Neotropical Region by 23 species, several of which are unfortunately known only from the type series.

The current classification of the Hydrometridae and its approximately 125 species (Polhemus and Polhemus 2008) still follows almost entirely that proposed by Andersen (1977), with a subdivision in three subfamilies: Heterocleptinae, Limnobatodinae, and Hydrometrinae. Both Heterocleptinae and Limnobatodinae are represented on the Neotropical Region by a single species each, respectively, *Veliometra schuhi*, from northern Brazil, and *Limnobatodes paradoxus* Hussey, recorded from Honduras, Guyana, French Guiana, Brazil, and Peru (Schuh and Slater 1995; Moreira et al. 2011b; Cianferoni 2014).

As for the Hydrometrinae, it is the most diverse subfamily of Hydrometridae and holds *Hydrometra*, the only genus of the family known to occur outside the tropics.

This is represented on the Neotropical Region by 39 species ranging from Mexico to Argentina (Drake and Lauck 1959; Mychajliw 1961; Smith 1988a; Perez-Goodwyn 2001; Moreira and Barbosa 2013). Besides *Hydrometra*, other Neotropical Hydrometrinae are divided between the genera *Bacillometra*, with one species from French Guiana and Brazil, and *Bacillometroides* Polhemus and Polhemus, with three species from Peru and Bolivia that were previously held on the former genus (Schuh and Slater 1995; Polhemus and Polhemus 2010).

The greatest diversity of the family Veliidae is known from the Neotropical Region, especially because of the immense number of described species of *Rhagovelia* (Polhemus and Polhemus 2008; Padilla-Gil and Moreira 2013). About 300 out of the more than 900 species of the family have been described from the Neotropical Region, a number that will still increase much because many species of Microveliinae remain undescribed (Polhemus and Polhemus 2007, 2008).

Despite being shown by Damgaard (2008a) that Veliidae is paraphyletic because of the positioning of two of its subfamilies closer to Gerridae than to its other subfamilies, the group still has a generally accepted status of family. Some of its subfamilies and genera are probably also non-monophyletic, but an updated classification has not yet been proposed, and that summarized by Andersen (1982) and analyzed by Andersen and Weir (2004) is still followed. According to them, Veliidae is divided into the subfamilies Perittopinae, Rhagoveliinae, Veliinae, Microveliinae, Haloveliinae, and Ocelloveliinae.

Out of the six currently accepted subfamilies of Veliidae, Perittopinae, Haloveliinae, and Ocelloveliinae are not represented on the Neotropical Region. The first two are recorded only from the Oriental Region and gather one and five genera, respectively. The last one is monotypic and has a controversial positioning, with two species from South Africa that had been previously placed in the Mesoveliidae and Macroveliidae (Schuh and Slater 1995).

Two genera are currently accepted in Rhagoveliinae, and the cosmopolitan *Rhagovelia* is the only one with Neotropical representatives, because the species previously allocated on the genus *Trochopus* Carpenter were transferred by Polhemus (1997) to *Rhagovelia* under the *salina* group. Besides this synonymy, the same author reorganized the species of the genus in 17 groups, which formed five complexes (monophyletic) and a grade (non-monophyletic). After that, Moreira et al. (2012) proposed a new group for two species previously placed on the *robusta* group. A total of 184 species are distributed throughout the Americas, of which 97 % have a tropical distribution, and South America is the geographical area with the highest diversity (Padilla-Gil and Moreira 2013).

The subfamily Veliinae is currently divided into nine genera, out of which *Oiovelia*, *Paravelia*, *Platyvelia*, *Steinovelia* Polhemus and Polhemus, *Stridulivelia*, and *Veloidea* Gould occur on the Neotropical Region (Polhemus and Polhemus 1993; Schuh and Slater 1995). More genera of Neotropical Veliinae will surely be described in the future, because *Paravelia* is known to be a paraphyletic entity.

For a long period, *Oiovelia* was only represented by the type species from Venezuela, until Spangler (1986), Moreira et al. (2010), and Rodrigues et al. (2014) described seven additional species from South America. *Paravelia*, which is by far the most diverse of the subfamily, has 59 described species and used to gather even more before Polhemus and Polhemus (1993) transferred nine species to *Platyvelia* and five to *Steinovelia*. When described, these genera were fully composed by species previously allocated on *Paravelia*, and no new species have been described afterward.

Stridulivelia was divided by Polhemus (1979) into the subgenera *Aenictovelia* for the primarily Mesoamerican species that lack a stridulatory mechanism and *Stridulivelia* s.str. for the South American species with a stridulatory apparatus. Currently, five species are assigned to the subgenus *Aenictovelia* and ten to *Stridulivelia* s.str. (Polhemus and Spangler 1995). Finally, *Veloidea* is an exclusively Neotropical genus, now with five species distributed from Mexico to Honduras in Central America and on Colombia, Venezuela, and Ecuador in South America (Buzzetti and Zettel 2008).

Microveliinae is probably the most problematic subfamily of Veliidae on the Neotropics. Besides the small size of most species, the lack of diagnostic morphological characteristics makes the taxonomic study of the group very complicated, especially regarding the *Microvelia*. Many genera of Microveliinae are known only from Africa or from the Oriental region, and the lack of revisions focusing on the Neotropical fauna makes the number of genera recorded from tropical America be relatively small, despite of the large amount of described species.

Most of them (86 species from Mexico to Argentina) are held on the genus *Microvelia*, represented on the region by the subgenera *Kirkaldya* and *Microvelia* s.str. A few others are allocated on the genera *Euvelia* Drake (four species from Brazil, Peru, and Bolivia), *Husseyella* (one from Florida and the Caribbean, two from Brazil), *Xiphovelia* (one from central Brazil), and *Aegilipsovelia* (three from Mexico) (Polhemus 1970, 1977; Polhemus and Polhemus 1984a; Smith 1988b; Moreira et al. 2011b).

Lastly, following the classification summarized by Andersen (1982), Gerridae is divided into eight subfamilies: Charmatometrinae, Cylindrostethinae, Eotrechinae, Gerrinae, Halobatinae, Ptilomerinae, Rhagadotarsinae, and Trepobatinae. Out of these, Eotrechinae and Ptilomerinae are the only ones not occurring on the Neotropical Region, being known respectively from India and the Orient and from Madagascar to New Guinea (Schuh and Slater 1995).

Charmatometrinae is restricted to the Neotropics and comprised by three genera: *Brachymetra*, with nine species from Guatemala to Argentina (Harris and Drake 1945; Hungerford and Matsuda 1957; Drake 1957; Moreira et al 2011b); *Charmatometra*, with a single species from Venezuela, Colombia, and Ecuador (Mazzucconi et al. 2009); and *Eobates* Drake and Harris, with only one species from the island of Hispaniola and Peru (Aristizábal 2002).

Cylindrostethinae is also composed by three genera: the pantropical *Cylindrostethus*, represented on the Neotropics from Colombia and Trinidad and Tobago to Argentina by 10 of its 19 species (Floriano 2013; Floriano and Cavichioli 2013); *Platygerris*, with three species from Mexico to Colombia (Aristizábal 2002);

and *Potamobates*, with 19 described species that range from Mexico to Peru (Padilla-Gil and Damgaard 2011; Morales-Castaño et al. 2013).

Gerrinae is a cosmopolitan subfamily, currently divided into the tribes Gerrini and Tachygerrini. The former is cosmopolitan and represented on the Neotropical Region by the genera *Aquarius*, with two predominantly temperate species extending into Mexico and Guatemala and one species occurring on Chile and Peru, out of its 17 described species; *Gerris*, with a mainly Nearctic species reaching Mexico and one species endemic to the same country, from its 30 known species; *Limnogonus*, with seven species and subspecies ranging from Mexico to Argentina, out of a total of 28 species; and *Neogerris*, with eight species that range from the Caribbean to Argentina, out of a diversity of 13 species. The tribe Tachygerrini, restricted to Tropical America, is composed by *Eurygerris*, with nine species that range from Mexico to the Andean countries of South America; and *Tachygerris*, with 13 species distributed from Mexico to South America, except Chile, Argentina, and Uruguay (Damgaard et al. 2014).

Halobatinae is subdivided into the pantropical Halobatini and the Paleotropical Metrocorini. The former is composed by the genera *Asclepios* Distant and *Halobates*, out of which only marine species of the later are represented on the Neotropical Region. *Halobates robustus* Barber, from the Galápagos Islands, is the only coastal species occurring on the area; the remaining four are found on open ocean, and *H. micans* is the only species of the genus recorded from the Atlantic Ocean (Andersen and Cheng 2004).

Rhagadotarsinae is composed by only two genera, the Paleotropical *Rhagadotarsus* Breddin and the American *Rheumatobates*. According to Westlake et al. (2000), Polhemus and Westlake (still unpublished) recognized 37 taxa in *Rheumatobates*: 32 species, three undescribed species, and two subspecies, out of which 31 species and subspecies are Neotropical, being recorded from Mexico to Argentina. To that number, a species described from Colombia by Castro-Vargas and Morales-Castaño (2011) can be added.

Finally, the trepobatine water striders are currently organized on the tribes Naboandelini, Stenobatini, Metrobatini, and Trepobatini, of which only the last two are present on the Neotropical Region. Metrobatini is represented on the area only by 12 species of the genus *Metrobates*, which is exclusive of the New World. As for the Trepobatini, its Neotropical representatives include the single described species of *Lathriobatoides* Polhemus, from Northern Brazil; the four species of the exclusively South American *Halobatopsis* Bianchi; the eight species of *Ovatametra* Kenaga, ranging from Panama to Argentina; the six species of *Telmatometra* Bergroth, distributed from Mesoamerica to Brazil and Peru; the only known species of *Telmatometroides* Polhemus, recorded from Costa Rica, Panama, Colombia, and Ecuador; 7 out of the 12 described species of *Trepobates* Uhler, which is exclusive of the New World; and by the single species of *Trepobatoides* Hungerford and Matsuda, recorded from Bolivia and Peru (Polhemus and Polhemus 2002).

6.4.1 *Identification Key to the Families and Genera of Gerromorpha from the Neotropical Region*

1 Apterous, tarsal claws of forelegs preapical, tarsal claws of middle and hind legs apical, inhabitants of coral reefs..........................Hermatobatidae (*Hermatobates*)
1′ Apterous, brachypterous, or macropterous; positioning of tarsal claws equal on all legs; habitat variable..2
2 Tarsal claws apical (Fig. 6.7a, c, d, e) 3
2′ Tarsal claws preapical (Fig. 6.7b, f–k)...15
3 Body long and almost always slender (Fig. 6.1e); eyes located far from the the anterior margin of thorax; antennae and legs long and very slender (Fig. 6.1e) ....Hydrometridae (part) 4
3′ Body, legs, and antennae not as above; eyes located closer to anterior margin of thorax...7
4 Body length 3–5 mm; antennomere I much longer than II, distinctly passing apex of head..*Veliometra*
4′ Body length at least 6 mm; antennomere I subequal to or shorter than II, generally only slightly passing apex of head...5
5 Meso- and metasternum without longitudinal grooves........................*Hydrometra*
5′ Meso- and metasternum with longitudinal grooves...6
6 Antennomere IV bulbous distally; coxae equally spaced..................*Bacillometra*
6′ Antennomere IV filamentous, not expanded distally; middle coxae farther from hind coxae than fore coxae..*Bacillometroides*
7 Body compact and robust (Fig. 6.1d); ventral surface of head with a deep rostral groove (Fig. 6.5b); tarsal formula 2-2-2...Hebridae 8
7′ Body delicate, elongated, and suboval (Fig. 6.1b, c); ventral surface of head without deep rostral groove; tarsal formula 3-3-3................................10
8 Antenna shorter than maximum pronotum width, antennomeres thick, antennomeres I and IV with similar lengths..*Merragata*
8′ Antenna as long as maximum pronotum width, antennomeres slender, antennomere IV longer than I...9
9 Antenna appearing 5 segmented because of a false membranous joint present on antennomere IV (Fig. 6.12)..*Hebrus*
9′ Antenna not appearing 5 segmented; false joint absent...................*Lipogomphus*
10 Eyes slightly removed from anterior margin of thorax; bucculae anteriorly produced, laterally covering base of rostrum; pronotum posteriorly produced, covering all mesonotum (Fig. 6.1b).........................Macroveliidae (*Chepuvelia*)
10′ Eyes located closer to anterior margin of thorax; bucculae not produced, not covering base of rostrum; pronotum at most partially covering mesonotum (Fig. 6.1c)...Mesoveliidae (part) 11
11 Macropterous...*Mesovelia* (part)

11′ Apterous..12
12 Body length 1.2–1.3 mm; eyes vestigial (Fig. 6.2a)...........................*Cryptovelia*
12′ Body length at least 1.5 mm; eyes well-developed, globose...........................13
13 Mesonotum prolonged on middle, distinctly longer than pronotum ....*Mesovelia* (part)
13′ Mesonotum not prolonged, subequal in length or shorter than pronotum........14
14 Body shorter than 2 mm; ventral surface of head with a pair of prominent longitudinal carina behind base of pronotum.................................*Darwinivelia*
14′ Body length at least 2.5 mm, ventral surface of head without prominent longitudinal carina...*Speovelia*
15 Head elongated; eyes located far from the anterior margin of thorax ....Hydrometridae (*Limnobatodes*)
15′ Head not elongated; eyes located closer to anterior margin of thorax..............16
16 Middle coxae located much closer to hind coxae than to fore coxae (Fig. 6.13c); hind femur long, generally distinctly passing apex of abdomen..Gerridae (17)
16′ Middle coxae approximately equally distant from fore and hind coxae; hind femur shorter, usually at most slightly passing apex of abdomen (except for males of *Microvelia longipes*)..36
17 Eyes approximately reniform, emarginated on posterior portion of inner margin (Fig. 6.14a)...18
17′ Posterior portion of inner margin of eye not emarginated (Fig. 6.14b)............24
18 Antenna as long as or longer than body; antennomere IV distinctly longer than others..*Tachygerris*
18′ Antenna shorter than body; antennomere IV not longer than others................19
19 Body length greater than four times its width............................*Cylindrostethus*
19′ Body length smaller than four times its width..20
20 Tarsomere I of fore leg with about half the length of tarsomere II...................21
20′ Tarsomeres I and II of fore leg with subequal lengths.....................................23
21 Rostrum short, not reaching base of the prosternum........................*Potamobates*
21′ Rostrum longer, extending at least until the mesosternum..............................22
22 Pronotum with a large central light macula on anterior portion; middle tarsi without claws...*Neogerris*
22′ Pronotum with two relatively small elongated light maculae on anterior portion; middle tarsi with claws...*Limnogonus*
23 Pronotum not covering mesonotum; connexives without spines on posterior ends...*Eurygerris*
23′ Pronotum covering mesonotum; connexives with small spines on posterior ends...*Aquarius*
24 Tibia and tarsomere I of middle leg with a fringe of long erect setae; marine species...*Halobates*
24′ Tibia and tarsomere I of middle leg without fringe of long erect setae (except for some species of *Trepobates*); freshwater species..........................25

25 Antennomere II longer than III; antennomeres II–III of males distally widened........*Metrobates*
25′ Antennomere II subequal in length or shorter than III; antennomeres II–III of males not distally widened........26
26 Tarsomere I of fore leg longer than II........*Charmatometra*
26′ Tarsomere I of fore leg not longer than II........27
27 Tarsomere II of fore leg not longer than two times the length of I........28
27′ Tarsomere II of fore leg longer than two times the length of I........29
28 Pronotum yellowish with several distinct black stripes........*Eobates*
28′ Pronotum orange to reddish brown, usually only with lateral margins darkened, but sometimes with a median black stripe........*Brachymetra*
29 Middle femur longer than the middle tibia and than the hind femur, which can be modified in males........*Rheumatobates*
29′ Middle femur shorter than the middle tibia and than the hind femur........30
30 Antennomere III 10–40 % longer than I........31
30′ Antennomere III with 40–80 % the length of I........33
31 Antennomere III longer than two times the length of II........*Telmatometra*
31′ Antennomere III shorter than two times the length of II........32
32 Interocular space with a dark longitudinal stripe........*Telmatometroides*
32′ Interocular space without dark longitudinal stripe........*Lathriobatoides*
33 Antennomere I much longer than II–III together........*Trepobatoides*
33′ Antennomere I not distinctly longer than II–III together........34
34 Middle tibia distinctly shorter than body length........*Ovatametra*
34′ Middle tibia at least as long as body........35
35 Eyes in lateral view not extending beyond half of propleura; hind tibia distinctly shorter than two times the length of hind tarsus........*Trepobates*
35′ Eyes in lateral view extending beyond half of propleura; hind tibia longer than two times the length of hind tarsus........*Halobatopsis*
36 Head dorsum without impressed longitudinal line or pair of indentations near base; coxae of the same pair of legs inserted close to each other (Fig. 6.13a); male fore tibia without grasping comb........Mesoveliidae [*Mesoveloidea*]
36′ Head dorsum with an impressed longitudinal line and a pair of indentations near base; coxae of the same pair of legs well separated between them (Fig. 6.13b); male fore tibia usually with an apical grasping comb (Fig. 6.15)........Veliidae 37
37 Tarsal formula 1-2-2........38
37′ Tarsal formula 2-2-2 or 3-3-3........43
38 Middle tarsus with four leaflike structures subapically (modified claws and bipartite ventral arolium) (similar to Fig. 6.7i)........39
38′ Middle tarsus with at most three leaflike structures........40

39 Tarsomere II of middle leg longer than I, widened and deeply cleft subapically .... *Euvelia*
39′ Tarsomere II of middle leg shorter than I, not widened and deeply cleft subapically........ *Husseyella*
40 Antennomere I long and slender, subequal to or longer than head width between eyes........41
40′ Antennomere I much shorter and stouter........42
41 Antennomere I at least twice as long as distance from base of antenna to apex of head; tarsal claws of middle and hind legs very long, subequal in length to tarsomere I........ *Aegilipsovelia*
41′ Antennomere I and tarsal claws shorter........ *Microvelia* (*Kirkaldya*)
42 Middle tarsus with three leaflike structures subapically (modified claws and ventral arolium)........ *Xiphovelia*
42′ Middle tarsus without leaflike structures........ *Microvelia* (*Microvelia*)
43 Tarsal formula 2-2-2 or 3-3-3; middle tarsus deeply cleft, with claws and arolia modified in a swimming fan (Fig. 6.7j)........ *Rhagovelia*
43′ Tarsal formula 3-3-3; middle tarsus unmodified........44
44 Middle and hind tarsi each with four leaflike structures (modified claws and arolia) arising from a short cleft at middle of last article (Fig. 6.7i) .... *Veloidea*
44′ Middle and hind tarsi without leaflike structures........45
45 Body surface with several glabrous depressed structures, rounded or elongated; abdominal sternites with lateral shining glabrous grooves (Fig. 6.16) .... *Stridulivelia*
45′ Body surface not as above; abdominal sternites without lateral grooves........46
46 Last tarsomere laterally expanded, suboval (Fig. 6.7h); tarsomere II of middle leg with at most 1.5 times the length of I........ *Oiovelia*
46′ Last tarsomere not expanded, subcylindrical; tarsomere II of midleg with at least two times the length of I........47
47 Tubercles of venter of thorax, if present, centrally located (in at least one species the mesoacetabula are slightly expanded centrally, but do not form distinct tubercles) (Fig. 6.17a)........ *Paravelia*
47′ Metasternum with pair of anterolateral tubercles, near mesoacetabula........48
48 Mesoacetabula prolonged into tubercles, positioned in front of the tubercles of the metasternum (Fig. 6.17b)........ *Platyvelia*
48′ Mesoacetabula not prolonged into tubercles, only metasternal tubercles present (Fig. 6.17c)........ *Steinovelia*

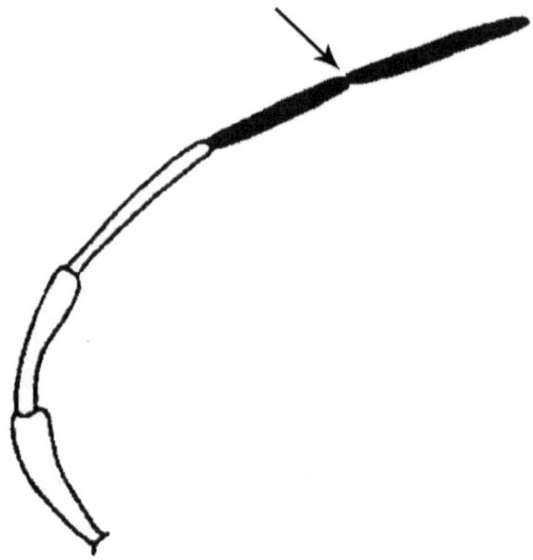

Fig. 6.12 Antenna of *Hebrus* (Hebridae) showing the superficial constriction present on the fourth article (Modified from Mazzucconi et al. 2009)

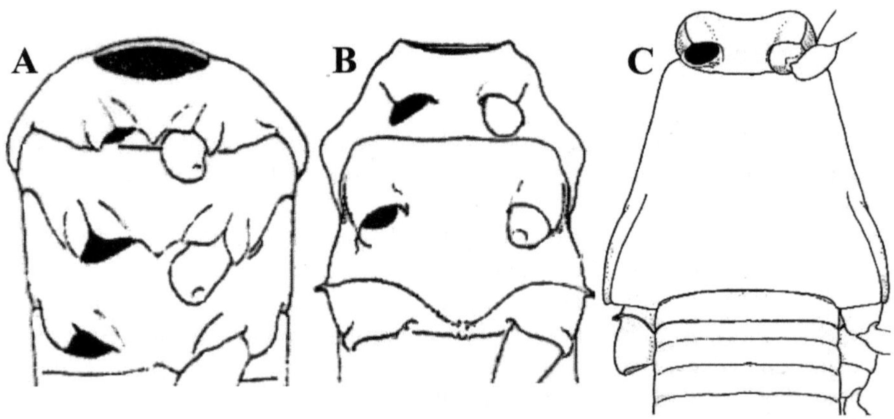

Fig. 6.13 Insertion of the coxae on different Gerromorpha. (**a**) *Mesovelia* (Mesoveliidae). (**b**) *Rhagovelia* (Veliidae). (**c**) *Charmatometra* (Gerridae) (Modified from Andersen 1982)

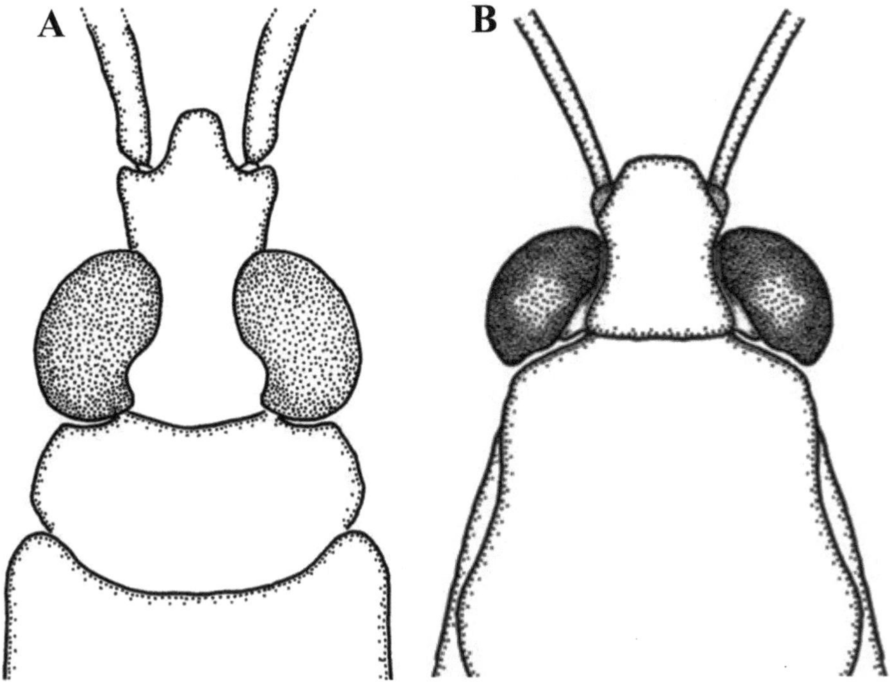

Fig. 6.14 Different eye shapes on Gerridae. (**a**) *Cylindrostethus*. (**b**) *Brachymetra* (Modified from Moreira et al. 2011a)

Fig. 6.15 Detail of the tibial grasping comb of *Velia*, a structure present on many males of the family Veliidae (*GC* grasping comb; modified from Cianferoni and Santini 2013)

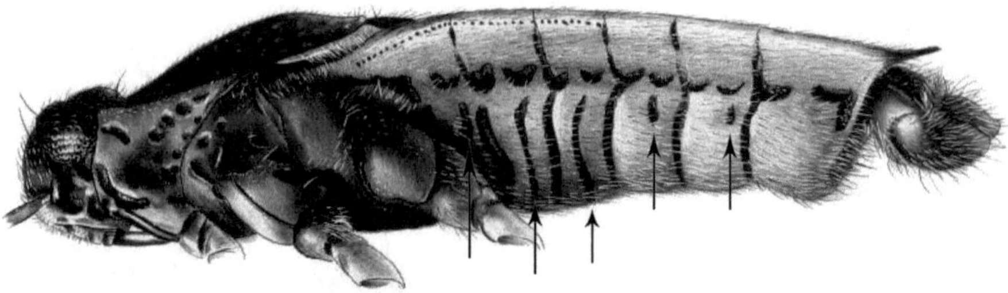

Fig. 6.16 *Stridulivelia* (Veliidae), lateral view of body showing grooves of abdominal sternites (Modified from Drake and Menke 1962). The number of grooved sternites varies from species to species

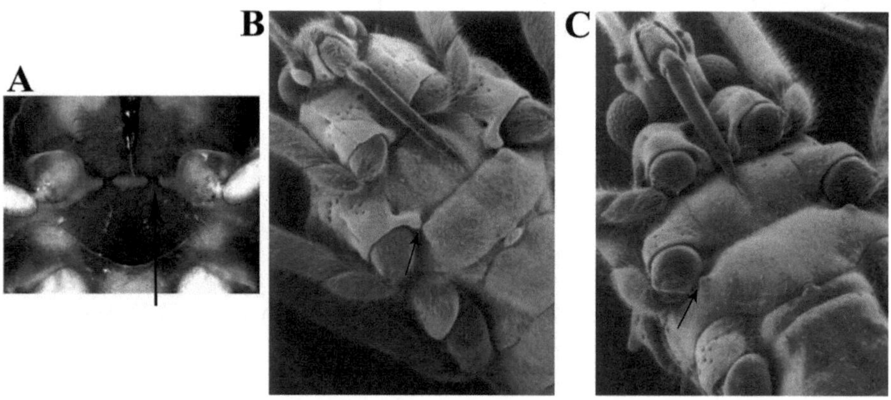

Fig. 6.17 Distribution of thoracic tubercles on different genera of Neotropical Veliinae (Veliidae). (**a**) *Paravelia*. (**b**) *Platyvelia*. (**c**) *Steinovelia* (Modified from: **a** – Moreira and Barbosa 2012; **b**, **c** – Polhemus and Polhemus 1993)

6.5 Concluding Remarks

On the last three decades, the Neotropical Gerromorpha have been target to many faunistical and taxonomic studies, and several species are being described as new or recorded from previously unknown localities. This effort serves as base for phylogenetic and biogeographical studies to be developed in the future, which will contribute to the understanding of the evolution and distributional patterns of the semiaquatic bugs.

Despite of that fact, there is still much to be researched concerning the Neotropical gerromorphans, and more than a hundred undescribed species from the region are known to exist (Polhemus and Polhemus 2008). The same authors pointed the Guiana Shield and the eastern slopes of the Andes as areas of high diversity and endemicity of gerromorphans and considered a very large area of the Atlantic coastal rainforests of Brazil extending from Rio de Janeiro north to the mouth of the

Amazon as "black holes" (under-collected areas) for the aquatic and semiaquatic bugs.

Finally, the formation on Neotropical countries of specialists in all fields of research with interest on the semiaquatic bugs is fundamental for the development of biological, ecological, and systematic studies regarding the group, which are currently extremely scarce when compared with those published concerning Nearctic and Palearctic species.

References

Andersen NM (1973) Seasonal polymorphism and developmental changes in organs of flight and reproduction in bivoltine pondskaters (Hem. Gerridae). Ent Scand 4:1–20

Andersen NM (1975) The *Limnogonus* and *Neogerris* of the Old World with character analysis and a reclassification of the Gerrinae (Hemiptera: Gerridae). Ent Scand Supp 7:1–96

Andersen NM (1976) A comparative study of locomotion on the water surface in semiaquatic bugs (Insecta, Hemiptera, Gerromorpha). Vidensk Meddr Dansk Naturh Foren 139:337–396

Andersen NM (1977) A new and primitive genus and species of Hydrometridae (Hemiptera, Gerromorpha) with a cladistic analysis of relationships within the family. Ent Scand 8:301–316

Andersen NM (1978) A new family of semiaquatic bugs for *Paraphrynovelia* Poisson with a cladistic analysis of relationships (Insecta, Hemiptera, Gerromorpha). Steenstrupia 4:211–225

Andersen NM (1979) Phylogenetic inference as applied to the study of evolutionary diversification of semiaquatic bugs (Hemiptera: Gerromorpha). Syst Zool 28:554–578

Andersen NM (1981) Semiaquatic bugs: phylogeny and classification of the Hebridae (Heteroptera: Gerromorpha) with revisions of *Timasius*, *Neotimasius* and *Hyrcanus*. Syst Entomol 6:377–412

Andersen NM (1982) The semiaquatic bugs (Hemiptera, Gerromorpha). Phylogeny, adaptations, biogeography and classification. Entomonograph 3:1–455

Andersen NM (1998) Water striders from the Paleogene of Denmark with a review of the fossil record and evolution of semiaquatic bugs (Hemiptera, Gerromorpha). Biol Skr 50:1–157

Andersen NM (1999) *Cryptovelia stysi* sp. n. from Borneo with a reanalysis of the phylogeny of the Mesoveliidae (Heteroptera: Gerromorpha). Acta Soc Zool Bohem 63:5–18

Andersen NM, Cheng L (2004) The marine insect *Halobates* (Heteroptera: Gerridae): biology, adaptations, distribution, and phylogeny. Oceanogr Mar Biol Annu Rev 42:119–180

Andersen NM, Polhemus JT (1976) Water-striders (Hemiptera: Gerridae, Veliidae, etc.). In: Cheng L (ed) Marine insects. North-Holland Publishing Company and American Elsevier Publishing Company, Amsterdam-Oxford/New York, pp 187–224

Andersen NM, Polhemus JT (1980) Four new genera of Mesoveliidae (Hemiptera, Gerromorpha) and the phylogeny and classification of the family. Ent Scand 11:369–392

Andersen NM, Polhemus DA (2003) A new genus of terrestrial Mesoveliidae from the Seychelles (Hemiptera: Gerromorpha). J New York Entomol Soc 111:12–21

Andersen NM, Weir TA (2000) The coral treaders, *Hermatobates* Carpenter (Hemiptera: Hermatobatidae), of Australia and New Caledonia with notes on biology and ecology. Inv Tax 14:327–345

Andersen NM, Weir TA (2004) Mesoveliidae, Hebridae, and Hydrometridae of Australia (Hemiptera: Heteroptera: Gerromorpha), with a reanalysis of the phylogeny of the semiaquatic bugs. Inv Syst 18:467–522

Anderson LD (1932) A monograph of the genus *Metrobates* (Hemiptera, Gerridae). Univ Kansas Sci Bull 20:297–311

Aristizábal HG (2002) Los hemípteros de la película superficial del agua en colombia. Parte 1, Familia Gerridae. Academia Colombiana de Ciencias Exactas, Físicas y Naturales, Bogotá
Armúa de Reyes C, Estévez AL, Schnack JA, Brataszczuk EM (2005) Importancia de los belostomátidos como potenciales controladores biológicos en los ambientes acuáticos de la Provincia de Corrientes. Com Cien Tecnol Univ Nac Nordeste 2005(B-006):1–2
Arnqvist G, Mäki M (1990) Infection rates and pathogenicity of trypanosomatid gut parasites in the water strider *Gerris odontogaster* (Zett.) (Heteroptera: Gerridae). Oecologia 84:194–198
Ashmole NP, Ashmole MJ (1967) Comparative feeding ecology of sea birds of a tropical oceanic island. Bull Peabody Mus Nat Hist 24:1–131
Bacon JA (1956) A taxonomic study of the genus *Rhagovelia* (Hemiptera, Veliidae) of the Western Hemisphere. Univ Kansas Sci Bull 38:695–913
Benjamin RK (1967) Laboulbeniales on semi-aquatic Hemiptera. Laboulbenia Aliso 6:111–136
Brinkhurst RO (1959) Alary polymorphism in the Gerroidea (Hemiptera-Heteroptera). J Anim Ecol 28:211–230
Brinkhurst RO (1960) The British distribution of the water-bug *Velia saulii* Tamanini with some notes on alary polymorphism. Proc R Entomol Soc Lond 35A:91–92
Butler EA (1923) A biology of British Hemiptera-Heteroptera. Witherby, London
Buzzetti FM, Zettel H (2008) The largest species of Veliidae (Hemiptera: Heteroptera) in the world: *Veloidea aequatoriana* sp. n. from Ecuador. Tijdschr Entomol 151:187–191
Cabette HSR, Giehl NFS, Dias-Silva, K, Juen L, Batista JD (2010) Distribuição de Nepomorpha e Gerromorpha (Insecta: Heteroptera) da Bacia Hidrográfica do Rio Suiá-Miçú, MT: riqueza relacionada à qualidade da água e do hábitat. In: Gestão e educação ambiental – água, biodiversidade e cultura, vol 2. RiMa Editora, São Paulo, pp 113–137
Callahan JR (1974) Observations on *Gerris incognitus* and *Gerris gillettei* (Heteroptera: Gerridae). Proc Entomol Soc Wash 76:15–21
Carayon J (1971) Notes et documents sur l'appareil odorant métathoracique des hémiptères. Ann Soc Entomol Fr (N S) 7:737–770
Castro-Vargas MI, Morales-Castaño IT (2011) The *Rheumatobates* Bergroth, 1892 (Hemiptera: Heteroptera: Gerridae) of Colombia, including the description of *R. Plumipes* n. sp. and a key to represented species. Zootaxa 3040:1–18
Champion GC (1898) Hemiptera-Heteroptera. Biol Centrali-Americana, Rhynchota 2:1–416
Cheng L (1985) Biology of *Halobates* (Heteroptera: Gerridae). Annu Rev Entomol 30:111–135
Cheng L, Fernando CH (1971) Life history and biology of the riffle bug *Rhagovelia obesa* Uhler (Heteroptera: Veliidae) in Southern Ontario. Can J Zool 49:435–442
Cheng L, Douek M, Goring DAI (1978) UV absorption by gerrid cuticles. Limnol Oceanogr 23:554–556
China WE (1955) The evolution of the water bugs. Bull Natl Inst Sci India 7:91–103
China WE (1963) Hemiptera-Heteroptera collected by the Royal Society Expedition to South Chile 1958–1959. Ann Mag Nat Hist 13:705–723
Cianferoni F (2014) A new record for *Limnobatodes paradoxus* Hussey, 1925 from Guyana (Hemiptera: Heteroptera: Hydrometridae: Limnobatodinae). Zootaxa 3811:398–400
Cianferoni F, Santini G (2013) Comparative description of the hairy structures in two endemic *Velia* species (Insecta: Hemiptera: Heteroptera: Veliidae): *V. currens* (Fabricius, 1794) and *V. gridellii* Tamanini, 1947. Acta Zool 95:347–357
Cobben RH (1960) The Heteroptera of the Netherlands Antilles – I. Foreword, Gerridae, Veliidae, Mesoveliidae (water striders). Stud Fauna Curaçao Other Carrib Isl 11:1–34
Cobben RH (1965) Egg-life and symbiont transmission in a predatory bug, *Mesovelia furcata* Ms & Rey (Heteroptera, Mesoveliidae). Proceedings of the 12th international congress of Entomology, London, 1964, pp 166–168
Cobben RH (1968) Evolutionary trends in Heteroptera part I. Eggs, architecture of the shell, gross embryology and eclosion. Centre for Agricultural Publishing and Documentation, Wageningen
Cobben RH (1978) Evolutionary trends in Heteroptera part II. Mouthpart-structures and feeding strategies. H. Veenman & Zonen B.V, Wageningen
Cooper SD (1984) The effects of trout on water striders in stream pools. Oecologia 63:376–379

Corbet PS (1959) Notes on the insect food of the Nile crocodile in Uganda. Proc R Entomol Soc Lond 34A:17–22

Damgaard J (2008a) Phylogeny of the semiaquatic bugs (Hemiptera-Heteroptera, Gerromorpha). Insect Sys Evol 39:431–460

Damgaard J (2008b) Evolution of the semi-aquatic bugs (Hemiptera: Heteroptera: Gerromorpha) with a re-interpretation of the fossil record. Acta Entomol Mus Natl Pragae 48:251–268

Damgaard J, Moreira FFF, Hayashi M, Weir TA, Zettel H (2012) Molecular phylogeny of the pond treaders (Insecta: Hemiptera: Heteroptera: Mesoveliidae), discussion of the fossil record and a checklist of species assigned to the family. Insect Syst Evol 43:175–212

Damgaard J, Moreira FFF, Weir TA, Zettel H (2014) Molecular phylogeny of the pond skaters (Gerrinae), discussion of the fossil record and a checklist of species assigned to the subfamily (Hemiptera: Heteroptera: Gerridae). Insect Syst Evol 45:251–281, S1–S43

de la Torre-Bueno JR (1908) The broken hemelytra in certain Halobatinae. Ohio Nat 9:389–392

de la Torre-Bueno JR (1917) Life history of the northern *Microvelia* – *Microvelia borealis* Bueno (Hem., Het.). Entomol News 28:354–359

Drake CJ (1957) New Neotropical Halobatinae (Hemiptera: Gerridae). Bull Brooklyn Entomol Soc 52:128–130

Drake CJ, Chapman HC (1958) New Neotropical Hebridae, including a catalogue of the American species (Hemiptera). J Wash Acad Sci 48:317–326

Drake CJ, Cobben RH (1960) The Heteroptera of the Netherlands Antilles – II. Hebridae. Stud Fauna Curaçao Other Carrib Isl 11:35–43

Drake CJ, Lauck DR (1959) Descriptions, synonymy, and check-list of American Hydrometridae (Hemiptera: Heteroptera). Great Basin Nat 19:43–52

Drake CJ, Menke AS (1962) Water-striders of the subgenus *Stridulivelia* from Mexico, Central America, and the West Indies (Hemiptera: Veliidae). Proc U S Natl Mus 113:413–419

Drake CJ, van Doesburg PH (1966) Water-striders of the American genus *Trochopus* (Hemiptera: Veliidae). Stud Fauna Suriname Other Guyanas 8:65–76

Dufour L (1833) Recherches anatomiques et physiologiques sur les Hémiptères, acompagnées de considérations relatives a l'histoire naturelle et a la classification de ces insectes. Bachelier, Paris

Ekblom T (1926) Morphological and biological studies of the Swedish families of Hemiptera-Heteroptera. Part I. The families Saldidae, Nabidae, Lygaeidae, Hydrometridae, Veliidae and Gerridae. Zool Bidr Uppsala 10:29–179

Ekblom T (1930) Morphological and biological studies of the Swedish families of Hemiptera-Heteroptera. Part II. The families Mesoveliidae, Coreidae and Corixidae. Zool Bidr Uppsala 12:113–150

Estévez AL, Schnack JA (1980) Las ninfas de *Merragata lacunifera* (Berg) (Hemiptera, Hebridae). Neotropica 26:29–33

Floriano CFB (2013) Taxonomia de Cylindrostethus Mayr, 1865 (Hemiptera: Heteroptera: Gerridae) da América do Sul. Dissertation, Universidade Federal do Paraná

Floriano CFB, Cavichioli RR (2013) A new species of *Cylindrostethus* Mayr (Hemiptera, Gerromorpha, Gerridae) for the Neotropical Region, with an identification key for the species of Group 1 *sensu* Drake 1952. Zootaxa 3702:187–192

Frick KE (1949) The biology of *Microvelia capitata* Guerin, 1857, in the Panama Canal Zone and its role as a predator on Anophelinae larvae. Ann Entomol Soc Am 42:77–100

Frick MG, Williams KL, Bolten AB et al (2009) Foraging ecology of oceanic-stage loggerhead turtles *Caretta caretta*. Endanger Species Res 9:91–97

Galbreath JE (1973) Diapause in *Mesovelia mulsanti* (Hemiptera: Mesoveliidae). J Kansas Entomol Soc 46:224–233

Grimaldi D, Engel MS (2005) Evolution of the insects. Cambridge University Press, Cambridge

Harrington RW, Harrington ES (1972) Food of female marsh killifish, *Fundulus confluentus* Goode and Bean, in Florida. Amer Midland Nat 87:492–502

Harris HM, Drake CJ (1945) A new *Brachymetra* from Peru with a list of known species (Hemiptera, Gerridae). Proc Entomol Soc Wash 47:211–212

Haverschmidt F (1962) Notes on the feeding habits and food of some hawks from Surinam. Condor 64:154–158

Heckman CW (2011) Encyclopedia of South American aquatic insects: Hemiptera – Heteroptera. Illustrated keys to known families, genera, and species in South America. Springer, New York

Henriquez NP, Spence JR (1993) Host location by the gerrids egg parasitoid *Tiphodytes gerriphagus* (Marchal) (Hymenoptera: Scelionidae). J Insect Behav 6:455–466

Herring JL (1949) A new species of *Rheumatobates* from Florida (Hemiptera, Gerridae). Fla Entomol 32:160–165

Herring JL (1955) A new American genus of Veliidae (Hemiptera). Fla Entomol 38:21–25

Hoffmann WE (1924) The life histories of three species of gerrids (Heteroptera, Gerridae). Ann Entomol Soc Am 17:419–430

Hoffmann WE (1932) The biology of three North American species of *Mesovelia* (Heteroptera-Mesoveliidae). Can Entomol 64:88–134

Hoyt RD (1970) Food habits of the silverjaw minnow, *Ericymba buccata* Cope, in an intermittent stream in Kentucky. Amer Midland Nat 84:226–236

Hungerford HB (1917) The life-history of *Mesovelia mulsanti* White. Psyche 24:73–84

Hungerford HB (1920) The biology and ecology of aquatic and semiaquatic Hemiptera. Kansas Univ Sci Bull 11:3–341

Hungerford HB, Matsuda R (1957) Descriptions of two species of the genus *Brachymetra* (Gerridae, Hemiptera). J Kansas Entomol Soc 30:19–25

Hynes HBN (1948) Notes on the aquatic Hemiptera-Heteroptera of Trinidad and Tobago, B.W.I., with a description of a new species of *Martarega* B. White (Notonectidae). Trans R Entomol Soc Lond 99:341–360

Iglesias MS, Crespo FA (2003) Crecimiento relativo en *Brachymetra albinerva albinerva* (Amyot & Serville, 1843) (Heteroptera: Gerridae). Physis 60B:5–10

Iglesias MS, Crespo FA (2008) Growth-pattern variation in two species of *Rheumatobates* (Heteroptera: Gerridae) with different degrees of sexual dimorphism. Can Entomol 140:380–384

Ignacimuthu S (2002) Biological control of insect pests. Curr Sci 82:1196–1197

Jamieson GS, Scudder GGE (1979) Predation in *Gerris* (Hemiptera): reactive distances and locomotion rates. Oecologia 44:13–20

Jordan KHC (1932) Zur Kenntnis der Eies und der Larven von *Microvelia schneideri* Schultz. Zeitschr Wissensch Insektenbiol 27:18–22

Jordan KHC (1951) Autotomie bei *Mesovelia furcata* Mls. R. (Hem. Het. Mesoveliidae). Zool Anz 147:205–209

Kaitala A (1991) Phenotypic plasticity in reproductive behaviour of water striders: trade-offs between reproduction and longevity during food stress. Funct Ecol 5:12–18

Konopko SA, Mazzucconi SA (2005) Los estadios larvales de *Trepobates taylori* (Hemiptera: Heteroptera: Gerridae). Rev Soc Entomol Arg 64:113–114

Konopko SA, Mazzucconi SA (2011) The immature stages of the genus *Trepobates* Uhler (Hemiptera: Heteroptera: Gerridae), with an identification key to instars and the description of the nymphs of *T. taylori* (Kirkaldy). Zootaxa 2733:1–15

Laird M (1956) Studies of mosquitoes and freshwater ecology in the South Pacific. Bull R Soc N Z 6:1–213

Lanciani CA (1985) Parasitism of nymphal *Mesovelia mulsanti* (Hemiptera: Mesoveliidae) by the water mite *Hydryphantes tenuabilis* (Acariformes: Hydryphantidae). Fla Entomol 68:352–354

Lanciani CA (1987) Rearing immature *Mesovelia mulsanti* (Hemiptera: Mesoveliidae) on a substratum of duckweed. Fla Entomol 70:286–288

Liche H (1936) Beobachtungen über das Verhalten der Wasseläufer (Gerridae, Hemiptera Heteroptera). Bull Int Acad Pol Sci Lett BII:525–546

Linsenmair KE, Jander R (1963) Das "Entspannungsschwimmen" von *Velia* und *Stenus*. Naturwissensch 50:231

Maier CT (1977) The behavior of *Hydrometra championiana* (Hemiptera: Hydrometridae) and resource partitioning with *Tenagogonus quadrilineatus* (Hemiptera: Gerridae). J Kansas Entomol Soc 50:263–271

Malipatil MB, Monteith GB (1983) One new genus and four new species of terrestrial Mesoveliidae (Hemiptera: Gerromorpha) from Australian and New Caledonia. Aust J Zool 31:943–955

Mazzucconi SA, Bachmann AO (1997) Los géneros *Oiovelia* y *Stridulivelia* de la Argentina (Heteroptera: Veliidae). Rev Soc Entomol Arg 56:62

Mazzucconi AS, López-Ruf ML, Bachmann AO (2009) Hemiptera-Heteroptera: Gerromorpha y Nepomorpha. In: Domínguez E, Fernández HR (eds) Macroinvertebrados bentônicos sudamericanos. Sistemática y biologia. Fundación Miguel Lillo, Tucumán, pp 167–231

McPherson JE (1988) Life history of *Mesovelia mulsanti* (Hemiptera: Mesoveliidae) in Southern Illinois. Great Lakes Entomol 21:19–23

McPherson JE, Taylor SJ (2006) Observations on the field life history of *Steinovelia stagnalis* (Hemiptera: Veliidae) in Southern Illinois, U.S.A., with a survey of the biological literature. Entomol News 117:399–405

Menke AS (1979) The semiaquatic and aquatic Hemiptera of California (Heteroptera: Hemiptera). Bull Calif Insect Surv 21:1–166

Meyer HW (1971a) Visuelle Schlüsselreize für die Auslösung der Beutefanghadlung beim Bachwasserläufer *Velia caprai* (Hemiptera, Heteroptera). 1. Untersuchung der räumlichen und zeitlichen Reizparameter mit formverschiedenen Attrapen. Zeitsch Vergl Physiol 72:260–297

Meyer HW (1971b) Visuelle Schlüsselreize für die Auslösung der Beutefanghadlung beim Bachwasserläufer *Velia caprai* (Hemiptera, Heteroptera). 1. Untersuchung der Wirkung zeitlicher Reizmuster mit Flimmerlicht. Zeitsch Vergl Physiol 72:298–342

Miyamoto S (1953) Biology of *Microvelia diluta* Distant, with descriptions of its brachypterous form and larval stages. Sieboldia 1:113–133

Molano-Rendón F, Morales-Castaño IT, Serrato-Hurtado C (2008) Clasificación y hábitats de Gerridae (Heteroptera – Gerromorpha) en Colombia. Acta Biol Colomb 13:41–60

Morales-Castaño IT, Molano-Rendón F, Castro-Vargas MI (2013) Nueva especie de *Potamobates* Champion, 1898 (Insecta: Hemiptera: Heteroptera: Gerridae: Cylindrostethinae) del Pacífico Colombiano. Bol Cient Centro Mus 17:189–195

Moreira FFF, Barbosa JF (2012) Two new species of *Paravelia* Breddin, 1898 and distributional notes concerning the Veliidae from Minas Gerais State, Brazil (Insecta: Hemiptera: Heteroptera: Gerromorpha). Zootaxa 3354:58–68

Moreira FFF, Barbosa JF (2013) A new *Hydrometra* (Hemiptera: Heteroptera: Hydrometridae) from northeastern Brazil, with a key to the species recorded from the country. Zootaxa 3619:70–74

Moreira FFF, Ribeiro JRI (2009) Two new *Rhagovelia* (Heteroptera: Veliidae) and new records for twelve species in southeastern Brazil. Aquat Insects 31:45–61

Moreira FFF, Nessimian JL, Rúdio JA, Salles FF (2010) New species and new records of Veliidae from Espírito Santo State and adjacent Minas Gerais State, Brazil, with notes on nomenclature (Insecta: Heteroptera: Gerromorpha). J Nat Hist 44:2761–2801

Moreira FFF, Alecrim VP, Ribeiro JRI, Nessimian JL (2011a) Identification key to the Gerridae (Insecta: Heteroptera: Gerromorpha) from the Amazon River floodplain, Brazil, with new records for the Brazilian Amazon. Zoologia 28:269–279

Moreira FFF, Barbosa JF, Ribeiro JRI, Alecrim VP (2011b) Checklist and distribution of semiaquatic and aquatic Heteroptera (Gerromorpha and Nepomorpha) occurring in Brazil. Zootaxa 2958:1–74

Moreira FFF, Barbosa JF, Ribeiro JRI (2012) Veliidae (Insecta, Heteroptera, Gerromorpha) from southeastern Brazil: three new species from Rio de Janeiro State, a new species group for Neotropical *Rhagovelia* Mayr, and notes on distribution and synonymy. Rev Bras Entomol 56:147–158

Muñoz SR, Moreira FFF, Naranjo CL (2010) Checklist, distribution, and habitat of the semiaquatic and aquatic bugs from Cuba (Hemiptera: Heteroptera: Dipsocoromorpha, Leptopodomorpha, Gerromorpha and Nepomorpha). Zootaxa 2562:1–23

Mychajliw S (1961) Four new species of *Hydrometra* from the New World (Hemiptera: Hydrometridae). J Kansas Entomol Soc 34:27–33

Nakasuji F, Dyck VA (1984) Evaluation of the role of *Microvelia douglasi atrolineata* (Bergroth) (Heteroptera: Veliidae) as predator of the brown planthopper *Nilaparvata lugens* (Stål) (Homoptera: Delphacidae). Res Popul Ecol 26:134–149

Nieser N, Melo AL (1997) Os heterópteros aquáticos de Minas Gerais: guia introdutório com chave de identificação para as espécie de Nepomorpha e Gerromorpha. Editora UFMG, Belo Horizonte

Nieser N, Melo AL (1999) A new species of *Halobatopsis* (Heteroptera: Gerridae) from Minas Gerais (Brazil), with a key to the species. Ent Ber 59:97–102

Nummelin M (1997) Wing dimorphism of the water strider *Limnogonus franciscanus* (Stål) (Heteroptera: Gerridae) in a seasonal tropical climate. Entomol Fennica 8:167–170

Padilla-Gil DN, Arcos OP (2011) Estadios y variación temporal de *Eurygerris fuscinervis* (Heteroptera: Gerridae) en los Andes de Colombia. Rev Colomb Entomol 36:300–303

Padilla-Gil DN, Damgaard J (2011) A new species of *Potamobates* Champion from Colombia with a re-analysis of phylogenetic relationships (Hemiptera: Gerridae). Zootaxa 2879:41–49

Padilla-Gil DN, Moreira FFF (2013) Checklist, taxonomy and distribution of the *Rhagovelia* Mayr, 1865 (Hemiptera: Heteroptera: Veliidae) of the Americas. Zootaxa 3640:409–424

Padilla-Gil DN, Pacheco-Chaves B (2012) New records of *Rheumatobates* Bergroth (Hemiptera: Heteroptera: Gerridae) from the Pacific coast of Colombia and Costa Rica, with a key to males of *Rheumatobates* in the Eastern Tropical Pacific. Zootaxa 3427:33–46

Perez-Goodwyn PJ (2001) A new *Hydrometra* species from Argentina (Heteroptera: Hydrometridae). Fla Entomol 84:127–130

Poisson RA (1957) Hétéroptères aquatiques. Faune Fr 61:1–263

Polhemus JT (1969) A new *Rheumatobates* from Mexico (Hemiptera: Gerridae). J Kansas Entomol Soc 42:509–511

Polhemus JT (1970) A new genus of Veliidae from Mexico (Hemiptera). Proc Entomol Soc Wash 72:443–448

Polhemus JT (1974) The *austrina* group of the genus *Microvelia* (Hemiptera: Veliidae). Great Basin Nat 34:207–217

Polhemus JT (1975) New estuarine and intertidal water striders from Mexico and Costa Rica (Hemiptera: Gerridae, Mesoveliidae). Pan-Pacific Entomol 51:243–247

Polhemus JT (1977) Type-designations and other notes concerning Veliidae (Insecta: Hemiptera). Proc Entomol Soc Wash 79:637–648

Polhemus JT (1979) A new species of *Stridulivelia* from Mexico, and new subgenus from Middle America (Hemiptera: Veliidae). Pan-Pacific Entomol 55:46–50

Polhemus JT (1990) Surface wave communication in water striders; field observations of unreported taxa (Heteroptera: Gerridae, Veliidae). J New York Entomol Soc 98:383–384

Polhemus JT (1994) Stridulatory mechanisms in aquatic and semiaquatic Hemiptera. J New York Entomol Soc 102:270–274

Polhemus DA (1997) Systematics of the genus Rhagovelia Mayr (Heteroptera: Veliidae) in the Western Hemisphere (exclusive of the angustipes complex). Entomological Society of America, Lanham

Polhemus JT, Polhemus DA (1984a) Notes on Neotropical Veliidae (Hemiptera) VI. Revision of the genus *Euvelia* Drake. Pan-Pacific Entomol 60:55–62

Polhemus JT, Polhemus DA (1984b) Studies on Neotropical Veliidae (Hemiptera) VII. Descriptions of four new species of *Paravelia* Breddin. Amazoniana 8:339–349

Polhemus JT, Polhemus DA (1988) Family Hebridae Amyot and Serville, 1843. The velvet water bugs. In: Henry TJ, Froeschner RC (eds) Catalog of the Heteroptera, or true bugs, of Canada and the Continental United States. E. J. Brill, Leiden/New York/København/Köln, pp 152–155

Polhemus JT, Polhemus DA (1991) A review of the veliid fauna of bromeliads, with a key and description of a new species (Heteroptera: Veliidae). J New York Entomol Soc 99:204–216

Polhemus JT, Polhemus DA (1993) Two new genera for New World Veliinae (Heteroptera: Veliidae). J New York Entomol Soc 101:391–398

Polhemus JT, Polhemus DA (2002) The Trepobatinae (Gerridae) of New Guinea and surrounding regions, with a review of the World fauna. Part 6. Phylogeny, biogeography, World checklist, bibliography and final taxonomic addenda. Insect Syst Evol 33:253–290

Polhemus JT, Polhemus DA (2007) Global trends in the description of aquatic and semiaquatic Heteroptera species, 1758–2004. Tijdschr Entomol 150:271–288

Polhemus JT, Polhemus DA (2008) Global diversity of true bugs (Heteroptera; Insecta) in freshwater. Hydrobiology 595:379–391

Polhemus JT, Polhemus DA (2010) *Bacillometroides*, a new genus of Hydrometridae (Heteroptera) for three previously described species from South America. Entomol Am 116:58–63

Polhemus JT, Spangler PJ (1995) A review of the genus *Stridulivelia* Hungerford and two new species (Heteroptera: Veliidae) from South America. Proc Entomol Soc Wash 97:128–152

Popov YA (1971) Origin and main evolutionary trends of Nepomorpha bugs. Proc 13th Int Congr Entomol 1:282–283

Porter TW (1950) Taxonomy of the American Hebridae and the natural history of selected species. PhD thesis, University of Kansas

Rensing L (1962) Beiträge zur vergleichenden Morphologie, Physiologie und Ethologie der Wasserläufer. Zool Beitr N F 7:447–485

Riley CFC (1925) Some aspects of the general ecology and behavior of the water-strider, *Gerris rufoscutellatus* Latreille. Entomol Rec J Var 37:65–72, 86–93, 107–115

Riley MK (2006) Comparison of food preference and behavior of two waterstriders *Halobates hawaiiensis* and *Limnogonus luctuosus* (Hemiptera: Gerridae) in Moorea, French Polynesia. Water Resour Center Arch ESPM 107:1–12

Rodrigues HDD, Melo AL, Ferreira-Kepler RL (2014) Taxonomic revision of the Neotropical genus *Oiovelia* (Hemiptera: Heteroptera: Veliidae). Acta Entomol Mus Natl Pragae 54:65–98

Rodríguez-Castro VA, Quiroz-Martinez H, Solis-Rojas C, Tejada LO (2006) Mass rearing and egg release of *Buenos scimitar* Bare as biocontrol of larval *Culex quinquefasciatus*. J Am Mosq Control Assoc 22:123–125

Rowe L, Arnqvist G (2011) Sexual selection and the evolution of genital shape and complexity in water striders. Evolution 66:40–54

Rowe L, Westlake KP, Currie DC (2006) Functional significance of elaborate secondary sexual traits and their evolution in the water strider genus *Rheumatobates*. Can Entomol 138:568–577

Schuh RT, Slater JA (1995) True bugs of the world (Hemiptera: Heteroptera). Classification and natural history. Cornell University Press, Ithaca

Scudder GGE (1959) The female genitalia of the Heteroptera: morphology and bearing on classification. Trans R Entomol Soc London 111:405–467

Shaalan EAS, Canyon DV (2009) Aquatic insect predators and mosquito control. Trop Biomed 26:223–261

Silvey JKG (1931) Observations on the life-history of *Rheumatobates rileyi* (Berg.) (Hemiptera-Gerridae). Papers Mich Acad Sci Arts Lett 13:433–447

Smith CL (1988a) Family Hydrometridae Billberg, 1820. The marsh traders. In: Henry TJ, Froeschner RC (eds) Catalog of the Heteroptera, or true bugs, of Canada and the Continental United States. E. J. Brill, New York, pp 156–158

Smith CL (1988b) Family Veliidae Amyot and Serville, 1843. The marsh traders. In: Henry TJ, Froeschner RC (eds) Catalog of the Heteroptera, or true bugs, of Canada and the Continental United States. E. J. Brill, New York, pp 734–742

Smith BP (1989) Impact of parasitism by larval *Limnochares aquatica* (Acari: Hydrachnida; Limnocaridae) on juvenile *Gerris comatus*, *Gerris alacris*, and *Gerris buenoi* (Insecta: Hemiptera; Gerridae). Can J Zool 67:2238–2243

Southwood TRE, Leston D (1959) Land and water bugs of the British Isles. Frederick Warne & Co., London
Spangler PJ (1986) Two new species of water-striders of the genus *Oiovelia* from the Tepui Cerro de la Neblina, Venezuela (Hemiptera: Veliidae). Proc Entomol Soc Wash 88:438–450
Spence JR, Andersen NM (1994) Biology of water striders: interactions between systematics and ecology. Annu Rev Entomol 39:101–128
Spence JR, Spence DH, Scudder GGE (1980) Submergence behavior in *Gerris*: underwater basking. Amer Midland Nat 103:385–391
Sprague IB (1956) The biology and morphology of *Hydrometra martini* Kirkaldy. Univ Kansas Sci Bull 38:579–693
Steiniger F (1933) Die erscheinungen der Katalepsie bei Stabheuschrecken und Wasserläufern. Zeitsch Morph Okol Tiere 26:591–708
Svensson AM, Danielsson I, Rydell J (2002) Avoidance of bats by water striders (*Aquarius najas*, Hemiptera). Hydrobiology 489:83–90
Takahashi R (1921) Observations on Hydrometridae (2). Insect World 25:8–12
Taylor SJ, McPherson JE (1999) Morphological variation and polyvoltinism of *Microvelia pulchella* (Heteroptera: Veliidae) in southern Illinois, USA. Acta Soc Zool Bohem 63:237–249
Taylor SJ, McPherson JE (2000) Comparison of two population sampling methods used in filed life history studies of *Mesovelia mulsanti* (Heteroptera: Gerromorpha: Mesoveliidae) in southern Illinois. Great Lakes Entomol 33:223–230
Taylor SJ, McPherson JE (2003) Voltinism and laboratory rearing of *Microvelia hinei* (Heteroptera: Gerromorpha: Veliidae). Great Lakes Entomol 36:1–9
Teran AF, Vogt RC, Gomez MFS (1995) Food habits of an assemblage of five species of turtles in the Rio Guaporé, Rondônia, Brazil. J Herpetol 29:536–547
Vepsäläinen K, Kaitala A, Kaitala V (1985) Reproductive traits of the waterstrider *Gerris thoracicus* in unpredictable environments: a simulation study. Oikos 45:266–272
Vermette R, Fairbairn DJ (2002) How well do mating frequency and duration predict paternity success in the polygynandrous water strider *Aquarius remigis*? Evolution 56:1808–1820
Villagra CA, Villalobos CA, Tapia DH, Rodriguez-Auad K (2001) Sexual dimorphism and behaviour in the water strider *Gerris chilensis* (Berg) (Hemiptera: Gerridae). Rev Chil Entomol 28:87–93
Warren A (1915) Dragonflies and their food. Proc Entomol Soc Hawaii 3:72–82
Wesenberg-Lund C (1943) Biologie der Süsswasserinsekten. Gyldendalske Boghandel-Nordisk Forlag, København
Westlake KP (1998) Sexual conflict and the evolution of male elaboration in the water strider genus Rheumatobates (Heteroptera: Gerridae). PhD dissertation, University of Toronto
Westlake KP, Rowe L, Currie DC (2000) Phylogeny of the water strider genus *Rheumatobates* (Heteroptera: Gerridae). Syst Entomol 25:125–145
Wharton R (1981) Diving by the veliid *Trochopus plumbeus* (Uhler) (Hemiptera). Fla Entomol 64:194
Wilcox RS (1979) Sexual discrimination in *Gerris remigis*: role of a surface wave signal. Science 206:1325–1327
Williams FX (1944) Biological studies in Hawaiian water-loving insects part V. Heteroptera or bugs. Proc Hawaiian Entomol Soc 12:186–196
Wilson CA (1958) Aquatic and semiaquatic Hemiptera of Mississippi. Tulane Stud Zool 6:115–170
Zettel H (1999) *Nieserius* gen.n., a new genus of the subfamily Hyrcaninae (Heteroptera: Hebridae) from Thailand, Laos, and Nepal, with the first known subaquatic species of Gerromorpha. Aquat Insects 21:39–52

Part IV
Infraorder Nepomorpha

Chapter 7
The True Water Bugs (Nepomorpha)

Julianna F. Barbosa and Higor D.D. Rodrigues

Abstract The infraorder Nepomorpha is one of the most specialized groups of heteropterans, with most of its species spending the entire life cycle within the water. True water bugs include ten families, 52 genera, and approximately 730 species in the Neotropical region, which shelters about 30 % of the global diversity of the group. Despite this fact, little is known about their biology, ecology, and geographical distribution in the Neotropics. Several genera need to be revised and many species remain with outdated descriptions. The systematics of the infraorder, although has received little attention in recent years, remains with various unresolved questions, and no phylogenetic study was conducted for any endemic group of this region.

7.1 Introduction

The most traditional classification of the Heteroptera was proposed by Latreille (1810), who used the Linnaean term Hemiptera to refer to the Rhyngota group of Fabricius (1803). In Latreille's classification, heteropterans were divided in two groups based on the lifestyle, namely, Hydrocorisae (currently Nepomorpha) for aquatic species and Geocorisae for terrestrial species. Subsequently, Dufour (1833) divided Geocorisae, recognizing a third group named Amphibicorisae, referring to the semiaquatic species living on the water surface (currently Gerromorpha). This classification, although based on behavioral characters, is highly typological and does not reflect their evolutionary history (Andersen 1979, 1982; Schuh and Slater 1995).

J.F. Barbosa (✉)
Instituto de Biologia, Departamento de Zoologia, Laboratório de Entomologia,
Universidade Federal do Rio de Janeiro, Avenida Carlos Chagas Filho,
373, CCS, Bloco A, Sala 107 Cidade Universitária, Rio de Janeiro, RJ 21941-971, Brazil
e-mail: julianna.freires@gmail.com

H.D.D. Rodrigues
Museu de Zoologia da Universidade de São Paulo,
Avenida Nazaré, 481, Ipiranga, São Paulo, SP 04263-000, Brazil
e-mail: higorddr@gmail.com

A.R. Panizzi, J. Grazia (eds.), *True Bugs (Heteroptera) of the Neotropics*,
Entomology in Focus 2, DOI 10.1007/978-94-017-9861-7_7

Fieber (1861) introduces Gymnocerata (= Geocorisae) and Cryptocerata (= Hydrocorisae), names created in reference to the length of the antennae and used until recently in the literature (e.g., Mahner 1993). Leston et al. (1954) introduced the suffix "morpha" in their classification, proposing the infraorders Cimicomorpha and Pentatomomorpha. Popov (1968) suggested for the first time the name Nepomorpha for the aquatic bugs, and since then the suffix has been used for naming heteropteran infraorders (Štys and Kerzhner 1975). However, Mahner (1993) regarded Nepomorpha as an ambiguous name and advocated the usage of the old name Cryptocerata. In his review of Mahner's work, Andersen (1995) promoted the disuse of the old name to maintain the nomenclatural stability of infraordinal names.

Nepomorpha comprises more than 2,400 valid species distributed in all biogeographical regions, except Antarctica (Polhemus and Polhemus 2008a). This is a group of specialized true bugs, adapted to aquatic habitats, with species living in lentic environments, such as small pools and lakes, and lotic environments, ranging from small streams to large rivers. Species belonging to families Ochteridae and Gelastocoridae display riparian habits, occupying the margins of aquatic ecosystems, although they are commonly collected at considerable distances from water (Nieser and Melo 1997). Besides living in different aquatic habitats, many species are able to colonize degraded environments, which make this group of insects viable for biomonitoring programs of water quality. A few species are able to colonize saline (Naucoridae) or acidified waters (Corixidae) (e.g., La Rivers 1951; Henrikson and Oscarson 1981). Species of Corixidae and Notonectidae are the main colonizers of temporary waters, due to their good dispersion ability and adaptations to water level fluctuations (Williams 2006).

7.2 General Characteristics and Diagnosis

Aquatic heteropterans have a high morphological diversity, with body shape ranging from ovate and flattened dorsoventrally to globular and rounded, convex on the dorsal surface and straight ventrally, elongated, or elongate ovoid. Most species show a uniform brown color, varying from yellow to black. However, species with distinctly colored body, featuring different patterns of spots, are also common. Besides the variation of forms and colors, aquatic heteropterans are widely diverse in total body length, with some species measuring a little more than 1 mm (Pleoidea), whereas others are more than 100 mm long (Belostomatidae) (Schuh and Slater 1995).

In general, nepomorphans are mainly characterized by the reduced antennae, usually hidden under the compound eyes; the unique exception is the Oriental genus *Fischerotrephes* Zettel, 1994 (Helotrephidae), in which brachypterous specimens do not have antennae (Zettel et al. 2011). Antennae vary in number of segments (1–4) and are filiform, thickened, or flattened and may or may not have lateral projections. Compound eyes are well developed, occupying a large part of the dorsolateral margins of the head. In Ochteroidea, it has an internal emargination on the dorsal surface. Ocelli are absent in most groups, except in Ochteroidea and the subfamily Diaprepocorinae (Corixidae). Cephalic trichobothria are always absent (Popov 1971). The rostrum has three or four segments, generally being small and robust, except in Ochteridae, Aphelocheiridae, and some genera of Helotrephidae, where it can reach the middle coxae. Corixidae have unsegmented rostrum, usually triangular, tapering apically, with transverse sutures.

The pronotum is long, covering the meso- and metanotum. Scutellum is always exposed, but not reaching the middle length of the forewings. In some subfamilies of Corixidae, the pronotum almost entirely covers the scutellum. Forewings of various groups, traditionally named hemelytra, are divided into an anterior coriaceus part (clavus, corium, embolium) and a membranous posterior part. However, hemelytra are not shared by all families – species of Pleidae and Helotrephidae have forewings similar to elytra, entirely leathery in both brachypterous and macropterous forms (Bachmann and López-Ruf 1994).

Wing polymorphism occurs in several families of Heteroptera, usually being associated with dispersion, which can be a consequence of habitat instability, increasing population density (Muraji and Nakasuji 1988), hormonal changes (Southwood 1961), or photoperiod (Vepsäläinen 1971, 1974). According to Schuh and Slater (1995), nepomorphans generally have the following five types of wings: staphylinoid (covering less than half of the abdomen, usually ending at abdominal segment III; clavus and corium not delimited; with vestigial embolium or not; posterior margin of the wing truncated; hind wings may be absent or reduced), brachypterous (usually not covering abdominal tergites VI–VII; clavus and corium fused or not, generally with at most a fine suture present; embolium generally present and well defined; hind wings usually reduced), coleopteroid (usually covering the whole abdomen; clavus and corium elongated, generally fused; embolium developed or not; membrane reduced or inconspicuous, providing a uniform appearance to wing, similar to a coleopteran elytra), submacropterous (exposing only a small part of the posterior abdominal tergites, generally by reduction of part of the membrane; clavus, corium, embolium, and membrane present; hind wings reduced or developed), and macropterous (clavus, corium, embolium, and membrane developed; hind wings fully developed). Usually only the macropterous form has flight muscles developed (Schuh and Slater 1995). Despite the existence of these designations for wing types in Heteroptera, researchers working with aquatic heteropterans often use only the terms brachypterous and macropterous, with all other designations being included in brachyptery. However, staphylinoid and brachypterous individuals with at least three exposed abdominal tergites are rare, only occurring in Aphelocheiridae, and Naucoridae genera *Cryphocricos* Signoret and *Limnocoris* Stål. Intraspecific polymorphism is important in the taxonomy of these species, because some morphological characters are directly related to the polymorphic state (Polhemus and Polhemus 2008b). In some species of *Limnocoris*, the shape of the posterolateral angle of the pronotum and embolium varies considerably between brachypterous and macropterous individuals. Therefore, the correct identification of certain species can be hindered when this polymorphism is not well documented.

Forelegs of Nepomorpha are generally modified and may be raptorial (e.g., Belostomatidae, Gelastocoridae, Nepidae, Naucoridae) with forefemora enlarged and usually grooved at inner surface for reception of tibiae or scoop shaped with the foretarsi unsegmented and margined by a fringe of long setae (Corixidae). In some genera, foretarsi are fused with tibiae and claws may be reduced or absent. In others, such as Ochteridae and Pleidae, all legs are apparently cursorial. Hind legs are generally fringed with golden setae and in some groups can be slightly flattened, which helps in swimming. Tarsal formula varies among families and genera, being, for example, 1-1-1 (all Nepidae), 1-2-2 (all Potamocoridae), 2-2-3 (all Ochteridae), or 3-3-3 (some Pleidae). Dorsal and ventral arolia are usually present in adults and nymphs (Schuh and Slater 1995).

Some groups may have ventromedian carinae on thoracic sterna (e.g., *Belostoma* Latreille), on abdominal sternites (e.g., Notonectidae), or on the thorax and abdomen [e.g., Helotrephidae and *Limnocoris* (Naucoridae)]. Sexual dimorphism is observed in some families, especially regarding the relative length, presence of stridulatory structures, different tarsal formula, and symmetry of abdominal sternites. Abdominal tergite VIII is modified as a pair of siphons in Belostomatidae (retractable) and Nepidae (nonretractable). Male genital segments vary from symmetrical (e.g., Belostomatidae, Nepidae, Potamocoridae) to asymmetrical (e.g., Corixidae, Gelastocoridae, most Naucoridae and Notonectidae), including modifications such as 90° rotation of the genital capsule in Helotrephidae (Nieser and Chen 2002), Corixidae, Gelastocoridae and Ochteridae. Female ovipositor is generally not developed. However, it can be relatively well developed and possess robust spines in some groups, such as the Pleidae.

7.3 General Biology and Ecology

As stated above, wing polymorphism is very frequent in most families, with micropterous, brachypterous, staphylinoid, coleopteroid, submacropterous, and macropterous individuals. Wing reduction in the Nepomorpha was long thought not to exist except where there was an adaptation for underwater respiration (Schuh and Slater 1995). According to Popham (1960), once a plastron had been acquired, the need for gas store replenishment and for the hemelytra to collect air from the surface no longer exists. The loss of wings, flight muscles, and reduction of the median thoracic tracheal trunks can be explained as adaptations to permanent submergence in fast-flowing streams or rivers. Schuh and Slater (1995) argue that the reduction of the wings in the Nepomorpha is not necessarily related to adaptations to underwater respiration, because some species that use oxygen from the atmosphere (some Nepidae) also have reduced hind wings.

Little is known about the mechanisms responsible for generating forms that are unable to fly, and the only feature that seems to be common to most of these species is the stability of the habitat. Brown (1951) made this connection in Corixidae, observing that those species normally found on temporary habitats migrate a lot, while those found in permanent habitats do it less frequently. According to him, those species of Corixidae whose migration rate was restricted by brachypterism were also confined to permanent habitats.

For most individuals, wings perform an important role on respiration. Therefore, the full absence of wings is not observed on Nepomorpha, although their strong reduction is present. On first instars, respiration is cutaneous, and the wings and pterothorax having participation on respiration from the third instar on (Popham 1960). The main air supply of adult aquatic bugs is atmospheric, with replenishment of air stores. The secondary source is the oxygen dissolved in the water, which the insects obtain by means of the "physical gill" effect of their air stores (Parsons 1970). On fast-flowing streams or rivers, surface replenishment of these stores is not possible. Individuals living on this kind of habitats acquired a more specialized type of respiration through the plastron – a "gas store" communicating with the tracheal system and usually in the form of a thin film of constant volume and large surface area, retained by hydrofuge hairs, and therefore not liable to loss by diffusion (Thorpe 1950). On Neotropical bugs, this system can be observed in Neotrephinae (Helotrephidae), Cryphocricinae, and Laccocorinae (Naucoridae) (Parsons and Hewson 1974; Hinton 1976).

In Nepidae, the water scorpions, all air stores are normally concealed from the water (Parsons 1972), and there is a closure of the abdominal spiracles, associated with loss of the ventral gas stores and physical gill respiration (Popham 1960). In most representatives of Naucoridae and other families of aquatic bugs, air stores act as physical gills, with loss of air by diffusion and necessity of replenishment at the water surface. In Corixidae, Naucoridae, and Pleidae, secretion from scent glands is used on a grooming behavior related to prevention of contamination by microorganisms on ventral hydrofuge hairs (Kovac and Maschwitz 1989, 1991). Such contamination would lead to the wetting of the respiratory region, resulting in the loss of the ventral air bubble and drowning of the insects (Kovac and Maschwitz 1991). These glands are absent in Belostomatinae (Belostomatidae) (Staddon 1971), Anisopinae (Notonectidae) (Staddon and Thorne 1974), Nepidae and Nerthrinae (Gelastocoridae) (Staddon and Thorne 1979). Other common behavioral traits found among certain water bugs include grooming and cleaning of mouthparts and legs, especially after feeding, and production of unpleasant odors from the well-developed scent glands, the latter most probably as a defense mechanism (McCafferty 1981).

Aquatic bugs are observed in a great variety of habitats, including acidified lakes, saline ponds, and hot springs (Henrikson and Oscarson 1985; Papáček and Kovac 2001). Ochteridae and Gelastocoridae comprise riparian saltatorial bugs. Ochterids, the velvety shore bugs, are terricolous or littoral that live along the shores of ponds or streams, but they also run on aquatic plants (Papáček 2001). Two Neotropical species of *Ochterus* Latreille occur in salt marshes: *O. aeneifrons* (Champion) and *O. perbosci* (Guérin-Méneville) (Polhemus 1976). Commonly called toad bugs, individuals of Gelastocoridae can be observed jumping on margins of aquatic habitats, remaining motionless when disturbed. With pruinose surface, individuals of *Gelastocoris* Kirkaldy are perfectly camouflaged with sand grains, being also found on mud (Todd 1955). *Nerthra* Say, on the other hand, shows uniform color pattern and is more associated with leaf litter, sometimes being found far away from the margins of water bodies. They seem to spend much of their time burrowing in the soil, also being found on algae on rocky sea coasts, roots of orchids, leaf mold, and under debris of heaps of dry grass (Todd 1955, 1957, 1965). Estévez and López-Ruf (2006) and Estévez et al. (2010) observed adults and nymphs of *Nerthra ranina* (Herrich-Schäeffer) and *N. gaucha* Estévez & Schnack submersed and swimming underwater.

Corixids (water boatmen) are bottom dwellers. They are found in most stagnant or nearly stagnant aquatic habitats where there is a bottom of relatively fine soil particles (Macan 1938). Midlegs are used for clinging to the bottom or other substrate while feeding or resting (Lauck 1979). Belostomatidae (toe-biters or electric-light bugs) and Nepidae are constantly associated with vegetation, where they cling on with the tip of the abdomen in contact with the surface film (Hungerford 1919; Keffer 1996). In general, these bugs are observed in stagnant waters or small streams with slow current (Menke 1963; Keffer 1996). Until recently, no members of *Horvathinia* Montandon had been collected in the aquatic habitat. All of its species were described based on specimens attracted to the light. Schnack et al. (2006) examined two specimens collected in a shallow subtropical permanent pond. However, there is almost nothing published on their ecology or reproductive behavior (Estévez and Ribeiro 2011).

Limnocoris is a genus of benthic naucorids (creeping water bugs), while *Cryphocricos* has some species living under conditions of very strong current (Nieser and López-Ruf 2001). Specimens of *Ambrysus* Stål can be found on rocky streams with moderate to fast currents, clinging to the branches and stones on margins, and beneath leaves on bedrock streams (Usinger 1946; Nieser 1975; Rodrigues et al. 2012b). Very little is known about the life history and habitats of Potamocoridae. Most specimens have apparently been collected at lights (Schuh and Slater 1995), although on Macaé River (southeastern Brazil) individuals of this family were collected mainly on sandy and gravel bottom. In Costa Rica, *Potamocoris* Hungerford specimens were collected in environments with a certain level of contamination in a region of agricultural activity (Herrera and Springer 2012).

Notonectidae, Helotrephidae, and Pleidae are pleuston inhabitants, swimming on their backs and propelling the body with the hind legs. Helotrephids, or beetle backswimmers, have been found on holes in the rocky bottom, filled with soil and covered with a thin layer of mud (China 1936), and in large or small, stable or temporary, quiet, and running waters, rock pools, waterfalls, seeping rocks, hot springs, water tanks, and outdoor ceramic bowls (Papáček 2001). Pleids, or pygmy backswimmers, are found swimming actively both in normal and upside position, on clear or turbid waters (Bachmann 1968, 1998). These bugs live in tangles of dense aquatic vegetation which they use for grasping and maintaining their position on the water column of stagnant ponds, swamps, bogs, sluggish drainage ditches, secluded coves in lakes, and the more sluggish parts of slowly moving streams (Drake and Chapman 1953; Gittelman 1974). Notonectids, or backswimmers, live in both lotic and lentic waters, staying in the water column most of the time and being able to maintain their position without clinging on the vegetation. In *Buenoa* Kirkaldy this hability is related to the presence of hemoglobin (Miller 1964). Representatives of *Martarega* White have been observed jumping over the water surface when disturbed.

7.3.1 Eggs and Oviposition

The oviposition of this group of insects is very diverse, with females that deposit their eggs individually or grouped, submerged or slightly submerged above the water surface, or even distant from the water. Eggs may be deposited in different ways: in a horizontal position on the substrate, perpendicular to the substrate (exophytic oviposition), or inserted in the substrate, which are usually plant tissues (endophytic oviposition) or sediment present at the bottom of aquatic ecosystems (Mazzucconi et al. 2009).

Nepids can deposit their eggs glued to plants or algae masses, where part of the egg is submerged, and respiratory horns remain projected above the water surface. Some species of *Ranatra* F. can also deposit eggs far below the water surface (Hinton 1981). In some genera of Belostomatidae, more specifically the subfamily Belostomatinae, females lay eggs on the hemelytra of males, which carry them until the moment of eclosion (Hungerford 1925). On the other hand, females of Lethocerinae lay their eggs on substrates (rocks, logs, plants) that are slightly above the surface of water, leaving the tasks of moisturizing and protection from predators to the males (Ichikawa 1988).

In Corixidae, eggs are generally glued to substrates below the water surface and may or may not be attached by an elongated stalk. The females of Naucoridae also glue eggs on plants and rocky substrates, although endophytic oviposition is known for the Palearctic species *Ilyocoris cimicoides* (L.) (Cobben 1968). A series of articles describing American naucorid eggs has been published, and López-Ruf (1989) found that the disposition of aeropylar system is a taxonomic character at the genus level. In addition, eggs have specific taxonomic characters, with the chorionic sculpturing differing interspecifically (Sites and Nichols 1999). The number of micropyles ranges between 2 and 3 in *Ambrysus*, *Limnocoris*, and *Pelocoris* Stål (Sites and Nichols 1999, 2001) and 3 and 5 in *Cryphocricos hungerfordi* Usinger (Sites and Nichols 1993).

Females of Ochteroidea oviposit out of water. Those of Ochteridae usually deposit their eggs in sandy environments, remains of plant material, or wet soil. As for gelastocorids, oviposition occurs in sandy places (mainly *Gelastocoris*), and eggs may be partly or completely buried, or in small holes in the mud under stones, sometimes distant from the banks of water (mainly *Nerthra*) (Hinton 1981). Hungerford (1922) reared *Gelastocoris oculatus* (F.) in laboratory and observed that the number of eggs laid by a female may vary from 1 to 13. Brown and McPherson (1994) also reared the same species in laboratory and observed that females deposited their eggs singly or in groups of two or three, on the substrate or glued to gravel. Females of Notonectidae usually oviposit in plant tissues, individually or in rows; in *Notonecta* L. eggs can be deposited exposed on the surface of rocks or plants (Hinton 1981). In Pleoidea, oviposition is also endophytic, although Cobben (1968) suggested an entirely exposed deposition in some genera of Helotrephidae due to egg shape.

7.3.2 Nymph Feeding and Development

Nepomorphans are paurometabolous insects, in which nymphs develop gradually until they reach the adult stage, with the same lifestyle on both stages. There are generally five nymphal instars, although a few species have only four. Nymphs are predaceous and similar to adults in color and body shape, differing by the smaller size, absence of wings, and non-developed genitalia. Immatures seem to live in shallower sites and in general have more gregarious habits than adults. Usually in tropics and/or subtropics, species or populations can be multivoltine, with three or more generations annually (Saulich and Musolin 2007).

Little is known about the life history of Neotropical species of Nepomorpha. Most of the research and rearing was developed upon North American species, some of them occurring from Canada to Mexico. The development of these species is univoltine, typical of temperate regions. Hungerford (1922) wrote about the development of *Gelastocoris oculatus*, and as stated by Kevan (1942) for *Nerthra nepaeformis* (F.) from Trinidad, nymphs of both species rest upon the sand for molting. Under laboratory conditions, Brown and McPherson (1994) fed nymphs of *G. oculatus* with *Tenebrio molitor* L. and *Drosophila melanogaster* Meigen.

Nymphs of *Ochterus banksi* Barber (another North American species) construct small individual cells in the moist sand in which they molt. Nymphal stages move slower than adults and carry sand grains on their bodies, presumably for the purpose of concealment (Bobb 1951).

Konopko (2010, 2012, 2013), Konopko and Mazzucconi (2011), Konopko and Melo (2009), and Konopko et al. (2010, 2011) provided a series of descriptions of nymphs of the Corixidae from Argentina: *Ectemnostega montana* (Lundblad), *E. quechua* (Bachmann), *E. stridulata* (Hungerford), *Sigara schadei* (Hungerford), *S. denseconscripta* (Breddin), *Tenagobia incerta* Lundblad, *T. schadei* Lundblad, and *Trichocorixa mendozana* Jaczewski. Detritus, algae, cladocerans, and larvae of Chironomidae were supplied as food, the nymphs being fed solely with algae. López-Ruf and Bachmann (1994) and Estévez et al. (2010) described the nymphal instars of *Neoplea absona* Drake & Chapman, *N. maculosa* (Berg) (Pleidae), and *Nerthra gaucha* Estévez & Schnack (Gelastocoridae).

Descriptions of the immature stages were conducted for some Naucoridae species. Life cycle, including voltinism, eggs, and all immature stages of three species from the northernmost Neotropics were described: *Ambrysus l. lunatus* Usinger (Sites and Nichols 1990), *Cryphocricos hungerfordi* Usinger (Sites and Nichols 1993), and *Limnocoris lutzi* La Rivers (Sites and Nichols 2001). Regarding the fauna of Central and South America, Sites (1991) described eggs and nymphs of *Pelocoris poeyi* (Guérin-Méneville), nymphs of 3–5 instars of *Limnocoris ovatulus* Montandon (López-Ruf 1993), 3–5 instars of *Placomerus micans* La Rivers (López-Ruf 1996), 3–5 instars of *Ctenipocoris schadei* (De Carlo) (López-Ruf 2004), and 1, 3–5 instars of *Ambrysus attenuatus* Montandon (López-Ruf and Hernández 2007). For *Ambrysus lunatus* Usinger (Naucoridae), Sites and Nichols (1990) provided larvae of *Prosimulium* Roubaud. First instars of *Curicta scorpio*

Stål from Texas (a species that occurs up to Argentina) were fed in lab with *D. melanogaster* and the third instar with corixids or small notonectids (Keffer et al. 1994). Except for Sites and Nichols (1990, 1993, 2001), these works are usually taxonomic and lack data about life history.

The first nymphs of *Notonecta* and probably all of the Anisopinae feed to a large extent upon microcrustaceans in the water (Hungerford 1933). Pleidae feed as well on small crustaceans, probably *Cladocera* Latreille (Bachmann 1968).

Most of the features that allow distinguishing the genera of Nepomorpha are not complete developed on nymphs, with sexual and alar characteristics being absent. In Notonectidae, it is possible to identify the different genera based on the labrum, pronotum, and anteapical spur. In Belostomatidae, the number of tarsomeres and claws on the anterior leg and shape of the posterior femora and abdomen are reliable for genera identification. The shape of the pronotum and anterior legs are useful for distinguish genera of Gelastocoridae. This differentiation is not so simple for Corixidae. The pronotum of the nymphs are not developed, and the pala and tibia of the front legs still have a single segment (Hungerford 1948).

7.3.3 Adult Feeding and Reproduction

Except Corixidae, all aquatic Heteroptera are strictly predaceous. Anterior legs of Corixidae are used for collecting food, with the tarsi highly specialized for this purpose (Popham 1961). Its customary food consists of the matter found in deposits on the bottom of the pool and upon the dead leaves that have lodged in its shallow waters. It can include organic matter; diatoms; desmids; oscillatoria; sometimes threads of live (more often dead) filaments of *Zygnema* Agardh, Euglena, *Paramecium* Müller, and *Chlamydomonas* Ehrenberg; spores of various algae; unicellular plants; and animals (Hungerford 1917). There are, indeed, some species of Corixidae in which the palae are not flattened and spoon shaped and which may be predaceous (Hungerford 1948).

In general, nepomorphans feed on crustaceans, dipteran larvae, and other insects, sometimes even on small vertebrates, such as fishes and tadpoles. Belostomatidae, Gelastocoridae, Naucoridae, Nepidae, and Potamocoridae have raptorial forelegs, with enlarged femora and cylindrical tibiae. *Lethocerus* Mayr has frequently been observed feeding on animals many times larger than themselves: fishes, frogs, salamanders, and freshwater snakes (Menke 1963). In the laboratory, adults of *Lethocerus* killed and digested a large *Hydrophilus* Geoffrey and those of *Belostoma malkini* Lauck, the freshwater shrimp *Palaemonetes* Heller, the large tree frog *Hypsiboas crepitans* (Wied-Neuwied), and the fishes *Poecilia* Bloch & Schneider and *Rivulus* Poey (Cullen 1969). Individuals of *Ranatra* wait for their prey hiding between aquatic plants in a more vertical position, with the head directed downward (Nieser 1975). In Gelastocoridae, prey consist of small insects, especially the larvae of those with which they live in association (Todd 1955).

In Notonectidae, Pleidae, and Ochteridae, the forelegs are not modified, although important on predation. Specimens of *Buenoa* feed largely upon small crustaceans and mosquito larvae, which they hold in the crib formed by the bristles arming the forelegs (Truxal 1953). Pleids can feed on small crustaceans, small larvae of aquatic insects, annelids, and animals that fall onto the water surface, such as Collembola and small Diptera (Papáček 2001). The hind legs of *Neoplea striola* (Fieber) are sometimes used for crawling and prey handling, while other backswimmers use these legs strictly for swimming (Gittelman 1977).

Cannibalism is often observed in *Notonecta* and can occur regularly even under normal conditions, although it is more frequent among starved individuals (Zalom 1978) preying upon young nymphs (Fox 1975). There are also records of cannibalism in Pleidae (Papáček 2001), between nymphs of *Abedus* Stål (Belostomatidae), if their prey are not abundant (Hungerford 1919), and in *Ambrysus occidentalis* La Rivers, as a result of intraspecific aggression (Constantz 1974). Hungerford (1922, 1923) observed females of *Gelastocoris oculatus* feeding upon their own offspring and cannibalistic behavior in the first instar of *Cymatia* Flor (Holarctic/Oriental genera of Corixidae).

Nepomorphans generally mate in the water, whereas surface and shore bugs mate in their respective habitats (McCafferty 1981). Stridulation is quite common in these bugs, with an important role on mating. In *Buenoa*, the courting maneuver includes male stridulation, with a chirping sound by rubbing the front legs against the base of the rostrum as they approach the females (Hungerford 1933). Stridulation through anterior femoral ridges and anterior coxal pegs is also known in some species (Wilcox 1975). Male backswimmers stay below the female when mating, with the genital segments reaching around the right side of the female and the front and middle legs employed in the mating clasp (Hungerford 1924). A stridulatory mechanism was cited by Polhemus (1994) for males and females of an undescribed species of *Martarega* – denticles on costal margin of hemelytra/basal ridge on hind femur. Denticles on hemelytral costal margin are observed in females of *M. hungerfordi* Truxal; *M. nieseri* Barbosa, Ribeiro, & Nessimian; *M. oriximinaensis* Barbosa, Ribeiro, & Ferreira-Keppler; and *M. siolii* Barbosa, Nessimian, & Takiya, with no ridges on hind femora.

Males of Corixidae have distinctly asymmetrical abdomen. On mating, male and female are held together thanks to conical pegs of male anterior legs. The male darts onto the back of the female, which he holds by pressing his head against her pronotum and using the distal pegs of the pala to attach under the curved longitudinal ridge of the lateral flange on the female hemelytra (Popham 1961). There was a general conception that the strigil in this family has a stridulatory function; however, Popham et al. (1984) demystified and stated that this structure is in fact responsible for maintaining the subelytral gas stores, while surfacing during mating. In the subfamily Micronectinae, stridulation is better known in *Micronecta* Kirkaldy, an Old World genus, although the same mechanism is observed in *Tenagobia* Bergroth, and *Monogobia* Nieser & Chen with ridges present on abdominal segment 8 (genital capsule; right lobe in *Tenagobia*) or basally on the right paramere (Jansson 1989;

Jansson and Meyer-Rochow 1990; Nieser and Chen 2006). In Corixinae, anterior legs along with the head are involved in stridulation (Aiken 1985).

As in male corixids, the abdomen of male gelastocorids and ochterids is asymmetrical. The male mounts the female, grasping her with the middle pair of legs, the first pair flexed beneath him, and the abdomen somewhat to the left side, due to the asymmetrical structure of the male genitalia (Hungerford 1922; Polhemus 1976). Stridulation is known only in males of *Nerthra*, being generated by ridges on abdomen and proctiger (Polhemus and Lindskog 1994). Males of *Pelocoris femoratus* (Palisot) (Naucoridae) are said to also mounting females, with forelegs grasping the prothorax/mesothorax junction (Brewer and Sites 1994). In this family, stridulation is known in *Limnocoris*, produced by ridges on the abdominal sternites (at least on 2 and 3) and distal part of the hind femur (Aiken 1985; Polhemus 1994; Polhemus and Lindskog 1994).

Males of *Curicta* Stål (Nepidae), in their turn, approach a female, usually from her side, and grasp her with his three adjacent legs. The male twists the distal half of his abdomen ventrally under the female, which considerably deforms the male's abdomen and places his last abdominal tergum obliquely underneath the female's genital operculum (Keffer and McPherson 1993). In *Ranatra* (Nepidae), the male stridulation is produced by rubbing the coxae and femora of the forelegs (Torre Bueno 1905b).

Although sound production has been recorded in Belostomatidae, no stridulatory mechanism was described (Aiken 1985). This family is famous for its parental care. Belostomatinae are back-brooders: females lay eggs on the backs of the males, where they remain until the hatching of larvae. Male brood care is obligatory in the Neotropical belostomatine species of *Abedus* and *Belostoma* (Estévez and Ribeiro 2011). This behavior involves added risks, reduces predatory efficiency, and precludes additional mating for brooding males (Smith 1976, 1979).

Mating on Neotropical representatives of Helotrephidae and Pleidae is poorly known, and there are no records of stridulation in these families.

7.4 Classification and Diversity

Historically, the first studies made in an attempt to systematically classify the Nepomorpha began with Esaki and China (1927) and China (1933). In these classifications, they considered the Corixidae as the most basal group of the lineage. China (1955) proposed that the greater the specialization for aquatic environment, the greater the derivation on a scale of evolutionary time. This led him to consider Ochteroidea as the most basal family of water bugs, being a generalized group. Subsequently, Popov (1971), in his extensive study based on extant and fossil species, proposed that the Nepomorpha have arisen from a lineage of proto-heteropterans, whose nearest ancestor would be the proto-ochterids. In turn, this last lineage gave rise to three major groups in the mid-Triassic period: Nepoidea, Corixoidea, and Ochteroidea (which initially also included Naucoroidea, Notonectidae, and

Pleoidea). In his classification, for the first time, Nepoidea was considered as the basal group of Nepomorpha, followed by Corixidae, Ochteroidea, Naucoroidea, Notonectidae, and Pleoidea.

Rieger (1976) proposed his classification based on morphological characters, also recognizing Nepoidea as the basal group, followed by Ochteroidea, Corixidae, Naucoroidea, Aphelocheiridae, Notonectidae, and Pleoidea, the last six groups forming a monophyletic group based mainly on the clypeus and food pump structures. In addition, Naucoroidea was recovered as the sister group to Aphelocheiridae + Notonectoidea (Notonectidae and Pleidae + Helotrephidae). Later, Mahner (1993) presented one of the most comprehensive phylogenetic frameworks for the group so far, which was based on morphological characters (although not using a formal phylogenetic analysis). He also considered Nepoidea as the basal group, and unlike Rieger (1976), Corixidae was considered more closely related to Ochteroidea. However, the main critical to Mahner's work is the controversial use of the rules of classification and nomenclature. He created several suprageneric names, such as Tripartita and Cibariopectinata, characterized respectively by a tripartite food pump and possession of so-called cibariopectines in the food pump. In addition, he resurrected the old names Cryptocerata and Gymnocerata, justifying the hypothesis that aquatic true bugs are a sister group to the rest of the heteropterans.

Hebsgaard et al. (2004) proposed a classification using mostly the morphological characters of Rieger (1976) and Mahner (1993), in addition to molecular data (mitochondrial gene 16S and nuclear gene 28S), which were analyzed both separately and in a combined analysis. The final hypothesis was obtained by combined molecular and morphological data analysis, which supported the monophyly of all nepomorphan superfamilies and families, except Naucoroidea, that showed to be paraphyletic. The data strongly supported the relationship of Potamocoridae and Aphelocheiridae, and thus, the authors created the superfamily Aphelocheiroidea to allocate these two families; Naucoroidea was left with only the family Naucoridae.

Since then, Nepomorpha has always been considered a monophyletic group. The main synapomorphies of the infraorder, according to Mahner (1993), are (1) presence of three pairs of tympanal organs in the two pterothoracic and first abdominal segments, (2) particular structure of the midgut epithelium and ileum, (3) the spiral katatrepsis of the embryos, and (4) the presence of spiracular sieve plates and structure of the mesepimeron. However, more recently, Hua et al. (2009), based on an analysis of mitochondrial genomes from nine nepomorphan families, proposed a different hypothesis of relationship to this group of insects, with Nepomorpha being a paraphyletic group. The superfamily Pleoidea was elevated to the infraordinal level and considered the sister group of the Nepomorpha that consists Corixoidea + ((Naucoroidea + Notonectidae) + (Ochteroidea + Nepoidea)). They also recognized the old superfamily Naucoroidea (Aphelocheiridae + Naucoridae) as monophyletic, grouping Aphelocheiroidea *sensu* Hebsgaard et al. (2004) again within Naucoroidea. As Weirauch and Schuh (2011) mentioned, this great change in the systematics of the true water bugs still seems premature, because of the few taxa examined and the lack of a combined analysis using morphological characters.

Analyzing only morphological characters of the mouthparts, Brożek (2013) conclusions concur with Popov (1971), Mahner (1993), and Hebsgaard et al. (2004), in the placement of Nepoidea and Corixoidea as a sister group with respect to the remaining nepomorphan families, but differ in the placement of the Potamocoridae. This family is recognized as a sister group to Naucoridae, and they together form the superfamily Naucoroidea, with Aphelocheiroidea including only Aphelocheiridae. Li et al. (2014) reanalyzed the mitochondrial genomes of Hua et al. (2009) with new sampling strategies. Their analysis also recovered Corixoidea as a basal group, although it supports Nepomorpha as monophyletic and the close relationship between Notonectoidea and Pleoidea.

There are more than 2,400 species of Nepomorpha worldwide, divided into 11 families: Aphelocheiridae, Belostomatidae, Corixidae, Gelastocoridae, Helotrephidae, Naucoridae, Nepidae, Notonectidae, Ochteridae, Pleidae, and Potamocoridae (Štys and Jansson 1988). Polhemus and Polhemus (2008a) mentioned 732 species from the Neotropical region, divided into ten families and 52 genera. The only family that does not occur in the Neotropics is Aphelocheiridae, which is monotypic and recorded mainly from the Afrotropical, Oriental, and Australian regions, with a few species in the Palearctic. The genus *Aphelocheirus* Westwood has approximately 80 valid species (Zettel et al. 2008).

Currently, some authors consider Micronectinae and Diaprepocorinae as distinct families from Corixidae (Mahner 1993; Nieser 2002; Chen et al. 2005), with this family being restricted to Corixinae, Cymatiainae, Heterocorixinae, and Stenocorixinae. They are all treated as subfamilies of Corixidae here, since there are no robust studies to corroborate with the newly proposed classification. Corixinae is the most diverse subfamily, including ten of the 17 genera of Corixidae, about 60 % of them with less than ten species. The tribe Graptocorixini has two genera, both with Neotropical representatives: *Neocorixa* Hungerford, with two species, and *Graptocorixa* Hungerford with 12 species (Hungerford 1948; Jansson 1978). The other genera belong to the tribe Corixini, with *Centrocorisa* Lundblad (2), *Corisella* Lundblad (7), *Krizousacorixa* Hungerford (3), *Neosigara* Lundblad (7), *Orocorixa* (1),[1] *Ramphocorixa* Abbot (2), and *Trichocorixella* (1) being the least diverse (Hungerford 1948; Jansson 1979; Nieser and Padilla-Gil 1992; Nieser and Chen 2006; Tinerella and Polhemus 2006). The Neotropical genus *Ectemnostegella* Lundblad was downgraded by Bachmann (1979) to subgenus of *Ectemnostega* Enderlein and has currently 12 species; the New World *Trichocorixa* Kirkaldy and *Morphocorixa* Jaczewski have 13 species each (Hungerford 1948; Nieser 1969; Roback and Nieser 1980; Bachmann 1981; Jansson and Polhemus 1987). The cosmopolitan genus *Sigara* F. has some problems of delimitation, with great variation between species. At the moment, the genus includes 32 Neotropical species, held mainly in the subgenus *Tropocorixa* Hutchinson (Hungerford 1948; Bachmann 1961, 1979, 1981; Bernardo et al. 2012).

[1] The recently described and still monotypic genus *Orocorixa* is very poorly known and feebly differentiated from *Neosigara*, being a little larger, with a smaller lower peg row on male pala.

Heterocorixinae includes solely the genus *Heterocorixa* White, which is exclusively Neotropical, and bears 20 species (Hungerford 1948; Jaczewski 1950; Nieser 1970). Micronectinae has three representatives in the Neotropical region: the monotypic *Monogobia* (Micronectini stat. nov.), with *M. elongata* Nieser & Chen, and *Synaptogobia* Nieser & Chen (Synaptogobiini stat. nov.), with two species, both known only from northern Brazil (Nieser & Chen 2006). The genus *Tenagobia* (Micronectini) has a widespread distribution on the Western Hemisphere, with 28 Neotropical species (Nieser 1977; Bachmann 1979; Polhemus and Nieser 1997; Nieser and Chen 2008).

Naucoridae is the second most diverse family of Nepomorpha in terms of Neotropical genera and the first in number of species (ca. 180) (Polhemus and Polhemus 2008a). The family is subdivided into five subfamilies, Cheirochelinae, Cryphocricinae, Laccocorinae, Limnocorinae, and Naucorinae (Štys and Jansson 1988), and only the first one does not occur in the Neotropics. Regarding the relationships among these subfamilies, only the study of Mahner (1993) was carried out so far. In this work, he proposed Laccocorinae as basal group and Limnocorinae as sister group of Cryphocricinae, considering both as monophyletic taxa. However, more studies are needed to corroborate or refute Mahner's hypothesis.

Cryphocricinae is endemic to the New World, with three tribes and five genera, all occurring in the Neotropical region. Cryphocricini has two genera: the widespread Neotropical *Cryphocricos*, with 13 species (De Carlo 1940; Usinger 1947; Bachmann 1998), and *Procryphocricos* Polhemus from Venezuela and Colombia, with four species (Polhemus 1991; Sites and Camacho 2014; Sites and Shepard 2015).

Ambrysini is currently composed of the genera *Ambrysus* and *Carvalhoiella* De Carlo. The former has 79 valid species, being most diverse of the family, distributed mainly in Mexico and southwestern USA; 29 species have been recorded from South America (La Rivers 1953a, b, 1958, 1962a, b; Polhemus and Polhemus 1983; Davis 1986; Bachmann 1998; Moreira et al. 2011; Rodrigues et al. 2012b; Sites and Shepard 2015). As for *Carvalhoiella*, it has only three species restricted to Brazil and Suriname (Nieser et al. 1999). Originally, this last genus was described in Ambrysinae but then moved to Naucorinae by Nieser (1975) due to the very close generic limits between it and *Pelocoris*. Later on, Nieser et al. (1999) moved it back to Ambrysinae (*sensu* Usinger 1941). Cataractocorini bears only the genus *Cataractocoris* Usinger, with three species ranging from south of Mexico to El Salvador (Sites 2004; Sites et al. 2013).

Although the subfamily Laccocorinae has been downgraded to tribe of Naucorinae by Popov (1970), the status of Montandon (1897) for this taxon was maintained in the literature, including a series of poorly diverse genera. *Decarloa* is monotypic, with *D. darlingtoni* La Rivers, known only from Haiti (Štys and Jansson 1988). *Interocoris* La Rivers, with only *I. mexicanus* (Usinger) endemic to Mexico, was originally described as a subgenus of *Heleocoris* Stål but then raised to genus by Polhemus and Polhemus (2008b). As for *Heleocoris*, all of its South American species were transferred to *Ctenipocoris* Montandon (Polhemus 1987; Polhemus and Polhemus 2008b), a genus that also occurs on Afrotropical and Oriental regions (Štys and Jansson 1988). The only exception is *Heleocoris plaumanni* De Carlo, which was not transferred to *Ctenipocoris* because Polhemus and Polhemus (2008b) could not examine material of this species. Moreover, they argued that according to De Carlo's (1968b) figure of the foreleg, this species is certainly very different from the other Neotropical *Ctenipocoris*. Despite that fact, some authors (e.g., López-Ruf 2004; Herrera 2013) consider this species as part of *Ctenipocoris*, which now has six species in the Neotropics, recorded from Central and South America.

Limnocorinae is currently monotypic, with only the valid genus *Limnocoris* being considered valid. The other three genera described within the subfamily (*Borborocoris* Stål, *Usingerina* La Rivers, and *Sattleriella* De Carlo) were regarded as groups of species with isolated modifications and synonymized (Nieser and López-Ruf 2001). *Limnocoris* is endemic to the New World and has a wide distribution throughout the Americas, occurring from the USA to Argentina. Most of its 73 valid species are South American, whereas only 16 have been recorded from Central America or southern USA (Nieser et al. 2013).

The subfamily Naucorinae is represented in the Neotropics by two genera. *Placomerus* was monotypic, with *P. micans* La Rivers, known from Argentina, Brazil, Bolivia, and Paraguay (Moreira et al. 2011). *Placomerus obscuratus* Sites & Camacho was recently described from Venezuela and Guyana. *Pelocoris*, on its turn, is a New World genus with 13 Neotropical species (Nieser 1969; Davis 1986; Manzano et al. 1995; Polhemus and Sites 1995).

Despite being treated as a subfamily of Naucoridae by Van Doesburg (1984), Potamocoridae was formally raised to family by Cobben (1978). It is monotypic and the sole family of Nepomorpha occurring exclusively on Neotropical region. The genus *Coleopterocoris* Hungerford was recently synonymized with *Potamocoris* by Polhemus and Carrie (2013). *Potamocoris* now includes 11 species from Belize, Costa Rica, Suriname, Brazil, Ecuador, Peru, Paraguay, and northwestern Argentina (Polhemus and Polhemus 1983; Longo et al. 2005; Mazzucconi et al. 2009; Moreira et al. 2011; Herrera and Springer 2012, 2014; Polhemus and Carrie 2013).

Belostomatidae currently comprises three subfamilies: Belostomatinae, Horvathiniinae, and Lethocerinae (Lauck and Menke 1961; Štys and Jansson 1988). Belostomatinae, the most diverse one, is represented in the Neotropics by three genera: *Abedus*, *Belostoma*, and *Weberiella* De Carlo. The first one is subject of continuous questionings, being cited by some authors as a subgenus of *Belostoma* (De Carlo 1968a; Schnack and Estévez 1990). If considered a genus, it includes 11 species from the Neotropical region, distributed along Central America and recently reported for Colombia (Menke 1960; Bogan et al. 2013; Romero and Noriega 2013). *Belostoma* is the most diverse genus, with 77 Neotropical species ranging from Mexico to Argentina (Lauck 1959; Lanzer de Souza 1980, 1992; Estévez and Polhemus 2001, 2007; Estévez and Armúa de Reyes 2003; Ribeiro and Alecrim 2008; Ribeiro and Estévez 2009). Finally, *Weberiella* is exclusively Neotropical, with *Weberiella rhomboides* (Menke), occurring in northern and central-western Brazil and French Guiana (Estévez and Ribeiro 2011).

Horvathiniinae, with only one genus, had all species synonymized with *Horvathinia pelocoroides* Montandon by Schnack and Estévez (2005), except for *H. lenti* De Carlo. The genus has records from Argentina, Bolivia, Brazil, and Paraguay (Estévez and Ribeiro 2011). Lethocerinae was recently reviewed by Perez-Goodwyn (2006). *Benacus* Stål, originally described as genus, was downgraded to a subgenus of *Lethocerus* by Lauck and Menke (1961). However, many authors continued treating it as a distinct genus (Brooks and Kelton 1967; Usinger 1968; Popov 1971), and it was formally reerected by Perez-Goodwyn (2006). It only includes one species, *B. griseus* (Say), that occurs on Mexico, Honduras, Cuba, Jamaica, and Colombia (Estévez and Ribeiro 2011). *Lethocerus* is a cosmopolitan genus, with 15 Neotropical species (Perez-Goodwyn 2006).

Nepidae is formed by two subfamilies: Nepinae and Ranatrinae. Nepinae is represented by two genera in the Neotropical region. *Curicta* is a New World genus with 22 species recorded from the Neotropical region (Keffer 1996). *Telmatotrephes* Stål occurs in the Neotropical and Oriental regions, with four species on the former (Lansbury 1972). Ranatrinae is represented by the cosmopolitan genus *Ranatra*, with 63 Neotropical species (Kuitert 1949; De Carlo 1972; Lansbury 1974; Roback and Nieser 1974; Nieser 1975; Menke 1979; Bachmann 1998; Nieser and Burmeister 1998; Naranjo et al. 2010; Moreira et al. 2011).

Notonectidae is represented in the Neotropics by its two subfamilies, with about a hundred species (Polhemus and Polhemus 2008a). Notonectinae includes the tribes Notonectini and Nychiini. *Enitharoides* Brooks (Notonectini), exclusively Neotropical, was created as a subgenus of *Enithares* Spinola to include its Neotropical species, being posteriorly raised to genus by Štys and Říha (1975). *Notonecta* (Notonectini) is the only cosmopolitan genus of Notonectidae, with about 40 Neotropical species (Hungerford 1933; Mazzucconi et al. 2009; Barbosa and Nessimian 2013). The genus was reviewed by Hungerford (1933), and few species were described since then. Truxal (1949) reviewed the New World genus *Martarega* (Nychiini), currently with 22 species. The subfamily Anisopinae is represented in the New World by the genus *Buenoa*, with 56 Neotropical species (Truxal 1949, 1953; Roback and Nieser 1974; Barbosa et al. 2010a, b, 2015; Padilla-Gil 2010, 2012; Moreira et al. 2011; Barbosa and Rodrigues 2013).

Gelastocoridae has two subfamilies. Gelastocorinae is monotypic, with *Montandonius* Melin synonymized with *Gelastocoris* by Todd (1955), although Nieser (1975) still considered them distinct genera. *Gelastocoris* occurs in New World, with 23 Neotropical species (Todd 1955; Estévez and Schnack 1980; Boulard and Jauffret 1984). Nerthrinae is also monotypic, with the cosmopolitan genus *Nerthra* and about 30 Neotropical species (Todd 1955; Schnack and Estévez 1979; Polhemus and Lindskog 1994).

Besides Potamocoridae, Ochteridae, Helotrephidae, and Pleidae are the families with the lowest number of Neotropical records, with less than 20 recorded species for each one. Ochteridae is represented by the genera *Ocyochterus* Drake & Gómez-Menor, with two species, *O. victor* (Bolívar) endemic to Ecuador (Cianferoni 2012) and *O. irmae* Polhemus & Polhemus described from Ecuador and Colombia (Polhemus and Polhemus 2014), and the cosmopolitan *Ochterus*, whose New World representatives were reviewed by Schell (1943a, b) and belong to the subgenus *Neochterus* Mahner (Cianferoni 2012), including 18 species (Drake 1952; Kormilev and De Carlo 1952; Nieser 1975; Polhemus and Polhemus 1976; Nieser and Chen 1992; Cordeiro et al. 2014).

The subfamily Neotrephinae (Helotrephidae) is endemic to the Neotropical region and includes two genera. *Paratrephes* is monotypic, with *P. hintoni* China occurring mainly in northern South America, with one record from southeastern Brazil (Rodrigues et al. 2012a). *Neotrephes* China has nine species (Papáček and Zettel 2004) recorded mainly from southeastern and southern Brazil and from northeastern Argentina (Bachmann 1998; Nieser and Chen 2002; Rodrigues et al. 2012a). Pleidae is a small family with most diversity occurring in the Neotropical region. *Neoplea* Esaki & China is a New World genus, with 12 Neotropical species, and *Paraplea* Esaki & China is cosmopolitan, with one species recorded from the Americas, *P. puella* (Barber). Finally, *Heteroplea* is monotypic, with *H. stictosoma* Cook being restricted to Venezuela (Cook 2011).

7.5 Economic Importance

Nepomorphans, although having no direct economic importance to humans, play a significant role on the environment. They are generally abundant in lentic or lotic, preserved or not, aquatic ecosystems developing a key part in the energy cycle of these environments. Individuals of all families are predators (not exclusively in the case of Corixidae) and certainly have great influence upon other organisms (Stout 1981, 1982; Sites and Willig 1991; Papáček 2000). Species of all families can be considered as potential biological control agents of invertebrates of medical importance, such as the larvae of mosquitoes and aquatic gastropods.

Immatures of some families of hematophagous dipterans that transmit diseases to humans have been reared in the laboratory to serve as food to species of aquatic heteropterans. For example, Sites and Nichols (1990) used larvae of Simuliidae to feed *Ambrysus lunatus*; Consoli et al. (1989) reared *Belostoma micantulum* (Stål) (Belostomatidae) using larvae and pupae of *Aedes fluviatilis* (Lutz) (Culicidae) as food; and Pereira et al. (1991) also reared *B. micantulum*, but using larvae of *Culex quinquefasciatus* Say (Culicidae). Many are the studies of predation on invertebrates of medical importance by nepomorphans under laboratory conditions, especially involving Belostomatidae, Naucoridae, and Nepidae, which can be seen in the list provided by Jenkins (1964).

In contrast, some species of Notonectidae can cause damage to cultures of several fish species (Torre Bueno 1905a). A single specimen of *Notonecta* can feed on more than 2,500 fish larvae/day. Corixidae also cause damage to various pisciculture tanks, by feeding on fish eggs and larvae (Papáček 2000). On the other hand, eggs, immatures, and adults of corixids and notonectids are used as human food in parts of Mexico (Hungerford 1948; Hutchinson 1993). Additionally, representatives of corixids can be dehydrated and used as food for ornamental fish and domestic birds (Hutchinson 1993).

The use of this group of insects as bioindicators of water quality is still incipient in the Neotropics. Additionally, considering that most of the published works have taken into account only generic identifications and that many nepomorphan genera occur in preserved, degraded, or modified environments, such studies might turn out uninformative. Despite the difficulty to identify specimens at specific level, this should be the path to be followed in order to reliably use aquatic Heteroptera in evaluations of water quality.

7.6 Key for Families and Genera of the Neotropical Nepomorpha

Modified from Hungerford (1948), De Carlo (1963), Nieser (1975), Nieser and Chen (2006), Cook (2011), Estévez and Ribeiro (2011), Cianferoni (2012), and Polhemus and Polhemus (2014).

1a. Ocelli present (if obsolete or absent, then head transverse and eyes pedunculate or subpedunculate) (Fig. 7.2A); species usually found on sandy shores of water bodies 2
1b. Ocelli absent; aquatic species 5
2a (1a). Antennae shorter than the head, not observed dorsally and hidden in a groove beneath head (Fig. 7.2C); rostrum short, reaching anterior coxae; scutellum tumid; fore femora enlarged with a groove for reception of the tibia (Fig. 7.2B) .... 3 (Gelastocoridae)
2b. Antennae a little longer than the head (Fig. 7.3A), observed dorsally; rostrum long, reaching at least posterior coxae (Fig. 7.3C); scutellum flat; all legs cursorial (anterior femora not enlarged) .... 4 (Ochteridae)
3a (2a). Pronotum rectangular, generally with anterior and posterior margin with the same width (Fig. 7.1A); fore femur subtriangular, very broad at base, about as wide as long; foretarsus fused with the tibia, bearing one claw (Fig. 7.1B); color commonly uniform .... *Nerthra* Say
3b. Pronotum with the anterior margin distinctly shorter than posterior margin (Fig. 7.2C); fore femur moderately incrassate, about twice as long as wide; foretarsus not fused with the tibia, bearing two claws (Fig. 7.2B); color heterogeneous, pruinose .... *Gelastocoris* Kirkaldy
4a (2b). Lateral expansions of pronotum as wide as an eye; vertex of head without channels parallel to inner margins of eyes; body length 6.75–8.20 mm .... *Ocyochterus* Drake & Gómez-Menor
4b. Lateral expansions of the pronotum distinctly narrower than an eye (Fig. 7.3B); vertex of head with channels parallel between the eyes; body length 3.30–6.30 mm.. .... *Ochterus* Latreille
5a (1b). Forelegs raptorial (femur enlarged, with a groove for reception of the tibia) (Figs. 7.4 and 7.7A) 6
5b. Forelegs not raptorial 26
6a (5a). Hemelytral membrane with venation; respiratory siphon present at the end of the body (Figs. 7.6B and 7.8D) 7
6b. Hemelytral membrane without venation; respiratory siphon absent from the end of the body (Figs. 7.11B and 7.12A) 15
7a (6a). All tarsi 1-segmented; hind legs without fringe of developed bristles; siphon long and not retractile (Fig. 7.6B) .... 8 (Nepidae)
7b. Tarsi 2- or 3-segmented, rarely foretarsus 1-segmented; hind legs with a fringe of developed bristles; siphon short and retractile (Fig. 7.8D).... 10 (Belostomatidae)
8a (7a). Body cylindrical; anterolateral corners of pronotum not wider than head (including eyes) (Fig. 7.4) .... *Ranatra* F.

8b. Body flattened; anterolateral corners of pronotum wider than head (including eyes) 9
9a (8b). Body and pronotum elongate; pronotum longer than wide, ratio length/larger width equal to 5–6 (Fig. 7.5) .... *Curicta* Stål
9b. Body oval; pronotum quadrate, ratio length/larger width equal to or smaller than 3 (Fig. 7.6A, B) .... *Telmatotrephes* Stål
10a (7b). Segment 1 of rostrum short, thicker than long and obviously shorter than segment 2; abdominal sternites 5 and 6 subdivided laterally by suture-like fold (Fig. 7.8A); spiracles located on or adjacent to mesal margins of connexivum 11
10b. Segment 1 of rostrum much longer than greatest thickness, subequal in length to segment 2; abdominal sternites not subdivided by a suture; spiracles located near center of connexivum 13
11a (10a). Body elongate (Fig. 7.8D); foretarsus 3-segmented (often appearing 2-segmented externally) and bearing one long claw (Fig. 7.8B); tibia and tarsus of hind leg thinly compressed, much more dilated than middle tibia and tarsus (Fig. 7.8C); genital operculum of females with spines and acutely rounded at apex 12
11b. Body oval (Fig. 7.7B); foretarsus 2-segmented (often appearing 1-segmented), bearing two very short, vestigial claws (Fig. 7.7A); tibia and tarsus of middle and hind legs similar, narrow, flattened, but not broadly dilated; genital operculum of females fringed with hairs, rounded at apex .... *Horvathinia* Montandon
12a (11a). Inner pad of setae of fore femur with two symmetrical furrows; pads of hind femur with just one; external borders of ventral laterotergites 2 and 3 narrowed, almost straight .... *Lethocerus* Mayr
12b. Inner pad of setae of fore femur without any trace of furrow; pads of hind femur with just a deep cleft, but not divided; external borders of ventral laterotergites 2 and 3 not narrowed, arcuate, following the border line of abdomen .... *Benacus* (Say)
13a (10b). Membrane of hemelytra well developed, with its greatest width larger than that of clavus; most of its cells equal in length, in the form of long, narrow rectangles (Fig. 7.9) .... *Belostoma* Latreille
13b. Membrane of hemelytron distinctly reduced, with its greatest width equal to or smaller than that of clavus; most of its cells usually small or feebly defined, or short and rectangular, of various lengths (Fig. 7.10) 14
14a (13b). Lateral margins of abdomen not smooth, interrupted at borders between segments (scalelike laterotergites) (Fig. 7.10) .... *Weberiella* De Carlo
14b. Lateral margins of abdomen smooth, not scalelike .... *Abedus* Stål
15a (6b). Antenna long, extending beyond the lateral margin of the head; fore femur can be expanded or not; male genitalia always symmetric; small bugs, body length less than 4 mm (Fig. 7.11A, B) .... *Potamocoris* Hungerford (Potamocoridae)
15b. Antenna short, not exceeding the lateral margin of the head in dorsal view; fore femur always expanded (Fig. 7.13B, 16A, 18C); male genitalia usually asymmetric; body length equal to or exceeding 4 mm .... 16 (Naucoridae)

16a (15b). Pro-pleurae covering the posterior part of the prosternum (Fig. 7.13B) 17
16b. Prosternum exposed posteriorly (Fig. 7.18B) 19
17a (16a). Ventral surface of abdomen pubescent almost or quite to lateral margins (Fig. 7.12B) .... *Ambrysus* Stål (Fig. 7.12A) (*Ambrysus montandoni* La Rivers has the lateral margins of the abdominal sternites naked, without pubescence. This species shares with those of *Carvalhoiella*, in addition to this feature, the anterior margin of pronotum slightly concave behind interocular space; absence of the male accessory genitalic process of tergite VI; and the shape of the divergent pseudoparameres of tergite VIII)
17b. Ventral surface margined by a glabrous area, especially on the abdomen (Fig. 7.13B, C) 18
18a (17b). Anterior margin of pronotum deeply emarginate behind interocular space (as observed in *Ambrysus* sp. Fig. 7.12A) .... *Cataractocoris* Usinger
18b. Anterior margin of pronotum straight or somewhat concave behind interocular space (Fig. 7.13A) .... *Carvalhoiella* De Carlo
19a (16b). Foreleg pretarsus with 2 claws, often minute and closely appressed, sometimes resembling a single structure (Fig. 7.14C); middle and hind femora with two longitudinal rows of conspicuous bristles or spinelike setae on ventral face, in addition to 2 usual rows on posterior face (Fig. 7.14B) 20
19b. Foreleg pretarsus with or without 1 min claw (Fig. 7.18C); middle and hind femora without additional rows of bristles or spiniform setae on ventral surfaces, or with just an indication of the upper row, or with some scattered additional setae (Fig. 7.15B) 22
20a (19a). Posterior angles of pronotum pointing obtusely but distinctly backward; hemelytra separately pointed .... *Decarloa* La Rivers
20b. Posterior angles of pronotum not pointing backward; hemelytra with fully developed membranes and covering the abdomen 21
21a (20b). Female with two tarsal segments on foreleg; male phallotheca symmetrical .... *Ctenipocoris* Montandon (Fig. 7.14A, B)
21b. Female with one tarsal segment on foreleg; male phallotheca asymmetrical .... *Interocoris* La Rivers
22a (19b). Body broadly oval to subcircular, flattened (Fig. 7.15A, B); meso- and metasterna with prominent, broad, laterally expanded median longitudinal carinae bearing foveae or otherwise excavate (Fig. 7.15C, D); inner margins of eyes (dorsal view) distinctly diverging anteriorly (Fig. 7.15A) .... *Limnocoris* Stål
22b. Body more elongate, robust, dorsum moderately convex (Fig. 7.18A, B); meso- and metasternal longitudinal carinae inconspicuous, thin, platelike, or absent; inner margins of the eyes in dorsal view usually distinctly converging anteriorly, infrequently parallel sided to slightly diverging (Fig. 7.18A) 23
23a (22b). Anterior margin of pronotum deeply emarginated behind interocular space (Fig. 7.16A) 24
23b. Anterior margin of pronotum straight behind interocular space (Fig. 7.17A) 25

24a (23a). Lateral margins of pronotum crenulate (Fig. 7.16A); abdominal tergites distinct (Fig. 7.16B); male abdominal tergite 7 asymmetrical .... *Cryphocricos* Signoret

24b. Lateral margins of pronotum smooth; abdominal tergites fused; male abdominal tergite 7 symmetrical .... *Procryphocricos* Polhemus

25a (23b). Mesofemur greatly flattened dorsoventrally, with its dorsoposterior margins extended into thickened, flap-like projections and equipped with fine brush (Fig. 7.17B); tarsi of meso- and metalegs longer than their corresponding tibiae .... *Placomerus* La Rivers (Fig. 7.17A)

25b. Mesofemur not flattened or margined as described above (Fig. 7.18B); tarsi of meso- and metalegs shorter than their corresponding tibiae .... *Pelocoris* Stål (Fig. 7.18A)

26a (5b). Body flattened (Fig. 7.19A); rostrum apparently not segmented, with transverse striations (Fig. 7.20B); foretarsi modified, spatulate, with a fringe of long bristles and without distinct claws (Fig. 7.22) .... 27 (Corixidae)

26b. Body convex (Fig. 7.36); rostrum distinctly segmented, without transverse striations (Fig. 7.33A); foretarsi not modified, with distinct claws (Fig. 7.34C) 42

27a (26a). Pronotum not covering the scutellum (Fig. 7.19A) 28

27b. Pronotum almost entirely covering the scutellum (Fig. 7.21B) 30

28a (27a). Mesosternum with a distinct carina (Fig. 7.19C) .... *Synaptogobia* Nieser & Chen (Fig. 7.19A, B)

28b. Mesosternum without carina (Fig. 7.20B) 29

29a (28b). Antennae three-segmented; metaxyphus short (Fig. 7.20B) .... *Tenagobia* Bergroth (Fig. 7.20A, B)

29b. Antennae one-segmented; metaxyphus long .... *Monogobia* Nieser & Chen

30 (27b). Infraocular portion of genae very broad, the lower margin of the eye concave, the hypo-ocular suture arising near the subacute production of the inferior angle of the eye (Fig. 7.21A) .... *Heterocorixa* White (Fig. 7.21B)

30b. Infraocular portion of gena not broad, or if broad, the hypo-ocular suture when present arising about midway along the ventral margin of the eye (Fig. 7.31) 31

31a (30b). With rather thick, well-developed apical claw on foretarsus of both sexes; pala of both sexes narrowly digitiform; males with row of stout short pegs along the upper edge of the palm (Fig. 7.22); usually with a mat of hairlike setae on frons in both sexes 32

31b. With apical claw on the foretarsus spinelike, usually resembling the spines along the lower margin of the palm; pala not narrowly digitiform, of various shapes, male with row of pegs usually well above the fringe of setae along the upper edge of the palm (Fig. 7.25); setae on frons not mat-like, often absent 33

32a (31a). Male abdomen sinistral, strigil absent; female face slightly concave; female abdomen slightly asymmetrical .... *Neocorixa* Hungerford

32b. Male abdomen dextral, strigil present; female face not concave and abdomen normal .... *Graptocorixa* Hungerford
33a (31b). Male foretibia not produced over pala (Fig. 7.23B 34
33b. Male foretibia distinctly produced over pala (Fig. 7.30 38
34a (33a). Vertex of males conically produced anterad (acuminate), frons deeply concave (Fig. 7.23A); pala of male forelegs nearly divided by deep transverse groove on dorsal surface (Fig. 7.23B); costal margin of elytra in females with an elongate-ellipsoid, polished thickening at basal third of length .... *Ramphocorixa* Abbott
34b. Vertex of males not produced, if produced, rounded with frons flat (Fig. 7.24A); pala of male forelegs without deep transverse groove on dorsal side; costal margin of females not thickened at basal third (Fig. 7.24B) 35
35a (34b). Inner posterior angle of eye acutely produced (Fig. 7.24A); pronotum and hemelytra smooth; anterior femur of male greatly inflated (Fig. 7.24B) .... *Krizousacorixa* Hungerford
35b. Inner posterior angle of eye not acutely produced; pronotum and hemelytra rastrated; anterior femur of male not inflated (Fig. 7.25) 36
36a (35b). Foretibiae with apical comb of spines (Fig. 7.25); males with dextral asymmetry, without strigil .... *Morphocorixa* Jaczewski
36b. Foretibiae without comb of spines (Fig. 7.26); males with sinistral asymmetry and without strigil, or with dextral asymmetry and with strigil 37
37a (36b). Male pala with one row of pegs (Fig. 7.26); metaxyphus short, triangular .... *Sigara* F.
37b. Male pala with two rows of pegs (Fig. 7.27); metaxyphus long, apically rounded .... *Neosigara* Lundblad
38a (33b).Inner posterior angle of eye sharply right angulate, occasionally slightly produced (Fig. 7.28) .... *Trichocorixella* Jaczewski
38b. Inner posterior angle of eye obtuse angulate 39
39a (38b). Ventral surface of hind femur pubescent only on basal third of the front margin 40
39b. Ventral surface of hind femur pubescent at least for one-half of the front margin 41
40a (39a).Apices of clavi not exceeding a line drawn through nodal furrows of the hemelytra in female and slightly exceeding in males; tibiae strongly produced apically over the pala (Fig. 7.29); males with sinistral asymmetry, strigil on left .... *Trichocorixa* Kirkaldy
40b. Clavi distinctly exceeding a line drawn through nodal furrows of the hemelytra in both sexes; tibiae slightly produced over the pala (Fig. 7.30); males with dextral asymmetry, strigil on right .... *Ectemnostega* Enderlein
41a (39b). Males with dextral asymmetry, without strigil .... *Centrocorisa* Lundblad
41b. Males with dextral asymmetry, with strigil .... *Corisella* Lundblad

42a (26b). Body elongate, without punctuations on the dorsal surface (Fig. 7.34A); hind legs longer than fore- and middle legs; hind tibiae and tarsi with distinct fringe of setae; hind tarsal claws apparently absent .... 43 (Notonectidae)

42b. Body globose, with dorsal surface punctuate (Fig. 7.36); legs almost symmetrical; setae of hind tibiae and tarsi do not form a fringe; hind tarsal claws normal 46

43a (42a). Labrum rounded (Fig. 7.32A); hemelytral commissure with definite hairlined pit at the anterior end (Fig. 7.32B); males with stridulatory comb at foretibia (Fig. 7.32C); spine at caudosinistral margin of seventh abdominal tergite (Fig. 7.32D) .... *Buenoa* Kirkaldy

43b. Labro triangular (Fig. 7.33A); hemelytral commissure without definite hairlined pit at the anterior end (Fig. 7.33B); males without stridulatory comb at foretibia and without spine at caudosinistral tergite 44

44a (43a). Eyes connected dorsally (Fig. 7.33B); hemelytral process present at the end of the clavus (Fig. 7.33C); mesofemora without anteapical spur .... *Martarega* White

44b. Eyes separated (Fig. 7.34A); hemelytral process absent at the end of the clavus; mesofemora with anteapical spur (Fig. 7.34C) 45

45a (44b). Lateral margins of pronotum foveate (Fig. 7.34B); genital capsule asymmetrical .... *Enitharoides* Brooks

45b. Lateral margins of pronotum not foveate (Fig. 7.35); genital capsule symmetrical .... *Notonecta* L.

46a (42b). Head fused with the pronotum (Fig. 7.36); antennae one-segmented in brachypterous forms and two-segmented in macropterous forms; rostrum four-segmented .... 47 (Helotrephidae)

46b. Head not fused with the pronotum (Fig. 7.38); antennae and rostrum three-segmented .... 48 (Pleidae)

47a (46a). Rostrum long, extending to the middle of the hind coxae, fourth segment nearly four times as long as the third; cephalonotal suture virtually straight behind the eyes (Fig. 7.36); lateral margins of pronotum feebly sinuate, punctuation weak .... *Paratrephes* China

47b. Rostrum short, not reaching the middle coxae, third and fourth segments with subequal length (Fig. 7.37); cephalonotal suture in brachypterous form distinctly curved backward behind the eyes; lateral margins of pronotum strongly sinuate, punctuation strong .... *Neotrephes* China

48a (46b). Head with distinct callus on vertex (Fig. 7.38); tarsal formula 3-3-3 .... *Heteroplea* Cook

48b. Head without callus on vertex; tarsal formula 3-2-3 or 3-2-2 49

49a. Tarsal formula 3-2-3 (Fig. 7.39); median carinae on segments 2–5 of venter .... *Neoplea* Esaki & China

49b. Tarsal formula 2-2-3; median carinae on segments 2–6 of venter ....
.... *Paraplea* Esaki & China

Figs. 7.1–7.3 1A–B: *Nerthra* sp. – A. *habitus* dorsal; B. *habitus* ventral. 2A–C: *Gelastocoris* sp. – A. head; B. frontal view of head and forelegs; C. *habitus* dorsal. 3A–C: *Ochterus* spp. A. head and pronotum; B. *habitus* dorsal of *O. aeneifrons surinamensis* Nieser; C. *habitus* lateral. *AT* antenna, *OC* ocelli (Photos by HDD Rodrigues, 2A by JFBarbosa)

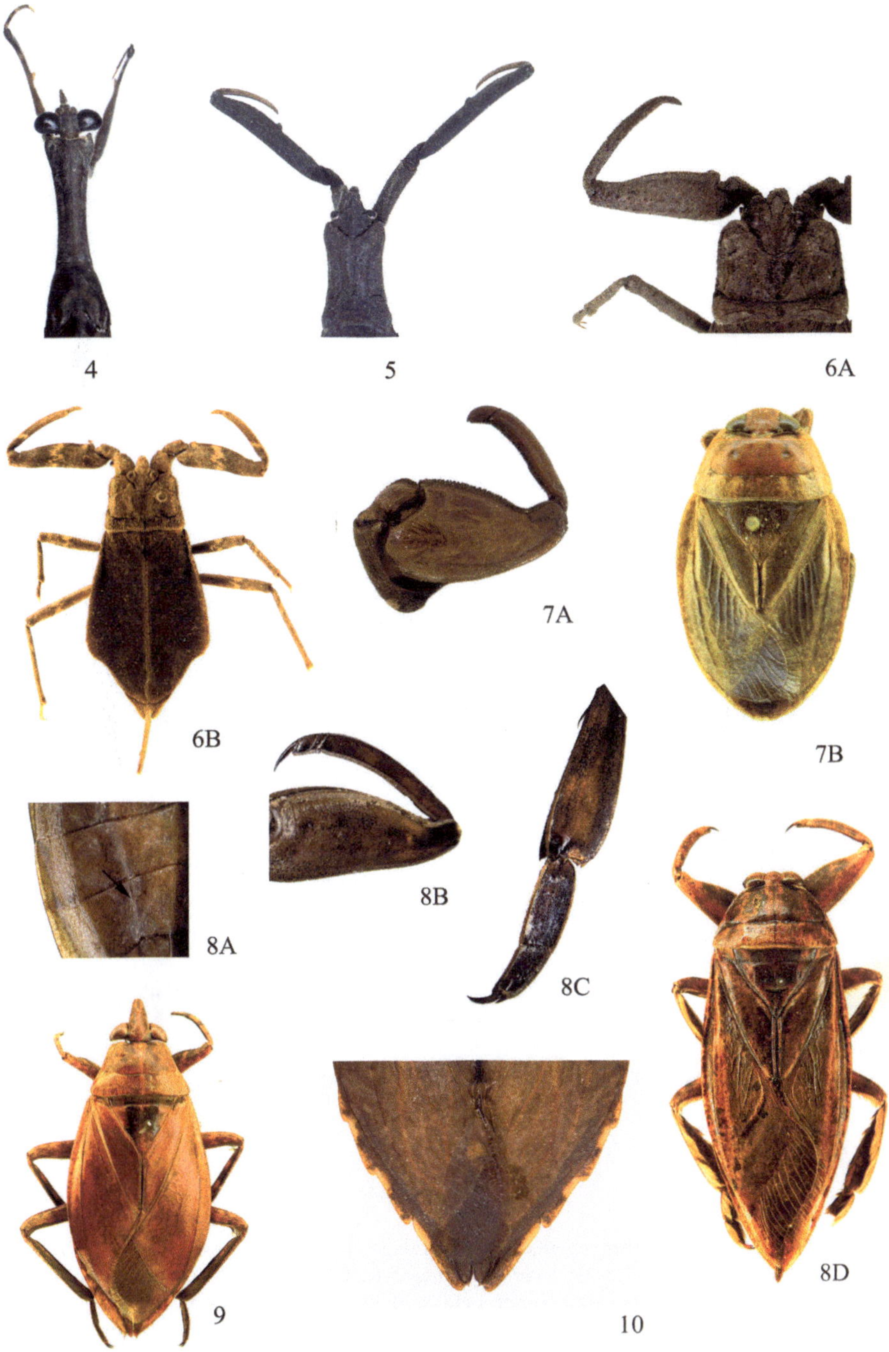

Figs. 7.4–7.10 4–5: head and pronotum – 4. *Ranatra* sp.; 5. *Curicta* sp. 6A–B: *Telmatotrephes* spp. – A. head and pronotum of *Telmatotrephes carvalhoi* De Carlo; B. *habitus* dorsal of *Telmatotrephes grandicollis* Kuitert. 7A–B: *Horvathinia pelocoroides* Montandon – A. foreleg; B. *habitus* dorsal. 8A–D: *Lethocerus* sp. – A. suture-like fold on the abdomen; B. foreleg; C. hind tibia and tarsi; D. *habitus* dorsal. 9. *Habitus* dorsal of *Belostoma* sp. 10. Abdomen with scalelike laterotergites of *Weberiella rhomboides* (Menke) (Photos by HDD Rodrigues)

Figs. 7.11–7.14 11A–B: *Potamocoris nelsoni* (Longo, Ribeiro, & Nessimian) – A. *habitus* ventral; B. *habitus* dorsal. 12A–B: *Ambrysus usingeri* La Rivers – A. *habitus* dorsal; B. *habitus* ventral. 13A–C: *Carvalhoiella stysi* Nieser, Pelli & Melo – A. *habitus* dorsal; B. *habitus* ventral; C. ventral view of abdomen. 14A–C: *Ctenipocoris* sp. – A. *habitus* dorsal; B. *habitus* ventral; C. foreleg with detail of tarsi with two tarsomeres and two claws (Photos by HDD Rodrigues)

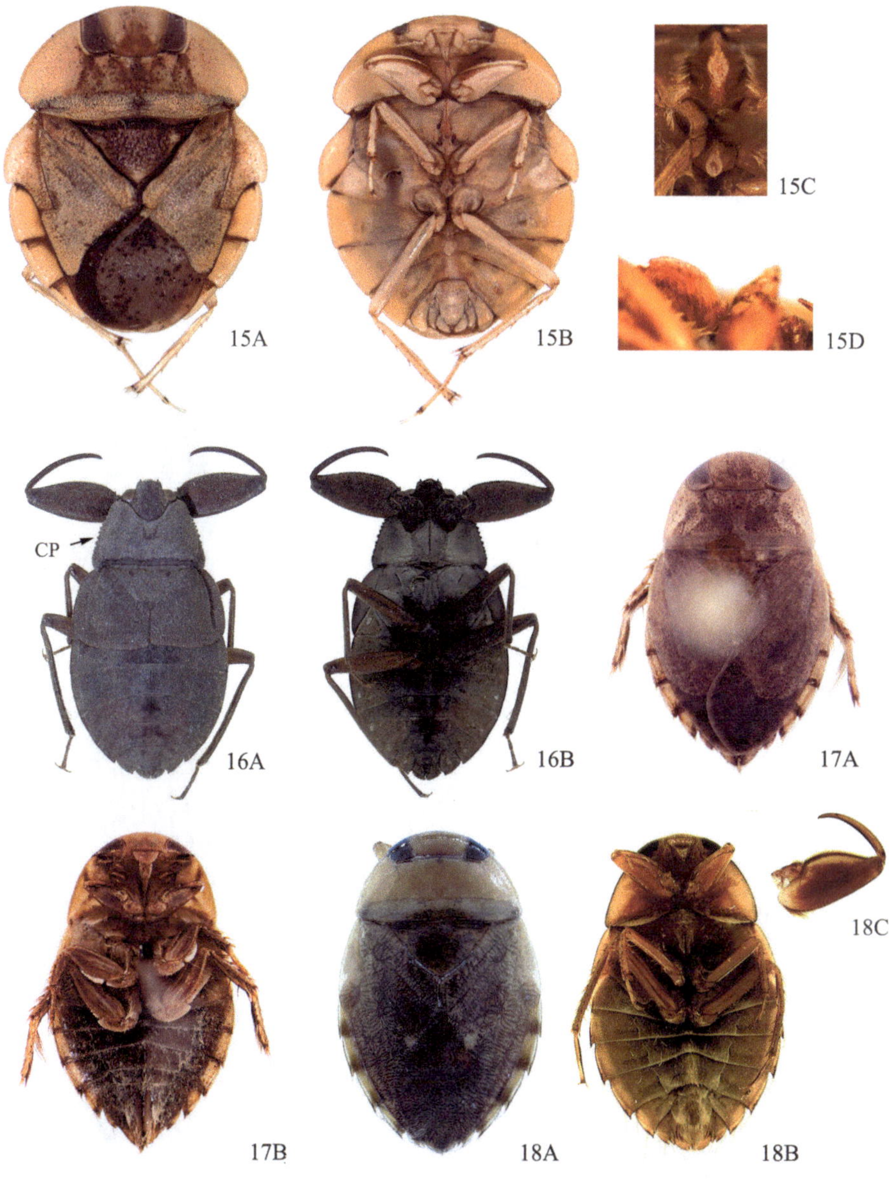

Figs. 7.15–7.18 15A–B: *Limnocoris submontandoni* La Rivers – A. *habitus* dorsal; B. *habitus* ventral. 15C–D: *Limnocoris bruchi* De Carlo – C. ventral view of longitudinal carinae on meso- and metasternum; D. lateral view of longitudinal carina on mesosternum. 16A–B: *Cryphocricos* sp. – A. *habitus* dorsal; B. *habitus* ventral. 17A–B: *Placomerus micans* La Rivers. A. *Habitus* dorsal; B. *habitus* ventral. 18A–C: *Pelocoris bipunctulus* (Herrich-Schäffer). A. *Habitus* dorsal; B. *habitus* ventral; C. foreleg. *CP* crenulate pronotum (Photos by HDD Rodrigues)

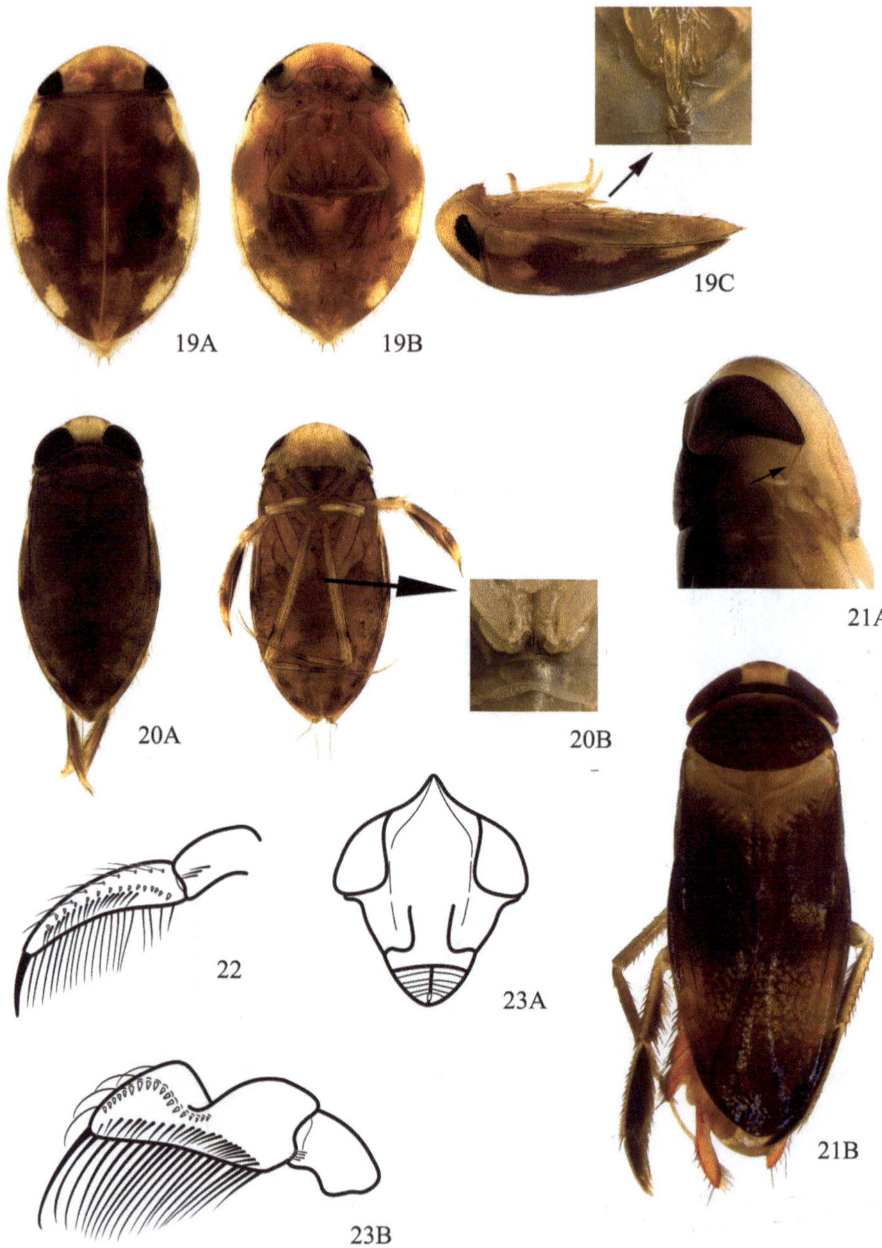

Figs. 7.19–7.23 19A–C: *Synaptogobia* sp. – A. *habitus* dorsal; B. *habitus* ventral; C. *habitus* lateral, with detail of carina on mesosternum. 20A–B: *Tenagobia* sp. – A. *habitus* dorsal; B. metaxyphus. 21A–B: *Heterocorixa* sp. – A. lateral view of the gena with hypo-ocular suture (*arrow*); B. *habitus* dorsal. 22. Pala of *Graptocorixa* sp. 23A–B: *Ramphocorixa* sp. – A. ventral view of the head; B. pala (Photos 19 to 21A by HDD Rodrigues, 21B–23B by JF Barbosa, illustrations modified from Hungerford 1948 and Bachmann 1981)

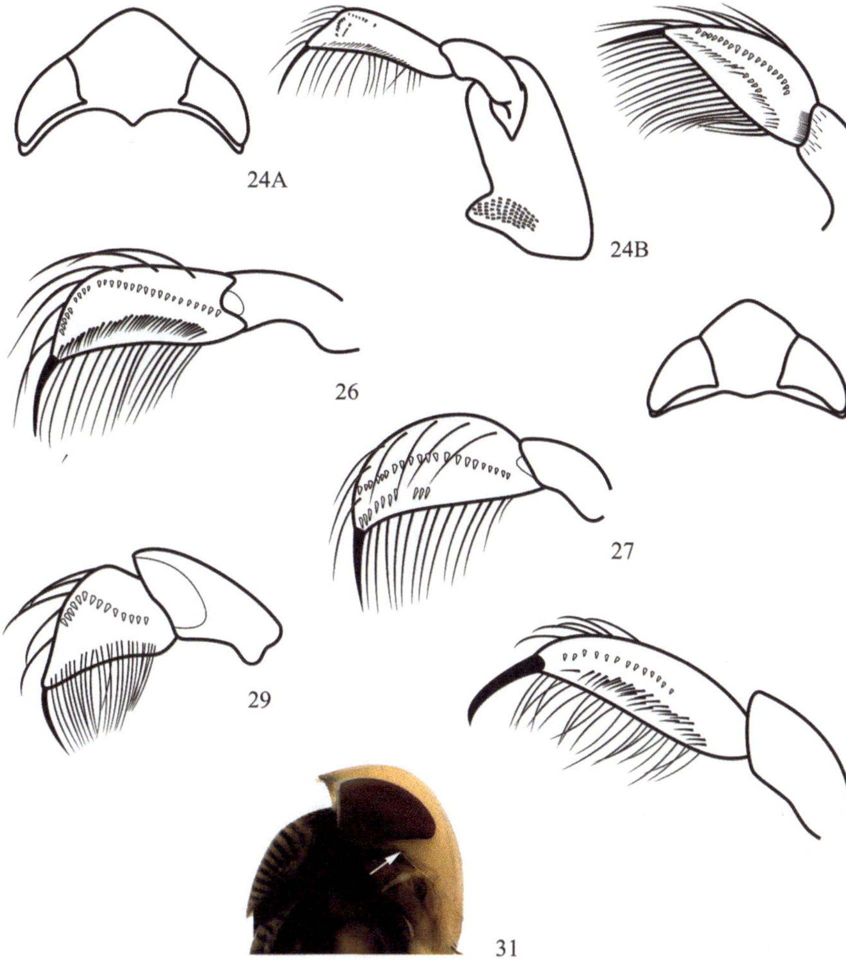

Figs. 7.24–7.31 24A–B: *Krizousacorixa* sp. – A. dorsal view of the head; B. foreleg. 25–27, 29, 30: pala – 25. *Morphocorixa* sp.; 26. *Sigara* sp.; 27. *Neosigara* sp.; 29. *Trichocorixa* sp.; 30. *Ectemnostega* sp. 28: dorsal view of the head of *Trichocorixella* sp. (Illustrations modified from Hungerford 1948 and Bachmann 1981). 31: lateral view of the gena of *Centrocorisa* sp. with hypoocular suture (*arrow*) (Illustrations modified from Hungerford 1948 and Bachmann 1981 by JF Barbosa; Photo 31 by HDD Rodrigues)

Figs. 7.32–7.35 32A–D: *Buenoa* sp. – A. ventral view of the head; B. hairlined pit at anterior end of hemelytra; C. foreleg; D. lateral view of abdomen. 33A–C: *Martarega* sp. – A. ventral view of the head; B. *habitus* dorsal; C. hemelytra. 34A–C: *Enitharoides* sp. – A. *habitus* dorsal; B. lateral view of head and pronotum (*arrow* indicates the pronotum foveate); C. middle leg with detail of anteapical spur. 35: Dorsal view of the pronotum of *Notonecta* sp. *LB* labrum, *SA* stridulatory area, *SP* spine, *HP* hemelytral process (Photos 32A–C by JF Barbosa, 32D–35 by HDD Rodrigues)

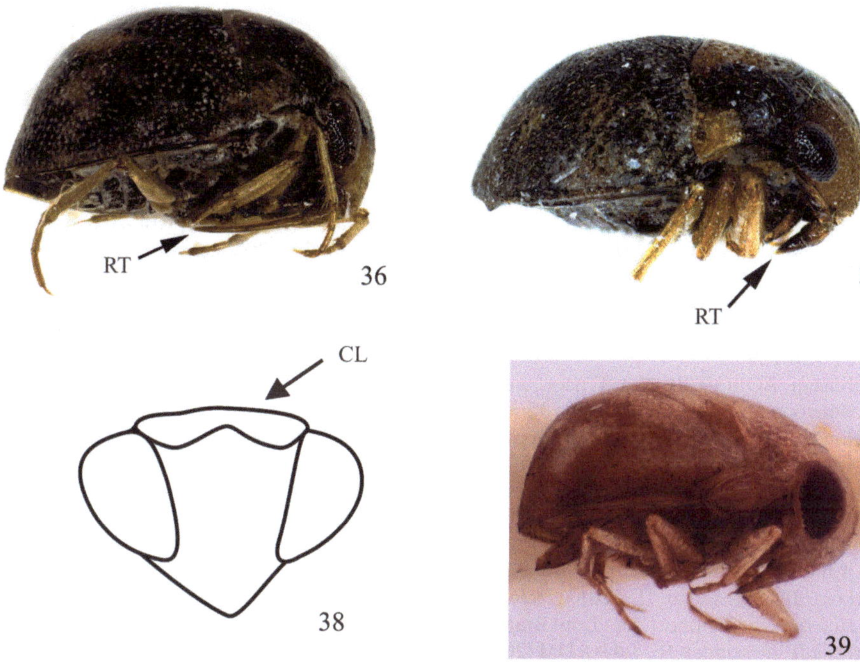

Figs. 7.36–7.39 36, 37, 39: *Habitus* lateral – 36. *Paratrephes hintoni* China (Modified from Rodrigues et al. 2012a); 37. *Neotrephes lanemeloi* Nieser & Chen; 39. *Neoplea* sp. 38: Frontal view of the head of *Heteroplea* sp. (Modified from Cook 2011). *CL* callus, *RT* rostrum (Illustration modified from Cook 2011 by JF Barbosa; Photos by HDD Rodrigues)

7.7 Concluding Remarks

Many Neotropical countries lack records or surveys concerning the infraorder Nepomorpha. Most taxonomic knowledge about the group is very restricted, mainly to Costa Rica, Cuba, Colombia, Venezuela, Suriname, Ecuador, Brazil, and Argentina. Polhemus and Polhemus (2008a) estimated as 107 the number of undescribed species of Nepomorpha in the Neotropical region. This number is probably higher when considering large holes on surveys. If the taxonomy is not well developed, knowledge about life histories and behavior, for example, become huge gaps for the most part of the region. Besides that, phylogenetic relationships of the subfamilies and genera of Nepomorpha are almost unknown. Despite all this lack of knowledge about Neotropical fauna, the number of researchers, or researchers in training process with this group, is increasing, which makes possible that part of this knowledge gap can be filled or enhanced in the coming years.

References

Aiken RB (1985) Sound production by aquatic insects. Biol Rev 65:163–211

Andersen NM (1979) Phylogenetic inference as applied to the study of evolutionary diversification of semiaquatic bugs (Hemiptera: Gerromorpha). Syst Zool 28(4):554–578

Andersen NM (1982) The semiaquatic bugs (Hemiptera, Gerromorpha). Phylogeny, adaptations, biogeography and classification. Entomonograph 3:1–455

Andersen NM (1995) Phylogeny and classification of aquatic bugs (Heteroptera, Nepomorpha). An essay review of Mahner's 'Systema Cryptoceratum Phylogeneticum'. Ent Scand 26:159–166

Bachmann AO (1961) Notas sobre Corixidae (Hemiptera) (2ª série). Neotropica 7:19–24

Bachmann AO (1968) Las Pleidae de la Republica Argentina (Hemiptera). Rev Soc Entomol Arg 30:121–129

Bachmann AO (1979) Notas para una monografia de las Corixidae Argentinas (Insecta, Heteroptera). Acta Zool Lilloana 35:305–350

Bachmann AO (1981) Insecta, Hemiptera, Corixidae. Fauna Agua Dulce Rep Arg 35:1–270

Bachmann AO (1998) Heteroptera acuáticos. In: Coscarón S, Morrone JJ (eds) Biodiversidad de Artrópodos Argentinos. Una Perspectiva Biotaxonómica. Ediciones Sur, La Plata, pp 163–180

Bachmann AO, López-Ruf ML (1994) Los Pleoidea de la Argentina (Insecta: Hemiptera). Fauna Agua Dulce Rep Arg 35:1–32

Barbosa JF, Nessimian JL (2013) New species and new records of *Notonecta* (Hemiptera: Heteroptera: Notonectidae) from Brazil. Zoologia 30:692–696

Barbosa JF, Nessimian JL, Takiya DM (2015) Description of two new species of *Martarega* White, 1879 (Heteroptera: Nepomorpha: Notonectidae), one based on the mixed type-series of *M. oriximinaensis* Barbosa, Ribeiro & Ferreira-Keppler, 2010. Zootaxa 3947(3):417–424

Barbosa JF, Nessimian JL, Ferreira-Keppler RL (2010a) Two new species of *Buenoa* Kirkaldy, 1904 (Hemiptera: Heteroptera: Notonectidae) from the States of Amazonas and Roraima, Brazil. Zootaxa 3682:534–540

Barbosa JF, Ribeiro JRI, Nessimian JL (2010b) A new species of *Buenoa* Kirkaldy (Hemiptera, Heteroptera, Notonectidae) from Rio de Janeiro, Brazil. Rev Bras Entomol 54:560–564

Barbosa JF, Rodrigues HDD (2013) A new species of *Martarega* White, 1879, with new distributional records of Notonectidae (Hemiptera: Heteroptera: Nepomorpha) from Brazil. Zootaxa 3682:534–540

Bernardo LP, Ribeiro JRI, Stenert C, Maltchik L (2012) Uma nova espécie de *Sigara* Fabricius (Hemiptera, Heteroptera, Corixidae) e redescrição das espécies do gênero com registro no Estado do Rio Grande do Sul, Brasil. Rev Bras Entomol 56:1–24

Bobb ML (1951) Life history of *Ochterus banksi* Barber (Hemiptera: Ochteridae). Bull Brooklyn Entomol Soc 46:92–100

Bogan MT, Gutierrez-Ruacho OG, Alvarado-Castrado JA, Lytle DA (2013) New records of *Martarega*, *Graptocorixa*, and *Abedus* (Heteroptera: Notonectidae, Corixidae, Belostomatidae) from Northwestern Mexico and Arizona, including the first record of *Graptocorixa emburyi* in the United States. Southwest Nat 58:494–497

Boulard M, Jauffret P (1984) Description d'un *Gelastocoris* nouveau du Brésil et précisions sur les genitalia mâles et femelles de ce genre d'hémiptères (Gelastocoridae). Nouvelle Rev Entomol 1:7–18

Brewer DW, Sites RW (1994) Behavioral Inventory of *Pelocoris femoratus* (Hemiptera: Naucoridae). J Kansas Entomol Soc 67:193–198

Brooks AR, Kelton LA (1967) Aquatic and semiaquatic Heteroptera of Alberta, Saskatchewan, and Manitoba (Hemiptera). Mem Entomol Soc Can 51:1–92

Brown ES (1951) The relation between migration rate and type of habitat in aquatic insects, with special reference to certain species of Corixidae. Proc Zool Soc London 121:539–545

Brown LN, McPherson JE (1994) Life history and laboratory rearing of *Gelastocoris oculatus oculatus* (Fabricius) (Hemiptera: Gelastocoridae) with descriptions of immature stages. Proc Entomol Soc Wash 96:516–526
Brożek J (2013) Deliberations on the external morphology and modification of the labial segments in the Nepomorpha (Heteroptera: Insecta) with notes on the phylogenetic characteristics. Sci World J 790343:1–49
Chen PP, Nieser N, Zettel H (2005) The aquatic and semi-aquatic bugs (Heteroptera: Nepomorpha & Gerromorpha) of Malesia, Fauna Malesiana handbooks, 5. Brill, Leiden/Boston
China WE (1933) A new family of Hemiptera-Heteroptera with notes on the phylogeny of the suborder. Ann Mag Nat Hist 10:180–196
China WE (1936) The first genus and species of Helotrephidae (Hemiptera) from the New World. Ann Mag Nat Hist 10:527–538
China WE (1955) XLIV – A reconsideration of the systematic position of the family Joppeicidae Reuter (Hemiptera–Heteroptera), with notes on the phylogeny of the suborder. Ann Mag Nat Hist 8:353–370
Cianferoni F (2012) The Ochteridae of Ecuador, with new records and an identification key (Hemiptera: Heteroptera: Nepomorpha). Zootaxa 3260:62–68
Cobben RH (1968) Evolutionary trends in Heteroptera. Part I. Eggs, architecture of the shell, gross embryology and eclosion. Centre of Agricultural Publishing and Documentation, Wageningen
Cobben RH (1978) Evolutionary trends in Heteroptera. Part II. Mouthpart-structures and feeding strategies. LandbHoogesch, Wageningen
Consoli RAGB, Pereira MH, Melo AL, Pereira LH (1989) *Belostoma micantulum* Stal 1858 (Hemiptera: Belostomatidae) as a predator of larvae and pupae of *Aedes fluviatilis* (Diptera: Culicidae). Bull World Health Organ 30:5–150
Constantz GD (1974) The mating behavior of a creeping water bug, *Ambrysus occidentalis* (Hemiptera: Naucoridae). Am Midl Nat 92:234–239
Cook JL (2011) A new genus and species of Pleidae (Hemiptera) from Venezuela, with notes on the genera of Pleidae. Zootaxa 3067:26–34
Cordeiro IRS, Moreira FFF, Silva FAC (2014) A new *Ochterus* (Hemiptera: Heteroptera: Ochteridae) from northeastern Brazil, with a key to the species recorded from the country. Zootaxa 3860:493–497
Cullen MJ (1969) The biology of giant water bugs (Hemiptera: Belostomatidae) in Trinidad. Proc R Entomol Soc Lond (A) 44:123–136
Davis JR (1986) New distributional records of Naucoridae (Hemiptera) from the United States and Mexico, with morphological and taxonomic notes. Southwest Nat 31:535–539
De Carlo JA (1940) I) Descripcion de tres especies nuevas del genero *Cryphocricus* Signoret. II) Una nueva especie del genero *Heleocoris* Stål (Hemiptera – Naucoridae). Rev Soc Entomol Arg 10:426–433
De Carlo JA (1963) Un nuevo genero y una nueva especie de la subfamilia Ambrysinae (Hemiptera: Naucoridae). Rev Soc Ent Arg 24:9–11
De Carlo JA (1968a) Estudio comparativo del aparato genital macho de *Belostoma* sp. y *Abedus* sp. (Hemiptera, Belostomatidae). Rev Soc Entomol Arg 31:115–119
De Carlo JA (1968b) Tres especies nuevas del genero *Coleopterocoris* y una especie nueva del genero *Heleocoris* (Hemiptera, Naucoridae). Physis 28:193–197
De Carlo JA (1972) Clave de las especies americanas descriptas del genero *Ranatra* Fabricius. Rev Soc Entomol Arg 34:177–185
Drake CJ (1952) Concerning American Ochteridae (Hemiptera). Fla Entomol 35:72–75
Drake CJ, Chapman HC (1953) Preliminary report on the Pleidae (Hemiptera) of the Americas. Proc Biol Soc Wash 66:53–59
Dufour L (1833) Recherches anatomiques et physiologiques sur les Hémiptères accompagnées de considérations relatives à l'historie naturelle et à la classification de ces insectes. Mém Savants Etrang Acad Sci, Paris, pp 3–333
Esaki T, China WE (1927) A new family of aquatic Heteroptera. Trans Entomol Soc Lond 75:279–295

Estévez AL, Armúa de Reyes AC (2003) Una nueva especie de *Belostoma* (Heteroptera: Belostomatidae) de la laguna Ibera (Corrientes, Argentina). FACENA 19:123–127
Estévez AL, López Ruf ML (2006) Separación de subfamilias en la familia de chinches Gelastocoridae (Hemiptera). Rev Biol Trop (Int J Trop Biol) 54:1–4
Estévez AL, López Ruf ML, Hernández EP (2010) Description of the preimaginal instars of *Nerthra gaucha* Estévez-Schnack (Hemiptera: Heteroptera: Gelastocoridae). Zootaxa 2444:61–64
Estévez AL, Polhemus JT (2001) The small species of *Belostoma* (Heteroptera, Belostomatidae). I. Key to species groups and a revision of the *denticolle* group. Iheringia Sér Zool 91:151–158
Estévez AL, Polhemus JT (2007) The small species of *Belostoma* (Heteroptera: Belostomatidae): revision of *plebejum* group. Rev Biol Trop (Int J Trop Biol) 55:147–155
Estévez AL, Ribeiro JRI (2011) *Weberiella* De Carlo, 1966 (Insecta: Heteroptera: Belostomatidae) revisited: redescription with a key to the genera of Belostomatidae and considerations on back-brooding behavior. Zool Anzeiger 250:46–54
Estévez AL, Schnack JA (1980) Subfamilia Gelastocorinae Champion. Revision de las especies de America del Sur (Hemiptera, Gelastocoridae). Obra Centenario Museo La Plata 6:75–89
Fabricius JC (1803) Systema Rhyngotorum. Apud Carolum Reichard, Brunsvigiae, pp vi–314
Fieber FX (1861) Die europäischen Hemiptera. Halbflügler. (Rhynchota Heteroptera). Carl Gerold's Sohn, Vienna, p 444
Fox LR (1975) Cannibalism in natural populations. Annu Rev Ecol Syst 6:87–106
Gittelman SH (1974) The habitat preference and immature stages of *Neoplea striola* (Hemiptera: Pleidae). J Kansas Entomol Soc 47:491–503
Gittelman SH (1977) Leg segment proportions, predatory strategy and growth in backswimmers (Hemiptera: Pleidae, Notonectidae). J Kansas Entomol Soc 50:161–171
Hebsgaard MB, Andersen NM, Damgaard J (2004) Phylogeny of true water bugs (Nepomorpha: Hemiptera–Heteroptera) based on 16S and 28S rDNA and morphology. Syst Entomol 29:488–508
Henrikson L, Oscarson HG (1981) Corixids (Hemiptera – Heteroptera), the new top predators in acidified lakes. Verh. Int. Verein. Theor Angew Limnol 21:1616–1620
Henrikson L, Oscarson HG (1985) Waterbugs (Corixidae, Hemiptera-Heteroptera) in acidified lakes: habitat selection and adaptations. Ecol Bull 37:232–238
Herrera F (2013) First record of the genus *Ctenipocoris* (Heteroptera: Naucoridae) in Central America, with a preliminary key to the American species and description of a new species. Zootaxa 3731:338–344
Herrera F, Springer M (2012) First record of the family Potamocoridae (Hemiptera: Heteroptera) in Costa Rica and of *Coleopterocoris* Hungerford, 1942 in Central America. Zootaxa 3333:66–68
Herrera F, Springer M (2014) New species of *Potamocoris* Hungerford (Heteroptera, Potamocoridae) from Costa Rica and a key to the species. Zootaxa 3884:492–496
Hinton HE (1976) Plastron respiration in bugs and beetles. J Insect Physiol 22:1529–1550
Hinton HE (1981) Biology of insect eggs, vol 1–3. Pergamon Press, Oxford
Hua J, Li M, Dong P, Cui Y, Xie Q, Bu W (2009) Phylogenetic analysis of the true water bugs (Insecta: Hemiptera: Heteroptera: Nepomorpha): evidence from mitochondrial genomes. BMC Evol Biol 9:1–11
Hungerford HB (1917) Food habits of Corixids. J NY Entomol Soc 25:1–5
Hungerford HB (1919) The biology and ecology of aquatic and semiaquatic Hemiptera. Kansas Univ Sci Bull 11:1–328
Hungerford HB (1922) The life history of the toad bug *Gelastocoris oculatus* Fabr. (Gelastocoridae). Kansas Univ Sci Bull 14:145–171
Hungerford HB (1924) Stridulation of *Buenoa limnocastoris* Hungerford and systematic notes on the *Buenoa* of the Douglas Lake region of Michigan, with the description of a new form (Notonectidae-Hemiptera). Ann Entomol Soc Am 17:223–226

Hungerford HB (1925) Notes on the giant water bugs (*Lethocerus* and *Benacus* – Belostomatidae Hemiptera). Psyche 32:88–92
Hungerford HB (1933) The genus *Notonecta* of the world (Notonectidae-Hemiptera). Univ Kansas Sci Bull 21:1–195
Hungerford HB (1948) The Corixidae of the Western Hemisphere (Hemiptera). Univ Kansas Sci Bull 32:5–827
Hutchinson GE (1993) A treatise on limnology IV. The Zoobenthos. Wiley, New York
Ichikawa N (1988) Male brooding behaviour of the giant water bug *Lethocerus deyrollei* Vuillefroy (Hemiptera: Belostomatidae). J Ethol 6:121–127
Jaczewski T (1950) Notes on Corixidae. XXIV–XXVIII. Polskie Pismo Entomologiczne (Bull Entomol Pologne) 19:103–113
Jansson A (1978) A new species of *Graptocorixa* from Mexico. Pan-Pac Entomol 54:207–209
Jansson A (1979) A new species of *Krizousacorixa* from Mexico (Heteroptera, Corixidae). Pan-Pac Entomol 55:258–260
Jansson A (1989) Stridulation of Micronectinae (Heteroptera: Corixidae). Ann Ent Fenn 55:161–175
Jansson A, Meyer-Rochow VB (1990) Sound production in *Synaptonecta issa* (Heteroptera: Corixidae, Micronectinae) – an Asian bug that turned up in a New Zealand aquarium. Appl Ent Zool 25:328–331
Jansson A, Polhemus JT (1987) Revision of the genus *Morphocorixa* Jaczewski (Heteroptera, Corixidae). Ann Entomol Fenn 53:105–118
Jenkins DW (1964) Pathogens, parasites and predators of medically important arthropods: annotated list and bibliography. Bull World Health Organ 30(suppl):1–150
Keffer SL, McPherson JE (1993) Curictan copulation and waterscorpion higher classification (Heteroptera: Nepidae). Proc Entomol Soc Wash 95:74–78
Keffer SL (1996) Systematics of the New World waterscorpion genus *Curicta* Stål (Heteroptera: Nepidae). J NY Entomol Soc 104:117–215
Keffer SL, Taylor SJ, McPherson JE (1994) Laboratory rearing and descriptions of immature stages of *Curicta scorpio* (Heteroptera: Nepidae). Ann Entomol Soc Am 87:17–26
Kevan DKM (1942) Some observations on *Mononyx nepaeformis* (Fabricius). Proc R Entomol Soc Lond (A) 17:109–110
Konopko SA (2010) Description of the nymphs of *Ectemnostega (Ectemnostegella) stridulata* (Hungerford 1948) (Hemiptera: Heteroptera: Corixidae). Zootaxa 2639:19–34
Konopko SA (2012) Description of the immature stages of *Sigara (Tropocorixa) schadei* (Hungerford) (Hemiptera: Heteroptera: Corixidae). Zootaxa 3487:41–57
Konopko SA (2013) The nymphs of *Sigara (Tropocorixa) denseconscripta* (Hemiptera, Heteroptera, Corixidae). Iheringia Sér Zool 103:302–312
Konopko AS, Mazzucconi SA (2011) Morphometry and chaetotaxy of the nymphs of *Ectemnostega (Ectemnostegella) quechua* (Bachmann 1961) (Insecta: Hemiptera: Heteroptera: Corixidae). J Nat Hist 45:1995–2014
Konopko SA, Melo MC (2009) Larval morphology of *Ectemnostega (Ectemnostegella) montana* (Lundblad 1928) (Hemiptera: Heteroptera: Corixidae: Corixinae), with an emphasis on chaetotaxy. Zootaxa 2315:1–18
Konopko SA, Mazzucconi AS, Bachmann AO (2010) Description of the nymphs of *Tenagobia (Incertagobia) incerta* Lundblad 1929 and *Tenagobia (Schadeogobia) schadei* Lundblad 1929 (Hemiptera: Heteroptera: Micronectidae), with emphasis on morphometry and chaetotaxy. Zootaxa 2511:39–58
Konopko SA, Mazzucconi AS, Bachmann AO (2011) Description of the immature stages of *Trichocorixa mendozana* Jaczewski (Hemiptera: Heteroptera: Corixidae). Zootaxa 3060:47–61
Kormilev NA, De Carlo JA (1952) Una especie nueva del genero *Ochterus* Latreille (1807) del Paraguay (Hemiptera, Ochteridae). An Soc Cient Arg 153:155–159

Kovac D, Maschwitz U (1989) Secretion-grooming in the water bug *Plea minutissima*: a chemical defence against microorganisms interfering with the hydrofuge properties of the respiratory region. Ecol Entomol 14:403–411

Kovac D, Maschwitz U (1991) The function of the metathoracic scent gland in corixid bugs (Hemiptera, Corixidae): secretion-grooming on the water surface. J Nat Hist 25:331–340

Kuitert LC (1949) Some new *Ranatra* from the Americas. J Kansas Entomol Soc 22:24–34

Lansbury I (1972) A revision of the genus *Telmatotrephes* Stål (Hemiptera-Heteroptera, Nepidae). Zool Scripta 1:271–286

Lansbury I (1974) Notes on *Ranatra* (*Amphischizops*) *compressicollis* Montandon with a review of its systematic position within the America *Ranatra* (Hemiptera-Heteroptera, Nepidae). Zool Scripta 3:23–30

Lanzer de Souza ME (1980) Inventário da distribução geográfica da família Belostomatidae Leach, 1815, (Hemiptera, Heteroptera) na região neotropical. Iheringia Ser Zool 55:43–86

Lanzer de Souza ME (1992) Nova espécie do gênero *Belostoma* Latreille, 1807 (Heteroptera, Belostomatidae). Iheringia Ser Zool 72:147–150

La Rivers I (1951) A revision of the genus *Ambrysus* in the United States (Hemiptera: Naucoridae). Univ Calif Publ Entomol 8:277–338

La Rivers I (1953a) A new species and subspecies of *Ambrysus* from Guatemala (Hemiptera: Naucoridae). Pan-Pac Entomol 29:138–142

La Rivers I (1953b) Two new Naucorid bugs of the genus *Ambrysus*. Proc US Nat Mus 103(3311):1–7

La Rivers I (1958) New *Ambrysus* records for Mexico (Hemiptera, Naucoridae). Proc Ent Soc Wash 60:71–74

La Rivers I (1962a) A new species of *Ambrysus* from Costa Rica (Hemiptera, Naucoridae). Pan-Pac Entomol 38:234–236

La Rivers I (1962b) Another Mexican *Ambrysus* (Hemiptera: Naucoridae). Entomol News 73:152–155

Latreille PA (1810) Considérations générales sur l'ordre naturel des animaux. Chez F Schoell, Paris

Lauck DR (1959) Three new species of *Belostoma* from Mexico and Central America (Hemiptera: Belostomatidae), with a list of North American Species. Bull Chicago Acad Sci 11:1–9

Lauck DR (1979) Family Corixidae/ water boatmen. In: Menke AS (ed) The semiaquatic and aquatic Hemiptera of California (Heteroptera: Hemiptera), Bulletin of the California Insect Survey 21. University of California Press, Berkeley, pp 1–166

Lauck DR, Menke AS (1961) The higher classification of the Belostomatidae (Hemiptera). Ann Entomol Soc Am 54:644–657

Leston D, Pendergrast JG, Southwood TRE (1954) Classification of the terrestrial Heteroptera (Geocorisae). Nature 174:91–92

Li T, Hua J, Wright AM, Cui Y, Xie Q, Bu W, Hillis DM (2014) Long-branch attraction and the phylogeny of true water bugs (Hemiptera: Nepomorpha) as estimated from mitochondrial genomes. BMC Evol Biol 14:1–12

Longo RS, Ribeiro JRI, Nessimian JL (2005) A new species of *Coleopterocoris* Hungerford from Southeastern Brazil, with notes on *C. hungerfordi* De Carlo and *C. kleerekoperi* Hungerford (Hemiptera: Heteroptera: Potamocoridae). Zootaxa 1016:39–47

López-Ruf ML (1989) Los huevos de algunas especies de los generos *Pelocoris* y *Ambrysus* (Heteroptera Limnocoridae). Limnobios 2:720–724

López-Ruf ML (1993) Descripcion de ninfas de *Limnocoris ovatulus* Montandon (Heteroptera: Limnocoridae). Rev Soc Entomol Arg 52:17–20

López-Ruf ML (1996) Descripcion de las ninfas 3, 4 y 5 de *Placomerus micans* (Heteroptera: Naucoridae). Rev Soc Entomol Arg 55:73–77

López-Ruf ML (2004) Redescripción de *Ctenipocoris schadei* y descripción de las larvas III y V (Hemiptera: Naucoridae). Rev Soc Entomol Arg 63:19–24

López-Ruf ML, Bachmann AO (1994) Descripcion de las ninfas de *Neoplea absona* (Drake y Chapman) y *Neoplea maculosa* (Berg) (Heteroptera: Pleidae). Rev Soc Entomol Arg 53:25–31

López-Ruf ML, Hernández P (2007) Redescripción de *Ambrysus (Ambrysus) attenuatus* Montandon y descripción de las larvas V, IV, III y I (Hemiptera: Naucoridae). Lundiana 8:3–7

Macan TT (1938) Evolution of aquatic habitats with special reference to the distribution of Corixidae. J Anim Ecol 7:1–19

Mahner M (1993) Systema cryptoceratum phylogeneticum (Insecta, Heteroptera). Zoologica 48:1–303

Manzano MR, Nieser N, Caicedo G (1995) Lista preliminas de heterópteros acuáticos en la Isla Gorgona y Llanura del Pacífico. Biblioteca José Jerónimo Triana 11:47–72

Mazzucconi SA, López-Ruf L, Bachmann AO (2009) Hemiptera – Heteroptera: Gerromorpha y Nepomorha. In: Domínguez E, Fernández HR (eds) Macroinvertebrados Bentónicos Sudamericanos. Sistemática y Biología. Fundación Miguel Lillo, Tucumán, pp 167–231

McCafferty WP (1981) Water bugs (Order Hemiptera). In: Aquatic entomology. The fishermen's and ecologist's illustrated guide to insects and their relatives. Science Books International, Boston, pp 168–188

Menke AS (1960) A taxonomic study of the genus *Abedus* Stål (Hemiptera, Belostomatidae). Univ Calif Publ Entomol 16:393–439

Menke AS (1963) A review of the genus *Lethocerus* in North and Central America, including the West Indies (Hemiptera: Belostomatidae). Ann Entomol Soc Am 56:261–267

Menke AS (1979) Family Nepidae. In: Menke AS (ed) The semiaquatic and aquatic Hemiptera of California (Heteroptera: Hemiptera), Bulletin of the California Insect Survey 21. University of California Press, Berkeley, pp 1–166

Miller PL (1964) The possible role of haemoglobin in *Anisops* and *Buenoa* (Hemiptera: Notonectidae). Proc R Entomol Soc Lond (A) 30:166–175

Montandon AL (1897) Hemiptera cryptocerata. Fam. Naucoridae. – Sousfam. Laccocorinae. Verhandlungen der Zoologisch-Botanischen Gesellschaft in Wien 47:435–454 (reprint paginated 1–20)

Moreira FFF, Barbosa JF, Ribeiro JRI, Alecrim VP (2011) Checklist and distribution of semiaquatic and aquatic Heteroptera (Gerromorpha and Nepomorpha) occurring in Brazil. Zootaxa 2958:1–74

Muraji M, Nakasuji F (1988) Comparative studies on life history traits of three wing dimorphic water bugs, *Microvelia* spp. Westwood (Heteroptera: Veliidae). Res Popul Ecol 30:315–327

Naranjo C, Riviaux SM, Moreira FFF, Court RC (2010) Taxonomy and distribution of aquatic and semiaquatic Heteroptera (Insecta) from Cuba. Rev Biol Trop (Int J Trop Biol) 58:897–907

Nieser N (1969) The Heteroptera of the Netherlands Antilles – VII. Corixidae. Stud Fauna Curação Other Caribb Isl 28:135–164

Nieser N (1970) Corixidae of Suriname and the Amazon with records of other Neotropical species. Stud Fauna Suriname Other Guyanas 12:43–70

Nieser N (1975) The water bugs (Heteroptera: Nepomorpha) of the Guyana Region. Stud Fauna Suriname Other Guyanas 16:1–103

Nieser N (1977) A revision of the genus *Tenagobia* Bergroth (Heteroptera: Corixidae). Stud Neotrop Fauna Environ 12:1–56

Nieser N (2002) Guide to aquatic Heteroptera of Singapore and Peninsular Malaysia. IV. Corixoidea. Raffle Bull Zool 50:263–274

Nieser N, Burmeister EG (1998) Two new species of *Ranatra* from Brazil (Insecta, Heteroptera, Nepidae). Spixiana 21:159–163

Nieser N, Chen PP (1992) Notes on Gelastocoridae and Ochteridae (Heteroptera) with the description of five new species. Storkia 1:2–13

Nieser N, Chen PP (2002) Six new species of *Neotrephes* China, 1936 (Heteroptera: Helotrephidae) from Brazil, with a key to Neotropical Helotrephidae. Lundiana 3:31–40

Nieser N, Chen PP (2006) Two new genera and new subfamily of Micronectidae (Heteroptera, Nepomorpha) from Brazil. Denisia 19:523–534

Nieser N, Chen PP (2008) A new species of *Tenagobia* Bergroth from Bolivia, with notes on the subgenus *Fasciagobia* Nieser, 1977 (Heteroptera, Nepomorpha: Micronectidae). Folia Entomol Hungarica 69:5–13

Nieser N, López-Ruf M (2001) A review of *Limnocoris* Stål (Heteroptera: Naucoridae) in Southern South America East of the Andes. Tijdschr Entomol 144:261–328

Nieser N, Melo AL (1997) Os heterópteros aquáticos de Minas Gerais: guia introdutório com chave de identificação para as espécies de Nepomorpha e Gerromorpha. Editora UFMG, Belo Horizonte pp 1–180

Nieser N, Padilla-Gil N (1992) Three new species of Corixidae from Colombia (Heteroptera). Ent Ber Amst 52:38–46

Nieser N, Chen PP, Melo AL (2013) A new species and new synonymy in *Limnocoris* (Hemiptera: Heteroptera: Naucoridae) from Brazil. Acta Musei Moraviae Sci Biol 98:335–346

Nieser N, Pelli A, Melo AL (1999) Two new Ambrysinae (Heteroptera: Naucoridae) from Minas Gerais, Brazil. Acta Soc Zool Bohem 63:157–163

Padilla-Gil DN (2010) Two new species of *Martarega* (Hemiptera: Heteroptera: Notonectidae) and a new species of *Tachygerris* (Hemiptera: Heteroptera: Gerridae) from Colombia. Zootaxa 2560:61–68

Padilla-Gil DN (2012) Two new species of *Buenoa* (Hemiptera: Heteroptera: Notonectidae) from Colombia, with a key to the species in Colombia. Zootaxa 3316:50–56

Papáček M (2000) Small aquatic bugs (Nepomorpha) with slight or underestimated economic importance. In: Schaefer CW, Panizzi AR (eds) Heteroptera of economic importance. CRC Press, Boca Raton, pp 591–600

Papáček M (2001) Small aquatic and ripicolous bugs (Heteroptera: Nepomorpha) as predators and prey: the question of economic importance. Eur J Entomol 98:1–12

Papáček M, Kovac D (2001) Three new species of Limnotrephini (Heteroptera, Nepomorpha, Helotrephidae) from Peninsular Malaysia and Thailand. Linzer Biologische Beiträge 33:1047–1055

Papáček M, Zettel H (2004) Helotrephidae of the world (Hemiptera: Heteroptera: Nepomorpha) checklist and bibliography. Acta Soc Zool Bohem 68:99–108

Parsons MC (1970) Respiratory significance of the thoracic and abdominal morphology of the three aquatic bugs *Ambrysus*, *Notonecta* and *Hesperocorixa* (Insecta, Heteroptera). Z Morph Tiere 66:242–298

Parsons MC (1972) Respiratory significance of the thoracic and abdominal morphology of *Belostoma* and *Ranatra*. Z Morph Tiere 73:163–194

Parsons MC, Hewson RJ (1974) Plastral respiratory devices in adult *Cryphocricos* (Naucoridae: Heteroptera). Psyche 8:510–527

Pereira MH, Melo AL, Pereira LH (1991) Laboratory rearing of *Belostoma micantulum* Stål, 1858 (Hemiptera, Belostomatidae). Rev Bras Biol 51:603–606

Perez-Goodwyn PJ (2006) Taxonomic revision of the subfamily Lethocerinae Lauck & Menke (Heteroptera: Belostomatidae). Stuttgarter Beitr Naturk Ser A 695:1–71

Polhemus DA (1987) *Heleocoris brasiliensis* De Carlo is a *Ctenipocoris* (Hemiptera: Naucoridae). Pan-Pac Entomol 63:370

Polhemus DA, Carrie RH (2013) A new species of *Potamocoris* (Heteroptera: Potamocoridae) from Belize, and synonymy of the genus *Coleopterocoris*. Tijdschr voor Entomol 156:141–149

Polhemus DA, Polhemus JT (2014) *Ocyochterus irmae*, a beautiful new species of Ochteridae (Heteroptera) from the Northern Andes. Tijdschr voor Entomol 157:163–171

Polhemus JT (1976) Chapter 9: Shore bugs (Hemiptera: Saldidae, etc.). In: Cheng L (ed) Marine insects. North-Holland Publishing Company, Amsterdam/Oxford, pp 225–262

Polhemus JT (1991) A new and primitive genus of Cryphocricinae (Heteroptera: Naucoridae). Pan-Pac Entomol 67:119–123

Polhemus JT (1994) Stridulatory mechanisms in aquatic and semiaquatic Heteroptera. J N Y Entomol Soc 102:270–274

Polhemus JT, Lindskog P (1994) The stridulatory mechanism of *Nerthra* Say, a new species, and synonymy (Heteroptera: Gelastocoridae). J NY Entomol Soc 102:242–248

Polhemus JT, Nieser N (1997) New species of *Tenagobia* (Heteroptera: Corixidae) from Venezuela. Entomol News 108:379–381

Polhemus JT, Polhemus DA (1976) Aquatic and semiaquatic Heteroptera of the Grand Canyon (Insecta: Hemiptera). Great Basin Nat 36:221–226

Polhemus JT, Polhemus DA (1983) Notes on Neotropical Naucoridae II. A new species of *Ambrysus* and review of the genus *Potamocoris* (Hemiptera). Pan-Pac Entomol 58:326–329

Polhemus JT, Polhemus DA (2008a) Global diversity of true bugs (Heteroptera; Insecta) in freshwater. Hydrobiology 595:379–391

Polhemus JT, Polhemus DA (2008b) Intraspecific morphological polymorphism in Naucoridae (Hemiptera: Heteroptera) with notes on nomenclature and synonymy. Acta Entomol Mus Nat Pragae 48:289–298

Polhemus JT, Sites RW (1995) The identity of *Pelocoris biimpressus* Montandon and synonymy of *Pelocoris* species in the Southwestern United States (Heteroptera: Naucoridae). Proc Entomol Soc Wash 97:654–658

Popham EJ (1960) On the respiration of aquatic Hemiptera Heteroptera with special reference to the Corixidae. Proc Zool Soc London 135:209–242

Popham EJ (1961) The function of paleal pegs of Corixidae (Hemiptera Heteroptera). Nature 190:742–743

Popham EJ, Bryant MT, Savage AA (1984) The function of the abdominal strigil in male corixid bugs. J Nat Hist 18:441–444

Popov YA (1968) Origin and main evolutionary trends of Nepomorpha bugs. In: Proceedings of the 13th international congress of entomology, Moscow, vol 1. Nauka, Leningrad, pp 282–283

Popov YA (1970) Notes on the classification of the recent Naucoridae (Heteroptera, Nepomorpha). Bull Acad Polonaise Sci 18:93–98

Popov YA (1971) Historical development of the Hemiptera infraorder Nepomorpha (Heteroptera). Trudy Paleontol Inst Acad Sci 129:1–230

Ribeiro JRI, Alecrim VP (2008) Duas novas espécies de *Belostoma* Latreille, 1807 (Hemiptera: Heteroptera: Belostomatidae) do grupo *plebejum sensu* Nieser, 1975. Acta Amazon 38:179–188

Ribeiro JRI, Estévez AL (2009) The small species of *Belostoma* Latreille (Heteroptera, Belostomatidae). III. A revision of *oxyurum* group, with a new species from Brazil and description of the male of *B. noualhieri* Montandon. Rev Bras Entomol 53:207–215

Rieger C (1976) Skelett und Muskulatur des Kopfes und Prothorax von *Ochterus marginatus* Latreille. Zoomorphologie 83:109–191

Roback SS, Nieser N (1974) Aquatic Hemiptera (Heteroptera) from the Llanos of Colombia. Proc Acad Nat Sci Phila 126:29–49

Roback SS, Nieser N (1980) IV. Aquatic Hemiptera (Heteroptera). Proc Acad Nat Sci Phila 132:196–198

Rodrigues HDD, Melo AL, Ferreira-Keppler RL (2012a) Macropterous form of *Paratrephes* China, 1940, with new distributional records of Neotrephinae from Brazil (Hemiptera: Heteroptera: Helotrephidae). Zootaxa 3483:82–88

Rodrigues HDD, Melo AL, Sites RW (2012b) Redescription of *Ambrysus montandoni* La Rivers, 1963 (Hemiptera: Heteroptera: Naucoridae), with the first records for Brazil. Zootaxa 3357:63–68

Romero I, Noriega JA (2013) Chinches acuáticas de la superfamilia Nepoidea (Hemiptera: Nepomorpha) de Colombia: nuevos registros para suramérica y ampliación de su distribución en el país. Biota Colomb 14:92–107

Saulich AH, Musolin DL (2007) Seasonal development of aquatic and semiaquatic true bugs (Heteroptera). St. Petersburg University Press, St. Petersburg (in Russian, with extended 6p English Summary). ISBN 978-5-288-04332-1

Schell DV (1943a) The Ochteridae (Hemiptera) of the Western Hemisphere. J Kansas Entomol Soc 16:29–36

Schell DV (1943b) The Ochteridae (Hemiptera) of the Western Hemisphere (Continued). J Kansas Entomol Soc 16:37–47

Schnack JA, Estévez AL (1979) Monografia sistematica de los Gelastocoridae de la Republica Argentina (Hemiptera). Limnobios 1:346–380

Schnack JA, Estévez AL (1990) On the taxonomic status of *Abedus* Stål (Hemiptera, Belostomatidae). Rev Bras Ent 34:267–269

Schnack JA, Estévez AL (2005) On the taxonomic status of the genus *Horvathinia* Montandon (Hemiptera: Belostomatidae). Zootaxa 1016:21–27

Schnack JA, Estévez AL, Armúa-de-Reyes AC (2006) Laguna Don Blanco, Argentina: first record of *Horvathina* (Hemiptera: Belostomatidae) as a wetland dweller. Entomol News 117:197–202

Schuh RT, Slater JA (1995) True bugs of the world (Hemiptera: Heteroptera). Classification and natural history. Cornell University Press, Ithaca

Sites RW (1991) Egg ultrastructure and descriptions of nymphs of *Pelocoris poeyi* (Guerin Méneville) (Hemiptera: Naucoridae). J NY Entomol Soc 99:622–629

Sites RW (2004) Distributional records of *Cataractocoris* (Heteroptera: Naucoridae) in Mesoamerica. Entomol News 115:173–174

Sites RW, Camacho J (2014) Neotropical genera of Naucoridae (Hemiptera: Heteroptera: Nepomorpha): New species of *Placomerus* and *Procryphocricos* from Guyana and Venezuela. Zootaxa 3753:469–482

Sites RW, Nichols BJ (1990) Life history and descriptions of immature stages of *Ambrysus lunatus lunatus* (Hemiptera: Naucoridae). Ann Entomol Soc Am 83:800–808

Sites RW, Nichols BJ (1993) Voltinism, egg structure, and descriptions of immature stages of *Cryphocricos hungerfordi* (Hemiptera: Naucoridae). Ann Entoml Soc Am 86:80–90

Sites RW, Nichols BJ (1999) Egg architecture of Naucoridae (Heteroptera): internal and external structure of the chorion and micropyle. Proc Entomol Soc Wash 101:1–25

Sites RW, Nichols BJ (2001) Voltinism, egg structure, and descriptions of immature stages of *Limnocoris lutzi* (Heteroptera: Naucoridae). Ann Entomol Soc Am 94:26–32

Sites RW, Shepard WD (2015) Neotropical genera of Naucoridae (Hemiptera: Heteroptera: Nepomorpha): new species of *Ambrysus* and *Procryphocricos* from Belize and Peru. Ann Entomol Soc Am doi:10.1093/aesa/sav027

Sites RW, Willig MR (1991) Microhabitat associations of three sympatric species of Naucoridae (Insecta: Hemiptera). Environ Entomol 20:127–134

Sites RW, Reynoso-Velasco D, Novelo-Gutiérrez R (2013) Revision of the Mesoamerican Genus *Cataractocoris* Usinger (Hemiptera: Heteroptera: Nepomorpha: Naucoridae). Zootaxa 3682:475–484

Smith RL (1976) Male brooding behavior of the water bug *Abedus herberti* (Hemiptera: Belostomatidae). Ann Entomol Soc Am 69:740–747

Smith RL (1979) Paternity assurance and altered roles in the mating behaviour of a giant water bug, *Abedus herberti* (Heteroptera: Belostomatidae). Anim Behav 27:716–725

Southwood TRE (1961) A hormonal theory of the mechanism of wing polymorphism in Heteroptera. Proc R Entomol Soc Lond 36:63–66

Staddon BW (1971) Metasternal scent glands in Belostomatidae (Heteroptera). J Entomol 46:69–71

Staddon BW, Thorne MJ (1974) Observations on the metathoracic scent gland system of the back-swimmer *Notonecta glauca* L. (Heteroptera: Notonectidae). J Entomol (Ser A) 48:223–227

Staddon BW, Thorne MJ (1979) The metathoracic scent gland system in Hydrocorisae (Heteroptera: Nepomorpha). Syst Entomol 4:239–250

Stout RJ (1981) How abiotic factors affect the distribution of two species of tropical predaceous aquatic bugs (Family Naucoridae). Ecology 62:1170–1178
Stout RJ (1982) Effects of a harsh environment on the life history patterns of two species of tropical aquatic Hemiptera (Family Naucoridae). Ecology 63:75–83
Štys P, Jansson A (1988) Check-list of recent family-group and genus-group names of Nepomorpha (Heteroptera) of the world. Acta Entomol Fenn 50:1–44
Štys P, Kerzhner I (1975) The rank and nomenclature of higher taxa in recent Heteroptera. Acta Ent Bohemoslov 72:64–79
Štys P, Říha P (1975) Studies on tertiary Notonectidae from Central Europe (Heteroptera). Acta Univ Carolinae 1973:163–184
Thorpe WH (1950) Plastron respiration in aquatic insects. Biol Rev 25:344–390
Tinerella PP, Polhemus JT (2006) A new species of *Neosigara* Lundblad (Heteroptera: Corixidae) from Colombia with a revised key to adults. Russ Entomol J 15:221–222
Todd EL (1955) A taxonomic revision of the family Gelastocoridae (Hemiptera). Univ Kansas Sci Bull 37:277–475
Todd EL (1957) Five new species of Gelastocoridae with comments on other species (Hemiptera). Proc Entomol Soc Wash 59:145–162
Todd EL (1965) A note on the occurrence of *Nerthra rugosa* (Desj.) in Brazil (Hemiptera: Gelastocoridae). Proc Entomol Soc Wash 67:111
Torre Bueno JR (1905a) The genus *Notonecta* in America North of Mexico. J NY Entomol Soc 13:143–167
Torre Bueno JR (1905b) The tonal apparatus of *Ranatra quadridentata*, Stål. Can Entomol 37:85–87
Truxal FS (1949) A study of the genus *Martarega* (Hemiptera, Notonectidae). J Kansas Entomol Soc 22:1–24
Truxal FS (1953) A revision of the genus *Buenoa* (Hemiptera: Notonectidae). Univ Kansas Sci Bull 35:1351–1517
Usinger RL (1941) Key to the subfamilies of Naucoridae with a generic synopsis of the new subfamily Ambrysinae (Hemiptera). Ann Entomol Soc Am 34:5–16
Usinger RL (1946) Notes and descriptions of *Ambrysus* Stal with an account of the life history of *Ambrysus mormon* Montd. (Hemiptera: Naucoridae). Univ Kansas Sci Bull 31:185–210
Usinger RL (1947) Classification of the Cryphocricinae (Hemiptera: Naucoridae). Ann Entomol Soc Am 40:329–343
Usinger RL (1968) Aquatic Hemiptera. In: Usinger RL (ed) Aquatic insects of California, with keys to North American genera and California species, 3rd edn. University of California Press, Berkeley/Los Angeles, pp 182–228
Van Doesburg PH (1984) A new species of *Potamocoris* Hungerford, 1941 from Suriname (Heteroptera: Naucoridae). Zool Mededelingen 59:19–25
Vepsäläinen K (1971) The roles of photoperiodism and genetic switch in alary polymorphism in *Gerris* (Het., Gerridae) (a preliminary report). Acta Entomol Fenn 28:101–102
Vepsäläinen K (1974) Determination of wing length and diapause in water-striders (*Gerris* Fabr., Heteroptera). Hereditas 77:163–175
Weirauch C, Schuh RT (2011) Systematics and evolution of Heteroptera: 25 years of progress. Annu Rev Entomol 56:487–510
Wilcox RS (1975) Sound-producing mechanisms of *Buenoa macrotibialis* Hungerford (Hemiptera: Notonectidae). Int J Insect Morphol Embryol 4:169–182
Williams DD (2006) The biology of temporary waters. Oxford University Press, Oxford/New York
Zalom FG (1978) A comparison of predation rates and prey handling times of adult *Notonecta* and *Buenoa* (Hemiptera: Notonectidae). Ann Entomol Soc Am 71:143–148
Zettel H, Lane DJW, Moore S (2008) *Aphelocheirus* (s.str.) *bruneiensis* sp.n., a new benthic water bug (Hemiptera: Heteroptera: Aphelocheiridae) from Brunei, and a key to Bornean *Aphelocheirus*. Zootaxa 1920:61–68
Zettel H, Papáček M, Kovac D (2011) Guide to the aquatic Heteroptera of Singapore and Peninsular Malaysia: VII. Family Helotrephidae. Raffle Bull Zool 59:171–179

Part V
Infraorder Leptopodomorpha

Chapter 8
Shore Bugs (Saldidae)

Alan L. de Melo

Abstract The Saldidae is one of the most ancient and extant families of insects belonging to the infraorder Leptopodomorpha that includes small (2–8 mm), oval-shaped insects known as shore bugs. They can be found in several habitats from marine intertidal places to edges of streams and lakes or far away from water in dry environments. Aspects of general characteristics, biology, and diagnosis are presented, and keys for the four extant families and for the subfamilies, genera, and known species of Saldidae occurring in the Neotropical region, mainly South America, are included.

8.1 Introduction

The infraorder Leptopodomorpha Popov includes about 400 species of small oval-shaped insects from four extant families: Aepophilidae with only known species from the marine intertidal zone in Palearctic region (Europe and maybe Mediterranean region of North Africa); Leptopodidae, with several species occurring worldwide, but only one species found in Ecuador and Colombia (Schuh and Slater 1995); Omaniidae with a few species of the marine intertidal zone in the Indo-Pacific region; and Saldidae with about 340 known species worldwide and from three extinct families (Archegocimicidae, Mesolygaeidae, Palaeoleptidae) (Popov 1989; Popov et al. 1994). Recently, some authors, after new fossil discoveries, accepted more families (Ryzhkova 2012, 2013). They are known as shore bugs and can be found in several habitats from marine intertidal places to edges of streams and lakes or be found far away from water in dry environments (Henry 2009).

Members of this group present some typical characteristics that can, with few exceptions, be easily recognized and can aid to separate individuals in different families; shore bugs are relatively uncommon in insect collections because they are easy to be overlooked by nonspecialists.

A.L. de Melo (✉)
Departamento de Parasitologia, Instituto de Ciências Biológicas, Universidade Federal de Minas Gerais, Av. Antonio Carlos, 6627, Belo Horizonte, MG 31270-901, Brazil
e-mail: alan@ufmg.br

A.R. Panizzi, J. Grazia (eds.), *True Bugs (Heteroptera) of the Neotropics*, Entomology in Focus 2, DOI 10.1007/978-94-017-9861-7_8

Some fossils mainly from the Miocene period and belonging to this infraorder were found in amber from several localities in different countries. Fossils of the Cretaceous and Jurassic periods have been found, but some authors did not accept these species as members of the Leptopodomorpha (Cobben 1980, 1987; McKellar and Engel 2014; Ryzhkova 2013; Zhang et al. 2011, 2012).

There are a few records of parasites (protozoa, mites, and entomogenous worms) occurring in these intertidal insects (Poinar 1975, 2012; Polhemus and Chapman 1979; Stock and Lattin 1976).

8.2 General Characteristics and Diagnosis

Most shore bugs (Fig. 8.1) are typically small (2–8 mm) and ovate, and some tiny species morphologically resemble terrestrial species rather than members of Nepomorpha and Gerromorpha; however, they possess morphological and physiological attributes including parts of the body covered by fine setae that repel water and bind layers of air to the body indicating that they are able to survive and adapt to periods of submergence during high tides or other aquatic perturbation.

Fig. 8.1 A typical specimen of Saldidae (scale bar = 1 mm) (Photo A. L. de Melo)

All the apterous, brachypterous, and macropterous forms have visible antennae (filiform and longer than the head) with all segments of nearly equal diameter: segment 1 short and segments 2–4 longer. Compound eyes are always present and usually very large (sometimes reniform) and occupying the side of the head that is usually relatively short and broad and present at least three pairs of trichobothria dorsally. The labium is short or long, with three or four segments inserted ventrally and extending backwards between the legs; ocelli are usually present. Some species present the forewings undifferentiated or reduced being part of the body covered by a minute setae; others present the hemelytra covering the abdomen divided into a clavus, endocorium, exocorium, and membrane, and its veins form three, four, or five closed cells in macropterous forms. In Saldidae, wing polymorphism is common (Polhemus and Chapman 1979). Legs are usually relatively short, slender, or longer. The hind coxae are large, transverse with broad coxal cavity in thoracic pleurae, and adapted for hopping or jumping. Femora are sometimes armed with spines; all tarsi are usually three segmented with apical claws. Male genitalia are symmetrical; parameres are hook-shaped with a distinct processus; ovipositor valvulae are developed or reduced and platelike; spermatheca is present.

The Saldidae is one of the most ancient and the best-known families of insects belonging to Leptopodomorpha with about 340 species in two subfamilies: Chiloxanthinae and Saldinae (Popov 1971; Polhemus and Chapman 1979; Rasnitsyn and Quicke 2002). This cosmopolitan family is recorded in all continents (except Antarctic) being more speciose in Northern Hemisphere (Schuh et al. 1987; Schuh and Polhemus 2009).

8.3 General Biology and Ecology

The habits of various species are yet poorly known, and few of them that were studied do not allow generalizations concerning their life cycles. The lack of economic importance prevents more detailed studies in several countries.

Saldidae usually copulate side by side (Polhemus and Chapman 1979), and according to Cobben (1957), the female's forewing and male abdominal segments show some modifications to facilitate this copulatory position.

Eggs are laid in or on the vegetation or other substrates. They are pearly white oblong cylinders about 1 mm long. Maturation of most saldid eggs requires approximately a week.

Five nymphal instars are known. Egg to adult developmental time varies according to species and temperature and varies from 3 weeks to a month (Wiley 1922; Jordan and Wendt 1938). There are some cold-adapted species and others that can overwinter in the egg or adult stages, and depending on the temperature, they are uni- or multivoltine (Wroblewski 1966). Parental care is not well known in Saldidae.

Most members of Saldidae, which do not reach 2 mm, live along the edges of beaches, lake shores, and stream banks (including muddy or sandy shores), and all species of this group (adults and nymphs) are predaceous, feeding on small insects (Brooks and Kelton 1967), mainly Diptera larvae, each other (cannibalism) (Wroblewski 1966), and their own eggs (Rimes 1951). Some species can adapt to become scavenger or carnivore (Merritt et al. 2008). It is possible that some saldid species can find their prey submerged or on the surface by chemoreceptors located on the antenna and by visual means (Polhemus 1976). After catching the prey, the rostrum is inserted, and protease enzymes are injected to immobilize and liquefy the body contents before they can be as reported in several other predaceous Heteroptera sucked up (Picado 1937, 1939; Cheng 1967; Pereira and Melo 1998).

8.4 Classification and Diversity

The tentative classification of Saldidae was proposed by Cobben (1959), and since that time, works by Polhemus (1985a, b), Schuh and Polhemus (1980), and Schuh and Slater (1995) presented important contributions. The phylogenetic analyses for Saldidae were carried by Polhemus (1977) and Schuh and Polhemus (2009) with very interesting information.

The keys for Saldidae are adapted mainly from Heckman (2011) and several other sources cited in the references, and only the representatives for Neotropical region, mainly South America, are included.

8.4.1 Key to the Families

1. Labium short, not reaching beyond the apex of the fore coxa..........Leptopodidae
– Labium long, reaching at least beyond the hind coxa...2
2. Compound eyes small, not covering the head; ocelli absent. Wings reduced to small pads...Aepophilidae
– Compound eyes larger, covering most of the surface of the head. Wings cover most of the abdomen...3
3. Length 2 mm or shorter. Eyes covering about 1/3 of the lateral surface of the pronotum reaching beyond its collar...Omaniidae
– Length 2.2 mm or more. Eyes did not reach the pronotal collar; ocelli present. Wings cover the abdomen...Saldidae

8.4.2 *Key to the Genera of Saldidae*

1. Antennal segments 3 and 4 swollen, dorsum of pronotum with two distinct conical projections...*Saldoida* Osborn.
This genus is easily recognized by the two sharply conical projections of the pronotum, and antennal segments 3 and 4 are swollen. *Saldoida* is Nearctic and Oriental in distribution and contains few described species
– Antennal segments 3 and 4 not swollen, dorsum of pronotum without distinct conical projections...2
2. Embolar fracture longer, reaching at least as far as the hind end of the claval suture...3.
– Embolar fracture short, not reaching the hind end of the claval suture.................4.
3. Hemelytral membrane with four terminal closed cells..*Paralosalda* Polhemus & Evans.
One species, *Paralosalda innova* Polhemus & Evans, inhabits the rocky intertidal zone along the Pacific coasts of Costa Rica, Panama, and Colombia
– Hemelytral membrane with five terminal closed cells.................*Pentacora* Reuter
The five elongated cells of the membrane readily identify *Pentacora*. It is a relatively warm-adapted genus that is found in saline and rocky habitats. It is distributed mainly throughout North and Central America. There are records of *Pentacora sphacelata* (Uhler) in the Galapagos Islands and Spain
4. Head with a transverse swelling..5.
– Head without transverse swelling but covered with short setae................*Oiosalda* Drake & Hoberlandt.
One species, *Oiosalda caboti* Drake and Hoberlandt, is known only from Colombia
5. Hemelytral membrane with five cells..............................*Pseudosaldula* Cobben.
The presence of a transverse swelling on the head and the five closed cells of the hemelytral membrane identify this genus
– Hemelytral membrane with four cells...6.
6. Membrane with first (inner) cell protruding anteriorly half its length beyond second cell...*Salda* F.
The members of this genus are identified by the four elongated cells of the membrane and are large (up to 6 mm) dark coloration. Only few species are described and at least one was recorded from Florida. This genus is Holarctic and temperate to arctic adapted
– Membrane with all cells of subequal length or with first cell protruding anteriorly much less than half its length beyond the second cell..7.
7. Hypocostal ridge simple; without secondary hypocostal ridge..*Rupisalda* Polhemus.
The four closed cells of the membrane of approximately equal length and the simple structure of the hypocostal ridge identify species in this genus
– Hypocostal ridge complex, with secondary hypocostal ridge..............................8.

8. Corium with two distinct veins...*Saldula* Van Duzee. The four closed cells of the membrane readily identify *Saldula*, and it can be distinguished from *Micracanthia* by the distal veins of the corium (best viewed from the underside of the hemelytra). This genus is cosmopolitan
– Corium without veins..*Micracanthia* Reuter. Very similar to *Saldula* and is distinguished by the apparent lack of a forked vein in the corium, or the apex on the median vein is not forked. This is best viewed from the underside of the hemelytra. It is mainly Holarctic (at least one species reaching Alaska) in distribution. Several species are known from Asia, Europe, Mexico, the USA and from the Caribbean islands including Costa Rica and at least one from South America (Polhemus 1977)

8.4.3 Key to the Species of Pseudosaldula

1. First antennal segment with black marking or lateroventral stripe......................2.
– First antennal segment usually uniform, without black mark..............................4.
2. Fore and middle tibiae and sometimes the hind tibia with a defined dark stripe on the dorsal surface..3.
– All tibiae lacking distinct stripes on the dorsal surface..*Pseudosaldula penai* Schuh & Polhemus (Bolivia, Chile, Argentina)
3. Dorsum coated by golden setae; clavus and corium usually black, with pruinose areas. First antennal segment with black stripes on ventrolateral and dorsomedial surfaces...................................*Pseudosaldula chilensis* (Blanchard) (Argentina, Chile)
– Dorsum coated by black setae; corium with pale markings without pruinose areas. First antennal segment with a black longitudinal stripe on the ventrolateral surface..*Pseudosaldula salina* Schuh & Polhemus (Argentina)
4. Hemelytra with dull surfaces except along the overlapping borders of the claval commissure and sometimes on limited areas along the edge of the corium........5.
– Hemelytra with shiny surfaces on the corium, embolar area, veins, and sometimes along the claval commissure and some areas of the membrane..........8.
5. Dorsum with long, dark, and dull setae. Hemelytral setae twice as long as the diameter of hind tibia...........................*Pseudosaldula pilosa* Schuh & Polhemus (Argentina, Chile)
– Dorsum with short and recumbent setae. Hemelytral setae equal or shorter than the diameter of a hind tibia..6.
6. Pronotum campanulate, with slightly to strongly concave margins; irregular ivory-colored macula on the subapical surface of the exocorium...*Pseudosaldula saxicola* Schuh & Polhemus (Argentina, Peru)

– Pronotum flattened with slightly to strongly convex margins. Only a pale mark on the subapical surface of the exocorium..7.

7. Corium coated by shiny, golden, recumbent setae; corium, clavus, and membrane with several purplish, pruinose areas and numerous long setae on the membranes veins...............................*Pseudosaldula vulgaris* Schuh & Polhemus (Bolivia, Peru)

– Corium not coated by shiny, golden, recumbent setae; corium, clavus, and membrane with few purplish, pruinose areas and scattered setae on the membrane veins..*Pseudosaldula bergi* (Haglund) (Argentina, Chile)

8. Anterior part of exocorium and radial vein shiny; the rest of the corium is dull...*Pseudosaldula huamachuco* Schuh & Polhemus (Peru)

– Posterior part of exocorium shiny and variable..9.

9. Exocorium and part of the posterior half of the endocorium with polish and shiny areas..10.

– Exocorium at the embolar area with polish and shiny areas. If this area reaches the posterior half, then the radial vein is also polished.......................................11.

10. Exocorium and some area of the posterior part of the endocorium and entire membrane of brachypterous specimens with shiny pattern (macropterous forms only basal third of the membrane polished)......................................*Pseudosaldula andensis* (Distant) (Ecuador, Peru)

– Exocorium and some area of the posterior part of the endocorium with shiny pattern interrupted by a dull area along the radial vein, basal area, and veins of membrane polished in brachypterous forms (macropterous forms have only basal third of the membrane polished).............................*Pseudosaldula bruesi* (Drake) (Peru)

11. Pronotum campanulate, with slightly to strongly concave margins; dorsum with black setae without pruinose areas..12.

– Pronotum not campanulate with slightly to strongly convex margins, dorsum with both black and shiny golden setae and pruinose areas...............................13.

12. Dorsum with erect and long setae.....*Pseudosaldula yungas* Schuh & Polhemus (Bolivia, Peru)

– Dorsum with erect and of moderate length setae..*Pseudosaldula antioquia* Schuh & Polhemus (Colombia)

13. Embolar area polished at posterior half; radial vein smooth and shiny..*Pseudosaldula aurea* Schuh and Polhemus (Argentina)

– Embolar area polished and extends for its entire length; radial vein is matt...*Pseudosaldula perula* (Drake) (Argentina, Peru)

8.4.4 *Key to the Species of* Rupisalda

1. Pubescence dense and yellowish brown; first antennal segment stout and yellowish...*Rupisalda verdica* (Drake & Hottes) (Brazil: Santa Catarina)
– Pubescence not dense and golden or grayish; first antennal segment slender or darker..2.
2. Legs testaceous, with whitish marks..............................*Rupisalda zelena* (Drake) (Peru)
– Legs light brown to pale, without markings..3.
3. First antennal segment brownish.................................*Rupisalda icolana* (Drake) (Brazil: Santa Catarina)
– First antennal segment lighter at apical part, with a broad whitish or yellowish ring...*Rupisalda ventralis* (Stål) (Brazil: Mato Grosso, Minas Gerais, Rio de Janeiro)

8.4.5 *Key to the Species of* Saldula

1. Lateral margins of pronotum, pale...2.
– Lateral margins of pronotum dark or fuscous brown..3.
2. Legs yellowish, with brown markings.................................*Saldula coxalis* (Stål) (North, South, and Central America including the Caribbean islands)
– Legs testaceous and sometimes marked with brown....................*Saldula doeringi* Drake & Carvalho (Argentina)
3. Head, pronotum, and scutellum black, not dull, rugulose...................................4.
– Head, pronotum, and scutellum polished, dull, or coated by pubescence, not rugulose...5.
4. Outer margin of male paramere with prominent swelling and a slight swelling along the inner surface..*Saldula humilis* (Say) (North America, Caribbean islands, Surinam, Brazil—Santa Catarina)
– Outer margin of male paramere without swelling, with prominent tubercle along the inner surface...*Saldula dentulata* (Hodgden) (Central America, Caribbean islands, Ecuador, Panama, Peru)
5. Lateral margins of pronotum concave or straight...6.
– Lateral margins of the pronotum convex...9.
6. Dorsal surfaces of head and thorax black, dull, and with a silvery or golden pubescence and a few longer black setae on thorax..*Saldula peruviana* (Cobben) (Peru)
– Dorsal surfaces of the head and thorax black, shiny, with a sparse, pale pubescence..7.

7. Pubescence on hemelytra is moderately long and dull velvet black with moderately long, golden; semi-recumbent pilosity..*Saldula scitula* Drake & Hottes (Brazil: Santa Catarina)
– Pubescence on hemelytra is long and dark brownish or short and silvery pilosity..8.
8. Head, pronotum, and scutellum with grayish brown pubescence and erect dark setae; dorsal surface of hemelytra with long, erect, dark brownish setae.............. ....*Saldula hodgdeni* (Drake) (Bolivia)
– Head, pronotum, and scutellum with pale and short pubescence and few longer black setae; dorsal surface of hemelytra with short silvery pubescence..*Saldula chartoscirtoides* Cobben (Brazil: Santa Catarina)
9. Shape of the female is broadly ovate; hemelytra fuscous brown...*Saldula lynchi* Drake & Carvalho (Argentina)
– Shape not broadly ovate; hemelytra with different pattern................................10.
10. Length about 4.35 mm................................*Saldula differata* Drake & Carvalho (Argentina)
– Length is usually up to 4 mm...11.
11. Hemelytra with pruinose or pubescent areas...12.
– Hemelytra without pruinose areas...*Saldula pallipes* (F.) (Holarctic Region, Mexico, Central America, Caribbean islands, South America, and Africa)
12. Corium with two or three dark spots..........*Saldula penningtoni* Drake and Carvalho [Argentina, Brazil (Minas Gerais, Santa Catarina), Chile, Paraguay, Peru, Uruguay]
– Corium with bluish white pruinose area on the inner part...*Saldula galapagosana* (Polhemus) (Galapagos Islands). Remark: The record of *Saldula saltatoria* (L.), a Holarctic species from Colombia (Polhemus and Chapman 1979) needs confirmation

8.5 Concluding Remarks

Many species of Leptopodomorpha have been misidentified mainly by nonspecialists, and apart from scarce references from the older literature that sometimes are difficult to obtain, there are today few specialists studying this group of insects, and for a better knowledge of this group, sampling must be improved. So, more people are needed to collect more intensively in faunal studies regarding shore bugs biology, including habitats, ecological traits, phylogenetics, historical biogeography, altitude, and temperature gradients of this interesting but little-known insects that can be found in almost all zoogeographic regions worldwide. Yet, fascinating questions concerning speciation and evolution, habitat disturbance, or fragmentation of natural habitats by rapid and continued anthropogenic changes arise. Shore bugs' relationship with the extinct Leptopodomorpha or with natural preys and predators needs be improved or in the future, with more collected specimens, studies could be extended further to include their parasites.

References

Brooks AR, Kelton LA (1967) Aquatic and semiaquatic Heteroptera of Alberta, Saskatchewan, and Manitoba (Hemiptera). Mem Ent Soc Can 51:3–92

Cheng L (1967) Studies on the biology of Gerridae (Hem., Heteroptera). I: Observations on the feeding of *Limnogonus fossarum* (F.). Entomol Mon Mag 102:121–129

Cobben RH (1957) Beitrag zur Kenntnis der Uferwanzen (Hem. Het. Fam. Saldidae). Ent Berichten 17:245–257

Cobben RH (1959) Notes on the classification of Saldidae with the description of a new species from Spain. Zool Mededelingen 36:303–316

Cobben RH (1980) The Saldidae of the Hawaiian archipelago (Hemiptera: Heteroptera). Pac Insect 22:1–34

Cobben RH (1987) New African Leptopodomorpha (Heteroptera: Saldidae, Omaniidae, Leptopodidae), with an annotated checklist of Saldidae from Africa. I. New species of the genus *Saldula* (Saldidae). Rev Zool Afr 100:399–421

Heckman CW (2011) Encyclopedia of South American aquatics insects: Hemiptera – Heteroptera. Springer, London

Henry TJ (2009) Biodiversity of the Heteroptera. In: Foottit RG, Adler PH (eds) Insect biodiversity: science and society. Wiley-Blackwell, Oxford, pp 223–226

Jordan KHC, Wendt A (1938) Zur Biologie von *Salda litoralis* L. (Hem. Het.). Stettiner Ent Zeitng 99:273–292

McKellar RC, Engel MS (2014) The first Mesozoic Leptopodidae (Hemiptera: Heteroptera: Leptopodomorpha), from Canadian Late Cretaceous amber. Hist Biol 26(6):702–709. doi:10.1080/08912963.2013.838753

Merritt RW, Cummins KW, Berg MB (2008) An introduction to the aquatic insects of North America. Kendall/Hunt Publ. Co., Dubuque

Pereira MH, Melo AL (1998) Influência do tipo de presa no desenvolvimento e na preferência alimentar de *Belostoma anurum* (Herrich-Schaffer, 1848) e *B. plebejum* (Stal, 1858) (Heteroptera, Belostomatidae). Oecologia Brasiliensis 5:41–49

Picado TC (1937) Estudo experimental sobre o veneno de *Lethocerus delpontei* (De Carlo) (Hemiptera-Belostomidae). Mem Inst Butantan 10:305–310

Picado TC (1939) Étude experimentale du venin de *Lethocerus delpontei* (De Carlo). Trav Stn Zool Wimereux 13:533–562

Poinar GO Jr (1975) Entomogenous nematodes. A manual and host list of insect nematode associations. Brill, Leiden

Poinar GO Jr (2012) *Halophilanema prolata* n. gen., n. sp. (Nematoda: Allantonematidae) a parasite of the intertidal bug, *Saldula laticollis* (Reuter) (Hemiptera: Saldidae) on the Oregon coast. Par Vector 5:24

Polhemus JT (1976) Shore bugs (Hemiptera: Saldidae, etc.). In: Cheng L (ed) Marine insects. North-Holland Publ. Co, New York, pp 225–261

Polhemus JT (1977) The biology and systematics of the Saldidae of Mexico and Middle America. Ph.D. thesis, University of Colorado, Boulder, CO, USA

Polhemus JT (1985a) Shore bugs (Heteroptera, Hemiptera; Saldidae): a world overview and taxonomy of Middle American forms. The Different Drummer, Englewood

Polhemus JT (1985b) Nomenclatural changes for North American Saldidae. Proc Entomol Soc Wash 87:893

Polhemus JT, Chapman HC (1979) Family Saldidae/shore bugs. In: Menke AS (ed) The semi-aquatic and aquatic Hemiptera of California (Heteroptera: Hemiptera), Bulletin of the California Insect Survey 21. University of California Press, Berkeley, pp 16–33

Popov YA (1971) The historical development of Hemiptera infraorder Nepomorpha (Heteroptera). Trudy Paleontol Institut Moskva (Nauk Akademii CCCP) 129:1–230 (in Russian)

Popov YA (1989) Some aspects of systematic of Leptopodoidea. Acta Biol Silesiana 13:63–68

Popov YA, Dolling WR, Whalley PES (1994) British Upper Triassic and Lower Jurassic Heteroptera and Coleorrhyncha (Insecta: Hemiptera). Genus 5:307–347

Rasnitsyn AP, Quicke DLJ (2002) History of insects. Kluwer Academic Publishers, Dordrecht

Rimes GD (1951) Some new and little-known shore-bugs (Heteroptera-Saldidae) from the Australian region. Trans R Soc S Aust 74:135–145

Ryzhkova OV (2012) New Saldoid bugs of the family Enicocoridae (Hemiptera: Heteroptera: Leptopodomorpha) from the Lower Cretaceous of Mongolia. Paleontol J 46:485–494 (original in Russian)

Ryzhkova OV (2013) New saldoid bug of the family Archegocimicidae (Hemiptera: Heteroptera: Leptopodomorpha) from the Middle Jurassic of Eastern Siberia. Paleontol J 47:180–184 (original in Russian)

Schuh RT, Polhemus JT (1980) *Saldolepta kistnerorum*, new genus and new species from Ecuador (Hemiptera, Leptopodomorpha), the sister group of *Leptosalda chiapensis*. Am Mus Novit 2698:1–5

Schuh RT, Polhemus JT (2009) Revision and analysis of *Pseudosaldula* Cobben (Insecta: Hemiptera: Saldidae): a group with a classic Andean distribution. Bull Am Mus Nat Hist 323:1–102

Schuh RT, Slater JA (1995) True bugs of the world (Hemiptera: Heteroptera). Classification and natural history. Comstock, Ithaca

Schuh RT, Galil B, Polhemus JT (1987) Catalog and bibliography of Leptopodomorpha (Heteroptera). Bull Am Mus Nat Hist 185:243–406

Stock MW, Lattin JD (1976) Biology of intertidal *Saldula palustris* (Douglas) on the Oregon coast (Heteroptera: Saldidae). J Kans Entomol Soc 49:313–326

Wiley GO (1922) Life history notes on two species of Saldidae (Heteroptera) found in Kansas. Kans Univ Sci Bull 14:301–311

Wroblewski A (1966) Shore bugs (Heteroptera, Saldidae) of Poland. Polskie Pismo Ent 36:219–302

Zhang WT, Yao YZ, Ren D (2011) New shore bug (Hemiptera, Heteroptera, Saldidae) from the Early Cretaceous of China with phylogenetic analyses. Zookeys 130:185–198

Zhang WT, Yao YZ, Ren D (2012) A new Early Cretaceous shore bug (Hemiptera: Heteroptera: Saldidae) from China. Zootaxa 3647:585–592

Part VI
Infraorder Cimicomorpha

Chapter 9
Minute Pirate Bugs (Anthocoridae and Lyctocoridae)

Diego L. Carpintero

Abstract A brief historical overview of the study of Anthocoridae and Lyctocoridae (Hemiptera: Heteroptera) found in the Neotropical region is presented. Dichotomic keys are provided for the recognition of the subfamilies to the subtribal level and a taxonomic list of all species known to date as well as images for guidance. Comments on most important species and a summary of the main literature are included.

9.1 Introduction

The first and most extensive taxonomic revision of the Anthocoridae sensu lato was made by O. M. Reuter, in 1884, describing many new taxa and reviewing the known taxa until then. Knowledge of Neotropical fauna, little studied until then except for the work of Stål (1860, 1873), was widely enriched by Reuter's monograph. Subsequently, this author described some new taxa (1909).

Another essential work for understanding the Neotropical fauna of anthocorids was made by Champion (1900), describing new taxa and with the particularity of providing excellent drawings of many of the species described by him. Also Poppius (1909, 1913) provided a number of new taxa for the continental fauna.

Other important taxonomical and faunistic contributions were chronologically made by Herring (1966a, b, 1976), especially the latter, with an excellent key using metapleural scent gland as a discriminatory character for genera. Péricart (1972) made an excellent monograph, and the same author (1996) wrote the catalog of Palaearctic fauna, very useful to show us his very authoritative point of view regarding the taxonomy of this group. Kelton (1978) made an excellent work on the fauna of Canada, very useful for those who work with Neotropical fauna. Later, Froeschner (1981, 1985) and Carpintero et al. (1997) made regional catalogs. Schuh and Slater

D.L. Carpintero (✉)
Museo Argentino de Ciencias Naturales Bernardino Rivadavia,
Avenida Angel Gallardo 470, Buenos Aires 1405, Argentina
e-mail: dcarpint@macn.gov.ar

A.R. Panizzi, J. Grazia (eds.), *True Bugs (Heteroptera) of the Neotropics*,
Entomology in Focus 2, DOI 10.1007/978-94-017-9861-7_9

(1995) and Lattin (2000) presented an overview of the family worldwide; the latter concentrated his review in managed ecosystems; Carpintero (2002) wrote the first catalog devoted to Neotropical species. Carpintero and Dellapé (2006, 2012) revised and described new Neotropical taxa.

Monophyly of the family Anthocoridae, after the excellent work of Carayon (1972), was challenged by posterior analysis by Schuh and Štys (1991) and Schuh et al. (2009) who studied the phylogenetic relationships of this group among the Cimicomorpha. Ying et al. (2008) and Min et al. (2012) followed this criterion, but they did not use the "Lasiochilidae" for their molecular analysis. Carpintero and Dellapé (2008) described and analyzed a series of new characters of the tribe Cardiastethini, highlighting the need for more extensive studies of the morphology of this group, as a previous step toward a best comprehension of their taxonomy.

With respect to morphology, biology, ecology, and importance of these bugs, it should be mentioned the above works of Carayon (1972), Péricart (1972), and Schuh and Slater (1995) and also the work on the biology of the Heteroptera by Miller (1971). Slater (1982) made an excellent characterization of the family.

The taxonomic criteria used in this chapter, which will give validity to family status of the Lyctocoridae, considered their morphological characters, in addition to their ecology, warrant the use of such taxonomic status. As for lasiochilines, they continue to be considered among the Anthocoridae. A review of this subfamily is still in arbitration for publication, which restore its status as a subfamily of Anthocoridae.

9.2 General Characteristics and Diagnosis

Anthocoridae are small to minute (1.15–5 mm) insects, somewhat flattened, with porrect head and ocelli always present in macropterous forms. First instar nymphs have five ommatidia. The labium is four segmented (but appears to be three segmented). The labrum is wide and flap-like. The metathoracic scent gland is deeply bilobed behind, and a median accessory gland is present. The forewing has an elongated, median chorial groove, with a short ventral fold, which does not reach the cuneus along the costal margin. In macropterous forms, a cuneus is present (except in *Amphiareus*), and the membrane has 3–4 not very evident veins. The hind wing has or lacks a hamus. Abdominal terga II and III have the dorsal laterotergites separated from the mediotergites, but tergum IV to the posterior end is entire (II and III are entire in a few brachypterous species). Nymphs have dorsal abdominal scent gland openings present between terga III–IV, IV–V, and V–VI (a vestigial gland opening is sometimes present between VI and VII). The tibiae have spongy fossae either well developed or vestigial; foretibiae in males are, in most of genera, armed with a row of teeth or spines in the inner margin; the tarsi have three segments. Male genitalia are always asymmetrical. Hemocoelic insemination is present in Anthocorinae. It occurs by the aedeagus that becomes a membranous tube, sliding along a groove in the left paramere. The coupling mode of Lasiochilinae is controversial, regarding the "normal" coupling suggested by Carayon (1957, 1972). A parietovaginal gland is present (Lasiochilinae). The paragenital system has either advanced seminal conceptacles or copulatory tubes. Fertilization takes place in the vitellarium. There are two to seven testes follicles. The ovipositor is usually developed or sometimes reduced. The principal salivary gland is bilobed, and the accessory gland is elongated (modified from Slater 1982).

Lyctocoridae are characterized by having males with left paramere flat, with a single-sheet form, and without median sulcus. The *phallus* is tubular and well visible, with *ductus seminis* striated transversely and extended apically by a *processus gonopori* called acus and with the shape of an injection needle. The female has genital apophyses in the middle of the anterior region of sternite VII.

Most species of Anthocoridae have functional wings and are capable of flight. In a few taxa, it occurs in both long-winged (macropterous) and short-winged (brachypterous) forms, e.g., species of *Xylocoris* and *Lasiochilus*, and some other are known only for brachypterous species as *Astemmocoris* and some Lasiochilinae and Cardiastethini.

It is noteworthy that, for a correct taxonomic characterization of this family, at least until the generic level, the shape of metathoracic scent gland is a character constant enough to be used successfully.

9.3 General Biology and Ecology

Anthocoridae are chiefly predaceous bugs that occur on a variety of habitats where they feed on different arthropods, including mites and insects. A few are plant feeders, at least in part, chiefly on pollen [e.g., *Orius insidiosus* (Say), *O. tristicolor* (White), and *Paratriphleps laeviusculus* (Champion)]. A European species occurs in ant nests [(*Xylocoris formicetorum* (Boheman)]. Little is known about this myrmecophilus species, but what is known suggests that other anthocorids may be found in such sites, perhaps even in termite colonies. Other species occurs in the litter layer (*Lasiochilus* spp., *Xylocoris* spp.), on forbs (*Orius* spp.), on shrubs (*Anthocoris* spp., *Orius* spp.), and on different parts of trees (*Anthocoris* spp., *Brachysteles* sp., *Lyctocoris* spp., and most of Scolopini Scolopina).

Long-distance movement of species of Anthocoridae seems to involve trivial movement combined with movements by air currents. Individuals in the litter layer may be picked up by wind disturbances and moved considerable distances. Southwood (1960) provided an extensive review of the flight activity of the Heteroptera. He reviewed captures of Anthocoridae by means of light and suction traps and airplanes. Glick (1957) did early studies on the arthropod distribution in the air column. One anthocorid that is well represented in his samples was *Orius insidiosus* (Say), a common species in a variety of agroecosystems. Southwood felt that the small size and limited flight capabilities of these bugs made them more dependent on long-range movement via upper air layers. Herring (1958, 1966a) related evidence of Heteroptera that has been moved in this way, specifically via hurricanes.

In view of many continental species of Anthocoridae found on the Caribbean Islands (e.g., *Lasiochilus*, *Cardiastethus*, *Brachysteles*, *Buchananiella*, and *Orius*), long-range movements over water seem to be the most reasonable way for these small insects to have originally colonized these islands. A similar phenomenon was noted on the Seychelles Islands by Distant (1913).

9.3.1 Eggs and Oviposition

The egg is ovoid or elongated, with a pale sculptured chorion and a usually circular operculum. There are neither fertilization tubes nor micropyles, but numerous pseudomicropyles are present.

The characters most useful for egg systematics, apart from the general shape and size, include the appearance of the chorial surface (smooth or covered with bumps, cross-linked or not), the shape and the cross-linking of the operculum, and the flange.

Cobben (1968) studied the embryonic development of various Anthocoridae (*Anthocoris*, *Orius*, *Lyctocoris*). The duration of embryonic development after spawning depends greatly on the temperature. The total duration of the hatching takes few minutes.

Oviposition varies according to the different developmental levels of the ovipositor of the females.

9.3.2 Nymph Feeding and Development

The nymphs do not differ greatly from the adult, and growth is driven by shedding or ecdysis. The last one, called fledging, is accompanied by full development of wings (for winged forms) and sexual organs. The normal number of nymphal stages is five, but it is possible that some species have four stages.

All nymphs from 1st to 5th instars are equipped with three dorsal abdominal odoriferous glands, more or less circular; they appear reddish by transparency in the early stages and come out by two posterior holes on the intersegmental membranes III–IV, IV–V, and VI–V. Nymphs do not bear ocelli.

The wing development appears only gradually in nymphs. Completely absent in the 1st instar, they are sometimes detectable in the 2nd nymph as slight posterior edges of the mesonotum and metanotum. In the 3rd instar, these wing pads are always visible and present in the form of small-posterior lateral extensions of the meso- and metanotal plates. In the 4th instar they are more developed, but hemelytra do not completely cover the hind wings. At the last 5th instar, hemelytral and wing extensions usually hide the hind wings (except on the inner edge). Tarsi show two segments at all nymphal stages. The sexual organs are externally invisible except at the 5th instar, where ovipositor of the female is apparent.

Nymphal food does not differ substantially from that of adults. The young Anthocoridae, after hatching, are voracious predators of small arthropods, and they attack the same prey that of adults, preferring only individuals proportional to their size (eggs and young larvae). The number of prey consumed can reach or exceed, in *Anthocoris* and *Orius*, several hundred mites or a hundred aphids during nymphal development of a single individual, and predatory activity is most important during the 5th instar. It seems at least one meal is needed between two molts, and the main meal precedes the ecdysis. Growth is sometimes compatible with a not predatory feeding habit (Péricart 1972).

Key for determining the age of nymphs (modified from Péricart 1972):

1. Without or with barely distinguishable wing pads..2
– Wing pads distinguishable..4
2. Eyes with only five well-developed ommatidia..3
– Eyes clearly with more than five ommatidia..2nd instar
3. Body length substantially less than 1.7 times that of the egg..................1st instar
– Body length substantially greater than 1.7 times that of the egg............2nd instar
4. The hemelytral and hind wing pads are very small. The first did not arrive at the basis of the second...3rd instar
– The hemelytral and hind wing pads are larger. The first, covering at least part of the second..5
5. The hemelytral pads do not reach the apex of functional wings.............4th instar
– The hemelytral pads arrive beyond the apex of functional wings............5th instar

9.3.3 Adult Feeding and Reproduction

Much of the Anthocoridae live on plants, on leaves, on branches, on flowers, or under the bark, where they seek their prey: Sternorrhyncha (aphids, Coccoidea, and psyllids), other Heteroptera (eggs or small nymphs), Thysanoptera (preferred preys of many Oriini), Lepidoptera (eggs or small caterpillars), Psocoptera and Collembola (which are also attacked), and Coleoptera, small species living under bark. They also feed on mites (*Tetranychus*) and oribatida living on plants (Péricart 1972). *Xylocoris* was found in poultry farms feeding on Diptera larvae and eggs, which were developed in the deposits of their feces. Others are subcortical predators of bark beetles (e.g., *Lyctocoris*, *Scoloposcelis*, and *Xylocoris*).

The Lasiochilinae inhabit the soil, but they also live on plants. Many species were reported as introduced in the USA on flowers and fruits (Horton et al. 2013). It is also noted that *Lyctocoris campestris* has the blood-sucking habit, and its presence in bird nests is very common (Aramburú et al. 2009). A tendency toward hematophagy is also manifested occasionally in many species of Anthocoridae. Biting of *Orius* and *Anthocoris* in humans has been reported many times in the Old World (Péricart 1972).

Predatory habits of most Anthocoridae are well recognized; the plant-feeding habits of some are less well known and documented (Lattin 1999).

It is very important to note that many species of this family are polyphagous, feeding on prey and also on plant juices, which determines their success to develop in different environments, both natural and anthropized.

The Anthocoridae sensu lato have in common with other Cimicoidea families (Cimicidae, Polyctenidae, Plokiophilidae) and Nabidae a peculiar mode of insemination which distinguish them from other Hemiptera. In these groups, insemination is not done by pouring the sperm in the female genital tract but injecting it in the hemocoel or in organs secondarily developed for this purpose and not associated with the genital vias, which are subsequently reached. This phenomenon is called "traumatic insemination."

9.4 Classification and Diversity

The family Anthocoridae sensu lato contains between 400 and 600 species distributed worldwide (Lattin 1999) including two families: Anthocoridae and Lyctocoridae. In the Neotropical region, there are 39 genera and 128 known species of Anthocoridae and two genera and six species of Lyctocoridae.

Anthocoridae Fieber *(sensu lato)*
- Anthocorinae Fieber (*sensu* Schuh and Štys 1991)
  - Anthocorini Fieber
  - Oriini Carayon
  - Scolopini Carayon
  - Almeidini Carayon (not present in the Neotropics)
  - Blaptostethini Carayon
  - Cardiastethini Carayon
  - Xylocorini Carayon
- Lasiochilinae Carayon

Lyctocoridae Reuter

Key for suprageneric taxa of Anthocoridae (sensu lato) present in the Neotropics (modified from Carayon 1972; Péricart 1972; Herring 1976; Kelton 1978; Schuh and Slater 1995; and Carpintero 2014):

1. Male with left paramere flat, single-sheet form, without median sulcus. *Phallus* tubular, well visible, with *ductus seminis* striated transversely and extended apically by a processus gonopori called acus, with the shape of an injection needle. Female with genital apophyses in the middle of the anterior region of sternite VII..Lyctocoridae

– Male with left paramere with different shapes but always with a median sulcus throughout much of its length. *Phallus* in membranous tube shape, or with lateral membranous lobes, apically never as an acus. Females without a genital apophysis in sternite VII..Anthocori dae...2

2. Metapleural scent gland curved or straight and directed backward and never prolonged by a carina....................Lasiochilinae
– Metapleural scent gland curved forward, sometimes prolonged by a carina. If curved backward, it is prolonged by a carina and directed forward........................ ....Anthocorinae....................3
3. Antennal segments II–IV of equal thickness; segment III and IV fusiforms, with erected hairs never longer than twice the diameter of the segment.......................4
– Antennal segments III and IV always thinner than II, with erected hairs longer than twice the diameter of the segment....................5
4. Claws with large pulvilli; anterior male tibiae with a row of teeth in their inner margin; left paramere spiraled....................Oriini
– Claws without pulvilli; anterior male tibiae without a row of teeth in their inner margin; left paramere falciform....................Anthocorini
5. Males with a brush of hair on the left side of the fourth abdominal sternum. Females with two short copulatory tubes opening medially in the intersegmental membranes VII–VIII, and each formed of an anterior segment and a posterior segment....................Blaptostethini
– Males without the above brush of hairs. Females, if having copulatory tubes, are never with these features....................6
6. Males with a unique glandular opening on sternum IV or V. Metapleural scent gland is short and gently curved forward and whose apex is located near the center of the metapleura.......................................Scolopini...............................7
– Males without glandular opening on sternum IV or V. Metapleural scent gland is long and strongly curved forward and whose apex reaches the anterior margin of the metapleura. If the apex of the scent gland does not reach the anterior margin of the metapleura that extends to the anterior margin by a thin carina.................8
7. The body is oval, with long and thin legs. Anterior femora are not thickened nor provided with teeth or spines. The corium is punctate. Metasternum with a median longitudinal carina....................Calliodina
– Body depressed dorsoventrally, with subparallel sides. Legs robust. Femora, especially the anterior and posterior ones, are very thick and provided with spines or teeth on the ventral edge. The corium is smooth. Metasternum without a median carina....................Scolopina
8. Ovipositor well developed; anterior tibiae widened at its apex, with a wide *fossula spongiosa*. Metapleural scent gland is never prolonged by a carina. Externally angular....................Xylocorini
– Ovipositor vestigial; anterior tibiae thin, with *fossula spongiosa* vestigial or very small. Metapleural scent gland is prolonged by a carina and directed forward. If there is no carina, it is externally rounded..Cardiastethini

Plate 9.1

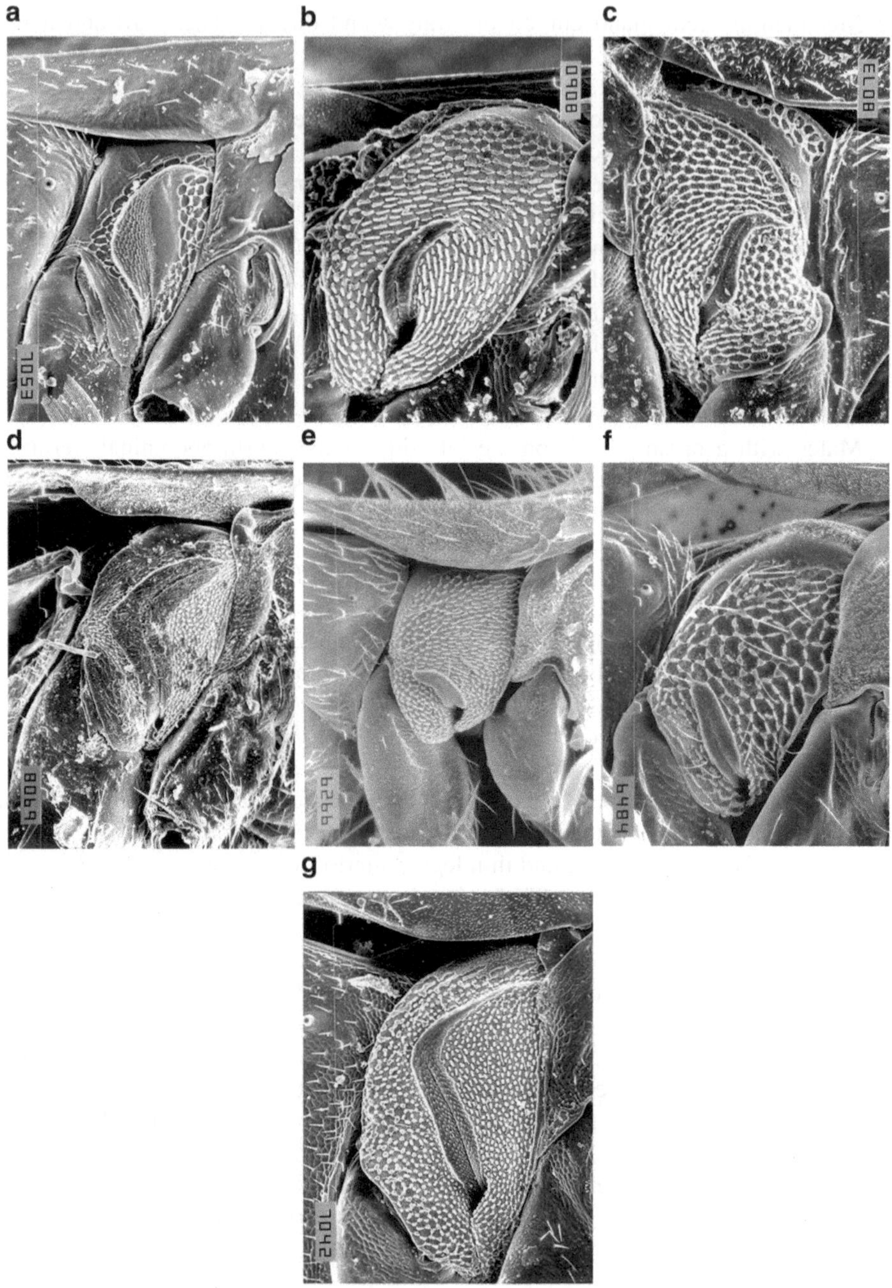

Plate 9.1 (**a**) Genus *Orius* (Oriini), metapleura. (**b**) Genus *Calliodis* (Scolopini), metapleura. (**c**) Genus *Buchananiella* (Cardiastethini), metapleura. (**d**) Genus *Xylocoris* (Xylocorini), metapleura. (**e**) Genus *Lasiochilus* (Lasiochilinae), metapleura. (**f**) Genus *Plochiocoris* (Lasiochilinae), metapleura. (**g**) Genus *Lyctocoris* (Lyctocorinae), metapleura (Photos by D Carpintero)

9.4.1 Main Species

Several species of Anthocoridae are used in biocontrol, some with great success in recent years, such as species of the genus *Orius*, *Anthocoris*, *Montandoniola*, and *Xylocoris*, to control important agricultural pests such as thrips, aphids, and lepidopteran larvae (Schuh and Slater 1995).

The Neotropical species of the tribes of Anthocoridae, most used as biocontrol agents, are listed below as well as their distribution:

Abbreviations: CA = Central America; CI = Caribbean Islands; PC = Pacific Islands; SA = South America

Family Anthocoridae Fieber
Subfamily Anthocorinae Fieber
Tribe Xylocorini Carayon
Xylocoris (Arrostelus) flavipes (Reuter)
Dist. SA
X. (Proxylocoris) afer (Reuter)
Dist. CA, CI, SA
X. (P.) albonotatus (Champion)
Dist. CA
X. (P.) bimaculatus (Champion)
Dist. CA
X. (P.) cacti Carayon
Dist. CA
X. (P.) sordidus (Reuter)
Dist. CA, CI, PC, SA
X. (X.) vicarius (Reuter)
Dist. SA

Remarks: Yamada et al. (2006) Ballal et al. (2003), and Sing and Arbogast (2008) demonstrated the use of several species of this genus to control, especially of, stored grain pests.

Tribe Anthocorini Fieber
Anthocoris albiger Reuter
Dist. CA
A. antevolens Buchanan-White
Dist. CA, SA
A. dentipes Champion
Dist. CA
A. fulvipennis Reuter
Dist. CA
A. nigripes Reuter
Dist. CA

A. obsoletus (Blanchard)
Dist. SA
A. rufotinctus Champion
Dist. CA
A. variicornis Champion
Dist. CA
A. variipes Champion
Dist. CA
Dufouriellus ater (Dufour)
Dist. SA
Elatophilus (E.) antennatus Kelton
Dist. CA
Macrotrachelia albovittata Champion
Dist. CA
M. elongata Champion
Dist. CA
M. nigronitens (Stål)
Dist. CA, SA
M. nitida Champion
Dist. CA
M. opacipennis Champion
Dist. CA
M. thripiformis Champion
Dist. CA
Melanocoris nigricornis Champion
Dist. CA
M. obovatus Champion
Dist. CA

Remarks: Species of this tribe are used in biocontrol of Homoptera (Mullois-Juan and Bravaccini 2009; Péricart 1972).

Tribe Oriini Carayon
Macrotracheliella conica (Blanchard)
Dist. SA
M. laevis Champion
Dist. CA, SA
Orius alcides Herring
Dist. SA
O. brevicollis (Blanchard)
Dist. SA

O. championi Herring
Dist. CA, SA
O. euryale Herring
Dist. CA
O. flaviceps (Poppius)
Dist. CA, CI
O. florentiae Herring
Dist. SA
O. fuscus (Reuter)
Dist. CA, SA
O. insidiosus (Say)
Dist. CA, CI, SA
O. ixionides Herring
Dist. CA
O. jasiones Herring
Dist. CA
O. lesliae Herring
Dist. SA
O. pallidus (Poppius)
Dist. SA
O. pele Herring
Dist. SA
O. perpunctatus (Reuter)
Dist. CA, CI, SA
O. pumilio (Champion)
Dist. CA CI
O. thyestes Herring
Dist. SA
O. tristicolor (Buchanan-White)
Dist. CA, CI, PC, SA
Paratriphleps laeviusculus Champion
Dist. CA, CI, SA
P. pallidus (Reuter)
Dist. CI

Remarks: Probably this group includes the anthocorids which are more used as biocontrol. They are particularly used to control thrips populations in cultivated plants (Péricart 1972; Kelton 1978; Lattin 1999).

Plate 9.2

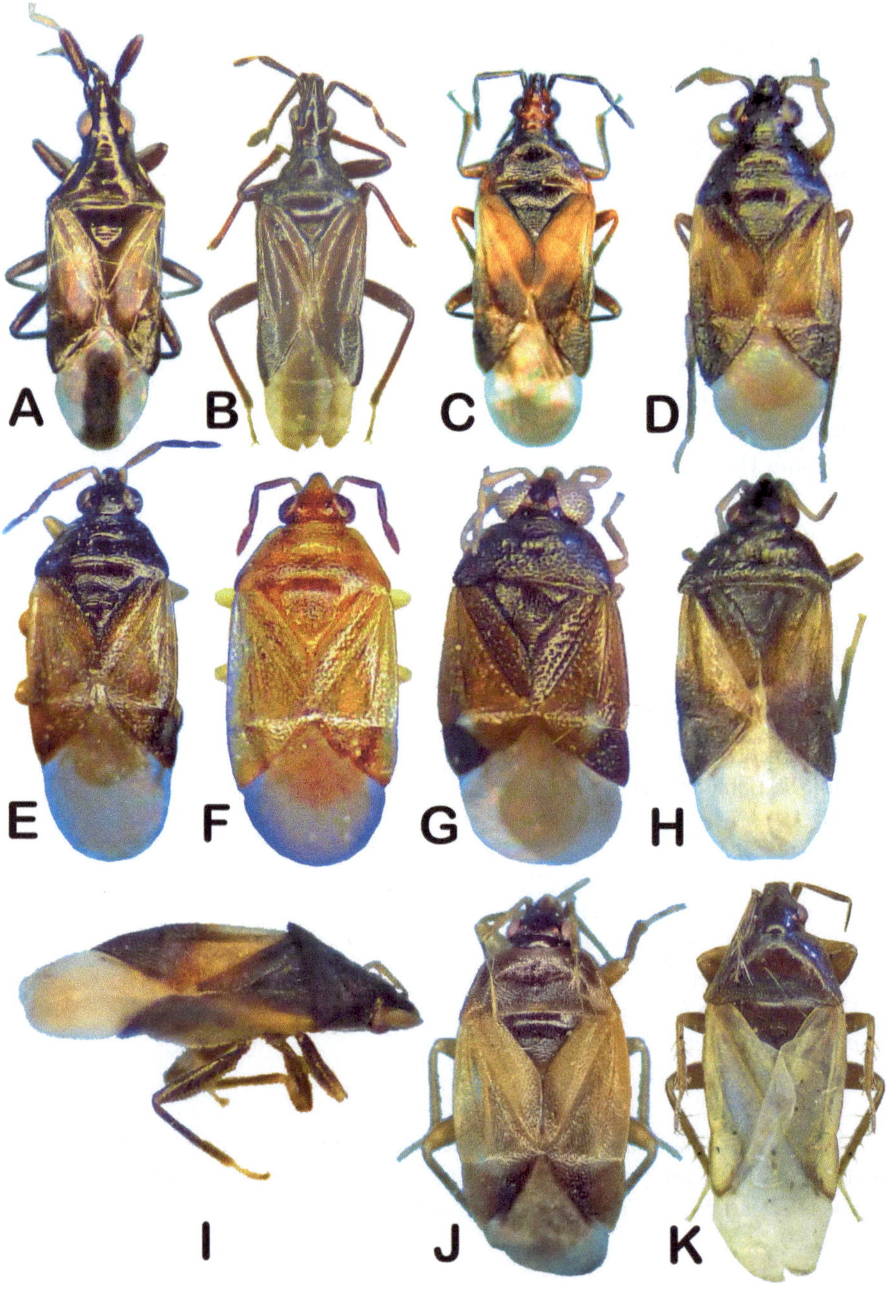

Plate 9.2 (**a**) *Montandoniola moraguesi* (Oriini), habitus. (**b**) *Macrotrachelia nigronitens* (Anthocorini), habitus. (**c**) *Anthocoris nemoralis* (Anthocorini), habitus. (**d**) *Orius insidiosus* (Oriini), habitus. (**e**) *Orius reedi* (Oriini), habitus. (**f**) *Orius pallidus* (Oriini), habitus. (**g**) *Orius perpunctatus* (Oriini), habitus. (**h**) *Orius tristicolor* (Oriini), habitus. Dorsal view. (**i**) *Orius tristicolor* (Oriini), habitus. Lateral view. (**j**) *Lyctocoris campestris* (Lyctocoridae), habitus. (**k**) *Xylocoris flavipes* (Xylocorini), habitus (Photos by D Carpintero)

There are also three known introduced species in the Neotropics: *Montandoniola moraguesi* (Puton) (Anthocorinae: Oriini).

Remarks: Palearctic distribution. Introduced in Bermuda and Hawaii for biocontrol of thrips (Dobbs and Boyd 2006). Expanded to Central and South America. First record for Brazil and Argentina.

Blaptostethus pallescens Poppius (Anthocorinae: Blaptostethini)
Remarks: This species is distributed in Asia and Africa, introduced in Brazil (first record), and studied as a biocontroller of the stored rice grain moth, *Corcyra cephalonica* Stainton (Ballal et al. 2003).

Anthocoris nemoralis (F.) (Anthocorinae: Anthocorini)
Remarks: Palearctic distribution. This species is introduced in Canada, first recorded in the Neotropics of Chile, and used in Europe for biocontrol of *Cacopsylla pyri*, a pest of pears (Carpintero 2014).

9.4.2 Secondary Species

Family Lyctocoridae Reuter
Astemmocoris cimicoides Carayon & Usinger
  Dist. SA
Lyctocoris (L.) *campestris* (F.)
  Dist. CA, PC, SA
L. (L.) latus Poppius
  Dist. SA
L. (L.) mexicanus Kelton
  Dist. CA
L. (Metriosteles) signoreti Reuter
  Dist. SA
L. (M.) spanbergii Reuter
  Dist. SA

Plate 9.3

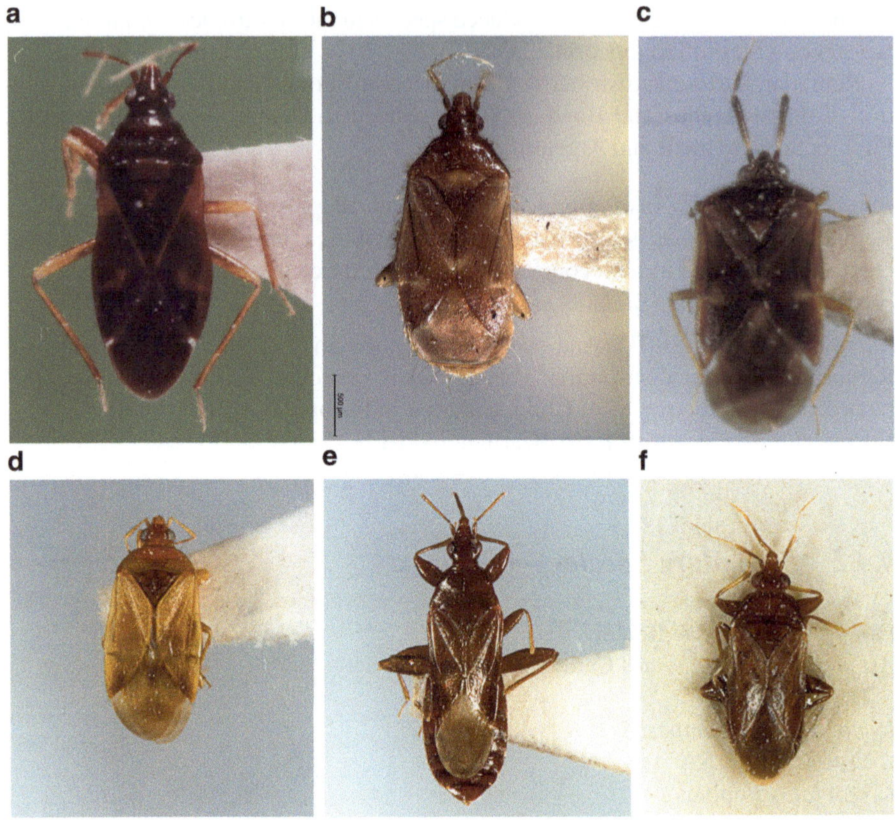

Plate 9.3 (**a**) *Lasiocolpus minor* (Lasiochilinae), habitus. (**b**) *Dilasia microps* (Lasiochilinae), habitus. (**c**) *Buchananiella continua* (Cardiastethini), habitus. (**d**) *Calliodis pallescens* (Scolopini Calliodina), habitus. (**e**) *Scolopocoris gracilicornis* (Scolopini Scolopina), habitus. (**f**) *Solenonotus* sp. (Scolopini), habitus (Photos by D Carpintero)

Family Anthocoridae Fieber
Subfamily Lasiochilinae Carayon
(Modified from Carpintero 2002)
Dilasia ashlocki (Herring)
  Dist. PC
D. basalis (Reuter)
  Dist. CA, CI, SA
D. colludens (Buchanan-White)
  Dist. SA
D. foveicollis (Champion)
  Dist. CA, CI
D. fuscula Reuter
  Dist. CA, CI

D. reuteri (Champion)
Dist. CA
D. unicolor (Reuter)
Dist. CA, CI, SA
Eusolenophora divisa (Champion)
Dist. CA, CI
E. testacea Poppius
Dist. SA
Lasiochilus galateae Reuter
Dist. SA
L. hirtellus Drake & Harris
Dist. CA, CI
L. microps Champion
Dist. CA, CI, SA
L. mirificus Drake & Harris
Dist. CA, CI
L. pallidulus Reuter
Dist. CA, CI, PC, SA
L. punctipennis Champion
Dist. CA, SA
L. sulcatus Champion
Dist. CA, SA
L. sulcicollis Reuter
Dist. SA
L. varicolor Uhler
Dist. CA, CI, SA
Lasiocolpus biguttatus Poppius
Dist. SA
L. elegans Reuter
Dist. SA
L. minor Champion
Dist. CA
L. rostralis (Poppius)
Dist. SA
L. sinuaticollis Reuter
Dist. CA
L. unicolor Poppius
Dist. SA
Plochiocoris longicornis Champion
Dist. CA, CI, SA
P. pilosus (Reuter)
Dist. SA

Tribe Cardiastethini Carayon
Alofa sodalis (Buchanan-White)
Dist. CI, PC, SA
Amphiareus constrictus (Stål)
Dist. CA, PC, SA
Buchananiella continua (Buchanan-White)
Dist. SA
B. devia Bergroth
Dist. PC
Cardiastethus aequinoctialis Poppius
Dist. PC, SA
C. cubanus Poppius
Dist. CI
C. elegans (Blanchard)
Dist. SA
C. flavus Poppius
Dist. CI
C. fraterculus Van Duzee
Dist. CA, CI
C. rugicollis Champion
Dist. CA, CI
C. testaceus (Blanchard)
Dist. SA
C. tropicalis Champion
Dist. CA, CI
C. uhleri China
Dist. CA, CI
Dasypterus assimilis Reuter
Dist. CA CI
D. discifer (Stål)
Dist. SA
D. limbatellus (Stål)
Dist. CA, PC, SA
D. ophthalmicus (Reuter)
Dist. CA, SA
Dolostethus pubescens Henry & Herring
Dist. SA
Pehuencoris gurguliophagus Carpintero & Dellapé
Dist. SA
Physopleurella floridana Blatchley
Dist. CI, SA

Tribe Scolopini Carayon
Subtribe Calliodina Carayon
Calliodis bifasciatus (Champion)
Dist. CA, CI
C. clarus (Buchanan-White)
Dist. SA
C. coloratus (Poppius)
Dist. SA
C. crawfordi (Poppius)
Dist. CA
C. maculipennis (Reuter)
Dist. CI
C. nebulosus (Uhler)
Dist. CA, CI, PC
C. pallescens (Reuter)
Dist. CA, CI, SA
C. picturatus Reuter
Dist. SA
C. pictus (Uhler)
Dist. CA, CI
C. punctatostriatus (Reuter)
Dist. CA, SA
C. signatus (Poppius)
Dist. CI
Eulasiocolpus megalops Champion
Dist. CA, SA
Guayascoris foreroi Carpintero & Dellapé
Dist. SA
Lasiocolpoides ciliatus Champion
Dist. CA
Lepidonannella opaca (Poppius)
Dist. SA
Nidicola aglaia Drake & Herring
Dist. CA
N. engys Drake & Herring
Dist. CA
N. etes Drake & Herring
Dist. CA
N. marginata Harris & Drake
Dist. CA

N. mazda Herring
Dist. PC
N. mitra Drake & Herring
Dist. CA
Opisthypselus punctaticollis Reuter
Dist. CA, SA
Zopherocoris armatus (Stål)
Dist. CA, SA

Subtribe Scolopina Carayon
Ameroscolopa flavicornis (Reuter)
Dist. CA, GT, MX
A. basilicus (Drake & Harris)
Dist. NA
Lasiochiloides denticulatus Champion
Dist. CA
L. socialis Drake & Harris
Dist. CA
Scolopa wygodzinskyi Carayon
Dist. SA
Scolopella brasiliensis Carayon
Dist. SA
Scolopocoris gracilicornis (Poppius)
Dist. SA

Remarks: Many of these species could potentially be used as biocontrol, using their versatility in the selection of their prey (Lyctocoridae), their wide diversity and large populations in the Neotropics (Calliodina), the diversity of environments that colonize (Cardiastethini), and their resistance (Lasiochilinae), allowing them to be transported over long distances and colonize new areas.

9.5 Concluding Remarks

Lattin (1999) mentioned that "colonization of non-native species is a worldwide problem that affects natural biological diversity and alters species richness whenever it occurs."

There are three species of Anthocoridae that have been introduced into Neotropical areas, accidentally or deliberately. Among them, only *Montandoniola moraguesi* (Puton) was intentionally introduced for biocontrol purposes and appears to have become established. Probably there are other nonindigenous species that await discovery and recognition.

Over recent years we have received inquiries about the identity of anthocorid species found in both wild and cultivated areas. The presence of alien species found in these samples, plus the purchase by producers of *Orius insidiosus* from abroad, should be of alert to those responsible for plant health and quarantine. We do not believe it is necessary to introduce new foreign species into our fauna for biocontrol, because the Neotropics has many species of anthocorids potentially suitable for such purpose.

References

Aramburú RM, Calvo G, Carpintero DL, Cicchino AC (2009) Artrópodos presentes en nidos de cotorra *Myiopsitta monacha* (Aves: Psittacidae). Rev Mus Arg Cienc Nat (NS) 11:1–5

Ballal CR, Singh SP, Poorani J, Gupta T (2003) Biology and rearing requirements of an anthocorid predator, Blaptostethus pallescens Poppius (Heteroptera: Anthocoridae). J Biol Control 17:29–33

Carayon J (1957) Introduction á l'étude des Anthocoridae omphalophores (Hemiptera: Heteroptera). Ann Soc Entomol France 126:158–197

Carayon J (1972) Caractères Systématiques et classification des Anthocoridae (Hemipt.). Ann Soc Entomol France (NS) 8:309–349

Carpintero DL (2002) Catalogue of the Neotropical Anthocoridae (Heteroptera). Rev Soc Entomol Arg 61:25–44

Carpintero DL (2014) Cimicoidea y Naboidea. In: Roig-Juñent SA,Claps LE, Morrone JJ (eds) Biodiversidad de artrópodos argentinos, vol 3. Nat Univ Tucumán, San Miguel de Tucumán, Argentina (in press), pp 365–376

Carpintero DL, Dellapé PM (2006) *Pehuencoris:* new genus of Cardiastethini (Heteroptera: Anthocoridae) from southern Argentina and Chile (Patagonia). Zool Sci 23:1039–1042

Carpintero DL, Dellapé PM (2008) *Rajburicoris*, a new genus of Cardiastethini, and discussion of the systematic position of *Dufouriellus* (Heteroptera: Heteroptera: Anthocoridae). Acta Entomol Mus Nat Pragae 48:503–509

Carpintero DL, Dellapé PM (2012) Neotropical Scolopini (Hemiptera: Heteroptera: Anthocoridae): new taxa, diagnostic characters and a key to the genera of the tribe. Acta Entomol Mus Nat Pragae 52:49–66

Carpintero DL, Maës JM, Coscarón MC (1997) Catálogo de los Anthocoridae (Heteroptera) de Nicaragua. Rev Nica Entomol 41:23–26

Champion GC (1900) Anthocoridae Insecta Rhynchota Hemiptera-Heteroptera, 2. In: Godman FD, Salvin O (eds) Biologia Centrali-Americana. Taylor & Francis, Boca Raton, pp 306–336

Cobben RH (1968) Part 1. Eggs, architecture of the shell, gross embryology, and eclosion. In: Evolutionary trends in Heteroptera. Centre for Agricultural Publishing and Documentation, Wageningen

Distant WL (1913) Reports of the Percy Sladen Trust Expedition to the Indian Ocean in 1905. Trans Linn Soc Lond 16:139–191

Dobbs TT, Boyd DW Jr (2006) Status and distribution of *Montandoniola moraguesi* (Hemiptera: Anthocoridae) in the continental United States. Fla Entomol 89:41–46

Froeschner RC (1981) Family Anthocoridae Amyot & Serville. Heteroptera or true bugs of Ecuador: a partial catalog. Smithson Control Zool 122:12–13

Froeschner RC (1985) Anthocoridae Amyot and Serville. Synopsis of the Heteroptera or true bugs of the Galápagos Islands. Smithson Control Zool 407:7–10

Glick PA (1957) Collecting insects by airplane in southern Texas. US Dep Agric Tech Bull 1158:1–28

Herring JL (1958) Evidence for hurricane transport and dispersal of aquatic Hemiptera. Pan-Pac Entomol 34:174–175

Herring JL (1966a) The Anthocoridae of the Galápagos and Cocos Islands (Hemiptera). Proc Entomol Soc Wash 68:127–130

Herring JL (1966b) The genus *Orius* of the Western Hemisphere (Hemiptera: Anthocoridae). Ann Entomol Soc Am 59:1093–1109

Herring JL (1976) Keys to genera of Anthocoridae of America North of Mexico, with description of a new genus (Hemiptera: Heteroptera). Fla Entomol 59:143–150

Horton DR, Lewis TM, Dobbs TT (2013) Interception of Anthocoridae, Lasiochilidae and Lyctocoridae at the Miami Plant Inspection Station (Hemiptera: Heteroptera). Fla Entomol 96:482–497

Kelton LA (1978) The Anthocoridae of Canada and Alaska (Heteroptera: Anthocoridae). In: The insects and Arachnids of Canada, Part 4. Publication 1639. Ottawa Agriculture Canada, Ottawa

Lattin JD (1999) Bionomics of Anthocoridae. Annu Rev Entomol 44:207–231

Lattin JD (2000) Minute pirate bugs (Anthocoridae). In: Schaefer CW, Panizzi AR (eds) Heteroptera of economic importance. CRC Press, Boca Raton, pp 607–637

Miller NCE (1971) Anthocoridae (Amyot & Serville) 1843. In: The biology of the Heteroptera. EW Classey Ltd, Hampton, pp 145–147

Min L, Ying T, Ying Z, Wenjun B (2012) Higher level phylogeny and the first divergence time estimation of Heteroptera (Insecta: Hemiptera) based on multiple genes. PLoS ONE 7:1–17

Mullois-Juan L, Bravaccini F (2009) *Anthocoris nemoralis*, depredador natural de la Psila. Hort Internaci 68:28–31

Péricart J (1972) Anthocoridae, Cimicidae et Microphysidae de l'Ouest- Paléarctique. 7. Hemiptera. In: Fauna de l'Europe et du Bassin Méditerranéen. Masson ed, Paris

Péricart J (1996) Family Anthocoridae Fieber, 1836 y Family Cimicidae Latreille, 1802. In: Aukema B, Rieger C (eds) Catalogue of the Heteroptera of the Palaearctic Region, vol 2. The Netherlands Entomological Society, Amsterdam, pp 108–144

Poppius BR (1909) Beiträge zur Kenntnis der Anthocoriden. Acta Soc Sci Fenn 37:1–45

Poppius BR (1913) Beiträge zur Anthocoriden-Fauna von Central- und Nord Amerika. Ann Soc Entomol Belgique 57:11–15

Reuter OM (1884) Monographia Anthocoridarum orbis terrestris. Acta Soc Sci Fenn 14:555–758

Reuter OM (1909) Anthocoridae novae descriptae. Ofv Finsk Vet Soc Förh 51:1–7

Schuh RT, Slater JA (1995) True bugs of the world (Hemiptera: Heteroptera). Classification and natural history. Cornell University Press, Ithaca

Schuh RT, Štys P (1991) Phylogenetic analysis of Cimicomorphan family relationships (Heteroptera). J NY Entomol Soc 99:298–350

Schuh RT, Weirauch C, Wheeler WC (2009) Phylogenetic relationships within the Cimicomorpha (Hemiptera: Heteroptera): a total-evidence analysis. Syst Entomol 34:15–48

Sing SE, Arbogast RT (2008) Optimal *Xylocoris flavipes* (Reuter) (Hemiptera: Anthocoridae) density and time of introduction for suppression of bruchid progeny in stored legumes. Environ Entomol 37:131–142

Slater JA (1982) Hemiptera. In: Parker SP (ed) Synopsis and classification of living organisms, vol 2. McGraw-Hill, New York, pp 417–447

Southwood TRE (1960) The flight activity of Heteroptera. Trans R Entomol Soc Lond 112:173–220

Stål C (1860) Bidrag till Rio de Janeiro-traktens Hemipter-fauna, 1. Kung Sv Vet Handl (NF) 2:1–84

Stål C (1873) Enumeratio Hemipterorum. 3. Kung Sv Vet Handl (NF) 11:1–163

Yamada K, Yasunaga T, Nakatani Y, Hirowatari T (2006) The minute pirate-bugs genus *Xylocoris* Dufour (Hemiptera: Heteroptera: Anthocoridae) from the rice mills in Thailand. Proc Entomol Soc Wash 108:525–553

Ying T, Weibing Z, Min L, Qiang X, Wenjun B (2008) Influence of data conflict and molecular phylogeny of major clades in Cimicomorphan true bugs (Insecta: Hemiptera: Heteroptera). Mol Phylogenet Evol 47:581–597

Chapter 10
Plant Bugs (Miridae)

Paulo S.F. Ferreira, Thomas J. Henry, and Lívia A. Coelho

Abstract The Miridae, or plant bugs, with more than 1,300 genera and 11,130 described species, represent the largest and most diverse family of Heteroptera. Although many plant bugs are phytophagous and may cause serious damage to crops, a great many are predatory and may be important in biological control programs. In the Neotropics, about 561 genera and more than 3,400 species are included in seven of the world's eight recognized subfamilies. In this chapter, we give an overview of the classification, general life history, and economic importance and provide identification keys to the Neotropical subfamilies and tribes. For each subfamily, we present a brief diagnosis, an overview of the respective tribes and subtribes, and key literature.

10.1 Introduction

The classification of the Miridae was first outlined by Reuter (1910), who recognized nine subfamilies and 18 divisions. Van Duzee (1917), in his catalog of the Hemiptera of North America north of Mexico, modified Reuter's names and arrangement and listed seven subfamilies. Knight (1923) recognized nine subfamilies based on studies of the arolia (parempodia), pretarsi, and genitalic structures, a grouping similar to that used by Van Duzee (1917). Carvalho (1952) published a synopsis of the higher classification of the family in which he listed six subfamilies, not including Isometopinae, and 24 tribes. Schuh (1976), in evaluating

P.S.F. Ferreira (✉)
Departamento de Entomologia, Universidade Federal de Viçosa (UFV),
Av. Peter Henry Rolfs s/n, Viçosa, MG 36570.900, Brazil
e-mail: pfiuza@ufv.br

T.J. Henry
Systematic Entomology Laboratory, Agricultural Research Service, United States Department of Agriculture, c/o Smithsonian Institution, National Museum of Natural History, MRC 168, P.O. Box 37012, Washington, DC 20013-7012, USA
e-mail: thomas.henry@ars.usda.gov

L.A. Coelho
Museu Regional de Entomologia, Universidade Federal de Viçosa (UFV),
Viçosa, MG 36570.900, Brazil
e-mail: liviaguiar04@gmail.com

A.R. Panizzi, J. Grazia (eds.), *True Bugs (Heteroptera) of the Neotropics*,
Entomology in Focus 2, DOI 10.1007/978-94-017-9861-7_10

pretarsal structures and femoral trichobothria, presented a cladistic analysis that supported a novel six-subfamily arrangement, including the Psallopinae and Isometopinae, but with the Deraeocorinae merged with the Mirinae and the Orthotylinae merged with the Phylinae. Although generally accepted by European hemipterists, Schuh's inclusion of the Isometopidae as a subfamily of Miridae marked the first time that this arrangement had been accepted in the Western Hemisphere. Schuh et al. (2009) recognized seven subfamilies, omitting the Psallopinae, and Jung and Lee (2012), likewise, supported a seven-family classification (minus Psallopinae) based on molecular sequence data, although they considered the Bryocorinae polyphyletic with the bryocorines (Bryocorini) and dicyphines (Dicyphini) having widely separated positions on their cladogram. Carvalho's (1955a) keys to the genera of the world and first modern world catalog (Carvalho 1957, 1958a, b, 1959, 1960) stimulated major interest in the family, and Schuh's (1995) much-needed update was followed by an invaluable online version (Schuh 2013) that has given unpreceded access to information on the Miridae throughout the world. Currently, eight subfamilies (Bryocorinae, Cylapinae, Deraeocorinae, Isometopinae, Mirinae, Orthotylinae, Phylinae, and Psallopinae) are recognized (Henry 2009; Schuh 2013).

The Miridae, commonly called plant bugs (United States and Canada), capsids (Britain), *Blindwanzen* (Germany), and *blomstertaeger* (Denmark), represent the largest family of Heteroptera (Henry and Wheeler 1988; Schuh and Slater 1995; Wheeler 2000a; Henry 2009) and one of the 20 most diverse families of Insecta (Cassis and Schuh 2012), with more than 11,130 described species (Schuh 2013). Henry and Wheeler (1988) suggested that the world fauna will approach 20,000 once the Neotropical species are fully studied. Plant bugs are found in all major zoogeographic regions of the world, except the Antarctic, with centers of diversity in tropical and Mediterranean ecosystems (Wheeler 2001; Cassis and Schuh 2012; Schuh 2013).

The Miridae form a monophyletic group (Wheeler et al. 1993; Wheeler 2001; Schuh et al. 2009; Jung and Lee 2012) characterized by the presence of femoral trichobothria, the loss of nymphal dorsal abdominal scent glands between segments 4 and 5, asymmetrical parameres, two- or three-segmented tarsi, small pulvilli attached near the bases of the claws, a triangular cuneus at the apex of the corium, and a hemelytral membrane with one or two closed cells (Schuh et al. 2009).

The fossil Miridae are insufficiently studied, and although a large number of species have been named (Wheeler 2001), most of them need revision (Zherikhin 2002; Grimaldi and Engel 2005). The earliest fossil mirid is recorded from the Upper Jurassic (Kukalová-Peck 1991; Shcherbakov and Popov 2002; Yao et al. 2007; Jung and Lee 2012). Leston (1979) suggested that the primary speciation of the Miridae took place during the Upper Cretaceous or Lower Cenozoic, following the rise of the angiosperms to dominance (Crepet 1979; Wheeler 2001). Based on molecular data, using the oldest available fossils as calibration points, Jung and Lee (2012) hypothesized that the earliest mirids may have evolved as early as the Permian.

10.2 General Characteristics and Diagnosis (Fig. 10.1)

Many plant bugs are often brightly colored red, orange, and yellow, often with spots and stripes, but the majority are more cryptically colored, often mottled green, gray, brown, or black, blending in remarkably well with their surroundings (Henry and Wheeler 1988; Wheeler and Henry 2004; Henry 2009). Myrmecomorphy is common in the family (Henry 2009), especially in the subfamilies Mirinae (e.g., Herdoniini), Orthotylinae (e.g., Ceratocapsini), and Phylinae (e.g., Cremnorrhini and Pilophorini). The adults vary considerably in size and appearance. They range

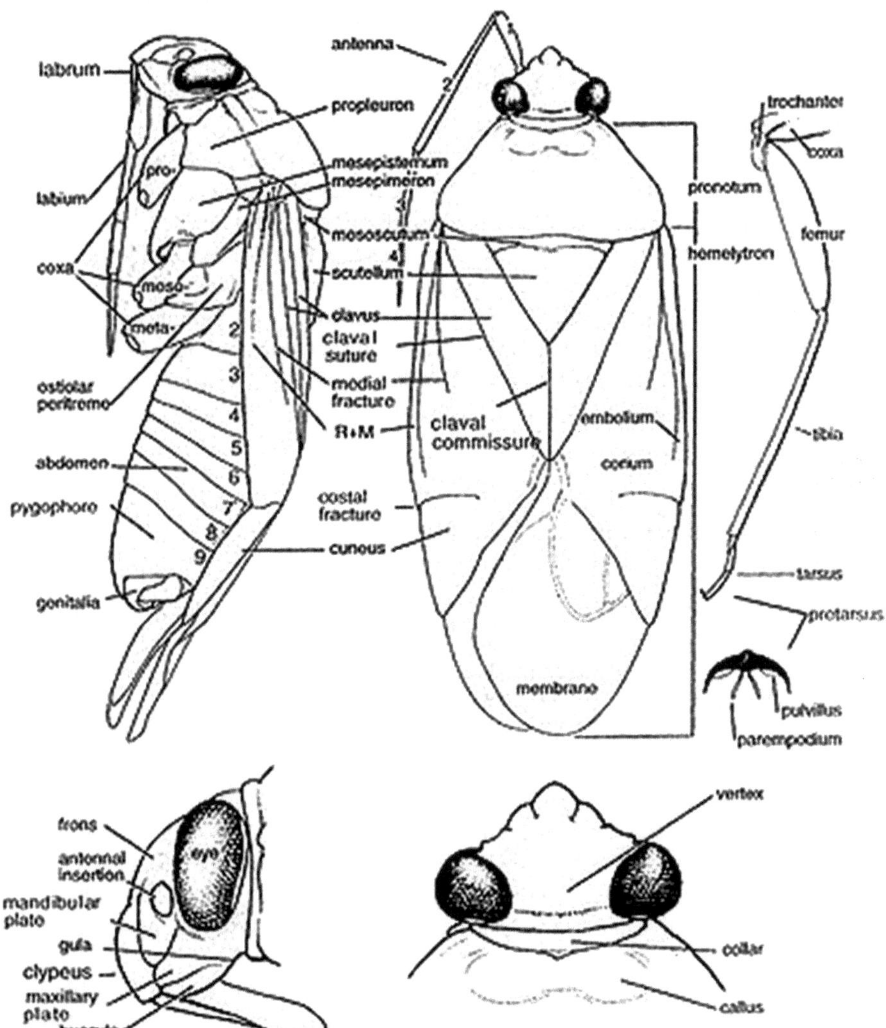

Fig. 10.1 Miridae general morphology (After Schwartz and Foottit 1992)

from little more than one mm, as in certain species of *Tytthus* Fieber, to 15 mm or more in some Neotropical Restheniini. They are elongate to ovoid. The head is triangular and usually prognathous. The eyes are usually large, and ocelli, considered plesiomorphic, are present only in the subfamily Isometopinae. The antennae and labium are four segmented. Most species are fully winged (macroptery), though submacroptery, brachyptery, microptery, and coleoptery are widespread, as are intraspecific wing polymorphism and sexual dimorphism. The hemelytra are distinctive, with the corium divided distally into a triangular cuneus and an apical membrane with one or two unequal cells divided by a short longitudinal vein, a character that is apomorphic for mirids. Trochanters of all legs are divided. The middle and hind femora possess two to 10 trichobothria on the ventral and lateral surfaces. The tarsi are most often three segmented. The pretarsus includes the claws, which can be simple or toothed, with setiform or fleshy parempodia and commonly with fleshy pulvilli and pseudopulvilli. The metathoracic scent glands are paired, and the external efferent or evaporative system is usually well developed. The evaporatorium occupies less than half of the metepisternum with a variably shaped and often diagnostic ostiolar peritreme. The metathoracic spiracle is either exposed and teardrop shaped or recessed. The male genitalia are asymmetrical, including the parameres and endosoma. The endosoma is partially membranous, often inflatable, and is confined by a sclerotized phallotheca. The female ovipositor is laciniate; the sperm storage organ is a large, saclike structure. Male and female genitalia have characters important in inferring higher-group relationships and identifying species (Davis 1955; Kelton 1959; Schuh 1975, 1976; Fontes 1981, 1989, 1996; Henry and Wheeler 1988; Schuh and Slater 1995; Wheeler 2001; Cassis and Schuh 2012).

10.3 General Biology and Ecology

Plant bugs are ecologically successful because of their high species diversity, large populations, and wide geographic ranges (Wheeler 2001). Various mirids have distributions that correspond with long-recognized biogeographic regions and intercontinental patterns (e.g., Gondwanan or Paleotropical) (Schuh and Slater 1995). Biogeographic studies have identified centers of endemism, interrelationships, and areas of species richness correlated with areas with high endemism (Wheeler 2001).

Mirids are abundant across a variety of habitats, ranging from mountain tops to salt marshes and coastal dunes (Southwood and Leston 1959; Marples 1966; Denno 1977; Wheeler 2001). Many species specialize in glandular-hairy and even carnivorous plants (Wheeler 2001), and species of *Ranzovius* Reuter are among the few heteropterans that live as commensals in spider webs (Wheeler and McCaffrey 1984; Henry 1999).

The habits and biology of the Miridae have been documented in numerous, widely scattered publications (Henry 2009). Wheeler (2000a, b, 2001) synthesized data on the biology, hosts, and habits, significantly increasing knowledge of the family. Much of what we know about courtship and oviposition behavior, fecundity, longevity, and voltinism is the result of studies involving species of agricultural importance. The life histories of most plant bugs, however, are unknown.

Miridae may have a significant economic impact on agricultural crops. Many species are major pests (Wheeler 2000a, 2001; Henry 2009; Cassis and Schuh 2012), whereas others are predatory and considered beneficial and have great potential as biological control agents (Wheeler 2000b; Sanchez et al. 2004; Cassis and Schuh 2012). Because even small mirid populations can cause severe injury, applied entomologists frequently underestimate the importance of plant bugs. Feeding symptoms in the Miridae probably are more diverse than in any other heteropteran family (Wheeler 2000a). Most crop plants have at least one mirid species among their complement of pests. Changes in agricultural practices have been responsible for increasing the host range of many plant bugs, and the introduction of numerous exotic plants has increased the supply of new hosts and has allowed the establishment of nonnative taxa around the world (Wheeler 2001).

Mirids often are not recognized as predators. Species with predatory habits do not exhibit the typical specialized morphological characteristics for capturing prey. Predation by mirids on arthropod eggs and early instars of other arthropods make them potentially important in biological control (Wheeler 2000b). Even plant bugs that are significant crop pests can, under certain conditions, be considered beneficial because they are able to switch from phytophagy to zoophagy (Wheeler 2000a, b, 2001). A mirid that is normally a key crop pest can become an important predator of other pests on the same crop. Some species have been described either as pests or as useful natural enemies, depending on where they occur, seasonality, hosts, control measures, and availability of alternative food sources (Wheeler 1976, 2000b, 2001).

Their piercing-sucking mouthparts and the presence of powerful salivary secretions allow mirids to explore nutrient resources, such as dried carrion and feces (Wheeler 2001). Most plant bugs are oligophagous, occurring on members of one plant family or on plants belonging to a few closely related families (Wheeler 2000a). Only a few species are considered monophagous (Wheeler 2001). For example, Pires et al. (2007, 2008) showed that *Platyscytus decempunctata* (Carvalho) feeds only on *Solanum cernuum* Vell. (Solanaceae). In contrast, some species are strongly polyphagous, such as *Lygus lineolaris* (Palisot de Beauvois), which may feed on more than 300 plants (Young 1986).

10.4 Life History

Mirids often have life cycles that last no more than six weeks. They usually have five nymphal instars, although the number may be as low as three or as high as six (Schuh and Slater 1995; Wheeler 2001). They reach the adult stage in 15–30 days. Eggs are usually inserted into plant tissue. In most temperate regions, mirids are univoltine and overwinter in the egg stage. Others, however, such as *Lygus* spp., have multiple generations per year and often overwinter as adults. Eclosion occurs in early to late spring in phytophagous species, which are associated with new growth and flower development of the host plant. Predatory species often have life cycles synchronized with those of their preferred prey and usually appear as adults later in the season than phytophagous species. The life histories of most tropical Miridae, however, are poorly studied or are unknown (Schuh and Slater 1995; Wheeler 2001).

Plant bugs are remarkable for their genitalic evolution (Kelton 1959; Cassis and Schuh 2012), including traumatic insemination and sexual antagonistic coevolution (Tatarnic et al. 2006; Tatarnic and Cassis 2010; Cassis and Schuh 2012). Parthenogenesis occurs in a few mirid species, such as the predatory *Campyloneura virgula* (Herrich-Schaeffer), where males are rare (Wheeler and Henry 1992; Henry 2012a).

The Miridae have significant potential as indicators of ecological changes, mainly because they are susceptible to many pesticides and are vulnerable to habitat disturbances (Fauvel 1999; Wheeler 2001). Brown (1991) presented a list of mirids as potential bioindicators in conservation biology studies. Taking into account their great species diversity, wide range of feeding habitats, and the damage they inflict on fruits, field crops, and ornamental plants, the Miridae are among the most important of all insect groups (Henry and Wheeler 1988; Wheeler 2001; Cassis and Schuh 2012).

10.5 Classification and Diversity

The Miridae, placed (with Thaumastocoridae and Tingidae) in the superfamily Miroidea, are currently separated into eight subfamilies: Bryocorinae, Cylapinae, Deraeocorinae, Isometopinae, Mirinae, Orthotylinae, Phylinae, and Psallopinae (Schuh 2013). More than 11,130 described species are included in 44 tribes worldwide (Schuh 2013; Schuh and Menard 2013).

All mirid subfamilies represented in the Neotropics (considering all of Mexico, Central and South America, and the West Indies), except the Psallopinae (Wolski and Henry 2015). About 3,429 species and 561 genera are included in 25 tribes, representing 27 % of the world mirid fauna. Mexico has about 621 species; Central America, 440; South America, 2,008; and the West Indies, 406. Tables 10.1 and 10.2 show countries of the Neotropical Region with their subfamilies, tribes, number of genera, and species represented. The countries with the largest number of genera and species are Brazil (280 genera, 1,081 species), Mexico (213, 621), Argentina (163, 524), Peru (124, 327), and Panama (112, 234). For the largest islands of the West Indies (Hernández and Henry 2010), Cuba has 57 genera and 105 species recorded; Puerto Rico, 40 and 63; Jamaica, 36 and 48; Hispaniola, 22 and 28; and Grenada, 15 and 23. Mexico has 20 tribes; Central America, 23; and South America, 22. The majority of the genera are represented by only a few species (Fig. 10.2); 49 % percent have only one species; 62 %, one or two species; and 75 %, one to four species.

The Neotropical genera with the greatest numbers of species in descending order are *Prepops* Reuter (Mirinae), *Phytocoris* Fallen (Mirinae), *Eccritotarsus* Stål (Bryocorinae), *Ceratocapsus* Reuter (Orthotylinae), *Taedia* Distant (Mirinae), *Orthotylus* Fieber (Orthotylinae), *Pycnoderes* Guérin-Menèville (Bryocorinae), *Lopidea* Uhler (Orthotylinae), *Fulvius* Stål (Cylapinae), and *Dagbertus* Distant (Mirinae).

The percentages of Neotropical species by subfamily (Fig. 10.3) in descending order are Orthotylinae (33 %), Mirinae (28 %), Deraeocorinae (14 %), Bryocorinae (10 %), Cylapinae (7.4 %), Phylinae (7 %), and Isometopinae (0.50 %). Ninety percent of the Miridae in the Neotropical Region are represented in the following tribes: Orthotylini (34 %) (Orthotylinae), Mirini (19 %) (Mirinae), Eccritotarsini (10 %) (Bryocorinae), Phylini (8 %) (Phylinae), Deraeocorini (8 %) (Deraeocorinae), Herdoniini (5 %) (Mirinae), Restheniini (3 %) (Mirinae), and Clivinematini (3 %) (Deraeocorinae).

Most of the literature supporting the study of Neotropical Miridae is found primarily in taxonomic papers, including keys; catalogs; reviews of subfamilies, tribes, and genera; and descriptions of genera and species (e.g., Carvalho 1955a, 1957–1960; Schuh 1995, 2013; Carvalho and Costa 1997; Ferreira 1980; Ferreira and Henry 2011).

Maldonado's (1969) Miridae of Puerto Rico and Hernández and Henry's (2010) Miridae of Cuba contain diagnoses and keys to the subfamilies, tribes, genera, and species that are useful in recognizing many of the more widely distributed taxa. Studies on insect-plant relationships, biogeography, and integrated pest management are few, but see literature cited in Ferreira et al. (2001), Ferreira and Henry (2011), Schuh (1991, 2014), and Wheeler (2000a, b, 2001). Carvalho and Carpintero (1993) and Carpintero et al. (2006) provided a checklist of the Miridae of Argentina. Although treating mostly North American taxa, H. H. Knight's (1923, 1941, 1968) comprehensive works on the Miridae of Connecticut, Illinois, and the Nevada Test Site are still useful for keying subfamilies, tribes, and genera.

There also are few Neotropical mirid studies hypothesizing phylogenetic relationships. Ferreira (1998) revised the tribe Clivinematini and gave a phylogenetic analysis of the tribe; Schuh (1991) treated the Pilophorini; Schwartz (1987, 2008), the Stenodemini; and Tatarnic and Cassis (2012), the Halticini. Forero (2008) revised and provided an analysis of the *Hadronema* Uhler group; Henry and Kim (1984), *Neurocolpus* Reuter; Henry (1991,1999, 2012b), *Keltonia* Knight, *Pseudatomoscelis* Reuter, *Ranzovius* Distant, and *Tytthus* Fieber; Sadowska-Woda et al. (2008), *Fulvius* Stål (based on molecular data); Schuh and Schwartz (1985), *Rhinacloa* Reuter; and Schwartz (1984, 2011), *Irbisia* Reuter and *Slaterocoris* Wagner.

One cannot consider the Miridae without mentioning José C. M. Carvalho. He published a majority of the taxonomic work on the Neotropical Miridae in an amazing, career-long series of nearly 500 papers in which he described almost 400 genera and more than 2,100 species (Schuh 2013). Carvalho and Froeschner (1987, 1990, 1994) listed the genera and species described by Carvalho from 1943 to January 1993, along with their type depositories.

Table 10.1 Number of genera and species by tribe in Central America and the West Indies

Mexico and Central America		MEX		GUA		BEL		ELS		HON		NIC		CRI		PAN		CUB		JAM		DOR		WIN	
Subfamily	Tribe	GN	SP	GN	SP	GN	SP	GN	SP	GN	SP	GN	SP	GN	SP	GN	SP	GN	SP	GN	SP	GN	SP	GN	SP
Bryocorinae	Bryocorini	1	1	0	0	0	0	1	1	0	0	1	2	1	1	0	0	1	1	1	1	0	0	2	3
Bryocorinae	Dicyphini	5	9	4	8	0	0	1	1	0	0	5	11	4	6	2	4	3	5	0	0	1	1	0	0
Bryocorinae	Eccritotarsini	24	74	18	39	0	0	10	15	5	6	12	28	17	29	23	56	9	13	3	5	1	1	6	9
Cylapinae	Cylapini	1	1	2	2	0	0	0	0	1	2	0	0	0	0	2	8	0	0	0	0	0	0	0	0
Cylapinae	Fulviini	3	21	2	6	0	0	2	3	1	1	2	15	2	7	6	18	1	3	2	2	0	0	2	5
Cylapinae	Vaniini	0	0	1	1	0	0	0	0	0	0	0	0	1	2	1	1	0	0	0	0	0	0	0	0
Deraeocorinae	Clivinematini	11	33	5	9	0	0	1	1	1	1	2	5	5	7	5	7	1	1	1	1	2	2	2	2
Deraeocorinae	Deraeocorini	6	6	4	4	0	0	0	0	0	0	2	2	2	2	6	9	1	1	0	0	1	1	0	0
Deraeocorinae	Hyaliodini	4	8	4	6	1	1	2	2	0	0	6	9	5	7	8	13	3	4	0	0	2	3	1	2
Deraeocorinae	Surinamellini	1	3	0	0	0	0	0	0	0	0	1	1	0	0	1	1	1	2	0	0	0	0	0	0
Deraeocorinae	Termatophylini	4	4	1	1	0	0	0	0	1	1	0	0	1	1	0	0	1	1	1	1	0	0	1	2
Isometopinae	Diphlebini	0	0	0	0	0	0	0	0	0	0	0	0	0	0	0	0	1	2	0	0	0	0	0	0
Isometopinae	Myiommini	0	0	0	0	0	0	0	0	0	0	0	0	0	0	0	0	0	0	0	0	0	0	0	0

Mirinae	Herdoniini	10	16	3	3	0	0	0	0	0	0	0	0	1	1	5	6	0	0	0	0	0	0	0	0
Mirinae	Mirini	33	121	23	40	0	0	13	19	12	17	22	40	13	25	26	60	12	27	7	10	1	1	9	12
Mirinae	Restheniini	17	49	4	7	0	0	2	3	2	2	3	5	1	6	4	12	1	8	1	1	1	1	1	2
Mirinae	Stenodemini	7	15	4	6	1	1	3	3	3	3	3	4	2	2	4	4	4	4	0	0	0	0	0	0
Orthotylinae[a]	Ceratocapsini	1	1	2	3	0	0	0	0	0	0	2	4	0	0	1	2	2	6	1	2	0	0	1	3
Orthotylinae	Halticini	1	3	1	2	0	0	1	2	0	0	1	2	1	1	0	0	1	2	0	0	0	0	0	0
Orthotylinae	Orthotylini	53	179	6	15	0	0	6	6	0	0	13	14	9	14	7	10	7	9	9	11	1	2	10	12
Phylinae[b]	Hallodapini	2	5	0	0	0	0	0	0	0	0	0	0	0	0	1	1	1	1	0	0	0	0	0	0
Phylinae	Phylini	27	65	7	7	1	1	5	5	6	9	10	25	4	6	9	21	6	13	7	7	5	7	8	11
Phylinae	Pilophorini	2	7	1	1	1	1	0	0	0	0	1	2	1	1	0	0	2	5	1	2	1	1	1	1
	Total	213	621	92	160	4	4	47	61	32	42	86	169	70	118	112	233	58	108	34	43	16	20	44	64

MEX Mexico, *GUA* Guatemala, *BEL* Belize, *ELS* El Salvador, *HON* Honduras; *NIC* Nicaragua, *CRI* Costa Rica, *PAN* Panama, *CUB* Cuba, *JAM* Jamaica, *DOR* Dominican Republic; Caribbean or West Indies (not included: Cuba, Dominican Republic, and Jamaica) Antigua and Barbuda, Bahamas, Barbados, Cayman Islands, Dominica, Grenada, Haiti, St. Kitts and Nevis, Montserrat, Saint Christopher-Nevis-Anguilla, Saint Lucia, Saint Vincent and the Grenadines, Trinidad and Tobago, and Turks and Caicos Islands; *GN* genus number, *SP* species number

[a]Excluding Coridromini

[b]Phylini (*sensu lato*), prior to Schuh and Menard (2013) and Menard et al. (2014)

Table 10.2 Number of genera and species by tribe in countries of South America

South America															
		COL		VEN		GUY		SUR		FGU		ECU		PER	
Subfamily	Tribe	GN	SP	GN	SP	GN	SP	GN	SP	GN	SP	GN	SP	GN	SP
Bryocorinae	Bryocorini	0	0	0	0	0	0	0	0	0	0	1	1	1	1
Bryocorinae	Dicyphini	3	9	2	3	1	2	3	5	1	2	4	12	6	13
Bryocorinae	Eccritotarsini	7	22	15	24	2	2	6	8	2	3	9	19	18	51
Cylapinae	Cylapini	0	0	2	2	0	0	0	0	0	0	2	2	1	6
Cylapinae	Fulviini	2	6	2	9	2	3	2	4	2	6	2	7	2	5
Cylapinae	Vaniini	1	1	0	0	0	0	0	0	0	0	0	0	0	0
Deraeocorinae	Clivinematini	2	3	3	4	0	0	0	0	0	0	1	1	3	4
Deraeocorinae	Deraeocorini	4	6	2	3	0	0	0	0	1	1	1	1	3	7
Deraeocorinae	Hyaliodini	4	12	5	6	1	1	1	1	0	0	6	7	10	16
Deraeocorinae	Surinamellini	0	0	1	1	0	0	1	1	0	0	1	1	1	1
Deraeocorinae	Termatophylini	0	0	1	2	1	1	1	1	0	0	0	0	0	0
Isometopinae	Diphlebini	0	0	0	0	0	0	0	0	0	0	0	0	0	0
Isometopinae	Myiommini	0	0	0	0	0	0	0	0	0	0	0	0	0	0
Mirinae	Herdoniini	1	1	1	1	0	0	1	1	0	0	2	2	7	7
Mirinae	Mirini	20	46	19	39	3	3	9	10	12	28	25	68	38	96
Mirinae	Restheniini	7	17	4	9	0	0	1	3	1	2	5	14	7	51
Mirinae	Stenodemini	3	8	3	3	0	0	1	1	0	0	4	6	5	9
Orthotylinae	Ceratocapsini	0	0	1	3	1	1	1	2	0	0	1	1	1	3
Orthotylinae	Halticini	0	0	0	0	0	0	1	2	0	0	1	2	1	2
Orthotylinae	Orthotylini	7	13	7	14	1	1	1	1	3	3	8	11	10	20
Phylinae	Hallodapini	1	1	0	0	0	0	0	0	0	0	0	0	0	0
Phylinae	Phylini	7	8	8	15	0	0	5	6	1	1	10	22	9	32
Phylinae	Pilophorini	0	0	0	0	1	1	0	0	1	1	0	0	1	3
	Total	69	153	76	138	13	15	34	46	24	47	83	177	124	327

Subfamily	Tribe	BOL		PAR		ARG		URU		CHI		BRA	
		GN	SP	GN	SP	GN	SP	GN	SP	GN	SP	GN	SP
Bryocorinae	Bryocorini	0	0	1	1	1	1	0	0	0	0	1	3
Bryocorinae	Dicyphini	1	4	4	4	4	10	1	2	2	3	5	27
Bryocorinae	Eccritotarsini	12	29	8	13	18	53	3	5	2	2	34	152
Cylapinae	Cylapini	2	2	0	0	1	1	0	0	0	0	12	23
Cylapinae	Fulviini	2	3	2	9	3	15	0	0	1	2	9	32
Cylapinae	Vaniini	1	1	0	0	0	0	0	0	0	0	1	1
Deraeocorinae	Clivinematini	0	0	0	0	6	9	0	0	0	0	4	10
Deraeocorinae	Deraeocorini	0	0	1	1	4	7	0	0	4	6	11	23
Deraeocorinae	Hyaliodini	6	6	3	4	7	13	0	0	2	2	11	38
Deraeocorinae	Surinamellini	0	0	0	0	1	2	0	0	0	0	4	9
Deraeocorinae	Termatophylini	0	0	0	0	0	0	0	0	0	0	1	1
Isometopinae	Diphlebini	0	0	0	0	0	0	0	0	0	0	0	0
Isometopinae	Myiommini	0	0	0	0	3	7	0	0	0	0	1	3
Mirinae	Herdoniini	2	2	7	11	10	27	1	1	0	0	20	54
Mirinae	Mirini	20	34	20	39	31	123	5	5	10	44	55	277
Mirinae	Restheniini	6	19	7	22	15	96	2	5	3	6	17	134
Mirinae	Stenodemini	2	2	3	3	10	26	3	4	1	2	10	20
Orthotylinae	Ceratocapsini	1	4	1	4	1	17	0	0	0	0	1	47
Orthotylinae[a]	Halticini	0	0	0	0	1	3	1	1	0	0	1	3
Orthotylinae	Orthotylini	8	13	8	13	30	79	1	1	10	33	55	145
Phylinae[b]	Hallodapini	0	0	0	0	1	1	0	0	0	0	1	1
Phylinae	Phylini	4	6	3	4	15	31	0	0	8	16	24	72
Phylinae	Pilophorini	0	0	1	1	1	3	0	0	0	0	1	7
	Total	67	125	69	129	163	524	17	24	43	116	279	1,082

COL Colombia, *VEM* Venezuela, *GUY* Guyana, *SUR* Suriname, *FGU* French Guiana, *ECU* Ecuador, PER Peru, *BOL* Bolivia, *PAR* Paraguay, *ARG* Argentina, *URU* Uruguay, *CHI* Chile, *BRA* Brazil

[a]Excluding Coridromini

[b]Phylini (*sensu lato*), prior to Schuh and Menard (2013) and Menard et al. (2014)

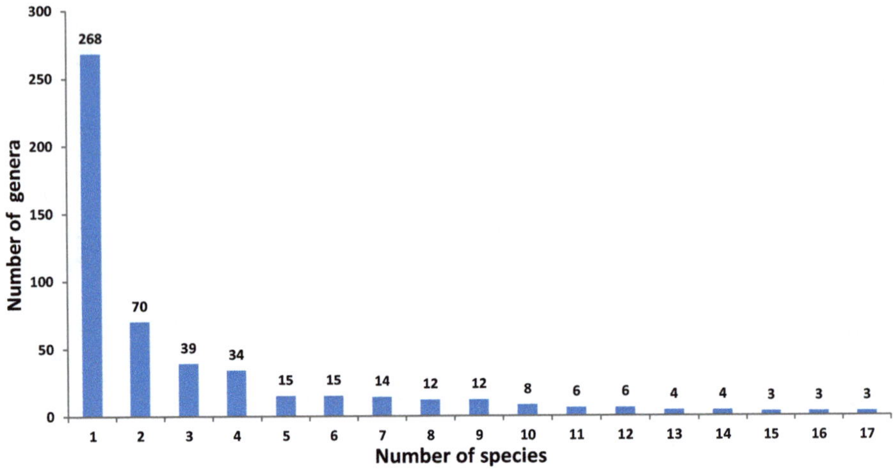

Fig. 10.2 Number of genera in relation to number of species of Neotropical Miridae

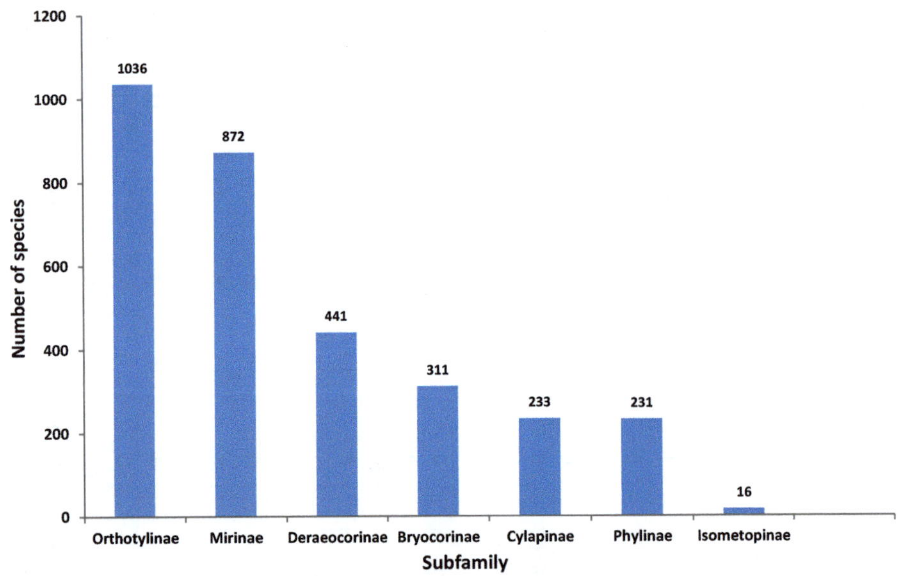

Fig. 10.3 Number of species by subfamily of Neotropical Miridae

In the following section, we provide keys to the subfamilies and tribes of Neotropical Miridae, modified from Schuh and Slater (1995), Carvalho and Costa (1997), and Hernández and Henry (2010), and other references as noted, followed by a diagnosis of the subfamily and a highlight of the most significant papers containing substantial numbers of new taxa and/or identification keys. Only seven of the eight recognized subfamilies are included. Previously, the Neotropical genus

Isometocoris Carvalho and Sailer had been considered the only New World representative of the subfamily Psallopinae (Henry and Maldonado 1982). In a forthcoming paper, however, Wolski and Henry (2015) document that *Isometocoris* is not a psallopine; thus, this subfamily is omitted from the key.

Key to the Neotropical subfamilies of Miridae

1. Tarsi two segmented, the first segment shorter than the second; ocelli present .... Isometopinae
– Tarsi two or three segmented; ocelli absent 2
2. Parempodia large, fleshy, apically convergent or divergent, arising between claws; phallotheca never cone shaped 3
– Parempodia setiform, if large, fleshy, and convergent, then with a cone-shaped phallotheca nested within the left paramere and protruding posteriorly from genital capsule 4
3. Parempodia divergent at apices, usually thickened; a prominent collar present, separated from anterior margin of pronotum by a deep furrow present .... Mirinae
– Parempodia convergent at apices, usually more slender; pronotal collar absent, if present, not separated from pronotum by a deep suture .... Orthotylinae
4. Pulvilli present, free, or connected along inner surface of claws, sometimes difficult to see, in which case a collar is always absent 5
– Pulvilli absent; collar present or absent 6
5. Pulvilli arising from base or inner margin of claw; hemelytral membrane with two cells; tarsi uniformly slender .... Phylinae
– Pulvilli arising from ventral surface of claw; hemelytral membrane with one or two cells; tarsi thickened distally or slender with tiny pretarsi .... Bryocorinae
6. Claws distinctly toothed or thickened at bases; never toothed apically .... Deraeocorinae
– Claws not toothed or thickened basally; apex almost always with a fine subapical tooth .... Cylapinae

10.5.1 Subfamily Bryocorinae

This subfamily is recognized by the three-segmented tarsi, the often large and fleshy pulvilli (Eccritotarsini) attached to the bases of the claws and the often distally thickened (Eccritotarsini) or sometimes slender (Dicyphini) tarsal segments, with tiny indistinct claws. The Bryocorini and Eccritotarsini have only one closed cell on the hemelytral membrane, whereas the Dicyphini (Dicyphina and Monaloniina) have two.

Carvalho and Ferreira (1995) keyed 66 of the Neotropical genera of Bryocorinae (Eccritotarsini), and Ferreira and Henry (2011) provided a synopsis and keys to the four tribes, 24 genera, and 56 species of Minas Gerais, Brazil.

The following key is modified from Schuh and Slater (1995), Hernández and Henry (2010), and Ferreira and Henry (2011).

Key to the Neotropical tribes of Bryocorinae

1. Hemelytral membrane with two closed cells; tarsi uniformly slender; delicate slender species, sometimes spotted dorsally; genital capsule often bearing tubercules or processes on the right side; pretarsal structures very small, pulvilli small; parempodia present (includes Dicyphina and Monaloniina) .... Dicyphini
– Hemelytral membrane with only one closed cell; tarsi dilated distally; oval, robust species, without dorsal spots; genital capsule without tubercules on the right side; pulvilli large, distinct; parempodia present or absent.................. 2
2. Claws without true pulvilli, but with pseudopulvilli arising from the unguitractor plate, recurved and convergent apically; parempodia absent; metathoracic scent efferent system with bulbous peritremal disc and distinct ostiolar canal .... Bryocorini
– Claws with large and flattened pulvilli, arising or continuing from the inner claw surface, without pseudopulvilli; parempodia asymmetrical; metathoracic scent efferent system usually with a narrow tongue-shaped peritremal disc and weakly developed ostiolar canal ....Eccritotarsini

Bryocorini (Fig. 10.4)

The Bryocorini are the smallest of the three tribes, containing four genera in the world, with only *Monalocoris* Dahlbom occurring in the Neotropics. These small, oval bugs are characterized by the rounded pronotal collar, reduced number of femoral trichobothria, and their specialized feeding on ferns (Pterophyta). Only seven Neotropical species are known (Schuh 2013).

Dicyphini (Figs. 10.5 and 10.6)

The Neotropical Dicyphini are recognized by their slender body form, elongate head, slender antennae, trapeziform pronotum with a distinct narrow collar, wide mesoscutum, translucent hemelytra, often reduced auricles and evaporative area, and respiratory horns on the egg.

The subfamily is separated into three subtribes, of which only the Dicyphina Monaloniina occur in the Neotropics. Cassis (1984) revised and keyed Dicyphina and gave generic status to many of the previous subgeneric names, such as *Engytatus* Reuter, *Nesidiocoris* Kirkaldy, and *Tupiocoris* China and Carvalho. Carvalho and Ferreira (1994b) keyed the Neotropical genera.

The widespread *Engytatus modestus* (Distant) and *Nesidiocoris tenuis* (Reuter) are well-known pests of various crop plants, including tobacco, tomato, and eggplant

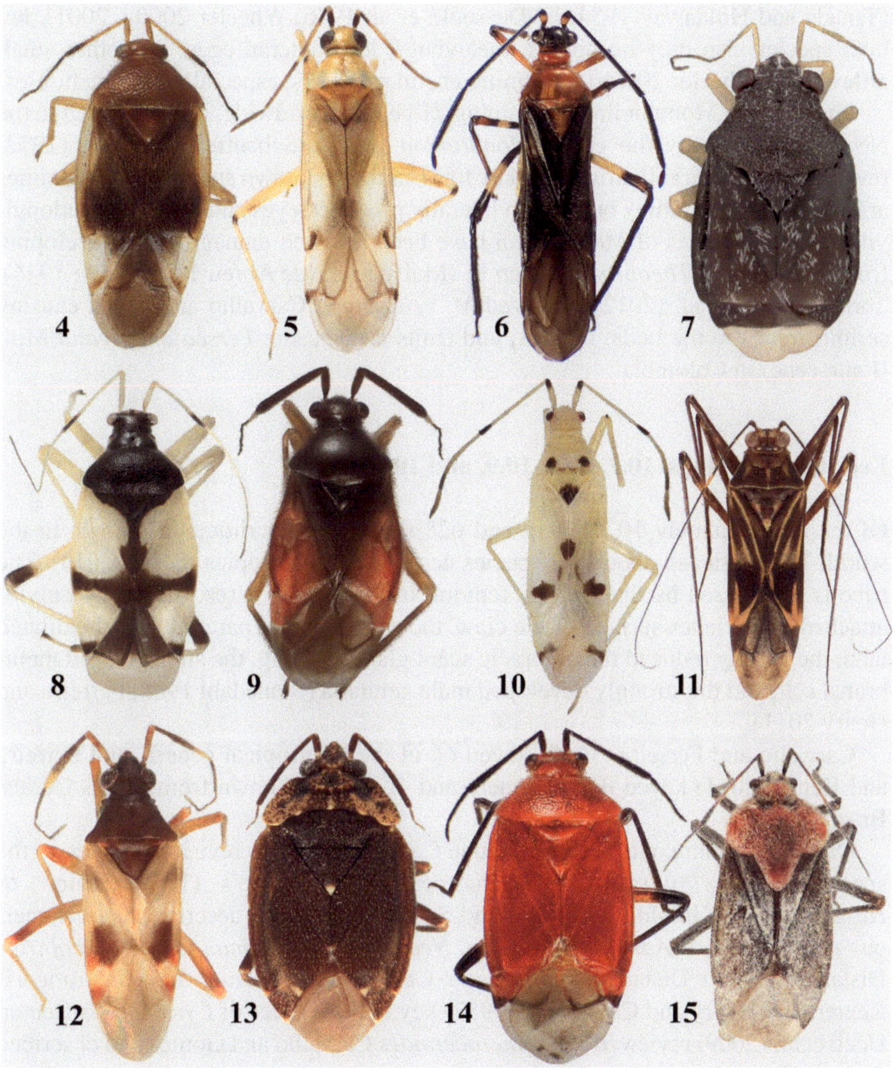

Figs. 10.4–10.15 Neotropical Miridae. 4, *Monalocoris pallidiceps* (Reuter) (Bryocorinae: Bryocorini). 5, *Nesidiocoris tenuis* (Reuter) (Bryocorinae: Dichypini: Dicyphina). 6, *Monalonion annulipes* Signoret (Bryocorinae: Dicyphini: Monaloniina). 7, *Cyrtocapsus femoralis* Reuter (Bryocorinae: Eccritotarsini) 8, *Eccritotarsus brevicuneatus* Carvalho (Bryocorinae: Eccritotarsini). 9, *Pachymerocerus fairmairei* (Stål) (Bryocorinae: Eccritotarsini). 10, *Chius maculatus* Distant (Bryocorinae: Eccritotarsini). 11, *Cylapus striatus Reuter* (Cylapinae: Cylapini). 12, *Fulvius anthocoroides* Stål (Cylapinae: Fulviini). 13, *Peritropis izyai* Wolski and Henry (Cylapini: Fulviini). 14, *Ambracius dufouri* Stål (Deraeocorinae: Clivinematini). 15, *Clivinema regalisimilis* Carvalho (Deraeocorinae: Clivinematini) (Photos by G Ouellette)

(Tanada and Holdaway 1954; El-Dessouki et al. 1976; Wheeler 2000a, 2001), but both species also prey on aphids, mealybugs, lepidopteran eggs, and other small arthropods (Wheeler 2000b), becoming useful predators, especially in greenhouses.

The subtribe Monaloniina, containing 21 genera worldwide, is represented in the New World only by the genus *Monalonion* Herrich-Schaeffer. Carvalho (1972) reviewed the genus and provided a key to 11 of the 16 known species. Monaloniines are relatively large, shiny bugs, with laterally produced eyes and distinct pseudopulvilli. Several species of *Monalonion* have been reported damaging the developing fruits and pods of *Theobroma cacao* L. (Malvaceae) (de Abreu 1977; Wille 1944). Torres Jaimes et al. (2012) observed *M. velezangeli* Carvalho and Costa causing serious injury to the buds, flowers, and fruits of avocado, *Persea americana* Mill. (Lauraceae) in Colombia.

Eccritotarsini (Figs. 10.7, 10.8, 10.9, and 10.10)

Of the approximately 107 genera and 628 species of Eccritotarsini known in the world, 70 genera and about 452 species occur in the Neotropics (Schuh 2013). The tribe is recognized by the swollen femoral trichobothrial bases, the fleshy pulvilli attached to the inner surface of the claw, the asymmetrical parempodia, the dilated tarsi, the greatly reduced metathoracic scent gland opening, the single closed membranal cell, and the strongly developed male genitalia (Stonedahl 1988; Ferreira and Henry 2011).

Carvalho and Ferreira (1995) keyed 66 of the Neotropical genera, and Ferreira and Henry (2011) keyed the 18 genera and 42 species known from Minas Gerais, Brazil.

Other important contributions include Carvalho's (1953) revision and key to the species of *Eccritotarsus* Stål; Schaffner and Carvalho's (1981) review of *Hesperolabops* Kirkaldy; Carvalho and Schaffner's (1985) descriptions of numerous new species of *Adneella* Carvalho, *Neella* Reuter, *Sysinas* Distant, *Neofurius* Distant, *Neosilia* Distant, *Nototremates* Carvalho and China, and *Dichroocoris* Reuter; and Henry and Carvalho's (1987) key to the species of *Cyrtocapsus* Reuter. Uceli et al. (2009) reviewed *Pachymeroceroides* Carvalho and Gomes and described the peculiar male hemelytra.

This subfamily contains a number of economically important taxa. *Pycnoderes quadrimaculatus* Guérin-Meneville, often called the bean capsid, is a well-known pest of beans, squash, sweet potatoes, and garden crops, where it causes chlorosis, mottling, leaf drop, and premature aborting of fruits (Wheeler 2000a). Feeding by the Neotropical *Tenthecoris orchidearum* (Reuter) may produce chlorotic spots or even kill the foliage on orchids, and, in heavy infestations, the entire plants may die (Gimingham 1928; Wheeler 2001). In the early twentieth century, species of *Tenthecoris* were considered serious pests of orchids grown in greenhouses in the Britain (Carvalho and Leston 1952), France (Denis 1908), and the United States

(Weiss 1917), but with the advent of modern control practices, these eccritotarsines seldom become a problem (Wheeler 2001).

10.5.2 Subfamily Cylapinae

Of the cylapine tribes recognized by Cassis and Schuh (2012), only Cylapini (12 genera and 42 species), Fulviini (15 genera and 88 species), and Vaniini (one genus and four species) occur in the Neotropics (Gorczyca 2000, 2006; Schuh 2013). This subfamily is recognized by the long labium, often extending well beyond the hind coxae; the long, slender antennae; the membrane having two closed cells; and the long, slender legs, setiform parempodia, lack of pulvilli, and long, slender claws, usually with a distinct subapical tooth. Members of the Cylapini and Vaniini have relatively broad heads, often with stylate or protruding eyes (e.g., *Cylapus* Say), whereas the Fulviini (e.g., *Fulvius* Stål and *Peritropis* Uhler) have more elongate heads, with smaller, nonprotruding eyes.

Gorczyca (2006) cataloged the subfamily for the world, and Gorczyca (2000) and Cassis and Monteith (2006) studied the phylogeny of the group. Gorczyca (2006) recognized four tribes, whereas Cassis and Monteith (2006) and Cassis and Schuh (2012) listed five, with the addition of Vaniini. Carvalho and Ferreira (1994a) provided a key to the 20 Neotropical genera.

The following key is modified from Gorczyca (2000), with information from Cassis and Schuh (2012) incorporated.

Key to the tribes of Cylapinae

1. Body elongate oval; head horizontal, elongate, tapering anteriorly; eyes relatively small, not bulging or stylate; antenna usually shorter than body, second segment longest .... Fulviini
– Body elongate; head vertical, somewhat flattened in front; eyes bulging, often stylate; antenna usually longer than the body, third segment longest 2
2. Labium short, relatively stout, extending to or only slightly beyond hind coxae; parempodia flattened .... Vaniini
– Labium long, slender, extending well beyond hind coxae onto abdomen; parempodia rounded .... Cylapini

Cylapini (Fig. 10.11)

Members of the tribe are best recognized by their stalked eyes, long slender labium, often extending to the genital segments, auriculate evaporatorium, and toothed claws with setiform parempodia.

Twelve genera and 42 species of Cylapini are known in the Neotropics (Schuh 2013). Carvalho and Fontes (1968a) provided a key to six genera of the *Cylapus* complex (*Amapacylapus* Carvalho and Fontes, *Cylapocerus* Carvalho and Fontes, *Cylapus* Say, *Peltidocylapus* Poppius, *Trichocylapus* Poppius, and *Valdasus* Stål), Carvalho (1987a) added two new genera (*Cylapinus* Carvalho, *Tucuruisca* Carvalho) six new species, and Carvalho (1989a) described five genera (*Adcylapocoris* Carvalho, *Cylapocorella* Carvalho, *Cylapocorisca* Carvalho, *Microcylapus* Carvalho, *Valdasoides* Carvalho) and two species.

Fulviini (Figs. 10.12 and 10.13)

Fifteen genera and 88 species of Fulviini are known from the Neotropical Region (Schuh 2013). Carvalho and Costa (1994) revised and keyed the species *Fulvius*, one of the larger Neotropical mirid genera; Ferreira and Henry (2002) added two new species from Brazil, and Hernández and Henry (2010) described one from Cuba; Sadowska-Woda et al. (2008) presented a preliminary phylogenetic analysis, and Wolski and Henry (2012, 2013) revised *Peritropis* Uhler, describing 17 new species. Carvalho (1989a) described a number of new species (*Adcylapocoris* Carvalho, *Cylapocorella* Carvalho, *Cylapocorisca* Carvalho), and Henry and Silva de Paula (2004) described *Rhyparochromomiris femoratus*, a peculiar ant-mimetic species resembling a herdoniine mirine with greatly swollen front femora similar to those found in certain Rhyparochromidae (Lygaeoidea). Although Schuh (2013) included *R. femoratus* in the Cylapini, the long second antennal segment and male parameres suggest that this peculiar cylapine belongs in the Fulviini.

Vaniini

Vaniines are often brightly colored with orange and red markings. Their protruding eyes and flattened parempodia will distinguish them from most other Cylapinae. Gorczyca (1997) treated the *Vannius* complex (*Cylapus* Say, *Valdasus* Stål, *Vannius* Distant) and cataloged the world fauna. Cassis and Monteith (2006) presented a phylogenetic analysis of the *Vannius* complex, and Cassis and Schuh (2012) provided a summary of relationships.

Little is known of the habits of most cylapines. Species of *Fulvius* are often taken under the loose bark of dead trees, and most are thought to be predatory (Wheeler 2001; Henry et al. 2011), though a few observations suggest that some may feed on fungi (Schuh 1976). *Fulvius anthocoroides* (Reuter) has been observed feeding on mites and insects in stored nuts, and the North American *F. imbecilis* Say is documented feeding on Diptera and Coleoptera larvae and other arthropods (Henry et al. 2011). Herring (1976) considered his new genus and species, *Trynocoris lawrencei*, a predator of ciid beetle larvae associated with fungi growing on forest trees. Though long thought to be predaceous, Wheeler and Wheeler (1994) observed fungal spores and conidia in the gut contents of the North American *Cylapus tenuicornis* (Say) (and a Neotropical *Cylapus* sp.), providing evidence for mycophagous feeding habits in the genus. Cassis and Monteith (2006) noted that their new vaniine genus and species, *Kauakamiris krypton*, described from New Caledonia, was most easily collected in moldy cut-branch traps comprised of no particular plant species, suggesting it is mycetophagous.

10.5.3 Subfamily Deraeocorinae

Members of this subfamily can be distinguished by the distinctly punctate, often polished dorsum and the basally toothed claws, although certain other deraeocorines (e.g., Termatophylini) may lack distinct dorsal punctures and have a dull coloration (Schuh and Slater 1995; Cassis 1995; Hernández and Henry 2010).

The following key is adapted from Carvalho (1955a) and Hernández and Henry (2010).

Key to Tribes of Deraeocorinae

1. Head elongate, pointed apically, with anthocorid-like facies; eyes large, placed far from basal angle of the head .... Termatophylini
– Head not elongate, without anthocorid-like facies; eyes variable, placed near basal angle of the head 2
2. Pronotum with a distinct impressed line running from anterolateral corner to margin of calli, sometimes strongly punctate .... Clivinematini
– Pronotum without an impressed line 3
3. Hemelytra hyaline; emboliar margin of corium wide; males and females with an elongate anal tube .... Hyaliodini
– Hemelytra not hyaline; emboliar margin of corium narrow; males and females without an anal tube 4
4. Species ant mimetic; antennal segments linear; vertex depressed, sometimes punctate; male vesica with two membranous lobes apically .... Surinamellini
– Species not ant mimetic; antennal segment I thicker than remaining segments, segment II clavate at apex; vertex smooth, not depressed; male vesica without two membranous lobes apically .... Deraeocorini

Carvalho's (1955a) now outdated keys to the world genera include the most useful key to the Neotropical deraeocorine genera. Maldonado (1969) keyed the genera and species of Puerto Rico, and Hernández and Henry (2010) treated the Cuban genera.

Clivinematini (Figs. 10.14 and 10.15)

Seventeen genera and about 55 species of Neotropical clivinematines are known (Schuh 2013). Henry (1979a) reviewed *Bothynotus* Fieber, and Ferreira and Henry (2010) revised *Ambracius* Stål. Ferreira (1993, 1996a, b, 1998, 2001) provided a comprehensive treatment of the Clivinematini, describing four new genera and providing information on their biogeography and biology, a cladistic analysis, and a key to the 17 Neotropical genera. Ferreira and Schaffner (2004) reviewed *Admetus* Distant, and Kerzhner and Schuh (1998) provided the replacement name *Admetomiris* for the preoccupied *Admetus*. Carpintero et al. (2008) added the new genus *Valdesiana* Carpintero and Dellapé and provided a key to the six genera and nine species of Clivinematini from Argentina.

Based on a few observations, most clivinematines are thought to prey on species of Sternorrhyncha, especially those in the family Ortheziidae. Knight (1928) reported *Clivinema sericea* Knight feeding on a species of *Orthezia* (Hemiptera: Ortheziidae) in New Mexico, Miller and Schuh (1995) observed *C. coalina* Bliven preying on *O. annae* Cockerell on *Atriplex polycarpa* (Torr.) Wats. in California, Ferreira (1998) reported *Hemicerocoris bicolor* Carvalho feeding on soft scales on guava in Mexico, and Ferreira and Henry (2010) documented records of *Ambracius dufouri* Stål preying on *Orthezia* sp. in Brazil.

Deraeocorini (Fig. 10.16)

Twenty-seven genera and 62 species of the Deraeocorini are recorded from the Neotropical Region (Schuh 2013). Stonedahl et al. (1997) keyed the North American genera, which include a few of the Neotropical taxa. Most work on the nominotypical tribe has been descriptive. Papers containing keys to species include Carvalho and Maldonado's (1982) study of *Lundiella* and Carvalho and Costa's (1990c) review of *Diplozona* Van Duzee. Hernández and Stonedahl (1999) most recently added the three new genera *Acutifromiris*, *Cephalomiroides*, and *Scutellograndis* from Chile.

Nearly all members of this tribe are considered zoophagous. In North America, Wheeler et al. (1975) showed that *Deraeocoris nebulosus* (Uhler) was an effective predator of various ornamental pests, including mites, aphids, scales, whiteflies, psyllids, and lace bugs. Wheeler (2001) summarized the predatory habits of numerous species of *Deraeocoris* occurring mostly in the Nearctic and Palearctic regions.

Hyaliodini (Figs. 10.17, 10.18, and 10.19)

Twenty genera and about 111 species of Hyaliodini are known from the Neotropics (Schuh 2013). The most important works, including keys to species, are Carvalho and Schaffner (1977) for *Annona* Distant, Carvalho (1982) for *Antias* Distant, Carvalho (1974d) for *Auchus* Distant, and Henry and Ferreira (2003) for *Paracarniella* Henry and Ferreira. Several of the larger hyaliodine genera, particularly *Hyaliodes* Reuter, *Hyaliodocoris* Knight, *Knightonia* Carvalho and Drake, and *Paracarnus* Distant, are in need of revision.

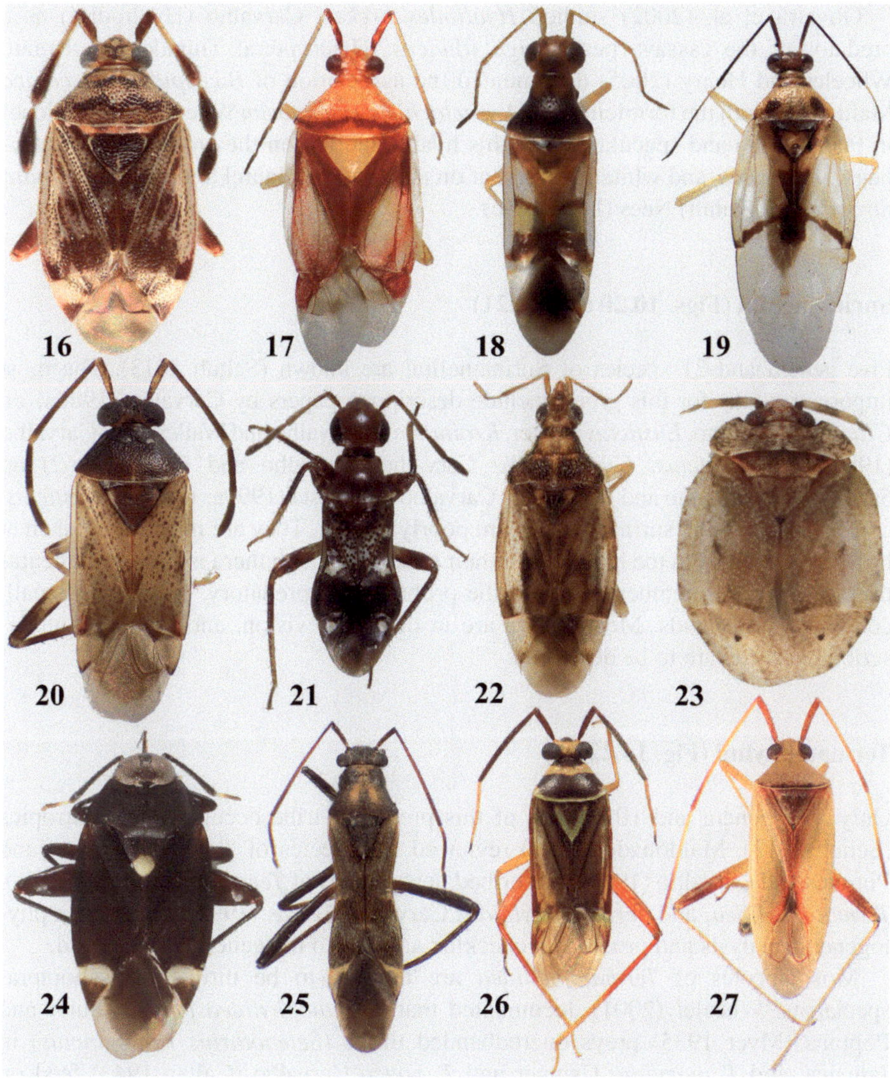

Figs. 10.16–10.27 Neotropical Miridae. 16, *Diplozona collaris* Van Duzee (Deraeocorinae: Deraeocorini). 17, *Brasiliocarnus fraudans* (Carvalho) (Deraeocorinae: Hyaliodini). 18, *Bicuspidatiella conica* Maldonado (Deraeocorinae: Hyaliodini). 19, *Knightonia knighti* (Carvalho) (Deraeocorinae: Hyaliodini). 20, *Eustictus brunneus* Maldonado (Deraeocorinae: Surinamellini). 21, *Surinamella doesburgi* Carvalho and Rosas (Deraeocorinae: Surinamellini). 22, *Termatophylidea brunnea* Maldonado (Deraeocorinae: Termatophylini). 23, *Diphleps unica* Bergroth (Isometopinae: Diphlebini). 24, *Myiomma mexicanum* Henry (Isometopinae: Myiommini). 25, *Adxenetus petiolatus* (Stål) (Mirinae: Herdoniini). 26, *Calondas fasciatus* Distant (Mirinae: Mirini). 27, *Creontiades rubrinervis* (Stål) (Mirinae: Mirini) (Photos by G Ouellette)

Oliveira et al. (2002) studied *Hyaliodes beckeri* Carvalho (Hyaliodini) as a predator of the cassava pest *Vatiga illudens* (Heteroptera: Tingidae) in Brazil. Wheeler and Henry (2005) documented the association of *Bicuspidatiella conica* Maldonado with the formicine ant *Myrmelachista ramulorum* Wheeler (Formicidae) in Puerto Rico and speculated that this hyaliodine fed on the ant-attended mealybugs, soft scales, and whiteflies present on the host tree, laurel amarillo, *Nectandra turbacensis* (Kunth) Nees (Lauraceae).

Surinamellini (Figs. 10.20 and 10.21)

Five genera and 21 species of Surinamellini are known (Schuh 2013). The most important works for this group include descriptive papers by Carvalho (1988a) on *Craoiella* Carvalho, *Eustictus* Reuter, *Krainacoris* Carvalho and Wallerstein; Carvalho (1989b) on *Eustictus*, *Guapimirella* Carvalho; Carvalho and Rosas (1962) on *Surinamella* Carvalho and Rosas; and Carvalho and Costa (1990a, 1991) on *Eustictus*.

The habits of the surinamellines are poorly known. They are most often taken at lights or beaten from the branches of their hosts. Although there is little documentation, most or all members of this tribe probably are predatory, preying on small, coexisting arthropods. Most genera are in of need revision, and numerous undescribed taxa remain to be described.

Termatophylini (Fig. 10.22)

Only four genera and 10 species of this predatory tribe occur in the Neotropics (Schuh 2013). Maldonado (1970) reviewed the species of *Termatophylidea*, and Poppius and Carvalho (1955b) described new species of *Termatophylella* Carvalho, *Termatophylidea*, and *Termatophyloides* Carvalho. Cassis (1995) presented a phylogenetic analysis and provided a checklist and key to the genera of the world.

Most species of *Termatophylidea* are thought to be thrips (Thysanoptera) specialists. Wheeler (2001) documented that *Termatophylidea pilosa* Reuter and Poppius (Myer 1935) preys on redbanded thrips (*Selenothrips rubrocinctus*) in Jamaica, and *T. maculata* Usinger and *T. opaca* Carvalho (Callan 1943) feed on redbanded thrips and grass thrips (*Caliothrips insularis*), respectively.

10.5.4 Subfamily Isometopinae (Figs. 10.23 and 10.24)

The Isometopinae are recognized by the overall round to elongate-oval body form; the trapeziform pronotum with carinate lateral margins; the strongly modified head, often with holoptic eyes, and distinct ocelli between the eyes; the sexually dimorphic antennae with segment II longer and more thickened in males than females; and the small size, ranging from 2.00 to 3.00 mm.

Herczek (1993) discussed the systematic relationships and position of the subfamily and recognized the two tribes Diphlebini and Myiommini in the New World. For the latter, he further recognized three subtribes, two of which occur in the Neotropical Region: Myiommina and Plaumanocorina.

Nine genera and 31 species are known from the Neotropics (Schuh 2013). Ghauri and Ghauri (1983) keyed the genera of the world, but overlooked several Western Hemisphere genera. Henry (1980) provided a key to the New World genera, reviewed the genera *Lidopus* and *Wetmorea*, and described the new genera *Brailovskiocoris*, *Lidopiella*, and *Myiopus*. Henry and Herring (1979) revised *Corticoris* McAtee and Malloch, Henry (1979c) described new species and provided a key to the New World species of *Myiomma* Uhler, and Henry and Carpintero (2012) treated the Isometopinae from Argentina, parts of Brazil, and Paraguay and described nine new species.

The genus *Isometocoris* Carvalho and Sailer (1954), first placed in Isometopinae, was transferred to Psallopinae by Henry and Maldonado (1982). Wolski and Henry (2015) document the placement of this genus in the Cylapinae.

With the apparent exception of one Oriental species, evidence indicates that all Isometopinae are scale predators (Wheeler and Henry 1978a; Henry 1984b; Wheeler 2001).

10.5.5 Subfamily Mirinae

The Mirinae is recognized by the distinct pronotal collar, three-segmented tarsi, divergent fleshy parempodia, inflatable endosomal membrane, and the large, ringed secondary gonoporal opening.

The following key is modified from Hernández and Henry (2010).

Key to the Tribes of Mirinae

1. First segment of hind tarsus as long as second and third combined; antennae usually very long and slender; head porrect; frons usually with a distinct median sulcus; coloration brown or green .... Stenodemini
– First segment of hind tarsus shorter than second and third combined; antennae relatively short; frons without a median sulcus; coloration variable 2
2. Species mostly dull black, marked with red, orange, or yellow on head, pronotum, and/or scutellum, dorsal surface velvety; metathoracic evaporative area small; pronotal collar wide mesally .... Restheniini
– Species variably colored, usually green, brown, or black with variable markings, dorsal surface dull or shiny, but never velvety; metathoracic evaporative area prominent; pronotal collar more narrow 3
3. Strongly myrmecomorphic; abdomen constricted at base in most taxa, hemelytra usually constricted through middle, corium usually with pruinose patches and often with a white spot near middle margin .... Herdoniini
– Not strongly myrmecomorphic; abdomen never constricted at base; hemelytra usually subparallel or convex, usually without pruinose patches and white spots .... Mirini

Herdoniini (Fig. 10.25)

Twenty-seven genera and 78 species of this group of strikingly ant-mimetic taxa are known from the Neotropics (Schuh 2013). Carvalho (1973a) provided a list of the genera, and Carvalho (1973b) keyed the genera of the world. In a series of well-illustrated papers, Carvalho and Ferreira (Carvalho and Ferreira 1973a, b, c, d, e, f, g, h, i, j, k, l) treated most of the herdoniine genera, usually accompanied by a key to species: *Adxenetus* Carvalho and Ferreira, *Allommatisca* Carvalho and Ferreira, *Allommatus* Reuter, *Barberiella* Poppius, *Camponotidea* Reuter, *Cearana* Carvalho and Ferreira, *Dacerla* Bergroth, *Guarania* Carvalho and China, *Haarupia* Poppius, *Herdonisca* Carvalho and Ferreira, *Herdonius* Stål, *Laurinia* Reuter, *Lepidoxenetus* Poppius, *Paradacerla* Carvalho and Usinger, *Paraxenetus* Reuter, *Proxenetus* Carvalho and Ferreira, *Sphinctothorax* Stål, *Xenetomorpha* Poppius, *Xenetopsis* Poppius, *Xenetus* Distant, *Zacynthus* Distant, and *Zosippus* Distant.

Herdoniines are remarkable in their resemblance to ants, with most having the abdomen constricted at the base and hemelytra narrowed through the middle. Wheeler and Henry (1980) studied *Barberiella formicoides* (Poppius) in the eastern United States and documented its habits and intimate association with three different species of ants. Little other information is available on the habits and life history of this fascinating group.

Mirini (Figs. 10.26, 10.27, 10.28, and 10.29)

Eighty-seven genera and more than 600 species of Mirini occur in the Neotropics. Carvalho's (1955a) key remains the only attempt to identify all genera of Mirini.

Phytocoris Fallén, the largest genus within the Miridae (Stonedahl 1988b), is represented in the Neotropics by 94 species (Schuh 2013), the second largest number after *Prepops* Reuter (Restheniini). Stonedahl (1988b) revised and keyed the western North American species, many of which range into Mexico, but few studies treat the Neotropical fauna much beyond describing new species. Other useful works on this genus include Maldonado (1969), Carvalho and Ferreira (1969), Carvalho and Gomes (1970), Carvalho (1990b), Carvalho and Costa (1990b, d, 1995), Carpintero and Chérot (2008, 2011), and Hernández and Henry (2010).

Important revisions of the largest genera, with keys to species, include Carvalho (1986a) for 12 species of *Calocorisca* Distant; Carvalho (1986b) for 12 species of *Chrysodasia* Carvalho; Carvalho and Fontes (1983) for 38 species of *Dagbertus* Distant; Carvalho and Gomes (1980), for 26 species of *Derophthalma* Berg; Carvalho (1986c) for 13 species of *Euchilocoris* Reuter; Carvalho (1976a) for 34 species of *Horcias* Distant; Carvalho and Jurberg (1976) for 12 species of *Horciasinus* Carvalho and Jurberg; Carvalho and Fontes (1972) and Carvalho (1985c) for 29 species of *Neostenotus* Reuter; Carvalho and Ferreira (1971) and Fontes (1981) for 19 of *Notholopus* Bergroth; Carvalho (1975c) for 18 species of *Poeas* Distant; Carvalho and Gomes (1969) and Ferreira (1980) for 22 species of *Polymerus* Hahn; Carvalho and Costa (1989) for 20 species (South American spp.

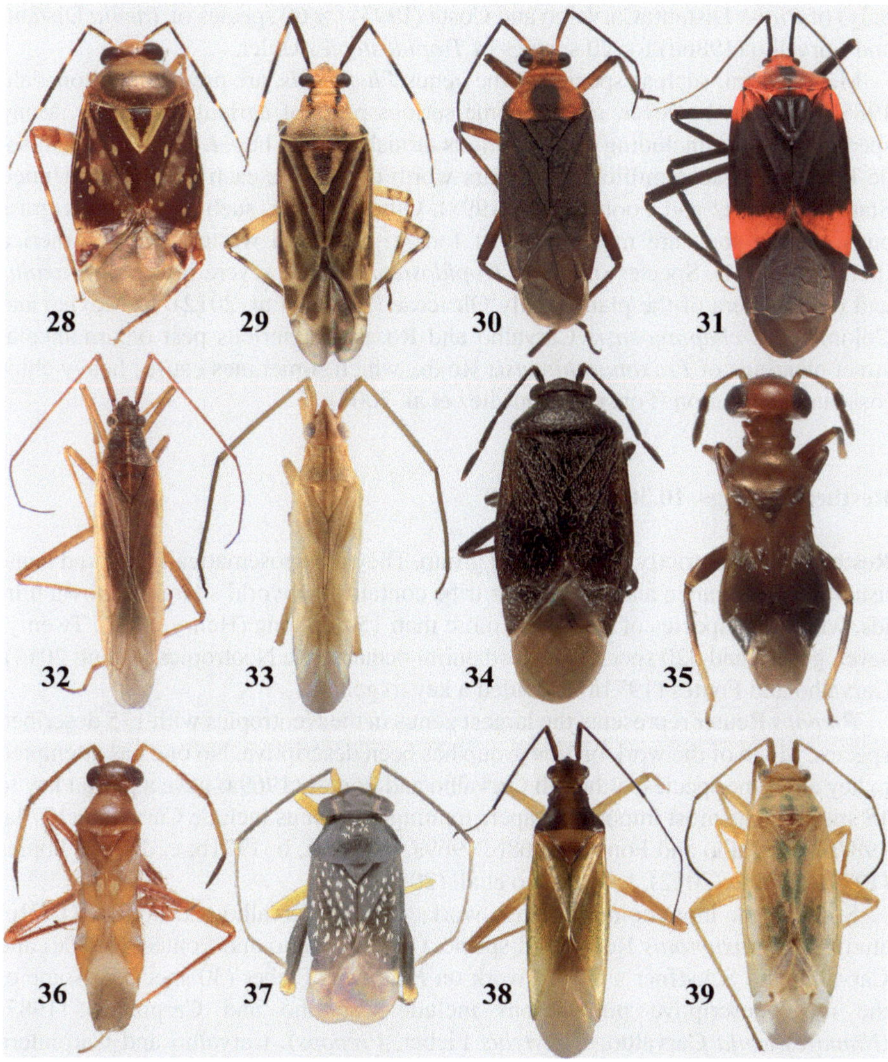

Figs. 10.28–10.39 Neotropical Miridae. 28, *Horcias nobilellus* (Berg) (Mirinae: Mirini). 29, *Piasus cribricollis* (Stål) (Mirinae: Mirini). 30, *Prepops latipennis* (Stål) (Mirinae: Restheniini). 31, *Resthenia scutata* Spinola (Mirinae: Restheniini). 32, *Stenodema dohrni* (Stål) (Mirinae: Stenodemini). 33, *Trigonotylus tenuis* (Reuter) (Mirinae: Stenodemini). 34, *Izyacapsus kerzhneri* Henry (Orthotylinae: Ceratocapsini). 35, *Zanchisme mexicanus* Carvalho and Schaffner (Orthotylinae: Ceratocapsini). 36, *Sericophanes ornatus* (Berg) (Orthotylinae: Ceratocapsini). 37, *Microtechnites bractatus* (Say) (Orthotylinae: Halticini). 38, *Carvalhomiris bifurcatus* Forero (Orthotylinae: Orthotylini). 39, *Chileria colla* Carvalho and Carpintero (Orthotylinae: Orthotylini) (Photos by G Ouellette)

only) of *Proba* Distant; Carvalho and Costa (1993) for 60 species of *Taedia* Distant; and Carvalho (1986d) for 20 species of *Tropidosteptes* Uhler.

Many Mirini, such as species of the genus *Phytocoris*, are predatory (Stonedahl 1988b). Others, however, may become serious pests of agricultural crops. Many species of *Lygus*, including the ubiquitous tarnished plant bug, *L. lineolaris* (Palisot de Beauvois), cause millions of dollars worth of damage each year in the United States (Schwartz and Foottit 1992, 1998). Other mirines, such as the black grass bugs, *Irbisia* spp., are major pests of forage grasses in western North America (Schwartz 1984). Species of genus *Tropidosteptes* cause severe injury to *Fraxinus* and other genera of the plant family Oleaceae (Henry et al. 2012). In Mexico and Colombia, *T. chapingoensis* Carvalho and Rosas is a serious pest of ornamental street plantings of *Fraxinus chinensis* Roxb., which sometimes causes heavy chlorosis and defoliation (Fonseca-González et al. 2007).

Restheniini (Figs. 10.30 and 10.31)

Restheniines are strictly a New World group. They are aposematically colored bugs, usually red or orange and black. This tribe contains the world's largest known mirids, with some species of *Resthenia* more than 15 mm long (Henry 2009). Twenty-seven genera and 320 species of Restheniini occur in the Neotropics (Schuh 2013). Carvalho and Fontes (1971a) provided a key to genera.

Prepops Reuter represents the largest genus in the Neotropics with 175 described species. Much of the work on this group has been descriptive. No one has attempted to key all of the species, although Carvalho and Fontes (1969a) gave a partial key to 18 species. The most important papers treating this genus include Carvalho (1974a, 1987b), Carvalho and Fontes (1968b, 1969a, b, 1970a, b, 1971b, c, d, e, f), Fontes (1989), Coelho (2012), and Coelho et al. (2012).

Some of the more comprehensive works include Carvalho and Fontes' (1971e) study of *Chiloxionotus* Reuter (28 species) and Carvalho and Fontes' (1970c) and Carvalho and Schaffner's (1975) work on *Platytylus* Fieber (30 species). Some of the most descriptive publications include Carvalho and Carpintero (1987) (*Nanniresthenia* Carvalho, *Platytylus* Fieber, *Prepops*), Carvalho and Carpintero (1989) (*Carpinteroa* Carvalho, *Chiloxionotus*, *Mimoncopeltus* Kirkaldy, *Nanniresthenia*, *Platytylus*, *Prepops*), Carvalho and Ferreira (1968) (*Lampsophorus* Reuter, *Mabelia* Kirkaldy), and Carvalho (1974b, c) [*Lampsophorus*, *Mabelia*, *Prepops*, *Prepopsella* Carvalho, *Resthenia* Spinola, *Restheniella* Carvalho]; Carvalho (1975a) [*Chiloxinotus*, *Opistheurista* Carvalho, *Prepops*]; Carvalho (1987b, c, d) [*Carpinteroa*, *Eurylomata* Reuter, *Fontesius* Carvalho, *Guaranimiris* Carvalho, *Oncerometopus* Reuter, *Platytylus*, *Prepops*, *Seabracoris* Carvalho]; Carvalho (1988b, c) [*Chiloxionotus*, *Prepops*, *Resthenia*]; Carvalho (1989c) [*Carpinteroa*, *Mimoncopeltus*, *Platytylus*, *Prepops*, *Resthenia*]).

Although the hosts and feeding habits of most restheniines are unknown, several species may cause economic damage to crops. *Opistheurista clandestina* (Van Duzee) causes extensive foliar spotting and blotching to beans, cowpeas, and sweet potatoes in the southern United States and Central America (Wheeler 2001), resulting in injury resembling that caused by mites and thrips (Jones 1921). *Prepops latipennis* (Stål), another Neotropical species, produces similar injury to beans and potatoes (King and Saunders 1984).

Stenodemini (Figs. 10.32 and 10.33)

Schwartz (2008) revised the Stenodemini, provided a key to the genera of the world, and presented a cladistic analysis supporting the monophyly of this tribe of grass-feeding bugs. Thirteen genera and about 46 species are known from the Neotropics (Schuh 2013).

Carvalho and Fontes (1981) reviewed the Neotropical species of *Collaria* Provancher, Carvalho (1975b) keyed the species of *Stenodema* Laporte, and Carvalho and Fontes (1969c) treated *Neotropicomiris* Carvalho and Fontes. Other mostly descriptive contributions include Carpintero and Estevez (2001) (*Cynodonmiris* Carpintero and Estevez, *Spartinomiris* Carpintero and Estevez), Carvalho and Fontes (1969c) (*Collaria*, *Dolichomiris* Reuter, *Kuscheliana* Carvalho, *Neotropicomiris*, *Ophthalmomiris* Berg, *Porpomiris* Berg, *Stenodema*, *Trigonotylus* Fieber), and Carvalho (1975b) [*Dolichomiris*, *Megaloceroea* Fieber, *Stenodema*, *Trigonotyliscus* Carvalho (a junior synonym of *Trigonotylus*), *Trigonotylus*].

Losses to prairie grasses, important graze for cattle and milk production in Colombia, have been attributed to stenodemines. *Collaria scenica* (Stål), the predominant pest species in that region, has caused losses up to 25 % of dry matter, the decrease of 10 % grass digestibility, and the reduction of milk production by up to 5 l/animal/day, resulting in an economic loss of 20 % to producers (Martinez and Barreto 1998; Duarte et al. 1998; Barreto 2011). Its biology has been studied on wheat, including adult morphometry (Carlessi et al. 1999).

Collaria oleosa (Distant) is a potential pest of wheat in Brazil (Da Silva et al. 1994) and is known to reduce sorghum yields in Cuba (Ryder et al. 1968). Ferreira et al. (2001) evaluated the economic importance of *Collaria oleosa* (Distant), *Dolichomiris linearis* Reuter, *Neotropicomiris costalis* Carvalho and Fontes, *N. longirostris* Carvalho and Fontes, *Opisthocoris carmelitanus* Carvalho and Costa, and *Trigonotylus tenuis* Reuter on grasses in Minas Gerais, Brazil.

10.5.6 Subfamily Orthotylinae

The Orthotylinae are distinguished by the usually complex male genitalia, especially the endosoma, often with multiple branching spicules, and parameres; the relatively large, open genital aperture; and the fleshy, convergent parempodia between the claws that are similar to those found in pilophorine Phylinae (distinguished, however, by the strap-like vesica similar to all other phylines).

The following key is modified from Hernández and Henry (2010).

Key to the Tribes of Orthotylinae

1. Small, oval, compact species with saltatorial hind femora 2
– Usually larger, more elongate, medium-sized species, hind femora not saltatorial 3
2. Black species; head strongly dorsoventrally elongate; dorsum usually with patches of white, scalelike setae; parameres small, simple .... Halticini
– Pale yellow to brown species; head not strongly dorsoventrally elongate; dorsum without scalelike setae; left paramere, long, sickle shaped .... Coridromini
3. Head truncate posteriorly, with a distinct basal carina; species often ant mimetic but, if not, antennal segments III and IV as thick or thicker than antennal segment II; male parameres sickle shaped to complex and branching; vesica without or, at most, with one simple spiculum; phallotheca readily visible within genital capsule when viewed caudally .... Ceratocapsini
– Head not truncate posteriorly, without a distinct basal carina; species not ant mimetic; antennal segments III and IV more slender than segment II; male parameres relatively simple; vesica with multiple, often branching or complex spiculi; phallotheca not visible within genital capsule .... Orthotylini

Ceratocapsini (Figs. 10.34, 10.35, and 10.36)

The Ceratocapsini are an exclusively New World group, comprising more than 10 genera (Henry 1979b, 1994, 2006, 2013) and about 200 species, about half of which occur in the Neotropics. Many ceratocapsines are remarkably myrmecomorphic and are frequently collected in association with ants (Wheeler 1991; Henry 1994). Carvalho (1958b), in his world catalog, synonymized the tribe with the nominate tribe Orthotylini. Since then, however, Carvalho et al. (1983) and Henry (e.g., 1994, 2006, 2013) have given evidence for its monophyly, and, more recently, Schuh (2013) recognized the tribe in his online catalog.

Carvalho et al. (1983) reviewed the species of *Ceratocapsus* Reuter for the Neotropical Region, described 45 new species, and provided a key to distinguish them. Carvalho and Costa (1988) reviewed the genus *Sericophanes* Reuter and provided a key to the Neotropical species, and Hernández and Henry (1999) treated the Cuban species of *Ceratocapsus*. Since then, most papers have been primarily descriptive. More recently, Henry (2006) added the new genus *Izyacapsus* to accommodate two new species from Mexico, Henry (2013) added the new genus *Joseocoris* to include two new species from Argentina and Brazil, and Henry (2015) revised the *Renodaeus* group of species, comprising eight genera and 48 species, including 26 new species, many of which are Neotropical.

Ceratocapsines are largely predatory. Wheeler and Henry (1978b) documented *Ceratocapsus modestus* (Uhler) preying on grape phylloxera, *Daktulosphaira viti-*

foliae (Fitch), in North America. Carvalho et al. (1983) provided an overview of the feeding habits of the group, including a report of *C. dispersus* Carvalho and Fontes feeding on the tobacco budworm *Heliothis virescens* (F.) in Peru. Gravena and Pazetto (1987) observed *C. dispersus* Carvalho and Fontes and *C. mariliensis* Carvalho and Fontes preying on the eggs and larvae of the cotton leafworm *Alabama argillacea* (Hübner) and tobacco budworm in Brazil. Wheeler (2000b, 2001) provided a summary of ceratocapsine feeding habits.

Coridromini

This newly recognized tribe was established for the Old World genus *Coridromius* Signoret (Tatarnic and Cassis 2012), based on hypodermic genitalia formed by the left paramere and aedeagus, the reduced female genitalia, the short round body, and strongly saltatorial hind femora (Tatarnic et al. 2006).

Only the adventive *Coridromius chenopoderis* Tatarnic and Cassis (2008) occurs in Neotropical Region. In addition to the Australian Region, it is now known from Hawaii, the southwestern United States (and Florida), and Mexico, where it feeds on plants in the family Chenopodiaceae (Tatarnic and Cassis 2008).

Halticini (Fig. 10.37)

The halticines are a group of mostly small, shiny, black mirids, with a dorsoventrally elongate head, often with silvery scalelike dorsal setae set in scattered patches, simple parameres, and a mostly membranous endosoma with two sclerotized spiculi (Tatarnic and Cassis 2012).

Only the genus *Microtechnites* Berg and five species are currently recognized from the Neotropics (Schuh 2013). Tatarnic and Cassis (2012) transferred most New World species of *Halticus* Hahn to *Microtechnites* and placed *Cafayatina* Carvalho and Carpintero as a junior synonym of it. Henry (1983) reviewed the New World *Halticus*, several of which are now in *Microtechnites*, and provided a key to distinguish species.

The ubiquitous garden fleahopper, *Microtechnites bractatus* (Say) (widely known in the literature under the combination *Halticus bractatus*), occurs throughout much of the Western Hemisphere from Canada to Argentina and Brazil (Henry and Wheeler 1988), where it may cause significant injury to agricultural crops. Beyer (1921) and Cagle and Jackson (1947) studied the life history. This species may cause severe foliar chlorosis or bleaching, often resulting in heavy crop losses. It has a wide host range, including small grains and many forage and garden crops; major losses to alfalfa, clover, and tomatoes have been recorded (Beyer 1921; Morrill 1925).

Orthotylini (Figs. 10.38, 10.39, 10.40, and 10.41)

Carvalho (1985b) provided a key to genera of Neotropical Orthotylinae.

Other important contributions to the study of this subfamily in the Neotropics are Forero (2008) (*Aoplonema* Knight, *Aoplonemella* Forero, *Daleapidea* Knight, *Hadronema* Uhler, *Hadronemidea* Reuter, *Lopidea* Uhler, *Origonema* Forero, *Scutomiris* Forero, *Tupimiris* Carvalho and Schaffner); Forero and Schwartz (2009) (*Aoplonema*); Forero (2009) (*Chileria* Carvalho, *Orthotylus* Fieber); Henry (1978, 2001) (*Hyalochloria* Reuter); Knight and Schaffner (1968) (*Lopidea*, *Mayamiris* Knight and Schaffner), 1975 (*Lopidea*), 1976 (*Ilnacora* Reuter); Maldonado and Ferreira (1971) (*Carvalhomiris* Maldonado and Ferreira); Maldonado (1980) (*Jobertus* Distant); Schaffner (1969) (*Lopidella* Knight); and Schaffner and Schwartz (2008) (*Ficinus* Distant, *Jornandes* Distant).

The habits of the Orthotylinae are widely varied. Numerous species may become pests of various crops and ornamentals. For example, Wheeler and Henry (1976) documented the life history and damage by the honey locust plant bug, *Blepharidopterus chlorionis* (Say), on ornamental honey locust, *Gleditsia triacanthos* L. (Fabaceae), in the United States. Species of the onion plant bug genus *Labopidea* Uhler may seriously injure commercial onions and garlic (Henry 1982), and *Falconia intermedia* (Distant) is known to discolor *Lantana camara* L. (Verbenaceae) in Mexico (Palmer and Pullen 1998), making it a potential biological control agent. Ferreira et al. (2001) and Pires et al. (2007, 2008) highlighted the feeding habits and life history of *Platyscytus decempunctatus* on its only known host plant, *Solanum cernuum* Vell. (Solanaceae).

A great number of other orthotylines are predatory and have potential as biocontrol agents. The Holarctic *Blepharidopterous angulatus* (Fallén), known as the black-kneed capsid, is a well-documented predator of aphid and mites in orchards and ornamental plantings (Wheeler and Henry 1992). Hsiao (1945) described *Hyalochloria denticornis* from specimens taken on cotton and beans heavily infested with a species of *Empoasca* (Cicadellidae), suggesting that it was attracted to the these leafhoppers as prey, and Beingolea (1959, 1960) observed this mirid feeding on *Anomis texana* Riley (Lepidoptera) and *Aphis gossypii* Glover (Aphididae) on cotton in Peru. Henry (2001) speculated that members of this genus are opportunistic predators, tracking their prey across multiple and often unrelated hosts. Numerous other members of this tribe, such as *Heterotoma planicornis* (Pallas), *Orthotylus ramus* Knight, and *O. marginalis* Reuter, have been observed preying on aphids, scales, and other small arthropods (Wheeler 2001), making it plausible that many of the poorly known Neotropical taxa should have similar predatory tendencies.

Figs. 10.40–10.51 Neotropical Miridae. 40, *Hyalochloria bispina* Henry (Orthotylinae: Orthotylini). 41, *Jobertus chryselectrus* Distant (Orthotylinae: Orthotylini). 42, *Hyalopsallus diaphanus* (Reuter) (Phylinae: Exaeretini). 43, *Moissonia cuneata* (Stål) (Phylinae: Exaeretini). 44, *Cyrtopeltocoris albofasciatus* Reuter (Phylinae: Hallodapini). 45, *Spanagonicus albofasciatus* (Reuter) (Phylinae: Nasocorini). 46, *Reuteroscopus hamatus* Kelton (Phylinae: Phylini: Keltoniina). 47, *Anomalocornis peyreti* Couturier and Costa (Phylinae: Phylini: Phylina). 48, *Platyscytus binotatus* Reuter (Phylinae: Phylini: Phylina). 49, *Sthenaridea vulgaris* (Distant) (Phylinae: Pilophorini). 50, *Semium subglaber* Knight (Phylinae: Semiini: Semiina). 51, *Tytthus neotropicalis* Carvalho (Phylinae: Semiini: Semiina) (Photos by G Ouellette)

10.5.7 Subfamily Phylinae

This subfamily is recognized by the hairlike parempodia in most taxa (except Pilophorini and some Hallodapini), the fleshy pulvilli often found along the ventral surface of each claw, the mittlike left paramere that usually cups the phallotheca, and the strap-like vesica of the male genitalia. Schuh (1974) noted that males are always macropterous, whereas females sometimes may be brachypterous.

Schuh (1995), in his catalog of the Miridae of the world, recognized five phyline tribes (Auricillocorini, Hallodapini, Leucophoropterini, Phylini, and Pilophorini) and, subsequently, listed six tribes in his online catalog (Schuh 2013), following Wyniger's (2010) resurrection of Knight's tribe Pronotocrepini. These studies were followed by Schuh and Menard (2013), who significantly revised the classification and proposed eight tribes, based on a combination of molecular and morphological evidence, and Menard et al. (2014), who further redefined the classification in a "total-evidence phylogenetic analysis," proposed the tribes Coatonocapsini, Cremnorrhini, Decomiini, Hallodapini, Leucophoropterini, Nasocorini, Phylini, Pilophorini, and Semiini.

Only six of these nine phyline tribes are known from the Neotropics. Unfortunately, several of the newly proposed taxa lack meaningful morphological support, so that they cannot be readily keyed or defined without supporting molecular data, and genera can be placed in their respective tribes and subtribes only by following Schuh and Menard's (2013) listing. As a consequence, we provide an overview of the tribes as defined by Schuh and Menard (2013) and Menard et al. (2014), but are unable to provide a key to distinguish them until these taxa are defined morphologically.

Cremnorrhini

Schuh and Menard (2013) and Menard et al. (2014) recognized this tribe based exclusively on molecular data, indicating that unifying morphological characters are yet to be defined. Two subtribes within this tribe were recognized (Schuh and Menard 2013), the Old World Coatonocapsina and the mostly north temperate Cremnorrhina, the latter of which includes all taxa occurring in the Western Hemisphere. Subsequently, Menard et al. (2014) elevated the Old World Coatonocapsini to tribal status.

Most members of this tribe occur in the western United States and Canada, with only six species of *Coquillettia* Uhler, one of *Teleorhinus* Uhler, and one species of the large Holarctic genus *Macrotylus* Fieber recorded from Mexico (Henry and Wheeler 1988; Wyniger 2010, 2012). Most are recognized by their elongate, often ant-like body, the frequently brachypterous females, the oblique head with the clypeus visible dorsally, the pronotum with an upturned collar-like area, the enlarged free pulvilli, and the distinctive male and female genitalia (Wyniger 2010; Schuh and Menard 2013). Wyniger (2010, 2012) revised and provided identification keys to the New World taxa under the tribal name Pronotocrepini, now treated as a synonym of Cremnorrhina by Schuh and Menard (2013).

Exaeretini (Figs. 10.42 and 10.43)

Only three Neotropical genera and about seven species are included in this tribe (Schuh and Menard 2013), including *Gonoporomiris* Henry and Schuh, *Hyalopsallus* Carvalho and Schaffner, and the circumtropical genus *Moissonia* Reuter. Schuh and Menard (2013) defined this tribe based on the relatively long labium extending beyond the hind coxae, the elongate claws without or with only very small pulvilli, a large twisted gonopore, a twisted S-shaped endosoma composed of two straps connected by a membrane, and an elevated posterior lobe on the left paramere.

Gonoporomiris contains two species (Henry and Schuh 2002; Schuh and Schwartz 2004), *Hyalopsallus* one (Henry and Wheeler 1982; Carvalho 1990a), and *Moissonia* four Neotropical species (Schuh 1995). Little is known of their habits, but they are probably phytophagous. Henry and Wheeler (1982) recorded adults and nymphs of *H. diaphanus* (Reuter) from *Crotalaria incana* L. (Fabaceae) in Florida, and Henry and Schuh (2002) reported adults of *Gonoporomiris mirificus* (Distant) from the flowers of cabbage palmetto, *Sabal* sp. (Arecaceae) on Grand Bahama Island.

Hallodapini (Fig. 10.44)

The Hallodapini are mostly Old World and include taxa previously placed in the Auricillocorini, a tribe recently synonymized with it by Schuh and Menard (2013). Hallodapines are recognized by the uniquely shaped flattened pronotal collar, projecting frons and clypeus, and the twisted endosoma. *Cyrtopeltocoris* Reuter, the only hallodapine genus represented in the Western Hemisphere, occurs primarily in the western United States, with one species known from Cuba and two ranging into Baja California, Mexico. Most species of *Cyrtopeltocoris* are shiny, dark brown, with distinct white maculae or spots on the hemelytra. Knight (1968) provided a key to distinguish eight of the eleven known species; Hernández and Henry (2010) redescribed *C. cubanus* Poppius.

Little is known of their habits. Females are brachypterous and extremely antlike. Males are most frequently taken at lights.

Nasocorini (Fig. 10.45)

Ten genera and about 63 species of this tribe occur in the Neotropics (Schuh and Menard 2013), including the widespread genera *Atractotomus* Fieber, *Rhinacloa* Reuter, *Chlamydatus* Curtis, and *Spanagonicus* Berg. Schuh and Menard (2013) defined the Nasocorini based on the flattened vertex of many species; the short, wide hind femora, often with a row of dorsal spicules on the hind femur; the sericeous or scalelike setae; and the cushion-like pulvilli covering the undersurface of the claws.

Recent revisions and keys to species of the Holarctic genera *Atractotomus* (Stonedahl 1990) and *Chlamydatus* (Schuh and Schwartz 2005), and the New World genera *Rhinacloa* (Schuh and Schwartz 1985) and *Spanagonicus* Menard (in press) will allow recognition of most species.

Members of this tribe are primarily phytophagous, but many have predatory tendencies. Stonedahl (1990) discussed the feeding habitus of the species of *Atractotomus*, noting that members of the *magnicornis* group probably were partially predacious. Wheeler (2001) summarized the predatory habits of *A. magnicornis* (Fallén) and *A. mali* (Meyer-Dür), both of which are well-known predators of aphids, scales, mites, and other small arthropods.

Phylini

As noted by Schuh and Menard (2013), this tribe "continues to be an amalgam of genera lacking easily characterized unifying characters," making it impossible to provide a diagnosis that will allow others to recognize the taxa included in this group. Schuh and Menard (2013) and Menard et al. (2014) recognized the subtribes Exocarpocorina (not in the Neotropics), Keltoniina, Oncotylina, and Phylina.

Keltoniina (Fig. 10.46)

Keltoniina contains one Afrotropical genus and three New World genera (Schuh and Menard 2013). This subtribe is recognized by the somewhat elongate head, small eyes; multiple types of vestiture, including long, erect, almost bristlelike setae, and clumps or patches of white sericeous setae; and the generally, stout, C-shaped endosoma.

The genus *Pseudatomoscelis* Reuter contains four species found from the southern United States to Panama and the West Indies, and *Keltonia* Knight includes 13 species, ranging from the southern United States and south to Colombia and Venezuela (Henry 1991). Henry (1991) revised and keyed the species *Keltonia* and *Pseudatomoscelis*, the latter of which includes the cotton fleahopper, *P. seriata* (Reuter), a well-known pest of cotton in the southern United States, and Henry (2002) added a new species of *Pseudatomoscelis* from the Dominican Republic.

The large genus *Reuteroscopus* Kirkaldy includes 54 species found from the United States to South America. The bright yellow body, marked with dark brown, is characteristic of most species, but the male endosoma with peculiar "feathered" appendages is unique in the Miridae. Kelton (1964) and Knight (1965) described and keyed most of the species, but the species limits, especially those described from females, are still unclear, reflecting the great need for revision.

Oncotylina

Although molecular data unites the taxa included in this subtribe, no morphological characters define them (Schuh and Menard 2013; Menard et al. 2014).

Schuh and Menard (2013) placed nine Neotropical genera in this subtribe, with most ranging only into Mexico: *Hamatophylus* Weirauch (one species), *Knightophylinia* Schaffner (1978) (one species), *Lineatopsallus* Henry (two species), *Maculimiris* Weirauch (two species), *Phymatopsallus* Knight (five species), *Schaffneropsallus* (one species), and *Viscacoris* Weirauch (four species).

Ranzovius Distant, with 10 described species occurring from the United States to southern Brazil (Henry 1999), is one of the more interesting phyline genera. All species are obligate spider commensals. Wheeler and McCaffrey (1984) studied the life history and association of the apparently kleptoparasitic *R. clavicornis* (Knight) (as *R. contubernalis* Henry) with the theridiid spider *Anelosimus studiosus* (Hentz); the bug appeared to feed mainly on small entrapped insects too small to attract the spider's attention and occasionally on dry cadavers. Henry (1984a, 1999) revised the genus and provided a key to species.

Phylina (Figs. 10.47 and 10.48)

The only morphological characters uniting this subtribe are the vaguely diagnostic long labium extending beyond the hind coxae and the absence of pronotal calli (Menard et al. 2014). Nevertheless, molecular evidence apparently always groups these taxa (Menard et al. 2014). Because of the lack of a morphological definition, all Neotropical genera placed in Phylina are treated as *incertae sedis* by Schuh and Menard (2013), indicating that much work remains to clarify the position of the taxa included in this group.

Thirteen genera and about 49 Neotropical species are placed in Phylina (Schuh and Menard 2013). Some of the included genera and the most important papers treating them are *Anomalocornis* Carvalho and Wygodzinsky (Couturier and Costa 2002), *Bicurvicoris* Carvalho and Schaffner (Carvalho and Schaffner 1973), *Crassicornus* Carvalho (Carvalho 1984), *Parafulvius* Carvalho (Costa and Couturier 2000), *Platyscytisca* Costa and Henry (Henry and Costa 2003), and *Platyscytus* Reuter (Maldonado and Carvalho 1981).

Pilophorini (Fig. 10.49)

This worldwide tribe, containing many myrmecomorphic taxa, is recognized by the fleshy convergent parempodia between the claws; the white sericeous or silky setae, often occurring in patches or distinct bands; and the carinate and upturned posterior margin of the vertex. Only three genera and about 11 species are known from the Neotropics.

Schuh and Schwartz (1988) treated the New World species of *Pilophorus* and the nine mostly Neotropical species of *Sthenaridea* Reuter, most of which are associated with grasses (Poaceae) and sedges (Cyperaceae). Schuh (1991) provided a phylogenetic, biogeographic, and host plant analysis of this tribe. Hernández and Henry (2010) described the new species *Pilophorus cubanus*, which is associated with *Pinus* (Pinaceae). Most species of *Pilophorus* are, at least partially predacious, feeding on aphids and other arthropods (Schuh and Schwartz 1988); Wheeler (2001) summarized the known hosts and feeding habits for some of the Nearctic and Palearctic species. More recently, Henry (2012c) described the peculiar new genus and species *Randallophorus schuhi* from Paraguay, which apparently is most closely related to *Sthenaridea* based on male genitalia.

Semiini (Figs. 10.50 and 10.51)

The tribe Semiini, as defined by Schuh and Menard (2013), contains two subtribes, the Exocarpocorina, with four Neotropical genera and six species restricted to Argentina and Chile, and the Semiina, with three genera and 16 species ranging from the United States to South America. The tribe is defined by the eyes being parallel to the anterior margin of the pronotum, the reduced scent gland opening, the sclerotized ventral sack in females, and the simple, elongate endosoma, with or without a well-developed secondary gonopore or a broad, flattened, T-shaped endosoma with a well-developed secondary gonopore.

Exocarpocorina

Only four genera [*Ampimpacoris* Weirauch (one species) and Schuh, *Araucanophylus* Carvalho (two species), *Chiloephylus* Carvalho (one species), and *Gonzalezinus* Carvalho (two species)] and six poorly known species of the subtribe are known in the Neotropics, all of which occur only in Argentina or Chile. All remaining genera are found in the Australian Region, making this the only transantarctic distribution in the Phylinae (Schuh and Menard 2013). These taxa are grouped based on unique asymmetrical sclerites of the female vestibulum, the elaborate posterior wall, the L-shaped phallotheca, and the expanded central portion of the left paramere (Schuh and Menard 2013; Menard et al. 2014).

Semiina

Some of the better-known genera of the subtribe Semiina include *Semium* Reuter, with five Neotropical species (Kelton 1974), and *Tytthus* Fieber, with 11 species (Henry 2012b). All species of *Semium* are associated with species of *Chamaesyce* (Euphorbiaceae; older records from *Euphorbia* sp. apply to this genus) (Kelton

1974). Species of the genus *Tytthus* prey exclusively on the eggs of grass-feeding plant hoppers (Delphacidae), making them important biological control agents on agricultural monocots, such as rice and sugarcane (Henry 2012b).

10.6 Concluding Remarks

The Miridae, with more than 11,000 species, represent one of the 20 most hyperdiverse insect groups (Cassis and Schuh 2012). As their common name implies, many plant bugs are phytophagous and often become pests of agricultural crops and ornamental plants. Lygus bugs (*Lygus* spp.), the cotton fleahopper (*Pseudatomoscelis seriatus*), the cocoa capsids (*Monalonion* spp.), the orchid bugs (*Tenthecoris* spp.), the garden fleahopper (*Microtechnites bractatus*), and the bean capsid (*Pycnoderes quadrimaculatus*) are just a few examples of mirid plant pests that bring notoriety to the family through their feeding (Wheeler 2000a), which may be expressed as chlorosis, cankers, shot holing, stunting, fruit drop, yellowing and wilting foliage, and defoliation (Wheeler 2001). Many others have the potential to cause enormous economic losses even when symptoms are not readily apparent (Wheeler 2001).

Though mirids have not evolved the obvious morphological adaptations for capturing prey, such as raptorial front legs found in phymatine reduviids or stout, piercing-sucking mouthparts found in asopine pentatomids and most Reduviidae, many mirids, nevertheless, are successful predators. Most mirids will feed on the eggs, early instar larvae, nymphs, and dead or dying invertebrates (Wheeler 2001). Some taxa possess powerful salivary enzymes that immobilize their prey (Cohen 1996). Most members of the subfamily Deraeocorinae and many species of Mirinae (e.g., *Phytocoris* spp.), Orthotylinae (e.g., *Blepharidopterus angulatus*), Phylinae (e.g., *Campylomma verbasci*), and Bryocorinae (e.g., *Campyloneura virgula*, *Engytatus modestus*) have strong predaceous tendencies.

Despite their great importance both as plant pests and as predators, many new taxa remain to be described, especially in the Neotropical Region, and the biology, host plants, and phylogenetic relationships of most are unknown, except for the most agriculturally important taxa. Henry and Wheeler's (1988) speculation that the world fauna will approach 20,000 species once the Neotropical species are more fully studied likely is an underestimation. We hope this brief overview will stimulate much-needed additional taxonomic and biological work on this highly diverse, speciose group of economically important bugs.

Acknowledgments We are grateful to Michael D. Schwartz (Canadian Collection of Insects, Arachnids, and Nematodes, Ottawa, Ontario, Canada) for permission to reproduce the schematic illustration of *Lygus* sp. showing morphological structures (from Schwartz, M.D. and R. G. Foottit, 1992, Lygus bugs on the prairies. Biology, systematics, and distribution. Technical Bulletin 1992-4E, Agriculture Canada, Ottawa. 44 pp) and Gary Ouellette (Systematic Entomology Laboratory, ARS, USDA, c/o National Museum of Natural History, Washington, DC) for the color plates used in the chapter. We also thank Conselho Nacional de Desenvolvimento Científico e Tecnológico (CNPq) for financial support to Lívia Aquiar Coelho. A. G. Wheeler, Jr. (School of Agricultural, Forest, and Environmental Sciences, Clemson University, Clemson, South Carolina) kindly reviewed the manuscript.

References

Barreto TN (2011) Desarrollo de un sistema de manejo y alerta temprana para la chinche de los pastos Collaria scenica Stål, en relación con la variabilidad y el cambio climático en el altiplano cundiboyacense. Informe Técnico Final, Corporación Colombiana de Investigación Agropecuaria, CI Tibaitatá, Mosquera Cundinamarca

Beingolea GO (1959) Notas sobre a *Hyalochloria denticornis* Tsai Yu-Hsiao (Hemip.: Miridae) predator de los huevos de *Anomis texana* Riley (Lepidop.: Noctuidae). Rev Peruana Entomol Agric 2:51–59

Beingolea GO (1960) Notas adicionales sobre *Hyalochloria denticornis* Tsai Yu-Hsiao (Hemipt.: Miridae), predator de los huevos de *Anomis texana* Riley (Lep.: Noctuidae). Rev Peruana Entomol Agric 3:1–5

Beyer AH (1921) Garden flea-hopper in alfalfa and its control. US Dept Agric Bur Entomol Bull 964:1–27

Brown KS Jr (1991) Conservation of Neotropical environments: insects as indicators. In: Collins NM, Thomas JA (eds) The conservation of insects and their habitats. Academic, San Diego, pp 349–404

Cagle LR, Jackson HW (1947) Life history of the garden fleahopper. Virginia Agric Exp Stn Tech Bull 107:1–27

Callan EM (1943) Natural enemies of the cacao thrips. Bull Entomol Res 34:313–321

Carlessi LRG, Corseiul E, Salvadori JR (1999) Aspectos biológicos e morfométricos de *Collaria scenica* (Stal) (Hemiptera: Miridae) em trigo. An Soc Entomol Brasil 28:65–73

Carpintero DL, Chérot F (2008) Sur quelques nouveaux *Phytocoris* Fallén, 1814 provenant de l'Argentine et du Chili. I (Heteroptera: Miridae). Nouv Rev Entomol (N S) 24:221–258

Carpintero DL, Chérot F (2011) Una nueva especie de *Phytocoris* Fallén, 1814 (Hemiptera: Heteroptera: Miridae). Heteropterus Rev Entomol 11:209–214

Carpintero DL, Estevez AL (2001) Two new genera of Stenodemini, *Spartinomiris* and *Cynodonmiris*, from Argentina (Heteroptera: Miridae). Neotropica 47:25–32

Carpintero DL, Dellapé PM, Cheli G (2008) *Valdesiana curiosa*: a remarkable new genus and species of Clivinematini (Hemiptera: Miridae: Deraeocorinae) from Argentina, with a key to the Argentinean genera and species. Zootaxa 1672:61–68

Carpintero DL, Dellapé PM, Melo MC (2006) New records of Heteroptera (Hemiptera) from Argentina. Zootaxa 1129:1–29

Carvalho JCM (1952) On the major classification of the Miridae (Hemiptera). (With keys to subfamilies and tribes and a catalogue of the world genera.). An Acad Bras Cienc 24:31–110

Carvalho JCM (1953) Neotropical Miridae, LV: the genus *Eccritotarsus* Stål with descriptions of fourteen new species (Hemiptera). An Acad Bras Cienc 25:469–493

Carvalho JCM (1955a) Keys to the genera of Miridae of the world (Hemiptera). Bol Mus Paraense Emílio Goeldi Zool 11:1–151

Carvalho JCM (1955b) Neotropical Miridae, LXV: New genera and species of bugs of the tribe Termatophylini (Hemiptera: Deraeocorinae). Proc U S Nat Mus 104:641–649

Carvalho JCM (1957) A catalogue of the Miridae of the world. Part I. Subfamilies Cylapinae, Deraeocorinae, Bryocorinae. Arq Mus Nac, Rio de Janeiro 44:1–158

Carvalho JCM (1958a) A catalogue of the Miridae of the world. Part II. Subfamily Phylinae. Arq Mus Nac, Rio de Janeiro 45:1–216

Carvalho JCM (1958b) A catalogue of the Miridae of the world. Part III. Subfamily Orthotylinae. Arq Mus Nac, Rio de Janeiro 47:1–161

Carvalho JCM (1959) A catalogue of the Miridae of the world. Part IV. Subfamily Mirinae. Arq Mus Nac, Rio de Janeiro 48:1–384

Carvalho JCM (1960) A catalogue of the Miridae of the world. Part V. Bibliography and general index. Arq Mus Nac, Rio de Janeiro 51:1–194

Carvalho JCM (1972) Mirídeos neotropicais, CXLVI: Gênero *Monalonion* H.-S., 1853 (Hemiptera). Anais Acad Brasil Ciencis 44:119–143

Carvalho JCM (1973a) Mirídeos neotropicais, CLXXVII: Estudos sobre a tribo Herdoniini Distant. XV – Lista dos gêneros e espécies conhecidas até o presente (Hemiptera). Rev Brasil Biol 33(suppl):195–196

Carvalho JCM (1973b) Neotropical Miridae, CLXXVIII: Studies on the tribe Herdoniini Distant XVI: key to the world genera (Hemiptera). Rev Brasil Biol 33(suppl):197–200

Carvalho JCM (1974a) Neotropical Miridae, CLXXXV: Descriptions of three new species of *Prepops* Reuter, tribe Resthenini (Hemiptera). Rev Brasil Biol 34:467–472

Carvalho JCM (1974b) Mirídeos neotropicais, CLXXXI: Estudos sobre a tribo Resthenini Reuter. VIII: descrição de *Prepopsella* n. gen. e seis espécies novas (Hemiptera). An Acad Bras Cienc 46:303–310

Carvalho JCM (1974c) Neotropical Miridae, CLXXXIV: *Restheniella* n. gen. (Resthenini) and *Veramiris* n. gen. (Herdoniini) from Brazil and Mexico (Hemiptera). Rev Brasil Biol 34:461–466

Carvalho JCM (1974d) Mirídeos neotropicais, CLXXXVI: Revisão dos gêneros *Auchus* Distant e *Piestotomus* Bergroth (Hemiptera). Rev Brasil Biol 34:495–504

Carvalho JCM (1975a) Mirídeos neotropicais, CXCV: Gênero *Opistheurista* Carvalho e descrições de novas espécies da tribo Resthenini (Hemiptera). Rev Brasil Biol 35:553–560

Carvalho JCM (1975b) Neotropical Miridae, CLXXXVIII: On the genera *Dolichomiris* Reuter, *Megaloceroea* Fieber, *Stenodema* Laporte, *Trigonotyliscus* n. gen. and *Trigonotylus* Fieber (Hemiptera). Rev Brasil Biol 35:121–140

Carvalho JCM (1975c) Mirídeos neotropicais, CLXXXIX: Descrições de espécies novas de *Poeas* e *Taedia* (Hemiptera). Rev Bras Biol 35:167–206

Carvalho JCM (1976a) Mirídeos neotropicais, CC: Revisão do gênero *Horcias* Distant, come descrições de espécies novas (Hemiptera). Rev Brasil Biol 36:429–472

Carvalho JCM (1976b) Mirídeos neotropicais, CCVI: Revisão do gênero *Horciasinus Carvalho & Jurberg (Hemiptera)*. Rev Brasil Biol 36:811–834

Carvalho JCM (1982) Mirídeos neotropicais, CCXXIX: Gênero *Antias* Distant com descrições de novas espécies (Hemiptera). Rev Brasil Biol 42:325–334

Carvalho JCM (1984) Mirídeos neotropicais, CCLII: Descrições de novos gêneros e espécies da tribo Phylini Douglas & Scott (Hemiptera). Bol Mus Paraense Emílio Goeldi 1:143–206

Carvalho JCM (1985a) Mirídeos neotropicais, CCL: Gêneros e espécies do complexo *Pseudocarnus-Antias-Carmelus* (Hemiptera). Rev Brasil Biol 44:377–386

Carvalho JCM (1985b) Mirídeos neotropicais, CCLIII: Descrições de novos gêneros e espécies da tribo Orthotylini Van Duzee (Hemiptera). Rev Brasil Biol 45:249–298

Carvalho JCM (1985c) Mirídeos neotropicais, CCLVII: Revisão de algumas espécies descritas por O. M. Reuter e correções taxonômicas (Hemiptera). Rev Brasil Biol 45:489–497

Carvalho JCM (1986a) Mirídeos neotropicais, CCLXI: Gênero *Calocorisca* Distant com descrições de espécies novas (Hemiptera). Rev Brasil Biol 46:55–77

Carvalho JCM (1986b) Mirídeos neotropicais, CCLXVI: Gênero *Chrysodasia* Reuter com descrições de novas espécies (Hemiptera). Rev Brasil Biol 46:217–229

Carvalho JCM (1986c) Mirídeos neotropicais, CCXVII: Genero *Euchilocoris* Reuter com descrições de novas espécies (Hemiptera). Rev Brasil Biol 46:257–271

Carvalho JCM (1986d) Mirídeos neotropicais, CCLXX: Espécies regionais do gênero *Tropidosteptes* Uhler (Hemiptera). Rev Brasi Biol 46:401–413

Carvalho JCM (1987a) Mirídeos neotropicais, CCLVI: Dois gêneros e seis espécies novas da tribo Cylapini (Hemiptera). Acta Amazonica 16(17):589–598

Carvalho JCM (1987b) Mirídeos neotropicais, CCLXXXIX: Adições e notas sobre a tribo Resthenini Reuter (Hemiptera). An Acad Bras Cienc 59:393–401

Carvalho JCM (1987c) Neotropical Miridae, CCXC: Two new genera of the tribe Resthenini Reuter (Hemiptera). An Acad Bras Cienc 59(4):405–410

Carvalho JCM (1987d) Neotropical Miridae, CCXCI: Seventeen new species of the genus *Prepops* Reuter (Hemiptera). An Acad Bras Cienc 59:411–431

Carvalho JCM (1988a) Mirídeos neotropicais, CCLXXXII: Um gênero e três espécies novas da tribo Surinamellini Carvalho & Rosas (Hemiptera). Bol Mus Paraense Emílio Goeldi 4:9–16

Carvalho JCM (1988b) Mirídeos neotropicais, CCCIII: Ilustrações das espécies de Resthenini descritas por Reuter 1905 da Venezuela (Hemiptera). An Acad Bras Cienc 60:245–250
Carvalho JCM (1988c) Mirídeos neotropicais, CCLXXXVIII: Novas espécies notas e correções taxonômicas sobre Resthenini (Hemiptera). Rev Brasil Biol 48:349–369
Carvalho JCM (1989a) Mirídeos neotropicais, CCCVI: Novos gêneros e espécies da tribo Cylapini Kirkaldy (Hemiptera). Bol Mus Paraense Emílio Goeldi 5:79–94
Carvalho JCM (1989b) Mirídeos neotropicais, CCCXII: Novo gênero e espécies de Surinamellini (Hemiptera). Bol Mus Paraense Emílio Goeldi 5:213–224
Carvalho JCM (1989c) Mirídeos neotropicais, CCCII: Descrições e nota sobre Resthenini Reuter (Hemiptera). Rev Brasil Biol 49:435–442
Carvalho JCM (1990a) Mirídeos neotropicais, CCCXVII: Sôbre as espécies colecionadas por E. P. Van Duzee, descritas por Reuter, originarias da Jamaica (Hemiptera). Rev Brasil Biol 50:191–220
Carvalho JCM (1990b) Mirídeos neotropicais, CCCXXII: Novas espécies de *Phytocoris* Fallen e correção taxonômica em *Eurychilella* Reuter (Hemiptera). Rev Brasil Biol 50:765–767
Carvalho JCM, Carpintero DJ (1987) Mirídeos neotropicais, CCLXXII: Descrições de sete espécies novas da tribo Resthenini Reuter (Hemiptera). Rev Bras Zool 4:19–31
Carvalho JCM, Carpintero DL (1989) Mirídeos neotropicais, CCCXI: Novas espécies da tribo Resthenini Reuter, colecionadas na Argentina e Peru (Hemiptera). Rev Brasil Biol 49:1109–1123
Carvalho JCM, Carpintero DL (1993) An annotated list of the Miridae of the Argentine Republic (Hemiptera). Rev Brasil Biol 53:397–420
Carvalho JCM, Costa LAA (1988) Neotropical Miridae, CCXCII: Revision of the genus *Sericophanes* Reuter in South and Central America (Hemiptera). Rev Brasil Biol 48:897–909
Carvalho JCM, Costa LAA (1989) Mirídeos neotropicais, CCCV: Espécies sul-americanas do gênero Proba Distant (Hemiptera). An Acad Brasil Cienc 60:333–346
Carvalho JCM, Costa LAA (1990a) Mirídeos neotropicais, CCCXXXV: Novas espécies do gênero *Eustictus* Reuter (Hemiptera). Bol Mus Paraense Emílio Goeldi 6:181–189
Carvalho JCM, Costa LAA (1990b) Mirídeos neotropicais, CCCXXXVI: Novas espécies dos gêneros *Phytocorisca* e *Phytocoris* colecionados no sul do Brasil. An Acad Bras Cienc 62:299–304
Carvalho JCM, Costa LAA (1990c) Mirídeos neotropicais, CCCXIV: Gênero *Diplozona* Van Duzee com descrição de espécies novas (Hemiptera). Rev Brasil Biol 50:174–186
Carvalho JCM, Costa LAA (1990d) Mirídeos neotropicais, CCCXXIX: Novas espécies de *Phytocoris* e *Adphytocoris* da América Central e América do Sul (Hemiptera). Rev Brasil Biol 50:769–781
Carvalho JCM, Costa LAA (1991) Mirídeos neotropicais, CCCXXXV: Novas espécies do gênero *Eustictus* Reuter (Hemiptera). Bol Mus Para Emilio Goeldi, Zool 6:181–189
Carvalho JCM, Costa LAA (1993) Mirídeos neotropicais, CCCLIV: Novas espécies de *Taedia* Distant do Brasil (Hemiptera). Rev Brasil Biol 53:217–239
Carvalho JCM, Costa LAA (1994) The Genus *Fulvius* from the Americas (Hemiptera: Miridae). An Inst Biol Univ Nac Autón México, Ser Zool 65:63–135
Carvalho JCM, Costa LAA (1995) Mirídeos neotropicais, CDIII: Três espécies novas do gênero *Phytocoris* Fallén da América Central (Hemiptera). Rev Brasil Biol 55:577–581
Carvalho JCM, Costa LAA (1997) Chaves taxonômicas de subfamilias e tribos de Miridae Hahn, 1831 do mundo (Insecta, Heteroptera). Arq Museu Nac, Rio de Janeiro 57:1–49
Carvalho JCM, Ferreira PSF (1968) Mirídeos neotropicais, CII: Sobre os gêneros "*Lampsophorus*" Reuter e "*Mabelia*" Kirkaldy com descrição de espécie nova (Hemiptera). Rev Brasil Biol 28:207–213
Carvalho JCM, Ferreira PSF (1969) Mirídeos neotropicais, CXI: Gênero *Phytocoris* Fallén na fauna Chilena (Hemiptera). An Acad Bras Cienc 41:601–623
Carvalho JCM, Ferreira PSF (1971) Mirídeos neotropicais, CXXXI: Gênero *Notholopus* Bergroth (Hemiptera). Rev Brasil Biol 31:165–173
Carvalho JCM, Ferreira PSF (1973a) Mirídeos neotropicais, CLXIII: Estudos sobre a tribo Herdoniini Distant. I – Gênero *Allommatus* Reuter (Hemiptera). Rev Brasil Biol 33:87–104

Carvalho JCM, Ferreira PSF (1973b) Mirídeos neotropicais CLXIV: Estudos sobre a tribo Herdoniini Distant. II – Gênero *Allommatisca* n. gen. (Hemiptera). Rev Brasil Biol 33:105–107

Carvalho JCM, Ferreira PSF (1973c) Mirídeos neotropicais, CLXVI: Estudos sobre a tribo Herdoniini Distant. IV – *Adxenetus* gen. nov. (Hemiptera). Rev Brasil Biol 33:115–124

Carvalho JCM, Ferreira PSF (1973d) Mirídeos neotropicais, CLXVII: Estudos sobre a tribo Herdoniini Distant. V – Gênero *Barberiella* Poppius 1914 (Hemiptera). Rev Brasil Biol 33:125–132

Carvalho JCM, Ferreira PSF (1973e) Mirídeos neotropicais, CLXVIII: Estudos sobre a tribo Herdoniini Distant. VI – Descrições de dois novos gêneros (Hemiptera). Rev Brasil Biol 33:133–136

Carvalho JCM, Ferreira PSF (1973f) Mirídeos neotropicais, CLXIX: Estudos sobre a tribo Herdoniini Distant. VII – Gênero *Haarupia* Poppius (Hemiptera). Rev Brasil Biol 33:137–146

Carvalho JCM, Ferreira PSF (1973g) Mirídeos neotropicais, CLXX: Estudos sobre a tribo Herdoniini Distant. VIII – Gênero *Herdonius* Stål (Hemiptera). Rev Brasil Biol 33:147–154

Carvalho JCM, Ferreira PSF (1973h) Mirídeos neotropicais, CLXXI: Estudos sobre a tribo Herdoniini Distant. IX - Gênero *Lepidoxenetus* Poppius (Hemiptera). Rev Brasil Biol 33:155–158

Carvalho JCM, Ferreira PSF (1973i) Mirídeos neotropicais, CLXXII: Estudos sobre a tribo Herdoniini Distant. X - Gênero *Paraxenetus* Reuter (Hemiptera). Rev Brasil Biol 33:159–174

Carvalho JCM, Ferreira PSF (1973j) Mirídeos neotropicais, CLXXIII: Estudos sobre a tribo Herdoniini Distant. XI - Gênero *Proxenetus* n. gen. (Hemiptera). Rev Brasil Biol 33:175–182

Carvalho JCM, Ferreira PSF (1973k) Mirídeos neotropicais, CLXXV: Estudos sobre a tribo Herdoniini Distant. XIII - Gênero *Zacynthus* Distant (Hemiptera). Rev Brasil Biol 33:183–190

Carvalho JCM, Ferreira PSF (1973l) Mirídeos neotropicais, CLXXIV: Estudos sobre a tribo Herdoniini Distant. XII - Gêneros *Xenetus* Distant e *Xenetopsis* Poppius (Hemiptera). Rev Brasil Biol 33:179–182

Carvalho JCM, Ferreira PSF (1994a) Mirídeos neotropicais, CCCLXXXVIII: Chave para os gêneros de Cylapinae Kirkaldy, 1903 (Hemiptera). Rev Ceres 41:327–334

Carvalho JCM, Ferreira PSF (1994b) Mirídeos Neotropicais, CCCXCIII: Chave para os gêneros de Dicyphinae Reuter 1883 (Hemiptera). Rev Ceres 41:669–675

Carvalho JCM, Ferreira PSF (1995) Mirídeos neotropicais, CCCXC: Chave para os gêneros neotropicais de *Bryocorinae* Baerensprung, 1860 (Heteroptera). Rev Ceres 42:469–496

Carvalho JCM, Fontes AV (1968a) Mirídeos neotropicais, CI: Revisão do complexo *Cylapus* Say, com descrições de gêneros e espécies novos (Hemiptera). Rev Brasil Biol 28:273–282

Carvalho JCM, Fontes AV (1968b) Mirídeos neotropicais, C: Três novas espécies da tribo Resthenini (Hemiptera). Rev Brasil Biol 28:29–34

Carvalho JCM, Fontes AV (1969a) Mirídeos neotropicais CVI: Estudos sobre o gênero *Prepops* Reuter (I), (Hemiptera). An Acad Bras Cienc 41:259–275

Carvalho JCM, Fontes AV (1969b) Mirídeos Neotropicais, XCI: Estudos sobre o gênero *Prepops* Reuter- (II) (Hemiptera). An Acad Bras Cienc 41:575–581

Carvalho JCM, Fontes AV (1969c) Mirídeos neotropicais, CX: Sôbre a posição sistemática de alguns gêneros da tribo Stenodemini China (Hemiptera). Rev Brasil Biol 29:329–350

Carvalho JCM, Fontes AV (1970a) Mirídeos neotropicais, CXVI: Estudos sobre o gênero *Prepops* Reuter (III) (Hemiptera). An Acad Bras Cienc 42:367–388

Carvalho JCM, Fontes AV (1970b) Mirídeos neotropicais, CXVIII: Estudos sobre o gênero *Prepops* Reuter (IV) (Hemiptera). An Acad Bras Cienc 42:793–814

Carvalho JCM, Fontes AV (1970c) Mirídeos neotropicais, CXXII. Estudos sôbre o gênero *Platytylus* Fieber, 1858 (Hemiptera). Stud Entomol 13:441–462

Carvalho JCM, Fontes AV (1971a) Mirídeos neotropicais, CXIII: Descrição de *Platytyloides* n. gen., na tribo Resthenini Reuter (Hemiptera). Rev Brasil Biol 31:17–22

Carvalho JCM, Fontes AV (1971b) Mirídeos neotropicais, CXXIX: Chave sistemática para os gêneros da tribo Resthenini Reuter (Hemiptera). Rev Brasil Biol 31:141–144

Carvalho JCM, Fontes AV (1971c) Mirídeos neotropicais, CXV: Adições e correções a tribo Resthenini Reuter (Hemiptera). Papeis Avulsos Zool 24:117–122

Carvalho JCM, Fontes AV (1971d) Mirídeos neotropicais, CXXXIII: Estudos sobre o gênero *Prepops* Reuter (V) (Hemiptera). Rev Brasil Biol 31:319–326

Carvalho JCM, Fontes AV (1971e) Mirídeos neotropicais, CXXVI: Sôbre o gênero *Chiloxionotus* Reuter (Hemiptera). An Acad Brasi Cienc 43:671–685

Carvalho JCM, Fontes AV (1971f) Mirídeos neotropicais, CXXXIV: Estudos sobre o gênero *Prepops* Reuter (VI) (Hemiptera). An Acad Bras Cienc 43:791–802

Carvalho JCM, Fontes AV (1972) Mirídeos neotropicais, CXLI: Gênero *Alda* Reuter, com descrições de novas espécies (Hemiptera). Rev Brasil Biol 32:85–96

Carvalho JCM, Fontes AV (1981) Mirídeos neotropicais, CCXXV: Revisão do gênero *Collaria* Provancher no continente Americano (Hemiptera). Experientiae 27:11–46

Carvalho JCM, Fontes AV (1983) Mirídeos neotropicais, CCXXXIII: Gênero *Dagbertus* Distant. – Descrições de espécies e revisão das que ocorrem na região (Hemiptera). Rev Brasil Biol 43:157–176

Carvalho JCM, Froeschner RC (1987) Taxonomic names proposed in the insect order Heteroptera by José Candido de Melo Carvalho from 1943 to January 1985, with type depositories. J New York Entomol Soc 95:121–224

Carvalho JCM, Froeschner RC (1990) Taxonomic names proposed in the insect order Heteroptera by José Candido de Melo Carvalho from January 1985 to January 1989, with type depositories. J New York Entomol Soc 98:310–346

Carvalho JCM, Froeschner RC (1994) Taxonomic names proposed in the insect order Heteroptera by José Candido de Melo Carvalho from January 1989 to January 1993. J New York Entomol Soc 102:481–508

Carvalho JCM, Gomes IP (1969) Mirídeos neotropicais, CXII: Espécies do gênero *Polymerus* Hahn que ocorrem no Chile (Hemiptera). Rev Brasil Biol 29:477–486

Carvalho JCM, Gomes IP (1970) Mirídeos neotropicais, CXIV: Algumas espécies de *Phytocoris* Fallén da Argentina, Brasil, Colômbia e Paraguai (Hemiptera). Rev Brasil Biol 30:115–136

Carvalho JCM, Gomes IP (1980) Mirídeos neotropicais, CCXVIII: Revisão do gênero *Derophthalma* Berg, 1883 (Hemiptera). Experientiae 26:93–146

Carvalho JCM, Jurberg J (1976) Mirídeos neotropicais, CCVI: Revisão do gênero *Horciasinus* Carvalho & Jurberg (Hemiptera). Rev Brasil Biol 36:811–834

Carvalho JCM, Leston D (1952) The classification of the British Miridae (Hem.), with keys to genera. Entomol Mon Mag 88:231–251

Carvalho JCM, Maldonado J (1982) Neotropical Miridae, CCXXVIII: Genus *Lundiella* Carvalho with descriptions of new species (Hemiptera). Rev Brasil Biol 42:55–61

Carvalho JCM, Rosas AF (1962) Mirídeos neotropicais, XCI: Uma tribo e dois gêneros novos (Hemiptera). Rev Brasil Biol 22:427–432

Carvalho JCM, Sailer RI (1954) A remarkable new genus and species of Isometopidae from Panama (Hemiptera: Isometopinae). Entomol News 65:85–88

Carvalho JCM, Schaffner JC (1973) Neotropical Miridae, CLIV: *Bicurvicoris*, *Hyalopsallus* and *Nigrimiris*, new genera of Phylini (Hemiptera). Rev Brasil Biol 33:17–22

Carvalho JCM, Schaffner JC (1975) Neotropical Miridae, CXCVIII: Review of the genera *Callichila* Reuter and *Platytylus* Fieber (Hemiptera). Rev Brasil Biol 35:705–736

Carvalho JCM, Schaffner JC (1977) Neotropical Miridae, CCX: Review of the genus *Annona* Distant (Hemiptera). Rev Brasil Biol 37:247–265

Carvalho JCM, Schaffner JC (1985) Neotropical Miridae, CCLIV: Descriptions of new species and one previously described species of Bryocorinae from Mexico, Central America and Venezuela (Hemiptera). Folia Entomol Mex 64:3–32

Carvalho JCM, Fontes AV, Henry TJ (1983) Taxonomy of the South American species of *Ceratocapsus*, with descriptions of 45 new species (Hemiptera: Miridae). USDA Technical Bull 1676:1–58

Cassis G (1984) A systematic study of the subfamily Dicyphinae (Heteroptera: Miridae). PhD dissertation, Oregon State University, Corvallis

Cassis G (1995) A reclassification and phylogeny of the Termatophylini (Heteroptera: Miridae: Deraeocorinae), with a taxonomic revision of the Australian species, and a review of the tribal classification of the Deraeocorinae. Proc Entomol Soc Wash 97:258–330

Cassis G, Monteith GB (2006) A new genus and species of Cylapinae from New Caledonia with re-analysis of the *Vannius* complex phylogeny (Heteroptera: Miridae). Mem Queensland Mus 52:13–26

Cassis G, Schuh RT (2012) Systematics, biodiversity, biogeography, and host associations of the Miridae (Insecta: Hemiptera: Heteroptera: Cimicomorpha). Annu Rev Entomol 57:377–404

Coelho LA (2012) Contribuição à taxonomia e biogeografia do gênero Prepops Reuter, 1905 (Hemiptera: Miridae). Dissertation, Universidade Federal de Viçosa, Brazil

Coelho LA, Ferreira PSF, Costa LAA (2012) *Prepops schuhi*: a distinctive new species of Miridae (Hemiptera: Heteroptera: Mirinae: Resthenini) from Rio Grande do Sul, Brazil. Entomol Am 118:145–151

Cohen AC (1996) Plant feeding by predatory Heteroptera: Evolutionary and adaptational aspects of trophic switching. In: Alomar O, Wiedenmann RN (eds) Zoophytophagous Heteroptera: implications for life history and integrated pest management. Thomas Say Publ Entomol. Entomological Society of America, Lanham, pp 1–17

Costa LAA, Couturier G (2000) Mirídeos neotropicais: Uma nova espécie do gênero *Parafulvius* Carvalho (Heteroptera: Miridae: Phylinae). Rev Française d'Entomol (ns) 22:119–122

Couturier G, Costa LAA (2002) Mirideos Neotropicais: uma nova species do genero *Anomalocornis* Carvalho and Wygodzinsky, 1945 [Heteroptera, Miridae, Phylinae]. Revue francaise d' Entomologie (n.s.) 24:193–196

Crepet WL (1979) Insect pollination: A paleontological perspective. BioScience 29:102–108

Da Silva DB, Alves RT, Ferreira PSF, Camargo AJA (1994) *Collaria oleosa* (Distant, 1883) (Heteroptera: Miridae), uma praga potencial na cultura do trigo na região dos cerrados. Pesq Agropec Bras 29:2007–2012

Davis NT (1955) Morphology of the female organs of reproduction in the Miridae (Hemiptera). Ann Entomol Soc Am 48:132–150

de Abreu JM (1977) Mirídeos neotropicais associados ao cacaueiro. In: Lavabre EM (ed) Les Mirides Du Cacaoyer. Institut Français du Café et du Cacao, Paris, pp 85–106

Denno RF (1977) Comparison of the two assemblages of sap-feeding insects (Homoptera-Hemiptera) inhabiting two structurally different salt marsh grasses in the genus *Spartina*. Environ Entomol 6:359–372

Denis F (1908) An orchid parasite. Gard Chron (Lond) 3(43):313

Duarte AO, Castillo ST, Gómez F, Rey A, Aragón R (1998) El chinche de los pastos: Efectos de su ataque y estrategias para su control en fincas lecheras de Cundinamarca y Boyacá. Corpoica, Tibaitata, 18 p

El-Dessouki SA, El-Kifl AH, Helal HA (1976) Life cycle, host plants and symptoms of damage of the tomato bug, *Nesidiocoris tenuis* Reut. (Hemiptera: Miridae), in Egypt. Z Pflanzenkr Pflanzenshutz 83:204–220

Fauvel G (1999) Diversity of Heteroptera in agroecosystems: Role of sustainability and bioindication. Agric Ecosyst Environ 74:275–303

Ferreira PSF (1980) Taxonomia das espécies do genero *Polymerus* Hahn, 1831 (Hemiptera, Miridae), da região cisandina, América do Sul. Experientiae 26:329–386

Ferreira PSF (1993) Descrição do macho e primeiro registro para o Brasil de *Termatophylidea opaca* Carvalho (Heteroptera: Miridae: Termatophylini). An Soc Entomol Brasil 22:485–490

Ferreira PSF (1996a) Newly recognized synonyms, combinations and status change in the tribe Clivinematini (Heteroptera: Miridae: Deraeocorini). An Soc Entomol Brasil 25:269–274

Ferreira PSF (1996b) Descriptions of four new genera of the tribe Clivinematini (Heteroptera: Miridae: Deraeocorini). An Soc Entomol Brasil 25:259–268

Ferreira PSF (1998) The tribe Clivinematini: Cladistic analysis, geographic distribution and biological considerations (Heteroptera, Miridae). Rev Brasil Entomol 42:53–57

Ferreira PSF (2001) Diagnoses and description of the world genera of the tribe Clivinematini (Heteroptera: Miridae). Stud Neotrop Fauna 36:227–240

Ferreira PSF, Henry TJ (2002) Description of two new species of *Fulvius* Stål (Heteroptera: Miridae: Cylapinae) from Brazil, with biological and biogeographic notes on the genus. Proc Entomol Soc Wash 104:56–62

Ferreira PSF, Henry TJ (2010) Revision of the genus *Ambracius* Stål, 1860 (Heteroptera: Miridae: Deraeocorinae: Clivinematini), with the description of three new species. Zootaxa 2485:1–15

Ferreira PSF, Henry TJ (2011) Synopsis and keys to the tribes, genera, and species of Miridae (Hemiptera: Heteroptera) of Minas Gerais, Brazil. Part I: Bryocorinae. Zootaxa 2920:1–41

Ferreira PSF, Schaffner JC (2004) A review of the genus *Admetus* Distant, 1883 (Heteroptera: Miridae: Clivinematini). Stud Neotrop Fauna 39:117–121

Ferreira PSF, Da Silva ER, Coelho LBN (2001) Miridae (Heteroptera) fitófagos e predadores de Minas Gerais, Brasil, com ênfase em espécies com potencial econômico. Iheringia Ser Zool 91:159–169

Fonseca-González J, Cibrián-Tovar D, Villanueva-Morales A, Lomelí-Flores JR (2007) Descripción y ciclo de vida de la chinche del fresno *Tropidosteptes chapingoensis* Carvalho & Rosas (Hemiptera: Miridae). Ra Ximhai 3:443–459

Fontes AV (1981) Estudos comparativos da genitália da fêmea no gênero *Notholopus* Bergroth, 1922 (Hemiptera: Miridae). Arq Mus Nac, Rio de Janeiro 56:137–183

Fontes AV (1989) Contribuição ao estudo da genitália da fêmea de algumas espécies de *Prepops* Reuter, 1905 (Hemiptera, Miridae). Bol Mus Nac (n s) (Zool) 330:1–31

Fontes AV (1996) Contribuição ao estudo da genitália de fêmeas de *Polymerus* Hahn, 1831 (Hemiptera, Miridae). Rev Brasil Entomol 40:137–141

Forero D (2008) Revision and phylogenetic analysis of the *Hadronema* group (Miridae: Orthotylinae: Orthotylini), with descriptions of new genera and new species, and comments on the Neotropical genus *Tupimiris*. Bull Am Mus Nat Hist 312:1–172

Forero D (2009) Description of one new species of *Chileria* and three new species of *Orthotylus*, with nomenclatural and distributional notes on Neotropical Orthotylinae (Heteroptera: Miridae: Orthotylini). Am Mus Novitates 3642:1–50

Forero D, Schwartz MD (2009) Description of a new species of *Aoplonema* (Hemiptera: Miridae: Orthotylinae) with a biogeographic analysis and reassessment of the phylogeny of the genus. Entomol Am 115:67–76

Ghauri MSK, Ghauri FYK (1983) A new genus and new species of Isometopidae from North India, with a key to world genera (Heteroptera). Reichenbachia 21:19–25

Gimingham CT (1928) An introduced capsid injurious to orchids. Entomol Mon Mag 64:272–274

Gorczyca J (1997) Revision of the *Vannius*-complex and its subfamily placement (Hemiptera: Heteroptera: Miridae). Genus 8:517–553

Gorczyca J (2000) A systematic study on Cylapinae with a revision of the Afrotropical Region (Heteroptera, Miridae). Sydawnictwo Uniwersytetu Śląkiego, Katowice, 174 p

Gorczyca J (2006) The catalog of the subfamily Cylapinae Kirkaldy, 1903 of the world (Hemiptera, Heteroptera, Miridae). Monogr Upper Silesian Mus 5:1–100

Gravena S, Pazetto JA (1987) Predation and parasitism of cotton leafworm eggs, *Alabama argillacea* [Lep.: Noctuidae]. Entomophaga 32:241–248

Grimaldi D, Engel MS (2005) Evolution of the insects. Cambridge University Press, New York

Henry TJ (1978) Review of the Neotropical genus *Hyalochloria*, with descriptions of ten new species (Hemiptera: Miridae). Trans Am Entomol Soc 104:69–90

Henry TJ (1979a) Review of the New World species of *Bothynotus* Fieber (Hemiptera: Miridae). Florida Entomol 62:232–244

Henry TJ (1979b) Review of the *Ceratocapsus lutescens* group, with descriptions of seven new species from the eastern United States (Hemiptera: Miridae). Proc Entomol Soc Wash 81:401–423

Henry TJ (1979c) Review of the New World species of *Myiomma* with descriptions of eight new species (Hemiptera: Miridae: Isometopinae). Proc Entomol Soc Wash 81:552–569

Henry TJ (1980) Review of *Lidopus* Gibson and *Wetmorea* McAtee and Malloch, descriptions of three new genera and two new species, and key to New World genera (Hemiptera: Miridae: Isometopinae). Proc Entomol Soc Wash 82:178–194

Henry TJ (1982) The onion plant bug genus *Labopidicola* (Hemiptera: Miridae): Economic implications, taxonomic review, and description of new species. Proc Entomol Soc Wash 84:1–15

Henry TJ (1983) The garden flea hopper genus *Halticus* (Hemiptera: Miridae): Resurrection of an old name and key to species of the Western Hemisphere. Proc Entomol Soc Wash 85: 607–611

Henry TJ (1984a) Revision of the spider-commensal plant bug genus *Ranzovius* Distant (Heteroptera: Miridae). Proc Entomol Soc Wash 86:53–67

Henry TJ (1984b) New species of Isometopinae (Hemiptera: Miridae) from Mexico, with new records for previously described North America species. Proc Entomol Soc Wash 86:337–345

Henry TJ (1991) Revision of *Keltonia* and the cotton fleahopper genus *Pseudatomoscelis*, with the description of a new genus and an analysis of their relationships (Heteroptera: Miridae: Phylinae). J New York Entomol Soc 99:351–404

Henry TJ (1994) Revision of the myrmecomorphic plant bug genus *Schaffneria* Knight (Heteroptera: Miridae: Orthotylinae). Proc Entomol Soc Wash 96:701–712

Henry TJ (1999) The spider-commensal plant bug genus *Ranzovius* (Heteroptera: Miridae: Phylinae) revisited: Three new species and a revised key, with the description of a new sister genus and phylogenetic analysis. Acta Soc Zool Bohem 63:93–115

Henry TJ (2001) Review of the orthotyline plant bug genus *Hyalochloria*, with a key and descriptions of four new species (Hemiptera: Heteroptera: Miridae). J New York Entomol Soc 109:235–262

Henry TJ (2002) New species of the plant bug genera *Keltonia* Knight and *Pseudatomoscelis* Reuter (Heteroptera: Miridae: Phylinae). Proc Entomol Soc Wash 104:97–105

Henry TJ (2006) *Izyacapsus* (Heteroptera: Miridae: Orthotylinae), a new ceratocapsine plant bug genus established to accommodate two new species from Mexico. Russian Entomol J 15:163–170

Henry TJ (2009) Biodiversity of the Heteroptera. In: Foottit RG, Adler PH (eds) Insect biodiversity: science and society. Wiley-Blackwell Publishing Ltd., Oxford, UK, pp 223–263

Henry TJ (2012a) First eastern North American records of *Campyloneura virgula* (Hemiptera: Heteroptera: Miridae: Bryocorinae). Proc Entomol Soc Wash 114:159–163

Henry TJ (2012b) Revision of the plant bug genus *Tytthus* (Hemiptera, Heteroptera, Miridae, Phylinae). ZooKeys 220:1–114

Henry TJ (2012c) *Randallophorus schuhi*, a new Neotropical genus and species of Pilophorini (Hemiptera: Miridae: Phylinae). Entomol Am 118:177–182

Henry TJ (2013) *Joseocoris*, new genus and two new species of Ceratocapsini (Heteroptera: Miridae: Orthotylinae) from Argentina and Brazil. Rev Soc Entomol Argentina 72:147–153

Henry TJ (2015) Revision of the *ceratocapsine Renodaeus* group: *Marinonicoris*, *Pilophoropsis*, *Renodaeus*, and *Zanchisme*, with descriptions of four new genera (Heteroptera, Miridae, Orthotylinae). ZooKeys 490:1–156

Henry TJ, Carpintero DL (2012) Review of the jumping tree bugs (Hemiptera: Heteroptera: Miridae: Isometopinae) of Argentina and nearby areas of Brazil and Paraguay, with descriptions of nine new species. Zootaxa 3345:41–58

Henry TJ, Carvalho JCM (1987) A peculiar case history: *Hemisphaerodella mirabilis* Reuter is the nymphal stage of *Cyrtocapsus caligineus* (Stål) (Heteroptera: Miridae: Bryocorinae). J New York Entomol Soc 95:290–293

Henry TJ, Costa LAA (2003) Review of the Neotropical phyline genus *Platycytisca* (Hemiptera: Heteroptera: Miridae), with the description of a new species from Bahia, Brazil. J New York Entomol Soc 111:120–125

Henry TJ, Ferreira PSF (2003) Three new genera and three new species of Neotropical Hyaliodini (Hemiptera: Heteroptera: Miridae: Deraeocorinae), with new combinations and new synonymy. J New York Entomol Soc 111:96–119

Henry TJ, Herring JL (1979) Review of the genus *Corticoris* with descriptions of two new species from Mexico (Hemiptera: Miridae: Isometopinae). Proc Entomol Soc Wash 81:82–96

Henry TJ, Kim KC (1984) Genus *Neurocolpus* Reuter (Heteroptera: Miridae): Taxonomy, economic implications, hosts and phylogenetic review. Trans Am Entomol Soc 110:1–75

Henry TJ, Maldonado J (1982) The four "ocelli" of the isometopine genus *Isometocoris* Carvalho and Sailer (Hemiptera: Miridae). Proc Entomol Soc Wash 84:245–249

Henry TJ, Schuh RT (2002) Two new genera to accommodate two North American plant bugs (Heteroptera: Miridae: Phylinae). Proc Entomol Soc Wash 104:211–220

Henry TJ, Silva de Paula A (2004) *Rhyparochromomiris femoratus*, a remarkable new genus and species of Cylapinae (Hemiptera: Heteroptera: Miridae) from Ecuador. J New York Entomol Soc 112:176–182

Henry TJ, Wheeler AG Jr (1982) New United States records for six Neotropical Miridae (Hemiptera) in southern Florida. Florida Entomol 65:233–241

Henry TJ, Wheeler AG (1988) Family Miridae Hahn, 1833. In: Henry TJ, Froeschner RC (eds) Catalog of the Heteroptera, or true bugs, of Canada and the Continental United States. E. J. Brill, Leiden, pp 251–507

Henry TJ, Caldwell DL, Halbert SE (2012) *Tropidosteptes forestierae* (Hemiptera: Heteroptera: Miridae: Mirinae): A new species of plant bug injuring ornamental Florida swamp privet, *Forestiera segregata* (Oleaceae), in South Florida. Insecta Mundi 0240:1–10

Henry TJ, Hoffman RL, Wolski A (2011) First North American record of the Old World cylapine *Fulvius subnitens* Poppius (Hemiptera: Heteroptera: Miridae) from Virginia, with descriptions and a key to the U.S. species of *Fulvius*. Proc Entomol Soc Wash 113:127–136

Herczek A (1993) Systematic position of Isometopinae Fieb. (Miridae, Heteroptera) and their interrelationships. Prace Naukowe Uniw Slaskiego 1357:1–86

Hernández LM, Henry TJ (1999) Review of the *Ceratocapsus* of Cuba, with descriptions of three new species and a neotype designation for *C. cubanus* Bergroth (Heteroptera: Miridae: Orthotylinae). Caribb J Sci 35:201–214

Hernández LM, Henry TJ (2010) The plant bugs or Miridae (Hemiptera: Heteroptera) of Cuba. Pensoft Series Faunistica, Sofia/Moscow

Hernández LM, Stonedahl GM (1999) Three new genera of Deraeocorini (Heteroptera: Miridae: Deraeocorinae) from South America. J New York Entomol Soc 107:14–27

Herring JL (1976) A new genus and species of Cylapinae from Panama (Hemiptera: Miridae). Proc Entomol Soc Wash 78:91–94

Hsiao T-Y (1945) A new plant bug from Peru, with note on a new genus from North America (Miridae: Hemiptera). Proc Entomol Soc Wash 47:24–27

Jones TH (1921) *Opisthuria clandestina* var. *dorsalis* Knight injurious to legumes. J Econ Entomol 14:405

Jung S, Lee S (2012) Molecular phylogeny of the plant bugs (Heteroptera: Miridae) and the evolution of feeding habits. Cladistics 28:50–79

Kelton LA (1959) Male genitalia as taxonomic characters in the Miridae (Hemiptera). Canadian Entomol Suppl 11:1–72

Kelton LA (1964) Revision of the genus *Reuteroscopus* Kirkaldy 1905 with descriptions of eleven new species (Hemiptera: Miridae). Canadian Entomol 96:1421–1433

Kelton LA (1974) A new species of *Semium* (Heteroptera: Miridae) from Mexico, with new records on distribution for two other species. Canadian Entomol 105:1583–1584

Kerzhner IM, Schuh RT (1998) Replacement names for junior homonyms in the family Miridae (Heteroptera). Zoosystematica Rossica 7:171–172

King ABS, Saunders JL (1984) The invertebrate pests of annual food crops in Central America: A guide to their recognition and control. Overseas Development Administration, London

Knight HH (1923) Family Miridae (Capsidae). In: Britton WE (ed) Guide to the insects of Connecticut. Part IV. The Hemiptera or sucking insects of Connecticut, vol 34. State Connecticut Geological and Natural History Survey Bulletin, Hartford, pp 422–658

Knight HH (1928) Key to the species of *Clivinema* with descriptions of seven new species (Hemiptera, Miridae). Proc Biol Soc Wash 41:31–36

Knight HH (1941) The plant bugs, or Miridae, of Illinois. Bull Illinois Nat Hist Surv 22:1–234

Knight HH (1965) A new key to species of *Reuteroscopus* Kirk. with descriptions of new species (Hemiptera, Miridae). Iowa State J Sci 40:101–120

Knight HH (1968) Taxonomic review: Miridae of the Nevada Test Site and the western United States. Brigham Young Univ Sci Bull, Biol Ser 9:1–282

Knight HH, Schaffner JC (1968) *Lopidea* Uhler: New species and records from Mexico and southwestern United States, with *Mayamiris*, related new genus from Mexico (Hemiptera, Miridae). Iowa State J Sci 43:71–81

Knight HH, Schaffner JC (1975) Additional species of *Lopidea* Uhler from Mexico and Guatemala (Hemiptera, Miridae). Iowa State J Res 49:413–422

Knight HH, Schaffner JC (1976) New and old species of the genus *Ilnacora* Reuter (Hemiptera, Miridae). Iowa State J Res 50:399–407

Kukalová-Peck J (1991) Fossil history and the evolution of hexapod structures. In: CSIRO (ed) The insects of Australia. A textbook for students and research workers, vol I, 2nd edn. Cornell University Press, Ithaca, New York pp 141–179

Leston D (1979) The eversible rectal organ of certain Miridae (Hemiptera) and its function. Florida Entomol 62:409–411

Maldonado Capriles J (1969) The Miridae of Puerto Rico (Insecta, Hemiptera). Univ Puerto Rico Agr Exp Sta Tech Pap 45:1–133

Maldonado J (1970) Descriptions of new species of the genus *Termatophylidea* with a key to the known species. Proc Entomol Soc Wash 72:119–126

Maldonado J (1980) The genus *Jobertus* Distant, 1884 (Hemiptera: Miridae: Orthotylinae). J Agri Univ Puerto Rico 64:304–309

Maldonado J, Carvalho JCM (1981) A new species of *Platyscytus* Reuter and a key to the species of the genus (Hemiptera: Miridae). Rev Brasil Biol 41:385–388

Maldonado J, Ferreira PSF (1971) *Carvalhomiris brachypterus*, a new mirid genus and species from Colombia (Hemiptera, Miridae). Rev Brasil Biol 31:345–347

Marples TG (1966) A radionuclide tracer study of arthropod food chains in a *Spartina* salt marsh ecosystem. Ecology 47:270–277

Martinez GE, Barreto TN (1998) La chinche de los pastos *Collaria scenica* Stål en la Sabana de Bogotá. Boletín de Investigación Corporación, Corpoica, 66 p

Menard KL (in press) A review of the genus *Spanagonicus* Berg (Miridae: Phylinae: Nasocorini) with the description of novel antennal characters, the description of a new species from Central America, and a key to currently known taxa. *Zootaxa*

Menard KL, Schuh RT, Woolley JB (2014) Total-evidence phylogenetic analysis and reclassification of the Phylinae (Insecta: Heteroptera: Miridae), with the recognition of new tribes and subtribes and a redefinition of Phylini. Cladistics 30:391–427

Miller RS, Schuh RT (1995) Predation by *Clivinema coalinga* Bliven (Heteroptera: Miridae: Deraeocorinae: Clivinematini) of *Orthezia annae* Cockerell (Sternorrhyncha: Ortheziidae). J New York Entomol Soc 102:383–384

Morrill AW (1925) Commercial entomology on the west coast of Mexico. J Econ Entomol 18:707–716

Myer JG (1935) Notes on cocoa-beetle and cocoa-thrips. Trop Agric 12:22

Oliveira MSS, Fialho JF, Icuma IM (2002) Ocorrência do mirídeo *Hyaliodes beckeri* predando o percevejo-de-renda *Vatiga illudens* em mandioca no Distrito Federal. Ministério da Agricultura, Pecuária e Abastecimento, Comunicado Técnico 67, Brasilia

Palmer WA, Pullen KR (1998) The host range of *Falconia intermedia* (Distant) (Hemiptera: Miridae): A potential biological control agent for *Lantana camara* L. [Verbenaceae]. Proc Entomol Soc Wash 100:633–635

Pires EM, Ferreira PSF, Guedes RNC, Serrão JE (2007) Morphology of the phytophagous bug *Platyscytus decempunctatus* (Carvalho) (Heteroptera: Miridae). Neotrop Entomol 36:510–513

Pires EM, Guedes RNC, Serrão JE, Ferreira PSF (2008) Seasonal and interpopulational morphometry variation of *Platyscytus decempunctatus* (Carvalho 1945) (Heteroptera: Miridae). Biota Neotrop 8:39–46

Reuter OM (1910) Neue Beiträge zur Phylogenie und Systematik der Miriden nebst einleitenden Bemerkungen über die Phylogenie der Heteropteren-Familien. Mit einer Stammbaumstafel. Acta Soc Sci Fenn 37:1–167

Ryder WD, Neyra M, Chong O (1968) Grain sorghum insects in Cuba, and some effects of DDT high-volume emulsion sprays on their abundance and yield. Rev Cubana Cienc Agric 2:245–252

Sadowska-Woda I, Cherot F, Malm T (2008) A preliminary phylogenetic analysis of the genus *Fulvius* Stål (Hemiptera: Miridae: Cylapinae) based on molecular data. Insect Syst Evol 39:407–417

Sanchez JA, Gillespie DR, Gillespie RR (2004) Plant preference in relation to life history traits in the zoophytophagous predator *Dicyphus hesperus*. Entomol Exp Appl 112:7–19

Schaffner JC (1969) Review of the genus *Lopidella* with descriptions of two new species (Hemiptera: Miridae). J Kansas Entomol Soc 42:387–391

Schaffner JC (1978) *Knightensis* and *Knightophylinia*, new genera of the tribe Phylini (Heteroptera: Miridae). J Kansas Entomol Soc 51:441–448

Schaffner JC, Carvalho JCM (1981) Additional cactus plant bugs of the genus *Hesperolabops* (Miridae, Hemiptera). Folia Entomol Mex 47:69–80

Schaffner JC, Schwartz MD (2008) Revision of the Mexican genera *Ficinus* Distant and *Jornandes* Distant with the description of 21 new species (Heteroptera: Miridae: Orthotylinae: Orthotylini). Bull Am Mus Nat Hist 309:1–87

Schuh RT (1974) The Orthotylinae and Phylinae (Hemiptera: Miridae) of South Africa with a phylogenetic analysis of the ant-mimetic tribes of the two subfamilies for the world. Entomol Am 47:1–332

Schuh RT (1975) The structure, distribution, and taxonomic importance of trichobothria in the Miridae (Hemiptera). Am Mus Novitates 2585:1–26

Schuh RT (1976) Pretarsal structure in the Miridae (Hemiptera) with a cladistic analysis of relationships within the family. Am Mus Novitates 2601:391–439

Schuh RT (1991) Phylogenetic, host and biogeographic analyses of the Pilophorini (Heteroptera: Miridae: Phylinae). Cladistics 7:157–189

Schuh RT (1995) Plant bugs of the world (Insecta: Heteroptera: Miridae). Systematic catalog, distributions, host list and bibliography. New York Entomological Society, New York

Schuh RT (2013) On-line systematic catalog of plant bugs (Insecta: Heteroptera: Miridae). http://research.amnh.org/pbi/catalog/ (2003–2013). Accessed 1 Aug 2014

Schuh RT, Menard KL (2013) A revised classification of the Phylinae (Insecta: Heteroptera: Miridae): Arguments for the placement of genera. Am Mus Novitates 3785:1–72

Schuh RT, Schwartz MD (1985) Revision of the plant bug genus *Rhinacloa* Reuter with a phylogenetic analysis (Hemiptera: Miridae). Bull Am Mus Nat Hist 179:382–469

Schuh RT, Schwartz MD (1988) Revision of the New World Pilophorini (Heteroptera: Miridae: Phylinae). Bull Am Mus Nat Hist 187:101–201

Schuh RT, Schwartz MD (2004) New genera, new species, new synonyms, and new combinations in North American and Caribbean Phylinae (Heteroptera: Miridae). Am Mus Novitates 3436: 1–36

Schuh RT, Schwartz MD (2005) Review of the North American *Chlamydatus* Curtis species, with new synonymy and the description of two new species (Heteroptera: Miridae: Phylinae). Am Mus Novitates 3471:1–55

Schuh RT, Slater JA (1995) True bugs of the world (Hemiptera: Heteroptera), classification and natural history. Cornell University Press, New York

Schuh RT, Weirauch C, Wheeler WC (2009) Phylogenetic relationships within the Cimicomorpha (Hemiptera: Heteroptera): a total-evidence analysis. Syst Entomol 34:15–48

Schwartz MD (1984) A revision of the black grass bug genus *Irbisia* Reuter (Heteroptera: Miridae). J New York Entomol Soc 92:193–306

Schwartz MD (1987) Phylogenetic revision of the Stenodemini with a review of the Mirinae (Heteroptera: Miridae). Dissertation, City University of New York

Schwartz MD (2008) Revision of the Stenodemini with a review of the included genera (Hemiptera: Heteroptera: Miridae: Mirinae). Proc Entomol Soc Wash 110:1111–1201

Schwartz MD (2011) Revision and phylogenetic analysis of the North American genus *Slaterocoris* Wagner with new synonymy, the description of five new species and a new genus from Mexico, and a review of the genus *Scalponotatus* Kelton (Heteroptera: Miridae: Orthotylinae). Bull Am Mus Nat Hist 354:1–290

Schwartz MD, Foottit RG (1992) Lygus bugs on the prairies: Biology, systematics, and distribution. Agric Can Tech Bull 4E:1–46

Schwartz MD, Foottit RG (1998) Revision of the Nearctic species of the genus *Lygus* Hahn, with a review of the Palaearctic species (Heteroptera: Miridae), vol 10. Mem Entomol Int 10:1–428. Associated Publishers, Gainesville, Florida

Shcherbakov DE, Popov YA (2002) Superorder Cimicidea Laicharting, 1781 Order Hemiptera Linné, 1758. The bugs, cicadas, plantlice, scale insects, etc. (=Cimicida Laicharting, 1781, =Homoptera Leach, 1815+Heteroptera Latreille, 1810). In: Rasnitsyn AP, Quicke DLJ (eds) History of insects. Kluwer Academic Publishers, Dordrecht, pp 143–157

Southwood TRE, Leston D (1959) Land and water bugs of the British Isles. Frederick Warne & Co., London

Stonedahl GM (1988) Revisions of *Dioclerus*, *Harpedona*, *Mertila*, *Myiocapsus*, *Prodromus*, and *Thaumastomiris* (Heteroptera: Miridae, Bryocorinae, Eccritotarsini). Bull Am Mus Nat Hist 187:1–99

Stonedahl GM (1990) Revision and cladistic analysis of the Holarctic genus *Atractotomus* Fieber (Heteroptera: Miridae: Phylinae). Bull Am Mus Nat Hist 198:1–88

Stonedahl GM, Lattin JD, Razafimahatratra V (1997) Review of the *Eurychilopterella* complex of genera, including the description of a new genus from Mexico (Heteroptera: Miridae: Deraeocorinae). Am Mus Novitates 3198:1–33

Tanada Y, Holdaway FG (1954a) Feeding habits of the tomato bug, *Cyrtopeltis* (*Engytatus*) *modestus* (Distant), with special reference to the feeding lesion on tomato. Hawaii Agric Exp Stn Tech Bull 24:1–40

Tanada Y, Holdaway FG (1954b) Feeding habits of the tomato bug, *Cyrtopeltis* (*Engytatus*) *modestus* (Distant), with special reference to the feeding lesion on tomato. Hawaii Agricultural Experiment Station, University of Hawaii, Honolulu, 40 p

Tatarnic NJ, Cassis G (2008) Revision of the plant bug genus *Coridromius* Signoret (Insecta: Heteroptera: Miridae). Bull Am Mus Nat Hist 315:1–95

Tatarnic NJ, Cassis G (2010) Sexual coevolution in the traumatically inseminating plant bug genus *Coridromius*. J Evolution Biol 23:1321–1326

Tatarnic NJ, Cassis G (2012) The Halticini of the world (Insecta: Heteroptera: Miridae: Orthotylinae): Generic reclassification, phylogeny, and host plant associations. Zool J Linn Soc London 164:558–658

Tatarnic NJ, Cassis G, Hochuli DF (2006) Traumatic insemination in the plant bug genus *Coridromius* Signoret (Heteroptera: Miridae). Biol Lett 2:58–61

Torres Jaimes LF, Correa Londoño GA, Cartagena Valenzuela JR, Monsalve García DA, Londoño Zuluaga ME (2012) Relationship of *Monalonion velezangeli* Carvalho & Costa (Hemiptera: Miridae) with the phenology of avocado (*Persea americana* Mill., cv Hass). Rev Fac Nal Agr Medellin 65:6659–6665

Uceli LF, Ferreira PSF, Pirovani VD (2009) Review of *Pachymeroceroides* genus with the description of a new species (Hemiptera: Heteroptera: Miridae: Bryocorinae). Zootaxa 2174:37–44

Van Duzee EP (1917) Catalogue of the Hemiptera of America north of Mexico (excepting the Aphididae, Coccidae, and Aleurodidae). Univ California Pub Tech Bull Entomol 2:1–902

Weiss HB (1917) Some unusual orchid insects (Hem., Lep., Dip., Col.). Entomol News 28:24–29

Wheeler AG (1976) Lygus bugs as facultative predators. In: Scott DR, O'Keeffe LE (eds) Lygus bug: Host plant interactions. University of Idaho Press, Moscow, pp 28–35

Wheeler AG (1982) New United States records for six neotropical Miridae (Hemiptera) in southern Florida. Fla Entomol 65:233–241

Wheeler AG (1991) Plant bugs of *Quercus ilicifolia*: Myriads of mirids (Heteroptera) in pitch pine-scrub oak barrens. J New York Entomol Soc 99:405–440

Wheeler AG (2000a) Plant bugs (Miridae) as plant pests. In: Schaefer CW, Panizzi AR (eds) Heteroptera of economic importance. CRC Press, Boca Raton, Florida, pp 37–83

Wheeler AG (2000b) Predacious plant bugs (Miridae). In: Schaefer CW, Panizzi AR (eds) Heteroptera of economic importance. CRC Press, Boca Raton, pp 657–693

Wheeler AG (2001) Biology of the plant bugs (Hemiptera: Miridae). Pests, predators, opportunists. Cornell University Press, Ithaca, New York

Wheeler AG, Henry TJ (1976) Biology of the honeylocust plant bug, *Diaphnocoris chlorionis*, and other mirids associated with ornamental honeylocust. Ann Entomol Soc Am 69:1095–1104

Wheeler AG, Henry TJ (1978a) Isometopinae (Hemiptera: Miridae) in Pennsylvania: Biology and descriptions of fifth instars, with observations of predation on obscure scale. Ann Entomol Soc Am 71:607–614

Wheeler AG, Henry TJ (1978b) *Ceratocapsus modestus* (Hemiptera: Miridae), a predator of grape phylloxera: Seasonal history and description of fifth instar. Melsheimer Entomol Ser No 25:6–10

Wheeler AG, Henry TJ (1980) Seasonal history and host plants of the ant mimic *Barberiella formicoides* Poppius, with description of the fifth-instar (Hemiptera: Miridae). Proc Entomol Soc Wash 82:269–275

Wheeler AG, Henry TJ (1992) A synthesis of the Holarctic Miridae (Heteroptera): Distribution, biology, and origin, with emphasis on North America. Thomas Say Foundation, Entomological Society of America, Lanham, Maryland

Wheeler AG, Henry TJ (2004) Plant bugs (Hemiptera: Miridae). In: Capinera J (ed) Encyclopedia of entomology. Kluwer Academic Publishers, Oxford, pp 1737–1742

Wheeler AG, Henry TJ (2005) Description of the adult and fifth instar of a myrmecomorphic plant bug, *Bicuspidatiella conica* Maldonado (Hemiptera: Miridae: Deraeocorinae), with notes on habits. Proc Entomol Soc Wash 107:209–213

Wheeler AG, McCaffrey JP (1984) *Ranzovius contubernalis*: Seasonal history, habits, and description of fifth instar, with speculation on the origin of spider commensalism in the genus *Ranzovius* (Hemiptera: Miridae). Proc Entomol Soc Wash 86:1063–1068

Wheeler QD, Wheeler AG (1994) Mycophagous Miridae? Associations of Cylapinae (Heteroptera) with pyrenomycete fungi (Euascomycetes: Xylariaceae). Proc Entomol Soc Wash 102:114–117

Wheeler AG, Stinner BR, Henry TJ (1975) Biology and nymphal stages of *Deraeocoris nebulosus* (Hemiptera: Miridae), a predator of arthropod pests on ornamentals. Ann Entomol Soc Am 68:1063

Wheeler WC, Schuh RT, Bang R (1993) Cladistic relationships among higher groups of Heteroptera: congruence between morphological and molecular data sets. Entomol Scand 24:121–137

Wille JE (1944) Insect pests of cacao in Peru. Trop Agric 21:143

Wolski A, Henry TJ (2012) Revision of the New World species of *Peritropis* Uhler (Heteroptera: Miridae: Cylapinae). Insect Syst Evol 43:213–270

Wolski A, Henry TJ (2013) Corrigendum to: revision of the new world species of *Peritropis* Uhler (Hemiptera: Miridae: Cylapinae). Insect Syst Evol 44:107–109

Wolski A, Henry TJ (2015) A new subfamily placement for the genus *Isometocoris* Carvalho and Sailer (Hemiptera: Heteroptera: Miridae), with the description of a new species from Brazil. Proc Entomol Soc Wash 117:407–418

Wyniger D (2010) Resurrection of the Pronotocrepini, with revisions of the Nearctic genera *Orectoderus* Uhler, *Pronotocrepis* Knight, and *Teleorhinus* Uhler, and comments on the Palearctic *Ethelastia* Reuter (Heteroptera: Miridae: Phylinae). Am Mus Novitates 3703:1–67

Wyniger D (2012) Revision of the Nearctic genus *Coquillettia* Uhler with a transfer to the tribe Phylini, the description of 14 new species, a new synonymy, and the description of two new Nearctic genera *Leutiola* and *Ticua* and two new species (Heteroptera: Miridae: Phylinae). Entomol Am 117:134–211 [2011]

Yao Y, Wanzhi C, Dong R (2007) The oldest known fossil plant bug (Hemiptera: Miridae), from Middle Jurassic of Inner Mongolia, China. Zootaxa 1442:37–41

Young OP (1986) Host plants of the tarnished plant bug, *Lygus lineolaris* (Heteroptera: Miridae). Ann Entomol Soc Am 79:747–762

Zherikhin VV (2002) Ecological history of terrestrial insects. In: Rasnitsyn AP, Quicke DLJ (eds) History of insects. Kluwer Academic Publishers, Dordrecht, pp 331–388

Chapter 11
Damsel Bugs (Nabidae)

Maria del C. Coscarón, S. Kristine Braman, and Marcela Cornelis

Abstract Damsel bugs, Nabidae, are valued predators in agricultural and urban systems. Worldwide, there are considered to be 31 genera and 386 species known, with 11 genera and 83 species from the Neotropical region. While little bionomic information is available for Neotropical nabid species, highlighting the need for future research, the biology of *Nabis (Tropiconabis) capsiformis* Germar, is representative of the family. *N. (T.) capsiformis* demonstrated a type I functional response when feeding on bollworm eggs. Under laboratory conditions, total generation time was 51.6 days. Nabids insert their eggs into plant tissue with only the operculum remaining visible. *N. (T.) capsiformis* averaged 105.3 eggs/female. Much remains to be discovered concerning this fascinating and important group of predatory heteropterans. Knowledge about biology and taxonomy of this group is fundamental to developing future programs in integrated pest control.

11.1 Introduction

The Nabidae, or damsel bugs, are regarded as beneficial insects, appreciated for their predatory potential in suppressing insect pests in crops such as alfalfa, soybean, and cotton (Braman 2000). Nabids have been known to probe plants for moisture (Stoner 1972), occasionally carry plant pathogens (Burgess et al. 1983), and even

M.C. Coscarón (✉)
División Entomología, Museo de La Plata,
Paseo del bosque s/n, La Plata 1900, Buenos Aires, Argentina
e-mail: mcoscaron@fcnym.unlp.edu.ar

S.K. Braman
Department of Entomology, UGA Griffin Campus,
1109 Experiment Street, Griffin, GA 30223-1797, USA
e-mail: kbraman@uga.edu

M. Cornelis
Facultad de Ciencias Exactas y Naturales, Universidad Nacional de La Pampa,
Uruguay 151 L6300CLB, Santa Rosa, La Pampa, Argentina
e-mail: cornelismarcela@gmail.com

A.R. Panizzi, J. Grazia (eds.), *True Bugs (Heteroptera) of the Neotropics*,
Entomology in Focus 2, DOI 10.1007/978-94-017-9861-7_11

occasionally bite humans (Faúndez and Carvajal 2011). They feed on small insects, many of which are considered pests, yet also feed on insects regarded as beneficial, e.g., anthocorids. Nabids also serve as prey for spiders, assassin bugs, birds, etc. Nabids possess raptorial front legs with thickened spines and have purportedly been given their common name from their habit of holding the front legs up like a damsel holding a skirt hem up while dancing. Nabids overwinter as adults or late-stage nymphs although some have been reported to overwinter as eggs (Harris 1928).

The majority of information published on biology and ecology and impact of nabids have focused on those inhabiting agroecosystems, with little known about the bionomics of the many species occurring outside managed ecosystems (Wheeler 2001). Characters used in identifying species include external morphological characters: size and shape of male claspers, interocular distances, trichobothria, length of antennal segments, degree of fuscous markings, spines on the legs, and size and shape of internal seminal depository of females. While there are considered to be 31 genera and 386 species known worldwide (Henry 2009), for the Neotropical region, 11 genera and 83 species are known (Volpi and Coscarón 2010; Coscarón and Volpi 2013). Neotropical Nabidae are not well known at present, but a key to the nabids of Argentina (Cornelis and Coscarón 2013) provides descriptions and identification characters for five genera and 14 species (Coscarón, in press). The biology, ecology, economic importance, and susceptibility to insecticides for the two best-studied species in the region, *Nabis (Tropiconabis) capsiformis* Germar and *Nabis punctipennis*, are provided herein. There continues to be a particular need for biological information on neotropical nabid species especially in defining potential and actual impact on known pest species.

11.2 General Characteristics and Diagnosis

11.2.1 General Characteristics

The Nabidae are a small group of insects ranging from 5 to 15 mm. Many are elongate and of drab coloration, whereas others are more stout bodied and occasionally possess distinctive red and black color patterns (Fig. 11.1a–d). The general aspect in different views can be seen in Fig. 11.2a, b.

11.2.2 Diagnosis

Labium flexible and usually curving, reaching onto pro- or mesothorax, with 4 distinct segments (Fig. 11.2a, b, e); antennal prepedicellite present, variable in length (Figs. 11.1b and 11.2e); fossula spongiosa present on fore and middle tibiae of most taxa (Fig. 11.2d, e); costal fracture present or absent; membrane with 2 or 3 elongate

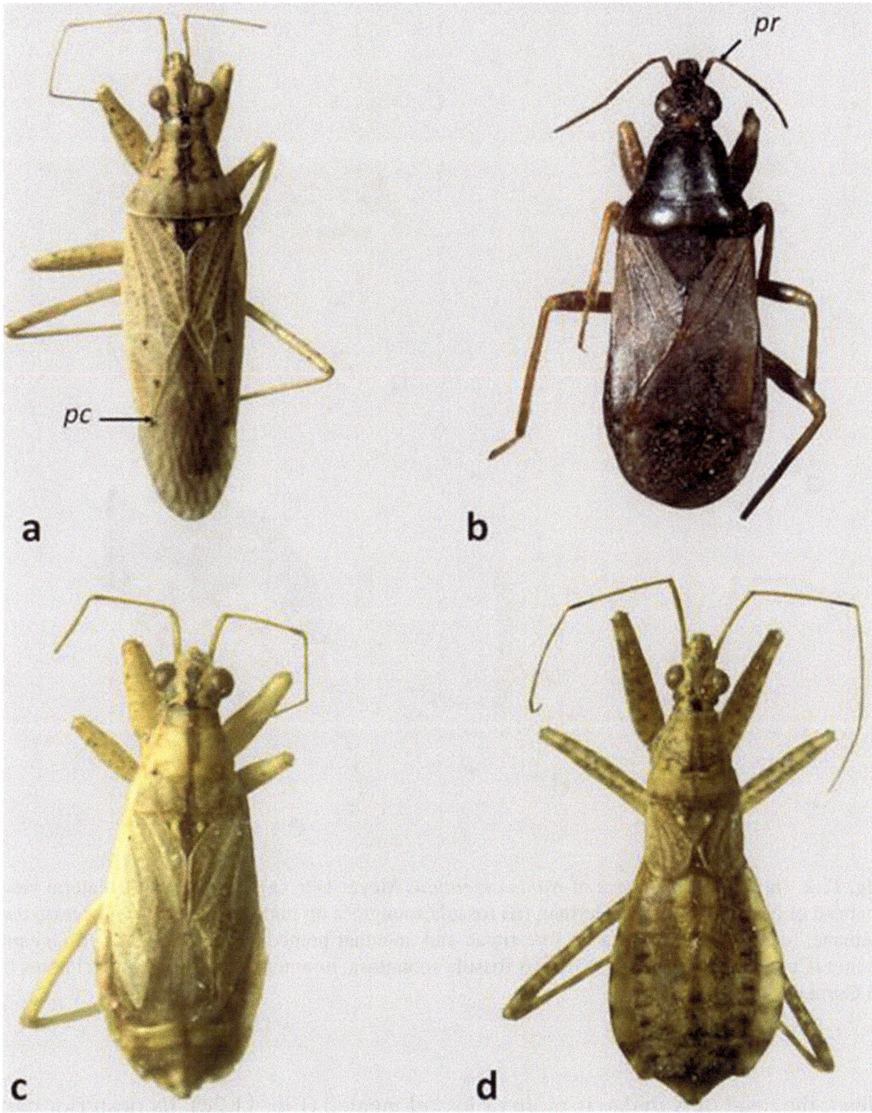

Fig. 11.1 Different color patterns of Nabidae, (**a**) *Nabis punctipennis* Blanchard (macropterous form), (**b**) *Pagasa fuscipennis* Reuter (macropterous form) (Cornelis and Coscarón 2013), (**c**) *Nabis faminei* Stål (brachypterous form), (**d**) *Hoplistoscelis* sp. Reuter (micropterous form) (*pc* processus corial, *pr* antennal prepedicellite) (Photos by M Cornelis)

cells, usually with emanating veins and a stub (processus corial) (Fig. 11.1a); abdominal spiracles 2–8 present, located either on laterosternites or medio-sternites; parastigmal pits present in most taxa; Ekblom's organ usually present in males (Fig. 11.3a, d, e).

Fig. 11.2 (**a, b**) General aspect of *Nabis argentinus* Meyer-Dür, (**a**) dorsal view, (**b**) lateral view, (**c**) head of *Nabis capsiformis* Germar, (**d**) fossula spongiosa on middle tibiae of *Nabis mexicanus* Remane, (**e**) fossula spongiosa on fore tibiae and antennal prepedicellite of *Pagasa fuscipennis* Reuter (Cornelis and Coscarón 2013) (*fs* fossula spongiosa, *pr* antennal prepedicellite) (Photos by M Cornelis)

Head: the head of Nabidae is more or less elongated (Fig. 11.2c). Its posterior part, cylindrical, engaged in the collar of the prothorax with a neck.

Thorax: it is more or less bell shape (Fig. 11.1a–d).

Legs: according to the type of prey, foretibiae and forefemora are simple; others are enlarged and armed with heavy spines, forming a formidable apposable grasping apparatus.

Wings: polymorphism is common in many species (Fig. 11.1a–d) particularly in those living at higher latitudes (Asquith and Lattin 1990). According to Kerzhner (1983), *Nabis (T.) capsiformis*, "pale damsel bugs," are long-winged bugs that fly well and are the most widespread species in the Nabidae.

Abdomen: trichobothria are present on abdomen in some taxa.

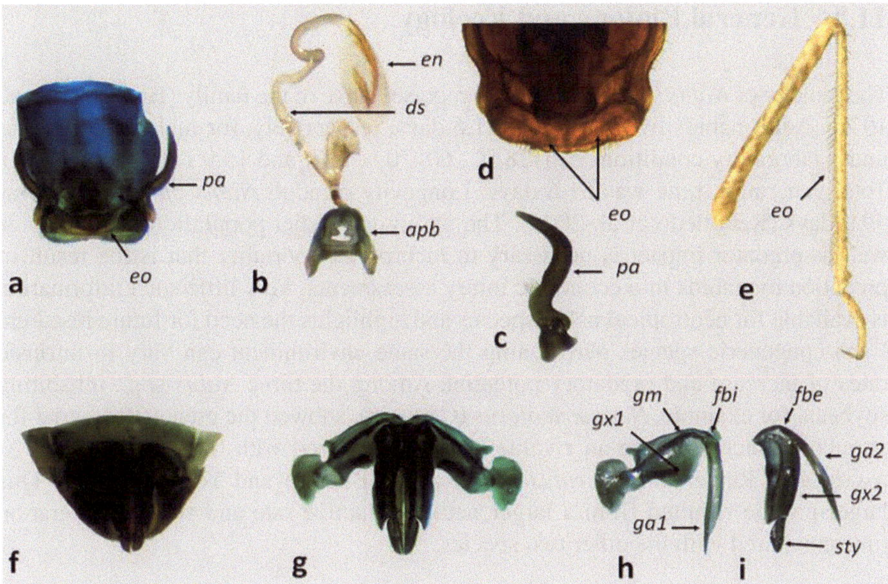

Fig. 11.3 (**a–c**) Male genitalia *Nabis faminei* Stål, (**a**) pygophore, (**b**) aedeagus, (**c**) paramere (Cornelis and Coscarón 2013), (**d**) detail of Ekblom's organ *Nabis argentinus* Meyer-Dür, (**e**) hind tibiae of male *Nabis argentinus* Meyer-Dür showing spines of Ekblom's organ, (**f–i**) female genitalia *Pagasa fuscipennis* Reuter, (**f–g**) genital segment, (**h**) first gonapophysis and gonocoxite 1, (**i**) second gonapophysis and gonocoxite 2 (Cornelis and Coscarón 2013) (*apb* articulatory apparatus, *ds* ductus seminis, *en* endosoma, *eo* Ekblom's organ, *fbe* external fibula, *fbi* internal fibula, *ga1* and *ga2* gonapophysis 1 and 2, *gm* gonangulum, *gx1* and *2* gonocoxites 1 and 2, *pa* paramere, *sty* styloid) (Photos by M Cornelis)

Male genitalia (Fig. 11.3a–c) are usually symmetrical, sometimes parameres (some Prostemmatinae) or phallus asymmetrical.

Female genitalia (Fig. 11.3f–i) are ovipositor laciniate reduced in some cases: spermatheca in the form of vermiform gland.

11.2.3 Specialized Morphology

Some Nabidae possess a structure known as Ekblom's organ Kerzhner (1981). It consists of two diagonal grooves surrounded by specialized setae, situated behind the posterior foramen of the pygophore (Fig. 11.3a, d), and a group of specialized setae at the posterodistal margin of the hind tibia (Fig. 11.3e) that are rubbed across the pygophoral portion of the organ to distribute attractant pheromones from rectal glands (Carayon 1970). The parastigmal pits are small depressions containing a concentration of apparently secretory setae, located adjacent to the spiracles on one or more of the ventral laterotergites of abdominal segments 3–7 or in the sternum 3 of others (Carayon 1948, 1950). Their function is unknown.

11.3 General Biology and Ecology

The biology of *Nabis (T.) capsiformis* is representative of the family (Hormchan et al. 1976). Adult nabids lived 14.9 and 21.6 days, respectively, for males and females under laboratory conditions (26–28 °C, 60–70 % RH, and 15:9 L:D photoperiod). Total generation time was 51.6 days. Longevity of adult *Nabis punctipennis* was 59.9 days (Rebolledo et al. 2005). The ability to predict population phenology as well as predator impact is necessary to incorporate mortality that is the result of predation by nabids into economic injury assessments. Very little such information is available for neotropical nabid species and highlights the need for future research. Even congeneric species partitioning the same environment can vary in intrinsic rates of increase and predatory potential. Among the three *Nabis* spp., inhabiting soybean, for example, *N. americoferus* (Carayon), showed the greatest potential for population increase with an r value of 0.72 compared with 0.50 and 0.44 for *N. roseipennis* Reuter and *N. rufusculus* Reuter (Braman and Yeargan 1988). This larger r value resulted from a larger net reproductive rate and shorter generation time compared with the other two species.

11.3.1 Eggs and Oviposition

Nabids insert their eggs into plant tissue with only the operculum remaining visible (Braman 2000). Embryonic development takes place after the eggs are laid. Eggs with 1 micropyle are shown in Fig. 11.4a. Eggs of *Nabis (T.) capsiformis* required 7.6 days to hatch under laboratory conditions described above (Hormchan et al. 1976).

11.3.2 Nymph Feeding and Development

Nymphal nabids are important predators in agroecosystems, yet few identification aids are available to assist the pest management practitioner. Elvin and Sloderbeck (1984) developed a key to the nymphs of four species of Nabidae from the southeastern USA: *Nabis roseipennis, N. americoferus, N. (T.) capsiformis,* and *Hoplistoceslis deceptivus* (Harris). Cornelis et al. (2012) provide a description of instars III–V of *N. (T.) capsiformis* collected in Argentina. Nymphal dorsal abdominal scent glands are present between terga 4/5, 5/6, and 6/7 (Fig. 11.4d). Trichobothria are present on the abdomen of nymphs of some taxa. The third, fourth, and fifth instars can be seen in Fig. 11.4b–d.

Péricart (1987) gave a general contribution to the knowledge of the immature stages of Nabidae. Currently, contributions on instars included three species of *Nabis*: *N. alternatus* Parshley (Taylor 1949), *N. roseipennis*, and *N. rufusculus* (Mundinger 1922); several species of *Nabicula*, *N. americolimbata* Carayon, *N. fla-*

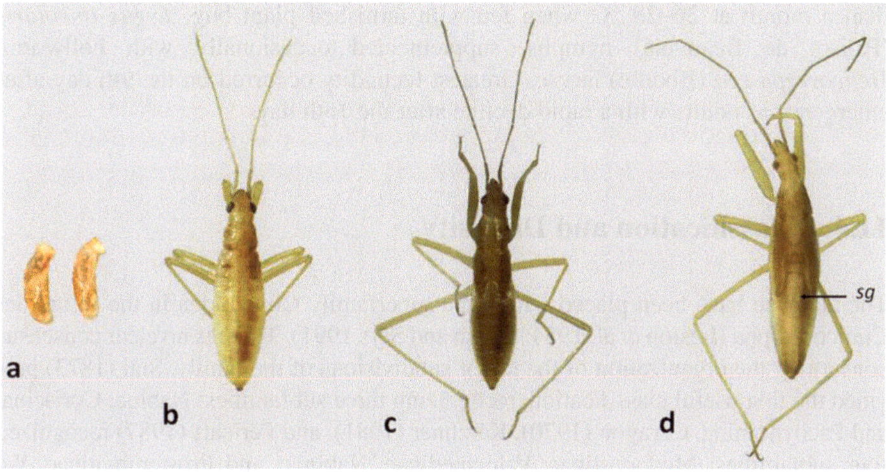

Fig. 11.4 (**a**) Eggs of *Pagasa fuscipennis* Reuter (Cornelis and Coscarón 2013), (**b–d**) nymphs of *Nabis capsiformis* Germar, (**b**) third instar, (**c**) fourth instar, (**d**) fifth instar (Cornelis et al. 2012) (*sg* scent glands) (Photos by M Cornelis)

vomarginata Scholtz, *N. limbata* Dahlbom, *N. nigrovittata nearctica* Kerzhner, *N. propinqua* Reuter, *N. subcoleoptrata* Kirby, and *N. vanduzeei* Kirkaldy (Larivière 1994); *Lasiomerus annulatus* Reuter (Larivière 1993); *Arachnocoris albomaculatus* Scott (Myers 1925) (Nabinae); and *Alloeorhynchus trimacula* Stein (Cervantes-Peredo 2004) (Prostemmatinae).

Ojeda-Peña (1971) described the nymphs, eggs, and biology of *Nabis (T.) capsiformis* Germar in Peru. *N. punctipennis* is an important predator of several insect pests of cotton in the central coastal region; *N. (T.) capsiformis* appears to play a similar role in the northern coastal region (Ojeda-Peña 1971). The egg and five nymphal instars averaged 9.3 and 17.8 days, respectively, at 27 °C and 75 % R.H. when fed with larvae of *Bucculatrix thurberiella* Busck. The adults lived for 3–8 days, females laying an average of 112.4 eggs each. A generation lasted 26–35 days in summer. Hormchan et al. (1976) determined that male *N. (T.) capsiformis* develop faster than females, 18 versus 22.4 days, respectively, at 26–28 °C on a diet of first instar tarnished plant bugs and bollworms.

11.3.3 Adult Feeding and Reproduction

All Nabidae practice vaginal copulation, with fertilization taking place in the mesodermal oviducts near the base of the ovarioles or near the pedicels. Insemination can be normal or hemocoelic. Others practice traumatic intravaginal insemination (Carayon 1977). *Nabis (T.) capsiformis* averaged 105.3 eggs per female (Hormchan et al. 1976) with 78 % viability under laboratory conditions. Adults survived less than a month at 26–28 °C when fed with tarnished plant bug, *Lygus lineolaris* (Palisot de Beauvois) nymphs, supplemented occasionally with bollworm, *Helicoverpa zea* (Boddie) larvae. Greatest fecundity occurred on the 9th day after emergence as adults with a rapid decline after the 15th day.

11.4 Classification and Diversity

The Nabidae have been placed within the superfamily Cimicoidea in the infraorder Cimicomorpha (Leston et al. 1954; Schuh and Stys 1991). There is no clear consensus concerning the organization of the major subdivisions of the family. Stål (1873) provided the first useful classification, recognizing three subfamilies: Nabina, Coriscina, and Pachynomina. Carayon (1970), Kerzhner (1981), and Péricart (1987) recognized four subfamilies: Medocostinae, Velocipedinae, Nabinae, and Prostemmatinae. We follow the subfamily classification of Schuh and Stys (1991) in defining the Nabidae as including only the Nabinae and Prostemmatinae. They defined the nabids on the basis of 4-segmented labium, membrane venation, and the presence of Ekblom's organ; Kerzhner's (1981) classification is followed for the tribes.

The subfamily Nabinae is classified into four tribes, Arachnocorini, Carthasini, Gorpini, and Nabini, which are united by the presence of moderately long labial segments, the lack of scutellar trichobothria, and opening of pygophore oriented dorsad. The subfamily Prostemmatinae is divided into the Phorticini and Prostemmatini, each with only two genera, both of which possess scutellar trichobothria and have the posterior foramen of the pygophore in a caudal or ventral position. Prostemmatini is considered to be more primitive (Kerzhner 1981, 1996).

For the world, 31 genera and 386 species are known (Henry 2009); for the Australian region, 7 genera and 22 species; for the Palearctic region, 10 genera and 112 species (Kerzhner 1996); and for Canada and the continental USA, 10 genera and 41 species (Henry and Lattin 1988); Kerzhner and Henry 2008) and for the Neotropical region (Volpi and Coscarón 2010; Coscarón and Volpi 2013), 11 genera and 83 species. There are only partial catalogs for Argentina, two genera and seven species (Pennington 1920–1921); five genera and 14 species (Cornelis and Coscarón 2013); Ecuador, two genera and four species (Froeschner 1981); Galápagos, one genus and three species (Froeschner 1985); and Panama, 10 genera and 17 species (Froeschner 1999).

11.4.1 Main Species

Family Nabidae Costa
Subfamily Nabinae Costa
Tribe Arachnocorini Reuter

Genus *Arachnocoris* Scott
Arachnocoris alboannulatus Costa Lima
Distribution: Brazil

Arachnocoris albomaculatus Scott
Distribution: Brazil

Arachnocoris berytoides Uhler
Distribution: Guadeloupe. Grenada. Martinique. Puerto Rico

Arachnocoris dispar Scott
Distribution: Brazil

Arachnocoris eberhardi Kerzhner
Distribution: Costa Rica

Arachnocoris myersi China
Distribution: Brazil

Arachnocoris panamensis (Distant)
Distribution: Panama

Arachnocoris setosus Kerzhner
Distribution: Costa Rica

Arachnocoris simoni Bergroth
Distribution: Venezuela

Arachnocoris thesauri López
Distribution: French Guyana

Arachnocoris torquatus Bergroth
Distribution: Venezuela

Arachnocoris trinitatis Bergroth
Distribution: Trinidad

Arachnocoris varius López
Distribution: French Guyana

Tribe Carthasini Blatchley
Genus *Praecarthasis* Kerzhner
Praecarthasis froeschneri Kerzhner
Distribution: Brazil (Mato Grosso). Ecuador (Pastaza). Peru (Junin)

Praecarthasis gibbus Kerzhner
Distribution: Ecuador (Pastaza). Panama (Barro Colorado). Peru (Cuzco, Junin)

Praecarthasis nigrescens Kerzhner
Distribution: Brazil (Mato Grosso, Pará). Peru (Junin)

Praecarthasis panamensis (Harris)
Distribution: Panama (Porto Bello)

Praecarthasis paprzyckii Kerzhner
Distribution: Peru (Junin, Huánuco)

Praecarthasis pusillus Kerzhner
Distribution: Brazil (Rio de Janeiro)

Genus *Carthasis* Champion

Carthasis championi Harris
Distribution: Panama (Chiriqui)

Carthasis rufonotatus Champion
Distribution: Panama (Chiriqui)

Tribe Gorpini Reuter

Genus *Neogorpis* Barber

Neogorpis spinicollis Kerzhner
Distribution: Panama (Barro Colorado, Canal Zone, Colon)

Tribe Nabini Costa

Genus *Hoplistoscelis* Reuter

Hoplistoscelis confusa Kerzhner and Henry
Distribution: Colombia. Costa Rica. Dominican Republic. El Salvador. Grenada. Grenadines. Guatemala. Haiti. Honduras. Jamaica. México (Chiapas, Guerrero, Michoacan, Morelos, Oaxaca, San Luis Potosi, Sinaloa, Tabasco, Veracruz). Panama. Puerto Rico. Venezuela

Hoplistoscelis dentipes (Harris)
Distribution: México

Hoplistoscelis nigriventris (Stål)
Distribution: México

Hoplistoscelis pallescens (Reuter)
Distribution: Central America. Costa Rica. Venezuela. West Indies

Hoplistoscelis sericans (Reuter)
Distribution: Guatemala (Guatemala, Quetzaltenango, Sacatepequez). México (Guerrero, Jalisco, México City, Morelos, Veracruz)

Hoplistoscelis sordida (Reuter)
Distribution: Argentina (Entre Ríos). Brazil. Central and South America from central part of México to Argentina. Costa Rica (Cartago). Eastern North American. Guatemala (Sacatepequez, Vera Paz). México (Guerrero, Jalisco, Morelos, Tabasco, Vera Cruz). Panama (Chiriqui). West Indies

Genus *Lasiomerus* Reuter

Lasiomerus andabata Kerzhner
Distribution: Guatemala. México (Tabasco). Panama (Chiriqui)

Lasiomerus constrictus (Champion)
Distribution: Argentina (Buenos Aires). Guatemala. Honduras. México. Panama (Chiriqui). México to Panama

Lasiomerus signatus (Uhler)
Distribution: Grenada. México (Tabasco). Panama (Chiriqui)

Lasiomerus spinicrus (Reuter)
Distribution: Antilles. Argentina (Corrientes). Bolivia. Brazil. Cuba. Ecuador. Grenada. Guadeloupe. Guyana. Haití. México. Panama (Chiriqui). Peru. México to Brazil

Genus *Metatropiphorus* Reuter

Metatropiphorus alvarengai Kerzhner
Distribution: Argentina (Buenos Aires). Brazil (Bahia, Mato Grosso, Santa Catarina). Surinam (Marowijne)

Genus *Nabis* Latreille

Nabis argentinus Meyer-Dür
Distribution: Argentina (Buenos Aires, Catamarca, Córdoba, Chaco, La Pampa, Mendoza, Río Negro, Salta, San Luis). Uruguay (Artigas, Canelones, Colonia, Durazno, Maldonado, Montevideo, Paysandú)

Nabis crassipes Reuter
Distribution: Guatemala (Guerrero, Morelos). México (México City)

Nabis faminei Stål
Distribution: Argentina (Buenos Aires, Patagonia, Tierra del Fuego). Chile

Nabis ferus (L.)
Distribution: Guatemala (Quetzaltenango). México (Durango, Guerrero, Veracruz)

Nabis galapagoensis Kerzhner
Distribution: Ecuador (Galápagos Islands)

Nabis paranensis Harris
Distribution: Argentina (Buenos Aires, Jujuy, La Pampa, Misiones). Brazil (Parana)

Nabis punctipennis Blanchard
Distribution: Argentina (Chubut, Mendoza, Neuquén, Río Negro). Chile (Osorno). Continental Chile and Archipielago Juan Fernández

Nabis reductus Kerzhner
Distribution: Ecuador (Galápagos Islands)

Nabis roripes Stål
Distribution: Argentina (Misiones). Brazil (Mato Grosso). Colombia (Cundinamarca). Peru (Loreto, San Juan)

Nabis seticrus Harris
Distribution: Argentina (Salta). Brazil (Mato Grosso, Rio de Janeiro)

Nabis tandilensis Berg
Distribution: Argentina (Buenos Aires)

Subgenus *Nabis* Latreille

Nabis (Nabis) mexicanus Remane
Distribution: Guatemala. México

Subgenus *Tropiconabis* Kerzhner

Nabis (Tropiconabis) capsiformis Germar
Distribution: Argentina (Buenos Aires, Catamarca, Córdoba, Corrientes, Jujuy, La Pampa, Mendoza, Misiones, Salta, Santiago del Estero). Brazil (Mato Grosso do Sul, Rio de Janeiro, Pará). British Guyana. Chile (Arica). Continental Chile and Easter Island. México. Peru (Lima). Uruguay (Montevideo)

Nabis (Tropiconabis) consimilis Reuter
Distribution: Chile. Colombia. Ecuador (Galápagos Islands). Peru

Nabis (Tropiconabis) latior Kerzhner and Henry
Distribution: Bahamas

Subfamily Prostemmatinae Reuter

Tribe Prostemmatini Reuter

Genus *Alloeorhynchus* Fieber

Alloeorhynchus alayoi Kerzhner
Distribution: Cuba

Alloeorhynchus delicatus Harris
Distribution: Panama

Alloeorhynchus jamaicensis Kerzhner
Distribution: Jamaica

Alloeorhynchus maldonadoi Kerzhner
Distribution: Puerto Rico

Alloeorhynchus moritzii (Stein)
Distribution: Antigua and Barbuda (Saint John's). Argentina (Formosa, Chaco, Corrientes, Misiones). Grenada

Alloeorhynchus slateri Kerzhner
Distribution: Jamaica

Alloeorhynchus trimacula (Stein)
Distribution: Guatemala. México (Oaxaca). Panama

Alloeorhynchus vittativentris Stål
Distribution: Argentina (Misiones). Colombia. Panama

Genus *Pagasa* Stål

Pagasa luctuosa Van Duzee
Distribution: Colombia

Subgenus *Lampropagasa* Reuter

Pagasa (Lampropagasa) cobbeni Kerzhner
Distribution: Curaçao Island (Lesser Antilles). Netherlands Antilles

Pagasa (Lampropagasa) confusa Kerzhner
Distribution: Costa Rica. Guatemala. México. Panama. Puerto Rico

Pagasa (Lampropagasa) flavipennis Kerzhner
Distribution: Bolivia

Pagasa (Lampropagasa) fusca (Stein)
Distribution: Guatemala. México. Panama

Pagasa (Lampropagasa) fuscipennis Reuter
Distribution: Argentina (Buenos Aires, Chaco, Chubut, Córdoba, La Pampa, Misiones, Río Negro, Salta, San Luis, Santa Fé). Brazil (Goiás, Minas Gerais, Santa Catarina, São Paulo). Paraguay. Uruguay (Montevideo, Colonia, Soriano)

Pagasa (Lampropagasa) similis Poppius
Distribution: Brazil (Amazonas, Mato Grosso). Surinam

Subgenus *Pagasa* Stål

Pagasa (Pagasa) aenescens Stål
Distribution: Brazil. Surinam

Pagasa (Pagasa) amazonica Poppius
Distribution: Brazil (Amazonas, Maranhão). Paraguay. Surinam

Pagasa (Pagasa) bimaculata Harris
Distribution: Brazil (Mato Grosso). Panama. Paraguay. Surinam

Pagasa (Pagasa) brailovskyi Kerzhner
Distribution: México. Panama

Pagasa (Pagasa) brunneipes Kerzhner
Distribution: Brazil

Pagasa (Pagasa) costalis Reuter
Distribution: Argentina (Buenos Aires, Salta). Ecuador. Paraguay. Surinam (Saramacca). Uruguay (Montevideo)

Pagasa (Pagasa) doesburgi Kerzhner
Distribution: Surinam

Pagasa (Pagasa) henryi Kerzhner
Distribution: Brazil. Panama. Peru

Pagasa (Pagasa) luteiceps (Walker)
Distribution: Argentina (Corrientes, Entre Ríos, Formosa, Santa Fe, Santiago del Estero) Brazil (Mato Grosso, Rio de Janeiro). Colombia (Magdalena). México (Vera Cruz). Panama (Barro Colorado Island). Peru. Surinam. Tobago Island. Uruguay. Venezuela

Pagasa (Pagasa) margaritae Kerzhner
Distribution: Surinam

Pagasa (Pagasa) pallidiceps (Stål)
Distribution: Brazil

Pagasa (Pagasa) pallipes Stål
Distribution: Colombia. Costa Rica. Cuba. Honduras. México. Panama (Chiriqui). Venezuela

Pagasa (Pagasa) planipes Harris
Distribution: Peru (Junin)

Pagasa (Pagasa) prostemmatoides Kerzhner
Distribution: México. Salvador

Pagasa (Pagasa) ruficeps (Walker)
Distribution: Brazil

Pagasa (Pagasa) signatipennis Reuter
Distribution: Argentina (Formosa). Bolivia. Brazil (Mato Grosso). Colombia. Paraguay. Surinam. Venezuela

Tribe Phorticini Kerzhner

Genus *Phorticus* Stål

Phorticus collaris Stål
Distribution: México

Phorticus speciosus Harris
Distribution: Panama

Phorticus viduus Stål
Distribution: Brazil (Rio de Janeiro)

Phorticus obscuripes Stål
Distribution: Brazil (Rio de Janeiro)

11.4.2 Fossils and Genomic Data

Asquith and Lattin (1990) suggested a Late Cretaceous age for the genus, predicting a much older age for the group as a whole. The extinct *Metatropiphorus succini* (Jordan) from the Late Eocene Baltic amber (40–35 Ma) belongs to the genus considered the most primitive in the subfamily Nabinae (Kerzhner 1981). Li et al. (2012a, b) presented the complete mt genome of *Alloeorhynchus bakeri*, a representative of Prostemmatinae, and provided analyses of the nucleotide composition, codon usage, compositional biases, and RNA secondary structure. They evaluated the phylogenetic position of Nabidae in Heteroptera based on the sequences of PCGs. This represented the first sequenced mitochondrial genome within the family Nabidae. Further, a study with additional species found that the relationship of three tribes from two subfamilies of Nabidae was well represented by the mitogenome data and was consistent with current phylogeny of the family Nabidae constructed mainly from morphological traits (Li et al. 2012a).

11.4.3 Economic Importance

Adult female *Nabis (Tropiconabis) capsiformis* were found to be potentially very effective predators of *Heliothis* spp. eggs and larvae (Samson and Blood 1980). *N. (T.) capsiformis* was the major consumer of small velvetbean caterpillar, *Anticarsia gemmatalis* Hübner larvae, in a Florida study (Godfrey et al. 1989). Ribeiro and Castiglioni (2008) reported *N. (T.) capsiformis* among the predators collected in soybean in Uruguay contributing to predation of red-shouldered stink bug, *Piezodorus guildinii* (Westwood) eggs. *Nabis (T.) capsiformis* demonstrated a type I functional response when feeding on bollworm eggs (Parajulee et al. 2006). *Nabis punctipennis* adults consumed a daily average of 12.4 aphids, *Acyrthosiphon pisum* (Harris) (Rebolledo et al. 2005). Nymphs consumed an average of 9.1 aphids daily. Adults and nymphs preferred adult aphids rather than nymphs.

Nabis spp. are among the main predators collected in soybean in Brazil (Correa et al. 1977; Campos et al. 1997; Thomazini 2001; Fernandes et al. 2010). Abundance of predators, including nabids, appeared not to be affected by cultivar when soybeans were sampled by sweeping or beat cloth (Belorte et al. 2004). Suction sampling and visual observation were equally effective sampling methods for nabids in tomatoes (Raga et al. 1990).

An assessment of selective insecticides targeting lepidopterous pests of soybean for potential nontarget effects on *Nabis (T.) capsiformis* indicated that both methoxyfenozide and indoxacarb had fewer nontarget effects than pyrethroid, organophosphate, or carbamate insecticides (Baur et al. 2003). *Nabis (T.) capsiformis* populations were reduced by lambda-cyhalothrin and monocrotophos applications targeting *A. gemmatalis* on soybean (White et al. 1992). Among four insecticides tested, spinosad had the least effect on *Nabis punctipennis* (Romero et al. 2009). Biopesticides did not affect nabid intrinsic rate of increase, net reproductive rate, or generation time for predators fed with healthy *Anticarsia gemmatalis* larvae and predators fed with larvae infected by a formulation of a nuclear polyhedrosis virus (NPV) of this host (Watanabe et al. 1999).

11.4.4 Tritrophic Interactions

Nabids are an important component of the natural enemy community impacting pest populations with plant feeding also important to development and persistence in the habitat (Braman 2000). Nymphal nabids screened from Bt corn in the field indicated that these natural enemies were among the three orders (Araneae, Coleoptera, and Heteroptera) found to contain Cry1Ab endotoxins above the threshold of 0.5 ng Cry1Ab per gram of fresh weight (Harwood et al. 2005). However, in field studies, the abundance of predators, including nabids, across 3 years in cotton fields with standard grower practices failed to exhibit any negative impact of Bt cotton on predator populations (Torres and Ruberson 2005). Nabids were among the arthropod faunae in orchard ground cover where plant species richness was a major driver for the fauna communities, both for the arthropod and pollinator assemblages (García and Miñarro 2013). Beneficial arthropods including nabids were higher in the presence of cover crops in an orchard under organic production (Fernandez et al. 2008). Nabid species were most abundant in hairy vetch in a study evaluating interplanting crops on cotton and effects on predatory hemipterans (Loya-Ramirez et al. 2003). Conservation biological control strategies that utilize non-crop plants to enhance populations of natural enemies and their persistence require careful selection of plant species to conserve nabids and other predatory heteropterans and serve as a source rather than a sink (Perdikis et al. 2011 and references therein).

11.5 Concluding Remarks

Much remains to be discovered concerning this fascinating and important group of predatory heteropterans. Taxonomic determination is still a difficult task, because characters of relevant importance, such as the external morphological and genital structures, have been inconsistently used. Therefore, the number of valid species may be different from the ones described so far. Knowledge about biology and taxonomy of this group is fundamental to developing future programs in integrated pest control.

References

Asquith A, Lattin JD (1990) *Nabicula (Limnonabis) propinqua* (Reuter) (Heteroptera: Nabidae): dimorphism, phylogenetic relationships and biogeography. Tijdschr Entomol 133:3–16

Baur ME, Ellis J, Hutchinson K, Boethel D (2003) Contact toxicity of selective insecticides for non-target predaceous hemipterans in soybeans. J Entomol Sci 38:269–277

Belorte LCC, Ramiro ZA, Faria AM (2004) Occurrence of predators in five soybean cultivars [*Glycine max* (L.) Merrill, 1917] district of Araçatuba, São Paulo state, Brazil. Arq Inst Biol (São Paulo) 71:45–49

Braman SK (2000) Damsel bugs (Nabidae). In: Schaefer CW, Panizzi AR (eds) Heteroptera of economic Importance. CRC Press, Boca Raton, pp 639–656

Braman SK, Yeargan KV (1988) Comparison of developmental and reproductive rates of *Nabis americoferus*, *N. roseipennis*, and *N. rufusculus* (Hemiptera: Nabidae). Ann Entomol Soc Am 81:923–930

Burgess L, Dueck J, McKenzie DL (1983) Insect vectors of the yeast *Nematospora coryli* in mustard, *Brassica juncea*, crops in southern Saskatchewan. Can Entomol 115:25–30

Campos OR, Campos AR, Lara FM (1997) Seasonal occurrence of pest insects and entomophagous predators in two varieties of soybean [*Glycine max* (L.) Merrill], in the region of Ilha Solteira – SP. Cult Agron 6:1–11

Carayon J (1948) Les organes parastigmatiques des Hémiptères Nabidae. C R Acad Sci Fr 227:864–866

Carayon J (1950) Caractères anatomiques et position systématique des Hémiptères Nabidae (note préliminaire). Bull Mus Natl Hist Nat Paris Ser 2(22):95–101

Carayon J (1970) Étude des *Alloeorhynchus* d'Afrique centrale avec quelques remarques sur la classification des Nabidae (Hemiptera). Ann Soc Entomol Fr 6:899–931

Carayon J (1977) Insemination extra-genitale traumatique. In: Grasse PP (ed) Traite de zoologie. Masson, Paris, pp 351–390

Cervantes-Peredo L (2004) *Alloeorhynchus trimacula* (Stein) (Heteroptera: Nabidae: Prostemmatinae), a predator of Rhyparochromidae (Lygaeoidea) associated with figs in Mexico. Proc Entomol Soc Wash 106:346–351

Cornelis M, Coscar6n MC (2013) The Nabidae (Insecta, Hemiptera, Heteroptera) of Argentina. ZooKey 333:1–30

Cornelis M, Quirán E, Coscar6n MC (2012) Description of some immature stages of *Nabis (Tropiconabis) capsiformis* (Hemiptera: Nabidae). Rev Mex Biodiv 83:1009–1012

Correa BS, Panizzi AR, Newman GG, Turnipseed SG (1977) Geographical distribution and seasonal abundance of the main insect pests of soybean and their predators. Ann Soc Entomol Brasil 6:40–50

Coscar6n MC (in press) Nabidae. In: Catalogue of the Heteroptera or true bugs of Argentina. Zootaxa

Coscar6n MC, Volpi LN (2013) Nomenclatural and bibliographic corrections to the Catalog of Nabidae (Hemiptera: Heteroptera) for the Neotropical Region. Zootaxa 3646:93–96

Elvin MK, Sloderbeck PE (1984) A key to the nymphs of selected species of Nabidae (Hemiptera) in the southeastern USA. Fla Entomol 67:269–273

Faúndez EI, Carvajal MA (2011) A human case of biting by *Nabis punctipennis* (Hemiptera: Heteroptera: Nabidae) in Chile. Acta Entomol Musei Nat Pragae 51:407–409

Fernández DE, Cich6n LI, Sánchez EE, Garrido SA, Gittins C (2008) Effect of different cover crops on the presence of arthropods in an organic apple (Malus domestica Borkh) orchard. J Sustain Agric 32:197–211

Fernandes FL, Picanço MC, Fernandes MES, Xavier VM, Martins JC, Silva VF (2010) Natural biological control of pests and ecological interactions with predators and parasitoids in bean crop. Biosci J 26:6–14

Froeschner RC (1981) Heteroptera or true bugs of Ecuador: a partial catalog. Smithson Contrib Zool 322:55–57

Froeschner RC (1985) Synopsis of the Heteroptera or true bugs of the Galápagos Islands. Smithson Contrib Zool 407:39–40

Froeschner RC (1999) True bugs (Heteroptera) of Panama: a synoptic catalog as a contribution to the study of Panamanian biodiversity. Am Entomol Inst 61:1–393

García RR, Miñarro M (2013) Role of floral resources in the conservation of pollinator communities in cider-apple orchards. Agric Ecosyst Environ 183:118–126

Godfrey KE, Whitcomb WH, Stimac JL (1989) Arthropod predators of velvetbean caterpillar, *Anticarsia gemmatalis* Hübner eggs and larvae. Environ Entomol 18:118–123

Harris HM (1928) A monographic study of the hemipterous family Nabidae as it occurs in North America. Entomol Am 9:1–97

Harwood JD, Wallin WG, Obrycki JJ (2005) Uptake of Bt endotoxins by nontarget herbivores and higher order arthropod predators: molecular evidence from a transgenic corn agroecosystem. Mol Ecol 14:2815–2823

Henry TJ (2009) Biodiversity of Heteroptera. In: Foottit R, Adler P (eds) Insect biodiversity: science and society. Wiley-Blackwell, New York, pp 223–263

Henry TJ, Lattin JD (1988) Family Nabidae Costa, 1853. The damsel bugs. In: Henry TJ, Froeschner RC (eds) Catalog of the Heteroptera, or true bugs, of Canada and the Continental United States. Brill, Leiden, pp 508–520

Hormchan P, Schuster MF, Hepner LW (1976) Biology of *Tropiconabis capsiformis*. Ann Entomol Soc Am 69:1016–1018

Kerzhner IM (1981) Fauna of the USSR. Bugs. Heteroptera of the family Nabidae. Acad Sci USSR Zool Inst Nauka 13(2):326, Leningrad (in Russian)

Kerzhner IM (1983) Airborne *Nabis capsiformis* (Heteroptera: Nabidae) from the Atlantic, Indian, and Pacific Oceans. Int J Entomol 25:273–275

Kerzhner IM (1996) Family Nabidae. In: Aukema B, Rieger C (eds) Catalogue of the Heteroptera of the Palaearctic Region, vol 2. Netherlands Entomological Society, Amsterdam, pp 84–107

Kerzhner IM, Henry TJ (2008) Three new species, notes and new records of poorly known species, and an updated checklist for the North American Nabidae (Hemiptera: Heteroptera). Proc Entomol Soc Wash 110:988–1011

Larivière M (1993) *Lasiomerus annulatus* (Reuter) (Hemiptera: Nabidae) in Southern Canada: ecology and phenology, with a description of the last-instar nymph. Can Entomol 125:513–519

Larivière M (1994) Biodiversity of *Nabicula* Kirby species (Hemiptera: Nabidae) in Canada: faunistic review, bioecology, biogeography. Can Entomol 126:327–378

Leston D, Pendergrast JG, Southwood TRE (1954) Classification of the terrestrial Heteroptera (Geocorisae). Nature 174:91–92

Li H, Liu H, Cao L, Shi A, Yang H, Cai W (2012a) The complete mitochondrial genome of the damsel bug *Alloeorhynchus bakeri* (Hemiptera: Nabidae). In J Biol Sci 8:93–107

Li H, Liu H, Song F, Shi A, Zhou X, Cai W (2012b) Comparative mitogenomic analysis of damsel bugs representing three tribes in the family Nabidae (Insecta: Hemiptera). PLoS ONE 7:e45925. doi:10.1371/journal.pone.0045925

Loya-Ramírez JG, García-Hernández JL, Ellington JL, Thompson DV (2003) The impact of interplanting crops on the density predation of Hemiptera predators. Interciencia 28:415–420

Mundinger FG (1922) The life history of two species of Nabidae (Hemiptera: Heteroptera). NY State Coll For Publ 16:149–167

Myers JG (1925) Biological notes on *Arachnocoris albomaculatus* Scott (Hemiptera: Nabidae). J NY Entomol Soc 33:136–146

Ojeda-Peña D (1971) Biología y hábitos de *Nabis capsiformis* Germar (Hemip.: Nabidae). Rev Per Entom 14:297–303

Parajulee MN, Shrestha RB, Leser JF, Wester DB, Blanco CA (2006) Evaluation of the functional response of selected arthropod predators on bollworm eggs in the laboratory and effect of temperature on their predation efficiency. Environ Entomol 35:379–386

Pennington MS (1920–1921) Lista de los hemípteros heterópteros de la República Argentina. Segunda Parte, Buenos Aires, pp 17–28

Perdikis D, Fantinou A, Lykouressis D (2011) Enhancing pest control in annual crops by conservation of predatory Heteroptera. Biol Control 59:13–21

Péricart J (1987) Hémiptères Nabidae d'Europe Occidentale et du Maghreb. Faune France 71:1–185

Raga V, Gravena S, Bortoli SA, Arai J, Wassano GN (1990) Insect survey and arthropod predator activity in tomato crop of determined growth. Ann Soc Entomol Brasil 19:253–271

Rebolledo R, Villegas G, Klein C, Aguilera A (2005) Fluctuación poblacional, capacidad depredadora y longevidad de *Nabis punctipennis* Blanchard (Hemiptera: Nabidae). Agric Tec (Chile) 65:442–446

Ribeiro A, Castiglioni E (2008) Caracterización de las poblaciones de enemigos naturales de *Piezodorus guildinii* (Westwood) (Hemiptera: Pentatomidae). Agrociencia 12:48–56

Romero C, Araya JE, Guerrero MA, Curkovic T (2009) Effects of carbaril, metamidofos, lambda cyhalothrin and spinosad on nymph of *Nabis punctipennis* Blanchard (Hemiptera: Nabidae). Bol Sanidad Veg Plagas 35:371–376

Samson PR, Blood PRB (1980) Voracity and searching ability of *Chrysopa signata* (Neuroptera: Chrysopidae), *Micromus tasmaniae* (Neuroptera: Hemerobiidae) and *Tropiconabis capsiformis* (Hemiptera: Nabidae). Aust J Zool 28:575–580

Schuh RT, Stys P (1991) Phylogenetic analysis of cimicomorphan family relationships (Heteroptera). J NY Entomol Soc 99:298–350

Stål C (1873) Enumeratio Hemipterorum. Bidrag till en förteckning öfver aller hittills kända Hemiptera, jemte systematiska meddelanden. III. Kongl Svenska Vet-Akad Handl 11:1–163

Stoner A (1972) Plant feeding by *Nabis*, a predaceous genus. Environ Entomol 1:557–558

Taylor EJ (1949) A life history study of *Nabis alternatus*. J Econ Entomol 42:991

Thomazini MJ (2001) Insects associated with soybean in the State of Acre, Brazil. Acta Amazon 31:673–681

Torres JB, Ruberson JR (2005) Canopy- and ground-dwelling predatory arthropods in commercial Bt and non-Bt cotton fields: patterns and mechanisms. Environ Entomol 34:1242–1256

Volpi LN, Coscarón MC (2010) Catalog of Nabidae (Hemiptera: Heteroptera) for the Neotropical Region. Zootaxa 2513:50–68

Watanabe MA, DeNardo EAB, Maia AHN (1999) Efeito do nucleopoliedrovirus de *Anticarsia gemmatalis* Hübner (Lepidoptera: Noctuidae) na biologia de *Nabis* sp. (Heteroptera: Nabidae). Ann Soc Entomol Brasil 28:461–468

Wheeler AG Jr (2001) Additional Florida records of the seldom-collected *Lasiomerus andabata* (Hemiptera:Nabidae). Fla Entomol 84:723

White JS, Brown RA, Bettencourt AL, Soares CLS (1992) Lambda-cyhalothrin: effects on natural pest control in Brazilian soyabeans. Proc Brighton Crop Prot Conf Pest Dis 1:811–816

Chapter 12
Assassin Bugs (Reduviidae Excluding Triatominae)

Hélcio R. Gil-Santana, Dimitri Forero, and Christiane Weirauch

Abstract Reduviidae is the largest family of predaceous terrestrial Heteroptera, with about 7,000 described species in 25 subfamilies, and is one of the three most speciose families within Hemiptera. A general overview on Neotropical members of this family is furnished, with an updated account on the taxonomy for each subfamily. Keys to genera of almost all subfamilies are presented.

12.1 Introduction

Reduviidae is the largest family of predaceous terrestrial Heteroptera (Henry 2009). They are voracious predators and hence are also called "assassin bugs" (Ambrose 2000). About 25 subfamilies are recognized, although there is a lack of consensus on some of these concepts (Maldonado 1990; Schuh and Slater 1995; Weirauch et al. 2014). The family comprises nearly 7,000 described species and is therefore one of the three most speciose families within Hemiptera (Forero 2008; Henry 2009).

Reduviids show remarkable morphological diversity and range in size from relatively small and extremely delicate insects of few millimeters to very large insects (Schuh and Slater 1995; Weirauch 2008). With exception of Triatominae that feed on vertebrate blood, all reduviids are considered to be arthropod predators (e.g., Miller 1953; Readio 1927).

H.R. Gil-Santana (✉)
Laboratório de Diptera, Fundação Instituto Oswaldo Cruz (Fiocruz), Av. Brasil, 4365, Manguinhos, Rio de Janeiro, RJ 21040-360, Brazil
e-mail: helciogil@uol.com.br

D. Forero
Laboratorio de Entomología, Unidad de Ecología y Sistemática, Pontificia Universidad Javeriana, Carrera 7 No.43-82, Bogotá, Colombia
e-mail: forero-i@javeriana.edu.co

C. Weirauch
Department of Entomology, University of California, Riverside, CA 92521, USA
e-mail: christiane.weirauch@ucr.edu

A.R. Panizzi, J. Grazia (eds.), *True Bugs (Heteroptera) of the Neotropics*, Entomology in Focus 2, DOI 10.1007/978-94-017-9861-7_12

12.2 General Characteristics and Diagnosis

The great majority of Reduviidae are medium to large true bugs and often show elongate or ovoid body shapes and long legs with stout fore femora. Overall, Reduviidae is one of the most morphologically diverse groups of Heteroptera ranging from delicate and elongate Emesinae (Wygodzinsky 1966) and Bactrodinae (Coscarón and Melo 2003) to large and robust Hammacerinae (Coscarón and Giacchi 1987b). Reduviidae can also mimic other groups of Heteroptera (Stride 1954), as well as other insects, such as Hymenoptera, with a number of species participating in Müllerian mimicry complexes (Haviland 1931; Hogue 1993).

The diagnostic characters include compound eyes usually large, ocelli often present (exceptions noted in the subfamily treatments); second antennal segment with trichobothria; labium with three visible segments (except in Centrocnemidinae and Hammacerinae that have four), usually stout and curved, but sometimes relatively thin and straight; the distal segment of labium with plectrum and prosternum with a stridulatory groove; membrane of hemelytra usually with two or three elongated cells; many taxa with a fossula spongiosa at the apex of the fore and mid-tibiae, present in one or more pairs of legs or absent; Brindley's glands between the metathorax and the first abdominal segment; female genitalia with lateral spermathecae; males with the eighth abdominal segment telescoped largely into the seventh segment and usually with symmetrical genitalia; and presence of three pairs of dorsal abdominal scent glands in nymphs (which can persist in adults) (Schuh and Slater 1995; Weirauch 2008).

12.3 General Biology and Ecology

Although many Reduviidae appear to feed on a wide variety of arthropods, assassin bugs in some groups show prey preferences or even prey specializations (see summary in Hwang and Weirauch 2012). For example, Ectrichodiinae seem to be specialized to prey on millipedes (Cachan 1952; Forthman and Weirauch 2012), several taxa are termite predators, e.g., Salyavatinae (McMahan 1983), and one species of Harpactorinae-Apiomerini, *Micrauchenus lineola* (F.) (Bérenger and Pluot-Sigwalt 2009); species of Holoptilinae prey on ants (Jacobson 1911) and certain Emesinae and Harpactorinae on spiders (Wignall and Taylor 2008; Jackson et al. 2010). Additional preferences, some of them tentative, are summarized by Ambrose (2000).

12.4 Phylogenetic Relationships

Recent phylogenetic work at the suprageneric level in Reduviidae has advanced our understanding of the relationships among the main lineages of reduviids. Weirauch (2008) was the first to present a morphology-based phylogenetic analysis of the

Reduviidae that incorporated species as terminal taxa, thus avoiding a ground plan approach and testing the monophyly of the included subfamilies. Representatives of 21 subfamilies were part of the analyses for which the main results can be summarized as follows: Hammacerinae is considered the sister group of all remaining Reduviidae; the Phymatine Complex (Phymatinae, Holoptilinae, and Centrocnemidinae) is monophyletic; Harpactorinae is the sister group of the remaining "higher Reduviidae"; Ectrichodiinae and Tribelocephalinae are sister groups; Emesinae, Saicinae, and Visayanocorinae are grouped in a clade; and Salyavatinae is paraphyletic with respect to Sphaeridopinae. The polyphyletic nature of Reduviinae was documented for the first time, although the extent of this problem remained largely unknown.

Weirauch and Munro (2009) reassessed the phylogeny of Reduviidae using molecular markers (~3.3 Kbp) of mitochondrial and nuclear ribosomal genes. Hammacerinae were not recovered as sister group to all remaining Reduviidae, but instead as sister to the Phymatine Complex. Reduviidae are therefore split into two main clades, the Hammacerinae + Phymatine Complex and the "higher Reduviidae." Peiratinae, not Harpactorinae, were found to be the sister group to the remaining higher Reduviidae, placing Harpactorinae in a more derived position. Other results corroborated relationships proposed in the morphological analysis (Weirauch 2008), most importantly the close relationship of Triatominae to some large Neotropical Reduviinae and to the Stenopodainae.

More recently, Hwang and Weirauch (2012) presented the most comprehensive molecular analysis of Reduviidae to date (178 terminals, 18 subfamilies) that emphasized additional investigation of the polyphyly of Reduviinae. The authors found 11 or 14 (depending on analyses) distinct clades of Reduviinae, the composition, relationships, and classification of which now warrant additional systematic study. Other noteworthy results were the paraphyly of Triatominae with respect to the reduviine genus *Opisthacidius* and the polyphyly of Cetherinae.

The classification of Reduviidae will be impacted by these and future phylogenetic results. In particular, investigating the phylogenetic positions of certain Neotropical taxa, most importantly the Bactrodinae, Elasmodeminae, and Phimophorinae, is the top priority.

12.5 Classification, Diversity, and Biology

Twenty-five subfamilies are currently considered as valid (Weirauch et al. 2014). Recent and future phylogenetic hypotheses will provide arguments for modifications of this classification that will recognize a monophyletic Reduviinae and accommodate other clades currently classified as Reduviinae in existing or new subfamilies. The following outline of subfamilies that occur in the Neotropics follows the subfamilies recognized by Schuh and Slater (1995), with addition of the Tribelocephalinae that were only recently documented to also occur in the New World (Weirauch 2010). Keys to assassin bug subfamilies that are distributed in the

Neotropics were presented by Forero (2004). The keys published by Schuh and Slater (1995) and Weirauch et al. (2014) are for all subfamilies of Reduviidae worldwide.

12.5.1 Bactrodinae

Bactrodinae are exclusively Neotropical and include only the genus *Bactrodes* Stål (McAtee and Malloch 1923; Maldonado 1990). *Bactrodes* was recently revised by Coscarón and Melo (2003), who recognized five species. Species in the genus are distributed from Mexico to Argentina (Maldonado 1990; Coscarón and Melo 2003).

Nothing much is known about the biology of *Bactrodes*. Forero (2006) reported *Chaetocnema* sp. (Coleoptera: Chrysomelidae) as field prey for *B. femoratus* (F.). This species has also been found on the plant species *Aciotis laxa* (Melastomataceae) in French Guiana (Bérenger and Pluot-Sigwalt 1997). Several species of *Bactrodes* appear to be associated with Melastomataceae species with sticky trichomes that the assassin bugs evade aided by their strongly modified, asymmetrical pretarsus (Weirauch and Zhang personal observation).

12.5.2 Cetherinae

Cetherinae (Fig. 12.1) are represented in the Neotropics only by the genus *Eupheno* Gistel, which includes three described species (Maldonado 1990). Champion (1898) presented a key to separate two of the three known species.

Species of *Eupheno* are found on tree trunks or under the bark of decaying trees (Champion 1898; Haviland 1931; Hwang and Weirauch 2012), where they prey upon different arthropods including beetles (Champion 1898) and termites (Haviland 1931).

12.5.3 Chryxinae

This small and rarely collected subfamily is composed of four genera and five species (Gil-Santana et al. 2007b; Weirauch 2012).

Very little is known about the biology of this secretive group of Reduviidae. *Lentia corcovadensis* has been collected in soft soil next to large trees and collected together with *Fulvius quadristillatus* Stål (Miridae) (Wygodzinsky 1946). Wygodzinsky (1946) remarked the similarity in coloration and movements of *Lentia* to some Anthocoridae and Miridae. *Petasolentia goellnerae* was collected at a mercury vapor light in Peru (Weirauch 2012).

Fig. 12.1 *Eupheno pallens* (Laporte) (Cetherinae), nymph, live specimen, covered with debris as camouflage

Key to genera and species of Chryxinae (based on Gil-Santana et al. 2007b; Weirauch 2012):

1. Total length 8.0–9.0 mm; veins on corium indistinct; connexivum with uniform clear coloration.. *Wygodzinskyella travassosi* (Lent & Wygodzinsky)
1′. Total length 3.1–5.3 mm; veins on corium distinct, at least basally; connexivum with clear and dark alternate colors..2
2. Head with process on frons.............................*Petasolentia goellnerae* Weirauch
2′. Head without process on frons..3
3. with ocelli and an acute process on its ventral surface; corium of hemelytra with a small costal cell..*Lentia corcovadensis* Wygodzinsky
3′. Head without ocelli or an acute process on its ventral surface; corium of hemelytra withoutasmallcostalcell..................................*Chryxus*Champion......................4
4. Male of 5 mm in length; femora with uniform coloration; hemelytra with corium on basal half yellowish, apical half dark, membrane whitish with a large blackish spot inside discal cell and another blackish spot in the distal region; veins clear... ....*Chryxus tomentosus* Champion
4′. Male of 3.6 mm in length; femora with darkened rings; hemelytra with corium whitish, with a subbasal dark spot, apical half dark whitish area, giving the impression of a transverse pale band; membrane brownish with two whitish spots; veins darkened...............*Chryxus bahianus* Gil-Santana, Costa & Marques

12.5.4 Ectrichodiinae

There are 22 genera and more than 100 described species of Ectrichodiinae (Figs. 12.2, 12.3, 12.4, and 12.5) in the New World (Maldonado 1990; Carpintero and Maldonado 1996; Bérenger and Gil-Santana 2005; Baena and Susín 2007; Gil-Santana and Baena 2009). Dougherty (1995) recognized 19 genera for the New World, including five described as new (*Cryptonannus*, *Doblepardocoris*, *Schuhella*, *Sinchocoris*, and *Wygodzinskyocoris*), whereas Carpintero and Maldonado (1996) listed only 17. She also proposed the following synonymies: *Anapothea* Miller, *Jorgcoris* Carpintero, and *Parapothea* Carpintero as junior synonyms of *Pothea* Amyot & Serville; *Pseudoracelda* Carpintero as a junior synonym of *Racelda* Signoret; and *Santainezia* Miller as a junior synonym of *Rhiginia* Stål. Carpintero and Maldonado (1996) either did not recognize or overlook these changes proposed by Dougherty (1995), with the exception of the synonymies regarding *Anapothea* and *Santainezia*, which had already been proposed by Wygodzinsky (1959) and Carpintero and Maldonado (1988), respectively (Gil-Santana et al. 2013a). Until a taxonomic revision and a phylogenetic analysis of all these genera are produced, we are accepting Carpintero and Maldonado (1996) arrangement with the addition of the new taxa proposed by Dougherty (1995) and Bérenger and Gil-Santana (2005) and the renewal of *Parapothea* as junior synonym of *Pothea* by Gil-Santana (2014).

Eleven of the Neotropical genera are currently monotypic, and we here list the included species: *Borgmeierina dentata* Wygodzinsky, *Cryptonannus punctella* Dougherty, *Jorgcoris aczeli* (Wygodzinsky), *Margacoris thaumastos* Carpintero, *Pseudopothea luacesi* (Bruner & Fracker), *Pseudoracelda macrocephala* Carpintero, *Schuhella hermosa* Dougherty, *Travassocoris pulchra* Wygodzinsky, *Wygodzinskyocoris nigripes* Dougherty, and *Xarada inca* Carpintero (Carpintero 1980; Maldonado 1990; Dougherty 1995; Carpintero and Maldonado 1996).

Brontostoma currently includes more than 20 species (Maldonado 1990; Dougherty 1995; Gil-Santana et al. 2004, 2005b, 2013a; Gil-Santana and Baena 2009), although Maldonado (1990) and Dougherty (1995) disagreed on the validity of some of them (see Gil-Santana et al. 2005b for a discussion). A taxonomic revision of *Brontostoma* is urgently needed. Keys to species by Wygodzinsky (1951) and Gil-Santana et al. (2004, 2005b, to Brazilian species) are of limited use due to intraspecific variation, misidentifications, and problematic species delimitations. For instance, Wygodzinsky (1951) misidentified *Brontostoma trux* Stål (Fig. 12.2) as *B. rubrovenosum* (Stål), an error that was carried over into later keys (Gil-Santana et al. 2004, 2005b; but see Gil-Santana et al. 2013a for correction).

Daraxa contains 15 species (Dougherty 1995; Carpintero and Maldonado 1996; Bérenger and Gil-Santana 2005), mostly described by Carpintero (1980). The monotypic genus *Cryptonannus* is extremely similar to *Daraxa* and may be congeneric (Forero 2004).

Pothea was subdivided into two subgenera by Carpintero (1980) and currently includes 33 species (Carpintero and Maldonado 1996; Gil-Santana and Costa 2005; Gil-Santana 2014). The Argentinean species were revised by Carpintero (1978).

Fig. 12.2 *Brontostoma trux* (Stål) (Ectrichodiinae), female, live specimen

Fig. 12.3 *Pothea jaguaris* (Carpintero) (Ectrichodiinae), female, dorsal view

Fig. 12.4 *Rhiginia lateralis* (Lepeletier & Serville) (Ectrichodiinae), male, live specimen

Fig. 12.5 *Racelda monstrosa* Carpintero (Ectrichodiinae), female, dorsal view. Scale 5.0 mm

Carpintero and Maldonado (1990) published a key to species in the subgenus *Pothea* (*Brachypothea*) Carpintero.

Baena and Susín (2007) provided an up-to-date checklist of the 19 species of *Rhiginia*, but outlined that the taxonomic status of some species is under discussion. There is no key to identify the species of *Rhiginia*.

The remaining genera of Ectrichodiinae currently comprise between two and six species. The four species of *Cricetopareis* can be identified with the key by Carpintero and Maldonado (1991). Dougherty (1995) provided a key for the two described species of *Doblepardocoris*. *Ectrichodiella* has two species, *E. minima* (Valdés) from Cuba and *E. rafaeli* (Gil-Santana and Coletto-Silva) from Brazil (Gil-Santana et al. 2013a). *Pseudodaraxa* also contains two described species, *P. albidops* Carpintero and *P. littoralis* Carpintero, both from Argentina (Carpintero 1980). Bérenger and Gil-Santana (2005) provided a key to the three species in the genus *Pseudozirta* that was described by them. *Racelda* has six species (Maldonado 1990; Carpintero and Maldonado 1996; Bérenger and Gil-Santana 2005). In this genus, the females are apterous (Dougherty 1995; Carpintero and Maldonado 1996) (Fig. 12.5). The two species of *Sinchocoris* were described and keyed by Dougherty (1995). The five species included in *Zirta* (Dougherty 1995, Carpintero & Maldonado 1996) were revised by Dougherty (1986).

Ectrichodiinae are frequently found in leaf litter (Miller 1953; Louis 1974; Hwang and Weirauch 2012). The scarce biological data suggest that Ectrichodiinae might be specialized millipede predators (reviewed in Forthman and Weirauch 2012), a behavior that was documented among Neotropical taxa for species of *Brontostoma* and *Rhiginia* (Carpintero and Maldonado 1996, Forthman and Weirauch 2012).

The Ectrichodiinae differ from most Reduviidae by the high frequency of wing length polymorphism, ranging from macropterous to apterous conditions, in which females commonly exhibit wing reduction (Dougherty 1995) (Figs. 12.2 and 12.5).

Key to the New World genera of Ectrichodiinae (modified from Gil-Santana et al. 2013a):

1. Antennal insertion shielded laterally by a small sclerite. Scutellum with two midlateral projections and an apical blunt tip. Tarsi two segmented. Fore and mid-tibia without spongy fossa.................................*Ectrichodiella* Fracker & Bruner

1′. Antennal insertion with at most a small process on the antennifer. Scutellum with two distal prongs. Tarsi three segmented. Fore and mid-tibia with spongy fossa 2

2. Antennal insertion shielded laterally by a small process on the antennifer; vertex elevated, ocellar tubercle conical ....*Jorgcoris* Carpintero

2′. Antennal insertion not shielded by an antennifer process; vertex not elevated, ocellar tubercle conical or rounded.. 3

3. Four antennal segments... 4

3′. Six or more [apparent] antennal segments... 5

4. Ocelli not raised on an ocellar tubercle; abdominal sternites with heavy punctuation.. *Schuella* Dougherty

4′. Ocelli raised on an ocellar tubercle; abdominal sternites without heavy punctuation..*Zirta* Stål

5. Fore femur with a ventral cleft...6

5′. Fore femur without ventral cleft, although it may be armed on ventral surface...9

6. Coloration uniformly black..........................*Wygodzinskyocoris* Dougherty 1995

6′. Coloration with a combination of dark and light brown....................................7

7. Abdominal sternites heavily punctuated........................*Cryptonannus* Dougherty

7′. Abdominal sternites lacking heavy punctuation...8

8. Head elongated in lateral view, i.e., head length greater than the head height....... ....*Sinchocoris* Dougherty

8′. Head subtriangular in lateral view, i.e., head length and height subequal............ ....*Doblepardocoris* Dougherty

9. Fore femora with a row of large dentiform processes ventrally .... *Borgmeierina* Wygodzinsky

9′. Fore femora unarmed or at most with a series of minute denticles or stiffened hairs ventrally...10

10. Postocular region with a pair of blunt elevations; seven antennal segments; fore and mid-femora incrassate, with a ventral carina and a row of setigerous and dentiform tubercles; length 9–9.5 mm..................................*Xarada* Carpintero

10′. Postocular region without a pair of elevations; seven or eight antennal segments; fore and mid-femora incrassated or not incrassated, without a ventral carina and arowofsetigerousanddentiformtubercles...11

11. Seven antennal segments; anterior pronotal lobe with a pair of paramedial carinated lobes, ocellar tubercle conical; prongs of scutellum close basally, divergent distally, spiniform..................................*Travassocoris* Wygodzinsky

11′. Seven or eight antennal segments; anterior pronotal lobe without a pair of paramedial carinated lobes; ocellar tubercle not conical: prongs of scutellum separated basally, subparallel...12

12. Robust species of 15 to more than 40 mm length; fore femora thickened, sometimes strongly so; mid-femora less frequently thickened, both with blunt tubercles or sharp and dentiform processes set on areas with short stiff hairs; fore and mid-trochanters with similar armature; fore and mid-tibiae slightly or strongly thickening toward the apex, with spongy fossa well developed......... ....*Brontostoma* Kirkaldy

12′. Smaller and/or less robust species; femora slender or slightly thickened; another setofcharacters...13

13. Head longer than wide..14

13′. Head length as long as or shorter than the width..19

14. First (visible) labial segment elongate, longer than second and third (visible) together; pronotum smooth and shiny........................*Pothea* Amyot & Serville

14′. First (visible) labial segment shorter than or at most subequal to second and third (visible) together; pronotum opaque, typically rugose, seldom smooth and shiny...15

15. Second (visible) labial segment longer than first; anterior pronotal lobe with distinct sculpture, posterior pronotal lobe rugose; metasternum with two transverse carinae..*Margacoris* Carpintero
15′. Second (visible) labial segment subequal in length to first segment; different set ofcharacters..16
16. First (visible) labial segment shorter than second and third together; second segment subequal to first, at most slightly longer or shorter; body red orange and black,rarelybrownishspecies..17
16′. First (visible) labial almost as long as or longer than second and third together; second distinctly shorter than first; body dark brown, brownish, blackish, at mostwithyellowishmarkingsonspecies...18
17. Longitudinal sulcus of the anterior pronotal lobe well developed anteriorly, but not reaching transverse sulcus; pronotum often rugose on anterior lobe, opaque, or moderately shiny; length 10–26 mm..*Rhiginia* Stål
17′. Longitudinal sulcus of the anterior pronotal lobe deep medially, but not reaching anterior or posterior margins of lobe; pronotum shiny, smooth; length 12–15 mm..*Pseudozirta* Bérenger & Gil-Santana
18. First antennal segment approximately half as long as the head; median longitudinal sulcus on anterior pronotal lobe obsolete; length 9–11.2 mm.......... ....*Pseudopothea* Wygodzinsky
18′. First antennal segment about as long as the head; median longitudinal sulcus well developed on anterior pronotal lobe and extending onto posterior lobe continuously; length 8–17 mm...*Racelda* Signoret
19. Body not flattened dorsoventrally..20
19′. Body flattened dorsoventrally..21
20. With ventrolateral elevations posterior to eyes; eyes, ocellar tubercles, and ocelli large to very large; legs slender, ventrally without spines or carinae; spongy fossa very small, less than 1/5 length of fore and 1/10 length of mid-tibiae; length 14–25 mm..*Cricetopareis* Breddin
20′. Without ventrolateral elevations posterior to eyes; eyes, ocellar callus conical or flattened; fore and mid-legs strongly carinated below, femora with setigerous tubercles and dentiform spines; spongy fossa on fore and mid-tibiae moderately developed, extending to between 1/5 and 1/3 length of segment; body length 6–13 mm...*Daraxa* Stål
21. Longitudinal sulcus of anterior pronotal lobe reduced to a fovea; anteocular region longer than postocular; head elongated; fore and mid-femora slightly enlarged, fusiform, ventrally carinated with setigerous tubercles ....*Pseudodaraxa* Carpintero
21′. Longitudinal sulcus of pronotum extending across both lobes; anteocular region much shorter than postocular; head hemispherical, vertical; fore femora enlarged basally, narrowing at the apex, curved, thinly carinated ventrally, on basal 2/3, with setigerous and teeth-like tubercles; mid- and hind femora similar, slender, straight, without carinae.................*Pseudoracelda* Carpintero

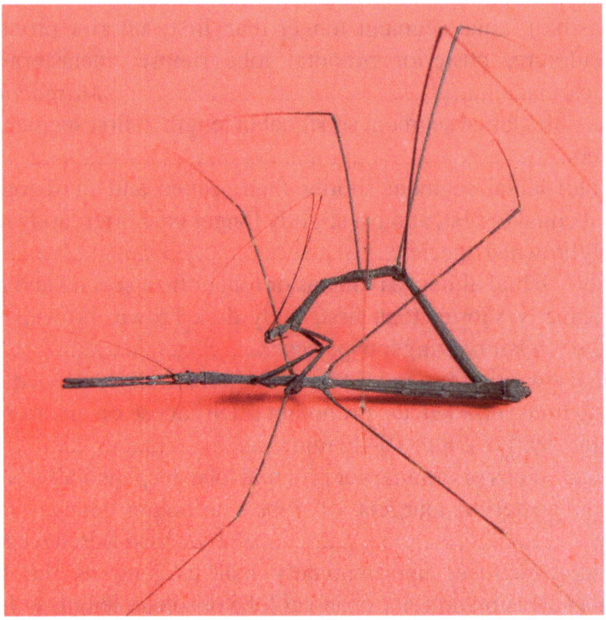

Fig. 12.6 *Ghinallelia brasiliensis* (Dohrn) (Emesinae), male and female in copula

12.5.5 Elasmodeminae

This subfamily includes only the genus *Elasmodema* Stål that contains three species known from Argentina, Brazil, and Paraguay (Wygodzinsky 1949a). Species can be identified using information in Wygodzinsky (1944) and Kormilev (1948). The biology of *Elasmodema* is essentially unknown. Wygodzinsky (1944) reported *E. setigerum* from a bird nest and mentioned that *E. erichsoni* was commonly collected under bark of wooden fence posts that were not occupied by ants.

12.5.6 Emesinae

Nearly 30 genera of Emesinae (Fig. 12.6) classified in five tribes occur in the Neotropics (Wygodzinsky 1966; Maldonado 1990). The monograph of Emesinae by Wygodzinsky (1966) remains the main source for study of this group, including keys to genera and species, although several new species and minor taxonomical changes have been proposed by diverse authors since then.

Wygodzinsky (1966) reviewed biological and ecological data for Emesinae. In addition to habitats shared with many other reduviids, some species have adapted to specific environments such as bird nests, caves, and spider and psocid webs

Fig. 12.7 *Microtomus conspicillaris* (Drury) (Hammacerinae), female, dorsal view

(Wygodzinsky 1966; Gil-Santana et al. 2007a). Especially, the association with spider webs has resulted in various anatomical modifications of body and legs (Wygodzinsky 1966; Soley et al. 2011). *Emesa mourei* Wygodzinsky has been found in both the webs of spiders and caterpillars (Gil-Santana and Jurberg 2003). Additional biological data on a few Neotropical species were published by Santiago-Blay and Maldonado (1988), Gil-Santana et al. (2005a), and Forero (2007).

12.5.7 Hammacerinae

This subfamily (Fig. 12.7) includes the genera *Homalocoris* Perty and *Microtomus* Illiger, which both are restricted to the New World (Maldonado 1990).

The eight species of *Homalocoris* can be identified using information in Maldonado (1987, 1996) and Maldonado and Santiago-Blay (1991). *Microtomus* includes 12 valid species (Maldonado 1990; Coscarón et al. 2003). *Microtomus* was first revised by Stichel (1926), and Costa Lima (1935) provided a now outdated key to species. About half of the currently recognized species have been treated in more recent taxonomic papers (Coscarón and Giacchi 1987a, b; Giacchi and Coscarón 1986, 1992; Melo and Coscarón 2004).

Little is known about the biology of Hammacerinae. They are usually found under bark of decaying trees (Champion 1898; Readio 1927), and *Homalocoris* has also been collected using light traps (Forero personal observation).

Key to the genera of Hammacerinae (adapted from Forero 2004):

1. Second antennal segment with 8–18 pseudosegments; total length less than 13 mm (except *H. nermini* Maldonado that measures ~20.5 mm); abdominal sternites of males without setose areas...*Homalocoris*

1′. Second antennal segment with 23–28 pseudosegments; total length more than 13 mm; abdominal sternites of males with setose areas....................*Microtomus*

12.5.8 Harpactorinae

This subfamily contains the greatest number of genera and species in the Neotropical region and worldwide (Schuh and Slater 1995; Forero 2011). Only two of the seven currently recognized tribes (Weirauch et al. 2014) occur in the New World, the Apiomerini and Harpactorini.

Key to New World tribes of Harpactorinae (based on Schuh and Slater 1995):

1. Ocelli lateral, more widely separated from one another than from eyes; labium straight; claws entire..Apiomerini

1′. Ocelli less widely separated from one another than from eyes; labium curved; claws with a basal tooth..Harpactorini

In addition, all Apiomerini except *Micrauchenus* Amyot & Serville have a sulcus on the apex of the fore tibia to receive the tarsus. The fore tibiae is densely pubescent, a feature shared with Harpactorinae tribes Diaspidiini and Ectinoderini from Old World (Forero 2004; Forero et al. 2011).

Apiomerini

There are 12 extant genera in Apiomerini (Gil-Santana et al. 2002, 2003; Bérenger 2006; Forero et al. 2011). With the exception of *Apiomerus* Hahn, which has several Nearctic species, Apiomerini (Fig. 12.8) are exclusively found in the Neotropical region (Maldonado 1990; Gil-Santana et al. 2003).

Seven genera of Apiomerini are monotypic, and their species are listed in the following: *Agriocoris flavipes* (F.), *Beharus cylindripes* (F.), *Calliclopius nigripes* (L.), *Fourcartus parecisi* Bérenger, *Micrauchenus lineola* (F.), *Ponerobia bipustu-*

Fig. 12.8 *Apiomerus nigricollis* Stål (Harpactorinae, Apiomerini), preying on a curculionid beetle

lata (F.), and *Sphodrolestes vittaticollis* Stål (Maldonado 1990; Gil-Santana et al. 2003; Bérenger 2006). The eight species of *Agriocleptes* Stål can be identified following Forero (2002).

Apiomerus comprises more than 100 described species (Maldonado 1990; Forero et al. 2010; Berniker et al. 2011). The first attempt to monograph the genus was that of Costa Lima et al. (1951). Later studies have improved our understanding of the taxonomy of the group (e.g., Gil-Santana and Milano 2007; Forero et al. 2010; Berniker et al. 2011), and additional revisionary work is underway (Forero and Weirauch personal communication).

Heniartes Spinola comprises 31 species (Maldonado 1990, Gil-Santana and Forero 2009), which can be identified using publications by Wygodzinsky (1947, 1953). *Manicocoris* includes two species that were keyed by Gil-Santana et al. (2002).

Apiomerini, particularly species of *Apiomerus*, have been observed to prey on bees, hence the common name "bee killer assassin bugs" (Marques et al. 2003; Silva and Gil-Santana 2004; Gil-Santana and Forero 2010), although their overall prey spectrum is not restricted to Hymenoptera. Some species of *Apiomerus* resemble meliponine bees (Hogue 1993) and are suspected to mimic them (Gil-Santana et al. 2003). *Apiomerus pictipes* Herrich-Schäffer has been speculated to attract *Trigona* bees using a chemical lure (Weaver et al. 1975). On more than one occasion, individuals of *Apiomerus nigricollis* Stål (Fig. 12.8) in the Southeast Atlantic Forest in Brazil have been noticed to exude a scent similar to flowers while being perched – with extended forelegs and waiting for prey – on top of leaves (Gil-Santana, personal observation). Almost all species of Apiomerini studied so far are polyphagous (Gil-Santana et al. 2003) except *Micrauchenus lineola* (F.) that feeds exclusively on termites (Bérenger and Pluot-Sigwalt 2009). Some species, mainly of *Apiomerus*, have been considered for use as biological control agents (Gil-Santana 2002).

Apiomerus species collect plant resins with their forelegs that may facilitate prey capture in immature and adults. In females, this resin is particularly important during oviposition (Eisner 1988; Choe and Rust 2007; Forero et al. 2011): stored on the ventral surface of the abdomen that in females of *Apiomerus* features distinct setose patches, the resin coating is transferred to the egg mass after oviposition using specialized leg structures, the metatibial combs; first instar immatures then use this resin from the egg mass to sticky-coat their own fore tibia, greatly facilitating prey capture (Eisner 1988; Forero et al. 2011). Other Apiomerini species are believed to gather resins as well, but direct observations for taxa other than *Manicocoris* (Melo et al. 2005) and *Beharus* Amyot & Serville (Uyttenboogaart 1901) are lacking.

Key to Apiomerini genera [based on Gil-Santana et al. (2003) and Bérenger (2006)]:

1. Head elongated, straight in lateral view; ocelli far to each other; antenna with all segments cylindrical....................2
1′. Head short, bent posteriorly in lateral view, with a conspicuous spiniform tubercle in the gular region; ocelli close to each other; antennal segments third and fourth enlarged and laterally compressed....................*Sphodrolestes* Stål
2. Head much longer than wide; postocular region longer than anteocular; anterolateral angles of pronotum acute; scutellum triangular, enlarged at the apex in some species; fore tibiae laterally compressed....................3
2′. Head barely longer than wide; postocular region longer, as long or shorter than anteocular; anterolateral angles of pronotum blunt; scutellum of variable structure; fore tibiae cylindrical....................4
3. Head about as long as the pronotum; first antennal segment somewhat thicker than the others; membrane of hemelytra darkened; apex of mid-tibiae curved....................*Heniartes* Spinola
3′. Head longer than pronotum; first antennal segment about as thick as the other segments; hemelytra with hyaline patches on corium and membrane; apex of mid-tibiae not curved....................*Agriocleptes* Stål
4. Anterior acetabula very prominent; scutellum generally triangular or rectangular; sulcus at the apex of mid-tibia absent....................5
4′. Anterior acetabula not as prominent; scutellum with the apex somewhat rectangular; sulcus at the apex on mid-tibia present....................10
5. First antennal segment more than twice as long as the head; antennal insertion very close together....................*Micrauchenus* Amyot & Serville
5′. First antennal segment shorter than the head; antennal insertion not very close together or distinctly separated....................6
6. Postocular region twice as long as the anteocular or longer; hemelytra much longer than abdomen....................7
6′. Postocular region about as long as the anteocular or slightly longer; hemelytra not surpassing or barely surpassing the abdomen....................8
7. Head shorter than pronotum; interocular region with a concavity; fore and hind femora of approximately the same length; fore tibiae densely pilose, incrassate in the distal half....................*Calliclopius* Stål

7′. Head about as long as the pronotum; interocular region flat or slightly depressed; fore femora somewhat shorter than hind femora; fore tibiae not as densely pilose, slightly thickened in midportion, tapering distally .... *Manicocoris* Stål
8. Head oval, much shorter than pronotum; antenna short, with second segment twice as long as the first segment, which is expanded apically .... *Ponerobia* Amyot & Serville
8′. Head elongate, a little shorter than pronotum; antenna long, with second segment about as long as the first, which is cylindrical, although thicker than the others... 9
9. Anterolateral angle of pronotum acute; fore femora longer than hind femora, the latter slightly thickened on basal half; median process of pygophore emarginate apically.. *Amauroclopius* Stål
9′. Anterolateral angle of pronotum blunt; fore femora approximately as long as the hind femora, of uniform thickness; median process of pygophore truncate apically.. *Beharus* Amyot & Serville
10. First (visible) labial segment shorter than anteocular region; scutellum posteriorly flat .... *Apiomerus* Hahn
10′. First (visible) labial segment approximately as long as the anteocular region; scutellum depressed posteriorly.. 11
11. First antennal segment three times as long as the anteocular region; hind tarsi three segmented; female fully winged... *Agriocoris* Stål
11′. First antennal segment 1.3 times as long as the anteocular region; hind tarsi two segmented; female micropterous .... *Fourcartus* Bérenger

Harpactorini

Harpactorini (Figs. 12.9 and 12.10) is the most speciose tribe of Harpactorinae and Reduviidae as a whole (Schuh and Slater 1995; Zhang and Weirauch 2014) that comprises 51 genera in the Neotropical region alone (Forero 2011, 2012; Swanson 2012). The only published key to separate the genera is by Stål (1872) and now badly outdated (Forero 2011). A key that includes all Neotropical genera is being prepared (D. Swanson, personal communication). Many taxonomical changes including the description of new genera have occurred since the late nineteenth century, rendering identifications difficult without consultation of original descriptions and a well-identified reference collection (Forero 2011). Publications containing new descriptions and revisions within Harpactorini were referenced by Forero et al. (2008) and Forero (2011). The latter paper also supplied a checklist of Neotropical Harpactorinae genera.

Members of the Harpactorini are commonly associated with plant crops, and some have been studied as biological control agents of crop pests (Grundy 2007; Grundy and Maelzer 2003). Several observations show that certain Harpactorini may be associated with particular plants (summarized in Bérenger and Pluot-Sigwalt 1997). These polyphagous predators often live on a single plant species, not only as adults but sometimes also as immature instars and eggs. While seeking prey,

Fig. 12.9 *Arilus carinatus* (Forster) (Harpactorinae, Harpactorini), nymph, live specimen, preying on a termite

Fig. 12.10 *Zelus versicolor* (Herrich-Schäffer) (Harpactorinae, Harpactorini), female on flower of *Bidens rubifolia* Kunth (Asteraceae), preying on a meliponine bee

some Harpactorini apparently found certain plants to have other advantages and became adapted to visiting these plants species exclusively. In addition to hosting the arthropod prey species the assassin bugs feed on, these plants provide additional resources such as sugary or proteinaceous secretions produced in extrafloral nectaries or other structures (Bérenger and Pluot-Sigwalt 1997; Gil-Santana and Alves 2011). While Habib (1976) noted that several cultivated tree species in Brazil seem to be inhabited by *Zelus leucogrammus* (Perty), Gil-Santana and Alves (2011) observed an association between *Zelus versicolor* (Herrich-Schäffer) and *Bidens rubifolia* Kunth (Asteraceae) (Fig. 12.10), in the same country. Bérenger and Pluot-Sigwalt (1997) observed in French Guiana that *Zelus araneiformis* Haviland may complete its postembryonic development feeding exclusively on glandular formations of *Cecropia obtusa* Tréc. (Cecropiaceae). Harpactorini have also evolved a unique predation strategy using sticky traps that is restricted to species in the New World (Zhang and Weirauch 2013). Other than Apiomerini that rely on external supplies of resin that can be used as sticky traps, a large proportion of Neotropical species of Harpactorini produce sticky substances themselves (Zhang and Weirauch 2013, 2014). Derived from integumental glands mostly on the fore and mid-tibiae, these sticky secretions serve as sticky traps to facilitate prey capture.

Remarkable examples of mimicry occur between certain groups of Hymenoptera and several taxa of Harpactorini. Species of *Notocyrtus* are recognized as mimics of meliponine bees, which they resemble mostly because of the angular and inflated pronotum (Haviland 1931; Jackson 1973; Gil-Santana 2008). Two mimetic species pairs have so far been identified: *Trigona fulviventris* Guérin and *Ptilotrigona lurida* (Smith) are models of *Notocyrtus dorsalis* (Gray) and *N. colombianus* Carvalho & Costa, respectively (Jackson 1973; Gil-Santana 2008). Several species in different genera resembles wasps, especially ichneumonid wasps (Maldonado and Lozada 1992). Species of *Hiranetis* Spinola, *Graptocleptes* Stål, and several additional, closely related and hard to delineate genera mimetic Braconidae and/or Ichneumonidae. This group of genera is in need of a comprehensive revision in order to clarify their systematics (Gil-Santana et al. 2013b).

12.5.9 Holoptilinae

Neolocoptiris villiersi Wygodzinsky & Usinger, described from a female from Guyana is the only extant species of Holoptilinae known from the New World (Wygodzinsky and Usinger 1963; Maldonado 1990). Poinar (1991) described one additional species based on a Dominican amber fossil. Old World species are known to prey on ants (e.g., Jacobson 1911), but the biology of New World species is unknown, except one observation that found an early instar immature underneath tree bark (G. Zhang, personal communication).

12.5.10 Peiratinae

There are 11 genera of Peiratinae in the Neotropics (Maldonado 1990; Cai and Taylor 2006; Melo 2012). Among them, *Sirthenea* Spinola is exceptional in being circumtropical, thus comprising species from the Old and New Worlds (Maldonado 1990; Willemse 1985).

Four of the Neotropical genera are monotypic and comprise: *Froeschnerisca vittata* (Coscarón), *Lentireduvius brasiliensis* Cai & Taylor, *Synnamarinus rasahusoides* Maldonado & Bérenger, and *Zeraikia novafriburguensis* Gil-Santana & Costa (Coscarón 1995, 1997; Bérenger et al. 1996; Gil-Santana and Costa 2003; Cai and Taylor 2006; Melo 2012).

Eidmannia was revised by Coscarón (1986b), who recognized six South American species. According to Coscarón and Carpintero (1994), eight species of *Melanolestes* are found in the Neotropical region. *Phorastes* includes only two species, *P. femoratus* (Degeer) and *P. incognitus* van Doesburg (1981). *Rasahus* was revised by Coscarón (1983), followed by the description of additional species (Coscarón 1986a; Coscarón and Maldonado 1988), and more than 20 Neotropical species are currently recognized (Maldonado 1990). The 12 species of *Sirthenea* that occur in the Neotropics were revised by Willemse (1985). *Thymbreus* includes three species which were studied by Coscarón (1994). *Tydides* and its four included species were revised by Lent and Jurberg (1967).

Peiratinae are ground-dwelling reduviids (Miller 1953; Louis 1974) which feed on other arthropods (Ambrose 1987; Readio 1927). Little detail is known on the biology of the Neotropical species; specimens of various genera are typically collected at lights and are known to inflict very painful bites (Forero and Weirauch personal communication).

Key to Peiratinae genera [based on Lent and Jurberg (1967); Froeschner (1981); Coscarón (1986b, 1995); Bérenger et al. (1996); Gil-Santana and Costa (2003); Cai and Taylor (2006); Melo (2012)]:

1. Head with preocellar transverse groove deeply impressed; fore coxa elongated, apical third to half-extended caudad of the prosternal process............................2
1′. Head with obsolete or shallowly impressed transverse groove; fore coxa short, the apex not or only slightly extending caudad of prosternal process................9
2. Mid-tibia without spongy fossa..*Sirthenea* Spinola
2′. Mid-tibia with spongy fossa occupying a fifth or more of tibial length..............3
3. Metapleural sulcus straight..4
3′. Metapleural sulcus curved..5
4. Anteocular region distinctly longer than the postocular portion; connexivum uniformly colored..*Tydides* Stål
4′. Anteocular and postocular regions subequal in length; connexivum bicolored..*Sinnamarynus* Maldonado & Bérenger
5. Fore tibia with spongy fossa restricted to its apical third; hemelytra, when present, uniformly fuscous to black...*Melanolestes* Stål

5′. Fore tibia with spongy fossa occupying more than the apical third; hemelytra, when present, with at least two colors 6
6. Fore tibia with spongy fossa occupying the distal half; body robust, length about three times its width .... *Lentireduvius* Cai & Taylor
6. Fore tibia with spongy fossa occupying the distal three fourths or more of its length, if shorter, then eyes small, about half as wide as the length of the interocular space in dorsal view (*Eidmannia*); body slender, three times longer than wide 7
7. Eyes small, about half as wide as the length of the interocular space in dorsal view, not reaching dorsal or ventral margins or the head in lateral view; scutellum with apical prolongation distinctly upcurved, apex erect .... *Eidmannia* Taeuber
7′. Eyes large, about half as wide or wider than the length of the interocular space in dorsal view, reaching dorsal and ventral margins of the head in lateral view; scutellum with apical prolongation horizontal or oblique, apex not erect 8
8. Scutellum with posterior process acuminate; male pygophore subrectangular, basal plate of phallus complex; female with 10th tergite with a projection .... *Froeschnerisca* Coscarón
8′. Scutellum with posterior process short, rounded, not acuminate; male pygophore quadrangular or rounded, basal plate of phallus simple; female with 10th tergite without a projection .... *Rasahus* Amyot & Serville
9. Posterior lobe of the head abruptly converging from the eyes to neck; metapleural sulcus nearly straight, except near lateral margin of supporting sclerite .... *Thymbreus* Stål
9′. Posterior lobe of the head abruptly converging to collar from a point considerably posterior to eyes; metapleural sulcus curved 10
10. Head distinctly shorter than the length of the anterior pronotal lobe; space between the hind coxae narrower than the hind coxal diameter .... *Phorastes* Kirkaldy
10′. Head somewhat longer than the anterior pronotal lobe; space between the hind coxae wider than the hind coxal diameter .... *Zeraikia* Gil-Santana & Costa

12.5.11 Phimophorinae

The monotypic genus *Phimophorus* Bergroth is the only representative of the subfamily Phimophorinae in the Neotropical region. *Phimophorus spissicornis* Bergroth was described based on a single specimen from Brazil (Bergroth 1886). Usinger and Wygodzinsky (1964) redescribed the species, and Maldonado (1990) also reported it from Ecuador.

Phimophorus spissicornis has frequently been found on palms in Colombia (Chaverra-Rodriguez et al. 2010), in microhabitats shared with various Triatominae species (Lent and Wygodzinsky 1979; Chaverra-Rodriguez et al. 2010). In addition, *P. spissicornis* has also been found in Bromeliaceae and bananas in Brazil (Lent and Jurberg 1977).

12.5.12 Phymatinae

Although treated as a separate family by some authors, Carayon et al. (1958) convincingly argued that the so-called ambush bugs should be regarded as part of the Reduviidae, a result that has been corroborated by all recent phylogenetic analyses (e.g., Hwang and Weirauch 2012). Two of the four recognized tribes occur in the Neotropical region, the Macrocephalini and Phymatini. Kormilev (1962) and Froeschner and Kormilev (1989) revised and cataloged, respectively, the group. Delimitation of species and even certain genera is difficult, due to widespread sexual dimorphism, color polymorphism, and the lack of comprehensive, well-illustrated taxonomic revisions. The group, and especially the speciose genus *Phymata* Latreille, is in severe need of a modern revision.

Phymatinae are sit-and-wait ambush predators that mostly hunt on flowers (Balduf 1941; Dodson and Marshall 1984). Little is known about the particular natural history of Neotropical species. Nearctic species have been shown to engage in mate guarding (Dodson and Marshall 1984) and have become models to investigate sexual selection (Punzalan et al. 2008).

Key to the tribes of Phymatinae [based on Froeschner and Kormilev (1989) and Forero (2004)]:

1. Head above the eye and propleura just ventral to the lateral margin each with a distinct longitudinal groove for reception of antennae at rest................Phymatini
1′. Head and propleura without such a longitudinal groove for reception of antennae...Macrocephalini

Macrocephalini

Two of the Neotropical genera of this tribe are monotypic and include *Extraneza nasuta* Barber from Puerto Rico and *Kormilevida iviei* (Kormilev & Doesburg) from Dominican Republic (Froeschner and Kormilev 1989; Kormilev and van Doesburg 1991). *Lophoscutus* Kormilev comprises approximately 65 Neotropical species (Froeschner and Kormilev 1989; Kormilev and van Doesburg 1991), although a clear distinction from *Macrocephalus* Swederus may not be possible in all cases. Keys are available for species occurring in North and Central America (Kormilev 1987) and the Caribbean (Kormilev and van Doesburg 1991), but not for South America. The 14 Neotropical species of *Macrocephalus* were keyed by Froeschner and Kormilev (1989).

Key to the American genera of Macrocephalini [modified from Kormilev and van Doesburg (1991) and Forero (2004)]:

1. Fore tibia without tarsi..........2
1′. Fore tibia with tarsi..........3
2. Pale spot on scutellum large, shaped as a spearhead or lancet, sometimes three-pronged posteriorly; parameres with a subapical branch.......... *Macrocephalus* Swederus
2′. Pale spot on scutellum small, restricted to the base of median carina; carina thin, reaching the tip of the scutellum, or lateral borders of abdominal segments III and IV produced into rounded lobes; parameres hook-shaped, lacking a subapical branch.......... *Lophoscutus* Kormilev
3. Second antennal segment 1.5 times as long as the first; buccula forming canals for reception of antennae at rest; abdominal segments II–IV with connexivum very large laterally.......... *Kormilevida* Doesburg
3′. Second antennal segment shorter than the first; buccula does not form such canals; abdominal segments II–IV with connexivum not very large laterally.......... *Extraneza* Barber

Phymatini

Three Neotropical genera of the tribe Phymatini are monotypic and include *Anthylla nervosopunctata* (Signoret) from Chile, *Kelainocoris farri* Kormilev from Jamaica, and *Paraphymata saileri* from Haiti (Froeschner and Kormilev 1989).

The three species belonging to *Neoanthylla* Kormilev, two from Brazil and one from Peru, were keyed by Froeschner and Kormilev (1989). *Phymata* Latreille has the highest number of species in the subfamily, 84 of which occur in the Neotropical region (Froeschner and Kormilev 1989). Kormilev (1962) monographed the genus and is the main source to identify species in this genus. Similar to the situation in Harpactorini, a well-identified reference collection is instrumental for identifying many taxa.

Key to the genera of Phymatini [modified from Froeschner and Kormilev (1989)].

1. Mid- and hind tibiae with upper side carinate laterally and sulcate medially.......... 2
1′. Mid- and hind tibiae convex on the upper side, neither carinate nor sulcate..........3
2. Fore femur subtriangular, more or less swollen.......... *Phymata* Latreille
2′. Fore femur elongately ovate, gently constricted near midlength, subapically with a carina forming a ring on exterior surface.......... *Paraphymata* Kormilev
3. First (visible) labial segment twice as long as the second; sutures between sternites II–V indistinct.......... *Kelainocoris* Kormilev

3′. First (visible) labial segment at most as long as the second; sutures between sternites II–V generally distinct..4
4. Fore femur subtriangular, exterior surface convex, granulated, dull; sutures between abdominal sternites II and III clearly visible.......................*Anthylla* Stål
4′. Fore femur elongately ovate, external face flat, smooth, glossy, polished; sutures between abdominal sternites II and III indistinct...
....*Neoanthylla* Kormilev

12.5.13 Physoderinae

Two monotypic genera of Physoderinae are known from the New World that include *Cryptophysoderes fairchildi* Wygodzinsky & Maldonado, known from Panama and Venezuela (Wygodzinsky and Maldonado 1972; Carcavallo and Tonn 1976), and *Leptophysoderes orellana* Weirauch that was described from Ecuador (Weirauch 2006a). Forero and Weirauch (2005) synonymized the putative physoderine *Harpinoderes cicheroi* Martinez & Carcavallo under the reduviine *Aradomorpha crassipes* Champion. Additional, undescribed Physoderinae occur in the Neotropical region (W.S. Hwang personal communication).

Little is known on the biology of Neotropical Physoderinae, other than they are thought to be cryptic (Wygodzinsky and Maldonado 1972). *Cryptophysoderes* was observed preying on larvae of Calliphoridae (Diptera) in a dung trap (Martínez and Carcavallo 1989) and collected in a hollow tree (Wygodzinsky and Maldonado 1972). *Leptophysoderes* was collected by canopy fogging (Weirauch 2006a).

12.5.14 Reduviinae

In Reduviinae (Fig. 12.11), 14 Neotropical genera are currently included (Maldonado 1990; Melo 2012; Gil-Santana et al. 2013a). Four genera are monotypic and include the following species: *Corupaia brasiliensis* Lent & Wygodzinsky, *Namapa caroli* Wygodzinsky & Lent, *Patago patagonicus* Kirkaldy, and *Peregrinator biannulipes* (Montrouzier & Signoret) (Maldonado 1990). *Pseudozelurus* Lent & Wygodzinsky is not monotypic, but only *P. superbus* (Champion) occurs in the Neotropical region (Lent and Wygodzinsky 1959; Maldonado 1990).

Aradomorpha includes three species (Maldonado 1990), which can be recognized using the key by Wygodzinsky (1949b). *Leogorrus* was revised by Melo (2007), who recognized and keyed 14 species, although additional species await description (M.C. Melo personal communication). *Microlestria* has three described species (Maldonado 1990), two of which were keyed by Champion (1898). Seven of the ten species of *Nalata* were keyed by Champion (1898). *Neivacoris* has three species, which were discussed by Lent and Wygodzinsky (1947). The eight species

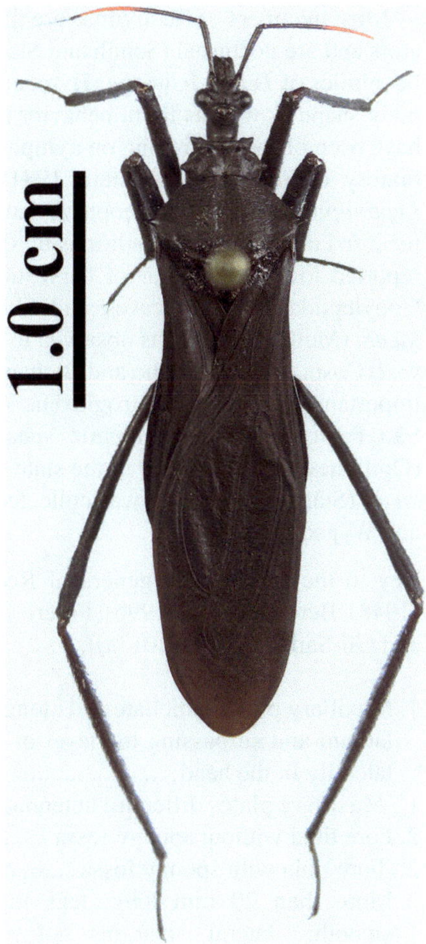

Fig. 12.11 *Zelurus lepeletierianus* (Kirkaldy) (Reduviinae), female, dorsal view

of *Opisthacidius* were studied and keyed by Lent and Wygodzinsky (1947, 1956). The two species of *Pantopsilus*, *P. bosqui* Lent & Wygodzinsky, and *P. longipes* (Berg) are known only from Argentina and were studied by Costa Lima (1940) and Lent and Wygodzinsky (1947). *Zeluroides* has two species, *Z. americanus* from the USA and Mexico and *Z. mexicanus* from Mexico (Maldonado 1990). Lent and Wygodzinsky (1959) recognized three subspecies for *Z. americanus* and two for *Z. mexicanus* and provided a key to distinguish them. With approximately 130 species and several subspecies, *Zelurus* is the largest genus of Neotropical Reduviinae (Maldonado 1990). The revision of Costa Lima (1940) included a key to the species. This key has been outdated for some time because of numerous taxonomical changes including the description of new species (e.g., Lent and Wygodzinsky 1945, 1951, 1955, 1966, 1968).

Most members of Reduviinae are thought to be general insect or arthropod predators and are nocturnal (Schuh and Slater 1995). Many species of *Zelurus* appear to be mimics of *Pepsis* F. wasps (Hymenoptera, Pompilidae) with respect to color and body shape as well as flight behavior (Costa Lima 1940). Some species of *Zelurus* have been observed preying on nymphs of Triatominae, both under natural and laboratory conditions (Carpintero 1981). In Brazil, predation on two species of *Leptinopterus* Hope (Coleoptera, Lucanidae) by *Zelurus angularis* (Stål) was recorded under natural conditions by Grossi et al. (2012). A species of *Zelurus* was reported to be a predator of harvestmen (*Goniosoma* sp., Arachnida, Opiliones, Gonyleptidae) in most caves studied by Machado et al. (2003). *Goniosoma spelaeum* (Mello-Leitão) was observed to be the prey of cave-dwelling *Zelurus travassosi* (Costa Lima) (Trajano and Bichuette 2010). *Zelurus travassosi* is also the most important predator of the trogloxene species *G. spelaeum* in the Brazilian state of São Paulo and the troglophile species *Daguerreia inermes* Soares & Soares (Opiliones, Gonyleptidae) in the state of Paraná (Machado et al. 2003). *Neivacoris steini* (Stål) (Reduviinae) was collected in termite nests on several occasions (Lent and Wygodzinsky 1947).

Key to the Neotropical genera of Reduviinae [based on Lent and Wygodzinsky (1948), Bérenger et al. (1996), Forero (2004), Gil-Santana and Coletto-Silva (2005), and Gil-Santana et al. (2013a)]:

1. Maxillary plates lamellate and elongated, including between them the base of the labium and surpassing the level of the first antennal segment; antenna inserted laterally in the head .... *Aradomorpha* Champion

1′. Maxillary plates different; antennae inserted dorsally in the head 2

2. Fore tibia without spongy fossa 3

2′. Fore tibia with spongy fossa 5

3. More than 20 mm long; legs long and slender, dorsal surface of femora smooth; lateral margins of scutellum with a pair of subbasal tubercles .... *Patago* Bergroth

3′. Less than 10 mm long; legs short, dorsal surface of femora strongly granulated; lateral margins of scutellum without tubercles 4

4. Fore trochanter ventrally with a stout spiniform process; corium and membrane of hemelytra sharply detached from each other, their relative dimensions as usual .... *Nalata* Stål

4′. Fore trochanter without a spiniform process; corium of hemelytra reduced to a narrow external band, not sharply detached from the membrane, the latter very large, elongated forward to reach the hind border of pronotum .... *Microlestria* Stål

5. Apex of all femora at ventral surface with a distinct pair of dentiform processes .... *Leogorrus* Stål

5′. Apex of all femora at ventral surface without such processes 6

6. Disc of anterior pronotal lobe without tubercles or spines 7

6′. Disc of anterior pronotal lobe with tubercles or spines 11

7. Fore and mid-femora with teeth along the ventral surface .... *Pantopsilus* Berg

7′. Fore and mid-femora without teeth processes on the ventral surface 8
8. Smaller species, 6–8 mm in length 9
8′. Larger species, over 10 mm in length 10
9. Body covered with long fine hairs; posterior angles of connexivum not salient .... *Peregrinator* Kirkaldy
9′. Body not covered with long fine hairs; posterior angles of connexivum with denticular lateral processes .... *Namapa* Wygodzinsky & Lent
10. Anterior femora slightly sulcate longitudinally on ventral surface, with two longitudinal rows of very numerous short bristles; the two membrane cells of equal width .... *Corupaia* Lent & Wygodzinsky
10′. Anterior femora not sulcate ventrally, with simple hairs only; the two membrane cells with different dimensions......... *Pseudozelurus* Lent & Wygodzinsky
11. Pronotum distinctly granulated 12
11′. Pronotum not granulated 13
12. Disc of anterior pronotal lobe with four tubercles; fore and mid-femora more incrassate than the hind femora .... *Opisthacidius* Berg
12′. Disc of anterior pronotal lobe with a pair of tubercles or short spines; fore and median femora slender, about the same diameter as the hind femora .... *Zeluroides* Lent & Wygodzinsky
13. Mandibular plates swollen, prominent, reaching or surpassing the apex of the head; clypeus vertical .... *Neivacoris* Lent & Wygodzinsky
13′. Mandibular plates less developed, not reaching the apex of the head; clypeus never vertical .... *Zelurus* Hahn

12.5.15 Saicinae

In this subfamily (Fig. 12.12), 10 genera are known from the Neotropical region (Gil-Santana and Costa 2009; Gil-Santana 2011). Five genera are currently monotypic, with the included species being *Bagriella ornata* McAtee & Malloch, *Buninotus niger* Maldonado, *Caprilesia almirantiana* Gil-Santana, Marques & Costa, *Kiskeyana palassaina* (Weirauch & Forero), and *Saicireta correntina* Melo & Coscarón (Maldonado 1990; Gil-Santana et al. 2006; Weirauch and Forero 2007a, b; Melo and Coscarón 2005).

Oncerotrachelus Stål has 14 New World species, with 13 of them occurring in the Neotropics (Gil-Santana 2013). McAtee and Malloch (1923) and Barber (1931) provided keys to four and five species of this genus, respectively. The two species of *Paratagalis* Monte, *P. spinosus* Monte, and *P. zikani* Gil-Santana & Costa (Fig. 12.12) can be identified by Gil-Santana and Costa (2009). Two species of *Pseudosaica* Blinn, *P. panamaensis* Blinn, and *P. florida* (Barber) occur in the Neotropical region (Blinn 1990; Gil-Santana and Marques 2005), and Blinn (1990) allows for their identification. *Saica* Amyot & Serville is the second largest genus of

Fig. 12.12 *Paratagalis zikani* Gil-Santana & Costa (Saicinae), male holotype, dorsal view

Saicinae in New World, with 13 recognized species (Maldonado 1990; Blinn 1994; Gil-Santana and Marques 2005). Champion (1898) keyed five species and McAtee and Malloch (1923) six. *Tagalis* Stål has seven described species (Gil-Santana 2011; Melo 2008), but a few more await description (Forero, personal observation). In the meantime, the known species can be identified by the key provided by Gil-Santana (2011).

Little is known of the biology and natural history of Saicinae. Specimens have been most commonly collected at lights (Schuh and Slater 1995) or swept and beaten from vegetation. Gil-Santana et al. (2010) included a synopsis of the biological and ecological information available for New World Saicinae, but also recorded *Tagalis evavilmae* Gil-Santana, Gouveia & Zeraik as an inhabitant of birds' nests, a first for Saicinae. *Kiskeyana palassaina* was found in leaf litter next to a stream and under rocks (Weirauch and Forero 2007a). Given that most Saicinae possess a scopula, a hairy attachment structure on the tarsi of all pairs of legs, it has been speculated that they can dwell on smooth plant surfaces (Weirauch and Forero 2007a).

Key to the New World genera of Saicinae (modified from Gil-Santana and Costa 2009):

1. Foreleg without spines, at most with erect setae 2
1′. Fore femur with two rows of spines, fore tibiae either with setae or with spines 5
2. Posterior pronotal lobe with upward projecting spines or tubercles; mesoscutum (scutellum) and metanotum apically with vertical spines or tubercles 3

2′. Pronotum generally unarmed, but sometimes with humeral spines; apex of mesoscutum produced into a long horizontal tapering spine, metanotum without spine or tubercle...*Oncerotrachelus* Stål
3. Opposed surfaces of labium and head with spinelike setae or bristles; forewing with two to three cells; metapleura without a tubercle near coxal cavity 4
3′. Opposed surfaces of labium and head without spinelike setae or bristles; forewing with four cells; metapleura with a tubercle near coxal cavity .... *Saicireta* Melo & Coscarón
4. Process on lower anterior angle of prothorax acute to subacute; second antennal segment about half as long as the first antennal segment; medial process of male pygophore bifurcate; posterior margin of seventh abdominal sternite in females vertical or subvertical...*Saica* Amyot & Serville
4′. Process on the lower anterior angle of the prothorax subconical; second antennal segment about ¾ as long as the first antennal segment; medial process of male pygophore a single, erect barbless spine; posterior margin of seventh abdominal sternite in females sloping ventrocephalad........... *Pseudosaica* Blinn
5. Fore tibia with a row of setae directed mediad.......*Kiskeyana* Weirauch & Forero
5′. Fore tibia with a ventral row of spines.. 6
6. Humeral angles of pronotum without processes, rounded...................*Tagalis* Stål
6′. Humeral angles of pronotum with spinelike processes 7
7. Fore coxae and anterior pronotal lobe unarmed......*Bagriella* McAtee & Malloch
7′. Fore coxae spined, anterior pronotal lobe with four spines or rounded humps 8
8. Anterior pronotal lobe with four spines; two labial segments spined .... *Paratagalis* Monte
8′. Anterior pronotal lobe with four humps; one or all three labial segments spined 9
9. Only the first (visible) labial segment spined; forewings with four closed cells...*Buninotus* Maldonado
9′. All three (visible) labial segments spined; forewings with two closed cells....................................... *Caprilesia* Gil-Santana, Marques & Costa

12.5.16 Salyavatinae

The only New World representative of the Salyavatinae is the genus *Salyavata* Amyot & Serville that was recently revised by van Doesburg and Forero (2012), who recognized and keyed 11 species.

Species of *Salyavata* feed on termites, particularly *Nasutitermes* species, using a bait and capture strategy (McMahan 1982, 1983). Different species of *Salyavata* might feed on different species of termites (van Doesburg and Forero 2012). Immatures of *Salyavata* are known to glue material of the termites' nests onto their integument, presumably to camouflage themselves (McMahan 1982, 1983). Integumental glands and trichomes involved in this camouflaging behavior were studied by Weirauch (2006b).

Fig. 12.13 *Sphaeridops aurantius* Gil-Santana, Costa & Zeraik (Sphaeridopinae), male paratype, dorsal view

12.5.17 Sphaeridopinae

The subfamily Sphaeridopinae (Fig. 12.13) currently includes three exclusively Neotropical genera and six species (Gil-Santana and Alencar 2001; Forero 2004). *Sphaeridops* comprises three species and *Veseris* two, which were keyed by Gil-Santana et al. (2000) and Gil-Santana and Alencar (2001), respectively. *Volesus* contains only a single species, *V. nigripennis* Champion (Maldonado 1990; Forero 2004).

Immatures of Sphaeridopinae share a unique type of integumental glands and trichomes with Salyavatinae, and it was therefore suspected that they might display camouflaging behavior similar to the one observed in Salyavatinae (Weirauch 2006b; McMahan 1982, 1983). In addition, Sphaeridopinae have been found in association with termites and fed on them in captivity (P. Wygodzinsky personal communication in: McMahan 1982) indicating that they might also share their prey specialization on termites with the closely related Salyavatinae.

Key to the genera of Sphaeridopinae [modified from Forero (2004)].

1. Interocular space larger than the width of an eye; antennifer apically straight .... *Volesus* Champion
1′. Interocular space shorter than the width of an eye; antennifer variable 2
2. Antennifer bifurcated; prosternum not excavated medially ....*Sphaeridops* Amyot & Serville
2′. Antennifer straight apically or only with lateral projection, never bifurcated; prosternum excavated medially... *Veseris* Stål

12.5.18 Stenopodainae

In this subfamily Stenopodainae (Fig. 12.14), 18 genera are recognized in the Neotropical region (Wygodzinsky and Giacchi 1994). The key to New World genera of Stenopodainae by Wygodzinsky and Giacchi (1994) separates only Neotropical genera and excludes the recently described Nearctic genus *Arenaeocoris* Blinn (Blinn 2012).

The three species of *Achillas* Torre Bueno were keyed by Bérenger (2001). Five species of *Apronius* Stål were keyed by Maldonado (1986a) and two additional species described by Maldonado and Lozada (1991) and Bérenger (2001), amounting to seven valid species in this genus. *Ctenotrachelus* Stål has 21 species (Maldonado 1990, 1995a; Bérenger 2001), with partial keys presented by Barber (1930) (11 spp.) and Maldonado (1995a) (four spp.). The six species of *Diaditus* Stål were revised and keyed by Giacchi (1982). The three species of *Gnathobleda* (*Gnathobleda*) Stål were revised by Giacchi (1977), and three species have been included in *Gnathobleda* (*Pnohirmus*) Stål (Maldonado 1990). Wygodzinsky and Giacchi (1986) keyed five of the species included in both subgenera. *Kodormus* Barber has three species, which can be identified using Bérenger and Maldonado (1996). *Narvesus* Stål and its two species, *N. carolinensis* Stål and *N. minor* Barber, were revised and keyed by Giacchi (1974). *Ocrioessa* Bergroth has two species, *O. cornutulus* (Berg) and *O. lobuliventris* Bergroth. The genus and *O. cornutulus* were redescribed by Giacchi (1985). A revision of the American species of *Oncocephalus* Klug was presented by Giacchi (1984), who recognized ten American species in this genus. *Otiodactylus* Pinto is monotypic, including only *O. signatus* Pinto. *Pnirontis* Stål includes two subgenera: *Pnirontis* (*Pnirontis*) Stål and *Pnirontis* (*Centromelus*) Fieber (Wygodzinsky and Giacchi 1994). The species formerly included in *Rutuba* Torre Bueno were transferred to *Pnirontis* (*Pnirontis*) by Giacchi (1988a), who considered the genus *Rutuba* as a junior synonym of *Pnirontis*. Maldonado (1990) cataloged 31 Neotropical species in this genus (including three as *Rutuba* spp.). Maldonado (1994a) and Giacchi (1996) further described two more species. Barber (1930) keyed 18 species; Maldonado (1986b) keyed 12 species which he considered

Fig. 12.14 *Rhyparoclopius aokiae* Gil-Santana (Stenopodainae), male holotype, dorsal view

as belonging to the "*subinermis*" group (species with first segment unarmed below), and Maldonado (1994a) presented a key to species of *Pnirontis* that measure 16 mm or more in length. *Podormus* Stål is monotypic, including only *P. granulatus* Stål, which was redescribed by Bérenger (2001). Maldonado (1990) cataloged six Neotropical species of *Pygolampis* Germar; Barber (1930) keyed four species, followed by the description of two other species by Costa Lima and Campos-Seabra (1945). The four species of *Rhyparoclopius* Stål were keyed by Gil-Santana (2012) (Fig. 12.14). Maldonado (1994b) provided a key to five species of *Seridentus* Osborn, followed by the description of the sixth species in this genus by Giacchi (1998). A revision and additional notes with a key to the 11 species of *Stenopoda* Laporte were published by Giacchi (1969, 1988b). *Stenopodessa* Barber has two species, *S. lanei* Costa Lima & Campos-Seabra and *S. piligera* Barber (Maldonado 1990).

Little is known about the biology of Stenopodainae. Species are collected under logs, on the ground in open grassy areas, in leaf litter, while sweeping and beating vegetation, and, most frequently, at lights (D. Forero and C. Weirauch personal observation).

Key to Neotropical genera of Stenopodainae [modified from Wygodzinsky and Giacchi (1994)]:

1. First antennal segment produced beyond insertion of the second segment; distinct hemelytral cell present proximad of the basal cell 2

1′. First antennal segment not produced beyond insertion of the second segment; distinct hemelytral cell proximad of basal cell absent 3

2. Antennifer armed with a distinct, nearly erect spine; first segment of mid- and hind tarsi as long as the second; posterior angles of connexival segments salient or strongly salient .... *Pnirontis* (*Pnirontis*) Stål
2′. Antennifer unarmed or provided with minute spines; first segment of mid- and hind tarsi shorter than the second; posterior angles of connexival segments varied .... *Pnirontis* (*Centromelus*) Fieber
3. First (visible) labial segment approximately twice as long as the second and third combined.. *Pygolampis* Germar
3′. First (visible) labial segment subequal to, or shorter than, the second and third combined.. 4
4. Prosternum distinctly elongate behind the fore coxae as long as, or longer than, coxae 5
4′.Prosternum shorter behind coxae than length of coxae, or coxae inserted at the hind margin of prosternum.. 8
5. First (visible) labial segment about as long as the second; anterolateral angles of collar and humeri acutely spinous; disc of anterior pronotal lobe with paired distinct spinelike tubercles; scutellum with long, erect spine; fore coxae elongate cylindrical, about twice as long as wide; hemelytral apical cu-pcu cross-vein obsolescent.. *Ocrioessa* Bergroth
5′. First (visible) labial segment much longer than second; anterolateral angles of collar and humeri bluntly rounded; disc of anterior pronotal lobe lacking spinelike tubercles; scutellum with short, blunt process; fore coxae shorter, less than twice as long as wide; apical cu-pcu cross-vein normally developed......... 6
6. Posterior angles of connexival segments 2–5 with spined foliaceous lobes; angles of seventh male abdominal segment long, acutely produced................................ .... *Achillas* Torre Bueno
6′. Posterior angles of connexival shortly produced; angles of seventh male abdominal segment not as above 7
7. Pronotum at least twice longer than wide; anterior prosternal processes short; sides of abdomen parallel or subparallel; last abdominal segment ending in two moderately developed, not acute apical lobes........................ *Ctenotrachelus* Stål
7′. Pronotum a little longer than wide; anterior prosternal processes strongly produced, curved apically; sides of abdomen subparallel anteriorly and diverging apically; last male abdominal segment ending in two flaring, widely developed, apically acute lobes.. *Seridentus* Osborn
8. Body and appendages with dense, adpressed pubescence and numerous tiny, erect bristles; basal discal cell of the hemelytra fused to M, in many cases only for a comparatively short distance... 9
8′. Body glabrous or variously pubescent but not as above; basal cell fused to M for a considerable distance or not in direct contact with M at all.......................... 11
9. Fore tarsi two segmented; fore femora strongly incrassate, fusiform; humeri rounded; first antennal segment much shorter than head....... *Rhyparoclopius* Stål
9′. Fore tarsi three segmented; fore femora only feebly incrassate, subcylindrical; humeri pointed; first antennal segment about as long as the head or longer than the head 10

10. Anterior tibiae with elongate spongy fossa; postocular region of the head nearly parallel-sided in dorsal view, abruptly constricted at neck .... *Stenopoda* Laporte
10′. Anterior tibiae lacking spongy fossa; postocular region of the head gently converging behind eyes to neck.. *Stenopodessa* Barber
11. Body elongate fusiform, often five times (or more) as long as the maximum width; head subcylindrical, anteocular and postocular regions of equal length; antennae long and slender, first segment always longer than anteocular region; pronotum narrow, its length along midline at least equal to width across humeri; male genitalia, when in situ, with cuplike posterior extension of pygophore completely covering parameres; female genital area narrowly pointed posteriorly 12
11′. Body not elongate fusiform, broader, always less than five times as long as the maximum width; head of varied shapes, anteocular region longer than postocular; antennae with first segment generally shorter than, only rarely as long as the anteocular region; pronotum wider across humeri than along midline; male genitalia, when in situ, with parameres not covered, clearly visible; female genital area not narrowly pointed posteriorly 13
12. Head with distinct, often branched, setigerous processes ventrolaterally behind eyes; maxillary plates shortly produced as rounded lobes, or strongly produced as acute lobes, on either side of base of labium .... *Gnathobleda* (*Gnathobleda*) Stål
12′. Head with or without distinct setigerous processes ventrolaterally behind eyes; maxillary plates not produced as described above.. *Gnathobleda* (*Pnohirmus*) Stål)
13. Fore femora strongly incrassate, at least twice as thick as the mid- and hind femora 17
13′. Fore femora slender, less than twice as thick as the mid- and hind femora.... 14
14. Mandibular plates produced anteriad between antennae, subcylindrical, parallel, rounded apically, projecting well beyond the apex of the head; apex of antennifer attaining the apex of the head; second (visible) labial segment only about half as long as the first one.. *Diaditus* Stål
14′. Mandibular plates not as above; apex of antennifer considerably short, not attaining the apex of the head; second (visible) labial segment not, or only very slightly shorter than the first segment....................................... 15
15. Head, thorax, abdomen ventrally, legs and base of hemelytra with numerous uniformly dense, small setigerous tubercles, metapleura with conspicuous sooty black evaporatory area .... *Podormus* Stål
15′. Setigerous tubercles when present not as above; metapleura with or without sooty black evaporatory area 16
16. Fore femora and to a lesser degree mid- and hind femora conspicuously tuberculate on ventral and dorsal surfaces; postocular region, in dorsal view, longer than eyes; vein emanating from basal discal cell conspicuously curved... *Nitornus* Stål

16′. Femora not tuberculate; postocular region shorter than eyes; vein emanating from inner basal cell almost straight .... *Narvesus* Stål
17. Fore tibiae with conspicuous spongy fossa, approximately twice as long as the diameter of the tibia .... *Apronius* Stål
17′. Fore tibiae without distinct spongy fossa, with spongy fossa as long or shorter than the diameter of the tibia 18
18. Head posterolaterally behind eyes with ramose setigerous processes; eyes distinctly though shortly pilose; fore and hind tibiae distinctly curved; connexival margins lobulate or denticulate 19
18′. Head posterolaterally behind eyes at most with simple setiferous spines; eyes not pilose; fore and hind tibiae almost straight; connexival margins entire .... *Oncocephalus* Klug
19. Disc of anterior pronotal lobe with paired conspicuous tubercles; evaporatory area of metapleura large, sooty black .... *Kodormus* Barber
19′. Disc of anterior pronotal lobe without conspicuous tubercles; evaporatory area of metapleura reduced in size, stramineous .... *Otiodactylus* Pinto

12.5.19 Tribelocephalinae

Until recently, no Tribelocephalinae were known to occur in the New World (Maldonado 1990; Weirauch 2010). A recently described new genus and species, *Tribelocodia ashei* Weirauch, from French Guyana, is currently classified to be part of the Tribelocephalinae. This species shares characters with both, Ectrichodiinae and the closely related Tribelocephalinae, but based on the results of a morphology-based cladistic analysis, it is currently placed in the Tribelocephalinae (Weirauch 2010). Nothing is known about the biology of *Tribelocodia.*

12.5.20 Vesciinae

The four Neotropical genera of the Vesciinae that accommodate 16 valid species can be distinguished based on the key provided below that is modified from Maldonado (1995b).

Microvescia Wygodzinsky comprises only *M. costalimai* Wygodzinsky (Wygodzinsky 1943; Maldonado 1990, 1995b). *Mirambulus* includes two species, *M. morio* Breddin and *M. niger* Breddin (Maldonado 1990, 1995b). The five species included in *Pessoaia* (Maldonado 1990, 1995b) can be identified using keys and descriptions by Wygodzinsky (1943) and Abalos (1945). *Vescia* includes eight species, five of which were keyed by Wygodzinsky (1943).

Mirambulus morio has been found in disturbed habitats in secondary forests, and *Pessoaia limai* was collected in Malaise traps and light traps in tropical dry forests in Colombia (Forero 2006).

Key to the genera of Vesciinae [modified from Maldonado (1995b)]:

1. Anteocular region shorter than the postocular; first antennal segment longer than the anteocular region 2
1′. Anteocular region longer than the postocular; first antennal segment shorter than the anteocular region 3
2. Spine between antennifers absent; ocelli present.................*Mirambulus* Breddin
2′. Spine between antennifers present; ocelli absent .... *Vescia* Stål
3. Pronotum glabrous and head nearly so, some species with scattered setae; the anterior pronotal lobe with short spinules and fine corrugations; the humeral angle spined or not; apex of prosternum reaching mesosternum............ ....*Pessoaia* Costa Lima
3′. Head and pronotum densely pilose; anterior lobe of pronotum without spinules and corrugations; the humeral angle always spined; apex of prosternum not reaching mesosternum ....*Microvescia* Wygodzinsky

12.6 Concluding Remarks

With this synopsis of the Neotropical Reduviidae, we make it clear that although a taxonomic framework exists, further work is needed to understand the full extent of the Neotropical biota. Future phylogenetic hypotheses, in particular at the species level, will help accommodate a wealth of natural history observations and let us ask deep evolutionary questions. It is also clear that many new taxa are waiting to be discovered and described in years to come, but this is only going to be accomplished with thorough fieldwork. We also have highlighted what we know with regard to basic biology and natural history of reduviids. Future reduviid researchers have an ample field to explore in both the taxonomic and biological aspects of the Reduviidae in the Neotropical region.

References

Abalos JW (1945) *Pessoaia alvaradoi* n. sp. (Hemiptera, Reduviidae). Anal Inst Med Reg Tucumán 1:229–237

Ambrose DP (1987) Biological, behavioral, and morphological tools in the biosystematics of Reduviidae (Insecta, Heteroptera, Reduviidae). Proc Indian Acad Sci Anim Sci 96:499–508

Ambrose DP (2000) Assassin bugs (Reduviidae excluding Triatominae). In: Schaefer CW, Panizzi AR (eds) Heteroptera of economic importance. CRC Press, Boca Raton, pp 695–712

Baena M, Susín J (2007) *Rhiginia guentheri* – a new species from Peru (Heteroptera: Reduviidae: Ectrichodiinae). Mainzer naturwissenschaftliches Archiv/Beiheft 31:113–119

Balduf WV (1941) Life history of *Phymata pennsylvanica americana* Melin (Phymatidae, Hemiptera). Ann Entomol Soc Am 34:204–214

Barber HG (1930) Essay on the subfamily Stenopodinae of the New World. Entomol Am 10:149–238

Barber HG (1931) A new *Oncerotrachelus* from Cuba (Hemiptera; Reduviidae). Bull Brooklyn Entomol Soc 26:185–186

Bérenger J-M (2001) Hétéroptères Reduviidae nouveaux ou peu connus de Guyane française. I. Stenopodainae). Bull Soc Entomol Fr 106:5–18

Bérenger J-M (2006) Um nouveau genre d'Apiomerini du Brésil (Heteroptera, Reduviidae, Harpactorinae). Nouv Rev d'Entomol 22:369–375

Bérenger J-M, Gil-Santana H (2005) Nouveau genre et nouvelles espèces d'Ectrichodiinae d'Amérique du Sud (Heteroptera, Reduviidae). Bull Soc Entomol Fr 110:509–516

Bérenger J-M, Maldonado CJ (1996) Une nouvelle espèce du genre Néotropical *Kodormus* Barber [Heteroptera, Reduviidae, Stenopodainae]. Revue Fran Ent (n s) 18:35–37

Bérenger J-M, Pluot-Sigwalt D (1997) Relations privilégiées de certains Heteroptera Reduviidae prédateurs avec les végétaux. Premier cas connu d'un Harpactorinae phytophage. C R Acad Sci Paris Sci de la vie 320:1007–1012

Bérenger J-M, Pluot-Sigwalt D (2009) Notes sur *Micrauchenus lineola* (Fabricius 1787), espèce termitophile et termitophage (Heteroptera: Reduviidae: Harpactorinae, Apiomerini). Ann Soc Entomol Fr (n s) 45:129–133

Bérenger J-M, Maldonado CJ, Pluot-Sigwalt D (1996) Un nouveau genre de Reduviinae de Guyane et notes sur la sous-famille (Heteroptera, Reduviidae). Bull Soc Entomol Fr 101:251–256

Bergroth E (1886) Zur kenntniss der Aradiden. Verh Zool Bot Ges 36:53–60

Berniker L, Szerlip S, Forero D, Weirauch C (2011) Revision of the *crassipes* and *pictipes* species groups of *Apiomerus* Hahn (Hemiptera: Reduviidae: Harpactorinae). Zootaxa 2949:1–113

Blinn RL (1990) *Pseudosaica panamaensis*, a new genus and species of assassin bug from Panama (Heteroptera: Reduviidae: Saicinae). J NY Entomol Soc 98:347–351

Blinn RL (1994) Synopsis of the Saicinae (Heteroptera: Reduviidae) of America North of Mexico, with the description of a new species of *Saica* from the eastern United States. J NY Entomol Soc 102:62–66

Blinn RL (2012) *Arenaeocoris enervatus* (Hemiptera: Heteroptera: Reduviidae: Stenopodainae), a new genus and species from de Southeastern United States. Zootaxa 3478:105–110

Cachan P (1952) Etude de la prédation chez les Réduvides de la région Éthiopienne. 1. La prédation en groupe chez *Ectrichodia gigas* H-Sch. Phys Comp Oecol 2:378–385

Cai W, Taylor SJ (2006) *Lentireduvius*, a new genus of Peiratinae from Brazil, with a key to the New World genera (Hemiptera: Reduviidae). Zootaxa 1360:51–60

Carayon J, Usinger RL, Wygodzinsky P (1958) Notes on the higher classification of the Reduviidae, with the description of a new tribe of the Phymatinae (Hemiptera-Heteroptera). Rev Zool Bot Afr 57:256–281

Carcavallo RU, Tonn RJ (1976) Clave gráfica de Reduviidae (Hemiptera) hematófagos de Venezuela. Bol Direc Malar San Amb 16:244–265

Carpintero DJ (1978) Revisión del género *Pothea* Amyot & Serville, 1843.Sus espécies en la Argentina (Hemiptera, Reduviidae, Ectrichodiinae). Revta Museo Argentino Ci Nat Bernardino Rivadavia 5:177–286

Carpintero DJ (1980) Nuevos Ectrichodiinae americanos (Insecta – Hemiptera – Reduviidae). Acta Scientifica 14:3–33

Carpintero DJ (1981) Sobre Reduviidae predatores de Triatominae. Com Museo Argen Ci Nat Bernardino Rivadavia 6:83–92

Carpintero DJ, Maldonado CJ (1988) Contributions to the knowledge of American Ectrichodiinae I. Notes about *Rhiginia* (Hemiptera: Reduviidae). J Agric Univ Puerto Rico 72:251–254

Carpintero DJ, Maldonado CJ (1990) Contributions to the knowledge of American Ectrichodiinae II. Notes about *Rhiginia* and *Pothea* (Hemiptera: Reduviidae). J Agric Univ Puerto Rico 74:449–456

Carpintero DJ, Maldonado CJ (1991) Contributions to the knowledge of American Ectrichodiinae III. The genus *Cricetopareis* Breddin, 1903 (Hemiptera: Reduviidae). J Agric Univ Puerto Rico 75:81–86

Carpintero DJ, Maldonado CJ (1996) Diagnostic characters and key to the genera of American Ectrichodiinae (Heteroptera, Reduviidae). Caribb J Sci 32:125–141
Champion GC (1898) Insecta Rhynchota. Hemiptera-Heteroptera. In: Godman FD, Salvin O (eds) Biologia Centrali Americana, vol II. Taylor & Francis, London
Chaverra-Rodriguez D, Forero D, Jaramillo ON, Triana Ch O (2010) New record and ecological notes of *Phimophorus spissicornis* (Hemiptera: Heteroptera: Reduviidae: Phimophorinae) in Colombia. Rev Col Entomol 36:176–178
Choe D-H, Rust MK (2007) Use of plant resin by a bee assassin bug, *Apiomerus flaviventris* (Hemiptera: Reduviidae). Ann Entomol Soc Am 100:320–326
Coscarón MC (1983) Revisión del género *Rasahus*. Revta Mus La Plata (n s) Secc Zool 13:75–138
Coscarón MC (1986a) Nueva espécie del gênero *Rasahus* Amyot & Serville, 1843 del Brasil (Heteroptera, Reduviidae, Piratinae). Revta Bras Ent 30:303–305
Coscarón MC (1986b) Revisión del gênero *Eidmannia* Taeuber, 1934 (Heteroptera, Reduviidae). Revta Bras Entomol 30:311–322
Coscarón MC (1994) Systematics and phylogenetic analysis of *Thymbreus* Stål (Heteroptera: Reduviidae: Peiratinae). Zool Med 68:221–230
Coscarón MC (1995) A new neotropical genus of Peiratinae (Reduviidae, Heteroptera). Revta Bras Entomol 39:453–457
Coscarón MC (1997) *Froeschnerisca* nom. nov. for the junior homonym of *Froeschneriellia* Coscarón (Heteroptera: Reduviidae). Physis Secc C 54:49
Coscarón MC, Carpintero DL (1994) Revision of the genus *Melanolestes* Stål (Heteroptera: Reduviidae, Peiratinae). Ent Scand 24:361–381
Coscarón MC, Giacchi JC (1987a) Revisión de la subfamilia Microtominae (Hemiptera, Reduviidae). I. *Microtomus lunifer* (Berg). Rev Soc Entomol Arg 44:243–250
Coscarón MC, Giacchi JC (1987b) Revisión de la subfamilia Microtominae. III. *Microtomus conspicillaris* (Drury, 1782) (Heteroptera: Reduviidae). Phys Secc C 45:59–65
Coscarón MC, Maldonado CJ (1988) Contribution to the knowledge of *Rasahus* Amyot & Serville, 1843 (Hemiptera: Reduviidae); description of a new species. J Agric Univ Puerto Rico 72:247–250
Coscarón MC, Melo MC (2003) Revision of the subfamily Bactrodinae (Heteroptera, Reduviidae), with a phylogenetic analysis of *Bactrodes*. Zootaxa 304:1–15
Coscarón MC, Melo MC, Cuello N (2003) Synonymizing *Microtomus sticheli* Costa Lima, 1935 under *M. tibialis* Stichel, 1926 and redescription of the species (Heteroptera: Reduviidae). Zootaxa 288:1–6
Costa Lima A (1940) Sobre as especies de *Spiniger* (Hemiptera: Reduviidae). Mem Inst Oswaldo Cruz 35:1–123
Costa Lima A, Campos-Seabra CA (1945) Stenopodinae da coleção do Instituto Oswaldo Cruz (Hemiptera: Reduvioidea: Reduviidæ) (2ª nota). Mem Inst Oswaldo Cruz 42:287–292
Costa Lima A, Campos-Seabra CA, Hathaway CR (1951) Estudo dos Apiômeros (Hemiptera: Reduviidae). Mem Inst Oswaldo Cruz 49:273–442
Costa Lima A (1935) Genero *Microtomus* Illiger, 1807. Ann Acad Bras Sci 7:315–322
Dodson G, Marshall L (1984) Mating patterns in an ambush bug *Phymata fasciata* (Phymatidae). Am Midl Nat 112:50–57
Dougherty V (1986) Review of the South American genus *Zirta* (Hemiptera: Reduviidae: Ectrichodiinae), with description of three species. Ann Entomol Soc Am 79:691–699
Dougherty V (1995) A review of the New World Ectrichodiinae genera (Hemiptera: Reduviidae). Trans Am Entomol Soc 121:173–225
Eisner T (1988) Insekten als fursorgliche Eltern. Verh Deut Zool Gesel 81:9–17
Forero D (2002) An up-to-date key to the species of the genus *Agriocleptes* (Heteroptera: Reduviidae: Apiomerinae). Rev Colomb Entomol 28:207–209

Forero D (2004) Diagnosis de los géneros neotropicales de la família Reduviidae (Hemiptera:Heteroptera), y su distribución en Colombia (excepto Harpactorinae). In: Fernández F, Andrade G, Amat G (eds) Insectos de Colombia, vol 3. Universidad Nacional de Colombia, Bogotá, pp 128–275

Forero D (2006) New records of Reduviidae (Hemiptera: Heteroptera) from Colombia and other Neotropical countries. Zootaxa 1107:1–47

Forero D (2007) Description of a new species of *Liaghinella* (Hemiptera: Heteroptera: Reduviidae: Emesinae) from the Colombian Andes, with notes on its feeding habits and conservation status. Zootaxa 1502:55–68

Forero D (2008) The systematics of the Hemiptera. Rev Colomb Ent 34:1–21

Forero D (2011) Classification of Harpactorinae assassin bugs (Hemiptera: Heteroptera: Reduviidae). Bol Mus Entomol Francisco Luis Gallego 3:9–24

Forero D (2012) *Pronozelus*, a new Neotropical harpactorine genus and species from Colombia (Hemiptera: Heteroptera: Reduviidae: Harpactorinae). Entomol Am 118:278–284

Forero D, Weirauch C (2005) Synonymy of *Harpinoderes cicheroi* Martinez & Carcavallo, 1989 with *Aradomorpha crassipes* Champion, 1899 (Hemiptera: Heteroptera: Reduviidae). Zootaxa 950:1–4

Forero D, Gil-Santana HR, van Doesburg PH (2008) Redescription of the Neotropical genus *Aristathlus* (Heteroptera, Reduviidae, Harpactorinae). In: Grozeva S, Simov N (eds) Advances in Heteroptera research: festschrift in honor of 80th anniversary of Michail Josifov. Pensoft, Sofia/Moscow, pp 85–103

Forero D, Berniker L, Szerlip S (2010) A polychromatic new species of *Apiomerus* (Hemiptera: Reduviidae: Harpactorinae) from Central America. Zootaxa 2522:44–60

Forero D, Choe DH, Weirauch C (2011) Resin gathering in Neotropical resin bugs (Insecta: Hemiptera: Reduviidae): functional and comparative morphology. J Morphol 272:204–229

Forthman M, Weirauch C (2012) Toxic associations: a review of the predatory behaviors of millipede assassin bugs (Hemiptera: Reduviidae: Ectrichodiinae). Eur J Entomol 109:147–153

Froeschner RC (1981) Heteroptera or true bugs of Ecuador: a partial catalog. Smithson Control Zool 322:1–147

Froeschner RC, Kormilev NA (1989) Phymatidae or ambush bugs of the world: a synonymic list with keys to species except *Lophoscutus* and *Phymata* (Hemiptera). Entomography 6:1–76

Giacchi JC (1969) Revision del genero *Stenopoda* Laporte, 1833 (Hemiptera, Reduviidae, Stenopodainae). Phys Secc C 34:1–26

Giacchi JC (1974) Revision de los Stenopodainos americanos. III. El genero *Narvesus* Stal, 1859 (Heteroptera, Reduviidae). Phys Secc C 33:53–64

Giacchi JC (1977) Revision de los Stenopodainos americanos. IV. El genero *Gnathobleda* Stal, 1859 (Heteroptera- Reduviidae). Phys Secc C 37:261–274

Giacchi JC (1982) Revision de los Stenopodainos americanos. V. El genero *Diaditus* Stal, 1859 (Heteroptera, Reduviidae). Phys Secc C 41:9–27

Giacchi JC (1984) Revision de los Stenopodainos americanos. VI. Las especies americanas del genero *Oncocephalus* Klug, 1830 (Heteroptera- Reduviidae). Phys Secc C 42:39–62

Giacchi JC (1985) Revision de los Stenopodainos americanos. VII. Redescripcion de los gêneros *Pnirontis* Stal, 1859; *Pnohirmus* Stal 1859; *Ctenotrachelus* Stal 1868; *Ocrioessa* Bergroth, 1918 y *Kodormus* Barber, 1930 (Heteroptera: Reduviidae). Phys Secc C 43:61–70

Giacchi JC (1988a) Revision de los Stenopodainos americanos. VIII. Estudio preliminar acerca del status genérico de *Pnirontis* Stål, 1859 y *Rutuba* Torre Bueno, 1914 (Heteroptera, Reduviidae). Phys Secc C 46:5–13

Giacchi JC (1988b) Revision de los Stenopodainos americanos. IX. El genero *Stenopoda* Laporte, 1833 (Heteroptera: Reduviidae): enmiendas y notas adicionales. Phys Secc C 46:47–59

Giacchi JC (1996) Una nueva especie del subgénero *Pnirontis* (*Pnirontis*) Stal, 1868 (Stenopodainae: Reduviidae: Heteroptera). Phys Secc C 52:15–17

Giacchi JC (1998) Una nueva especie del género *Seridentus* SOsborn, 1904 (Stenopodainae, Reduviidae, Heteroptera). Phys Secc C 56:31–32

Giacchi JC, Coscarón MC (1986) Revisión de la subfamilia Microtominae. II. *Microtomus purcis* (Drury, 1782); *M. luctuosus* (Stål, 1854) y *M. cinctipes* (Stål, 1858) (Heteroptera: Reduviidae). Phys Secc C 44:103–112

Giacchi JC, Coscarón MC (1992) Revisión de la subfamilia Microtominae. IV. *Microtomus gayi* (Spinola, 1852) y *M. pessoai* Lent y Suarez, 1956 (Heteroptera, Reduviidae). Phys Secc C 47:67–71

Gil-Santana HR (2002) Predação de *Lagria villosa* Fabricius, 1783 (Coleoptera: Lagriidae) por *Apiomerus nigrilobus* Stål, 1872 (Hemiptera: Reduviidae: Apiomerinae) em Cabo Frio, Estado do Rio de Janeiro, Brasil. Entomol Vect 9:201–208

Gil-Santana HR (2008) New records, and nomenclatural and biological notes on Reduviidae (Hemiptera: Heteroptera) from Bolivia and Brazil. Zootaxa 1785:43–53

Gil-Santana HR (2011) Three new species of *Tagalis* Stål (Hemiptera: Heteroptera: Reduviidae: Saicinae) from Brazil. Zootaxa 2996:33–48

Gil-Santana HR (2012) *Rhyparoclopius aokiae* sp. nov., a remarkable Stenopodainae (Hemiptera: Heteroptera: Reduviidae) from Mato Grosso do Sul, Brazil, with taxonomical notes on other species of *Rhyparoclopius* Stål. Zootaxa 3478:93–104

Gil-Santana HR (2013) Two new species of *Oncerotrachelus* Stål (Hemiptera: Heteroptera: Reduviidae: Saicinae), from Amazonas, Brazil, with taxonomical notes. Zootaxa 3626:268–278

Gil-Santana HR (2014) *Pothea berengeri* sp. nov. from Brazil, with taxonomical notes on *Pothea furtadoi* Gil-Santana & Costa and *Pothea jaguaris* (Carpintero) and reinstatement of *Parapothea* Carpintero as junior synonym of *Pothea* Amyot & Serville (Hemiptera: Heteroptera: Reduviidae: Ectrichodiinae). Zootaxa 3826:497–516

Gil-Santana HR, Alencar J (2001) Sobre o gênero *Veseris* Stål 1865, com *Eurylochus* Torre-Bueno, 1914, como sinônimo novo e chaves para identificação (Hemiptera, Reduviidae, Sphaeridopinae). Entomol Vect 8:95–104

Gil-Santana HR, Alves RJV (2011) Association between *Zelus versicolor* (Herrich-Schäffer) (Hemiptera, Reduviidae, Harpactorinae) and *Bidens rubifolia* Kunth (Asterales, Asteraceae). Entomol Brasilis 4:30–32

Gil-Santana HR, Baena M (2009) Two new species of *Brontostoma* Kirkaldy (Hemiptera: Heteroptera: Reduviidae: Ectrichodiinae) from Bolivia, with description of the male genitalia of two other species of the genus, and description of the female of *B. doughertyae* Gil-Santana, Lopes, Marques & Jurberg. Zootaxa 1979:41–52

Gil-Santana HR, Coletto-Silva A (2005) *Berengeria rafaeli*, gen. nov., sp. nov. de Reduviinae da Reserva Ducke, Estado do Amazonas, Brasil (Hemiptera-Heteroptera, Reduviidae). Acta Amazon 35:153–156

Gil-Santana HR, Costa LAA (2003) Um novo gênero e espécie de Peiratinae de Nova Friburgo, Rio de Janeiro, Brasil (Hemiptera, Heteroptera, Reduviidae). Rev Bras Zool 20:3–8

Gil-Santana HR, Costa LAA (2005) *Pothea furtadoi* sp. nov. (Hemiptera, Heteroptera, Reduviidae, Ectrichodiinae) de Mato Grosso e Minas Gerais, Brasil. Rev Bras Zool 22:400–404

Gil-Santana HR, Costa LAA (2009) A new species of *Paratagalis* Monte from Brazil with taxonomical notes and a key to New World genera of Saicinae (Hemiptera: Heteroptera: Reduviidae: Saicinae). Zootaxa 2197:20–30

Gil-Santana HR, Forero D (2009) A new species of *Notocyrtus*, a new synonym of *Coilopus*, and new records and notes on other Harpactorini (Hemiptera: Heteroptera: Reduviidae: Harpactorinae) from South America. Zootaxa 2148:55–67

Gil-Santana HR, Forero D (2010) Taxonomical and biological notes on Neotropical Apiomerini (Hemiptera: Heteroptera: Reduviidae: Harpactorinae). Zootaxa 2331:57–68

Gil-Santana HR, Jurberg J (2003) Sobre a ocorrência de *Emesa mourei* Wygodzinsky, 1945 (Hemiptera, Reduviidae, Emesinae) em teias biológicas. Entomol Vect 10:61–66

Gil-Santana HR, Marques OM (2005) Primeiro registro de *Saica apicalis* Osborn & Drake para o Brasil e *Pseudosaica florida* (Barber), com notas taxonômicas e chave para os gêneros de Saicinae do Brasil (Hemiptera, Reduviidae). Rev Bras Zool 22:405–409

Gil-Santana HR, Milano P (2007) Descrição de Apiomerus costai sp. nov. do Mato Grosso, com notas taxonômicas sobre Apiomerus Hahn (Hemiptera, Heteroptera: Reduviidae, Harpactorinae, Apiomerini). Neotr Entmol 36:314–316

Gil-Santana HR, Costa LAA, Zeraik SO (2000) Espécie nova de *Sphaeridops* Amyot & Serville, 1843 (Hemiptera, Reduviidae, Sphaeridopinae). Bol Mus Nac N S Zool 423:1–6

Gil-Santana HR, Costa LAA, Zeraik SO (2002) Sinonimização de *Paramanicocoris* Lima, Hathaway & Seabra, 1948 e *Manicocoris* Stål, 1866, com redescrição de *M. rubroniger* (Lima, Hathaway & Seabra, 1948), comb. nov. (Hemiptera, Reduviidae, Harpactorinae, Apiomerini). Bol Mus Nac N S Zool 490:1–7

Gil-Santana HR, Costa LAA, Forero D, Zeraik SO (2003) Sinopse dos Apiomerini, com chave ilustrada para os gêneros (Hemiptera-Heteroptera, Reduviidae, Harpactorinae). Publ Avuls Mus Nac 97:3–21

Gil-Santana HR, Costa LAA, Marques OM (2004) *Brontostoma bahiensis* sp. nov. de Ectrichodiinae (Hemiptera, Heteroptera, Reduviidae) do Estado da Bahia, Brasil. Rev Bras Zool 21:127–130

Gil-Santana HR, Costa LAA, Baena M, Zeraik SO (2005a) *Ghinallelia brasiliensis* (Dohrn): notas biológicas e taxonômicas (Hemiptera-Heteroptera: Reduviidae: Emesinae). Bol SEA 37:257–265

Gil-Santana HR, Lopes CM, Marques OM, Jurberg J (2005b) Descrição de *Brontostoma doughertyae* sp. nov. e estudo morfológico comparativo com *B. rubrum* (Amyot & Serville, 1843) (Hemiptera: Heteroptera: Reduviidae: Ectrichodiinae). Entomol Vect 12:75–94

Gil-Santana HR, Marques OM, Costa LAA (2006) *Caprilesia almirantiana* gen. nov. and sp. nov. of Saicinae from Brazil (Hemiptera, Reduviidae). Rev Bras Zool 23:392–394

Gil-Santana HR, Alves VR, Barrett TV, Costa LAA (2007a) A new species of *Phasmatocoris* Breddin (Hemiptera: Reduviidae: Emesinae) from the Amazon, Brazil. Zootaxa 1642:43–51

Gil-Santana HR, Costa LAA, Marques OM (2007b) Sinopse dos Chryxinae (Hemiptera, Reduviidae). Rev Bras Zool 24:77–83

Gil-Santana HR, Gouveia FBP, Zeraik SO (2010) *Tagalis evavilmae* sp. nov. (Hemiptera: Reduviidae: Saicinae), an inhabitant of birds' nests in Amazonas, Brazil, with taxonomical notes and a key to the species of *Tagalis* Stål. Zootaxa 2721:1–14

Gil-Santana HR, Baena M, Grillo H (2013a) *Berengeria* Gil-Santana & Coletto-Silva, a junior synonym of *Ectrichodiella* Fracker & Bruner, with new records and taxonomic notes on Ectrichodiinae from Brazil, and keys to Ectrichodiinae and Reduviinae genera of the New World (Hemiptera: Heteroptera: Reduviidae). Zootaxa 3652:60–74

Gil-Santana HR, Davranoglou L-R, Neves JA (2013b) A new synonymy of *Graptocleptes bicolor* (Burmeister), with taxonomical notes (Hemiptera: Heteroptera: Reduviidae: Harpactorinae: Harpactorini). Zootaxa 3700:348–360

Grossi PC, Koike RM, Gil-Santana HR (2012) Predation on species of *Leptinopterus* Hope (Coleoptera, Lucanidae) by three species of Reduviidae (Hemiptera, Heteroptera) in the Atlantic Forest, Brazil. Entomol Brasilis 5:88–92

Grundy PR (2007) Utilizing the assassin bug, *Pristhesancus plagipennis* (Hemiptera: Reduviidae), as a biological control agent within an integrated pest management programme for *Helicoverpa* spp. (Lepidoptera: Noctuidae) and *Creontiades* spp. (Hemiptera: Miridae) in cotton. Bull Entomol Res 97:281–290

Grundy PR, Maelzer DA (2003) Towards the on-farm conservation of the assassin bug *Pristhesancus plagipennis* (Walker) (Hemiptera: Reduviidae) during winter using crop plants as refuges. Aust J Entomol 42:153–158

Habib MEM (1976) Estudos biológicos sobre *Zelus leucogrammus* Perty, 1834 (Hemiptera, Reduviidae, Zelinae). An Soc Entomol Brasil 5:120–129

Haviland MD (1931) The Reduviidae of Kartabo Bartica District, British Guiana. Zoologica 7:129–154

Henry TJ (2009) Biodiversity of Heteroptera. In: Foottit RG, Adler PH (eds) Insect biodiversity: science and society. Blackwell Publishing Ltd., Oxford, pp 223–263

Hogue CL (1993) Latin American insects and entomology. University of California Press, Los Angeles

Hwang WS, Weirauch C (2012) Evolutionary history of assassin bugs: insights from divergence dating and ancestral state reconstruction. PLoS ONE 7:e45523

Jackson JF (1973) Mimicry of *Trigona* bees by a reduvid (Hemiptera) from British Honduras. Fla Entomol 56:200–202

Jackson RR, Salm K, Nelson XJ (2010) Specialized prey selection behavior of two East African assassin bugs, *Scipinnia repax* and *Nagusta* sp. that prey on social jumping spiders. J Insect Sci 10(82):1–19

Jacobson E (1911) Biological notes on the hemipteron *Ptilocerus ochraceus*. Tijdschr Entomol 54:175–179

Kormilev NA (1948) Una especie nueva de la familia Elasmodemidae Leth.& Sev. (1896) de la Republica Argentina (Hemiptera-Heteroptera, Reduvioidea). Rev Soc Entomol Argent 14:141–147

Kormilev NA (1962) Revision of the Phymatinae (Hemiptera, Phymatidae). Phillipine J Sci 89:287–486

Kormilev NA (1987) Notes on North and Central America *Lophoscutus* spp. (Hemiptera, Phymatidae, Macrocephalinae). Proc Entomol Soc Wash 89:701–705

Kormilev NA, van Doesburg PH (1991) Notes on Caribbean Phymatidae (Hemiptera: Heteroptera). Zool Med Leiden 65:277–285

Lent H, Jurberg J (1967) Revisão dos Piratinae americanos. III: as espécies do gênero "*Tydides*" Stal, com um estudo sobre a genitália (Hemiptera, Reduviidae). Atas Simp Biota Amaz (Zool) 5:337–363

Lent H, Jurberg J (1977) As ninfas de quinto estádio de *Phimophorus spissicornis* Bergroth, 1886 (Hemiptera, Reduviidae). Rev Bras Biol 37:521–524

Lent H, Wygodzinsky P (1945) Contribuição ao conhecimento do gênero *Zelurus* Hahn (*Spiniger* auct.) (Reduviidae, Hemiptera). Mem Inst Oswaldo Cruz 43:205–269

Lent H, Wygodzinsky P (1947) Contribuição ao conhecimento dos "Reduviinae" americanos (Reduviidae, Hemiptera). Rev Bras Biol 7:341–368

Lent H, Wygodzinsky P (1948) On two new genera of American "Reduviinae", with a key and notes on others (Reduviidae, Hemiptera). Rev Bras Biol 8:43–55

Lent H, Wygodzinsky P (1951) Espécies do gênero "*Zelurus*" Hahn (Hemiptera, Reduviidae). Rev Bras Biol 11:173–179

Lent H, Wygodzinsky P (1955) Espécies de *Zelurus* Hahn e *Opisthacidius* Berg de Colômbia, Ecuador e Peru (Reduviidae, Hemiptera). Mem Inst Oswaldo Cruz 53:135–142

Lent H, Wygodzinsky P (1956) Situação atual do gênero "*Opisthacidius*" Berg, 1879 (Hemiptera, Reduviidae). Rev Bras Biol 16:327–334

Lent H, Wygodzinsky P (1959) Sobre los gêneros "*Pseudozelurus*" Lent & Wygod.y "*Zeluroides*" Lent & Wygod. (Reduviinae, Reduviidae, Hemiptera). Rev Bras Biol 19:351–365

Lent H, Wygodzinsky P (1966) Os tipos de G. Breddin de "*Spiniger*" (= "*Zelurus*" Hahn) (Hemiptera, Reduviidae). Rev Bras Biol 26:145–164

Lent H, Wygodzinsky P (1968) Situação atual das espécies de "*Zelurus*" Hahn do grupo "*formosus*" (Hemiptera, Reduviidae). Rev Bras Biol 28:317–326

Lent H, Wygodzinsky P (1979) Revision of the Triatominae (Hemiptera, Reduviidae), and their significance as vectors of Chagas disease. Bull Am Mus Nat Hist 163:123–520

Louis D (1974) Biology of Reduviidae of cocoa farms in Ghana. Am Midl Nat 91:68–89

Machado SF, Ferreira RL, Martins RP (2003) Aspects of the population ecology of *Goniosoma* sp. (Arachnida, Opiliones, Gonyleptidae) in limestone caves in southeastern Brazil. Trop Zool 16:13–31

Maldonado CJ (1986a) Two new species of *Apronius* Stål with notes on the genus (Heteroptera: Reduviidae: Stenopodainae). J NY Entomol Soc 94:174–179

Maldonado CJ (1986b) The genus *Pnirontis*, the *subinermis* group of species (Hemiptera: Reduviidae: Stenopodainae). J Agric Univ Puerto Rico 70:9–18

Maldonado CJ (1987) *Homalocoris punctatus* n. sp. and key to the species in the genus (Reduviidae, Microtominae). J Agric Univ Puerto Rico 71:249–253

Maldonado CJ (1990) Systematic catalogue of the Reduviidae of the world. Caribb J Sci (Spec Publ) 1:1–694

Maldonado CJ (1994a) *Pnirontis grandis* n. sp. (Heteroptera: Reduviidae: Stenopodainae). Proc Entomol Soc Wash 96:713–716

Maldonado CJ (1994b) A new species of *Seridentus* Osborn, 1904 and a key to the species (Heteroptera: Reduviidae: Stenopodainae). Proc Entomol Soc Wash 96:717–722

Maldonado CJ (1995a) A new *Ctenotrachelus* and notes about other species in the genus (Heteroptera: Reduviidae: Stenopodainae). Caribb J Sci 31:281–283

Maldonado CJ (1995b) Notes about Vesciinae, a key to the World genera, and description of two new *Chopardita* (Heteroptera: Reduviidae). Proc Entomol Soc Wash 97:626–633

Maldonado CJ (1996) Reduviidae (Heteroptera) collected by fogging the forest canopy in Peru. Proc Entomol Soc Wash 98:233–236

Maldonado CJ, Lozada RPW (1991) A new species of *Apronius* Stal (Stenopodainae: Reduviidae) from Peru. J Agric Univ Puerto Rico 75:407–409

Maldonado CJ, Lozada RPW (1992) Key to the group of Neotropical wasp-mimetic harpactorine genera and the description of a new species (Hemiptera: Reduviidae). Proc Entomol Soc Wash 94:162–165

Maldonado CJ, Santiago-Blay JA (1991) Classification of *Homalocoris* (Heteroptera: Reduviidae: Hammacerinae), with the description of a new species. Proc Entomol Soc Wash 93:703–708

Marques OM, Gil-Santana HR, Magalhães ACA, Carvalho CAL (2003) Predação de *Apiomerus lanipes* (Fabricius, 1803) (Hemiptera: Reduviidae) sobre *Apis mellifera* (Linnaeus, 1758), no estado da Bahia, Brasil. Ent Vect 12:419–429

Martínez A, Carcavallo RU (1989) Physoderinae Neotropicales (Hemiptera-Reduviidae). Chagas 5:11–18

McAtee WL, Malloch JR (1923) Notes on American Bactrodinae and Saicinae (Heteroptera: Reduviidae). Ann Entomol Soc Am 15:247–254

McMahan EA (1982) Bait-and-capture strategy of a termite-eating assassin bug. Insect Soc 29:346–351

McMahan EA (1983) Adaptations, feeding preferences, and biometrics of a termite-bating assassin bug (Hemiptera: Reduviidae). Ann Entomol Soc Am 76:483–486

Melo MC (2007) Revision of the Neotropical genus *Leogorrus* Stål (Hemiptera: Reduviidae). Insect Syst Evol 38:51–92

Melo MC (2008) New records of Peruvian Reduviidae (Heteroptera), with the description of a new species of *Tagalis* Stål 1860 (Saicinae). Zootaxa 1763:55–62

Melo MC (2012) On the taxonomic placement of the genus *Sinnamarynus* (Hemiptera: Heteroptera: Reduviidae), and a new record of *S. rasahusoides* from Peru. Check List 8:540–541

Melo MC, Coscarón MC (2004) Redescription of *Microtomus reuteri* Berg (Heteroptera: Reduviidae: Hammacerinae) from South America. Entomol News 115:249–254

Melo MC, Coscarón MC (2005) *Saicireta correntina*, a new genus and species of assassin bug from Argentina (Heteroptera, Reduviidae, Saicinae) with a key to the New World genera. Dtsch Entomol Z 52:245–249

Melo MC, Berkov A, Coscaron MC (2005) Redescription of *Manicocoris rufipes* (Fabricius 1787), including nymphs I, II, III, and V (Reduviidae: Harpactorinae: Apiomerini), and its association with *Clusia* fruits. Stud Neotrop Fauna E 40:55–64

Miller NCE (1953) Notes on the biology of the Reduviidae of southern Rhodesia. Trans Zool Soc Lond 27:541–672

Poinar G (1991) *Praecoris dominicana* gen. n., sp. n. (Hemiptera: Reduviidae: Holoptilinae) from Dominican amber, with an interpretation of past behavior based on functional morphology. Insect Syst Evol 22:193–199

Punzalan D, Rodd FH, Rowe L (2008) Sexual selection mediated by the thermoregulatory effects of male colour pattern in the ambush bug *Phymata americana*. Proc R Entomol Soc Lond B 275:483–492

Readio PA (1927) Studies on the biology of the Reduviidae of America north ofMexico. Kansas Univ Sci Bull 17:1–291

Santiago-Blay JA, Maldonado CJ (1988) Observation on the true bug *Emesa tenerrima*, a possible spider mimic, and *Ghilianella borincana* (Hemiptera: Reduviidae: Emesinae) from Puerto Rico. Proc Entomol Soc Wash 90:369–372

Schuh RT, Slater JA (1995) True bugs of the world (Hemiptera: Heteroptera): classification and natural history. Cornell University Press, Ithaca

Silva AC, Gil-Santana HR (2004) Predation of *Apiomerus pilipes* (Fabricius) (Hemiptera, Reduviidae, Harpactorinae, Apiomerini) over Meliponinae bees (Hymenoptera, Apidae), in the state of Amazonas, Brazil. Rev Bras Zool 21:769–774

Soley F, Jackson R, Taylor P (2011) Biology of *Stenolemus giraffa* (Hemiptera: Reduviidae), a web invading, araneophagic assassin bug from Australia. New Zeal J Zool 38:297–316

Stål C (1872) Enumeratio Reduviinorum Americae. In: Enumeratio Hemipterorum. Kongliga Svenska Vetenskaps-Akademiens Handlingar 10:66–128

Stichel W (1926) Die Gattung *Microtomus* Illiger (Reduviidae). Dtsch Entomol Z 1926:179–190

Stride GO (1954) On the specific status of *Phonoctonus subimpictus* Stål. Rev Zool Bot Afr 50:13–16

Swanson DR (2012) A new synonym in the Harpactorinae of the New World (Heteroptera: Reduviidae). Proc Entomol Soc Wash 114:250–254

Trajano E, Bichuette ME (2010) Diversity of Brazilian subterranean invertebrates, with a list of troglomorphic taxa. Subterr Biol 7:1–16

Usinger RL, Wygodzinsky P (1964) Description of a new species of *Mendanocoris* Miller, with notes on the systematic position of the genus (Reduviidae, Hemiptera, Insecta). Am Mus Novit 2204:1–13

Uyttenboogaart DL (1901) Nadere mededeeling over *Beharus lunatus*. Tijdschr Entomol 45:36–37

van Doesburg PH (1981) Nomenclature in the genus *Phorastes* Kirkaldy, 1900 (Heteroptera: Reduviidae: Peiratinae). Entomol Ber 41:183–185

van Doesburg PH, Forero D (2012) Revision of the genus *Salyavata* Amyot & Audinet-Serville, 1843 (Heteroptera: Reduviidae: Salyavatinae). Zool Med Leiden 86:515–568

Weaver EC, Clarke ET, Weaver N (1975) Attractiveness of an assassin bug to stingless bees. J Kans Entomol Soc 48:17–18

Weirauch C (2006a) New genus and species of Physoderinae (Heteroptera: Reduviidae) from the New World with a revised diagnosis of Physoderinae Miller. Am Mus Novit 3510:1–9

Weirauch C (2006b) Anatomy of disguise: camouflaging structures in nymphs of some Reduviidae (Heteroptera). Am Mus Novit 3542:1–18

Weirauch C (2008) Cladistic analysis of Reduviidae (Heteroptera: Cimicomorpha) based on morphological characters. Syst Entomol 33:229–274

Weirauch C (2010) *Tribelocodia ashei*, new genus and new species of Reduviidae (Insecta: Hemiptera), has implications on character evolution in Ectrichodiinae and Tribelocephalinae. Insect Syst Evol 41:103–122

Weirauch C (2012) *Petasolentia goellnerae* gen. nov., sp. nov., a new genus and species of Chryxinae (Heteroptera: Reduviidae). Entomol Z 122:119–122

Weirauch C, Forero D (2007a) *Kiskeya palassaina*, new genus and species of Saicinae (Heteroptera: Reduviidae) from the Dominican Republic. Zootaxa 1468:57–68

Weirauch C, Forero D (2007b) *Kiskeyana* new replacement name for the assassin bug *Kiskeya* (Hemiptera: Heteroptera: Reduviidae) from the Dominican Republic. Zootaxa 1530:68

Weirauch C, Munro JB (2009) Molecular phylogeny of the assassin bugs (Hemiptera: Reduviidae), based on mitochondrial and nuclear ribosomal genes. Mol Phylogenet Evol 53:287–299

Weirauch C, Bérenger JM, Berniker L, Forero D, Forthman M, Frankenberg S, Freedman A, Gordon E, Hoey-Chamberlain R, Hwang WS, Michael A, Udah O, Watson C, Zhang G, Zhang J (2014) An illustrated identification key to assassin bug subfamilies and tribes (except Emesinae). Can J Arthropod Identif 26:1–115

Wignall AE, Taylor PW (2008) Biology and life history of the araneophagic assassin bug Stenolemus bituberus including a morphometric analysis of the instars (Heteroptera, Reduviidae). J Nat Hist 42:59–76

Willemse L (1985) A taxonomic revision of the New World species of *Sirthenea* (Heteroptera: Reduviidae: Peiratinae). Zool Verhand 215:1–67

Wygodzinsky P (1943) Contribuição ao conhecimento da subfamília Vesciinae (Hemiptera, Reduviidae). Rev Bras Biol 3:203–223

Wygodzinsky P (1944) Contribuição ao conhecimento do gênero "Elasmodema" Stal, 1860 (Elasmodemidae, Reduvioidea, Hemiptera). Rev Bras Biol 4:193–213

Wygodzinsky P (1946) Sôbre um novo gênero e uma nova espécie de Chryxinae e considerações sôbre a subfamília (Reduviidae, Hemiptera). Rev Bras Biol 6:173–180

Wygodzinsky P (1947) Contribuição ao conhecimento do gênero *Heniartes* Spinola, 1837 (Apiomerinae, Reduviidae, Hemiptera). Arq Mus Nac 41:11–64

Wygodzinsky P (1949a) Elenco sistemático de los reduviiformes americanos. Instituto de Medicina Regional de la Universidad Nacional de Tucumán, Monografia 1:1–102

Wygodzinsky P (1949b) Notas sobre Reduviidae argentinos (Hemiptera). Anal Inst Med Reg Tcumán 2:335–340

Wygodzinsky P (1951) Notas sobre Ectrichodiinae Neotropicales (Hemipt. Reduviidae). Rev Soc Entomol Argent 15:35–52

Wygodzinsky P (1953) Contribuição ao conhecimento dos gêneros "Heniartes" Spindola, 1837 e "Agriocleptes" Stal (Apiomerinae, Reduviidae, Hemiptera). Rev Bras Biol 13:369–380

Wygodzinsky P (1959) Notas y descripciones de Reduviidae bolivianas (Hemiptera). Acta Zool Lill 17:293–320

Wygodzinsky P (1966) A monograph of the Emesinae (Reduviidae, Hemiptera). Bull Am Mus Nat Hist 133:1–614

Wygodzinsky P, Giacchi JC (1986) A preliminary study about the generic status of Gnathobleda Stal, 1859 and Pnohirmus Stal, 1859 (Reduviidae: Stenopodainae). Phys Secc C 44:141–145

Wygodzinsky P, Giacchi JC (1994) Key to the genera of Stenopodainae of the New World (Insecta, Heteroptera, Reduviidae). Phys Secc C 49:5–9

Wygodzinsky P, Maldonado CJ (1972) Description of the first genus of physoderine assassin bugs (Reduviidae, Hemiptera) from the New World. Am Mus Novit 2504:1–7

Wygodzinsky P, Usinger RL (1963) Classification of the Holoptilinae and description of the first representative from the New World (Hemiptera: Reduviidae). Proc R Entomol Soc Lond 32:47–52

Zhang G, Weirauch C (2013) Sticky predators: a comparative study of sticky glands in harpactorine assassin bugs (Insecta: Hemiptera: Reduviidae). Acta Zool 94:1–10

Zhang G, Weirauch C (2014) Molecular phylogeny of Harpactorini (Insecta: Reduviidae): correlation of novel predation strategy with accelerated evolution of predatory leg morphology. Cladistics 30:339–351

Chapter 13
Hematophagous Bugs (Reduviidae, Triatominae)

José Jurberg, Cleber Galvão, Christiane Weirauch, and Felipe F.F. Moreira

Abstract The members of the subfamily Triatominae (Heteroptera, Reduviidae) are vectors of *Trypanosoma cruzi* (Chagas), the causative agent of Chagas disease or American trypanosomiasis. As important vectors, triatomine bugs have attracted ongoing attention, and thus, various aspects of their systematics, biology, ecology, biogeography, and evolution have been studied for decades. In the present chapter, the authors summarize the current knowledge on the systematics, phylogeny, evolution, and biology of these vectors and discuss the implications for human health.

13.1 Introduction

The Triatominae are bloodsucking bugs, which since the discovery of Carlos Chagas (1909) have been recognized as actual or potential vectors of Chagas disease. The first triatomine bug species was described in 1773, by C. De Geer (Fig. 13.1), but according to Lent and Wygodzinsky (1979), the first report on aspects and habits dated back to 1590 by Reginaldo de Lizárraga. While traveling to inspect convents in Peru and Chile, this priest noticed the presence of large hematophagous insects that attacked at night. In subsequent reports, various travelers and naturalists also mentioned the presence of these insects in South America. One of the most celebrated reports was by Charles Darwin, during his trip to South America in the H.M.S. Beagle in 1835, when he wrote: "One which I caught at Iquique, (for they

J. Jurberg (✉) • C. Galvão • F.F.F. Moreira
Laboratório Nacional e Internacional de Referência em Taxonomia de Triatomíneos, Instituto Oswaldo Cruz, Fiocruz, Av. Brasil, 4365, Manguinhos, Rio de Janeiro, RJ 21040-360, Brazil
e-mail: jjurberg@ioc.fiocruz.br; galvao@ioc.fiocruz.br; felipento@hotmail.com

C. Weirauch
Department of Entomology, University of California, Riverside, CA 92521, USA
e-mail: christiane.weirauch@ucr.edu

A.R. Panizzi, J. Grazia (eds.), *True Bugs (Heteroptera) of the Neotropics*, Entomology in Focus 2, DOI 10.1007/978-94-017-9861-7_13

Fig. 13.1 Male of *Triatoma rubrofasciata* (formerly *Cimex rubrofasciatus*), the first formally described species of Triatominae (From Jurberg and Galvão 2006)

are found in Chile and Peru,) was very empty. When placed on a table, and though surrounded by people, if a finger was presented, the bold insect would immediately protrude its sucker, make a charge, and if allowed, draw blood. No pain was caused by the wound. It was curious to watch its body during the act of sucking, as in less than ten minutes it changed from being as flat as a wafer to a globular form…" (Darwin 1845; Lent and Wygodzinsky 1979).

American trypanosomiasis or Chagas disease was discovered in 1909. Two years before, the Brazilian physician Carlos Chagas (1879–1934) was sent by Oswaldo Cruz to Lassance, a small village in the state of Minas Gerais, Brazil, to conduct an antimalaria campaign in the region where a railway was being built. Chagas stayed in the area for about 1 year, during which time a railroad engineer, Cantarino Mota, alerted him to the presence of hematophagous insects referred to locally as "barbeiros" (the local term for triatomines) (Fig. 13.2). Alerted to the presence of these insects inside the human dwellings, the doctor decided to investigate the possibility that they might transmit some parasite to humans, since besides malaria he had detected clinical signs that were difficult to interpret.

The local population complained about an uncomfortable feeling referred as palpitation, and Chagas observed arrhythmia and signs of heart failure among local

Fig. 13.2 *Panstrongylus megistus*, the species of Triatominae in which C. Chagas found *T. cruzi* (From Jurberg and Galvão 2006)

residents, along with reports of unexplained sudden death. Upon dissecting the triatomines he found flagellates in their intestinal tract. Believing that they belonged to *Trypanosoma minasense*, which infected black tufted-ear marmosets (*Callithrix penicillata*) in the same region, he sent several triatomine specimens to Oswaldo Cruz in Rio de Janeiro to feed on uninfected marmosets. Several weeks later he returned to Rio de Janeiro to discover a new trypanosome in the blood of one of the animals. As a tribute to his mentor, he named the species *Trypanosoma cruzi*. He then returned to Minas Gerais in an attempt to identify the parasite's vertebrate host. After numerous negative blood samples, he found an infected cat. Some 30 days later he returned to the house where he had discovered the infected animal and found a little girl named Berenice, just 2 years old, who was febrile and presented circulating forms of *T. cruzi* in her bloodstream.

From 1909 to 1912, Chagas described a new disease, its causative agent, natural reservoirs, and the vector, a set of discoveries unparalleled in international medicine to this day. As a result of his work, in 1912 Chagas received the Schaudinn Award from the Institute of Tropical Diseases in Hamburg, Germany (Chagas Fo 1968). Thus, one and the same researcher, in inverse order and in a short space of time, discovered a new disease that would later bear his last name, first recognizing the vector, next the parasite and reservoirs, and finally the clinical disease in humans (Chagas 1909). Chagas' discovery was overlooked by the Brazilian scientific community, as represented by the National Academy of Medicine, and was treated with disbelief for more than 20 years, because some scientists questioned the very existence of the disease. It was in Argentina in 1935 that Salvador Mazza submitted studies on the disease to the Annual Meeting of the Argentine Society of Tropical Medicine and together with Cecílio Romaña gave a new dimension and credibility to the problem (Chagas Fo 1968).

For more than one century, since the first description by De Geer, triatomines were studied merely from a descriptive point of view. However, beginning in 1909, when Chagas discovered the disease, and due to its newly acquired relevance to human health, studies began on the clinical form of the disease, the protozoan, and the vertebrate hosts, as well as the vector biology and transmission mechanisms. Advances in vector taxonomy began with Arthur Neiva, one of the most important scientists in this phase, who, in 1911, began describing various species, culminating with the publication of his dissertation "Revisão do gênero *Triatoma* Lap." in 1914. Important monographs were published subsequently by Pinto (1925) and Del Ponte (1930), in addition to other extensive studies published by Neiva and Lent (1936, 1941), Usinger (1944), Abalos and Wygodzinsky (1951), and Ryckman (1962), culminating in the grand works by Lent and Wygodzinsky (1979) and Carcavallo et al. (1998/1999). Relevant information concerning the triatomines was also compiled by Jurberg and Galvão (2006), which served as a basis for the preparation of the present chapter.

13.1.1 Chagas Disease

Currently there is no vaccine or effective cure for chronic Chagas disease, primarily a zoonosis, a parasitic disease of wild animals transmitted by sylvatic species of triatomine bugs. The adaptation of some triatomine species to human dwellings was secondary, as was the parasite's domiciliary cycle. American trypanosomiasis is now an endemic disease affecting mostly Latin America, primarily in rural populations of Central and South America, where it is an important public health problem. According to the World Health Organization, an estimated 28 million people are at risk of the disease, with 15 million currently infected (Guhl and Lazdins-Helds 2007).

Animals infected with *T. cruzi* are always mammals, as the parasite cannot develop in the blood of birds, reptiles, or amphibians. A triatomine bug that has sucked the blood of a mammal (including a human) infected with *T. cruzi* acquires the infection and the protozoan then reproduces, multiplying in the insect's digestive tract and producing the infective forms. Infection remains in the insect throughout its lifespan and can occur both in the nymphs and adults. Only the eggs are not affected, so the second generation remains uninfected until its first ingestion of infected blood (i.e., there is no transovarian transmission). Therefore, triatomine bugs reared in the laboratory with blood from uninfected animals can be used safely in experiments.

When the triatomine bug bites, it defecates during or right after bloodsucking, eliminating infective forms of the parasite in its feces. These forms can penetrate actively either through the bite's orifice, the mucosa, or small wounds and scratches on the skin. Upon entering the bloodstream the parasites are transported to the muscles or organs (mainly the heart and digestive tract), where they multiply and cause lesions. Penetration of *Trypanosoma cruzi* through the skin can cause a local reaction known as *chagoma*, and another important sign is unilateral bipalpebral swelling, leaving the patient's eye practically closed. This is the so-called Romaña's sign, named after its discoverer, an Argentine physician.

There are three distinct phases in the disease: acute, indeterminate, and chronic. In the acute phase (3–4 weeks), the infection varies from an asymptomatic to a severe and fatal form, the latter mainly in children or debilitated individuals, characterized by high fever, while other symptoms like diarrhea and vomiting can appear when the digestive tract is affected. The indeterminate phase is characterized by low parasitemia without clinical signs, which can persist or evolve into a chronic disease. The chronic form normally appears 10–15 years after the acute phase, and Chagasic cardiopathy is the most common manifestation, the digestive form producing visceromegalies, especially megaesophagus and megacolon.

13.1.2 Epidemiological Importance and Implications for the Human Health

All triatomine bugs must be regarded as potential vectors of *T. cruzi*, which infects a wide variety of sylvatic and domestic mammals, but so far only a few species have become well adapted to living in human dwellings, thereby acquiring epidemiological importance for humans (Jurberg and Galvão 2006). The most important vector species are *Triatoma infestans* (Klug), *T. brasiliensis* Neiva, *T. dimidiata* (Latreille), *Panstrongylus megistus* (Burmeister), and *Rhodnius prolixus* (Stål). Heavy domestic infestations of triatomines can be highly stressful, because these populations can reach several hundred individuals of different stages, representing a high daily blood loss.

According to the World Health Organization, Chagas disease is the third most important parasitic disease next to malaria and schistosomiasis, based on the resulting disability and work limitations. In Latin America, it is the fourth most important disease, following respiratory diseases, diarrhea, and AIDS (Schofield 1994). Each year about 500,000 people are infected with the disease, which is still incurable. The existing drugs are only partially effective in the acute phase, which mainly attacks the heart and digestive tract. Approximately 10 % of the infected individuals develop clinical signs and symptoms of chronic Chagas disease.

Occurrence of the classic form of the disease in a given area depends on three basic factors: presence of *T. cruzi* (the etiological agent), domiciliated triatomine bugs (vectors), and humans and other mammals (hosts) inhabiting the domiciliary environment. In addition to the classic infection model through contaminated triatomine feces, accounting for 80 % of the infections, other mechanisms contribute to cases of Chagas disease, such as transfusion of contaminated blood and blood products (16 %), congenital transmission from infected mothers (2 %), and the rest from organ transplants, infection by the oral route through ingestion of contaminated food, and laboratory accidents (Schofield 1994).

13.1.3 Control

After the successful campaign to control domestic triatomine populations in some South American countries, the new target of studies should be the species invading controlled areas. In recent years, there has been increased reporting of sylvatic species invading human dwellings and peridomiciliary environment in South America (Valente et al. 1998; Coura et al. 1999; Valente 1999; Almeida et al. 2000; Sandoval et al. 2000, 2004; Galvão et al. 2001b; Vivas et al. 2001; Wolff and Castillo 2002; Sousa et al. 2004; Sousa and Galvão 2004). The majority of these findings are adult insects and flight represents an important form of dispersion to previously controlled areas. Studies on triatomine flight capacity can facilitate identification of species with the tendency to invade dwellings and to allow the application of adequate vector surveillance. Effective surveillance is obviously important to avoid reinfestation after control, or resurgence of any vector population. Chagas disease control should be based on various independent but complementary work fronts and has become a public health priority in the affected countries due to epidemiological relevance and the high financial costs for the economy.

Use of Insecticides

Vector control uses residual-action insecticides, which must be applied both on the inside and outside parts of houses and outbuildings. One problem is that the insecticides do not affect eggs laid in inaccessible places like cracks and crevices in the buildings. The vector species that are subject to control should be considered, since the biological cycle should be the basis for systematizing the intervals and number of insecticide applications over the course of the year.

Housing Improvement

This should be a primary goal, because typical poor rural dwellings are made of mud and wattle, with thatched palm roofs and packed earthen floors, and have domestic and wild animals living in the same environment, thus facilitating massive infestations by triatomines. Anti-triatomine measures include construction of simple housing with measures like smooth walls without cracks, ceramic tile roofs protected below with ceilings, and well-organized furniture and utensils to avoid the formation of refuges for insects.

Health Education and Research

Despite extensive research on Chagas disease, little information has been generated to teach rural communities about it. The most effective control measure is a health education project targeting rural communities and health professionals. A population that is well informed about the disease is better prepared to prevent the entry and persistence of vectors and to notify the health authorities about existing problems. Information about the disease and vectors and the means to control them should be provided to the entire population, in the school system, on radio and television, and in community centers through leaflets, posters, and films. Unfortunately this is still a distant reality in Latin American countries.

When Carlos Chagas discovered the medical importance of triatomine bugs in 1909, only 33 species were known, and it was up to Arthur Neiva (through the Oswaldo Cruz Institute) to launch taxonomical studies on this group by creating the embryo for the Triatomines Laboratory that has been operating nonstop for more than a century. In 1989, the Triatomines Laboratory was transformed into the current *Laboratório Nacional e Internacional de Referência em Taxonomia de Triatomíneos* (LNIRTT; National and International Reference Laboratory for Triatomine Taxonomy) (Jurberg 1999), housing the world's largest collection of triatomines, with about 24,000 dry specimens, consisting of the Herman Lent Collection with 9,000 specimens and the Rodolfo Carcavallo Collection with 15,000 specimens. The laboratory has a large-scale insectary and is currently rearing 43 species in approximately 150 crystallizers, in addition to a collection kept in alcohol, with 45 species. The collections remain open and continue to receive material and make donations, and the laboratory is open to new scientific collaboration.

13.2 General Characteristics and Diagnosis

Most triatomine species can be identified on the basis of their external morphology. Their general appearance is similar to that of other reduviids, and the general external and internal morphologies are shown in Fig. 13.3. Detailed schemata of morphological traits were published by Lent and Wygodzinsky (1979) and several scanning electron microscopy illustrations were provided by Carcavallo et al. (1998/1999). The length of most species varies from 20 to 28 mm, but *Dipetalogaster maxima* (Uhler) can reach 44 mm in length, and the smallest species is *Alberprosenia goyovargasi* (Martínez & Carcavallo), with 5 mm. The color pattern (Figs. 13.4, 13.5, 13.6, 13.7, 13.8, and 13.9) varies, with an overall black or piceous color and spotted patterns of yellow, brown, orange, or red (Jurberg et al. 2004, 2012, 2014).

Adults differ from nymphs by the presence of ocelli and well-developed external genitalia and wings, with the exception of two species of *Mepraia* Mazza, Gajardo & Jörg 1940, which display wing polymorphism (Jurberg et al. 2002). Females are larger than males and have pointed or truncate abdominal apex, whereas in males

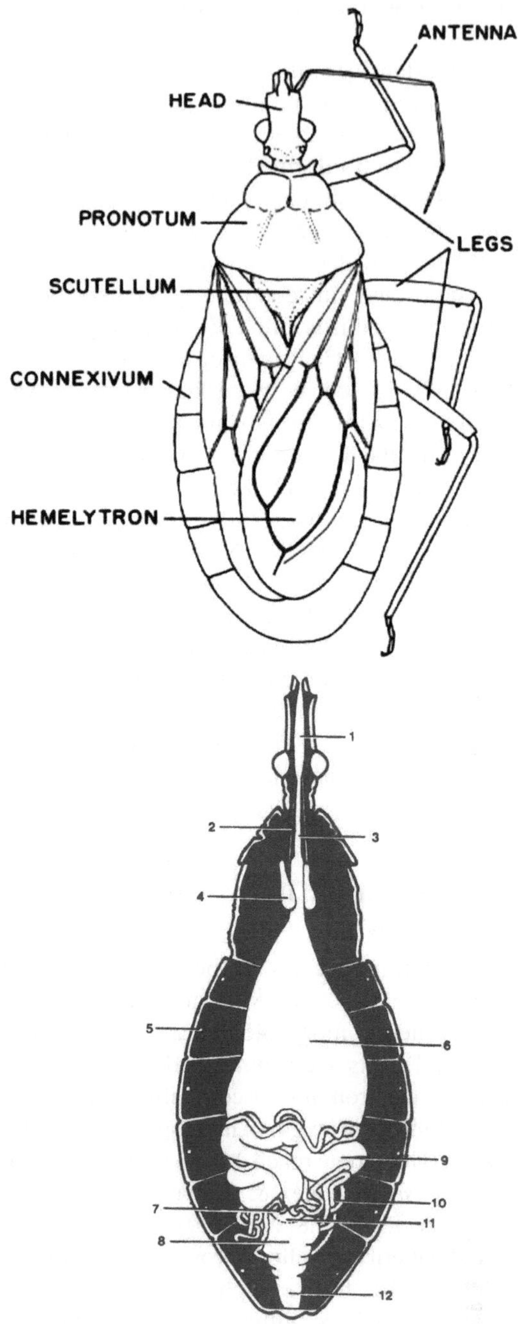

Fig. 13.3 General external and internal morphologies of the Triatominae (1, pharynx; 2, salivary duct; 3, esophagus; 4, salivary gland; 5, third respiratory spiracle; 6, promesenterium; 7, basal ampoule; 8, rectal ampoule; 9, postmesenterium; 10, Malpighian tubules; 11, light zone; 12, rectum) (From Lent and Wygodzinsky 1979 and Ramírez 1985)

Fig. 13.4 Color pattern of triatomine bugs: *Rhodnius brethesi* (From Jurberg and Galvão 2006)

Fig. 13.5 Color pattern of triatomine bugs: *Rhodnius nasutus* (From Jurberg and Galvão 2006)

Fig. 13.6 Color pattern of triatomine bugs: *Triatoma tibiamaculata* (From Jurberg and Galvão 2006)

Fig. 13.7 Color pattern of triatomine bugs: *Triatoma lecticularia* (From Jurberg and Galvão 2006)

Fig. 13.8 Color pattern of triatomine bugs: *Triatoma carcavalloi* (From Jurberg and Galvão 2006)

Fig. 13.9 Color pattern of triatomine bugs: *Triatoma brasiliensis* (From Jurberg and Galvão 2006)

the apex is rounded. The morphology of triatomine eggs and nymphs (Fig. 13.10) has been the focus of several studies, and a summary of these works was provided by Galvão et al. (2005).

Besides the external structure, size, proportions, and coloration, genital features and the internal morphology also provide insights to the taxonomy and systematics of the Triatominae. The morphology of the salivary glands and proportions between the testicular tubes aid on the generic differentiation, and the structure of male genitalia can be used for the specific diagnosis, as demonstrated in Lent and Wygodzinsky (1979), Gonçalves et al. (1987), and Santos et al. (1997). These characters, in association with the study of live specimens when available, and the use of molecular, cytogenetic, morphometric, and histochemical tools provide support for the discovery of cryptic species and for the development of phylogenetic analysis (Jurberg 2003; Galvão 2003).

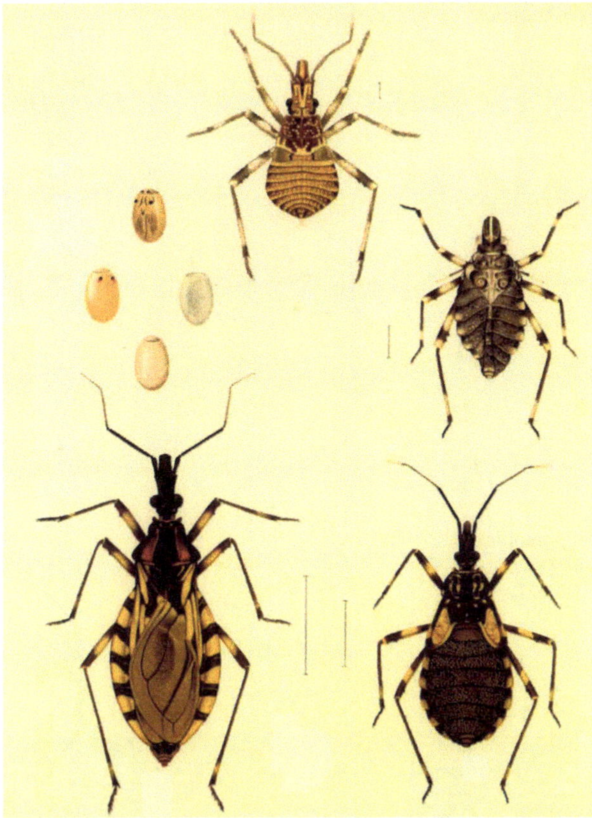

Fig. 13.10 Eggs, nymphs, and female of *Triatoma brasiliensis* (From Jurberg and Galvão 2006)

13.2.1 Salivary Glands

The salivary complex is an important characteristic for the differentiation of the genera of Triatominae. The structures forming the complex are homogeneous among the genera, but display variable shapes (Figs. 13.11 and 13.12). Del Ponte (1921) was the first to demonstrate the importance of the salivary complex of the triatomines for the understanding of the taxonomy of the group, using *Triatoma infestans* as a model. Other comparative studies have been performed using different species of the genera *Triatoma* Laporte, *Mepraia*, *Panstrongylus* Berg, *Psammolestes* Bergroth, and *Rhodnius* Stål (Barreto 1922; Baptist 1941; Barth 1954).

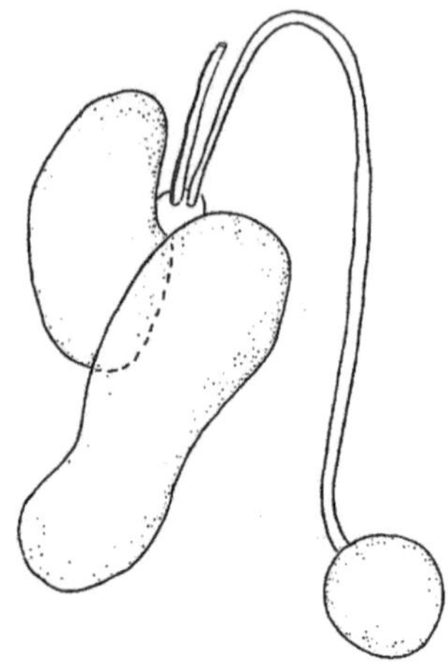

Fig. 13.11 Salivary complex of *Triatoma infestans* (From Santos et al. 1997)

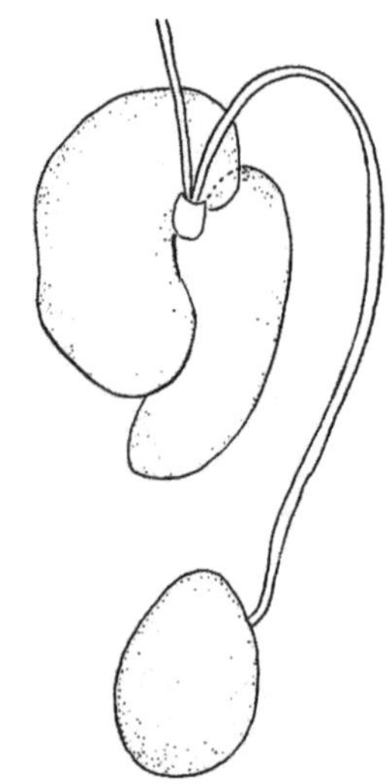

Fig. 13.12 Salivary complex of *Panstrongylus megistus* (From Santos et al. 1997)

13.2.2 Phallic Structure

Opposite to the females, in which the genital plates are simple and extremely homogeneous, the male genital complex of the Triatominae is composed of approximately 15 features, which are highly variable and can be used for generic and specific differentiation (Singh-Pruthi 1925; Dupuis 1955, 1963; Lent and Jurberg 1969, 1975) (Figs. 13.13 and 13.14). This variation was first studied by Lent and Jurberg (1965), who used the comparative analysis of the phallic structures as a tool for the identification of a new species in the genus *Psammolestes* (*P. tertius* Lent & Jurberg), at that time composed by only two species.

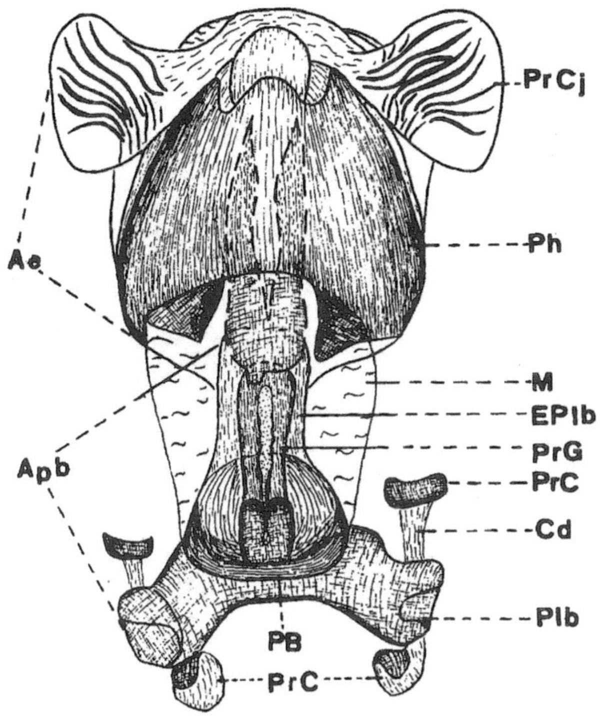

Fig. 13.13 Phallus of *Mepraia spinolai* in ventral view (*Ae* aedeagus, *Apb* articular apparatus, *Cd* dorsal connection, *EPlb* median extension of the basal plate, *M* membrane, *PB* basal bridge, *PrC* capitate process, *PrCj* conjunctive process, *Ph* phallosome, *PrG* process of the gonopore) (From Lent and Jurberg 1967)

Through the continuous use of this technique, phallic structures have been investigated in 16 genera and a great amount of species by the above cited authors, which culminated with the construction of a comparative table of presence/absence (+/–) and shape of the genital structure of the triatomines, which was presented in Carcavallo et al (1998/1999) and herein reproduced on Fig. 13.15. It is important to notice on the board that *Rhodnius stali* Lent, Jurberg & Galvão and *R. pictipes* Stål share some features with species of *Triatoma* that are not present on other species of the genus, which is an issue that needs to be further investigated (Figs. 13.16 and 13.17).

13.2.3 Testicular Tubes

Schreiber et al. (1968) noticed that the triatomines present a fixed number of seven testicular tubes (Fig. 13.18), but with different composition according to the genera, which proportioned taxonomists a new tool for characterizing and confirming

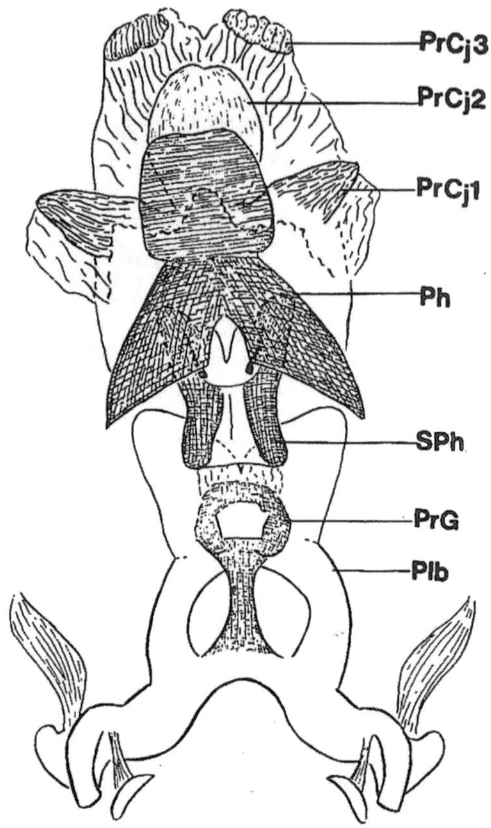

Fig. 13.14 Phallus of *Rhodnius stali* in ventral view (*Ph* phallosome, *PlB* basal plate, *PrCj 1, 2, 3* processes of the conjunctive 1, 2, 3, *SPh* support of the phallosome) (From Lent et al. 1993)

generic statuses. An example is the revalidation by Lent et al. (1994) of the genus *Mepraia*, in which there is a very short tube, two short tubes, two medium tubes, and one long tube. Males of *Rhodnius* present five short and two long; those of *Triatoma* have three short, two medium, and two long; and in *Panstrongylus* there are seven tubes of uniform size.

13.3 General Biology and Ecology

Knowledge of triatomine biology in the natural habitat is scarce, and for several species the natural ecotopes have not been described. Triatominae species are found in almost any habitat offering a degree of climatic stability and access to a blood source. Most of them tolerate a range of air humidity between 30 and 80 %, and temperatures of 24–28 °C are satisfactory. Their development is usually slow at temperatures below 16 °C, whereas temperatures above 40 °C are lethal. During the rigorous winters of Patagonia, they undergo lethargy, and on strong summers they increase the number of blood meals in order to compensate the higher loss of water.

	PB	PrG	EPlb	Ph	PrPh	SPh	PrCj	PrEn	V	PrP	
Cavernicola	–	curto +	curto +	estreito curto	–	curto +	impar apical +	1+1 +	–	+	CAVERNICOLINI
Psammolestes	+	+	+	+	+	–	–	impar pequeno +	–	+	
Rhodnius	+	+	+	+	+	–	vários +	+	–	+	RHODNIINI
R. stali *R. pictipes*	+	+	+	cone +	–	+	+	+	–	duplo +	
Triatoma	+	+	+	+	–	+	–	1+1 +	+	+	
Hermanlentia	+	muito longo +	muito longo +	ápice digitiforme +	–	+	1+1com dentes +	–	–	+	
Mepraia	+	+	+	+	–	–	1+1 aliforme	–	arco +	+	
Eratyrus	+	+	+	+	–	+	–	+	+	+	
Dipetalogaster	+	+	+	+	–	+	–	impar pequeno +	+	+	TRIATOMINI
Panstrongylus	+	+	+	+	–	+	–	1+1 +	+	+	
Paratriatoma	+	+	+	+	–	+	–	1+1 +	+	+	
Linshcosteus	+	+	+	+	–	+	–	1+1 +	+	+	
Bolbodera	–	muito longo +	muito longo +	+	–	+	–	–	–	+	
Belminus	+	muito longo	muito longo	+	–	+	–	–	+	triangular +	
Microtriatoma	+	curto estreito +	curto +	+	–	base+ larga +	1 apical 1+1 +	impar grande +	–	ápice truncado +	BOLBODERINI
Parabelminus	+	curto largo +	curto +	quase inapar +	–	base+ larga +	1+1 apical ventral +	1+1 +	–	ápice truncado +	
Alberprosenia	–	+	+	+	–	+	–	–	–	+	ALBERPROSENIINI

Fig. 13.15 Comparative scheme of phallic structures on Triatominae genera (*PB* basal bridge, *PrG* process of the gonopore, *EPlb* median extension of the basal plate, *Ph* phallosome, *PrPh* process of the phallosome, *SPh* support of the phallosome, *PrCj* conjunctive process, *PrEn* process of the endosoma, *V* vesica, *PrP* process of the pygophore) (From Carcavallo et al. 1998/1999)

In general, most triatomines are nocturnal and during the day they remain in their resting places, although they may sometimes go out to suck blood during the day under adverse conditions. In colonies reared in the laboratory, the bugs seek food sources in broad daylight.

Most species of the subfamily are sylvatic, living in bird nests or animal dens, under tree bark, or in tree hollows, bromeliads, palm trees, and other ecotopes, feeding on various animals. Gaunt and Miles (2000) summarized the habitats of representatives of the triatomine genera *Rhodnius* and *Triatoma* showing that most species in the former genus live in or associated with palm trees, while most species in the latter genus live in or associated with rocky/terrestrial habitats. However, dur-

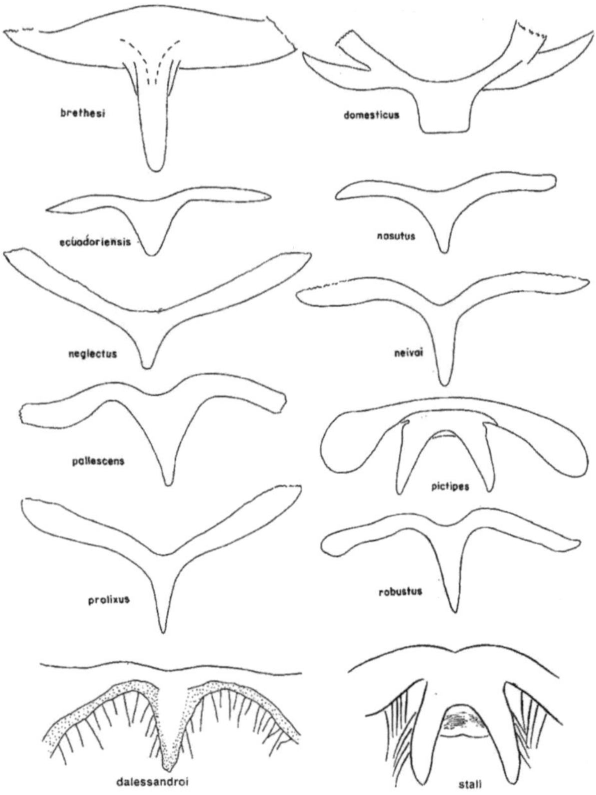

Fig. 13.16 Differences between the median processes of the pygophores of 11 species of the genus *Rhodnius* (From Carcavallo et al. 1998/1999)

ing their evolutionary process some species acquired the capacity to colonize man-made structures near human dwellings, like chicken coops and pigsties, and are thus referred to as peridomiciliated. Others can colonize the interior of human dwellings and are thus called domiciliated. The latter are epidemiologically more important, since they account for transmission of the disease to humans (Carcavallo et al. 1998/1999). Many rural inhabitants live in dwellings with straw roofs and mud walls, and these provide various forms of shelter for the triatomines (Fig. 13.19).

13.3.1 Locomotory Activity

According to Browne (1975), there are two forms of locomotory activity, one spontaneous (circadian), apparently without interference from external stimuli, and another oriented by external stimuli. Triatomines display various processes with temporal modulation, and spontaneous locomotor activity is one of these processes

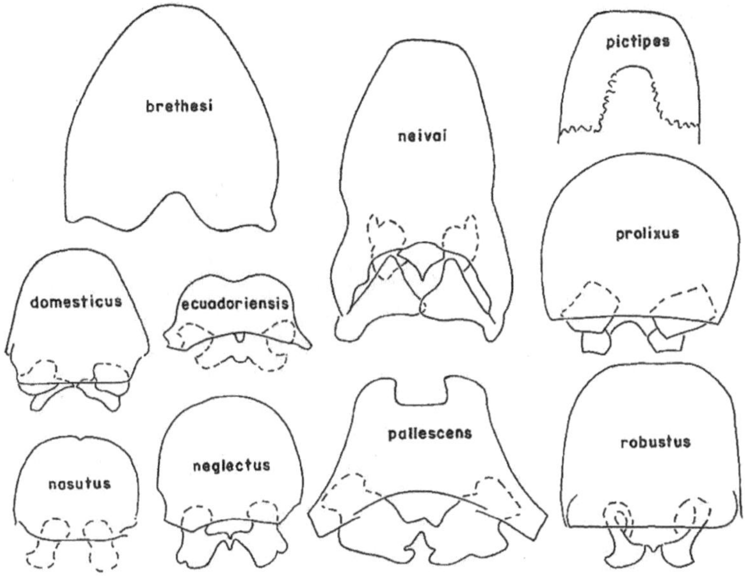

Fig. 13.17 Differences between the phallosomes of ten species of the genus *Rhodnius* (From Carcavallo et al. 1998/1999)

that can be observed individually and can be considered a true circadian rhythm. Various authors used actographic records to demonstrate that locomotor activity in both adults and nymphs is intensified during the early hours of the nocturnal period (Schofield 1976; Núñez 1982; Settembrini 1984; Lazzari 1992).

According to Lazzari (1992), the circadian rhythm of spontaneous locomotor activity is divided into two well-demarcated moments: the search for food occurs in late afternoon/dusk and the search for shelter occurs at dawn. This hypothesis was tested subsequently by Lorenzo and Lazzari (1998), who filmed the locomotor activity of insects in an arena containing refuges and concluded that *Triatoma infestans* demonstrated greater activity and motivation to feed in the early hours of the evening than at the end of this period.

13.3.2 Aggregation

Aggregative behavior in triatomines is mediated principally by the response to the presence of chemical signals (Velázquez-Antich 1968). Studies performed with *T. infestans* and *Rhodnius prolixus* demonstrated the existence of gregarious behavior as a response to volatile substances found in the feces (Schofield and Patterson 1977; Ondarza et al. 1986). Aggregation mediated by chemical substances contained in feces was recently demonstrated by Pires et al. (2002) in *Panstrongylus megistus* and by Vitta et al. (2002) in *Triatoma pseudomaculata* Corrêa & Espínola.

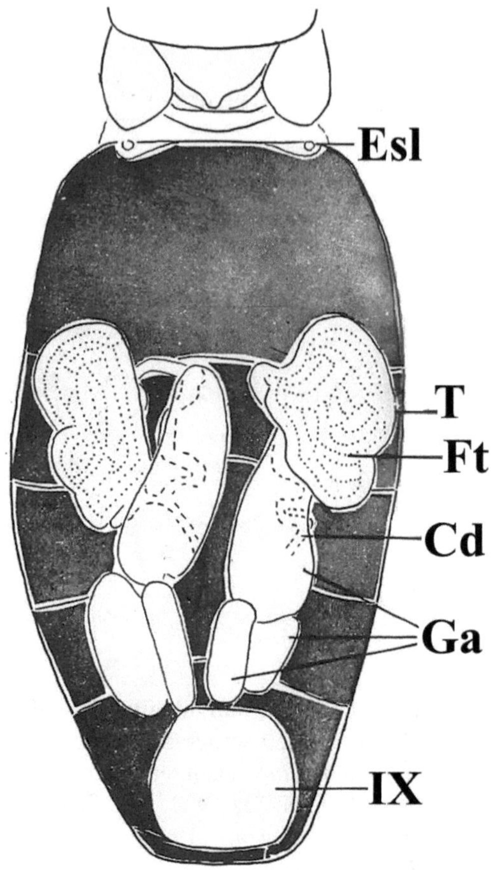

Fig. 13.18 Dissected abdomen of a male *Mepraia spinolai* showing structure of the sexual organs (*Cd* deferent canal, *Esl* spiracle, *Ft* testicular follicles, *Ga* accessory gland, *IX* pygophore or ninth abdominal segment, *T* testicle) (From Gonçalves et al. 1987)

Despite the various attempts of analysis, the nature of the chemical signal present in feces is still not well known, but it is known that feces act as signalers for the insects' refuges. These results show the need for future research aimed at the characterization of an interspecific gregarious chemical compound, in order to aid the control and monitoring of triatomine populations.

13.3.3 Camouflage

The ability of *Triatoma dimidiata* nymphs to camouflage themselves with particles of dust from the soil was described in detail by Hase (1940). Zeledón et al. (1969) observed the same species in both the field and the laboratory and called attention to this behavior's epidemiological importance. According to Zeledón et al. (1973), this behavior is present in various triatomine species, but to variable degrees, and it appears to be completely absent in others. One can clearly establish a correlation between the habitats where some species live (in contact with as opposed to distant

Fig. 13.19 Rural dwelling subject to infestation by triatomine bugs (From Carcavallo et al. 1998/1999)

from the ground) and the presence or absence of this behavior. This behavior's epidemiological implications deserve further investigation, since it interferes directly with the efficiency of control measures, and prior knowledge of the species presenting this behavior is necessary in order to develop entomological surveillance methodologies.

13.3.4 Dispersion

Dispersion in triatomines can occur passively, i.e., involuntarily by humans, or actively, through the adult insects' flight (Galvão et al. 2001b). Knowledge is still limited concerning the mechanisms involved in dispersion by flight. The insects apparently respond directly to external conditions, but not to an internal clock. This is an extremely relevant aspect because areas that have been chemically treated and are free of triatomine foci can be recolonized by flying specimens. In *Triatoma infestans*, the mean flight distance is about 200 m (Schofield and Matthews 1985), but flights of more than 1 km have been observed in the field (Schweigmann et al. 1988). Triatomine flight capacity has been studied both in the field and the laboratory primarily based on releasing and recapturing the insects and has furnished important information on the most flight-capable species. However, no experiments have been done so far focusing on observations of flight behavior itself (Lehane and Schofield 1981; Schofield et al. 1991, 1992; Galvão et al. 2001b).

Fig. 13.20 Triatomine eggs: agglutinated eggs of *Rhodnius domesticus* (From Jurberg and Galvão 2006)

Fig. 13.21 Triatomine eggs: free eggs of *Triatoma brasiliensis* (From Jurberg and Galvão 2006)

13.3.5 Eggs and Oviposition

In general, the eggs of the majority of triatomine species are deposited free in the environment, although some species have an adhesive substance that makes the eggs stick to the substrate (Figs. 13.20 and 13.21). *Triatoma infestans*, a domiciliated species that originally inhabited rodent nests, is still found in the Bolivian altiplano and lays its eggs loose in the sites it inhabits. *Triatoma platensis* Neiva and *T. delpontei* Romaña & Abalos, ornithophilous species, stick their eggs to the substrate of the bird nests in which they live. Species of *Rhodnius* also stick their eggs to the substrate. *Rhodnius prolixus*, which inhabits the nest of *Mycteria americana* L., a migratory bird, can have its eggs spread when they are adhered to the bird's feathers. *Psammolestes arthuri* (Pinto), on its turn, oviposits in clusters, like members of the reduviid subfamilies Harpactorinae and Apiomerinae (Lent and Wygodzinsky 1979).

Oviposition normally occurs 10–30 days after copulation, which can be repeated several times during the adult's lifespan, although a female fecundated just once remains permanently fertilized. The number of eggs varies by species and principally as a function the female's degree of feeding. Recently laid eggs are whitish or pearly in color and change color over time until they become pinkish or brownish, near the day of hatching. During this phase, with the eggs transparent, one can see the black ocular spots, as the 1st stage nymphs emerge from the egg leaving the exuviae appearing (protonymph).

13.3.6 Feeding and Development

Triatomine bugs are obligatory hematophagous insects in all stages of their development and in both sexes, requiring numerous blood meals to complete their development. The amount of blood ingested varies according to the species as well as in relation to the life stage, and generally the 4th and 5th instar nymphs are the ones that feed the most. When they finish a meal, their body changes appearance, since the volume of blood ingested is so great that the abdomen dilates considerably, giving them a globose appearance by stretching the intersegmental and connexival membranes.

The vast majority of species feed on the blood of mammals or birds, but some can feed on reptile or amphibian blood (Carcavallo et al. 1998/1999). Coprophagy (ingestion of feces), kleptohematophagy (sucking blood already ingested by another triatomine), and hemolymphagy (sucking hemolymph from other arthropods) have also been reported by various authors (Lafont 1912; Brumpt 1914; Lent and Martins 1940; Wood 1941; Ryckman 1951; Sandoval et al. 2000). Lorosa et al. (2000) and Ruas-Neto et al. (2001) showed that hemolymphagy can be an important survival strategy in nature, observing *T. circummaculata* (Stål) and *T. rubrovaria* (Blanchard) sucking blattids (cockroaches) in natural rock piles. Laboratory experiments by these authors demonstrated that the biological cycle of both species can be completed exclusively with hemolymph from these insects; however, after the imaginal molt, the males and females of both species could not survive on hemolymph. According to some authors, this aspect could be correlated with studies on the evolution of these two species, which still show this ancestral predatory characteristic.

Some triatomines are considered stenophagic, that is, adapted to feeding only on given hosts, but the vast majority have eclectic feeding habits. Carcavallo et al. (1998/1999) published an extensive list correlating each species to its respective food source, which can be identified through techniques like the precipitin test (Siqueira 1960; Freitas et al. 1960). The results, however, may reflect much more the predominance of a given host or hosts in an area than a true food preference, and it is common to detect the presence of blood from various hosts in a single insect, which can lead to different interpretations. Could it mean great species mobility? Host mobility? Absence of preference? Opportunism? These questions can only be answered safely by associating these results with those obtained through ethological studies.

Feeding behavior depends on various types of stimuli in order to induce hematophagy. Various studies have demonstrated that heat and carbon dioxide are the principal stimuli involved in the search for and biting of the host (Botto-Mahan et al. 2002), but heat appears to act only on the search for food and does not interfere in the feeding itself. The thermoreceptors are concentrated mainly in the antennae, which perform characteristic movements in the presence of a heat source, and bilateral antennectomy results in the impossibility of locating a heat source. According to Lazzari and Núñez (1989), it is possible to make *Triatoma infestans* nymphs suck cold blood, as long as the bite is induced by thermal stimulation of the antennae. Other types of observations, aimed to characterize the species with greatest vector potentials, are to check the number of bites per meals, duration of the meal, and defecation site. These aspects are highly epidemiologically relevant, because with more contacts between vectors and hosts, greater is the probability of infection or transmission of *Trypanosoma cruzi* (Rocha et al. 1997).

Resistance to long periods of fasting and the fact that many species are generalists favor their survival in nature. This capacity of the triatomines has been known for decades. In 1926 Uribe reported the survival of a 3rd instar nymph of *Rhodnius prolixus* for 5 months. The literature shows discrepancies in the survival periods of various species studied according to the methodology employed, which can show variation in the feeding sources, relative humidity, and temperature, as well as the stress the insect suffers during handling (Wood 1951; Friend and Smith 1977; Mascarenhas 1990; Galvão et al. 2001a; Dias-Lima and Sherlock 2002, Martínez-Ibarra et al. 2003). Resistance to fasting can vary between and within species. Among the different stages, the 4th and 5th are normally most resistant, because of their higher capacity to ingest blood. Various authors have focused their laboratory studies on the resistance period to fasting among the various species, including Galvão et al. (1996, 1999a) and Jurberg and Costa (1989a, b).

In the domestic environment, these insects generally suck the blood of their victims at night while they are asleep, and the bites generally occur on the uncovered parts of the body. That is why the most popular name for the triatomine bugs in Brazil is *barbeiro* (meaning barber in Portuguese; for a list of vernacular names, see Table 13.1). The sleeping individuals are unaware of the bugs, because the bites are generally painless due to the anesthetic and anticlotting action of the saliva, although rare cases of hypersensitivity can occur. Some species may defecate while feeding, while others defecate soon afterwards or even abandon the food source and defecate far from the sucking site, a fact characterizes them as either good or bad transmitters of the Chagas disease.

The triatomines are parasites with slow development, whose life cycle from egg to adult includes five nymphal stages (Fig. 13.22), ranging from 3 to 4 months in *Rhodnius prolixus* up to 2 years in *Panstrongylus megistus*. This life span can vary considerably according to the temperature to which the bugs are submitted, a fact that was early studied by Neiva (1913), who observed that high temperatures accelerated the embryonic period of *Triatoma infestans*. Carcavallo and Martínez (1972) obtained shorter cycles in specimens of three species of *Triatoma* reared at high temperatures as compared to those reared at variable temperatures. Silva (1985,

Table 13.1 Examples of vernacular names for Triatominae

Region	Names
USA	Kissing bug, conenose bug, big bedbug (note that "assassin bug" refers to predatory Reduviidae)
	China bug (refers to *T. protracta* on the Pacific coast, once assumed to come from the orient)
	Red-banded conenose (refers to *T. rubrofasciata* and/or *T. sanguisuga*)
Mexico	Chinche besucona (Spanish—"kissing bug")
	Chinche hosicona [Spanish—"trunked bug" (trunked, as in elephant)]
	Chinche picuda (Spanish—"biting bug")
	Chinchona (Spanish—"big bug")
	Pik (Nahuatl, probably refers to the sound made when flying bug strikes wall)
Belize	Bush chinche (implies the absence of domestic Triatominae in Belize)
Central America	Chinche besucona (or just chinche; note that "telepate" more usually refers to cimicid bedbugs)
	Talaje (Spanish—"cutting bug," but mainly applied to cimicid bedbugs)
	Chuluyu (NE Honduras—usually applied specifically to *R. prolixus*, probably derives from *chuzu*, "needle" or "spear,"+*uyu*, "alive")
	Polvoso (Spanish—"dusty"; in NE Honduras usually applied specifically to *T. dimidiata*, probably refers to the habit of nymphs to cover themselves with dust)
	Chinche bebe sangre (Costa Rica, Spanish—"blood-drinking bug")
Cuba	Sangrejuela (Spanish—"bloodstealer," also refers to leeches)
Colombia	Pito (Spanish—"whistle" or "horn," but may also refer to the bug's extended mouthparts)
	Chinche picuda
Venezuela	Chipo (colloquial Spanish—"little bug")
	Îipi (Macuxi Indians; refers specifically to *T. maculata*)
Ecuador	Chinchorro (Spanish+Quechua—"large bug")
Peru	Chirimacha (Quechua—"bug that dislikes the cold" or "drunk with the cold")
Paraguay	Chichá guazú (Guarani, probably derived from Spanish *chinche*, "bug,"+Guarani *guazú*, "big")
	Itchajuponja (Ayoreo, probably derived from Spanish *chinche*, "bug,"+*jupona*, "sucker")
	Sham bui tá (Aché—"insect that does harm by its dejections" mainly used for blister beetles)
	Timbucú (in the Guarani of the Chiriguano and Tapieté groups, signifies "long beak")
Bolivia	Vinchuca (Quechua—"bug that lets itself fall")
	Uluchi (Quechua—"bug without wings"; refers to nymphal stages)
	Timbucú
Chile	Vinchuca
Argentina	Vinchuca
Uruguay	Vinchuca

(continued)

Table 13.1 (continued)

Region	Names
Brazil	Barbeiro (Portuguese—"barber" or "shaver")
	Furão (Portuguese—"big piercing bug")
	Chupão (Portuguese—"big sucking bug")
	Bicudo (Portuguese—"beaked bug")
	Fincão [Portuguese (Brazil-RS)]—"big piercing bug")
	Cascudo [Portuguese (Brazil-MG)—"thick-skinned bug," used mainly for nymphs]
	Chupança [Portuguese (NE Brazil)—"sucking bug"]
	Procotó (or brocotó, borocotó) [Portuguese (NE Brazil)—"bug that hides in cracks"]
	Gigolô [Portuguese (NE Brazil)—"exploiter, esp. of women"]
	Percevejo (or bicho) de parede [Portuguese (NE Brazil)—"wall bedbug"]
	Gaudério (or bicho gaudério) [Portuguese (Brazil-PE)—"indigent thief"] rondão [Portuguese (NE Brazil)—"big bug that observes from hiding"]
	Percevejão (Portuguese—"big bedbug")
	Percevejo do sertão (Portuguese—"bedbug from Sertão," sertão = interior of Brazil)
	Percevejo das pedras (Brazil-RGS) (Portuguese—"bedbug among the stones"; *T. rubrovaria*)
	Piolho de piassava (Brazil-Amazon region) (Portuguese—"louse from the piassaba palm"; *R. brethesi*)
	Vunvum (Portuguese—probably onomatopoeic for the sound of bug flight)
	Josipak [Matacos Indians (Roraima)]
	Îipi [Macuxi Indians (Roraima and Venezuela); refers specifically to *T. maculata*]

From Schofield and Galvão (2009)

1988, 1989a, b, c, 1990a, b, 1992), Silva and Silva (1988a, b, c, d, e, 1989, 1990a, b, 1991, 1993), and Silva et al. (1995) compared the developmental times of several species reared at 25 and 30 °C, showing a reduction of approximately 30 days in three species of *Rhodnius* and from 40 to 60 days in *Triatoma* species. Various other authors have demonstrated acceleration in the developmental period of different species submitted to increased temperature (Galvão et al. 1995, 1999b; Rocha et al. 1994, 2001a, b).

According to most authors, when associated with low relative humidity, the cycle is shortened by metabolic alteration and dehydration, increasing the number of meals to balance the energy budget and water loss. These results support the hypothesis that higher temperatures and lower relative humidity, as possible consequences of global warming, could accelerate the life cycle of triatomines. The result is a change in the population dynamics of some Chagas disease vectors, extending their geographic distribution towards more temperate regions as well increasing the density of some populations.

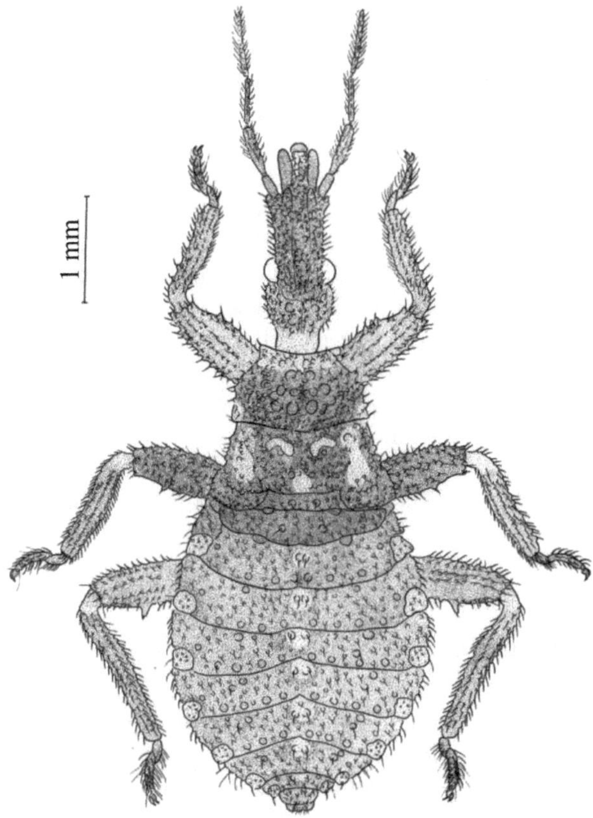

Fig. 13.22 Fourth instar nymph of *Belminus herreri* (From Rocha et al. 2002)

The longevity of nymphs and adults varies by species and ambient conditions. Laboratory experiments generally use what are considered ideal conditions for most species, with a mean temperature close to 28 °C and relative humidity around 70 %, blood feeding on adequate sources for each species, and a photoperiod of 12 h. Of course in nature the insect is influenced by various factors that are difficult to control in the laboratory. Results obtained in the laboratory are thus merely an approximation of what occurs in nature, but knowledge of the biological cycles and population dynamics allows an estimation of the species' growth and colonization capacity, principally for anthropophilic species, but also for so-called secondary species with a tendency towards domiciliation (Carcavallo et al. 1998/1999).

Laboratory Rearing

Important information on triatomine biology has been obtained through rearing and observation in the laboratory and it is not difficult to establish and maintain colonies of these insects. Some appropriate requirements include the control of air temperature and relative humidity. In the LNIRTT, Oswaldo Cruz Institute, Rio de Janeiro, colonies are maintained in glass crystallizers (20 cm high by 20 cm in diameter), covered with a reduced-mesh nylon screen to avoid the escape of 1st instar nymphs and the entry of predators (microhymenoptera, ants, and spiders). Strips of filter paper are placed inside the crystallizers to help absorb the humidity and increase the circulating area. A wooden stand placed inside each crystallizer serves as a support for the hosts (pigeons and mice), which are anesthetized and immobilized before being offered as the food source. Black paper strips are placed outside each crystallizer to limit the amount of light striking the recipient.

Success in maintaining the colonies depends on adherence to the abovementioned items, taking into consideration that offering the same food source repeatedly for long periods of time leads to deficiencies in the insect's development. When cleaning the recipients, the strips of filter paper, which become soiled with the insects' feces, are changed so the environment does not become overloaded, but some strips impregnated with feces are left in the crystallizers to allow recently hatched nymphs to have contact with their natural digestive tract symbionts, which aid in digesting blood.

13.3.7 Reproduction

The first information on copulation in triatomines was published by Neiva (1914), who noted that *Panstrongylus megistus* females appeared to copulate only once, maintaining the eggs fertile throughout their lifespan. Courting is not complex in this group of insects, and copulation in some species has been observed in the laboratory. Prior to copulation, the male approaches the female attempting to immobilize her with the three legs on one side of the body in a dorsolateral position (Fig. 13.23) (Abalos and Wygodzinsky 1951; Hack and Bar 1979; Lent and Wygodzinsky 1979; Lima et al. 1986; Rojas et al. 1990; Manrique and Lazzari 1994). Copulation itself can last 15–30 min, and the male fertilizes the female by depositing spermatophores in her vagina. The spermatophores then burst and release the spermatozoa that migrate to the spermathecae, where they remain protected while awaiting the passage of the successive ova to fertilize.

The presence of sexual pheromones in triatomines has been the target of research for many years, and evidence of chemical attraction between males and females was found in *Rhodnius prolixus* by Velazquez-Antich (1968) and in *Triatoma infestans* and *Panstrongylus megistus* by Neves and Paulini (1981). According to Baldwin et al. (1971), sexual pheromones are released during copulation, leading to an aggregation of males around the couple. Similar conclusions were obtained by Manrique and Lazzari (1994), studying *Triatoma infestans*. On the other hand, the complete absence of sexual attraction was demonstrated by Schofield and Moreman (1979), Hack and Bar (1979), and Lima and MacCord (1994), which can probably be explained by the differing methodologies applied.

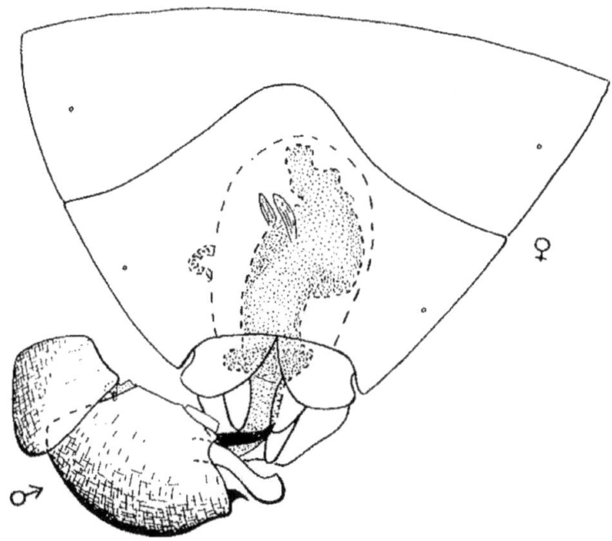

Fig. 13.23 Last abdominal segments of a male and a female of *Rhodnius prolixus* during copulation (From Carcavallo et al. 1998/1999)

13.4 Classification and Diversity

The subfamily Triatominae includes 146 extant species (Galvão et al. 2003; Schofield and Galvão 2009; Rosa et al. 2012; Gonçalves et al. 2013; Jurberg et al. 2013), and its current classification that recognizes five tribes is based mainly on the revision by Lent and Wygodzinsky (1979), the most important systematic study concerning the group. Since that revision, there has been considerable work, including the descriptions of several new taxa. Therefore, a summary of the recent classification and a list of Neotropical species updated from Galvão et al. (2003) and Schofield and Galvão (2009) are herein provided (Tables 13.2 and 13.3). Regarding the identification of the genera and species, extensive dichotomous keys were provided by Lent and Wygodzinsky (1979), Lent et al. (1994), and Carcavallo et al. (1998/1999).

13.4.1 Phylogeny and Evolution

Triatominae has long been classified as a subfamily of the predominantly predatory Reduviidae (Usinger 1943; Maldonado-Capriles 1990; Schuh and Slater 1995). The subfamily is diagnosed, among other features, by the slender and straight labium of its members and their blood-feeding habits (Lent and Wygodzinsky 1979; Clayton 1990; Weirauch 2008). However, phylogenetic relationships of Triatominae are not

Table 13.2 Current systematic classification of the subfamily Triatominae

Subfamily	Tribes	Genera	Number of valid names
Triatominae	Alberproseniini	*Alberprosenia*	2
	Bolboderini	*Belminus*	8
		Bolbodera	1
		Microtriatoma	2
		Parabelminus	2
	Cavernicolini	*Cavernicola*	2
	Rhodniini	*Psammolestes*	3
		Rhodnius	18
	Triatomini	*Dipetalogaster*	1
		Eratyrus	2
		Hermanlentia	1
		Linshcosteus	6
		Meccus	3
		Mepraia	3
		Nesotriatoma	3
		Panstrongylus	14
		Paratriatoma	1
		Triatoma[1]	71

Updated from Galvão et al. (2003) and Schofield and Galvão (2009)
[1]Excluding *T. dominicana*, a fossil species

Table 13.3 Checklist of the Neotropical species of Triatominae, with their geographical distribution

Species and author	Countries or territories
Alberprosenia goyovargasi Martínez & Carcavallo	Venezuela
A. malheiroi Serra, Atzingen & Serra	Brazil
Belminus costaricensis Herrer, Lent & Wygodzinsky	Costa Rica, Mexico
B. herreri Lent & Wygodzinsky	Colombia, Panama
B. laportei Lent, Jurberg & Carcavallo	Brazil
B. peruvianus Herrer, Lent & Wygodzinsky	Peru
B. pittieri Osuna & Ayala	Venezuela
B. rugulosus Stål	Colombia, Venezuela
Bolbodera scabrosa Valdés	Cuba
Microtriatoma borbai Lent & Wygodzinsky	Brazil
M. trinidadensis (Lent)	Brazil, Bolivia, Colombia, Peru, Trinidad and Tobago, Venezuela
Parabelminus carioca Lent	Brazil
P. yurupucu Lent & Wygodzinsky	Brazil
Cavernicola lenti Barrett & Arias	Brazil

(continued)

Table 13.3 (continued)

Species and author	Countries or territories
C. pilosa Barber	Brazil, Colombia, Panama, Peru, Venezuela
Psammolestes arthuri (Pinto)	Colombia, Venezuela
P. coreodes Bergroth	Argentina, Bolivia, Brazil, Paraguay
P. tertius Lent & Jurberg	Brazil
Rhodnius amazonicus Almeida, Santos & Sposina	Brazil, French Guiana
Rhodnius brethesi Matta	Brazil, Colombia, Venezuela
R. colombiensis Mejia, Galvão & Jurberg	Colombia
R. dalessandroi Carcavallo & Barreto	Colombia
R. domesticus Neiva & Pinto	Brazil
R. ecuadoriensis Lent & León	Ecuador, Peru
R. milesi Carcavallo, Rocha, Galvão & Jurberg	Brazil
R. montenegrensis Rosa, Rocha, Gardim, Pinto, Mendonça, Ferreira-Filho, Carvalho, Camargo, Oliveira, Nascimento, Cilense & Almeida	Brazil
R. nasutus Stål	Brazil
R. neglectus Lent	Brazil
R. neivai Lent	Colombia, Venezuela
R. pallescens Barber	Belize, Colombia, Costa Rica, Panama
R. paraensis Sherlock, Guitton & Miles	Brazil
R. pictipes Stål	Belize, Brazil, Colombia, Ecuador, Guyana, French Guiana, Peru, Suriname, Trinidad and Tobago, Venezuela
R. prolixus Stål	Bolivia, Brazil, Colombia, Costa Rica, El Salvador, Ecuador, Guatemala, Guyana, French Guiana, Honduras, Mexico, Nicaragua, Panama, Suriname, Trinidad and Tobago, Venezuela
R. robustus Larrousse	Bolivia, Brazil, Colombia, Ecuador, French Guiana, Peru, Venezuela
R. stali Lent, Jurberg & Galvão	Bolivia, Brazil
R. zeledoni Jurberg, Rocha & Galvão	Brazil
Dipetalogaster maxima (Uhler)	Mexico
Eratyrus cuspidatus Stål	Colombia, Ecuador, Guatemala, Mexico, Panama, Peru, Venezuela
E. mucronatus Stål	Bolivia, Brazil, Colombia, Ecuador, Guatemala, Guyana, French Guiana, Panama, Peru, Suriname, Trinidad and Tobago, Venezuela

(continued)

Table 13.3 (continued)

Species and author	Countries or territories
Hermanlentia matsunoi (Fernández-Loayza)	Peru
Meccus bassolsae (Alejandre Aguilar, Nogueda Torres, Cortéz Jimenez, Jurberg, Galvão & Carcavallo)	Mexico
M. longipennis (Usinger)	Mexico
M. mazzottii (Usinger)	Mexico
M. pallidipennis (Stål)	Mexico
M. phyllosomus (Burmeister)	Mexico
M. picturatus (Usinger)	Mexico
Mepraia gajardoi Frias, Henry & Gonzalez	Chile
M. spinolai (Porter)	Chile
Nesotriatoma bruneri Usinger	Cuba
N. flavida (Neiva)	Cuba
N. obscura Maldonado & Farr	Jamaica
Paratriatoma hirsuta Barber	Mexico, USA
Panstrongylus chinai (Del Ponte)	Ecuador, Peru, Venezuela
P. diasi Pinto & Lent	Bolivia, Brazil
P. geniculatus (Latreille)	Argentina, Bolivia, Brazil, Colombia, Costa Rica, Ecuador, Guatemala, Guyana, French Guiana, Mexico, Nicaragua, Panama, Paraguay, Peru, Suriname, Uruguay, Trinidad and Tobago, Venezuela
P. guentheri Berg	Argentina, Bolivia, Paraguay, Uruguay
P. howardi (Neiva)	Ecuador
P. humeralis (Usinger)	Panama
P. lenti Galvão & Palma	Brazil
P. lignarius (Walker)	Brazil, Peru, Guyana, Suriname, Venezuela
P. lutzi (Neiva & Pinto)	Brazil
P. megistus (Burmeister)	Argentina, Bolivia, Brazil, Paraguay, Uruguay
P. mitarakaensis Bérenger & Blanchet	French Guiana
P. rufotuberculatus (Champion)	Argentina, Bolivia, Brazil, Colombia, Costa Rica, Ecuador, Mexico, Panama, Peru, Venezuela
P. tupynambai Lent	Brazil, Uruguay
T. arthurneivai Lent & Martins	Brazil
T. baratai Carcavallo & Jurberg	Brazil
T. barberi Usinger	Mexico
T. bolivari Carcavallo, Martínez & Pelaez	Mexico
T. boliviana Martínez, Chávez, Sossa, Aranda, Vargas & Vidaurre	Bolivia
T. brailovskyi Martínez, Carcavallo & Pelaez	Mexico

(continued)

Table 13.3 (continued)

Species and author	Countries or territories
T. brasiliensis Neiva	Brazil
T. breyeri Del Ponte	Argentina
T. carcavalloi Jurberg, Rocha & Lent	Brazil
T. carrioni Larrousse	Ecuador, Peru
T. circummaculata (Stål)	Brazil, Uruguay
T. costalimai Verano & Galvão	Brazil
T. deaneorum Galvão, Souza & Lima	Brazil
T. delpontei Romaña & Abalos	Argentina, Bolivia, Brazil, Uruguay, Paraguay
T. dimidiata (Latreille)	Belize, Colombia, Costa Rica, El Salvador, Ecuador, Guatemala, Honduras, Mexico, Nicaragua, Panama, Peru, Venezuela
T. dispar Lent	Colombia, Costa Rica, Ecuador, Panama
T. eratyrusiformis Del Ponte	Argentina
T. garciabesi Carcavallo, Cichero, Martínez, Prosen & Ronderos	Argentina, Bolivia
T. gerstaeckeri (Stål)	USA, Mexico
T. gomeznunezi Martínez, Carcavallo & Jurberg	Mexico
T. guasayana Wygodzinsky & Abalos	Argentina, Bolivia, Paraguay
T. guazu Lent & Wygodzinsky	Brazil
T. hegneri Mazzotti	Mexico
T. incrassata Usinger	USA, Mexico
T. indictiva Neiva	USA, Mexico
T. infestans (Klug)	Argentina, Bolivia, Brazil, Chile, Ecuador, Paraguay, Peru, Uruguay
T. jatai Gonçalves, Teves-Neves, Santos-Mallet, Carbajal-de-la-Fuente & Lopes	Brazil
T. jurbergi Carcavallo, Galvão & Lent	Brazil
T. klugi Carcavallo, Jurberg, Lent & Galvão	Brazil
T. lecticularia (Stål)	USA, Mexico
T. lenti Sherlock & Serafim	Brazil
T. limai Del Ponte	Argentina
T. maculata (Erichson)	Aruba, Brazil, Bonaire, Curaçao, Colombia, Guyana, French Guiana, Suriname, Venezuela
T. matogrossensis Leite & Barbosa	Brazil
T. melanocephala Neiva & Pinto	Brazil
T. mexicana (Herrich-Schaeffer)	Mexico
T. neotomae Neiva	USA, Mexico

(continued)

Table 13.3 (continued)

Species and author	Countries or territories
T. nigromaculata (Stål)	Colombia, Peru, Venezuela
T. nitida Usinger	Costa Rica, Guatemala, Honduras, Mexico
T. oliveirai (Neiva, Pinto & Lent)	Brazil
T. patagonica Del Ponte	Argentina, Uruguay
T. peninsularis Usinger	Mexico
T. petrochiae Pinto & Barreto	Brazil
T. pintodiasi Jurberg, Cunha & Rocha	Brazil
T. platensis Neiva	Argentina, Paraguay, Uruguay
T. protracta (Uhler)	USA, Mexico
T. pseudomaculata Corrêa & Espínola	Brazil
T. recurva (Stål)	USA, Mexico
T. rubida (Uhler)	USA, Mexico
T. rubrofasciata (De Geer)	Andaman Islands, Angola, Argentina, Azores, Bahamas, Brazil, Burma, Cambodia, Caroline Islands, China, Comoros, Congo, Cuba, Dominican Republic, Formosa, French Guiana, Goa, Grenada, Guadeloupe, Haiti, Hawaii, Hong Kong, India, Indonesia (Borneo, Java, Sumatra), Jamaica, Madagascar, Malaysia, Martinica, Mauritius Islands, New Guinea, Okinawa, Philippines, Saint Croix, Saint Vincent, Saudi Arabia, Seychelles, Sierra Leone, Singapore, South Africa, Sri Lanka, Thailand, USA, Venezuela, Vietnam, Zanzibar
T. rubrovaria (Blanchard)	Argentina, Brazil, Uruguay
T. ryckmani Zeledón & Ponce	Costa Rica, Guatemala, Honduras
T. sanguisuga (Leconte)	USA, Mexico
T. sherlocki Papa, Jurberg, Carcavallo, Cerqueira & Barata	Brazil
T. sinaloensis Ryckman	Mexico
T. sordida (Stål)	Argentina, Bolivia, Brazil, Paraguay, Uruguay
T. tibiamaculata (Pinto)	Brazil
T. vandae Carcavallo, Jurberg, Rocha, Galvão, Noireau & Lent	Brazil
T. venosa (Stål)	Bolivia, Colombia, Costa Rica, Ecuador, Peru
T. vitticeps (Stål)	Brazil
T. williami Galvão, Souza & Lima	Brazil
T. wygodzinskyi Lent	Brazil

Updated from Galvão et al. (2003)

resolved, with problems persisting in three areas. First, it is unclear if Triatominae are actually monophyletic, i.e., derived from a single ancestor that gave rise to the entire clade of blood-feeding assassin bugs (Hypsa et al. 2002; Schofield and Galvão 2009; Weirauch and Munro 2009; Patterson and Gaunt 2010; Hwang and Weirauch 2012). Second, different groups of predatory Reduviidae have been proposed to be the closest relatives of the Triatominae. Third, relationships and evolutionary history of clades within the Triatominae are not well understood.

Alternative phylogenetic hypotheses show Triatominae to be monophyletic (Hypsa et al. 2002; Weirauch and Munro 2009; Patterson and Gaunt 2010), paraphyletic (Hwang and Weirauch 2012), or polyphyletic (de Paula et al. 2005; Schofield and Galvão 2009). This directly impacts our understanding of the number of times blood feeding evolved within the Reduviidae, as well as the ecological circumstances under which this transition occurred. Triatominae were considered a natural group or clade until Schofield (1994) suggested that they may in fact represent a polyphyletic assemblage of multiple blood-feeding lineages derived independently from predatory ancestors. This hypothesis stimulated a number of phylogenetic studies that attempted to clarify origin and relationships of Triatominae. Several analyses were hampered by poor sampling of in-group (Triatominae) and/or out-group (remaining Reduviidae) taxa or were based on insufficient character data, e.g., Weirauch (2008; 162 morphological characters; only four spp. of Triatominae), Weirauch and Munro (2009; 3,300 bp of 16S, 28S, and 18S rDNA; only five spp. of Triatominae), Hypsa et al. (2002; 57 spp. of Triatominae, but mostly based on ~500 bp of 16S rDNA), or de Paula et al. (2005; 57 of Triatominae and 15 out-group spp.; 16S rDNA only).

Two recent analyses are derived from more substantial datasets: Patterson and Gaunt (2010) analyzed one ribosomal nuclear gene (D2 region of 28S rDNA) and four protein-coding mitochondrial loci (~2,700 bp), 45 out-group taxa, and 13 species of Triatominae representing the tribes Bolboderini, Triatomini, and Rhodniini. Despite a substantial amount of missing data, their analyses showed strong support for the monophyly of Triatominae. The dataset of Hwang and Weirauch (2012) comprised 178 spp. of Reduviidae and out-groups and included 13 spp. of Triatominae (Cavernicolini, Rhodniini, and Triatomini) and five molecular markers (rDNA: 16S, 18S, 28S D2, 28S D3–5; nuclear protein coding: wingless; ~3,700 bp). Triatominae were rendered paraphyletic by the reduviine genus *Opisthacidius* Berg that formed the sister group to the *Rhodnius*+*Cavernicola* clade. These current conflicting hypotheses ask for further testing using comprehensive molecular and morphological datasets and should aim on including representatives of all five triatomine tribes.

Even with the question of triatomine monophyly unresolved, recent analyses converge on placing kissing bugs in close relationship with a small number of predatory Reduviidae. Earlier authors proposed either an assemblage comprising Salyavatinae, Sphaeridopinae, and Reduviinae (Usinger 1943), Physoderinae (Lent and Wygodzinsky 1979), Stenopodainae (Clayton 1990; some trees only), Harpactorinae or Reduviinae (Hypsa et al. 2002), or Harpactorinae, Ectrichodiinae, Reduviinae, and Salyavatinae (de Paula et al. 2005) as closest relatives of the Triatominae. Subsequent analyses, based on molecular and morphological data sets,

Fig. 13.24 Lateral view of *Linshcosteus karupus*

found a sister group relationship of large-bodied Reduviinae (different combinations of the genera *Zelurus* Hahn, *Centrogonus* Bergroth, and *Opisthacidius*) with the Triatominae, with the Stenopodainae forming the sister group to that clade (Weirauch 2008; Weirauch and Munro 2009; Patterson and Gaunt 2010; Hwang and Weirauch 2012).

Given that the larger picture of triatomine relationships is only beginning to emerge, it is not surprising that lower-level relationships among kissing bugs are largely obscure. It is unclear if the five tribes are monophyletic and even the monophyly of the genera *Triatoma* and *Rhodnius* is not well established (e.g., Lent and Wygodzinsky 1979; Garcia and Powell 1998; Sainz et al. 2004; de Paula et al. 2005). The Oriental distribution of the *T. rubrofasciata* complex and the morphologically distinctive, genus *Linshcosteus* Distant that is classified within the Triatomini (Hypsa et al. 2002) has long puzzled researchers. Patterson and Gaunt (2010) found (based on a fixed molecular clock model) that Triatominae originated around 107 million years ago, before the Gondwanan breakup. Under this scenario, the disjunct distribution of Triatominae could be attributed to vicariance. Based on a (generally thought to be more reliable) relaxed clock model using fossil calibration points, Hwang and Weirauch (2012) showed Triatominae to be much younger (~32 million years), making dispersal from South America the currently better supported hypothesis.

13.4.2 Geographical Distribution

The vast majority of the triatomines is found only in the New World, although a few species occur in East Asia and the coast of Australia. In the Neotropical and Nearctic regions, species are found between 42°N and 46°S. One species, *Triatoma rubrofasciata*, is tropicopolitan, and additional seven species of *Triatoma* occur in Southern and Southeastern Asia and Northern Australia. *Linshcosteus* (Fig. 13.24) is the only genus restricted to the Old World, specifically to the Indian subcontinent (Galvão et al. 2003). Carcavallo et al. (1998/1999) provided several maps showing the geographical distribution and altitudinal/latitudinal dispersion of all American species of Triatominae.

13.5 Concluding Remarks

More than a century after the discovery of the medical importance of the triatomines, there are still millions of people infected or living at risk of contracting Chagas disease, of which they are vectors. In addition to the classical vectorial transmission of the disease by the direct contact with the contaminated feces of these insects due to their hematophagous habits, alternative ways of transmission emerge as serious public health issues. In some regions of northern Brazil, for example, the ingestion of food contaminated by infected triatomines, especially the locally appreciated *açaí* pulp, is responsible for outbreaks of acute Chagas disease (Nóbrega et al. 2009). Also, due to massive human migration, the disease has been more often reported from non-endemic areas, mainly due to blood transfusions or organ transplants from infected, but not symptomatic, immigrants from Latin America (Kirchhoff 1993). Considering these facts, it becomes clear that studies concerning the taxonomy, evolution, biology, and ecology of triatomine bugs are extremely urgent in order to support the control of a highly neglected disease.

References

Abalos JW, Wygodzinsky P (1951) Las Triatominae Argentinas (Reduviidae, Hemiptera). Publ Inst Med Reg 601:1–179

Almeida CE, Vinhaes MC, Almeida JR, Silveira AC, Costa J (2000) Monitoring the domiciliary and peridomiciliary invasion process of *Triatoma rubrovaria* in the state of Rio Grande do Sul, Brazil. Mem Inst Oswaldo Cruz 95:761–768

Baldwin WF, Knigth AG, Lynn KR (1971) A sex pheromone in the insect *Rhodnius prolixus* (Hemiptera: Reduviidae). Can Entomol 103:18–22

Baptist BA (1941) The morphology and physiology of the salivary glands of Hemiptera, Heteroptera. Q J Microsc Sci 83:91–139

Barreto AL (1922) Estudos sobre anatomia do gênero *Triatoma*. Ap. salivar. Mem Inst Oswaldo Cruz 15:127–130

Barth R (1954) Estudos anatômicos e histológicos sobre a sub-família Triatominae (Heteroptera, Reduviidae) IV. O complexo das glândulas salivares de *Triatoma infestans*. Mem Inst Oswaldo Cruz 52:517–585

Botto-Mahan C, Cattan PE, Canals M (2002) Field tests of carbon dioxide and conspecifics as baits for *Mepraia spinolai*, wild vector of Chagas disease. Acta Trop 82:377–380

Browne ELB (1975) Regulatory mechanisms in insect feeding. Adv Insect Physiol 11:1–116

Brumpt E (1914) Importance du cannibalisme et coprophagie chez les Réduvidés hématophàges (*Rhodnius* et *Triatoma*) pour la conservation des trypanosomas pathogenes en dehors de l'hote vertebré. Bull Soc Pathol Exot 7:702–705

Carcavallo RU, Martínez A (1972) Life cycles of some species of *Triatoma* (Hemiptera. Reduviidae). Can Entomol 104:699–704

Carcavallo RU, Galíndez-Girón I, Jurberg J, Lent H (1998/1999) Atlas of Chagas disease vectors in the Americas. Editora Fiocruz, Rio de Janeiro

Chagas C (1909) Nova Tripanozomíaze humana. Estudos sôbre a morfologia e o ciclo evolutivo do *Schizotrypanum cruzi* n. gen., n. sp.; agente etiolójico de nova entidade mórbida do homem. Mem Inst Oswaldo Cruz 1:159–218

Chagas Fo C (1968) Histórico sobre a doença de Chagas. In: Cançado R (ed) Doença de Chagas. Editora de Cultura Médica, Belo Horizonte, pp 5–21

Clayton RA (1990) A phylogenetic analysis of the Reduviidae (Hemiptera: Heteroptera) with redescription of the subfamilies and tribes. Doctor's thesis, The George Washington University, USA

Coura JR, Junqueira ACV, Bóia MN, Fernandes O (1999) Chagas disease: from bush to huts and houses. Is it the case of the Brazilian Amazon? Mem Inst Oswaldo Cruz 94(Suppl 1):379–384

Darwin CR (1845) Journal of researches into the natural history and geology of the countries visited during the voyage of H.M.S. Beagle round the world, under the Command of Capt. Fitz Roy, 2nd edn. J. Murray, London

De Geer C (1773) Mémoires pour servir à l'historie des insectes. Pierre Hesselberg, Stockholm

de Paula AS, Diotaiuti L, Schofield CJ (2005) Testing the sister-group relationship of the Rhodniini and Triatomini (Insecta: Hemiptera: Reduviidae: Triatominae). Mol Phylogenet Evol 35:712–718

Dias-Lima AG, Sherlock IA (2002) Resistência ao jejum em diferentes espécies de Triatomíneos (Hemiptera, Reduviidae, Triatominae). Entomol Vect 9:47–70

Dupuis C (1955) Les genitalia des Hemipteres. Mem Mus Nat Hist Nat Ser A Zool 6:183–278

Dupuis C (1963) Progrés récents de l'étude des genitalia des Hétéroptères (etude bibliographique critique). Doctor's thesis, Museum National d'Histoire Naturelle, Paris, France

Freitas JLP, Siqueira AF, Ferreira AO (1960) Investigações epidemiológicas sobre triatomíneos de hábitos domésticos e silvestres com auxílio da reação de precipitinas. Rev Inst Med Trop 2:90–99

Friend WG, Smith JJB (1977) Factors affecting bloodsucking insects. Annu Rev Entomol 22:309–331

Galvão C (2003) A sistemática dos triatomíneos (Hemiptera, Reduviidae) de De Geer ao DNA. Entomol Vect 10:511–530

Galvão C, Jurberg J, Cunha V, Mello RP (1995) Biologia do *Triatoma nitida* Usinger, 1939 em laboratório (Hemiptera, Reduviidae). Mem Inst Oswaldo Cruz 90:657–663

Galvão C, Jurberg J, Lent H (1996) Resistência ao jejum de *Triatoma nitida* Usinger, 1939 em laboratório (Hemiptera, Reduviidae, Triatominae). Mem Inst Oswaldo Cruz 91:639–640

Galvão C, Rocha DS, Jurberg J, Carcavallo RU (1999a) Resistência ao jejum de *Triatoma melanosoma* Martínez, Olmedo & Carcavallo, 1987 (Hemiptera, Reduviidae) em laboratório. Entomol Vect 6:705–779

Galvão C, Rocha DS, Cunha V, Presgrave OAF, Jurberg J, Carcavallo RU (1999b) Influência da temperatura no ciclo de vida de *Triatoma melanosoma* Martínez, Olmedo & Carcavallo, 1987 (Hemiptera, Reduviidae). Mem Inst Oswaldo Cruz 94:851–854

Galvão C, Rocha DS, Cunha V, Jurberg J, Carcavallo RU (2001a) Tempo de alimentação e defecação das ninfas de *Triatoma melanosoma* Martínez, Olmedo & Carcavallo, 1987 (Hemiptera, Reduviidae) em diferentes condições de temperatura e umidade. Rev Bras Zool 18:233–242

Galvão C, Rocha DS, Jurberg J, Carcavallo RU (2001b) Início da atividade de vôo em *Triatoma infestans* (Klug, 1834) e *T. melanosoma* Martínez, Olmedo & Carcavallo, 1987 (Hemiptera, Reduviidae). Mem Inst Oswaldo Cruz 96:137–140

Galvão C, Carcavallo RU, Rocha DS, Jurberg J (2003) A checklist of the current valid species of the subfamily Triatominae Jeannel, 1919 (Hemiptera, Reduviidae) and their geographical distribution, with nomenclatural and taxonomic notes. Zootaxa 202:1–36

Galvão C, McAloon FM, Rocha DS, Schaefer CW, Patterson J, Jurberg J (2005) Description of eggs and nymphs of *Linshcosteus karupus* Galvão, Patterson, Rocha, & Jurberg, 2002 (Hemiptera: Reduviidae: Triatominae). Ann Entomol Soc Am 98:861–872

Garcia BA, Powell JR (1998) Phylogeny of species of *Triatoma* (Hemiptera: Reduviidae) based on mitochondrial DNA sequences. J Med Entomol 35:232–238

Gaunt M, Miles M (2000) The ecotypes and evolution of triatomine bugs (Triatominae) and their associated trypanosomes. Mem Inst Oswaldo Cruz 95:557–565

Gonçalves TCM, Lent H, Almeida JR (1987) Estudo anatômico e morfométrico dos folículos testiculares de algumas espécies de Triatominae (Hemiptera, Reduviidae). Mem Inst Oswaldo Cruz 82:543–555

Gonçalves TCM, Teves-Neves SC, Santos-Mallet JR, Carbajal-de-la-Fuente AL, Lopes CM (2013) *Triatoma jatai* sp. nov. in the State of Tocantins, Brazil (Hemiptera: Reduviidae: Triatominae). Mem Inst Oswaldo Cruz 108:429–437

Guhl F, Lazdins-Helds JK (2007) Reporte sobre la enfermedad de Chagas. World Health Organization, Brussels

Hack WH, Bar ME (1979) Comportamento sexual de *Triatoma infestans* en laboratório. Facena 3:105–117

Hase A (1940) Ueber *Triatoma dimidiata* (Hemiptera, Triatomidae) I Teil. Zeitschr Parasitenk 11:419–429

Hwang WS, Weirauch C (2012) Evolutionary history of Assassin Bugs (Insecta: Hemiptera: Reduviidae): insights from divergence dating and ancestral state reconstruction. PLoS One 7(9), e45523

Hypsa V, Tietz DF, Zrzavý J, Rego ROM, Galvão C, Jurberg J (2002) Phylogeny and biogeography of Triatominae (Hemiptera: Reduviidae): molecular evidence of a New World origin of the Asiatic clade. Mol Phylogenet Evol 23:447–457

Jurberg J (1999) Noventa anos da descoberta da doença de Chagas e a criação do Centro de Referência em Taxonomia de Triatomíneos. Entomol Vect 6:315–322

Jurberg J (2003) Ferramentas usadas em taxonomia de triatomíneos – o uso múltiplo. Entomol Vect 10:497–509

Jurberg J, Costa JM (1989a) Estudos sobre a resistência ao jejum e aspectos nutricionais de *Cavernicola lenti* Barret & Arias, 1985 (Hemiptera, Reduviidae, Triatominae). Mem Inst Oswaldo Cruz 84:129–137

Jurberg J, Costa JM (1989b) Estudos sobre a resistência ao jejum e aspectos nutricionais de *Triatoma lecticularia* (Stål, 1859) (Hemiptera, Reduviidae, Triatominae). Mem Inst Oswaldo Cruz 84:393–399

Jurberg J, Galvão C (2006) Biology, ecology and systematics of Triatominae (Heteroptera, Reduviidae) vectors of Chagas disease, and implications for human health. Denisia 19:1096–1116

Jurberg J, Lent H, Carcavallo RU, Rocha DS, Galvão C, Lasserre DF (2002) Estudo morfológico da genitália externa masculina de *Mepraia gajardoi* Frias, Henry & Gonzalez, 1998 (Hemiptera, Reduviidae), com comentários de suas relações filogenéticas. Entomol Vect 9:559–577

Jurberg J, Galvão C, Noireau F, Carcavallo RU, Rocha DS, Lent H (2004) Uma iconografia dos triatomíneos. Entomol Vect 11:457–494

Jurberg J, Galvão C, Rocha DS, Dale C, Cunha V (2012) Vetores da doença de Chagas no Brasil. Instituto Oswaldo Cruz, Rio de Janeiro

Jurberg J, Cunha V, Cailleaux S, Raigorodschi R, Lima MS, Rocha DS, Moreira FFF (2013) *Triatoma pintodiasi* sp. nov. do subcomplexo *T. rubrovaria* (Hemiptera, Reduviidae, Triatominae). Rev Pan-Amaz Saúde 4:43–56

Jurberg J, Rodrigues JMS, Moreira FFF, Dale C, Cordeiro IRS, Lamas VD Jr, Galvão C, Rocha DS (2014) Atlas iconográfico dos triatomíneos do Brasil (Vetores da doença de Chagas no Brasil). Instituto Oswaldo Cruz, Rio de Janeiro

Kirchhoff LV (1993) American Trypanosomiasis (Chagas' disease) – a tropical disease now in the United States. New Engl J Med Res 329:639–644

Lafont A (1912) Trypanosomide d'un reduvide (*Conorhenus rubrofasciatus*) inoculable au rat et à la sourris. Ann Inst Pasteur 26:893–922

Lazzari CR (1992) Circadian organization of locomotion activity in the haematophagous bug *Triatoma infestans*. J Insect Physiol 38:895–903

Lazzari CR, Núnez JA (1989) Blood temperature and feeding behavior in *Triatoma infestans* (Heteroptera: Reduviidae). Entomol Genet 14:183–188

Lehane MJ, Schofield CJ (1981) Field experiments on dispersive flight by *Triatoma infestans*. Trans R Soc Trop Med Hyg 75:399–400

Lent H, Jurberg J (1965) O gênero *Psammolestes* Bergroth, 1911 com um estudo sobre a genitália das espécies. Rev Brasil Biol 25:349–376

Lent H, Jurberg J (1967) Algumas informações sobre *Triatoma spinolai* Porter, 1934, com um estudo sobre as genitálias externas (Hemiptera, Reduviidae). Rev Brasil Biol 27:273–288

Lent H, Jurberg J (1969) O gênero *Rhodnius* Stål, 1859 com um estudo sobre a genitália das espécies (Hemiptera, Reduviidae, Triat.). Rev Brasil Biol 29:487–560

Lent H, Jurberg J (1975) O gênero *Panstrongylus* Berg, 1879 com um estudo sobre a genitália externa das espécies (Hemiptera, Reduviidae, Triatominae). Rev Brasil Biol 35:379–438

Lent H, Martins AV (1940) Estudo sobre os Triatomideos do Estado de Minas Gerais com descrição de uma espécie nova. Rev Entomol 11:877–886

Lent H, Wygodzinsky P (1979) Revision of Triatominae (Hemiptera, Reduviidae) and their significance as vectors of Chagas disease. Bull Am Mus Nat Hist 163:123–529

Lent H, Jurberg J, Galvão C (1993) *Rhodnius stali* n. sp., afim de *Rhodnius pictipes* Stål, 1872 (Hemiptera, Reduviidae, Triatominae). Mem Inst Oswaldo Cruz 88:605–614

Lent H, Jurberg J, Galvão C (1994) Revalidação do gênero *Mepraia* Mazza, Gajardo, Jörg, 1940 (Hemiptera, Reduviidae, Triatominae). Mem Inst Oswaldo Cruz 89:347–352

Lima MM, MacCord JR (1994) Possible absence of attraction to odor in *Panstrongylus megistus* (Hemiptera: Reduviidae) under laboratory conditions. Mem Inst Oswaldo Cruz 89:271–274

Lima MM, Jurberg P, Almeida JR (1986) Behavior of triatomines (Hemiptera: Reduviidae) vectors of Chagas disease. I Courtship and copulation of *Panstrongylus megistus* (Burm., 1835) in the laboratory. Mem Inst Oswaldo Cruz 81:1–5

Lorenzo MG, Lazzari CR (1998) Activity pattern in relation to refuge exploitation and feeding in *Triatoma infestans* (Hemiptera: Reduviidae). Acta Trop 70:163–170

Lorosa ES, Jurberg J, Souza ALA, Vinhaes MC, Núñes IM (2000) Hemolinfa de Dictyoptera na manutenção do ciclo biológico silvestre de *Triatoma rubrovaria* (Blanchard, 1843) e *Triatoma circummaculata* (Stål, 1859) (Hemiptera, Reduviidae, Triatominae). Entomol Vect 7:287–296

Maldonado-Capriles J (1990) *Systematic catalogue of the Reduviidae of the world (Insecta: Heteroptera)*. University of Puerto Rico, Mayaguëz

Manrique G, Lazzari CR (1994) Sexual behaviour and stridulation during mating in *Triatoma infestans* (Hemiptera: Reduviidae). Mem Inst Oswaldo Cruz 89:629–633

Martínez-Ibarra JA, López MN, Robles MRH, Guillén YG (2003) Influence of the blood meal source on the biology of *Meccus picturatus* Usinger 1939 (Hemiptera: Reduviidae:Triatominae) under laboratory conditions. Mem Inst Oswaldo Cruz 98:227–232

Mascarenhas BM (1990) Triatomíneos da Amazônia. Sobre o ciclo evolutivo de *Rhodnius brethesi* Matta, 1919 (Hemiptera, Reduviidae, Triatominae). Bol Mus Paraense Emílio Goeldi Ser Zool 6:191–202

Neiva A (1913) Informações sobre a biologia da vinchuca *Triatoma infestans* Klug. Mem Inst Oswaldo Cruz 5:24–31

Neiva A (1914) Revisão do gênero Triatoma Lap. Doctor's thesis, Faculdade de Medicina, Rio de Janeiro, RJ, Brasil

Neiva A, Lent H (1936) Notas e commentarios sobre triatomideos. Lista de especies e sua distribuição geographica. Rev Entomol 6:153–190

Neiva A, Lent H (1941) Sinopse dos Triatomideos. Rev Entomol 12:61–92

Neves DP, Paulini E (1981) Atração sexual em *Panstrongylus megistus* e *Triatoma infestans* (Hemiptera, Reduviidae) por feromônio. Rev Bras Entomol 25:301–306

Nóbrega AA, Garcia MH, Tatto E, Obara MT, Costa E, Sobel J, Araujo WN (2009) Oral transmission of Chagas disease by consumption of açaí palm fruit, Brazil. Emerg Insect Dis 15:653–655

Núnez JA (1982) Food source orientation and activity in *Rhodnius prolixus* Stål (Hemiptera: Reduviidae). Bull Entomol Res 72:253–262

Ondarza RN, Gutiérrez-Martínez A, Malo EA (1986) Evidence for the presence of sex and aggregation pheromones from *Triatoma mazzottii* (Hemiptera: Reduviidae). J Econ Entomol 79:688–692

Patterson JS, Gaunt MW (2010) Phylogenetic multi-locus codon models and molecular clocks reveal the monophyly of haematophagous reduviid bugs and their evolution at the formation of South America. Mol Phylogenet Evol 56:608–621

Pinto C (1925) Ensaio monographico dos Reduvideos hematophagos ou "barbeiros". Doctor thesis, Faculdade de Medicina, Rio de Janeiro, Brasil

Pires HHR, Lorenzo MG, Diotaiuti L, Lazzari CR, Lorenzo-Figueiras AN (2002) Aggregation behaviour in *Panstrongylus megistus* and *Triatoma infestans*: inter and intraspecific responses. Acta Trop 81:47–52

Ponte D (1921) Contribución al estudio del género *Triatoma* Laporte. Anatomia interna. Rev Inst Bacteriol Dep Nac Hig 3:133–196

Ponte D (1930) Catálogo descriptivo de los géneros *Triatoma* Lap., *Rhodnius* Stål, e *Eratyrus* Stål. Rev Inst Bacteriol Dep Nac Hig 5:855–937

Ramírez JP (1985) Chipos de Venezuela. J. Ramírez Pérez, Villa de Cura

Rocha DS, Galvão C, Jurberg J (1994) Biologia do *Rhodnius pictipes* Stål, 1872 em condições de laboratório (Hemiptera, Reduviidae, Triatominae). Mem Inst Oswaldo Cruz 89:265–270

Rocha DS, Fonseca AH, Costa FA, Jurberg J, Galvão C (1997) Desenvolvimento de *Rhodnius pictipes* Stål, 1872 alimentado através de membrana de silicone e em camundongos (Hemiptera, Reduviidae, Triatominae). Mem Inst Oswaldo Cruz 92:553–558

Rocha DS, Jurberg J, Carcavallo RU, Presgrave OAF, Cunha V, Galvão C (2001a) Influência da temperatura e umidade no desenvolvimento ninfal de *Rhodnius robustus*. Rev Saúde Publ 35:400–406

Rocha DS, Jurberg J, Carcavallo RU, Cunha V, Galvão C (2001b) Influência da temperatura e umidade na biologia de *Rhodnius neglectus* Lent, 1954 em laboratório (Hemiptera, Reduviidae, Triatominae). Rev Soc Bras Med Trop 34:357–363

Rocha DS, Patterson JS, Sandoval CM, Jurberg J, Ângulo VM, Esteban EA, Galvão C (2002) Description and ontogenetic morphometrics of nymphs of *Belminus herreri* Lent & Wygodzinsky (Hemiptera: Reduviidae: Triatominae). Neotrop Entomol 34:491–497

Rojas JC, Malo EA, Gutierrez-Martínez A (1990) Mating behavior of *Triatoma mazzottii* Usinger (Hemiptera: Reduviidae) under laboratory conditions. Ann Entomol Soc Am 83:598–602

Rosa JA, Rocha CS, Gardim S, Pinto MC, Mendonça VJ, Ferreira-Filho JCR, Carvalho EOC, Camargo LMA, Oliveira J, Nascimento JD, Cilense M, Almeida CE (2012) Description of *Rhodnius montenegrensis* n. sp. (Hemiptera: Reduviidae: Triatominae) from the State of Rondônia, Brazil. Zootaxa 3478:62–76

Ruas-Neto AL, Corseuil E, Cavalieri A (2001) Development of rupestrian triatomines (Hemiptera: Reduviidae: Triatominae) following hemolymphagy on blaberids (Blattodea: Blaberidae) in Rio Grande do Sul State, Brazil. Entomol Vect 8:205–216

Ryckman RE (1951) Recent observations of cannibalism in *Triatoma* (Hemiptera, Reduviidae). J Parasitol 37:43–437

Ryckman RE (1962) Biosystematics and hosts of the *Triatoma protracta* complex in North America. *Univ.* Calif Publ Entomol 27:93–239

Sainz AC, Mauro LV, Moriyama EN, García BA (2004) Phylogeny of triatomine vectors of *Trypanosoma cruzi* suggested by mitochondrial DNA sequences. Genetica 121:229–240

Sandoval CM, Joya M, Gutierrez R, Angulo VM (2000) Cleptohaemathophagia of the triatome bug *Belminus herreri*. Med Vet Entomol 14:100–101

Sandoval CM, Duarte R, Gutierrez R, Rocha DS, Angulo VM, Esteban L, Reyes M, Jurberg J, Galvão C (2004) Feeding sources and natural infection of *Belminus herreri* (Hemiptera, Reduviidae, Triatominae) from dwellings in Cesar, Colombia. Mem Inst Oswaldo Cruz 99:137–140

Santos CM, Jurberg J, Galvão C, Lent H (1997) Análise morfológica comparativa do complexo salivar dos Triatominae (Hemiptera, Reduviidae). Entomol Vect 4:155–162

Schofield CJ (1976) An actograph for the study of circadian rhythms of locomotory activity of triatomine bugs. Trans R Soc Trop Med Hyg 70:277

Schofield CJ (1994) Triatominae biologia y control. Eurocommunica Publications, Bognor Regis

Schofield CJ, Galvão C (2009) Classification, evolution and species groups within the Triatominae. Acta Trop 110:88–100

Schofield CJ, Matthews JNS (1985) Theoretical approach to active dispersal and colonization of houses by *Triatoma infestans*. J Trop Med Hyg 88:211–222

Schofield CJ, Moreman K (1979) Apparent absence of a sex attractant in adult *Triatoma infestans* (Klug), vector of Chagas' disease. Trans R Soc Trop Med Hyg 70:165–166

Schofield CJ, Patterson JW (1977) Assembly pheromone of *Triatoma infestans* and *Rhodnius prolixus* nymphs (Hemiptera: Reduviidae). J Med Entomol 13:727–734

Schofield CJ, Lehane MJ, Mcewan P, Catalá SS, Gorla DE (1991) Dispersive flight by *Triatoma sordida*. Trans R Soc Trop Med Hyg 85:676–678

Schofield CJ, Lehane MJ, Mcewan P, Catalá SS, Gorla DE (1992) Dispersive flight by *Triatoma infestans* under natural climatic conditions in Argentina. Med Vet Entomol 6:51–56

Schreiber G, Penalva F, Carvalho HC (1968) Morfologia comparada dos túbulos testiculares e sistemática dos Triatominae (Hemiptera, Reduviidae). Cienc Cult 20:640–641

Schuh RT, Slater JA (1995) True bugs of the world (Hemiptera: Heteroptera): classification and natural history. Cornell University Press, Ithaca

Schweigmann N, Vallve S, Muscio O, Ghilini M, Alberti A, Wisnivesly-Colli C (1988) Dispersal flight by *Triatoma infestans* in an arid area of Argentina. Med Vet Entomol 2:401–404

Settembrini BP (1984) Circadian rhythms of locomotion activity in *Triatoma infestans* (Hemiptera, Reduviidae). J Med Entomol 21:204–212

Silva IG (1985) Influência da temperatura na biologia de triatomíneos. I. *Triatoma rubrovaria* (Blanchard, 1843) (Hemiptera, Reduviidae). Rev Goiana Med 31:1–37

Silva IG (1988) Influência da temperatura na biologia de triatomíneos. VII. Rhodnius prolixus Stål, 1859 (Hemiptera, Reduviidae). Rev Trop 17:145–155

Silva IG (1989a) Influência da temperatura na biologia de triatomíneos. VIII – Triatoma matogrossensis Leite & Barbosa, 1953 (Hemiptera, Reduviidae). An Soc Entomol Brasil 18:91–94

Silva IG (1989b) Influência da temperatura na biologia de triatomíneos. XII – *Triatoma platensis* Neiva, 1913 (Hemiptera, Reduviidae). Rev Pat Trop 18:15–18

Silva IG (1989c) Influência da temperatura na biologia de triatomíneos. XIV – *Triatoma protracta* (Uhler, 1894) (Hemiptera, Reduviidae). Rev Pat Trop 18:77–80

Silva IG (1990a) Influência da temperatura na biologia de triatomíneos. XIII. *Dipetalogaster maximus* (Uhler, 1894) (Hemiptera, Reduviidae). An Soc Entomol Brasil 19:111–114

Silva IG (1990b) Influência da temperatura na biologia de triatomíneos. VI. *Triatoma brasiliensis* Neiva, 1911 (Hemiptera, Reduviidae). Rev Bras Entomol 34:307–311

Silva IG (1992) Influência da temperatura na biologia de triatomíneos. XVII. *Triatoma tibiamaculata* (Pinto, 1926) (Hemiptera, Reduviidae). Rev Pat Trop 21:27–31

Silva IG, Silva HHG (1988a) Influência da temperatura na biologia de Triatomíneos. II. *Rhodnius neglectus* Lent, 1954 (Hemiptera, Reduviidae). Rev Goiana Med 34:29–37

Silva IG, Silva HHG (1988b) Influência da temperatura na biologia de Triatomíneos. IV. *Triatoma infestans* (Klug, 1834) (Hemiptera: Reduviidae). An Soc Entomol Brasil 17:443–454

Silva IG, Silva HHG (1988c) Influência da temperatura na biologia de triatomíneos. V. *Triatoma sordida* (Stål, 1859) (Hemiptera, Reduviidae). Rev Pat Trop 17:133–144

Silva IG, Silva HHG (1988d) Influência da temperatura na biologia de triatomíneos. X. *Triatoma vitticeps* Stål, 1859 (Hemiptera, Reduviidae). Rev Goiana Med 34:39–45

Silva IG, Silva HHG (1988e) Influência da temperatura na biologia de Triatomíneos. XI. *Rhodnius robustus* Larrousse, 1927 (Hemiptera, Reduviidae). Rev Goiana Med 34:145–154

Silva IG, Silva HHG (1989) Influência da temperatura na biologia de triatomíneos. IX. Rhodnius nasutus Stål, 1859 (Hemiptera, Reduviidae). Mem Inst Oswaldo Cruz 84:377–382

Silva IG, Silva HHG (1990a) Influência da temperatura na biologia de triatomíneos. XV. *Rhodnius ecuadoriensis* Lent & León, 1958 (Hemiptera, Reduviidae). Rev Goiana Med 36:49–53

Silva IG, Silva HHG (1990b) Influência da temperatura na biologia de Triatomíneos. XIV. Rhodnius pictipes Stål, 1872 (Hemiptera, Reduviidae). Rev Pat Trop 19:151–157

Silva IG, Silva HHG (1991) Influência da temperatura na biologia de triatomíneos. XVI. Panstrongylus herreri Wygodzinsky, 1948 (Hemiptera, Reduviidae). An Soc Entomol Brasil 20:277–281

Silva IG, Silva HHG (1993) Influência da temperatura na biologia de Triatomíneos. III. *Panstrongylus megistus* (Burmeister, 1835) (Hemiptera, Reduviidae). Rev Bras Entomol 37:489–496

Silva IG, Fernandes FF, Silva HHG (1995) Influência da temperatura na biologia de triatomíneos. XX. *Triatoma maculata* (Erichson, 1848) (Hemiptera, Reduviidae). Rev Pat Trop 24:49–64

Singh-Pruthi H (1925) The morphology of the male genitalia in Rhynchota. Trans R Entomol Soc 1:127–267

Siqueira AF (1960) Estudos sobre a reação de precipitina aplicada e identificação de sangue ingerido por triatomíneos. Rev Inst Trop São Paulo 2:41–53

Sousa LC, Galvão C (2004) Colonização do *Panstrongylus lutzi* Neiva & Pinto, 1923 (Hemiptera: Reduviidae: Triatominae) em ecótopos artificiais no Ceará. Rev Soc Bras Med Trop 37:35

Sousa LC, Bastos AJ, Sousa J, Galvão C (2004) Infestação domiciliar por *Panstrongylus lutzi* Neiva & Pinto, 1923 e outras espécies de Triatomíneos (Hemiptera: Reduviidae: Triatominae) na região Noroeste do Ceará. Rev Soc Bras Med Trop 37:181–182

Usinger RL (1943) A revised classification of the Reduvioidea with a new subfamily from South America (Hemiptera). Ann Entomol Soc Am 36:602–618

Usinger RL (1944) The Triatominae of North and Central America and West Indies and their public health significance. US Publ Health Serv Publ Health Bull 288:1–83

Valente VC (1999) Potential for domestication of *Panstrongylus geniculatus* (Latreille, 1811) (Hemiptera, Reduviidae, Triatominae) in the municipality of Muaná, Marajó Island, State of Pará, Brazil. Mem Inst Oswaldo Cruz 94:399–400

Valente VC, Valente SAS, Noireau F (1998) Chagas disease in the Amazon Basin: association of *Panstrongylus geniculatus* (Hemiptera: Reduviidae) with domestic pigs. J Med Entomol 35:99–103

Velázquez Antich A (1968) Atracción por olor en ninfas y adultos de *Rhodnius prolixus*. Rev Inst Med Trop Sao Paulo 10:242–246

Vitta ACR, Lazzari CR, Diotaiuti L, Lorenzo MG (2002) Aggregation mediated by faeces and footprints in *Triatoma pseudomaculata* (Heteroptera: Reduviidae), a Chagas disease vector. Mem Inst Oswaldo Cruz 97:865–867

Vivas AS, Barazarte H, Molina de Fernandez D (2001) Primer registro de *Eratyrus mucronatus* Stål, 1959 (Hemiptera: Reduviidae) en el ambiente domiciliário em Venezuela. Entomotropica 16:215–217

Weirauch C (2008) Cladistic analysis of Reduviidae (Heteroptera: Cimicomorpha) based on morphological characters. Syst Entomol 33:229–274

Weirauch C, Munro JB (2009) Molecular phylogeny of the assassin bugs (Hemiptera: Reduviidae), based on mitochondrial and nuclear ribosomal genes. Mol Phylogenet Evol 53:287–299

Wolff M, Castillo C (2002) Domiciliation trend of *Panstrongylus rufotuberculatus* in Colombia. Mem Inst Oswaldo Cruz 97:297–300

Wood SF (1941) Notes on the distribution and habits of reduviid vectors of Chagas' disease in the Southwestern United States. Pan-Pac Entomol 17:85–94

Wood SF (1951) Importance of feeding and defecation times of insects vectors in transmission of Chagas'disease. J Econ Entomol 44:52–54

Zeledón R, Zúniga A, Swartzelder JC (1969) The camouflage of *Triatoma dimidiata* and the epidemiology of Chagas' disease in Costa Rica. Bol Chileno Parasitol 24:106–108

Zeledón R, Valerio C, Valerio JE (1973) The camouflage phenomenon in several species of Triatominae (Hemiptera: Reduviidae). J Med Entomol 10:209–211

Chapter 14
Lace Bugs (Tingidae)

Marcus Guidoti, Sara I. Montemayor, and Éric Guilbert

Abstract The members of the Tingidae family are commonly known as "lace bugs" because of the lacelike appearance of their pronotum and hemelytra. The family is distributed worldwide and is constituted by three subfamilies, Tinginae, Cantacaderinae, and Vianaidinae; all of them represented in the Neotropical Region being the last one endemic of this region. There are around 300 genera and 2,500 species in this family, but for the Neotropical Region, there are known approximately 70 genera and 600 species; most of them belong to the subfamily Tinginae. All lace bugs are phytophagous and host specific, with a few exceptions. Among them, there are some that have economic importance as crop pests and others that are useful for controlling harmful weeds in field crops. Herein we characterize and diagnose the family, we mention some of the most relevant aspects of their biology and ecology, and we provide a summary of its taxonomy with some comments about its phylogeny. The economically most relevant species of the Neotropical Region are listed; for all of them, we provide information about their host plants, distributions, and related literature. We also developed a key to the most economically important Neotropical genera.

M. Guidoti (✉)
Programa de Pós-graduação em Biologia Animal, Instituto de Biociências, Universidade Federal do Rio Grande do Sul (UFRGS), Av. Bento Gonçalves 9500, prédio 43435, Bairro Agronomia, Porto Alegre, RS 91501-970, Brazil
e-mail: marcus.guidoti@gmail.com

S.I. Montemayor
Divisiõn Entomología, Museo de La Plata,
Paseo del bosque s/n, La Plata 1900, Buenos Aires, Argentina
e-mail: smontemay@gmail.com

É. Guilbert
UMR7205 MNHN/CNRS/EPHE/UPMC, Muséum National d'Histoire Naturelle,
45 rue Buffon, CP 50, Paris F-75 005, France
e-mail: guilbert@mnhn.fr

A.R. Panizzi, J. Grazia (eds.), *True Bugs (Heteroptera) of the Neotropics*,
Entomology in Focus 2, DOI 10.1007/978-94-017-9861-7_14

14.1 Introduction

Tingidae is a family of small hemimetabolous, plant-feeding bugs usually called "lace bugs" due to the lacelike aspect of their hemelytra and pronotum of most of its species (Schuh and Slater 1995) (Fig. 14.1). They are unique among the Heteroptera because it is the single family with a high specific richness in which all its members are exclusively phytophagous. They are worldwide distributed and comprise about 2,500 species arranged in 300 genera. The family is composed of three subfamilies: Vianaidinae, Cantacaderinae, and Tinginae (Schuh et al. 2006); all of them are represented in the Neotropical Region, Vianaidinae being endemic of this region (though there are fossil records from the Nearctic Region). Approximately 600 species distributed in around 70 genera are currently known for this region.

The size ranges from 2 to 8 mm, and besides their highly complex morphology, they are poorly colored, generally varying in shades of brown. Tingids are characterized by a low flight activity presenting gregarious and sedentary habits and by being generally host specific (though they feed on a great number of botanical families). Adults and nymphs are usually found on the undersides of leaves (with some exceptions), where they feed on the sap of living plants by piercing the epidermis with their very slender stylets. These stylets are protrusive and retractile and can

Fig. 14.1 Tingids on the leaves of their host plants: (**a**) *Acanthocheila* sp., (**b**) *Phymacysta* sp., (**c**) *Leptopharsa* sp., and (**d**) *Gargaphia* sp. (Photos by C. D'Haese)

easily penetrate the cellular tissue to extract the sap. Their feeding activities may cause great injury and plasmolysis of the foliage. Many cultivated and wild plants of prime importance in agriculture and horticulture are seriously affected by the feeding activities of these insects (Drake and Ruhoff 1965). The entire life cycle is completed on the same host plant species and on the very same part of the plant in most of the cases.

The members of this family are adapted to a wide range of habitats; some species occur on the ground, associated with the upper parts of the roots or lower parts of stems, some are subterranean, some live in middle height or on the plant canopy, and some non-Neotropical genera induce galls and feed on pollen and moss (Péricart 1983). Some are described as associated with ants, because they were found in ant nests or collected together (Van Doesburg 1977). They are small and difficult to see due to their hyaline structures and then remain poorly known, despite the number of species already described.

14.2 General Characteristics and Diagnosis

Most species are characterized by the lacelike network of areolae (Fig. 14.2a–c), but some species can present punctuations instead (Fig. 14.2d, e). The head is armed with a maximum of nine spines (Fig. 14.3). These spines are extremely relevant for tingid taxonomy, and they are named according to their position. The antennae have four segments as well as the labium which is inserted ventrally in the head. In Vianaidinae, compound eyes can be absent or extremely reduced (Fig. 14.4a), and the ocelli are absent in all but the macropterous forms of this subfamily. The bucculae are well developed (Fig. 14.4b), areolated, and sometimes joined anteriorly. The dorsum and the hemelytra are slightly or heavily punctuated and/or areolated. The pronotum is convex or flat and composed of a collar, a pronotal disk with or without carinae, and a pair of lateral expansions (paranota), with or without a posterior projection covering the scutellum (Fig. 14.2). The paranota and the carinae can be well developed or absent, with different degrees of reflection and numbers of rows of areolae. The collar can be developed forming a hood, which can fully or partially cover the head (Fig. 14.5). The rostral channel forms a groove which is framed with sternal laminae (Fig. 14.6). The dorsal laterotergites are present, but the ventral ones are absent. The abdominal spiracles 2–8 are located on abdominal sterna. The male genitalia are symmetrical; the phallus has some sclerotized eversible structures (Lee 1969). The parameres are usually directed backwards; the ovipositor is laciniate; the connection between the first valvula and first valvifer is lost, as in the Miridae (Schuh and Slater 1995). The spermatheca is absent and the pseudospermathecae present, but it seems to be unrelated with the sperm storage function (Marchini et al. 2010).

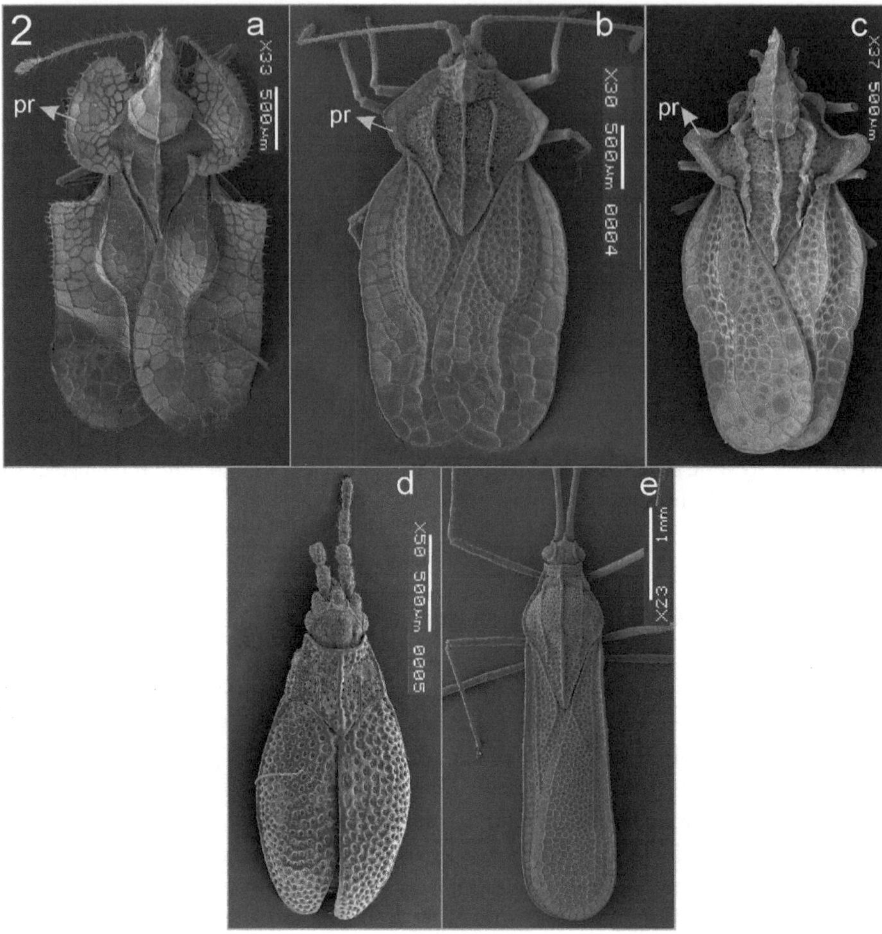

Fig. 14.2 Dorsal habitus, pr: paranota: (**a**) *Corythucha* sp., (**b**) *Gargaphia* sp., (**c**) *Corythaica* sp., (**d**) *Coleopterodes* sp., and (**e**) *Tigava* sp. (Photos by S. Montemayor)

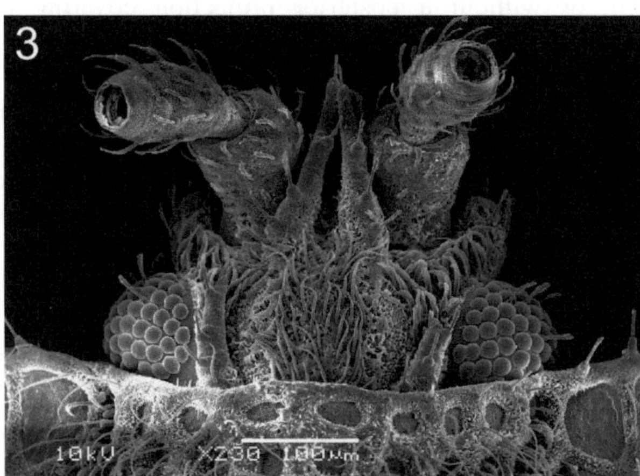

Fig. 14.3 Dorsal view of head of *Tingis americana*, presenting four spines (Photo by M. Guidoti)

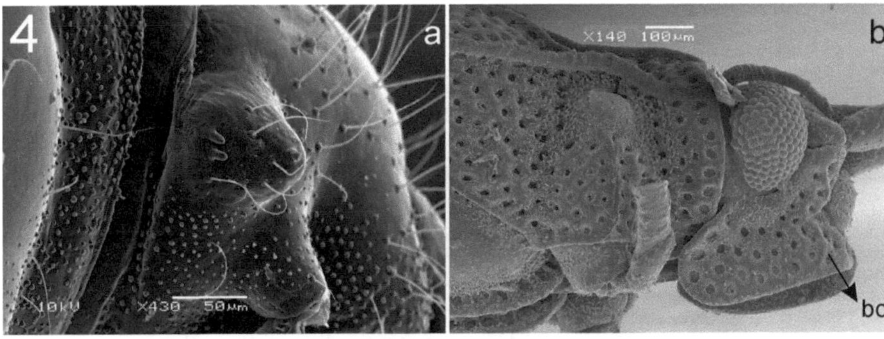

Fig. 14.4 Lateral view of the eye; bc, buccula: (**a**) *Thaumamannia vanderdrifti* Doesburg, 1977 (Vianaidinae), and (**b**) *Tigava* sp. (Photos by M. Guidoti and S. Montemayor)

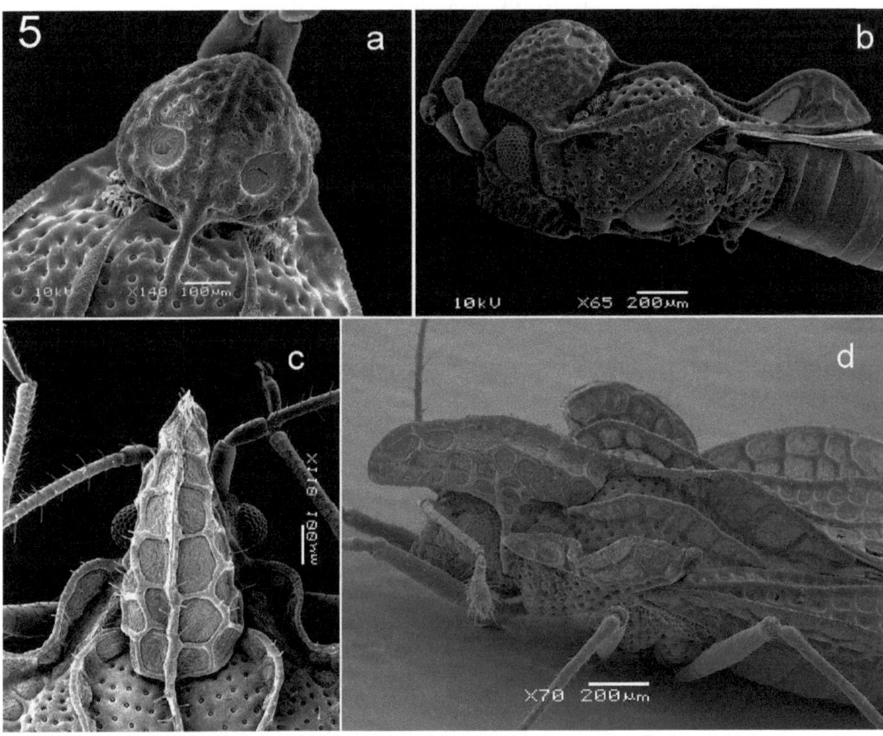

Fig. 14.5 Dorsal and lateral view of the hood: (**a**, **b**) *Sphaerocysta globifera* Stål, (**c**, **d**) *Corythaica* sp. (Photos by M. Guidoti and S. Montemayor)

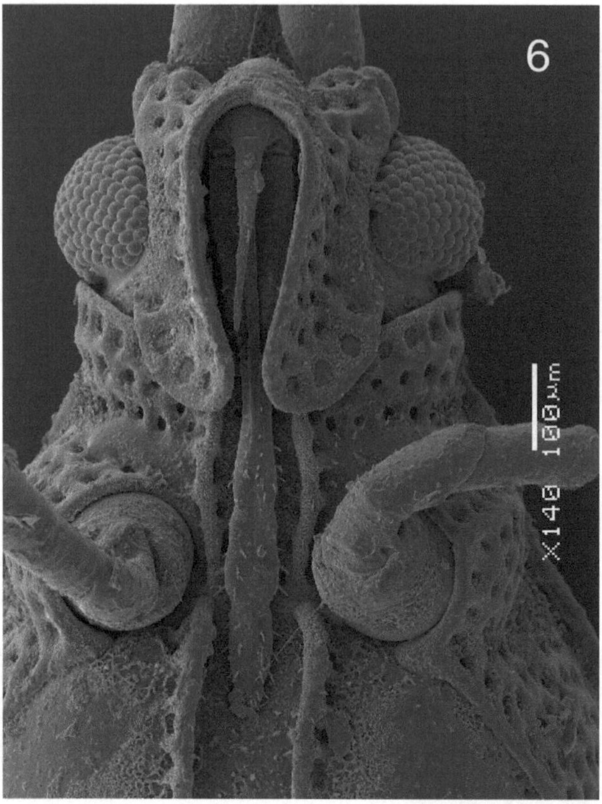

Fig. 14.6 Rostral channel of *Tigava* sp. (Photo by S. Montemayor)

14.3 General Biology and Ecology

14.3.1 Reproduction, Eggs, and Oviposition

The reproductive behavior has been described for very few species [e.g., Kogan 1960 – *Corythaica cyathicollis* (Costa)], and no courtship behavior has been reported. Reproductive parameters, such as fertility and fecundity, could vary in great scale between species, and most of the studies were carried with species from the Palearctic Region (e.g., Stusak 1968) and only a few with Neotropical species (e.g., Cividanes et al. 2004). The development of male and female reproductive organs was described in detail by Eguagie (1976), for *Tingis ampliata* (Herrich-Schaeffer), a non-Neotropical species of a cosmopolitan and probably paraphyletic genera. Generally, there are one or two generations per year (univoltine or bivoltine), although multivoltine species are known (e.g., some *Corythucha* Stål species – Neal and Douglas 1990). The mechanisms that regulate the number of generations in Tingidae are photoperiodic induction of diapause and noncleidoic eggs. The first one was reported in *Corythucha* (Neal et al. 1992) and perhaps is present in genera that overwinter as adults (Neal and Schaefer 2000). The last one was reported for *Stephanitis pyrioides* (Scott) and consists in a dependency on imbibing water from the host for the completion of the embryonic development (Neal and Schaefer 2000). Thus, during development, the absorption of moisture results in an increase in egg size and weight. As a possible consequence of the moisture requirement, seasonal fluctuations in the moisture availability could also affect voltinism (Neal and Schaefer 2000). And a third mechanism present in *Acalypha* Westwood species, a non-Neotropical genus, is to overwinter as nymphs (Pericart 1983).

Eggs have been poorly studied. Their size is around 0.4–0.85 mm (Péricart 1983), and their structure is similar to the one of the other Cimicomorphan families, particularly the Miridae. The posterior part is hemispherical; the sides are slightly compressed laterally and the axis slightly curved; the anterior part is concave and closed by an opercle. The chorion which constitutes the external envelop of the egg is often whitish or clear colored and sometimes brownish at the extremities. The egg shape and structure are quite homogeneous within tingid species.

The oviposition site is variable as there are species that drop their eggs on roots, stems, bud flowers, and/or leafs. The eggs can be inserted on the spongy mesophyll with only the operculum outside the vegetal tissue (endophytic oviposition), partially inserted in the vegetal tissue (pseudo-endophytic oviposition), or placed at the surface of the vegetal tissue (exophytic oviposition). They are usually oviposited in the abaxial surface of the leaf and can be arranged in small or big groups, or they can be isolated, in the midvein or lateral veins or on the leaf surface in several places. Generally, when oviposited on the leaf's veins, they are found on the veins' axillae. The same species can have more than one site for oviposition, so it is difficult to characterize an oviposition strategy at specific level. In some *Gargaphia* Stål species, more than one female oviposits on the same site, forming clusters of eggs from different females. Two Palearctic genera, *Copium* Thunberg and *Paracopium* Distant, are known as gall inducers, and the time and method of laying and fixation of the egg to the corolla are connected with floral cecidogenesis.

14.3.2 Nymph Development and Maternal Care

Most of the species have five instars although four-instar cycle was reported for *Stephanitis rhododendri* Horváth (Johnson 1936). The nymphal morphology varies from simple to highly complex, presenting impressive spinelike structures on the head and on the posterior margins of thoracic and abdominal segments (Fig. 14.7). Some of these cuticular structures present glandular activity, secreting chemical compounds that are related to osmoregulatory functions (Livingstone 1978) and sensorial functions (Aldrich et al. 1991) or have defensive properties (Mason et al. 1991; Scholze 1992). These structures have been described for species of several

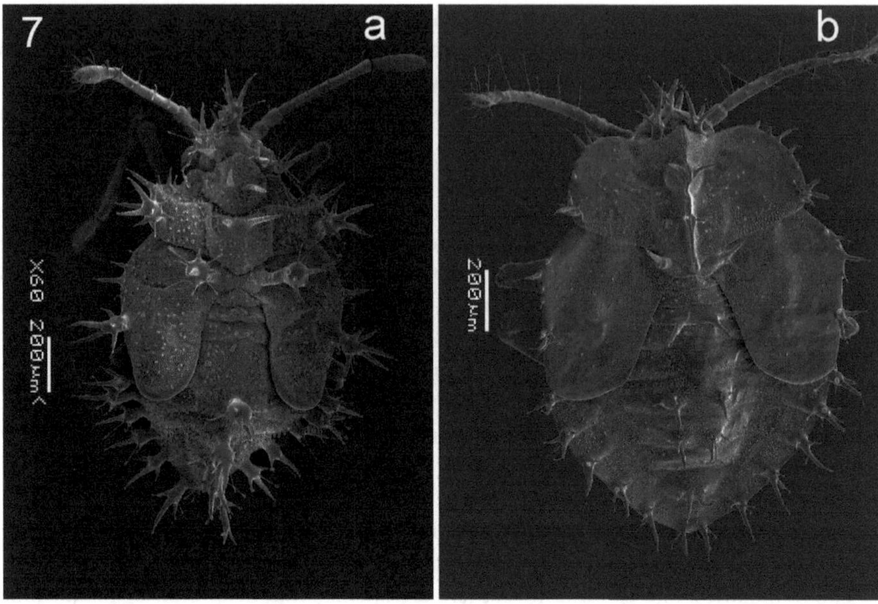

Fig. 14.7 Dorsal habitus of fifth instar nymphs: (**a**) *Corythaica* sp. and (**b**) *Corythucha* sp. (Photos by S Montemayor)

genera and are a reliable source of taxonomic characters (Guilbert 2004a, 2005; Lee 1969). Nymphs also have one pair of abdominal scent glands, which in the subfamilies Tinginae and Cantacaderinae are disposed on the center of the posterior margin of IV and V abdominal segments and in Vianaidinae they are laterally disposed in the posterior margin of the IV segment.

First and second instars are very similar among species; the interspecific differences are more evident from the third instar onwards. Wing pads become conspicuously developed at the fourth or fifth instar, but they are already visible as a small lateral development of the posterior margin of mesonotum in the second instar. Some species belonging to different genera present parental care of nymphs and eggs, sometimes with tricky and aggressive protective behaviors (Tallamy and Denno 1981; Tallamy and Iglay 2004). Very often, nymphs exhibit a gregarious habit; when disturbed and scattered, they quickly group together again. They usually feed and develop to adults on the same plant and even on the same leaf. Some species have all their nymphs fully studied and described, and most of them are from the Palearctic Region (e.g., Livingstone 1976). But only recently (e.g., Guilbert and Montemayor 2010; Guidoti and Barcellos 2013; Montemayor 2009; Montemayor and Dellapé 2010; Montemayor et al. 2011), Neotropical nymphs started to figure in the specialized literature. It has been reported, for example, that an oligophagous Neotropical species, *Tingis americana* Drake, could present different fitness and nymphal development rates in different host plants, due to unequal biological parameters (Moreira et al. 2013). The nymphal morphology is well known for some genera, but little is known about its functionality and evolution (Scholze 1992; Guilbert 2004b). Scholze (1992) studied its functionality, whereas Guilbert (2004b) its evolution and hypothesized nymphal outgrowths as adaptations rather than merely hyperthelic structures. The evolutionary pattern of the development of these structures was studied, and the heterochronic events involved were classified as generally peramorphic, where the next stage (instar) has overdeveloped structures and, in some cases, paedomorphic, where the next stage has underdeveloped structures (Guilbert et al. 2008).

Maternal care of offspring, a rare behavior within Cimicomorpha's families, has been described for three tropical genera, two of them distributed in the Neotropics (*Gargaphia* and *Leptobyrsa* Stål). Chemical compounds were isolated from nymphs and adults, and they seem to play an important role in communication and also as predatory protection strategy (e.g., Aldrich et al. 1991; Mason et al. 1991; Scholze 1992). In *Gargaphia solani* Heidemann, one female takes care of groups of eggs oviposited by several other females until this brood becomes adult (Tallamy and Denno 1981). This probably occurs because the maternal care induces a great survivor rate for the species, but also exposes the female involved in danger. Thus, the smaller is the number of females exposed to predatory activity, the higher is the number of females that could survive (Tallamy 1985). *Leptobyrsa decora* Drake, presents some communal effort to protect the brood (Loeb and Bell 2006), and there are a couple of biological studies where the effect of such maternal care on the brood success through the nymphal development is clear (Melksham 1984). In the Afrotropical species *Compseuta picta* Schouteden, adult females push away the aggressor from the brood and move the wings against the aggressor (Tallamy and Iglay 2004). Besides the mentioned examples, it is expected that more tingids, including Neotropical genera, present these ethological traits.

14.4 Classification and Diversity

The classification of tribes, genera, and species is mostly based on external morphological characters, especially the lacy outgrowths of the pronotum and of the hemelytra. Systematics and taxonomy of the Tingidae are a hard and paradoxal task, due to such particular outgrowths. In contrast to what happens in other families, characters of the genitalia are not frequently used since Drake and Davis (1960) noted that the male genitalia were not of taxonomic value, in spite of Lee's (1969) observation of useful variations of the phallys for the oriental species of the genus *Stephanitis* Stål. Tegumentary structures used to identify the species are not always constant. Today, some genera remain unclearly defined, especially in tropical groups. Drake (with his many coauthors) is the single most prolific author of the group. Drake's latest works were illustrated with habitus views of various tingids, although only few of his works contain aids to identification.

For almost a century, the placement of Piesmatidae within Tingidae has been a case of argument. Tullgren (1918) recognized that Tingidae lacked abdominal trichobothria, whereas Piesmatidae possessed them, and Reuter (1910) classified Piesmatidae and Tingidae together in the superfamily Tingitoidea. Finally, Leston et al. (1954) and Drake and Davis (1960) have clearly shown that these two families are not closely related. Piesmatidae were consistently divorced from the Tingidae and placed in the infraorder Pentatomomorpha. Drake and Davis (1960) proposed the superfamily Miroidea Hahn, based on morphological structures from a global concept, to hold the families Tingidae Laporte and Miridae Hahn, and then systematized the Tingidae into the subfamilies Cantacaderinae Stål, Tinginae Laporte, and Vianaidinae Kormilev. This concept was followed in the catalog of Drake and Ruhoff (1965). Stys and Kerzhner (1975), following Scudder's (1959) conclusion, which was based on a study of the female genitalia of the Heteroptera, treated it as a superfamily including two families: Tingidae (with two subfamilies, Cantacaderinae and Tinginae) and Vianaididae. Owing to its distinct general facies and unique characters, the Vianaidinae were treated as a family by Kormilev (1955) in its original description and later by Scudder (1959), Carayon (1962), and Stys and Kerzhner (1975). Schuh and Stys (1991) and Schuh and Slater (1995) followed Drake and Davis (1960) and Drake and Ruhoff (1965) and treated this taxon as a subfamily of the Tingidae.

On the basis of the first phylogenetic analysis of the Tingidae, Lis (1999) raised the subfamilies Cantacaderinae sensu Drake & Davis and Tinginae sensu Drake & Davis into families: the Cantacaderinae and the Tingidae grouped into the superfamily Tingoidea together with the Vianaididae. Cantacaderinae was divided into two subfamilies, Cantacaderinae and Carldrakeaninae, whereas Phatnomatini sensu Drake & Davis was raised as a subfamily and transferred into the newly raised Tingidae, together with the Tinginae (Tingidae sensu Drake & Davis). In a different phylogenetic analysis, Guilbert (2001) found the Cantacaderinae monophyletic but included into the Tinginae. In a third phylogenetic analysis, including larval characters, Guilbert (2004b) retrieved Cantacaderinae sensu Drake & Davis and Tinginae sensu Drake & Davis. Schuh et al. (2006), on the basis of Lis' study, presented evidences to justify the placement of Vianaidinae as a sister group of Tinginae + Cantacaderinae. Guilbert (2012) validated the groups within Cantacaderinae proposed by Lis (1999), but showed a different relationship within these groups. Guilbert et al. (2014) presented a phylogenetic hypothesis based upon molecular and morphological characters in which Vianaidinae is included in Tingidae, as a sister group of Cantacaderinae + Tinginae, with the Phatnomatini included in Tinginae. In the same contribution, the phylogenetic status of the Tinginae tribes is discussed (Guilbert et al. 2014), as they were in previous contributions (Guilbert 2001, 2004b).

14.5 Economic Importance

14.5.1 Main Genera/Species

***Amblystira* Stål**

This American genus is constituted by 20 species, most of them distributed in the Neotropics. The only member of the genus that has been recorded as a pest is *A. machalana* Van Duzee. It is commonly known as "black lace bug" and feeds on cassava, *Manihot esculenta* Crantz, an important crop of South America. Arias and Belloti (2003) studied the life cycle, behavior, and damage caused by *A. machalana* on cassava. This species is distributed in Venezuela, Colombia, Ecuador, and Peru (Arias and Belloti 2003; Drake and Ruhoff 1965). The only key to species available for the genus is restricted to the South American *Amblystira* (Montemayor 2010a).

***Acanthocheila* Stål**

This is a genus originally constituted by 16 species. Froeschner (1995) reviewed the genus and synonymized some of its species and presented a key to identify them. Currently, it is constituted by seven species distributed in the entire continent, though most of them are known from South and Central America. *Acanthocheila armigera* (Stål) is frequently mentioned because it causes damages in plants of the genera *Nicotiana*, *Ouratea*, and *Pisonia* (Stonedahl et al. 1992; Neal and Schaeffer 2000). It is known from the USA, Mexico, Guatemala, Cuba, El Salvador, Honduras, Haiti, Costa Rica, Panama, Puerto Rico, Trinidad, Venezuela, Colombia, Ecuador, Peru, Bolivia, Brazil, and Argentina (Froeschner 1995).

***Carvalhotingis* Froeschner**

Froeschner (1995) described this Neotropical genus to accommodate five species originally described as *Acanthocheila*, and, in this same contribution, he provides a key to species. The two species with economic importance are *C. visenda* (Drake & Hambleton) and *C. hollandi* (Drake). *Carvalhotingis visenda* was the first biological control agent approved for release against cat's claw creeper *Macfadyena unguis-cati* (L.) (Bignoniaceae) in Australia (Dhileepan et al. 2010), and it has also been released in South Africa (King et al. 2011). Cat's claw creeper, a climbing woody vine, is native from Mexico through Central America to tropical South America including Trinidad and Tobago. In Australia, cat's claw creeper is a major environmental weed in coastal Queensland and New South Wales, where it is a major threat to biodiversity in riparian and rain forest communities (Dhileepan et al. 2007). *Carvalhotingis visenda* was originally distributed in Brazil, Peru, and

Argentina. *Carvalhotingis hollandi* (Drake), which is also being used as a biological control agent of *Macfadyena unguis-cati* in South Africa (King et al. 2011), was originally distributed in Brazil, Paraguay, and Argentina.

Corythaica Stål

This is an American genus comprised of 21 species, 14 of which are distributed in the Neotropical Region. Hurd (1945) made a comprehensive review of the genus that includes a key to species, redescriptions of all the species known at that moment, and drawings of them. Montemayor and Melo (2012) made a review of the Argentinean *Corythaica* where they describe new species, provide a key to the Argentinean *Corythaica*, and raised from synonymy *C. passiflorae* (Berg). *Corythaica cyathicollis* Costa is frequently mentioned in the literature as a major pest for several solanaceous crops (Kogan 1960; Neal and Schaefer 2000; Stonedahl et al. 1992). Montemayor and Melo (2012) discussed the identity of *C. cyathicollis* Costa and mentioned that probably almost all the citations of *C. cyathicollis* in the literature do not refer to Costa's species and most likely pertain to *C. monacha* (Stål) or *C. passiflorae* (Montemayor & Melo 2012). Because of this identity confusion, the distribution and host plants of these three species must be reevaluated (Montemayor and Melo 2012).

Corythucha Stål

This is a large genus with more than 75 species. It is widely distributed in the Americas, and it has also been accidentally introduced into Europe and Asia. The great majority of its species are from North America (Montemayor 2009). Despite the number of species in this genus, there is a remarkable uniformity among them. One striking feature of the host plants of this genus is that a large number of trees are included, in contrast to the herbaceous plants which predominate as hosts for the majority of tingids (Hurd 1946). Gibson (1918) made a review of the genus in which he described several species and provided a key to 56 species (among them *C. gossypii* F. and *C. ciliata* Say known from the Neotropical Region and with economic importance); this is the last comprehensive study of the genus. *Corythucha gossypii* is a serious pest of beans and cotton, hence the common name "cotton or bean lace bug." However, it has been reported in another 24 host plants including castor bean (*Ricinus communis* L.) (Herney-Varón et al. 2010) and sunflower (*Helianthus annuus* L.). Cotton lace bug damage can be observed on the plant's foliage, ranging from some leaf stippling from considerable yellowing to bleaching. This can reduce plant vigor, and as a consequence, fruit production can be affected or even prevent fruit from forming when infestations are heavy (Miller and Nagamine 2005). All immature stages of this species are described and illustrated by Lopez-Montes et al. (1982). It is known from almost all the New World. *Corythucha ciliata*, commonly known as the sycamore bug, is a pest of the popular ornamental tree *Platanus occidentalis* L., sycamore, and of its hybrids. These trees are frequently used as street-side trees in urban areas. In cases of severe infestations, host trees have been observed defoliated in late summer. Several consecutive years of severe *C. ciliata* damage, combined with other stress factors, may kill the trees. It is suspected that *C. ciliata* may serve as vector of two fungi, *Ceratocystis fimbriata* Hell. et Halsted forma *specialis platani* Walter and *Apiognomonia* (= *Gnomonia*) *veneta* (Sacc. and Speg) (Ruiting et al. 2009). It is believed that the pest came from North America, from where it has spread throughout Europe and parts of Asia (Õszi et al. 2005). In the Neotropical Region, it has been only reported for Chile (Prado 1990).

Dictyla Stål

This genus is constituted by over 60 species distributed worldwide. There are 17 known species from America and 13 from the Neotropical Region. Brailovsky and Torres (1986) provided a redescription of the genus and mentioned its possible Neotropical origin. The most mentioned species in the literature is *D. monotropidia* (Stål), a Neotropical species that has been frequently reported on *Cordia* spp. (Drake and Ruhoff 1965) as well as on cotton in Argentina (Fenton 1934) and on orchids from Mexico to the boarder of Texas (Swezey 1945; Drake and Ruhoff 1965; Neal and Schaefer 2000). The feeding activities of this insect damage the leaves that fall prematurely causing low flowering and fruiting and a developmental delay of the plants (Arguedas and Fallas 1993). There are some studies about the biology of the species (e.g., Fallas et al. 1993; Martínez et al. 2012). *Dictyla monotropidia* is known from Mexico, Cuba, Costa Rica, El Salvador, Guatemala, Haiti, Honduras, Jamaica, Panama, Puerto Rico, Trinidad, Venezuela, Colombia, Peru, Brazil, Bolivia, Paraguay, and Argentina. Up to now, the known host plants are *Cordia alliodora* (Ruiz & Pav.) Cham., *C. curassavica* (Jacq.) Roem. & Schult., *C. gerascanthus* L., *C. tomentosa* Cham. & Schltdl., *C. trichotoma* (Vell.) Arráb. ex Steud, and *Gossypium* sp. Guilbert and Montemayor (2010) provided a description and illustrations of the fifth instar as well as some comments on the fourth and third instars. There is no key to recognize the species of *Dictyla*, and because of the great number of species that constitute it and its morphological diversity, it is not easy to properly identify them.

Gargaphia Stål

This is a large American genus constituted by over 50 species, most of them distributed in the Neotropical Region. The interrupted rostral channel at meso-metasternal suture with transverse laminae converging medially is a diagnostic character of the genus. It includes several species which are economically important as plant feeders. Drake (1917) developed a key for the Nearctic species of *Gargaphia*. Some years later, Gibson (1919a) made a review of the genus when it was constituted by 25 species and provided a key to 16 of them (none of the Neotropical species with economic importance were included), and this was the last comprehensive study of the genus. The Neotropical species most mentioned as pests are *G. lunulata* (Mayr), *G. sanchezi* Froeschner, and *G. torresi* Costa Lima. *Gargaphia lunulata* feeds on many important South American plants from several families: Euphorbiaceae (*Euphorbia* sp., *Manihot* sp., *Ricinus* sp.), Leguminosae [*Cassia* sp., *Glycine max* (L.), *Phaseolus* spp.], and Malvaceae (*Gossypium* sp., *Hibiscus* sp., and *Urena* sp.) (Stonedahl et al. 1992; Neal and Schaefer 2000). It is also a pest of *Passiflora caerulea* L., a climbing plant with ornamental, medic, and alimentary interest (Ajmat et al. 2003). It is distributed in Colombia, Brazil, Paraguay, Argentina, and Uruguay. *Gargaphia sanchezi* is known from Colombia and attacks beans, *Phaseolus vulgaris* L. The biology of this species was studied by Schoonhoven et al. (1975), and usually the populations of *G. sanchezi* are very high, causing great damage on the plants. Finally, *G. torresi* feeds on several economically important plants such as cotton, sweet potato, bean, corn, and sunflower as well as other plants (*Canavalia* sp., *Gossypium* sp., *Ipomoea* sp., *Helianthus* sp., *Hibiscus* sp., *Parthenium* sp., *Phaseolus*, *Sida* sp., *Triumfetta* sp., *Xanthium* sp., and *Zea* sp.). It has some natural enemies such as *Xylocoris* sp. (Hemiptera: Anthocoridae), *Franklinothrips* sp. (Thysanoptera: Aeolothripidae), and *Bochartia* sp. (Acari: Erythraeidae) (Silva and Barbosa 1986). *Gargaphia torresi* is known from Brazil, Bolivia, Paraguay, and Argentina. The two most important studies that have been performed on the biology of *G. torresi* are related to the effects of temperature on their development, fecundity, and longevity (Domingues-da-Silva 2004) and to different aspects of its life cycle (Arce-de-Hamity et al. 2006).

Leptobyrsa Stål

This genus was described by Stål for *Tingis steini* Stål, and later Champion (1897) redefined the genus to include five more Central American species. Since Champion's modification, 11 more species were added. Drake and Poor (1937) reviewed the genus, and from the 17 species that constituted the genus, only eight were left, all of them Neotropical. Unfortunately, there is no key for the genus. Among the *Leptobyrsa* spp., *L. decora* Drake has importance as a control agent of the weed *Lantana camara* L. to which it causes severe defoliation (Day and Zalucki 2009). Harley and Kassulke (1971) studied the life history of this species as a potential biological control agent of *L. camara*. Melksham (1984) compared the oviposition and maternal care of two populations from Colombia and Peru, and Thomas and Ellison (2000) listed the countries where *L. decora* has been released as a biocontrol agent. This species was described by Drake from Colombia and Ecuador, and later it was reported from Peru (Harley and Kassulke 1971) and the Galapagos Islands (Henry and Wilson 2004). *Leptobyrsa decora* has been introduced widely around the world, including Australia, the Cook Islands, Fiji, Guam, Ghana, Hawaii, Palau, South Africa, and Zambia, to help control *Lantana camara* L. (Thomas and Ellison 2000; Henry and Wilson 2004). A photograph of this species is also available in Henry and Wilson (2004).

Leptocysta Stål

This genus was erected to accommodate *L. sexnebulosa* (Stål), first described in the genus *Tingis*. Currently, the genus includes six species, all distributed in South America. Monte (1946) provided the first key for the genus, when it was constituted by three species, and excellent drawings of these species. Several years later, Montemayor (2010b) provided another revision of the genus with the description of new species and a key to all the currently known species. The only species mentioned feeding on economically important plants is *L. sexnebulosa* (Stonedahl et al. 1992), and the host plants are *Antennaria* sp., *Ipomoea batatas*, *Mikania* sp., and *Parthenium*, and *Vernonia* sp. *Leptocysta sexnebulosa* is distributed in Venezuela, Colombia, Peru, Brazil, Paraguay, and Argentina (Montemayor 2010b). For the remaining species of the genus, there are no records of host plants.

Leptodictya Stål

This genus is confined to the Americas; it is constituted by more than 50 species, most of them distributed in the Neotropical Region. The *Leptodictya* mentioned as pest is *L. tabida* (Herrich-Schaeffer), commonly known as the sugarcane lace bug because it feeds on sugarcane (*Saccharum* spp. hybrids) as well as other species of Poaceae (Setamou et al. 2005). High populations of *L. tabida* reduce plant vigor, cause leaves to senesce prematurely, and reduce the total area of photosynthesis (Neal and Schaefer 2000). It is known from the USA (Florida, Hawaii, and Texas), Mexico, Cuba, Costa Rica, El Salvador, Guatemala, Nicaragua, Panama, and Venezuela (Chang 1986). This author provided a good review with respect to the life cycle, bionomics, damage, and control along with a morphological description of adults and nymphs of *L. tabida*.

Leptopharsa Stål

This is a very large and diverse genus highly variable in its morphology. Through the years, several genera have been described on the base of species originally described as *Leptopharsa*. It is constituted by over 100 species, most of them from America, except for four African ones and two Australian ones. Most of its members are distributed in the Neotropical Region. There are two Neotropical species well known for their economic importance: *L. gibbicarina* Froeschner and *L. heveae* Drake & Poor. Adults and nymphs of *L. gibbicarina* cause important yield losses in the crops of the African oil palm (*Elaeis guineensis* Jacquin). The lesions caused by the feeding activities of this tingid allow the infestation of the plant by a fungi complex that causes a severe disease called *Pestalotiopsis* or gray leaf blight (Escalante et al. 2010). *Leptopharsa gibbicarina* is distributed in Colombia. The original description (Froeschner 1976) provides very good drawings of the species. The rubber tree lace bug, *L. heveae*, is known from Brazil, and it was described as a pest of *Hevea brasiliensis* Muell. Arg. known as the Pará rubber tree, sharinga tree, or, most commonly, the rubber tree. This tree is of major economic importance because the milky latex extracted from it is the primary source of natural rubber. *Leptopharsa heveae* occurs in high populations in rubber tree plantations, and it is a limiting factor in rubber production due to the loss of photosynthetic tissue (Santos and De-Freitas 2008). The control of the pest has been made mainly with chemical products, which cause environmental contamination. The alternative would be the use of biological control agents; however, information about *L. heveae* natural enemies is scarce. The entomopathogenic fungus Sporothrix insectorum (Hoog & Evans) (Neal and Schaeffer 2000) and the parasitoid *Erythmelus tingitiphagus* (Soares) have been found naturally on eggs of the rubber tree lace bug (Santos and De-Freitas 2008).

Pleseobyrsa Drake & Poor

This is a genus that belongs to the tribe Tingini and was erected to place four species, three previously assigned to the genus *Leptobyrsa* Stål and a new species. After the description of *Pleseobyrsa*, five other species were described for the genus. Until 1969, *Pleseobyrsa* was constituted by nine species, but Froeschner (1969) described a new genus belonging to the tribe Litadeini Drake and Ruhoff, *Stragulotingis*, to place two *Pleseobyrsa* species. In a later publication, Froeschner transferred two other species from *Pleseobyrsa* to *Stragulotingis*, described a new *Pleseobyrsa*, and provided a key to the species of both genera (Froeschner 1991). Some years later, another *Pleseobyrsa* was described, and in this contribution, an updated key to the species of *Pleseobyrsa* was provided as well as illustrations and photographs of adult and fifth nymph of *P. persea* (Montemayor et al. 2011). The genus includes currently seven species, and three feed on avocado tree, causing damage: *P. boliviana* (Drake & Poor), *P. chiriquensis* (Champion), and *P. persea. Pleseobyrsa boliviana* is known from Bolivia; *P. chiriquensis* from Costa Rica, Panama, Venezuela, and Colombia; and *P. persea* from Costa Rica.

Pseudacysta Blatchley

This genus only includes *P. perseae* (Heidemann), first described as *Acysta* Champion, and is commonly known as the avocado bug. As its vernacular name indicates, it is harmful for *Persea americana* P. Miller (avocado). *Pseudacysta persea* destroys leaf cells (Moznette 1922) and causes chlorosis and browning (Medina-Gaud et al. 1991). Other host plants mentioned in the literature are *Comphora officinalis*, *Persea carolinensis*, and *Persea gratissima* (Heidemann 1908). In the original description of the species, egg and nymphs were also described, and drawings of the adult, fifth nymphal instar, and eggs were provided. Hurd (1946) redescribed the genus and discussed some other aspects such as resemblance with *Acysta* and distribution. The known distribution of *P. perseae* includes the USA, Mexico, Cuba, Dominican Republic, Guatemala, Jamaica, Puerto Rico, St. Lucia, St. Thomas, St. John, St. Croix, French Guiana, and Venezuela (Humeres et al. 2009).

Stephanitis Stål

This is a large genus constituted by around 60 species. Hórvath (1912) divided it into four subgenera: *Stephanitis* Stål, *Menodora* Hórvath, *Norba* Hórvath, and *Omoplax* Hórvath; only the first one is present in America (Hurd 1946). One of the species with economic importance is *S. pyrioides* (Scott), an important pest of azaleas (*Rhododendron* spp.). The azalea lace bug has become of increasing importance as azaleas have been more and more popular as ornamentals. Its origin is thought to be Japan, from where it was first described, and it has been accidentally introduced in several countries. The azalea lace bug is considered one of the most important problems affecting azaleas since its introduction (Shakunthala and Braman 2012). Illustrations of *S. pyrioides* are available at Dickerson and Weiss (1917), Weiss and Headlee (1918), Shen et al. (1985), and Gomez-Menor (1954). Shakunthala and Braman (2012) provided an excellent review of all the information available related to the azalea lace bug.

Teleonemia Costa

Teleonemia Costa is a taxonomically complex genus with a high interspecific variation. It is constituted by over 80 species distributed worldwide. *Teleonemia elata* Drake, *T. harleyi* Froeschner, *T. prolixa* (Stål), and *T. scrupulosa* Stål are Neotropical important species because they feed on *Lantana camara* (Winder & Harley 1983) and therefore have been used for the biocontrol of this weed which is considered to be among the world's ten worst invasive weeds. *Teleonemia elata* has been released in Australia, Cook Islands, South Africa, Uganda, and Zambia; *T. harleyi* and *T. prolixa* in Australia; and *T. scrupulosa* in Ascension Island, Australia, Fiji, Ghana, Guam, Hawaii, India, Indonesia, Kenya, Madagascar, Micronesia, New Caledonia, Niue, Northern Mariana Islands, Palau, Papua New Guinea, Samoa, Solomon Islands, South Africa, St. Helena, Tanzania, Tonga, Uganda, Vanuatu, Zambia, Zanzibar, and Zimbabwe (Thomas and Ellison 2000). Photographs of adults and nymphs of *T. elata* as well as comments on the species are provided by Harley and Kassulke (1971). The original distribution of this species is Brazil, Chile, Paraguay, and Peru (Drake and Ruhoff 1965). Harley and Kassulke (1973) provided SEM photographs of adults, fifth instar, and egg of *T. harleyi*. This species is distributed in Trinidad. *Teleonemia prolixa* has not only been considered for the biological control of *L. lantana* but also for *Mikania micrantha* Kunth, a Neotropical vine that is causing great damage to crops of tea, teak, rubber, oil palm, and coconut in Southeast Asia (Cock 1982). It is distributed in the Neotropical Region in Guatemala, Jamaica, Panama, Surinam, Trinidad and Tobago, British Guiana, Venezuela, Colombia, Ecuador, Peru, Brazil, Bolivia, Paraguay, and Argentina. *Teleonemia scrupulosa* eggs, nymphs, and adults are described and illustrated by Roonwal (1952), and more recently, SEM photographs and redescriptions of the nymphs were provided (Guidoti and Barcellos 2013). It is distributed in the Neotropical Region in the following countries: Cuba, Costa Rica, Guatemala, Haiti, Jamaica,

Netherlands Antilles (Aruba, Curacao, Klein Bonaire, Bonaire), Panama, Trinidad, Windward Islands (Grenada, St. Vincent), French Guiana, British Guiana, Venezuela, Peru, Brazil, Colombia, and Paraguay.

Vatiga Drake & Hambleton

The species of the genus *Vatiga* are native to the Neotropics and exhibit a decided preference for plants of the genus *Manihot* Miller, mainly *Manihot esculenta* Crantz, commonly known as cassava and one of the most important crops in Africa, Central and South America, and Asia. Froeschner (1993) reviewed the genus, provided a key to species, synonymized some species, and elevated to species status a subspecies, so as a result of Froeschner's study, the genus is currently constituted by five species. *Vatiga illudens* (Drake) is one of the species most frequently mentioned in the literature and also most frequently misidentified. It is mentioned from Brazil, Colombia, Cuba, Dominican Republic, Ecuador, Guyana, Haiti, Jamaica, Lesser Antilles, Puerto Rico, Reunion Island, Trinidad and Tobago, USA, and Venezuela. Oliveira et al. (2001) studied the biology of this species in experimental conditions, and Fialho et al. (2009) evaluated the economic damage caused in cassava root and foliage yield. *Vatiga manihotae* (Drake) is the second most mentioned species, and it is also the most widely distributed. According to Froeschner (1993), it is known from Cuba, Trinidad, Venezuela, Colombia, Peru, Brazil, Paraguay, and Argentina. The other species of the genus have more restricted distributions, such as *V. cassiae* (Drake & Hambleton), only known from Brazil; *V. pauxilla* (Drake & Poor), from Argentina; and *V. varianta* (Drake), restricted to Brazil and Colombia.

Phatnoma Fieber

This genus occurs in several geographic regions of the world. Most of the 27 species of the genus are similar in structure and intraspecific variability and therefore are difficult to separate (Froeschner 1996). For this genus, only a very outdated partial key exists (Gibson 1919b). *Phatnoma* is distributed in South and Central America, Africa, Asia, and Oceania. For the Neotropics, the species that has been reported feeding on plants with economic importance is *P. marmorata* Champion. It feeds on cacao and pineapple, and it is distributed in Costa Rica, Honduras, Panama, Trinidad, Brazil, and Ecuador. Its fifth nymph has been described by Guilbert (2005); however, very little is known about its biology.

14.5.2 Key to the Most Economically Important Neotropical Genera

The key below is valid for the genera of economic importance, and it may lead to erroneous results if used for other genera.

1. Seven cephalic spines, anterolateral paranotal angle projecting as a distinct angle or spiniform process .... *Phatnoma* Fieber
– Less than seven cephalic spines, paranota otherwise 2
2. Hood absent or scarcely developed 3
– Hood well developed 10
3. Radio-cubitus vein "C" shaped, paranota folded over pronotum .... *Dictyla* Stål
– Radio-cubitus vein not "C" shaped, paranota not folded over pronotum 4
4. Paranota lacking or poorly developed 5
– Paranota well developed 6
5. None or two rudimentary cephalic spines present (occipital pair), paranota represented by a small pale earlike appendage at each lateral angle, discoidal area not closed behind .... *Pseudacysta* Blatchley
– Two cephalic spines present (occipital pair), paranota absent or very slender carina like, discoidal area closed behind .... *Amblystira* Stål
6. Paranota projected forward, much broader anteriorly .... *Pleseobyrsa* Drake & Poor
– Paranota not projected forward, same width in all its length or posteriorly broader 7
7. Paranota armed with spines, lateral carinae absent or only posteriorly developed 8
– Paranota without spines, lateral carinae fully developed 9
8. Without occipital spines, paranota armed with 7–10 long, stout spines .... *Carvalhotingis* Froeschner
– With occipital spines, paranota armed with 5–10 long, stout spines .... *Acanthocheila* Stål
9. Generally five cephalic spines (sometimes less), elytra broadening from base, discoidal area not surpassing middle of hemelytra .... *Leptopharsa* Stål
– Between two and five cephalic spines; elytra elongated, parallel margined; discoidal area generally surpassing middle of hemelytra .... *Teleonemia* Costa
10. Hood not surpassing length of head 11
– Hood surpassing length of head 14
11. Mesosternal rostral laminae deeply constricted at mesosternum .... *Vatiga* Drake & Hambleton
– Mesosternal rostral laminae subparallel 12
12. Rostral channel interrupted at meso-metasternal suture by the metasternal laminae that converge and contact to each other .... *Gargaphia* Stål
– Rostral channel not interrupted at meso-metasternal suture 13
13. Hood tectiform shaped, paranota folded over pronotum, posterior process fully developed, elytra and paranota .... *Leptodictya* Stål
– Hood globose or subglobose, paranota not folded, projected forward in front, posterior process abbreviated, elytra and paranota armed with hairs and/or spines .... *Leptobyrsa* Stål
14. Paranota not projected forward 15
– Paranota projected forward 16

15. Hood gradually narrowing toward apex, tip surpassing antennal segment I; paranota with basal folds wider at the callus .... *Corythaica* Stål
– Hood tectiform, not surpassing antennal segment I; paranota without basal fold, produced evenly ....*Leptopharsa* Stål
16. Hood compress, slightly wider backwards, narrow through all its length; discoidal area not raised, surpassing middle of hemelytra ....*Leptocysta* Stål
– Hood globose or subglobose, much wider backwards, narrower anteriorly; discoidal area raised, not surpassing middle of hemelytra 17
17. Hemelytra rectangular shaped abruptly widened at base, broadest anteriorly, margins parallel or slightly concave .... *Corythucha* Stål
– Hemelytra gradually widening toward apex, broadest apically, margins rounded .... *Stephanitis* Stål

14.6 Concluding Remarks

Tingidae are exclusive plant feeders and usually monophagous, features that make them particularly interesting from an economical point of view. However, there is a noticeable lack of information about this group of insects. The economic importance of lace bugs will continue increasing as species emigrate (e.g., as concealed eggs) and as minor crops gain importance and expand to serve a burgeoning world population (Neal and Schaefer 2000).

Among the Neotropical lace bugs are some potential pests for different economically important crops, though usually only large populations cause serious damage. It is not easy to control and manage lace bug populations because of their different reproductive strategies (e.g., oviposition strategies), the annual number of generations, and the habit of living in the underside of the leaf. The ecology and behavior of most of the species remain unknown, and the knowledge of these different aspects of their biology is fundamental for the establishment of management strategies of their populations. Another little known aspect of these complex insects is their chemical communication, which seems to play an important role in intraspecific interactions, particularly between females and offspring. Predator and parasitoid information are scarce, and it could be useful when establishing control strategies for Tingidae populations.

The taxonomy of Tingidae is mostly based on external morphological characters which exhibit a high diversity of shapes. As such, relationships within Tingidae subfamilies and tribes are far from known. Suprageneric taxa have been currently discussed without highly supported hypothesis proposed so far. Immature forms seem to present very important characters that should be more explored in a taxonomic and phylogenetic framework. Morphological studies on unexplored structures could reveal important source of information. The tingid taxonomy and systematics are still in need of further studies to improve our knowledge on biological, ecological, ethological, and evolutionary questions.

Behavior and ecological parameters are the keys to understand economic problems raised by tingid species. As such, they will help provide the adequate solution. Considering that this chapter focused on Neotropical tingids, a more comprehensive work should be done since the last worldwide contribution was published 13 years ago (Neal and Schaeffer 2000).

References

Ajmat MV, Coviella MA, Pannunzio MJ (2003) Aspectos morfológicos, biológicos y daño de *Gargaphia lunulata* (Mayr) 1865 (Heteroptera: Tingidae) sobre *Passiflora caerulea* L. (Passifloraceae). Bol Sanidad Veg Plagas 29:339–346

Aldrich JR, Neal JW, Oliver JE, Lusby WR (1991) Chemistry via-à-vis maternalism in lace bugs (Heteroptera: Tingidae): alarm pheromones and exudate defense in *Corythucha* and *Gargaphia* species. J Chem Ecol 17:2307–2323

Arce-de-Hamity M, Zamar MI, Neder-de-Román L (2006) Tabla de vida y fecundidad de *Gargaphia torresi* Costa Lima (Hemiptera: Tingidae) sobre girasol. Idesia 24:37–40

Arguedas M, Fallas E (1993) La chinche de encaje del laurel *Dictyla monotropidia*. Serie Plagas Enferm 7:1–4

Arias BV, Belloti AC (2003) Ciclo biológico, comportamiento e importancia económica de *Amblystira machalana* (Heteroptera: Tingidae) en el cultivo de la yuca (*Manihot esculenta* Crantz). Rev Colomb Entomol 29:143–148

Brailovsky H, Torres L (1986) Hemiptera – Heteroptera de México XXXVI. Revisión genérica de la Familia Tingidae Laporte. Ann Inst Biol UNAM 56:869–932

Carayon J (1962) Observations sur l'appareil odorifique des Hétéroptères, particulièrement celui des Tingidae, Vianaididae et Piesmatidae. Cahiers Nat 18:1–16

Champion GC (1897) Rhynchota. In: Salvin O, Godman FDC (eds) Biologia Centrali-Americana. R.H. Porter, Londres, pp 1–32

Chang VCS (1986) The sugarcane lacebug: a new insect pest in Hawaii. Hawaiian Sugar Tech. In: Proceedings of the 44th annual conference report, Hawaiian Sugar Technologists, pp A27–A29

Cividanes FJ, Fonseca FS, Santos TM (2004) Distribuição de *Leptopharsa heveae* em seringal no Estado de São Paulo. Pesq Agropec Brasil 39:1053–1056

Cock MJW (1982) Potential biological control agents for *Mikania micrantha* HBK from the Neotropical Region. Trop Pest Manage 28:242–254

Day MD, Zalucki MP (2009) *Lantana camara* Linn. (Verbenaceae). In: Rangaswamy M, Gadi VPR, Anantanarayanan CR (eds) Biological control of tropical weeds using arthropods. Cambridge University Press, New York, pp 211–246

Dhileepan K, Trevino M, Snow L (2007) Specificity of *Carvalhotingis visenda* (Hemiptera: Tingidae) as a biological control agent for cat's claw creeper *Macfadyena unguis-cati* (Bignoniaceae) in Australia. Biol Control 41:283–290

Dhileepan K, Treviño M, Bayliss D, Saunders M, McCarthy J, Shortus M, Snow EL, Walter GH (2010) Introduction and establishment of Carvalhotingis visenda (Hemiptera: Tingidae) as a biological control agent for cat's claw creeper Macfadyena unguis-cati (Bignoniaceae) in Australia. Biol Control 55:58–62

Dickerson EL, Weiss HB (1917) The azalea lace-bug, *Stephanitis pyrioides* (Scott) (Tingitidae, Hemiptera). Entomol News 28:101–105

Domingues-da-Silva CA (2004) Efeitos da temperatura no desenvolvimento, fecundidade e longevidade de *Gargaphia torresi* Lima (Hemiptera, Tingidae). Rev Brasil Entomol 48:547–552

Drake CJ (1917) Key to the Nearctic species of *Gargaphia* with the description of a new species (Hem., Heter.). Entomol News 28:227–228

Drake CJ, Davis NT (1960) The morphology, phylogeny, and higher classification of the family Tingidae, including the description of a new genus and species of the subfamily Vianaidinae (Hemiptera: Heteroptera). Entomol Am 39:1–100

Drake CJ, Poor ME (1937) Concerning the genus *Leptobyrsa* Stål (Hemiptera). Proc Biol Soc Wash 50:163–166

Drake CJ, Ruhoff FA (1965) Lacebugs of the world: a catalog (Hemiptera: Tingidae). US Nat Mus Bull 243:1–634

Eguagie WE (1976) Changes in body weight and reproductive organs of *Tingis ampliata* H.-S. (Heteroptera: Tingidae). J Nat Hist 10:157–166

Escalante M, Damas D, Marquez D, Gelvez W, Chacón H, Díaz A, Monreno B (2010) Diagnóstico y evaluación de *Pestalotiopsis* e insectos inductores, en plantaciones de palma aceitera al sur del lago Maracaibo, Venezuela. Bioagro 22:211–216

Fallas EM, Arguedas MA, Briceño MD (1993) Dispersión y métodos de cría de *Dictyla monotropidia* (Hemiptera: Tingidae). Rev Biol Tropical 41:509–513

Fenton FA (1934) Tingitoidea affecting cotton. Can Entomol 66:198–199

Fialho J, Vieira EA, Paula-Moraes SV, Silva MS, Junqueira NTV (2009) Economic damage caused by lacebug upon cassava root and foliage yield. Sci Agrár 2:151–155

Froeschner RC (1969) Zoogeographic and systematic notes on the lace bug Litadeini, with the description of the new genus *Strangulotingis* (Hemiptera: Tingidae). Great Basin Nat 29:129–132

Froeschner RC (1976) Description of a new species of lace bug attacking the oil palm in Colombia (Hemiptera: Tingidae). Proc Entomol Soc Wash 78:104–107

Froeschner RC (1991) The lace bug genera *Pleseobyrsa* and *Stragulotingis*: reviews, keys and description of one new species in each (Heteroptera: Tingidae: Tinginae). Proc Biol Soc Wash 93:767–771

Froeschner RC (1993) The Neotropical lace bugs of the genus *Vatiga* (Heteroptera: Tingidae), pests of cassava: new synonymies and key to species. Proc Biol Soc Wash 95:457–462

Froeschner RC (1995) Review of the new world lace bug genera *Acanthocheila* Stål and *Carvalhotingis* new genus (Heteroptera: Tingidae). Proc Entomol Soc Wash 92:331–339

Froeschner RC (1996) Lace bug genera of the world, I: introduction, subfamily Cantacaderinae (Heteroptera: Tingidae). Smiths Control Zool 574:1–43

Gibson EH (1918) The genus *Corythucha* Stål (Tingidae; Heteroptera). Trans Am Entomol Soc 44:69–104

Gibson EH (1919a) The genus *Gargaphia* Stål (Tingidae; Heteroptera). Trans Am Entomol Soc 45:187–201

Gibson EH (1919b) The genus *Phatnoma* Fieber (Tingidae; Heteroptera). Trans Am Entomol Soc 45:181–185

Gomez-Menor J (1954) Tingides que viven sobre el peral. Bol Patol Veg Entomol Agric 20:369–392

Guidoti M, Barcellos A (2013) On the nymphs of lantana lace bug *Teleonemia scrupulosa* Stål (Hemiptera: Heteroptera: Tingidae: Tinginae): ontogenetic features of integumentary structures highlighted. Zootaxa 3613:289–296

Guilbert É (2001) Phylogeny and evolution of exaggerated traits among the Tingidae (Heteroptera, Cimicomorpha). Zool Scripta 30:313–324

Guilbert É (2004a) Immature stages of New Caledonian Tingidae (Heteroptera): description and development. Eur J Entomol 101:261–271

Guilbert É (2004b) Do larvae evolve the same way as adults in Tingidae (Insecta: Heteroptera)? Cladistics 20:139–150

Guilbert É (2005) Morphology and evolution of larval outgrowths of Tingidae (Insecta, Heteroptera), with description of new larvae. Zoosystema 27:95–113

Guilbert É (2012) Phylogeny of Cantacaderinae (Heteroptera: Tingidae) revisited after the description of a new genus and new species from New Caledonia. Eur J Entomol 109:111–116

Guilbert É, Montemayor SI (2010) Tingidae (Insecta, Heteroptera) from the Argentinan Yungas: new records and descriptions of selected fifth instars. Zoosystema 32:549–565
Guilbert É, Desutter-Grandcolas L, Grandcolas P (2008) Heterochrony in Tingidae (Insecta: Heteroptera): paedomorphosis and/or peramorphosis? Biol J Linn Soc 93:71–80
Guilbert E, Damgaard J, D'Haese C (2014) Phylogeny of the lacebugs (Insecta: Heteroptera: Tingidae) using morphological and molecular data. Syst Entomol 39:431–441
Harley KLS, Kassulke RC (1971) Tingidae for biological control of *Lantana camara* (Verbenaceae). Entomophaga 16:389–410
Harley KLS, Kassulke RC (1973) The suitability of *Teleonemia harleyi* for biological control of *Lantana camara* in Australia. Entomophaga 18:343–347
Heidemann O (1908) Two new species of North American Tingitidae. Proc Biol Entomol Wash 10:103–108
Henry JT, Wilson MR (2004) First records of eleven true bugs (Hemiptera: Heteroptera) from the Galapagos Islands, with miscellaneous notes and corrections to published reports. J NY Entomol Soc 112:75–86
Herney-Varón E, Moreira MD, Corredor JP (2010) Efecto de *Corythucha gossypii* sobre las hojas de higuerilla: Criterios para su muestreo y control con insecticidas. Corp Colomb Invest Agropec 11:41–47
Horváth G (1912) Species generis *Tingitidarum stephanitis*. Ann Historico-naturales Musei Nat Hungar 10:319–339
Humeres EC, Morse JG, Stouthamer R, Roltsch W, Hoddle MS (2009) Evaluation of natural enemies and insecticides for control of *Pseudacysta perseae* (Hemiptera: Tingidae) on avocados in Southern California. Fla Entomol 92:35–42
Hurd MP (1945) A monograph of the genus *Corythaica* Stål (Hemiptera: Tingidae). Iowa State Coll J Sci 20:79–99
Hurd MP (1946) Generic classification of North American Tingoidea (Hemiptera-Heteroptera). Iowa State Coll J Sci 20:429–493
Johnson CG (1936) The biology of *Leptobyrsa rhododendri* Horvath (Hemiptera, Tingitidae), the Rhodoendron lacebug. I. Introducion, bionomics and life history. Ann Appl Biol 23:342–368
King AM, Williams HE, Madire LG (2011) Biological control of cat's claw creeper, *Macfadyena unguis-cati* (L.) A.H.Gentry (Bignoniaceae), in South Africa. Afr Entomol 19:366–377
Kogan M (1960) *Corythaica cyathicollis* (Costa, 1864), aspectos sistemáticos, biológicos e econômicos (Hemiptera, Tingidae). Mem Instituto Oswaldo Cruz 58:59–88
Kormilev NA (1955) A new myrmecophil family of Hemiptera from the delta of Rio Parana, Argentina. Rev Ecuat Entomol Parasitol 2:465–477
Lee CE (1969) Morphological and phylogenic studies on the larvae and male genitalia of the East Asiatic Tingidae (Heteroptera). J Fac Agric 15:138–256
Leston D, Pendergrast JG, Southwood TRE (1954) Classification of the terrestrial Heteroptera (Geocorisae). Nature 174:91
Lis B (1999) Phylogeny and classification of Cantacaderini [= Cantacaderidae stat. nov.] (Hemiptera: Tingoidea). Ann Zool 49:157–196
Livingstone D (1976) On the functional anatomy of the egg and the description of the nymphal instars of *Dasytingis rudis* Drake & Poor (Heteroptera: Tingidae), a sap sucker on *Vitex negundo* (Verbinaceae). J Nat Hist 10:529–544
Livingstone D (1978) On the body outgrowths and the phenomenon of 'sweating' in the nymphal instars of Tingidae (Hemiptera: Heteroptera). J Nat Hist 12:377–394
Loeb MLG, Bell LK (2006) Distribution of care-giving effort in a communally breeding lace bug: fair guarding without coercion. J Insect Behav 19:19–30
Lopez-Montes AJ, Villa-Machado B, Madrigal-Cardeno A (1982) Ciclo de vida de la chinche de encaje *Corythucha gossypii* (F.) (Hemiptera: Tingidae) en girasol (*Helianthus annuus* L.). Rev Colomb Entomol 8:19–27
Marchini D, Del-Bene G, Dallai R (2010) Functional morphology of the female reproductive apparatus of *Stephanitis pyrioides* (Heteroptera, Tingidae): a novel role for the pseudospermathecae. J Morphol 271:473–482

Martínez HE, Ospina CM, Montoya EC, Constantino LM, Machado PB (2012) Aspectos biológicos de *Dictyla monotropidia* (Hemiptera: Tingidae), en nogal cafetero *Cordia alliodora* (Boraginaceae). Rev Colomb Entomol 38:306–313

Mason JR, Neal J, Oliver JE, Lusby WR (1991) Bird-repellent properties of secretions from nymphs of the Azalea lace bug. Ecol Appl 1:226–230

Median-Gaud S, Segarra-Carmona AE, Franqui RA (1991) The avocado lacewing bug, Pseudacysta perseae (Heidemann) (Hemiptera: Tingidae). J Agric Univ P R 75:185–188

Melksham JA (1984) Colonial oviposition and maternal care in two strains of *Leptobyrsa decora* Drake (Hemiptera: Tingidae). J Aust Entomol Soc 23:205–210

Miller LT, Nagamine WT (2005) First record of *Corythucha gossypii* (Hemiptera: Tingidae) in Hawaii, including notes on host plants. Proc Hawaii Entomol Soc 37:85–88

Monte O (1946) Revisão do genero "*Lepocysta*" Stål. Rev Brasil Biol 6:325–331

Montemayor SI (2009) Description of a new *Corythucha* Stål from Argentina (Hemiptera: Heteroptera: Tingidae), with a description of its life cycle. Zootaxa 2170:61–68

Montemayor SI (2010a) Description of a new *Amblystira* (Hemiptera: Heteroptera: Tingidae) from Argentina with a key to the South American species of the genus. Zootaxa 2675:65–68

Montemayor SI (2010b) Review of the genus *Leptocysta* Stål with descriptions of two new species (Hemiptera: Heteroptera: Tingidae) from Argentina. Zootaxa 2641:62–68

Montemayor SI, Dellapé PM (2010) On the identity of *Gargaphia subpilosa* Berg, 1879, *G. bergi* Monte, 1940 and *G. penningtoni* Drake, 1928 (Insecta, Hemiptera, Heteroptera, Tingidae), with the description of immatures of *G. bergi*. Zoosystema 32:155–162

Montemayor SI, Melo MC (2012) Synopsis of the genus *Corythaica* Stål (Insecta, Heteroptera, Tingidae), with the description of three new species from Argentina. Stud Neotrop Fauna Environ 47:119–130

Montemayor SI, González-Herrera A, Villalobos K (2011) Description of a new *Pleseobyrsa* (Heteroptera: Tingidae) from Costa Rica. Rev Mex Biodiversidad 82:475–480

Moreira DC, Redaelli LR, Guidoti M, Barcellos A (2013) Compared nymphal development of *Tingis americana* (Hemiptera, Tingidae) in two *Handroanthus* species (Bignoniaceae) and reproductive parameters in seedlings of *Handroanthus heptaphyllus*. Iheringia Sér Zool 103:195–199

Moznette GF (1922) The avocado, its insect enemies and how to combat them. USDA Farmer Bull 1261:1–31

Neal JW Jr, Douglas LW (1990) Seasonal dynamics and the effect of temperature in *Corythucha cydoniae* (Heteroptera: Tingidae). Environ Entomol 19:1299–1304

Neal JW Jr, Schaefer CW (2000) Lace bugs (Tingidae). In: Schaefer CW, Panizzi AR (eds) Heteroptera of economic importance. CRC Press, Boca Raton, pp 85–137

Neal JW Jr, Tauber CA, Tauber MJ (1992) Photoperiodic induction of reproductive diapause in *Corythucha cydoniae* (Heteroptera: Tingidae). Environ Entomol 21:1400–1418

Oliveira MAS, Fialho JF, Alves RT, Oliveira JNS, Gomes AC (2001) Dinâmica populacional do percevejo-de-renda (*Vatiga illudens*) na cultura da mandioca no Distrito Federal. Bol Pesq Desenv 3:1–15

Õszi B, Lanányi M, Hufnagel L (2005) Population dynamics of the sycamore lace bug in Hungary. Appl Ecol Environ Res 4:135–150

Péricart J (1983) Hémiptères Tingidae Euro-Méditerranéens. Faune Fr 69:1–620

Prado E (1990) Presencia en Chile de *Corythucha ciliata* (Say) (Hemiptera: Heteroptera: Tingidae). Rev Chile Entomol 18:53–55

Reuter OM (1910) Neue Beiträge zur Phylogenie und Systematik der Miriden nebst einleitenden Bemerkungen über die Phylogenie der Heteropteren-Familien. Acta Soc Sci Fenn 37:1–169

Roonwal JL (1952) The natural establishment and dispersal of an imported insect in India the lantana bug *Teleonemia scrupulosa* Stål (= *lantanae* Distant) (Hemiptera, Tingidae), with a description of its egg, nymphs and adult. J Zool Soc India 4:1–16

Ruiting J, Feng LYW, Yuzhou D (2009) Spread of and damage by an exotic lacebug, *Corythuca ciliata* (Say, 1832) (Hemiptera: Tingidae), in China. Entomol News 120:409–414

Santos RS, De-Freitas S (2008) Parasitismo de *Erythmelus tingitiphagus* (Soares) (Hymenoptera: Mymaridae) em ovos de *Leptopharsa heveae* Drake & Poor (Hemiptera: Tingidae), em plantios de seringueira (*Hevea brasiliensis* Müell. Arg.). Biol Control 37:571–576

Scholze W (1992) Sekretschwitzen bei Netzwanzenlarven: Zur Morphologie, Chemie und biologischen Bedeutung von Integumentbildungen mit sekretorischer Funktion bei Netzwanzenlarven (Heteroptera, Tingidae). Inaugural-dissertation zur Erlangung des Doktorgrades der Fakulta¨ t Biologie, Chemie und Geowissenschaften, Universita¨ t Bayreuth, Bayreuth

Schoonhoven A, Burbano F, Arenas R (1975) Notes on the biology of the lace bug *Gargaphia sanchezi* (Hemiptera, Tingidae) a pest of beans (*Phaseolus vulgaris*). Turrialba 25:327

Schuh RT, Slater JA (1995) True bugs of the world (Hemiptera: Heteroptera). Classification and natural history. Cornell University Press, Ithaca

Schuh RT, Stys P (1991) Phylogenetic analysis of Cimicomorphan family relationships (Heteroptera). J NY Entomol Soc 99:298–350

Schuh RT, Cassis G, Guilbert É (2006) Description of the first recent macropterous species of Vianaidinae (Heteroptera: Tingidae) with comments on the phylogenetic relationships of the family within the Cimicomorpha. J NY Entomol Soc 114:38–53

Scudder GGE (1959) The female genitalia of the Heteroptera: morphology and bearing on classification. Trans R Entomol Soc Lond 111:405–467

Setamou M, Showler AT, Reagan TE, Jones WA, Bernal JS (2005) *Leptodictya tabida* (Hemiptera: Tingidae): a potential threat to sugarcane production in lower Rio Grande valley of Texas. J Econ Entomol 98:1018–1023

Shakunthala N, Braman SK (2012) A scientific review on the ecology and management of the azalea lace bug *Stephanitis pyrioides* (Scott) (Tingidae: Hemiptera). J Entomol Sci 47:247–263

Shen HW, Wu WJ, Yang PS (1985) The biology of the azalea lacebug, *Stephanitis pyrioides* (Scott) I. The morphology of the azalea lacebug, *Stephanitis pyrioides* (Scott). Mem Coll Agric Nat Taiwan Univ 25:143–154

Silva CCAD, Barbosa SMDL (1986) An outbreak of *Gargaphia torresi* in bean crop in Alagoas state. Pesq Agropec Brasil 21:1003–1004

Stonedahl GM, Dolling WR, Heaume GJ (1992) Identification guide to common tingid pests of the world (Heteroptera: Tingidae). Trop Pest Manage 38:438–449

Stusák JM (1968) Notes on the bionomics and immature stages of *Tingis stachydis* (Fieber) (Heteroptera, Tingidae). Acta Entomol Bohemosl 65:412–421

Stys P, Kerzhner IM (1975) The rank and nomenclature of higher taxa in recent Heteroptera. Acta Entomol Bohemsl 72:65–79

Swezey O (1945) Insects associated with orchids. Proc Hawaiian Entomol Soc 12:343–403

Tallamy DW (1985) "Egg dumping" in lace bugs (*Gargaphia solani*, Hemiptera: Tingidae). Behav Ecol Sociobiol 17:357–362

Tallamy DW, Denno RF (1981) Maternal care in *Gargaphia solani* (Hemiptera: Tingidae). Anim Behav 29:771–778

Tallamy DW, Iglay RB (2004) Maternal care in *Compseuta picta*, an African lace bug (Heteroptera: Tingidae). J Insect Behav 17:247–249

Thomas SE, Ellsion CA (2000) A century of classical control of *Lantana camara*: can pathogens make a significant difference? In: Spencer NR (ed) Proceedings of the X international symposium on biological control of weeds. Montana State University, Bozeman, pp 97–104

Tullgren A (1918) Zur Morphologie und Systematik der Hemipteren. Entomol Tidskrift 39:113–132

Van Doesburg PH (1977) A new species of *Thaumamannia* from Surinam (Heteroptera, Tingidae, Vianaidinae). Zool Mededelingen 52:187–189

Weiss HB, Headlee TJ (1918) Some new insect enemies of greenhouse and ornamental plants in New Jersey. NJ Agric Exp Stn Circ 100:3–9

Winder JA, Harley KLS (1983) The phytophagous insects on Lantana in Brazil and their potential for biological control in Australia. J Entomol 29:346–362

Part VII
Infraorder Pentatomomorpha

Chapter 15
Flat Bugs (Aradidae)

Maria del C. Coscarón and Eugenia F. Contreras

Abstract Aradidae or "flat bugs" have a flattened shape and somber colors well adapted for a lifestyle on or under the bark of dead trees. The pronounced elongation of the stylets is a remarkable adaptation to exploit juices of fungi during feeding. For the Neotropical Region, 80 genera and 509 species are recognized. General characteristics, diagnosis, biology, and ecology of Neotropical aradids are presented.

15.1 Introduction

Aradids are commonly known as "flat bugs." Their flattened form and somber colors are adaptations for a life on or under the bark of dead trees. They have succeeded in utilizing the mycelia of various fungi. This ability has been accomplished by a remarkable elongation of the stylets, which lie in a vast coil within the head of the bug when it is not feeding and extend for a long distance into the slender threads of the fungus when it is feeding. Many flat bugs are mycophagous, as suggested by the fact that the majority of the species live either under the bark of decaying trees or under bark chips, twigs, or debris on the floor of moist forests; probably others feed on the sap of dying or living trees. Macropterous species tend to live on a restricted number of host trees; some species feed on the phloem, cambium, and xylem of living trees and may affect the growth of these trees. Some aradids may be associated with termites and occasionally they are observed infected by mites.

M.C. Coscarón (✉)
División Entomología, Museo de La Plata,
Paseo del bosque s/n, La Plata 1900, Buenos Aires, Argentina
e-mail: mcoscaron@fcnym.unlp.edu.ar

E.F. Contreras
Instituto de Biología de la Altura, Universidad Nacional de Jujuy,
Av. Bolivia 1661, San Salvador de Jujuy 4600, Argentina
e-mail: eugecon@yahoo.com.ar

A.R. Panizzi, J. Grazia (eds.), *True Bugs (Heteroptera) of the Neotropics*,
Entomology in Focus 2, DOI 10.1007/978-94-017-9861-7_15

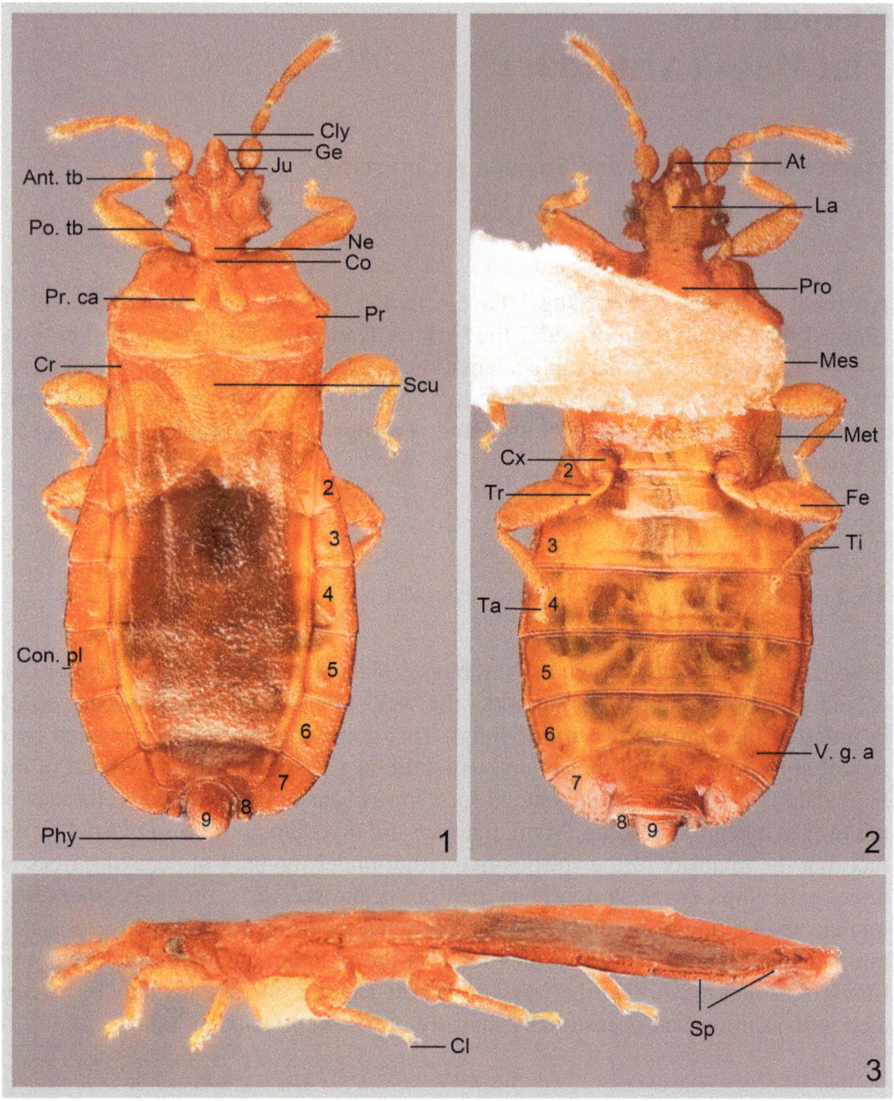

Plate 15.1 *Aneurus patriciae* Picchi. Fig. 1, dorsal view: Ant. tb: antenniferous tubercles; Cly, clypeus; Co, collar; Con. pl, connexival plate; Cr, corium; Ge, gena; Ju, jugum; Ne, neck; Phy, pygophore; Po. tb, postocular tubercles; Pr, pronotum; Pr. ca, pronotal callosities; Scu, scutellum. Fig. 2, ventral view: At, atrium; Cx, coxa; Fe, femur; La, labium; Mes, mesosternum; Met, metasternum; Pro, prosternum; Ta, tarsus; Ti, tibia; Tr, trochanter; V.g.a, ventral glabrous areas. Fig. 3, lateral view: Cl, claws; Sp, spiracles

Aradids are generally somber in color, black or brown, and elliptical, oval, or rectangular in shape. They range from 3 to 11 mm in length. Many tropical species are wingless and the dorsal surface is frequently appearing granular or rugose.

The family Aradidae exhibits a variety of wing development, macroptery, brachyptery, and aptery. There is camouflage, plus a rough surface that holds layers of dirt and debris which are important antipredator survival adaptations in these litter dwelling species.

Many species never go out of the bark, spending their whole life under it, whereas some tropical species have flights of dispersion during the spring.

15.2 General Characteristics and Diagnosis

Flat bugs are easily identified by bearing mandibular and maxillary stylets extremely elongated and coiled. Ocelli are absent, labia usually short and stout with four distinct segments, trochanters sometimes fused with femora, and tarsi bisegmented. Figures 1, 2, and 3 (dorsal, lateral, and ventral views, respectively) show the main characters used for the description of species.

The head (Figs. 4, 5, and 6) is of the prognathous type, and it is highly specialized to accommodate the mycetophagous sucking mouth parts; mouth parts have a general similar pattern among species, but they differ strikingly in degree of development of the various parts. The insertion of the rostrum is distinctive for each subfamily. They have elongated maxillary and mandibular stylets that are coiled within the head. In the anterior part of the head (clypeus), the bugs have coiled setae composed of mandibular and maxillary stylets. In the resting position, both the stylets are coiled up to form a compact double spiral (Weber 1930), and when the bug is feeding, the length of the uncoiled stylets may be as much as five to six times that of the insect body length.

The thorax (Fig. 1) bears the wings which are well developed and are diverse in development and venation. The wings may fall off, and this character is usually dimorphic, females usually being brachypterous and males macropterous; occasionally, females might be in both forms, and rarely males are dimorphic (Schuh and Slater 1995). Wing reduction is known to occur on half of the known species, and it can be extreme, leaving no trace of wings. Miller (1938) recognized that the apterous tropical flat bug that was considered a nymph was, in fact, an adult. Micropterous forms are not easily distinguishable from truly apterous forms; the latter shows more or less extreme reduction and fusions of the thoracic sclerites, especially the scutellum, and may also be covered with coarse incrustations that almost obscure the body texture and shape. Monteith (1969) demonstrated that forms that were considered new genera were in fact macropterous and apterous morphs of the same species.

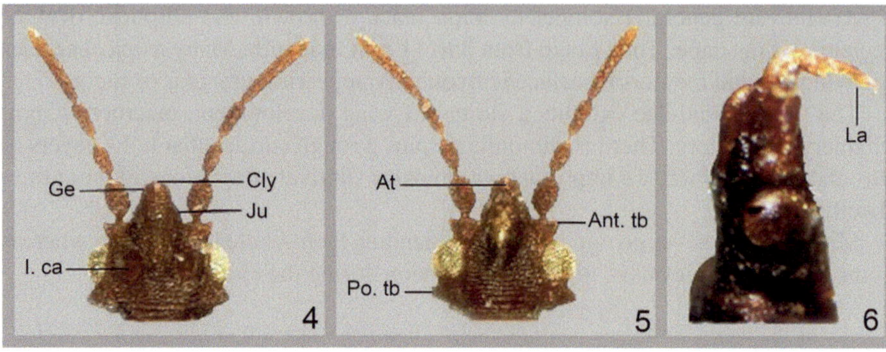

Plate 15.2 *Aneurus mexicanus* Kormilev. Fig. 4, head dorsal view: Cly, clypeus; Ge, gena; Ju, jugum; I. ca, interocular callosities. Fig. 5, head ventral view: Ant. tb, antenniferous tubercle; At, atrium; Po. tb, postocular tubercles. *Isodermus gayi* Erichson. Fig. 6, head lateral view: La, labium

The metathoracic scent glands are well developed (Carayon 1955), and the canals and evaporating areas are variously and often strikingly developed. The dorsal scent glands may persist in adults.

The legs in Aradidae vary considerably. The trochanters are commonly distinct from the femora; in some subfamilies those of the front and middle legs are fused with scarcely a trace of the dividing suture; in other cases, trochanters are completely fused. The front tibiae have an apical comb on the inner side; in a few groups this structure is preapical, spine-like, and directed inward at right angles to the main axis of the tibia. Tarsi are typically two segmented. Arolia occur in most subfamilies and in a majority of the genera of the world. The legs are variously modified for stridulation, the femora with scrapers, and the tibiae with files.

The abdomen (Figs. 1 and 2) shows the connexivum as a part of the tergum, since the spiracles are considered to mark the lower limit of the tergum. The connexiva of the first segment or the lateral parts of the first tergite are absent. The spiracles are clearly recognizable from the 2nd to the 8th segments in both sexes in all subfamilies.

The female genital segments (Fig. 7) begin with the 7th segment. The median longitudinal split of the 7th sternum is common to all subfamilies, except Aneurinae. Male genital segments (Fig. 8) begin with the 8th segment.

Stridulatory structures are present in the ventral surface of the abdomen. These take the form of a comb on the hind margin of the 4th segment on either side, a file-like curved elevation on the 3rd segment, and long longitudinal knife-like carinae on the 4th segment. The file or comb-like structures are opposed by carinae on the femora and knife-like surfaces of the hind tibiae.

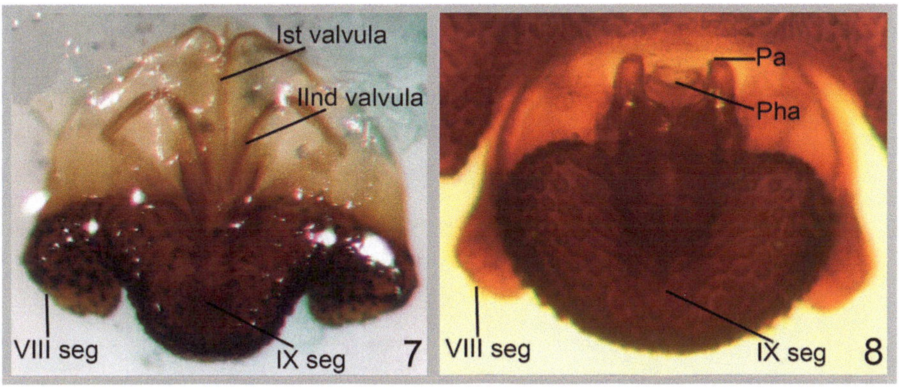

Plate 15.3 *Neuroctenus subandinus* Kormilev. Fig. 7, female genitalia: Ist valvula, first valvula; IInd valvula, second valvula; VIII seg, eighth segment; IX seg, ninth segment. Fig. 8, male genitalia: Pa, paramere; Pha, phallus; VIII seg, eighth segment; IX seg, ninth segment

15.3 General Biology and Ecology

Flat bugs are inhabitants of the wet rainforest floor, where they find the food sources (fungi) to feed on; the loss of wings is due to the no need for dispersal flights. However, the macropterous species, which live subcortically feeding on ephemerous food sources, are believed to use wings for a dispersal flight to colonize new wood sources. Several bark flat bugs species are known to live on small-diameter branches (Heiss and Péricart 2007) where fungi are generally more diversified (Heilmann-Clausen and Christensen 2004; Nordén et al. 2004).

Most of the available information on their biology and ecology is based on occasional observations (Heiss and Péricart 2007). Gossner et al. (2007) studied habitat requirements (e.g., habitat continuity, host and resource availability, and shade tolerance). According to Marchal et al. (2012), aradids vary in their habitat preferences and can be sensitive to different types of exploitations. Some species need large senescent trees, while others depend on small fresh deadwood and, therefore, are more sensitive to fuelwood harvesting.

Flat bugs sometimes live with termites in nests (Hendee 1933; Kormilev and Froeschner 1987; Usinger 1936). Aradids have been reported living in nest of birds (Vásárhelyi 1979) and rodents (Blatchley 1934) and in galleries of wood-boring beetles. According to Usinger and Matsuda (1959), they hibernate under any bark that is loose, and many records are not necessarily related to their feeding habits.

McClure (1932) reported prenatal care and stated that females leave the egg mass, and presumably males move on top of the eggs and remain there for 2 weeks or until the eggs hatch and even remain with young nymphs for 1 or 2 days. Taylor

Plate 15.4 Cryptic: *Mezira paragranuliger* Kormilev Figs. 9 and 10

(1988) suggests the occurrence of paternal care rather than maternal care. Although paternal care is known in a few Heteroptera (e.g., Ralston (1977), Ridley (1978), Smith (1980)) most notably in the Belostomatidae, maternal care is by far more typical.

Aradids are cryptical (Figs. 9 and 10) and move quickly when disturbed. They are difficult to see when resting on the surface of the bark but more often, and especially in the nymphal stage, are conspicuous against the white surface of the fungus when feeding.

15.3.1 Eggs and Oviposition

Copulation in aradids takes place in their natural habitats on fungi, beneath the bark. During copulation, the male positions below the female. Eggs are deposited in frass beneath the bark or in rotten logs or stumps or glued to the surface of the bark or fungi. Some species lay their eggs on the substrate free, while others lay their eggs with a sticky substance to adhere to the substrate. Some eggs are found partially embedded in the softwood in the galleries of termites and also laid in a regular manner attached to pieces of bark. Eggs are elongate-oval in shape, without a cap but with several micropylar processes at the anterior end. Hexagonal reticulations are a characteristic feature of these eggshells after hatching. A single female may lay from 14 to 45 eggs. In general, aradids have one or two annual generations and are represented in all the biogeographical regions of the world (Usinger and Matsuda 1959).

15.3.2 Nymph Feeding and Development

Nymphs (Figs. 11, 12, 13, and 14) are commonly found with the adults under a loose bark, and in these circumstances, it is not known whether this represents a concentration on or about a limited source of food or whether a genuine gregarious instinct is involved (Usinger and Matsuda 1959). Nymphal abdominal scent glands are present between terga 4/5 and 5/6 or between 4/5 and 5/6 or, less common, between terga 3/4. They are prominent and functional (Schuh and Slater 1995).

In general, aradids are uni- or bivoltine (i.e., one or two generations per year). However, the species *Aradus cinnamomeus* Panzer has a 2-year life span in most parts of Europe; in northern Europe, the life span is 3 years (Brammanis 1975; Heliövaara and Väisänen 1987). Eggs are laid in May; nymphs hatch in June and reach the 4th instar during the first summer. After hibernation, at the base of the tree or in the surrounding litter (Brammanis 1975; Heliövaara 1982), the bugs become adult in July–August of the following year, after which they hibernate. It is not until the following spring that the bugs mate and begin laying eggs (Tropin 1949; Brammanis 1975; Heliövaara 1984; Vásárhelyi and Böröcz 1987).

15.3.3 Adult Feeding and Reproduction

Most of the species are subcortical and mycetophagous. They live on or under the bark of dead trees and the coiled setae are remarkable adaptations for feeding on juices of fungi. They are mostly found on branches that have died recently and are not totally desiccated. When the tree becomes too dry, they apparently abandon it for another branch. They are, however, weak fliers and have a low vagility (Jacobs 1986).

The association of aradids with a particular species of trees has been summarized by Kiritshenko (1913), Stichel (1957), and Usinger and Matsuda (1959) for Europe and by Parshley (1921) for the Nearctic Region. Aradids are generally found in quiescent state, feeding, resting, or in hibernation. According to Yonke (1991), a particular aradid may be suitable for specific fungal growth.

Several species of aradids have been collected using the Berlese funnel on ground forest litter or by sweep netting the dry foliage or by means of using malaise or light traps (Table 15.1).

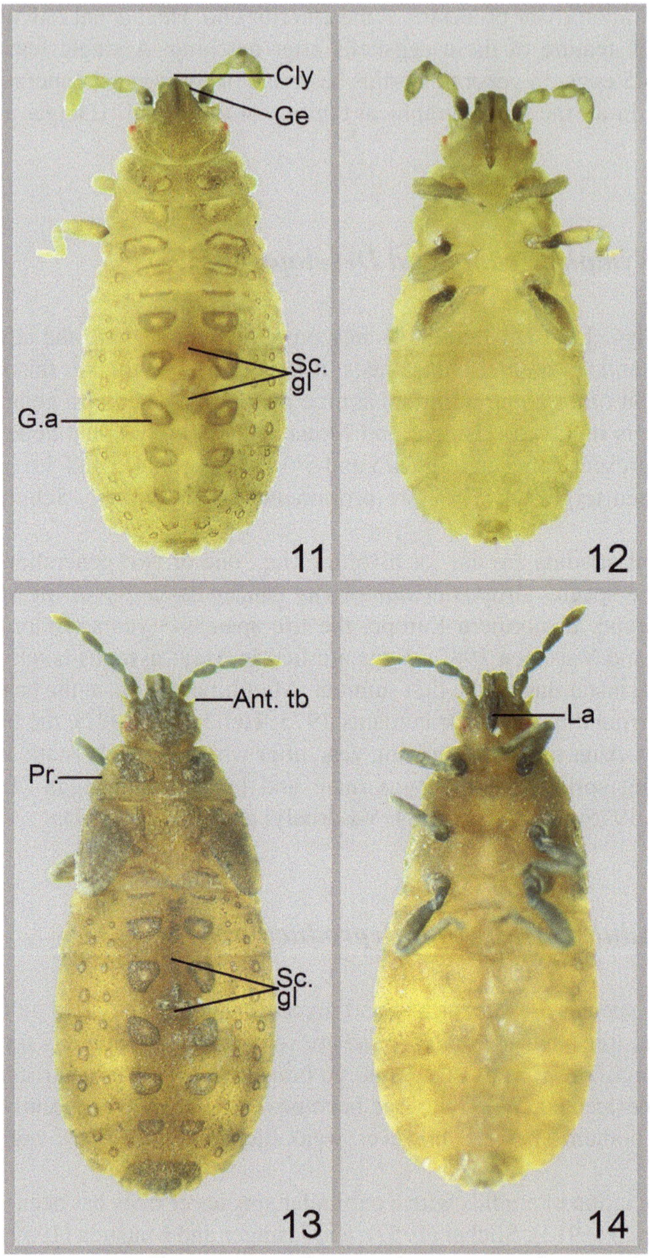

Plate 15.5 *Neuroctenus subandinus* Kormilev. Fig. 11, first immature stage, dorsal view: Cly, clypeus; G. a, glabrous area; Ge, gena; Sc gl, scent glands. Fig. 12, first immature stage, ventral view. Fig.13, fourth immature stage, dorsal view: Ant. tb, antenniferous tubercles; Pr, pronotum; Sc gl, scent glands. Fig. 14, fourth immature stage, ventral view: La, labium

Table 15.1 Collecting methods used to sample aradids

Species	Collecting methods
Eretmocoris disparis Drake & Maldonado	Berlese funnel from ground forest litter (Drake and Maldonado 1955)
Peggicoris zeteki Drake	Berlese funnel from forest litter (Drake 1956)
Acaricoris barroanus Drake & Kormilev; *Acaricoris clausus* Drake & Kormilev	Berlese funnel from forest litter (Drake and Kormilev 1958)
Proxius (Neoproxius) nicaraguensis Kormilev	Swept from dry foliage (Kormilev 1982)
Aradus pericarti Heiss; *Aphleboderris comata* Champion; *Aphleboderris pubescens* (Walker); *Dysodius magnus* Heiss; *Helenus hesiformis* White; *Hesus acuminatus* (F.); *Hesus cordatus* (F.); *Hesus flaviventris* (Burmeister); *Hesus humeralis* Heiss; *Mezira cf. boliviana* Kormilev; *Mezira barberi* Kormilev; *Mezira handlirschi* (Bergroth); *Mezira laeviventris* (Champion); *Neuroctenus longulus* Bergroth; *Neoproxius gypsatus* (Bergroth); *Phyllotingis eximia* (Haglund); *Phyllotingis cf. triangula* Heiss; *Placogenys cockerelli* Usinger & Matsuda; *Santaremia robusta* Kormilev	Interception trap, malaise trap, light trap (Heiss and Moragues 2009)

15.4 Classification and Diversity

The family Aradidae has been placed within the infraorder Pentatomomorpha (Leston 1954). Today, flat bugs are grouped into 8 subfamilies: Aneurinae, Aradinae, Calisiinae, Carventinae, Chinamyersiinae, Isoderminae, Mezirinae, and Prosympiestinae (Schuh and Slater 1995). The subfamily Chinamyersiinae is classified in two tribes, Tretocorini and Chinamyersiini, and the subfamily Prosympiestinae is divided in two tribes, Llaimacorini and Prosympiestini. Usinger and Matsuda (1959) provided a systematic account of the family and Kormilev and Froeschner (1987) a world catalog. Vásárhelyi (1987) and Grozeva and Kerzhner (1992) provided a hypothesis on the relationship among the Aradidae.

The family consists of about 1,931 species distributed in approximately 233 genera in the world. For the Nearctic Region, 10 genera and 123 species are known (Henry 2009); for the Neotropical Region, 80 genera and 509 species are referred (Table 15.2) (Coscarón and Contreras 2012); the Palearctic Region is represented by 28 genera and 204 species, and for the Australian Region, there are 38 genera and 143 species cited (Henry 2009).

Table 15.2 List of species of aradids from the Neotropical Region and their geographical distribution

Subfamily	Genera	Species	Distribution
Aneurinae	*Aneurillus*	*A. doesburgui* (Kormilev)	Suriname: Albina
	Aneurosoma	*A. dissimile* (Bergroth)	Argentina: Misiones; Chaco. Brazil: Amazonas; Rio Grande do Sul Colombia and Guadeloupe
			Panama: Chiriqui. Paraguay: Villa Rica
	Aneurus	*A. aterrimus* Kormilev	Costa Rica: Puntarenas
		A. barberi Kormilev	Saint Lucia, British West Indies
		A. bosqui Kormilev	Argentina: Misiones. Brazil: Rio Grande do Sul. Paraguay: Villarica
		A. brailovskyi Kormilev	Mexico: Veracruz
		A. bucki Kormilev	Brazil: Rio Grande do Sul
		A. burmeisteri Bergroth	Colombia
		A. championi Kormilev	Guatemala: Sacatepequez. Mexico: Veracruz
		A. equatoriensis Kormilev	Ecuador
		A. fiskei Heidemann Kormilev	Cuba: Cienfuegos. Mexico: Nueva León
		A. froeschneri Kormilev	Mexico: Veracruz. Panama: El valle
		A. guanacastensis Kormilev	Costa Rica: Guanacaste
		A. haitiensis Kormilev	Haiti
		A. herediensis Kormilev	Costa Rica: Guanacaste; Heredia
		A. hispaniolensis Picchi	Haiti
		A. hrdyi Ŝtys	Cuba
		A. maryae Picchi	Mexico: Michoacan
		A. mexicanus Kormilev	Mexico: Veracruz
		A. minutus Bergroth	Guatemala. Panama: Chiriqui
		A. montanus Champion	Guatemala: Totonicapan
		A. nasutus Kormilev	Guadeloupe
		A. neojamaicensis Picchi	Jamaica

		A. patriciae Picchi	Jamaica. Cuba: Villa Clara.
		A. pisoniae Kormilev	Puerto Rico: Maricao
		A. pusillus Kormilev	Guatemala: Alta Verapaz. Panama
		A. schuhi Kormilev	Brazil: Amazonas
		A. slateri Picchi	The Bahamas
		A. usingeri Picchi	Costa Rica: Guanacaste. Guatemala: Escuintla
			Mexico: Veracruz. Panama: Chiriqui
		A. vauriei Kormilev	Guadeloupe. Jamaica
		A. veracruzensis Picchi	Mexico: Veracruz
		I. aibonitensis (Kormilev)	Puerto Rico: Aibonito
		I. bergi (Kormilev)	Argentina: Corrientes
		I. bispinosus (Kormilev)	Brazil: Santa Catarina; Rio Grande do Sul
		I. bolivianus (Kormilev)	Bolivia
		I. carioca (Kormilev)	Brazil: Rio de Janeiro
		I. costaricensis (Kormilev)	Costa Rica: Guanacaste
		I. flavomaculatus (Distant)	Ecuador: Pichincha
		I. fritzi (Kormilev)	Bolivia. Peru
		I. leptocerus (Hussey)	Guatemala: Baja Verapaz. Mexico: Chiapas
			Panama. Chiriqui
		I. longicornis (Kormilev)	Costa Rica
		I. marginalis (Walker)	Bolivia. Colombia
		I. monrosi (Kormilev)	Argentina: Salta
		I. plaumanni (Kormilev)	Brazil: Santa Catarina
		I. politus (Say)	Cuba. Guatemala. French Guiana: St Laurent du Maroni
		I. sahlbergi (Bergroth)	Brazil: Minas Gerais; São Paulo

(continued)

Table 15.2 (continued)

Subfamily	Genera	Species	Distribution
		I. simulans (Walker)	Brazil: Rio de Janeiro
		I. subdipterus (Burmeister)	Argentina: Misiones; Buenos Aires; Chaco; Córdoba; Formosa; Jujuy; Formosa; Salta; Santiago del Estero; Tucumán. Brazil
		I. tenuis (Champion)	Panama: Chiriqui. Mexico: Guerrero. Nicaragua: Rivas
		I. wygodzinskyi (Picchi)	Jamaica
Aradinae	*Aradiolus*	*A. chemsaki* Kormilev	Mexico
		A. paradoxus Kormilev	Mexico: Oaxaca
		A. angustellus (Blanchard)	Argentina: Buenos Aires; Córdoba; Chaco; Santa Cruz; Rio Negro. Chile. Peru: Cuzco
		A. brasiliensis Usinger	Argentina: Misiones. Brazil: Mato Grosso; Santa Catarina; São Paulo. Paraguay: San Bernardino
		A. compressicornis Stål	Colombia. Venezuela: Trujillo
		A. falleni Stål	Brazil: Chapada; Rio de Janeiro. Guatemala: Alta Verapaz
			Panama. Mexico: Guerrero; Chiapas. Belize: Cayo
		A. fronterana Drake	Costa Rica: Guanacaste. Mexico: Tabasco
		A. gracilicornis Stål	Cuba. Mexico: Morelos; Chiapas
		A. granuliger Kormilev	Mexico
		A. mexicanus Usinger	Argentina: Misiones; Tucumán. Brazil: Espirito Santo
			Mexico: Temascaltepec
		A. pallidicornis Stål	
		A. penningtoni Drake	Argentina: Buenos Aires; Corrientes; Chaco; Formosa; Jujuy; Misiones; Mendoza; Santa Fé
			Bolivia: Santa Cruz. Brazil: Chapada
			Paraguay: Horqueta

		A. pericarti Heiss	Suriname: Marowijne. French Guiana
		A. testaceus Kormilev	Mexico
Calisiinae	*Calisiopsis*	*C. ampliceps* Champion	Brazil: Porto Alegre; Santa Catarina. Panama
		C. brasiliensis Kormilev	Brazil: Rio Grande do Sul; Santa Catarina
		C. brodzinskyorum Froeschner	Dominican Republic
		C. kormilevi Froeschner	Peru: Madre de Dios
		C. minuta Kormilev	Mexico: Tampico
		C. nigrotuberculata (Froeschner)	Mexico: Texas
		C. planiceps Kormilev	Brazil: Santa Catarina
	Calisius	*C. affinis* Barber	The Bahamas
		C. bilobatus Kormilev	Bolivia
		C. brasiliensis Kormilev	Brazil: Santa Catarina
		C. confusus Kormilev	Brazil: Santa Catarina
		C. contubernalis Bergroth	Guadeloupe
		C. elegantulus Bergroth	Guadeloupe
		C. farri Kormilev	Jamaica: Saint Thomas
		C. ferox Champion	Panama
		C. gracilis Kormilev	Guatemala
		C. insignis Kormilev	Guatemala
		C. longiventris Kormilev	Panama
		C. major Bergroth	Venezuela: La Guaira
		C. pallipes Stål	Brazil
		C. pangoaesnsis Kormilev	Peru
		C. placidus Horváth	Brazil: Goiáz; São Paulo
		C. pulcher Kormilev	Brazil: Minas Gerais

(continued)

Table 15.2 (continued)

Subfamily	Genera	Species	Distribution
Carventinae	*Acaricoris*	*A. austeris* Drake & Kormilev	Guadeloupe
		A. barroanus Drake & Kormilev	Panama
		A. clausus Drake & Kormilev	Puerto Rico: Mayagüez
		A. haitiensis Kormilev	Haiti
		A. lattini Heiss	Dominican Republic
	Aglaocoris	*A. clarkei* Drake	Dominica in Leeward West Indies Islands
		A. comes Drake	Guadeloupe
		A. cubanus Drake	Cuba
		A. drakei Kormilev	Haiti: Port-au-Prince
		A. invisus Drake	Guadeloupe Island.
		A. natalii Drake & Maldonado	Puerto Rico: Yauco; Mayaguez
		A. orientalis Grillo Ravello	Cuba
		A. rectangularis Usinger & Matsuda	Santo Domingo
		A. vicinus Drake	Guadeloupe
	Alayocoris	*A. hirsutus* Grillo Ravelo	Cuba: La Habana
	Aparilocoris	*A. mexicanus* (Kormilev)	Mexico: Oaxaca
		A. venezuelanus Kormilev	Venezuela: Aragua
	Apterocoris	*A. surinamensis* Kormilev & Doesburg	Suriname: Brokopondo
	Atactocoris	*A. farri* Kormilev	Jamaica
		A. perneri Heiss	Jamaica
	Baracoris	*B. alayoi Grillo Ravelo*	Cuba
	Carpocoris	*C. cubanus* Grillo Ravelo	Cuba
	Carventus	*C. chilensis* Kormilev	Chile: Concepción
		C. mexicanus Bergroth	Colombia: Cundinamarca. Panama: Chiriqui
	Cubanocoris	*C. alayoi* Grillo Ravelo	Cuba

	Dihybogaster	*D. incrustatus* Kormilev	Brazil: Santa Catarina
		D. plana Kormilev	Brazil: Santa Catarina
		D. plaumanni Kormilev	Brazil: Santa Catarina
		D. wygodzinskyi Usinger & Matsuda	Brazil: Rio de Janeiro
	Eretmocoris	*E. cubensis* Usinger & Matsuda	Cuba
		E. dominicus Kormilev	Dominica in Leeward West Indies Islands
		E. gigas Usinger & Matsuda	Puerto Rico
		E. longicornis Usinger & Matsuda	Haiti
		E. minimus Usinger & Matsuda	Cuba: Cienfuegos
		E. productus Usinger & Matsuda	Guadeloupe
		E. prominens Usinger & Matsuda	Puerto Rico
		E. tatei Harris & Drake	Puerto Rico
	Glyptocoris	*G. annulatus* Kormilev	Brazil: Santa Catarina
		G. confusus Kormilev	Brazil: Santa Catarina
		G. espiritosantensis Wygodzinsky	Brazil: Espirito Santo
		G. fluminensis Wygodzinsky	Brazil: Rio de Janeiro
		G. insularis Drake	Guadeloupe
		G. milleri Wygodzinsky	Brazil: Rio de Janeiro
		G. minutus Kormilev	Brazil: Santa Catarina
		G. plaumanni Kormilev	Brazil: Mato Grosso
		G. sejunctus Harris & Drake	Brazil
		G. spinosus Usinger & Matsuda	Brazil: Parana; Rio de Janeiro
		G. verus Drake	Guadeloupe in Leeward West Indies Islands
	Kolpodaptera	*K. minuta* Kormilev	Guatemala
		K. panamensis Usinger & Matsuda	Panama

(continued)

Table 15.2 (continued)

Subfamily	Genera	Species	Distribution
		K. prominens Usinger & Matsuda	Guadeloupe
		K. rugosa Kormilev	Cuba
	Laevsoicoris	*L. lissonotus* Grillo Ravelo	Cuba
	Lepidocoris	*L. garcesi* Grillo Ravelo	Cuba: Camagüey
	Neoproxius	*N. amazonicus* (Kormilev)	Brazil: Amazonas
		N. carioca (Kormilev)	Brazil: Rio de Janeiro
		N. costariquensis (Kormilev)	Costa Rica: Turrialba
		N. gypsatus (Bergroth)	Bolivia: Santa Cruz. Brazil: Pará; São Paulo
			French Guiana. Guatemala: Alta Verapaz. Panama. Peru. Venezuela
		N. incaicus (Kormilev)	Peru
		N. lindemannae (Kormilev)	Brazil: Amazonas
		N. magdalenae (Kormilev)	French Guiana. Nicaragua
		N. nicaraguensis (Kormilev)	Nicaragua
		N. palliatus (Champion)	Guatemala. Panama. Peru
		N. panamensis (Kormilev)	Panama
		N. personatus (Champion)	Guatemala: Alta Verapaz. Panama: Chiriqui
		N. peruvianus (Kormilev)	Peru
	Peggicoris	*P. zeteki* (Drake)	Panama
	Planocoris	*P. orientalis* Grillo Ravelo	Cuba
	Pokoldaptera	*P. humicola* Grillo Ravelo	Cuba
	Protokolpodaptera	*P. habanense* Grillo Ravelo	Cuba: La Habana
	Proxius	*P. incrustatus* Stål	Brazil: Rio de Janeiro
		P. schwarzii Heidemann	Cuba: Camagüey; La Habana
	Psorosoma	*P. forficulinum* Champion	Panama
	Reeceicus	*R. saileri* Drake	Peru

	Rhysocoris	*R. cubanus* Kormilev	Cuba: Cienfuegos
		R. disparis (Drake & Maldonado)	Puerto Rico: Mayagüez
		R. jamaicensis Kormilev	Jamaica
		R. rugosus Usinger & Matsuda	Haiti
	Tainocoris	*T. alayoi* Grillo Ravelo	Cuba
	Yvacoris	*Y. litoralis* Grillo Ravelo	Cuba
	Zayasocoris	*Z. trinitarius* Grillo Ravelo	Cuba
Isoderminae	*Isodermus*	*I. gayi* (Spinola)	Chile. Argentina: Neuquén; Río Negro; Santa Cruz; Tierra del Fuego
Mezirinae	*Allelocoris*	*A. dryadis* Drake & Harris	Brazil: Rio de Janeiro
	Aphleboderrhis	*A. alata* Kormilev	Brazil
		A. comata Champion	Brazil: Pará; Rio Grande do Sul. French Guiana
			Panama: Chiriqui. Peru
		A. hirticornis (Bergroth)	Brazil
		A. pilosa Stål	Argentina: Misiones; Jujuy; Tucumán. Paraguay. Peru
		A. procerula (Bergroth)	Brazil
		A. pubescens (Walker)	Brazil: Pará. French Guiana. Guatemala: Alta Verapaz
			Mexico. Panama: Chiriqui. Peru. Suriname: Paramaribo
		A. tomentosa (Bergroth)	Brazil
	Artagerus	*A. crispatus* Stål	Bolivia. Brazil: Pará; Rio de Janeiro. Panama: Chiriqui. Peru
		A. hispidus Champion	Brazil: Amazonas; Para. Panama: Chiriqui. Peru: Huánuco
		A. histricus Stål	Brazil. Mexico: Veracruz
		A. martinezi Kormilev	Bolivia: Cochabamba. Brazil: Amapá; Amazonas. Peru

(continued)

Table 15.2 (continued)

Subfamily	Genera	Species	Distribution
		A. montandoni Bergroth	Brazil: Amazonas
		A. plaumanni Kormilev	Brazil: São Paulo; Santa Catarina
		A. setosus Stål	Colombia: Bogotá. Panama
	Asterocoris	*A. australis* Drake & Harris	Brazil: Rio de Janeiro
		A. schubarti Wygodzinsky	Brazil: Espirito Santo; São Paulo
	Banksiessa	*B. pubescens* Usinger & Matsuda	Panama
	Bergrothiessa	*B. confusa* Kormilev	Brazil: Parana
		B. grossa Kormilev	Brazil: Parana
		B. intermediaria (Kormilev)	Brazil: Santa Catarina
		B. paranensis Kormilev	Brazil: Parana
		B. plaumanni Kormilev	Brazil: Santa Catarina; Minas Gerais
		B. rufa Kormilev	Brazil: Parana
		B. usingeri (Wygodzinsky)	Brazil: Rio de Janeiro
	Chapadia	*C. alata* Kormilev	Brazil
	Cinyphus	*C. alatus* Kormilev	Mexico: Xalapa
		C. amazonicus Kormilev	Bolivia: Santa Cruz
		C. andinus Kormilev	Peru
		C. armillatus Bergroth	Bolivia. Colombia
		C. emarginatus (Stål)	Guatemala: Alta Verapaz. Mexico: Chiapas
		C. lutosus Champion	Panama. Peru
		C. meziroides Kormilev	Peru
		C. minutus Blöte	Bolivia: Coroico
		C. ovatus Kormilev	Brazil. Peru
		C. peruvianus Kormilev	Peru
		C. saileri Kormilev	Mexico: Veracruz

		C. squalidus Champion	Nicaragua: Chontales. Panama: Chiriqui. Venezuela
		C. subtruncatus Bergroth	Nicaragua: Chontales. Panama: Chiriqui. Venezuela
		C. terminalis Kormilev	Venezuela: Aragua
		C. venezuelanus Kormilev	Venezuela: Aragua
	Coloborrhinus	*C. ovatus* (Kormilev)	Brazil: Minas Gerais. Peru. Suriname
		C. peruvianus Kormilev	Peru
		C. pumilio Champion	Brazil: Minas Gerais. Panama
	Ctenoneurus	*C. necopinatus* Bergroth	Saint Thomas in Leeward West Indies Islands
	Delnocoris	*D. micropterus* Kormilev	Costa Rica: Heredia
	Dysodius	*D. ampliventris* Bergroth	Brazil: Amazonas. Colombia. Panama. Suriname
		D. brailovskyi Heiss	Mexico
		D. brevipes Bergroth	Costa Rica: Guanacaste. Mexico: Chiapas; Guerrero; Oaxaca
		D. crenulatus (Stål)	Colombia. Costa Rica: Guanacaste; Puntarenas
			Guatemala: Alta Verapaz; Baja Verapaz. Panama: Chiriqui
			Mexico
		D. equatorianus Kormilev	Ecuador: Napo-Pastaza. French Guiana: Saint-Laurent
			du Maroni. Suriname: Brokopondo
		D. lunatus F.	Argentina: Chaco. Bolivia: Cochabamba; Coroico; Chapare; Santa Cruz. Brazil: Amazonas; Mato Grosso; Pará; São Paulo; Santa Catarina. Colombia
			Costa Rica. Ecuador: Napo; Pichincha. French Guiana:
			Saint-Laurent du Maroni. Guatemala: Alta Verapaz. Honduras: Cortés. Mexico: Campeche; Quintana Roo; Oaxaca; Veracruz; Yucatan. Nicaragua: Chontales. Panama: Chiriqui. Paraguay. Peru: Cuzco; Madre de Dios. Suriname: Brokopondo. Venezuela: Caracas. Trinidad and Tobago

(continued)

Table 15.2 (continued)

Subfamily	Genera	Species	Distribution
		D. magnus (Heiss)	Bolivia: Coroico. Brazil: Santa Catarina; Mato Grosso.
			Ecuador: Napo. British Guiana. French Guiana: Saint-Laurent du Maroni. Peru: Amazonas; Madre de Dios. Suriname: Paramaribo
	Emydocoris	*E. montanus* Drake	Brazil
		E. stali Kormilev	Brazil: Rio de Janeiro
		E. testudinatus Usinger	Brazil: Rio de Janeiro.
	Eunotoplocoris	*E. ruckesi* Kormilev	Peru
	Forficulassa	*F. lobulata* Kormilev	Brazil: Pará
	Helenus	*H. hesiformis* White	Brazil: Amazonas. French Guiana: Saint-Laurent du Maroni. Peru: Amazonas. Suriname
		H. hirsutus Champion	Panama
	Hesus	*H. acuminatus* (F.)	Bolivia: Chapare; Cochabamba. Brazil: Chapada; Amazonas; Pará. Colombia. Ecuador: Napo; Sucumbios. French Guiana: Saint-Laurent du Maroni. Peru
		H. cordatus (F.)	Brazil: Pará; Chapada; Amazonas. French Guiana: Saint-Laurent du Maroni. Nicaragua: Chontales. Panama: Chiriqui. Peru: Amazonas. Surinam: Paramaribo
		H. humeralis Heiss	Brazil: Amazonas; Rondônia. Colombia. Ecuador: Napo; Sucumbios. French Guiana. Peru: Amazonas; Huánuco; Madre de Dios
		H. flaviventris (Burmeister)	Bolivia: Santa Cruz. Brazil: Amazonas; Chapada; Pará. Colombia. French Guiana: Saint-Laurent du Maroni. Nicaragua. Panama: Chiriqui. Peru: Madre de Dios. Venezuela
		H. mexicanus Kormilev	Guatemala. Honduras. Mexico: Chiapas; Yucatan
			Panama. Trinidad and Tobago. Venezuela

		H. subarmatus Stål	Bolivia: Cochabamba. Ecuador: Napo; Sucumbios.
			Colombia. Suriname: Paramaribo. Brazil: Rio de Janeiro; Rondônia. Peru: Amazonas
	Illibius	*I. laticeps* Stål	Colombia. Costa Rica: Heredia. Panama
	Kelaino	*K. kjellanderi* Kormilev	Colombia
	Kormilevia	*K. aberrans* Kormilev	Colombia
		K. dureti (Kormilev)	Argentina: Misiones
		K. gerali Kormilev	Brazil: Santa Catarina
		K. montrouzieri (Kormilev)	Brazil: Santa Catarina
		K. plaumanni (Kormilev)	Brazil: Santa Catarina
		K. setifera Usinger & Matsuda	Brazil
		K. teresopolitana (Wygodzinsky)	Brazil: Rio de Janeiro
	Limonocoris	*L. jolyi* Kormilev	Venezuela
	Lobocara	*L. oblonga* Bergroth	Argentina: Misiones. Bolivia: Coroico. Brazil: Amazonas. French Guiana
		L. ovata Bergroth	Brazil: Amazonas. British Honduras. French Guiana.
			Guatemala. Nicaragua: Chontales
	Mapiri	*M. paradoxa* Kormilev	Bolivia
	Melanosterphus	*M. spinosus* (F.)	Brazil: Amazonas; Pará; Chapada. French Guiana
	Mezira	*M. abdominalis* (Stål)	Cuba. Puerto Rico
		M. amazonica Kormilev	Peru: Amazonas
		M. americana (Spinola)	Argentina: La Rioja; Neuquén; Río Negro. Chile: Los Ríos
		M. andina Kormilev	Bolivia: Cochabamba. Brazil: Amazonas. Peru
		M. angustata (Champion)	Panama: Chiriqui

(continued)

Table 15.2 (continued)

Subfamily	Genera	Species	Distribution
		M. argentinensis Kormilev	Argentina: Córdoba; Corrientes; Entre Ríos; Santa Fé. Paraguay: Caaguazu. Brazil: Santa Catarina. Peru
		M. armata Kormilev	Peru
		M. arnaudi Kormilev	Peru
		M. auripilosa Kormilev	Mexico
		M. barberi Kormilev	Brazil: Amazonas. Venezuela. Peru: Huánuco. French Guiana
		M. birabeni Kormilev	Argentina: Cordoba
		M. boliviana Kormilev	Bolivia: Cochabamba, Chapare
		M. bonaerensis Kormilev	Argentina: Buenos Aires
		M. bouvieri (Bergroth)	Colombia
		M. brachyptera Kormilev	Jamaica
		M. brasiliensis Kormilev	Brazil: Santa Catarina
		M. bridarollii Kormilev	Bolivia: Chapare
		M. bruchi Kormilev	Argentina: Cordoba
		M. carinata Usinger	Mexico
		M. carioca Kormilev	Brazil: Rio de Janeiro
		M. championi Kormilev	Mexico: Veracruz
		M. chemsaki Kormilev	Costa Rica: Cartago
		M. cimicoides Walker	Brazil
		M. constricta (Champion)	Guatemala
		M. costalimai Kormilev	Brazil. Paraguay
		M. crenulata Kormilev	Venezuela
		M. cubana Kormilev	Cuba
		M. divisa (Champion)	Guatemala

		M. doesburgi (Kormilev & Froeschner)	Suriname
		M. dybasi Kormilev	Colombia: Villavicencio
		M. emarginata (Say)	Mexico
		M. equatoriana Kormilev	Ecuador: Esmeralda
		M. eurycephala Kormilev	Brazil: Rio Grande do Sul
		M. flavicans (Stål)	Brazil: Rio de Janeiro. Suriname
		M. formosa Kormilev	Argentina: Corrientes; Formosa
		M. fritzi Kormilev	Brazil: Santa Catarina
		M. gracilis Kormilev	Brazil: Amazonas
		M. gradata (Bergroth)	Brazil: Rio Grande do Sul
		M. granulata (Say)	Cuba: Guantanamo
		M. granuliger (Stål)	Argentina: Buenos Aires; Entre Ríos; Misiones; Salta; Santa Fé; Tucumán
			Brazil: Rio Grande do Sul; Chapada; Rio de Janeiro; Espirito Santo. Paraguay: Caaguazú
		M. guanacastensis Kormilev	Costa Rica: Guanacaste
		M. guianensis Kormilev	British Guiana
		M. halaszfyae Kormilev	Peru
		M. handlirschi (Bergroth)	Brazil: Amazonas; Pará. French Guiana: Saint-Laurent du Maroni. Panama: Chiriqui. Suriname
		M. hondurensis Kormilev	Honduras
		M. horvathi (Bergroth)	Brazil
		M. hyperlobata Kormilev	Brazil: Amazonas
		M. inca Kormilev	Peru
		M. incrustata Kormilev	Brazil
		M. jamaicensis Bergroth	Jamaica

(continued)

Table 15.2 (continued)

Subfamily	Genera	Species	Distribution
		M. kjellanderi Kormilev	Brazil: Amazonas. Colombia. Panama. Suriname
		M. laeviventris (Champion)	Brazil: São Paulo. Ecuador: Napo-Pastaza
			French Guiana. Panama: Chiriqui
		M. lata (Champion)	Guatemala: Alta Verapaz
		M. ligneola (Bergroth)	Brazil. Peru
		M. lobata (Say)	Mexico: Chiapas; Oaxaca
		M. longipilis (Champion)	Mexico: Guerrero
		M. luteonotata Kormilev	Brazil: Chapada
		M. maculata Kormilev	Mexico: Zacatecas
		M. maculiventris (Champion)	Guatemala: Alta Verapaz
		M. mexicana Kormilev	Mexico: Veracruz; Nuevo Leon
		M. minor Kormilev	Brazil
		M. moesta Stål	Mexico
		M. nana (Champion)	Guatemala
		M. nasalis Kormilev	Colombia: Tolima. Mexico: Veracruz
		M. neonigripennis Kormilev	Venezuela
		M. neonigripennis neonigripennis Kormilev	Argentina: Salta
		M. neonigripennis misionensis Kormilev	Argentina: Misiones
		M. neotropicalis (Champion)	Mexico: Chiapas; Michoacan. Guatemala: Alta Verapaz
		M. nigripennis Usinger	Argentina: Chaco; Jujuy; Misiones; Salta; Tucumán. Brazil: Santa Catarina; Río Grande do Sul. Paraguay: Horqueta
		M. oblonga Kormilev & Heiss	Brazil: Rio de Janeiro
		M. obscura (Distant)	Ecuador: Chimborazo
		M. occidentalis Kormilev	Mexico: Veracruz
		M. paraangustata Kormilev	Mexico

		M. paraensis Kormilev & Heiss	Brazil: Pará
		M. paragranuliger Kormilev	Argentina: Chaco; Jujuy; Salta; Misiones; Tucumán. Brazil: Santa Catarina; Espirito Santo; Rio de Janeiro
		M. paraguayensis Kormilev	Paraguay: Horqueta
		M. paralata Kormilev	Costa Rica: Limon
		M. paratropicalis Kormilev	Mexico: Chiapas
		M. parvicornis Kormilev	Bolivia
		M. pauperula Kormilev	Brazil: Rio de Janeiro. French Guiana
		M. peruviana Kormilev	Peru
		M. pilosa Kormilev	Mexico
		M. placida placida Kormilev	Puerto Rico
		M. placida haitiensis Kormilev	Haiti: Port-au-Prince
		M. plaumanni Kormilev	Brazil: Rio Grande do Sul
		M. proseni Kormilev	Argentina: Salta. Peru
		M. proxima Kormilev	Peru.
		M. punctiventris (Stål)	Colombia
		M. pusilla Kormilev	Guatemala: Alta Verapaz
		M. regularis (Champion)	Argentina: Tucumán. Costa Rica
			Guatemala. Mexico: Veracruz
		M. reuteri (Bergroth)	Argentina: Buenos Aires; Santa Fé; Misiones. Brazil
		M. romani Kormilev	Brazil: Amazonas
		M. rugicornis (Champion)	British Honduras: Belize. Mexico: Oaxaca
		M. rugiventris (Champion)	Guatemala. Peru
		M. saltensis Kormilev	Argentina: Salta; Tucumán. Brazil
		M. sangabrielensis Kormilev	Brazil
		M. sanmartini Kormilev	El Salvador. Peru. Venezuela: Bolivar

(continued)

Table 15.2 (continued)

Subfamily	Genera	Species	Distribution
		M. scheveni Heiss	Dominican Amber
		M. sinuata (Champion)	Mexico: Guerrero
		M. spissigrada Kormilev	Argentina: Misiones
		M. surinamensis Blöte	Suriname
		M. tartagalensis Kormilev	Argentina: Salta
		M. timida Kormilev	Brazil: Mato Grosso
		M. trinidadensis Kormilev	Trinidad and Tobago
		M. tropicalis Kormilev	Mexico: Jalisco
		M. variegata Kormilev	Mexico: Veracruz
		M. venezuelana Kormilev	Venezuela
		M. veracruzensis Kormilev	Mexico: Veracruz
		M. vianai Kormilev	Argentina: Cordoba
		M. vicina Kormilev & Froeschner	Peru
		M. vulcanica Kormilev	Ecuador: Chimborazo
		M. yucatana (Champion)	Mexico: Yucatan
	Miorrhynchus	*M. angulatus* Kormilev	Peru
		M. bolivianus Kormilev	Bolivia: Cochabamba. Peru
		M. brasiliensis Kormilev	Brazil: Mato Grosso; Santa Catarina
		M. championi Kormilev	Bolivia: Santa Cruz. Peru
		M. jatahyensis Kormilev	Brazil: Goiás
		M. longicornis Kormilev	Peru: Madre de Dios
		M. longipes Champion	Panama: Chiriqui
		M. paraguayensis Kormilev	Brazil: Santa Catarina. Paraguay: Guairá
		M. peruvianus Kormilev	Ecuador: Napo. Peru: Pachitea
		M. plaumanni Kormilev	Brazil: Mato Grosso

		M. proseni Kormilev	Bolivia: Cochabamba. Brazil: Amazonas. Colombia
		M. schuhi Kormilev	Peru: Junin
		M. undulatus Kormilev	Peru
		M. usingeri Kormilev	Panama
	Mystilocoris	*M. pubescens* Usinger & Matsuda	Colombia
	Nannium	*N. bituberculatum* Champion	Guatemala
		N. brasiliense Kormilev	Brazil: Santa Catarina
		N. elongatum Bergroth	Peru. Venezuela
		N. kormilevi Doesburg	French Guiana. Suriname
		N. parvum Bergroth	Costa Rica: Guanacaste. Guatemala. Panama: Chiriqui. Peru
		N. subovatulum Bergroth	Brazil
	Neormenocoris	*N. costaricensis* (Kormilev)	Costa Rica: Heredia
	Neuroctenus	*N. amazonicus* Kormilev	Brazil: Amazonas; Mato Grosso. Colombia. Peru
			Venezuela: Portuguesa
		N. amplus Champion	Mexico: Guerrero
		N. andrei (Kormilev)	Colombia
		N. aztequi Kormilev	Mexico
		N. bergrothi Champion	Guatemala: Alta Verapaz. Mexico: Chiapas; Oaxaca. Peru
		N. centralis (Berg)	Argentina: Buenos Aires; Corrientes; Catamarca; Córdoba; Jujuy; La Rioja; Salta; Santa Fé; Tucumán
			Brazil: Santa Catarina
		N. chilensis Kormilev	Chile
		N. colombianus Kormilev	Colombia: Villavicencio
		N. dilatatus Bergroth	Costa Rica. Guatemala. Mexico: Veracruz

(continued)

Table 15.2 (continued)

Subfamily	Genera	Species	Distribution
		N. discrepans Kormilev	Mexico
		N. distanti Bergroth	Mexico
		N. insignis Kormilev	Brazil: Santa Catarina
		N. litigiosus (Stål)	Bolivia: Santa Cruz. Guatemala: Alta Verapaz. Costa Rica
			Panama: Chiriqui. Mexico
		N. longissimus Kormilev	Brazil: Pará
		N. longiusculus Kormilev	Brazil: Santa Catarina
		N. longiventris Kormilev	Bolivia: Coroico. Colombia: Cauca. Panama. Peru
		N. longulus Bergroth	Brazil: Amazonas; Mato Grosso. Peru
		N. mexicanus (Champion)	Colombia. Mexico: Oaxaca
		N. niger Bergroth	Guatemala. Mexico: Morelos
		N. ovatus Stål	Mexico
		N. papyrinus Bergroth	Mexico: Morelos
		N. proseni Kormilev	Bolivia: Santa Cruz
		N. punctulatus (Burmeister)	Argentina: Corrientes; Misiones. Brazil: Rio Grande do Sul
			Colombia. Guatemala. Nicaragua: Chontales. Panama: Chiriqui. Paraguay: Caaguazú
		N. robustus Kormilev	Colombia. Peru
		N. rossi Kormilev	Ecuador: Napo-Pastaza
		N. rubiginosus Bergroth	Bolivia: Coroico. Colombia. Cuba. Mexico. Peru
		N. schlingeri Kormilev	Ecuador: Napo-Pastaza. Peru
		N. simplex (Uhler)	Cuba

		N. subandinus Kormilev	Argentina: Catamarca; Jujuy; Mendoza; Tucumán; Salta. Bolivia
		N. subparallelus Champion	Guatemala: Alta Verapaz. Mexico: Veracruz
		N. substitutus Kormilev	Mexico: Chihuahua
		N. surinamensis Kormilev	Suriname
		N. terginus (Stål)	Brazil: Rio Grande do Sul; Rio de Janeiro
			Colombia. Guatemala. Venezuela.
		N. trigonus Bergroth	Guatemala
		N. uhleri Bergroth	Mexico: Michoacan
		N. uruguayensis Kormilev	Uruguay
		N. vanduzeei Kormilev	Panama
	Notapictinus	*N. amazonicus* Kormilev	Brazil: Amazonas
		N. angulatus Kormilev	Peru
		N. araguaensis Kormilev	Venezuela: Aragua
		N. beckeri (Kormilev)	Brazil: Rio Grande do Sul; Santa Catarina
		N. bimaculatus Kormilev	Brazil: Santa Catarina
		N. brachypterus (Drake & Kormilev)	Brazil: Santa Catarina
		N. breviceps (Champion)	Panama: Chiriqui
		N. christae Kormilev	Brazil: Amazonas
		N. denticollis (Champion)	Panama: Chiriqui
		N. derivatus (Kormilev)	Brazil: Mato Grosso
		N. diharpagus Kormilev	Peru
		N. dissimilis Kormilev & Heiss	Brazil
		N. dollingi Kormilev	Costa Rica
		N. dominicus (Usinger)	Dominican Republic
		N. dyscritus Kormilev	Peru

(continued)

Table 15.2 (continued)

Subfamily	Genera	Species	Distribution
		N. equatoriensis Kormilev	Ecuador
		N. hoguei Kormilev	Brazil: Rio de Janeiro
		N. incaicus Kormilev	Peru
		N. kjellanderi Kormilev	Peru
		N. luteoincrustatus (Kormilev)	Bolivia
		N. maculatus (Kormilev)	Peru
		N. martinezi (Kormilev)	Argentina: Salta
		N. micropterus Kormilev	Brazil: Parana
		N. nanus (Kormilev)	Paraguay
		N. notatus Kormilev	Suriname
		N. ornatus Kormilev	Brazil: Rio de Janeiro
		N. paramaculatus Kormilev	Brazil: Chapada
		N. parviceps (Champion)	Panama: Chiriqui
		N. parvulus Kormilev	Panama
		N. piliger Kormilev	Colombia: Cauca
		N. platyceps Kormilev	Brazil: Chapada
		N. plaumanni Kormilev	Brazil: Santa Catarina
		N. quadraticeps (Champion)	Ecuador: Napo-Pastaza. Panama: Chiriqui
		N. rutilus (Kormilev)	Brazil: Santa Catarina
		N. sanmigueli (Kormilev)	Argentina: Tucumán; San Luis
		N. similis Kormilev & Heiss	Brazil: Rio de Janeiro
		N. subparallelus Kormilev	Brazil: Amazonas
		N. surinamensis Kormilev	Suriname
		N. terminalis Kormilev	Brazil: Santa Catarina
		N. testaceus Kormilev & Heiss	Brazil: Rio de Janeiro
		N. uruguayensis Kormilev	Uruguay

		N. varelai Kormilev	Venezuela: Bolívar
		N. venezuelanus Kormilev	Venezuela
	Notoplocoris	*N. guanabarensis* Kormilev	Brazil
		N. leptocerus Kormilev	Brazil
		N. mendesi Wygodzinsky	Brazil: Rio de Janeiro
		N. montei Usinger	Brazil: Rio de Janeiro
		N. ovatus Kormilev	Brazil: Parana
		N. potensis Drake & Harris	Brazil: Rio de Janeiro
		N. robustus Kormilev	Brazil: Espirito Santo
		N. sobrali Wygodzinsky	Brazil: Rio de Janeiro
		N. triangulatus Kormilev	Brazil: Santa Catarina
		N. usingeri Kormilev & Heiss	Brazil: São Paulo
	Ormenocoris	*O. stylatus* Usinger & Matsuda	Ecuador: Napo
	Parahesus	*P. truncatus* Kormilev	Bolivia
	Paramezira	*P. meziroides* (Kormilev)	Brazil
	Phyllotingis	*P. eximia* (Haglund)	Brazil: Amazonas; Rondônia. Colombia: Caquetá. French Guiana
		P. interjecta (Bergroth)	Bolivia. Brazil. Colombia. Panama
		P. lanceolata (F.)	Brazil: Amazonas. Peru: Amazonas; Iquitos
			Cayenne, French Guiana. Suriname: Brokopondo
		P. reducta Heiss	Trinidad
		P. triangula Heiss	Brazil: Mato Grosso. French Guiana
	Pictinus	*P. armatus* Champion	Panama
		P. bechynei Kormilev	Brazil: Amazonas
		P. brasiliensis (Wygodzinsky)	Brazil: Rio de Janeiro

(continued)

Table 15.2 (continued)

Subfamily	Genera	Species	Distribution
		P. carioca Kormilev	Brazil: Rio de Janeiro
		P. cinctipes Stål	Bogotá
		P. fictus Kormilev	Brazil: Santa Catarina
		P. fronto Bergroth	Brazil: Santa Catarina
		P. granuliferus Kormilev	Brazil: Parana
		P. pilosulus Kormilev	Brazil: Parana
		P. rhombocarinatus Kormilev	Brazil: Santa Catarina
		P. simulans (Walker)	Brazil: Rio de Janeiro
		P. spiniger Champion	Panama: Chiriqui
		P. stali Kormilev	Peru
		P. stolidus Kormilev	Brazil: Santa Catarina
		P. venezuelanus Kormilev	Venezuela: Bolivar
		P. wittmeri Kormilev	Brazil: São Paulo
	Placogenys	*P. brachyptera* (Kormilev)	Brazil: Santa Catarina; Minas Gerais
		P. clarkei Kormilev	Argentina: Tucumán.
		P. cockerelli Usinger & Matsuda	Brazil: Rondônia; Pará. British Guiana
			French Guiana: Saint-Laurent du Maroni
		P. constricta Kormilev	Ecuador
		P. explanata (Kormilev)	Brazil: Amapá; Pará. Peru
		P. parva Kormilev	Brazil: Espirito Santo
	Pseudopictinus	*P. dispar* Kormilev	Brazil: Parana
	Rhynchomirus	*R. brachypterus* Kormilev	Peru: Junin
	Saileriessa	*S. stigmata* Usinger & Matsuda	Panama
	Santaremia	*S. robusta* Kormilev	Brazil: Pará. French Guiana
Prosympiestinae	*Llaimocoris*	*L. penai* Kormilev	Chile: Cautin

In the Neotropical Region, for Argentina four genera and five species were reported (Pennington 1921); Coscarón (in press) increased these numbers to 14 genera and 41 species; records from Ecuador comprise nine genera and 15 species (Froeschner 1981); from Panama 30 genera and 60 species were reported (Froeschner 1999), and from Chile six genera and six species were referred (Prado 2008).

Regional keys for the Neotropical Region were only made for Argentina by Contreras and Coscarón (2012) including diagnosis of genera, geographic distribution, and additional distributional records (Contreras et al. 2011).

Aradids were found in the following substrates: under the bark of a rotten log (Wygodzinsky 1948); on the bark during night (Heiss and Moragues 2009), under the bark (Kormilev 1982); under the bark of dead trees (Wygodzinsky 1948); beneath the bark of decaying branches of sourwood (*Oxydendrum*) and beneath the loose bark scales on dry and decaying branches of trees (Heidemann 1904); *Zanthoxylum americanum*, *Oxydendrum*, *Platanus*, *Quercus*, *Carya*, *Rhus*, beneath the bark of *Pinus montezumae*, elms, and oaks (Usinger and Matsuda 1959); on orchid plants, under *Schinopsis lorentzii* and on *Pisonia subcordata* (Kormilev 1953, 1958, 1964); in larval galleries of Cerambycidae abandoned of *Celtis tala* (Kormilev 1956); and in deserted galleries of a small cerambycid in dead branches of sumac (Heidemann 1904).

In this family an important forest pest is the species *Aradus cinnamomeus* Panzer (usually called bark bugs); it feeds on the phloem, cambium, and xylem of living *Pinus* sp. and *Larix* sp. (Pinaceae) (Kiritshenko 1913) and causes stunting of the growth of these trees (Strawinski 1925). It has a Palearctic distribution (Helioövaara 2000). This bug forms groups in large colonies preferably on young pine trees, 7–12 cm in diameter, sucking the leaves and twigs.

Records of aradids from ambar include registers from Birmanian ambar (100 million years ago) (Heiss and Grimaldi 2001, 2002), registers from the Baltic dated 40–50 million years ago (Heiss 1997, 1998, 2002, 2012) also Dominican amber (Heiss 2000).

15.5 Concluding Remarks

The understanding of Aradidae diversity from a systematic, phylogenetic, biological, ecological, and economic viewpoint, including their role in conservation biology and global warming, is fundamental. An enormous amount of important and biological, biogeographical, and host information useful to other disciplines remains to be discovered in the world and especially in the Neotropical Region.

References

Blatchley WS (1934) Notes on a collection of Heteroptera taken in the vicinity of Los Angeles California. Trans Am Entomol Soc 60:1–16

Brammanis L (1975) Die Kiefernrindenwanze, *Aradus cinnamomeus* Panz. (Hemiptera: Heteroptera). Ein Beitrag zur Kenntnis der Lebensweise und der forstlichen Bedeuntung Stud For Suecica 123:1–81

Carayon J (1955) Quelques Caractères anatomiques des Hémiptères Aradides. Rev Fr Entomol 22:169–180

Contreras E, Coscarón MC (2012) The Aradidae (Insecta, Hemiptera, Heteroptera) of Argentina. Zootaxa 3500:1–35

Contreras E, Neder L, Coscarón MC (2011) Contribution to the knowledge of flat bugs (Hemiptera: Heteroptera) from the temperate South America. Mun Entomol Zool 6:764–768

Coscarón MC, Contreras EF (2012) Catalog of Aradidae (Hemiptera: Heteroptera) for the Neotropical Region. Zootaxa 3466:1–103

Coscarón, M.C. (in press) Aradidae. Catalogue of the Heteroptera or true bugs of Argentina. Zootaxa

Drake CJ (1956) New neotropical genera and species of apterous Aradidae (Hemíptera). J Wash Acad Sci 46:322–327

Drake CJ, Maldonado J (1955) New Apterous Aradids from Puerto Rico. J Wash Acad Sci 45:289–294

Drake CJ, Kormilev NA (1958) Concerning the Apterous Aradidae of the Americas (Hemiptera). Ann Entomol Soc Am 51:241–247

Froeschner RC (1981) Heteroptera of true bugs of ecuador: a partial catalog. Smithson Contrib Zool 322:1–147

Froeschner RC (1999) True bugs (Heteroptera) of Panamá: a symoptic catalog as a contribution to the study of Panamanian diversity. American Entomological Institute, Gainesville

Gossner MH, Engel H, Blaschke M (2007) Factors determining the occurrence of flat bugs (Aradidae) in beech dominated forests. Waldoekologie 4:59–89

Grozeva SM, Kerzhner IM (1992) On the phylogeny of Aradid Subfamilies Heteroptera: Aradidae). Act Zool Hung 38:199–205

Heidemann O (1904) Notes on a few Aradidae occurring north of the Mexican boundary. Proc Entomol Soc Wash 6:229–233

Heilmann-Clausen J, Christensen M (2004) Does size matter? On the importance of various dead wood fractions for fungal diversity in Danish beech forests. For Ecol Manag 201:105–117

Heiss E (1997) Erstnachweis einer Aneurinae aus dem Baltischen Bernstein: *Aneurus ancestralis* n. sp. (Heteroptera, Aradidae). Wiss Mitteil Carolinea 55:111–112

Heiss E (1998) Revision der Familie Aradidae des Baltischen Bernsteins I: Bisher beschriebene Taxa der Gattung *Aradus* und zwei neue Arten (Insecta, Heteroptera). Mitt aus dem Geol-Paläontol Inst der Univ Ham 81:251–268

Heiss E (2000) First record of a Mezirinae flat bug from Dominican amber: *Mezira scheveni* n. sp (Heteroptera, Aradidae). Entomol Nach 7:6–10

Heiss E (2002) Aradidae in Baltischem Bernstein – Beispiel eines über 50 Millionen Jahre alten erfolgreichen Insektenbauplanes. Heteroptera 15:8–9

Heiss E (2012) Revision der Aradidae des Baltischen Bernsteins VII. Eine neue *Aneurus*-Art (Heteroptera: Aradidae). Entomol Zeit Stutt 122:109–110

Heiss E, Grimaldi D (2001) *Archearadus burmensis* gen. n., n. sp., a remarkable Mesozoic Aradidae in Burmese amber (Heteroptera: Aradidae). Carolinea 59:99–102

Heiss E, Grimaldi D (2002) The first known female of *Archearadus burmensis* Heiss and Grimaldi, 2001, in Cretaceous Burmese amber (Heteroptera: Aradidae). Zeit der Arb Öst Entom 54:55–59

Heiss E, Péricart J (2007) Hémipteres Aradidae Piesmatidae et Dipsocoromorphes euro-méditerranéens. Faune France 91:1–509

Heiss E, Moragues G (2009) Flat bugs of French Guyana- a preliminary faunal list (Heteroptera, Aradidae). Linzer Biol Beitr 41(2):1659–1675

Heliövaara K (1982) Overwintering sites of the pine bark-bug, *Aradus cinnamomeus* (Heteroptera, Aradidae). Ann Entomol Fenn 48:105–108

Heliövaara K (1984) Ecology of the pine bark bug, *Aradus cinnamomeus* (Heteroptera, Aradidae). A forest entomological approach. Univ Helsinski Dept Agric For Zool Rep 7:1–38
Heliövaara K (2000) Flat bugs (Aradidae). In: Schaefer CW, Panizzi AR (eds) Heteroptera of economic importance. CRC Press, Boca Raton, pp 513–517
Heliövaara K, Väisänen R (1987) Geographic variation in the life-history of *Aradus cinnamomeus* and breakdown mechanism of the reproductive isolation of allochronic bugs (Heteroptera, Aradidae). Silva Fenn 17:351–357
Hendee EC (1933) The association of the termites *Kalotermes minor*, Reticulitermes hesperus and Zootermopsis angusticollis with fungi. Univ Calif Pubs Zoöl 39:111–134
Henry TJ (2009) Biodiversity of Heteroptera. In: Foottit R, Adler P (eds) Insect biodiversity: science and society. Wiley-Blackwell, Hoboken, pp 223–263
Jacobs DH (1986) Morphology and taxonomy of sub-saharan *Aneurus* species with notes on their phylogeny, biology and cytogenetics (Heteroptera: Aradidae: Aneurinae). Entomol Mem Dep Agric Water Supply Rep S Afr 64:1–29
Kiritschenko NA (1913) Aradidae et Dysodiidae. Faun Russia 6:11–301
Kormilev NA (1953) Aradidae (Hemiptera) Argentinae II. Acta Zool Lil 13:207–259
Kormilev NA (1956) Notas sobre Aradidae Neotropicales VI (Hemiptera). Anal Soc Cient Arg 162:148–159
Kormilev NA (1958) Notes on the Aradidae in the U.S. national museum (Hemiptera). Proc Un Stat Nat Mus 109:209–222
Kormilev NA (1964) Notes on Aradidae in the U.S. national museum III. Subfamily Mezirinae (Hemiptera). Proc Un Stat Nat Mus 115(3483):245–258
Kormilev NA (1982) Records and descriptions of central American Aradidae (Hemiptera). Was J Biol 40(1–2):26–44
Kormilev NA, Froeschner RC (1987) Flat bugs of the world. A synonymic list (Heteroptera: Aradidae). Entomology 5:1–246
Leston D (1954) Classification of the terrestrial Heteroptera (Geocorisae). Nature 174:91–92
Marchal L, Paillet Y, Guilbert E (2012) Habitat characteristics of Aradidae (Insecta: Heteroptera) in two French deciduous forest. J Insect Conserv 17:269–278
McClure HE (1932) Incubation of bark bug eggs. Entomol News 43:188–189
Miller NCE (1938) A new subfamily of Malaysian Dysodiidae (Rhynchota). Ann Mag Nat Hist 5:498–510
Monteith GB (1969) A remarkable case of alary dimorphism in the Aradidae (Hemiptera) with a generic synonymy and a new species. J Aust Entomol Soc 8:87–94
Nordén B, Ryberg M, Gotmark F, Olausson B (2004) Relative importance of coarse and fine woody debris for the diversity of wood inhabiting fungi in temperate broadleaf forests. Biol Conserv 117:1–10
Parshley HM (1921) Essay on the American species of *Aradus* (Hemiptera). Trans Entomol Soc 47:1–106
Pennington MS (1921) Lista de los Hemípteros Heterópteros de la República Argentina. Segunda parte: Familia Aradidae, Liedidae, Ligaeidae, Phyrrhocoridae, Tigididae, Enicocephalidae, Phimatidae, Reduviidae, Nabidae, Cimicidae, Miridae (parte), 17–28; Belostomatidae, Gelastocoridae, Corixidae, 29–47. Buenos Aires
Prado EC (2008) Conocimiento actual de Hemiptera-Heteroptera de Chile con lista de especies. Bol Mus Nac Hist Nat Chile 57:31–75
Ralston JS (1977) Egg guarding by male assassin bugs of the genus *Zelus* (Hemiptera: Reduviidae). Psyche 84:103–107
Ridley M (1978) Parenteral care. Anim Behav 26:904–932
Schuh RT, Slater JA (1995) True bugs of the world (Hemiptera: Heteroptera): classification and natural history. Cornell University Press, Ithaca
Smith RL (1980) Evolution of exclusive postcopulatory paternal care in the insects. Fla Entomol 63:65–78
Stichel W (1956–1962) Verzeichnis der Paläarktischen Hemiptera: Heteroptera, pts 1–4. Hermsdorf, Berlin

Strawinski K (1925) Historja naturalna Korowca sosnowego *Aradus cinnamomeus* Pnz. (Hemiptera-Heteroptera). Roczn Nauk Rolnicz Lesn 13:644–693

Taylor SJ (1988) Observations on parental care in the family Aradidae (Heteroptera). Great Lakes Entomol 21:159–161

Tropin IV (1949) The pine bark bug and its control. Goslesbumizdat, Moscow

Usinger RL (1936) Studies in the American Aradidae with descriptions of new species (Hemiptera). Ann Entomol Soc Am 29:490–516

Usinger RL, Matsuda R (1959) Classification of the Aradidae. British Museum, London

Vásárhelyi T (1979) Three new species of the New Guinea genus *Biroana* (Heteroptera, Aradidae). Pac Insect 21:90–94

Vásárhelyi T (1987) On the relationships of the eight aradid subfamilies (Heteroptera). Acta Zool Acad Sci Hung 33:263–267

Vásárhelyi T, Böröcz P (1987) Studies on the life-history of *Aradus cinnamomeus* in Hungary (Heteroptera: Aradidae). Folia Entomol Hung 48:233–239

Weber H (1930) Biologie der Hemipteren. Biologische Studienbücher, XI. J. Springer, Berlin

Wygodzinsky P (1948) Studies on some Apterous Aradidae from Brazil (Hemiptera). Bol do Mus Nac Rio Janeiro 86:1–23

Yonke TR (1991) Order Hemiptera. In: Stehr FW (ed) Immature insects, vol 2. Kendall/Hunt Publishing, Dubuque, pp 22–65

Chapter 16
The Big-Eyed Bugs, Chinch Bugs, and Seed Bugs (Lygaeoidea)

Thomas J. Henry, Pablo M. Dellapé, and Alexandre S. de Paula

Abstract The Lygaeoidea, representing the second largest superfamily in the infraorder Pentatomomorpha, are one of the most diverse groups of Heteroptera with about 700 genera and more than 4,200 species in the world. In the Neotropics, about 184 genera and 836 species are included in 12 of the world's 16 families. For each family, we provide a diagnosis; an overview of the classification; information on the general life history, ecology, and economic importance; and comprehensive keys to subfamilies, tribes, and genera for the Neotropical Region, including Mexico, Central and South America, and the West Indies.

16.1 Introduction

The Lygaeoidea represent the second largest superfamily within the Pentatomomorpha with more than 4,200 species worldwide (Henry 2009). Though the superfamily currently is recognized by most contemporary workers, its status has fluctuated. For example, Štys (1961, 1967) grouped the Lygaeoidea and Pyrrhocoroidea with the Coreoidea but kept the superfamily Piesmatoidea; Henry and Froeschner (1988) accepted the Lygaeoidea but retained Piesmatoidea as a separate superfamily;

T.J. Henry (✉)
Systematic Entomology Laboratory, Agricultural Research Service, United States Department of Agriculture, c/o Smithsonian Institution, National Museum of Natural History, MRC 168, P.O. Box 37012, Washington, DC 20013-7012, USA
e-mail: thomas.henry@ars.usda.gov

P.M. Dellapé
División Entomología, Museo de La Plata, Paseo del bosque s/n, La Plata B1900FWA, Buenos Aires, Argentina
e-mail: pdellape@gmail.com

A.S. de Paula
Departamento de Ciências Biológicas, Instituto de Ciências Exatas e Biológicas, Universidade Federal de Ouro Preto (UFOP), Campus Morro do Cruzeiro Bauxita, Ouro Preto, MG 35400-000, Brazil
e-mail: hetalex@terra.com.br

A.R. Panizzi, J. Grazia (eds.), *True Bugs (Heteroptera) of the Neotropics*, Entomology in Focus 2, DOI 10.1007/978-94-017-9861-7_16

Schaefer (1993) also recognized Lygaeoidea but included the Piesmatidae and Idiostolidae; and Henry (1997a) recognized six superfamilies within the infraorder based on a cladistic analysis, merging the Piesmatoidea with the Lygaeoidea and giving nearly all subfamilies within the Lygaeidae (sensu lato) family status. More recently, Li et al. (2005), in using molecular sequence data, concluded that the Berytidae and Piesmatidae nested within the Pyrrhocoroidea, making the Lygaeoidea paraphyletic. Given the strong morphological support for a monophyletic Lygaeoidea Henry (1997a), including the Berytidae and Piesmatidae, the acceptance of such a novel hypothesis needs further investigation (Forero 2008).

We recognize 12 of the 16 families within the Lygaeoidea as occurring in the Neotropical Region, following Henry (1997a). For each family, we give a diagnosis, an overview of the classification, and keys to all of the subfamilies, tribes, and genera. In addition, we provide a brief overview of the general life history, habits, ecology, and economic importance of the major lygaeoid species.

The Artheneidae are all Old World, except for two Palearctic species, *Chilacis typhae* (Perrin) (Wheeler and Fetter 1987) and *Holcocranum saturejae* (Kolenati) (Hoffman and Slater 1995), recently established in the United States. *Polychisme poecilus* (Spinola), a South American species previously placed in Artheneidae (Slater and Brailovsky 1986), has been shown to belong in the lygaeid subfamily Ischnorhynchinae (Kerzhner 1997). The Cryptorhamphidae, containing only two genera and four species, are restricted to the Australian Region (Hamid 1971; Cassis and Gross 2002), and the Malcidae, with three genera and about 29 species, are restricted to the Oriental and Palearctic regions (Štys 1967; Kerzhner 2001). The Meschiidae, the most recently recognized family of Heteroptera, was described from Australia and India to accommodate the genus *Meschia* Distant, containing four species, the new genus *Neomeschia* Malipatil, and the new species *N. queenslandicus* Malipatil (Malipatil 2014).

Our classification closely aligns with Sweet's (2000a) interpretation of the Lygaeoidea, with the exception of his proposal to elevate the lygaeid (sensu stricto) subfamilies Ischnorhynchinae and Orsillinae to family status, but without giving specific character information to support his hypothesis. As a consequence, we feel that the strong character support for a monophyletic Lygaeidae (Henry 1997a), which includes these two subfamilies and the Lygaeinae, precludes any argument to separate them until new evidence is offered.

Slater (1964a) and Slater and O'Donnell (1995) cataloged the Lygaeidae (sensu lato) of the world; Ashlock and Slater (1988) cataloged the Nearctic species, which includes numerous taxa also occurring in the Neotropics; Schuh and Slater (1995) provided a good overview of the subfamilies and tribes; and Henry (1997a) provided a key to help distinguish 15 of the 16 recognized families of Lygaeoidea. There have been a number of outstanding treatments of the Lygaeoidea with keys, including Froeschner's (1981) list and keys to the Ecuadorian Heteroptera and (Froeschner 1985) synopsis of the Galapagos fauna; Slater and Baranowski's (1990) Lygaeidae of Florida; Slater and Brailovsky's (2000) Lygaeidae of Mexico, with keys to tribes and a checklist of species; Peck's (2001) list and keys to the Heteroptera of the Galapagos Islands; Baranowski and Slater's (2005) Lygaeidae of the West Indies; and Dellapé's (2014) Lygaeoidea of Argentina, with keys to genera and a checklist of species. More regional checklists include Paula and Ferreira's (1998, 2000) list for Minas Gerais, Brazil; Cervantes and Brailovsky's (2011) list for Veracruz, Mexico; and the lists for Argentina by Melo et al. (2004) (Corrientes), Melo et al. (2011) (Chaco National Park), and Dellapé and Carpintero (2012) (Buenos Aires Hills). Also, although aimed primarily at the North American fauna, this author went by Torre-Bueno's (1946) synopsis and keys and Slater and Baranowski's (1978) *How to Know the True Bugs* are useful for many taxa ranging into Mexico, Central America, and the West Indies.

Sweet (1964a, b) detailed the ecology and feeding habits of many widespread New World Lygaeoidea. Sweet (2000a, b) provided an especially thorough overview of the economic importance of many of the most important taxa in the superfamily.

16.2 Life History, Ecology, and Economic Importance

The Lygaeoidea are a diverse, highly successful group of true bugs found in all zoogeographic regions. Important reviews of the habits, life history, and economic importance include Sweet (1960, 1964a, b, 2000a) and Schuh and Slater (1995) and the many papers cited in these works. Many lygaeoids have attracted considerable attention as model species for the study of insect physiology (Slater and O'Donnell 1995) and agriculture (Sweet 2000a). More recently, Burdfield-Steel and Shuker (2014) provided an overview emphasizing their behavior, evolution, and ecology.

Lygaeoid nymphs typically are associated with the adults (Schuh and Slater 1995) and are often gregarious (Aller and Caldwell 1979). Although parental care is widespread in the Heteroptera, no evidence has been found for this behavior within the Lygaeoidea. A number of species in temperate areas, including *Lygaeus equestris* (L.) and *L. simulans* Deckert, show reproductive diapause and migratory capabilities affected by temperature and photoperiod (Solbreck 1979; Dingle et al. 1980). These adaptations allow them to avoid or survive low temperatures during the winter months, as well as to migrate by following seasonal patterns and host plant abundance (Dingle et al. 1980; Attisano et al. 2013).

Wing modifications within the Lygaeoidea fall into four categories (Slater 1977): (1) aptery, or the complete absence of wings; (2) sub-brachyptery, where the forewings extend only to the end of the fifth abdominal tergite; (3) brachyptery, where the forewings are reduced and do not cover the sixth and seventh abdominal terga and the hind wings are reduced but usually not flaplike; and (4) macroptery, where the clavus and corium are distinct, the membrane is well developed, and the hind wings are elongate. Flightless morphs are favored by habitat permanency (Slater 1977).

Slater (1977) and Slater and Baranowski (1990) recognized three major lygaeoid habitats: arboreal, geophilic, and laminophilic. Many arboreal species live on plants above ground level. These species are fully winged and usually readily fly when disturbed. Geophiles live on the ground in the litter layer, where they feed on fallen seeds. A number of the geophilous species may climb plants to feed on mature seeds, but most spend most of their lives on the ground. Laminaphiles live between the sheaths of leaves and stems of grasses, sedges, and rushes. Unlike arboreal species, they often are brachypterous (Slater and Baranowski 1990).

The feeding habits of seed bugs were extensively discussed by Sweet (1960, 1964a, b, 2000a). Most lygaeids feed on seeds or plant sap, whereas only a few are predatory. They have piercing-sucking mouthparts, which are made up of the mandibles and maxillae modified to form needlelike stylets lying within a grooved labium (Schuh and Slater 1995). Feeding methods can be divided into two "types": "stylet-sheath" feeders and "lacerate-flush" feeders (Schuh and Slater 1995; Sweet 2000a). The majority of Lygaeoidea are lacerate-flush feeders, a method commonly used by Miridae and other heteropterans that feed on portions of the plant rich in nutrients, such as seeds (Schuh and Slater 1995; Wheeler 2001), whereas the families Blissidae, Malcidae, and Colobathristidae are predominantly sap feeders.

16.3 General Characteristics and Diagnoses of the Lygaeoidea

Henry (1997a) documented the paraphyly of Lygaeidae as treated by previous authors (e.g., Slater 1964a; Slater and O'Donnell 1995; Schuh and Slater 1995). That the Berytidae, Colobathristidae, Malcidae, and Piesmatidae nested within the Lygaeidae in his analysis provided sufficient evidence to either reduce these well-defined families to subfamily status within the Lygaeidae (sensu lato) or to elevate many of the subfamilies within the paraphyletic Lygaeidae to family level. Henry (1997a) chose the latter as the more informative strategy to reflect his hypothesized phylogeny, based on three synapomorphies, the reduced venation nearly always lacking closed cells on the hemelytral membrane, and the incrassate fore femora found in all basal taxa but lost in a number of distal groups. Thus, 16 families are now recognized in the superfamily Lygaeoidea.

The following key modified from Henry (1997a) and Dellapé (2014) will distinguish the Neotropical lygaeoid families.

Key to the Neotropical families of Lygaeoidea

1. Abdominal spiracles on segment II ventral 2
– Abdominal spiracles on segment II dorsal 4
2. Trichobothria present on head; ovipositor, at most, dividing abdominal sternite VII; suture between abdominal sterna IV and V usually curving anteriorly and usually ending before attaining lateral abdominal margin, except in Plinthisinae .... Rhyparochromidae
– Trichobothria never present on head; ovipositor dividing at least sternites VI and VII; suture between abdominal sterna IV and V straight, always attaining lateral abdominal margin 3

3. Profemora weakly incrassate, little thicker than metafemora; base of hemelytral membrane with a distinct closed cell .... Heterogastridae
– Profemora strongly incrassate, much thicker than metafemora; base of hemelytral membrane without a closed cell .... Pachygronthidae
4. Tarsi two segmented; ocelli present or absent; trichobothria lacking on abdominal segments IV and V, often lacking on other segments as well .... Piesmatidae
– Tarsi three segmented; ocelli present; trichobothria present on abdominal segments IV and V 5
5. Each ocellus nearly encircled by a distinct groove 6
– Each ocellus not encircled by a groove.. 11
6. Abdominal spiracles on segments V and VI ventral .... Colobathristidae
– Abdominal spiracles on segments V and VI dorsal... 7
7. Connexiva on abdominal segments V to VI produced into conspicuous dentate lobes .... Malcidae (Old World only)
– Connexiva on abdominal segments V to VII simple, never produced into conspicuous lobes 8
8. First antennal segment long, slender, often apically clavate, subequal in thickness to and nearly always longer than segments II and III .... Berytidae
– First antennal segment short, stout, barrel shaped, much shorter and thicker than segments II and III 9
9. Hemelytra impunctate or with only a few indistinct punctures, corium hyaline to translucent beyond constricted base; apex of scutellum bifid; head broad, eyes substylate, vertex wider than anterior width of pronotum .... Ninidae
– Hemelytra distinctly punctate on clavus and corium, corium opaque throughout, never constricted; apex of scutellum rounded or acute; head not broadened, eyes never substylate, vertex always narrower than anterior width of pronotum 10
10. Buccula short, not extending posteriorly beyond bases of antennae; abdominal trichobothria present on sternites II to VII.. Cymidae
– Buccula long, extending posteriorly to base of head; abdominal trichobothria present only on sternites V and VI .... Cryptorhamphidae (Old World only)
11. Abdominal spiracles on segments III and IV ventral...................................... 12
– Abdominal spiracles on segments III and IV dorsal 13
12. Lateral pronotal margin explanate or with a wide flattened carina; female abdomen rounded caudally; male abdominal sternite VII without clusters or combs of setae ventrally...
.... Artheneidae (Old World, except for two introduced Nearctic species)
– Lateral pronotal margin rounded or, at most, weakly carinate; female abdomen often rounded caudally; male abdominal segment VII with transverse combs or clusters of setae ventrally .... Oxycarenidae
13. Abdominal spiracles on segments V to VI ventral; sutures between tergites 4/5 and 5/6 curving forward through middle .... Geocoridae
– Abdominal spiracles on segments V to VI dorsal; all abdominal tergites transverse, sutures never curving forward 14

14. Abdominal spiracles on segment VII dorsal; each pronotal callus with an impressed, transverse, usually shiny groove; scutellum usually with a cross-shaped carina...Lygaeidae
– Abdominal spiracles on segment VII ventral; pronotal calli without impressed grooves; scutellum without a cross-shaped carina.................................Blissidae

16.4 Family Berytidae

16.4.1 General Characteristics and Diagnosis

Berytidae, commonly referred to as stilt bugs, comprise a small group of morphologically diverse lygaeoids. Typical stilt bugs, such as the metacanthine genera *Jalysus* Stål and *Metacanthus* Costa, are elongate, slender insects, with long slender legs and antennae, often as long as or longer than the body. Other taxa, however, deviate from this general appearance. Members of the genus *Hoplinus* Stål have spindle-shaped bodies and shorter and stouter legs and antennae and often are armed with spines on the head, pronotum, and hemelytra, species of *Parajalysus* Distant are more robust and armed with three long, erect spines on the pronotum, and the recently described *Cuscohoplininus pagoreni* Dellapé and Carpintero has only a single spine on the middle of the anterior pronotal lobe (Dellapé and Carpintero 2007). Other taxa, such as *Pronotacantha* Uhler, have spines on the pronotum and scutellum and distinctly banded appendages, those of *Phaconotus* Harris are adorned with pearl-like tubercles on the anterior lobe of the pronotum, and the genus *Diabolonotus* Henry has two anteriorly directed "devil-like," pronotal horns.

Synapomorphies defining the family are the elongate bilobed head; basally tapered buccula; long, slender, apically clavate first antennal segment; basally narrowed scutellum; subparallel to basally constricted hemelytra; grooved metasternum; dentate claws; grooved, quadrate abdominal segment II; undivided abdominal segment VII in females; hidden dorsal spiracles; and midlateral position of the trichobothria on abdominal segment III (Henry 1997b, c).

16.4.2 Classification and Diversity

Thirty-seven genera and about 174 species of Berytidae are known in the world (Henry and Froeschner 1998; Henry 2002, 2007; Dellapé and Carpintero 2007; Cai et al. 2011, 2013). The family is separated into three subfamilies and six tribes: the Berytinae (and Berytini and Berytinini), Gampsocorinae (and Gampsocorini and Hoplinini), and Metacanthinae (and Metacanthini and Metatropini) (Henry 1997b).

All three subfamilies but only four of six tribes are known from the Neotropical Region. The Neotropical Berytinae are represented only by the widespread North American *Neoneides muticus* (Say), which also occurs in Baja California, Mexico. The Neotropical Gampsocorinae are represented only by the genus *Gampsocoris* Fuss and four species in the nominate tribe Gampsocorini and nine genera and 36 species in the Hoplinini (Henry and Froeschner 1998; Henry 2002; Dellapé and Carpintero 2007). Two genera and 12 species of Neotropical Metacanthinae are known, all of which are placed in the nominate tribe Metacanthini (Henry and Froeschner 1998; Henry 2007).

Most early work on Neotropical berytids was limited to descriptions of a few new genera and species. Stål (1874) provided the first synopsis of the family with the first keys to genera and the species of *Jalysus*. Most subsequent papers were mostly descriptive (e.g., Distant 1880–1893; Horváth 1905), until McAtee's (1919) review of the Nearctic fauna, which included several Neotropical genera. Harris (1943) added additional South American genera and species but little was provided to aid identification until berytid specialist J. M Štusák (e.g., 1967, 1968, 1971, 1973, 1977) provided a series of well-illustrated descriptions of new genera and species and clarification of previously confused taxa. Other important works include Štusák and Cobben's (1975) keys to the Antillean species and Froeschner's (1981) checklist and keys to the Ecuadorian genera and species. More recently, Henry (1997c) monographed the family for the Western Hemisphere, treating 13 genera and 52 species, including 12 genera and 49 species from the Neotropics. Subsequently, Henry (2002) reviewed the genus *Hoplinus*, clarified the identity of the type species, *H. spinosissimus* Signoret and described the new species *H. paulai* from Brazil; Henry (2007) described the new species of *Jalysus ossesae* from Brazil; and Dellapé and Carpintero (2007) described the new hoplinine genus and species *Cuscohoplininus pagoreni* from Peru.

The following keys to the Neotropical subfamilies, tribes, and genera of Berytidae are modified from Henry (1997c).

Key to the Neotropical subfamilies of Berytidae

1. Head, pronotum, and undersurface of thorax bordering rostral sulcus always clothed with appressed sericeous or woolly pubescence; ventral surface of abdomen deeply punctate (except Old World genus *Yemmatropis*) .... Berytinae
– Head, pronotum, and undersurface of thorax along the rostral sulcus without appressed sericeous or woolly pubescence; ventral surface of abdomen never punctate 2
2. Metathoracic scent channel smooth, extended onto an elongate spout or digitiform spine (except Old World genus *Metatropis*) .... Metacanthinae
– Metathoracic scent channel lined with overlapping scalelike plates, extended onto a pouchlike structure, or scent channel and spout completely absent, at most, with a blunt to elongate tubercle comprised of honeycombed chambers .... Gampsocorinae

Subfamily Berytinae Puton

Only one genus and species of this subfamily, belonging to the tribe Berytini Puton, occurs in the Neotropical Region. The widespread *Neoneides muticus* (Say) occurs throughout much of the United States and Canada and into northern Mexico (Henry 1997c).

Subfamily Metacanthinae

Only the nominate tribe Metacanthini occurs in the Neotropical Region.

Key to the genera of Neotropical Metacanthini

1. Ostiolar spout ending in an acutely produced apical spine (Fig. 2).... *Jalysus* Stål
– Ostiolar spout without an apical spine, instead apically rounded and weakly recurved near the level of the hemelytra .... *Metacanthus* Costa

Subfamily Gampsocorinae Southwood and Leston

Key to the Neotropical Tribes of Gampsocorinae

1. Ostiolar scent channel distinct, set within a pouchlike spout lined with overlapping scalelike plates; each side of pronotal collar with an erect spine or tubercle ....Gampsocorini
– Ostiolar scent channel absent, never with a pouchlike spout; pronotum often variously armed with spines or tubercles, but never with a distinct collar having a spine or tubercle on either side ....Hoplinini

Tribe Gampsocorini

Gampsocoris Fuss (Fig. 1) is the only Neotropical genus belonging to this tribe. Henry (1997c) provided a key to the four known species.

Tribe Hoplinini

The nine genera included in this tribe are restricted to the Western Hemisphere. Henry (1997c) revised the group and included keys to species.

Key to the Neotropical genera of Hoplinini

1. Posterior lobe of pronotum armed with erect spines 2
– Posterior lobe of pronotum without spines 4
2. Head with five or more median spines... *Hoplinus* Stål

Plate 16.1 Figs. 1–12. 1, *Gampsocoris tuberculatus* Štusák (Berytidae: Gampsocorinae). 2, *Jalysus sobrinus* Stål (Berytidae: Metacanthinae). 3, *Blissus parasigaster* Drake (Blissidae). 4, *Patritius grossus* Haglund (Blissidae). 5, *Colobasiastes similis* Horváth (Colobathristidae). 6, *Cymodema breviceps* Stål (Cymidae: Cyminae). 7, *Geocoris callosulus* Berg (Geocoridae: Geocorinae). 8 *Isthmocoris imperialis* Distant (Geocoridae: Geocorinae). 9, *Epipolops frondosus* Herrich-Schaeffer (Geocoridae: Pamphantinae). 10, *Cephalocattarus waorani* Slater and Henry (Geocoridae: Pamphantinae). 11, *Kleidocerys virescens* (F.) (Lygaeidae: Ischnorhynchinae). 12, *Polychisme ferruginosus* (Stål) (Lygaeidae: Ischnorhynchinae) (Photos by G Ouellette)

– Head without median spines, at most, with a small rounded tubercle on vertex or a slender spine on middle of posterior lobe 3
3. Pronotum with three rows of long, slender spines, one row along each lateral margin and one along meson; posterior lobe of head with six white tubercles, the middle one often spinelike .... *Pronotacantha* Uhler
– Pronotum with four stout spines, one on anterior lobe and three along basal margin of posterior lobe; posterior lobe of head without tubercles .... *Parajalysus* Distant
4. Base of pronotal disc with three high, acute ridges; head and pronotum with a well-developed median carina; legs and antennae with pilose setae three or more times the diameter of the segments .... *Metajalysus* Štusák
– Base of pronotal disc often convex, but never with three high ridges; appendages with or without pilose setae three or more times diameter of segment.................. 5
5. Scutellum hidden by a strongly explanate basal edge of pronotum; body and appendages with erect, bristlelike or long, simple setae; metafemora distinctly bowed 6
– Scutellum clearly visible and armed with a distinct spine or tubercle; body and appendages without bristlelike or pilose setae; metafemora not bowed 7
6. Anterior lobe of pronotum with two long, stout, anteriorly directed, blunt spines or tubercles; side of pronotum without a swollen protuberance visible from dorsal aspect; appendages with long, pilose setae, setae on tibiae three or four times longer than the diameter of the respective segments .... *Diabolonotus* Henry
– Anterior lobe of pronotum unarmed, without stout spines; side of pronotum with a swollen protuberance visible from dorsal aspect; appendages sometimes with erect, bristlelike setae, but those on tibiae at most subequal to diameter of respective segments.. *Xenoloma* Harris
7. Anterior lobe of pronotum with an erect spine at middle between calli .... *Cuscohoplininus* Dellapé and Carpintero
– Anterior lobe of pronotum lacking an erect spine.. 8
8. Anterior lobe of pronotum with 6–11 white, rounded tubercles; posterior edge of pronotum unarmed; posterior lobe of head with a short white tubercle between ocelli and often with a white tubercle on either side of each ocellus .... *Phaconotus* Harris
– Anterior lobe of pronotum with only five white, rounded tubercles restricted to narrow collar region (two rounded ones on either side and one in between two erect, white spines); posterior edge of pronotum with four upturned tubercles; posterior lobe of head without tubercles .... *Oedalocanthus* Henry

16.5 Family Blissidae

16.5.1 General Characteristics and Diagnosis

The Blissidae, or chinch bugs, are elongate to broadly oval, often flattened to access leaf sheaths of their hosts, and range in size from less than 3 mm to more than 15 mm. They are recognized by the often entirely or partially pruinose surfaces of head and pronotum, a tubercle on the genital capsule, the peculiarly winged sperm reservoirs in many taxa, and the dorsal position of the abdominal spiracles and by being the only lygaeoid group to feed entirely on plant sap, rather than seeds. Blissids range from fully winged to frequently short winged or brachypterous, having only remnant wing pads as in the genera *Blissus* Burmeister and *Ischnodemus* Fieber or even the complete loss of wing pads as in the genera *Aulacoblissus* Slater and *Howdenoblissus* Štys.

16.5.2 Classification and Diversity

The Blissidae comprise about 51 genera and 436 species worldwide (Cassis and Gross 2002; Henry 2009). Only 16 genera and about 109 species are known from the Neotropical Region (Slater 1979; Slater and O'Donnell 1995; Dellapé and Montemayor 2009). Of these, about 72 % are in the genera *Blissus* (12 spp.), *Ischnodemus* (34 spp.), *Patricius* Distant (12 spp.), and *Toonglasa* Distant (21 spp.; see discussion below). Slater (1979) monographed the world fauna and provided keys to all genera and most species except for the genus *Blissus*. Although Leonard (1968a) studied the species of *Blissus* for eastern North America and described a few new species (Leonard 1968b, 1970), members of the genus remain difficult to identify and are in great need of revision. Slater and Brailovsky (1983, 1990), in revising the primarily Neotropical genus *Toonglasa* Distant, synonymized *Extarademus* Slater and Wilcox (1966) and provided keys to the 21 known species. Slater (1986b) established the genus *Aulacoblissus* to accommodate a new micropterous species from Venezuela, Štys (1991) described *Howdenoblissus slateri*, a similar-appearing apterous species from Colombia, and Brailovsky and Barrera (2012) added *Napoblissus foreroi*, also with greatly abbreviated hemelytra from Ecuador. Dellapé and Montemayor (2009) described the most recently recognized species of *Ischnodemus* and provided descriptions of the male and immature stages of *I. subflavus* Slater and Wilcox. Henry (1997a) elevated Blissinae to family status.

The following key to genera was created in part from Slater (1979) and in part based on our original research. In our opinion, the genus *Toonglasa* is not monophyletic as now interpreted. As a consequence, until additional revisionary studies can be conducted, we are reestablishing *Toonglasa* as a monotypic genus, containing only the type species *T. forficuloides* Distant and resurrecting *Extarademus* to accommodate the remaining 20 species, including its type species, *Macropes collaris* Signoret, as designated by Slater and Wilcox (1966).

Key to the Neotropical genera of Blissidae

1. Fore coxal cavities open behind 2
– Fore coxal cavities closed behind 9
2. Pronotum at least partially pruinose 3
– Pronotum shiny, without pruinose areas 5
3. Pronotum entirely pruinose (Fig. 3) .... *Blissus* Burmeister
– Anterior half of pronotum shiny; posterior half pruinose 4
4. Membrane entirely white; posterior margin of pronotum tan, contrasting with a darker anterior area .... *Caveloblissus* Slater and Wilcox
– Membrane dark brown, with a subbasal pale band; posterior pronotal lobe entirely dark .... *Praeblissus* Barber
5. Macropterous to micropterous (with at least short wing pads) 6
– Apterous (without wing pads) 7
6. Abdominal venter with a prominent stridulitrum on segments 3 to 6; metathoracic scent-gland auricle simple and rounded; only fore femur with one large and several small spines .... *Heteroblissus* Barber
– Abdominal venter lacking a stridulitrum; metathoracic scent-gland auricle strongly produced anteriorly; all femora with spines .... *Praetorblissus* Slater
7. Fore femur lacking spines .... *Howdenoblissus* Štys
– Fore femur with one or two spines 8
8. Fore femur with only one small spine; scutellum lacking a median elevation .... *Aulacoblissus* Slater
– Fore femur with two spines, one large and one small; scutellum with a median elevation 9
9. Labium short, not reaching fore coxae; pronotum pruinose .... *Patritiodemus* Slater and Ahmad
– Labium longer, always reaching fore coxae or beyond; pronotum shiny or in part pruinose 10
10. Fore femur lacking spines .... *Ischnodemus* Fieber
– Fore femur with one or more spines 11
11. Fore femur with only one spine 12
– Fore femur with two or more spines 14
12. Membrane composed of numerous small reticulate cells .... *Reticulatodemus* Slater and Wilcox
– Membrane without small reticulate cells 13
13. Broad flattened species; fore femur strongly incrassate; pro- and mesosternum grooved to receive labium; each side of seventh abdominal segment in males with a prominent posteriorly directed projection; scent-gland auricle relatively broad, slightly curving forward distally, and distinctly raised above the evaporative surface .... *Toonglasa* Distant, revised status
– Slender, elongate species; fore femur not strongly incrassate; pro- and mesosternum not grooved to receive labium; seventh abdominal segment in males without projections; scent-gland auricle slender, curving forward, and flat against evaporative surface .... *Extarademus* Slater and Wilcox

14. Fore femur with two spines; head and pronotum with scalelike setae .... *Xenoblissus* Barber
– Fore femur with three or more spines 15
15. Relatively slender species; only fore femur multispinose, middle and hind femora without spines .... *Procelladernus* Slater and Wilcox
– All femora with multiple spines 16
16. Body broad and strongly flattened; head shiny; pruinose areas of pronotum confined to the area around collar and sometimes a narrow line across the transverse impression .... *Riggiella* Kormilev
– Body elongate and not flattened; head pruinose; pronotum usually at least in part pruinose (Fig. 4) .... *Patritius* Distant

16.6 Family Colobathristidae

16.6.1 General Characteristics and Diagnosis

The Colobathristidae (Fig. 5) comprise a small group of tropical bugs with about half the known genera found in the Neotropical Region. They range from about 6.0 mm to over 20 mm. They are characterized by their slender elongate bodies, long slender legs and antennae, bulging eyes, quadrate bilobed thorax, slender often spined scutellum, hyaline hemelytra with few hardly visible membranal veins, and slender elongate abdomen constricted at the base. Colobathristids also possess characters appearing in part coreoid and lygaeoid (Henry 1997a). Štys (1966) indicated they have a platelike ovipositor similar to those in the Coreoidea or Pyrrhocoroidea but internal genital structures similar to those in the malcid line. Cobben (1968) indicated that the eggs were most similar to the Coreoidea. Kumar (1968), however, considered the three-lobed salivary glands and the fingerlike gastric ceca on the midgut "decidedly lygaeid features" (sensu lato), as well as the aedeagus, the arrangement of the trichobothria, and the position of the spiracles. In Henry's (1997a) phylogenetic analysis, the colobathristids nest within the Lygaeoidea, forming a sister-group relationship with the Berytidae.

16.6.2 Classification and Diversity

Twenty-five genera and more than 80 species of Colobathristidae (Kerzhner 2001; Štys and Exnerová 2012) are placed in two subfamilies, the nominate Colobathristinae and the monogeneric Dayakiellinae, containing only two Indonesian species of *Dayakiella* Horváth (Štys 1966). Štys and Exnerová (2012) provided a key to the 13 Old World genera. Thirteen genera and about 40 species are recorded from the Neotropics. *Phaenacantha saileri* Kormilev, described from Guatemala, is the only representative of an otherwise exclusively Old World genus. Horváth (1904) monographed the family and Kormilev (1949a, 1949b, 1951) described additional new Neotropical taxa. Carvalho and Henry (1986) described the new genus *Parathristes* to accommodate *P. carajaensis* from Pará, Brazil, the largest species of the family yet discovered. Štys and Henry (2015) added the new genus *Neolabradoria* and new species *N. inexpectata* from Peru, provided the replacement name *Bradaloria* for the preoccupied *Labradoria* Kormilev, and gave a revised key to the Neotropical genera.

Kormilev (1951) provided a key to the Neotropical genera, modified from Horváth (1904), and Carvalho and Costa (1989) provided a well-illustrated update, with the addition of *Parathristes*. The following key is modified from Štys and Henry (2015). The genus *Curupira* Distant (1888), tentatively placed as a synonym of *Colobathristes* Burmeister by Carvalho and Costa (1989), is not included until type material is studied and its identity clarified.

Key to the genera of Neotropical Colobathristidae

1. Side of head with a distinct nearly straight to lunate stridulitrum 2
– Side of head without a stridulitrum 7
2. Scutellum with a long, erect, spine, usually longer than the scutellum 3
– Scutellum without an erect spine, apex spiniform, horizontal or subhorizontal, usually shorter than the scutellum 6
3. Distance between ocelli greater than space from an ocellus to an eye; sternite IV as long as III .... *Diascopoea* Horváth
– Distance between ocelli much less than distance between an ocellus and an eye 4
4. Antennal segment IV with a distinct white ring at base .... *Neocolobrathristes* Kormilev
– Antennal segment IV without a white ring at base 5
5. Scutellar spine nearly glabrous; length of antennal segments III and IV subequal; vertex with one groove .... *Calliseidus* Horváth
– Scutellar spine with numerous long setae; antennal segment IV distinctly longer than III; vertex with two grooves .... *Trichocentrus* Horváth
6. Anterior lobe of pronotum bulbous or horn shaped, higher than head; scutellar spine horizontal or subhorizontal; undersurface of fore tibia with distinct denticles .... *Peruda* Distant
– Anterior lobe of pronotum not bulbous, lower than the head and posterior lobe; scutellum without horizontal spine, upper surface with a small dent or impression; fore tibia without distinct denticles .... *Perudella* Kormilev
7. Scutellum with a long, erect, spine, usually longer than scutellum 8
– Scutellum without an erect spine, apex spiniform, horizontal or subhorizontal, usually shorter than scutellum 11
8. Head and anterior lobe of pronotum almost glabrous, with only thin, simple setae 9
– Head and anterior lobe of pronotum densely coated with adpressed, woolly, silvery pubescence 10

9. Rostrum extending only to bases of fore coxae; anterior lobe of pronotum much longer than posterior lobe .... *Parathristes* Carvalho and Henry
– Rostrum extending nearly to middle coxae; anterior lobe of pronotum shorter than posterior lobe.. *Phaenacantha* Horváth
10. Distance from an ocellus to an eye about twice the distance between ocelli; rostral segment IV longer than III; ventral edge of genital capsule with a sharp, protruding process... *Carvalhoia* Kormilev
– Distance from an ocellus to an eye subequal or slightly greater than the distance between ocelli; rostral segments III and IV equal in length; ventral edge of genital capsule lacking a sharp process .... *Colobathristes* Burmeister
11. Distance from an ocellus to an eye less than distance between ocelli; vertex with a median groove in front of ocelli 12
– Distance from an ocellus to an eye three times greater than distance between ocelli; vertex in front of ocelli with two shallow grooves............................... 13
12. Distance between ocelli slightly greater than distance from an ocellus to an eye; scutellum subtriangular, with a slender, pointed, horizontal spine (meso- and metanotum covered); hemelytra narrower than abdomen.................................. .... *Bradaloria* Štys and Henry
– Distance between ocelli subequal to distance from an ocellus to an eye; scutellum transverse, oval, with only a low dentiform tubercle (medial parts of meso- and metanotum exposed); hemelytra as wide as abdomen .... *Neolabradoria* Štys and Henry
13. Scutellum with subhorizontal spine visible in lateral aspect; fore femora lacking or with only a few tiny denticles on distal third; antennal segment III short, only about one-third of body length.. *Piptocentrus* Horvath
– Scutellum with horizontal spine, not visible in lateral aspect; fore femur with numerous small denticles over entire length; antennal segment III long, longer than half the body length.. *Colobasiastes* Breddin

16.7 Family Cymidae

16.7.1 General Characteristics and Diagnosis

New World cymids are small, punctate, usually yellowish-brown bugs, measuring from 3.0 to nearly 5.0 mm. Members of this family have the ocelli nearly encircled by a groove, a short barrel-shaped antennal segment I, a short buccula not extending posteriorly beyond the level with the bases of the antennae, spiracles II to VI dorsal, and the dorsal abdominal scent-gland scars appearing singly between terga 4/5 (*Cymodema* and a few *Cymus*), in twos between terga 3/4 and 4/5 (*Cymus*), and threes between terga 3/4, 4/5, and 5/6 (Ontiscinae).

16.7.2 Classification and Diversity

Nine genera and about 54 species known worldwide (Hamid 1975; Henry 2009) are placed in two subfamilies, the Cyminae and Ontiscinae, of which only the former occurs in the New World. In the Neotropical Region, only two genera and eight species are known. *Cymodema* Spinola contains only two species, with *C. breviceps* (Stål) (Fig. 6) the most widespread, occurring from the United States to Argentina and Brazil, and *C. barberi* Hamid, known only from Mexico. *Cymus* Hahn contains six species, with one known only from Brazil, four recorded from Mexico, and one from Guatemala and Mexico. Hamid (1975) monographed the group as a subfamily and provided keys to the genera and species of the world. Henry (1997a) gave Cyminae family status (minus the Ninini) and hypothesized it as the sister group to the remainder of the "malcid line," including in sequence the Ninidae, Malcidae, Colobathristidae, and Berytidae.

The following key will separate the only two Neotropical cymid genera.

Key to the Neotropical genera of Cymidae

1. First antennal segment not exceeding apex of clypeus; with two dorsal abdominal scent-gland scars between terga 3/4 and 4/5; seventh abdominal spiracle ventral .... *Cymus* Hahn
– First antennal segment exceeding the apex of clypeus; with only one dorsal abdominal scent-gland scar between terga 4/5; seventh abdominal spiracle dorsal (Fig. 6) .... *Cymodema* Signoret

16.8 Family Geocoridae

16.8.1 General Characteristics and Diagnosis

Members of this subfamily are readily recognized by their kidney-shaped to stylate eyes, extending laterally beyond the anterior margins of the pronotum; the broad heads; the relatively stout ovoid bodies; and the posteriorly curved abdominal sutures between terga 4/5 and 5/6. In addition, the abdominal spiracles on segments II, III, and IV are dorsal and those on V, VI, and VII usually are ventral (Henry 2009), except for members of the Australian tribe Australocorinae, in which all abdominal spiracles (II–VII) are dorsal (Malipatil 2012).

16.8.2 Classification and Diversity

The Geocoridae, or big-eyed bugs, are a worldwide group comprising 27 genera and about 280 species (Henry 2009, 2013; Malipatil 2012; Rengifo-Correa et al. 2013). The family is separated into five subfamilies: the Australocorinae Malipatil,

Bledionotinae Reuter, Geocorinae Stål, Henestarinae Douglas and Scott, and Pamphantinae Barber and Bruner (Slater 1999; Henry 1997a, 2009, 2013; Malipatil 2012). The Australocorinae, including one genus and four species restricted to Australia (Malipatil 2012); the Bledionotinae, containing one genus and species restricted to the Palearctic (Slater 1964a; Péricart 2001); and the Henestarinae, with three genera, are restricted to the Old World, with the exception of the problematic *Coriantipus inopinatus* Bergroth described from Argentina. The Geocorinae occur in all zoogeographic regions and the Pamphantinae are found only in the New World tropics (Henry 2013; Rengifo-Correa et al. 2013), with the exception of one genus and species described from Queensland, Australia (Slater 1981).

The Neotropical Geocorinae are represented by only four genera and about 25 species (Slater 1964a; Slater and O'Donnell 1995), which represents only about 10 % of the world fauna (Readio and Sweet 1982). The eastern US species of *Geocoris* Fallén and *Isthmocoris* McAtee were revised by Readio and Sweet (1982). Brailovsky (2013) provided a key to the fourteen Neotropical species of *Ninyas* Distant. The monobasic geocorine genus *Stenogeocoris* and the species *S. horvathi* were described by Montandon (1913) based on one specimen from Córdoba Province, Argentina; the type specimen is lost and there have been no additional records since Montandon's description.

The Pamphantinae, comprising ten genera and 48 species (Henry 2013; Rengifo-Correa et al. 2013), are separated into three tribes, the Cattarini Slater, the Epipolopini Slater, and the nominate Pamphantini (Slater 1999). Brailovsky (1989b) described one new genus and two new species of Pamphantini and provided a key to the genera. Slater and Henry (1999) reviewed the cattarine genus *Cattarus*, with four new species, and described the remarkable ant-mimetic *Cephalocattarus waorani* from Ecuador. Baranowski and Slater (2005) gave a key to the three genera and 13 species known from the West Indies. Henry (2006) revised the stalk-eyed genus *Epipolops*, described five new species, and provided a key to distinguish the 14 known species, and Henry (2013) described the new genus and species *Cymapamphantus valentineorum* from the British Virgin Islands and provided a checklist and keys to the tribes and genera of the Pamphantinae. Rengifo-Correa et al. (2013) described two additional species of *Epipolops* and provided a revised key to species and a phylogenetic analysis of the genus.

The following key to the genera of Geocorinae is modified from Readio and Sweet (1982) and Baranowski and Slater (2005). The keys to tribes and genera of the Pamphantinae are from Henry (2013).

Key to the Neotropical subfamilies of Geocoridae

1. Sutures on abdominal sterna II, III, and IV fused and without lateral trichobothria; body elongate, antlike .... Pamphantinae
– Sutures on abdominal sterna II, III, and IV entire and with distinct, lateral trichobothria; body more stout, not antlike .... Gecorinae

Key to the Neotropical genera of Geocorinae

1. Claval commissure well developed; clypeus with a slight sulcus; basal three abdominal sterna often fused .... *Ninyas* Distant
– Claval commissure not present; clypeus with a complete median sulcus; basal three abdominal sterna not fused 2
2. Rostral segment II longer than III; eyes stylate, remote from anterior angles of pronotum; ocular suture incomplete, extending laterally from the ocellus around base of eye stalk to apical margin of head, ending at anterior margin of eye (Fig. 8) .... *Isthmocoris* McAtee
– Rostral segment II shorter than III; eyes semistylate, sometimes near or in contact with anterior angles of pronotum; ocular suture absent or suture complete, extending laterally from the ocellus, around base of eye stalk to apical margin of head and posteriorly back to ocellus (Fig. 7) 3
3. Body elongate, about three times longer than wide; pronotum subquadrangular, wider at the level of and slightly constricted behind the calli .... *Stenogeocoris* Montandon
– Body short and stout, never more than twice as long as wide. Pronotum subquadrangular .... *Geocoris* Fallén

Key to the Neotropical tribes of Pamphantinae

1. Males and females with a distinct lunate stridulitrum on side of head below eyes and a plectron on inner face of fore femur; male abdomen with a distinct tubercle on each side of segment III; anterior and posterior pronotal lobes separated by a deep transverse impression .... Cattarini
– Males and females without a stridulitrum on head or a plectron on fore femur; male abdominal segment III without a tubercle, though sometimes slightly swollen; pronotal lobes usually not separated by a deep transverse impression . 2
2. Eyes strongly stylate, extending laterally well beyond outer margin of head; pronotum usually with lobes and/or spines along lateral margins; includes only *Epipolops* (Fig. 9) .... Epipolopini
– Eyes not stylate, never extending beyond outer margin of head; pronotum entire, without lateral lobes or spines .... Pamphantini

Key to the Neotropical genera of Cattarini

1. Lateral margins of pronotal lobes unarmed .... *Cattarus* Stål
– Lateral margin of anterior pronotal lobe with a broad winglike process and each humeral angle of posterior lobe with a spinelike projection (Fig. 10) .... *Cephalocattarus* Slater and Henry

Key to the Neotropical genera of Pamphantini

– Anterior femur without a distinct subapical spine 2
– Anterior femur with one or two distinct subapical spines 3

2. Anterior and posterior lobes of pronotum little separated, without a deep transverse impression; posterior lobe of pronotum almost flat, without swellings .... *Parapamphantus* Barber
– Anterior and posterior lobe of pronotum separated by a deep transverse impression; posterior lobe with a distinct swelling or blunt tubercle on each side .... *Tropicoparapamphantus* Brailovsky
3. Head lacking ocelli 4
– Head with distinct ocelli between eyes 5
4. Anterior pronotal lobe greatly swollen or globose; narrow posterior lobe with a long, slender, curving spine arising at each humeral angle .... *Abpamphantus* Barber
– Anterior pronotal lobe not greatly swollen, two and half times as long and only slightly wider than posterior lobe; humeral angles unarmed .... *Cymapamphantus* Henry
5. Eyes large and substylate, inner margin of eye extending past anterior angle of pronotum; distance between ocelli less than to subequal to the distance from an ocellus to an eye; posterior half of hemelytra convex or rounded; profemur with two spines (apical spine broken on three specimens examined) .... *Neopamphantus* Barber and Bruner
– Eyes prominent, but not substylate, inner margin of eye not extending past anterior angle of pronotum; each ocellus closer to eye than to each other; hemelytra subparallel throughout; profemur with only one spine .... *Pamphantus* Stål

16.9 Family Heterogastridae

16.9.1 General Characteristics and Diagnosis

This family is recognized by the ventral abdominal spiracles, the lack of trichobothria on the head, the weakly incrassate fore femora, the distinct closed cell at the base of the hemelytral membrane, and the deeply inserted ovipositor often extending to abdominal segment V (Scudder 1962a; Henry 1997a, 2009).

16.9.2 Classification and Diversity

The Heterogastridae are primarily an Old World group comprising about 24 genera and 100 species (Cassis and Gross 2002; Henry 2009), with only two native (Ashlock and Slater 1988) and one introduced (Wheeler and Hoebeke 2013) species of *Heterogaster* Schilling occurring in North America. Only *H. behrensii* (Uhler) gets into northern Mexico as the sole Neotropical member of the family. Scudder

(1962a) provided a key to the genera of the world and Henry (1997a) elevated Heterogastrinae to family status.

16.10 Family Lygaeidae

16.10.1 General Characteristics and Diagnosis

The Lygaeidae are recognized by the impressed, often shiny, transverse line across the calli; the Y-shaped pattern on the scutellum; and the dorsal position of abdominal spiracles II through VII. Members of the subfamily Ischnorhynchinae are dull brown to reddish brown, elongate-oval, nearly glabrous bugs having the clavus punctate and the base of the pronotum entire and not depressed on either side of the scutellum. The Lygaeinae are usually moderate to large, often aposematically colored, orange and red bugs having an impunctate clavus, the base of the pronotum depressed on either side of the scutellum, and a distinct subcosta on the hemelytra. The Orsillinae are relatively small, dull, yellowish brown, often pubescent bugs having a depressed area at the base of the pronotum and lacking punctures on the clavus and a subcosta.

16.10.2 Classification and Diversity

The Lygaeidae (sensu stricto) comprise about 102 genera and 970 species worldwide (Slater and O'Donnell 1995; Henry 2009). The family is separated into three subfamilies, the Ischnorhynchinae, Lygaeinae, and Orsillinae (Henry 1997a). Sweet (2000a) suggested that the Ischnorhynchinae and Orsillinae should be given family status. He did not, however, provide information to support his hypothesis; thus, we maintain the family as documented by Henry (1997a). Brailovsky (1982a) and A. ("A" used to distinguish Alex Slater from James A. Slater – i.e., Slater without initials throughout text) Slater (1992) provided keys to the Neotropical genera; Baranowski and Slater (2005) monographed the Lygaeoidea of the West Indies and gave keys to all of the genera and species. Henry (1997a) accorded the Lygaeidae (sensu stricto), including the Ischnorhynchinae, Lygaeinae, and Orsillinae, family status.

The Ischnorhynchinae comprise about 15 genera and 77 species worldwide, with only four genera and nine species known from the Neotropics (Slater 1964a; Slater and O'Donnell 1995; Baranowski and Slater 2005). The primarily Holarctic genus *Kleidocerys* Stephens contains four Neotropical species. *Neokleidocerys* Scudder, with three Neotropical species and previously treated as a subgenus of *Kleidocerys*, was elevated to generic rank by Slater and Brailovsky (1989). *Polychisme* Kirkaldy, with only one species, previously was placed in the family Artheneidae (Slater and Brailovsky 1986) but was returned to Ischnorhynchinae by Kerzhner (1997).

Syzygitis Bergroth, previously considered a synonym of *Polychisme* (Slater 1967), was reinstated by Slater and Brailovsky (1986). Scudder (1962a) revised the subfamily and provided a key to the world genera.

Lygaeinae is largest of the three subfamilies, with about 57 genera and 500 species worldwide. Twenty-two genera and about 175 species are known in the Neotropics (A. Slater 1992; A. Slater and Baranowski 2001). Brailovsky (1982a) and Baranowski and Slater (2005) provided keys to all or a substantial number of the Neotropical genera, and A. Slater (1992) revised the subfamily and provided keys to all New World genera and a checklist of species. Slater and O'Donnell (1995) summarized the many new combinations resulting from A. Slater's (1992) work. A. Slater and Baranowski (2001) described the most recent Neotropical genus, *Melanopleuroides*, from the Dominican Republic. Other important works include Brailovsky's revisions of *Lygaeus* F. (1978), *Craspeduchus* Stål (1979), *Acroleucus* Stål (1980), *Ochrimnus* Stål (1982b), and *Torvochrimnus* Stål (1983), among other important papers. Many lygaeines, such as the genera *Lygaeus* F. and *Oncopeltus* Stål, are aposematically colored red and orange. The vast literature base involving mostly laboratory studies of *Oncopeltus fasciatus* (Dallas), commonly called the large milkweed bug, was compiled by Slater and O'Donnell (1995).

The Orsillinae comprise about 30 genera and 250 species, separated into four tribes (Ashlock 1967a; Henry 2009). In the Neotropics, three tribes, nine genera, and about 46 species are known. Ashlock and Lattin (1963) established *Xyonysius* for ten species previously placed mostly in *Nysius* Dallas, Ashlock (1967a) revised and provided a key to the orsilline genera of the world, and Hamilton (1983) established *Neortholomus* for the New World species of *Ortholomus* Stål. Barber (1947) revised the North America species of *Nysius*, but this work is badly outdated and the Neotropical species remain confused and difficult to identify.

We provide keys to the subfamilies and keys to the genera of Ischnorhynchinae modified from Scudder (1962b), Lygaeinae modified from A. Slater (1992) and Baranowski and Slater (2005), and the Orsillinae modified from Ashlock (1967a).

Key to the Neotropical subfamilies of Lygaeidae

1. Clavus punctate; base of pronotum entire, not depressed .... Ischnorhynchinae
– Clavus impunctate; base of pronotum bordering either side of scutellum depressed or flattened 2
2. Hemelytra with a distinct subcosta and membrane lacking intervannal veins; coloration often in large part red or orange; most species nearly glabrous .... Lygaeinae
– Hemelytra lacking a subcosta and membrane with intervannal veins; coloration dull gray to yellowish brown; species often pubescent .... Orsillinae

Key to the Neotropical genera of Ischnorhynchinae

1. Lateral margin of pronotum lacking a distinct carina; corium with a short row of punctures along claval suture near inner angle and a complete row extending the entire length; middle of hemelytra largely hyaline 2

– Lateral margin of pronotum distinctly carinate; corium with only a single row or two or more entire rows of punctures; middle of corium more opaque............... 3
2. Anterior half of pronotum narrowed and elongate; corium with apical margin and apex with a rectangular or L-shaped macula......................................
....*Neokleidocerys* Slater and Brailovsky
– Anterior half of pronotum not narrowed and elongate; corium at most with two small maculae apically (Fig. 11)..*Kleidocerys* Stephens
3. All abdominal spiracles dorsal; basal two-thirds of lateral margin of hemelytra broadly explanate, broader at middle than diameter of antennal segment I...........
....*Syzygitis* Bergroth (Chile)
– Spiracles on abdominal segments III to VII ventral; basal two-thirds of lateral margin of hemelytra only narrowly explanate, subequal at middle to diameter of antennal segment II (Fig. 12)..*Polychisme* Kirkaldy

Key to the Neotropical genera of Lygaeinae

1. Eyes on stalks...*Nicuesa* Distant
– Eyes not on stalks...2
2. Callus depressed on either side of median carina; hemelytral membrane dark with apical margin or rounded macula clear or hyaline........................*Acroleucus* Stål
– Callus not depressed or interrupted medially; hemelytral membrane not dark with a hyaline apical area...3
3. Pronotum with four transverse depressions behind the calli......... *Ochrimnus* Stål
– Pronotum without four transverse depressions behind the calli 4
4. Scutellum swollen, raised above hemelytra; basal margin of pronotum on either side of scutellum flattened and produced posteriorly (Fig. 14) ....*Oncopeltus* Stål
– Scutellum not swollen; basal margin of pronotum not produced posteriorly 5
5. Posterior pronotal lobe, at most, as high mesally as lateral margins; disc flat or nearly flat; median carina on basal third distinct 6
– Posterior pronotal lobe higher mesally than at lateral margins; disc distinctly convex; median carina absent 9
6. Clavus pale, with a dark vein on inner and outer margin.....................................
....*Neacoryphus* Scudder
– Clavus unicolorous...7
7. Dorsal pubescence short, usually semierect, and dense; membrane uniformly dark ....*Dalmochrimnus* Brailovsky
– Dorsal pubescence short, recumbent, and sparse; membrane usually marked with white...8
8. Larger species, length 6 mm or more; membrane dark with narrow lateral margin white or entirely white, with dark veins; brachypters uncommon........................
.... *Melacoryphus* A. Slater

– Smaller species, length usually less than 6 mm; membrane usually with a large, central white spot, if largely white with dark veins, then basal third dark; brachypters common .... *Lygaeospilus* Barber
9. Pronotum coarsely punctate, punctures extending onto disc .... *Oxygranulobaphus* Brailovsky
– Pronotum finely punctate, punctures restricted to depressed areas just before and after calli 10
10. Pronotum dark, at most with anterior angles paler 11
– Pronotum extensively pale, at least with more than anterior angles pale 12
11. Head with pale spots; posterior metapleural margin nearly straight (Fig. 13) .... *Melanopleurus* Stål
– Head without pale spots; posterior metapleural margin distinctly concave (Dominican Republic)......................... *Melanopleuroides* Slater and Brailovsky
12. Claval veins either paler or darker than surrounding area 13
– Claval veins concolorous with surrounding area 14
13. Lateral margin of the pronotum dark brown; veins on corium and membrane darker than surrounding areas (Fig. 15).................... *Torvochrimnus* Brailovsky
– Lateral margin of pronotum pale; veins on corium paler than surrounding area; veins on membrane concolorous with surrounding area.................................... .... *Hadrosomus* A. Slater
14. Claval margin bordering scutellum pale, contrasting with dark margin bordering corium or a semicircular, submedian spot 15
– Clavus unicolorous or pale basally and dark apically or with only the commissural margin pale.. 16
15. Apical margin of corium red or orange with a narrow outer black border .... *Anochrostomus* A. Slater
– Apical margin of the corium yellow 17
16. Postcallar impression interrupted on either side of median line forming a short accessory branch .... *Craspeduchus* Stål
– Postcallar impression entire, not interrupted on either side of median line .... *Ochrostomus* Stål
17. Pronotum longer medially than wide at anterior margin; hemelytral membrane brown with a pale median line .... *Ektyphonotus* A. Slater
– Pronotum distinctly shorter medially than wide anteriorly; hemelytral membrane not brown with a pale median line 18
18. Costal margin of corium pale apically and basally, dark brown to black between; dark markings on corium either restricted to the costal margin and a small discal spot or a triangular spot with the longest side at costal margin; thoracic pleura almost entirely gray or black... *Lygaeus* Fabricius
– Costal margin of corium usually entirely pale, if interrupted, then dark markings of corium forming a subrectangular spot with anterior and posterior edges straight; thoracic pleura with wide pale margins 19
19. Length more than 10 mm; pronotum pale orange with a small, transverse, brown spot on either side of basal margin and a dark quadrate spot on each callus .... *Biblochrimnus* Brailovsky

Plate 16.2 Figs. 13–24. 13, *Melanopleurus bistriangularis* (Say) (Lygaeidae: Lygaeinae). 14, *Oncopeltus sexmaculatus* Stål (Lygaeidae: Lygaeinae). 15, *Torvochrimnus poeyi* Guérin-Méneville (Lygaeidae: Lygaeinae). 16, *Neortholomus jamaicensis* (Dallas) (Lygaeidae: Orsillinae). 17, *Nysius irroratus* (Spinola) (Lygaeidae: Orsillinae). 18, *Cymoninus notabilis* (Distant) (Ninidae). 19, *Crophius convexus* Barber (Oxycarenidae). 20, *Oxycarenus hyalinipennis* (Costa) (Oxycarenidae). 21, *Pachygrontha compacta* Distant (Pachygronthidae: Pachygronthinae). 22, *Phlegyas annulicrus* Stål (Pachygronthidae: Teracriinae). 23, *Parapiesma cinereum* (Say) (Piesmatidae). 24, *Plinthisus parvioculatus* Slater (Rhyparochromidae: Plinthisinae) (Photos by G Ouellette)

– Length 7 mm or less; pronotum with pale orange to red, but with at least a large dark quadrate spot on the posterior lobe 20
20. Pronotum uniformly dark brown, with only median line pale .... *Achlyosomus* A. Slater
– Anterior pronotal lobe orange to red; posterior lobe dark, with median line pale 21
21. Head slightly swollen; ocelli lower than vertex when viewed laterally .... *Pseudacroleucoides* Brailovsky
– Head slightly less swollen; ocelli about level with vertex when viewed laterally .... *Latochrimnus* Brailovsky

Key to the Neotropical genera of Orsillinae

1. Costal margin of hemelytron straight to at least level with apex of clavus; fore femur sometimes spined; connexivum often exposed (Orsillini) 2
– Costal margin of hemelytron straight only to level with apex of scutellum; fore femur never spined; connexivum not exposed 4
2. Mesopleuron and propleuron evenly meeting, not overlapping (widespread) (Fig. 16)... *Neortholomus* Hamilton
– Mesopleuron appearing to overlap the propleuron 3
3. Vertex with a carina extending from each ocellus to antenniferous tubercle; head short, anteocular length less than two times length of an eye; labium extending to hind coxae (Brazil)... *Aborsillus* Barber
– Vertex lacking carinae; head elongate, anteocular length more than two times length of an eye; labium long, extending to apex of abdomen (North America and Mexico) .... *Belonochilus* Uhler
4. Hemelytron without a complete row of punctures on either side of claval suture; buccula impunctate (Nysiini) (widespread) (Fig. 17) .... *Nysius* Dallas
– Hemelytron with a distinct row of punctures on either side of claval suture; buccula usually punctate (Metrargini) 5
5. Buccula not extending beyond level with middle of eye (widespread) .... *Xyonysius* Ashlock and Lattin
– Buccula nearly reaching base of head.. 6
6. Antenniferous tubercle acute or quadrate; explanate costal margin of hemelytra with alternating light and dark spots 7
– Antenniferous tubercle not produced; costal margin of hemelytra without alternating light and dark spots 8
7. Scutellum longer than prothorax; veins of corium raised; hemelytral membrane reduced (Juan Fernandez Islands)............................... *Robinsonocoris* Kormilev
– Scutellum shorter than prothorax; veins of corium not raised; hemelytral membrane fully developed (Galapagos Islands)................. *Darwinysius* Ashlock
8. Apex of scutellum rounded; macropterous (brachypters unknown) (Argentina) .... *Balionysius* Ashlock
– Apex of scutellum acute; hemelytra coleopteriform; hemelytral membrane greatly abbreviated (Colombia).. *Coleonysius* Ashlock

16.11 Family Ninidae

16.11.1 General Characteristics and Diagnosis

Most Ninidae range from about 3.0 to 4.0 mm long and are recognized by the broad head, the stylate eyes, the broad vertex, the transverse sulcus before the ocelli, the mostly hyaline and often basally constricted hemelytra, the bifid apex of the scutellum, and the dorsal position of abdominal spiracles II–IV (Scudder 1957; Henry 1997a). The head and pronotum of several taxa are coated with a powdery white residue.

16.11.2 Classification and Diversity

Ninids are a small family comprising five genera and 16 species worldwide (Slater 1964a; Slater and O'Donnell 1995). Scudder (1957) revised the Ninini as a tribe of the Cyminae and provided a key to genera and species. Only three genera and six species of Ninidae occur in the Neotropics. *Cymoninus notabilis* (Distant) is the most widespread New World species, ranging from the southern United States, throughout the West Indies, and into much of South America (Scudder 1957), and *C. wilcoxae* Brailovsky is known only from Mexico (Brailovsky 1975). *Neoninus illustris* Distant occurs from Mexico to Brazil, *N. argentinus* Kormilev is known only from Argentina, and *N. montanellus* Brailovsky occurs in Brazil and Venezuela (Brailovsky 1989a). The monotypic genus *Paraninus* Scudder is represented by *P. gracilis* Scudder, described from Brazil and British Guiana. This small family previously was included as a tribe of the Cyminae prior to its elevation to family status (Henry 1997a). The following key is modified from Scudder (1957).

Key to the Neotropical genera of Ninidae

1. Apical half of first labial segment slender................................*Paraninus* Scudder
– Apical half of first labial segment swollen..2
2. Antennal segments II and III more slender than segment IV; distal half of clavus clear or transparent (Fig. 18).. *Cymoninus* Breddin
– Antennal segments II and III as thick as segment IV; distal half of clavus pruinose and opaque ....*Neoninus* Distant

16.12 Family Oxycarenidae

16.12.1 General Characteristics and Diagnosis

Oxycarenids are characterized by the punctate porrect head; hyaline often explanate hemelytra; abdominal spiracles III to VII ventral and II dorsal; lack of lateral trichobothria on sterna III, IV, and V; a median trichobothrium on sternum V; abdominal sutures complete to lateral margins (segments III, IV, and V often fused); a truncate female abdomen; and a transverse comb of glandular setae on the male abdomen (Henry 1997a, 2009; Henry and Dellapé 2009). Previously treated as a subfamily of the Lygaeidae (sensu lato), the group was accorded the group family status by Henry (1997a).

16.12.2 Classification and Diversity

The Oxycarenidae include about 24 genera and 150 species, most of which occur in the Old World (Péricart 2001; Henry 2009; Henry and Dellapé 2009; Brailovsky and Cervantes 2011). Five genera and 22 species are found in the Neotropics (Slater 1964a; Dellapé and Cheli 2007; Henry and Dellapé 2009; Brailovsky and Cervantes 2011). The largest genus *Crophius* Stål, comprising 19 species, including eight from the Neotropics (Slater 1964a; Dellapé and Cheli 2007), was synonymized by Hoberlandt (1987) under the Palearctic *Anomaloptera* Amyot and Serville, an action followed by a number of subsequent authors (e.g., Slater and O'Donnell 1995; Péricart 1998, 2001; Dellapé and Cheli 2007). Henry et al. (2015), however, have shown that *Crophius* is distinct from *Anomaloptera*, as is *Mayana* Distant (containing the two species *M. costatus* Distant and *M. diruptus* Distant), also previously placed in synonymy with *Crophius* (Van Duzee 1910). The genus *Dycoderus* Uhler (1901) is represented by the species *D. picturatus* (Uhler), known only from the western United States; *Neaplax* Slater (1974) contains two species, *N. mexicanus* Slater and *N. baja* Brailovsky and Cervantes (2011), both known only from Mexico; and *Notocoderus* Henry and Dellapé (2009) was described to accommodate *N. argentinus* Henry and Dellapé from Argentina. The large Palearctic genus *Oxycarenus* Costa is represented in the New World by the introduced *O. hyalinipennis* (Costa) (Fig. 20), a pest of cotton first detected in Brazil as early as 1917, now known to occur throughout much of South America and the West Indies (Slater and Baranowski 1994). More recently, it was detected in the Florida Keys on naturalized wild cotton, *Gossypium* sp. (Nagoshi et al. 2012), but apparently it has been eradicated (NAPPO 2014).

The following key is modified from Henry et al. (2015).
Key to the Neotropical Genera of Oxycarenidae

1. Ocelli absent.......... 2
– Ocelli present 3
2. Head strongly globose in front of eyes; pronotum trapeziform, not separated into two lobes, but with a wide collar-like area; costal margin of hemelytra explanate .... *Neaplax* Slater
– Head not strongly globose; pronotum lacking a collar and separated into two distinct lobes, with anterior lobe two or more times longer than posterior lobe; costal margin of hemelytra not explanate.......... *Notocoderus* Henry and Dellapé
3. Fore femora with only one spine; labium usually extending only to middle coxae, if longer, never onto abdomen.......... *Crophius* Stål
– Fore femora with two or more spines; labium extending to or beyond middle coxae 4
4. Fore femora with two spines; labium extending only to middle coxae or slightly beyond; hemelytral membrane with numerous closed cells .... *Mayana* Distant
– Fore femora with four spines; labium extending beyond hind coxae, well onto abdomen; hemelytral membrane without closed cells (Fig. 19) .... *Oxycarenus* Costa

16.13 Family Pachygronthidae

16.13.1 General Characteristics and Diagnosis

Members of this family are separated into two subfamilies, the Pachygronthinae and the Teracriinae (Slater 1955). Pachygronthines are distinguished by the elongate bodies, strongly incrassate spined fore femora, and the frequently unusually long first antennal segment. Teracriines are more stout bodied, have shorter antennae, and are held together as a subfamily by the ventral position of the abdominal spiracles (Henry 1997a).

16.13.2 Classification and Diversity

The Pachygronthidae comprise about 13 genera and 78 species worldwide (Cassis and Gross 2002; Henry 2009). Only the two pachygronthine genera *Oedancala* Amyot and Serville, with 14 species, and *Pachygrontha* Germar, with six species, are known from the Neotropics (Slater 1955, 1964a; Slater and O'Donnell 1995). The teracriine genus *Phlegyas* Stål contains only three species, two of which occur in the Neotropical Region. *Phlegyas annulicrus* Stål is known from the western

United States and Mexico and *P. patruelis* Berg has been reported from Argentina, Paraguay, and Uruguay. All members of the family feed on monocots of the families Cyperaceae (sedges), Juncaceae (rushes), and Poaceae (grasses) (Slater 1955). Slater (1955) revised the family for the world and gave keys to genera and species, and Henry (1997a) gave the group family status. The following key is modified from Slater (1955) and Dellapé (2014).

Key to the Neotropical subfamilies and genera of Pachygronthinae

1. Antennal segment I not or barely attaining apex of clypeus, always much shorter than antennal segments II–IV (Teracriinae) (Fig. 22)...................... *Phlegyas* Stål
– Antennal segment I greatly exceeding apex of clypeus, usually as long or longer than antennal segments II–IV (Pachygronthinae)... 2
2. Eye as wide as or wider than long; distance between base of the antenna and anterior margin of eye as long as or longer than length of eye; antennal segment I clavate; antennal segment IV shorter than either segment II or III (Fig. 21) ....*Pachygrontha* Germar
– Eye longer than wide; distance between base of antenna and anterior margin of eye less than length of eye; antennal segment I gradually swollen to apex; antennal segment IV longer than either segment II or III..................................... .... *Oedancala* Amyot and Serville

16.14 Family Piesmatidae

16.14.1 General Characteristics and Diagnosis

The Piesmatidae, often called ash-gray plant bugs, are recognized by the small size (5 mm or less); the dull yellowish-gray coloration; the broad head; the produced juga often surpassing the apex of the clypeus; the widely placed ocelli (often reduced or absent in brachypters); the uni-, bi-, or tricarinate pronotum; the reticulate or areolate hemelytra; the dorsal spiracle on abdominal segment II; the absence of trichobothria on abdominal segments IV and V; and the two-segmented tarsi.

16.14.2 Classification

The Piesmatidae superficially resemble certain Tingidae because of the reticulate hemelytra and the two-segmented tarsi. Drake and Davis (1958), however, showed that they belonged in the infraorder Pentatomomorpha based on the presence of abdominal trichobothria, a true spermatheca, and a differentiated endophallus. Their superfamily placement has fluctuated from being included in their own superfamily, the Piesmatoidea, to their current position in the Lygaeoidea (Henry 1997a). Worldwide nine genera and about 45 species are separated into two subfamilies, the widespread Piesmatinae and the Afrotropical Psamminae (Henry 1997a, 2009; Dellapé 2014).

Drake and Davis (1958) treated the world genera and Heiss and Péricart (1983) revised the Palearctic fauna. Péricart (1974) established two new subgenera, *Afropiesma* and *Parapiesma*, for the widely distributed genus *Piesma* Le Peletier and Serville, and Heiss and Péricart (1997) later gave these subgenera generic status. As a consequence, most of the New World species previously placed in *Piesma* now belong in *Parapiesma* Péricart, including the widespread *P. cinereum* (Say). Two genera and two species occur in the Neotropics. *Miespa reedi* (Drake) is known only from Chile, and *Parapiesma cinereum* (Say) occurs throughout much of the New World.

Key to the genera and species of Neotropical Piesmatidae

1. Spiracles VI and VII subventral laterally; pronotum tricarinate (Fig. 23) .... *Parapiesma* Péricart
– Spiracle VI dorsal, only VII subventral laterally; pronotum with only a single median carina .... *Miespa* Drake

16.15 Family Rhyparochromidae

16.15.1 General Characteristics and Diagnosis

The Rhyparochromidae are commonly called seed bugs because most feed on mature seeds (Sweet 1964a, b). They are recognized by the presence of head trichobothria; a fused suture between sterna IV and V, usually curving forward anterolaterally from the midline of the sternum and not reaching the dorsal margin of the abdomen (except in Plinthisinae); and the fore femora usually incrassate and armed below with spines (Schuh and Slater 1995; Henry 1997a; Dellapé 2014). Most are dull brown to black, often with contrasting pale or white marks or mottled patterns on the hemelytra. They range from very small (e.g., Antillocorini, Lilliputocorini) to large (e.g., some Lethaeini and Myodochini).

Rhyparochromids have a broad range of feeding habits. Many species live on the ground, feeding on fallen seeds. Other species climb vegetation when mature seeds are available, some live on weedy vegetation (Slater 1977; Slater and Baranowski 1990), and others are arboreal and live in forests canopies (Slater et al. 2009; Dellapé and Henry 2010). The Cleradini, which feed on vertebrate blood, are the exception to the seed-feeding habit within the family.

Numerous species are myrmecomorphic, and although many are not extremely similar to ants in their morphology, adults and especially nymphs of some genera (such as the myodochines *Neopamera* Harrington, *Pseudopachybrachius* Malipatil, and *Heraeus* Stål) mimic ants in their movements (Slater and Baranowski 1990).

16.15.2 Classification and Diversity

The Rhyparochromidae are the most diverse group of the Lygaeoidea, comprising more than 2,000 species in 14 tribes or about a half the world's fauna. According to Henry's (1997a) phylogenetic classification, the family, defined by the presence of head trichobothria, contains the subfamilies Plinthisinae and Rhyparochrominae. The Plinthisinae, with only the nominate tribe, have complete abdominal sutures reaching the dorsal margins of the abdomen, whereas most other rhyparochromines have abdominal sutures IV–V abbreviated and not reaching the dorsal margin of abdomen. The apparent presence of a complete abdominal sternal suture on segments IV–V in at least 11 tribes of Rhyparochrominae suggests multiple independent origins of this character as discussed by Slater et al. (2009). We note (personal observation), however, that in most of these exceptions segments IV–V are modified in some way, indicating that the previous oversimplified presence or absence interpretation of the abbreviated abdominal sutures needs to be redefined and expanded.

Key to the subfamilies of Rhyparochromidae

1. Males with a stridulatory mechanism on abdominal segment I and hind wing; females with a conjunctival membrane between abdominal sterna 4/5; pronotum wider across anterior one-third than across humeral angles .... Plinthisinae
– Males without abdominal and hind-wing stridulatory mechanisms; both sexes usually with abdominal sterna 4/5 fused; pronotum variable, usually wider across humeral angles than across anterior lobe ....Rhyparochrominae

16.15.3 Subfamily Plinthisinae

The Plinthisinae possess head trichobothria, similar feeding habits, and an overall general appearance but lack the fused suture between the abdominal terga 4/5 found in most rhyparochromids.

Plinthisines usually have the pronotum expanded across the anterior lobe; the fore femora are heavily incrassate and spined; the wings are often greatly reduced or staphylinoid; males have a stridulatory mechanism on abdominal segment I and the hind wing; and females have a conjunctival membrane between sterna 4/5 (Slater and Woodward 1982; Schuh and Slater 1995; Slater and Brailovsky 2000). Putshkov (1958) argued that plinthisines merit subfamily status. According to Slater and Woodward (1982), the Plinthisini represent the basal group of the family, and Henry (1997a) corroborated the monophyly of the group based on the head trichobothria and maintained its subfamily status.

The subfamily includes only two genera, the monotypic *Bosbequius* Distant from Myanmar and the large genus *Plinthisus* Stephens (Fig. 24), with most of the species distributed in the Palearctic, a few species in North America, and only three Neotropical species described by Slater (1971).

16.15.4 Subfamily Rhyparochrominae[1]

Key to the tribes of Rhyparochrominae

1. Abdominal spiracles II, III, and IV dorsal 2
– All abdominal spiracles ventral.. 3
2. Inner laterotergites absent; lateral pronotal margins almost always rounded ....Myodochini
– Inner laterotergites present; lateral pronotal margins variable from rounded to carinate .... Udeocorini
3. Posterior pair of trichobothria on abdominal sternum V positioned one above the other 4
– Posterior pair of trichobothria on abdominal sternum V positioned one in front of the other in a linear series on segments IV and V 7
4. Ocelli lateral and behind eyes; suture between abdominal sterna IV and V attaining a lateral connexival margin; abdominal tergum III usually not sclerotized; labial segment II usually not attaining the base of the head .... Cleradini
– Ocelli between and slightly posterior to eyes; suture between abdominal sterna IV and V usually not attaining lateral connexival margin and usually markedly curving anteriorly from venter dorsally; labium variable, but usually with segment II reaching or exceeding base of head.. 5
5. Medium sized, length usually more than 5 mm; apical corial margin straight .... Ozophorini
– Very small to minute, length usually less than 3.5 mm; apical corial margin usually concave 6
6. Inner laterotergites absent; metathoracic scent-gland auricle strongly curved anteriorly; tarsi two segmented .... Lilliputocorini
– Inner laterotergites present; metathoracic scent-gland auricle straight or curved posteriorly; tarsi three segmented 7
7. Apical corial margin deeply concave; inner laterotergites present; head lacking iridescent areas; abdominal scent-gland scars between terga III/IV, IV/V, and V/VI ....Antillocorini
– Apical corial margin straight; inner laterotergites absent; head with iridescent areas basally; abdominal scent-gland scars between terga V/VI minute or absent .... Lethaeini

Tribe Antillocorini

The Antillocorini are a cosmopolitan group, with about 100 species occurring in tropical and subtropical regions of the world. Seventeen of the 34 currently recognized genera and 34 species occur in the Neotropics (Slater 1964a; Slater and O'Donnell 1995; Brambila 2000).

[1] In the proof stage of this chapter, we discovered that we omitted the tribe Rhyparochromini and the adventive genera *Dieuches* Dohrn (see Henry and Froeschner 1993) and *Elasmolomus* Stål (see Slater 1972) from the keys.

The Antillocorini are characterized by the bucculae joined by a carina well behind the labium; a deeply concave apical corial margin; the ventral abdominal spiracles; the linear trichobothria on abdominal sternum V; the presence of inner laterotergites; and immatures with a double or "troughed" suture between terga 3/4 and 4/5 and with a field of spines or tubercles laterally, with three pairs of scent glands between abdominal terga 3/4, 4/5, and 5/6 (Ashlock 1964; Sweet 1977; Slater and Woodward 1982; Schuh and Slater 1995; Slater and Brailovsky 2000).

Slater (1980) described several new taxa, discussed the phylogenetic relationships, and provided a key to the genera of the Western Hemisphere. Some Neotropical species lack a deeply concave apical corial margin and the linear arrangement of abdominal trichobothria. Schuh and Slater (1995) stated that the group may not be monophyletic.

The following key is modified from Slater's (1980) key.

Key to the Neotropical genera of Antillocorini

1. Apical corial margin straight or, at most, very slightly sinuate near inner angle 2
– Apical corial margin with inner third deeply and conspicuously concave 5
2. Suture between abdominal sterna 4/5 straight, reaching dorsal margin .... *Caeneusia* Strand
– Suture between abdominal sterna 4/5 curving anteriorly, not reaching dorsal margin 3
3. Lateral pronotal margins rounded; metathoracic scent-gland auricle weakly curved posteriorly .... *Paradema* Slater
– Lateral pronotal margins bluntly calloused or subcarinate; metathoracic scent-gland auricle curved posteriorly 4
4. Body elongate and slender, more than 3 mm long, macropterous .... *Schuhocoris* Slater
– Body small to minute, less than 2 mm long, robust, coleopteroid .... *Branstettocoris* Brailovsky
5. Lateral pronotal margins rounded or faintly calloused 6
– Lateral pronotal margins calloused or carinate 7
6. Body above and below polished and shining .... *Acolhua* Distant
– Head and abdominal sternum shining, constrasting with dull pruinose or subpruinose surface on remainder of body .... *Bathydema* Uhler
7. Trichobothria on abdominal sternum V linear, the posterior trichobothrium directly below or well behind the spiracle 8
– Trichobothria on abdominal sternum V not linear; if linear then the posterior trichobothrium anterior to the spiracle 9
8. Fore femur mutic; posterior trichobothrium on abdominal sternum V well behind spiracle .... *Arimacoris* Baranowsky and Slater
– Fore femur with two rows of spines, larger in males; posterior trichobothrium on abdominal sternum V directly below or slightly caudad of spiracle (Fig. 26) .... *Valeris* Brambila
9. Trichobothria on abdominal sternum V linear, anterior to spiracle 10

Plate 16.3 Figs. 25–36. 25, *Bathydema jamaicensis* Slater and Baranowski (Rhyparochromidae: Rhyparochriminae: Antillocorini). 26, *Valeris subcavicola* (Scudder) (Rhyparochromidae: Rhyparochriminae: Antillocorini). 27, *Clerada apicicornis* Signoret (Rhyparochromidae: Rhyparochriminae: Cleradini). 28, *Cistalia signoretii* (Guérin-Méneville) (Rhyparochromidae: Rhyparochriminae: Lethaeini). 29, *Paragonatas divergens* (Distant) (Rhyparochromidae: Rhyparochriminae: Lethaeini). 30, *Distingphyses insignis* Distant (Rhyparochromidae: Rhyparochriminae: Myodochini). 31, *Myodocha froeschneri* A. Slater (Rhyparochromidae: Rhyparochriminae: Myodochini). 32, *Prytanes confusus* (Barber) (Rhyparochromidae: Rhyparochriminae: Myodochini). 33, *Ptochiomera nodosa* Say (Rhyparochromidae: Rhyparochriminae: Myodochini). 34, *Xenydrium formiciforme* Bergroth (Rhyparochromidae: Rhyparochriminae: Myodochini). 35, *Ozophora costaricensis* Slater and O'Donnell (Rhyparochromidae: Rhyparochriminae: Ozophorini). 36, *Tempyra biguttula* Stål (Rhyparochromidae: Rhyparochriminae: Udeocorini) (Photos by G Ouellette)

– Trichobothria on abdominal sternum V not linear, somewhat dorsoventral relative to one another, usually below or behind spiracle 11
10. Prosternum with a deep median groove .... *Cligenes* Distant
– Prosternum lacking a deep median groove .... *Botocudo* Kirkaldy
11. Dorsal surface almost or entirely shiny and polished 12
– Dorsal surface either completely pruinose or with head shiny, contrasting with the pruinose pronotum 14
12. Dorsal surface completely shiny, not pruinose 13
– Dorsal surface almost entirely shiny and polished, pruinose only narrowly across the base of head between ocelli .... *Germacoris* Baranowsky and Slater
13. Fore femora with 8–9 sharp spines; head and anterior pronotal lobe smooth and impunctate .... *Terenocoris* Slater
– Fore femora with only 2–3 spines; head and pronotum with numerous, irregular, coarse punctures .... *Trachinocoris* Slater
14. Lateral pronotal margins sharply and acutely carinate .... *Paurocoris* Slater
– Lateral pronotal margins with a blunt calloused ridge, but lacking a sharp, acute edge 15
15. Dorsal surface with strongly variegated coloration; metathoracic scent-gland auricle broad, lobate, not strongly curved caudad; evaporative area with grooves .... *Antillodema* Slater
– Dorsal surface dull yellow, brown or chestnut, without bright variegated contrasting coloration; metathoracic scent-gland auricle slender and strongly curved posteriorly; evaporative area lacking conspicuous grooves 16
16. Outer margin of metathoracic scent-gland evaporative area elevated into a raised subcarinate ridge, only slightly removed from outer lobe of scent-gland auricle; head pruinose, similar in texture to that of pronotum; spiracle of abdominal segment IV below lateral shelf; paramere with an elongate, acutely pointed inner projection .... *Scythinus* Distant
– Outer margin of metathoracic scent-gland evaporative area not strongly elevated into a ridge, well removed from outer lobe of scent-gland auricle; head shiny, strongly contrasting with texture of pronotum; spiracle on abdominal segment IV on lateral shelf; paramere lacking a well-developed inner projection .... *Antillocoris* Kirkaldy

Tribe Cleradini

The Cleradini, with 20 genera and 54 species, are confined to the Old World tropics, except for *Clerada apicicornis* Signoret (Fig. 27) introduced into the Western Hemisphere (Schuh and Slater 1995). This species has been recorded from Cuba, the Dominican Republic, Grenada, Haiti, Jamaica, Puerto Rico, St. Vincent, St. Thomas, and the Virgin Islands in the West Indies and Brazil, Colombia, and

Venezuela from South America (Slater 1964a; Morales et al. 1969; Torres et al. 2000; Baranowski and Slater 2005).

Members of the Cleradini have laterally placed ocelli; a short antennal segment III and a short labium not exceeding the base of the head; an unarmed, slender fore femur; the abdomen with all spiracles ventral and without inner laterotergites; an expanded connexival membrane; a secondary longitudinal abdominal suture; complete abdominal sutures between sternal suture IV–V; bifurcate parameres; and eggs with a pseudoperculum (Slater and Woodward 1982; Schuh and Slater 1995; Slater and Brailovsky 2000; Slater et al. 2009).

Tribe Lethaeini

The tribe Lethaeini, with 38 genera and more than 160 species worldwide (Li et al. 2011), has its greatest diversity in the tropical and subtropical parts of the Afrotropical, Oriental, and Australian regions (Slater 1986a). It is one of the most diverse tribes in the Neotropics, with 13 genera and 36 species known.

Lethaeines are recognized by the linear placement of the trichobothria on abdominal sternum V; a rounded buccular groove joined immediately behind the labium; a carinate juga; a trichobothrium at each anterior corner of the pronotum; the abdomen with all spiracles ventral and with or without inner laterotergites; the absence of a Y chromosome; the extreme modification of the sperm reservoir; the immatures with a double or "troughed" suture between terga 3/4, 4/5, and 5/6 and with a field of spines or tubercles laterally; the reduced scent gland between abdominal terga 5/6; and the presence of an iridescent area or areas on the head (Ashlock 1964; Slater and Woodward 1982; O'Donnell 1991).

The following key is modified from Baranowski and Slater's key (2005) to the West Indian lethaeines.

Key to the Neotropical genera of Lethaeini

1. Pronotal trichobothria absent; eyes large, occupying more than half the length of the head; juga reduced; densely setose species .... *Lipostemmata* Berg
– Pronotum with a pair of trichobothria on anterolateral region; eyes small; juga well developed; less setose species 2
2. Dorsal surface of body strongly shiny 3
– Dorsal surface of body only partly shiny 5
3. Pronotum trapeziform, narrowing anteriorly; head with two iridescent patches; fore femur slightly incrassate 4
– Pronotum subquadrangular; head with one iridescent patch; fore femur strongly incrassate and compressed .... *Rhaptus* Stål
4. Body with long setae dorsally; eyes with two long, forward-curving setae .... *Xestocoris* Van Duzee
– Body almost glabrous or with only a few long, scattered setae dorsally; eyes without two long, forward-curving setae .... *Bubaces* Distant

5. Head with two iridescent patches 6
– Head with one iridescent patch 8
6. Inner laterotergites absent; body with short setae dorsally .... *Stictolethaeus* O'Donnell
– Inner laterotergites present; body with long setae dorsally 7
7. Dorsal surface weakly punctate; evaporative area extensive; macropterous or brachypterous species .... *Valtissius* Barber
– Dorsal surface strongly punctate; evaporative area reduced, restricted to areas surrounding auricle; coleopteroid species .... *Esuris* Stål
8. Transverse pronotal impression distinct, dividing pronotum into anterior and posterior lobes; lateral pronotal margins broadly explanate and sinuate; collar triangular, well separated from the pronotum by a row a punctures .... *Neopetissius* O'Donnell
– Transverse pronotal impression indistinct; lateral pronotal margins carinate and narrowly explanate; collar not differentiated 9
9. Evaporatory area extended anteriorly over the mesopleura, reaching or nearly reaching dorsal margin 10
– Evaporatory area not extended anteriorly, distant from dorsal margin of the mesopleura .... *Cryphula* Stål
10. Antennal segment I with a pale annulus (Fig. 28) .... *Cistalia* Stål
– Antennal segment I without a pale annulus 11
11. Larger, length over 6 mm .... *Petissius* Distant
– Smaller, length less than 5 mm 12
12. Lateral margin of pronotum explanate .... *Gonatoides* Slater
– Lateral margin of pronotum rounded (Fig. 29) .... *Paragonatas* Distant

Tribe Lilliputocorini

The Lilliputocorini are circumtropical and comprise the nominotypical genus *Lilliputocoris* Slater and Woodward and ten species (Slater and O'Donnell 1995). The only record from the Neotropics is *L. neotropicalis* Slater and Woodward, described from a female from Pará, Brazil.

The Lilliputocorini are minute bugs, less than 2 mm long, recognized by the clavate antennal segments II and III; the reduced and modified ovipositor; the two-segmented tarsi; the uniquely formed metathoracic scent-gland auricle and adjacent evaporative area; the abdomen with all spiracles ventral and without inner laterotergites; the immatures with a double or "troughed" suture between terga 3/4, 4/5, and a field of spines or tubercles laterally; and the reduction or loss of the scent gland between abdominal terga 5/6 (Slater and Woodward 1982; Schuh and Slater 1995; Slater and Brailovsky 2000).

Tribe Myodochini

The tribe Myodochini is among the most diverse groups of Rhyparochromidae (Dellapé and Henry 2010), comprising 75 genera and 307 species worldwide, with 36 genera and 117 species occurring in the Neotropics.

The Myodochini usually possess a deeply incised transverse impression across the pronotum and a rounded anterior pronotal lobe; the abdomen has the spiracles on segments II, III, and IV dorsal and lacks inner laterotergites; and immatures have a Y-suture between abdominal terga 3/4 and three pairs of scent glands between abdominal terga 3/4, 4/5, and 5/6 (Slater and Woodward 1982; Schuh and Slater 1995; Slater and Brailovsky 2000).

Three Neotropical species now assigned to the genera *Pseudocnemodus* Barber, *Ptochiomera* Say, and *Sisamnes* Distant are misplaced. *Pseudocnemodus martinezi* Brailovsky from Paraguay probably should be included in *Bergicoris* Dellapé and probably is a synonym of *B. multifarious* Berg; the identity and position of the Chilean species in *Ptochiomera* are doubtful; and *Sisamnes annulicollis* (Berg) from Argentina probably belongs in *Neopamera* Harrington. Clarification of these taxa will be presented in a forthcoming paper by PMD.

Some myodochines, such as species in the genera *Neopamera*, *Pseudopachybrachius*, and *Prytanes*, are abundant and commonly collected in disturbed habitats and generally climb vegetation, whereas other taxa are strict geophiles, presumably in specialized habitats, and are poorly represented in collections. Most genera have fewer than ten species, and in the case of genera with more species (e.g., *Paromius*), only a few are Neotropical. The actual diversity of this group is much higher than present numbers indicate, and many new taxa remain to be described.

Harrington (1980) revised Myodochini and provided a generic key and a cladistic and biogeographic analysis. Nine Neotropical genera have been described since her work. *Ashlockobius* Slater and A. Slater (1999) was synonymized with *Villalobosothignus* Brailovsky 1984 by Dellapé and Montemayor (2011). The following key is modified from the Harrington's (1980) key to the myodochine genera of the world.

Key to the Neotropical genera of Myodochini

1. Interocular distance less than postocular distance 2
– Interocular distance equal to or exceeding postocular distance 4
2. Posterior margins of ocelli posterior to posterior margins of eyes; eyes round (Fig. 31) .... *Myodocha* Latreille
– Posterior margins of ocelli anterior to posterior margins of eyes; eyes elongate-oval 3
3. Fore femur incrassate, with strong spines; ocelli sessile; antennal segment I not extending beyond clypeus (tylus) .... *Pephysena* Distant
– Fore femur weakly incrassate, slender, with only a few medium-sized and minute spines; ocelli protuberant; antennal segment I extending beyond clypeus .... *Dushinckanus* Brailovsky

4. Ocelli absent...................................... *Andercnemodus* Brailovsky and Cervantes
– Ocelli present 5
5. Crescent-shaped, striated stridulitrum present ventrolaterally on anterior portion of abdomen.. 6
– Striated stridulitrum lacking on anterior portion of abdomen 10
6. Head prolonged into a distinct neck; stridulitrum confined to abdominal sterna II and III; plectrum on hind femur consisting of two chisel-like projections ....*Erlacda* Signoret
– Head not prolonged into a neck; stridulitrum extending onto sternum IV; plectrum on hind femur consisting of a line or scattered field of minute tubercles............ 7
7. Pronotum shiny; length of antennal segment I greater than interocular distance; pronotum shiny, not pruinose, and never deeply punctate; male fore tibia typically armed with a spine or spines ....*Pseudomera* Distant
– Pronotum pruinose; length of antennal segment I less than interocular distance; pronotum pruinose, in some cases very deeply punctate; male fore tibia unarmed 8
8. Stridulitrum diffuse, not readily apparent; pronotum and head deeply punctate; abdomen with a band of long silvery setae covering much of sternum IV in lateral view ....*Stridulocoris* Harrington
– Stridulitrum readily apparent; pronotum and head not deeply punctate; abdomen of uniform vestiture, without a band of long silvery setae.................................. 9
9. Lateral margin of posterior pronotal lobe angled posterolaterad at approximately a 45 degree angle; pronotum markedly bilobed, with a deeply incised, transverse impression; always macropterous; antennal segment IV with a pale proximal band; fore femoral spines clearly in two rows (present on both the inner and outer edge of ventral surface)..*Froeschneria* Harrington
– Lateral margin of the posterior pronotal lobe angled posterolaterad at less than a 45 degree angle; transverse impression not deeply incised; often submacropterous; antennal segment IV usually uniformly dark; if with a pale band, then fore femoral spines in a single row (present only along inner edge of ventral surface) .... *Ligyrocoris* Stål
10. Evaporative area reduced, occupying much less than half of metapleural area; claval punctation usually in three rows; pronotum shiny, never pruinose............11
– Evaporative area not reduced, occupying much more than half of metapleural area; claval punctation in more than three rows; pronotum dull, shagreened, or pruinose.. 14
11. Antennae with enlarged segments; segment III strongly swollen, distally of much greater diameter than segment I (Fig. 33) ....*Ptochiomera* Say
– Antennae filiform; segment III generally filiform, if somewhat clavate, diameter no greater distally and often more slender than segment I 12
12. Collar not apparent on the anterior pronotal lobe; ventral surface of head not grooved; buccular juncture not near labial insertion, occurring at level of eyes; pronotum dorsoventrally compressed, lateral margins subcarinate 13

– Collar apparent on anterior pronotal lobe but not demarked posteriorly; ventral surface of head grooved; buccular juncture U shaped near labial insertion; lateral margins of both pronotal lobes rounded (Fig. 32) .... *Prytanes* Distant
13. Buccular juncture broadly U shaped; fore femur with spines only along the inner edge of the ventral surface; anterior margin of abdominal sternum II scalloped .... *Kolenetrus* Barber
– Buccular juncture elongate, terminating in a V-shaped end; fore femur with spines not confined to a single row; anterior margin of abdominal sternum II not scalloped .... *Scintillororis* Slater and Brailovsky
14. Pronotum dorsoventrally compressed with a distinct lateral carina on both lobes .... *Megacholula* Harrington
– Pronotum not compressed; lateral pronotal margins ecarinate 15
15. Both lobes of pronotum and head uniformly and coarsely punctate; head broad, posterior margin of eyes touching anterior pronotal angles; sternal scalloping prolonged and clearly visible on sternum II; a very narrow anterior pronotal collar vaguely indicated but never demarked posteriorly by a groove............. 16
– Head and anterior pronotal lobe impunctate or only vaguely punctate; head generally with eyes removed from anterior pronotal angles; sterna scalloping usually lacking, if present, largely hidden under metapleuron and not prolonged; in most cases, with a distinct anterior pronotal collar demarked posteriorly by a distinct groove... 18
16. Clavus with three regular rows of punctures; apex of scutellum rounded, elevated, and distinctly pale; buccula with a posterior projection .... *Paracholula* Harrington
– Clavus often with more than three rows of punctures; scutellum unicolorous; buccula without a posterior projection.. 17
17. Abdomen with a large glabrous area or diffuse stridulitrum on lateral portions of sterna II and III; hind femur with a plectrum of scattered spines on basal half; abdomen often with a band of long silvery setae covering much of sternum IV in lateral view; lateral pronotal margins ecarinate........................... .... *Stridulocoris* Harrington
– Abdomen without a stridulitrum or band of long setae as above; hind femur devoid of spines; lateral margin of posterior pronotal lobe with a blunt carina, sometimes forked over humeral angles....................................... *Cholula* Distant
18. Mesacetabulum with mesepimeron emergent (or barely) from between the meso- and metepisternum 19
– Mesepimeron enclosed by metepisternum; touching mesepisternum 28
19. Head elongate, often with a distinct neck; postocular distance equal to or greater than distance between ocelli 20
– Head less elongate, never with a distinct neck; postocular distance less than distance between ocelli 24

20. Eyes elongate-oval; posterior margin of ocelli anterior to posterior margin of the eyes 21
– Eyes rounded; posterior margin of ocelli posterior to posterior margin of eyes 22
21. Head with a clearly defined cylindrical neck; vertex of head convex; anterior pronotal lobe pruinose.......... *Pephysena* Distant
– Head with a much less well-defined neck; vertex of head flat, depressed between the eyes; anterior pronotal lobe shiny .... *Tenuicoris* Slater and Harrington
22. Head essentially flat when viewed laterally, not showing a gradual rounded constriction from eyes to insertion of head; fore coxa without a spine .... *Catenes* Distant
– Head when viewed laterally showing gradual rounded constriction from eyes to insertion of head; fore coxa armed with a spine 23
23. Ventral portion of collar produced anteriorly.......... *Heraeus* Stål
– Ventral portion of collar not produced anteriorly.......... *Paisana* Dellapé
24. Juga rounded; pygophore with anterior margin of dorsal aperture with small denticles; aedeagus with large, stout spines on conjunctiva and vesica .... *Myodacanthus* Dellapé
– Juga forming a ridge; pygophore with anterior margin of dorsal aperture smooth; aedeagus spined or unspined.......... 25
25. Eyes not protruding or surpassing dorsal margin of head in lateral view 26
– Eyes protruding, inserted high, and surpassing dorsal margin of head in lateral view.......... 27
26. Head and anterior pronotal lobe lower than posterior pronotal lobe in lateral aspect; fore coxal spine poorly developed or absent; color predominantly blackish brown, posterior pronotal lobe with characteristic orange areas along lateral margins and paired orange maculae medially adjacent to transverse impression.......... *Orthaea* Dallas
– Dorsal surface of head and both lobes of pronotum essentially in the same plane; fore coxal spine(s) well developed; posterior pronotal lobe lacking orange markings along lateral margins.......... *Neopamera* Harrington
27. Profemur elongate and slender, with a few spines restricted to distal end; mesepimeron clearly emergent; aedeagus spined.......... *Neomyocoris* Dellapé and Montemayor
– Profemur incrassate with two rows of spines ventrally; mesepimeron barely emergent; aedeagus without spines.......... *Thoraea* Dellapé and Montemayor
28. Head broad, interocular distance equal to or greater than width of pronotal collar 29
– Head not broad, interocular distance less than width of pronotal collar 31
29. Strikingly ant mimetic, with lateral margin of head between eye and insertion of antenna expanded into a platelike curving ridge; lateral margin of pronotal collar with a rounded but distinct spine (Fig. 34) .... *Xenydrium* Poppius and Bergroth

– Not ant mimetic, lateral margin of head between eye and base of antenna not expanded into a ridge; lateral margin of pronotal collar without a spine 30

30. Head elongate behind eyes, forming a short cylindrical neck; V-shaped buccular juncture with a strong midventral carina; antennal segment I short, remote from apex of head; scutellum pruinose on basal half, shiny on black tumid central area and apex .... *Acrolophyses* Dellapé and Henry

– Head with a shorter neck; midventral carina of buccular junction less developed; antennal segment I longer, almost attaining apex of head; scutellum completely pruinose (Fig. 30) .... *Distingphyses* Scudder

31. Head and anterior pronotal lobe, including collar, impunctate, or collar with only a few faint punctures; pronotal collar narrow and ringlike .... *Perigenes* Distant

– Pronotal collar usually broad and distinctly punctate; rest of anterior pronotal lobe and head also often punctate, though sometimes with indistinct, minute punctures 32

32. Pronotum tapering cephalad with anterior lobe flattened and weakly convex; anterior lobe distinctly lower than posterior lobe in lateral view; collar with a characteristic median depression to posterior margin; abdomen equal to or longer than combined length of head and pronotum .... *Paromius* Fieber

– Anterior pronotal lobe usually strongly convex, not lower than posterior lobe in lateral view; collar without median depression; abdomen shorter than combined length of head and pronotum 33

33. Small, length 4.0 mm or less; ant mimetic, head strongly swollen in lateral view; eyes small, rounded, protruding, and beadlike .... *Bacacephalus* Harrington

– Larger than above; if small, then not ant mimetic with a swollen head and beadlike eyes 34

34. Large, usually greater than 6 mm long; head broad and jugal ridge above antennal segment I distinct 35

– Small, generally less than 5 mm long; jugal ridge above antennal segment I usually narrow and poorly developed .... *Pseudopachybrachius* Malipatil

35. Hemelytra with a broad, dark, transverse band at level of corial apex; male fore tibia unarmed; aedeagus with serrated projections on vesica and two strong spines on conjunctiva .... *Pseudoparomius* Harrington

– Hemelytra without a broad, dark, transverse band at level of corial apex; male fore tibia with or without spines; aedeagus spined or unspined but without serrate projections 36

36. Antennal segment IV with apex paler; anterior pronotal lobe elongate, at least 1.5 times as long as posterior pronotal lobe, but never globose; male fore tibia not strongly curved, with only a few short, sharp spines; aedeagus unspined .... *Villalobosothignus* Brailovsky

– Antennal segment IV with or without a broad yellowish band subbasally, but never with apex paler; anterior pronotal lobe elongate and globose; male fore tibia curved on anterior half and straight distally, with a large spine medially; aedeagus spined .... *Bergicoris* Dellapé

Tribe Ozophorini

The Ozophorini have a worldwide distribution, with most taxa occurring in the Neotropical and Oriental regions (Slater 1986a; Schuh and Slater 1995; Cassis and Gross 2002). Twenty-eight genera and 194 species are currently included in the tribe, with 13 genera and about 87 species known from Neotropics, mostly in the large genus *Ozophora* Uhler.

Ozophorines usually have a porrect, grooved head; the hind wing lacks a hamus and secondary veins; the abdomen lacks inner laterotergites and all spiracles are ventral; and the immatures possess a Y-suture between abdominal terga 3/4 and three pairs of scent glands between abdominal terga 3/4, 4/5, and 5/6 (Slater and Woodward 1982; Schuh and Slater 1995; Slater and Brailovsky 2000). The genus *Ozophora* is one of the most speciose taxa of Rhyparochromidae in the Western Hemisphere. Slater (1995) provided a key to the 51 mainland Neotropical species.

The following key is based on Ashlock and Slater's (1982) key to the New World Ozophorini and Ashlock's (1985) key to the *Bergidea* group.

Key to the Neotropical genera of Ozophorini

1. Fore femur unarmed beneath, at most with a patch of 20 or more minute spicules 2
– Fore femur beneath with one or more stout, compound spines set on the tubercles in both sexes 5
2. Ocelli present 3
– Ocelli absent 4
3. Fore femur with a patch of 20 or more minute spicules, more evident in males; hemelytra with clavus and corium differentiated, macropterous .... *Pamozophora* Ashlock and Slater
– Fore femur unarmed; hemelytra with clavus and corium indistinguishably fused and nearly uniformly, coarsely punctate .... *Allotrophora* Slater and Brailovsky
4. Body short, stout, elliptical; vertex of head strongly convex; eyes large; pronotum extremely short and broad, width more than two times median length, disc strongly convex, lateral margin broadly explanate .... *Icaracoris* Slater
– Body elongate-oval; vertex of head flattened, eyes smaller, dorsoventrally elongate; pronotum nearly quadrate, width two times median length, disc less convex, lateral margins not explanate .... *Longinischus* Brailovsky
5. Basal segments of abdomen laterally with a cross-striated, crescent-shaped stridulitrum; plectrum consisting of scattered small tubercles on base of hind femur .... *Lygofuscanellus* Scudder
– Basal segments of abdomen laterally without stridulatory structures 6
6. Lateral edge of corium (viewed laterally) with a finely striated stridulitrum; hind femora with plectrum (with a regular and dense field of tubercles on apical half); lateral margins of pronotum sharply and broadly explanate, gently narrowing to humeral angles, without a posterior notch .... *Balboa* Distant
– Lateral edge of corium without a stridulitrum, at most with minute crenulations on narrow lateral edge; hind femora without a plectrum; lateral margins of pronotum bluntly carinate or, if explanate (some species of *Ozophora*), then ending abruptly to form a notch at humeral angles 7

7. Pronotal collar and pronotal lobes usually distinct 8
– Pronotal collar and pronotal lobes always indistinct 10
8. Coleopteroid; ocelli absent 9
– Usually macropterous, if hemelytra modified, then not extremely so, or if coleopteroid (*Ozophora ovalis*, Colombia), ocelli present (Fig. 35) .... *Ozophora* Uhler
9. Eyes protruding laterally on short stalks; costal margin of corium clearly emarginate; claval commissure shorter than length of head and pronotum combined.. *Pseudomenotelus* Brailovsky and Cervantes
– Eyes large, sessile, not protruding laterally; costal margin of corium weakly emarginate; claval commissure longer than length of head and pronotum combined... *Pseudomenus* Ashlock and Slater
10. Coleopteroid, claval commissure longer than pronotum; ocelli absent; metapleural evaporative area covering less than one-third of metapleuron..... 11
– Often submacropterous or brachypterous, but never coleopteroid, claval commissure shorter than pronotum; ocelli present; metapleural evaporative area occupying at least one-half of metapleuron (Chile and Argentinean Patagonia) .... *Bergidia* Breddin
11. Body short, elliptical, strongly convex; hemelytra without any trace of membrane, claval commissure longer than head and pronotum taken together .... *Brailovskyocoris* Slater
– Body elongate oval; hemelytra with membrane reduced, claval commissure shorter than head and pronotum taken together.. 12
12. Antennal segment III longer than segment I and more than half as long as segment II; pronotum and hemelytra with surface, including punctures, dull, not sculptured or rugose; texture of anterior lobe of pronotum and base of scutellum not distinct from remainder of dorsum; evaporative area covering mesepimeron (Juan Fernandez Islands).. *Micrymenus* Bergroth
– Antennal segment III shorter than segment I and less than half as long as segment II; pronotum and hemelytra with surface, especially punctures, shiny, obviously rugose and sculptured; anterior lobe of pronotum and basal area of scutellum with a finely pebbled texture, strongly contrasting with surrounding areas; evaporative area covering only posterior part of mesepimeron (Juan Fernandez Islands) .... *Rugomenus* Ashlock

Tribe Udeocorini

The Udeocorini, comprising 17 genera and 38 species, are most abundant in the Australian Region (Slater 1964a; Slater and O'Donnell 1995). Only three genera and five species have been recorded from the Neotropics. The presence of the Nearctic *Tempyra biguttula* Stål in Argentina probably represents an introduction. Slater (1986a) argued for a Gondwanan origin of the group.

Members of the tribe have rounded, carinate, or explanate pronotal margins; the abdomen has dorsal spiracles on segments II, III, and IV and inner laterotergites; and the immatures possess a Y-suture between abdominal terga 3/4 and have three pairs of scent glands between abdominal terga 3/4, 4/5, and 5/6 (Sweet 1967; Slater and Woodward 1982; Schuh and Slater 1995; Slater and Brailovsky 2000).

Key to the Neotropical genera of Udeocorini

1. Body surface shiny and uniformly dark .... *Astemmoplitus* Spinola
– Body surface dull and not uniformly dark 2
2. Hemelytra covering abdomen in dorsal view; corium with a pale, round macula distally (Fig. 36).. *Tempyra* Stål
– Hemelytra narrower than abdomen, connexiva visible in dorsal view; corium with a pale, elongate macula distally .... *Bathycles* Distant

16.16 Main Species

The Lygaeoidea include many species of economic importance. Sweet (2000a) gave a detailed account of the most important ones, which are concentrated in the families Blissidae, Geocoridae, Lygaeidae, and Oxycarenidae, though the diverse family Rhyparochromidae has a large number of economically important seed-feeding species (e.g., Henry and Froeschner 1993) and numerous nuisance pests that often cluster in and around homes and commercial buildings (Henry and Adamski 1998; Henry 2004).

The Blissidae, also known as chinch bugs, are probably the most economically important group of Lygaeoidea (Sweet 2000a; Samuels et al. 2002). The chinch bugs are sap suckers specialized for feeding on monocotyledonous plants (Slater 1976). Many blissids are among the most important pests of grasses (Poaceae), including barley, corn (maize), millets, oats, rice, rye, sorghum, and wheat (Sweet 2000a). *Blissus leucopterus* (Say) is considered the most important New World species (Slater 1976; Sweet 2000a). Pereira and da Silva (1988) reported *B. leucopterus* (Say) attacking *Brachiaria radicans* Napper in Minas Gerais, Brazil. Valério et al. (1999) documented *Blissus antillus* Leonard damaging more than 800 acres of Tangola grass pastures in southern Mato Grosso.

Members of the family Geocoridae are in large part predaceous, although they also feed on seeds and foliage of plants, especially as early instars (Sweet 1960; Tamaki and Weeks 1972). As noted by Sweet (2000b), geocorids have complex nutritional requirements and evidently require plant food for optimal development. Species such as *Geocoris uliginosus* (Say) may be considered omnivorous (Carstens et al. 2008). Nevertheless, their importance as biological control agents is well documented (Naranjo and Gibson 1996; Coll and Ruberson 1998; Hagen et al. 1999). A list of the prey for species of *Geocoris* in the United States was compiled by Readio and Sweet (1982).

The family Lygaeidae is separated into three subfamilies, the Ischnorhynchinae, Lygaeinae, and Orsillinae. The life cycles and habitats of most of the species of Ischnorhynchinae are not known. Scudder (1962b) gave available host records, Wheeler (1976) studied the seasonal history and summarized hosts of the Holarctic birch catkin bug, *Kleidocerys resedae* (Panzer) and Cervantes and Baez (2010) presented the life histories of *K. punctatus* Distant and *K. virescens Fabricius*. Only a few species of Lygaeinae are recorded as pests on agricultural crops (Slater 1964b). Aposematically colored members of the genus *Oncopeltus* are restricted to milkweeds and other plants of the family Asclepiadaceae (Wheeler 1983), from which they sequester toxic cardenolides, making them unpalatable to potential predators (Scudder and Duffy 1972).

Other species, such as *Lygaeus kalmii* Stål (Wheeler 1983) and *Neacoryphus bicrucis* (Say) (Solbreck and Pehrson 1979), have been found feeding on numerous plant families, even though asclepiads remain their preferred hosts (Wheeler 1983). The Orsillinae include species that feed on seeds, as well as on vascular tissues. Much of the recorded damage occurs when large populations migrate from wild hosts to crop plants, especially during times of water stress (Ashlock 1967b). Economically important population of *Nysius* develop when the seed supply of the preferred host becomes insufficient for a large population, and the bugs move to a new host, often one of agricultural importance. *Nysius simulans* Stål has been reported from São Paulo state, attacking corn, cotton, and rice, and *Neopamera bilobata* (Say), from Minas Gerais, attacking tobacco (Costa Lima 1940).

The Oxycarenidae feed on both seeds and sap, chiefly of the plant family Malvaceae. The introduced *Oxycarenus hyalinipennis* (Costa) often is of major economic importance on cotton and other Malvaceae, sometimes developing large numbers and feeding between the fibers of the cottonseeds, causing staining of the cotton bolls (Annecke and Moran 1982). Almost nothing is known of the feeding habits of the largest New World genus, *Crophius*, though (Blatchley 1926) reported *C. disconotus* (Say) in numbers on goldenrod, *Solidago* sp. (Asteraceae).

Only a scattering of records are available for the economic species of Rhyparochromidae (Sweet 2000a). Species such as *Neopamera bilobata* (Say), *Pseudopachybrachius vincta* (Say), and their relatives sporadically damage crops, especially strawberries (Slater and Baranowski 1990). Sweet (1964b) found *Cryphula trimaculata* (Distant) inhabiting old fields, feeding on seeds of perennial bunch grasses, such as *Schizachyrium scoparium* J. Presl, *Festuca rubra* L., and *Panicum* sp. *Elasmolomus sordidus* (F.) and *Dieuches armatipes* (Walker) are potential pest of peanuts (Slater 1972; Henry and Froeschner 1993). Though documented as feeding on the blood of small rodents (Harrington 1983, 1990), *Clerada apicicornis* Signoret apparently prefers feeding on other insects, including the blood-feeding reduviid *Rhodnius prolixus* Stål (Torres et al. 2000).

Although the Berytidae were long thought to be primarily phytophagous (Wheeler and Schaefer 1982; Péricart 1984), many have strong predatory tendencies (Henry 1997a, 2000). *Parajalysus spinosus* Distant is said to cause serious damage on cacao, though a later study considered it an important predator of *Heliothis virescens* (F.) eggs (Wille 1951). Other species, such as *Parajalysus andinus* (Horváth), may play an important role in pollinating cacao (Henry 1997b).

Henry and Froeschner (1998) documented the known hosts, and Henry (2000) provided an overview of the feeding habits in the family.

The Colobathristidae and Piesmatidae are also sap feeders (Sweet and Schaefer 1985; Heiss and Péricart 1983; Sweet 2000a). Colobathristids, found in the Neotropical and the Oriental-Australian regions, all feed on grasses, including bamboos. No economically important colobathristids have been reported in the Neotropics, although *Phaenacantha* (*Anorygma*) *saccharicida* (Karsch) is a pest of sugar cane in Australia (Cassis and Gross 2002). Many piesmatids specialize on plants of the family Chenopodiaceae, including the widespread *Parapiesma cinereum* (Say) (Schaefer 1981; Heiss and Péricart 1983).

16.17 Concluding Remarks

The Lygaeoidea, though predominantly seed feeders, include some of the most economically important species of Heteroptera, such as the sap-feeding chinch bugs. A lesser number of others, such as the geocorids, are in large part predatory. The group represents the second largest superfamily of the infraorder Pentatomomorpha, with 16 families, more than 700 genera, and 4,200 species worldwide (Henry 2009). The Neotropical fauna treated in this chapter comprises about 184 genera and 836 species, representing 26 % of the world's genera and only 20 % of the species, suggesting that the number of new taxa awaiting description is immense. That nearly 30 new species of *Heraeus* Stål are being described in a forthcoming revision (Dellapé et al. in review) further illustrates the amount of work that remains to be done in the Neotropics. We hope this review of the superfamily, which includes information on hosts, habits, economic importance, classification, and keys to all Neotropical genera in 12 families, will stimulate interest in this large, diverse group of taxonomically and ecologically important insects.

Acknowledgments We are grateful to Gary Ouellette (Systematic Entomology Laboratory, Agricultural Research Service, US Department of Agriculture, c/o the National Museum of Natural History, Washington, DC) for preparing the color plates included in this chapter and to Pavel Štys (Charles University, Prague, Czech Republic) and Alfred G. Wheeler (Clemson University, Clemson, South Carolina, USA) for their careful reviews of the manuscript and suggestions for its improvement.

References

Aller T, Caldwell RL (1979) Investigation of the possible presence of an aggregation pheromone in the milkweed bugs, *Oncopeltus fasciatus* and *Lygaeus kalmii*. Physiol Entomol 4:287–290

Annecke DP, Moran VG (1982) Insects and mites of cultivated plants in South Africa. Butterworths, Durban

Ashlock PD (1964) Two new tribes of Rhyparochrominae: a re-evaluation of the Lethaeini (Hemiptera-Heteroptera: Lygaeidae). Ann Entomol Soc Am 57:414–422

Ashlock PD (1967a) A generic classification of the Orsillinae of the world (Hemiptera-Heteroptera: Lygaeidae). University of California Press, Berkeley

Ashlock PD (1967b) New records and name changes of North American Lygaeidae (Hemiptera: Heteroptera: Lygaeidae). Proc Entomol Soc Wash 79:575–582

Ashlock PD (1985) A revision of the *Bergidea* group: a problem in classification and biogeography (Hemiptera-Heteroptera: Lygaeidae). J Kansas Entomol Soc 57:675–688

Ashlock PD, Lattin JD (1963) Stridulatory mechanisms in the Lygaeidae, with a new American genus of Orsillinae (Hemiptera: Heteroptera). Ann Entomol Soc Am 56:693–703

Ashlock PD, Slater JA (1982) A review of the genera of Western Hemisphere Ozophorini with two new genera from Central America (Hemiptera-Heteroptera: Lygaeidae). J Kansas Entomol Soc 55:737–750

Ashlock PD, Slater JA (1988) Family Lygaeidae Schilling, 1829. Infericornes Amyot and Serville, 1843; Myodochidae Kirkaldy, 1899; Geocoridae Kirkaldy, 1902. In: Henry TJ, Froeschner RC (eds) Catalog of the Heteroptera, or true bugs, of Continental United States. E. J. Brill, Leiden/New York, pp 167–245

Attisano A, Tregenza T, Moore AJ, Moore PJ (2013) Oosorption and migratory strategy of the milkweed bug, *Oncopeltus fasciatus*. Anim Behav 86:651–657

Baranowski RM, Slater JA (2005) The Lygaeidae of the West Indies. Florida Agric Exp Station Bull 402:1–266

Barber HG (1947) Revision of the genus *Nysius* in the United States and Canada (Hemiptera: Heteroptera: Lygaeidae). J Acad Sci Washington 37:354–366

Blatchley WS (1926) Heteroptera or true bugs of Eastern North America, with especial reference to the faunas of Indiana and Florida. Nature Publishing Company, Indianapolis

Brailovsky H (1975) Contribución al estudio de los Hemiptera-Heteroptera para México IV. Una nueva especie de *Cymonimus* Breddin (Lygaeidae-Cyminae-Ninini). Rev Soc Mexicana Hist Nat 36:177–181

Brailovsky H (1978) Estudio de género *Lygaeus* Fabricius 1794, del Nuevo Mundo, con descripción de cinco nuevas especies (Hemiptera-Heteroptera-Lygaeidae-Lygaeinae). An Inst Biol Univ Nac Aut México Ser Zool 49:123–166

Brailovsky H (1979) Revisión de género *Craspeduchus* Stal con descripción de dos nuevas especies (Hemiptera-Heteroptera-Lygaeidae-Lygaeinae). An Inst Biol Univ Nac Aut México Ser Zool 50:205–226

Brailovsky H (1980) Revisión del género *Acroleucus* Stal (Hemiptera-Heteroptera-Lygaeidae-Lygaeinae). Folia Entomol Mexicana 44:39–120

Brailovsky H (1982a) Un nuevo arreglo nomenclatorial y descripción de tres nuevos géneros y dos nuevas especies americanas de la subfamilia Lygaeinae (Hemiptera-Heteroptera-Lygaeidae). An Inst Biol Univ Nac Aut México Ser Zool 52:259–276

Brailovsky H (1982b) Revisión del complejo *Ochrimnus*, con descripción de nuevas especies y nuevo géneros (Hemiptera-Heteroptera-Lygaeidae-Lygaeinae). Folia Entomol Mexicana 51:1–163

Brailovsky H (1983) Revisión del género *Torvochrimnus* Brailovsky (Hemiptera-Heteroptera-Lygaeidae-Lygaeinae). An Inst Biol Univ Nac Aut México Ser Zool 53:285–320

Brailovsky H (1984) Un nuevo género y una nueva especie sudamericanos de la tribu Myodochini (Hemiptera: Heteroptera: Lygaeidae). An Inst Biol Univ Nac Aut México Ser Zool 54:79–85

Brailovsky H (1989a) Una especie nueva del género *Neoninus* Distant (Hemíptera-Heteroptera-Lygaeidae-Cyminae-Ninini) de Sudamérica. An Inst Biol Univ Nac Aut México Ser Zool 59:153–158

Brailovsky H (1989b) Un género y dos especies nuevas de hemipteros (Lygaeidae, Bledionotinae, Pamphantinae) del Brasil. An Inst Biol Univ Nac Aut México Ser Zool 59:193–202

Brailovsky H (2013) Description of four new species of *Ninyas* from Venezuela, a key to the known species and some new records (Hemiptera: Heteroptera: Geocoridae). Acta Mus Moraviae Sci Biol (Brno) 98:395–406

Brailovsky H, Barrera E (2012) A remarkable new micropterous Blissidae (Hemiptera, Heteroptera, Lygaeoidea) from South America. Dtsch Entomol Z 59:43–45

Brailovsky H, Cervantes Peredo L (2011) A second species of the genus *Neaplax* Slater 1974 from Mexico (Heteroptera: Lygaeidae: Oxycarenidae). Proc Entomol Soc Wash 110:1–6

Brambila J (2000) A review of *Cligenes* with the description of a new genus, *Valeris* (Hemiptera: Rhyparochromidae: Antillocorini). Florida Entomol 83:303–315

Burdfield-Steel E, Shuker DM (2014) The evolutionary ecology of the Lygaeidae. Ecol Evol 4:2278–2301

Cai B, Dang K, Bu W (2011) *Paraberytus* Štusák, a new record genus from China, with description of a new species (Hemiptera, Heteroptera, Berytidae, Berytini). Acta Zootax Sinica 36:241–245

Cai B, Ye Z, Bu W (2013) A review of *Yemmalysus* Štusak, 1972 from China, with description of one new species (Hemiptera: Heteroptera: Berytidae). Zootaxa 3736:338–344

Carstens JD, Baxendale FP, Heng-Moss TM, Wright RJ (2008) Predation of the Chinch Bug, *Blissus occiduus* Barber (Hemiptera: Blissidae) by *Geocoris uliginosus* (Say) (Hemiptera: Lygaeidae). J Kansas Entomol Soc 81:328–338

Carvalho JCM, Costa LAA (1989) Chave para identificaçno dos gLneros neotrópicos da família Colobathristidae (Hemiptera). Rev Brasil Biol 49:271–277

Carvalho JCM, Henry TJ (1986) Sobre um gênero novo peculiar da família Colobathristidae (Hemiptera) da regino de Carajás (Pará, Brasil). Bol Museu Paraense Emilio Goeldi Zool 2:85–91

Cassis G, Gross GF (2002) Hemiptera: Heteroptera (Pentatomomorpha). In: Houston WK, Maynard GV (eds) Zoological catalogue of Australia, vol 27.3B. CSIRO, Melbourne, pp 1–737

Cervantes Peredo L, Baez MB (2010) Life histories of the seed bugs, *Kleidocerys punctatus* and *Kleidocerys virescens*. J Insect Sci 10:91; insectscience.org/10.91

Cervantes Peredo LM, Brailovsky H (2011) Chinches: Lygaeoidea (Insecta: Heteroptera). In: Cruz-Angón A, Lorea-Hernández F, Hernández-Ortiz V, Morales-Mavil JE (eds) La biodiversidad en Veracruz. Estudio de estado. Diversidad de especies: conocimiento actual, vol 2. CONABIO, Gobierno del Estado de Veracruz, Universidad Veracruzana, Instituto de Ecología, A. C. Distrito Federal, pp 327–337

Cobben RH (1968) Evolutionary trends in Heteroptera. Part I. Eggs, architecture of the shell, gross embryology, and eclosion. Centre for Agricultural Publishing and Documentation, Wageningen

Coll M, Ruberson JR (1998) Predatory Heteroptera: their ecology and use in biological control. Thomas Say Publications in Entomology, Lanham

Costa Lima A (1940) Insetos do Brasil – Hemípteros – 2° Tomo. Escola Nacional de Agronomia, Rio de Janeiro

Dellapé PM (2014) Lygaeoidea. In: Roig-Juñent L, Claps E, Morrone JJ (eds) Biodiversidad de artrópodos Argentinos, vol 3. Editorial INSUE, Argentina, pp 89–106

Dellapé PM, Carpintero D (2007) *Cuscohoplininus pagoreni*: a new genus and species of Hoplinini stilt bug from Peru (Heteroptera: Berytidae). Rev Biol Trop 55:673–676

Dellapé PM, Carpintero D (2012) Relevamiento de los Heteroptera (Insecta: Hemiptera) de las sierras de Tandil, provincia de Buenos Aires, Argentina. Rev Museo Arg Cien Nat 14:125–134

Dellapé PM, Cheli GH (2007) A new species of *Anomaloptera* Amyot & Serville from Patagonia (Hemiptera: Lygaeoidea: Oxycarenidae). Zootaxa 1528:65–68

Dellapé PM, Henry TJ (2010) *Acrolophyses*, a new seed bug genus and two new species (Hemiptera: Heteroptera: Rhyparochromidae: Myodochini) from forest-canopy fogging in Ecuador and Peru. Insect Syst Evol 41:75–89

Dellapé PM, Melo MC, Henry TJ. A phylogenetic review of the true bug genus *Heraeus* Stål 1862 (Hemiptera: Rhyparochromidae: Myodochini), with the description of two new genera and 30 new species. Zool J Linnean Soc (in review)

Dellapé PM, Montemayor SI (2009) Description of a new species of *Ischnodemus* from Peru, and the male and immature stages of *I. subflavus* (Hemiptera: Heteroptera: Lygaeoidea: Blissidae). Rev Mexicana Biodiv 80:687–691

Dellapé PM, Montemayor SI (2011) On the identity of *Ashlockobius* Slater & Slater and *Villalobosothignus* Brailovsky (Hemiptera: Heteroptera: Rhyparochromidae: Myodochini), with the description of a new arboreal species from Ecuador. Zootaxa 2748:47–52
Dingle H, Alden BM, Blakley NR, Kopec D, Miller ER (1980) Variation in photoperiodic response within and among species of milkweed bugs (*Oncopeltus*). Evolution 34:356–370
Distant WL (1880–1893) Insecta. Rhychota. Hemiptera-Heteroptera. In Goodman FD, Salvin O (eds) Biologia Centrali-Americana, vol I. London. x + 462 p, 39 pls (1880: 1–88; 1881: 89–168; 1882: 169–224; 1883: 225–264; 1884: 265–304; 1889: 305–328; 1893: i–xx + 329–462)
Distant WL (1888) Enumeration of the Van Volxem Collection of Rhynchota contained in the Brussel's Museum. Part II. In: Comptes-Rendus des Séances. Serie III, No. 95. Bull Soc Entomol Belgique 33:VII–XIII
Drake CJ, Davis NT (1958) The morphology and systematics of the Piesmatidae (Hemiptera), with keys to world genera and American species. Ann Entomol Soc Am 51:567–581
Forero D (2008) The systematics of the Hemiptera. Rev Colomb Entomol 34:1–21
Froeschner RC (1981) Heteroptera or true bugs of Ecuador: A partial catalog. Smithsonian Contr Zool 322:1–147
Froeschner RC (1985) Synopsis of the Heteroptera or true bugs of the Galápagos Islands. Smithsonian Contr Zool 407:1–84
Hagen KS, Mills NJ, Gordh G, McMurtry JA (1999) Terrestrial arthropods predators of insect and mite pests. In: Bellows TS, Fisher TW (eds) Handbook of biological control: principles and applications of biological control. Academic, San Diego, pp 383–503
Hamid A (1971) A revision of the Cryptorhamphinae (Heteroptera: Lygaeidae) including the description of two new species from Australia. J Australian Entomol Soc 10:163–174
Hamid A (1975) A systematic revision of the Cyminae (Heteroptera: Lygaeidae) of the world with a discussion of the morphology, biology, phylogeny and zoogeography. Occas Pap Entomol Soc Nigeria 14:1–179
Hamilton SW (1983) *Neortholomus*, a new genus of Orsillini (Hemiptera–Heteroptera: Lygaeidae: Orsillinae). Univ Kansas Sci Bull 52:197–234
Harrington BJ (1980) A generic level revision and cladistic analysis of the Myodochini of the world (Hemiptera, Lygaeidae, Rhyparochrominae). Bull Am Mus Nat Hist 167:49–116
Harrington BJ (1983) A new species of Cleradini (Hemiptera, Lygaeidae, Rhyparochrominae) from the Central–African – Republic and Ghana. J New York Entomol Soc 91:63–67
Harrington BJ (1990) Detecting evidence of hematophagy in dry museum specimens of *Clerada apicicornis* (Hemiptera: Lygaeidae: Rhyparochrominae). Ann Entomol Soc Am 83:545–548
Harris HM (1943) Art. XVI. New Neididae (Hemiptera) from South America, with notes on some little-known species. Ann Carnegie Mus 29:443–450
Heiss E, Péricart J (1983) Revision of Palaearctic Piesmatidae (Heteroptera). Mitteilungen Münchner Entomol Gesellschaft 73:61–171
Heiss E, Péricart J (1997) Revised taxonomic status of some Old World Piesmatidae (Heteroptera). Zeitschrift Arbeitsgeminschaft Österreichischer Entomol 49:119–120
Henry TJ (1997a) Phylogenetic analysis of family groups with the infraorder Pentatomomorpha (Hemiptera: Heteroptera), with emphasis on the Lygaeoidea. Ann Entomol Soc Am 90:275–301
Henry TJ (1997b) Cladistic analysis and revision of the stilt bug genera of the world (Heteroptera: Berytidae). Contr Am Entomol Inst 30:1–100
Henry TJ (1997c) Monograph of the stilt bugs, or Berytidae (Heteroptera), of the Western Hemisphere. Mem Entomol Soc Washington 19:1–149
Henry TJ (2000) Stilt bugs (Berytidae). In: Schaefer CW, Panizzi AR (eds) Heteroptera of economic importance. CRC Press, Boca Raton, pp 725–735
Henry TJ (2002) Review of the stilt bug genus *Hoplinus* with the description of a new species and notes on other Hoplinini (Heteroptera: Berytidae: Gampsocorinae). J New York Entomol Soc 110:182–191
Henry TJ (2004) *Raglius alboacuminatus* (Goeze) and *Rhyparochromus vulgaris* (Schilling) (Lygaeoidea: Rhyparochromidae): Two Palearctic bugs newly discovered in North America. Proc Entomol Soc Wash 106:513–522

Henry TJ (2006) Revision of the New World lygaeoid genus *Epipolops* (Heteroptera: Geocoridae: Pamphantinae: Epipolopini), with descriptions of five new species. Can Entomol 138:504–530

Henry TJ (2007) A newly discovered Brazilian species of the stilt bug genus *Jalysus* (Hemiptera: Heteroptera: Berytidae) associated with myrmecophytic plants. Proc Entomol Soc Wash 109:324–330

Henry TJ (2009) Biodiversity of Heteroptera. In: Foottit R, Adler P (eds) Insect biodiversity: science and society, 1st edn. Blackwell Publishing, Oxford, UK, pp 223–263

Henry TJ (2013) *Cymapamphantus valentineorum*, a new genus and species of Pamphantinae (Heteroptera: Lygaeoidea: Geocoridae) from the British Virgin Islands, with a checklist of the species and keys to the tribes and genera of the subfamily. Proc Entomol Soc Wash 115:392–401

Henry TJ, Adamski D (1998) *Rhyparochromus saturnius* (Rossi) (Heteroptera: Lygaeoidea: Rhyparochromidae), a Palearctic seed bug newly discovered in North America. J New York Entomol Soc 106:132–140

Henry TJ, Dellapé PM (2009) A new genus and species of Oxycarenidae (Hemiptera, Heteroptera, Lygaeoidea) from Argentina. ZooKeys 25:49–59

Henry TJ, Froeschner RC (eds) (1988) Catalog of the Heteroptera, or true bugs, of Canada and the Continental United States. E. J. Brill, Leiden

Henry TJ, Froeschner RC (1993) *Dieuches armatipes* (Walker) (Heteroptera: Lygaeidae) newly discovered in the Western Hemisphere. Proc Entomol Soc Wash 95:449–452

Henry TJ, Froeschner RC (1998) Catalog of the stilt bugs, or Berytidae, of the world (Insecta: Hemiptera: Heteroptera). Contr Am Entomol Inst 30:1–72

Henry TJ, Dellapé PM, Scudder GGE (2015) Resurrection of the genera *Crophius* Stål and *Mayana* Distant from synonymy with *Anomaloptera* Amyot and Serville, and a key to the New World genera (Hemiptera: Heteroptera: Oxycarenidae). Proc Entomol Soc Wash 117: 367–380

Hoberlandt L (1987) Results of the Czechoslovak-Iranian Entomological Expeditions to Iran 1970,1973 and 1977. Heteroptera, Lygaeidae, Oxycareninae. Acta Entomol Musei Nat Pragae 42:12–29

Hoffman RL, Slater JA (1995) *Holcocranum saturejae*, a Palearctic cattail bug established in eastern United States and tropical Africa (Heteroptera: Lygaeidae: Artheneinae). Banisteria 26:12–15

Horváth G (1904) Monographia Colobathristinarum. Ann Musei Nat Hungarici 2:117–172

Horváth G (1905) Berytidae novae a Dre G. Horváth descriptae. Ann Musei Nat Hungarici 3:56–60

Kerzhner IM (1997) East Palaearctic species of the genus *Artheneis* (Heteroptera: Lygaeidae). Zoosyst Rossica 6:213–232

Kerzhner IM (2001) Family Malcidae. In: Aukema B, Rieger C (eds) Catalogue of Heteroptera of the Palaearctic Region. Netherlands Entomological Society, Amsterdam, pp 227–229

Kormilev NA (1949a) Notas sobre los Colobathristidae de Bolivia con la descripción de un género y una especies nuevos (Hemiptera). Univ Nac La Plata Not Museo La Plata 14(Zool 124):167–176

Kormilev NA (1949b) La Familia Colobathristidae Stal en la Argentina. Acta Zool Lilloana 7:359–383

Kormilev NA (1951) Notas sobre Colobathristidae neotropicales (Hemiptera), con la descripción de tres géneros y siete especies nuevas. Rev Brasil Biol 11:63–84

Kumar R (1968) Aspects of the morphology and relationships of the superfamilies Lygaeoidea, Piesmatoidea and Pyrrhocoroidea (Hemiptera: Heteroptera). Entomol Mon Mag 103:251–261

Leonard DE (1968a) A revision of the genus *Blissus* (Heteroptera: Lygaeidae) in eastern North America. Ann Entomol Soc Am 61:239–250

Leonard DE (1968b) Three new species of *Blissus* from the Antilles (Heteroptera: Lygaeidae). Proc Entomol Soc Wash 70:150–153

Leonard DE (1970) A new North American species of *Blissus* (Heteroptera: Lygaeidae). Can Entomol 102:1531–1533

Li H-M, Den R-Q, Wang J-W, Chen Z-Y, Jia F-L, Wang X-Z (2005) A preliminary phylogeny of the Pentatomomorpha (Hemiptera: Heteroptera) based on nuclear 18S rDNA and mitochondrial DNA sequences. Mol Phylogen Evol 37:313–326

Li J, Gao C, Bu W (2011) Review of the tribe Lethaeini Stål (Hemiptera: Heteroptera: Lygaeoidea: Rhyparochromidae) from China, with a key to Chinese genera and species. Zootaxa 3126:28–38

Malipatil MB (2012) Australocorinae, a new subfamily of Geocoridae (Hemiptera: Heteroptera, Lygaeoidea) from Australia, with descriptions of a new genus and two new species. Zootaxa 3554:75–88

Malipatil MB (2014) Meschiidae, a new family of Lygaeoidea (Hemiptera: Heteroptera) from India and Australia, with descriptions of a new genus and two new species. Zootaxa 3815:233–248

McAtee WL (1919) Key to the Nearctic genera and species of Berytidae (Heteroptera). J New York Entomol Soc 27:79–92

Melo MC, Dellapé PM, Carpintero DL, Coscarón MC (2004) Reduviidae, Miridae, y Lygaeoidea colectados en Colonia Carlos Pellegrini (Esteros del Iberá, Corrientes, Argentina). Rev Soc Entomol Argentina 63:59–67

Melo MC, Dellapé PM, Carpintero DL, Montemayor SI (2011) Heteroptera (Hemiptera) from the Chaco National Park (Argentina). Zootaxa 2999:1–19

Montandon AL (1913) Nouvelles formes de Geocorinae appartenant aux collections du Muséum national Hongrois. Ann Musei Nat Hungarici 11:211–219

Morales A, Corredor-Arjona A, Osorno-Mesa E, Parra-Marquez J (1969) Infección natural de *Mus musculus* con *Trypanosoma cruzi*, en una región de Colombia, S.A. Rev Acad Colomb Cienc Exactas Fisicas Nat 13:375–377

NAPPO (2014) Official pest report: cotton seed bug (*Oxycarenus hyalinipennis*) eradicated from Florida. North American Plant Protection Organization (NAPPO), Phytosanitary Alert System http://www.pestalert.org/oprDetail.cfm?oprID=577

Naranjo SE, Gibson RL (1996) Phytophagy in predaceous Heteroptera: effects on life history and population dynamics. In: Alomar O, Wiedenmann RN (eds) Zoophytophagous Heteroptera: implications for life history and integrated pest management. Thomas Say Publications in Entomology, Lanham, pp 57–93

Nagoshi RN, Paraiso O, Brambila J, Kairo MT (2012) Assessing the usefulness of DNA barcoding to identify *Oxycarenus hyalinipennis* (Hemiptera: Oxycarenidae) in Florida, a potentially invasive pest of cotton. Florida Entomol 95:1174–1181

O'Donnell JE (1991) A new coleopteroid Lethaeini from southern South America (Hemiptera: Lygaeidae: Rhyparochrominae). J New York Entomol Soc 99:87–96

Paula A, Ferreira PSF (1998) Fauna de Heteroptera de la "Mata do Córrego do Paraíso", Viçosa, Minas Gerais, Brasil. I. Riqueza y diversidad específicas. An Inst Biol Univ Nac Aut Mex Ser Zool 69:39–51

Paula A, Ferreira PSF (2000) Fauna de Heteroptera de la "Mata do Corrego do Paraiso", Viçosa, Minas Gerais, Brasil. II. Patrones temporales. Distribucion anual y estacionalidad. An Inst Biol Univ Nac Aut Mex Ser Zool 71:7–19

Peck SB (2001) Smaller orders of insects of the Galapagos Islands, Ecuador: Evolution, ecology, and diversity. NRC Research Press, Ottawa

Pereira RP, da Silva SC (1988) Ocorrência de *Blissus leucopterus* em pastagens do estado do Rio de Janeiro. Pesq Agropec Bras 23:551–553

Péricart J (1974) Subdivision du genre *Piesma* [Hem. Piesmatidae] et remarques diverses. Ann Soc Entomol France 10:51–58

Péricart J (1984) Hémiptères Berytidae Euro–Méditerranéens. Faune de France. France et régions limitrophes, 70th edn. Féderation Française des Sociétés de Sciences Naturelles, Paris

Péricart J (1998) Hémiptères Lygaeidae Euro-Méditerranéens. Vol. 2. Systématique: Seconde Partie. Oxycareninae, Bledionotinae, Rhyparochrominae (1). Faune France 84B:I–III, 1–453

Péricart J (2001) Family Lygaeidae Schilling, 1829 seed-bugs. In: Aukema B, Rieger C (eds) Catalogue of the Heteroptera of the Palaearctic Region. Pentatomomorpha I, vol 4. The Netherlands Entomological Society, Amsterdam

Putshkov VG (1958) Larvae of Hemiptera-Heteroptera. I. Lygaeidae. Revue d'Entomol l'URSS 37:392–413
Readio J, Sweet MH (1982) A review of the Geocorinae of the United States east of the 100th Meridian (Hemiptera: Lygaeidae). Misc Publ Entomol Soc Am 12:1–91
Rengifo-Correa L, Brailovsky H, Henry TJ, Morrone JJ (2013) Phylogenetics and evolutionary morphology of the Neotropical true bug genus *Epipolops* (Hemiptera: Heteroptera: Geocoridae). Syst Entomol 39:127–140 DOI:. doi:10.1111/syen.12039
Samuels RI, Coracini DLA, dos Santos CAM, Gava CAT (2002) Infection of *Blissus antillus* (Hemiptera: Lygaeidae) eggs by the entomopathogenic fungi *Metarhizium anisopliae* and *Beauveria bassiana*. Biol Control 23:269–273
Schaefer CW (1981) Improved cladistic analysis of the Piesmatidae and consideration of the known host plants. Ann Entomol Soc Am 74:536–539
Schaefer CW (1993) The Pentatomomorpha (Hemiptera: Heteroptera): an annotated outline of its systematic history. Europ J Entomol 90:105–122
Schuh RT, Slater JA (1995) True bugs of the world (Hemiptera: Heteroptera). Classification and natural history. Cornell University Press, Ithaca
Scudder GGE (1957) A revision of Ninini (Hemiptera-Heteroptera, Lygaeidae) including the description of a new species from Angola. Publ Cult Comp Diamantes Angola 34:91–108
Scudder GGE (1962a) New Heterogastrinae (Hemiptera) with a key to the genera of the world. Opuscula Entomol 27:117–127
Scudder GGE (1962b) The Ischnorhynchinae of the world (Hemiptera: Lygaeidae). Trans Royal Entomol Soc London 114:163–194
Scudder GGE, Duffy SS (1972) Cardiac glycosides in the Lygaeinae (Hemiptera: Lygaeidae). Can J Zool 50:35–42
Slater JA (1955) A revision of the subfamily Pachygronthinae of the world (Hemiptera: Lygaeidae). Philip J Sci 84:1–160
Slater JA (1964a) A catalogue of the Lygaeidae of the world. 2 vols. University of Connecticut, Storrs
Slater JA (1964b) Results of the Lund University expedition in 1950–1951. Chapter II. Hemiptera (Heteroptera): Lygaeidae. South Afric Anim Life 10:15–228
Slater JA (1967) Synonymy in the Lygaeidae (Hem.). Proc Entomol Soc Wash 69:244–245
Slater JA (1971) The first Neotropical records of the genus *Plinthisus* with the description of three new species. J Kansas Entomol Soc 44:377–384
Slater JA (1972) The occurrence of *Elasmolomus sordidus* (F), a potential pest of peanuts, in Brazil (Hemiptera: Lygaeidae). O Biologico 38:394–395
Slater JA (1974) *Neaplax*, a new genus of Oxycareninae from the Western Hemisphere (Hemiptera: Lygaeidae). J Kansas Entomol Soc 47:517–522
Slater JA (1976) Monocots and chinch bugs: a study of host plant relationships in the lygaeid subfamily Blissinae (Hemiptera: Lygaeidae). Biotropica 8:143–165
Slater JA (1977) The incidence and evolutionary significance of wing polymorphism in lygaeid bugs with particular reference to those of South Africa. Biotropica 9:217–229
Slater JA (1979) The systematics, phylogeny, and zoogeography of the Blissinae of the world (Hemiptera, Lygaeidae). Bull Am Mus Nat Hist 165:1–180
Slater JA (1980) Systematic relationships of the Antillocorini of the Western Hemisphere (Hemiptera: Lygaeidae). Syst Entomol 5:199–226
Slater JA (1981) Two new genera of Lygaeidae from northern Australia including the first member of the Pamphantini from the Eastern Hemisphere (Hemiptera: Heteroptera). Australian J Entomol 20:111–118
Slater JA (1986a) A synopsis of the zoogeography of the Rhyparochrominae (Heteroptera: Lygaeidae). J New York Entomol Soc 94:262–280
Slater JA (1986b) *Aulacoblissus*, a new genus of micropterous Blissinae from Venezuela (Hemiptera: Lygaeidae). Florida Entomol 69:661–665
Slater A (1992) A genus level revision of Western Hemisphere Lygaeinae (Heteroptera: Lygaeidae) with keys to species. Univ Kansas Sci Bull 55:1–56

Slater JA (1995) Fifteen new species of *Ozophora* from Central and South America with a key to mainland Neotropical species (Hemiptera: Lygaeidae). Am Mus Novit 3135:1–31

Slater JA (1999) The systematic position of the Pamphantinae with description of two new tribes and a new species of *Cattarus* (Hemiptera: Lygaeoidea: Geocoridae). Acta Soc Zool Bohemicae 63:199–208

Slater JA, Baranowski RM (1978) How to know the true bugs (Hemiptera–Heteroptera). Wm. C. Brown Co. Publ, Dubuque

Slater JA, Baranowski RM (1990) The Lygaeidae of Florida (Hemiptera: Heteroptera). Arthropods of Florida and neighboring lands areas. Florida Dep Agric Consumer Serv 14:1–211

Slater JA, Baranowski RM (1994) The occurrence of *Oxycarenus hyalinipennis* (Costa) (Hemiptera: Lygaeidae) in the West Indies and new Lygaeidae records for the Turks and Caicos Islands of Providenciales and North Caicos. Florida Entomol 77:495–497

Slater A, Baranowski RM (2001) *Melanopleuroides dominicanus*, a new lygaeine genus and species from the Dominican Republic (Heteroptera: Lygaeidae). Florida Entomol 84:131–132

Slater JA, Brailovsky H (1983) Review of the Neotropical genus *Toonglasa* (Hemiptera: Lygaeidae). Ann Entomol Soc Am 76:523–535

Slater JA, Brailovsky H (1986) The first occurrence of the subfamily Artheneinae in the Western Hemisphere with the description of a new tribe (Hemiptera: Lygaeidae). J New York Entomol Soc 94:409–415

Slater JA, Brailovsky H (1989) El género *Neokleidocerys* (Scudder) *status nov*. y descripción de una especie nueva (Hemiptera-Heteroptera-Lygaeidae-Ischnorhynchinae. An Inst Biol Univ Nac Aut México Ser Zool 59:181–191

Slater JA, Brailovsky H (1990) A further contribution to the systematics of the genus *Toonglasa* (Hemiptera: Lygaeidae: Blissinae). J New York Entomol Soc 98:406–423

Slater JA, Brailovsky H (2000) Lygaeidae (Hemiptera). In: Llorente-Bousquets JE, González-Soriano E, Papavero N (eds) Biodiversidad, taxonomía y biogeografía de artrópodos de México: Hacia una síntesis de su conocimiento, vol II. Univ Nac Aut México, México, pp 319–333

Slater JA, Henry TJ (1999) Notes and descriptions of new Pamphantinae, including four new species of *Cattarus* and a remarkable new myrmecomorphic genus and species (Heteroptera: Lygaeoidea: Geocoridae). J New York Entomol Soc 107:304–330

Slater JA, O'Donnell JE (1995) A catalogue of the Lygaeidae of the world (1960–1994). New York Entomological Society, New York

Slater JA, Slater A (1999) *Ashlockobius*, a new genus of Myodochini from Venezuela (Hemiptera: Lygaeoidea: Rhyparochromidae: Myodochini). Proc Entomol Soc Wash 101:138–142

Slater JA, Wilcox DB (1966) An analysis of three new genera of Neotropical Blissinae (Hemiptera: Lygaeidae). Ann Entomol Soc Am 59:61–76

Slater JA, Woodward TE (1982) Lilliputocorini, a new tribe with six new species of *Lilliputocoris*, and a cladistic analysis of the Rhyparochrominae (Hemiptera, Lygaeidae). Am Mus Novit 2754:1–23

Slater JA, Schuh RT, Cassis G, Johnson CA, Pedraza-PeZaloza P (2009) Revision of *Laryngodus* Herrich-Schaeffer, an *Allocasuarina* feeder, with comments on its biology and the classification of the family (Heteroptera: Lygaeoidea: Rhyparochromidae). Invertebr Syst 23:111–133

Solbreck C (1979) Induction of diapause in a migratory seed bug, *Neacoryphus bicrucis* (Say) (Heteroptera, Lygaeidae). Oecologia 43:41–49

Solbreck C, Pehrson I (1979) Relations between environment, migration and reproduction in a seed bug, *Neacoryphus bicrucis* (Say) (Heteroptera, Lygaeidae). Oecologia 43:51–62

Stål C (1870–1876) Enumeratio Hemipterorum: Bidrag till en Företeckning öfver alla hittills kända Hemiptera, jemte systematiska meddelanden. Parts 1–5. Kongliga Svenska Vetenskaps-Akademiens Handlingar, 1870, part 1, 9(1): 1–232; 1872, part 2, 10: 1–159; 1873, part 3, 11: 1–163; 1874, part 4, 12: 1–186; 1876, part 5, 14: 162

Štusák JM (1967) New stilt bugs from the tropics (Heteroptera, Berytidae). Acta Entomol Musei Nat Pragae 37:279–295

Štusák JM (1968) A new genus of Neotropical stilt-bugs (Hemiptera: Berytidae). J New York Entomol Soc 76:2–8

Štusák JM (1971) A new species of *Parajalysus* Distant from Brazil (Heteroptera, Berytinidae). Acta Entomol Bohemosl 68:149–152

Štusák JM (1973) A new species of *Pronotacantha* Uhler from Mexico (Heteroptera, Berytinidae). Acta Entomol Bohemosl 70:45–48

Štusák JM (1977) A new Neotropical stilt bug – *Metajalysus horvathi* gen. et sp. n. (Heteroptera: Berytidae). Acta Zool Acad Scient Hungaricae 23:421–426

Štusák JM, Cobben RH (1975) The Heteroptera of the Netherlands Antilles – X Berytinidae (stilt bugs). Stud Fauna Curação Caribbean Islands 159:63–78

Štys P (1961) Morphology of the abdomen and female ectodermal genitalia of the trichophorous Heteroptera and bearing on their classification. Proc XXth Int Congr Entomol 1:37–43

Štys P (1966) Revision of the genus *Dayakiella* Horv. and notes on its systematic position (Heteroptera; Colobathristidae). Acta Entomol Bohemosl 63:27–39

Štys P (1967) Monograph of Malcinae, with consideration of morphology and phylogeny of related groups (Heteroptera, Malcidae). Acta Entomol Musei Nat Pragae 37:351–516

Štys P (1991) First apterous genus and species of Lygaeidae: Blissinae (Heteroptera). Acta Entomol Bohemosl 88:265–271

Štys P, Henry TJ (2015) A new genus and species of Colobathristidae (Hemiptera: Heteroptera) from Peru, with a replacement name for the preoccupied *Labradoria* Kormilev, and a revised key to the Neotropical genera. Proc Entomol Soc Wash 117:27–35

Štys P, Exnerová A (2012) A new genus and species of Oriental Colobathristidae (Hemiptera: Heteroptera) with a key to the Eastern Hemisphere genera and morphological and functional considerations. Entomol Am 118:53–65

Sweet MH (1960) The seed bugs: a contribution to the feeding habits of the Lygaeidae (Hemiptera: Heteroptera). Ann Entomol Soc Am 53:317–321

Sweet MH II (1964a) The biology and ecology of the Rhyparochrominae of New England (Heteroptera: Lygaeidae). Part I. Entomol Am 43:1–124

Sweet MH II (1964b) The biology and ecology of the Rhyparochrominae of New England (Heteroptera: Lygaeidae). Part II. Entomol Am 44:1–201

Sweet MH II (1967) The tribal classification of the Rhyparochrominae (Heteroptera: Lygaeidae). Ann Entomol Soc Am 60:208–226

Sweet MD (1977) The systematic position of the seedbug genus *Neosuris* Barber, 1924 (Hemiptera: Lygaeidae) with a discussion of the zoogeographical significance of the genus and notes on the distribution and ecology of *N. castanea* (Barber, 1911) and *N. fulgida* (Barber, 1918). J Kansas Entomol Soc 50:569–574

Sweet MH II (2000a) Seed and chinch bugs (Lygaeoidea). In: Schaefer CW, Panizzi AR (eds) Heteroptera of economic importance. CRC Press, Boca Raton, pp 143–264

Sweet MH II (2000b) Economic importance of predation by big-eyed bugs (Geocoridae). In: Schaefer CW, Panizzi AR (eds) Heteroptera of economic importance. CRC Press, Boca Raton, pp 713–724

Sweet MH, Schaefer CW (1985) Systematic status and biology of *Chauliops fallax* Scott, with a discussion of the phylogenetic relationships of the Chauliopinae (Hemiptera: Malcidae). Ann Entomol Soc Am 76:526–536

Tamaki G, Weeks RE (1972) Biology and ecology of two predators, *Geocoris pallens* Stål and *G. bullatus* (Say). USDA Tech Bull 1446:1–46

Torre-Bueno JR de al (1946) A synopsis of the Hemiptera-Heteroptera of America north of Mexico. Part III. Family XI–Lygaeidae. Entomol Am 26:1–141

Torres M, Cárdenas E, Pérez S, Morales A (2000) Haematophagy and cleptohaematophagy of *Clerada apicicornis* (Hemiptera: Lygaeidae), a potential biological control agent of *Rhodnius prolixus* (Hemiptera: Reduviidae). Mem Inst O Cruz 95:131–133

Uhler PR (1901) Some new genera and species of North American Hemiptera. Proc Entomol Soc Wash 4:507–515

Valério JR, Vieira JM, Valle LCS (1999) Ocorrência de *Blissus antillus* Leonard (Hemiptera: Lygaeidae: Blissinae) em Pastagem no Estado de Mato Grosso do Sul. An Soc Entomol Brasil 28:527–529

Van Duzee EP (1910) Monograph of genus *Crophius* Stal. Bull Buffalo Soc Nat Sci 9:389–398

Wheeler AG Jr (1976) Life history of *Kleidocerys resedae* on European white birch and ericaceous shrubs. Ann Entomol Soc Am 69:459–463

Wheeler AG Jr (1983) The small milkweed bug, *Lygaeus kalmii* (Hemiptera, Lygaeidae): milkweed specialist or opportunist? J New York Entomol Soc 91:57–62

Wheeler AG Jr (2001) Biology of the plant bugs (Hemiptera: Miridae). Pests, predators, opportunists. Cornell University Press, Ithaca, 506 pp

Wheeler AG Jr, Fetter JE (1987) *Chilacis typhae* (Heteroptera: Lygaeidae) and the subfamily Artheneinae new to North America. Proc Entomol Soc Wash 89:244–249

Wheeler AG Jr, Hoebeke ER (2013) Establishment of the Palearctic *Heterogaster urticae* (F.) (Hemiptera: Lygaeoidea: Heterogastridae) in North America, with new British Columbia records of the native *H. behrensii* (Uhler). Proc Entomol Soc Wash 115:189–196

Wheeler AG Jr, Schaefer CW (1982) Review of stilt-bug (Hemiptera: Berytidae) host plants. Ann Entomol Soc Am 75:498–506

Wille JE (1951) Biological control of certain cotton insects and the application of new organic insecticides in Peru. J Econ Entomol 44:13–18

Chapter 17
Cotton Stainers (Pyrrhocoridae) and Bordered Plant Bugs (Largidae)

Carl W. Schaefer

Abstract The Pyrrhocoridae and the Largidae, making up the superfamily Pyrrhocoroidea, are moderately sized insects, often with bright and contrasting colors. Several genera of Largidae resemble ants, either in color or in form. These insects range from 7 to 18 mm (Pyrrhocoridae) and from 5 to 16 mm (Largidae). They feed on seeds, on the ground, or in the fruits. However, for most largids, we do not know much at all about their host plants or biology. *Dysdercus* (the only Neotropical genus in Pyrrhocoridae) can be harmful, feeding on cotton seeds; however, this damage occurs rarely in the Neotropics and more often in the Paleotropics.

17.1 Introduction

The superfamily Pyrrhocoroidea contains two families, the Largidae and the Pyrrhocoridae. Pyrrhocoroidea is grouped with the trichophora (which is not a formal name), together with the Lygaeoidea, Coreoidea, and Pentatomoidea; each of these superfamilies has regularly placed trichobothria (sensory hairs) on their abdominal sternites (the third through the seventh). These families constitute the infraorder Pentatomomorpha, to which some have included the superfamily Aradoidea [but see Sweet (2006), who treats it as a separate infraorder].

17.2 Classification

The Pyrrhocoridae are of course most closely related to the Largidae, sharing several characters such as the lack of ocelli, the (usually) distinct division of the pronotum into anterior and posterior regions, a characteristic pattern of trichobothria (long sensory hairs arranged symmetrically on abdominal sterna three through

---

Carl W. Schaefer: Author deceased at the time of publication

A.R. Panizzi, J. Grazia (eds.), *True Bugs (Heteroptera) of the Neotropics*, Entomology in Focus 2, DOI 10.1007/978-94-017-9861-7_17

seven), and more arcane characteristics of the male and female genitalia (Schaefer 1964, 1966). Indeed, the two families were in the past considered subfamilies of a larger Pyrrhocoridae until Hussey (1929) suggested that each be raised to family level; China (1954) raised Largidae to family level, and Southwood (1956) created the formal term "Pyrrhocoroidea." Two families, Pyrrhocoridae and Largidae, in a superfamily Pyrrhocoroidea, are now the accepted classification.

Of the two families, the Largidae may be the more primitive, a view put forth by Schaefer (1964, 1966), partly because the largid ovipositor is laciniate (designed for cutting into plant tissue for the inserting of eggs) like that of the presumably ancestral Lygaeoidea and unlike the flattened broadened ovipositor of the Pyrrhocoridae (designed for the laying of eggs onto surfaces); for an opposing view, see Kumar (1968). In addition, as mentioned above, the forefemora of the Pyrrhocoridae are often (not always) swollen, and the family's metathoracic scent gland apparatus lacks a raised "auricle"; both may be advanced features, and the unswollen (or slightly swollen) condition of the forefemora in Largidae and the presence in this family of an auricle are possibly primitive features.

The superfamily Pyrrhocoroidea is distinguished by the absence of ocelli, thus making it easy to identify. So also is the Miridae, but mirids are smaller and easily distinguished in many other features. See also Schaefer and Stehlík (2013) for an account of the Pyrrhocoroidea of the Caribbean Sea Region.

A key to families of the Pyrrhocoroidea (this key will work for New and Old World Pyrrhocoroidea) is given below.

Forefemora not or slightly swollen, terete, sometimes distinctly sulcate; metathoracic scent gland complex, usually auriculate; terminus of 7th venter in female medially cleft; ovipositor laciniate; male's aedeagus without thecal appendage and vesica with distinct coils.............................Largidae

Forefemora usually distinctly swollen, neither terete nor sulcate; metathoracic scent gland complex but not auriculate; terminus of 7th venter in female entire (not cleft); ovipositor platelike; aedeagus with or without thecal appendage and vesica straight or slightly coiled .... Pyrrhocoridae

Because only the single genus of Pyrrhocoridae occurs in the New World, the key separating Largidae and Pyrrhocoridae will perforce identify *Dysdercus*.

17.3 Pyrrhocoridae

There are at least 30 genera in the family Pyrrhocoridae (Hussey 1929), but only *Dysdercus* occurs both in the New World and the Old. The other genera, some with many species, occur in the Paleotropics [e.g., *Dindymus*, with 78 species (Stehlík 2013b)], but only *Dysdercus* has reached the Neotropics. About 65–70 species occur worldwide in this genus (van Doesburg 1968; Beccari and Gerini 1970), and nearly 40 species and about 25 subspecies of New World *Dysdercus* occur in the Neotropics and the sub-Neotropics, extending up to the southern United States (van Doesburg 1968; Henry 1988a). These Neotropical species and subspecies of *Dysdercus* have been ably revised and keyed by van Doesburg (1968). Costa Lima et al. (1962) also has a key, but it is less conclusive.

The Pyrrhocoridae have been tentatively divided into two subfamilies, a division which I believe has not been published. However, because only one genus occurs in the New World, this controversy is here moot. This single genus, *Dysdercus*, occurs also throughout the Old World tropics [including Australia, but no Pyrrhocoroidea occur in New Zealand (Lariviére and Larochelle 2004, p. 191, 2014)].

At least 25 species are Paleotropical (Hussey 1929, Freeman 1947), and in the Paleotropics, they are far more diverse. Nevertheless, the actual numbers of species are more in the Neotropics (about 40 species) than in the Paleotropics (about 25–30 species). It seems likely that the New World species are all descended from a single species (or closely related members) of one of two subgenera of subgenera found only in Africa (the only other subgenus is mostly Oriental and Australasian) (see Freeman 1947; Stehlík 1965a, b; van Doesburg 1968; Schaefer 2013). The greater diversity of *Dysdercus* species in the Old World, as well as the far greater diversity of pyrrhocorid genera, supports the idea that the New World *Dysdercus* are derived from an Old World group and not the reverse. More recently, Zrzavý and Nedvěd (1997) show cladistically that the New World *Dysdercus* are all closely related, that they came from an Afrotropical group, and that the Afrotropical group did not come from the Neotropical group (see also van Doesbug 1968). The phylogenetic relationships of the New World species of *Dysdercus* have also been ably if tentatively discussed by van Doesburg (1968).

In the Galapagos Islands, there are two species of *Dysdercus* (*D. concinnus* Stål and *D. lunulatus* Uhler). These are not closely related, so Froeschner (1985) believes these two originated from two separate invasions of these islands. The first is on Santa Cruz and the second is on San Cristóbal Island.

17.3.1 Diagnosis

Species of *Dysdercus* are neither particularly large nor particularly small, ranging from 8 to 18 mm long; they are elongate oval, and their bodies are somewhat flattened (Figs. 17.1 and 17.2). Nearly all species are aposematic (warningly colored), red, yellow, or orange, and these colors may be contrastingly patterned with black; ventrally, each thoracic and abdominal segment is often vividly striped with white or yellow and a darker color—red, orange, deep yellow, or black; many species have a contrastingly black or white spot dorsally. The result is a striking and quite lovely

Fig. 17.1 Pyrrhocoridae: *Dysdercus andreae* (L.) (Dominican Republic) (length: = 9 mm) (Photo by J O'Donnell)

Fig. 17.2 Pyrrhocoridae: *Dysdercus ruficeps* (Perty) (Panama) (length = 15 mm) (Photo by J O'Donnell)

insect, whether viewed from above or below. They often feed on common plants (mostly members of the order Malvales) and often occur in aggregations, so they are frequently seen.

17.3.2 Characterization

Like all hemipterans, *Dysdercus* species have slender elongated sucking mouthparts that reach from the front of the head well onto the middle of the thorax and occasionally onto the abdomen, and, like most heteropterans, their hind wings are membranous, but their forewings are partly membranous and partly sclerotized. Characteristic of Pyrrhocoroidea, the superfamily to which they (and Largidae) belong, is the presence of two closed cells in the forewing's membrane and a lack of ocelli; in addition, the pronotum is clearly divided into an anterior third and a posterior two thirds, these sometimes of different colors.

Pyrrhocoridae can be distinguished from Largidae most readily by the presence in the females of the latter of a median split in sternum seven, indicative of an elongate or laciniate ovipositor; the ovipositor of Pyrrhocoridae is platelike, and the female's seventh sternum is entire. Other differences between members of the two families are as follows: in Pyrrhocoridae, the forefemora are at the most slightly swollen (usually slender), and the metathoracic scent gland lacks an auricle (a raised ridge surrounding the opening); the forefemora of Largidae are often swollen, and the metathoracic scent gland has an auricle. In *Dysdercus* itself, the metathoracic scent gland apparatus is somewhat reduced, and the associated evaporative area is much reduced [as seems to be true generally of aposematic bugs related to Pyrrhocoroidea (Schaefer 1972)]. No region of the body is densely covered with setae or hairs, nor does any part bear a thick coating of tiny hairs (as does occur in some New World Largidae). Also, most species of *Dysdercus* are larger than the New World species of Largidae; those *Dysdercus* species that are smaller are far more brightly and contrastingly colored.

17.3.3 Natural History

Members of *Dysdercus* live for the most part on or near the ground, often on fairly low plants upon which they feed. Preferring members of the dicotyledon order Malvales, *Dysdercus* can become serious pests of cotton, a member of that order (see below). Like other pyrrhocorids—and indeed like the bugs in related families—*Dysdercus* feeds upon the ripening seeds, or secretions on the seeds, but not other parts of the plant. Because such food is seasonal, some species of *Dysdercus* migrate during periods when food is scarce (Schaefer and Ahmad 2000). These movements are in some cases correlated with the bugs' reproductive cycle. In at least one Neotropical species, *Dysdercus fulvoniger* (De Geer), the flight muscles of females become greatly reduced, presumably to provide additional nutrients to developing eggs (Davis 1975).

That the adults and nymphs are warningly colored strongly suggests that the bugs are bad tasting and may warn away prospective predators (Svádová et al. 2010 and references therein), who studied a common European species, *Pyrrhocoris apterus* (L.). These bugs also often live in groups and perhaps thus collectively warn off vertebrate predators, which can see the brightly colored assemblages from a distance. There are some isolated records of vertebrates feeding on *Dysdercus* and some records of other arthropods doing so. The scattered nature of these records and particularly their paucity suggest predators are not important in population control. Also, of little importance are parasites (on predators and parasites, see Schaefer and Ahmad 2000). It appears thus that populations of *Dysdercus* are held in check by availability of food.

Despite their occasional economic importance, their wide distribution, and their high visibility to humans, these bugs' biologies and ecology remain poorly known. An excellent account of the Neotropical *Dysdercus* species may be found in van Doesburg (1968) and a more general summary of pyrrhocorid biology and ecology in Schaefer and Ahmad (2000).

17.3.4 Economic Importance

Dysdercus species feed mostly on malvalean plants, a group which includes cotton. In some parts of the tropics, *Dysdercus* can do serious harm to cotton crops, but they do much less harm in the New World than in the Old. However, at least one New World *Dysdercus*, *D. peruvianus* (Guérin), feeds in Peru on wild malvalean plants in the spring and then moves to cotton in the summer and can cause damage (Anonymous 1961).

In the West Indies, *Dysdercus* species feed on four families of the Malvales (Malvaceae, Sterculiaceae, Bombaceae, and Tiliaceae), but prefer Malvaceae. These species prefer wetter islands (Grenada, St. Vincent) rather than drier ones (Montserrat). On these islands, wild malvaceous hosts have nearly been eradicated and, in general, *Dysdercus* is no longer a threat to cotton production (Squire 1939); however, this was written many years ago.

Like many other heteropterans, *Dysdercus* sucks juices from the reproductive parts of plants and, in so doing, may weaken or, when abundant, even kill young cotton bolls. However, unique to *Dysdercus* is another form of damage: the microorganisms that cause internal boll disease can be transmitted mechanically from infected cotton boll to uninfected boll. Luckily, this infection seems to occur less often and less seriously in the New World than in the Old. The excreta of the bugs, and the symptoms of boll disease, discolor the cotton fibers, rendering them unfit for cloth production. For more details and a review of published work, see Schaefer and Ahmad (2000).

17.3.5 Movements and Changes in Food Plants

Dysdercus species feed mostly on seeds, a food rich in nutrients, but one less readily available than other plant tissues. The seeds of a given plant species ripen, fall, and remain available and edible over a relatively brief period. An insect subsisting on seeds must either tie its life cycle closely to that of its host or feed on the seeds of several plants whose phenologies do not greatly overlap. Usually, of course, these plants will be fairly closely related (being fed on by the same insects) and share similar defenses against predators. Such an insect breaching the defenses of one plant species will more easily breach those of that species' relatives. As different populations of the insect species, in different parts of its range, become adapted to different species of plants, speciation of the insects may occur endemically if the original host should be lost. Speciation may be hastened if newly attached species of plants develop new defenses in response to this attack, defenses which the new attacker breaches but which its conspecifics cannot.

Something of the opposite may also occur if a population of a species which feeds generally on members of a plant group be carried somehow beyond its range. Encountering there other species of its plant group, the insect may flourish and, now allopatric, speciate. Both phenomena probably explain both the Pyrrhocoroidea itself and the genus *Dysdercus*.

Some *Dysdercus* feed fairly generally on plants of the Malvales, but prefer certain species: for example, *D. mimus* (Say) on *Sida* spp. and *D. fernaldi fuscofasciatus* Blőte on *Hibiscus bifurcatus* (van Doesburg 1968). Other species, such as those with narrow isolated ranges (van Doesburg 1968), may have arrived as a small population (perhaps a gravid female) and, adapted to living on some malvalean plants at home, continued to live on others abroad. Indeed, the original colonizing stock of *Dysdercus* in the New World (from Africa: see van Doesburg 1968; Stehlík 1965a, b) survived on one side of the Atlantic on plants it had grown accustomed to on the other.

The ability to feed on several members of a related group of plants has economic consequences, if of course the plant group contains species of economic importance. It appears that the most economically important species of *Dysdercus* are those whose fairly broad food-plant range allows them to feed on non-crop reservoir plants until crop monocultures (usually cotton) become available. These reservoirs are the more important the more evanescent the food source is: and seeds are relatively short lived. *Dysdercus* species thus resemble *Leptocorisa* (Alydidae: Micrelytrinae: Leptocorisini) and *Clavigralla* (Coreidae: Pseudophloeinae) species, which damage river and pulse crops, respectively [references and brief discussion in Schaefer and Mitchell (1983)].

Of the twenty New World species listed by van Doesburg (1968) as feeding on cotton, twelve are also recorded from at least three other plants (excluding species of doubtful identification); it is likely, moreover, that several of those apparently restricted to cotton actually feed on other malvalean plants as well, the biology of most of those species not having been studied.

Dysdercus cingulatus (F.) can withstand several days' starvations, and indeed, there is some evidence this period can be extended through selection (Ahmad and Khan 1980). This being so, and other species doing the same thing, it increases the likelihood that species which feed on one plant species may, in time, feed on another.

Dysdercus migrate as adults and, during this time, may gather water and energy from nectar of various flowers and from probing fruits, both sometimes of plant species not usually used as hosts (Myers 1927; van Doesburg 1968); even other insects may be so used (Schouteden 1912; Kamble 1978). Citrus seems to be such a favored plant (Moizant and Téran 1970, Riley and Howard *in* Myers 1927) and indeed Myers quote a West Indian source that fresh orange juice is a good bait for *Dysdercus*. The fondness for citrus may however only reflect the likelihood that citrus groves often lay near cotton fields.

Because seasonal oviposition follows migration, the amino acids gathered from nectar may be used for egg production as well as for quick energy; a larger source of nitrogen is of course the histolyzed flight muscles in *Dysdercus* (Davis 1975).

This habit of *Dysdercus* (and other pyrrhocorids) may have led to the adoption of new hosts. Srivastava and Gupta (1971) suggest a population of *D. cingulatus* may move from cotton to wheat in this way; nymphs damaged the ears of the latter in a field where the former had grown shortly before. This switch seems unlikely to be permanent, for feeding on wheat can probably not yield reproducing adults. Nevertheless, if migrating females deposit more eggs upon nonhost plants visited for a brief sustenance, and if these plants are not too dissimilar chemically from the usual host (as cotton and wheat are), populations of pyrrhocorids may develop upon new hosts. If the process is repeated, each new host visited by a migrating female, being different from its predecessor, a "chain" of populations may develop the host of whose last link is a plant species markedly different from that of the first.

This explanation may account for the choices by *Dysdercus* species which, although each preferring a difference malvaceous host, will when that host senesces move to another less preferred (Dingle and Arora 1973; Robertson 1977; Kamble 1978). As a result, several species of bug may occur on one host, although for only one species is that host both the primary preferred one; for other bugs, that host is a temporary necessity. Thus, *Dysdercus koenigii* (F.) much prefers *Althea rosea* to *Hibiscus* or cotton and *Hibiscus* slightly over cotton (Shukla and Upadhaya 1972) [but see Chatterjee and Raychaudhuri (1976)], who found cotton preferred in the field by this species).

It appears then that *Dysdercus* leaves the preferred malvaceous host when it becomes unsuitable and flies to a less preferred plant, malvaceous or not. It is not clear from the literature if [as Janzen (1972) believes] a non-malvaceous secondary host merely provides water to a bug on the way to a malvaceous plant, or if the now-unsuitable primary host is always forsaken for a non-malvaceous plant by some species of *Dysdercus* (e.g., those studied by Myers in Cuba). More usually—perhaps always—a malvaceous secondary host is sought (as by the bugs Robertson studied in Tanzania), and this explains why *Dysdercus* species are so widespread as pest of cotton, even though several species do not prefer it (Shukla and Padhyaya 1972). Indeed, *Hibiscus esculentus* has been suggested as a trap plant in India.

As populations of *Dysdercus* move from host to host, different preferences will build up, depending upon how long different hosts—or the same hosts in different regions—are available. It is quite likely that *D. koenigii* in Uttar Pradesh does prefer *Althea* to cotton (Shukla and Padhyaya 1972), whereas in West Bengal the reverse is true (Chatterjee and Raychaudhuri 1960); the differences may lie in the different relative abundances and availabilities of these two plants in these two Indian states.

It is unfortunate that these ideas have not really been applied to the Neotropics, especially as the New World *Dysdercus* are all more closely related to each other than are the Old World *Dysdercus*. Perhaps because the pest status of New World *Dysdercus* is far less than that in the Old World, these speculations on the movement of these insects are merely speculative and, for now, of small importance. Note that the Indian subcontinent species *Dysdercus koenigii* has been confused with other species, for many years (see also Hussey 1929). Qadri and Ahmed (sic: Ahmad) (2001) have cleared much of this confusion. However, with global warming, *Dysdercus* may move to the north into cotton fields (and other malvaceous crops) and even to the south in southern South America.

17.4 Largidae

We recognize two subfamilies in Largidae: Physopeltinae, which is exclusively Old World, and Larginae, whose 14 genera occur only in the New World. Within Larginae, we recognize the tribes Largini (six genera) and Arhaphini (four genera and Largulini (four genera)). Here, we key the subfamilies and tribes of the World Largidae and the genera of Neotropical Largidae.

We have been unable to discover how it was and why Largidae have been called "the bordered plant bugs." They do feed, in general, on plants. But why *bordered*? Some *Largus* and many *Arhaphe* species do have horizontal stripes across their pronotum or wings. However, in some species of *Largus*, there may be a horizontal stripe on the pronotum, but the stripe may be absent. Perhaps these stripes are the reason for "bordered plant bugs." But so of course do many *Dysdercus*, as well as other Pyrrhocoridae of the Old World.

A key to the subfamilies and tribes of Largidae (this key will work for New and Old World Largidae) is presented below.

1. Eyes pedunculate or sessile; head never with depression behind bucculae; labium not reaching hind coxae; forefemur not sulcate. Metathoracic scent gland apparatus with small, rounded, or elongated, sessile auricle (New World) Larginae 2
   Eyes sessile; head with shallow depression behind bucculae; labium reaching at least to hind coxae; forefemur distinctly sulcate. Metathoracic scent gland apparatus with large, semicircular, partly free-standing auricle (Old World) Physopeltinae 3
2. (a) Eyes pedunculate or sessile. Never ant mimetic. Occipital suture present. Hemelytra usually fully developed. Auricle of metathoracic scent gland apparatus rounded. Without conspicuously slender antennae. Callar and pronotal lobes separated from one another. Genital capsule without parandria nor with a high lamellar, broken ventral rim (New World) .... Larginae: Largini
   (b) Eyes sessile. Never ant mimetic. Occipital suture present. Hemelytra fully developed. Auricle of metathoracic scent gland apparatus rounded. "With conspicuously slender antennae, callar and pronotal lobes almost unseparated." Genital capsule "with massive parandria or with a high lamellar, broken ventral rim" [from Stehlík and Jindra's (2007) description of Largulini] (Jamaica, Dominican Republic) .... Larginae: Largulini Stehlík and Jindra (2007)
   (c) Eyes sessile (but bulging in *Thaumastaneis* and *Vasarhelyecoris*). Always ant mimetic. Occipital suture absent. Hemelytra usually reduced. Auricle of metathoracic scent gland apparatus rounded (*Vasarhelyecoris*) or elongate (New World) .... Larginae: Arhaphini
3. Antenna very long, first segment longer than head plus prothorax. Sexually dimorphic (male's abdomen longer). One species: *Macrocheraia grandis* (Gray) (Asia) .... Physopeltinae: Lohitini
   Antenna not very long, first segment never as long as head plus prothorax. Not sexually dimorphic (Old World) .... Physopeltinae: Physopeltini

17.4.1 Subfamily Larginae (Figs. 17.3, 17.4, 17.5, 17.6, 17.7, and 17.8)

This exclusively New World subfamily is in general smaller than *Dysdercus* and not so warningly colored. Except for *Largus*, the species are considerably longer than wide; in *Largus*, the species are wider. Several species, especially in the genus *Largus*, are pale reddish, but no Larginae appear to warn away predators by tasting bad. In Arhaphini, protection is accomplished by appearing antlike, either in color (*Arhaphe* and *Pararhaphe*) or in structure (*Thaumastaneis* and *Vasarhelyecoris*). In Largini, a few species may mimic velvet ants.

The distribution of Larginae is in the New World, including southern (and occasionally northern) United States through South America.

Tribe Largini Amyot and Serville

This tribe consists of six species, of which *Largus* is the larger and stouter. The key to genera of Largini [modified from van Doesburg's (1966)] is presented below. The distributions are taken from Stehlík (2013a).

Fig. 17.3 Largidae: Larginae: Largini: *Largus convivus* (Stål) (Mexico) (length = 15 mm) (Photo by J O'Donnell)

Fig. 17.4 Largidae: Larginae: Largini: *Rosaphe stylophthalmum* (Stål) (French Guiana) (length = 12 mm) (Photo by J O'Donnell)

Fig. 17.5 Largidae: Larginae: Largini: *Acinocoris lunaris* (Gmelin) (Bolivia) (length = 10 mm) (Photo by J O'Donnell)

Fig. 17.6 Largidae: Larginae: Largulini: *Largulus parallelus* Hussey (Jamaica) (length = 6 mm) (Photo by J O'Donnell)

Fig. 17.7 Largidae: Larginae: Arhaphini: *Arhaphe argula* (Bliven) (Mexico) (length = 10 mm) (Photo by J O'Donnell)

Fig. 17.8 Largidae: Larginae: Arhaphini: *Vasarhelyecoris ophthalmicus* Brailovsky & Barrera (Peru) (length = 18 mm) (Photo by J O'Donnell)

1. Eyes pedunculate (on stalks), eyestalks at least as long as the eyes are seen from above 2
   Eyes barely or not at all pedunculate 4
2. Anterior lobe of pronotum longer than pronotal posterior disk (Mexico to Brazil and Paraguay) .... *Fibrenus*
   Anterior lobe shorter than disk 3
3. Posterior corners of pronotum rounded; head and pronotum without many long black setae (Costa Rica to Argentina) .... *Acinocoris* (Fig. 17.5)
   Posterior borders of pronotum acute; head and pronotum with many long black seta (Venezuela and Guyana to Argentina) .... *Rosaphe* (Fig. 17.4)
4. Body broad, its length less than 2.5 times its length (northern United States to Argentina) .... *Largus* (Fig. 17.3)
4. Body slender, its length at least three times its width 6
5. Length of pronotum about 1.1 times its width; eyes large [Honduras to Argentina (Dellape and Melo 2012)] .... *Theraneis*
6. Length of pronotum about 1.3 times its width; eyes small (southern United States to Brazil) .... *Stenomacra*

Distribution is New World, southern (and occasionally northern) United States through South America.

These six genera are not antlike and in general are narrow. *Largus* is the largest genus (about 60 species); the others range from a few species up to 16 (*Acinocoris*) and 21 (*Theraneis*) (Stehlík 2013a). The five genera are generally brown, light brown, or yellow brown; *Largus* is an exception. One species, *Rosaphe stylophthalmum* Stål (Fig. 17.4) has a red corium, and others, *Fibrenus gibbicollis* Stål and some *Acinocoris* (Fig. 17.5), have a pattern reminiscent of *Dysdercus*—a yellow or black corium with two broad black or yellow stripes or dots across it. Some species of *Theraneis* and of *Rosaphe* (= *Astemma*) (Stehlík 2007) are thought to mimic velvet ants (Hymenoptera: Mutillidae) (Costa Lima 1940).

Largus species (Fig. 17.3) are also brown or dark brown, but there are some species with black and/or black and red. They are larger than most other genera of Largini and are wider.

Of these genera, a few species of *Largus* extend further north, into the United States: *L. cinctus* Herrich-Schaeffer and *L. succinctus* (L.) occur in some northern states, but these species also are far more abundant in the southern United States and Mexico (Henry 1988a, b).

The genus *Wupatkius*, described by Bliven (1956), is quite clearly a synonym of *Largus*, and Stehlík (2013a) mentions this as well. Halstead (1972a) synonymized this genus. Nevertheless, Bliven (1973) continued to recognize the genus *Wupatkius*. Since then, no formal description of the synonymy has occurred or any discussion of it. Therefore, the name still stands.

Tribe Largulini Stehlík and Jindra

The key is simplified from Stehlík and Brailovsky (2011), and the distributions are from Stehlík (2013a).

1. Anterior lobe of pronotum swollen (Dominican Republic) .... *Neolargulus*
   Anterior lobe of pronotum not swollen 2
2. Body larger (♂♂ 8.75–9 mm; ♀♀ 10–11 mm) (Jamaica)................*Armilargulus*
   Body smaller (♂♂ 7–8 mm; ♀♀ 8–10 mm)... 3
3. Pronotum pale, with two black spots anteriorly (Jamaica).......................*Largulus*
   Pronotum pale, no dark spots (Jamaica) .... *Paralargulus*

Three of the genera (and species) are found only in Jamaica; the fourth is found in the Dominican Republic. It is uncertain if these species will be found elsewhere. The biogeography will be interesting.

The tribe Largulini was raised from the tribe Largini by Stehlík and Jindra (2007), largely because of the peculiar shape of the male genitalia, "with massive parandria or with a lamellar broken ventral surface" (p. 14). The tribe consists of four genera (above), each with only one species.

Tribe Arhaphini Bliven

The distribution of the Arhaphini goes from southern (occasionally northern) United States into Brazil.

A note on Bliven's odd movement of this group into the family Alydidae: Bliven (1973) erected the subfamily Arhaphinae and moved it from Pyrrhocoroidea into Alydidae; at the same time, he moved the alydid subfamily Micrelytrinae (=Leptocorisini and Micrelytrini) from Alydidae to Coreidae, without discussion [for a current view of the higher classification of Alydidae, see Schaefer (1999)]. Bliven was impressed by several morphological features shared by Arhaphinae and Alydinae, particularly "general body form," type of pubescence, partial pronotal carina, the resemblance of nymphal alydines and adult arhaphines to ants, and the stridulitrum-plectrum complex. The last of these similarities (which Bliven considered the most important) does not prove phylogenetic affinity (Schaefer and Pupedis 1981), particularly as most alydines (contrary to Bliven's belief) lack it (Schaefer et al. 1989). The other similarities occur too generally within Heteroptera to warrant including Arhaphinae within Alydidae. Indeed, some alydid micrelytrines (e.g., *Dulichius*, *Trachelium*, some *Cydamus*), which Bliven moved into Coreidae, are ant mimetic and look rather like arhaphines. So also do two Old World genera of Pyrrhocoridae, *Myrmoplasta* Gerstaecker, and *Courtesius* Distant.

As mentioned, Bliven (1973) believed that similarities between the two pulled Arhaphinae into Alydidae (in his restricted sense, *sans* Micrelytrinae). To quote him: "there was a successful mingling of genes between a *Tollius*-like (a genus in Alydinae) form and a largid." He also believed differences between Arhaphinae and other largids pushed this subfamily out from Largidae; but his only evidence is the undocumented statement that "these insects (i.e., arhaphines) have long been recognized as aberrant within the family Largidae" (Bliven 1973, p. 125).

This removal of Arhaphinae to Alydidae was at best curiosity; Henry (1988a, b) was quite correct to mention it and then to restore Arhaphinae to Largidae. See another account of these movements of Bliven in Stehlík (2011).

In creating the higher-group category Arhaphinae, Bliven (1973) did not consider the Old World forms in Largidae. Having done so, we believe (as did Bliven) that *Arhaphe* and its relatives form a group equivalent to *Largus* and its relatives, but, *contra* Bliven, we believe that each group should be of tribal rank (see the key above).

A key to the genera of Arhaphini is given below, with the distributions from Stehlík (2013a).

1. Body not elongated; body black or reddish brown, with some white, yellow, or orange markings; pronotal humeri without sublateral spines; metathoracic scent glands elongated 2
   Body elongated, all black or dark brown; pronotal humeri with sublateral spines; metathoracic scent glands elongated or not 3

2. Head shorter than pronotum (United States north to Illinois and Virginia and into Central America .... *Arhaphe* Herrich-Schaeffer (Fig. 17.7).
   Head longer than pronotum (Guatemala) .... *Pararhaphe* Henry
3. Scutellum with blunt spine; eyes in lateral view not higher than vertex of head; metathoracic scent glands elongated (South America, including Bolivia (Schaefer 2000) and Paraguay (Arnold 2013)...........*Thaumastaneis* Kirkaldy and Edwards
   Scutellum without blunt spine; eyes in lateral view higher than vertex of head; metathoracic scent glands rounded (Peru) .... *Vasarhelyecoris* Brailovsky & Barrera (Fig. 17.8).

The bodies of the Arhaphini are brachypterous, elongate, black or reddish brown, or black with white-to-orange markings. In *Arhaph*e and *Pararhaphe*, the hemelytra are pale orange to white, with a black medial portion and a subterminal band of black to brown. The other two genera are black and white. The head is very prominent, sometimes with a distinct neck, and their eyes are also prominent, but never stylate. The species range in size from 5–10 mm (*Arhaphe* and *Pararhaphe*) to 8–13 mm (*Thaumastaneis*, females larger) and 16–17 mm (*Vasarhelyecoris*).

Members of all four genera more or less resemble ants. Members of *Arhaphe* and *Pararhaphe* are less antlike than are those of *Thaumastaneis* and *Vasarhelyecoris*. The last two are shaped like ants, whereas *Arhaphe* and *Pararhaphe* achieve their antlike appearances through color pattern: contrastingly pale spots or bands so placed as to resemble the head-thorax constriction and thus the petiole of ants.

Thaumastaneis and *Vasarhelyecoris* also have pronotal spines, which the other two genera lack. The former bears a spine also on the scutellum; similar spines are found elsewhere, in ant-mimicking micrelytrines (Alydidae) and colobathristids. A similar structure occurs in the immature (and myrmecomorphic) forms of the larguline (Largulini) *Largulus parallelus* Hussey, but not in the (nonmyrmecomorphic) adults (Schaefer 2000).

17.4.2 Diagnosis

Arhaphe and *Pararhaphe* may most easily be distinguished from *Thaumastaneis* and *Vasarhelyecoris* by having some white or yellow structural (not setal) coloration, and by lacking the "neck" of the last two genera, *Vasarhelyecoris* most closely resembles *Thaumastaneis*, but is larger and lacks the scutellar spine.

*Pararhaph*e: Henry (1988b) changed the more commonly used name *Japetus*, because of the homonymy of Stål's name (1863, of a Hemiptera: Fulgoridae) with the more recent *Japetus* of Distant (1883).

17.5 Habitats and Food Plants of Larginae

Nearly nothing is known about habitats, and even less about food plants, particularly for the tribe Arhaphini. But some isolated, and often unsubstantiated, comments have been published on Arhaphini. There are more, and somewhat better substantiated, comments on *Largus* (Largini), a genus of more species.

Arhaphe species are not uncommon, scurrying like ants in and under dried leaves (Barber 1910, 1911; de la Torre-Bueno 1942; Halstead 1972b; Brailovsky 1981), occasionally with water nearby; not much else is known about them. In particular, nearly nothing is known about food plants. de la Torre-Bueno (1942) wrote that he brought four adults of *Arhaphe arguta* (Bliven 1956) onto Bermuda grass (*Cynodon dactylon,* Gramineae), "on which they proceeded to feed, as I carefully noted." However, they may have been only getting water because, usually on the ground, it is quite possible arhaphines actually feed on fallen seeds.

It is probably significant that, although abundant locally, there is no mention of the Arhaphini in Ahmad and Schaefer (1987), which is a list of food plants of Pyrrhocoroidea. There are some references to the other tribe, Largini. In general, several *Largus* appear to feed on seeds, in fruits, or on the ground. *Largus* may also feed on dead insects (Myers 1927, P. L. Mitchell personal communication) and even dog food (P. L. Mitchell personal communication). *Largus* eggs are laid on the ground. *Acinocoris* and *Fibrenus* appear to feed on seeds on plants (van Doesburg 1966).

Although eggs of *Largus* are laid on the ground, the nymphs seem usually to be upon plants and usually on flowering or fruiting plants. This may also be the case with other Largini.

Like other phytophagous heteropterans, many of these largines probably get water and, especially, proteins and salts from bird droppings and dead bodies (and dog food) (Schaefer 1980; Schaefer and Mitchell 1983).

17.6 Concluding Remarks

Overall, these bugs are of a size and conspicuousness that many are easily collected. As a result, the taxonomy of Pyrrhocoroidea has been well worked out, thanks to van Doesburg (1968) (*Dysdercus*) and several papers by Brailovsky and Stehlík (Largidae). However, we need to know the biologies, especially for the Largidae. For *Dysdercus*, it would be very useful to know more about how many species, and where and when, they feed on cotton, which was a worse problem years ago than now (Schaefer and Ahmad 2000). With global warming, this problem may occur again.

Acknowledgments I thank heartily Dr. Jane O'Donnell for the lovely photographs and also heartily to Ms Virge Kask, who did fine and recondite work on them.

References

Ahmad M, Khan NH (1980) Chemical repellants for *Dysdercus koenigii* (Fabr.). Indian J Entomol 42:820–821

Ahmad I, Schaefer CW (1987) Food plants and feeding biology of the Pyrrhocoroidea (Hemiptera). Phytophaga 1:75–92

Anonymous (1961) Insects not known to occur in the United States. Peruvian cotton stainer (*Dysdercus peruvianus* Guérin). Crop Econ Insect Rep US Dept Agric 11:1–2

Arnold K (2013) Fragmenta Heteroptera Neotropica XIV (Insecta: Hemiptera: Heteroptera). Edessana 3:25–59

Barber HG (1910) Some Mexican Hemiptera-Heteroptera new to the fauna of the United States. J NY Entomol Soc 18:34–39

Barber HG (1911) Descriptions of some new Hemiptera-Heteroptera. J NY Entomol Soc 19:23–31

Beccari F, Gerini V (1970) Catalogo delle specie appartenenti al genere *Dysdercus* Boisduval (Rhynchota, Pyrrhocoridae). Riv Agric Subtrop Trop 64:1–67

Bliven BP (1956) New Hemiptera from the western states with illustrations of previously described species and new synonymy on the Psyllidae. B. P. Bliven, Eureka (privately published)

Bliven BP (1973) A third paper on Hemiptera associated with Pyrrhocoridae. Occident Entomol 1:123–133

Brailovsky H (1981) *Arhaphe* H.S., descripcion de nuevas especies (Hemiptera: Heteroptera: Largidae). Folia Entomol Mex 48:81–109

Chatterjee NB, Raychaudhuri DN (1960) Bionomics of *Dysdercus koenigii* (Fabr.). Indian Agric 4:104–112

Chatterjee NB, Raychaudhuri DN (1976) Some aspects of host-specificity in the cotton stainer *Dysdercus koenigii* Fabr. (Pyrrhocoridae: Heteroptera). In: Ananthakrishnan TN (ed) Insects and host specificity. Symposium on problems of host-specificity in insects. Loyola College, Madras, pp 47–49

China WE (1954) Notes on the nomenclature of the Pyrrhocoridae (Hemiptera Heteroptera). Entomol Mon Mag 110:188–189

Costa Lima AM (1940) Insetos do Brasil. 2° Tomo. Capítulo 22. Hemipteros. Escola Nacional de Agronomia, Série Didática N. 3, Rio de Janeiro, Brasil

Costa Lima AM, Guitton N, Ferreira OV (1962) Sobre as espécies americanas do gênero *Dysdercus* Boisduval (Hemiptera, Pyrrhocoridae, Pyrrhocorinae). Mem Inst Oswaldo Cruz 60:21–58

Davis NT (1975) Hormonal control of flight muscle histolysis in *Dysdercus fulvoniger*. Ann Entomol Soc Am 68:710–714

de la Torre-Bueno J (1942) Notes on *Arhaphe cicindeloides* and Walker and *Japetus mimeticus* Barber. Bull Brooklyn Entomol Soc 37:68–69

Dellape PM, Melo MC (2012) *Theraneis vittata* Spinola, 1837 (Hemiptera: Heteroptera: Largidae): first record from Argentina. Check List 8:237–238

Dingle H, Arora G (1973) Experimental studies of migration in bugs of the genius *Dysdercus*. Oecologia 12:119–140

Distant WL (1883) Insecta. Rhynchota. Hemiptera-Heteroptera, vol 1. Biologia Centrali-Americana, London

Freeman P (1947) A revision of the genus *Dysdercus* Boisduval (Hemiptera, Pyrrhocoridae), excluding the American species. Trans R Entomol Soc 98:373–424

Froeschner RC (1985) Synopsis of the Heteroptera or true bugs of the Galapagos Islands. Smithson Contrib Zool 407:1–84

Halstead TF (1972a) A review of the genus *Arhaphe* Herrich-Schäffer (Hemiptera: Largidae). Pan-Pac Entomol 48:1–7

Halstead TF (1972b) Notes and synonymy in *Largus* Hahn with a key to United States species (Hemiptera: Largidae). Pan-Pac Entomol 48:246–248

Henry TJ (1988a) Family Pyrrhocoridae Fieber, 1860, the cotton stainers. In: Henry TJ, Froeschner RC (eds) Catalog of the Heteroptera, or true bugs, of Canada and the Continental United States. E.J. Brill, Leiden, pp 613–615

Henry TJ (1988b) Family Largidae Amyot and Serville, 1843, the largid bugs. In: Henry TJ, Froeschner RC (eds) Catalog of the Heteroptera, or true bugs, of Canada and the Continental United States. E.J. Brill, Leiden, pp 159–165

Hussey RF (1929) General catalogue of Hemiptera. Fascicle III. Pyrrhocoridae. Smith College, Northampton, pp 1–144

Janzen DH (1972) Escape in space by *Sterculia apetala* seeds from the bug *Dysdercus fasciatus* in a Costa Rican deciduous forest. Ecology 53:350–361

Kamble ST (1978) Bionomics of *Dysdercus koenigii* Fabr. (Hemiptera: Pyrrhocoridae). J NY Entomol Soc 79:154–157

Kumar R (1968) Aspects of the morphology and relationships of the superfamilies Lygaeoidea, Piesmatoidea and Pyrrhocoroidea (Hemiptera: Heteroptera). Entomol Mon Mag 103:251–261 (dated 1967, published 1968)

Lariviére MC, Larochelle A (2004) Fauna of New Zealand (Ko te Aitanga Pepeke o Aotearoe). Heteroptera (Insecta: Hemiptera): catalogue. Menaaki Whenue Press, Lincoln 50:1–330

Lariviére MC, Larochelle A (2014) Checklist of the New Zealand Heteroptera (Insecta: Hemiptera): an update based on the 2004 to 2013 literature. Zootaxa 3755:347–367

Moizant RC, Téran JB (1970) *Dysdercus maurus* Distant (Hemiptera, Pyrrhocoridae) sobre *Citrus* spp. Agron Trop (Maracay) 20:267–269

Myers JG (1927) Ethological observations on some Pyrrhocoridae of Cuba (Hemiptera-Heteroptera). Ann Entomol Soc Am 20:279–300

Qadri SS, Ahmed I [sic: Ahmad] (2001) Redescription of the only red cotton stainer's [sic] species *Dysdercus koenigii* (F.) (Hemiptera: Pyrrhocoridae) found in Pakistan with reference to its genitalia and their bearing on its phylogenetic relationships. Pak J Entomol 16:25–28

Robertson IAD (1977) Records of insects taken at light taps in Tanzania. IV – Seasonal changes in catches and effect of the lunar cycle on insects of the genus *Dysdercus* (Heteroptera: Pyrrhocoridae). Ministry of Overseas Development, Centre Overseas Pest Res, London, UK, Misc Rep 30, 8 pp

Schaefer CW (1964) The morphology and higher classification of the Coreoidea (Hemiptera-Heteroptera): Parts I and II. Ann Entomol Soc Am 57:670–684

Schaefer CW (1966) An abdominal anomaly in a coreid (Hemiptera: Heteroptera), with some phylogenetic notes. Bull Brooklyn Entomol Soc 49(50):77–80

Schaefer CW (1972) Degree of metathoracic scent-gland development in the trichophorous Heteroptera (Hemiptera). Ann Entomol Soc Am 65:810–821

Schaefer CW (1980) The host plants of the Alydinae, with a note on heterotypic feeding aggregations (Hemiptera: Coreoidea: Alydidae). J Kansas Entomol Soc 53:115–122

Schaefer CW (1999) The higher classification of the Alydidae (Hemiptera: Heteroptera). Proc Entomol Soc Wash 101:94–98

Schaefer CW (2000) Systematic notes on Larginae (Hemiptera: Largidae). J NY Entomol Soc 108:130–145

Schaefer CW (2013) A new species of *Dysdercus*: *Dysdercus stehliki* sp.nov. (Hemiptera: Heteroptera: Pyrrhocoridae) from Brazil. Acta Musei Moraviae 98:381–390

Schaefer CW, Ahmad I (2000) Cotton stainers and their relatives (Pyrrhocoroidea: Pyrrhocoridae and Largidae). In: Schaefer CW, Panizzi AR (eds) Heteroptera of economic importance. CRC Press, Boca Raton, pp 271–308

Schaefer CW, Mitchell PL (1983) Food plants of the Coreoidea (Hemiptera: Heteroptera). Ann Entomol Soc Am 76:591–615

Schaefer CW, Pupedis RJ (1981) A stridulatory device in certain Alydinae (Hemiptera: Heteroptera: Alydidae). J Kansas Entomol Soc 54:143–152

Schaefer CW, Stehlík JR (2013) Caribbean Sea Region Pyrrhocoroidea (Hemiptera: Pyrrhocoridae, Largidae). Neotrop Entomol 42:372–383

Schaefer CW, Schaffner JC, Ahmad I (1989) The *Alydus*-group, with notes on the alydine genital capsule (Hemiptera: Heteroptera: Alydidae). Ann Entomol Soc Am 82:500–507

Schouteden H (1912) Les hémiptères parasites des cotonniers en Afrique. Rev Zool Africaine 1:297–321

Shukla GS, Upadhyaya VK (1972) Relative food preferences of *Dysdercus koenigii* (Hemiptera: Pyrrhocoridae). Ann Entomol Soc Am 65:762–763

Southwood TRE (1956) The structure of the eggs of the terrestrial Heteroptera and its relationship to the classification of the group. Trans R Entomol Soc 108:163–220

Squire FA (1939) Observations on cotton stainers (*Dysdercus*) in the West Indies. Bull Entomol Res 30:289–291

Srivastava RP, Gupta RS (1971) *Dysdercus cingulatus* Fabr., a new pest of wheat—a possible case of host cross-over. Indian J Entomol 33:354

Stehlík JL (1965a) Mission Zoologique de l'I.R.S.C. en Afrique Orientale (P. Basilewsky-N. Leleup, 1957) Pyrrhocoridae (Het.). Acta Musei Moravia (Scientia Naturales) 50:211–252

Stehlík JL (1965b) Pyrrhocoridae and Largidae collected by E.S. Brown on Solomon Islands (Heteroptera). Acta Musei Moraviae (Scientia Naturales) 50:253–292

Stehlík JL (2007) New combinations in the genus *Rosaphe* Kirkaldy et Edwards, 1902 (=*Astemma* auct., *nec* Le Peletier et Serville, 1825) (Heteroptera: Largidae: Larginae). Acta Musei Moraviae, Scientiae Biologicae (Brno) 92:111–113

Stehlík JL (2011) Redescription of *Pararhaphe* and review of *Arhaphe* (Hemiptera: Heteroptera: Largidae) of America north of Mexico. Zootaxa 3058:35–54

Stehlík JL (2013a) Review and reclassification of the Old World genus *Phyospelta* (Hemiptera: Heteroptera: Pyrrhocoridae). Acta Entomol Musei Nationalis Pragae 53:505–584

Stehlík JL (2013b) A description of four new species of *Dindymus* (Hemiptera: Heteroptera: Pyrrhocoridae) from south-eastern Asia. Acta Musei Moraviae Scientiae Biol 98:41–51

Stehlík JL, Brailovsky H (2011) Two new genera of the tribe Largulini (Hemiptera: Heteroptera: Largidae) from Greater Antilles. Acta Entomol Mus Nat Pragae 51:449–456

Stehlík JL, Jindra Z (2007) Largulini—a new tribe of Larginae from Jamaica (Heteroptera, Largidae). Entomol Basiliensa et Collectionis Frey 29:13–20

Svádová KH, Exnervová A, Kopečová M, Štys P (2010) Predator dependent mimetic associations: do passerine birds avoid Central European red-and-black Heteroptera? Eur J Entomol 107:349–355

Sweet MH (2006) Justification for the Aradimorpha as an infraorder of the suborder Heteroptera (Hemiptera, Prosorrhyncha) with special reference to the pregenital abdominal structure. Denisea 19:225–248

van Doesburg Jr PH (1966) Heteroptera of Suriname: I. Largidae and Pyrrhocoridae. Stud Fauna Suriname Guyanas 9:1–60

van Doesburg PH Jr (1968) A revision of the New World species of *Dysdercus* Guérin Méneville (Heteroptera, Pyrrhocoridae). Zool Med 97:1–215

Zrzavý J, Nedvěd O (1997) Phylogeny of the New World *Dysdercus* (Insecta: Hemiptera: Pyrrhocoridae) and evolution of their colour patterns. Cladistics 13:109–213

Chapter 18
Broad-Headed Bugs (Alydidae)

Antônio R. Panizzi and Carl W. Schaefer

Abstract The broad-headed bugs (Alydidae) are divided into two subfamilies, Alydinae and Micrelytrinae, each divided into two tribes, Daclerini and Alydini, and Micrelytrini and Leptocorisini, respectively. The family has 53 genera and about 250 species; in the Neotropics, there are 21 genera. Alydids are small (8–20 mm), slender, with a triangular head; nymphs of alydines mimic ants, the adults of some Micrelytrini also mimic ants. The most studied species in the Neotropics is the alydine *Neomegalotomus parvus* (Westwood), usually associated with legumes, and may be a pest on soybean. Other common genera include *Hyalymenus* Amyot & Serville, *Stenocoris* Burmeister, *Cydamus* Stål, and *Trachelium* Herrich-Schäffer. Studies on taxonomy and bioecology on alydids of the Neotropics are needed.

18.1 Introduction

Alydidae Amyot and Serville, 1843, were treated as a subfamily of the family Coreidae and even as a tribe (Schaffner 1964); now it has been treated as a family, together with Coreidae, Rhopalidae, Hyocephalidae, and Stenocephalidae, in the superfamily Coreoidea (Schaefer 1964).

This family contains 53 genera and approximately 250 species, mostly tropical or subtropical, in all regions of the world. There are only two genera that span both the Old and the New World, *Alydus* and *Megalotomus*. These genera are Holarctic, but *Alydus* extends from Alaska through Canada into Mexico (Brailovsky and Flores 1979; Froeschner 1988; Maw et al. 2000).

The genera of Alydinae have been revised by Schaffner (1964; 22 species worldwide); the world genera of the subfamily Micrelytrinae, tribe Leptocorisini, were

Carl W. Schaefer: Author deceased at the time of publication

A.R. Panizzi (✉)
Laboratório de Entomologia, Embrapa Trigo, Caixa Postal 3081, Passo Fundo, RS 99001-970, Brazil
e-mail: antonio.panizzi@embrapa.br

A.R. Panizzi, J. Grazia (eds.), *True Bugs (Heteroptera) of the Neotropics*, Entomology in Focus 2, DOI 10.1007/978-94-017-9861-7_18

revised by Ahmad (1965; seven species worldwide). There has been no list of the subfamily Micrelytrinae, tribe Micrelytrini; our data came from Dolling's "Catalogue of the Palaearctic Region" (2006) (eight genera, Palearctic), which may be undercounted, especially for tropical Asian and African genera. The Alydinae contains two tribes, the Alydini and the Daclerini, the latter with only one genus, *Daclera*; this latter tribe has not been accepted by some workers and is included in Schaffner (1964) as a genus. *Acestra* has also been considered a Micrelytrinae tribe (Li and Zheng 1993); both suggested tribes (Daclerini and Acestrini) are from the Old World.

Schaefer (1999) worked out the higher classification of the family Alydidae. It now has two subfamilies (Alydinae and Micrelytrinae). The Alydinae contains two tribes (Daclerini and Alydini); Micrelytrinae also has two tribes, the Micrelytrini and the Leptocorisini, both distributed worldwide.

Within the Neotropics, there are eight genera of Alydinae and 13 genera of Micrelytrinae in the tribes Micrelytrini (11 species) and Leptocorisini (2 species). These genera have been keyed in Schaefer (2004), as well as the subfamilies, tribes, and subtribes (of Leptocorisini).

18.2 General Characteristics and Diagnosis

Species of Alydidae are fairly small, ranging from 8 to 20 mm. Alydids are narrow, i.e., they are much longer than wide, especially in many of the Leptocorisini and some of the Micrelytrini. The dorsa of the head and thorax lack ridges or bumps (except for spines on some Micrelytrini) and are flattened. The head is triangular and resembles the heads of ants. Some of the adults of Micrelytrini (genera *Cydamus* and *Trachelium*), as well as the nymphs of Alydinae, are antlike, presumably to ward off predators. This ant mimicry (myrmecomorphy) has been reported for several species of Alydidae by several authors for the past 80+ years (Nicholson 1927; Mathew 1935; Costa Lima 1940; Kormilev 1953; Kumar 1966; Schaefer 1972; Elzinga 1978; Sisson 1980; Oliveira 1985). The family's common name, broad-headed bugs, actually refers only to the Alydinae: their "heads are indeed broader between the eyes than are those of other coreoids" (Schaefer 2004).

Additional diagnostic characters include bucculae very short, antennae dorsally inserted with segment 1 not constricted at base, ocelli not placed on elevations, corium elongated on costal margin, metathoracic scent glands auricles well developed, and tibia nonsulcate (Schuh and Slater 1995).

18.3 Classification and Diversity

The family Alydidae was divided into three subfamilies [Leptocorisinae, Alydinae, and Micrelytrinae (Ahmad 1965)]. More recently, the family was divided into two subfamilies (Alydinae and Micrelytrinae), this last containing two tribes, Leptocorisini (a former subfamily of Ahmad's classification) and Micrelytrini

(Schaefer 1999). A key for the two subfamilies, including the tribes for the last, is presented below according to Schaefer (2004), with slight modifications.

1. Hind femur bearing spines; trichobothria of abdominal sternum five arranged in a row lateral or anterior to spiracle .... Alydinae
– Hind femur without spines; trichobothria of abdominal sternum five arranged in a triangle posterior to spiracle.. Micrelytrinae (2)
2. Second rostral segment shorter than 3rd and 4th together; 3rd rostral segment more than half long as 4th; evaporative area of metathoracic scent gland smooth.. Leptocorisini
– Second rostral segment longer than 3rd and 4th together; 3rd rostral segment less than half long as 4th; evaporative area of metathoracic scent gland ridged .... Micrelytrini

The complete higher classification of Alydidae based on Schaefer (1999) is shown below:

Family Alydidae Amyot and Serville

Subfamily Alydinae Amyot and Serville

Tribe Alydini (New and Old World)
Tribe Daclerini (Old World)

Subfamily Micrelytrinae Stål

Tribe Micrelytrini Stål (New and Old World)
Tribe Leptocorisini Stål

Subtribe Leptocorisidi Stål (New and Old World)
Subtribe Noliphidi Ahmad (Old World)

The most comprehensive taxonomic study of the Alydidae family is by Schaffner (1964), Ahmad's revision of the Leptocorisini (1965), and Kormilev's (1953) revision of part of the Micrelytrini: these are the only family-group taxonomic revisions. Much work needs to be done on revisions of other family groups and genera and also on alydids' biology. For the Neotropics (South America), Froeschner (1981) keyed the subfamilies and genera.

18.3.1 Alydinae

This subfamily contains individuals in all major zoogeographic zones. The best-known genera include *Alydus* F., *Hyalymenus* Amyot and Serville, *Megalotomus* Fieber, *Neomegalotomus* Schaffner and Schaefer, and *Riptortus* Stål. In the Neotropics, *Hyalymenus* and *Neomegalotomus* include the most common and known species (Grazia et al. 2012).

The Alydinae prefer legumes (Schaefer 1972, 1980; Schaefer and Mitchell 1983; Panizzi 1988; Santos and Panizzi 1998a, b), and some species are considered pests on leguminous crops (see section 4 on main species). There are also several references to different species of alydines being attracted to and feeding on carrion, vertebrate fecal matter, and cow urine and ammonia (Schaefer 1980; Adler and Wheeler 1984; Ventura and Panizzi 2000; Silva et al. 2010). These nitrogen-rich materials attract alydines, and it has been speculated that these bugs might require higher concentration of nitrogen than do other bugs, based on this behavior and their association with nitrogen-rich plants (i.e., legumes) (Panizzi et al. 2000).

18.3.2 Micrelytrinae

In this subfamily, several genera (*Cydamus*, *Darmistus*, *Esperanza*) reach the southwestern states of the USA. *Protenor* occurs throughout the USA and into Canada (Froeschner 1988). *Esperanza texana* Barber occurs farther north and its range may be expanding (Johnston 1927; Hussey 1948; Froeschner 1980). With global warming, *Esperanza*'s distribution is expanding, not only to the north but to the south. It now occurs as far south as Costa Rica (Schaefer 2003).

In the Neotropics, the main genera in Micrelytrinae (Micrelytrini) are *Stenocoris* Burmeister, *Cydamus* Stål, and *Trachelium* Herrich-Schäffer (Grazia et al. 2012). We do not know what Micrelytrini feed on. Leptocorisini feed on grasses; in the Orient, some species may become pests on rice.

18.4 General Biology

Data on the biology of alydids are available for those species with major or minor economic importance, such as species of *Leptocorisa* [*L. acuta* (Thunberg) and *L. oratorius* (F.)], associated with rice throughout Asia; species of *Riptortus* [*R. dentipes* (F.), *R. linearis* (F.), *R. pedestris* (F.), and *R. serripes* (F.)] associated with legume crops in Asia, Africa, and Australia; and *Alydus pilosulus* Herrich-Schäffer and *Megalotomus quinquespinosus* Say, associated with legumes in the Nearctic region (Canada and USA) (references in Panizzi et al. 2000).

For the Neotropical alydids, not much data is available on their biology. Most information has been published on the species which have some economic importance, such as *Neomegalotomus parvus* (Westwood), pest of legume crops, particularly in Brazil (section 5.1); other species studied in greater detail are those on the genus *Hyalymenus* which show mimetic association with ants (Oliveira 1985).

18.5 Main Species

Most members of Alydidae in the world are not pests, and in general their biology, ecology, and host plants are largely unknown. There is only one species which is considered common and is of somewhat economic importance in the Neotropics. The second species in this genus is less common, and they are presented below.

18.5.1 Neomegalotomus parvus (Westwood)

The new genus *Neomegalotomus* Schaffner & Schaefer was erected to accommodate the Neotropical species formerly included in the genus *Megalotomus* (Schaffner & Schaefer 1998; Schaefer & Panizzi 1998). *N. simplex* (Westwood), *N. latifascia* (Berg), and *N. pallescens* Stål were all synonymized with *N. parvus* (Schaefer & Ahmad 2008).

N. parvus (Westwood) occurs in the Neotropics between 24° N and 30° S longitude; the northernmost are from central Mexico on the west and St. Vincent and Barbados on the east; the southernmost distribution is northern Argentina, southern Brazilian states, and Uruguay (Schaefer and Ahmad 2008); these authors do not include Rio Grande do Sul, the southernmost state of Brazil, where specimens of *N. parvus* have been collected recently (in Passo Fundo, 28 ° S, AR Panizzi, unpublished).

N. parvus is, by far, the most studied species of alydid in the Neotropics, particularly in Brazil. It was first reported on *Crotalaria* sp. in Rio de Janeiro state (Costa Lima 1919), and other hosts include common bean, soybean, cotton, tomato, lupin, pigeon pea, and lablab (Panizzi 1988; Chandler 1989; Santos and Panizzi 1998a).

Eggs (Fig. 18.1) are laid singly or in groups. On soybean, eggs are laid on the upper third of the plant, preferably on the lower side of leaves, close to the midrib (Panizzi et al. 1996). On pigeon pea, *Cajanus cajan* (L.), eggs are laid preferably in-between seeds (crevices) of mature pods (Ventura and Panizzi 2000, 2003). Interesting to mention that during maintenance of *N. parvus* colony in the laboratory, in seven occasions, females laid eggs on the body of conspecifics, and on three events, eggs remained on the insect body until they hatched (Panizzi and Santos 2001).

Nymphs mimic ants (Fig. 18.2) and are darkish. Adult males are pale brown, with a whitish band along both sides of the thorax; females are entirely darkish brown (Fig. 18.3). Body length is ca. 10 mm (Costa Lima 1919, Paradela Fo et al. 1972).

Several studies on the nymphal and adult biology, population dynamics, oviposition, and feeding preferences of *N. parvus* on selected cultivated and noncultivated legumes (e.g., soybean, pigeon pea, lablab, green bean, indigo, lupin, and cowpea) have been published (Panizzi 1988; Ventura and Panizzi 1997; Santos and Panizzi 1998a; Ventura et al. 2000a, b; Ventura and Panizzi 2003, 2004, 2005). In general,

Fig. 18.1 Eggs of *Neomegalotomus parvus* on a soybean stem (**a**) and on crevices of a pigeon pea pod mature (**b**) (Courtesy of JJ Silva)

Fig. 18.2 Ant-mimic nymph of *Neomegalotomus parvus* on a soybean pod (Courtesy of JJ Silva)

Fig. 18.3 Adult (female) of *Neomegalotomus parvus* on green bean pods (Courtesy of JJ Silva)

this alydid is most associated with mature soybean and pigeon pea plants. In the laboratory, colonies of *N. parvus* are easily kept using mature seeds of pigeon pea, to allow the use of this bug for research and teaching purposes (Ventura and Panizzi 1997).

Although considered a minor pest, this alydid damages common beans by transmitting the yeast spot disease caused by *Nematospora coryli* Peglion and by causing seedling mortality (Paradela Fo et al. 1972; Chandler 1984, 1989). It also can cause reduction in seed vigor and viability in soybean (Santos and Panizzi 1998b). In Brazil, *N. parvus* adults are parasitized by at least three species of tachinids (Santos and Panizzi 1997).

18.5.2 Neomegalotomus rufipes (Westwood)

This second species in the genus *Neomegalotomus*, *N. rufipes*, occurs from southeastern Florida south to Central America and the West Indies islands in the Caribbean, including Antigua, Bahamas, Cuba, Dominican Republic, Grand Cayman, Grenada, Guadeloupe, Jamaica, Martinique, Puerto Rico, St. Lucia, and St. Martin; despite its affinity to legumes, apparently it is not a pest in these islands (Schaefer and Ahmad 2008). These authors provide a key to species of *Neomegalotomus*, shown below:

1a. Metathoracic scent gland auricle usually somewhat flattened and separation between anterior and posterior parts of auricle shallow; median protuberance of ventral rim of male's genital capsule pointing medially .... *Neomegalotomus parvus* (Westwood)
1b. Metathoracic scent gland auricle rounded, convex, separation between anterior and posterior parts deep; median protuberance of male's genital capsule pointing .... *Neomegalotomus rufipes* (Westwood)

The reference to the occurrence of *Neomegalotomus rufipes* in São Paulo state in Brazil on cotton, on sunn hemp, *Crotalaria juncea* L., on the legume *Macroptilium heterophyllum* (Humb. and Bonpl. ex Willd.), and on the ornamental plant *Asclepias curassavica* L. mentioned by Silva et al. (1968) is probably a mistake, because this species is not known to occur in this area.

18.6 Secondary Species

There are several genera/species of alydids in the Neotropics. Froeschner (1981) keyed the genera of South American alydids, and he included the following: in Alydinae, *Apidaurus* Stål, *Alydus* F., *Hyalymenus* Amyot and Serville, *Burtinus* Stål, and *Megalotomus* Fieber (now *Neomegalotomus* Schaffner and Schaefer); in Micrelytrinae (Leptocorisini), *Stenocoris* Burmeister and *Lyrnessus* Stål; and in Micrelytrinae (Micrelytrini), *Calamocoris* Breddin, *Cydamus* Stål, *Trachelium* Herrich-Schäffer, *Bactrophya* Breddin, and *Bactrocoris* Kormilev.

Among these genera, species worth of mention are *Hyalymenus pulcher* (Stål), *Stenocoris americanus* Ahmad, *S. fabricii* Ahmad, *S. tipuloides* (De Geer), *Bactrophya aequatoriana* Breddin, *Calamocoris erubescens* Breddin, *C. nigrolimbatus* Breddin, and *Cydamus inauratus* Distant. *Cydamus pictipes* (Stål) is reported on grasses and on castor bean, *Ricinus communis* L. in São Paulo state, Brazil (Silva et al. 1968).

In Brazil, two species of *Hyalymenus* have been studied in detail regarding myrmecomorphy (ant mimicry): *Hyalymenus limbativentris* Stål, which is usually associated with solanaceous plants, feeding on fruits, and *Hyalymenus tarsatus* (F.), which is found on plants of several (five) families, feeding on flower and/or fruits (Oliveira 1985). This last species is reported on the legume *Cassia occidentalis* in Brazil (Silva et al. 1968).

18.7 Concluding Remarks

Alydids in the Neotropical region are largely unknown, and their biology, except for *Neomegalotomus parvus* (Westwood), which has occasional a pest status, is little studied. Data in the literature are scant, and most information in the literature is restrained to lists of species (catalog type of publication) and taxonomic keys for subfamilies and genera (e.g., Schaffner 1964; Froeschner 1981; Schaefer 2004; Schaefer and Ahmad 2008). Moreover, information (published and not) seems to be confused, such as the reference to *Leptocorisa filiformis* (F.) occurring in coffee plantations in São Paulo state, Brazil (Silva et al. 1968), and *Leptocorisa* sp. occasionally infesting soybean and rubber plant, *Hevea brasiliensis* (Willd. ex Adr. de Juss.) also in São Paulo, Brazil (AM Faria, pers comm to ARP). The genus *Leptocorisa* is distributed in the Orient and Australia (Ahmad 1965, Livermore et al. http://Coreoidea.SpeciesFile.org) and is not supposed to occur in the Neotropics. This example, clearly illustrates the strong need to revise the taxonomic status of the species of alydids in the Neotropics. In addition, research work on their biology, particularly on their life history in nature, to reveal their association with host plants either cultivated or not is needed.

References

Adler PH, Wheeler AG Jr (1984) Extra-phytophagous food sources of Hemiptera-Heteroptera: bird droppings, dung, and carrion. J Kansas Entomol Soc 57:21–27

Brailovsky H, Flores RZ (1979) Contribución al estudio de los Hemiptera- Heteroptera de México: XVII. Revision de la familia Alydidae Amyot y Serville. An Inst Biol Univ Nac Autón México (Ser Zool) 50:255–339

Ahmad I (1965) The Leptocorisinae (Heteroptera: Alydidae) of the world. Bull British Museum Nat Hist Entomol Suppl 5:1–156

Chandler L (1984) Crop life table studies of the pests of beans (*Phaseolus vulgaris* L.) at Goiânia, Goiás. Rev Ceres 31:284–298

Chandler L (1989) The broad-headed bug, *Megalotomus parvus* (Westwood) (Hemiptera: Alydidae), a dry season pest of beans in Brazil. Annu Rep Bean Improv Coop 32:84–85

Costa Lima AM (1919) Nota sobre o mimetismo da nympha do *Alydus* (*Megalotomus*) *pallescens* com formiga e considerações relativa à espécie *Galeottus formicarius* (Hemiptera-Coreidae). Arch Esc Sup Agric Niterói 4:5–8

Costa Lima AM (1940) Insetos do Brasil. Hemiptera, vol 2. Ministerio Da Agricultura, Rio de Janeiro

Dolling WR (2006) Family Alydidae Amyot and Serville, 1843. In: Aukema B, Rieger C (eds) Catalogue of the Heteroptera of the Palaearctic Region, vol 5. Netherland Entomological Society, Amsterdam, pp 28–42

Elzinga RJ (1978) Fundamentals of entomology. Prentice-Hall, Englewood Cliffs

Froeschner RC (1980) Is *Esperanza texana* Barber (Hemiptera: Alydidae) extending its range? Entomol News 91:92

Froeschner RC (1981) Heteroptera or true bugs of Ecuador: a partial catalog. Smithson Contr Zool 322:1–147

Froeschner RC (1988) Family Alydidae Amyot and Serville, 1843. The broad-headed bugs. In: Henry TJ, Froeschner RC (eds) Catalog of the Heteroptera, or true bugs, of Canada and the Continental United States. E.J. Brill, Leiden, pp 4–11

Grazia J, Cavichioli RR, Wolff VRS, Fernandes JAM, Takiya DM (2012) Hemiptera Linnaeus, 1758. In: Rafael JA, Melo GAR, Carvalho CJB, Casari AS, Constantino R (eds) Insetos do Brasil diversidade e taxonomia. Holos Editora, Ribeirão Preto, pp 347–405

Hussey RF (1948) Esperanza Texana [sic] in Florida (Hemiptera, Coriscidae). Bull Brooklyn Entomol Soc 43:115

Johnston HG (1927) *Esperanza texana* Barber found in Louisiana (Hemiptera Coreidae). Bull Brooklyn Entomol Soc 22:221

Kormilev NA (1953) Revisión de Micrelytrinae Stål de la Argentina, com descripción de um género y siete espécies nuevos de Argentina, Brasil y Bolivia (Hemiptera, Coreidae). Rev Soc Entomol Arg 16:49–66

Kumar R (1966) Studies on the biology, immature stages, and relative growth of some Australian bugs of the superfamily Coreoidea (Heteroptera). Austr J Zool 14:895–991

Li X-Z, Zheng L-Y (1993) Preliminary study on the phylogeny of Alydidae (Hemiptera, Coreoidea). Acta Zootax Sinica 18:330–343

Livermore LJR, Lemaître VA, Dolling WR, Webb MD. Coreoidea species file online. Version 5.0/5.0. [26/07/2014]. http://Coreoidea.SpeciesFile.org

Mathew AP (1935) Tansformational deceptive resemblance as seen in the life-history of a plant bug (*Riptortus pedestris*), and of a mantis (*Evantissa pulchra*). J Bombay Nat Hist Soc 37:803–813

Maw HEL, Foottit RG, Hamilton KGA, Scudder GGE (2000) Checklist of the Hemiptera of Canada and Alaska. NRC Research Press, Ottawa

Nicholson AJ (1927) A new theory of mimicry in insects. Austr Zool 5:10–104

Oliveira PS (1985) On the mimetic association between nymphs of *Hyalymenus* spp. (Hemiptera: Alydidae). Zool J Linn Soc 83:371–384

Panizzi AR (1988) Biology of *Megalotomus parvus* (Hemiptera: Alydidae) on selected leguminous food plants. Insect Sci Appl 9:279–285

Panizzi AR, Santos CH (2001) Unusual oviposition on the body of conspecifics by phytophagous heteropterans. Neotrop Entomol 30:471–472

Panizzi AR, Hirose E, Oliveira EDM (1996) Egg allocation by *Megalotomus parvus* (Westwood) (Heteroptera: Alydidae) on soybean. An Soc Entomol Brasil 25:537–543

Panizzi AR, Schaefer CW, Natuhara Y (2000) Broad-headed bugs (Alydidae). In: Schaefer CW, Panizzi AR (eds) Heteroptera of economic importance. CRC Press, Boca Raton, pp 321–336

Paradela Fo O, Rossetto CJ, Pompeu AS (1972) *Megalotomus parvus* (Westwood) (Hemiptera: Alydidae), vector de *Nematospora coryli* Peglion em feijoeiro. Bragantia 31:5–10

Santos CH, Panizzi AR (1997) Tachinid parasites of adult *Megalotomus parvus* West. (Hemiptera: Alydidae). An Soc Entomol Brasil 26:577–578

Santos CH, Panizzi AR (1998a) Danos qualitativos causados por *Neomegalotomus parvus* (Westwood) em sementes de soja. An Soc Entomol Brasil 27:387–393

Santos CH, Panizzi AR (1998b) Nymphal and adult performance of *Neomegalotomus parvus* (Hemiptera: Alydidae) on wild and cultivated legumes. Ann Entomol Soc Am 91:445–451

Schaefer CW (1972) Clades and grades in the Alydidae. J Kansas Entomol Soc 45:135–141

Schaefer CW (1980) The host plants of the Alydinae, with a note on heterotypic feeding aggregations (Hemiptera: Coreoideae; Alydidae). J Kansas Entomol Soc 53:115–122

Schaefer CW (1999) The higher classification of the Alydidae (Hemiptera: Heteroptera). Proc Entomol Soc Wash 101:94–98

Schaefer CW (2003) A revision of *Darmistus* Stål (Hemiptera: Alydidae: Micrelytrinae). Proc Entomol Soc Wash 105:950–966

Schaefer CW (2004) Key to the genera of New World Alydidae (Hemiptera: Heteroptera). Proc Entomol Soc Wash 106:280–287

Schaefer CW, Ahmad I (2008) A revision of *Neomegalotomus* (Hemiptera: Alydidae). Neotrop Entomol 37:30–44

Schaefer CW, Mitchell PL (1983) Food plants of the Coreoidea (Hemiptera: Heteroptera). Ann Entomol Soc Am 76:591–615

Schaefer CW, Panizzi AR (1998) The correct name of "Megalotomus" pests of soybean (Hemiptera: Alydidae). An Soc Entomol Brasil 27:669–670

Schaffner JC (1964) A taxonomic revision of certain genera of the tribe Alydini (Heteroptera: Coreidae). PhD dissertation, Iowa State University, IO, USA

Schaffner JC, Schaefer CW (1998) *Neomegalotomus* new genus (Hemiptera: Alydidae: Alydinae). Ann Entomol Soc Am 91:395–396

Schuh RT, Slater JA (1995) True bugs of the world (Hemiptera: Heteroptera): classification and natural history. Cornell University Press, Ithaca

Silva AG D'A, Gonçalves CR, Galvão DM, Gonçalves AJL, Gomes J, Silva MN, Simoni L (1968) Quarto catálogo dos insetos que vivem nas plantas do Brasil – seus parasitas e predadores, Part II, vol I. Ministerio Da Agricultura, Rio de Janeiro

Silva JJ, Arruda-Gatti IC, Mikami AY, Pissinati A, Panizzi AR, Ventura MU (2010) Attraction of *Neomegalotomus parvus* (Westwood) to cow urine and ammonia. Sci Agric 67:84–86

Sisson RF (1980) Deception: formula for survival. Nat Geogr Mag 157:394–415

Ventura MU, Panizzi AR (1997) *Megalotomus parvus* West. (Hemiptera: Alydidae): Inseto adequado para experimentação e didática entomológica. An Soc Entomol Brasil 26:579–581

Ventura MU, Panizzi AR (2000) Oviposition behavior of *Neomegalotomus parvus* (West.) (Hemiptera: Alydidae): daily rhythm and site choice. An Soc Entomol Brasil 29:391–400

Ventura MU, Panizzi AR (2003) Population dynamics, gregarious behavior and oviposition preference of *Neomegalotomus parvus* (Westwood) (Hemiptera: Heteroptera: Alydidae). Braz Arch Biol Technol 46:33–39

Ventura MU, Panizzi AR (2004) Responses of *Neomegalotomus parvus* (Hemiptera: Alydidae) to color and male-lured traps. Braz Arch Biol Technol 47:531–535

Ventura MU, Panizzi AR (2005) Morphology of olfactory sensilla and its role in host plant recognition by *Neomegalotomus parvus* (Westwood) (Heteroptera: Alydidae). Braz Arch Biol Technol 48:589–597

Ventura MU, Montalván R, Panizzi AR (2000a) Feeding preferences and related types of behaviour of *Neomegalotomus parvus*. Entomol Exp Appl 97:309–315

Ventura MU, Silva JJ, Panizzi AR (2000b) Phytophagous *Neomegalotomus parvus* (Westwood) (Hemiptera: Alydidae) feeding on carrion and feces. An Soc Entomol Brasil 29:841–843

Chapter 19
Leaf-Footed Bugs (Coreidae)

José Antônio Marin Fernandes, Paula Levin Mitchell, Laurence Livermore, and Malin Nikunlassi

Abstract The family Coreidae is distributed worldwide, but these phytophagous bugs are most abundant in the tropics and subtropics. In the Neotropical region, all of the subfamilies and 16 tribes are represented. In tropical ecosystems, these bugs feed on herbs and shrubs in open areas of forests as well as at the forest edge. Some species are spectacularly colored, and unusual expansions of antennae, humeral angles, femora, or tibiae occur in many groups. Some of them move lazily even when disturbed and hardly fly to escape; others are extremely nimble, fast flying away when disturbed. They are frequently encountered in crops, representing important pests in several commodities. No one common name is universally accepted for the family, and none of the frequently used names (e.g., squash bug, leatherbug, leaf-footed bug, Randwanzen) are collectively appropriate for all members of the family.

J.A.M. Fernandes (✉)
Instituto de Ciências Biológicas, Universidade Federal do Pará,
Rua Augusto Corrêa n°1, Belém, PA 66075-110, Brazil
e-mail: joseamf@ufpa.br

P.L. Mitchell
Department of Biology, Winthrop University, Rock Hill, SC 29733, USA
e-mail: mitchellp@winthrop.edu

L. Livermore
Biodiversity Informatics, Insects Division, The Natural History Museum,
Cromwell Road, London SW7 5BD, UK
e-mail: l.livermore@nhm.ac.uk

M. Nikunlassi
Biodiversity Informatics, The Natural History Museum,
Cromwell Road, London SW7 5BD, UK
e-mail: malin.nikunlassi@gmail.com

A.R. Panizzi, J. Grazia (eds.), *True Bugs (Heteroptera) of the Neotropics*,
Entomology in Focus 2, DOI 10.1007/978-94-017-9861-7_19

19.1 Introduction

The family was established by Leach (1815), but Stål (1867, 1870) was the first author to propose a suprageneric structure to Coreidae. Surprisingly not much has been done at this level since then, and much of the suprageneric organization is still the same. The most comprehensive work at this level since Stål was presented by Schaefer (1964, 1965). For many years, the literature on Coreidae was fragmentary; keys are still lacking for several important groups. During the last few decades, Brailovsky and coworkers have revised or contributed to the knowledge of several Neotropical genera of Coreidae (see throughout the chapter). Only recently, a print catalog for the Neotropical Coreidae (Packauskas 2010) and an online catalog for Coreoidea of the world (CoreoideaSF Team 2015) were compiled. These recent revisions and catalogs will allow future works on Coreidae, especially on relationships among taxa.

19.2 General Characteristics and Diagnosis

Coreids are usually heavy-bodied insects, elongate or elliptical in shape. Species of the genus *Thasus* are bulky insects reaching 30–40 mm, and species of *Vilga* are delicate 6–10 mm bugs. Many have the hind femora enlarged, with spines and tubercles; hind tibiae curved, with spines or leaflike expansions; and third and fourth antennal segments dilated or tumid (Plate 19.1). Most of the species have dull colors like brown, gray, and yellow, but some have bright colors like red, green, orange, or sometimes a metallic coloring. Head smaller than pronotum; antennae with four segments and inserted above a line drawn through the eye; membrane of forewing with several veins; femora of hind legs frequently incrassate; hind tibiae terete or slightly to strongly dilated; abdominal spiracles ventral; 3 trichobothria on abdominal segments III–VI, 2 on segment VII; nymphal dorsal abdominal scent gland openings between terga 4/5 and 5/6; genital plates flattened; spermatheca only with proximal pump flange; eggs with pseudoperculum except in Pseudophloeinae.

19.3 General Biology and Ecology

Coreids are plant feeders; hosts include gymnosperms and angiosperms and monocots and dicots. However, dicots are far more common as hosts than monocots, and few coreid species are exclusive feeders on gymnosperms (Schaefer and Mitchell 1983). Opportunistic feeding on fecal matter, carrion, and insect eggs has been reported (Adler and Wheeler 1984; Menezes-Netto et al. 2012) but is never the primary source of nutrients. Plant preference spans a wide spectrum from specialization on a single genus to extreme polyphagy; reported host plants of one African

Leptoglossus zonatus
PHOTO: *W.O. Ree, Jr.*

Pachylis sp.
PHOTO: *José A.M. Fernandes*

Pachylis sp.
PHOTO: *José A.M. Fernandes*

Holhymenia sp.
PHOTO: *José A.M. Fernandes*

Eubule farinosa.
PHOTO: *José A.M. Fernandes*

Crinocerus sanctus
PHOTO: *Arthur Anker*

Spartocera sp.
PHOTO: *Andreas Kay*

Spartocera sp.
PHOTO: *Andreas Kay*

Plate 19.1 Field images of Coreidae

mictine species, *Anoplocnemis curvipes* (F.), represent >30 plant families. Legumes are the primary host plants of Pseudophloeinae and several tribes of Coreinae (Acanthocerini, Mictini, Nematopodini); this may be a primitive trait of Coreoidea (Schaefer and Mitchell 1983). Some genera or tribes may be characterized by their host plant preferences (e.g., *Chelinidea* spp. on Cactaceae), but most cannot; a single genus (e.g., *Leptoglossus* Guerin) may include polyphagous species and those with highly restricted feeding habits (Mitchell 2006).

Most coreids are winged, but members of the Australian tribe Agriopocorini are usually wingless (Schuh and Slater 1995), and the single described species of Barreratalpini is brachypterous (Packauskas 1994). Both macropterous and brachypterous forms are known in the Neotropical pseudophloeine *Vilga mexicana* Distant (Dolling 1977). At least one coreid has been described as migratory (Aldrich et al. 1982), but most movement involves shorter distances. Coreids move seasonally among host plants and between hosts and overwintering areas in temperate regions, occasionally becoming a problem in houses. Dispersal may occur in the nymphal stage, but in polyphagous species such as *L. phyllopus*, most movement occurs when recently molted winged adults disperse from deteriorating nymphal hosts (Mitchell 2006). Vast geographic range expansion has occurred through commercial movement of plants. For example, the conifer-feeding *Leptoglossus occidentalis* Heidemann, formerly restricted to western North America, spread eastward as far as Ontario and Connecticut and then, following accidental introductions, has spread through most of Europe, most recently being reported from Turkey (Hizal 2012) and parts of Asia, most recently Korea (Ahn et al. 2013). The Neotropical *Leptoglossus zonatus* (Dallas), long known as a highly polyphagous crop pest in Central and South America and Mexico, has also expanded its range recently, moving across the southern United States from Texas to Florida, becoming a problem in satsuma oranges and other citrus (Xiao and Fadamiro 2010).

A number of coreids are economic pests; grain legumes, tree nuts, rice, fruits, and garden vegetables are among the crops attacked. Several species pose problems in forest management (conifers and eucalypts). Only a small proportion of coreid species damage crops, but those few can cause serious losses (Mitchell 2000). Worldwide, legume feeders are likely the most important economically, especially in combination with alydids and pentatomids. Pests in the Neotropics include *Anasa* spp. on cucurbits, *Spartocera* spp. on Solanaceae and tobacco, *Phthiacnemia picta* (Drury) on tomatoes and *Leptoglossus* spp. on a variety of crops. Direct damage to grains and developing seeds within pods and seed heads is most common; pod abortion or reduced seed production may result from feeding damage. Fruits may become malformed, pithy, or corky, or develop blotches, spots, depressed areas or lesions. Wilting due to vascular damage can also result in losses, and the role of coreids and other Heteroptera in transmission of diseases is only recently becoming appreciated (Mitchell 2004). No coreids are known to transmit viruses or phytoplasmas, but 13 species are associated with fungal pathogens (Mitchell 2004) and 30 species representing 17 primarily Neotropical genera harbor trypanosomatids (adapted from Camargo and Wallace 1994 and Godoi et al. 2002). A survey of Amazonian Heteroptera showed coreids to far outnumber other families in frequency of trypanosomatid infection; >80 % of Coreidae were infected, many with the plant parasite *Phytomonas*, compared with <10 % of other phytophagous families examined (Godoi et al. 2002). *Phthiacnemia picta* not only harbor phytomonads but have been shown to acquire and infect tomatoes with these flagellates (Jankevicius et al. 1989; Fiorini et al. 1993). *Leptoglossus zonatus* (Dallas) can transmit a trypanosomatid to corn kernels (Jankevicius et al. 1993). In the United States, transmission by *Anasa tristis* De Geer of the fastidious phloem-colonizing bacterium causing cucurbit yellow vine disease has recently been documented and extensively studied (Bruton et al. 2003). Although a few coreid species have been investigated as potential biocontrol agents (Mitchell 2000), none have been successfully adopted, either due to failure to disperse or insufficient host specificity.

Natural enemies of coreids include parasitic tachinid flies; several species of egg parasitoids in the families Encyrtidae, Eurytomidae, and Platygastridae; and both vertebrate and invertebrate predators, particularly reduviids, spiders, and ants. Guarding of eggs by females has been reported for one species (Hemmingsen 1947). Scent glands, present in both adults and nymphs, provide effective protection from some predators. Blackbirds feeding on *Coreus marginatus* (L.) and various pentatomid species will wipe bugs along on the ground, or hit them repeatedly against the ground, to avoid consuming the defensive secretions (Schlee 1992). Adult scent glands are located ventrally on the metathorax, whereas nymphal scent glands are on the dorsum of the abdomen. Compounds produced by these glands differ between juveniles and adults. Chemical composition of the scent fluid was reviewed by Aldrich (1988).

19.3.1 Eggs and Oviposition

Egg shapes vary among coreid species. Ovoid or elliptical eggs are most common, but hemicylindrical, cuboidal, globular, and oblong eggs are known (Cobben 1968; Moulet 1995). The portion attached to the substrate may be flattened, producing an egg that is triangular in cross-section (e.g., *Acanthocephala terminalis* (Dallas), Yonke and Medler 1969b). However, in pseudophloeines, the side of the egg facing the substrate is convex (Cobben 1968). The hemicylindrical eggs of *Leptoglossus*, *Phthia s.l.*, and *Narnia* (as well as several Old World genera) are laid in long chains, end to end (Amaral 1981a; Mitchell and Mitchell 1986; Mitchell 2000), but other types may be oviposited singly, in small groups, adjacent rows, large clusters, or even atop one another as in the Old World pseudophloeine, *Ceraleptus gracilicornis* H.-S. (Cobben 1968). Most eggs are laid on host plants, generally on the undersides of leaves, but the egg chains of *Leptoglossus phyllopus* (L.) may be placed along virtually any linear surface—host or non-host, living or nonliving (Mitchell and Mitchell 1986). Females of *Phyllomorpha laciniata* (Villers) may choose to place eggs on leaves of the host plant or on the backs of males. Such oviposition blocks wing movement so that individuals carrying eggs cannot fly and are subject to higher predation (Reguera and Gomendio 1999). Paternal care was found to be higher in a population subjected to heavy egg parasitism; eggs laid on males suffered less parasitism than those laid on plants (Gomendio et al. 2008).

Cobben (1968) provides detailed descriptions of the eggs of several species. A pseudoperculum is present in most Coreinae in the shape of a round cap, but not in Pseudophloeinae. In the latter subfamily, the extensive eclosion fracture may entirely separate the two halves of the eggshell. The number and placement of aero-micropyles also varies, from just a few to 60 in the species studied. They are most commonly arranged in a ring which crosses or more rarely coincides with the circular pseudoperculum. The chorion may be up to 100 µm thick, and the egg burster is triangular with a sclerotized ridge (Cobben 1968).

Eggs are cemented to the substrate by the female, and it has been suggested that trypanosomatid cysts may survive in this material, thereby providing a source of infection (Camargo and Wallace 1994). However, *Burkholderia* bacterial symbionts are not acquired from the eggs in *Thasus neocalifornicus* Brailovsky & Barrera and must be acquired de novo, presumably from the soil, by the early instars (Olivier-Espejel et al. 2011).

19.3.2 Nymph Feeding and Development

Five stadia occur between hatching and adult emergence. First instars often remain in the vicinity of the egg mass and appear to need only a source of water for successful molt to the second stadium (Cook and Neal 1999). Probing into plant tissue may occur, but the tissue selected is always vegetative (tendrils, leaves, terminal buds) (Rodrigues et al. 2007a; Wheeler and Miller 1990). Under laboratory conditions, survivorship of first instars provided with only a water wick was equivalent to that of first instars provided green bean and sunflower seed, and only the bean was used by the nymphs when food was provided (Mitchell 1980a).

Second through fifth instars require a nutrient source and feed similarly to the adults in most cases, although the shorter stylets may prevent access to deeper plant tissues. Younger instars are generally gregarious; research with *Phthia picta* (Amaral 1981a) has shown that grouped nymphs develop faster than isolated nymphs. Experimental studies of nymphal social groupings in the cactus bug *Chelinidea vittiger aequoris* McAtee (Miller et al. 2012) indicate that females reared alone are less likely to forage socially as adults than nymphs reared in a group; no similar effect of nymphal environment on adult behavior was found for males. Aggregation of juvenile bugs has been attributed to a number of factors, including both abiotic (humidity) and biotic (protection from predation) (Panizzi 2004a). Coreid nymphs are often highly visible—brightly colored with flattened hind tibiae and/or antennal segments—and all have dorsal abdominal scent glands. Aposematic feeding aggregations of nymphs of a Costa Rican coreid species responded to disturbance by pulsating, spraying anal fluid, and releasing scent fluid (Aldrich and Blum 1978). The various secretions of the dorsal abdominal glands function in defense against natural enemies as repellents, deterrents, and/or toxins (Noge et al. 2012) and as an alarm pheromone. Nymphs will also respond to the alarm pheromones of adults in *L. zonatus* (Leal et al. 1994), but not in *T. neocalifornicus* (Prudic et al. 2008). Interestingly, the chemistry of coreid nymphal secretions differs from that of pentatomids, in that aldehyde chains are shorter and little or no *n*-tridecane is present. The short chain keto-aldehyde of coreids (4-oxo-(E)-2-hexenal) is potent against ants, whereas the longer chain compounds of pentatomid nymphs are less potent but are synergized by *n*-tridecane (Eliyahu et al. 2012).

19.3.3 Adult Feeding and Reproduction

Coreids feed on a range of plant parts, including vegetative tissue, phloem, buds, fruits, developing seeds, and mature seeds (Mitchell 2000). Such preferences are often specific at the tribal level; bugs are either associated with vegetative and vascular tissue or with reproductive plant parts. Kumar (1966) separated Australian coreids into those that fed exclusively on sap or those that fed on sap and fruit, and this early observation has held for other groups studied subsequently (Schaefer and Mitchell 1983). Like all Pentatomomorpha, coreids produce two types of saliva, gelling saliva to line the path of the stylets, producing a salivary sheath, and watery saliva containing enzymes (Miles 1972; Cobben 1978). The mode of feeding in several species of Australian coreids has been described as an "osmotic pump." In this process, a short salivary sheath extends into parenchyma associated with vascular tissue. Sucrase in the watery saliva induces phloem unloading and discharge of parenchyma cell contents into the spaces between cells, and nutrients are thereby obtained from the extracellular spaces without direct penetration of the membranes of individual cells. The resulting effect on the plant may be a necrotic lesion or wilting of shoots beyond the feeding site (Miles and Taylor 1994). This feeding mode has been associated with species from Amorbini, Mictini, and Dasynini but may be more widespread within the family. True phloem feeding—similar to that of aphids, in which stylets penetrate phloem sieve elements—is assumed for the very few coreid species (*Hygia cliens* Dolling, *Notobitus affinis* (Dallas), and *Cloresmus* sp.) in which ant-tending for honeydew has been observed (Maschwitz et al. 1987). No coreids feed exclusively from xylem, but hydration from xylem certainly occurs. Stylet sheaths terminating in xylem are reported for *Leptoglossus phyllopus* (L.), which feeds primarily on developing seeds (Mitchell 2006), and wilting of squash leaves has been attributed to xylem blockage by the gelling saliva of *A. tristis* (Neal 1993).

The number of generations per year varies with latitude and as a function of crop cycles, host plant phenology, and host plant specificity. In temperate zones, the overwintering stage is usually the adult. Adults are long-lived; mark-release-recapture studies indicate maximal lifespan of over 50 days (Mitchell 2006). Courtship and mating have been described for several species and involve antennation, touching with the legs, or other body contact prior to copulation (Eberhard 1998; Miller 2011; Tatarnic and Spence 2013). Both males and females may mate multiple times. Duration of mating varies within and between species but may be as long as 6–7 h (McLain et al. 1993; Tatarnic and Spence 2013). Mating often occurs on fruits, and feeding by one or both partners during copulation is common. The most frequent mating position is end to end, with the male facing the opposite direction from the female. Copulatory success appears to correlate with size of male (McLain et al. 1993; Tatarnic and Spence 2013).

In several coreid tribes, a pronounced sexual dimorphism occurs in the size and armament of the hind femur, although the expression of this trait may vary within and across populations (Miller and Emlen 2010). This trait has been linked to male-male combat and territoriality or female defense polygyny in at least 8 species (Fujisaki 1980, 1981; Mitchell 1980b; Aldrich et al. 1982; Miyatake 1995, 1997; Eberhard 1998; Miller and Emlen 2010; Miller 2013). However, large mixed gender aggregations occur in other species (Mitchell 1980b; Solomon and Froeschner 1981), and male-specific attractant compounds have been identified for several species (Millar 2005). These species-specific volatile compounds are released from epidermal cells, ventral abdominal glands, or the lateral lobe of the metathoracic scent gland, depending on species (Aldrich et al. 1982, 1993; Soldi et al. 2012), and may serve in short-range communication, as attractants for females only or as aggregation pheromones.

19.4 Classification and Diversity

Although distributed worldwide, coreids are most abundant in the tropics. A recent catalog of New World Coreidae (Packauskas 2010) lists 158 genera that contain species with distribution records from the Neotropical region. All three subfamilies—Coreinae, Meropachyinae, and Pseudophloeinae—are represented in this region. A detailed study of the species richness of the coreoid fauna of relict pluvial subtropical forest in southern Brazil revealed 20 species of Coreidae present, with one species, *Cebrenis supina* Brailovsky, dominating beat-sheet collections (Barcellos et al. 2008). A number of economically important species have been extensively studied, but for the majority of Neotropical species, unfortunately little or nothing is known about their biology. Here we present keys to the subfamilies and the tribes of Coreinae of the Neotropical region; in addition, we present information on the biology of these groups, insofar as it is known. Species of major economic importance, and others that have been studied in depth, are treated separately in Sect. 19.5.

Key to subfamilies of the Neotropical Region Coreidae (adapted from Packauskas 1994).

1. Tibiae not sulcate on outer surface; metathoracic scent gland orifices with a single large or two nearly fused auricles; meso- and metasterna sulcate; posterior coxae contiguous or nearly so; all species less than 12 mm long............................Pseudophloeinae
– Tibiae clearly or slightly sulcate on outer surface; other characteristics different from above; length variable...................................2
2. Metathoracic scent gland orifices deep between coxae; hind coxae far separate; hind femora strongly clavate and curved; hind tibiae with an apical tooth or spine...................Meropachyinae
– Scent gland lateral to coxae; hind coxae not so separate; hind femora not or rarely strongly clavate and usually straight; hind tibiae unarmed at apex..Coreinae

19.4.1 Pseudophloeinae Stål (Plate 19.2)

Pseudophloeinae species are known to feed on legumes (Fabaceae) that grow in open woodlands or grasslands with scattered trees; apparently dense forests are not occupied by most of the species (Dolling 1986).

Pseudophloeinae is composed of two tribes: Clavigrallini and Pseudophloeini. The former apparently is monophyletic and the latter paraphyletic according to Dolling (1986). Pseudophloeinae is represented in the New World by three genera: *Coriomeris* Westwood, *Ceraleptus* Costa, and *Vilga* Stål, all belonging to Pseudophloeini. The first two genera are Holarctic in distribution, but one species of *Coriomeris* is found in the mountainous areas of Oaxaca, Mexico. Only *Vilga* Stål is registered to the Neotropical region.

Packauskas (1994) in the key to subfamilies of the New World Coreidae characterized Pseudophloeinae by the tibiae not sulcate; metathoracic scent gland orifices with a single large or two nearly fused auricles; meso- and metasternum sulcate; males with a bilobed genital capsule opening filled by the clearly visible parameres. Dolling (1986, 1987) and Schuh and Slater (1995) presented a more complete diagnosis for Pseudophloeinae including: antennae inserted at sides of head; antennifers with porrect or deflexed process at outer apical angles; rostrum at rest reaching metasternum; posterior coxae separated by about the width of a coxa or less; femora moderately to strongly clavate; posterior femora typically with two or more large subapical spines beneath on anterior side with some smaller spines, tubercles, or granules between them and a terminal series of about four spines diminishing in length toward apex of femur; tibiae without apical spines; membrane of hemelytron with a compound vein near base almost parallel with apical margin of corium; posterolateral angle of sternite VII almost always and of sternites III–VI commonly produced into teeth or spines; female paratergite VIII without functional spiracle; first valvula of ovipositor without peg-like setae; spermatheca with bulb lunate and duct devoid of prominent flange; egg not operculate or pseudoperculate, opening by a transverse eclosion rent. Pseudophloeines are easy to recognize because they do not

Plate 2 - Pseudophloeinae Stål, 1868 (Scalebars = 5mm): 1. *Ceraleptus denticulatus* Froeschner, 1963 [USNM]; 2. *Ceraleptus pacificus* Barber, 1914 [USNM]. 3. *Ceraleptus probolus* Froeschner, 1963 [USNM]; 4. *Coriomeris humilis* (Uhler, 1872) [BMNH]; 5. *Coriomeris insularis* Dolling & Yonke, 1976 [BMNH]; 6. *Coriomeris occidentalis* Dolling & Yonke, 1976 [CAS]; 7. *Vilga (Echinovilga) dallasi* Distant, 1881 [BMNH]; 8. *Vilga (Laevivilga) divaricata* Distant, 1893 [BMNH]; 9. *Vilga (Platyvilga) peruviana* Dolling, 1977 [BMNH]; 10. *Vilga (Trichovilga) mexicana* Distant, 1892 [BMNH]; 11. *Vilga (Vilga) westwoodi* (Kolenati, 1845) [BMNH]; 12. *Vilga (Vilgula) dissimilis* Distant, 1893 [BMNH].

Plate 19.2 Coreinae plate—Pseudophloeinae

have the dorsal surface of the tibiae sulcate; a tribe of Coreinae—Hydarini—also lack these sulci, but do not occur in the Neotropical region. All remaining Coreinae taxa present tibiae sulcate.

The genus *Vilga* is composed of 14 species divided into 6 subgenera (Dolling 1977). This genus has a striking feature, probably apomorphic, grouping all species and not presented in any other genera of the subfamily—the freely projecting lobe on the seventh tergum of the male. This taxon is also characterized by the absence of an antevannal vein ("glochis") in the hind wing, a feature used by Stål (1870) to characterize Pseudophloeinae. Dolling (1977) noted that this absence can be a plesiomorphic state shared with the Old World *Hoplolomia* Stål and *Risbecocoris* Izzard. This genus was revised and keyed by Dolling (1977); he also included in his key a pair of couplets to separate *Vilga* from *Ceraleptus* and *Coriomerus*. No host plants are known for this genus, but legumes are likely based on the feeding habits of other Pseudophloeini. One species, *Vilga mexicana* Distant, is unusual in having both macropterous and brachypterous forms (Dolling 1977).

19.4.2 Meropachyinae Stål (Plate 19.3)

Meropachyinae has a striking characteristic that is the hind tibiae with an apical spine or tooth used in keys to separate Meropachyinae from other subfamilies. Other characteristics used are the small head and narrow thorax; metathoracic scent gland opening deep between coxae, opening anteriorly; projections of peritreme fused; evaporatorium moderate and without ridges; veins of the forewing arising from a vein close to corial border; widely separated hind coxae; hind femur curved and strongly incrassate; second through fifth abdominal sterna fused in both sexes; plus several characteristics from male and female genitalia.

This subfamily has a history of name changes that started with Stål (1867) erecting it under Meropachydida afterwards changed to Merocorina (Stål 1870). Further authors used both names to designate the same group of genera; thus, Meropachydinae was indicated as the correct name for the subfamily by Schaefer (1964) followed by Froeschner (1981), Henry and Froeschner (1988), Schuh and Slater (1995), and Brailovsky (1998a, 1999a) among others. Merocorinae was used by Van Duzzee (1916), Torre-Bueno (1941) and Kormilev (1954). Packauskas (1994) explained that Meropachyinae should be used rather than Meropachydinae, and more recently Brailovsky (2009a) and CoreoideaSF Team (2015) also use this name.

Kormilev (1954) revised Meropachyinae and proposed three tribes: Merocorini Stål, Meropachyini Stål, and Spathophorini Kormilev. After that, only keys and lists of regional fauna were presented till Brailovsky (1998a) started a series of works dealing with the entire subfamily. Additional keys to the tribes may be found in Packauskas (1994). The subfamily currently includes 27 genera, most of them monotypic, comprising about 60 species (CoreoideaSF Team 2015).

Plate 3 - Meropachyinae Stål, 1867 (Scalebars = 5mm): 1. *Diariptus hexacanthus* Stål, 1860 [ZMHB]; 2. *Flavius lineaticornis* Stål, 1862 [NHMW]; 3. *Gracchus integer* (Burmeister, 1835) [ZMHB]; 4. *Himellastella conica* Brailovsky & Barrera, 1998 [ZMHB]; 5. *Hirilcus gracilis* (Burmeister, 1835) [ZMHB]; 6. *Hirilcus variolosus* (Burmeister, 1835)[ZMHB]; 7. *Larraldiella terminalis* Brailovsky, 1999 [UNAM]; 8. *Lycambes andicola* Breddin, 1903 [DEI]; 9. *Lycambes sargi* Breddin, 1904 [DEI]; 10. Lycambes varicolor Stål, 1862 [NHMW]; 11. *Spathophora biclavata* (Fabricius, 1803) [ZMUC]; 12. *Paralycambes pronotalis* Brailovsky, 1998 [UNAM].

Plate 19.3 Meropachyinae plate

Merocorini Stål

This tribe can be identified by the small size of the species; scutellum triangular, short, not surpassing base of claval commissure, the apex of scutellum elevated as a small knob and not hiding claval commissure; abdominal spiracles circular and located almost on the anterior third of each sternite (Kormilev 1954). Only one genus is currently included—*Merocoris* Perty—divided in two subgenera. The nominal subgenus has five species spread from Argentina to Central Part of Brazil; this group can be recognized by the scutellar disk with a high conical tubercle. The subgenus *Corynocoris* Mayr has three species, but one is registered only to North America, the other two distributed in Central America and Great Antilles; this group can be recognized by the scutellar disk almost flat. Brailovsky and Barrera (2009a) described a new species from Brazilian Cerrado areas collected on grassland, and presented a key to subgenera and species of *Merocoris*. Other *Merocoris* spp. are reported to feed on milkworts (*Polygala* sp.) (Schaefer and Mitchell 1983).

Meropachyini Stål

This tribe is easily recognized by the elongate scutellum extending beyond the distal end of the clavus; hind acetabulae laterally projected, visible in dorsal view; posterior tibiae broadly curved distally. The hind coxae widely separated looking like thoracic appendages, and the hind coxae strongly incrassate and long produce a unique facies remarkable among coreids.

This tribe contains 18 genera so far with 30 species chiefly South American and only a few species registered to Central America. Brailovsky (1999a) started the study with Meropachyini describing several genera and presenting a key to the known genera. Further contributions (Brailovsky and Barrera 2001; Brailovsky and Luna 2000; Brailovsky 2009a) included several new genera and species plus an updated key. The large old genera—*Hirilcus* Stål, *Marichisme* Kirkaldy, *Meropachys* Burmeister, and *Peranthus* Stål—are still to be revised. Little is known of the biology of this group. A key to genera can be found in Brailovsky (1999a, 2009a) and Brailovsky and Barrera (2001).

Spathophorini Kormilev

Spathophorini is a tribe recognizable by the triangular scutellum, flat, shorter than clavus; hind femora not attaining the apex of the abdomen; antennal segment IV usually slender; and abdominal spiracles elliptical and positioned between the anterior and middle third of the segment. This tribe is quite neglected because after Kormilev (1954), not much was added to it, except works by Brailovsky (1998a,

2000, 2008) and Brailovsky and Barrera (1998a). *Lycambes varicolor* Stål is reported to feed on *Lonchocarpus guatemalensis* and *L. cruentus* (Peredo 1999). A key to genera can be found in Brailovsky (1998a, 2000, 2008).

19.4.3 Coreinae Leach

Most of the coreids are included here. Extremely diverse in shape, color, and size; distance from eye to ocellus smaller than that between ocelli; anterolateral opening of the metathoracic scent gland well developed, peritreme with developed projections, auricle usually occupying half of the size of evaporatorium; corial margins straight or slightly sinuate; membranal veins of forewing arising from a transverse vein very close, or touching, corial margin; abdominal terga 1–2 and 3–7 fused; genital capsule of male without lateral prolongations; articulation of female genital plates usually membranous; gonangulum usually flat and folded.

These are the most common and striking coreids both in the field and in collections. This is by far the largest subfamily, with a mostly tropical worldwide distribution. This is also the subfamily in which discussions about taxonomical levels of certain groups are concentrated. Neotropical tribes are not much affected by such discussions; thus, they will not be treated here (see Schuh and Slater 1995 for more on suprageneric changes). Hydarini was raised by Ahmad (1970) to subfamily level, but recent works (i.e., Schuh and Slater 1995; Brailovsky 2010a) have considered this taxon as a tribe, which is the position followed here. Coreinae comprises 32 tribes, but Neotropical fauna comprises only 11 tribes, presented below. Several authors worked with Coreinae, but the subfamily still needs a phylogenetic analysis to raise hypotheses about relationships among genera. Packauskas (1994) presented a key to the Neotropical tribes, which is a keystone to understanding Coreinae.

Key of Coreinae (adapted from Packauskas 1994). Packauskas used Coreini to refer to the set of genera we are considering under Hypselonotini according to the Coreoidea catalog (CoreoideaSF Team 2015).

1. Tylus projecting conspicuously beyond juga as a distinct knob or strongly compressed plate; juga strongly deflex..2
– Tylus not or only vaguely projecting beyond juga; juga variable.........................3
2. Tylus projected as a distinct knob; ventral surface of the body with several small black spots; hind femur of the males not or slightly incrassate..Placoscelini
– Tylus well projected as a strongly compressed plate; ventral surface of the body with ventral color greatly variable but not as above; hind femur of the males greatly incrassate................................Acanthocephalini

3. With the combination of hind femora spinose or tuberculate; strongly incrassate in males; head not over half as long as pronotum and not extending past antenniferous tubercles or very slightly so; antenniferous tubercles occupying most of anterior head width, space between them rarely wider than one tubercle....4
– Not with this combination of characters....5
4. Metathoracic scent gland with a single fused auricle (sometimes bilobed); ocellar tubercles large; usually with a spine present on antenniferous tubercles; parameres membranous....Acanthocerini
– Metathoracic scent gland with two completely separated auricles (area between them depressed); ocellar tubercles small; antenniferous tubercles never armed; parameres sclerotized....Nematopodini
5. Head porrect; juga and tylus extended forward beyond antenniferous tubercles; bucculae never reaching midlength of head; all femora spinose below, rostrum almost reaching or extending onto abdomen....Anisoscelini
– Not with this combination of characters....6
6. Juga extended and exceeding or equal to tylus in length....7
– Juga deflexed or at least not exceeding nor equal to tylus in length....8
7. Head shorter than pronotum; antennae terete; brachypterous....Barreratalpini
– Head equal in length to pronotum; antennae triangular in cross-section; macropterous....Chelinideini
8. Distance between hind coxae equal to distance from coxa to lateral margin; antenniferous tubercles prominent, subcontiguous above deflexed tylus and juga....Spartocerini
– Distance between hind coxae much narrower than distance from coxa to lateral margin, if distances nearly equal then head elevated between antenniferous tubercles; tylus and juga variable....9
9. Third antennal segment expanded, other segments terete; tylus and juga deflexed; not filling space between antenniferous tubercles; all femora armed distally....Chariesterini
– Third antennal segment never expanded, if appearing expanded, then other segments also appearing expanded; tylus and juga usually extended anteriorly between antenniferous tubercles; femora rarely armed....10
10. Spiracles situated twice as far from posterior border of middle abdominal segments (III–VI) as from anterior border....Discogastrini
– Spiracles equidistant from anterior and posterior borders of middle abdominal segments (III–VI)....11
11. Fourth antennal segment flattened; third antennal segment over twice as long as second....Hydarini
– Fourth antennal segment not flattened; third antennal segment usually shorter than second, never twice as long as second....Hypselonotini

Acanthocephalini Stål (Plate 19.4)

This tribe groups species with medium to large size and interesting color patterns that are found in well-preserved forest. The tribe was established by Stål (1870) as Acanthocephalina, but no description was given. Schaefer (1964) also did not present a description of the tribe but mentioned the striking spinelike tylus. The most significant features of this tribe are the tylus, which projects conspicuously beyond the juga as a distinct acute plate, juga deflexed, abdominal spiracles close to anterior margin of the sternum, hind femur of males greatly incrassate, all femora of both sexes spinose ventrally, and hind tibia slightly or strongly expanded in both sexes (Packauskas 1994; Brailovsky 1997a). Several genera were described and revised or new species added in the last decades mainly by Brailovsky and coauthors (Brailovsky 1991, 1999b; Brailovsky and Barrera 1999, 2005). The largest genus *Acanthocephala*—one of the most common and conspicuous—still needs revision; Brailovsky (2006b) reviews the Mexican species, and McPherson et al. (2011) treats the species of America north of Mexico. No key to the entire tribe has been published so far.

This tribe is distributed across the Americas, but the largest number of species is found in the North and Central parts of South America, especially Peru and Ecuador; only a few are distributed across Central America and Nearctic region. Acanthocephalini includes 15 genera: *Acanthocephala* Laporte (27 species, three of them invading Nearctic region and three exclusively Nearctic), *Cervantistellus* Brailovsky & Barrera (two), *Cleotopetalops* Brailovsky (three), *Ctenomelynthus* Breddin (five), *Empedocles* Stål (two), *Ichilocoris* Brailovsky & Barrera (two), *Laminiceps* Costa (18), *Leptopetalops* Breddin (two), *Lucullia* Stål (one), *Meluchopetalops* Breddin (two), *Petalops* Amyot & Serville (12), *Placophyllopus* Blöte (one), *Salapia* Stål (17), *Stenometapodus* Breddin (five), and *Thymetus* Stål (two) (CoreoideaSF Team 2015).

The natural history and behavior of *Acanthocephala* spp. have been studied in depth by several authors. The incrassate hind femora of males are used as a weapon in aggressive encounters in two species. Eberhard (1998) provides a detailed description of feeding, aggression, and mating behavior in *Acanthocephala declivis guatemalensis* [which may be *Acanthocephala alata* Burmeister; see McPherson et al. (2011) and Brailovsky (2006a, b) and similar aggressive behavior occurs in *A. femorata* (Mitchell 1980b)]. The trunks and branches of *Cordia* sp. (Boraginaceae) trees were the feeding site for *A. declivis*, whereas *A. femorata* males defended territories on the stalks of cultivated sunflower. McPherson et al. (2011) provide host plant records for two species whose distribution extends into the Neotropics: *Acanthocephala femorata* (F.) and *Acanthocephala declivis* (Say). Vegetative plant parts (stems, shoots, and branches) appear to be the preferred sites for species in this genus, but more information is needed regarding other genera before a generalization may be made regarding the tribe.

Plate 4 - Acanthocephalini Stål, 1870 (Scalebars = 5mm): 1. *Acanthocephala alata* (Burmeister, 1835) [ZMHB]; 2. *Cleotopetalops bicolor* Brailovsky, 1999 [CAS]; 3. *Ctenomelynthus brunneiventris* Breddin, 1903 [ISNB]; 4. *Laminiceps superbus* Breddin, 1903 [USNM]; 5. *Leptopetalops neotropicalis* Brailovsky, 1991 [RNHL]; 6. *Lucullia flavovittata* Stål, 1865 [BMNH]; 7. *Meluchopetalops banausus* Breddin, 1903 [DEI]; 8. *Petalops napoensis* Brailovsky, 1991 [BMNH]; 9. *Placophyllopus colthurnatus* Blöte, 1938 [BMNH]; 10. *Salapia nigra* Brailovsky, 1992 [CAS]; 11. *Stenometapodus impictus* Breddin, 1903 [ISNB]; 12. *Thymetus ochropterus* (Stål, 1860) [BMNH].

Plate 19.4 Coreinae plate—Acanthocephalini

Acanthocerini Bergroth (Plate 19.5)

Species belonging to this tribe can be found in open areas on herbs, shrubs, and small trees but also in crops. Bugs belonging to Acanthocerini have the head sub-quadrate; ocelli placed on prominent tubercles; blunt tylus slightly exceeding the jugae; antenniferous tubercles often armed laterally with a spine; antennal segment four robust; metathoracic scent gland opening placed almost ventrally, with a single, sometimes double, fused auricle; femora incrassate especially in males, sometimes armed subapically on ventral surface; hind tibiae lacking foliaceous expansions; spermatheca with the bulb elongate; parameres membranous (O'Shea 1980a; Packauskas 1994). This tribe has received little attention since O'Shea (1980a) for a group with species easy to find and catch, very common in collections, and with a certain economic importance. Brailovsky (1987a, 1988a) and Brailovsky and Barrera (2003a) described new genera and added important information to identify some genera.

Acanthocerini includes 17 genera (most of them with very few species) and 50 species: *Acanthocerus* Palisot de Beauvois (three species), *Athaumastus* Mayr (8), *Beutelspacoris* Brailovsky (two), *Camptischium* Amyot & Serville (three), *Crinocerus* Burmeister (one), *Dersagrena*, Kirkaldy (three); *Elachisme* Kirkaldy (one), *Euthochtha* Mayr (one), *Golema* Amyot & Serville (two), *Lybindus* Stål (four), *Machtima* Amyot & Serville (three), *Moronopelios* Brailovsky (two), *Rondoneva* Brailovsky & Barrera (one), *Sagotylus* Mayr (one), *Schaeferocoris* O'Shea (one), *Thlastocoris* Mayr (two), and *Zoreva* Amyot & Serville (12) (CoreoideaSF Team 2015). This group is basically South American, but genera like *Acanthocerus* and *Euthochtha* have a North American and Antillean distribution, and *Sagotylus* can be found in the northern part of the Neotropical region and United States.

The natural history of *Athaumastus haematicus* (Stål) has been described in Brazil (Merti 1940, cited in O'Shea 1980a); Schaefer and Mitchell (1983), Schaefer and O'Shea (1979), and Bosq (1937) also give host plant records for this minor crop pest. On pineapple, *Lybindus dichrous* Stål adults and nymphs cluster on the peduncle of the fruit and suck plant juices; plants without fruits are not attacked (Mariconi 1953). Other minor economic pests include *Thlastocoris laetus* Mayr on pineapple; *Machtima crucigera* (F.) and *Euthochtha galeator* (F.) on flowers; *Dersagrena flaviventris* (Berg) on cotton; *Camptischium clavipes* (F.) on castor bean, ice cream bean, and the medicinal plant, guaco; and *Sagotylus confluens* (Say) on cowpea (Mead 2014; Mitchell 2000; Schaefer and O'Shea 1979, and references therein). The immature stages of *E. galeator* in North America were described by Yonke and Medler (1969a), and combat between males of this species was noted by Mitchell (1980b). *Crinocerus sanctus* (F.), a major pest of cowpea in Brazil, is treated below (section "*Crinocerus* Burmeister"). A key to the genera of Acanthocerini can be found in O'Shea (1980a).

Plate 5 - Acanthocerini Bergroth, 1913 (Scalebars = 5mm): 1. *Acanthocerus lobatus* (Burmeister, 1835) [BMNH]; 2. *Athaumastus haematicus* (Stål, 1860) [ZMHB]; 3. *Beutelspacoris sanchezi* Brailovsky, 1987 [UNAM}; 4. *Camptischium clavipes* (Fabricius, 1803) [BMNH]; 5. *Dersagrena subfoveolata* (Berg, 1892) [BMNH]; 6. *Elachisme mundula* (Stål, 1859) [BMNH]; 7. *Euthochtha galeator* (Fabricius, 1803) [USNM]; 8. *Golema histrio* (Burmeister, 1835) [ZMHB]; 9. *Lybindus rufocinctus* Stål, 1860 [ZMHB]; 10. *Moronopelios vespiformis* Brailovsky, 1988 [CAS]; 11. *Sagotylus confluens* (Say, 1832) [BMNH]; 12. *Zoreva lacerna* Brailovsky & Barrera, 1982 [CAS];

Plate 19.5 Coreinae plate—Acanthocerini

Anisoscelini Laporte (Plate 19.6)

Since Stål (1868), several authors have used erroneously the name Anisocelidini to designate this tribe. Species belonging to Anisoscelini are very characteristic because of the foliaceous expansion of the hind tibiae, bright colors, and abundance in agricultural and urban environments. Body elongate and narrow, ratio of body length to width greater or equal to three, less than four. Head usually longer than scutellum. Head usually porrect; tylus and mandibular plates never deflexed; both usually extend beyond antenniferous tubercles. Distance between antenniferous tubercles greater than width of one tubercle. Antennal segment IV never the shortest and longer than segment III. Antennal segment I usually longer than head. Bucculae short extending or not to the level of anterior margin of the eye. Rostrum extending to at least metacoxae, usually onto abdomen. Pronotum steeply declivent; with distinct collar, punctate, occasionally rugose, never tuberculate; humeral angles acute to obtusely angled; posterior angles obtuse to rounded; callar region raised. Usually lacking metapleural supracoxal spine. Metapleural scent gland opening separated, with two auricles; anterior auricle larger. Abdomen rarely wider than thorax. Spiracle of abdominal segment IV closer to lateral edge of segment than to anterior or posterior edge. Abdominal segments never spined posterolaterally. Connexivum VI separated from dorsum or slightly fused at apex; connexivum VII fused to dorsal plate. Scutellum longer than wide or subequal, rarely wider than long. All femora spined below, at least distally. Hind femora rarely incrassate or spined above.

This tribe is one of the largest in the Neotropics and the subject of recent revisions: taxonomic and nomenclatural changes led Leptoscelini to be considered part of Anisoscelini; important revisions affected *Leptoglossus* and *Phthia*, genera that contain several species quite common in human-altered environments and causing damage to a number of crops; other studies revealed several new genera and species (Packauskas 1994, 2008; Packauskas and Schaefer 2001; Allen 1969; Osuna 1984; Gibson and Holdridge 1918; Brailovsky 1989, 1990a, 1990b, 1996, 1997b, 2001a; Brailovsky and Barrera 1994, 1998b; Brailovsky and Sánchez 1983a).

Anisocelini comprises 27 genera: *Anisoscelis* Latreille (11 species); *Baldus* Stål (one); *Bellamynacoris* Brailovsky (two); *Belonomus* Uhler (one); *Chondrocera* Laporte (two); *Coribergia* Casini (one); *Dalmatomammurius* Brailovsky (six); *Diactor* Perty (two); *Holhymenia* Lepeletier & Serville (eight); *Kalinckascelis* Brailovsky (one); *Leptoglossus* Guérin-Méneville (57); *Leptopelios* Brailovsky (one); *Leptoscelis* Laporte (25); *Leptostellana* Brailovsky (one); *Malvana* Stål (three); *Malvanaioides* Brailovsky (five); *Narnia* Stål (seven, mostly Nearctic with four species invading northern boundaries of Neotropical region); *Onoremia* Brailovsky (one); *Phthia* Stål (four); *Phthiacnemia* Brailovsky (one); *Phthiadema* Brailovsky (three); *Phthiarella* Brailovsky, 2009 (4); *Plunentis* Stål, 1860 (2); *Rhytidophthia* Brailovsky, 2009 (1); *Sephinioides* Brailovsky (two); *Tarpeius* Stål (two); and *Ugnius* Stål (two) (CoreoideaSF Team 2015). Most of the species belonging to this tribe are found in tropical areas, but a few are found in North America (e.g., species of *Leptoglossus*). Osuna (1984) made several changes,

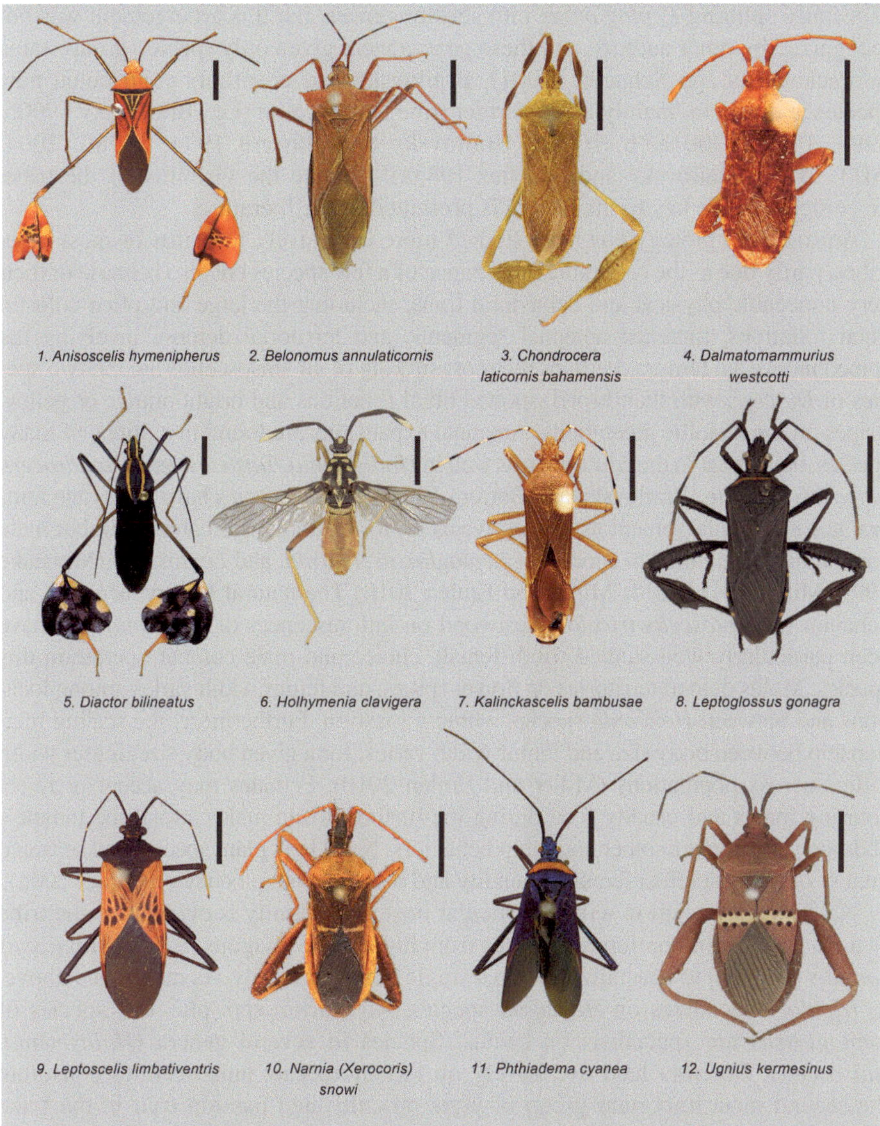

Plate 6 - Anisoscelini Laporte, 1832 (Scalebars = 5mm): 1. *Anisoscelis hymenipherus* Westwood, 1840 [OXUM}; 2. *Belonomus annulaticornis* Uhler, 1869 [USNM}; 3. *Chondrocera laticornis bahamensis* Brailovsky & Barrera, 1994 [USNM}; 4. *Dalmatomammurius westcotti* Brailovsky, 1982 [UNAM]; 5. *Diactor bilineatus* (Fabricius, 1803) [MNHN]; 6. *Holhymenia clavigera* (Herbst, 1784) [BMNH]; 7. *Kalinckascelis bambusae* Brailovsky, 1990 [UNAM]; 8. *Leptoglossus gonagra* (Fabricius, 1775) [BMNH]; 9. *Leptoscelis limbativentris* Breddin, 1901 [DEI]; 10. *Narnia (Xerocoris) snowi* Van Duzee, 1906 [CAS]; 11. *Phthiadema cyanea* (Signoret, 1862) [NHMW]; 12. *Ugnius kermesinus* (Linnaeus, 1758) [BMNH].

Plate 19.6 Coreinae plate—Anisoscelini

especially splitting *Leptoglossus* into several genera, but this arrangement was not recognized by other authors, and these genera were given only species-group status by Packauskas and Schaefer (2001). Brailovsky and coauthors contributed new species and genera, mainly in the former tribe Leptoscelini (i.e., Brailovsky 1990b, 1993, 1997b, 2001a, b, 2009b; Brailovsky and Barrera 1994, 1998b, 2004, 2011, 2013; Brailovsky and Sánchez 1983a). Despite the visibility of the tribe, no comprehensive key to the genera is presented in the literature.

Anisosceline biology has been studied more extensively than that of most other tribes, partly due to the economic importance of a few species but also because of their very noticeable physical and behavioral traits, including the large and often colorful tibial foliations, flattened antennal segments, and territorial defense involving the spined incrassate femora. Perhaps the most striking of all anisoscelids are the two species of *Diactor*, with their broad, spotted tibial foliations and bright orange or yellow stripes on the metallic green body. Antennal expansions are found in nymphs of many species, but appear in the adult stage as well in *Holhymenia*, *Tarpeius*, and *Chondrocera laticornis* Laporte (Brailovsky and Barrera 1994). Not all species have incrassate femora, nor is the enlargement as impressive as in, for example, Nematopodini, but male combat has been noted in species of *Leptoglossus*, *Narnia*, and *Leptoscelis* (Miyatake 1997; Miller 2011, 2013; Miller and Emlen 2010) The natural history, ecology, and behavior of *Leptoscelis tricolor* Westwood on inflorescences of *Heliconia* spp. have been particularly well-studied. Both female choice and male combat operate in this species. Males defend territories on flower spikes, and femur width varies among locations and between *Heliconia* species within a location. Furthermore, the scaling relationship between body size and femur width varies; for a given body size, femur width differs across populations (Miller and Emlen 2010). Females may accept or reject courting males and usually feed during the mating, while males stroke the female's abdomen and perform other courtship behaviors. Natal host plant species and seasonal quality of the host affect female fecundity and male courtship behavior (Miller 2008).

No strong association with a particular host plant family is evident for the tribe as a whole. Feeding preference ranges from highly polyphagous [e.g., *Leptoglossus zonatus* (Dallas)] to specialization on a single genus or family. As mentioned above, *L. tricolor* specializes on *Heliconia* species. All *Narnia* spp. plus two species of *Leptoglossus* are specialists on cactus. Species in several genera (*Holhymenia*, *Anisoscelis*, *Diactor*) feed exclusively on Passifloraceae; anisoscelines constitute the second most important group of pests on cultivated passion fruit in the Lake Maracaibo region of Venezuela (Aguiar-Menezes et al. 2002). The latter authors provide information on the life history of *Diactor bilineatus* (F.).

Most if not all anisoscelines seem to be associated with reproductive plant parts. Feeding on vascular tissue (primarily xylem) and nectar is reported, but xylem represents a source of hydration (Rodrigues et al. 2007a; Mitchell 2006). Nymphs are usually observed on buds or fruits. When two species of oligophagous anisoscelines were offered a mixed diet of *Passiflora* spp., food preference (species and tissue) changed with instar (Rodrigues et al. 2007b). Polyphagous species [e.g., *Leptoglossus phyllopus* (L.)] may move generationally among host plants, tracking plant reproductive structures (Mitchell 2006).

Host plant records for various species are given by Brailovsky and Sánchez (1983a) and Schaefer and Mitchell (1983). Species and genera of major importance, including targets of ecological and behavioral research as well as serious economic pests, are treated separately in Sect. 19.5.2 (*Anisoscelis*, *Holhymenia*, *Leptoglossus Narnia*, and *Phthia s.l.*).

Barreratalpini Brailovsky (Plate 19.7)

This tribe can be recognized by body brown, medium size, dorsoventrally flattened; head longer than wide, shorter than pronotum; mandibular plates longer than tylus but not contiguous in front of it; antennae terete, first antennal segment strong, fourth the shortest; brachypterous; ventral surface of the femora with two short distal spines; tibiae terete; abdomen enlarged; connexival angles not developed.

This tribe was proposed by Brailovsky (1988b) for the new genus and species—*Barreratalpa bambusii*—collected on bamboo in Mexico. After the initial description, only Packauskas (1994) mentioned this tribe. Brailovsky (1988b) provides a short but useful discussion justifying the separation of the new tribe from others.

Chariesterini Stål (Plate 19.7)

This small tribe is formed by four genera and 24 species, most living in Central America and the northern part of South America. No formal description has been made, but species can be identified by the small size; body narrow with lateral margins almost parallel; third antennal segment expanded, other segments terete; tylus and juga deflexed not filling space between antenniferous tubercles; all femora armed distally; distance between hind coxae much narrower than distance from coxa to lateral margin, hind tibiae never expanded. The single species of *Chariesterus* registered to Brazil—*C. armatus*—can be collected on Euphorbiaceae (*Euphorbia* spp.). The Caribbean species *Chariesterus gracilicornis* (Stål) is reported from amaranth and cucurbits in addition to euphorbs (Schaefer and Mitchell 1983). The two most comprehensive works are the revision of *Chariesterus* by Ruckes (1955) and the description of *Ruckesius* by Yonke (1972) that also presented a key to the genera.

Chelinideini Blatchley (Plate 19.7)

Species of Chelinideini belong to a single genus—*Chelinidea* Uhler. They are known as cactus bugs because they live on these plants causing damage to certain species. Hamlin (1924) revised *Chelinidea* Uhler and gave a more comprehensive description of the genus. Herring (1980) revised *Chelinidea* again adding new species, a key to the species, and new information about distribution, biology, and host plants. They can be recognized by the body short somewhat broader, oval. Head as

Plate 7 - Barreratalpini Brailovsky, 1988 (Scalebars = 5mm): 1. *Barreratalpa bambusii* Brailovsky, 1988 [UNAM]; **Chariesterini Stål, 1868:** 2. *Chariesterus moestus* Burmeister, 1835 [ZMHB]; 3. *Chariesterus robustus* Distant, 1892 [BMNH]; 4. *Plapigus abdominalis* (Dallas, 1852) [BMNH]; 5. *Ruckesius kapleri* Yonke, 1972 [USNM]; 6. *Staluptus marginalis* (Burmeister, 1835) [ZMHB]; **Chelinideini Blatchley, 1926:** 7. *Chelinidea tabulata* (Burmeister, 1835) [ZMHB]; **Discogastrini Stål, 1868:** 8. *Cnemomis ticinensis* Brailovsky & Barrera, 1986 [RNHL]; 9. *Savius carayoni* Brailovsky, 1986 [RNHL]; 10. *Savius jurgiosus nigroclavatus* Brailovsky, 1986 [UNAM] 11. *Scamurius abnormis* Brailovsky, 1986 [RNHL]; 12. *Schuhgaster formidabilis* Brailovsky, 1993 [UNAM].

Plate 19.7 Coreinae plate—Barreratalpini-Chariesterini-Chelinideini-Discogastrini long as pronotum. Tylus appearing conical from above. Juga acutely pointed and level with or surpassing tylus. Eyes small, subcircular, and closer to base of the head than to antennal tubercle. First three antennal segments prismatic; basal segment flattened and crested on inner edge; second and third subequal; distal rounded and tapered toward either extremity. Bucculae about one-third length of the basal segment of rostrum. Rostrum just attaining the abdomen; first segment slightly longer than the head; second subequal to first; third the shortest and fourth two-thirds the length of the second. Pronotum nearly twice as long as wide; anterior margin forming a narrow collar; anterolateral margin entire straight to slightly rounded; humeral angles rounded to obtuse and not developed. Sterna sulcated to receive the rostrum. Connexivum bent upward and distinctly dilated. Posterior femora enlarged and armed with two rows of teeth on under surface; anterior and middle femora armed with smaller and fewer teeth on under surface (adapted from Hamlin 1924).

The genus *Chelinidea* Uhler is formed by five species found in Nearctic region but also reaching or crossing Neotropical northern boundaries. Only *C. tabulata* (Burmeister) has a larger distribution in the Neotropical region, reaching Venezuela. *Chelinidea staffilesi* and *C. tabulata* nymphs were studied and described by Brailovsky et al. (1994). Several species, including *C. tabulata*, have been considered as biological control agents for cacti (*Opuntia* spp.) but have either failed to disperse or were never released due to the potential to damage other plants (Mitchell 2000 and references therein).

Discogastrini Stål (Plate 19.7)

Species belonging to this tribe are usually medium in size; elongate; light green to green. Head is square and shorter than pronotum. Tylus longer than mandibular plates but not projected in a spine, both usually anteriorly extended between antenniferous tubercles. Rostrum short not reaching mesosternum. Pronotal disk rough and transversely sulcate. Pronotal angles rounded or subacute but never developed. Evaporatorium slightly developed and sulcate. Femora and tibiae terete. Spiracles on anterior third of each abdominal segment. Spine of the sternite VII bifid.

This tribe is formed by eight genera [*Cnemomis* Stål (eight species), *Coryzoplatus* Spinola (two), *Discogaster* Burmeister (five), *Karnaviexallis* Brailovsky (one), *Lupanthus* Stål (two), *Savius* Stål (nine), *Scamurius* Stål (11), and *Schuhgaster* Brailovsky (one)]. Most of the species are found in South America with only a few specimens living in the northern part of the Neotropical region. Only *Cnemomis* and *Scamurius* were revised (Brailovsky and Barrera 1986a; Brailovsky 1987b), but Brailovsky added valuable information to four other genera (Brailovsky 1986a, b, c; Brailovsky and Barrera 2003b) and described two others (Brailovsky 1984, 1993). Information on biology is limited; *Savius jurgiosus jurgiosus* Stål was reported from leaves and pods of *Prosopis laevigata* and leaves of *Ziziphus pedunculata* (Brailovsky et al. 1995) and from *Buddleja sessiliflora* (Brailovsky 1986b) in Mexico; immatures of this subspecies are described by Brailovsky et al. (1995). *Savius* sp. was also reported from the petioles and shoots of Texas ebony, *Ebenopsis*

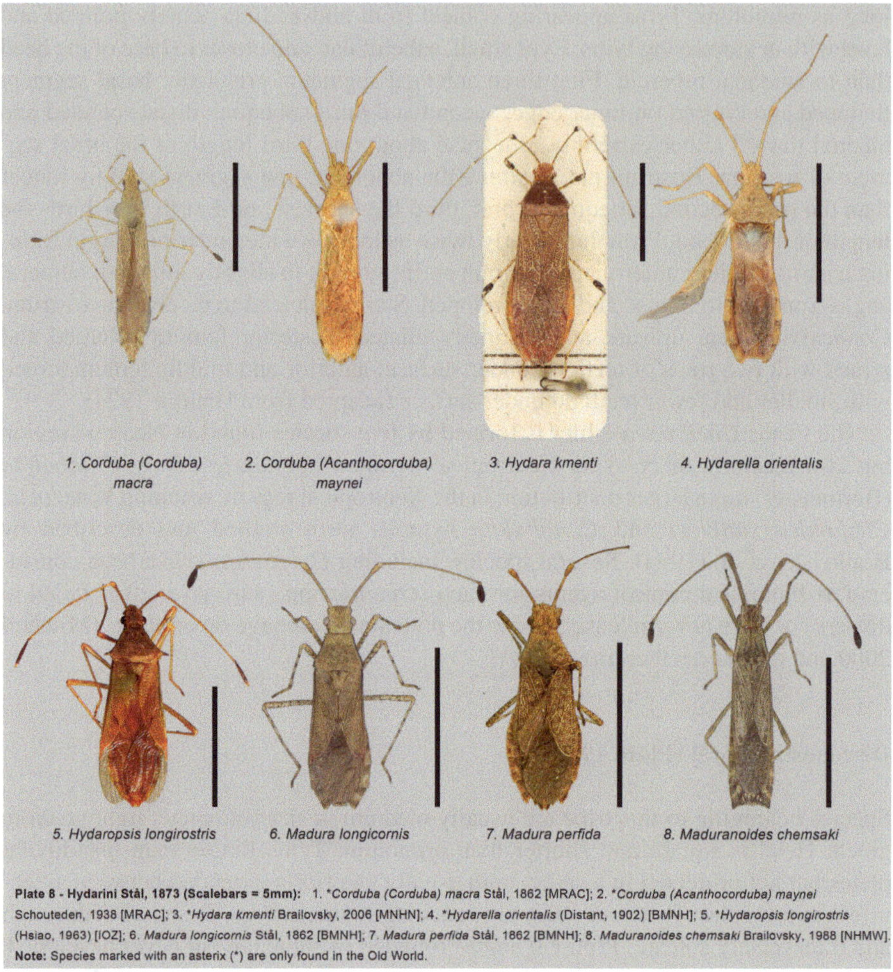

Plate 19.8 Coreinae plate—Hydarini

ebano (Berl.) Barneby & Grimes (as *Pithecellobium*) in southern Texas (Schaefer and Mitchell 1983). A key to the genera included in Discogastrini can be found in Brailovsky (1984, 1993).

Hydarini Stål (Plate 19.8)

This tribe was always considered quite apart from other taxa (Stål 1873; Schaefer 1964) and Ahmad (1970) raised this taxon to subfamily level, but later authors (with the exception of (CoreoideaSF Team 2015) are still considering Hydarini as a tribe. The tribe is characterized by the lack of a dorsal tibial sulcus; paramere with a subdistal projection; broad direct articulation of the first valvifer and valvula; and lack of a spermathecal bulb. Head squared to rectangular and rough; first antennal segment slightly swollen distally; antennal segment III more than twice as long as segment II, and antennal segment IV short and flattened; spur of the third metathoracic axillary sclerite well developed; the anterior and posterior lobes of metathoracic peritreme completely separated; evaporatorium without ridges; connexivum ending at sixth tergite; trichobothria of third sternum level with spiracles; anterior spur of the seventh sternum double; trichobothria of fourth sternum anterior to spiracle; all femora slightly swollen anteriorly; and the hind tibiae not expanded.

This small tribe is formed by nine genera and 23 species; most of the genera and species are found in Africa and Asia. The Neotropical fauna is composed of *Hydarellamixia* (one species), *Hydascelis* (one), *Madura* (three), and *Maduranoides* (one). These species are distributed from Argentina to the United States. Brailovsky (1988c, 2010a) provides the necessary information to identify genera. *Madura perfida* Stål is reported from grasses and dry corn (references in Brailovsky and Rivera 1979) and from *Annona cherimola* Mill. (Peña and Bennett 1995). A key to the genera known from Neotropical region can be found in Brailovsky (2010a).

Hypselonotini Bergroth (Plates 19.9 and 19.10)

This tribe was first mentioned by Bergroth (1913) as Hypselonotaria, which was also used by Blöte (1935). No description or diagnosis for the tribe was given by Bergroth or Blöte, so a study of the relationships among these genera is badly needed and a careful evaluation of the separation between Hypselonotini and Coreini considered. Several genera belonging to this tribe were the subject of recent revisions, and when mentioned, Coreini is considered as the valid tribe for them (Brailovsky 1983a, 1985, 1988d, 1988e, 1989, 1990c, 1992a, 1995a; Brailovsky and Cadena 1992; Brailovsky and Garcia 1987; Brailovsky and Melendez 1989; Whitehead 1974). Species belonging to this tribe are distributed across the Americas, including Chile, with a considerable number living in subtropical areas. No key for the genera of the tribe is known.

This tribe comprises 34 genera and 352 species: *Acanonicus* Westwood (three species), *Acidomeria* Stål (11), *Althos* Kirkaldy (24), *Anasa* Amyot & Serville (76), *Arioge* Stål (one), *Astygiton* Berg (one), *Catorhintha* Stål (32), *Cebrenis* Stål (25), *Cebreniscella* Brailovsky (two), *Cebrenistella* Brailovsky (three), *Cimolus* Stål (four), *Collatia* Stål (five), *Daphnasa* Brailovsky (four), *Eldarca* Signoret (two), *Encedonia* Stål (two), *Hypselonotus* Hahn (12), *Microbasis* Dallas (one), *Namacus* Amyot & Serville (four), *Nirovecus* Stål (three), *Nisoscolopocerus* Barber (two), *Paryphes* Burmeister (28), *Petersitocoris* Brailovsky (one), *Petersitocoroides* Brailovsky (four), *Rochrosoma* Reed (one), *Scolopocerus* Uhler (four), *Sethenira* Spinola (five), *Sphictyrtus* Stål (20), *Stenoprasia* Horváth (five), *Stiroptus* Stål (one), *Sundarus* Amyot & Serville (31), *Vazquezitocoris* Brailovsky (14), *Viklundocoris* Brailovsky (one), *Villasitocoris* Brailovsky (1), and *Zicca* Amyot & Serville (19) (CoreoideaSF Team 2015).

Plate 9 - Hypselonotini Bergroth, 1913 (Scalebars = 5mm): 1. *Acidomeria sp.* [BMNH]; 2. *Althos pericarti* Brailovsky, 1990 [BMNH]; 3. *Anasa litigiosa* Stål, 1862 [NHMW]; 4. *Arioge germainii* (Signoret, 1863) [NHMW]; 5. *Catorhintha guttula* (Fabricius, 1794) [RNHL]; 6. *Cebrenis cauta* Brailovsky, 1995 [MNHN]; 7. *Cimolus luteus* Brailovsky, 2001 [MNHN]; 8. *Daphnasa mucronata* (Distant, 1881) [MNHN]; 9. *Encedonia mutica* (Signoret, 1863) [NHMW]; 10. *Hypselonotus lineatus* Stål, 1862 [NHMW]; 11. *Namacus prominulus* (Stål, 1860) [ZMHB]; 12. *Nirovecus claviger* Stål, 1860 [BMNH].

Plate 19.9 Coreinae plate—Hypselonotini

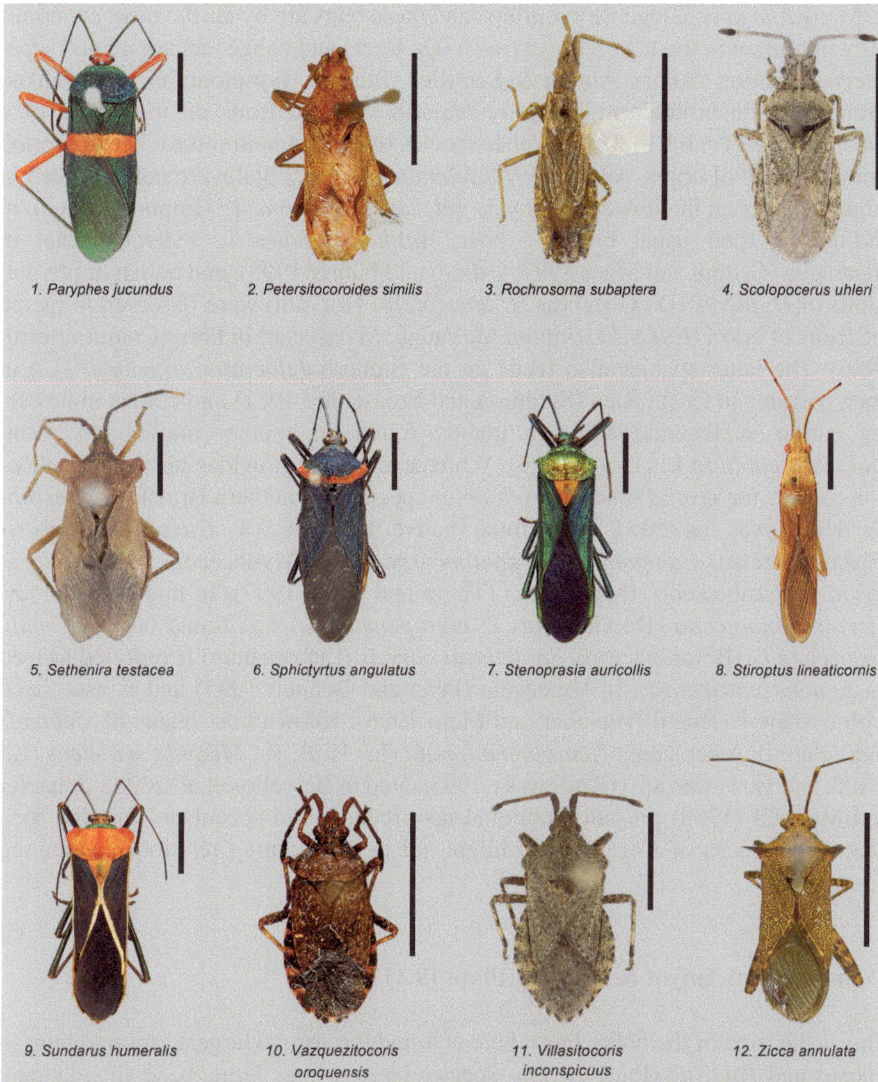

Plate 10- Hypselonotini Bergroth, 1913 (Scalebars = 5mm): 1. *Paryphes jucundus* Horváth, 1913 [HNHM]; 2. *Petersitocoroides similis* Brailovsky, 1992 [RNHL]; 3. *Rochrosoma subaptera* Reed, 1899 [USNM]; 4. *Scolopocerus uhleri* Distant, 1881 [BMNH]; 5. *Sethenira testacea* Spinola, 1837 [MRSN]; 6. Sphictyrtus angulatus Horváth, 1913 [HNHM]; 7. *Stenoprasia auricollis* Horváth, 1913 [HNHM]; 8. *Stiroptus lineaticornis* Stål, 1860 [ZMHB]; 9. *Sundarus humeralis* Horváth, 1913 [HNHM]; 10. *Vazquezitocoris oroquensis* Brailovsky, 1990 [UNAM]; 11. *Villasitocoris inconspicuus* (Herrich-Schäffer, 1842) [BMNH]; 12. *Zicca annulata* (Burmeister, 1835) [NHMW].

Plate 19.10 Coreinae plate—Hypeslonotini 2

Many *Anasa* spp. feed on cucurbits, and these bugs are by far the most economically important in the tribe; *Anasa tristis* (De Geer) and congeners are treated separately in section "*Anasa* Amyot & Serville". But the Hypselonotini is not linked strongly to Cucurbitaceae; *Paryphes blandus* Horváth feeds on the leaves of a cucurbit vine (Young 1980), but other species of Hypselonotini have been reported from a variety of crops. Adults of *Hypselonotus lineatus* Stål were noted to feed in large numbers on the flowers of physic nut, *Jatropha curcas* L. (Euphorbiaceae), in addition to their usual breeding host, *Waltheria indica* L. (Sterculiaceae) in Nicaragua (Grimm and Maes 1997; Grimm and Führer 1998), and both nymphs and adults of *H. fulvus* (De Geer) (as *H. lanceolatus* Horváth) were observed to pierce the fruits of arazá (*Eugenia stipitata* McVaugh, Myrtaceae) in Peru (Couturier et al. 1996). The latter species also feeds on the euphorb *Julocroton argenteus* (L.) in open pastures in Costa Rica (Solomon and Froeschner 1981) and comes to flowering cotton in Trinidad although mainly found on another malvaceous plant, *Malachra capitata* L. (Urich 1916). Whitehead (1974) provides host plant collection records for several other *Hypselonotus* species. In southern Brazil, *H. interruptus* Hahn was recorded from three native tree species: *Casearia sylvestris* (Flacourtiaceae), *Campomanesia xanthocarpa* Berg (Myrtaceae), and *Myrciaria rivularis* Cambessedes (Myrtaceae) (Thum and Costa 1997). In this same survey, *Zicca nigropunctata* (De Geer) (as *Z. nigropunctulata*) was found on *Patagonula americana L.* (Boraginaceae). *Sphictyrtus chrysis* (Lichtenstein) is reported to feed on *Annona muricata* L. in Venezuela (Peña and Bennett 1995) and is associated with cashew in Brazil (Bleicher and Melo 1996). Known host plants of *Cebrenis supina* are in Asteraceae (*Neurolaena lobata* (L.) R. B. R., *Mikania scandens* (L.) Willd, and *Verbesina* sp.) (Brailovsky 1995, cited in Barcellos et al. 2008). Schaefer and Mitchell (1983) present additional host records and speculated that the food plant preferences of Coreini (s.l.) might aid in determining relationships among genera.

Nematopodini Amyot & Serville (Plate 19.11)

This is the tribe of the bulky bugs; here is found the second largest group of bugs—*Thasus* and *Pachylis* (Plate 19.11). Species large to very large, head subquadrate; tylus projecting slightly beyond juga; antenniferous tubercles unarmed, occupying most of anterior head, ocellar tubercles small; metathoracic scent gland opening placed almost ventrally, peritreme with two completely separated lobes and area between them depressed; femora incrassate, posterior femora markedly incrassate especially in male and armed ventrally, anterior femora distally armed with two ventral spines; tibiae sometimes dilated; abdominal venter unarmed; paramere heavily sclerotized (O'Shea 1980b; Packauskas 1994).

Nematopodini and Acanthocerini were part of the Mictini until O'Shea and Schaefer (1978) reinstated them, separating the Neotropical species of the first two

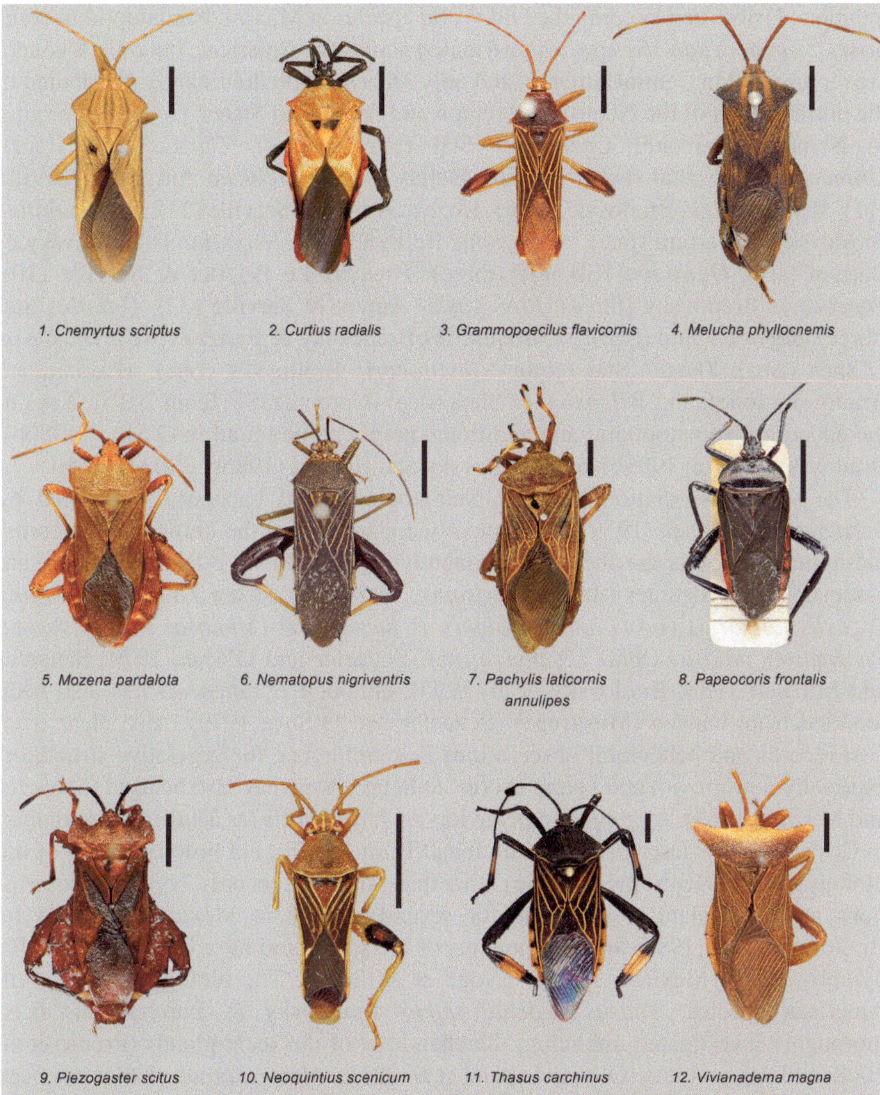

Plate 11 - Nematopodini Amyot & Serville, 1843 (Scalebars = 5mm): 1. *Cnemyrtus scriptus* (Burmeister, 1835) [ZMHB]; 2. *Curtius radialis* Brailovsky, 1986 [NHMW]; 3. *Grammopoecilus flavicornis* (Fabricius, 1803) [ZMUC]; 4. *Melucha phyllocnemis* (Burmeister, 1835) [ZMHB]; 5. *Mozena pardalota* Brailovsky & Barrera, 2001 [UNAM]; 6. *Nematopus nigriventris* Breddin, 1904 [DEI]; 7. *Pachylis laticornis annulipes* Blöte, 1938 [RNHL]; 8. *Papeocoris frontalis* Brailovsky, 2002 [UNAM]; 9. *Piezogaster scitus* Brailovsky & Barrera, 1984 ; 10. *Neoquintius scenicum* (Brailovsky & Barrera, 1986) [RNHL]; 11. *Thasus carchinus* Brailovsky & Barrera, 1994 [UNAM]; 12. *Vivianadema magna* Brailovsky, 1987 [AMNH].

Plate 19.11 Coreinae plate—Nematopodini

mentioned tribes and keeping the Old World species in Mictini. Nematopodini comprises 22 genera and 161 species distributed across the Americas, but only 8 genera are represented in Central America and only *Mozena* is predominantly distributed in the northern part of the Neotropical region and the United States. Genera belonging to Nematopodini are: *Cnemyrtus* Stål (two species), *Curtius* Stål (four), *Grammopoecilus* Stål (two), *Himella* Dallas (three), *Melucha* Amyot & Serville (11), *Meluchamixia* Brailovsky (one), *Mozena* Amyot & Serville (31), *Nectoquintius* Brailovsky & Barrera (one), *Nematopus* Berthold (31), *Neoquintius* Brailovsky & Barrera (six), *Ouranion* Kirkaldy (three), *Pachylis* Le Peletier & Serville (10), *Papeocoris* Brailovsky (three), *Piezogaster* Amyot & Serville (33), *Quintius* Stål (three), *Saguntus* Stål (two), *Stentoquintius* Brailovsky & Barrera (two), *Thasopsis* O'Shea (two), *Thasus* Stål (eight), *Tovarocoris* Brailovsky (one), *Vivianadema* Brailovsky (one), and *Wilcoxina* O'Shea (one) (CoreoideaSF Team 2015). Keys to the genera of Nematopodini and additional notes can be found in O'Shea (1980b), Brailovsky (1987a, 1995b), and Brailovsky and Barrera (1986b, 2002, 2009b).

The strong association between Nematopodini and Fabaceae was noted by Schaefer and O'Shea (1979) and is clearly apparent from the known host records; mesquite and acacia are the most frequently reported hosts. All *Mozena* spp. are associated with legumes (*Acacia*, *Prosopis*, *Schrankia*), as are *Thasus* (*Prosopis*), *Pachylis laticornis* (F.) (*Pithecellobium*), *P. hector* Stål (*Mimosa*, *Prosopis*), and *Piezogaster odiosus* (Stål) (*Pachyrrhizus*) (Schaefer and O'Shea 1979; Schaefer and Mitchell 1983; Brailovsky et al. 1995), although *P. laticornis* has also been reported from banana (Musaceae) (Schaefer and O'Shea 1979). Also clear from host records and behavioral observations is a preference for vegetative structures, especially new growth and young shoots, although pods may also be used (Schaefer and Mitchell 1983). Aggregation behaviors of *P. laticornis* (as *Thasus acutangulus*) in Costa Rica are described by Aldrich and Blum (1978), but little is known of the biology of other Neotropical species other than host plant records. Nymphal descriptions and host plant associations for several species in Mexico are given by Brailovsky et al. (1995) and components of the scent fluid have been identified for nymphs of one Mexican species (Noge et al. 2012). The biology of the North American species, *Thasus neocalifornicus* Brailovsky & Barrera, has been thoroughly investigated, including the chemistry of the scent glands (Prudic et al. 2008) and the gut biota (Olivier-Espejel et al. 2011). *Mozena obtusa* Uhler has been extensively studied in Texas as a potential biological control agent for mesquite in Australia and South Africa (Cuda and DeLoach 1998).

Placoscelini Stål (Plate 19.12)

This tribe was proposed by Schaefer (1968) as Stenoscelideini, but only Packauskas (2006) presented a more complete description and the synapomorphies that support this taxon (mostly from the internal genitalia). Brailovsky and Barrera (2012) presented a diagnosis of the tribe and a key for the genera, both reproduced below. Dolling in the Coreoidea Species File (CoreoideaSF Team 2015) commented that

Plate 12 - Placoscelini Stål, 1868 (Scalebars = 5mm): 1. *Nyttum beckeri* Dolling & Casini, 1988 [BMNH]; 2. *Plaxiscelis plebeja* Stål, 1860 [BMNH]; 3. *Stenoeurilla pellucida* (Brailovsky & Barrera, 1997) [BMNH]; 4. *Stenoscelidea morona* Brailovsky & Barrera, 1996 [USNM]; **Spartocerini Amyot & Serville, 1843:** 5. *Euagona diana* Dallas, 1852 [BMNH]; 6. *Eubule rugulosa* Brailovsky, 1992 [UNAM]; 7. *Menenotus diminutus* Walker, 1871 [BMNH]; 8. *Molchina xantha* Brailovsky, 2006 [UNAM]; 9. *Sephina quintanarooana* Brailovsky, 1987 [UNAM]; 10. *Spartocera gigantea* Distant, 1892 [BMNH]; **Incertae sedis:** 11. *Amblyomia foreroi* Brailovsky, 1998 [NHMW]; 12. *Mamurius mopsus* Stål, 1862 [UNCB].

Plate 19.12 Coreinae plate—Placoscelini Spartocerini

Placoscelini Stål (as Placoscelida) has priority over Stenoscelideini Schaefer, because *Plaxiscelis* Spinola (= *Placoscelis* Agassiz) is included in this tribe. Head with postclypeal sulcus single or double (with internal ridge); clypeus raised as a short tubercle above deflexed juga; distance between antenniferous tubercles slightly less than width of one tubercle; antennal segment I longer than maximal length of head; antennal segment IV the longest and never longer than II and III together; rostrum reaching posterior margin of mesosternum or anterior third of metasternum; rostral segment III the shortest. Pronotum slightly declivent, bilobate; callar region slightly raised; collar present; humeral angles obtusely rounded or with a tiny spinous projection; metathoracic scent gland peritreme in form of two separate auricles with distinct bridge between them. Macropterous, hemelytra extending beyond apex of last abdominal segment. Connexival segments raised above terga; posterior angle unarmed, never spined postero-apically; abdominal spiracles circular.

This tribe comprises only four genera: *Nyttum* Spinola (10 species), *Plaxiscelis* Spinola (eight), *Stenoeurilla* Brailovsky & Barrera (eight), and *Stenoscelidea* Westwood (13) (CoreoideaSF Team 2015). These taxa are composed of species found basically in South America and previously placed in Acanthocephalini, except those belonging to *Stenoeurilla*. *Stenoeurilla aenescens* (Stål) (as *Stenoscelidea*) was reported by Solomon and Froeschner (1981) to feed on *Lonchocarpus costaricensis* Pittier. A key can be found in Brailovsky and Barrera (2012).

Spartocerini Amyot & Serville (Plate 19.12)

This tribe was called erroneously Corecorini by several authors that followed van Duzee (1916) considering *Corecoris* Hahn as the valid name instead of *Spartocera* Laporte. Specimens small to large with body robust. Head rectangular and declivent. Antenniferous tubercles close together, unarmed and projected beyond tylus. Antennae short with segments stout. Rostrum short just surpassing procoxae. Pronotum steeply declivent with anterior margin tuberculate or crenulate. Metacoxae separate from each other by the same distance of the metacoxa to the lateral margin. Posterior wing with cubital vein strongly curved anteriorly and hamus reaching the base of the wing. Spur of the third alar sclerite not developed. Trichobothria of the third abdominal segment located before the spiracle. Inner laterotergites of the connexivum fused.

This tribe is exclusively Neotropical and represented by six genera: *Euagona* Dallas (three species), *Eubule* Stål (13), *Menenotus* Laporte (two), *Molchina* Amyot & Serville (six), *Sephina* Amyot & Serville (24), and *Spartocera* Laporte (18). No key for all genera has been produced, and information is spread out in the literature. Only *Euagona* was revised recently (Brailovsky 2010b), but other genera were treated to some extent—*Eubule* (Brailovsky 1992b), *Sephina* (Brailovsky 1983b, 1987a, 2001b; Brailovsky and Sánchez 1983b), and *Molchina* (Brailovsky 2006a, key to the species) (CoreoideaSF Team 2015). Two species of *Sephina* are reported to feed on *Metastelma* spp. (Apocynaceae). *Eubule sculpta* (Perty) is reported from *Morrenia* sp. (Apocynaceae) but also from a cucurbit host (Schaefer and Mitchell 1983). The brightly colored bull's horn bugs, *Euagona* spp., exhibit an unusual trait, in that the dorsal coloration of the abdomen of nymphs appears to resemble that of the wings of the adults (W.R. Dolling, pers. comm.). Only one genus in this tribe is of major economic importance: *Spartocera* spp. are covered in detail in section "*Spartocera* Laporte".

Incertae sedis (Plate 19.12) Two genera of Neotropical Coreidae were considered *Incertae sedis* by CoreoideaSF Team(2015). The genus *Amblyomia* Stål is composed by three species revised by Brailovsky (1998b) and considered in Leptoscelidini. Species are known from Mexico and Colombia and apparently feed on bromeliads. Packauskas (2008) removed *Amblyomia* from this tribe and considered it *Incertae sedis* based on an uncoiled vesica and absence of a posttylar sulcus.

The genus *Mamurius* Stål is composed by two northern Neotropical species resembling a small *Piezogaster* (Nematopodini). Some diagnostic characteristics were mentioned by Brailovsky (1982) comparing this genus to *Dalmatomammurius*. O'Shea (1979) considered *Mamurius* out of Nematopodini due to the small size, tylus much projected anteriorly and genitalia not similar to those of the nematopodines.

19.5 Main Species

Here we discuss some genera of major importance for which the biology, ecology, behavior, natural enemies, and/or pest status were not covered in detail in Sect. 19.4. Information on many of these economically important species was summarized in Mitchell (2000); rather than repeat information already available in the literature, we refer the reader to *Heteroptera of Economic Importance* (Schaefer and Panizzi 2000) and focus here primarily on more recent research and updates to our knowledge.

19.5.1 Acanthocerini Bergroth

Crinocerus Burmeister

The single species in this genus, *Crinocerus sanctus* (F.) (Plate 19.1), has a quite recognizable dorsal pattern described by O'Shea (1980a) as "a light brown St. Andrew's cross on a darker brown background." Aspects of the life history and photographs of immature stages and adults are provided by Amaral (1986). Photographs of an adult male, illustrating the incrassate femora, declivent pronotum, and distinctive markings, may also be found in Pall and Coscarón (2013).

These bugs have a wide host range, including leaves, green stems, and nuts of cashew (Bleicher and Melo 1996); oranges, limes, cotton, and leaves and fruit of guava (Amaral 1986); and new shoots and green fruit of acerola (*Malpighia* spp.) (Soglia et al. 1998). On the latter host, wilting of shoots and deformation of fruit are reported. However, *C. sanctus* is most commonly a pest of legumes, including *Canavalia ensiformis*, guandu bean, *Dolichos* sp., and particularly cowpea, on which bugs cause serious losses, feeding on stems, new foliage, and pods. (Bosq 1937; Amaral 1986; Daoust et al. 1985). In northern Brazil (Pará, Amazonas, Acre), this species is considered a principal pest of cowpea, forming part of a pod-sucking bug complex that includes the pentatomid *Piezodorus guildinii* (Westwood) (Daoust et al. 1985). Laboratory rearing studies using cowpea plus acerola or guandu bean leaves plus cowpea have shown nymphal development to range from 56 days at 27 ± 2 °C (Soglia et al. 1998) to 58.41 ± 1.25 days at 22 ± 2 °C (Amaral 1986).

19.5.2 Anisoscelini Laporte

Anisoscelis Latreille

Bugs in this genus feed on *Passiflora* spp.; flower buds and fruits of cultivated passion fruit damaged by feeding may wither or drop from the vine (Polaszek and Kolberg 2008). *Anisoscelis foliaceus* (F.) has been the most extensively studied, particularly in southern Brazil where it overlaps and shares host plants with *Holhymenia clavigera* Herbst (Rodrigues and Moreira 2005). Eggs are ovoid, golden initially but later turning brown; nymphs are green in the early instars, darkening to ochre in the fifth instar. The expansion of the hind tibia becomes visible in third instars and continues to enlarge during development. Eggs and juveniles of both species were described and compared (Rodrigues and Moreira 2005) and were found to be remarkably similar through the fourth instar except in size and tibial foliation; *H. clavigera* are larger and lack the broad, leaflike expansions characteristic of Anisoscelini. Rearing experiments showed the wild host, *P. suberosa*, to be optimal for growth and development compared with cultivated passion fruit, *P. edulis*. A mix of vegetative and reproductive structures is used, and plant part preference changes over the course of development (Rodrigues et al. 2007b, 2008). Detailed histological studies of feeding sites showed mainly penetration to xylem in mature leaves, whereas several sites within the fruit were targeted, including pericarp, aril (pulp), endosperm, and embryo (Rodrigues et al. 2007a). *Anisoscelis flavolineatus* Blanchard is listed as a pest of cultivated passion fruit in Columbia (Urueta-Sandino 1975), but the genus is not mentioned for Venezuela (Aguiar-Menezes et al. 2002). Only low densities of *A. foliaceus* were found on cultivated passion vines in southeastern Brazil (Caetano et al. 2000, cited in Rodrigues et al. 2007b). Egg parasitoids of *A. foliaceus* include *Gryon chrysolaum* Walker and *G. vitripenne* Masner (Polaszek and Kolberg 2008).

Holhymenia Lepeletier & Serville

These unusual-looking bugs, with black and white markings and clear wings, are thought to be Batesian mimics of ichneumonid wasps (Rodrigues and Moreira 2005) (Plate 19.1). Two very similar species, *Holhymenia histrio* (F.) and *H. clavigera* (Plate 19.6, #6), cause extensive damage to cultivated passion fruit by feeding on stems, floral buds, and fruits. Feeding and consequent damage by *Holhymenia* spp. resembles that of *A. foliaceus*; xylem is probed on leaves, but feeding occurs on pericarp, aril, and seeds. First instars fed on tendrils of *P. edulis* and terminal buds and leaves of the native host, *P. suberosa*; fruit was preferred by later instars. Adults preferred green fruits over purple fruits (Rodrigues et al. 2007a, 2008). The egg and immatures of *H. clavigera* were described in detail by Rodrigues and Moreira (2005) and closely resemble those of *A. foliaceus* until the fifth instar, when the foliated antennal segments of *H. clavigera* develop a brown margin and the legs are strikingly banded. Two platygastrid parasitoids have been reared from eggs of *H. rubiginosus* Breddin: *G. chrysolaum* and *G. vitripenne* (Polaszek and Kolberg 2008).

Leptoglossus Guérin-Menéville

Primarily Neotropical with some Nearctic species, this large genus includes some of the more damaging and most extensively studied of the coreid economic pests. The majority of species are primarily found in South America (Allen 1969), but two species have achieved considerably wider distribution. The range of *Leptoglossus gonagra* (F.) extends well beyond the Neotropics, to Africa, Asia, and Australia. The North American conifer pest, *L. occidentalis* Heidemann, recently invaded Europe, Turkey, and East Asia; it represents a potential threat to commercial seed orchards and European alpine forest ecosystems (Lesieur et al. 2014). In the Neotropics, however, the most damaging species is unquestionably the highly polyphagous *L. zonatus* (Dallas), distributed from Argentina north to the southern United States.

Leptoglossus spp. are characterized by an elongate body shape, head porrect, humeral angles of the pronotum slightly to greatly expanded, hind femora slightly incrassate in males, and hind tibiae foliated, although the shape of the inner and outer dilations varies among species. Sexual dimorphism of the tibial dilations has been reported for a few species (Schaefer et al. 2008). A transverse fascia may be present on the corium; its color varies from bright white to pale, and shape varies in width and in the appearance of the margin: straight or zigzag (Allen 1969). Presently, 61 species are placed in the genus (CoreoideaSF Team 2015), although information on biology is available for relatively few species, primarily those with economic importance or unusual feeding habits.

Some species are host specific at the genus level; others are highly polyphagous, but there appear to be no evolutionary patterns of relatedness among the preferred host plants. Two species are associated with Amazonian palms, *L. hesperus*

Brailovsky in Brazil and *L. lonchoides* Allen in Peru (Couturier et al. 1991; Brailovsky and Couturier 2003). In North America, two species breed exclusively on mistletoe, and two are associated with conifers (Schaefer and Mitchell 1983). The conifer-feeding species are related to one another and to a more broadly feeding species that includes gymnosperms in the diet; together they constitute the *corculus* group of Allen (1969) and belong to the *zonatus* group of Packauskas and Schaefer (2001). Similarly, the mistletoe feeders are closely related (Allen 1969). However, the three specialists on Magnoliaceae (*L. katiae* Schaefer & Packauskas, *L. fulvicornis* (Westwood), and *L. dilaticollis* Guérin) are divided between two species groups (Schaefer et al. 2008). Similarly, the various cactus feeders (*L. subauratus* Distant, *L. dentatus* Berg, and *L. cinctus* Herrich-Schäffer) are divided among three species groups (Packauskas and Schaefer 2001); thus, host plant relationships do not always reflect taxonomic relationships.

The more generalist feeders, pests on a variety of crops, are primarily in the *zonatus* group. Species with exceptionally broad host plant ranges include *L. phyllopus* (L.) (Mitchell 2006), *L. zonatus* (Dallas) (Fernandes and Grazia 1992), *L. stigma* (Herbst) (Amaral and Cajueiro 1977), and *L. gonagra* (F.) (Mitchell 2000). Other Neotropical species mentioned as being crop pests are *L. conspersus* Stål on tomato and passion fruit (Urueta-Sandino 1975; Mitchell 2000); *L. balteatus* (L.) on guava, loofah, oranges, tomatoes, and legumes (Barber and Bruner 1947; Schaefer and Mitchell 1983); *L. chilensis* (Spinola) on various fruits; and *L. cinctus* on various fruits in addition to cacti (Faúndez and Carvajal 2011; Schaefer and Mitchell 1983). The latter species has the distinction of being the only coreid ever reported to bite a human, presumably to acquire fluids (Faúndez and Carvajal 2011). Allen (1969) notes that damage to guava, cashew, lychee, and annatto attributed to *L. stigma* probably refers to the very similar *L. concolor* (Walker).

Several species of *Leptoglossus* are known to transmit or facilitate the transmission of plant diseases, including fungal pathogens and fruit-inhabiting trypanosomatids. Two fungal diseases of pistachio, botryosphaeria panicle and shoot blight and stigmatomycosis, are transmitted by *L. clypealis* Heidemann in North America. Among the Neotropical species, *L. gonagra* transmits citrus diseases, *L. zonatus* transmits stigmatomycosis from pomegranate to citrus, and puncture wounds on fruit caused by *L. stigma* facilitate the entry of fungi. *Phytomonas* spp. are carried internally (salivary glands, alimentary canal) in *L. gonagra*, *L. ingens* (Mayr), and *L. stigma* (Mitchell 2004 and references therein). The ability of *L. zonatus* to harbor and transmit *Phytomonas mcgheei* to corn seeds has been studied experimentally; this species is considered a potential trypanosomatid vector because trypanosomes from bug salivary glands could infect corn in laboratory studies (Jankevicius et al. 1993).

Mitchell (2000) reviews the biology, life history, and distribution of *L. stigma*. This species, similar to *L. concolor* (Walker) but with a more steeply declivent pronotum and smooth pronotal calli, ranges from Mexico through South America east of the Andes, as far south as Paraguay and Brazil (Allen 1969; Packauskas 2010). An irregular fascia is usually, but not always, present on the corium. Food plants include pumpkin, guava, starfruit, cashew, oranges, mango, pomegranate, and tangerine. Adults attack the fruits and buds of guava (Amaral and Cajueiro 1977 and references therein). The development and behavior of *L. stigma* (as *Veneza stigma*) was observed and documented in the laboratory (Amaral and Cajueiro 1977). Nymphs could be reared on a diet of guava, pumpkin, and chayote, although mortality was high. Durations of the egg stage and nymphal stadia were determined and adult reproductive behavior described.

The distribution of *L. phyllopus* extends from the United States south through much of Central America, but it has only been considered to be of economic importance in the United States. These bugs have a broad, straight transverse fascia on the corium and bicolored hind wings (Allen 1969). The natural history, seasonal dispersal, host plant preference, feeding behavior, and tissue preference were examined by Mitchell (2006), using mark-release-recapture, field censuses, laboratory rearing experiments, and histological techniques. Stylets penetrated to xylem on vegetative tissue and to developing seeds and xylem on pods of green bean and tomato fruit. Nymphs developed equally well on reproductive structures from plants from different families, but could not survive on vegetative tissue alone. In the field, host-switching occurred frequently; teneral adults moved from the nymphal host to another plant species shortly after emergence, suggesting that the observed polyphagy is sequential rather than concurrent and the ability to develop on a variety of plants allows versatility in locating hosts in the appropriate reproductive condition.

Allen (1969) confirmed the synonymy of *L. membranaceus* (F.) with *L. australis* (F.). Baranowski and Slater (1986) then synonymized the Eastern Hemisphere *L. australis* with the Neotropical *L. gonagra s.s.*, thereby expanding the range of this species nearly worldwide. There is great variability in tibial dilations and humeral angles, but the orange-red to yellow lunate transverse fascia on the pronotal disk and the variegated antennae are quite recognizable in this species (Plate 19.6, #8) (Allen 1969; Baranowski and Slater 1986). Mitchell (2000) summarized the Eastern Hemisphere literature under *australis* and that of the Western Hemisphere under *gonagra*. In the Neotropics, *L. gonagra* damages citrus throughout its range, with damage reported in Florida, Puerto Rico, Argentina, and Brazil; however, this insect is highly polyphagous, and other crops attacked include pumpkin, chayote, loofah, cucumber, bitter gourd, watermelon, guava, araçá, passion fruit, papaya, mango, pomegranate, tobacco, physic nut, and corn. However, breeding hosts are more limited; bitter gourd, loofah, watermelon, and citron support nymphal populations (Mitchell 2000). Damage to citrus fruit has been described in detail: fruit drop, necrotic spots and lesions, dried, damaged, and bitter pulp, and secondary fungal infections are characteristic of *L. gonagra* infestations (Calza et al. 1964; Albrigo and Bullock 1977).

Extensive studies of the biology of *L. gonagra*, including egg stage duration and hatching success, nymphal development on leaves and fruit of pumpkin, and adult reproductive cycle were conducted in the lab under varying conditions of temperature and humidity (Amaral and Storti 1976). High mortality occurred during the second stadium on both pumpkin and bitter gourd, the wild host. Duration of the total life cycle was 80.7 ± 1.12 days. Development times (in days) were 8.2 for eggs and 3.6, 13.9, 10.3, 11.2, and 15.4 for the five nymphal instars, respectively.

Nymphal development and growth under controlled laboratory conditions were compared among five species of *Passiflora* (Caetano and Boiça 2000). Nymphs could not complete development on fruits alone. When provided branches, buds, flowers, and fruits, nymphs failed to complete development on *P. nitida* and showed the lowest adult weight on *P. giberti*. *P. laurifolia* was judged to be the most susceptible species, based on development time, adult weight, and longevity.

The literature on *L. zonatus* has grown substantially in recent years, due to increased interest in its economic importance on a variety of crops in South America and its spread across the southern United States. This species is typically characterized by two large whitish-yellow spots on the pronotal disk and a wide irregular fascia across the corium (Plate 19.1), but the fascia may vary in width or be missing entirely (Allen 1969). The eggs and immatures are described and illustrated by Fernandes and Grazia (1992), with observations on adult behavior. Mitchell (2000) summarizes older information in the literature on biology, host plants, economic damage, and control.

In the southern United States, satsuma mandarin oranges are damaged directly by bug feeding and indirectly by transmission of the yeast, *Nematospora coryli*; *L. zonatus* is considered to be a key pest (Henne et al. 2003; Xiao and Fadamiro 2010). The latter authors evaluate damage to satsumas (2010) and provide developmental data for nymphs and adult life history and reproductive parameters (2009). Preferences were tested between ripened and unripened fruit and among tomato, satsumas, kumquats, lemons, and peach; tomato was highly preferred, but *L. zonatus* can survive and develop on a diet entirely composed of satsuma fruit. Fruit maturity did not affect preference (Xiao and Fadamiro 2009). Among the wild hosts in the United States is Chinese tallow, an invasive euphorb (Johnson and Allain 1998).

Development of *L. zonatus* and the reproductive parameters of adults have also been studied on sorghum and corn (Matrangolo and Waquil 1994) and on corn, soybean, and green bean (Panizzi 1989). Corn seeds were far preferable for rearing than pods of either legume, and this insect is abundant in corn fields in Brazil (de Souza and Amaral 1999a; Panizzi 2004b). An African tree widely planted in greenbelts, *Spathodea campanulata* Beauv (Bignoniaceae), supports adults and nymphs and may serve as an alternate host (de Souza and Amaral 1999a). The entire life cycle can also be completed on physic nut; instar duration data, nymphal measurements, and reproductive parameters have been determined for this host plant (Grimm and Somarriba 1999). *L. zonatus* is the second most common heteropteran on this crop in Nicaragua (Grimm and Maes 1997) and also occurs on physic nut in Mexico (Tepole-García et al. 2012). Damage to pomegranate (Raga et al. 1995) and guava (Pires et al. 2013) has also been described, and occurrence of this insect on *Annona* spp. (Peña and Bennett 1995), *Malpighia emarginata* Sessé & Moc, *Morus nigra* L., and *Mangifera indica* L. (Pires et al. 2013) has been recorded.

Omnivory has also been observed in *L. zonatus*; egg masses of fall armyworm, *Spodoptera frugiperda*, were consumed by adults in maize fields in Brazil (Menezes-Netto et al. 2012). *L. zonatus* has also been observed to orient toward and land on humans and on plastic cylinders hung from poles in corn fields, possibly representing either territorial or recognition behavior (Panizzi 2004b). Natural enemies include *Trichopoda pennipes* F. and egg parasitoids including *Gryon* sp. (de Souza and Amaral 1999b). Control using entomopathogens (*Beauveria bassiana* (Bals.) Vuil. and *Metarhizium anisopliae* (Metsch.) Sorok) appears promising in laboratory and field trials and was more effective than treatment with neem seed (Grimm and Guharay 1998).

Narnia Stål

Narnia spp. inhabit primarily the desert areas of North America; however, the distribution of *N. femorata* Stål extends to Guatemala (Brailovsky 1975). Various cacti are hosts for *Narnia* spp., including *Opuntia* spp. (Brailovsky 1975), but unlike the cactus pad specialists in Chelinideini, these bugs preferentially feed on fruits. Brailovsky et al. (1994) provide detailed descriptions of the egg, nymphal stadia, and life cycle of *N. femorata* in Mexico. Eggs are laid on the spines and pads, and the life cycle is dependent on the fruiting and flowering cycle of the host. First instars are gregarious, but do not feed; late instars disperse. Recent behavioral studies of *Narnia femorata* Stål have added substantially to our understanding of coreid feeding and mating. Adults reared exclusively on pads survive, but development is delayed and body size is smaller compared with individuals with access to fruits (Nageon de Lestang and Miller 2009). In the presence of cactus fruit, females prefer the odor of males who developed on fruit although no preference is exhibited in the absence of fruit odors (Addesso et al. 2014). Both male-male competition and female choice operate in this species; enlarged femora in males are used in combat (Miller 2013) to defend territories on cactus, but male body size and leg area only strongly affect the outcome of male-male competition in the presence of females (Procter et al. 2012).

Phthia Stål

The genus *Phthia* Stål was recently revised by Brailovsky (2009c) with the erection of four new genera. Of the 13 species originally in *Phthia s.l.*, *Phthia picta* (Drury) has been placed alone in the genus *Phthiacnemia* Brailovsky, while *P. lunata* (F.) remains, along with three other species, in *Phthia s.s.* Three additional species were placed in *Phthiadema* Brailovsky, four in *Phthiarella* Brailovsky, and one in *Rhytidophthia* Brailovsky. Brailovsky (2009c) provides a key to genera, which may be distinguished from one another by the development of the humeral angles and the extent of metallic iridescence on the head, clavus, and corium. Although all are Neotropical, only the widely distributed *P. picta* has received any mention in the economic literature; this species may be distinguished by the dull, non-metallic body surface and the truncate (non-acute) humeral angles. A narrow orange-red transverse band may be present on the posterior portion of the pronotum (see illustration in Brailovsky 2009c), and other orange markings may be variably present (Baranowski and Slater 1986), but unlike *Phthia s.s.*, no yellow transverse fasciae occur on the anterior lobe or on the corium (Brailovsky 2009c). Literature records for this species may be found under *Dallacoris* Osuna in addition to *Phthia* and *Phthiacnemia*.

Host plant records are available for *P. picta*, *P. lunata*, and *Rhytidophthia splendens* (Valdés). The latter, endemic to Cuba, was collected from *Taonabo* (=*Ternstroemia*) *parviflora* Krug & Urban (Theaceae) (Barber and Bruner 1946). *P. lunata* is reported from Cucurbitaceae (*Citrullus* and *Cucurbita*) (Maes and Goellner-Scheiding 1993). Published host plant records for *P. picta* are summarized by Schaefer and Mitchell (1983), Baranowski and Slater (1986), and Mitchell (2000). Serantes (1973) lists the known hosts in Argentina. *Solanum nigrum* L. is the common wild breeding host in Florida, and occurrence on tomato has been reported throughout the range of *P. picta*, from the Caribbean to Argentina (Baranowski and Slater 1986; Serantes 1973). However, the species is not always considered an economically important pest (Mitchell 2000), and despite the common names, "chinche de tomate" and "percevejo do tomate," it is not restricted to Solanaceae. This species is in fact highly polyphagous, feeding on cultivated cucurbits and legumes in addition to tomato and eggplant; plants from 11 families are reported as feeding or breeding hosts (Schaefer and Mitchell 1983). Feeding punctures cause direct damage resulting in deformed fruit; indirect damage includes entry pathways for fungal pathogens (Amaral 1980) and transmission of a trypanosomatid of tomatoes, *Phytomonas serpens* (Gibbs) (Jankevicius et al. 1989; Freymuller et al. 1990; Fiorini et al. 1993). Flagellates of *P. serpens* were found in the digestive tract of field-collected insects and appear in hemocoel and salivary glands 1 week following acquisition in the laboratory. Transmission in both directions (from *P. picta* to tomato fruit and from tomato fruit to insect) was documented. The nature of the relationship between tomato and *P. serpens* is not known with certainty; fruit-dwelling trypanosomatids may not actually be pathogenic although phloem-inhabiting trypanosomatids, also transmitted by bugs, cause serious disease and economic losses (e.g., hartrot of coconut transmitted by *Lincus lobuliger* Breddin (Sgrillo et al. 2005).

The eggs of *P. picta*, like those of *Leptoglossus* and *Narnia*, are hemicylindrical and laid in a single row, end to end. The juvenile stages are described by Serantes (1973) with illustrations of the egg and first and fifth instars. Detailed measurements, descriptions, and illustrations of all five instars are provided by da Silva et al. (2001). Younger instars, brightly colored and gregarious, develop faster when reared in aggregations (Amaral 1981a). Life history and behavioral observations are given by Serantes (1973) for Argentine populations and Amaral (1980, 1981a, b) for Brazil and reviewed by Mitchell (2000) and da Silva and Carvalho (2001). More recent studies of development under controlled laboratory conditions (da Silva and Carvalho 2001) showed total nymphal development time to be 34.39 ± 0.67 days on tomato at 26 ± 1 °C, 70 ± 10 % RH, and 14: [L:D] photoperiod, with 56 % mortality.

For females, the pre-oviposition period was 62.83 ± 9.86 days, with 9.37 ± 1.48 days between egg batches. Data are also provided on clutch size, adult longevity, and development time for eggs and each instar.

19.5.3 Hypselonotini Bergroth

Anasa Amyot & Serville

This large genus comprises 76 species (CoreoideaSF Team 2015) with distributions ranging from Canada to northern Argentina and Uruguay (Brailovsky 1985; Packauskas 2010), although the majority of species are Neotropical. These medium-sized bugs are elongate-oval, with the pronotum strongly declivent; the connexivum widely exposed; the antennae, femora, and tibiae slender; and the beak extending at least to the middle coxae (Blatchley 1926). Although several species have been noted as injurious to cucurbit crops, only *Anasa tristis* (De Geer) is currently considered a serious economic pest. Reports of other *Anasa* spp. feeding on cultivated cucurbits include *A. trilineata* Stål in Venezuela; *A. incompta* Brailovsky in Colombia; *A. costalis* Stål , *A. litigiosa* Stål, *A. maculipes* Stål, *A. ruficornis* Stål, and *A. uhleri* Stål in Mexico; and *A. guttifera* Berg in Argentina (Brailovsky 1985; Brailovsky et al. 1994). The widely distributed *A. scorbutica* (F.) is reported from loofah, squash, gourds, maize, tomato, and physic nut in various parts of its range (Mitchell 2000 and references therein), but cucurbits are the only hosts on which breeding was noted (Barber and Bruner 1947). *A. andresii* (Guérin-Méneville), distributed from the southern United States and the Caribbean to Colombia and Ecuador (Packauskas 2010), is reported from cucurbits, including *C. pepo* (L.) and *C. ficifolia* Bouché, cotton, tomatoes, and bromeliads (Brailovsky 1985) and was described in Cuba to be "at times a pest on squash" (Barber and Bruner 1947). On squash, these bugs cluster on the leaf underside, causing wilting and death, and may also feed on fruits (Mitchell 2000 and references therein). Interestingly, a number of *Anasa* spp. have been collected from bromeliads: *A. litigiosa* (Plate 19.9, #3) occurs on *Tillandsia* in addition to wild and cultivated cucurbits (Brailovsky 1985), and *A. costalis* has been reported from both *C. pepo* and two species of *Tillandsia* (Brailovsky 1985; Brailovsky et al. 1994), whereas *A. capaneodes* Stål and *A. impictipes* Stål may be more strictly associated with *Tillandsia* spp. (Brailovsky 1985). Brailovsky (1985) compiled a thorough list of all known host plants and habitats for *Anasa* spp., adding more Mexican host plants subsequently (Brailovsky et al. 1994). Of the 17 species for which host plants are known, 13 feed on cucurbits (although not always exclusively), with stem, leaf, and fruit all mentioned as feeding sites. With the exception of *A. tristis* and two strictly North American species (*A. armigera* (Say) and *A. repetita* Heidemann), virtually nothing is known of the natural history of *Anasa* species (Brailovsky 1985).

The notorious "squash bug," *A. tristis*, occurs from Canada to Brazil and is a pest of squash and pumpkin in the United States and Mexico. A vast amount of research has been dedicated to understanding the biology of *A. tristis*, the damage it causes, and potential control methods. Mitchell (2000) reviewed this primarily North American literature, including life cycle, behavior, feeding and oviposition preferences, histological studies of feeding damage, insecticidal control methods, cultural controls, natural enemies, and tritrophic interactions. More recent studies have revealed that, in addition to direct damage to vascular tissues, the squash bug transmits a phloem-colonizing bacterium, *Serratia marcescens* Bizio, the causative agent of cucurbit yellow vine disease (Bruton et al. 2003). This disease was first described in 1988 in Texas and Oklahoma and subsequently spread eastward in the United States. The pathogen can be acquired in the nymphal stage and transmitted after the bug reaches adulthood; transmission appears to be propagative rather than noncirculative. No bacteria are found in the cibarium, suggesting that the likely location for retention of the bacterium is the hemocoel (Wayadande et al. 2005). Overwintering infected adults can transmit *S. marcescens* to cucurbit crops the following spring (Pair et al. 2004). The pathway of transmission from insect to phloem sieve tube cells is not yet understood (Wayadande et al. 2005; Maskey 2010); wilting caused by *A. tristis* has been attributed to xylem blockage (Neal 1993), but stylets do penetrate to the phloem tissue. However, the diameter of the stylets is too large to permit direct insertion into phloem sieve tube elements in the manner of aphid or leafhopper feeding (A. Wayadande, personal communication), and despite several studies of squash bug feeding using electrical penetration graphing (Bonjour et al. 1991; Cook and Neal 1999; Maskey 2010), the mode of feeding and the mode of pathogen transmission in *A. tristis* remain uncertain (Maskey 2010). Nonetheless, the economic importance of the squash bug, and possibly other *Anasa* spp. as well, has increased with the recognition of its vector competence.

19.5.4 Spartocerini Amyot & Serville

Spartocera Laporte (Plate 19.1)

Much of the older literature on this group, especially the species of economic importance, was published under the synonym *Corecoris* Hahn. These bugs are characteristically large and ovoid, with a broad abdomen and prominent connexivum. Of the 18 species distributed from the southern United States through South America, six have been noted as injurious to crops. In the United States, *Spartocera diffusa* (Say) damages potato; Chittenden (1927) provides nymphal descriptions and information on natural history and host plants. A preference for solanaceous crops also characterizes the Neotropical species, but sweet potato (*Ipomoea batatas* (L.), Convolvulaceae) is also attacked. *Solanum nigrum* L. is reported as a wild host for several pest species, including *S. diffusa*, *Spartocera brevicornis* (Stål), *Spartocera batatas* (F.), and *Spartocera fusca* (Thunberg) (Chittenden 1927; Bosq

1937; Costa Lima 1940; Halbert 2003). Little is known of the natural history of species that pose no economic threat, with one exception. *Spartocera lativentris* Stål feeds on the leaves and stems of *Solanum sisymbriifolium* L., with two generations per year in southern Brazil. No feeding on fruits was observed, and first instars were not observed to feed (Becker and Prato 1982). These authors also provide detailed information on the eggs of this species, including a description, oviposition sites, egg clusters, female reproductive organs, and egg mortality, hatching success, and parasitism.

Spartocera batatas (F.) feeds on sweet potato in Jamaica, Cuba, Puerto Rico, and other Caribbean islands; it is currently considered to reach minor pest status only in Puerto Rico. One infestation in Florida has also been reported (Halbert 2003 and references therein). Adults are parasitized by the tachinid fly, *Trichopoda pennipes* (F.) in the Caribbean and in Brazil (Amaral and Vieira 1978). Both *S. brevicornis* and *S. fusca* are associated with various cultivated solanaceous crops; tobacco and tomato, respectively, are mentioned by Costa Lima (1940) for Brazil. However, *S. fusca* has been noted elsewhere in its broad range to feed on a variety of hosts including *Capsicum annuum* (chili peppers) and *Physalis* sp. and has been collected from citrus. In Florida, it breeds on *Solanum americanum* P. Mill (Blatchley 1926; Brailovsky and Sánchez 1983b; Baranowski and Slater 1986). *Spartocera granulata* Stål is also registered in Brazil as a pest of cultivated solanaceous crops (Becker and Prato 1982).

By far the most damaging species in the genus, and the best studied, is *Spartocera dentiventris* Berg, a pest of tobacco in Brazil. Feeding causes the tobacco leaves to twist and wilt, causing economic losses. Detailed information on the biology of this species is provided by Caldas et al. (1998, 1999, 2000) including a description of the immatures, reproductive biology, longevity of adults, and cage studies of life history in tobacco. Mark-release-recapture in tobacco fields (Jesus et al. 2002) documented the population dynamics of the bug over the tobacco growing season, showing an aggregated pattern of distribution and predation by the reduviid *Cosmoclopius nigroannulatus* Stål on the second generation of nymphs Canto-Silva & Romanowski (2003). This assassin bug shows a type II functional response when feeding on first instar *S. dentiventris* (da Rocha and Redaelli 2004). Reported parasitoids of both *S. dentiventris* and *S. lativentris* include the eurytomid wasp *Neorileya ashmeadi* Crawford and the platygastrid *Gryon gallardoi* (Brèthes) (Becker and Prato 1982; Santos et al. 2001). Extensive studies have been conducted on the behavior of *G. gallardoi* as a potential biological control agent for *S. dentiventris* (da Rocha et al. 2007, 2008).

19.6 Concluding Remarks

Two recently compiled catalogs (Packauskas 2010; CoreoideaSF Team 2015) provide an invaluable resource for continued and expanded study of the Neotropical Coreidae. Catalogs of the North American fauna (e.g., Baranowski and Slater 1986; Henry and Froeschner 1988) have been available for some time, but only recently has information on the entire New World Coreidae become easily accessible. As Packauskas (2010) noted, studies of New World coreids, and especially the Neotropical genera, languished for over 100 years. The higher classification of the Coreoidea has received attention (e.g., Schaefer 1964) and Stål's (1867) key to New World subfamilies and tribes has been updated (Packauskas 1994), but no modern comprehensive key to genera exists. The voluminous contributions of Brailovsky and his collaborators (see References for an appreciation of the extent of these works) include keys to genera for five Neotropical tribes discussed herein; papers by Yonke (1972) and O'Shea (1980a, b) provide information and keys for another three tribes. However, keys are lacking for some of the largest Neotropical coreid tribes and those with the most economically damaging members (e.g., Hypselonotini). Furthermore, keys to species exist for fewer than half the described genera in most tribes (see listings in Packauskas 1994). The Coreidae of the Neotropical region form an impressive and complex assemblage, with striking, often bizarre, characters and spectacular coloration. It is hoped that the advent of the newly launched Coreoidea Species File Online (CoreoideaSF Team 2015) will stimulate further, much needed, work on these elegant bugs.

Acknowledgments PLM thanks Lisonéia Fiorentini Smaniotto and JAMF thanks Murilo Victor Silva Bitar for invaluable assistance with literature searches. The majority of specimens were photographed as part of Leverhulme Trust Project F/00 696 and through research supported by the SYNTHESYS Project which was financed by European Community Research Infrastructure Action under the FP6 "Structuring the European Research Area" Programme. Specimens from the BMNH were prepared by Mick Webb and photographed by Tristan Bantock, Holly Dawson, and Laurence Livermore. All other specimen photographs were taken by Laurence Livermore. The image editing procedure and design templates were developed by Malin Nikunlassi. Image editing was done by Elizabeth Livermore, Laurence Livermore, and Rose Stainthorp. Copyright of individual specimen images remains that of their host collections which are marked by brackets in specimen plate captions. All specimen images are also available from http://coreoidea.speciesfile.org/.

We would like to thank the following photographers who provided field images: Arthur Anker (https://www.flickr.com/photos/artour_a/), Andreas Kay (https://www.flickr.com/photos/75374522@N06/), and W.O. Ree, Jr. (Texas A&M AgriLife Extension, Bryan, Texas, USA). We would also like to thank our colleagues who provided access and assistance to the following collections: AMNH, USA, New York, New York, American Museum of Natural History; CAS, USA, California, San Francisco, California Academy of Sciences; BMNH, United Kingdom, London, The Natural History Museum; DEI, Germany, Müncheberg, Senckenberg Deutsches Entomologisches Institut; HNHM, Hungary, Budapest, Hungarian Natural History Museum; IOZ, China, Beijing, Chinese Academy of Sciences, Institute of Zoology; ISNB: Belgium, Brussels, Institut Royal des Sciences Naturelles de Belgique; MNHN, France, Paris, Muséum National d'Histoire Naturelle; MRAC, Belgium, Tervuren, Musee Royal de l'Afrique Centrale; MRSN, Italy, Torino, Museo Regionale di Scienze Naturali; NHMW, Austria, Wien, Naturhistorisches Museum Wien; OXUM, United Kingdom, Oxford, University Museum of Natural History; RNHL, Netherlands, Leiden, Naturalis Biodiversity Centre; UNAM, Mexico, Mexico D.F., Universidad Nacional Autonoma de Mexico; UNCB, Colombia, Bogotá, Universidad Nacional de Colombia, Insituto de Ciencias Naturales de la Universidad Nacional; USNM, USA, Washington D.C., National Museum of Natural History; ZMHB. Germany, Berlin, Museum für Naturkunde der Humboldt-Universität; ZMUC, Denmark, Copenhagen, University of Copenhagen, Zoological Museum.

References

Addesso KM, Short KA, Moore AJ, Miller CW (2014) Context-dependent female mate preferences in leaf-footed cactus bugs. Behaviour 151:479–492

Adler PH, Wheeler AG Jr (1984) Extra-phytophagous food source of Hemiptera-Heteroptera: bird droppings, dung, and carrion. J Kans Entomol Soc 57:21–27

Aguiar-Menezes EL, Menezes EB, Cassino PCR, Soares MA (2002) Passion fruit. In: Peña JE, Sharp JL, Wysoki M (eds) Tropical fruit pests and pollinators: biology, economic importance, natural enemies, and control. CAB International, Wallingford, pp 360–390

Ahmad I (1970) Some aspects of the female genitalia of *Hygia* Uhler 1861(Coreidae: Colpurinae) and their bearing in classification. Pak J Sci 2:235–243

Ahn SJ, Son D, Choo HY, Park CG (2013) The first record on *Leptoglossus occidentalis* (Hemiptera: Coreidae) in Korea, a potential pest of the pinaceous tree species. J Asia-Pacific Entomol 16:281–284

Albrigo LG, Bullock RC (1977) Injury to citrus fruits by leaffooted bug and citron plant bugs. Proc Fl State Hort Soc 90:63–67

Aldrich JR (1988) Chemical ecology of the Heteroptera. Annu Rev Entomol 33:211–238

Aldrich JR, Blum MS (1978) Aposematic aggregation of a bug (Hemiptera: Coreidae): the defensive display and formation of aggregations. Biotropica 10:58–61

Aldrich JR, Kochansky JP, Lusby WR, Dutky SR (1982) Volatile male-specific natural products of a coreid bug (Hemiptera: Heteroptera). J Chem Ecol 8:1369–1376

Aldrich JR, Waite GK, Moore C, Payne JA, Lusby WR, Kochansky JP (1993) Male-specific volatiles from nearctic and Australasian true bugs (Heteroptera: Coreidae and Alydidae). J Chem Ecol 19:2767–2781

Allen RC (1969) A revision of the genus *Leptoglossus* Guerin (Hemiptera: Coreidae). Entomol Am 45:35–140

Amaral FBF (1980) Observações biológicas em adultos acasalados e não acasalados de um coreídeo praga – *Phthia picta* (Drury, 1770). An Soc Entomol Bras 9:75–79

Amaral FBF (1981a) Efeito de dietas naturais e de fatores ambientais na biologia de *Phthia picta* (Drury, 1770) sob condições de laboratório (Hemiptera, Coreidae). Rev Bras Biol 41:845–853

Amaral FBF (1981b) Aspectos comportamentais de *Phthia picta* (Drury, 1770) em condições de laboratório (Hemiptera, Coreidae). Rev Bras Biol 41:441–446

Amaral FBF (1986) Observações sobre o ciclo biológico de *Crinocerus sanctus* (Fabricius, 1775) (Hemiptera; Coreidae) sob condições de laboratório. An Soc Entomol Bras 15:5–18

Amaral FBF, Cajueiro IVM (1977) Observações sobre o ciclo biológico de *Veneza stigma* (Herbest [sic], 1784) Osuna 1975 (Hemiptera, Coreidae) em laboratório. An Soc Entomol Bras 6:164–172

Amaral FBF, Storti AF (1976) Estudos biológicos sobre *Leptoglossus gonagra* (Fabricius, 1775), (Coreidae, Hemiptera) em laboratório. An Soc Entomol Bras 5:130–137

Amaral FBF, Vieira AOS (1978) Ocorrência de *Trichopoda pennipes* (Fabricius, 1794) (Diptera, Tachinidae) parasitando *Corecoris batatas* (Fabricius, 1798) (Hemiptera Coreidae) no Brasil. An Soc Entomol Bras 7:67–68

Baranowski RM, Slater JA (1986) Coreidae of Florida (Hemiptera: Heteroptera). Arthropods of Florida and neighboring land areas, vol 12, no 630. Florida Department of Agriculture and Consumer Services, the Division of Plant Industry, Gainesville

Barber HG, Bruner SC (1946) Records and descriptions of miscellaneous Cuban Hemiptera. Bull Brooklyn Entomol Soc 16:52–61

Barber HG, Bruner SC (1947) The Coreidae of Cuba and the Isle of Pines with the description of a new species (Hemiptera – Heteroptera). Mem Soc Cubana Hist Nat 19:77–88

Barcellos A, Schmidt LS, Brailovsky H (2008) Abundance and species richness of Coreoidea (Hemiptera: Heteroptera) from Parque Estadual do Turvo, southern Brazil. Neotrop Entomol 37:406–412

Becker M, Prato MD (1982) Natality and natural mortality of *Spartocera lativentris* Stal, 1870 (Heteroptera: Coreidae) in the egg stage. An Soc Entomol Bras 11:269–281
Bergroth E (1913) Supplementum Catalogi hemipterorum bruxellensis. II. Coreidae, Pyrrhocoridae, Colobathristidae, Neididae. Mém Soc Entomol Belg 22:125–183
Blatchley WS (1926) Heteroptera or true bugs of eastern North America. Nature Publ Co, Indianapolis
Bleicher E, Melo QMS (1996) Artrópodes associados ao cajueiro no Brasil. 2ª. ed. Fortaleza: EMBRAPA/CNPAT. Documentos 9:1–35
Blöte HC (1935) Catalogue of the Coreidae in the Rijksmuseum van Natuurlijke Historie. Part II. Coreinae, first part. Zool Mededeelingen 18:181–227
Bonjour EL, Fargo WS, Webster JA, Richardson PE, Brusewitz GH (1991) Probing behavior comparisons of squash bugs (Heteroptera: Coreidae) on cucurbit hosts. Environ Entomol 20:143–149
Bosq JM (1937) Lista preliminar de los hemipteros (Heteropteros), especialmente relacionados con la agricultura nacional. Rev Soc Entomol Arg 9:111–134
Brailovsky H (1975) Distribucion de las especies de Narnia Stål (Coreidae-Coreinae-Anisoscelini) y desdcription de una nueva especie. Rev Soc Mex Hist Nat 26:169–176
Brailovsky H (1982) Hemiptera-Heteroptera de México XXIV. Nuevos registros de la tribu Mictini y descripción de un nuevo género y dos nuevas especies (Coreidae). An Inst Biol Univ Nac Aut México Ser Zool 52:277–288
Brailovsky H (1983a) El género *Acidomeria* Stål con descripción de una nueva especie (Hemiptera-Heteroptera-Coreidae-Coreini). Fol Entomol Mexic 56:3–20
Brailovsky H (1983b) El género *Sephina* Amyot y Serville en Venezuela con descripción de una nueva especie (Hemiptera-Heteroptera-Coreidae-Spartocerini). An Inst Biol Univ Nac Aut México Ser Zool 53:277–283
Brailovsky H (1984) Descripción de un nuevo género y una nueva especie de la tribu Discogastrini (Hemiptera: Heteroptera-Coreidae). Fol Entomol Mexic 62:17–26
Brailovsky H (1985) Revisión del género *Anasa* Amyot-Serville (Hemiptera-Heteroptera-Coreidae-Coreini). Monogr Inst Biol Univ Nac Aut México 2:1–266
Brailovsky H (1986a) Una especie nueva del género *Coryzoplatus* Spinola (Hemiptera-Heteroptera-coreidae-Discogastrini). An Inst Biol Univ Nac Aut México Ser Zool 57:113–117
Brailovsky H (1986b) Hemiptera-Heteroptera de México XXXV. Revisión de la familia Coreidae Leach. I. Tribu Discogastrini Stal. An Inst Biol Univ Nac Aut México Ser Zool 56:401–422
Brailovsky H (1986c) Un nouveau Discogastrini d' Amérique du Sud (Hemiptera. Coreidae). Ann Soc Entomol Fr 22:211–214
Brailovsky H (1987a) Three new genera and six new species of Neotropical Coreidae (Heteroptera). J N Y Entomol Soc 95:518–530
Brailovsky H (1987b) Revisión del género *Scamurius* Stål (Hemiptera-Heteroptera-Coreidae-Discogastrini). Fol Entomol Mexic 70:25–52
Brailovsky H (1988a) Un nuevo género y dos nuevas especies de Coreidos Neotropicales (Hemiptera-Heteroptera-Coreidae Acanthocerini). An Inst Biol Univ Nac Aut México Ser Zool 58:171–178
Brailovsky H (1988b) Hemiptera-Heteroptera de México XXXIX. Descripción de una tribu nueva, un género nuevo y una especie nueva de coreidos recolectados en bambu (*Bambusa* sp.) (Coteidae-Coreinae). An Inst Biol Univ Nac Aut México Ser Zool 58:155–164
Brailovsky H (1988c) La tribu Hydarini Stål, en el continente Americano con descripción de dos nuevos géneros, una nueva especie y una nueva subespecie (Hemiptera-Heteroptera-Coreidae). An Inst Biol Univ Nac Aut México Ser Zool 58:623–649
Brailovsky H (1988d) Dos nuevas especies del género *Sethenira* Spinola y nuevos arreglos nomenclatoriales dentro de *Acidomeria* Stål (Hemiptera-Heteroptera-Coreidae-Coreini). An Inst Biol Univ Nac Aut México Ser Zool 58:179–198
Brailovsky H (1988e) Revisión del género *Sundarus* Amyot-Serville (Hemiptera-Coreidae-Coreini) para el continente Americano. An Inst Biol Univ Nac Aut México Ser Zool 58:561–622

Brailovsky H (1989) Revisión del género *Scolopocerus* Uhler con descripción de una especie nueva (Hemiptera-Heteroptera-Coreidae-Coreini) de Norteamérica. An Inst Biol Univ Nac Aut México Ser Zool 59:65–76

Brailovsky H (1990a) Nuevos arreglos nomenclatoriales y descripción de dos especies nuevas del género *Dalmatomammurius* Brailovsky (Hemiptera-Heteroptera-Coreidae-Leptoscelidini). An Inst Biol Univ Nac Aut México Ser Zool 61:343–355

Brailovsky H (1990b) Géneros nuevos y especies nuevas de Coreidos Neotropicales (Hemiptera-Heteroptera-Coreidae: Acanthocerini, Leptoscelidini y Anisoscelidini). An Inst Biol Univ Nac Aut México Ser Zool 61:107–123

Brailovsky H (1990c) Revisión del complejo "*Althos*" (Hemiptera-Heteroptera-Coreidae). Publ Espec Univ Aut México 5:1–156

Brailovsky H (1991) Descripción de la segunda especie del género *Leptopetalops* (Hemiptera-Heteroptera: Coreidae: Acanthocephalini). An Inst Biol Univ Nac Aut México Ser Zool 62:97–102

Brailovsky H (1992a) Un género y tres especies nuevas de Coreidos Neotropicales (Hemiptera-Heteroptera-Coreidae-Coreinae-Coreini). An Inst Biol Univ Nac Aut México Ser Zool 63:185–199

Brailovsky H (1992b) El género *Eubule* con descripción de cuatro especies y algunos registros nuevos (Hemiptera-Heteroptera-Coreidae-Spartocerini). An Inst Biol Univ Nac Aut México Ser Zool 63:201–213

Brailovsky H (1993) Género nuevo y especies nuevas de coreidos neotropicales (Hemiptera-Heteroptera-Coreidae: Acanthocerini, Chariesterini, Coreini, Discogastrini, Leptoscelidini, y Nematopodini). An Inst Biol Univ Nac Aut México Ser Zool 64:109–127

Brailovsky H (1995a) Review of the "*Cebrenis*" complex (Hemiptera-Heteroptera-Coreidae-Coreinae-Coreini).). Publ Espec Univ Aut México 15:1–124

Brailovsky H (1995b) New genera and new species of Neotropical Coreidae (Hemiptera: Heteroptera). Pan-Pacific Entomol 71:217–226

Brailovsky H (1996) A new genus and a new species of Leptoscelidini (Heteroptera: Coreidae) from Costa Rica. Proc Entomol Soc Wash 98:215–220

Brailovsky H (1997a) One new genus and two new species of Acanthocephalini (Hemiptera: Coreidae: Coreinae). Ann Entomol Soc Am 90:550–554

Brailovsky H (1997b) *Leptostellana parva* gen. nov. et spec. nov. from Costa Rica (Insecta: Heteroptera: Coreidae: Leptoscelidini). Reichenbachia 32:19–24

Brailovsky H (1998a) A new species of *Paralycambes* Kormilev from South America (Hemiptera: Heteroptera: Coreidae: Meropachydinae). J New York Entomol Soc 106:57–63

Brailovsky H (1998b) A review of the genus *Amblyomia* Stål (Heteroptera: Coreidae: Coreidae: Leptoscelini). Fla Entomol 81:475–482

Brailovsky H (1999a) The tribe Merophachydini with descriptions of five new genera, synonymical notes, and a key to the genera (Heteroptera: Coreidae: Meropachydinae). Proc Entomol Soc Wash 101:573–594

Brailovsky H (1999b) One new genus and three new species of Acanthocephalini (Hemiptera: Heteroptera: Coreidae: Coreinae). J New York Entomol Soc 107:247–255

Brailovsky H (2000) A new genus and three new species of the tribe Spathophorini (Heteroptera: Coreidae). Entomol News 111:97–106

Brailovsky H (2001a) A new genus and a new species of Leptoscelidini (Heteroptera: Coreidae: Coreinae) from Brazil. Proc Entomol Soc Wash 103:917–921

Brailovsky H (2001b) Five new species of Neotropical Coreidae (Insecta: Hemiptera). Reichenbachia 31:67–79

Brailovsky H (2006a) A new species of *Molchina* Amyot and Serville, 1843 from Ecuador (Heteroptera: Coreidae: Spartocerini). Fla Entomol 89:233–238

Brailovsky H (2006b) A review of the Mexican species of *Acanthocephala* Laporte, with description of one new species (Heteroptera, Coreidae, Coreinae, Acanthocephalini). Denisia 50:249–268

Brailovsky H (2008) Revision of the genus *Diariptus* Stål (Hemiptera: Heteroptera: Coreidae: Meropachydinae: Spathophorini). Zootaxa 1936:59–65

Brailovsky H (2009a) New genera and new species of Meropachyini (Hemiptera: Heteroptera: Coreidae: Meropachyinae) from South America. Proc Entomol Soc Wash 111:555–563

Brailovsky H (2009b) Two new species of *Malvanaioides* Brailovsky and key to the known taxa (Hemiptera: Heteroptera: Coreidae: Coreinae: Leptoscelini). Proc Entomol Soc Wash 111:365–369

Brailovsky H (2009c) Revision of the *Phthia* generic complex with a description of four new genera (Hemiptera: Heteroptera: Coreidae: Coreinae: Leptoscelini). Acta Entomol Musei Nat Pragae 49:59–74

Brailovsky H (2010a) New genus and new species of Hydarini (Hemiptera, Heteroptera, Coreidae) from South America. Deutsche Entomol Zeitscrift 57:85–88

Brailovsky H (2010b) Review of the Neotropical genus *Euagona* Dallas (Hemiptera: Heteroptera: Coreidae: Spartocerini) with description of one new species. Proc Entomol Soc Wash 112:1–9

Brailovsky H, Barrera E (1986a) El género *Cnemomis* Stål y descripción de cuatro nuevas especies (Hemiptera-Heteroptera-Coreidae-Discogastrini). An Inst Biol Univ Nac Aut México Ser Zool 57:119–136

Brailovsky H, Barrera E (1986b) El género *Quintius* Stål con descripción de un subgénero nuevo y tres especies nuevas (Hemiptera-Heteroptera-Coreidae-Nematopodini). An Inst Biol Univ Nac Aut México Ser Zool 56:437–452

Brailovsky H, Barrera E (1994) Descripción de cuatro especies y una subespecie nuevas de la Tribu Anisoscelidini (Hemiptera-Heteroptera-Coreidae). An Inst Biol Univ Nac Aut México Ser Zool 65:45–62

Brailovsky H, Barrera E (1998a) A new genus, two new species, and synonymical notes on the tribe Spathophorini (Heteroptera: Coreidae: Meropachydinae), with a key to genera of the tribe. Proc Entomol Soc Wash 100:746–754

Brailovsky H, Barrera E (1998b) A review of the Costa Rican species of *Leptoglossus* Guerin, with descriptions of two new species (Hemiptera: Heteroptera: Coreidae: Coreinae: Anisoscelini). Proc Calif Acad Sci 50:167–184

Brailovsky H, Barrera E (1999) An analysis of the genus *Salapia* Stål with description of six new species, and some taxonomic rearrangements (Hemiptera : Heteroptera : Coreidae : Acanthocephlalini). Pan-Pacific Entomol 75:130–146

Brailovsky H, Barrera E (2001) A further contribution to the systematics of the tribe Meropachyini (Heteroptera: Coreidae: Meropachyinae). Fla Entomol 84:613–627

Brailovsky H, Barrera E (2002) New genera and new species of neotropical Nematopodini (Hemiptera: Heteroptera: Coreidae: Coreinae). Pan Pac Entomol 78(4):265–275

Brailovsky H, Barrera E (2003a) A new genus, one new species, and synonymical notes on the tribe Acanthocerini (Heteroptera: Coreidae: Coreinae). Proc Entomol Soc Wash 105:883–888

Brailovsky H, Barrera E (2003b) Three new species and new distributional records of Discogastrini from Central and South America (Hemiptera: Heteroptera: Coreidae: Coreinae). J N Y Entomol Soc 111:22–28

Brailovsky H, Barrera E (2004) Six new species of *Leptoglossus* Guérin (Hemiptera: Heteroptera: Coreidae: Coreinae: Anisoscelini). J N Y Entomol Soc 112:56–74

Brailovsky H, Barrera E (2005) *Cervantistellus*: a new genus of Acanthocephalini from Mexico (Hemiptera: Coreidae: Coreinae). Ann Entomol Soc Am 98:48–54

Brailovsky H, Barrera E (2009a) New species of *Merocoris* (*Merocoris*) perty from Brazil, with keys to known subgenera and species of the tribe Merocorini (Hemiptera: Heteroptera: Coreidae: Meropachyinae). Fla Entomol 92:134–138

Brailovsky H, Barrera E (2009b) A revision of the genus *Neoquintius* nov. status, with the description of four new species and analysis of the *Quintius* complex (Hemiptera: Heteroptera: Coreidae: Nematopodini). Rev Mex Biodiv 80:411–418

Brailovsky H, Barrera E (2011) Two new species of *Leptoscelis* from South America (Hemiptera: Heteroptera: Coreidae: Leptoscelini). Het Rev Entomol 11:195–199

Brailovsky H, Barrera E (2012) Review of the *Stenoscelidea* complex and key to the known genera of Stenoscelideini (Hemiptera: Heteroptera: Coreidae). Acta Entomol Musei Nat Pragae 52:355–370

Brailovsky H, Barrera E (2013) New species of *Narnia* (Hemiptera: Heteroptera: Coreidae: Coreinae: Anisoscelini) from Mexico and key to the known species of the genus. Zootaxa 3736:285–890

Brailovsky H, Cadena A (1992) Revisión del género *Zicca* (Hemiptera-Heteroptera-Coreidae-Coreinae-Coreini). Publ Espec Univ Aut México 9:1–101

Brailovsky H, Couturier G (2003) A new species of *Leptoglossus* (Heteroptera: Coreidae: Anisoscelini) associated with the Amazonian palm *Mauritia flexulosa* (Arecaceae: Lepidocaryeae) in Peru. Entomol News 114:18–22

Brailovsky H, Garcia M (1987) Revisión del género *Catorhintha* Stål (Hemiptera-Heteroptera-Coreidae-Coreinae-Coreini). Monogr Inst Biol Univ Nac Aut Mexico 4:1–148

Brailovsky H, Luna A (2000) A new Neotropical genus of Meropachyni (Heteroptera: Coreidae: Meropachydinae). J New York Entomol Soc 108:163–170

Brailovsky H, Melendez V (1989) Revisión del género *Sphictyrtus* Stål (Hemiptera-Heteroptera-Coreidae-Coreinae-Coreini). An Inst Biol Univ Nac Aut México Ser Zool 60:1–76

Brailovsky H, Rivera L (1979) Revisión del género *Madura* Stål, con descripción de una nueva especie (Hemiptera-Heteroptera-Coreidae-Coreini). An Inst Biol Univ Nac Aut México Ser Zool 50:241–250

Brailovsky H, Sánchez C (1983a) Hemiptera-Heteroptera de México XXIX. Revisión de la familia Coreidae Leach. Parte 4. Tribu Anisoscelidini Amyot-Serville. An Inst Biol Univ Nac Aut México Ser Zool 53:219–275

Brailovsky H, Sánchez C (1983b) Hemiptera-Heteroptera de México XXVI. Revisión de la familia Coreidae Leach. Parte 3. Tribu Spartocerini Amyot-Serville. An Inst Biol Univ Nac Aut México Ser Zool 53:181–203

Brailovsky H, Barrera E, Mayorga C, Ortega Leon G (1994) Estadios ninfales de los Coreidos del Valle de Tehuacán, Puebla. (Hemiptera-Heteroptera) I. *Chelinidea staffilesi*, *C. tabulata* y *Narnia femorata*. An Inst Biol Univ Nac Aut México Ser Zool 65:241–264

Brailovsky H, Mayorga C, Ortega León G, Barrera E (1995) Estadios ninfales de los Coreidos del Valle de Tehuacán, Puebla. (Hemiptera-Heteroptera) II. Especies asociadas a huizacheras (*Acacia* spp.) y mezquiteras (*Prosopis* spp.): *Mozena lunata*, *Pachylis hector*, *Savius jurgiosus jurgiosus* y *Thasus gigas*. An Inst Biol Univ Nac Aut México Ser Zool 66:57–80

Bruton BD, Mitchell F, Fletcher J, Pair SD, Wayadande A, Melcher U, Brady J, Bextine B, Popham TW (2003) *Serratia marcescens*, a phloem-colonizing, squash bug-transmitted bacterium: causal agent of cucurbit yellow vine disease. Plant Dis 87:937–944

Caetano AC, Boiça AL Jr (2000) Desenvolvimento de *Leptoglossus gonagra* Fabr. (Heteroptera: Coreidae) em espécies de maracujazeiro. An Soc Entomol Brasil 29:353–359

Caldas B-HC, Redaelli LR, Dieffenbach LMG (1998) Descrição dos estágios imaturos de *Corecoris dentiventris* Berg (Hemiptera: Coreidae). An Soc Entomol Brasil 27:405–412

Caldas B-HC, Redaelli LR, Dieffenbach LMG (1999) Parâmetros reprodutivos de *Corecoris dentiventris* Berg (Hemiptera:Coreidae) em cultura de fumo (*Nicotiana tabacum*). An Soc Entomol Brasil 28:595–600

Caldas B-HC, Redaelli LR, Dieffenbach LMG (2000) Biology of *Corecoris dentiventris* Berg, 1884 (Hemiptera, Coreidae) in tobacco culture (*Nicotiana tabacum*). Rev Brasil Biol 60:173–178

Calza R, Orlando A, Rossetti V, Nakadaira JT (1964) O percevejo *Leptoglossus gonagra* (Fabr. 1775) em citros, no estado de São Paulo. O Biológico 30:188–197

Camargo EP, Wallace FG (1994) Vectors of plant parasites of the genus *Phytomonas*. Adv Dis Vector Res 10:333–359

Canto-Silva CR, Romanowski HP (2003) Population fluctuation, immature mortality and adult longevity of *Spartocera dentiventris* (Berg) (Hemiptera: Coreidae) on *Nicotiana tabacum* (Solanaceae). Neotrop Entomol 32:399–406

Chittenden FH (1927) The Florida potato plant-bug. State Plant Board Fla Quarterly Bull 11:115–118

Cobben RH (1968) Evolutionary trends in Heteroptera. Part I. Eggs, architecture of the shell, gross embryology and eclosion. Agric Res Reports, Wageningen 707:1–475

Cobben RH (1978) Evolutionary trends in Heteroptera. Part II. Mouth-part structures and feeding strategies. Meded Landbouwhogeschool Wageningen 78:1–407

Cook CA, Neal JJ (1999) Feeding behavior of larvae of *Anasa tristis* (Heteroptera: Coreidae) on pumpkin and cucumber. Environ Entomol 28:173–177

CoreoideaSF Team (2015) Coreoidea species file online. Version 5.0/5.0 [14 Jan 2015]. http://Coreoidea.SpeciesFile.org

Costa Lima A (1940) Insetos do Brasil, vol 2, Hemipteros. Escola Nacional de Agronomia. Série didática no. 3, Rio de Janeiro

Couturier G, Clement CR, Viana Filho P (1991) *Leptoglossus lonchoides* Allen (Heteroptera, Coreidae), causante de la caída de los frutos de *Bactris gasipaes* (Palmae) en la Amazonia central. Turrialba 41:293–298

Couturier G, Quiñones L, González R, Riva R, Young F (1996) Los insectos plaga de las Myrtaceae frutales en Pucallpa, Amazónia Peruana. Rev Per Entomol 39:125–130

Cuda JP, DeLoach CJ (1998) Biology of *Mozena obtusa* (Hemiptera: Coreidae), a candidate for the biological control of mesquite, *Prosopis* spp. (Fabaceae). Biol Contr 13:101–110

da Rocha LD, Redaelli LR (2004) Functional response of *Cosmoclopius nigroannulatus* (Hem.: Reduviidae) to different densities of *Spartocera dentiventris* (Hem.: Coreidae) nymphae. Braz J Biol 64:309–316

da Rocha LD, Kolberg R, de Mendonça Jr MS, Redaelli LR (2007) Body size variation in *Gryon gallardoi* related to age and size of the host. Biol Control 52:161–173

da Rocha LD, Sant Ana J, Redaelli LR (2008) Discrimination of *Spartocera dentiventris* (Berg, 1884) (Hemiptera: Coreidae) eggs by *Gryon gallardoi* (Brèthes, 1913) (Hymenoptera: Scelionidae). Braz J Biol 68:161–167

da Silva RA, Carvalho GS (2001) Aspectos biológicos de *Phthia picta* (Drury, 1770) (Hemiptera: Coreidae) em tomateiro sob condições controladas. Ciência Rural 31:381–386

da Silva RA, Flores PS, Carvalho GS (2001) Descrição dos estágios imaturos de *Phthia picta* (Drury) (Hemiptera: Coreidae). Neotrop Entomol 30:253–258

Daoust RA, Roberts DW, Das Neves BP (1985) Distribution, biology, and control of cowpea pests in Latin America. In: Singh SR, Rachie KO (eds) Cowpea research, production and utilization. Wiley, Chichester, pp 249–264

de Souza CEP, Amaral F (1999a) Nova planta hospedeira de *Leptoglossus zonatus* (Dallas) (Heteroptera: Coreidae). An Soc Entomol Brasl 28:753–756

de Souza CEP, Amaral F (1999b) Ocorrência natural de parasitóides de *Leptoglossus zonatus* (Dallas) (Heteroptera: Coreidae). An Soc Entomol Bras 28:757–759

Dolling WR (1977) A revision of the neotropical genus *Vilga* Stål (Hemiptera: Coreidae). Syst Entomol 2:27–44

Dolling WR (1986) The tribe Pseudophloeini (Hemiptera: Coreidae) in the Old World Tropics with a discussion on the distribution of the Pseudophloeinae. Bull Br Mus Nat Hist Entomol 53:151–212

Dolling WR (1987) A minetic coreid bug and its relatives (Hemiptera: Coreide). J Nat Hist 21:1259–1271

Eberhard WG (1998) Sexual behavior of *Acanthocephala declivis guatemalana* (Hemiptera: Coreidae) and the allometric scaling of their modified hind legs. Ann Entomol Soc Am 91:863–871

Eliyahu D, Ceballos RA, Saeidi V, Becerra JX (2012) Synergy versus potency in the defensive secretions from nymphs of two pentatomomorphan families (Hemiptera: Coreidae and Pentatomidae). J Chem Ecol 38:1358–1365

Faúndez EI, Carvajal MA (2011) Primer registro de una picadura de *Leptoglossus chilensis* (Spinola, 1852) (Hemiptera: Heteroptera: Coreidae) en un ser humano. Bol Biodivers Chile 6:1–21

Fernandes JAM, Grazia J (1992) Estudo dos estágios imaturos de *Leptoglossus zonatus* (Dallas, 1852) (Heteroptera: Coreidae). An Soc Entomol Bras 21:179–188

Fiorini JE, Silva PMF, Brazil RP, Attias M, Esteves MJG, Angluster J (1993) Axenic cultivation of a pathogenic *Phytomonas* species isolated from tomato fruit, and from its phytophagic insect vector, *Phthia picta* (Hemiptera: Coreidae). Cytobios 75:163–170

Freymuller E, Milder R, Jankevicius JV, Jankevicius SI, Camargo EP (1990) Ultrastructural studies on the trypanosomatid *Phytomonas serpens* in the salivary glands of a phytophagous hemipteran. J Protozool 37:225–229

Froeschner RC (1981) Heteroptera or true bugs of Ecuador: a partial catalog. Smithsonian Contrib Zool 322:1–147

Fujisaki K (1980) Studies on the mating system of the winter cherry bug, *Acanthocoris sordidus* Thunberg (Heteroptera: Coreidae) I. Spatio-temporal distribution patterns of adults. Res Popul Ecol 21:317–331

Fujisaki K (1981) Studies on the mating system of the winter cherry bug, *Acanthocoris sordidus* Thunberg (Heteroptera: Coreidae) II. Harem defense polygyny. Res Popul Ecol 23:262–279

Gibson EH, Holdridge A (1918) The genus *Narnia* Stål and a key to the genera of Anisoscelini A. and S. (Coreidae: Heteroptera). Psyche 25:1–4

Godoi MMI, Serrano MG, Teixeira MMG, Camargo EP (2002) A PCR-based survey on *Phytomonas* (Euglenozoa: Trypanosomatidae) in phytophagous hemipterans of the Amazon region. J Eukaryot Microbiol 49:275–279

Gomendio M, García-González P, Reguera P, Rivero A (2008) Male egg carrying in *Phyllomorpha laciniata* is favoured by natural not sexual selection. Anim Behav 75:763–770

Grimm C, Führer E (1998) Population dynamics of true bugs (Heteroptera) in physic nut (*Jatropha curcas*) plantations in Nicaragua. J Appl Entomol 122:515–521

Grimm C, Guharay F (1998) Control of leaf-footed bug *Leptoglossus zonatus* and shield-backed bug *Pachycoris klugii* with entomopathogenic fungi. Biocontrol Sci Technol 8:365–376

Grimm C, Maes J-M (1997) Insectos asociados al cultivo de tempate (*Jatropha curcas*) en el Pacifico de Nicaragua III Coreoidea (Heteroptera). Rev Nica Entomol 42:16–34

Grimm C, Somarriba A (1999) Suitability of physic nut (*Jatropha curcas* L.) as single host plant for the leaf-footed bug *Leptoglossus zonatus* Dallas (Het., Coreidae). J Appl Entomol 123:347–350

Halbert S (2003) Giant sweetpotato bug, *Spartocera batatas* (Fabricius) (Insecta: Hemiptera: Coreidae). University of Florida IFAS Extension Document EENY-305 (IN582). http://edis.ifas.ufl.edu/pdffiles/IN/IN58200.pdf

Hamlin JC (1924) A review of the genus *Chelinidea* (Hemiptera-Heteroptera) with biological data. Ann Entomol Soc Am 17:193–208

Hemmingsen AM (1947) Plant bug guarding eggs and offspring and shooting anal jets (*Physomerus grossipes* F., Coreidae). Entomol Medd 25:200

Henne DC, Johnson SJ, Bourgeois WJ (2003) Pest status of leaf-footed bugs (Heteroptera: Coreidae) on citrus in Louisiana. Proc Fla Hort Soc 116:240–241

Henry TJ, Froeschner RC (1988) Catalog of the Heteroptera, or true bugs, of Canada and the continental United States. E.J. Brill, Leiden

Herring JL (1980) A review of the cactus bugs of the genus *Chelinidea* with the description of a new species (Hemiptera: Coreidae). Proc Entomol Soc Wash 82:237–251

Hizal E (2012) Two invasive alien insect species, *Leptoglossus occidentalis* (Heteroptera: Coreidae) and *Cydalima perspectalis* (Lepidoptera: Crambidae), and their distribution and host plants in Istanbul province, Turkey. Fla Entomol 95:344–349

Jankevicius JV, Jankevicius SI, Campaner M, Conchon I, Maeda LA, Teixeira MMG, Freymuller E, Camargo EP (1989) Life cycle and culturing of *Phytomonas serpens* (Gibbs), a trypanosomatid parasite of tomatoes. J Protozool 36:265–271

Jankevicius SI, de Almeida ML, Jankevicius JV, Cavazzana M Jr, Attias M, De Souza W (1993) Axenic cultivation of trypanosomatids found in corn (*Zea mays*) and in phytophagous hemipterans (*Leptoglossus zonatus* Coreidae) and their experimental transmission. J Eukar Microbiol 40:576–581

Jesus CR, Redaelli LR, Romanowski HP (2002) Population dynamics and spatial distribution of *Spartocera dentiventris* (Berg) (Hemiptera: Coreidae) adults on *Nicotiana tabacum* L. (Solanaceae). Neotrop Entomol 31:541–549

Johnson SR, Allain LK (1998) Observations on insect use of Chinese tallow [*Sapium sebiferum* (L.) Roxb.] in Louisiana. Castanea 63:188–189

Kormilev N (1954) Notas sobre Coreidae Neotropicales II; (Hemiptera) Mericorinae de la Argentina y países limítrofes. Rev Ecuat Entomol Parasitol 2:153–185

Kumar R (1966) Studies on the biology, immature stages and relative growth of some Australian bugs of the superfamily Coreoidea (Hemiptera: Heteroptera). Austr J Zool 14:895–991

Leach WE (1815) Coreidae. In: Edinburgh encyclopaedia conducted by David Brewster with the assistance of gentlemen eminent in science and literature, vol 9. Blackwood, Edinburgh, pp 121–183

Leal WS, Panizzi AR, Niva CC (1994) Alarm pheromone system of leaf-footed bug *Leptoglossus zonatus* (Heteroptera: Coreidae). J Chem Ecol 20:1209–1216

Lesieur V, Yart A, Guilbon S, Lorme P, Auger-Rozenberg M-A, Roques A (2014) The invasive *Leptoglossus* seed bug, a threat for commercial seed crops, but for conifer diversity? Biol Invasions. doi:10.1007/s10530-013-0630-9

Maes J-M, Goellner-Scheiding U (1993) Catalogo de los Coreoidea (Heteroptera) de Nicaragua. Rev Nica Entomol 25:1–19

Mariconi FAM (1953) O "percevejo do abacaxi" (*Lybindus dichrous* Stål, 1859). O Biológico 19:155–162

Maschwitz U, Fiala B, Dolling WR (1987) New trophobiotic symbioses of ants with south east Asian bugs. J Nat Hist 21:1097–1107

Maskey K (2010) Comparison of electrical penetration graph waveforms of squash bug feeding on watermelon and its relatives. Thesis, Oklahoma State University

Matrangolo WJR, Waquil JM (1994) Biologia de *Leptoglossus zonatus* (Dallas)(Hemiptera: Coreidae) alimentados com milho e sorgo. An Soc Entomol Bras 23:419–423

McLain DK, Burnette LB, Deeds DA (1993) Within season variation in the intensity of sexual selection on body size in the bug *Margus obscurator* (Hemiptera Coreidae). Ethol Ecol Evol 5:75–86

McPherson JE, Packauskas RJ, Sites RW, Taylor SJ, Bundy CS, Bradshaw JD, Mitchell PL (2011) Review of *Acanthocephala* (Hemiptera: Heteroptera: Coreidae) of America north of Mexico with a key to species. Zootaxa 2835:30–40

Mead FW (2014) Coreid bug, leaf-footed bug, *Euthochtha galeator* (Fabricius) (Insecta: Hemiptera: Coreidae). EENY293 UF/IFAS Extension. https://edis.ifas.ufl.edu/in570

Menezes-Netto AC, Varella AC, Fernandes OA (2012) Maize-dwelling insects omnivory in *Spodoptera frugiperda* (J.E. Smith) (Lepidoptera: Noctuidae) egg masses. Braz Arch Biol Technol 55:97–100

Miles PW (1972) The saliva of Hemiptera. Adv Ins Physiol 9:183–255

Miles PW, Taylor GS (1994) 'Osmotic pump' feeding by coreids. Entomol Exp Appl 73:163–173

Millar JG (2005) Pheromones of true bugs. Top Curr Chem 240:37–84

Miller CW (2008) Seasonal effects on offspring reproductive traits through maternal oviposition behavior. Behav Ecol 19:1297–1304

Miller CW (2011) The heliconia bug, *Leptoscelis tricolor*. Featured creatures. http://entnemdept.ufl.edu/creatures/misc/bugs/heliconia_bug.htm

Miller CW (2013) Sexual selection: male-male competition. In: Losos J (ed) The Princeton guide to evolution. Princeton University Press, Princeton, pp 641–646

Miller CW, Emlen DJ (2010) Across and within population differences in the size and scaling relationship of a sexually-selected trait in *Leptoscelis tricolor* (Hemiptera: Coreidae). Ann Entomol Soc Am 103:209–215

Miller CW, Fletcher RJ Jr, Anderson BD, Nguyen LD (2012) Natal social environment influences habitat selection later in life. Anim Behav 83:473–477

Mitchell PL (1980a) Host plant utilization by leaf-footed bugs: an investigation of generalist feeding strategy. Dissertation, University of Texas at Austin

Mitchell PL (1980b) Combat and territorial defense of *Acanthocephala femorata* (Hemiptera, Coreidae). Ann Entomol Soc Am 73:404–408

Mitchell PL (2000) Leaf-footed bugs. In: Schaefer CW, Panizzi AR (eds) Heteroptera of economic importance. CRC Press, Boca Raton, pp 337–403

Mitchell PL (2004) Heteroptera as vectors of plant pathogens. Neotrop Entomol 33:519–545

Mitchell PL (2006) Polyphagy in true bugs: a case study of *Leptoglossus phyllopus* (L.) (Hemiptera, Heteroptera, Coreidae). Denisia 19:1117–1134

Mitchell PL, Mitchell FL (1986) Parasitism and predation of leaffooted bug (Hemiptera: Heteroptera: Coreidae) eggs. Ann Entomol Soc Am 79:854–860

Miyatake T (1995) Territorial mating aggregation in the bamboo bug, *Notobitus meleagris*, Fabricius (Heteroptera: Coreidae). J Ethol 13:185–189

Miyatake T (1997) Functional morphology of the hind legs as weapons for male contests in *Leptoglossus australis* (Heteroptera: Coreidae). J Insect Behav 10:727–735

Moulet P (1995) Hémiptères Coreoidea euro-méditerranéens. Faune de France 81:1–336

Nageon de Lestang F, Miller CW (2009) Effects of diet on development and survivorship of *Narnia femorata* nymphs (Hemiptera: Coreidae). Fla Entomol 92:511–512

Neal JJ (1993) Xylem transport interruption by *Anasa tristis* feeding causes *Cucurbita pepo* to wilt. Entomol Exp Appl 69:195–200

Noge K, Prudic KL, Becerra JX (2012) Defensive roles of (E)-2-alkenals and related compounds in Heteroptera. J Chem Ecol 38:1050–1056

O'Shea R (1979) Redescriptions of three Neotropical coreid genera of uncertain tribal placement (Heteroptera). Entomol News 90:45–50

O'Shea R (1980a) A generic revision of the Acanthocerini (Hemiptera:Coreidae: Coreinae). Stud Neotr Fauna Environ 15:57–80

O'Shea R (1980b) A generic revision of the Nematopodini (Hemiptera:Coreidae: Coreinae). Stud Neotr Fauna Environ 15:197–225

O'Shea R, Schaefer CW (1978) The Mictini are not monophyletic (Hemiptera, Coreidae: Coreinae). Ann Entomol Soc Am 71:776–784

Olivier-Espejel S, Sabree ZL, Noge K, Becerra JX (2011) Gut microbiota in nymph and adults of the giant mesquite bug (*Thasus neocalifornicus*)(Heteroptera: Coreidae) is dominated by *Burkholderia* acquired de novo every generation. Environ Entomol 40:1102–1110

Osuna EA (1984) Monografia de la tribu Anisoscelidini (Hemiptera: Heteroptera Coreidae). 1. Revisión genérica. Bol Entomol Venez (NS) 3:77–145

Packauskas RJ (1994) Key to the subfamilies and tribes of the New World Coreidae (Hemiptera) with a checklist of published keys to genera and species. Proc Entomol Soc Wash 96:44–53

Packauskas RJ (2006) On the re-erection of the tribe Stenoscelideini Schaefer (Heteroptera, Coreidae, Coreinae). In: Rabitsch W (ed) Hug the bug- For love of true bugs. Festschrift zum 70. Geburtstag von Ernst Heiss, pp 539–542. *Denisia* 19:1–1184

Packauskas RJ (2008) The position of the Leptoscelini and other taxonomic changes within the family Coreidae (Hemiptera: Heteroptera). Zootaxa 1669:63–68

Packauskas R (2010) Catalog of the Coreidae, or leaf-footed bugs, of the new world. Fort Hays Stud 5:1–270

Packauskas RJ, Schaefer CW (2001) Clarifaction of some taxonomic problems in Anisoscelini and Leptoscelini (Hemiptera: Coreidae: Coreinae). Proc Entomol Soc Wash 103:249–256

Pair SD, Bruton BD, Mitchell F, Fletcher J, Wayadande A, Melcher U (2004) Overwintering squash bugs harbor and transmit the causal agent of cucurbit yellow vine disease. J Econ Entomol 97:74–78

Pall JL, Coscarón MC (2013) Synopsis of Acanthocerini (Hemiptera, Coreidae) from Argentina. ZooKeys 305:33–53

Panizzi AR (1989) Desempenho de ninfas e adultos de *Leptoglossus zonatus* (Dallas, 1852) (Hemiptera: Coreidae) em diferentes alimentos. An Soc Entomol Bras 18:375–379

Panizzi AR (2004a) Adaptive advantages for egg and nymph survivorship by egg deposition in masses or singly in seed-sucking Heteroptera. In: Gujar GT (ed) Contemporary trends in insect science. Campus Books International, Delhi, pp 60–73

Panizzi AR (2004b) A possible territorial or recognition behavior of *Leptoglossus zonatus* (Dallas) (Heteroptera, Coreidae). Rev Brasil Entomol 48:577–579

Peña JE, Bennett FD (1995) Arthropods associated with *Annona* spp. in the Neotropics. Fla Entomol 78:329–349

Peredo LC (1999) Description of the immature stages, adult morphology and biology of *Pantochlora vivida* Stål (Heteroptera: Pentatomidae: Edessinae). J N Y Entomol Soc 107:372–385

Pires EM, Nogueira RM, da Silva CJ, Pelissari F, Ferreira JAM, Soares MA (2013) New sucking coreid species in *Psidium guajava*. Scient Electron Arch 4:31–35

Polaszek A, Kolberg R (2008) Rediscovery of *Gryon chrysolaum* (Walker)(Hymenoptera: Platygastridae)—an egg parasitoid of *Holhymenia rubiginosa* Breddin and *Anisoscelis foliacea* Fabricius (Heteroptera: Coreidae). Zootaxa 1968:39–44

Procter DS, Moore AJ, Miller CW (2012) The form of sexual selection arising from male–male competition depends on the presence of females in the social environment. J Evol Biol 25:803–812

Prudic KL, Noge K, Becerra JX (2008) Adults and nymphs do not smell the same: the different defensive compounds of the giant mesquite bug (*Thasus neocalifornicus*: Coreidae). J Chem Ecol 34:734–741

Raga A, de Piza CT Jr, de Souza Fo MF (1995) Ocorrência e danos de *Leptoglossus zonatus* (Dallas) (Heteroptera: Coreidae) em romã, *Punica granatum* L., em Campinas, São Paulo. An Soc Entomol Bras 24:183–185

Reguera P, Gomendio M (1999) Predation costs associated with parental care in the golden egg bug *Phyllomorpha laciniata* (Heteroptera: Coreidae). Behav Ecol 10:541–544

Rodrigues D, Moreira GRP (2005) Comparative description of the immature stages of two very similar leaf footed bugs, *Holymenia clavigera* (Herbst) and *Anisoscelis foliacea marginella* (Dallas) (Hemiptera, Coreidae, Anisoscelini). Rev Bras Entomol 49:7–14

Rodrigues D, Sampaio DS, Isaias RMDS, Moreira GRP (2007a) Xylem and seed feeding by two passion vine leaffooted bugs, *Holymenia clavigera* and *Anisoscelis foliacea marginella* (Hemiptera: Coreidae: Anisoscelini), with notes on mouthpart morphology. Ann Entomol Soc Am 100:907–913

Rodrigues D, Duarte LS, Moreira GRP (2007b) Performance consequences of food mixing in two passion vine leaf-footed bugs, *Holymenia clavigera* Herbst, 1784 and *Anisoscelis foliacea marginella* Dallas, 1852 (Hemiptera: Coreidae). Braz J Biol 67:91–99

Rodrigues D, Soares GLG, Moreira GRP (2008) Feeding preference of *Holymenia clavigera* and *Anisoscelis foliacea marginella* (Hemiptera: Coreidae: Anisoscelini) in relation to intra- and interspecific features of host plants (Passifloraceae). Environ Entomol 37:1323–1331

Ruckes H (1955) The genus *Chariesterus* de Laporte (Heteroptera, Coreidae). Am Museum Novit 1721:1–16

Santos RSS, Redaelli LR, Dieffenbach LMG (2001) Ocorrência de parasitismo em ovos de *Spartocera dentiventris* (Berg) (Hemiptera: Coreidae) em cultura de fumo. Neotrop Entomol 30:731–733

Schaefer CW (1964) The morphology and higher classification of the Coreoidea (Hemiptera-Heteroptera): parts I and II. Ann Entomol Soc Am 57:670–684

Schaefer CW (1965) The morphology and higher classification of the Coreoidea (Hemiptera-Heteroptera): parts III. Misc Publs Entomol Soc Am 5:1–76

Schaefer CW (1968) The morphology and higher classification of the Coreoidea (Hemiptera: Heteroptera) part IV. The Acanthocephala-group and the position of *Stenoscellidea* Westwood (Coreidae). Occ Pap Univ Conn (Biol Sci) 1:153–199

Schaefer CW, Mitchell PL (1983) Food plants of the Coreoidea (Hemiptera: Heteroptera). Ann Entomol Soc Am 76:591–615

Schaefer CW, O'Shea R (1979) Host plants of three coreine tribes (Hemiptera: Heteroptera: Coreidae). Ann Entomol Soc Am 72:519–523

Schaefer CW, Panizzi AR (eds) (2000) Heteroptera of economic importance. CRC Press, Boca Raton

Schaefer CW, Packauskas RP, Bu K (2008) A new species of *Leptoglossus* (Hemiptera: Coreidae: Coreinae: Anisoscelini) from Brazil. Proc Entomol Soc Wash 110:61–66

Schlee MA (1992) La prédation des hétéroptères par les oiseaux. Trav Laborat Biol Évol Insectes Hémipteroidea 5:87–96

Schuh RT, Slater JA (1995) True bugs of the world (Hemiptera: Heteroptera). Classification and natural history. Cornell University Press, Ithaca

Serantes HE (1973) Biologia de *Phthia picta* (Drury) (Hemiptera: Coreidae). Fitotec Lat Am 9:1–9

Sgrillo RB, Moura JIL, Sgrillo KRPA (2005) Simulation model for *Phytomona* epidemics in coconut trees. Neotrop Entomol 34:527–538

Soglia MCM, Sá WMS, Nascimento AS (1998) Aspectos bioecológicos do percevejo *Crinocerus sanctus* (Fabr., 1775) (Heteroptera, Coreidae), praga da aceroleira (*Malpighia puncifolia*). Abstr XVII Congr Bras Entomol, Rio de Janeiro

Soldi RA, Rodrigues MACM, Aldrich JR, Zarbin PHG (2012) The male-produced sex pheromone of the true bug, *Phthia picta*, is an unusual hydrocarbon. J Chem Ecol 38:814–824

Solomon JC, Froeschner RC (1981) Notes on food resources and behavior of the family Coreidae (Hemiptera) in a semi-deciduous tropical forest. Proc Entomol Soc Wash 83:428–431

Stål C (1867) Bidrag till hemipterernas systematic. Öfvers K Vet Akad Förhandl 24:491–560

Stål C (1868) Hemiptera Fabriciana. Kong Sven Vet Akad Handl 7:1–148

Stål C (1870) Hemiptera insularum Philippinarum – Bidrag till Philippinska oarnes Hemipter-fauna. Ofv Kong Vet Akad Forh 7:607–776

Stål C (1873) Enumeratio Hemipterorum. 3. Kong Sven Vet Akad Handl 11:1–163

Tatarnic NJ, Spence JR (2013) Courtship and mating in the crusader bug, *Mictis profana* (Fabricius). Austr J Entomol 52:151–155

Tepole-García RE, Pineda-Guillermo S, Martínez-Herrera J, Castrejón-Gómez VR (2012) Records of two pest species, *Leptoglossus zonatus* (Heteroptera: Coreidae) and *Pachycoris klugii* (Heteroptera: Scutelleridae), feeding on the physic nut, *Jatropha curcas*, in Mexico. Fla Entomol 95:208–210

Thum AB, Costa EC (1997) Coreidae (Heteroptera) associados a espécies florestais. Cienc Flor 7:27–31

Torre-Bueno JR (1941) A synopsis of the Hemiptera-Heteroptera of America north of Mexico. Part II. Families Coreidae, Alydidae, Corizidae, Neididae, Pyrrhocoridae and Thaumastotheriidae. Entomol Am 21:41–122

Urich FW (1916) Insects affecting the cotton plant in Trinidad. Trinidad Tobago Bull XV:18–19

Urueta-Sandino EJ (1975) Plagas de los cultivos de badea, curuba, maracuya, papaya, y vid en el occidente Antioqueño. Sec Agric Fomento, Medellín

Van Duzzee EP (1916) Check list of the Hemiptera (Excepting the Aphididae, Aleurodidae and Coccidae) of America, North of Mexico. New York Entomological Society, New York

Wayadande A, Bruton B, Fletcher J, Pair S, Mitchell F (2005) Retention of cucurbit yellow vine disease bacterium *Serratia marcescens* through transstadial molt of vector *Anasa tristis* (Hemiptera: Coreidae). Ann Entomol Soc Am 98:770–774

Wheeler AG Jr, Miller GL (1990) *Leptoglossus fulvicornis* (Heteroptera: Coreidae), a specialist on magnolia fruits: seasonal history, habits, and descriptions of immature stages. Ann Entomol Soc Am 83:753–765

Whitehead DR (1974) Variation and synonymy in *Hypselonotus* (Heteroptera: Coreidae). J Wash Acad Sci 64:223–233

Xiao YF, Fadamiro H (2009) Host preference and development of *Leptoglossus zonatus* (Hemiptera: Coreidae) on satsuma mandarin. J Econ Entomol 102:1908–1914

Xiao YF, Fadamiro HY (2010) Evaluation of damage to satsuma mandarin (*Citrus unshiu*) by the leaffooted bug, *Leptoglossus zonatus* (Hemiptera: Coreidae). J Appl Entomol 134:694–703

Yonke T (1972) A new genus and two new species of Neotropical Chariesterini (Hemiptera: Coreidae). Proc Entomol Soc Wash 74:283–287

Yonke TR, Medler JT (1969a) Description of immature stages of Coreidae. 1. *Euthochtha galeator*. Ann Entomol Soc Am 62:469–473

Yonke TR, Medler JT (1969b) Description of immature stages of Coreidae. 2. *Acanthocephala terminalis*. Ann Entomol Soc Am 64:474–476

Young AM (1980) Notes on the interaction of the Neotropical bug *Paryphes blandus* Horvath (Hemiptera: Coreidae) with the vine *Anguria warscewiczii* Hook F. (Cucurbitaceae). Brenesia 17:27–42

Chapter 20
Scentless Plant Bugs (Rhopalidae)

Trevor M. Fowles, Maria del C. Coscarón, Antônio R. Panizzi, and Scott P. Carroll

Abstract Rhopalidae are closely related to the much more speciose Coreidae, and like them are plant feeders, although more associated with reproductive tissues and seeds. The two rhopalid subfamilies, Rhopalinae and Serinethinae, occur worldwide, with the latter more restricted to tropical latitudes. About 38 % of the ca. 210 rhopalid species and 38 % of the 21 genera have at least partially Neotropical distributions. The serinethine genus *Jadera* Stål is particularly associated with the Neotropics and like other members of the subfamily feeds on seeds of Sapindaceae (soapberry family). Most Neotropical rhopalid species are unstudied, and few appear to have major economic importance. However, serinethines are important models in insect reproductive life history, behavior, and human-induced evolution. Rhopalids are not scentless, and a more accurate family name is needed.

20.1 Introduction

The rhopalids are a small worldwide family of approximately 200 species of plant-feeding bugs in 21 genera and 209 species (Henry 2009). Formerly treated as a subfamily of the Coreidae, usually with the name Corizinae, Schaefer (1965) ended a century of taxonomic controversy with his definitive elevation of these insects to family Rhopalidae, with subfamilies Rhopalinae and Serinethinae. Chopra (1967)

T.M. Fowles (✉) • S.P. Carroll
Department of Entomology, University of California, Davis, CA 95616, USA
e-mail: tmfowles.work@gmail.com; spcarroll@ucdavis.edu

M.C. Coscarón
División Entomología, Museo de La Plata,
Paseo del bosque s/n, La Plata 1900, Buenos Aires, Argentina
e-mail: mcoscaron@fcnym.unlp.edu.ar

A.R. Panizzi
Laboratório de Entomologia, Embrapa Trigo, Caixa Postal 3081,
Passo Fundo, RS 99001-970, Brazil
e-mail: antonio.panizzi@embrapa.br

A.R. Panizzi, J. Grazia (eds.), *True Bugs (Heteroptera) of the Neotropics*,
Entomology in Focus 2, DOI 10.1007/978-94-017-9861-7_20

Table 20.1 Summary of the known number of Rhopalidae genera and species for the Australian (1), Nearctic (2), Neotropical (3), Palearctic (4) and regions and the world (5)

Australian		Nearctic		Neotropical		Palearctic		World	
Genus	Species	Genus	Species	Genus	Species	Genus	Species	Genus	Species
2	6	10	39	9	81	14	69	21	209

(1) Based on Cassis and Gross (2002)
(2) Based on chapters in Henry (1988)
(3) Based on this chapter
(4) Based on chapters in Aukena and Rieger (1995–2006)
(5) Based on Henry (2009)

extensively revised the tribal classification of the family and included keys to subfamilies, tribes, and genera. Göllner-Scheiding (1983) published a world catalog of the family. In Rhopalinae there are approximately 150 species in six tribes and 17 genera, while the Serinethinae (soapberry bugs) consist of about 65 species in three genera.

For the Neotropical region, four of the six rhopaline tribes are present (Table 20.1), with none of these being exclusive to the New World. Likewise, two of the three serinethine genera are present, and the genus *Jadera* Stål is exclusive to the New World.

Several partial catalogs have been developed for South America: Argentina (five species in two genera, Pennington 1920–1921, and 35 species in five genera, Coscarón 2014), mainland Ecuador (10 species in four genera, Froeschner 1981), the Galápagos Islands (three species in three genera, Froeschner 1985), Nicaragua (12 species in six genera, Maes and Göllner-Scheiding 1993), and Panama (14 species in four genera, Froeschner 1999). In addition, for Brazil, Grazia et al. (2012) refer to five genera with 26 species, namely, the rhopalines *Niesthrea* Spinola (six spp.), *Liorhyssus* Stål (one sp.), *Harmostes* Burmeister (eight spp.), and *Xenogenus* Berg (two spp.), and the serinethine *Jadera* (nine spp.).

Only a few species of Neotropical rhopalids have been studied in any detail, and scientific knowledge of the general biology and ecology of this fauna is largely incomplete. In-depth information is available for a few *Jadera* species, though mainly at the margins of the Neotropical region in southern Florida and southern Brazil (e.g., Carroll et al. 2003a, b; Panizzi et al. 2005). To encourage further study, Carroll and Loye (2012) reviewed host associations of *Jadera* with plant species in the soapberry family (Sapindaceae). Much remains to be learned about the systematics of Neotropical rhopalids, as well as their potential economic importance (Gonzáles 1989). While there are no comprehensive keys for identification of the Neotropical rhopalid species, Pall and Coscarón (2012) have provided one for the Argentinian taxa, along with photographs and range maps. The Web site "Soapberry Bugs of the World" (www.soapberrybugs.org) is a center for information on serinethines.

20.2 General Characteristics and Diagnosis

Rhopalids have four-segmented mouthparts and antennae. They may resemble coreids but are generally narrower and smaller bodied (although a few Neotropical serinethines reach or exceed 15 mm in body length). Most rhopalines are much smaller than most serinethines and generally dull brownish in color. In contrast, many serinethines have contrasting coloration with bright red, orange, and black. Rhopalids are often misidentified, and other insects are likewise mistaken for rhopalids. The distinct paired ocelli between the eyes (Fig. 20.1a) distinguish serinethines from Pyrrhocoridae such as *Dysdercus* Guérin-Méneville, which lack ocelli. The rhopalid forewing membrane has close to 20 slender parallel veins arising from a transverse basal vein (Fig. 20.1a), further distinguishing serinethines from Pyrrhocoridae and the superficially similar lygaeine Lygaeidae and rhopalines from the similar Orsillinae Lygaeidae, all of which have just 4–5 major veins in the membrane. Adult rhopalids lack the scent glands on the thorax between the middle and hind coxae present in Coreidae and Lygaeidae and are also unusual in retaining functional dorsal abdominal glands between the 4–5 and 5–6 terga (Fig. 20.1b).

The commonly applied name for this family, "scentless plant bugs," reflects the absence of metathoracic scent glands in serinethines. However, this name is misleading and inappropriate because these and other rhopalids commonly produce redolent compounds from scent glands (Aldrich et al. 1990a). Members of the more diverse subfamily, Rhopalinae, in fact possess fully functional metathoracic scent glands as adults, in common with other hemipterans (Aldrich et al. 1990a; Davidova-

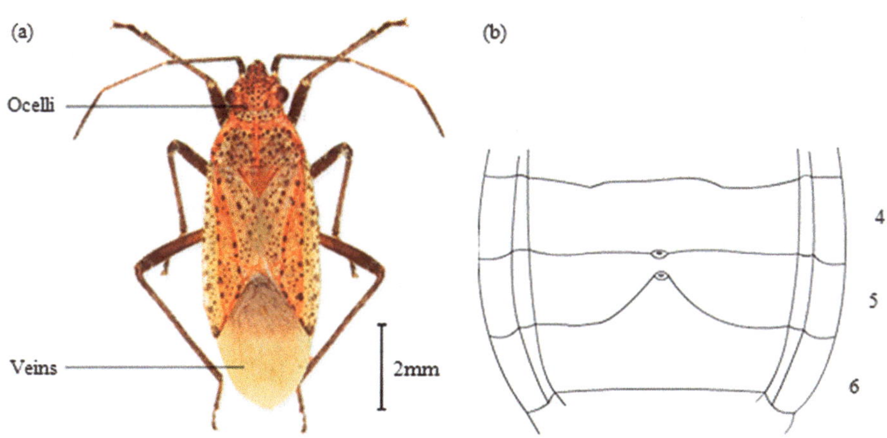

Fig. 20.1 (**a**) *Jadera choprai* Göllner-Scheiding; the ocelli distinctly raised. The front wing membrane has many veins of similar size. (**b**) *Liorhyssus hyalinus*. Image adapted from Davidova-Vilímová et al. (2000). The dorsal abdominal glands between 4–5 and 5–6 ; the terga are unusual among Hemiptera because they remain functional in the adult stage

Vilímová et al. 2000). Moreover, both subfamilies retain the anterior dorsal abdominal gland usually found only in immature Heteroptera (Davidova-Vilímová et al. 2000), and adult males also secrete fragrances from glands in the genitalia (Aldrich 1988; Aldrich et al. 1990a). If the vestigial metathoracic scent glands of Serinethinae indeed indicate a reduced capacity for synthesis of de novo chemical defense, it may have evolved in association with their conversion of toxic cyanogens from the seeds of their host plants (Braekman et al. 1982). They can secrete, excrete, and bleed these toxins to defend against predators (Aldrich et al. 1990b). Ribeiro (1989) found that predatory birds and toads rejected *J. haematoloma* (Herrich-Schäffer) after initially ingesting them. However, glandular secretions may have an additional function in rhopalids. Schwartz et al. (2009) reported that the odorous monoterpene secretions from the dorsal abdominal glands of *Boisea* Kirkaldy, which the insects actively spread across their cuticles with the legs, inhibit the germination and halt the growth of the fungal pathogen *Beauveria bassiana* (Bals.), which is potentially lethal to serinethines (Reinert et al. 1999). Accordingly, the redolent glandular products and vascular toxins of rhopalids likely serve multiple important functions in these insects.

20.3 General Biology and Ecology

To varying extents, rhopalids specialize on particular host plant taxa as food resources for growth and reproduction. In consequence, many aspects of the general biology and ecology may be predicted from knowledge of host plant range, phenology, morphology, and chemical characteristics. In some cases there is evidence of host–rhopalid coevolution, as in the variably inflated fruits of balloon vines coupled with variation in the mouthpart lengths of their specialist serinethine seed predators (e.g., Carroll 1988). Even specialists may take water or nutriment from other sources, however, including flowers, sweet fruits, and dead insects, which may sometimes lead to confusion about the definitive host on which they depend for growth and reproduction. In addition, probably well under 1 % of museum specimens have reliable host records, making it challenging to determine host relations without focused field studies (Carroll and Loye 2012). In this section we treat development and reproduction in further detail, organized around subfamily, tribe, and host relations.

20.3.1 Eggs and Oviposition

Rhopalids typically lay their eggs on or near host plants. The most extensive analysis of rhopalid eggs was undertaken by Vilímová and Rohanová (2010). They found that the differences in egg characteristics between tribes supported Schaefer and Chopra's (1982) treatment of Rhopalidae. Detailed studies of eggs and oviposition have not been made for most Neotropical rhopalids, so that we have included observations from temperate and subtropical congeners and conspecifics to describe some key attributes.

All rhopaline eggs have two micropylar processes, and this conformation is regarded as the ancestral state by Vilímová and Rohanová (2010). Tribe Rhopalini, with characters such as poorly defined pseudopercula, appears to be the most primitive. The widely distributed rhopaline *Liorhyssus hyalinus* (F.) typically lays eggs on host sepals or nearby stems and leaf midribs.

Until now, contributions on instars included *Arhyssus hirtus* (Torre-Bueno) (Wheeler and Henry 1984), *A. lateralis* (Say) (Paskewitz and McPherson 1983), *Esperanza texana* Barber (Wheeler and Henry 1984), *Harmostes* (*Harmostes*) *reflexulus* (Say) (Yonke and Walker 1970a, b), *Liorhyssus hyalinus* (F.) (Cornelis et al. 2012), *Niesthrea louisianica* Sailer (Wheeler 1977), *Rhopalus* (*Brachycarenus*) *tigrinus* (Schilling) (Wheeler and Hoebeke 1988) and *R.* (*Rhopalus*) *parumpunctatus* Schill. (Stroyan 1954), and *Xenogenus gracilis* Reed (Diez and Coscarón 2015).

Eggs may be parasitized by Hymenoptera in the *Telenomus podisi* (Ashmead) group (Cornelis et al. 2012).

In tribe Harmostini, the attachment stalk is not developed. *Harmostes reflexulus* (Say) lays one to four eggs on the floral pappi of Asteraceae. Eggs require about 8 days to hatch over an average month-long oviposition period (Yonke and Walker 1970a, b).

The ovoid eggs of Niesthreini are deposited in small loose clusters (Vilímová and Rohanová 2010) and have hexagonal surface reticulation on the chorion (Paskewitz and McPherson 1983). Within this group the pseudoperculum ranges from well defined (e.g., *Niesthrea louisianica* Sailer) to poorly defined [e.g., *Arhyssus* Stål and *N. sidae* (F.)]. Each egg is attached by a thin, short strand protruding from the dorsal side; whether this is homologous with egg stalks in Chorosomatini is uncertain (Paskewitz and McPherson 1983; Vilímová and Rohanová 2010). Overwintered *N. louisianica* oviposit on the undersides of leaves in spring; successive generations oviposit on flower pedicels as buds develop and fruits and seeds become more abundant. As many as 1,000 eggs may be laid by a single female (Wheeler 1977).

Vilímová and Rohanová (2010) regard the egg characteristics of subfamily Serinethinae as derived relative to the Rhopalinae. Rather than just two micropylar processes, 6–45 are present, arranged in one or two broad circles. At least two *Jadera* species, *J. choprai* Göllner-Scheiding and *J. haematoloma* (Herrich-Schaeffer), exhibit a rare oviposition behavior, burying their nearly round eggs in a hole they dig in the soil using their forelegs (Carroll 1988, 1991; Panizzi et al. 2002). *J. choprai* nymphs are able to emerge even when extra soil is deposited on top of the eggs (Panizzi et al. 2002). These authors speculated that burying may deter predation and prevent desiccation. Burying eggs may be more widespread within the genus, but is not universal. For example, *J. obscura* (Westwood) and *J. coturnix* (Burmeister) feeding on host vines in the canopy drop their eggs to the ground. This oviposition coincides with the dehiscence of host seeds from the canopy to the forest floor [Wolda and Tanaka 1987; Tanaka and Wolda 1987 (as *J. aeola*)].

20.3.2 Nymphal Feeding and Development

Little is known about the biology of juvenile tropical rhopalines. In the Nearctic, bivoltine *Arhyssus lateralis* (Say) develops during the summer months and requires about 40 days to mature (Paskewitz and McPherson 1983). *Niesthrea louisianica* matures in about 25 days (Wheeler 1977), and *Harmostes reflexulus* requires 22 days to eclose after hatching, with most time in the 1st and 5th instars (4.7 and 6.7 days, respectively; Yonke and Walker 1970a, b).

More is known about nymphal feeding and development in Neotropical serinethines. Nymphs of *Jadera choprai* in southern Brazil feed and develop on the sapindaceous weedy balloon vine, *Cardiospermum halicacabum* (L.) (Panizzi and Hirose 2002; Panizzi et al. 2002, 2005), which is common among row crops (Lorenzi 2000). In the laboratory more than 75 % of nymphs complete development on mature balloon vine seeds, with nymphs reared in groups surviving better. Group rearing also speeds development time (average 35 days in groups, 40 days alone; Panizzi et al. 2005). Often, groups of nymphs are found feeding on a single seed, forming a cluster around the seed; the many stylet sheaths left behind after each feeding session are visible on the seed surface (Fig. 20.2a, b). These authors also provide a detailed description of the eggs and nymphs of this species. Similarly, *J. obscura* on Barro Colorado Island in Panama develop over a 5–6-week span, feeding on seeds of sapindaceous canopy and light gap lianas in aggregations that may exceed a million individuals in number (Wolda and Tanaka 1987; Carroll and Loye 1987). Wolda and Tanaka (1987) described nymphal development of this species in detail.

Jadera haematoloma, which ranges from the United States to South America and has been accidentally introduced to Hawaii and Taiwan, develops on ten or more native and introduced sapinds (Carroll and Loye 2012). Development time is similar

Fig. 20.2 Rhopalid bug nymphs and adults feeding on a single mature host fruit or seed; (**a**) nymphs of *Jadera choprai* feeding on a mature seed of the balloon vine *Cardiospermum halicacabum*; (**b**) mature seed of *C. halicacabum* showing stylet sheaths left on the seed integument after each feeding session (Photos AR Panizzi); (**c**) nymphs and adults of *Jadera haematoloma* on the dehiscent fruit of the balloon vine *Cardiospermum corundum*. Clusters of feeding nymphs may completely cover a seed, forming a bright red "ball of bugs" around it (Photo SP Carroll)

to that reported for other *Jadera*. In Florida, USA, this native insect, after adopting an invasive Asian ornamental tree as a host in Florida in the mid-twentieth century, has evolved to a remarkable degree in many traits, including egg size, developmental ability, and the length of mouthparts used in feeding on host seedpods (Carroll and Boyd 1992; Carroll et al. 1997, 1998, 2003a). Experiments rearing bugs from both hosts on either native or nonnative host seeds indicate that the offspring of early colonists from the native balloon vine host probably survived and developed poorly on the introduced tree at the time they first colonized it but now exploit that host with an efficiency similarly to which their ancestors from the native host showed on the native host. The derived population on the tree has, however, lost its ability to perform well on the native host (Carroll et al. 1998). Ironically, the most detailed description of the nymphal stadia of this species is in a beautifully complete review of the species by Tsai et al. (2013), based on observations of the recently introduced population in Taiwan. Nymphs and adults of *J. haematoloma* cluster on seeds of the balloon vine *Cardiospermum corundum* (L.) forming a bright red "ball of bugs" around it (Fig. 20.2c).

20.3.3 Adult Feeding and Reproduction

In contrast to rhopalines, fewer serinethines enter temperate latitudes, although the two New World *Boisea* are largely temperate (Carroll and Loye 2012). *Boisea trivittata* (Say) enters tropical latitudes in eastern Mexico, where its patchy highland distribution matches that of its otherwise north temperate host tree, *Acer negundo* L. Globally, sapinds in the widely distributed balloon vine genus *Cardiospermum* L. are the most commonly used, though many others among the over 150 sapind genera and close to 2,000 species are also exploited, particularly in the Paleotropics, by the third serinethine genus *Leptocoris* Hahn (Carroll and Loye 2012). The Neotropical sapind flora is distinctive in that the 500 species of sapind lianas in the five genera *Cardiospermum*, *Paullinia* L., *Serjania* Mill., *Thinouia* Planch. & Triana, and *Urvillea* Kunth constitute 60 % of regional family diversity and >25 % of the world family diversity (Acevedo-Rodríguez et al. 2011). The majority of the native host records for *Jadera* (the principal Neotropical serinethine) are from these genera, suggesting that much of the diversification of *Jadera* has occurred in association with these host radiations (Carroll and Loye 2012).

Adult feeding and reproduction of *J. choprai* has been extensively studied. In the laboratory, on mature seeds of the balloon vine *C. halicacabum*, males lived on average for 86.6 days, while females lived for 47.7 days; average fecundity (number of eggs/female) was 146.4 eggs and egg hatchability 85.3 % (Panizzi et al. 2005). *J. haematoloma* in Florida has similar vital statistics, but lays about twice as many eggs in captivity (Carroll et al. 1998, 2003b). Both species are flight and wing polymorphic. In *J. haematoloma*, there are four types, which in females are long-winged flyers that retain the flight muscles after mating (and presumably feeding), long-winged flyers that histolyze the muscles after feeding and mating, long-winged bugs that do not develop flight muscles, and short-winged bugs that do not develop flight muscles (Dingle and Winchell 1997). The flight morphs differ in life history, with flying individuals, which tend to be larger, having a much later age of first reproduction, while flightless bugs become reproductive within a few days of eclosion if food is present. Histolyzing bugs represent an intermediate state in terms of reproductive options, and flight morph frequencies appear to be evolving in populations on introduced hosts that differ from the native hosts in the spatial and temporal availability of seeds (Carroll et al. 2003b).

A risk of flightlessness in *J. haematoloma* and likely other *Jadera* is that flightless individuals are not capable of physiological diapause in the absence of food (Carroll et al. 2003b). Thus, the reproductive life span of congenitally flightless individuals depends on the persistence of seeds within a highly circumscribed locale. Temperate populations of this species in the central United States enter a photoperiodic diapause as day length declines in advance of the winter season. In subtropical Florida, seeds of *Cardiospermum* are available most of the year but are locally sporadic depending on plant phenology and the activities of the multi-taxon community of seed predators (Carroll et al. 2003b; Carroll and Loye 2006). Flight-capable adults there do not diapause in response to shortening day length but, in addition to being able to fly in search of more productive sites, are also able to enter a starvation-based diapause that will be advantageous during periods of low seed availability at local and regional scales (sensu Dingle 1972; Dingle and Arora 1973).

On Barro Colorado Island in Panama, *J. obscura* and *J. coturnix* diapause during much of the year due to rather synchronous seed predation in their primary liana hosts (Wolda and Tanaka 1987; Tanaka and Wolda 1987). During this period, *J. obscura*, which is entirely volant, travels the understory in swarms that may exceed 100,000 individuals, taking moisture from the leaves of small nonhost trees. *J. coturnix* (which is often referred to by the invalid name *J. aeola*) similarly diapauses and then produces two generations during the late dry season (Tanaka and Wolda 1987). The first generation is produced by volant adults coming out of a 10-month dormancy. They aggregate on newly mature host seeds and oviposit around those hosts. The developing offspring then consume much of that year's seed crop, and upon emerging as adults many fly to sheltered locations, aggregating beneath understory leaves to await for the next year's seeds. A subset of the bugs however emerges as small, short-lived, but highly fecund flightless and short-winged individuals that will neither migrate nor diapause, but instead quickly feed on remnant seeds, mate, and oviposit. All resulting offspring in laboratory trials were long winged, suggesting that the risky tactics of producing a subset of obligatorily reproductive individuals that are incapable of diapausing are coordinated to produce diapausing offspring (Tanaka and Wolda 1987). Circumstances permitting lineages that succeed in completing this extra cycle could potentially greatly out-reproduce lineages constrained to the more conservative tactic with a single annual generation.

In summary, diapausing aggregations that may reach enormous numbers are a common feature of volant adult serinethines, whether the diapause is induced by photoperiod in cold winter areas, by predictable food shortages when host seed production is highly seasonal, or during periods of food shortage that are not as predictable (Fig. 20.3). Both New World *Jadera* and Old World *Leptocoris* commonly cluster beneath the green leaves of hosts or other nearby plants for weeks to many months in advance of the next period of seed production (Tanaka and Wolda 1987; Wolda and Tanaka 1987; Tanaka et al. 1987; Carroll et al. 2005; Zych 2010). Densities may reach up to 500 bugs/leaf (Wolda and Tanaka 1987). These and especially the third soapberry bug genus, *Boisea*, are renowned nuisance pests in many regions based on their propensity to aggregate and to enter homes in large numbers (e.g., Schowalter 1986, Fig. 20.3). This behavior of serinethines may not have been reported in the Neotropics.

While diapause is less well studied in the subfamily Rhopalinae, there is evidence of diapause in temperate populations of rhopalines with partially Neotropical ranges. In the Niesthreini, this includes *N. louisianica* Wheeler (1977) and *A. lateralis*, Readio (1928), and in the Harmostini *H. reflexulus* (Yonke and Walker 1970a).

Fig. 20.3 Adult *Boisea trivittata* aggregate during nonreproductive periods and may inhabit buildings in large numbers (Photo courtesy of Jodi Green under Creative Commons https://creativecommons.org/licenses/by-nc-nd/2.0/)

20.4 Classification and Diversity

A key to the subfamilies, tribes, and genera of Rhopalidae in the Neotropics is shown below (Chopra (1967), revised by Froeschner (1981) and used by Pall and Coscarón (2012); authors have added *Stictopleurus*).

1 Lateral margin of the pronotum with a distinct notch delimiting the collar posteriorly. Suture between visible abdominal segments III and IV as strong and deep as suture between sternites IV and V. Subfamily Serinethinae .... *Jadera* Stål

Lateral margins of the pronotum not notched immediately posterior to collar. Suture between visible abdominal segments III and IV much shallower and weaker than other sutures. Subfamily Rhopalinae Amyot and Serville................ 2

2(1) Posterior femur thicker than the median or anterior femora, ventrally with several strong spines.. 3

Posterior femur not thicker than the anterior or median femora, no spines ventrally.. 4

3(2) Anterolateral angle of the pronotum produced anteriorly as an acute tooth. Clypeus acutely projecting beyond the juga. Tribe Harmostini Stål .... *Harmostes* Burmeister

Anterolateral angle of the pronotum not produced. Clypeus bluntly rounded apically, not or slightly (less than own width) surpassing the juga. Tribe Chorosomini Douglas and Scott.. *Xenogenus* Berg

4(3). Phragma at junction of the first and second abdominal terga poorly developed in males, absent in females; the apodeme of the seventh sternum very small in males, absent in females. Phallus with one dorsomedian lobe on theca. Rhopalini Amyot and Serville.. 5

Phragma at junction of the first and second abdominal terga moderately developed in males, absent in females; the apodeme of the seventh sternum well developed in males, absent in females. Phallus with a pair of dorsolateral lobes on the theca. Niesthreini Chopra.. 6

5(4) Metapleura divided and metathoracic scent gland openings not reduced. Three pairs of asymmetrical, sclerotized, lateral, conjunctival appendages*Liorhyssus* Stål

Metapleura undivided and reduced metathoracic scent gland openings, absence of the left distal lateral conjunctival appendage..................... *Stictopleurus* Stål

6(5) Last connexival segment dorsally with an inconspicuous oblique suture setting off lateral basal angle (generally, the membrane must be lifted to see this suture). Labium not surpassing the posterior coxae..................... *Arhyssus* Stål

Last connexival segment without a dividing suture. Labium reaching or surpassing the third abdominal sternite................................ *Niesthrea* Spinola

The number of rhopalid species in the different ecoregions is referred on Table 20.1 (Göllner-Scheiding 1997). For the Neotropics, the following eight genera and 80 species are recorded (Table 20.2 lists references and resident countries for each species below):

Table 20.2 References and resident countries for nine genera and 81 species and three subspecies of rhopalids from the Neotropics

Subfamily: Rhopalinae
Tribe: Chorosomatini (Fieber)
Genus: *Xenogenus* Berg
gracilis (Reed)
Distribution: Argentina, Chile, South America
References: Göllner-Scheiding (1983), Pall and Coscarón (2012), Pall et al. (2013), Diez and Coscarón (2015)
picturatum Berg http://heteroptera.myspecies.info
Distribution: Argentina, Bolivia, Cuba, Mexico, Nicaragua, Puerto Rico, Saint Vincent, Uruguay, Central and South America
References: Göllner-Scheiding (1983), Maes and Göllner-Scheiding (1993), Coscarón (1998), Carpintero and De Biase (2011), Dellapé and Carpintero (2012), Pall and Coscarón (2012), Pall et al. (2013), Diez and Coscarón (2015)
Tribe: Harmostini
Genus: *Aufeius* Stål
impressicollis Stål
Distribution: Guatemala, Mexico
References: Göllner-Scheiding (1983)
Genus: *Harmostes* Burmeister
Subgenus: *Harmostes* Burmeister
angustatus Van Duzee
Distribution: Mexico
References: Göllner-Scheiding (1983), Henry (1988)
bilobatus Melo & Montemayor
Distribution: Argentina
References: Melo and Montemayor (2011)
brevispinus Blöte
Distribution: Argentina, Bolivia
References: Göllner-Scheiding (1983), Pall and Coscarón (2012)
disjunctus Barber
Distribution: Ecuador
References: Göllner-Scheiding (1983)
dorsalis Burmeister
Distribution: Argentina, Guatemala, Mexico, Nicaragua, Peru, from British Columbia to Argentina/Peru

(continued)

Table 20.2 (continued)

References: Blöte (1934), Göllner-Scheiding (1983), Maes and Göllner-Scheiding (1993), Pall and Coscarón (2012)
enatus Brailovsky & Soria
Distribution: Mexico
References: Brailovsky and Soria (1981)
formosus Distant
Distribution: Mexico
Reference: Göllner-Scheiding (1983)
fraterculus (Say)
Distribution: Mexico, Central and South America
References: Göllner-Scheiding (1983), Henry (1988)
fusiformis Harris
Distribution: Peru
References: Göllner-Scheiding (1983)
gravidator (F.)
Distribution: Argentina, Mexico, Central and South America (Argentina, Bolivia, Brazil, Colombia, Peru, Venezuela)
References: Göllner-Scheiding (1983), Pall and Coscarón (2012)
incisuratus Distant
Distribution: Brazil, Chile, Colombia, Peru
Göllner-Scheiding (1983)
nebulosus Stål
Distribution: El Salvador, Guatemala, Mexico, Nicaragua, Panama, Central America
References: Göllner-Scheiding (1983), Maes and Göllner-Scheiding (1993)
parafraterculus Göllner-Scheiding
Distribution: Argentina, Brazil
References: Göllner-Scheiding (1983), Pall and Coscarón (2012)
paraprolixus Göllner-Scheiding
Distribution: Bolivia
References: Göllner-Scheiding (1998)
prolixus Stål
Distribution: Argentina, Bolivia, Brazil, Mexico, Paraguay, Peru, Uruguay, South America (Argentina, Bolivia, Brazil, Paraguay, Peru, Uruguay)
References: Göllner-Scheiding (1983), Melo et al. (2011), Dellapé and Carpintero (2012), Pall and Coscarón (2012), Pall et al. (2013)
reflexulus (Say)
Distribution: Cuba, Mexico
References: Göllner-Scheiding (1983)
rubrum Melo & Montemayor http://heteroptera.myspecies.info
Distribution: Argentina
References: Melo and Montemayor (2011)
serratus (F.)

(continued)

Table 20.2 (continued)

Distribution: Argentina, Bolivia, Brazil, Costa Rica, Cuba, Dominican Republic, Ecuador, El Salvador, Guatemala, Honduras, Jamaica, Lesser Antilles, Mexico, Nicaragua, Panama, Paraguay, Puerto Rico, Venezuela, West Indies, Central and South America
References: Blöte (1934), Göllner-Scheiding (1983), Maes and Göllner-Scheiding (1993), Melo et al. (2011), Dellapé and Carpintero (2012), Pall and Coscarón (2012)
signoreti Reed
Distribution: Argentina, Chile
References: Göllner-Scheiding (1983), Pall and Coscarón (2012)
splendens Harris
Distribution: Argentina, Bolivia, Brazil, Paraguay
References: Göllner-Scheiding (1983) (Melo and Montemayor, personal communication)
Subgenus: *Neoharmostes* Göllner-Scheiding
apicatus Stål
Distribution: Mexico, South America (Argentina, Bolivia, Brazil, Chile, Paraguay, Uruguay)
References: Göllner-Scheiding (1983), Pall and Coscarón (2012)
bergi Göllner-Scheiding
Distribution: Argentina, Bolivia
References: Göllner-Scheiding (1998), Melo and Montemayor (personal communication)
confinis Harris 1942
Distribution: Chile
References: Göllner-Scheiding (1983)
corazonus Distant
Distribution: Argentina, Chile, Ecuador, Peru
References: Göllner-Scheiding (1983), Pall and Coscarón (2012)
corizoides Jensen-Haarup
Distribution: Argentina, Uruguay
References: Göllner-Scheiding (1983), Pall and Coscarón (2012)
imitabilis Harris
Distribution: Argentina
References: Göllner-Scheiding (1983), Pall and Coscarón (2012)
insitivus Harris
Distribution: Argentina, Chile
References: Göllner-Scheiding (1983), Pall and Coscarón (2012)
marmoratus (Blanchard)
Distribution: Argentina, Bolivia, Chile, Peru
References: Göllner-Scheiding (1983), Pall and Coscarón (2012)
minor (Spinola)
Distribution: Argentina, Chile
References: Göllner-Scheiding (1983), Pall and Coscarón (2012)
petulans Harris
Distribution: Argentina, Bolivia, Peru, Uruguay

(continued)

Table 20.2 (continued)

References: Göllner-Scheiding (1983), Pall and Coscarón (2012)
procerus Berg http://heteroptera.myspecies.info
Distribution: Argentina, Brazil, Peru, Uruguay, Patagonia
References: Göllner-Scheiding (1983), Coscarón (1998), Marrero et al. (2008), Carpintero (2009), Carpintero and De Biase (2011), Melo et al. (2011), Cava et al. (2012), Dellapé and Carpintero (2012), Pall and Coscarón (2012)
Tribe: Niesthreini
Genus: *Arhyssus* Stål
confusus Chopra
Distribution: Mexico
References: Göllner-Scheiding (1983)
lateralis (Say)
Distribution: Guatemala, Mexico
References: Göllner-Scheiding (1983), Henry (1988)
parvicornis (Signoret)
Distribution: Costa Rica, Mexico, Paraguay, Utah to Costa Rica
References: Nickel (1958), Göllner-Scheiding (1983), Henry (1988)
peruensis Chopra, 1968
Distribution: Peru
References: Göllner-Scheiding (1983)
pilosus (Signoret)
Distribution: Venezuela
References: Göllner-Scheiding (1983)
punctatus (Signoret)
Distribution: Guatemala, Honduras, Mexico, Nicaragua, Panama, Colorado to Honduras
References: Göllner-Scheiding (1983), Henry (1988), Maes and Göllner-Scheiding (1993)
slateri Chopra
Distribution: Mexico
References: Göllner-Scheiding (1983)
tricostatus (Spinola)
Distribution: Argentina, Chile
References: Göllner-Scheiding (1983), Melo and Montemayor (personal communication)
validus (Uhler)
Distribution: Mexico
References: Göllner-Scheiding (1983)
Genus: *Niesthrea* Spinola
agnes Chopra
Distribution: Argentina
References: Göllner-Scheiding (1983), Pall and Coscarón (2012)
ashlocki Froeschner

(continued)

Table 20.2 (continued)

Distribution: Ecuador
References: Froeschner (1989), Henry and Wilson (2004)
brevicauda Chopra
Distribution: Peru
Göllner-Scheiding (1983)
dentatus Chopra
Distribution: Brazil
References: Göllner-Scheiding (1983)
dignus Chopra
Distribution: Brazil
References: Göllner-Scheiding (1983)
fenestratus (Signoret)
Distribution: Chile
References: Göllner-Scheiding (1983)
josei Göllner-Scheiding
Distribution: Argentina
References: Pall and Coscar0n (2012)
louisianica Sailer
Distribution: Mexico
References: Göllner-Scheiding (1983)
pictipes (Stål)
Distribution: Argentina, Brazil, Ecuador, Mexico, Paraguay
References: Froeschner (1981), Göllner-Scheiding (1983), Carpintero and De Biase (2011), Dellapé and Carpintero (2012), Pall and Coscar0n (2012)
pictipes casinii Göllner-Scheiding
Distribution: Argentina, Uruguay
References: Göllner-Scheiding (1984)
sidae (F.)
Distribution: Columbia, Ecuador, Mexico, Nicaragua, Panama, Puerto Rico, Venezuela, Caribbean, Central America, South America, and West Indies
References: Blöte (1934), Göllner-Scheiding (1983), Maes and Tellez (1988), Henry (1988), Maes and Göllner-Scheiding (1993), Henry and Wilson (2004)
similis Chopra
Distribution: Argentina, Brazil
References: Göllner-Scheiding (1983), Melo et al. (2011), Pall and Coscar0n (2012)
ventralis (Signoret)
Distribution: Guatemala, Mexico, Nicaragua, Texas to Guatemala
References: Göllner-Scheiding (1983), Maes and Göllner-Scheiding (1993)
vincentii (Westwood)
Distribution: Argentina, Brazil, Paraguay, Venezuela, West Indies
References: Blöte (1934), Göllner-Scheiding (1983), Pall and Coscar0n (2012)
Tribe: Rhopalini
Genus: *Liorhyssus* Stål

(continued)

Table 20.2 (continued)

hyalinus (F.)
Distribution: Argentina, Chile, Cuba, Guatemala, Mexico, Nicaragua, Venezuela, Patagonia, all the world
References: Dallas (1852) as *Rhopalus ruber* Dallas, Göllner-Scheiding (1983), Maes and Göllner-Scheiding (1993), Hernandez (1994), Cermeli et al. (2004), Werner (2009), Carpintero and De Biase (2011), Dellapé and Carpintero (2012), Pall and Coscarón (2012)
kaltenbachi Göllner-Scheiding
Distribution: Bolivia, Brazil, Paraguay
References: Göllner-Scheiding (1983)
lineatoventris (Spinola)
Distribution: Argentina, Chile, Ecuador
References: Göllner-Scheiding (1983), Pall and Coscarón (2012)
Genus: *Stictopleurus* Stål
knighti Harris
Distribution: Mexico
References: Göllner-Scheiding (1983)
punctiventris (Dallas)
Distribution: Mexico
References: Henry (1988)
Subfamily: Serinethinae
Genus: *Jadera* Stål
aeola (Dallas)
Distribution: Brazil, Antilles, from Mexico to Argentina
References: Göllner-Scheiding (1983), Barcellos et al. (2008), Pall and Coscarón (2012)
Note: see *coturnix*
aeola aeola (Dallas)
Distribution: Argentina, Guatemala, Mexico, Nicaragua, Caribbean, from Mexico to Argentina
References: Göllner-Scheiding (1979), no distributional information Göllner-Scheiding (1983), Maes and Göllner-Scheiding (1993), Carpintero and De Biase (2011), Melo et al. (2011), Pall and Coscarón (2012)
aeola rufoculis (Kirby, 1890)
Distribution: Brazil
References: Göllner-Scheiding (1983)
antica (Walker)
Distribution: Dominican Republic, West Indies
References: Göllner-Scheiding (1983)
bayardae Göllner-Scheiding
Distribution: British Honduras, Guatemala, Mexico, Panama, from Mexico to Panama
References: Göllner-Scheiding (1983), Henry (1988), Colombia SPC personal observation (2011)
choprai Göllner-Scheiding
Distribution: Argentina, Bolivia, Brazil, Paraguay, Uruguay, South America

(continued)

Table 20.2 (continued)

References: Göllner-Scheiding (1979), Göllner-Scheiding (1983), Carpintero and De Biase (2011), Pall and Coscarón (2012)
coturnix (Burmeister)
Distribution: Brazil, Uruguay
Note: According to Göllner-Scheiding (1979) *coturnix* (Burmeister) is a synonym of *aeola* (Dallas). Henry (1988) lists each as separate species; we have followed Henry (1988) and the Coreoidea Species File which consider *coturnix* a valid species
decipiens Göllner-Scheiding
Distribution: Argentina, Brazil, West Indies
References: Göllner-Scheiding (1983), Pall and Coscarón (2012)
diaphona Göllner-Scheiding
Distribution: Mexico, Nicaragua, Panama
References: Göllner-Scheiding (1983), Maes and Göllner-Scheiding (1993)
golbachi Göllner-Scheiding
Distribution: Argentina, Bolivia, Ecuador
References: Göllner-Scheiding (1983), Pall and Coscarón (2012)
haematoloma (Herrich-Schäffer)
Distribution: Argentina, Belize, Cuba, Colombia, Guatemala, Mexico, Nicaragua, Uruguay, from Central America to Colombia and Venezuela, West Indies
References: Göllner-Scheiding (1983), Maes and Téllez (1988), Maes and Göllner-Scheiding (1993), Bressa et al. (2001), Pall and Coscarón (2012)
harrisi Göllner-Scheiding
Distribution: Bolivia, Brazil
References: Göllner-Scheiding (1979)
obscura (Westwood)
Distribution: Argentina, Brazil, Costa Rica, from Mexico to Argentina/Peru
References: Blöte (1934), Göllner-Scheiding (1983), Carroll and Loye (1987), Pall and Coscarón (2012)
parapectoralis Göllner-Scheiding
Distribution: Argentina, Brazil
References: Göllner-Scheiding (1983), Schaefer (2001), Carpintero and De Biase (2011), Pall and Coscarón (2012)
pectoralis Stål
Distribution: Bolivia, Brazil, Uruguay
References: Göllner-Scheiding (1983)
peruviana Göllner-Scheiding
Distribution: Peru
References: Göllner-Scheiding (1983)
pyrrholoma Stål
Distribution: Bolivia, Brazil, Colombia, Ecuador
References: Göllner-Scheiding (1983)
sanguinolenta (F.)
Distribution: Argentina, Belize, Cuba, Ecuador, Nicaragua, Paraguay, Puerto Rico, St. Croix, Culebra, St. John, St. Thomas, Uruguay
References: Heidemann (1901), Barber (1923) as *Jadera rubrofusca*, Göllner-Scheiding (1983), Maes and Göllner-Scheiding (1993), Bressa et al. (2001), Carpintero (2009), Pall and Coscarón (2012), personal observation Scott Carroll (1987)

(continued)

Table 20.2 (continued)

schuhi Göllner-Scheiding
Distribution: Venezuela
References: Göllner-Scheiding (1983)
silbergliedi Froeschner
Distribution: Galápagos Islands
References: Froeschner (1985)
similaris Göllner-Scheiding
Distribution: Ecuador
References: Göllner-Scheiding (1983)
Genus: *Boisea* Stål
trivittata (Say)
Distribution: Mexico
References: Scott Carroll personal observation (2011) (new record for the Neotropical region)

Note: Geographic distribution was taken into account from Göllner-Scheiding (1983) and posterior publications. The classification follows Coreoidea Species File (http://Coreoidea.SpeciesFile.org/)

Family: Rhopalidae
Subfamily: Rhopalinae
Tribe: Chorosomatini (Fieber)
Genus: *Xenogenus* Berg
gracilis Reed
picturatum Berg
Tribe: Harmostini
Genus: *Aufeius* Stål
impressicollis Stål
Genus: *Harmostes* Burmeister
Subgenus: *Harmostes* Burmeister
angustatus Van Duzee
bilobatus Melo & Montemayor
brevispinus Blöte
disjunctus Barber
dorsalis Burmeister
enatus Brailovsky & Soria
formosus Distant
fraterculus (Say)
fusiformis Harris
gravidator (F.)
incisuratus Distant
nebulosus Stål
parafraterculus Göllner-Scheiding

(continued)

paraprolixus Göllner-Scheiding
prolixus Stål
reflexulus (Say)
rubrum Melo & Montemayor
serratus (F.)
signoreti Reed
splendens Harris
Subgenus: *Neoharmostes* Göllner-Scheiding
apicatus Stål
bergi Göllner-Scheiding
confinis Harris
corazonus Distant
corizoides Jensen-Haarup
imitabilis Harris
insitivus Harris
marmoratus (Blanchard)
minor (Spinola)
petulans Harris
procerus Berg
Tribe: Niesthreini
Genus: *Arhyssus* Stål
confusus Chopra
lateralis (Say)
parvicornis (Signoret)
peruensis Chopra
pilosus (Signoret)
punctatus (Signoret)
slateri Chopra
tricostatus (Spinola)
validus (Uhler)
Genus: *Niesthrea* Spinola
agnes Chopra
ashlocki Froeschner
brevicauda Chopra
dentatus Chopra
dignus Chopra
fenestratus (Signoret)
josei Göllner-Scheiding
louisianica Sailer
pictipes (Stål)
pictipes casinii Göllner-Scheiding
sidae (F.)
similis Chopra
ventralis (Signoret)

(continued)

vincentii (Westwood)
Tribe: Rhopalini
Genus: *Liorhyssus* Stål
hyalinus (F.)
kaltenbachi Göllner-Scheiding
lineatoventris (Spinola)
Genus: *Stictopleurus* Stål
punctiventris (Dallas)
knighti Harris
Subfamily: Serinethinae
Genus: *Jadera* Stål
aeola (Dallas)
aeola aeola (Dallas)
aeola rufoculis (Kirby)
antica (Walker)
bayardae Göllner-Scheiding
choprai Göllner-Scheiding
coturnix (Burmeister)
decipiens Göllner-Scheiding
diaphona Göllner-Scheiding
golbachi Göllner-Scheiding
haematoloma (Herrich-Schäffer)
harrisi Göllner-Scheiding
obscura (Westwood)
parapectoralis Göllner-Scheiding
pectoralis Stål
peruviana Göllner-Scheiding
pyrrholoma Stål
sanguinolenta (F.)
schuhi Göllner-Scheiding
silbergliedi Froeschner
similaris Göllner-Scheiding
Genus: *Boisea* Stål
trivittata (Say)

20.4.1 Main Species

***Harmostes reflexulus* (Say)**

This is considered the most common *Harmostes* in the Nearctic by Blatchley (1926). In Göllner-Scheiding's (1978) revision of the genus, she lists its range as extending into Mexico and Cuba.

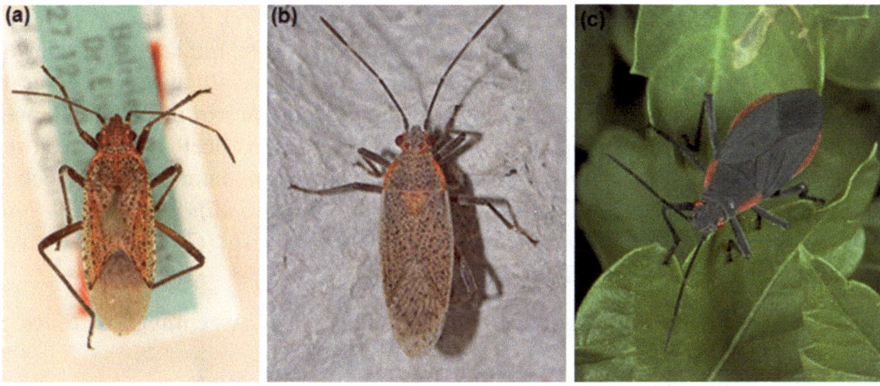

Fig. 20.4 (**a**) Adult of *Jadera choprai*, (**b**) *Jadera coturnix*, and (**c**) *Jadera haematoloma*

In Missouri it is a bivoltine species that feeds on many Asteraceae, and its seasonal migration follows a procession of maturing blooms throughout the summer (Yonke and Walker 1970a). Of their field-captured adults, 15.7 % were parasitized by the fly *Leucostoma acirostre* Reinhard.

Relevant literature: Van Duzee (1909) distribution in Florida, Yonke and Walker (1970a) life history, and Yonke and Walker (1970b) description of the egg and nymphs.

Jadera choprai Göllner-Scheiding (Fig. 20.4a)

An abundant species in the Neotropics. Many museum specimens have been misidentified as *J. sanguinolenta*. About 8–12 mm in length. They are larger and more orange in color than *J. sanguinolenta*, *J. decipiens*, and *J. goldbachi*. They are recognized by ocher-orange dorsal sides marked with irregular spots. The head and scutellum are orange (the head sometimes brown). The ventral side is ocher. Appendages are dark brown.

J. choprai preys on the seeds of *Cardiospermum corindum* as well as *Cardiospermum halicacabum* var. *halicacabum*. Like *J. haematoloma*, they have been observed carrying seeds (Panizzi and Hirose 2002).

Distribution: Argentina, Bolivia, Brazil, Paraguay, and Uruguay.

Relevant literature: Göllner-Scheiding (1979) description, Panizzi et al. (2002) oviposition, and Panizzi and Hirose (2002) seed-carrying behavior.

Jadera coturnix (Burmeister) (Fig. 20.4b)

Most twentieth-century literature refers to this species as *aeola* after being designated by Kirby in 1852. However, Burmeister had already named it *coturnix* in 1835, giving the latter name priority. Göllner-Scheiding considers this species to be the most widely distributed *Jadera* (Göllner-Scheiding 1979). *J. coturnix* varies in size and color, typically brown with numerous dark spots and red lateral margins of the head and pronotum. Spots may form vertical lines, especially near the center of the head. Spots may be bigger on veins of hemelytra and on part of the membrane. About 8–13 mm in length. Appendages are brown. *J. coturnix* ocelli are much closer together than those of *J. hinnulea. J. coturnix* is a seed predator of sapindaceous vines and has been observed feeding on *Cardiospermum grandiflorum* L., *Paullinia sessiliflora* Radlk., *P. turbacensis* Kunth, *Serjania decaplueria* (Croat), *S. mexicana* L., and *Thinouia myriantha* (Tr. & Planch.).

Distribution: Mexico, Nicaragua, Panama, Ecuador, Brazil, Argentina, Puerto Rico, and Virgin Islands.

Relevant literature: Göllner-Scheiding (1979) description, (1983) general catalog, Tanaka and Wolda (1987) flight and life history polymorphism, and Carroll and Loye (2012) host records.

Jadera haematoloma (Herrich-Schäffer) (Fig. 20.4c)

This rhopalid is about 10–14 mm in length. It has a charcoal dorsal ground color with broad red lateral stripes on the sides of the head and pronotum (most distinct on the pronotum); the eyes are also red. The ventral side is dark except for margins of the abdomen and posterior segment, which are red. Appendages are black.

The red lateral margins on the pronotum and black dorsum resemble *J. diaphona* and *J. pyrrholoma*. The head of *J. haematoloma* is relatively wider than *J. pyrrholoma. J. haematoloma* is intermediate in size between these two species. Its charcoal and red abdomen help distinguish it from the uniformly orange-red abdominal venter of *J. diaphona*. The best diagnostic feature is the male genitalia, illustrations and descriptions for which are provided by Göllner-Scheiding (1979).

In addition to feeding on seeds, *J. haematoloma* also feeds opportunistically on vulnerable conspecifics (Carroll and Loye 1987). Comprehensive tables of host records are provided by (Carroll and Loye 2012). *J. haematoloma* aggregates in large numbers around hosts, competing for food and mates. Males guard their mates, by copulating until the onset of oviposition. During oviposition a male may stand on top of a female, fending off other males and re-copulating after all eggs are laid (Carroll 1988). Some populations are rapidly evolving in adaptation to introduced host plants (discussed above).

Distribution: Mexico, Guatemala, Nicaragua, Belize, Panama, Colombia, Venezuela, Uruguay, Argentina, Cuba, Puerto Rico, United States, Hawaii, and Taiwan (Tsai et al. 2013).

Relevant literature: Schaefer (1965), 10 (morphology, genitalia, figures); Göllner-Scheiding (1979), 57 (redescription, lectotype, genitalia, figures, distribution); Schaefer and Chopra (1982), 226 (morphology, host plants); Schaefer and Mitchell (1983), 593 (host plants); Mead (1985) (diagnostic characters, wing polymorphism, larva, photos, host plant, distribution, map, phenology, aggregation, impact on human, control); Carroll and Loye (1987), 373 (host plants, feeding, coevolution with host plants, aggregation, ecology, distribution, map); Carroll (1988), 54 (records, distribution, host plants, phenology, development, reproductive behavior, and ecology); Ribeiro (1989), 466 (records, host plant, aggregation, aposematism, development); Aldrich et al. (1990a), 200 (laboratory rearing, chemical ecology); Carroll (1991), 510 (reproductive behavior and ecology); Carroll and Boyd (1992), 1,053 (intraspecific variability, evolution); Carroll (1993), 156 (reproductive ecology); Carroll and Dingle (1996), 210 (records, host plants, feeding, intraspecific variability, evolution); Dingle and Winchell (1997), 365 (genetic and physiological control of wing polymorphism); Carroll et al. (1998), 956 (records, host plants, reproductive ecology, adaptation); Reinert et al. (1999), 469 (pest status, biological control); Winchell et al. (2000), 1,365 (wing polymorphism, physiology); Carroll et al. (2003b), S80 (genetic variation, selection); Carroll et al. (2003a), 135 (host plants, records, polymorphism); Dingle et al. (2009), 2,031 (genetic architecture, intraspecific variability, selection); Carroll and Loye (2012), 675 (host plants); Zych et al. (2012) (stridulation); and Tsai et al. (2013) (the most comprehensive review of the species and the source of this redacted reference list).

Liorhyssus hyalinus (F.) (Fig. 20.5a)

Worldwide and throughout the Neotropics (Göllner-Scheiding 1983), this species is distinguished by a hyaline membrane of the hemelytra extending beyond the abdomen. Its length is 5.5–6.5 mm and width 1.8–2.5 mm. Eggs are usually deposited in groups of 20–30 (Vilímová and Rohanová 2010). Early instars aggregate. Cornelis et al. (2012) detail nymphs, parasites, and hosts.

All life stages of *L. hyalinus* feed on *Sorghum vulgare* Pers., damaging crops in Venezuela (Cermeli et al. 2004). Cage experiments showed no significant damage to lettuce seed yield and germination until unnatural numbers of insects (400) and their nymphs fed on a single plant (Carlson 1959).

Distribution: Mexico, Guatemala, Nicaragua, Colombia, Venezuela, Ecuador, Chile, Argentina, Cuba, Grenada, and Puerto Rico.

Relevant literature: Readio (1928) (eggs, life history, host plants), Cermeli et al. (2004) (sorghum crop damage), Cornelis et al. (2012) (life history, nymphs, host plants, parasites), Davidova-Vilímová et al. (2000) (scent glands, systematics), and Carlson 1959 (effects on lettuce seed production).

Harmostes serratus F. (Fig. 20.5b)

Primarily a Neotropical species with range extending into the Southern United States (Göllner-Scheiding 1978, 1983). When viewed laterally it can be distinguished from other Harmostes by ventrally indented antenniferous tubercles. It can be distinguished from other *Harmostes* by the crenulate to serrate lateral margins of the thorax. The head is long with lateral spinous processes between the eyes and antennae, size 7–8 mm. long, and width 2.3–2.5 mm (Gibson 1917).

Fig. 20.5 (**a**) Adult of *Liorhyssus hyalinus* (Photo courtesy of L. Brehmer and E. Flores-Guevara), (**b**) *Harmostes serratus*, and (**c**) *Niesthrea sidae* (Courtesy of Hannah Mason)

H. serratus has been found feeding on *Dahlia* Cav. in several locations in Puerto Rico (Wolcott 1948).

Distribution: Mexico, Guatemala, El Salvador, Honduras, Nicaragua, Costa Rica, Panama, Colombia, Ecuador, Brazil, Peru, Bolivia, Paraguay, Argentina, Cuba, Dominican Republic, Puerto Rico, Jamaica, and Trinidad and Tobago.

Relevant literature: Göllner-Scheiding (1978) (revision of the genus) and Gibson (1917) (the genus *Harmostes* Berm).

***Niesthrea sidae* (F.) (Fig. 20.5c)**

This is the most widespread species in the genus and widely distributed in the Neotropics (Henry and Wilson 2004). Sailer (1961) first identified *N. sidae* and established it as the type specimen for the genus. He noted many museum specimens, and descriptions in the literature erroneously identify with *N. louisianica*, to which it is very similar. Chopra (1973) considers it a complex of highly variable local populations, which may contribute to the confusion. It is distinguished from *N. ashlocki* by the shorter parameres, by the less deeply convex apex of the medioventral lobe, and by the longer, denser pubescence of the head, pronotum, and legs (Sailer 1961; Henry and Wilson 2004).

It has been observed feeding on the malvaceous plants *Abutilon* Mill. and *Sida* L. (Wheeler 1977; Schaefer and Chopra 1982), as well as the fruit of the euphorb *Jatropha curcas* L. (Grimm and Führer 1998). Collections of *N. sidae* have been made on the leaves of *Lantana* L. and *Prosopis* L. (Palmer and Pullen 1995; Smith and Ueckert 1974), though these may be coincidental landings and not hosts.

Distribution: Mexico, Nicaragua, Panama, Colombia, Venezuela, Ecuador, Galápagos Islands, Brazil, Peru, Bolivia, Paraguay, Puerto Rico, and Grenada.

Relevant literature: Sailer (1961) designated lectotype, Chopra (1973) revised genus, Henry and Wilson (2004) description, and Galápagos record.

20.5 Concluding Remarks

Rhopalids belong to a small but widespread family that is important in both applied and basic entomology. Yet very little is known about the biology of most of its approximately 200 species. Ironically, while members the subfamily Serinethinae are commonly noted as nuisance pests because many species form large aggregations in domestic settings (e.g., Mead 1985), they are little studied from a practical perspective and instead are more commonly the subject of theoretically motivated studies in evolution and ecology (reviewed by Tsai et al. 2013). Genus *Jadera* is speciose and common throughout the Neotropics but has not been actively studied at low latitudes since the 1980s. In addition to the value that might derive from more pragmatically motivated studies of serinethines, investigations of contemporary evolution on introduced sapindaceous host plants, for which *Jadera* is well known in North America, could be logically expanded to South America.

Subfamily Rhopalinae is more diverse than Serinethinae at higher levels of classification (Chopra 1967), and four of the six rhopaline tribes occur in the Neotropics: Harmostini, Chorosomini, Rhopalini, and Niesthreini. Many of the species feed on reproductive tissues of Asteraceae, and others are associated with Malvaceae and Poaceae. In contrast to Serinethinae, none of the Rhopalinae are known to use sapindaceous plants as hosts.

The negative economic impacts of rhopalids appear to be limited. In Serinethinae they are related mainly to the nuisance of diapausing aggregations entering structures and occasional fruit spotting. Panizzi et al. (2002) suggested that the seed predation by *Jadera* species on weedy *Cardiospermum halicacabum* in cultivated soybean could be beneficial, and Carroll (2011) suggested that adaptation to this same plant in northern Australia wildlands by the serinethine *Leptocoris tagalicus* (Burmeister) could supply genotypes for the control of the congeneric Neotropical environmental weed *C. grandiflorum* in eastern Australia. Such considerations would benefit from further development. In Rhopalinae, *Liorhyssus* has been recognized as a pest of sorghum in Venezuela (Cermeli et al. 2004). Many rhopalines are small, cryptically colored, and furtive. It is possible that their presence on crops is underappreciated, and this too deserves additional consideration in field agriculture in the Neotropics.

References

Acevedo-Rodriguez P, Van Welzen PC, Adema F, Van Der Ham RWJM (2011) Sapindaceae. In: Flowering plants. Eudicots. Springer, Heidelberg, pp 357–407

Aldrich JR (1988) Chemical ecology of the Heteroptera. Annu Rev Entomol 33:211–238

Aldrich JR, Carroll SP, Oliver JE, Lusby WR, Rudmann AA, Waters RM (1990a) Exocrine secretions of scentless plant bugs; *Jadera, Boisea, Niesthrea* species (Hemiptera: Heteroptera: Rhopalidae). Biochem Syst Ecol 18:369–376

Aldrich JR, Carroll SP, Lusb WR, Thompson MJ, Kochansky JP, Waters RM (1990b) Sapindaceae, cyanolipids, and bugs. J Chem Ecol 16:199–210

Aukema B, Rieger C (eds) (1995–2006) Catalogue of the Heteroptera of the Palaearctic Region, vol 4 and 5, Pentatomomorpha. Netherlands Entomological Society, Amsterdam

Barber HG (1923) A preliminary report on the Hemiptera-Heteroptera of Porto Rico collected by the American Museum of Natural History. Am Mus Novit 75:1–13

Barcellos A, Schmidt LS, Brailovsky H (2008) Abundance and richness of Coreoidea (Hemiptera: Heteroptera) from Parque Estadual do Turvo, Southern Brazil. Neotrop Entomol 37:406–412

Blatchley WS (1926) Heteroptera or true bugs of Eastern North America, with especial reference to the faunas of Indiana and Florida. Nature Publishing, Indianapolis

Blöte HC (1934) Catalogue of the Coreidae in the Rijksmuseum van Natuurlijke Historie: Part I. Corizinae, Alydinae. Zool Meded 17:253–285

Braekman JC, Daloze D, Pasteels JM (1982) Cyanogenic and other glucosides in a neo-guinean bug *Leptocoris isolata*: possible precursors in its host-plant. Biochem Syst Ecol 10:355–364

Brailovsky H, Soria F (1981) Contribución al estudio de los Hemiptera-Heteroptera de México: XVIII. Revisión de la tribu Harmostini Stål (Rhopalidae) y descripción de una nueva especie. An Inst Biol Univ Nac Auton Mex (Zool) 51:123–168

Bressa MJ, Papeschi AG, Mola LI, Larramendy ML (2001) Autosomal univalents as a common meiotic feature in *Jadera haematoloma* (Heteroptera: Rhopalidae: Serinethinae). Eur J Entomol 98:151–157

Carpintero DL (2009) Insecta: Hemiptera: Heteroptera. In: Athor J (ed) Parque Costero del Sur: Naturaleza, conservación y patrimonio cultural. Fundación de Historia Natural "Félix de Azara", Ciudad Autónoma de Buenos Aires, pp 295–305

Carpintero DL, De Biase S (2011) Los Hemiptera Heteroptera de la Isla Martín García (Buenos Aires, Argentina). Historia Nat Tercera Serie 1:27–47

Carroll SP (1988) Contrasts in reproductive ecology between temperate and tropical populations of *Jadera haematoloma*, a mate-guarding hemipteran (Rhopalidae). Ann Entomol Soc Am 81:54–63

Carroll SP (1991) The adaptive significance of mate guarding in the soapberry bug, *Jadera haematoloma* (Hemiptera: Rhopalidae). J Insect Behav 4:509–530

Carroll SP (1993) Divergence in male mating tactics between two populations of the soapberry bug: guarding versus nonguarding. Behav Ecol 4:156–164

Carroll SP (2011) Conciliation biology: the eco-evolutionary management of permanently invaded biotic systems. Evol Appl 4:184–199

Carroll SP, Boyd C (1992) Host race radiation in the soapberry bug: natural history with the history. Evolution 46:1052–1069

Carroll SP, Dingle H (1996) The biology of post-invasion events. Biol Conserv 78:207–214

Carroll SP, Loye JE (1987) Specialization of *Jadera* (Hemiptera: Rhopalidae) on the seeds of Sapindaceae (Sapindales), and coevolutionary responses of defense and attack. Ann Entomol Soc Am 80:373–378

Carroll SP, Loye JE (2006) Invasion, colonization, and disturbance; historical ecology of the endangered Miami blue butterfly. J Insect Conserv 10:13–27

Carroll SP, Loye JE (2012) Soapberry bug (Hemiptera: Rhopalidae: Serinethinae) native and introduced host plants: biogeographic background of anthropogenic evolution. Ann Entomol Soc Am 105:671–684

Carroll SP, Dingle H, Klassen SP (1997) Genetic differentiation of fitness-associated traits among rapidly evolving populations of the soapberry bug. Evolution 51:1182–1188

Carroll SP, Klassen SP, Dingle H (1998) Rapidly evolving adaptations to host ecology and nutrition in the soapberry bug. Evol Ecol 12:955–968

Carroll SP, Dingle H, Famula TR (2003a) Rapid appearance of epistasis during adaptive divergence following colonization. Proc R Soc Lond B (Suppl) 270:580–583

Carroll SP, Marler M, Winchell R, Dingle H (2003b) Evolution of cryptic flight morph and life history differences during host race radiation in soapberry bug *Jadera haematoloma*. Ann Entomol Soc Am 96:135–143

Carroll SP, Loye JE, Dingle H, Mathieson M, Zalucki MP (2005) Ecology of *Leptocoris* Hahn (Hemiptera: Rhopalidae) soapberry bugs in Australia. Aust J Entomol 44:344–353

Carlson EC (1959) The effect of *Lygus* and Hyaline grass bugs on lettuce seed production. J Econ Entomol 52:242–244

Cassis G, Gross GF (2002) Hemiptera: Heteroptera (Pentatomomorpha). In: Houston WWK, Wells A (eds) Zoological catalog of Australia. CSIRO, Melbourne, pp 1–737

Cava MB, Corronca JA, Coscarón MC (2012) Composition and structure of Heteroptera communities (Hemiptera) in high altitude habitats of a national protected area in Argentina. Mun Entomol Zool 7:1214–1226

Cermeli M, Sánchez J, Morales P, Godoy F (2004) *Liorhyssus hyalinus* (F.)(Hemiptera: Rhopalidae) nueva plaga del sorgo en Venezuela. Entomotropica 19:101–103

Chopra NP (1967) The higher classification of the family Rhopalidae (Hemiptera). Trans R Entomol Soc Lond 119:363–399

Chopra NP (1973) A revision of the genus Niesthrea Spinola (Rhopalidae: Hemiptera). J Nat Hist 7(4):441–459

Cornelis M, Quirán E, Coscarón M (2012) The scentless plant bug, *Liorhyssus hyalinus* (F.) (Hemiptera: Heteroptera: Rhopalidae): description of immature stages and notes on its life history. Zootaxa 3525:83–88

Coscarón MC (1998) Los ejemplares tipo de Coreoidea (Heteroptera) depositados en la colección del Museo de La Plata. SerTéc y Did, Museo de La Plata 34:1–5

Coscarón MC (2014) Rhopalidae. In: Catalogue of the Heteroptera or true bugs of Argentina. Zootaxa (in press)

Dallas WS (1852) List of the specimens of hemipterous insects in the collection of the British Museum, vol 2. Taylor & Francis Incorporated, London

Davidova-Vilímová J, Nejedla M, Schaefer CW (2000) Dorso-abdominal scent glands and metathoracic evaporatoria in adults of central European Rhopalidae (Hemiptera: Heteroptera), with a discussion of phylogeny and higher systematics. Eur J Entomol 97:213–222

Dellapé PM, Carpintero DL (2012) Relevamiento de los Heteroptera (Insecta: Hemiptera) de las sierras de Tandil, provincia de Buenos Aires, Argentina. Rev Mus Argentino Cienc Nat Nova Serie 14:125–134

Diez F, Coscarón MC (2015) Contribution to the knowledge of Patagonia, Argentina: redescription of the genus Xenogenus Berg 1883 (Hemiptera: Heteroptera: Rhopalidae) and description of immature stages of *Xenogenus gracilis* Reed, 1899. Zootaxa 3919(3):573–582

Dingle H, Carroll SP, Famula TR (2009) Influence of genetic architecture on contemporary local evolution in the soapberry bug, *Jadera haematoloma*: artificial selection on beak length. J Evol Biol 22:2031–2040

Dingle H (1972) Migration strategies of insects. Science 175:1327–1335

Dingle H, Arora G (1973) Experimental studies of migration in bugs of the genus *Dysdercus*. Oecologia 12:119–140

Dingle H, Winchell R (1997) Juvenile hormone as a mediator of plasticity in insect life histories. Arch Insect Biochem Physiol 35:359–373

Froeschner RC (1981) Heteroptera or true bugs of Ecuador: a partial catalog. Smithson Contrib Zool 322:55–57

Froeschner RC (1985) Synopsis of the Heteroptera or true bugs of the Galápagos Islands. Smithson Contrib Zool 407:39–40

Froeschner RC (1989) A small collection of Heteroptera from the Galapagos Islands, with the description of the new species *Niesthrea* ashlocki and a list of *Niesthrea* species (Rhopalidae). Proc Biol Soc Wash 102:609–612

Froeschner RC (1999) True bugs (Heteroptera) of Panama: a synoptic catalog as a contribution to the study of Panamanian biodiversity. Am Entomol Inst 61:1–393

Gibson EH (1917) The genus *Harmostes* Burm. Entomol News 28:439–450

Göllner-Scheiding U (1978) Revision der Gattung *Harmostes* Burm., 1835 (Heteroptera, Rhopalidae) und einige Bemerkungen zu den Rhopalinae. Mitt Mus Nat kd Berl Zool Reihe Mitt Zool Mus Berlin 54:257–311

Göllner-Scheiding U (1979) Die Gattung *Jadera* Stål, 1862 (Heteroptera, Rhopalidae). Dtsch Entomol Z 26:47–75

Göllner-Scheiding U (1983) General-kataloge der Familie Rhopalidae (Heteroptera). Mitt Zool Mus Berlin 59:37–189

Göllner-Scheiding U (1984) Ergänzungen zu den Gattungen Liorhyssus STAL, 1870, Niesthrea SPINOLA, 1837, und Rhopalus SCHILLING, 1827 (Heteroptera, Rhopalidae). Mitt zool Mus Berlin 60:115–121

Göllner-Scheiding U (1997) Die Rhopalidae der afrotropischen Region unter besonderer Berücksichtigung der Fauna der Republik Namibia (Insecta: Heteroptera, Coreoidea). Mitt Mus Nat kd Berl Zool Reihe 73:291–308

Göllner-Scheiding U (1998) Three new species of Rhopalidae: *Harmostes (Harmostes) paraprolixus* spec. nov. and *Harmostes (Neoharmostes) bergi* spec. nov. from Bolivia, and *Peliochrous thomasi* spec. nov. from Uganda (Insecta: Heteroptera: Coreoidea). Reichenbachia 32:175–179

González RH (1989) Insectos y ácaros de importancia agrícola y cuarentenaria en Chile. Editora Ograma, Santiago

Grazia J, Cavichioli RR, Wolff VRS, Fernandes JAM, Takiya DM (2012) Hemiptera. In: Rafael JA, Melo GAR, Carvalho CJKB, Casari AS, Constantino R (eds) Insetos do Brasil. Diversidade e Taxonomia. Ed. Holos, Ribeirão Preto, pp 347–405

Grimm C, Führer E (1998) Population dynamics of true bugs (Heteroptera) in physic nut (*Jatropha curcas*) plantations in Nicaragua. J Appl Entomol 122:515–521

Harris HM (1942) Notes on *Harmostes*, with descriptions of some new species (Hemiptera: Corizidae). J Wash Acad Sci 32:28–32

Heidemann O (1901) Papers from the Hopkins Stanford Galapagos Expedition, 1898–1899.1. Entomological results (1): Hemiptera. Proc Wash Acad Sci 3:363–370

Henry TJ (1988) Family Rhopalidae Amyot and Serville, 1843. In: Henry TJ, Froeschner RC (eds) Catalog of the Heteroptera, or true bugs, of Canada and the continental United States. Brill, Leiden, pp 652–664

Henry TJ (2009) Biodiversity of Heteroptera. In: Foottit R, Adler P (eds) Insect biodiversity: science and society. Blackwell, Chichester

Henry TJ, Wilson MR (2004) First records of eleven true bugs (Hemiptera: Heteroptera) from the Galápagos Islands, with miscellaneous notes and corrections to published reports. J N Y Entomol Soc 112(1):75–86

Hernandez LM (1994) Hemiptera Heteroptera of the sugarcane agroecosystem in Cuba Hemiptera Heteroptera del agroecosistema de cana de azucar en Cuba. Cienc Biol Acad Cienc Cuba 27:12–17

Livermore LJR, Lemaître VA, Dolling WR, Webb MD (2014) Coreoidea species file online. Version 5.0/5.0. [retrieval date]. http://Coreoidea.SpeciesFile.org

Lorenzi H (2000) Plantas daninhas do Brasil, 3rd edn. Instituto Plantarum de Estudos da Flora Ltda, Nova Odessa

Maes PJ, Göllner-Scheiding U (1993) Catálogo de los Coreoidea (Heteroptera). Rev Nica Entomol 25:1–19

Maes PJ, Tellez RJ (1988) Catalogo de los insectos y artropodos terrestres asociados a las principales plantas de importancia económica en Nicaragua. Rev Nica Entomol 5:1–95

Marrero HJ, Carpintero DL, Zalba SM (2008) Relevamiento de la diversidad de Heteroptera (Hemiptera) terrestres del cerro Cura Malal, provincia de Buenos Aires, Argentina. Rev Museo Argentino Ciencias Naturales, Nueva Serie 10:29–35

Mead FW (1985) *Jadera* scentless plant bugs in Florida (Hemiptera: Rhopalidae). Florida Department of Agriculture and Consumer Service, Entomology Circular 277, Gainesville, FL, USA

Melo MC, Montemayor SI (2011) Two new species of *Harmostes* (*Harmostes*) Burmeister (Hemiptera: Heteroptera: Rhopalidae) from Argentina. Stud Neotrop Fauna Environ 46:131–136

Melo MC, Dellapé PM, Carpintero DL, Montemayor SI (2011) Heteroptera (Hemiptera) from the Chaco National Park (Argentina). Zootaxa 2999:1–19

Nickel JL (1958) Agricultural Insects of the Paraguayan Chaco 1. J Econ Entomol 51:633–637

Pall JL, Coscarón MC (2012) The Rhopalidae (Hemiptera: Heteroptera) of Argentina. J Nat Hist 46:1441–1465

Pall JL, Quirán E, Coscarón MC (2013) New records of the family Rhopalidae (Heteroptera) in the province of La Pampa (Argentina). Mun Ent Zool 8:382–386

Palmer WA, Pullen KR (1995) The phytophagous arthropods associated with *Lantana camara*, *L. hirsuta*, *L. urticifolia*, and *L. urticoides* (Verbenaceae) in North America. Biol Contib 5:54–72

Panizzi AR, Hirose E (2002) Seed-carrying and feeding behavior of *Jadera choprai* Göllner-Scheiding (Heteroptera: Rhopalidae). Neotrop Entomol 31:327–329

Panizzi AR, Hirose E, Chocorosqui VR (2002) Unusual oviposition behavior by a seed feeding bug (Heteroptera: Rhopalidae). Neotrop Entomol 31:477–479

Panizzi AR, Schaefer CW, Hirose E (2005) Biology and descriptions of nymphal and adult *Jadera choprai* (Hemiptera: Rhopalidae). Ann Entomol Soc Am 98:515–526

Paskewitz SM, McPherson JE (1983) Life history and laboratory rearing of *Arhyssus lateralis* (Hemiptera: Rhopalidae) with descriptions of immature stages. Ann Entomol Soc Am 76:477–482

Pennington MS (1920) Lista de los Hemípteros Heterópteros de la República Argentina. Printed by the author at Brasil street, # 1045, Buenos Aires, Argentina

Readio PA (1928) Studies on the biology of the genus *Corizus* (Coreidae, Hemiptera). Ann Entomol Soc Am 21:189–201

Smith LL, Ueckert DN (1974) Influence of insects on mesquite seed production. J Range Manag 27:61–65

Sailer RI (1961) The identity of *Lygaeus sidae* Fabricius, type species of the genus *Niesthrea*. Proc Entomol Soc Wash 63:293–299

Schwarz J, Gries R, Hillier K, Vickers N, Gries G (2009) Phenology of semiochemical-mediated host foraging by the western boxelder bug, *Boisea rubrolineata*, an aposematic seed predator. J Chem Ecol 35:58–70

Reinert JA, Knauf TA, Maranz SJ, Bishr M (1999) Effect of *Beauveria bassiana* fungus on the boxelder and red shouldered bugs (Hemiptera: Rhopalidae). Fla Entomol 82:469–474

Ribeiro ST (1989) Group effects and aposematism in *Jadera haematoloma* (Hemiptera: Rhopalidae). Ann Entomol Soc Am 82:466–475

Schaefer CW (2001) Variation in *Jadera parapectoralis* Göllner-Scheiding (Hemiptera: Rhopalidae). Neotrop Entomol 30:473–474

Schaefer CW (1965) The morphology and higher classification of the Coreoidea (Hemiptera-Heteroptera). Part III. The families Rhopalidae, Alydidae, and Coreidae. Misc Publ Entomol Soc Am 5:1–76

Schaefer CW, Mitchell PL (1983) Food plants of the Coreoidea (Hemiptera: Heteroptera). Ann Entomol Soc Am 76:591–615

Schaefer CW, Chopra NP (1982) Cladistic analysis of the Rhopalidae, with a list of food plants. Ann Entomol Soc Am 75:224–233

Schowalter TD (1986) Overwintering aggregation of *Boisea rubrolineatus* (Heteroptera: Rhopalidae) in western Oregon. Environ Entomol 15:1055–1056

Stroyan HLG (1954) Notes on the early stages of *Rhopalus parumpunctatus* Schill. (Hemiptera: Coreidae). Proc R Entomol Soc Lond Ser A Gen Entomol 29:32–38

Tanaka S, Wolda H (1987) Seasonal wing length dimorphism in a tropical seed bug: ecological significance of the short-winged form. Oecologia 73:559–565

Tanaka S, Wolda H, Denlinger DL (1987) Seasonality and its physiological regulation in three neotropical insect taxa from Barro Colorado Island, Panama. Insect Sci Appl 8:507–514

Tsai JF, Hsieh YX, Rédei D (2013) The soapberry bug, *Jadera haematoloma* (Insecta, Hemiptera, Rhopalidae): first Asian record, with a review of bionomics. ZooKeys 297:1–41

Van Duzee EP (1909) Observations on some Hemiptera taken in Florida in the spring of 1908. Bull Buffalo Soc Nat Sci 9:149–230

Vilímová J, Rohanová M (2010) The external morphology of eggs of three Rhopalidae species (Hemiptera: Heteroptera) with a review of the eggs of this family. Acta Entomol Musei Natioalis Pragae 50:75–95

Werner DJ (2009) Nachweise von *Liorhyssus hyalinus* (Heteroptera: Rhopalidae) in Schleswig-Holstein und anderswo. Heteropteron 31:35–36

Wheeler AG (1977) Life history of *Niesthrea louisianica* (Hemiptera: Rhopalidae) on rose of Sharon in North Carolina. Ann Entomol Soc Am 70:631–634

Wheeler AG Jr, Henry TJ (1984) Host plants, distribution, and description of fifth-instar nymphs of two little-known Heteroptera, *Arhyssus hirtus* (Rhopalidae) and *Esperanza texana* (Alydidae). Fla Entomol 67:521–529

Wheeler AG Jr, Hoebeke RE (1988) Biology and seasonal history of *Rhopalus* (Brachycarenus) *tigrinus*, with descriptions of immature stages (Heteroptera: Rhopalidae). J NY Entomol Soc 96:381–389

Winchell R, Dingle H, Moyes CD (2000) Enzyme profiles in two wing polymorphic soapberry bug populations (*Jadera haematoloma*: Rhopalidae). J Insect Phys 46:1365–1373

Wolcott GN (1948) Insects of Puerto Rico. J Agric Univ Puerto Rico 32:749–975

Wolda H, Tanaka S (1987) Dormancy and aggregation in a tropical insect *Jadera obscura* (Hemiptera: Rhopalidae). Proc Koninklijke Nederlandse Akademie van Wetenschappen Series C Biol Med Sci 90:351–356

Yonke TR, Walker DL (1970a) Field history, parasites, and biology of *Harmostes reflexulus* (Say) (Hemiptera: Rhopalidae). J Kans Entomol Soc 4:444–450

Yonke TR, Walker DL (1970b) Description of the egg and nymphs of *Harmostes reflexulus* (Hemiptera: Rhopalidae). Ann Entomol Soc Am 63:1749–1754

Zych AF (2010) Seasonal group behavior of adult *Jadera haematoloma* (Hemiptera: Rhopalidae) in central Florida following host seed exhaustion. Fla Entomol 93:644–646

Zych AF, Mankin RW, Gillooly JF, Foreman E (2012) Stridulation by *Jadera haematoloma* (Hemiptera: Rhopalidae): production mechanism and associated behaviors. Ann Entomol Soc Am 105:118–127

Chapter 21
Burrower Bugs (Cydnidae)

Cristiano F. Schwertner and Cristiane Nardi

Abstract Burrower bugs are typical pentatomoid insects, small to medium in size, recognized by the morphological adaptations for digging. However, some cydnids live aboveground or on vegetation, feeding on falling seeds or even plant tissues.

The family has a worldwide distribution, being well represented in tropical and temperate regions, known in the fossil record since at least the Late Cretaceous. Cydnidae includes more than 750 species in 93 genera, divided in six subfamilies. In the Neotropical region, 145 species are recorded and included in the subfamilies Amnestinae (38 spp.), Cephalocteinae (eight), Cydninae (97), and Sehirinae (one). The monophyly of the Cydnidae and phylogenetic relationships with other pentatomoid families are still controversial. Among the subfamilies, only Cephalocteinae has a strong support in a phylogenetic context. Burrower bugs are phytophagous and seem to be polyphagous, and although the group has been considered of little economic importance, damage to crops in the Neotropical region has been growing in the last 15 years. Because of the burrowing habits and small size of most of the species, people unnoticed its presence, limiting taxonomic characterization, host plants records, as well as damage and symptoms to cultivated plants. In this chapter, an overview of the group in the Neotropical region is presented, including an identification key for subfamilies and genera with the characterization of the commonest species.

C.F. Schwertner (✉)
Departamento de Ciências Biológicas, Instituto de Ciências Ambientais, Químicas e Farmacêuticas, Universidade Federal de São Paulo (UNIFESP), Rua São Nicolau, 210 Centro, Diadema, SP 09913-030, Brazil
e-mail: schwertner@unifesp.br

C. Nardi
Departamento de Agronomia (DEAGRO), Universidade Estadual do Centro-Oeste (UNICENTRO), Rua Camargo Varela de Sá, 03, Guarapuava, PR 85040-080, Brazil
e-mail: cnardi@unicentro.br

A.R. Panizzi, J. Grazia (eds.), *True Bugs (Heteroptera) of the Neotropics*, Entomology in Focus 2, DOI 10.1007/978-94-017-9861-7_21

21.1 Introduction

The burrower bugs are typical pentatomoid insects, small to medium in size, and with a unique life habit among true bugs. Most of the species are recognized by the morphological adaptations for digging: wide and flattened head, legs with strong and long spines, and modified tibia and tarsi. These characteristics are more conspicuous in the species with fossorial habitats that feed on roots. However, some cydnids live aboveground or on vegetation, feeding on falling seeds or even plant tissues (Schuh and Slater 1995; Schaefer 2009).

The family has a worldwide distribution, being well represented in tropical and temperate regions (Froeschner 1960; Lis 1999a, 2002). It includes more than 750 species in 93 genera, divided in six subfamilies (Table 21.1) (Grazia et al. 2008; Pluot-Sigwalt and Lis 2008).

Cydnids were considered to be primitive pentatomoids (Schaefer 1968, 1981), despite the derived features associated with fossorial habits. Fossil Cydnidae have been assigned from the Mesozoic to Lower Miocene (Shcherbakov and Popov 2002; Grimaldi and Engel 2005). At least one extant cydnid subfamily, Amnestinae, is known from the Late Cretaceous and Early Jurassic (Yao et al. 2007) and represents part of the dominant land bug fauna during the Cretaceous (Shcherbakov and Popov 2002).

The monophyly of the group has been questioned (Grazia et al. 2008; Pluot-Sigwalt and Lis 2008), and the phylogenetic position among pentatomoids is still uncertain (Lis 2010). Some cydnids show maternal care of eggs and young (Agrawal et al. 2001, 2004; Nakahira and Kudo 2008; Filippi-Tsukamoto et al. 2008), a behavior also found in other related families (Filippi-Tsukamoto et al. 1995; Tallamy and Schaefer 1997).

Burrower bugs are phytophagous, and most of the species seems to be polyphagous (Froeschner 1960; Lis et al. 2000). Although the group has been considered of

Table 21.1 Number of extant genera and species of Cydnidae by subfamilies and tribes for the world and Neotropical region

	World		Neotropical (total[a])		Neotropical (endemic[b])	
Subfamily/Tribe	Gen	sp	Gen	sp	Gen	sp
Amaurocorinae	3	5	0	0	0	0
Amnestinae	2	38	2	38	1	35
Cephalocteinae	8	32	2	8	1	7
Cephalocteini	2	4	0	0	0	0
Scaptocorini	6	28	2	8	1	7
Cydninae	65	576	11	97	5	76
Cydnini	11	114	1	1	0	0
Geotomini	54	462	11	97	5	78
Garsauriinae	3	17	0	0	0	0
Sehirinae	12	72	1	1	0	0
Total	93	742	16	131	7	118

[a]Includes species with Nearctic distribution
[b]Includes species found only in the Neotropical region

little economic importance (Lis et al. 2000), damage to crops in the Neotropical region has been growing in the last 15 years (Oliveira et al. 2000, 2013). Lis et al. (2000) reviewed the species of economic importance, and since then, several studies have added new knowledge about the biology of the species associated to crops (e.g., Chapin and Thomas 2003; Oliveira and Malaguido 2004; Riis et al. 2005b; Chapin et al. 2006; Nardi et al. 2008; Vivan et al. 2013).

21.2 General Characteristics and Diagnosis

Size varying from 2 to 25 mm, color black or brown, usually with a glossy or shining surface (Figs. 21.1, 21.2, 21.3, 21.4, 21.5, and 21.6). Body ovoid and convex. Head quadrate or semicircular, wide, and often explanate (Figs. 21.7, 21.8, 21.9, and 21.10). Antenna 5-segmented. Scutellum never surpassing the middle of the

Fig. 21.1 *Amnestus pusio* (Stål) (Cydnidae: Amnestinae), dorsal view (Photo CF Schwertner)

Fig. 21.2 *Amnestus pusio* (Stål) (Cydnidae: Amnestinae), frontal view (Photo CF Schwertner)

Fig. 21.3 *Scaptocoris castanea* Perty (Cydnidae: Cephalocteinae: Scaptocorini), dorsal view (Photo CF Schwertner)

Fig. 21.4 *Scaptocoris castanea* Perty (Cydnidae: Cephalocteinae: Scaptocorini), lateral view (Photo CF Schwertner)

Fig. 21.5 *Cyrtomenus mirabilis* (Perty) (Cydnidae: Cydninae: Geotomini), dorsal view (Photo CF Schwertner)

Fig. 21.6 *Sehirus cinctus cinctus* (Palisot de Beauvois) (Cydnidae: Sehirinae), dorsal view (Photo CF Schwertner)

Fig. 21.7 *Amnestus pusio*, detail of the head (frontal view) (Photo CF Schwertner)

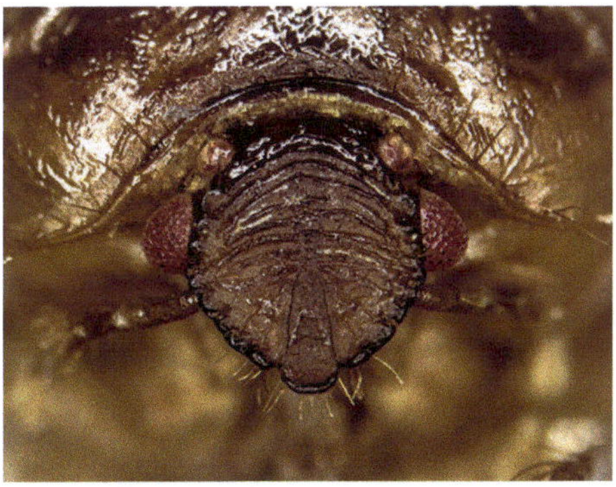

Fig. 21.8 *Scaptocoris castanea*, detail of the head (frontal view) (Photo CF Schwertner)

Fig. 21.9 *Cyrtomenus mirabilis*, detail of the head (dorsal view) (Photo CF Schwertner)

Fig. 21.10 *Sehirus cinctus cinctus*, detail of the head (frontal view) (Photo CF Schwertner)

Fig. 21.11 *Scaptocoris castanea*, median tibia (Photo CF Schwertner)

abdomen. Distal margins of coxae with a row of setae or bristles, known as coxal combs. Tibia with strong and long spines in all extension (Figs. 21.11, 21.12, 21.13, 21.14, and 21.15), anterior tibia usually compressed (Figs. 21.12, 21.14, 21.15);

Fig. 21.12 *Cyrtomenus teter*, anterior tibia (Photo CF Schwertner)

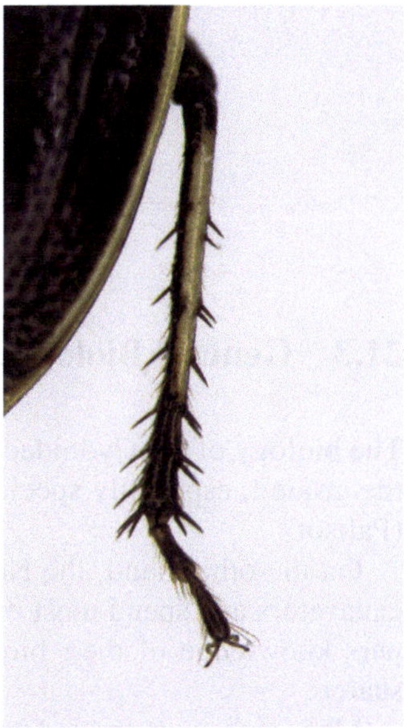

Fig. 21.13 *Sehirus cinctus cinctus*, posterior tibia (Photo CF Schwertner)

tarsi 3-segmented; in Cephalocteinae, anterior and posterior tibiae are strongly modified, and tarsi may be reduced or absent (Figs. 21.14, 21.15, 21.29). Abdominal trichobothria on sterna 3–7 oblique, longitudinal, or transversal to the spiracle line, usually 2+2. Nymphal scent glands present between abdominal terga 3/4, 4/5, and 5/6.

Fig. 21.14 *Atarsocoris giselleae* (Carvalho), anterior tibia (Photo CF Schwertner)

Fig. 21.15 *Scaptocoris castanea*, anterior tibia (Photo CF Schwertner)

21.3 General Biology and Ecology

The biology of the Cydnidae is relatively well known for species that live above the ground, especially species of the subfamily Sehirinae, e.g., *Sehirus cinctus* (Palisot).

On the other hand, the Neotropical species of the family are predominantly excavators and spend most of their lives in the soil. The subterranean habit hampers knowledge of their biology, ecology, and behavior, and such studies are scarce.

Most of the nymphs and adults of Cephalocteinae and Cydninae are sap-feeding through the roots of the hosts, although some species, e. g., *Pangaeus bilineatus* (Say), were observed sucking the content of ground pods of *Arachis hypogaea* L. (Chapin et al. 2006). For Sehirinae and Amnestinae, feeding occurs in aboveground structures of the host plants. The species *Sehirus cinctus* consumes mature fallen seeds from several host plants (Froeschner 1960; Sites and McPherson 1982), while *Amnestus* spp. were found feeding on fruits and seeds of *Ficus colubrinae* Standl. in Mexico (Mayorga and Cervantes 2001).

21.3.1 Eggs and Oviposition

The females of Cephalocteinae and Cydninae lay single eggs, belowground (García and Bellotti 1980; Riis et al. 2005b). In Sehirinae, females lay egg masses in shallow cracks on the soil surface (Sites and McPherson 1982). At least one species of Amnestinae, *Amnestus ficus* Mayorga and Cervantes, lays eggs inside the fruits of *Ficus* (Moraceae) (Mayorga and Cervantes 2001).

Maternal care is observed in those species that lay egg masses, i.e., *Sehirus cinctus cinctus* and *Adomerus triguttulus* (Motschulsky). The females guard the eggs until the nymphs leave the place after hatching (Southwood and Hine 1950; Nakahira et al. 2013). This behavior was not observed in species of other subfamilies than Sehirinae.

When oviposition occurs in the soil, eggs are usually deposited near the host plant. This behavior probably increases the efficiency of food location by the offspring, which typically have limited ability to move. The moisture and soil texture seem to determine the depth of the oviposition site (Willis and Roth 1962; Riis and Esbjerg 1998; Riis et al. 2005b). In times of prolonged drought in Brazilian Cerrado, eggs of *Scaptocoris* spp. were found about 1.5 m deep (Nardi et al. 2008).

Eggs of Cydnidae are characterized by corium smooth, uniform creamy coloration and no conspicuous projections (Fig 21.16a) (García and Bellotti 1980; Mayorga and Cervantes 2001; Vivan et al. 2013). Incubation time can vary from one week (García and Bellotti 1980; Sites and McPherson 1982; Riis et al. 2005b) up to 4 weeks (Sales and Medeiros 2001).

21.3.2 Nymph Feeding and Development

The nymphal stage of the Neotropical species of Cephalocteinae and Cydninae lives within the soil, feeding in the roots of the hosts. Typically polyphagous, it is likely that nymphs feed on plants closer to the site hatching.

The newly hatched nymphs measure from 1.0 to 2.5 mm in length (García and Bellotti 1980; Sites and McPherson 1982; Riis et al. 2005b) and move around in the soil alone. Like most of the terrestrial Heteroptera, postembryonic development occurs in five instars (Fig. 21.16), visually characterized by the morphology of the individuals. The last three instars have wing pads, which become more evident in the 5th instar (Fig. 21.16f). Time of postembryonic development remains unknown for most of the species, especially those that do not have established methods for laboratory rearing. Laboratory studies found that adults of *Cyrtomenus bergi* Froeschner, under 23.9 ± 1.1 °C and gravimetric moisture of 24 %, emerge in 112 days on average (García and Bellotti 1980; Riis et al. 2005a). For other species of Cydnidae, the development period can be shorter, as in *S. cinctus cinctus* (54 days) (Sites and McPherson 1982), although many still remain unknown. Field experiments

Fig. 21.16 Immature stages of *Scaptocoris carvalhoi*. (**a**) Eggs; (**b**) first instar; (**c**) second instar; (**d**) third instar; (**e**) fourth instar; (**f**) fifth instar (Photos C Nardi)

in pasture areas in Mato Grosso State, Brazil, conducted by Sales Jr. and Medeiros (2001) verified that egg-adult period of *Scaptocoris carvalhoi* Becker is 150–180 days.

As for insects in general, immature stage in cydnids is extremely sensitive to temperature changes, humidity, population density, and food availability (Tauber et al. 1986). For some nymphs (e.g., *Scaptocoris* spp.), dry periods may determine their location at greater depths in the soil, apparently in search for higher humidity (Oliveira and Malaguido 2004; Nardi et al. 2007; Pereira et al. 2012).

It remains unknown the occurrence of diapause in Neotropical populations. However, *P. bilineatus* from Texas (USA) showed diapause during the winter, when overwinter individuals remained inactive at 6–8 inches belowground (Cole 1988).

21.3.3 Adult Feeding and Reproduction

Most of the biological aspects of Cydnidae are difficult to be determined, but it is known that adults may live from 100 to 300 days (García and Bellotti 1980; Sales and Medeiros 2001). Observations from Riis et al. (2005b) and García and Bellotti (1980) showed that the preoviposition in *C. bergi* is 10.5 days and the female can lay at least 250 eggs. Similar reports for other species of Neotropical cydnids are unknown from the literature.

As for the nymphal stage, adults of most Cydnidae (subfamilies Cephalocteinae and Cydninae) occur in the soil, where they remain during much of life. The structure and physical conditions of the soil are extremely important for their survival. Generally, the burrower bug exhibits preference for sandy soils, to the detriment of clay soils (Riis and Esbjerg 1998; Riis et al. 2005a; Oliveira and Malaguido 2004), probably due to the lower water accumulation and lower compression, features that facilitate excavation (Willis and Roth 1962). In fact, the moisture seems to be determinant in the distribution of insects in the soil, since in times of drought, adults of Cephalocteinae seem to avoid dry areas and the surface of the soil, and individuals can be found more than 2 m deep where the soil is wetter (Nardi et al. 2007). The search for ideal conditions of temperature and soil moisture also has been related to persistence of some underground Cydnidae as tenants in nests of *Atta* spp., i.e., *Atarsocoris giselleae* (Carvalho) and *Amnestus* spp. (Froeschner 1960, 1975; Moser 1963; Soares et al. 2006).

Most of the Neotropical Cydnidae is polyphagous, and plants of several families have been reported as hosts (Becker 1967; Riis et al. 2005b; Timonin 1958; Mayorga and Cervantes 2001; Chapin et al. 2006) (Table 21.2). Host plants include small size plants, such as *A. hypogaea*, to larger-sized *Eucalyptus* spp. Very often these insects are reported as agricultural pests, and, depending on the population level, the symptoms of their occurrence can be checked on the shoots, whose leaves turn yellow and dry (Gallo et al. 2002; Riis et al. 2005b; Oliveira and Malaguido 2004; Pereira et al. 2012).

Burrower bugs have the ability to leave the ground during a period of adult life. Typically, this behavior is associated with flight capacity, which allows individuals to reach long distances and colonize new environment. The ability to fly and leave the soil is associated with the development of wings, which may vary between species and even between individuals of a population. According to Froeschner (1988), adults of *Tominotus caecus* (Van Duzee) are brachypterous, which prevents these individuals to leave the soil by flying. Nardi et al. (2008) studied *S. carvalhoi* of the Brazilian Cerrado and described the occurrence of brachypterous and macropterous individuals. The brachypterous adults are unable to move the wings and have no fly ability, while macropterous ones have high mobility and ability to fly. The flight behavior of aggregates of *S. carvalhoi* is called swarm, which occurs at the end of the day, usually after rain events. During these swarms, the insects fly about 3 ft high and can reach distances of several kilometers, often attracted by light. However, contrary to what occurs in other groups of insects, flight activity does not seem to be related to mating and is only associated with colonization of new areas, location of food and finding new sexual partners (Willis and Roth 1962; Oliveira and

Table 21.2 Host plants related to Neotropical species of Cephalocteinae and Cydninae

Species	Host plant family	Host plants	References
Scaptocoris buckupi	Asteraceae	*Senecio brasiliensis* Less	Becker (1967)
	Poaceae	*Brachiaria* spp.	Oliveira et al. (2003)
Scaptocoris carvalhoi	Fabaceae	*Glycine max* L.	Becker (1967)
		Phaseolus vulgaris L.	Becker (1996)
	Malvaceae	*Gossypium* sp.	Souza Fo et al. (1997)
	Poaceae	*Zea mays* L.	
		Pastures (several species)	
	Solanaceae	*Nicotiana tabacum* L.	
Scaptocoris castanea	Arecaceae	*Cocos nucifera* L.	Moreira (1923)
	Asteraceae	*Helianthus annuus* L.	Costa Lima (1940)
	Euphorbiaceae	*Manihot esculenta* Crantz	Hayward (1943)
	Fabaceae	*Arachis hypogaea* L.	Andrade and Puzzi (1953)
		Glycine max L.	Silva et al. (1968)
		Lupinus albus L.	Brewer (1972)
		Medicago sativa L.	Brisolla et al. (1985)
		Phaseolus vulgaris L.	Salvadori (1999)
		Pisum sativum L.	Oliveira et al. (2000)
	Malvaceae	*Gossypium* sp.	Matias et al. (2011)
	Meliaceae	*Azadirachta indica* A. Juss.	
	Myrtaceae	*Eucalyptus* spp.	
	Musaceae	*Musa* sp.	
	Poaceae	*Brachiaria* spp.	
		Oryza sativa L.	
		Panicum sp.	
		Pennisetum glaucum L.	
		Saccharum officinarum L.	
		Sorghum bicolor L.	
		Triticum vulgare Vill.	
		Zea mays L.	
	Portulacaceae	*Portulaca oleracea* L.	
	Rubiaceae	*Coffea* sp.	
	Solanaceae	*Capsicum annuum* L.	
		Lycopersicon esculentum Mill.	
		Nicotiana tabacum L.	
Scaptocoris divergens	Musaceae	*Musa* sp.	Roth (1961)
Scaptocoris talpa	Musaceae	*Musa* sp.	Timonin (1961a, b)
Atarsocoris giselleae	Amaranthaceae	*Telanthera maritima* (Mart.) Moq.	Becker (1967)
	Chenopodiaceae	*Chenopodium ambrosioides* L.	
	Palmae	*Diplothemium maritimum* Mart.	
	Poaceae	Several species	

(continued)

Table 21.2 (continued)

Species	Host plant family	Host plants	References
Cyrtomenus bergi	Euphorbiaceae	*Manihot esculenta* Crantz	CIAT (1981, 1989)
Cyrtomenus mirabilis	Fabaceae	*Arachis hypogaea* L.	Cividanes et al. (1981)
		Phaseolus vulgaris L.	Lacerda (1983)
		Pisum sativum L.	Herrera (1988)
	Liliaceae	*Allium cepa* L.	
		Allium fistulosum L.	
	Poaceae	*Saccharum officinarum* L.	
		Sorghum bicolor (L.) Moench	
		Zea mays L.	
		Pastures (several species)	
		Weeds (several species)	
	Palmae	*Elaeis guineensis* Jacq.	
	Solanaceae	*Solanum tuberosum* L.	
	Rubiaceae	*Coffea arabica* L.	
	Umbelliferae	*Coriandrum sativum* L.	
Pangaeus bilineatus	Amaranthaceae	*Spinacia* sp.	Gould (1931)
	Fabaceae	*Arachis hypogaea* L.	Cole (1988)
	Malvaceae	*Gossypium* sp.	Smith and Pitts (1974)
	Rosaceae	*Fragaria vesca* L.	Otten (1956)
	Solanaceae	*Capsicum* sp.	Froeschner (1988)
			Lis et al. (2000)
Tominotus communis	Fabaceae	*Arachis hypogaea* L.	Smith and Pitts (1974)
Sehirus cinctus cinctus	Anacardiaceae	*Toxicodendron radicans* (L.) Kuntze	Stoner (1920)
	Apiaceae	*Chaerophyllum procumbens* (L.) Crantz	Hart (1919)
	Apocynaceae	*Apocynum cannabinum* L.	Froeschner (1941)
		Asclepias spp.	Blatchley (1926)
		Ageratina altissima (L.) King and H.E. Rob.	McPherson and Mohlenbrock (1976)
		Antennaria plantaginifolia (L.) Richards.	
		Erigeron philadelphicus L.	
		Helianthus spp.	
		Solidago spp.	
		Verbesina alternifolia (L.)	
	Ebenaceae	*Diospyros virginiana* L.	

(continued)

Table 21.2 (continued)

Species	Host plant family	Host plants	References
	Lamiaceae	*Lamium purpureum* L.	
		Melilotus spp.	
		Monarda punctata L.	
		Perilla frutescens (L.) Britt.	
		Physostegia virginiana (L.) Benth.	
		Stachys spp.	
		Teucrium canadense L.	
	Poaceae	*Elymus virginicus* L.	
		Phleum pratense L.	
		Poa spp.	
	Polygonaceae	*Rumex crispus* L.	
	Rosaceae	*Prunus* spp.	
		Rubus spp.	
	Scrophulariaceae	*Verbascum thapsus* L.	

Malaguido 2004; Nardi et al. 2008). Swarming behavior is also mentioned for other species of the Neotropical Scaptocorini (Becker 1967).

Although little is known about the reproduction of Neotropical cydnids, for those species that live in the soil, mating occurs in this habitat, as showed for *Scaptocoris divergens* Froeschner (Willis and Roth 1962) and *S. carvalhoi* (Nardi 2005). During copulation males and females remain connected by the genitalia and in opposite directions (Willis and Roth 1962). Whereas Cydnidae present solitary behavior, finding sexual partners may be difficult, since the environment where these bugs live restricts large-scale communication between individuals.

Like other Heteroptera, it is likely that copulation in Cydnidae is mediated by chemicals and sound signals. Studies by Gogala et al. (1974), Gogala (1984), and Cokl et al. (2006) show that the morphology of the stridulatory apparatus of males and females is different in some species (Fig. 21.17); consequently, sound production also differs between the sexes. However, there are no detailed studies on the implications of these differences in the encounter between sexual partners. Vibrational signals have a low propagation capacity, reaching distances of a few centimeters. Thus, structures to amplify such signals and allow it to spread to greater distances would be necessary; the root of the host plants would be one of those structures (Drašlar and Gogala 1976; Cokl and Virant-Doberlet 2003).

With respect to chemical signals in Cydnidae, few studies have been conducted. Pluot-Sigwalt (2008) identified basi-abdominal glands in males of tribes Cydnini and Geotomini (Cydninae) that produced pheromones. This pheromone is probably associated with the location of sexual partners, courtship, and copulation, since the secretion of the gland is associated with increased sexual activity. However, this characteristic appears to be unique to the subfamily Cydninae; basi-abdominal glands are absent in male of Amaurocorinae, Cephalocteinae, and Sehirinae (Pluot-Sigwalt 2008). No species of Amnestinae and Garsauriinae was analyzed to date with respect to this characteristic.

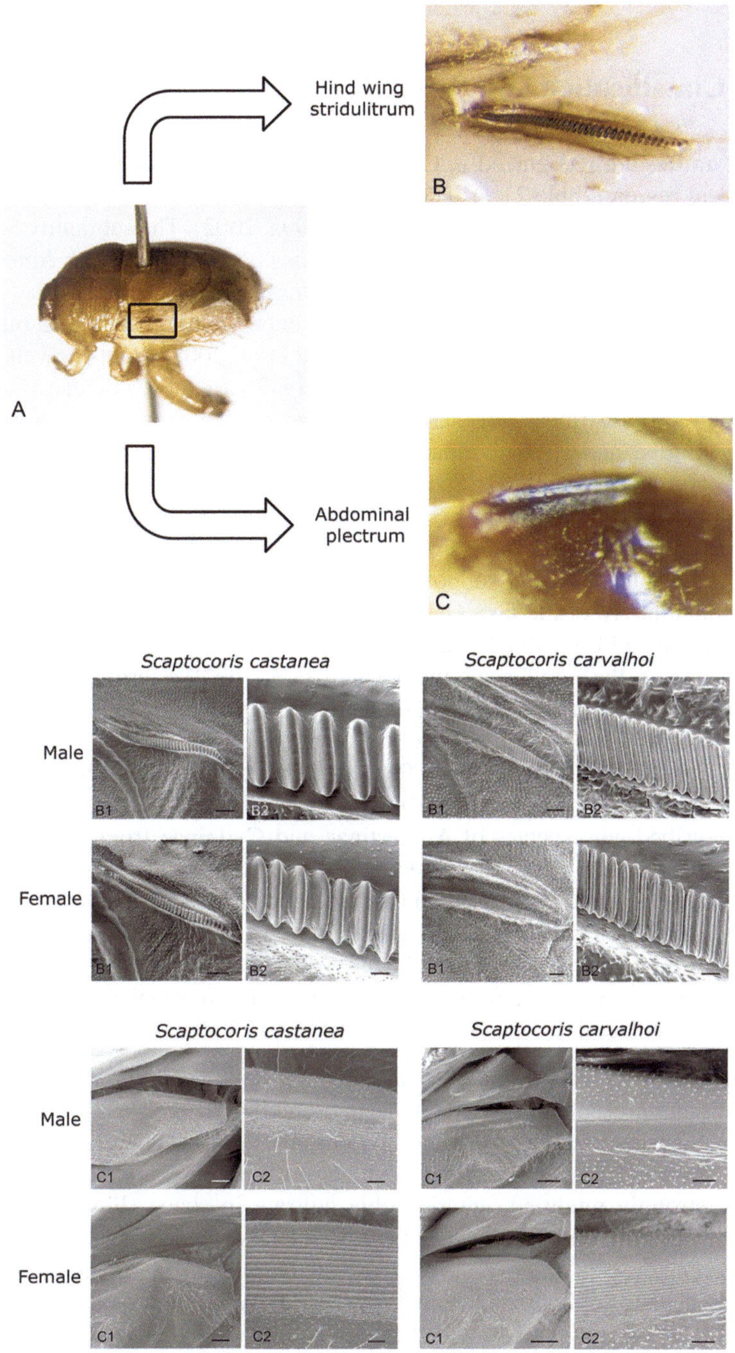

Fig. 21.17 Male and female stridulatory apparatus of burrower bugs, *Scaptocoris castanea* and *Scaptocoris carvalhoi*. (**a**) Adult body showing the stridulatory apparatus location; (**b**, **b1**) stridulitrum situated on the postcubital vein of the hind wings (scale bar = 100 μm); (**b2**) middle third of the stridulitrum, showing teeth detail (scale bar = 10 μm); (**c**, **c1**) stridulatory plectrum situated on both lateral sides of the frontal edge of the tergal plate (scale bar = 100 μm); (**c2**) middle third of stridulatory plectrum (scale bar = 20 μm) (Photos C Nardi)

21.4 Classification and Diversity

Six subfamilies are recognized in Cydnidae, and four of them are recorded in the Neotropical region (Table 21.1). The subfamilies Amaurocorinae and Garsauriinae are restricted to the Eastern hemisphere (Lis 1999a, 2002). The subfamily Sehirinae is predominantly Palearctic, with only one species occurring in North America and reaching the neotropics (Froeschner 1960; Mayorga 2002).

The monophyly of the Cydnidae and phylogenetic relationships with other pentatomoid families are still controversial (Grazia et al. 2008; Pluot-Sigwalt and Lis 2008; Lis 2010). Two of the typical characteristics of the Cydnidae (i.e., coxal combs and spinose tibiae) are shared with the related families Parastrachiidae, Thaumastellidae, and Thyreocoridae, sometimes included as subfamilies in a more inclusive taxon (Dolling 1981; Schuh and Slater 1995). The classification of this group of families has been intensively studied (Pluot-Sigwalt and Lis 2008; Lis 2010), and general conclusions remain unsettled. Among the subfamilies, only Cephalocteinae has a strong support in a phylogenetic context (Lis 1999b), while the non-monophyly of the taxa Geotomini (Cydninae) and Sehirinae is suggested (Pluot-Sigwalt and Lis 2008; Lis 2010).

Froeschner (1960) reviewed the Cydnidae in the Western Hemisphere and included identification keys to all genera and species known at that time to the neotropics. Becker (1967) monographed the Scaptocorini of the Neotropical region, described new species of *Scaptocoris* Perty, and proposed the new genus *Atarsocoris* to include *A. giselleae*. Froeschner (1975) and Mayorga and Cervantes (2001, 2005, 2009) described new species of Amnestinae and Cydninae from the Neotropical region, while Eger (2008) described a new genus and three new species for Amnestinae. Grazia et al. (2004) updated the taxonomy of the Cephalocteinae, and Marco and Coscar6n (2011) summarized the literature of the group in the Neotropical region after the publication of Froeschner (1960).

Identification key to subfamilies of Cydnidae of the Neotropical region (based on Froeschner 1960, 1981):

1. Scutellum short, not surpassing the middle length of the abdomen, claval commissure present, almost as long as the scutellum (Fig. 21.22)..... Amnestinae
– Scutellum usually surpassing the middle length of the abdomen, claval commissure never present 2
2. Tibiae strongly modified, anterior tibia cultrate in shape (Figs. 21.14, 21.15, 21.29), posterior tibia club shaped, posterior tarsi always absent (Figs. 21.18, 21.19) .... Cephalocteinae
– Tibiae usually cylindrical, sometimes strongly compressed and modified (Fig. 21.12), but never as described above 3
3. Pronotum with a lateral, submarginal row of setigerous punctures (Fig. 21.5); tarsal segment II subequal in diameter than I and III (Fig. 21.20).......... Cydninae
– Pronotum without a row of setigerous punctures; tarsal segment II distinctly narrower than I and III (Fig. 21.21) .... Sehirinae

Fig. 21.18 *Atarsocoris giselleae* (Carvalho), posterior tibia (Photo CF Schwertner)

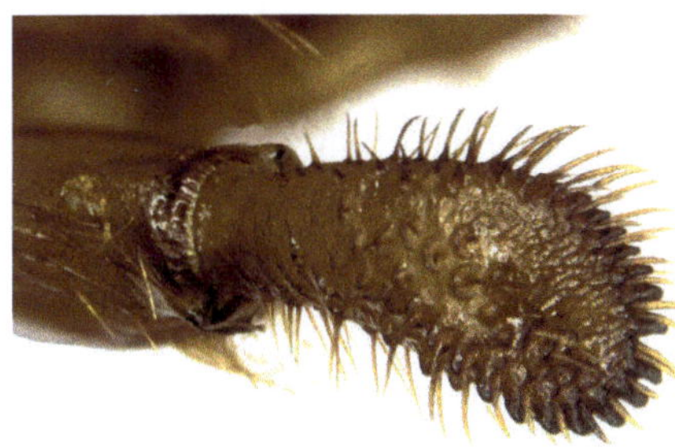

Fig. 21.19 *Scaptocoris castanea*, posterior tibia (Photo CF Schwertner)

Fig. 21.20 *Cyrtomenus mirabilis*, anterior tarsus (Photo CF Schwertner)

Fig. 21.21 *Sehirus cinctus cinctus*, anterior tarsus (Photo CF Schwertner)

Fig. 21.22 *Amnestus pusio*, scutellum and hemelytra in detail (Photos CF Schwertner)

21.4.1 Subfamily Amnestinae

These very small bugs (1.6–4.5 mm) are reddish brown to blackish brown in color and easily recognized by the presence of the claval commissure (Fig. 21.22). Other unique characteristics include clypeus and juga with marginal pegs (Fig. 21.7), posterior margin of pronotum sinuate laterally near humeral angles, short scutellum, and anterior tibia compressed. Amnestinae is predominantly Neotropical (Table 21.1), with two species endemic to the Nearctic region. One species, *Amnestus pusillus* Uhler, is well established in the Iran (Province of Gilan), probably introduced by man (Mayorga et al. 2012).

Species of Amnestinae are among the oldest fossils of true bugs, dated at least from the Cretaceous/Jurassic (Shcherbakov and Popov 2002; Yao et al. 2007). Distribution of this subfamily was widespread during the Mesozoic, and species were found in localities included today in Neotropical and Palearctic regions

(Shcherbakov and Popov 2002). Other Amnestinae fossils were recorded in amber from Mexico (Chiapas) and the Dominican Republic (Thomas 1988, 1994).

The two extant genera of Amnestinae can be separated by the development of the hemelytral membrane and the presence of ocelli (Eger 2008). In *Amnestus*, the hemelytral membrane is developed, occupying 1/3 of the hemelytral length (Fig. 21.1), and the ocelli are present; in *Lattinestus*, the hemelytral membrane is reduced, occupying 1/7 or less of the hemelytral length, and the ocelli are absent (Eger 2008).

Adults of Amnestinae were recorded in litter of humid forest (Eger 2008; Mayorga and Cervantes 2001), low vegetation in moist areas (Van Duzee 1894; Blatchley 1926 in McPherson and Mohlenbrock 1976), bluegrass on sandy knolls (*Andropogon furcatus* Muhl. consortia) (McPherson and Mohlenbrock 1976), and beach drift (Torre-Bueno 1915; Froeschner 1960). Nymphs and adults of *A. ficus* were found belowground, 10 cm deep in the soil, or inside of the *Ficus* fruit, feeding on freshly fallen fruits and seeds of *Ficus cotinifolia* H.B.K., *F. insipida* Willd, *F. trigonata*, and *F. colubrinae* Standley (Mayorga and Cervantes 2001). Furthermore, Caudell (1924, in Mayorga and Cervantes 2001) found *Amnestus subferrugineus* (Westwood) associated with guano in bat caves, suggesting that this species feed on the seeds contained in the excrements.

The species of Amnestinae are poorly studied, and most of the recent publications include descriptions of new species, morphology, and host plant records (Lis 1998; Lis and Hohol-Kilinkiewicz 2001; Lis and Pluot-Sigwalt 2002; Yao et al. 2007; Pluot-Sigwalt and Lis 2008; Eger 2008; Lis 2010; Mayorga et al. 2012; Mayorga-Martinez and Brailovsky 2012). The most complete work about *Amnestus* biology and behavior was done by Mayorga and Cervantes (2001). These authors showed that female of *A. ficus* laid single eggs inside the fruits of *Ficus* or inside the soil. Nymphs and adults are diurnal and disperse from one tree to another, probably attracted to fresh fruit volatiles. According to Froeschner (1960), these insects are attracted to light, although Mayorga and Cervantes (2001) collected only a few specimens with light traps. Life cycle of *A. ficus* takes ca. 25 days, and in this short period, individuals need to find suitable fruits and seeds before rain falls and competitors (e. g., ants and crabs) remove the seeds (Laman 1996; Mayorga and Cervantes 2001).

21.4.2 Subfamily Cephalocteinae

Besides the strongly modified tibiae, the species of Cephalocteinae are also recognized by having globular and strongly convex body (Figs. 21.3 and 21.4), coloration varying from yellow amber to dark brown, and length between 4 and 10 mm (Schuh and Slater 1995; Lis 1999b; Grazia et al. 2004). Two tribes are recognized, and their distribution includes all biogeographical regions (Lis 1999b). Two genera are found in the neotropics: *Atarsocoris* Becker (1967) (monotypic) and *Scaptocoris* Perty (1833) with seven species. Species of Scaptocorini are related to several host plants, including many crops (Becker 1967, 1996; Souza Fo et al. 1997; Oliveira et al. 2003; Grazia et al. 2004) (Table 21.2).

The genus *Atarsocoris* is characterized by the apex of tylus bifid (Figs. 21.23 and 21.24), ostiole exposed (Fig. 21.26), and anterior tarsi absent. In the genus

Scaptocoris, the apex of the tylus is entire (Figs. 21.8, 21.25, and 21.27), the ostiole covered (Fig. 21.28), and the anterior tarsi present or reduced, but always with tarsal insertion present (Fig. 21.29; see also Grazia et al. 2004). Lis and Hohol-Kilinkiewicz (2001) showed that the number and arrangement of trichobothria can also be used to differentiate *Atarsocoris* and *Scaptocoris* (one trichobothria on each side of the abdominal segment in the first, two in the second).

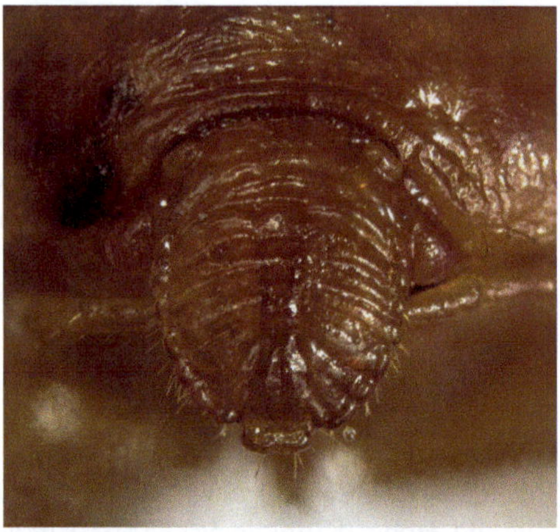

Fig. 21.23 *Atarsocoris giselleae*, head in frontal view (Photo CF Schwertner)

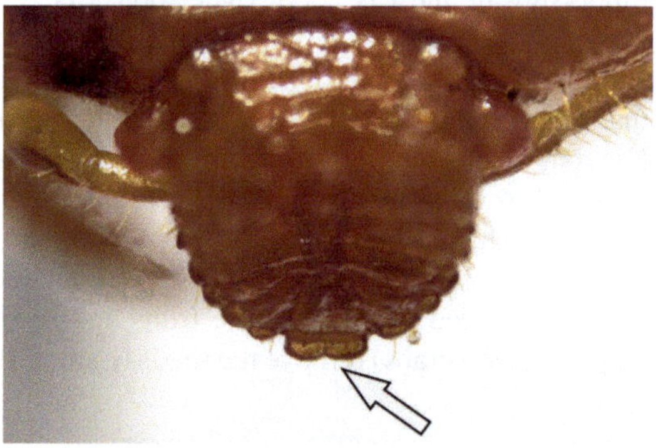

Fig. 21.24 *Atarsocoris giselleae*, detail of the head in dorsal view, showing apex of tylus (Photo CF Schwertner)

Fig. 21.25 *Scaptocoris carvalhoi*, head in dorsal view (Photo CF Schwertner)

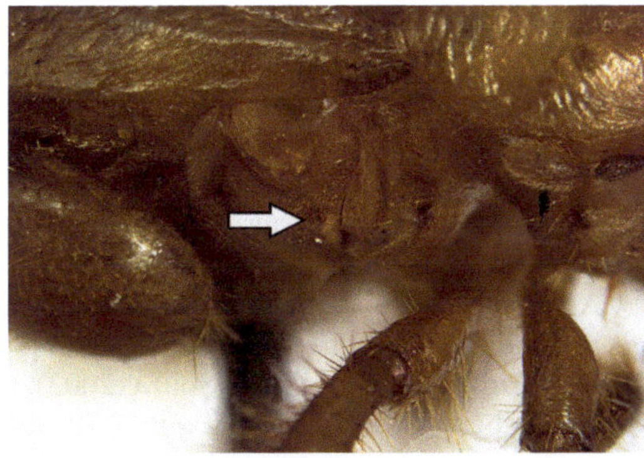

Fig. 21.26 *Atarsocoris giselleae*, evaporatorium (Photo CF Schwertner)

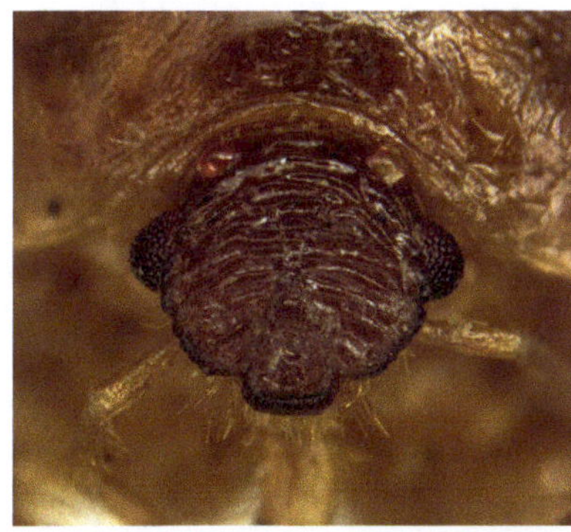

Fig. 21.27 *Scaptocoris divergens*, head in dorsal view (Photo CF Schwertner)

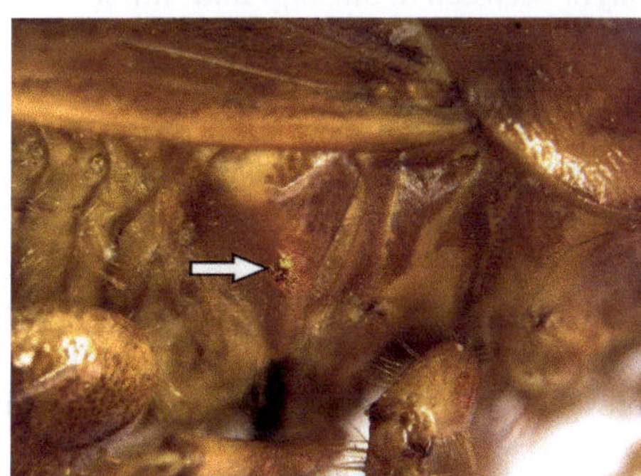

Fig. 21.28 *Scaptocoris minor*, evaporatorium (Photo CF Schwertner)

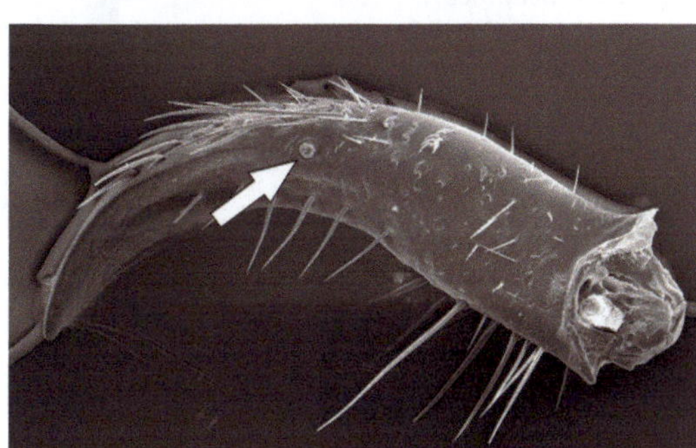

Fig. 21.29 *Scaptocoris carvalhoi*, anterior tibia (Photo C Nardi)

In the Western Hemisphere, Scaptocorini is distributed from the USA to Argentina, although it is more common in Brazil, where large populations cause significant damage to crops. *Scaptocoris buckupi* Becker, *S. carvalhoi*, and *S. castanea* Perty are the main species associated to damage (Froeschner 1960; Becker 1967; Grazia et al. 2004) and can be differentiated by the shape and arrangement of corbicles present in the posterior tibiae and presence of tarsus or tarsal insertion in the anterior and median leg (Grazia et al. 2004).

Damage caused by brown burrower bugs in crops and pastures is due to the feeding of adults and nymphs, sucking the sap of plants through the roots (Souza 2002; Gallo et al. 2002). According to Souza (2002), the initial symptom of the attack is a slight number of nearby plants with yellowish leaves, withered and dry aspect. However, during severe infestations, these symptoms may occur in plants distributed in several acres.

21.4.3 Main Species

***Scaptocoris carvalhoi* Becker (= *Atarsocoris brachiariae* syn. jr.) (Figs. 21.16, 21.17, 21.25, 21.29, 21.30, and 21.31a)**

Adults and nymphs (Fig. 21.16) of *S. carvalhoi* have anterior tarsi absent, but tarsal insertion present, like a small knob (Fig. 21.29), and the discal tubercles of the corbicle (Fig. 21.30) distinctly elongated (Becker 1967; Grazia et al. 2004). The synonym between *S. buckupi* and *Atarsocoris brachiariae* Becker was defined recently (Grazia et al. 2004).

Vivan et al. (2013) published the first detailed morphological record of immatures belonging to *Scaptocoris*, describing *S. carvalhoi*. Eggs of *S. carvalhoi* are elliptical and have 1.60 mm length and 0.90 mm width, a smooth chorion surface, and no visible aero-micropylar processes (Fig. 21.16a). Nymphal stage of *S. carvalhoi* presents five instars, an elliptical body, and fossorial scythe-like forelegs. The tarsi are absent as in adults, and the prototarsal insertion region becomes visible only in the 4th instar. Nymphs from the 1st to 4th instar of *S. carvalhoi* showed the presence of 1 + 1 trichobothria in urosternites III to VII, close to the anterior margin

Fig. 21.30 *Scaptocoris carvalhoi*, posterior tibia (Photo CF Schwertner)

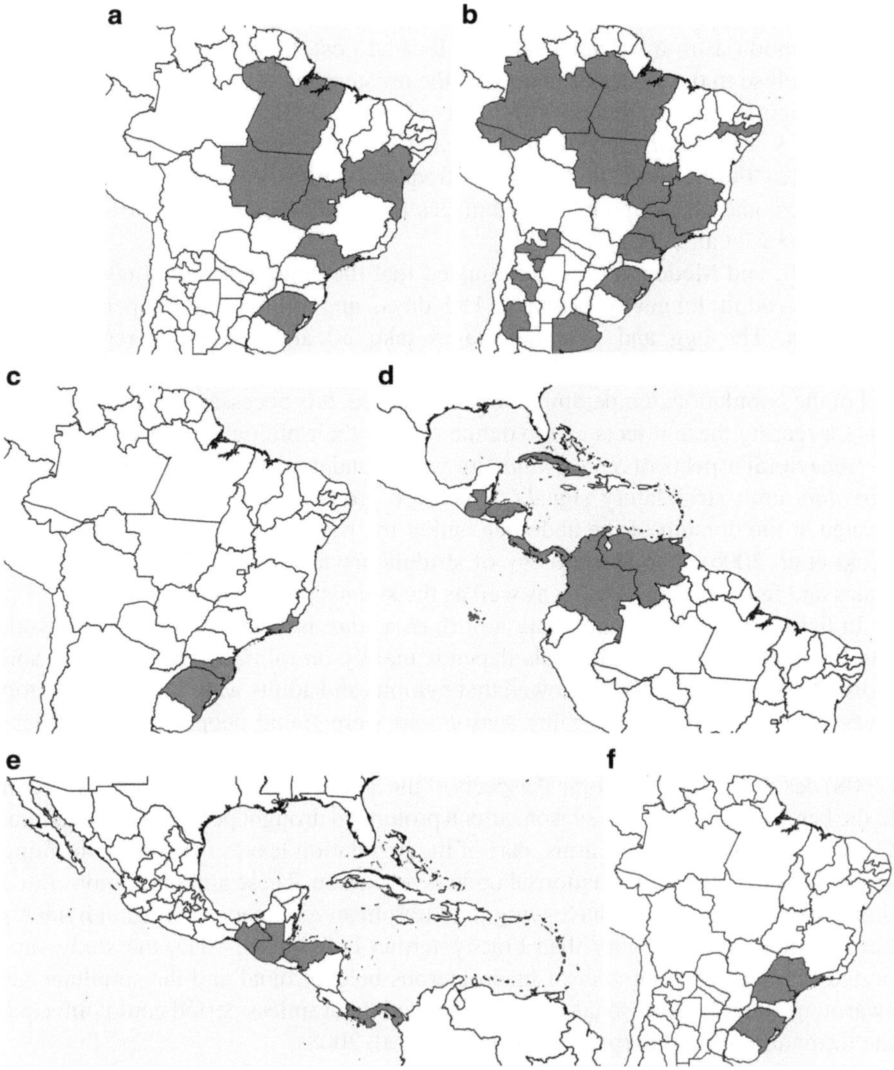

Fig. 21.31 Distribution maps of Cephalocteinae in the Neotropical region. (**a**) *Scaptocoris carvalhoi*; (**b**) *Scaptocoris castanea*; (**c**) *Scaptocoris buckupi*; (**d**) *Scaptocoris divergens*; (**e**) *Scaptocoris talpa*; (**f**) *Atarsocoris giselleae*

and inside the spiracles; besides these trichobothria, fifth instars presented 1+1 pre-trichobothria in urosternites III to V located posteriorly, almost in the row of spiracles close to the posterior margin of the urosternites.

Although frequently related with plants of Poaceae (Becker 1967), nymphs and adults of *S. carvalhoi* are polyphagous and occur on plants of several families (Table 21.2). Furthermore, this species is frequently mistaken with other brown burrower bugs, and all symptoms and damages to crops are generally attributed to *S. castanea* (Lis et al. 2000).

Sales Jr. and Medeiros (2001) estimated that the egg-adult period takes 150 to 180 days, adult longevity is about 150 days, and preoviposition period takes 18.2 days. The egg and nymphal stages take 30 and 145 days, respectively. Nevertheless, these biological data are approximate and obtained without any control of the population, temperature, or soil moisture. It is necessary to develop methods for rearing these insects and to define exactly their biological aspects.

Behavioral aspects of *S. carvalhoi* are poorly understood. Male and female of *S. carvalhoi* emit stridulatory signals, which are produced by a stridulatory system located at the dorsum of the abdomen and at the basal area of the hind-wing vein (Cokl et al. 2006). The morphology of stridulatory apparatus is different between males and females (Fig. 21.17), as well as the signals produced by them.

In field conditions, the population of *S. carvalhoi* is found stratified in the soil, and the distribution of individuals depends mainly on rainfall conditions and soil moisture. Nardi et al. (2007) showed that nymphs and adults were located in the top layer of the soil during the rainy season and were found deeper during rainless period in the Brazilian Cerrado region (16° 51′ 50″ S, 50° 33′ 43″ W). Nardi et al. (2008) described some ecological aspects of the adults of *S. carvalhoi*, which swarm in the beginning of the rainy season, after a prolonged drought period in the Brazilian Cerrado region. In these swarms, part of the population leaves the soil, performing long flights to locations with more abundant vegetation. These authors demonstrated that adults of *S. carvalhoi* have wing polymorphism and macropterous individuals have greater flight capacity than brachypterous individuals. Also, the study supported that increased number of macropterous belowground and the simultaneous swarming could be a seasonal-related event and that rainless period could influence the formation of polymorphic adults (Nardi et al. 2008).

Since *S. carvalhoi* is pointed as an important species of crop plants in Brazil, biological and ecological data are extremely important to develop methods for integrated management of these insects. According to Nakano et al. (2001), *S. carvalhoi* feeds through the roots of the hosts, causing developmental delay, wilting, and death of plants. In areas where the soil is degraded, the damage caused by these insects is more severe (Oliveira et al. 2000).

Scaptocoris castanea Perty (Figs. 21.3, 21.4, 21.8, 21.11, 21.15, 21.19, 21.31b)

Adults of *S. castanea* are characterized by the presence of anterior and median tarsi; median tibia with dorsal area flat and glabrous; surface of corbicle flattened (Fig. 21.19); discal tubercles arranged in irregular row, extending from the base to near the apex of the corbicle.

Cokl et al. (2006) demonstrated that males and females of *S. castanea* emit stridulatory signals, which were registered when the adults were on the roots of soybean. Males and females also show distinct stridulatory tergum system, with one ridge in male and 13 ridges in female. Furthermore, differences between male and female emissions were expressed significantly in syllable repetition time, in the repetition time of the first and second pulse train, and in the interval between them, in the number of pulses per pulse train, and in their repetition rate.

Although cited as the most frequent species in Brazil and Argentina, Becker (1967) and Lis et al. (2000) commented that possibly many reports of occurrence assigned to species are mistaken. *Scaptocoris castanea* cause damage in a wide range of host plants, and apparently populations occur in areas previously cultivated with pastures (Puzzi and Andrade 1957).

The damage caused by nymphs and adults of *S. castanea* occurs by sucking the sap of the plants through the roots, causing withering, drying, and even the death of plants (Oliveira et al. 2000). In the Southeast and Midwest of Brazil, population peak of these insects occurs in late spring and summer, while the greatest number of adults in the soil is observed in summer (from December to February) (Oliveira and Malaguido 2004). Nymphs and adults remain in the upper soil layers (above 20 cm) during rainy seasons, going down deep when water availability is low. Based on this, Oliveira and Malaguido (2004) suggest that sampling for population estimates should be made at depths of 30–50 cm in order to reach the population throughout the year. Furthermore, these authors suggested that the best time for the application of control measures is the period when the bugs are more accessible, i.e., closer to the soil surface.

The population density of *S. castanea* also appears to vary with the cultural agricultural practices. Pereira et al. (2012) showed that *S. castanea* population is larger in integrated crop-livestock system (corn associated with *Brachiaria decumbens* Stapf), with no tillage. Both nymphs and adults of *S. castanea* were more numerous in areas where corn was cultivated for two consecutive years and where pasture was renewed every 2 years. Moreover, lower insect population densities were observed in plots that remained only with *Brachiaria* pasture.

Despite the importance of *S. castanea* in agriculture, few results have been obtained with the management of the populations in the field, due to the difficulty to target these insects by conventional control. Thus, we emphasize the need for more basic studies, aiming to recognize aspects of their biology, ecology, and behavior, to improve the strategies for their management.

Secondary Species

Scaptocoris buckupi Becker (Figs. 21.31c)

Adults of *S. buckupi* are characterized by the presence of tarsal insertion in the anterior leg (shared with *S. carvalhoi*) and median tibiae with the dorsal area uniformly convex and bristly, and tubercles of the corbicle in a median regular row, except for two of them in the middle third, arranged obliquely (Becker 1967).

Scaptocoris buckupi is understudied with respect to their biology and behavior. Since the description of the species by Becker (1967), the occurrence of nymphs and adults associated with *Senecio brasiliensis* (Spreng.) is the only available information about these insects (treated as *Atarsocoris* sp. in Oliveira et al. 2003). Grazia et al. (2004) reviewed the taxonomy of the species associated to pastures in Brazil and identified them as *S. buckupi*.

This insect is more abundant at the base of the weed plants where they probably found the favorable conditions to survive (Oliveira et al. 2003). Moreover, populations of *S. buckupi* seem to follow the general behavior of other *Scaptocoris*, living in sandy-textured, low-level organic matter soil.

Scaptocoris divergens (Figs. 21.27 and 21.31d)

S. divergens is found in Guatemala, Honduras, Costa Rica, Panama, Trinidad, Colombia and Venezuela. This species is characterized by having the clypeus distinctly surpassing the apex of jugae, greatly expanding forward (Fig. 21.27) (Becker 1967). Host plant records include roots of banana (Roth 1961) (Table 21.2). Willis and Roth (1962) commented on the occurrence of swarms of *S. divergens* in Colombia, during six or seven nights for a month, with several individuals collected in light traps. In this study, the authors illustrate the mating behavior in *Scaptocoris* for the first time, an extremely difficult task because this behavior occurs belowground (Willis and Roth 1962; Nardi 2005).

Roth (1961) described the morphology of the scent gland of *S. divergens* and identified the secretion as a compound of seven aldehydes, two furans, and two quinones. These compounds showed fungistatic and fungicidal activity against *Fusarium* and repellent activity against ants (*Pheidole* sp.).

Scaptocoris talpa (Fig. 21.31e)

Scaptocoris talpa is the largest species within *Scaptocoris*, reaching 8.8 mm in length, with corbicle tubercles flattened and with approximately 25 setigerous punctures in the lateral margin of the pronotum (Becker 1967). This species was found feeding on the roots of banana (Timonin 1958) and sugarcane (Champion 1900 cited in Becker 1967) in Honduras and Guatemala, respectively.

Timonin (1961a, b) showed that the scent gland volatiles released by these insects reduced the population of four isolates of *Fusarium oxysporum* Schlecht. f. sp. *cubense* and 14 other isolated soil-inhabiting fungi, associated with banana cultivation. These compounds proved to be important for the protection of banana and tomato plants against soil fungi and pathogenic nematodes (Timonin 1961a).

Atarsocoris giselleae (Figs. 21.14, 21.18, 21.23, 21.24, 21.26, and 21.31f)

This is the only species included the genus *Atarsocoris* Becker (Grazia et al. 2004). According to Becker (1967), individuals of this species were collected in coastal dune vegetation, where prevailed grasses, shrubs, and small trees. Label records indicated that this species also swarms in some periods of the year. No other information is available about this peculiar species.

21.4.4 Subfamily Cydninae

This group represents the most speciose subfamily of Cydnidae (Table 21.1). The size ranges from 3 to 20 mm, and body coloration is usually black or dark brown (Figs. 21.5, 21.38, and 21.39). Despite the morphological variation found among the subfamily, species can be recognized by the more compressed body, presence of setigerous punctures in the lateral margins of pronotum and hemelytra, anterior tibia never cultrate, and tarsi always arising at or near the apex of tibia (Froeschner 1960, 1981). In the Neotropical region, only species of the tribe Geotomini are indigenous. The species *Cydnus aterrimus* (Foster) (Cydnini) found in the neotropics is considered an accidental introduction in the Western Hemisphere (Froeschner 1960; Lis 1996); *C. aterrimus* was recorded in the USA and Tobago by a single specimen in each locality (Froeschner 1960).

The Neotropical fauna of Cydninae, Geotomini, is predominantly endemic, with some species reaching the Nearctic region (Table 21.3). Only the genus *Microporus* Uhler is found outside the Western Hemisphere, with species also found in the

Table 21.3 Species of subfamily Cydninae by genera in the Neotropical region

Taxon	World	Neotropical (total)	Neotropical (endemic)
Cydnini	114	1	0
Cydnus[a]	7	1	0
Geotomini	443	96	78
Cyrtomenus[b]	8	7	6
Dallasiellus[b]	31	27	23
Ectinopus	3	3	3
Melanaethus	18	10	7
Microporus[a]	16	3	2
Onalips	3	3	3
Pangaeus[b]	23	22	17
Prolobodes	3	3	3
Rhytidoporus[b]	6	6	4
Tominotus[b]	14	14	10
Total	557	97	78

[a]Genera with species outside the Western Hemisphere
[b]Genera predominantly Neotropical, with species reaching the Nearctic region

Afrotropical, Australian, Oriental, and Palearctic regions (Lis 1999a, 2002). Fossil of the extant genus *Dallasiellus* is known from Dominican amber of Lower Miocene (Thomas 1994).

Economically important species in the Neotropical region are included in the genera *Cyrtomenus* Amyot and Serville and *Pangaeus* Stål. Nymphs and adults of these taxa feed on roots, tubers (Riis et al. 2005b), and ground pods (Riis et al. 2005b; Chapin et al. 2004, 2006), reducing productivity and facilitating infection by soil pathogens such as *Fusarium*, *Aspergillus*, *Pythium*, and *Genicularia* (Riis et al. 2005b), as well as the development of aflatoxins (Chapin et al. 2004). At least one species of *Tominotus* is recorded feeding on peanuts (Lis et al. 2000), but this species was never studied in detail, and the damage level was unknown. Biological information about other genera of Neotropical Cydninae is still limited.

21.4.5 Key to Some of the Genera of Cydninae: Geotomini of the Neotropical Region (Modified From Froeschner 1960).

We include in the key those genera with species discussed in this chapter (*Cyrtomenus*, *Dallasiellus*, *Pangaeus*, *Prolobodes*, and *Tominotus*). All of them have the anterior part of the peritreme without an enlarged, differentiated apical structure (Fig. 21.32) and were separated by Froeschner (1960) from the remaining genera in the first step of his key. To identify those genera with the anterior part of the peritreme modified (*Ectinopus* Dallas, *Macroporus* Uhler, *Melanaethus* Uhler, *Microporus* Uhler, and *Onalips* Signoret), the works of Froeschner (1960, 1981) and Mayorga (2002) are appropriate references:

1. Pronotum anteriorly with deep, sharply impressed line paralleling anterior margin from side to side (Figs. 21.33a and 21.38)..*Pangaeus*
– Pronotum anteriorly without this impressed line, rarely with a partial, vague line laterally, sometimes with a row of punctures in the same area.................... 2
2. Posterior tibia conspicuously compressed (Fig. 21.34), anterior and posterior faces glabrous, not spined; spines of the posteroventral margin conspicuously longer, thinner, and more tapering than those of dorsal margin...........................3
– Posterior tibia not or only weakly compressed (Figs. 21.35); dorsal and ventral spines equally developed.. 4
3. Labial segment II with a large, semicircular, foliaceous lobe (Fig. 21.36)...*Prolobodes*
– Labial segment II without large foliaceous lobe.................................*Cyrtomenus*
4. Margin of each jugum with a complete row (extending from the eye to the apex of jugum) of coarse, more or less contiguous punctures giving rise to numerous long hairs and usually also to a row of pegs (Fig. 21.39) ....*Tominotus*
– Margin of each jugum without a complete row of coarse setigerous punctures, which can be absent or, if present, not extending more than ¾ of way to apical angle of jugum; pegs never present (Fig. 21.37)................................*Dallasiellus*

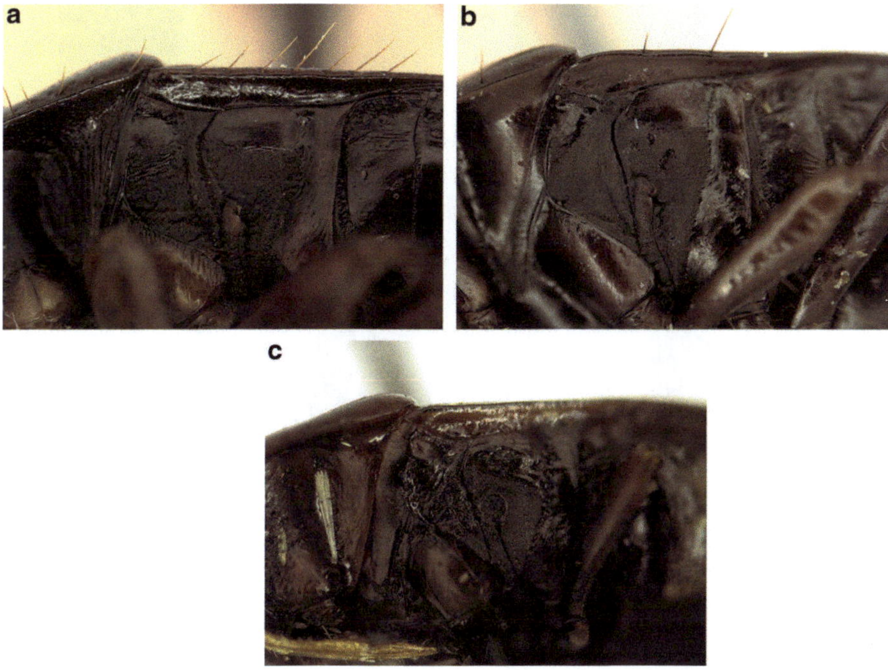

Fig. 21.32 Evaporatoria in Cydninae. (**a**) *Cyrtomenus mirabilis*; (**b**) *Pangaeus neogeus*; (**c**) *Melanaethus spinolae* (Photos CF Schwertner)

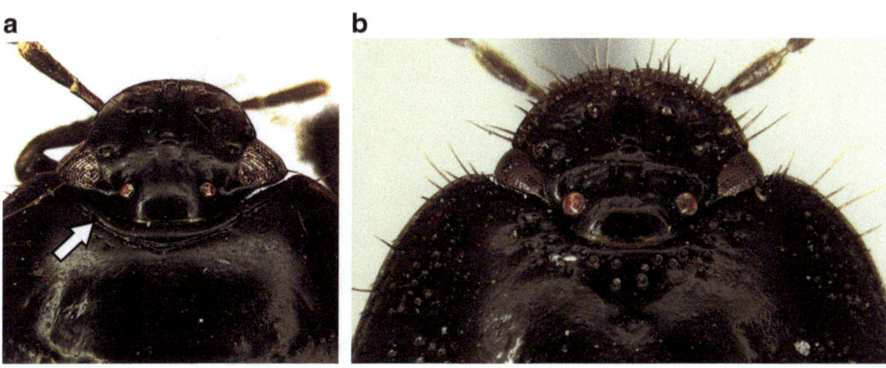

Fig. 21.33 Detail of pronotum in dorsal view. (**a**) *Pangaeus neogeus*; (**b**) *Cyrtomenus mirabilis* (Photos CF Schwertner)

Fig. 21.34 Posterior tibia, *Cyrtomenus mirabilis*. (**a**) Lateral view; (**b**) dorsal view (Photos CF Schwertner)

Fig. 21.35 Posterior tibia, *Dallasiellus longulus* (Photo CF Schwertner)

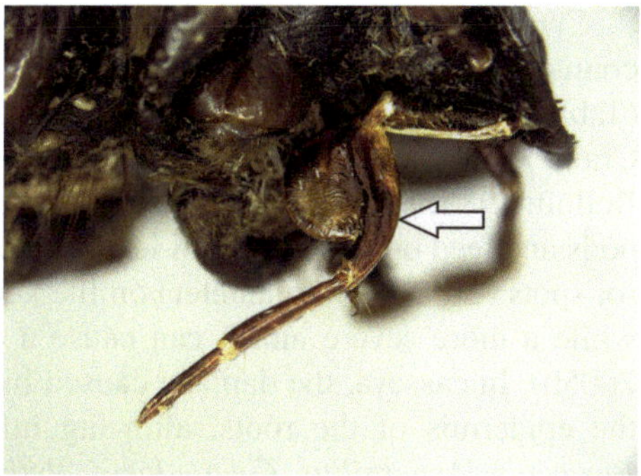

Fig. 21.36 Rostrum of *Prolobodes giganteus* (second segment in detail) (Photo CF Schwertner)

Fig. 21.37 *Dallasiellus longulus*, head in dorsal view (Photo CF Schwertner)

21.4.6 Main Species

Cyrtomenus bergi Froeschner (Fig. 21.40a) and *Cyrtomenus mirabilis* (Perty) (Figs. 21.5, 21.9, 21.20, 21.32a, 21.33b, 21.34, and 21.40a)

The species *Cyrtomenus bergi* and *C. mirabilis* were included in the subgenus *Cyrtomenus* by Froeschner (1960) that can be differentiated from the subgenus *Syllobus* by the mesopleural evaporatorium entire, not posteriorly interrupted by a shining, submarginal band (Fig. 21.32a). Among species of the subgenus, the absence of a transverse row of prominent setigerous punctures in the urosternites IV to VI and posterior tibia not so compressed as the anterior tibia (Figs. 21.12 and 21.34) allow the recognition of both species. Becker and Galileo (1982) suggested that *C. bergi* and *C. mirabilis* may be the same species; indeed, they are morphologically identical, and the characters proposed to separated them (Froeschner 1960) are difficult to ascertain (CF Schwertner, unpublished data).

Eggs of *C. bergi* have a cream hyaline coloration, are oval shaped, and have an average size of 1.35 mm (length) and 0.92 mm (width); the incubation time is 13.6 days; at hatching, nymphs have whitish-yellowish coloration and are about 1.7 mm and at the fifth instar reach 6.2 mm on average (García and Bellotti 1980).

Cyrtomenus bergi is polyphagous and considered an agricultural pest in several countries, causing damage in plants of numerous families (Riis et al. 2005b) (Table 21.2). Most of the damage was reported in cassava (*Manihot esculenta* Crantz) and peanuts (*A. hypogaea*) (CIAT 1989; García and Bellotti 1980; Arias and Bellotti 1985; Bellotti et al. 1999; Riis et al. 2005b). In peanuts, *C. bergi* pierce the pods and feed on the kernels. A less severe attack will cause yellow to brownish dry rot spots (ca. 1.5 mm diameter) on the kernels of both mature and immature pods, while a more severe attack can cause a complete loss of the harvest (Riis et al. 2005b). In cassava, the damage caused by *C. bergi* is due to the stylet insertion in the epidermis of the roots, allowing infection by soil microorganisms such as *Fusarium*, *Aspergillus*, *Genicularia*, *Pythium*, *Diplodia*, and *Phytophthora* (CIAT 1989), deteriorating the root system of the plant, reducing the content of starch, and affecting the quality of cassava. According to Arias and Bellotti (1985), 20–30 % of roots affected result in 100 % commercial loss.

Biological parameters of *C. bergi* may vary according to the food consumed by nymphs and adults (García and Bellotti 1980, Riis et al. 2005b). In cassava, egg to adult development time is ca. 90 days, whereas in peanut was ca. 60 days; longevity of females reared on peanuts was significantly higher (316.1 days) than females reared on maize (199.1 days) or cassava (111.7 days) (Riis et al. 2005b).

Cyrtomenus bergi is one of the most studied species of Cydnidae. Basic biological aspects are well known (García and Bellotti 1980; Riis et al. 2005b), as well as information about behavior, population dynamics, and food preference (Riis 1997; Riis et al. 2005a, b). Chemical, biological, and cultural control was discussed in several works (e.g., Bellotti and Riis 1994; Calcedo and Bellotti 1994; Sánchez and Bellotti 1997; Barberena and Bellotti 1998; Riis et al. 2003; Jaramillo et al. 2005).

Cyrtomenus mirabilis is considered an important species in cultivated peanuts in Peru, Paraguay, Argentina, and Brazil (Froeschner 1960; Zucchi et al. 1993; Gallo et al. 2002). Damage is the same as described for *C. bergi*. In Brazil, Waquil et al. (2003) reported the occurrence of individual feeding on roots of *Sorghum bicolor* L.

Pangaeus bilineatus (Say) (Figs. 21.38 and 21.40b)

Within the genus, *P. bilineatus* can be recognized by the presence of three or more submarginal setigerous punctures on each jugum and the absence of two key characters used to diagnose other species of *Pangaeus*: ventral surface of posterior femora without tubercles and subapical line of the pronotum without punctures (Froeschner 1960). *Pangaeus bilineatus* lives in the soil and feeds on the roots or underground pods. Females lay single eggs in the soil near plant structures, where newly hatched immatures can rapidly access the roots (Smith and Pitts 1974). Nymphs and adults of this species are considered pest of peanuts in the USA (Chapin and Thomas 2003; Chapin et al. 2006) and have been recorded in other crops (Gould 1931; Otten 1956; Smith and Pitts 1974; Lis et al. 2000) (Table 21.2).

Fig. 21.38 *Pangaeus bilineatus*, habitus dorsal view (Photo CF Schwertner)

In spinach crops, *P. bilineatus* attack newly germinated seeds (Gould 1931; Smith and Pitts 1974), and sucking the stem results in plant wilting and dying (Smith and Pitts 1974).

In peanut, feeding of *P. bilineatus* in the kernels reduces the flavor and the oil quality besides the increment of seed contamination by aflatoxin (Chapin et al. 2006). The damage of nymphs and adults of this species is called "pitting," which is recognized by light yellow to dark brown lesions as a result of this burrower-bug feeding. According to Chapin et al. (2004), 98 % of all aflatoxin contamination in peanut seeds was related to *P. bilineatus* feeding. Overwintering adults were present in conservation tillage of peanut field before planting, and two to three subsequent generations were observed (Chapin and Thomas 2003). These authors demonstrated that peanuts strip-tilled into corn or wheat residue had greater *P. bilineatus* populations and kernel feeding than conventional tillage or strip tillage into rye residue. The control of this species has been carried out by chemical methods, besides conventional tillage practices and crop rotation system (Smith and Pitts 1974; Chapin and Thomas 2003).

In peanut crop areas, Highland and Lummus (1986) used white light traps and captured numerous individuals of *P. bilineatus* (Say). According to these authors, peaks of adult catches can be closely related to field infestations, indicating that light traps are useful tools in an integrated pest management program for this species.

Although the species had been pointed as a vector of certain plant virus disease in the USA (Sailer 1958 in McPherson and Mohlenbrock 1976), there is no detailed account about plant-virus-insect interaction.

Secondary Species

Tominotus communis (Uhler) (Figs. 21.39 and 21.40c)

T. communis is distributed in Nicaragua, British West Indies, Cuba, Haiti, the Dominican Republic, Puerto Rico, the Bahamas, and the USA. The genus *Tominotus* Mulsant and Rey is distributed from the USA to Argentina, but knowledge about the species is scarce. *T. communis* is widespread in Central and North America (Fig. 21.40c) and can be recognized by the narrower scutellar apex and polished dorsum (Froeschner 1960).

This species was recorded in peanut crops associated with a *P. bilineatus* population in Texas. In these areas, nymphs and adults of *T. communis* caused decrease of seed quality, deforming the kernels and turning them badly flavored (Smith and Pitts 1974).

21.4.7 Subfamily Sehirinae

Species of the subfamily Sehirinae are easily recognized among Neotropical Cydnidae by the absence of the row of setigerous punctures in the lateral margin of pronotum and the shape of the tarsal segment II distinctly narrower than I and III. Froeschner (1960) also called attention to the presence of narrow, creamy white margins of the pronotum, corium, and abdomen (Fig. 21.6), a unique pattern among cydnids in the Neotropical region.

Fig. 21.39 *Tominotus communis*, habitus dorsal view (Photo CF Schwertner)

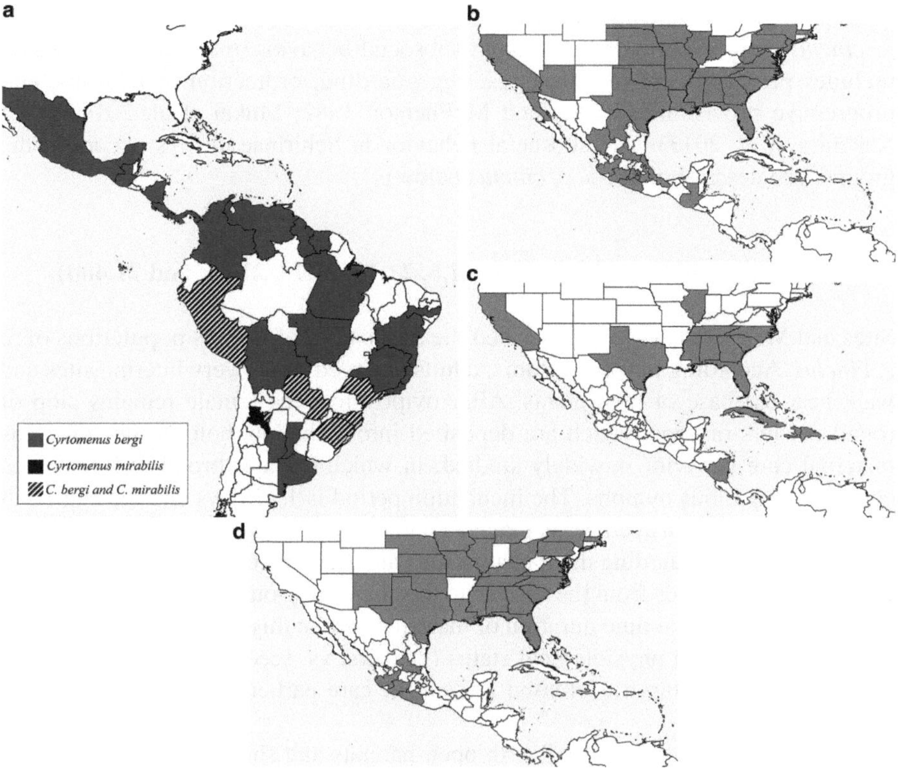

Fig. 21.40 Distribution maps of Cydninae (**a**, **b**, and **c**) and Sehirinae (**d**) in the Neotropical region. (**a**) *Cyrtomenus bergi* and *C. mirabilis*; (**b**) *Pangaeus bilineatus*; (**c**) *Tominotus communis*; (**d**) *Sehirus cinctus cinctus*

Sehirinae is predominantly Palearctic (Lis 1999a, 2002); only the genus *Sehirus* Amyot and Serville has been found in the Western Hemisphere (Table 21.1). The single endemic New World species, with three recognized subspecies, range from southern Canada to Mexico (Froeschner 1960). *S. cinctus cinctus* is well distributed in Mexico (Fig. 21.40d), reaching the Neotropical region.

Although most of the cydnids seem to live close to the soil or belowground, individuals of Sehirinae are commonly collected in aboveground parts of host plants. Nymphs and adults of *Sehirus* are often caught in various herbs, especially mint species (Lamiaceae). McDonald (1968) reared *S. cinctus albonotatus* Dallas on *Stachys palustris* L. (Lamiaceae), while Rider (2012) found this species in *Physostegia parviflora* Nutt. (Lamiaceae). Besides the phytophagous habit of *Sehirus* spp., individuals of this species were found feeding on the beebread discarded from honeybee colonies. According to Van Engelsdorp and Donovall (2009), several adults of *S. cinctus* were found associated with the beebread, probably feeding on a mixture of pollen and honey stored in the colony cells. These authors also observed *S. cinctus* nectaring on *Physostegia virginiana* (L.) Bentham (Lamiaceae).

Like the related group Parastrachiidae, burrower bugs of the Sehirinae (i.e., *S. cinctus* and *Adomerus* spp.) present subsocial behavior (maternal care), which includes production of trophic eggs, egg guarding, protection of nymphs, and progressive provisioning (Sites and McPherson 1982; Mukai et al. 2010, 2012; Nakahira et al. 2013). The subsocial behavior in Sehirinae always occurs in the ground (see description for *S. c. cinctus* below).

***Sehirus cinctus cinctus* Palisot (Figs. 21.6, 21.10, 21.13, 21.21, and 21.40d)**

Sites and McPherson (1982) described the life history of Illinois populations of *S. c. cinctus*. According to these authors, adults emerged from overwintering sites and walk near the base of host plants. After oviposition, the female remains atop of round clusters of eggs, which are deposited into a shallow hole in the soil. This maternal care behavior is widely studied, in which females brood and provision eggs and gregarious nymphs. The incubation period is 10.2 days and the egg-adult period is 53.2 days, in average. Females remain with the eggs and early instar larvae for several weeks, guarding them against ants and other predators and feeding them by carrying mint seeds from the surface to the chamber (Southwood and Hine 1950; McDonald 1968). The time duration of maternal care in this species can be affected by maternal age and physiological status (i.e., first vs. second reproduction), with older females showing a decreased amount of care earlier than younger females (Kligh 1997).

Populations of *S. c. cinctus* live in open habitats and show aggregate behavior, feeding on leaves or fallen seeds of mint and other species of plants (Sites and McPherson 1982; Van Engelsdorp and Donovall 2009) (Table 21.2). Although *S. c. cinctus* has been related to some economic important plants (*Prunus* spp., *Helianthus* spp.) (Table 21.2), this species is not considered an agricultural pest in the Neotropical region.

21.5 Concluding Remarks

The cydnids are a diverse and widely distributed group of bugs and in the Neotropical region represent an important part of the true bug fauna. Remarkable features include the unique burrower habit and morphological adaptations for digging, subsocial behavior, important fossil records, and damage to cultivated plants.

Because to the burrowing habits and small size of most of the species, people unnoticed its presence, limiting taxonomic characterization, host plants records, as well as damage and symptoms to cultivated plants.

More comprehensive investigations must be conducted about the biology, ecology, and behavior of cydnids. It is essential to establish methodologies to study this group of heteropterans in natural habitats and under laboratory conditions. Furthermore, the influence of environmental conditions on field populations, dispersal, and seasonal adaptations may also be useful for improving integrated management.

Although the taxonomy of the Cydnidae is considered well understood (Schuh and Slater 1995), still much work on the Neotropical taxa needs to be done. For some species, several reports have been mistaken [e.g., occurrence and damage attributed to *S. castanea* are actually due to the attack of *S. carvalhoi*, Lis et al. (2000)], and a correct identification is essential to define and delineate studies about each species, as well as control and management measures. In some cases, the taxonomy at the species level still awaits revision (e.g., *Cyrtomenus bergi* and *C. mirabilis*), and it is probable that new species will be found (*Amnestus* spp.) (Mayorga and Cervantes 2001, 2005). In addition to the species already reported as economically important, other cydnids might become agricultural pests, and monitoring natural populations of these insects may help to understand such potential.

Finally, studies on the evolution of these bugs are still lacking, and a phylogenetic classification of the group awaits better resolution. Investigations on the phylogenetic relationships of the Neotropical taxa will be crucial to test current classification and evolutionary hypothesis, helping to improve knowledge on this diverse group of true bugs.

References

Agrawal AF, Brodie ED, Brown J (2001) Parent-offspring coadaptation and the dual genetic control of maternal care. Science 292:1710–1712

Agrawal AF, Brown JM, Brodie ED III (2004) On the social structure of offspring rearing in the burrower bug, *Sehirus cinctus* (Hemiptera: Cydnidae). Behav Ecol Sociobiol 57:139–148

Andrade AC, Puzzi D (1953) Experiências com inseticidas orgânicos para controlar o "percevejo castanho" (*Scaptocoris castaneus*) em cana-de-açúcar. O Biológico 19:187–189

Arias B, Bellotti AC (1985) Aspectos ecológicos de manejo de *Cyrtomenus bergi* F. Chinche de la viruela en el cultivo de la Yuca, *Manihot esculenta* C. Rev Colomb Entomol 11:42–44

Barberena MF, Bellotti AC (1998) Parasitismo de dos razas de nemátodos *Heterorhabditis bacteriophora* sobre chinche *Cyrtomenus bergi* (Hemiptera: Cydnidae) en el laboratorio. Rev Colomb Entomol 24:7–11

Becker M (1967) Estudos sobre a família Scaptocorinae na região neotropical (Hemiptera: Cydnidae). Arq Zool 15:291–325

Becker M (1996) Uma nova espécie de percevejo-castanho (Heteroptera: Cydnidae: Scaptocorinae). Ann Soc Entomol Bras 25:95–102

Becker M, Galileo MHM (1982) A genitália de macho em cinco gêneros neotropicais da subfamília Cydninae (Heteroptera: Cydnidae). Rev Bras Biol 42:21–30

Bellotti AC, Riis L (1994) Cassava cyanogenic potential and resistance to pests and diseases. Acta Hortic 375:141–151

Bellotti AC, Smith L, Lapointe S (1999) Recent advances in cassava pest management. Annu Rev Entomol 44:343–370

Blatchley WS (1926) Heteroptera or true bugs of Eastern North America with special reference to the faunas of Indiana and Florida. Nature Publishing Company, Indiana

Brewer M (1972) *Scaptocoris castaneus* Perty, chinche dañina a raíces de alfafa (Hemiptera-Cydnidae). Idia 194:27–28

Brisolla AD, Furtado EL, Cardim DF, Kawasoto OS (1985) Ocorrência do percevejo castanho – *Scaptocoris castaneus* Perty, 1930 – em bananal na região litorânea do Estado de São Paulo. O Biológico 51:135–137

Calcedo AM, Bellotti AC (1994) Evaluación del potencial del nemátodo entomógeno *Steinernema carpocapsae* w. (Rhabditida: Steinernematidae) para el control de *Cyrtomenus bergi* F. (Hemiptera: Cydnidae) en condiciones de laboratorio. Rev Colomb Entomol 20:241–246

Centro Internacional de Agricultura Tropical CIAT (1981) Cassava program annual report for 1980. Cali, Colombia

Centro Internacional de Agricultura Tropical CIAT (1989) Cassava program annual report. Centro Internacional de Agricultura Tropical CIAT, Cali

Chapin JW, Thomas JS (2003) Burrower bugs (Heteroptera: Cydnidae) in peanut: seasonal species abundance, tillage effects, grade reduction effects, insecticide efficacy, and management. J Econ Entomol 96:1142–1152

Chapin JW, Dorner JW, Thomas JS (2004) Association of a burrower bug (Heteroptera: Cydnidae) with aflatoxin contamination of peanut kernels. J Entomol Sci 39:71–83

Chapin JW, Sanders TH, Dean LO et al (2006) Effect of feeding by a burrower bug, *Pangaeus bilineatus* (Say) (Heteroptera: Cydnidae), on peanut flavor and oil quality. J Entomol Sci 41:33–39

Cividanes FJ, Silveira S, Machado PS (1981) Flutuação populacional de cidnídeos coletados em regiões canavieiras de São Paulo. Científica 9:241–247

Cokl A, Virant-Doberlet M (2003) Communication with substrate-borne signals in small plant-dwelling insects. Annu Rev Entomol 48:29–50

Cokl A, Nardi C, Bento JMS, Hirose E, Panizzi AR (2006) Transmission of stridulatory signals of the burrower bugs, *Scaptocoris castanea* and *Scaptocoris carvalhoi* (Heteroptera: Cydnidae) through the soil and soybean. Physiol Entomol 31:371–381

Cole CL (1988) Stratification and survival of diapausing burrower bugs. Southwest Entomol 13:243–246

Costa Lima AM (1940) Insetos do Brasil, 2, Hemipteros. Série Didática 3, Escola Nacional de Agricultura, Rio de Janeiro, RJ, Brasil

Dolling WR (1981) A rationalized classification of the burrower bugs (Cydnidae). Syst Entomol 6:61–67

Drašlar K, Gogala M (1976) Structure of stridulatory organs of insects from fam. Cydnidae (Heteroptera). Biol Vestn 24:175–200

Eger JE (2008) A new genus and three new species of burrowing bugs (Hemiptera: Heteroptera: Cydnidae: Amnestinae). Proc Entomol Soc Wash 110:940–947

Filippi-Tsukamoto L, Nomakuchi S, Kuki K, Tojo S (1995) Adaptiveness of parental care in *Parastrachia japonensis* (Hemiptera: Cydnidae). Ann Entomol Soc Am 88:374–383

Filippi-Tsukamoto L, Baba N, Inadomi K, Yanagi T, Hironaka M, Nomakuchi S (2008) Pre- and post-hatch trophic egg production in the subsocial burrower bug, *Canthophorus niveimarginatus* (Heteroptera: Cydnidae). Die Naturwissenschaften 96:201–211

Froeschner RC (1941) Contributions to a synopsis of the Hemiptera of Missouri. Pt. 1, Scutelleridae, Podopidae, Pentatomidae, Cydnidae, Thyreocoridae. Am Midl Nat 26:122–146

Froeschner RC (1960) Cydnidae of the western hemisphere. Proc U S Nat Mus 111:337–680

Froeschner RC (1975) Three new species of burrowing bugs found in association with ants in Brazil (Hemiptera: Cydnidae). J Kansas Entomol Soc 48:105–110

Froeschner RC (1981) Heteroptera or true bugs of Ecuador: a partial catalog. Smithson Contrib Zool 322:1–147

Froeschner RC (1988) Family Cydnidae Billberg, 1820, burrowing bugs. In: Henry TJ, Froeschner RC (eds) Catalog of the Heteroptera, or true bugs of Canada and the continental United States, 1st edn. E. J. Brill, Leiden, pp 119–129

Gallo D, Nakano O, Silveira-Neto S, Carvalho RPL, Baptista GC, Berti Fo E, Parra JRP, Zucchi RA, Alves SB, Vendramim JD, Marchini LC, Lopes JRS, Omoto C (2002) Entomologia agrícola. Fealq, Piracicaba

García CA, Bellotti AC (1980) Estudio preliminar de la biología y morfología de *Cyrtomenus bergi* F. Nueva plaga de la yuca. Rev Colomb Entomol 6:55–61

Gogala M (1984) Vibration producing structures and songs of terrestrial Heteroptera as systematic character. Biol Vestn 32:19–36

Gogala M, Čokl A, Drašlar K, Blaževič A (1974) Substrate-borne sound communication in Cydnidae (Heteroptera). J Comp Physiol 94:25–31

Gould CS (1931) *Pangaeus uhleri*, a pest of spinach. J Econ Entomol 24:484–486

Grazia J, Schwertner CF, Silva EJE (2004) Arranjos taxonônicos e nomenclaturais em Scaptocorini (Hemiptera: Cydnidae, Cephalocteinae). Neotrop Entomol 33:511–512

Grazia J, Schuh RT, Wheeler WC (2008) Phylogenetic relationships of family groups in Pentatomoidea based on morphology and DNA sequences (Insecta: Heteroptera). Cladistics 24:932–976

Grimaldi D, Engel MS (2005) Evolution of the insects. Cambridge University Press, Cambridge, MA

Hart CA (1919) The Pentatomoidea of Illinois with keys to the nearctic genera. Ill Nat Hist Surv Bull 13:157–223

Hayward KJ (1943) Memoria annual del año 1942. Rev Ind Agric 33:66–84

Herrera MG (1988) Reconocimiento y manejo de la chinche subterránea Cyrtomenus bergi Froeschner., en cultivos de "Cebolla de Rama" en Pereira. Instituto Colombiano Agropecuario (ICA), Pereira, Colombia

Highland BH, Lummus PF (1986) Use of light traps to monitor flight activity of the burrowing bug, *Pangaeus bilineatus* (Hemiptera: Cydnidae), and associated field infestations in peanuts. J Econ Entomol 79:523–526

Jaramillo J, Borgemeister C, Ebssa L, Gaigl A, Tobón R, Zimmermann G (2005) Effect of combined applications of *Metarhizium anisopliae* (Metsch.) Sorokin (Deuteromycotina: Hyphomycetes) strain CIAT 224 and different dosages of imidacloprid on the subterranean burrower bug *Cyrtomenus bergi* Froeschner (Hemiptera: Cydnidae). Biol Control 34:12–20

Klight SL (1997) Factors influencing maternal behaviour in a burrower bug, *Sehirus cinctus* (Heteroptera: Cydnidae). Anim Behav 53:105–112

Lacerda JI (1983) Dano causados au dendê (*Elais guineesis*) por ação do *Cyrtomenus bergi* (Froeschner, 1960) (Hemiptera: Cydnidae). Rev Floresta 14:59–60

Laman TG (1996) The impact of seed harvesting ant (*Pheidole* sp nov.) on *Ficus* establishment in the canopy. Biotropica 28:777–781

Lis JA (1996) Taxonomy and phylogeny of the genus *Cydnus* F., 1803 (Hemiptera: Heteroptera: Cydnidae). Ann Soc Entomol Fr 32:403–409

Lis JA (1998) *Amnestus raunoi* n. sp. from Iran, the first Old World representative of the New World subfamily Amnestinae (Hemiptera : Heteroptera: Cydnidae). Entomol Fen 9:5–7

Lis JA (1999a) Burrower bugs of the Old World – a catalogue (Hemiptera: Heteroptera: Cydnidae). Genus 10:165–249

Lis JA (1999b) Taxonomy and phylogeny of Cephalocteinae with a reference to their historical biogeography (Hemiptera: Heteroptera: Cydnidae). Pol Pis Entomol 68:111–131

Lis JA (2002) Burrower bugs described after the Old World catalogue of the family (Hemiptera: Heteroptera: Cydnidae). Pol Pis Entomol 71:7–17

Lis JA (2010) Coxal combs in the Cydnidae *sensu lato* and three other related "cydnoid" families – Parastrachiidae, Thaumastellidae, Thyreocoridae (Hemiptera: Heteroptera): Functional, taxonomic, and phylogenetic significance. Zootaxa 2476:53–64

Lis JA, Hohol-Kilinkiewicz A (2001) Abdominal trichobothrial pattern and its taxonomic and phylogenetic significance in Cephalocteinae (Hemiptera: Heteroptera: Cydnidae). Ann Zool 51:293–297

Lis JA, Pluot-Sigwalt D (2002) Nymphal and adult cephalic chaetotaxy of the Cydnidae (Hemiptera: Heteroptera), and its adaptive, taxonomic and phylogenetic significance. Eur J Entomol 99:99–109

Lis JA, Becker M, Schaefer CW (2000) Burrower bugs (Cydnidae). In: Schaefer CW, Panizzi AR (eds) Heteroptera of economic importance. CRC Press, Boca Raton, pp 405–419

Marco AM, Coscarón MDC (2011) Burrower bugs described after the Cydnidae of the Western–Hemisphere catalogue 1960 (Hemiptera: Heteroptera: Cydnidae). Zootaxa 2766:57–63

Matias FI, Sampaio MV, Coelho L, Grazia J (2011) Occurrence of *Scaptocoris castanea* Perty (Hemiptera: Cydnidae) damaging *Azadirachta indica* (Meliaceae) seedlings in Brazil. Neotrop Entomol 40:288–289

Mayorga MC (2002) Revision generica de la familia Cydnidae (Hemiptera-Heteroptera) en Mexico, con un listado de las especies conocidas. An Inst Biol Univ Nac Auton Mex Zool 73:57–192

Mayorga CM, Brailovsky H (2012) Four new species of *Lattinestus* (Hemiptera: Heteroptera: Cydnidae: Amnestinae). Rev Mex Biodiv 83:1013–1024

Mayorga MC, Cervantes PL (2001) Life cycle and description of a new species of *Amnestus* Dallas (Hemiptera-Heteroptera: Cydnidae) associated with the fruit of several species of *Ficus* (Moraceae) in Mexico. J N Y Entomol Soc 109:392–402

Mayorga CM, Cervantes LP (2005) Description of six new species of *Amnestus* Dallas (Hemiptera: Heteroptera: Cydnidae) from Mexico. J N Y Entomol Soc 113:159–173

Mayorga CM, Cervantes LP (2009) Two new species of *Amnestus* from Guatemala, with new records for some other Guatemalan burrower bugs (Hemiptera: Heteroptera: Cydnidae). Zootaxa 2311:19–37

Mayorga MC, Cervantes-Peredo L, Lis JA (2012) *Amnestus raunoi* J.A. Lis, the only representative of the Amnestinae in the Old World: a synonym of *Amnestus pusillus* Uhler (Hemiptera: Heteroptera: Cydnidae). Zootaxa 3445:67–68

Mcdonald FJD (1968) Some observations on *Sehirus cinctus* (Palisot de Beauvois) (Heteroptera: Cydnidae). Can J Zool 46:855–858

McPherson JE, Mohlenbrock RH (1976) A list of the Scutelleroidea of the La Rue-Pine Hills Ecological Area with notes on biology. Gr Lakes Entomol 9:125–169

Moreira C (1923) Insetos nocivos aos aroozaes e seu controle. Alm Agric Bras 1923:193–194

Moser JC (1963) Contents and structure of *Atta texana* nest in summer. Ann Entomol Soc Am 56:286–291

Mukai H, Hironaka M, Baba N, Yanagi T, Inadomi K, Filippi L, Nomakuchi S (2010) Maternal-care behaviour in *Adomerus variegatus* (Hemiptera: Cydnidae). Can Entomol 142:52–56

Mukai H, Hironaka M, Tojo S, Nomakuchi S (2012) Maternal vibration induces synchronous hatching in a subsocial burrower bug. Anim Behav 84:1443–1448

Nakahira T, Kudo S (2008) Maternal care in the burrower bug *Adomerus triguttulus*: defensive behavior. J Insect Behav 21:306–316

Nakahira T, Tanaka KD, Kudo SI (2013) Maternal provisioning and possible joint breeding in the burrower bug *Adomerus triguttulus* (Heteroptera: Cydnidae). Entomol Sci 16:151–161

Nakano O, Romano FCB, Pessini MM (eds) (2001) Pragas de solo. ESALQ/USP, Piracicaba

Nardi C (2005) Percevejos castanhos (Hemiptera, Cydnidae, Scaptocoris): aspectos morfológicos, ecológicos e comportamentais. Dissertação, Universidade de São Paulo-Escola Superior de Agricultura Luiz de Queiroz, Piracicaba, SP, Brasil

Nardi C, Fernandes PM, Rodrigues OD, Bento JMS (2007) Flutuação populacional e distribuição vertical de *Scaptocoris carvalhoi* Becker (Hemiptera: Cydnidae) em área de pastagem. Neotrop Entomol 36:107–111

Nardi C, Fernandes PM, Bento JMS (2008) Wing polymorphism and dispersal of *Scaptocoris carvalhoi* (Hemiptera: Cydnidae). Ann Entomol Soc Am 101:551–557

Oliveira LJ, Malaguido AB (2004) Flutuação populacional dos percevejos castanhos da raiz, *Scaptocoris castanea* Perty (Hemiptera: Cydnidae), no perfil do solo em áreas produtoras de soja nas regiões centro-oeste e sudeste do Brasil. Neotrop Entomol 33:283–291

Oliveira LJ, Malaguido AB, Nunes J Jr, Corso IC, DeAngelis S, Farias LC, Hoffmann-Campo CB, Lantmann A (2000) Percevejos castanhos da raiz em sistemas de produção de soja. Embrapa-Soja, Londrina

Oliveira EDM, Pasini A, Fonseca ICB (2003) Association of the soil bug *Atarsocoris* sp. (Hemiptera: Cydnidae) with the weed *Senecio brasiliensis* Less. Neotrop Entomol 32:155–157

Oliveira LJ, Roggia S, Salvadori JR, Ávila CJ, Fernandes PM, Oliveira CM (2013) Insetos que atacam raízes e nódulos da soja. In: Hoffman-Campo CB, Correia-Ferreira BS, Moscardi F (eds) Soja: manejo integrado de insetos e outros artrópodes-praga. Embrapa, Brasília, pp 75–144

Otten E (1956) Heteroptera, Wanzen, Halbflüger. In: Blunck H (ed) Tierische Schädlinge an Nutzpflanzen, Lieferung. Heteroptera, Homoptera I, 3 edn. Parey, Berlin, Germany, pp 1–149

Pereira MF, Peres RM, Borges RS (2012) Population of *Scaptocoris castanea* Perty (Hemiptera: Cydnidae) in a crop-livestock integration system. Neotrop Entomol 41:409–413

Pluot-Sigwalt D (2008) A pair of basi-abdominal sex pheromone glands in the male of some burrower bugs (Hemiptera: Heteroptera: Cydnidae). Acta Entomol Mus Nat Pragae 48:511–522

Pluot-Sigwalt D, Lis JA (2008) Morphology of the spermatheca in the Cydnidae (Hemiptera : Heteroptera): Bearing of its diversity on classification and phylogeny. Eur J Entomol 105:279–312

Puzzi D, Andrade AC (1957) O "percevejo castanho" – *Scaptocoris castaneus* (Perty) – no Estado de São Paulo. O Biológico 23:157–163

Rider DA (2012) The Heteroptera (Hemiptera) of North Dakota I: Pentatomomorpha: Pentatomoidea. Gr Lakes Entomol 45:312–380

Riis L (1997) *Behaviour and population growth of the burrower bug, Cyrtomenus bergi Froeschner: Effects of host plants and abiotic factors*. Department of Ecology and Molecular Biology Royal Veterinary and Agricultural University, Copenhagen

Riis L, Esbjerg P (1998) Movement, distribution, and survival of *Cyrtomenus bergi* (Hemiptera: Cydnidae) within the soil profile in experimentally simulated horizontal and vertical soil water gradients. Environ Entomol 27:1175–1181

Riis L, Bellotti AC, Bonierbale M, O'Brien GM (2003) Cyanogenic potential in cassava and its influence on a generalist insect herbivore *Cyrtomenus bergi* (Hemiptera: Cydnidae). J Econ Entomol 96:1905–1914

Riis L, Belotti AC, Arias B (2005a) Bionomics and population growth statistics of *Cyrtomenus bergi* (Hemiptera: Cydnidae) on different host plants. Fla Entomol 88:1–10

Riis L, Esbjerg P, Bellotti AC (2005) Influence of temperature and soil moisture on some population growth parameters of *Cyrtomenus bergi* (Hemiptera: Cydnidae). Fla Entomol 88:11–22

Roth LM (1961) A study of the odoriferous gland of *Scaptocoris divergens* (Hemiptera: Cydnidae). Ann Entomol Soc Am 54:900–911

Sales O Jr, Medeiros MO (2001) Percevejos castanhos da raiz em pastagens. In: Anais da Reunião Sul-Brasileira Sobre Pragas De Solo. Embrapa Soja, Londrina

Salvadori JR (1999) Pragas de solo em culturas graníferas. Comunicado Técnico 26. EMBRAPA, Passo Fundo

Sánchez D, Bellotti AC (1997) Patogenicidad de hongos hyphomicetes sobre *Cyrtomenus bergi* F. (Hemiptera: Cydnidae) chinche subterráneo de la yuca. Rev Colomb Entomol 23:31–37

Schaefer CW (1968) The homologies of the female genitalia in the Pentatomoidea (Hemiptera-Heteroptera). J N Y Entomol Soc 76:87–91

Schaefer CW (1981) The land bugs (Hemiptera: Heteroptera) and their adaptive zones. Rostria 33:67–83

Schaefer CW (2009) *Prosorrhyncha* (Heteroptera and Coleorrhyncha). In: Resh VH, Cardé RT (eds) Encyclopedia of insects, 2nd edn. Academic, Amsterdam, pp 839–855

Schuh RT, Slater JA (1995) True bugs of the world (Hemiptera: Heteroptera). Classification and natural history. Cornell University Press, Ithaca

Shcherbakov DE, Popov YA (2002) Superorder Cimicidea Laicharting, 1781. Order Hemiptera Linne, 1758. The bugs, cicadas, plant lice, scale insects, etc. In: Rasnitsyn AP, Quicke DLJ (eds) History of insects. Kluwer, Dordrecht, pp 143–157

Silva AGD, Gonçalves CR, Galvão DM, Gonçalves AJL, Gomes J, Silva MN, Simoni L (1968) Quarto catálogo dos insetos que vivem nas plantas do Brasil: seus parasitos e predadores. Ministério da Agricultura, Rio de Janeiro

Sites RW, McPherson JE (1982) Life history and laboratory rearing of *Sehirus cinctus cinctus* (Hemiptera: Cydnidae), with descriptions of immature stages. Ann Entomol Soc Am 75:210–215

Smith JW, Pitts JT (1974) Pest status of *Pangaeus bilineatus* attacking peanuts in Texas. J Econ Entomol 67:111–113

Soares IMF, Della Lucia TMC, Santos AA, Nascimento IC, Delabie JHC (2006) Caracterização de ninhos e tamanho de colônia de *Acromyrmex rugosus* (F. Smith) (Hymenoptera, Formicidae, Attini) em restingas de Ilhéus, BA, Brasil. Rev Bras Entomol 50:128–130

Sousa CR (2002) Composição populacional e mobilidade no solo do percevejo castanho Atarsocoris brachiariae (Hemiptera: Cydnidae). Dissertação, Universidade Federal de Goiás, GO, Brasil

Southwood TRE, Hine DJ (1950) Further notes on the biology of *Sehirus bicolor* (L.) (Hemiptera, Cydnidae). Entomol Mon Mag 86:299–301

Souza Fo MF, Ramiro ZA, Raga A, Thomazini MJ (1997) Ocorrência de Scaptocoris castanea e Atarsocoris brachiariae (Heteroptera: Cydnidae) na cultura da soja no Estado de São Paulo. In: Anais da Reunião de Pesquisa de Soja da Região Central do Brasil. Instituto Biológico, Jaboticabal

Stoner D (1920) The Scutelleroidea of Iowa. Univ Iowa Stud Nat Hist 8:1–140

Tallamy DW, Schaefer CW (1997) Maternal care in the Hemiptera: ancestry, alternatives, and current adaptive value. In: Choe JC, Crespi BJ (eds) The evolution of social behavior in insects and arachnids. Cambridge University Press, Cambridge, pp 94–115

Tauber JM, Tauber AC, Masaki S (eds) (1986) Seasonal adaptations of insects. Oxford University Press, New York

Thomas DB (1988) Fossil Cydnidae (Heteroptera) from the Oligo-Miocene amber of Chiapas, Mexico. J N Y Entomol Soc 96:26–29

Thomas DB (1994) Fossil Cydnidae (Heteroptera) in the Dominican amber. J N Y Entomol Soc 102:303–309

Timonin MI (1958) *Scaptocoris talpa* Champ. on the roots of banana and other plants. Plant Prot Bull FAO 6:1–2

Timonin MI (1961a) The interaction of plant, pathogen, and *Scaptocoris talpa* Champ. Can J Bot 39:695–703

Timonin MI (1961b) Effect of volatile constituents of *Scaptocoris talpa* Champ. on the growth of soil fungi. Plant Soil 14:323–334

Torre-Bueno JR (1915) Heteroptera in beach drift. Entomol News 26:274–279

Van Duzee EP (1894) A list of the hemiptera of buffalo and vicinity. Buffalo Soc Nat Sci Bull 5:167–204

Van Engelsdorp D, Donovall LR (2009) New plant host record and first record of the burrower bug *Sehirus cinctus* (Palisot de Beauvois) (Hemiptera: Cydnidae) associating with honey bees, *Apis mellifera* L. (Hymenoptera: Apidae). Proc Entomol Soc Wash 111:903–906

Vivan LM, Nardi C, Grazia J, Bento JM (2013) Description of the immatures of *Scaptocoris carvalhoi* Becker (Hemiptera: Cydnidae). Neotrop Entomol 42:288–292

Waquil JM, Viana PA, Cruz I (2003) Manejo de pragas na cultura do sorgo. Circular Técnica 22, Embrapa, MAPA, Brasília

Willis ER, Roth LM (1962) Soil and moisture relations of *Scaptocoris divergens* Froeschner (Hemiptera: Cydnidae). Ann Entomol Soc Am 55:21–33

Yao YZ, Cai WZ, Ren D (2007) The first fossil Cydnidae (Hemiptera: Pentatomoidea) from the late Mesozoic of China. Zootaxa 1388:59–68

Zucchi RA, Silveira Neto S, Nakano O (1993) Guia de identificação de pragas agrícolas. FEALQ, Piracicaba

Chapter 22
Stink Bugs (Pentatomidae)

Jocélia Grazia, Antônio R. Panizzi, Caroline Greve, Cristiano F. Schwertner, Luiz A. Campos, Thereza de A. Garbelotto, and José Antônio Marin Fernandes

Abstract The Pentatomidae, representing the fourth largest family within Heteroptera, are one of the most diverse groups with about 800 genera and more than 4,700 species in the world. In the Neotropics, about 230 genera and 1,400 species are included in seven subfamilies of the world's nine subfamilies; four subfamilies are exclusively of the Neotropics. In this chapter, for each subfamily, a diagnosis, an overview of the classification, and information on life history, ecology, and economic importance are given. Comprehensive keys and diagnosis to the subfamilies, tribes, and genera for the Neotropical Region, including Mexico, Central and South America, and the West Indies, are also given.

J. Grazia (✉) • C. Greve • L.A. Campos
Departamento de Zoologia, Instituto de Biociências, Universidade Federal do Rio Grande do Sul (UFRGS), Av. Bento Gonçalves 9500, prédio 43435, Bairro Agronomia,
Porto Alegre, RS 91501-970, Brazil
e-mail: jocelia@ufrgs.br; carolinegreve@gmail.com; luiz.campos@ufrgs.br

A.R. Panizzi
Laboratório de Entomologia, Embrapa Trigo, Caixa Postal 3081,
Passo Fundo, RS 99001-970, Brazil
e-mail: antonio.panizzi@embrapa.br

C.F. Schwertner
Departamento de Ciências Biológicas, Instituto de Ciências Ambientais,
Químicas e Farmacêuticas, Universidade Federal de São Paulo (UNIFESP),
Rua São Nicolau, 210 Centro, Diadema, RS 09913-030, Brazil
e-mail: schwertner@unifesp.br

T. de A. Garbelotto
Programa de Pós-Graduação em Biologia Animal, Instituto de Biociências,
Universidade Federal do Rio Grande do Sul (UFRGS), Av. Bento Gonçalves 9500,
prédio 43435, Bairro Agronomia, Porto Alegre, RS 91501-970, Brazil
e-mail: therezagarbelotto@hotmail.com

J.A.M. Fernandes
Instituto de Ciências Biológicas, Universidade Federal do Pará,
Rua Augusto Corrêa n°1, Belém, PA 66075-110, Brazil
e-mail: joseamf@ufpa.br

A.R. Panizzi, J. Grazia (eds.), *True Bugs (Heteroptera) of the Neotropics*,
Entomology in Focus 2, DOI 10.1007/978-94-017-9861-7_22

22.1 Introduction

Pentatomidae is the fourth most numerous family within Heteroptera including more than 4,700 species in more than 800 genera (Table 22.1). Pentatomids are represented in all zoogeographical regions although the tropical and subtropical faunas are more diverse; they are commonly known as stink bugs. Rolston and McDonald (1979), Rolston (1981, 1984), and Rolston et al. (1980) provided keys to the identification of subfamilies, tribes, and genera of the Western Hemisphere. The subfamilies Asopinae, Cyrtocorinae, Discocephalinae, Edessinae, and Pentatominae are widely represented in the Neotropical Region, while Podopinae and Stirotarsinae only have one species each. The subfamilies Cyrtocorinae, Discocephalinae, and Stirotarsinae are exclusively of the Neotropics. The Edessinae, with the exception of two species found in the Southern United States of America (McPherson 1982), occur in the Neotropics.

The asopines have predacious habits and the incrassate rostrum is the morphological feature common to the group as an adaptation to predatory behavior. They are found in all zoogeographical regions.

Table 22.1 Number of genera and species in the subfamilies and tribes of Pentatomidae of the Neotropical region

	World		Neotropics	
Subfamily/tribe	Genera	Species	Genera	Species
Asopinae	63	357	23	100
Cyrtocorinae	4	11	4	11
Discocephalinae	76	315	76	315
Discocephalini	43	192	43	192
Ochlerini	33	123	33	123
Edessinae	7	>300	7	>300
Pentatominae	380	>2,803	116	>667
Carpocorini	106	>800	53	>223
Catacanthini	7	53	5	32
Halyini	82	>390	2	13
Mecideini	1	17	1	4
Menidini	27	>100	2	23
Nezarini	20	150	3	>70
Pentatomini	65	>900	30	>200
Piezodorini	4	>25	1	1
Procleticini	11	33	11	23
Sciocorini	11	>150	1	3
Strachiini	14	>100	1	13
Unplaced	32	92	6	53
Podopinae	62	255	1	1
Stirotarsinae	1	1	1	1
Total			228	>1,400

Cyrtocorinae is a small subfamily, cryptically colored with the tree bark substratum where they live; the scutellum with a median, long, and tough spine is a morphological feature found in all species. They are distributed from Mexico to Argentina.

Discocephalinae includes two tribes, Discocephalini and Ochlerini, with more than 300 species almost all medium sized and dark colored. The morphology of the immatures and the biology within the group are poorly known. Exceptions are found in Discocephalini, the genus *Antiteuchus* Dallas which is a natural host of cacao (*Theobroma cacao*) in the Amazon region and shows maternal care habits, and Ochlerini, the genera *Lincus* Stål and *Macropygium* Spinola of which some species are important vectors of *Phytomonas* spp. (Trypanosomatidae) in palms in South America.

Edessinae includes the most colored and largest stink bugs, with great diversity in the Amazon region, and more than 300 known species. They have in common the strong development of the metasternum. For more than a century, the group was treated as a tribe of Pentatominae but studies in the last two decades raised the taxon to subfamily rank and increased the number of genera to seven.

The subfamily Pentatominae is the most diverse within the Pentatomidae and includes species recorded in all the zoogeographical regions in numbers higher than 3,000 distributed in about 600 genera. About 1/4 of this diversity is present in the Neotropics. They are all phytophagous and many species are economically important as pests of cultivated crops. Pentatominae is not a monophyletic group and does not have a definition based on phylogenetic studies. Different authors consider the classification within Pentatominae as having from 8 to more than 40 tribes. Following Rider's tentative classification, 11 tribes are registered in the Neotropics: Catacanthini Atkinson, Carpocorini Mulsant & Rey, Halyini Amyot & Serville, Mecideini Distant, Menidini Atkinson, Nezarini Atkinson, Pentatomini Leach, Piezodorini Atkinson, Procleticini Pennington, Sciocorini Amyot & Serville, and Strachiini Mulsant & Rey.

The subfamily Podopinae has worldwide distribution but is poorly represented in the Neotropics, with only one species. The subfamily Stirotarsinae was recently erected to include *Stirotarsus abnormis* Bergroth, only known from the northern region of South America.

22.2 Subfamily Asopinae

22.2.1 Introduction

Species of the subfamily Asopinae are predaceous, a secondary condition among pentatomids. More than 350 species and 60 genera are known worldwide (Thomas 1992, 1994). Asopines are easily recognized by the incrassate rostrum (Fig. 22.1), and some species are useful as biocontrol agents (De Clercq 2000, 2008).

Fig. 22.1 *Apateticus lineolatus* (Herrich-Schaeffer), lateral showing detail of the incrassate rostrum (Photo by CF Schwertner)

22.2.2 General Characteristics and Diagnosis

The general morphology of the species is similar to other groups of pentatomids. Diagnostic characteristics include modifications of the head associated to the feeding habits: rostrum strongly incrassate, insertion of the labium very close to the base of labrum, and posterior margins of buccula merged (Gross 1975; Gapud 1991). The group also shared the presence of the genital plates of the male pygophore and presence of thecal shield (Gapud 1991; Gapon and Konstantinov 2006).

22.2.3 General Biology and Ecology

Asopines predate insects and other small organisms, usually slow and soft-bodied individuals, i.e., larvae of Lepidoptera, Diptera, and Hymenoptera (Schuh and Slater 1995; De Clercq 2008). Most species seems to be generalists, but some may have a more restricted diet (i.e., oligophagous) or may be related to specific habitats (De Clercq 2000, 2008). Nymphs and adults are observed sucking plant fluids and water, probably to acquire the basic needs during periods of prey scarcity. However, asopines do not develop unless they have access to nutrients of animal origin, required from the 2nd instar on. Early nymphs (2nd and 3rd instars) tend to attack and feed together, while the late instars and adults tend to attack alone; nymphs and adults may show cannibalism when preys are not available (De Clercq 2000, 2008).

Life cycle of the asopines is similar to that of the phytophagous pentatomids. Adults emerge and start to feed and reproduce using different preys; some species may live up to 3 months. They use visual, chemical, and tactile cues to locate and recognize their prey. Mating behavior is similar to phytophagous species, with males starting courtship by antennal movement. Eggs are laid in masses, on different parts of plants used by adults or on nearby structures. The number of eggs per clutch varies between species, between individuals, or between clutches of the same female. There are five nymphal instars. First instars are not predatory and only need moisture for survival. From the 2nd instar on, they need to feed on prey for survival. Initial nymphs are gregarious and tend to attack in groups; gregarious behavior decreases with the development, and nymphs disperse in search for prey. Incubation period, nymphal development time, prereproductive period, and longevity are quite variable; influence of food and abiotic conditions (temperature and humidity) impact the expression of these traits. Egg parasitoids (Hymenoptera: Scelionidae) are recorded as natural enemies.

Although highlighted as potential agents for pest control, the use of asopines in biological control programs is still limited. Only about 10 % of the species have been the focus of such studies (De Clercq 2008). In the Neotropical Region, species of the genera *Alcaeorrhynchus* Bergroth, *Apateticus* Dallas, *Euthyrhynchus* Dallas, *Brontocoris* Thomas, *Perillus* Stål, *Podisus* Herrich-Schäffer, *Stiretrus* Laporte, *Supputius* Distant, and *Tylospilus* Stål have been the focus of recent studies. Two species, *Podisus maculiventris* (Say) and *Podisus nigrispinus* (Dallas), show great potential in controlling populations of phytophagous insects (Saini 1994; De Clercq 2000). *P. maculiventris* is the only asopine commercially available for augmentative biological control (De Clerq 2008).

22.2.4 Classification and Diversity

In the Neotropical Region, 23 genera and 100 species are known (Table 22.2). Taxonomy of the subfamily Asopinae was revised by Thomas (1992), which includes identification keys to all genera and species of the Western Hemisphere.

Table 22.2 Checklist of genera and species of Neotropical Asopinae (distribution as available in the literature)

Species	Distribution
Alcaeorrhynchus grandis (Dallas)	USA, MEX, TRI, CO, BR, UR, ARG, ECU, BOL, CU, JAM, CR, HON, VEN
Alcaeorrhynchus phymatophorus (P. de B.)	PR, HT, DOR, CU, USA (Florida Keys)
Andrallus spinidens (F.)	USA, MEX, NIC, CR, CU, HON, DOR
Apateticus lineolatus (Herrich-Schaeffer)	USA, MEX, CR, ESAL, PAN, HON, CO, VEN, NIC, ECU,
Apateticus marginiventris (Stål)	USA, MEX
Apoecilus invarius (Walker)	USA, MEX
Brontocoris nigrolimbatus (Spinola)	CHI
Brontocoris tabidus (Signoret)	CHI, PAR, ARG, BR
Colpothyreus flavolineatus (Blanchard)	BOL
Comperocoris roehneri (Philippi)	CHI, ARG
Coryzorhaphis leucocephala Spinola	BR
Coryzorhaphis carneolus Erichson	GUY, BR, PER, BOL, ECU, FG
Coryzorhaphis cruciata Stål	MEX, BEL, HON, PAN, CO, VEN, BR

(continued)

Table 22.2 (continued)

Species	Distribution
Coryzorhaphis superba Breddin	PER, BOL, ECU
Coryzorhaphis egeri Thomas	MEX, CR, HON
Coryzorhaphis dollingi Thomas	ECU
Discocera caynnensis Laporte	BR, PER, SUR, FG
Discocera coccínea (F.)	BR, PER, COL, ECU, SUR, PAR, BOL
Euthyrhynchus floridanus (L.)	USA, MEX, GUA, ESAL, HON, CR, PAN, ECU, BR, BOL, NIC
Heteroscelis servillei Laporte	BR, PER, BOL, ARG, PAR, VEN, GUY, FG, SUR, TRI
Heteroscelis lepida (Stål)	USA, MEX, GUA, PAN, CO, VEN, FG, ECU, HON
Heteroscelis bimaculata (Walker)	BOL, BR, PER, PAN
Heteroscelis robustus Thomas	BR, PAR
Marmessulus nigricornis Bergroth	BR, ARG, URU
Marmessulus brasilianus Schouteden	BR
Oplomus dichrous (Burmeister)	USA, MEX, HON
Oplomus catena (Drury)	BR, FG, ARG, URU, PAR, ECU
Oplomus cruentus (Burmeister)	ARG, BR, PER, URU
Oplomus mutabilis Stål	MEX, GUA, HON
Oplomus ebulinus (Herrich-Schaeffer)	BR, PER, BOL, PAR, ECU
Oplomus salamandra (Burmeister)	COL, VEN, SUR, ECU, BOL, PER, BR
Oplomus marginalis (Westwood)	BR, PAN, COL, ECU, PER, BOL, ARG, VEN, SUR, PAR
Oplomus mundus Stål	USA, MEX, GUA, CR, PAN, NIC, HON
Oplomus pulcher Dallas	MEX, CR, PAN, HON
Oplomus annotatus Uhler	CU
Oplomus punctatus Montandon	ARG
Oplomus pulchiventris Horvath	ARG
Ornithossoma rivieri Kormilev	ARG
Parajalla sanguineosignata (Spinola)	CHI
Perillus confluens (Herrich-Schaeffer)	USA, MEX, GUA, ELS, CR, HON
Perillus circumcintus Stål	USA, CAN, MEX
Perillus bioculatus (F.)	USA, CAN, MEX
Perillus splendidus (Uhler)	USA, MEX
Podisus aenescens (Stål)	MEX, GUA, HON, CR, PAN, CO, ECU, BOL, PER, ARG, BR, FG, PAR
Podisus sculptus Distant	CR, PAN, PER, BOL
Podisus formosus Costa Leite	BR
Podisus mucronatus Uhler	USA (Florida), CU, PUR, DOR
Podisus falcatus Distant	GUA, HON, CR
Podisus semialbus (Walker)	BR, VEN
Podisus curvispina Bergroth	BR

(continued)

Table 22.2 (continued)

Species	Distribution
Podisus cornutus (Dallas)	CO, BR
Podisus gundlachii (Guérin-Ménéville)	CU
Podisus borinquensis Barber	PUR
Podisus congrex (Stål)	MEX, GUA, CR, ESAL, PAN, VEN, ECU, HON
Podisus crassimargo (Stål)	BR, ECU, COL, VEN, CHI, PAN
Podisus volxemi Distant	BR, ECU
Podisus tinctus (Dallas)	COL, ECU
Podisus subferrugineus Barber & Bruner	CU, JAM
Podisus sagitta (F.)	USA (Texas), MEX, GUA, ESAL, HON, NIC, CR, PAN, VEN, JAM, GRE, DRE, HAI, DOR, PUR, TRI, Curacao, CU, BAH
Podisus nigrispinus (Dallas)	BR, PAN, CR, BOL, ARG, PER, ECU, SUR, COL, GUY, PAR
Podisus distinctus (Stål)	ECU, BOL, COL, BR, VEN, FG, PAR, ARG, PER
Podisus maculiventris (Say)	USA, MEX, CAN, HAI, DRE, BAH
Podisus neglectus Westwood	USA, MEX
Podisus affinis Distant	MEX, GUA, HON, PAN
Podisus mexicanus Distant	MEX
Podisus nigriventris Distant	MEX, GUA, NIC, HON, CR, PAN, COL
Podisus insignis Distant	GUA
Podisus trucidatus Thomas	MEX, HON, CR, PAN
Podisus ventralis (Dallas)	VEN, BR, ECU, PER, ARG
Podisus sordidus (Stål)	ECU (Galapagos)
Podisus rostralis (Stål)	BR, BOL
Podisus pallipes (Dallas)	VEN, BR, ECU, ARG
Podisus mactans Thomas	ECU
Podisus serieventris Uhler	CAN, USA, MEX
Stiretrus quinquepunctatus Germar	HAI, DOR
Stiretrus decemguttatus (Lepeletier & Serville)	BRA, SUR, BOL, PAR, URU, ARG, PER
Stiretrus decastigmus (Herrich-Schaeffer)	ARG, BRA, BOL, PAR
Stiretrus erythrocephalus (Lep. & Serville)	ARG, BRA, BOL, PAR, URU
Stiretrus anchorago (F.)	USA, MEX, GUA, ESAL, CR, HON, PAN, NIC
Stiretrus bifrenatus Breddin	COL, ECU, PER, BRA, URU
Stiretrus loratus Germar	BR, BOL
Stiretrus cinctellus Germar	BR, ARG
Supputius typicus Distant	MEX, CR, PAN, VEM, COL, ECU, BR, HON
Supputius cincticeps (Stål)	BR, VEN, PR, CHI, BOL, ARG
Supputius pulchricornis (Stål)	MEX
Tylospilus armatus Thomas	ARG
Tylospilus cloelia (Stål)	MEX, HON, PAN, COL, BOL, BR, PAR, ARG
Tylospilus chilensis (Spinola)	CHI, ARG

(continued)

Table 22.2 (continued)

Species	Distribution
Tylospilus nigrobinotatus (Berg)	ARG, BRA, URU
Tylospilus acutissimus Stål	USA, MEX, NIC, COL, ANT, DOR
Tylospilus megaspilus (Walker)	BR
Tylospilus distans Bergroth	BR, CHI
Tylospilus peruvianus (Horvath)	PERU, SUR, BR
Tynacantha marginata Dallas	BR, ARG, BOL, PER, COL, ECU, PAR, URU
Tynacantha splendes Distant	PAN, CR
Tyrannocoris rex Thomas	BR, VEN
Tyrannocoris nigriceps Thomas	BR, ARG
Tyrannocoris rideri Thomas	PUR
Tyrannocoris jole (Stål)	MEX, HON, CU, HAI, RDO
Zicrona americana Thomas	USA, MEX

Since then, studies about morphology, biology, and distribution records of some of the Neotropical species have been published (see Sects. 2.5 and 2.6).
Key to the genera of subfamily Asopinae with species included in biological control studies in the Neotropical Region (adapted from Thomas 1992). Number in superscript indicates Neotropical genera that are keyed in the same step (see list at the end of the key).

1. Scutellum greatly enlarged, covering most of the abdomen, apex surpassing coria in length, almost attaining apex of abdomen (Fig. 22.2) .... *Stiretrus* Laporte[1,2]
1′. Scutellum not covering most of the abdomen, usually shorter than coria (Figs. 22.3 and 22.4) 2
2. Apical part of the scutellum enlarged, equal in length or longer than the frenal margin of the scutellum, and about as wide as the corium (Figs. 22.3 and 22.5) .... 3[3]
2′. Apical part of the scutellum small, shorter than the frenal margin and much narrower than the corium (Fig. 22.4) 4
3. Frenal margin of the scutellum shorter than the postfrenal scutellum; protibia usually expanded.......... *Oplomus* Spinola (Fig. 22.5)
3′. Frenal margin of the scutellum longer than postfrenal scutellum; protibia always prismatic, never expanded.......... *Perillus* Stål (Fig. 22.3)
4. Scent gland with elevated peritreme, surrounded by evaporatoria (Fig. 22.6) 5
4′. Ostiole of scent gland without elevated peritreme, terminating in short or long sulcus not surrounded by evaporatoria (Fig. 22.7) .... *Alcaeorrhynchus* Bergroth[4] (Fig. 22.8)
5. Protibiae dilated, expanded.......... *Euthyrhynchus* Dallas[5]
5′. Protibiae not dilated.......... 6

6. Base of the abdomen with anteriorly directed tubercle or spinous process (Fig. 22.9)..7
6′. Base of the abdomen may be prominent but without forwardly directed tubercle or spinous process..................*Supputius* Distant (in part)[6] (Fig. 22.10)
7. Rostral segment II distinctly longer than III and IV combined ....*Supputius* Distant (in part)[7]
7′. Rostral segment II shorter or subequal in length to III and IV combined.........8[8]
8. Abdominal spine long, extending to mesocoxae; posterior tibiae terete; most or all of the dorsal surface of the head with pale colorless punctuation..................9
8′. Abdominal spine usually short, not surpassing metacoxae; posterior tibia usually sulcate; dorsal surface of the head densely punctuate.....................................10
9. Scent gland peritreme long, reaching more than half way to pleural edge; anterolateral pronotal margin dentate to spinulose............... *Brontocoris* Thomas
9′. Scent gland peritreme short, reaching less than halfway to the pleural edge; anterolateral pronotal margin rugulose, smooth (Fig. 22.11) ....*Tylospilus* Stål (Fig. 22.12)
10. Inner margins of jugae distinctly convergent (Fig. 22.13); males with glandular patches of silk hairs on abdominal venter...........*Apateticus* Dallas[9] (Fig. 22.14)
10′. Inner margins of jugae parallel or concave but never convergent (Fig. 22.15); males without glandular patches of silk hairs on abdominal venter ....*Podisus* Herrich-Schäffer (Figs. 22.4, 22.16, and 22.17).

1, *Colpothyreus* Stål; 2, *Discocera* Laporte; 3, *Heteroscelis* Laporte, *Coryzorhaphis* Spinola, *Comperocoris* Stål, *Marmessulus* Bergroth; 4, *Parajalla* Distant, *Rhacognathus* Fieber, *Zicrona* Amyot & Serville; 5, *Eocanthecona* Bergroth; 6, *Ornithosoma* Kormilev, *Andrallus* Bergroth; 7, *Picromerus* Amyot & Serville; 8, *Thynacantha* Dallas, *Tyrannocoris* Thomas; 9, *Apoecilus* Stål

22.2.5 Main Species

De Clercq (2000) reviewed the most studied species of asopines (main and secondary) in a world basis and discussed information available for each one. In this and the next section, we present a brief characterization for some of the common Neotropical species that have been the focus of recent studies.

Fig. 22.2 *Stiretrus anchorago* (F.), dorsal view (Courtesy of G. Fauske)

Fig. 22.3 *Perillus bioculatus* (F.), dorsal view (Courtesy of G. Fauske)

Fig. 22.4 *Podisus maculiventris* (Say), dorsal view (Photo by CF Schwertner)

Brontocoris tabidus **(Signoret)**

The genus *Brontocoris* included only two species, and *B. tabidus* differs from *B. nigrolimbatus* (Spinola) by the anterolateral pronotal margin crenulate and connexivum with a black spot in each anterolateral and posterolateral angles (Thomas 1992; Rider 2014); in *B. nigrolimbatus* the pronotal margin is spinulose and the connexivum uniformly pale yellow. Considering a promising biological agent in Brazil (Zanuncio et al. 1994; De Clercq 2000), it has been the focus of several studies (e.g., Oliveira et al. 2005; Zanuncio et al. 2006; Lemos et al. 2009, 2010; DeMenezes et al. 2013; Pires et al. 2011).

Perillus bioculatus **(F.) (Fig. 22.3)**

Among the species included in the genus, *P. bioculatus* can be recognized by the following suite of characters: presence of an ante-apical spine or tubercle in the profemur, pronotum tumid and without lunate black mark on the anterior half, abdominal spiracles enclosed within black area, and dorsal punctuation widely spaced (Thomas 1992). The two-spotted stink bug is one of the main predators of the Colorado potato beetle (*Leptinotarsa decemlineata*, Chrysomelidae); the biology and other aspects of

Fig. 22.5 (**a**) *Oplomus catena* (Drury), dorsal view (Photo by CF Schwertner). (**b**) *Oplomus dichrous* (Burmeister), dorsal view (Courtesy of G. Fauske)

Fig. 22.6 *Supputius cincticeps* (Stål), ventral view showing detail of the evaporatorium (Photo by CF Schwertner)

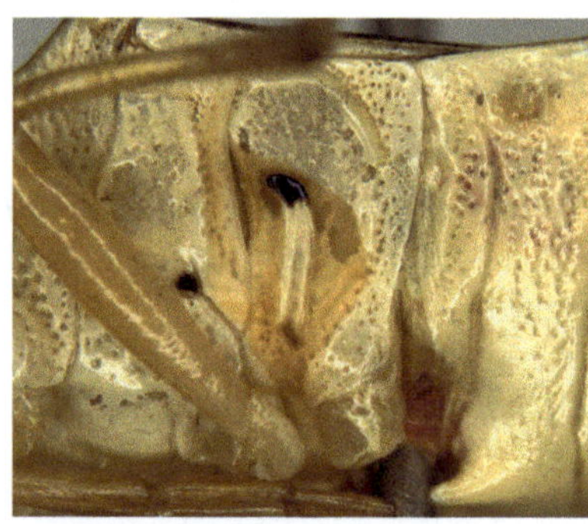

Fig. 22.7 *Alcaeorrhynchus grandis* (Dallas), ventral view showing detail of the evaporatorium (Photo by CF Schwertner)

Fig. 22.8 *Alcaeorrhynchus grandis* (Dallas), dorsal view (Photo by CF Schwertner)

Fig. 22.9 *Apateticus lineolatus* (Herrich-Schaeffer), ventral view (Photo by CF Schwertner)

Fig. 22.10 *Supputius cincticeps* (Stål), dorsal view (Photo by CF Schwertner)

Fig. 22.11 *Tylospilus distans* Bergroth, head in dorsal view (Photo by CF Schwertner)

this asopine have been investigated by several authors (reviewed in De Clercq 2000; see also Adams 2000, 2001; Wittmeyer and Coudron 2001; Coudron et al. 2002; Coudron and Kim 2004; Matlock 2005; Greenstone et al. 2010).

Podisus maculiventris (Say) (Figs. 22.4 and 22.15)

Thomas (1992) recognized several groups of species within *Podisus*, including the *maculiventris* group of species. Species included in this group can be recognized by the dorsal surface of pronotum without *calli*; anterolateral margins of the pronotum

a b

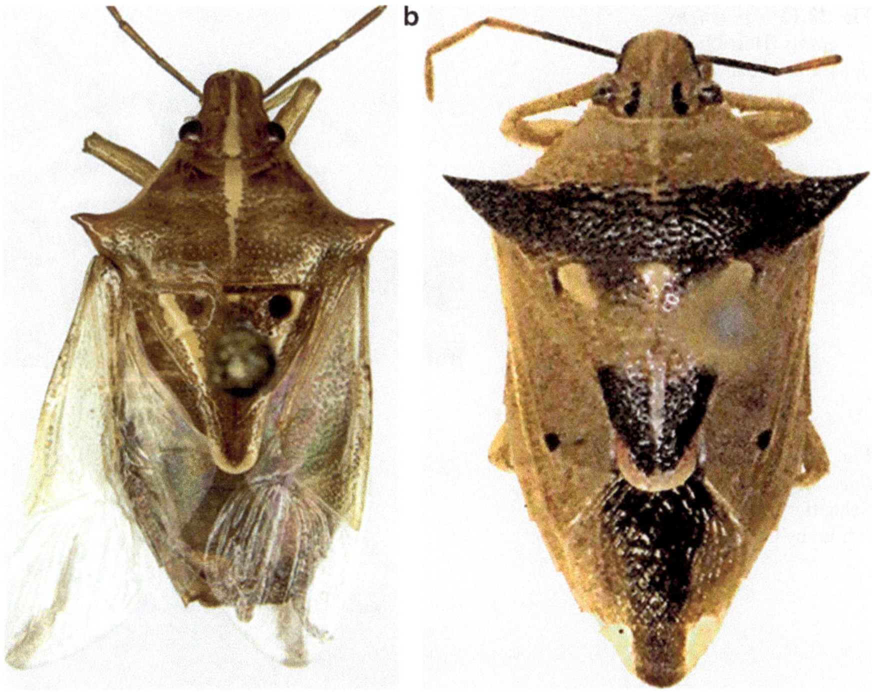

Fig. 22.12 (**a**) *Tylospilus distans* Bergroth, dorsal view (Photo by CF Schwertner). (**b**) *Tylospilus acutissimus* Stål, dorsal view (Courtesy of G. Fauske)

not inflated and dentate, crenulate, or rugulose; and humeral angles simple and directed laterad, not elongately produced (Thomas 1992). Within the group, *P. maculiventris* is the only species with humeral angles laterally produced in acuminate angles (Fig. 22.4) and with the abdominal spine reaching between metacoxae. It is the most common species in North America, also known as spined soldier bug, and extensive literature about its biology, role, and potential use in biological control programs has been published (De Clercq 2000, 2008).

Podisus nigrispinus (Dallas) (Fig. 22.16)

Species included in the *sagitta* group, which is similar to the *maculiventris* group (see above diagnosis of *P. maculiventris*), however, have the humeral angles distinctly bifid or emarginate (Fig. 22.16) (Thomas 1992). Within the group, *P. nigrispinus* can be recognized by the last rostral segment darker than the others and by the anterolateral margins of pronotum paler than the pronotal disk. One of the most common species in the Neotropical Region, *P. nigrispinus* has been found attacking several pest insects (De Clercq 2000). Several studies has been carried out regarding

Fig. 22.13 *Apateticus lineolatus* (Herrich-Schaeffer), head in dorsal view (Photo by CF Schwertner)

Fig. 22.14 *Apateticus lineolatus* (Herrich-Schaeffer), dorsal view (Photo by CF Schwertner)

Fig. 22.15 *Podisus maculiventris* (Say), head in dorsal view (Photo by CF Schwertner)

Fig. 22.16 *Podisus nigrispinus* (Dallas), dorsal view (Photo by CF Schwertner)

the biology, predation potential in different agroecosystems, and rearing methods of this species, also known as the Brazilian soldier spined bug (De Clercq 2000; Vivan et al. 2003; Lemos et al. 2003, De Medeiros et al. 2004; Sousa-Souto et al. 2006; Torres et al. 2006; Neves et al. 2009; Rodrigues et al. 2009; De Bortoli et al. 2011; Soares et al. 2011; Laumann et al. 2013; Vacari et al. 2013).

Fig. 22.17 *Podisus sagitta* (F.), dorsal view (Photo by CF Schwertner)

***Supputius cincticeps* (Stål) (Figs. 22.6 and 22.10)**

The genus *Supputius* include three species, and *S. cincticeps* can be recognized by rostral segment II as long as III and IV combined, apex of scutellum concolorous with the disk and abdominal tubercle strongly produced, reaching between metacoxae (Thomas 1992). Although not so common as *P. nigrispinus*, this species is also found attacking several pests in South America (De Clercq 2000; Zanuncio et al. 2004, 2005a, b; De Castro et al. 2013).

22.2.6 Secondary Species

The species included in this section are far less studied than the species included in the previous section, and little is known about them besides their distribution and morphological aspects. However, they represent common Neotropical asopines and potential control agents in biological control programs.

***Alcaeorrhynchus grandis* (Dallas) (Figs. 22.7 and 22.8)**

This is the most common and widespread species of *Alcaeorrhynchus*, the only one found in the continental Americas (Thomas 1992). The genus include only two species, which can be identified by the direction of the spined humeral angle: it is bent somewhat forward in *A. grandis* and directed laterad in *A. phymatophorus*. The species *A. grandis* is the largest Neotropical asopine and is associated to several cultivated ecosystems, attacking lepidopterous larvae (Malaguido and Panizzi 1998a; De Clercq 2000).

Apateticus lineolatus (Herrich-Schaeffer) (Figs. 22.1, 22.13, and 22.14)

The genus *Apateticus* include two species, and *A. lineolatus* can be recognized by the posterior angle of the pronotum without a tooth and the basal abdominal spine robust, protruding between metacoxae (Thomas 1992). There are few studies about this species (De Clercq 2000); in Mexico, *A. lineolatus* was found feeding in larvae of the beetles *Epilachna varivestis* (Coccinellidae) and *Leptinotarsa* spp. (Chrysomelidae), pests of important crops as beans and potatoes.

Euthyrhynchus floridanus (L.)

This genus is monotypic, despite the high color polymorphism found in *E. floridanus*; Thomas (1992) mentioned that North and South American populations represent distinct lineages (forms) of this species. A very conspicuous species, with red and metallic blue coloration, with extensive literature on bionomics under laboratory conditions is available (De Clercq 2000; Briceño 2014). However, only a few notes about its life history on the field are known (Avila-Núñez et al. 2009).

Oplomus dichrous (Burmeister) (Fig. 22.5b)

The genus *Oplomus* includes 12 species, and *O. dichrous* can be recognized by two unique characters: prosternum expanded between procoxae and juga longer than tylus, contiguous and deflexed anteriorly in males (Thomas 1992). A common species within it distribution range, it seems to be an important predator of *L. decemlineata* Mexico (De Clercq 2000), but laboratory and field tests showed that *O. dichrous* has little potential as control agent in cool climates.

Podisus spp.

Podisus sagitta (F.) (Fig. 22.17) and *P. distinctus* (Stål)

Both species belong to the *sagitta* group (Thomas 1992), together with *P. nigrispinus* (see diagnose for *P. nigrispinus* above). They can be separated by the coloration of the rostral segments: the last rostral segment darker than the three preceding segments in *P. sagitta* and the first rostral segment distinctly paler than the three succeeding segments in *P. distinctus*. Besides, both species have the anterolateral margin of pronotum concolor, which allows separating these species from *P.*

nigrispinus. The biology of *Podisus* spp. is similar, and *P. sagitta* and *P. distinctus* have been studied as control agents of different pests in North and South America (De Clercq 2000; Matos Neto et al. 2004).

Podisus rostralis (Stål)

This species is included in the *maculiventris* group, together with *P. maculiventris* (see diagnosis above) and another eight species of *Podisus* (Thomas 1992). The group is widely distributed in the American continent; however *P. rostralis* has a more restricted distribution (Table 22.2). It can be recognized among other species of *Podisus* by a unique set of characters: dorsum of the head densely punctuated, hemelytral membrane transparent with a mesial dark blotch, and basal rostral segment contrastingly pale to three dark succeeding segments. In Brazil, *P. rostralis* was found attacking pests of *Eucalyptus* spp. and information about its biology under laboratory conditions are available (Molina-Rugama et al. 1998; Matos Neto et al. 1999; Lemos et al. 2005).

Stiretrus spp.

The genus includes eight species; *S. anchorago* (F.) (Fig. 22.2) and *S. decemguttatus* (Lepeletier & Serville) can be recognized by the protibia distinctly dilated, anterolateral margins of the pronotum concavely sinuate to subrectilineal (never convex), and basal abdominal spine long, protruding between mesocoxae (Thomas 1992). The species can be separated by the length of the ostiole: *S. anchorago* has the ostiole long, reaching at least halfway to the pleural margin; in *S. decemguttatus* it is short, reaching one-third the distance to the pleural margin.

The distribution of both species does not overlap, and *S. anchorago* is the only species within the genus that occurs in the north of Panama (all the other species of *Stiretrus* are known from South America). More data on the bionomics and life history of *S. anchorago* are available (De Clercq 2000), but its effectiveness in agroecosystems still needs more attention. More recently, Paleari (2013) studied the development, the polymorphism, and some ecological aspects of *S. decemguttatus*.

Tylospilus acutissimus (Stål) (Fig. 22.12b)

This species can be recognized within the genus *Tylospilus* by humeral angle spinose, often slightly directed forward and without a post-apical tooth on the posterior edge; base of the scutellum with only indistinct calluses at basal angles, never elongated; ostiole extending halfway to the pleural margin; and basal abdominal spine reaching almost to the mesocoxae (Thomas 1992). Although considered a common species in some parts of the United States, this species has received minor attention and only scattered data about its biology is known (De Clercq 2000).

22.2.7 Remarks

The subfamily Asopinae includes a diverse group of species, with worldwide distribution and relatively well-known taxonomy (Thomas 1992, 1994). However, studies in a phylogenetic context are still lacking for the group. The monophyly of the subfamily and included genera was never questioned, but its position within Pentatomidae and the phylogenetic relationships within the subfamily are unresolved.

The unique feeding habits within the superfamily Pentatomoidea make the group an exciting subject of study in many distinct research areas such as evolutionary biology, comparative physiology, and crop protection.

Knowledge about the Neotropical asopines is mostly restricted to species related to agroecosystems. Details about species distribution, life history traits, comparative morphology, and molecular data are still scarce and will be crucial to understand the diversity and evolution of the group in the Neotropical Region.

22.3 Subfamily Cyrtocorinae

22.3.1 Introduction

This exclusively Neotropical taxon was revised by Packauskas and Schaefer (1998); it includes four genera and 11 species. Besides the record of *Cyrtocoris trigonus* (Germar) from California (Banks 1910; Horváth 1916; Brailovsky et al. 1988), Packauskas and Schaefer (1998) agreed with Henry and Froeschner (1988) that the lack of any subsequent discovery of *Cyrtocoris* White in the United States makes Bank's record suspect. According to Packauskas and Schaefer (1998), Kormilev (1955) appears to have been the first author to present evidence for raising Cyrtocorinae to family rank as distinct from the Pentatomidae allying Cyrtocoridae with Cydnidae.

22.3.2 General Characteristics and Diagnosis

Length 6–10 mm. Body black to brown in ground color, often covered with brown-tan and/or white scalelike setae, giving the bug a lighter brown color dorsally and ventrally. Ventral covering usually variegated with areas of deep brown and yellow-white. Body broadly rectangular in dorsal view, in lateral view convex above, and nearly flat below; trapezoidal to rectangular in posterior view. Head subquadrate, angles rounded in front of the eyes; mandibular plates developed, flattened, and depressed on either side of clypeus. Apex of clypeus at base of rostrum expanded laterally beyond bucculae. Bucculae extending past eyes. Antennae arising from below eyes; V segment longest, slightly swollen distally; II shortest (very short). Rostrum and antennae fitting into midventral sulcus on thoracic sterna. Pronotum broadly trapezoidal; humeri greatly expanded and directed downward anterolaterally. Scutellum broadly U-shaped, extending to end of the abdomen; most of corium and sometimes part of membrane exposed. Tarsi 2-segmented; 2nd segment 2–4 times as long as 1st. Metathoracic scent gland opening sustained by an auricle peritreme. Abdominal segments II–IV exposed dorsally, strongly extended laterally and angled ventrally. Each abdominal spiracle closer to preceding segment than to lateral edge of its own segment or to posterior segment; 2nd abdominal spiracle on membranous anterior of segment. Abdominal sterna III–VII laterally with one pair of trichobothria side by side, those on anterior sterna lateral to spiracle; trichobothria all of equal size.

22.3.3 General Biology and Ecology

Cyrtocorines are of minor economic importance and they are not commonly found. As a consequence little is known about their biology. *C. trigonus* were found feeding on stems of the weed plant arrowleaf sida, *Sida rhombifolia* L. (Malvaceae), in Londrina, Paraná State, Brazil. Despite egg deposition in the laboratory, nymphs did not develop on this plant. Nymphs and adults of *Cyrtocoris egeris* Packauskas & Schaefer were observed feeding on immature soybean pods in Argentina (Reconquista, Santa Fé Province). Under laboratory conditions, nymphs of neither species could be reared beyond the 3rd instar (Schaefer et al. 2005). Adult and nymphal *C. egeris* fed gregariously on the stems of *Acalypha diversifolia* Jacquin (Euphorbiaceae) and moved to the leaves when these became available; adults could also feed on *Piper* sp. (Piperaceae) (Brailovsky et al. 1988). *Cyrtocoris gibbus* (F.) is reported to feed on stems of other legume plants such as the trees *Acacia* sp. and "bracatinga" (*Mimosa scabrella* Bentham), in Rio de Janeiro (Costa Lima 1940; Silva et al. 1968). Adults of *C. trigonus* feed on *Syngonium podophyllum* Schott (Araceae), *Acalypha* sp. (Euphorbiaceae), *Piper* sp., and *Pothomorpha umbellatum* (L.) Miquel (Piperaceae) (Brailovsky et al. 1988, as *C. obtusus* Horváth).

Not many studies have been published focusing on Cyrtocorinae immatures. Brailovsky et al. (1988) described the egg and nymphs of *C. egeris* [as *Cyrtocoris trigonus* (Germar)] from Mexico, including notes on the biology of the species. Schaefer et al. (1998) described the last three instars of *C. egeris* from Ecuador. Bianchi et al. (2011) studied under SEM the egg and the 1st and 5th instars of *C. egeris*; their results corroborate the hypothesis of Schaefer et al. (1998) on the absence of spiracles on urosternite I of nymphs and the trichobothrial arrangement (1 + 1 trichobothria on 1st instar and 2 + 2 trichobothria on following instars) contrary to Brailovsky et al. (1988). Also, the placement of the trichobothria relative to the spiracle on different sterna, becoming more medial posteriorly, described for the later instars and adult of *C. egeris* in Schaefer et al. (1998) were confirmed by Bianchi et al. (2011). The presence of three irregular rows of aero-micropylar processes in the egg stage and the large amount of dorsal abdominal sclerotization in the nymphal stage, as stated by Schaefer et al. (1998), seem to be distinctive features of Cyrtocorinae, comparable to the degree of modification of the adult stage. On the other hand, barrel-shaped eggs and the pattern of the nymphal morphology are shared by Cyrtocorinae and other subfamilies of the Pentatomidae (Bianchi et al. 2011).

22.3.4 Classification and Diversity

Cyrtocorinae includes four genera, three of them monotypic and one with eight species: *Ceratozigum* Horváth, monotypic with *C. horridum* (Germar) (Fig. 22.18a, b) registered for Brazil, Panama, Peru, and Venezuela; *Cyphothyrea* Horváth, monotypic, with *C. erosa* Horváth, from Peru and Brazil; *Cyrtocoris* White, the most diverse genus, including eight species – *C. andicola* Horváth (Peru), *C. egeris* Packauskas & Schaefer (Mexico through Brazil), *C. gibbus* (F.) (widely distributed), *C. montanus* Horváth (Peru), *C. obtusus* Horváth (Brazil), *C. paraensis* Pirán (Southern Brazil), *C. simplex* Horváth (Brazil), and *C. trigonus* (Germar) (Mexico through Argentina) – and *Pseudocyrtocoris* Jensen-Haarup, monotypic with *P. laceratus* (Herrich-Schäffer) from Argentina and Brazil.

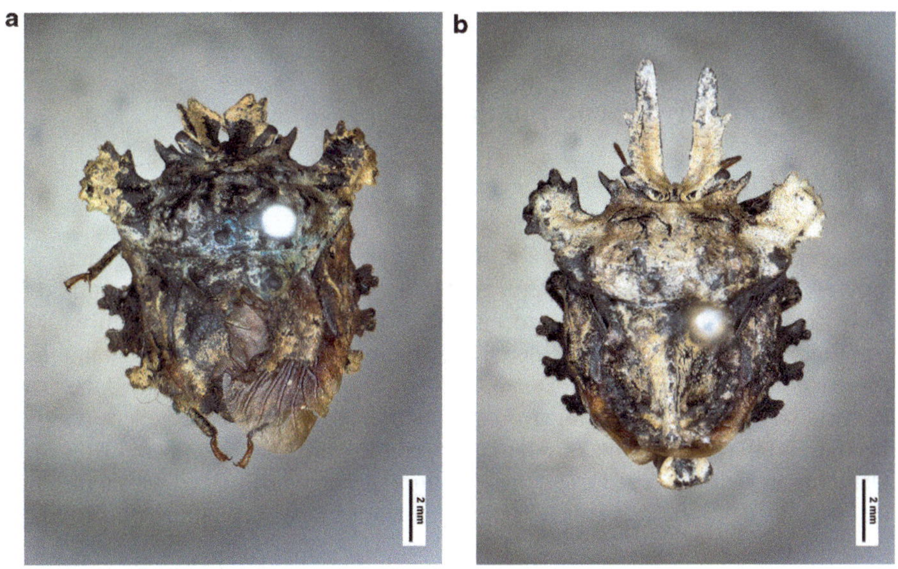

Fig. 22.18 *Ceratozigum horridum* (Germar): (**a**) female; (**b**) male (Photos by A Ferrari)

Packauskas and Schaefer (1998) provided detailed redescriptions of the taxa and keys to separate the genera and the *Cyrtocoris* species, beside the intraspecific variability found in this genus; they also discussed the peculiar sexual dimorphism found in *Ceratozigum horridum* (Germar). There are two forms of the head in the males themselves, one form resembling the female mandibular plates (greatly produced anteriorly) and another form (but in no female) where the mandibular plates extend far forward.

Kormilev (1955) appears to have been the first author to present evidence for raising Cyrtocorinae to family rank as distinct from the Pentatomidae allying Cyrtocoridae with Cydnidae. Kormilev's arguments persuaded Rolston and McDonald (1979) to raise Cyrtocorinae, and for the most part Cyrtocoridae has been recognized since. Gapud (1991) separated Cyrtocorinae (as a pentatomid subfamily) from the rest of Pentatomidae by the absence of a triangulin, 2nd valvifers with a distinct median fusion line, and male phallotheca relatively flexible. Packauskas and Schaefer (1998) considered the presence of a triangulin, 2nd valvifers completely fused, and a rigid phallotheca as apomorphies of Pentatomidae minus Cyrtocoridae. Gapud (1991) placed the Cyrtocorinae + Pentatomidae sensu stricto as the most apical taxa, "strongly separated from the rest of Pentatomoidea" by the loss of first valvulae, the absence of the gonangulum, the invagination and dilation on the spermathecal duct, the retention of membranous flaps of the 2nd valvulae, and the presence of an anteroposterior pair of basal sclerites on the spermathecal base. In Grazia et al. (2008) the morphological characters that consistently support the recognition of a broadly conceived Pentatomidae [Cyrtocorinae (Aphylinae + Pentatomidae sensu stricto)] are the loss of gonapophyses 8 and the first *rami*, gonapophyses 9 reduced and fused to gonocoxites 9, gonangulum absent, and the ductus receptaculi dilated and invaginated, forming three distinct walls.

22.4 Subfamily Discocephalinae

22.4.1 *Introduction*

Discocephalinae (Fig. 22.19) is a group of phytophagous species (77 genera; 307 species) restricted to the Neotropical Region. This subfamily was proposed by Fieber (1860) as the family "Discocephalida." It was treated as a subfamily for the first time by Stål (1868) and considered by Kirkaldy (1909) as a tribe of Pentatominae. More recently, Rolston and McDonald (1979) followed Stål and regarded the group as a subfamily, providing diagnosis and including 23 American genera removed from Halyini (Pentatominae). These 23 genera were later grouped in the tribe Ochlerini by Rolston (1981), and the remaining genera were placed within the nominal tribe Discocephalini. This tribe arrangement of Discocephalinae was followed by subsequent authors.

Fig. 22.19 Representative species of Discocephalinae; (**a**) *Discocephala marmorea* Laporte (Discocephalini); (**b**) *Ochlerus signoreti* Breddin (Ochlerini). Scale bars = 1 mm (Photos by LA Campos & TA Garbelotto)

22.4.2 General Characteristics and Diagnosis

This subfamily includes species usually with flattened body, small to medium sized, mottled brown, and black or fuscous. They can be diagnosed primarily by the insertion of the labium, usually arising on or posterior to the anterior limit of the eyes (Fig. 22.20), and the pairs of abdominal trichobothria usually laterad of the imaginary line connecting the spiracles (Fig. 22.21) – in other subfamilies the trichobothria are usually in line with spiracles (Rolston and McDonald 1979; Rolston 1981).

The two tribes of Discocephalinae, Discocephalini and Ochlerini, can be distinguished from one another by the general aspect, color, and the structure of the dorsal surface of the third tarsal segment of the hind legs of females, convex in Discocephalini and shallowly depressed or flattened in Ochlerini (Fig. 22.22) (Rolston 1981).

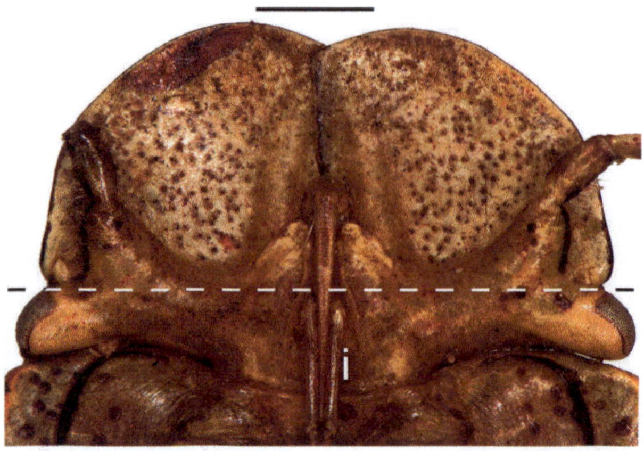

Fig. 22.20 Ventral view of the head of *Tetragonotum megacephalum* Ruckes, depicting the insertion of the 1st rostral segment (*i*) placed posterior to the anterior limit of eyes (*dashed line*). Scale bar = 1 mm (Photo by LA Campos & TA Garbelotto)

Fig. 22.21 Ventral view of part of the abdomen of *Opophylax extenebratus* Bergroth, depicting the trichobothria (*t*) placed laterad of imaginary line (*dashed line*) connecting spiracles (*s*). Scale bar = 1 mm (Photo by LA Campos & TA Garbelotto)

22.4.3 General Biology and Ecology

Discocephalinae are all phytophagous, considered to be phloem feeders (Waldkircher et al. 2004; Guerra et al. 2011). Some species of *Antiteuchus* Dallas (Discocephalini) are natural hosts of cacao (*Theobroma cacao*) in the Amazon region and in the Brazilian state of Bahia, and some species of *Lincus* Stål and *Macropygium* Spinola

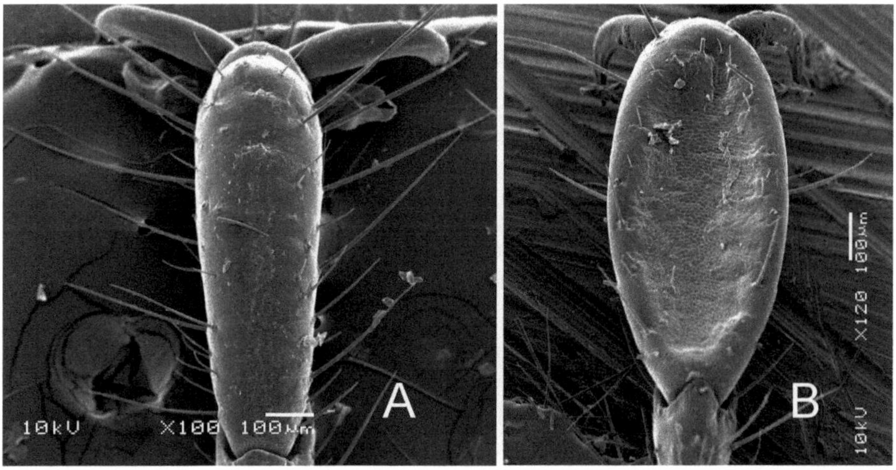

Fig. 22.22 Scanning electron micrographs, third tarsomere of female hind legs of representative species of Discocephalinae, dorsal; (**a**) *Antiteuchus tripterus* (F.); (**b**) *Schraderiellus cinctus* (Ruckes) (Photos by LA Campos & T Roell)

(Ochlerini) are important vectors of *Phytomonas staheli* (Trypanosomatidae) in palms in tropical South America (Costa Lima 1940; Desmier-De-Chenon 1984; Couturier and Kahn 1989; Llosa et al. 1990; Dollet et al. 1993).

The morphology of the immatures and the biology within the group is poorly known; a few studies mention immatures without describing them. Maternal care is recorded in Discocephalini for species of *Antiteuchus* (Eberhard 1975; Santos and Albuquerque 2001a, b) with female bugs protecting both egg masses and immatures; trophobiosis with ants is recorded for an unidentified species of *Lincus* (Stahel 1964) and for *Eurystethus* (*H.*) *microlobatus* Ruckes (Guerra et al. 2011). Field observations by LA Campos and TA Garbelotto indicate that parental care and trophobiosis with ants also occur in *Dinocoris rufitarsus* Ruckes and *Catulona pensa* Rolston, respectively.

22.4.4 Classification and Diversity

Systematics of Discocephalinae is yet in a very early stage. Although a first cladistic approach is available for Ochlerini (Fig. 22.23) (Campos and Grazia 2006), hypotheses of relationship among genera of Discocephalini are based on morphological interpretation by classic authors (e.g., Ruckes, Rolston), although lacking at all for several genera. The relationship among Discocephalinae and other pentatomid subfamilies is also obscure. Campos and Grazia (2006) recovered the monophyly of the subfamily sustained by two exclusive synapomorphies, first segment of labium reaching the prosternum and the dorsal surface of the male proctiger membranous

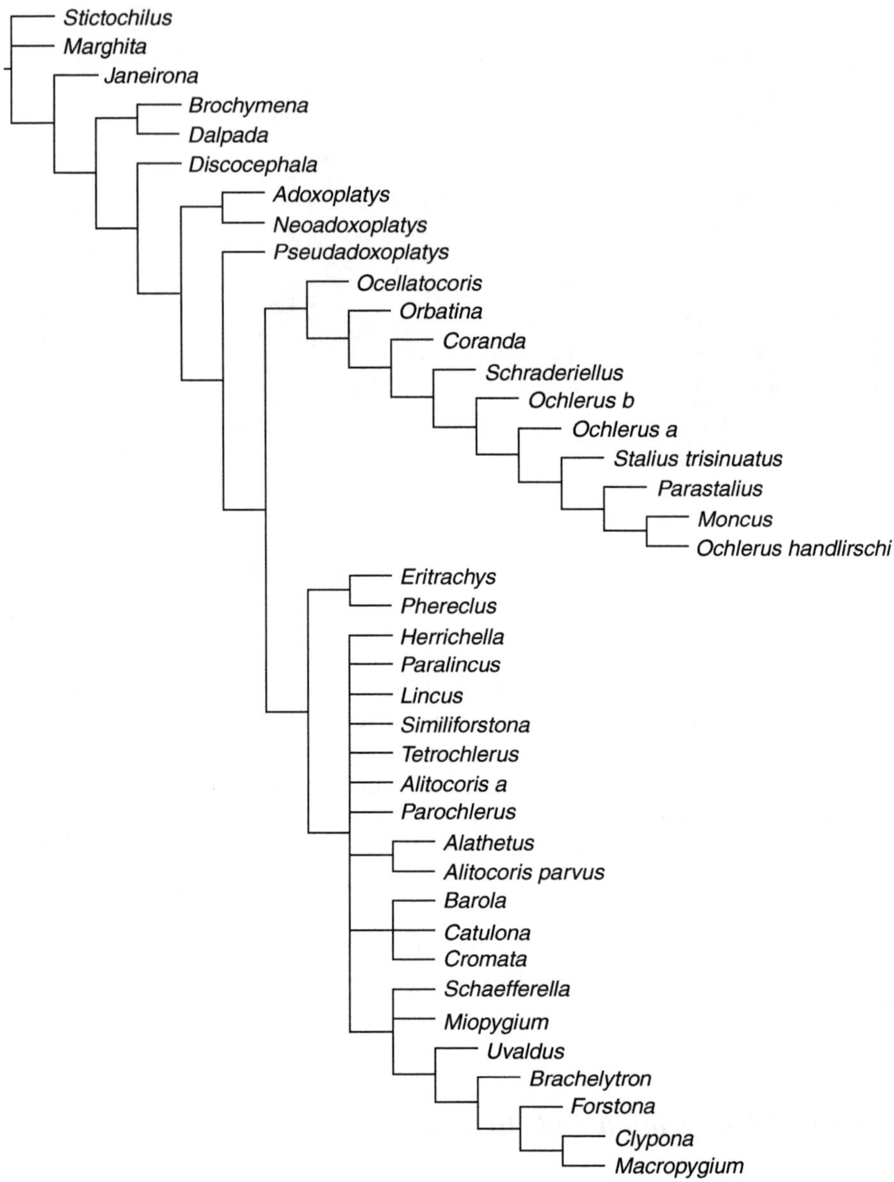

Fig. 22.23 Phylogeny of Ochlerini depicting the relationship between genera; strict consensus tree after successive weighting (Redrawn from Campos and Grazia (2006))

at basal third and one homoplastic synapomorphy, metasternum with median carina. But, as mentioned before, the tribe arrangement adopted for Discocephalinae (with two tribes, Discocephalini and Ochlerini) is the same since Rolston and McDonald (1979) and Rolston (1981).

Discocephalini

Discocephalini comprises 43 genera and 192 species (Table 22.3), including the monotypic fossil genus *Acantocephalonotum* Petrulevičius & Popov. Members of this tribe are brown, often mottled with black or shiny black (Figs. 22.19a and 22.24), although a formal diagnosis for this tribe has never been published. A phylogeny for

Table 22.3 Checklist of genera and species of Discocephalini (distribution as available in the literature)

Genera	Species
Abascantus Stål	*A. grandis* Becker – BRA
	A. lobatus Stål (type) – BRA
	A. pubescens Becker – PER
Ablaptus Stål	*A. amazonus* Stål (type) – BRA
	A. bolivianus Becker & Grazia – BOL
	A. brevirostrum Rolston – COL; CRI; PAN
	A. costaricensis Grazia & Zwetsch – CRI
	A. phoenix Grazia & Zwetsch – CRI
	A. simillimus Becker & Grazia – BOL
	A. tavakiliani Rolston – GUF
	A. varicornis (Walker) – BRA
Acanthocephalonotum Petrulevičius & Popov[a]	*A. martinsnetoi* Petrulevičius & Popov – ARG[a]
Acclivilamna Ruckes	*A. vicina* (Signoret) (type) – BRA
Agaclitus Stål	*A. australis* Becker & Grazia – BRA
	A. dromedarius Stål (type) – BOL; BRA; GUF; PER
	A. fallenii Stål – BRA?; GUF
Alcippus Stål	*A. reticulatus* (Stål) (type) – BRA
Allinocoris Ruckes	*A. nubilus* (Dallas) (type) – BRA
Alveostethus Ruckes	*A. latifrons* (Dallas) (type) – BRA
	A. politus (Signoret) – BRA; COL; PER; VEN
	A. pseudopolitus (Ruckes) – PRY
	A. regulosus (Fieber) – BRA
Antiteuchus Dallas	*A. amapensis* Fernandes & Grazia – BRA
	A. amplus (Walker) – BOL; BRA; ECU; GUF; PER

(continued)

Table 22.3 (continued)

Genera	Species
	A. bartletti Ruckes – GUY
	A. beckerae Fernandes & Grazia – BOL; BRA
	A. confinium Ruckes – ECU; PER
	A. costaricensis Ruckes – CRI; PAN
	A. cuspidatus Ruckes – COL; PAN
	A. doesburgi Fernandes & Grazia – BRA; SUR
	A. exiguus Fernandes & Grazia –BRA
	A. fulvescens Ruckes – PER
	A. geometricus Engleman – BRA
	A. graziae Engleman – BOL; BRA; COL; ECU; PER
	A. guianensis Ruckes – BRA; GUF; GUY
	A. innocens Engleman – BLZ; GTM; MEX
	A. kerzhneri Rider –PER
	A. ledeburi Fernandes & Grazia – BRA
	A. macraspis (Petry) – BRA; COL; CRI; GUF; PAN; PER; SUR; VEN;
	A. maculosus Ruckes – ECU
	A. marini Fernandes & Grazia – BRA
	A. marmoratus (Erichson, 1848) – GUY
	A. melanicus Fernandes & Grazia – PER
	A. melanoleucus (Westwood) – BOL; BRA; GUY; PER; SUR; VEN
	A. mimeticus Ruckes – BRA; PER
	A. mixtus (F.) (type) – ARG; BOL; BRA; GUF; GUY; SUR; TTO; VEN
	A. nebulosus Ruckes – BRA
	A. nigricans Ruckes – COL; ECU
	A. pallescens Stål – BRA; GUY; SUR
	A. panamensis (Ruckes) – PAN
	A. peruensis Ruckes – PER
	A. pictus Ruckes – ECU; PER
	A. punctissimus Ruckes – BRA
	A. radians Ruckes – BRA
	A. rideri Rolston – BRA
	A. rolstoni Engleman – COL; ECU
	A. ruckesi Rolston – PER
	A. rufitasus Rolston – VEN
	A. schuhi Engleman – BRA; ECU
	A. sepulcralis (F.) – BOL; BRA; SUR; TTO; VEN
	A. simulatus Fernandes & Grazia – BRA
	A.tatei (Ruckes) (*species inquirenda*) – VEN
	A. tesselatus (Westwood) – BRA

(continued)

Table 22.3 (continued)

Genera	Species
	A. tripterus (F.) – ARG; BOL; BRA; COL; ECU; PAN; PRY; TTO; VEN
	A. variegatus Dallas – BOL; BRA; PER
Callostethus Ruckes	*C. guattatopunctatus* (F.) (type) – BRA; PER
	C. flavolineatus Fernandes & Grazia – BRA
Cataulax Spinola	*C. annulicornis* Walker (*insertae sedis*) – BRA
	C. eximius (Stål) (type) – BRA *C. froeschneri* Grazia, Campos & Becker – BRA
	C. pudens (Distant) – PAN; VEN *C. punctipes* Walker (*insertae sedis*) – BRA
	C. radians Grazia, Campos & Becker – PER
	C. subtiliterconspersus Grazia, Campos & Becker – BRA; GUF *C. subvittatus* Walker (*insertae sedis*) - BRA
Colpocarena Stål	*C. complanata* (Burmeister) (type) – BRA; SUR; VEN
Coriplatus White	*C. depressus* White – BRA; COL; CUB; GUF; GUY; VEN
Dinocoris Burmeister	
Dinocoris Burmeister	*D.* (*D.*) *antennatus* (Dallas) – BOL; BRA; PER
	D. (*D.*) *corrosus* (Herrich-Schäffer) – ARG; BRA; PAR
	D. (*D.*) *fabricii* Becker & Grazia – unknown
	D. (*D.*) *gibbosus* (Fallou) – BRA; PAN; VEN
	D. (*D.*) *gibbus* (Dallas) – ARG; BRA; PRY VEN
	D. (*D.*) *histrio* (L.) – GUY
	D. (*D.*) *maculatus* Laporte – BRA; COL; PER; VEN
	D. (*D.*) *nigroantennatus* Becker & Grazia – COL; PER
	D. (*D.*) *ramosus* (Walker) – MEX
	D. (*D.*) *reticulatus* Becker & Grazia – BRA
	D. (*D.*) *rufitarsus* Ruckes – BRA; COL; HND; NIC; PAN
	D. (*D.*) *variolosus* (L.) (type) – GUF; PAN; TTO
Praedinocoris Becker & Grazia	*D.* (*P.*) *lineatus* (Dallas) (type) – BRA; GUF; GUY; PER; SUR; VEN
	D. (*P.*) *nigrodecoratus* Becker & Grazia – BRA
	D. (*P.*) *prolineatus* Becker & Grazia – ARG; BRA; PRY; URY
Discocephala Laporte	*D. carvalhoi* Becker & Grazia – BRA
	D. deplanata Walker – BRA
	D. marmorea Laporte (type) – BRA
Discocephalessa Kirkaldy	*D. andina* (Breddin) – ECU
	D. humilis (Herrich-Schäffer) – COL
	D. notulata (Stål) (type) – CRI; MEX
	D. sordida (Walker) – BRA; GUF
	D. terminalis (Walker) – BOL; BRA

(continued)

Table 22.3 (continued)

Genera	Species
Dryptocephala Laporte	*D. asperula* Perty – ?BRA
	D. brullei Laporte (type) –BRA
	D. crenata Ruckes – PER
	D. cydnoides (Perty) – ?BRA
	D. dentata Fieber – BRA; URY
	D. dentifrons (Latreille) –PER
	D. integra Walker – BRA
	D. latiloba Stål – BRA
	D. lipoloba Ruckes – BRA
	D. lurida (Erinchson) – ARG; BRA; COL; GUY
	D. maculosa Ruckes – BRA
	D. nigricornis Ruckes – BRA
	D. obtusiceps Stål – BRA; COL; ECU; PER
	D. punctata Amyot & Serville – BRA; PER; VEN
	D. spinosa Mayr – BRA
Eurystethus Mayr	*E. deplanatus* Becker – BRA
	E. goianensis Becker – BRA
Eurystethus Mayr	*E.* (*E.*) *ellipsoidalis* Ruckes – PAN
	E. (*E.*) *macroconus* Ruckes – BRA
	E. (*E.*) *nigropunctatus* Mayr (type) – BRA
	E. (*E.*) *ornatus* Ruckes – BRA
	E. (*E.*) *ovalis* Ruckes – BOL; PER
	E. (*E.*) *pallescens* Ruckes – SUR
	E, (*E.*) *parvulus* Ruckes –BRA
	E. (*E.*) *sordidus* Ruckes – GUF
	E. (*E.*) *spurculus* Ruckes – GUF
Hispidisoma Ruckes	*E.* (*H.*) *fulvescens* Ruckes – BRA
	E. (*H.*) *microlobatus* Ruckes – BRA; SUR
	E. (*H.*) *nigricornis* Ruckes – ECU
	E. (*H.*) *nigroviridis* Ruckes – BRA
	E. (*H.*) *punctissimus* Ruckes – BRA
	E. (*H.*) *sacculatus* Ruckes – BRA
	E. (*H.*) *variegatus* Ruckes (type) – BOL; BRA
Glyphuchus Stål	*G. sculpturatus* Stål (type) – BRA
Grassatorama Rider	*G. nigroventris* (Ruckes) (type) – CRI
	G. reticulatus (Ruckes) – VEN
	G. sinuatus (Ruckes) – CRI
Harpogaster Kormilev	*H. willineri* Kormilev (type) – ARG; BRA
Ischnopelta Stål	*I. luteicornis* (Walker) – BRA
	I. oblonga (Fieber) – BRA
	I. scutellata (Signoret) (type) – BRA; PRY; VEN
Lineostethus Ruckes	*L. clypeatus* (Stål) (type) – MEX
	L. graziae Hildebrand & Becker – MEX

(continued)

Table 22.3 (continued)

Genera	Species
	L. marginellus (Stål) – MEX; USA
	L. tenebricornis (Ruckes) – MEX; USA
Mecistorhinus Dallas	*M. amplus* (Walker) – BRA
	M. complanatus (Distant) – PAN
	M. coralium Ruckes – PER
	M. guatemalensis (Distant) – GTM
	M. josephi (Stål) – GUF
	M. obscurus (Dallas) – MEX
	M. rufescens Dallas (type) – BRA
	M. semilugens Bergroth – GUF
	M. tesselatus (Westwood) – BRA
	M. tibialis Ruckes – CRI; PAN
	M. variegatus Ruckes – ECU
Oncodochilus Fieber	*O. integer* Breddin – BRA
Oncodochilus Fieber	*O.* (*O.*) *aradiformis* (Herrich-Shäffer) (type) – BRA
	O. (*O.*) *patruelis* (Stål) –BRA
Oncoeochilus Breddin	*O.* (*O.*) *taschenbergi* (Breddin) (type) – BRA
	O. (*O.*) *cruciatulus* Breddin –BOL
Opophylax Bergroth	*O. extenebratus* Bergroth (type) – BRA
	O. signoreti (Distant) – PAN
Pandonotum Ruckes	*P. bergrothi* Becker – BRA
	P. punctiventris Ruckes (type) – BRA
Paralcippus Becker & Grazia	*P. dimidiatus* (Ruckes) (type) –ECU
Parantiteuchus Ruckes	*P. hemitholus* Ruckes (type) – GUF
Parvamima Ruckes	*P. bicolor* Ruckes (type) – PAN
	P. mexicana Rolston – MEX
Patronatus Ruckes	*P. binotatus* Ruckes (type) – COL; PAN
	P. flavierus Ruckes – VEN
	P. punctissimus Ruckes – GUY
Pelidnocoris Stål	*P. haglundi* Ruckes – BRA
	P. majusculus Ruckes – PAN
	P. stalii Haglund (type) – COL; CRI; PAN; MEX
Phineus Stål	*P. fucopunctatus* Stål (type) – MEX
Phoeacia Stål	*P. erubescens* (Distant) – GTM; PAN
	P. gibba (Fieber) – BRA
	P. lineaticeps (Stål) (type)– BRA
Placidocoris Ruckes	*P. bivittatus* Ruckes (type) – ARG
Platycarenus Fieber	*P. umbraculatus* (Fieber) (type) – BRA; COL; ECU;GUF; GUY; PAN; PER; SUR; VEN
Priapismus Distant	*P. costaricensis* Rolston – CRI
	P. ecuadorensis Rolston – ECU

(continued)

Table 22.3 (continued)

Genera	Species
	P. foveatus Distant (type) – CRI; PAN
	P. maculatus Distant – PAN
	P. pini Rolston – HND
Psorus Bergroth	*P. cassidiformis* Bergroth (type) –BRA; PER; TTO
	P. paraensis Fernandes, Grazia & Lobo – BRA
	P. manauara Fernandes, Grazia & Lobo – BRA
Ruckesiocoris Rider	*R. nitidus* (Ruckes) (type) – PAN
Sympiezorhincus Spinola	*S. punctipes* Dallas – BRA
	S. tristis Spinola (type) – BRA; GUF; PRY
Tetragonotum Ruckes	*T. megacephalum* Ruckes (type) – BRA
Uncinala Ruckes	*U. tau* Ruckes (type) – BRA
Uncicrus Ruckes	*U. kollarii* (Fieber) (type) – BRA
Insertae sedis and *species inquirenda*	*Empicoris marmoreus* Spinola (see Fernandes and Grazia 2006)

[a]Fossil
[b]The monotypic genera *Anhanga* Distant and *Braunus* Distant (previously included in Discocephalini) are being transfered to Pentatominae (TdeA Garbelotto, personal communication, and Barão et al. in press)

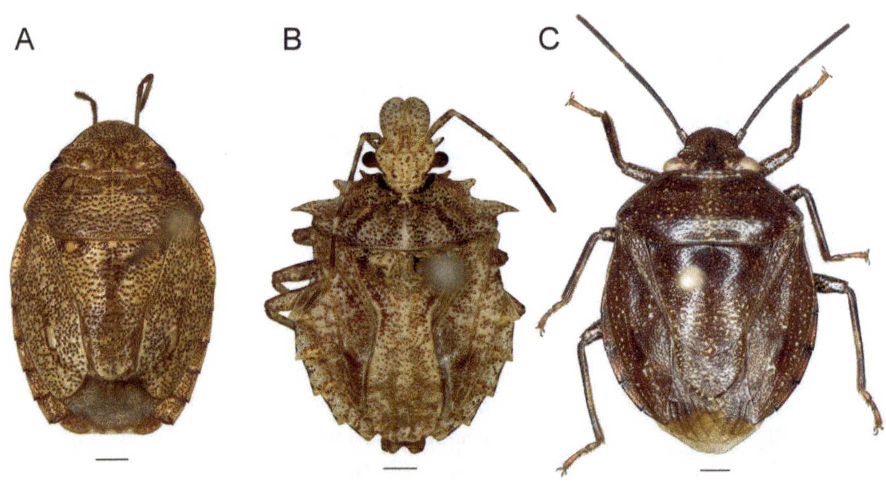

Fig. 22.24 Representative species of Discocephalini; (**a**) *Acclivilamna vicina* (Signoret); (**b**) *Coriplatus depressus* White; (**c**) *Antiteuchus rideri* Rolston. Scale bars = 1 mm (Photos by LA Campos & TA Garbelotto)

this group is being prepared by TdeA Garbelotto, but groups of genera were recognized by some authors based on morphological similarity, as detailed below.

The genera *Abascantus* Stål, *Coriplatus* White, *Eurystethus* Mayr, *Paralcippus* Becker & Grazia, and *Pelidnocoris* Stål shared the form and proportions of the metasternum and the spatial relationships of the mesocoxae and metacoxae – where a metasternum broadly hexagonal, somewhat wider than long (Fig. 22.25), causes

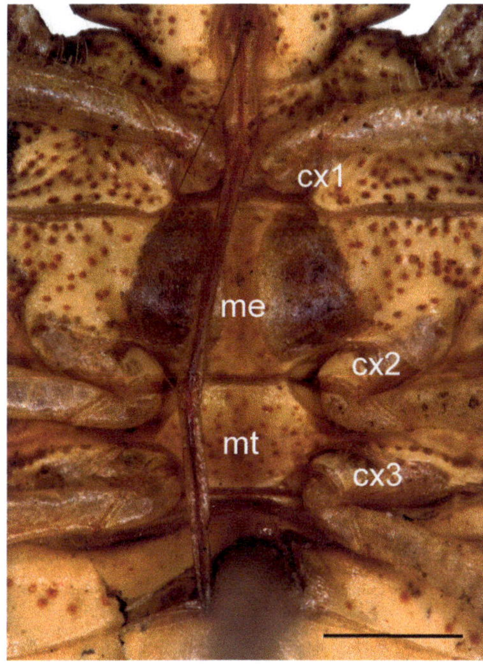

Fig. 22.25 Ventral view of thoracic sternites of *Paralcippus dimidiatus* (Ruckes). *cx1* procoxae, *cx2* mesocoxae, *cx3* metacoxae, *me* mesosternum, *mt* metasternum. Scale bar = 1 mm (Photos by LA Campos & TA Garbelotto)

the metacoxae to be placed much farther apart from each other than they are distant from the respective mesocoxae, the shape of pronotum, and the widely exposed connexivum (last two not in *Eurystethus* Mayr) (Ruckes 1966a; Ruckes and Becker 1970; Becker 1977; Becker and Grazia 1986). Also, *Abascantus* Stål, *Coriplatus* White, and *Pelidnocoris* Stål have in common the shape of scutellum and gonocoxites VIII of external female genitalia much enlarged, covering reduced laterotergites IX. *Paralcippus* Becker & Grazia and *Eurystethus* Mayr share a shorter scutellum (Becker 1977; Becker and Grazia 1986). The genus *Alcippus* Stål shares with *Paralcippus* Becker & Grazia and *Eurystethus* Mayr the presence of the laterotergites IX and the shorter scutellum, and though the general facies of *Alcippus* is similar to that of *Paralcippus* it has the meso and metacoxae mutually equidistant (Becker and Grazia 1986, 1989). *Pandonotum* Ruckes was also placed near *Eurystethus* by the shape of the head and scutellum, the labium with an intercalary unit between the two basal segments, and the reticulate venation of hemelytral membrane (Ruckes 1965).

Ablaptus Stål, *Agaclitus* Stål, *Cataulax* Spinola, *Sympiezorhincus* Spinola, and *Uncinala* Ruckes are considered related especially because of the sexual dimorphic hemelytral membrane, with sclerotized areas and a transversal furrow in males (Fig. 22.26), less clearly differentiated in *Cataulax*. Species of these genera except *Ablaptus* also have the apical angle of corium reaching the apex of the abdomen (Ruckes 1965; Becker and Ruckes 1969; Becker and Grazia 1989, 1992, 1995; Grazia et al. 2000).

Fig. 22.26 Hemelytral membranes of male of *Uncinala tau* Ruckes. *s* sclerotized areas, *t* transversal furrow. Scale bar = 1 mm (Photo LA Campos & TA Garbelotto)

The larger group of genera within Discocephalini is known as "broadheaded discocephalines"; it includes 14 genera whose species have the interocular width equal to or greater than the length of the head (Figs. 22.19a and 22.24a), the basal segment of the rostrum projecting onto the prosternum, and the base of abdominal venter unarmed (Ruckes 1966b; Becker 1977; Rolston 1990).

Keys for two of these groups of genera are available, one for the broadheaded discocephalines (Rolston 1990) and another to separate *Antiteuchus* Dallas of similar genera (Ruckes 1964). Keys for species of several genera are also available (Ruckes 1966a for *Eurystethus*, 1966b for *Alveostethus* and *Lineostethus*, 1966c for *Dryptocephala*, 1966d for *Pelidnocoris*; Rolston 1984 for *Priapismus*; Becker and Grazia 1989 for *Ablaptus*, 1992 for *Agaclitus*; Grazia et al. 2000 for *Cataulax*; Grazia and Zwetsch 2000 for *Ablaptus*; Fernandes and Grazia 2006 for *Antiteuchus*), but some of them (Ruckes 1966a, b for *Lineostethus*, Rolston 1984, and Fernandes and Grazia 2006) are out to date since new species were described after the publication of the keys.

Ochlerini

Ochlerini was raised by Rolston (1981) and presently includes 32 genera and 115 species (Table 22.4) (Garbelotto et al. 2013; Cervantes-Peredo and Ortega-León 2014; Simões and Campos 2014). Species of Ochlerini are, with few exceptions, recognizable by their dull black or fuscous coloration, and primarily by the shallowly excavated or flattened superior surface of the third tarsal segment of the hind legs in females (Fig. 22.22b) and sometimes in males as well (Rolston 1981).

Table 22.4 Checklist of genera and species of Ochlerini (distribution as available in the literature)

Genera	Species
Adoxoplatys Breddin	*A. brasiliensis* Kormilev – BRA
	A. bridarollii Kormilev – BOL
	A. comis Breddin – PER; BOL
	A. gallardoi Kormilev – ARG
	A. giaii Kormilev – ARG
	A. minax Breddin (type) – PER; BOL
	A. panamensis Kormilev – PAN
	A. willineri Kormilev – BOL
Alathetus Dallas	*A. haitiensis* Rolston – HAI
	A. rufitarsis Dallas (type) – JAM
Alitocoris Sailer	*A. grandis* Garbelotto & Campos – PAN
	A. lateralis Garbelotto & Campos – CRI
	A. ornatus Garbelotto & Campos – PAN
	A. schraderi Sailer (type) – CRI; GTM; HND; PAN
	A. maculosus Sailer (*sedis mutabilis*)[a] – GTM; HND
	A. brunneus Sailer (*sedis mutabilis*)[a]
Barola Rolston	*B. farfala* Rolston (type) – PAN
Brachelytron Ruckes	*B. angelicus* Ruckes (type) – BRA
Catulona Rolston	*C. apaga* Rolston – BRA
	C. lucida Campos & Grazia – BRA
	C. pensa Rolston (type) – BRA
Clypona Rolston	*C. aerata* Rolston (type) – ARG
Coranda Rolston	*C. castana* Rolston – ECU
	C. picipes (Stål) (type) – COL; PER
Cromata Rolston	*C. ornata* Rolston (type) – BRA
	C. graziae Campos – VEN; BRA
Eritrachys Ruckes	*E. bituberculata* Ruckes (type) – CRI; PAN
	E. brailovskyi Ortega-León & Thomas – ECU
Forstona Rolston	*F. speciosa* Rolston (type) – BRA
Herrichella Distant	*H. thoracica* Distant (type) – COL
Hondocoris Thomas	*H. cavei* Thomas (type) – HND
Lincus Stål	*L. anulatus* Rolston – PAN; BRA
	L. apollo Dolling – FRG
	L. armiger Breddin – BOL
	L. bipunctatus (Spinola) – GUY; FRG; BRA
	L. breddini Rolston – SUR
	L. convexus Rolston – PER
	L. dentiger Breddin – PER; ECU
	L. discessus (Distant) – CRI; GUY
	L. fatigus Rolston – SUR
	L. hebes Rolston – PER
	L. incisus Rolston – SUR

(continued)

Table 22.4 (continued)

Genera	Species
	L. lamelliger Breddin – COL; FRG; SUR
	L. laminatus Rolston – PER
	L. lethifer Dolling – ECU
	L. leviventris Rolston – PER
	L. lobuliger Breddin – BRA
	L. malevolus Rolston – PER
	L. manchus Rolston – BOL
	L. modicus Rolston – ECU
	L. operosus Rolston – VEN
	L. parvulus (Ruckes) – PER; BRA
	L. repizcus Rolston – PER
	L. rufospilotus (Westwood) (type) – COL; GUY; SUR; PER; BRA
	L. securiger Breddin – BOL; PER; BRA
	L. singularis Rolston – PER
	L. sinuosus Rolston – PER
	L. spathuliger Breddin – PER
	L. spurcus Rolston – PER
	L. styliger Breddin – COL; PER
	L. substyliger Rolston – COL
	L. subuliger Breddin – COL; VEN
	L. tumidifrons Rolston – PAN; TTO; VEN
	L. vallis Rolston – PER
	L. vandoesburgi Rolston – SUR
	L. varius Rolston – PER
Macropygium Spinola	*M. reticulare* (F.) (type) – BRA[b]
Miopygium Breddin	*M. cyclopeltoides* Breddin (type) – BRA
	M. grossa Ruckes – BRA
Moncus Stål	*M. obscurus* (Dallas) (type) – BRA
	M. monachus Bergroth – FRG
Neoadoxoplatys Kormilev	*N. haywardi* Kormilev – ARG
	N. longirostra Ruckes – COL
	N. saileri Kormilev (type) – MEX
	N. thomasi Cervantes & Ortega – MEX
Ocellatocoris Campos & Grazia	*O. dasys* Campos & Grazia (type) (MNRJ) – BRA
Ochlerus Spinola	*O. bergrothi* Breddin – UNKNOWN
	O. bistillatus Breddin – PER; BOL
	O. cinctus Spinola (type) – ?BRA
	O. coriaceus Herrich-Schäffer – COL; VEN
	O. circummaculatus Stål – BRA
	O. communis Breddin – UNKNOWN
	O. cotylophorus Breddin – COL; ?PER

(continued)

Table 22.4 (continued)

Genera	Species
	O. dentijugis Breddin – PER
	O. handlirschi Breddin – MEX; HND
	O. incisulus Breddin – VEN
	O. lutosus Herrich-Schäffer – BRA
	O. notatulus Breddin – BRA
	O. profanus Breddin – PER
	O. rusticus Breddin – BRA
	O. signoreti Breddin – FRG
	O. stylulatus Breddin – PER
	O. tenuicornis Breddin – VEN
Orbatina Ruckes	*O. fuligina* Ruckes (type) – PAN; COL; BOL
Paralincus Rolston	*P. bimaculatus* (Ruckes) – GUY
	P. silvae Rolston – SUR; BRA
	P. terminalis (Walker) (type) – BRA
	P. sordidus (Herrich-Shäffer) (*nomen dubium*) – BRA
Parastalius Matesco, Grazia, & Campos	*P. trisinuatus* (Rolston) (type) – CRI; PAN
	P. rolstoni Matesco, Grazia & Campos – CRI
Parochlerus Breddin	*P. latus* Breddin (type) – PER; BRA
Phereclus Stål	*P. pluto* Stål (type) – COL
	P. punctatus (Dallas) – COL
	P. antennatus Distant – COL
Pseudadoxoplatys Rolston	*P. mendacis* Rolston (type) – PER; BOL
Schaefferella Spinola	*S. fusca* Rolston – ECU; PER
	S. incisa (Herrich-Schäffer) (type) – BRA
Schraderiellus Rider	*S. cinctus* (Ruckes) (DBTC) – CRI
	S. hughesae (Ruckes) (type) – CRI
Similiforstona Campos & Grazia	*S. bella* Campos & Grazia (type) – BRA
Stalius Rolston	*S. castaneus* (Distant) – NIC
	S. leonae Ortega – MEX, GUA
	S. scutellatus Thomas – CRI
	S. tartareus (Stål) (type) – MEX; HND; COL
Tetrochlerus Breddin	*T. fissiceps* Breddin (type) – ECU
Uvaldus Rolston	*U. concolor* Rolston (type) – BRA
Xynocoris Garbelotto & Campos	*X. calosus* Garbelotto & Campos – BOL
	X. crinitus Garbelotto & Campos – BRA
	X. cupreus Garbelotto & Campos – BRA
	X. egeri Garbelotto & Campos – CRI, BRA
	X. insularis Garbelotto & Campos – TTO
	X. lucidus Garbelotto & Campos – BRA
	X. meridionalis Garbelotto & Campos – BRA, PAR

(continued)

Table 22.4 (continued)

Genera	Species
	X. parvus (Distant) (type) – CRI, PAN, VEN, SUR, BRA, ECU
	X. recavus Garbelotto & Campos – VEN, BRA
	X. tuberculatus Garbelotto & Campos – VEN, BRA

[a]See Garbelotto et al. (2013)
[b]There are several undescribed species of *Macropygium* occurring in different countries, from Mexico to Argentina

The most comprehensive systematic work for Ocherini was made by Rolston (1992a), including the description of nine genera, diagnosis for the tribe, and key and diagnosis for its genera. The phylogeny of genera of Ochlerini and its sister-group relationship to Discocephalini were recently the theme of a cladistic study (Campos and Grazia 2006) (Fig. 22.23). Unambiguous synapomorphies of Discocephalinae are the long first segment of labium, reaching prosternum, and the male segment X largely membranous at least on its basal third (Campos and Grazia 2006, characters 13 and 37). Unambiguous synapomorphy of Ochlerini is the presence of a dorsal depression of the third tarsal segment of the hind legs of females (Campos and Grazia 2006, character 30).

The most basal genera within Ochlerini are *Adoxoplatys* Breddin, *Neoadoxoplatys* Kormilev, and *Pseudadoxoplatys* Rolston, whose species can be easily distinguished from other ochlerines by a usually long and flattened body, light-brown colored and of somewhat glossy aspect (Fig. 22.27a). All other genera were grouped into two clades, named *Ocellatocoris* and *Herrichella* clades by Campos and Grazia (2006). The first includes the type genus *Ochlerus* Spinola and seven other genera (Fig. 22.19b), and the later includes the remaining 20 genera known by that time (Fig. 22.27b), except *Hondocoris* Thomas that was not included in the cladistic analysis. The *Herrichella* clade was subject of another cladistic analysis that aimed to give a phylogeny-based classification of the genus *Alitocoris* Sailer and related species (Fig. 22.28) (Garbelotto et al. 2013).

The checklist for genera and species of Ochlerini provided by Campos and Grazia (2006) is updated here, with the inclusion of taxonomic information from several sources (Arismendi and Thomas 2003; Matesco et al. 2007; Ortega-León and Thomas 2010; Garbelotto et al. 2011, 2013; Simões and Campos 2014; Cervantes-Peredo and Ortega-León 2014) (Table 22.4).

22.4.5 Remarks

The study of Discocephalinae remains a challenging issue, since species of most genera are poorly represented in scientific collections and their natural history is virtually unknown. The description of life cycles and morphology of immatures are

Fig. 22.27 Representative species of Ochlerini; (**a**) *Adoxoplatys comis* Breddin; (**b**) *Xynocoris parvus* (Distant). Scale bars = 1 mm (Photos by LA Campos & TA Garbelotto)

strongly needed. Behaviors registered for a few species such as parental care and trophobiosis with ants are interesting research themes, as well as feeding relationship with plants especially of Ochlerini. Phylogeny and systematics of Discocephalinae are also in need for investigation, focusing the review and modern taxonomic treatment of genera, use of molecular data in cladistic analyses, and relationship with other subfamilies of Pentatomidae.

22.5 Subfamily Edessinae

22.5.1 Introduction

Edessinae is one of the largest subfamilies of Pentatomidae with about 300 species (Schuh and Slater 1995; da Silva et al. 2013) distributed only in Neotropical Region, except for two species found in the Southern United States of America [*Edessa bifida* (Say) and *E. florida* Barber] (McPherson 1982). This subfamily includes seven genera without tribal division. Edessinae was until recently treated as a tribe of Pentatominae but was raised to subfamily by Rolston and McDonald (1979). This taxon has well-established limits but they are quite similar to the limits of the genus *Edessa*, causing a lot of confusion and turning the latter a depot of species.

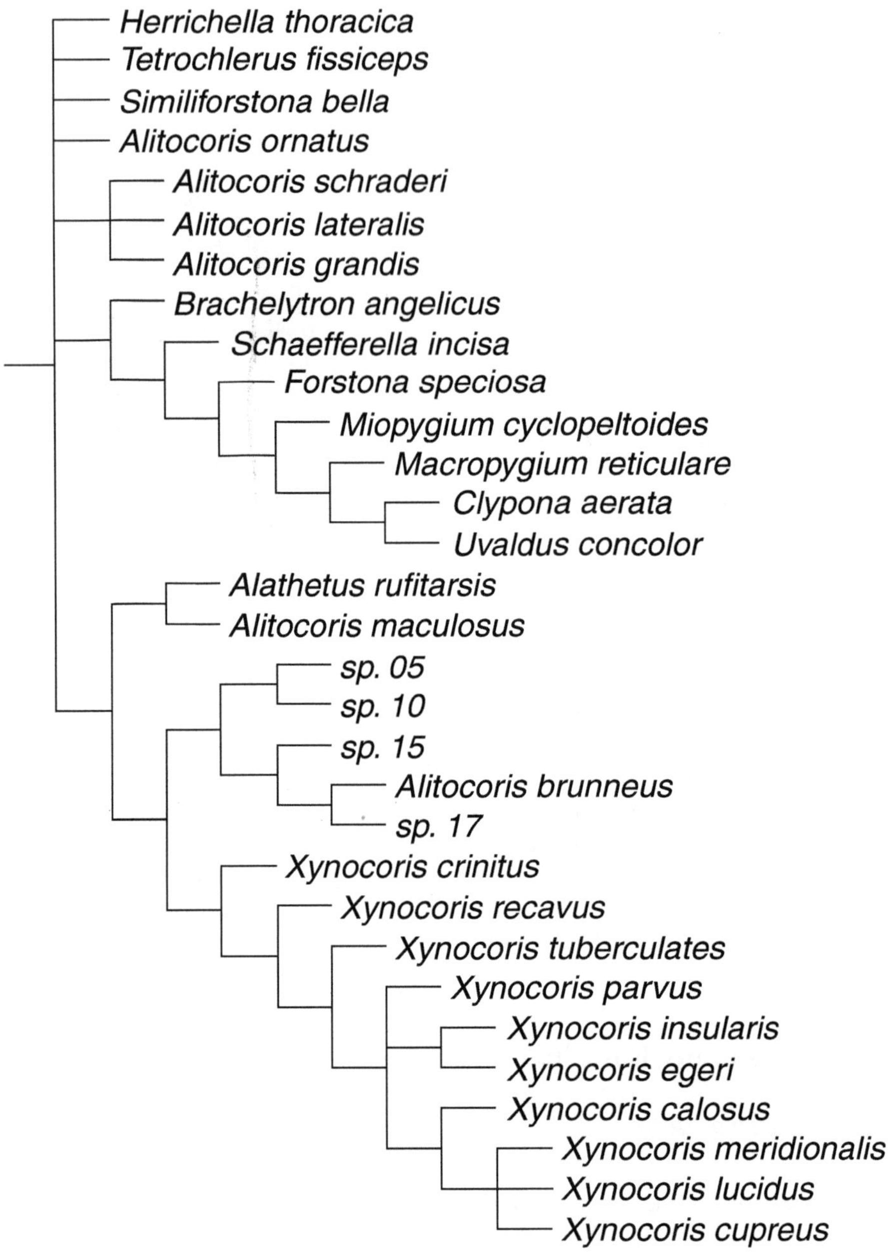

Fig. 22.28 Phylogeny of the *Herrichella* clade depicting the relationship between species of the *Alitocoris* group; strict consensus tree after implied weighting (Redrawn from Garbelotto et al. (2013))

22.5.2 *General Characteristics and Diagnosis*

Medium to large size bugs (1–3 cm long). Very attractive bugs with color ranging from light green to greenish-brown with highly contrasting connexivum and ventral surface. Head short and wide. Antennae 4 or 5 segmented. Humeral angles ranging from not to highly developed conical or flat, apex rounded, or acute. Metasternum strongly produced anteriorly onto mesosternum (prosternum in *Pantochlora* Stål) and usually bifid except for *Pantochlora* and *Doesburgedessa* that show a single projection. First rostral segment short, reaching anterior notched part. Posterior part of metasternal process also notched, receiving medial abdominal tubercle (Stål 1872; Rolston and McDonald 1979; Schuh and Slater 1995). Distal angle of the abdominal segment VII range from before the level of genital plates to surpassing them by far (Stål 1872). Dorsal rim of pygophore with a superior process. Ventral surface of the body usually lighter than dorsal with dark stripes at least on abdominal intersegmentar areas (Figs. 22.29, 22.30, 22.31, 22.32, 22.33, and 22.34).

Female internal genitalia: Description is almost the same of Pentatominae but capsula seminalis with 3 digitiform processes and a beak-like projection on the thickening of the vaginal intima (Barcellos and Grazia 2003b).

Fig. 22.29 *Edessa oxyacantha* Breddin (Photo by F Carvalho-Filho)

Fig. 22.30 *Edessa laticornis* Stål: (**a**) adult; (**b**) 5th instar nymph (Photo by F Carvalho-Filho)

Fig. 22.31 *Edessa bifida* (Say) (Photo by F Carvalho-Filho)

Fig. 22.32 *Edessa elaphus* Breddin (Photo by F Carvalho-Filho)

Fig. 22.33 *Edessa dolichocera* Burmeister, lateral view (Photo by F Carvalho-Filho)

Fig. 22.34 *Edessa dolichocera* Burmeister, frontal view (Photo by F Carvalho-Filho)

Male internal genitalia: The phallus of Edessinae is quite simple; phalloteca is strongly esclerotized, tubular to pear-like, and elongated; conjunctiva absent; vesica small, elliptical in lateral view; and *ductus seminis distalis* short, not or slightly surpassing the apex of vesica.

22.5.3 General Biology and Ecology

Edessinae are all plant feeder bugs, and only one species is economically important as pests of cultivated crops – *Edessa meditabunda* (F.) (Fig. 22.35), particularly on soybean and on tobacco (Silva et al. 1968; Rizzo 1971; Lopes et al. 1974; Panizzi et al. 2000). Another species is recognized as a secondary pest – *Edessa rufomarginata* (De Geer) – that feeds on Leguminosae but is more often found on Solanaceae, causing damage to tomato (Silva et al. 1968; Rizzo and Saini 1987; Panizzi et al. 2000). The species mentioned before are very common throughout South and Central Americas due to their capability to invade areas modified by man and are well studied with several registers over time. On the other hand, others may have local importance and are barely or not studied so far– e.g., a not described species that is an important pest in camu camu (Myrtaceae) crops in Pucallpa, Peru (Iannacone et al. 2007). Several other species can be found on cultivated or native plants of economical or ornamental interest but cause no apparent damage: *Edessa scabriventris* Stål on guava and Brazilian cherry (Lima et al. 2010) and *Paraedessa stolida* (L.) on acerola (JAM Fernandes, personal observation).

Only a few works concentrated on economically important species were produced so far. Rizzo (1971) presented detailed information on the biology, immature stages, field ecology, and host plants of *Edessa meditabunda*. This species has been the subject of studies regarding its association with and damaging potential to soy-

Fig. 22.35 *Edessa meditabunda* (F.) (Courtesy of JJ Silva)

bean (Silva et al. 2012) and sunflower (Malaguido and Panizzi 1998b) and nymph development and feeding behavior (Panizzi and Machado-Neto 1992); also, the behavior of newly hatched nymphs on the corions has been recently studied in detail (Calizotti and Panizzi 2014).

Rizzo and Saini (1987) presented biology data on *Edessa rufomarginata*. This species was subject of another study focused on the description of the immature stages (Fortes and Grazia 1990) under controlled laboratorial conditions. The same species was observed and studied in the field revealing several interesting ecological data including trophobiotic interaction between nymphs of *E. rufomarginata* and ants (Silva and Oliveira 2010). A species not recognized as pest – *Edessa scabriventris* – had some ecological and biological characteristics revealed in Lima et al. (2010). A probably new species considered pest in camu camu crop had its biology, immature stages, and field ecology studied (Iannacone et al. 2007).

22.5.4 Classification and Diversity

The subfamily Edessinae Amyot & Serville has seven genera: *Edessa* F. (about 280 species), *Brachystethus* Laporte (10 species), *Peromatus* Amyot & Serville (seven species), *Olbia* Stål (three species), *Pantochlora* Stål (one species), *Doesburgedessa* Fernandes (five species), and *Paraedessa* Silva & Fernandes (nine species). The genus *Brachystethus* is recognized by the short metasternal process and anterior apex slightly notched. *Peromatus* has only four antennal segments. *Olbia* has a pair of spines on the posterior margin of the pronotum, near the scutellum base.

Pantochlora is characterized by the anterior apex of the metasternal process not bifurcate, projected between procoxae, laterally flattened, and attached to the sternum. *Doesburgedessa* is recognized by the metasternal process with a single conical anterior projection which is detached from the mesosternum. *Paraedessa* has a reduction of the genital plates of the females plus a remarkable esclerotization of the gonapophyses 8; males show a lateral expansion of the proctiger and a dark crest covering base of the paramere. *Edessa* is composed of several groups of species with the metasternal process clearly bifurcating anteriorly but variable in many other characteristics (Fernandes 2010). Edessinae is a monophyletic group Barcellos and Grazia 2003a) supported by several characteristics, but its composition at supra-specific level is still not clear due to *Edessa* that is hiding the real diversity of the subfamily. Studies are revealing different groups of species (e.g., Fernandes and Van Doesburg 2000; Fernandes and Campos 2011; da Silva and Fernandes 2012) and also new genera *Doesburgedessa* (Fernandes 2010) and *Paraedessa* (da Silva et al. 2013). The heterogeneity of *Edessa* can be confirmed by its confusing taxonomical and nomenclatural history that culminates with Kirkaldy's catalog (1909) where he listed the subgenera but considered himself unable to include the species in them (footnote p. 153). Genus *Edessa* despite its size still has hundreds of new species to be described. New taxonomical studies probably will split *Edessa* in different taxa. A phylogenetic study is also important to present a hypothesis of relationship among taxa and confirm or not that *Edessa* is a paraphyletic taxon. In this book the genera *Lopadusa*, *Mediocampus*, *Neopharnus*, *Pharnus*, *Platistocoris*, and *Praepharnus* are included in Pentatomini (Pentatominae).

Key to genera of Edessinae and species group of *Edessa*. The concept of Stål (1872) dividing the genus in species groups is adopted here; genus in italic and group of species in bold.

1. Body rounded; apex of metasternal process level with mesocoxae, anteriorly raised, and sulcated but not clearly bifurcated .... *Brachystethus* Laporte
– Body with different shapes but usually elongated, lateral margins subparallel, metasternal process not as above 2
2. Metasternal process anteriorly clearly bifurcated 4
– Metasternal process projected anteriorly in a simple process 3
3. Metasternal process long, reaching mesocoxae, laterally compressed, and fused with sterna, and apex widely rounded; humeral angle slightly projected and acute; found in Central America .... *Pantochlora* Stål
– Metasternal process short, not reaching procoxae, conical, not in contact with sterna, apex truncate; humeral angle projected laterally or dorsolaterally, conical, apex sharp; found in Amazon region ....*Doesburgedessa* Fernandes
4. Four segmented antennae; second antennal segment very long, at least three times longer than the first; humeral angle truncate or globose; paramere robust and large; specimens usually bigger than 2 cm........*Peromatus* Amyot & Serville
– Five segmented antennae; second antennal segment short, at most twice longer than the first; humeral angle and paramere strongly variable in shape; specimens usually smaller than 2 cm 5

5. Posterior margin of the pronotum with a large pair of spines overlapping basal angles of scutellum; humeral angles long, sinuous, and laterally directed; pronotal disk with green and yellow stripes connecting humeral angles .... *Olbia* Stål
– Posterior margin of the pronotum without large spines or other projections; humeral angles variable in shape; pronotal disk not as above (*Edessa*) 6
6. Humeral angles rounded or sharp but not laterally developed more than the distance between the eyes 7
– Humeral angles variable in shape and laterally developed more than the distance between the eyes 11
7. Species elongate; abdominal segment VII projected posteriorly and contrastingly dark at least in distal half 8
– Species elliptical; abdominal segment VII not posteriorly projected and concolorous with connexivum 10
8. Corium brown with yellow veins; dark rounded spots below the spiracles; seventh abdominal segment mostly contrastingly dark dorsally, ventrally concolorous, posterior projection at least 1.5 times longer than wide at base .... **Pygoda** Amyot & Serville
– Corium brown variegate with whitish and blackish spots; without dark spots below spiracles; abdominal segment VII contrastingly dark dorsally only in distal half, ventrally with spots or stripes contrastingly dark, posterior projection at most 1.2 times longer than wide at base 9
9. Posterior part of proctiger expanded laterally; behind this part bristles short and sparse; pygophore with a rough ridge on each side of proctiger; gonocoxites 8 slightly to strongly reduced, leaving visible part of gonapophyses 8; gonapophyses 8 strongly developed and sclerotized .... *Paraedessa* Silva & Fernandes
– Posterior part of proctiger not expanded laterally at most with a small carina and dense tuft of long bristles; pygophore without ridge close to proctiger; gonocoxites 8 not reduced and leaving visible a tiny part of gonapophyses 8 at most; gonapophyses 8 not developed and membranous .... *Hypoxys* Amyot & Serville
10. Pronotal disk tumid; costal margin of corium uniformly colored and punctured; connexivum with punctures dark .... *Ascra* Say
– Pronotal disk not tumid. Costal margin of corium with punctures and usually color distinct from the rest; connexivum with punctures concolor .... **Aceratodes** Amyot & Serville
11. Humeral angle long, apex acute. Abdominal segment VII with apex twisted, strongly projected posteriorly, surpassing considerably the genital plates. Pygophore with a yellow spot on the base of each genital cup process .... **Dorypleura** Amyot & Serville
– Humeral angle long or short; apex acute, truncate, or globose, black or yellow contrasting with general coloration. Abdominal segment VII with apex not twisted projected posteriorly but not or slightly surpassing the genital plates. Pygophore without any spot on the base of genital cup process .... *Edessa* F.

22.5.5 Remarks

Edessinae was treated as a tribe of Pentatominae and was raised to subfamily by Rolston and McDonald (1979). This taxon has well-established limits; nevertheless its concept is quite similar to the genus *Edessa*. They are well represented in scientific collections, and modern taxonomical studies probably will split *Edessa* in different taxa. Recent studies are revealing different groups of species and also new genera within *Edessa*, a probably a paraphyletic taxon, deserving a phylogenetic study to establish the relationships within the subfamily and among the Pentatomidae.

22.6 Subfamily Pentatominae

22.6.1 Introduction

Pentatominae is the most diverse subfamily of Pentatomidae and includes species distributed in all the regions of the world (Schuh and Slater 1995; Grazia et al. 1999). The exact number of groups (tribes, genera, and species) gathered in Pentatominae is uncertain, since the definition of the subfamily is still under discussion. Nonetheless, it is possible to estimate that between 400 and 600 genera and 2,700 and 3,300 species are included in Pentatominae (Grazia et al. 1999; Cassis and Gross 2002; Rider 2014). In the Neotropical Region, approximately 120 genera and about 680 species occur.

22.6.2 General Characteristics and Diagnosis

Extremely variable in color, form, and size. First segment of labium arising anterior to the line of anterior margin of the eyes, with variable length in relation to the bucculae. Humeral angles frequently with process (acute or not); scutellum not reaching the apex of the abdomen, with frenal region extending from 2/5 to 2/3 the length of the scutellum. Metasternum rarely produced anteriorly. Abdominal segments III to VII with two pairs of trichobothria, near the spiracular line (Rolston and McDonald 1979; Schuh and Slater 1995) (Figs. 22.36, 22.37, 22.38, 22.39, 22.40, 22.41, 22.42, and 22.43).

Female internal genitalia: vesicular region of ductus receptaculi membranous, invaginated into a sclerotized rod; capsula seminalis with anterior and posterior annular flanges, bearing or not processes of variable number (Rolston and McDonald 1979; Schuh and Slater 1995). Male internal genitalia: conjunctiva usually membranous, sometimes absent; vesica usually as a sclerotized tube, with variable length (Rolston and McDonald 1979; Schuh and Slater 1995).

Fig. 22.36 *Oenopiella punctaria* Stål (Photo by A Ferrari)

Fig. 22.37 *Agroecus scabricornis* Herrich-Schaeffer (Photo by L Barros)

22.6.3 General Biology and Ecology

Pentatominae are all phytophagous and many species are economically important as pests of cultivated crops (Grazia et al. 1999; Panizzi et al. 2000). In the Neotropics the Pentatominae cause damage mainly in Leguminosae and Gramineae crops. *Nezara viridula* (L.) (Fig. 22.40), *Piezodorus guildinii* (Westwood) (Fig. 22.44), and *Euschistus heros* (F.) (Fig. 22.45) are important pests of soybean (Panizzi et al. 2000),

Fig. 22.38 *Euschistus irroratus* Bunde, Grazia, & Mendonça Jr. (Photo by J Grazia)

Fig. 22.39 *Pellaea stictica* (Dallas) (Courtesy of CSRibeiro-Costa)

Fig. 22.40 *Nezara viridula* (L.) (Photo by A Ferrari)

Fig. 22.41 *Phalaecus* sp. (Photo by J Grazia)

Fig. 22.42 *Roferta marginalis* (Herrich-Schaeffer) (Photo by JAM Fernandes)

Fig. 22.43 *Stictochilus tripunctatus* Bergroth (Photo by R Lupoli)

and species of *Arvelius* Spinola, *Loxa* Amyot & Serville, *Agroecus* Dallas, *Dichelops* Spinola, and *Chinavia* Orian also occur in this crop, sometimes causing damage (Panizzi and Slansky 1985; Panizzi et al. 2000; Schwertner and Grazia 2007).

Mormidea v-luteum (Lichtenstein), *Oebalus poecilus* (Dallas), *Oebalus ypsilon-griseus* (De Geer) (Fig. 22.46), and *Tibraca limbativentris* Stål (Fig. 22.47) cause

Fig. 22.44 *Piezodorus guildinii* (Westwood) (Courtesy of JJ Silva)

Fig. 22.45 *Euschistus heros* (F.) (Courtesy of JJ Silva)

damage to rice and wheat (Panizzi et al. 2000). Also, *Glyphepomis* Berg species are rice pests (Campos and Grazia 1998; Farias et al. 2012) and *Dichelops melacanthus* (Dallas) (Fig. 22.48) is considered an important problem in wheat (Chocorosqui and Panizzi 2004; Manfredi-Coimbra et al. 2005) and corn crops (Ávila and Panizzi 1995) in Brazil.

Loxa spp. and *Arvelius albopunctatus* (De Geer) (Fig. 22.49) cause damage in Solanaceae plants in Brazil (Panizzi et al. 2000). The last one is also reported to feed in tobacco in Brazil (Marques et al. 2006).

Pentatominae are, in general, highly polyphagous, and their life history includes several host plant sequences. For some species this has been investigated in details in the Neotropics, as illustrated by the studies conducted by Panizzi (1997).

Fig. 22.46 *Oebalus ypsilongriseus* (De Geer) (Courtesy of H. Prando)

Fig. 22.47 *Tibraca limbativentris* Stål (Courtesy of H. Prando)

Fig. 22.48 *Dichelops melacanthus* (Dallas) (Courtesy of JJ Silva)

Fig. 22.49 *Arvelius albopunctatus* (De Geer) (Courtesy of JJ Silva)

22.6.4 Classification and Diversity

Pentatominae is not a monophyletic group (Gapud 1991; Grazia et al. 2008) and does not have a definition based on phylogenetic studies. Therefore, authors diverge about composition of the subfamily (Cassis and Gross 2002). Schuh and Slater (1995), for example, included eight tribes in Pentatominae, from which only Pentatomini is registered in the Neotropical Region. Rider (2014) considers 42 tribes in Pentatominae, 11 registered in Neotropics: Catacanthini Atkinson, Carpocorini Mulsant & Rey, Halyini Amyot & Serville, Mecideini Distant, Menidini Atkinson, Nezarini Atkinson, Pentatomini Leach, Piezodorini Atkinson, Procleticini Pennington, Sciocorini Amyot & Serville, and Strachiini Mulsant & Rey. The relationship among tribes is unknown; also most of the tribes do not have diagnostic characteristics to define them, which make the establishment of the exact composition of these groups difficult. Six genera of the Neotropical Region are considered as unplaced in Rider's classification (Table 22.5).

Table 22.5 Checklist of Neotropical Pentatominae genera (distribution as available in the literature)

Carpocorini	
Acledra Signoret	BRA, CHI, ARG, URU
Agroecus Dallas	PER, ECU, BOL, GUY, FGU, COL, VEZ, BRA, PER, BOL, ARG, PAR
Amauromelpia Fernandes & Grazia	CR, FGU, BRA, BOL, COL, PER
Berecynthus Stål	MEX, CR, PAN, COL, VEZ, SUR, BRA, PER
Braunus Distant	CR, VEZ, COL, ECU, PER, BOL
Bucerocoris Mayr	BRA
Caonabo Rolston	BRA
Caribo Rolston	BAH, JAM, IS. VIR
Caracia Stål	VEZ
Copeocoris Mayr	BRA, ARG
Cosmopepla Stål	USA, MEX, ELS, CR, VEZ, COL, ECU
Cradia Bergroth	BRA
Curatia Stål	BRA
Dichelops Spinola	PAN, BRA, ECU, PAR, VEZ, BRA, PER, BOL, PAR, ARG, URU
Epipedus Spinola	BRA
Euschistus Dallas	USA, MEX, NIC, GTM, ELS, HON, CR, PAN, COL, VEZ, W.I. (ALL), GUY, BRA, ECU, BOL, FGU, PER, CUB, JAM, CURAÇAO, BLZ, ARG, URU,
Galedanta Amyot & Serville	CR, GUY, COL, BRA
Glyphepomis Berg	BRA, ARG, URU
Hypanthracos Grazia & Campos	BRA, URU
Hypatropis Bergroth	HON, NIC,CR, PAN, TTO, COL, VEZ, SUR, BRA, ARG, URU
Lattinidea Rider & Eger	CR, PAN, COL, VEZ, FGU, BRA, ECU, PER
Lattinellica Rider & Eger	COL, BRA, ECU, PER
Ladeaschistus Rolston	BRA, BOL, PAR, ARG, URU
Luridocimex Grazia, Fernandes & Schwertner	BRA
Mathiolus Distant	GTM
Mecocephala Dallas	BRA, ARG, URU
Menudo Thomas	PUR
Mormidea Amyot & Serville	CAN, USA, MEX, NIC, PUR, CUB, HIS, DRE, COL, GUY, FGU, PER, BOL, SUR, BRA, ECU, TRI, PAN, ARG, URU, VEZ, JAM, PAR, GRE, TRI

(continued)

Table 22.5 (continued)

Carpocorini	
Oebalus Stål	USA, MEX, W.I. (ALL), SUR, BRA, PER, PAR, ARG, URU, TRI, COL, GUY, ECU, BOL, HIS, PUR, CUB
Oenopiella Bergroth	BRA, CHI, ARG, URU
Ogmocoris Mayr	BRA
Padaeus Stål	USA, MEX, GTM, NIC, HON, CR, ECU, COL?
Parahypatropis Grazia & Fernandes	BRA, ARG
Paramecocephala Benvegnú	SUR, GUY, GUF, BRA, ECU, ARG, URU
Paratibraca Campos & Grazia	PAN, TTO, SUR, COL, BRA, PER, BOL
Parentheca Berg	BRA, PAR, ARG, URU
Pedinonotus Fernandes & Grazia	BRA
Pentatomiana Grazia & Barcellos	BRA (RJ, SP, SC)
Poriptus Stål	TTO, COL, BRA, BOL, ARG
Prionotocoris Kormilev	BOL, ARG
Proxys Spinola	USA, MEX, DRE, CUB, GTM, NIC, FGU, BLZ, CR, GRE, PAN, COL, VEZ, BRA, ARG, URU
Sibaria Stål	MEX, ALL CENTRAL AM., COL, VEZ, FGU, GUY, SUR, TTO, BRA, ECU, PER, BOL, PAR, ARG
Spinalanx Rolston & Rider	COL, PER, ECU, BOL
Stysiana Grazia, Fernandes & Schwertner	COL, SUR, TTO, GUY, BRA, PER, ARG, PAR
Tibraca Stål	DRE, CUB, CR, COL, VEZ, BRA, PER, BOL, ARG, URU
Trichopepla Stål	CAN, USA, MEX
Catacanthini	
Arocera Spinola	MEX, BLZ, HON, CR, PAN, CUB, JAM, PUR, DRE, TTO, VEZ, FGU, GTM, CR, COL, GUY, SUR, FGU, BRA, ECU, BRA, BOL, ARG
Boea Walker	CR, BRA
Rhyssocephala Rider	MEX, BLZ, GTM, CR, PAN, JAM, COL, VEZ, ECU, BRA, PER, BOL
Runibia Stål	USA, JAM, Virgen Is, COL, VEZ, ECU, BRA, PER, BOL, GUF, ARG, PAR, URU
Vulsirea Spinola	USA, MEX, HON, HAI, JAM, CR, CUB, PAN, COL, VEZ, BRA, BOL
Halyini	
Brochymena Amyot & Serville	USA, MEX, HON, CR, BAH, CUB
Parabrochymena Larivière	USA, MEX, GTM, CR
Mecideini	
Mecidea Dallas	USA, MEX, NIC, HON, CR, PUR, ANTIGUA, ARG

(continued)

Table 22.5 (continued)

Carpocorini	
Menidini	
Elanela Rolston	SUR, BRA, PER
Rio Kirkaldy	MEX, GTM, ELS, CR, PAN, VEZ, SUR, BRA, ARG
Nezarini	
Chinavia Orian	CAN, USA, MEX, BLZ, CR, CUB, DRE, PUR, JAM, Grenada, Guadeloupe, Dominica, St. Croix, VIR IS., GTM, NIC, HAI, HON, BAH, COL, VEZ, SUR, FGU, GUY, BRA, ECU, PER, BOL, CHI, PAR, ARG, URU
Chlorochroa Stål	CAN, USA, MEX
Nezara Amyot & Serville	MEX, NIC, HON, CUB, DRE, PAN, VEZ, BRA, CHI, ARG, URU
Roferta Rolston	BLZ, PUR, PAN, TRI, CUB, BRA, ARG
Pentatomini	
Adevoplitus Grazia & Becker	MEX, GTM, CR, PAN, VEZ
Arvelius Spinola	DRE, GUP, MAR, BAR, SVI, USA, MEX, ELS, GTM, HON, CUB, DRE, PUR, VIR, ANT, BAH, NIC, JAM, CR, PAN, COL, VEZ, SUR, GUY, TTO, HAI, ECU, BRA, BOL, PAR, ARG, URU, GUY
Banasa Stål	USA, MEX, ALL CENTRAL AM., ALL W.I., COL, ECU, VEZ, PER, BOL, BRA, PAR, SUR, GUY, GUF, ARG, URU,
Brasilania Jensen-Haarup	BRA
Chlorocoris Spinola	USA, MEX, BLZ, NIC, HON, GTM, JAM, CR, PAN, COL, VEZ, SUR, TRI, ECU, BRA, PER, BOL, PAR, ARG
Chloropepla Stål	BRA, BOL, VEZ, CR, PER, GUY, FGU,
Disderia Bergroth	MEX, GTM, BLZ
Elsiella Froeschner	ECU
Eludocoris Thomas	CR
Evoplitus Amyot & Serville	BRA
Fecelia Stål	PUR, HAI, DRE
Glaucioides Thomas	HON, CR, PAN, VEZ, SUR, GUY, BRA
Grazia Rolston	MEX, CUB, DRE, PAN, VEZ, BRA, PAR
Janeirona Distant	BRA, PER, PAR, ARG
Kermana Rolston	USA, MEX, CR, HON, PAN, BRA, ARG, URU
Lopadusa Stål	PAN, VEZ, GUY, BRA, ECU, PER, BOL, PAR, ARG
Loxa Amyot & Serville	USA, MEX, CUB, DRE, BAH, PUR, DOMINICA, GUADALOUPE, CURAÇAO, GRENADINIES, ST. LUCIA, JAM, HON, NIC, PAN, COL, GUY, VEZ, SUR, FGU, BRA, PER, ARG
Marghita Ruckes	BRA, ARG
Mayrinia Horváth	NIC, CR, COL, VEZ, GUY, BRA, PER, BOL, PAR, ARG
Mediocampus Thomas	DOR
Modicia Stål	CUB
Myota Spinola	BRA

(continued)

Table 22.5 (continued)

Carpocorini	
Neopharnus Van Duzee	EUA, CUB
Neotibilis Grazia & Barcellos	MEX, CR, PAN, COL, VEZ, GUY, GUF, BRA, ECU, PER, BOL
Nocheta Rolston	SUR, BRA
Pallantia Stål	MEX, GTM, PAN, VEZ, PER, BRA, PAR, ARG
Paratibilis Ruckes	MEX, PER
Pellaea Stål	USA, MEX, NIC, HON, CR, PAN, COL, VEZ, GUY, BRA, ECU, ARG, URU
Phalaecus Stål	GUY, SUR, GUF, BRA, COL
Pharnus Stål	CUB
Pharypia Stål	MEX, HON, GTM, CR, PAN, COL, VEZ, FGU, BRA, ARG
Placocoris Mayr	BRA, ARG
Platistocoris Rider	BRA
Praepharnus Barber & Bruner	CUB
Pseudevoplitus Ruckes	VEZ, BRA, PER, BOL, ARG
Ramosiana Kormilev	MEX, CR, PAN, PER, BOL
Rhyncholepta Bergroth	PAN, VEZ, GUF, VEZ, BRA, BOL
Rideriana Grazia & Frey-da-Silva	BRA, PER, BOL
Serdia Stål	CR, PAN, VEZ, BRA, PER, BOL, PAR, ARG
Similliserdia Fortes & Grazia	BRA
Stictochilus Bergroth	ARG, BRA
Taurocerus Amyot & Serville	MEX, GTM, CR, PAN, COL, GUY, BRA, PER, BOL
Tibilis Stål	BRA, ECU, PER, BOL
Vidada Rolston	PER
Piezodorini	
Piezodorus Fieber	USA, TRI, PUR, GTM, HON, NIC, JAM, CUB, DRE, St. Vincent, Grenada, PAN, VEZ, BRA, ECU, PAR, ARG
Procleticini	
Aleixus McDonald	BRA
Brepholoxa Van Duzee	USA, MEX, W.I.
Dendrocoris Bergroth	USA, MEX, GTM,
Lobepomis Berg	ARG
Neoderoploa Pennington	BOL, ARG
Odmalea Bergroth	GUF, TTO, PAN, VEZ, COL, BRA, BOL, ARG, URU
Parodmalea Rider	BRA
Procleticus Berg	ARG
Pseudobebaeus Distant	BRA, PER
Terania Pirán	ARG
Thoreyella Spinola	BRA, PAR, ARG, URU

(continued)

Table 22.5 (continued)

Carpocorini	
Strachiini	
Murgantia Stål	USA, MEX, GTM, ANT, CR, PAN, BRA, PER, BOL
Sciocorini	
Trincavellius Distant	ECU, CHI, PER
Unplaced	
Capivaccius Distant	MEX, HON, PAN, BRA, ARG
Cyptocephala Berg	USA, CENTRAL AM. (ALL), BAH, CUB, DRE, JAM, VIR IS., DRE, PUR, BRA, PER, BOL, ARG, URU
Patanius Rolston	BRA
Senectius Rolston	PER
Tepa Rolston & McDonald	USA, CURAÇAO
Thyanta Stål	USA, MEX, GTM, BLZ, ELS, HON, NIC, CR, PAN, BAH, CUB, JAM, DRE, HAI, PUR, VIR IS., BARBADO, BRITISH W.I., FRENCH W.I., CURAÇAO, TTO, VEZ, COL, SUR, FGU, BRA, PER, ECU, CHI, BOL, PAR, ARG, URU

Catacanthini was established by Atkinson (1888) to include *Catacanthus* Spinola (15 spp.) and *Chalcocoris* Dallas (two spp.), both genera not known in the Neotropical Region. More five Neotropical genera were added since then (Table 22.5). Some of the most colorful species of Pentatominae are included in Catacanthini. Keys for identification of species are available for *Arocera* (Rider 1992), *Rhyssocephala* (Rider 1991), and *Runibia* (Zwetsch and Grazia 2001). A study of the relationships within Catacanthini using morphologic characters is under development (J. Grazia, unpublished).

Carpocorini was first defined by Mulsant and Rey (1866), including *Carpocoris* Kolenati, *Dryocoris* Mulsant & Rey (junior synonym of *Holcostethus*), and *Peribalus* Mulsant & Rey (junior synonym of *Holcostethus*). This tribe is one of the most diverse in the subfamily (Cassis and Gross 2002) and includes many of the species of economic importance as *Euschistus heros* (F.), *Oebalus poecilus* (Dallas), and *Tibraca limbativentris* Stål. In the Neotropical Region, 53 genera are included in Carpocorini. A phylogenetic analysis based on morphology and DNA sequences is under development (J. Grazia, unpublished).

The tribe Halyini was established by Amyot and Serville (1843) to include a heterogeneous group of 16 genera, which currently included different families of Pentatomoidea and subfamilies and tribes of Pentatomidae. Dallas (1851) considered the group as a separate family; however the Halyini as currently circumscribed share all synapomorphies of the Pentatomidae. The tribe includes 82 genera with worldwide distribution, with great diversity in the Afrotropical and Australian regions (Linnavuori 1982; Gross 1975). Two genera are recorded in the Western Hemisphere (Table 22.5), with species found in the northern Neotropical Region.

Distant (1902) propose Mecideini to include *Aenaria* Stål and *Mecidea* Dallas. The proposal was questioned by Bergroth (1905), and *Mecidea* was grouped with other seven genera today included in the tribe Diemeniini Kirkaldy. Mecideini was accepted as a separated tribe (i.e., Rolston and McDonald 1979) but remains monotypic. The genus *Mecidea* includes 16 species distributed in all biogeographic regions except the Australian region (Sailer 1952). In the Neotropics, *Mecidea* has a disjunct distribution, with 4 described species (Sailer 1952; Thomas 2000; Grazia and Schwertner 2008).

Menidini was described to include four genera: *Cresphontes* Stål, *Antestia* Stål (currently Antestiini), *Apines* Dallas, and *Menida* Motschulsky (Atkinson 1888). The definition of the tribe and relationship among the genera, however, are in need of phylogenetic studies (Rider 1998). In the Neotropical Region, only two genera occur from the 28 actually assigned to the tribe (Table 22.5). Keys to the identification of the species of the Neotropical genera, *Rio* Kirkaldy and *Elanela* Rolston are available (Fortes and Grazia 2000; Grazia and Greve 2011) and *Rio* was revised (Grazia and Fortes 1995).

The tribe Nezarini is represented by four genera in the Neotropical Region: *Chinavia*, *Chlorochroa* Stål, *Roferta* Rolston, and *Nezara* Amyot & Serville. The last one is represented only by *N. viridula* (L.), which has a cosmopolitan distribution (Ferrari et al. 2010). The group was proposed to include species of *Nezara* and *Acrosternum* Fieber (Atkinson 1888). Cladistic analyses in the genus and group of species level were published recently (Ferrari et al. 2010; Genevcius et al. 2012; Genevcius and Schwertner 2014); however, the tribe lacks a formal definition, based on phylogenetic studies. A pictorial key for the Brazilian species of *Chinavia*, which include some secondary pests in South America, is presented in Schwertner and Grazia (2007).

Pentatomini is the most diverse tribe of Pentatominae. The original description brings as the only character defining the group, the scutellum not covering the wings or elytra (Leach 1815). No further definition was elaborated and the scattered phylogenetic evidences indicate that Pentatomini is not a monophyletic group (Gapud 1991; Hassan and Kitching 1993; Campos and Grazia 2006). Nonetheless, several studies using cladistic methodology at genus level for Neotropical groups put some light on to the relationships, especially among groups of species and genus level (Grazia 1997; Fortes and Grazia 2005; Simões et al. 2012; Greve et al. 2013). A phylogenetic analysis based in morphology and DNA sequences testing the *Chlorocoris* group was recently developed (Greve, personal communication), and the results derived from the total evidence analysis and the morphological analysis alone are not in complete agreement. The differences among the two hypotheses of relationship prevent the proposal of a unified classification for the taxa studied, not supporting the group as a monophyletic clade.

In the Neotropical Region, Piezodorini is represented only by the genus *Piezodorus* Fieber, with the species *P. guildinii* (Westwood) being a major pest of soybean in South America (Panizzi et al. 2000). The original description was based in species of *Piezodorus* and *Ambiorix* Stål and defines the tribe by "tibiae generally rounded, rarely furnished above with a narrow and obsolete furrow, or flat and emarginate; venter spinose at the base, spine sometimes extended to the head; apical angles of the sixth abdominal segment sometimes produced in a large acute tooth" (Atkinson 1888).

The tribe Procleticini was described to include two monotypic genera: *Lobepomis* Berg and *Procleticus* Berg (Pennington 1920). Currently it includes 11 genera (Table 22.5) almost exclusively Neotropical (Rider 1994; Bernardes et al. 2009; Schwertner and Grazia 2012). Only *Brepholoxa* and *Dendrocoris* have species occurring in the Nearctic region (Rider 1994). The group is considered monophyletic and is supported by metasternum sulcate, ventral rim of pygophore with a medial U-shaped emargination and with 1 + 1 process with different degrees of development, and gonocoxites 8 smaller than laterotergites 9 (Schwertner and Grazia 2012).

Amyot and Serville (1843) proposed the group Sciocorides to include genera currently in different subfamilies of Pentatomidae. A more restricted concept of the taxon was applied by Stål (1876), who restricted the group to *Dyroderes* Spinola, *Mennacarus* Amyot & Serville, *Pododus* Amyot & Serville, and *Sciocoris* Fállen. The tribe includes currently 11 genera distributed in all biogeographic regions, although poorly represented in the Neotropics, with only one genus (Table 22.5).

The group Strachiaries (Mulsant and Rey (1866)) included originally only the genus *Strachia* Hahn. Stål (1876) expanded the group to include *Agaeus* Dallas, *Bagrada* Stål, *Cinxia* Stål, *Eurydema* Laporte, and *Stenozygum* Dallas. Currently the tribe contains 14 genera worldwide, with only the genus *Murgantia* with representatives in the Neotropical Region (Brailovsky and Barrera 1889).

Other six genera treated as unplaced in Rider (2014) are registered in the Neotropical Region (*Capivacius* Distant, *Cyptocephala* Berg, *Patanius* Rolston, *Senectius* Rolston, *Tepa* Tolston & McDonald, and *Thyanta* Stål). With the exception of *Cyptocephala* (4 spp.) and *Thyanta* (27 spp.), the remaining genera are monotypic.

Key to the genera of Pentatominae of economic importance (modified from Rolston 1974 and Rolston and McDonald 1984):

1. Third urosternite with medial process.. 2
1′. Third urosternite without process.. 5
2. Specimens not longer than 9 mm, yellowish brown in color *Piezodorus* Fieber
2′. Specimens with at least 10 mm, green in life, or reddish brown or pale yellow in dry preserved.. 3
3. Mandibular plates surpassing clypeus in length; mandibular plates apex acute; lateral margin of pronotum partially crenulated .... *Arvelius* Spinola
3′. Mandibular plates not surpassing clypeus, rounded at apex; lateral margin of pronotum not ornamented.. 4
4. Wide colored band in the margins of the head, pronotum, or basal third of hemelytra; spiracle sometimes surrounded by yellow callus, without any other macula.. *Chinavia* Orian
4′'. Margins of the head, pronotum, and basal third of hemelytra concolor with the body or with a thin yellow band; spiracle without callus and green macula may be present.. *Nezara* Amyot & Serville
5. Distal apex of the femora bearing a strong spine; specimens green in vivo (can be reddish brown or pale yellow in dry preserved specimens) .... *Loxa* Amyot & Serville

5′. Distal apex of the femora unarmed, specimens usually brown or dark 6
6. Two rows of small tubercles present on inferior surface of femora .... *Agroecus* Dallas
6′. Inferior surface of the femora lacking tubercles 7
7. Small specimens, at most10 mm length, usually not reaching this length 8
7′. Specimens length over 10 mm 10
8. Specimens' dorsum uniform brown to dark in color, without ornamentation of different colors; insects shorter than 7 mm .... *Glyphepomis* Berg
8′. Pronotum and scutellum frequently with ivory callous; insects around 10 mm in length 9
9. Apex of first rostral segment projecting past bucculae .... *Mormidea* Amyot & Serville
9′. First rostral segment lying entirely between bucculae .... *Oebalus* Stål
10. Mandibular plates conspicuously longer than clypeus, with apices usually acute .... *Dichelops* Spinola
10′. Mandibular plates nearly equal to clypeus in length, sometimes a little longer or shorter 11
11. Peritreme long, evanescent .... *Thyanta* Stål
11′. Peritreme short 12
12. First antennal segment reaching and sometimes slightly surpassing the apex of the head; anterolateral margins of pronotum distinctly undulating, denticulate, or vertically rugose .... *Euschistus* Dallas
12′. First antennal segment not attaining the apex of the head; anterolateral margins entire .... *Tibraca* Stål

22.6.5 Remarks

The Pentatominae is the most speciose subfamily within Pentatomidae and are represented in all geographic regions. New characters should be investigated and analyzed in all Pentatominae groups, in order to help in the clarification of infra-subfamily relations. Cladistics studies at genus and species levels in the subfamily are needed.

22.7 Subfamilies Podopinae and Stirotarsinae

22.7.1 Introduction

The subfamilies Podopinae and Stirotarsinae are represented by only one species each in the Neotropical Region. The subfamily Podopinae is distributed worldwide; most of its diversity can be found in the Afrotropical, Oriental, and Palearctic regions (Davidová-Vilímová and Štys 1994; Rider 2014). The Stirotarsinae was erected recently (Rider 2000) to include *Stirotarsus abnormis* Bergroth, a species known only by a female specimen collected in Peru (Fig. 22.50a, b).

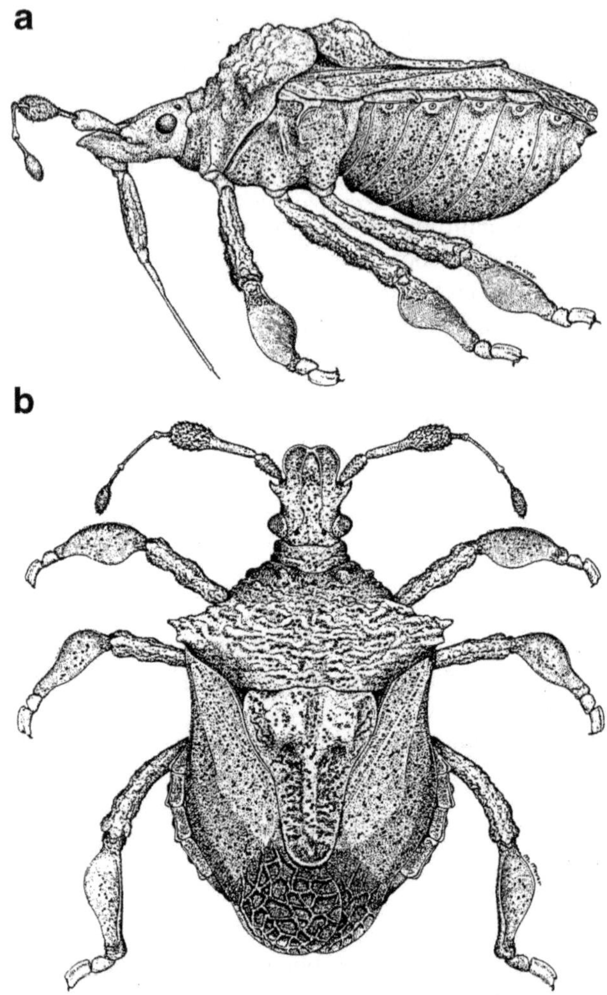

Fig. 22.50 *Stirotarsus abnormis* Bergroth, dorsal view (**a**) and ventral view (**b**) (From Rider (2000), with permission of the Annals of Entomological Society of America)

22.7.2 General Characteristics and Diagnosis

Characteristics of Podopinae include antenniferous tubercle visible in dorsal view; scutellum usually elongated, surpassing 2/3 the abdominal length; lateral rim of pygophore well developed, projecting on the ventral rim; ventral wall of pygophore with a median depression; lateral lobes of the conjunctiva strongly sclerotized (Schaefer 1981; Gapud 1991; Gapon and Konstantinov 2006).

The subfamily Stirotarsinae can be recognized by the antennal segments I, II, and V distinctly inflated; rostrum three-segmented, without apparent segmentation and anterior half enlarged; tarsi two-segmented, enlarged and with a longitudinal carina; tibiae foliate, dorsally concave. Other diagnostic characteristics, like the general body surface wrinkly, head longer than wide, sternum sulcated, and peritreme and evaporatorium reduced, are found in several groups of Pentatomidae and Pentatomoidea (i.e., Cyrtocorinae and some Podopinae).

22.7.3 Classification and Diversity

The Podopinae was first recognized by Stål (1876) as a group of genera. Lethierry and Severin (1893) gave status of subfamily to the group. The current classification was established by Davidová-Vilímová and Štys (1994), with five tribes: Brachycerocorini, Deroploini, Graphosomatini, Podopini, and Tarisini (Rider 2014). The monotypic genus *Neoleprosoma* was described by Kormilev and Pirán (1952) to include the Neotropical species *N. argentinensis* Kormilev & Pirán (Fig. 22.51). The authors considered *Neoleprosoma* related to *Leprosoma* Baerensprung and other similar genera of Podopinae from the Palearctic region, although the classification of the species was never reviewed. Davidová-Vilímová and Štys (1994) included *Neoleprosoma* in the tribe Graphosomatini. Distribution records of *N. argentinensis* include Argentina (Buenos Aires, Córdoba, and Santa Fé) and Uruguay (Artigas), in the southern region of South America.

Fig. 22.51 *Neoleprosoma argentinensis* Kormilev & Pirán (Photo by CF Schwertner)

The monotypic *Stirotarsus* was proposed by Bergroth (1911) to include *S. abnormis* from Peru, but because of its aberrant features, the classification among Pentatomidae was unsettled (Rider 2000). Based on the presence of a *ductus receptaculi* dilated and invaginated with three distinct walls in the female spermatheca and the unique set of characters found in *S. abnormis*, Rider (2000) proposed to place the genus in a separate subfamily. Poorly represented in collections (only one female), this taxon is probably endemic to the northern region of South America. No biological information is available; however previous hypothesis placed this taxon related to the asopines.

22.8 Concluding Remarks

The Pentatomidae, though predominantly herbivorous, includes predacious species and some economically important species of Heteroptera, such as the green stink bugs and the brown bugs. The group represents the second largest family of the infraorder Pentatomomorpha, with nine subfamilies, about 850 genera, and 4.700 species worldwide. The Neotropical fauna treated in this chapter comprises about 230 genera and more than 1,400 species, in seven subfamilies, representing 23 % of the world's genera and about 33 % of the species, suggesting that the number of new taxa awaiting description is immense. We hope this review of the family, which includes information on hosts, habits, economic importance, classification, checklists of species and/or genera, and keys to many of the Neotropical genera, will provide information to stimulate new interest in this large and diverse group of taxonomically and ecologically important insects.

Acknowledgments To David Rider (North Dakota State University, Fargo, USA) for allowing access to the collection at the NDSU and providing most of the Asopinae specimens that illustrate the group in this chapter. To Gerald Fauske (North Dakota State University, Fargo, USA), who kindly took the pictures of the asopines from the NDSU collection.

References

Adams TS (2000) Effect of diet and mating on oviposition in the twospotted stink bug *Perillus bioculatus* (F.) (Heteroptera: Pentatomidae). Ann Entomol Soc Am 93:1288–1293

Adams TS (2001) Morphology of the internal reproductive system of the male and female two-spotted stink bug, *Perillus bioculatus* (F.) (Heteroptera: Pentatomidae) and the transfer of products during mating. Invert Repr Develop 39:45–53

Amyot CJB, Serville A (1843) Histoire Naturelle des Insectes. Hémiptères. Librairie Encyclopedique de Roret ed., Paris. lxxvi and 675pp

Arismendi N, Thomas DB (2003) Pentatomidae (Heteroptera) of Honduras: a checklist with description of a new ochlerine genus. Insecta Mundi 17:219–236

Atkinson ET (1888) Notes on Indian Rhynchota, Heteroptera, no 4. J Asiatic Soc Bengal 57:118–184

Ávila C, Panizzi AR (1995) Occurrence and damage by *Dichelops* (*Neodichelops*) *melacanthus* (Dallas)(Heteroptera: Pentatomidae) on corn. An Soc Entomol Brasil 24:193

Avila-Núñez JL, Ortega LDO, Pisarelli MPC (2009) Un caso de depredación de adulto de *Gonodonta pyrgo* Cramer 1777 (Lepidoptera: Noctuidae) por *Euthyrhynchus floridanus* (Linnaeus 1767)(Heteroptera: Pentatomidae: Asopinae). Entomotropica 23:173–175

Banks N (1910) Catalogue of the Nearctic Hemiptera-Heteroptera. American Entomological Society, Philadelphia

Barão KR, Garbelotto TA, Campos LA, Grazia J (in press) Unusual looking pentatomids: reassessing the taxonomy of *Braunus* distant and *Lojus* McDonald (Heteroptera: Pentatomidae). Zootaxa

Barcellos A, Grazia J (2003a) Cladistics analysis and biogeography of *Brachystethus* Laporte (Heteroptera, Pentatomidae, Edessinae). Zootaxa 256:1–14

Barcellos A, Grazia J (2003b) Revision of *Brachystethus* (Heteroptera, Pentatomidae, Edessinae). Iheringia, Sér Zool 93:413–446

Becker M (1977) A review of the genus *Colpocarena* Stål (Heteroptera, Pentatomidae, Discocephalinae). Rev Brasil Biol 37:367–373

Becker M, Grazia J (1986) A new genus of Discocephaline Pentatomid from Ecuador (Heteroptera, Pentatomidae, Discocephalinae). Rev Bras Biol 46:453–459

Becker M, Grazia J (1989) Novas contribuições ao gênero *Ablaptus* Stål, 1864 (Heteroptera, Pentatomidae, Discocephalinae). Mem Inst O Cruz 84:57–68

Becker M, Grazia J (1992) Revisão do gênero *Agaclitus* Stål (Heteroptera, Pentatomidae, Discocephalinae). Rev Bras Entomol 36:831–842

Becker M, Grazia J (1995) The bug genus *Uncinala* Ruckes, 1965 (Heteroptera, Pentatomidae, Discocephalinae). Rev Bras Entomol 39:709–716

Becker M, Ruckes H (1969) A review of the genus *Sympiezorhincus* Spinola (Heteroptera, Pentatomidae, Discocephalinae). Am Mus Novit 2391:1–17

Bergroth E (1905) Rhynchota neotropica. I Revue d'Entomologie 24:104–112

Bergroth E (1911) Zur Kenntnis der neotropischen Arminen (Hem. Het.). Wien Entomol Zeitg 30:117–130

Bernardes JLC, Schwertner CF, Grazia J (2009) Cladistic analysis of *Thoreyella* and related genera (Hemiptera: Pentatomidae: Pentatominae: Procleticini). Zootaxa 1984:1–23

Bianchi FM, Matesco VC, Campos LA, Grazia J (2011) External morphology of the egg and the first and fifth instars of *Cyrtocoris egeris* Packauskas & Schaefer (Hemiptera: Heteroptera: Pentatomidae: Cyrtocorinae). Zootaxa 2991:29–34

Brailovsky H, Barrera E (1989) El Genero *Murgantia* Stal, com description de cuatro espécies nuevas y algunos registros nuevos (Hemiptera, Heteroptera, Pentatomidae, Pentatomini) de America Latina. An Inst Biol Univ Nal Auton México, Ser Zool 59:219–244

Brailovsky H, Cervantes-Peredo L, Mayorga C (1988) Hemiptera-Heteroptera de Mexico XL: La familia Cyrtocoridae Distant en la Estacion de Biologia Tropical "Los Tuxtlas" (Pentatomoidea). An Inst Biol Univ Nac Autón Méx (Zool) 58(1987):537–560

Briceño RD (2014) Caracterización del comportamiento y las señales vibratorias en *Euthyrhynchus floridanus* (Hemiptera: Pentatomidae) durante el cortejo y la cópula. Rev Biol Trop 62:95–108

Calizotti GS, Panizzi AR (2014) Behavior of first instar nymphs of *Edessa meditabunda* (F.) (Hemiptera: Pentatomidae) on the egg mass. Fla Entomol 97:277–280

Campos LA, Grazia J (1998) Revisão de *Glyphepomis* Berg, 1891 (Heteroptera, Pentatomidae). Rev Bras Entomol 41:203–212

Campos LA, Grazia J (2006) Análise cladística e biogeografia de Ochlerini (Hemiptera, Pentatomidae, Discocephalinae). Iheringia Sér Zool 96:147–163

Cassis G, Gross G (2002) Hemiptera: Heteroptera (Pentatomomorpha). In: Houston W, Wells A (eds) Zoological catalogue of Australia. Csiro Publishing, Melbourne, xiv + 737 p

Castro AAD, Canevari GDC, Pikart TG, Ribeiro RC, Serrão JE, Zanuncio TV, Zanuncio JC (2013) Salivary gland histology of the predator *Supputius cincticeps* (Heteroptera: Pentatomidae). Ann Entomol Soc Am 106:273–277

Cervantes-Peredo L, Ortega-León G (2014) Description of a new species of *Neoadoxoplatys* and immature stages of *Neoadoxoplatys saileri* Kormilev (Heteroptera: Pentatomidae) associated with Bamboo. Neotr Entomol 43:236–244

Chocorosqui VR, Panizzi AR (2004) Impact of cultivation systems on *Dichelops melacanthus* (Dallas)(Heteroptera: Pentatomidae) population and damage and its chemical control on wheat. Neotr Entomol 33:487–492

Costa Lima AM (1940) Insetos do Brasil. Tomo 2. Capitulo XXII. Hemipteros. Escola Nacional de Agronomia, Rio de Joneiro

Coudron TA, Kim Y (2004) Life history and cost analysis for continuous rearing of *Perillus bioculatus* (Heteroptera: Pentatomidae) on a zoophytogenous artificial diet. J Econ Entomol 97:807–812

Coudron TA, Wittmeyer J, Kim Y (2002) Life history and cost analysis for continuous rearing of *Podisus maculiventris* (Say) (Heteroptera: Pentatomidae) on a zoophytophagous artificial diet. J Econ Entomol 95:1159–1168

Couturier G, Kahn F (1989) Bugs of *Lincus* spp. vectors of Marchitez and Hartrot (oil palm and coconut diseases) on *Astrocaryum* spp., Amazonian native palms. Principes 33:9–20

Dallas WS (1851) List of the specimens of hemipterous insects in the collection of the British museum. Part 1. Trustees of the British Museum, London, 368 pp

da Silva VJ, Fernandes JAM (2012) A new species group in *Edessa* Fabricius, 1803 (Heteroptera: Pentatomidae: Edessinae). Zootaxa 3313:12–22

da Silva VJ, Nunes DM, Fernandes JAM (2013) *Paraedessa*, a new genus of Edessinae (Hemiptera: Heteroptera: Pentatomidae. Zootaxa 3716:395–416

Davidová-Vilímová J, Štys P (1994) Diversity and variation of trichobothrial patterns in adult Podopinae (Heteroptera: Pentatomidae). Acta Univ Carolinae Biol 37(1993):33–72

De Bortoli SA, Otuka AK, Vacari AM, Martins MI, Volpe HX (2011) Comparative biology and production costs of *Podisus nigrispinus* (Hemiptera: Pentatomidae) when fed different types of prey. Biol Contr 58:127–132

De Clercq P (2000) Predaceous stinkbugs (Pentatomidae: Asopinae). In: Schaefer CW, Panizzi AR (eds) Heteroptera of economic importance. CRC Press, Boca Raton, pp 737–789

De Clercq P (2008) Predatory Stink Bugs (Hemiptera: Pentatomidae, Asopinae). In: Capinera JL (ed) Encyclopedia of insects, 2nd Ed. Kluwer Academic Publishers, Dordrecht, pp 3042–3045

DeMenezes CWG, Soares MA, De Assis SL, De Menezes SJMC, Dos Santos JB, Zanuncio JC (2013) *Brontocoris tabidus* (Heteroptera: Pentatomidae) preying on *Podalia walkeri* (Lepidoptera: Megalopygidae) on eucalypt plants in Brazil. Fla Entomol 96:261–263

Desmier-De-Chenon R (1984) Recherches sur le genre *Lincus* Stål, Hemiptera Pentatomidae Discocephalinae, et son rôle éventuel dans la transmission de la Marchitez du palmier à huile et du Hart – Rot du cocotier. Oléagineux 39:1–6

Distant WL (1902) The fauna of British India, including Ceylon and Burma, vol 1. Published under the authority of the secretary of state for India in council, London. xxxviii + 438 pp

Dollet, M, Alvanil F, Diaz A, Louvet C, Gargani D, Segur C, Muller E, Julia JF, Calvache H, Renard JL, Maldes JM (1993) Les pentatomides vecteurs des trypanosomes associes au Hartrot du cocotier et Marchitez du palmier. In: ANPP (ed) Conférence Internationale sur les Ravageurs en Agriculture, 3ème, Montpellier, Annales… Tome III

Eberhard WG (1975) The ecology and behavior of a subsocial pentatomid bug and two scelionid wasps: strategy and counterstrategy in a host and its parasites. Smithson Contrib Zool 205:1–39

Farias P, Klein J, Sant'Ana J, Redaelli L, Grazia J (2012) First records of *Glyphepomis adroguensis* (Hemiptera, Pentatomidae) and its parasitoid, *Telenomus podisi* (Hymenoptera, Platygastridae), on irrigated rice fields in Rio Grande do Sul, Brazil. Rev Bras Entomol 56:383–384

Fernandes JAM (2010) A new genus and species of Edessinae from Amazon Region (Hemiptera: Heteroptera: Pentatomidae). Zootaxa 2662:53–65

Fernandes JAM, Campos LD (2011) A new group of species of *Edessa* Fabricius, 1803 (Hemiptera: Heteroptera: Pentatomidae). Zootaxa 3019:63–68

Fernandes JAM, Grazia J (2006) Revisão do gênero *Antiteuchus* Dallas (Heteroptera, Pentatomidae, Discocephalini). Rev Bras Entomol 50:165–231

Fernandes JAM, van Doesburg PH (2000) The *E. cervus*-group of *Edessa* Fabricius, 1903 (Heteroptera: Pentatomidae: Edessinae). Zool Med Leiden 74:151–165

Ferrari A, Schwertner CF, Grazia J (2010) Review, cladistic analysis and biogeography of *Nezara* Amyot & Serville (Hemiptera: Pentatomidae). Zootaxa 2424:1–41

Fieber FX (1860) Die europäischen Hemiptera. Halbflüger (Rynchota Heteroptera): Nach der analytischen Methode bearbeitet. Gerold, Wien

Fortes NDF, Grazia J (1990) Estudo dos estágios imaturos de *Edessa rufomarginata* (De Geer, 1773) (Heteroptera-Pentatomidae). An Soc Entomol Brasil 19:191–200

Fortes NDF, Grazia J (2000) Novas espécies de *Rio* Kirkaldy, 1909 (Heteroptera, Pentatomidae). Iheringia Sér Zool 88:67–102

Fortes NDF, Grazia J (2005) Review and cladistic analysis of *Serdia* Stål (Heteroptera, Pentatomidae, Pentatomini). Rev Bras Entomol 49:294–339

Gapon DA, Konstantinov FV (2006) On the structure of the aedeagus in shield bugs (Heteroptera, Pentatomidae): III. Subfamily Asopinae. Entomol Rev 86:806–819

Gapud VP (1991) A generic revision of the subfamily Asopinae, with consideration on its phylogenetic position in the family Pentatomidae and superfamily Pentatomoidea (Hemiptera-Heteroptera). Philippines Entomol 8:865–961

Garbelotto TA, Campos LA, Grazia J (2011) A new synonymy in *Alitocoris* Sailer (Hemiptera: Pentatomidae: Discocephalinae). Rev Bras Entomol 55:138–140

Garbelotto TA, Campos LA, Grazia J (2013) Cladistics and revision of *Alitocoris* with considerations on the phylogeny of the *Herrichella* clade (Hemiptera, Pentatomidae, Discocephalinae, Ochlerini). Zool J Linn Soc 168:452–472

Genevcius BC, Schwertner CF (2014) Review and phylogeny of the *geniculata* group, genus *Chinavia* (Heteroptera: Pentatomidae), with notes on biogeography and morphological evolution. Zootaxa 3847:33–56

Genevcius BC, Grazia J, Schwertner CF (2012) Cladistic analysis and revision of the *obstinata* group, genus *Chinavia* Orian (Hemiptera: Pentatomidae). Zootaxa 3434:1–30

Grazia J (1997) Cladistics analysis of the *Evoplitus* genus group of Pentatomini (Heteroptera, Pentatominae). J Comp Biol 2:115–129

Grazia J, Fortes NDF (1995) Revisão do gênero *Rio* Kirkaldy, 1909 (Heteroptera, Pentatomidae). Ver Bras Entomol 39:409–430

Grazia J, Greve C (2011) Contributions to the knowledge of *Elanela*: *Elanela jordi* sp. nov., from Amazonas, Brazil (Hemiptera: Heteroptera: Pentatomidae). Heterop Rev Entomol 11:261–266

Grazia J, Schwertner CF (2008) Pentatomidae e Cyrtocoridae. In: Claps LE, Debandi G, Roig-Juñent S (eds) Biodiversidad de Artrópodos Argentinos, vol 2, Tucumán. Sociedad Entomológica Argentina, Argentina, pp 223–234

Grazia J, Zwetsch A (2000) Duas novas espécies de *Ablaptus* Stål (Heteroptera, Pentatomidae, Discocephalinae). Rev Brasil Zool 17:989–994

Grazia J, Fortes NDF, Campos LA (1999) Pentatomoidea. In: Brandão C, Cancello E (eds) Biodiversidade do Estado de São Paulo, Brasil: síntese do conhecimento ao final do século XX. FAPESP, São Paulo, pp 101–112

Grazia J, Campos LA, Becker M (2000) Revision of *Cataulax* Spinola, with *Architas* Distant as a new synonymy (Heteroptera: Pentatomidae: Discocephalini). An Soc Entomol Bras 29:475–488

Grazia J, Schuh RT, Wheeler WC (2008) Phylogenetic relationships of family groups in Pentatomoidea based on morphology and DNA sequences (Insecta: Heteroptera). Cladistics 24:932–976

Greenstone MH, Szendrei Z, Payton ME, Rowley DL, Coudron TC, Weber DC (2010) Choosing natural enemies for conservation biological control: use of the prey detectability half-life to rank key predators of Colorado potato beetle. Entomol Exp Appl 136:97–107

Greve C, Schwertner CF, Grazia G (2013) Cladistic analysis and synopsis of *Chloropepla* Stål (Hemiptera: Heteroptera: Pentatomidae) with the description of three new species. Insect Syst Evol 44:1–43

Gross GF (1975) Plant-feeding and other bugs (Hemiptera) of South Australia. Heteroptera, Part I. A.B. James, Adelaide

Guerra TJ, Camarota F, Castro FS, Schwertner CF, Grazia J (2011) Throphobiosis between ants and *Eurystethus microlobatus* Ruckes 1966 (Hemiptera: Heteroptera: Pentatomidae) a cryptic, gregarious and subsocial stinkbug. J Nat Hist 45:1101–1117

Hassan SA, Kitching IJ (1993) A cladistic analysis of the tribes of the Pentatomidae (Heteroptera). Jap J Entomol 61:651–669

Henry TJ, Froeschner RC (1988) Catalog of the Heteroptera or true bugs of Canada and the continental United States. Brill, Leiden

Horváth G (1916) Revisio Cyrtocorinarum. Ann Musei Nat Hung 14:219–224

Iannacone J, Perez D, Tueros A (2007) Ciclo de vida y aspectos poblacionales de *Edessa* aff. *aulacosterna* Stal, 1872 (Heteroptera: Pentatomidae) chinche del fruto del camu camu (Myrtaceae) en zona de restinga, Ucayali, Perú. Acta Amazon 37:635–642

Kirkaldy GW (1909) Catalogue of the Hemiptera (Heteroptera) Vol. I: Cimicidae. Feliz L. Dames, Berlin

Kormilev NA (1955) La subfamilia Cyrtocorinae Distant en la Argentina, (Hemiptera Pentatomoidae [sic]). *Rev*. Ecuat Ent Parasit 2:321–334

Kormilev NA, Pirán AA (1952) La subfamilia Graphosomatinae (Jokow.) en la Argentina, con la descripcion de un genero y una especie nuevos. An Soc Cient Argent 153:212–218

Laumann RA, Kavčič A, Moraes MC, Borges M, Čokl A (2013) Reproductive behaviour and vibratory communication of the neotropical predatory stink bug *Podisus nigrispinus*. Physiol Entomol 38:71–80

Leach WE (1815) Order VIII. Hemiptera. In: Brewster D (ed) The Edinburgh Encyclopedia. Entomology, Edinburgh, pp 120–126

Lemos WP, Ramalho FS, Serrão JE, Zanuncio JC (2003) Effects of diet on development of *Podisus nigrispinus* (Dallas) (Het., Pentatomidae), a predator of the cotton leafworm. J Appl Entomol 127:389–395

Lemos WDP, Ramalho FDS, Serrão JE, Zanuncio JC (2005) Morphology of female reproductive tract of the predator *Podisus nigrispinus* (Dallas) (Heteroptera: Pentatomidae) fed on different diets. Braz Arch Biol Technol 48:129–138

Lemos WP, Zanuncio JC, Ramalho FS, Serrão JE (2009) Fat body of the zoophytophagous predator *Brontocoris tabidus* (Het.: Pentatomidae) females: Impact of the herbivory and age. Micron 40:635–638

Lemos WP, Zanuncio JC, Ramalho FS, Zanuncio VV, Serrão JE (2010) Herbivory affects ovarian development in the zoophytophagous predator *Brontocoris tabidus* (Heteroptera, Pentatomidae). J Pest Sci 83:69–76

Lethierry L, Severin G (1893) Catalogue général des Hémiptères. Bruxelles, Pentatomidae

Lima MS, Fernandes JAM, Lima IMM (2010) Record of *Edessa scabriventris* Stăl (Hemiptera: Pentatomidae) associated to *Eugenia uniflora* (Brazilian-Cherry) and *Psidium guajava* (Guava) (Myrtaceae), in north-northeastern Brazil. Neotr Entomol 39:674–675

Linnavuori RE (1982) Pentatomidae and Acanthosomatidae (Heteroptera) of Nigeria and the Ivory Coast, with remarks on species of the adjacent countries in West and Central Africa. Acta Zool Fenn 163:1–176

Llosa JF, Couturier G, Kahn F (1990) Notes on the ecology of *Lincus spurcus* and *L. malevolus* (Heteroptera: Pentatomidae: Discocephalinae) on Palmae in Forests of Peruvian Amazonia. Ann Soc Entomol Fr (N S) 26:249–254

Lopes OJ, Link D, Basso IV (1974) Pentatomídeos de Santa Maria – lista preliminar de plantas hospedeiras. Rev Cent Cien Rur 4:317–322

Malaguido AB, Panizzi AR (1998a) *Alchaeorrhynchus grandis* (Dallas): An eventual predator of *Chlosyne lacinia saundersii* Doubleday & Hewitson on sunflower in Northern Paraná state. An Soc Entomol Brasil 27:671–674

Malaguido AB, Panizzi AR (1998b) Pentatomofauna associated with sunflower in northern Paraná state, Brazil. An Soc Entomol Bras 27:473–475

Manfredi-Coimbra S, Silva JJ, Chocorosqui VR, Panizzi AR (2005) Danos do percevejo barriga-verde *Dichelops melacanthus* (Dallas) (Heteroptera: Pentatomidae) em trigo. Ciên Rur 35:1243–1247

Marques O, Coutinho M, Gil-Santana H, Schmidt C (2006) *Arvelius albopunctactus* (DeGeer) (Hemiptera: Pentatomidae) in tobacco (*Nicotiana tabacum* L.) in Brazil. Magistra 18:110–111

Matesco VC, Grazia J, Campos LA (2007) Description of new genus and species of Ochlerini from Central America (Hemiptera: Pentatomidae: Discocephalinae). Zootaxa 1565:63–68

Matlock RB Jr (2005) Impact of prey size on prey capture success, development rate, and survivorship in *Perillus bioculatus* (Heteroptera: Pentatomidae), a predator of the Colorado potato beetle. Environ Entomol 34:1048–1056

Matos Neto FC, Zanuncio JC, Freitas LC, Gomes BMR (1999) Nymphal development of the predator *Podisus rostralis* (Heteroptera: Pentatomidae) fed with three alternative prey. Brenesia 50:72–77

Matos Neto FC, Oliveira HN, Zanuncio JC, Holtz AM, Oliveira I, Fialho MC (2004) Weight gain of the predator *Podisus distinctus* (Heteroptera: Pentatomidae) with combinations of the preys *Tenebrio molitor* (Coleoptera: Tenebrionidae) and *Musca domestica* (Diptera: Muscidae). Rev Biol Trop 52:101–108

McPherson JE (1982) The Pentatomoidea (Hemiptera) of northeastern North America with emphasis of the fauna of Illinois. Southern Illinois University Press, Carbondale

Medeiros RSD, Silva AMCD, Zanuncio JC, Ramalho FDS, Serrão JE, Cecon PR (2004) Oviposition pattern of the predator *Podisus nigrispinus* (Heteroptera: Pentatomidae) under different temperatures. Biocontrol Sci Technol 14:487–498

Menezes CWG, Soares MA, De Assis SL, De Menezes SJMC, Dos Santos JB, Zanuncio JC (2013) *Brontocoris tabidus* (Heteroptera: Pentatomidae) preying on *Podalia walkeri* (Lepidoptera: Megalopygidae) on eucalypt plants in Brazil. Fla Entomol 96:261–263

Molina-Rugama AJ, Zanuncio JC, Zanuncio TV, Oliveira ML (1998) Reproductive strategy of *Podisus rostralis* (Stal)(Heteroptera: Pentatomidae) females under different feeding intervals. Biocontrol Sci Technol 8:583–588

Mulsant É, Rey C (1866) Histoire naturelle des punaises de France, vol II, Pentatomides. Savy et Deyrolle, Paris

Neves RC, Torres JB, Vivan LM (2009) Reproduction and dispersal of wing-clipped predatory stinkbugs, *Podisus nigrispinus* in cotton fields. BioControl 54:9–17

Oliveira I, Zanuncio JC, Serrão JE, Zanuncio TV, Pinon TBM, Fialho MDCQ (2005) Effect of female weight on reproductive potential of the predator *Brontocoris tabidus* (Signoret, 1852) (Heteroptera: Pentatomidae). Braz Arch Biol Technol 48:295–301

Ortega-León G, Thomas DB (2010) Especie nueva de *Eritrachys* (Hemiptera: Pentatomidae: Ochlerini) de Ecuador. Rev Mex Biodivers 81:61–64

Packauskas RJ, Schaefer CW (1998) Revision of the Cyrtocoridae (Hemiptera: Pentatomoidea). Ann Entomol Soc Am 91:363–386

Paleari LM (2013) Developmental biology, polymorphism and ecological aspects of *Stiretrus decemguttatus* (Hemiptera, Pentatomidae), an important predator of cassidine beetles. Rev Bras Entomol 57:75–83

Panizzi AR (1997) Wild hosts of pentatomids: ecological significance and role in their pest status on crops. Annu Rev Entomol 42:99–122

Panizzi AR, Machado-Neto E (1992) Development of nymphs, and feeding habits of nymphal and adult *Edessa meditabunda* (Heteroptera: Pentatomidae) on soybean and on sunflower. Ann Entomol Soc Am 85:477–482

Panizzi A, Slansky F Jr (1985) Review of phytophagous pentatomids (Hemiptera: Pentatomidae) associated with soybean in the Americas. Fla Entomol 68:184–214

Panizzi AR, McPherson JE, James DG, Javahery M, McPherson RM (2000) Stink bugs (Pentatomidae). In: Schaefer CW, Panizzi AR (eds) Heteroptera of economic importance. CRC Press, Boca Raton, pp 421–474

Pennington M (1920) Lista de los Hemiptera Heterόptera de la República Argentina. Primera parte, Buenos Aires, pp 29–47

Pires EM, Zanuncio JC, Serrão JE (2011) Cannibalism of *Brontocoris tabidus* and *Podisus nigrispinus* during periods of pre-release without food or fed with *Eucalyptus cloeziana* plants. Phytoparasitica 39:27–34

Rider DA (1991) *Rhyssocephala*, new genus, with the description of three new species (Heteroptera: Pentatomidae). J N Y Entomol Soc 99:583–610

Rider DA (1992) Revision of *Arocera* Spinola, with the description of two new species (Heteroptera: Pentatomidae). J N Y Entomol Soc 100:99–136

Rider DA (1994) A generic conspectus of the Tribe Procleticini Pennington (Heteroptera, Pentatomidae), with the description of *Parodmalea rubella*, new genus and species. J N Y Entomol Soc 102:193–221

Rider DA (1998) Nomenclatural changes in the Pentatomoidea (Hemiptera-Heteroptera: Pentatomidae, Tessaratomidae). III. Generic level changes. Proc Entomol Soc Wash 100:504–510

Rider DA (2000) Stirotarsinae, new subfamily for *Stirotarsus abnormis* Bergroth (Heteroptera: Pentatomidae). Ann Entomol Soc Am 93:802–806

Rider DA (2014) Pentatomoidea home page. North Dakota State University, Fargo

Rizzo HFE (1971) Aspectos morfologicos y biologicos de *Edessa meditabunda* (F.) (Hemiptera, Pentatomidae). Rev Per Entomol 14:272–281

Rizzo HF, Saini ED (1987) Aspectos morfologicos y biologicos de *Edessa rufomarginata* (De Geer) (Hemiptera, Pentatomidae). Rev Fac Agron 8:51–63

Rodrigues ARS, Torres JB, Siqueira HAA, Teixeira VW (2009) *Podisus nigrispinus* requer cópulas longas para o sucesso reprodutivo. Neotr Entomol 38:746–753

Rolston LH (1974) Revision of the genus *Euschistus* in Middle America (Hemiptera, Pentatomidae, Pentatomini). Entomol Am 48:1–102

Rolston LH (1981) Ochlerini, a new tribe in Discocephalinae (Hemiptera: Pentatomidae). J N Y Entomol Soc 89:40–42

Rolston LH (1984) A revision of the genus *Priapismus* Distant (Hemiptera: Pentatomidae). J Kansas Entomol Soc 87:119–126

Rolston LH (1990) Key and diagnoses for the genera of 'broadheaded' discocephalines (Hemiptera: Pentatomidae). J N Y Entomol Soc 98:14–31

Rolston LH (1992) Key and diagnoses for the genera of Ochlerini (Hemiptera: Pentatomidae: Discocephalinae). J N Y Entomol Soc 100:1–41

Rolston LH, McDonald FJD (1979) Keys and diagnoses for the families of Western Hemisphere Pentatomoidea, subfamilies of Pentatomidae and tribes of Pentatominae (Hemiptera). J N Y Entomol Soc 87:189–207

Rolston LH, McDonald FJD (1984) A conspectus of Pentatomini of the Western Hemisphere. Part 3 (Hemiptera: Pentatomidae). J N Y Entomol Soc 92:69–86

Rolston LH, McDonald FJD, Thomas DB (1980) A conspectus of Pentatomini genera of the Western Hemisphere. Part I (Hemiptera: Pentatomidae). J N Y Entomol Soc 88:120–132

Ruckes H (1964) The genus *Antiteuchus* Dallas, with descriptions of new species (Heteroptera, Pentatomidae, Discocephalinae). Bull Am Musf Nat Hist 127:47–102

Ruckes H (1965) Several new genera and species of discocephaline pentatomids (Heteroptera: Pentatomidae). New York Entomol Soc LXXIII:114–134

Ruckes H (1966a) The genus *Eurystethus* Mayr, with the descriptions of new species (Heteroptera, Pentatomidae, Discocephalinae). Am Mus Novitates 2254:1–37

Ruckes H (1966b) An analysis and a breakdown of the genus *Platycarenus* Fieber (Heteroptera, Pentadomidae, Discocephalinae). Am Mus Novitates 2255:1–42

Ruckes H (1966c) The genus *Dryptocepha* Laporte (Heteroptera, Pentadomidae, Discocephalinae). Am Mus Novitates 2256:1–31

Ruckes H (1966d) A review of the bug genus *Pelidinocoris* Stål (Heteroptera, Pentadomidae, Discocephalinae). Am Mus Novitates 2257:1–8

Ruckes H, Becker M (1970) The bug genus *Coriplatus* White (Heteroptera, Pentatomidae, Discocephalinae). Am Mus Novit 2409:1–11

Sailer RI (1952) A review of the stink bugs of the genus *Mecidea*. Proc US Natl Mus 102:471–505

Saini ED (1994) Aspectos morfologicos y biologicos de *Podisus connexivus* Bergroth (Heteroptera, Pentatomidae). Rev Soc Entomol Argent 53:35–42

Santos AV, Albuquerque GS (2001a) Custos ecofisiológicos do cuidado maternal em *Antiteuchus sepulcralis* (Fabricius) (Hemiptera: Pentatomidae). Neotr Entomol 30:105–111

Santos AV, Albuquerque GS (2001b) Eficiência do cuidado maternal de *Antiteuchus sepulcralis* (Fabricius) (Hemiptera: Pentatomidae) contra inimigos naturais do estágio de ovo. Neotr Entomol 30:641–646

Schaefer CW (1981) Genital capsules, trichobothria, and host plants of the Podopinae (Pentatomidae). Ann Entomol Soc Am 74:590–601

Schaefer CW, Packauskas RJ, Eger JE (1998) Nymphs of *Cyrtocoris egeris* (Hemiptera: Pentatomoidea: Cyrtocoridae). Ann Entomol Soc Am 91:452–457

Schaefer CW, Panizzi AR, Coscarón MC (2005) New records of plants fed upon by the uncommon heteropterans *Cyrtocoris egeris* Packauskas & Schaefer and *C. trigonus* (Germar) (Hemiptera: Cyrtocoridae) in South America. Neotr Entomol 34:127–129

Schuh RT, Slater JA (1995) True bugs of the world (Hemiptera: Heteroptera): classification and natural history. Cornell University Press, Ithaca

Schwertner CF, Grazia J (2007) O gênero *Chinavia* Orian (Hemiptera, Pentatomidae, Pentatominae) no Brasil, com chave pictórica para os adultos. Rev Bras Entomol 51:416–435

Schwertner CF, Grazia J (2012) Review of the Neotropical genus *Aleixus* McDonald (Hemiptera: Heteroptera: Pentatomidae: Procleticini), with description of a new species and cladistic analysis of the tribe Procleticini. Entomol Am 118:252–262

Silva DP, Oliveira PS (2010) Field biology of *Edessa rufomarginata* (Hemiptera: Pentatomidae): phenology, behavior, and patterns of host plant use. Environ Entomol 39:1903–1910

Silva AGD'A, Gonçalves CR, Galvão DM, Gonçalves AJL, Gomes J, Silva MN, Simoni L (1968) Quarto Catálogo dos Insetos que Vivem nas Plantas do Brasil – Seus Parasitas e Predadores. Parte II, vol I. Ministry of Agriculture, Rio de Janeiro

Silva FAC, Silva JJ, Depieri RA, Panizzi AR (2012) Feeding activity, salivary amylase activity and superficial damage to soybean seed by adult *Edessa meditabunda* (F.) and *Euschistus heros* (F.) (Hemiptera: Pentatomidae). Neotr Entomol 41:386–390

Simões FL, Campos LA (2014) Taxonomic notes on *Ochlerus*: revisiting Herrich-Schäffer's species (Hemiptera: Pentatomidae: Discocephalinae: Ochlerini). Zootaxa 3774:496–500

Simões FL, Ferrari A, Grazia J (2012) Is *Elsiella* Froeschner, 1981 a valid genus? (Hemiptera: Heteroptera: Pentatomidae: Pentatominae). Zootaxa 3238:39–48

Soares MA, Batista JD, Zanuncio JC, Lino-Neto J, Serrão JE (2011) Ovary development, egg production and oviposition for mated and virgin females of the predator *Podisus nigrispinus* (Heteroptera: Pentatomidae). Acta Sci Agron 33:597–602

Sousa-Souto L, Júnior E, Santos W, Lima ER, Zanuncio JC, Fonseca MDG (2006) Sperm depletion: a cost for single mated females of *Podisus nigrispinus* (Dallas, 1851) (Heteroptera: Pentatomidae). Braz Arch Biol Technol 49:923–926

Stahel G (1964) Die Siebröhrenkrankheit (Phloemnekrose, Flagellatose) des Kaffebaumes. Neth J Agric Sci 4:260–264

Stål C (1868) Hemiptera Fabriciana. Kong Sv Vet-Ak Handl 7:1–148

Stål C (1872) Enumeratio Hemipterorum. Bildrag till en förteckning öfver alla hittilis kända Hemiptera, jemte systematiska meddelanden. 2. Bidrag till. Kong Sv Vet-Ak Handl 10:1–159

Stål C (1876) Enumeratio Hemipterorum. Bidrag till en Förteckning öfver alla hittills kända Hemiptera, Jemte Systematiska Meddelanden. Kong Sv Vet-Ak Handl 14:1–162

Thomas DB (1992) Taxonomic synopsis of the Asopinae Pentatomidae (Heteroptera) of the Western Hemisphere, Monographs 16. The Thomas Say Foundation, ESA, Lanham, pp 1–156

Thomas DB (1994) Taxonomic synopsis of the Old World asopine genera Pentatomidae: Heteroptera). Insecta Mundi 8:145–212

Thomas DB (2000) Pentatomidae (Hemiptera). In: Bousquets JEL, Soriano EG, Papavero N (eds) Biodiversidad, taxonomia y biogeografía de artrópodos de México; hacia una síntesis de su conocimiento, vol II. Universidad Nacional Autónoma de México, Instituto de Biologia, México, pp 335–352

Torres JB, Zanuncio JC, Moura MA (2006) The predatory stinkbug *Podisus nigrispinus*: biology, ecology and augmentative releases for lepidopteran larval control in *Eucalyptus* in Brazil. Biocontrol News Info 27:1–18

Vacari AM, De Bortoli SA, Goulart RM, Volpe HXL, Otuka AK, Veiga ACP (2013) Comparison of eggs, larvae, and pupae of *Plutella xylostella* (Lepidoptera: Plutellidae) as prey for *Podisus nigrispinus* (Hemiptera: Pentatomidae). Ann Entomol Soc Am 106:235–242

Vivan LM, Torres JB, Veiga AF (2003) Development and reproduction of a predatory stinkbug, *Podisus nigrispinus* in relation to two different prey types and environmental conditions. BioControl 48:155–168

Waldkircher G, Webb MD, Maschwitz U (2004) Description of a new shieldbug (Hemiptera: Plataspidae) and its close association with a species of ant (Hymenoptera: Formicidae) in Southeast Asia. Tijdschr Entomol 147:21–28

Wittmeyer JL, Coudron TA (2001) Life table parameters, reproductive rate, intrinsic rate of increase, and estimated cost of rearing *Podisus maculiventris* (Heteroptera: Pentatomidae) on an artificial diet. J Econ Entomol 94:1344–1352

Zanuncio JC, Alves JB, Zanuncio TV, Garcia JF (1994) Hemipterous predators of eucalypt defoliator caterpillars. For Ecol Manage 65:65–73

Zanuncio JC, Lacerda MC, Junior JSZ, Zanuncio TV, Da Silva AMC, Espindula MC (2004) Fertility table and rate of population growth of the predator *Supputius cincticeps* (Heteroptera: Pentatomidae) on one plant of *Eucalyptus cloeziana* in the field. Ann Appl Biol 144:357–361

Zanuncio TV, Zanuncio JC, Serrão JE, Medeiros RS, Pinon T, Sediyama CA (2005a) Fertility and life expectancy of the predator *Supputius cincticeps* (Heteroptera: Pentatomidae) exposed to sublethal doses of permethrin. Biol Res 38:31–39

Zanuncio JC, Beserra EB, Molina-Rugama AJ, Zanuncio TV, Pinon TBM, Maffia VP (2005b) Reproduction and longevity of *Supputius cincticeps* (Het.: Pentatomidae) fed with larvae of *Zophobas confusa*, *Tenebrio molitor* (Col.: Tenebrionidae) or *Musca domestica* (Dip.: Muscidae). Braz Arch Biol Technol 48:771–777

Zanuncio JC, Lemos WP, Lacerda MC, Zanuncio TV, Serrao JE, Bauce E (2006) Age-dependent fecundity and fertility life tables of the predator *Brontocoris tabidus* (Heteroptera: Pentatomidae) under field conditions. J Econ Entomol 99:401–407

Zwetsch A, Grazia J (2001) Revision of the genus *Runibia* (Heteroptera, Pentatomidae, Pentatomini). Iheringia Sér Zool 91:5–28

Chapter 23
Shield Bugs (Scutelleridae)

Joseph Eger Jr., Aline Barcellos, and Luciana Weiler

Abstract Scutelleridae Leach includes three subfamilies, 25 genera, and almost 100 species in the Neotropics. Both subfamilies Scutellerinae and Elvisurinae are represented by a single genus, and Pachycorinae contains the majority of the species. Neotropical shield bugs are usually dull colored, but polymorphisms and sexual dimorphisms have caused a considerable confusion in their taxonomy. Several genera need revision, and many new taxa are awaiting formal descriptions. Biological and ecological data are scarce, with a few species reported as minor pests.

23.1 Introduction

The family Scutelleridae Leach is global in distribution and quite diverse in size and appearance. These bugs have typically been referred to as "shield bugs," but recent papers on Old World representatives have used the term "jewel bugs" (Cassis and Vanags 2006; Tsai et al. 2011) because of their bright colors. However, this term does not accurately describe all scutellerids as many are rather dull colored. The Palearctic Scutelleridae are fairly well known and were recently catalogued by Göllner-Scheiding (2006). Australian and Taiwanese representatives have been revised by Cassis and Vanags (2006) and by Tsai et al. (2011), respectively. Neotropical shield bugs tend to be rather cryptic in coloration and poorly known

J. Eger Jr. (✉)
Dow AgroSciences LLC, 2606 S. Dundee Street, Tampa, FL 32629, USA
e-mail: jeeger@dow.com

A. Barcellos
Museu de Ciências Naturais, Fundação Zoobotânica do Rio Grande do Sul,
Rua Dr. Salvador França, 1427, Porto Alegre, RS 90690-000, Brazil
e-mail: alinebar.fzb@gmail.com

L. Weiler
Programa de Pós-Graduação em Biologia Animal, Instituto de Biociências,
Universidade Federal do Rio Grande do Sul (UFRGS), Av. Bento Gonçalves 9500,
prédio 43435, Bairro Agronomia, Porto Alegre, RS 91501-970, Brazil
e-mail: weiler.luciana@gmail.com

A.R. Panizzi, J. Grazia (eds.), *True Bugs (Heteroptera) of the Neotropics*,
Entomology in Focus 2, DOI 10.1007/978-94-017-9861-7_23

taxonomically and biologically. Species numbers are probably greatly underestimated. Polymorphism in several species has contributed to a considerable degree of nomenclatural confusion as well. Overall, these little studied bugs are in need of taxonomic, ecological, and biological studies in the Neotropics.

Contributing to the lack of research on Neotropical Scutelleridae is the fact that there are few economically important species in this area. In the Middle East, species of *Eurygaster* Laporte are major pests of wheat, while *Calidea* spp. and *Tectocoris diophthalmus* (Thunberg) are pests of cotton in Africa and Australia, respectively (Javahery et al. 2000). A number of other species are minor pests in the Old World, but New World species tend to be minor pests of minor crops.

23.2 General Characteristics and Diagnosis

Shield bugs represent a diverse pentatomoid family, with 80 genera and about 500 species worldwide (Tsai et al. 2011). They are recognized by the beetle-like appearance, due to the great development of mesoscutellum (scutellum), which covers almost the entire abdomen. An enlarged scutellum occurs in a number of other pentatomoid families, including Aphylidae and Lestoniidae which are found only in Australia. In the New World, Thyreocoridae, Cyrtocoridae, and some Pentatomidae (primarily in the subfamilies Asopinae and Podopinae) have an enlarged scutellum. The tibiae of thyreocorids possess strong spines, much like cydnids, whereas scutellerids have strong setae at most. Cyrtocorids have a quadrate body shape and a spinose to quadrate projection on the scutellum which is lacking in scutellerids, while asopine pentatomids have an enlarged rostrum not found in scutellerids. Finally, the trichobothria of podopines are single while those of scutellerids are paired. Scutellerids vary greatly in shape, color, and size, but they are usually medium-sized pentatomoids (10–20 mm). Most of the species diversity occurs in the tropics (Lattin 1964).

Besides the enlarged scutellum, other diagnostic, but not always present, characters are rostrum attaining at least mesocoxae, prosternal keel well developed, hemelytra weakly sclerotized, legs usually unarmed, paired trichobothria, external genitalia visible or concealed by seventh sternite, spermathecal bulb with simple wall, genital capsule with a ringlike 8th segment, phallus often with two or three pairs of well-developed conjunctival projections, and parameres commonly uncinate (Schuh and Slater 1995; Tsai et al. 2011; Barcellos et al. 2014).

The location of the metathoracic ostiole, the presence or absence of the ostiolar peritreme, and the length and shape of the latter are diagnostic characters useful for identifying the different genera of Neotropical scutellerids and may be helpful in the separation of species in a limited number of genera. The external genitalia within Scutelleridae are usually simple but may be diagnostic for separation of species. Specifically, the shape of the posterior margin of the genital cup, the parameres, and the female genital plates may offer valuable taxonomic characters at the species level, depending on the genus. Internal genitalia, especially the development and shape of the conjunctival projections and phallotheca, are usually diagnostic. Variations in the spermatheca, including the shape of the spermathecal bulb, and distal and proximal flanges are important as described by authors such as McDonald (1966), Kumar (1965), and Candan et al. (2007, 2010).

Unlike the Afrotropical and Oriental shield bugs, those from Neotropics are usually dull brown, with only a few aposematic, and frequently polymorphic, species (e.g., species of *Agonosoma* Laporte, *Augocoris* Burmeister, *Pachycoris* Burmeister, and *Tiridates* Stål). Sexually dimorphic species are also found at least in three genera (*Agonosoma*, *Lobothyreus* Mayr, and *Crathis* Stål). Both polymorphism and sexual dimorphism have caused many synonymies and considerable confusion in the taxonomy of this group.

In the Neotropics, 25 genera are recorded, some of them also distributed in Nearctic region along with several undescribed genera. About 112 species are recorded in this region; however, as most of the genera need revision, this number is probably underestimated.

23.3 General Biology and Ecology

For Neotropical fauna, there are studies on the biology of *A. flavolineata* Laporte (Paleari 1992b), *P. klugii* Burmeister (Cervantes-Peredo 2002), *P. torridus* (Scopoli) (Rodrigues et al. 2011), and *Dystus puberulus* Stål (Cervantes-Peredo 2004).

Scutellerids are generally considered to be exclusively phytophagous, although some authors consider these bugs generalists (Javahery et al. 2000). Host plant records in Neotropical shield bugs are scarce (Table 23.1). Chérot et al. (1998) reported *Chrysocoris* (*Fitha*) *fascialus* (White) feeding on a decomposing snake in Thailand, while Eger et al. (2015) have collected large numbers of New World scutellerids at butterfly traps baited with putrefied fish carrion.

Maternal care has been described for *A. flavolineata* (Paleari 1992b), *P. torridus* (Hussey 1934; Santos et al. 2005), *P. klugii* (Cervantes-Peredo 2002), and also *P. stallii* Uhler (Williams et al. 2005). Females guard their eggs and first instar nymphs and may transmit aggregation semiochemicals to first instar nymphs. Nymphs become less aggregated as they develop, and fifth instars disperse totally.

Aposematism is usually associated with acquisition of toxic compounds by shield bugs from their host plants, providing protection against predation. Such strategy has been described for *Pachycoris*, whose species are known to feed preferentially on euphorbs. Williams et al. (2001) studied the chemical compounds of exocrine glands of *P. stallii*, a shield bug occurring in northwestern Mexico. This bug feeds on *Croton californicus* Muell. Arg., a Euphorbiaceae toxic to vertebrates. *Pachycoris stallii* uses toxic compounds obtained from this host plant to prevent predation by other invertebrates. The mostly South American *P. torridus* has the euphorb *Cnidoscolus pubescens* Pohl as its supposed preferential host plant (Santos et al. 2005) although there are several other records of host plants for this species. *Agonosoma flavolineata*, *Chelycoris haglundi* (Montandon), *C. lethierryi*

Table 23.1 Distribution and biology of Neotropical Scutelleridae[a]

Species	Distribution	Biology and notes
Scutellerinae		
Augocoris ehrenbergii Germar	Mexico, Guatemala, Venezuela*, Brazil*, Argentina*	
Augocoris gomesii Burmeister	USA (Arizona) and Mexico, South to Brazil	Attracted to light (Torre-Bueno 1914)
Augocoris illustris (F.)	USA (Florida), Mexico South to Argentina, and Brazil	Host plants: Phyllanthaceae, *Phyllanthus epiphyllanthus* L.; Sapotaceae, *Chrysophyllum cainito* L., *C. oliviforme* L., *Manilkara zapota* (L.) van Royen, *Mimusops elengi* L., *Sideroxylon foetidissimum* Jacq. (Barber and Bruner 1932; Wolcott 1923, 1936; Bruner et al. 1945).
		Attracted to light (Callan 1948; Hussey 1956)
Augocoris nigripennis Dallas	Venezuela	
Augocoris poeyi Uhler	Cuba	
Augocoris rufus Dallas	Colombia	
Augocoris rugulosus Herrich-Schaeffer	Brazil	
Pachycorinae		
Acantholomidea porosa (Germar)	Southern USA, Cuba, Mexico, Guatemala, Colombia	Host plant: Fabaceae, *Phaseolus vulgaris* L. (Maes 1994)
Agonosoma bicolor (Westwood)	Brazil	
Agonosoma flavolineata Laporte	Grenada, French Guyana, Brazil, Paraguay	Host plants: Convolvulaceae, *Ipomoea batatas* (L.) Lam.; Euphorbiaceae, *Croton glandulosus* L., *Jatropha gossypifolia* L.; Malvaceae, *Gossypium hirsutum* L. (Quintanilla et al. 1976; Paleari 1992b; Smith and Heard 2003; Rider 2014).
		Life history: Paleari (1992b)
Agonosoma trivittata (Panzer)	Grenada, Panama, Colombia, Venezuela, French Guyana, Brazil	Host plant: Euphorbiaceae, *Cnidoscolus urens* (L.) Arthur (Hallman 1979)
Ascanius atomarius (Germar)	Mexico*, Brazil, Bolivia, Argentina	Host plant: Rosaceae, *Ascanius* sp. collected on *Rubus erythrocladus* Mart. (Barcellos et al. 2014)
Ascanius hirtipes (Herrich-Schaeffer)	Brazil, Paraguay, Argentina	

(continued)

Table 23.1 (continued)

Species	Distribution	Biology and notes
Camirus brevilinea (Walker)	Brazil	
Camirus conicus (Germar)	Mexico and Trinidad, S. to Colombia, Venezuela, Guyana, Brazil	
Camirus divergens (Walker)	Brazil, "Amazon region"	
Camirus impressicollis Stål	Brazil	
Camirus moestus (Stål)	USA (Arizona), Mexico, Nicaragua	Host plants: Asteraceae, *Ambrosia ambrosioides* (Cav.) Payne; Malvaceae, *Gossypium* sp. (Maes 1994; Rider 2014)
Camirus socius (Stål)	Mexico	
Chelycoris haglundi (Montandon)	Brazil, Peru, Paraguay, Argentina	Host plant: Euphorbiaceae, *Croton* sp. (Barcellos et al. 2014)
Chelycoris lethierryi (Montandon)	Brazil, Peru, Argentina, Uruguay	Host plant: Euphorbiaceae, *Croton* sp. (Barcellos et al. 2014)
Chelycoris scitulus (Walker)	Brazil, Peru, Bolivia, Paraguay, Argentina	
Chelycoris vittatus Distant	Brazil, Paraguay	
Chelyschema incarnata (Germar)	Brazil	
Chelyschema leucotelus (Walker)	Brazil	
Chelyschema trinotata (Walker)	Guatemala, French Guyana, Brazil	
Coptochilus ferrugineus Amyot & Serville	Ecuador, Brazil, Bolivia, Argentina	
Coptochilus lentiginosus Berg	Brazil, Paraguay, Argentina	
Coptochilus Neotropicsis Distant	French Guyana, Brazil	Collected at light (Eger et al. 2015)
Crathis ansata (Distant)	Belize, Costa Rica, Panama, Bolivia*	
Crathis longifrons Stål	French Guyana, Colombia, Brazil, Peru	Collected at fish carrion-baited butterfly traps (Eger et al. 2015)
Diolcus cassidoides (Walker)	Hispaniola	

(continued)

Table 23.1 (continued)

Species	Distribution	Biology and notes
Diolcus disjunctus Barber	Puerto Rico	
Diolcus irroratus (F.)	USA (Florida), Cuba, Hispaniola, Puerto Rico, Jamaica, Antigua, Saint Vincent	Host plants: Combretaceae, *Conocarpus erectus* L.; Malvaceae, *Gossypium* sp. (Eger et al. 2015, Wolcott 1941)
Diolcus pusillus Berg	Argentina, Uruguay	
Diolcus variegatus (Herrich-Schaeffer)	Cuba, Hispaniola, Jamaica, adventive in Florida	Host plant: Malvaceae, *Malvastrum corchorifolium* (Desc.) Britt., *Sida acuta* Burm. (cited as *Sida carpinifolia* L.) (Bruner et al. 1945; Eger and Baranowski 2002)
Dystus puberulus Stål	Mexico, Costa Rica, Brazil, Bolivia	Host plant: Moraceae, *Ficus* spp. (Cervantes-Peredo 2004)
		Life history, immatures (Cervantes-Peredo 2004)
Ephynes brevicollis Stål	Brazil	
Galeacius crowleyi Distant	Brazil	
Galeacius martini Schouteden	French Guyana, Brazil	Host plant: Melastomataceae, *Miconia sellowiana* Naudin (Bianchi et al. 2011).
		Immatures described (Bianchi et al. 2011)
Galeacius simplex Breddin	Peru	
Galeacius tessellatus Distant	Panama	
Homaemus proteus Stål	Southern United States, Mexico, Jamaica*, Costa Rica, Colombia	
Lobothyreus brasiliensis Schouteden	Brazil	
Lobothyreus breviceps Breddin	Peru	
Lobothyreus illex Bergroth	French Guyana, Brazil,	
Lobothyreus lobatus (Westwood)	Colombia, Brazil	
Misippus spinolae (Signoret)	Brazil, Chile, Argentina, Uruguay	Host plant: Poaceae, *Oryza sativa* L. and *Sorghum bicolor* (L.) (Quintanilla et al. 1976)

(continued)

Table 23.1 (continued)

Species	Distribution	Biology and notes
Orsilochides glirina (Bergroth)	Brazil	
Orsilochides leucoptera (Germar)	Brazil, Peru, Bolivia, Paraguay, Argentina	Host plants: Euphorbiaceae, *Croton* sp.; Malvaceae, *Gossypium hirsutum* L. Moench; Poaceae, *Sorghum bicolor* (L.) Moench (Quintanilla et al. 1976; Barcellos et al. 2014)
Orsilochides scurrilis (Stål)	Mexico, Guatemala, El Salvador, Nicaragua, Bolivia*	
Orsilochides stictica (Dallas)	USA (Texas), Mexico, Guatemala, Honduras	
Orsilochides variabilis (Herrich-Schaeffer)	Mexico, Belize, Guatemala, Nicaragua, Costa Rica, Colombia, Argentina*, Uruguay*	
Pachycoris chrysomelinus Walker	Ecuador	
Pachycoris fabricii (L.)	Caribbean, Mexico* to Brazil*	Host plants: Boraginaceae, *Cordia nitida* Vahl.; Euphorbiaceae, *Croton discolor* (Willd.) Raf., *Croton humilis* L., *Croton* sp.; Verbenaceae, *Lantana involucrata* L., *Lantana* sp. (Barber 1939; Wolcott 1923, 1941)
Pachycoris klugii Burmeister	Mexico, northern Central America, Colombia*, Brazil*, Paraguay*	Host plant: Euphorbiaceae, *Jatropha curcas* L. (Cervantes-Peredo 2002; Martínez-Herrera et al. 2010; Tepole-García et al. 2012)
		Immatures described, maternal care, field observations (Cervantes-Peredo 2002)
Pachycoris nitidiventris Breddin	Costa Rica	
Pachycoris obscuratus Herrich-Schaeffer	South America	
Pachycoris stallii Uhler	Mexico	Host plant: Euphorbiaceae, *Croton californicus* Muell. Arg. (Williams et al. 2005)
		Immatures described, maternal care, life history, egg parasite *Telenomus pachycoris* (Costa Lima) (Hymenoptera: Scelionidae) and predator *Dorymyrmex bicolor* (Hymenoptera: Formicidae) Wheeler (Williams et al. 2005)

(continued)

Table 23.1 (continued)

Species	Distribution	Biology and notes
Pachycoris torridus (Scopoli)	Mexico and California* south to Argentina and Brazil	Host plants: Anacardiaceae, *Anacardium occidentale* L., *Mangifera indica* L., *Schinus terebinthifolius* Raddi; Aquifoliaceae, *Ilex paraguariensis* St. Hil.; Boraginaceae, *Cordia* sp.; Euphorbiaceae, *Aleurites fordii* Hemsley, *Cnidoscolus pubescens* Pohl, *Croton californicus* Müll. Arg., *Jatropha curcas* L., *Jatropha* sp., *Manihot esculenta* Crantz, *Sapium haematospermum* Müll. Arg.; Malpighiaceae, *Malpighia glabra* L.; Myrtaceae, *Eucalyptus* sp., *Psidium guajava* L., *P. guineense* Sw. (cited as *P. araca* Raddi); Poaceae, *Oryza sativa* L.; Rubiaceae, *Coffea* sp. (Hussey 1934; Bosq 1937, 1940; Costa Lima 1940; Callan 1948; Silva et al. 1968; Maes 1994; Sánchez-Soto and Nakano 2002; Sánchez-Soto et al. 2004; Santos et al. 2005; Michelotto et al. 2006; Rodrigues et al. 2011; Souza et al. 2012; Rider 2014)
		Egg parasitoid, *Paratelenomus pachycoris* Costa Lima (Costa Lima 1928) (Hymenoptera: Scelionidae); parasitoids, *Trichopoda pennipes* (F.), *T. pilipes* (F.) (Diptera: Tachinidae) (Guimarães 1977); parental care, ecology, field observations (Hussey 1934; Santos et al. 2005). Collected at fish carrion-baited butterfly traps (Eger et al. 2015)
Polytes bicolor Distant	Peru	
Polytes bimaculatus Eger	Peru	Collected at fish carrion-baited butterfly traps (Eger et al. 2015)
Polytes bullocki Eger	Peru	
Polytes confusus Eger	Peru	
Polytes discrepans (Uhler)	Peru	
Polytes fenestra Breddin	Colombia, Peru, Bolivia, Argentina	
Polytes granulatus (Walker)	Colombia, Ecuador, Peru, Bolivia, Argentina	
Polytes lattini Eger	Peru	
Polytes leopardinus Distant	Peru	
Polytes lineolatus (Dallas)	Costa Rica, Panama, Colombia, Venezuela, Ecuador, Peru, Bolivia	

(continued)

Table 23.1 (continued)

Species	Distribution	Biology and notes
Polytes obscurus (Dallas)	Colombia, Brazil*, Peru, Bolivia, Argentina	
Polytes nigrovittatus Eger	Peru	
Polytes propinquus (Walker)	Venezuela	
Polytes rubromaculatus Distant	Ecuador	
Polytes ruth (Breddin)	Peru, Bolivia	
Polytes similis Eger	Peru	
Polytes speculiger Breddin	Peru	
Polytes tigrinus (Vollenhoven)	Colombia, Peru	
Polytes velutinus (Dallas)	Colombia, Venezuela, Ecuador, Peru, Bolivia	
Sphyrocoris obliquus (Germar)	Southern USA throughout the Caribbean and Central America into Colombia and Venezuela	Host plants: Asteraceae, *Bidens bipinnata* L., *Bidens* sp.; Fabaceae, *Glycine max* L. Merrill; Lamiaceae, *Salvia mellifera* Greene; Malvaceae, *Gossypium barbadense* L. (Callan 1948; Eger 2012)
		Egg parasitoid, *Trissolcus trinidadensis* Crawford (Hymenoptera: Scelionidae) (Crawford 1913; Callan 1948)
Symphylus affinis (F.)	South America, Antilles*	
Symphylus albomaculatus Distant	Panama	
Symphylus caribbeanus Kirkaldy	USA (Texas, Florida), Mexico, Caribbean, Central America	Host plants: Anacardiaceae, *Mangifera indica* L.; Clusiaceae, *Clusia major* L. (as *C. rosea* Jacq.); Fabaceae, *Phaseolus lathyroides* L.; Lamiaceae, *Salvia mellifera* Greene; Myricaceae, *Myrica cerifera* L.; (Blatchley 1926; Bruner et al. 1945; Callan 1948; Lattin 1964)
		Attracted to light (McPherson 1977)
Symphylus cyphonoides (Walker)	Mexico, Panama, Ecuador, Brazil, Bolivia,	Collected at fish carrion-baited butterfly traps (Eger et al. 2015)
Symphylus deplanatus (Herrich-Schaeffer)	Mexico, Guatemala, Belize, Jamaica, Saint Vincent, Grenada, Trinidad y Tobago, Panama, Ecuador, Brazil, Paraguay, Argentina	Host plant: Boraginaceae, *Cordia macrostachya* (Jacquin) Roemer & Schultes (Rider 2014)
		Attracted to light (Uhler 1894; Torre-Bueno 1914)

(continued)

Table 23.1 (continued)

Species	Distribution	Biology and notes
Symphylus devexus Walker	Brazil	
Symphylus enac Breddin	Bolivia	
Symphylus infamatus Uhler	Mexico	
Symphylus lativittatus Breddin	Ecuador	
Symphylus leucospilus (Walker)	Amazon Region, Brazil	
Symphylus luminosus (Germar)	Mexico	
Symphylus modestus Distant	Guatemala, Panama	
Symphylus obtusus Dallas	Mexico, Panama, Colombia	
Symphylus poecilus Dallas	Brazil	
Symphylus ramivitta Walker	Belize, Ecuador, Brazil, Peru, Bolivia, Paraguay, Argentina	Collected at fish carrion-baited butterfly traps (Eger et al. 2015)
Symphylus rivulosus (Walker)	Mexico, Guatemala, Nicaragua, Trinidad y Tobago, Panama, Brazil	
Symphylus signoretti Distant	Mexico	
Testrina wolfii (Fabricius)	Brazil, South America	
Tetyra antillarum Kirkaldy	USA (Florida), Mexico, Antilles	Attracted to light (Wolcott 1941)
Tetyra pinguis (Germar)	Mexico, Guadalupe, Guatemala, Jamaica, Honduras, Colombia, French Guyana, Brazil, Bolivia	Host plant: Rubiaceae, *Coffea* sp. (Maes 1994)
Tetyra poecila Berg	Brazil, Argentina	Host plants: Fabaceae, *Acacia caven* (Mol.) Mol. (cited as *A. cavenia*), *A. farnesiana* (L.) Willd., *Gleditsia triacanthos* L.; Solanaceae, *Solanum bonariense* L. (Berg 1879)
		Immatures described (Berg 1879)
Tiridates mexicanus (Herrich-Schaeffer)	Mexico	

(continued)

Table 23.1 (continued)

Species	Distribution	Biology and notes
Tiridates rubrocinctus rubrocinctus (Herrich-Schaeffer)	Mexico	
Tiridates rubrocinctus schaffneri Eger	Mexico, Honduras	
Elvisurinae		
Nesogenes boscii (Fabricius)	Greater Antilles, Saint Vincent	

[a]Species treated in this table are those treated as valid by Kirkaldy (1909) unless more recent work has modified the status of these species. Subspecies treated by Kirkaldy are not treated here. *Localities followed by an asterisk are probably the result of a misidentification and are considered questionable

(Montandon), and *Orsilochides leucoptera* (Germar) were observed feeding on *Croton* spp. in Brazil (Eger et al. 2015).

Adult diapause has been reported for *A. flavolineata* during the dry season in southeastern Brazil (Paleari 1992b). Also, adults of *Pachycoris torridus* were found, apparently in diapause, inside terrestrial bromeliads during the winter in southern Brazil (Barcellos et al. 2014).

23.3.1 Eggs and Oviposition

Scutellerid eggs are spherical or barrel-shaped, similar to those of other pentatomoids, with small aero-micropylar processes variable in number among species, and hatching occurs through a triangular or T-shaped egg burster. The eggs of the Neotropical *Dystus puberulus* have a smooth surface, with 6–7 aero-micropylar processes (Cervantes-Peredo 2004). In *Galeacius martini* Schouteden, eggs are reddish brown, with the surface of the chorion translucent, nearly smooth under light stereomicroscope, and with a mean of 24 aero-micropylar processes (Bianchi et al. 2011). Studies employing scanning electron microscopy (SEM) for Neotropical fauna are available for *Pachycoris stallii* (Williams et al. 2005) and *Galeacius martini* (Bianchi et al. 2011).

Scutellerid eggs are laid in two (more often) or more rows, on several parts of the host plant. Egg masses vary in number, averaging from 14 (*Agonosoma flavolineata*, Paleari 1992b; *D. puberulus*, Cervantes-Peredo 2004) to almost a hundred eggs (75–92) in *P. torridus* (Santos et al. 2005) and 81.4 on average in *P. klugii* (Cervantes-Peredo 2002). Under laboratory conditions, individuals of *P. torridus* reared on the physic nut, *Jatropha curcas* L., showed a preoviposition period twice as long as, and laid more eggs than, those fed on strawberry guava – *Psidium cattleianum* Sabine (Myrtaceae) (Borges Filho et al. 2013).

23.3.2 Nymph Feeding and Development

A few Neotropical species have had their biological cycle studied. Life cycle is highly variable in length. For *Agonosoma flavolineata* reared on *Croton glandulosus* L. (Euphorbiaceae), a mean total development time of about 42 days was recorded, including a 7-day incubation period (Paleari 1992b). Cervantes-Peredo (2004) observed a total life cycle of 60 days in *D. puberulus* reared on fruits of *Ficus* spp. The nymphal stage of *P. torridus* is about 55 days long when fed on *J. curcas* or strawberry guava, under laboratory conditions (Borges Filho et al. 2013). First stadium nymphs, as in other pentatomoids, do not feed, using only reserves from embryonic stage. Apparently, this fact contributes to the low mortality observed in this instar, as food availability is not a constraint. Conversely, the role of aggregation is considered crucial for the survival of the first instar.

23.3.3 Adult Feeding and Reproduction

Shield bugs are broadly phytophagous, although there are some observations of necrophagy or at least attraction to carrion as mentioned previously (Chérot et al. 1998, J. Eger et al. 2015). Stems, leaves, and flowers may be used as food although fruit and seeds seem to be most commonly used. Data on the known host plants are still scarce (Table 23.1), but some patterns may be pointed out. *Pachycoris torridus* has a wide range of reported host plants of different botanical families, but shows preference for euphorbs, similar to its congeners. *Symphylus* spp. are recorded mostly on Malvaceae and Lamiaceae. *Sphyrocoris obliquus* (Germar) also seems to be polyphagous.

The sex ratio in collections is sometimes puzzlingly biased. Eger (1990) observed that specimens of *Polytes* Stål in collections were overwhelmingly males. Genera such as *Crathis* and *Lobothyreus* have many more males than females sampled, and we do not know the reason for that. However, some scutellerids attracted to fish carrion are primarily males (Eger et al. 2015), so collection method may have more to do with sex ratio differences in collections and this bias may not be indicative of a bias in natural populations.

23.4 Classification and Diversity

The higher classification of scutellerids has not been well studied historically, but recent work has helped clarify the subfamily status of various taxa. Cassis and Vanags (2006) provide a detailed discussion of the infrafamilial groups. There are currently eight subfamilies, only three of which are further divided into tribes. The subfamilies Tectocorinae McDonald and Cassis, Odontoscelinae Amyot and

Serville, and Hoteinae Carapezza are exclusively Old World. New World representatives of Odontotarsinae Mulsant and Rey and Eurygastrinae Amyot and Serville occur only in the Nearctic region, leaving the Neotropics with three subfamilies, Elvisurinae Stål, Pachycorinae Amyot and Serville, and Scutellerinae Leach.

Pachycorinae was historically characterized by stridulatory areas on the abdominal sterna associated with a tibial plectrum and also by the presence of a single interclaval vein in the hind wing (Fischer 2001). Included in the subfamily were three genera with Old World distribution (*Deroplax* Mayr, *Hotea* Amyot and Serville, and *Tectocoris* Hahn) (Kumar 1965). Lattin (1964) and others contested the monophyly of Pachycorinae, as the stridulatory areas appeared to have evolved more than once within the family (Lattin 1964). McDonald and Cassis (1984) placed *Tectocoris* in a monotypic subfamily, Tectocorinae, based primarily on genitalic characters and the presence of abdominal sternal glands in males. Carapezza (2009) described genitalic differences between the Old World genera *Hotea* and *Deroplax* and the New World genera of Pachycorinae and erected a new subfamily, Hoteinae, for these two genera. He also pointed out that species of *Ellipsocoris* Mayr (Odontotarsinae) possess stridulatory patches on the abdominal sterna. Cassis and Vanags (2006) mentioned that stridulatory areas occur in *Nesogenes* (Elvisurinae) as well (see discussion below). Moreover, the interclaval vein is also shared with Odontotarsinae and Eurygastrinae (Czaja 2007). The removal of *Deroplax*, *Hotea*, and *Tectocoris* from the subfamily makes the Pachycorinae exclusively New World in distribution. Based on these characters, associated with the absence of a unique pattern of genitalia morphology, Tsai et al. (2011) suggested a non-monophyly of Pachycorinae, a hypothesis that is currently under investigation (Weiler et al. in prep.).

Scutellerinae is primarily an Old World subfamily which contains two tribes, Sphaerocorini Stål and Scutellerini Leach. The Scutellerinae are represented in this hemisphere by only a single genus, *Augocoris*, which belongs in the tribe Scutellerini. The subfamily Elvisurinae is also represented by a single genus in the Neotropics. Horvath (1921) placed his new genus *Nesogenes* in the Elvisurinae based on the developed thoracic sterna. The type and only species is *Cimex boscii* F. which was previously placed in the genus *Diolcus* Mayr. This species is distributed in the Greater Antilles, where most of the species of *Diolcus* are found, and the thoracic sterna are developed into relatively thick carinae (Fig. 23.53), not at all like the thin platelike lobes of the Elvisurinae. *Nesogenes* appears to be more closely related to species of *Diolcus* than to any of the Elvisurini and probably should be removed from Elvisurini and placed back in the Pachycorinae, if not back into the genus *Diolcus* (Eger et al. 2015).

The Neotropical scutellerid fauna, the poorest known of the zoogeographic regions, currently includes three subfamilies, Pachycorinae with 23 genera, Scutellerinae, represented in the region only by *Augocoris*, and Elvisurinae represented by the probably misplaced genus, *Nesogenes*. With the exception of *Stethaulax* Uhler, all remaining pachycorine genera are recorded in the Neotropics, with a few occurring in both the Nearctic and Neotropical Regions (*Acantholomidea*, *Diolcus*, *Homaemus*, *Orsilochides*, *Sphyrocoris*, and *Tetyra*). However, it is likely that some species currently assigned to *Symphylus* may actually belong to *Stethaulax*, which would extend the range of this genus into the Neotropics as well (Lattin 1964). There are 112 recorded species, but this number is certainly underestimated.

The scutellerine genus *Augocoris* includes six species, of which *A. illustris* and *A. gomesii* are the best known in Neotropics. Both are polymorphic, ranging from almost entirely white dorsally to black or blue with red markings (Lattin 1964).

The taxonomic knowledge of the Neotropical pachycorine fauna is incipient. There are revisions only for *Agonosoma* (Paleari 1992a), *Tiridates* (Eger 1987), *Polytes* (Eger 1990), and *Sphyrocoris* Mayr (Eger 2012) although Eger and Lattin (1995) reviewed the generic placement and synonymy of types of New World scutellerids found in the British Museum. Most of the remaining genera are known only from their original descriptions. Some of them are particularly problematic. *Symphylus* is a "catch-all" genus which probably represents a non-monophyletic group in need of revision. Some species currently placed in *Symphylus* may belong to related genera (*Ephynes* Stål, *Galeacius* Distant, and *Stethaulax*) or may require the establishment of new genera and there appear to be a large number of undescribed species in this group. Additionally, there are many apparently undescribed species in genera such as *Camirus* Stål, *Diolcus*, and *Tetyra* F.

23.4.1 Main Species, Economic Aspects

Pachycoris torridus (Scopoli) (Fig. 23.27) and *P. klugii* Burmeister, both highly polymorphic species, are reported as pests of *J. curcas* (Grimm and Maes 1997; Cervantes-Peredo 2002; Rodrigues et al. 2011; Souza et al. 2012), damaging the seeds. *P. torridus* occurs from southern Mexico to Argentina, but is more abundant in South America and is the only species considered to be of economic importance in Brazil. Besides *J. curcas*, *P. torridus* is reported on several host plants, causing damage to strawberry guava and tung, *Vernicia fordii* (Hemsl.) (Euphorbiaceae) (Borges Filho et al. 2013). *Pachycoris klugii* is distributed from Mexico into Central America.

Augocoris illustris (F.) (Fig. 23.2) – this representative of the only scutellerine genus in the Neotropics – is highly variable in color, species varying from light, almost white with variable dark spots, to black with red markings. Barber and Bruner (1932) provided illustrations of the variability in this species. Reported hosts are *Chrysophyllum oliviforme* L. and other species of Sapotaceae.

Augocoris gomesii Burmeister (Fig. 23.1) – this species is a little larger and more ovate than the preceding but also shows a great deal of variability in color. Host plants are not known, but it can be abundant at lights in some parts of South America.

Agonosoma flavolineata (Laporte) (Fig. 23.13) and *A. trivittata* (Panzer) – these species are widely distributed in South America. Females of the former species and both sexes of the latter are polymorphic (Paleari 1994). *Agonosoma trilineatum* was introduced to Australia as a biological control agent for the introduced weed *Jatropha gossypifolia* L. (Smith and Heard 2003).

Symphylus deplanatus (Herrich-Schaeffer) – this species is very common in collections and is a medium-sized (10–12 mm) shield bug. It is also polymorphic, showing different patterns of black to brown markings. There is one synonym and two subspecies of this species, but based on genitalia, there is probably a complex of species that may be misidentified in collections.

Tetyra pinguis Germar – described from Brazil, this species is recorded for all Neotropics. This wide distribution, together with variable size and color patterns, might indicate that it represents a complex of several species.

23.4.2 Secondary Species

Crathis ansata (Distant) – this is a Central American species which differs from its congeners by the anterior part of scutellum conspicuously more elevated than pronotum.

Crathis longifrons Stål (Fig. 23.9) – this species is widely distributed in South America, and the southern limit of its range is Central Brazil. Curiously, there are large series of male specimens contrasting with very few females in collections. It has been collected at fish carrion-baited traps.

Tetyra poecila Berg – this species is very commonly found in collections from southern Brazil and Argentina. It was recorded on Fabaceae and Mimosaceae.

Symphylus spp. (Fig. 23.17) – the genus as currently conceived contains a large number of mostly brown mottled species that are frequently collected at light and at fish carrion-baited traps.

Symphylus ramivitta Walker (Fig. 23.19) – its color pattern, consisting of longitudinal stripes, is very distinctive among its congeners. This species is broadly distributed in the Neotropics.

Symphylus cyphonoides (Walker) (Fig. 23.18) – this is one of the smallest species within the genus, is yellowish brown, has rounded shape, and distributed from Central America to Brazil. Adults have been collected on fruits of *Cervantesia* sp. (Santalaceae) in Peru.

Galeacius martini Schouteden (Fig. 23.23) – this bug is a very distinctive pachycorine, with humeri strongly produced. The immature stages were described by Bianchi et al. (2011).

Camirus spp. (Fig. 23.4) – these brown nondescript species are frequently collected when sweeping secondary growth along trails or roadways.

Homaemus spp. (Fig. 23.15) and *Sphyrocoris obliquus* (Germar) (Fig. 23.16) – these are also dull brown insects with some darker patterning. As with *Camirus* spp., they are commonly swept from vegetation in disturbed areas, the latter taxa being reported from *Bidens* spp. (Eger 2012). Distribution for both species ranges from the United States into northern South America.

Natural enemies include invertebrate predators such as the ant *Dorymyrmex bicolor* Wheeler on *P. stallii* (Williams et al. 2001); egg parasitoids such as the platygastrid wasps *Trissolcus trinidadensis* Crawford and *Telenomus pachycoris*

Figs. 23.1–23.4 Dorsal habitus. 1: *Augocoris gomesii*; 2: *Augocoris illustris*; 3: *Acantholomidea porosa*; 4: *Camirus* sp. Dimensional lines equal 1.0 mm (Photos by J Eger)

(Costa Lima) (as *Pseudotelenomus pachycoris*) (Gabriel et al. 1988), respectively, on *Sphyrocoris obliquus* (Germar) and *P. torridus*; and adult parasitoids such as the tachinids *Trichopoda pennipes* (F.) on *P. klugii* (Cervantes-Peredo 2002) and *T. pilipes* (F.) and *T. pilipes* on *P. torridus* (Guimarães 1977).

Key to Neotropical genera of Scutelleridae [adapted from Schouteden (1904), Lattin (1964), and Barcellos et al. (2014)]

Figs. 23.5–23.8 Dorsal habitus. 5: *Ascanius hirtipes*; 6: *Chelycoris haglundi*; 7: *Coptochilus ferrugineus*; 8: *Chelyschema* sp. Dimensional lines equal 1.0 mm (Photos by J Eger)

Figs. 23.9–23.12 Dorsal habitus. 9: *Crathis longifrons*; 10: *Lobothyreus brasiliensis*; 11: *Diolcus chrysorrhoeus*; 12: *Nesogenes boscii*. Dimensional lines equal 1.0 mm (Photos by J Eger)

Figs. 23.13–23.16 Dorsal habitus. 13: *Agonosoma flavolineata*; 14: *Tiridates rubrocinctus*; 15: *Homaemus proteus*; 16: *Sphyrocoris obliquus*. Dimensional lines equal 1.0 mm (Photos by J Eger)

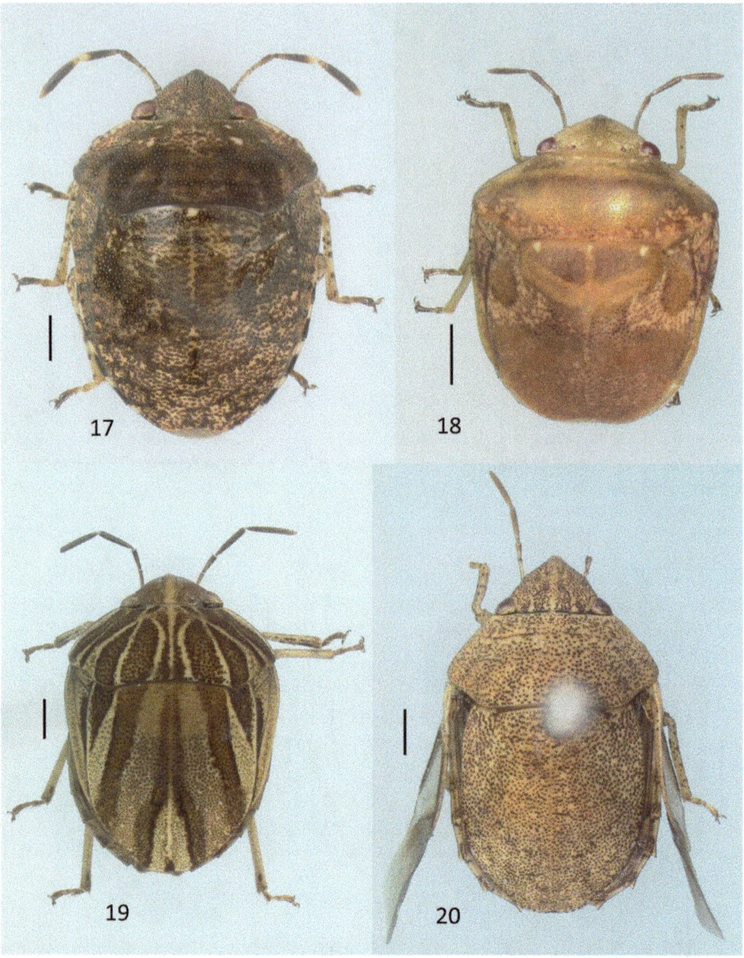

Figs. 23.17–23.20 Dorsal habitus. 17: *Symphylus* sp.; 18: *Symphylus cyphonoides*; 19: *Symphylus ramivitta*; 20: *Ephynes brevicollis*. Dimensional lines equal 1.0 mm (Photos by J Eger)

Figs. 23.21–23.24 Dorsal habitus. 21: *Polytes fenestra*; 22: *Polytes bimaculatus*; 23: *Galeacius martini*; 24: *Misippus spinolae*. Dimensional lines equal 1.0 mm (Photos by J Eger)

Figs. 23.25–23.28 Dorsal habitus. 25: *Orsilochides leucoptera*; 26: *Tetyra* sp.; 27: *Pachycoris torridus*; 28: *Dystus puberulus*. Dimensional lines equal 1.0 mm (Photos by J Eger)

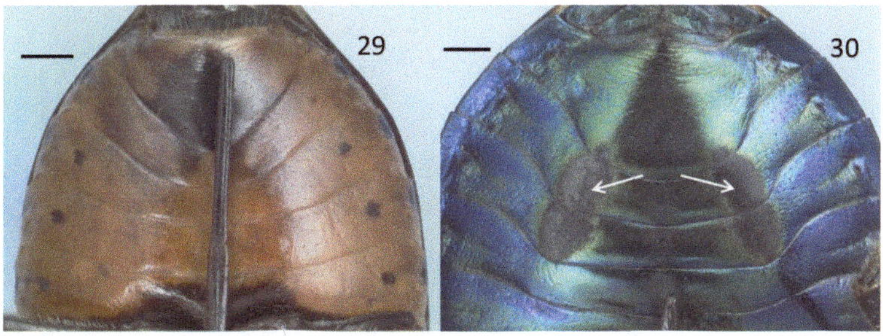

Figs. 23.29–23.30 Abdominal venter. 29: *Augocoris gomesii*; 30: *Pachycoris torridus*, *arrows* indicate striated areas. Dimensional lines equal 1.0 mm (Photos by J Eger)

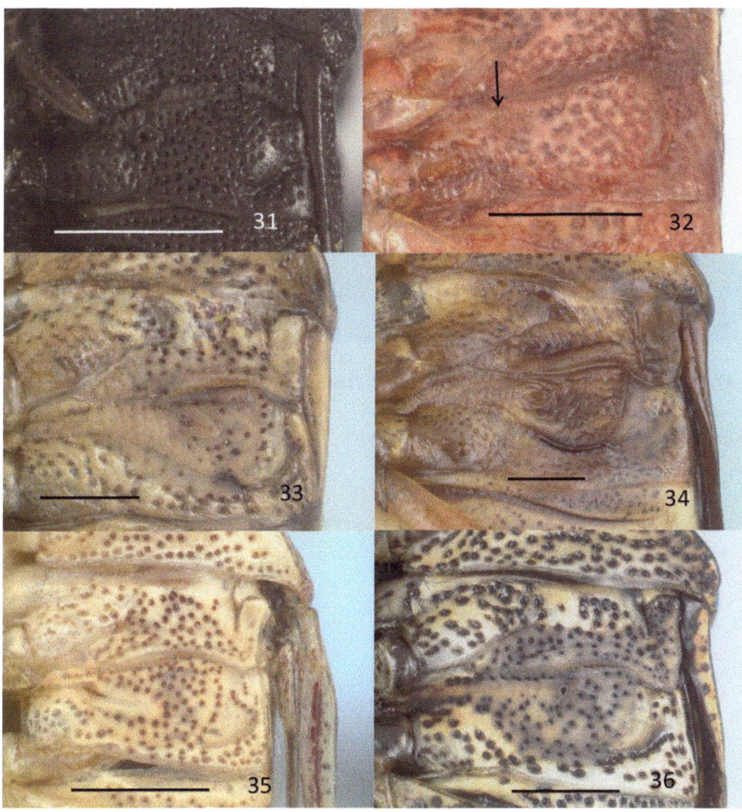

Figs. 23.31–23.36 External efferent system. 31: *Acantholomidea porosa*; 32: *Misippus spinolae*, *arrow* indicates location of ostiole; 33: *Chelycoris lethierryi*; 34: *Chelyschema* sp.; 35: *Ephynes brevicollis*; 36: *Polytes obscurus*. Dimensional lines equal 1.0 mm (Photos by J Eger)

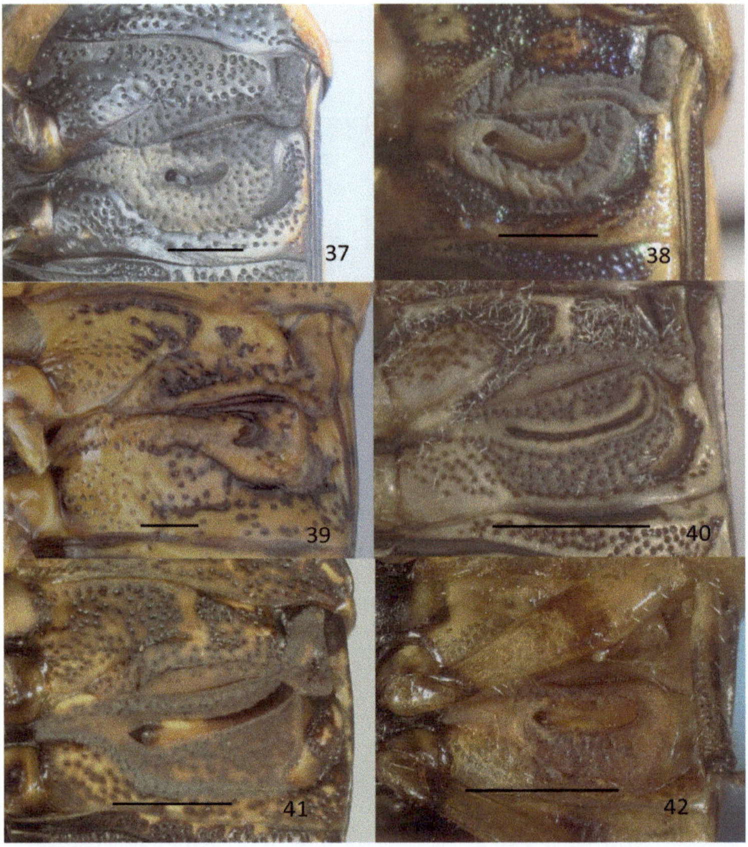

Figs. 23.37–23.42 External efferent system. 37: *Tiridates rubrocinctus*; 38: *Agonosoma flavolineata*; 39: *Tetyra* sp.; 40: *Homaemus proteus*; 41: *Crathis longifrons*; 42: *Galeacius martini*. Dimensional lines equal 1.0 mm (Photos by J Eger)

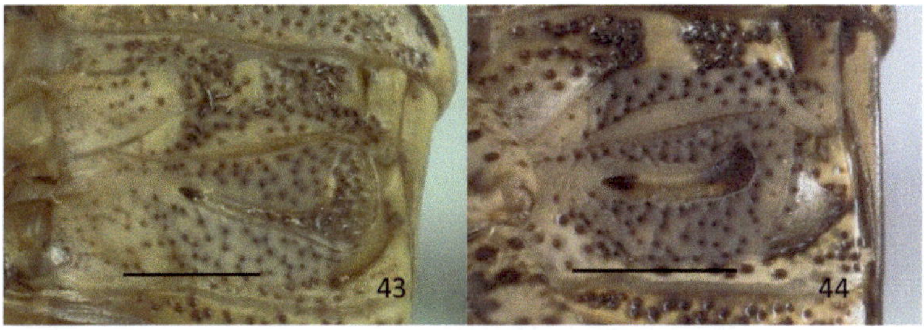

Figs. 23.43–23.44 External efferent system. 43: *Sphyrocoris obliquus*; 44: *Symphylus caribbeanus*. Dimensional lines equal 1.0 mm (Photos by J Eger)

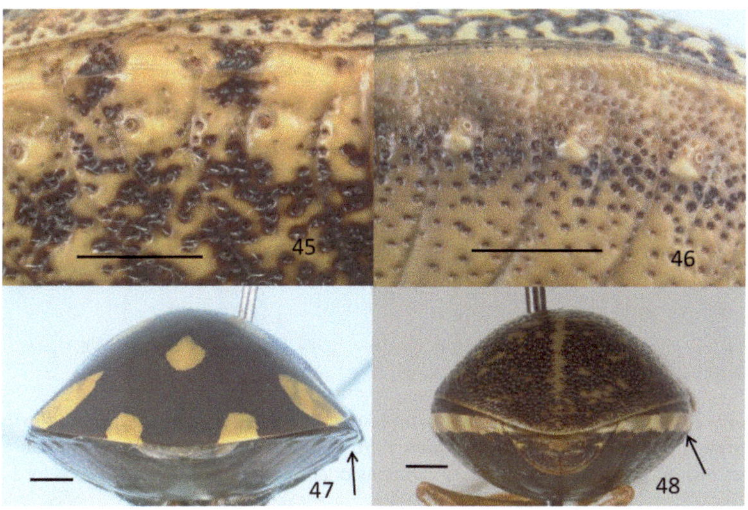

Figs. 23.45–23.48 45–46: Lateral view of abdomen showing spiracles. 45: *Ascanius hirtipes*; 46: *Chelycoris lethierryi*. 47–48: Caudal view of abdomen. 47: *Pachycoris torridus*; 48: *Orsilochides guttata*. *Arrows* indicate lateral compression in *P. torridus* and lateral convexity in *O. guttata*. Dimensional lines equal 1.0 mm (Photos by J Eger)

Figs. 23.49–23.54 49: *Acantholomidea porosa*, lateral view; 50: *Coptochilus ferrugineus*; 51–52: Lateral view of head. 51: *Acantholomidea porosa*; 52: *Camirus* sp. 53: Thoracic sterna of *Nesogenes boscii*; 54: Proleg of *Tetyra* sp. Dimensional lines equal 1.0 mm (Photos by J Eger)

1. Stridulatory areas absent on abdominal venter (Fig. 23.29); antennae three-segmented (Scutellerinae) .... *Augocoris* (Figs. 23.1, 23.2)
1'. Stridulatory areas present on both sides of midline on each of at least abdominal sternites 5–6 (Fig. 23.30); antennae five-segmented (Pachycorinae) 2
2. Head with anterior margins and anterior portion of lateral margins elevated (Fig. 23.50).. *Coptochilus* (Fig. 23.7)
2'. Head with anterior and lateral margins not elevated 3
3. Distance from metacoxa to ostiole shorter than distance from ostiole to lateral margin of metapleuron (Figs. 23.31, 23.32, 23.34, 23.35, 23.37, 23.38, 23.40, 23.41, 23.42, 23.43, and 23.44) 4
3'. Distance from metacoxa to ostiole longer than or subequal to distance from ostiole to lateral margin of metapleuron (Figs. 23.33, 23.36, 23.39) 19
4. Pronotum with a median transverse impression, this usually conspicuous (Fig. 23.49); head usually declivent (Figs. 23.51, 23.52); coloration brown to black, frequently mottled; punctures usually provided with short, decumbent setae.. 5
4'. Pronotum not impressed transversely; head usually porrect; coloration variable; punctures rarely provided with setae and if so, setae generally long and erect 6
5. Head strongly declivent; bucculae with distinct lobe anteriorly, lobe strongly truncate posteriorly (Fig. 23.51); coloration black (Fig. 23.3); stridulatory areas not extending onto seventh sternite .... *Acantholomidea*
5'. Head moderately to slightly declivent; bucculae smoothly rounded, not truncate posteriorly (Fig. 23.52); coloration brown, usually mottled (Fig. 23.4); stridulatory areas frequently extending onto seventh sternite .... *Camirus*
6. Ostiole not attended by peritreme or attending structure less than twice as long as the width of ostiole (Figs. 23.32, 23.34) 7
6'. Ostiole attended by peritreme that is much more than twice as long as width of ostiole (Figs. 23.35, 23.37, 23.38, 23.40, 23.41, 23.42, 23.43, and 23.44).. 10
7. Ostiole less than one third of distance from metacoxa to lateral margin of metapleuron (Fig. 23.32) (southern South America)....... *Misippus* (Fig. 23.24)
7'. Ostiole distinctly more than one third of distance from metacoxa to lateral margin of metapleuron (Fig. 23.34).. 8
8. Located in South America into southern Central America ....
.... *Chelyschema* (Fig. 23.8)
8'. Located in the Caribbean and North America into Mexico 9
9. Thoracic sterna sulcate, sulcus bordered by well-developed carina (Fig. 23.53) ...
.... *Nesogenes* (Fig. 23.12)
9'. Thoracic sterna only shallowly sulcate, carina not developed............................
.... *Diolcus* (Fig. 23.11)
10. Ostiolar peritreme reaching anterolateral corner of metathoracic evaporative area or separated from anterolateral corner by less than its width at that point (Figs. 23.40, 23.41, 23.43)... 11

10'. Apex of ostiolar peritreme separated from anterolateral corner of metathoracic evaporative area by at least its width at that point (Figs. 23.35, 23.37, 23.38, 23.42, 23.44) 14
11. Peritreme greatly expanded apically (Fig. 23.43) .... *Sphyrocoris* (Fig. 23.16)
11'. Peritreme roughly parallel sided, groove-like (Figs. 23.40, 23.41) 12
12. Peritreme scar-like, its sides elevated (Fig. 23.40) .... *Homaemus* (Fig. 23.15)
12'. Peritreme an elevated groove (Fig. 23.41) 13
13. Anterolateral pronotal margins straight or slightly sinuous (Fig. 23.10) .... *Lobothyreus*
13'. Anterolateral pronotal margins concave (Fig. 23.9) .... *Crathis*
14. Length of peritreme less than ½ distance from ostiole to lateral margin of metapleuron (Fig. 23.37) 15
14'. Length of peritreme more than ½ distance from ostiole to lateral margin of metapleuron (Figs. 23.35, 23.38, 23.42, 23.44) 16
15. Posterior margin of last abdominal sternite convex, covering male and female genitalia; antennae mostly black .... *Tiridates* (Figs. 23.14, 23.37)
15'. Posterior margin of last abdominal sternite concave mesially, exposing male and female genitalia; antennae dark with white annulations .... *Tetyra antillarum*
16. Peritreme spout shaped, curved anterad (Fig. 23.38)..... *Agonosoma* (Fig. 23.13)
16. Peritreme relative straight, may be curved anterad only at apex 17
17. Pronotum twice as wide as long, anterolateral margins strongly developed; posterior margin of scutellum strongly concave (Fig. 23.23) .... *Galeacius*
17'. Pronotum less than twice as wide as long, anterolateral margins rounded, not developed; posterior margin of scutellum rounded, truncate, or shallowly concave (Figs. 23.17–23.20) 18
18. Posterolateral angles of last three abdominal sternites strongly developed into distinct spines (Fig. 23.20); ostiolar peritreme angled posterad (Fig. 23.35) .... *Ephynes*
18'. Posterolateral angles of last three abdominal sternites developed at most into small spines (Figs. 23.17–23.19); ostiolar peritreme extended laterad, slightly curved anterad (Fig. 23.44) .... *Symphylus*
19. Ostiole not attended by ruga, sulcus, or auricle or attending structure shorter than the diameter of ostiole (Fig. 23.36) 20
19'. Ostiole attended by distinct ruga, sulcus, or auricle at least as long as the diameter of ostiole (Figs. 23.33, 23.39) 21
20. Head broadly rounded anteriorly; anterolateral pronotal margins moderately developed .... *Testrina*
20'. Head usually narrowing anteriorly, mostly triangular; anterolateral pronotal margins not developed (Figs. 23.21, 23.22) .... *Polytes*
21. Antennal segments 4 and 5 distinctly flattened; dorsum with conspicuous erect setae (Fig. 23.28) .... *Dystus*
21'. Antennal segments 4 and 5 cylindrical or, at most, slightly flattened; dorsum usually lacking setae or at most with short, decumbent setae 22

22. Each abdominal spiracle within or subtended by callus; tylus surpassing juga by at least diameter of second antennal segment (Figs. 23.45,23.46)................... 23
22'. Abdominal spiracles not within or subtended by calli.................................... 24
23. Each abdominal spiracle located in dorsal edge of concolorous callus (Fig. 23.45); body shape oval (Fig. 23.5)..*Ascanius*
23'. Each abdominal spiracle subtended by or adjacent to pale callus (Fig. 23.46); body shape elongate oval (Fig. 23.6)..*Chelycoris*
24. Distance from metacoxa to ostiole about 1.5 times the distance from ostiole to lateral margin of metapleura (Fig. 23.39); if less than that, femora provided with black annulus or ring of black macules (Fig. 23.54).............. *Tetyra* (Fig. 23.26)
24'. Distance from metacoxa to ostiole subequal to or slightly greater than distance from ostiole to lateral margin of metapleuron; femora uniformly colored..... 25
25. Abdominal sternites compressed near lateral margins (Fig. 23.47).................... ....*Pachycoris* (Fig. 23.27)
25'. Abdominal sternites not compressed laterally (Fig. 23.48) .... *Orsilochides* (Fig. 23.25)

23.5 Concluding Remarks

The knowledge on Neotropical shield bugs is still an open field of investigation. We need to make serious efforts to study their taxonomy, systematics, and bioecology. Phylogenies of the group, at any taxonomic level, should include larger samples, if possible, with representatives of all biogeographic areas, especially those regarding molecular characters. Polymorphism also deserves further studies, from the molecular and phylogeographical approaches. For that, collections in sampled areas should be emphasized with special attention to under explored microhabitats.

References

Barber HG (1939) Insects of Porto Rico and the Virgin Islands – Hemiptera – Heteroptera (excepting the Miridae and Corixidae). Sci Surv Puerto Rico 14:263–441

Barber HG, Bruner SC (1932) The Cydnidae and Pentatomidae of Cuba. J Dep Agric Univ Puerto Rico 16:231–284

Barcellos A, Eger J Jr, Grazia J (2014) Scutelleridae. In: Roig-Juñent S, Claps LE, Morrone JJ (eds) Biodiversidad de Artrópodos Argentinos, vol 3. Sociedad Entomologica Argentina, Mendoza, pp 409–415

Berg C (1879) Hemiptera Argentina enumeravit speciesque novas descripsit. Ann Soc Cient Argent 5:297–314

Bianchi FL, Garbelotto TA, Campos LA (2011) Description of immatures of *Galeacius martini* Schouteden (Hemiptera: Heteroptera: Scutelleridae). Zootaxa 3110:37–45

Blatchley WS (1926) Heteroptera or true bugs of Eastern North America with special reference to the faunas of Indiana and Florida. The Nature Publishing, Indianapolis

Borges Filho RC, Pratissoli D, Nava DE, Monte FG, Guidoni AL, Silva SDA, Polanczyk RA (2013) Development of *Pachycoris torridus* (Hemiptera: Scutelleridae) on *Jatropha curcas* (Euphorbiaceae), *Psidium cattleianum* (Myrtaceae) and *Aleurites fordii* (Euphorbiaceae). Fla Entomol 96:1149–1157

Bosq JM (1937) Lista preliminar de los Hemipteros (Heteropteros) especialmente relacionados con la agricultura national. Rev Soc Entomol Argent 9:111–134

Bosq JM (1940) Lista preliminar de los Hemipteros (Heteropteros) especialmente relacionados con la agricultura national (continuación). Rev Soc Entomol Argent 10:399–417

Bruner SC, Scaramuzza LC, Otera AR (1945) Catalogo de los insectos que atacan a las plantas economicas de Cuba. Cuba Est Exp Agron Bol 63:1–246

Callan EMC (1948) The Pentatomidae, Cydnidae and Scutelleridae of Trinidad. B W I Proc R Entomol Soc Lond (B) 17:115–124

Candan S, Suludere Z, Erbey M (2007) Morphology of eggs and spermatheca of *Odontotarsus purpureolineatus* (Heteroptera, Scutelleridae). Biol (Bratislava) 62:763–769

Candan S, Suludere Z, Güllü M (2010) Description of spermatheca and eggs of *Eurygaster austriaca* (Schrank, 1778) (Heteroptera: Scutelleridae), based on optical and scanning electron microscopy. Turk J Zool 35:653–662

Carapezza A (2009) On some Old World Scutelleridae (Heteroptera). Nouv Rev Entomol (NS) 25:197–212

Cassis G, Vanags L (2006) Jewel bugs of Australia (Insecta, Heteroptera, Scutelleridae). In: Rabitsch W (ed) Hug the bug – for love of true bugs. Festschrift zum 70. Geburtstag von Ernst Heiss. Denisia 19:275–398

Cervantes-Peredo L (2002) Description, biology, and maternal care of *Pachycoris klugii* (Heteroptera: Scutelleridae). Fla Entomol 85:464–473

Cervantes-Peredo L (2004) *Dystus puberulus* Stål (Heteroptera: Scutelleridae) a shield bug associated with figs in Mexico. Fla Entomol 87:528–532

Chérot F, Pauwels OSG, Eger JE, Chimsunchart C (1998) Unusual feeding behavior in a scutellerid: a case of scavenging on snake. Nat Hist Bull Siam Soc 46:207–208

Costa Lima A (1928) *Pseudotelenomus pachycoris* (n. g., n. sp.), parasite des oeufs de *Pachycoris torridus* (Scop.). Comptes Rendu Soc Biol 99:880–883

Costa Lima A (1940) Insetos do Brasil. 2° Tomo, Hemípteros. Escola Nacional de Agronomia, Rio de Janeiro

Crawford JC (1913) Descriptions of new Hymenoptera, No. 8. Proc U S Natl Mus 46(2029):343–352

Czaja J (2007) The hindwing venation and its taxonomic value in Afrotropical Scutelleridae (Hemiptera: Heteroptera). Ann Zool 57:307–318

Eger JE Jr (1987) A review of the genus *Tiridates* Stål (Heteroptera: Pentatomoidea: Scutelleridae). Fla Entomol 70:339–350

Eger JE Jr (1990) Revision of the genus *Polytes* Stål (Heteroptera: Scutelleridae). Ann Entomol Soc Am 83:115–141

Eger JE Jr (2012) The genus *Sphyrocoris* Mayr (Heteroptera: Scutelleridae: Pachycorinae). Gt Lakes Entomol 45:235–250

Eger JE Jr, Baranowski RM (2002) *Diolcus variegatus* (Heteroptera: Pentatomoidea: Scutelleridae), a Caribbean species established in south Florida. Fla Entomol 85:267–269

Eger JE Jr, Lattin JD (1995) Generic placement and synonymy of some New World Scutelleridae (Hemiptera: Heteroptera) in the British Museum (Natural History). J N Y Entomol Soc 103:412–420

Eger JE Jr, Brailovsky H, Henry TJ (2015) Heteroptera attracted to butterfly traps baited with fish or shrimp carrion. Fla Entomol (in press)

Fischer C (2001) Ein Beitrag zum Grundmuster, phylogenetischen System und zur Verwandtschaft der Scutelleridae (Heteroptera, Pentatomidae). Dissertation, Freien Universität Berlin, Germany

Gabriel D, Calcagnolo G, Tancini RS, Dias Netto N, Petinelli A Jr, Araújo JBM (1988) Estudo com o percevejo *Pachycoris torridus* (Scopoli, 1772) (Hemiptera, Scutelleridae) e seu inimigo natural *Pseudotelenomus pachycoris* Lima, 1928 (Hymenoptera, Scelionidae) em cultura do pinhão manso *Jatropha* spp. Biológico 52:17–20

Göllner-Scheiding U (2006) Scutelleridae. In: Aukema B, Rieger C (eds) Catalogue of the Heteroptera of the Palaearctic region, vol 5, Pentatomomorpha II. Netherlands Entomological Society, Wageningen, pp 190–227

Grimm C, Maes J-M (1997) Insectos asociados al cultivo de tempate (*Jatropha curcas* L.) (Euphorbiaceae) en el Pacifico de Nicaragua. I. Scutelleridae (Heteroptera). Rev Nicar Entomol 39:13–26

Guimarães JH (1977) Host-parasite and parasite-host catalogue of South American Tachinidae (Diptera). Arq Zool (São Paulo) 28:1–131

Hallman G (1979) Importancia de algunas relaciones naturales plantas-artropodos en la agricultura de la zona calida del Tolima central. Rev Colomb Entomol 5:19–26

Horvath G (1921) Genera duo nova Scutelleridarum. Ann Mus Natl Hung 18:145–146

Hussey RF (1934) Observations on *Pachycoris torridus* (Scop.), with remarks on parental care in other Hemiptera. Bull Brooklyn Entomol Soc 29:133–145

Hussey RF (1956) Additions to the United States list of Hemiptera. Fla Entomol 39:88

Javahery M, Schaefer CW, Lattin JD (2000) Shield bugs (Scutelleridae). In: Schaefer CW, Panizzi AR (eds) Heteroptera of economic importance. CRC, Boca Raton, pp 475–503

Kumar R (1965) Contributions to the morphology and relationships of Pentatomoidea (Hemiptera: Heteroptera). Pt I Family Scutelleridae. J Entomol Soc Qld 4:41–55

Lattin JD (1964) The Scutellerinae of America North of Mexico (Hemiptera: Heteroptera: Pentatomidae). Thesis, University of California, CA, USA

Maes J-M (1994) Catalogo de los Pentatomoidea (Heteroptera) de Nicaragua. Rev Nicar Entomol 28:1–29

Martínez-Herrera J, Martinez-Ayala AL, Makkar H, Francis G, Becker K (2010) Agroclimatic conditions, chemical and nutritional characterization of different provenances of *Jatropha curcas* L. from Mexico. Eur J Sci Res 39:396–407

McDonald FJD (1966) The genitalia of North American Pentatomoidea (Hemiptera: Heteroptera). Quaest Entomol 2:7–150

McDonald FJD, Cassis G (1984) Revision of the Australian Scutelleridae Leach (Hemiptera). Aust J Zool 32:537–572

McPherson JE (1977) The first report of the occurrence of *Symphylus caribbeanus* (Hemiptera: Scutelleridae) in Texas. J Kans Entomol Soc 50:529–530

Michelotto MD, Silva RA, Busoli AC (2006) Stinkbugs (Hemiptera: Heteroptera) collected on Barbados cherry plants (*Malpighia glabra* L.) in Jaboticabal, SP, Brazil. Arq Inst Biol Sao Paulo 73:123–125

Paleari LM (1992a) Revisão do gênero *Agonosoma* Laporte, 1832 (Hemiptera, Scutelleridae). Rev Bras Entomol 36:505–520

Paleari LM (1992b) Biologia de *Agonosoma flavolineata* Laporte, 1832 (Hemiptera, Scutelleridae). Rev Bras Entomol 36:521–526

Paleari LM (1994) Variação sazonal de freqüência e abordagem genética dos padrões de cor de *Agonosoma flavolineata* Laporte, 1832 (Hemiptera, Scutelleridae). Rev Bras Entomol 38:47–56

Quintanilla RH, Margheritis AE, Rizzo HF (1976) Catalogo de hemípteros hallados en la Provincia de Corrientes (Argentina). Rev Soc Entomol Argent 35:115–133

Rider DA (2014) Pentatomoidea home page. North Dakota State University. Available at http://www.ndsu.nodak.edu/ndsu/rider/Pentatomoidea/. Accessed Jan 2014.

Rodrigues SR, Oliveira HN, Santos WT, Abot AR (2011) Aspectos biológicos e danos de *Pachycoris torridus* em pinhão–manso. Bragantia 70:356–360

Sánchez-Soto S, Nakano O (2002) Ocorrência de *Pachycoris torridus* (Scopoli) (Hemiptera: Scutelleridae) em acerola (*Malpighia glabra* L.) no Brasil. Neotrop Entomol 31:481–482

Sánchez-Soto S, Milano P, Nakano O (2004) Nova planta hospedeira e novos padrões cromáticos de *Pachycoris torridus* (Scopoli) (Hemiptera: Scutelleridae) no Brasil. Neotrop Entomol 33:109–111

Santos JC, Silveira FAO, Almeida FVM, Fernandes GW (2005) Ecology and behavior of *Pachycoris torridus* (Hemiptera: Scutelleridae): new host plant, color polymorphism, maternal care and parasitism. Lundiana 6:107–111

Schouteden H (1904) Heteroptera. Fam. Pentatomidae. Subfam. Scutellerinae. Genera Insectorum 24:1–98

Schuh RT, Slater JA (1995) True bugs of the world (Hemiptera: Heteroptera). Classification and natural history. Cornell University Press, Ithaca

Silva AGA, Gonçalves CR, Galvão DM, Gonçalves AJL, Gomes J, Silva MN, Simoni L (1968) Quarto Catálogo dos insetos que vivem nas plantas do Brasil, seus parasitos e predadores. Parte II, Tomo I. Insetos, hospedeiros e inimigos naturais. Ministério da Agricultura, Rio de Janeiro

Smith K, Heard T (2003) Biological control of the bellyache bush. Pestic Outlook 14:145

Souza GK, Pikart TG, Oliveira HN, Serrão JE, Zanuncio JC (2012) Color polymorphism in *Pachycoris torridus* (Hemiptera: Scutelleridae) and its taxonomic implications. Rev Chil Hist Nat 85:357–359

Tepole-García RE, Pineda-Guillermo S, Martínez-Herrera J, Castrejón-Gómez VR (2012) Records of two pest species, *Leptoglossus zonatus* (Heteroptera: Coreidae) and *Pachycoris klugii* (Heteroptera: Scutelleridae), feeding on the physic nut, *Jatropha curcas*, in Mexico. Fla Entomol 95:208–210

Torre-Bueno JR (1914) Phototropism in Heteroptera. Bull Brooklyn Entomol Soc 9:90–96

Tsai J-F, Rédei D, Yeh GF, Yang MM (2011) Jewel bugs of Taiwan (Heteroptera: Scutelleridae). National Chung Hsing University, Taichung

Uhler PR (1894) On the Hemiptera-Heteroptera of the island of Grenada, West Indies. Proc Zool Soc London 1894:167–224

Williams L, Evans PE, Bowers WS (2001) Defensive chemistry of an aposematic bug. *Pachycoris stallii* Uhler and volatile compounds of its host plant *Croton californicus* Muell-Arg. J Chem Ecol 27:203–215

Williams L, Coscarón MC, Dellapé PM, Roane TM (2005) The shield-backed bug, *Pachycoris stallii*: description of immature stages, effect of maternal care on nymphs, and notes on life history. J Insect Sci 5:1–13

Wolcott GN (1923) "Insectae Portoricensis." A preliminary annotated check-list of the insects of Porto Rico, with descriptions of some new species. J Dep Agric Univ Puerto Rico 7:241–256

Wolcott GN (1936) "Insectae Borinquensis." A revised annotated check-list of the insects of Puerto Rico. J Dep Agric Univ Puerto Rico 20:1–627

Wolcott GN (1941) A supplement to "Insectae Borinquensis". J Dep Agric Univ Puerto Rico 25:33–158

Chapter 24
Negro Bugs (Thyreocoridae)

Viviana C. Matesco and Jocélia Grazia

Abstract The Thyreocoridae, usually known as "negro bugs," includes 212 species in 12 genera distributed in two subfamilies Thyreocorinae (seven species in three genera) which are not represented in the neotropics and Corimelaeninae (205 species in nine genera) with a worldwide distribution. Relatively small and dark colored, often shiny, this is a poorly studied family within Pentatomoidea and its relationships still being debated. They are exclusively herbivorous, and, at least, 23 Mexican and Neotropical species have some record of associated plant. Besides that, in the literature, there is no information about damage caused by these bugs. Immature stages are known only for eight species occurring in Mexico and Neotropical region, including descriptions of eggs and/or nymphs and/or life history. The only known key for the Neotropical representatives of the family is that found in McAtee and Malloch (Ann Carnegie Mus 21:191–411, 1933).

24.1 Introduction

The Thyreocoridae Amyot & Serville (=Corimelaenidae Uhler) is a common but poorly studied group in the Pentatomoidea. It includes the Thyreocorinae, with seven species in three genera, and Corimelaeninae, with 205 species in nine genera (McAtee and Malloch 1933; Sailer 1940, 1941; Kormilev 1956a, b; McPherson and Sailer 1978; Stys and Davidová 1979; Dolling 1981; Ahmad and Moizuddin 1982; Lis 2006). Previous authors considered the Thyreocorinae (= Corimelaeninae) as a subfamily of Cydnidae (i.e., Gapud 1991) by the presence of coxal fringes and tibial spines. Grazia et al. (2008), in a cladistic analysis of the Pentatomoidea, suggested

V.C. Matesco
Programa de Pós-Graduação em Biologia Animal, Instituto de Biociências, Universidade Federal do Rio Grande do Sul (UFRGS), Av. Bento Gonçalves 9500, prédio 43435, Bairro Agronomia, Porto Alegre, RS 91501-970, Brazil
e-mail: vimatesco@yahoo.com.br;

J. Grazia (✉)
Departamento de Zoologia, Instituto de Biociências, Universidade Federal do Rio Grande do Sul (UFRGS), Av. Bento Gonçalves 9500, prédio 43435, Bairro Agronomia, Porto Alegre, RS 91501-970, Brazil
e-mail: jocelia@ufrgs.br

A.R. Panizzi, J. Grazia (eds.), *True Bugs (Heteroptera) of the Neotropics*, Entomology in Focus 2, DOI 10.1007/978-94-017-9861-7_24

that the Parastrachiinae Schaefer, Dolling, and Tachikawa should be treated as part of a more broadly conceived Thyreocoridae.

Only the subfamily Corimelaeninae is represented in the Neotropical region. Basic information (e.g., life cycle, geographical distribution) for many species is either incomplete, inaccurate, or lacking (McPherson 1974), especially in the neotropics. Keys to the genera, subgenera, and species are found in McAtee and Malloch (1933). *Galgupha* Amyot & Serville is the most diverse Neotropical genus with 155 species in 15 subgenera. Table 24.1 shows the number of Thyreocoridae species registered in the world, in Mexico, and in the Neotropical region. Table 24.2 provides the number of Mexican and Neotropical species in each genus and subgenus of the Corimelaeninae.

24.2 General Characteristics and Diagnosis

Negro bugs (Figs. 24.1, 24.2, 24.3, 24.4, 24.5, 24.6, 24.7, and 24.8) have a size from 3 to 8 mm, body oval-elongate, dorsally convex, and ventrally flat. Dark color (castaneous or black), usually shiny. Head declivent, subtriangular in dorsal view. Antennae five segmented. Scutellum strongly convex recovering most of hemelytra. Corium reduced, some species with the visible part of corium in yellow. In the Corimelaeninae, hind wings with jugal lobe perforated. Tibiae with setae and spines in all their extension; tarsi 3-segmented.

24.2.1 Immatures

Thyreocoridae immatures have been broadly overlooked, probably due to their minute size, somewhat secretive habits, and lack of knowledge about host plants, which is required for nymphal development (Yonke 1991). General features of the eggs and nymphs of thyreocorid species have been briefly described (Southwood 1956; Cobben 1968; Hinton 1981; Yonke 1991). Immature stages are known only for eight species occurring in Mexico and Neotropical region, including descriptions of eggs and/or nymphs and/or life history (Table 24.3).

Southwood (1956) and Cobben (1968) discussed the egg characters of two thyreocorid species, *Corimelaena* sp. and *Thyreocoris scarabaeoides* (L.), respectively. At least five Nearctic Corimelaeninae species have their immature described

Table 24.1 Number of genera and species of the subfamilies of the Thyreocoridae (Hemiptera) occurring in the world, Mexico, and Neotropical region

	World		Mexico		Neotropics	
	Genera	Species	Genera	Species	Genera	Species
Thyreocoridae	12	212	6	42	9	159
Corimelaeninae	9	205	6	42	9	159
Thyreocorinae	3	7	0	0	0	0

Table 24.2 List of genera and subgenera included in Thyreocoridae: Corimelaeninae and its respective number of species occurring in Mexico and Neotropical region

Genus	Subgenus	Total number of species	Number of Mexican species	Number of Neotropical species
Alkindus Distant		2	1	2
Amyssonotum Horváth		1	1	1
Corimelaena White		31	16	13
Cydnoides Malloch		9	4	3
	Cydnoides Malloch	3	2	0
	Sayocoris McAtee & Malloch	3	2	0
	Cosmarioides McAtee & Malloch	3	0	3
Eumetopia Westwood		1	0	1
Galgupha Amyot & Serville		155	19	133
	Acritophleps McAtee & Malloch	1	0	1
	Acrotmetus Horváth	6	1	6
	Astiroderma Horváth	2	0	2
	Bonaria McAtee & Malloch	1	0	1
	Charoda McAtee & Malloch	1	0	1
	Ctenopoda McAtee & Malloch	2	0	2
	Euryscytus Horváth	38	4	34
	Galgupha Amyot & Serville	8	4	1
	Gyrocnemis McAtee & Malloch	37	5	33
	Microcompsus Horváth	3	0	3
	Nothocoris McAtee & Malloch	27	4	22
	Orocoris McAtee & Malloch	1	0	0
	Psestophleps McAtee & Malloch	18	1	17
	Pteronomos McAtee & Malloch	9	0	9
	Trepocnemis McAtee & Malloch	1	0	1
Godmania Horváth		1	1	1
Pericreps Horváth		4	0	4
Pruhleria McAtee & Malloch		1	0	1

McAtee and Malloch (1933), Sailer (1940, 1941), Kormilev (1956a, b), McPherson and Sailer (1978), and Dolling (1981)

Fig. 24.1 *Eumetopia fissipes* Westwood, adult, *dorsal view*

Fig. 24.2 *Alkindus crassicosta* Horváth, adult, *dorsal view*

Fig. 24.3 *Corimelaena lateralis* (F.), adult, *dorsal view*

Fig. 24.4 *Eumetopia fissipes* Westwood, female, head and pronotum, *dorsal view*

Fig. 24.5 *Eumetopia fissipes* Westwood, male, head and pronotum, *dorsal view*

Fig. 24.6 *Corimelaena lateralis* (F.), corium, *dorsolateral view*

Fig. 24.7 *Corimelaena lateralis* (F.), female, abdomen, *ventral view*

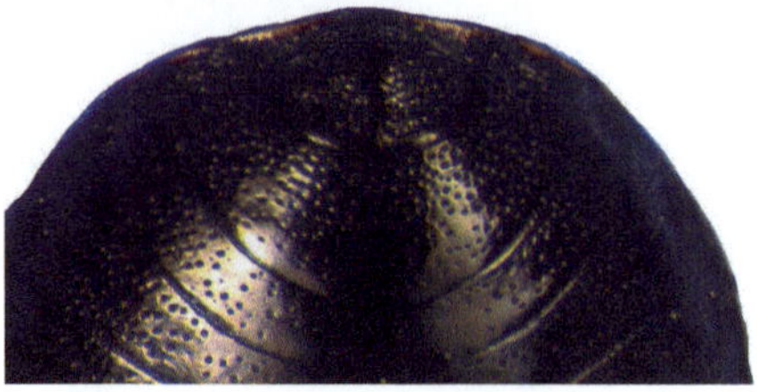

Fig. 24.8 *Corimelaena lateralis* (F.), male, abdomen, *ventral view*

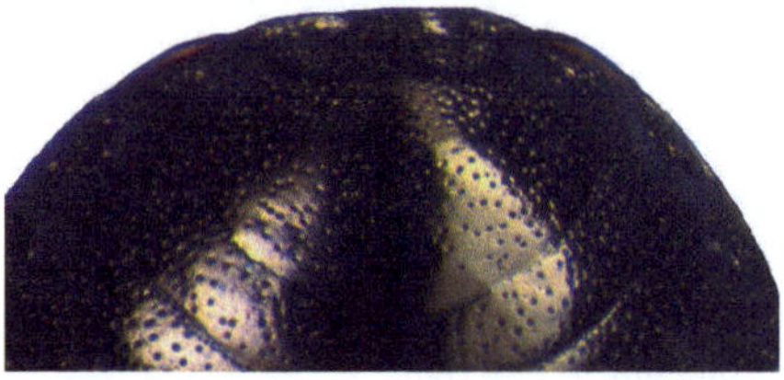

Table 24.3 List of thyreocorid species occurring in Mexico and Neotropical region with described immature stages (eggs and/or nymphs) and/or immature biology

	Eggs			
Species	SEM[a]	SM[a]	Nymphs	Biology
Corimelaena (*Corimelaena*) *lateralis* (F.)		McPherson (1972)	McPherson (1972)	McPherson (1971, 1972)
Corimelaena (*Parapora*) *extensa* Uhler		Lung and Goeden (1982)	Lung and Goeden (1982)	Lung and Goeden (1982)
Corimelaena (*P.*) *incognita* (McAtee & Malloch)	Bundy and McPherson (2009)		Bundy and McPherson (2009)	Bundy and McPherson (2009)
Galgupha (*Euryscytus*) *difficilis* (Breddin)	Matesco et al. (2012)			
Galgupha (*Galgupha*) *atra* Amyot & Serville		Javahery (1994)		
Galgupha (*G.*) *ovalis* Hussey		Biehler and McPherson (1982)	Biehler and McPherson (1982)	Biehler and McPherson (1982)
Galgupha (*Gyrocnemis*) *fossata* McAtee & Malloch	Matesco et al. (2012)			

[a]*SEM* scanning electron microscopy, *SM* stereomicroscopy

(McPherson 1972; Biehler and McPherson 1982; Lung and Goeden 1982; Bundy and McPherson 1997, 2009). Matesco et al. (2012) described the egg ultrastructure of two Neotropical species *Galgupha* (*Euryscytus*) *difficilis* (Breddin) and *Galgupha* (*Gyrocnemis*) *fossata* McAtee & Malloch.

The eggs are elongate with rounded poles. Corium is thin, apparently smooth, but wrinkled when observed in scanning electron microscopy (Bundy and McPherson 1997, 2009; Matesco et al. 2012). Initially white, the eggs become reddish or dark red with the embryo development; the translucent chorion allows viewing the structures inside the egg. Aero-micropylar processes around the apical pole are pedunculated, curved inward, varying in numbers from 4 to 10, and differing in the peduncle size. Operculum is absent and the eclosion line irregular and transversal (Javahery 1994). Bundy and McPherson (2009) described the egg burster of *Corimelaena incognita* (McAtee & Malloch) as rhomboid, with a short median process; Southwood (1956) and Cobben (1968) mentioned the egg burster is reduced in both subfamilies. In thyreocorid eggs, no pattern has been recognized yet at the generic level. Egg characters in the family are scarce – the egg structure is simple – and basic information is available for only a few species.

Matesco et al. (2012), based on the egg morphology of ten Corimelaeninae species, outlined a consistent pattern: eggs are laid singly on parts of the host plants; each egg is cylindrical with rounded ends and white before embryonic development; the chorion is thin and shiny with a few (2–10) short, stalked aero-micropylar processes around the anterior pole; lengths of eggs and micropylar processes are 0.65–1.08 and 0.03–0.05 mm, respectively.

Nymphs are similar to the adults in body and head shapes. In general, nymph morphological characters are similar to the other Pentatomoidea families. Most of the body is dark, black, or castaneous; the abdomen varies from yellow to reddish always with dark punctures. The body is strongly convex with punctures uniformly distributed. Tibiae spines are present. Median dorsal abdominal plates present in all abdominal segments, segments I and II with paired plates; scent glands ostiole paired and placed between segments III–IV, IV–V, and V–VI; and lateral abdominal plates subquadrangular. Other characters, apparently unique to this family, are metanotum partially fused on 1st to 3rd instars; posterior and anterior margins of the dorsal abdominal plates of segments IV and V, respectively, are contiguous.

24.3 General Biology and Ecology

Negro bugs are exclusively herbivorous, generally associated to bushes, feeding and living in the host plant reproductive parts as flowers and developing or ripe fruits. McPherson (1971, 1972) registered the preference of *Corimelaena lateralis* (F.) for mature plants of *Daucus carota* L. (Apiaceae), instead of inflorescences of the same plant. *Corimelaena pulicaria* (Germar) does not show this preference (McPherson 1972). The latter species is registered in distinct host plants up to 10 different plant families (Table 24.2). Nevertheless, all species studied used, along their life cycle in the same area, only one (i.e., Biehler and McPherson 1982) or two (i.e., Bundy and McPherson 1997) plant families as preferential sites for feeding, mating, and oviposition.

Schaefer (1988) compiled data on the host plants for several Pentatomoidea families, including Thyreocoridae. The data do not point to any preference for host plant at subfamily or genus level. Most species feed on the reproductive parts, rich in nitrogen (especially *Corimelaena* spp.); a few occur over the soil (especially *Cydnoides* spp.) (Schaefer 1988). Rider (2013) provided a list of several published records of host plants of Thyreocoridae. At least, 23 Mexican and Neotropical species have some record of associated plant in the literature (Table 24.4).

All studies of thyreocorids at field were developed in the northern hemisphere; the life cycle corresponds in general to the pentatomid life cycle (Grazia and Schwertner 2008). They could be uni-, bi-, or multivoltine: in low latitudes, a higher number of generations per year occur, whereas in higher latitudes, only one generation per year is possible (McPherson 1972; Lung and Goeden 1982; Bundy and McPherson 1997, 2009). The adults show dormancy, and during this period, they could be found under litter, soil, and also rocks; they leave the dormancy sites at late winter or early spring and then move to the host plants and start feeding. Precopula behavior was described by Bundy and McPherson (1997): the male stands behind the female and starts the contact by antennation; after the female acceptance, the copula takes place with the individuals positioned in opposite directions, just touching genitalic structures; copulation can last hours.

Thyreocorids lay eggs singly and glued laterally to the substrate, often in the host plant reproductive parts, less frequently in branches or leaves. Egg development varies between eight and 11 days, depending on the species and also the temperature and humidity. First instar nymphs are actives, differing from other Pentatomoidea families' nymphs, feeding and not showing gregarious behavior. Nymphs' development varies from 30 to 45 days, probably influenced by the diet and also by the local temperature and humidity. Tachinidae (Diptera) and Scelionidae (Hymenoptera) egg parasitoids are the major natural enemies recorded.

24.4 Classification and Diversity

The evolution of Thyreocoridae within Pentatomoidea has been discussed (Gapud 1991; Grazia et al. 2008; Lis et al. 2012, and Matesco et al. in prep.), usually suggesting a close relationship with the Cydnidae.

The hypothesis of Gapud (1991) of the relationships within Pentatomoidea based on morphology recognizes that the Cydnidae and Thyreocoridae are strongly related by the presence of coxal fringes and tibial spines. However, the Cydnidae can be separated from the Thyreocoridae by setigerous punctures in the head and pronotum. Thyreocoridae shares weak characters with other families, such as the broad scutellum, covering the abdomen; short frenum; and laterotergites nine contiguous.

Table 24.4 List of species of Thyreocoridae: Corimelaeninae occurring in Mexico and Neotropical region and its respective potential host plants and geographical distribution

Species	Plant family	Plant species	Distribution
Alkindus atratus Distant	Fabaceae	*Glycine* sp.[24]	Mexico and Neotropical (Guatemala[27], Honduras[24], El Salvador[24], Nicaragua, Aruba, Curaçao, Venezuela, Costa Rica, Colombia, Panama, Brazil[27])
		Phaseolus sp.[24]	
	Malvaceae	*Gossypium* sp.[24]	
		Sida sp.[24]	
	Moraceae	*Ficus* sp.[27]	
	Musaceae	*Musa* sp.[24]	
	Myrtaceae	*Eucalyptus* sp.[24]	
	Poaceae	*Cenchrus* sp.[24]	
		Echinochloa sp.[24]	
		Oryza sp.[24]	
		Sorghum sp.[24]	
		Zea mays L.[2]	
	Rubiaceae	*Coffea* sp.[24]	
	Sterculiaceae	*Theobroma* sp.[24]	
A. crassicosta Horváth	Poaceae	*Andropogon bicornis* L.[20]	Neotropical (Brazil)
Amyssonotum rastratum (Stål)			Mexico and Neotropical (Belize, Guatemala, Costa Rica, Trinidad, Colombia, Panama, Suriname, Brazil, Peru, Bolivia)
Corimelaena alticola (Horváth)			Mexico
C. barberi (McAtee & Malloch)			Mexico and Neotropical (Costa Rica)
C. championi (Distant)			Neotropical (Belize, Guatemala)
C. cognata (Van Duzee)	Loasaceae	*Eucnide* sp.[47]	Mexico and Neotropical (Cuba, Jamaica, Nicaragua, Colombia, Venezuela[2])

(continued)

Table 24.4 (continued)

Species	Plant family	Plant species	Distribution
C. contrasta (McAtee & Malloch)			Mexico
C. corallina (McAtee & Malloch)			Mexico and Neotropical (Guatemala)
C. digitata (McAtee & Malloch			Mexico
C. elegans (McAtee & Malloch)			Mexico and Neotropical (Guatemala, Costa Rica, Panama, Trinidad)
C. extensa Uhler	Asteraceae	*Baccharis salicifolia* (Ruiz & Pav.) Pers.[9]	Mexico[16]
	Scrophulariaceae	*Linaria vulgaris* Mill.[40]	
	Solanaceae	*Nicotiana glauca* Graham[23]	
C. incognita (McAtee & Malloch)	Chenopodiaceae	*Atriplex* sp.[6]	Mexico
	Scrophulariaceae	*Leucophyllum frutescens* (Berland.) I.M.Johnst.[10]	
	Solanaceae	*Nicotiana glauca* Graham[19]	
		Nicotiana obtusifolia Martens & Galeotti[19]	
C. interrupta Malloch			Mexico and Neotropical (Guatemala, Honduras, Costa Rica)
C. lateralis (F.)	Apiaceae	*Chaerophyllum procumbens* (L.) Crantz[31]	Mexico and Neotropical (Guatemala[16])
		Chaerophyllum tainturieri Hook[42]	
		Daucus carota L.[29, 30, 31, 42]	
		Torilis japonica (Houtt.) DC.[42]	

	Asteraceae	*Erigeron philadelphicus* L.[31]	
		Solidago altissima L.[30]	
	Geraniaceae	*Geranium carolinianum* L.[31, 42]	
		Geranium maculatum L.[25]	
	Juglandaceae	*Juglans nigra* L.[33]	
	Poaceae	*Bromus tectorum* L.[42]	
		Elymus hystrix L.[31]	
	Ranunculaceae	*Ranunculus macranthus* Scheele[19]	
	Rubiaceae	*Galium aparine* L.[1, 42]	
	Vitaceae	*Parthenocissus quinquefolia* (L.) Planch.	
C. micans (McAtee & Malloch)			Neotropical (Guatemala)
C. minuta Uhler			Neotropical (Cuba, Haiti, Dominican Republic, Puerto Rico, Jamaica)
C. minutissima Malloch			Mexico
C. nigra Dallas	Ranunculaceae	*Ranunculus* sp.[22]	Mexico
C. palmeri (McAtee & Malloch)			Mexico
C. parana McAtee & Malloch			Neotropical (Brazil)

(continued)

Table 24.4 (continued)

Species	Plant family	Plant species	Distribution
C. pulicaria (Germar)	Anacardiaceae	*Rhus glabra* L.[15]	Mexico[13] and Neotropical (Guatemala, Martinique, Venezuela[2])
	Apiaceae	*Daucus carota* L.[30, 31, 42]	
		Chaerophyllum procumbens (L.) Crantz[31]	
		Torilis japonica (Houtt.) DC.[42]	
	Apocynaceae	*Apocynum cannabinum* L.[31, 42]	
	Asclepiadaceae	*Asclepias incarnata* L.[31]	
	Asteraceae	*Antennaria plantaginifolia* (L.) Richardson[31]	
		Baccharis neglecta Britton[34]	
		Cirsium arvense (L.) Scop.[28]	
		Erigeron philadelphicus L.[31]	
		Erigeron strigosus Muhlenberg[33, 42]	
		Eupatorium sp.[43]	
		Iva frutescens L.[21]	
	Geraniaceae	*Geranium maculatum* L.[25]	
	Juglandaceae	*Juglans nigra* L.[32, 33]	
	Liliaceae	*Uvularia* sp.[15]	
	Plantaginaceae	*Veronica peregrina* L.[14, 44]	
	Poaceae	*Sorghum halepense* L. (Pers.)[42]	
	Polygonaceae	*Polygonum* sp.[15]	
	Rhamnaceae	*Ceanothus americanus* L.[14, 44]	
	Rubiaceae	*Galium verum* L.[1]	
	Scrophulariaceae	*Linaria vulgaris* Mill.[28, 40]	
		Verbascum thapsus L.[31]	

C. signoretii (McAtee & Malloch)			Neotropical (Brazil)
C. tibialis (F.)			Mexico and Neotropical (Jamaica, Costa Rica, Panama, Colombia, Saint Vincent and the Grenadines, Grenada, Venezuela, Trinidad, Guiana, Suriname, French Guiana, Brazil, Ecuador, Bolivia)
Cydnoides (*Cydnoides*) *ciliatus* (Uhler)	Euphorbiaceae	*Euphorbia* sp.[44]	Mexico
	Fabaceae	*Cassia marilandica* L.[8, 44]	
C. (*C.*) *confusus* McAtee & Malloch	Euphorbiaceae	*Euphorbia* sp.	Mexico
C. (*Sayocoris*) *obtusus* (Uhler)	Euphorbiaceae	*Euphorbia polycarpa* Benth.[44]	Mexico[16]
C. (*S.*) *peregrinus* McAtee & Malloch			Mexico[44]
C. (*Cosmarioides*) *nitens* McAtee & Malloch			Neotropical (Brazil)
C. (*C.*) *reticulatus* McAtee & Malloch			Neotropical (Brazil)
C. (*C.*) *setiventris* McAtee & Malloch			Neotropical (Bolivia, Paraguay)
Eumetopia fissipes Westwood			Neotropical (Brazil)
Galgupha (*Acritophleps*) *luteomarginata* McAtee & Malloch			Neotropical (Brazil, Bolivia)
G. (*Acrotmetus*) *brasilianus* (Jensen-Haarup)			Neotropical (Brazil, Bolivia, Paraguay)
G. (*A.*) *crassa* McAtee & Malloch			Neotropical (Brazil)

(continued)

Table 24.4 (continued)

Species	Plant family	Plant species	Distribution
G. (*A.*) *grossa* McAtee & Malloch			Neotropical (French Guiana, Brazil, Paraguay[17])
G. (*A.*) *minuenda* McAtee & Malloch			Neotropical (Brazil)
G. (*A.*) *schulzii* (F.)	Rubiaceae	*Spermacoce verticillata* L.[11]	Mexico e Neotropical (Panama, Trinidad[11], French Guiana, Brazil, Argentina)
	Solanaceae	*Solanum tuberosum* L.[36]	
G. (*A.*) *tucumanus* (Horváth)			Neotropical (Brazil, Argentina)
G. (*Astiroderma*) *albipennis* (Eschsholtz)			Neotropical (Guatemala, French Guiana, Brazil, Argentina, Uruguay)
G. (*A.*) *breddini* McAtee & Malloch			Neotropical (Brazil, Argentina[17])
G. (*Bonaria*) *longirostris* (Berg)			Neotropical (Argentina)
G. (*Charoda*) *simplex* McAtee & Malloch			Neotropical (Brazil, Chile, Argentina[17])
G. (*Ctenopoda*) *castor* McAtee & Malloch			Neotropical (Brazil)
G. (*C.*) *maculosa* (Berg)			Neotropical (Brazil, Paraguay, Argentina[3])
G. (*Euryscytus*) *aliena* McAtee & Malloch			Neotropical (Brazil)
G. (*E.*) *apicata* McAtee & Malloch			Neotropical (Suriname, French Guiana)
G. (*E.*) *assimilis* McAtee & Malloch			Neotropical (Brazil)
G. (*E.*) *australis* McAtee & Malloch			Neotropical (Brazil, Bolivia, Paraguay, Argentina)
G. (*E.*) *basalis* (Germar)	Moraceae	*Ficus* sp.[12]	Neotropical (Jamaica[46], Brazil)

G. (*E.*) *caudata* McAtee & Malloch			Neotropical (Argentina)
G. (*E.*) *caudiculata* McAtee & Malloch			Neotropical (Argentina)
G. (*E.*) *contra* McAtee & Malloch			Neotropical (Argentina)
G. (*E.*) *contracta* McAtee & Malloch			Neotropical (Brazil)
G. (*E.*) *corvina* (Horváth)			Neotropical (Brazil, Argentina)
G. (*E.*) *cruenta* (Horváth)			Mexico
G. (*E.*) *curvata* McAtee & Malloch			Neotropical (Brazil, Bolivia[17])
G. (*E.*) *curvatula* McAtee & Malloch			Neotropical (Brazil)
G. (*E.*) *difficilis* (Breddin)	Scrophulariaceae	*Buddleja Brasiliensis* Jacq. ex Spreng.[26]	Neotropical (Brazil, Argentina)
G. (*E.*) *foveolata* (Jensen-Haarup)			Neotropical (Brazil)
G. (*E.*) *insula* McAtee & Malloch			Neotropical (Paraguay, Argentina)
G. (*E.*) *jaczewskii* McAtee & Malloch			Neotropical (Brazil, Argentina[17])
G. (*E.*) *glabrata* McAtee & Malloch			
G. (*E.*) *laevis* (Stål)			Neotropical (Brazil)
G. (*E.*) *lucretia* McAtee & Malloch			Mexico e Neotropical (Guatemala)
G. (*E.*) *mammula* McAtee & Malloch			Neotropical (Brazil)

(continued)

Table 24.4 (continued)

Species	Plant family	Plant species	Distribution
G. (*E.*) *nitida* McAtee & Malloch			Neotropical (Panama)
G. (*E.*) *obesa* McAtee & Malloch			Neotropical (Paraguay)
G. (*E.*) *opacifrons* McAtee & Malloch			Neotropical (Brazil)
G. (*E.*) *opercula* McAtee & Malloch			Mexico
G. (*E.*) *parae* McAtee & Malloch			Neotropical (Brazil)
G. (*E.*) *parallela* McAtee & Malloch			Neotropical (Brazil, Bolivia)
G. (*E.*) *punctata* McAtee & Malloch			Mexico
G. (*E.*) *rasilis* (Horváth)			Neotropical (Colombia, Guiana, Suriname, French Guiana, Brazil, Bolivia)
G. (*E.*) *regularis* McAtee & Malloch			Neotropical (Argentina)
G. (*E.*) *romana* McAtee & Malloch			Neotropical (Brazil)
G. (*E.*) *rostrata* McAtee & Malloch			Neotropical (Brazil)
G. (*E.*) *sahlbergi* McAtee & Malloch			Neotropical (Brazil)
G. (*E.*) *signatipennis* (Germar)			Neotropical (Brazil, Paraguay, Argentina)
G. (*E.*) *smidtii* (F.)			Neotropical (Brazil)

G. (*E.*) *valvata* McAtee & Malloch			Neotropical (French Guiana)
G. (*E.*) *variipennis* McAtee & Malloch			Neotropical (Brazil)
G. (*E.*) *vulnerata* (Horváth)			Neotropical (Brazil, Paraguay)
G. (*Galgupha*) *atra* Amyot & Serville	Asteraceae	*Helianthus decapetalus* L.[31]	Mexico
	Euphorbiaceae	*Euphorbia maculata* L.[48]	
	Juglandaceae	*Juglans nigra* L.[33]	
	Plantaginaceae	*Plantago aristata* Michaux[41, 44]	
	Rosaceae	*Rubus allegheniensis* Porter[31]	
	Scrophulariaceae	*Castilleja sessiliflora* Pursh[41]	
		Verbascum thapsus L.[31]	
G. (*G.*) *denudata* (Uhler)			Mexico
G. (*G.*) *monostira* Horváth			Mexico
G. (*G.*) *ovalis* Hussey	Asteraceae	*Vernonia interior* Small[39]	Mexico and Neotropical (Guatemala)
	Juglandaceae	*Juglans nigra* L.[32]	
	Plantaginaceae	*Plantago aristata* Michaux[7]	
G. (*Gyrocnemis*) *acuta* McAtee & Malloch			Neotropical (Brazil)
G. (*G.*) *basigrapha* (Horváth)			Neotropical (Peru, Bolivia[17])
G. (*G.*) *bicornis* McAtee & Malloch			Neotropical (Brazil)
G. (*G.*) *concava* McAtee & Malloch			Neotropical (Brazil, Bolivia, Paraguay, Argentina)
G. (*G.*) *cruralis* (Stål)	Aquifoliaceae	*Ilex paraguariensis* A. St. Hil.[37]	Neotropical (Brazil, Bolivia, Argentina)

(continued)

Table 24.4 (continued)

Species	Plant family	Plant species	Distribution
	Poaceae	*Sorghum bicolor* L. Moench[36]	
		Oryza sativa L.[36]	
		Zea mays L.[5]	
	Rutaceae	*Citrus* sp.[36]	
	Solanaceae	*Solanum tuberosum* L.[4,5]	
G. (*G.*) *cydnoidea* McAtee & Malloch			Neotropical (Colombia, Venezuela)
G. (*G.*) *differentialis* McAtee & Malloch			Neotropical (Brazil, Peru, Bolivia, Paraguay)
G. (*G.*) *dimorpha* McAtee & Malloch			Neotropical (Bolivia, Paraguay)
G. (*G.*) *diversa* McAtee & Malloch			Neotropical (Paraguay, Argentina)
G. (*G.*) *durionei* Kormilev			Neotropical (Bolivia[17])
G. (*G.*) *fossata* McAtee & Malloch	Scrophulariaceae	*Buddleja brasiliensis* Jacq. ex Spreng.[26]	Neotropical (Brazil, Paraguay, Argentina)
G. (*G.*) *geayi* McAtee & Malloch			Neotropical (Venezuela)

G. (*G.*) *guttiger* (Stål)	Arecaceae	*Elaeis* sp.[24]	Mexico and Neotropical (Cuba, Haiti, Dominican Republic, Jamaica, Belize[13], Guatemala, Honduras, Nicaragua, Costa Rica, Panama, Colombia, Venezuela, Trinidad, Bolivia)
	Brassicaceae	*Brassica* sp.[24]	
	Cucurbitaceae	*Cucurbita* sp.[24]	
	Fabaceae	*Cajanus* sp.[24]	
		Phaseolus sp.[24]	
	Lauraceae	*Persea* sp.[24]	
	Malvaceae	*Gossypium* sp.[24]	
		Sida sp.[24]	
	Nyctaginaceae	*Boerhavia* sp.[24]	
		Bougainvillea sp.[24]	
	Pedaliaceae	*Sesamum* sp.[24]	
	Poaceae	*Oryza* sp.[24]	
		Sorghum sp.[24]	
		Zea sp.[24]	
	Rubiaceae	*Coffea* sp.[24]	
	Solanaceae	*Lycopersicon* sp.[24]	
		Nicotiana sp.[24]	
		Solanum sp.[24]	
G. (*G.*) *impressa* (Horváth)			Neotropical (Venezuela, Brazil, Peru, Bolivia, Argentina)
G. (*G.*) *inaequalis* McAtee & Malloch			Neotropical (Colombia, Venezuela, Bolivia)
G. (*G.*) *intermedia* McAtee & Malloch			Neotropical (Brazil, Argentina[17])
G. (*G.*) *lineata* McAtee & Malloch			Neotropical (Brazil, Argentina)
G. (*G.*) *maculipennis* (Germar)			Neotropical (Brazil[3], Paraguay, Argentina)

(continued)

Table 24.4 (continued)

Species	Plant family	Plant species	Distribution
G. (*G.*) *morbiloci* McAtee & Malloch			Mexico
G. (*G.*) *nitens* (Breddin)			Neotropical (Brazil, Colombia, Peru, Bolivia)
G. (*G.*) *nitidipennis* McAtee & Malloch			Neotropical (Venezuela)
G. (*G.*) *notha* McAtee & Malloch			Neotropical (Argentina)
G. (*G.*) *novatra* McAtee & Malloch			Neotropical (Argentina)
G. (*G.*) *novella* McAtee & Malloch			Neotropical (Brazil)
G. (*G.*) *occulta* McAtee & Malloch			Neotropical (Brazil)
G. (*G.*) *omissa* McAtee & Malloch			Neotropical (Brazil)
G. (*G.*) *parva* McAtee & Malloch			Neotropical (Brazil, Paraguay)
G. (*G.*) *punctifer* McAtee & Malloch			Mexico and Neotropical (Belize, Guatemala, Costa Rica, Panama, Colombia, Bolivia)
G. (*G.*) *quadrisignata* (Stål)	Verbenaceae	*Lantana* sp.[35]	Mexico and Neotropical (Belize, Guatemala, Honduras, El Salvador, Costa Rica)
G. (*G.*) *reinhardtii* (Jensen-Haarup)			Neotropical (Venezuela, Brazil)
G. (*G.*) *singularis* McAtee & Malloch			Neotropical (Argentina)
G. (*G.*) *soror* McAtee & Malloch			Mexico

G. (*G.*) *torresi* Kormilev			Neotropical (Argentina[17])
G. (*G.*) *triconcava* McAtee & Malloch			Neotropical (Brazil, Bolivia, Paraguay, Argentina)
G. (*G.*) *unica* McAtee & Malloch			Neotropical (Colombia)
G. (*Microcompsus*) *daldorfii* (Fabricius)			Neotropical (Brazil)
G. (*M.*) *haywardi* Kormilev			Neotropical (Argentina[18])
G. (*M.*) *vinculata* (Germar)	Rubiaceae	*Spermacoce verticillata* L.[11]	Neotropical (Colombia, Venezuela, Trinidad, Suriname, French Guiana, Brazil, Bolivia, Paraguay, Argentina)
G. (*Nothocoris*) *akarna* McAtee & Malloch			Neotropical (Argentina)
G. (*N.*) *alutacea* McAtee & Malloch			Neotropical (Brazil)
G. (*N.*) *boliviano* McAtee & Malloch			Neotropical (Brazil, Bolivia)
G. (*N.*) *brevis* McAtee & Malloch			Neotropical (Brazil)
G. (*N.*) *brunnipennis* (Germar)			Neotropical (Brazil, Paraguay)
G. (*N.*) *caracasana* McAtee & Malloch			Neotropical (Venezuela)
G. (*N.*) *chilocoroides* (Walker)			Mexico and Neotropical (French Guiana, Brazil)
G. (*N.*) *chrostowski* McAtee & Malloch			Neotropical (Brazil)
G. (*N.*) *coccineloides* Horváth			Neotropical (French Guiana, Brazil, Bolivia, Paraguay, Argentina)

(continued)

Table 24.4 (continued)

Species	Plant family	Plant species	Distribution
G. (*N.*) *fossula* McAtee & Malloch			Mexico
G. (*N.*) *grenadensis* McAtee & Malloch			Neotropical (Colombia)
G. (*N.*) *inornata* McAtee & Malloch			Neotropical (Argentina)
G. (*N.*) *juno* McAtee & Malloch			Neotropical (Argentina)
G. (*N.*) *magna* Sailer			Mexico[38]
G. (*N.*) *marginalis* McAtee & Malloch			Neotropical (Colombia)
G. (*N.*) *marginicollis* Horváth			Neotropical (Colombia, Venezuela, Grenada)
G. (*N.*) *microphthalma* McAtee & Malloch			Neotropical (Brazil)
G. (*N.*) *nitiduloides* (Wolff)	Asteraceae	*Helenium* sp.	Mexico and Neotropical (Guatemala, Nicaragua[24], Costa Rica, Brazil)
	Fabaceae	*Glycine* sp.[24]	
	Malvaceae	*Gossypium* sp.	
	Plantaginaceae	*Plantago patagonica* Jacq.[44,45]	
	Poaceae	*Saccharum* sp.	
	Verbenaceae	*Lantana* sp.[35]	
G. (*N.*) *oculata* McAtee & Malloch			Neotropical (Paraguay)
G. (*N.*) *parvula* Horváth			Neotropical (Peru)
G. (*N.*) *repetita* McAtee & Malloch			
G. (*N.*) *semilimbata* Horváth			Neotropical (Brazil, Paraguay)

G. (*N.*) *stramineipennis* McAtee & Malloch			Neotropical (Brazil)
G. (*N.*) *surda* McAtee & Malloch			Neotropical (Brazil)
G. (*N.*) *terminalis* (Walker)			Neotropical (Brazil, Bolivia, Paraguay, Uruguay)
G. (*Psestophleps*) *altera* McAtee & Malloch			Neotropical (Brazil)
G. (*P.*) *bergiana* McAtee & Malloch			Neotropical (Brazil, Peru, Paraguay, Argentina)
G. (*P.*) *bisignata* McAtee & Malloch			Neotropical (Brazil, Argentina)
G. (*P.*) *carbonata* McAtee & Malloch			Neotropical (Argentina)
G. (*P.*) *casta* McAtee & Malloch			Neotropical (Argentina)
G. (*P.*) *costomaculata* McAtee & Malloch			Neotropical (Brazil, Paraguay)
G. (*P.*) *denieri* Kormilev			Neotropical (Argentina[17])
G. (*P.*) *fimbriata* McAtee & Malloch			Neotropical (Venezuela, Suriname)
G. (*P.*) *imitans* McAtee & Malloch			Neotropical (Venezuela[2], Brazil)
G. (*P.*) *inops* McAtee & Malloch			Neotropical (Brazil)
G. (*P.*) *mayana* McAtee & Malloch			Neotropical (Honduras)
G. (*P.*) *media* McAtee & Malloch			Neotropical (Guiana, Suriname)

(continued)

Table 24.4 (continued)

Species	Plant family	Plant species	Distribution
G. (*P.*) *mexicana* McAtee & Malloch			Mexico
G. (*P.*) *neobisignata* McAtee & Malloch			Neotropical (French Guiana, Brazil, Paraguay, Argentina, Uruguay)
G. (*P.*) *obovata* McAtee & Malloch			Neotropical (Argentina)
G. (*P.*) *pallipennis* McAtee & Malloch			Neotropical (Brazil)
G. (*P.*) *porcata* Horváth			Neotropical (Brazil, Paraguay, Argentina)
G. (*P.*) *tabellula* McAtee & Malloch			Neotropical (Suriname)
G. (*Pteronomos*) *amitta* McAtee & Malloch			Neotropical (Argentina)
G. (*P.*) *china* McAtee & Malloch			Neotropical (Brazil, Peru, Bolivia, Argentina)
G. (*P.*) *fritzi* Kormilev			Neotropical (Chile[17])
G. (*P.*) *meridiana* McAtee & Malloch			Neotropical (Paraguay, Argentina)
G. (*P.*) *oblonga* McAtee & Malloch			Neotropical (Brazil, Paraguay, Argentina)
G. (*P.*) *punctifrons* McAtee & Malloch			Neotropical (Argentina)
G. (*P.*) *ruficornis* (Dallas)			Neotropical (Colombia)
G. (*P.*) *vianai* Kormilev			Neotropical (Argentina[18])
G. (*P.*) *vittifrons* McAtee & Malloch			Neotropical (Bolivia, Argentina)

G. (*Trepocnemis*) *anomala* McAtee & Malloch			Neotropical (Argentina)
Godmania aterrima Horváth			Mexico and Neotropical (Guatemala)
Pericrepis afer McAtee & Malloch			Neotropical (Argentina)
P. bergi Kormilev			Neotropical (Argentina[18])
P. bosqui Kormilev			Neotropical (Argentina[18])
P. callosula Horváth			Neotropical (Brazil, Paraguay[17], Argentina)
Pruhleria incerta (Uhler)	Musaceae	*Musa* sp.	Neotropical (Cuba, Nicaragua[13])

All data were taken from McAtee and Malloch (1933), except when indicated by the superscript numbers that correspond to references

Referências: 1. Batra (1984), 2. Becker and Grazia-Vieira (1971), 3. Berg (1891). 4. Bertels (1962), 5. Bertels and Baucke (1966), 6. Bibby (1961), 7. Biehler and McPherson (1982), 8. Blatchley (1926), 9. Boldt and Robbins (1990), 10. Bundy and McPherson (2009), 11. Callan (1948), 12. Costa Lima (1936), 13. Distant (1880), 14. Forbes (1905), 15. Froeschner (1941), 16. Froeschner 1988, 17. Kormilev (1956a), 18. Kormilev (1956b), 19. Jones (1993), 20. Klein et al. (2013), 21. Lago and Testa (2000), 22. Linsley and MacSwain (1959), 23. Lung and Goeden (1982), 24. Maes (1994), 25. Martin (1965), 26. Matesco et al. (2012), 27. Matesco and Grazia (2013), 28. Maw (1976), 29. McPherson (1971), 30. McPherson (1972), 31. McPherson and Mohlenbrock (1976), 32. McPherson and Weber (1990), 33. Nixon et al. (1975), 34. Palmer (1987), 35. Palmer and Pullen (1995), 36. Quintanilla et al. (1976), 37. Quintanilla et al. (1981), 38. Sailer (1941), 39. Schwitzgebel and Wilbur (1942), 40. Smith (1959), 41. Stoner (1920), 42. Tecic and McPherson (2005), 43. Uhler (1894), 44. Torre-Bueno 1939, 45. van Duzee (1904), 46. van Duzee (1907), 47. van Duzee (1923)

Grazia et al. (2008) analyzed the relationships within the Pentatomoidea, using 135 terminal taxa, 57 morphological characters, and sequence data from genetic regions of rDNA 18S, rDNA 16S, rDNA 28S, and COI. The study supports the monophyly of the Corimelaenidae (=Thyreocoridae), represented by *Thyreocoris* Schrank+*Allocoris* McAtee & Malloch (=*Corimelaena* White) based on seven homoplasious synapomorphies. Those are posterior and humeral angles of pronotum not developed; hind wings with A1 stridulitrum; prosternum deeply sulcate; prosternum strongly carinate; coxae with a fringe of setae, bristles, or scales; fore tibiae with a row of stout setae on lateral margin; and gonocoxites 9 joined medially by membrane.

In the morphological analysis of Grazia et al. (2008), Corimelaenidae shows up as a basal group within Cydnidae. In some trees of the molecular analysis, *Allocoris* is the sister group of Parastrachiidae (*Parastrachia* Distant+*Dismegistus* Amyot & Serville). Grazia et al. (2008) suggest that Parastrachiidae should be given a subfamily rank and treated as part of a broadly conceived Thyreocoridae.

The analysis of Lis et al. (2012) tests the monophyly of the Dinidoridae and its systematic position within Pentatomoidea, based on sequence data of rDNA 12S and 16S. Although the analysis of Thyreocoridae was not one of the authors' aims and the results about the family were not discussed, three terminal taxa were included: *Galgupha difficilis* (Breddin), *Strombosoma impictum* (Stål), and *Thyreocoris scarabaeoides* (L.). In any of the analysis (rDNA 12S, rDNA 16S, and the combined analysis), Thyreocoridae was not recovered as monophyletic. In the combined analysis, *S. impictum* and *T. scarabaeoides* show up as sister groups (supporting Thyreocorinae), and *G. difficilis* comes on a distantly related branch.

Matesco et al. (in prep.) made considerations on the phylogeny of the Thyreocoridae, emphasizing the relationships among species of *Corimelaena*. They included 40 terminal taxa and 114 morphological characters of the adult and immature stages. According to the consensus tree, *Corimelaena* has been recovered as monophyletic, but not its subgenera. The monophyly of Thyreocoridae, Corimelaeninae, *Strombosoma* Amyot & Serville, and the subgenus *Galgupha* was also corroborated. The monophyly of the Thyreocorinae and the genera *Alkindus* Distant and *Galgupha*, however, was not supported by the results. The authors suggested the maintenance of family status to Thyreocoridae (not including the Parastrachiidae) and the abandonment of the subgenera of *Corimelaena*.

In Matesco et al. (in prep.), the monophyly of Thyreocoridae is supported by six synapomorphies, which are a high density of the head punctation, presence of punctation on the area between the eye and the ocellus, presence of the pseudoperitreme on mesopleuron, anterior margin of the sternite VII angulate in the male, absence of the carina on the internal face of the dorsal rim of the pygophore, and parameres partially exposed.

Thyreocoridae was first established by Amyot and Serville (1843) as Thyréocorides. Uhler (1872) suggested the name Corimelaenidae to include the genera *Galgupha* and *Corimelaena*. Lethierry and Severin (1893) considered the subfamily Corimelaenidae within the Pentatomidae, removing it from the subfamily Cydnidae. Horváth

(1919) treated the group as the subfamily Thyreocorinae of the Cydnidae and recognized two tribes: Canoparia and Thyreocoraria. McAtee and Malloch (1928) excluded the Canoparia from Thyreocorinae. McAtee and Malloch (1933) considered Thyreocorinae as a subfamily of Pentatomidae, being equivalent to the current family status within Pentatomoidea (Rolston and McDonald 1979; Ahmad and McPherson 1990). Froeschner (1960), following McAtee and Malloch (1933), separated Thyreocoridae from the Cydnidae, giving it a status of family. Dolling (1981) recognized Thyreocorinae and Corimelaeninae as two independent subfamilies within the Cydnidae. Thyreocoridae is currently given a family status by Stys and Davidová (1979). Lis (2006) considered both Thyreocorinae and Corimelaeninae as subfamilies of Thyreocoridae.

The status (family vs. subfamily) and the name of the group (Thyreocoridae vs. Corimelanidae) have been intensely disputed (Leston 1953; Rolston and McDonald 1979; Dolling 1981; Ahmad and Moizuddin 1982; Froeschner 1988), but its composition has remained almost unchanged (McAtee and Malloch 1933; Dolling 1981; Lis 2006). Within the Corimelaeninae, the only genus whose position has been disputed is *Eumetopia* Westwood, excluded from Thyreocoridae by McAtee and Malloch (1933), but later included in the Corimelaeninae by Dolling (1981).

Taxonomic studies of the Neotropical thyreocorids are scarce. Only the genus *Alkindus* (Matesco and Grazia 2013) has been revised after the broad revision of the group by McAtee and Malloch (1933). The genus is associated with at least 15 species of plants. Its distribution is restricted to forest areas in the Neotropical region, and there is a large gap between the distributions of both species: *Alkindus atratus* Distant occurs from Mexico to northern Brazil, and *A. crassicosta* Horváth is restricted to southern and southeastern Brazil (Matesco and Grazia 2013).

In diversity studies and directed searches, bugs of this family are collected in abundance (McPherson 1974). In recent diversity analysis in southern Brazil, Thyreocoridae is an especially rich and diverse group within Pentatomoidea, following Pentatomidae (Barcellos 2006; Schmidt and Barcellos 2007; Mendonça et al. 2009).

The current state of knowledge of thyreocorid species, however, prevents the identification of all collected specimens, hindering both the analysis of diversity and the accumulation of information about these insects in nature. Mendonça et al. (2009) attribute the impossibility of identification to the species level to the current lack of knowledge of the group in the Neotropical region.

In the neotropics, several regional studies on the Thyreocoridae are available, for the fauna of Cuba (Barber and Bruner 1932), Puerto Rico (Barber 1939), Jamaica (Van Duzee 1907), Nicaragua (Maes 1994), Grenada (Uhler 1894), Trinidad (Callan 1948), Venezuela (Becker and Grazia-Vieira 1971), French Guiana (Becker and Grazia-Vieira 1977), Brazil (McAtee and Malloch 1928; Grazia et al. 1999; Barcellos 2006; Schmidt and Barcellos 2007; Mendonça et al. 2009; Grazia and Schwertner 2011), Ecuador (Froeschner 1981), Chile (Porter 1933), and Argentina (Berg 1879, 1884, 1891; Grazia et al. in press).

24.5 Economic Importance

Although several thyreocorid species have been recorded in crops, there is no information about damage caused by these bugs (Lis et al. 2000). Schuh and Slater (1995) just mentioned that some species of Corimelaeninae could be secondary pests to orchards and flowers. There are scattered reports of noticeable damage to cultivated and ornamental plants; more often, reference is made to the unpleasant taste given to the small berries in which these insects occur (Froeschner 1988). However, none of thyreocorid species are mentioned among the Heteroptera of economic importance (Schaefer and Panizzi 2000).

References

Ahmad I, McPherson JE (1990) Male genitalia of the type species of *Corimelaena* White, *Galgupha* Amyot & Serville, and *Cydnoides* Malloch (Hemiptera: Cydnidae: Corimelaeninae) and their bearing on classification. Ann Entomol Soc Am 83:162–170

Ahmad I, Moizuddin M (1982) A new species of the subfamily Thyreocorinae (Pentatomoidea: Cydnidae) from Pakistan with a note on the systematic position of the subfamily. Annot Zool Bot 148:1–8

Amyot CJB, Serville A (1843) Histoire Naturelle des Insectes. Hémiptéres. Librairie Encyclopedique de Roret ed, Paris

Barcellos A (2006) Hemípteros terrestres. In: Becker FG, Ramos RA, Moura LA (eds) Região da Lagoa do Casamento e dos Butiazais de Tapes, Planície Costeira do Rio Grande do Sul. Ministério do Meio Ambiente, Brasil, Brasília, pp 198–209

Barber HG (1939) Insects of Porto Rico and the Virgin Islands – Hemiptera-Heteroptera (excepting the Miridae and Corixidae). Scientific Survey of Porto Rico and the Virgin Islands. New York Acad Sci 14:263–441

Barber HG, Bruner SC (1932) The Cydnidae and Pentatomidae of Cuba. J Dep Agric Puerto Rico 16:231–284

Batra SWT (1984) Phytophages and pollinators of *Galium* (Rubiaceae) in Eurasia and North America. Environ Entomol 13:1113–1124

Becker M, Grazia-Vieira J (1971) Contribuição ao conhecimento da superfamília Pentatomoidea na Venezuela (Heteroptera). Iheringia (Zool) 40:3–25

Becker M, Grazia-Vieira J (1977) The Pentatomoidea [Heteroptera] collected in French Guiana by the expedition of the Muséum National d'Histoire Naturelle. Ann Soc Entomol Fr 13:53–67

Berg C (1879) Hemiptera Argentina enumeravit speciesque novas descripsit. Pauli E. Coni, Bonariae

Berg C (1884) Addenda et emmendanda ad Hemiptera Argentina. Paul E. Coni, Bonariae

Berg C (1891) Nova Hemiptera faunarum Argentinae et Uruguayensis. Ann Soc Ci Argent 32:164–175

Bertels A (1962) Insetos – hóspedes de solanáceas. Iheringia 25:1–11

Bertels A, Baucke O (1966) Segunda relação das pragas das plantas cultivadas no Rio Grande do Sul. Pesq Agropec Bras 1:17–46

Bibby FF (1961) Notes on miscellaneous insects of Arizona. J Econ Entomol 54:324–333

Biehler JA, McPherson JE (1982) Life history and laboratory rearing of *Galgupha ovalis* (Hemiptera: Corimelaenidae), with descriptions of immature stages. Ann Entomol Soc Am 75:465–470

Blatchley WS (1926) Heteroptera or true bugs of Eastern North America with special reference to the faunas of Indiana and Florida. The Nature Publ. Co., Indianapolis

Boldt PE, Robbins TO (1990) Phytophagous and flower-visiting insect fauna of *Baccharis salicifolia* (Asteraceae) in the southwestern United States and Northern Mexico. Environ Entomol 19:515–523

Bundy CS, McPherson JE (1997) Life history and laboratory rearing of *Corimelaena obscura* (Heteroptera: Thyreocoridae) with descriptions of immature stages. Ann Entomol Soc Am 90:20–27

Bundy CS, McPherson JE (2009) Life history and laboratory rearing of *Corimelaena incognita* (Hemiptera: Heteroptera: Thyreocoridae), with descriptions of immature stages. Ann Entomol Soc Am 102:1068–1076

Callan EMC (1948) The Pentatomidae, Cydnidae and Scutelleridae of Trinidad, B.W.I. Proc R Entomol Soc Lond (B) 17:115–124

Cobben RH (1968) *Evolutionary trends in Heteroptera.* Part I Eggs, architecture of the shell, gross embryology and eclosion. Agricultural Research Reports, Wageningen

Costa Lima AM (1936) Terceiro catálogo dos insectos que vivem nas plantas do Brasil. Ministério da Agricultura, Departamento Nacional da Producção Vegetal, Escola Nacional de Agronomia, Rio de Janeiro

Distant WL (1880) Insecta. Rhynchota, Hemiptera-Heteroptera. In: Godman FD, Salvin O (eds) Biologia Centrali – Americana, vol 1. England, London, pp 1–88

Dolling WR (1981) A rationalized classification of the burrower bugs (Cydnidae). Syst Entomol 6:61–76

Forbes SA (1905) Twenty-third report of the state entomologist on the noxious and beneficial insects of the state of Illinois. R.R. Donnelley & Sons Co., Chicago

Froeschner RC (1941) Contributions to a synopsis of the Hemiptera of Missouri, Pt. 1. Scutelleridae, Podopidae, Pentatomidae, Cydnidae, Thyreocoridae. Am Midl Nat 26:122–146

Froeschner RC (1960) Cydnidae of the western hemisphere. Proc U S Nat Mus 111:337–680

Froeschner RC (1981) Heteroptera or true bugs of Ecuador: a partial catalog. Smithson Control Zool 322:1–147

Froeschner RC (1988) Family Thyreocoridae Amyot & Serville, 1843. The Negro Bugs. In: Henry TJ, Froeschner RC (eds) Catalog of the Heteroptera, or true bugs, of Canada and the Continental United States. E. J. Brill, New York, pp 698–707

Gapud VP (1991) A generic revision of the subfamily Asopinae, with consideration of its phylogenetic position in the family Pentatomidae and superfamily Pentatomoidea (Hemiptera-Heteroptera). Philipp Entomol 8:865–961

Grazia J, Fortes NDF, Campos LA (1999) Superfamília Pentatomoidea. In: Brandão CRF, Cancello EM (eds) Invertebrados Terrestres – Biodiversidade do Estado de São Paulo. Síntese do conhecimento ao final do século XXI. Vol. 5. FAPESP, , pp 101-112

Grazia J, Matesco VC, Schwertner CF. Hemiptera: Thyreocoridae. In: Roig-Juñent S, Claps LE, Morrone JJ (eds) Biodiversidad de Artrópodos Argentinos. Vol. 3. Sociedad Entomológica Argentina (in press)

Grazia J, Schuh RT, Wheeler WC (2008) Phylogenetic relationships of family groups in Pentatomoidea based on morphology and DNA sequences (Insecta: Heteroptera). Cladistics 24:932–976

Grazia J, Schwertner CF (2008) Pentatomidae e Cyrtocoridae. In: Claps LE, Debandi G, Roig-Juñent R (dirs) Biodiversidad de Artrópodos Argentinos, vol 2. Sociedad Entomológica Argentina, Tucumán, pp 223–234

Grazia J, Schwertner CF (2011) Checklist dos percevejos-do-mato (Hemiptera: Heteroptera: Pentatomoidea) do Estado de São Paulo, Brasil. Biota Neotrop 11:1–12

Hinton HE (1981) Biology of insect eggs. Pergamon Press, Oxford

Horváth G (1919) Analecta ad cognitionem Cydnidarum. Annls Mus Nat Hung 17:205–273

Javahery M (1994) Development of eggs in some true bugs (Hemiptera-Heteroptera). Part I. Pentatomoidea. Can Entomol 126:401–433

Jones WA Jr (1993) New host and habitat associations for some Arizona Pentatomoidea and Coreidae. Southwest Entomol suppl 16:1–29

Klein JT, Redaelli LR, Barcellos A (2013) *Andropogon bicornis* (Poales, Poaceae): a hibernation site for Pentatomoidea (Hemiptera: Heteroptera) in a rice-growing region of Southern Brazil. Neotrop Entomol 42:240–245

Kormilev NA (1956a) Notas sobre Pentatomoidea neotropicales IV (Hemiptera). Acta Sci Inst Invest San Miguel 3:1–13

Kormilev NA (1956b) Notas sobre Pentatomoidea neotropicales V (Hemiptera). Acta Sci Inst Invest San Miguel 4:3–7

Lago PK, Testa SIII (2000) The terrestrial Hemiptera and auchenorrhynchous Homoptera of Point Clear Island and surrounding marshlands, Hancock County, Mississippi. J Miss Acad Sci 45:184–193

Leston D (1953) The suprageneric nomenclature of the British Pentatomoidea (Hemiptera). Entomol Gaz 4:13–25

Lethierry L, Severin G (1893) *Catalogue général des Hémiptéres – Pentatomidae*. Bruxelles, F. Hayez, Belgium

Linsley EG, MacSwain JW (1959) Ethology of some *Ranunculus* insects with emphasis on competition for pollen. Univ Calif Publ Entomol 16:1–46

Lis JA (2006) Family Thyreocoridae Amyot & Serville, 1843 – negro bugs. In: Aukema B, Rieger C (eds) Catalogue of the Heteroptera of the Palaearctic Region, vol 5. Entomological Society, Wageningen, pp 148–149

Lis JA, Becke M, Schaefer CW (2000) Burrower bugs (Cydnidae). In: Schaefer CW, Panizzi AR (eds) Heteroptera of economic importance. CRC Press, Boca Raton, pp 505–512

Lis JA, Lis P, Ziaja DJ, Kocorek A (2012) Systematic position of Dinidoridae within the superfamily Pentatomoidea (Hemiptera: Heteroptera) revealed by the Bayesian phylogenetic analysis of the mitochondrial 12S and 16S rDNA sequences. Zootaxa 3423:61–68

Lung KYH, Goeden RD (1982) Biology of *Corimelaena extensa* on tree tobacco, *Nicotiana glauca*. Ann Entomol Soc Am 75:177–180

Maes JM (1994) Catalogo de los Pentatomoidea (Heteroptera) de Nicaragua. Rev Nicar Entomol 28:1–29

Martin MC (1965) An ecological life history of *Geranium maculatum*. Am Midl Nat 73:111–149

Matesco VC, Grazia J (2013) Revision of the genus *Alkindus* Distant (Hemiptera: Heteroptera: Thyreocoridae: Corimelaeninae). Zootaxa 3750:57–70

Matesco VC, Bianchi FM, Campos LA, Grazia J (2012) Egg ultrastructure of two species of *Galgupha* Amyot & Serville, with a discussion of the eggs and oviposition patterns of thyreocorid and allied groups (Hemiptera: Heteroptera: Pentatomoidea: Thyreocoridae). Zootaxa 3247:43–51

Matesco VC, Schwertner CF, Grazia J Cladistic analysis of *Corimelaena* White based on morphology, with considerations on the phylogeny of the Thyreocoridae (Hemiptera: Heteroptera: Pentatomoidea) (*in prep*)

Maw MG (1976) An annotated list of insects associated with Canada thistle (*Cirsium arvense*) in Canada. Can Entomol 108:235–244

McAtee WL, Malloch JR (1928) Thyreocorinae from the state of Paraná, Brazil (Hemiptera: Pentatomidae) (From the scientific results of the Polish Zoological Expedition to Brazil in the years 1921–1924). Ann Mus Zool Polon 7:32–44

McAtee WL, Malloch JR (1933) Revision of the subfamily Thyreocorinae of the Pentatomidae (Hemiptera-Heteroptera). Ann Carnegie Mus 21:191–411

McPherson JE (1971) Notes on the laboratory rearing of *Corimelaena lateralis lateralis* (Hemiptera: Corimelanidae) on wild carrot. Ann Entomol Soc Am 64:313–314

McPherson JE (1972) Life history of *Corimelaena lateralis lateralis* (Hemiptera: Thyreocoridae) with descriptions of immature stages and list of other species of Scutelleroidea found with it on wild carrot. Ann Entomol Soc Am 65:906–911

McPherson JE (1974) Three negro bug state records for Illinois (Hemiptera: Corimelaenidae). Trans Il Acad Sci 67:361–363

McPherson JE, Mohlenbrock RH (1976) A list of the Scutelleroidea of the La Rue-Pine Hills Ecological Area with notes on biology. Gt Lakes Entomol 9:125–169
McPherson JE, Sailer RI (1978) A new species of *Corimelaena* (Hemiptera: Thyreocoridae) from America North of Mexico. J Kansas Entomol Soc 51:516–520
McPherson JE, Weber BC (1990) Seasonal flight patterns of Hemiptera (excluding Miridae) in a southern Illinois black walnut plantation. Gt Lakes Entomol 23:105–120
Mendonça MS, Schwertner CF, Grazia J (2009) Diversity of Pentatomoidea (Hemiptera) in riparian forests of southern Brazil: taller forests, more bugs. Rev Bras Entomol 53:121–127
Nixon PL, McPherson JE, Cuda JP (1975) A list of the Scutelleroidea (Hemiptera) collected on immature black walnut trees in Southern Illinois with some notes on biology. Trans Ill Acad Sci 68:409–413
Palmer WA (1987) The phytophagous insect fauna associated with *Baccharis halimifolia* L. and *B. neglecta* Britton in Texas, Louisiana, and northern Mexico. Proc Entomol Soc Wash 89:185–199
Palmer WA, Pullen KR (1995) The phytophagous arthropods associated with *Lantana camara, L. hirsuta, L. urticifolia,* and *L. urticoides* (Verbenaceae) in North America. Biol Control 5:54–72
Porter CE (1933) Acerca de los Thyreocorinae chilenos. Rev Chil Hist Nat 37:198–200
Quintanilla RH, Margheritis AE, Rizzo HF (1976) Catalogo de hemipteros hallados en la Provincia de Corrientes (Argentina). Rev Soc Entomol Argent 35:115–133
Quintanilla RH, Rizzo HF, Núñez AS (1981) Catalogo preliminar de Hemipteros hallados en la provincia Misiones (Argentina). Rev Fac Agron 2:145–161
Rider DA (2013) *Plant Host Records – Thyreocoridae*. Available on: http://www.ndsu.nodak.edu/ndsu/rider/Pentatomoidea/Hosts/plant_Corimelaenidae.htm. Access 16/XII/2013
Rolston LH, McDonald JD (1979) Keys and diagnoses for the families of Western Hemisphere Pentatomoidea, subfamilies of Pentatomidae and tribes of Pentatominae (Hemiptera). J N Y Entomol Soc 87:189–207
Sailer RI (1940) A new species of Thyreocorinae. J Kansas Entomol Soc 13:62–63
Sailer RI (1941) A new species of Thyreocorinae from Mexico. J Kansas Entomol Soc 14:90–91
Schaefer CW (1988) The food plants of some "primitive" Pentatomoidea (Hemiptera: Heteroptera). Phytophaga 2:19–45
Schaefer CW, Panizzi AR (2000) Heteroptera of economic importance. CRC Press, Boca Raton
Schmidt LS, Barcellos A (2007) Abundância e riqueza de espécies de Heteroptera (Hemiptera) do Parque Estadual do Turvo, sul do Brasil: Pentatomoidea. Iheringia 97:73–79
Schuh RT, Slater JA (1995) True bugs of the world (Hemiptera: Heteroptera): classification and natural history. Cornell University Press, Ithaca
Schwitzgebel RB, Wilbur DA (1942) Lepidoptera, Hemiptera and Homoptera associated with ironweed, *Vernonia interior* Small in Kansas. Trans Kans Acad Sci 45:195–202
Smith JM (1959) Notes on insects, especially *Gymnaetron* spp. (Coleoptera: Curculionidae), associated with toadflax, *Linaria vulgaris* Mill. (Scrophulariaceae), in North America. Can Entomol 91:116–121
Southwood TRE (1956) The structure of the eggs of the terrestrial Heteroptera and its relationship to the classification of the group. Trans R Entomol Soc Lond 108:163–221
Stoner D (1920) The Scutelleroidea of Iowa. Univ Iowa Stud Nat Hist 8:1–140
Stys P, Davidová J (1979) Taxonomy of *Thyreocoris* (Heteroptera, Thyreocoridae). Annot Zool Bot 134:1–40
Tecic DL, McPherson JE (2005) Resurvey of the Pentatomoidea (Heteroptera) of the La Rue-Pine Hills Research Natural Area in Union County, Illinois. Gt Lakes Entomol 37:30–70
Torre-Bueno JR (1939) A synopsis of the Hemiptera-Heteroptera of America North of Mexico. Part I, Families Scutelleridae, Cydnidae, Pentatomidae, Aradidae, Dysodiidae and Termitaphididae. Entomol Am 19:141–206
Uhler PR (1872) Notices of the Hemiptera of the Western Territories of the United States, chiefly from the surveys of Dr. F. V. Hayden. In: Hayden FV (ed) Preliminary report of the United

States geological survey of Montana and portions of adjacent territories, being a fifth annual report of progress. Government Printing Office, Washington, DC, pp 392–423
Uhler PR (1894) On the Hemiptera-Heteroptera of the Island of Grenada, West Indies. Proc Zool Soc Lond 1894:167–224
Van Duzee EP (1904) Annotated list of the Pentatomidae recorded from America, North of Mexico, with descriptions of some new species. T Am Entomol Soc 30:1–80
Van Duzee EP (1907) Notes on Jamaican Hemiptera: a report on a collection of Hemiptera made on the island of Jamaica in the spring of 1906. Bull Buffalo Soc Nat Sci 8:1–79
Van Duzee EP (1923) A rearrangement of our North American Thyreocorinae (Hemip.). Entomol News 34:302–305
Yonke TR (1991) Order Hemiptera. In Stehr FW *Immature insects*, vol. 2. Dubuque, Kendall/Hunt Publishing Co, Dubuque, pp 22–65

Chapter 25
Less Diverse Pentatomoid Families (Acanthosomatidae, Canopidae, Dinidoridae, Megarididae, Phloeidae, and Tessaratomidae)

Cristiano F. Schwertner and Jocélia Grazia

Abstract The superfamily Pentatomoidea comprises about 7,000 species worldwide divided in 15 families and shows a remarkable diversity. In the Neotropical region, more than 1,900 species in 10 families are recorded. In this chapter, the knowledge about the families Acanthosomatidae, Canopidae, Dinidoridae, Megarididae, Phloeidae, and Tessaratomidae, which include 66 species in 28 genera in the region, is summarized. Although representing less diverse lineages within the superfamily Pentatomoidea, studies about these families will be fundamental to understand the evolution and natural history of the true bugs in the Neotropical region. Some of them are endemic to the Neotropics (i.e., Canopidae and Megarididae) or have relict distribution (i.e., Acanthosomatidae, Phloeidae, and Tessaratomidae). Characteristics like cryptic coloration (Phloeidae) and maternal care behavior (Acanthosomatidae) make these bugs good models for evolutionary and ecological studies. However, little information is available for most of the species. Identification keys for all families of Pentatomoidea found in the Neotropics as well as characterization of all families and genera treated in this chapter are provided.

25.1 Introduction

The superfamily Pentatomoidea includes mainly phytophagous bugs, recognized by the developed scutellum, 2+2 abdominal trichobothria lateral to the spiracle line, opening of the genital capsule in males (the pygophore) directed posteriorly, and

C.F. Schwertner (✉)
Departamento de Ciências Biológicas, Instituto de Ciências Ambientais, Químicas e Farmacêuticas, Universidade Federal de São Paulo (UNIFESP), Rua São Nicolau, 210 Centro, Diadema, SP 09913-030, Brazil
e-mail: schwertner@unifesp.br

J. Grazia
Departamento de Zoologia, Instituto de Biociências, Universidade Federal do Rio Grande do Sul (UFRGS), Av. Bento Gonçalves 9500, prédio 43435, Bairro Agronomia, Porto Alegre, RS 91501-970, Brazil
e-mail: jocelia@ufrgs.br

A.R. Panizzi, J. Grazia (eds.), *True Bugs (Heteroptera) of the Neotropics*, Entomology in Focus 2, DOI 10.1007/978-94-017-9861-7_25

eggs barrel shaped (Schuh and Slater 1995; Grazia et al. 2008). The group comprises about 7,000 species worldwide and shows a remarkable morphological diversity (Fig. 25.1), separated in 15 families (Grazia et al. 2008). In the Neotropical region, more than 1,900 species in ten families are recorded (Table 25.1).

Fig. 25.1 Morphological diversity of Pentatomoidea families. (**a**) *Canopus* sp. (Canopidae); (**b**) *Dinidor mactabilis* Perty (Dinidoridae); (**c**) *Phloea subquadrata* Spinola (Phloeidae), male and female in copula (Photos: (a) D Ellyahu (b) CF Schwertner (c) JC Bernardes)

Table 25.1 Number of genera and species of Pentatomoidea families for the world and for the Neotropical Region

		World		Neotropical Region	
Family	Subfamily	Genera	Species	Genera	Species
Acanthosomatidae		57	183	22	34
	Acanthosomatinae	16	132	1	1
	Blaudusinae	25	30	12	21
	Ditomotarsinae	16	21	9	12
Canopidae	[a]	1	8	1	8
Cydnidae	[b]	93	742	16	131
Dinidoridae		13	115	1	6
	Dinidorinae	10	68	1	6
	Megymeninae	3	17	0	0
Lestoniidae	[a]	1	2	0	0
Megarididae	[a]	1	16	1	16
Parastrachiidae	[a]	2	4	0	0
Pentatomidae	[b]	900	4.700	240	>1,500
Phloeidae	[a]	3	4	2	3
Plataspididae	[a]	59	560	0	0
Scutelleridae	[b]	81	450	25	112
Tessaratomidae		55	240	1	3
	Natalicolinae	8	14	0	0
	Oncomerinae	14	60	1	3
	Tessaratominae	32	166	0	0
Thyreocoridae	[b]	12	212	9	159
Thaumastellidae	[a]	1	3	0	0
Urostylididae	[a]	4	80	0	0

[a]Without classification at subfamilial or tribal level

[b]For the classification and diversity at subfamilial/tribal level, see corresponding chapter in this book

In this chapter, the knowledge about the families Acanthosomatidae, Canopidae, Dinidoridae, Megarididae, Phloeidae, and Tessaratomidae in the Neotropical region is summarized, which include 66 species in 28 genera in the region (Table 25.2). Canopidae and Megarididae are exclusively found in the Neotropics (McAtee and Malloch 1928, 1933), while Phloeidae has a relic distribution, with one species found in the Australian region (Grazia et al. 2008). To help identify the families treated here, an identification key to all families of Pentatomoidea found in the Neotropics is provided.

Table 25.2 Checklist of the species and country distribution of the less diverse families of the Pentatomoidea in the Neotropical Region

Family/Subfamily	Species	Distribution
Acanthosomatidae		
Acanthosomatinae	*Elasmucha lateralis* (Say)	CAN, USA, MEX
Blaudusinae		
Blaudusini	*Bebaeus punctipes* Dallas	CO, PAR, VEM, BOL
	Blaudus ruficornis Stål	COL
Lanopini	*Acrophyma bicallosa* (Stål)	COL, BRA
	Acrophyma cumingii (Westwood)	CHI, ARG
	Cochabambia martinezi Pirán	BOL
	Ea australis Distant	CHI, ARG
	Ea septentrionalis Carvajal, Fáundez, and Rider	CHI
	Hellica johni Froeschner	ARG, URU
	Hellica johnpolhemi Froeschner	BOL, ARG
	Hellica nitida Haglund	BR (AM?, SC, RS), ARG
	Lanopis algescens Bergroth	ARG
	Lanopis chubuti Distant	ARG
	Lanopis rugosus Signoret	CHI, ARG
	Lanopis splendens Distant	ARG
	Mazanoma variada Rolston and Kumar	CHI
	Phorbanta variabilis (Signoret)	CHI, ARG
	Pseudosinopla canaliculus (Reed)	CHI
	Sinopla humeralis Signoret	CHI
	Sinopla perpunctatus Signoret	CHI, ARG
	Sniploa obsoletus Signoret	CHI
	Sniploa shajovskoii Kormilev	ARG
Ditomotarsinae		
Ditomotarsini	*Archaeoditomotarsus* Fáundez, Carvajal, and Rider	CHI

(continued)

Table 25.2 (continued)

Family/Subfamily	Species	Distribution
	Cylindrocnema plana Mayr	CHI, ARG
	Ditomotarsus punctiventris Spinola	CHI, ARG
	Hyperbius geniculatus Signoret	CHI, ARG
	Nopalis sulcatus Signoret	CHI
	Planois patagonus Distant	ARG
	Planois gayi (Spinola)	CHI, ARG
	Praesus incarnatus Stål	COL
	Rolstonus rolstoni Froeschner	COL
	Tolono decoratus Rolston and Kumar	COL, ECU
	Tolono confusus Carvajal, Rider, and Faúndez	COl, ECU
	Tolono pallidus Carvajal, Rider, and Faúndez	COL
Canopidae	*Canopus andinus* Horváth	PER
	Canopus burmeisteri McAtee and Malloch	BRA (AM), BOL
	Canopus caesus (Germar)	BR (PA)
	Canopus fabricii McAtee and Malloch	CR, PAN, BR
	Canopus germari McAtee and Malloch	CR, PAN, COL, ECU
	Canopus globosus Horváth	BRA, PER, BOL
	Canopus impressus F.	BR
	Canopus orbicularis Horváth	GUI, BRA
Dinidoridae		
Dinidorinae	*Dinidor braziliensis* Durai	BRA, PAR
	Dinidor impicticollis Stål	COL, ECU
	Dinidor mactabilis Perty	BRA, ARG
	Dinidor pulsator (Schouteden)	PER
	Dinidor rufocinctus Stål	PAN, COL, BRA, ECU, BOL
	Dinidor saucius Stål	BRA
Megarididae		
	Megaris antennata McAtee and Malloch	GUA
	Megaris atratula Stål	BLZ
	Megaris constricta McAtee and Malloch	GUA
	Megaris hemisphaerica McAtee and Malloch	GUY, VEN, BRA
	Megaris laevicollis Stål	BRA, PAR, ARG, VEN?
	Megaris longula McAtee and Malloch	BRA
	Megaris majuscula McAtee and Malloch	CUB

(continued)

Table 25.2 (continued)

Family/Subfamily	Species	Distribution
	Megaris nigritula Stål	BRA
	Megaris peruviana Horváth	PER
	Megaris puertoricensis Barber	PUR
	Megaris punctulata Horváth	TRI
	Megaris rotunda McDonald	BRA
	Megaris semiamicta McAtee and Malloch	GUA, PAN
	Megaris stalii McAtee and Malloch	BRA
	Megaris trinotata Distant	PAN
	Megaris vianai Kormilev	ARG
Phloeidae		
	Phloea corticata (Drury)	BRA
	Phloea subquadrata Spinola	BRA
	Phloephana longirostris (Spinola)	BRA
Tessaratomidae		
Oncomerinae		
Piezosternini	*Piezosternum subulatum* (Thunberg)	MEX, NIC, CUB, RDO, PUR, GDE (Guadeloupe), MAR, PAN, VEM, COL, FGU, GUY, SUR, BRA, ECU
	Piezosternum thunbergi Stål	BRA, BOL, PAR, ARG, URU
	Piezosternum venezolanum Pirán	VEN

Key to the families of Neotropical Pentatomoidea (modified from Rolston and McDonald 1979):

1. Antennae IV- or V-segmented; MTG opening distant from the lateral margin of metapleura....2
– Antennae III-segmented; MTG opening near the lateral margin of metapleura (Fig. 25.2) .... Phloeidae
2. Scutellum enlarged covering most of hemelytra (Fig. 25.3)....8
– Scutellum leaving most of hemelytra exposed even when scutellum attains the apex of abdomen....3
3. Trichobothria in urosternites III–VII on large callus located mesad of the adjacent spiracle (Fig. 25.4) ....Dinidoridae
– Trichobothria not on large callus, both rarely mesad of spiracles on all urosternites 4
4. Pronotum extending over the base of scutellum (Fig. 25.43) .... Tessaratomidae
– Pronotum ending at the base of scutellum 5
5. Tibial spines if present are confined to the apex of tibiae....6
– Tibial spines along all tibiae in addition to setae (Fig. 25.5) 7

6. Tarsi 2-segmented; urosternite VIII exposed in males (Fig. 25.6); Pendergrast's organs (Fig. 25.7) usually present in females .... Acanthosomatidae
– Tarsi usually 3-segmented; urosternite VIII always in males concealed; Pendergrast's organs in females always absent................................ Pentatomidae
7. Fore tibiae usually expanded (Fig. 25.8a), sometimes cultrate with tarsi inserted midway of length bearing a lateral row of especially stout spines unless cultrate; the apex of scutellum usually narrowly rounded ....Cydnidae
– Fore tibiae subcylindrical; spines on the lateral margin not notably stout (Fig. 25.9); the apex of scutellum broadly rounded (Fig. 25.10) ....Thyreocoridae
8. Tarsi 2-segmented ....Megarididae
– Tarsi 3-segmented 9
9. Tibial spines numerous, along all tibiae in addition to setae; if weakly spined, lateral margins of the pronotum fringed with long setae ....Thyreocoridae
– Tibial spines if present confined to the apex 10
10. Sutures of abdominal venter complete, reaching lateral margins (Fig. 25.11); the second antennal segment much longer than the diameter..........................11
– Sutures of the abdominal venter obsolete laterad of spiracles; the second antennal segment subequal to the diameter ....Canopidae
11. Trichobothria paired (Fig. 25.12); frena lacking............................ Scutelleridae
– Trichobothria single; short frena present .... Pentatomidae

Fig. 25.2 *Phloeophana longirostris* (Spinola) (Phloeidae), metathoracic gland (MTG) opening (Photo CF Schwertner)

Fig. 25.3 *Galgupha coccineloides* Horváth (Thyreocoridae) dorsal view (Photo CF Schwertner)

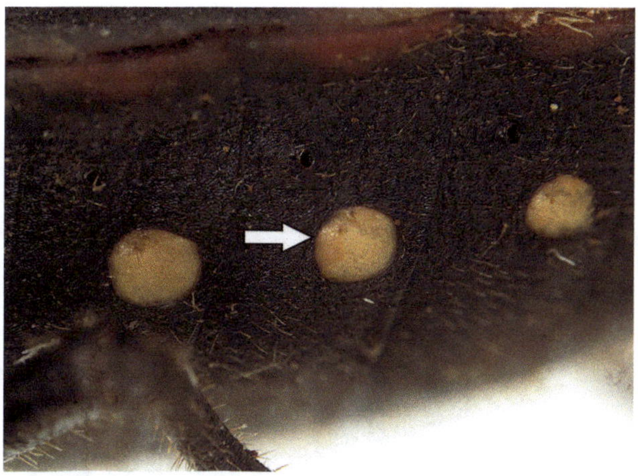

Fig. 25.4 *Dinidor mactabilis* Perty (Dinidoridae), detail of the *lateral view* of abdomen (Photo CF Schwertner)

Fig. 25.5 *Cyrtomenus mirabilis* Berg (Cydnidae), posterior tibia in *dorsal view* (Photo CF Schwertner)

Fig. 25.6 *Planois gayi* (Spinola) (Acanthosomatidae), detail of ventral abdomen showing urosternite VIII exposed in male (Photo CF Schwertner)

Fig. 25.7 *Planois gayi* (Spinola) (Acanthosomatidae), detail of *lateral view* of abdomen showing Pendergrast's organ (Photo CF Schwertner)

Fig. 25.8 Fore tibia in Cydnidae, lateral view. (**a**) *Cyrtomenus teter* (Spinola), Cydnidae; (**b**) *Scaptocoris castanea Perty*, Cephalocteinae (Photos CF Schwertner)

Fig. 25.9 *Galgupha* sp. (Thyreocoridae), fore tibiae (Photo CF Schwertner)

Fig. 25.10 *Galgupha* sp. Thyreocoridae, *lateral view* (Photo CF Schwertner)

Fig. 25.11 *Tetyra* sp. (Scutelleridae), partial *ventral view* showing sutures of abdominal venter (Photo CF Schwertner)

Fig. 25.12 *Tetyra* sp. (Scutelleridae), detail of the *ventral view* (Photo CF Schwertner)

25.2 Acanthosomatidae

Species included in the family Acanthosomatidae resemble median to large pentatomids (i.e., Fig. 25.15) and may be recognized by the shorter scutellum, 2-segmented tarsi, male abdominal segment 8 large and exposed, and Pendergrast's organ present in most of the females. The family includes more than 200 species in 57 genera (Grazia et al. 2008; Faúndez 2014; Carvajal et al. 2015) (Table 25.1), distributed mainly in temperate regions or at high altitudes in the subtropics. All studied species are phytophagous; maternal care is registered for some species (Tallamy and Schaefer 1997).

25.2.1 General Characteristics and Diagnosis

Adults are recognized for having 2-segmented tarsi and short scutellum, never covering the corium; size ranging from 5 to 20 mm in length; antenna 5-segmented, antenniferous tubercles not visible dorsally in most species; mesosternum usually with a conspicuous carina (subfamily Acanthosomatinae); urosternite III often with spine anteriorly developed; 1 + 1 trichobothria in abdominal urosternites III to VII, transversal; males with abdominal segment VIII exposed; females usually with Pendergrast's organs (glandular abdominal organs), rounded or oval, located in the urosternites V to VII (only in the VII in some species); and posterior margin of VII deeply emarginated.

Immatures. Southwood (1956) and Putshkov (1959) provided diagnoses for eggs and nymphs of several families and species of Heteroptera distributed in Europe, including Acanthosomatidae. Four Nearctic species in two genera of the subfamily Acanthosomatinae were described by Cobben (1968), who also discusses the data available so far for the whole group. At least two species have the nymphs formally described (Carter and Hoebeke 2003; Martinez et al. 2003; Faúndez 2007b) and allow us to infer that the immatures of the acanthosomatids share the general aspects of other families of Pentatomoidea (i.e., Grazia and Schwertner 2008).

Characteristics of the immatures include elliptical eggs with narrower anterior pole, smooth corium when observed in optical microscopy, and cup-shaped aero-micropylar processes ("aero-micropylar cup" in Cobben 1968); nymphs with oval body (Fig. 25.13); and anterior openings of scent glands farther apart than the median and posterior openings. In *Sinopla perpunctatus* Signoret, the dorsal plates in the third, fourth, and fifth instars are divided longitudinally, forming paired plates (Martinez et al. 2003, Faúndez 2007a). This feature seems to be unique among immature pentatomoids (Putshkov 1959).

Fig. 25.13 *Sinopla perpunctatus* Signoret, fifth instar nymphs (From Faúndez 2007b. Used with permission from Boletin de la Sociedad Entomológica Aragonesa)

25.2.2 General Biology and Ecology

Few species of Acanthosomatidae have the biology studied in detail to date; most of the investigations were done with Palaearctic species, especially those that present maternal care (Kudo and Nakahira 1993; Kudo 2000). Biological data of Neotropical species include mostly field observations and label records. More recently, studies about the biology of the Neotropical species have been published (Faúndez and Osorio 2010b).

All known acanthosomatids are phytophagous, usually recorded in more than one host, which includes trees and shrubs (Kumar 1974; Schaefer and Ahmad 1987; Faúndez 2007c, 2009). Feeding sites include young tissues of the host and reproductive parts (Schaefer and Ahmad 1987; Faúndez 2007a, c). Casual records of feeding on decaying organic matter and predation including cannibalism are known (Miller 1971), probably related to the shortage of suitable host plants (Schaefer and Ahmad 1987).

Polyphagy seems to be the rule, but specializations can occur at different levels (Schaefer and Ahmad 1987; Faúndez 2007a, 2009). For instance, *Sinopla perpunctatus* is registered only in *Nothofagus* spp. in Chile and is considered a specialized species (Faúndez 2007a; Osorio 2009). Otherwise, *Acrophyma cumingii*

(Westwood) and *Ditomotarsus punctiventris* Spinola are considered generalists (Faúndez 2007b, 2009). Acanthosomatidae species are found feeding in several different host plant families, including Aceraceae, Anacardiaceae, Cornaceae, Apiaceae, Aquifoliaceae, Araliaceae, Asteraceae, Betulaceae, Caprifoliaceae, Casuarinaceae, Cupressaceae, Cyperaceae, Fabaceae, Juncaceae, Moraceae, Myrtaceae, Pinaceae, Poaceae, Proteaceae, Rhamnaceae, Rosaceae, Salicaceae, Nothofagaceae, Polygonaceae, Tiliaceae, and Ephedraceae (Schaefer and Ahmad 1987; Schaefer et al. 2000; Faúndez 2007a, 2008; Osorio 2009). In a survey conducted in riparian forests of the Pampas in southern Brazil (Mendonça et al. 2009), *Hellica nitida* Haglund was one of the most abundant pentatomoids, indicating preference of this species by this type of vegetation.

Life cycle is similar to that described for Pentatomidae (Grazia and Schwertner 2008). Faúndez (2007b) gives a brief description of the life cycle of *D. punctiventris*, providing details of the mating behavior (Faúndez 2007b; Faúndez et al. 2009). Females lay eggs in clutches, corresponding more or less with the number of ovarioles (usually 10–14). Egg and nymphal development time, prereproductive females, and adult longevity have not been studied in any of the Neotropical species.

Maternal care is recorded in species of the subfamily Acanthosomatinae (Schuh and Slater 1995) distributed mainly in Australian, Nearctic, Oriental, and Palearctic regions, including *Elasmucha lateralis* (Say, 1831) which is also recorded in the Neotropical region (Table 25.1). Faúndez and Osorio (2010a, b) described maternal care behavior for the first time in *Sinopla perpunctatus* (Signoret 1864), a species distributed exclusively in the Neotropics. The authors also reported coloration change in the female associated to the reproductive period and guard of eggs and nymphs.

Schaefer et al. (2000) reviewed the species of acanthosomatids with economic importance, and no records of Neotropical species damaging cultivated plants were found in the literature. More recently, *Phorbanta variabilis* (Signoret 1863) was recorded on Magellan barberry, *Berberis microphylla*, Berberidaceae, and *Ditomotarsus punctiventris* (Spinola 1852) is considered as a quarentenary species for the USA (E Fáundez & M Carvajal, personal communication 2014).

25.2.3 Classification and Diversity

Acanthosomatidae was recognized as a higher taxon by Signoret (1864) and treated as a subfamily or tribe of Pentatomidae by different authors (Schuh and Slater 1995). The current classification includes three subfamilies: Acanthosomatinae, Blaudusinae (tribes Blaudusini and Lanopini), and Ditomotarsinae (tribes Ditomotarsini and Laccophorellini) (Kumar 1974; Kment 2005).

Fischer (1994) and Grazia et al. (2008) defined the monophyly of the family. However, relationships of Acanthosomatidae with other Pentatomoidea are still unsettled (Kment 2005; Carvajal and Faúndez 2013). According to Gapud (1991), the family is related to Dinidoridae, Tessaratomidae, and Scutelleridae, in a relatively basal position within Pentatomoidea. Grazia et al. (2008) defined the family in different positions according to the analyses performed: the morphological evidence pointed to a basal position, while molecular data suggest the family as a derived taxon and related to Pentatomidae. Both analyses showed a sister-group relationship between Acanthosomatidae and Lestoniidae, the latter occurring exclusively in Australia.

The distribution of the Acanthosomatidae occurs predominantly in the Southern Hemisphere of the World, including Argentina, Australia, Chile, and South Africa. Only few genera (i.e., *Acanthosoma*, *Elasmucha*, and *Elasmostethus*) have representatives in the Northern Hemisphere. In the Neotropics, at least 22 genera and 34 species are described, and the faunas of Argentina and Chile are more diverse (Table 25.2).

Classical contributions with descriptions of new taxa were made in the nineteenth and early twentieth centuries by Westwood (1837), Spinola (1850, 1852), Signoret (1864), Mayr (1864), Stål (1867, 1872), Haglund (1868), Distant (1911), and Bergroth (1917). The generic world fauna was monographed by Kumar (1974). Keys to the genera found in the Western Hemisphere were provided by in Rolston and Kumar (1975), updated below. More recent studies on the Neotropical fauna include Froeschner (1997, 2000), Faúndez (2007a, b, c, 2009), Faúndez et al. (2009, 2014), Faúndez and Osorio (2010a, b), Carvajal and Faúndez (2013), Carvajal et al. (2014), Faúndez (2014), and Faúndez et al. (2014).

Knowledge about the diversity and the classification of the group in the Neotropical region is currently under revision, including the description of several new taxa (M Carvajal and E Faúndez, personal communication 2014).

Key to the genera of Acanthosomatidae of the Neotropical region (modified from Rolston and Kumar 1975 and Faúndez 2014).

1. Median tubercle or spine present at the base of the abdominal venter 11
1′ Base of the abdominal venter smoothly convex 2
2(1) Distal end of the first antennal segment clearly surpassing the apex of head 3
2′ Distal end of the first antennal segment reaching little if any beyond the apex of head 6
3(2) Longitudinal sulcus on the prosternum before coxae as deep as the diameter of the rostrum, little wider; the distal diameter of the first antennal segment usually about twice the basal diameter .... *Cylindrocnema* Mayr
3′ Longitudinal sulcus on the prosternum absent or much broader than the diameter of the rostrum; the first antennal segment subcylindrical 4
4(3) The length of the first antennal segment more than 4/5 the length of the head measured dorsally .... *Planois* Signoret (Fig. 25.14)
4′ The length of the first antennal segment less than 3/5 the length of the head measured dorsally 5
5(4) The prosternum shallowly depressed lengthwise; humeral angles developed .... *Nopalis* Signoret (Fig. 25.15)

5′ The prosternum transversely convex; humeral angles not developed .... *Ditomotarsus* Spinola (Fig. 25.16)
6(2) Juga far surpassing the anteclypeus, usually contiguous before the anteclypeus 7
6′ Juga not or scarcely surpassing the anteclypeus... 8
7(6) The anterior preocular part of the head strongly deflexed, prosternum a strong, mediolongitudinal groove, measuring 8–10 mm.. .... *Mazanoma* Rolston and Kumar (Fig. 25.17)
7′ The anterior preocular part of the head not strongly deflexed, prosternum without mediolongitudinal groove, measuring 5–6 mm .... *Rolstonus* Froeschner (Fig. 25.18)
8(6) Ostiolar peritreme reaching more than halfway from inner margin of the ostiole to the lateral margin of the metapleuron 10
8′ Ostiolar peritreme short.. 9
9 The pronotum with anterolateral margins slightly sinuous; mesosternum weakly carinate; female with one pair of Pendergrast's organs; body coloration black, shiny, and marked with shades of yellow, with pronotum having a crescent curving ivory mark, from posterolateral margins to the anterior submargin .... *Tolono* Rolston and Kumar (Fig. 25.19)
9′ The pronotum with anterolateral margins straight; the mesosternum without carina; female with two pairs of Pendergrast's organs; body coloration different from above, pronotum predominantly red .... *Praesus* Stål (Fig. 25.20)
10(8) Juga strongly concave laterally; coloration blackish .... *Hyperbius* Stål (Fig. 25.21)
10′ Juga not strongly concave laterally; coloration not blackish.............................. .... *Archaeoditomotarsus* Fáundez, Carvajal, and Rider
11(1) The mesosternal carina greatly produced, extending anteriorly beyond procoxae; the abdominal spine appressed to the right side of the posterior portion of the mesosternal carina 12
11′ The mesosternal carina weakly developed or absent....................................... 13
12(11) Ostiolar peritreme reaching a little more than halfway from the inner margin of the ostiole to the lateral margin of the metapleuron .... *Elasmucha* Stål (Fig. 25.22)
12′ Ostiolar peritreme reaching about three-fourth distance from the inner margin of the ostiole to the lateral margin of the metapleuron .... *Elasmostethus* Fieber (Fig. 25.23)
13(11) The abdominal spine surpassing mesocoxae... 14
13′ The abdominal spine not reaching mesocoxae 15
14(13) Lateroposterior margins of the pronotum not produced into any spines or processes; the ostiole reaching more than halfway from the inner margin of the ostiole to the lateral margin of the metapleuron; the abdominal spine attaining procoxae .... *Blaudus* Stål (Fig. 25.24)
14′ Humeral angles of the pronotum produced into flat processes; the ostiole reaching about one-third distance from the inner margin of the ostiole to the lateral margin of the metapleuron; the abdominal spine attaining the head .... *Bebaeus* Dallas (Fig. 25.25)

15(13) Juga contiguous before the anteclypeus .... *Sniploa* Signoret
15′ Juga not surpassing the anteclypeus or, if longer than the anteclypeus, neither markedly convergent nor contiguous 16
16(15) Ostiolar peritreme extending much farther than halfway from the inner margin of the ostiole to the lateral margin of the metapleuron .... *Phorbanta* Stål (Fig. 25.26)
16′ Ostiolar peritreme extending halfway or less from the inner margin of the ostiole to the lateral margin of the metapleuron 17
17(16) Ostiolar peritreme reaching halfway from the inner margin of the ostiole to the lateral margin of the metapleuron.............. *Lanopis* Signoret (Fig. 25.27)
17′ Ostiolar peritreme reaching not more than one-third distance from the inner margin of the ostiole to the lateral margin of the metapleuron........................ 18
18(17) The spine at the base of the abdominal venter clearly extending onto the metasternum 19
18′ The base of the abdominal venter tuberculate, with the tubercle not or scarcely surpassing the posterior margin of the metasternum 20
19(18) The apex of head broad, with the anteclypeus and each paraclypeus individually rounded and sides scarcely concave before the eyes .... *Ea* Distant
19′ The apex of the head a narrow smooth parabola, sides distinctly concave before the eyes... *Acrophyma* Bergroth (Fig. 25.28)
20(18) The first antennal segment reaching or slightly surpassing the apex of head 21
20′ The mesosternum without carina; the first antennal segment not reaching the apex of head .... *Hellica* Stål* (Fig. 25.29)
21(20) The ostiolar peritreme narrow and not fleshy (Fig. 25.30)................................ .... *Sinopla* Signoret (Fig. 25.31)
21′ The ostiolar peritreme wide and fleshy .... *Pseudosinopla* Faúndez

* Following Faúndez et al. (2014), the genus *Cochabambia* Pirán will key with *Hellica*. The genus *Cochabambia* from *Hellica* can be separated by more elongated body punctuation denser and shallower, and pronotum trapezoidal in shape.

25.3 Canopidae

Canopids are exclusively distributed in the Neotropical region (Table 25.2), with all eight known species belonging to genus *Canopus* F. that are medium sized (mm), with almost rounded body, dorsally convex and ventrally flat (Fig. 25.32) Coloration black and shiny, with purple and green reflexes. Scutellum enlarged, covering entirely the abdomen and most part of hemelytra (Grazia et al. 2012). The wide scutellum is shared with other pentatomoid families (Aphylidae, Canopidae,

Fig. 25.14 *Planois gayi* (Spinola), *dorsal view* (Photo CF Schwertner)

Fig. 25.15 *Nopalis sulcatus* Signoret, *dorsal view* (Photo CF Schwertner)

Fig. 25.16 *Ditomotarsus punctipes* Spinola, *dorsal view* (Photo CF Schwertner)

Fig. 25.17 *Mazanoma varia* Rolston and Kumar, *dorsal view* (Photo CF Schwertner)

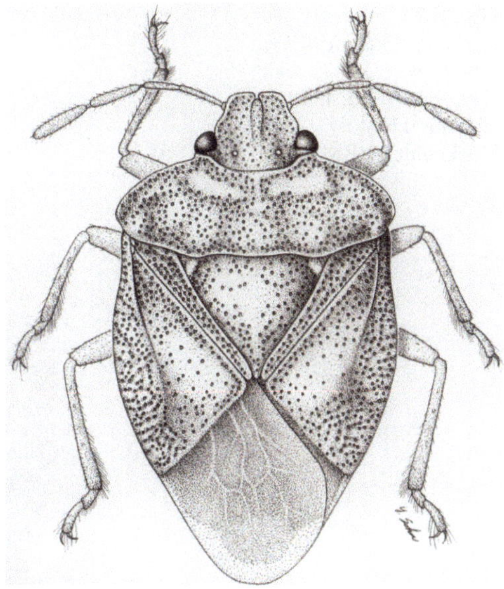

Fig. 25.18 *Rolstonus rolstoni* Froeschner (From Froeschner 2000. Used with permission of Entomologica Americana)

Fig. 25.19 *Tolono decoratus* Rolston and Kumar, *dorsal view* (Photo modified from the original available at http://ihs.myspecies.info/taxonomy/term/94, Copyright American Museum of Natural History licensed under a Creative Commons Attribution CC BY Licence, http://creativecommons.org/licenses/by/3.0/)

Fig. 25.20 *Praesus incarnatus* Stål, *dorsal view* (Photo: Bert Gustafsson. Copyright Swedish Museum of Natural History, Stockholm – NRM)

Fig. 25.21 *Hyperbius geniculatus* Signoret, *dorsal view* (Photo CF Schwertner)

Fig. 25.22 *Elasmucha lateralis* (Say), dorsal view (Photos CF Schwertner)

Fig. 25.23 *Elasmostethus placidus* (Walker), *dorsal view* (Photo CF Schwertner)

Fig. 25.24 *Blaudus ruficornis* Stål, *dorsal view* (Photo: Bert Gustafsson. Copyright Swedish Museum of Natural History, Stockholm – NRM)

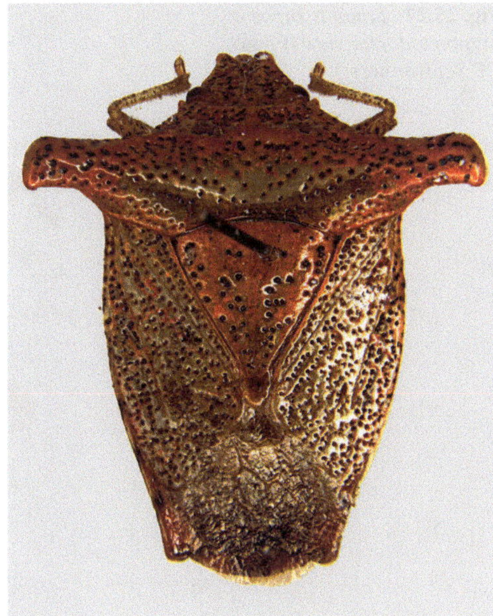

Fig. 25.25 *Bebaeus punctipes* Dallas, *dorsal view* (Photo CF Schwertner)

Fig. 25.26 *Phorbanta variabilis* (Signoret), *dorsal view* (Photo CF Schwertner)

Fig. 25.27 *Lanopis rugosus* Signoret, *dorsal view* (Photo CF Schwertner)

Fig. 25.28 *Acrophyma cumingii* (Westwood), *dorsal view* (Photo CF Schwertner)

Fig. 25.29 *Hellica nitida* Haglund, *dorsal view* (Photo CF Schwertner)

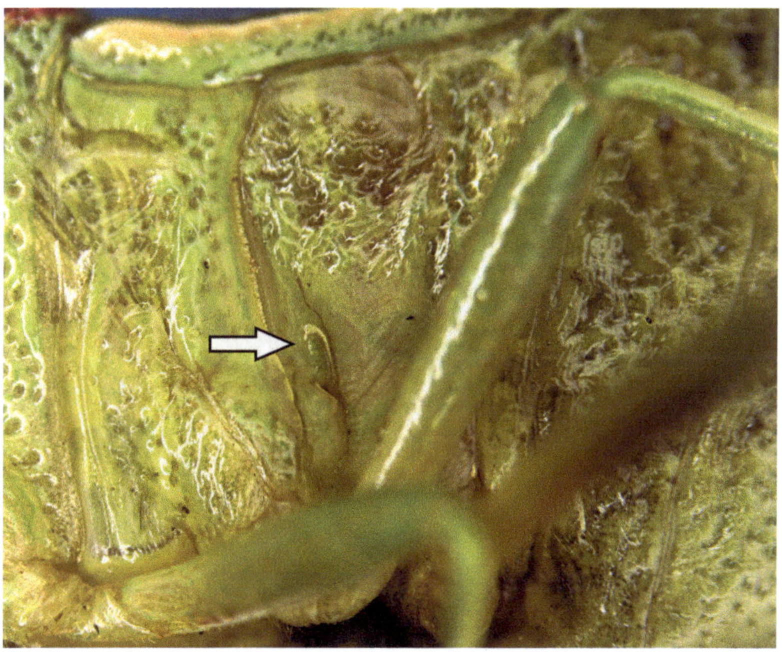

Fig. 25.30 *Sinopla perpunctatus* Signoret, (**b**) *Lanopis rugosus*, detail of ventral view showing ostiolar peritreme (Photo CF Schwertner)

Fig. 25.31 *Sinopla perpunctatus* Signoret, *dorsal view* (Photo CF Schwertner)

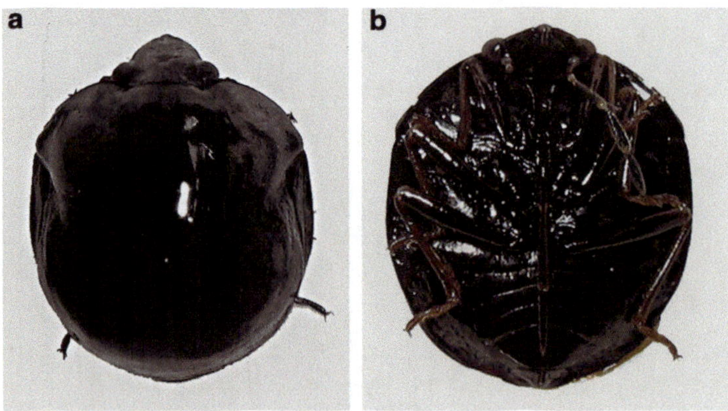

Fig. 25.32 *Canopus* sp. (**a**) *dorsal view*, (**b**) *ventral view* (Photos CF Schwertner)

Lestoniidae, Plataspididae, and Scutelleridae), but the similarity among these families is only superficial (McDonald 1979; Grazia et al. 2008). The family is poorly represented in collections, probably due to the small size and difficult sampling, sometimes being mixed up with Coleoptera. It was recognized as a subfamily of Pentatomidae by McAtee and Malloch (1928).

25.3.1 General Characteristics and Diagnosis

The canopids do not have a common name; they are medium sized (5–7 mm) with obovate outline, slightly narrower posteriorly than anteriorly, strongly convex dorsally, and flattened ventrally; the head is short, length of the head anterior to the eyes almost as long as the eye; margins of mandibular plates slightly reflexed; the prosternal sulcus and strongly laminate propleural carinae present; antennae V-segmented, with segment II reduced, subequal to the diameter; the cutellum covering the abdomen and almost all hemelytra; hemelytra elongated, twice as long as the abdomen with line of weakness for folding at the end of the costa, membrane with at least five parallel veins; hind wings with lobate posterior margins; tibiae setose, tarsi III-segmented; abdomen with obsolete sutures laterad to the spiracles; and trichobothria on sterna III–VII placed longitudinally mesad of the spiracular line. The nymphs are strongly convex and sclerotized, with three pairs of dorsal abdominal scent gland openings between terga III–IV, IV–V, and V–VI, the anterior gland openings twice the width of the other two; nymphal sterna II and III are divided mesally.

25.3.2 General Biology and Ecology

Biology and ecology of the canopids are poorly known. McHugh (1994) registered nymphs and adults of *Canopus* spp. on fungi and found spores in the digestive system, confirming the fungivorous habits of these insects.

25.3.3 Classification and Diversity

A single genus, *Canopus* F., and eight species are known (Table 25.2). The first three species described in the genus were based in nymphs and were not recognized by McAtee and Malloch (1928) in their review of the group. Horváth (1919) proposed the tribe Canoparia and added five new species. McAtee and Malloch (1928) treated Canopinae as a subfamily of Pentatomidae, transferred two previously described species to *Canopus*, synonymized two species described by Horváth (1919), and described three new species, totalizing the eight species so far.

McDonald (1979) recognized Canopidae as a family, describing the morphology of the male and female genitalia of *C. caesus*, *C. orbicularis*, and *C. impressus* and the female genitalia of *C. burmeisteri* and *C. fabricii*. The development of the scutellum may suggest the canopids being related to the Megarididae and Plataspididae, but McDonald (1979) considered them quite clearly as separate families. The only key to the species is found in McAtee and Malloch (1928). Schaefer (1980, 1988) asserted that the families Canopidae, Cydnidae, Cyrtocoridae, Lestoniidae, Megarididae, Plataspididae, and Thaumastellidae consist of a primitive group within the Pentatomoidea, because of the plesiomorphic characters they have in common.

Gapud (1991) considered the Canopidae related to the Scutelleridae. The characters defining the group and the presence of a prosternal sulcus and prosternal carina strong laminate are shared with Megarididae and Thyreocoridae (Grazia et al. 2008). The relationships of the Canopidae with the Cydninae (Cydnidae) were also pointed out by Grazia et al. (2008) in the combined cladistic analysis using morphology and molecular data, but these results are in need of further tests.

25.4 Dinidoridae

Despite its relatively large size (up to 30 mm) and its aposematic coloration, these bugs are poorly represented in collections. The family was reviewed by Durai (1987), which divided it into two subfamilies and four tribes (Table 25.1). Its species are predominantly found in the Afrotropical and Oriental regions. In the Neotropics, only one genus, *Dinidor* Latreille, is recorded, with six species distributed in South America (Table 25.2).

25.4.1 General Characteristics and Diagnosis

Adults have a large size (10–30 mm) body that is ovoid and robust (Fig. 25.33); general color usually dark brown or black, with red/yellow stripes and maculae usually present; head carinate laterally and buccula short and elevated; with 4–5 segments antennae; and rostrum reaching meso and metacoxa. Humeral angles never developed. Scutellum triangular, width subequal to the length, never covering the corium, and apex usually blunt. Hemelytral with venation reticulate. Tarsi 2–3 segments.

Eggs or nymphs of any species of the Neotropical Dinidoridae have not been formally described. Knowledge is restricted to species from other regions (Malipatil and Kumar 1975; Danielczok and Kocorek 2003). Information about the eggs is also provided by Southwood (1956) and Cobben (1968). Eggs and nymphs of *D. mactabilis* were illustrated in Schwertner and Grazia (2014).

Fig. 25.33 *Dinidor mactabilis* Perty, *dorsal view* (Photo CF Schwertner)

In general, the immatures of dinidorids follow the pattern described for those of other families of Pentatomoidea (Cobben 1968; Schuh and Slater 1995; Grazia and Schwertner 2008). Unique characteristics include the shape of the eggs and of the aero-micropylar processes, the aspect of the chorion surface, and the development of some nymph structures.

Eggs are cylindrical, deposited in rows, and horizontally positioned to the substrate (a unique condition among pentatomoids). The chorion can be smooth or ornamented, usually thicker than in other families of Pentatomoidea. Aero-micropylar processes reduced, eccentrically arranged around the egg, with a number of processes variable from 30 to 50. Operculum not apparent, a semicircular line along the upper region, opposite to the region in contact with the substrate. Nymphs with body oval and robust. Openings of the dorsal abdominal glands of nymphs present only between tergites 4/5 and 6/5 (when present between tergites 3/4, the openings are reduced).

25.4.2 General Biology and Ecology

Only a few species of the family Dinidoridae have their biology studied (Schaefer et al. 2000). Most of the available data is limited to label records or other field observations. Schaefer and Ahmad (1987) compiled records of known host plants for four families of Pentatomoidea, including Dinidoridae. Schaefer et al. (2000) reviewed and discussed the data available for the species of economic importance. None of these studies include Neotropical species.

All species studied so far are exclusively phytophagous, feeding both in reproductive and vegetative parts of their hosts. The data indicate that polyphagy is widespread, but monophagy in some species is likely. Schafer and Ahmad (1987) and Schaefer et al. (2000) pointed out that some species can show preference for certain plant families (e.g., *Aspongopus* spp. for plants of the family Cucurbitaceae).

All species studied are univoltine and usually use more than one host plant throughout its life cycle (Schaefer et al. 2000). Some species may exhibit gregarious behavior during part of the year, being found in large quantities together on the host. Dinidoridae species oviposit egg masses averaging 14–28 eggs. Fertility is considered low, not exceeding 1–2 clutches/female (Schaefer et al. 2000).

One Neotropical species, *Dinidor mactabilis* (Dinidorinae), was recorded in *Smilax japecanga* Grisebach (Smilacaceae) in southern Brazil (Grazia et al. 2012). An interesting biological data on this species is the large number of ovipositions grouped on branches of the host plant.

There are no records of crop damage by dinidorids in the Neotropical region. Species in these families considered pests are found in the Oriental and Palearctic regions (*Aspongopus* spp.), associated to cucurbit crops.

25.4.3 Classification and Diversity

Durai (1987) recognized the subfamilies Dinidorinae (with tribes Dinidorini and Thalmini) and Megymeninae (with tribes Megymenini and Eumenotini). Kocorek and Lis (2000) reviewed the Megymeninae, proposed a new tribe (Byrsodepsini), and treated Eumenotini as junior synonym of Megymenini. Keys for identification of all genera and species can be found in Durai (1987). Rolston et al. (1996) summarized the current classification for the family. In the Neotropical region, only the endemic genus *Dinidor* is found, including six species distributed in Central and South Americas (Table 25.2).

Gapud (1991) considered Dinidoridae and Tessaratomidae sister-groups, based on two synapomorphies: presence of spiracles of abdominal segment II, partially exposed, and laterotergites 9 very large in females. Grazia et al. (2008) found similar results, although Dinidoridae may be paraphyletic or monophyletic according to the analyses performed.

25.5 Megarididae

Megaridids are small (5 mm or less in length) and coleopteroid in shape. Strongly convex dorsally, the scutellum is enlarged and covering the abdomen and wings (Fig. 25.34). The distribution is exclusively Neotropical, ranging from Mexico to southern Brazil; one species is endemic to Cuba and another to Puerto Rico. It was recognized as a subfamily of Pentatomidae by McAtee and Malloch (1928) and elevated to family by McDonald (1979). The family is poorly represented in collections, probably due to the small size and difficult sampling.

25.5.1 General Characteristics and Diagnosis

Adults have a body that is ovoid and strongly convex dorsally; dark in color and polished; anterior margins of the head and pronotum carinate and bucculae undeveloped; antennae 4-segmented, with many setae as long as the diameter of segments in females, much longer in males; scutellum almost completely covering hemelytra; forewing longer than the abdomen with thin areas about the middle costa, adapting the wing for folding; membrane with one or none parallel veins; and tibia without spines, tarsi 2-segmented.

Information about immatures are scarce (McAtee and Malloch 1928; Schuh and Slater 1995), and there is no formal description of eggs. General body of nymphs similar to adults in shape and color and heavily sclerotized, with abdominal terga not distinguished.

Fig. 25.34 *Megaris* sp., *dorsal view* (Photo VC Matesco)

25.5.2 General Biology and Ecology

Biology and ecology of the megaridids are completely ignored; it is accepted that all species are exclusively phytophagous. There are records of *Megaris puertoricensis* Barber and *M. semiamicta* McAtee and Malloch em *Eugenia* spp. (Myrtaceae), feeding on flowers of its hosts (references in Schuh and Slater 1995).

25.5.3 Classification and Diversity

A single extant genus, *Megaris* Stål, and 16 extant species are known (McDonald 1979) (Table 25.2). The only key to the species is found in McAtee and Malloch (1928), which should be complemented with the descriptions of Barber (1939), Kormilev (1956), and McDonald (1979). More recently, Poinar and Heiss (2013) described a new genus and species from Dominican amber fossil of Tertiary (age estimated between 45 and 15 mya).

25.6 Phloeidae

Phloeidae is a family of true bugs with peculiar aspect and with unique cryptic morphology. The species are flattened, with the margins of the head, thorax, and abdominal segments expanded into large lobes (Figs. 25.1c, 25.35, 25.36), which make these bugs to be mistaken with the bark of trees they live (Lent and Jurberg 1965; Grazia et al. 2012). They are large in size (20–25 mm). The family is distributed mainly in the Neotropical region and has been better studied since the mid-1960s (Lent and Jurberg 1965). More recently, thorough studies have been conducted on this exquisite group of bugs (Guilbert 2003; Bernardes et al. 2005; Salomão et al. 2012).

25.6.1 General Characteristics and Diagnosis

The body is depressed, with external margins of juga, pronotum, base of corium, and abdomen broadly foliate. Eyes divided into dorsal and ventral portion. The antenna 3-segmented, only partially visible dorsally, hidden below expanded juga; segment 1 very long and segment 3 curved. Peritreme opening near the lateral margin of the pleuron. Hemelytral membrane reticulate. Tarsi 3-segmented. Abdominal sterna 3–7 with trichobothria arranged longitudinally to the spiracular

Fig. 25.35 *Phloeophana longirostris* (Spinola), *dorsal view* (Photo CF Schwertner)

line. Spiracle 2 present and partially exposed. Abdominal scent gland of nymph openings present between terga 3/4, 4/5, and 5/6 (the latter with only one opening).

25.6.2 General Biology and Ecology

The life cycle of these bugs occurs entirely on the trunks of the host (Salomão et al. 2012). Females protect eggs and early nymphs; the latter attach themselves to the abdomen of the mother (Lent and Jurberg 1965; Guilbert 2003) (Fig. 25.37). Nymphs and adults feed on the vascular system of the hosts (Bernardes et al. 2005). Host plants include *Cecropia* sp. (Urticaceae), *Croton floribundus* Spreng (Euphorbiaceae), *Ficus* spp. (Moraceae), *Enterolobium maximum* Ducke (Mimosacea), *Eugenia cauliflora* de Berg (Myrtaceae), *Miconia cubatensis* Hoehne (Melastomataceae), *Machaerium* spp. (Fabaceae), *Myrcia* sp. (Myrtaceae), *Parkia*

Fig. 25.36 *Phloea subquadrata* Spinola, *dorsal view* (Photo CF Schwertner)

multijuga Benth, *Psidium* sp. (Myrtaceae), *Terminalia catappa* L. (Rosaceae), *Sucurinega guaraiva* Kuhlmann (Euphorbiaceae), and *Parkia multijuga* Benth (Mimosaceae). Salomão et al. (2012) gave an excellent account about the group and summarized several studies that have been carried on in the Serra do Japi, São Paulo, Brazil, where the three species occur sympatrically.

25.6.3 Classification and Diversity

The family includes four species in three genera (Grazia et al. 2008), with disjunct distribution in the Southern Hemisphere: *Serbana* Distant (one species) occurs only in Borneo (Leston 1953), while species of the genera *Phloea* Spinola (two species) and *Phloeophana* Leston (one species) occur only in South America, their distribution restricted to Brazil from Pará to Rio Grande do Sul and restricted to Amazon and Atlantic Rainforest regions (Lent and Jurberg 1965). The relationship of the Phloeidae among Pentatomoidea families is still equivocal (Grazia et al. 2008).

Fig. 25.37 *Phloea subquadrata* (Spinola), female ventral view with nymphs attached in the abdomen (Photo CF Schwertner)

25.7 Tessaratomidae

This family of large and robust bugs, sometimes exceeding 40 mm in length, is mainly distributed in the Old World tropics, restricted to the Southern Hemisphere (Schuh and Slater 1995; Grazia et al. 2012). About 45 genera and 235 species are known worldwide (Table 25.1). One genus, *Piezosternum* Amyot and Serville, has wide distribution and includes three endemic species to the Neotropics (Table 25.2).

Similar to large pentatomids, tessaratomids can be distinguished from them by a very small head; antenna usually 4-segmented; short rostrum; metasternum produced anteriorly, generally into a large blade-like projection reaching the anterior coxae; and pronotum extending over the base of the scutellum.

25.7.1 General Characteristics and Diagnosis

Adults are large (10–40 mm) and have a body that is ovoid to elongated and robust (Fig. 25.38). Varied coloration, but the Neotropical species are predominantly green. The head is laterally carinate, relatively small in comparison to the body size;

Fig. 25.38 *Piezosternum thunbergi* Stål, *dorsal view* (Photo CF Schwertner)

buccula is very short. Antennae 4- or 5-segmented (usually 4). The rostrum is short, reaching but not surpassing the anterior coxa. Humeral angles may be developed, a condition found in all Neotropical species; the posterior margin of the pronotum reaching up on the base of the scutellum. The scutellum triangular, longer than broad, never covering the corium, apex usually pointed. Hemelytral membrane without reticulated veins. Hind wing with hamus. The metasternum produced laterad and anteriorly, reaching the anterior coxae; the posterior margin of the metasternum truncate at the junction with the abdomen. Tarsi with 2- or 3-segments.

Knowledge about eggs and nymphs of these families is restricted to species from other regions (Southwood 1956; Cobben 1968; Kumar 1969; McDonald 1969; Malipatil and Kumar 1975; Magnien et al. 2008).

Eggs are spherical, deposited perpendicular to the substrate. Chorion smooth, thinner than in Dinidoridae. Aero-micropylar processes are reduced and knob like, circulating the middle region of the egg, in variable numbers (30 to 60). Operculum not apparent, line of operculum dorsolateral extending basally. Nymphs of the 1st, 2nd, and 3rd instars with oval, flattened body; late instars elongated and more robust body. Fifth instar nymph of *P. thunbergi* is illustrated in Schwertner and Grazia (2014).

25.7.2 General Biology and Ecology

Only a few species of tessaratomids have their biology studied (McDonald 1969; Malipatil and Kumar 1975; Schaefer et al. 2000; Dzerefos et al. 2009); most of the available data is limited to label data or other types of field observation. Schaefer and Ahmad (1987) refer to known host plants, and Schaefer et al. (2000) reviewed and discussed the available data of the species related to crops. None of these studies provided information about the Neotropical species.

All species studied so far are exclusively phytophagous, feeding on both reproductive and vegetative parts of their host. Polyphagy seems to be the rule, but as for Dinidoridae, some species may show host preference [e.g., *Musgraveia sulciventris* (Stål) feeds primarily on plants of the family Rutaceae] (Schaefer and Ahmad 1987; Schaefer et al. 2000).

All species are univoltine and usually use more than one host plant throughout its life cycle (Schaefer et al. 2000; Dzerefos et al. 2009), with biological aspects very similar to Dinidoridae (see item 3.3). Depending on the species, females can lay in masses of four rows (3-4-4-3 formula), with an average of 14–28 eggs per clutch. Fertility is considered low, not exceeding 1–2 clutches/female (Schaefer et al. 2000). Eggs of *Encosternum delegorguei* Spinola took an average of 18 ± 9 days to hatch (temperatures ranging from 11 to 25 °C), while the nymphs took four months to reach adulthood (Dzerefos et al. 2009). Adults of this species overwinter from autumn to late winter. Maternal care has been described for some tessaratomids of the subfamily Oncomerinae in the Australian region (Gogala et al. 1998; Monteith 2006). Three genera show a similar behavior to that described for species of Phloeidae, where the nymphs are carried on the modified body of the female for a period of time after hatching (Monteith 2006).

The Neotropical species *Piezosternum subulatum* (Thunberg) was recorded in *Cucurbita* sp. (Cucurbitaceae), *Coffea* sp. (Rubiaceae), *Hibiscus* sp. (Malvaceae), *Lycopersicon* sp. (Solanaceae), and *Psidium* sp. (Myrtaceae) in Nicaragua (Maes 1994). There are no records of damage to crops in the Neotropical region. A single species of this family is considered pest in the Australian region, *Musgraveia sulciventris* Stål (Schaefer et al. 2000).

25.7.3 Classification and Diversity

In a cladistic analysis, Sinclair (1989) suggested that Tessaratomidae is polyphyletic and raised the Oncomerinae to family status, but recently it was restored as a subfamily (Sinclair 2000). Rolston et al. (1993) catalogued the world fauna. Schuh and Slater (1995) recognized the three subfamilies mentioned above, the Tessaratominae subdivided into five tribes (Eusthenini, Platytatini, Prionogastrini, Sepinini, and Tessaratomini), and the Oncomerinae into two tribes (Oncomerini and Piezosternini). Grazia et al. (2008) found different results concerning the monophyly of the family, according to the analyses performed: the group appears as monophyletic or paraphyletic, but always related to the Dinidoridae. The intrafamiliar classification was recently discussed by Kment and Vilímová (2010). No phylogenetic hypothesis concerning the relationships within the family is available. Identification key to the Neotropical species is available in Pirán (1971).

25.8 Concluding Remarks

The families treated in this chapter represent less diverse lineages of the superfamily Pentatomoidea, although with great importance to understand the evolution and natural history of the true bugs in the Neotropical region. Most of the species found in the Neotropics are endemic taxa (i.e., Canopidae and Megarididae) or have a relict distribution (i.e., Acanthosomatidae, Phloeidae, and Tessaratomidae), and studies in these groups will certainly be helpful to have a better picture of the fauna of the region.

The species found in the Neotropics show remarkable characteristics, like cryptic coloration (Phloeidae) and maternal care behavior (Acanthosomatidae), and may be good models for evolutionary and ecological studies. For some families, like Canopidae and Megarididae, there is little information available despite scattered distribution records. Future research in these families should include more detailed studies on the distribution of the species, phylogenetic relationships, morphology and natural history.

Acknowledgments The authors would like to thank Eduardo Faúndez and Mariom Carvajal (North Dakota State University, Fargo, USA) for invaluable information and kind assistance with Acanthosomatidae part, David Rider (North Dakota State University, Fargo, USA) for kindly allowing access to the collection at the NDSU, and Tom J. Henry (ARS/USA, National Museum of Natural History, Washington D.C.) for providing most of the Acanthosomatidae specimens used to illustrate the group in the chapter.

References

Barber HG (1939) Scientific survey of Puerto Rico and the Virgin Islands, Vol. XIV (3). Insects of Puerto Rico and the Virgin Islands-Hemiptera-Heteroptera (excepting the Miridae and Corixidae). New York Acad Sci 14:263–441

Bergroth E (1917) New and little-known heteropterous Hemiptera in the United States National Museum. Proc USA Natl Museum 51:215–239

Bernardes JL, Grazia J, Barcellos A, Salomão AT (2005) Descrição dos estágios imaturos e notas sobre a biologia de *Phloea subquadrata* (Heteroptera, Phloeidae). Iheringia Sér Zool 95:415–420

Carter ME, Hoebeke ER (2003) Biology and seasonal history of *Elasmostethus atricornis* (Van Duzee) (Hemiptera: Acanthosomatidae), with descriptions of the immature stages and notes on pendergrast organs. Proc Entomol Soc Wash 105:525–534

Carvajal MA, Faúndez EI (2013) Rediscovery of *Sinopla humeralis* Signoret, 1864 (Hemiptera: Heteroptera: Acanthosomatidae). Zootaxa 3637:190–196

Carvajal MA, Faúndez EI, Rider DA (2014) New data on the genus *Ea* Distant, 1911 (Hemiptera: Heteroptera: Acanthosomatidae), with description of a new species from Chile. An Inst Patagonia 42:53–58

Carvajal MA, Rider DA, Faúndez EI (2015) Revision of the Genus *Tolono* Rolston and Kumar, 1975 (Hemiptera: Heteroptera: Acanthosomatidae), with Descriptions of Two New Species. Proc Entomol Soc Wash 117:7–13

Cobben RH (1968) Evolutionary trends in Heteroptera. Part I: Eggs, architecture of the shell, gross embriology and eclosion. Wagen Centre Agric Publ Docum, Wageningen

Danielczok T, Kocorek A (2003) External morphology of eggs of four African species of *Coridius* Ill. (Heteroptera : Pentatomoidea : Dinidoridae). Pol J Entomol 72:63–73

Distant WL (1911) Rhynchotal notes, LIII. Ann Mag Nat Hist 8:242–258

Durai PSS (1987) A revision of the Dinidoridae of the world (Heteroptera: Pentatomoidea). Orient Insects 21:163–360

Dzerefos CM, Witkowski ETF, Toms R (2009) Life-history traits of the edible stinkbug, *Encosternum delegorguei* (Hem., Tessaratomidae), a traditional food in southern Africa. J Appl Entomol 133:749–759

Faúndez EI (2007a) Asociación críptica entre *Sinopla perpunctatus* Signoret, 1863 (Acanthosomatidae: Hemiptera) y el Ñirre *Nothofagus antarctica* (G. Forster) Oersted (Fagaceae) en la Región de Magallanes (Chile). Bol Soc Entomol Aragonesa 40:563–564

Faúndez EI (2007b) Lista de las especies de Acanthosomatidae (Insecta: Hemiptera) de la región de Magallanes (Chile). An Ins Patagonia 35:75–78

Faúndez EI (2007c) Notes on the biology of *Ditomotarsus punctiventris* Spinola, 1852 (Hemiptera: Acanthosomatidae) in the Magellan region, and comments about the crypsis in Acanthosomatidae. An Ins Patagonia 35:67–70

Faúndez EI (2008) A new record of Mazanoma variada Rolston & Kumar, 1974 (Hemiptera: Heteroptera: Acanthosomatidae). Bol Soc Entomol Aragon 43:499–500

Faúndez EI (2009) Contribution to the knowledge of the genus *Acrophyma* Bergroth, 1917 (Hemiptera: Heteroptera: Acanthosomatidae). Zootaxa 2137:57–65

Faúndez EI (2014) A new genus for a Chilean species of Acanthosomatidae (Hemiptera: Heteroptera). Zootaxa 3768:596–600

Faúndez EI, Osorio GA (2010a) Contribution to the knowledge of *Cylindrocnema plana* Mayr, 1864 (Hemiptera: Heteroptera: Acanthosomatidae). An Ins Patagonia 38:159–161

Faúndez EI, Osorio GA (2010b) New data on the biology of *Sinopla perpunctatus* Signoret, 1864 (Hemiptera: Heteroptera: Acanthosomatidae). Bol Biodivers Chile 3:24–31

Faúndez EI, Osorio GA, Bahamondez CP, Monsalve EA (2009) Comportamiento gregario en los hábitos sexuales de *Ditomotarsus punctiventris* Spinola, 1852 (Hemiptera: Heteroptera: Acanthosomatidae). Bol Biodivers Chile 1:15–18

Faúndez EI, Rider DA, Carvajal MA (2014) *Cochabambia* Pirán, 1959 (Hemiptera: Heteroptera: Acanthosomatidae), a senior homonym of *Cochabambia* Marcuzzi, 1985 (Coleoptera: Tenebrionidae), with notes on the placement of *Cochabambia* Pirán. Zootaxa 3793:595–596

Fischer C (1994) Das Pendergrast-Organ der Acanthosomatidae (Heteroptera, Pentatomoidea): Schutz des Eigeleges vor Räubern und Parasiten? Sitz Ges Natur Fr Berlin 33:129–142

Froeschner RC (1997[1995]) *Rolstonus rolstoni*, new genus and new species of Acanthosomatidae from Argentina (Heteroptera: Pentatomoidea: Ditomotarsini). J N Y Entomol Soc 103:360–363

Froeschner RC (2000[1999]) Revision of the South American genus *Hellica* Stål (Heteroptera: Acanthosomatidae). J N Y Entomol Soc 107:164–170

Gapud VP (1991) A generic revision of the subfamily Asopinae, with consideration of its phylogenetic position in the family Pentatomidae and superfamily Pentatomoidea (Hemiptera-Heteroptera). Philipp Entomol 8:865–961

Gogala M, Yong H-S, Brühl C (1998) Maternal care in *Pygoplatys* bugs (Heteroptera: Tessaratomidae). Eur J Entomol 95:311–315

Grazia J, Schwertner CF (2008) Pentatomidae e Cyrtocoridae In: Claps LE, Debandi G, Roig-Juñent S (eds) Biodiversidad de artrópodos Argentinos, vol 2, Tucumán, Sociedad Entomológica Argentina, pp 223–234

Grazia J, Schuh RT, Wheeler WC (2008) Phylogenetic relationships of family groups in Pentatomoidea based on morphology and DNA sequences (Insecta: Heteroptera). Cladistics 24:932–976

Grazia J, Cavicchioli RR, Wolff VRS, Fernandes JAM, Takiya DM (2012) Hemiptera. In: Rafael JA, Melo GAR, Carvalho CJB, Casari S (eds) Os insetos do Brasil: Diversidade e taxonomia. Editora Holos, Ribeirão Preto, pp 348–405

Guilbert ERIC (2003) Habitat use and maternal care of *Phloea subquadrata* (Hemiptera: Phloeidae) in the Brazilian Atlantic forest (Espirito Santo). Eur J Entomol 100:61–63

Haglund JCE (1868) Hemiptera nova. Stettin Entomol Zeitung 29:150–163

Horváth G (1919) Ergebnisse einer mit Unterstützung der Kais. Akademie der Wissenschaften in Wien ausgeführten zoologischen Forschungsreise von weiland Prof. Dr. Franz Tölg. nach Kleinasien (Amanus Gebirge). V. Rhynchota. Archiv Naturgeschichte 85:146–147

Kment P (2005) Revision of *Mahea* Distant, 1909, with a review of the Acanthosomatidae (Insecta: Heteroptera) of Madagascar and Seychelles. Acta Entomol Musei Nat Pragae 45:21–50

Kment P, Vilímová J (2010) Thoracic scent efferent system of the Tessaratomidae *sensu lato* (Hemiptera: Heteroptera: Pentatomoidea) with implication to the phylogeny of the family. Zootaxa 2363:1–59

Kocorek A, Lis JA (2000) A cladistic revision of the Megymeninae of the world (Hemiptera: Heteroptera: Dinidoridae). Polskie Pismo Entomol 69:7–30

Kormilev NA (1956) Notas sobre Pentatomoidea neotropicales IV (Hemiptera). Acta Sci Inst Invest San Miguel 3:1–13

Kudo S-I (2000) The guarding posture of females in the subsocial bug *Elasmucha dorsalis* (Heteroptera: Acanthosomatidae). Eur J Entomol 97:137–139

Kudo S-I, Nakahira T (1993) Brooding behavior in the bug *Elasmucha signoreti* (Heteroptera: Acanthosomatidae). Psyche 100:121–126

Kumar R (1969) Morphology and relationships of the Pentatomoidea (Heteroptera). IV. Oncomerinae (Tessaratomidae). Aust J Zool 17:553–606

Kumar R (1974) A revision of world Acanthosomatidae (Heteroptera: Pentatomidae). Key to and description of subfamilies, tribes and genera, with designation of types. Aust J Zool Supl Ser 34:1–60

Lent H, Jurberg J (1965) Contribuição ao conhecimento dos Phloeidae Dallas, 1851, com um estudo sôbre genitália (Hemiptera, Pentatomoidea). Rev Bras Biol 25:123–144

Leston D (1953) "Phloeidae" Dallas: systematics and morphology, with remarks on the phylogeny of "Pentatomoidea" Leach and upon the position of "Serbana" Distant (Hemiptera). Rev Bras Biol 13:121–140

Maes J-M (1994) Catálogo de los Pentatomoidea (Heteroptera) de Nicaragua. Rev Nicar Entomol 28:1–29

Magnien P, Smets K, Pluot-Sigwalt D, Constant J (2008) A new species of *Pygoplatis* Dallas (Heteroptera, Tessaratomidae) from the Damar agroforests in Sumatra: description, immatures and biology. Nouv Rev d'Entomol 24:99–112

Malipatil MB, Kumar R (1975) Biology and immature stages of some Queensland Pentatomomorpha (Hemiptera: Heteroptera). J Aust Entomol Soc 14:113–128

Martinez P, Dellapé PM, Coscarón MDelC, Giganti H (2003) Immature stages of *Sinopla perpunctatus* (Heteroptera: Acanthosomatidae) from Argentina. Entomol News 114:147–151

Mayr GL (1864) Diagnosen neuer Hemipteren. Verh Zool-Bot Ges Wien 14:903–914

McAtee WL, Malloch JR (1928) Sinopsis of pentatomid bugs of the subfamilies Megaridinae and Canopinae. Proc USA Nat Mus 72:1–21

McAtee WL, Malloch JR (1933) Revision of the subfamily Thyreocorinae of the Pentatomidae (Hemiptera-Heteroptera). Ann Carnegie Museum 21:191–411

McDonald FJD (1969) Life cycle of the bronze orange bug *Musgraveia sulciventris* (Stål) (Hemiptera: Tessaratomidae). Aust J Zool 17:817–820

McDonald FJD (1979) A new species of *Megaris* and the status of the Megarididae McAtee & Malloch and Canopidae Amyot & Serville (Hemiptera: Pentatomoidea). J N Y Entomol Soc 87:42–54

McHugh JV (1994) On the natural history of Canopidae (Heteroptera: Pentatomoidea). J N Y Entomol Soc 102:112–114

Mendonça MS Jr, Schwertner CF, Grazia J (2009) Diversity of Pentatomoidea (Hemiptera) in riparian forests of Bagé, RS, southern Brazil: Taller forests, more bugs. Rev Bras Entomol 53:121–127

Miller NCE (1971) The biology of the Heteroptera, 2nd edn. E.W. Classey Ltd., Hampton, 206p

Monteith GB (2006) Maternal care in Australian oncomerine shield bugs (Insecta, Heteroptera, Tessaratomidae). Denisia 19:1135–1152

Osorio GA (2009) Primer registro para la región del Maule de *Sinopla perpunctatus* Signoret, 1864 (Hemiptera: Heteroptera: Acanthosomatidae). Bol Biodivers Chile 1:35–37

Pirán AA (1971) La família Tessaratomidae (Hemiptera-Heteroptera) en la región neotropical. Acta Zool Lilloana 26:197–208

Poinar G Jr, Heiss E (2013) *Minysporops dominicanus* gen. n., sp. n. (Hemiptera: Pentatomoidea: Megarididae), a megaridid in Dominican amber. Hist Biol 25:95–100

Putshkov VG (1959) Larvae [sic] of Hemiptera-Heteroptera belonging to the superfamily Pentatomoidea. I. Identification keys of the families Pentatomoidea and of the species Acanthosomidae, Cydnidae and Scutelleridae. Zool Zhurnal 38:1190–1206 [in Russian]

Rolston LH, Kumar R (1975[1974]) Two new genera and two new species of Acanthosomatidae (Hemiptera) from South America, with a key to the genera of the Western Hemisphere. J N Y Entomol Soc 82:271–278

Rolston LH, McDonald FJD (1979) Keys and diagnoses for the families of Western Hemisphere Pentatomoidea, subfamilies of Pentatomidae and tribes of Pentatominae (Hemiptera). J N Y Entomol Soc 87:189–207

Rolston LH, Aalbu RL, Murray MJ, Rider DA (1993) A catalogue of the Tessaratomidae of the world. Papua New Guinea J Agric For Fish 36:36–108

Rolston LH, Rider DA, Murray MJ, Aalbu RL (1996) Catalog of the Dinidoridae of the World. Papua New Guinea J Agric For Fish 39:22–101

Salomão AT, Postali TC, Vasconcellos-Neto J (2012) Bichos-cascas na Serra do Japi: história natural dos percevejos Phloeidae (Hemiptera). In: Vasconcellos-Neto J, Polli PR, Penteado-Dias AM (eds) Novos olhares, novos saberes sobre a Serra do Japi: ecos de sua biodiversidade. Editora CRV, Curitiba, pp 321–337

Schaefer CW (1980) The sound-producing structures of some primitive Pentatomoidea (Hemiptera: Heteroptera). J N Y Entomol Soc 88:230–235

Schaefer CW (1988) The food plants of some "primitive" Pentatomoidea (Hemiptera: Heteroptera). Phytophaga 2:19–45

Schaefer CW, Ahmad I (1987) The food plants of four pentatomoid families (Hemiptera: Acanthosomatidae, Tessaratomidae, Urostylidae, and Dinidoridae). Phytophaga 1:21–34

Schaefer CW, Panizzi AR, James DG (2000) Several small pentatomoid families (Cyrtocoridae, Dinidoridae, Eurostylidae, Plataspidae, and Tessaratomidae). In: Schaefer CW, Panizzi AR (eds) Heteroptera of economic importance. CRC Press, Boca Raton, pp 505–512

Schuh RT, Slater JC (1995) True bugs of the world (Hemiptera: Heteroptera): Classification and natural history. Cornell University Press, Ithaca

Schwertner CF, Grazia J (2014) Dinidoridae, Megarididae e Tessaratomidae. In: Roig-Juñent S, Claps LE, Morrone JJ (eds) Biodiversidad de Artrópodos Argentinos, vol 3. INSUE – Universidad Nacional de Tucumán, Tucumán

Signoret V (1864[1863]) Revision des Hémiptères du Chili. Ann Soc Entomol France 3:541–588

Sinclair DP (1989) A cladistic, generic revision of the Oncomeridae Stål n. stat. and Tessaratomidae Schilling n. stat. (Hemiptera: Heteroptera: Pentatomoidea). PhD dissertation, University of Sydney, Australia

Sinclair DP (2000) A generic revision of the Oncomerinae (Heteroptera: Pentatomoidea: Tessaratomidae). Mem Qld Museum 46:307–329

Southwood TRE (1956) The structure of the eggs of the terrestrial Heteroptera and its relationship to the classification of the group. Trans Entomol Soc Lond 108:163221

Spinola M (1850) Di alcuni generi d'Insetti Artroidignati nuovamente proposti. Mem Soc Italiana Sci Res Modena 25:61–178
Spinola M (1852) Hemípteros. In Gay C, Historia física y política de Chile. Zoología 7:113–320
Stål C (1867) Bidrag till Hemipterernas systematik. Öfvers K Vetensk-akad forh 24:491–534
Stål C (1872) Enumeratio Hemipterorum. 2. *K. Svenska vetensk-akad.Handl.*10:1–159
Tallamy DW, Schaefer CW (1997) Maternal care in the Hemiptera: Ancestry, alternatives, and current adaptive value. In: Choe JC, Crespi BJ (eds) The evolution of social behavior in insects and arachnids. Cambridge University Press, Cambridge/New York, pp 94–115
Westwood JO (1837) In: Hope FW, A catalogue of Hemiptera in the collection of the Rev. F. W. Hope, M. A. with short Latin diagnoses of the new species. London, Pt. 1, 46 p

Index

A.R. Panizzi, J. Grazia (eds.), *True Bugs (Heteroptera) of the Neotropics*,
Entomology in Focus 2, DOI 10.1007/978-94-017-9861-7

B

H

O

T

U

V

MIX
Papier aus verantwortungsvollen Quellen
Paper from responsible sources
FSC® C105338
FSC
www.fsc.org

If you have any concerns about our products,
you can contact us on
ProductSafety@springernature.com

In case Publisher is established outside the EU,
the EU authorized representative is:
Springer Nature Customer Service Center GmbH
Europaplatz 3, 69115 Heidelberg, Germany

Printed by Libri Plureos GmbH
in Hamburg, Germany